International Table of Atomic Weights*†

Name	Symbol	Atomic Number	Atomic Weight
Actinium	Ac	89	(227)
Aluminum	Al	13	26.981539
Americium	Am	95	(243)
Antimony	Sb	51	121.75
Argon	Ar	18	39.948
Arsenic	As	33	74.92159
Astatine	At	85	(210)
Barium	Ba	56	137.327
Berkelium	Bk	97	(247)
Beryllium	Be	4	9.012182
Bismuth	Bi	83	208.98037
Boron	B	5	10.811
Bromine	Br	35	79.904
Cadmium	Cd	48	112.411
Calcium	Ca	20	40.078
Californium	Cf	98	(251)
Carbon	C	6	12.011
Cerium	Ce	58	140.115
Cesium	Cs	55	132.90543
Chlorine	Cl	17	35.4527
Chromium	Cr	24	51.9961
Cobalt	Co	27	58.93320
Copper	Cu	29	63.546
Curium	Cm	96	(247)
Dysprosium	Dy	66	162.50
Einsteinium	Es	99	(252)
Erbium	Er	68	167.26
Europium	Eu	63	151.965
Fermium	Fm	100	(257)
Fluorine	F	9	18.9984032
Francium	Fr	87	(223)
Gadolinium	Gd	64	157.25
Gallium	Ga	31	69.723
Germanium	Ge	32	72.61
Gold	Au	79	196.96654
Hafnium	Hf	72	178.49
Helium	He	2	4.002602
Holmium	Ho	67	164.93032
Hydrogen	H	1	1.00794
Indium	In	49	114.82
Iodine	I	53	126.90447
Iridium	Ir	77	192.22
Iron	Fe	26	55.847
Krypton	Kr	36	83.80
Lanthanum	La	57	138.9055
Lawrencium	Lr	103	(260)
Lead	Pb	82	207.2
Lithium	Li	3	6.941
Lutetium	Lu	71	174.967
Magnesium	Mg	12	24.3050
Manganese	Mn	25	54.93805
Mendelevium	Md	101	(258)
Mercury	Hg	80	200.59
Molybdenum	Mo	42	95.94
Neodymium	Nd	60	144.24
Neon	Ne	10	20.1797
Neptunium	Np	93	(237)
Nickel	Ni	28	58.69
Niobium	Nb	41	92.90638
Nitrogen	N	7	14.00674
Nobelium	No	102	(259)
Osmium	Os	76	190.2
Oxygen	O	8	15.9994
Palladium	Pd	46	106.42
Phosphorus	P	15	30.973762
Platinum	Pt	78	195.08
Plutonim	Pu	94	(244)
Polonium	Po	84	(209)
Potassium (Kalium)	K	19	39.0983
Praseodymium	Pr	59	140.90765
Promethium	Pm	61	(145)
Protactinium	Pa	91	231.03588
Radium	Ra	88	(226)
Radon	Rn	86	(222)
Rhenium	Re	75	186.207
Rhodium	Rh	45	102.90550
Rubidium	Rb	37	85.4678
Ruthenium	Ru	44	101.07
Samarium	Sm	62	150.36
Scandium	Sc	21	44.955910
Selenium	Se	34	78.96
Silicon	Si	14	28.0855
Silver	Ag	47	107.8682
Sodium	Na	11	22.989768
Strontium	Sr	38	87.62
Sulfur	S	16	32.066
Tantalum	Ta	73	180.9479
Technetium	Tc	43	(98)
Tellurium	Te	52	127.60
Terbium	Tb	65	158.92534
Thallium	Tl	81	204.3833
Thorium	Th	90	232.0381
Thulium	Tm	69	168.93421
Tin	Sn	50	118.710
Titanium	Ti	22	47.88
Tungsten	W	74	183.85
Unnilennium	Une	109	(266)
Unnilhexium	Unh	106	(263)
Unniloctium	Uno	108	(265)
Unnilpentium	Unp	105	(262)
Unnilquadium	Unq	104	(261)
Unnilseptium	Uns	107	(262)
Uranium	U	92	238.0289
Vanadium	V	23	50.9415
Xenon	Xe	54	131.29
Ytterbium	Yb	70	173.04
Yttrium	Y	39	88.90585
Zinc	Zn	30	65.39
Zirconium	Zr	40	91.224

* Based on relative atomic mass of $^{12}C = 12$.

† The values given in the table apply to elements as they exist in materials of terrestrial origin and to certain artificial elements. Values in parentheses are the mass number of the isotope of the longest half-life.

Chemistry: *Principles and Practice*

CHEMISTRY:

SAUNDERS GOLDEN SUNBURST SERIES
Saunders College Publishing
Harcourt Brace Jovanovich College Publishers

Principles & Practice

Daniel L. Reger
Professor of Chemistry
University of South Carolina

Scott R. Goode
Associate Professor of Chemistry
University of South Carolina

Edward E. Mercer
Professor of Chemistry
Associate Dean of the College of Science and Mathematics
University of South Carolina

Illustrated by
George V. Kelvin

Fort Worth Philadelphia San Diego New York Orlando Austin
San Antonio Toronto Montreal London Sydney Tokyo

This book is dedicated to our wives,
Cheryl, Regis, and Carol,
our families, and our students.

Text Typeface: Times Roman
Compositor: Progressive Typographers, Inc.
Acquistions Editor: John Vondeling
Associate Editor: Nanette Kauffman
Managing Editor: Carol Field
Project Editor: Sally Kusch
Copy Editor: Jay Freedman
Manager of Art and Design: Carol Bleistine
Art Director: Christine Schueler
Art and Design Coordinator: Caroline McGowan
Text Designer: Rebecca Lemna
Cover Designer: Lawrence R. Didona
Text Artwork: George V. Kelvin
Computer-generated structures: TRIPOS Associates
Photo Research Editor: Dena Digilio-Betz
Layout Artist: Dorothy Chattin
Director of EDP: Tim Frelick
Production Manager: Charlene Squibb
Marketing Manager: Marjorie Waldron

Cover Credit: Chemistry is based on the results of laboratory experiments. (Larry Cameron)

Printed in the United States of America

CHEMISTRY: PRINCIPLES AND PRACTICE

ISBN 0-03-73333-2

Library of Congress Catolog Card Number: 92-063004
3456 069 987654321

Contents Overview

Contents

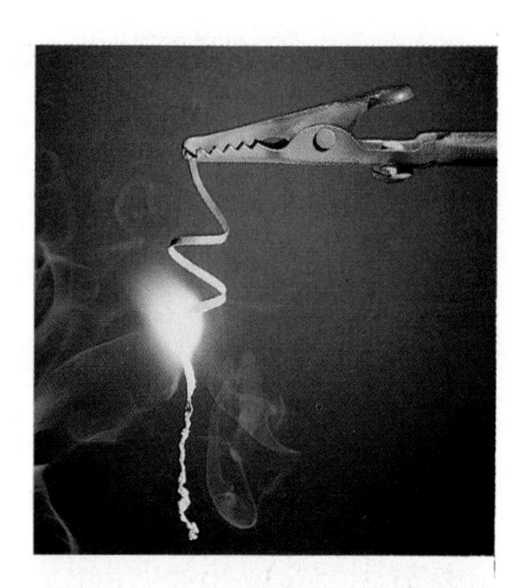

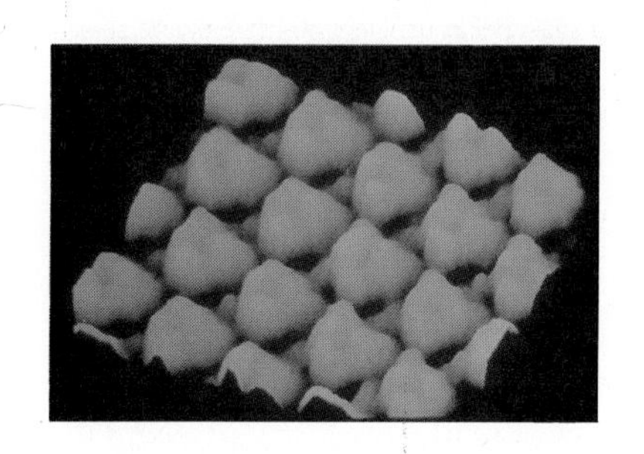

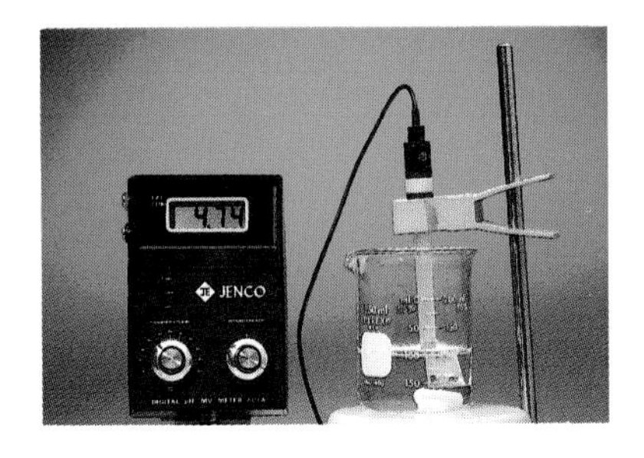
JENCO

Preface

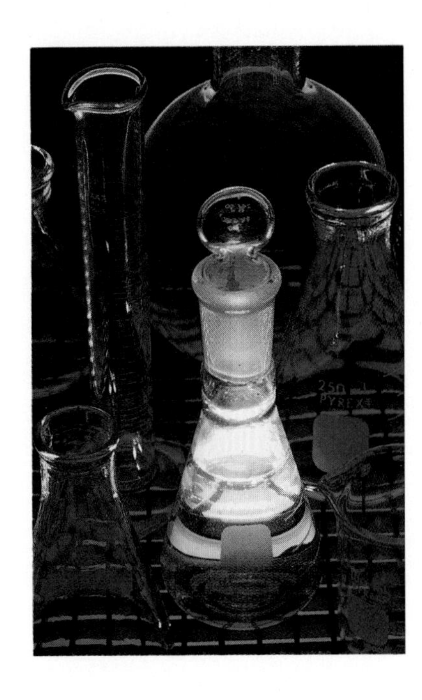

Philosophy and Goals

Our main goal in writing a general chemistry text is to bring the excitement of chemistry to our students and convince them that chemistry is a living, changing science, one in which people discover new facts and develop new concepts every day. We present the principles of chemistry in this context to avoid the impression that chemistry comes from textbooks rather than from experiments. Chemistry is first and foremost an experimental science, and the facts and theories that are its foundation have come from many years of experimentation.

Organization

The overall organization of the material in this text follows a general order that has been established over the years. This textbook is designed to be used by students who are interested in further study in chemistry and related areas, such as biology, engineering, geology, and the medical professions. The material is presented with numerous concrete examples, stressing logical, problem-solving approaches rather than rote learning. We have streamlined the presentation of the material by developing only those concepts needed to fully explain main principles. A new topic is introduced after consideration of why it is important and only when the students' current knowledge base will allow them to truly understand the concept. Topics are brought up when they can be explained clearly and completely. The language of chemistry is carefully developed; new terms are defined when they are first introduced. Also, the objectives of each new section are clearly stated; we believe that new concepts are learned more readily when the student is given the opportunity to know the reasons why the material is important.

We have refined the general presentation of the key topics. The first two chapters introduce the student to the basic concepts and language of chemistry. For programs with well-prepared students, these chapters are designed such that they can be made assigned reading. Stoichiometry is the first main topic, and a general method for executing calculations based on chemical equations is developed. The method is not "plug into this formula," but rather a sequential reasoning process that is applicable to a whole series of calculations ranging from mass-mass conversions, to reactions in solution, to heat flow in chemical equations, and to reactions involving gases. Example problems in the text are complemented with flow diagrams to help students visualize the problem-solving process. After this treatment, students will understand these calculations on a conceptual basis as well as obtain the correct answers. This approach fosters critical-thinking skills by helping students to develop a strategy rather than to memorize rote operations.

Early in the first stoichiometry chapter (Chapter 3) we clearly demonstrate how molar mass is a conversion factor that relates grams to moles. Empirical and molecular formulas, balancing equations, and the use of chemical equations in stoichiometry calculations starting with mass data are presented in the first stoichiometry chapter. Chapter 4 covers solution stoichiometry and heat in chemical reactions. Enthalpy is introduced as part of the chemical equation. This approacn to thermochemistry is unique, presenting thermochemistry as a natural progression of equation stoichiometry. The concepts of enthalpy as a state function and Hess's law are also developed at this early stage. Gases are covered next (Chapter 5) because we believe that early placement of this material is helpful for the first-semester laboratory, although the chapter can be taught after structure and bonding. A unique unified approach to gas laws is presented, and again, reaction stoichiometry is emphasized. In all of these chapters, the concepts are illustrated with important, real-life chemical reactions in the many worked-example problems. We believe our integration of descriptive chemistry throughout the text, in worked examples, in featured topics, and in exercises helps to solidify the concepts of chemical reactivity.

Chapters 6 through 9 develop atomic and molecular structure. The models and theories are developed as a natural progression from experimental observations. We emphasize the periodic table as a tool to help learn electron configurations, as well as trends in ionization energies and the sizes of atoms and ions. The presentation of bonding and shapes of molecules is supported by high-quality drawings that picture atoms and orbitals in proper perspective. The molecular orbital section is presented at two levels to help meet the needs of different instructors. The first section covers homonuclear diatomic molecules, and the second introduces heteronuclear diatomic molecules and delocalizing bonding. The organization allows the instructor to omit, teach a basic introduction, or defer molecular orbital theory to a later time in the course. Chapters 10 and 11 on liquids and solids reinforce the concepts of interactions between molecules and ions.

A systematic approach to equilibria is presented in Chapters 12 through 14. Many students of general chemistry find this topic difficult, but we clarify the material by introducing a strategy that works for all equilibrium systems. The introduction to equilibria uses simple gas-phase reactions, which provide a review of stoichiometry, and also serve as examples of equilibria.

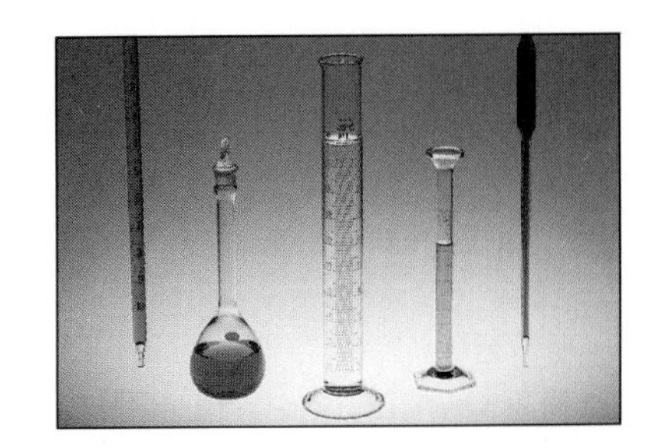

Solubility equilibria and the common ion effect are introduced at this point so that relevant and descriptive chemical problems can be treated early. Chapter 13 presents the concepts of solution equilibria to acid-base reactions. We present strong and weak electrolytes in this equilibrium chapter. The equilibria of aqueous solutions of acids and bases, as well as the reactions of acids with bases (titration equilibria), are all presented from the same orientation. In Chapter 14, the systematic treatment continues through acid-base titration curves. These three chapters can be taught in the first semester or may be moved to later in the second semester, depending on the needs of individual courses.

The material on equilibria is followed by complete chapters on thermodynamics and kinetics. In both of these chapters, experimental data are used to introduce the concepts. The chapter on thermodynamics integrates stoichiometry and concepts such as the principle of Le Chatelier. The chapter on kinetics builds on the concepts of thermodynamics to present activation energy and to describe reaction mechanisms.

A comprehensive redox/electrochemistry chapter follows. We defer oxidation numbers and redox equations to this point, because early introduction of oxidation numbers can be done only by rote memorization. Introduction of these concepts in the context of electrochemistry allows students to appreciate why the material is important, increasing their conceptual understanding of the topic. The text is completed with chapters on metallurgy and transition metal chemistry, main group chemistry, nuclear chemistry and a combined organic chemistry and biochemistry chapter. Partial or complete coverage of this material will be appropriate for individual courses.

Overall, the design of the text enables students with different backgrounds and different methods of learning to master the mixture of material that constitutes a general chemistry course. Importantly, they will also leave the course with an appreciation for chemistry—its principles and its practices.

Supporting Materials

Printed Test Bank/Computerized Test Bank by Reger, Goode and Mercer. Computerized Test Bank is available for IBM $5\frac{1}{4}''$, IBM $3\frac{1}{2}''$ and MacIntosh.

Instructor's Manual by Reger, Goode, Mercer, and Gabrielli offers suggestions for organization of the course and provides carefully worked-out solutions to the end-of-chapter exercises *that are not* numbered in blue in the text.

Student Solutions Manual by John DeKorte of Northern Arizona University. With an emphasis on accuracy and clarity, this meticulously prepared manual presents fully worked-out solutions to the end-of-chapter exercises with exercise numbers printed in blue. Informative and helpful, the manual refers students to any pertinent text, tables, and art in the book that would enhance understanding of the problem to be solved, and, where appropriate, also briefly notes information to clarify the problem-solving.

Study Guide by Geoffrey Davies and Edward Witten of Northeastern University. Developed to complement the approach of the textbook, the Study Guide is an interactive way for the student to review objectives by section, terminology of the chapter, and the math used in the chapter. Opening with a

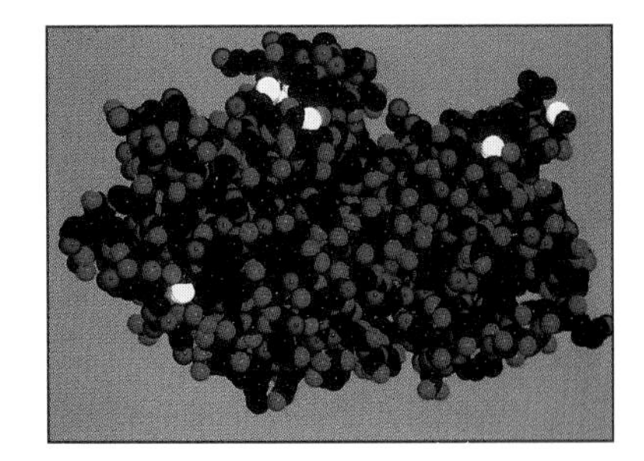

Self Test and closing with a Chapter Test, each chapter of the Study Guide gives the student ample opportunity to practice taking examinations. Plenty of Practice Exercises are provided for problem-solving mastery. Answers to the Self Test, Chapter Test and Practice Exercises are given at the end of each Study Guide chapter.

Experimentation and Analysis in the Chemistry Laboratory by Daniel L. Reger, Eugene R. Weiner (University of Denver) and William R. Gilkerson (University of South Carolina). Classic experiments are made modern with a clean visual presentation and meaningful graphics and safety icons. Emphasizes clear, precise instructions and safety in the laboratory. Forty experiments. Instructor's Manual available.

Lecture Outline by Ronald Ragsdale (University of Utah). Helps the student organize the material in the textbook. In addition, a **Problems Book** by Professor Ragsdale contains over 1200 multiple choice questions to give students additional practice with problem solving and test taking.

Color Overhead Transparencies. A collection of more than 100 full-color images from the book is available to the adopter.

Saunders Chemistry Update Newsletter Published quarterly, the Newsletter features high-interest chemistry issues and research based on readings from various academic and industry publications. Each of the five to ten brief articles is keyed to specific locations within Reger, Goode and Mercer: *Chemistry: Principles and Practice,* exposing students to current, real chemistry in relation to the principles covered in the textbook.

Computer Programs and Video Materials Because our offerings in technology—software, video and videodisc—for the general chemistry course are constantly being improved and updated, we ask that you contact your local Saunders College Publishing representative for the latest information, or call the friendly folks in the Saunders/Harcourt Brace Jovanovich Customer Relations Department. Dial 1-800-776-2606.

Shakhashiri Videotapes by Bassam Shakhashiri (University of Wisconsin, Madison). Fifty commonly performed, 3- to 5-minute classroom experiments.

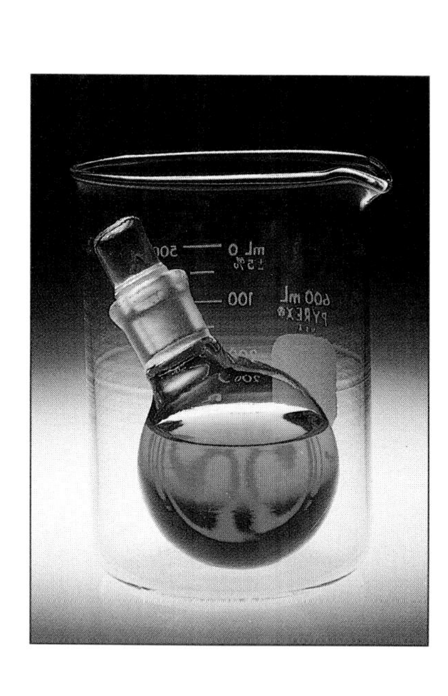

Saunders Chemistry Videodisc and Remote Control/Barcode Guide Contains all of the Shakhashiri demonstrations and over 600 images from various Saunders sources. Remote Control/Barcode Guide allows for convenient movement through the disc.

Periodic Table Videodisc and Remote Control/Barcode Guide by Alton J. Banks (Southwest Texas State University). Also published by JCE: Software. A visual compilation of information about chemical elements, their properties, their uses, and their reactions with air, water, acids, and bases. This videodisc is especially useful for demonstrating chemical reactions in a large lecture room.

KC? Discoverer Software A computer database explores 48 different properties of the elements. Includes numerical and nonnumerical information, finds all the elements in a specified range of values, graphs any numeric property against any other, lists in order of atomic number up to 4 properties of the elements, and sorts elements in increasing alphabetic or numeric order. For IBM or MacIntosh. Works alone or can interact with the Periodic Table Videodisc.

Acknowledgments

A book is very much a team project. We are truly indebted to our colleagues and our reviewers, who have patiently explained chemistry, worked problems, provided their best examples, discussed strategies, and looked for errors. Among the many people who helped were Joe Emily, Scott Mason, and Chris Thomas. Mike McClure and Barton Hawkins performed many hours of research to help bring life to the project in the form of unique *Insights into Chemistry*.

People outside the team of authors who taught part or all of the material in the text included Dave Garza, Steve Morgan and Regis Goode. These people provided a great number of insights and suggestions. We particularly thank our classes of general chemistry students, especially the 1991–92 classes, who helped us test the material in draft form and provided valuable information. Students are harsh critics.

In the course of writing the text, several of our colleagues agreed to more than just review the book, but to check the accuracy of numerical values and the viability of the chemistry, and even to coauthor ancillary material. Especially instrumental were John DeKorte and Alan Gabrielli.

Our photographic team, Larry Cameron and Bob Philp, brought a wonderful sense of design, photography, and chemistry to our book. George Kelvin, our immensely skilled and literate scientific illustrator, improved the presentation with a unique ability to simplify art and distill drawings to their essence. Pat Chernovicz helped provide unique computer-generated art.

Our publisher, John Vondeling, and his experience in the field, were crucial in shifting the equilibrium position of the project towards completion. The team at Saunders College Publishing, particularly Dena Digilio-Betz, Jay Freedman, Nanette Kauffman, Sally Kusch, and Christine Schueler, were not only helpful and competent, they provided support, guidance, and therapy, as needed.

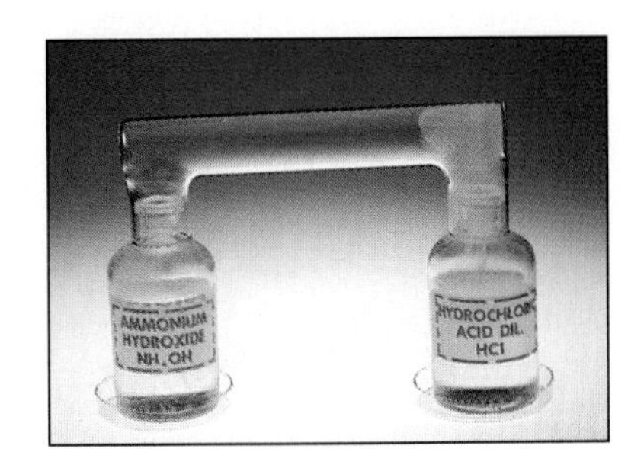

Reviewers

We would also like to acknowledge the reviewers of the book. They provided knowledge, insight, and plain common sense to help guide us during a sometimes arduous development path.

Toby Block, *Georgia Institute of Technology*
Robert S. Bly, *University of South Carolina*
Lawrence Brown, *Appalachian State University*
Juliette Bryson, *Chabot College*
Allan Colter, *University of Guelph*
Ernest Davidson, *Indiana University, Bloomington*
Geoffrey Davies, *Northeastern University*
John DeKorte, *Northern Arizona University*
Grover Everett, *University of Kansas, Lawrence*
David Garza, *University of South Carolina*
Michael Golde, *University of Pittsburgh*
Frank Gomba, *United States Naval Academy*
Robert Gordon, *Queen's University*
Henry Heikkinen, *University of Northern Colorado*

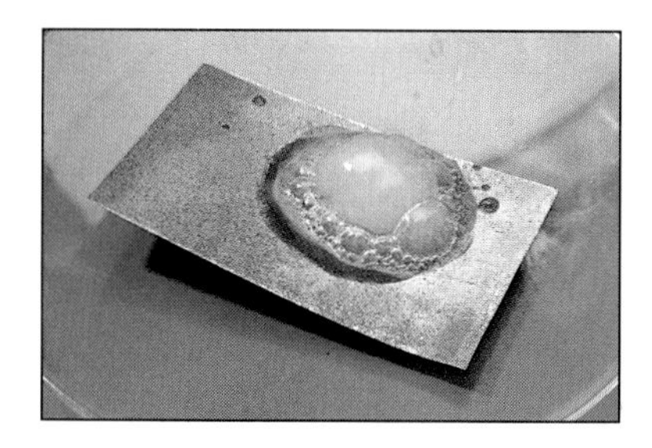

James Holler, *University of Kentucky*
Thomas Huang, *Eastern Tennessee University*
Colin Hubbard, *University of New Hamsphire*
Wilbert Hutton, *Iowa State University*
Philip Lamprey, *University of Massachusetts, Lowell*
Bruce Mattson, *Creighton University*
Hector McDonald, *University of Missouri, Rolla*
Jack McKenna, *St. Cloud State University*
Jennifer Merlic, *Santa Monica College*
Stephen L. Morgan, *University of South Carolina*
Gardiner Myers, *University of Florida*
George Pfeffer, *University of Nebraska*
Robert H. Philp, Jr., *University of South Carolina*
David Pringle, *University of Northern Colorado*
J.M. Prokipcak, *University of Guelph*
Ronald Ragsdale, *University of Utah*
Robert Richman, *Mt. St. Mary's College*
Eugene Rochow, *Harvard University*
Dennis Rushforth, *University of Texas, San Antonio*
James Sodetz, *University of South Carolina*
Helen Stone, *Ben L. Smith High School*
Ronald Strange, *Fairleigh Dickinson University*
Raymond Trautman, *San Francisco State University*
Eugene R. Weiner, *University of Denver*
Edward Wong, *University of New Hampshire*

Daniel L. Reger
Scott R. Goode
Edward E. Mercer
December 1992

Features

A number of the features in this book have been designed to make the study of chemistry a logical and enjoyable process.

Chapter Four

Stoichiometry II: Chemical Reactions in Solution and Thermochemistry

For a chemical reaction to occur, individual molecules must collide. For this reason, chemical reactions are usually carried out in solution because molecules in solution are in constant motion and readily collide. Since reaction stoichiometry and yields are based on the *number of moles* of reactants in the solution and not the volume of the solution, we need a method to calculate the number of moles of a substance in a particular volume of solution.

4.1 Solutions and Molarity

Objectives

- To explain the relationship between amount of solute, volume of solution, and concentration of solution expressed as molarity
- To describe how to prepare solutions of known molarity from weighed samples or from concentrated solutions

A common type of solution is made by dissolving a solid in a liquid, frequently water. The liquid is the **solvent,** the component that has the same

◀ Most chemical reactions are carried out in solution.

Introduction of Topics

Each chapter begins with a brief introduction to the material, and each section starts with a series of learning objectives (these should be read "The student should be able to. . .") that are formulated to help the student focus on the principles or concepts that need to be mastered. A clear, practical explanation of *why* the topic of the section is important generally follows the learning objectives.

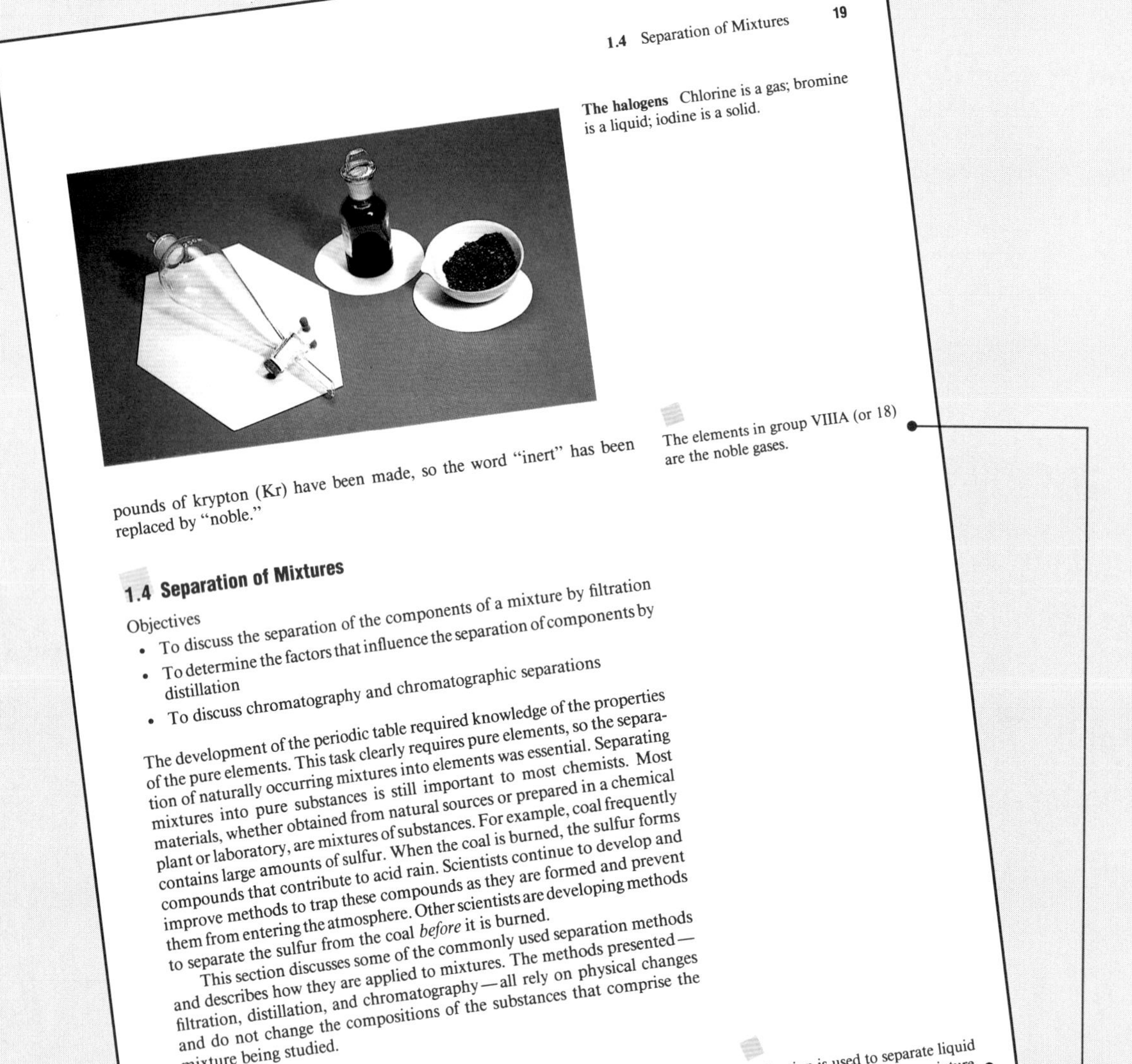

1.4 Separation of Mixtures 19

The halogens Chlorine is a gas; bromine is a liquid; iodine is a solid.

pounds of krypton (Kr) have been made, so the word "inert" has been replaced by "noble."

The elements in group VIIIA (or 18) are the noble gases.

1.4 Separation of Mixtures

Objectives

- To discuss the separation of the components of a mixture by filtration
- To determine the factors that influence the separation of components by distillation
- To discuss chromatography and chromatographic separations

The development of the periodic table required knowledge of the properties of the pure elements. This task clearly requires pure elements, so the separation of naturally occurring mixtures into elements was essential. Separating mixtures into pure substances is still important to most chemists. Most materials, whether obtained from natural sources or prepared in a chemical plant or laboratory, are mixtures of substances. For example, coal frequently contains large amounts of sulfur. When the coal is burned, the sulfur forms compounds that contribute to acid rain. Scientists continue to develop and improve methods to trap these compounds as they are formed and prevent them from entering the atmosphere. Other scientists are developing methods to separate the sulfur from the coal *before* it is burned.

This section discusses some of the commonly used separation methods and describes how they are applied to mixtures. The methods presented—filtration, distillation, and chromatography—all rely on physical changes and do not change the compositions of the substances that comprise the mixture being studied.

Filtration

A mixture that consists of part solid and part liquid is usually easy to separate into these components. We can perform a **filtration,** separating the solid

A filtration is used to separate liquid from solid components in a mixture.

Margin Notes

Margin notes emphasize and reinforce concepts. As each objective of a section poses a problem or topic, its margin notes often pose a one-sentence solution or answer. Each margin note is carefully chosen to emphasize and/or summarize an important concept. Margin notes have a narrow purpose and are designed to help the student focus on the key concept without detracting from the information flow of the text.

Example Problems

Example problems are numerous and worked in a straightforward and logical fashion. The strategy is presented, the problem solved, and warnings of potential errors and pitfalls provided. Each worked-example problem is followed by a similar "Understanding" problem and answer so that the students can test their comprehension of the topic. Practical descriptive chemistry is incorporated into the example problems.

Example 17.5 Titration Using $KMnO_4$

A 1.225 g sample of iron ore is dissolved in acid and treated to convert all of the metal to iron(II). The titration of this sample in acid solution with 0.0180 *M* $KMnO_4$ solution requires 45.30 mL to reach the endpoint. What is the mass percentage of iron in the sample of ore?

Solution In order to find the mass percentage of iron in the sample, we must solve a reaction stoichiometry problem to find the mass of iron. The strategy we will employ is shown in the diagram.

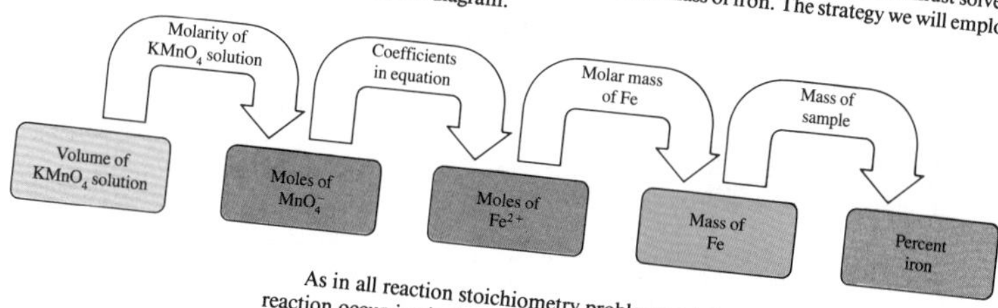

As in all reaction stoichiometry problems, a balanced equation is needed. The reaction occurring in this example is described by the equation

$$5Fe^{2+} + MnO_4^- + 8H^+ \longrightarrow 5Fe^{3+} + Mn^{2+} + 4H_2O$$

The first step is the calculation of the number of moles of the known reactant, in this case the permanganate ion.

$$\text{moles } MnO_4^- = 45.30 \text{ mL } KMnO_4 \left(\frac{1 \text{ L}}{1000 \text{ mL}}\right)\left(\frac{0.0180 \text{ mol}}{1 \text{ L}}\right)$$
$$= 8.15 \times 10^{-4} \text{ mol } MnO_4^-$$

The reaction stoichiometry is used to find the number of moles of iron. Since 5 mol Fe^{2+} reacts with 1 mol $KMnO_4$,

$$\text{moles of Fe} = 8.15 \times 10^{-4} \text{ mol } MnO_4^- \left(\frac{5 \text{ mol Fe}}{1 \text{ mol } MnO_4^-}\right) = 4.08 \times 10^{-3} \text{ mol Fe}$$

The quantity of iron must be expressed in grams for calculating the mass percentage in the sample, so we multiply by the molar mass of iron:

$$\text{mass of Fe} = 4.08 \times 10^{-3} \text{ mol Fe} \left(\frac{55.85 \text{ g}}{1 \text{ mol}}\right) = 0.228 \text{ g Fe}$$

The percentage of iron in the sample is calculated from the mass of iron, as determined in the titration, and the mass of the sample.

$$\text{percentage of Fe} = \frac{0.228 \text{ g Fe}}{1.225 \text{ g sample}} \times 100\% = 18.6\% \text{ Fe}$$

Understanding A 20.00 mL sample of an iron(II) solution requires 24.30 mL of a 0.0192 *M* $KMnO_4$ solution to react completely. What is the molar concentration of iron(II) in the sample solution?

Answer: 0.117 *M* iron(II)

Flow Diagrams: Conceptual Problem-Solving

Flow diagrams are used in many problems that involve mathematical calculations. The flow diagrams show the starting point in the problem and the operations needed to get to the solution. The flow diagrams are carefully formulated to aid the student in developing problem-solving thought processes and strategies.

Zinc reacts with hydrochloric acid to give off bubbles of hydrogen gas.

Example 5.8 Using Volumes of Gases in Equations

Hydrogen gas is frequently prepared in the laboratory by the reaction of zinc and hydrochloric acid. The other product is $ZnCl_2$. Calculate the volume of hydrogen produced at 744 torr pressure and 27 °C from the reaction of 32.2 g of zinc and 500 mL of 2.20 molar HCl.

Solution The strategy of this example is interesting because three different methods are used in calculations with the numbers of moles of the three different substances. (1) The molar mass is used to calculate the number of moles from the mass of zinc. (2) The molarity and the volume of solution are used to calculate the number of moles of HCl. (3) The ideal gas law is used to convert the number of moles of hydrogen gas into volume of hydrogen gas. As always, the chemical equation is used to relate the number of moles of one substance to moles of another.

First, write the chemical equation.

$$Zn + 2HCl \longrightarrow ZnCl_2 + H_2$$

Second, use the information given in the problem to calculate the numbers of moles of zinc and hydrochloric acid. The amounts of two reactants are given in this example, so this is a limiting-reactant problem. We need to calculate the number of moles of hydrogen gas produced by complete consumption of each reactant.

Mass of Zn → Molar mass of Zn → Moles of Zn → Coefficients in chemical equation → Moles of H_2

Volume of HCl solution → Molarity of HCl solution → Moles of HCl → Coefficients in chemical equation → Moles of H_2

Choose smaller amount → Ideal gas equation → Volume of H_2 gas

$$\text{moles Zn} = 32.2 \text{ g Zn}\left(\frac{1 \text{ mol Zn}}{65.39 \text{ g Zn}}\right) = 0.492 \text{ mol Zn}$$

$$\text{moles HCl} = 0.500 \text{ L HCl soln}\left(\frac{2.20 \text{ mol HCl}}{1 \text{ L HCl soln}}\right) = 1.10 \text{ mol HCl}$$

Use the coefficients in the equation to calculate the amount of hydrogen one could obtain from each of the reactants.

$$\text{moles } H_2 \text{ based on Zn} = 0.492 \text{ mol Zn}\left(\frac{1 \text{ mol } H_2}{1 \text{ mol Zn}}\right) = 0.492 \text{ mol } H_2$$

$$\text{moles } H_2 \text{ based on HCl} = 1.10 \text{ mol HCl}\left(\frac{1 \text{ mol } H_2}{2 \text{ mol HCl}}\right) = 0.550 \text{ mol } H_2$$

The zinc yields the smaller amount of hydrogen and is therefore the limiting reactant. The problem is completed by using the ideal gas law.

$$P = \frac{744}{760} \text{ atm} = 0.979 \text{ atm} \qquad V = ?$$

$$n = 0.492 \text{ mol } H_2 \qquad T = 300 \text{ K}$$

$$V = \frac{nRT}{P} = \frac{(0.492 \text{ mol } H_2)(0.0821 \text{ L atm/mol K})(300 \text{ K})}{0.979 \text{ atm}}$$

$$= 12.4 \text{ L } H_2$$

Understanding Many scientists believe that when the earth's atmosphere evolved, some of the oxygen gas came from the decomposition of water induced by solar radiation.

$$2H_2O \xrightarrow{\text{light}} 2H_2 + O_2$$

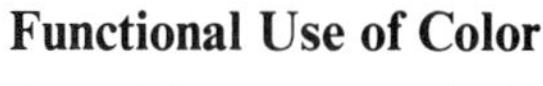

Functional Use of Color

Quantities are placed in functional color-coded boxes. Operations that are used to interconvert quantities are placed in process arrows.

- Volume, Equilibrium Constant, Rate Parameters
- Moles, Chemical Equations, Empirical Formulas
- Concentration, Density, Pressure, % Yield, Number of Atoms, pH
- Thermodynamic quantities—Enthalpy, Entropy, Free Energy
- Mass
- Operations

Insights Into Chemistry

Insights into Chemistry are written and graphically designed to intrigue the intellectually curious student. Whether the topic provides historical perspective, biography, closer treatment of a subject, or an application, the discussion is upbeat and interesting. These *Insights* provide a break from problem solving and calculations. Newspaper style headings create excitement for chemistry.

INSIGHTS into CHEMISTRY

Science, Society, and Technology

Modern Alloys Are Widely Used. Techniques Evolved from the Ancients

Stainless steels are mostly iron, but chromium and nickel are added to make the metal resistant to rust. The concentrations of these and other alloying elements determine many of the properties of the alloy. Three common alloys are described here.

Type 301 stainless is commonly used to make cooking utensils, sinks, and commercial kitchen counters and cabinets. It is sometimes called "18/8" stainless steel, referring to the percentage of chromium and nickel in the alloys.

Type 316L is easy to weld and highly resistant to attack by chemicals, including concentrated acids such as sulfuric acid. It is widely used in the chemical industry to build reactors and other vessels.

Type 420 is a heat-treatable alloy. It is used by knifemakers, who will grind a blade, heat it in a furnace, and then quickly cool it in water or oil. The resulting blade is extremely hard, but also very brittle. (The blade might shatter if dropped on the ground at this stage.) A second heating step, called annealing, makes the blade less brittle. The knife blade is heated to about 500 °C and allowed to cool slowly to room temperature.

The following elements are added to iron to form these stainless steel alloys, in percentage by weight:

	301	316L	420
C	0.15 max	0.03 max	0.15 max
Mn	2.00 max	2.00 max	1.00 max
P	0.045 max	0.045 max	0.040 max
S	0.030 max	0.030 max	0.030 max
Si	1.00 max	1.00 max	1.00 max
Cr	16.00–18.00	16.00–18.00	12.00–14.00
Ni	6.00–8.00	10.00–14.00	
Mo		2.00–3.00	

Damascus swords were made from a steel that contained a relatively high concentration of carbon. While most carbon steels are too brittle to be used for a sword, the steelmakers of Damascus discovered that if they heated the steel to the right temperature, added carbon powder, hammered the metal, and repeated the process, they were rewarded with an extraordinarily strong and sharp knife or sword. The physical processes of heating and hammering were just as important to making Damascus steel as was adding the carbon.

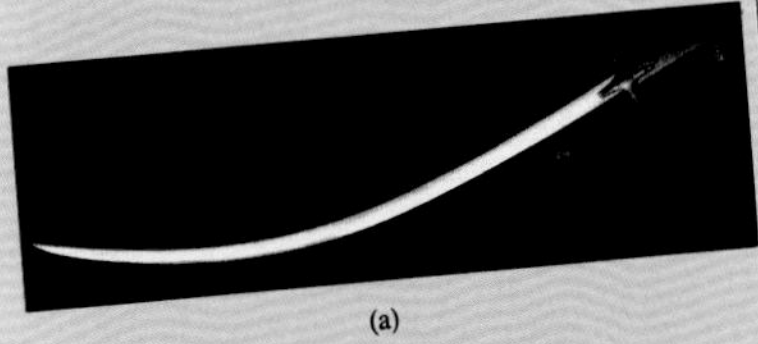

(a)

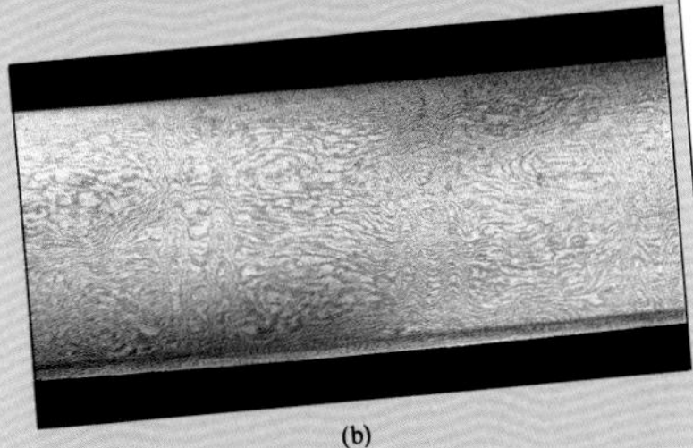

(b)

Damascus sword (a) Damascus swords were renowned for their strength and sharpness. (b) The uneven distribution of carbon produced a unique striated pattern in the steel.

neous mixtures such as liquid solutions or metal alloys. Alloys of copper and nickel can range between 0% and 100% of either metal. Two of these alloys are constantan (60% copper, 40% nickel) and coinage nickel (75% copper and 25% nickel).

Not only is the composition of any compound identical for all samples, but each of its intensive properties is also the same. The melting point is an important property that distinguishes a substance from a mixture. The temperature of a pure substance remains constant as it melts; most mixtures melt over a range of temperatures.

All samples of a compound have the same composition and intensive properties.

Full-Color Photography

Over 200 new, original photographs bring the chemistry alive. These photographs are carefully integrated into the text and are used to showcase descriptive chemistry, as well as to provide concrete examples of chemical principles.

9.3 Valence Bond Theory 331

Figure 9.13 Mixing paint as an analogy to the formation of hybrid orbitals The mathematical mixing of atomic orbitals to form hybrid orbitals is similar to mixing different colors of paint to obtain a new color. After the paint is mixed, there are still the same number of cups of paint, but the color has changed.

pictorially (Figure 9.14). The *sign* of the wave function's amplitude in one lobe of a *p* orbital is opposite to the sign in the other lobe (indicated by shading in the figure). When the *p* orbital is added to the *s* orbital, there is reinforcement of the wave amplitude on one side of the nucleus and cancellation on the other, forming an *sp* hybrid orbital that has one large lobe and

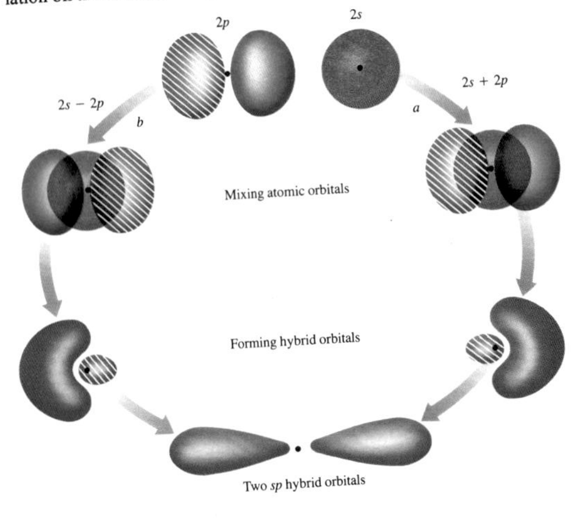

Figure 9.14 *sp* Hybrid orbital The *sp* hybrid orbitals are formed from the addition, *a*, and subtraction, *b*, of one *s* and one *p* atomic orbital. In *a*, the amplitudes of the wave functions reinforce on the right side of the nucleus and in *b* on the left side. The bottom drawing shows the two new hybrid orbitals about the nucleus, with the shape of the orbitals elongated and the smaller lobes omitted for clarity.

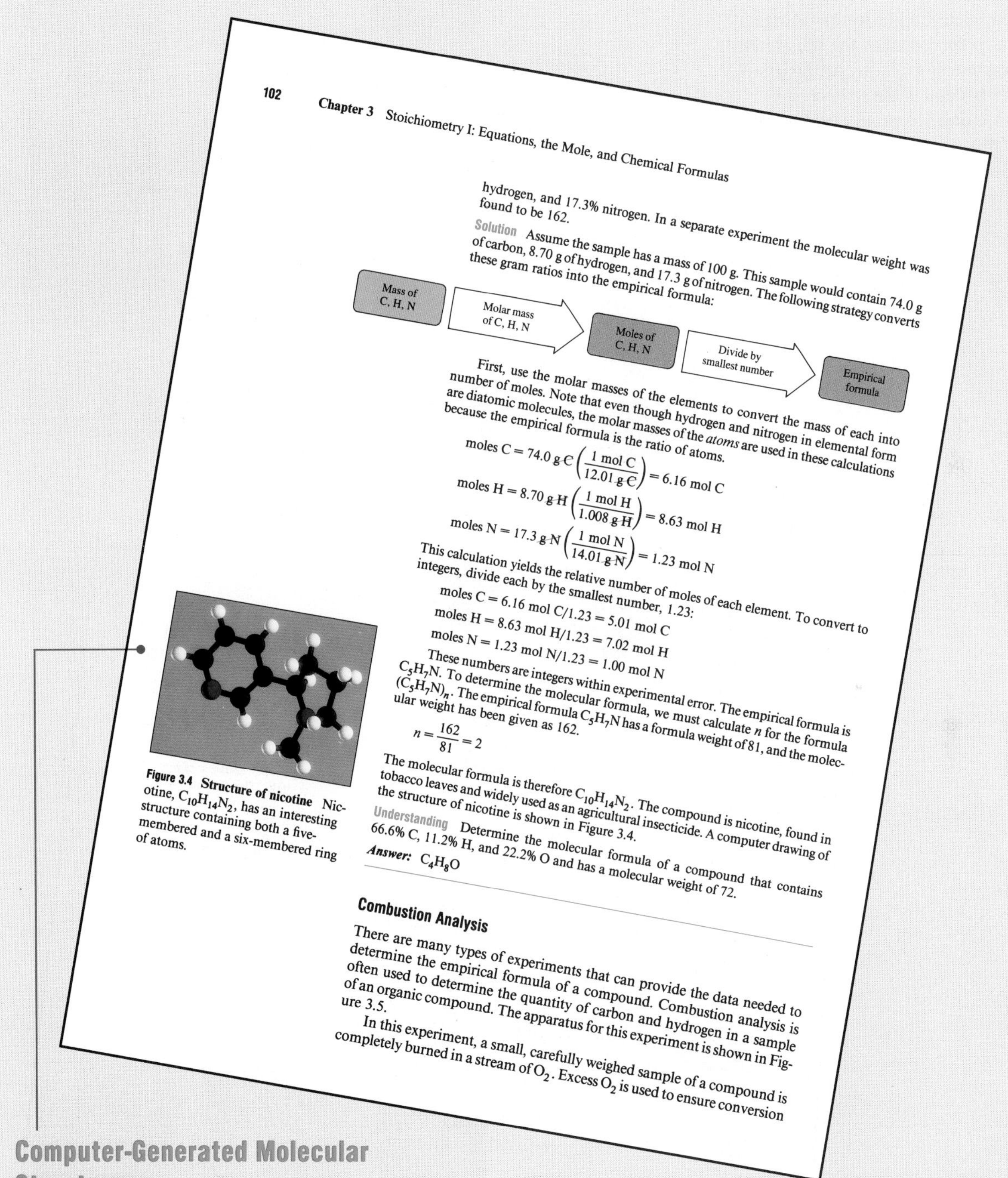

102 **Chapter 3** Stoichiometry I: Equations, the Mole, and Chemical Formulas

hydrogen, and 17.3% nitrogen. In a separate experiment the molecular weight was found to be 162.

Solution Assume the sample has a mass of 100 g. This sample would contain 74.0 g of carbon, 8.70 g of hydrogen, and 17.3 g of nitrogen. The following strategy converts these gram ratios into the empirical formula:

First, use the molar masses of the elements to convert the mass of each into number of moles. Note that even though hydrogen and nitrogen in elemental form are diatomic molecules, the molar masses of the *atoms* are used in these calculations because the empirical formula is the ratio of atoms.

$$\text{moles C} = 74.0\ \text{g C}\left(\frac{1\ \text{mol C}}{12.01\ \text{g C}}\right) = 6.16\ \text{mol C}$$

$$\text{moles H} = 8.70\ \text{g H}\left(\frac{1\ \text{mol H}}{1.008\ \text{g H}}\right) = 8.63\ \text{mol H}$$

$$\text{moles N} = 17.3\ \text{g N}\left(\frac{1\ \text{mol N}}{14.01\ \text{g N}}\right) = 1.23\ \text{mol N}$$

This calculation yields the relative number of moles of each element. To convert to integers, divide each by the smallest number, 1.23:

$$\text{moles C} = 6.16\ \text{mol C}/1.23 = 5.01\ \text{mol C}$$

$$\text{moles H} = 8.63\ \text{mol H}/1.23 = 7.02\ \text{mol H}$$

$$\text{moles N} = 1.23\ \text{mol N}/1.23 = 1.00\ \text{mol N}$$

These numbers are integers within experimental error. The empirical formula is C_5H_7N. To determine the molecular formula, we must calculate n for the formula $(C_5H_7N)_n$. The empirical formula C_5H_7N has a formula weight of 81, and the molecular weight has been given as 162.

$$n = \frac{162}{81} = 2$$

The molecular formula is therefore $C_{10}H_{14}N_2$. The compound is nicotine, found in tobacco leaves and widely used as an agricultural insecticide. A computer drawing of the structure of nicotine is shown in Figure 3.4.

Understanding Determine the molecular formula of a compound that contains 66.6% C, 11.2% H, and 22.2% O and has a molecular weight of 72.

Answer: C_4H_8O

Figure 3.4 Structure of nicotine Nicotine, $C_{10}H_{14}N_2$, has an interesting structure containing both a five-membered and a six-membered ring of atoms.

Combustion Analysis

There are many types of experiments that can provide the data needed to determine the empirical formula of a compound. Combustion analysis is often used to determine the quantity of carbon and hydrogen in a sample of an organic compound. The apparatus for this experiment is shown in Figure 3.5.

In this experiment, a small, carefully weighed sample of a compound is completely burned in a stream of O_2. Excess O_2 is used to ensure conversion

Computer-Generated Molecular Structures

Sophisticated, mainframe chemistry software and modern high-resolution imaging are used to produce many of the more complex molecular structures in the text. The result is a truer representation of structures at the molecular level.

Illustrations

Renowned scientific illustrator George Kelvin has prepared all of the illustrative art, leading to a stimulating art program that skillfully depicts the science. All drawings are done in proper geometric perspective.

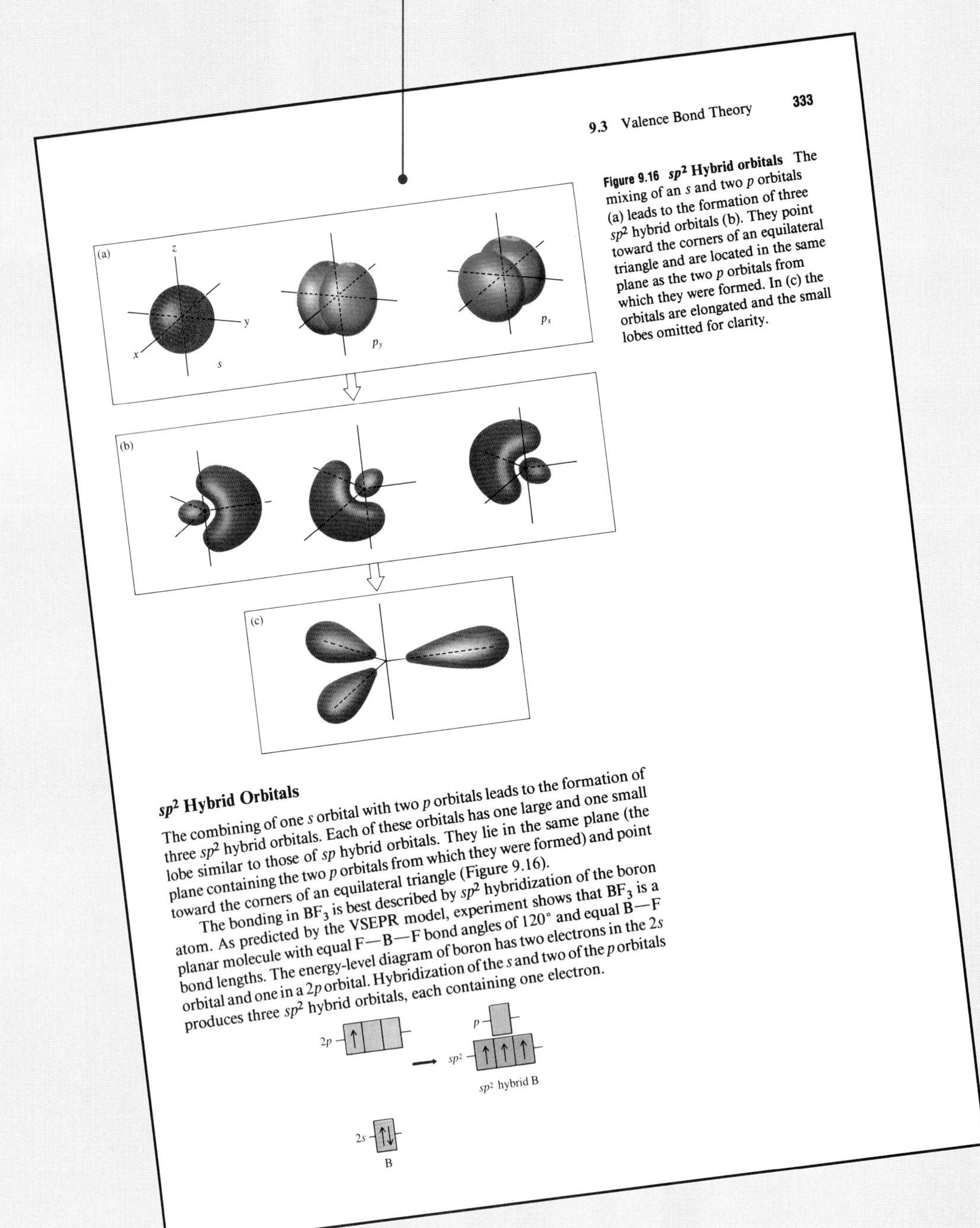

Figure 9.16 ***sp*² Hybrid orbitals** The mixing of an *s* and two *p* orbitals (a) leads to the formation of three sp^2 hybrid orbitals (b). They point toward the corners of an equilateral triangle and are located in the same plane as the two *p* orbitals from which they were formed. In (c) the orbitals are elongated and the small lobes omitted for clarity.

sp^2 Hybrid Orbitals

The combining of one *s* orbital with two *p* orbitals leads to the formation of three sp^2 hybrid orbitals. Each of these orbitals has one large and one small lobe similar to those of *sp* hybrid orbitals. They lie in the same plane (the plane containing the two *p* orbitals from which they were formed) and point toward the corners of an equilateral triangle (Figure 9.16).

The bonding in BF_3 is best described by sp^2 hybridization of the boron atom. As predicted by the VSEPR model, experiment shows that BF_3 is a planar molecule with equal F—B—F bond angles of 120° and equal B—F bond lengths. The energy-level diagram of boron has two electrons in the 2*s* orbital and one in a 2*p* orbital. Hybridization of the *s* and two of the *p* orbitals produces three sp^2 hybrid orbitals, each containing one electron.

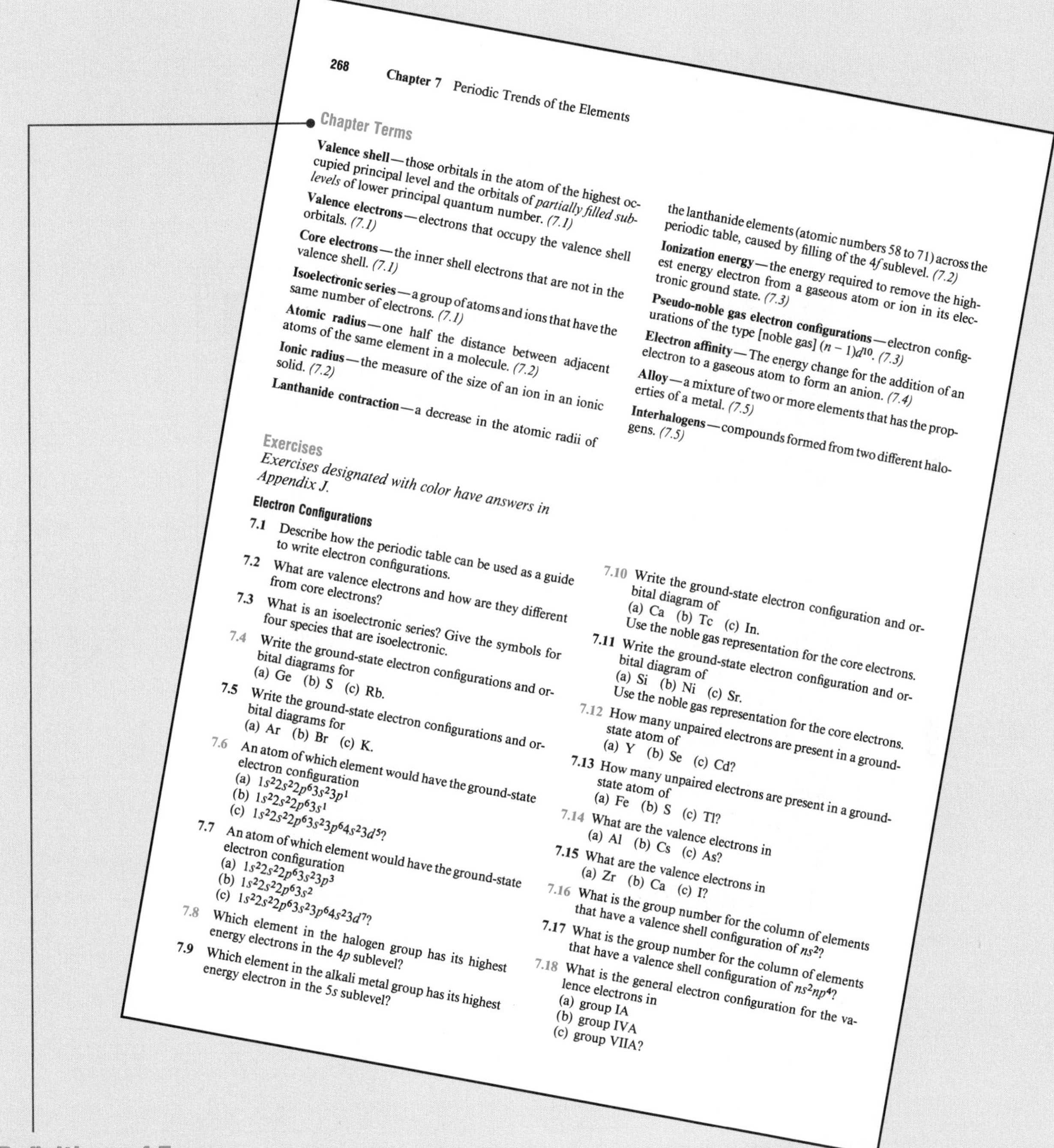

268 Chapter 7 Periodic Trends of the Elements

Chapter Terms

Valence shell—those orbitals in the atom of the highest occupied principal level and the orbitals of *partially filled sublevels* of lower principal quantum number. *(7.1)*

Valence electrons—electrons that occupy the valence shell orbitals. *(7.1)*

Core electrons—the inner shell electrons that are not in the valence shell. *(7.1)*

Isoelectronic series—a group of atoms and ions that have the same number of electrons. *(7.1)*

Atomic radius—one half the distance between adjacent atoms of the same element in a molecule. *(7.2)*

Ionic radius—the measure of the size of an ion in an ionic solid. *(7.2)*

Lanthanide contraction—a decrease in the atomic radii of the lanthanide elements (atomic numbers 58 to 71) across the periodic table, caused by filling of the 4*f* sublevel. *(7.2)*

Ionization energy—the energy required to remove the highest energy electron from a gaseous atom or ion in its electronic ground state. *(7.3)*

Pseudo-noble gas electron configurations—electron configurations of the type [noble gas] $(n-1)d^{10}$. *(7.3)*

Electron affinity—The energy change for the addition of an electron to a gaseous atom to form an anion. *(7.4)*

Alloy—a mixture of two or more elements that has the properties of a metal. *(7.5)*

Interhalogens—compounds formed from two different halogens. *(7.5)*

Exercises

Exercises designated with color have answers in Appendix J.

Electron Configurations

7.1 Describe how the periodic table can be used as a guide to write electron configurations.

7.2 What are valence electrons and how are they different from core electrons?

7.3 What is an isoelectronic series? Give the symbols for four species that are isoelectronic.

7.4 Write the ground-state electron configurations and orbital diagrams for
(a) Ge (b) S (c) Rb.

7.5 Write the ground-state electron configurations and orbital diagrams for
(a) Ar (b) Br (c) K.

7.6 An atom of which element would have the ground-state electron configuration
(a) $1s^22s^22p^63s^23p^1$
(b) $1s^22s^22p^63s^1$
(c) $1s^22s^22p^63s^23p^64s^23d^5$?

7.7 An atom of which element would have the ground-state electron configuration
(a) $1s^22s^22p^63s^23p^3$
(b) $1s^22s^22p^63s^2$
(c) $1s^22s^22p^63s^23p^64s^23d^7$?

7.8 Which element in the halogen group has its highest energy electrons in the 4*p* sublevel?

7.9 Which element in the alkali metal group has its highest energy electron in the 5*s* sublevel?

7.10 Write the ground-state electron configuration and orbital diagram of
(a) Ca (b) Tc (c) In.
Use the noble gas representation for the core electrons.

7.11 Write the ground-state electron configuration and orbital diagram of
(a) Si (b) Ni (c) Sr.
Use the noble gas representation for the core electrons.

7.12 How many unpaired electrons are present in a ground-state atom of
(a) Y (b) Se (c) Cd?

7.13 How many unpaired electrons are present in a ground-state atom of
(a) Fe (b) S (c) Tl?

7.14 What are the valence electrons in
(a) Al (b) Cs (c) As?

7.15 What are the valence electrons in
(a) Zr (b) Ca (c) I?

7.16 What is the group number for the column of elements that have a valence shell configuration of ns^2?

7.17 What is the group number for the column of elements that have a valence shell configuration of ns^2np^4?

7.18 What is the general electron configuration for the valence electrons in
(a) group IA
(b) group IVA
(c) group VIIA?

Definitions of Terms

Each new term is clearly indicated in bold type and defined when first introduced in the text. The terms and definitions are listed at the end of each chapter; in addition, an alphabetical glossary with definitions and section references is given in Appendix I.

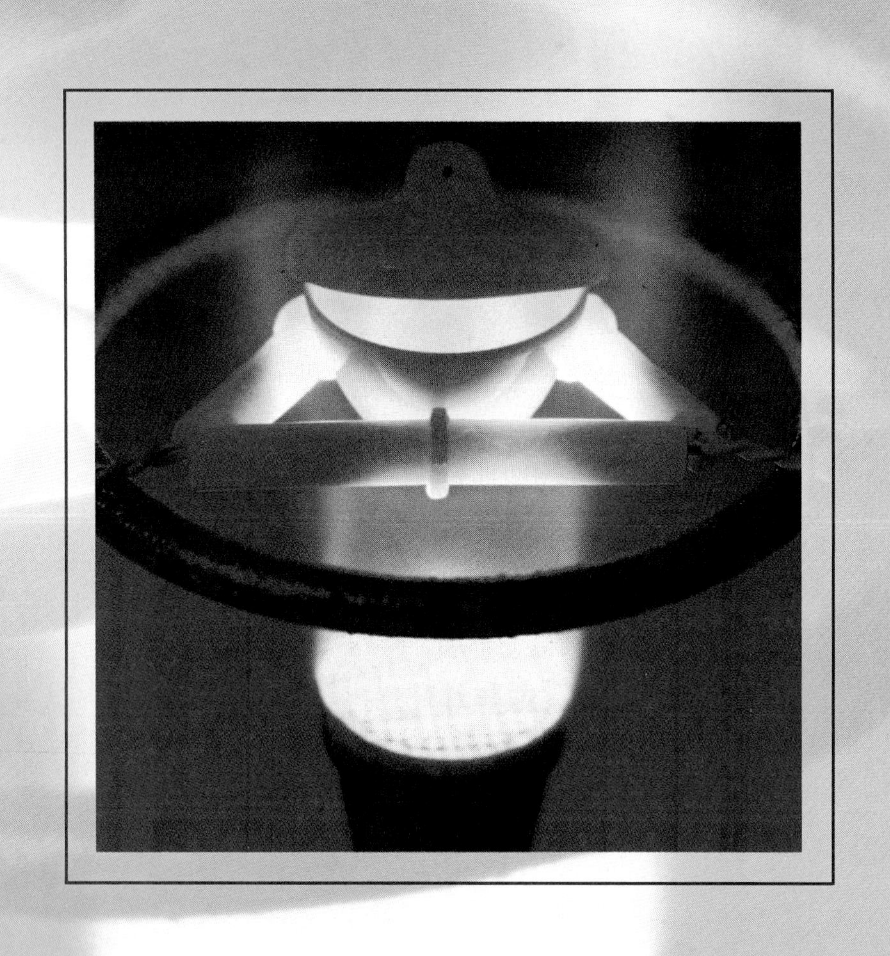

Chapter One

Introduction to Chemistry

Chemistry is the study of matter and its interactions with other matter and with energy. Everything that we see, touch, and feel is matter. Chemistry is used not only by scientists, but by all people, since it describes everyday occurrences as well as those in a test tube. Any description of chemistry, however, is just words, and does not convey the wide variety of projects that chemists work on, the urgency of many chemical problems, and the excitement of the search for solutions.

Chemists investigate many different aspects of their science as they try to answer questions such as:

Why does bread smell so good when it is baking?
How can acid rain be neutralized?
What, if anything, do the chemicals that slow the growth of cancer have in common?
Can we manufacture a theater curtain that will not burn even when exposed to an open flame?

Many of the advantages enjoyed by our society originate in chemistry. Modern advances like wrinkle-free clothing have freed us from drudgery and allowed us the time for other pursuits, including leisure activities. The production of a modern worker in a 40-hour week has never been greater. As society has exploited chemistry, however, serious and complex problems have developed. Pollution, waste management, and environmental cancer are problems for which solutions must be found. The solutions will come from a coalition of scientists, including chemists, life scientists, social scientists, and others. Chemistry is very much an ongoing, living science, not

Chemistry is based on results from laboratory experiments.

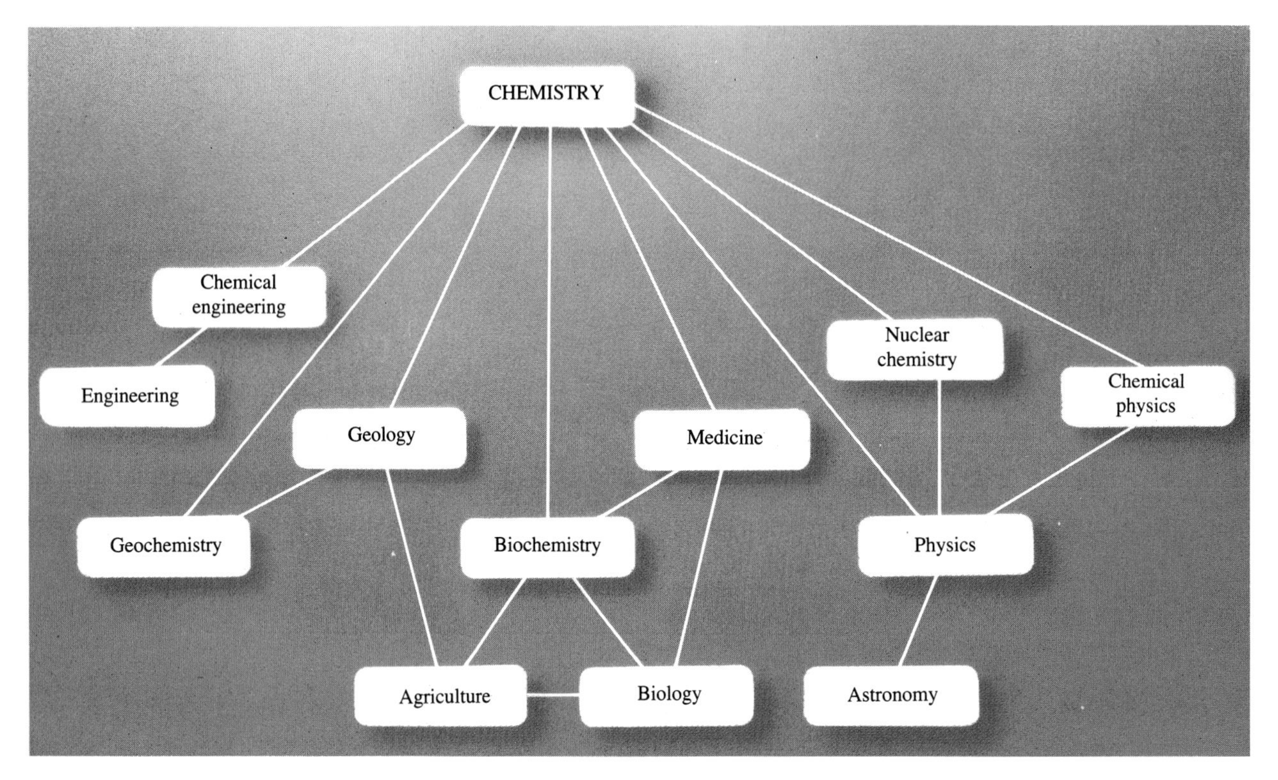

Figure 1.1 Chemistry and the natural sciences.

static material mastered by long-dead scholars. New advances and discoveries are made daily.

Chemistry is at the core of our scientific knowledge. All of the natural sciences (physical science, life science, and earth science) explore different relationships between materials, their interactions with each other, and their interactions with energy. The relationships between chemistry and other natural sciences are illustrated in Figure 1.1.

We feel the impact of chemistry each and every day of our lives. It is difficult to name more than a handful of issues that affect society without noticing the presence and the central role of chemistry. The need for abundant pure water, the use of petroleum, the fight against disease, and proposed trips to the stars all involve chemistry. Much of the subject of chemistry has been well studied, and the results of these studies have led to important knowledge about the universe in which we live and products that improve the way we live. Other facets of chemistry are unknown or not completely understood. Chemistry is a living, evolving, experimental science.

1.1 The Nature of Chemistry

Objectives

- To define the science of chemistry
- To identify thc proccsses in the scientific method of investigation

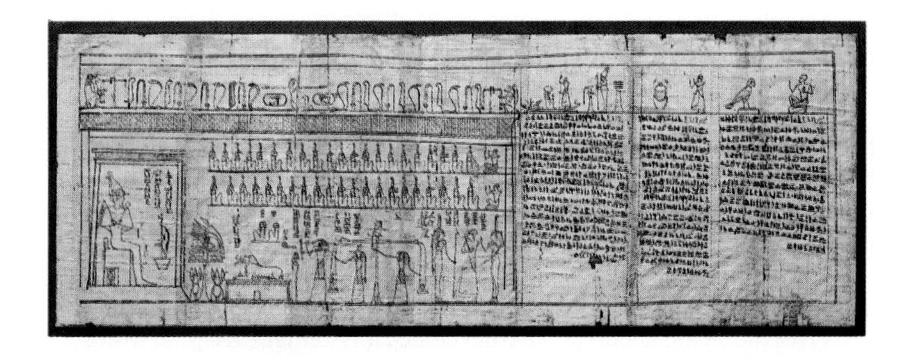

Ancient Egyptian record of chemical reaction

The study of matter and its interactions is called **chemistry.** Chemistry is an experimental science. The knowledge of chemistry comes from centuries of careful observations. Archaeologists have discovered records left by chemists employed by the pharaohs of Egypt; these records detail their observations of the properties of materials and of the reactions that transform them into other materials. All chemists make observations, suggest some basic principles to explain their results, and then use these principles to predict the results of other experiments, perhaps reactions that have not yet been observed. If the predictions are correct, the principles are tested with more experiments. If the predictions fail, though, the principles are modified to include the new results.

There are some very practical reasons to learn the principles of chemistry. There are literally millions of chemicals, and they can undergo billions of reactions. Rather than memorize each of these reactions, it is best to try to develop a few models that predict them. We use the term model to describe both a mathematical relationship ("the volume increases as temperature increases . . .") and a qualitative picture that helps to predict the outcome of an experiment. There are good models that predict the behavior of gases, for example. In contrast, we are just beginning to identify the chemicals and processes associated with the changes that occur as people grow old, and the models that describe aging are quite tentative.

Chemistry is a laboratory science that is based on observations of changes that occur when different materials are mixed.

The Scientific Method

Advances in chemistry require both experimental data and theoretical explanations. One cannot advance without the other. Chemists have limited time and resources, and they need guidance to choose which of the many possible experiments are likely to yield the most information. Experimentation that is guided by theory and past experiments has a name—the **scientific method.** Many different experimental approaches qualify—there is not just one "scientific" method.

The word "science" has an interesting origin. The ancient Romans thought that learning could be divided into two disciplines. The first was art (from the Latin root *ar*, meaning "to join"). Art was a way of arranging or joining details to achieve a satisfying result. The second discipline is science, derived from the Latin *scientia,* meaning "knowledge." Science is guided by knowledge.

There is no single "scientific" method; the term refers to experiments that are guided by knowledge.

Scientists first draw on a large body of experience in formulating the ideas for their experiments. They use both experimental data and theory for direction. A chemist who is trying to design a drug to fight cancer will first review the results that have been published in the scientific literature. Perhaps one drug was effective in preventing the cancer from spreading, but also had dangerous side effects when used on humans. The chemist might try modifying the drug to eliminate its toxicity without changing its effectiveness, perhaps first using computer programs that relate a chemical structure to its properties. Improving, modifying, and extending our knowledge are all components of scientific investigations.

Over the years, scientists have developed a systematic usage of words, or nomenclature, to describe their investigations. The word **hypothesis** is used to describe a possible explanation for an event even if the explanation is an untested assumption. If we can summarize a large number of observations with a statement (or equation), we call the statement a **law.** The French chemist Jacques Charles performed careful measurements of the volume of a gas as it was heated or cooled. He explained his observations with what is now called Charles's law: the volume of a gas changes in proportion to the temperature. (Charles's law is discussed in detail in Chapter 5.)

A law summarizes the observations, but it provides no explanation. Scientists try to explain the laws of nature, and the explanation is called a **theory.** Scientists know that a gas expands as it is heated, as stated by Charles's law. But more importantly, these observations can be explained by a relatively simple theory, the kinetic theory of gases. Please note that scientists reserve the word "theory" for an explanation of the laws of nature, a very narrow usage for the word. This contrasts sharply with day-to-day usage: "I have a theory about why the basketball team lost last night. I think. . . ." Most people would define a theory as an educated hunch or guess, but a scientist carefully distinguishes between theory, hypothesis, and guess.

Again, for emphasis: remember that chemistry is first and foremost an experimental science. Theories are always subject to modification, extension, and perhaps rejection, on the basis of new discoveries. Theories are not perfect, but rather the best understanding of current knowledge. Scientists accept theories, but they are prepared to modify them as new data become available. Sometimes theory suggests that the results of an experiment were incorrect, so the experiment is repeated, usually under more carefully controlled conditions. In fact, many of the very best experiments subject current theories to vigorous tests as scientists try to develop more exact and complete theoretical descriptions of chemistry.

Chemists have chosen a career in chemistry not because of any hypotheses, laws, or theories, but because they desire to know what happens when some substances interact. They have questions about the nature of matter, and they want answers. Chemists turn to their knowledge of chemistry, its models, laws, and theories, but in the end, they recognize that chemistry is an experimental science. If experimental results disagree with theory, there are only three possible approaches to follow. We can change the experimental data, we can suspect that the experiment was not carried out correctly, or we can modify the theory. Changing data is called dishonesty; improving the experiment and modifying the theory are called science.

1.2 Matter

Objectives

- To define matter and its properties
- To identify the properties of matter as intensive or extensive and physical or chemical
- To classify a sample of matter based on its properties and composition

As we look around us, everything we see (and much of what we cannot see) is composed of matter. **Matter** is defined as anything that has mass and occupies space. The food we eat, the air we breathe, and the books we read are all samples of matter. Few subjects are more important than the study of matter and its properties.

In the definition of matter, we have used the term mass. **Mass** is a measure of the quantity of matter in an object. **Weight,** the force that results from the attraction between matter and the earth (or any other large object), is the most familiar property of matter. The weight of an object will vary from one location to another, but the mass is always the same. The weight of the famous moon rocks increased sixfold when they were brought to the earth, because of the earth's larger gravitational attraction. The mass of these rocks, however, did not change. Mass can be measured with a *balance,* such as the one shown in Figure 1.2. A balance measures the mass of an object by comparing it to objects of known mass. The balance determines the mass rather than the weight, because both the standard mass used for the comparison and the object being measured experience the same gravitational attraction. A device like a bathroom scale, however, measures the weight of an object; its reading will depend on gravitational attractions (Figure 1.3).

Figure 1.2 **Laboratory balance**

Matter has mass and occupies space.

Properties of Matter

Anything we can observe or measure about a sample of matter is called a **property**. We all strive to understand matter and its properties; whether we speak about chemicals in a beaker or the food we eat, we are still talking about matter. Our attempts to understand matter and how it behaves lead us to many different measurements of the properties of matter. As a result of observations made through the centuries, many ways of classifying properties have been developed.

One classification scheme divides properties into intensive and extensive groups. **Extensive properties** are those that depend on the size of the sample; they measure *how much* matter is in a particular sample. Mass and volume are typical extensive properties. **Intensive properties** are those that are independent of the size of the sample; these properties define *what* the sample is, and not how much of it is present. Color, physical state (solid, liquid, or gas) at a given temperature, and melting point (the temperature at which a solid is converted into a liquid) are all examples of intensive properties—none of these depend on the size of the sample. Extensive properties are used to measure the quantity of matter; intensive properties are used to compare different samples and determine whether they contain

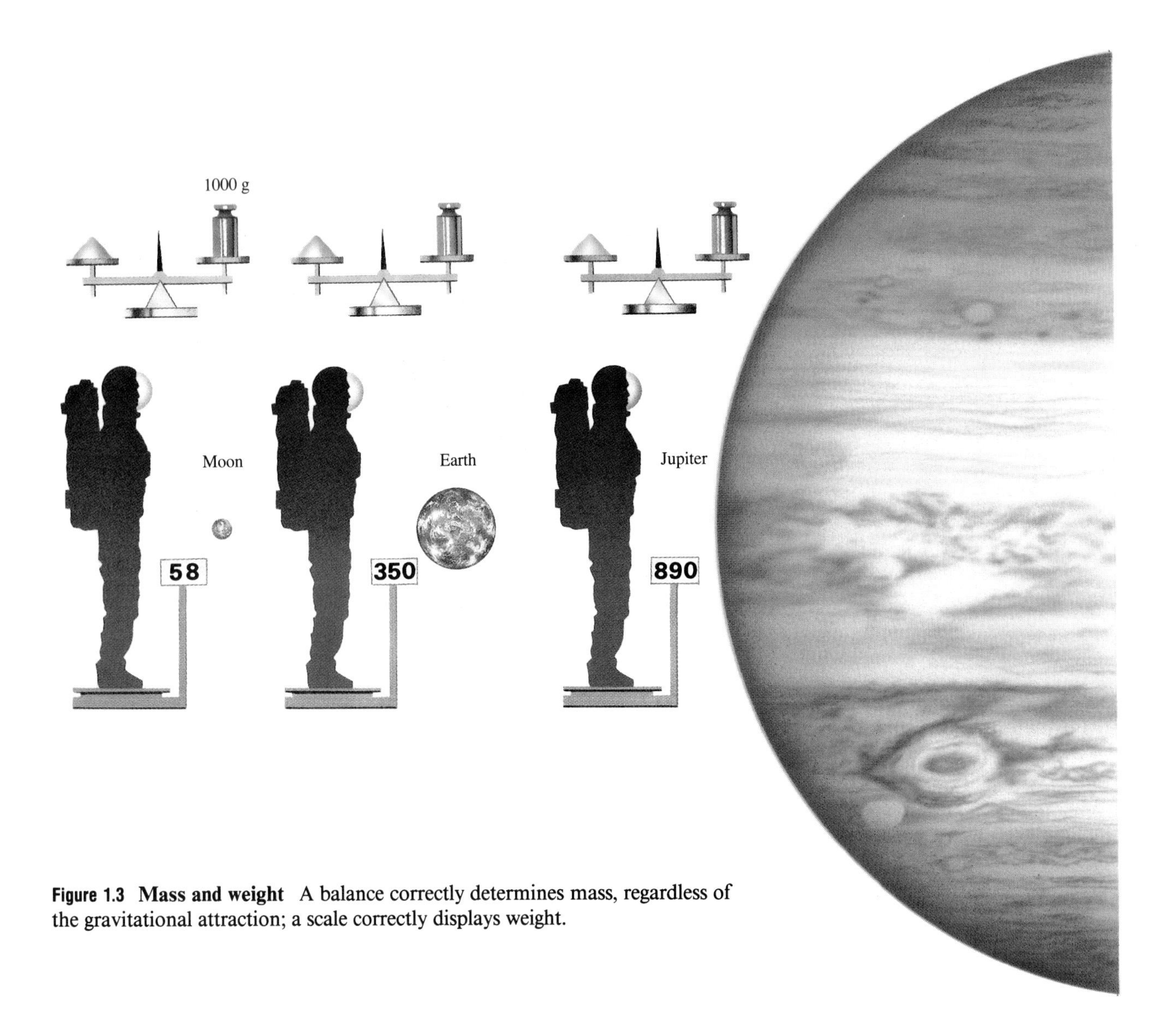

Figure 1.3 Mass and weight A balance correctly determines mass, regardless of the gravitational attraction; a scale correctly displays weight.

Extensive properties measure how much of the matter is in a particular sample. Intensive properties define the identity of the sample.

the same or different materials. If a large number of the intensive properties of two samples are identical, then it is reasonable to assume that both samples are composed of the same material. Thus, by reporting the intensive properties of particular samples of matter, chemists in all parts of the world are able to compare their results.

The examples of intensive properties given above might suggest that these properties are rather limited in number. However, it is possible to define additional intensive properties by taking the ratio of two extensive properties. The most familiar example of this is the density of matter. **Density** is defined as the mass-to-volume ratio:

$$\text{density} = \frac{\text{mass}}{\text{volume}}$$

Both the mass and the volume are measures of the quantity of matter present, but their ratio, the density, is independent of the size of the sample. A 10.0-cm^3 sample of aluminum has a mass of 27.0 g, while 20.0 cm^3 of aluminum has a mass of 54.0 g. The density for both of these samples is the same, 2.70 g/cm^3:

$$d = \frac{27.0\ \text{g}}{10.0\ \text{cm}^3} = 2.70\ \text{g/cm}^3 \qquad \frac{54.0\ \text{g}}{20.0\ \text{cm}^3} = 2.70\ \text{g/cm}^3$$

Other examples of intensive properties are solubility (typically the mass of a material that can dissolve in a specified volume of water) and specific electrical conductance.

Properties can also be classified as physical or chemical. **Physical properties** are those that can be measured without changing the kind of matter in the sample (its composition). The mass of a sample, the volume it occupies, and its color can all be observed without changing the composition of the sample; mass, volume, and color are all physical properties. **Chemical properties** describe the tendency of a material to react, forming new and different substances. Explosiveness and flammability are examples of chemical properties.

Physical properties are those that can be measured without changing the composition of the sample. Chemical properties describe the tendency of a material to react, forming new and different substances.

Changes in the properties of a substance can also be classified as physical or chemical changes. A **physical change** can be observed without changing the composition of the substance; for example, freezing is a physical change. In a **chemical change,** at least part of the substance is converted into a different kind (or kinds) of matter. The rusting of iron and the burning of wood both produce new kinds of matter with properties entirely different from those of the initial sample; these are examples of chemical changes. The failure of a sample of matter to undergo chemical change is also considered to be a chemical property. The fact that gold does not react with air is a chemical property of that material. Most chemical changes are also accompanied by physical changes, like changes in color or temperature.

When matter undergoes a physical change, the composition does not change. In a chemical change, some matter is converted into a different kind of matter.

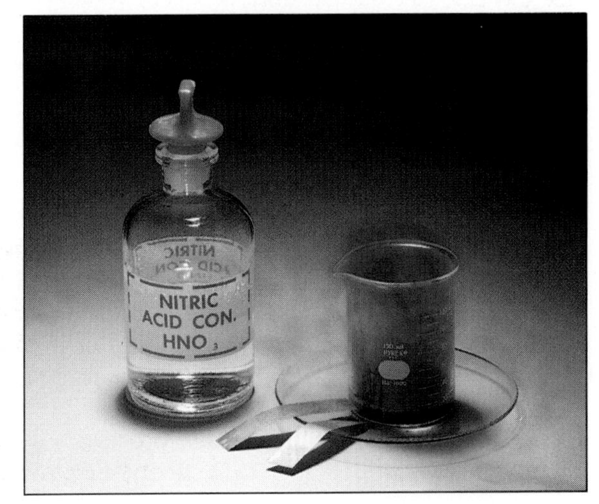

Making a statue by melting metals such as brass or bronze is a physical change. The reaction with certain acids is a chemical change.

Example 1.1 Properties of Matter

In each part a property or change is underlined. Classify each as intensive or extensive and chemical or physical.

(a) The color of sulfur is yellow.
(b) The rusting of iron requires oxygen.
(c) Ice melts at 32 °F.
(d) The density of water is 1.00 grams per cubic centimeter.
(e) A new pencil is 10 inches long.

Solution

(a) intensive, physical
(b) intensive, chemical
(c) intensive, physical
(d) intensive, physical
(e) extensive, physical

Classification of Matter

As a first step in understanding how the properties of matter depend on the kind of matter, we try to classify properties by using common characteristics. One such classification scheme might be by physical state, *i.e.,* solid, liquid, or gas. From the point of view of the chemist, the most useful classification of matter is that shown in Figure 1.4.

Matter can be broadly divided into pure substances and mixtures. A **mixture** is matter that can be separated into two or more substances by using the differences in physical properties of the components. An important characteristic of mixtures is that their composition can vary over a wide

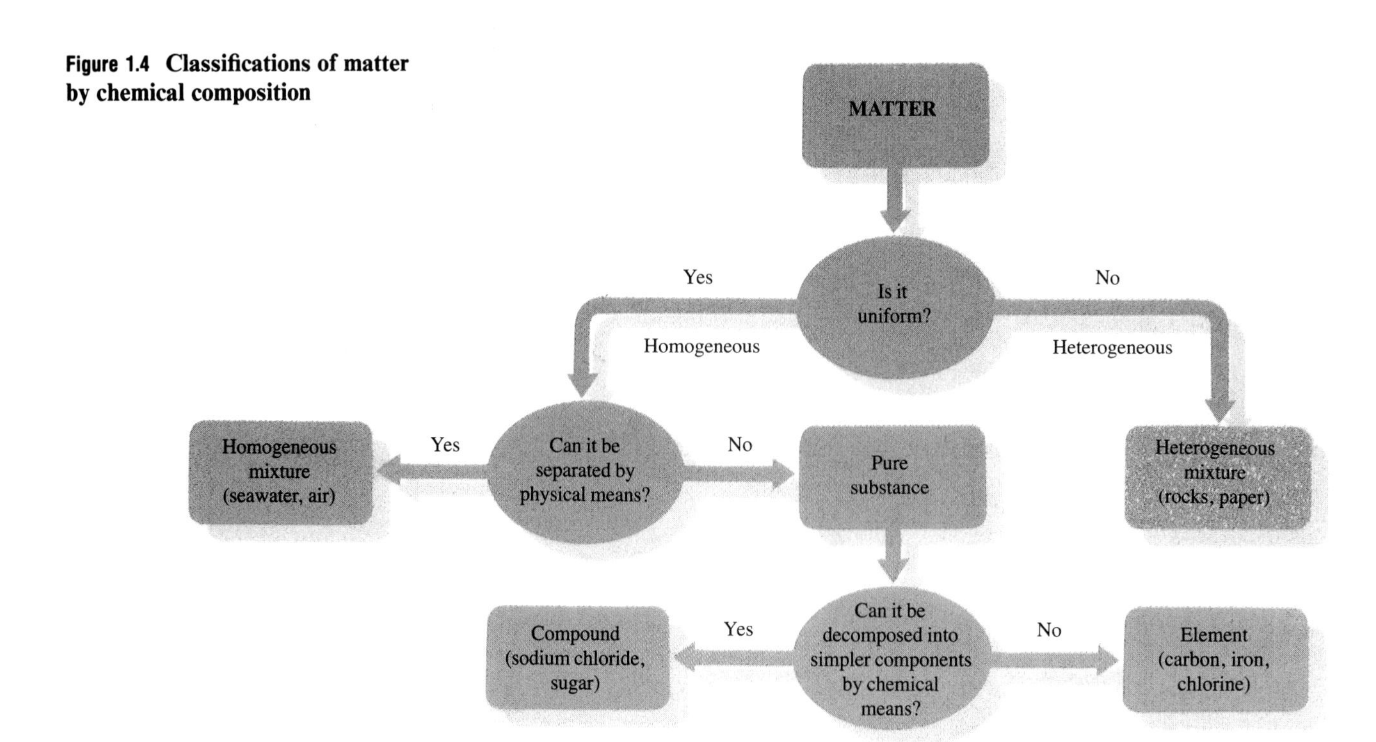

Figure 1.4 Classifications of matter by chemical composition

range. Mixtures can be homogeneous (uniform) or heterogeneous. The first step in classifying matter is to determine whether the matter is uniform throughout. If some parts of the sample have properties different from those of other parts, then it is a **heterogeneous mixture.** If it is uniform, it could be a **homogeneous mixture,** or it could be a **pure substance,** one that cannot be separated into component parts by a physical process. To the chemist, whenever the word "substance" is used alone, the modifier "pure" is understood.

Mixtures can be homogeneous (everywhere the same) or heterogeneous (in which the composition differs at different places).

Mixtures

Most of the matter that we encounter in our everyday lives is composed of mixtures of substances. Even a loaf of bread or a glass of milk is actually a complex mixture of several substances (Figure 1.5). In some cases it is quite obvious that a sample is a mixture, while in other cases careful examination and complicated equipment might be needed to determine whether the sample is a pure substance or a mixture.

Black gunpowder is an example of a mixture that consists of three solid substances: carbon, sulfur, and potassium nitrate. These three components can be separated by taking advantage of their different solubilities in two liquids: water and carbon disulfide. Potassium nitrate dissolves in water but not in carbon disulfide. Sulfur is soluble in carbon disulfide but not in water. Carbon does not dissolve in either of these liquids. If the gunpowder is treated with water, the liquid will contain the potassium nitrate, while the solid is a mixture of carbon and sulfur. After separating the liquid and solid, we can dissolve the sulfur by adding carbon disulfide to the solid and sepa-

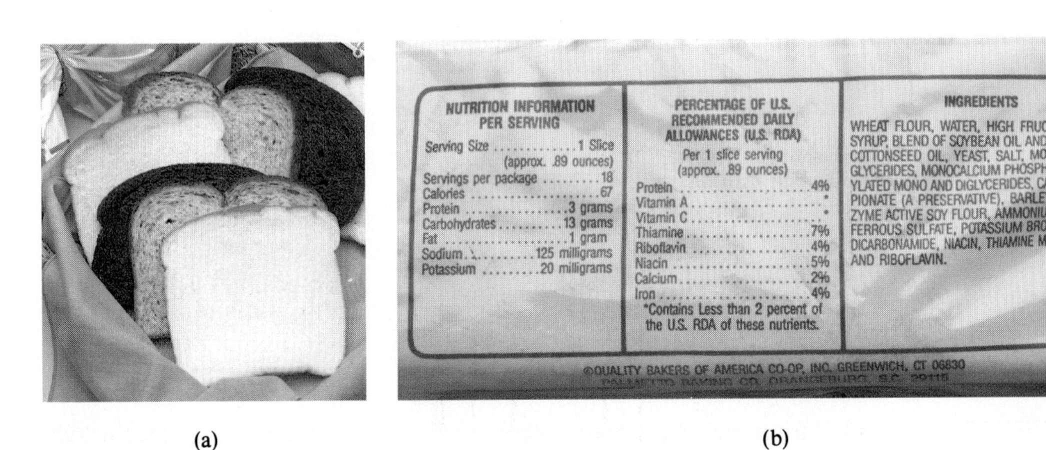

(a) (b)

Figure 1.5 Common mixtures of chemicals Bread might look homogeneous, but it is actually a mixture of several different substances.

Recently, some consumer advocates tested bread wrappers (not the bread, but the plastic bags in which the bread was wrapped) and found that some of them contained lead-based inks and dyes. While lead is not banned from use in printer's inks, many people reuse the bread wrappers to store food. Sometimes, the bread wrappers become reversed and food is stored directly against the lead-containing ink. Since there *is* a legitimate worry about lead poisoning, most manufacturers have decided to remove lead-containing inks from their products.

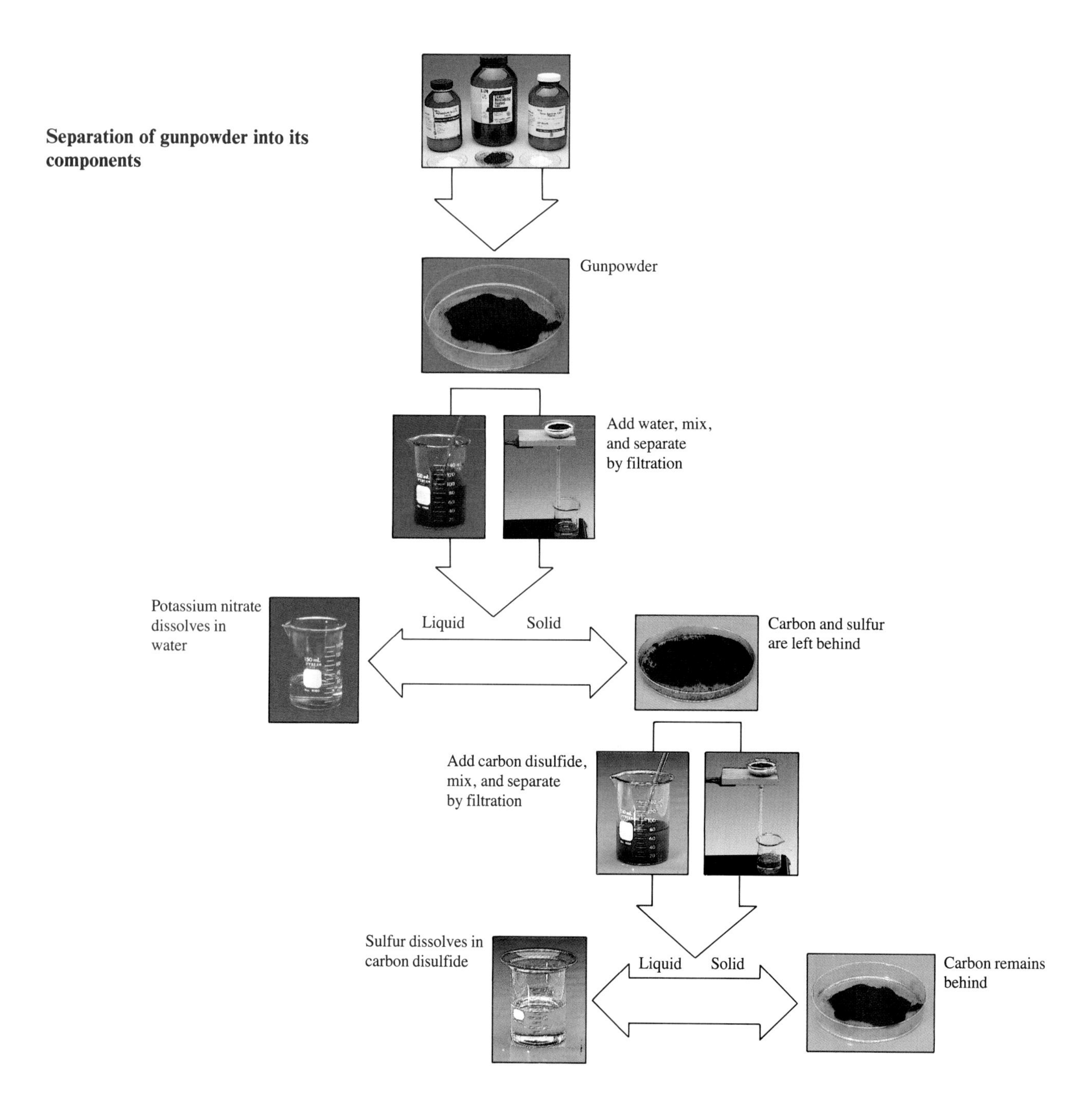

Gunpowder
Add water, mix, and separate by filtration
Potassium nitrate dissolves in water
Liquid
Solid
Carbon and sulfur are left behind
Add carbon disulfide, mix, and separate by filtration
Sulfur dissolves in carbon disulfide
Liquid
Solid
Carbon remains behind

Separation of gunpowder into its components

Table 1.1 Composition of Dry Air

Substance	Concentration (% by volume)
nitrogen	78.084
oxygen	20.946
argon	0.934
carbon dioxide	0.033
other	0.003

rate it from the solid carbon. When we finally evaporate the liquids, the two dissolved materials (sulfur and potassium nitrate) are recovered.

Four different mixtures are involved in the separation of black gunpowder into its components: the gunpowder, the water solution, the mixture of carbon and sulfur, and the carbon disulfide solution. If the carbon residue contains small amounts of potassium nitrate and sulfur, then it is also a mixture. Gunpowder is an example of a heterogeneous mixture, because microscopic examination reveals that some of the particles are black, others are yellow, and others are colorless crystals. The solution of potassium nitrate in water and the solution of sulfur in carbon disulfide are homogeneous because every drop of each of these liquids is identical in all respects. *Homogeneous mixture* and **solution** are two names for the same thing.

Solutions are homogeneous mixtures. Solutions can be solids, liquids, or gases.

Not all solutions are liquids. Many **alloys,** mixtures of a metal and another substance (usually another metal), are solid solutions. Bronze is a common alloy, a homogeneous mixture of copper and tin. Bronze is used to make statues because it is easy to cast and it resists weathering very well. Since all gaseous mixtures are homogeneous, they are all solutions. Air is a solution that contains nitrogen, oxygen, and many other gases; the composition of dry air is listed in Table 1.1.

Bronze statuary The Romans cast bronze statues that have survived two thousand years. This statue depicts a Roman prince, either Gaius or Lucius Caesar, both of the first century BC.

Substances

In common usage, the word substance is generally synonymous with matter, but most chemists use the word to mean "pure substance," one that cannot be separated into its components by physical means like filtration. When two different substances combine to form one or more new substances, the properties of the product or products are generally different from those of the starting materials. Table 1.2 lists some of the properties of sodium, chlorine, and sodium chloride (the product that results when these two substances combine). Sodium chloride is table salt.

Another very important difference between homogeneous mixtures and substances is that a mixture can exhibit variable composition, while a substance does not. A mixture such as a solution of sugar in water might contain a teaspoon or a tablespoon of sugar in a cup of water, or might be made in many other proportions. However, all samples of a *substance* such as sodium chloride are the same, whether they are made in the laboratory by the combination of sodium and chlorine, mined in Michigan, or separated from seawater.

There are millions of known substances, and more are being discovered every day. Of these substances, there are a few (about 107) that are different from the others. These substances are called **elements,** and they differ in that they cannot be decomposed into simpler substances by normal chemical means. All other substances are **compounds,** ones that can be decomposed into simpler substances or into their elements by chemical processes. The relationship of the elements and compounds is shown in Figure 1.4.

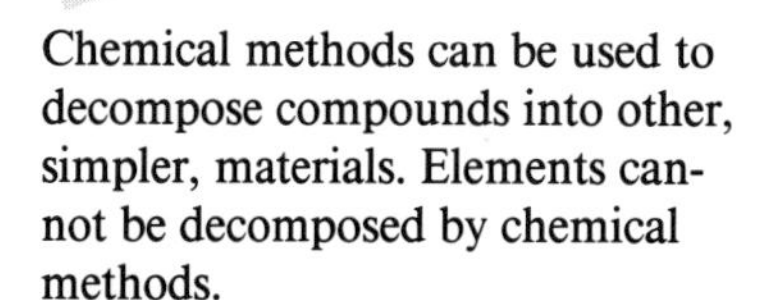

Chemical methods can be used to decompose compounds into other, simpler, materials. Elements cannot be decomposed by chemical methods.

Elements and Symbols

Each of the elements has a unique name. Many elements, such as the metals silver, gold, and copper, have been known since ancient times. Other naturally occurring elements have been isolated in pure form only during the past

Table 1.2 Properties of Sodium, Chlorine, and Sodium Chloride

Property	Sodium	Chlorine	Sodium Chloride
physical state at room temperature	solid	gas	solid
appearance	metallic silvery	green	clear
melting point	97.8 °C	−103 °C	800 °C

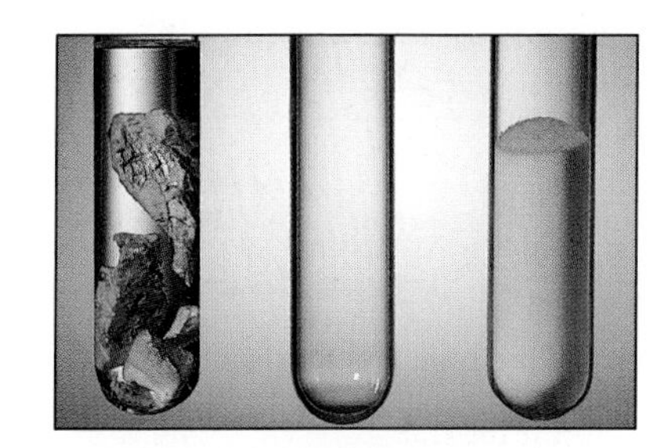

Photographs of sodium, chlorine, and sodium chloride.

Table 1.3 The Names and Symbols of Several Common Elements

Name	Symbol	Name	Symbol	Name	Symbol
aluminum	Al	fluorine	F	nitrogen	N
barium	Ba	gold	Au	oxygen	O
bromine	Br	hydrogen	H	phosphorus	P
calcium	Ca	iodine	I	potassium	K
carbon	C	iron	Fe	silicon	Si
chlorine	Cl	magnesium	Mg	silver	Ag
chromium	Cr	mercury	Hg	sodium	Na
copper	Cu	nickel	Ni	sulfur	S

fifty years. The most recently discovered elements do not occur in nature, but have been produced by the techniques of high-energy physics.

The **symbols** of the elements are abbreviations for their names. These consist of one or two letters,* with the first letter always capitalized and the second one lower case. For many of the elements the symbol is an obvious abbreviation of its name; for example, C is carbon, N is nitrogen, Ca is calcium, and Te is tellurium. In other cases, particularly for elements that have been known since ancient times, the symbol is an abbreviation of the ancient name of the element, usually in Latin. Examples include Na for sodium (natrium), Pb for lead (plumbum), Au for gold (aurum), and Sn for tin (stannum). An alphabetical list of the elements, along with their symbols and some other important information, is given inside the front cover. Table 1.3 lists the names and symbols of several common elements that will be used frequently in this text. You should become familiar with these elements and their symbols.

Compounds

The majority of pure substances are compounds. A compound can be decomposed into simpler substances, and eventually into its constituent elements, by normal chemical methods. Since all compounds are composed of two or more elements, the names of compounds are based upon the names of the elements. This topic will be discussed later in more detail. For the present, we will consider some important characteristics of compounds, those used to distinguish compounds from mixtures.

Unlike a mixture, a compound always has the same elements present in the same proportion. In all samples of sodium chloride, 39.3% of the mass is the element sodium and 60.7% is chlorine. Water consists of 11.2% hydrogen and 88.8% oxygen. Such definite composition is not observed for homoge-

* Recently, the scientific community has decided that the elements above 103 will be temporarily named in a systematic manner and designated by symbols that contain three letters. For example, element 104 is unnilquadium (Unq), from the Latin for one *(unis)* zero *(nihil)* four *(quattuor)*. After a new element is isolated and confirmed independently, the discoverers will assign a name and symbol. Since all of the most recently discovered elements are very unstable (they exist for only fractions of a second before decomposing), they are not discussed in this text.

INSIGHTS into CHEMISTRY

Science, Society, and Technology

Modern Alloys Are Widely Used. Techniques Evolved from the Ancients

Stainless steels are mostly iron, but chromium and nickel are added to make the metal resistant to rust. The concentrations of these and other alloying elements determine many of the properties of the alloy. Three common alloys are described here.

Type 301 stainless is commonly used to make cooking utensils, sinks, and commercial kitchen counters and cabinets. It is sometimes called "18/8" stainless steel, referring to the percentage of chromium and nickel in the alloys.

Type 316L is easy to weld and highly resistant to attack by chemicals, including concentrated acids such as sulfuric acid. It is widely used in the chemical industry to build reactors and other vessels.

Type 420 is a heat-treatable alloy. It is used by knifemakers, who will grind a blade, heat it in a furnace, and then quickly cool it in water or oil. The resulting blade is extremely hard, but also very brittle. (The blade might shatter if dropped on the ground at this stage.) A second heating step, called annealing, makes the blade less brittle. The knife blade is heated to about 500 °C and allowed to cool slowly to room temperature.

The following elements are added to iron to form these stainless steel alloys, in percentage by weight:

	301	316L	420
C	0.15 max	0.03 max	0.15 max
Mn	2.00 max	2.00 max	1.00 max
P	0.045 max	0.045 max	0.040 max
S	0.030 max	0.030 max	0.030 max
Si	1.00 max	1.00 max	1.00 max
Cr	16.00–18.00	16.00–18.00	12.00–14.00
Ni	6.00–8.00	10.00–14.00	
Mo		2.00–3.00	

Damascus swords were made from a steel that contained a relatively high concentration of carbon. While most carbon steels are too brittle to be used for a sword, the steelmakers of Damascus discovered that if they heated the steel to the right temperature, added carbon powder, hammered the metal, and repeated the process, they were rewarded with an extraordinarily strong and sharp knife or sword. The physical processes of heating and hammering were just as important to making Damascus steel as was adding the carbon.

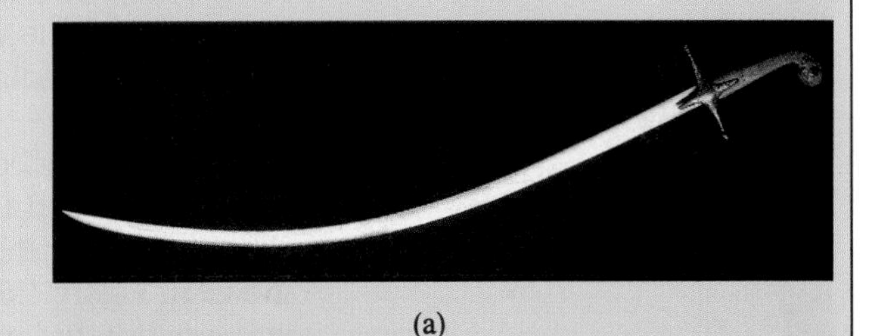

(a)

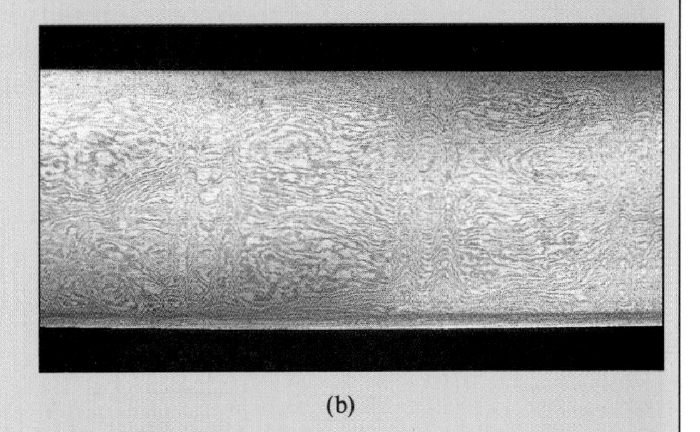

(b)

Damascus sword (a) Damascus swords were renowned for their strength and sharpness. (b) The uneven distribution of carbon produced a unique striated pattern in the steel.

neous mixtures such as liquid solutions or metal alloys. Alloys of copper and nickel can range between 0% and 100% of either metal. Two of these alloys are constantan (60% copper, 40% nickel) and coinage nickel (75% copper and 25% nickel).

Not only is the composition of any compound identical for all samples, but each of its intensive properties is also the same. The melting point is an important property that distinguishes a substance from a mixture. The temperature of a pure substance remains constant as it melts; most mixtures melt over a range of temperatures.

All samples of a compound have the same composition and intensive properties.

1.3 The Periodic Table

Objectives

- To discuss the periodic table of the elements and how to organize the elements into groups and periods
- To describe the properties of elements in several common groups

By the middle of the 19th century, chemists had isolated many of the elements, and they began a systematic investigation of their properties. As the chemical and physical properties of the elements were determined, scientists noted that some elements were quite similar to others. For example, lithium, sodium, and potassium have similar chemical properties, and the density of sodium is approximately halfway between that of lithium and potassium. The same is true of the group of elements chlorine, bromine, and iodine. The two groups, however, have very different properties. Grouping and classifying elements is the first step toward understanding their properties; the most widely used classification scheme, the periodic table, is discussed in this section.

Working independently, the Russian chemist Dimitri Mendeleev (1834–1907) and the German physicist Lothar Meyer (1830–1895) proposed arranging the elements in a table, the modern version of which is shown in Figure 1.6 and the inside cover of this book. This table, known as the **periodic table,** arranges elements (shown by their symbols) into rows and places elements with similar chemical and physical properties in the same columns. The lighter elements are at the top of the column and the heavier elements at the bottom.

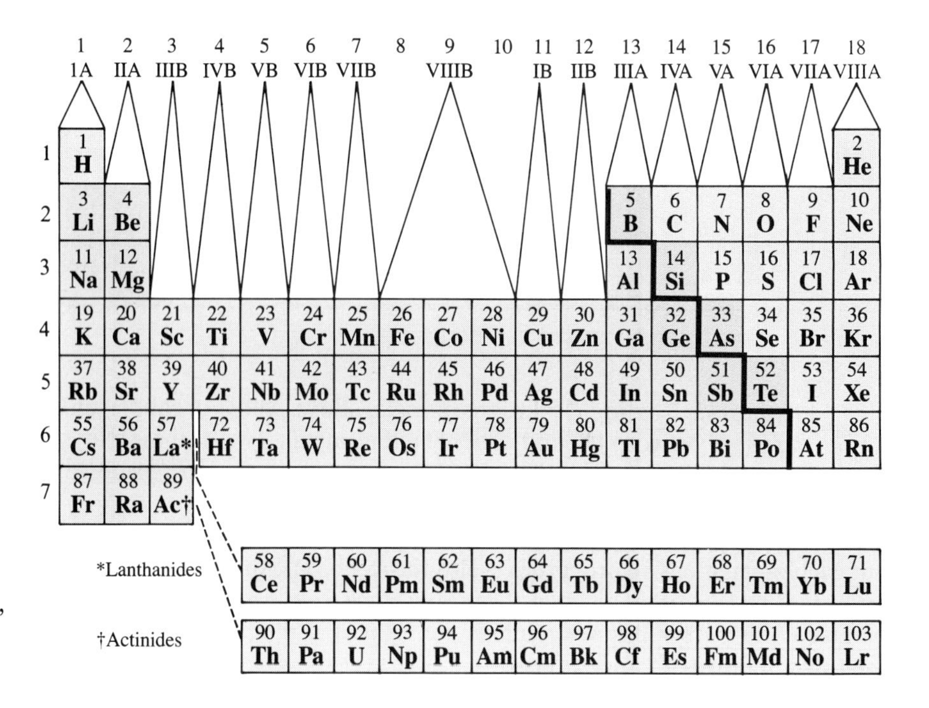

Figure 1.6 Periodic table of the elements Metals are shown in blue, metalloids in green, and nonmetals in yellow.

(a)

(b)

The periodic table (a) Dimitri Mendeleev. (b) Lothar Meyer. (c) Mendeleev's "original" periodic table, and (d) the English translation. The English translation of the periodic table was published in 1871. The group headings at the top describe the products observed when the elements react with hydrogen and oxygen. Mendeleev left space under aluminum for gallium, then unknown. Elements in parentheses are continuations from previous periods.

(c)

TABELLE II

Reihen	Gruppe I. — R^2O	Gruppe II. — RO	Gruppe III. — R^2O^3	Gruppe IV. RH^4 RO^2	Gruppe V. RH^3 R^2O^5	Gruppe VI. RH^2 RO^3	Gruppe VII. RH R^2O^7	Gruppe VIII. — RO^4
1	H=1							
2	Li=7	Be=9,4	B=11	C=12	N=14	O=16	F=19	
3	Na=23	Mg=24	Al=27,3	Si=28	P=31	S=32	Cl=35,5	
4	K=39	Ca=40	—=44	Ti=48	V=51	Cr=52	Mn=55	Fe=56, Co=59, Ni=59, Cu=63.
5	(Cu=63)	Zn=65	—=68	—=72	As=75	Se=78	Br=80	
6	Rb=85	Sr=87	?Yt=88	Zr=90	Nb=94	Mo=96	—=100	Ru=104, Rh=104, Pd=106, Ag=108.
7	(Ag=108)	Cd=112	In=113	Sn=118	Sb=122	Te=125	J=127	
8	Cs=133	Ba=137	?Di=138	?Ce=140	—	—	—	— — — —
9	(—)	—	—	—	—	—	—	
10	—	—	?Er=178	?La=180	Ta=182	W=184	—	Os=195, Ir=197, Pt=198, Au=199.
11	(Au=199)	Hg=200	Tl=204	Pb=207	Bi=208	—	—	
12	—	—	—	Th=231	—	U=240	—	— — — —

(d)

Row	Group I — R_2O	Group II — RO	Group III — R_2O_3	Group IV RH_4 RO_2	Group V RH_3 R_2O_5	Group VI RH_2 RO_3	Group VII RH R_2O_7	Group VIII — RO_4
1	H = 1							
2	Li = 7	Be = 9.4	B = 11	C = 12	N = 14	O = 16	F = 19	
3	Na = 23	Mg = 24	Al = 27.3	Si = 28	P = 31	S = 32	Cl = 35.5	
4	K = 39	Ca = 40	— = 44	Ti = 48	V = 51	Cr = 52	Mn = 55	Fe = 56, Co = 59, Ni = 59, Cu = 63
5	(Cu = 63)	Zn = 65	— = 68	— = 72	As = 75	Se = 78	Br = 80	
6	Rb = 85	Sr = 87	?Yt = 88	Zr = 90	Nb = 94	Mo = 96	— = 100	Ru = 104, Rh = 104, Pd = 106, Ag = 108
7	(Ag = 108)	Cd = 112	In = 113	Sn = 118	Sb = 122	Te = 125	I = 127	
8	Cs = 133	Ba = 137	?Di = 138	?Ce = 140				
9								
10			?Er = 178	?La = 180	Ta = 182	W = 184		Os = 195, Ir = 197, Pt = 198, Au = 199
11	(Au = 199)	Hg = 200	Tl = 204	Pb = 207	Bi = 208			
12				Th = 231		U = 240		

In the original table proposed by Mendeleev, not all of the spaces were filled with known elements. He proposed the existence of several elements that would fill these spaces, and predicted their expected properties. The discovery four years later of gallium, with properties very similar to those predicted by Mendeleev for one such element, showed the usefulness of this arrangement.

Each row of the table is called a **period** and is numbered. The properties of the elements change regularly going across the period. The elements in each column of the table have similar properties. Each column is called a **group,** and the groups are numbered across the top. The traditional method (in North America) for numbering uses a combination of Roman numerals and the letters A or B to label each group. Recently, numbering the groups 1 through 18 has been adopted; this scheme is also shown in Figure 1.6. The older labelling scheme will generally be used in this text, but the new notation will frequently be shown in parentheses.

The elements in a column, or a group, have similar chemical properties.

One important way to classify elements is to divide them into metals and nonmetals. A **metal** is broadly defined as a material that has luster and is a good electrical conductor. A majority of the elements, those in the center and left side of the table, are metals. **Nonmetals,** elements that typically do not conduct an electrical current, include the elements in the top right part of the table. Many periodic tables, such as the one in Figure 1.6, show a line dividing the metals from the nonmetals. The elements along the line show some properties of both metals and nonmetals, and are called **metalloids.** A particularly interesting feature of metalloids is that they are **semiconductors,** weak conductors of electricity, a property that makes them extremely useful in solid-state electronics. Hydrogen, the lightest element, is generally listed in group 1A (1), but is a nonmetal and is sometimes also shown in group VIIA (17).

The elements can be divided into metals, metalloids, and nonmetals.

The elements in groups labeled A (1, 2, 13 – 18) are historically called the **representative elements** or **main group elements.** The metals in the center part of the table, the B groups (3 – 12), are called the **transition metals.** Two series of heavier elements are set off at the bottom of the table to save space. The first series, the **lanthanides** (cerium, Ce, through lutetium, Lu) are the elements that follow lanthanum (La) in period 6. The second series, the **actinides** (thorium, Th, through lawrencium, Lr), fit in period 7 after actinium (Ac). These two series of elements, also known as the **inner transition metals,** do not belong to any of the groups labeled at the top of the table. Most of the actinide elements do not occur in nature, but have been made in nuclear physics laboratories.

Table 1.4 Abundances of the Major Elements in the Earth's Crust

Element	Abundance (% by weight)
oxygen	49.5
silicon	25.7
aluminum	7.4
iron	4.7
calcium	3.4
sodium	2.6
potassium	2.4
magnesium	1.9
hydrogen	0.9
titanium	0.6
chlorine	0.2
phosphorus	0.1
manganese	0.1
carbon	0.1
sulfur	0.1

Important Groups of Elements

Several important groups of elements have specific names and very characteristic properties. It is important for you to be familiar with these common elements.

The elements in group IA (1) are known as the **alkali metals** (Figure 1.7). These metals are soft and have low melting points. They are very reactive, with the reactivity generally increasing down the group. Their high reactivity toward water and many other substances requires that they be handled with

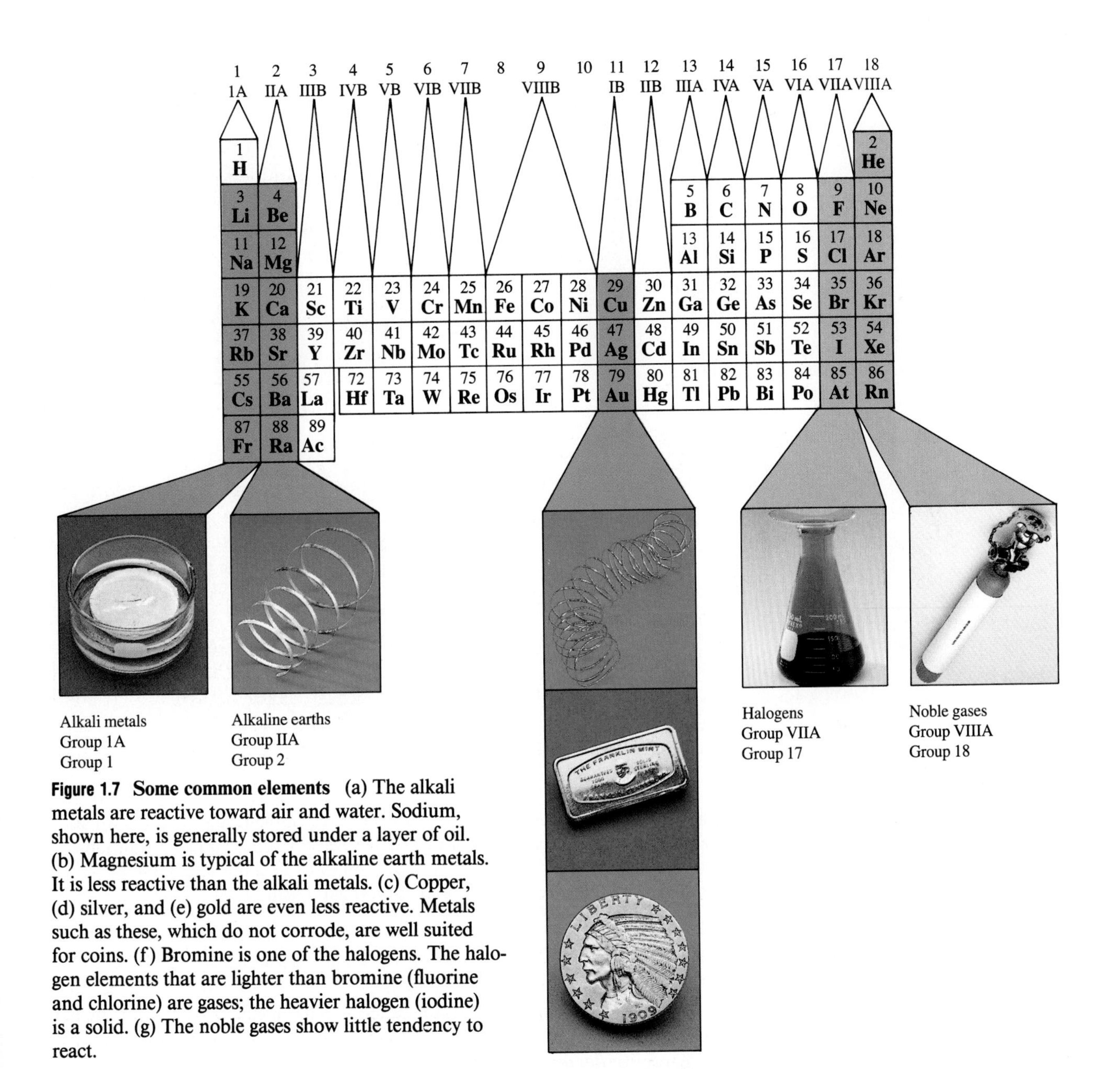

Figure 1.7 Some common elements (a) The alkali metals are reactive toward air and water. Sodium, shown here, is generally stored under a layer of oil. (b) Magnesium is typical of the alkaline earth metals. It is less reactive than the alkali metals. (c) Copper, (d) silver, and (e) gold are even less reactive. Metals such as these, which do not corrode, are well suited for coins. (f) Bromine is one of the halogens. The halogen elements that are lighter than bromine (fluorine and chlorine) are gases; the heavier halogen (iodine) is a solid. (g) The noble gases show little tendency to react.

extreme care in the laboratory. Figure 1.8 shows what happens on mixing sodium with water.

Group IA (or 1) elements are the alkali metals.

Sodium (Na) and potassium (K) are abundant elements in the earth's crust, and always occur in compounds. Table 1.4 shows the elements that are most abundant in the crust of the earth.

Elements in group IIA (2) are known as the **alkaline earth metals.** These elements are also reactive, but less reactive than the alkali metals. Magnesium (Mg) and calcium (Ca) are very abundant in the earth's crust, and calcium is an important constituent of bones, sea shells, and coral reefs.

The elements in group IIA (or 2) are called alkaline earth metals.

(a) (b) (c) (d)

Figure 1.8 Mixing sodium with water An energy-releasing reaction occurs when sodium metal is added to water. A flammable gas is produced and the solution turns a special dye red. The quantity of sodium must be quite small; if large amounts of sodium were used, an explosion might result.

The halogens are the elements in group VIIA (or 17).

Group VIIA (17) elements are called the **halogens,** a word that means "salt-formers." The halogens are among the most reactive nonmetals, with the reactivity decreasing down the column. Fluorine is probably the most reactive of all the elements. At room temperature and pressure, fluorine is a yellow gas, chlorine is a greenish yellow gas, bromine is a red liquid and iodine is a nearly black, metallic-looking solid. Chlorine is the most abundant element in this group. It is a major constituent of table salt and is found in seawater.

The elements of group VIIIA (18), on the right side of the table, are known as the **noble gases.** None had been discovered when the periodic table was first proposed, but they were easily inserted as an additional group. The name "inert gases" was applied to these elements because they are all gases at room temperature, and prior to 1962 all were thought to be completely unreactive. Now, a number of compounds of xenon (Xe) and a few com-

The halogens Chlorine is a gas; bromine is a liquid; iodine is a solid.

pounds of krypton (Kr) have been made, so the word "inert" has been replaced by "noble."

The elements in group VIIIA (or 18) are the noble gases.

1.4 Separation of Mixtures

Objectives

- To discuss the separation of the components of a mixture by filtration
- To determine the factors that influence the separation of components by distillation
- To discuss chromatography and chromatographic separations

The development of the periodic table required knowledge of the properties of the pure elements. This task clearly requires pure elements, so the separation of naturally occurring mixtures into elements was essential. Separating mixtures into pure substances is still important to most chemists. Most materials, whether obtained from natural sources or prepared in a chemical plant or laboratory, are mixtures of substances. For example, coal frequently contains large amounts of sulfur. When the coal is burned, the sulfur forms compounds that contribute to acid rain. Scientists continue to develop and improve methods to trap these compounds as they are formed and prevent them from entering the atmosphere. Other scientists are developing methods to separate the sulfur from the coal *before* it is burned.

This section discusses some of the commonly used separation methods and describes how they are applied to mixtures. The methods presented—filtration, distillation, and chromatography—all rely on physical changes and do not change the compositions of the substances that comprise the mixture being studied.

Filtration

A filtration is used to separate liquid from solid components in a mixture.

A mixture that consists of part solid and part liquid is usually easy to separate into these components. We can perform a **filtration,** separating the solid

Separation of coffee from the grounds The ground coffee is separated from the liquid by a paper filter. Coffee filters are much more porous than the laboratory filter paper shown in Figure 1.9. The flow rate through coffee filters is much faster and the liquid extracts only the aromatic coffee flavors, leaving the bitter components behind. If laboratory filter paper is used, the filtration takes much longer and the coffee is quite bitter.

Figure 1.9 Filtration Different types of filter paper are used, depending mostly on the size of the solid particles.

Water in a municipal treatment plant is passed through a bed of sand to remove any solids from the water stream.

from the liquid by passing the mixture through a porous barrier. A typical laboratory filtration is shown in Figure 1.9. The mixture is poured through paper or cloth that allows the liquid to pass but stops the solid. Many chemical reactions are designed so that the desired substance is a solid and the other components remain in solution. Reactions of this type will be covered in detail in Chapter 4.

Distillation

A liquid solution can often be separated into components by **distillation,** a separation process based on differences in boiling points. The distillation consists of two steps: (1) the mixture is heated to convert at least part of the liquid into a vapor, and (2) the vapor is condensed to a liquid in a different container. The apparatus used for these operations, shown in Figure 1.10, is called a still. For example, pure water can be separated from a salt-water solution by distillation. The salt-water solution is heated to its boiling point so the liquid water is converted into steam. The steam is condensed and collected as pure water. The salt in the solution does not vaporize at the boiling point of the solution and remains in the original flask. While simple, this process cannot generally supply drinking and irrigation water on a large scale because it requires a large (and expensive) use of energy to boil the solution.

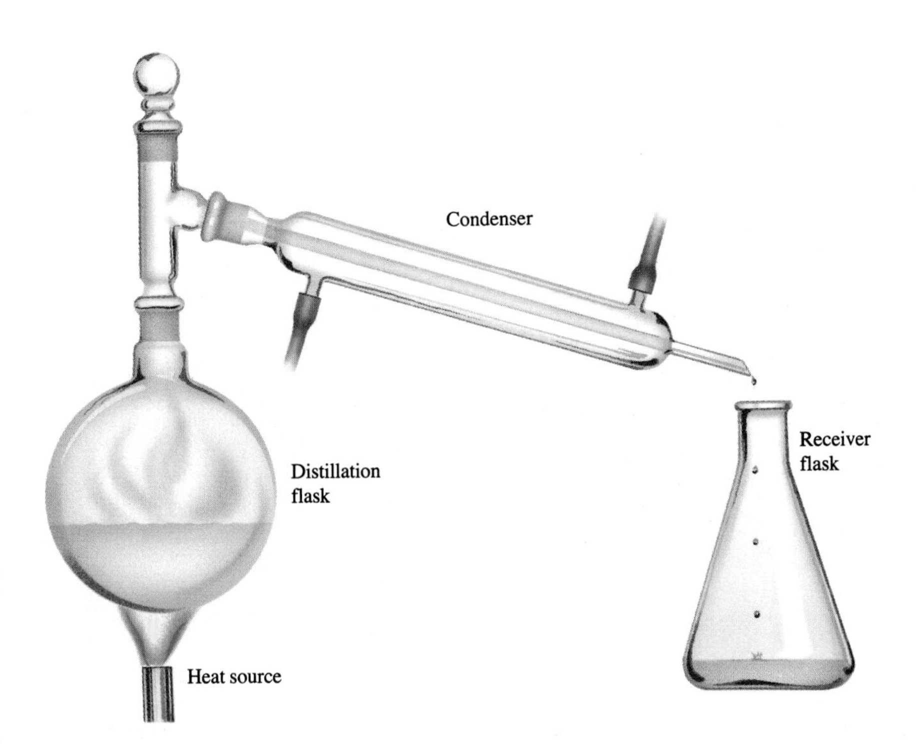

Figure 1.10 Distillation apparatus Heat is applied to the distillation flask to convert some of the liquid into a vapor. The vapor travels to the condenser, which is cooled with circulating water, and the condensed vapor is collected in a receiver flask.

Distillation can also be used to separate a mixture of two or more liquids that boil at different temperatures. An important example is the distillation of crude oil to separate components such as gasoline, kerosene, and diesel fuel. This distillation is more difficult than that of salt water because as the mixture of liquids is heated, all of the components form vapors, the word used to describe the gas-phase substance formed at temperatures below the boiling point. The liquid with the lowest boiling point (most volatile) will vaporize to a greater extent than other, higher boiling, liquids, but the vapor contains some of each of the components. The composition of the vapor will be determined by the relative volatility and the amount of each component in the liquid mixture. If the apparatus in Figure 1.10 is used, the first liquid collected will be enriched in the most volatile component. The collected liquid could be distilled again, and now the vapor would be enriched further because the starting liquid for the second distillation contained more of the lowest boiling component than was present in the original liquid.

A distillation separates a mixture by the differences in volatility of the components.

A better way to separate liquid mixtures is to use the apparatus shown in Figure 1.11. A long column (often filled with a solid material such as glass beads) is placed between the distillation flask and the condenser. As the vapors travel up the column, they condense but are revaporized as the column is heated by the ascending vapors. The vapors at the top of the column are enriched in the most volatile component and can be condensed

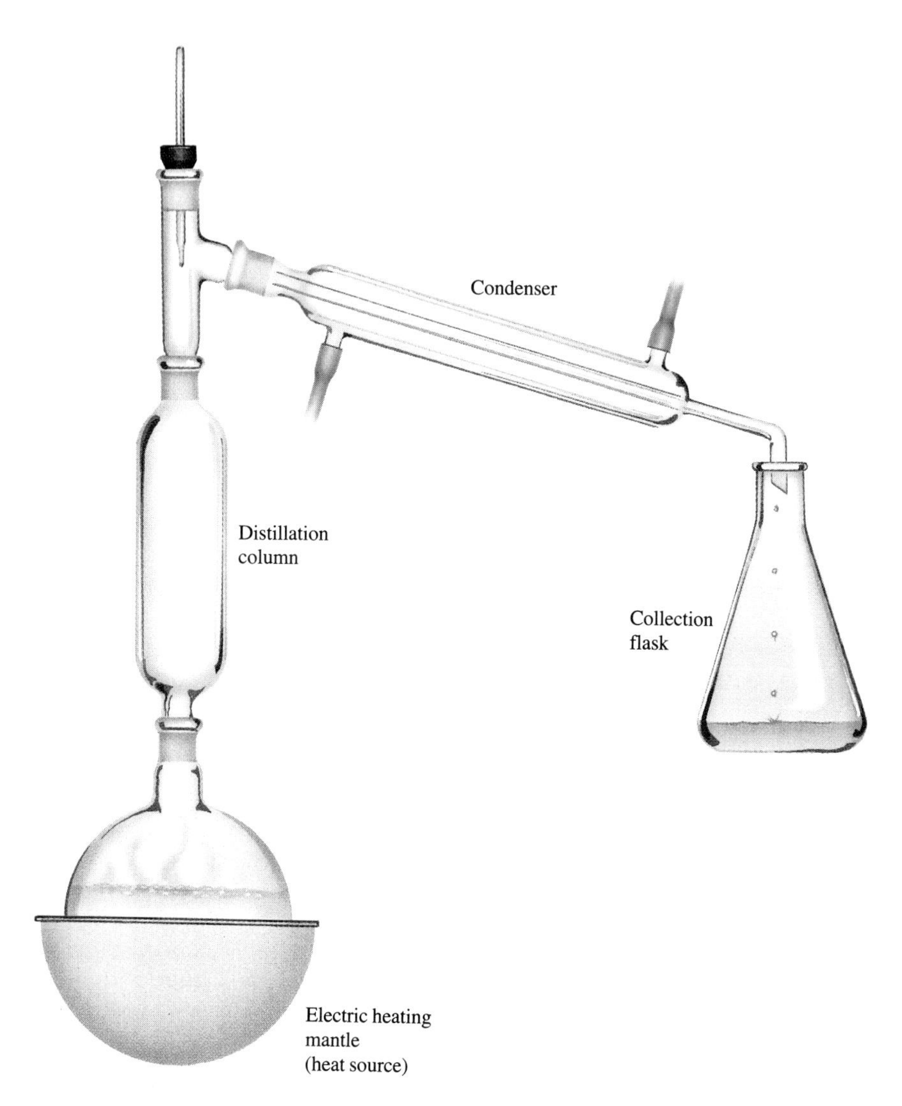

Figure 1.11 Distillation apparatus A distillation column is used to separate a mixture of volatile liquids. The heat is supplied by an electrical heater—open flames cannot be used safely in the presence of many materials.

Figure 1.12 Distillation of petroleum Large distillation towers are used in a petroleum refinery. This particular column is located at a refinery in Catlettsburg, Kentucky, that is operated by Ashland Oil, Inc. The refinery can process more than 200,000 barrels of crude oil per day. This computer-controlled refinery can be switched from the production of fuels to lubricants to petrochemicals quickly and efficiently.

(cooled by water) and collected in the collection flask. Next, the second most volatile component will reach the top, and it can be collected in a second receiver flask. The temperature increases during the distillation because the second component has a higher boiling point than the first. The towers that are seen at petroleum and chemical plants (Figure 1.12) are distillation columns, built on the large scale needed to process large volumes of materials.

Chromatography

Mixtures can be separated into component substances based on differences in attraction to other materials. A common example of these attractive forces is seen when certain foods are spilled on clothing. The coloring agents in grape jelly, for example, are strongly attracted to many kinds of fabric. Such stains are difficult to remove. Other stains, equally unsightly, are easily

(a) (b) (c)

Figure 1.13 Paper chromatography of ink A dot of ink is placed near the bottom of a piece of paper, and the edge of the paper is immersed in a liquid. The separation improves as the liquid moves farther up the paper. (a) Before the experiment; (b) after 1 minute; (c) after 5 minutes.

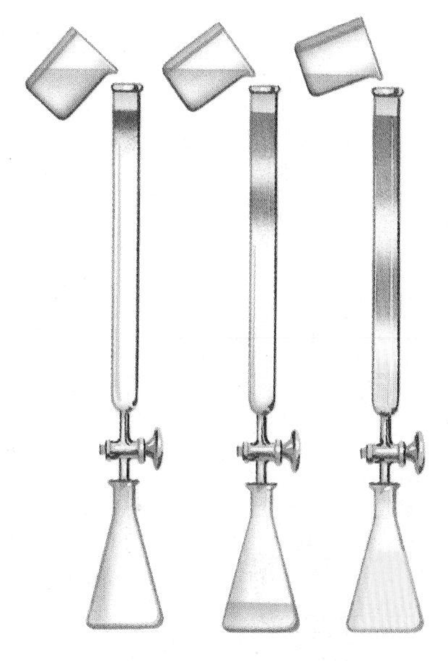

Figure 1.14 Column chromatography The separation of the components of chlorophyll, the coloring agent in green leaves, was one of the first applications for column chromatography.

removed by washing with water. The attractive forces between the grape jelly pigment and the fabric could provide the basis for a separation technique that removes the color from grape jelly.

Chromatography is a technique that employs two materials, one moving and the other stationary, and is used to separate a mixture into its components. A typical system might use a liquid (the mobile phase) sweeping past a solid (stationary phase). The solid stationary phase attracts some of the components more strongly than others, slowing down their passage through the column. The division of the components between the stationary and mobile phases is the basis for a chromatographic separation.

In *paper chromatography,* illustrated in Figure 1.13, the paper is the stationary phase, and a liquid such as acetone or alcohol is the mobile phase. The mixture of materials is placed near the bottom of a piece of paper. The edge of the paper is placed in the mobile phase liquid, which rises up the paper by capillary attraction. Those components of the mixture that are not strongly held to the paper will move up with the liquid; those that are attracted more strongly by the paper will move more slowly, causing their separation. Figure 1.13 shows the chromatographic separation of the components of one kind of ink. The colors that are strongly attracted by the paper remain near the bottom while the others move further up the paper.

A similar separation can be done by column chromatography shown in Figure 1.14. In this case, the mixture is dissolved in a liquid and placed at the top of a column packed with a solid material. Additional liquid is added to the column and the components of the mixture separate just as in the paper chromatography experiment. As each component of the mixture reaches the bottom of the column, it is collected as a solution of the pure substance.

A chromatographic separation employs two phases, one moving and one stationary. Compounds that prefer the stationary phase move slowly; those that prefer the moving phase move quickly.

INSIGHTS into CHEMISTRY

Science, Society, and Technology

Gas Chromatography, a Modern Separation Technique

Gas chromatography (GC) is one of the most versatile and powerful separation techniques used by chemists today. Many important compounds become gases when heated; gas chromatography is useful for the separation and analysis of these compounds. A mixture can be separated into components even when the components are quite similar to each other (boiling points that differ by only 0.01 °C, for example). The compositions of complex mixtures can be determined by this technique, even when the samples contain hundreds of compounds. Typical of this class of mixtures are petroleum products.

Another advantage to chemical analysis by GC is that extremely small samples are required. The sensitivity makes the technique particularly useful for the analysis of evidence from a crime scene, for industrial applications such as flavor chemistry, and for environmental studies.

Separation by GC occurs as the moving phase, a gas, passes over the stationary phase (often a liquid coated on the walls of a glass column). The column is generally a coil of capillary glass tubing. The basic components of a gas chromatograph are shown in Figure 1.15.

The steps in the separation process for a typical liquid sample mixture follow: The sample to be analyzed (a small fraction of a drop) is drawn into a syringe and injected into a heated zone, where it immediately vaporizes. A stream of an inert gas, usually helium or nitrogen, carries the sample into the column. The column is enclosed in an oven to keep the sample in the gaseous state. In the column, the components of the sample travel at speeds that are determined by their interactions with the stationary phase that has been chosen. If the proper stationary phase has been selected and the column is long enough, the components will be completely separated before they emerge from the column.

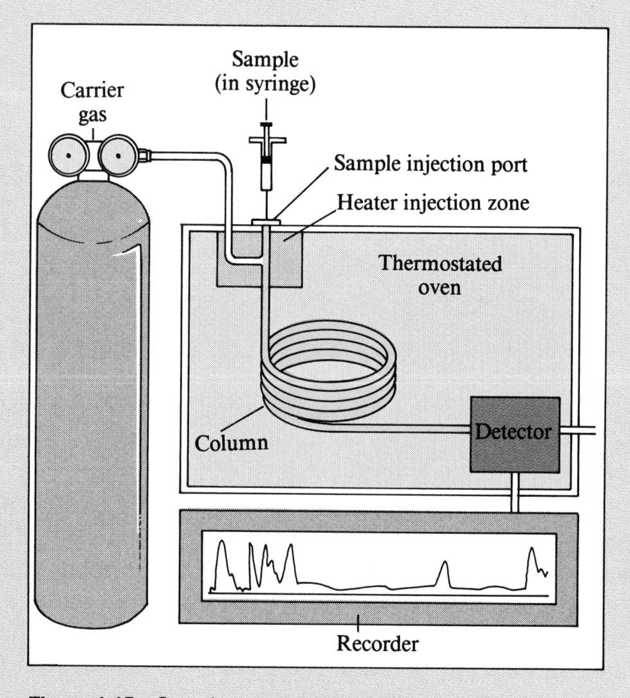

Figure 1.15 Gas chromatograph This block diagram represents a typical gas chromatograph.

1.5 Measurement and Units

Objectives

- To describe the SI units of measurement
- To derive unit conversion factors
- To convert measurements from one set of units to another
- To develop conversion factors derived from equivalent quantities

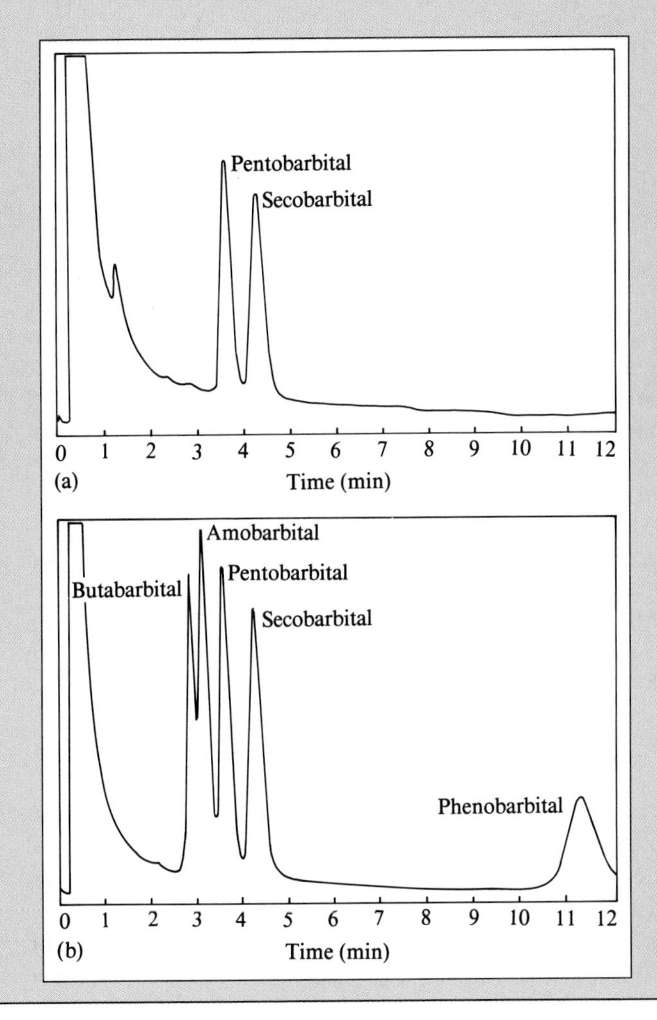

As each component emerges, it enters a detector where some property is measured (density and thermal conductivity have been used) and recorded on a moving piece of graph paper. This recording is called a chromatogram. The chromatogram is a plot of the detector response against time. The time required for a component to emerge from the column is called its retention time and is a useful identifying characteristic of the substance. Figure 1.16 shows how retention time can be used to tentatively identify two barbiturates by comparison to a known mixture of barbiturates.

The detector response is generally proportional to the quantity of each component present in the sample (an extensive property). Thus, not only can you identify the substance, but you can often determine its concentration.

Although gas chromatography has existed as an analytical method for about 40 years, important advances continue to be made. The gas chromatography is combined with computers and advanced detectors to help scientists identify biologically important chemicals such as cancer-fighting agents. Identifying environmental pollutants is another area of use for these instruments because of the very small quantities that can be detected.

Figure 1.16 Gas chromatograms The retention times of five barbiturates are determined from a chromatogram of a mixture of known composition. (b) The presence of pentobarbital and secobarbital, two drugs, can be inferred from the chromatogram of a sample of unknown composition (a).

Scientific progress is based on gathering and interpreting careful observations. These observations can be expressed either in a qualitative way, like "That car is going fast," or in a quantitative manner: "That car is going 95 miles per hour." Qualitative descriptions involve terms such as hot, cold, fast, slow, heavy, and light. Quantitative descriptions, those that use a number to describe a property, communicate much more information. The development of modern chemistry would not have been possible without the quantitative data gathered over the past several centuries.

Table 1.5 The SI Base Units

Quantity	Unit	Abbreviation
length	meter	m
mass	kilogram	kg
time	second	s
temperature	kelvin	K
amount	mole	mol
electric current	ampere	A
luminous intensity	candela	cd

For measurements to be meaningful, there must be accepted standards of comparison, so that the numbers reported for quantities like distance, time, volume, and mass have the same meaning for everyone. When we say that a pen is six inches long, we mean that the pen is six times as long as the length that has been defined as one inch. **Units** are standards that are used for quantitative comparison between measurements of the same type of quantity. The scientific community has adopted the SI units *(Le Système International d'Unités)* to express measurements. These units are an outgrowth of the metric system that was created during the French Revolution; the system originated when the French rejected anything related to the deposed monarchy.

While most SI units have been accepted by the scientific community, a few have not; some nonsystematic units are still commonly used for certain measurements.

Base Units

There are seven types of quantities in which all measurements can be expressed: length, mass, time, temperature, amount of substance, electrical current, and luminous intensity. The SI defines a **base unit** for each of these; they are given in Table 1.5.

All other physical quantities can be expressed as combinations of base units, called **derived units.** Area has units of length squared or m^2 in SI units; volume is expressed in m^3. Density is the ratio of mass to volume and has units of kg/m^3.

Quantities are usually expressed in units that avoid very large or very small numbers. Units of length in the English system vary from inches to miles to light-years. It is more convenient and meaningful to express the size of a candy bar as 1.0 ounce rather than 0.000031 ton. Unfortunately, dealing with these units requires conversion factors (like the number of ounces in a ton) that are often hard to remember. Conversion among SI units is generally simpler.

The SI creates units of different sizes by attaching a prefix that moves the decimal point. The prefix *kilo* (used in the base unit for mass) means 1000 or 10^3. An object that has a mass of one kilogram has *exactly* the same mass as 1000 g. These prefixes and their meanings and abbreviations are given in Table 1.6. The prefixes that will be used most frequently in this text are shown in bold type.

Table 1.6 Prefixes Used with SI Units*

Prefix	Abbreviation	Meaning
exa	E	10^{18}
peta	P	10^{15}
tera	T	10^{12}
giga	G	10^9
mega	M	10^6
kilo	k	10^3
hecto	h	10^2
deka	d	10^1
deci	d	10^{-1}
centi	c	10^{-2}
milli	m	10^{-3}
micro	μ	10^{-6}
nano	n	10^{-9}
pico	p	10^{-12}
femto	f	10^{-15}
atto	a	10^{-18}

* Students who are unfamiliar with exponential notation should refer to Appendix A for an explanation.

Unit Conversion Factors

Changes are easily made among the SI units because the meanings of the prefixes provide conversion factors. A **unit conversion factor** is a fraction in which the numerator is a quantity that is equal or equivalent to the quantity in the denominator, but expressed in different units. For example, a factor for converting units of kilograms to grams can be derived from the definition of the prefix. Since *kilo* means 10^3,

The prefixes for SI units indicate the power of ten by which the base unit is multiplied.

$$1 \text{ kg} = 1 \times 10^3 \text{ g}$$

The conversion factor is:

$$\text{unit conversion factor} = \left(\frac{10^3 \text{ g}}{1 \text{ kg}}\right)$$

A conversion between units is simply a multiplication process in which the quantity to be converted is multiplied by the appropriate unit conversion factor. *A unit conversion factor is equal to one* because the quantity in the numerator of the fraction is identical in size to the quantity in the denominator. Consider converting a mass of 0.251 kilograms to grams. The strategy is to multiply 0.251 kg by the unit conversion factor that gives the number of grams in a kilogram.

$$\text{mass} = 0.251 \text{ kg}\left(\frac{1000 \text{ g}}{1 \text{ kg}}\right) = 251 \text{ g}$$

Note that the units of kg for the given mass were cancelled by the units of kg in the denominator of the conversion factor. The mass is not changed by multiplying by the unit conversion factor, only the units in which it is expressed. The conversion factor is placed in parentheses and has a numeral "1" behind it, simply for emphasis.

Multiplying by the unit conversion factor does not change the quantity, just the units in which it is expressed.

If you make a mistake, say by writing the unit conversion factor upside down, the units will not cancel.

$$\text{mass} = 0.251 \text{ kg}\left(\frac{1 \text{ kg}}{1000 \text{ g}}\right) = 0.000251 \text{ kg}^2/\text{g}$$

Even if you neglected the units and thought that 0.251 kg was correctly converted to 0.000251 g, the application of some common sense would tell you otherwise. The answer is totally unreasonable, because it is a smaller number expressed in a smaller unit. It is always a good idea to check that the answer is logical.

This approach to changing units of measure is often called the "factor-label method," since the numbers provide a "factor" with the units forming a "label." The method is further illustrated in Example 1.2.

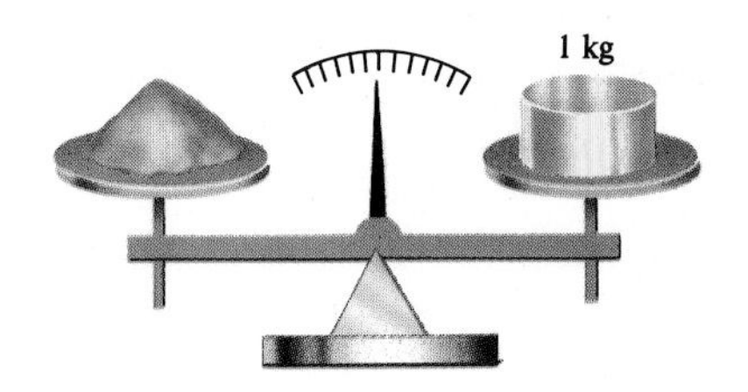

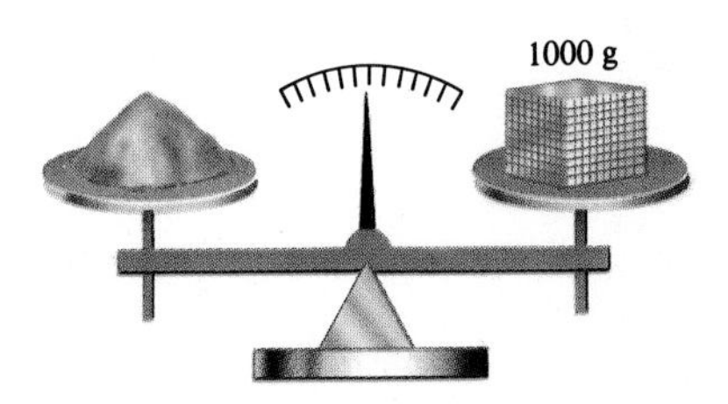

Kilo means 1000.

Example 1.2 Conversion of Units

Convert each of the following measurements into the standard SI base unit: (a) 454 g (b) 105 pm. The strategy is shown in the flow diagram:

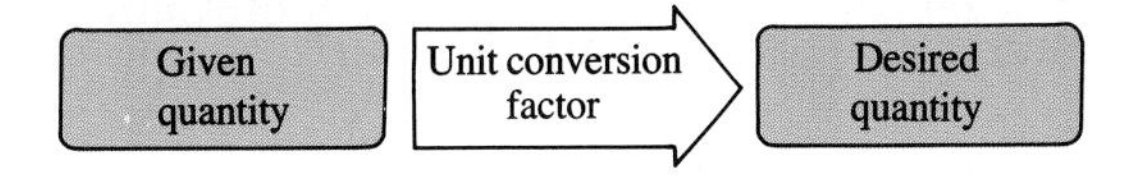

Solution

(a) The SI base unit of mass is the kilogram, with *kilo* meaning 1000. Thus, 1 kg = 1000 g, and so the needed conversion factor is

$$\text{unit conversion factor} = \left(\frac{1\ \text{kg}}{1000\ \text{g}}\right)$$

Notice that the g-to-kg conversion factor contains the same information as the kg-to-g conversion factor. Each relationship allows us to convert from one of the units to the other.

The mass of 454 g is expressed in kilograms by multiplying by the unit conversion factor.

$$\text{mass} = 454\ \text{g}\left(\frac{1\ \text{kg}}{1000\ \text{g}}\right) = 0.454\ \text{kg}$$

(b) The prefix pico means 10^{-12}, so 1 pm = 1×10^{-12} m.

$$\text{unit conversion factor} = \left(\frac{10^{-12}\ \text{m}}{1\ \text{pm}}\right)$$

We get the answer by multiplying the given length by the unit conversion factor:

$$\text{length} = 105\ \text{pm}\left(\frac{10^{-12}\ \text{m}}{1\ \text{pm}}\right) = 1.05 \times 10^{-10}\ \text{m or } 0.000000000105\ \text{m}$$

Understanding A 10-km race is popular among runners. Express this distance in the SI base unit.

Answer: 10,000 m

Conversion Among Derived Units

Unit conversion among derived units is performed by similar methods. The unit conversion factors are formed from the definitions, but sometimes operations like squaring and cubing are required to make the unit conversion factors contain the desired units.

Volume

The volume of a rectangular box is the product of its length times width times height. A volume is always a product of three lengths, so the standard unit of volume is a cube with dimensions equal to the base unit: $1\ \text{m} \times 1\ \text{m} \times 1\ \text{m} = 1\ \text{m}^3$. A cubic meter is an inconveniently large volume for laboratory scale experiments—a cubic meter of water weighs about a ton. Volume measurements in cubic centimeters are much more common. Conversions between these volume units are similar to those of length, except that volume is length cubed. The relationship between the lengths (m and cm) can be written first, and then the relationship between the volumes can be determined by cubing the equivalent lengths.

identical lengths: $100\ \text{cm} = 1\ \text{m}$

identical volumes: $(100\ \text{cm})^3 = (1\ \text{m})^3$

$10^6\ \text{cm}^3 = 1\ \text{m}^3$

$$\text{unit conversion factor} = \left(\frac{10^6\ \text{cm}^3}{1\ \text{m}^3}\right)$$

A nonsystematic unit used to express volumes is the liter, abbreviated as L. There are exactly 1000 L in 1 m^3. The milliliter (10^{-3} L) is also commonly used. One mL is identical to one cm^3.

$$1\ \text{m}^3 = 10^3\ \text{L} = 10^6\ \text{cm}^3$$

$$1\ \text{L} = 1000\ \text{mL} = 1000\ \text{cm}^3$$

$$1\ \text{cm}^3 = 1\ \text{mL}$$

Example 1.3 illustrates conversions among these units of volume.

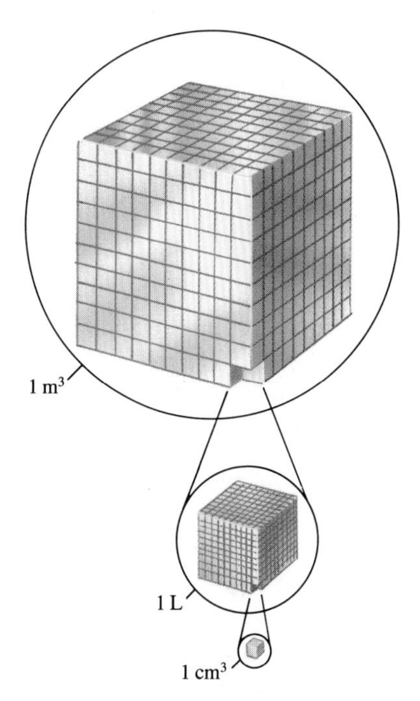

The drawings show the relative sizes of 1 cm^3, 1 L, and 1 m^3.

Example 1.3 Conversion of Volume Units

Express a volume of 322 mL in units of:
(a) liters (b) cm^3 (c) m^3

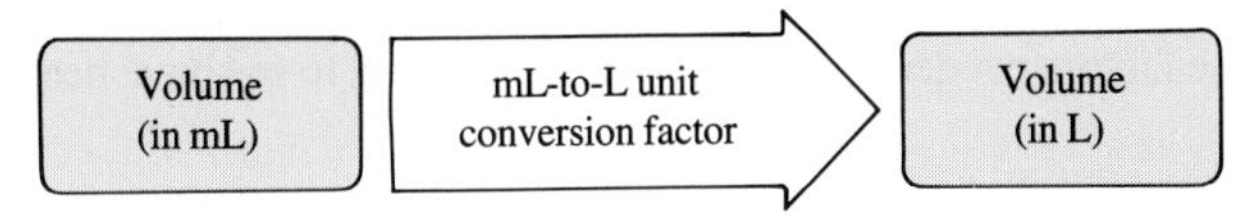

Solution

(a) The unit conversion factor can be derived from the meaning of the prefix *milli*.

$$\text{unit conversion factor} = \left(\frac{1\ \text{L}}{1000\ \text{mL}}\right)$$

$$\text{volume} = 322\ \text{mL}\left(\frac{1\ \text{L}}{1000\ \text{mL}}\right) = 0.322\ \text{L}$$

(b) The milliliter and the cubic centimeter represent the same volume, so the conversion factor is

$$\text{unit conversion factor} = \left(\frac{1\ \text{cm}^3}{1\ \text{mL}}\right)$$

Obviously the numerical value of our volume will not change, just the units:

$$\text{volume} = 322\ \text{mL}\left(\frac{1\ \text{cm}^3}{1\ \text{mL}}\right) = 322\ \text{cm}^3$$

(c) From the answer to part (b) and the knowledge that 1 m^3 is the same as 10^6 cm^3, a volume in mL can be expressed in m^3:

$$\text{volume} = 322\ \text{cm}^3\left(\frac{1\ \text{m}^3}{10^6\ \text{cm}^3}\right) = 3.22 \times 10^{-4}\ \text{m}^3$$

Density

Many intensive properties (mentioned in Section 1.2) that are used for identifying substances are defined by a ratio of two extensive properties. One of the most frequently used of these is density, defined as the mass per unit volume. Using the base units in the International System, density would have the units kg/m^3, or kg m^{-3}. For most substances, this unit is an inconvenient size, so the densities of solids and liquids are expressed in g/cm^3 (the same as g/mL); gas densities are generally expressed in g/L. A unit conversion in which two units change (kg to grams *and* m^3 to cm^3) requires two conversion factors. The factors can be applied separately or together. These two processes are presented in the next example.

Example 1.4 Conversion between Density Units

Express a density of 8.4 g/cm^3 in the SI base units.

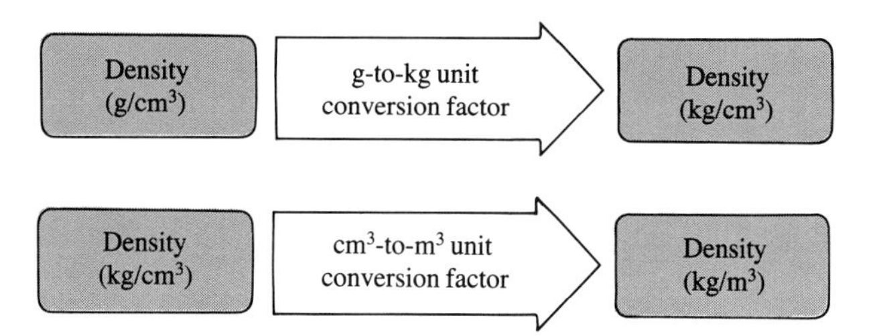

Solution The factors needed to convert g to kg and cm^3 to m^3 have been previously derived.

$$\text{density} = 8.4 \frac{\text{g}}{\text{cm}^3}\left(\frac{1 \text{ kg}}{1000 \text{ g}}\right) = 8.4 \times 10^{-3} \frac{\text{kg}}{\text{cm}^3}$$

$$8.4 \times 10^{-3} \frac{\text{kg}}{\text{cm}^3}\left(\frac{10^6 \text{ cm}^3}{\text{m}^3}\right) = 8.4 \times 10^{3} \frac{\text{kg}}{\text{m}^3}$$

The calculation in which more than one conversion factor is used is sometimes called a *chain calculation.* Most calculators can perform these calculations quite efficiently, but instructions differ from one particular device to the next.

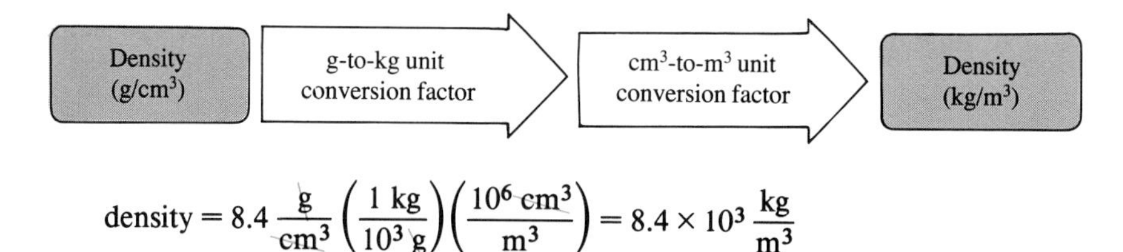

$$\text{density} = 8.4 \frac{\text{g}}{\text{cm}^3}\left(\frac{1 \text{ kg}}{10^3 \text{ g}}\right)\left(\frac{10^6 \text{ cm}^3}{\text{m}^3}\right) = 8.4 \times 10^{3} \frac{\text{kg}}{\text{m}^3}$$

When performing chain calculations, it is especially important to evaluate your answer to make sure that it is sensible.

Understanding The density of a gas is 1.05 g/L. Convert to the SI base units.

Answer: The density of the gas is 1.05 kg/m^3.

English System

Most people in the United States are more familiar with the English system of measurements than with SI. Table 1.7 summarizes the relationships between the SI and the English system. This table, taken from a more complete presentation in Appendix C, demonstrates the simplicity of the SI units and prefixes and also provides the information needed to make conversions from one system to the other.

Table 1.7 Relations in the SI and English System

SI Units	English Units	SI-English Equivalents
length		
1 km = 10^3 m	1 ft = 12 in	1 in = 2.54 cm
1 cm = 10^{-2} m	1 yd = 3 ft	1 m = 39.37 in
1 mm = 10^{-3} m	1 mile = 5280 ft	1 mi = 1.609 km
1 nm = 10^{-9} m		
volume		
1 m^3 = 10^6 cm^3	1 gal = 4 qt	1 L = 1.057 qt
1 cm^3 = 1 mL	1 qt = 57.75 in^3	1 qt = 0.946 L
mass		
1 kg = 10^3 g	1 lb = 16 ounces	1 lb = 453.6 g
1 mg = 10^{-3} g	1 ton = 2000 lb	1 ounce = 28.35 g

Example 1.5 Conversions between SI and English System

Perform the following conversions:

(a) Express 5.000 pounds of sugar in kilograms.
(b) A bottle of wine contains 750 mL. How many quarts is this volume?

Solution

(a) From Table 1.7, the connection between the SI and English system for units of mass is 1 pound = 453.6 g. This relationship forms the unit conversion factor.

$$\text{mass of sugar} = 5.000\ \text{lb}\left(\frac{453.6\ \text{g}}{1\ \text{lb}}\right) = 2268\ \text{g}$$

To obtain the mass in the desired units of kg requires a second conversion.

$$\text{mass of sugar} = 2268\ \text{g}\left(\frac{1\ \text{kg}}{1000\ \text{g}}\right) = 2.268\ \text{kg}$$

The calculations could have been chained together, as shown in (b).

(b) $$\text{volume of wine} = 750\ \text{mL}\left(\frac{1\ \text{L}}{1000\ \text{mL}}\right)\left(\frac{1\ \text{qt}}{0.946\ \text{L}}\right) = 0.793\ \text{qt}$$

Understanding What is the mass in grams of a candy bar that weighs 1.40 ounces?

Answer: 39.7 g

Temperature Conversion Factors

Temperature is a familiar quantity to most of us. In the scientific community, and in much of the world, temperatures are generally measured in the units of degrees Celsius, abbreviated °C. The Celsius scale is defined by assigning the freezing point of water a value of 0 °C and the boiling point of water as 100 °C. The difference between melting and boiling is 100 degrees in the Celsius scale. The scale was formerly called the centigrade scale because the interval between freezing and boiling was divided into 100 equal units.

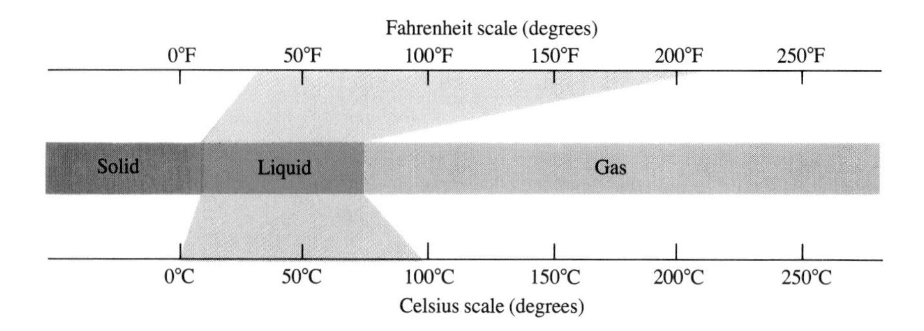

Figure 1.17 The freezing and boiling points of water The freezing and boiling points of water are shown on the Fahrenheit and Celsius scales.

In the United States, the Fahrenheit scale (°F) is used to express temperature. The freezing and boiling points of water are assigned values of 32 °F and 212 °F on this scale, so the difference is 180 °F. Figure 1.17 shows the relation between these two scales.

A unit conversion factor can be derived because a difference of 180 °F is the same as a difference of 100 °C.

$$\text{unit conversion factor} = \left(\frac{180\ °\text{F}}{100\ °\text{C}}\right) = \left(\frac{1.8\ °\text{F}}{1.0\ °\text{C}}\right)$$

Conversion between these temperature scales involves more than just multiplying by the unit conversion factor because 0 °C represents a different temperature than 0 °F. In fact, 0 °C is equal to 32 °F, so 32 °F must be added when converting from Celsius to Fahrenheit.

$$T_F = T_C\left(\frac{1.8\ °\text{F}}{1.0\ °\text{C}}\right) + 32\ °\text{F}$$

Unit conversions can require operations other than multiplication.

where T_F and T_C are the temperatures on the Fahrenheit and Celsius scales, respectively.

Many years after the Celsius temperature scale had been defined, scientists discovered that no temperature below −273.15 °C could be obtained. This lowest possible temperature is referred to as absolute zero. The SI temperature scale uses the same degree as the Celsius scale, but starts at absolute zero. The scale is named after Lord Kelvin. The relationship between Celsius and Kelvin temperatures is

$$T_K = T_C + 273.15$$

T_K is the temperature on the Kelvin scale; T_C is the Celsius temperature. The SI unit of temperature is the kelvin, abbreviated K. SI units do not use "degrees"; a temperature of 100 °C (the boiling point of water) corresponds to 373 K. Figure 1.18 shows a comparison of the three scales.

Since the size of the unit is the same on the Kelvin and Celsius temperature scales, a change in temperature of 12 °C is also a change of 12 K.

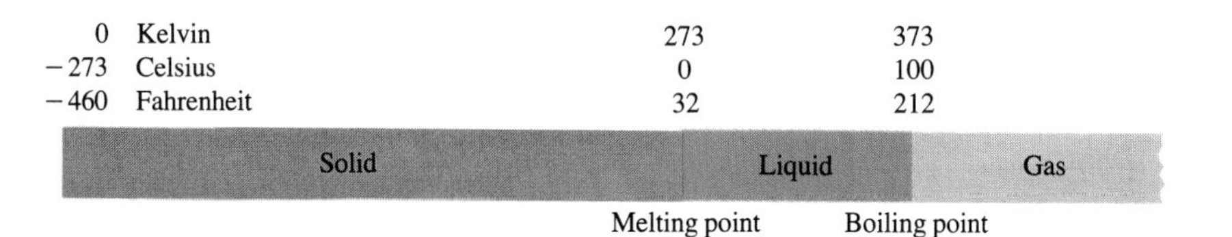

Figure 1.18 Temperature scales and the phase changes of water The three temperature scales that are most commonly used, and the relation of each to the melting and boiling points of water. The Kelvin scale is used to define the SI unit of temperature.

Example 1.6 Conversion among Temperature Scales

The densities of solids and liquids are usually measured at a temperature of 25 °C. Express this temperature on the Fahrenheit and Kelvin scales.

Solution The relationship needed to convert between the Fahrenheit and Celsius scales is

$$T_F = 25\ °\text{C}\left(\frac{1.8\ °\text{F}}{1.0\ °\text{C}}\right) + 32\ °\text{F} = 77\ °\text{F}$$

To find the temperature on the Kelvin scale, we simply change the zero point by adding 273:

$$T_K = 25 + 273 = 298\ \text{K}$$

Understanding The boiling point of benzene is 80 °C. Express this temperature in °F and in K.

Answer: 176 °F, 353 K

Equivalencies and Conversion Factors

Up to this point we have used the factor-label unit conversion method for calculations in which both units represent the same kind of quantity. This method can also be used to change from one type of measurement to another. For example, if we know the mass of a sample and the density of the material, the factor-label method can be used to find the volume occupied by that sample. When we say that the density of copper is 8.92 g/cm^3, we mean that a volume of 1 cm^3 is equivalent to a mass of 8.92 grams; that is, each 1 cm^3 of the sample has a mass of 8.92 g. We shall use the symbol $\simeq$ to represent "is equivalent to." For copper,

$$1\ \text{cm}^3\ \text{Cu} \simeq 8.92\ \text{g Cu}$$

Density of copper A cube of copper that is 1.0 cm on a side (a volume of 1 cm^3) has a mass of 8.92 g.

Equivalencies form the basis for conversion factors. The two conversion factors derived from the density of copper are

$$\left(\frac{8.92\ \text{g Cu}}{1\ \text{cm}^3\ \text{Cu}}\right) \quad \text{and} \quad \left(\frac{1\ \text{cm}^3\ \text{Cu}}{8.92\ \text{g Cu}}\right)$$

The use of this type of conversion factor is illustrated in Example 1.7.

Equivalent methods of expressing the same quantity can be used to derive conversion factors.

Example 1.7 Conversions between Equivalent Units

What is the volume occupied by 25.0 g of aluminum? The density of aluminum is 2.70 g cm^{-3}.

Solution Since the mass of this sample is known, and we want to find the volume, we must multiply the mass by a conversion factor that has units of grams in the denominator and volume units in the numerator. The equivalency based on the density of this material is

$$1\ \text{cm}^3\ \text{Al} \simeq 2.70\ \text{g Al}$$

The needed conversion factor is

$$\text{unit conversion factor} = \left(\frac{1\ \text{cm}^3\ \text{Al}}{2.70\ \text{g Al}}\right)$$

Applying this conversion factor, we obtain the desired volume:

$$\text{volume of aluminum} = 25.0 \text{ g Al} \left(\frac{1 \text{ cm}^3 \text{ Al}}{2.70 \text{ g Al}}\right) = 9.26 \text{ cm}^3 \text{ Al}$$

Understanding What is the mass of a 20.0 mL sample of mercury at 25 °C? The density of mercury is 13.5 g/mL at this temperature.

Answer: 270 grams

In the last example, the factor-label method was used to change one type of measurement into an equivalent measurement. Conversion factors based upon known *chemical* relationships will be used frequently throughout this text. These relationships will enable us to predict the amount of materials formed in a lab-scale reaction. These same relationships enable chemists to determine how much gasoline can be refined from a barrel of oil, how much limestone is consumed by acid rain, and how much heat can be produced by a cubic foot of natural gas.

1.6 Uncertainty in Measurements

Objectives

- To distinguish between accuracy and precision
- To use the convention of significant digits for expressing the uncertainty of measurements
- To express the results of calculations to the correct number of significant digits

Many automobile drivers have tried to avoid fines in traffic court (usually unsuccessfully) by claiming that the police made an error in measuring their speed. Police, on the other hand, rarely issue a speeding ticket unless the automobile exceeds the speed limit by several miles per hour because drivers cannot read their speedometers exactly. These situations illustrate two limitations that we encounter in making measurements. The driver in the first situation said that the measurement of speed was not accurate. **Accuracy** is the term used to express the agreement of the measured value with the true value of the same quantity. The policeman in the second situation referred to the precision of the measurement of speed. **Precision** expresses the agreement among repeated measurements. A high-precision measurement is one that produces the same results, time after time. On many autos, the width of the speedometer needle covers two miles per hour on the scale, so the driver might read the same speed differently at different times.

Accuracy and Precision

Table 1.8 shows four sets of data for the measurement of the mass of a nickel whose true mass is 5.11 g; the table illustrates both precision and accuracy. In each set of data the same nickel was measured four times, but on a different balance. The average value for each set is taken as the best value, and the range of values found (range is defined as the difference between the largest and smallest values) is a measure of the agreement among the individual

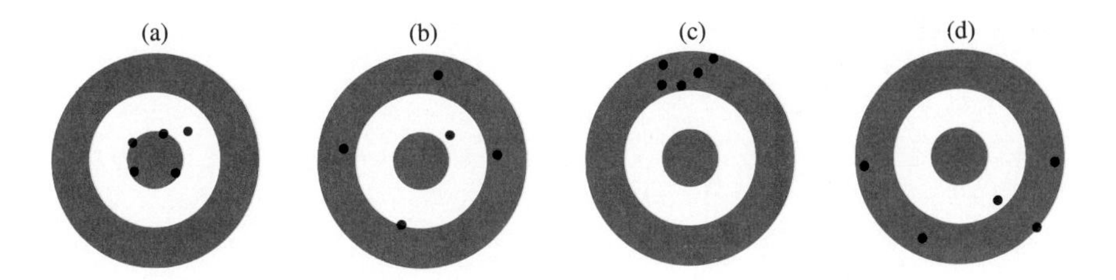

Accuracy and precision Accuracy and precision do not necessarily go hand-in-hand. Data can be (a) accurate and precise, (b) accurate but not precise, (c) precise but not accurate, or (d) neither accurate nor precise.

determinations. The determination of mass on Balance 1 is both precise and accurate, since the range of values is small and the average agrees with the true value. Balance 2 provides an accurate value, but it is not precise, because the range of individual measurements is large. Balance 3 is precise, but not accurate, possibly because it was improperly calibrated. Balance 4 is neither precise nor accurate. Accurate numbers have a small **error,** while precise numbers have a small **uncertainty.**

Accuracy expresses how close a number is to the true value. Precision expresses the repeatability, or how close duplicate values are to each other.

Table 1.8 Accuracy and Precision: Repetitive Weighing of an Object (True Mass = 5.11 g) on Several Balances

	Measured Mass (g)			
	Balance 1	Balance 2	Balance 3	Balance 4
	5.10	5.02	5.22	5.35
	5.12	5.20	5.21	5.10
	5.11	5.25	5.21	5.40
	5.11	4.97	5.20	5.15
average	5.11	5.11	5.21	5.25
range	0.02	0.28	0.02	0.25

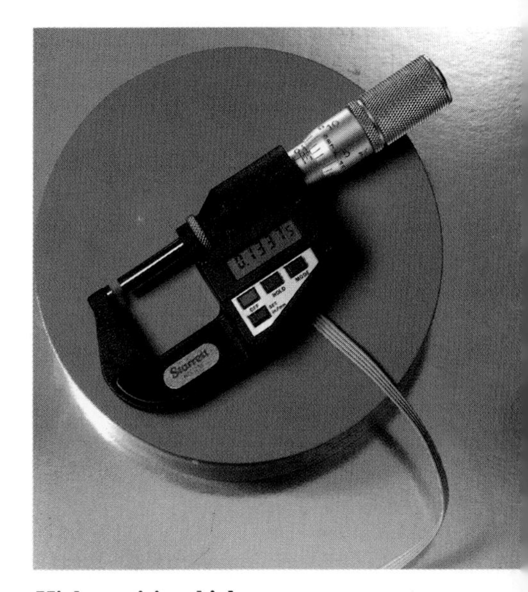

High-precision, high-accuracy measurements A digital micrometer can measure dimensions of items several centimeters long with high precision.

Absolute and Relative Precision

All measured quantities have some uncertainty, which often depends on the measuring device used. For example, we could measure the length of a playing field by counting the number of paces it takes to walk from one end to the other. An uncertainty of about one foot might be expected for this measurement. If we use a steel tape that is graduated in inches, the uncertainty would be a fraction of an inch. Laser measuring devices can reduce the uncertainty to a few hundredths of an inch. From this simple example, we can see that much of the uncertainty in any measurement will depend upon the measuring device used. In addition, no matter how good the measuring device, it must be used properly to minimize error and uncertainty.

When a measurement is made and reported, the precision of the measurement must also be reported. This precision is sometimes given explicitly by using the symbol ±. In this notation the length of a pencil might be reported as 21.3 ± 0.2 cm, if several measurements fell between 21.1 and 21.5 cm. These ± numbers are also called the **absolute uncertainties;** they

Figure 1.19 Devices that measure the volume of liquids Several devices are used to measure the volume of liquids. Considerations include ease of use as well as accuracy and precision. From left to right: the blue liquid is in a 10-mL graduated pipet that can dispense liquids with an accuracy of ±0.06 mL; the reddish liquid is in a volumetric flask that will contain 100.00 ±0.08 mL when filled to the mark; the yellow liquid is in a 100-mL graduated cylinder that measures volumes to ±0.4 mL; the green liquid is in a 10-mL graduated cylinder that measures volumes to ±0.06 mL; the red liquid at the right is in a 10-mL transfer pipet that delivers the liquid with an accuracy of ±0.04 mL.

The absolute uncertainty expresses the precision in the same units as the number itself.

express the uncertainty in the same units as the measurement. Figure 1.19 shows several pieces of laboratory equipment that are used to measure the volume of liquids, along with the typical uncertainty in the measurement.

Another way of expressing precision is the **relative uncertainty,** the ratio of the absolute uncertainty to the measured value, often stated as a percentage:

$$\%\text{ relative uncertainty} = \left(\frac{\text{absolute uncertainty}}{\text{measured value}}\right) 100\%$$

Two masses given as (a) 10.00 ± 0.03 grams and (b) 50.00 ± 0.03 grams both have the same absolute uncertainty of 0.03 grams, but the second number has a smaller relative uncertainty:

The relative uncertainty expresses the precision as a fraction (usually a percentage) of the number.

$$\text{(a) } \%\text{ relative uncertainty} = \frac{\pm 0.03\text{ g}}{10.00\text{ g}} 100\% = \pm 0.3\%$$

$$\text{(b) } \%\text{ relative uncertainty} = \frac{\pm 0.03\text{ g}}{50.00\text{ g}} 100\% = \pm 0.06\%$$

Significant Digits

Often, a different method of reporting uncertainty is used. Although the absolute and relative uncertainties can be calculated, a simpler method provides estimates of the uncertainties that are nearly as good as those obtained with detailed calculations.

In the **significant digit** convention, the uncertainty is presumed to be ±1 in the last digit reported. (Many scientists use "significant figure" as a synonym for "significant digit.") If a volume is reported as 12.3 mL, there is an

	12.00	0.1005	0.000532	15
Number of significant digits	4	4	3	2
Implied uncertainty	±0.01	±0.0001	±0.000001	±1
Relative uncertainty	$\frac{\pm 0.01}{12.00}$	$\frac{\pm 0.0001}{0.1005}$	$\frac{\pm 0.000001}{0.000532}$	$\frac{\pm 1}{15}$
% relative uncertainty	±0.08%	±0.10%	±0.19%	±6.7%

Figure 1.20 Significant digits The arrow above the number points to the first significant digit, and that below the number shows the last significant digit.

implied uncertainty of ±0.1 mL; the volume could be as small as 12.2 mL or as large as 12.4 mL. If the uncertainty of the measurement of this volume were ±0.01 mL, then the volume would be recorded as 12.30 mL.

The decimal place of the last digit determines the absolute uncertainty. The **number of significant digits** (or the number of significant figures) is the number of digits from the first nonzero digit to the last digit. This number provides a measurement of the relative uncertainty. The numbers 0.1, 0.0020, 0.0103, and 12.00 have one, two, three, and four significant figures, respectively. Note that the zeros preceding the first nonzero digit are not counted in the number of significant digits. They are simply being used to locate the decimal point. A zero is significant only if it is preceded somewhere by a nonzero digit. Some examples are given in Figure 1.20.

Numbers are presumed to have an absolute uncertainty of ±1 in the last digit.

Example 1.8 Determining the Number of Significant Digits

How many significant digits are present in each of the following measured quantities, and what is the uncertainty indicated by the number given?

(a) A package of candy has a mass of 103.42 g.
(b) The mass of a milliliter of gas is 0.003 g.
(c) The volume of a solution is 0.2500 L.

Solution

(a) Counting from the first nonzero digit in 103.42 to the last digit expressed, there are five significant figures. The uncertainty is ±0.01 grams.
(b) None of the zeros in 0.003 are significant; they show only the location of the decimal point. There is only one significant digit, and the uncertainty is ±0.001 g.
(c) The 2 is the first nonzero digit in 0.2500 L, so there are four significant digits, including the last two zeros. The uncertainty is ±0.0001 L.

Understanding How many significant digits are in the number 0.01020?

Answer: Four significant digits

When a measured number does not contain a decimal point and ends in one or more zeros, for example 2000, the number of significant digits is unclear. It is not known which of these zeros are used simply to locate the decimal point and which, if any, define precision. One way to remove this ambiguity is to express the measurement in **exponential notation,** also called **scientific notation.** This method expresses a quantity as the product of two numbers. The first is a number between 1 and 10, and the second is 10 raised to some whole number power. When exponential notation is used, the uncertainty is expressed more clearly. If we write 2000 as 2.00×10^3, the uncertainty is 0.01×10^3 or 10. Exponential notation is also a space-saving way to represent very small and very large numbers. For example it is much more compact to express the number 0.00000000431 as 4.31×10^{-9}. Exponential notation is reviewed in Appendix A.

Uncertainty in Calculations

Measurement of mass The mass of the material is determined by subtracting the mass of the empty container from the mass of the container plus its contents.

In many experiments the quantities of interest must be calculated from several measured values. For example, we must often determine the mass of a sample from the mass of the empty container and the total mass of the sample plus the container.

mass of sample = total mass − mass of container

The uncertainty in the mass of the sample depends on the uncertainty in each of the measurements.

The number of significant digits in a calculated value depends upon the uncertainties of the measurements *and* the type of mathematical operations used. An electronic calculator does not preserve the significant digit convention; it generally displays as many digits as fit across the calculator face. It is our responsibility to determine the number of significant digits in the result of any calculation, since the significant digits represent the uncertainty in the measurement.

Addition and Subtraction

The uncertainty in the result of an addition or subtraction is determined from the absolute uncertainties of the original numbers. The absolute uncertainty in the answer can be no smaller than the absolute uncertainty in any of the numbers from which it is derived. Consider the difference between two numbers.

$$25.34 - 24.0 = 1.34$$

The calculator will read 1.34, but if we were to express the answer to two decimal places, we would imply an uncertainty of ± 0.01 ($\pm$ one digit in the last place). This level of uncertainty is not justified because one of the numbers (24.0) has an uncertainty of ± 0.1. The same uncertainty is present in the result, so it should be expressed to one decimal place.

$$25.34 - 24.0 = 1.3$$

The *uncertainty* in the result of an *addition* or *subtraction* is determined from the *decimal place of the numbers that are used to derive the answer.* If, for

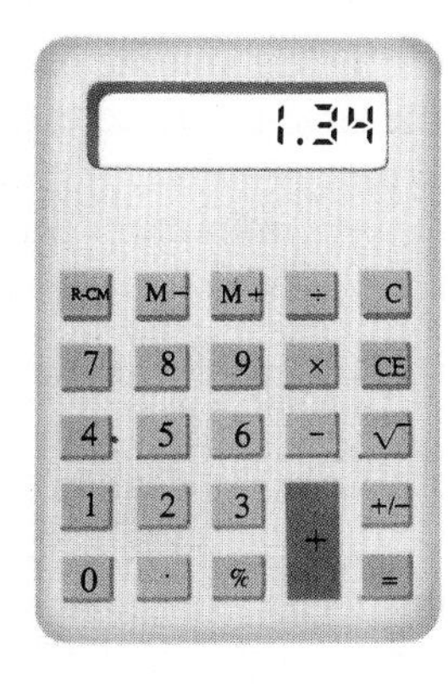

Significant digits The operator must report the proper number of significant digits—the calculator will not.

example, a number with uncertainty in the third decimal place is added to a number with uncertainty in the fourth place, the result will have uncertainty in the third decimal place.

An electronic calculator can also provide too few digits.

$$28.39 - 6.39 = \underset{\text{(calculator)}}{22} = \underset{\text{(correct)}}{22.00}$$

The answer is derived from numbers with two decimal places, so the answer should also be expressed to two decimal places. In this case, many calculators display only two figures, instead of the four that are significant.

When the result of a calculation has too many digits, we must *round* the number, to reflect the proper number of significant digits, rounding a 5 to the even number. If a calculation of a quantity has three significant digits, and your calculator displays 12.35, then report 12.4; if the display is 12.25, then report 12.2.

The number of significant digits in the result of an addition or subtraction problem is determined by the *decimal place* of the least significant digit.

Multiplication and Division

The precision in a multiplication or division process is determined by the relative uncertainty of the combined numbers. The *number of significant digits* in a product or quotient is the same as the *fewest* number of significant digits in the original numbers. For example, to determine the density of a substance, the mass and the volume of the sample must be measured. For a sample with a mass of 7.311 g and a volume of 7.7 cm^3, the density is

$$d = \frac{7.311\text{ g}}{7.7\text{ cm}^3} = \underset{\text{(calculator)}}{0.9494805\text{ g/cm}^3} = \underset{\text{(correct)}}{0.95\text{ g/cm}^3}$$

The density is expressed to two significant digits, the same as the measured volume. Note that although the implied uncertainty in the volume is ± 0.1, the implied absolute uncertainty in the density is ± 0.01.

The rationale for these operations becomes clearer if we calculate the minimum and maximum values that the density can have. The mass, 7.311 g, is in the range from 7.310 g to 7.312 g. The volume of the sample may be measured in a graduated cylinder. The measured volume is 7.7 cm^3; it may be as low as 7.6 or as high as 7.8 cm^3. The ratio of the smallest mass to

$$\begin{array}{r} 12.516 \\ 3.15 \\ 24.1 \\ \hline 39.766 \end{array}$$

39.8

Round the results shown on the calculator to 39.8 to report the proper number of significant digits.

the largest volume provides the lowest estimate of the density. The ratio of the largest mass to the smallest volume provides the highest estimate.

$$\text{minimum density} = \frac{7.310\text{ g}}{7.8\text{ cm}^3} = 0.937\text{ g/cm}^3$$

$$\text{maximum density} = \frac{7.312\text{ g}}{7.6\text{ cm}^3} = 0.963\text{ g/cm}^3$$

$$\text{density range} = 0.937\text{ to }0.963\text{ g/cm}^3$$

$$\text{rounded density range} = 0.94\text{ to }0.96\text{ g/cm}^3$$

$$\text{density} = 0.95 \pm 0.01\text{ g/cm}^3$$

The number of significant digits in the result of a multiplication or division is determined by the smallest number of significant digits in the numbers from which it is calculated.

The range of densities derived from the calculations of minimum and maximum is more involved, but provides the same result. *The product or quotient should have no more significant digits than the fewest number in any of the quantities from which it is derived.*

If the product 0.5000×6.0000 is evaluated on a calculator, the display shows 3 as the result. The component numbers have four and five significant figures; therefore, the result must have four significant figures, so 3.000 is the correct representation.

Example 1.9 Combining Measured Quantities

Express the result of the calculation to the correct number of significant digits.

(a) $0.092 \times 25.32/27.31$
(b) $55.8752 - 56.533$
(c) $0.198 \times 10.012937 + 0.8021 \times 11.009305$
(d) $2.334 \times 10^{-2} + 3.1 \times 10^{-3}$

Solution

(a) This calculation involves only multiplication and division. Two of the three numbers have four significant digits, but the third has only two. The result of the calculation will only have two significant digits.

$$\frac{0.092 \times 25.32}{27.31} = \underset{\text{(calculator)}}{0.085296228} = \underset{\text{(correct)}}{0.085}$$

(b) Only subtraction is involved in this calculation. We can identify the last significant digit more easily by writing the numbers in a column:

$$\begin{array}{rl} 55.8752 & \\ -\ 56.533 & \\ \hline -\ 0.6578 & \text{(calculator)} \\ -\ 0.658 & \text{(correct)} \end{array}$$

There is uncertainty in the third decimal place of the second number, so the result should be expressed to three decimal places.

(c) First evaluate the two products, then add the results. The first product has three significant digits, and the second product has four. The sum is reported to two decimal places.

$$\begin{array}{rcrcr} 0.198 \times 10.012937 & = & 1.98256 & = & 1.98 \\ +\ 0.8021 \times 11.009305 & = & 8.83056 & = & 8.831 \\ \hline & & 10.81312 & = & 10.81 \end{array}$$

The numbers are shown rounded in the last column so that the number of significant digits is clear. In general, round the answer only in the last step, preserving the full number of digits in intermediate steps.

(d) The problem is simplified if both numbers are expressed in exponential notation with the same power of ten. Choosing the second number to change,

$$3.1 \times 10^{-3} = 0.31 \times 10^{-2}$$

Now the uncertainty in the addition operation is easy to interpret.

$$\begin{array}{r} 2.334 \times 10^{-2} \\ +0.31 \times 10^{-2} \\ \hline 2.64 \times 10^{-2} \end{array}$$

Understanding Express the results of the calculation to the correct number of significant digits: 1.33/55.494 + 10.00.

Answer: 10.02

The concept of significant digits applies only to measured numbers, or quantities calculated from measured numbers. Three kinds of numbers never limit significant digits. These numbers are:

1. Counted numbers or tallies. There are exactly five fingers on a hand, or 24 students in a class. There is no uncertainty in the numbers 5 and 24.
2. Defined numbers. The meanings of the prefixes in the *Système International* are defined quantities. There are exactly 100 cm in a meter, so there is no uncertainty at all in the unit conversion factor 100 cm/m.
3. The power of ten, when exponential notation is used, is an exact number and will never limit the number of significant digits.

Summary

Chemistry is central to all branches of science. Chemistry has advanced by careful experimentation guided by the *scientific method.* The results of a large number of experimental observations are summarized into *laws.* Scientists then try to formulate *theories* that explain these laws of nature.

Matter is anything that has *mass* and occupies space. *Extensive properties* tell us how much of a given type of matter that we have in a given sample, while *intensive properties* are characteristic of the type of matter and do not depend on the size of the sample. *Density,* the ratio of mass to volume, is an important intensive property that is derived from the ratio of two extensive properties. *Physical properties* are observed without changing the composition of the sample, whereas *chemical properties* describe how the sample undergoes *chemical change.*

The broadest division of matter is into *mixtures* and *pure substances.* Mixtures can be either *homogeneous,* the same throughout, or *heterogeneous,* in which different parts have different properties. Pure substances cannot be separated by physical means and are divided into *compounds* and *elements.* Compounds can be decomposed into elements by chemical means.

A variety of techniques can be used to separate the components of a mixture. A *filtration* is used to separate a solid from a liquid. A *distillation* is used to separate volatile liquids from a mixture. A mixture of liquids can be separated by a fractional distillation. *Chromatography* is used to separate many different types of mixtures.

The *periodic table* arranges the elements in approximate order of increasing mass in rows, known as *periods,* and columns, called *groups.* The elements in any group have similar chemical properties. The groups are numbered for identification. Most chemists use a system of Roman numerals (I to VIII) and the letters A or B; some use Arabic numerals from 1 to 18. The group on the left side of the table, labeled IA (or 1), is the *alkali metal* group. The elements of the next group, labeled IIA (2), are the *alkaline earth metals.* The elements on the right side of the table are called the *noble gases* (VIIIA or 18) because they are not very reactive. To the left of the noble gases are the *halogens* (VIIA or 17). The *metals* are the elements in the center and left side, and the *nonmetals* are in the top right of the periodic table. Metals have luster and can conduct electrical current; nonmetals have neither property. A few elements that lie between the metals and nonmetals on the periodic table are called the *metalloids.* The groups of elements labeled as A on the periodic table are called the *representative* or *main group* elements, and those labeled as B are the *transition metals.* The *lanthanide* and *actinide* metals are set off at the bottom of the table to save space.

Scientists express measurements using the *SI* units. In this system seven *base* units are defined, and all other units of measure are derived from these. *Unit conversion factors* are very useful in changing from one unit type to another. The use of conversion factors to change the units of a quantity is known as the factor-label method. These conversion factors may be based upon definitions or equivalencies.

Both the Celsius and Kelvin scales are used to measure temperature. These scales have the same size unit of measure, but the Kelvin scale is based on an absolute zero.

All measurements have an uncertainty associated with them. *Significant digits* are used to indicate the uncertainty in measurements. The proper number of significant figures must be determined by the uncertainty in the measurements and how they are combined. In addition and subtraction, the number of significant figures in the answer is determined from the absolute uncertainty of the numbers from which the answer is derived. In multiplication and division calculations, the number of significant figures in the answer is the same as in the quantity with the fewest number of significant figures.

Chapter Terms

Chemistry—the study of matter and its interactions with other matter and with energy. *(1.1)*

Scientific method—investigations that are guided by theory. *(1.1)*

Hypothesis—a possible explanation for observed results. *(1.1)*

Law—a statement or equation that summarizes a large number of observations. *(1.1)*

Theory—an explanation of the laws of nature. *(1.1)*

Matter—anything that has mass and occupies space. *(1.2)*

Mass—a measure of the quantity of matter in a sample or object. *(1.2)*

Weight—the force of attraction between two objects. The weight of an object will change from one location to another but the mass is always the same. *(1.2)*

Property—anything that can be observed or measured. *(1.2)*

Extensive properties—those that depend upon the specific sample that is under observation. Examples are mass and volume. *(1.2)*

Intensive properties—those that are identical in any sample of a particular substance. Examples are color, density, and melting point. *(1.2)*

Density—the ratio of mass to volume. *(1.2)*

Physical property—one that can be observed without changing the substances present in the sample. *(1.2)*

Chemical property—the tendency to react and form new substances. *(1.2)*

Physical change—one that can be observed without changing the substance(s) present. *(1.2)*

Chemical change—a process in which one or more new substances are produced. *(1.2)*

Mixture—two or more substances that can be separated by taking advantage of different physical properties of the substances. *(1.2)*

Substance—matter that cannot be separated into component parts by a physical process. *(1.2)*

Heterogeneous mixture—a mixture in which different parts have different properties and composition. *(1.2)*

Homogeneous mixture—a mixture in which all parts of the sample exhibit identical properties. *(1.2)*

Solution—another name for a homogeneous mixture. *(1.2)*

Alloy—a mixture of a metal and one or more additional elements, often a second metal. *(1.2)*

Element—a substance that cannot be decomposed into a simpler substance by normal chemical means. *(1.2)*

Compound—a substance that can be decomposed into its elements by chemical processes. *(1.2)*

Symbol—an abbreviation for an element that consists of one or two letters, usually related to the name of the element. *(1.2)*

Periodic table—an arrangement of the elements (shown by their symbols) into rows and columns so that elements with similar chemical properties are in the same column. *(1.3)*

Period—a row of the periodic table. *(1.3)*

Group—a column of the periodic table. *(1.3)*

Metal—a material that has luster and is a good conductor of electricity. Metallic elements are located in the center and left side of the periodic table. *(1.3)*

Nonmetal—a material that lacks the characteristics of a metal. Nonmetallic elements are in the top right part of the periodic table. *(1.3)*

Metalloid—an element with properties intermediate between those of a metal and a nonmetal. The elements along the dividing line between metals and nonmetals in the periodic table. *(1.3)*

Semiconductor—a weak conductor of electricity. *(1.3)*

Representative elements (also called **main group elements**)—the elements in groups labeled A (IA to VIIIA) or groups 1 to 2 and 13 to 18. *(1.3)*

Transition metals—elements in groups labeled B (IIIB to VIIIB, IB to IIB) or groups to 3 to 12. *(1.3)*

Lanthanides—the elements in period 5 between lanthanum (La) and hafnium (Hf), cerium (Ce) through lutetium (Lu). *(1.3)*

Actinides—the elements in period 6 following actinium (Ac), thorium (Th) through lawrencium (Lr). *(1.3)*

Inner transition elements—the lanthanides and the actinides. The inner transition elements are placed at the bottom of the periodic table. *(1.3)*

Alkali metals—the elements in group IA (1): Li, Na, K, Rb, Cs, Fr. *(1.3)*

Alkaline earth metals—the elements in group IIA (2): Be, Mg, Ca, Sr, Ba, Ra. *(1.3)*

Halogens—the elements in group VIIA (17): F, Cl, Br, I, At. *(1.3)*

Noble gases—the elements of group VIIIA (18). They were formerly known as the "inert gases": He, Ne, Ar, Kr, Xe, Rn. *(1.3)*

Filtration—the process of separating a mixture of a solid and liquid by passing it through a porous barrier. *(1.4)*

Distillation—the separation of components of a liquid mixture based on differences in volatility. It consists of two steps: (1) the mixture is heated to convert the liquid into a vapor, its gaseous form, and (2) the vapor is condensed to a liquid in a different container. *(1.4)*

Chromatography—a technique that employs two materials, one moving and one stationary, to separate a mixture into its components. *(1.4)*

Units—standards used for quantitative comparison between measurements of the same type of quantity. *(1.5)*

Base unit—any of the defined units in the SI. *(1.5)*

Derived unit—a unit that is composed of combinations of base units. *(1.5)*

Unit conversion factor—a fraction in which the numerator and the denominator express the same quantity in different units. *(1.5)*

Equivalency—a relationship that expresses the same quantity in two different types of units. Equivalent quantities are represented by the symbol ≏. *(1.5)*

Accuracy—the agreement of the measured value with a true or accurately known value of the same quantity. *(1.6)*

Precision—agreement among repeated measurements. *(1.6)*

Error—the difference between the measured result and the true value. *(1.6)*

Uncertainty—related to precision. Measurements of high precision have small uncertainty. *(1.6)*

Absolute uncertainty—a measure of the expected range in the measurement. *(1.6)*

Relative uncertainty—the ratio of the absolute uncertainty of a number to the number itself, often expressed as a percentage. *(1.6)*

Significant digit—all digits in a measurement from the first nonzero digit through the first digit that is uncertain. Also called "significant figure." *(1.6)*

Number of significant digits—the number of digits from the first nonzero digit to the last significant digit in the quantity. *(1.6)*

Exponential notation—a quantity expressed as the product of a number between 1 and 10 multiplied by a power of 10. Also called "scientific notation." *(1.6)*

Exercises

Exercises designated with color have answers in Appendix J.

The Nature of Science

1.1 Compare the use of the words "theory" and "hypothesis" by scientists and by the general public.

1.2 Go to the library to examine some current thoughts about topics like "What caused the ice age?" or "Why did dinosaurs become extinct?" Try to separate the factual data from the writer's opinions and determine whether *you* would classify the material as theory, hypothesis, etc.

1.3 Draw a diagram similar to Figure 1.1 that places the following words in the proper relationships: theory, hypothesis, hunch, guess, model, data.

Properties

1.4 Define matter, mass, and weight.

1.5 Give examples of homogeneous and heterogeneous mixtures.

1.6 Is the light from an electric bulb an intensive or extensive property?

1.7 In each part there is an underlined property. Classify the property as intensive or extensive, and as chemical or physical.
(a) An apple is red.
(b) Sulfuric acid converts sugar into carbon and steam.
(c) One pound of butter.
(d) Sugar is soluble in water.
(e) Wood burns in air forming carbon dioxide and water.

1.8 In each part there is an underlined property. Classify the property as intensive or extensive, and as chemical or physical.
(a) Mercury is a silvery liquid.
(b) A ball is a spherical object.
(c) Sodium and chlorine react to form table salt.
(d) A sample of water has a mass of 28 grams.
(e) The density of aluminum is 2.70 g/cm^3.

1.9 Classify the following as chemical changes or physical changes:
(a) melting of ice
(b) burning a piece of paper
(c) leaves changing color
(d) iron rusting
(e) sugar dissolving in water
(f) a firecracker exploding
(g) magnetizing an iron nail
(h) baking a cake

1.10 Which of the following describe physical changes and which describe chemical changes?
(a) milk souring
(b) water evaporating
(c) forming copper wire from a bar of copper
(d) frying an egg
(e) shattering a glass window
(f) distillation of alcohol
(g) mixing an Alka-Seltzer tablet with water
(h) producing light from an incandescent light bulb

1.11 In the following description of the element aluminum, identify which of the properties are chemical and which are physical. "Aluminum is a silver colored metal that dissolves in an alkaline solution with the evolution of a gas. It can be isolated from a molten mixture of alumina and cryolite by passing an electric current through the melt. The metal has a density of 2.70 g cm^{-3}, melts at 660 °C, and boils at 2467 °C. Aluminum is vaporized in a vacuum and deposited on a smooth glass surface to make mirrors for telescopes."

1.12 In the following description of the element fluorine, identify which of the properties are chemical and which are physical. "Fluorine is a pale yellow corrosive gas that reacts with practically all organic and inorganic substances. Finely divided metals, glass, ceramics, carbon, and even water burn in fluorine with a bright flame. Small amounts of compounds of this element in drinking water and toothpaste prevent dental cavities. The free element has a melting point of −219.6 °C, and boils at −188.1 °C. Fluorine is one of the few elements that form compounds with the element xenon."

Elements, Compounds, and Mixtures

1.13 Are all alloys homogeneous solutions?

1.14 Explain the difference between substances, compounds, and elements.

1.15 Classify each of the following as an element, compound, or mixture. Identify mixtures as homogeneous or heterogeneous.
(a) water
(b) window cleaner
(c) 14 karat gold
(d) copper
(e) air
(f) sugar

1.16 Classify each of the following as an element, compound, or mixture. Identify mixtures as homogeneous or heterogeneous.
(a) cough syrup
(b) green leaf
(c) helium
(d) a muddy river
(e) a window glass

1.17 For each of the symbols, give the name of the element.
(a) O
(b) Na
(c) Si
(d) P
(e) Cl
(f) Pb

1.18 Give the symbol used for each of the following elements.
(a) silver
(b) carbon
(c) sulfur
(d) bromine
(e) helium

The Periodic Table

1.19 Cadmium is frequently found as an impurity in zinc minerals. Explain this fact.

1.20 Define group and period. Develop a method that you can use to convert group numbers between the Roman numeral system popular in the United States and the Arabic numerals that are used in many other countries.

1.21 Provide the name and symbol for an element that is in the same group with the element given.
(a) sodium (d) neon
(b) Fe (e) Ge
(c) bromine (f) Ca

1.22 Provide the name and symbol for an element that is in the same group with the element given.
(a) Ti (d) Ba
(b) oxygen (e) argon
(c) fluorine (f) N

1.23 From the position in the periodic table, identify each of the following elements as a representative, a transition, or an inner transition element.
(a) silicon (d) Np
(b) Cr (e) barium
(c) magnesium

1.24 From the position in the periodic table, identify each of the following elements as a representative, a transition, or an inner transition element.
(a) sulfur (d) europium
(b) iron (e) Br
(c) K

1.25 Give the symbol and name for each of the following elements:
(a) The alkaline earth element in the same period as sulfur.
(b) A noble gas in the same period as potassium.
(c) The heaviest alkali metal.
(d) A halogen in the same period with tin (Sn).

1.26 Give the symbol and name for each of the following elements:
(a) The alkali metal in the same period as chlorine.
(b) A halogen in the same period as magnesium.
(c) The heaviest alkaline earth metal.
(d) A noble gas in the same period with carbon.

1.27 How many elements are in
(a) the third period?
(b) group IVA?
(c) the alkaline earth family?
(d) the first transition series?
(e) the chromium group?

1.28 Which pair of the following elements would you expect to show the greatest similarity in physical and chemical properties? Te, Sr, P, Na, Kr, Ra

Separation of Mixtures

1.29 Suggest ways of separating the following mixtures:
(a) salt and water
(b) the dyes in blue ink
(c) sugar and carbon black
(d) gasoline and water
(e) gasoline and kerosene
(f) sawdust and sugar

1.30 Suggest ways to separate the following mixtures:
(a) potassium nitrate and water
(b) sand and salt
(c) alcohol and water
(d) sulfur and carbon
(e) kerosene and water

Units and Conversions

1.31 What are the *base* SI units used to express each of the following quantities?
(a) the mass of a bag of sugar
(b) the distance from the earth to the moon
(c) the temperature of a sunny August day
(d) the time it takes to run 15 kilometers

1.32 What are the *base* SI units used to express each of the following quantities?
(a) the mass of a person
(b) the distance from London to Paris
(c) the boiling point of water
(d) the length of a chemistry lecture

1.33 Perform the conversions needed to fill in the blanks. Use exponential notation if appropriate.
(a) 3.01 cm = ______ m = ______ mm = ______ nm
(b) 50.0 cm^3 = ______ dm^3 = ______ mL
= ______ L = ______ m^3
(c) 23.1 g = ______ mg = ______ kg
(d) 72 °F = ______ °C = ______ K
(e) 45 s = ______ ms = ______ minutes

1.34 The 1500-meter race is sometimes called the "metric mile." Express this distance in miles.

1.35 A standard sheet of paper in the United States is 8.5 × 11 inches. Express the area of this sheet of paper in cm^2.

1.36 Wine is sold in 750-mL bottles. How many quarts of wine are there in a case of 12 bottles?

1.37 The speed limit on limited-access roads in Canada is 100. km per hour. How fast is this in miles per hour? In meters per second?

1.38 Derive an equation, including units, to convert between temperature scales:
(a) Fahrenheit to Celsius
(b) Kelvin to Fahrenheit
(c) Fahrenheit to Kelvin

1.39 Express the boiling point of benzene, 80 °C, in degrees Fahrenheit and in kelvins.

1.40 Helium has the lowest boiling point of any substance. It boils at 4.21 K. Express this temperature in °C and °F.

1.41 At what temperature will a Celsius thermometer give the same numerical reading as a Fahrenheit thermometer?

1.42 An irregularly shaped piece of metal with a mass of 47.8 g is placed into a graduated cylinder containing 30.0 mL of water. The water level rises to 48.5 mL. What is the density of the metal in g/cm^3?

1.43 A 50.0-mL graduated cylinder has a mass of 52.1 g. When it is filled with an unknown liquid to the 40.0 mL mark, the cylinder and liquid have a combined mass of 88.5 g. Calculate the density of the liquid in g/cm^3.

1.44 The density of benzene at 25.0 °C is 0.879 g/cm^3. What volume, in liters, will 2.50 kg of benzene occupy?

1.45 What is the radius of a copper sphere (density = 8.92 g/cm^3) whose mass is 3.75×10^3 g? The volume of a sphere is given by the equation $V = (4/3)\pi r^3$.

1.46 Lead has a density of 11.4 g/cm^3. What is the mass of a lead brick measuring 8.50 inches by 5.10 inches by 3.20 inches?

1.47 Acetone, the solvent in nail polish remover, has a density of 0.791 g/cm^3. What will be the volume of 25.0 g of acetone?

Significant Digits

1.48 Explain the relationship between the number of significant digits and the precision of a measurement.

1.49 Describe a computation in which *your* calculator makes an error in the number of significant digits.

1.50 How many significant digits does each of the numbers contain?
(a) 1.5003
(b) 0.0070
(c) 5.7
(d) 2.00×10^7
(e) 15.9994
(f) 0.010003

1.51 How many significant digits does each of the numbers contain?
(a) 88
(b) 4.104
(c) 5×10^3
(d) 5.0005
(e) 22.9898
(f) 0.0040

1.52 Perform the calculations indicated and express the answer to the correct number of significant digits. Use exponential notation where needed.
(a) 17.2×12.55
(b) $1.4 \times 1.11/42.33$
(c) 18.33×0.0122
(d) $25.7 - 25.25$
(e) $19.5 + 2.35 + 0.037$
(f) $2.00 \times 10^3 - 1.7 \times 10^1$

1.53 Perform the calculations indicated and express the answer to the correct number of significant digits. Use exponential notation where needed.
(a) 1.88×36.35
(b) $1.04 \times 3.114/42$
(c) $15/25.69$
(d) $45.2 - 37.25$
(e) $28.5 + 4.43 + 0.073$
(f) $3.10 \times 10^3 - 5.1 \times 10^1$

1.54 The following calculations involve multiplication/division and addition/subtraction operations of measured values in the same problem. Evaluate each and give the answer to the correct number of significant digits.

(a) $\dfrac{(25.12 - 1.75) \times 0.01920}{(24.339 - 23.15)}$

(b) $\dfrac{55.4}{(26.3 - 18.904)}$

(c) $(0.9221 \times 27.977) + (0.0470 \times 28.976) + (0.309 \times 29.974)$

1.55 The following calculations involve multiplication/division and addition/subtraction operations of measured values in the same problem. Evaluate each and give the answer to the correct number of significant digits.

(a) $\dfrac{(48.35 - 35.18) \times 0.12}{(33.792 - 31.426)}$

(b) $\dfrac{48.33}{(35.2 - 29.0)}$

(c) $(0.0742 \times 6.01512) + (0.9258 \times 0.190100)$

Unit Conversions

1.56 Compare and contrast equivalent quantities with equal quantities.

1.57 Draw a block diagram that explains how to convert km/hr to m/s.

1.58 The speed of sound in air at sea level is 340 m/s. Express this speed in miles per hour.

1.59 A package of aluminum foil with an area of 75 ft^2 weighs 12 ounces. Use the density of aluminum, 2.70 g/cm^3, to find the average thickness of this foil in nanometers. (volume = thickness × area)

1.60 A light-year, the distance light travels in one year, is a unit used by astronomers to measure the large distances between stars. Calculate the distance in miles represented by one light-year. Assume that the length of a year is 365.25 days and that light travels at a rate of 3.00×10^8 m/s.

1.61 The radius of a sodium atom is 1.86×10^{-10} m. Assuming that the sodium atom is spherical, how many atoms placed side-by-side would be needed to equal the length of a standard sheet of paper, 11.0 inches?

1.62 How many square meters will 1.00 gallon of paint cover if it is applied to a uniform thickness of 8.00×10^{-2} millimeters?

1.63 The distance between the centers of the two oxygen atoms in an oxygen gas molecule is 1.21×10^{-10} m. Calculate this distance in feet.

1.64 Gold leaf, which is used for many decorative purposes, is made by hammering pure gold into very thin sheets. Assuming that a sheet of gold leaf is 1.27×10^{-5} cm thick, how many square feet of gold leaf could be obtained from 28.35 g of gold? The density of gold is 19.3 g/cm^3.

1.65 The speed of light is 3.00×10^8 m/s. Assuming that the distance from the earth to the sun is 93,000,000 miles, how many minutes does it take for light to reach the earth from the sun?

Additional Exercises

1.66 The mass of a piece of metal is 134.412 g. It is placed into a graduated cylinder that contains 12.35 mL of water. The volume of the metal and water in the cylinder is found to be 19.40 mL. Calculate the density of the metal.

1.67 Consider two liquids, liquid A with a density of 0.98 g/mL and liquid B with a density of 1.03 g/mL. Notice that one density is known to two significant digits and the other to three. Calculate the volume of liquid A in a sample that weighs 9.9132 g; be sure to express your result to the proper number of significant digits. Calculate the volume of the same mass of liquid B, again making sure that you have the appropriate number of significant digits.

Recording the number of significant digits is only one way to estimate the uncertainty. Repeat the calculations of volume by using the minimum and maximum values of density to calculate a maximum and minimum volume. The range between the two is also a measure of uncertainty.

Compare the estimated uncertainties in the two liquids as measured by the two techniques. Do all estimates give the same answer? Should they? Explain any disagreements.

Chapter Two

Atoms, Molecules, and Ions

For thousands of years, scientists have tried to learn why one material is different from another. For example, at room temperature and pressure, helium is a colorless gas, chlorine is a greenish yellow gas, bromine is a red liquid, sulfur is a yellow solid, mercury is a silver-gray liquid, and copper is a shiny reddish brown solid (Figure 2.1).

The physical characteristics of these elements are different and so are their chemical behaviors. For example, bromine will react violently with many substances, but helium is unreactive under all conditions. This chapter will lay the foundation to develop concepts and models that can explain these observations.

2.1 Dalton's Atomic Theory

Objectives

- To describe modern atomic theory and to show how the theory explains the laws of constant composition, multiple proportions, and conservation of mass
- To interpret how molecular formulas represent combinations of atoms

More than 2300 years ago, Greek philosophers asked whether a sample of matter divided into smaller and smaller pieces would retain the bulk properties of the substance. In other words, is matter "continuous," or is it com-

Scanning tunneling microscope view of iodine atoms.

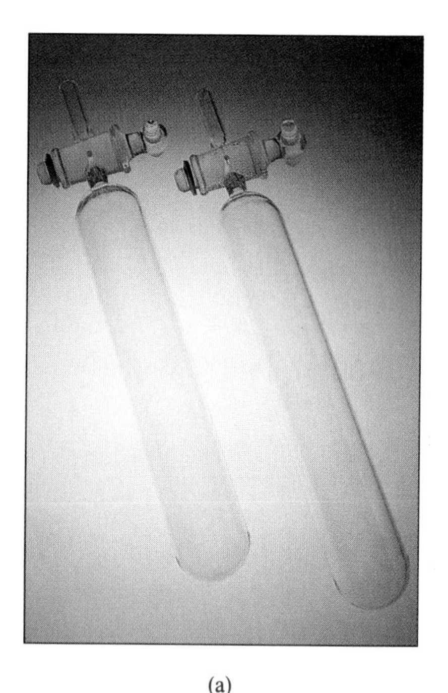

(a)

(b)

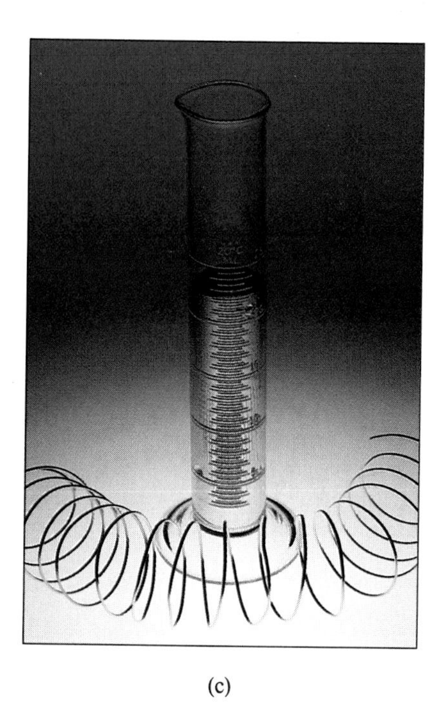

(c)

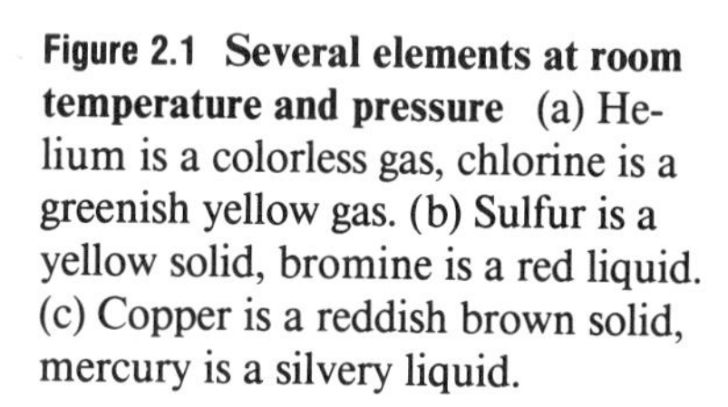

Figure 2.1 Several elements at room temperature and pressure (a) Helium is a colorless gas, chlorine is a greenish yellow gas. (b) Sulfur is a yellow solid, bromine is a red liquid. (c) Copper is a reddish brown solid, mercury is a silvery liquid.

posed of small indivisible particles that if further subdivided do not retain the bulk properties of the sample ("discontinuous" matter)? A proponent of the idea that matter is discontinuous, Democritus (ca. 460–370 BC), proposed that matter was composed of very small particles that he called atoms. This idea was widely debated for centuries, but no conclusion was reached because the existence of atoms could not be proved. It was not possible to resolve this issue until scientists developed the techniques necessary to perform experiments that could differentiate "continuous" from "discontinuous" matter.

Building on the earlier work of other scientists of the time, John Dalton (1766–1844) proposed that matter is indeed made of small individual particles (is discontinuous). **Dalton's atomic theory** as expressed in modern terms states that:

John Dalton

1. Matter is composed of small indivisible particles called *atoms.* **Atoms** are the smallest units of an element that enter into chemical combination.
2. An *element* is composed entirely of one type of atom. The properties of atoms of one element are different from those of any other element.
3. A *compound* contains atoms of two or more different elements. The relative number of atoms of each element in a given compound is always the same.
4. Atoms do not change in chemical reactions. Chemical reactions involve changing the way in which the atoms are joined together.

Dalton's theory enabled him to explain the results of many experiments. Years of experiments had shown that the mass ratio of elements in a pure substance was always the same. For example, chemical analysis of water

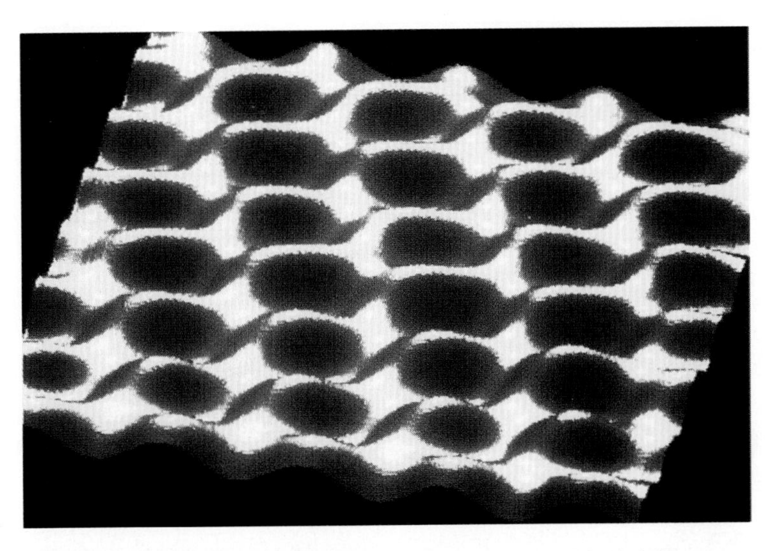

Scanning tunneling microscope view of carbon atoms The bright areas mark the location of individual carbon atoms in graphite.

Compound A
1.00 g carbon 0.08 g hydrogen

Compound B
1.00 g carbon 0.16 g hydrogen

Compound C
1.00 g carbon 0.24 g hydrogen

Law of multiple proportions The masses of hydrogen that combine with one gram of carbon in these compounds are in the ratios of 1 : 2 : 3.

always finds a ratio of 8.0 g of oxygen for every 1.0 g of hydrogen. This observation is known as the **law of constant composition:** all samples of a pure substance contain the same elements in the same proportions by mass. Dalton's third postulate explains this law because it states that a given compound is always made up of the same types of atoms in the same ratios.

The law of constant composition is explained by the fact that the relative number of atoms of each element in a given compound is always the same.

All samples of a particular compound contain the same elements in the same proportions, but sometimes more than one compound can be formed from the same elements. The compositions of these compounds reveal an important relationship that is called the **law of multiple proportions:** the masses of one element (in each compound) that combine with a fixed mass of a second element are always in a ratio of small whole numbers. For example, there are two common compounds that contain only carbon and oxygen. In one, 1.33 g of oxygen combine with 1.00 g of carbon; in the other, 2.66 g of oxygen combine with 1.00 g of carbon. Thus, the ratio of the masses of oxygen that combine with one gram of carbon is 2.66/1.33 or two-to-one. It is now known that one compound contains one carbon atom for each oxygen atom (carbon monoxide), and the other contains one carbon atom for every two oxygen atoms (carbon dioxide). Carbon dioxide has twice as many oxygen atoms per carbon atom as does carbon monoxide, so the ratio of the masses of oxygen that combine with one gram of carbon should be two-to-one, as observed experimentally. This law is explained by the existence of indivisible atoms as formulated by Dalton.

The existence of atoms explains the law of multiple proportions.

The fourth postulate explains the **law of conservation of mass:** there is no detectable loss or gain in mass when a chemical reaction occurs. In a chemical reaction, the way in which the atoms are combined is changed, but neither the number of atoms nor the type of atoms changes. Because the numbers and types of atoms do not change during a reaction, the mass cannot change.

The existence of atoms explains the law of conservation of mass.

INSIGHTS into CHEMISTRY

Development of Chemistry

Dalton's Atomic Theory Provides Foundation for Chemistry

Today, John Dalton is regarded as the father of modern atomic theory. However, Dalton was not the first person to propose that all matter is composed of particles called atoms. For instance, Isaac Newton published statements that indicated his belief in the concept of atoms. What made Dalton's theory so significant in the history of chemistry?

One important factor is that Dalton's theory was firmly founded on the results of scientific experiments. Earlier atomic theories were based more on philosophical arguments and speculations than on physical evidence. Dalton's ideas about the structure of matter grew from his interest in the weather and the composition of the atmosphere (he kept daily records of the weather for 57 years), which led him to investigate the physical and chemical properties of gases. All accounts of Dalton agree that he was a rather poor experimentalist, but fortunately he also had the published results of others who were skilled in the laboratory. In fact, much of the data upon which Dalton based the law of multiple proportions had been available in the chemical literature for at least 15 years before he published his theory. Dalton's ability to recognize and interpret the relationships among experimental data was one of his greatest assets. Using Proust's law of constant composition, Lavoisier's law of conservation of mass, and his own law of multiple proportions, Dalton formed the major ideas of his atomic theory. These ideas helped to explain many of the experimental results that had accumulated, and encouraged others to perform experiments to test their validity. In 1808, the same year that Dalton published a much more detailed explanation of his atomic theory, the experimental work of Thomas Thomson and William Hyde Wollaston provided confirmation of the law of multiple proportions and thus increased the general acceptance of Dalton's theory.

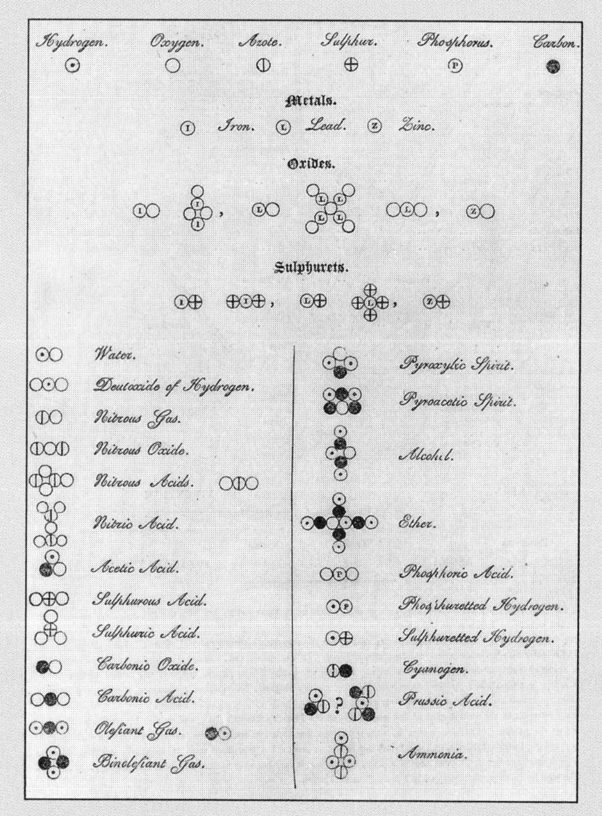

Dalton's proposed atomic symbols

Dalton also emphasized the use of visual aids to help explain his theory. Since he imagined an atom to be a tiny, indivisible sphere, it was natural for him to use cir-

It is important to realize that these three "laws" are based on careful laboratory experiments. These experiments have been repeated on numerous occasions.

Molecules

In many pure substances, both elements and compounds, the atoms are grouped into small clusters called molecules. A **molecule** is a combination of atoms joined tightly together so that they behave as a single particle. If all of the atoms in the molecule are the same, the substance is an element; if atoms

Dalton's scale of relative atomic weights

cles to represent atoms. Each element was represented as a specific type of circle, with each circle standing for an individual atom. His symbols for compounds are very similar to our use of structural formulas today. Again, Dalton was not the first to use symbols to represent the elements—alchemists had used symbols for hundreds of years. Dalton's innovation was that each symbol represented an individual atom of an element.

Probably the most important factor in the acceptance of Dalton's work was his use of quantitative measurements of mass to describe chemical reactions. Dalton believed that all of the atoms of a particular element had the same size and weight, while atoms of different elements had different weights. Dalton's explanation of the laws of constant composition and multiple proportions certainly implied that elements had these characteristics. Dalton felt that it was extremely important to determine the atomic weight of each element by experiment. Of course, direct measurement was impossible, but he felt that relative weights could be determined from the published data on composition of substances such as water and the oxides of carbon. In order to determine the relative weights, Dalton needed the correct chemical formula for the combined elements. Since there were no definitive experiments to guide him, Dalton decided to make some assumptions. He assumed that, unless there was a specific reason to think otherwise, the elements in a simple compound combined on the basis of one atom to one atom. Other combinations of the same elements were in the simplest ratios possible, such as one to two. This became known as Dalton's rule of simplicity. As an example, Dalton assumed the formula for water to be HO. Because the data for water showed that 8 g of oxygen (the experimental value in Dalton's time was closer to 7 g) combined with 1 g of hydrogen, Dalton assigned a relative weight of 1 for hydrogen, the element he chose as his reference standard, and a relative weight of 8 for oxygen. Using such reasoning, Dalton published the first table of the relative weights of atoms.

The publication of a table of relative atomic weights was not without problems. Inaccuracies in experimental data and assumptions of formulas caused errors in Dalton's work. Unfortunately, Dalton was unwilling to correct his mistakes. Even when the work of other scientists such as Joseph Louis Gay-Lussac and Amedeo Avogadro suggested ways to correct his chemical formulas and weights, Dalton chose to ignore these people or to label their work as incorrect. Because of Dalton's opposition to corrections of his work, inaccuracies and uncertainties about molecular formulas and relative weights of atoms persisted for almost 50 years after his theory was published. In spite of its flaws, Dalton's table of relative weights was an important milestone in the development of chemistry as a quantitative science. It introduced the importance of weight as a characteristic of an element, and it encouraged other scientists to perform quantitative experiments to determine more accurate values for the weights of the atoms.

of different elements combine to form a molecule, the substance is a compound. The simplest molecules are **diatomic;** that is, they are composed of two atoms. The stable elemental forms of hydrogen, oxygen, nitrogen, and the halogens are diatomic molecules. Although the elements oxygen and hydrogen form diatomic molecules, just one atom of oxygen combines with two atoms of hydrogen to form a molecule of water (Figure 2.2).

A **molecular formula** gives the number of every type of atom in a molecule. It is an abbreviated way of describing the composition of a molecule. In the formula, the elements that are present are identified by their chemical symbols. Each symbol is followed by a numerical *subscript* to indicate the

A molecular formula gives the symbols of the elements in the molecule and the number of atoms of each element that are present.

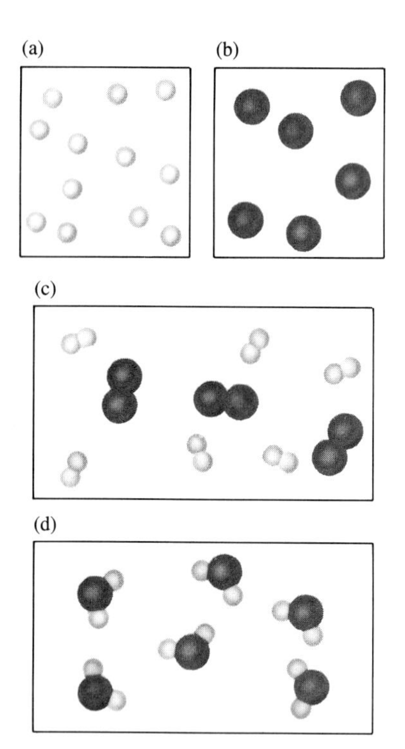

Figure 2.2 Atoms and molecules of hydrogen, oxygen and water Atoms of hydrogen (a) and oxygen (b) form diatomic molecules (c). (d) Water is composed of hydrogen and oxygen atoms, combined in a 2 : 1 ratio.

1 1A	2 IIA	3 IIIB	4 IVB	5 VB	6 VIB	7 VIIB	8	9 VIIIB	10	11 IB	12 IIB	13 IIIA	14 IVA	15 VA	16 VIA	17 VIIA	18 VIIIA
1 **H**																	2 **He**
3 **Li**	4 **Be**											5 **B**	6 **C**	7 **N**	8 **O**	9 **F**	10 **Ne**
11 **Na**	12 **Mg**											13 **Al**	14 **Si**	15 **P**	16 **S**	17 **Cl**	18 **Ar**
19 **K**	20 **Ca**	21 **Sc**	22 **Ti**	23 **V**	24 **Cr**	25 **Mn**	26 **Fe**	27 **Co**	28 **Ni**	29 **Cu**	30 **Zn**	31 **Ga**	32 **Ge**	33 **As**	34 **Se**	35 **Br**	36 **Kr**
37 **Rb**	38 **Sr**	39 **Y**	40 **Zr**	41 **Nb**	42 **Mo**	43 **Tc**	44 **Ru**	45 **Rh**	46 **Pd**	47 **Ag**	48 **Cd**	49 **In**	50 **Sn**	51 **Sb**	52 **Te**	53 **I**	54 **Xe**
55 **Cs**	56 **Ba**	57 **La***	72 **Hf**	73 **Ta**	74 **W**	75 **Re**	76 **Os**	77 **Ir**	78 **Pt**	79 **Au**	80 **Hg**	81 **Tl**	82 **Pb**	83 **Bi**	84 **Po**	85 **At**	86 **Rn**
87 **Fr**	88 **Ra**	89 **Ac†**															

*Lanthanides	58 **Ce**	59 **Pr**	60 **Nd**	61 **Pm**	62 **Sm**	63 **Eu**	64 **Gd**	65 **Tb**	66 **Dy**	67 **Ho**	68 **Er**	69 **Tm**	70 **Yb**	71 **Lu**
†Actinides	90 **Th**	91 **Pa**	92 **U**	93 **Np**	94 **Pu**	95 **Am**	96 **Cm**	97 **Bk**	98 **Cf**	99 **Es**	100 **Fm**	101 **Md**	102 **No**	103 **Lr**

Diatomic elements The elements colored in yellow exist as diatomic molecules.

number of atoms of that element that are present in the molecule (the absence of a subscript means one atom). Molecular hydrogen is written as H_2 and water is written as H_2O. As stated in Dalton's third postulate, the molecular formula of a substance is always the same. In subsequent chapters, we will learn how the formula of a compound is determined by experiment.

Molecular formula	Cl_2	CH_4	C_2H_6
Structural formula	Cl—Cl	H | H—C—H | H	H H | | H—C—C—H | | H H
Model of Molecule			

Figure 2.3 Molecular and structural formulas

Figure 2.3 shows different ways of writing the formulas of substances. The first line shows the molecular formula. The second line shows the **structural formula,** indicating how the atoms are connected in the molecule. The figure also shows models that help us visualize the shapes of molecules.

2.2 Structure of Atoms

Objectives

- To describe the three main particles that make up an atom, and the structure of the atom
- To define isotopes, and to interpret the symbol that represents each isotope
- To describe the process by which atoms can lose or gain electrons to form ions, and to differentiate this process from radioactivity

Although atoms were initially viewed as indivisible, a number of experiments—performed or interpreted long after Dalton proposed his theory—have shown that atoms are composed mainly of three types of particles. They are:

1. **Electrons:** An electron is a particle that has a mass of 9.109×10^{-31} kg and a charge of -1.602×10^{-19} coulombs (the coulomb is the SI unit for electric charge).
2. **Protons:** A proton is a particle that has a mass of 1.673×10^{-27} kg and a charge of $+1.602 \times 10^{-19}$ coulombs.
3. **Neutrons:** A neutron is a particle that has a mass of 1.675×10^{-27} kg and no charge.

Particle	Mass	Relative Mass	Relative Charge
electron	9.109×10^{-31} kg	0	1−
proton	1.673×10^{-27} kg	1	1+
neutron	1.675×10^{-27} kg	1	0

Notice two important properties of these particles:

1. The charges of the electron and proton are exactly equal, but opposite in sign. Atoms are electrically neutral species, and must contain equal numbers of electrons and protons. Experiments show that charges on all particles are multiples of the charge on electrons and protons, so we generally refer to an electron as having a relative charge of 1− and a proton as having a relative charge of 1+.
2. The masses of the proton and neutron are nearly the same, but the mass of the electron is much less (the ratio of the mass of a proton to that of an electron is 1836). The electrons provide only a small fraction of the total mass of an atom—nearly all of the mass of the atom is accounted for by the protons and neutrons.

The mass of an atom arises from the protons and neutrons.

Discovery of the Electron

The electron was the first component of the atom to be identified. The discovery of the electron and its properties resulted from many years of experiments, each of which added a bit more knowledge about the relationship between electricity (electrons) and matter. Ultimately, they led to our modern theory of atomic structure.

The first evidence of a connection between electrical and chemical properties came from the work of Sir Humphrey Davy (1778–1829). He passed currents through molten compounds, causing the process called electrolysis (discussed in Chapter 17) to occur. In this way Davy isolated a series of new elements, including sodium, potassium, calcium, magnesium, strontium, and barium. The work was continued by Michael Faraday (1791–1867), who had been Davy's assistant and protégé. Faraday found that there was a quantitative relationship between the amount of electrical charge passed through the molten compounds and the mass of the element produced. On the basis of his observations, he formulated Faraday's laws of electrolysis (which will be discussed in detail in Chapter 17). These laws, which established the connection between electricity and chemistry, also indicated that an electrical current consisted of definite units of electrical charge. A number of investigators worked during the 1800s to understand the nature of electricity and its relationship to matter.

Toward the end of the 1800s, scientists began to investigate the flow of electricity in devices called gas discharge tubes. These were glass tubes that had a metal electrode at each end, and that contained gas at low pressure. When high voltages were applied across the electrodes, an electrical discharge—a flow of electricity—occurred and the gas began to glow (modern neon signs and fluorescent lights work the same way). Experiments were conducted to try to understand the emission of light from a gas. They revealed that even when the gas was almost completely removed, a glow was observed as long as an appropriate electrical voltage was applied. These tests also indicated that the negative electrode was the source of some unusual emission. Because these emissions came from the negative electrode, called the cathode, they were soon called cathode rays. In further experiments, scientists found that cathode rays traveled in straight lines, heated a metal foil placed in their path, and could be deflected by electrical and magnetic fields. One group of scientists believed cathode rays to be some sort of light or energy, while another group believed that cathode rays were electrically charged particles.

The controversy was settled in 1897 by a British physicist, J. J. Thomson (1856–1940), by a series of experiments with specially prepared gas discharge tubes. Thomson found that, by applying carefully controlled magnetic and electrical fields to the cathode rays, he could determine the charge-

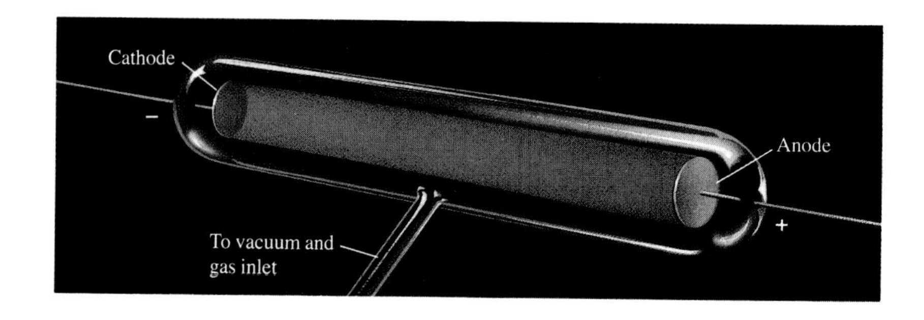

Gas discharge tube When a high voltage is applied across a partially evacuated tube, the tube begins to glow.

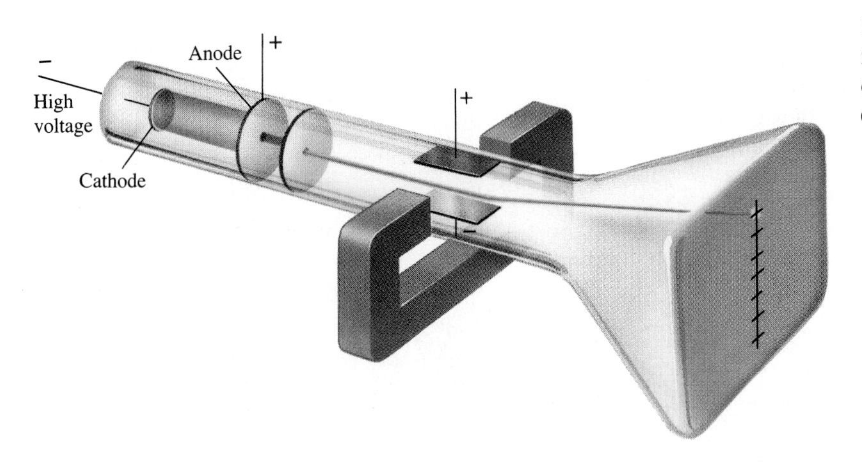

Effect of a magnetic field on a cathode ray Application of a magnetic field causes a cathode ray to be deflected. An electric field also deflects the cathode ray.

to-mass ratio (often expressed as e/m, where e stands for the charge and m stands for mass) of the cathode ray.

This measurement established beyond a doubt that cathode rays were electrically charged particles; the direction of their deflection indicated that they were negatively charged. The value for e/m that Thomson obtained was about -1.0×10^{11} coulombs/kilogram. (The currently accepted value is -1.76×10^{11} coulombs/kilogram.) Thomson also discovered that the same value for the e/m ratio was obtained regardless of the cathode material or the gas in the tube. These observations suggested that cathode ray particles were fundamental particles of matter and were therefore constituents of all atoms. The particles were called electrons, a name suggested years earlier for the particle that theoretically carried electricity. In 1906, Thomson received the Nobel Prize in physics for his work on the electron.

Having established a value for the e/m ratio, Thomson and his students at the Cavendish Laboratory at Cambridge University set out to develop a method to measure the value of e. They soon developed a technique known as the "cloud method." This method consisted of measuring the rate at which an electrically charged cloud of water mist fell under the influence of gravity, and then the modified rate when the cloud was subjected to an electrical field. The difference between the two rates could be used to calculate the value for e. The method had serious inconsistencies and, even with improvements, gave unsatisfactory and imprecise results.

In 1909, Robert A. Millikan (1868–1953), an American physicist teaching at the University of Chicago, improved the "cloud method" to provide a reliable measurement of the charge on the electron. He found that when the cloud of water mist was subjected to a much stronger electrical field than Thomson used, the cloud dissipated and left only a few individual droplets in view. He realized that it would be much more accurate to work with a single drop than with the numerous droplets in a cloud. By observing the rate of free fall of a single uncharged drop, he could calculate its mass. He then exposed the drop to a source of ionizing radiation to give it an electrical charge and switched on the electrical field between the two plates. Millikan measured the electrical force needed to keep the drop from falling because of gravity. Knowing the mass of the drop and the strength of the field, he could calculate the total electrical charge on the drop.

Millikan oil drop experiment Oil drops are formed by the injector and are charged by interaction with x rays. Millikan adjusted the electrical field between the charged plates so as to suspend the charged oil drops (they never fell due to gravity or rose due to the electric field), and calculated the charges on the oil drops from these data.

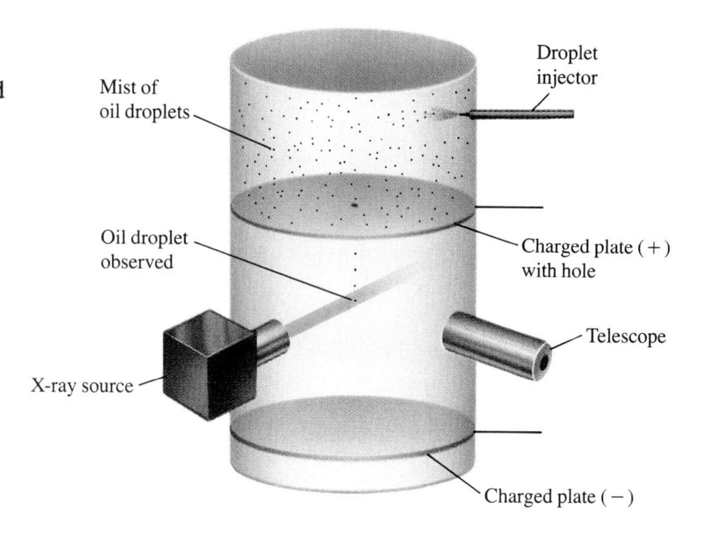

Millikan observed that the charge on any particular drop was always an integral multiple of a single quantity, which he took to be the charge carried by a single electron. Although he published a value for e in 1909, Millikan was somewhat troubled about his results because of the evaporation of the water drops. He solved this problem by substituting oil for water. The rise and fall of an oil drop could be observed for up to four and one-half hours at a time. Millikan carefully repeated his measurements and, in 1913, published a new value for e equal to -1.60×10^{-19} coulombs. Using Millikan's value for e and Thompson's value for e/m, scientists could calculate the value for m, the mass of the electron. To three significant figures, the modern value is 9.11×10^{-31} kilograms. Millikan was awarded the Nobel Prize in 1923.

The Nuclear Atom

Scientists initially described the structure of atoms on the basis of simple electrostatic forces between protons and electrons. These forces obey **Coulomb's law:** the force between two charged objects is directly proportional to the product of the charges and inversely proportional to the square of the distance between them. In equation form,

$$F = \frac{kQ_1Q_2}{r^2}$$

where k is a constant, Q_1 and Q_2 are the charges on the particles, and r is the distance between the charged particles. The force, F, is repulsive if the sign of the charge of both particles is the same and attractive if the charges are of opposite sign. That is, like-charged particles repel and oppositely charged particles attract. Electrostatic forces explain many phenomena discussed in this book.

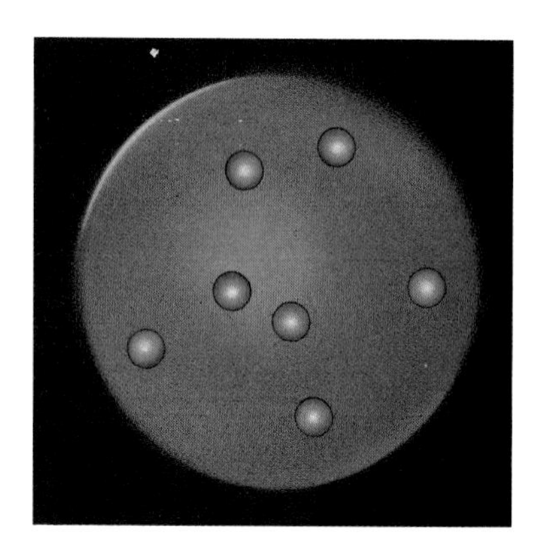

Figure 2.4 Thomson's model of an atom Thomson's model of the atom had electrons embedded in a sphere of positive charge.

In the late 1800s, J. J. Thomson proposed a model of an atom based on Coulomb's law. In Thomson's model, the positive charge and the mass were uniformly distributed and the electrons were spread out to minimize electrostatic repulsions (Figure 2.4). This description of an atom, referred to as the "plum pudding" model, was shown to be incorrect by an important experiment performed in the laboratory of Ernest Rutherford. In this experiment, shown in Figure 2.5, high energy alpha particles (particles with a positive charge of 2+ and a mass four times as great as that of a proton) were allowed to hit a thin gold target and their deflections were measured. Most of the alpha particles went through the gold with no deflection, just as if they were traveling through empty space. The surprising result of the experiment was that a few of the alpha particles were deflected through large angles and, in fact, some were deflected back in the direction from which they came. If the positive charge and mass of an atom were spaced out evenly, as presumed by the Thomson model, the observed deflection of the relatively heavy alpha particles would be impossible. Rutherford said, "It was almost as incredible as if you fired a 15-inch shell at a piece of tissue paper and it came back and hit you."

The Thomson model of the atom was not consistent with the results, so Rutherford proposed a new model for the atom. In Rutherford's model, the positive charge and nearly all of the mass of the atom are in a central core, with the electrons at a relatively large distance from this core. Calculations based on the experimental data showed that the central core, which Rutherford called the **nucleus,** is extremely small, even in comparison to the size of the atom itself. Most of the volume of the atom is thus occupied by the electrons, which move rapidly about the nucleus. Although he could not explain why the protons were close together, an unlikely situation in view of Coulomb's law, Rutherford knew that his experimental results could not be explained without proposing that all of the protons in the atom were contained in this dense, positively charged nucleus. It is now known that the nucleus contains both the protons and neutrons; there are forces called *strong nuclear binding forces,* which are stronger than those described by Coulomb's law, that hold these particles together in the nucleus. In general, the ratio of protons to neutrons found in the nuclei of atoms is between 0.7 and 1.

Atoms contain protons and neutrons in a central core, called the nucleus, which is surrounded by electrons.

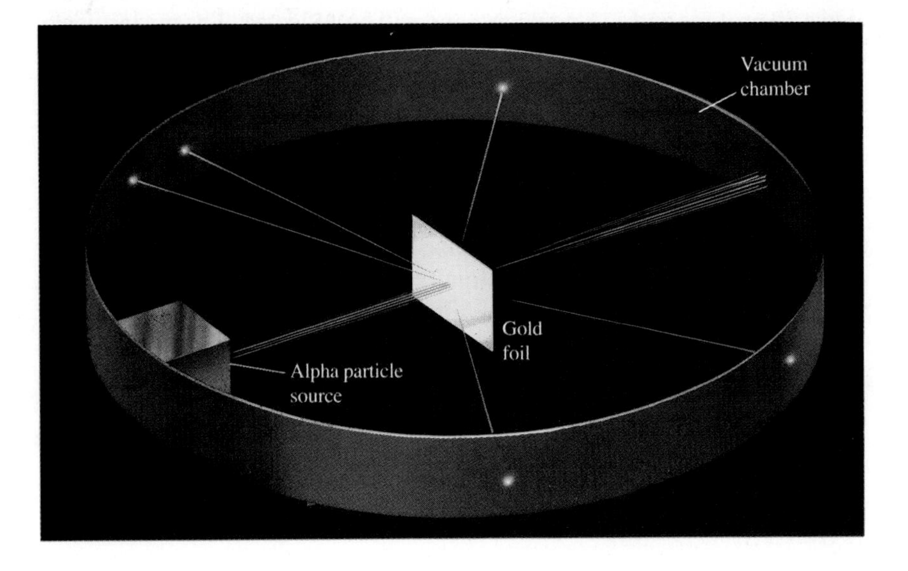

Figure 2.5 The Rutherford experiment Alpha particles are directed toward a very thin piece of gold foil inside a vacuum chamber. Detectors indicate that while most of the particles go through the foil, some are deflected through large angles and some are deflected back in the direction from which they came.

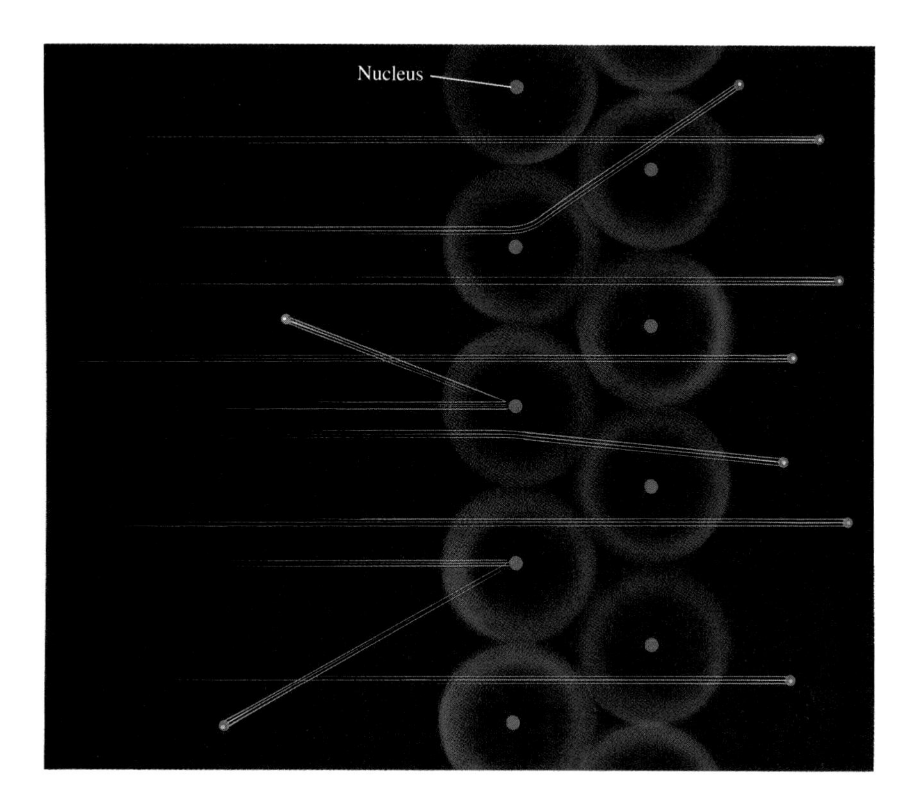

Figure 2.6 Deflection of alpha particles by the nucleus In the Rutherford model of the atom, most alpha particles will pass through the gold foil, but some will come close to a massive, highly positively charged nucleus and be deflected.

This model explains the results of Rutherford's experiment. The volume occupied by atoms is mostly occupied by the electrons. Electrons have a low relative mass compared to alpha particles and will not measurably deflect them. Thus, most of the alpha particles will not come close to a gold nucleus and will travel through the gold foil without being deflected. A few will come close to a dense, highly positively charged gold nucleus (it is now known that it has 79 protons and 118 neutrons) and will be deflected. The magnitude of the deflection is determined by how close the alpha particle comes to the nucleus. The alpha particles aimed directly at the gold nucleus are bounced backwards (Figure 2.6).

Atomic Number and Isotopes

The nuclear model allows us to describe how the atoms of an element differ from those of another element. Atoms of different elements have different numbers of protons in the nucleus. The **atomic number** (represented by the letter Z) is the number of protons in the nucleus of an atom. Each element has its own unique atomic number. In fact, the identity of an element is defined by its atomic number. Hydrogen is the element with one proton in its nucleus; its atomic number is 1. Helium has two protons, so its atomic number is 2. The atomic number of lithium is 3, and so on. The atomic number is listed above the symbol for each element in the periodic table (Figure 2.7).

1	2	3	4	5	6	7	8	9	10	11	12	13	14	15	16	17	18
1A	IIA	IIIB	IVB	VB	VIB	VIIB		VIIIB		IB	IIB	IIIA	IVA	VA	VIA	VIIA	VIIIA
1 **H**																	2 **He**
3 **Li**	4 **Be**											5 **B**	6 **C**	7 **N**	8 **O**	9 **F**	10 **Ne**
11 **Na**	12 **Mg**											13 **Al**	14 **Si**	15 **P**	16 **S**	17 **Cl**	18 **Ar**
19 **K**	20 **Ca**	21 **Sc**	22 **Ti**	23 **V**	24 **Cr**	25 **Mn**	26 **Fe**	27 **Co**	28 **Ni**	29 **Cu**	30 **Zn**	31 **Ga**	32 **Ge**	33 **As**	34 **Se**	35 **Br**	36 **Kr**
37 **Rb**	38 **Sr**	39 **Y**	40 **Zr**	41 **Nb**	42 **Mo**	43 **Tc**	44 **Ru**	45 **Rh**	46 **Pd**	47 **Ag**	48 **Cd**	49 **In**	50 **Sn**	51 **Sb**	52 **Te**	53 **I**	54 **Xe**
55 **Cs**	56 **Ba**	57 **La***	72 **Hf**	73 **Ta**	74 **W**	75 **Re**	76 **Os**	77 **Ir**	78 **Pt**	79 **Au**	80 **Hg**	81 **Tl**	82 **Pb**	83 **Bi**	84 **Po**	85 **At**	86 **Rn**
87 **Fr**	88 **Ra**	89 **Ac†**															

*Lanthanides	58 **Ce**	59 **Pr**	60 **Nd**	61 **Pm**	62 **Sm**	63 **Eu**	64 **Gd**	65 **Tb**	66 **Dy**	67 **Ho**	68 **Er**	69 **Tm**	70 **Yb**	71 **Lu**
†Actinides	90 **Th**	91 **Pa**	92 **U**	93 **Np**	94 **Pu**	95 **Am**	96 **Cm**	97 **Bk**	98 **Cf**	99 **Es**	100 **Fm**	101 **Md**	102 **No**	103 **Lr**

Figure 2.7 Periodic table
Atomic numbers are shown on the periodic table above the symbol for the element.

The mass of an atom is determined by the number of protons and neutrons. The **mass number** (represented by the letter A) is the total number of protons and neutrons in an atom. The mass number does not identify a specific element; the atomic number does that. In fact, different atoms of the same element can differ in the numbers of neutrons. For example, there are three types of hydrogen atoms. Most atoms of hydrogen have no neutrons in the nucleus, a few have one neutron, and a very few have two neutrons (Figure 2.8). All hydrogen atoms have one proton and one electron, and virtually the same chemical properties, but they can have different mass numbers.

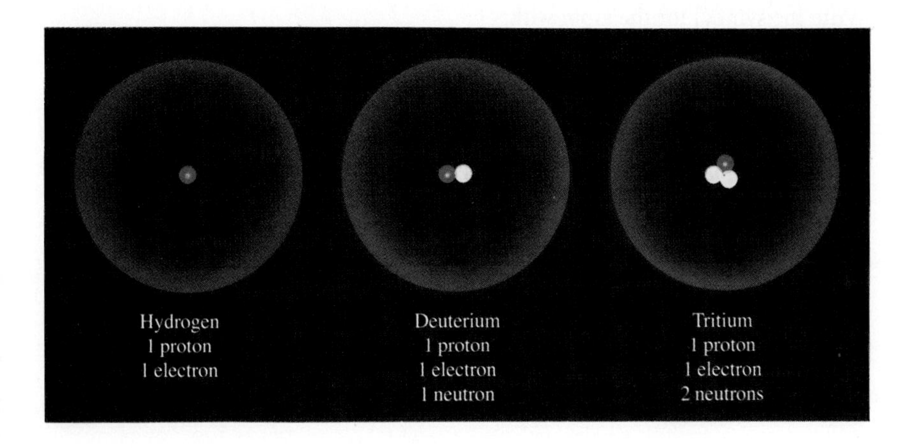

Figure 2.8 Isotopes of hydrogen
There are three isotopes of hydrogen. Each kind of atom has one proton and one electron, but the number of neutrons ranges from zero to two. The isotope with one neutron is called deuterium, and the isotope with two neutrons is called tritium.

Atoms of the same element can have different numbers of neutrons in their nuclei.

Isotopes are atoms of the same element that have different numbers of neutrons. Isotopes have the same atomic number but different mass numbers. All isotopes of the same element have the same number of electrons, which is equal to the number of protons in the nucleus (to maintain electrical neutrality). The mass numbers and relative abundances of the isotopes of an element can be determined by mass spectroscopy, an experimental method described in *Insights into Chemistry* on the next page.

The existence of isotopes is an interesting phenomenon. About 75% of the naturally occurring elements have more than one stable isotope. Titanium and nickel, for example, each have five stable isotopes, whereas copper and chlorine each have only two. Fluorine (mass number = 19) and phosphorus (mass number = 31) are examples of elements that have just one stable isotope. The different isotopes of elements are used by chemists for important experiments, such as the dating of fossils and other geological samples.

A shorthand notation is used to designate isotopes of each element. Specific isotopes of an element are often designated using the form

$${}^{A}_{Z}X$$

where X is the symbol of the element, A is the mass number, and Z is the atomic number. The three isotopes of hydrogen are

$${}^{1}_{1}H \qquad {}^{2}_{1}H \qquad {}^{3}_{1}H$$

Notice that the atomic number is the same for all three of these isotopes; if the atomic number is not 1, the atom is not hydrogen. Oxygen also has three isotopes.

$${}^{16}_{8}O \qquad {}^{17}_{8}O \qquad {}^{18}_{8}O$$

Including the atomic number with the symbol is optional in the notation, because either identifies the element. The oxygen isotopes are often written as

$${}^{16}O \qquad {}^{17}O \qquad {}^{18}O$$

Example 2.1 Symbols of Atoms

Write the symbol for the atom with:

(a) 6 protons and 6 neutrons
(b) 13 protons and 14 neutrons.

Solution

(a) The element with 6 protons is carbon. The mass number is the sum of the numbers of protons and neutrons (6 + 6 = 12).

$${}^{12}_{6}C \quad \text{or} \quad {}^{12}C$$

(b) Aluminum is the element that has 13 protons; the mass number is 27.

$${}^{27}_{13}Al \quad \text{or} \quad {}^{27}Al$$

Understanding Write the symbol for the atom with 19 protons and 20 neutrons.

Answer: ${}^{39}_{19}K$ or ${}^{39}K$

INSIGHTS into CHEMISTRY

Science, Society, and Technology

Mass Spectrometers Are Now Used by Scientists to Measure Isotope's Mass and Abundance

A mass spectrometer measures the mass and relative abundances of the isotopes present in a sample of an element. One type of mass spectrometer is pictured here. A curved tube is evacuated with a vacuum pump, and a sample of the element is then introduced as a gas into one end. The gas is exposed to a beam of high-energy electrons which convert the atoms of the element into cations. The high voltage between the plates accelerates these cations through a slit so that they travel down the tube.

The curved section of the tube has a magnetic field perpendicular to the direction of the ions. This magnetic field deflects the ions into a curved path. The degree of curvature of the path depends on the mass and charge of the ion, and the sizes of the accelerating voltage and the magnetic field strength. The charge-to-mass ratio of the ions that reach the detector can be determined from the known magnetic field strength, which is varied to bring each set of ions to the detector. The output for a sample of neon is shown. The position of the signal indicates the mass of each isotope, and the strength of the signal (shown as signal height in the drawing) indicates the relative abundance. For neon, the main isotope is ^{20}Ne, with minor isotopes of ^{21}Ne and ^{22}Ne.

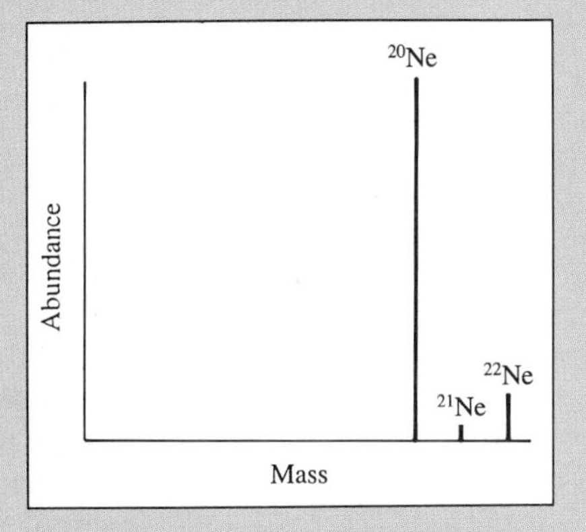

The mass spectrum of neon The mass spectrum of neon shows three isotopes. The most abundant is ^{20}Ne, but some ^{21}Ne and ^{22}Ne atoms are also observed.

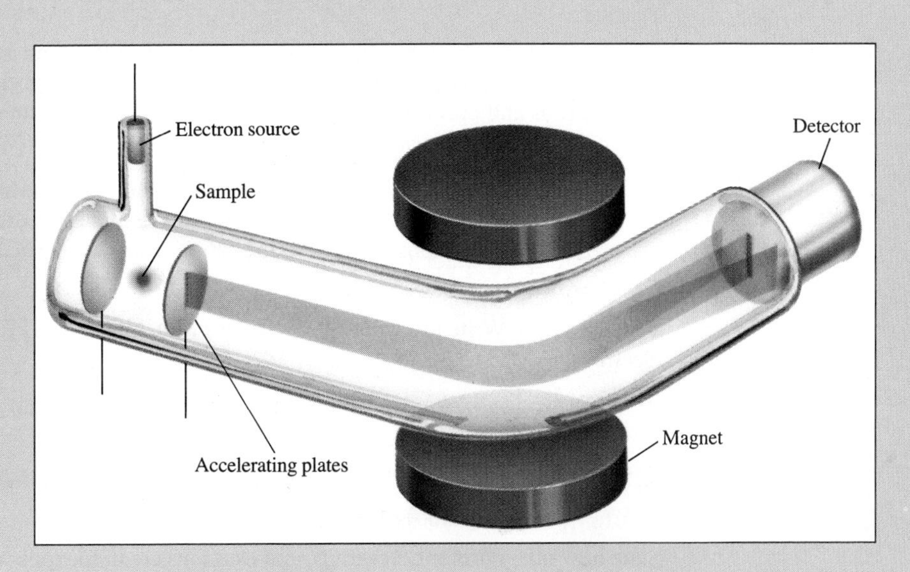

Diagram of a mass spectrometer

Ions

Atoms can gain or lose electrons to form charged particles called ions.

Atoms can gain or lose electrons, producing charged particles called **ions.** A **cation** is an ion that has a positive charge; an **anion** has a negative charge. Cations have fewer electrons than protons, whereas anions have more electrons than protons. Ions are formed by the gain or loss of electrons by neutral atoms; the number of protons in the nucleus never changes in a chemical process.

Ion	Composition	Charge
cation	more protons than electrons	+
anion	more electrons than protons	−

The charge on an ion is indicated by a superscript following the symbol. A sodium cation with a net charge of 1+ is written as Na^+; an anion of oxygen with a 2− charge is written as O^{2-}. This notation can be combined with that for isotopes to indicate specific charged isotopes of elements. The symbol $^{37}Cl^-$ represents the isotope of chlorine that contains 17 protons (all isotopes of chlorine contain 17 protons), 20 neutrons, and 18 electrons (one more electron than protons to give the ion the 1− charge). The isotope of magnesium that contains 12 protons, 13 neutrons, and 10 electrons is written as $^{25}Mg^{2+}$.

Example 2.2 Symbols of Atoms and Ions

(a) Write the symbol for a species with eight protons, nine neutrons, and eight electrons.

(b) Write the symbol for a species with 20 protons, 20 neutrons, and 18 electrons.

Solution

(a) The eight protons define the atomic number as 8, so the atom is oxygen. The sum of the numbers of protons and neutrons is 17, the mass number. The numbers of electrons and protons are equal, so no charge is needed. The symbol is

$$^{17}_{8}O \quad \text{or} \quad ^{17}O$$

(b) This species has an atomic number of 20, so it is calcium. The mass number is 40, the sum of the numbers of protons and neutrons. Since there are two more protons than electrons, the species is a cation with a charge of 2+. The symbol is

$$^{40}_{20}Ca^{2+} \quad \text{or} \quad ^{40}Ca^{2+}$$

Understanding Write the symbol for a species with 23 protons, 28 neutrons, and 21 electrons.

Answer: $^{51}_{23}V^{2+}$ or $^{51}V^{2+}$

Example 2.3 Symbols for Ions

State how many protons, neutrons, and electrons are in the ions with the following symbols:

(a) $^{23}_{11}Na^+$ (b) $^{81}_{35}Br^-$

Solution

(a) This sodium cation has 11 protons, 12 neutrons (mass number of 23 − atomic number of 11) and 10 electrons, one fewer than the number of protons. Remember, the number of electrons is calculated from the number of protons and the overall charge.

(b) This anion of bromine has 35 protons, 46 neutrons, and 36 electrons, one more than the number of protons.

Understanding How many protons, neutrons, and electrons are in the ion with the symbol $^{39}K^{+}$?

Answer: 19 protons, 20 neutrons, and 18 electrons

Example 2.4 Components of Atoms and Ions

Fill in the blanks in the following table.

	(a)	(b)
symbol	$^{15}N^{3-}$	—
atomic number	—	—
mass number	—	24
charge	—	2+
number of protons	—	12
number of electrons	—	—
number of neutrons	—	—

Solution

(a) The symbol given specifies the other items in the column. The atomic number subscript has been omitted, but from the periodic table we know the element nitrogen has an atomic number of 7. The mass number is the superscript of 15 before the symbol, and the charge of 3− is the superscript that follows. The number of protons is 7, the atomic number. The number of electrons is the number of protons plus three (the negative charge) = 10, and the number of neutrons is the mass number (15) minus the atomic number (7) = 8.

(b) The ion must be Mg since this is the element with 12 protons in the nucleus, and the atomic number is 12. The mass number is 24 and the charge 2+, making the correct symbol $^{24}Mg^{2+}$. The number of electrons is two less than the number of protons, or 10, and the number of neutrons is the mass number minus the atomic number = 12. In table form:

	(a)	(b)
symbol	$^{15}N^{3-}$	$^{24}Mg^{2+}$
atomic number	7	12
mass number	15	24
charge	3−	2+
number of protons	7	12
number of electrons	10	10
number of neutrons	8	12

Understanding Write the symbol for the ion that has a mass number of 79, an atomic number of 34, and 36 electrons.

Answer: $^{79}Se^{2-}$

Radioactivity

An atom can gain or lose electrons to form ions, but the number of protons and neutrons in a stable nucleus does not change in chemical reactions. However, the nuclei of certain isotopes of some elements are unstable; an unstable nucleus spontaneously transforms into a more stable nucleus with the emission of a small particle and energy. This phenomenon is known as **radioactivity.** The three most common types of emissions are called alpha (α) particles, beta (β) particles, and gamma (γ) rays. The alpha particles are positively charged helium nuclei ($^4He^{2+}$). These were the particles used in Rutherford's experiment described earlier. Beta particles are high-energy electrons. Gamma rays are a form of high-energy electromagnetic radiation. Radioactivity will be covered in some detail in Chapter 20.

Radioactivity refers to changes in the nucleus of an atom, not chemical reactions.

2.3 Ionic Compounds

Objectives

- To write the empirical formulas of compounds formed from cations in groups IA and IIA and anions in groups VIA and VIIA
- To specify the names, formulas and charges of the important polyatomic ions
- To compare the physical properties expected for ionic compounds to those of molecular compounds

We saw earlier that atoms can gain or lose electrons to form ions. An **ionic compound** is composed of cations and anions joined to form a neutral species. The resulting compounds are held together by electrostatic forces described by Coulomb's law, and adopt structures that maximize the attraction of oppositely charged species and minimize the repulsive forces between charged species with charges of the same sign. *Ionic compounds generally form from the combination of metals with nonmetals.* An example of an ionic compound is sodium chloride. It is made up of equal numbers of sodium cations (Na^+) and chlorine anions (Cl^-). The structure of this compound is shown in Figure 2.9. In this structure, each Na^+ ion is surrounded by six Cl^- ions and, in turn, each Cl^- ion is surrounded by six Na^+ ions. This arrangement forms an extended three-dimensional array.

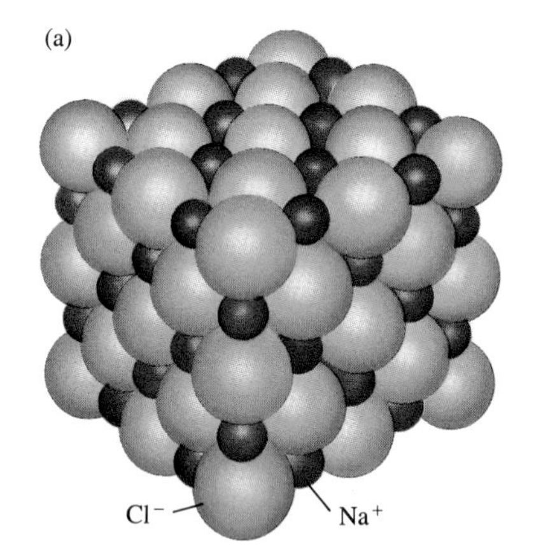

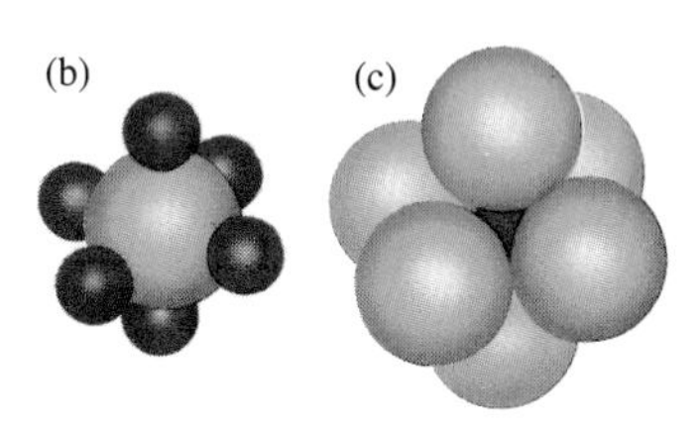

Figure 2.9 Structure of NaCl (a) The three-dimensional array of ions in solid NaCl. (b) Each Cl^- anion is surrounded by six Na^+ cations, and (c) each Na^+ cation is surrounded by six Cl^- anions.

The subscripts in the formulas of ionic compounds have a slightly different meaning from those in molecules. The molecular formula gives the actual numbers and types of atoms in a molecule. However, the exact number of ions in the three-dimensional array of an ionic substance depends on the size of the sample. Instead, the formula of an ionic compound is an **empirical formula** that gives the relative numbers of different elements in a substance, using the smallest whole numbers for subscripts. The formula of sodium chloride is NaCl (it is customary to write the cation first). Each grain of salt contains a very large number of sodium cations and chlorine anions, but they are present in a ratio of one to one. Because the substance is electrically neutral, the sum of the charges contributed by the cations and anions in the formula of an ionic compound must be zero.

The periodic table can be used to help predict the expected charges on many ions. In general, the metallic elements form cations while the nonme-

INSIGHTS into CHEMISTRY

Chemical Reactivity

With Their Different Properties, Heavier Hydrogen Isotopes Are Used to Study Many Important Reactions

While many elements occur as mixtures of isotopes, hydrogen is the only element that has isotopes that differ in mass by a factor of two and three, a fact that has significant effects on the physical properties of the compounds formed by the isotopes. The isotope with a mass number of 1 is by far the most abundant, and it is called hydrogen or, very rarely, protium. The isotope with a mass number of 2 is called deuterium and is often denoted by the symbol D. The isotope with a mass number of 3 is called tritium and is often denoted by the symbol T. Tritium is radioactive, while the other two isotopes are stable. In a naturally occurring sample of hydrogen, there is one atom of deuterium for every 7000 hydrogen atoms, and one atom of tritium for every 10^{18} hydrogen atoms. Because it occurs naturally in such minute amounts, tritium is obtained from nuclear power plants or specially designed reactors, where it is produced by a variety of nuclear reactions. One use of tritium is as a component of some nuclear weapons.

Deuterium can be obtained from natural sources. The usual source is one of its compounds, deuterium oxide, D_2O, also known as heavy water. The properties of heavy water are compared with those of H_2O in the following table. Because of the difference in boiling points, fractional distillation can be used to separate H_2O and D_2O. Samples of water in which 99.5+% of the hydrogen atoms present are deuterium can be obtained by careful fractional distillation.* Pure deuterium, D_2, can then be obtained from the heavy water.

Properties of H_2O and D_2O

	H_2O	D_2O
melting point	0.0 °C	3.8 °C
boiling point	100.0 °C	101.4 °C
density at 4 °C	1.000 g/cm^3	1.108 g/cm^3

Deuterium and tritium have become valuable tools for studying the reactions of compounds containing hydrogen. A compound can be "labeled" by replacing one or more ordinary hydrogen atoms with deuterium or tritium atoms. The resulting compound is chemically nearly identical to the original compound. As the compound reacts, the path taken by the heavier isotopes can be monitored: deuterium by mass spectrographic analysis and tritium by counting its radioactive decay. (The isotopes used in this manner are also called tracers.) Many important reactions, including digestion and body metabolism, can be studied with this technique.

* A sample of water that contains both hydrogen and deuterium will be a mixture of H_2O, HDO, and D_2O in a statistical ratio. As the H_2O is removed by the distillation, some HDO is converted into H_2O and D_2O to keep the mixture statistical (see Chapter 12).

---

tallic elements, especially those closest to the right side of the periodic table (excluding the noble gases) form anions.

Elements generally gain or lose the number of electrons that will give them the same number of electrons as the nearest noble gas. Main group metals form cations with a charge equal to the group number. Group IA (1) elements form cations with a 1+ charge, group IIA (2) elements form cations with a 2+ charge, and groups IIIB (3) and IIIA (13) metals form cations with a 3+ charge. Main group nonmetals form anions with a charge equal to the number of electrons that need to be added to the group number in order to equal eight. This number of electrons will give the anion the same number of electrons as the noble metals. Group VIIA (17) elements form anions with a 1− charge, group VIA (16) elements form anions with a 2− charge, and nitrogen from group VA (15) forms an anion with a 3− charge. These common ions are shown in Table 2.1.

The charges on many ions can be determined from their group number.

If the charges on the ions in a compound are known, we can write the formula of the ionic compound by adjusting the subscripts so that the sum of

Table 2.1 Charges on Common Ions

IA	IIA	IIIB	IIIA	VA	VIA	VIIA
Li^+	Be^{2+}			N^{3-}	O^{2-}	F^-
Na^+	Mg^{2+}		Al^{3+}		S^{2-}	Cl^-
K^+	Ca^{2+}	Sc^{3+}	Ga^{3+}		Se^{2-}	Br^-
Rb^+	Sr^{2+}	Y^{3+}	In^{3+}		Te^{2-}	I^-
Cs^+	Ba^{2+}					

The empirical formulas of ionic compounds balance the charges of the ions.

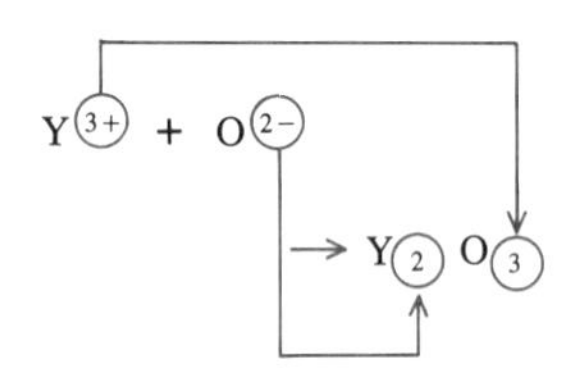

the charges is zero. For cases in which the ions have equal but opposite charges, the subscripts will always be 1 because empirical formulas are expressed as the smallest whole number ratio. The formula of the compound formed by Ca^{2+} and O^{2-} is CaO (remember that the subscript 1 is not written). An example of a case in which the charges are not the same is the ionic compound formed by Mg^{2+} and Cl^-. It takes two 1− charged chloride ions to balance the 2+ charge on the magnesium cation. The formula is $MgCl_2$. Note in this formula that the subscript of the cation is the absolute value of the charge of the anion and the subscript of the anion is the absolute value of the charge on the cation. The formula of the salt made from Y^{3+} and O^{2-} is Y_2O_3. The overall charge is balanced by using the 3 charge of Y as the subscript for O and the 2 charge of O as the subscript for Y.

Example 2.5 Empirical Formulas of Ionic Compounds

Write the empirical formulas of compounds made up from:
(a) Ca^{2+} and Br^- (b) Mg^{2+} and S^{2-} (c) K^+ and O^{2-}

Solution

(a) Two Br^- anions balance the charge of one Ca^{2+} cation. Thus, the empirical formula is $CaBr_2$. Another way to arrive at this answer is to use the numerical charge of 1 on Br as the subscript for Ca and the charge of 2 on Ca as the subscript for Br.
(b) Both ions have the same number for the charge. In the formula MgS the sum of the charges is zero. In this case, the method of using the charge of the other ion as a subscript would yield the formula Mg_2S_2, which is reduced to MgS because the empirical formula uses the smallest whole numbers for the subscript.
(c) Two K^+ cations balance the charge of one O^{2-} anion. The formula is K_2O.

Understanding What is the empirical formula of the compound made up from Na^+ and Se^{2-}?

Answer: Na_2Se

Polyatomic Ions

So far, we have considered only **monatomic ions,** ions formed by the loss or gain of electrons by single atoms. Ions can also be formed by *groups* of atoms joined together by the same kinds of forces that occur in molecules. In ions of this type, such as NH_4^+ (ammonium ion) and OH^- (hydroxide ion), the total numbers of protons and electrons in the entire group of atoms are not equal.

Table 2.2 Polyatomic Anions

Name	Formula	Name	Formula
acetate	$CH_3CO_2^-$	nitrate	NO_3^-
carbonate	CO_3^{2-}	nitrite	NO_2^-
chlorate	ClO_3^-	perchlorate	ClO_4^-
chromate	CrO_4^{2-}	permanganate	MnO_4^-
cyanide	CN^-	phosphate	PO_4^{3-}
dichromate	$Cr_2O_7^{2-}$	sulfate	SO_4^{2-}
hydroxide	OH^-	sulfite	SO_3^{2-}

Charged species consisting of more than one atom are known as **polyatomic ions.** In ionic solids, ammonium is the only important polyatomic cation, but there are many polyatomic anions. Table 2.2 is a short list of polyatomic anions that are used in the next several chapters. A more extensive list of ions is given in Appendix D.

The empirical formula of an ionic compound formed from polyatomic ions can also be predicted from the ionic charges. Treat each polyatomic ion as an inseparable group of atoms with the total charge given in Table 2.2, and write the empirical formula that yields a neutral compound. For compounds in which the subscript of the polyatomic ion is greater than 1, place parentheses around the entire polyatomic group, as in $(NH_4)_3PO_4$.

Polyatomic ions are groups of atoms with a fixed charge.

Example 2.6 Empirical Formulas of Ionic Compounds

Write the formula for the compound made up of

(a) Ba^{2+} and NO_3^-
(b) sodium cation and hydroxide anion
(c) potassium cation and dichromate anion

Solution

(a) Two NO_3^- anions are needed to balance the charge of one Ba^{2+} cation. The formula is $Ba(NO_3)_2$. Note that the parentheses around the NO_3^- group mean that there are two complete NO_3^- groups for every Ba^{2+}. It is incorrect to write the formula as BaN_2O_6.
(b) The charge on the sodium cation is 1+ because it is in group IA (1), and the hydroxide anion is OH^-. The formula is NaOH.
(c) The potassium cation has a 1+ charge and the dichromate anion is $Cr_2O_7^{2-}$. The formula is $K_2Cr_2O_7$.

Understanding Write the formula for the compound made up of the ammonium cation and the sulfate anion.

Answer: $(NH_4)_2SO_4$

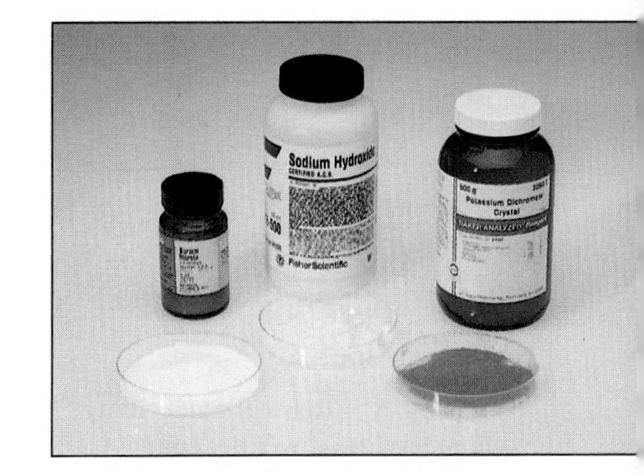

Photograph of (from left to right) barium nitrate, sodium hydroxide, and potassium dichromate.

Physical Properties of Ionic and Molecular Compounds

Ionic and molecular substances usually have very different physical properties. Ionic solids have three-dimensional structures that are held together by strong electrostatic forces because each ion is surrounded by ions of the

Ionic compounds are generally hard crystalline solids, whereas small molecular compounds are generally liquids or gases at room temperature.

opposite charge. In general, ionic materials form hard crystalline solids that must be heated to high temperatures before they melt and to extremely high temperatures before they vaporize.

Even though the solid is made up of ions, the solid does not conduct electrical current because the ions, which would carry electrical charge, cannot move (Figure 2.10a). In the liquid state, ionic substances conduct electrical current because the ions are now free to move (Figure 2.10b). Most

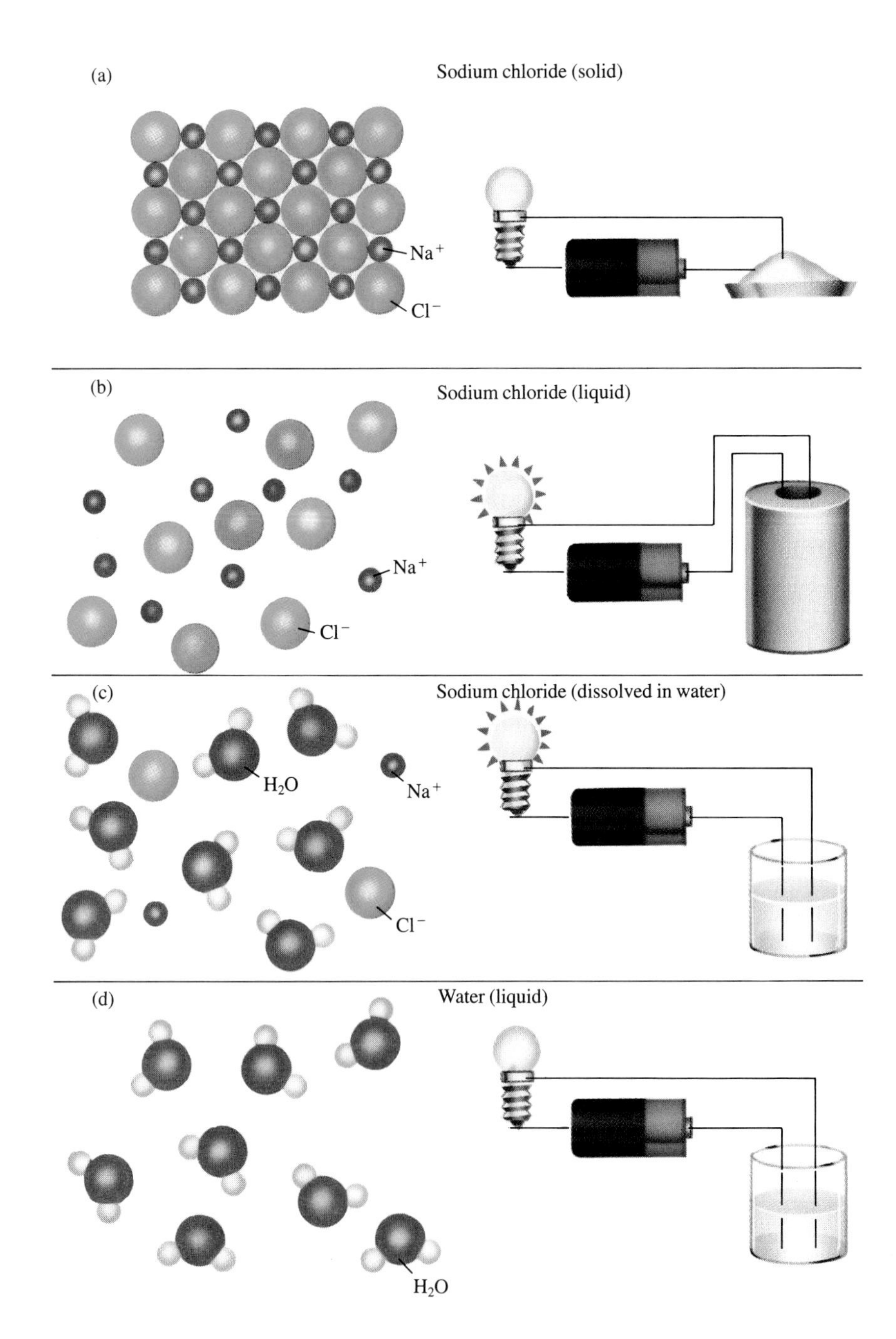

Figure 2.10 Electrical conductivity Ionic solids (a) will not conduct electrical current, but will conduct as molten liquids (b) or when dissolved in water (c). Pure water (d) will not conduct electrical current. Note that tap water contains ions and will conduct an electrical current, so it is important to keep electrical appliances away from bathtubs.

ionic substances will dissolve in water; the solid structure breaks up and separates into individual cations and anions (surrounded by the water molecules), a process that is called **dissociation.** For example, when NaCl dissolves in water it separates into Na^+ cations and Cl^- anions (Figure 2.10c). An ionic substance when dissolved in water conducts an electrical current because the ions in solution are free to move. Pure water does not conduct electricity because there are no charge carriers present (Figure 2.10d).

Small molecular compounds like water (H_2O) and molecular bromine (Br_2) have properties that are very different from those of ionic compounds. Substances consisting of small molecules generally form low-melting solids, liquids, or gases at room temperature (molecular substances that contain many atoms have different properties). In contrast to the strong electrostatic forces that hold together ionic solids, the forces that hold one molecule to another are weak (these forces will be covered in Chapter 10). Figure 2.11 depicts bromine molecules in all three phases. In the solid, the molecules are in fixed positions. The solid is easy to melt—it becomes a liquid at −7 °C and the liquid boils at 59 °C. In the liquid, the molecules are still in close contact, but are free to move. In the gas phase the molecules are in motion and are well separated. In all phases, the individual Br_2 molecules remain intact. None of these phases of bromine conduct electrical current because there are no charged species present.

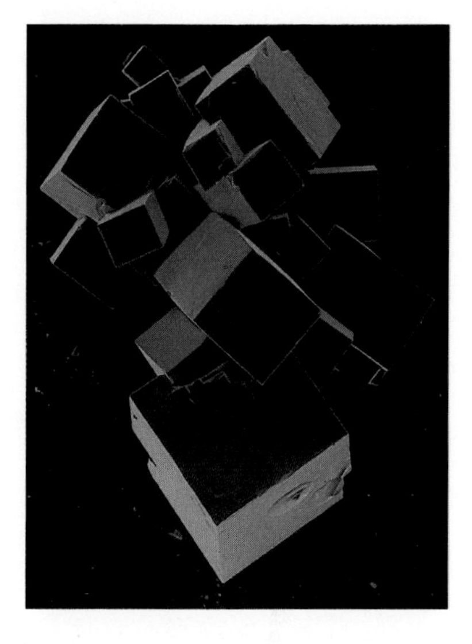

NaCl Color-tinted scanning electron micrograph of small, carefully grown crystals of sodium chloride.

2.4 Chemical Nomenclature

Objective

- To name simple ionic and molecular compounds

Chemists use names as well as formulas to identify compounds. In the early development of chemistry, many different methods were used to name substances. Millions of different compounds have been isolated and many more are being prepared daily, so an organized naming system is needed. **Chemical nomenclature** is the organized system for the naming of compounds. This section will outline the methods used to name ionic and simple molecular compounds.

Ionic Compounds

Ionic compounds composed of monatomic ions are named by using the name of the element present as the cation (generally a metal) followed by a modified name of the element that forms the anion (generally a nonmetal). The anions are named by adding the suffix -ide to the first part of the element name. Table 2.3 gives important examples of the names of monatomic anions.

Binary compounds, compounds composed of only two elements, are easily named. Table salt, NaCl, is sodium chloride. Magnesium bromide is the name for $MgBr_2$. Note that the numbers of ions in the empirical formula are inferred from the known charges of the ions and are not given in the name.

Figure 2.11 Three phases of bromine Bromine solid melts at room temperature. When the liquid warms, red bromine gas forms.

Table 2.3 Names of Common Monatomic Anions

Anion	Name	Anion	Name
H^-	hydride	F^-	fluoride
N^{3-}	nitride	Cl^-	chloride
O^{2-}	oxide	Br^-	bromide
S^{2-}	sulfide	I^-	iodide

Binary ionic compounds are named with the cation name first, followed by the first part of the element name of the anion with an *-ide* ending.

Example 2.7 Naming Binary Ionic Compounds

Name the compounds:
(a) BaI_2 (b) MgO (c) Na_2S

Solution In each case write the name of the cation followed by the name of the anion as it appears in Table 2.3:

(a) barium iodide,
(b) magnesium oxide,
(c) sodium sulfide.

Understanding Name the compound $CaCl_2$.

Answer: calcium chloride

Polyatomic ions are named as given in Table 2.2. Ammonium sulfide is the name for $(NH_4)_2S$. The formula of magnesium nitrate is $Mg(NO_3)_2$.

Example 2.8 Formulas of Ionic Compounds of Polyatomic Ions

Write the formulas for the following compounds:

(a) ammonium chromate,
(b) barium perchlorate,
(c) sodium sulfate.

Solution

(a) An ammonium ion has a 1+ charge and chromate, CrO_4^{2-}, has a 2− charge, so two ammonium cations are needed to balance the charge. The ammonium ion is placed in parentheses to separate the subscripts: $(NH_4)_2CrO_4$.
(b) Two polyatomic perchlorate anions, each with a 1− charge, are needed to balance the 2+ barium ion, so the formula is $Ba(ClO_4)_2$.
(c) Two sodium 1+ cations balance the charge of the 2− sulfate anion, giving the formula Na_2SO_4. No parentheses are needed for the sulfate because the subscript 1 is not written.

Understanding Write the formula for the compound sodium carbonate.

Answer: Na_2CO_3

Naming Transition Metal Compounds

The charges on ions of metals in groups I, II, and III are readily predicted from the periodic table. However, this cannot be done for most of the other metals, especially the transition metals—they can have several different

Table 2.4 Naming Transition Metal Compounds

Compound	Modern Name	Older Name
$FeCl_2$	iron(II) chloride	ferrous chloride
$FeCl_3$	iron(III) chloride	ferric chloride
Cu_2O	copper(I) oxide	cuprous oxide
CuO	copper(II) oxide	cupric oxide
$CrCl_2$	chromium(II) chloride	chromous chloride
Cr_2S_3	chromium(III) sulfide	chromic sulfide

charges. For example, iron can form two different ionic compounds with chlorine, having the formulas $FeCl_2$ and $FeCl_3$. The charge of the iron ion in $FeCl_2$ must be 2+ and in $FeCl_3$ it must be 3+. In the modern system of nomenclature for these compounds, a large Roman numeral is written in parentheses after the name of the metal to specify the charge. The compound $FeCl_2$ is iron(II) chloride, read "iron two chloride." The compound $FeCl_3$ is iron(III) chloride.

Roman numerals in parentheses are used to indicate the charge on a transition metal ion.

An older system of nomenclature also exists for some of the more common ions. This system uses the suffixes -ous and -ic to designate the lower and higher charged cations, respectively. The Latin name for the element is used as the root with some metals. For example, ferrous and ferric are the names of Fe^{2+} and Fe^{3+}, and cuprous and cupric are the names of Cu^+ and Cu^{2+}, respectively. This system will not be used in this text, but does appear in the older chemical literature. Examples of names of transition metal compounds are given in Table 2.4.

Example 2.9 Names of Transition Metal Compounds

Write the modern names of the following compounds:
(a) $CoBr_2$ (b) $Cr_2(SO_4)_3$ (c) $CuNO_3$ (d) $Fe(OH)_3$

Solution Each of these compounds is formed from a metal that can have more than one charge. The charge on the anion is known.

(a) The 2− total charge from the two 1− charged bromide anions is balanced by a 2+ charge on cobalt, so the compound is cobalt(II) bromide.
(b) Each of the three polyatomic sulfate anions has a 2− charge. In order to balance this 6− total negative charge, each chromium cation must have a 3+ charge. The compound is chromium(III) sulfate.
(c) The nitrate group has a 1− charge, so this is copper(I) nitrate.
(d) Hydroxide has a single negative charge, so the compound is iron(III) hydroxide.

Understanding Write the modern name of $Co(CH_3CO_2)_2$.

Answer: cobalt(II) acetate

Molecular Compounds

Many important molecular compounds have common names that are not systematic in nature. Thus, H_2O is water, NH_3 is ammonia, and CH_4 is methane—the name is not related to the formula. However, the majority of

Table 2.5 Prefixes for Naming Molecular Compounds

Number	Prefix	Example	Name
one	mono-	CO	carbon monoxide
two	di-	CO_2	carbon dioxide
three	tri-	SO_3	sulfur trioxide
four	tetra-	SF_4	sulfur tetrafluoride
five	penta-	PF_5	phosphorus pentafluoride
six	hexa-	SF_6	sulfur hexafluoride

binary molecular compounds have systematic names that are determined by methods similar to those used in naming ionic compounds. Because molecular compounds are generally formed from two or more nonmetals, two general rules are needed to decide which element should appear first in the name and in the formula (analogous to the cation in naming salts).

1. The element further to the left in the periodic table appears first.
2. The element closer to the bottom within any group appears first.

Hydrogen, which frequently is shown in two places on the periodic table, has its own rules. Hydrogen is given as the second element in the compounds it forms with elements in groups IA to VA, and first in its compounds with the group VIA and VIIA elements. Oxygen is also special and appears last in all cases, except in OF_2. These rules create the following ordering scheme: Al, B, Si, C, As, P, N, **H,** Se, S, I, Br, Cl, **O,** F. Hydrogen bromide is written HBr (hydrogen is before bromine on the list).

In certain cases, more than one compound can be formed from the same elements. We have already seen two combinations of carbon and oxygen, CO and CO_2. When more than one compound can be formed, we use prefixes to indicate the number of atoms of each element in the molecule. These prefixes, along with an example for each, are shown in Table 2.5. The prefix mono- (for one) is used only with the second element. Also, it is common practice to drop the last letter of a prefix that ends in *a* or *o* before anions that begin with a vowel, especially "oxide" (CO is carbon monoxide).

Prefixes indicate the number of atoms of each element present in a molecule of a compound.

This nomenclature is particularly useful for the oxides of nitrogen because six of them are well characterized. Two are dinitrogen monoxide, N_2O, and nitrogen dioxide, NO_2. It would be a terrible mistake to confuse these because N_2O is a relatively nontoxic material sometimes used as an anesthetic (laughing gas), and NO_2 is extremely toxic. Nomenclature can be very important!

Example 2.10 Names of Molecular Compounds

Write the formulas for the following compounds:

(a) selenium trioxide,
(b) dinitrogen tetroxide,
(c) phosphorus pentachloride,
(d) carbon tetrachloride.

Solution

(a) The prefix tri- means three: SeO_3.
(b) The prefixes di- and tetra- stand for two and four: N_2O_4. Note that in the name of this compound the *a* of tetra- is dropped.
(c) Penta- stands for five: PCl_5
(d) CCl_4

Understanding Write the formula for sulfur tetrachloride.

Answer: SCl_4

Example 2.11 Names of Molecular Compounds

Give the names for the following compounds:
(a) N_2O_5 (b) AsI_3 (c) XeF_6

Solution

(a) This name is another example that drops the last letter of the prefix: dinitrogen pentoxide.
(b) This name is arsenic triiodide.
(c) Xenon hexafluoride, an interesting example of a compound of a "noble gas."

Understanding Give the name for the compound IF_3.

Answer: iodine trifluoride

Summary

Dalton's atomic theory describes matter as being composed of small individual particles called *atoms.* The existence of atoms explains the law of constant composition (all samples of a substance contain the same elements in the same proportions), the law of multiple proportions (for compounds formed from the same elements, the masses of one element that combine with a fixed mass of the other are in the ratio of small whole numbers) and the law of conservation of mass (there is no loss or gain in mass when a chemical reaction takes place). Atoms are nature's building blocks, and *molecules* are combinations of atoms joined tightly together. Molecules are generally combinations of atoms of nonmetals. *Molecular formulas* are used to describe the number of every type of atom in molecules.

Atoms contain three different kinds of particles: (1) *protons* that have a relative charge of 1+ and relative mass of 1; (2) *neutrons* that have no charge and a relative mass of 1: (3) *electrons* that have a relative charge of 1− and relative mass of nearly 0. The protons and neutrons are closely packed in a central core of the atom called the *nucleus,* and the electrons are spaced at a relatively large distance around this core. The *atomic number* is the number of protons in the nucleus, and it defines the type of atom (that is, the element). The *mass number* is the sum of the numbers of protons and neutrons in the atomic nucleus. Atoms that have the same atomic number but different mass numbers are known as *isotopes.* Atoms can gain or lose electrons to form *ions. Cations* are positively charged ions, and *anions* are negatively charged ions. The nucleus of the atom does not change when ions are formed. In contrast, certain nuclei of some elements are unstable and transform into other nuclei with the emission of a small particle and energy. This phenomenon is known as *radioactivity.*

Cations and anions aggregate to form neutral species known as *ionic compounds.* The ions in ionic compounds are held together by electrostatic attractive forces (coulombic forces). An ionic compound is generally formed from a cationic metal and anionic nonmetal. The *empirical formula* of an ionic compound gives the relative number of ions. The formula of an ionic compound can be written by balancing the charges of its ions. *Polyatomic ions* are groups of atoms that have a net charge. Ionic compounds are generally hard crystalline solids whereas most substances consisting of small molecules form gases, liquids or low-melting solids.

Chemical nomenclature is a method for systematically naming compounds. Binary ionic compounds are named by first naming the cation followed by the name of the anion. The name of a monatomic anion consists of the first part of the element name plus an -ide suffix. The charges on cations of the transition metals can be different in different compounds and are indicated in the name by a Roman numeral in parentheses. Molecular compounds are named in a similar manner, with the element more to the left of the periodic table generally named first. In cases where more than one compound can be made from the same elements, prefixes are required that indicate the number of atoms of each element present in the molecule.

Chapter Terms

Atom—the smallest unit of an element that enters into a chemical combination. *(2.1)*

Law of constant composition—all samples of a pure substance contain the same elements in the same proportions by mass. *(2.1)*

Law of multiple proportions—if two elements unite to form more than one compound, the masses of one element (in each compound) that combine with a fixed mass of the second element are in a ratio of small whole numbers. *(2.1)*

Law of conservation of mass—there is no detectable loss or gain in mass when a chemical reaction occurs. *(2.1)*

Molecule—a combination of atoms joined tightly together so that they behave as a single particle. *(2.1)*

Diatomic molecule—a molecule that is composed of two atoms. *(2.1)*

Molecular formula—gives the number and type of every atom in a molecule. *(2.1)*

Structural formula—indicates how the atoms are connected in the molecule. *(2.1)*

Electron—a very small particle that has a mass of 9.11×10^{-31} kg and a charge of -1.602×10^{-19} coulombs. The relative charge is 1− and the relative mass is approximately 0. *(2.2)*

Proton—a small particle that has a mass of 1.673×10^{-27} kg and a charge of $+1.602 \times 10^{-19}$ coulombs. The absolute charge is the same as that of the electron but opposite in sign. The relative charge is 1+ and the relative mass is 1. *(2.2)*

Neutron—a small particle that has a mass of 1.675×10^{-27} kg and no charge. The relative mass is 1. *(2.2)*

Coulomb's law—the force between two charged objects is directly proportional to the product of the charges and inversely proportional to the square of the distance between them. *(2.2)*

$$\text{force} = \frac{kQ_1Q_2}{r^2}$$

Nucleus—the small, heavy, positively charged core of an atom. *(2.2)*

Atomic number—the number of protons in the nucleus of an atom. It has the symbol *Z*. *(2.2)*

Mass number—the total number of protons and neutrons in an atom. It has the symbol *A*. *(2.2)*

Isotopes—atoms of the same element that have different numbers of neutrons. *(2.2)*

Ion—a charged particle formed by the addition or removal of electrons from an atom or group of atoms. *(2.2)*

Cation—a positively charged ion. *(2.2)*

Anion—a negatively charged ion. *(2.2)*

Radioactivity—a spontaneous nuclear reaction that transforms a relatively unstable nucleus into a more stable nucleus with the emission of a small particle and energy. *(2.2)*

Ionic compound—a compound composed of cations and anions joined to form a neutral species. *(2.3)*

Empirical formula—gives the relative numbers of different elements in a substance, using the smallest whole numbers for subscripts. *(2.3)*

Monatomic ion—an ion formed by the loss or gain of electrons by a single atom. *(2.3)*

Polyatomic ion—charged species made up of more than one atom. *(2.3)*

Dissociation—the separation of an ionic solid into individual anions and cations when dissolved in solution (generally water). *(2.3)*

Chemical nomenclature—the organized system for the naming of compounds. *(2.4)*

Binary compound—a compound composed of only two elements. *(2.4)*

Exercises

Exercises designated with color have answers in Appendix J.

Atomic Theory

2.1 In one of the compounds that contain only oxygen and nitrogen, 1.75 g of nitrogen combine with 1.00 g of oxygen. In another, 0.875 g of nitrogen combine with 1.00 g of oxygen. Show that these compounds obey the law of multiple proportions.

2.2 In one binary sulfur-oxygen compound 1.0 g of sulfur combines with 1.0 g of oxygen, and in another compound 0.67 g of sulfur combine with 1.0 g of oxygen. Show that these compounds obey the law of multiple proportions.

2.3 Identify the postulate of Dalton's atomic theory that explains
(a) the law of conservation of mass.
(b) the law of constant composition.

2.4 How does Dalton's atomic theory explain
(a) the fact that a sample of pure NaCl obtained from a mine in the United States contains the same elements in the same ratio as NaCl obtained from a mine in France?
(b) the fact that the mass of the hydrogen peroxide molecule, H_2O_2, is the same as the sum of the masses of the hydrogen, H_2, and oxygen, O_2, molecules from which it was formed?

Structure of Atoms

2.5 Compare and contrast the terms molecule, atom, element and compound. Give examples of each. Some of your examples will fit more than one term, so make it clear which term fits each example.

2.6 Compare the masses and charges of the important particles that make up atoms.

2.7 Describe Thomson's model of the structure of atoms.

2.8 How did the Rutherford experiment disprove the Thomson model of the atom?

2.9 What difference might you expect if aluminum foil had been used in the Rutherford experiment in place of gold foil?

2.10 Describe the modern view of the structure of an atom.

2.11 Define the terms
(a) atomic number,
(b) mass number,
(c) isotope.
Discuss how the atomic number and mass number are related in various isotopes of elements.

2.12 Write the symbol that describes each of the following isotopes:
(a) an atom that contains seven protons and eight neutrons.
(b) an atom that contains 31 protons and 39 neutrons.
(c) an atom that contains 18 protons and 22 neutrons.

2.13 Write the symbol that describes each of the following isotopes:
(a) an atom that contains five protons and six neutrons.
(b) an atom that contains 25 protons and 30 neutrons.
(c) an atom that contains 14 protons and 14 neutrons.

2.14 Give the numbers of protons, neutrons and electrons in the following species:
(a) $^{79}_{33}As$
(b) $^{51}_{23}V$
(c) $^{128}_{52}Te$

2.15 Give the numbers of protons, electrons and neutrons in the following species:
(a) $^{32}_{16}S$ (b) $^{24}_{12}Mg^{2+}$ (c) $^{37}_{17}Cl^-$

2.16 Write the symbol for the ion with
(a) eight protons, 10 electrons, and eight neutrons.
(b) 34 protons, 36 electrons, and 45 neutrons.
(c) 28 protons, 26 electrons, and 31 neutrons.

2.17 Write the symbol for the ion with
(a) four protons, two electrons, and five neutrons.
(b) 32 protons, 30 electrons, and 40 neutrons.
(c) 35 protons, 36 electrons, and 44 neutrons.

2.18 Given the partial information in each column in the following table, fill in the blanks.

symbol	___	$^{40}Ca^{2+}$	___	___
atomic number	11	___	___	___
mass number	___	___	81	___
charge	___	___	1−	2−
number protons	___	___	35	52
number electrons	10	___	___	___
number neutrons	12	___	___	76

2.19 Given the partial information in each column in the following table, fill in the blanks.

symbol	$^{127}I^-$	___	___	___
atomic number	___	___	___	17
mass number	___	16	___	___
charge	___	0	2+	1−
number protons	___	8	38	___
number electrons	___	___	___	___
number neutrons	___	___	50	20

2.20 Define radioactivity. Compare radioactivity with the process of removing an electron from an atom to form an ion.

Ionic Compounds

2.21 Describe the types of elements that generally combine to form ionic compounds and the types that combine to form molecular compounds.

2.22 How are the properties of ionic compounds different from those of molecular compounds?

2.23 Write the symbol for the monatomic ion expected for each of the following elements:
(a) potassium (b) bromine (c) barium (d) sulfur

2.24 Write the symbol for the monatomic ion expected for each of the following elements:
(a) iodine (b) magnesium (c) oxygen (d) sodium

2.25 What are the empirical formulas for compounds made from
(a) Li^+ and I^- (b) Cs^+ and O^{2-} (c) Y^{3+} and Cl^-?

2.26 What are the empirical formulas for compounds made from
(a) Ca^{2+} and S^{2-} (b) Mg^{2+} and N^{3-} (c) Fe^{2+} and F^-?

2.27 Write the empirical formulas for ionic compounds made from the following elements:
(a) magnesium and fluorine
(b) sodium and oxygen
(c) scandium and selenium
(d) barium and nitrogen

2.28 Write the empirical formulas for ionic compounds made from the following elements:
(a) calcium and chlorine
(b) rubidium and sulfur
(c) lithium and nitrogen
(d) yttrium and selenium

2.29 Write the formulas and charges of
(a) chromate ion (b) carbonate ion (c) sulfate ion

2.30 Write the formulas and charges of
(a) hydroxide ion (b) chlorate ion (c) permanganate ion

2.31 Write the formulas of
(a) sodium nitrate
(b) beryllium hydroxide
(c) ammonium acetate
(d) potassium sulfite

2.32 Write the formulas of
(a) magnesium nitrite
(b) lithium phosphate
(c) barium cyanide

2.33 Explain why most ionic compounds are hard solids at room temperature whereas many molecular substances such as H_2O and O_2 are liquids or gases.

Nomenclature

2.34 Write the names of the following ionic compounds:
(a) LiI (b) Mg_3N_2 (c) Na_3PO_4 (d) $Ba(ClO_4)_2$

2.35 Write the names of the following ionic compounds:
(a) NH_4Br (b) $BaCl_2$ (c) K_2O (d) $Sr(NO_3)_2$

2.36 Write the names of the following transition metal compounds:
(a) $CoCl_3$ (b) $FeSO_4$ (c) CuO

2.37 Write the names of the following transition metal compounds:
(a) $RhBr_2$ (b) CuCN (c) $V(NO_3)_3$

2.38 Write the formulas of
(a) manganese(III) sulfide
(b) iron(II) cyanide
(c) potassium sulfide
(d) mercury(II) chloride

2.39 Write the formulas of
(a) calcium nitride
(b) chromium(III) perchlorate
(c) tin(II) fluoride
(d) potassium permanganate

2.40 Write the formulas for the following molecular compounds:
(a) sulfur tetrafluoride
(b) nitrogen trichloride
(c) dinitrogen pentoxide
(d) chlorine trifluoride

2.41 Write the formulas for the following molecular compounds:
(a) sulfur difluoride
(b) silicon tetrachloride
(c) gallium trichloride
(d) dinitrogen trioxide

2.42 Write the names for the following molecular compounds:
(a) PBr_5 (b) SeO_2 (c) B_2Cl_4 (d) S_2Cl_2

2.43 Write the names for the following molecular compounds:
(a) HI (c) SO_2
(b) NF_3 (d) N_2Cl_4

2.44 Potassium chloride is used as a substitute for sodium chloride by some people who have hypertension or heart problems. What is the formula of potassium chloride?

2.45 The compound MnO is added to glass during manufacture to improve the color of the glass. Write the name of MnO.

Additional Problems

2.46 The common name for a slurry of $Mg(OH)_2$ in water is milk of magnesia. Give the proper name of this compound.

2.47 The common name of KNO_3 is saltpeter, and $(NH_4)_2CO_3$ is smelling salts (it gives off ammonia). Give the proper names of these two compounds.

2.48 Write the symbol, including atomic number, mass number, and charge, for each of the following species:
(a) A halogen with a mass number of 35 and a 1− charge.
(b) An alkali metal with 18 electrons, 20 neutrons, and a 1+ charge.

2.49 Write the symbol, including atomic number, mass number, and charge, for each of the following species:
(a) A neutral noble gas element with 21 neutrons in its nucleus.
(b) An alkaline earth metal with a mass number of 40 and a 2+ charge.

2.50 Name the following compounds and indicate whether they are ionic or molecular.
(a) NO (c) Na_2O
(b) $Y_2(SO_4)_3$ (d) NBr_3

2.51 Write the formulas of the following compounds and indicate whether they are ionic or molecular.
(a) calcium phosphate
(b) germanium dioxide
(c) iron(III) sulfate
(d) phosphorus tribromide

2.52 Given the partial information in each column in the following table, fill in the blanks.

symbol	___	___	___	$^{28}Si^{2-}$
atomic number	___	___	49	___
mass number	70	103	___	___
charge	___	3+	1+	___
number protons	31	___	___	___
number electrons	28	42	___	___
number neutrons	___	___	65	___

2.53 Plutonium was first isolated by Glenn Seaborg and coworkers in 1940 as the ^{238}Pu isotope; they made it by deuterium bombardment of uranium. Give the number of protons, neutrons, and electrons in an atom of this isotope of plutonium.

2.54 From the list of elements Li, Ca, Fe, Al, Cl, O, C, and N, write the formulas and names of compounds with the following characteristics:
(a) An ionic compound with the formula MX_2, where M is an alkaline earth metal and X is a nonmetal.
(b) A molecular substance with the formula AB_2, where A is a group IVA element and B is a group VIA element.
(c) A compound with the formula M_2X_3, where M is a transition metal and X is a nonmetal.

2.55 Describe the three isotopes of hydrogen. Write the symbol and give the name of each isotope.

2.56 Write the symbol for each of the following species:
(a) A cation with a mass number of 23, an atomic number of 11, and a charge of 1+.
(b) A member of the nitrogen group (group V) that has a 3+ charge, 48 electrons, and 70 neutrons.
(c) A noble gas with no charge and 48 neutrons.

2.57 Write the formulas for the following compounds:
(a) sodium selenide
(b) nickel(II) bromide
(c) dinitrogen pentoxide
(d) copper(II) sulfate
(e) ammonium sulfite

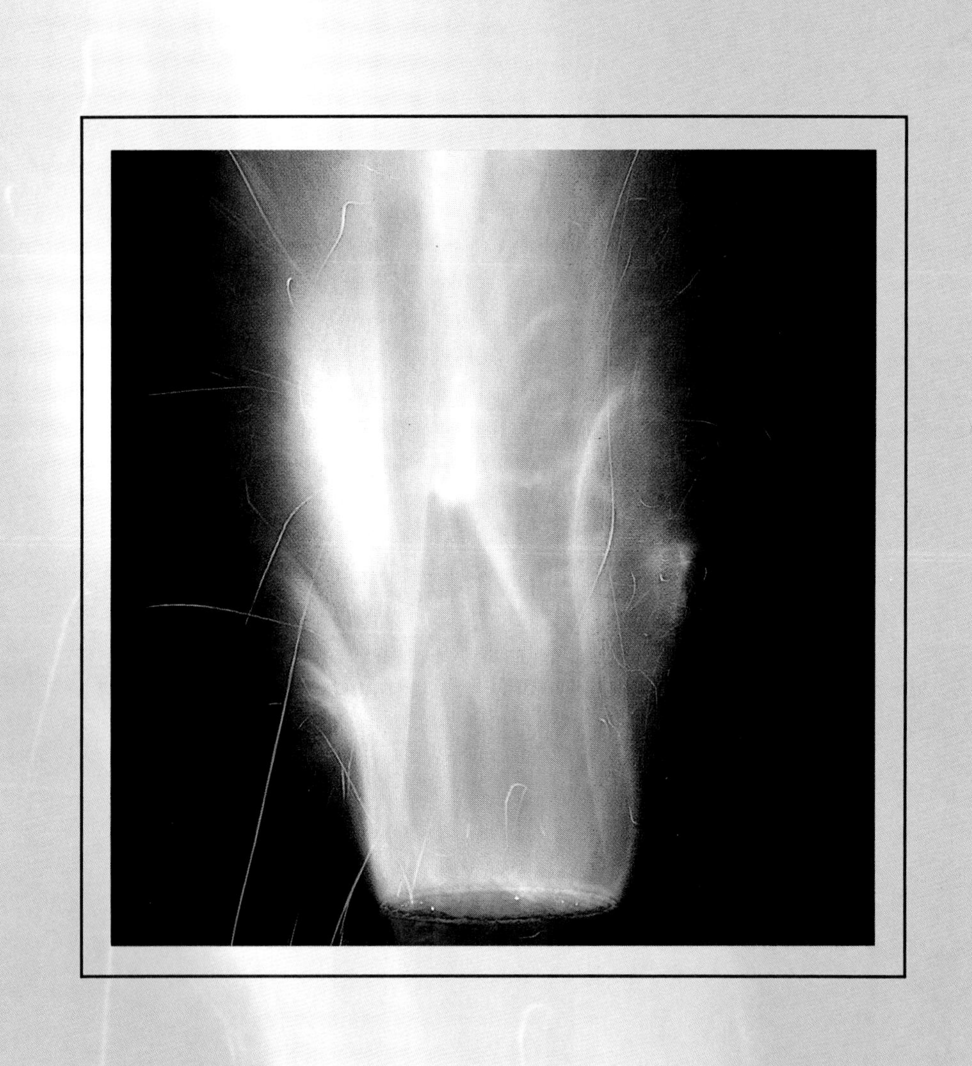

Chapter Three

Stoichiometry I: Equations, the Mole, and Chemical Formulas

Chemistry progressed from an art to a science when chemists began to measure how much of each substance was used up by a chemical reaction and how much of the resulting substances was produced. In this chapter and the next, we will discuss **stoichiometry,** the study of *quantitative* relationships involving substances and their reactions. We will also develop a method for determining the formulas of the new substances produced in these reactions.

3.1 Chemical Equations

Objectives

- To balance a chemical expression
- To write chemical equations for (a) the combustion of an organic compound in oxygen and (b) a neutralization reaction

Some mixtures of substances are unreactive under almost any conditions; other mixtures react violently. Knowledge of how different substances react is central to the science of chemistry and is needed to understand the world around us. Chemists have been doing experiments for centuries and know the outcomes of many reactions. New reactions are always being investigated in laboratories all over the world. Frequently, the outcome of a new

The combustion of zinc dust in a flame.

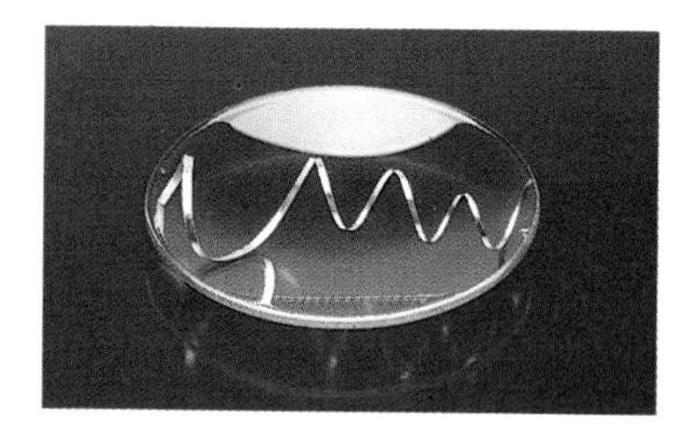

(a)

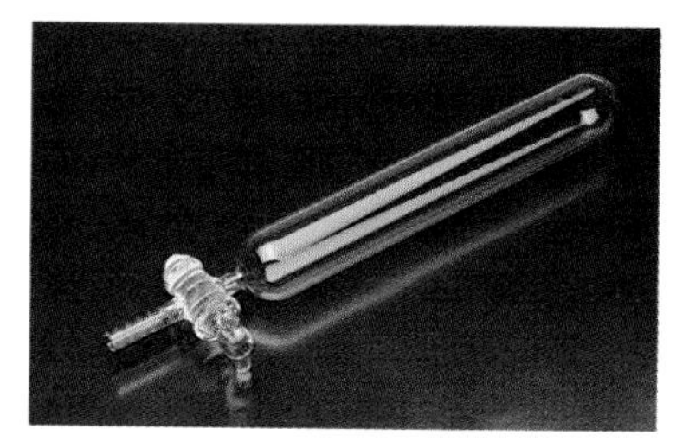

(b)

(c)

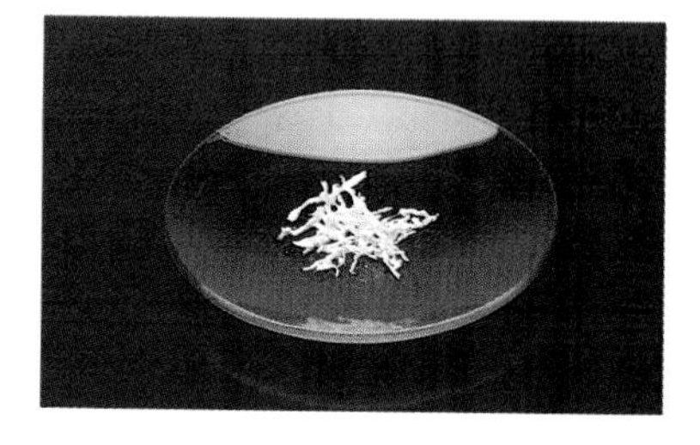

(d)

Magnesium burning in oxygen Magnesium (a) reacts with oxygen (b). The reaction (c) is accompanied by a bright light and magnesium oxide (d) is formed as the product.

reaction can be predicted from knowledge about other reactions that have been studied previously.

Chemical changes can be compactly described by equations. Rather than using an equal sign when writing equations, chemists generally use an arrow that is read as "yields." For example, the reaction of magnesium and oxygen to form magnesium oxide is written as

$$2Mg + O_2 \longrightarrow 2MgO$$

The magnesium and oxygen are called the **reactants,** the substances that are consumed, and the magnesium oxide is called the **product,** the substance that is formed. The arrow points from the reactants to the products. The **chemical equation** describes the identities and relative amounts of reactants and products in a chemical reaction.

Each side of the preceding equation contains two magnesium atoms and two oxygen atoms. The equation is said to be *balanced* because it follows the fourth postulate of Dalton's atomic theory, that matter is conserved.

In general, two steps are needed to write a chemical equation. First, determine the identities (the formulas) of the reactants and products in the chemical reaction, and write each formula on the appropriate side of an arrow. At this point, because it is not necessarily balanced, we call what we have written a **chemical expression.** Identifying the formulas of reactants and products can be a difficult task in some cases; methods that are useful for these determinations are described in later sections. An example of a chemical expression is shown below for the reaction of the diatomic molecules H_2 and Cl_2 to form HCl.

$$H_2 + Cl_2 \longrightarrow HCl$$

This chemical expression is *not* a balanced equation, because there are different numbers of hydrogen atoms and chlorine atoms on the left and right sides of the arrow. The second step in writing an equation is to adjust the number that precedes each formula to balance the numbers of atoms of each type on each side of the arrow. These numbers are the **coefficients,** the number of units of each substance involved in the chemical reaction. This example is easy to balance; just place a coefficient of 2 in front of the HCl.

$$H_2 + Cl_2 \longrightarrow 2HCl$$

The equation now shows two atoms of hydrogen and two atoms of chlorine on each side of the arrow; the equation is balanced.

It is incorrect to balance the chemical expression by changing the formula of any of the substances. *Do not alter the subscripts in any of the substances in order to balance the expression.* It is known from experiments that HCl is the correct molecular formula of the product. Balancing the expression by changing the product to H_2Cl_2 is incorrect because one molecule of HCl does not contain two atoms of each element. Using a coefficient of 2 to balance the equation is correct; two molecules of HCl are formed in the reaction, each of them containing one hydrogen atom and one chlorine atom. Figure 3.1 shows a schematic representation of this reaction.

Sodium metal reacts with chlorine gas to produce sodium chloride. In this case the chemical expression is

$$Na + Cl_2 \longrightarrow NaCl$$

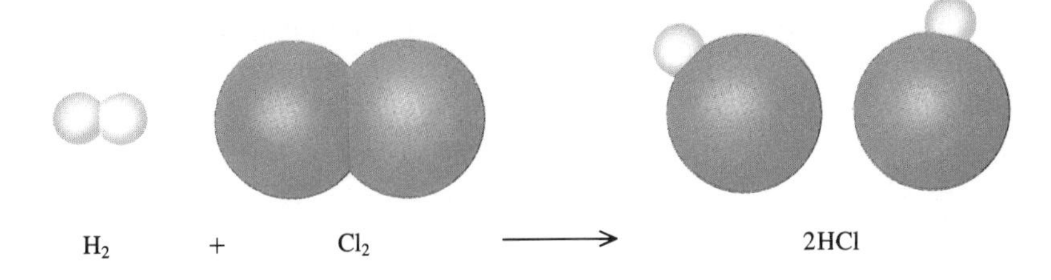

Figure 3.1 Reaction of hydrogen and chlorine The reaction of one molecule of hydrogen with one molecule of chlorine forms two molecules of hydrogen chloride. Both sides of the equation contain the same numbers of hydrogen and chlorine atoms.

This chemical expression is not an equation because there are two chlorine atoms on the left and only one on the right. Placing a coefficient of 2 before the sodium chloride brings the chlorine atoms into balance, but takes the sodium out of balance. Also placing a coefficient of 2 before the sodium produces the balanced equation.

$$2Na + Cl_2 \longrightarrow 2NaCl$$

Writing Balanced Equations

The most common method used to convert a chemical expression into an equation is to adjust the coefficients in a systematic manner until the expression is brought into balance. The best way to learn to balance equations is by practice. If your first effort to balance the equation has not worked after a few minutes, begin again using a new starting point. An attempt that is going to be successful will generally yield the equation quickly. If two or three attempts fail, or if you are using very large coefficients, check that the formulas of the reactants and products are correct. You must start with the correct formulas in the chemical expression!

Equations are balanced by adjusting the coefficients.

One effective way to begin balancing the chemical expression is to assume that the balanced equation contains one molecule of the most complicated substance. Bring the atoms of this substance into balance by adjusting the coefficients of the substances on the other side of the equation. In addition, it is particularly useful to balance elements that appear only once on

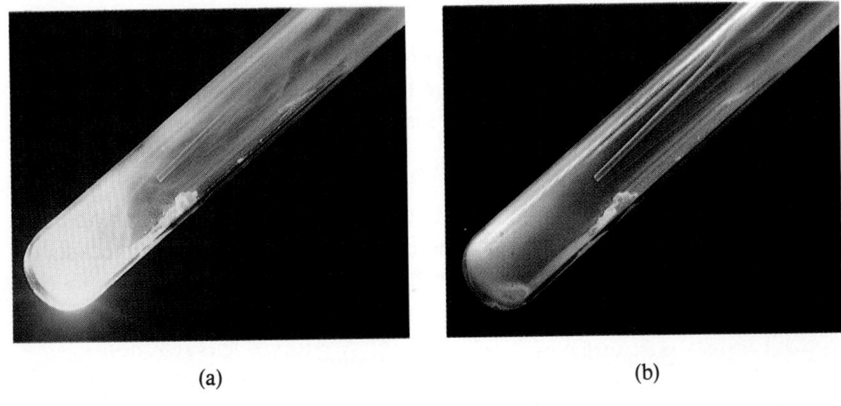

(a) (b)

Sodium reacting with chlorine Sodium metal and chlorine gas react (a) to form sodium chloride (b).

each side of the equation. Consider, for example, the chemical expression for the burning of propane, C_3H_8, (a fuel that is used in many areas for home heating and cooking):

$$C_3H_8 + O_2 \longrightarrow CO_2 + H_2O$$

The C_3H_8 is the most complicated substance, so assume that its coefficient is 1 and proceed to balance its atoms on the product side. Each carbon dioxide molecule contains one carbon atom, so placing a coefficient of 3 in front of the CO_2 will balance carbon. Each water molecule contains two hydrogen atoms, so four water molecules will contain eight hydrogen atoms. Placing a coefficient of 4 before the H_2O balances hydrogen.

$$C_3H_8 + O_2 \longrightarrow 3CO_2 + 4H_2O$$

Next, the number of oxygen atoms on the left side of the equation must be made equal to the number on the right. There are six oxygen atoms in three CO_2 molecules (3 molecules × 2 oxygen atoms/molecule) and four oxygen atoms in four molecules of H_2O (4 molecules × 1 oxygen atom/molecule), giving a total of 10 atoms of oxygen on the product side. Since each molecule of oxygen contains two atoms, the coefficient of O_2 is 10/2, or 5.

$$C_3H_8 + 5O_2 \longrightarrow 3CO_2 + 4H_2O$$

The important final step is to check that your answer is correct. In this case, each side of the equation contains three carbon, eight hydrogen, and ten oxygen atoms.

Balancing the reaction of oxygen and butane, C_4H_{10}, another compound that is used as a portable fuel, requires an additional step at the end. The chemical expression is

$$C_4H_{10} + O_2 \longrightarrow CO_2 + H_2O$$

Starting as before, assume that the reaction involves one molecule of butane, and place a 4 in front of the CO_2 and a 5 in front of the H_2O to bring the carbon and hydrogen into balance. The product side has 13 oxygen atoms present [$(4 \times 2) + (5 \times 1) = 13$], requiring a fraction, 13/2, before the O_2 to make the expression balance.

$$C_4H_{10} + \tfrac{13}{2}O_2 \longrightarrow 4CO_2 + 5H_2O$$

While these operations produce a correct chemical equation, chemical equations are most often written with the smallest correct set of whole number coefficients. To eliminate the fraction, multiply the coefficients of all substances in the equation by 2. This leaves the equation still in balance.

$$2C_4H_{10} + 13O_2 \longrightarrow 8CO_2 + 10H_2O$$

In checking, we find 8 carbon, 20 hydrogen, and 26 oxygen atoms on each side of the equation. Note that the coefficients are the smallest correct set of whole numbers.

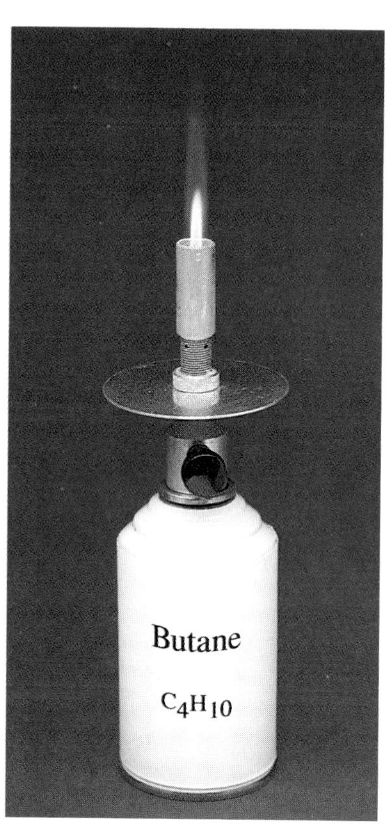

Burning butane gas Butane burns in air to produce carbon dioxide and water. It is a good fuel for a portable burner.

Example 3.1 Balancing Chemical Expressions

Balance the following chemical expression:

$$NaOH + Al + H_2O \longrightarrow H_2 + NaAlO_2$$

Solution In this case, the $NaAlO_2$ is the most complicated species. Its atoms are in balance without adding any coefficients to the reactant side. Only the hydrogen atoms are out of balance. They come into balance with a 3/2 coefficient for H_2.

$$NaOH + Al + H_2O \longrightarrow \tfrac{3}{2}H_2 + NaAlO_2$$

This yields a balanced equation, but both sides of the equation should be multiplied by 2 to remove the fraction.

$$2NaOH + 2Al + 2H_2O \longrightarrow 3H_2 + 2NaAlO_2$$

Check the final answer (2 Na, 2 Al, 4 O, and 6 H atoms on each side).

Understanding Balance the following expression, using whole number coefficients.

$$Al + HCl \longrightarrow AlCl_3 + H_2$$

Answer: $2Al + 6HCl \longrightarrow 2AlCl_3 + 3H_2$

Figure 3.2 Drain cleaners Drano is a solid that contains sodium hydroxide (lye) and a small quantity of aluminum. The cleaning is provided by the reaction of sodium hydroxide with materials like food and hair. The aluminum reacts to form hydrogen gas, which stirs the mixture.

In many cases, the physical state of each substance in the equation must be specified. We use the symbols (s) for solid, (ℓ) for liquid, (g) for gas, and (aq) for substances dissolved in water (aqueous solution). The reaction balanced in Example 3.1, taking place in water, is written

$$2NaOH(aq) + 2Al(s) + 2H_2O(\ell) \longrightarrow 3H_2(g) + 2NaAlO_2(aq)$$

The NaOH and $NaAlO_2$ dissolve in water, the Al is a solid, the H_2O is a liquid, and the H_2 is a gas at the temperature and pressure of the reaction (Figure 3.2). A mixture of sodium hydroxide and aluminum is used in several commercial drain cleaners. Although the major effect of the cleaner comes from the reaction of the sodium hydroxide (lye) with grease, hair, and so forth in the drain, the gaseous hydrogen that forms in the reaction causes agitation and helps unclog the drain.

Example 3.2 Writing Balanced Equations

(a) The reaction of sodium carbonate (Na_2CO_3) with nitric acid (HNO_3) yields water, carbon dioxide, and sodium nitrate. Write the equation.

(b) The highly toxic gas phosphine, PH_3, is useful in the synthesis of the semiconductor indium phosphide, InP. It can be prepared by the reaction of Ca_3P_2 (calcium phosphide) and water. Calcium hydroxide is also produced in this reaction. Write the equation.

Solution

(a) First write the chemical expression, using the correct formula for each compound.

$$Na_2CO_3 + HNO_3 \longrightarrow H_2O + CO_2 + NaNO_3$$

1. Assume a coefficient of 1 for Na_2CO_3; the carbon is then in balance as written.
2. Put a 2 in front of the $NaNO_3$ to balance Na.

$$Na_2CO_3 + HNO_3 \longrightarrow H_2O + CO_2 + 2NaNO_3$$

3. Put a 2 in front of the HNO_3 to balance NO_3^-. Since the polyatomic nitrate group appears on both sides of the equation, it is best to treat it as a group for balancing the chemical expression.
4. The coefficient of water is found to be correct by checking the H and O balance.

$$Na_2CO_3 + 2HNO_3 \longrightarrow H_2O + CO_2 + 2NaNO_3$$

(b) The first thing needed is the chemical expression.

$$Ca_3P_2 + H_2O \longrightarrow PH_3 + Ca(OH)_2$$

Start by placing a 3 before $Ca(OH)_2$ and a 2 before PH_3 to bring calcium and phosphorus into balance.

$$Ca_3P_2 + H_2O \longrightarrow 2PH_3 + 3Ca(OH)_2$$

Now balance hydrogen. On the product side there are six hydrogen atoms in $2PH_3$ and six more in $3Ca(OH)_2$. Place a 6 before the water.

$$Ca_3P_2 + 6H_2O \longrightarrow 2PH_3 + 3Ca(OH)_2$$

The equation is in balance because six oxygen atoms are on each side. Remember that the $(OH)_2$ part of $Ca(OH)_2$ means there are two atoms of oxygen and two atoms of hydrogen in every unit of $Ca(OH)_2$. The parentheses are used rather than O_2H_2 because hydroxide ion, OH^-, is one of the polyatomic anions discussed in Chapter 2.

Understanding Bromine (Br_2) reacts with sodium iodide to yield sodium bromide and iodine (I_2). Write the chemical equation.

Answer: $Br_2 + 2NaI \longrightarrow I_2 + 2NaBr$

Combustion reaction An oil well burning out of control is an example of a combustion reaction.

Types of Chemical Reactions

Chemists have recognized and classified many types of reactions. Identification of a particular type of reaction is often useful in writing chemical equations. Two types of reactions are particularly important to learn at this point because they are frequently encountered.

Combustion Reactions

Organic compounds, compounds made up of carbon atoms in combination with other elements such as hydrogen, oxygen, and nitrogen, react with oxygen in combustion reactions. A **combustion reaction** is the process of burning. Examples are the burning of wood and natural gas. Unless told differently, assume that the products of the combustion of organic compounds containing only carbon, hydrogen, and oxygen are always CO_2 and H_2O.

Organic compounds react with excess oxygen to produce CO_2 and H_2O—a combustion reaction.

Example 3.3 **Combustion Equations**

(a) Write the equation for the combustion of ethanol, also called ethyl alcohol (C_2H_5OH).
(b) Write the equation for the combustion of benzene (C_6H_6).

Solution

(a) The combustion of an organic compound yields CO_2 and H_2O. The chemical expression is

$$C_2H_5OH + O_2 \longrightarrow CO_2 + H_2O$$

In order to achieve balance with a coefficient of 1 for the most complicated species, place a 2 before CO_2 and a 3 before H_2O.

$$C_2H_5OH + O_2 \longrightarrow 2CO_2 + 3H_2O$$

Formulas of organic compounds are often written to emphasize their structures, not in the most compact form. Note that a C_2H_5OH molecule contains six H atoms.

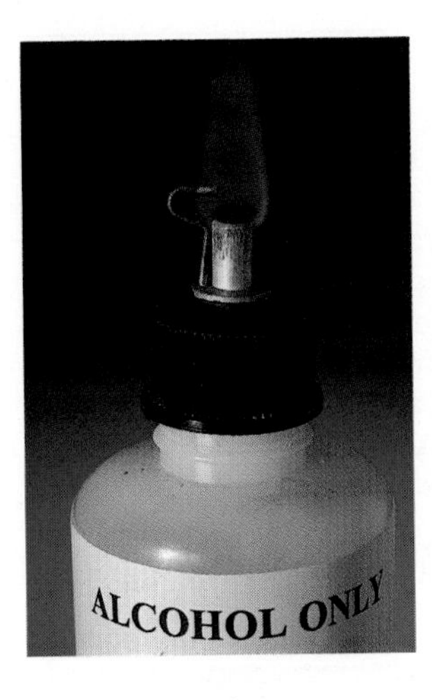

(a)

(b)

(c)

Combustion reactions of alcohols
(a) Small ethanol heating lamp used by a jeweler. (b) Sterno is a brand of "jellied" methanol used in cooking. (c) Some racing cars use alcohol fuel because it is less likely to explode in a collision than gasoline.

There are seven oxygen atoms on the product side. Since on the reactant side there is one oxygen atom in the ethanol, a 3 in front of O_2 will bring the expression into balance.

$$C_2H_5OH + 3O_2 \longrightarrow 2CO_2 + 3H_2O$$

(b) The chemical expression is

$$C_6H_6 + O_2 \longrightarrow CO_2 + H_2O$$

A 6 before the CO_2 and a 3 before the H_2O are needed to balance the carbon and hydrogen atoms from benzene.

$$C_6H_6 + O_2 \longrightarrow 6CO_2 + 3H_2O$$

To balance the O_2, a 15/2 coefficient is needed.

$$C_6H_6 + \tfrac{15}{2}O_2 \longrightarrow 6CO_2 + 3H_2O$$

Multiply both sides of the equation by 2 to obtain whole number coefficients.

$$2C_6H_6 + 15O_2 \longrightarrow 12CO_2 + 6H_2O$$

Benzene has been used extensively in many chemical processes. Recently, its use has been restricted since studies have shown that it can cause a number of health problems. It is still used in the laboratory, but it must be handled properly by trained individuals.

Understanding Write the equation for the combustion of methanol (CH_3OH).

Answer: $2CH_3OH + 3O_2 \longrightarrow 2CO_2 + 4H_2O$

Neutralization Reactions

Another important type of reaction is a **neutralization reaction,** the reaction of an acid and a base to yield water and a salt. The simplest definition of an **acid** is any substance that provides the hydrogen cation, H^+, for reaction in water solution. An example is HNO_3.

$$HNO_3(\ell) \xrightarrow{H_2O} H^+(aq) + NO_3^-(aq)$$

(Text continues on p. 90.)

INSIGHTS into CHEMISTRY

Chemical Reactivity

Nitric and Sulfuric Acids Are Culprits in Acid Rain: No Easy Answers

In recent years, acid rain has been the subject of a great deal of debate and concern. The harmful effects of acid rain have been highly publicized—lakes that can no longer support aquatic life, forests that are dying, and buildings and monuments that are literally dissolving away with every rainfall. The problem is not restricted to the highly industrialized nations—it is a worldwide problem. The causes are fairly well understood, but the solutions to the problem are often expensive or generate problems of their own. Difficult decisions will have to be made because acid rain is a problem that must be solved.

Many people assume that acid rain is a recent phenomenon caused by atmospheric pollution. In fact, all rain can be called acid rain. Rainwater is normally slightly acidic because dissolved carbon dioxide from the atmosphere reacts to form carbonic acid:

$$CO_2(g) + H_2O(\ell) \longrightarrow H_2CO_3(aq)$$

However, acids other than carbonic acid are found in the rain falling in many parts of the world, especially the northeastern United States and western Europe.

The greater acidity of rain is primarily caused by oxides of nitrogen and sulfur, which are present in the atmosphere from a variety of sources. Nitrogen monoxide, NO, is formed in the atmosphere by electrical storms, as well as in combustion processes, particularly in automobile engines. Nitrogen monoxide then reacts with oxygen in the atmosphere to produce nitrogen dioxide, NO_2, which reacts with more oxygen and water in clouds and rain to form a solution of nitric acid.

$$N_2(g) + O_2(g) \longrightarrow 2NO(g)$$

$$2NO(g) + O_2(g) \longrightarrow 2NO_2(g)$$

$$4NO_2(g) + 2H_2O(\ell) + O_2(g) \longrightarrow 4HNO_3(aq)$$

The nitrogen oxides produced by automobiles, besides contributing to acid rain, also are major components of photochemical smog. Nitrogen dioxide is responsible for the characteristic brown color of this form of smog.

The sulfur oxides that are major contributors to acid rain have a variety of sources. Volcanic eruptions may spew thousands of tons of sulfur dioxide, SO_2, into the atmosphere; other natural sources are forest fires and the bacterial decay of organic matter.

In addition to these natural sources, the activities of humans generate large amounts of sulfur dioxide. The main sources are the burning of sulfur-containing coal and other fossil fuels and the roasting of metal sulfides in the production of metals such as zinc and copper. The sulfur dioxide can be converted into sulfur trioxide, SO_3, by several reactions, one of which is

$$SO_2(g) + NO_2(g) \longrightarrow SO_3(g) + NO(g)$$

Volcanic eruption The eruption of a volcano can spew large amounts of $SO_2(g)$ into the atmosphere.

Statue damaged by acid rain Acid rain causes significant damage to marble statues.

The sulfur trioxide dissolves in water forming a sulfuric acid solution.

$$SO_3(g) + H_2O(\ell) \longrightarrow H_2SO_4(aq)$$

Every year, acid rain causes millions of dollars of damage to stone buildings and monuments, especially those made of marble. Greek and Roman monuments that have lasted for centuries are now rapidly deteriorating. Acid rain is also extremely harmful to plant and animal life. Many lakes and rivers now have water too acidic for fish or other life. Acid rain can also dissolve metal compounds from the soil, which are then washed into lakes where they may poison the fish. Metal compounds often become concentrated in fish, and eating those fish may poison people. Acid rain can also destroy the ability of many trees and plants to resist disease and bacteria, by dissolving the waxy coating on their leaves.

Lakes in limestone areas can neutralize acid rain because the limestone reacts directly with the acids (the following reaction is a simplified version):

$$\underset{\text{limestone}}{CaCO_3(s)} + H_2SO_4(aq) \longrightarrow CaSO_4(aq) + CO_2(g) + H_2O(\ell)$$

Lakes that are now too acidic could be neutralized by adding crushed limestone or lime, CaO. Some lakes in Sweden are being treated in this manner. However, this measure can be at best a temporary solution. Besides being costly and impractical for large bodies of water, this procedure treats the results of acid rain while the causes remain untreated.

What can be done to solve the problem of acid rain? An obvious answer would be to prevent the escape of the nitrogen and sulfur oxides into the atmosphere. Because automobile emissions are the major contributor of nitrogen oxides, some states have enacted very tough emission control standards for cars sold and operated within their borders. California has been a leader in the fight for higher standards. The auto manufacturers have lobbied strongly against tougher standards, claiming that the cost to consumers would be too high and that the standards would be too difficult to meet.

A decrease of sulfur oxides also presents difficult choices for the public. The removal of sulfur from fossil fuels is extremely expensive and technically difficult to accomplish. A cheaper but less efficient method is to remove the SO_2 gas after it has been formed in the combustion of these fuels. The technology for this process is fairly simple, but the installation costs and the disposal of the resulting solid waste present other problems. Alternate energy sources such as solar, geothermal, and nuclear would help to solve the problem, but these alternatives are expensive and unpopular with certain groups. As stated earlier, the public is going to have to make some difficult choices to reduce acid rain. The problem is not going to "wash" away.

Table 3.1 Common Acids

Acid	Name
HF	hydrofluoric acid
HCl	hydrochloric acid
HBr	hydrobromic acid
HCN	hydrocyanic acid
HNO_2	nitrous acid
HNO_3	nitric acid
H_2SO_3	sulfurous acid
H_2SO_4	sulfuric acid
$HClO_4$	perchloric acid
H_3PO_4	phosphoric acid

The H_2O over the arrow and the (aq) after the ion mean that the species are dissolved in water. When HNO_3 dissolves, it separates into ions, a process known as **ionization.** Another way to write $H^+(aq)$ ion is H_3O^+ (hydronium ion), indicating that the hydrogen cation is associated with a water molecule.

$$HNO_3(\ell) + H_2O(\ell) \longrightarrow H_3O^+(aq) + NO_3^-(aq)$$

Hydrogen ions do associate strongly with water molecules, but there is experimental evidence for $H_5O_2^+$ and $H_9O_4^+$ as well as H_3O^+. The H_3O^+ formulation is particularly useful in certain problems considered in later chapters. In considering reaction stoichiometry, the $H^+(aq)$ formulation is preferred because it makes equations simpler to balance.

Table 3.1 lists several compounds that are acids in water solution. A more complete list is found in Appendix F. Often a different name is used for these compounds when they are in the pure state rather than dissolved in water. For example, pure HCl exists as a gas at room temperature and pressure, and is named hydrogen chloride. It is an acid when dissolved in water, and is referred to as a hydrochloric acid solution.

The simplest definition of a **base** is any substance that provides hydroxide anion (OH^-) in water. The most common bases are the hydroxides of group IA and IIA. When dissolved in water, these compounds dissociate into ions. Two examples are NaOH and $Ca(OH)_2$.

$$NaOH(s) \xrightarrow{H_2O} Na^+(aq) + OH^-(aq)$$

$$Ca(OH)_2(s) \xrightarrow{H_2O} Ca^{2+}(aq) + 2OH^-(aq)$$

The equations for acids and bases dissolving in water are the first ones encountered so far in this text that show charged species. When you write an equation that contains charged species, *the net charge on both sides of the equation must be the same,* as well as the number of atoms of each element. Although the net charge must be in balance, the charges on each side of the equation do *not* have to equal zero.

When you mix an acid and a base, their reaction always yields H_2O and the respective *salt.* A **salt** is a compound made up of the cation from a base and the anion from an acid. Be careful to write the formula of the salt correctly.

$$HCl(aq) + NaOH(aq) \longrightarrow H_2O(\ell) + NaCl(aq)$$

$$H_2SO_4(aq) + Ca(OH)_2(aq) \longrightarrow 2H_2O(\ell) + CaSO_4(s)$$

Notice how $H^+(aq)$ ions from the acids combine with $OH^-(aq)$ ions from the bases to produce H_2O. Both acids and bases are corrosive while many salts are not, so the formation of water and salt from an acid and a base is called a **neutralization reaction.**

$$H^+(aq) + OH^-(aq) \longrightarrow H_2O(\ell)$$

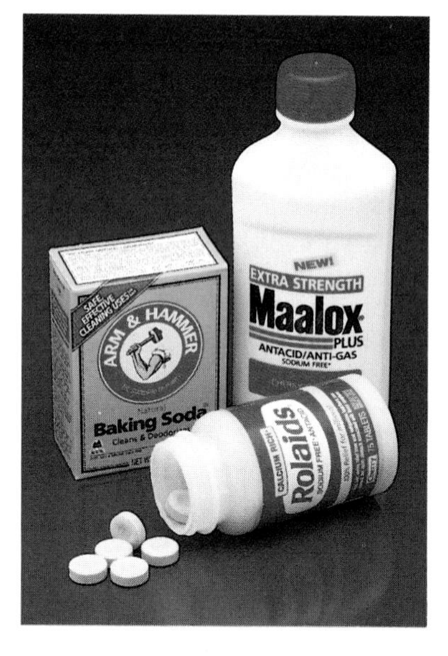

Antacids Products that neutralize acid to cure "heartburn" are bases.

Example 3.4 Neutralization Reactions

Write the equation for the reaction of nitric acid with magnesium hydroxide.

Solution This reaction is an example of a neutralization reaction. The chemical expression is

$$HNO_3 + Mg(OH)_2 \longrightarrow H_2O + Mg(NO_3)_2$$

To balance this chemical expression, assume a coefficient of 1 for $Mg(NO_3)_2$. The Mg is in balance, and the nitrate group can then be balanced by placing a 2 in front of HNO_3. Since the left side now has a total of four hydrogen atoms, two water molecules are needed to bring the equation into balance.

$$2HNO_3 + Mg(OH)_2 \longrightarrow 2H_2O + Mg(NO_3)_2$$

In balancing any acid-base reaction, the number of hydrogen ions contributed by the acid and the number of hydroxide ions contributed by the base are both equal to the coefficient of the H_2O formed. The reaction shown in this example is similar to the one by which milk of magnesia neutralizes excess stomach acid (HCl).

Each $H^+(aq)$ provided by the acid is neutralized by one $OH^-(aq)$ from the base, forming one H_2O molecule.

Understanding Write the equation for the reaction of hydrochloric acid with calcium hydroxide.

Answer: $2HCl + Ca(OH)_2 \longrightarrow 2H_2O + CaCl_2$

3.2 The Mole

Objectives

- To describe the concept of the mole
- To explain the atomic weight scale
- To determine the molar mass of any element or compound
- To interconvert between mass, moles, and numbers of atoms or molecules

Sulfur, which occurs most commonly as S_8 molecules, burns in oxygen to form sulfur dioxide.

$$S_8(s) + 8O_2(g) \longrightarrow 8SO_2(g)$$

This equation states that one molecule of S_8 combines with eight molecules of O_2 to produce eight molecules of SO_2. While this statement is certainly true, the small sizes and masses of these molecules make it impossible to observe the reaction on this scale. A laboratory-scale reaction might use several grams of material—quantities that are convenient to measure and handle. An industrial-scale reaction would be much larger. The chemical equation provides information about the number of molecules involved in a reaction, but this information is not directly useful to the chemist in the laboratory, because individual molecules are too small to count. Instead, chemists usually measure the mass of each reactant and product.

A similar problem is faced by a shopper needing 100 nails. Since nails are priced by mass, the clerk will ask how many pounds or kilograms are needed. In order to buy the right number of nails, the shopper must know how many nails of the desired size are in one kilogram. A conversion between amount (number of nails) and mass (kg of nails) is needed. In a similar manner, the chemist needs to convert the number of molecules (the type of data used in an equation) to mass (which can readily be measured), and *vice versa*.

Reaction of sulfur with oxygen Sulfur burns in oxygen with a blue flame.

What unit is used to count the numbers of atoms? If we were selling eggs, we would use the unit of dozen, or if selling cans of soda, we might use the unit of six-pack. Because of the extremely small size of atoms, a unit that contains many more entities is needed.

The mole is the SI unit for amount and contains 6.022×10^{23} units, such as atoms or molecules.

The SI system defines the unit of *amount* of substance as the *mole* (abbreviated as mol). One **mole** is the amount of substance that contains as many entities as there are atoms in exactly 12 grams of the ^{12}C isotope of carbon. The mole is the *unit* of the quantity "amount of a substance," analogous to the meter being the unit of the quantity "length." The number of atoms present in 12 g of ^{12}C has been determined experimentally to be 6.022×10^{23} atoms/mol (to four significant figures) and is known as **Avogadro's number,** named to honor the nineteenth-century physicist Amedeo Avogadro. The number is extremely large. For example, many scientists believe the earth is over four billion years old. In order to count out one mole of atoms over this time period, one would have to count out almost five million atoms per second. Figure 3.3 shows photographs of one-mole quantities.

Amedeo Avogadro Amedeo Avogadro (1776–1856) studied the behavior of gases. His work related the number of molecules to the volume of a gas.

Atomic Mass Unit

The **atomic mass unit** (u) has been established to compare the mass of any atom to that of ^{12}C. The mass of an atom of ^{12}C is defined as exactly 12 u, and all other atoms are compared to this standard. The atomic mass unit is extremely small by laboratory standards:

$$1 \text{ u} = 1.66 \times 10^{-27} \text{ kg}$$

The choice of carbon for the standard is somewhat arbitrary, but notice that the 12 u is numerically equal to the mass number of ^{12}C. Thus, the mass of a proton or neutron is about 1 u. On this scale, ^{24}Mg has an atomic mass of 24 u; that is, one atom of ^{24}Mg has about twice the mass of one atom of ^{12}C. ^{4}He has an atomic mass of 4 u, so three ^{4}He atoms have the same mass as one ^{12}C atom. Rounded to whole numbers, these values are the same as the mass

Figure 3.3 One mole of elements One mole of each of several elements: iron as a rod, liquid mercury in the cylinder, copper wire, sodium metal pictured under oil to protect it from reaction with water in the air, aluminum as a powder, and nitrogen gas in the balloons.

number of the atom, but when expressed more precisely they differ slightly from whole numbers ($^{24}Mg = 23.98504$ u, $^{4}He = 4.00260$ u) because the mass of an atom is determined by other factors besides the number of protons and neutrons in the nucleus.

Isotopic Distributions and Atomic Weight

Most elements have several isotopes, and the isotopic composition of the naturally occurring elements is generally constant and independent of the origin of the sample. Since the isotopes of a given element have different numbers of neutrons, they have different atomic masses. For example, naturally occurring lithium is a mixture of two isotopes: 7.42% of the atoms are ^{6}Li (mass = 6.015 u) and 92.58% are ^{7}Li (mass = 7.016 u). An average mass based on this isotopic distribution must be used in calculations involving the mass of lithium. If natural lithium were 50% ^{6}Li and 50% ^{7}Li, then the average mass would be about 6.5 u, but natural lithium is mainly ^{7}Li so the average is much closer to 7 u. A *weighted average* that takes into account the natural abundance of each isotope must be calculated. In this calculation, 7.42% and 92.58% must be expressed as the decimal fractions 0.0742 and 0.9258, respectively.

$$\text{average mass Li} = 0.0742 \times 6.015 \text{ u} + 0.9258 \times 7.016 \text{ u}$$
$$= 6.941 \text{ u}$$

This number appears under Li in the periodic table. It is shown without units and is (for historical reasons) called the **atomic weight,** the average mass in atomic mass units of an atom of the element. The term **atomic mass** is technically correct, but atomic weight is the more commonly used term. Remember that atomic weight is an *average* that reflects the natural isotopic distribution of the element.

The atomic weight of an element is the average mass of an atom in a natural sample of the element.

Example 3.5 Atomic Weight Calculation

Chlorine has two stable isotopes: ^{35}Cl with a natural abundance of 75.77% and a mass of 34.97 u, and ^{37}Cl with a natural abundance of 24.23% and an atomic mass of 36.97 u. Calculate the atomic weight of a natural sample of chlorine.

Solution The atomic weight is the weighted average of the two isotopes:

$$\text{atomic weight Cl} = 0.7577 \times 34.97 \text{ u} + 0.2423 \times 36.97 \text{ u}$$
$$= 35.45 \text{ u}$$

Understanding Boron has two stable isotopes: ^{10}B with a natural abundance of 19.9% and an atomic mass of 10.01 u, and ^{11}B with a natural abundance of 80.1% and an atomic mass of 11.01 u. Calculate the atomic weight of a natural sample of boron.

Answer: 10.81 u

Molar Mass

The atomic weight scale is based on the masses of atoms. In the laboratory, chemists work with moles of atoms or molecules. The **molar mass** of any substance is the mass (in grams) of one mole of that substance. The molar mass of an element is numerically equal to its atomic weight, but the molar mass is the mass of one mole and has the units of grams/mole. For example, the molar mass of a naturally occurring sample of lithium is its atomic weight in grams, 6.941 g/mol. Since one mole of lithium is equivalent to 6.941 g, we can write the following relationship:

$$1 \text{ mol Li} \triangleq 6.941 \text{ g Li}$$

We use this equivalency to convert from mass to moles and *vice versa.* A second equivalency, which allows the interconversion of moles and number of atoms, is derived from Avogadro's number:

$$1 \text{ mol Li} \triangleq 6.022 \times 10^{23} \text{ atoms of Li}$$

These are the kinds of equivalencies that are needed for most chemical problems because they apply to an actual sample of the element rather than to a single isotope. The periodic table displays the atomic weight of each element, which is also the molar mass of an element when expressed in grams/mole.

Using the Mole and Avogadro's Number with Elements

The definition of the mole solves both problems outlined at the start of this section, since it allows us to interconvert mass (in grams or kilograms) and amount (in moles) on a scale large enough to be useful. The molar mass, shown numerically on the periodic table, allows the interconversion of mass and moles; Avogadro's number allows the interconversion of moles and number of atoms.

Example 3.6 **Conversion Between Mass, Moles, and Number of Atoms**

(a) How many moles of Ar are present in 4.4 g of that element?
(b) How many atoms are present in 4.4 g of Ar?

Solution

(a) This problem requires conversion from grams (mass) into number of moles (amount). The molar mass of a substance gives the mass, in grams, of one mole. From the periodic table we see that the molar mass of argon is 39.9 g/mol.

$$1 \text{ mole Ar} \triangleq 39.9 \text{ g Ar}$$

This relationship is used to calculate the number of moles of argon, given the mass. Always write the units with each number. While writing units will take a little extra time, the cancellation of units provides a valuable check.

Mass of Ar → Molar mass of Ar → Moles of Ar

$$\text{moles Ar} = 4.4 \text{ g Ar} \left(\frac{1 \text{ mol Ar}}{39.9 \text{ g Ar}}\right) = 0.11 \text{ mol Ar}$$

In review, the mass of argon is converted into number of moles by using the molar mass of argon. Since 4.4 g is about 10% of the mass of one mole of argon, the answer of 0.11 mol is reasonable.

(b) Avogadro's number is the number of atoms in one mole of argon.

$$1 \text{ mol Ar} \simeq 6.02 \times 10^{23} \text{ atoms Ar}$$

In part (a) we calculated that 4.4 g is 0.11 mol of argon. Now use Avogadro's number to calculate the number of atoms in that number of moles:

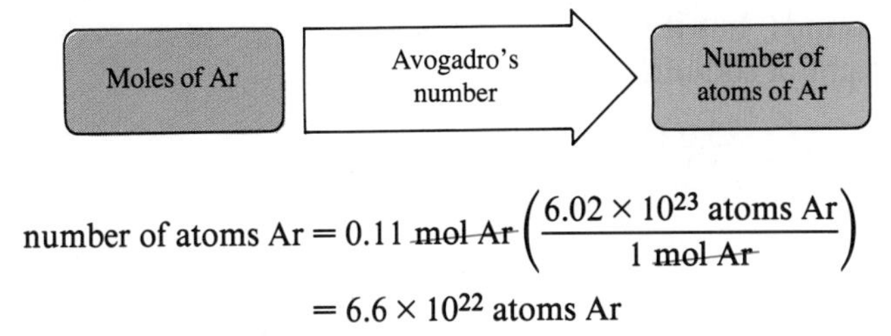

$$\text{number of atoms Ar} = 0.11 \cancel{\text{mol Ar}} \left(\frac{6.02 \times 10^{23} \text{ atoms Ar}}{1 \cancel{\text{mol Ar}}} \right)$$

$$= 6.6 \times 10^{22} \text{ atoms Ar}$$

In these two calculations, three significant figures are used for the molar mass and Avogadro's number. In calculations where experimental data are given, use one more significant figure for these numbers than in the least accurate number in the data given in the problem. This process avoids round-off errors.

Understanding How many moles of Sn are in 10.1 g of that substance? Also, calculate the number of atoms in 10.1 g of Sn.

Answer: 0.0851 mol Sn, 5.12×10^{22} atoms Sn

Molecular Weight

Since a molecule is a combination of atoms, the weight of a molecule can be determined from the weight of each atom in the molecule. The **molecular weight** is the sum of the atomic weights of all atoms present in the molecular formula, expressed in u. The *molar mass* of the compound is this same number, but expressed in grams/mole. The molecular weight of CO_2 can be calculated from the atomic weights given in the periodic table, taking into account the subscripts in the molecular formula. The formula indicates that one molecule of CO_2 contains one atom of carbon and two atoms of oxygen.

$$1 \cancel{\text{atom C}} \left(\frac{12.01 \text{ u}}{1 \cancel{\text{atom C}}} \right) = 12.01 \text{ u}$$

$$2 \cancel{\text{atom O}} \left(\frac{16.00 \text{ u}}{1 \cancel{\text{atom O}}} \right) = 32.00 \text{ u}$$

$$\text{molecular weight } CO_2 = 44.01 \text{ u}$$

Since CO_2 has a molecular weight of 44.01 u, the molar mass of CO_2 is 44.01 g/mol and there are 6.022×10^{23} molecules of CO_2 in 44.01 g of CO_2.

The number of significant figures to use in this type of calculation is sometimes arbitrary. If you are asked for the molecular weight of a compound, use either one or two digits after the decimal point, depending on the desired precision. Thus, the molecular weight of carbon dioxide can be correctly stated as 44.0 u or 44.01 u. In calculations where experimental data are given, as indicated in Example 3.6, use one more significant figure in the molar mass than the least accurate number in the data given in the problem.

The molar mass of a compound is the molecular or formula weight in grams.

Ionic compounds, described in Section 2.3, do not exist as individual molecules. The term molecular weight does not apply. The formula of an ionic substance gives the relative numbers of anions and cations in simplest form (the empirical formula). The **formula weight** is the sum of the atomic weights of the atoms in any formula. The term molar mass is still used for ionic compounds, but it is really the formula weight in grams/mole. The formula weight of sodium carbonate, Na_2CO_3, is

$$2\ \text{atom Na}\left(\frac{23.0\ \text{u}}{1\ \text{atom Na}}\right) = 46.0\ \text{u}$$

$$1\ \text{atom C}\left(\frac{12.0\ \text{u}}{1\ \text{atom C}}\right) = 12.0\ \text{u}$$

$$3\ \text{atom O}\left(\frac{16.0\ \text{u}}{1\ \text{atom O}}\right) = 48.0\ \text{u}$$

$$\text{formula weight } Na_2CO_3 = 106.0\ \text{u}$$

Using this formula weight, the molar mass of Na_2CO_3 is 106.0 g/mol.

The correct determination of the molar masses of compounds is important in many chemical calculations. Do each one systematically. These first two examples show the complete calculation using the conversion factor method. A more convenient form is

$$\begin{array}{lll} 2(\text{Na}) & 2 \times 23.0 = & 46.0 \\ 1(\text{C}) & 1 \times 12.0 = & 12.0 \\ 3(\text{O}) & 3 \times 16.0 = & 48.0 \\ \hline \end{array}$$

$$\text{formula weight } Na_2CO_3 = 106.0$$

As shown, molecular weights and formula weights are frequently written without the units of u.

Example 3.7 Calculation of Molar Mass

Hydrazine, N_2H_4, is a fuel that has been used as a rocket propellant. What is the molar mass of hydrazine?

Solution Work the problem systematically, using the data given in the periodic table.

$$\begin{array}{lll} 2(\text{N}) & 2 \times 14.01 = & 28.02 \\ 4(\text{H}) & 4 \times 1.01 = & 4.04 \\ \hline \end{array}$$

$$\text{molecular weight } N_2H_4 = 32.06$$

The molar mass is 32.06 g/mol.

Understanding What is the molar mass of calcium bromide?

Answer: 199.9 g/mol

Using the Mole and Avogadro's Number with Compounds

The molar mass of an element or compound is used to interconvert mass and number of moles.

Calculations that involve the masses of molecules are performed in the same way as calculations involving elements. First determine the molar mass of the compound, and then use it to interconvert gram and mole information.

Avogadro's number allows the interconversion of moles and number of molecules or formula units.

Example 3.8 Conversion of Mass to Moles

(a) How many moles of N_2H_4 are present in 14.2 g?

(b) How many moles of nitrogen atoms and how many moles of hydrogen atoms are present in the sample?

Solution

(a) The molar mass of N_2H_4, 32.06 g/mol, was calculated in Example 3.7. Since 14.2 g is about half the mass of one mole, we can estimate that our final answer will be about half a mole of N_2H_4. The molar mass (32.06 g/mol N_2H_4) is used for the exact conversion of grams into moles.

Mass of N_2H_4 → Molar mass of N_2H_4 → Moles of N_2H_4

$$\text{moles } N_2H_4 = 14.2 \text{ g } N_2H_4 \left(\frac{1 \text{ mol } N_2H_4}{32.06 \text{ g } N_2H_4}\right) = 0.443 \text{ mol } N_2H_4$$

(b) The formula N_2H_4 states that there are two moles of nitrogen atoms and four moles of hydrogen in every one mole of N_2H_4.

$$2 \text{ mol N} \simeq 1 \text{ mol } N_2H_4 \quad \text{and} \quad 4 \text{ mol H} \simeq 1 \text{ mol } N_2H_4$$

These equivalencies are used to calculate the number of moles of *atoms* present.

Moles of N_2H_4 → Formula of N_2H_4 → Moles of N and H atoms

$$\text{moles N} = 0.443 \text{ mol } N_2H_4 \left(\frac{2 \text{ mol N}}{1 \text{ mol } N_2H_4}\right) = 0.886 \text{ mol N}$$

$$\text{moles H} = 0.443 \text{ mol } N_2H_4 \left(\frac{4 \text{ mol H}}{1 \text{ mol } N_2H_4}\right) = 1.77 \text{ mol H}$$

Understanding How many moles are present in a 20.4 g sample of sodium nitrate?

Answer: 0.240 mol

Example 3.9 Conversion of Mass to Number of Molecules

How many molecules are present in 14.2 g of N_2H_4?

Solution From Avogadro's number, we know that 1 mol $N_2H_4 \simeq 6.022 \times 10^{23}$ molecules of N_2H_4. In Example 3.8 it was found that 14.2 g of N_2H_4 is equivalent to 0.443 mol N_2H_4.

Moles of N_2H_4 → Avogadro's number → Molecules of N_2H_4

$$\text{molecules } N_2H_4 = 0.443 \text{ mol } N_2H_4 \left(\frac{6.022 \times 10^{23} \text{ molecules } N_2H_4}{1 \text{ mol } N_2H_4}\right)$$

$$= 2.67 \times 10^{23} \text{ molecules } N_2H_4$$

This answer, about half of Avogadro's number, is reasonable since about half a mole of N_2H_4 was present.

Understanding How many molecules are present in 10.6 g of N_2O?

Answer: 1.45×10^{23} molecules of N_2O

Example 3.10 Conversion of Moles to Mass

(a) Ethylene, C_2H_4, is used for such diverse applications as the ripening of citrus fruits and the preparation of plastics. What is the mass of 3.22 mol of ethylene?
(b) What is the mass of 0.331 mol of LiCl?

Solution

(a) First, we find the molar mass of C_2H_4:

$$\begin{array}{lr} 2(\text{C}) \quad 2 \times 12.01 = & 24.02 \\ 4(\text{H}) \quad 4 \times 1.008 = & 4.03 \\ \hline \text{molecular weight } C_2H_4 = & 28.05 \end{array}$$

The molar mass of C_2H_4, 28.05 g/mol, is used to calculate the mass of C_2H_4.

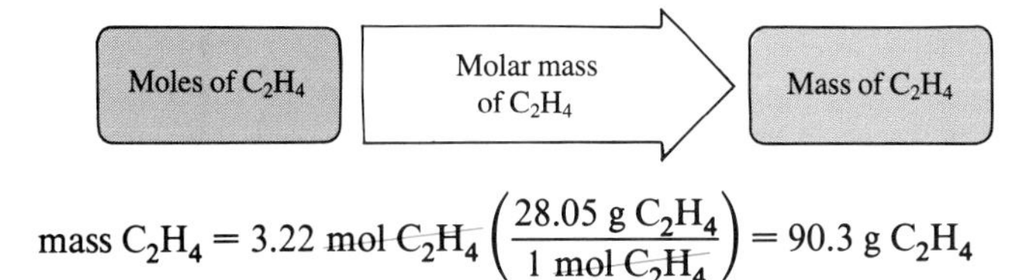

$$\text{mass } C_2H_4 = 3.22 \text{ mol } C_2H_4 \left(\frac{28.05 \text{ g } C_2H_4}{1 \text{ mol } C_2H_4} \right) = 90.3 \text{ g } C_2H_4$$

(b) The formula weight of LiCl is

$$\begin{array}{lr} 1(\text{Li}) \quad 1 \times 6.94 = & 6.94 \\ 1(\text{Cl}) \quad 1 \times 35.45 = & 35.45 \\ \hline \text{formula weight LiCl} = & 42.39 \end{array}$$

The molar mass is 42.39 g/mol, and is used to calculate the mass of LiCl.

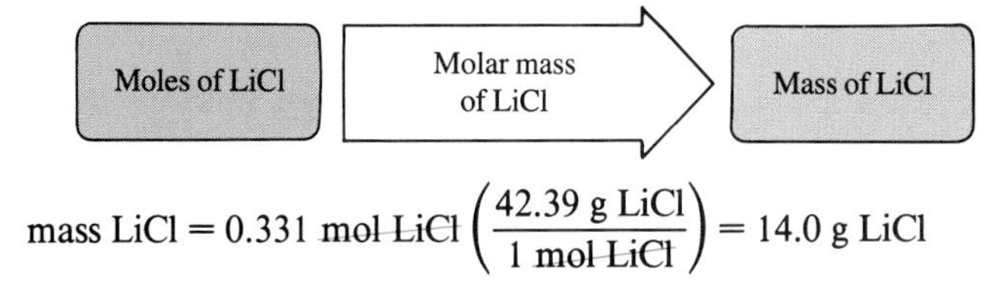

$$\text{mass LiCl} = 0.331 \text{ mol LiCl} \left(\frac{42.39 \text{ g LiCl}}{1 \text{ mol LiCl}} \right) = 14.0 \text{ g LiCl}$$

Understanding What is the mass of 43.1 mol of phosphorus pentachloride?

Answer: 8.97×10^3 g or 8.97 kg

3.3 Empirical Formulas

Objectives

- To calculate the mass percentage of any element in a compound
- To determine the empirical formula from gram ratios, mass percentage, or data from a combustion analysis
- To determine the molecular formula from the empirical formula and the molar mass

When a new compound is prepared or isolated, one of the first experimental tasks is to find its molecular formula. A molecular formula represents numerical information. For example, from the formula of benzene, C_6H_6, we know that one molecule of benzene consists of six carbon and six hydrogen atoms. We also know that one mole of benzene contains six moles of carbon and six moles of hydrogen atoms. This information can be converted into mass data as outlined in Section 3.2. For benzene,

$$\begin{array}{ll} 6(\text{C}) & 6 \times 12.01 = 72.06 \\ \underline{6(\text{H})} & \underline{6 \times \ \ 1.01 = \ \ 6.06} \\ \multicolumn{2}{l}{\text{molecular weight } C_6H_6 = 78.12} \end{array}$$

From this molecular weight determination, we know that one mole of benzene has a total mass of 78.12 g that comes from 72.06 g of carbon and 6.06 g of hydrogen. This information can be used to calculate the percentage of the mass of the pure substance that is contributed by each element.

$$\text{mass percent C} = \frac{72.06 \text{ g C}}{78.12 \text{ g compound}} \times 100\% = 92.24\% \text{ C}$$

$$\text{mass percent H} = \frac{6.06 \text{ g H}}{78.12 \text{ g compound}} \times 100\% = 7.76\% \text{ H}$$

The same answer is obtained for any compound having the general formula C_nH_n. For example, the compound acetylene, C_2H_2, also is 92.24% carbon and 7.76% hydrogen by mass. Thus, the percentage composition of a compound can be based on its *empirical formula* (the *relative* numbers of atoms of the elements in a compound expressed as the smallest whole number ratio) as well as on its molecular formula. The percentage composition calculated from the empirical formula of both benzene and acetylene, CH, is the same as that calculated from the molecular formulas.

Mass percentages of elements in a compound can be calculated from the empirical or molecular formula.

Example 3.11 Mass Composition

Aspirin is a remarkable analgesic (pain killer) and is also used in treating heart conditions. Interestingly, scientists do not completely understand why aspirin reduces pain.

(a) From its chemical formula, $C_9H_8O_4$, calculate the percentage by mass of each element in aspirin.
(b) What mass of carbon is in 2.44 g of aspirin?

Solution

(a) First calculate the mass of each element present, and sum them to find the molecular weight.

$$\begin{array}{ll} 9(\text{C}) & 9 \times 12.01 = 108.09 \\ 8(\text{H}) & 8 \times \ \ 1.01 = \ \ \ 8.08 \\ \underline{4(\text{O})} & \underline{4 \times 16.00 = \ \ 64.00} \\ \multicolumn{2}{l}{\text{molecular weight } C_9H_8O_4 = 180.17} \end{array}$$

Thus, one mole of aspirin has a mass of 180.17 g coming from 108.09 g of carbon, 8.08 g of hydrogen, and 64.00 g of oxygen. Use these values to calculate the mass composition.

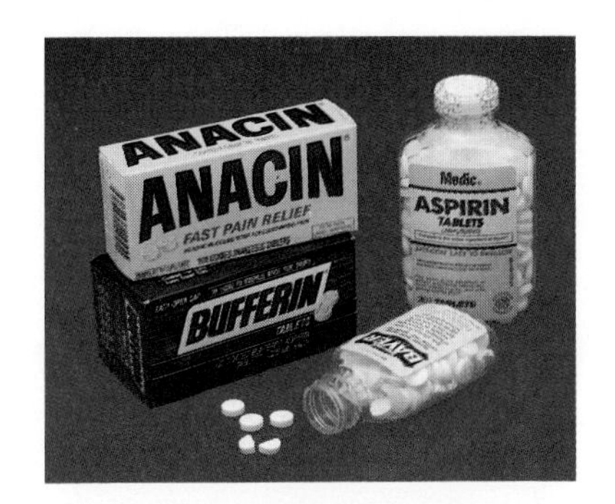

Aspirin is an important analgesic and, taken in low doses, appears to reduce chances of heart disease.

$$\text{mass percent C} = \frac{108.09 \text{ g C}}{180.17 \text{ g total}} \times 100\% = 59.99\% \text{ C}$$

$$\text{mass percent H} = \frac{8.08 \text{ g H}}{180.17 \text{ g total}} \times 100\% = 4.48\% \text{ H}$$

$$\text{mass percent O} = \frac{64.00 \text{ g O}}{180.17 \text{ g total}} \times 100\% = 35.52\% \text{ O}$$

(b) To find the mass of carbon in a 2.44 g sample of aspirin, we multiply this mass by the fraction of the mass (59.99/100) that is carbon.

$$\text{mass of carbon} = 2.44 \text{ g C} \times 0.5999 = 1.46 \text{ g C}$$

Understanding Calculate the percentage by mass of each element in $C_2H_2F_4$.

Answer: 23.54% C, 1.98% H, 74.48% F

Determination of Empirical Formula

The empirical formula of a compound can be calculated from mass percentage data by reversing the procedure just described. This calculation yields only the empirical formula, because the percentage composition by mass is based only on the relative numbers of atoms of each element in the compound. Additional experimental information (the molecular weight of the compound) is needed to convert the empirical formula into the molecular formula.

The results of an experiment (described later) show that 2.000 g of a particular compound consists of 1.713 g of carbon and 0.287 g of hydrogen. If we can calculate the relative numbers of moles of each element, we will also know the relative numbers of atoms, and thus the empirical formula. In order to calculate the relative numbers of moles, we need to convert the masses of each element into moles. The molar mass of each element is used for this conversion.

$$1 \text{ mol C} \simeq 12.011 \text{ g C}$$

$$1 \text{ mol H} \simeq 1.008 \text{ g H}$$

The masses of the elements (as determined in the experiment) are used to calculate the number of moles of carbon and hydrogen in this sample:

$$\text{moles C} = 1.713 \text{ g C}\left(\frac{1 \text{ mol C}}{12.011 \text{ g C}}\right) = 0.1426 \text{ mol C}$$

$$\text{moles H} = 0.287 \text{ g H}\left(\frac{1 \text{ mol H}}{1.008 \text{ g H}}\right) = 0.285 \text{ mol H}$$

This calculation provides the relative numbers of moles of carbon and hydrogen in this substance, but is not the final answer. A formula represents the ratio of atoms, and these must be whole numbers. While the formula $C_{0.1426}H_{0.285}$ is technically correct, it does not make any sense to refer to 0.1426 of a carbon atom. We need to adjust the mole ratio to whole numbers. A method that is usually successful for making this adjustment is to divide the number of moles of each element by the smallest number of moles found. This procedure will convert the smallest number to 1 and all other values to a number greater than 1, without changing their ratio.

$$\text{moles C} = 0.1426 \text{ mol C}/0.1426 = 1.000 \text{ mol C}$$

$$\text{moles H} = 0.285 \text{ mol H}/0.1426 = 2.00 \text{ mol H}$$

The empirical formula is CH_2.

The process for the determination of the empirical formula is represented by the following steps:

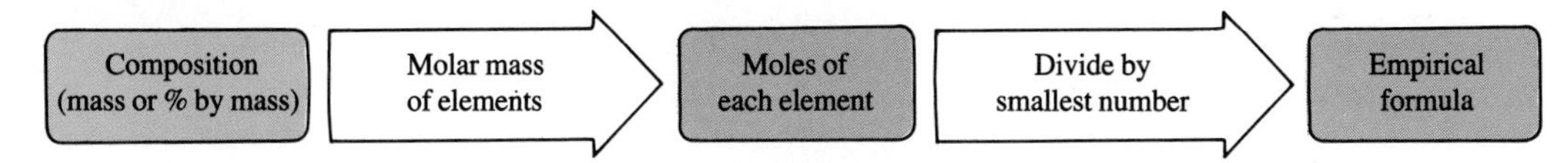

In the last example, the mass of hydrogen in the compound is much less than the mass of the carbon, but in the empirical formula there are two hydrogen atoms for every carbon atom. In trying to estimate empirical formulas, the large differences in the molar masses of different elements need to be taken into account, especially if hydrogen is present.

The molar masses of elements are used to convert between mass composition and empirical formulas.

Conversion of Empirical Formula into Molecular Formula

To convert the empirical formula to a molecular formula, we must know the molecular weight of the molecule. Several kinds of experiments (described later) can be used to determine molecular weight. For the compound just discussed, with an empirical formula of CH_2, the molecular weight was experimentally determined and found to be 42. The formula weight of the empirical formula, CH_2, is 14. The molecular formula must be a whole number multiple of the empirical formula, CH_2, C_2H_4, $C_3H_6 \cdots (CH_2)_n$, where n is the number of times the empirical formula occurs in the molecular formula. The value of n is calculated from

The molecular weight of the compound must be known to determine a molecular formula from the empirical formula.

$$n = \frac{\text{molecular weight}}{\text{empirical formula weight}}$$

In this case:

$$n = \frac{42}{14} = 3$$

The molecular formula is C_3H_6, three times the empirical formula.

In the first example in this section, the known formula of benzene was used to determine the percentage composition by mass. Empirical formula calculations are frequently based on the mass percent composition. If the composition is given as percentages, assume that a 100.00-g sample has been analyzed. The percentage of each element is then the number of grams of that element in the 100.00-g sample. For example, we calculated that benzene, C_6H_6, was 92.24% carbon and 7.76% hydrogen by mass. A 100.00-g sample of benzene will contain 92.24 g of carbon and 7.76 g of hydrogen. This procedure is illustrated in the next example.

The mass percentage of each element in a compound is the same as the number of grams of each element present in a 100-g sample.

Example 3.12 Empirical and Molecular Formulas

Calculate the empirical and molecular formulas of a compound extracted from tobacco. Chemical analysis shows that this substance contains 74.0% carbon, 8.70%

hydrogen, and 17.3% nitrogen. In a separate experiment the molecular weight was found to be 162.

Solution Assume the sample has a mass of 100 g. This sample would contain 74.0 g of carbon, 8.70 g of hydrogen, and 17.3 g of nitrogen. The following strategy converts these gram ratios into the empirical formula:

Mass of C, H, N → Molar mass of C, H, N → Moles of C, H, N → Divide by smallest number → Empirical formula

First, use the molar masses of the elements to convert the mass of each into number of moles. Note that even though hydrogen and nitrogen in elemental form are diatomic molecules, the molar masses of the *atoms* are used in these calculations because the empirical formula is the ratio of atoms.

$$\text{moles C} = 74.0\ \text{g C}\left(\frac{1\ \text{mol C}}{12.01\ \text{g C}}\right) = 6.16\ \text{mol C}$$

$$\text{moles H} = 8.70\ \text{g H}\left(\frac{1\ \text{mol H}}{1.008\ \text{g H}}\right) = 8.63\ \text{mol H}$$

$$\text{moles N} = 17.3\ \text{g N}\left(\frac{1\ \text{mol N}}{14.01\ \text{g N}}\right) = 1.23\ \text{mol N}$$

This calculation yields the relative number of moles of each element. To convert to integers, divide each by the smallest number, 1.23:

$$\text{moles C} = 6.16\ \text{mol C}/1.23 = 5.01\ \text{mol C}$$

$$\text{moles H} = 8.63\ \text{mol H}/1.23 = 7.02\ \text{mol H}$$

$$\text{moles N} = 1.23\ \text{mol N}/1.23 = 1.00\ \text{mol N}$$

These numbers are integers within experimental error. The empirical formula is C_5H_7N. To determine the molecular formula, we must calculate n for the formula $(C_5H_7N)_n$. The empirical formula C_5H_7N has a formula weight of 81, and the molecular weight has been given as 162.

$$n = \frac{162}{81} = 2$$

The molecular formula is therefore $C_{10}H_{14}N_2$. The compound is nicotine, found in tobacco leaves and widely used as an agricultural insecticide. A computer drawing of the structure of nicotine is shown in Figure 3.4.

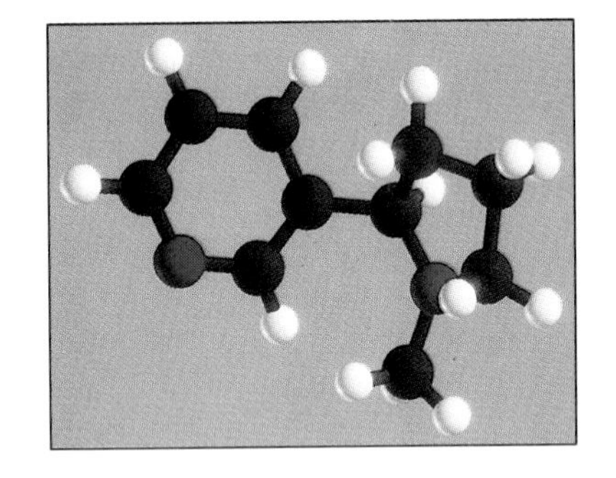

Figure 3.4 Structure of nicotine Nicotine, $C_{10}H_{14}N_2$, has an interesting structure containing both a five-membered and a six-membered ring of atoms.

Understanding Determine the molecular formula of a compound that contains 66.6% C, 11.2% H, and 22.2% O and has a molecular weight of 72.

Answer: C_4H_8O

Combustion Analysis

There are many types of experiments that can provide the data needed to determine the empirical formula of a compound. Combustion analysis is often used to determine the quantity of carbon and hydrogen in a sample of an organic compound. The apparatus for this experiment is shown in Figure 3.5.

In this experiment, a small, carefully weighed sample of a compound is completely burned in a stream of O_2. Excess O_2 is used to ensure conversion

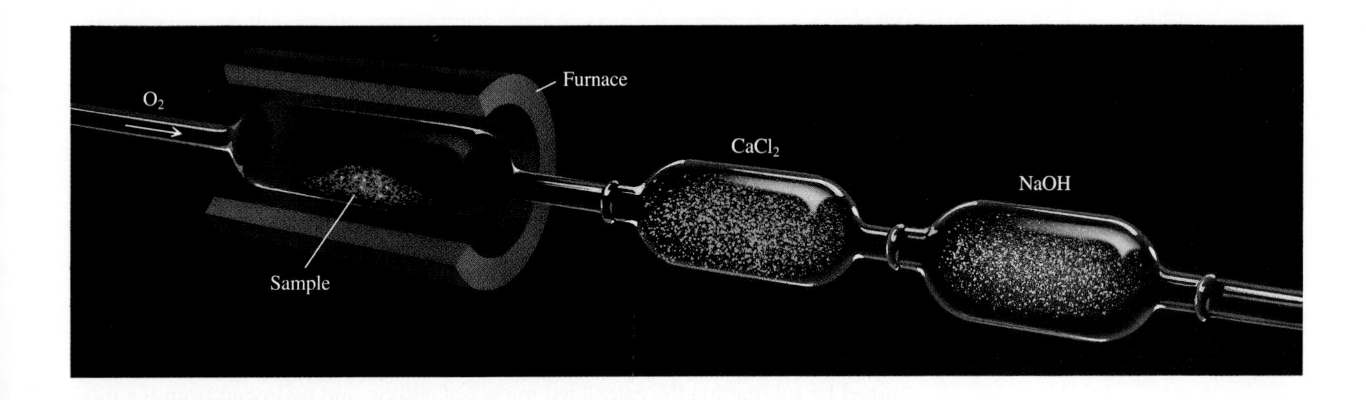

(a)

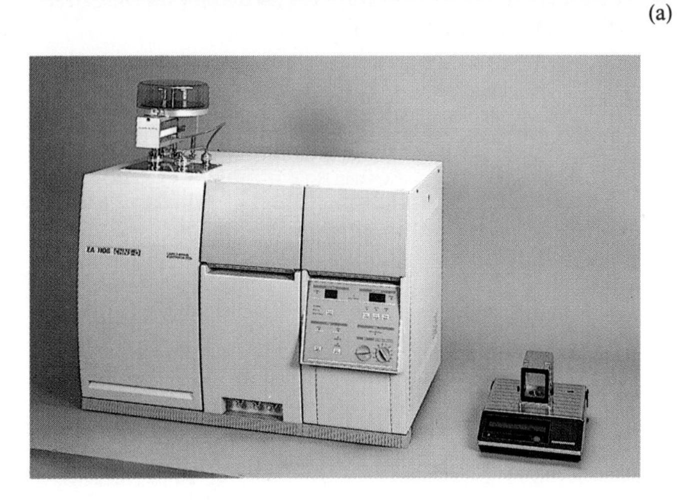

(b)

Figure 3.5 A combustion train (a) A combustion train is used to analyze for carbon and hydrogen in a compound. $CaCl_2$ is frequently used in the first trap because it absorbs the H_2O but not the CO_2. Sodium hydroxide is used in the second trap. It cannot be in the first trap because NaOH absorbs both H_2O and CO_2. (b) Photograph of a modern combustion analysis instrument. Only a few milligrams of sample are needed for the analysis.

of all the carbon into CO_2 and all the hydrogen into H_2O. The H_2O is collected in the first trap, and the CO_2 is collected in the second trap. The two traps are weighed before and after the combustion of the sample to determine the masses of the H_2O and CO_2 absorbed. Although the calculations would be simplified if pure hydrogen and carbon were generated and weighed, that experiment is quite difficult to perform and the results would be imprecise. To obtain the desired empirical formula, we must find the number of moles of hydrogen in the H_2O and the number of moles of carbon in the CO_2. The formula of water indicates that there are two moles of hydrogen atoms and one mole of oxygen atoms in every mole of water. Thus, two moles of hydrogen and one mole of oxygen are chemically equivalent to one mole of water.

Empirical formulas are determined in a combustion analysis by converting the weight of CO_2 and H_2O into moles of carbon and hydrogen.

$$2 \text{ mol H} \simeq 1 \text{ mol } H_2O \qquad \text{and} \qquad 1 \text{ mol O} \simeq 1 \text{ mol } H_2O$$

For CO_2,

$$1 \text{ mol C} \simeq 1 \text{ mol } CO_2 \qquad \text{and} \qquad 2 \text{ mol O} \simeq 1 \text{ mol } CO_2$$

The determination of an empirical formula by combustion analysis is illustrated in the next example.

Example 3.13 Combustion Analysis

The apparatus shown in Figure 3.5a is used to analyze the composition of a compound made up of only carbon, hydrogen, and oxygen. After combustion of a 0.1000 g sample, the mass of the first trap (collecting H_2O) increased by 0.0928 g and the mass of the second trap (collecting CO_2) increased by 0.228 g. What is the empirical formula of this compound? (Note that it is possible for the sum of the masses of the H_2O and CO_2 to be greater than the mass of the starting sample because some of the oxygen in the products comes from the added O_2.)

Solution The empirical formula requires the relative amounts of C, H, and O in the sample. The masses of CO_2 and H_2O must be converted into moles of CO_2 and H_2O, and then into moles of C and H, respectively. The amount of oxygen will be determined by subtracting the calculated masses of carbon and hydrogen from the total mass of the sample.

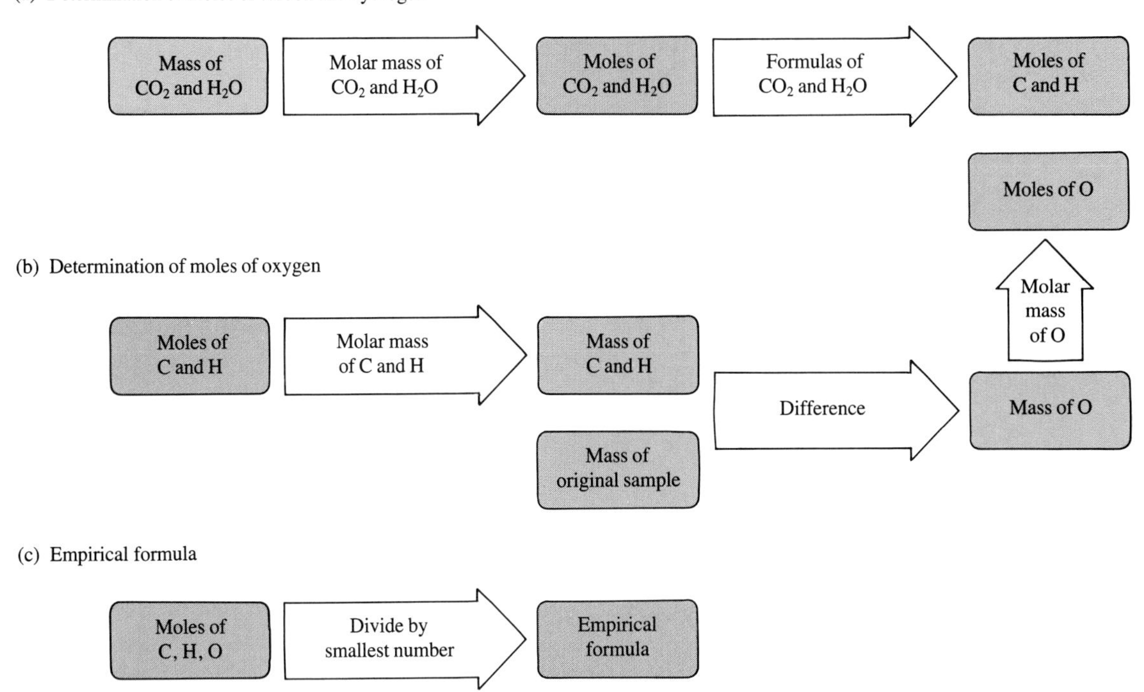

First we find the number of moles of CO_2 collected by the trap:

$$\text{moles } CO_2 = 0.228 \cancel{\text{g } CO_2}\left(\frac{1 \text{ mol } CO_2}{44.01 \cancel{\text{g } CO_2}}\right) = 0.00518 \text{ mol } CO_2$$

From the formula of carbon dioxide, 1 mol C $\simeq$ 1 mol CO_2, so the number of moles of carbon in the sample was

$$\text{moles C} = 0.00518 \cancel{\text{mol } CO_2}\left(\frac{1 \text{ mol C}}{1 \cancel{\text{mol } CO_2}}\right) = 0.00518 \text{ mol C}$$

Now we find the number of moles of water collected:

$$\text{moles } H_2O = 0.0928 \cancel{\text{g } H_2O}\left(\frac{1 \text{ mol } H_2O}{18.02 \cancel{\text{g } H_2O}}\right) = 0.00515 \text{ mol } H_2O$$

From the formula of water, 2 mol H $\hat{=}$ 1 mol H_2O, so the number of moles of hydrogen in the sample was

$$\text{moles H} = 0.00515\ \cancel{\text{mol H}_2\text{O}}\left(\frac{2\text{ mol H}}{1\ \cancel{\text{mol H}_2\text{O}}}\right) = 0.0103\text{ mol H}$$

The mass of oxygen in the sample cannot be determined directly in this experiment because excess O_2 is added to ensure complete combustion. Only part of the oxygen present in the products was present in the sample of the compound. However, the sample contains only carbon, hydrogen and oxygen, and from the law of conservation of mass we know that

$$\text{mass of sample} = \text{mass of C} + \text{mass of H} + \text{mass of O}$$

The mass of the sample was given and the masses of hydrogen and carbon can be calculated from the molar amounts. The mass of oxygen can then be calculated by difference:

$$\text{mass O} = \text{total mass of sample} - (\text{mass C} + \text{mass H})$$

$$\text{mass C} = 0.00518\ \cancel{\text{mol C}}\left(\frac{12.01\text{ g C}}{1\ \cancel{\text{mol C}}}\right) = 0.0622\text{ g C}$$

$$\text{mass H} = 0.0103\ \cancel{\text{mol H}}\left(\frac{1.008\text{ g H}}{1\ \cancel{\text{mol H}}}\right) = 0.0104\text{ g H}$$

$$\text{mass O} = 0.1000\text{ g} - (0.0622\text{ g} + 0.0104\text{ g}) = 0.0274\text{ g O}$$

The number of moles of hydrogen and carbon are known, and we now calculate the number of moles of oxygen:

$$0.0274\ \cancel{\text{g O}}\left(\frac{1\text{ mol O}}{16.00\ \cancel{\text{g O}}}\right) = 0.00171\text{ mol O}$$

Convert the numbers of moles of elements in the sample to whole-number ratios by dividing each by the smallest number.

$$\text{moles C} = 0.00518\text{ mol C}/0.00171 = 3.03\text{ mol C}$$

$$\text{moles H} = 0.0103\text{ mol H}/0.00171 = 6.02\text{ mol H}$$

$$\text{moles O} = 0.00171\text{ mol O}/0.00171 = 1.00\text{ mol O}$$

The empirical formula is C_3H_6O.

Understanding A compound contains only carbon, hydrogen, and nitrogen. After combustion of a 0.500 g sample, the mass of the first trap (collecting H_2O) increases by 0.698 g, and the mass of the second trap (collecting CO_2) increases by 0.977 g. What is the empirical formula of this compound?

Answer: C_2H_7N

The method of dividing through by the smallest number does not always yield whole numbers, but rather makes the smallest subscript equal to 1. Many empirical formulas, such as $C_3H_6O_2$, do not have any of the elements with a subscript of 1. An experiment to determine the empirical formula of this compound might yield numbers like C = 0.0282 mol, H = 0.0564 mol, and O = 0.0188 mol. Dividing by the smallest number, 0.0188, will yield the values C = 1.50 mol, H = 3.00 mol, and O = 1.00 mol. To find the correct formula, each of these numbers must be multiplied by the *same* smallest whole number (so the ratio is not changed) that will convert them to integers. In this case, multiplying each number of moles by 2 converts all of them to

INSIGHTS into CHEMISTRY

Development of Chemistry

Richards Measurements Cast Doubt About Atomic Weights: New Standards Lead to Nobel Prize

An important piece of information available in the periodic table is the atomic weight of each element. Atomic weight as a characteristic of an element was first introduced as part of John Dalton's atomic theory. Once the significance of an atomic weight scale was realized, the accurate determination of atomic weights became a goal of many nineteenth-century European chemists. The honor for the most accurate chemical determination of many atomic weights, however, belongs to an American chemist, Theodore W. Richards (1868–1928). Working primarily during the first decade of the twentieth century, Richards established standards for accuracy and precision that probably represent the limit for atomic weight determination by purely chemical methods.

When Richards began his work in 1887, the accepted values of the atomic weights were based upon the work of the Belgian chemist Jean Sarvais Stas (1813–1891). Stas, working with large quantities of materials, was highly regarded for his lengthy and careful procedures. His values were so well received that, until Richards began to have doubts, no one had seriously questioned them or attempted to check his work. Richards's interest in atomic weights began in graduate school at Harvard. His doctoral thesis problem was the accurate determination of the relative weights of hydrogen and oxygen by studying the exact composition of water. Richards's technique involved passing a weighed quantity of dry hydrogen gas over hot copper oxide and weighing the water formed. He could then calculate the mass of oxygen combined with the hydrogen and obtain the hydrogen-to-oxygen ratio. He measured a ratio of 1 to 15.869, which gave hydrogen a relative weight of 1.0082 compared to 16 for oxygen. This value differed slightly from the accepted value for hydrogen, a fact that caused Richards to begin doubting other accepted values. Using the same experiment, Richards obtained an atomic weight for copper that also did not agree with the accepted value. He then began to work with other metallic elements, and even confirmed that the atomic weight of cobalt is greater than that of nickel, although cobalt precedes nickel in the periodic table. During these studies, Richards perfected his analytical techniques, and introduced the practices of working with small quantities of material and of forming precipitates from dilute rather than concentrated solutions. He tried to think of every possible source of error and to account for it. Above all, he set rigorous standards for the purity of his samples. His excellent work led to his selection for studies in Europe, where he worked with some of the recognized leaders in physical and analytical chemistry. Upon his return, he accepted a teaching position in analytical chemistry at Harvard.

T. W. Richards accurately determined the atomic weights of many elements.

By 1905, Richards had become aware of serious errors in Stas's atomic weight values. His knowledge of physical chemistry and the latest theories of precipitation enabled him to point out the sources of Stas's errors and to make the proper corrections during his own analyses. Richards, along with his graduate students, redetermined the atomic weights of many of the elements previously studied by Stas; silver, nitrogen, chlorine, sodium, and potassium. In all cases, Richards's work produced significant changes in the accepted values. All in all, Richards and his students determined the atomic weights of 25 elements; those of an additional 30 elements were determined in the laboratories of two of his former students. Richards soon became recognized as the leading authority in this field of analytical chemistry, leading to a Nobel Prize in 1914.

In addition to his atomic weight determinations, Richards made contributions in the fields of thermochemistry and electrochemistry. He also made a comprehensive study of the atomic weight of lead obtained from radioactive sources. His finding in 1914 that lead from these sources had an atomic weight different from that of lead obtained from nonradioactive sources was one of the first experimental confirmations of the existence of isotopes. As his most lasting contribution, he established Harvard as a major center for physical and analytical chemistry research. More than 60 young scientists studied with him and became renowned chemists in their own right.

whole numbers: C = 3.00, H = 6.00, and O = 2.00. The empirical formula is $C_3H_6O_2$. The overall strategy is outlined below.

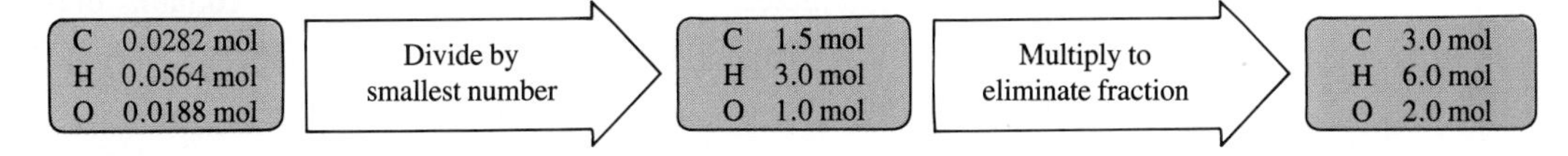

Example 3.14 Empirical and Molecular Formulas

Analysis of a sample of a compound shows that it contains 1.11 g carbon, 0.187 g hydrogen, and 0.889 g oxygen. What is the empirical formula? What is the molecular formula if the molecular weight is 354?

Solution The strategy is outlined in the following diagram.

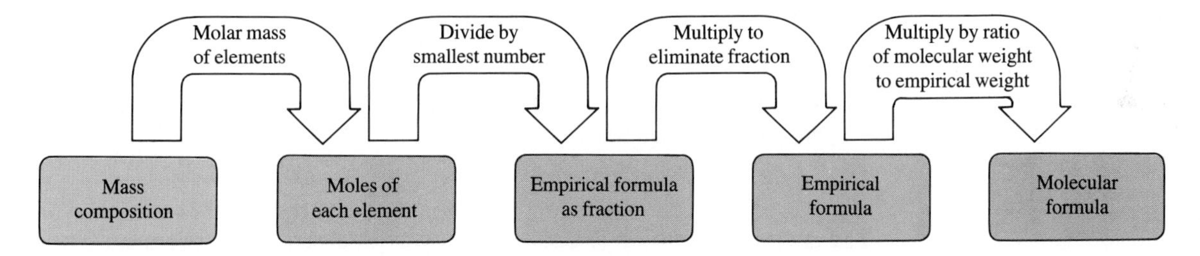

First, determine the number of moles of each element.

$$\text{moles C} = 1.11\ \cancel{\text{g C}}\left(\frac{1\ \text{mol C}}{12.01\ \cancel{\text{g C}}}\right) = 0.0924\ \text{mol C}$$

$$\text{moles H} = 0.187\ \cancel{\text{g H}}\left(\frac{1\ \text{mol H}}{1.008\ \cancel{\text{g H}}}\right) = 0.186\ \text{mol H}$$

$$\text{moles O} = 0.889\ \cancel{\text{g O}}\left(\frac{1\ \text{mol O}}{16.00\ \cancel{\text{g O}}}\right) = 0.0556\ \text{mol O}$$

Divide by the smallest number of moles found.

$$\text{moles C} = 0.0924\ \text{mol C}/0.0556 = 1.66\ \text{mol C}$$

$$\text{moles H} = 0.186\ \text{mol H}/0.0556 = 3.34\ \text{mol H}$$

$$\text{moles O} = 0.0556\ \text{mol O}/0.0556 = 1.00\ \text{mol O}$$

The data given in this problem are known to three significant figures. Clearly, we cannot round off 1.66 or 3.34 to the nearest integer. We need to multiply each by the same smallest number that will convert them into integers. In this case, multiply all of the molar quantities by 3.

$$\text{moles C} = 1.66\ \text{mol C} \times 3 = 4.98\ \text{mol C}$$

$$\text{moles H} = 3.34\ \text{mol H} \times 3 = 10.0\ \text{mol H}$$

$$\text{moles O} = 1.00\ \text{mol O} \times 3 = 3.00\ \text{mol O}$$

These are now integers within the accuracy of the measurement, and the empirical formula is $C_5H_{10}O_3$. The empirical formula weight is 118. The experimentally determined molecular weight of the compound is 354. Thus, the number of times the empirical formula occurs in the molecule $(C_5H_{10}O_3)_n$ is

$$n = \frac{354}{118} = 3$$

All of the subscripts in the empirical formula must be tripled to obtain the molecular formula of $C_{15}H_{30}O_9$.

Understanding What is the molecular formula of a substance that contains only 0.801 g of carbon and 0.101 g of hydrogen and has a molecular weight of 54?

Answer: C_4H_6

Chemical analysis is not limited to carbon and hydrogen. For example, the chlorine content in a substance can frequently be determined by reaction with $AgNO_3$. All of the chlorine in the substance is converted into solid AgCl, which can be collected and weighed. This method will be outlined in detail in Chapter 4.

3.4 Mass Relationships in Equations

Objectives

- To calculate the mass of a product formed or a reactant consumed from the chemical equation
- To determine percent yields

The chemical equation is a convenient way to describe any chemical reaction quantitatively (Section 3.1). The equation not only tells us what happens, it also expresses stoichiometric relationships among the species present. These relationships between numbers of molecules or moles allow us to derive conversion factors that are used to calculate the number of moles of one substance in an equation, given the number of moles of any other substance in the equation. For example, the following equation can be interpreted in terms of either molecules or moles.

$$2H_2 \quad + \quad O_2 \quad \longrightarrow \quad 2H_2O$$

$$2 \text{ molecules } H_2 + 1 \text{ molecule } O_2 \longrightarrow 2 \text{ molecules } H_2O$$

$$2 \text{ moles } H_2 \quad + \quad 1 \text{ mole } O_2 \quad \longrightarrow \quad 2 \text{ moles } H_2O$$

The coefficients in the equation give chemical equivalencies that can be used to express moles of any substance in the equation in the equivalent number of moles of any other substance in the equation. For the formation of water from hydrogen and oxygen, these equivalencies are

$$2 \text{ mol } H_2 \simeq 1 \text{ mol } O_2$$

$$2 \text{ mol } H_2 \simeq 2 \text{ mol } H_2O$$

$$1 \text{ mol } O_2 \simeq 2 \text{ mol } H_2O$$

Since we know how to convert mass data into moles, and *vice versa,* this knowledge can be combined with the stoichiometry of the chemical equation to answer questions such as "What mass of water is produced when 5.0 g of hydrogen is burned with excess oxygen?" Any time we need to determine information about one substance from given information about the quantity of another substance, we will need the chemical equation that relates the two substances of interest. The complete procedure for this type of operation is:

1. Write the balanced equation.
2. Calculate the number of moles of the species for which the mass is given, using the appropriate mass-mole conversion factor.
3. Use the coefficients from the balanced equation to convert the moles of the given substance into moles of the substance desired.
4. Calculate the mass of the desired species, using the appropriate mole-mass conversion factor.

Steps 2 and 4 involve unit conversions based on the molar masses calculated from the periodic table. Step 3 involves a unit conversion based on the coefficients in the equation written in Step 1. This procedure is summarized in the following diagram.

The coefficients in a chemical equation are used to calculate moles of one substance in the equation from the number of moles of a second substance.

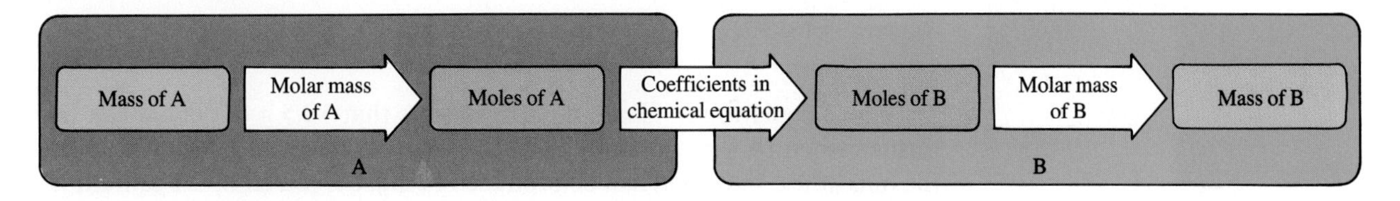

Example 3.15 Equation Stoichiometry

Determine the mass of Ga_2O_3 formed from the reaction of 14.5 g of gallium metal with excess O_2.

Solution We will use the four-step procedure just outlined. Write the balanced equation; convert mass of the given substance, gallium, into moles; use the chemical equation to convert moles of gallium into moles of Ga_2O_3; and finish the problem by converting moles of Ga_2O_3 into grams of Ga_2O_3.

First, write the equation.

$$4Ga(s) + 3O_2(g) \longrightarrow 2Ga_2O_3(s)$$

Second, calculate the number of moles of gallium that react. The molar mass of gallium is 69.72 g/mol, so 14.5 g of Ga is

$$\text{moles Ga} = 14.5 \text{ g Ga} \left(\frac{1 \text{ mol Ga}}{69.72 \text{ g Ga}}\right) = 0.208 \text{ mol Ga}$$

Third, use the coefficients of the equation to determine the conversion factor that relates moles of gallium to moles of Ga_2O_3. In this case,

$$4 \text{ mol Ga} \simeq 2 \text{ mol } Ga_2O_3$$

This equivalency is used to convert moles of gallium into moles of Ga_2O_3:

$$\text{moles } Ga_2O_3 = 0.208 \text{ mol Ga} \left(\frac{2 \text{ mol } Ga_2O_3}{4 \text{ mol Ga}}\right) = 0.104 \text{ mol } Ga_2O_3$$

Fourth, use the molar mass of Ga_2O_3 (187.4 g/mol) to calculate the mass of Ga_2O_3:

$$\text{mass } Ga_2O_3 = 0.104 \text{ mol } Ga_2O_3 \left(\frac{187.4 \text{ g } Ga_2O_3}{1 \text{ mol } Ga_2O_3}\right) = 19.5 \text{ g } Ga_2O_3$$

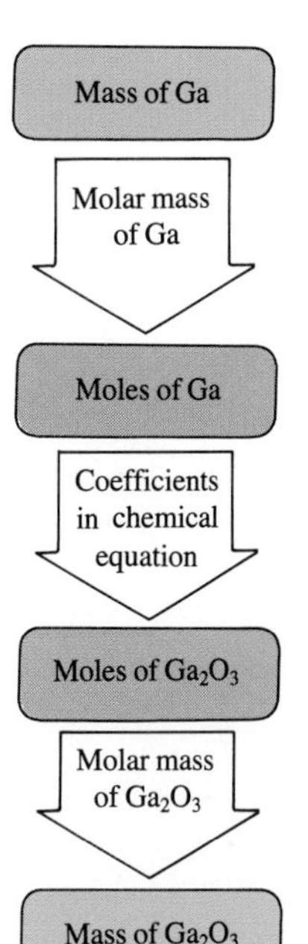

To summarize, the number of moles of gallium was calculated from the mass of gallium by using a conversion factor derived from atomic weights. Moles of gallium oxide were calculated from moles of gallium by using the coefficients of the balanced equation, and the problem was finished with a conversion of moles back into grams.

The answer is reasonable in that 14.5 g of gallium produced 19.5 g of Ga_2O_3. We expect the mass of the Ga_2O_3 produced to be somewhat greater than the mass of Ga that reacts, since it also includes some oxygen.

Note that it is possible to combine the unit conversion steps in this problem into a single chain calculation.

$$14.5\text{ g Ga}\left(\frac{1\text{ mol Ga}}{69.72\text{ g Ga}}\right)\left(\frac{2\text{ mol Ga}_2\text{O}_3}{4\text{ mol Ga}}\right)\left(\frac{187.4\text{ g Ga}_2\text{O}_3}{1\text{ mol Ga}_2\text{O}_3}\right) = 19.5\text{ g Ga}_2\text{O}_3$$

This combined procedure may be simpler, but be careful to include each of the unit conversion factors, and check that all units cancel except those desired for the answer of the problem.

Understanding What mass of SO_3 will form from the reaction of 4.1 g of SO_2 with excess O_2?

Answer: 5.1 g SO_3

The calculation in Example 3.15 tells us that 19.5 g of Ga_2O_3 can be formed when 14.5 g of gallium burns in excess of O_2. The 19.5 g of Ga_2O_3 is the maximum mass that can be produced because when all of the gallium is consumed, the reaction must stop. This calculated mass of product is the **theoretical yield,** the maximum quantity of product that can be obtained from a chemical reaction based on the amounts of starting materials. In Example 3.15 it is assumed that all of the O_2(g) needed for the reaction was present. We could also calculate the mass of O_2 consumed in the reaction.

Example 3.16 Equation Stoichiometry

Given the following equation:

$$2\text{PbS} + 3\text{O}_2 \longrightarrow 2\text{PbO} + 2\text{SO}_2$$

(a) what mass of O_2 will react with 4.10 g of PbS and
(b) what is the theoretical yield of PbO?

Solution

(a) The same series of steps used in Example 3.15 is used for this problem. The equation can be used to determine the amounts of reactants needed as well as the amounts of products formed in the reaction.

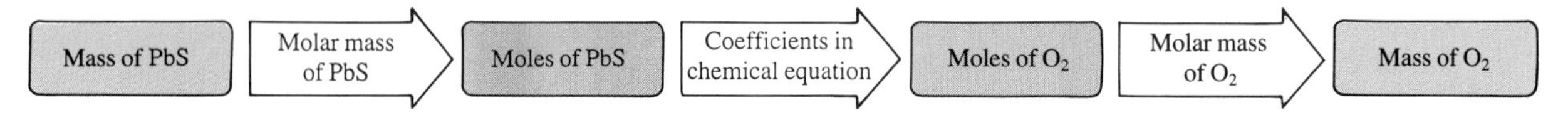

The balanced equation has been given. The second step is to calculate the number of moles of PbS that react (the molar mass of PbS is 239.3 g/mol).

$$\text{moles PbS} = 4.10\text{ g PbS}\left(\frac{1\text{ mol PbS}}{239.3\text{ g PbS}}\right) = 0.0171\text{ mol PbS}$$

From the equation, we know that three moles of O_2 are equivalent to two moles of PbS. The third step is to calculate the number of moles of O_2 needed to react with 0.0171 mol of PbS.

$$\text{moles O}_2 = 0.0171\text{ mol PbS}\left(\frac{3\text{ mol O}_2}{2\text{ mol PbS}}\right) = 0.0257\text{ mol O}_2$$

Fourth, finish the problem by calculating the mass of O_2 consumed, using the molar mass of O_2.

$$\text{mass } O_2 = 0.0257 \text{ mol } O_2 \left(\frac{32.00 \text{ g } O_2}{1 \text{ mol } O_2}\right) = 0.822 \text{ g } O_2$$

(b) The strategy is the same as in part (a), except that the equation is used to determine the amount of PbO rather than O_2.

Moles of PbS → Coefficients in chemical equation → Moles of PbO → Molar mass of PbO → Mass of PbO

The equation and the amount of PbS, 0.0171 moles, are known. Calculate the number of moles of PbO that will form.

$$\text{moles PbO} = 0.0171 \text{ mol PbS} \left(\frac{2 \text{ mol PbO}}{2 \text{ mol PbS}}\right) = 0.0171 \text{ mol PbO}$$

The theoretical yield is calculated by converting moles of PbO into grams.

$$\text{mass PbO} = 0.0171 \text{ mol PbO} \left(\frac{223.2 \text{ g PbO}}{1 \text{ mol PbO}}\right) = 3.82 \text{ g PbO}$$

The reaction in this example is commercially important. Many metal ores are mined as mixtures of sulfides and oxides. The first step in the refining process is to convert all of the ore into metal oxides by treating it with oxygen at high temperatures (Figure 3.6). This process is called "roasting." The oxide is then used in further chemical reactions to produce metal.

Understanding Given the following equation, calculate the mass of O_2 needed to react completely with 7.4 g of NO.

$$2NO + O_2 \longrightarrow 2NO_2$$

Answer: 3.9 g O_2

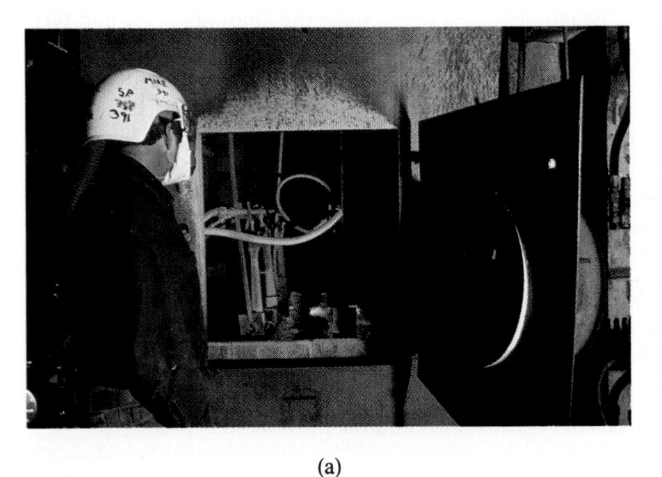

(a)

(b)

Figure 3.6 Roasting of PbS (a) Lead sulfide is converted into lead oxide in the high temperature "roasting" process. (b) The SO_2 gas is collected from the gas stream and used to prepare sulfuric acid.

Calculating the Reaction Yield

The procedure in Example 3.16b shows the calculation of the *theoretical* yield of PbO. However, it is often difficult to achieve the theoretical yield in the laboratory or an industrial process. For example, a reaction might produce a gas that is difficult to collect. Or if a solid is formed, some of it might stick to the walls of the reaction vessel and remain uncollected. Sometimes, reactions (called side reactions) other than the one described by the equation occur, consuming some starting material without forming the expected product. Thus, frequently not all of the product predicted by the stoichiometry calculation is isolated. The mass of product isolated from a reaction is known as the **actual yield.** It cannot, of course, exceed the theoretical yield. Chemists try to come as close to the theoretical yield as possible, and their success is presented as a **percent yield:**

The yield of a product isolated from a reaction is generally lower than the theoretical yield.

$$\text{percent yield} = \frac{\text{actual yield}}{\text{theoretical yield}} \times 100\%$$

Example 3.17 Determination of Percent Yield

Ammonia, NH_3, is an important compound used in the preparation of fertilizer. It is usually synthesized from nitrogen and hydrogen. The commercial production of ammonia from nitrogen and hydrogen is known as the Haber process, named after the German chemist who determined how to carry out the reaction on an industrial scale.

$$N_2 + 3H_2 \longrightarrow 2NH_3$$

(a) Calculate the mass of N_2 consumed when 45.3 g of H_2 reacts to form NH_3.
(b) What is the theoretical yield of NH_3?
(c) What is the percent yield if 102 g of NH_3 is isolated in this experiment?

Solution

(a and b) The balanced equation is used to calculate the amounts of N_2 and NH_3 equivalent to 45.3 g of H_2. The first step, the balanced equation, is given. Second, calculate the number of moles of H_2 that react:

$$\text{moles } H_2 = 45.3 \cancel{\text{ g } H_2}\left(\frac{1 \text{ mol } H_2}{2.016 \cancel{\text{ g } H_2}}\right) = 22.5 \text{ mol } H_2$$

Third, calculate the numbers of moles of N_2 and NH_3 equivalent to 22.5 mol of H_2, using the coefficients of the balanced equation.

$$\text{(a) moles } N_2 = 22.5 \cancel{\text{ mol } H_2}\left(\frac{1 \text{ mol } N_2}{3 \cancel{\text{ mol } H_2}}\right) = 7.50 \text{ mol } N_2$$

$$\text{(b) moles } NH_3 = 22.5 \cancel{\text{ mol } H_2}\left(\frac{2 \text{ mol } NH_3}{3 \cancel{\text{ mol } H_2}}\right) = 15.0 \text{ mol } NH_3$$

Fourth, calculate the mass of N_2 consumed and the mass of NH_3 produced.

$$\text{(a) mass } N_2 = 7.50 \cancel{\text{ mol } N_2}\left(\frac{28.01 \text{ g } N_2}{1 \cancel{\text{ mol } N_2}}\right) = 210 \text{ g } N_2$$

$$\text{(b) mass } NH_3 = 15.0 \cancel{\text{ mol } NH_3}\left(\frac{17.03 \text{ g } NH_3}{1 \cancel{\text{ mol } NH_3}}\right) = 255 \text{ g } NH_3$$

(c) The percent yield is calculated by dividing the mass of ammonia actually isolated in the reaction by the theoretical yield (times 100%).

$$\text{percent yield} = \frac{102\ \text{g}}{255\ \text{g}} \times 100\% = 40.0\%$$

The calculation shows that 45.3 g of H_2 will react with 210 g of N_2 and can produce a maximum of 255 g of NH_3. In this example, the chemist was able to collect only 102 g of the product, for a yield of 40.0%.

Understanding What is the percent yield if 2.4 g of NH_3 is obtained from the reaction of 0.64 g of H_2 with excess N_2?

Answer: 67%

Figure 3.7 Cobalt(II) chloride Many transition metal ions will combine with water molecules. The number of water molecules present changes the theoretical yield. The sample of cobalt(II) chloride on the left contains fewer water molecules per cobalt than the sample on the right. Not only does the sample on the right weigh more, but it can have different properties (such as color).

Laboratory workers occasionally observe an actual yield that is *greater* than the theoretical yield. A yield that is greater than 100% means that the product is not pure. For example, if a reaction takes place in solution and the desired product is a solid, this solid must be separated from the solution. If the solid is still wet after it is isolated, its mass will be greater than expected (Figure 3.7). To obtain the correct mass, the solid must be dried carefully. Also, the desired substance may be contaminated by other products or by excess reactants. In these cases, a purification procedure such as chromatography or distillation may be needed. Of course, both experimental and calculation errors can be made that will lead to incorrect yields. Any time the actual yield exceeds the theoretical yield, the chemist needs to look further to determine the source of the error.

3.5 Limiting Reactants

Objectives

- To identify the limiting reactant
- To calculate the yield based on a limiting reactant

The stoichiometry problems that we have solved so far have had one clearly identified reactant on which the calculation was based, and it was assumed that all other reactants were present in excess. In more typical situations faced in the laboratory, the masses of two or more of the reactants are known. Which of these reactants should be used in a stoichiometry calculation? The situation is analogous to a baker who needs six eggs and two cups of flour to make a cake. If the baker has 18 eggs and four cups of flour, only two cakes can be made. Each cake requires two cups of flour, and with only four cups available only two cakes can be made. Even though there are enough eggs to make three cakes, the baker is *limited* to making two cakes by the amount of flour.

As an example of a chemical reaction that is limited by one reactant, consider the reaction of mercury with oxygen. The coefficients tell us that two moles of Hg react for every one mole of O_2.

$$2Hg + O_2 \longrightarrow 2HgO$$

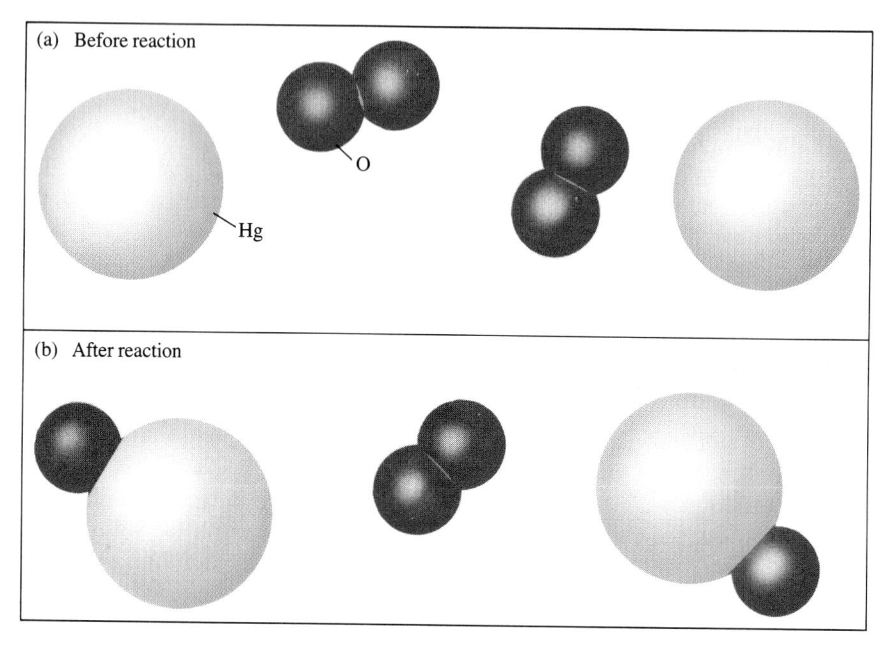

Figure 3.8 Limiting reactant If two atoms of mercury and two molecules of oxygen (a) react, two units of mercury oxide form (b), but one molecule of oxygen remains. Oxygen is present in excess, and mercury is the limiting reactant.

If the reaction begins with two moles of each reactant, all of the mercury will be consumed, but only one mole of the O_2 will be consumed (Figure 3.8). When the last atom of Hg is consumed, the reaction must stop. In this case, mercury is the **limiting reactant,** the reactant that is completely consumed when the chemical reaction occurs. When we calculate the amount of product formed, the calculation must be based on the limiting reactant, not other reactants that are present in excess.

The limiting reactant is completely consumed in a chemical reaction. The quantity of products formed in the reaction is determined by the quantity of the limiting reactant.

The limiting reactant is the one that yields the smallest amount of product, because the reaction ends when any one of the starting reactants is exhausted. Hence, a good approach to limiting-reactant problems is to calculate the number of moles of product formed from the given quantity of each reactant.

Example 3.18 Limiting Reactant

Calculate the mass, in grams, of H_2O produced from the reaction of 10.0 g of H_2 with 99.8 g of O_2.

Solution The approach to this problem is similar to the examples in Section 3.4, but the quantities of *both* reactants must be converted into quantity of products in order to determine the limiting reactant. The limiting reactant will be the one that produces less product.

The first step, the equation, is

$$2H_2 + O_2 \longrightarrow 2H_2O$$

Be careful at this point not to designate H_2 as the limiting reactant simply because fewer grams of it are present. A conversion into moles and proper use of the equation coefficients are needed to choose the limiting reactant.

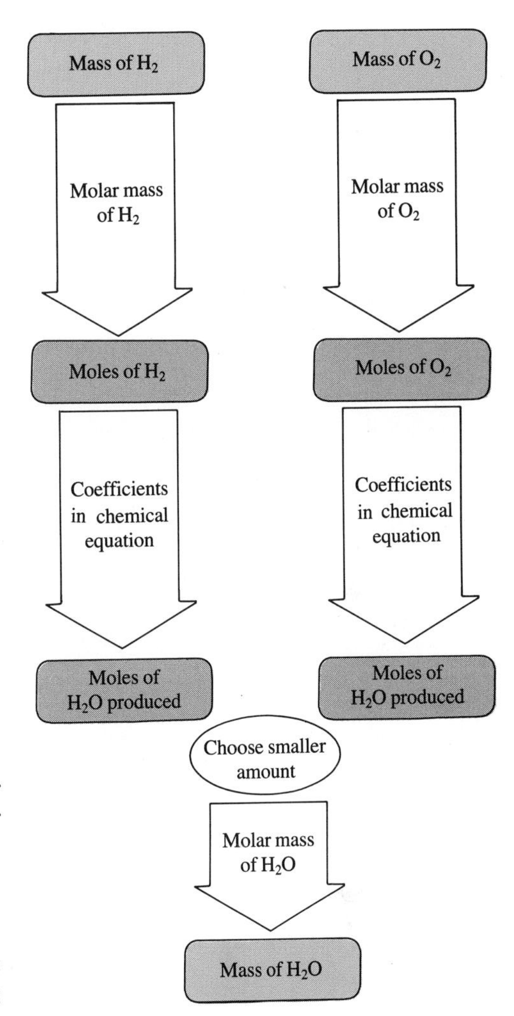

The second step is to calculate the numbers of moles of the reactants from their masses.

$$\text{moles } H_2 = 10.0 \text{ g } H_2 \left(\frac{1 \text{ mol } H_2}{2.016 \text{ g } H_2}\right) = 4.96 \text{ mol } H_2$$

$$\text{moles } O_2 = 99.8 \text{ g } O_2 \left(\frac{1 \text{ mol } O_2}{32.00 \text{ g } O_2}\right) = 3.12 \text{ mol } O_2$$

Keep in mind that the reactant present in the smaller number of moles is *not necessarily* the limiting reactant. Carry out the third step, using the coefficients in the equation, to calculate the number of moles of the desired substance (H_2O in this case) equivalent to the number of moles of each reactant.

$$\text{moles } H_2O \text{ based on } H_2 = 4.96 \text{ mol } H_2 \left(\frac{2 \text{ mol } H_2O}{2 \text{ mol } H_2}\right) = 4.96 \text{ mol } H_2O$$

$$\text{moles } H_2O \text{ based on } O_2 = 3.12 \text{ mol } O_2 \left(\frac{2 \text{ mol } H_2O}{1 \text{ mol } O_2}\right) = 6.24 \text{ mol } H_2O$$

This calculation shows that 10.0 g of H_2 can produce a maximum of 4.96 moles of H_2O if O_2 is present in excess. Similarly, 99.8 g of O_2 can produce 6.24 moles of H_2O if H_2 is present in excess. Clearly, H_2 is the limiting reactant because it produces less H_2O. The fourth step, the calculation of the mass of H_2O formed in the reaction, must be based on the number of moles of H_2.

$$\text{mass } H_2O = 4.96 \text{ mol } H_2O \left(\frac{18.02 \text{ g } H_2O}{1 \text{ mol } H_2O}\right) = 89.4 \text{ g } H_2O$$

Since there is an excess of O_2, some oxygen will remain at the end of the reaction. The quantity of O_2 remaining can be calculated by determining the mass of O_2 needed to react with 10.0 g of H_2, and subtracting it from the starting amount. It can also be calculated by assuming the conservation of mass. We started with 109.8 g of reactants and produced 89.4 g of H_2O. The 20.4 g difference must be the mass of O_2 remaining, because all of the H_2 is consumed.

Understanding Given the following equation, what mass (in grams) of $AlCl_3$ can be produced from 4.4 g of Al and 12 g of Cl_2?

$$2\,Al + 3Cl_2 \longrightarrow 2AlCl_3$$

Answer: 15 g $AlCl_3$

Many of the calculations necessary to study chemical reactions in a quantitative manner have now been presented. Of course, before you can actually do the experiments, you must learn many procedures regarding the proper handling of the chemicals and equipment used to carry out reactions. The following problem represents a situation that is likely to arise in the laboratory, and will serve to integrate many of the concepts you have learned up to this point.

Example 3.19 Equation Stoichiometry

Phosphorus trichloride reacts with oxygen to yield $POCl_3$. In an experiment performed in the laboratory, 11.0 g of PCl_3 and 1.34 g of O_2 were mixed, and 11.2 g of $POCl_3$ were isolated. What is the percent yield?

Solution To determine the percent yield, we must calculate the theoretical yield based on the limiting reactant. The strategy for the calculation of the theoretical yield is the same as in Example 3.18.

First, write the equation.

$$2PCl_3 + O_2 \longrightarrow 2POCl_3$$

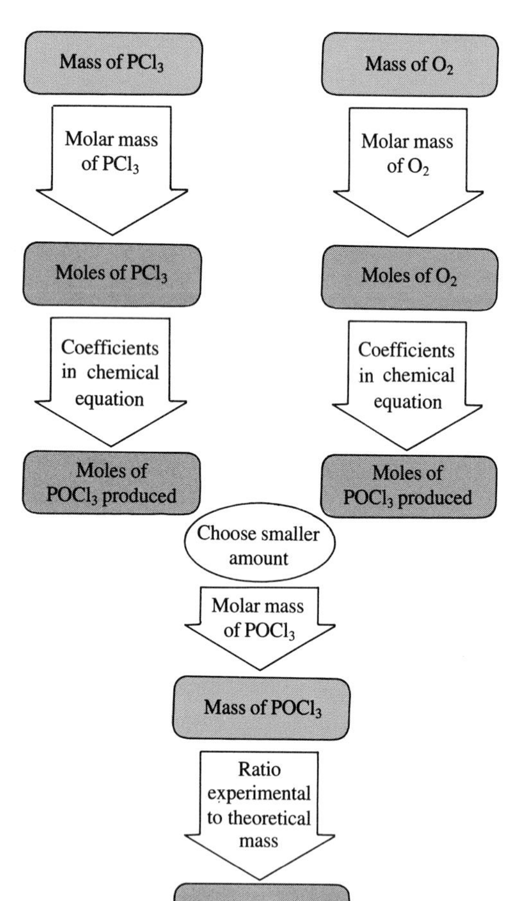

Second, convert the grams of both reactants into the equivalent number of moles.

$$\text{moles } PCl_3 = 11.0 \cancel{\text{ g } PCl_3} \left(\frac{1 \text{ mol } PCl_3}{137.3 \cancel{\text{ g } PCl_3}}\right) = 0.0801 \text{ mol } PCl_3$$

$$\text{moles } O_2 = 1.34 \cancel{\text{ g } O_2} \left(\frac{1 \text{ mol } O_2}{32.00 \cancel{\text{ g } O_2}}\right) = 0.0419 \text{ mol } O_2$$

Third, convert the moles of each reactant into moles of $POCl_3$. The reactant that yields the smaller number of moles of $POCl_3$ is the limiting reactant.

$$\text{moles } POCl_3 \text{ based on } PCl_3 = 0.0801 \cancel{\text{ mol } PCl_3} \left(\frac{2 \text{ mol } POCl_3}{2 \cancel{\text{ mol } PCl_3}}\right)$$
$$= 0.0801 \text{ mol } POCl_3$$

$$\text{moles } POCl_3 \text{ based on } O_2 = 0.0419 \cancel{\text{ mol } O_2} \left(\frac{2 \text{ mol } POCl_3}{1 \cancel{\text{ mol } O_2}}\right)$$
$$= 0.0838 \text{ mol } POCl_3$$

The PCl_3 is the limiting reactant, even though the mass and the number of moles of PCl_3 present at the beginning of the reaction are greater than those of oxygen. The number of moles of $POCl_3$ formed from PCl_3 is used in the fourth step to calculate the theoretical yield:

$$\text{mass } POCl_3 = 0.0801 \cancel{\text{ mol } POCl_3} \left(\frac{153.3 \text{ g } POCl_3}{1 \cancel{\text{ mol } POCl_3}}\right) = 12.3 \text{ g } POCl_3$$

The percent yield is the actual yield of the reaction divided by the theoretical yield, times 100%.

$$\text{percent yield} = \frac{11.2 \text{ g } POCl_3 \text{ isolated}}{12.3 \text{ g } POCl_3 \text{ theoretical}} \times 100\% = 91.0\% \text{ yield}$$

In summary, the theoretical yield when 1.34 g of O_2 reacted with 11.0 g of PCl_3 is 12.3 g of $POCl_3$. In practice, 11.2 g of $POCl_3$ were collected. The percent yield was 91.0%.

Understanding In the following reaction, 44 g of NH_3 reacted with 120 g of O_2, and 73 g of NO were isolated. What is the percent yield?

$$4NH_3 + 5O_2 \longrightarrow 4NO + 6H_2O$$

Answer: 94%

There are some practical reasons why a chemist would perform a reaction with some reactants in excess rather than with all present in stoichiometric amounts. Sometimes the result of a chemical reaction depends on an excess of one or more reactants being present. For example, the combustion of organic compounds will actually produce a mixture of CO, CO_2, and H_2O if oxygen is not present in excess. In many reactions, undesirable side products can be avoided by using an excess of one or more reactants. In other cases, an excess of certain reactants may be needed to increase the yield or shorten the length of time it takes for the reaction to occur.

Summary

Stoichiometry is the study of quantitative relationships involving substances and their reactions. The chemical equation gives a quantitative description of a chemical reaction. The first step in writing a chemical equation is to write the *chemical expression* that identifies all the *reactants* and *products* by their correct formulas. Balancing the expression is accomplished by adjusting the *coefficients* of the substances so that the number of atoms of each element is the same on both the product and reactant sides. Two important types of reactions are *combustion* and *neutralization* (acid-base) reactions.

The *mole* is the SI unit of *amount* and is used to express the quantitative relationships of matter. The mole is defined as the amount of a substance that contains as many units as there are atoms in 12 grams of carbon-12. This number is called *Avogadro's number* and has a value of 6.022×10^{23} units/mol. The mass in grams of one mole of any substance is its *molar mass.* Atomic weights, molecular weights, molar mass, and Avogadro's number are the key quantities in the important stoichiometric relationships in this chapter. These relationships are used to convert from the molecular to the molar scale and from mass into number of moles.

The percentage composition by mass of each element in a compound can be determined in the laboratory. The *empirical formula* of a substance is calculated from the mass composition and *vice versa.* The molecular weight of a compound is needed to find its molecular formula, given its empirical formula.

The coefficients in the chemical equation express not only the relative numbers of molecules involved in the chemical change, but also the number of moles of the substances consumed and produced by the reaction. All reaction stoichiometry problems should be solved by first converting the given quantities into numbers of moles. Then the coefficients in the equation are used to calculate the number of moles of the desired substances. The final solution might require another conversion of units. The most common reaction stoichiometry problems provide the masses of the reactants and ask for the masses of the products.

The quantities of products that are calculated from the chemical equation are *theoretical yields.* The *actual yields* are those obtained in the laboratory or factory. The degree of success in a chemical synthesis is often expressed as a *percent yield.*

When the quantity of more than one reactant is known, it is necessary to determine which is the *limiting reactant,* the reactant that is completely consumed in the reaction, and to base the stoichiometry calculation on this reagent.

Chapter Terms

Stoichiometry—the study of quantitative relationships involving substances and their reactions. *(3.1)*

Reactant—a substance that is consumed in a chemical reaction. *(3.1)*

Product—a substance that is formed in a chemical reaction. *(3.1)*

Chemical equation—an equation that describes the identities and relative amounts of reactants and products in a chemical reaction. *(3.1)*

Chemical expression—an expression that describes the identities of reactants and products in a chemical reaction, but is not necessarily balanced. *(3.1)*

Coefficient—the number of units of each substance in a chemical reaction. *(3.1)*

Organic compound—a compound made up of carbon atoms in combination with other elements such as hydrogen, oxygen, and nitrogen. *(3.1)*

Combustion reaction—the process of burning. Examples are the reactions of organic compounds with excess oxygen to yield carbon dioxide and water. *(3.1)*

Ionization—the separation of a molecular compound into individual cations and anions when dissolved (generally in water). *(3.1)*

Acid—a substance that provides the hydrogen cation in water solution. *(3.1)*

Base—a substance that provides the hydroxide anion in water. *(3.1)*

Salt—a compound made up of the cation from a base and the anion from an acid. *(3.1)*

Neutralization reaction—the reaction of an acid and a base to yield water and a salt. *(3.1)*

Mole—the amount of substance that contains as many entities as there are atoms in exactly 12 grams of ^{12}C (the isotope of carbon that contains six neutrons). *(3.2)*

Avogadro's number—the number of units in one mole, 6.022×10^{23} units/mol. *(3.2)*

Atomic mass unit—the base unit of a mass scale that defines one unit (u) as one twelfth the mass of a single atom of ^{12}C. *(3.2)*

Atomic weight (atomic mass)—the mass in atomic mass units of one atom of an element. It is an average mass that reflects the natural isotopic distribution of the element. *(3.2)*

Molar mass—the mass in grams of one mole of any substance; numerically the same as the molecular or formula weight. *(3.2)*

Molecular weight—the sum of the atomic weights of all atoms present in the molecular formula of a molecule. *(3.2)*

Formula weight—the sum of the atomic weights of all atoms in a formula. *(3.2)*

Theoretical yield—the maximum quantity of product that can be obtained in a chemical reaction, based on the amounts of starting materials. *(3.4)*

Actual yield—the amount of product isolated when a chemical reaction is carried out. *(3.4)*

Percent yield—the actual yield divided by the theoretical yield and multiplied by 100%. *(3.4)*

Limiting reactant—the reactant that is completely consumed in a chemical reaction. The amounts of products that can form and the amounts of the other reactants that can be consumed are determined by the amount of the limiting reactant present. *(3.5)*

Exercises

Exercises designated with color have answers in Appendix J.

Chemical Equations

3.1 What is the difference between the chemical expression and the chemical equation?

3.2 Balance the following chemical expressions.
(a) $C_5H_{12} + O_2 \longrightarrow CO_2 + H_2O$
(b) $NH_3 + O_2 \longrightarrow N_2 + H_2O$
(c) $KOH + H_2SO_4 \longrightarrow K_2SO_4 + H_2O$

3.3 Balance the following chemical expressions.
(a) $Mg_3N_2 + H_2O \longrightarrow NH_3 + Mg(OH)_2$
(b) $Fe + O_2 \longrightarrow Fe_2O_3$
(c) $Zn + H_3PO_4 \longrightarrow H_2 + Zn_3(PO_4)_2$

3.4 Balance the following chemical expressions.
(a) $N_2H_4 + N_2O_4 \longrightarrow N_2 + H_2O$
(b) $F_2 + H_2O \longrightarrow HF + O_2$
(c) $Na_2O + H_2O \longrightarrow NaOH$

3.5 Balance the following chemical expressions.
(a) $N_2O + NH_3 \longrightarrow N_2 + H_2O$
(b) $Cr_2S_3 + HCl \longrightarrow CrCl_3 + H_2S$
(c) $Fe_2S_3 + O_2 \longrightarrow Fe_2O_3 + SO_2$

3.6 Write an equation for the combustion (in excess oxygen) of each of the following compounds:
(a) C_6H_{12} (b) C_4H_8 (c) C_3H_6O (d) $C_4H_6O_2$

3.7 Write an equation for the combustion (in excess oxygen) of each of the following compounds:
(a) C_8H_{14} (b) C_5H_{10} (c) C_4H_8O (d) $C_5H_8O_2$

3.8 Write an equation for the reaction involving each of the following pairs of reactants:
(a) KOH and H_2SO_4
(b) calcium hydroxide and HCl
(c) HNO_3 and lithium hydroxide

3.9 Write an equation for the reaction involving each of the following pairs of reactants:
(a) $Mg(OH)_2$ and HF
(b) sodium hydroxide and H_3PO_4
(c) H_2SO_4 and magnesium hydroxide

3.10 The reaction of hydrazine, N_2H_4, with molecular oxygen is so violent that large quantities of gases and heat are given off rapidly. For this reason, hydrazine has been used as a rocket fuel. The products of the reaction are NO_2 and water. Write the equation.

3.11 The substance H_3PO_3 can be converted into H_3PO_4

and PH_3 by heating. Write the equation for this reaction.

3.12 Disulfur dichloride is used to vulcanize rubber. It is prepared by the reaction of elemental sulfur, S_8, and chlorine gas. Write the equation for this reaction.

3.13 Uranium dioxide reacts with carbon tetrachloride vapor at high temperatures, forming green crystals of uranium tetrachloride and phosgene ($COCl_2$), a poisonous gas. Write the equation for this reaction.

Atomic Weight and the Mole

3.14 Using a mass spectrometer, a sample of an element is found to contain 60.4% of atoms that have a mass of 68.93 u, while the remaining 39.6% of the atoms have a mass of 70.925 u. What is the atomic weight and identity of this element?

3.15 An element is found to have two isotopes that have masses of 62.9396 u and 64.9278 u, with 30.83% of the atoms being the heavier isotope. What is the atomic weight and identity of this element?

3.16 (a) How many moles are in 9.40 g of S?
(b) How many grams are in 3.3 mol of Al?
(c) How many grams are in 3.0×10^{25} atoms of Cl?

3.17 (a) How many moles are in 33.0 g of Se?
(b) How many atoms of N are in 5.6 g of N?
(c) What is number of moles in 2.14×10^{24} atoms of K?

3.18 What is the molar mass of each of the following compounds?
(a) NaOH (b) C_2H_4 (c) $Mg(OH)_2$

3.19 What is the molar mass of each of the following compounds?
(a) N_2O_4 (b) Na_2SO_4 (c) $C_6H_{10}O_2$

3.20 (a) How many moles are in 14.3 g of C_6H_6?
(b) Calculate the mass of 3.22×10^{22} molecules of SiH_4.

3.21 (a) What is the mass of 78.4 mol of CO_2?
(b) How many moles are in 192 g of $AgNO_3$?

3.22 Colchicine, $C_{22}H_{25}NO_6$, is a naturally occurring compound that has been used as a medicine since the time of the pharaohs in ancient Egypt. Although the reasons for its effectiveness are not yet clearly understood, it is still used to treat the inflammation in joints caused by a gout attack.
(a) What is the molar mass of colchicine?
(b) What is the mass, in grams, of 3.2×10^{22} molecules of colchicine?
(c) How many moles are in 326 g of colchicine?
(d) How many carbon atoms are in 50 molecules of colchicine?

3.23 Nickel tetracarbonyl, $Ni(CO)_4$, is a volatile, extremely toxic compound that can form when carbon monoxide gas is passed over finely divided nickel. Despite this toxicity, it has been used for more than a century in a method of making highly purified nickel.
(a) What is the mass of 1.00 mol of $Ni(CO)_4$?
(b) How many moles are in 3.22 g of $Ni(CO)_4$?
(c) How many molecules of $Ni(CO)_4$ are in 5.67 g?
(d) How many atoms of carbon are in 34 g of $Ni(CO)_4$?

3.24 Calculate the number of moles in each of the following samples:
(a) 2.2 g K_2SO_4
(b) 6.4 g $C_8H_{12}N_4$
(c) 7.13 g $Fe(C_5H_5)_2$

3.25 How many molecules are in 3.4 g of H_2?

3.26 Calculate the mass, in grams, of each of the following:
(a) 7.55 mol N_2O_4
(b) 9.2 mol $CaCl_2$
(c) 0.44 mol CO

3.27 (a) How many moles are in 48.0 g of H_2O_2?
(b) How many oxygen atoms are in this sample?

3.28 (a) What is the mass, in grams, of 3.50 mol of NO_2?
(b) How many molecules are in this sample?
(c) How many nitrogen and how many oxygen atoms are in the sample?

3.29 (a) How many moles are in 33.1 g of SO_3?
(b) How many molecules are in this sample?
(c) How many sulfur and how many oxygen atoms are in the sample?

Empirical Formulas

3.30 What is the percentage, by mass, of each element in the following substances?
(a) C_4H_8 (b) $C_3H_4N_2$ (c) Fe_2O_3

3.31 What is the percentage, by mass, of each element in the following substances?
(a) C_6H_{12} (b) $C_5H_{12}O$ (c) $NiCl_2$

3.32 What is the mass percent of carbon in CO? What is the mass of carbon in 4.9 g of CO?

3.33 What is the mass percent of each element in $C_4H_{10}O$? What is the mass of carbon in a 1.80 g sample of $C_4H_{10}O$?

3.34 What mass of carbon is in each of the following samples?
(a) 4.32 g of CO_2 (b) 2.21 g of C_2H_4
(c) 0.0443 g of CS_2

3.35 What is the mass of hydrogen in each of the following samples?
(a) 4.33 g H_2O (b) 1.22 g C_2H_2
(c) 4.44 g N_2H_4

3.36 What is the empirical formula of a compound that contains 0.139 g of hydrogen and 0.831 g of carbon?

3.37 What is the empirical formula of a substance made up of 0.80 g of carbon and 0.20 g of hydrogen?

3.38 What is the empirical formula of a compound made up of 0.571 g of carbon, 0.072 g of hydrogen, and 0.333 g of nitrogen?

3.39 What is the empirical formula of a compound made up of 0.152 g of nitrogen, and 0.348 g of oxygen?

3.40 A sample of a compound contains 1.11 g of carbon, 0.187 g of hydrogen, and 0.370 g of oxygen. What is the empirical formula?

3.41 A platinum compound named cisplatin is very effective in the treatment of certain types of cancer. Analysis shows that it contains 65.02% platinum, 2.02% hydrogen, 9.34% nitrogen, and 23.63% chlorine. What is its empirical formula?

3.42 What is the empirical formula of a substance that is 44.06% iron and 55.93% chlorine?

3.43 What is the empirical formula of a substance that contains only selenium and chlorine, and that is 52.7% selenium by mass?

3.44 A compound made up of titanium, carbon, and hydrogen is shown to contain 67.44% carbon and 5.67% hydrogen. What is the empirical formula?

3.45 Carvone is an oil isolated from caraway seeds that is used in perfumes and soaps. This compound contains 79.95% carbon, 9.40% hydrogen, and 10.65% oxygen. What is the empirical formula?

3.46 A 2.074-g sample containing only carbon, hydrogen, and oxygen is burned in excess O_2 to produce 3.80 g of CO_2 and 1.04 g of H_2O. What is the empirical formula?

3.47 A 2.000-g sample made up of Ca and Cl reacts with $AgNO_3$ to produce $Ca(NO_3)_2$ and 5.166 g of AgCl. What is the empirical formula of the starting material?

3.48 A compound is made up of C, H, N, and O. Combustion of a 1.48-g sample in excess O_2 yields 2.60 g of CO_2 and 0.799 g of H_2O. In a separate experiment it was shown that a 2.43 g sample contained 0.340 g of N. What is the empirical formula?

3.49 A compound with an empirical formula of HO has a molecular weight of 34. What is the molecular formula?

3.50 What are the molecular formulas of the following compounds?
(a) empirical formula = C_2H_4O; molecular weight = 132
(b) empirical formula = $C_3H_4NO_3$; molecular weight = 408

3.51 What is the molecular formula of a compound that has
(a) the empirical formula $C_5H_{10}O$ and a molecular weight of 258?
(b) the empirical formula PCl_3 and a molecular weight of 137.3?

3.52 A compound is made up of 62.0% carbon, 10.4% hydrogen, and 27.5% oxygen by mass and has a molecular weight of 174. What is its molecular formula?

3.53 The combustion of a 0.500-g sample of a compound made up of C, H, and N yields 0.338 g of H_2O and 1.42 g of CO_2. Its molecular weight is 186. What is the molecular formula?

3.54 Acetic acid gives vinegar its sour taste. Analysis of acetic acid shows it is 40.0% carbon, 6.71% hydrogen, and 53.3% oxygen. Its molecular weight is 60. What is its molecular formula?

3.55 Fructose, an important sugar, is made up of 40.0% carbon, 6.71% hydrogen, and 53.3% oxygen. Its molecular weight is 180. What is its molecular formula?

Mass Relationships in Equations

3.56 (a) Write the equation for the combustion of propylene, C_3H_6.
(b) What mass of CO_2 is produced from burning 2.45 g of C_3H_6?

3.57 (a) Write the equation for the combustion of C_4H_8O.
(b) What mass of O_2 is needed to react completely with a 5.33-g sample of C_4H_8O?

3.58 The reaction of P_4 (an elemental form of phosphorus) with Cl_2 yields PCl_5. What mass of Cl_2 is needed to react completely with 0.567 g of P_4?

3.59 What mass of NH_3 can form from reaction of 5.33 g of N_2 with excess H_2?

3.60 Aluminum metal reacts with sulfuric acid, H_2SO_4, to yield aluminum sulfate and hydrogen gas. What mass of aluminum metal is needed to produce 13.2 g of hydrogen?

3.61 Heating $CaCO_3$ yields CaO and CO_2. What mass of $CaCO_3$ is needed to produce 4.65 g of CaO?

3.62 In a reaction of 3.3 g of Al with excess HCl, 3.5 g of $AlCl_3$ is isolated (hydrogen gas also forms in this reaction). What is the percent yield of the aluminum compound?

3.63 In a reaction of HCl and NaOH, the theoretical yield of H_2O is 78.2 g. What is the theoretical yield of NaCl?

3.64 Lithium metal reacts with O_2 to form lithium oxide. What is the theoretical yield of lithium oxide if 0.45 g of lithium reacts with excess O_2?

3.65 A reaction of 43.1 g of CS_2 with excess Cl_2 yields 45.2 g of CCl_4 and 41.3 g of S_2Cl_2. What is the percentage yield of each?

Limiting Reactant

3.66 Hydrogen and nitrogen react to form ammonia, NH_3. What mass of NH_3 is produced from the reaction of 1.0 g H_2 and 14 g N_2?

3.67 What is the theoretical yield of P_4O_{10} formed from the reaction of 2.2 g of P_4 and 4.2 g of O_2? Express the yield in grams.

3.68 What mass of silver can be produced by the reaction of 3.22 g of zinc metal and 4.35 g of $AgNO_3$ according to the following equation?

$$Zn(s) + 2AgNO_3(aq) \longrightarrow 2Ag(s) + Zn(NO_3)_2(aq)$$

3.69 The reaction of 7.0 g of Cl_2 with 2.3 g of P_4 produces 7.1 g of PCl_5. What is the percent yield?

3.70 In the combustion of 33.5 g of C_3H_6, 16.1 g of H_2O is isolated. What is the percent yield?

3.71 The reaction of 55.3 g of I_2 with 2.23 g of H_2 yields 43.1 g of HI. What is the percent yield?

3.72 In the reaction of 23.1 g of NaOH with 21.2 g HNO_3, a 12.9-g sample of $NaNO_3$ is isolated.
(a) What is the percent yield?
(b) What mass of the reactant that was not limiting remains at the end of the reaction?

3.73 The reaction of salicylic acid, $C_7H_6O_3$, with acetic anhydride, $C_4H_6O_3$, forms the compound known as aspirin, $C_9H_8O_4$, and acetic acid, $C_2H_4O_2$: $C_7H_6O_3 + C_4H_6O_3 \rightarrow C_9H_8O_4 + C_2H_4O_2$.
(a) What mass of aspirin can form from a reaction containing 4.6 g of each starting material?
(b) What mass of the reactant that was not limiting remains at the end of the reaction?

Additional Exercises

3.74 Copper can be commercially obtained from an ore that is 10.0 mass percent chalcopyrite, $CuFeS_2$, as the only source of copper. How many tons of chalcopyrite are needed to produce 20.0 tons of 99.0% pure copper?

3.75 In_2S_3 can be converted into metallic indium by a two-step process. First, it is converted into In_2O_3 by reaction with oxygen, and then the metal is obtained by reaction of this oxide with carbon. Assume that the other product of the second reaction is carbon dioxide.
(a) Write the two equations for this process.
(b) What mass of indium can be produced from 35.7 kg of In_2S_3, assuming an excess of the other reactants?

3.76 Copper sulfate is generally isolated as its hydrate, $CuSO_4 \cdot x(H_2O)$. If a sample contains 25.5% Cu, 12.8% S, 57.7% O, and 4.04% H, what is the value of x?

3.77 The compound dinitrogen oxide, N_2O, is a nontoxic gas that is used as the propellant in cans of whipped cream. How many nitrogen *atoms* are in a 34.7 g sample of N_2O?

3.78 Morphine is a narcotic substance that has been used medically as a painkiller. Its use has been highly restricted because of its addictive nature. Morphine is 71.56% C, 6.71% H, 4.91% N, and 16.82% O, and its molecular mass is 285. What is the molecular formula of morphine?

3.79 Fill in the blanks in the following table.

Name	Empirical Formula	Molar Mass g/mol	Molecular Formula
dimethyl sulfoxide	C_2H_6SO	78	—
cyclopropane	—	—	C_3H_6
tryptamine	C_5H_6N	160	—
lactose	—	—	$C_{12}H_{22}O_{11}$

3.80 Heating $NaWCl_6$ at 300 °C converts it into Na_2WCl_6 and WCl_6. If the reaction of 5.64 g of $NaWCl_6$ leads to the isolation of 1.52 g of WCl_6, what is the percent yield?

3.81 The compound $K[PtCl_3(C_2H_4)]$ was first prepared by a Danish pharmacist around 1830. The structure of this complex has only recently been determined, and it is now known to be the first example of an important class of compounds known as organometallics. It can be prepared by the reaction of $K_2[PtCl_4]$ with ethylene, C_2H_4. The other product is potassium chloride.
(a) What mass of $K[PtCl_3(C_2H_4)]$ can be prepared from 45.8 g of $K_2[PtCl_4]$ and 12.5 g of ethylene?
(b) What mass of the reactant that was not limiting remains at the end of the reaction?

3.82 In many important chemical problems, conversions must be carried out by using two equations. An example is a process frequently used to analyze for copper. Copper salts dissolved in water will react with zinc metal to yield copper metal and a zinc salt:

$$CuSO_4(aq) + Zn(s) \longrightarrow Cu(s) + ZnSO_4(aq)$$

The metallic copper can be collected and weighed, while the $ZnSO_4$ stays dissolved in the water in which the reaction takes place. In order to ensure that all of the copper is converted to the metallic form, an excess of zinc must be added to the reaction. The excess zinc must be removed before the weight of the copper is determined, because it is also a solid. To remove the zinc, an excess of H_2SO_4 is added. This will react with the zinc metal but not the copper metal.

$$Zn(s) + H_2SO_4(aq) \longrightarrow ZnSO_4(aq) + H_2(g)$$

With the zinc thus removed, the copper metal can be weighed and used to find the percentage copper in the original compound. What mass of H_2 gas is produced when a solution containing 1.20 g of $CuSO_4$ is mixed with 1.20 g of zinc, followed by addition of excess H_2SO_4?

3.83 Molecular nitrogen can be converted into NO by the following reaction sequence. These reactions are the first steps in the industrially important conversion of atmospheric nitrogen into nitric acid. What mass of NO is formed from 100 g of N_2 with excess H_2 and O_2?

$$N_2 + 3H_2 \longrightarrow 2NH_3$$
$$4NH_3 + 5O_2 \longrightarrow 4NO + 6H_2O$$

Chapter Four

Stoichiometry II: Chemical Reactions in Solution and Thermochemistry

For a chemical reaction to occur, individual molecules must collide. For this reason, chemical reactions are usually carried out in solution because molecules in solution are in constant motion and readily collide. Since reaction stoichiometry and yields are based on the *number of moles* of reactants in the solution and not the volume of the solution, we need a method to calculate the number of moles of a substance in a particular volume of solution.

4.1 Solutions and Molarity

Objectives

- To explain the relationship between amount of solute, volume of solution, and concentration of solution expressed as molarity
- To describe how to prepare solutions of known molarity from weighed samples or from concentrated solutions

A common type of solution is made by dissolving a solid in a liquid, frequently water. The liquid is the **solvent,** the component that has the same

Most chemical reactions are carried out in solution.

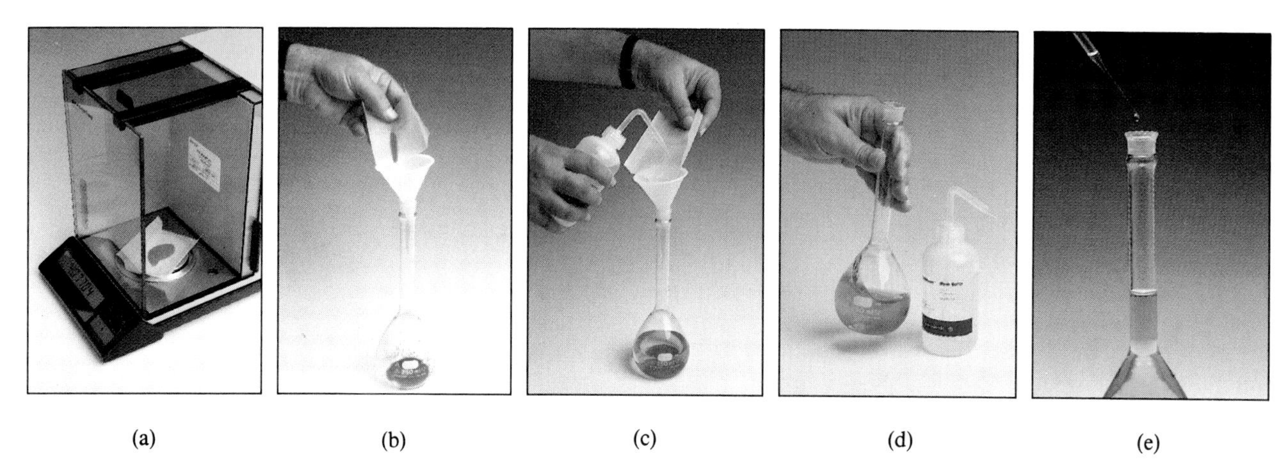
(a) (b) (c) (d) (e)

Figure 4.1 Preparation of a solution of known molarity (a) Weigh the sample of solute accurately. (b) Add the sample to the volumetric flask. (c) Wash the weighing paper with the solvent into the flask, and swirl the flask (d) to dissolve the solute. (e) Add more solvent until the level of solution is at the calibration mark on the neck of the flask.

state as the solution. The substance being dissolved is called the **solute.** The solute is often a solid, but can also be a gas or a liquid. When two liquids are mixed, the solvent is generally assumed to be the liquid present in larger quantity. In aqueous solutions, water is defined as the solvent.

The concentration of a solution expresses the quantity of solute in a given quantity of the solution. Several different units are used to express concentration. For the calculations involved in stoichiometry, the most useful unit of concentration is **molarity** (symbolized by M), the number of moles of solute in one liter of *solution.*

Molarity expresses concentration as moles of solute per liter of solution.

$$\text{molarity} = \frac{\text{moles of solute}}{\text{liters of solution}}$$

Figure 4-1 demonstrates one way to prepare a solution of known molar concentration. The solute is weighed and placed in a **volumetric flask** that has been calibrated to contain a known volume of liquid. A small amount of solvent is added and the solute is dissolved. The flask is then filled with more solvent to the calibration mark. Note that the molarity of a solution is based on the *total volume of solution* and not on the volume of added solvent.

Example 4.1 Calculating Molarity of a Solution

What is the molar concentration of a solution prepared by dissolving 35.2 g of NaCl in enough water to form 500 mL of solution?

Solution Molarity is *moles* of solute per *liter* of solution. To solve this problem, the mass of solute must be converted to moles, and the volume of solution must be expressed in liters.

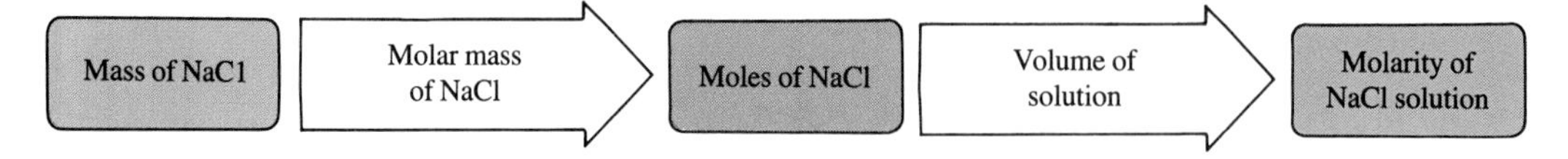

The molar mass of NaCl (58.44 g/mol) is used to calculate the number of moles of NaCl.

$$\text{moles NaCl} = 35.2 \text{ g NaCl}\left(\frac{1 \text{ mol NaCl}}{58.44 \text{ g NaCl}}\right) = 0.602 \text{ mol NaCl}$$

Second, the volume of solution (soln is used as an abbreviation for solution) is expressed in liters.

$$\text{liters NaCl soln} = 500 \text{ mL soln}\left(\frac{1 \text{ L}}{1000 \text{ mL}}\right) = 0.500 \text{ L soln}$$

The molarity is

$$\text{molarity NaCl} = \frac{0.602 \text{ mol NaCl}}{0.500 \text{ L soln}} = 1.20 \text{ mol NaCl/L soln}$$

In summation, dissolving 35.2 g of NaCl in water and diluting this mixture to exactly 500 mL yields a 1.20 *M* NaCl solution.

Understanding What is the molar concentration of a solution prepared by dissolving 32.1 g of KBr in enough water to form 3.2 L of solution?

Answer: 0.084 *M*

Molarity is a convenient concentration unit because the number of moles of solute in a known volume of solution can be easily calculated. This is important because stoichiometric calculations require knowledge of the number of moles of each reactant. Molar concentrations are used as conversion factors that *relate the volume of solution (expressed in liters) to the number of moles of solute present.* For a NaCl solution that is 1.20 *M*, one liter will contain 1.20 moles of NaCl.

$$1 \text{ L NaCl soln} \simeq 1.20 \text{ mol NaCl}$$

This equivalency can be used to convert volume of solution to number of moles or *vice versa.* Note that the concentration unit molarity allows the interconversion between volume of solution and number of moles in the same way that molar mass allows the interconversion between the mass of a sample and number of moles.

Example 4.2 Moles of Solute

How many moles of HCl are present in 2.4 L of a 0.33 *M* HCl solution?

Solution We know that for a 0.33 *M* HCl solution, 1 L HCl soln $\simeq$ 0.33 mol HCl. This equivalency is used to calculate the number of moles of HCl.

Volume of HCl solution → Molarity of HCl solution → Moles of HCl

$$\text{moles HCl} = 2.4 \text{ L HCl soln}\left(\frac{0.33 \text{ mol HCl}}{1 \text{ L HCl soln}}\right) = 0.79 \text{ mol HCl}$$

Understanding How many moles of NaCl are in 1.3 L of a 0.22 *M* NaCl solution?

Answer: 0.29 mol

Example 4.3 Mass of Solute

How many grams of potassium permanganate, $KMnO_4$, are present in 26 mL of a 0.33 *M* solution of $KMnO_4$?

Solution The volume and molarity of the solution are used to calculate the number of moles of $KMnO_4$; then the molar mass of $KMnO_4$ (158 g/mol) is used to convert the number of moles into grams.

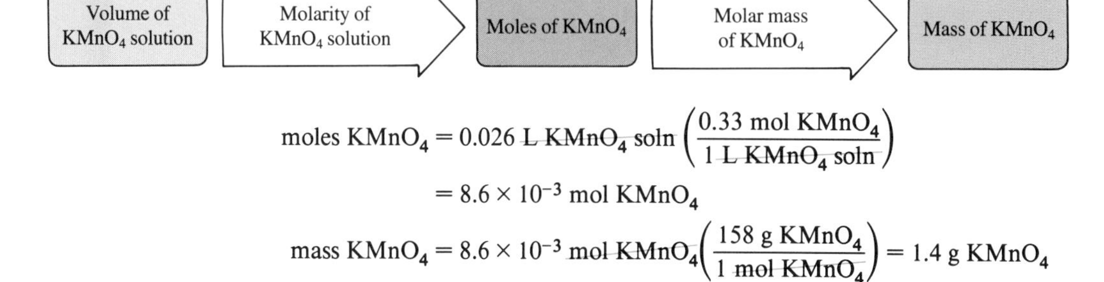

$$\text{moles } KMnO_4 = 0.026\ \cancel{\text{L } KMnO_4 \text{ soln}} \left(\frac{0.33 \text{ mol } KMnO_4}{1\ \cancel{\text{L } KMnO_4 \text{ soln}}}\right)$$
$$= 8.6 \times 10^{-3} \text{ mol } KMnO_4$$
$$\text{mass } KMnO_4 = 8.6 \times 10^{-3}\ \cancel{\text{mol } KMnO_4}\left(\frac{158 \text{ g } KMnO_4}{1\ \cancel{\text{mol } KMnO_4}}\right) = 1.4 \text{ g } KMnO_4$$

There are 1.4 g of $KMnO_4$ in 26 mL of 0.33 *M* $KMnO_4$ solution.

Understanding How many grams of calcium nitrate are present in 3.4 L of a 0.11 *M* solution of calcium nitrate?

Answer: 61 g

Dilution

Dilute solutions (ones of low concentration) can be prepared from more concentrated solutions. A dilute solution can be prepared by taking a small volume of the concentrated solution and adding pure solvent to it. This procedure is illustrated in Figure 4.2 and is similar to preparing a solution directly from a solid. The difference is that one measures a known *volume* of the concentrated solution rather than weighing a sample of the solute.

One device used to measure the volume of a solution accurately is a *pipet*. A **pipet** is a calibrated device designed to deliver an accurately known volume of liquid (Figure 4.2b). The liquid is drawn into the pipet by suction

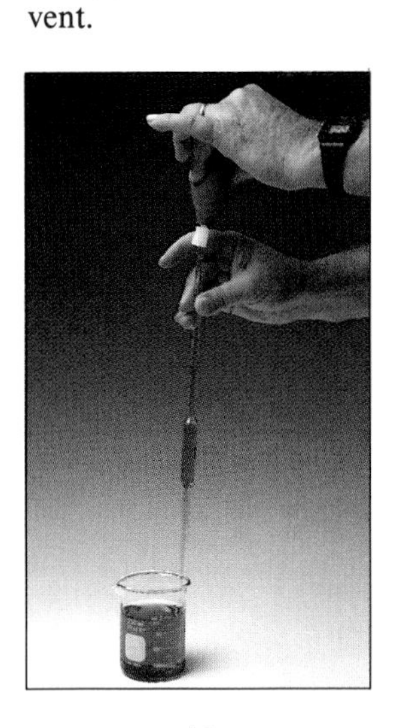

(a)

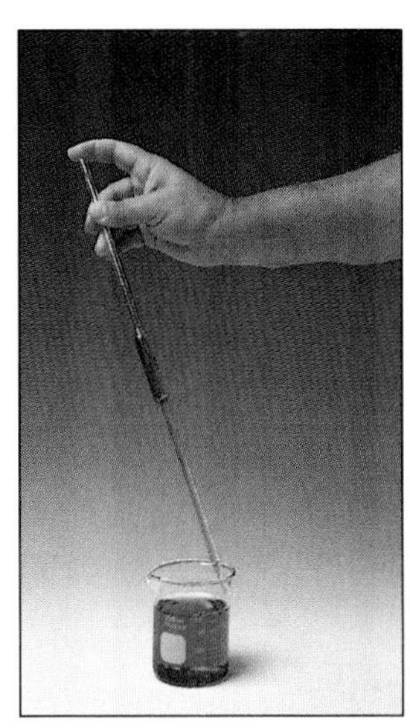

(b)

(c)

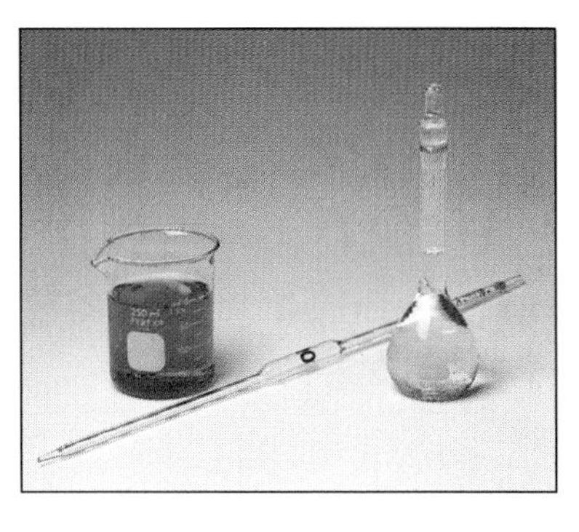

(d)

Figure 4.2 Preparation of a dilute solution from a concentrated solution (a) Draw the concentrated liquid into a pipet to a level just above the calibration mark. (b) Allow the liquid to settle down to the calibration line. Touch the tip of the pipet to the side of the container to remove any extra liquid. (c) Transfer this solution to a volumetric flask. Again, touch the pipet to the wall of the flask to ensure complete transfer, and (d) dilute to the mark with solvent.

from a rubber bulb. As with volumetric flasks, pipets of various sizes are available.

If 10 mL of a concentrated solution (often called the "stock solution") is diluted to 1.0 L with pure solvent, the concentration of the dilute solution can be determined from the volume and concentration of the stock solution and the total volume of the dilute solution. The key fact is that the number of moles of solute is not changed by dilution. The only difference between the two solutions is that more solvent is present in the dilute solution.

Solutions of known molarity can be prepared by dilution of more concentrated solutions.

$$\begin{matrix}\text{moles of solute in 10 mL} \\ \text{of concentrated solution}\end{matrix} = \begin{matrix}\text{moles of solute in 1.0 L} \\ \text{of dilute solution}\end{matrix}$$

Example 4.4 Dilution

What volume of a 3.24 *M* solution of HCl is needed to prepare 500 mL of 0.250 *M* HCl?

Solution In this problem, the molarities of both the concentrated and dilute solutions and the desired volume of the dilute solution are given. First, calculate the number of moles of HCl required in the dilute solution (dil) from its volume and molarity. The number of moles of HCl needed in the concentrated solution (conc) must be the same as the number in the dilute solution. Combine this information with the molarity of the concentrated solution to determine the required volume of the concentrated solution.

Volume of dilute HCl solution → [Molarity of dilute HCl solution] → Moles of HCl → [Molarity of concentrated HCl solution] → Volume of concentrated HCl solution

$$\text{moles HCl} = 0.500\ \text{L}\ \cancel{\text{HCl(dil) soln}}\left(\frac{0.250\ \text{mol HCl}}{1\ \cancel{\text{L HCl(dil) soln}}}\right)$$

$$= 0.125\ \text{mol HCl}$$

$$\text{volume HCl(conc)} = 0.125\ \cancel{\text{mol HCl}}\left(\frac{1\ \text{L HCl(conc) soln}}{3.24\ \cancel{\text{mol HCl}}}\right)$$

$$= 3.86 \times 10^{-2}\ \text{L HCl(conc)} = 38.6\ \text{mL HCl(conc)}$$

Thus, diluting 38.6 mL of a 3.24 molar HCl solution to a total volume of 500 mL yields a 0.250 molar solution. The answer is reasonable; the concentration dropped by a factor of a little more than ten (3.24 *M* to 0.250 *M*) while the volume increased by about the same factor (38.6 mL to 500 mL).

Understanding What volume of a 2.54 *M* solution of NaOH is needed to prepare 250 mL of 0.110 *M* NaOH?

Answer: 10.8 mL

In Example 4.4, the volume of a concentrated solution needed to prepare a dilute solution was calculated by the conversion factor method. The key to the calculation is that the numbers of moles of HCl in both the concentrated and dilute solutions are the same. The problem can also be solved by an algebraic method. Because the product of molarity (moles/volume) × volume yields moles, and the numbers of moles in the two solutions are equal, the following equation can be written:

$$\text{molarity(conc)} \times \text{volume(conc)} = \text{molarity(dil)} \times \text{volume(dil)}$$

$$M(\text{conc}) \times V(\text{conc}) = M(\text{dil}) \times V(\text{dil})$$

In using this equation to solve the problem in Example 4.4, simply solve the equation for the unknown quantity, the volume of the concentrated solution, and substitute the three values given in the problem:

$$M(\text{conc}) \times V(\text{conc}) = M(\text{dil}) \times V(\text{dil})$$

$$V(\text{conc}) = \frac{M(\text{dil}) \times V(\text{dil})}{M(\text{conc})}$$

$$= \frac{0.250\ \cancel{M\ \text{HCl}} \times 500\ \text{mL HCl}}{3.24\ \cancel{M\ \text{HCl}}}$$

$$= 38.6\ \text{mL HCl}$$

In this method of solving the problem, the units of volume of the concentrated solution are determined by the units of volume of the dilute solution, and the units of molarity cancel.

INSIGHTS into CHEMISTRY

Science, Society, and Technology

Volumetric Glassware Available in Wide Range of Accuracy and Precision

Several kinds of volumetric glassware are available for measuring liquids. Some glassware, for example the graduated cylinder, has a scale that provides only a limited precision. A graduated cylinder is easy to use and is the device of choice when an exact volume of solution is not crucial. When accuracy and precision are important, a *pipet* is used. Graduated pipets (pipets with multiple markings) are available, but are not as precise as *volumetric* pipets. A volumetric pipet is calibrated for a specific volume and is generally the most precise means of delivering accurately known volumes of solutions. For example, a 10-mL pipet has an uncertainty of 0.02 mL; that is, it will deliver between 9.98 and 10.02 mL of solution.

Pipets are labeled either "TD" or "TC." The TD pipets are calibrated "to deliver" a precise volume of solution. This type of pipet is used most often with aqueous solutions. After the solution is delivered, a small quantity of the solution remains in the tip. The manufacturer takes this remaining volume into account in the calibration. In contrast, a TC pipet is calibrated "to contain" a specific volume of solution. This type is designed to be blown out

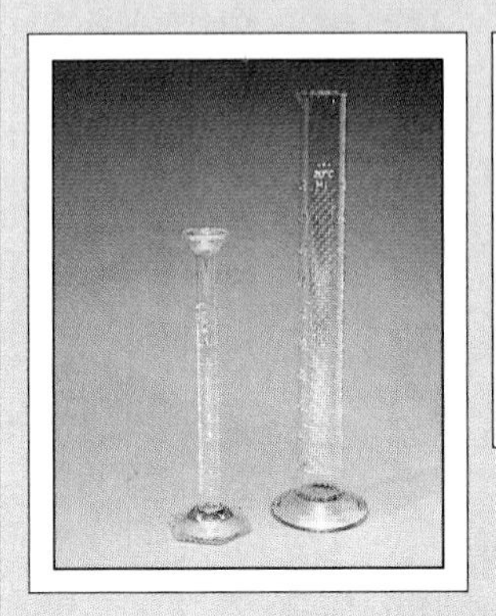

(a)

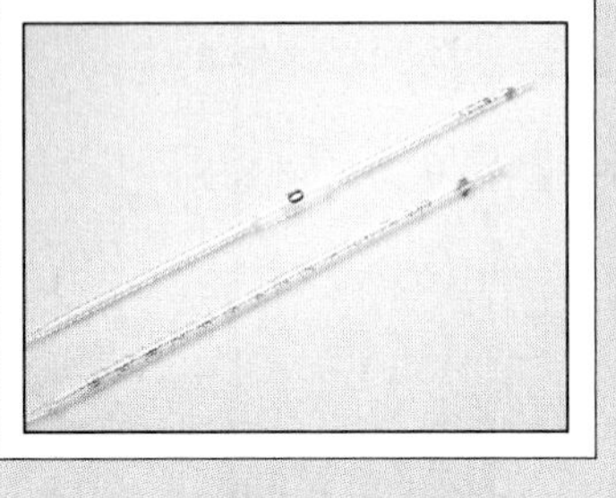

(b)

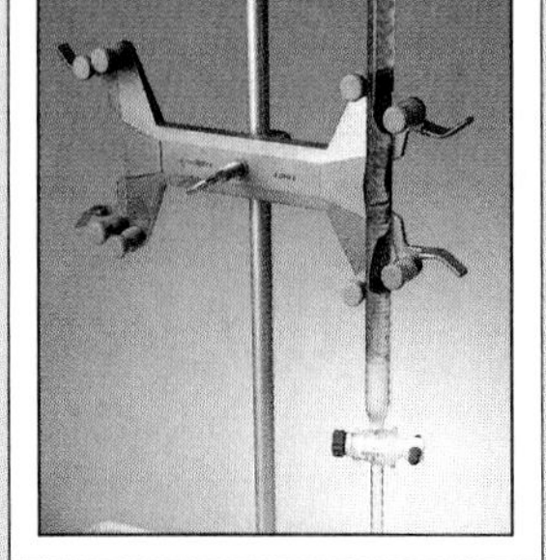

(c)

Volumetric glassware (a) Graduated cylinders. (b) Volumetric and graduated pipets. (c) Buret.

Example 4.5 Dilution

Calculate the molar concentration of a solution prepared by diluting 50 mL of 5.23 *M* NaOH to 2.0 L.

Solution First, calculate the number of moles of NaOH present from the volume and molarity of the concentrated solution. Then find the concentration of the dilute solution from the amount of NaOH and the volume of the dilute solution.

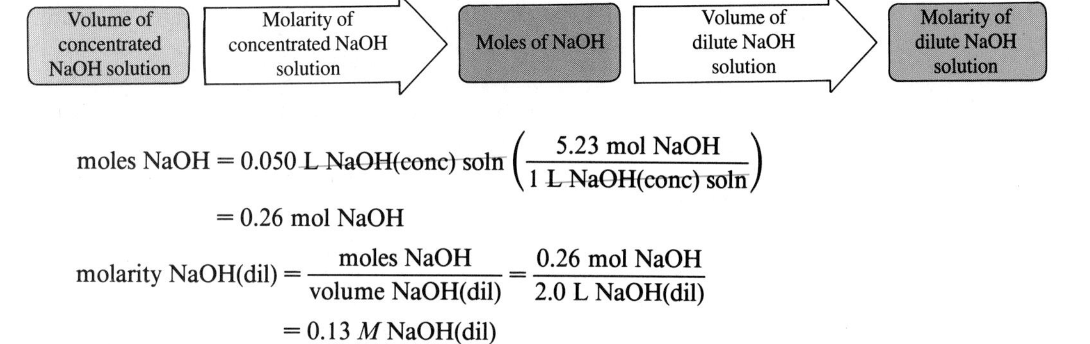

$$\text{moles NaOH} = 0.050\ \cancel{\text{L NaOH(conc) soln}} \left(\frac{5.23 \text{ mol NaOH}}{1\ \cancel{\text{L NaOH(conc) soln}}}\right)$$
$$= 0.26 \text{ mol NaOH}$$

$$\text{molarity NaOH(dil)} = \frac{\text{moles NaOH}}{\text{volume NaOH(dil)}} = \frac{0.26 \text{ mol NaOH}}{2.0 \text{ L NaOH(dil)}}$$
$$= 0.13\ M \text{ NaOH(dil)}$$

with air or rinsed with solvent; it is most often used with nonaqueous solutions.

A buret is similar to a graduated pipet, but the volume of solution delivered is controlled by a stopcock near the bottom of the cylinder. Burets are used to deliver reactants to reaction mixtures accurately.

Volumetric flasks are designed to contain a specific volume of solution. They are used to prepare a given volume of solution of known concentration starting from either a mass of a solute (usually a solid) or an accurately measured volume of a concentrated solution.

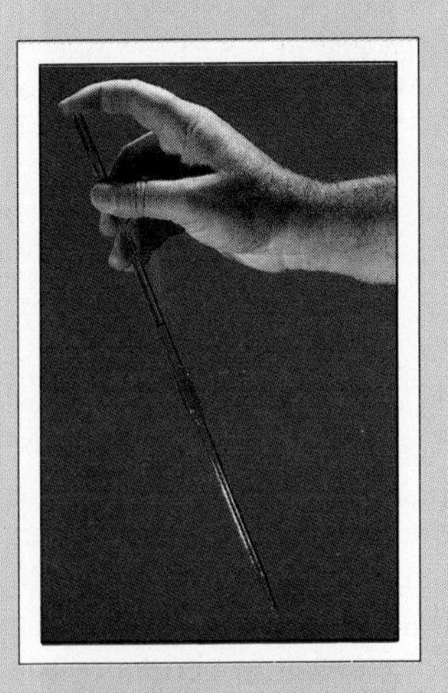

(a)

(b)

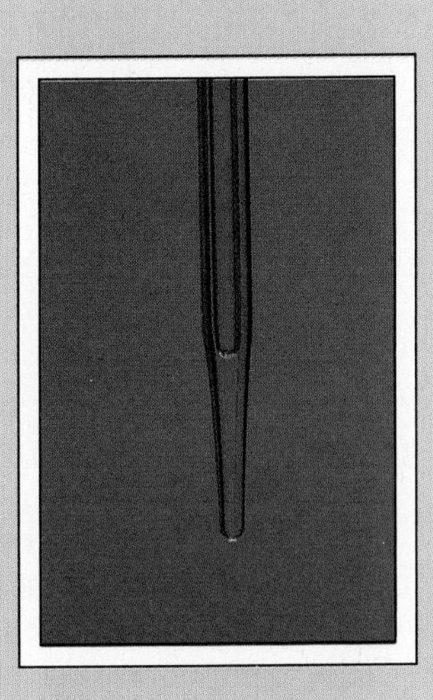

(c)

(a) A pipet is filled to the calibration mark. (b) After delivering the solution, a small amount of liquid (c) remains in the tip. The pipet is calibrated to take into account the liquid that remains in the tip.

This problem can also be solved by the algebraic method:

$$M(\text{conc}) \times V(\text{conc}) = M(\text{dil}) \times V(\text{dil})$$

$$M(\text{dil}) = \frac{M(\text{conc}) \times V(\text{conc})}{V(\text{dil})}$$

$$= \frac{5.23\ M\ \text{NaOH} \times 50\ \cancel{\text{mL NaOH}}}{2000\ \cancel{\text{mL NaOH}}}$$

$$= 0.13\ M\ \text{NaOH}$$

Understanding What is the molar concentration of a solution prepared by diluting 21 mL of 5.2 M HNO_3 to 0.50 L?

Answer: 0.22 M

4.2 Solution Stoichiometry

Objectives

- To perform stoichiometric calculations for systems that involve solutions
- To write net ionic equations from the full chemical equation
- To predict the water solubility of some common ionic substances

The number of moles of a reactant or product in a reaction can be calculated from the molarity and volume of solution.

Solutions of known concentration provide a fast and convenient way to deliver precise amounts of reactants into a reaction mixture. Stoichiometric calculations for reactions in solution are similar to those that we have already illustrated. In solution stoichiometry problems, the quantities of substances are expressed as the volumes of solutions with known concentrations, rather than in grams. The important similarity is that whether the mass or the solution volume of reactants or products is given or sought, the chemical equation is used to relate the number of moles of one substance to the number of moles of another.

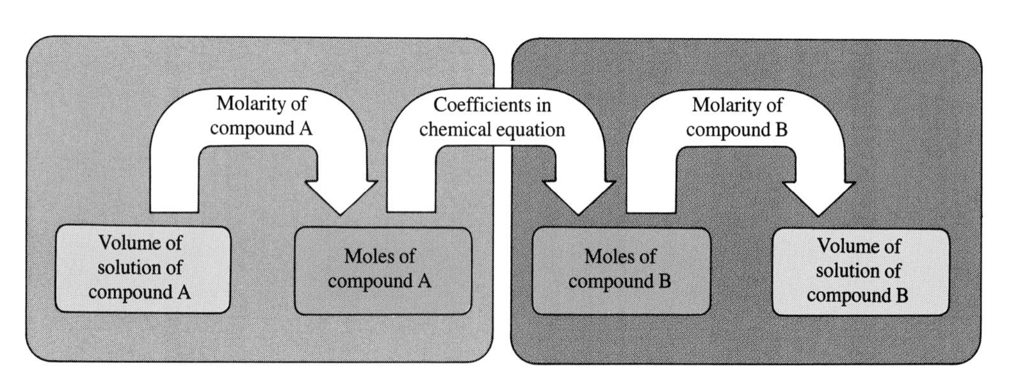

We now have two methods to convert from quantity to number of moles or *vice versa.* If the mass of a compound is given or needed, the *molar mass* of the compound generates the conversion factor. If the volume of solution of a compound is given or needed, the *molarity* of the solution generates the conversion factor. The stoichiometric relationships of the chemical equation are used to calculate the number of moles of any other substance in the equation in either case.

Example 4.6 Solution Stoichiometry

Calculate the mass, in grams, of $Al(OH)_3$ (molar mass = 78.00 g/mol) formed by the reaction of exactly 500 mL of 0.100 *M* NaOH with excess $Al(NO_3)_3$.

Solution The molarity of the solution is used to calculate the number of moles of NaOH from the volume of the NaOH solution. The coefficients of the equation are used to calculate the moles of $Al(OH)_3$. Moles of $Al(OH)_3$ are then converted to grams by using the molar mass.

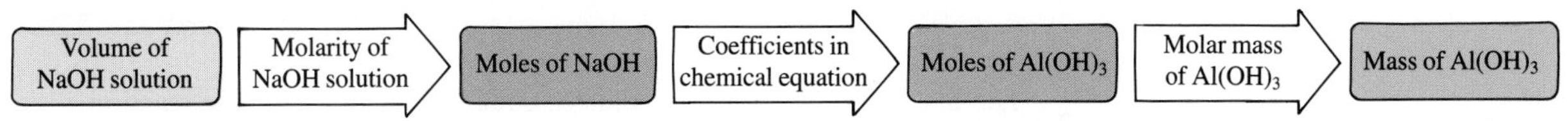

First, write the chemical equation.

$$Al(NO_3)_3(aq) + 3NaOH(aq) \longrightarrow 3NaNO_3(aq) + Al(OH)_3(s)$$

Second, calculate the number of moles of NaOH used in the reaction from the volume and concentration of the NaOH solution. NaOH is the limiting reactant because $Al(NO_3)_3$ is in excess.

$$1 \text{ L NaOH soln} \simeq 0.100 \text{ mol NaOH}$$

$$\text{moles NaOH} = 0.500 \text{ L } \cancel{\text{NaOH soln}} \left(\frac{0.100 \text{ mol NaOH}}{1 \text{ L } \cancel{\text{NaOH soln}}}\right)$$
$$= 0.0500 \text{ mol NaOH}$$

Third, use the stoichiometric equivalencies from the chemical equation to calculate the number of moles of $Al(OH)_3$ that are formed in the reaction.

$$\text{moles Al(OH)}_3 = 0.0500 \cancel{\text{ mol NaOH}} \left(\frac{1 \text{ mol Al(OH)}_3}{3 \cancel{\text{ mol NaOH}}}\right)$$
$$= 0.0167 \text{ mol Al(OH)}_3$$

Fourth, finish the problem by calculating the number of grams of $Al(OH)_3$, using the molar mass.

$$\text{mass Al(OH)}_3 = 0.0167 \cancel{\text{ mol Al(OH)}_3} \left(\frac{78.00 \text{ g Al(OH)}_3}{1 \cancel{\text{ mol Al(OH)}_3}}\right)$$
$$= 1.30 \text{ g Al(OH)}_3$$

Note that it is possible to combine the unit conversion steps in this problem into a single chain calculation.

$$\text{mass Al(OH)}_3 = 0.500 \text{ L } \cancel{\text{NaOH soln}} \left(\frac{0.100 \cancel{\text{ mol NaOH}}}{1 \text{ L } \cancel{\text{NaOH soln}}}\right) \left(\frac{1 \cancel{\text{ mol Al(OH)}_3}}{3 \cancel{\text{ mol NaOH}}}\right) \left(\frac{78.00 \text{ g Al(OH)}_3}{1 \cancel{\text{ mol Al(OH)}_3}}\right)$$
$$= 1.30 \text{ g Al(OH)}_3$$

Understanding What mass of AgCl will form in the reaction of exactly 500 mL of 1.30 *M* $CaCl_2$ with excess $AgNO_3$?

Answer: 186 g AgCl

In the previous example, the number of moles of product formed is calculated from the concentration and volume of NaOH and the coefficients in the balanced equation. The mass of product formed is calculated by using the molar mass of $Al(OH)_3$. It is often important to calculate the concentration or volume of a reactant or product from mass data, as illustrated in Example 4.7.

Example 4.7 Solution Stoichiometry

What volume of 0.20 M HNO_3 is needed to react completely with 37 g of $Ca(OH)_2$?

Solution The method used to solve this problem is similar to that in Example 4.6, except that in this case the number of moles of HNO_3 is calculated from the mass of $Ca(OH)_2$, and the molarity of HNO_3 is used to calculate the requested answer.

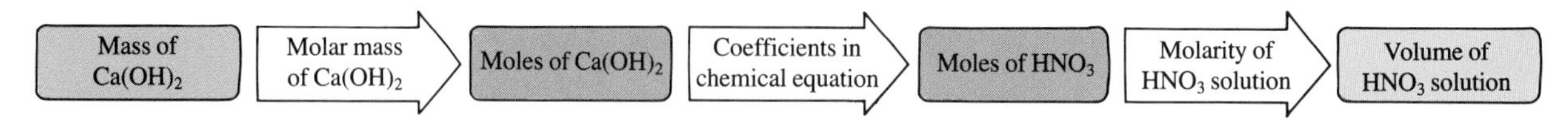

The first step is to write the balanced equation for the acid-base reaction.

$$2HNO_3(aq) + Ca(OH)_2(s) \longrightarrow 2H_2O(\ell) + Ca(NO_3)_2(aq)$$

Second, calculate the number of moles of $Ca(OH)_2$ (molar mass = 74.1 g/mol) that react.

$$\text{moles } Ca(OH)_2 = 37 \text{ g } Ca(OH)_2 \left(\frac{1 \text{ mol } Ca(OH)_2}{74.1 \text{ g } Ca(OH)_2}\right) = 0.50 \text{ mol } Ca(OH)_2$$

Third, use the coefficients of the chemical equation to calculate the equivalent number of moles of HNO_3.

$$\text{moles } HNO_3 = 0.50 \text{ mol } Ca(OH)_2 \left(\frac{2 \text{ mol } HNO_3}{1 \text{ mol } Ca(OH)_2}\right) = 1.0 \text{ mol } HNO_3$$

Fourth, finish the problem by calculating the volume of 0.20 M HNO_3 that contains 1.0 mol HNO_3.

$$\text{volume } HNO_3 \text{ soln} = 1.0 \text{ mol } HNO_3 \left(\frac{1 \text{ L } HNO_3 \text{ soln}}{0.20 \text{ mol } HNO_3}\right)$$

$$= 5.0 \text{ L } HNO_3 \text{ soln}$$

This is a relatively large volume of solution for a laboratory experiment. If we used a more concentrated HNO_3 solution, the volume of solution needed for complete reaction would be smaller.

Understanding What volume of 1.50 M HCl is needed to react completely with 4.50 g of $Mg(OH)_2$?

Answer: 103 mL

Net Ionic Equations

We have used notations such as NaCl(aq) and $Ca(NO_3)_2(aq)$ to represent ionic compounds in water solution. However, ionic compounds dissociate into ions when they dissolve in water. A more accurate representation of the solute in these solutions would be $Na^+(aq) + Cl^-(aq)$ and $Ca^{2+}(aq) + 2NO_3^-(aq)$. A chemical equation can be used to describe the process of compounds dissociating in water.

$$NaCl(s) \xrightarrow{H_2O} Na^+(aq) + Cl^-(aq)$$

$$Ca(NO_3)_2(s) \xrightarrow{H_2O} Ca^{2+}(aq) + 2NO_3^-(aq)$$

All ionic compounds that dissolve in water do so by dissociating into ions. (In thc next section we will see some rules for predicting the solubilities of ionic compounds.) In addition, a few molecular compounds such as HCl, HI, and HNO_3 ionize when dissolved in water. However, most molecular compounds are present as the neutral molecules when they dissolve in water. Sugar, $C_{12}H_{22}O_{11}$, is a common example of a compound that does not ionize when dissolved.

Let's examine the acid-base reaction that occurs when a solution of sodium hydroxide is mixed with a solution of hydrochloric acid. As in previous sections, we can write

$$HCl(aq) + NaOH(aq) \longrightarrow NaCl(aq) + H_2O(\ell)$$

This is called the **overall equation,** an equation that shows all of the reactants and products in undissociated form. However, all of the compounds except water exist in solution as ions [recall that HCl ionizes in water to produce $H^+(aq)$ and $Cl^-(aq)$], so a more accurate description of this reaction is

$$H^+(aq) + Cl^-(aq) + Na^+(aq) + OH^-(aq) \longrightarrow Na^+(aq) + Cl^-(aq) + H_2O(\ell)$$

This is the **complete ionic equation,** the equation that shows separately all species—both ions and molecules—as they are present in the solution. When we use this description for the substances in the reaction, the sodium ions and chloride ions are present in the same amounts on both sides of the equation. Since they undergo no change, they can be omitted from the equation (see Figure 4.3). The only chemical change is represented by the equation

$$H^+(aq) + OH^-(aq) \longrightarrow H_2O(\ell)$$

This equation is an example of a **net ionic equation,** an equation that shows only those species in the solution that actually undergo a chemical change. The ions that were omitted from the complete ionic equation are called **spectator ions** because they do not participate in the reaction. The net ionic

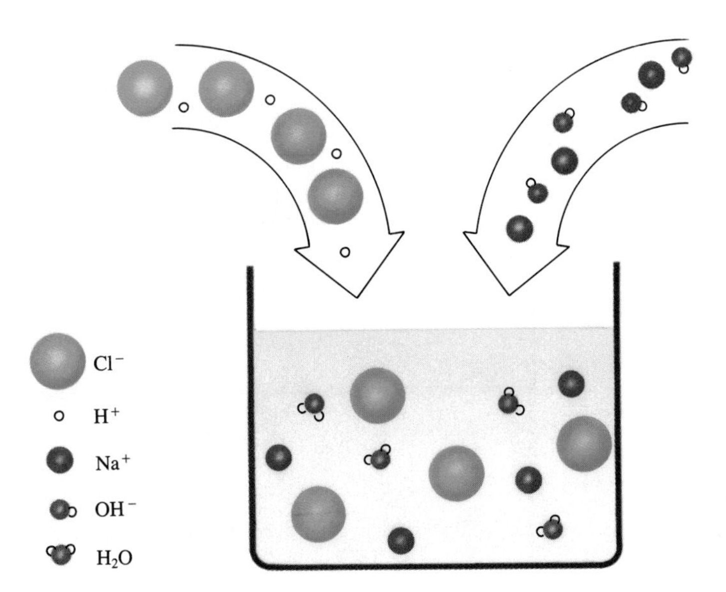

Figure 4.3 Net ionic equation The acid HCl and base NaOH in separate solutions are dissociated into ions. When mixed, the $Na^+(aq)$ and $Cl^-(aq)$ ions undergo no change, but the $H^+(aq)$ and $OH^-(aq)$ ions react to form more water.

The net ionic equation shows only species that are involved in the chemical change.

equation is a simpler description of the chemical reaction that is useful in certain situations because it helps focus on the species that undergo chemical change. Remember that even though the spectator ions do not participate in the reaction, they are present in the solution.

Example 4.8 Net Ionic Equation

Write the overall equation, the complete ionic equation, and the net ionic equation for the reaction of nitric acid (HNO_3) with KOH in water solution.

Solution The overall equation for this acid-base reaction is

$$HNO_3(aq) + KOH(aq) \longrightarrow H_2O(\ell) + KNO_3(aq)$$

Both of the reactants and the KNO_3 dissolve and separate into ions in water. The complete ionic equation is

$$H^+(aq) + NO_3^-(aq) + K^+(aq) + OH^-(aq) \longrightarrow H_2O(\ell) + NO_3^-(aq) + K^+(aq)$$

The K^+ and NO_3^- are present on both sides of the equation and do not undergo change. After we remove these spectator ions, the net ionic equation is

$$H^+(aq) + OH^-(aq) \longrightarrow H_2O(\ell)$$

Understanding Write the net ionic equation for the reaction in water of hydrobromic acid, HBr, and potassium hydroxide.

Answer: $H^+(aq) + OH^-(aq) \longrightarrow H_2O(\ell)$

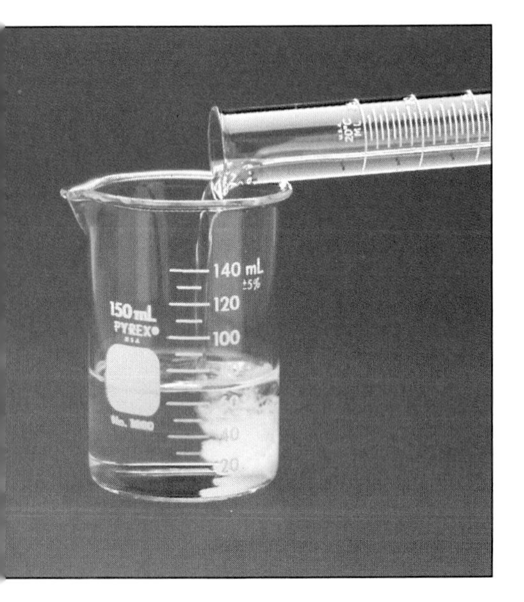

Figure 4.4 Precipitation of silver chloride Mixing solutions of sodium chloride and silver nitrate yields a precipitate of silver chloride.

Precipitation Reactions

A **precipitation reaction** involves the formation of an insoluble product or products from the reaction of soluble reactants. Precipitation reactions are conveniently described by net ionic equations. For example, mixing a solution of sodium chloride with a solution of silver nitrate produces solid silver chloride as a precipitate (Figure 4.4).

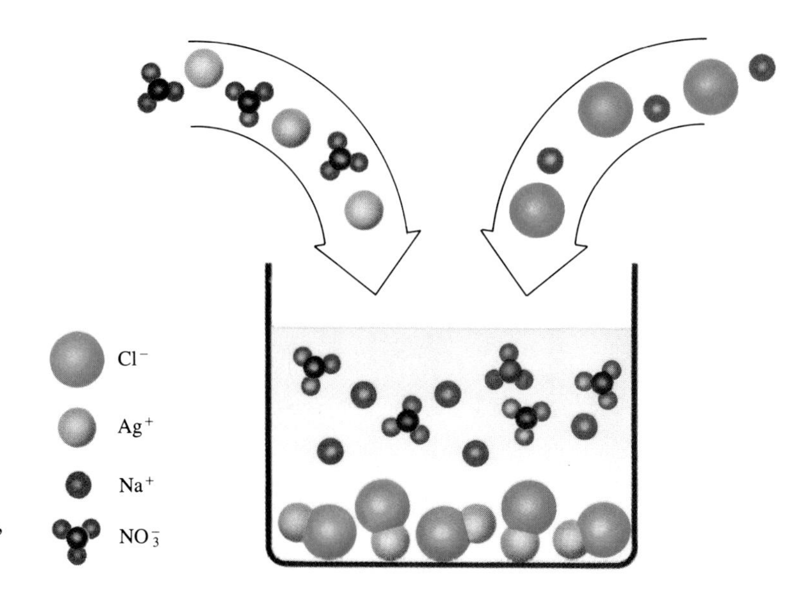

Figure 4.5 Precipitation reaction The compounds $AgNO_3$ and NaCl in separate solutions are dissociated into ions. When mixed, the $Na^+(aq)$ and $NO_3^-(aq)$ ions undergo no change, but the insoluble AgCl(s) precipitates.

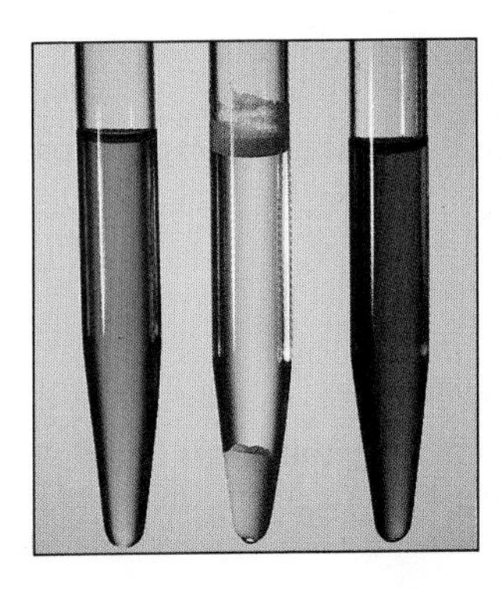

$NiCl_2$ Hg_2Cl_2 $CoCl_2$

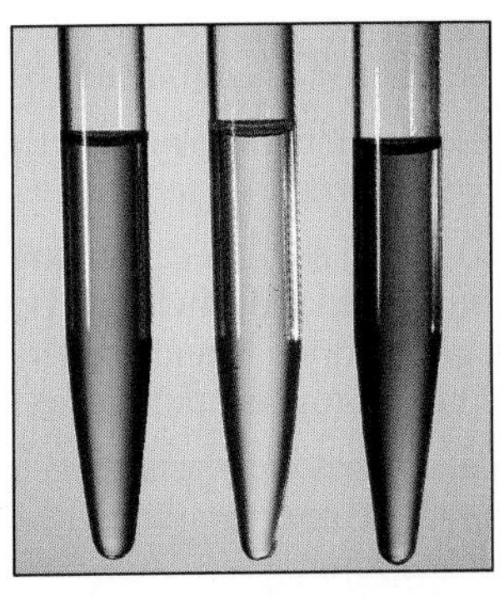

$Fe(NO_3)_3$ $NaNO_3$ $Cr(NO_3)_3$

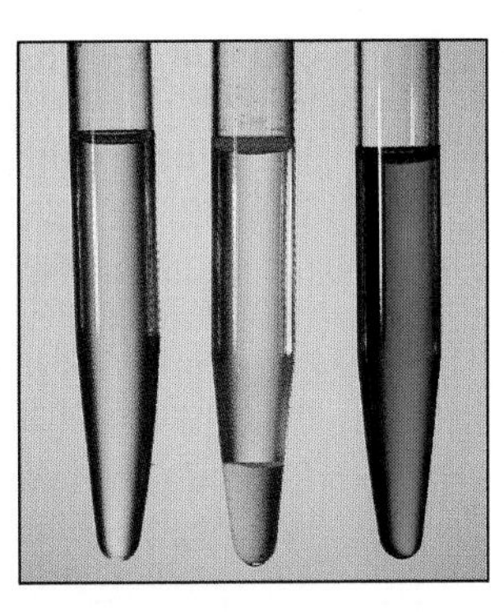

$FeSO_4$ $BaSO_4$ $CuSO_4$

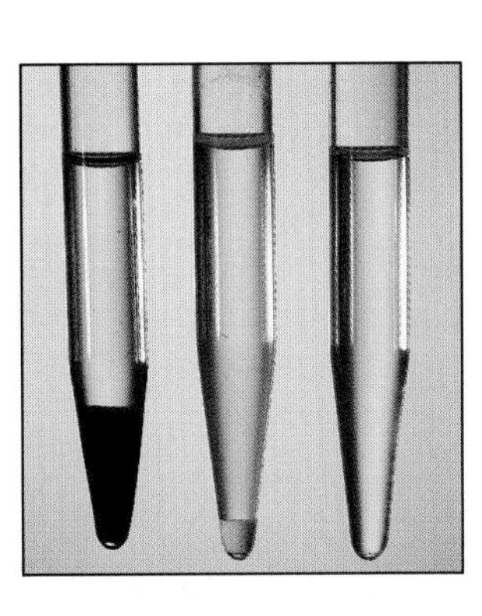

$Fe(OH)_3$ $Mg(OH)_2$ KOH

Figure 4.6 Determining solubility These tubes show the results of experiments in which ionic compounds are added to water. The results of experiments of this type determine the solubility rules in Table 4.1.

$$AgNO_3(aq) + NaCl(aq) \longrightarrow AgCl(s) + NaNO_3(aq)$$

Since all of the compounds in this reaction are soluble except for AgCl(s) and dissociate into ions in solution, the complete ionic equation is

$$Ag^+(aq) + NO_3^-(aq) + Na^+(aq) + Cl^-(aq) \longrightarrow AgCl(s) + Na^+(aq) + NO_3^-(aq)$$

The net ionic equation is (Figure 4.5)

$$Ag^+(aq) + Cl^-(aq) \longrightarrow AgCl(s)$$

Writing equations for precipitation reactions requires knowledge of which ionic compounds are soluble (will dissolve) and which are not. Unfortunately, there is no way to predict the solubility of compounds from general principles. The solubility of a compound must always be determined experimentally (Figure 4.6). From the results of such experiments, we can state some rules of thumb that are generally useful (although they have exceptions). Some simple solubility rules are given in Table 4.1.

The solubility of a compound is determined by experiment.

Table 4.1 Solubility Rules for Ionic Compounds in Water*

1. The common salts of the alkali metals (group IA (1)) and the ammonium ion (NH_4^+) are soluble.
2. Salts of nitrate (NO_3^-), chlorate (ClO_3^-), perchlorate (ClO_4^-), and acetate (CH_3COO^-) anions are soluble.
3. All chlorides, bromides, and iodides are soluble, except those of Ag^+, Pb^{2+}, and Hg_2^{2+} (the form in which Hg(I) occurs in water).
4. Sulfates (SO_4^{2-}) are soluble except those of Ba^{2+}, Pb^{2+}, Hg^{2+}, and Hg_2^{2+}.
5. Most metal hydroxides are insoluble. The exceptions are the hydroxides of the alkali metals and the heavier alkaline earth metals (Ca, Sr, Ba).
6. All carbonates (CO_3^{2-}) and phosphates (PO_4^{3-}) are insoluble, except those of the group IA metal and NH_4^+ ions.

* A concentration of 0.01 *M* or greater is called soluble.

Example 4.9 Equations for Precipitation Reactions

Using the solubility rules given in Table 4.1 as a guide, write the net ionic equations that describe any change that occurs upon mixing of

(a) solutions of $Pb(NO_3)_2$ and sodium carbonate.
(b) solutions of ammonium chloride and $AgClO_4$.
(c) solutions of potassium hydroxide and copper(II) chloride.
(d) solutions of ammonium bromide and nickel(II) sulfate.

Solution The rules indicate that the reactants in each part are soluble (it was given that they are solutions). All of the compounds are ionic; we must determine whether any insoluble products will form.

(a) Write the ions that are present from the reactants.

$$Pb^{2+}(aq) + 2NO_3^-(aq) + 2Na^+(aq) + CO_3^{2-}(aq)$$

The two possible products are sodium nitrate and lead carbonate. Sodium nitrate (rule 1) is soluble (which means that Na^+ and NO_3^- are spectator ions), but rule 6 in Table 4.1 indicates that lead carbonate will precipitate. The net ionic equation is

$$Pb^{2+}(aq) + CO_3^{2-}(aq) \longrightarrow PbCO_3(s)$$

(b) Write the ions that are present from the reactants.

$$NH_4^+(aq) + Cl^-(aq) + Ag^+(aq) + ClO_4^-(aq)$$

One of the two possible products, ammonium perchlorate, is soluble (so NH_4^+ and ClO_4^- are spectator ions), while silver chloride is not.

$$Ag^+(aq) + Cl^-(aq) \longrightarrow AgCl(s)$$

(c) One of the two possible products, KCl, is soluble, but the other, $Cu(OH)_2$, is insoluble (rule 5).

$$Cu^{2+}(aq) + 2OH^-(aq) \longrightarrow Cu(OH)_2(s)$$

(d) Ammonium sulfate (rules 1 and 4) and nickel bromide (rule 3) are both soluble, so no net reaction will occur.

Understanding Write the net ionic equation that describes the reaction that occurs when solutions of barium nitrate and sodium sulfate are mixed.

Answer: $Ba^{2+}(aq) + SO_4^{2-}(aq) \longrightarrow BaSO_4(s)$

The limited solubility of certain compounds is an important property that is exploited in many chemical applications. Example 4.10 illustrates a step in the recovery of excess silver ion from a used solution of photographic film developer.

Example 4.10 Solution Stoichiometry

Most photographic films, both color and black and white, contain silver compounds. Some of the silver is left in the film as part of the image, but much of it dissolves in the developing of the film. The silver is valuable and is generally recovered (Fig. 4.7). The concentration of the silver cations in the solution can be determined by adding a

Figure 4.7 Silver recovery unit This equipment can recover the silver from 480 gallons of liquid per day.

soluble chloride salt to precipitate them as silver chloride. The following calculation is based on data from this type of analysis.

A NaCl solution is added to 4.00 L of a solution of unknown silver ion concentration, producing 25.1 g of AgCl. Enough NaCl was added to ensure that all of the $Ag^+(aq)$ ions were precipitated. What was the molar concentration of Ag^+ in the photographic solution?

Solution We can calculate the number of moles of $Ag^+(aq)$ present in the solution from the number of moles of AgCl collected, by using the net ionic equation for the reaction.

$$Ag^+(aq) + Cl^-(aq) \longrightarrow AgCl(s)$$

Knowing the number of moles of $Ag^+(aq)$ and the volume of solution, we then calculate the molarity of the original solution. Note that the solution will contain spectator ions, but these ions do not affect the calculation because they are not measured in this problem.

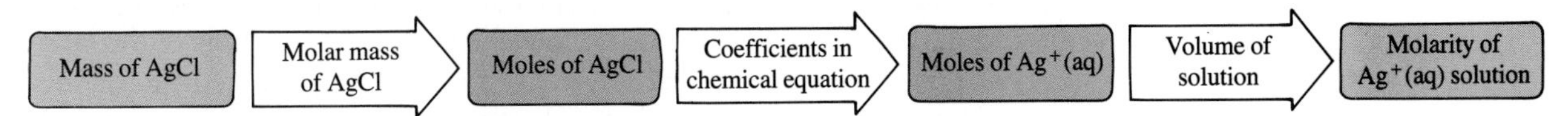

The number of moles of AgCl formed in the reaction is

$$\text{moles AgCl} = 25.1\ \cancel{\text{g AgCl}}\left(\frac{1\ \text{mol AgCl}}{143.3\ \cancel{\text{g AgCl}}}\right) = 0.175\ \text{mol AgCl}$$

The subscript 1 for silver in the formula of AgCl is used to convert moles of AgCl into moles of $Ag^+(aq)$.

$$\text{moles } Ag^+(aq) = 0.175\ \cancel{\text{mol AgCl}}\left(\frac{1\ \text{mol } Ag^+(aq)}{1\ \cancel{\text{mol AgCl}}}\right) = 0.175\ \text{mol } Ag^+(aq)$$

Since the original volume of the solution was 4.00 L, the molarity of $Ag^+(aq)$ is

$$\text{molar concentration } Ag^+(aq) = \frac{0.175\ \text{mol } Ag^+(aq)}{4.00\ \text{L soln}} = 0.0438\ M\ Ag^+(aq)$$

The concentration of the $Ag^+(aq)$ in the original solution was 0.0438 M.

Understanding What mass of solid can be isolated from the solution formed by mixing 100 mL of 0.22 *M* lead(II) nitrate with an excess of sodium chloride? What is the solid?

Answer: 6.1 g of $PbCl_2$

4.3 Chemical Analysis by Titration and Precipitation

Objectives

- To perform calculations using data from acid-base titrations
- To use data from gravimetric analyses to calculate concentration

Dying forest Acid rain is believed to contribute to the damage of many massive red spruce trees at Panther Gorge in the Adirondacks. Other trees in the forest do not appear to be damaged.

The identification of chemical species *(qualitative analysis)* and the determination of amounts or concentrations *(quantitative analysis)* are important not only in chemistry, but in fields such as medicine, agriculture, and law. Chemical analyses influence many economic and political decisions as well. Newspapers are filled with stories about the impact of chemicals on our lives. At the Olympic games and other sporting events, the news media report about performance-enhancing drugs such as steroids. Communities worry about the quality of their water and whether it has been contaminated by chemical waste. There is considerable debate about acid rain and its effects on lakes and forests.

Chemists are often at the center of these controversies because they are the people who perform the analyses necessary for rational action on many important issues. Many techniques have been designed to analyze substances, and new methods are being developed daily. Two of the most widely used methods are outlined in this section.

Acid-Base Titrations

To determine whether a lake has been damaged by an acid spill, the water must be analyzed. An obvious way to perform this analysis is to measure the amount of base needed to neutralize the acid present in a sample of the lake water. This analysis can be carried out by a **titration,** a procedure for the determination of the quantity of one substance by the addition of a measured amount of a second substance. The point at which the stoichiometrically equivalent amounts of the two reactants have been added is called the **equivalence point.** The reaction stoichiometry is used to determine the amount of acid in the sample from the measured amount of base added to reach the equivalence point. A very common way to detect the equivalence point in acid-base titrations is to add an **indicator,** a compound that changes color when an acidic solution becomes basic or *vice versa.* The point at which the indicator changes color is called the *end point* of the titration. For the analysis to be accurate, the analyst must select an indicator that changes color very close to the equivalence point.

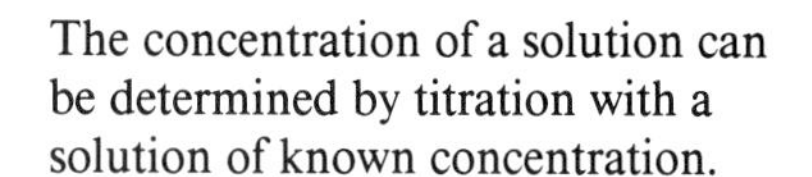

The concentration of a solution can be determined by titration with a solution of known concentration.

The analysis of the lake water involves several steps. A measured volume of the lake water is placed in a flask, and a few drops of indicator solution are added. A solution of base that has an accurately known concentration (called

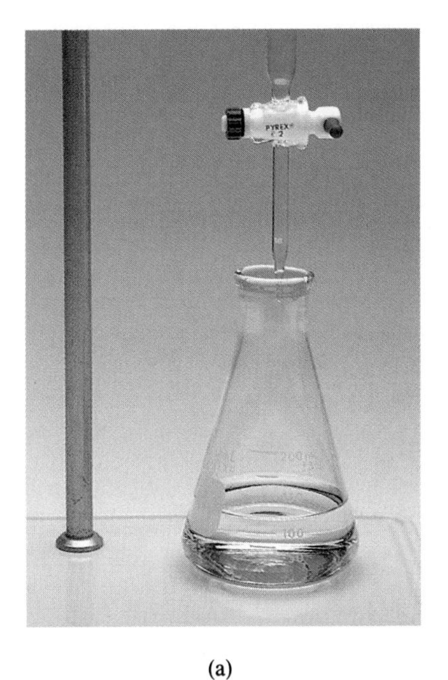

(a)

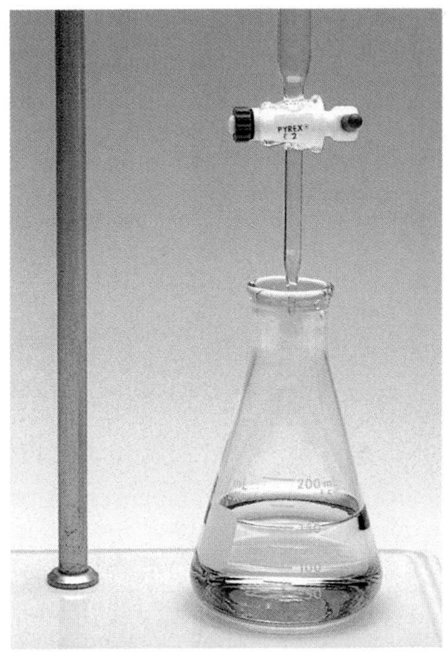

(b)

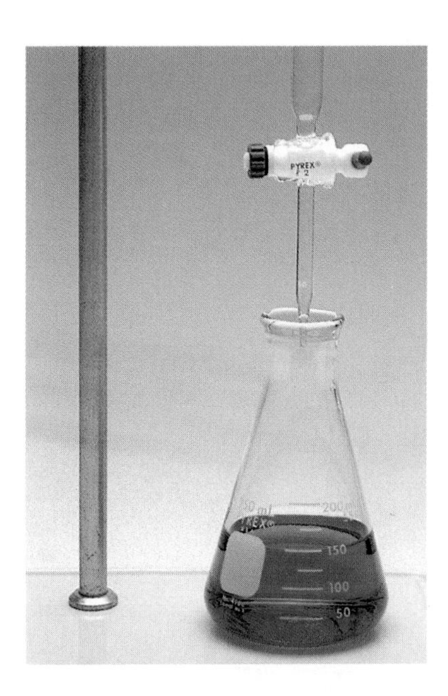

(c)

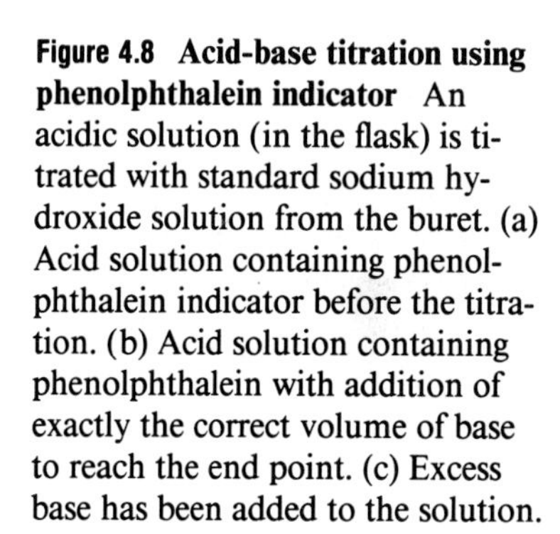

Figure 4.8 Acid-base titration using phenolphthalein indicator An acidic solution (in the flask) is titrated with standard sodium hydroxide solution from the buret. (a) Acid solution containing phenolphthalein indicator before the titration. (b) Acid solution containing phenolphthalein with addition of exactly the correct volume of base to reach the end point. (c) Excess base has been added to the solution.

a **standard solution**) is added from a buret. When the indicator changes color, the end point has been reached (Fig. 4.8), addition of the base is stopped, and the volume of solution delivered is read from the buret. The concentration of the acid in the sample is calculated from the data. We refer to chemical analysis in which the volume of a solution or substance is measured in the determination as **volumetric analysis.**

Example 4.11 Acid-Base Titration

A tank-car carrying concentrated acid overturns and spills into a small pond. An analysis of the pond must be performed before action is taken. A 2.00-L sample of water from the pond is titrated with a solution that is 0.0100 *M* in $OH^-(aq)$. What is the molar acid concentration in the lake water if 23.1 mL of the base solution is needed to reach the equivalence point?

Solution As in many stoichiometry problems, (1) a balanced equation is needed, (2) the information in the problem is used to determine the number of moles of one substance, (3) the coefficients in the equation are used to convert from moles of one substance into moles of another, and (4) the number of moles of the second substance calculated in step 3 is used to solve the problem.

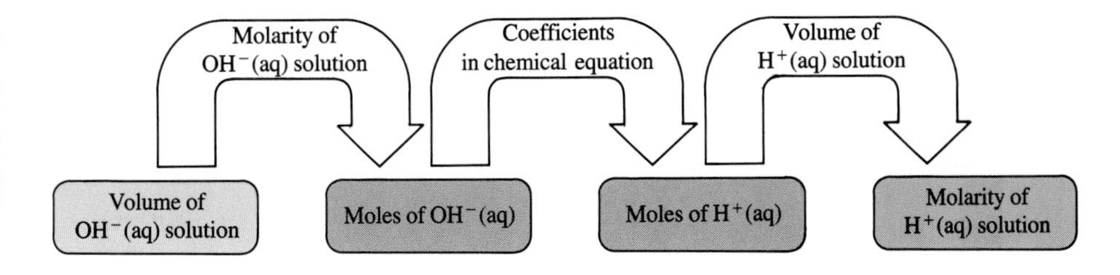

First, write the chemical equation. The net ionic equation for the reaction is

$$H^+(aq) + OH^-(aq) \longrightarrow H_2O(\ell)$$

Second, the number of moles of base added is determined from the volume (in liters) and concentration of the base solution.

$$\text{moles } OH^-(aq) = 0.0231 \text{ L } OH^-(aq) \text{ soln} \left(\frac{0.0100 \text{ mol } OH^-(aq)}{1 \text{ L } OH^-(aq) \text{ soln}}\right)$$

$$= 2.31 \times 10^{-4} \text{ mol } OH^-(aq)$$

Third, the number of moles of $H^+(aq)$ is the same as the number of moles of $OH^-(aq)$ because both ions have the same coefficient, 1, in the equation.

$$\text{moles } H^+(aq) = 2.31 \times 10^{-4} \text{ mol } OH^-(aq) \left(\frac{1 \text{ mol } H^+(aq)}{1 \text{ mol } OH^-(aq)}\right)$$

$$= 2.31 \times 10^{-4} \text{ mol } H^+(aq)$$

Fourth, this calculated amount of acid and the known volume of the sample are used to calculate the concentration of acid in the lake water.

$$\text{molarity } H^+(aq) = \frac{2.31 \times 10^{-4} \text{ mol } H^+(aq)}{2.00 \text{ L lake water}}$$

$$= 1.16 \times 10^{-4} \text{ molar } H^+(aq)$$

Understanding What is the molarity of NaOH in a solution if a 300-mL sample is neutralized by 55 mL of 1.33 *M* HCl solution?

Answer: 0.24 *M* NaOH

It is not always possible to prepare standard solutions by weighing a solid or by the dilution of a more concentrated standard solution as outlined in Section 4.1. Instead, titrations are frequently used to standardize solutions of unknown concentration. For example, hydrochloric acid solutions are frequently prepared by bubbling hydrogen chloride gas into a solvent. It is difficult to know accurately the mass of HCl added to the solvent. The concentration of the hydrochloric acid solution can be determined by titration of a measured mass of Na_2CO_3 with the acid (Example 4.12).

Example 4.12 Standardization of a Solution of HCl

What is the molar concentration of a solution of HCl if 25.9 mL is required to neutralize 0.210 g of pure Na_2CO_3? The equation for the reaction is

$$2HCl(aq) + Na_2CO_3(aq) \longrightarrow 2H_2O(\ell) + 2NaCl(aq) + CO_2(g)$$

Solution The strategy for this problem is

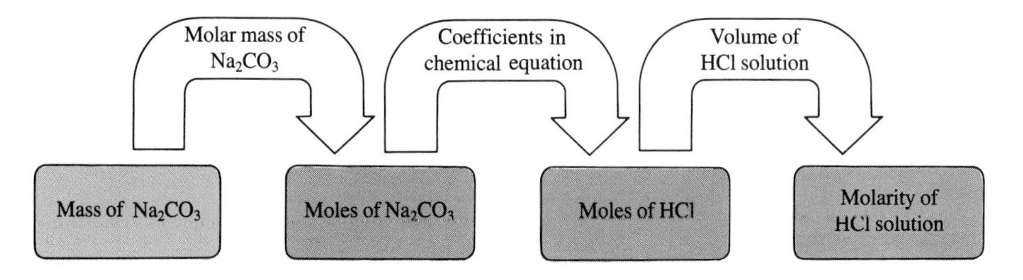

The equation is given. The second step is to determine the number of moles of Na_2CO_3 from the mass.

$$\text{moles } Na_2CO_3 = 0.210 \text{ g } Na_2CO_3 \left(\frac{1 \text{ mol } Na_2CO_3}{106.0 \text{ g } Na_2CO_3}\right)$$

$$= 1.98 \times 10^{-3} \text{ mol } Na_2CO_3$$

Third, the coefficients of the equation are used to calculate the number of moles of HCl.

$$\text{moles HCl} = 1.98 \times 10^{-3} \text{ mol } Na_2CO_3 \left(\frac{2 \text{ mol HCl}}{1 \text{ mol } Na_2CO_3}\right)$$

$$= 3.96 \times 10^{-3} \text{ mol HCl}$$

Now the concentration of the HCl solution may be computed from the volume of HCl solution (converted to liters) used in the titration.

$$\text{molarity HCl} = \frac{3.96 \times 10^{-3} \text{ mol HCl}}{0.0259 \text{ L HCl soln}}$$

$$= 0.153 \, M \text{ HCl}$$

Understanding What is the molarity of an HCl solution if 50.0 mL is neutralized by 10.0 mL of 0.23 *M* sodium hydroxide?

Answer: 0.046 *M*

Gravimetric Analysis

Precipitation reactions are also used for chemical analyses. If one component of a solution can be precipitated selectively, it can then be separated from solution, dried, and weighed. This procedure, analysis by mass, is known as **gravimetric analysis.** One of the most widely used gravimetric procedures is the determination of halides by adding silver nitrate to precipitate the silver halides. Another important gravimetric analysis is the determination of sulfate ion, SO_4^{2-}, by adding $BaCl_2$ to form insoluble $BaSO_4$ (Figure 4.9).

(a)

(b)

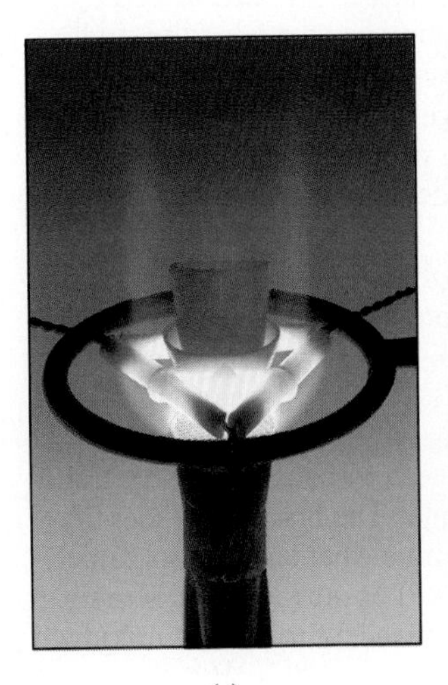

(c)

(d)

Figure 4.9 Gravimetric analysis (a) $BaSO_4$ precipitates upon addition of excess $BaCl_2$ to the solution containing SO_4^{2-}. (b) The solution is filtered to collect the $BaSO_4$. (c) The solid $BaSO_4$ is heated to remove all volatile impurities, cooled, and weighed (d) to determine the mass of $BaSO_4$.

INSIGHTS into CHEMISTRY

Chemical Reactivity

Mysterious Liquid in Vials from Civil War Easily Identified by Precipitation Reactions

Many students get an impression that professional chemists always use large, expensive instruments to perform their work. That's an incorrect impression—many chemical problems are addressed by the methods and techniques used as examples in this book.

Recently, historians found several small glass vials at the site of a Civil War explosion near Sumter, South Carolina. The vials were about 2 inches (50 mm) long and about $\frac{1}{2}$ inch (13 mm) in diameter, and contained a clear liquid. What was it?

Records from the Civil War indicated that Sumter was a center for Confederate stores and munitions. As the Union forces commanded by General William T. Sherman swept through the state in early 1865, the Confederate army burned the railroad bridges around Sumter, leaving a fully loaded munitions train hidden in the swamp. The Confederates were unable to spare any able-bodied men to guard the train, and Sherman heard rumors about the unguarded train. He dispatched Brigadier General Edward E. Potter to destroy the train and munitions. Potter set out on April 5, 1865, with about 2700 men supported by cavalry and artillery. The Confederacy was able to round up a force of 575 men from among the very young, the old, and the convalescent. A battle was fought on April 9. When the sole Confederate artillery piece was silenced, opposition was useless, and the road to Sumter was open. Ironically, April 9 was the day that Lee surrendered to Grant at the Appomattox courthouse. Potter's troops found the munitions train three miles into the swamp and blasted 3 locomotives and 35 cars loaded with military supplies.

The historians who found the vials near the old train tracks brought them to a chemist to find out what they contained. The vials were found packed in cotton, indicating importance (or explosiveness). The historians speculated that the contents might be chloroform (an anesthetic), an extract of opium, or nitroglycerine. Perhaps the vials might be levelling devices used to help aim a cannon.

The possibility of nitroglycerine, a widely used 19th-century explosive, was remote. Nitroglycerine is every bit as hazardous as people think—small vibrations can cause it to explode. Still, extreme caution was used when one of the vials was opened.

The contents, a water-clear, viscous fluid, were transferred to a screw-top container. The first step in identifying the sample was to determine whether it was organic (carbon-containing, like an oil) or inorganic (mineral in nature, ionic). The chemist mixed a drop of the unknown substance with water. Organic compounds do not generally mix with water, but many inorganic compounds do. The sample dissolved in water. The chemist noticed that adding a drop of the sample to some water in a test tube caused the solution to get very hot, so hot that it became uncomfortable to hold.

The acidity was checked next; the solution was extremely acidic, indicating that the unknown was a concentrated acid. The chemist next added barium chloride. A white precipitate formed. The only common anions that form precipitates with the barium cation are sulfate and phosphate. Next, some ammonium molybdate was added, a compound that would form a precipitate if phosphate was present. The lack of a precipitate indicated that the sample contained no phosphates. At this point it was clear that the sample was sulfuric acid, H_2SO_4, the strong acid that contains the sulfate anion. As additional confirmation, it is well known that sulfuric acid gets very hot when added to water.

A few relatively simple tests were all that was needed to determine the identity of the sample. The chemist had many years of experience, and this body of knowledge was important in making the determination quickly. However, the presence and absence of a precipitate, and not some complex and expensive test, was central to the procedure.

It is not as easy to determine the intended use of the vials of sulfuric acid as the chemical composition of the contents. Civil War historians note that nitroglycerine was discovered about a decade before the war, and it was often manufactured on the spot because it could not be transported. Perhaps the synthesis of nitroglycerine (or a different explosive) was the intended use of the sulfuric acid.

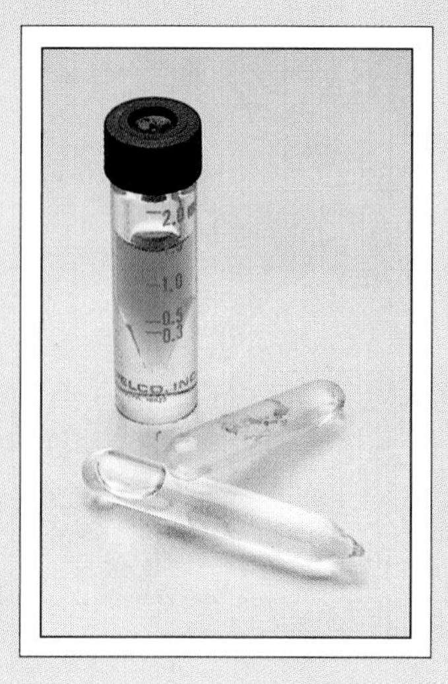

Vials found near Sumter, SC. One vial was opened and the contents transferred to the capped container for storage during the chemical analysis. The liquid in the capped container appears dark from contact with the cap.

Example 4.13 Analysis for Sulfate

Plaster of Paris, a white powder used to make casts that support broken bones, consists of $CaSO_4$ and a small amount of water. What is the percentage of $CaSO_4$ (molar mass = 136.1 g/mol) in a 0.511-g sample if reaction with an excess $BaCl_2$ solution yields 0.821 g of $BaSO_4$ (molar mass = 233.4 g/mol)?

Solution In this problem, we need to determine the mass of $CaSO_4$ in the 0.511-g sample from the mass of $BaSO_4$.

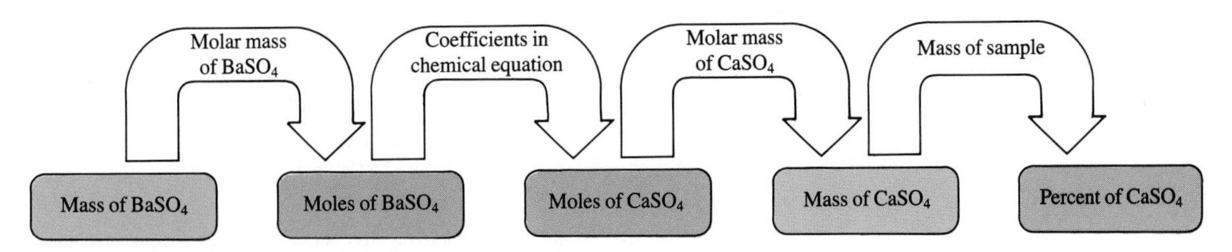

The chemical equation is

$$CaSO_4(aq) + BaCl_2(aq) \longrightarrow BaSO_4(s) + CaCl_2(aq)$$

Next, determine the number of moles of $BaSO_4$ formed in the precipitation reaction.

$$\text{moles } BaSO_4 = 0.821 \text{ g } BaSO_4 \left(\frac{1 \text{ mol } BaSO_4}{233.4 \text{ g } BaSO_4}\right)$$

$$= 0.00352 \text{ mol } BaSO_4$$

From the chemical equation, the number of moles of $CaSO_4$ is the same as the number of moles of $BaSO_4$.

$$\text{moles } CaSO_4 = 0.00352 \text{ mol } BaSO_4 \left(\frac{1 \text{ mol } CaSO_4}{1 \text{ mol } BaSO_4}\right)$$

$$= 0.00352 \text{ mol } CaSO_4$$

Next, calculate the mass of $CaSO_4$.

$$\text{mass } CaSO_4 = 0.00352 \text{ mol } CaSO_4 \left(\frac{136.1 \text{ g } CaSO_4}{1 \text{ mol } CaSO_4}\right) = 0.479 \text{ g } CaSO_4$$

The percentage $CaSO_4$ in the sample is calculated from this mass of $CaSO_4$ and the total mass of the sample.

$$\text{percentage } CaSO_4 = \frac{0.479 \text{ g } CaSO_4}{0.511 \text{ g sample}} \times 100\% = 93.7\%$$

Understanding What is the percentage of barium in an ionic compound of unknown composition if a 2.3-g sample of the compound dissolved in water produces 2.2 g of barium sulfate upon addition of an excess of sodium sulfate?

Answer: 56%

4.4 Enthalpy Changes in Chemical Reactions

Objectives

- To describe how heat flow in chemical reactions can be expressed as the change in enthalpy, ΔH

- To calculate the enthalpy change in a chemical reaction from the amount of reactants and the thermochemical equation

Up to this point, we have used chemical equations to calculate the quantities of substances consumed or produced. We have not yet considered another important part of chemical equations, the change in energy. Nearly all chemical reactions are accompanied by absorption or release of energy. The burning of wood or natural gas is a familiar example of a chemical reaction that gives off energy in the form of heat. Our everyday experience tells us that the quantity of heat produced by a fire depends on the amount and type of fuel that burns. A complete chemical equation includes a quantitative measure of this heat flow. This section presents the ideas needed to calculate the energy changes associated with chemical reactions. The energy changes in chemical reactions are quantitatively related to the amounts of reactants and products.

There are important practical reasons for considering these energy changes. A majority of the energy consumed in the United States is produced by chemical reactions. The most important of these reactions is the combustion of fossil fuels such as petroleum, natural gas, and coal. The study of the heat flow that accompanies a chemical reaction is called **thermochemistry.**

Some Basic Definitions

First, we must define some terms that are used in a special way in thermochemistry. The **system** is the matter of interest. Generally the system is composed of the substances of the reaction. The **surroundings** is all other matter, including the reaction container, the laboratory bench, and the person observing the reaction (Figure 4.10). In considering the heat flow in chemical reactions, we use the fact that energy is conserved. The **law of conservation of energy** states that the total energy of the universe (the system plus the surroundings) is constant during a chemical change.

Figure 4.10 The system The system is the matter of interest.

When a chemical reaction takes place, energy will either be transferred to or absorbed from the surroundings. That energy may take many forms, including light and sound, but in most reactions the energy is transferred as heat. Usually heat flow is measured by an increase or decrease in the temperature. A reaction is called **exothermic** if the system gives off heat to the surroundings. The combustion of natural gas to produce carbon dioxide and water is an example of an exothermic reaction. When an exothermic reaction occurs, heat must be removed from the reaction system to keep it at constant temperature. Therefore, heat can be written in the chemical equation as a product:

$$CH_4(g) + 2O_2(g) \longrightarrow CO_2(g) + 2H_2O(\ell) + \text{heat}$$

A reaction in which the system must absorb heat from the surroundings to keep the temperature constant is called **endothermic.** The formation of nitric oxide (NO) from the elements is an example of an endothermic reaction. In an endothermic reaction, heat must be added to the reaction system to keep it at constant temperature, so heat is written as a reactant:

$$N_2(g) + O_2(g) + \text{heat} \longrightarrow 2NO(g)$$

The most familiar measure of heat to most people is the calorie. A *calorie* (cal) was originally defined as the amount of heat needed to raise the temperature of one gram of water by one degree Celsius. The SI unit of heat is the **joule** (J), which is defined in terms of three of the base SI units as

$$1 \text{ J} = 1 \text{ kg m}^2/\text{s}^2$$

A good way to think about the joule is in comparison to the calorie:

$$1 \text{ cal} = 4.184 \text{ J}$$

Thus, it takes 4.184 J to raise the temperature of one gram of water from 14.5 °C to 15.5 °C (in the formal definition the temperature range must be specified).

Figure 4.11 Chemical reactions Most chemical reactions in the laboratory, such as this reaction of K_2CrO_4 and $Pb(NO_3)_2$ to yield solid $PbCrO_4$, are carried out under conditions of constant pressure.

Enthalpy and Thermochemical Equations

Chemical reactions are generally carried out in open containers (Figure 4.11). Under these conditions, the pressure (due to the atmosphere) varies little during the course of the reaction and can be considered constant. Under conditions of constant pressure, the quantity of heat *absorbed by the system* is called the **change in enthalpy,** and is represented by the symbol ΔH. When the symbol Δ (delta) precedes a letter or symbol, it means "change in." The symbol ΔH is pronounced as "delta H."

The direction of the heat flow determines the sign of the quantity ΔH. If the chemical reaction gives off heat (is exothermic), heat flows out of the system, and the sign of ΔH is negative. If the chemical reaction absorbs heat (is endothermic), heat flows into the system, and the sign of ΔH is positive (Figure 4.12).

A **thermochemical equation** is a chemical equation for which the value of ΔH is given. In the thermochemical equation, the change in enthalpy is written at the right, and the sign of ΔH used to indicate whether the reaction is exothermic or endothermic. This enthalpy change is determined by exper-

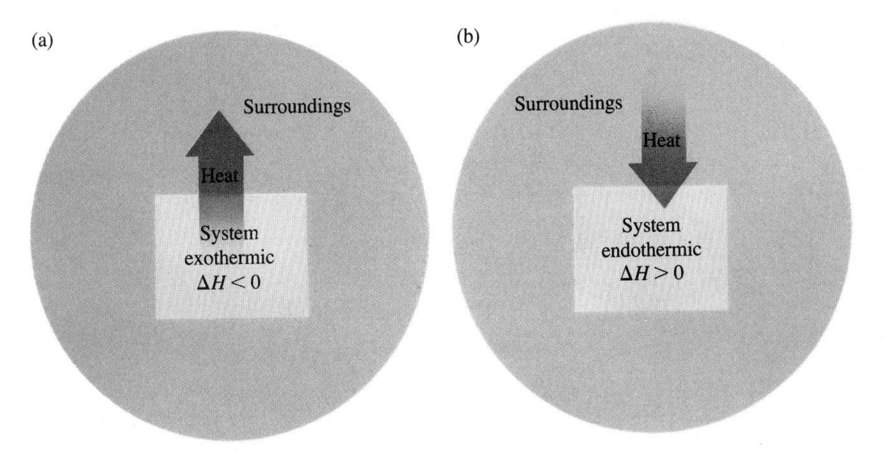

Figure 4.12 Direction of heat flow (a) In exothermic reactions, heat is transferred from the system to the surroundings. (b) In endothermic reactions, heat is transferred from the surroundings to the system.

iment (see Section 4.5). The thermochemical equation for an exothermic reaction, the combustion of methane, is shown in Equation 4.1; an endothermic reaction is shown in Equation 4.2.

$$CH_4(g) + 2O_2(g) \longrightarrow CO_2(g) + 2H_2O(\ell) \quad \Delta H = -890 \text{ kJ} \qquad [4.1]$$

$$N_2(g) + O_2(g) \longrightarrow 2NO(g) \quad \Delta H = +181.8 \text{ kJ} \qquad [4.2]$$

Change in enthalpy, ΔH, expresses heat flow in a chemical reaction that takes place at constant pressure.

The value of ΔH in a thermochemical equation assumes that the coefficients stand for moles, not molecules. The enthalpy change of −890 kJ in Equation 4.1 is observed when one mole of CH_4 and two moles of O_2 react to produce one mole of CO_2 and two moles of H_2O. Remember that a negative enthalpy change means that heat is given off by the system.

While it is good practice to include the physical states of the substances involved in any chemical equation, it is absolutely necessary to include the physical states when writing a thermochemical equation. From everyday experience we know that the melting of ice cools the surrounding matter and is therefore an endothermic process. The melting of ice can be described by a thermochemical equation. This reaction is an endothermic reaction, so ΔH is positive.

$$H_2O(s) \longrightarrow H_2O(\ell) \qquad \Delta H = +6.01 \text{ kJ}$$

Since energy is absorbed by the water when it melts, a mole of water in the liquid state has more energy than a mole of water in the solid state. Because the energy of a substance depends on its physical state, *the physical state of every substance must be included in all thermochemical equations.*

Endothermic reaction When ice melts, it absorbs heat from the surroundings.

Stoichiometry of Enthalpy Change in Chemical Reactions

Heat flow is a stoichiometric part of a thermochemical equation.

The enthalpy change is part of a thermochemical equation; the quantity ΔH is simply another stoichiometric quantity of the reaction. We can calculate the quantity of heat produced or absorbed in a reaction just as we have calculated the masses of products formed in a reaction. The thermochemical equation expresses the stoichiometric relationship between the number of moles of any substance in the equation and the quantity of heat produced or absorbed in the reaction. The thermochemical equation for the burning of ethane is

$$2C_2H_6(g) + 7O_2(g) \longrightarrow 4CO_2(g) + 6H_2O(\ell) \quad \Delta H = -3120 \text{ kJ} \qquad [4.3]$$

The thermochemical equivalencies are

$$2 \text{ mol } C_2H_6(g) \simeq -3120 \text{ kJ}$$

$$7 \text{ mol } O_2(g) \simeq -3120 \text{ kJ}$$

$$4 \text{ mol } CO_2(g) \simeq -3120 \text{ kJ}$$

$$6 \text{ mol } H_2O(\ell) \simeq -3120 \text{ kJ}$$

These equivalencies are used in the same way as are the mole-to-mole equivalencies introduced in Chapter 3. The following examples illustrate the process.

Example 4.14 Enthalpy as a Stoichiometric Quantity

Calculate the enthalpy change observed in the combustion reaction of 1.00 g of ethane using Equation 4.3.

Solution The same approach that was used previously for stoichiometric calculations is used here also. The mass of ethane is converted into moles and then the thermochemical equation is used to calculate ΔH.

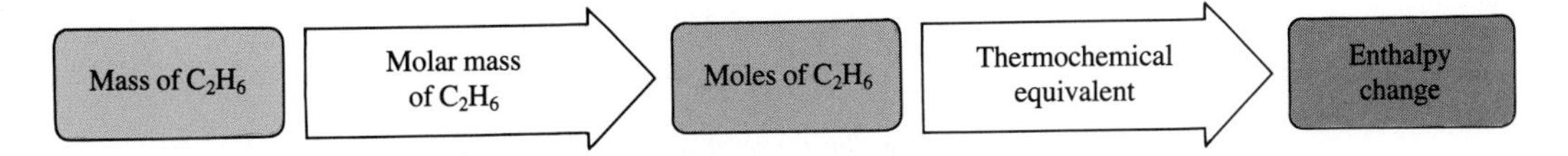

The equation is

$$2C_2H_6(g) + 7O_2(g) \longrightarrow 4CO_2(g) + 6H_2O(\ell) \qquad \Delta H = -3120 \text{ kJ}$$

The number of moles of C_2H_6 is calculated from the molar mass of C_2H_6 (30.07 g/mol).

$$\text{moles } C_2H_6 = 1.00 \text{ g } C_2H_6 \left(\frac{1 \text{ mol } C_2H_6}{30.07 \text{ g } C_2H_6}\right)$$

$$= 3.32 \times 10^{-2} \text{ mol } C_2H_6$$

The change in enthalpy given in the thermochemical equation gives us the equivalency relating mol C_2H_6 to ΔH:

$$2 \text{ mol } C_2H_6(g) \triangleq -3120 \text{ kJ}$$

Note that the 2 coefficient from the chemical equation is needed in this equivalence. Use it to determine the enthalpy change.

$$\text{enthalpy change} = 3.32 \times 10^{-2} \text{ mol } C_2H_6(g) \left(\frac{-3120 \text{ kJ}}{2 \text{ mol } C_2H_6(g)}\right)$$

$$= -51.8 \text{ kJ}$$

Remember that a negative sign for the change in enthalpy means that heat is given off by the system to the surroundings. We find that 51.8 kJ of heat is released for each gram of ethane consumed.

Understanding Calculate the enthalpy change observed when 5.0 g of O_2 is consumed completely by reaction with N_2 to form NO as given in Equation 4.2.

Answer: +28 kJ

Example 4.15 Enthalpy as a Stoichiometric Quantity

Propane (C_3H_8) is a gas that is often burned as a source of heat in places where natural gas is not available. If we need to produce 9.00×10^4 kJ of heat, what mass of propane must be burned? The thermochemical equation is

$$C_3H_8(g) + 5O_2(g) \longrightarrow 3CO_2(g) + 4H_2O(\ell) \qquad \Delta H = -2.22 \times 10^3 \text{ kJ}$$

Solution The strategy is the same as in Example 4.14, but in reverse.

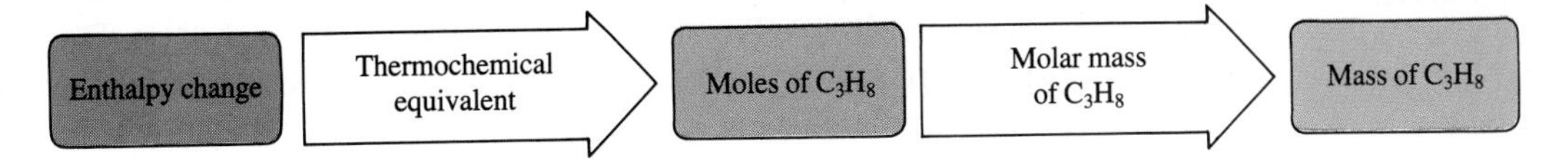

First, the thermochemical equation is used to calculate the number of moles of propane equivalent to the desired quantity of heat.

$1 \text{ mol } C_3H_8(g) \simeq -2.22 \times 10^3 \text{ kJ}$

$$\text{moles } C_3H_8 = -9.00 \times 10^4 \text{ kJ} \left(\frac{1 \text{ mol } C_3H_8}{-2.22 \times 10^3 \text{ kJ}}\right) = 40.5 \text{ mol } C_3H_8$$

We complete the problem by using the molar mass of propane to find the required mass.

$$\text{mass } C_3H_8 = 40.5 \text{ mol } C_3H_8 \left(\frac{44.10 \text{ g } C_3H_8}{1 \text{ mol } C_3H_8}\right) = 1.79 \times 10^3 \text{ g } C_3H_8$$

Understanding Determine the mass of methane that must be burned (Equation 4.1) to produce 4.00×10^3 kJ of heat.

Answer: 72.1 g

4.5 Calorimetry

Objective

- To describe how heat flow is measured by calorimetry

The change in enthalpy for a chemical reaction is determined by measuring the heat flow, a process called **calorimetry.** In a calorimetry experiment, the reaction of interest is carried out with known amounts of reactants, and the change in temperature is measured. The device in which the reaction takes place is known as a **calorimeter.** For the heat flow to be equal to the enthalpy change, the calorimeter must be operated at constant pressure. A convenient calorimeter for the experiment is a polystyrene coffee cup (Figure 4.13). An insulating device like the coffee cup is used because the heat change of the system due to the chemical reaction must be measured, avoiding any complications from heat exchanged with the surroundings. In this type of experiment, the reaction is usually performed in solution; the amount of solution must be measured because the observed temperature change depends on the amount of solution present.

Figure 4.13 Coffee cup calorimeter A coffee cup can be used as a calorimeter to determine ΔH for reactions carried out in solution.

The heat flow observed in a calorimeter is traditionally symbolized by the letter q. If the source of the heat is a chemical reaction performed under conditions of constant pressure, then q is the same as ΔH, the enthalpy change.

The change in temperature, ΔT, measured in the calorimeter experiment and the heat, q, given off or absorbed during the reaction are related by the heat capacity of the material undergoing the temperature change. The **heat capacity** of an object (such as the solution in a calorimeter) is defined as the quantity of heat required to raise the temperature of that object by 1 kelvin (or 1 °C). Heat capacity has units of J/K (or J/°C). The **specific heat** of a substance is the heat needed to raise the temperature of a one-*gram* sample of the substance by 1 kelvin, and it has units of J/g-K (or J/g-°C). The **molar heat capacity** of a substance is defined as the heat needed to raise the temperature of one mole of the substance by 1 kelvin. Table 4.2 shows the specific heats of several common substances. Note that water has a large

Table 4.2 Specific Heats*

Substance	Specific Heat (J/g-K)
$H_2O(\ell)$	4.184
CH_3CH_2OH (ethanol)	2.419
$CH_3CH_2OCH_2CH_3$ (ether)	2.320
Al(s)	0.900
Au(s)	0.129
$Hg(\ell)$	0.139
C(graphite)	0.720
MgO	0.92

* At 25 °C.

specific heat; it takes more energy to raise the temperature of one gram of water by 1 K than one gram of any of the other substances shown in the table.

If the mass of a sample and its specific heat are known, the relationship between heat (q) and change in temperature (ΔT) is given by

Heat flow is determined by measuring temperature change in a calorimeter.

$$q = mC_s\,\Delta T \qquad [4.4]$$

where q is the heat in joules, m is the mass in grams of the sample, C_s is the specific heat of the sample, and ΔT is $T_{\text{final}} - T_{\text{initial}}$.

Example 4.16 Heat Change Calculation

What quantity of heat must be added to a 20.0-g sample of aluminum to change its temperature from 23.0 °C to 34.0 °C?

Solution Use Equation 4.4 and the specific heat of aluminum (0.900 J/g-K) to calculate q. The temperature change is 34.0 °C − 23.0 °C = 11.0 °C (or 11.0 K).

$$q = mC_s\,\Delta T$$

$$q = 20.0\,\text{g} \times 0.900\,\frac{\text{J}}{\text{gK}} \times 11.0\,\text{K} = 198\ \text{J}$$

Understanding What quantity of heat is needed to raise the temperature of 120 g of water by 6.0 °C?

Answer: 3.0×10^3 J or 3.0 kJ.

The heat flow for a chemical reaction can be determined by measuring the temperature change when a reaction is carried out in solution in the coffee cup calorimeter. Since the pressure during the reaction is constant, the enthalpy change, ΔH, is directly related to the heat flow. The heat released by the reaction is transferred to the solution, causing an increase in its temperature. We assume that the coffee cup, the thermometer, and the stirrer do not influence the temperature change. As long as the solution is dilute, its specific heat will be close to that of the solvent, which is generally water. Frequently, q for the surroundings rather than for the system is measured in calorimetry experiments. When relating q and ΔH, any reaction causing temperature to increase will be exothermic.

Example 4.17 Calorimetry

A 50.0-mL sample of a dilute acid solution is added to 50.0 mL of a base solution in a coffee cup calorimeter. The temperature of the liquid increases from 18.20 °C to 21.30 °C. Calculate the heat flow for the reaction.

Solution Since the specific heat of the solution is not given, we assume that these dilute solutions have the same specific heat capacity as does water, 4.184 J/g-K. The heat evolved is calculated from the equation

$$q = mC_s\,\Delta T$$

where m is the mass of the solution and C_s is the specific heat of water. Since the two solutions total 100 mL, the mass will be quite close to 100 g because the density of water is 1.00 g/mL.

$$q = 100\ \text{g} \times 4.184\ \text{J/g-K} \times (21.30 - 18.20)\ \text{K}$$

$$q = 1.30 \times 10^3\ \text{J} = 1.30\ \text{kJ}$$

Understanding A chemical reaction releases enough heat to raise the temperature of 49.9 g of water from 17.82 °C to 19.72 °C. Calculate the heat evolved.

Answer: 397 J

Example 4.18 Heat Flow in Acid-Base Reactions

The reaction of an acid with a base is an exothermic process. In order to neutralize a 200-mL sample of an aqueous NaOH solution, 100 mL of 0.44 *M* aqueous HCl was needed. The temperature of the mixture increased by 1.96 K. Calculate the enthalpy change for the neutralization of one mole of hydrogen ions.

Solution The strategy is similar to the examples worked in Section 4.4, except that in this case the equivalency between heat and amount of acid is *measured* in the experiment and is used to *generate* the thermochemical equation. First the heat flow, q, is calculated. Since the reaction is at constant pressure, q can be used to calculate ΔH for one mole of H^+ ions.

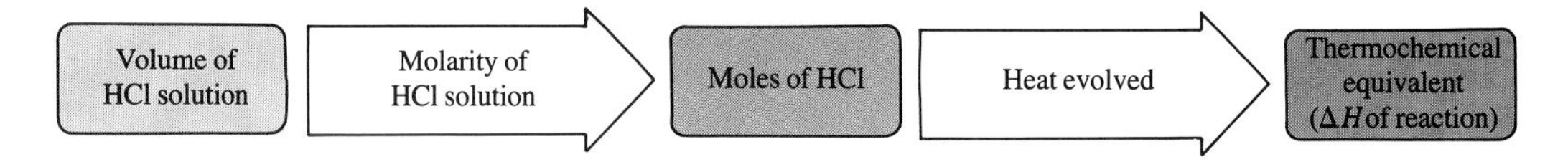

To calculate the heat flow,

$$q = mC_s\,\Delta T$$

$$q = 300\ \text{g} \times 4.184\ \text{J/g-K} \times 1.96\ \text{K}$$

$$q = 2.46 \times 10^3\ \text{J} = 2.46\ \text{kJ}$$

This heat flow is used to calculate ΔH for one mole of the reaction.

The complete equation for this acid-base reaction is

$$\text{HCl(aq)} + \text{NaOH(aq)} \longrightarrow \text{NaCl(aq)} + \text{H}_2\text{O}(\ell)$$

The Na^+ and Cl^- ions are spectator ions, making the net ionic equation

$$\text{H}^+\text{(aq)} + \text{OH}^-\text{(aq)} \longrightarrow \text{H}_2\text{O}(\ell)$$

Note that the physical state of each species must be given, since we are asked to derive the thermochemical equation. Calculate the number of moles of HCl that reacted.

$$\text{moles HCl} = 0.100 \cancel{\text{L HCl soln}} \left(\frac{0.44 \text{ mol HCl}}{1 \cancel{\text{L HCl soln}}}\right)$$

$$= 4.4 \times 10^{-2} \text{ mol HCl}$$

One mole of HCl yields one mole of $H^+(aq)$ in solution. We have determined that the neutralization of 4.4×10^{-2} mol $H^+(aq)$ generates 2.46 kJ of heat. The reaction is exothermic so ΔH is negative.

$$4.4 \times 10^{-2} \text{ mol } H^+(aq) \mathrel{\hat{=}} -2.46 \text{ kJ}$$

We need to determine the heat generated by one mole of acid in order to write the thermochemical equation for the neutralization reaction.

$$\Delta H_{\text{reaction}} = 1 \cancel{\text{mol } H^+(aq)} \left(\frac{-2.46 \text{ kJ}}{4.4 \times 10^{-2} \cancel{\text{mol } H^+(aq)}}\right) = -56 \text{ kJ}$$

The thermochemical equation is

$$H^+(aq) + OH^-(aq) \longrightarrow H_2O(\ell) \qquad \Delta H = -56 \text{ kJ}$$

Understanding The reaction of 0.440 g of magnesium with 400 g of hydrochloric acid solution causes the temperature of the solution to increase by 5.04 K. Calculate ΔH for the reaction.

$$Mg(s) + 2HCl(aq) \longrightarrow MgCl_2(aq) + H_2(g)$$

Answer: −466 kJ

The coffee cup calorimeter is not adequate for high-accuracy measurements. Scientists need to account for changes in temperature of the calorimeter, stirrer, and thermometer, as well as the water inside the cup. These problems can be solved, and temperature changes as small as 10 microdegrees (1 microdegree = 10^{-6} degree) can be measured.

4.6 Hess's Law

Objectives

- To define a state function
- To demonstrate that enthalpy is a state function
- To illustrate how changes in enthalpy are represented on an energy level diagram
- To calculate the enthalpy change (ΔH) of a reaction by using Hess's law

The enthalpy change for a chemical reaction is a very important and often necessary piece of information. For example, thermochemical information is essential in assessing the economics of fuels. When designing equipment for the commercial manufacture of compounds, the chemist or engineer must know whether heat must be supplied to or removed from the reaction mixture. For reactions that produce large quantities of heat, a new plant could literally blow up if the heat flow is not properly taken into account in the plant design. Unfortunately, it is difficult—and sometimes impossible—to determine experimentally the enthalpy change for a particular chemical reaction. A method is described in this section that can be used to calculate the enthalpy change for one reaction from experimentally determined enthalpy changes for other reactions.

Enthalpy is a state function; it does not depend on the history of the sample.

State Functions

A **state function** is any property of a system that is fixed by the present conditions and is independent of the system's previous history. The altitude of a climber on a mountain is an example of a state function. It does not make a difference whether the climber goes straight up the mountain or takes a trail that curves up the mountain (Figure 4.14). The altitude achieved by either path is the same, so it is a state function.

The value of a state function does not depend on how the state was achieved, but only on the conditions of the state itself. The mass of a sample is an example of a state function. The mass does not depend on how the sample was prepared or where it came from. The same is true of enthalpy; it is also a state function. Enthalpy does not depend on the history of the sample, but only on the conditions of pressure, temperature, chemical combination, and physical state.

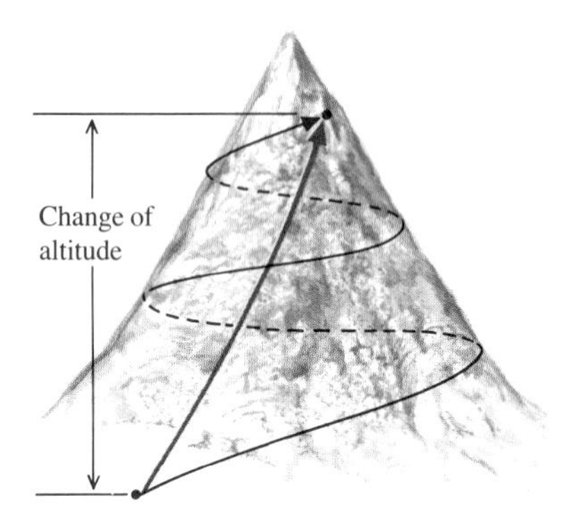

Figure 4.14 State function The altitude of a climber is a state function. The final altitude is the same going straight up the mountain or curving up the mountain.

Energy

The only kind of energy that we have considered so far is heat. Energy can take many forms; mechanical, electrical, and heat are just a few. All forms of energy fall into two categories: kinetic energy and potential energy. **Kinetic energy** is energy that matter possesses because it is in motion. The kinetic energy of an object depends on both its mass (m) and its speed (u).

$$\text{kinetic energy} = \tfrac{1}{2}mu^2$$

A moving automobile is an example of an object that possesses kinetic energy.

Potential energy is energy that matter possesses because of its position or condition. A brick on top of a building has more energy than one that is lying on the ground. This extra energy possessed by the brick at the top of the building is potential energy, since it depends on the elevated position of the brick. If the brick were dropped from the top of the building, its potential energy would be converted into kinetic energy as it fell. It is easy to under-

(a)

(b)

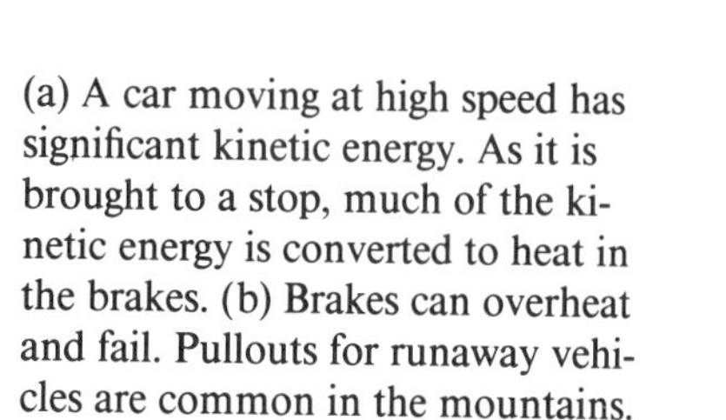

(a) A car moving at high speed has significant kinetic energy. As it is brought to a stop, much of the kinetic energy is converted to heat in the brakes. (b) Brakes can overheat and fail. Pullouts for runaway vehicles are common in the mountains.

stand why the potential energy of the brick on top of the building is a state function—it does not matter how the brick got to the top, only that it is there.

Compounds also possess energy as a result of the forces that hold the atoms together or, more formally, their state of chemical combination. This form of energy is called **chemical energy.** When a chemical reaction takes place, the energy stored in the reactants is not the same as that stored in the products. The difference in energy is often converted into heat (or comes from absorbed heat) as the chemical change takes place.

Thermochemical Energy Level Diagrams

The enthalpy change in a chemical reaction can be conveniently shown in diagram form. For example, from Equation 4.5 it is possible to say that the enthalpy of one mole of liquid water is 285.8 kJ less than the enthalpy of one mole of $H_2(g)$ plus one half mole of $O_2(g)$. (Fractions are commonly used as coefficients in thermochemical equations in order to express the heat change for one mole of a product.)

$$H_2(g) + \tfrac{1}{2}O_2(g) \longrightarrow H_2O(\ell) \qquad \Delta H = -285.8 \text{ kJ} \qquad [4.5]$$

Energy level diagrams graphically represent energy changes.

The enthalpy change for the formation of water from hydrogen and oxygen can be represented by an energy level diagram, as shown in Figure 4.15. Energy level diagrams are really one-dimensional graphs, in this case showing the relative enthalpies of the water and molecular hydrogen and oxygen. The energy units on the vertical axis are not absolute numbers because we do not know the absolute enthalpy of any given substance. They are relative numbers—we *do* know the *difference* in enthalpy between the reactants and the product.

Under certain conditions it is possible to reverse the direction of Equation 4.5 and decompose liquid water into the free elements:

$$H_2O(\ell) \longrightarrow H_2(g) + \tfrac{1}{2}O_2(g) \qquad \Delta H = +285.8 \text{ kJ} \qquad [4.6]$$

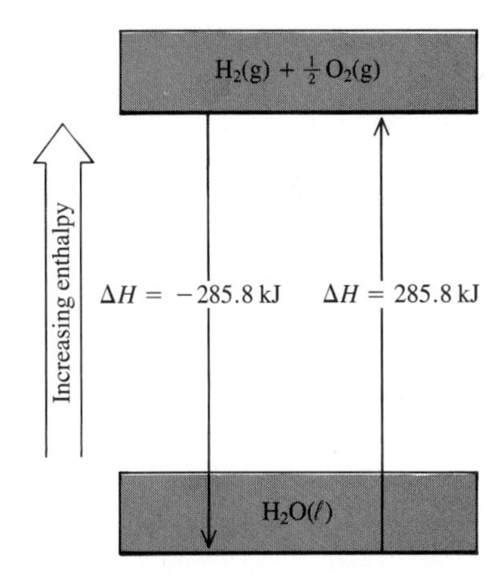

Figure 4.15 An energy level diagram shows two moles of hydrogen atoms and one mole of oxygen atoms in two different states of chemical combination; as one mole of $H_2(g)$ and one-half mole of $O_2(g)$, and as one mole of $H_2O(\ell)$. The conversion of one mole of $H_2(g)$ and one-half mole of $O_2(g)$ into one mole of water releases 285.8 kJ of energy, and the reverse reaction absorbs this same quantity of energy.

Since energy is released when water is formed from the diatomic elements, energy must be absorbed when the reverse reaction takes place. It is clear from the energy diagram that ΔH for Equation 4.6 must be numerically the same as the enthalpy change for Equation 4.5, but of opposite sign. That is, the reaction must *absorb* exactly as much heat from the surroundings as was *released* to the surroundings when written in the opposite direction. If this statement were not true, the law of conservation of energy would not be obeyed.

For certain reactions, it is difficult or impossible to measure the change in enthalpy directly. For instance, when carbon burns in a limited supply of oxygen, a mixture of both carbon monoxide and carbon dioxide is produced, so it is not possible to measure directly the enthalpy change for the reaction

$$C(s) + \tfrac{1}{2}O_2(g) \longrightarrow CO(g) \qquad [4.7]$$

Instead, an indirect method is used. We can measure the changes in enthalpy for carbon and carbon monoxide, each reacting with a large excess of oxygen to form CO_2. The thermochemical equations for these two reactions are

$$C(s) + O_2(g) \longrightarrow CO_2(g) \qquad \Delta H_1 = -393.5 \text{ kJ} \qquad [4.8]$$

$$CO(g) + \tfrac{1}{2}O_2(g) \longrightarrow CO_2(g) \qquad \Delta H_2 = -283.0 \text{ kJ} \qquad [4.9]$$

These two equations can be combined in such a way that the desired equation, 4.7, is obtained. Note that Equation 4.9 must be reversed (and the sign of ΔH changed) in order to have two equations that will sum to yield the desired equation.

$$C(s) + O_2(g) \longrightarrow CO_2(g) \qquad \Delta H = -393.5 \text{ kJ}$$

$$CO_2(g) \longrightarrow CO(g) + \tfrac{1}{2}O_2(g) \qquad \Delta H = +283.0 \text{ kJ}$$

$$C(s) + \tfrac{1}{2}O_2(g) \longrightarrow CO(g) \qquad \Delta H = -110.5 \text{ kJ}$$

The energy level diagram for this process is shown in Figure 4.16. First, one mole of carbon and one mole of oxygen react to form one mole of CO_2 in an exothermic step. The second reaction shows the endothermic step in which this CO_2 decomposes to form one mole of CO and half a mole of O_2. While we can directly measure the enthalpy change for the first reaction, we can experimentally measure only the reverse of the second reaction. As shown on the diagram, ΔH for the reverse reaction has the same magnitude but opposite sign from the reaction needed for the second step. Finally, ΔH for the desired reaction is calculated from the sum of the enthalpy changes for the two individual steps. Notice that half a mole of O_2 is on both the reactant and product sides of the two steps, so it is not in the calculated equation.

This example illustrates several properties of thermochemical equations. The rules that govern the combination of thermochemical equations

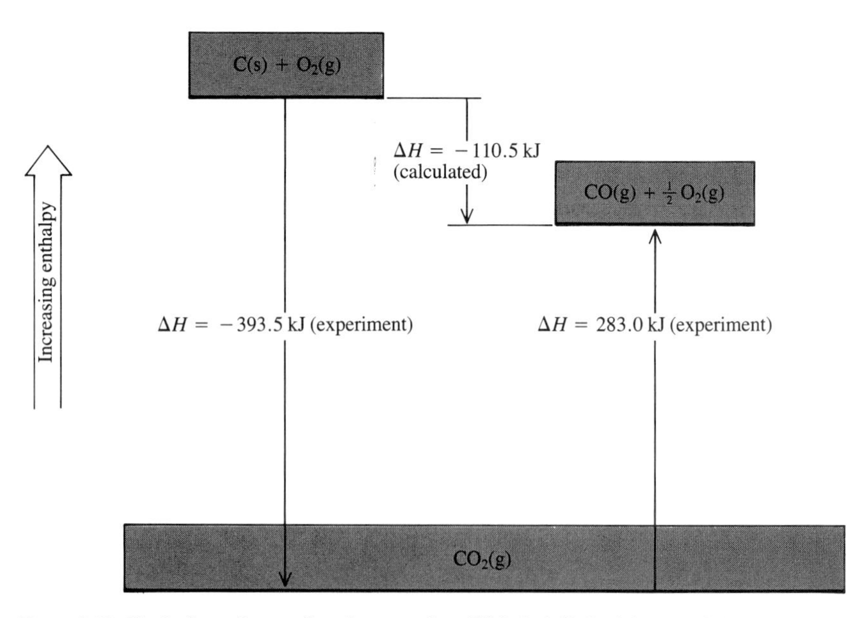

Figure 4.16 Enthalpy change for the reaction $C(s) + 1/2\ O_2(g) \longrightarrow CO(g)$ The diagram shows how the experimental measurements are combined to calculate the desired enthalpy change.

Table 4.3 Rules for Thermochemical Equations

1. The change in enthalpy for an equation obtained by adding two or more thermochemical equations is the sum of the enthalpy changes of the equations that have been added (Figure 4.16). This rule is known as **Hess's law.**
2. When a thermochemical equation is written in the reverse direction, the enthalpy change is numerically the same, but has the opposite sign (Figure 4.15).
3. Enthalpy change depends on the amount of substances in the reaction (it is an extensive property). For example, when the coefficients in a thermochemical equation are doubled, the enthalpy change must also double. This fact was assumed in the solutions of Examples 4.14 and 4.15.

are given in Table 4.3. These rules are a natural consequence of the law of conservation of energy and the fact that enthalpy is a state function.

Hess's law is a powerful tool for determining the enthalpy change that accompanies a reaction. It is not necessary to measure the enthalpy change directly for every reaction; Hess's law often lets us calculate the enthalpy change for one reaction, given the thermochemical equations for others.

Hess's law allows the ΔH of a reaction to be calculated from measured ΔH values of other reactions.

Problems that illustrate the use of Hess's law and the other properties of thermochemical equations are given in Examples 4.19 and 4.20.

Example 4.19 Hess's Law

Hydrogenation of hydrocarbons is an important reaction in the chemical industry. A simple example is the hydrogenation of ethylene to form ethane. Calculate the enthalpy change for this reaction.

$$\underset{\text{ethylene}}{C_2H_4(g)} + H_2(g) \longrightarrow \underset{\text{ethane}}{C_2H_6(g)} \qquad \Delta H = ?$$

Use the following thermochemical equations in obtaining your answer.

$$H_2(g) + \tfrac{1}{2}O_2(g) \longrightarrow H_2O(\ell) \qquad \Delta H = -285.8 \text{ kJ}$$
$$C_2H_4(g) + 3O_2(g) \longrightarrow 2CO_2(g) + 2H_2O(\ell) \qquad \Delta H = -1411 \text{ kJ}$$
$$C_2H_6(g) + \tfrac{7}{2}O_2(g) \longrightarrow 2CO_2(g) + 3H_2O(\ell) \qquad \Delta H = -1560 \text{ kJ}$$

Solution Ethylene and hydrogen are the reactants in the desired equation, and are also on the reactant side of the first two thermochemical equations that are given. The third thermochemical equation can be reversed to place the ethane on the product side, where it is found in the desired equation. When a thermochemical equation is written in the reverse direction, the sign of ΔH is changed. Adding these three thermochemical equations produces the desired equation. The overall enthalpy change is the sum of the enthalpy changes of each of the three equations.

$$H_2(g) + \tfrac{1}{2}O_2(g) \longrightarrow H_2O(\ell) \qquad \Delta H = -285.8 \text{ kJ}$$
$$C_2H_4(g) + 3O_2(g) \longrightarrow 2CO_2(g) + 2H_2O(\ell) \qquad \Delta H = -1411 \text{ kJ}$$
$$2CO_2(g) + 3H_2O(\ell) \longrightarrow C_2H_6(g) + \tfrac{7}{2}O_2(g) \qquad \Delta H = +1560 \text{ kJ}$$
$$\overline{C_2H_4(g) + H_2(g) \longrightarrow C_2H_6(g) \qquad \Delta H = -137 \text{ kJ}}$$

The net equation excludes substances that are common to both sides of the equation. Thus, in this example the $3\frac{1}{2}$ moles of oxygen, 2 moles of carbon dioxide, and 3 moles of water are not shown in the final equation.

Example 4.20 Hess's Law

Hydrocarbons of low molecular weight are converted into larger and more useful compounds by the chemical industry. Find the change in enthalpy for the synthesis of cyclohexane (C_6H_{12}) from ethylene.

$$3C_2H_4(g) \longrightarrow C_6H_{12}(\ell) \qquad \Delta H = ?$$

Use the information in Example 4.19 and the following thermochemical equation for the combustion of cyclohexane to solve this problem.

$$C_6H_{12}(\ell) + 9O_2(g) \longrightarrow 6CO_2(g) + 6H_2O(\ell) \qquad \Delta H = -3920 \text{ kJ}$$

Solution Since C_6H_{12} appears as a product in the desired equation, reverse the direction of the last equation and change the sign of ΔH.

$$6CO_2(g) + 6H_2O(\ell) \longrightarrow C_6H_{12}(\ell) + 9O_2(g) \qquad \Delta H = +3920 \text{ kJ}$$

We need to know the enthalpy change for the consumption of three moles of C_2H_4. The thermochemical equation, including the enthalpy change, for the reaction of ethylene with oxygen must be multiplied by three.

$$3C_2H_4(g) + 9O_2(g) \longrightarrow 6CO_2(g) + 6H_2O(\ell) \qquad \Delta H = 3 \times (-1411) \text{ kJ}$$

Adding these last two equations gives the desired reaction and the enthalpy change.

$$6CO_2(g) + 6H_2O(\ell) \longrightarrow C_6H_{12}(\ell) + 9O_2(g) \qquad \Delta H = +3920 \text{ kJ}$$
$$3C_2H_4(g) + 9O_2(g) \longrightarrow 6CO_2(g) + 6H_2O(\ell) \qquad \Delta H = 3 \times (-1411) \text{ kJ}$$
$$\overline{3C_2H_4(g) \longrightarrow C_6H_{12}(\ell) \qquad \Delta H = -313 \text{ kJ}}$$

Understanding Given the thermochemical equations

$$Sn(s) + Cl_2(g) \longrightarrow SnCl_2(s) \qquad \Delta H = -325 \text{ kJ}$$
$$SnCl_2(s) + Cl_2(g) \longrightarrow SnCl_4(\ell) \qquad \Delta H = -186 \text{ kJ}$$

what is ΔH for

$$Sn(s) + 2Cl_2(g) \longrightarrow SnCl_4(\ell) \qquad \Delta H = ?$$

Answer: -511 kJ

The importance of specifying the physical states of all substances in a thermochemical equation has been stressed already, but we can illustrate it better by an example. All changes of state are accompanied by enthalpy changes. The enthalpy change for a reaction will be different if the physical state of any of the substances is changed. Consider the combustion of methane as an example. When the reaction produces liquid water, the thermochemical equation is

$$CH_4(g) + 2O_2(g) \longrightarrow CO_2(g) + 2H_2O(\ell) \qquad \Delta H = -890 \text{ kJ} \quad [4.10]$$

The enthalpy change for the vaporization of one mole of liquid water has been determined experimentally as 44.0 kJ.

$$H_2O(\ell) \longrightarrow H_2O(g) \qquad \Delta H = 44.0 \text{ kJ} \quad [4.11]$$

By doubling Equation 4.11 and adding it to Equation 4.10 we obtain the equation for the combustion of methane to give steam, $H_2O(g)$.

$$CH_4(g) + 2O_2(g) \longrightarrow CO_2(g) + 2H_2O(\ell) \qquad \Delta H = -890 \text{ kJ}$$
$$2H_2O(\ell) \longrightarrow 2H_2O(g) \qquad \Delta H = +88.0 \text{ kJ}$$
$$\overline{CH_4(g) + 2O_2(g) \longrightarrow CO_2(g) + 2H_2O(g) \qquad \Delta H = -802 \text{ kJ}} \quad [4.12]$$

The difference between the ΔH values for Equations 4.10 and 4.12 is quite large, but the only difference is the H_2O produced in one reaction is in the liquid state, and in the other it is a gas. This example illustrates the importance of including the physical state in all thermochemical equations.

Summary

Many chemical reactions are carried out with one or more of the reactants dissolved in a *solvent.* To perform stoichiometric calculations, the *concentration* of the *solute* in the solution is necessary. *Molarity,* defined as the number of moles of solute in one liter of solution, is the most convenient unit of concentration for stoichiometry calculations. A solution of known concentration can be prepared by adding solvent to a weighed sample of solute and measuring the volume of the solution. It can also be prepared by dilution of a more concentrated solution of known concentration. In this case, a *pipet* is used to measure out a fixed volume of the concentrated solution that is diluted with pure solvent to form a known volume of dilute solution.

Stoichiometry calculations can be performed for reactions in solution. They generally require the calculation of number of moles from the molar concentrations and the volumes of solution. Calculations of reaction stoichiometry are simplified by using the *net ionic equation,* one that shows only those species in the reaction that undergo change.

Two types of reactions, acid-base and precipitation, are frequently employed for chemical analyses of solutions. Chemical analysis in which the volume of solution is measured is known as *volumetric analysis.* The determination of the concentration of a component in a solution by weighing a precipitate is known as *gravimetric analysis.* To predict which substances (if any) in a solution will precipitate, a series of solubility rules for ionic compounds have been presented.

A *titration* is used to determine the concentration of a solution of unknown concentration by adding an equivalent amount of a reactant solution of known concentration. A *buret* is used to measure accurately the volumes of added liquids. *Indicators* that change color at or near the *equivalence point* are used to determine when an equivalent amount of the solution of known concentration has been added.

Energy changes are also part of chemical equations. *Thermochemical equations* include information about heat flow. Heat flow at constant pressure is expressed by the *change in enthalpy,* ΔH. Reactions that give off heat are *exothermic* and have negative ΔH values, while reactions that absorb heat are *endothermic* and have positive ΔH values. According to the *law of conservation of energy,* all of the heat lost or gained by the *system* is transferred to or from the *surroundings.* Stoichiometry calculations that determine the heat flow in chemical reactions are performed by using the equivalencies given in the thermochemical equation.

Kinetic energy is the energy of motion and *potential energy* is the energy of position. *Chemical energy* is a form of potential energy arising from the forces that hold atoms together. Enthalpy, the measure of available chemical energy at constant pressure, is a *state function,* a property that is fixed by the state of the system. Since enthalpy is a state function, changes in enthalpy can be expressed in energy level diagrams. These diagrams are useful in demonstrating *Hess's law,* which states that the change in enthalpy for an equation obtained by adding two or more thermochemical equations is the sum of the enthalpy changes of the equations that have been added. Hess's law can be used to determine enthalpy changes of reactions for which they cannot be measured directly.

Chapter Terms

Solvent—the substance that has the same physical state as the solution. It is generally the component present in largest quantity; in aqueous solutions the solvent is water. *(4.1)*

Solute—any substance being dissolved to form a solution. Solutes are usually present in lesser quantity than the solvent. *(4.1)*

Molarity—the concentration unit defined as the number of moles of solute in one liter of solution. *(4.1)*

Volumetric flask—a container calibrated to hold an accurately known volume of liquid. *(4.1)*

Pipet—a device calibrated either to deliver or to contain a specific volume of liquid; it is used to measure accurately a fixed volume of solution. *(4.1)*

Overall equation—the equation that shows all of the reactants and products in undissociated form. *(4.2)*

Complete ionic equation—the equation that shows separately all species, both ions and molecules, as they are present in solution. *(4.2)*

Net ionic equation—the equation that shows only those species in the reaction that undergo change. *(4.2)*

Spectator ions—ions present in solution that do not undergo change. *(4.2)*

Precipitation reaction—the formation of an insoluble product or products from the reaction of soluble reactants. *(4.2)*

Titration—a procedure for the determination of the quantity of one substance by the addition of a measured amount of a second substance. *(4.3)*

Equivalence point—the point in a titration at which the reactants have been added in stoichiometrically equivalent amounts. *(4.3)*

Indicator—a compound that changes color at the *end point* of a titration. The indicator should be chosen so that the end point coincides as nearly as possible with the equivalence point. *(4.3)*

Standard solution—a solution of accurately known concentration. *(4.3)*

Volumetric analysis—a quantitative determination in which the volume of a solution or substance is measured. *(4.3)*

Gravimetric analysis—the selective precipitation of one component of a solution, followed by isolation and weighing of the precipitate, in order to determine the amount of that component. *(4.3)*

Thermochemistry—the study of the heat flow that accompanies chemical reactions. *(4.4)*

System—the matter of interest in a chemical reaction. *(4.4)*

Surroundings—all matter other than the system in a chemical reaction. *(4.4)*

Law of conservation of energy—the total energy of the universe (the system plus the surroundings) is constant during a chemical change. *(4.4)*

Exothermic reaction—a reaction that releases heat to the surroundings. *(4.4)*

Endothermic reaction—a reaction that absorbs heat from the surroundings. *(4.4)*

Joule (J)—the SI unit of heat, defined as

$$1 \text{ J} = 1 \text{ kg m}^2/\text{s}^2.$$

In comparison to the calorie, 1 calorie = 4.184 joules. *(4.4)*

Change in enthalpy, ΔH—the heat absorbed by the system under constant pressure. *(4.4)*

Thermochemical equation—a chemical equation that includes heat flow from the system to the surroundings. *(4.4)*

Calorimetry—the measurement of heat flow. *(4.5)*

Calorimeter—a device used to measure heat flow. *(4.5)*

Heat capacity—the quantity of heat required to raise the temperature of a sample by 1 kelvin (or 1 °C). *(4.5)*

Specific heat—the heat needed to raise the temperature of one *gram* of a substance by 1 kelvin; it has the units of J/g-K (or J/g-°C). *(4.5)*

Molar heat capacity—the heat needed to raise the temperature of one *mole* of a substance by 1 kelvin. *(4.5)*

State function—any property of a system that is fixed by the state of the system. *(4.6)*

Kinetic energy—the energy that matter possesses because of its motion. *(4.6)*

Potential energy—the energy that matter possesses because of its position or condition. *(4.6)*

Chemical energy—a form of potential energy derived from the forces that hold atoms together. *(4.6)*

Hess's law—the change in enthalpy for an equation obtained by adding two or more thermochemical equations is the sum of the enthalpy changes of the equations that have been added. *(4.6)*

Exercises

Exercises designated with color have answers in Appendix J.

Preparation of Solutions

4.1 Give a definition of molarity and describe how a solution of known molarity can be prepared from a solid or from a more concentrated solution.

4.2 Calculate the molarity of a solution prepared by dissolving 8.23 g of KOH in enough water to form 250 mL of solution.

4.3 Calculate the molarity of a solution prepared by dissolving 23.1 g of NaCl in enough water to form 500 mL of solution.

4.4 How many grams of $AgNO_3$ are needed to prepare 300 mL of a 1.00 *M* solution?

4.5 What mass, in grams, of Na_2SO_4 is needed to prepare 500 mL of a 3.50 *M* solution?

4.6 What volume of a 2.3 *M* HCl solution is needed to prepare 2.5 L of a 0.45 *M* HCl solution?

4.7 What volume of a 5.22 *M* NaOH solution is needed to prepare 1.00 L of a 2.35 *M* HCl solution?

4.8 What is the molarity of a glucose ($C_6H_{12}O_6$) solution prepared from 50 mL of a 1.0 *M* solution that is diluted with water to a final volume of 2.0 L?

4.9 What is the molarity of a solution of $Ca(OH)_2$ prepared from 100 mL of a 0.0234 *M* solution that is diluted with water to a final volume of 0.500 L?

4.10 What is the molarity of 2.0 L of solution prepared from
(a) 3.56 g of NaOH
(b) 25 mL of a 1.4 *M* NaOH solution?

4.11 What is the molarity of 250 mL of solution prepared from
(a) 0.12 g of sodium nitrate
(b) 0.75 mL of a 0.42 *M* NaOH solution?

4.12 How many moles of solute are in
(a) 33 mL of a 3.11 *M* HNO_3 solution
(b) 1.0 L of a 3.2 *M* HNO_3 solution?

4.13 How many moles of solute are in
(a) 0.22 L of a 1.2 *M* NaCl solution
(b) 500 mL of a 0.22 *M* solution of $AgNO_3$?

4.14 How many grams of solute are in
(a) 3.13 L of a 2.21 *M* HCl solution
(b) 1.5 L of a 1.2 *M* KCl solution?

4.15 How many grams of solute are in
(a) 0.113 L of a 1.00 *M* KBr solution
(b) 120 mL of a 2.11 *M* KNO_3 solution?

4.16 The substance KSCN is frequently used to test for iron in solution because a distinctive red color forms when it is added. As a laboratory assistant, you need to prepare 1.00 L of 0.20 *M* KSCN solution. What mass, in grams, of KSCN is needed?

4.17 Two liters of a 1.5 *M* solution of NaOH are needed for a laboratory experiment. A stock solution of 5.0 *M* NaOH is available. How is the desired solution prepared?

4.18 A 6.00-g sample of NaOH is added to a 1.00-L volumetric flask, and water is added to dissolve the solid and fill the flask to the mark. A 100-mL sample of this solution is added to a 5.00-L volumetric flask, and water is added to fill the flask to the mark. What is the concentration of NaOH in the second flask?

4.19 An 83.5-g sample contains NaCl contaminated with a substance that is not water-soluble. The sample is added to water, which is then filtered to form 234 mL of homogeneous solution. That solution is analyzed and found to be 3.22 *M* in NaCl. What is the percentage of NaCl in the original sample?

Solution Stoichiometry and Chemical Analysis

4.20 What mass of NaOH, in grams, is needed to neutralize 100 mL of 1.3 *M* HCl?

4.21 What mass of $BaSO_4$, in grams, forms in the reaction of 120 mL of 0.11 *M* $Ba(OH)_2$ with excess H_2SO_4?

4.22 What volume of 0.66 *M* HNO_3 is needed to neutralize 22 g of $Ca(OH)_2$?

4.23 What volume of 0.22 molar hydrochloric acid is needed to react completely with 2.5 g of magnesium hydroxide?

4.24 What is the molar concentration of a solution of HCl if 135 mL neutralizes 2.3 g of NaOH?

4.25 What is the molar concentration of a solution of $Ca(OH)_2$ if 255 mL neutralizes 33.4 mL of a 0.213-*M* solution of HCl?

4.26 What mass, in grams, of $BaSO_4$ will form in the reaction of 355 mL of 0.032 *M* H_2SO_4 with 266 mL of 0.015 *M* $Ba(OH)_2$?

4.27 Calculate the mass of magnesium hydroxide formed in the reaction of 1.2 L of a 5.5 *M* solution of sodium hydroxide and excess magnesium nitrate.

4.28 A 125 mL solution of $Ba(OH)_2$ was mixed with 75 mL of 0.10 *M* HCl. The resulting solution was still basic, and 35 mL of 0.012 *M* HCl was needed to neutralize the base. What is the molarity of the $Ba(OH)_2$ solution?

4.29 A solution was prepared by placing 14.2 g of KCl in a 1.00-L volumetric flask and adding water to dissolve the solid and fill the flask to the mark. What is the molarity of a $AgNO_3$ solution if 25 mL of the KCl solution will react exactly with 33.2 mL of the $AgNO_3$ solution?

4.30 Which of the following compounds dissolve in water?
(a) BaI_2 (c) Na_2CO_3
(b) $PbCl_2$ (d) $(NH_4)_2SO_4$

4.31 Which of the following compounds dissolve in water?
(a) Hg_2Cl_2 (c) KNO_3
(b) $CaBr_2$ (d) $AgClO_4$

4.32 Using the solubility rules given in Table 4.1, write the net ionic equations for any reaction that occurs on mixing each pair of solutions.
(a) solutions of sodium hydroxide and magnesium chloride

(b) solutions of sodium nitrate and magnesium bromide
(c) solutions of $Ba(ClO_4)_2$ and sodium carbonate

4.33 Using the solubility rules given in Table 4.1, write the net ionic equations for any reaction that occurs on mixing each pair of solutions.
(a) solutions of ammonium carbonate and magnesium chloride
(b) solutions of ammonium chloride and $AgClO_4$
(c) solutions of beryllium sulfate and sodium hydroxide

4.34 A water sample is known to contain one of the cations Pb^{2+} or Ba^{2+}. Treatment of the sample with NaCl produces a precipitate. Which of the metal cations does the solution contain?

4.35 A water sample is known to contain either Ag^+ or Mg^{2+} ions. Treatment of the sample with NaOH produces a precipitate, but treatment with KBr does not. Which of the metal cations does the solution contain?

4.36 A water sample is known to contain either Ca^{2+} or Ba^{2+} ions. Treatment of the sample with Na_2CO_3 produces a precipitate, but treatment with ammonium sulfate does not. Which of the metal cations does the solution contain?

4.37 A water sample is known to contain either Pb^{2+} or Ca^{2+} ions. Treatment of the sample with Na_2SO_4 produces a precipitate. Which of the metal cations does the solution contain?

4.38 A water sample is known to contain either Ca^{2+} or Hg^{2+} ions. How would you determine which ions are present?

4.39 A water sample is known to contain either Ag^+ or Ba^{2+} ions. How would you determine which ions are present?

4.40 Write the overall equation (including the physical states), the complete ionic equation, and the net ionic equation for the reaction that occurs when aqueous solutions of silver nitrate and sodium bromide are mixed. What mass of solid will precipitate if 345 mL of 0.33 *M* silver nitrate solution is mixed with 100 mL of a 1.3 *M* sodium bromide solution?

4.41 Write the overall equation (including the physical states), the complete ionic equation, and the net ionic equation for the reaction of aqueous solutions of sodium hydroxide and magnesium chloride. What mass of solid will form upon mixing 50 mL of 3.3 *M* sodium hydroxide with 35 mL of 1.0 *M* magnesium chloride?

4.42 Mixing excess potassium carbonate with 300 mL of a calcium chloride solution of unknown concentration yields 4.5 g of a solid. Give the formula of the solid and calculate the concentration of the calcium chloride solution.

4.43 What is the solid that precipitates, and how much of it will form, if an excess of sodium chloride solution is mixed with 10.0 mL of a 2.1 *M* silver nitrate solution?

4.44 What mass of lead(II) sulfate will precipitate upon mixing 20.0 mL of a 2.55 *M* solution of lead(II) acetate with an excess of sodium sulfate solution?

4.45 What volume of 0.112 *M* potassium carbonate is needed to precipitate all of the calcium ions in 50.0 mL of a 0.100 *M* solution of calcium chloride?

4.46 A solid forms when 21 mL of 3.5 *M* ammonium sulfate is mixed with excess barium chloride. Write the overall equation and calculate the mass of the precipitate.

4.47 A solid forms when 220 mL of 1.22 *M* sodium hydroxide is mixed with excess iron(II) chloride. Write the overall equation and calculate the mass of the precipitate.

4.48 What is the concentration of an HCl solution if a 100 mL sample is titrated with a 2.2 *M* solution of KOH, and 33.4 mL is needed to reach the equivalence point?

4.49 Potassium acid phthalate, $KHC_8H_4O_4$ (only one hydrogen will react with a base), is a crystalline solid that is available in a high state of purity, making it an excellent choice as a standard acid. What is the concentration of a NaOH solution if it takes 35 mL to neutralize a 0.022-g sample of the acid?

4.50 What is the percentage of calcium in an ionic compound of unknown composition if a 2.11-g sample of the compound is completely dissolved in water and produces 1.22 g of calcium carbonate upon addition of an excess of sodium carbonate?

4.51 What is the percentage of silver in an ionic compound of unknown composition if a 3.13-g sample of the compound is completely dissolved in water and produces 2.02 g of silver chloride upon addition of an excess of sodium chloride?

Thermochemistry

4.52 For the reaction

$$C(s) + O_2(g) \longrightarrow CO_2(g)$$

the enthalpy change is −393.5 kJ.
(a) Is energy released from or absorbed by the system in this reaction?
(b) What is the enthalpy change when 3.00 g of carbon is burned in an excess of oxygen?

4.53 The thermochemical equation for burning methane, the major component of natural gas, is

$$CH_4(g) + 2O_2(g) \longrightarrow CO_2(g) + 2H_2O(\ell) \qquad \Delta H = -890 \text{ kJ}$$

(a) Is this an endothermic or exothermic reaction?
(b) What is the enthalpy change when 1.00 g of methane is burned in an excess of oxygen?

4.54 The reaction of one mole of $O_2(g)$ and one mole of $N_2(g)$ to yield two moles of NO(g) is endothermic, with $\Delta H = +180$ kJ. What is the enthalpy change when 2.20 g of $N_2(g)$ reacts with an excess of oxygen?

4.55 The reaction of two moles of $H_2(g)$ with one mole of $O_2(g)$ to yield two moles of $H_2O(\ell)$ is exothermic, with $\Delta H = -572$ kJ. What is the enthalpy change when 10.0 g of $O_2(g)$ reacts with an excess of hydrogen?

4.56 The thermite reaction produces a great quantity of heat. Given the following thermochemical equation, what is the enthalpy change when 12.2 g of Al(s) reacts with an excess of $Fe_2O_3(s)$?

$$2Al(s) + Fe_2O_3(s) \longrightarrow Al_2O_3(s) + 2Fe(s) \qquad \Delta H = -852 \text{ kJ}$$

4.57 The formation of acetylene (C_2H_2), a gas used in welding, from its elements is endothermic. Given the following thermochemical equation, what is the enthalpy change when 33.2 g of C(s) reacts with an excess of $H_2(g)$?

$$2C(s) + H_2(g) \longrightarrow C_2H_2(g) \qquad \Delta H = +227 \text{ kJ}$$

4.58 The combustion of one mole of octane, C_8H_{18} (a component of gasoline), in excess oxygen is exothermic, giving 5.46×10^3 kJ of heat. What is the enthalpy change for burning 10.0 grams of octane?

4.59 The combustion of one mole of methanol, CH_3OH, in excess oxygen is exothermic, giving 727 kJ of heat. What is the enthalpy change for burning 10.0 grams of methanol? Compare this to the amount of heat produced by 10.0 g of octane, C_8H_{18}, a component of gasoline (see Exercise 4.58).

4.60 How much heat, in kilojoules, must be added to raise the temperature of a 300-gram sample of water from 22.5 °C to 39.1 °C? See Table 4.2 for the specific heat of water.

4.61 How much heat, in kilojoules, must be added to raise the temperature of a 500-gram sample of ethanol from 20.2 °C to 44.1 °C? See Table 4.2 for the specific heat of ethanol.

4.62 How much heat, in kilojoules, must be added to raise the temperature of a 20.0-g bar of aluminum from 22.5 °C to 34.2 °C? See Table 4.2 for the specific heat of aluminum.

4.63 How much heat, in kilojoules, must be added to raise the temperature of a 500-g ingot of gold from 21.4 °C to 45.9 °C? See Table 4.2 for the specific heat of gold.

4.64 Addition of a 7.11-g sample of NH_4NO_3 to 100 mL of water causes the temperature of the water to drop from 22.1 °C to 17.1 °C. Assuming that the mixture has the same specific heat as water and a mass of 107 g, calculate the heat flow in the process. Is the process exothermic or endothermic?

4.65 A 30-mL solution of a dilute acid is added to 70 mL of a base solution in a coffee cup calorimeter. The temperature of the solution increases from 22.3 °C to 24.1 °C. Assuming that the mixture has the same specific heat as water and a mass of 100 g, calculate the heat flow in the reaction. Is the process exothermic or endothermic?

4.66 A 100-mL solution of a dilute acid is added to 200 mL of a base solution in a coffee cup calorimeter. The temperature of the solution increases from 21.2 °C to 23.1 °C. Assuming that the mixture has the same specific heat as water and a mass of 300 g, calculate the heat flow in the reaction. Is the process exothermic or endothermic?

4.67 A 0.47-g sample of magnesium reacts with 200 g of dilute HCl in a coffee cup calorimeter to form $MgCl_2$ and H_2. The temperature increases by 10.9 °C as the magnesium reacts. Assuming that the mixture has the same specific heat as water, calculate the heat flow in the reaction. Is the process exothermic or endothermic?

4.68 Addition of 0.100 mol of $NH_4SCN(s)$ and 0.0500 mol of $Ba(OH)_2 \cdot 8H_2O$ to 500 mL of water causes the water temperature to drop from 22.8 °C to 19.7 °C. Assuming that the mixture has the same specific heat as water and a mass of 523 g, determine ΔH for the reaction

$$2NH_4SCN(s) + Ba(OH)_2 \cdot 8H_2O \xrightarrow{H_2O} Ba^{2+}(aq) + 2SCN^-(aq) + 2NH_3(aq) + 10H_2O(\ell)$$

4.69 A reaction that is used to propel rockets is

$$N_2O_4(\ell) + 2N_2H_4(\ell) \longrightarrow 3N_2(g) + 4H_2O(g)$$

This reaction has the advantage that neither product is toxic, so no dangerous pollution is released. When 10.0 g of liquid N_2O_4 is consumed in this reaction, 124 kJ of heat is released.

(a) Is the sign for the enthalpy change positive or negative?
(b) What is the value of ΔH for the chemical reaction as written?

4.70 When 2.00 g of octane, a component of gasoline, is burned in oxygen, the enthalpy change is −95.44 kJ. Calculate ΔH for the reaction

$$2C_8H_{18}(\ell) + 25O_2(g) \longrightarrow 16CO_2(g) + 18H_2O(\ell)$$

4.71 Ammonia is produced commercially by the direct reaction of the elements. The formation of 5.00 g of gaseous NH_3 by this reaction releases 13.56 kJ.
(a) What is the sign of the enthalpy change for this reaction?
(b) Calculate ΔH for the reaction

$$N_2(g) + 3H_2(g) \longrightarrow 2NH_3(g)$$

4.72 An important step in the manufacture of nitric acid is the reaction of nitrogen monoxide (NO) with oxygen. When 3.00 g of NO reacts with oxygen, the enthalpy change is −5.65 kJ. Calculate ΔH for the reaction

$$2NO(g) + O_2(g) \longrightarrow 2NO_2(g)$$

4.73 "Gasohol," a mixture of ethyl alcohol and gasoline, has been proposed as a fuel to help conserve our petroleum resources. This fuel is available on a limited basis. The thermochemical equation for the burning of ethyl alcohol is

$$C_2H_5OH(\ell) + 3O_2(g) \longrightarrow 2CO_2(g) + 3H_2O(\ell) \qquad \Delta H = -1366.8 \text{ kJ}$$

What is the enthalpy change for burning 2.00 g of ethyl alcohol? Compare this value to the information in Exercise 4.70. Which fuel gives more heat for a given weight?

4.74 The enthalpy change when one mole of methane (CH_4) is burned is −890 kJ. It takes 44.0 kJ to vaporize one mole of water. What mass of methane must be burned to provide the heat needed to vaporize 1.00 g of water?

4.75 Using the thermochemical equation in Exercise 4.73, calculate the mass of ethyl alcohol that must be burned to provide 500 J of heat.

4.76 Using the thermochemical equations

$$C_2H_6(g) + \tfrac{7}{2}O_2(g) \longrightarrow 2CO_2(g) + 3H_2O(\ell) \qquad \Delta H = -1560 \text{ kJ}$$

$$2C_2H_2(g) + 5O_2(g) \longrightarrow 4CO_2(g) + 2H_2O(\ell) \qquad \Delta H = -2599 \text{ kJ}$$

$$H_2(g) + \tfrac{1}{2}O_2(g) \longrightarrow H_2O(\ell) \qquad \Delta H = -286 \text{ kJ}$$

find ΔH for the reaction

$$C_2H_2(g) + 2H_2(g) \longrightarrow C_2H_6(g) \qquad \Delta H = ?$$

4.77 Using the thermochemical equations in Exercise 4.76 and in addition

$$CH_4(g) + 2O_2(g) \longrightarrow CO_2(g) + 2H_2O(\ell) \qquad \Delta H = -890 \text{ kJ}$$

$$C_2H_4(g) + 3O_2(g) \longrightarrow 2CO_2(g) + 2H_2O(\ell) \qquad \Delta H = -1411 \text{ kJ}$$

find ΔH for the reaction

$$C_2H_4(g) + 2H_2(g) \longrightarrow 2CH_4(g) \qquad \Delta H = ?$$

4.78 Calculate ΔH for the reaction

$$Zn(s) + \tfrac{1}{2}O_2(g) \longrightarrow ZnO(s) \qquad \Delta H = ?$$

given these equations:

$$Zn(s) + 2HCl(aq) \longrightarrow ZnCl_2(aq) + H_2(g) \qquad \Delta H = -152.4 \text{ kJ}$$

$$ZnO(s) + 2HCl(aq) \longrightarrow ZnCl_2(aq) + H_2O(\ell) \qquad \Delta H = -90.2 \text{ kJ}$$

$$2H_2(g) + O_2(g) \longrightarrow 2H_2O(\ell) \qquad \Delta H = -571.6 \text{ kJ}$$

4.79 Calculate ΔH for the reaction

$$Mg(s) + \tfrac{1}{2}O_2(g) \longrightarrow MgO(s) \qquad \Delta H = ?$$

given these equations:

$$Mg(s) + 2HCl(aq) \longrightarrow MgCl_2(aq) + H_2(g) \qquad \Delta H = -462 \text{ kJ}$$

$$MgO(s) + 2HCl(aq) \longrightarrow MgCl_2(aq) + H_2O(\ell) \qquad \Delta H = -146 \text{ kJ}$$

$$2H_2(g) + O_2(g) \longrightarrow 2H_2O(\ell) \qquad \Delta H = -571.6 \text{ kJ}$$

Additional Problems

4.80 A 5.3-g sample of NaOH is placed in a 1.00-liter volumetric flask, and water is added to the mark. A 100-mL sample of the resulting solution is placed in a 500-mL volumetric and additional water is added to the mark. What volume of the second sample is needed to neutralize 33 mL of 0.022 M H_2SO_4?

4.81 An aqueous solution of hydrazine, N_2H_4, can be prepared by the reaction of ammonia and sodium hypochlorite.

$$2NH_3(aq) + NaOCl(aq) \longrightarrow N_2H_4(aq) + NaCl(aq) + H_2O(\ell)$$

What mass of hydrazine, in grams, can be prepared from 50 mL of 1.22 M NH_3(aq) reacting with 100 mL of 0.44 M NaOCl(aq)?

4.82 An 80-mL sample of solution that is 2.33 M NaOH and 1.22 M KOH is evaporated to dryness. What mass of solid remains?

4.83 What is the concentration of hydroxide ion in a solution made by mixing 200 mL of 0.12 M NaOH with 300 mL of 0.15 M $Ba(OH)_2$? Assume that the final volume of solution is 500 mL.

4.84 What is the molar concentration of chloride ion in a solution formed by mixing 150 mL of 1.5 *M* NaCl with 200 mL of 2.0 *M* $CaCl_2$? Assume that the final volume of solution is 350 mL.

4.85 Sodium thiosulfate, $Na_2S_2O_3$, is used as a "fixer" in developing photographic film. The amount of $Na_2S_2O_3$ in a solution can be determined by a titration with I_2 according to the equation

$$2Na_2S_2O_3(aq) + I_2(aq) \longrightarrow Na_2S_4O_6(aq) + 2NaI(aq)$$

Calculate the concentration of $Na_2S_2O_3$ if 30.3 mL of a 0.112 *M* I_2 solution will react with a 100-mL sample of $Na_2S_2O_3$.

4.86 Tin(II) fluoride (stannous fluoride) is added to toothpaste as a convenient source of fluoride ion, which is known to help minimize tooth decay. The concentration of stannous fluoride in a particular toothpaste can be determined by precipitating the fluoride as PbClF.

$$SnF_2(aq) + 2Pb^{2+}(aq) + 2Cl^-(aq) \longrightarrow 2PbClF(s) + Sn^{2+}(aq)$$

The concentrations of Pb^{2+} and Cl^- are controlled so that $PbCl_2$ does not precipitate. If a sample of toothpaste that weighs 10.50 g produces a PbClF precipitate that weighs 0.105 g, what is the percentage of SnF_2 in the toothpaste?

4.87 A 0.3120-g sample of a compound made up of aluminum and chlorine yields 1.006 g of AgCl when mixed with enough $AgNO_3$ to react completely with all of the chlorine. What is the empirical formula of the compound?

4.88 Highly toxic NO gas can be prepared in the laboratory by the careful reaction of dilute sulfuric acid with an aqueous solution of sodium nitrite as shown below. What volume of 1.22 *M* sulfuric acid reacting with excess sodium nitrite is needed to prepare 2.44 g of NO?

$$3H_2SO_4(aq) + 3NaNO_2(aq) \longrightarrow 2NO + HNO_3(aq) + 3NaHSO_4(aq) + H_2O(\ell)$$

4.89 Although silver chloride is insoluble in water, addition of ammonia to a mixture of water and silver chloride causes the silver ions to dissolve as the $[Ag(NH_3)_2]^+$ ion. What is the concentration of $[Ag(NH_3)_2]^+$ ions that results from the addition of excess ammonia to a mixture of water and 0.022 g of silver chloride if the final volume of the solution is 150 mL?

4.90 Tetrachloroethane, $C_2H_2Cl_4$, is very effective at dissolving greases and oils. It is also nonflammable. These properties have made it an attractive solvent in industry. Unfortunately, long-term exposure can lead to jaundice and liver damage, so its use is being eliminated. Calculate ΔH for the reaction

$$C_2H_2(g) + 2Cl_2(g) \longrightarrow C_2H_2Cl_4(\ell) \qquad \Delta H = ?$$

given these equations:

$$2C(s) + H_2(g) \longrightarrow C_2H_2(g) \qquad \Delta H = +227 \text{ kJ}$$

$$2C(s) + H_2(g) + 2Cl_2(g) \longrightarrow C_2H_2Cl_4(\ell) \qquad \Delta H = +130 \text{ kJ}$$

4.91 Dissolving 6.00 g of $CaCl_2$ in 300 mL of water causes the temperature of the solution to increase by 3.43 °C. Assuming that the specific heat of the solution is 4.18 J/g-K, determine ΔH for the reaction

$$CaCl_2(s) \xrightarrow{H_2O} Ca^{2+}(aq) + 2Cl^-(aq)$$

mac

Chapter Five

The Gaseous State

Matter exists in three physical states: solid, liquid, and gas. In this chapter, the behavior and properties of gases are discussed. Our atmosphere is a sea of gases, mainly nitrogen and oxygen. A number of other gases are present in the atmosphere at low concentrations, and some of them are very important to life. Carbon dioxide, CO_2, is necessary for plants to survive, but many scientists are worried that an increase in the concentration of CO_2 in the air could contribute to global warming. Another important atmospheric gas is ozone, O_3. This is a highly toxic gas that is a source of air pollution at ground level, but at high altitudes it reduces the intensity of dangerous radiation from the sun.

Many gases are important in industrial processes. Both oxygen and nitrogen are separated from air and sold on a large scale. More than 50 billion pounds of nitrogen and 35 billion pounds of oxygen are produced in the United States each year. Most of the nitrogen is converted to another important gas, ammonia (NH_3), for use in fertilizers and to make plastics and fibers. Oxygen is used in the production of steel (Figure 5.1) and other metals, and for the propulsion of the NASA space shuttles.

Figure 5.1 Steel production Oxygen is used in the production of steel.

Natural gas, which is mainly methane (CH_4) formed by the decay of plants, is found trapped underground. It is used to heat homes, for cooking, and in the manufacture of hydrogen gas. Other gases are manufactured or separated from crude oil. An example is ethylene, C_2H_4, which is used in many ways, including the production of the plastic polyethylene. Hydrogen is produced from water as well as from natural gas; it is used widely in industry, and combines with oxygen for the propulsion of the NASA space shuttles.

In this chapter, the physical properties and behavior of gases will be discussed. We will develop a molecular model that explains the behavior of gases.

◀ Hot air balloons. The ideal gas law explains many of the properties of gases, including why hot air balloons rise.

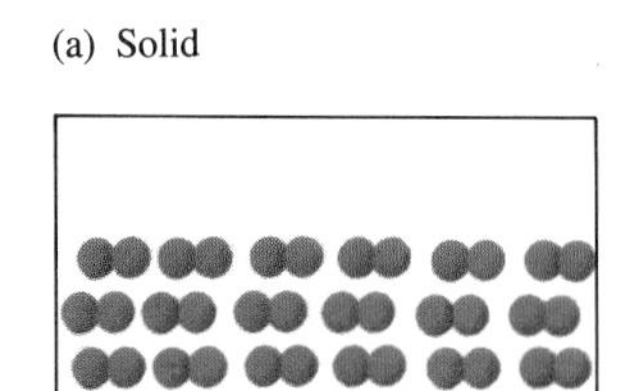

(b) Liquid

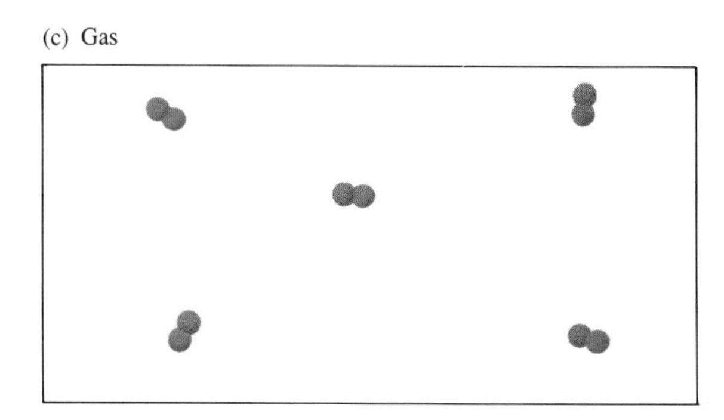

Figure 5.2 Molecules of bromine in the solid, liquid, and gas phases (a) Bromine in the solid phase has fixed shape and volume. (b) The liquid phase has a definite volume but not a fixed shape. (c) The gas phase has neither definite volume nor shape.

5.1 Properties and Measurements of Gases

Objectives

- To describe the characteristics of the three states of matter: solid, liquid, and gas
- To define the pressure of a gas and the units in which it is measured

A **gas** is a fluid that has no definite shape or volume. When placed in a container, a gas expands to fill the total volume of the container because the spaces between gas particles can change. For the same reason, a gas is also *compressible;* the volume occupied by a sample decreases with increases in pressure. A **liquid** is a fluid, a sample of which has a fixed volume but no definite shape. Like a gas, a sample of a liquid will take the shape of its container, but a liquid has definite volume and does not expand to fill the container. A sample of a **solid** has both fixed shape and volume. Both liquids and solids are **condensed phases,** phases that are resistant to volume changes because the spaces between the particles cannot readily change. Figure 5.2 shows how molecules of the element bromine are arranged in each of the three states.

As a consequence of the large distances between the molecules, the density of the gas phase is much lower than that of either of the condensed phases. Density is generally expressed in g/L for a gas, but the densities of liquids and solids are expressed in g/mL. A gas under atmospheric conditions is about 1000 times less dense than its liquid or solid state.

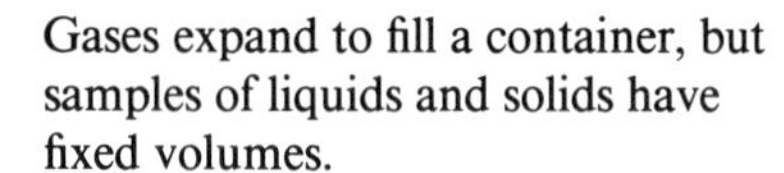

Gases expand to fill a container, but samples of liquids and solids have fixed volumes.

Pressure of a Gas

Pressure exerted on a surface is the force per unit area. The atmosphere of gas in which we live exerts a pressure due to the gravitational force of the earth attracting the gas molecules in the air. On the earth's surface, a sea of gas more than ten miles high presses down on us. We generally do not notice this pressure because it surrounds everything equally, but if you change altitude rapidly you can feel your ears "pop" because the pressure on the inner side of the eardrum changes more slowly than the outer pressure. At higher altitudes, on a mountain for example, the pressure of the atmosphere is less than at sea level because there are fewer gas molecules above you.

The pressure of the atmosphere is measured with a barometer, as shown in Figure 5.3. A long glass tube, sealed at one end, is filled completely with mercury and inverted into a dish of mercury. Gravitational attraction pulls down the column of mercury, producing a vacuum above the liquid in the tube. The column of liquid stops falling when the pressure caused by the mass of the mercury in the column is equal to the pressure exerted by the atmosphere on the surface of the mercury in the dish. Thus, the height of the mercury column is a measure of the atmospheric pressure. At sea level, the mercury column will be 760 mm high on an average day. If the mercury level in the barometer rises, the weather forecaster reports high pressure; if the atmospheric pressure is low the mercury level drops. Mercury is used in a barometer because it is a very high density liquid: 13.6 g/mL. When water is used in a barometer, the column of water is more than 10 meters high.

Pressure *differences* can be measured by using a manometer, one type of which is pictured in Figure 5.4. Mercury is placed in a U-shaped tube connected to the container of gas. The atmosphere exerts a pressure on the mercury surface on the open end of the tube, and the gas within the container exerts pressure on the other surface of the mercury. The difference between the heights of the two mercury surfaces is a measure of the difference between the pressure exerted by the gas in the container and the atmospheric pressure. The mercury column is lower on the end of the U-tube that experiences the higher pressure.

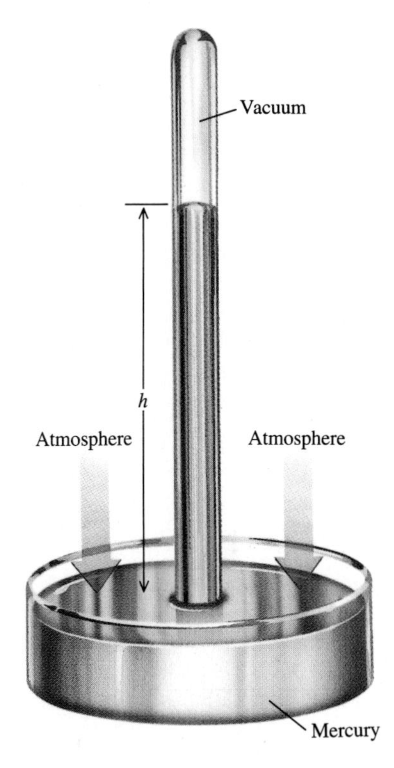

Figure 5.3 A barometer The pressure exerted by the atmosphere supports a column of mercury. The height of the column is used to measure the pressure of the atmosphere.

Units of Pressure Measurement

The SI unit of pressure is the pascal (Pa), named for the French scientist Blaise Pascal (1623–1662):

$$1 \text{ Pa} = 1 \text{ N/m}^2 = 1 \text{ kg/m s}^2$$

where N is the newton, the SI unit for force ($1 \text{ N} = 1 \text{ kg m/s}^2$), m is the meter, and s is the second. This unit of pressure is very small and has not been widely used by chemists.

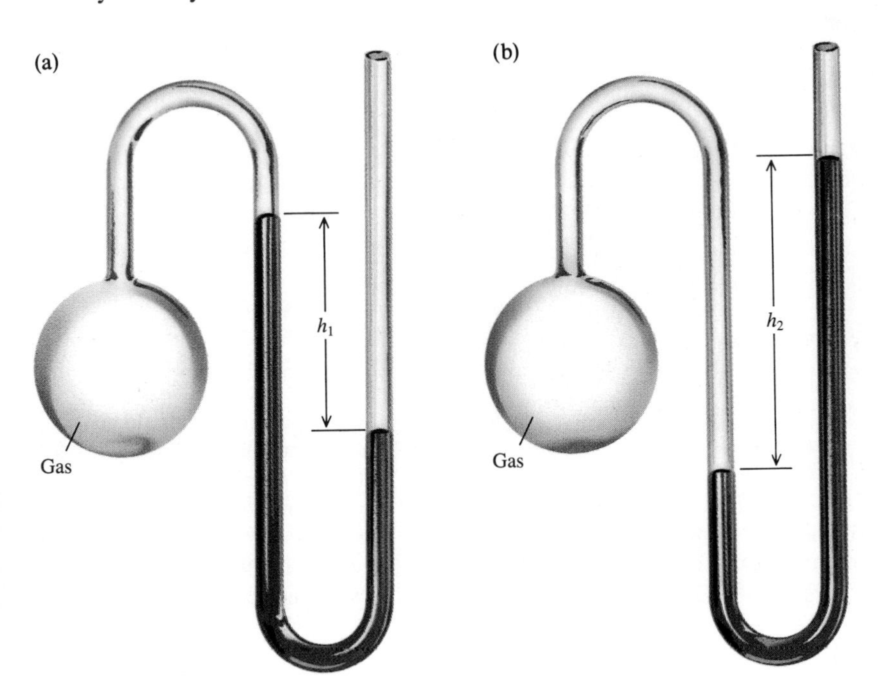

Figure 5.4 A manometer The difference between the heights of the mercury surfaces (h) in a U-tube measures the difference in pressure of a gas sample from atmospheric pressure. In (a) the pressure of the gas is less than atmospheric, and in (b) the pressure of the gas is greater than atmospheric.

INSIGHTS into CHEMISTRY

A Closer View

Plasmas — The Fourth State of Matter

A **plasma** is an ionized, electrically conductive gas. It is electrically neutral, containing enough free electrons to neutralize the charges on the positive ions present.

Plasmas are not rare; they are the most common form of matter in our universe. The most familiar is the sun, which is mainly a hydrogen plasma. In this hydrogen plasma, most of the hydrogen atoms have ionized, producing free electrons:

$$H \longrightarrow H^+ + e^-$$

The resultant plasma is electrically neutral, but ionic in nature. The mass of the sun is 1.99×10^{30} kg; the combined mass of the nine known planets is 2.57×10^{27} kg. These numbers tell us that our solar system is about 99.9% plasma, with the total of solids, liquids, and gases being only about 0.1%.

Plasmas are formed as the result of the interaction of energy with matter. When the energy input is high, ions and electrons are formed, and the matter becomes a plasma. The sun is an example of a nuclear fusion-powered plasma, but plasmas can also be formed by much less powerful sources.

Perhaps the most familiar plasmas are found in light sources. Common fluorescent lights, mercury and sodium vapor street lights, and the xenon arc lamps used in movie projectors are examples of plasma sources. The xenon arc lamp is fairly typical. A small spark between the two electrodes starts the ionization of the gas between them. Then a larger electric current is applied, and the ionized gas conducts the electrical current. The additional current causes more ionization, until a very energetic mixture of ions and electrons is produced:

$$Xe \longrightarrow Xe^+ + e^-$$

The electrons are attracted to the positive electrode; the cations move to the negative electrode. Many ion-electron collisions occur that result in the transfer of energy to the ions. The highly energetic ions emit radiation as part of a "cooling" mechanism.

The plasma can be used directly as a light source, or in combination with some other materials to tailor the light for a specific purpose. For example, light from the plasma in a fluorescent tube would be an unflattering blue-green, but the manufacturer coats the walls of the tube with materials called "phosphors." These phosphors, often compounds of the lanthanide series elements, absorb the blue-green radiation from the plasma and emit radiation that has about the same color as visible light.

Fluorescent tubes are available in a wide variety of colors. Tubes are available that mimic warm daylight (pinkish, near sunrise and sunset) or cold daylight (a north-facing window). Tubes are even available that concentrate the light in the colors needed by plants—"grow-lights" are heavy on blue and red but contain very little green.

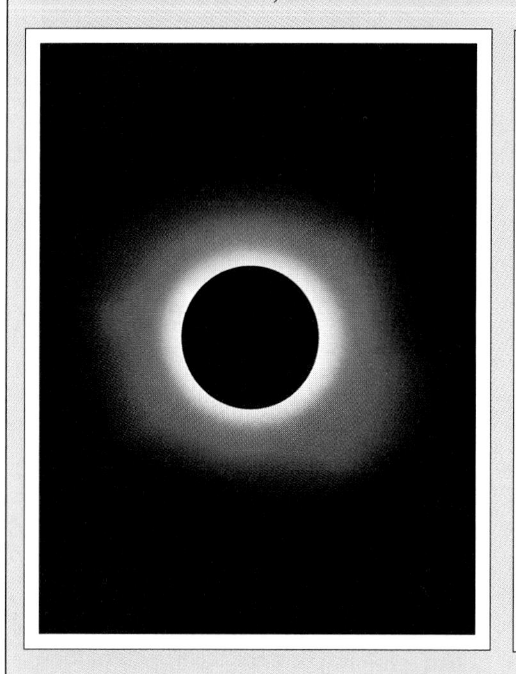

(a)

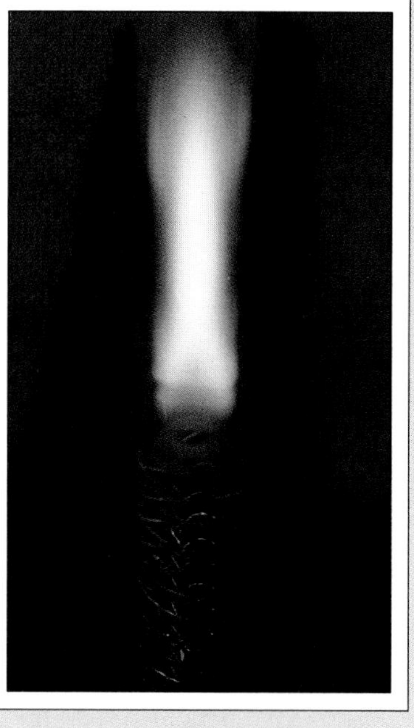

(b)

Plasmas (a) The sun is a plasma. Pictured here are solar flares, observable at a full eclipse. (b) A plasma can be easily prepared in the laboratory. In this photograph, the plasma is formed by the interaction of microwave energy and helium gas.

Table 5.1 Relationships Between Pressure Units

1 atm = 760 mm Hg	1 torr = 133.3 Pa
1 atm = 760 torr	1 atm = 14.7 psi
1 atm = 101.325 kPa	1 atm = 22.92 in. Hg
1 mm Hg = 133.3 Pa	

Two more convenient units for pressure are based on the mercury barometer and manometer. One standard atmosphere of pressure (1 atm) is the normal pressure at sea level, which is defined as the pressure exerted by a column of mercury exactly 760 mm high:

$$1 \text{ atm} = 760 \text{ mm Hg} = 101.325 \text{ kPa}$$

Another name for the unit "mm Hg" is the *torr,* so

$$1 \text{ torr} = 1 \text{ mm Hg}$$

The torr is named to recognize Evangelista Torricelli (1608–1647), an Italian student of Galileo who invented the barometer.

The English pressure unit, pounds per square inch (psi), is used in many engineering applications. This text will generally use atmospheres and torr to express pressure. Table 5.1 shows important relationships needed to convert between various units of pressure.

Example 5.1 Conversion of Pressure Units

Convert a pressure of 0.450 atm into units of:

(a) torr
(b) kPa

Solution

(a) The equality 1 atm = 760 torr is used to determine the pressure in torr.

$$\text{pressure torr} = 0.450 \text{ atm} \left(\frac{760 \text{ torr}}{1 \text{ atm}}\right) = 342 \text{ torr}$$

(b) The required conversion factor is obtained from the equation 1 atm = 101.3 kPa.

$$\text{pressure kPa} = 0.450 \text{ atm} \left(\frac{101.3 \text{ kPa}}{1 \text{ atm}}\right) = 45.6 \text{ kPa}$$

Understanding Express a pressure of 433 torr in atmospheres.

Answer: 0.570 atm

5.2 The Ideal Gas Law

Objectives

- To explain how the volume of a gas depends on pressure, temperature, and amount of sample
- To write the ideal gas law

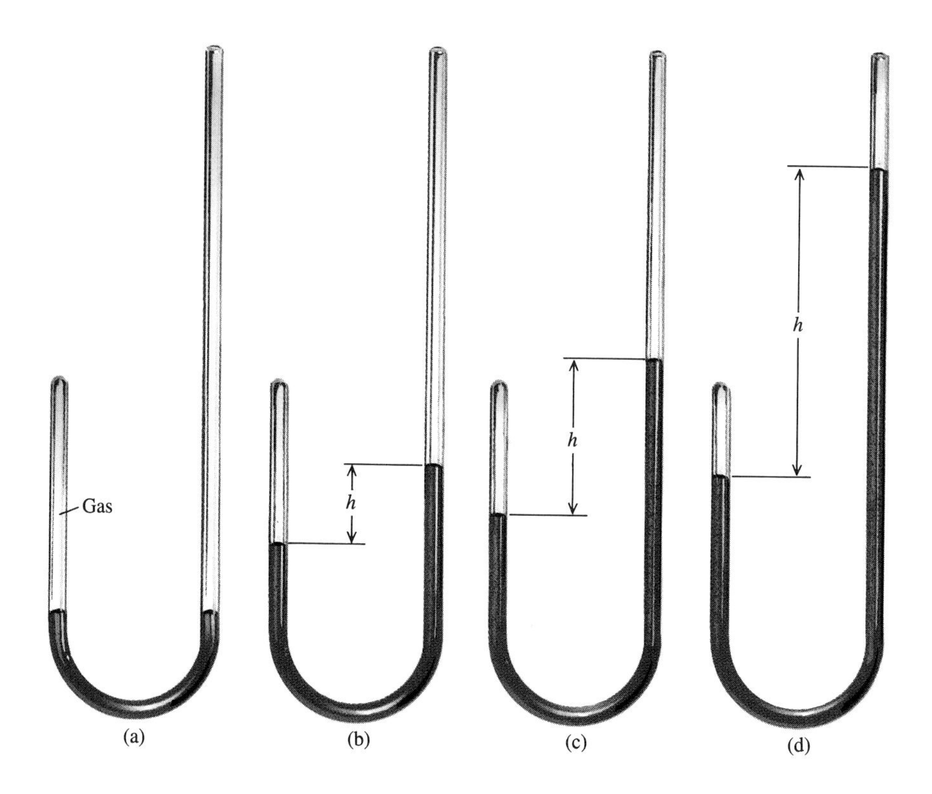

Figure 5.5 Change in volume of a gas with change in pressure From left to right, the increasing pressure on the sample of gas reduces the volume occupied by the sample.

The results of experiments performed over centuries demonstrate that four quantities are needed to describe the state of a gas: pressure (P), volume (V), temperature (T), and number of moles (n). A change in any one of these properties will influence the others. To illustrate these interrelationships, the change in volume of a gas will be examined as any one of the other three quantities varies, while holding the remaining two constant. Remember that these relationships apply to the gas phase only.

Volume and Pressure: Boyle's Law

Figure 5.5 shows an experiment that determines how pressure, measured with a manometer, influences volume of a gas sample. The pressure on the sample of gas is increased by adding mercury to the open end of the manometer. The pressure of the gas sample in the closed end of the tube is equal to atmospheric pressure plus the difference in the height (h) of the mercury surfaces. The experiment shows that the volume of the gas decreases as the pressure increases.

A plot of the volume measured in this experiment, as a function of the inverse of the pressure, is a straight line (Figure 5.6). The mathematical description of this relationship was first noted by an Irish chemist, Robert Boyle (1627–1691). **Boyle's law** states that at constant temperature the volume of a sample of gas is inversely proportional to the pressure. In equation form,

$$V = k_1 \times \frac{1}{P}$$

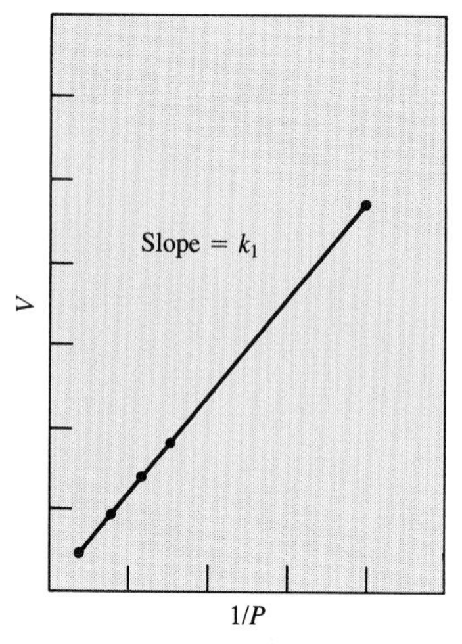

Figure 5.6 Plot of volume versus the inverse of pressure The volume of a gas is inversely proportional to the pressure.

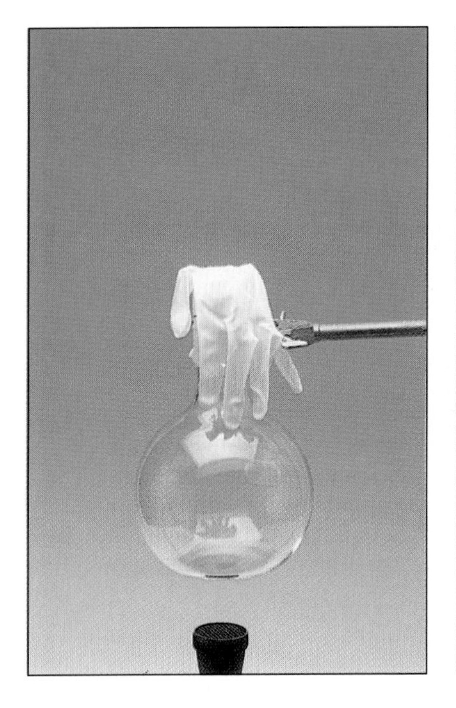

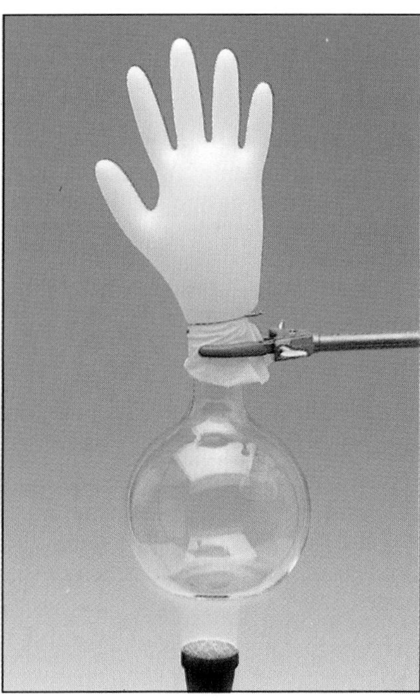

Figure 5.7 Heating a gas Heating a sample of gas causes the volume of the gas to increase when the pressure remains constant.

where k_1 is a constant (the slope of the line in Figure 5.6) that depends on the temperature and the amount of matter in the gas sample, the two quantities that were held constant in this experiment.

Volume and Temperature: Charles's Law

Figure 5.7 shows the effect of changing temperature on the volume of a gas, holding the pressure and amount of gas sample constant. Heating the gas increases the volume.

Figure 5.8 is a plot of the experimentally determined volumes of three different samples of gas as the temperature varies. When the Kelvin scale is used to measure temperature, doubling the temperature causes the volume of the gas to double. This relationship was determined by a French chemist and balloonist, Jacques Charles (1746–1823). **Charles's law** states that at constant pressure the volume of a fixed amount of gas is proportional to the absolute temperature, or

$$V = k_2 \times T$$

The graphs in Figure 5.8 give the experimental basis for the development of the Kelvin temperature scale and describe one of the first measurements to suggest that an absolute zero of temperature exists. Charles's law indicates that at absolute zero the volume of the gas must be zero. Does matter disappear at absolute zero? No—all gases liquefy before this temperature is reached. Since the basis for the graph is the measurement of the volume of a *gas,* Charles's law no longer applies once the sample becomes a liquid or a solid. Nevertheless, the graph can be extrapolated to zero volume, allowing the determination of the zero on the temperature scale. All three samples

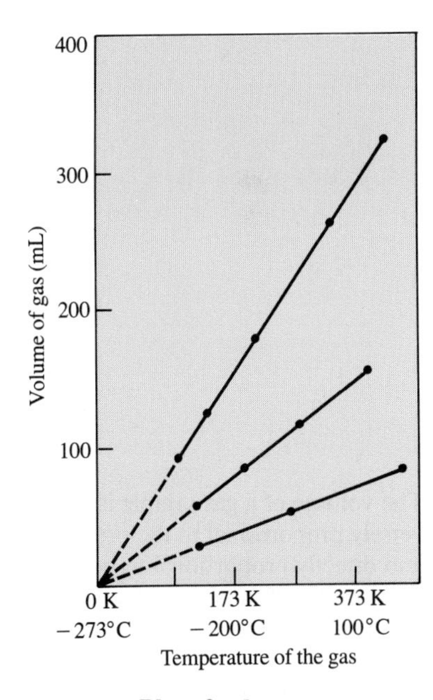

Figure 5.8 Plot of volume versus temperature The graph shows experimentally determined volumes of three gas samples as the temperature changes. The dotted lines are continuations of the experimental lines to lower temperatures. These extrapolations all reach zero volume at −273 °C.

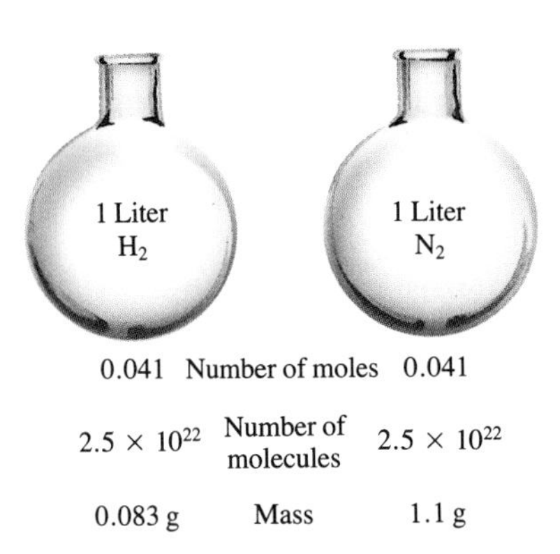

Figure 5.9 Flasks of two gases Equal-sized flasks of hydrogen and nitrogen gas at the same temperature and pressure contain the same numbers of molecules, but different masses.

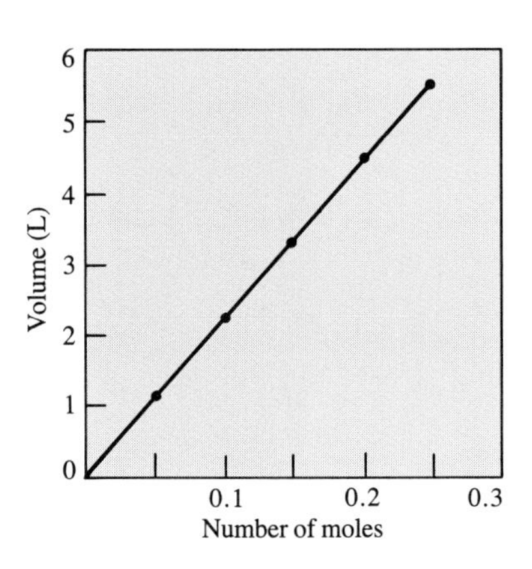

Figure 5.10 Plot of volume versus amount The volume of a gas at constant pressure and temperature is directly proportional to the number of moles of gas present.

reach a volume of zero at the same temperature. This temperature is called absolute zero and has the value −273.15 °C, the zero point of the Kelvin scale as outlined in Chapter 1.

Volume and Amount: Avogadro's Law

In 1811 Amedeo Avogadro proposed that at the same temperature and pressure, equal volumes of gases contain the same number of molecules. Over several decades, Avogadro's hypothesis was tested and found to be true within a few percent. The flasks in Figure 5.9 illustrate Avogadro's hypothesis for samples of hydrogen and nitrogen at normal temperature and pressure.

A modern statement of **Avogadro's law** is at constant pressure and temperature, the volume of a gas sample is proportional to the number of moles of gas present.

$$V = k_3 \times n$$

It is important to remember that the value of k_3 in Avogadro's law is the same at any given temperature and pressure regardless of the particular gas in the sample. Figure 5.10 shows Avogadro's law graphically.

The volume of a gas sample is inversely proportional to the pressure and directly proportional to both the number of moles and the temperature, in kelvins.

The Ideal Gas Law

The three laws, Boyle's, Charles's, and Avogadro's, state how volume changes with changes in pressure, temperature, and number of moles:

$$V = k_1 \times \frac{1}{P} \qquad \text{Boyle's law}$$

$$V = k_2 \times T \qquad \text{Charles's law}$$

$$V = k_3 \times n \qquad \text{Avogadro's law}$$

These three laws can be combined into a single equation known as the **ideal gas law:**

$$PV = nRT \quad [5.1]$$

where R is known as the ideal gas constant. The value of the constant R can be determined experimentally. The volume of *one mole* of an ideal gas at 273.1 K (0 °C) and 1.000 atm is 22.41 L. The conditions of 0 °C and 1 atm are known as **standard temperature and pressure (STP).** By substituting these values into the ideal gas equation, the value of R can be calculated:

$$R = \frac{PV}{nT} = \frac{(1.000 \text{ atm})(22.41 \text{ L})}{(1 \text{ mol})(273.1 \text{ K})} = 0.0826 \text{ L atm/mol K}$$

As shown in Table 5.2, the numerical value of R is dependent on the units used to measure pressure and volume. The ideal gas law is followed closely for all common gases, like H_2, O_2, and N_2, at normal temperatures and pressures.

Table 5.2 Values for the Ideal Gas Constant

R	Units
0.08206	L atm/mol K
8.314	J/mol K
8.314	kg m^2/s^2 mol K
1.987	cal/mol K

The ideal gas law expresses the interrelationships of volume, pressure, amount, and temperature.

5.3 The Influences of Changing Conditions on Gases

Objective

- To determine how a gas sample responds to changes in volume, pressure, and temperature

A problem commonly faced in the laboratory or chemical production plant is to determine how one of the four quantities that describe the state of a gas is influenced by changes in one or more of the other three. For example, if a methane tank is filled to a pressure of 3.2 atm on a winter day when the temperature is 2 °C, it is important to determine what the pressure will be on a summer day when the temperature has risen to 38 °C. If the pressure is going to increase substantially, a container that can withstand high pressures must be used.

The ideal gas law can be used to derive the equations that describe how any of the four quantities will vary, given changes in any of the other three. In problems of this type, we are given the state of a sample of gas under one set of conditions (the first set of conditions), and we need to find its state under a second set of conditions. In the case of the methane tank, the pressure is 3.2 atm at 2 °C (the first set of conditions), and we need to calculate the pressure of the sample at 38 °C (the second set of conditions). From the statement of the problem, it is clear that the number of moles of gas and the volume of the container do not change.

The ideal gas law is used to calculate changes in any of the four quantities, given changes in any others.

To solve the problem, we rearrange the ideal gas equation to place the variable quantities on the left side and the constant quantities on the right side.

State One	*State Two*
$P_1V = nRT_1$	$P_2V = nRT_2$
$\frac{P_1}{T_1} = \frac{nR}{V}$	$\frac{P_2}{T_2} = \frac{nR}{V}$

The right sides of both equations are the same; R is a constant, and in the experiment the volume and amount of sample did not change. Therefore, the left sides of both equations must be equal.

$$\frac{P_1}{T_1} = \frac{P_2}{T_2}$$

This equation expresses the relationship of the temperatures and pressures of a gas at two different conditions, with the volume and amount of the sample held constant. The data given in the problem can be arranged in tabular form.

State One	*State Two*
$P_1 = 3.2$ atm	$P_2 = ?$
$T_1 = 2$ °C	$T_2 = 38$ °C

In working any problem involving the gas laws, *temperature must be expressed in kelvins,* because it is the Kelvin scale that is based on absolute zero. In problems such as this one, pressure (and volume, in similar problems in which that is the variable) can be expressed in any appropriate units as long as both pressures (or volumes) are expressed in the same units. Converting the temperatures to kelvins ($T_K = T_C + 273$) yields

State One	*State Two*
$P_1 = 3.2$ atm	$P_2 = ?$
$T_1 = 275$ K	$T_2 = 311$ K

Rearrange the equation to place only the unknown quantity, P_2, on the left, and solve the equation by substituting the known values for P_1, T_1, and T_2.

$$\frac{P_1}{T_1} = \frac{P_2}{T_2}$$

$$P_2 = \frac{P_1 \times T_2}{T_1} = \frac{(3.2 \text{ atm})(311 \not{\text{K}})}{275 \not{\text{K}}} = 3.6 \text{ atm}$$

As with any problem, it is good practice to see whether the answer is reasonable. In the problem, the temperature increased. Pressure, like volume, is directly proportional to the absolute temperature (this relation is known as Amonton's law), and in the calculation the pressure did indeed increase from 3.2 to 3.6 atm as the temperature increased. The increase in pressure is modest because on the Kelvin scale the temperature changed from 275 to 311, just over 10%. The calculation indicates that because the increase in pressure going from winter to summer temperatures is modest, the original container for the methane gas is safe.

Example 5.2

In the lungs of a deep-sea diver ($V = 6.0$ L) at a depth of 100 meters, the pressure of the air is 9.7 atm. At a constant temperature of 37 °C, to what volume would the lungs expand if the diver was immediately brought to the surface (1.0 atm)?

Solution In this case the ideal gas equation already has the quantities that vary on the left side.

$$P_1V_1 = nRT \qquad P_2V_2 = nRT$$

The values of n, R, and T are constant, so the left sides of the two equations can be set equal.

$$P_1V_1 = P_2V_2$$

Set up a table of the known data.

$$P_1 = 9.7 \text{ atm} \qquad P_2 = 1.0 \text{ atm}$$
$$V_1 = 6.0 \text{ L} \qquad V_2 = ?$$

Rearrange the equation to place only the unknown quantity on the left, and solve the equation by substituting the known values.

$$V_2 = \frac{P_1V_1}{P_2} = \frac{(9.7 \text{ atm})(6.0 \text{ L})}{1.0 \text{ atm}} = 58 \text{ L}$$

Clearly, the diver needs to expel gas when rising to the surface.

Understanding At a pressure of 740 torr, a sample of gas occupies 5.00 L. Calculate the volume of the sample if the pressure is changed to 760 torr at constant temperature.

Answer: 4.87 L

Deep-sea diver A diver must rise from the bottom very slowly to allow time to expel the excess air from the lungs.

Example 5.3 Temperature and Volume Changes

A balloon filled with oxygen gas at 25 °C occupies a volume of 2.1 L. Assuming that the pressure remains constant, what will the volume be at 100 °C?

Solution This problem deals with Charles's law; the volume of a sample of gas will increase with increasing temperature at constant pressure. Rearrange the ideal gas law to place the variable quantities on the left side, and set the left sides of the two equations equal.

$$PV_1 = nRT_1 \qquad PV_2 = nRT_2$$

$$\frac{V_1}{T_1} = \frac{nR}{P} \qquad \frac{V_2}{T_2} = \frac{nR}{P}$$

$$\frac{V_1}{T_1} = \frac{V_2}{T_2}$$

Set up a table of the known data.

$$V_1 = 2.1 \text{ L} \qquad V_2 = ? \text{ L}$$
$$T_1 = 25 + 273 = 298 \text{ K} \qquad T_2 = 100 + 273 = 373 \text{ K}$$

Rearrange the equation to place only the unknown quantity on the left, and solve the equation by substituting the known values.

$$V_2 = \frac{V_1 \times T_2}{T_1} = \frac{(2.1 \text{ L})(373 \text{ K})}{298 \text{ K}} = 2.6 \text{ L}$$

As predicted by Charles's law, the volume increases as the temperature increases.

Understanding The volume of a sample of nitrogen gas at 27 °C changes from 0.440 L to 1.01 L when the sample is heated. What was the second temperature of the nitrogen sample?

Answer: 416 °C

Weather balloon Weather balloons are used to sample conditions in the upper atmosphere. They are not completely filled at launch because the helium expands as the balloon rises due to the decrease in pressure.

Example 5.4 Pressure, Volume, and Temperature Changes

A helium weather balloon is filled to a volume of 219 L on the ground, where the pressure is 754 torr and the temperature is 298 K. It is important to know how much the gas will expand as the balloon rises (pressure will fall) to ensure that the balloon can withstand the expansion. What will the volume be at an altitude of 10,000 meters (where the atmospheric pressure will be 210 torr) and a temperature of 230 K?

Solution Rearrange the ideal gas equation to place the variables on the left side, and set the left sides of the two equations equal.

$$P_1V_1 = nRT_1 \qquad P_2V_2 = nRT_2$$

$$\frac{P_1V_1}{T_1} = nR \qquad \frac{P_2V_2}{T_2} = nR$$

$$\frac{P_1V_1}{T_1} = \frac{P_2V_2}{T_2}$$

Make a table of the values given in the problem. The temperatures are already given in kelvins.

$P_1 = 754$ torr $P_2 = 210$ torr
$V_1 = 219$ L $V_2 = ?$
$T_1 = 298$ K $T_2 = 230$ K

Rearrange the equation to place only V_2 on the left side, and solve the equation by substituting the known values.

$$V_2 = \frac{P_1V_1T_2}{T_1P_2} = \frac{(754\ \cancel{\text{torr}})(219\ \text{L})(230\ \cancel{\text{K}})}{(298\ \cancel{\text{K}})(210\ \cancel{\text{torr}})} = 607\ \text{L}$$

The volume of the balloon will nearly triple as it rises.

Understanding The pressure of a sample of gas is 2.60 atm in a 1.54-L container at a temperature of 0 °C. Calculate the pressure exerted by this sample if the volume changes to 1.00 L and the temperature changes to 27 °C.

Answer: 4.40 atm

5.4 Calculations Using the Ideal Gas Law

Objectives

- To use the ideal gas law to calculate the pressure, volume, amount, or temperature of a gas, given values of the other three quantities
- To calculate the molar mass and the density of gas samples by using the ideal gas law

The ideal gas law is used to determine the value of any of the four quantities pressure, volume, amount, or temperature of a gas given values of the other three.

The ideal gas law can be used to calculate any of the four variables—pressure, volume, amount, or temperature—given the known values of the other three. In these calculations the value of R will be used, so it is necessary to convert the units of pressure into atmospheres and those for volume into liters.

The procedure can be illustrated by calculating the number of moles in a sample of argon gas that occupies a volume of 298 mL at a pressure of

351 torr and a temperature of 25 °C. First, the known values must be converted to match the units used in R.

For volume, 298 mL = 0.298 L.

Since 1 atm = 760 torr, the conversion of pressure to atmospheres is

$$\text{pressure in atm} = 351 \text{ torr}\left(\frac{1 \text{ atm}}{760 \text{ torr}}\right) = 0.462 \text{ atm}$$

For temperature,

$$T_K = T_C + 273 = 25 + 273 = 298 \text{ K}$$

Rearrange the ideal gas law to place the unknown, the number of moles, on the left, and solve the equation by substituting the known values with the necessary units.

$$PV = nRT$$

$$n = \frac{PV}{RT} = \frac{(0.462 \text{ atm})(0.298 \text{ L})}{(0.0821 \text{ L atm/mol K})(298 \text{ K})} = 5.63 \times 10^{-3} \text{ mol}$$

Note that the units cancel, leaving mol, the correct unit of the answer. Always carefully write and cancel units to ensure that you have combined the quantities correctly.

Example 5.5 Pressure of a Gas

Calculate the pressure of methane gas in a 3.3-L container that contains 1.2 mol of the gas at 25 °C.

Solution Make a table of the measured values with the necessary units.

$V = 3.3$ L $\quad n = 1.2$ mol $\quad T = 298$ K

Rearrange the ideal gas law with pressure on the left side, and solve the equation by substituting the known values.

$$PV = nRT$$

$$P = \frac{nRT}{V} = \frac{(1.2 \text{ mol})(0.0821 \text{ L atm/mol K})(298 \text{ K})}{3.3 \text{ L}} = 8.9 \text{ atm}$$

Understanding Calculate the temperature of a 350-mL container that contains 0.620 mol of an ideal gas at a pressure of 42.0 atm.

Answer: 289 K

Calculation of Molar Mass Using the Ideal Gas Law

Measuring the molar mass of gases is an important application of the ideal gas law. Before the development of mass spectroscopy, the molar masses of many substances were determined using the ideal gas law. The ideal gas law is used to calculate the number of moles (n) in a gas sample of known mass (m), then the molar mass ($\mathcal{M}$) is found using Equation 5.2, as shown in Example 5.6.

$$\mathcal{M} = \frac{m}{n} \qquad [5.2]$$

Example 5.6 Molar Mass

Calculate the molar mass of a gas if 0.495 g of the gas occupies 127 mL at 98 °C and 754 torr pressure.

Solution Make a table of the measured values with correct units.

$m = 0.495$ g $T = 371$ K

$P = 0.992$ atm $V = 0.127$ L

Use the ideal gas law to calculate the number of moles, n, of gas.

$$n = \frac{PV}{RT} = \frac{(0.992 \text{ atm})(0.127 \text{ L})}{(0.0821 \text{ L atm/mol K})(371 \text{ K})}$$

$$n = 4.14 \times 10^{-3} \text{ mol}$$

Use Equation 5.2 to convert the number of moles and mass of sample into molar mass.

$$\mathcal{M} = m/n = 0.495 \text{ g}/4.14 \times 10^{-3} \text{ mol}$$

$$\mathcal{M} = 120 \text{ g/mol}$$

Understanding Calculate the molar mass of a gas if a 19.2-g sample occupies 4.3 L at 27 °C and a pressure of 342 torr.

Answer: 244 g/mol

Often the molar mass is found from the density of the gas. Since the density is the mass of a one-liter sample of the gas, we solve for the molar mass by assuming a sample of one liter, as illustrated in Example 5.7.

Example 5.7 Molar Mass from the Density of a Gas

The density of a gas sample is 0.714 g/L at STP. What is the molar mass of this gas?

Solution The gas density tells us that a 1.00-L sample has a mass of 0.714 g. We therefore assume a sample of 1.00 L, and find the molar mass using the same strategy as in Example 5.6. From the definition of STP we construct a table of the data:

$m = 0.714$ g $T = 273$ K

$P = 1.00$ atm $V = 1.00$ L

First, from the pressure, temperature and volume, the number of moles in the sample is found by substituting into the ideal gas law.

$$n = \frac{PV}{RT} = \frac{(1.00 \text{ atm})(1.00 \text{ L})}{(0.0821 \text{ L atm/mol K})(273 \text{ K})} = 4.46 \times 10^{-2} \text{ mol}$$

The mass of this one-liter sample is known from the density, so Equation 5.2 is used to calculate the molar mass of the gas.

$$\mathcal{M} = \frac{0.714 \text{ g}}{4.46 \times 10^{-2} \text{ mol}} = 16.0 \text{ g/mol}$$

Understanding Calculate the molar mass of a gas that has a density of 1.64 g/L at 25 °C and a pressure of 763 torr.

Answer: 40.0 g/mol

Molar mass of a gas can be calculated from either the mass and the volume or the density, using the ideal gas law.

The approach to determining molar mass from either the mass and the volume of a gas sample or the density of the sample is the same.

5.5 Equation Stoichiometry Involving Gases

Objectives

- To perform stoichiometric calculations for reactions in which some or all of the reactants or products are gases
- To use volumes of gases in equation stoichiometry problems

The reactants and products in chemical reactions are frequently in the gas phase. The ideal gas law can be used to determine the number of moles, n, for use in problems involving reactions, in much the same way that molar mass is used for solids and molarity is used for compounds in solution. The equivalencies in the chemical equation are used (as in Chapters 3 and 4) to determine the conversion factors that relate moles of one substance to moles of another.

For example, we can determine the volume of hydrogen gas produced in a reaction of 4.21 g of sodium with excess water. The temperature, 25 °C, and the pressure, 0.993 atm, at which the reaction occurs must also be measured.

The strategy for the problem is similar to those for stoichiometric calculations carried out in Chapters 3 and 4.

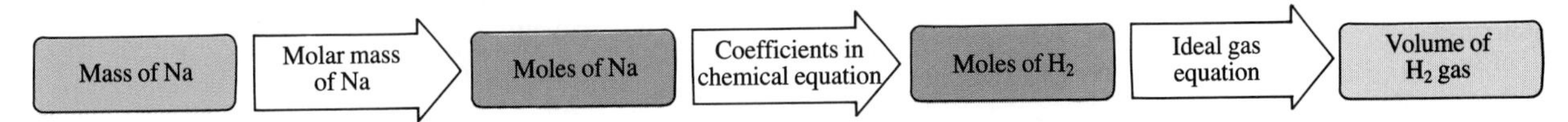

The first step, as always in stoichiometric calculations, is to write the equation.

$$2Na(s) + 2H_2O(\ell) \longrightarrow 2NaOH(s) + H_2(g)$$

Second, convert grams of sodium into moles.

$$\text{moles Na} = 4.21\ \text{g Na}\left(\frac{1\ \text{mol Na}}{22.99\ \text{g Na}}\right) = 0.183\ \text{mol Na}$$

Third, use the coefficients in the equation to calculate the number of moles of hydrogen gas equivalent to 0.183 mol of sodium.

$$\text{moles } H_2 = 0.183\ \text{mol Na}\left(\frac{1\ \text{mol } H_2}{2\ \text{mol Na}}\right) = 0.0915\ \text{mol } H_2$$

Use the ideal gas law to convert the moles of a gas sample into its equivalent volume.

Fourth, calculate the volume of hydrogen gas produced, by using the ideal gas law. The known values are

$P = 0.993$ atm $\qquad V = ?$

$n = 0.0915$ mol H_2 $\qquad T = 298$ K

Solve the ideal gas law for volume.

$$V = \frac{nRT}{P} = \frac{(0.0915\ \text{mol } H_2)(0.0821\ \text{L atm/mol K})(298\ \text{K})}{0.993\ \text{atm}}$$

$$= 2.25\ \text{L } H_2$$

Zinc reacts with hydrochloric acid to give off bubbles of hydrogen gas.

Example 5.8 Using Volumes of Gases in Equations

Hydrogen gas is frequently prepared in the laboratory by the reaction of zinc and hydrochloric acid. The other product is $ZnCl_2$. Calculate the volume of hydrogen produced at 744 torr pressure and 27 °C from the reaction of 32.2 g of zinc and 500 mL of 2.20 molar HCl.

Solution The strategy of this example is interesting because three different methods are used in calculations with the numbers of moles of the three different substances. (1) The molar mass is used to calculate the number of moles from the mass of zinc. (2) The molarity and the volume of solution are used to calculate the number of moles of HCl. (3) The ideal gas law is used to convert the number of moles of hydrogen gas into volume of hydrogen gas. As always, the chemical equation is used to relate the number of moles of one substance to moles of another.

First, write the chemical equation.

$$Zn + 2HCl \longrightarrow ZnCl_2 + H_2$$

Second, use the information given in the problem to calculate the numbers of moles of zinc and hydrochloric acid. The amounts of two reactants are given in this example, so this is a limiting-reactant problem. We need to calculate the number of moles of hydrogen gas produced by complete consumption of each reactant.

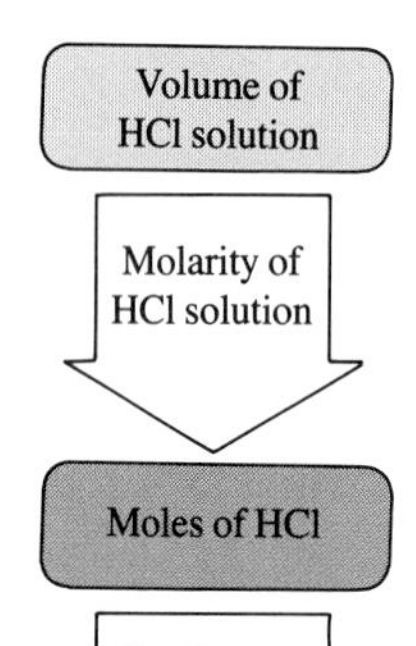

$$\text{moles Zn} = 32.2\ \text{g Zn}\left(\frac{1\ \text{mol Zn}}{65.39\ \text{g Zn}}\right) = 0.492\ \text{mol Zn}$$

$$\text{moles HCl} = 0.500\ \text{L HCl soln}\left(\frac{2.20\ \text{mol HCl}}{1\ \text{L HCl soln}}\right) = 1.10\ \text{mol HCl}$$

Use the coefficients in the equation to calculate the amount of hydrogen one could obtain from each of the reactants.

$$\text{moles } H_2 \text{ based on Zn} = 0.492\ \text{mol Zn}\left(\frac{1\ \text{mol } H_2}{1\ \text{mol Zn}}\right) = 0.492\ \text{mol } H_2$$

$$\text{moles } H_2 \text{ based on HCl} = 1.10\ \text{mol HCl}\left(\frac{1\ \text{mol } H_2}{2\ \text{mol HCl}}\right) = 0.550\ \text{mol } H_2$$

The zinc yields the smaller amount of hydrogen and is therefore the limiting reactant. The problem is completed by using the ideal gas law.

$$P = \frac{744}{760}\ \text{atm} = 0.979\ \text{atm} \qquad V = ?$$

$$n = 0.492\ \text{mol } H_2 \qquad T = 300\ \text{K}$$

$$V = \frac{nRT}{P} = \frac{(0.492\ \text{mol } H_2)(0.0821\ \text{L atm/mol K})(300\ \text{K})}{0.979\ \text{atm}}$$

$$= 12.4\ \text{L } H_2$$

Understanding Many scientists believe that when the earth's atmosphere evolved, some of the oxygen gas came from the decomposition of water induced by solar radiation.

$$2H_2O \xrightarrow{\text{light}} 2H_2 + O_2$$

What volume of oxygen at 754 torr and 40 °C is produced from the decomposition of 2.33 g of H_2O?

Answer: 1.68 L O_2

Volumes of Gases in Chemical Reactions

We have already seen that equal volumes of gases at the same temperature and pressure contain the same number of moles of each gas. *In chemical reactions, the volumes of gases will combine in the same proportions as the coefficients of the equation.* We can thus calculate directly the volume (rather than number of moles) of a gas produced by a reaction of gases as long as the pressure and temperature of both gases are the same. For example, ammonia gas is prepared by the reaction of hydrogen gas and nitrogen gas.

The coefficients in an equation can be interpreted as volumes of gases measured at the same temperature and pressure.

$$3H_2(g) + N_2(g) \longrightarrow 2NH_3(g)$$

$$3\text{ mol} + 1\text{ mol} \longrightarrow 2\text{ mol}$$

$$3\text{ L} + 1\text{ L} \longrightarrow 2\text{ L}$$

The equation states that three moles of hydrogen react with one mole of nitrogen to yield two moles of ammonia. It also states that three liters of hydrogen gas react with one liter of nitrogen gas to produce two liters of ammonia gas (Figure 5.11).

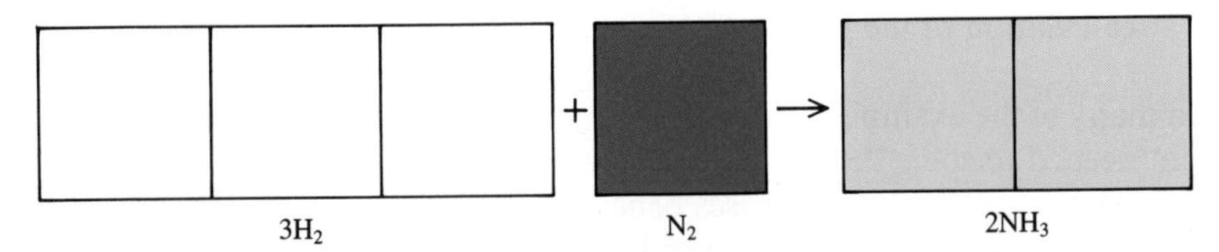

Figure 5.11 Volumes of gases in chemical reactions Three liters of hydrogen gas react with one liter of nitrogen gas to yield two liters of ammonia.

Example 5.9 Volumes of Gases in Chemical Reactions

Nitrogen monoxide, NO, is a pollutant given off by automobiles. It reacts with oxygen in the air to produce nitrogen dioxide, NO_2. Calculate the volume of NO_2 gas produced and the volume of O_2 gas consumed when 2.34 L of NO gas reacts with excess O_2. Assume all volumes are measured at the same pressure and temperature.

Solution Volumes of gases combine in the same proportions as the coefficients in the equation.

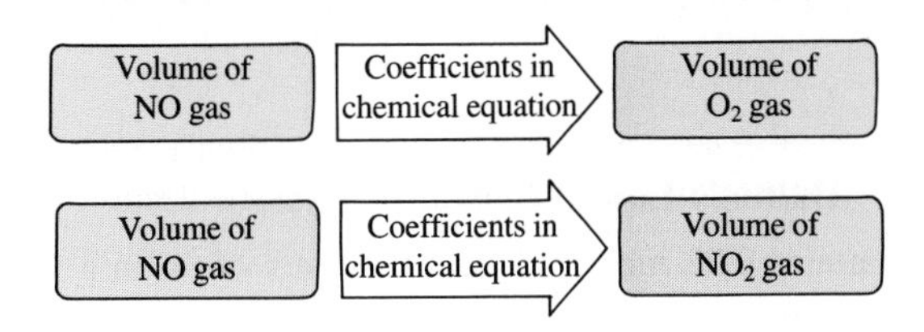

Air pollution Nitrogen oxides form in combustion reactions in the engines of automobiles and contribute to smog.

The equation is

$$2NO + O_2 \longrightarrow 2NO_2$$

Two volumes of NO are equivalent to one volume of O_2 and two volumes of NO_2. Using liters as the measure of volume,

$$2\ L\ NO \triangleq 1\ L\ O_2$$

$$2\ L\ NO \triangleq 2\ L\ NO_2$$

These equivalencies are used to calculate the volume of O_2 needed in the reaction and the volume of NO_2 produced.

$$\text{volume } O_2 = 2.34\ \cancel{L\ NO}\left(\frac{1\ L\ O_2}{2\ \cancel{L\ NO}}\right) = 1.17\ L\ O_2$$

$$\text{volume of } NO_2 = 2.34\ \cancel{L\ NO}\left(\frac{2\ L\ NO_2}{2\ \cancel{L\ NO}}\right) = 2.34\ L\ NO_2$$

Understanding Hydrogen, H_2, and chlorine, Cl_2, react to form hydrogen chloride, HCl. What volume of HCl forms from the reaction of 2.34 L of H_2 and 3.22 L of Cl_2?

Answer: 4.68 L HCl

5.6 Dalton's Law of Partial Pressure

Objectives

- To apply Dalton's law of partial pressure
- To define the mole fraction as a concentration unit and apply it to the calculation of the partial pressure of a gas in a mixture

In many of the examples in the previous sections, the identity of the gas was not needed to solve the example problems. There is frequently no need to specify the gas, because all gases behave the same at modest temperatures and pressures. In fact, we don't need to have a pure sample of gas to use the ideal gas law. If we fill a balloon with air and heat the air, it will expand to the same volume as if it were filled with pure nitrogen, argon, or any other gas. Many of the early experiments that led to the formulation of the gas laws were performed using samples of air rather than pure substances.

John Dalton was the first to realize that each gas in a mixture of gases exerts a pressure, which is the same as if it occupied the container by itself. This is called the component's partial pressure. His observations are summarized by **Dalton's law of partial pressure:** the total pressure of a mixture of gases is the sum of the partial pressures of all the components of the mixture. For a mixture of two gases A and B, the total pressure P_T is

The total pressure of a mixture of gases is the sum of the partial pressures exerted by each component.

$$P_T = P_A + P_B$$

where P_A and P_B are the partial pressures of gases A and B (Figure 5.12).

Example 5.10 Dalton's Law of Partial Pressure

A gas sample contains 0.22 mol of N_2 and 0.13 mol of O_2 in a 1.2-L container. Calculate the partial pressure of each gas and the total pressure at 50 °C.

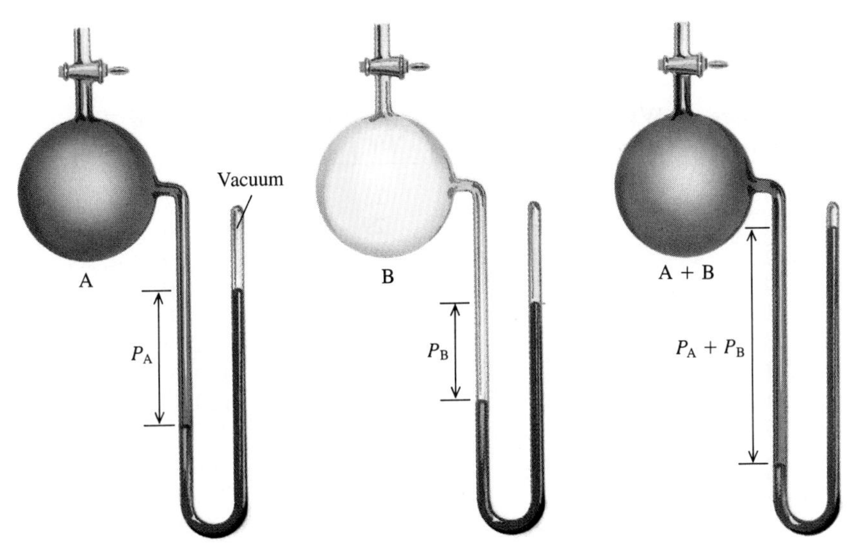

Figure 5.12 Pressure of a mixture of gases Mixing two gases produces a pressure that is the sum of the partial pressures of the individual gases.

Solution Use the ideal gas law to calculate the partial pressure of each gas in the container.

$P_{N_2} = ?$ $V_{N_2} = 1.2 \text{ L}$ $n_{N_2} = 0.22 \text{ mol}$ $T_{N_2} = 323 \text{ K}$

$P_{O_2} = ?$ $V_{O_2} = 1.2 \text{ L}$ $n_{O_2} = 0.13 \text{ mol}$ $T_{O_2} = 323 \text{ K}$

$$P_{N_2} = \frac{(n_{N_2})RT}{V_{N_2}} = \frac{(0.22 \text{ mol } N_2)(0.0821 \text{ L atm/mol K})(323 \text{ K})}{1.2 \text{ L}}$$

$$= 4.9 \text{ atm } N_2$$

$$P_{O_2} = \frac{(n_{O_2})RT}{V_{O_2}} = \frac{(0.13 \text{ mol } O_2)(0.0821 \text{ L atm/mol K})(323 \text{ K})}{1.2 \text{ L}}$$

$$= 2.9 \text{ atm } O_2$$

The total pressure is the sum of the partial pressures of the oxygen and nitrogen.

$$P_T = P_{N_2} + P_{O_2} = 4.9 \text{ atm} + 2.9 \text{ atm} = 7.8 \text{ atm}$$

Understanding Calculate the partial pressure of each gas and the total pressure in a 4.6-L container at 27 °C that contains 3.22 g of Ar and 4.33 g of Ne.

Answer: $P_{Ar} = 0.43$ atm, $P_{Ne} = 1.1$ atm, $P_T = 1.5$ atm

Collecting Gases by Displacement of Water

The apparatus shown in Figure 5.13 is frequently used to collect a gas produced in a chemical reaction. The volume of gas is measured by determining the volume of water displaced.

The gas sample collected by displacement of water is not pure because some water molecules are also present in the gas phase. Thus, the total pressure of the gas collected in the apparatus shown in Figure 5.13 is due to both the O_2 gas that was collected and the water vapor. As shown in Table

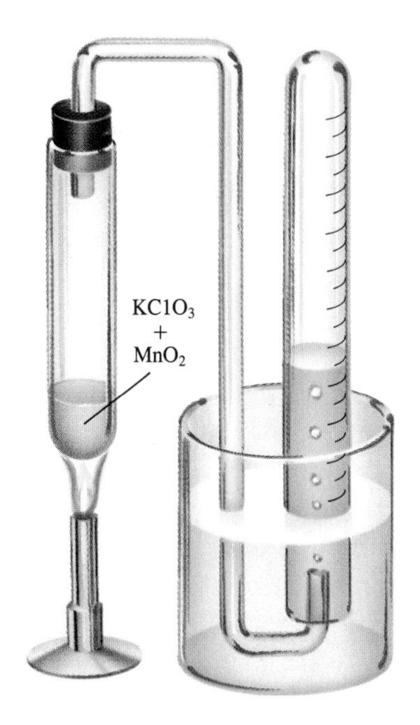

Figure 5.13 Collecting a gas over water The volume of gas produced in a chemical reaction can be measured by the displacement of water. The reaction shown is the thermal decomposition of $KClO_3$ (with MnO_2 added to speed up the reaction) to yield O_2 gas.

Table 5.3 Pressure of Water Vapor at Selected Temperatures

Temperature (°C)	Pressure of Water Vapor (torr)	Temperature (°C)	Pressure of Water Vapor (torr)
5	6.54	28	28.37
10	9.21	29	30.06
15	12.79	30	31.84
20	17.54	35	42.20
21	18.66	40	55.36
22	19.84	50	92.59
23	21.08	60	149.5
24	22.39	70	233.8
25	23.77	80	355.3
26	25.21	90	525.9
27	26.76		

5.3, the partial pressure of water present in the gas depends on the temperature of the water.

Example 5.11 Pressure of a Gas Collected Over Water

A sample of $KClO_3$ is heated and decomposes to produce 229 mL of gas, which was collected over water at 26 °C and a total pressure of 754 torr. How many moles of O_2 are formed?

Solution First, determine the partial pressure of the pure O_2 gas in the sample. Since we are interested in the amount of O_2, only the partial pressure of the O_2 gas must be used in the ideal gas law calculation. The partial pressure of water vapor at 26 °C is 25 torr (Table 5.3). From Dalton's law of partial pressure,

$$P_T = P_{O_2} + P_{H_2O}$$

$$P_{O_2} = P_T - P_{H_2O}$$

$$P_{O_2} = 754 \text{ torr} - 25 \text{ torr} = 729 \text{ torr } O_2$$

Calculate the amount of O_2 by using the ideal gas law.

$$P_{O_2} = 729 \text{ torr}\left(\frac{1 \text{ atm}}{760 \text{ torr}}\right) = 0.959 \text{ atm}$$

$$V = 0.229 \text{ L}$$

$$T = 299 \text{ K}$$

$$n = ?$$

$$n = \frac{PV}{RT} = \frac{(0.959 \text{ atm})(0.229 \text{ L})}{(0.0821 \text{ L atm/mol K})(299 \text{ K})}$$

$$= 8.95 \times 10^{-3} \text{ mol } O_2$$

Understanding Find the number of moles of hydrogen produced from a reaction of sodium and water. In the reaction, 1.3 L of gas is collected over water at 26 °C. The atmospheric pressure is 756 torr.

Answer: 0.051 mol H_2

Mole Fraction

A mixture of gases is a solution. Mole fraction is a convenient concentration unit to describe this gaseous mixture. **Mole fraction** is the number of moles of one component of a mixture divided by the total number of moles of all substances present in the mixture. The symbol χ (chi) is used to represent mole fraction. In a flask (Figure 5.14) containing 0.020 mol of argon and 0.060 mol of neon, the mole fraction of each gas is

$$\chi_{Ar} = \frac{0.020 \text{ mol Ar}}{0.080 \text{ mol total}} = 0.25$$

$$\chi_{Ne} = \frac{0.060 \text{ mol Ne}}{0.080 \text{ mol total}} = 0.75$$

The sum of the mole fractions of all components in the mixture is always one.

$$\chi_A + \chi_B + \chi_C \cdots \chi_n = 1$$

The relationship between the partial pressure and the mole fraction of a gas in a mixture can be derived from the ideal gas law. The pressure of any component of a mixture and the total pressure are expressed as

$$P_A = \frac{n_A RT}{V}$$

$$P_T = \frac{n_T RT}{V}$$

Dividing the first equation by the second yields

$$\frac{P_A}{P_T} = \frac{n_A RTV}{n_T RTV} = \frac{n_A}{n_T} = \chi_A$$

Thus, the partial pressure of a gas in a mixture can be calculated from its mole fraction and the total pressure.

$$P_A = \chi_A \times P_T$$

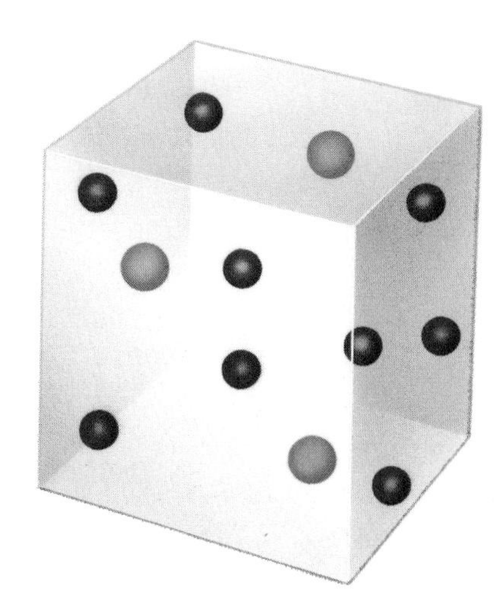

Figure 5.14 Mole fraction of a gas In a mixture of argon (yellow) and neon (red), the mole fractions are used to express the concentration of each gas.

Mole fraction is a convenient concentration unit for mixtures of gases.

Example 5.12 Partial Pressure of a Gas

What is the partial pressure of each gas in a flask that contains 0.20 mol of argon, 0.20 mol of nitrogen, and 0.30 mol of helium? The total pressure of gas is 4.2 atm.

Solution The partial pressure of each gas is its mole fraction times the total pressure. First calculate the total number of moles.

$$n_T = n_{Ar} + n_{N_2} + n_{He}$$
$$= 0.20 \text{ mol Ar} + 0.20 \text{ mol } N_2 + 0.30 \text{ mol He}$$
$$= 0.70 \text{ mol}$$

The partial pressures of argon and nitrogen are the same.

$$P_{Ar} = \chi_{Ar} \times P_T = \frac{0.20}{0.70} \times 4.2 \text{ atm} = 1.2 \text{ atm Ar}$$

$$P_{N_2} = \chi_{N_2} \times P_T = \frac{0.20}{0.70} \times 4.2 \text{ atm} = 1.2 \text{ atm } N_2$$

The partial pressure of helium is greater.

$$P_{He} = \chi_{He} \times P_T = \frac{0.30}{0.70} \times 4.2 \text{ atm} = 1.8 \text{ atm He}$$

The sum of the partial pressures of the three gases is, of course, equal to the total pressure.

Understanding What is the partial pressure of helium in a flask under a total pressure of 700 torr if it contains 10.2 mol of argon and 10.4 mol of helium?

Answer: 353 torr

5.7 Kinetic Molecular Theory of Gases

Objectives

- To outline the assumptions of the kinetic molecular theory
- To establish the relationships between molecular speed, energy, and the temperature of a gas
- To describe the Maxwell-Boltzmann distribution curve for the distribution of speeds of gas molecules
- To verify that the predictions of the kinetic molecular theory are consistent with experimental observations

The ideal gas law, as with all laws, was discovered experimentally. For example, scientists have shown that the volume of a gas at constant pressure is proportional to its temperature in kelvins. Chemists sought to understand *why* a single law can describe the physical behavior of all gases, regardless of the nature or size of the gas particles. The **kinetic molecular theory** describes the behavior of gas particles at the molecular level. The five assumptions of the theory are

1. Gases are made up of small particles that are in constant and random motion.
2. Collisions of gas particles with each other or the walls of the container are *elastic*—there is no loss in the total kinetic energy when the particles collide.
3. Gas particles are very small in size compared to the average distance that separates them.
4. There are no attractive or repulsive forces between gas particles.
5. The average kinetic energy of gas particles is proportional to the temperature on the Kelvin scale and is the same regardless of their identity.

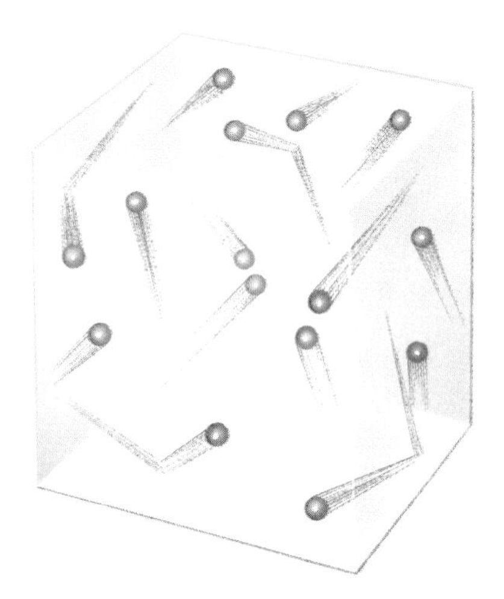

Figure 5.15 **Kinetic molecular theory of gases** Gas particles move rapidly in a container. The particles occupy only a small percentage of the total volume of the container. The collisions with the walls exert pressure.

This theory depicts gas particles as shown in Figure 5.15. The particles occupy only a small part of the volume of the box; most of it is empty space. The gas particles are in constant motion and collide with each other and the walls of the box. The direction and speed of the particles change when they collide, but the total energy of the gas does not change. The energy of the gas changes only if the temperature changes.

Pressure is the force per unit area. The kinetic molecular theory assumes that the pressure exerted by a gas comes from the collisions of the individual gas particles with the walls of the container. Obviously, pressure will increase if the force of the collisions or the number of collisions per second increases. The pressure of a gas is the same on all walls of its container.

The kinetic molecular theory of gases interprets on a molecular scale the observed behavior of gases.

Average Speed of Gas Particles

The fifth assumption of kinetic molecular theory is that the average kinetic energy of the gas particles is directly proportional to the temperature in kelvins. The *average* kinetic energy is considered because the gas particles are moving at different speeds, and the speed of each particle changes with each collision. A plot of the fraction of gas particles with a given speed versus that speed is shown in Figure 5.16. Some of the particles have very low speeds, while others move very rapidly. Plots of this type are known as *Maxwell-Boltzmann distribution curves.* Remember that the speed of each individual particle changes with each collision, but the distribution of the speeds will stay the same as in the figure. If the temperature is increased, the curve broadens and shifts to higher speeds, so the average speed increases.

A Maxwell-Boltzmann distribution describes the speed of gas molecules.

The average kinetic energy of a gas particle is related to the average squared speed by the relationship

$$\overline{KE} = \tfrac{1}{2}m\overline{u^2}$$

where the bars over kinetic energy (KE) and the squared speed (u^2) indicate average values, and m is the mass of the particle. The square root of the average squared speeds is called the **root mean square (rms) speed** and is indicated on Figure 5.16. The rms speed, u_{rms}, is the only kind of average speed we shall use.

Since the average kinetic energy of a gas particle is proportional to the temperature and the rms speed of the gas, it can be shown that

$$u_{rms} = \sqrt{\frac{3RT}{\mathcal{M}}} \qquad [5.3]$$

To obtain the rms speed of a molecule using Equation 5.3, we must express R as 8.314 J mol^{-1} K^{-1}, and the molar mass, $\mathcal{M}$, is expressed in *kilograms* per mole.

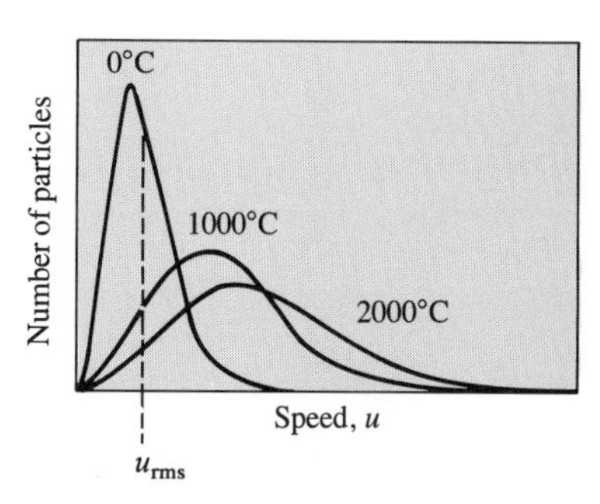

Figure 5.16 Maxwell-Boltzmann distribution The graph shows the number of particles that have a given speed versus the speed. The curve broadens and shifts to higher speeds as the temperature increases. The root mean square speed (u_{rms}) at 0 °C is shown on the graph.

Example 5.13 RMS Speed of Gas Particles

Calculate the root mean square speed of argon atoms at 27 °C.

Solution The molar mass of argon is 39.95 g/mol but must be converted to 0.03995 kg/mol before substituting in Equation 5.3. So units can be cancelled in the calculation, the joule has been expanded into its base unit equivalent of kg m^2 s^{-2}, when the value of R is substituted into Equation 5.3.

$$u_{rms} = \sqrt{\frac{3RT}{\mathcal{M}}} = \sqrt{\frac{3(8.314\ \text{kg m}^2\ \text{s}^{-2}\ \cancel{\text{mol}^{-1}}\ \cancel{\text{K}^{-1}})(300\ \cancel{\text{K}})}{0.03995\ \cancel{\text{kg}}\ \cancel{\text{mol}^{-1}}}}$$

$$= 433\ \text{m/s}$$

Understanding Calculate the root mean square speed of neon atoms at 27 °C.

Answer: 609 m/s

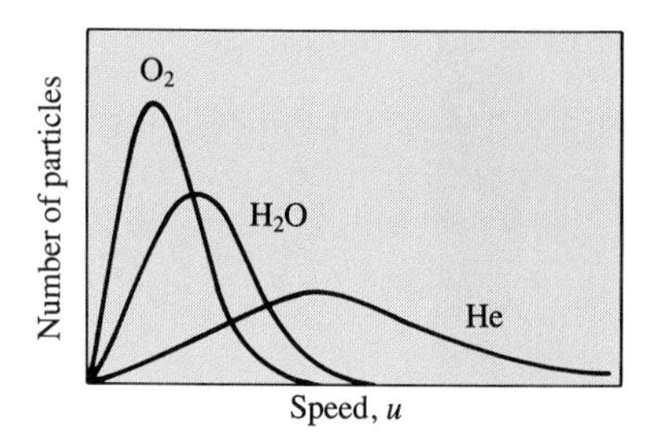

Figure 5.17 Distribution of speeds for particles of different masses The graph of the number of gas particles that have a given speed versus the speed for three different gases. At constant temperature, the rms speed of a gas increases as the molar mass decreases.

Equation 5.3 shows that the root mean square speed of a gas particle is proportional to the square root of temperature and inversely proportional to the square root of molar mass. We already saw in Figure 5.16 how the distribution of velocity changes with temperature. Figure 5.17 is a similar plot showing the speed distributions for three different gases at the same temperature. Gases with larger molar masses have lower rms speeds. This trend is also shown by the calculations in Example 5.13.

The observation that heavier particles have lower rms speeds is expected because the average kinetic energy of all gases is the same at the same temperature. At a given temperature, a heavier particle must be moving slower than a lighter particle if both are to have the same kinetic energy.

Comparison of Kinetic Molecular Theory and the Ideal Gas Law

It is important to compare the predictions of kinetic molecular theory with experimental observations of relationships between volume, pressure, temperature, and amount. For a theory to be useful it must be able to account for the experimental observations. We do the comparison only qualitatively, but calculations can show the quantitative relations are also correct.

Volume and Pressure—Compression of Gases

Kinetic molecular theory assumes that gas particles are small compared to the distances that separate them. Gases can expand to fill a larger container or compress into a smaller container because most of the volume of a gas is empty space. Solids and liquids are very different and do not readily compress because the particles are in close contact, and the particles themselves are not very compressible.

Boyle observed that the pressure of a gas increases when the volume decreases. This observation is explained by the model; as the size of a container decreases (at constant temperature), the number of collisions of the gas particles with the walls per unit area during any time interval will increase because the particles have less distance to travel between collisions with the walls. At constant temperature, the average force of each collision does not change, but in a smaller volume the same number of particles will strike a given area of the wall more often, so the pressure of gas in the container must increase (Figure 5.18).

Volume and Temperature

Equation 5.3 shows that increasing the temperature of a gas increases the rms speed of the gas particles. Two consequences of the increased speed of the gas particles are that each collision will exert a greater force on the walls and that the number of collisions per unit area per unit time will also increase. If pressure is to remain constant, the size of the container must increase to decrease the number of these more energetic collisions per unit area. The kinetic molecular theory thus predicts an increase in volume with an increase in temperature at constant pressure—Charles's law.

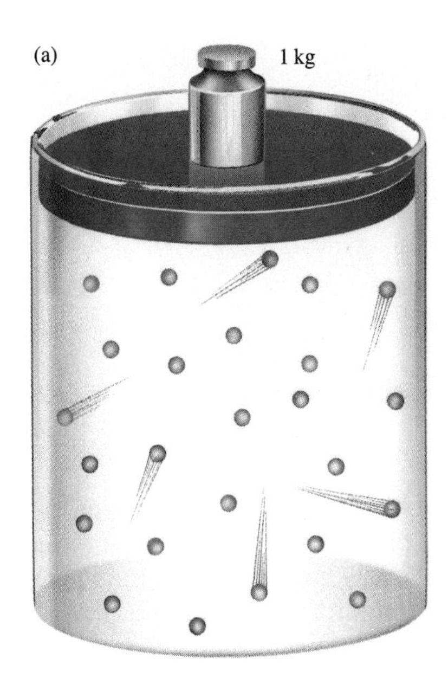

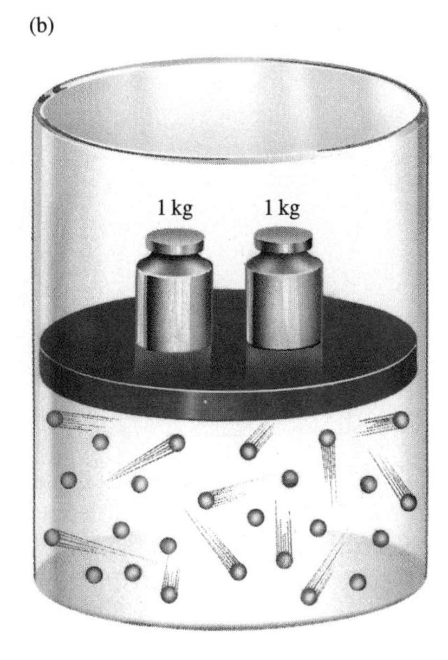

Figure 5.18 Changes in volume and pressure The pressure of the gas in (a) is less than that in (b) because there are fewer collisions per unit area per unit time with the walls in the larger space.

Volume and Amount

Increasing the number of gas particles in a container will increase the number of collisions with the walls per unit area per unit time. If the pressure is to remain constant, the volume of the container must increase. This phenomenon, of course, was stated years before kinetic molecular theory in Avogadro's law.

The kinetic molecular theory predicts the ideal gas law.

5.8 Diffusion and Effusion

Objectives

- To describe diffusion and effusion of a gas
- To calculate relative rates of effusion of two gases and use the data to calculate molar masses

The ability to predict the results of new observations is a required test of any theory. The laws governing the diffusion of gases are predicted by the kinetic molecular theory. **Diffusion** is the mixing of particles due to motion, such as the mixing of perfume molecules with air when the bottle is opened. The faster the molecular motion, the faster a gas will diffuse, although the rate of diffusion is always less than the rms speed of the gas because collisions change the direction of each particle many times.

Figure 5.19 shows an experiment that demonstrates relative rates of diffusion. Gaseous ammonia and hydrogen chloride react to form the white solid ammonium chloride.

$$NH_3(g) + HCl(g) \longrightarrow NH_4Cl(s)$$

The ammonia is introduced at one end of a tube at the same time the hydrogen chloride is added at the other end. The white ring of ammonium

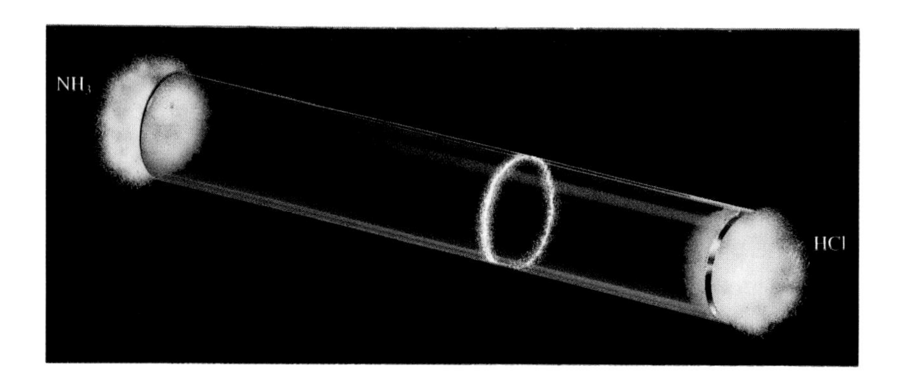

Figure 5.19 Diffusion of gases Ammonia diffuses faster than hydrogen chloride, forming a ring of ammonium chloride closer to the hydrogen chloride end of the tube.

chloride forms closer to the end of the tube where the hydrogen chloride was introduced. The kinetic molecular theory predicts this location of the ring because the rms speed of gas particles is inversely proportional to the square root of their molar mass (Equation 5.3). Ammonia has the lower molar mass, so it has the higher rms speed and thus diffuses down the tube faster.

The rates of diffusion and effusion are directly proportional to the square root of the temperature and inversely proportional to the square root of the molar mass.

Closely related to diffusion is effusion. **Effusion** is the passage of a gas through a small hole into an evacuated space. Thomas Graham (1805–1869) carefully measured the rates of effusion of a number of gases. **Graham's law** states that the rate of effusion of a gas is inversely proportional to the square root of its molar mass. The kinetic molecular theory predicts Graham's law because the rms speed of the gas particles is also inversely proportional to the square root of its molar mass. In comparing the rates of effusion of two gases, Graham's law is frequently written as

$$\frac{\text{rate of effusion of gas A}}{\text{rate of effusion of gas B}} = \sqrt{\frac{\mathcal{M}_B}{\mathcal{M}_A}} \qquad [5.4]$$

This expression can be used to compare the rates at which equimolar concentrations of helium and argon effuse through a small hole. In the same amount of time, more of the helium atoms effuse through the hole (Figure 5.20) because of their smaller atomic mass.

Argon Helium

Vacuum chamber

Figure 5.20 Relative rates of effusion of gases Atoms of helium will effuse through a small hole into a vacuum faster than will the heavier atoms of argon.

Graham's law can be used to calculate the difference between the rates of effusion of the two gases.

$$\frac{\text{rate of effusion of helium gas}}{\text{rate of effusion of argon gas}} = \sqrt{\frac{\mathcal{M}_{Ar}}{\mathcal{M}_{He}}}$$

$$= \sqrt{\frac{39.95 \text{ g/mol Ar}}{4.003 \text{ g/mol He}}} = 3.16$$

The lighter helium atoms effuse a little more than three times faster than argon atoms.

Gaseous diffusion (to which Graham's law also applies) is used to increase the fraction of ^{235}U in samples of that element. In an atomic bomb, the uranium atoms must be enriched substantially in ^{235}U. The natural abundance of ^{235}U is only 0.72%, with the remainder mostly ^{238}U. After testing a number of possible methods, scientists decided to convert the uranium to the gaseous compound UF_6, and use gaseous diffusion for the

Figure 5.21 UF_6 **diffusion separation plant**

separation. Graham's law can be used to calculate the ratio of the rates of diffusion of $^{235}UF_6$ and $^{238}UF_6$:

$$\frac{\text{rate of diffusion of } ^{235}UF_6 \text{ gas}}{\text{rate of diffusion of } ^{238}UF_6 \text{ gas}} = \sqrt{\frac{\mathcal{M}_{^{238}UF_6}}{\mathcal{M}_{^{235}UF_6}}}$$

$$= \sqrt{\frac{352.05 \text{ g/mol } ^{238}UF_6}{349.04 \text{ g/mol } ^{235}UF_6}} = 1.0043$$

The difference in rates of diffusion is very small, but the process can be repeated many times. To increase the concentration of the $^{235}UF_6$ to the required value of about 3%, a large separation plant was built in Oak Ridge, Tennessee. There the diffusion process is repeated successively to obtain the needed enrichment of ^{235}U (Figure 5.21).

Molar Mass Determinations by Graham's Law

Graham's law can be used to determine the molar mass of an unknown gas by measuring the time needed for equal volumes of a known gas and the unknown gas to effuse through the same small hole. Equation 5.4 relates the rate of effusion to molar mass. Gases with a higher rate of effusion will escape through the hole in a shorter time; the time it takes for a gas to effuse, t, is inversely proportional to the rate of effusion. Thus, Equation 5.4 becomes

The molar mass of a gas can be determined from relative rates of effusion.

$$\frac{\text{rate of effusion of gas A}}{\text{rate of effusion of gas B}} = \frac{t_B}{t_A} = \sqrt{\frac{\mathcal{M}_B}{\mathcal{M}_A}} \qquad [5.5]$$

Example 5.14 Determination of Molar Mass by Effusion

Calculate the molar mass of a gas if equal volumes of nitrogen and the unknown gas take 2.2 and 4.1 minutes, respectively, to effuse through the same small hole.

Solution Solve Equation 5.5 for the molar mass of the unknown gas (x) by squaring both sides and rearranging.

$$\frac{t_x}{t_{N_2}} = \sqrt{\frac{\mathcal{M}_x}{\mathcal{M}_{N_2}}}$$

$$\frac{(t_x)^2}{(t_{N_2})^2} = \frac{\mathcal{M}_x}{\mathcal{M}_{N_2}}$$

$$\mathcal{M}_x = \mathcal{M}_{N_2} \times \frac{(t_x)^2}{(t_{N_2})^2} = 28 \text{ g/mol} \times \frac{(4.1 \text{ min})^2}{(2.2 \text{ min})^2} = 97 \text{ g/mol}$$

Understanding Calculate the molar mass of a gas if equal volumes of oxygen gas and the unknown gas take 3.25 and 8.41 minutes, respectively, to effuse through a membrane.

Answer: 214 g/mol

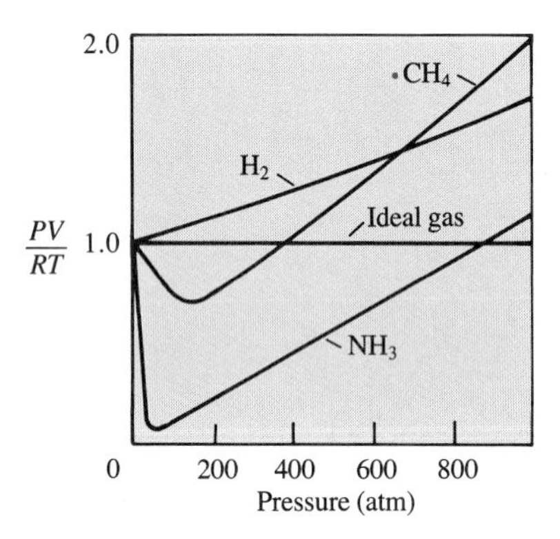

Figure 5.22 Behavior of a real gas Plot of PV/RT versus P for one mole of three gases. A gas that follows the ideal gas law will have a value of one at all pressures for PV/RT.

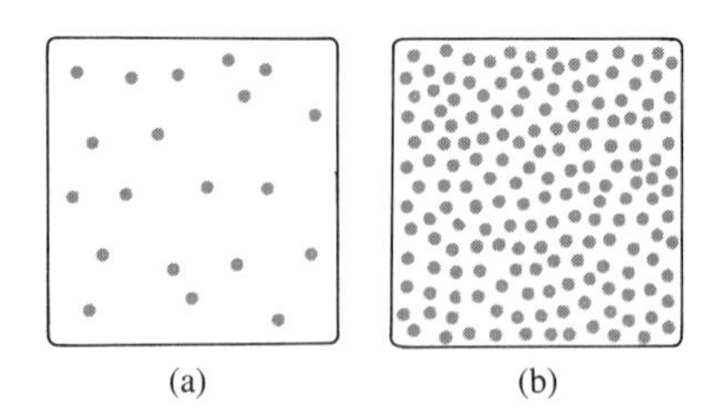

Figure 5.23 Gases at low and high pressures The size of the gas particles in container (a) at a low pressure is small in comparison to the volume occupied by the gas. The gas particles in (b) at a high pressure occupy a sizeable percentage of the volume of the gas.

5.9 Real Gases

Objectives

- To describe the conditions under which gases deviate from the ideal gas law
- To account for deviations from the ideal gas law by considering molecular volumes and the attractive forces between molecules

The ideal gas law was discovered by careful experimental observations. The kinetic molecular theory is a model that interprets the ideal gas law on the molecular level. As its name implies, the ideal gas law applies to an *ideal* gas, any gas that follows the five assumptions of the kinetic molecular theory. At pressures of about one atmosphere and temperatures well above the boiling point of the substance, most gases obey the ideal gas law quite closely.

Figure 5.22 shows a plot of measured values of PV/RT versus P for one mole each of three gases. For a gas that follows the ideal gas law, the measured values of PV/RT would follow the center line in Figure 5.22. At low pressures, all of the gases follow the ideal gas law, but as the pressure increases to high values (100 atm is a substantial pressure) deviations are seen.

Clearly, gases at high pressures do not behave as predicted by the ideal gas law. Does that mean that we should discard the law? No—it is useful at low pressures, the pressures at which chemists generally work with gases. What scientists do when experimental observations are at variance with a theory is to reevaluate both the theory and the experiments. In the case of deviations of gases from ideality, the assumptions made in the kinetic molecular theory are not justified under extreme conditions. When we examine deviations from the ideal gas law, we find that two of the assumptions of the kinetic molecular theory are not correct at high pressures. They are (a) that gas particles are small compared to the distances separating them, and (b) that there are no attractive forces between gas particles.

Each gas particle has an actual size. (Recall the definition of matter—it has mass and *occupies space.*) When many gas molecules are confined in a small volume (i.e., under high pressure), the volume occupied by the individual particles is no longer negligible compared to the volume of the gas sample (Figure 5.23). This results in the deviations for an ideal gas above the line in Figure 5.22 at high pressures.

Ammonia shows a significant deviation *below* the line at moderate pressures, and methane shows a slight deviation of this type. These deviations are caused by forces of attraction between the gas molecules (similar to the forces that hold molecules together in liquids, which will be discussed in Chapter 10). Gas molecules that are attracted to each other will not strike the wall as hard as predicted (Figure 5.24), reducing the pressure below that predicted by the ideal gas law. At high pressures, the molecules are forced closer together, making this attractive interaction more important. Ammonia dips most significantly below the line because, among the gases pictured, it has the strongest attractive forces between its molecules. Hydrogen is not observed to go below the line because its attractive forces are very small. At very high pressures, the volume factor dominates this attractive force, causing deviations above the line for all three gases.

Figure 5.25 shows a plot of PV/RT versus pressure for oxygen at three different temperatures. The behavior is more nearly ideal at higher temperatures. The deviations caused by the actual size of the gas particles and the attractive forces become less important in gas samples at higher temperatures (i.e., higher rms speeds of the molecules).

From the preceding discussion, two conclusions about the ideal gas law can be made: the ideal gas law is followed *best* at *low pressures and high temperatures*—when gases are far away from the conditions of temperature and pressure under which they will condense to a liquid. Consider SO_2, with a boiling point of −10 °C at one atmosphere. Below −10 °C, the forces between SO_2 molecules hold them closely together as a liquid. In SO_2 gas at temperatures just above −10 °C, the attractive forces between the molecules are sufficiently strong to cause considerable deviation from the ideal gas law. Measurable deviations of SO_2 gas from the ideal gas law occur even at room temperature, about 30 °C above its boiling point at one atmosphere. In comparison, nitrogen, with a boiling point of −196 °C, follows the ideal gas law very closely at normal temperatures and pressures. Usually, a gas will follow the ideal gas law under conditions of temperature and pressure that are more than 100° away from its condensation temperature (boiling point), as long as the pressure is not exceedingly high.

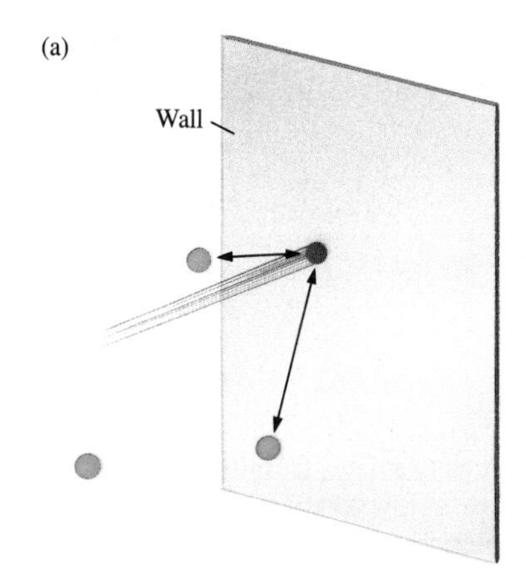

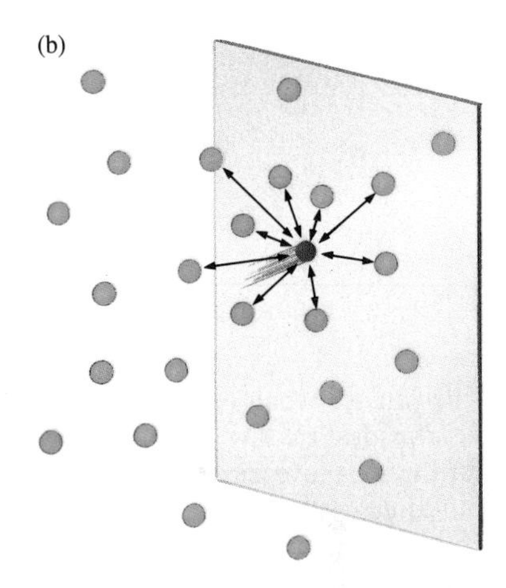

Figure 5.24 Forces of attraction in gases (a) At low pressures only a few molecules will attract a molecule about to hit the wall. (b) At higher pressures a large number of molecules will attract the molecule about to hit the wall, reducing the net force of each collision with the wall.

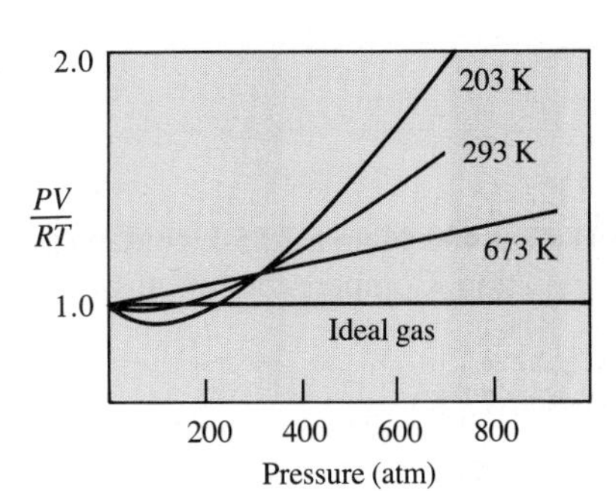

Figure 5.25 Behavior of gases with changes in temperature and pressure Gases follow the ideal gas law more closely at high temperatures.

Example 5.15 Obeying the Ideal Gas Law

Predict which gas is likely to follow the ideal gas law more closely.

(a) SO_2 gas at 0 °C compared to SO_2 at 100 °C.
(b) Nitrogen gas at 1 atm compared with N_2 at 100 atm, both at 25 °C.
(c) Oxygen gas and ammonia gas at −20 °C and 1 atm.

Solution

(a) The SO_2, like all gases, will follow the ideal gas law better at higher temperatures.
(b) Gases will follow the ideal gas law better at low pressures, so nitrogen at 1 atm follows the law better.
(c) At one atmosphere, oxygen boils at −183 °C and ammonia boils at −33 °C. The oxygen will follow the ideal gas law more closely because it is farther away from the temperature at which it will condense to a liquid.

A gas deviates from the ideal gas law at low temperatures and high pressures.

Understanding Which gas and set of conditions will follow the ideal gas law best: (a) N_2 at 25 °C and 1 atm, (b) SO_2 at 25 °C and 1 atm, or (c) N_2 at 25 °C and 100 atm?

Answer: (a)

van der Waals Equation

The van der Waals equation modifies the ideal gas law by accounting for the attractive forces between molecules and the volume the molecules occupy.

The ideal gas law can be extended to include the effects of attractive forces and the volume occupied by the particles. To correct for the volume occupied by the gas particles, we subtract the term nb from the volume, where n is the number of moles of gas and b is a constant that depends on the size of the gas particles. The volume term in the gas law then becomes $(V - nb)$. This corrected volume is the empty space, which is the only part of the sample that is compressed.

The pressure term can also be modified to correct for attractive forces by adding the term an^2/V^2 to the pressure, where a is a constant related to the strength of the attractive forces, and n and V are the number of moles and the gas volume. The pressure term in the gas law then becomes $(P + an^2/V^2)$. Substituting the new pressure and volume terms into the ideal gas law, we get the **van der Waals equation:**

$$\left(P + \frac{an^2}{V^2}\right)(V - nb) = nRT$$

The experimentally determined van der Waals constants are different for each gas; a few are given in Table 5.4.

Table 5.4 van der Waals Constants

Gas	a (atm L^2/mol^2)	b (L/mol)
H_2	0.244	0.0266
He	0.034	0.0237
Ne	0.211	0.0171
H_2O	5.46	0.0305
NH_3	4.17	0.0371
CH_4	2.25	0.0428
N_2	1.39	0.0391
O_2	1.36	0.0318
Ar	1.34	0.0322
CO_2	3.59	0.0427

Example 5.16 van der Waals Equation

Calculate the pressure of 2.01 mol of gaseous H_2O at 400 °C in a 2.55-L container, using the ideal gas law and the van der Waals equation. Compare the two answers.

Solution For the ideal gas law

$$P = \frac{nRT}{V} = \frac{(2.01\ \cancel{\text{mol}})(0.0821\ \cancel{\text{L}}\ \text{atm}/\cancel{\text{mol K}})(673\ \cancel{\text{K}})}{2.55\ \cancel{\text{L}}}$$

$$= 43.6\ \text{atm}$$

Rearranging the van der Waals equation to solve for pressure yields

$$P = \frac{nRT}{V - nb} - \frac{an^2}{V^2}$$

Substitute the measured values and the constants from Table 5.4 into this equation.

$$P = \frac{(2.01 \text{ mol})(0.0821 \text{ L atm/mol K})(673 \text{ K})}{2.55 \text{ L} - (2.01 \text{ mol})(0.0305 \text{ L/mol})} - \frac{(5.46 \text{ atm L}^2/\text{mol}^2)(2.01 \text{ mol})^2}{(2.55 \text{ L})^2}$$

$$= 44.6 \text{ atm} - 3.39 \text{ atm} = 41.2 \text{ atm}$$

Under these conditions, the ideal gas law and the van der Waals equation yield values that are fairly close. The correction for the volume (nb) increased the pressure, as expected, because it takes into account the actual size of the particles. The correction for attraction (an^2/V^2) decreased the pressure, as predicted earlier.

Understanding Calculate the pressure of 0.223 mol of ammonia gas at 30.0 °C in a 3.23-L container, using the ideal gas law and the van der Waals equation.

Answer: Ideal gas law = 1.72 atm, van der Waals = 1.70 atm. Under these conditions the correction is very small.

Summary

The state of a sample of gas is described by the volume, the pressure, the temperature, and the amount of the gas that is present. Experimental observations have shown that the volume of a gas sample is inversely proportional to pressure and directly proportional to temperature and amount of sample. These relationships are expressed by the *ideal gas law,*

$$PV = nRT$$

where R is the ideal gas constant and has been determined experimentally to be 0.0821 L atm/mol K.

The ideal gas law can be used to calculate changes in the state of a gas sample in which one or more of the four variables has changed. It can also be used to calculate any one of the four variables if three have been measured. The ideal gas law may also be used to determine the molar mass, given the volume and mass or the density of a sample.

The ideal gas law can be used with chemical equations to determine n, the number of moles of gas. In a reaction that involves two or more gases at the same temperature and pressure, the coefficients in the equation can be interpreted as volumes.

Dalton's law of partial pressure states that the total pressure of a mixture of gases is the sum of the partial pressures of the component gases. The partial pressure of a gas is the pressure it would exert if it alone occupied the container at the same temperature. When a gas is collected over water, the total pressure will be the sum of the pressure of the gas and the pressure of the water vapor that is also present. The partial pressures of the gases in a mixture are proportional to their *mole fractions.*

The *kinetic molecular theory* describes the behavior of gas particles at the molecular level. Gases are described as made up of small particles that are in constant and random motion. Collisions of gas particles with each other and the walls are *elastic,* and it is assumed that there are no attractive or repulsive forces between gas molecules. The average kinetic energy of the gas particles is proportional to the kelvin temperature. The *pressure* of a gas, which is the force per unit area exerted by the gas, comes from the particles rebounding from the walls of the container. The gas particles do not all move at the same speed, but display a distribution of speeds given by the Maxwell-Boltzmann distribution. The root mean square speed, u_{rms}, of a gas is proportional to the square root of temperature and inversely proportional to the square root of the molar mass. The assumptions of the kinetic molecular theory can be used to explain the ideal gas law.

Both *diffusion* and *effusion* are related to the speed of the gas molecules. *Graham's law* states that the rate of effusion is inversely proportional to the square root of the molar mass, and it can be used to determine the molar mass.

Gases behave as described by the ideal gas law at low pressures and high temperatures. Deviations from the law are observed at high pressures and low temperatures because both the attractive forces between molecules and the actual volume of the particles become important. The van der Waals equation describes the behavior of gases at high pressures more accurately than does the ideal gas law, because it contains terms to allow for attractive forces and for molecular size.

Chapter Terms

Gas—a fluid that has no definite shape or volume. *(5.1)*

Liquid—a fluid that has a fixed volume but no definite shape. *(5.1)*

Solid—a state of matter with a fixed shape and volume. *(5.1)*

Condensed phase—the solid and liquid states of matter. Any phase that is resistant to volume change. *(5.1)*

Pressure—the force per unit area exerted on a surface. *(5.1)*

Boyle's law—at constant temperature the volume of a gas sample is inversely proportional to the pressure ($P \times V = k_1$). *(5.2)*

Charles's law—at constant pressure the volume of a sample of a fixed amount of gas is proportional to the absolute temperature ($V = k_2 \times T$). *(5.2)*

Avogadro's law—at constant pressure and temperature the volume of a sample of gas is proportional to the number of moles of gas present ($V = k_3 \times n$). *(5.2)*

Ideal gas law—the equation that describes the state of a gas, $PV = nRT$, where P = pressure, V = volume, n = number of moles, R = ideal gas constant (0.0821 L atm/mol K), and T = temperature. *(5.2)*

Standard temperature and pressure (STP)—the conditions of 273 K (or 0 °C) and 1.00 atm. *(5.2)*

Dalton's law of partial pressure—the total pressure of a mixture of gases is the sum of the partial pressures of the component gases. *(5.6)*

Mole fraction—the number of moles of one component of a homogeneous mixture divided by the total number of moles of all substances present in the mixture. *(5.6)*

Kinetic molecular theory—a model that describes the behavior of gas particles at the atomic or molecular level. *(5.7)*

Root mean square (rms) speed—the square root of the average squared speeds of a collection of particles. *(5.7)*

Diffusion—the mixing of particles due to motion. *(5.8)*

Effusion—the passage of a gas through a small hole into an evacuated space. *(5.8)*

Graham's law—the rate of effusion of gases is inversely proportional to the square root of the molar mass. *(5.8)*

van der Waals equation—a description of the behavior of real gases, based on correcting the ideal gas law for particle size and attractive forces:

$$\left(P + \frac{an^2}{V^2}\right)(V - nb) = nRT$$

where a and b are constants experimentally determined for each gas. *(5.9)*

Exercises

Exercises designated with color have answers in Appendix J.

Properties of Matter

5.1 Describe the similarities and differences between the ways in which a gas and a liquid occupy a container.

5.2 Compare the densities of a substance as a solid, a liquid, and a gas.

5.3 Describe how atmospheric pressure is measured with a barometer and how pressure differences are measured with a manometer.

5.4 Make a drawing of an open-end manometer measuring 20 torr pressure on a sample of a gas.

5.5 Define three units used to describe pressure.

5.6 Express a pressure of
(a) 334 torr in atm
(b) 3944 Pa in atm
(c) 2.4 atm in torr.

5.7 Express a pressure of
(a) 3.2 atm in torr
(b) 54.9 atm in kPa
(c) 356 torr in atm.

5.8 The temperature terms for gas law problems must always be expressed in kelvins. Convert the following temperatures into kelvins:
(a) 45 °C (b) −28 °C (c) 230 °C.

Gas Laws

5.9 Describe the change in the volume of a gas that occurs when each of the following three quantities is increased with the other two held constant:
(a) pressure
(b) temperature
(c) amount.

5.10 A sample of gas at 1.02 atm of pressure and 39 °C is heated to 499 °C at constant volume. What is the new pressure in atm?

5.11 A 455-mL sample of argon gas at 0.330 atm is allowed to expand into a 9.36-L container. If the temperature remains constant, what is the new pressure of the argon gas in atm?

5.12 A 39.6-mL sample of gas is trapped in a syringe and heated from 27 °C to 127 °C. What is the new volume, in mL, in the syringe if the pressure is constant?

5.13 The quantity of gas in a 34-L balloon is increased from 3.2 mol to 5.3 mol at constant pressure. What is the new volume of the balloon at constant temperature?

5.14 The pressure of a balloon holding 166 mL of gas is increased from 399 torr to 1.00 atm. What is the new volume of the balloon, in mL, at constant temperature?

5.15 A sample of hydrogen gas is in a 2.33-L container at 745 torr and 27 °C. Express the pressure of hydrogen in atm after the volume is changed to 1.22 L and the temperature is changed to 100 °C.

5.16 Natural gas has been stored in expandable tanks that keep a constant pressure as gas is added or removed. The volume of the tank is 4.50×10^4 cubic feet when the tank contains 77.4 million moles of natural gas at −5 °C. What is the new volume of the tank if 5.3 million moles are used by consumers and the temperature rose to +7 °C?

5.17 Calculate the final temperature, in K, if a sample of gas at 358 K undergoes a pressure increase from 345 torr to 938 torr at constant volume.

5.18 The pressure of a 900-mL sample of helium is increased from 2.11 atm to 4.33 atm and the temperature is also increased from 0 °C to 22 °C. What is the new volume, in mL, of the sample?

5.19 A sample of methane gas at 22 °C is heated. The pressure increases from 1.54 atm to 2.13 atm, and the volume of the sample increases from 1.33 L to 1.89 L. To what temperature, in °C, is the methane heated?

5.20 A sample of argon occupies 3.22 L at 33 °C and 230 torr. How many moles of argon are present in the sample?

5.21 What is the temperature of a gas, in °C, if a 2.49-mole sample in a 24.0-L container is under a pressure of 2.44 atm?

5.22 A 3.00-L container is rated to hold a gas at a pressure no higher than 100 atm. Assuming the gas behaves ideally, what is the maximum number of moles of gas that this vessel can hold at 27 °C?

5.23 What is the pressure in atmospheres of 0.322 g of N_2 gas in a 300-mL container at 24 °C?

5.24 What is the volume, in liters, of a balloon that contains 82.3 mol of H_2 gas at 25 °C and 1.01×10^5 Pa?

5.25 What is the mass, in grams, of a sample of SO_2 if it occupies 200 mL at 27 °C and 1.22 atm?

5.26 What is the molar mass of a gas if a 0.550-g sample occupies 258 mL at a pressure of 744 torr and a temperature of 22 °C?

5.27 Calculate the molar mass of a gas if a 0.165-g sample at 1.22 atm occupies a volume of 34.8 mL at 50 °C.

5.28 The density of a gas sample is 2.41 g/L at STP. What is the molar mass of this gas?

5.29 The density of a gas sample is 1.43 g/L at STP. What is the molar mass of this gas?

Equation Stoichiometry with Gases

5.30 What volume, in milliliters, of hydrogen gas at 1.33 atm and 33 °C is produced by the reaction of 0.0223 g of lithium metal with excess water? The other product is LiOH.

5.31 The reaction of magnesium metal with HCl yields hydrogen gas and $MgCl_2$. What volume, in liters, of the gas is formed at 744 torr and 26 °C from 2.23 g of magnesium?

5.32 Heating potassium chlorate, $KClO_3$, yields oxygen gas and potassium chloride. What volume, in liters, of oxygen at 23 °C and 760 torr is produced from the decomposition of 4.42 g of potassium chlorate?

5.33 What volume of nitrogen gas, in liters, at 30 °C and 0.993 atm reacts with excess hydrogen to produce 4.22 g of ammonia?

5.34 What volume of hydrogen gas, in liters, is produced in the reaction of 1.33 g of zinc metal and 300 mL of 2.33 *M* H_2SO_4? The gas is collected at 1.12 atm of pressure and 25 °C.

5.35 What mass of water is formed from the reaction of 2.44 L of hydrogen gas and 3.11 L of oxygen gas? Both gases are at 734 torr pressure and 27 °C.

5.36 The "air" that fills the air-bags installed in automobiles is actually nitrogen produced by the decomposition of sodium azide, NaN_3. Assuming the other product to be metallic sodium, what volume, in liters, of nitrogen gas is released from the decomposition of 1.88 g of sodium azide? The pressure is 755 torr and the temperature is 24 °C.

5.37 What volume, in liters, of oxygen gas is consumed in the reaction with 3.22 L of hydrogen gas to form water? Assume that all gases are at the same pressure and temperature.

5.38 The gas hydrogen sulfide, H_2S, has the offensive smell associated with rotten eggs. It reacts slowly with the oxygen in the atmosphere to form sulfur dioxide and water. What volume of sulfur dioxide gas, in liters, is formed at constant pressure and temperature from 2.44 L of hydrogen sulfide, and what volume of oxygen gas is consumed?

5.39 There is considerable concern that an increase in the concentration of CO_2 in the atmosphere will lead to global warming. This gas is the product of the combustion of hydrocarbons used as energy sources. What volume of CO_2 gas, at constant temperature and pressure, is produced from the combustion of 2.00×10^3 L of CH_4 gas? What volume of oxygen gas is consumed?

Partial Pressure

5.40 What is the total pressure, in atm, in a container that holds 1.22 atm of hydrogen gas and 4.33 atm of argon gas?

5.41 What is the partial pressure of argon, in torr, in a container that also contains neon at 235 torr and is at a total pressure of 500 torr?

5.42 The pressure in a 3.11-L container is 4.33 atm. What is the new pressure in the tank when 2.11 L of gas at 2.55 atm is added to the container? All the gases are at 27 °C.

5.43 A 4.53-L sample of neon at 3.22 atm of pressure is added to a 10-L cylinder that contains argon. If the pressure in the cylinder is 5.32 atm after the neon is added, what was the original pressure of argon in the cylinder?

5.44 What is the pressure, in atm, in a 3.22-L container that holds 0.322 mol of oxygen and 1.53 mol of nitrogen? The temperature of the gases is 100 °C.

5.45 Calculate the partial pressure of oxygen, in atm, in a container that holds 3.22 mol of oxygen and 4.53 mol of nitrogen. The total pressure in the container is 7.32 atm.

5.46 A 10.5-g sample of hydrogen is added to a 30-L container that also holds argon gas at 1.53 atm. The gases are at 120 °C. What is the partial pressure of hydrogen gas in the mixture and what is the total pressure in the container?

5.47 Calculate the total pressure in a 5.00-L flask that contains 5.34 g of neon and 1.22 g of argon. The temperature of the gases is 30 °C.

5.48 What is the partial pressure of oxygen gas, in torr, collected over water at 26 °C if the total pressure is 755 torr (see Table 5.3)?

5.49 What is the total pressure, in torr, in a 1.00-L flask that contains 0.0311 mol of hydrogen gas that was collected over water (see Table 5.3)? The temperature is 25 °C.

5.50 Calculate the partial pressure of hydrogen gas, in atm, in a container that holds 0.220 mol of hydrogen and 0.432 mol of nitrogen. The total pressure is 5.22 atm.

5.51 What is the partial pressure of neon, in torr, in a flask under a total pressure of 209 torr if it contains 3.11 mol of neon and 1.02 mol of argon?

Kinetic Molecular Theory

5.52 Summarize the five assumptions of the kinetic molecular theory.

5.53 Discuss the origin of gas pressure in terms of the kinetic molecular theory.

5.54 Arrange the following gases in order of increasing rms speed of the particles at the same temperature: N_2, O_2, Ne.

5.55 Arrange the following gases in order of increasing rms speed of the particles at the same temperature: SO_2, H_2, Ar.

5.56 Arrange the following gases at the temperatures indicated in order of increasing rms speed of the particles: neon at 25 °C, neon at 100 °C, argon at 25 °C.

5.57 Arrange the following gases at the temperatures indicated in order of increasing rms speed of the particles: helium at 100 °C, neon at 50 °C, argon at 0 °C.

5.58 Calculate the rms speed of neon atoms at 100 °C.

5.59 Calculate the rms speed of SO_2 molecules at 127 °C. What is the rms speed if the temperature is doubled *on the Kelvin scale?*

5.60 Calculate the molar mass of a gas that has an rms speed of 518 m/s at 28 °C.

5.61 At what temperature, in kelvins, will neon atoms have an rms speed of 700 m/s?

5.62 Clearly define the terms diffusion and effusion.

5.63 Describe how the isotopes of uranium can be separated by a diffusion process.

5.64 Calculate the ratio of the rates of effusion of helium and neon gas.

5.65 Calculate the ratio of the rates of effusion of $^{238}UF_6$ and helium. ^{238}U has an atomic mass of 238.05 u.

5.66 Two identical balloons are filled, one with helium gas and the other with argon, at the same temperature and pressure. If the helium balloon loses 50 mL of volume by diffusion, how much will the argon balloon lose in the same length of time?

5.67 A container is filled with equal molar amounts of N_2 and SO_2 gas. Calculate the ratio of the rate of effusion of the two gases.

5.68 Calculate the molar mass of a gas if equal volumes of hydrogen and the unknown gas take 1.20 and 9.12 minutes, respectively, to effuse into a vacuum through a small hole.

5.69 Calculate the molar mass of a gas if equal volumes of oxygen and the unknown gas take 5.2 and 8.3 minutes, respectively, to effuse into a vacuum through a small hole.

Real Gases

5.70 Describe why gases at high pressures do not follow the ideal gas law.

5.71 Does the ratio PV/RT for a real gas have a value greater or less than one (a) if attractive forces between molecules are strong, and (b) if the volume of the gas particle becomes important relative to the total volume of the gas?

5.72 For the following pairs of gases at the given conditions, predict which one is *more likely* to follow the ideal gas law. Explain your choice.
(a) oxygen gas at −100 °C or at 100 °C
(b) nitrogen (boiling point = −196 °C) or xenon (boiling point = −107 °C) gas at −100 °C
(c) argon gas at 1 atm or at 50 atm of pressure

5.73 For the following pairs of gases at the given conditions, predict which one is *more likely* to follow the ideal gas law. Explain your choice.
(a) helium or sulfur dioxide, both at 25 °C
(b) nitrogen gas at −150 °C or at 100 °C
(c) helium gas at 1 atm or at 200 atm

5.74 Calculate the pressure, in atm, of 10.2 mol of argon at 530 °C in a 3.23-L container, using both the ideal gas law and the van der Waals equation.

5.75 Calculate the pressure, in atm, of 1.55 mol of nitrogen at 530 °C in a 3.23-L container, using both the ideal gas law and the van der Waals equation.

Additional Problems

5.76 It is important to check the pressure in car tires at the start of the winter because of the pressure drop. Calculate the pressure change in a tire inflated to 32 psi (pounds per square inch) at 90 °F if the temperature drops to 32 °F. Assume atmospheric pressure is 15 psi.

5.77 A sample of air was collected at the research station in the Antarctic to test for airborne pollutants. The sample was collected in a 1.00-L container at 764 torr and −20 °C. Calculate the pressure in the container when it was opened for analysis in a laboratory in South Carolina at a temperature of 22 °C.

5.78 A 2.8-L tank is filled with 0.24 kg of oxygen. What is the pressure in the tank at 20 °C?

5.79 The empirical formula of butane is C_2H_5. What is the molecular formula of butane if 1.26 g of butane gas occupies a volume of 544 mL at 27 °C and 744 torr?

5.80 To lose weight, we are told to exercise to "burn off the fat." Although fat is a complicated mixture, it has approximately the formula $C_{56}H_{108}O_6$. Calculate the volume of oxygen that must be consumed at 22 °C and 1.00 atm of pressure in order to "burn off" 5.0 pounds of fat. (Hint: start by writing the equation for the combustion of the fat.)

5.81 Three bulbs are connected by tubing, and the tubing is evacuated. The volume of the tubing is 22 mL. The first bulb has a volume of 50 mL and contains 2.00 atm of argon; the second bulb has a volume of 250 mL and contains 1.00 atm of neon; and the third bulb has a volume of 25 mL and contains 5.00 atm of hydrogen. If the stopcocks (valves) that isolate all three bulbs are opened, what is the final pressure of the whole system?

5.82 Calculate the mass of water produced in the reaction of 4.33 L of oxygen and 6.77 L of hydrogen gas. Both gases are at a pressure of 1.22 atm and a temperature of 27 °C.

5.83 Calculate the (a) rms speed of samples of hydrogen and nitrogen at STP, and (b) the average kinetic energies of the two gases under these conditions.

5.84 Lithium hydroxide is used to remove the CO_2 produced by the respiration of astronauts. An astronaut produces about 400 L of CO_2 at 24 °C and 1.00 atm of pressure every 24 hours. What mass of lithium hydroxide, in grams, is needed to remove the CO_2 for 24 hours? The equation is

$$2LiOH(s) + CO_2(g) \longrightarrow Li_2CO_3(s) + H_2O(\ell)$$

5.85 (a) Calculate the pressure, in atm, of 30.33 mol of hydrogen at 240 °C in a 2.44-L container using the van der Waals equation.
(b) Do the same calculation for methane under the same conditions.
(c) What is different about the two gases that causes the pressure in the containers to be different?

Chapter Six

Electrons in Atoms

Scientists have determined that an understanding of the way electrons are present in atoms is needed to explain many of the laws of chemical combination. For example, when hydrogen combines with oxygen it forms water, with each molecule containing two atoms of hydrogen and one atom of oxygen. In contrast, one atom of nitrogen combines with three atoms of hydrogen to form ammonia. Sodium forms a stable ion with a 1+ charge, calcium exists in compounds as a 2+ ion, and chlorine often combines as 1− ions. These are just a few of the many millions of facts that must be explained by a theory of chemical combination. The locations and energies of the electrons play a key role in the way atoms combine.

This chapter presents the modern model of atomic structure. Since much of the knowledge of the electronic structure of atoms is based upon observations of their interaction with light, we must first consider the nature of electromagnetic radiation. Keep in mind that the main goal in this chapter is to understand the nature of electrons in atoms.

6.1 The Nature of Light

Objectives

- To describe electromagnetic radiation in terms of wavelength, frequency, and energy
- To describe the models that are used to explain light
- To describe the dual nature of electromagnetic radiation

◀ The interaction of light with matter is used to probe the structures of atoms.

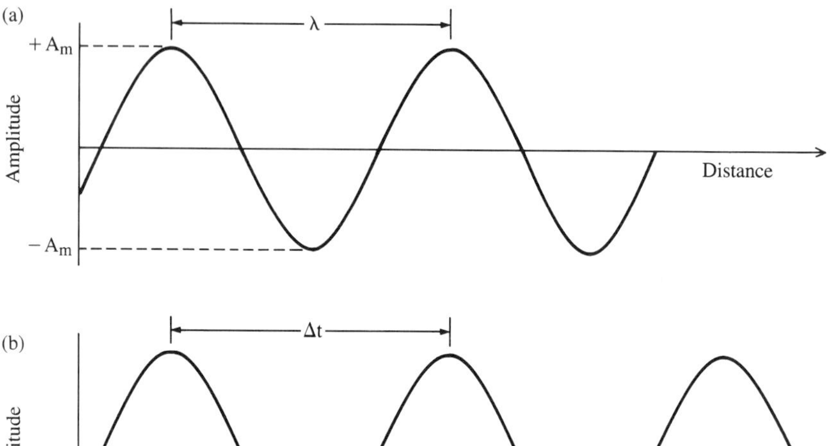

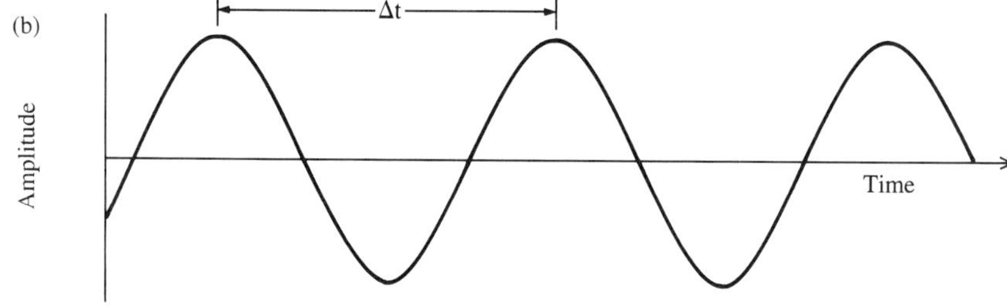

Figure 6.1 Typical waves (a) The wavelength, λ, is the distance between two successive peaks of the wave, and the amplitude is the vertical displacement. The amplitude, A, varies from $-A_m$ to A_m. (b) The length of time it takes for one complete wave to pass a point is Δt. The frequency of the wave, ν, is the number of wave maxima that pass a point in each second.

Waves The periodic nature of wave motion is not always easy to see.

The Wave Nature of Light

In the late 19th century, physicists knew that light could be described as waves that are similar to the waves that move through water. In water a disturbance produces an up-and-down motion of the surface. Although the crests of the waves move horizontally with time, the motion of the liquid is in the vertical direction. **Waves** are periodic in nature—they repeat at regular intervals of both time and distance.

Any wave is described by its amplitude, wavelength, and frequency, as shown in Figure 6.1. The maximum height of a wave is called its **amplitude;** the wave varies between $+A_m$ and $-A_m$. The amplitude of a light wave is related to the brightness of the source. The **wavelength** (λ, "lambda") is the distance between one peak and the next. In the SI system, wavelength is measured in meters. The **frequency** (ν, "nu") of a wave is the number of waves that pass a fixed point in one second. The SI unit for frequency is s^{-1}, and has been given the name **hertz,** which is abbreviated as Hz.

Light waves are called **electromagnetic radiation** because they consist of oscillating electric and magnetic fields, which are perpendicular to each other and perpendicular to the direction of motion, as shown in Figure 6.2. The periodic variations of the electric and magnetic fields of light are analogous to the vertical motion of the water particles.

In general, the speed at which a wave travels is the product of its wavelength and frequency. In a vacuum, all electromagnetic radiation travels at the same speed, $c = 3.00 \times 10^8$ m s^{-1}.

If either the wavelength or the frequency of electromagnetic radiation is known, the other can be calculated from the equation $c = \lambda\nu$.

$$c = \lambda\nu = 3.00 \times 10^8 \text{ m s}^{-1} \qquad [6.1]$$

From Equation 6.1, then, the longer the wavelength of light, the smaller is its frequency.

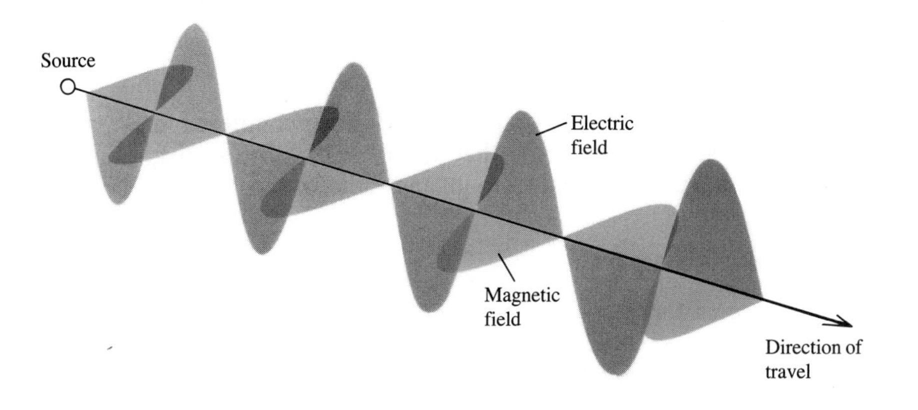

Figure 6.2 Electromagnetic radiation Light, or electromagnetic radiation, consists of oscillating electric and magnetic fields that are perpendicular to the direction of motion of the wave and to each other.

Example 6.1 Frequency and Wavelength

About 700 nm is the longest wavelength of light that can be detected by the human eye. What is the frequency of this electromagnetic wave?

Solution Rearrange Equation 6.1 to solve for the frequency. For the units of length to cancel, we must convert the wavelength into meters (1 nm = 10^{-9} m).

$$\nu = \frac{c}{\lambda}$$

$$\nu = \frac{3.00 \times 10^{8}\ \text{m s}^{-1}}{700\ \text{nm}}\left(\frac{1\ \text{nm}}{10^{-9}\ \text{m}}\right) = 4.28 \times 10^{14}\ \text{s}^{-1}$$

Understanding What is the frequency of radiation that has a wavelength of 3.00 meters? (This is in the range used for commercial FM radio transmission.)

Answer: 100 MHz or 100,000,000 waves per second.

The full range of electromagnetic radiation is very large. Only a very small part of this range, called **visible light,** can be detected by the human eye. Visible light has wavelengths from 400 nm ($\nu = 7.5 \times 10^{14}\ \text{s}^{-1}$) to 700 nm ($\nu = 4.3 \times 10^{14}\ \text{s}^{-1}$). The full range of electromagnetic radiation is shown in Figure 6.3, along with the common names used to identify different

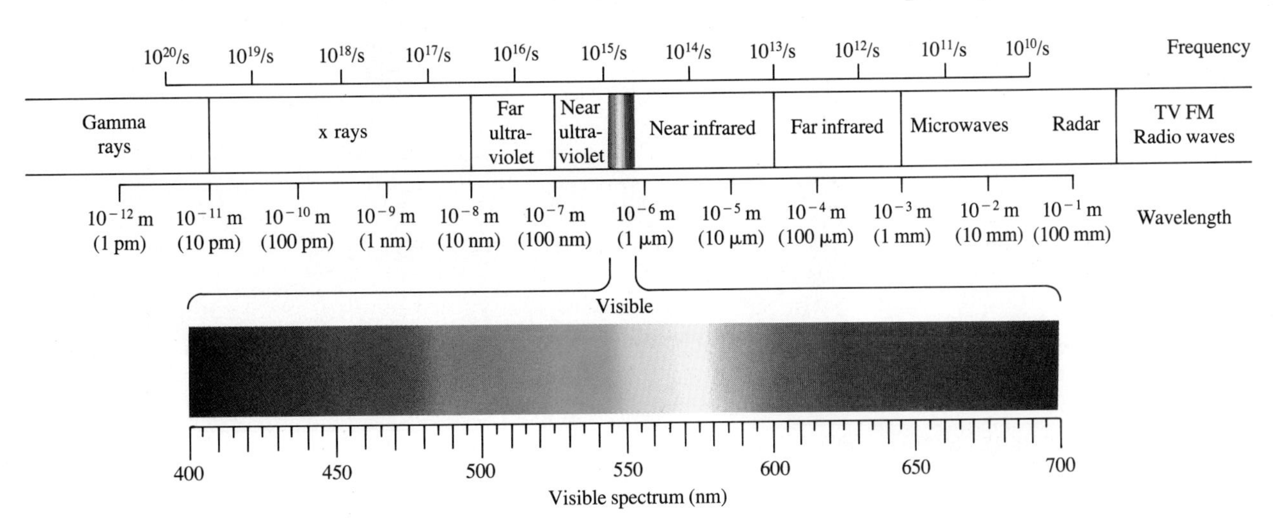

Figure 6.3 The electromagnetic spectrum The range of electromagnetic radiation is shown, and names commonly used to refer to different regions are identified. Divisions between the regions are not defined precisely.

ranges of wavelengths. Many of these names are encountered in everyday conversation, such as the x rays that are used for medical diagnosis, the microwaves that are used to heat food, and radio waves.

Quantization of Energy

At temperatures above 0 kelvin, matter emits electromagnetic radiation of all wavelengths. The white light from the common light bulb is produced by electrically heating a small tungsten wire to a high temperature.

Physicists of the 19th century, using the accepted model of electromagnetic radiation, were unable to explain the wavelength distribution of light

INSIGHTS into CHEMISTRY

A Closer View

Radiation Emitted by Heated Solids; Planck Solves "UV Catastrophe"

Prior to Planck's proposal of quantization, the accepted theories did not properly describe the distribution of light energy that is emitted by a heated object. While the light energy emitted at different temperatures was accurately predicted at long wavelengths, at short wavelengths the theory failed. The incorrect predictions at short wavelengths were called the "ultraviolet catastrophe."

The following figure shows the experimentally observed spectrum of light energy as a function of the temperature for a heated object. Not only does the total energy increase markedly with increasing temperature, but the wavelength at which the maximum energy is observed shifts to shorter wavelengths. The maximum in the curve at 5000 K occurs in the visible region of the spectrum,

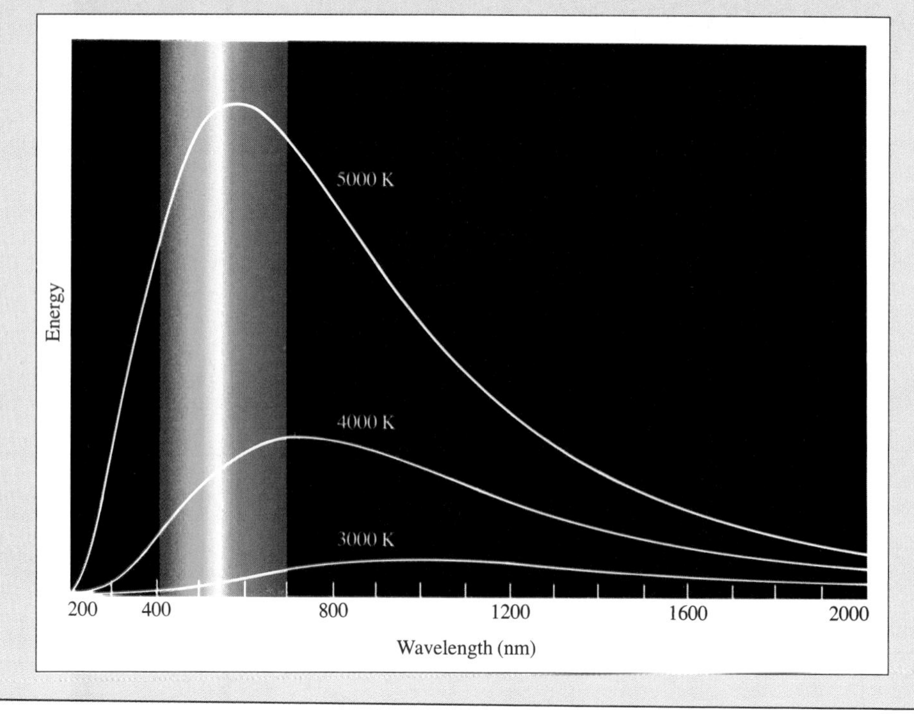

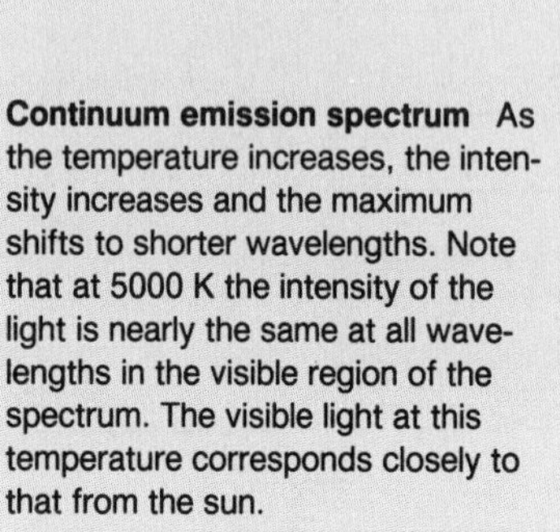

Continuum emission spectrum As the temperature increases, the intensity increases and the maximum shifts to shorter wavelengths. Note that at 5000 K the intensity of the light is nearly the same at all wavelengths in the visible region of the spectrum. The visible light at this temperature corresponds closely to that from the sun.

emitted by heated objects. In 1900, Max Planck (1858–1947) solved this problem with a brilliant assumption that violated the classical model of physics. He assumed that there was a smallest unit of energy that matter could absorb or emit, which he called a **quantum** of energy. Planck was able to show that the energy of a quantum is proportional to the frequency of the emitted light. In equation form,

$$\Delta E = h\nu$$

where h is a constant that has the value of 6.63×10^{-34} J s (joules seconds) and is called **Planck's constant.** The energy is written as ΔE to emphasize that the matter changes in energy when the light is absorbed or emitted.

The smallest amount of electromagnetic energy absorbed or emitted by matter is one quantum, $h\nu$.

and this curve coincides quite closely to the spectrum of light from the sun. Some scientists believe that the evolution of the human eye was influenced by the available light from the sun.

At temperatures above 600 °C enough visible light is produced so our eyes can detect it. As solids are heated, they first glow red, then yellow, and finally white (see the following figure). One method used to measure high temperatures is to compare the color of the light emitted by an object to a standard of known temperature.

At room temperatures practically no visible light is emitted by objects. However, even when there is insufficient visible light, images of an object can be produced by using the infrared radiation that it emits. The detected infrared radiation is amplified and electronically converted into a visible image.

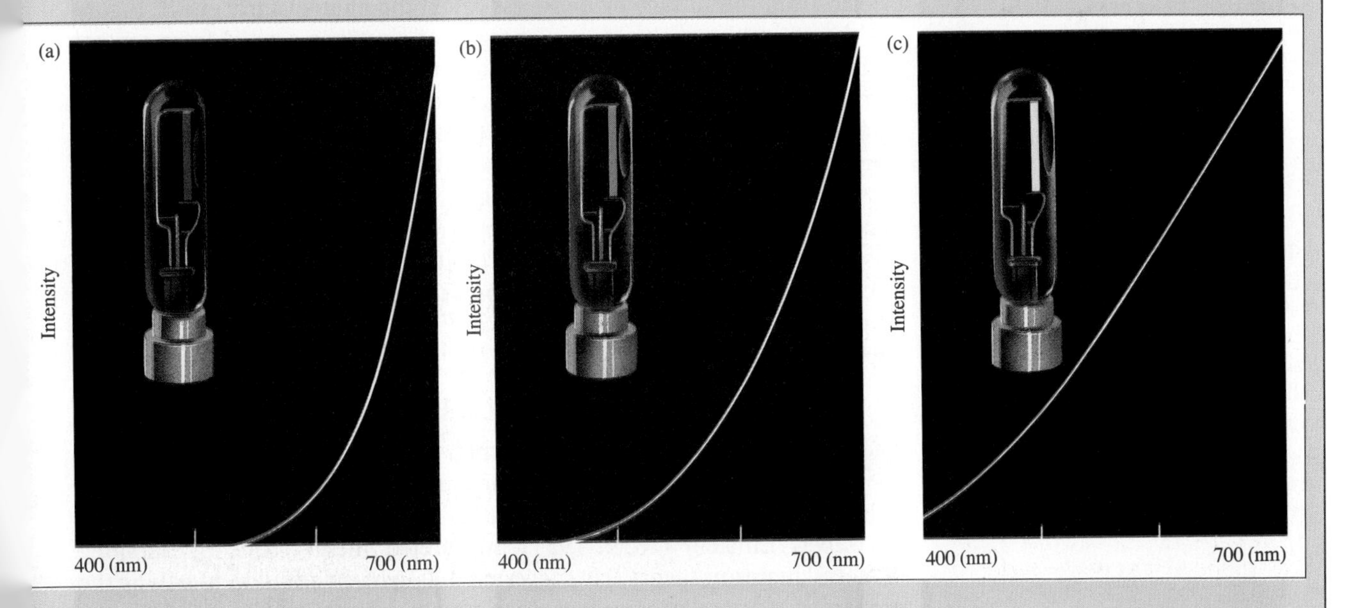

Light sources of different temperatures A calibrated tungsten lamp can be used to produce radiant energy with a spectral distribution that depends on the temperature. Each part shows a graph of the relative intensity of the visible spectrum and a tungsten lamp at that temperature. (a) 800 °C, the lamp glows a dull red. (b) 1500 °C, the lamp is brighter, but is still a red-yellow color. (c) 2500 °C, the light from the lamp appears white.

According to Planck, an object could only absorb or emit energy of $h\nu$, or $2h\nu$, $151h\nu$, or any other whole number multiple of $h\nu$. However, Planck's relationship forbids the sample to absorb or emit electromagnetic energy of $0.5h\nu$ or $12.6h\nu$. In other words, the electromagnetic energy absorbed or emitted by matter is *quantized.* This quantization of energy was an entirely new idea at the time. It was remarkably successful in accurately predicting the distribution of light energy as a function of the frequency of the light.

Example 6.2 The Quantum of Light Energy

What is the smallest quantity of light energy at a wavelength of 700 nm that is emitted by a heated object?

Solution According to Planck's equation, the minimum light energy is one quantum, or $h\nu$. The frequency of light with a wavelength of 700 nm was found in Example 6.1 to be $4.28 \times 10^{14}\ \text{s}^{-1}$. Now we use this frequency of the light to find the energy of one quantum.

$$\Delta E = h\nu = 6.63 \times 10^{-34}\ \text{J s} \times 4.28 \times 10^{14}\ \text{s}^{-1} = 2.84 \times 10^{-19}\ \text{J}$$

Understanding Calculate the energy of one quantum of light that has a frequency of 6.00×10^{15} Hz.

Answer: 3.98×10^{-18} J

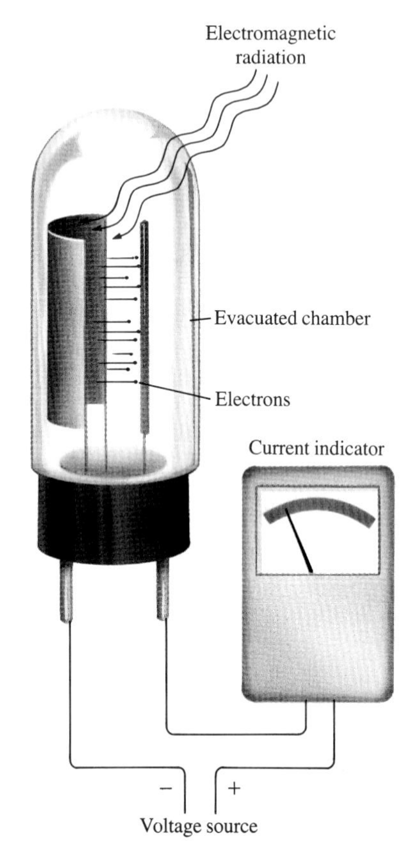

A photoelectric cell When light of high enough frequency strikes the metal surface in the tube, electrons are ejected. The electrons are attracted to the other electrode in the cell, producing an electric current in the external circuit. The electric current from a photoelectric cell is used to activate some automatic door openers.

Photoelectric Effect

When light strikes solid materials, particularly metals, electrons can be ejected from the surface by a process called the **photoelectric effect.** Figure 6.4 shows graphs of the kinetic energies of electrons ejected from several different metals. Each metal has a characteristic frequency (ν_0), called the **threshold frequency,** at which the kinetic energy of the emitted electrons is zero. Above the threshold frequency, the kinetic energy of the ejected electron increases linearly with the increase in the frequency of the light. Light of lower frequency than this threshold, no matter how intense, does not produce any photoelectrons. These observations contradict the predictions of classical physics. In the classical wave picture of light, electrons could be ejected by any frequency of light, as long as the light was bright enough. For all metals, the slope of the line given in Figure 6.4 is identical, and fits the equation

$$\text{kinetic energy} = h\nu - h\nu_0 = h(\nu - \nu_0) \tag{6.2}$$

where ν is the frequency of the absorbed light, and ν_0 is the threshold frequency, which is a characteristic of the particular metal. The product $h\nu_0$ is called the **work function** of the metal.

Albert Einstein (1879–1955) interpreted these results by applying Planck's quantum theory. He suggested that light, in addition to having the properties of waves, can be viewed as a stream of tiny particles, referred to in modern terms as **photons.** An electron is dislodged from the metal by collision with a single photon, having an energy of $h\nu$. Some of the energy, $h\nu_0$, must be used to overcome the attraction the solid has for the electrons, and the rest appears as the kinetic energy of the electron. If the energy of the absorbed photon is less than $h\nu_0$, no electron can be ejected, and the ab-

In the interpretation of the photoelectric effect, electromagnetic radiation is treated as particles of light (photons) instead of waves.

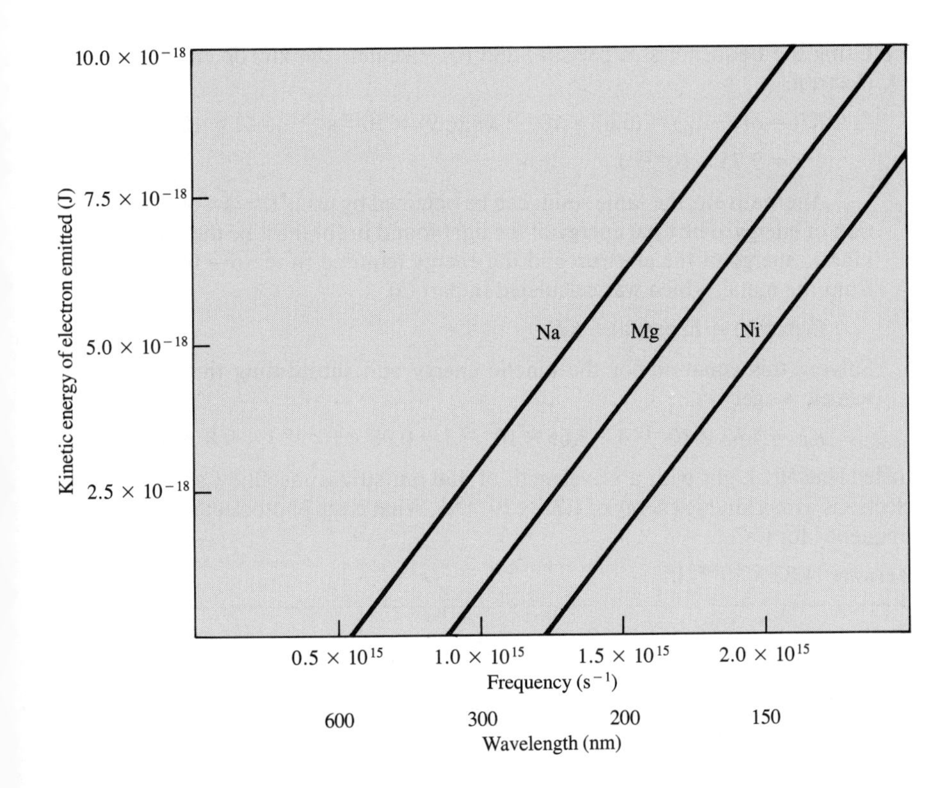

Figure 6.4 Electron energies from the photoelectric effect The threshold frequencies and corresponding wavelengths are: Na, $5.51 \times 10^{14}\ s^{-1}$ (544 nm); Mg, $8.89 \times 10^{14}\ s^{-1}$ (337 nm); Ni, $1.22 \times 10^{15}\ s^{-1}$ (245 nm). Note that the slopes of all three lines are identical and equal to Planck's constant, 6.63×10^{-34} J s.

sorbed energy simply heats the metal. Einstein's explanation of the photoelectric effect thus supported Planck's theory that energy is quantized and, more importantly, clarified that this quantization of energy is a property of the light. The energy of a photon of light is therefore given by the equation

$$E_{\text{photon}} = h\nu \qquad [6.3]$$

Example 6.3 The Photoelectric Effect

The threshold frequency (ν_0) that can dislodge an electron from metallic sodium is $5.51 \times 10^{14}\ s^{-1}$.

(a) What is the energy of a photon that has this frequency?
(b) What is the energy of a photon of light that has a wavelength of 430 nm?
(c) What is the kinetic energy of an electron that is ejected from sodium by light that has a wavelength of 430 nm?

Solution

(a) The energy of a photon is given by Equation 6.3:

$$E_{\text{photon}} = 6.63 \times 10^{-34}\ \text{J s} \times 5.51 \times 10^{14}\ \text{s}^{-1} = 3.65 \times 10^{-19}\ \text{J}$$

(b) First calculate the frequency of the photon by solving the equation $c = \lambda\nu$ for the frequency.

$$\nu = \frac{c}{\lambda} = \frac{3.00 \times 10^{8}\ \text{m s}^{-1}}{430\ \text{nm}}\left(\frac{1\ \text{nm}}{10^{-9}\ \text{m}}\right) = 6.98 \times 10^{14}\ \text{s}^{-1}$$

The energy of the photon can be calculated from Planck's equation.

$$E = h\nu = 6.63 \times 10^{-34}\ \text{J s} \times 6.98 \times 10^{14}\ \text{s}^{-1} = 4.63 \times 10^{-19}\ \text{J}$$

(c) Using the frequencies in parts (a) and (b), calculate the kinetic energy of the electron.

$$KE = h(\nu - \nu_0) = (6.63 \times 10^{-34} \text{ J s})(6.98 \times 10^{14} \text{ s}^{-1} - 5.51 \times 10^{14} \text{ s}^{-1})$$
$$= 9.75 \times 10^{-20} \text{ J}$$

Alternatively, the same result can be obtained by using the law of conservation of energy. The total energy of the light found in (b) must be the sum of the kinetic energy of the electron and the energy required to remove the electron from the metal, which was calculated in part (a).

$$E(\text{part b}) = E(\text{part a}) + KE$$

Solving this equation for the kinetic energy and substituting the calculated values, we get

$$KE = 4.63 \times 10^{-19} \text{ J} - 3.65 \times 10^{-19} \text{ J} = 0.98 \times 10^{-19} \text{ J} = 9.8 \times 10^{-20} \text{ J}$$

Understanding Light with a wavelength of 450 nm strikes metallic Cs, and ejects electrons with a kinetic energy of 1.22×10^{-19} J. What is the photoelectric threshold frequency for Cs?

Answer: 4.83×10^{14} Hz

The Dual Nature of Light

Many properties of light, such as the photoelectric effect, can be explained by treating light as a stream of particles, while other properties are more consistent with the interpretation of light as waves. These apparently conflicting models do not mean that light is two things, but that we have not yet been able to find a single model for describing light that is easily related to all of its properties. Describing light as both waves and particles is referred to as the **dual nature of light.** It is important to recognize that the dual nature of light originates in our models, and is not a true duality of the light. Up until his death in 1955, Albert Einstein had spent many years, without success, trying to develop a "unified model" that would resolve this duality.

6.2 Line Spectra and the Bohr Atom

Objectives

- To examine the origin of atomic line spectra
- To calculate the observed lines in the emission and absorption spectra of the hydrogen atom and ions that contain only one electron
- To relate the electronic energy levels in the hydrogen atom to the observed line spectrum and the Bohr model

When energy is added to a sample of gaseous atoms, a process called "excitation," the gas can emit light. Examination of the **spectrum** (the intensity of the light as a function of wavelength) reveals that the light from excited atoms is quite different from the light emitted by a heated solid. The heated solid produces a **continuum spectrum,** one in which all wavelengths of light are present. The atomic emission spectrum is called a **line spectrum,** because it contains light at discrete wavelengths separated by regions in which no light is emitted. Each element produces a line spectrum that is characteristic

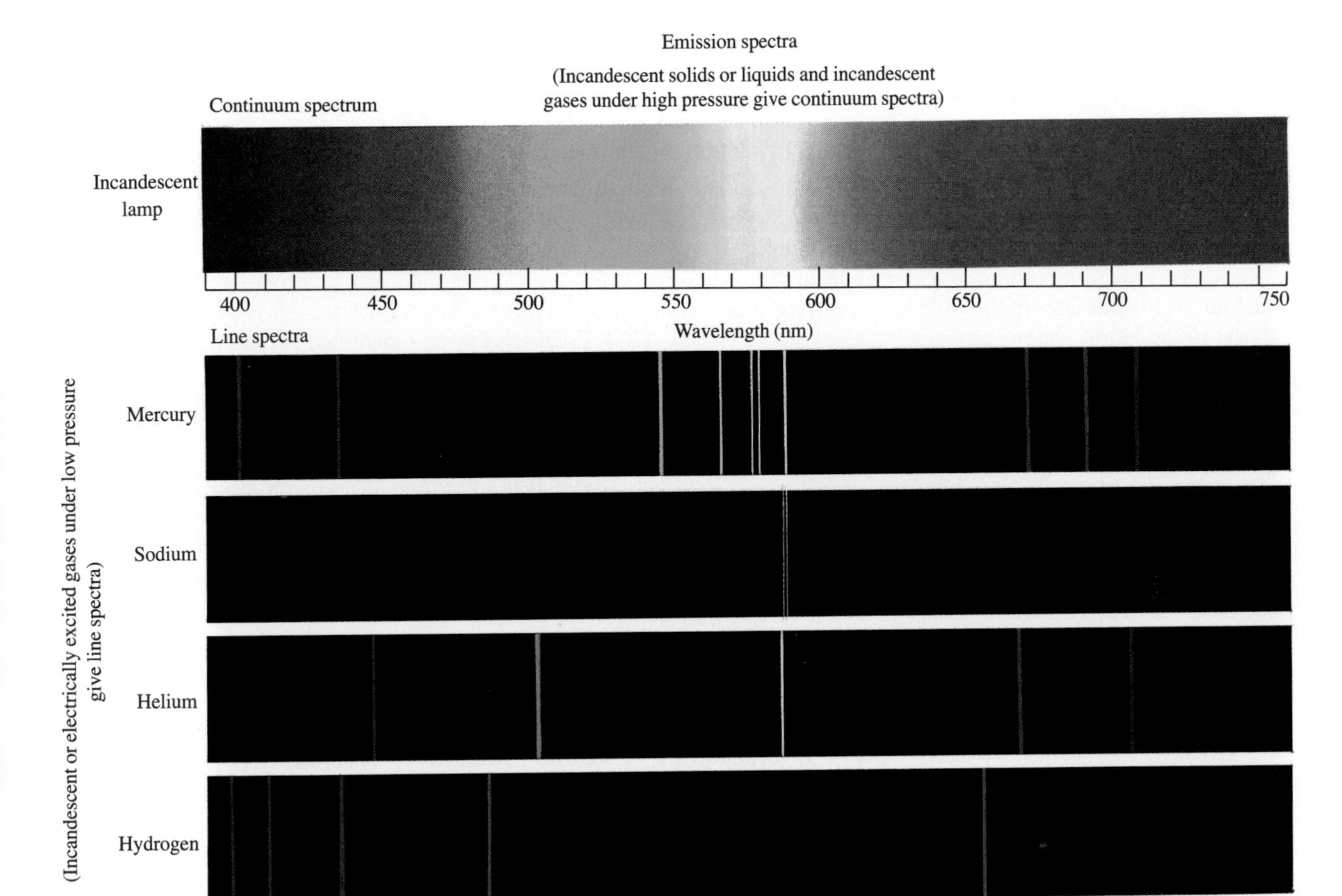

Figure 6.5 Emission spectra The emission spectra in the visible region for several elements.

of that element and is different from the spectrum of any other element. Long before the reason for this behavior was understood, line spectra were used to identify the elements present in samples of matter. In fact, in the 1860s the presence of unexpected emission lines observed in some samples of sodium and potassium led to the discovery of the elements cesium and rubidium. The line spectra of several elements are shown in Figure 6.5.

The spectrum of the simplest element, hydrogen, was studied in detail (see Figure 6.5). It was found that the wavelengths of the lines in the hydrogen spectrum could be calculated from a simple relationship, called the **Rydberg equation:**

$$\frac{1}{\lambda} = R_h\left(\frac{1}{n_1^2} - \frac{1}{n_2^2}\right) \qquad [6.4]$$

Here n_1 and n_2 are whole positive numbers with $n_1 < n_2$, and R_h is a constant, called the Rydberg constant, which has a value of $1.097 \times 10^7\ \text{m}^{-1}$. It is important to note that this equation was based upon the experimentally observed wavelengths of lines in the spectrum of the hydrogen atom. At the time there was no theoretical explanation for this correlation. The hydrogen atom spectrum consists of series of lines that are named after individuals. All lines in any series have the same value of n_1, with each line having a different

The Rydberg equation accurately predicts the wavelengths of all the observed lines in the spectrum of hydrogen atoms.

INSIGHTS into CHEMISTRY

Development of Chemistry

Scientists Predicted Hydrogen Spectrum; Modern Detectors Allow Its Observation

In 1885, a Swiss high-school teacher, Johann Balmer (1825–1898), published an empirical relationship that fit the four prominent lines of the hydrogen spectrum that appear in the visible region:

$$\lambda = 364.56 \text{ nm} \frac{n^2}{n^2 - 2^2}$$

where n is a whole number greater than 2. Several other lines predicted by this equation were identified by others. Balmer suggested in his paper that there might be other series of lines in which the 2^2 in his equation is replaced by 1^2, 3^2, 4^2, and so forth. Balmer's expression, in a slightly different form, is the Rydberg equation with $n_1 = 2$.

$$\frac{1}{\lambda} = 1.097 \times 10^7 \text{ m}^{-1} \left(\frac{1}{n_1^2} - \frac{1}{n_2^2}\right)$$

For many years there was no experimental verification of the other series of lines suggested by Balmer. In 1904, Friedrich Paschen identified two lines in the infrared region of the spectrum, for which n_1 was 3. Between 1906 and 1924, other series were identified in the ultraviolet region by Lyman ($n_1 = 1$), and in the infrared region by Brackett ($n_1 = 4$) and Pfund ($n_1 = 5$). The discovery of the Brackett and Pfund series had to wait until suitable detectors of far infrared radiation were developed. Each of these series of lines in the hydrogen spectrum was given the name of the person who identified it.

value for n_2. The names are the Lyman ($n_1 = 1$), Balmer ($n_1 = 2$), Paschen ($n_1 = 3$), Brackett ($n_1 = 4$), and Pfund ($n_1 = 5$) series.

Example 6.4 **Calculating Wavelength from the Rydberg Equation**

Calculate the wavelength (in nm) of the line in the hydrogen spectrum for $n_1 = 2$ and $n_2 = 4$.

Solution Substitute these values into the Rydberg equation.

$$\frac{1}{\lambda} = 1.097 \times 10^7 \left(\frac{1}{2^2} - \frac{1}{4^2}\right) \text{ m}^{-1} = 2.057 \times 10^6 \text{ m}^{-1}$$

Rearrange to solve for the wavelength, and convert the units to nanometers.

$$\lambda = \frac{1}{2.057 \times 10^6 \text{ m}^{-1}} = 4.861 \times 10^{-7} \text{ m} \left(\frac{1 \text{ nm}}{10^{-9} \text{ m}}\right) = 486.1 \text{ nm}$$

Light of this wavelength is blue.

Understanding Find the wavelength of the next line in the Balmer series, with $n_1 = 2$ and $n_2 = 5$.

Answer: 434.0 nm

The Bohr Model of the Hydrogen Atom

Once the relation between the energy of light and its frequency was discovered by Planck, the line spectra of atoms suggested that the energies of atoms are also quantized. By this we mean that an atom can exist only in certain allowed energy states. In 1911, Niels Bohr (1885–1962) proposed a model for the hydrogen atom which predicted that only certain energies of the atom are allowed. Bohr accepted Ernest Rutherford's recently proposed nuclear model for the atom. He assumed that the electron follows circular orbits around the nucleus, and he further assumed that the electron could have only certain values of angular momentum, *i.e.*, the angular momentum is quantized. Using his model, Bohr found that the allowed energies for the atom are also quantized. Each allowed state is described by a whole, positive number. The allowed energies are given by

$$E_n = -\left(\frac{2\pi^2 me^4}{h^2}\right)\left(\frac{1}{n^2}\right) \qquad [6.5]$$

where m is the mass of the electron, e is charge of the electron, and h is Planck's constant. The allowed energies, E_n, are found by using any whole positive number for n (1, 2, 3, . . .) in Equation 6.5, so many different energy states are possible. The light emitted by the atom must have an energy ($h\nu$) that is exactly equal to the difference between the energies of two of its allowed states:

$$h\nu = E_{n_2} - E_{n_1} = \left(\frac{2\pi^2 me^4}{h^2}\right)\left(\frac{1}{n_1^2} - \frac{1}{n_2^2}\right) \qquad [6.6]$$

Equation 6.6 is closely related to Rydberg's law. Equations 6.4 and 6.6 can be compared to show that the value of R_h, the Rydberg constant, is $2\pi^2 me^4/h^3c$. The value of the Rydberg constant calculated from the Bohr model is nearly identical to that found experimentally. The ability to calculate the experimental value of the Rydberg constant in terms of other physical constants was a major triumph of the Bohr model. Despite other deficiencies in Bohr's model for the hydrogen atom, the expression he derived for Rydberg's constant remains valid.

The existence of line spectra for the elements suggests that the energies of atoms are quantized.

The energy-level diagram for the hydrogen atom described by Bohr's theory is shown in Figure 6.6. The vertical lines show the electron's transitions between the allowed energy states of the atom. When an electron goes from one allowed energy state to a lower one, the difference in energy is released as radiation of a single wavelength. A hydrogen atom with its electron in the $n = 4$ state may return to the lowest energy state ($n = 1$) by emitting light in several ways. The electron can return to the $n = 1$ state in one step, by emitting a single photon with an energy equal to the energy difference between $n = 4$ and $n = 1$. Alternatively, the same energy change can occur by emission of as many as three photons, corresponding to the energies of the transitions from $n = 4$ to $n = 3$ to $n = 2$ to $n = 1$. For each transition, however, the energy must be emitted as a single photon. Each of the spectral series mentioned earlier corresponds to a set of transitions in which the final energy state is the same. For example, all transitions that end with the hydrogen atom having its electron in the $n = 2$ state belong to the Balmer series.

All of the energy released when an atom goes from one allowed energy state to a lower one is contained in a single photon of light.

If light energy is supplied to hydrogen atoms, the only photons absorbed are those with an energy that is the same as the difference between two allowed states of the electron. The **ground state** of an atom is its lowest allowed energy state. At normal temperatures nearly all of the atoms of hydrogen are present in the ground state, so nearly all of the absorption lines arise from transitions from the ground state ($n = 1$) to the excited states ($n > 1$). Thus the only lines observed in the absorption spectrum of hydrogen atoms are those in the Lyman series.

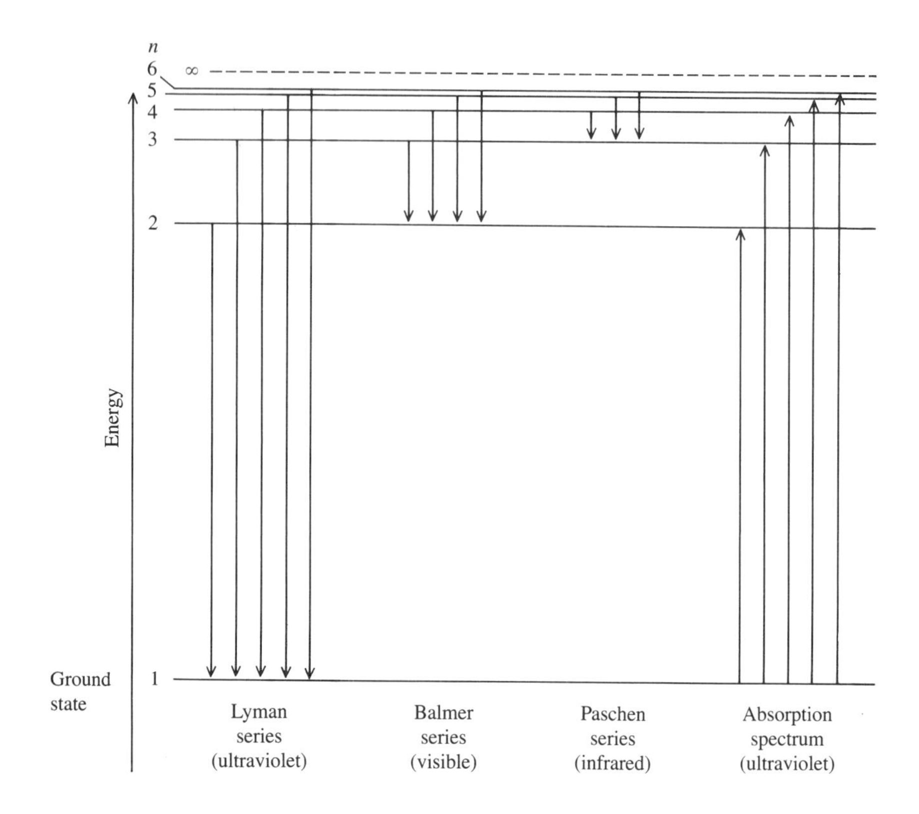

Figure 6.6 Transitions in the hydrogen atom The electron transitions that produce the lines in the Lyman, Balmer, and Paschen series in the emission spectrum of hydrogen are shown on the left. The absorption spectrum, shown on the right, contains only the lines in the Lyman series, since in a sample of hydrogen, nearly all the atoms are in the ground state.

While Bohr's model was a major advance in explaining the hydrogen spectrum, attempts to refine it and extend it to atoms other than hydrogen were unsuccessful. In addition, there are some fundamental theoretical problems with the Bohr model of the hydrogen atom that make it unacceptable. (See *Insights into Chemistry* on the Heisenberg uncertainty principle). An entirely different model was needed to account for the electronic structure of atoms.

6.3 Matter as Waves

Objectives

- To develop the wave model of matter
- To determine the wavelengths associated with particles of matter

In 1924, Louis de Broglie (1892–1987) proposed an entirely new way of considering matter in his doctoral dissertation. The success achieved by viewing electromagnetic radiation both as particles and as waves led de Broglie to suggest that the same duality may also apply to matter. He asked, "What if matter, in particular electrons, could also be described as waves?" To answer this question it was necessary to find some bridge that related typical wave properties, such as frequency or wavelength, to properties usually associated with particles of matter. A few years earlier Arthur Compton had performed experiments which showed that the momentum of a photon is given by the expression

$$\text{momentum} = p = h/\lambda \qquad [6.7]$$

INSIGHTS into CHEMISTRY

A Closer View

Heisenberg's Uncertainty Principle Limits Bohr's Atomic Model

Werner Heisenberg (1901–1976) postulated a very important principle of nature, one that limits the knowledge we may have about particles. This **Heisenberg uncertainty principle** states that it is not possible to know simultaneously both the position and momentum of a particle precisely. Expressed mathematically, the uncertainty principle is

$$\Delta x\, \Delta p \simeq \frac{h}{2\pi} = 1.06 \times 10^{-34}\ \text{kg m}^2\ \text{s}^{-1}$$

where Δx and Δp are the uncertainty in position and momentum, respectively, and h is Planck's constant.

The uncertainty principle makes it clear that the Bohr model of the atom is unacceptable. According to the Bohr model, an electron in the $n = 1$ orbit of the hydrogen atom follows a circular path having a radius of 53 pm, with a momentum of 1.99×10^{-24} kg m s^{-1}. If we assume that the uncertainty of the momentum is 1% of its value, or 1.99×10^{-26} kg m s^{-1}, then the uncertainty in its position is

$$\Delta x = \frac{1.06 \times 10^{-34}}{1.99 \times 10^{-26}} = 5.3 \times 10^{-9}\ \text{m} = 5300\ \text{pm}$$

The uncertainty in the position of the electron is a *hundred times* the radius of the Bohr orbit. The Bohr model thus calculates the position and the momentum of the electron more accurately than is possible within the limitations of the Heisenberg uncertainty principle.

Werner Heisenberg (1901–1976) German physicist. One of the pioneers in the field of quantum theory and discoverer of the uncertainty principle. He received the Nobel Prize in physics in 1932.

de Broglie suggested that the same relationship between wavelength and momentum of a photon might be used to relate the wave and particle properties of matter. The momentum of matter is the product of mass times velocity, so de Broglie proposed the use of Equation 6.8 to calculate the wavelength associated with an electron:

$$p = mu = h/\lambda \qquad [6.8]$$

Only a few years later, in 1928, Davisson and Germer performed an experiment in which electrons were observed to be diffracted by sodium chloride crystals. Diffraction is one of the properties of light that is very easily interpreted by using the wave model. The electron diffraction experiments confirmed that de Broglie's equation correctly calculated the wavelength of the electrons, and that small particles of matter do behave as waves.

Particles of matter exhibit the properties of waves under some circumstances.

Example 6.5 Calculating the Wavelength of an Electron

Find the wavelength of electrons that have a velocity of 3.00×10^6 meters per second.

Solution Solve de Broglie's Equation 6.8 for the wavelength, and substitute the known values, using the appropriate SI units, into the equation. The mass of the electron must be expressed in kg ($m = 9.11 \times 10^{-31}$ kg), and the velocity in m s^{-1}, when Planck's constant is expressed in joule-seconds (since 1 J = 1 kg m^2 s^{-2}).

$$\lambda = \frac{h}{mu} = \frac{6.63 \times 10^{-34}\ \text{kg m}^2\ \text{s}^{-1}}{(9.11 \times 10^{-31}\ \text{kg})(3.00 \times 10^6\ \text{m s}^{-1})} = 2.42 \times 10^{-10}\ \text{m}$$

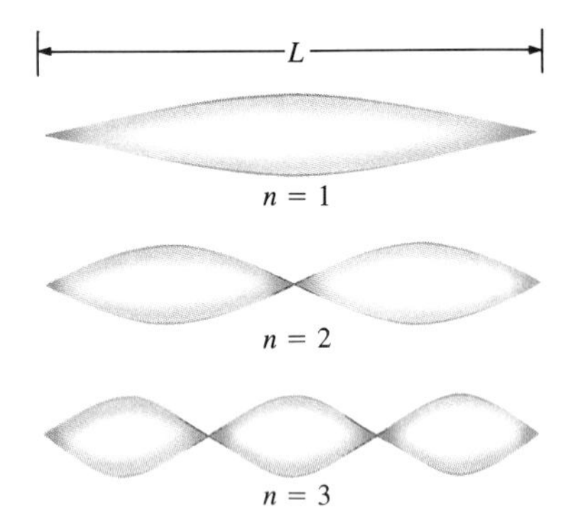

Figure 6.7 Standing waves The wavelengths at which a violin string will vibrate are those that satisfy the equation $L = n(\lambda/2)$. The waves with $n = 1$, 2, and 3 are shown.

This wavelength is comparable to that of x rays, which are also diffracted by crystalline solids.

Understanding What is the velocity (in m/s) of neutrons that have a wavelength of 0.200 nm? The mass of a neutron is 1.67×10^{-27} kg.

Answer: 1.98×10^3 m/s

de Broglie offered an explanation of the restricted energy states of the hydrogen atom by interpreting the electron as a standing or stationary wave. The vibration of a violin string is a simple example of a standing wave. When a violin string is plucked, its vibration is restricted to certain wavelengths, since the ends of the string cannot move. One-half the wavelength times a whole number must equal the length of the string (see Figure 6.7). de Broglie suggested that the circumference of a Bohr orbit must be a whole-number multiple of the electron's wavelength, so that a standing wave is produced (see Figure 6.8). If the electron were not a standing wave, it would partially cancel itself on each successive orbit, until its amplitude was zero and the electron (the wave) no longer existed!

The wave properties of matter are important in several modern applications. The diffraction of electrons and neutrons by molecules provides important information about the structure of molecules by allowing us to measure the distances between atoms. The electron microscope, which is capable of higher magnifications than can be achieved by a light microscope, is based upon the wave properties of electrons.

The Schrödinger Wave Equation

Shortly after de Broglie proposed that very small particles of matter might be described as waves, Erwin Schrödinger (1887–1961) applied the idea to describe the behavior of the electron in the hydrogen atom. A complete solution of his mathematical description is complicated, and is not needed in

Figure 6.8 Circular standing waves (a) The circumference of the circle is exactly five times the wavelength, so a stationary wave is produced. This is an allowed orbit. (b) The wave does not close on itself, since the circumference is 5.2 times the wavelength. This orbit is not allowed.

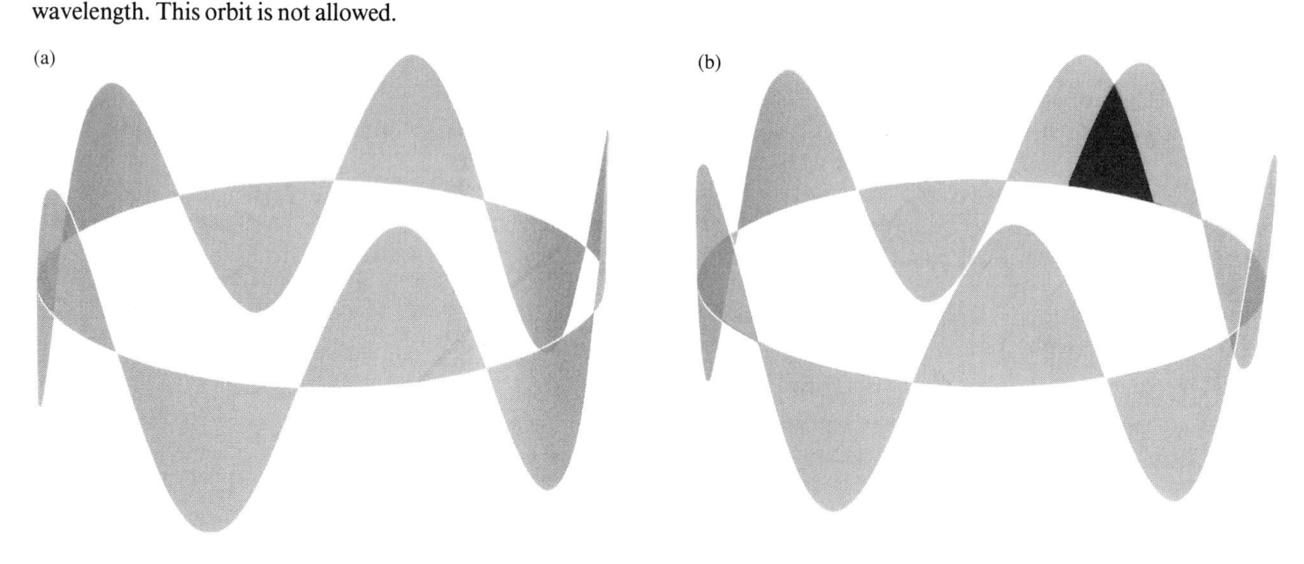

this book. However, the results are presented here because they are important to the understanding of the electronic structure of atoms.

1. The electron wave can be described by a mathematical function that gives the amplitude of the wave at any point in space. These amplitude functions are called **wave functions,** and are usually represented by the Greek letter ψ (psi).
2. The square of the wave function, ψ^2, gives the *probability* of finding the electron at any point in space. It is not possible to say exactly where the electron is located when we describe it as a wave. The wave model does not conflict with the Heisenberg uncertainty principle (p. 213), since it does not precisely define the location of the electron.
3. There are many wave functions that are acceptable descriptions of the electron wave in an atom. Each of these waves is characterized by a set of quantum numbers. The values of the quantum numbers are related to the shape and size of the electron wave and the location of the electron in three-dimensional space.
4. The energy of the electron can be calculated for each possible wave function. The electron energy is quantized and is identical to that predicted by the Bohr model of the hydrogen atom. However, while Bohr arbitrarily imposed this quantization of energy, it is a natural consequence of the solution to the wave equation. Schrödinger's approach to the description of matter as waves is consistent with the observation that energy is quantized and is one of the important tools of *quantum mechanics.*

There is no good analogy for the wave model of the atom as proposed by Schrödinger. Probably the best way to visualize the electrons in atoms is as a cloud of negative charge distributed about the nucleus of the atom, not as a rapidly moving small, hard particle. The electron cloud interpretation of wave functions is discussed in the next section.

6.4 Quantum Numbers in the Hydrogen Atom

Objectives

- To present the names and the allowed values and combinations of the quantum numbers in the hydrogen atom
- To use pictures that represent the electron in the hydrogen atom
- To present the notations used to represent the principal shells, the subshells, and the orbitals

In quantum mechanics matter is described as a wave. No wave can have a precise location, since it is defined by one complete period, which extends over a distance of one wavelength. Instead of a location for the electron, the wave model provides **quantum numbers** that describe the characteristics of the wave that is the electron. These quantum numbers are analogous to the coordinates used to locate the position of a particle. For example, the location of an airplane in flight is given by three numbers, the longitude, latitude, and altitude. Since the wave functions found by solving the Schrödinger equation are the quantum mechanical description of the electron's location in three-dimensional space, three kinds of quantum numbers must be specified. These three types of quantum numbers are represented by the symbols n, ℓ, and m_ℓ. The values of these quantum numbers give as much information about the location and the energy of the electron as is possible. In the solution of the Schrödinger equation the quantum numbers can have only certain allowed whole-number values. In addition, we find that the value of n restricts the values of ℓ, which in turn places restrictions on the values that m_ℓ may have. Each of these quantum numbers and its meaning is described in the following paragraphs.

The **principal quantum number** is called n. The allowed values for n are all positive whole numbers: $n = 1, 2, 3, \ldots$. The principal quantum number gives information about the *distance of the electron from the nucleus.* The larger the value of the principal quantum number, the greater is the average distance of the electron from the nucleus. Remember that the wave model does not provide a precise distance, and there is a small probability that any electron is very close to the nucleus as well as a small probability that it is very far from the nucleus, regardless of the value of n. As we shall see, several different wave functions can have the same value of n (except for $n = 1$). The term **principal shell** refers to all wave functions that have the same value of n, since they all have approximately the same average distance from the nucleus. The n quantum number is very important in determining the energy of the atom, since the distance of the electron from the nucleus is related to the energy of the atom. The further the electron is from the nucleus, the higher the energy.

The principal quantum number is designated by n and provides information about the distance of the electron from the nucleus.

The **angular momentum quantum number** is called ℓ. The allowed values of ℓ are 0, and all positive integers up to $(n - 1)$; $\ell = 0, 1, 2 \ldots (n - 1)$. Thus the ℓ quantum number must equal 0 for wave functions in the $n = 1$ shell. When $n = 4$, ℓ can have the values 0, 1, 2, or 3. The angular momentum quantum number, ℓ, can be associated with the *shape* that the electron probability function may have. Each value of the ℓ quantum number corresponds to a particular shape for the electron probability function. A **subshell** is the set of all the possible wave functions of the electron that have the same values of both the n and ℓ quantum numbers. Just as in the case of a principal shell, each subshell may consist of more than one wave function.

The angular momentum quantum number is ℓ. It describes the shape of the electron distribution.

To identify a subshell, values for both the principal and the angular momentum quantum numbers must be used. Lowercase letters are used to stand for different values of the ℓ quantum number, as follows:

angular momentum quantum number, ℓ	0	1	2	3	4
letter used	*s*	*p*	*d*	*f*	*g*

The first four letters, *s*, *p*, *d*, and *f*, were taken from the words formerly used to describe lines in atomic spectra: *sharp, principal, diffuse,* and *fundamental.* With the advent of quantum theory, this same terminology was applied. Using this notation, the subshell with $n = 3$ and $\ell = 1$ is called the $3p$ subshell.

Example 6.6 Allowed Combinations of Quantum Numbers

Give the notation used for each of the following subshells that is an allowed combination. If it is not an allowed combination, explain why.
(a) $n = 2, \ell = 0$ (b) $n = 1, \ell = 1$ (c) $n = 4, \ell = 2$

Solution

(a) $n = 2, \ell = 0$ is an allowed subshell. We use the letter *s* to express the value of $\ell = 0$, so the correct notation is 2*s*.
(b) Since ℓ must be at least one less than *n*, a value of $\ell = 1$ is not possible when $n = 1$.
(c) The letter *d* means that $\ell = 2$, so this subshell is referred to as 4*d*.

Understanding What is the notation for the subshell with $n = 3$ and $\ell = 1$?

Answer: 3*p*

The **magnetic quantum number** is called m_ℓ. Allowed values for m_ℓ are all whole numbers from $-\ell$ to $+\ell$. For example, if the ℓ quantum number is 2 (a *d* subshell), then m_ℓ may have the values −2, −1, 0, +1, and +2. The m_ℓ quantum number provides information about the *orientation in space* of the wave function. The term **atomic orbital** refers to a wave function that has specified values of *n*, ℓ, and m_ℓ. Each subshell consists of one or more orbitals. The number of orbitals in any given subshell is $(2\ell + 1)$, corresponding to the $(2\ell + 1)$ allowed values of the m_ℓ quantum number. An *s* subshell has only one orbital [2(0) + 1], a *p* subshell has three orbitals [2(1) + 1], a *d* subshell consists of 5 orbitals [2(2) + 1], and so on.

The magnetic quantum number is m_ℓ, and tells about the orientation of the electron distribution.

Once values are specified for these three quantum numbers, we have all the information that can be known about the location of the electron in three-dimensional space. We have described the size, the shape, and the orientation of the electron cloud. Table 6.1 shows the allowed combinations of these three quantum numbers, through the third principal shell.

Table 6.1 Allowed Combinations of the *n*, ℓ, and m_ℓ Quantum Numbers

Shell *n*	Subshell ℓ	Orbital m_ℓ
1	0 (1*s*)	0
2	0 (2*s*)	0
	1 (2*p*)	−1
		0
		1
3	0 (3*s*)	0
	1 (3*p*)	−1
		0
		1
	2 (3*d*)	−2
		−1
		0
		1
		2

Representations of Orbitals

The *square of the wave function* gives the *probability* that the electron will be found at any specific location in space. There are several common ways to draw pictures of the electron in an atom that portray the uncertainty of its location.

One method is to represent the probability of the electron with different densities of dots. At places where the probability is high, the concentration of dots is large. At locations where the probability is low, very few dots are present. Drawings like those in Figure 6.9a are often referred to as electron cloud or electron density representations, since the shading shows the electron as spread out over a region of space. The picture shows that the probability of finding the 1*s* electron near the nucleus is very high, decreasing at larger distances. The 1*s* orbital (in fact any *s* orbital) must be spherical, since the electron probability depends only on the distance of the electron from the nucleus, not on the direction.

All *s* orbitals have a spherical shape.

A second and more common way of representing an electron orbital is to use contour diagrams. In a contour diagram a surface is drawn that encloses

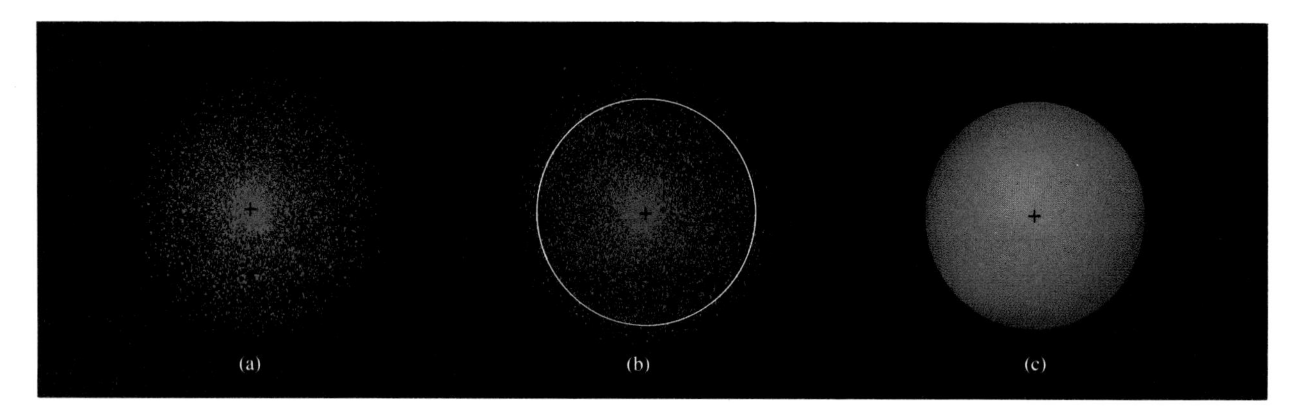

Figure 6.9 The hydrogen 1s orbital The nucleus is located at the origin of the coordinate system. (a) The concentration of the dots is proportional to the probability of the electron being found. (b) The circle encloses 90% of the dots. (c) Ninety percent of the electron's probability of being found lies within the contour surface.

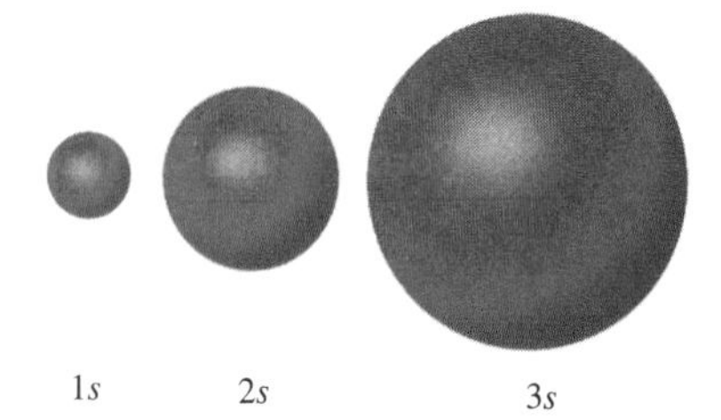

Figure 6.10 The size of *s* orbitals Contours for the *s* orbitals in the first three principal shells. The sizes are to scale.

some fraction of the electron probability, usually 90%. The value of ψ^2 is the same everywhere on the surface. Figure 6.9b shows the line that encloses 90% of the dots in the electron density diagram. The contour surface for the 1*s* orbital is represented in Figure 6.9c.

As the principal quantum number increases in value, the average distance of the electron moves farther from the nucleus, and the size of the contour surface increases. Figure 6.10 shows the contour surfaces for the 1*s*, 2*s*, and 3*s* orbitals. This representation for orbitals shows the shape and size of the electron cloud more clearly.

The *p* orbitals ($\ell = 1$) have a different shape from the *s* orbitals. They have two lobes, one on each side of the nucleus. Graphs of ψ and ψ^2 for a 2*p* orbital are given in Figure 6.11a and b. While the wave function itself has

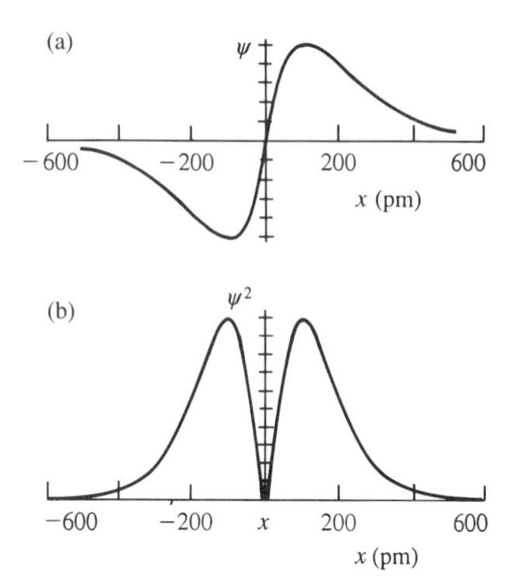

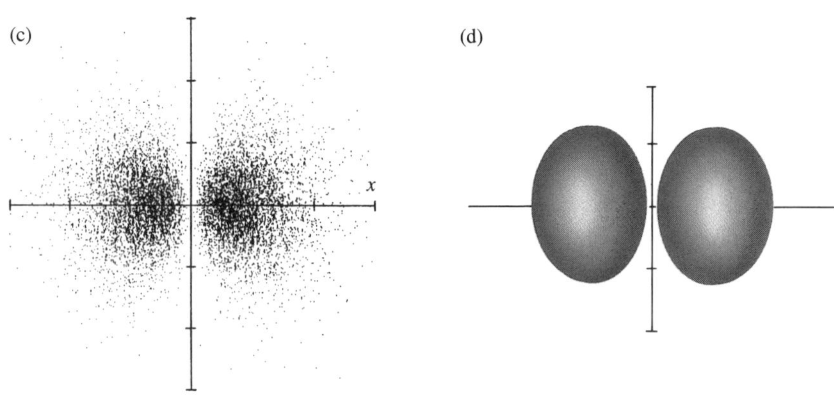

Figure 6.11 Representations of a 2*p* orbital (a) The graph of ψ for a 2*p* orbital directed along the *x* axis. (b) The square of the wave function is proportional to the electron probability. (c) The electron density is also represented by the distribution of dots. (d) The surface encloses 90% of the electron's probability.

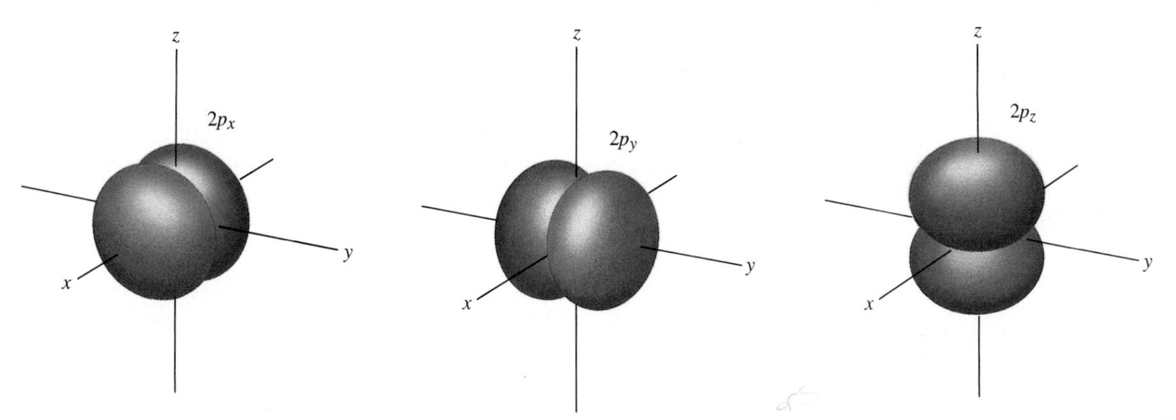

Figure 6.12 The three 2*p* orbitals The contour surfaces for the three 2*p* orbitals are identical in size and shape, but each is directed along a different axis.

different signs on opposite sides of the vertical axis, the square of ψ (the electron density) has the same pattern on both sides of the vertical axis. When we study the wave functions that describe the electrons in molecules (Chapters 8 and 9), the signs of the amplitude of the atomic wave functions become important. The electron density diagram and contour surface for the same 2*p* orbital are also shown in Figure 6.11d. All *p* orbitals, regardless of the value for the principal quantum number, have this same "dumbbell" shape, consisting of two lobes.

When the angular momentum quantum number is 1, there are three allowed values for the magnetic quantum number, so each *p* subshell must consist of three orbitals. In any principal shell the three different *p* orbitals have exactly the same size and shape, but different orientations. There is one *p* orbital directed along each of the three axes; these orbitals are referred to as the p_x, p_y, and p_z. Contours showing the relative orientation of the 2*p* orbitals are given in Figure 6.12. Each principal shell beyond the first has three *p* orbitals. Just as in the case of the *s* orbitals, the size of the contours for *p* orbitals increases as the value of the principal quantum number increases.

There are two lobes in *p* orbitals.

The contours for the five *d* orbitals ($\ell = 2$) are shown in Figure 6.13. Four of these have the same shape, with four identical lobes that point at the

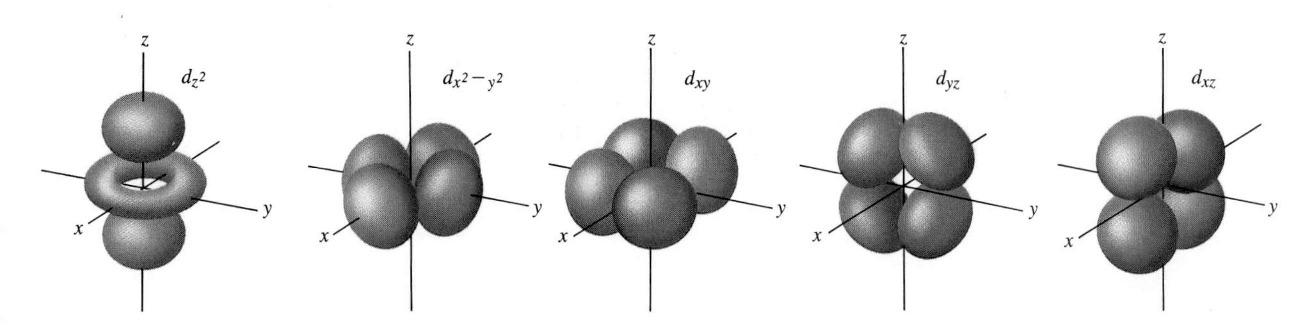

Figure 6.13 The contours for the five 3*d* orbitals Four of the five *d* orbitals have exactly the same shape, but differ in orientation. Although the d_{z^2} has a different appearance than the other four orbitals, it is mathematically equivalent.

corners of a square; these are labeled d_{xy}, d_{xz}, d_{yz}, and $d_{x^2-y^2}$. The remaining d orbital (d_{z^2}) looks different, but is mathematically equivalent to the other four.

The shapes of the seven f orbitals have also been calculated, but as you might suspect, they are more complex than those already shown. Since we won't need these in later chapters, they are not given here.

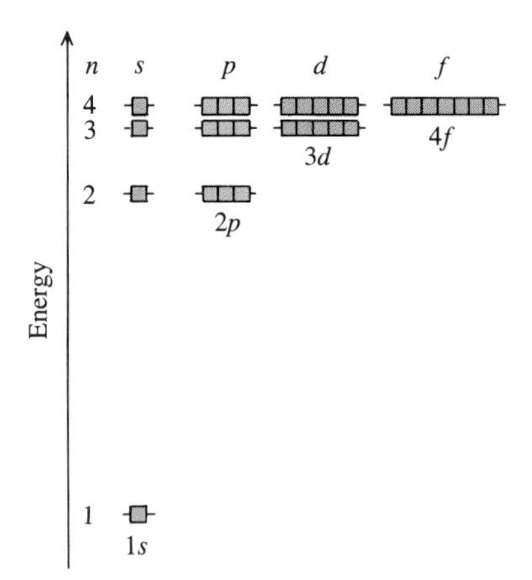

Figure 6.14 The energy level diagram for the hydrogen atom Each box represents one of the orbitals. The energy of each subshell is located by the short horizontal line at the center of the connected boxes. In hydrogen and other one-electron species (for example, He^+ or O^{7+}), all of the orbitals in any principal shell have identical energies.

Energies of the Hydrogen Atom

The energy of the hydrogen atom depends only upon the value of the principal quantum number of the electron's wave function. As the principal quantum number increases, the distance of the electron from the nucleus gets larger, so from Coulomb's law (Section 2.2), the energy of the atom increases with increasing n. The energy-level diagram for the hydrogen atom is given in Figure 6.14. If the principal quantum number is the same, no matter which orbital the electron occupies, the atom has exactly the same energy. Because the energy of the electron is inversely proportional to n^2, the energy separations between the shells decrease as n increases. This energy-level diagram is exactly the same as Bohr's (Figure 6.6), except that in Figure 6.14 the different subshells that make up each principal shell are identified as connected boxes.

If we generalize the quantum model to allow for different nuclear charges, the energy of a one-electron wave function is given by

$$E_n = -\frac{Z^2 R_h hc}{n^2} \qquad [6.9]$$

where Z is the nuclear charge, R_h is the Rydberg constant for hydrogen, and the other symbols have their usual meaning. With this equation we can calculate the spectrum of any ion that contains one electron—*e.g.*, He^+ or Li^{2+}. The one-electron spectrum of the oxygen(VII) ion has been used to identify the presence of oxygen in the atmosphere of the sun. At the very high temperatures on the surface of the sun, the elements have lost most of their electrons. By carefully studying the spectrum of light from the sun, it has been possible to identify the absorption lines for the O^{7+} ion.

The value of the principal quantum number determines the energy of any one electron wave function.

Example 6.7 Calculating Lines in the O^{7+} Spectrum

What is the wavelength of light required to raise an electron in the O^{7+} ion from the $n = 1$ shell to the $n = 2$ shell?

Solution Since the energy levels of all ions that contain only one electron are proportional to the square of the atomic number, we can change Rydberg's law to

$$\frac{1}{\lambda} = Z^2 R_h \left(\frac{1}{n_1^2} - \frac{1}{n_2^2}\right)$$

$R_h = 1.097 \times 10^7\ \text{m}^{-1}$, and for oxygen $Z = 8$. Using $n_1 = 1$ and $n_2 = 2$,

$$\frac{1}{\lambda} = 8^2 \times 1.097 \times 10^7\ \text{m}^{-1} \left(\frac{1}{1^2} - \frac{1}{2^2}\right)$$

$$= 64 \times 1.097 \times 10^7\ \text{m}^{-1} \left(\frac{3}{4}\right) = 5.266 \times 10^8\ \text{m}^{-1}$$

We can rearrange this equation to solve for λ and convert to nanometers:

$$\lambda = \frac{1 \text{ m}}{5.266 \times 10^8}\left(\frac{10^9 \text{ nm}}{1 \text{ m}}\right) = 1.90 \text{ nm}$$

This wavelength is in the x ray region of the electromagnetic spectrum.

Understanding Find the wavelength of the light emitted by an electron during a transition from the $n = 3$ to the $n = 1$ level in the C^{5+} ion.

Answer: 2.85 nm

Electron Spin

There is a fourth quantum number that does not come from the solution of the Schrödinger equation but is necessary to account for an important property of electrons. Scientists have observed that electrons act as small magnets when placed in a magnetic field. When a beam of electrons passes through a magnetic field, half of the electrons are deflected in one direction, and the other half are deflected in the opposite direction. The origin of this magnetic behavior is explained by visualizing the electron as a sphere with its charge on the surface, and postulating that it is allowed to spin only in a clockwise or counterclockwise direction (see Figure 6.15).

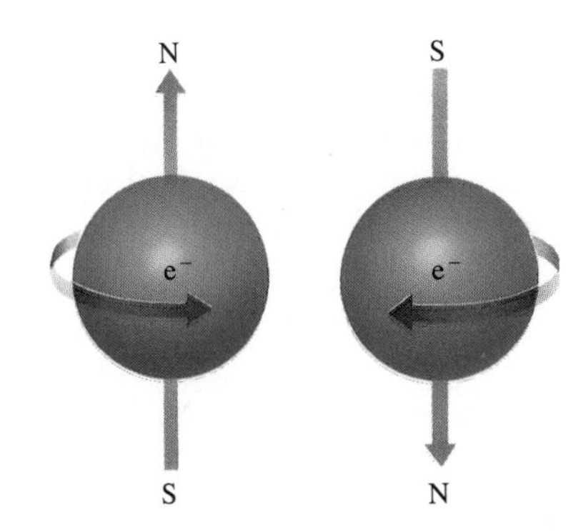

Figure 6.15 Electron spin When the charge of the electron spins in a counterclockwise or a clockwise sense, magnetic fields are generated in opposite directions.

In the quantum mechanical model of the hydrogen atom the magnetic behavior of the electron is described by the **electron spin quantum number,** which is represented by the symbol m_s. The allowed values of m_s are $+\frac{1}{2}$ and $-\frac{1}{2}$, corresponding to the two possible spin states for an electron. The electron spin does not depend on the values of any of the other quantum numbers. Two electrons that have the same spin are said to be *parallel,* while electrons with different spins (one $+\frac{1}{2}$ and the other $-\frac{1}{2}$) are called *opposed* or *paired.*

The electron spin quantum number, m_s, has only two allowed values, $+\frac{1}{2}$ and $-\frac{1}{2}$.

We can now summarize the wave description of the electron in the hydrogen atom. Four quantum numbers (n, ℓ, m_ℓ, and m_s) are needed to describe the electron in any hydrogen atom. Each of these quantum numbers provides some information about the probable location in space or the magnetic behavior of the electron. Remember, we must be satisfied with a probability distribution of the electron, since the exact location of a wave cannot be described.

6.5 Energy Levels for Multi-Electron Atoms

Objectives

- To explain the different effective nuclear charges experienced by electrons in different subshells
- To explain why the energy of an electron in a many-electron atom depends upon the ℓ quantum number as well as the n quantum number

The emission spectra of the other elements are not as easily interpreted as that for hydrogen. The spectra for the other elements are more complex because the energy of an electron depends on the angular momentum quantum number as well as the principal quantum number. In fact, the wavelengths of the emission lines are among the primary tools used to determine

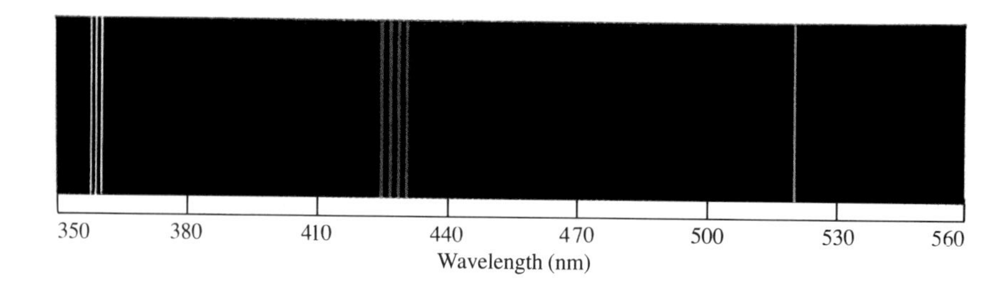

Figure 6.16 The emission spectrum of chromium The presence of these lines is used to identify the presence of chromium in a sample. Each element has a unique spectrum that may contain hundreds of lines.

the energy-level diagrams of atoms. Figure 6.16 shows the emission spectrum of chromium.

The electrons in many-electron atoms are described by the same four quantum numbers (n, ℓ, m_ℓ, m_s) that are used for the hydrogen atom. Unlike the hydrogen atom, the different subshells within the same shell do not have the same energy.

Effective Nuclear Charge

It is important to understand why subshells within the same principal shell differ in energy. From the distance term in Coulomb's law ($E = kQ_1Q_2/r$), we know that the energy of the atom decreases (it is in a more stable state) as the electron gets closer to the nucleus. For this reason, the energy of a 1s electron is lower than that of a 2s electron because the 1s electron is, on the average, closer to the nucleus. Experimental data show that for any atom that contains more than one electron, the energy is influenced by the ℓ quantum number, and the 2s subshell is lower in energy than the 2p subshell. A qualitative understanding of the dependence of the energy on the ℓ quantum number can be obtained by considering the electrostatic forces that act on the electrons in a multi-electron atom.

The single electron in the hydrogen atom, regardless of its location or the orbital it occupies, is attracted by the +1 charge of the nucleus. The situation in the lithium atom is more complex. Each electron is not only attracted by the +3 charge of the nucleus, but it is also repelled by the negative charges of the other two electrons. The electron-electron repulsions, known as *interelectronic repulsion forces,* reduce the effect of the positive charge of the nucleus on each electron, thus influencing their energies.

The net attraction of the nucleus for an electron at any distance r will be reduced, or shielded, by the repulsive forces from the electrons between it and the nucleus. This situation is represented schematically in Figure 6.17 for the lithium atom. The lowest energy state of lithium has two electrons in the 1s subshell, and one electron in the second principal shell. Because the two electrons in the 1s orbital are much closer to the nucleus than an electron in either the 2s or 2p subshell, most of the time the 1s electrons are between the nucleus and the third electron. The **effective nuclear charge** is the weighted average of the nuclear charge that affects any electron in the atom, after correcting for the interelectronic repulsions. The effective nuclear charge for the electron in the second shell will be considerably less than +3, since both 1s electrons are much closer to the nucleus than an electron in the second principal shell. The influence of interelectronic repulsions on the effective nuclear charge is frequently called **electron shielding.**

(a)

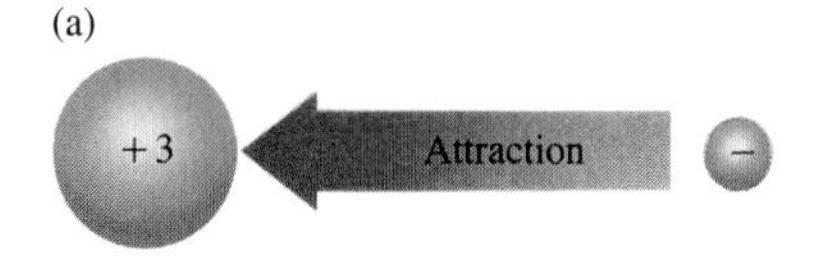

(b)

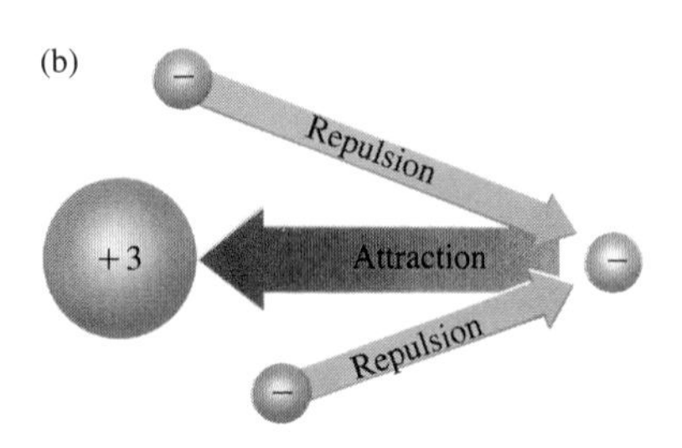

Figure 6.17 Effective nuclear charge (a) A single electron in Li^{2+} is subject to the full +3 charge of the nucleus. (b) The +3 attractive force of the nucleus on the outer electron is reduced, or shielded, by the repulsive forces from the inner electrons.

Detailed calculations of the effective nuclear charge for an electron in any given subshell must take into account the electron density distribution, as well as the average distance of the electron from the nucleus. To determine the effective nuclear charge for each electron, we need to know if the other electrons in the atom are between it and the nucleus. In the lithium atom, the $1s$ electrons are very close to the nucleus and experimental measurements show that the effective nuclear charge for them is close to +3. In the lithium atom's lowest energy state, the third electron occupies the second principal shell, and it has a very small probability of being closer to the nucleus than a $1s$ electron. The extent of shielding of the third electron by the $1s$ electrons depends on its distance from the nucleus. Close to the nucleus, where the electron has a small probability, it "feels" nearly all of the +3 nuclear charge. At large distances, the shielding by the $1s$ electrons is nearly complete and the electron experiences a nuclear charge of essentially +1, the charge of the nucleus minus the charge of the two $1s$ electrons. Since the third electron spends most of its time further from the nucleus than the $1s$ electrons, the effective nuclear charge will be a great deal smaller than +3.

The *effective* nuclear charge is the total nuclear charge corrected for the effect of interelectronic repulsions from other electrons.

Figure 6.18 shows plots of the electron probabilities (ψ^2) for the $2s$ and $2p$ orbitals, as a function of the distance from the nucleus. Although the average distances of the $2s$ and $2p$ electrons are about the same, the probability that the electron is close to the nucleus is larger for the $2s$ electron than for the $2p$ electron. Figure 6.18 shows that the $2s$ electron *penetrates* the electron density of the filled $1s$ shell more than does the $2p$ electron, so it experiences a slightly higher effective nuclear charge, which results in a lower energy state. Within any principal shell the penetration of the s orbital is always greater than that of the p orbitals, which in turn is greater than that of the d orbitals. This means that within any principal shell the subshells increase in energy in the order of increasing value of the ℓ quantum number. In the fourth principal shell, the greater penetration of electrons with lower values of the ℓ quantum number is reflected in the increasing order of energy for the subshells of $4s < 4p < 4d < 4f$.

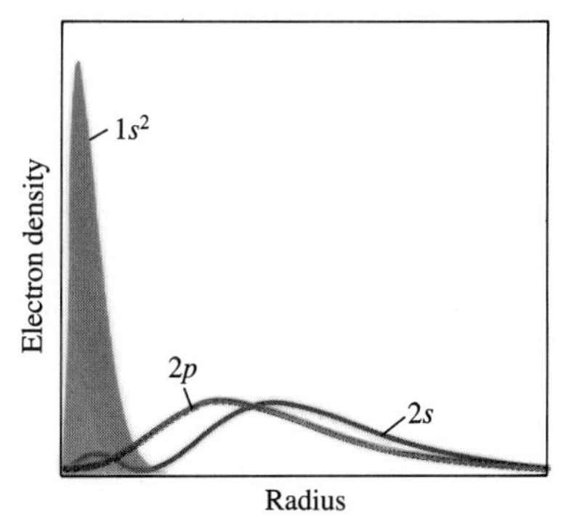

Figure 6.18 Probabilities of $2s$ and $2p$ electrons The electron probability for the $2s$ and $2p$ orbitals as a function of distance from the nucleus. The shaded area is the electron density from the two electrons in the $1s$ orbital. The greater penetration of the $2s$ electron causes it to be 179 kJ/mol more stable than an electron in the $2p$ subshell.

The greater the penetration of the inner shell by an electron, the greater is the effective nuclear charge that it experiences.

Energy Level Diagram of Multi-Electron Atoms

Electrons in different subshells experience different interelectronic repulsive forces, so the energy of an atom depends on which subshells are occupied. The energy level diagram for the magnesium atom, which takes into account interelectronic repulsions, is shown in Figure 6.19. The order in which the subshells are occupied is typical for atoms up to radon. We shall often speak of the "energy of the $3s$ electron" instead of the technically more correct phrase "the potential energy of an *atom* in which the $3s$ orbital is occupied."

In Figure 6.19, you can see that the energy of the $4s$ sublevel is lower than that of the $3d$ orbitals, because of the greater penetration of the inner electrons by the $4s$ orbital. There are several other cases in which electrons occupy one principal level, before all of the sublevels of lower principal quantum numbers have been used. As can be seen in Figure 6.19, the energy separation between sublevels gets quite small in the higher principal levels, so small changes in the shielding effects may cause the energy order to change from one element to the next. These situations will be examined in more

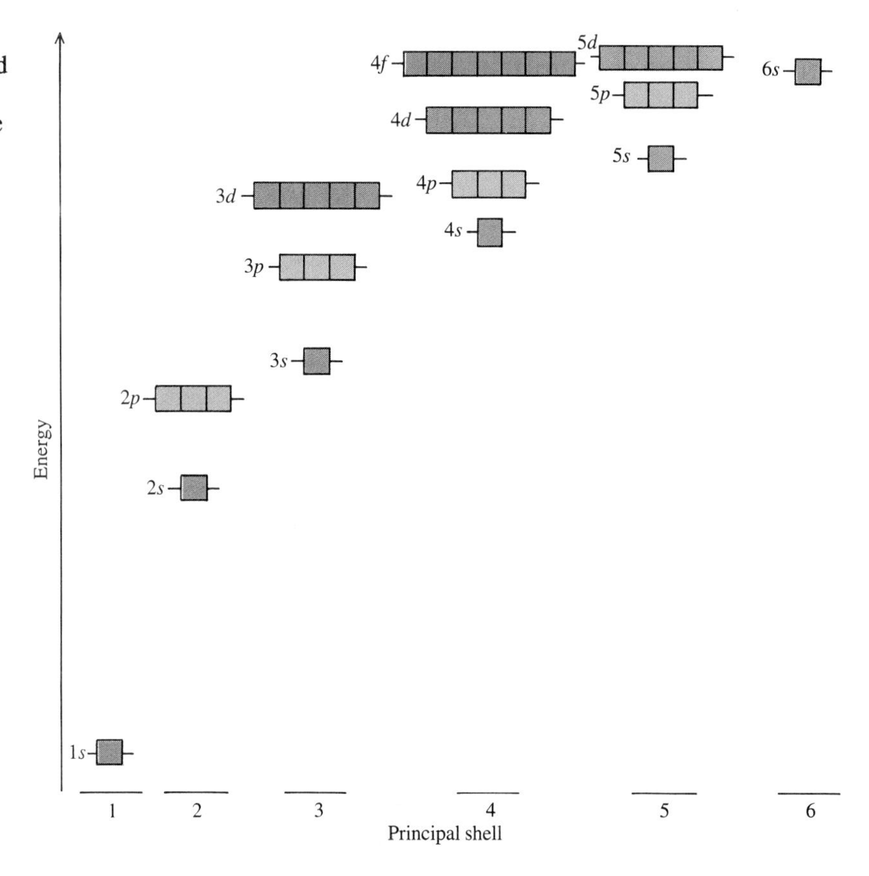

Figure 6.19 Energy level diagram for magnesium The energies of the first and second principal shells are not drawn to scale. The spacing for the energies of the sublevels above 3*s* is to scale.

detail in Chapter 7. Based on experimental observations, the subshells are usually occupied in the order:

$$1s < 2s < 2p < 3s < 3p < 4s < 3d < 4p < 5s < 4d < 5p < 6s < 4f < 5d < 6p < 7s < 5f < 6d$$

Since the energies of different orbitals depend only on the value of the n and ℓ quantum numbers, all of the orbitals in a subshell (designated by different values of the m_ℓ quantum number) have exactly the same energy. All of the orbitals in a sublevel have exactly the same energy, and are referred to as a **degenerate** set.

While the order of filling the sublevels can be memorized, Chapter 7 presents an explanation of how the periodic table is used to deduce the correct order of energies of the subshells.

6.6 Electrons in Multi-Electron Atoms

Objectives

- To use the Pauli exclusion principle to determine the maximum number of electrons in an orbital, subshell, or principal shell
- To represent the electron structure of an atom by writing the electron configuration

INSIGHTS into CHEMISTRY

A Closer View

Simple Methods to Help Remember Relative Energies of Sublevels

The order in which sublevels are occupied by electrons has been determined by experimental observation. A number of devices can be used to remember this order. All of them are simply empirical correlations that work, but they lack a firm theoretical basis. Here are a few:

1. The order of increasing energy for subshells can be predicted by using two simple rules that involve the values of the n and ℓ quantum numbers. These rules are:
   (a) The energy of subshells increases as the sum $(n + \ell)$ increases.
   (b) When two or more subshells have the same sum for $(n + \ell)$, the energy increases in order of increasing n.

   These rules are illustrated in the following table.

$(n+\ell)$	n	ℓ	Subshell
1	1	0	1*s*
2	2	0	2*s*
3	2	1	2*p*
	3	0	3*s*
4	3	1	3*p*
	4	0	4*s*
5	3	2	3*d*
	4	1	4*p*
	5	0	5*s*

   As the table shows, the application of these rules results in the experimentally observed order.
2. The second device involves writing all sublevels in a principal level on successive lines. On each horizontal line the order is that of increasing ℓ. A series of diagonal arrows, starting at the top and pointing down and to the left, results in the correct order for increasing energy.

```
1s ↙
2s ↙ 2p ↙
3s ↙ 3p ↙ 3d ↙
4s ↙ 4p ↙ 4d ↙ 4f ↙
5s ↙ 5p ↙ 5d ↙ 5f
6s ↙ 6p ↙ 6d
7s
```

   In fact, this diagram is a graphical representation of the first method presented.
3. The third method is to remember the following order:

$$ns \quad (n-2)f \quad (n-1)d \quad np$$

   where n is any whole positive number. Start with $n = 1$ and continue to increase it by one. The *f*, *d*, and *p* sublevels are included only if they are allowed. For example, when $n = 1$, there is no $-1f$, $0d$, or $1p$ subshell, so they are not included. For $n = 2$, only the *s* and *p* subshells are possible (there is no $0f$ or $1d$ sublevel). Proceeding in this manner, we obtain the normal order of increasing energy. When using this method, be sure to check for allowed combinations of n and ℓ, so that impossible sublevels will not be included.

While one or more of these empirical rules may be useful to you, we shall see in Chapter 7 that the structure of the periodic table is the most useful guide for obtaining the correct order of increasing energies of the subshells.

- To construct orbital diagrams and energy level diagrams for a given atom by using Hund's rule
- To predict the number of unpaired electrons in an atom

In a multi-electron atom the electrons are described by hydrogen-like wave functions. Each of these wave functions is described by the values assigned to the four quantum numbers (n, ℓ, m_ℓ, and m_s). These wave functions differ from those in the hydrogen atom because of the effect of interelectronic repulsions. Electrons in each subshell experience a different effective nuclear charge, so the energies of those electrons depend on both the n and the ℓ quantum numbers. The use of the quantum numbers to predict the shape and orientation of the electron clouds is the same as in the hydrogen atom.

The Pauli Exclusion Principle

By using the restrictions on the quantum numbers and the Pauli exclusion principle, you can easily determine the capacities of orbitals, subshells, and principal shells.

In 1925, Wolfgang Pauli (1900–1958) summarized the results of many experimental observations, with what is now known as the **Pauli exclusion principle:** *no two electrons in the same atom can have the same set of all four quantum numbers.* Only one of the four quantum numbers must have different values for two electrons to have a different set of quantum numbers. The Pauli exclusion principle is the quantum mechanical equivalent of saying that two objects cannot occupy the same space at the same time. (Recall the analogy between quantum numbers and coordinates that was made in Section 6.4.) Using the Pauli exclusion principle, we find that any orbital (described by the three quantum numbers n, ℓ, and m_ℓ) can have a maximum of two electrons in it, one with a spin of $+\frac{1}{2}$ and the other with a spin of $-\frac{1}{2}$. *Thus, the maximum number of electrons that can share a single orbital in an atom is two.* The Pauli exclusion principle also sets limits on the number of electrons that can be present in a subshell and a principal shell. These capacities are given in Table 6.2.

The Aufbau Principle

The quantum mechanical description of electrons in atoms can now be presented. We will construct an atom by a procedure called *aufbau,* the German word for "building up." In this procedure, electrons are added to the atom one at a time, until the proper number of electrons are present. As each electron is added, it will be assigned the quantum numbers of the lowest energy wave function that is available. The resulting electron configuration of the atom is its lowest energy state, which is called the ground state (Section 6.2). This is the state that is preferred by nature.

In the hydrogen atom there is only one electron, which occupies the $1s$ orbital in its ground state. The helium atom, with two electrons, has a ground state with both electrons in the $1s$ orbital ($n = 1$, $\ell = 0$, $m_\ell = 0$). These

Table 6.2 The Maximum Number of Electrons in Shells and Subshells

Capacity of Subshells				
subshell	s	p	d	f
number of orbitals $(2\ell + 1)$	1	3	5	7
number of electrons $2(2\ell + 1)$	2	6	10	14
Capacity of Principal Shells				
principal quantum number (n)	1	2	3	4
number of orbitals (n^2)	1	4	9	16
number of electrons ($2n^2$)	2	8	18	32

electrons must have opposite spins according to the Pauli exclusion principle. Electrons shown on an energy diagram are designated by arrows that represent the electron spin quantum number. An arrow points up if it has one spin quantum number, and down if it has the other. This notation has been used in Figure 6.20a to show the ground state of the helium atom. If one or more of the electrons is in any other allowed orbital of the diagram (see Figure 6.20b), the atom is in an **excited state.** The excited state is of higher energy and the atom tends to return to its ground state by losing energy, often by emitting a photon of light.

Two electrons in the same orbital must always have opposing spins, represented by "up" and "down" arrows.

Although the energy-level diagram is the most complete way to show the arrangement of electrons in atoms, chemists have developed a number of shorthand descriptions. **Orbital diagrams** are one way to show how the electrons are present in an atom. Each orbital is shown as a box with orbitals in the same subshell shown as connected boxes. The electrons in each orbital are represented by arrows pointing up or down for the two allowed values of the spin quantum number. Just as in the energy-level diagrams, if an orbital contains two electrons, the directions of the two arrows must be opposite to be consistent with the Pauli exclusion principle. The orbital diagram for the hydrogen atom is

[↑]

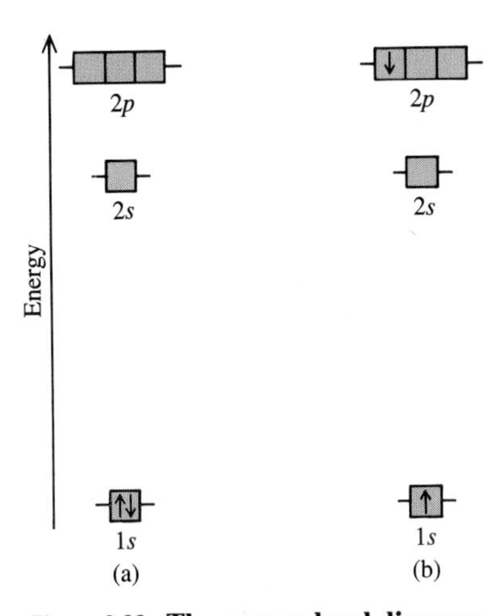

Figure 6.20 The energy-level diagram for the helium atom (a) The ground state electronic configuration. (b) An excited state configuration in which one electron occupies the 2*p* subshell. An excited atom returns to the ground state by losing energy. In any sample of atoms, nearly all of them are in the ground state; very high temperatures are needed to make the excited state populations large.

It would be equally correct to show the single electron as an arrow pointing down, but most chemists follow a convention of representing the first electron in an orbital with an "up" arrow. In orbital diagrams the electrons are represented as they are in energy-level diagrams, except that all of the orbitals are shown on a single line. The orbitals are shown in order of increasing energy, with gaps between boxes to indicate a difference in the energy of the orbitals.

An **electron configuration** lists the occupied subshells, using the usual notations (1*s* or 3*d*, for example), with the number of electrons in the subshell given by a superscript number. In this notation the electron configuration of a ground state hydrogen atom is $1s^1$. This is read as "one ess one."

There are two electrons in each atom of helium. The energy-level diagram of this atom has already been shown in Figure 6.20. The electron configuration and orbital diagram for helium are

		1*s*
He	$1s^2$	[↑↓]

The lithium atom, Li, contains three electrons, and the first two enter the 1*s* subshell with opposite spins. The third electron must go into the subshell with the next higher energy (2*s*), so that the Pauli exclusion principle is not violated. The electron configuration and orbital diagram are

		1*s*	2*s*
Li	$1s^22s^1$	[↑↓]	[↑]

Beryllium, with four electrons, completes the filling of the 2*s* subshell.

		1*s*	2*s*
Be	$1s^22s^2$	[↑↓]	[↑↓]

The orbitals are represented by boxes in both energy level diagrams and orbital diagrams.

The fifth electron in the boron atom must occupy the $2p$ subshell, which consists of three orbitals. The electron configuration and orbital diagram for boron are

		1s	2s	2p		
B	$1s^22s^22p^1$	$\uparrow\downarrow$	$\uparrow\downarrow$	$\uparrow$		

The three p orbitals are shown as connected boxes, to indicate that they form a degenerate set (all have the same energy). Any one of the three boxes could contain the electron, but by convention we usually place the electrons into the boxes proceeding from left to right.

The next element is carbon, which contains six electrons, and must have two electrons in the $2p$ subshell. There are several ways we could assign two electrons in the $2p$ subshell (fifteen ways, to be exact), but not all of these have the same energy. The magnetic properties of the carbon atom (see the following *Insight into Chemistry*) show that it contains two unpaired electrons. The second $2p$ electron must occupy a different orbital and have the same spin as the first electron, to be consistent with the observed magnetic properties of the atom.

We find whenever electrons are added to a subshell that contains more than one orbital, the electrons enter separate orbitals until there is one electron in each. These observations can be explained by the differences in interelectronic repulsions. Two electrons in the same orbital are closer together than if they are in separate orbitals, and therefore repel each other more strongly. Furthermore, experiments show that the spins of all the unpaired electrons are the same. This order is summarized by **Hund's rule:** In filling degenerate orbitals, one electron occupies each orbital and all have identical spins, before any two electrons are placed in the same orbital. Following Hund's rule, the electron configuration and orbital diagram for the carbon atom is

Degenerate orbitals are filled with one electron in each, before any electrons are paired.

		1s	2s	2p		
C	$1s^22s^22p^2$	$\uparrow\downarrow$	$\uparrow\downarrow$	$\uparrow$	$\uparrow$	

By using Hund's rule, we can easily write the electron configurations and orbital diagrams for the elements with atomic numbers 7 through 10. Note that the added electrons must form pairs starting with oxygen, since there are only three degenerate orbitals in the $2p$ subshell.

The electron configuration of an atom is compact, but it does not contain the detailed information about electron spins that is given in an orbital diagram.

		1s	2s	2p		
N	$1s^22s^22p^3$	$\uparrow\downarrow$	$\uparrow\downarrow$	$\uparrow$	$\uparrow$	$\uparrow$
O	$1s^22s^22p^4$	$\uparrow\downarrow$	$\uparrow\downarrow$	$\uparrow\downarrow$	$\uparrow$	$\uparrow$
F	$1s^22s^22p^5$	$\uparrow\downarrow$	$\uparrow\downarrow$	$\uparrow\downarrow$	$\uparrow\downarrow$	$\uparrow$
Ne	$1s^22s^22p^6$	$\uparrow\downarrow$	$\uparrow\downarrow$	$\uparrow\downarrow$	$\uparrow\downarrow$	$\uparrow\downarrow$

All of the important procedures for determining the electronic configurations of atoms have now been presented. The extension of the aufbau method to elements of higher atomic number is straightforward. Since the electronic configurations are related directly to the periodic table, further considerations of them are discussed in the next chapter.

INSIGHTS into CHEMISTRY

A Closer View

Chemists Are Interested in Magnetism to Study Spin Quantum Numbers of Electrons

The magnetic behavior of matter is important to the chemist, since it provides valuable information about the spin quantum numbers of the electrons present in a sample. A magnetic field exerts a force on all matter (see the following figure). **Diamagnetic matter** is repelled by a magnetic field, and **paramagnetic matter** is attracted into a magnetic field.

It is known that only materials containing unpaired electrons are paramagnetic; that is, they must have one or more orbitals that contain single electrons. The number of unpaired electrons in a molecule or ion is determined experimentally by measuring the force per unit mass exerted by its paramagnetism. The salts of many transition metal ions are paramagnetic.

Ferromagnetic materials behave as permanent magnets. All of the atoms that contain unpaired electrons line up so that their magnetic moments point in the same direction. Most ferromagnetic materials are metals. An alloy of aluminum, nickel, and cobalt called alnico is very ferromagnetic and is used in manufacturing permanent magnets that produce very strong magnetic fields.

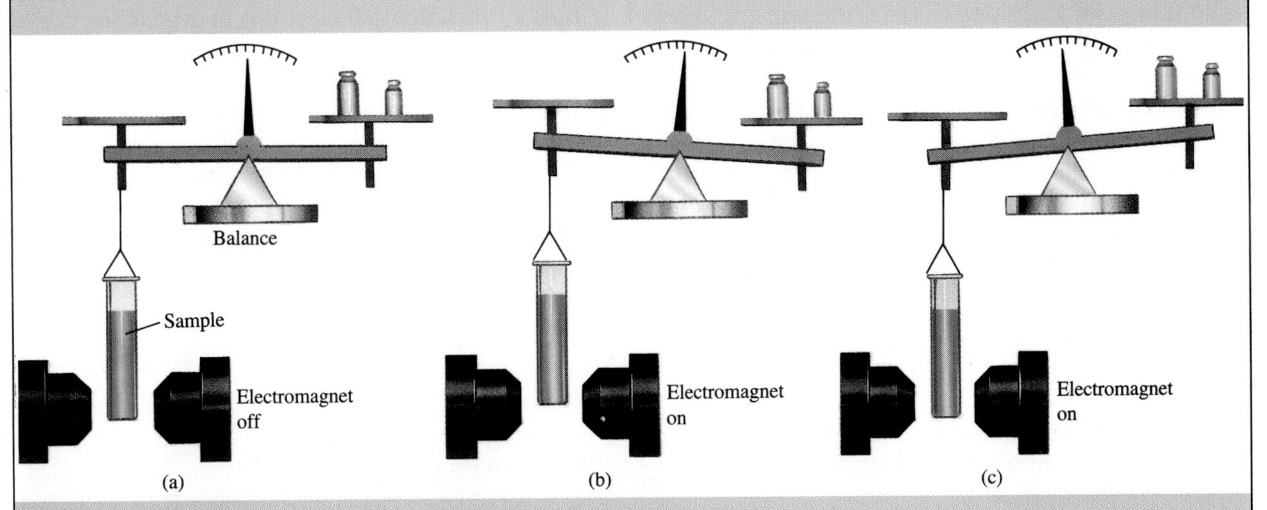

Measuring magnetic properties (a) Determine the weight of the sample in the absence of a magnetic field. Then weigh the sample in the presence of a magnetic field. (b) If the sample is diamagnetic, it is pushed weakly out of the field and has a smaller apparent weight. (c) If the sample is paramagnetic, it is attracted into the magnetic field and has a greater apparent weight.

Summary

Electromagnetic radiation can be described as waves that travel at a constant speed in a vacuum. The product of the *wavelength* and *frequency* of electromagnetic radiation always equals the speed of light, 3.00×10^8 m s^{-1}. Light can also be considered as a stream of *photons,* particles of light, each having an energy of $h\nu$, where h is *Planck's constant* and has the value of 6.63×10^{-34} J s. The particle nature of light is used to explain the *photoelectric effect* and the line spectra of the elements. Bohr was able to explain the lines in the hydrogen atom spectrum by assuming that the angular momentum of the electron was quantized.

de Broglie proposed that matter, normally viewed as particles, could be considered as waves with a wavelength of $\lambda = h/mu$. Schrödinger treated the electron in the hydrogen atom as a wave, and found a *wave function,* ψ, that contains three *quantum numbers,* n, ℓ, and m_ℓ. The square of the wave function, ψ^2, is the probability of finding the electron at any point in space. The value of each of the quantum numbers can be related to a characteristic of the wave function. The distance of the electron from the nucleus and its energy are determined by the value of n. The shape of the electron cloud is related to the ℓ quantum number, and its orientation is determined by the m_ℓ quantum number.

All wave functions with the same value of the n quantum number belong to the same *principal shell;* those in which both n and ℓ are the same are part of a *subshell;* when n, ℓ, and m_ℓ are all specified, the wave function is called an *orbital.* Subshells and orbitals are identified by giving the value of the principal quantum number (1, 2, 3, . .) followed by the letters *s, p, d,* or *f,* representing the values for the ℓ quantum number of 0, 1, 2, or 3, respectively. Thus, 3*p* refers to an orbital or subshell for which $n = 3$ and $\ell = 1$. In addition to the n, ℓ, and m_ℓ quantum numbers, which describe the location of the electron, a fourth quantum number, the *electron spin,* m_s, is needed to account for the magnetic properties of electrons and atoms.

Restrictions on the values of the quantum numbers mean that the following subshells are allowed.

1*s*
2*s*, 2*p*
3*s*, 3*p*, 3*d*
4*s*, 4*p*, 4*d*, 4*f*
.
.
.

The numbers of orbitals in the types of subshells are one for *s*, three for *p*, five for *d*, and seven for *f* subshells.

In an atom that contains two or more electrons, the energy of an electron depends on both the principal and angular momentum quantum numbers, because of interelectronic repulsions. The *Pauli exclusion principle* restricts the number of electrons that can be placed in an orbital to two, since there are only two allowed values for the electron spin quantum number. The Pauli exclusion principle determines the maximum number of electrons in any subshell as two for *s*, six for *p*, ten for *d*, and fourteen for *f* subshells. Furthermore, *Hund's rule* states that electrons in degenerate orbitals do not pair until there is one electron in each orbital of the set. The *ground state* electron configuration of an atom is built up by adding the appropriate number of electrons to the lowest energy subshells available, following the restrictions of the Pauli exclusion principle. The electron configuration ($1s^22s^2$, etc.) and the orbital diagram are two ways of representing electrons in atoms.

Chapter Terms

Wave—a periodic disturbance in a medium or in space which is described by specifying its amplitude, speed, wavelength, and frequency. *(6.1)*

Amplitude—the height of a wave. *(6.1)*

Wavelength—(λ, lambda) the distance from one peak of a wave to the next. In the SI system, wavelength is measured in meters. *(6.1)*

Frequency—(ν, nu) the number of waves that pass a fixed point in one second. The SI unit for frequency is s^{-1}, and has been given the name *hertz. (6.1)*

Hertz—(Hz) the SI unit of frequency. 1 Hz = 1 s^{-1}. *(6.1)*

Electromagnetic radiation—oscillating electric and magnetic fields that are both perpendicular to each other and to the direction of motion. *(6.1)*

Visible light—electromagnetic radiation in the wavelength range from 400 to 700 nm. Light in this wavelength range is seen by the human eye. *(6.1)*

Quantum—the smallest quantity of energy that is absorbed or emitted by matter as electromagnetic radiation. *(6.1)*

Planck's constant (h)—the proportionality constant that relates a quantum of energy to the frequency of the radiation absorbed or emitted. $h = 6.63 \times 10^{-34}$ J s. *(6.1)*

Planck's equation—$\Delta E = h\nu$. *(6.1)*

Photoelectric effect—the ejection of electrons from a solid by the absorption of a photon of light. *(6.1)*

Threshold frequency—(ν_0) the lowest frequency of light that can eject an electron from a solid. *(6.1)*

Work function ($h\nu_0$)—the minimum energy of light needed to eject an electron from a solid. *(6.1)*

Photon—a particle of light that possesses an energy of $h\nu$. *(6.1)*

Dual nature of light—electromagnetic radiation (light) is described both as waves and as particles. *(6.1)*

Spectrum—a graph of the intensity of light as a function of the wavelength or frequency. *(6.2)*

Continuum spectrum—a spectrum in which all wavelengths of light are present. This spectrum is characteristic of the light emitted by a heated solid. *(6.2)*

Line spectrum—a spectrum that contains light at discrete wavelengths separated by regions in which no light is emitted. Line spectra are produced by gaseous atoms of the elements. *(6.2)*

Rydberg equation—the equation that predicts the wavelengths of the lines in the hydrogen atom spectrum.

$$\frac{1}{\lambda} = R_h\left(\frac{1}{n_1^2} - \frac{1}{n_2^2}\right)$$

where n_1 and n_2 are whole positive numbers, with $n_1 < n_2$, and R_h is a constant, called the Rydberg constant, which has a value of 1.097×10^7 m^{-1}. *(6.2)*

Ground state—the state of the atom in which the electron configuration has the lowest possible energy. *(6.2)*

Heisenberg uncertainty principle—it is not possible to know simultaneously and precisely both the position and momentum of a particle. *(6.2)*

Wave function (ψ)—the equation of the wave that describes the location and energy of particles in the wave model of matter. *(6.3)*

Quantum numbers—numbers that describe the characteristics of wave functions, and are analogous to coordinates that describe the location of a particle. In the hydrogen atom, four quantum numbers are needed to describe the wave function of the electron. *(6.4)*

Principal quantum number (n)—the quantum number that contains information about the distance of an electron from the nucleus. It may have any positive integer value, and it affects the energy of the electron. Its value determines the energy of an electron in the hydrogen atom. *(6.4)*

Principal shell—the set of all wave functions in an atom that have the same value of n. *(6.4)*

Angular momentum quantum number (ℓ)—the quantum number that describes the shape of the electron probability wave in an atom. The allowed values of this quantum number are 0 and all positive whole numbers up to $(n - 1)$. *(6.4)*

Subshell—all the possible wave functions of electrons in an atom that have the same values for both the n and ℓ quantum numbers. Each principal shell consists of n subshells. *(6.4)*

Magnetic quantum number (m_ℓ)—the quantum number that describes the orientation of an electron wave function in an atom. The allowed values of this quantum number depend upon the value of the ℓ quantum number, and may have all integer values from $-\ell$ to $+\ell$. *(6.4)*

Atomic orbital—a wave function of an electron in an atom that has assigned values for all three of the quantum numbers n, ℓ, and m_ℓ. *(6.4)*

Electron spin quantum number (m_s)—the quantum number that represents one of the two allowed magnetic states of an electron. The allowed values of m_s are $+\frac{1}{2}$ and $-\frac{1}{2}$ only. *(6.4)*

Effective nuclear charge—the weighted average of the nuclear charge that influences any particular electron in an atom, after correcting for the effect of interelectronic repulsions. *(6.5)*

Electron shielding—the influence of interelectronic repulsions on the effective nuclear charge. *(6.5)*

Degenerate orbitals—all orbitals in an atom that have identically the same energy. In the hydrogen atom and one-electron ions, all wave functions that belong to the same principal shell form a degenerate set. In many-electron atoms, all orbitals in the same subshell form a degenerate set. *(6.5)*

Pauli exclusion principle—no two electrons in the same atom can have the same set of all four quantum numbers. *(6.6)*

Excited state—an atom in which one or more electrons occupy orbitals that leave lower energy orbitals partially or completely vacant. *(6.6)*

Orbital diagram—a drawing that represents the spins of electrons as "up" and "down" arrows placed in boxes that represent the orbitals. *(6.6)*

Electron configuration—a notation that describes the number of electrons in each subshell of an atom or ion; *e.g.*, $1s^22s^22p^2$ is the electron configuration of the carbon atom. *(6.6)*

Diamagnetic—matter that is repelled by a magnetic field. All electrons are paired in diamagnetic materials. *(6.6)*

Paramagnetic—matter that is attracted by a magnetic field. Only substances that contain unpaired electron spins are paramagnetic. *(6.6)*

Ferromagnetic—matter in which the magnetic moments of the atoms line up to produce a permanent magnet. *(6.6)*

Hund's rule—In filling a set of degenerate orbitals, each orbital is occupied by one electron, all with identical spins, before two electrons are placed in the same orbital. *(6.6)*

Exercises

Exercises designated with color have answers in Appendix J.

Electromagnetic Radiation

6.1 Two light sources have exactly the same color, but the second source has twice the brightness of the first. Which of the four quantities used to describe a light wave (amplitude, speed, frequency, wavelength) are the same and which are different for these two waves? Compare the energy of the photons and the total energy of the light from the two sources.

6.2 Two sources of light have wavelengths of 560 nm and 720 nm. Compare the velocity and frequency of light from these two light sources.

6.3 What is the frequency of light that has a wavelength of 400 nm? Name the spectral region for this light (ultraviolet, x ray, etc.).

6.4 Calculate the wavelength of electromagnetic radiation with a frequency of 6.00×10^{13} Hz. What is the name used for the spectral region of this radiation?

6.5 An FM radio station broadcasts at a frequency of 101.3 MHz. What is the wavelength of this radiation?

6.6 Gamma rays are electromagnetic radiation of very short wavelength emitted by the nuclei of radioactive elements. ^{91}Sr emits a gamma ray with a frequency of 2.47×10^{20} Hz. Express the wavelength of this radiation in picometers.

Energy of Light

6.7 The photoelectric effect for cadmium has a threshold frequency of 9.83×10^{14} Hz. For light of this frequency, find
(a) the wavelength,
(b) the energy of one photon (in joules), and
(c) the energy of one mole of photons (in kJ).

6.8 An argon ion laser emits light with a wavelength of 488 nm.
(a) What is the energy of one photon of light at this wavelength?
(b) If a particular argon ion laser produces 1.00 watt of power (1 watt = 1 joule/s), how many photons are produced each second by the laser?

6.9 What is the energy of one mole of photons that have a frequency of 2.50×10^{14} Hz?

6.10 The yellow light emitted by sodium vapor consists of photons with a wavelength of 589 nm. What is the energy change of a sodium atom that emits a photon with this wavelength?

6.11 The red color of neon signs is due to electromagnetic radiation with a wavelength of 640 nm. What is the change in the energy of a neon atom when it emits a photon of this wavelength?

The Photoelectric Effect

6.12 The photoelectric threshold frequency for carbon is $1.16 \times 10^{15}\ s^{-1}$.
(a) What is the longest wavelength of light that will eject electrons from a sample of solid carbon? What is the name of the region of the electromagnetic spectrum for light of this wavelength?
(b) Will electrons be ejected from carbon by any light in the visible region of the spectrum?

6.13 Electrons are ejected from sodium metal by light that has a wavelength less than 544 nm.
(a) What is the photoelectric threshold frequency for sodium metal?
(b) What is the kinetic energy (in joules) of the electrons ejected from sodium metal by light with a wavelength of 450 nm?
(c) How does the intensity of the light (its brightness) affect the kinetic energy of the electrons ejected? What does change when the intensity of the light is increased?

6.14 The charge of an electron is -1.602×10^{-19} coulombs. How many electrons must be ejected from the metal each second to produce an electric current of 1.0 microampere? (1 ampere = 1 coulomb s^{-1}). How many photons must be absorbed to produce this number of photoelectrons?

Line Spectra and the Bohr Atom

6.15 What is the wavelength (in nm) of the line in the spectrum of the hydrogen atom when the electron moves from the Bohr orbit with $n = 5$ to the orbit with $n = 2$? In what region of the electromagnetic spectrum (ultraviolet, visible, etc.) is this radiation observed?

6.16 What is the wavelength (in nm) for the line in the spectrum of Li^{2+} that arises from the electron moving from the Bohr orbit with $n = 6$ to that with $n = 4$? In what region of the electromagnetic spectrum (ultraviolet, visible, etc.) is this radiation observed?

6.17 The absorption spectra of ions have been used to identify the presence of the elements in the atmospheres of the sun and other stars. What is the wavelength of light that is absorbed by He^{+} ions, when they are excited from the Bohr orbit with $n = 3$ to the $n = 4$ state?

6.18 Why are many more lines observed in the emission spectra of hydrogen and other elements than are found in their absorption spectra?

Matter as Waves

6.19 What is the wavelength in nm for an electron that is moving at a velocity of 2.9×10^5 m s^{-1}?

6.20 The velocity of an electron in the first Bohr orbit is 2.19×10^6 m/s. What is the wavelength of this electron? The radius of the first Bohr orbit is 52.9 pm. Compare the wavelength of the electron to the circumference of the first Bohr orbit ($c = 2\pi r$).

6.21 Find the wavelength of an electron in the second Bohr orbit ($n = 2$) if its velocity is 1.094×10^6 m/s. The radius of the second Bohr orbit is 212 pm. Compare the wavelength of the electron to the circumference of the second Bohr orbit ($c = 2\pi r$).

6.22 Find the de Broglie wavelength of each of the following objects:
(a) a ball with a mass of 0.100 kg travelling at 40.0 m/s
(b) a 753-kg car travelling at 24.6 m/s (= 55 mph)
(c) a neutron (mass = 1.67×10^{-27} kg) with a velocity of 2.70×10^3 m/s. This is the rms speed of a neutron at normal room temperature.

6.23 What is the velocity of an electron that has a de Broglie wavelength of 1.00 nm?

6.24 For a diffraction experiment, neutrons (mass = 1.67×10^{-27} kg) with a wavelength of 0.150 nm are needed. What velocity must these neutrons have?

The Quantum Model of the Hydrogen Atom

6.25 How is the amplitude function, ψ, of electron waves related to the location of the electron in space?

6.26 Give the notation (1*s*, 2*s*, 2*p*, etc.) used to describe each of the following subshells. If it is not an allowed combination, state why.
(a) $n = 6, \ell = 1$
(b) $n = 3, \ell = 0$
(c) $n = 5, \ell = 2$
(d) $n = 4, \ell = 0$
(e) $n = 2, \ell = 3$

6.27 Give the notation (1*s*, 2*s*, 2*p*, etc.) used to describe each of the following subshells. If it is not an allowed combination, state why.
(a) $n = 5, \ell = 1$
(b) $n = 1, \ell = 1$
(c) $n = 3, \ell = 2$
(d) $n = 4, \ell = 3$
(e) $n = 7, \ell = 0$

6.28 Give the values of the n and ℓ quantum numbers for the subshells identified by
(a) 3*p*
(b) 5*d*
(c) 7*s*
(d) 4*f*
(e) 2*s*

6.29 Give the values of the n and ℓ quantum numbers for the subshells identified by
(a) 3*d*
(b) 5*p*
(c) 6*s*
(d) 5*f*
(e) 1*s*

6.30 Arrange the following orbitals for the hydrogen atom in order of increasing energy: $3p_x$, $2s$, $4d_{xy}$, $3s$, $4p_z$, $3p_y$, $4s$.

6.31 How does the electron spin quantum number affect the energy of the electron in the hydrogen atom?

6.32 In each part, a set of quantum numbers is given. If the set is an allowed combination of n, ℓ, m_ℓ, and m_s, give the subshell to which this wave function belongs (1*s*, 2*s*, 2*p*, etc.). If the combination of quantum numbers is not allowed, state why.
(a) $n = 2, \ell = 1, m_\ell = 0, m_s = -\frac{1}{2}$
(b) $n = 2, \ell = 2, m_\ell = 2, m_s = +\frac{1}{2}$
(c) $n = 3, \ell = 0, m_\ell = 0, m_s = +\frac{1}{2}$
(d) $n = 1, \ell = 0, m_\ell = 1, m_s = -\frac{1}{2}$
(e) $n = 3, \ell = 2, m_\ell = 2, m_s = +\frac{1}{2}$
(f) $n = 5, \ell = 0, m_\ell = 0, m_s = +\frac{1}{2}$

6.33 In each part, a set of quantum numbers is given. If the set is an allowed combination of n, ℓ, m_ℓ, and m_s, give the subshell to which this wave function belongs (1*s*, 2*s*, 2*p*, etc.). If the combination of quantum numbers is not allowed, state why.
(a) $n = 1, \ell = 1, m_\ell = 0, m_s = -\frac{1}{2}$
(b) $n = 4, \ell = 2, m_\ell = 2, m_s = +\frac{1}{2}$
(c) $n = 2, \ell = 0, m_\ell = 0, m_s = +\frac{1}{2}$
(d) $n = 3, \ell = 0, m_\ell = 1, m_s = -\frac{1}{2}$
(e) $n = 3, \ell = 2, m_\ell = 0, m_s = +\frac{1}{2}$
(f) $n = 6, \ell = 0, m_\ell = 0, m_s = +\frac{1}{2}$

6.34 (a) Which quantum number provides information about the magnetic orientation of the electron?
(b) Which quantum number is related to the average distance of the electron from the nucleus?
(c) What are the name and symbol used for the quantum number that determines the shape of the electron probability distribution?
(d) Which quantum number contains information about the orientation of the electron cloud?

6.35 (a) How many subshells are present in the $n = 3$ principal level?
(b) How many orbitals form the $4p$ sublevel?
(c) What is the maximum value of ℓ that is allowed in the shell with $n = 4$?
(d) What are the values of n and ℓ for a $3d$ subshell? Give all allowed values of the m_ℓ quantum number for this subshell.

6.36 In each part, sketch the contour surface for the orbital described.
(a) $n = 2, \ell = 0$
(b) $3p_x$
(c) $4d_{xy}$
(d) $n = 2, \ell = 1$
(e) $n = 1, \ell = 0$

6.37 Where is an electron in the p_z orbital most likely to be found? Where is the probability of finding this electron zero?

6.38 What is the designation for an orbital that has a spherical contour about the nucleus?

6.39 Sketch the shape of the contour surface for an electron with the quantum numbers $n = 3$, $\ell = 0$, $m_\ell = 0$, $m_s = -\frac{1}{2}$.

6.40 Sketch the orbital contour expected for an electron that has $n = 3$ and $\ell = 2$.

6.41 Show the shape of a contour for a p orbital.

Effective Nuclear Charge

6.42 In a carbon atom, does the $2s$ or $2p$ electron experience a higher effective nuclear charge? Explain.

6.43 Explain what is meant by "penetration" in the explanation of the dependence of electron energies on the ℓ quantum number.

Electron Configurations of Atoms

6.44 In each part, arrange the subshells in order of increasing energy in a many-electron atom.
(a) $5p$, $2p$, $3d$, $2s$, $3p$
(b) $1s$, $2p$, $3d$, $2s$, $4d$, $3s$
(c) $1s$, $2s$, $3s$, $2p$, $3p$, $4p$, $3d$

6.45 Write the electron configuration for
(a) all elements from lithium through neon that are not paramagnetic.
(b) the element with the largest paramagnetism among the first 10 elements in the periodic table.
(c) all elements with an atomic number less than 10 that have one unpaired electron.

6.46 Show the orbital diagram for each of the answers in Exercise 6.45.

6.47 Give the number of unpaired electrons present in the ground state electronic configuration of
(a) Li (b) He (c) F (d) B

6.48 Which of the following atoms are diamagnetic in the ground state? B, C, Ne, Be

6.49 What is the highest occupied subshell in each of these elements?
(a) hydrogen
(b) lithium
(c) nitrogen
(d) neon

6.50 What is the highest occupied subshell in each of these elements?
(a) helium
(b) beryllium
(c) carbon
(d) fluorine

6.51 Sketch the complete energy-level diagram for the nitrogen atom. Write the four quantum numbers assigned to each electron in the atom. (Some arbitrary choices will have to be made.)

6.52 Sketch the complete energy-level diagram for the carbon atom. Write the four quantum numbers assigned to each electron in the atom. (Some arbitrary choices will have to be made.)

6.53 Assign the four quantum numbers (n, ℓ, m_ℓ, m_s) to the highest energy electron in the boron atom.

6.54 Assign the four quantum numbers (n, ℓ, m_ℓ, m_s) to the highest energy electron in the lithium atom.

Additional Problems

6.55 The Lyman series of lines in the hydrogen atom spectrum arise from transition of the electron to the $n = 1$ state. Use the Rydberg equation to calculate the wave length (in nanometers) of the two lowest energy lines in the Lyman series.

6.56 The Paschen series of lines in the hydrogen atom spectrum arise from transitions to the $n = 3$ state. Use the Rydberg equation to calculate the wavelength (in nanometers) of the two lowest energy lines in the Paschen series.

6.57 According to both the Bohr model and the quantum mechanical model of the hydrogen atom, the energy of the atom can be calculated from the quantum number n using the equation

$$E_n = -\frac{hcR_h}{n^2}$$

Express the energy of the three lowest energy states of the hydrogen atom in joules, and in joules/mole.

6.58 Find the uncertainty in the position (in meters) of a 650-kg automobile that is moving at 55 miles per hour. Is this uncertainty in position significant?

6.59 Use the aufbau procedure to obtain the electron configuration and orbital diagram for each of the following:
(a) Be (b) B (c) Ne

6.60 Use the aufbau procedure to obtain the electron configuration and orbital diagram for each of the following:
(a) Li (b) F (c) O

6.61 An experiment uses single-photon counting techniques to measure light levels. If the wavelength of light emitted in an experiment is 589.0 nm, and the detector counts 1004 photons over a ten-second period, what is the power, in watts, striking the detector? (1 watt = 1 J/s.)

6.62 In extremely energetic systems like the sun, hydrogen emission lines can be seen from shells as high as $n = 40$. The spectrum emitted is quite striking, because the energy levels become spaced quite closely.
(a) Calculate the difference in energy between the $n = 2$ and 3 levels and compare to the difference in energy between the $n = 32$ and 33 levels.
(b) Calculate the largest energy difference that can be observed in the hydrogen atom.
(c) Explain why there is a limit to the energy difference.
(d) Is there a physical phenomenon that corresponds to this difference? Explain your answer.

6.63 A baseball weighs 220 g and a professional pitcher throws a fastball at a speed of 100 miles/hour and a curveball at 80 miles/hour. What wavelengths are associated with the motions of the baseball? If the uncertainty in the position of the ball is one half of a wavelength, which ball (fastball or curve) has a more precisely known position? Can the uncertainty in the position of a curveball be used to explain why batters frequently miss it?

6.64 The distance between layers of atoms in a crystal are measured by diffraction of waves with a wavelength comparable to the distance separating the atoms.
(a) What velocity must an electron have if a wavelength of 100 pm is needed for a diffraction experiment?
(b) Calculate the velocity of a neutron that has a wavelength of 100 pm. Compare this to the rms speed of a neutron at 300 K.

6.65 In each part identify the orbital diagram as the ground state, excited state, or an impossible state. If it is an impossible state tell why.

	$1s$	$2s$	$2p$		
(a)	↑	↑↓	↑		
(b)	↑↓	↑↓	↑↓		
(c)	↑↓	↑↓	↑	↑	↑

Chapter Seven

Periodic Trends of the Elements

The periodic table, described briefly in Section 1.3, was originally constructed by Mendeleev and Meyer on the basis of the known chemical and physical properties of the elements and their increasing atomic weights. For example, all the elements in group IA (1) are soft metals that react violently with water. Our knowledge of the basic structure of the atom now tells us that the atoms appear in the periodic table in order of increasing atomic number —each successive atom has one additional proton in its nucleus and one more electron around the nucleus. The known trends in group properties on which the periodic table was based can now be explained by the arrangement of electrons in atoms.

7.1 Electronic Structure and the Periodic Table

Objectives

- To correlate the positions of the elements in the periodic table with the arrangement of electrons in each element
- To distinguish valence electrons from core electrons for each element
- To write the electron configuration of an element based on its location in the periodic table
- To write the electron configuration of ions
- To identify an isoelectronic series

In Chapter 6 we examined the ground state electron configurations for the first ten elements. The electrons occupy the orbitals of lowest energy (aufbau

Potassium metal reacts violently with water.

principle); and when a subshell is being filled, the number of electrons with the same spin should be maximized (Hund's rule). Also, each electron must have a unique set of four quantum numbers (Pauli exclusion principle). Several methods to help remember the order of filling the orbitals were presented, but the information needed to write electron configurations can be found in the periodic table. The known chemical trends from which the periodic table was constructed correlate with the electron configurations of the outermost electrons.

Figure 7.1 shows how the periodic table can be divided into sections in which the same type of orbital is being filled. Also shown is the energy level diagram introduced in Chapter 6, with the subshells marked by the same color-coding scheme.

The arrangement of the elements in the periodic table reflects the order in which orbitals are filled on the energy level diagram. The first period has only two elements, hydrogen ($1s^1$) and helium ($1s^2$). The restrictions on the quantum numbers limit this first level to a $1s$ orbital that can hold only two electrons.

The third element, lithium ($1s^22s^1$), has an electron in the second principal level. The periodic table reflects the placement of an electron in the second shell by starting a second period. The second period (labeled to the left) contains the elements for which the outermost electrons are in the second principal level. It contains eight elements, two in which the $2s$ sublevel is being filled and six in which the $2p$ sublevel is being filled (the three $2p$ orbitals). The second period is six elements longer than the first because the p sublevel ($\ell = 1$) is allowed for the first time in the $n = 2$ level.

After filling the $2p$ sublevel, we fill the $3s$ sublevel—this starts the third period. *A new period starts on the periodic table each time electrons are placed in a new principal shell in the energy level diagram.* Each period always starts with the s subshell because, within any shell, the s subshell is always lowest in energy. Each period (except the first) ends with a filled p subshell.

The third period also contains eight elements, the elements formed by filling the $3s$ and $3p$ subshells. We might have expected more, since for $n = 3$ a d sublevel is also allowed. However, *the 4s sublevel is lower in energy than the 3d sublevel* for potassium and is filled first, as indicated on the energy level diagram in Figure 7.1b. The periodic table indicates that electrons go into the $n = 4$ level by starting the fourth period at the element potassium ($1s^22s^22p^63s^23p^64s^1$). After filling the $4s$ orbital, we next fill the $3d$ sublevel. The ten elements from scandium ($1s^22s^22p^63s^23p^64s^23d^1$) to zinc ($1s^22s^22p^63s^23p^64s^23d^{10}$) are located in the gap between groups IIA (2) and IIIA (13) on the periodic table. The gap exists because the d orbitals are not filled until the fourth period. The fourth period is completed by filling the $4p$ sublevel, and it contains a total of 18 elements.

The filling pattern of the fourth period repeats for the fifth period. In this period, the sublevels are filled in the order $5s$, $4d$, and then the $5p$. The f sublevel, allowed for the first time in the $n = 4$ shell, does not start to fill until after the $6s$ sublevel is filled. The elements that contain f electrons appear in the periodic table beginning at the sixth period, and are generally separated and placed at the bottom of the table to save space. Remember that the principal level for s and p subshells is the same as the period number,

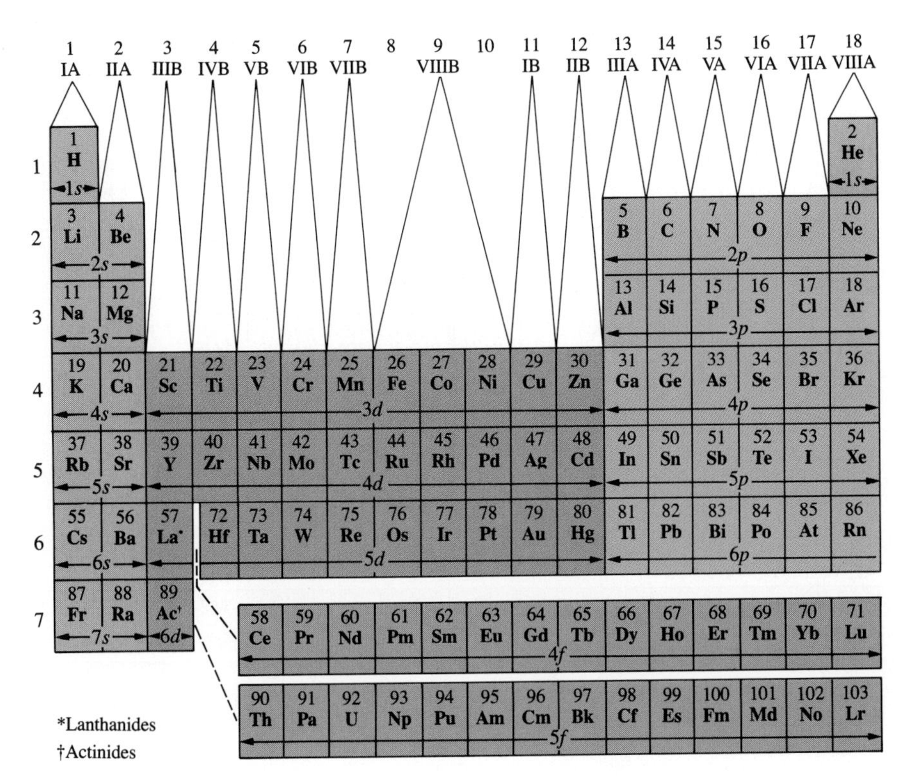

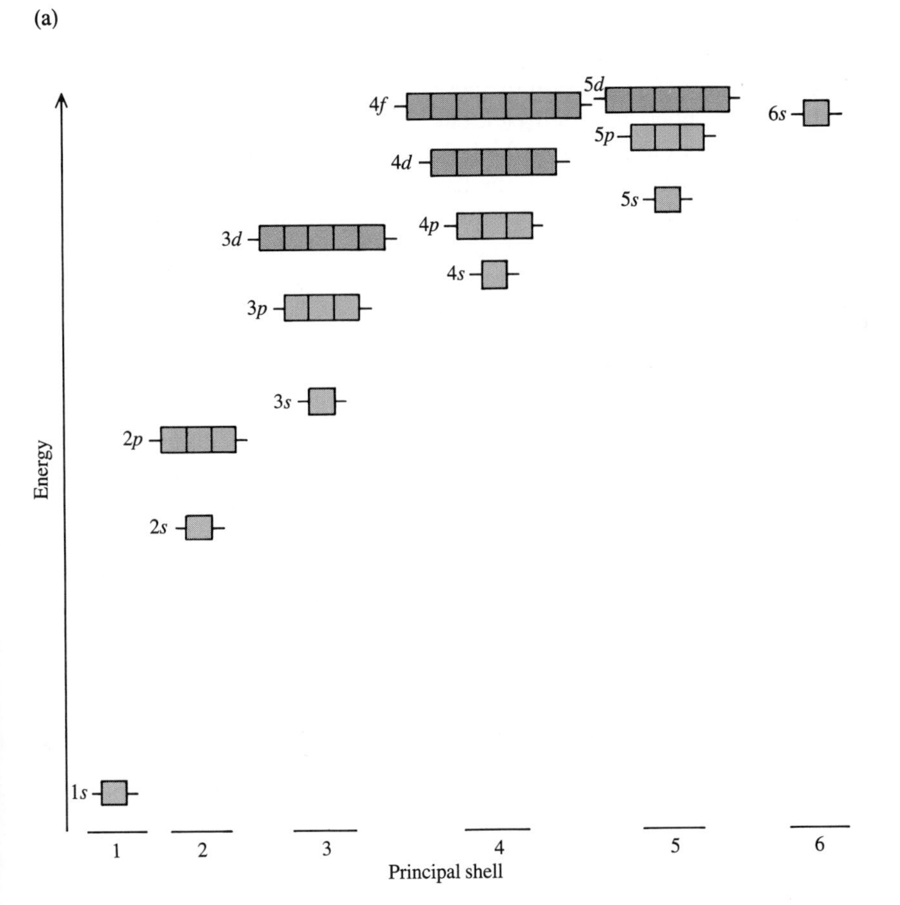

Figure 7.1 **(a) Periodic table indicating blocks of elements. (b) Energy level diagram appropriate for normal filling of sublevels.**

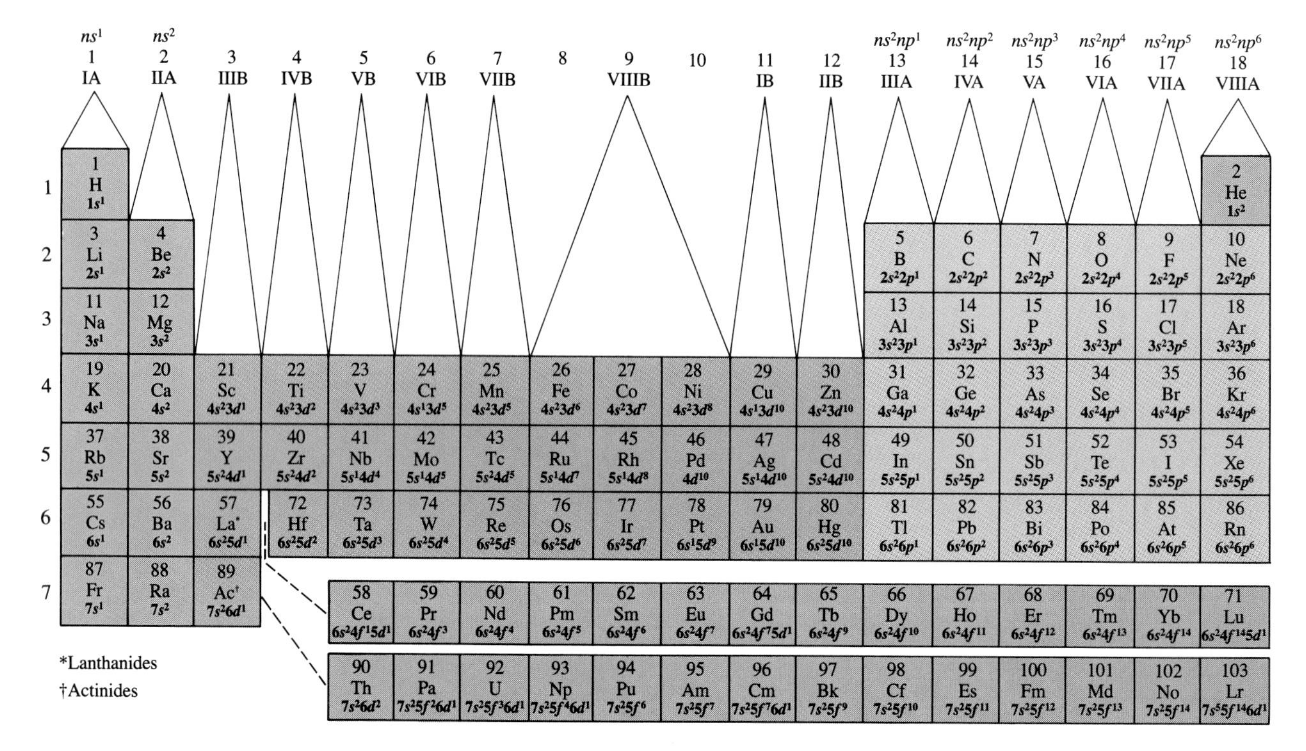

Figure 7.2 Periodic table showing electron configurations of the outermost electrons.

the principal level for a *d* subshell is one less than the period number, and the principal level for an *f* subshell is two less than the period number.

In summary, the colors on the periodic table in Figure 7.1a indicate the subshell being filled for each element. The two columns on the left are *s*-block elements, elements for which the highest energy electron occupies an *s* sublevel. The six columns on the right are *p*-block elements, the ten columns in the middle are *d*-block elements, and the *f*-block elements are at the bottom. The period number gives the value of *n* for the *s* or *p* subshells, the period number minus 1 gives the value of *n* for the *d* subshells, and the period number minus 2 gives the value of *n* for the *f* subshells.

The arrangement of elements in the periodic table can be explained by the order in which electrons occupy sublevels.

Valence Electrons

The elements in each group have similar electron configurations for the outermost electrons, although the values of the principal quantum number are different. The outermost electron configurations of the group IA elements can be described as ns^1, where n has the value 2 for lithium, 3 for sodium, 4 for potassium, etc. For the elements in group IIA, the electron configuration of the outer electrons is ns^2. For group IIIA, ns^2np^1 is the electron configuration, and so on across the periodic table. A periodic table giving these electron configurations is shown in Figure 7.2.

One obvious exception to this general rule (a few others will be discussed later) is found in group VIIIA; helium ($1s^2$) does not have the same outer electron configuration as the other members of the group (ns^2np^6). The chemical and physical properties of helium clearly show that it should be grouped with the other noble gases. Experimental evidence indicates that elements with the $1s^2$ and ns^2np^6 outermost electron configurations are unreactive.

An alternate method of writing electron configurations uses the symbol of a noble gas in brackets for the electrons in the noble gas electron configuration. Sodium can be written as either $1s^2 2s^2 2p^6 3s^1$ or [Ne]$3s^1$; both are equivalent. [Ne] is an abbreviated method for writing the electron configuration of neon, $1s^1 2s^2 2p^6$. Most chemists prefer the noble gas notation method—it is shorter, and it emphasizes the outermost electrons that determine the reactivity of the elements. This bracketed noble gas symbol can also be used in writing orbital diagrams of the elements. Table 7.1 shows a complete list of the electron configurations of the elements.

The similarity of the electron configurations and the chemical properties of elements in the same group correctly suggest that the outermost electrons in an atom determine its chemical behavior. For this reason we define the **valence shell** as those orbitals in the atom of the highest occupied principal level, and the orbitals of *partially filled sublevels* of lower principal quan-

Table 7.1 Ground State Electron Configurations of the Atoms

1	H	$1s^1$	37	Rb	[Kr] $5s^1$	72	Hf	[Xe] $6s^2 4f^{14} 5d^2$
2	He	$1s^2$	38	Sr	[Kr] $5s^2$	73	Ta	[Xe] $6s^2 4f^{14} 5d^3$
3	Li	[He] $2s^1$	39	Y	[Kr] $5s^2 4d^1$	74	W	[Xe] $6s^2 4f^{14} 5d^4$
4	Be	[He] $2s^2$	40	Zr	[Kr] $5s^2 4d^2$	75	Re	[Xe] $6s^2 4f^{14} 5d^5$
5	B	[He] $2s^2 2p^1$	41	Nb	[Kr] $5s^1 4d^4$	76	Os	[Xe] $6s^2 4f^{14} 5d^6$
6	C	[He] $2s^2 2p^2$	42	Mo	[Kr] $5s^1 4d^5$	77	Ir	[Xe] $6s^2 4f^{14} 5d^7$
7	N	[He] $2s^2 2p^3$	43	Tc	[Kr] $5s^2 4d^5$	78	Pt	[Xe] $6s^1 4f^{14} 5d^9$
8	O	[He] $2s^2 2p^4$	44	Ru	[Kr] $5s^1 4d^7$	79	Au	[Xe] $6s^1 4f^{14} 5d^{10}$
9	F	[He] $2s^2 2p^5$	45	Rh	[Kr] $5s^1 4d^8$	80	Hg	[Xe] $6s^2 4f^{14} 5d^{10}$
10	Ne	[He] $2s^2 2p^6$	46	Pd	[Kr] $4d^{10}$	81	Tl	[Xe] $6s^2 4f^{14} 5d^{10} 6p^1$
11	Na	[Ne] $3s^1$	47	Ag	[Kr] $5s^1 4d^{10}$	82	Pb	[Xe] $6s^2 4f^{14} 5d^{10} 6p^2$
12	Mg	[Ne] $3s^2$	48	Cd	[Kr] $5s^2 4d^{10}$	83	Bi	[Xe] $6s^2 4f^{14} 5d^{10} 6p^3$
13	Al	[Ne] $3s^2 3p^1$	49	In	[Kr] $5s^2 4d^{10} 5p^1$	84	Po	[Xe] $6s^2 4f^{14} 5d^{10} 6p^4$
14	Si	[Ne] $3s^2 3p^2$	50	Sn	[Kr] $5s^2 4d^{10} 5p^2$	85	At	[Xe] $6s^2 4f^{14} 5d^{10} 6p^5$
15	P	[Ne] $3s^2 3p^3$	51	Sb	[Kr] $5s^2 4d^{10} 5p^3$	86	Rn	[Xe] $6s^2 4f^{14} 5d^{10} 6p^6$
16	S	[Ne] $3s^2 3p^4$	52	Te	[Kr] $5s^2 4d^{10} 5p^4$	87	Fr	[Rn] $7s^1$
17	Cl	[Ne] $3s^2 3p^5$	53	I	[Kr] $5s^2 4d^{10} 5p^5$	88	Ra	[Rn] $7s^2$
18	Ar	[Ne] $3s^2 3p^6$	54	Xe	[Kr] $5s^2 4d^{10} 5p^6$	89	Ac	[Rn] $7s^2 6d^1$
19	K	[Ar] $4s^1$	55	Cs	[Xe] $6s^1$	90	Th	[Rn] $7s^2 6d^2$
20	Ca	[Ar] $4s^2$	56	Ba	[Xe] $6s^2$	91	Pa	[Rn] $7s^2 5f^2 6d^1$
21	Sc	[Ar] $4s^2 3d^1$	57	La	[Xe] $6s^2 5d^1$	92	U	[Rn] $7s^2 5f^3 6d^1$
22	Ti	[Ar] $4s^2 3d^2$	58	Ce	[Xe] $6s^2 4f^1 5d^1$	93	Np	[Rn] $7s^2 5f^4 6d^1$
23	V	[Ar] $4s^2 3d^3$	59	Pr	[Xe] $6s^2 4f^3$	94	Pu	[Rn] $7s^2 5f^6$
24	Cr	[Ar] $4s^1 3d^5$	60	Nd	[Xe] $6s^2 4f^4$	95	Am	[Rn] $7s^2 5f^7$
25	Mn	[Ar] $4s^2 3d^5$	61	Pm	[Xe] $6s^2 4f^5$	96	Cm	[Rn] $7s^2 5f^7 6d^1$
26	Fe	[Ar] $4s^2 3d^6$	62	Sm	[Xe] $6s^2 4f^6$	97	Bk	[Rn] $7s^2 5f^9$
27	Co	[Ar] $4s^2 3d^7$	63	Eu	[Xe] $6s^2 4f^7$	98	Cf	[Rn] $7s^2 5f^{10}$
28	Ni	[Ar] $4s^2 3d^8$	64	Gd	[Xe] $6s^2 4f^7 5d^1$	99	Es	[Rn] $7s^2 5f^{11}$
29	Cu	[Ar] $4s^1 3d^{10}$	65	Tb	[Xe] $6s^2 4f^9$	100	Fm	[Rn] $7s^2 5f^{12}$
30	Zn	[Ar] $4s^2 3d^{10}$	66	Dy	[Xe] $6s^2 4f^{10}$	101	Md	[Rn] $7s^2 5f^{13}$
31	Ga	[Ar] $4s^2 3d^{10} 4p^1$	67	Ho	[Xe] $6s^2 4f^{11}$	102	No	[Rn] $7s^2 5f^{14}$
32	Ge	[Ar] $4s^2 3d^{10} 4p^2$	68	Er	[Xe] $6s^2 4f^{12}$	103	Lr	[Rn] $7s^2 5f^{14} 6d^1$
33	As	[Ar] $4s^2 3d^{10} 4p^3$	69	Tm	[Xe] $6s^2 4f^{13}$	104	Unq	[Rn] $7s^2 5f^{14} 6d^2$
34	Se	[Ar] $4s^2 3d^{10} 4p^4$	70	Yb	[Xe] $6s^2 4f^{14}$	105	Unp	[Rn] $7s^2 5f^{14} 6d^3$
35	Br	[Ar] $4s^2 3d^{10} 4p^5$	71	Lu	[Xe] $6s^2 4f^{14} 5d^1$	106	Unh	[Rn] $7s^2 5f^{14} 6d^4$
36	Kr	[Ar] $4s^2 3d^{10} 4p^6$						

Valence electrons are the outermost electrons for each of the elements and determine their chemistry.

tum number. The **valence electrons** are the electrons that occupy the valence shell orbitals. All of the other electrons in an atom that are not in the valence shell are usually referred to as **core electrons** and are not directly involved in the chemical reactivities of the elements. The periodic table in Figure 7.2 shows the valence shell electron configurations for each element.

For the elements with partially filled *d* or *f* sublevels, those orbitals are a part of the valence shell. For example, the valence shell electron configuration for vanadium is $4s^23d^3$. However, for the *p*-block elements in the fourth and higher periods, such as gallium and iodine, the electrons in the completely filled *d* and *f* orbitals are not valence electrons. Experience has shown that electrons in completely filled *d* and *f* orbitals do not directly influence chemical changes. Thus, gallium has only three valence electrons, and iodine contains seven valence electrons, consistent with the other elements in the same groups. It is useful to note that with the exception of groups VIIIB and IB, *the number of valence electrons is equal to the group number.*

Writing Electron Configurations

The electron configuration of an element can be determined from its position in the periodic table.

The periodic table can be used as a guide in writing the electron configuration of any element. Just remember the four blocks of elements shown in Figure 7.1 and the number of each period. The period number gives the principal level of the *s*- and *p*-block orbitals; the principal level of the *d*-block elements is one less than the period number.

Example 7.1 Writing Electron Configurations

The elements calcium (Ca), tin (Sn), technetium (Tc), and samarium (Sm) are indicated in the following diagram in the framework of the periodic table. Write the valence electron configuration, the complete electron configuration, and the orbital diagram of each.

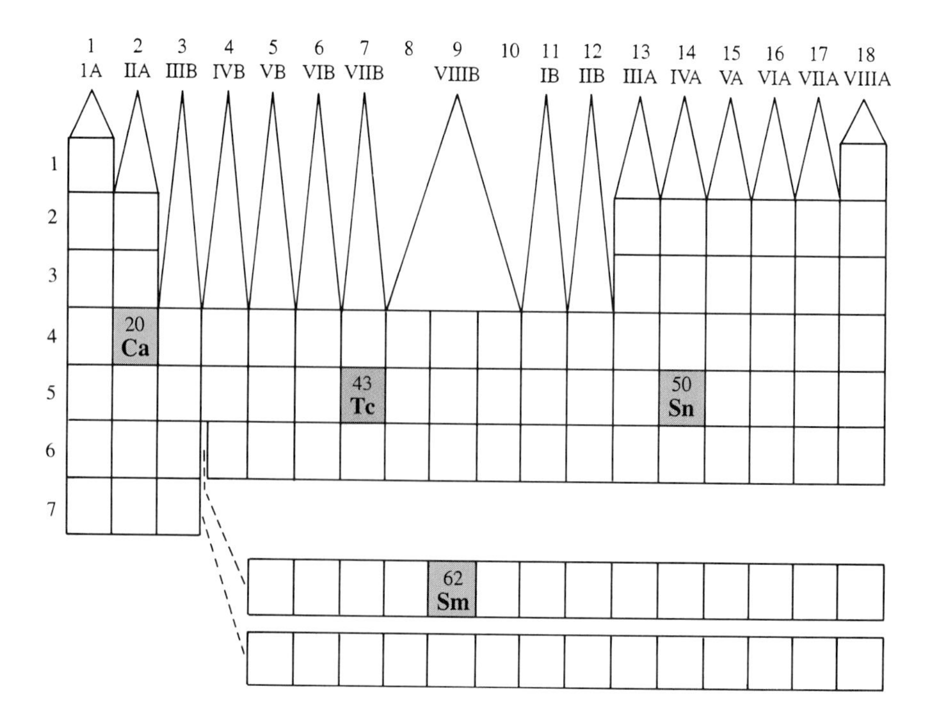

Solution *Calcium:* Calcium is the twentieth element and is in group IIA (2). It is an s-block element in the fourth period with two valence electrons. Its valence electron configuration is therefore $4s^2$. The core electrons that complete the electron configuration fill the sequence shown on the table starting with the first period, $1s^2$, the s-block of the second period, $2s^2$, the p-block of the second period, $2p^6$, the s-block of the third period, $3s^2$, and the p-block of the third period, $3p^6$. The electron configuration is thus $1s^22s^22p^63s^23p^64s^2$ or, using the noble gas notation, [Ar]$4s^2$. We can check the electron configuration by summing the number of electrons indicated (the superscript numbers); their sum is 20, the atomic number of calcium. The orbital diagram, using the noble gas notation, is

[Ar] ↑↓ 4s

Tin: Tin is in group IVA (14) and is a p-block element with four valence electrons. It is in the fifth period with the valence electron configuration $5s^25p^2$. The first 20 electrons fill orbitals in the same way as for calcium. The next electrons enter the d-block, first occupied with element 21, scandium. The principal level of the d-block elements is one less than the period number and there are ten elements in this section of the period, so we add $3d^{10}$. The p sublevel of the fourth period fills next, $4p^6$, the s sublevel of the fifth period, $5s^2$, the d-block of the fifth period, $4d^{10}$ (remember that the d sublevels have a principal quantum number of one less than the period number), and finally the last two electrons enter $5p^2$. The full configuration is $1s^22s^22p^63s^23p^64s^23d^{10}4p^65s^24d^{10}5p^2$ or [Kr]$5s^24d^{10}5p^2$. Note that the $4d^{10}$ electrons are *not* valence electrons, since the d sublevel is completely filled. The orbital diagram is

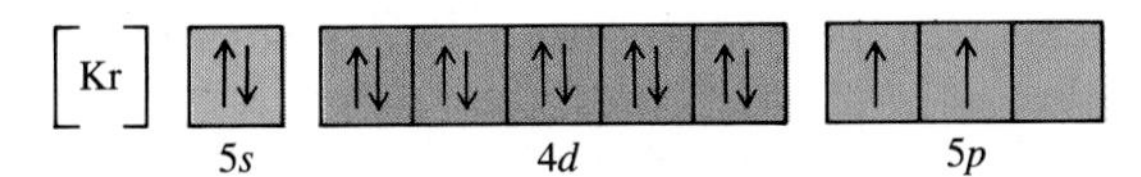

Technetium: Technetium is in group VIIB (7) and is a d-block element with seven valence electrons. Its valence electron configuration is $5s^24d^5$. The inner electrons are as outlined for tin, and the electron configuration is $1s^22s^22p^63s^23p^64s^23d^{10}4p^65s^24d^5$ or [Kr]$5s^24d^5$. The orbital diagram is

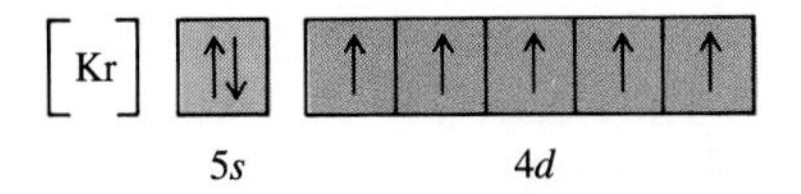

Samarium: Samarium is an f-block element. The first 54 electrons completely fill the subshells through $5p^6$, abbreviated as [Xe]. After xenon, the $6s$ level fills and then the $4f$ sublevel (the f level principal quantum number is two less than the period number). Samarium is the eighth element along the period (counting as if the table was not broken to save space), placing six electrons in the $4f$ sublevel. The valence electron configuration is $6s^24f^6$, and the electron configuration is [Xe]$6s^24f^6$. There are eight valence electrons. The orbital diagram is

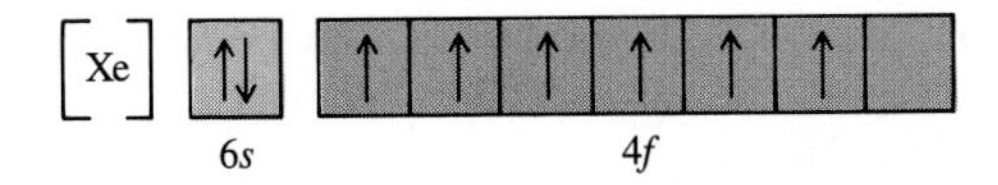

Understanding What are the electron configurations of nickel (Ni) and rubidium (Rb)?

Answer: Ni: [Ar]$4s^23d^8$; Rb: [Kr]$5s^1$

Anomalous Electron Configurations

While the electron configurations of most elements can be written after inspection of the periodic table, careful examination of Table 7.1 shows that there are some exceptions. The first two deviations from the expected filling order occur for chromium and copper, two important transition metals. For chromium, measurements show that the correct electron configuration is $[Ar]4s^13d^5$, not the expected $[Ar]4s^23d^4$. In a similar manner, the electron configuration for copper is $[Ar]4s^13d^{10}$ rather than $[Ar]4s^23d^9$. These exceptions are not surprising because $4s$ and $3d$ sublevels are close in energy. In filling these closely spaced sublevels, factors other than the simple aufbau principle can influence the most stable arrangement. In general, most of the exceptions produce half-filled or completely filled *d* or *f* subshells. The other two elements in group IB (9) are similar exceptions, as is molybdenum in group VIB (6).

Electron Configurations of Ions

Cations are formed by the loss of valence electrons from the orbitals with the highest n values.

Cations are formed from atoms by removing one or more valence electrons. *The electrons of highest n value are removed first; and for cases of the same n level, electrons are first removed from the sublevel of highest ℓ value.* Electrons in the highest principal level are the outermost electrons and are easiest to remove. Cations of elements in the *s* and *p* blocks are formed by removing the electrons that were added last. The electron configuration of beryllium is $1s^22s^2$, so the Be^{2+} ion has the electron configuration $1s^2$. For the *d*-block transition elements, *the ns electrons are lost before the (n − 1)d electrons.* An example is

$$\begin{array}{ccc} Fe & \longrightarrow & Fe^{2+} + 2e^- \\ [Ar]4s^23d^6 & \longrightarrow & [Ar]3d^6 \end{array}$$

Anions are formed by the addition of electrons to valence orbitals.

Anions are formed by adding electrons to atoms. The electron configurations are written using the same energy level diagram as used for atoms. The electron configuration of fluorine is $1s^22s^22p^5$ and that for fluoride anion, F^-, is $1s^22s^22p^6$. The electron configuration of O^{2-} is also $1s^22s^22p^6$. As in these two examples, monatomic anions complete the filling of the valence shell p orbitals.

Example 7.2 Electron Configurations of Ions

Write the electron configuration of:
(a) Na^+ (b) S^{2-} (c) Ni^{3+}.

Solution

(a) The sodium atom has the electron configuration $1s^22s^22p^63s^1$. Removing one electron from the highest n level, the $3s$, yields the electron configuration for Na^+ cation: $1s^22s^22p^6$ or [Ne].
(b) The S^{2-} anion has 18 electrons, two more than the sulfur atom: $[Ne]3s^23p^6$.
(c) The Ni atom has the electron configuration $[Ar]4s^23d^8$. Three electrons are removed to form the Ni^{3+} cation, but remember that the $4s$ electrons are removed before the $3d$: $[Ar]3d^7$.

Understanding Write the electron configuration of Co^{3+}.

Answer: $[Ar]3d^6$

Isoelectronic Series

The ions expected for the *s* and *p* block elements (the A group elements) can be determined from the group number. The elements of group IA (1) form cations with a 1+ charge, group IIA (2) form cations with a 2+ charge, group VIA (16) form anions with a 2− charge, and group VIIA (17) form anions with a 1− charge. The electron configurations of the common ions of oxygen, fluorine, sodium, and magnesium are

$$\underset{1s^2 2s^2 2p^4}{O} \xrightarrow{+2e^-} \underset{1s^2 2s^2 2p^6}{O^{2-}}$$

$$\underset{1s^2 2s^2 2p^5}{F} \xrightarrow{+1e^-} \underset{1s^2 2s^2 2p^6}{F^-}$$

$$\underset{1s^2 2s^2 2p^6 3s^1}{Na} \xrightarrow{-1e^-} \underset{1s^2 2s^2 2p^6}{Na^+}$$

$$\underset{1s^2 2s^2 2p^6 3s^2}{Mg} \xrightarrow{-2e^-} \underset{1s^2 2s^2 2p^6}{Mg^{2+}}$$

All four of these ions have the same electron configuration as neon.

An **isoelectronic series** is a group of atoms and ions that have the same number of electrons. The species O^{2-}, F^-, Ne, Na^+, and Mg^{2+} are isoelectronic—they all have 10 electrons and the electron configuration $1s^2 2s^2 2p^6$. The properties of atoms and ions in an isoelectronic series are often compared.

Species in an isoelectronic series have the same number of electrons.

Example 7.3 Electron Configurations of Isoelectronic Series

(a) Which elements or ions from among Ar, S^{2-}, Si^-, and Cl^{3+} are isoelectronic with P^+?

(b) Which ions from among Fe^{3+}, Ni^{3+}, and Co^{3+} are isoelectronic with Mn^{2+}?

Solution

(a) The P^+ cation has 14 electrons in the configuration $1s^2 2s^2 2p^6 3s^2 3p^2$. Ar and S^{2-} have 18 electrons, Si^- has 15 electrons, and Cl^{3+} has 14 electrons. Only Cl^{3+} is isoelectronic with P^+.

(b) The Mn^{2+} cation has 23 electrons and the electron configuration $[Ar]3d^5$ (remember to remove *first* the valence electrons with the highest *n* value, 4*s* in this case). The Fe^{3+} ion also has 23 electrons and the same electron configuration as Mn^{2+}; they are isoelectronic. The Ni^{3+} ion has 25 electrons ($[Ar]3d^7$) and Co^{3+} has 24 electrons ($[Ar]3d^6$). Neither is isoelectronic with Mn^{2+}.

Understanding Which elements or ions from among K, K^+, Ca^+, and Sc^{3+} are isoelectronic with Ar?

Answer: K^+ and Sc^{3+}

7.2 Sizes of Atoms and Ions

Objective

- To explain the periodic trends in sizes of atoms and ions

The detailed picture of the arrangement of electrons in atoms and ions can lead to an understanding of the chemical and physical properties of the elements. The relative sizes of atoms and ions have an important influence on chemical properties. The sizes of atoms and ions are determined by the electron cloud.

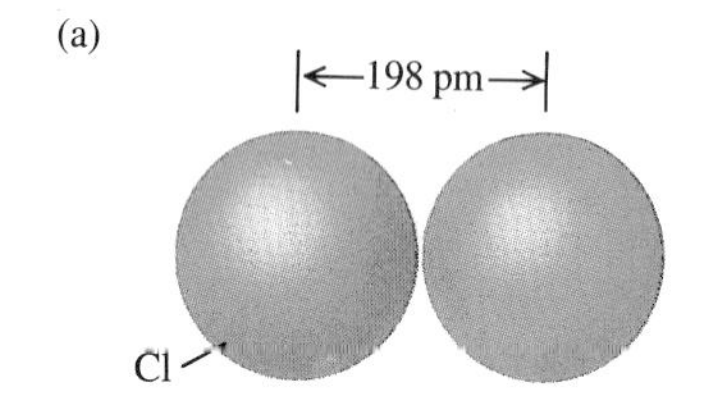

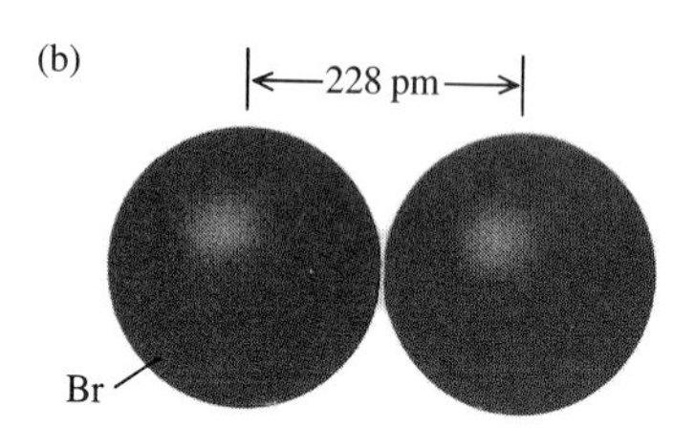

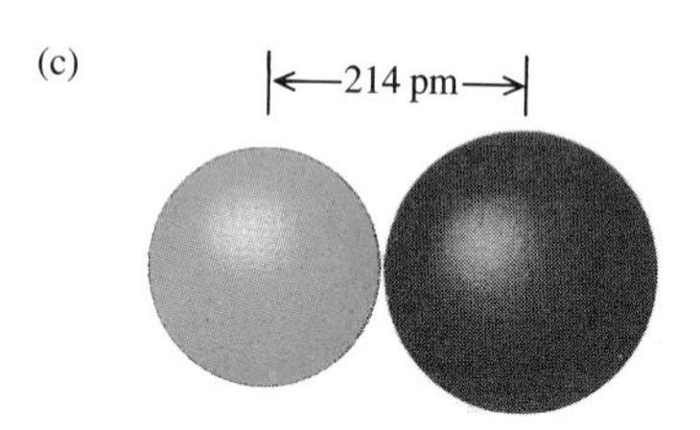

Figure 7.3 **Atomic radii** The atomic radii of chlorine and bromine are half the distance between nuclei in the homonuclear diatomic molecules. These values can be used to predict the distance between the nuclei in BrCl.

Measurement of Sizes of Atoms and Ions

How is the size of an atom or ion measured? We have already seen that the electron cloud does not have a fixed boundary; only a probability distribution can be determined. An **atomic radius** is one half the distance between adjacent atoms of the same element in a molecule. The atomic radius of an element can vary from one type of molecule to another, so a representative molecule must be chosen for the measurement. For example, the atomic radii of chlorine and bromine atoms can be determined by measuring the distance between the nuclei in the diatomic molecules. The distances between the nuclei in Cl_2 and Br_2 are 198 pm and 228 pm, respectively. The atomic radius of each can be defined as simply half of this distance, 99 pm for chlorine and 114 pm for bromine (Figure 7.3). These data allow us to predict that the distance between the chlorine and bromine nuclei in BrCl should be about 213 pm, close to the measured value of 214 pm.

Using somewhat similar methods, we can determine an approximate **ionic radius,** the measure of the size of an ion in an ionic solid. From the distances between the nuclei in ionic crystals, it is possible to assign consistent radii for stable cations and anions.

A number of general trends in the radii of atoms and ions have been identified. These general trends are related to how electron shielding contributes to the effective nuclear charge experienced by the valence electrons. The size of an atom or ion is determined by the electron cloud of the outermost electrons. The larger the effective nuclear charge for these valence electrons, the more strongly they are drawn toward the nucleus, resulting in a smaller size of the electron cloud. Two general features of electron shielding that are important in determining size are:

1. The electrons in any principal level are very effectively shielded by the electrons that occupy orbitals of lower n.
2. Electrons in the same sublevel and, to a lesser extent, in the same principal level do not shield each other very well.

We shall use these features repeatedly as we consider the relative sizes of atoms and ions.

Size Trends in Isoelectronic Series

The sizes of atoms and ions are important because they provide information about the structure and reactivity of molecules, metals, and ionic compounds. Size trends in an isoelectronic series will be considered first. Since

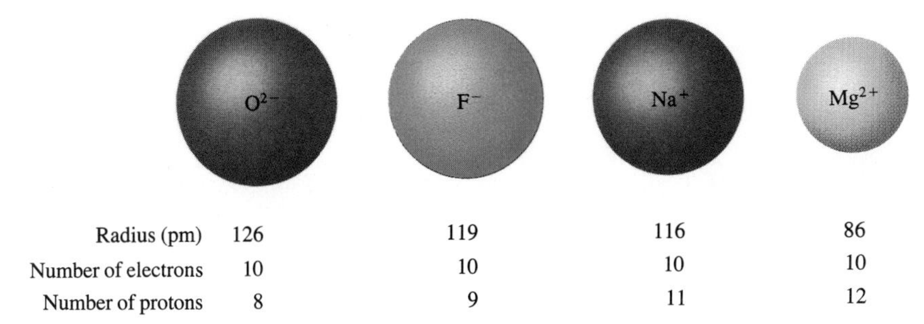

	O^{2-}	F^-	Na^+	Mg^{2+}
Radius (pm)	126	119	116	86
Number of electrons	10	10	10	10
Number of protons	8	9	11	12

Figure 7.4 Size trends for an isoelectronic series.

each member of an isoelectronic series has the same number of electrons, the sizes are determined by the number of protons in the nucleus. Figure 7.4 compares the sizes of four ions with the electron configuration $1s^22s^22p^6$.

The trend toward smaller radii as the atomic numbers increase in an isoelectronic series is easily understood by considering the effective nuclear charge experienced by the valence electrons. Since the number of electrons is constant throughout the series, the effective nuclear charge for the outermost electrons increases by approximately one for each succeeding element, causing the radius of the outermost electron clouds to decrease.

In an isoelectronic series, the species with the largest number of protons in its nucleus has the smallest radius.

Comparative Sizes of Atoms and Their Ions

The size of an atom can easily be compared to that of its own ions. Consider the sizes of the lithium atom and the lithium cation. The electron configuration of Li is $1s^22s^1$; for Li^+, it is $1s^2$. In this case, the number of protons in the nucleus is constant, but the lithium atom has an additional electron which is located in the $n = 2$ level. The cation has no electrons in the $n = 2$ level. In Chapter 6 we saw that the $1s$ electrons are very effective in shielding the $2s$ electron from the 3+ charge of the nucleus. Thus, the $2s$ electron is not tightly held to the nucleus, making the lithium atom much larger than its cation. This trend is general; *an atom is always larger than any of its cations* (Figure 7.5).

Figure 7.5 also shows the sizes of the iron atom and its 2+ and 3+ ions. The iron atom ($[Ar]4s^23d^6$) loses the two $4s$ electrons to form Fe^{2+} ($[Ar]3d^6$). This loss causes a large decrease in size, because the electrons in the $n = 4$ level have been removed from atomic iron, electrons that are effectively shielded from the nuclear charge by the $3d$ electrons. Loss of an additional electron to form Fe^{3+} ($[Ar]3d^5$) produces only a small decrease in radius, since electrons in the same subshell do not shield each other effectively from the nuclear charge.

Cations are smaller than the atoms from which they form.

By similar reasoning we explain the larger size of anions compared to their parent atoms. The addition of an electron to a fluorine atom ($1s^22s^22p^5$), forming a fluoride ion ($1s^22s^22p^6$), reduces the effective nuclear charge experienced by the outer electrons, producing a larger electron cloud. Again, this is general: *anions are larger than the atoms from which they are derived* (Figure 7.6).

Anions are larger than the atoms from which they form.

Figure 7.5 Sizes of atoms and cations (pm).

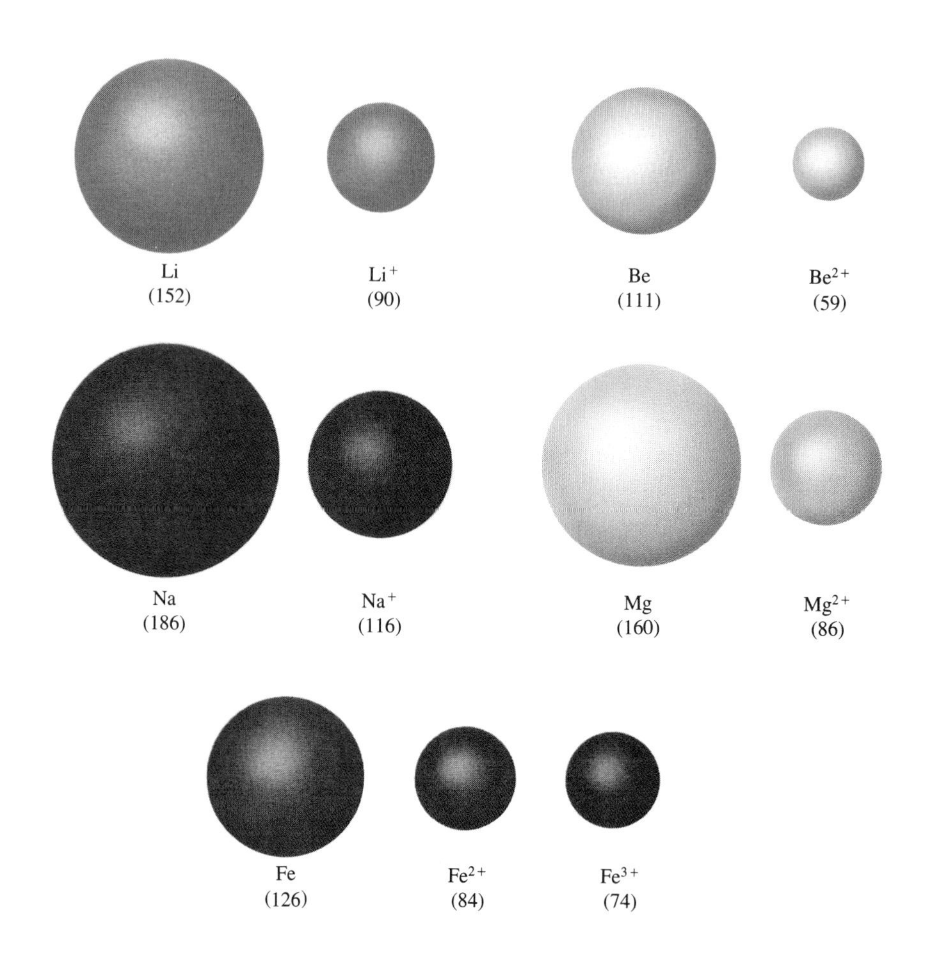

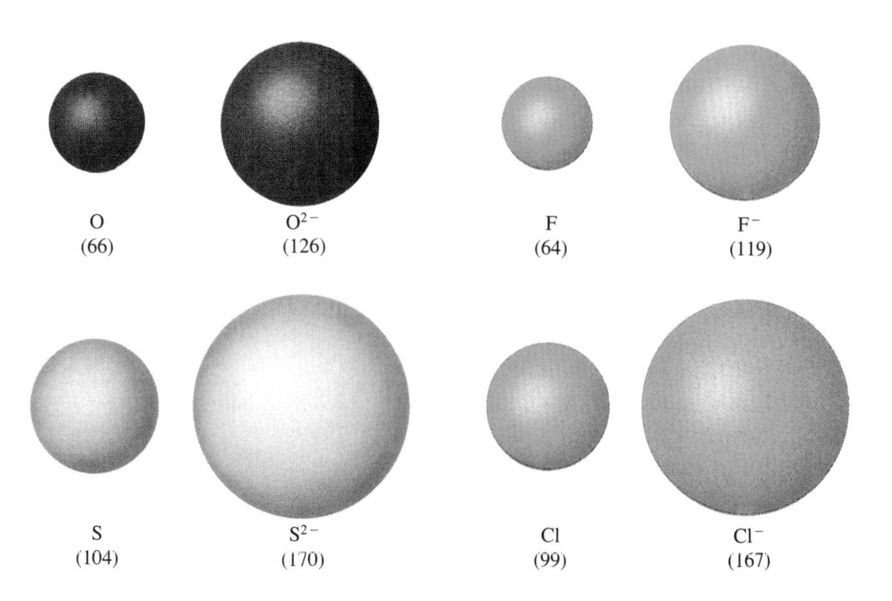

Figure 7.6 Sizes of atoms and anions (pm).

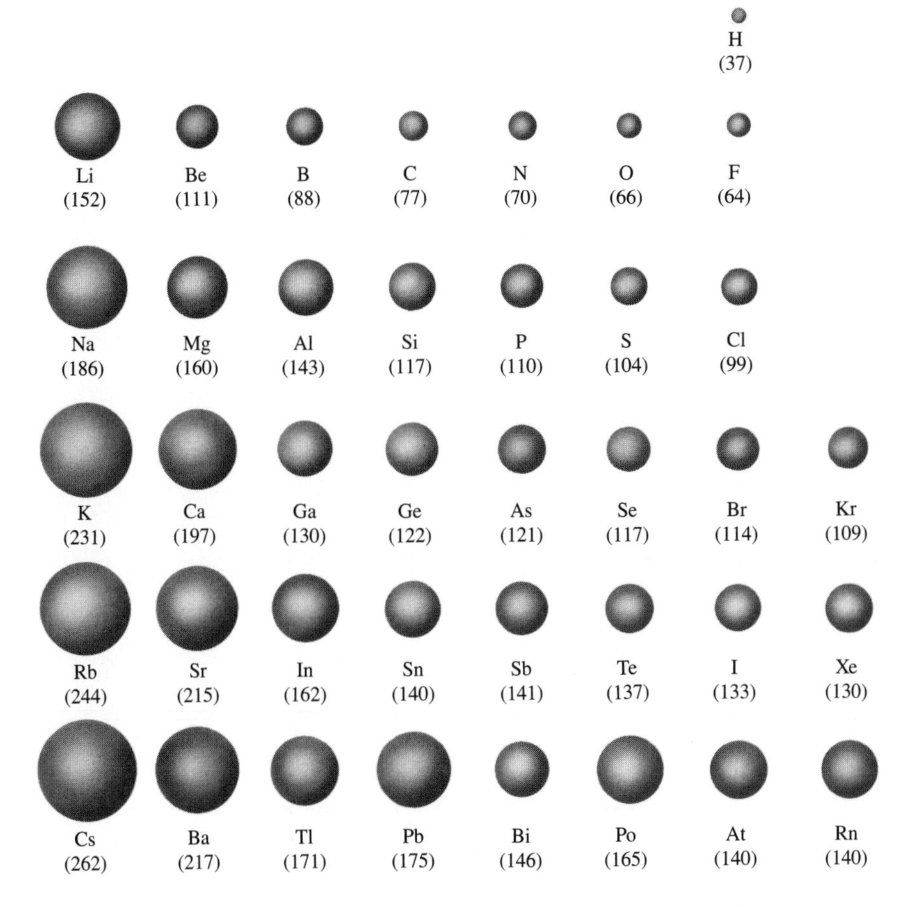

Figure 7.7 Atomic radii of the main group elements (pm).

Example 7.4 Sizes of Atoms and Ions

Identify the larger species of each of the following pairs:

(a) K or K^+
(b) S^{2-} or Cl^-
(c) Co^{2+} or Co^{3+}

Solution

(a) K has the electron configuration of $[Ar]4s^1$ and K^+ has the electron configuration of [Ar]. The electron in the 4*s* orbital makes K much larger than K^+.
(b) These two species have the same electron configuration ([Ar]). Since sulfur (Z = 16) has one less proton in its nucleus than does chlorine (Z = 17), the weaker nuclear attraction causes S^{2-} to be larger than Cl^-.
(c) Both Co^{2+} ($[Ar]3d^7$) and Co^{3+} ($[Ar]3d^6$) have valence electrons in the same sublevel, but Co^{2+} is larger because the additional shielding caused by the extra electron will expand the electron cloud.

Understanding Which is larger, Se^{2-} or Br^-?

Answer: Se^{2-}

Trends in the Sizes of Atoms

Figure 7.7 shows the atomic radii of the representative elements. These are measured values as described earlier, and values for helium, neon, and argon are not given because no compounds exist for these elements. Size trends in

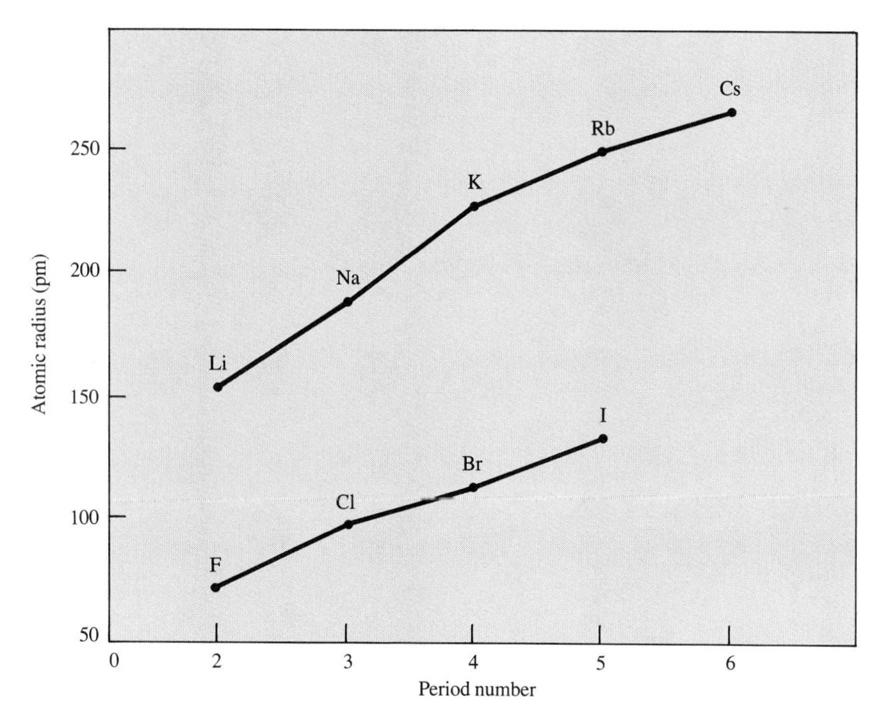

Figure 7.8 Plot of atomic radius vs. period number for the groups I and VIIA atoms.

The radii of atoms increase proceeding down any group because of the increase in the principal quantum number of the valence shell.

any group are easily predicted: the heavier the element in the group, the larger it will be. For example, we would expect that sodium, $1s^22s^22p^63s^1$, would be larger than lithium, $1s^22s^1$, since the $3s$ orbital is larger than the $2s$ orbital.

Shown in Figure 7.8 is a plot of atomic radius versus period number for the alkali metals (group IA) and the halogens (group VIIIA). The expected increase in size as we go down any group is clearly evident in these plots.

We need to consider the effective nuclear charge experienced by the valence electrons to determine the size trends of atoms in any given *period.* Compare, for example, atoms of lithium and beryllium. In both, the valence electrons are in a $2s$ orbital. Lithium has three protons, but the $2s$ electron is partially screened by the two electrons in the $1s$ orbital. The effective nuclear charge, Z_{eff}, for the $2s$ electron in lithium is about 1.3. Beryllium has an additional proton in its nucleus. Because the $2s$ electrons do not shield each other very effectively, Z_{eff} is about 2 for the $2s$ electrons in beryllium. The increased effective nuclear charge increases the electrostatic attraction of the nucleus for the outermost electrons, making beryllium *smaller* than lithium, even though it contains more electrons. This trend continues across the period. As electrons are added into the same principal level, the increasing effective nuclear charge for electrons in the outermost shell draws the electron cloud closer to the nucleus. The trends in sizes are summarized in the following periodic table. Elements at the bottom left of the table have the largest radii and those in the upper right are smallest.

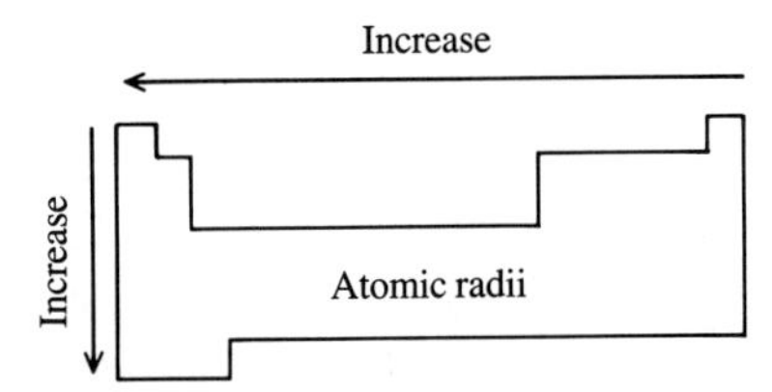

These trends in the atomic radii as a function of atomic number are shown graphically in Figure 7.9. The trends for the fourth period are expanded in Figure 7.10. The decrease in the size of the atoms across a period is clearly shown. There is a particularly large decrease in size on going from the group IA elements to the IIA elements in the same period, because electrons in an *ns* orbital do not shield each other effectively.

The radii of atoms decrease across a period because of the increase in effective nuclear charge.

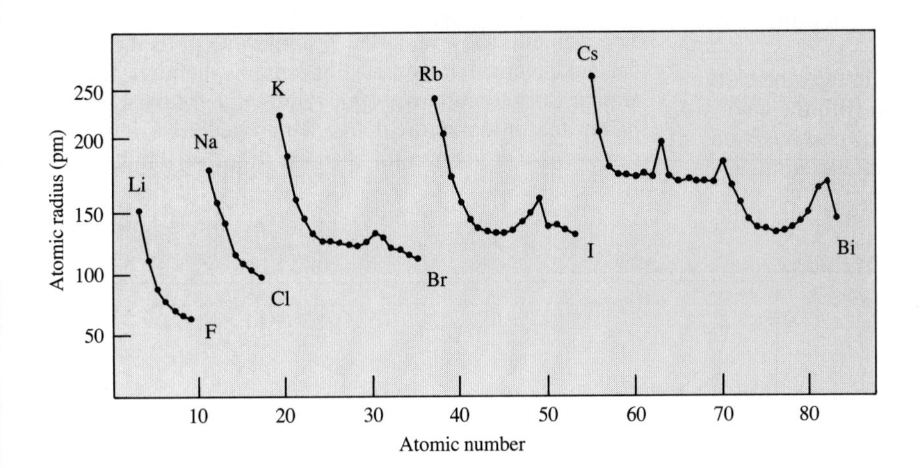

Figure 7.9 Plots of atomic radii of the elements.

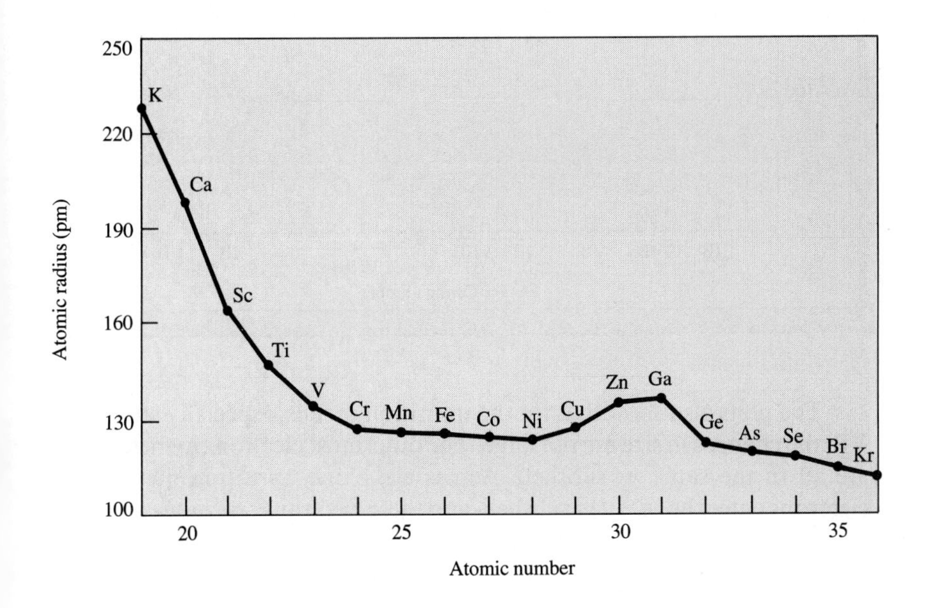

Figure 7.10 Atomic radii of the fourth period elements.

INSIGHTS into CHEMISTRY

A Closer View

The Lanthanide Contraction Interrupts the Periodic Table's Orderly Progression

The trends in atomic radii derived from the periodic table are of interest to chemists because many chemical and physical properties are determined in part by the size of the atom or ion of the element being considered. The trends in atomic radii of the main group elements are very regular. Across a period there is a decrease in atomic radii, while proceeding down a group there is an increase in the atomic radii. These trends are explained by considering the two major factors that affect the size—the effective nuclear charge and the principal quantum number of the outer electron orbital.

As already discussed, across a row in the *d*-block the decrease in size is not as pronounced as with the main group elements. The trends observed down a transition metal group show an even more unusual variation. There is an increase in radius between the fourth and fifth period transition metals, as expected from the increase in principal quantum number of the outermost electrons. In contrast, there is an unexpected similarity between the radii of the fifth and sixth period transition elements.

The explanation for this phenomenon has its origin in the lanthanide series of elements (from cerium, atomic number 58, to lutetium, atomic number 71), which appear between the elements lanthanum and hafnium in the sixth period. The electrons in the completely filled 4*f* sublevel in hafnium and the other elements of the sixth row transition metals do not completely shield the valence electrons, thus causing a larger effective nuclear charge for the outermost electrons. A result of this increase in effective nuclear charge is a reduction in the atomic radii, called the **lanthanide contraction.** This contraction nearly cancels the expected increase in size between the fifth and sixth period transition-metal elements.

An important consequence of the lanthanide contraction is that many of the fifth and sixth period transition elements show remarkable similarities in their physical and chemical properties. For example, hafnium is so similar to zirconium in atomic radius and chemical behavior that it took more than one hundred years after the discovery of zirconium for chemists to realize that haf-

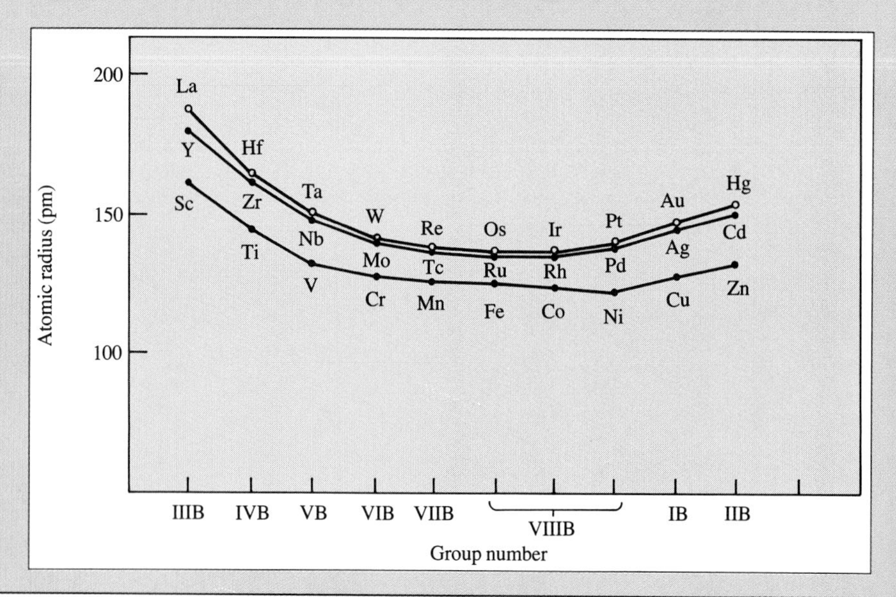

Sizes of the transition metals

The plots also show that for the transition metals, especially after group VB, the changes in size are not large. The outermost electrons in these metals are all in the same *ns* subshell. Across a *d*-block transition metal series, electrons enter the $(n-1)d$ subshell, and these electrons are quite effective in shielding the outermost *ns* electrons from the nuclear charge. Since Z_{eff} for the *ns* electrons is changing slowly in the transition elements, the sizes also change only slowly in the *d*-block. There is a small increase in size for the IIIA

nium was present as an impurity in every sample. Until 1923, when hafnium was finally identified, every published atomic weight for zirconium was wrong. All the physical constants that were published for zirconium actually applied to a naturally occurring mixture of zirconium and hafnium. Even with today's superior techniques, the two elements are difficult to separate from one another.

The lanthanide contraction has other consequences. One physical property that is directly influenced is the density of the sixth period elements. These elements have unusually high densities because their metallic radii are virtually the same as those of the fifth period elements in the same group, while their atomic masses are almost twice as large. Osmium and its neighbor iridium have the highest densities of any naturally occurring elements. The chemical activity of the sixth period elements is also influenced by the lanthanide contraction. Because of the high effective nuclear charge experienced by their valence electrons, sixth period elements such as platinum, gold, and mercury are relatively inert. Because of this chemical inactivity, platinum and gold are among the few metallic elements that occur in nature in the uncombined state.

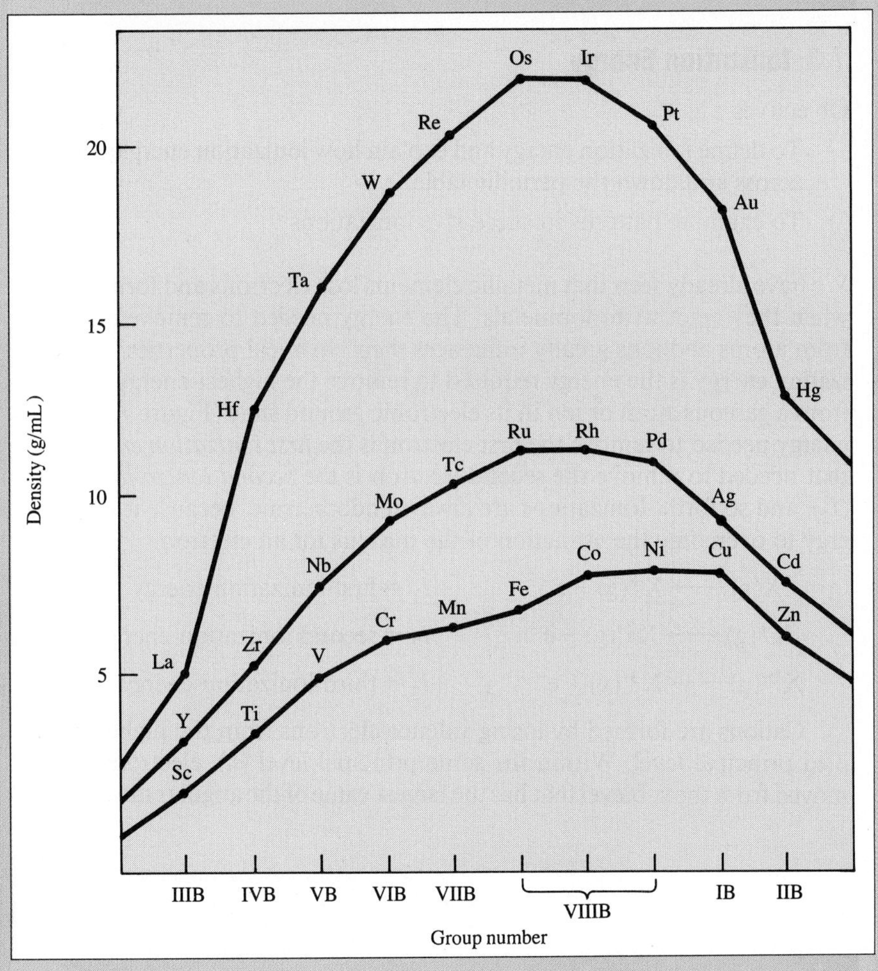

Densities of the transition metals

elements since the $(n-1)d$ sublevel is complete and the last electron occupies the np sublevel.

Example 7.5 Size Trends

List the following series of elements in order of increasing atomic radius: (a) Be, C, Mg (b) Rb, I, Br.

Solution

(a) Beryllium is in the same period with carbon and is to its left, so it is the larger of the two. Magnesium is below beryllium in the same group and is larger. The order of increasing size is C < Be < Mg.
(b) Rubidium is far to the left in the same period as iodine and is largest; bromine is above iodine and is the smallest: Br < I < Rb.

Understanding Order the elements O, S, and Al in increasing size.

Answer: O < S < Al

7.3 Ionization Energy

Objectives

- To define ionization energy and explain how ionization energies change across and down the periodic table
- To examine patterns in successive ionizations

We have already seen that metallic elements lose electrons and form cations when they react with nonmetals. The energy needed to remove electrons from atoms and ions greatly influences their chemical properties. The **ionization energy** is the energy required to remove the highest-energy electron from a gaseous atom or ion in its electronic ground state (Figure 7.11). The energy needed to remove the first electron is the *first ionization energy* (I_1), that needed to remove the second electron is the *second ionization energy* (I_2), and so forth. Ionizations are always endothermic, because it takes energy to overcome the attraction of the nucleus for an electron.

$$X(g) \longrightarrow X^{+}(g) + e^{-} \qquad I_1 = \text{first ionization energy}$$

$$X^{+}(g) \longrightarrow X^{2+}(g) + e^{-} \qquad I_2 = \text{second ionization energy}$$

$$X^{2+}(g) \longrightarrow X^{3+}(g) + e^{-} \qquad I_3 = \text{third ionization energy}$$

Cations are formed by losing valence electrons from the highest occupied principal level. Within the same principal level (n), electrons are removed from the sublevel that has the largest value of the angular momentum

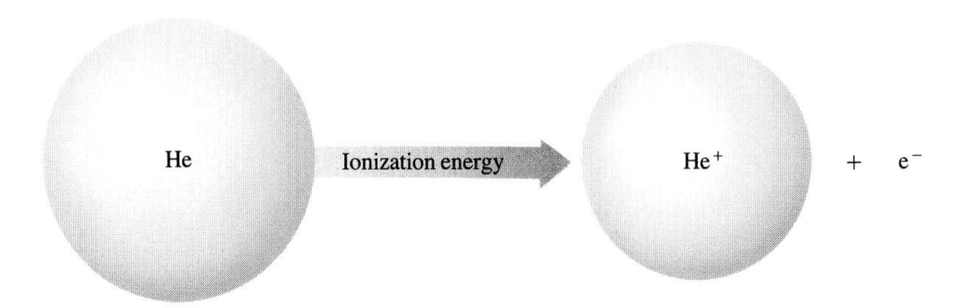

Figure 7.11 Ionization energy Ionization energy is the energy required to remove an electron from the ground state of a gaseous atom or ion.

quantum number (ℓ). The electron configurations of the ions formed by loss of the first three electrons from magnesium are shown here:

$$\underset{1s^22s^22p^63s^2}{Mg(g)} \longrightarrow \underset{1s^22s^22p^63s^1}{Mg^+(g)} + e^-$$

$$\underset{1s^22s^22p^63s^1}{Mg^+(g)} \longrightarrow \underset{1s^22s^22p^6}{Mg^{2+}(g)} + e^-$$

$$\underset{1s^22s^22p^6}{Mg^{2+}(g)} \longrightarrow \underset{1s^22s^22p^5}{Mg^{3+}(g)} + e^-$$

Trends in First Ionization Energies

The ionization energies are determined by how strongly the valence electrons are held by the nucleus. Electrons that are tightly held by the nucleus (those that experience a large effective nuclear charge, Z_{eff}) will be difficult to remove, whereas loosely held electrons (experience a low Z_{eff}) will be easily removed. Figure 7.12 shows a plot of first ionization energy versus atomic number. Figure 7.13 shows these trends in a periodic table for the representative elements.

The most obvious trend in the first ionization energies of the elements is a general increase across each period. We can explain this trend using the changes that occur in Z_{eff}. Consider the first ionization energy for the atoms

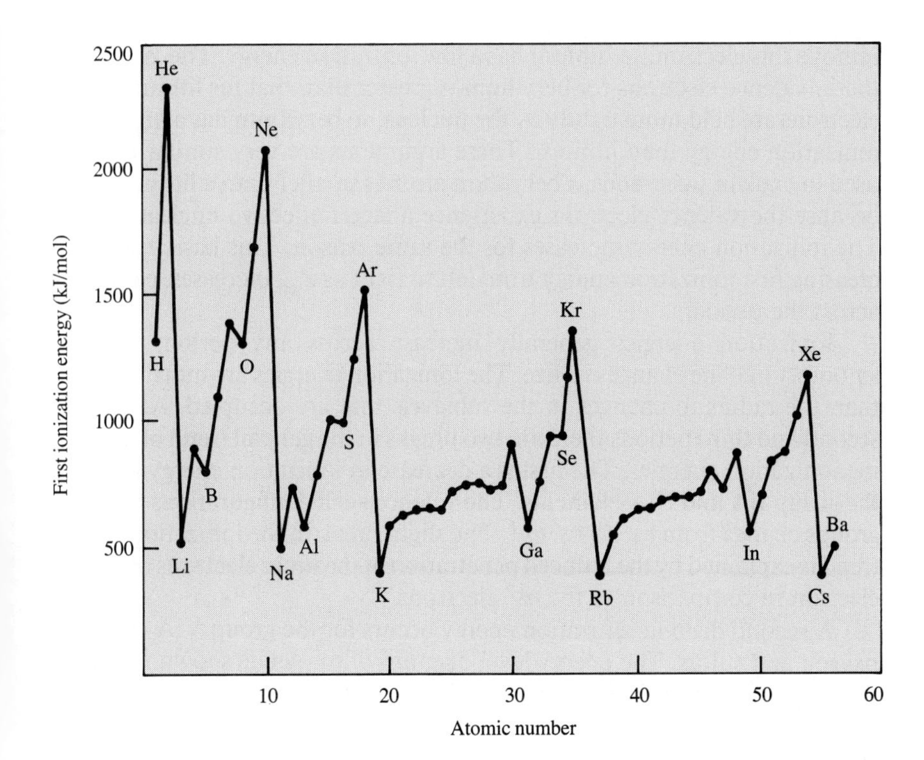

Figure 7.12 First ionization energies of atoms.

Figure 7.13 Three-dimensional plot of first ionization energies for the representative elements.

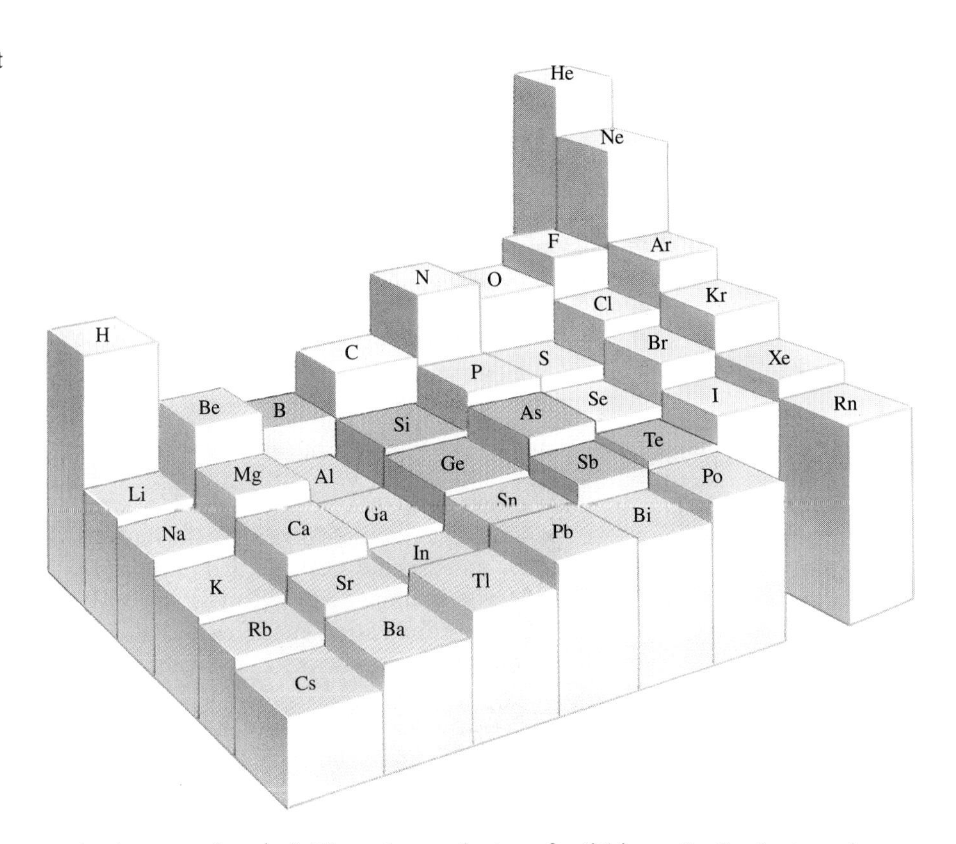

Ionization energies increase across a period because of the increase in effective nuclear charge.

in the second period. The valence electron for lithium, the 2*s* electron, is not tightly held by the nucleus. Only a small amount of energy is needed to remove this electron, so lithium has a low ionization energy. The Z_{eff} felt by the 2*s* valence electrons for beryllium is greater than that for lithium; these electrons are held more tightly by the nucleus, so beryllium has a *higher* first ionization energy than lithium. These arguments are very similar to those used to explain size trends; a beryllium atom is smaller than a lithium atom because the valence electrons experience a larger effective nuclear charge. The ionization energy increases for the same reason. This basic trend, increasing first ionization energy from left to right as Z_{eff} increases, continues across the period.

Ionization energies generally increase across any period, but not smoothly like the changes in size. The ionization energies are more sensitive than the radius to changes in the sublevels that are occupied. Across the second and third periods there are two breaks in the general trend of increasing ionization energies. The first is a decrease in ionization energy between the group IIA and IIIA elements. The valence shell configurations for these groups change from ns^2 to ns^2np^1. The slight break in the ionization energy trend is explained by the reduced penetration of the inner electrons by an np^1 electron in comparison to the ns^2 electrons.

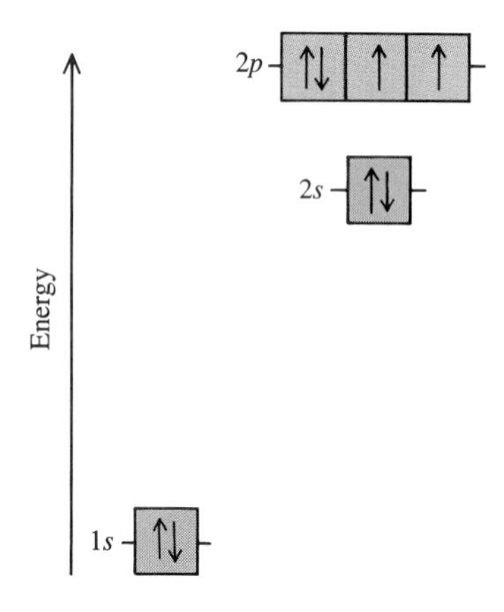

Figure 7.14 Energy level diagram of oxygen Electrons are paired in one of the 2*p* orbitals.

A second drop in ionization energy occurs for the group VIA elements oxygen and sulfur. The energy level diagram of oxygen is shown in Figure 7.14. The three *p* valence electrons of the group VA elements each occupy a separate *p* orbital. The fourth *p* valence electron that is present in oxygen and sulfur must be placed in an orbital that is already occupied. Two electrons in the same p orbital repel each other considerably more than two electrons in different p orbitals because they are in the same region of space, producing a decrease in the ionization energy. Note that both of the irregularities in ionization energies are small, and that the overall trends are dominated by the increase in the effective nuclear charge across a period while filling orbitals in the same principal shell.

Slight breaks in the increasing ionization energies across a period are observed at group IIIA, where electrons first enter the p subshell, and at group VIA, where electrons are first paired in the p subshell.

Within a group, the changes in ionization energies are not as large as the changes that occur within a period (Figure 7.15). In general, ionization energies decrease down a group. The decrease is caused by the fact that the valence electrons are in orbitals of increasing size, lowering coulombic attractions. The decrease is not as great as would be expected based on this factor alone because the effective nuclear charge felt by the valence electrons increases from the top to the bottom of the group, partially cancelling the size effect.

Ionization energies of representative elements decrease slightly going down a group.

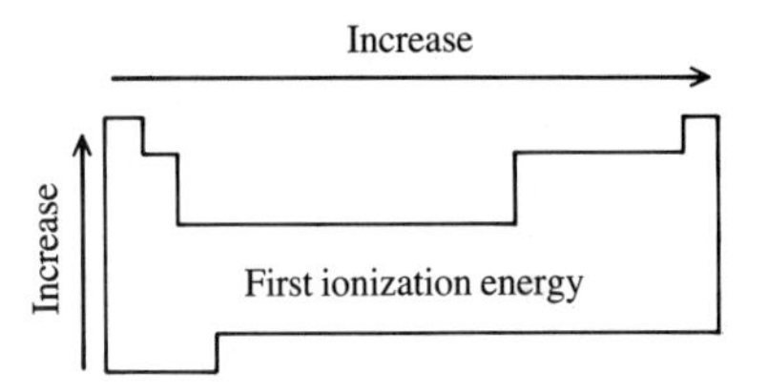

Ionization Energies of Transition Metals

For the transition elements (those in the d-block), *the ns electrons are lost before the (n − 1)d electrons.* Electrons in the highest principal shell are the outermost electrons and are the first removed by ionization.

Figure 7.12 shows that the changes in the ionization energies are small when progressing along the transition metals in any period. Across each series, an electron in the same ns orbital is being removed. With each successive element, the increase of one in the nuclear charge is almost cancelled by the shielding of the ns electrons by the additional $(n - 1)d$ electron. Thus, the effective nuclear charge experienced by the ns electrons increases only slowly, and this increase is reflected by the small rise in the ionization energies.

The chemistry of the transition metals is influenced by this slow change in ionization energy. Frequently, transition-metal elements of the same

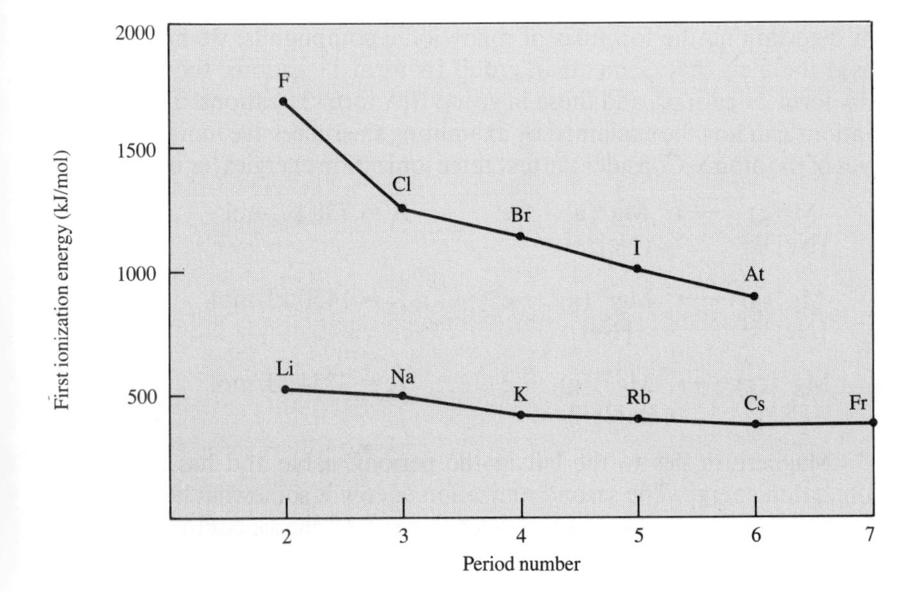

Figure 7.15 **First ionization energies down groups IA and VIIA.**

period form very similar compounds. Similar horizontal trends are not found with main group elements.

Ionization Energy Trends in an Isoelectronic Series

The trends in ionization energies along an isoelectronic series are easy to determine—the species with the greatest charge in the nucleus will have the *largest* first ionization energy. For example, Na^+ and Mg^{2+} both have the electron configuration $1s^22s^22p^6$. The higher nuclear charge and the smaller size of the Mg^{2+} ion both contribute to a higher ionization energy.

Example 7.6 Ionization Energy Trends

Predict which species in each of the following pairs has the higher ionization energy: (a) Mg or P (b) B or Cl (c) K^+ or Ca^{2+}.

Solution

(a) Magnesium and phosphorus are in the same period. Phosphorus is to the right and has the higher ionization energy.
(b) Boron and chlorine are in different periods, but chlorine is four groups to the right of boron. There is a small decrease in ionization energies on going from the second period to the third period, but this change is small compared to the increase in going from group IIIA to group VIIA. Chlorine has the higher ionization energy. Note that in this case the correct answer is clear. It is more difficult to compare closely positioned elements from different periods, such as boron and silicon, from the general trends among elements.
(c) The ions K^+ and Ca^{2+} are isoelectronic, but Ca^{2+} has one more proton in its nucleus and thus has the higher ionization energy.

Understanding Predict which species has the higher ionization energy: Al or Si.

Answer: Si

Charges of Cations

In determining the formulas of many ionic compounds, we have already used the facts that elements in group IA form 1+ cations, those in group IIA form 2+ cations, and those in group IIIA form 3+ cations. These observations can now be explained by examining the successive ionization energies of the atoms. Consider the first three ionization energies for magnesium:

$$\underset{[\text{Ne}]3s^2}{\text{Mg(g)}} \longrightarrow \underset{[\text{Ne}]3s^1}{\text{Mg}^+\text{(g)}} + e^- \qquad I_1 = 738 \text{ kJ/mol}$$

$$\underset{[\text{Ne}]3s^1}{\text{Mg}^+\text{(g)}} \longrightarrow \underset{[\text{Ne}]}{\text{Mg}^{2+}\text{(g)}} + e^- \qquad I_2 = 1450 \text{ kJ/mol}$$

$$\underset{[\text{Ne}]}{\text{Mg}^{2+}\text{(g)}} \longrightarrow \underset{[\text{He}]2s^22p^5}{\text{Mg}^{3+}\text{(g)}} + e^- \qquad I_3 = 7734 \text{ kJ/mol}$$

Magnesium lies to the left in the periodic table and has a low first ionization energy. The second ionization energy is somewhat larger, about double, because the ion that is produced has a 2+ charge compared to a 1+ charge. The third ionization energy is very large, because the third electron must be removed from the lower energy $n = 2$ level. These are not valence electrons and are tightly held by the nucleus. Thus, magnesium and all of the elements in group IIA lose two electrons to form cations with a charge of 2+. Similar arguments also apply to groups IA, IIIA, and IIIB; the valence electrons can be readily ionized, but not the core electrons. In each case, elements in these groups generally lose electrons until the cation formed has an electron configuration of a noble gas.

For the heavier elements in group IIIA, ionization of three electrons does not lead to a noble gas electron configuration. For gallium,

$$\underset{[Ar]4s^2 3d^{10} 4p^1}{Ga} \longrightarrow \underset{[Ar]3d^{10}}{Ga^{3+}} + 3e^-$$

The $3d$ electrons are *not* valence electrons and are not removed. The [noble gas]$(n-1)d^{10}$ electron configuration is known as a **pseudo-noble gas electron configuration** because several cations with this electron arrangement are stable. One other related trend for the heavier elements in groups IIIA is that two differently charged cations of the same element are found in stable compounds. For example, both TlCl and $TlCl_3$ are stable compounds. To form Tl^+, only the $6p^1$ electron is removed, leaving the [Xe]$6s^2 5d^{10}$ electron configuration. This electron configuration, as well as the pseudo-noble gas electron configuration, [Xe]$5d^{10}$, is found in compounds.

Representative metals can lose electrons until the cations attain a noble gas or a pseudo-noble gas electron configuration.

The charges on transition-metal ions are not as predictable. In general, the lowest positive charge found on a transition metal occurs when the ns electrons are ionized. A few other electron arrangements are particularly stable but depend on factors to be discussed later. The main point is that the transition metals form cations, but each metal can exist as more than one type of charged species. With the exception of the group IB elements, at least two electrons are normally lost from the metal in forming transition-metal compounds, consistent with the loss of both ns electrons.

Example 7.7 Ionization Energy Trends

Explain why the first ionization energy of lithium is less than that for beryllium, but the second ionization energy of beryllium is less than that for lithium.

Solution For both lithium and beryllium, the first electron is ionized from the $2s$ sublevel. Beryllium has one more proton in its nucleus and thus has a higher Z_{eff} for electrons in the $2s$ sublevel—its first ionization energy will be greater. For beryllium, the second electron also comes from the $2s$ sublevel, but for lithium it comes from an inner $1s$ sublevel. Removing an inner electron requires more energy than removing a valence electron, so the second ionization energy for lithium is larger than the second ionization energy for beryllium.

Understanding Does beryllium or boron have a larger *second* ionization energy?

Answer: boron

7.4 Electron Affinity

Objective

- To define electron affinity and explain the trends in electron affinities within the periodic table

Ionization energies provide information about how elements form cations. Many atoms, especially those of the elements on the right side of the periodic table, accept electrons to form anions. The **electron affinity** of an element is the energy change that accompanies the addition of an electron to a gaseous atom to form an anion.

$$A(g) + e^- \longrightarrow A^-(g)$$

Electron affinities are more difficult to measure than ionization energies. Representative values of electron affinities are given in Figure 7.16.

For most of the main-group elements, the electron affinity has a negative sign and thus the process is exothermic. As might be expected, the elements with the most exothermic (or most favorable) electron affinities are the group VIIA elements. The halogens are the elements with the highest effective nuclear charges that have a vacancy in the valence shell to hold an additional electron.

$$\begin{array}{lcl} Cl(g) + e^- & \longrightarrow & Cl^-(g) \\ [Ne]3s^2 3p^5 & & [Ne]3s^2 3p^6 = [Ar] \end{array}$$

Electron affinities are generally most favorable for elements with high ionization energies.

The group VIA elements also have electron affinities that are quite exothermic. The elements of groups VIA and VIIA also have high ionization energies. Thus, for these elements it is difficult to remove an electron, but energetically favorable to gain an electron. It is no surprise that these elements exist as anions in many compounds.

The electron affinity of nitrogen is 0, despite the fact that the electron affinities of the elements on either side of it in the periodic table are exothermic. The change in electron configuration of nitrogen upon addition of an electron is

$$\begin{array}{lcl} N(g) + e^- & \longrightarrow & N^-(g) \\ [He]2s^2 2p^3 & & [He]2s^2 2p^4 \end{array}$$

The nitrogen atom has an electron in each of its three *p* orbitals. The additional electron must pair with one of these electrons, causing larger electron-electron repulsions. These unfavorable interactions make the formation of a nitrogen anion less favorable than might have been expected. Nitrogen has a low electron affinity for the same reason that oxygen has a low ionization energy.

IA	IIA	IIIA	IVA	VA	VIA	VIIA
H −73						
Li −60	Be 241	B −27	C −122	N 0	O −141	F −328
Na −53	Mg 230	Al −43	Si −134	P −72	S −200	Cl −349
K −48	Ca 156	Ga −29	Ge −119	As −78	Se −195	Br −325
Rb −47	Sr 167	In −29	Sn −107	Sb −103	Te −190	I −295
Cs −46	Ba 52	Tl −19	Pb −35	Bi −91		

Figure 7.16 Electron affinities for selected main-group elements (kJ/mol).

Finally, electron affinities do not change dramatically down the groups. As with ionization energy trends down a group, the changes in size and Z_{eff} affect the electron affinities in opposite directions and largely cancel each other.

7.5 Trends in the Chemistry of Elements in Groups IA, IIA, and VIIA

Objective

- To explain the chemical trends of groups IA, IIA, and VIIA

IA	IIA
3 Li lithium	4 Be beryllium
11 Na sodium	12 Mg magnesium
19 K potassium	20 Ca calcium
37 Rb rubidium	38 Sr strontium
55 Cs cesium	56 Ba barium
87 Fr francium	88 Ra radium

The periodic table as developed by Mendeleev and Meyer arranges elements by chemical and physical properties. We now know that this arrangement groups together the elements that have the same number of valence electrons. The chemical properties of the elements are strongly influenced by factors such as sizes, ionization energies, and electron affinities. General trends of these properties have already been discussed. An important overall trend is that elements with low ionization energies are generally metallic and form cations in their compounds, while those with high ionization energies are nonmetals and form anions. We will consider in more detail the chemistry of elements in groups IA, IIA, and VIIA, which were discussed briefly in Chapter 1.

Group IA (1): The Alkali Metals

The group IA (1) metals, lithium, sodium, potassium, rubidium, cesium, and francium, all have a single electron in the *s* orbital of the valence shell. This electron is easily removed, so these elements are highly reactive, forming compounds that contain the metals as 1+ ions. In nature, these elements are found only in combination with other elements because they are too reactive to exist in the elemental state in the environment. Both sodium and potassium are abundant in nature, but lithium, rubidium, and cesium are very rare. Francium is exceedingly scarce, because all of its isotopes are radioactive. Hydrogen, which has a $1s^1$ electron configuration, is not a member of this group (even though it appears on top of this group on most periodic tables). Because it is the first element in the periodic table, hydrogen has unique chemical properties and is really in a group by itself.

The group IA elements are soft, silver-colored metals (Figure 7.17; cesium is golden) and they all melt at low temperatures compared to most other metals. Cesium, in fact, melts just above room temperature. The low melting point of sodium is one of the reasons it has been used as a cooling liquid in nuclear reactors.

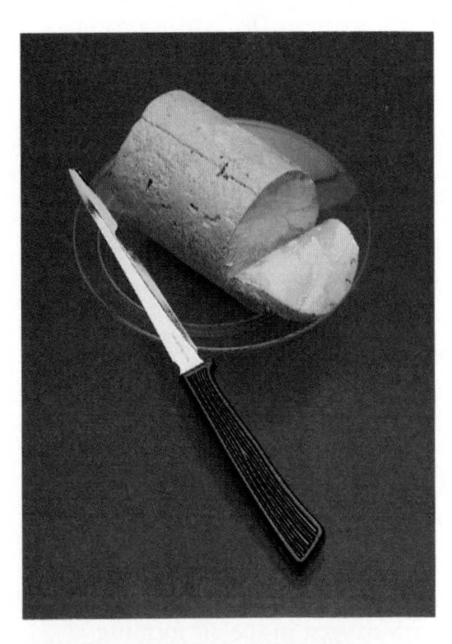

Figure 7.17 Sodium metal Sodium metal is soft and is easily cut by a knife. It is a silvery color when first cut, but loses its luster as it reacts rapidly with air.

Each of the alkali metals emits light of a characteristic color when a compound containing the metal is placed in a flame. Figure 7.18 shows the colors observed when salts of these elements are heated in a flame, a process known as a *flame test.* The colors of light emitted by these elements arise from their line spectra. Flame tests are used to determine whether these elements are present (qualitative analysis) in samples of unknown composi-

Figure 7.18 Flame test for alkali metal compounds The alkali metals lithium (a), sodium (b), and potassium (c) give off characteristic colors when heated in a flame. A drop of a solution that contains a salt of the metal is placed on a wire loop, and the loop is placed in a flame.

tion, and the intensity of these colors is used to determine the amount of these elements in the samples (quantitative analysis).

All of the alkali metals are very reactive, with the reactivity generally increasing down the group, as expected from the decrease in ionization energies. They all react with water to give similar products, hydrogen gas and solutions of the metal hydroxide. Lithium reacts only slowly, sodium reacts rapidly, potassium inflames, and rubidium and cesium react almost explosively with water.

$$2M(s) + 2H_2O(\ell) \longrightarrow 2MOH(aq) + H_2(g)$$

These reactions can be very dangerous because the hydrogen gas that is produced will react with oxygen in air to form water, another reaction that can be quite violent (Figure 7.19). Sodium, which is used commonly in the laboratory, needs to be treated very carefully, and is usually handled under oil to protect it from water vapor and oxygen in the air.

Figure 7.19 Reaction of potassium and water (a) Potassium reacts rapidly with water, giving off H_2 gas. (b) In air, flames are seen as the hydrogen gas produced reacts further with oxygen in the air forming water. (c) Under an atmosphere of argon (density greater than air), the potassium still reacts rapidly with water, but there is no flame because the argon excludes the oxygen.

INSIGHTS into CHEMISTRY

Chemical Reactivity

Blood Sodium and Potassium Levels Are Critical; Analyses Help Doctors Diagnose Patients

Sodium and potassium are two of the most important ions present in blood. Blood is a complex mixture of solids (mainly blood cells) and a liquid, called the *plasma.* Sodium is present in relatively high concentrations in the plasma. In contrast, only low concentrations of potassium are present in the plasma. *Inside* the blood cells the situation is reversed; the potassium concentration is high and the sodium concentration is low. Both ions play a role in the processes that control many functions of the body, such as muscle movement. When sodium and potassium levels are not in balance, these important processes do not function correctly. Concentrations of these ions in the blood can be tested for some medical conditions.

Physicians need to know sodium and potassium concentrations to make many important diagnoses. The chemical analysis for these elements is based on the emission spectra observed in a flame. Photographs of flames with sodium and potassium compounds added appear in Figure 7.18. Only 0.05 mL of blood, about a drop, is needed to measure the sodium and potassium content. The sample is diluted by adding 4.95 mL of solvent. The 5-mL diluted sample is sprayed slowly into a flame, where the light emission is observed. Both levels can be determined in the same sample by measuring the intensity of light at the wavelengths of characteristic lines in their atomic spectra.

The normal sodium concentration in blood is 0.135 to 0.148 *M*. Potassium concentrations are lower, in the range of 0.0040 to 0.0059 *M*. The development of an instrument that performs these analyses quickly, accurately, and inexpensively was an important improvement in hospital laboratory technology. Modern instruments provide readings with three significant digits in under a minute.

Sodium and potassium in blood This instrument can determine sodium and potassium concentrations in blood, using a sample of only 50 μL of blood. The instrument requires only a few seconds for a precise and accurate analysis.

Rapid determination of these important ionic concentrations can be very important to a sick person.

These metals also react with hydrogen and the halogens. Typical examples are

$$2K(s) + H_2(g) \longrightarrow 2KH(s)$$

$$2Na(s) + Cl_2(g) \longrightarrow 2NaCl(s)$$

Again, the heavier members of the group are generally more reactive. For example, lithium and sodium react slowly with liquid bromine; the other group IA elements react violently. The reaction of potassium with hydrogen forms KH. This ionic compound (it conducts electrical current above its

The reactivity of the group IA metals increases down the group. Their chemistry is dominated by the formation of M^+ ions.

melting point) is an example of a class of compounds called metal hydrides, compounds in which hydrogen is present as a negative ion.

The group IA metals react with molecular oxygen, but only lithium reacts to form the compound that we expect, Li_2O.

$$4Li(s) + O_2(g) \longrightarrow 2Li_2O(s)$$

Sodium reacts with excess O_2 to form a compound that contains the polyatomic *peroxide* anion, O_2^{2-}.

$$2Na(s) + O_2(g) \longrightarrow Na_2O_2(s)$$

The reaction is not stoichiometric, since the sodium peroxide is partially contaminated with some sodium oxide, Na_2O. Potassium, rubidium, and cesium react to form mixtures of three compounds. In addition to forming the oxide and peroxide, these metals also produce compounds that contain the polyatomic *superoxide* anion, O_2^-.

$$K(s) + O_2(g) \longrightarrow KO_2(s)$$

An unusual property of lithium is that it reacts with molecular nitrogen to form lithium nitride, a compound that contains the N^{3-} ion.

$$6Li(s) + N_2(g) \longrightarrow 2Li_3N(s)$$

Nitrogen is not a very reactive gas. In fact, nitrogen gas is used to protect materials that react with oxygen and water, a method that cannot be used for lithium. Although lithium is generally the least reactive of the group IA metals, it is the only one of these metals that reacts directly with molecular nitrogen.

Group IIA (2): The Alkaline Earth Metals

IA	IIA	IIIB
3 Li lithium	4 Be beryllium	
11 Na sodium	12 Mg magnesium	
19 K potassium	20 Ca calcium	21 Sc scandium
37 Rb rubidium	38 Sr strontium	39 Y yttrium
55 Cs cesium	56 Ba barium	57 La lanthanum
87 Fr francium	88 Ra radium	89 Ac actinium

The alkaline earth metals (group IIA), beryllium, magnesium, calcium, strontium, barium, and radium, have the valence shell configuration ns^2. These metals are reactive, but not as reactive as the group IA metals because they have higher ionization energies. The two valence electrons are readily removed; in most of their compounds they are present as 2+ cations. Both magnesium and calcium are very abundant in nature. Magnesium is found in many minerals, such as the magnesium limestone known as dolomite, $MgCO_3 \cdot CaCO_3$. Large land masses, such as the Dolomite Mountains in Italy, consist of this mineral. Large deposits of $CaCO_3$, formed from the fossilized remains of ancient life, are found in all parts of the world. Coral and seashells are also mainly composed of $CaCO_3$. Strontium and barium are moderately abundant, but beryllium and radium are rare. All of the isotopes of radium, as its name implies, are radioactive.

The metals have a silvery white appearance. They are softer than most metals, but harder and have higher melting points than the alkali metals. Like the alkali metals, the heavier metals of the alkaline earth group (calcium, strontium, and barium) show characteristic colors in a flame test (Figure 7.20). Beryllium and magnesium emit light when heated, but characteristic spectral lines are not in the visible region of the spectrum.

Magnesium has widespread industrial uses. As an **alloy,** a mixture of two or more elements that has the properties of a metal, it has properties similar

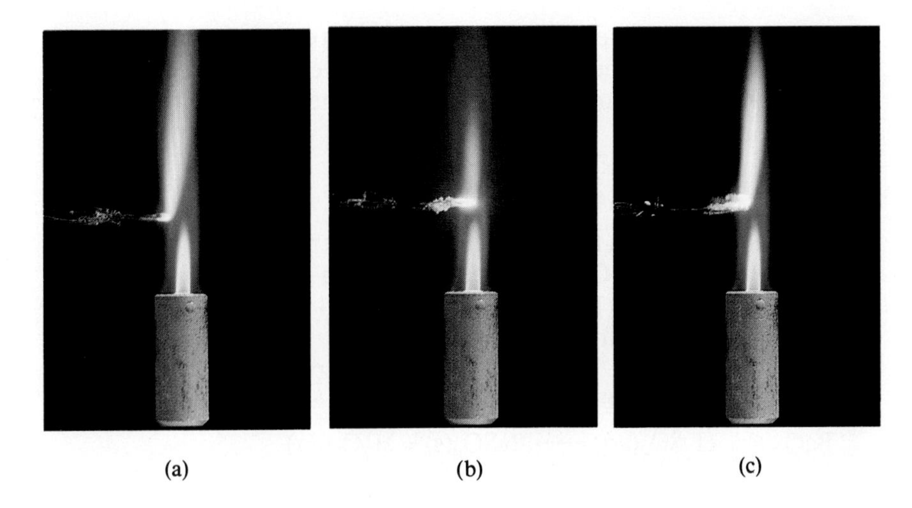

(a) (b) (c)

Figure 7.20 Flame test for group IIA metals The alkaline earth metals calcium (a), strontium (b), and barium (c) give off characteristic colors when heated in a flame.

to those of aluminum and is used as a construction material. Although magnesium is now about twice as expensive as aluminum, the use of magnesium might eventually increase, because it is easily isolated from seawater and has a density even lower than that of aluminum. Magnesium alloys are particularly useful in aeronautical applications, where low density and high strength are important. It is also added to aluminum to improve its mechanical properties. Mg-Al alloys can be machined much more easily than pure aluminum. Magnesium, in the form of Mg^{2+}, is an important constituent of the chlorophylls, compounds of great importance in photosynthesis.

As for the IA metals, the reactivity of the IIA metals increases down the group, a trend that follows the decrease in ionization energies. Beryllium does not react with water, even at elevated temperatures, nor with oxygen or the halogens below 600 °C. Magnesium reacts slowly with steam. Calcium, strontium, and barium react at room temperature with water to give the metal hydroxide and hydrogen gas.

$$M(s) + 2H_2O(\ell) \longrightarrow M(OH)_2(aq) + H_2(g) \qquad M = Ca, Sr, Ba$$

Reaction of calcium and water Calcium reacts with water to yield $Ca(OH)_2$ and H_2 gas.

The group IIA metals are not as reactive as the group IA metals. They form M^{2+} ions.

Beryllium and magnesium react with oxygen at elevated temperatures to form the oxide. The oxides are generally unreactive and act as coatings that protect the metal from further reaction. Calcium metal reacts with oxygen at room temperature.

$$2Ca(s) + O_2(g) \longrightarrow 2CaO(s)$$

Just as with the group IA (1) metals, the heavier elements of the group form salts of the peroxide ion, O_2^{2-}. Hydrogen peroxide is produced on a commercial scale by first forming barium peroxide by reaction of barium metal with oxygen, followed by treatment with sulfuric acid.

$$Ba(s) + O_2(g) \longrightarrow BaO_2(s)$$

$$BaO_2(s) + H_2SO_4(aq) \longrightarrow BaSO_4(s) + H_2O_2(aq)$$

Group VIIA (17): The Halogens

VIA	VIIA	VIIIA
8 O oxygen	9 F fluorine	10 Ne neon
16 S sulfur	17 Cl chlorine	18 Ar argon
34 Se selenium	35 Br bromine	36 Kr krypton
52 Te tellurium	53 I iodine	54 Xe xenon
84 Po polonium	85 At astatine	86 Rn radon

The chemistry of the halogens, fluorine, chlorine, bromine, iodine, and astatine, is dominated by the gain of one electron to attain a noble gas electron configuration. The halogens all exist as diatomic molecules, but are very reactive and occur in nature combined with other elements. Under standard conditions, fluorine is a light yellow gas and chlorine a deeper greenish-yellow gas, bromine is a deep red liquid (the only nonmetallic element that is a liquid at room temperature), and iodine is a shiny black solid.

Fluorine and chlorine are both abundant in nature—sodium chloride is one of the main components of seawater. Bromine is less abundant, but is found in sizeable quantities in certain inland seas such as the Dead Sea and the Great Salt Lake. Iodine is scarce, but is found in certain subterranean brine wells and in seaweed, where it was first discovered. The halogens are prepared industrially by electrolysis of these brine solutions (for fluorine, the molten salts are used directly). Astatine is very rare, because all of its isotopes are radioactive, and very little is known about the chemical behavior of the element.

Fluorine is the most reactive of all the elements. Because of its reactivity, it was not isolated until 1886, 60 years after the other halogens were prepared. It forms compounds with all other elements except helium, neon, and argon. It has a very interesting chemistry with xenon, reacting under different conditions to form XeF_2, XeF_4, and XeF_6. It also forms a special group of compounds called the **interhalogens,** compounds formed from two different halogens. The majority of the interhalogens have the general formula XF_n where X is one of the other halogens and $n = 1, 3, 5$, and even 7 in the compound IF_7.

The reactivity of the halogens decreases down the group. Their chemistry is dominated by the formation of X^- ions.

The reactivity of the halogens *decreases* down the group. For example, fluorine reacts explosively with hydrogen, but the reaction with hydrogen becomes less violent for the halogens further down the group and is very slow for iodine. In each case the hydrogen halide is formed.

$$X_2(g) + H_2(g) \longrightarrow 2HX(g) \qquad X = \text{group VIIA element}$$

The hydrogen halides all produce acidic solutions when they are dissolved in water. These acid solutions, particularly HF and HCl, are important industrially. HF is a very reactive, toxic material that can dissolve glass. It is used in the production of fluorocarbons, compounds used as refrigerants and aerosol propellents. Aqueous HCl is an inexpensive acid that is used in industry for a variety of purposes such as the synthesis of metal chlorides and other important chemicals.

Summary

The order for the filling of orbitals that is used in writing electron configurations of the elements can be determined from the periodic table. Each new period starts with the filling of an *ns* sublevel, where *n* is the period number. In the group IIIA (13) to VIIIA (18) elements, the *np* orbitals are being filled. An $(n-1)d$ sublevel is being filled across the transition-metal elements (groups IIIB (3) to IIB (12)). The $(n-2)f$ levels are being filled for the lanthanides and actinides, which are usually placed at the bottom of the table. Each time a *p* level is completely filled at group VIIIA (18), the electron configuration for a noble gas is achieved (helium has a $1s^2$ electron configuration). The symbol [noble gas symbol] is used to represent the electron configurations of the noble gas when writing electron configurations, a method that emphasizes the outermost or *valence electrons* in an atom or ion. The electron configurations of ions are written by starting with the electron configuration of the atoms and then adding or removing the correct number of electrons. In the formation of cations of the transition metals, the *ns* electrons are lost before the $(n-1)d$ electrons. Species that have the same number of electrons are *isoelectronic.*

The relative sizes of atoms and ions can be estimated from the electron configurations of the species and the trends in the effective nuclear charge, Z_{eff}, for the valence electrons. A cation will be smaller than its neutral atom; an anion will be larger. Along an isoelectronic series, the species with the largest number of protons in the nucleus will exert the strongest attraction for the electron cloud and will be smallest. When considering atoms along any given period, the valence electrons occupy the same principal level, but the number of protons in the nucleus is increasing. The Z_{eff} increases, drawing the outer electrons closer to the nucleus, causing the atoms to decrease in size along the period.

The trends in *ionization energy,* the energy required to remove the highest-energy electron from a gas-phase atom or ion, parallel the size trends. Along a period, the ionization energies *increase* because the effective nuclear charge is increasing and the radius is decreasing. There are two breaks in this general trend, at group IIIA when the *p* level is first occupied, and at group VIA when electrons are first paired in one of the *p* orbitals. Ionization energies decrease slightly down a group. Ionization energies change only slowly along a transition-metal series. In an isoelectronic series, the ionization energy will be largest for the element with the greatest number of protons. The energy needed to remove the second electron is always greater than the first, and the ionization energies for core electrons are very high.

The *electron affinity* is the energy change that accompanies the addition of an electron to a gas-phase atom or ion. In general, elements to the right side and top of the periodic table have exothermic (favorable) electron affinities. These are the elements that are generally observed to exist as anions in compounds with metals.

The *alkali metals,* group IA, all have the outer electron configuration of ns^1. Because of their low first ionization energies, they are very reactive, generally forming 1+ ions. They can be identified by the characteristic color given off when their salts are heated in a flame *(flame test).* The *alkaline earth metals,* group IIA, all have the outer electron configuration of ns^2. These metals are also reactive, forming 2+ cations, and the reactivity of both groups increases down the group. The *halogens,* group VIIA, are also very reactive (generally forming 1− ions), with the reactivity decreasing down the group.

Chapter Terms

Valence shell—those orbitals in the atom of the highest occupied principal level and the orbitals of *partially filled sublevels* of lower principal quantum number. *(7.1)*

Valence electrons—electrons that occupy the valence shell orbitals. *(7.1)*

Core electrons—the inner shell electrons that are not in the valence shell. *(7.1)*

Isoelectronic series—a group of atoms and ions that have the same number of electrons. *(7.1)*

Atomic radius—one half the distance between adjacent atoms of the same element in a molecule. *(7.2)*

Ionic radius—the measure of the size of an ion in an ionic solid. *(7.2)*

Lanthanide contraction—a decrease in the atomic radii of the lanthanide elements (atomic numbers 58 to 71) across the periodic table, caused by filling of the $4f$ sublevel. *(7.2)*

Ionization energy—the energy required to remove the highest energy electron from a gaseous atom or ion in its electronic ground state. *(7.3)*

Pseudo-noble gas electron configurations—electron configurations of the type [noble gas] $(n-1)d^{10}$. *(7.3)*

Electron affinity—The energy change for the addition of an electron to a gaseous atom to form an anion. *(7.4)*

Alloy—a mixture of two or more elements that has the properties of a metal. *(7.5)*

Interhalogens—compounds formed from two different halogens. *(7.5)*

Exercises

Exercises designated with color have answers in Appendix J.

Electron Configurations

7.1 Describe how the periodic table can be used as a guide to write electron configurations.

7.2 What are valence electrons and how are they different from core electrons?

7.3 What is an isoelectronic series? Give the symbols for four species that are isoelectronic.

7.4 Write the ground-state electron configurations and orbital diagrams for
(a) Ge (b) S (c) Rb.

7.5 Write the ground-state electron configurations and orbital diagrams for
(a) Ar (b) Br (c) K.

7.6 An atom of which element would have the ground-state electron configuration
(a) $1s^22s^22p^63s^23p^1$
(b) $1s^22s^22p^63s^1$
(c) $1s^22s^22p^63s^23p^64s^23d^5$?

7.7 An atom of which element would have the ground-state electron configuration
(a) $1s^22s^22p^63s^23p^3$
(b) $1s^22s^22p^63s^2$
(c) $1s^22s^22p^63s^23p^64s^23d^7$?

7.8 Which element in the halogen group has its highest energy electrons in the $4p$ sublevel?

7.9 Which element in the alkali metal group has its highest energy electron in the $5s$ sublevel?

7.10 Write the ground-state electron configuration and orbital diagram of
(a) Ca (b) Tc (c) In.
Use the noble gas representation for the core electrons.

7.11 Write the ground-state electron configuration and orbital diagram of
(a) Si (b) Ni (c) Sr.
Use the noble gas representation for the core electrons.

7.12 How many unpaired electrons are present in a ground-state atom of
(a) Y (b) Se (c) Cd?

7.13 How many unpaired electrons are present in a ground-state atom of
(a) Fe (b) S (c) Tl?

7.14 What are the valence electrons in
(a) Al (b) Cs (c) As?

7.15 What are the valence electrons in
(a) Zr (b) Ca (c) I?

7.16 What is the group number for the column of elements that have a valence shell configuration of ns^2?

7.17 What is the group number for the column of elements that have a valence shell configuration of ns^2np^4?

7.18 What is the general electron configuration for the valence electrons in
(a) group IA
(b) group IVA
(c) group VIIA?

7.19 What is the general electron configuration for the valence electrons in
(a) group IIA (b) group IIIA (c) group VIA?

7.20 Explain the fact that the first period contains two elements and the second period contains eight elements.

7.21 Why does the fourth period contain 18 elements?

7.22 Write the ground-state electron configuration of
(a) S^{2-} (b) Mn^{2+} (c) Ge^{2+}.

7.23 Write the ground-state electron configuration of
(a) Y^{3+} (b) Br^- (c) Rh^{3+}.

7.24 Write the symbol for a cation with a 1+ charge that has the electron configuration
(a) $1s^22s^22p^63s^23p^1$
(b) $1s^22s^22p^63s^1$
(c) $1s^22s^22p^63s^23p^64s^13d^6$?

7.25 Write the symbol for an anion with a 2− charge that has the electron configuration
(a) $[Ar]4s^23d^{10}4p^6$
(b) $1s^22s^22p^5$
(c) $1s^22s^22p^63s^23p^6$?

7.26 How many unpaired electrons are there in the ground state of
(a) Y^{3+} (b) Ni^{2+} (c) Cl^-?

7.27 How many unpaired electrons are there in the ground state of
(a) Co^{3+} (b) Sn^{2+} (c) Ru^{2+}?

7.28 Write the ground-state electron configurations for iron and chromium cations with a 3+ charge.

7.29 Which transition metal ion with a 3+ charge has the ground-state electron configuration $[Kr]4d^5$?

7.30 Write the symbol for a cation and an anion that are isoelectronic with Se.

7.31 Write the symbol for two cations and two anions that are isoelectronic with Kr.

7.32 Which cation(s), if any, of the second period elements with a 3+ charge have no unpaired electrons?

7.33 Which anion(s), if any, of the second period elements with a 2− charge have no unpaired electrons?

7.34 Which transition metal in the fourth period forms a 2+ ion with the greatest number of unpaired electrons?

7.35 Which transition metal in the fourth period forms a 3+ ion with the greatest number of unpaired electrons?

Size Trends

7.36 Which species in each of the following pairs is larger? Give an explanation for your answer.
(a) Na or Na^+ (b) O^{2-} or F^-
(c) Ni^{2+} or Ni^{3+}

7.37 Which species in each of the following pairs is larger? Give an explanation for your answer.
(a) Se^{2-} or Br^-
(b) Ru^{2+} or Ru^{3+}
(c) Mg^{2+} or Mg

7.38 Using only a periodic table as a guide, arrange each of the following series of atoms in order of increasing size.
(a) B, O, Li
(b) C, N, Si
(c) S, As, Sn

7.39 Using only a periodic table as a guide, arrange each of the following series of atoms in order of increasing size.
(a) Na, Be, Li
(b) P, N, F
(c) I, O, Sn

7.40 Using only a periodic table as a guide, arrange each of the following series of species in order of increasing size.
(a) Li, Be^{2+}, Be
(b) Cl, S, S^{2-}
(c) N, C, Si

7.41 Using only a periodic table as a guide, arrange each of the following series of species in order of increasing size.
(a) F, F^-, O^{2-}
(b) Al^{3+}, Mg, Na
(c) N, P, Si

7.42 Give an explanation for the fact that carbon atoms are larger than oxygen atoms even though oxygen has more electrons.

7.43 Why are sulfur atoms larger than oxygen atoms?

7.44 Explain why niobium atoms are considerably larger than vanadium atoms, but are about the same size as tantalum atoms.

Ionization Energy and Electron Affinity Trends

7.45 Define ionization energy.

7.46 Indicate which species in each pair has the higher ionization energy. Explain the reason for your answer.
(a) Si and Cl
(b) Na and Rb
(c) O^{2-} and F^-

7.47 Indicate which species in each pair has the higher ionization energy. Explain the reason for your answer.
(a) N and F
(b) Mg^{2+} and Na^+
(c) K and Si

7.48 Indicate which species in each pair has the higher ionization energy. Explain the reason for your answer.
(a) Ge and Cl
(b) B and F
(c) Al^{3+} and Na^+

7.49 Indicate which species in each pair has the higher ionization energy. Explain the reason for your answer.
(a) K and I
(b) Al and Al^+
(c) Cl^- and Ar

7.50 Using only a periodic table as a guide, arrange each of the following series of species in order of increasing first ionization energy.
(a) O, O^{2-}, F
(b) C, Si, N
(c) Te, Ru, Sr

7.51 Using only a periodic table as a guide, arrange each of the following series of species in order of increasing first ionization energy.
(a) S, Se^{2-}, O
(b) Fe, Br, F
(c) Cl, Cl^-, F

7.52 Even though ionization energies generally increase from left to right across the periodic table, the first ionization energy for aluminum is lower than that for magnesium. How can this be explained?

7.53 Explain why the first ionization energy of sodium is slightly lower than that for lithium.

7.54 Explain the fact that the first ionization energies of manganese, iron, and cobalt increase very slightly, whereas in the series gallium, germanium, and arsenic the first ionization energy increases considerably.

7.55 Explain why the second ionization energy for magnesium is about twice the first, but the third ionization energy is more than four times the second.

7.56 Which will be greater, the second ionization energy of boron or that of beryllium? Explain your answer.

7.57 Which will be greater, the second ionization energy of phosphorus or that of chlorine? Explain your answer.

7.58 Aluminum atoms are larger than silicon atoms, and the first ionization energy of silicon is larger than aluminum. Use changes in Z_{eff} to explain these trends.

7.59 Define electron affinity.

7.60 Which group on the periodic table would you expect to have large, negative electron affinities? Explain your answer.

7.61 How do the electron affinities vary down a group?

7.62 Explain why the electron affinity of lithium is slightly favorable (exothermic) whereas the electron affinity of beryllium is unfavorable (endothermic). Contrast these trends with the ionization energy trends of these two elements.

Chemistry of Groups IA, IIA, and VIIA

7.63 Describe the physical properties of the elements in group IA.

7.64 Describe the physical properties of the elements in group IIA.

7.65 Describe the physical properties of the elements in group VIIA.

7.66 What are the reactivity trends going down a group for the elements in group
(a) IA
(b) IIA
(c) VIIA?

7.67 Write the equation of the reaction, if any, of lithium with
(a) oxygen (b) nitrogen (c) chlorine (d) water.

7.68 Write the equation of the reaction, if any, of sodium with
(a) oxygen (b) nitrogen (c) chlorine (d) water.

7.69 Write the equation of the reaction, if any, of calcium with
(a) oxygen (b) nitrogen (c) water.

7.70 Write the equation of the reaction, if any, of barium with
(a) oxygen (b) nitrogen (c) water.

Additional Problems

7.71 Write the energy level diagram for titanium.

7.72 Write the energy level diagram for the 2+ cation of calcium.

7.73 Write the electron configuration of the copper atom and the 2+ cation of copper. Remember that the copper atom is an exception. Does the fact that the copper atom is an exception influence the electron configuration of Cu^{2+}?

7.74 Table 7.1 shows that palladium, with an electron configuration of $[Kr]4d^{10}$, is an exception. Write the electron configuration of the 2+ cation of palladium. Does the fact that the palladium atom is an exception influence the electron configuration of Pd^{2+}?

7.75 List the element for which the 2+ cation would have the electron configuration of $[Ar]3d^4$.

7.76 Chromium(IV) oxide is used to make high-quality magnetic recording tapes because of its magnetic properties. Write the formula of chromium(IV) oxide. Determine the number of unpaired electrons for a Cr^{4+} cation.

7.77 Which member of group VA has the (a) largest size, (b) the smallest ionization energy, (c) the greatest electron affinity?

7.78 Chlorine gas can be prepared by the electrolysis of NaCl. What volume of chlorine at STP can be prepared by the electrolysis of 2.44 g of NaCl?

7.79 Which ground-state electron configuration in the following series represents the element with (a) the largest size, (b) the smallest ionization energy, (c) the greatest electron affinity?
(1) $1s^22s^22p^1$
(2) $1s^22s^22p^4$
(3) $1s^22s^22p^5$

7.80 The mineral magnetite has the formula Fe_3O_4 and contains both Fe^{2+} and Fe^{3+}. Write the electron configurations of both cations of iron.

7.81 One of the cations Ga^{4+} and Mn^{4+} is not a stable species. Which one, and why?

7.82 Rank the following ions in terms of increasing sizes and increasing ionization energies: S^{2-}, K^+, Ca^{2+}.

7.83 Arrange the elements lithium, carbon, and oxygen in order of (a) increasing size, (b) increasing first ionization energy, (c) increasing second ionization energy, (d) number of unpaired electrons.

7.84 Using the pictures in Figures 7.18 and 7.20, suggest compounds to put in fireworks that would burn (a) red, (b) yellow, and (c) orange.

7.85 Fluorine proved to be one of the most difficult elements to isolate. In 1886, H. Moissan was able to isolate a sample of the gas by the electrolysis of KHF_2 in HF. For his effort, he received the Nobel Prize in chemistry in 1906. Why was fluorine so hard to isolate?

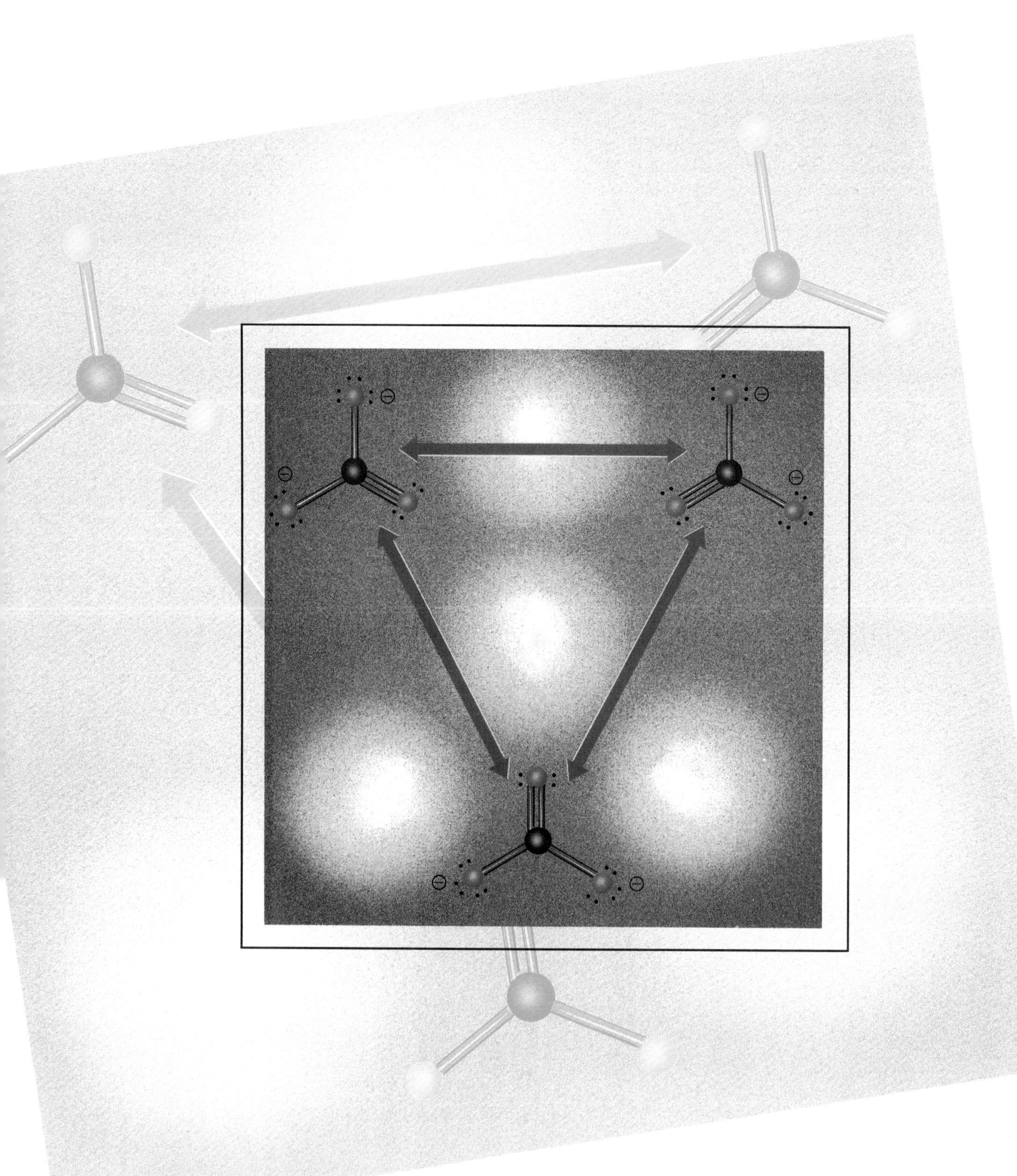

Chapter Eight

Chemical Bonds

A main goal for chemists is to be able to explain how and why elements combine to form compounds. Now that we have a model that relates the electronic structure of atoms and ions to other properties, this model can be extended to compounds. **Chemical bonds** are the forces that hold the atoms together in substances. Two classes (really limiting cases) of chemical bonds will be discussed that describe the forces that hold ionic and covalent compounds together, respectively.

8.1 Lewis Symbols

Objective

- To write Lewis electron-dot symbols for the elements and ions

Chemical bonds are formed by valence electrons. The inner electrons are held tightly by the nucleus and are not involved in bonding. A convenient and useful way to show the valence electrons that are used in bonding is by Lewis electron-dot symbols, first proposed by the American chemist G. N. Lewis (1875 – 1946). A **Lewis electron-dot symbol** consists of the symbol for the element surrounded by dots, one for each valence electron. Lewis electron-dot symbols for the representative elements to radium are shown in Figure 8.1. By convention, the first four electron dots are placed sequentially around the four sides of the element symbol, and additional electrons form pairs.

Lewis electron-dot symbols can also be written for ions. For many cations, the Lewis electron-dot symbol shows no electrons at all, because all of the valence electrons are removed upon formation of the ion.

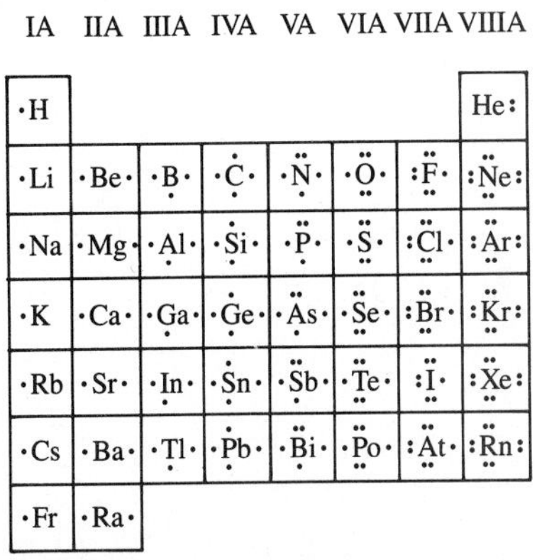

IA	IIA	IIIA	IVA	VA	VIA	VIIA	VIIIA
·H							He:
·Li	·Be·	·B·	·C·	·N·	·O·	:F·	:Ne:
·Na	·Mg·	·Al·	·Si·	·P·	·S·	:Cl·	:Ar:
·K	·Ca·	·Ga·	·Ge·	·As·	·Se·	:Br·	:Kr:
·Rb	·Sr·	·In·	·Sn·	·Sb·	·Te·	:I·	:Xe:
·Cs	·Ba·	·Tl·	·Pb·	·Bi·	·Po·	:At·	:Rn:
·Fr	·Ra·						

Figure 8.1 Lewis electron-dot symbols for the representative elements

There are three resonance structures for the carbonate ion.

$$Na\cdot \longrightarrow Na^+ + e^-$$

$$\cdot Ca\cdot \longrightarrow Ca^{2+} + 2e^-$$

For anions, electrons are added to the atom. Generally, the number of electrons added to form monatomic anions makes the ion isoelectronic with an atom of the next noble gas.

$$:\ddot{Cl}: + e^- \longrightarrow :\ddot{Cl}:^-$$

$$:\dot{O}: + 2e^- \longrightarrow :\ddot{O}:^{2-}$$

A Lewis electron-dot symbol shows the symbol of the element, the valence electrons as dots, and the charge, if any.

Both the chloride and oxide anions have eight valence electrons and the same electron configuration as a noble gas. The cations shown earlier are also isoelectronic with noble gases: Na^+ with the Ne atom and Ca^{2+} with the Ar atom.

Example 8.1 Lewis Electron-Dot Symbols

Write the Lewis electron-dot symbol for:
(a) fluorine atom (b) Be^{2+} (c) Br^-

Solution

(a) The fluorine atom is located in group VIIA, so it has seven valence electrons. They are placed around the symbol as three pairs and one single electron.

$$:\ddot{F}:$$

(b) The beryllium atom in group IIA has lost both of its valence electrons in forming the 2+ ion. The Lewis electron-dot symbol is the same as the symbol of the ion.

$$Be^{2+}$$

(c) The addition of an electron to the seven valence electrons of the bromine atom gives it eight.

$$:\ddot{Br}:^-$$

Understanding What is the Lewis electron-dot symbol for tin?

Answer: $\cdot\dot{Sn}\cdot$

8.2 Ionic Bonding

Objectives

- To use Lewis symbols to describe the formation of ionic compounds
- To describe how the charges on ions and the sizes of ions influence lattice energies

As outlined in Section 7.5, metals in groups IA and IIA tend to react with the nonmetals in groups VIA and VIIA to form ionic compounds. Electrons are easily removed from metals because metals have low ionization energies; electrons are easily added to nonmetals because their electron affinities are generally favorable. **Ionic bonding** is the bonding that results from the electrostatic attraction between positively charged cations and negatively charged anions. The formation of binary ionic compounds can be represented by using Lewis electron-dot symbols to show how electrons are lost from metals and gained by nonmetals.

$$\underset{[\text{He}]2s^1}{\text{Li}\cdot} + \underset{[\text{Ne}]3s^2 3p^5}{:\ddot{\underset{\cdot}{\text{Cl}}}:} \longrightarrow \underset{[\text{He}]}{\text{Li}^+} + \underset{[\text{Ar}]}{:\ddot{\underset{\cdot\cdot}{\text{Cl}}}:^-}$$

$$\underset{[\text{Ne}]3s^1}{\text{Na}\cdot} + \underset{[\text{He}]2s^2 2p^4}{:\dot{\underset{\cdot}{\text{O}}}:} + \underset{[\text{Ne}]3s^1}{\cdot\text{Na}} \longrightarrow \underset{[\text{Ne}]}{2\text{Na}^+} + \underset{[\text{Ne}]}{:\ddot{\underset{\cdot\cdot}{\text{O}}}:^{2-}}$$

Both lithium and sodium atoms lose one electron to form a cation with a 1+ charge, isoelectronic with a noble gas atom. Chlorine has seven valence electrons and gains one electron, filling its valence shell. Oxygen fills its valence shell, gaining two electrons to form the dinegative oxide ion. In forming Na_2O, two sodium atoms must transfer one electron each to supply the two electrons to oxygen.

Example 8.2 Lewis Electron-Dot Symbols of Ionic Compounds

Use Lewis electron-dot symbols to show the formation of (a) magnesium oxide and (b) calcium fluoride from the atoms.

Solution

(a) The magnesium atom has two valence electrons that it can readily lose. The oxygen atom has six valence electrons, and can gain two more electrons to fill its valence shell.

$$\cdot\text{Mg}\cdot + :\dot{\underset{\cdot}{\text{O}}}: \longrightarrow \text{Mg}^{2+} + :\ddot{\underset{\cdot\cdot}{\text{O}}}:^{2-}$$

(b) The calcium atom has two valence electrons to lose, but the fluorine atom with seven valence electrons can gain only one additional electron to fill its valence shell, so two fluorine atoms are needed.

$$\cdot\text{Ca}\cdot + :\ddot{\underset{\cdot}{\text{F}}}: + :\ddot{\underset{\cdot}{\text{F}}}: \longrightarrow \text{Ca}^{2+} + 2:\ddot{\underset{\cdot\cdot}{\text{F}}}:^-$$

Understanding Use Lewis electron-dot symbols to show the formation of lithium sulfide from the atoms.

Answer: $\text{Li}\cdot + \text{Li}\cdot + :\dot{\underset{\cdot}{\text{S}}}: \longrightarrow 2\text{Li}^+ + :\ddot{\underset{\cdot\cdot}{\text{S}}}:^{2-}$

Lattice Energy

The formation of ionic solids from the elements is usually a very exothermic process. For example

$$2\text{Li(s)} + \text{Cl}_2\text{(g)} \longrightarrow 2\text{LiCl(s)} \qquad \Delta H = -817 \text{ kJ}$$

A number of factors influence the size of such enthalpy changes. Two important factors that were presented in Chapter 7 are the ionization energy of the element that forms the cation and the electron affinity of the element that forms the anion. No matter which metal and nonmetal react to form the ionic solid, the sum of the ionization energy and the electron affinity is always an endothermic change. In the example of lithium chloride, the ionization energy of the lithium is +520 kJ/mol, and the electron affinity of chlorine is −349 kJ/mol, giving a sum of +171 kJ/mol.

The single most important contribution to the overall exothermic enthalpy change is the energy released when the oppositely charged ions are brought together in the ionic crystal lattice. When two electrically charged particles are brought together, the energy change is given by Coulomb's law:

$$E = \frac{kQ_1Q_2}{r}$$

where k is a constant, Q_1 and Q_2 are the charges on the two particles, and r is their final distance of separation. The energy is exothermic if the particles have opposite signs for their charges and endothermic if the charges are of the same sign. When a lithium ion and chloride ion come together to a distance equal to the sum of the ionic radii, the enthalpy change is approximately −490 kJ/mol.

Ionic solids are held together by strong coulombic forces.

The situation in forming an ionic solid from the gaseous ions is more complicated than the attraction of one cation to a single anion. In the ionic solid a very large number of cations and anions come together in an alternating arrangement of positive and negative ions. A number of arrangements are possible, depending on the sizes and charges of the ions. Figure 8.2 shows the arrangement of the ions in the sodium chloride and cesium chloride ionic structures. In solid sodium chloride each sodium cation is surrounded by six chloride anions, and each chloride is surrounded by six sodium cations. In the cesium chloride structure each cation and anion is surrounded by eight of the oppositely charged ions. These arrangements result in many attractions between the oppositely charged ions, but also many longer-range repulsions of the ions that have charges of the same sign. When the energy of all these attractive and repulsive interactions are added together, the net enthalpy change is always exothermic and proportional to the energy of attraction between a single cation and anion. The **lattice energy** is the energy required to separate one mole of an ionic crystal into the isolated gaseous ions (Figure 8.3).

Group II metals lose both valence electrons in order to maximize coulombic attractive forces, but the energy required to remove core electrons is too great.

The lattice energies of different ion pairs can be studied to explain the nature of ionic bonding forces. For instance, why does a metal such as magnesium lose two electrons rather than just one when it forms an ionic compound? We have already seen that the first ionization energy of magnesium is 738 kJ/mol and the second is 1450 kJ/mol. Even though it takes twice as much energy to remove the second electron, removing the second

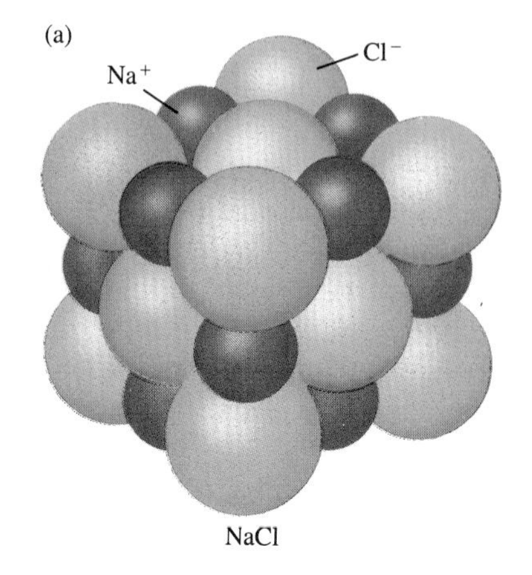

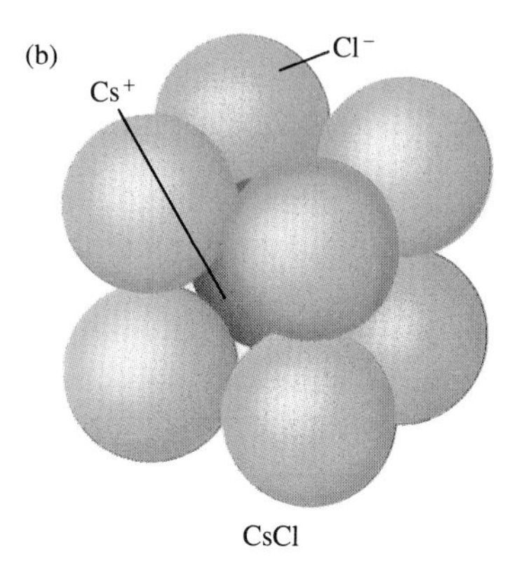

Figure 8.2 Ionic lattices of (a) NaCl and (b) CsCl In sodium chloride each cation is surrounded by six anions, but in cesium chloride each cation is surrounded by eight anions.

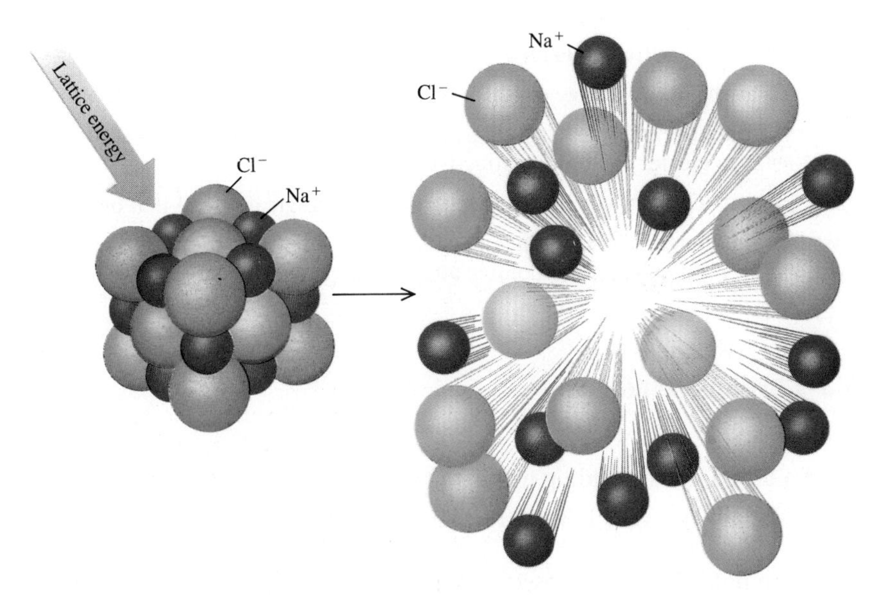

Figure 8.3 Lattice energy Lattice energy is the energy required to separate one mole of an ionic crystal solid into isolated gaseous ions.

electron doubles the charge (Q_1) on magnesium, thus increasing the electrostatic attraction substantially. The additional lattice energy caused by the 2+ ion more than compensates for the energy needed to ionize the second electron. As shown in Table 8.1, the lattice energies for MgX_2 (X = halide) compounds are considerably larger than for the group IA–halide compounds because of the larger charge on magnesium. These attractive forces would increase again if magnesium were to lose a third electron, but this does not happen because the third ionization energy is extremely high at 7734 kJ/mol. The energy gained by increasing the coulombic attractive forces would not be nearly as great as the energy needed to remove the third electron.

A similar situation exists for anions. Atoms of elements in the right side of the periodic table add electrons until their valence shell is filled. For example, chlorine will add one electron to form chloride, Cl^-, and oxygen will add two electrons to form oxide, O^{2-}. Oxygen atoms gain two electrons because they can accommodate them in their valence shell, and the higher charge causes the lattice energy to increase. Addition of an electron to chloride or oxide would be extremely endothermic because the electron cannot be accommodated by the valence shell.

Although it is less important than the charges, the r term in the potential energy equation affects lattice energies. If other factors are held constant, the lattice energy for ionic solids that consist of smaller ions is higher than that of larger ions. The lattice energy is greater, for example, in LiF than LiCl because the fluoride ion is smaller than the chloride ion. The charges are the same, but r is smaller for lithium fluoride. The lattice energy for MgO is the largest in Table 8.1 because of both the high charges of the ions and their small sizes.

Table 8.1 Lattice Energies

Compound	Lattice Energy (kJ/mol)
LiF	1036
LiCl	853
LiBr	807
LiI	732
NaF	923
NaCl	788
NaBr	736
NaI	686
KF	821
KCl	715
KBr	674
KI	632
MgF_2	2957
$MgCl_2$	2527
$MgBr_2$	2440
MgI_2	2327
Na_2O	2570
MgO	3938

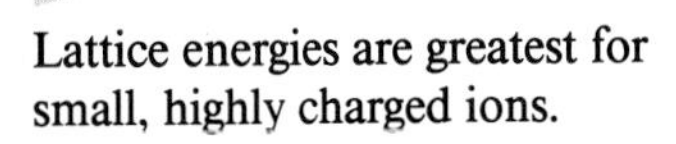
Lattice energies are greatest for small, highly charged ions.

Example 8.3 Lattice Energies

Explain why the lattice energy of Na_2O is considerably larger than that of NaF.

Solution The lattice energy of Na_2O is greater because the greater charge on O^{2-} than on F^- leads to greater coulombic attractive forces.

Understanding Explain why the lattice energy of KCl is greater than that of KI.

Answer: The charges of the ions in both of these compounds are the same, but Cl^- is smaller than I^-, making the coulombic attractive forces slightly greater for KCl.

8.3 Covalent Bonding

Objectives

- To describe the bonding forces in covalent compounds
- To be able to write Lewis structures of molecules and ions

Ionic compounds are composed of cations, generally from metals that have lost electrons, and anions, generally from nonmetals that have gained electrons. Ionic compounds are usually hard and have high melting points. Compounds (and elements) that contain only nonmetals usually consist of molecules with physical properties that are very different from ionic materials. For example, at room temperature, benzene, C_6H_6, is a liquid, and nitrogen dioxide, NO_2, is a gas that will condense into a liquid only at low temperatures (Figure 8.4). These dramatic differences in physical properties indicate that a second bonding model is needed to explain the bonding in molecules.

The bonding in the simplest molecule, H_2, is described by *sharing* of the two electrons. In this way, both electrons are attracted by each nucleus. *Each* nucleus is considered to have *gained* a share of both electrons and now has the helium noble-gas electron configuration. The sharing process is shown graphically in Figure 8.5. No new electrons have been added, but the two

(a)

(b)

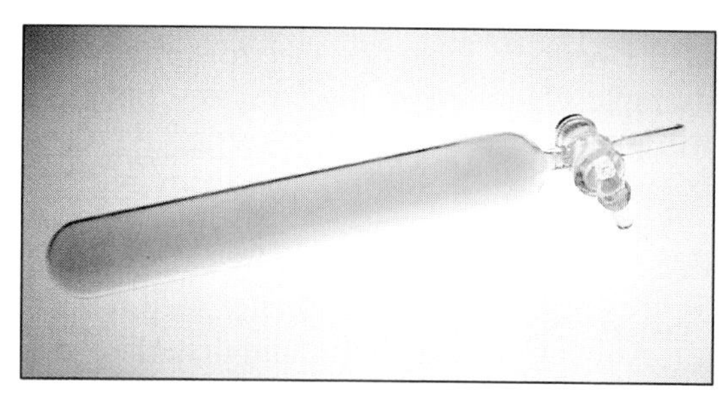

(c)

Figure 8.4 Ionic and covalent compounds Ionic compounds, such as $Ni(NO_3)_2 \cdot 6H_2O$ pictured here, are generally brittle solids (a). Many covalent compounds formed from nonmetallic elements are liquids (b, benzene) or gases (c, NO_2).

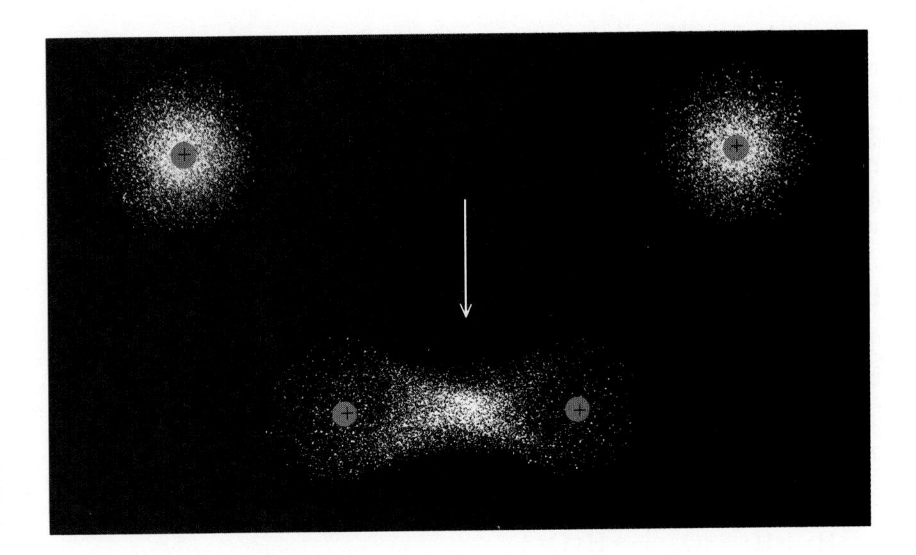

Figure 8.5 Sharing of electrons for H_2 The two nuclei in H_2 are held together by shared electrons.

shared electrons can be counted as being under the influence of both nuclear charges. A **covalent bond** is the bond that arises from atoms sharing electrons.

The diagram in Figure 8.6 shows how the energy of two hydrogen atoms changes as the distance between them varies. The energy is taken as zero when the atoms are far apart. As they move together, the electron clouds start to overlap. Both electrons are attracted by the positive charges of both nuclei. The overall energy of the atoms decreases; this is an exothermic process. The energy continues to decrease as the overlap increases until at close distances the nuclei come close enough that the repulsion between their *positive* charges becomes large. The low point of the curve is the internuclear distance at which the molecule is most stable. The **bond length** is the distance between

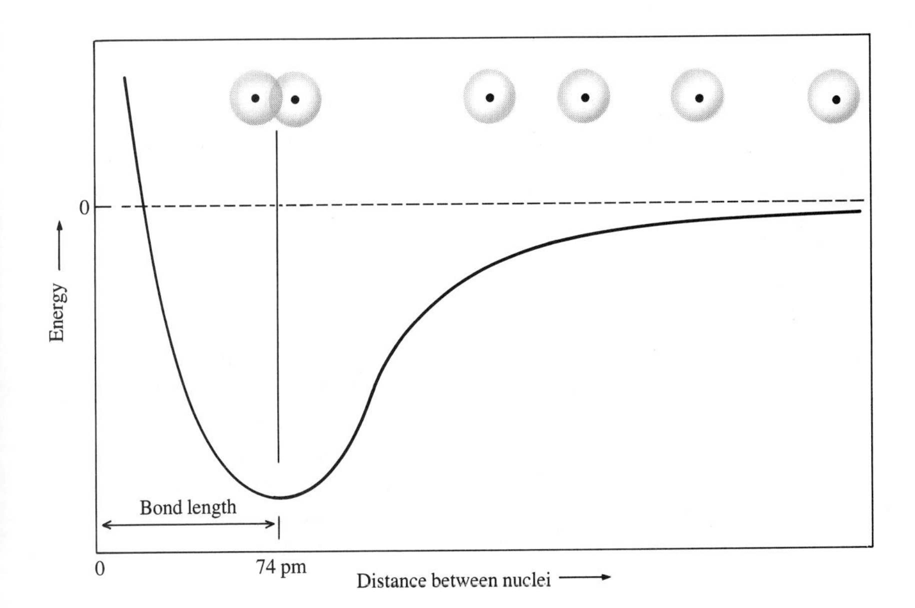

Figure 8.6 Potential energy curve for H_2 Two hydrogen atoms are most stable when their electron clouds overlap to form a covalent bond.

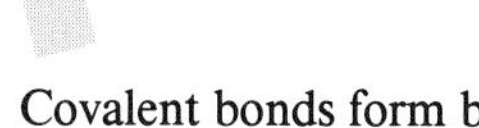

Covalent bonds form because two atoms sharing electrons are lower in energy than the isolated atoms.

the nuclei of two bonded atoms in a molecule. We would expect to find measured bond lengths to coincide with the energy minima in graphs such as Figure 8.6. For a covalent bond to form, nuclei must be close enough together that their electron clouds overlap and concentrate electron density between them. Since the electrons are occupying overlapping atomic orbitals, they must have opposite spins or the Pauli exclusion principle would be violated.

Lewis Structures

The Lewis electron-dot symbol notation can be used to describe the covalent bonds in molecules.

$$\mathrm{H\cdot + \cdot\ddot{\underset{\cdot\cdot}{Cl}}: \longrightarrow H:\ddot{\underset{\cdot\cdot}{Cl}}:}$$

$$\mathrm{:\ddot{\underset{\cdot\cdot}{F}}\cdot + \cdot\ddot{\underset{\cdot\cdot}{F}}: \longrightarrow :\ddot{\underset{\cdot\cdot}{F}}:\ddot{\underset{\cdot\cdot}{F}}:}$$

$$\mathrm{H\cdot + H\cdot + \cdot\ddot{\underset{\cdot}{O}}: \longrightarrow H:\ddot{\underset{\overset{\cdot\cdot}{H}}{O}}:}$$

Two types of pairs of electrons are shown in these structures. **Bonding pairs** of electrons are shared between two atoms. **Lone** or **nonbonding pairs** of electrons are on one atom and are not shared. Drawings of this type are known as **Lewis structures,** representations of covalent bonding using Lewis symbols that show shared electrons as dots (or lines) between atoms and unshared electrons as dots.

$$\mathrm{H:\ddot{\underset{\overset{\cdot\cdot}{H}}{O}}:}$$

bonding pairs · lone pairs

Frequently, a line is used to indicate a bonding pair in Lewis structures instead of the two dots.

$$\mathrm{H-\ddot{\underset{\overset{|}{H}}{O}}:}$$

Lewis structures are an easily understood picture that works well to account for the bonding in most molecules. Remember that *only valence electrons are shown in Lewis structures.*

Octet Rule

The noble gases are particularly unreactive, indicating that their electron arrangements are particularly stable. We have seen that the ions in ionic compounds frequently are isoelectronic with the noble gases. Similarly, covalent bonds form until the atoms achieve a noble gas electron configuration by *sharing* electrons. For hydrogen, only two electrons are needed to achieve a noble gas electron configuration, but for all other representative elements eight electrons are needed to fill an *s* and three *p* orbitals of the valence shell. The **octet rule** states that each atom in a molecule shares electrons until it is surrounded by eight valence electrons. Some of the electrons may be bonding electrons; some may be nonbonding, lone pair electrons. The octet rule is most useful for compounds of the second period. As presented later, there are many compounds of the elements in the third and higher periods for which the octet rule is not followed. Hydrogen, of course, will share only two electrons.

The octet rule is consistent with the fact that compounds formed from hydrogen with oxygen, nitrogen, and carbon have the formulas H_2O, NH_3, and CH_4.

```
                          H
   ..         ..          |
H—O:       H—N—H       H—C—H
   |          |           |
   H          H           H
```

Oxygen has six valence electrons and must share two more electrons with two hydrogen atoms to reach an octet. Nitrogen has five valence electrons and must share three more electrons, and carbon has four valence electrons and must share four more electrons to attain an octet. In most molecules that contain oxygen, each oxygen atom forms two covalent bonds to obtain an octet of valence electrons. In a similar manner, nitrogen usually forms three covalent bonds and carbon usually forms four covalent bonds.

Two atoms can share more than one pair of electrons between them. For example, consider molecular nitrogen, N_2. For both nitrogen atoms to attain an octet, each atom must share three additional electrons, a total of six for both.

$$:\dot{\underset{\cdot}{N}}\cdot + \cdot\dot{\underset{\cdot}{N}}: \longrightarrow :N:::N: \quad \text{or} \quad :N{\equiv}N:$$

The sharing of one pair of electrons is a **single bond,** the sharing of two pairs is a **double bond,** and the sharing of three pairs is a **triple bond.** An example of a compound that contains double bonds is CO_2. In this case each oxygen atom shares two electron pairs with the carbon atom. These four bonds also place an octet of electrons around the carbon atom.

$$:\dot{\underset{\cdot}{O}}: + \cdot\dot{\underset{\cdot}{C}}\cdot + :\dot{\underset{\cdot}{O}}: \longrightarrow \ddot{\underset{\cdot\cdot}{O}}::C::\ddot{\underset{\cdot\cdot}{O}} \quad \text{or} \quad \ddot{\underset{\cdot\cdot}{O}}{=}C{=}\ddot{\underset{\cdot\cdot}{O}}$$

The **bond order** is the number of electron pairs that are shared between two atoms. The bond order in each carbon-oxygen bond in CO_2 is two; the bond order in N_2 is three. As the bond order between two atoms increases, the average bond distance decreases. As an example, the average bond distances measured for compounds that contain single or double bonds between two nitrogen atoms and the triple bond distance in N_2 are

N—N	N=N	N≡N
147 pm	125 pm	110 pm

Writing Lewis Structures

The Lewis structure of a molecule is very important because it describes the bonding. To write a correct structure, we need to know the formula of the compound and which atoms are connected—that is, which atoms are sharing electrons. For water, H_2O, the two hydrogen atoms are bonded to the oxygen atom, but not to each other. For this simple molecule, this bonding arrangement is not hard to figure out because each hydrogen atom can make only one bond. In other cases, the connectivity is a more difficult problem that has to be solved by experiment. The **skeleton structure** shows which atoms are bonded to each other in a molecule. In the skeleton structure each connection represents *at least* a single bond. In this text the skeleton structure will generally be given, but experience and a few rules frequently make the assignment possible.

For simple molecules, a **central atom,** an atom bonded to two or more other atoms, is generally written first in the molecular formula. Hydrogen and fluorine must always be *terminal* (end) atoms, because these atoms form only one covalent bond. The same is generally true of the other halogens (exceptions are covered in Section 8.7). For compounds with more than one central atom, the molecular formula is frequently written to indicate the connectivity and help in writing the skeleton structure. For example, ethanol is generally written as CH_3CH_2OH to show that the skeleton structure is

```
    H   H
    |   |
H—C—C—O—H
    |   |
    H   H
```

Rules for Writing Lewis Structures

1. Write the skeleton structure.
2. Count the valence electrons of all of the atoms in the molecule. For polyatomic ions, subtract one electron for each positive charge or add one electron for each negative charge.
3. For each bond given in the skeleton structure, subtract two electrons from the total number of valence electrons. The result is the number of electrons that remain to complete the Lewis structure.
4. Calculate the number of unshared electrons needed to satisfy the octet rule for each atom (remember that hydrogen needs only two). If the number of electrons *needed* to satisfy the octet rule is equal to the number of electrons that still *remain* (step 3), finish the structure with lone pairs.
5. If there are fewer valence electrons remaining (step 3) than needed to satisfy the octet rule (step 4), multiple bonds are needed. Add one additional bond for every two additional electrons needed to satisfy the octet rule. Make the appropriate number of multiple bonds, and finish the structure with lone pairs.

Lewis structures are written by arranging the valence electrons as bond pairs and lone pairs to place an octet of electrons around each atom (two for hydrogen).

These rules give an organized method to arrange the available valence electrons so there is an octet about each atom. After writing a Lewis structure, check to make sure that the structure shows the *correct number of valence electrons.* Note that in step 5, using a pair of electrons as a bond pair rather than a lone pair effectively makes those two electrons count as four, since a shared pair of electrons is counted as part of the octet for *both* of the sharing atoms. In deciding where to place multiple bonds, be careful to place no more than eight electrons around an atom from the second period, and only two electrons on each hydrogen atom.

These rules can be applied to water to obtain the Lewis structure which was given earlier.

1. Only one skeleton structure is possible. The structure is pictured as "V"-shaped because, as will be shown in Section 9.1, that is the shape of the water molecule.

```
H—O
  |
  H
```

2. The total number of valence electrons is

$$\begin{array}{ll} 1(\text{O}) & 1 \times 6 = 6 \\ 2(\text{H}) & 2 \times 1 = \underline{2} \\ & \qquad\quad 8 \end{array}$$

3. The skeleton structure shows two bonds. Each bond uses two electrons for a total of four.

$$\begin{array}{r} \text{total number of valence electrons} = 8 \\ -\ \text{electrons used in skeleton structure} = \underline{4} \\ \text{remaining valence electrons} = 4 \end{array}$$

4. To obey the octet rule, the oxygen atom needs four unshared electrons.

```
H—O ←— needs 4e⁻ to complete octet
  |
  H
```

Four is the number remaining, so satisfy the octet rule with lone pairs.

```
   ..
H—O:
  |
  H
```

This Lewis structure is, of course, the same as that shown earlier.

Example 8.4 The Lewis Structure of Methanol

Write the Lewis structure of methanol, CH_3OH.

Solution

1. The skeleton structure is indicated by the way the formula has been written: three of the hydrogen atoms are bonded to carbon and one to oxygen, and these two central atoms are also bonded to each other.

```
  H
  |
H—C—O—H
  |
  H
```

2. The total number of valence electrons is

$$\begin{array}{ll} 1(\text{C}) & 1 \times 4 = \ \ 4 \\ 4(\text{H}) & 4 \times 1 = \ \ 4 \\ 1(\text{O}) & 1 \times 6 = \ \ \underline{6} \\ & \qquad\quad 14 \end{array}$$

3. The skeleton structure shows five bonds (3C—H, C—O, O—H). Each of these bonds uses two electrons, for a total of ten.

total number of valence electrons = 14
− electrons used in skeleton structure = 10
remaining valence electrons = 4

4. Calculate the number of electrons needed to satisfy the octet rule. The carbon atom has four bonds and thus already satisfies the octet rule. The oxygen atom has two bonds and requires two lone pairs to complete an octet.

```
      H
      |
  H—C—O—H
      |
      H
```

needs $0e^-$, $4e^-$ to complete octet

The four electrons needed are available, as calculated in step 3, and are used to finish the Lewis structure.

```
     H
     |   ..
  H—C—O—H
     |   ..
     H
```

Always check that each atom has an octet (hydrogen atoms need only two) and that the final Lewis structure has the correct number of valence electrons. For methanol, there are five bonding pairs and two lone pairs of electrons for a total of 14 electrons, the same as calculated in step 2.

Understanding Write the Lewis structure of dimethyl ether, CH_3OCH_3.

Answer:

```
     H       H
     |   ..  |
  H—C—O—C—H
     |   ..  |
     H       H
```

Example 8.5 Lewis Structure of Ethylene

Write the Lewis structure of ethylene, CH_2CH_2.

Solution

1. The skeleton structure is given by the formula

```
  H       H
   \     /
    C—C
   /     \
  H       H
```

2. The total number of valence electrons is

2(C)	2 × 4 =	8
4(H)	4 × 1 =	4
		12

3. The skeleton structure shows five bonds using two electrons each, for a total of ten.

total number of valence electrons = 12
− electrons used in skeleton structure = 10
remaining valence electrons = 2

4. Each carbon atom has three bonds and thus would need one lone pair each to finish the structure.

```
H       H
 \     /
  C—C
 /  ↑  ↑ \
H         H
```

needs $2e^-$, $2e^-$ to complete octet

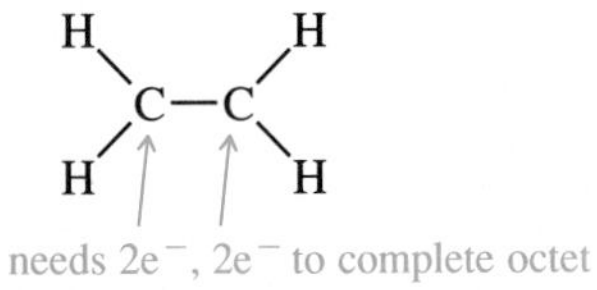

Four electrons are needed, but only two electrons remain in step 3. Multiple bonds are needed.

5. We have one pair of electrons too few to finish the structure with lone pairs. One double bond is needed to finish the structure. Clearly, it is placed between the carbon atoms.

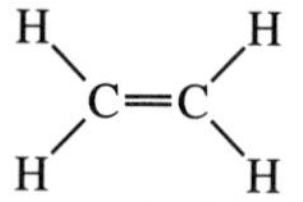

```
H       H
 \     /
  C=C
 /     \
H       H
```

The Lewis structure shows six bonds and accounts for all 12 valence electrons calculated in step 2. The hydrogen atoms each share two electrons and the carbon atoms each share eight electrons, so the structure is complete.

Multiple bonds are used in the Lewis structures of molecules that lack enough electrons to complete an octet around each atom using lone pairs.

Understanding Write the Lewis structure of acetone, C_3H_6O. The skeleton structure is

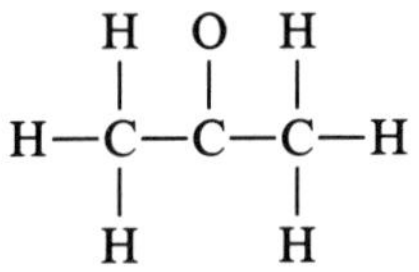

```
  H  O  H
  |  |  |
H—C—C—C—H
  |     |
  H     H
```

Answer:

```
  H :O: H
  |  ‖  |
H—C—C—C—H
  |     |
  H     H
```

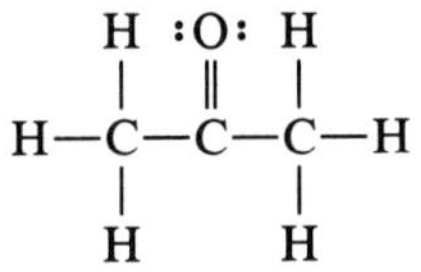

Example 8.6 Lewis Structure of Acetonitrile

Write the Lewis structure of CH_3CN.

Solution

1. The skeleton structure can be deduced from the formula.

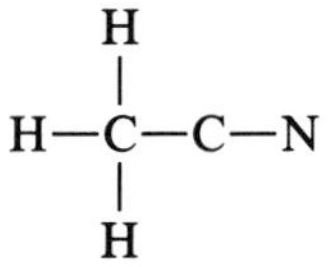

```
  H
  |
H—C—C—N
  |
  H
```

2. The total number of valence electrons is

$$\begin{array}{lll} 3(H) & 3 \times 1 = & 3 \\ 2(C) & 2 \times 4 = & 8 \\ 1(N) & 1 \times 5 = & \underline{5} \\ & & 16 \end{array}$$

3. The skeleton structure shows five bonds with two electrons each, for a total of ten.

$$\begin{array}{rr} \text{total number of valence electrons} = & 16 \\ -\text{ electrons used in skeleton structure} = & \underline{10} \\ \text{remaining valence electrons} = & 6 \end{array}$$

4. The first carbon atom has four bonds and needs no additional electrons for an octet. The second carbon atom has two bonds and would need two lone pairs, and the nitrogen atom has only one bond and would need three lone pairs to obtain an octet of electrons.

```
      H
      |
  H—C—C—N
      |
      H
```

needs $0e^-$, $4e^-$, $6e^-$ to complete octet

Thus, ten electrons (five lone pairs) are needed to finish the structure with lone pairs, but only six remain, so multiple bonds will be necessary.

5. We need two additional pairs of electrons to satisfy the octet rule, so two additional bonds are needed to complete the structure. Both additional bonds are placed between the carbon and nitrogen atoms, because any other arrangement would place more than an octet of electrons around one of the atoms.

```
      H
      |
  H—C—C≡N
      |
      H
```

Seven bonds are now present in the structure, using 14 of the 16 valence electrons. The remaining two electrons are placed on the nitrogen atom to complete its octet.

```
      H
      |
  H—C—C≡N:
      |
      H
```

Understanding Write the Lewis structure of allene, H_2CCCH_2.

Answer:

```
  H         H
   \       /
    C=C=C
   /       \
  H         H
```

8.4 Electronegativity

Objectives

- To define the electronegativity scale
- To use electronegativity differences to predict bond polarities

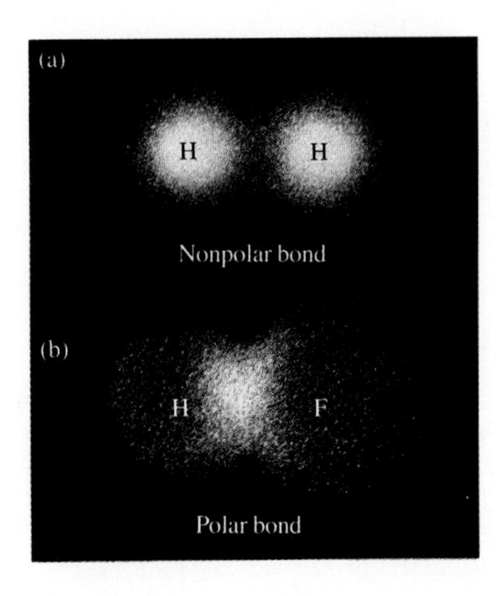

Figure 8.7 Electron density plots of the bonding electrons in H_2 and HF (a) The electrons in H_2 are shared equally. (b) The shared pair of electrons in HF is more strongly attracted to the fluorine atom.

The bonding description that we have given for ionic materials, in which electrons are transferred completely from one atom to another, and for molecular hydrogen, in which the two electrons are equally shared (Figure 8.7a), are really two extreme cases. When different atoms share electrons, the sharing is not exactly equal. The shared pair of electrons will spend more time around one nucleus than the other.

A **polar bond** is a covalent bond in which the bonding electrons are not shared equally by the two atoms. Figure 8.7b shows an electron density plot of the bonding pair of electrons in HF. The bonding pair of electrons spends more time, on average, around the fluorine atom than around the hydrogen atom because the fluorine atom attracts the shared electrons more strongly than does the hydrogen atom. This result is expected on the basis of the greater ionization energy of fluorine than of hydrogen.

The unequal sharing causes an imbalance in the electron distribution of the bonded atoms, which leaves a partial negative charge around the fluorine atom and a partial positive charge around the hydrogen atom. This *partial* charge is something less than the whole units of charge observed on ions. This charge separation is indicated as

$$\overset{\delta+}{\mathrm{H}}-\overset{\delta-}{\mathrm{F}}$$

where the δ indicates a partial charge. It is important to remember that the bond in HF is still covalent, but the sharing is not exactly even.

Electronegativity is a measure of the ability of an atom to attract the shared electrons in a chemical bond. Electronegativity of atoms correlates with properties such as ionization energy and electron affinity. Elements with a low ionization energy will have a low electronegativity and elements with a high ionization energy will have a high electronegativity. A number of methods have been suggested for calculating an electronegativity scale, which is a difficult problem because the electronegativity of an element differs slightly depending on the other elements to which it is bonded. Linus Pauling first introduced the concept of electronegativity and developed a widely used method for calculating a relative electronegativity scale. On this scale, fluorine, the most electronegative element, has a value of 4.0. Electro-

H 2.1						
Li 1.0	Be 1.5	B 2.0	C 2.5	N 3.0	O 3.5	F 4.0
Na 0.9	Mg 1.2	Al 1.5	Si 1.8	P 2.1	S 2.5	Cl 3.0
K 0.8	Ca 1.0	Ga 1.8	Ge 1.8	As 2.0	Se 2.4	Br 2.8
Rb 0.8	Sr 0.9	In 1.7	Sn 1.8	Sb 1.9	Te 2.1	I 2.5
Cs 0.7	Ba 0.9	Tl 1.8	Pb 1.9	Bi 1.9	Po 2.0	At 2.2

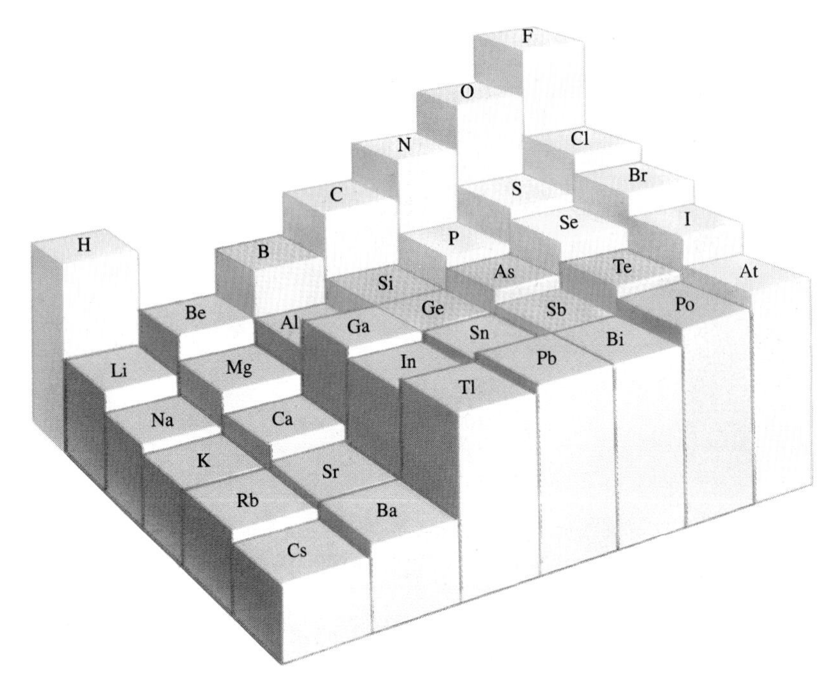

Figure 8.8 Electronegativities of representative elements

Elements with high ionization energies, those in the top right of the periodic table, have the highest electronegativities.

negativity values decrease down the periodic table and also to the left across the table (Figure 8.8). Oxygen has the second-highest electronegativity at 3.5, and nitrogen and chlorine tie for third at 3.0. Carbon and hydrogen have intermediate electronegativities of 2.5 and 2.1, respectively.

The difference in electronegativity between two covalently bonded atoms determines the *polarity,* the degree of charge separation, of the bond. The polarity in a bond is generally represented by an arrow pointing toward the more electronegative atom.

$$\overset{+\!\!\longrightarrow}{\text{H—F}} \qquad \overset{\leftrightarrow}{\text{H—Br}}$$

The HF bond is more polar than that in HBr because the electronegativity difference is larger. The length of the arrow indicates the electronegativity difference and thus the polarity of the bond. The base of the arrow is crossed to emphasize that this is the positive end.

Example 8.7 **Polarity of Bonds**

In each of the following pairs of bonds, which one is more polar? Show the direction of the polarity for the more polar bond.

(a) S—O and S—F
(b) H—C and H—N
(c) O—C and F—N

Solution

(a) Fluorine is more electronegative than oxygen, and both are more electronegative than sulfur. The S—F bond is more polar because of the greater difference in electronegativity. The polarity is

$$\overset{\longleftrightarrow}{\text{S—F}}$$

(b) Nitrogen is more electronegative than carbon. The H—N bond is more polar. The polarity is

$$\overset{\longleftrightarrow}{\text{H—N}}$$

(c) We need to look at Figure 8.8 to calculate the electronegativity difference in each bond. For O—C the electronegativity difference is $3.5 - 2.5 = 1.0$, and for F—N it is $4.0 - 3.0 = 1.0$. The bonds are of equal polarity. In a molecule, there may be slight differences due to the influence of other bonded atoms.

$$\overset{\longleftrightarrow}{\text{C—O}} \qquad \overset{\longleftrightarrow}{\text{N—F}}$$

Understanding Which bond is more polar, P—S or As—S?

Answer: As—S

Polar bonds are intermediate in nature between the two limiting cases of nonpolar bonds (such as H_2) and ionic bonds (such as CsF). The greater the difference in electronegativity between the bonded atoms, the greater the polarity of the bond. There is no clean break between molecular compounds that have polar covalent bonds and ionic compounds. It has been pointed out that ionic compounds generally are composed of a metal and a nonmetal, but compounds such as $AlCl_3$ (1.5 difference) and $SnCl_2$ (1.2 difference) containing the more electronegative metals have significant covalent character. Generally, if the difference in electronegativity is greater than 1.7, the bond has significant ionic character. However, experimental measurements of properties such as the electrical conductance of the melted compound are the only sure way to determine whether the compound is ionic.

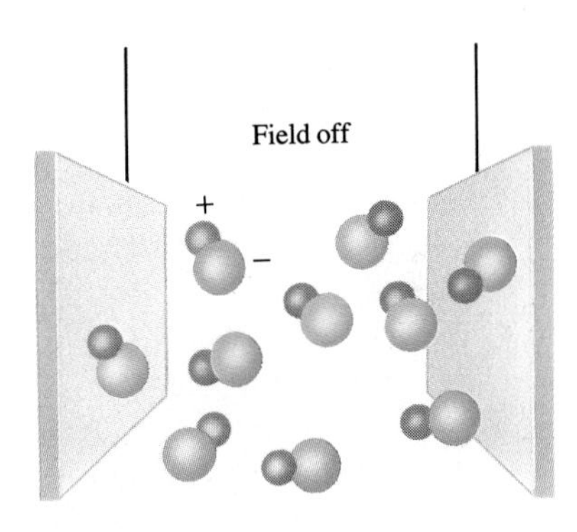

Dipole Moment

Polar diatomic molecules have positive and negative ends, and therefore are called dipoles. The **dipole moment** is the magnitude of the separated charge times the distance between the charges. The dipole moment is the measure of polarity.

The apparatus shown in Figure 8.9 is used to measure dipole moments. A sample is placed between two plates and an electric field is applied to the plates. Before the electric field is turned on, the sample molecules are randomly oriented. When the field is on, polar molecules will align their negative end toward the positive plate and their positive end toward the negative plate. This alignment increases the charge that the plates can hold, compared to the charge that can be held in the absence of polar molecules. This increase can be measured and the dipole moment calculated from it. The orientation of nonpolar molecules will not be affected by the field.

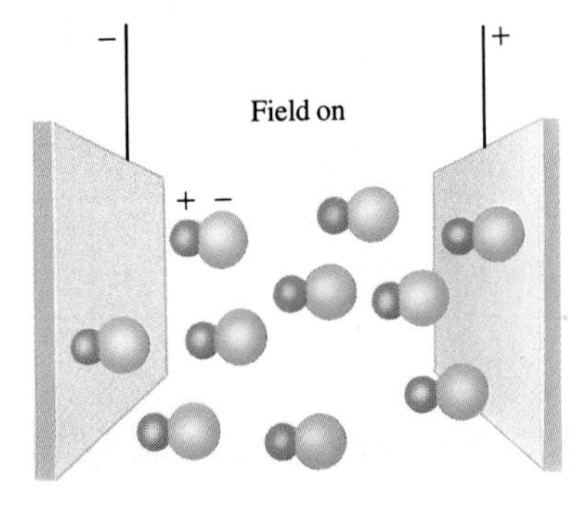

Figure 8.9 Measurement of dipole moments Polar molecules are randomly oriented when the field is off, but are aligned when the field is on.

INSIGHTS into CHEMISTRY

Development of Chemistry

Linus Pauling: Two Nobel Prizes for Work in Chemical Bonding and Nuclear Disarmament

Linus Carl Pauling (1901–) will be remembered as one of the most influential scientists of the twentieth century. His major achievements in chemistry have been in the fields of chemical bonding and molecular structure. He is one of the very few people who has received two Nobel Prizes, winning the prize in chemistry in 1954 and the Peace Prize in 1962. Throughout his career, he has been noted for his ability to take ideas and principles from one field of science and apply them to another.

Pauling was born in Portland, Oregon, on February 28, 1901. The son of a druggist, Pauling developed an early interest in chemistry. He graduated from Oregon State College in 1922 with a degree in chemical engineering. He then went to the California Institute of Technology in Pasadena, earning his Ph.D. in 1925. Like many of his generation, Pauling then went to Europe for postdoctoral studies. While there, he worked with Erwin Schrödinger and Niels Bohr and other pioneer scientists who were studying atomic and molecular structure and the new field of quantum mechanics. Upon his return to the United States in 1927, he accepted a faculty position at Cal Tech.

Linus Pauling

Pauling's early work was greatly influenced by his European studies. In 1931 he published a paper in which he applied quantum mechanics to explain how an electron-pair bond is formed by the interaction of two unpaired electrons, one from each of the two atoms. This was followed in 1935 by a book titled *Introduction to Quantum Mechanics,* which he co-authored with E. B. Wilson. He focused his attention on explaining and understanding the chemical bond. He introduced the concepts of electronegativity, resonance structures, and hybridization of atomic orbitals. These and other ideas were consolidated in his book *The Nature of the Chemical Bond,* published in 1939. Many consider this to have been the most influential chemistry book published in the twentieth century.

In the 1940s, Pauling turned his attention to biological systems. He applied his knowledge of molecular structure to proteins in the blood, particularly amino acids and polypeptides. He was among the first to propose that many proteins have structures held together by hydrogen bonds, giving them a helical shape. This same concept was used by Francis Crick and James Watson in their determination of the structure of DNA. Pauling's research group also discovered the abnormality in the molecular structure of hemoglobin associated with the genetic disease sickle cell anemia. In 1954 he was awarded the Nobel Prize in chemistry for his work on bonding and molecular structures.

After World War II, Pauling became concerned about the spread of nuclear weapons and their atmospheric testing. He worked actively for nuclear disarmament, and was a leader in obtaining a petition signed by more than 10,000 scientists urging an end to nuclear testing, which he presented to the United Nations. As a result of his efforts, he was awarded the Nobel Peace Prize in 1962. In recent years, Pauling has promoted the health benefits of large doses of vitamin C, especially in combatting the common cold and cancer. His views are controversial. He has published more than 1000 papers on science and politics, and is still actively pursuing his research in chemistry today (1992).

8.5 Formal Charges in Lewis Structures

Objectives

- To assign formal charges in Lewis structures
- To predict the relative stabilities of Lewis structures from formal charge

In all of the Lewis structures we have written, the bonds have been formed by the sharing of two electrons, with one electron coming from each atom.

Writing bonds in this manner has allowed the generalization that group VIA elements usually make two bonds to reach an octet of electrons, group VA elements make three bonds, and group IVA elements make four bonds. Not all Lewis structures can be completed by following this principle. An example is carbon monoxide, CO. Following the rules for writing the Lewis structure yields

:C≡O:

The structure places an octet of electrons about each atom and shows the correct number of valence electrons. The structure is different from those shown earlier, however, since the oxygen and carbon atoms make three bonds, not their normal numbers of two and four, respectively. This difference can be seen by separating the elements into their Lewis electron-dot symbols and assuming that the electrons in the bonds are divided equally between the two atoms.

·Ċ:⁻ ·Ȯ:⁺

Separating the shared electrons equally yields a Lewis symbol for carbon that has five valence electrons, two from the lone pair and three of the six electrons that formed the triple bond. Since carbon atoms have four valence electrons, this species is an anion. The Lewis symbol obtained for oxygen by separating the bonding electrons has only five valence electrons, compared to six for the atom, and is thus a cation. In order to indicate this bonding feature, the Lewis structure is written with **formal charge,** a charge assigned to atoms in Lewis structures by assuming the shared electrons are divided equally between the bonded atoms.

Formal charges are assigned to atoms in Lewis structures that make an unusual number of covalent bonds.

$\ominus$ $\oplus$
:C≡O:

The formal charges of neutral species must sum to zero. For charged species, the sum of the formal charges must be equal to the charge on the species.

Formal charges are assigned by assuming that bonding electrons are divided equally.

Most Lewis structures do not require any formal charges. Structures that require formal charges can be recognized by the number of bonds made by the elements. Carbon generally forms four bonds, nitrogen forms three bonds, oxygen forms two bonds, and fluorine forms one bond. If the Lewis structure requires more bonds than these numbers, the atom will have a positive formal charge. If it requires fewer bonds, then the atom will have a negative formal charge. For example, a nitrogen atom that forms four bonds will have a +1 formal charge, and one making only two bonds will have a −1 formal charge. Oxygen atoms with three bonds, as in CO, will have a +1 formal charge, and one making only one bond will have a −1 formal charge (Table 8.2).

The formal charge on an atom can also be calculated from the equation

formal charge = number of valence electrons of atom
− number of lone pair electrons − $\frac{1}{2}$ (number of shared electrons)

Applied to CO,

$$\text{formal charge of C} = 4 - 2 - \tfrac{1}{2}(6) = -1$$

$$\text{formal charge of O} = 6 - 2 - \tfrac{1}{2}(6) = +1$$

Table 8.2 Formal Charge

Species	Formal Charge	Species	Formal Charge
—N—	−1	—N— (four bonds)	+1
O—	−1	>O— (three bonds)	+1
—C— (three bonds)	−1		

Example 8.8 Formal Charge

Draw the Lewis structure of the ammonium cation, showing formal charges as needed.

Solution First, write the Lewis structures.

1. The skeleton structure of ammonium ion, NH_4^+, is

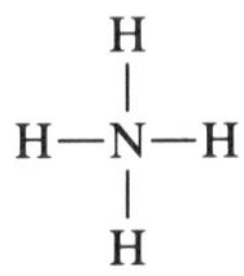

2. The total number of valence electrons, including the charge on the cation, is

$$\begin{array}{lll} 4(\text{H}) & 4 \times 1 = & 4 \\ 1(\text{N}) & 1 \times 5 = & 5 \\ 1+\text{ charge} & = & \underline{-1} \\ & & 8 \end{array}$$

3. The skeleton structure shows four bonds for a total of eight electrons. All of the valence electrons are used by the skeleton structure. The nitrogen atom has eight electrons around it and each hydrogen has the required two electrons—the Lewis structure is finished except for formal charge.

   Nitrogen has four bonds, so it has a +1 formal charge. Using the formula

   $$\text{formal charge} = 5 - 0 - \tfrac{1}{2}(8) = +1$$

The final Lewis structure is

```
     H
     | ⊕
  H—N—H
     |
     H
```

A final check shows the correct number of electrons around each atom and the correct number of valence electrons; finally, the formal charge is the charge of the ammonium ion.

Example 8.9 Formal Charge

Draw the Lewis structure of ozone, O_3, showing formal charges if needed.

Solution First write the Lewis structure.

1. The skeleton structure of ozone is

```
   O
  / \
 O   O
```

2. The total number of valence electrons is

   $$3(\text{O}) \quad 3 \times 6 = 18$$

3. The skeleton structure shows two bonds for a total of four electrons.

$$\begin{array}{rl} \text{total number of valence electrons} = & 18 \\ -\text{ electrons used in the skeleton structure} = & \underline{\ 4} \\ \text{remaining valence electrons} = & 14 \end{array}$$

4. To satisfy the octet rule, the center oxygen atom requires two lone pairs and those on the outside require three pairs each, for a total of 16 additional electrons.

O—O—O (O at apex bonded to two O atoms)

needs $6e^-$, $4e^-$, $6e^-$ to complete octet

Since only 14 electrons remain at step 3, one additional bond is needed.

5. The double bond can be placed between either of the pairs of bonded oxygen atoms.

O=O—O

The three bonds use six electrons. Complete the octet for each oxygen atom by using the remaining 12 electrons as lone pairs.

O(1)=O(2)—O(3) (with lone pairs)

Oxygen normally forms two bonds, and O(1) matches this description. However, O(2) forms three bonds and has a +1 formal charge, and O(3) forms one bond and has a −1 formal charge. Using the formula

$$\text{formal charge O(1)} = 6 - 4 - \tfrac{1}{2}(4) = 0$$

$$\text{formal charge O(2)} = 6 - 2 - \tfrac{1}{2}(6) = +1$$

$$\text{formal charge O(3)} = 6 - 6 - \tfrac{1}{2}(2) = -1$$

The correct Lewis structure is

O=O⊕—O⊖ (with lone pairs)

A final check shows that each oxygen atom has eight electrons, and the total number of electrons is the correct number of 18. The formal charges sum to the charge of the species, zero.

Understanding Assign formal charges to the atoms in N_2O, given the following Lewis structure.

:N=N=O:

Answer: :N⊖=N⊕=O:

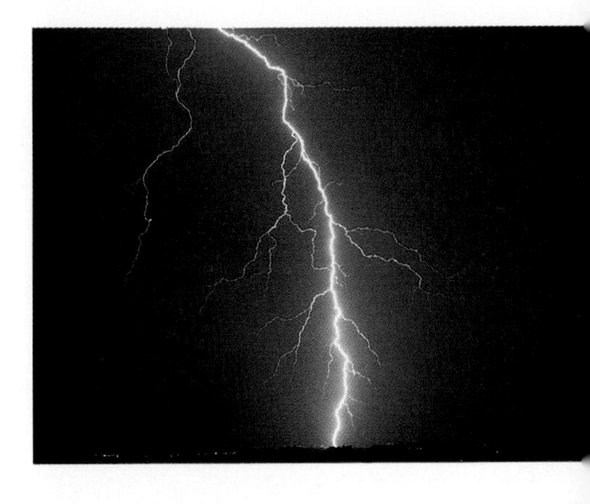

Ozone Ozone is formed by lightning and causes the odor associated with thunderstorms.

Formal Charges and Structure Stability

An important use of formal charges is to predict the relative stabilities of different Lewis structures. Comparisons of experimentally determined structures and Lewis structures allow several generalizations to be made.

1. Lewis structures that show the smallest formal charges are the most stable.
2. Lewis structures that have adjacent atoms with formal charges of the *same* sign are very unstable.
3. Lewis structures that place negative formal charges on the more electronegative atoms are more stable.
4. Opposite formal charges are usually on adjacent atoms.

Lewis structures that minimize formal charge are the best representations of molecular structures.

The use of these rules to predict the correct arrangement of atoms from the molecular formula is shown in Example 8.10.

Example 8.10 Structure of Hydrogen Cyanide

There are two possible ways to connect the atoms in hydrogen cyanide, HCN and HNC. Evaluate the formal charge to predict which is the more likely structure.

H—C—N   H—N—C

A   B

Solution The Lewis structure for arrangement A is

H—C≡N:

No formal charges are needed. The carbon atom forms four bonds and the nitrogen atom forms three.

The Lewis structure of B is

H—N⊕≡C⊖:

In this structure, a positive formal charge is assigned to the nitrogen atom because it forms four bonds, and a negative formal charge is on the carbon atom because it forms three bonds. Since Lewis structures that show the smallest formal charge are most stable, structure A is favored and is the structure actually observed for this molecule.

Understanding Use formal charge to predict which of the following structures is more likely to be correct for a molecule with the formula H_2CO.

A: H_2C—O (both H bonded to C)   B: H—C—O—H

Answer: A

8.6 Resonance in Lewis Structures

Objectives

- To write all possible resonance forms for a given molecule
- To predict the importance of different resonance structures

Some molecules or ions do not have one unique Lewis structure. An example is ozone, O_3, whose Lewis structure was illustrated in Example 8.9.

O(1)=O(2)⊕—O(3)⊖

When we wrote this Lewis structure, the placement of the double bond between atoms 1 and 2 was arbitrary. A second, equally correct structure places the double bond between atoms 2 and 3.

⊖O(1)—O(2)⊕=O(3)

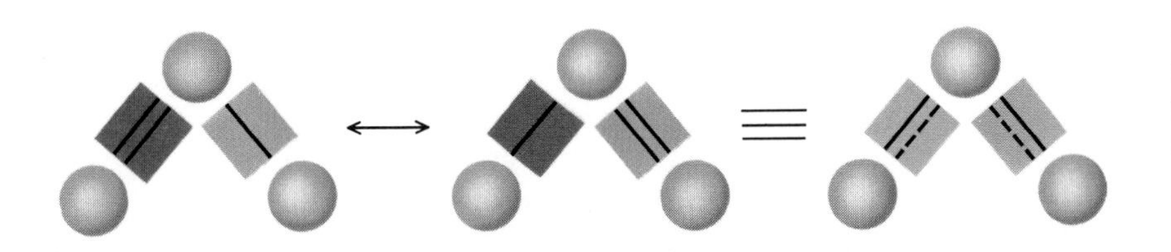

Figure 8.10 Averaging of resonance forms The actual bonding in cases with resonance is an average of the various resonance forms.

The two equally acceptable Lewis structures for O_3 are an example of resonance. **Resonance** is the use of two or more Lewis structures that differ only in the distribution of the valence electrons in representing the bonding in a species. The skeleton structure does *not* change; only the placement of bonds is different. We indicate resonance structures with a double-headed arrow.

$$\mathrm{O_{(1)}{=}\overset{\oplus}{O}_{(2)}{-}O_{(3)}^{\ominus} \longleftrightarrow {}^{\ominus}O_{(1)}{-}\overset{\oplus}{O}_{(2)}{=}O_{(3)}}$$

The two Lewis structures are called *resonance structures* or *resonance forms*. The two resonance forms for O_3 have the same types of bonds and are thus equivalent and of equal energy. Each resonance form has one O—O single bond and one O=O double bond, but each predicts different bond orders for the O—O bonds. Which one does experiment show is correct? The answer is *neither;* experiment shows that the O—O bond order is the same for both bonds and is about 1.5. Neither resonance form is correct by itself; the correct structure is an *average* of the two resonance forms. The bonding in molecules that have resonance forms does not bounce back and forth between the various resonance forms, but is the average of all forms as illustrated in Figure 8.10. Frequently an averaged structure is written in which the bond order of 1.5 is shown as a solid line and a dashed line.

More than one acceptable Lewis structure, known as resonance forms, can be written for many compounds.

One of the most important examples of resonance occurs in benzene, as shown in Example 8.11.

Example 8.11 Resonance Structures of Benzene

Write the resonance forms of benzene, C_6H_6.

Solution

1. The skeleton structure of this molecule is a six-membered ring of carbon atoms, with one hydrogen atom bonded to each carbon atom:

```
         H
         |
   H    C    H
    \ /   \ /
     C     C
     |     |
     C     C
    / \   / \
   H    C    H
        |
        H
```

2. The total number of valence electrons is

$$\begin{array}{ll} 6(\mathrm{C}) & 6 \times 4 = 24 \\ 6(\mathrm{H}) & 6 \times 1 = \underline{\ 6} \\ & \qquad\ \ 30 \end{array}$$

3. The skeleton structure has six C—C bonds and six C—H bonds, using 24 of these electrons.

$$\begin{aligned} \text{total number of valence electrons} &= 30 \\ -\text{ electrons used in the skeleton structure} &= \underline{24} \\ \text{remaining valence electrons} &= \ \ 6 \end{aligned}$$

4. In the skeleton structure of benzene, each carbon atom forms three bonds and thus shares six electrons. Each carbon atom would need an additional lone pair to satisfy the octet rule, a total of 12 electrons. Only six electrons remain.

5. Six additional electrons are needed to finish the structure with lone pairs, so three additional bonds are needed. There are two ways to put these bonds in the cyclic structure while giving each carbon atom four bonds:

```
        H                      H
        |                      |
   H    C       H         H    C       H
    \ //  \    /           \  /  \\   /
     C      C               C      C
     |      ||     <-->     ||     |
     C      C               C      C
    / \\  /  \             /  \  //  \
   H    C      H          H     C      H
        |                       |
        H                       H
```

Both structures complete the octet around each carbon atom and show the correct number of valence electrons. They are both equally satisfactory Lewis structures. The actual structure of the molecule is the average of these two resonance structures.

Understanding Write all the resonance forms for SO_2, a molecule with O—S—O connectivity.

Answer: $\ddot{\underset{..}{O}}=\overset{\oplus}{\underset{..}{S}}-\ddot{\underset{..}{O}}\!:^{\ominus} \longleftrightarrow {}^{\ominus}\!:\ddot{\underset{..}{O}}-\overset{\oplus}{\underset{..}{S}}=\ddot{\underset{..}{O}}$

Example 8.11 shows that there are two resonance forms for benzene, both of which are equivalent in energy. They differ in that the bonding between any two carbon atoms is a double bond in one structure and a single bond in the other. This structure is frequently written with a dotted circle to indicate that the true structure is an average of the two resonance forms.

```
        H                      H                      H
        |                      |                      |
   H    C       H         H    C       H         H    C       H
    \ //  \    /           \  /  \\   /           \  /  \    /
     C      C               C      C               C  .-.  C
     |      ||     <-->     ||     |       ≡       | (   ) |
     C      C               C      C               C  '-'  C
    / \\  /  \             /  \  //  \            /  \  /   \
   H    C      H          H     C      H         H    C      H
        |                       |                     |
        H                       H                     H
```

All of the C—C bond distances in benzene are found experimentally to be equal, and about the length expected for a bond order of 1.5.

Scanning tunneling microscope view of benzene The ring structure of benzene can be observed directly by a scanning tunneling microscope.

Species with Nonequivalent Resonance Forms

In the examples of resonance that have been shown so far, the resonance forms were of equal energy. Resonance forms are not always of equal energy, and it is important to determine which resonance form or forms best describe the actual bonding. Formal charge can be used to predict which resonance forms are the most important. The resonance forms of nitric acid, HNO_3, are a good example. There are three acceptable Lewis structures for this molecule, given the known skeletal structure. Two of the resonance forms are equivalent in energy (A and B), but the third form (C) is not.

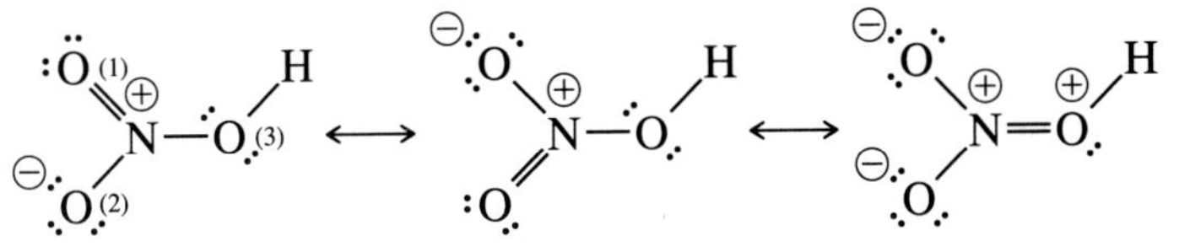

A B C

In all three of these structures, each atom has an octet of electrons (hydrogen has two), and each structure has the correct number of valence electrons, 24. Resonance forms A and B are equivalent in energy, but C is not. We can predict that structures A and B are better than C, because structure C shows more formal charges than A and B. Structure C also places formal charges of the same sign on adjacent atoms. We can say that resonance form C does not "contribute" significantly to the bonding; an average of resonance forms A and B best describes the bonding in nitric acid. This conclusion is verified by experiment, which shows that the N—O bond lengths are shorter to both O(1) and O(2) than O(3), consistent with a higher bond order.

Formal charge can be used to determine which resonance structures are most important.

Example 8.12 Resonance Structures of $(SCN)^-$

Write the resonance structures for thiocyanate anion, $(SCN)^-$, and indicate the relative contribution of each.

Solution

1. The connectivity is S—C—N.
2. The total number of valence electrons (remembering to add one for the overall negative charge) is 16.
3. The two bonds use four electrons, leaving 12 to finish the structure.
4. The nitrogen and sulfur atoms would each need 6 electrons and the carbon atom 4 electrons to satisfy the octet rule—a total of 16.

S—C—N

needs $6e^-$, $4e^-$, $6e^-$ to complete octet

5. Two additional bonds are needed to finish the structure. They can be placed in three different ways.

S=C=N or S≡C—N or S—C≡N

The Lewis structures for these possibilities are

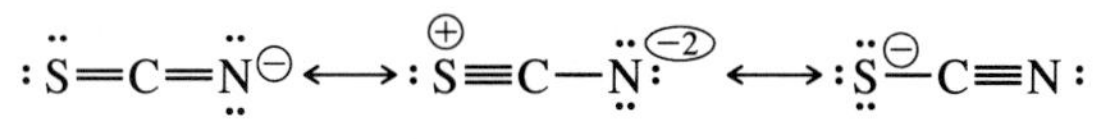

$:\ddot{S}=C=\ddot{N}^{\ominus} \longleftrightarrow :S^{\oplus}\equiv C-\ddot{N}:^{(-2)} \longleftrightarrow :\ddot{S}^{\ominus}-C\equiv N:$

Each of these structures has 16 valence electrons and an octet around each atom. Also, the sum of the formal charges is 1−, the charge on the thiocyanate anion, in each structure. Each of these resonance forms is different. The center structure with −2 and +1 formal charges is the least favorable because it does not minimize the formal charges. We can conclude that this resonance form does not make any significant contribution to the bonding in this anion. Both of the other two structures are stable and are important resonance forms for thiocyanate. However, the first structure is favored because the negative formal charge is on the more electronegative element, nitrogen. Thus, the first structure contributes more to the overall bonding than the third, but both are important.

Understanding Write all of the resonance forms for S_2O where one sulfur atom is the central atom. Indicate the relative contribution of each.

Answer: $\ddot{\underset{..}{S}}=\overset{\oplus}{\underset{..}{S}}-\ddot{\underset{..}{O}}:^{\ominus} \longleftrightarrow \overset{\ominus}{:}\ddot{\underset{..}{S}}-\overset{\oplus}{\underset{..}{S}}=\ddot{O}$ The first structure contributes more.

8.7 Molecules That Do Not Satisfy the Octet Rule

Objective

- To determine the types of molecules for which the octet rule is not obeyed

The Lewis structures of most species place eight electrons around each atom (two around hydrogen). There are three classes of molecules that do not obey the octet rule.

Electron Deficient Molecules

Elements of groups IIA (2) and IIIA (3) have only two and three valence electrons, respectively; not enough to make four electron-pair bonds. Many compounds of these elements, especially those of beryllium, boron, and aluminum, form compounds that are labeled as **electron deficient,** compounds for which the Lewis structures do not place eight electrons around the central atom. An example is BeH_2, an unstable molecule that has been observed in the gas phase. The Lewis structure of BeH_2 is

$$H-Be-H$$

Beryllium uses its two valence electrons to make electron-pair bonds with the two hydrogen atoms. Additional bonds cannot be made because the two bonds in the skeleton structure use all four of the available valence electrons. The same is true for BH_3, another unstable molecule that has been observed in the gas phase. The boron atom uses its three valence electrons, combined with the single electrons on the three hydrogen atoms, to form the three covalent bonds.

```
     H
    /
H—B
    \
     H
```

Elements in groups IIA and IIIA form compounds that are electron deficient because of a shortage of valence electrons.

The beryllium atom in BeH_2 and the boron atom in BH_3 are electron deficient, because there are only four and six electrons, respectively, around the central atom. The source of the electron deficiency is simply that elements in these groups do not have enough valence electrons to form four electron-pair bonds with other atoms.

Because these compounds are electron deficient, they generally associate with themselves or with other species. In the solid phase BeH_2 associates to form $(BeH_2)_n$, and the stable form of BH_3 is B_2H_6.

Example 8.13 Electron Deficient Molecules

Draw the Lewis structure of AlH_3. As with BH_3, this molecule has been observed at low pressure in the gas phase, but is generally observed in the associated form Al_2H_6.

Solution

1. The skeleton structure has to have the hydrogen atoms bonded to the central aluminum atom.
2. The total number of valence electrons is six.
3. The bonds of the skeleton structure use all six valence electrons, so the aluminum is electron deficient.

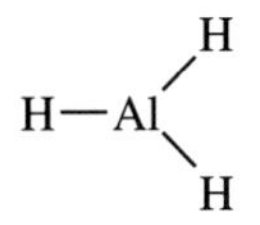

H—Al(—H)(—H)

The halides of beryllium, boron, and aluminum are also frequently given as examples of electron deficient molecules. The Lewis structure of BF_3 is typical.

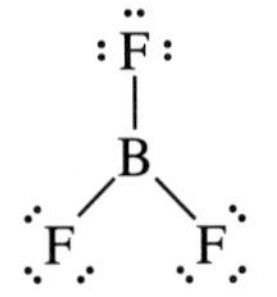

A

In this case, additional valence electrons (the fluorine lone pairs) are available, and following the rules for writing Lewis structures would give three equivalent resonance structures, each with one double bond.

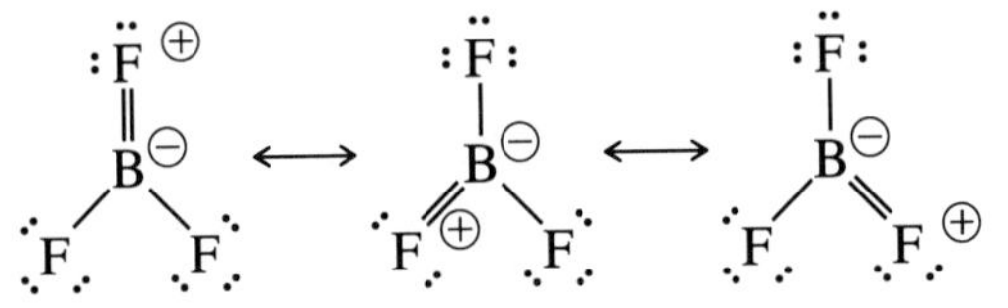

B ⟷ C ⟷ D

It has been argued that the electron deficient structure is preferred because these three resonance structures put a positive formal charge on fluorine, the more electronegative element. The best present experimental evidence is that all four resonance forms contribute.

Coordinate Covalent Bonds

So far, all covalent bonds have been formed by pairing an electron from one species with an electron from a second species. A **coordinate covalent bond** is a covalent bond in which both the electrons of the bonding pair come from one atom. A coordinate covalent bond is also frequently referred to as a *dative bond.* A coordinate covalent bond can be used to satisfy the octet rule for electron deficient molecules. For example, the lone pair on ammonia forms a coordinate covalent bond to boron in BH_3, forming a compound in which both the nitrogen and boron atoms have complete octets.

$$\mathrm{H_3\overset{\oplus}{N}{-}\overset{\ominus}{B}H_3}$$

Once formed, a coordinate covalent bond is no different from any other electron-pair bond, but it is a useful way to describe the formation of covalent bonds in certain cases. Coordinate covalent bonds will be very important in our later study of the chemistry of metals. An arrow pointing from the electron-pair donor to the acceptor is sometimes used to indicate a coordinate covalent bond.

$$\mathrm{H_3N{\rightarrow}BH_3}$$

Electron deficient compounds can attain an octet of electrons by forming coordinate covalent bonds.

If an arrow is used to indicate the coordinate covalent bond, no formal charges are shown.

Odd Electron Molecules

Any molecule that has an odd number of valence electrons must violate the octet rule. An example is NO, a molecule with 11 valence electrons. Two resonance forms for NO are

$$:\dot{N}{=}\ddot{O}: \longleftrightarrow :\overset{\ominus}{\ddot{N}}{=}\overset{\oplus}{\dot{O}}:$$

A B

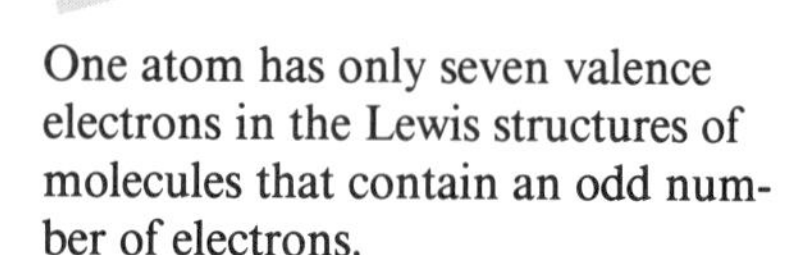

One atom has only seven valence electrons in the Lewis structures of molecules that contain an odd number of electrons.

Other resonance structures that place nine electrons about either atom are not important, because a second-period element has only four valence shell orbitals that can hold only eight electrons. The nitrogen atom in form A has only seven electrons, and the oxygen atom in form B has only seven electrons. Of the two resonance forms, A is the more important contributing form because it does not have any formal charge.

Odd electron molecules are often very reactive and form even electron species that obey the octet rule. For example, when gaseous NO is frozen into a solid, N_2O_2 forms with the Lewis structure

$$:\ddot{O}{=}\ddot{N}{-}\ddot{N}{=}\ddot{O}:$$

The formation of N_2O_2 from NO can be viewed as the pairing of the two unpaired electrons on the nitrogen atoms in resonance form A of two NO molecules.

$$:\ddot{O}{=}\ddot{N}\cdot + \cdot\ddot{N}{=}\ddot{O}: \longrightarrow :\ddot{O}{=}\ddot{N}{-}\ddot{N}{=}\ddot{O}:$$

Electron Rich Molecules

Many compounds exist in which the number of electrons available to draw the Lewis structure exceeds the number needed to satisfy the octet rule. In these cases, the excess electrons are placed on the central atom, giving it more than an octet of electrons. An example is PF_5. In this molecule, the five fluorine atoms are bonded to phosphorus, placing ten electrons about the phosphorus atom in the Lewis structure.

```
     ..
    :F: ..
     | / F:
 ..  |/  ''
:F — P
 ''  |\  ..
     | \ F:
    :F: ''
     ''
```

The molecule is **electron rich,** having more than eight electrons about an atom in a Lewis structure. A large class of compounds, in which the central atoms are electron rich, have the general formula YF_n where Y = P, S, Cl, As, Se, Br, Te, I, or Xe. In some cases, a fluorine atom can be replaced by another halogen or an oxygen atom. To handle writing Lewis structures for these compounds an additional rule, rule 6, is added to the five rules given in Section 8.3.

6. When *more* electrons are available than are needed to satisfy the octet rule for all atoms present, place the extra electrons around the central atom (it must be from the third or later row of the periodic table).

The octet rule can be exceeded for elements in the third and later periods, but not for elements in the second period. The octet rule is based on the idea that the valence *s* and *p* subshells can hold eight electrons. For atoms in the third and later periods, there are *d* orbitals that can hold additional electrons, exceeding an octet. In the second period there are no valence shell *d* orbitals, and the octet is never exceeded in stable species.

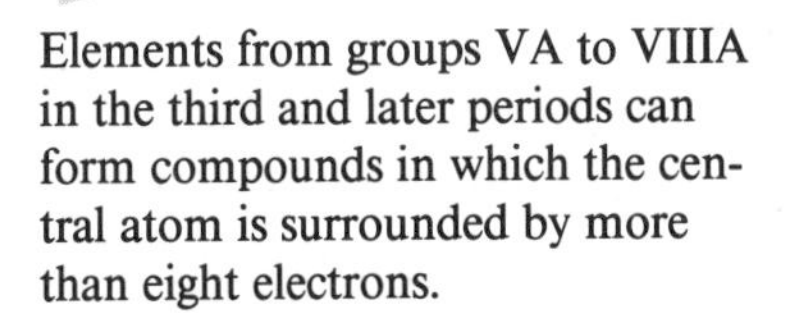
Elements from groups VA to VIIIA in the third and later periods can form compounds in which the central atom is surrounded by more than eight electrons.

Example 8.14 Electron Rich Compounds

Write the Lewis structure for XeF_4.

Solution Although xenon is a noble gas, it forms a number of compounds. In XeF_4, four of the eight xenon valence electrons are used to form the four bonds with the fluorine atoms, leaving two lone pairs on the central xenon atom:

```
 :F:   :F:
    \ /
    :Xe:
    / \
 :F:   :F:
```

The Lewis structure places a total of 12 electrons, four bond pairs and two lone pairs, about the xenon atom.

Understanding Write the Lewis structure for SF_6.

Answer:

```
        ..
   :F: :F: :F:
      \ | /
        S
      / | \
   :F: :F: :F:
        ''
```

Formal charges are not needed in the Lewis structures of PF_5, XeF_4, and SF_6. In each case, dividing the bonding electrons equally leads to Lewis symbols for the neutral atoms. Thus, for example, each of the five valence electrons on the phosphorus atom in PF_5 is used to make a bond to one of the five fluorine atoms.

Crystals of XeF_4

Oxides of *p* Block Elements from the Third and Later Periods

We can frequently write two types of Lewis structures for compounds with *p* block central atoms from the third or later periods bonded mainly to oxygen. An example is sulfuric acid, H_2SO_4.

```
        :Ö:⊖                          :O:
         |                             ‖
H—Ö—S(+2)—Ö—H  ⟷  H—Ö—S—Ö—H
         |                             ‖
        :Ö:⊖                          :O:

         A                              B
```

In structure A all atoms satisfy the octet rule, but a high formal charge of +2 is assigned to the sulfur atom. Lewis structure B has 12 electrons around the sulfur atom, but no formal charges are produced because the sulfur uses all six of its valence electrons to make six bonds. Resonance structure B is more important, because formal charges should be minimized. Experimental evidence supports the importance of structure B. The S—OH bond lengths are 157 pm, which can be taken as typical for a S—O single bond. The other two S—O bond lengths are shorter at 142 pm, indicating that there is substantial multiple bonding in these S—O bonds.

Example 8.15 Resonance Structures

Draw the important resonance structures for $HClO_3$.

Solution

1. The skeleton structure is

```
           O
          /
H—O—Cl
          \
           O
```

2. The total number of valence electrons is 26.
3. Eight electrons are used in the skeleton structure, leaving 18 to complete the structure with lone pairs.
4. The Lewis structure can be completed with 18 electrons.

```
             :Ö:⊖
            /
H—Ö—Cl:(+2)
            \
             :Ö:⊖
```

   This structure is valid, but it has considerable formal charges. A second structure can be written that eliminates the formal charges.

```
              Ö:
            //
H—Ö—Cl:
            \\
              O:
```

   This structure is favored because all formal charges are zero.

Understanding Write the important resonance forms for H_2SO_3 (both hydrogen atoms are bonded to oxygen atoms).

Answer:

```
        ˙O—H                  ˙O—H
  ..   /  ..            ..   /  ..
H—O—S:⊕      ⟷    H—O—S:
  ..   \               ..   \\
        :O:⊖                 O:
```

8.8 Bond Energies

Objectives

- To relate bond energies with bond strengths
- To calculate approximate enthalpies of reaction from bond energies

Covalent bonds form between many different elements. Clearly, not all types of covalent bonds are of equal strength; each involves different nuclei and electrons in different orbitals. Bond strengths are very important, because species with strong bonds are generally stable.

Bond dissociation energy or **bond energy** (D) is the energy required to break one mole of bonds in a gaseous species. The thermochemical equation that describes the bond dissociation for H_2 is

$$H_2(g) \longrightarrow H(g) + H(g) \qquad \Delta H = D(\text{H—H}) = 436 \text{ kJ/mol}$$

Bond energies are always endothermic and thus have a positive sign; it takes energy to break a bond.

Table 8.3 shows a number of important bond energies. For diatomic molecules, these numbers are measured exactly. A problem arises in measuring exact bond energies in polyatomic molecules because the energy required to break a bond is influenced by the other atoms. Consider the bond energy for *each* O—H bond in water.

$$H_2O(g) \longrightarrow OH(g) + H(g) \qquad \Delta H = 502 \text{ kJ/mol}$$

$$OH(g) \longrightarrow O(g) + H(g) \qquad \Delta H = 427 \text{ kJ/mol}$$

The values are not the same because the species in which we are breaking the bonds are not the same. In a similar manner, the O—H bond energy in CH_3OH is different (435 kJ/mol) from either of those given above. Since bond energies depend on the environment of the bonded atoms, the table gives *average* bond energies for all cases other than the diatomic molecules. While not exact, these numbers are fairly accurate since most bonds between the same two atoms are of similar strengths.

Bond energies measure the strengths of chemical bonds.

The values of bond energies span a fairly wide range. The C—H bond is more than twice as strong as the O—F bond. In comparing bond strengths between the same two atoms, double and triple bonds are stronger than single bonds, as expected, from their multiple sharing of electrons.

Bond Energies and Enthalpies of Reaction

The data in Table 8.3 can be used with Hess's law to calculate enthalpies of reaction. We calculate the energy required to break the correct number of moles of bonds of all the bonds in the reactants to form gaseous atoms, and

Table 8.3 Bond Energies (kJ/mol)*

Single Bonds			
C—H 414	N—H 389	O—H 463	F—F 159
C—C 348	N—N 163	O—O 146	Cl—F 253
C—N 293	N—O 201	O—F 190	Cl—Cl 242
C—O 351	N—F 272	O—Cl 203	Br—F 237
C—F 439	N—Cl 200	O—I 234	Br—Cl 218
C—Cl 328	N—Br 243		Br—Br 193
C—Br 276		S—H 339	I—Cl 298
C—I 238	H—H 436	S—F 327	I—Br 180
C—S 259	H—F 569	S—Cl 251	I—I 151
	H—Cl 431	S—Br 218	
Si—H 293	H—Br 368	S—S 266	
Si—Si 226	H—I 297		
Si—C 301			
Si—O 368			
Multiple Bonds			
C=C 611	O_2 498		
C≡C 837			
C=N 615	N=N 418		
C≡N 891	N≡N 946		
C=O 799			
C≡O 1072	S=O 523		
	S=S 418		

* The bond energies for the diatomic molecules can be measured directly; the other numbers are average bond energies.

we calculate the energy released when these atoms combine to form the new bonds of the products. The sum of these two numbers is the enthalpy of reaction. For the reaction of $H_2(g) + F_2(g) \rightarrow 2HF(g)$ the process is shown in Figure 8.11.

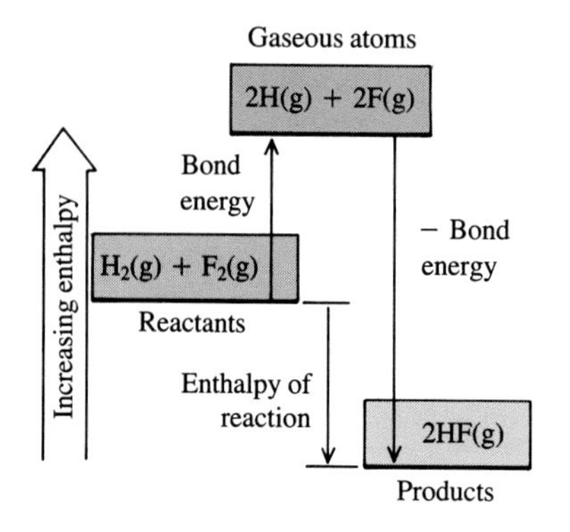

Figure 8.11 Calculation of enthalpy of reaction Energy-level diagram showing the use of bond energies to calculate the heat of reaction for $H_2(g) + F_2(g) \rightarrow 2HF(g)$.

In equation form, the calculation of enthalpies of reaction from bond energies is

$$\Delta H_{\text{reaction}} = \Sigma H(\text{bond energies of bonds broken}) - \Sigma H(\text{bond energies of bonds formed}) \quad [8.1]$$

The enthalpy of reaction can be calculated from the energy required to break all of the bonds in the reactants minus the bond energies of the new bonds formed in the products.

The negative sign in the equation indicates that we are making bonds in the products, an exothermic process, so the energy change is the opposite of a bond energy. Note also in the following calculation that one mole each of H_2 and F_2 is consumed, and *two* moles of HF are formed. The values given in Table 8.3 are for *one* mole of bonds.

$$\Delta H_{\text{reaction}} = [1\text{ mol} \times D(\text{H—H}) + 1\text{ mol} \times D(\text{F—F})] - [2\text{ mol} \times D(\text{H—F})]$$

$$\Delta H_{\text{reaction}} = [1\text{ mol} \times 436\text{ kJ/mol} + 1\text{ mol} \times 159\text{ kJ/mol}] - [2\text{ mol} \times 569\text{ kJ/mol}]$$

$$\Delta H_{\text{reaction}} = -543\text{ kJ}$$

The reaction is exothermic. Examination of the bond energies indicates that the big difference between molecules on the two sides of the equation is the weakness of the F—F bond, especially in comparison to the strong H—F bond. The bonds formed are stronger than the bonds broken, so the reaction is exothermic. We have already mentioned that F_2 is very reactive; the weak F—F bond is a major reason for this reactivity.

Calculations of this type yield correct $\Delta H_{\text{reaction}}$ values for reactions of diatomic molecules in the gas phase. For most other reactions, the calculated enthalpies of reaction are only approximately correct because *average* bond energies are used. The simplicity of being able to calculate ΔH values from a single small table, rather than measuring the result of each reaction experimentally, is the advantage of the method.

Example 8.16 Calculation of Enthalpies of Reaction

Use Table 8.3 to calculate an approximate enthalpy of reaction for

$$CH_4(g) + 2O_2(g) \longrightarrow CO_2(g) + 2H_2O(g)$$

Combustion of methane Methane gas burns in air in an exothermic reaction. Shown here is methane being flared in Saudi Arabia. The black smoke is evidence of incomplete combustion.

Solution Using Equation 8.1

$$\Delta H_{reaction} = [4 \text{ mol} \times D(\text{C—H}) + 2 \text{ mol} \times D(\text{O=O})] - [2 \text{ mol} \times D(\text{C=O}) + 4 \text{ mol} \times D(\text{O—H})]$$

Three important points are:

1. The Lewis structure of each compound in the reaction must be known, since bond energies depend on bond order. For example, both C—O bonds in CO_2 are double bonds.
2. In the calculation, the total number of moles of *bonds* being broken or made must be taken into account. Both the coefficients in the equation and the subscripts in the formulas need to be considered. One mole of CH_4 has four moles of C—H bonds; the two moles of H_2O contain four moles of O—H bonds.
3. Remember the minus sign in front of the *products* because formation of the products is the reverse of a bond energy.

Substitute the bond energy values from Table 8.3.

$$\Delta H_{reaction} = [4 \text{ mol} \times 414 \text{ kJ/mol} + 2 \text{ mol} \times 498 \text{ kJ/mol}] - [2 \text{ mol} \times 799 \text{ kJ/mol} + 4 \text{ mol} \times 463 \text{ kJ/mol}]$$

$$\Delta H_{reaction} = [2652 \text{ kJ}] - [3450 \text{ kJ}] = -798 \text{ kJ}$$

Again, the reaction is exothermic. In this case, the strong C=O bonds in CO_2 contribute the main difference in bond strengths between the reactants and products. Remember that enthalpies of reaction calculated from bond energies are only approximate; the experimentally measured value for this reaction is −802 kJ.

Understanding Calculate the enthalpy of reaction for the reaction of H_2 and N_2 to form two moles of NH_3, using data from Table 8.3.

Answer: −80 kJ

Summary

Lewis electron-dot symbols show the valence electrons (the outer electrons) as dots. In *ionic bonding,* some of the valence electrons are transferred from a metal to a nonmetal producing cations and anions that, for the representative elements, generally have noble-gas electron configurations. The main driving force for the formation of ionic solids is the *lattice energy,* the energy released when the ions are brought together to form an ionic crystaline solid. The lattice energy is greatest for small, highly charged ions.

Covalent bonds are formed by the sharing of valence electrons between atoms. *Lewis structures,* showing *bond pairs* and *lone pairs,* are used to represent covalent bonding. In Lewis structures, atoms (particularly those of the second period) share valence electrons to attain an *octet* of electrons (hydrogen atom has only two). The Lewis structure shows the number of valence electrons around each atom of the molecule or ion. Atoms can share one pair of electrons to form a *single bond,* two pairs to form a *double bond,* or three pairs to form a *triple bond.* Electrons are not shared equally between atoms of different elements, creating *polar bonds.* The *electronegativity* differences of the elements are used to determine the extent of this unequal sharing. Fluorine is the most electronegative element, and electronegativity decreases going down a group and to the left across a period. The polarity of the bond is measured by its *dipole moment.*

Formal charges are assigned to atoms in Lewis structures to indicate the charge that would result if the shared electrons were divided equally between the bonded atoms. Frequently, more than one Lewis structure can be written for a compound. Lewis structures that differ only in the placement of multiple bonds are called *resonance* forms. The actual bonding is an average of the bonding in the resonance forms, but not all resonance forms contribute equally in every case. Resonance forms with high formal charges or forms that place charges of the same sign on adjacent atoms do not contribute significantly to the bonding.

In certain cases, it may not be possible to write Lewis structures that obey the octet rule. A molecule in which an atom does not attain an octet is called *electron deficient;* such a molecule can accept an electron pair from another molecule or ion to form a *coordinate covalent bond* (dative bond). Species with central atoms from the third and later periods often exceed an octet around the central atom; such species are said to be *electron rich.*

Bond dissociation energy expresses the energy needed to break one mole of a particular type of covalent bond. In most cases, it is an average number because the energy needed to break a given type of bond may be different for different compounds. Bond energies can be used to calculate approximate values for the enthalpies of reactions.

Chapter Terms

Chemical bonds—the forces that hold the atoms together in substances. *(8.1)*

Lewis electron-dot symbol—the symbol of the element with one dot to represent each valence electron. *(8.1)*

Ionic bonding—the bonding that results from the electrostatic attraction between positively charged cations and negatively charged anions. *(8.2)*

Lattice energy—the energy required to separate one mole of an ionic solid into isolated gaseous ions. *(8.2)*

Covalent bond—the bond that arises from atoms sharing electron pairs. *(8.3)*

Bond length—the distance between the nuclei of two bonded atoms in a molecule. *(8.3)*

Bonding pairs—pairs of electrons shared between two atoms. *(8.3)*

Lone or nonbonding pairs—pairs of electrons that are not shared. *(8.3)*

Lewis structure—representation of covalent bonding, using Lewis symbols, that shows shared electrons as dots or lines between atoms and unshared electrons as dots. *(8.3)*

Octet rule—each atom in a molecule shares electrons until it is surrounded by eight valence electrons. The octet rule is most important for elements in the second period. *(8.3)*

Single bond—the bond formed by sharing one pair of electrons between two atoms. *(8.3)*

Double bond—the bond formed by sharing two pairs of electrons between two atoms. *(8.3)*

Triple bond—the bond formed by sharing three pairs of electrons between two atoms. *(8.3)*

Bond order—the number of electron pairs that are shared between two atoms. *(8.3)*

Skeleton structure—the drawing that shows which atoms are bonded to each other in a molecule. *(8.3)*

Central atom—an atom bonded to two or more other atoms. *(8.3)*

Polar bond—a covalent bond in which the bonding electrons are not equally shared by the two atoms. *(8.4)*

Electronegativity—a measure of the ability of an atom to attract the shared electrons in a chemical bond. *(8.4)*

Dipole moment—the magnitudes of separated charges times the distance between the charges. *(8.4)*

Formal charges—charges assigned to each atom in a Lewis structure, obtained by assuming that the shared electrons are divided equally between the bonded atoms. *(8.5)*

Resonance—the use of two or more Lewis structures that differ only in the distribution of the valence electrons in representing the bonding in a species. *(8.6)*

Electron deficient molecule—a molecule for which the Lewis structure has fewer than eight electrons around any atom. *(8.7)*

Coordinate covalent bond—a covalent bond in which both electrons of the bonding pair have come from one atom. A coordinate covalent bond is also frequently referred to as a *dative bond. (8.7)*

Electron rich molecule—a molecule that has more than eight electrons about an atom in its Lewis structure. *(8.7)*

Bond dissociation energy or bond energy—the energy required to break one mole of bonds in a gaseous species. *(8.8)*

Exercises

Exercises designated with color have answers in Appendix J.

Lewis Symbols and Ionic Bonding

8.1 What is a Lewis electron-dot symbol?

8.2 Write the Lewis symbol for

(a) sodium atom
(b) fluorine atom
(c) oxide 2− anion
(d) magnesium 2+ cation.

8.3 Write the Lewis symbol for
(a) sulfur atom
(b) iodide 1− anion
(c) beryllium atom
(d) gallium 2+ cation.

8.4 Identify the second period element A with the Lewis symbol
(a) ·A· (b) ·$\dot{\underset{\cdot}{A}}$· (c) ·$\dot{\underset{\cdot}{A}}$·$^-$.

8.5 Identify the third period element X with the Lewis symbol
(a) X^+ (b) ·$\dot{\underset{\cdot}{X}}$· (c) ·$\dot{\underset{\cdot}{X}}$·$^{2-}$.

8.6 Use Lewis electron-dot symbols to show the electron transfer during the formation of each compound from the appropriate atoms of (a) barium bromide, (b) potassium sulfide.

8.7 Use Lewis electron-dot symbols to show the electron transfer during the formation of each compound from the appropriate atoms of (a) beryllium oxide, (b) yttrium chloride.

8.8 What main factors control the magnitude of lattice energies? Which is most important?

8.9 Explain why (a) the lattice energy of NaI is greater than that of KI and (b) the lattice energy of $MgCl_2$ is greater than that of NaCl.

8.10 Explain why (a) the lattice energy of LiCl is greater than that of LiBr and (b) the lattice energy of Na_2O is greater than that of NaF.

8.11 Write the formulas of the ionic compounds that will form from the two pairs of elements given. With each pair, which will have the greater lattice energy? Explain your choice.
(a) potassium and sulfur; potassium and chlorine
(b) lithium and fluorine; rubidium and chlorine

8.12 Write the formulas of the ionic compounds that will form from the two pairs of elements given. With each pair, which will have the greater lattice energy? Explain your choice.
(a) lithium and oxygen; sodium and sulfur
(b) potassium and chlorine; magnesium and fluorine

8.13 For the ionic solids CaO and BaO, which has the greater lattice energy? Explain your choice.

8.14 Arrange the following ions in order of increasing size.
(a) K^+, Ca^{2+}, S^{2-}
(b) Na^+, F^-, O^{2-}

8.15 Which has the larger radius?
(a) S^{2-} or Se^{2-}
(b) Na^+ or Mg^{2+}

Lewis Structures

8.16 Describe the main difference between covalent and ionic bonding.

8.17 Define the octet rule and rationalize why the rule applies to most compounds made up of second period elements.

8.18 Write the formula of the simplest compound made from
(a) carbon and fluorine
(b) nitrogen and iodine
(c) oxygen and chlorine.

8.19 Write the formula of the simplest compound made from
(a) silicon and hydrogen
(b) nitrogen and chlorine
(c) oxygen and fluorine.

8.20 Write the Lewis structure for each of the following compounds. Label electrons as bond pairs or lone pairs.
(a) H_2S (b) H_2CO (c) PF_3

8.21 Write the Lewis structure for each of the following compounds. Label electrons as bond pairs or lone pairs.
(a) $SiCl_4$ (b) SF_2 (c) CS_2 (SCS)

8.22 What is the bond order between each pair of bonded atoms for the compounds in Exercise 8.20?

8.23 What is the bond order between each pair of bonded atoms for the compounds in Exercise 8.21?

8.24 Write the Lewis structure for
(a) AsH_3 (b) NO^+ (c) CF_3OH.

8.25 Write the Lewis structures for
(a) CN^-
(b) C_2H_6 (H_3CCH_3)
(c) N_2H_4 (H_2NNH_2).

8.26 Write the Lewis structures for each compound with the connectivity pictured.

(a) H_3C–C(=O)–H connectivity: H–C(H)(H)–C(O)–H
(b) H_2N–O–H
(c) H–C(H)(H)–C(H)–C(H)(H)
(d) H–N–N–H

8.27 Write the Lewis structure for each compound with the connectivity pictured.

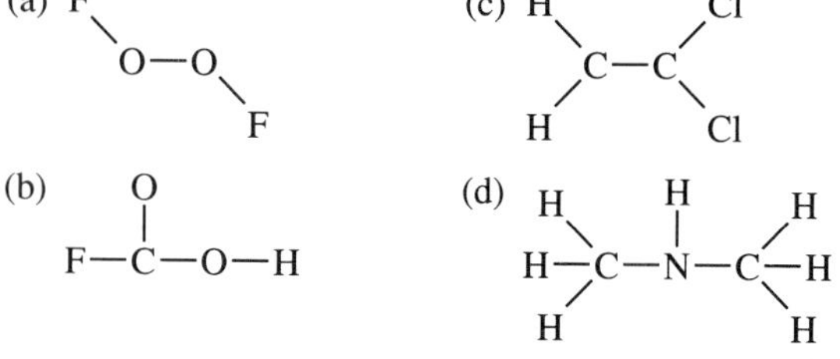

8.28 Write the Lewis structure for each compound with the connectivity pictured.

(a) H—C—C—H

(b) H—O—Cl

(c) H, H on C—C, with H on C and C—N

(d) H—O—O—H

8.29 Write the Lewis structure for each compound with the connectivity pictured.

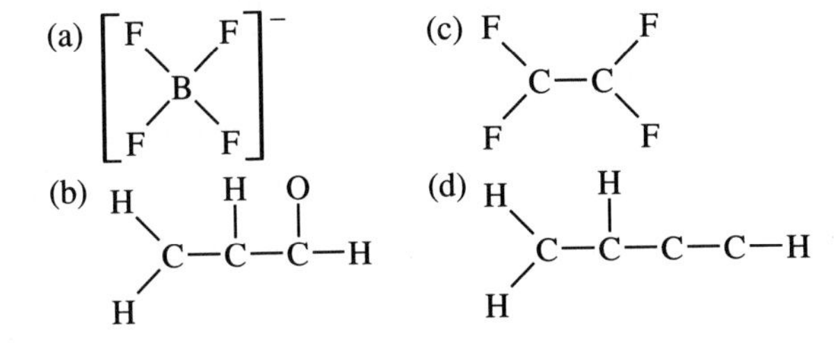

Electronegativity

8.30 What is a polar bond, and in what circumstances will it occur?

8.31 Outline the trends in the electronegativities of the representative elements.

8.32 Which atom in each of the following pairs has the higher electronegativity? Use only a periodic table.
(a) Br, I
(b) S, Cl
(c) C, N
(d) O, S
(e) K, Ge

8.33 List each of the following series of atoms in order of increasing electronegativity (use only a periodic table).
(a) F, S, Si
(b) Li, P, O
(c) Ga, B, S
(d) Br, Ca, Al

8.34 In each pair of bonds, indicate which has the greater polarity and show the direction of the dipole (use Figure 8.8 if needed).
(a) N—O, C—O
(b) Si—Ge, Ge—C
(c) S—H, O—H
(d) B—C, B—Si

8.35 In each pair of bonds listed, indicate which has the greater polarity and show the direction of the dipole (use Figure 8.8 if needed).
(a) N—F, F—Br
(b) F—N, S—P
(c) Cl—Br, S—Se
(d) C—N, B—F

8.36 Which of the following molecules will have the largest dipole moment: N_2, BrF, or ClF?

8.37 Which of the following molecules will have the largest dipole moment: O_2, BrCl, or ICl?

8.38 Explain the different behavior expected for HCl and Cl_2 when each is placed between two oppositely charged plates.

Formal Charge and Resonance

8.39 Explain how formal charges are assigned to atoms in Lewis structures.

8.40 The connectivity of H_3CN could be either H_2CNH or $HCNH_2$. Draw a Lewis structure for each and predict which connectivity is the more favorable arrangement.

8.41 Predict which connectivity, SCN^- or CSN^-, is the more favorable arrangement.

8.42 Write all possible resonance forms for the following species. Assign formal charges to each atom. Which resonance forms are the most important?
(a) NO_2^- (nitrogen is central)
(b) ClCN

8.43 Write all possible resonance forms for the species with the connectivity pictured. Assign formal charges to each atom. Which resonance forms are the most important?

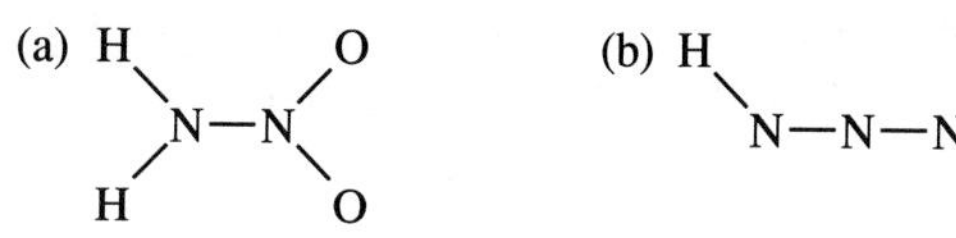

8.44 Write all possible resonance forms for the species with the connectivity pictured. Which resonance forms are the most important?

(a) H, H on C—N—N

(b) [O—C—N]⁻

8.45 Write all possible resonance forms for the species with the connectivity pictured. Which resonance forms are the most important?

(a) H, H—C (H) —C(O)—N(O)(O)

(b) [O—N(O)—O]⁻

8.46 Write all possible resonance forms for the species with the connectivity pictured. Which resonance forms are the most important?

(a) [N—N—N]⁻

(b) [O—C(O)(O)]²⁻

8.47 Write all resonance forms of chlorobenzene, C_6H_5Cl, a molecule with the same cyclic structure as benzene. In all answers, keep the C—Cl bond as a single bond.

8.48 Draw all resonance forms for methylisocyanate, CH_3NCO, a toxic gas used in the manufacturing of pesticides.

8.49 Write all resonance forms of the oxalate anion, $C_2O_4^{2-}$ (there are two terminal oxygen atoms on each carbon atom).

Exceptions to the Octet Rule

8.50 Write the Lewis structures of the following molecules and indicate whether they are odd electron species, electron deficient molecules, or electron rich molecules: (a) SeF_6, (b) BBr_3, (c) NO_2

8.51 Write the Lewis structures for the following species and indicate whether any of them are odd electron species, electron deficient species, or electron rich species: (a) IF_3, (b) ICl_4^-, (c) N_2^+

8.52 Give an example of a molecule that contains a coordinate covalent bond. How is this bond different from "normal" bonds?

8.53 Write the Lewis structure of IF_5.

8.54 Write all possible resonance forms for the species listed below with the connectivity pictured.
(a) O—Se—O (b) O—S—O with a third O bonded to S

8.55 For the species listed below, write the resonance form with an octet about the central atom and a second resonance form that minimizes formal charges.

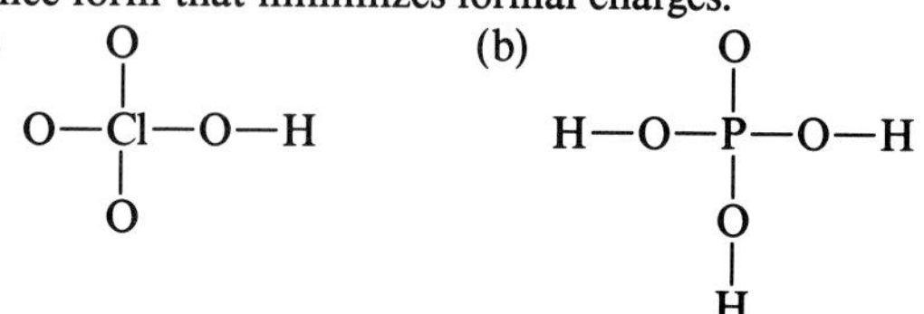
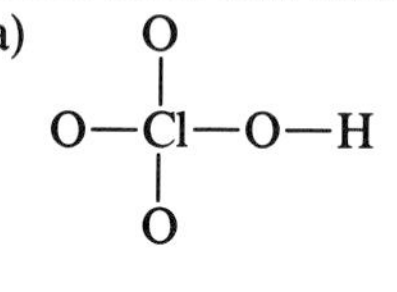

8.56 Write all possible resonance forms for the species with the connectivity pictured.
(a) $[O—Cl—O]^-$ (b) $[O—S—O—H]^-$ with a third O bonded to S

Bond Energies

8.57 Write the Lewis structures of HNNH and H_2NNH_2. Predict which molecule has the larger N—N bond energy.

8.58 Calculate the energy required to break all of the bonds (use Table 8.3) in one mole of
(a) NH_3 (b) CH_3OH

8.59 Calculate the energy required to break all of the bonds (use Table 8.3) in one mole of
(a) CH_2CF_2 (b) N_2H_4

8.60 Use Table 8.3 to calculate an approximate enthalpy change for (a) the reaction of molecular hydrogen and molecular oxygen in the gas phase to produce two moles of water vapor and (b) the reaction of carbon monoxide and molecular oxygen to form two moles of carbon dioxide.

8.61 Use Table 8.3 to calculate an approximate enthalpy change for
(a) the combustion of one mole of gaseous CH_3OH in excess molecular oxygen
(b) the reaction of one mole of carbon monoxide with molecular hydrogen to form gaseous methanol (CH_3OH).

8.62 Use Table 8.3 to calculate an approximate enthalpy change for
(a) the combustion of one mole of C_2H_4 in excess molecular oxygen to form gaseous water and CO_2
(b) the reaction of one mole of formaldehyde, H_2CO, with molecular hydrogen to form gaseous methanol (CH_3OH).

8.63 Use Table 8.3 to calculate an approximate enthalpy change for
(a) the reaction of H_2 and C_2H_2 to form one mole of C_2H_6
(b) the reaction of molecular hydrogen and molecular nitrogen to form one mole of ammonia.

Additional Problems

8.64 Given that the lattice energy of an ionic solid increases with the charges of the anion and cation, discuss why the formula of NaF is not NaF_2.

8.65 Use the octet rule to predict the element (E) from the second period that would be the central atom in
(a) EF_4^- and (b) EF_4^+.

8.66 The compound disulfur dinitrogen, S_2N_2, has a cyclic structure with alternating sulfur and nitrogen atoms. Draw two Lewis structures for S_2N_2 with different numbers of bonds.

8.67 The phosphorus oxyanion $[H_2P_2O_4]^{2-}$ contains a P—P bond with a hydrogen and two oxygen atoms bonded to each phosphorus. Draw a resonance form in which each phosphorus atom satisfies the octet rule and another in which they are electron-rich.

8.68 The compound SF_4CH_2 has an unusually short S—C distance of 155 pm. Use the Lewis structure to explain this short distance.

8.69 In the gas phase, the oxide N_2O_5 has a structure with an N—O—N core, with the other four oxygen atoms in terminal positions. In contrast, in the solid the stable form is $[NO_2]^+[NO_3]^-$. Draw one Lewis structure of the molecular form (with N—O—N single bonds) and all possible resonance forms of both ions observed in the solid. Remember that second period elements never exceed an octet.

8.70 The arrangement of the atoms in N_2O could be either N—N—O or N—O—N. Use the rules for the stability of Lewis structures with formal charge to predict the more likely arrangement.

8.71 Draw the Lewis structure and calculate the energy needed to break all of the bonds in one mole of CH_3NH_2.

8.72 Draw the Lewis structure of BrNO. Which is the more polar bond in the molecule?

8.73 Phosgene, Cl_2CO, is an extremely toxic gas that can be prepared by the reaction of CO with Cl_2. Calculate the approximate enthalpy change for this reaction, using data from Table 8.3.

8.74 Calculate an approximate enthalpy change (Table 8.3) for the reaction

$$HCN(g) + 2H_2(g) \longrightarrow H_3CNH_2(g)$$

8.75 Compare the trends in electronegativity and ionization energy within (a) a group and (b) a period. Explain any differences.

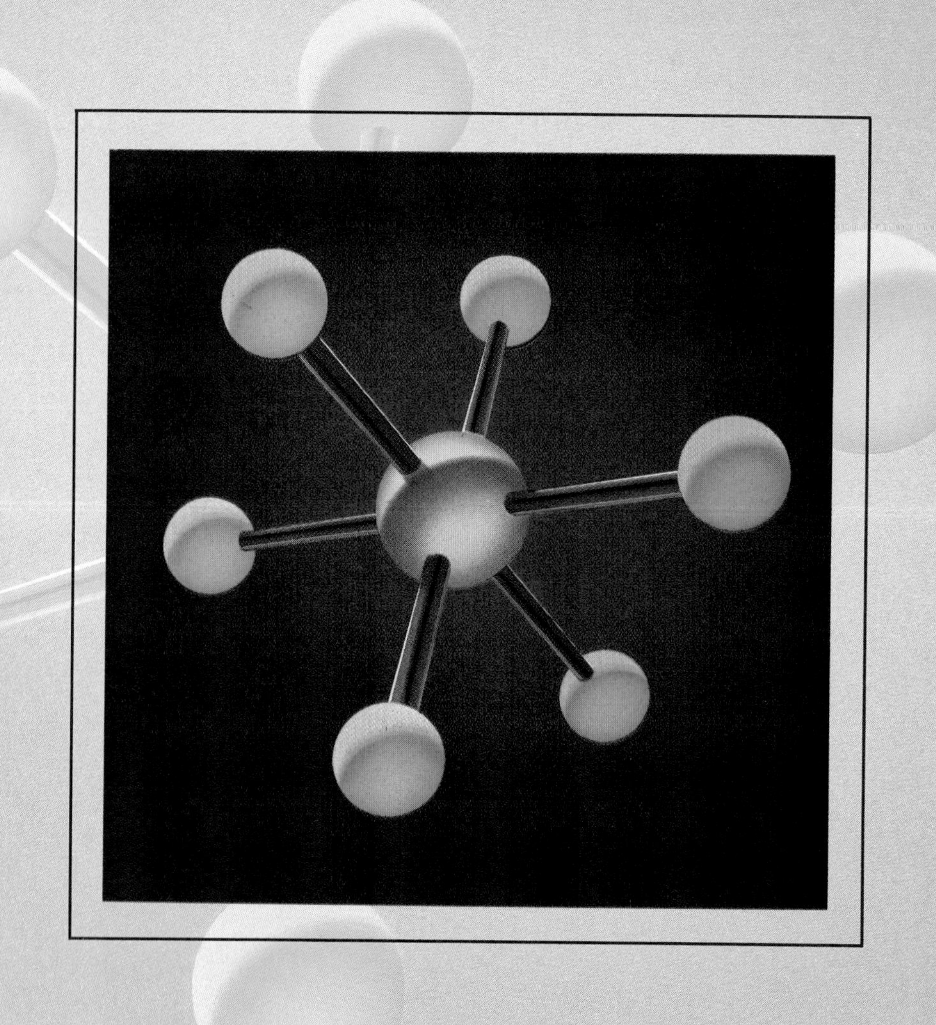

Chapter Nine

Molecular Structure and Bonding Theories

The shapes of molecules (that is, how the bonded atoms are arranged in space) exert a strong influence on their physical properties and chemical reactivities. On the basis of many experimentally determined molecular structures, chemists have developed a number of models that allow us to make fairly accurate predictions of molecular shapes. A simple model, based on Lewis structures, is extremely useful in predicting the shapes of many molecules, and is presented first. This model will be followed by two theories, valence bond theory and molecular orbital theory, that explain bonding in more detail. These two theories use the atomic orbitals to generate the wave functions for electrons in molecules.

9.1 Valence-Shell Electron-Pair Repulsion Model

Objectives

- To predict the shapes of molecules from the valence-shell electron-pair repulsion (VSEPR) model
- To distinguish between electron-pair arrangement and molecular geometry

The valence-shell electron-pair repulsion model (known as the VSEPR model) can predict the shapes of most molecules starting from the bonding description shown in a Lewis structure. The main premise of the model, based on the idea that the electron pairs about an atom repel each other, is stated in Rule 1.

The shapes of molecules, such as SF_6 pictured here, can be predicted by a simple model.

VSEPR Rule 1: Any molecule assumes a geometry that minimizes electrostatic repulsions between valence-shell electron pairs. The minimum repulsion occurs when the electron pairs are as far apart as possible.

Electron pairs are oriented as far apart as possible.

Starting with a Lewis structure, for each *central atom* (one that is bonded to at least two other atoms), we count the number of *lone pairs* plus the number of *bonded atoms.* This sum is used to determine the **electron-pair arrangement,** which is the arrangement adopted by the valence-shell electron pairs to maximize the distances between them. Figure 9.1 shows the arrangements that will keep the electron pairs at their maximum distance apart. These arrangements are given the names of the geometric solids that are formed by lines drawn to connect the electron pairs.

The basic idea of the model is easiest to see when there are only two pairs of electrons about the central atom, as in $BeCl_2$. A linear geometry places the two pairs of electrons on the central beryllium atom as far apart as possible. Any reduction of the Cl-Be-Cl angle below 180° will move the electron pairs closer together and increase the repulsion between them. Thus, for $BeCl_2$, a molecule in which the central atom has two bonded atoms and no lone pairs, the model predicts linear geometry and a Cl-Be-Cl bond angle of 180°.

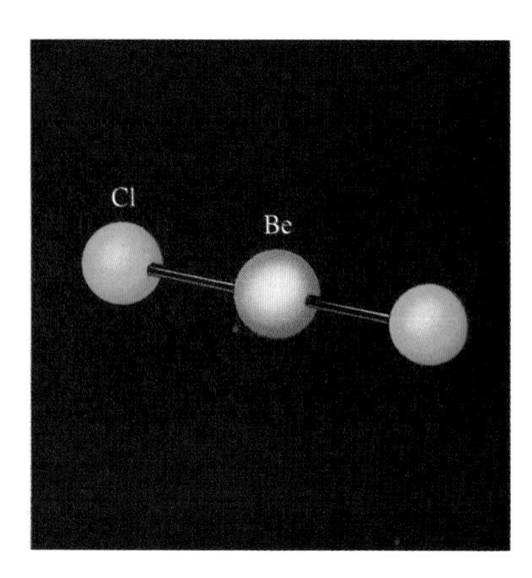

Hydrogen cyanide, HCN, is another molecule that is linear. The Lewis structure of HCN is

H—C≡N:

In this Lewis structure, the carbon atom forms four bonds, but in assigning the electron-pair arrangement we count the number of *bonded atoms* and lone pairs around the carbon atom. For multiple bonds, we can visualize that most of the shared electron density will be located between the bonded atoms. The effect of the repulsion of a multiple bond on the arrangement of the other electron pairs will be essentially the same as that of a single bond. The electron-pair arrangement for HCN is linear, the geometry for two pairs of electrons. Experiments confirm that HCN is linear.

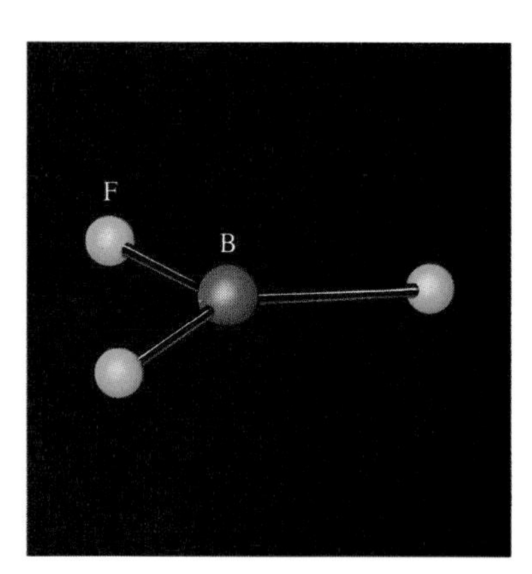

It is important to learn the shapes and bond angles associated with each electron-pair arrangement. For atoms with three pairs, as in BF_3, the electron-pair arrangement is trigonal planar. The bond angles in this molecule are all 120°, and the three fluorine atoms are in equivalent positions (the environment of each fluorine atom is the same). Note that all four atoms are in the same plane in this geometry.

Formaldehyde, H_2CO, is an example of a trigonal planar molecule involving a multiple bond.

H
 \
  C=O:
 /
H

The central carbon atom is bonded to two hydrogen atoms and one oxygen atom, and has no lone pairs. The electron-pair arrangement is trigonal planar, based on the three atoms bonded to the central carbon atom. In assigning the electron-pair arrangement, the electrons in the C—O double bond are counted the same as a single electron pair.

With four pairs, electron-pair repulsion is minimized by a tetrahedral arrangement, as pictured for CH_4. All angles between the electron pairs on the central atom in the tetrahedral arrangement are 109.5°. In this case, the

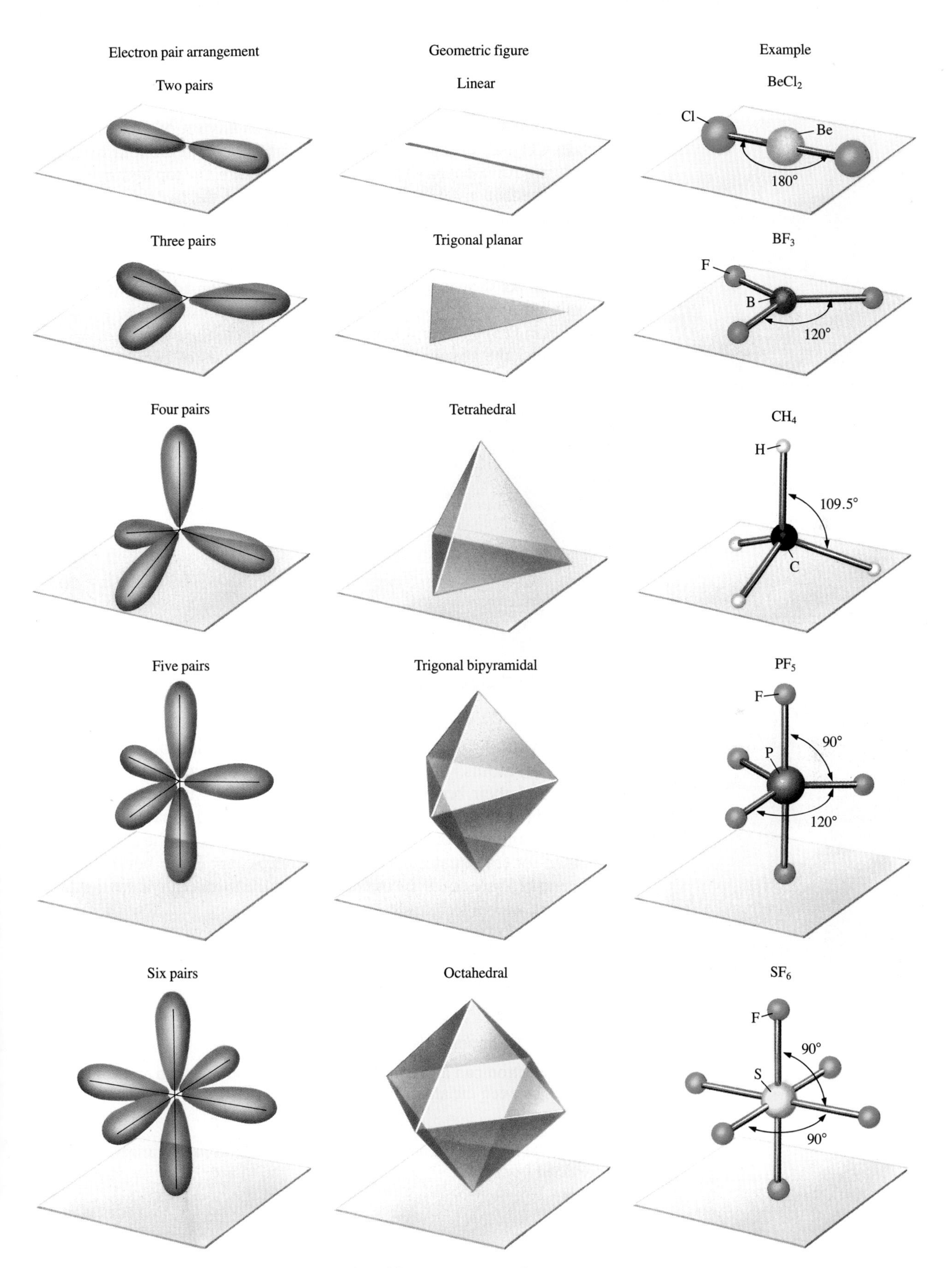

Figure 9.1 Geometric arrangements expected for different numbers of valence-shell electron pairs.

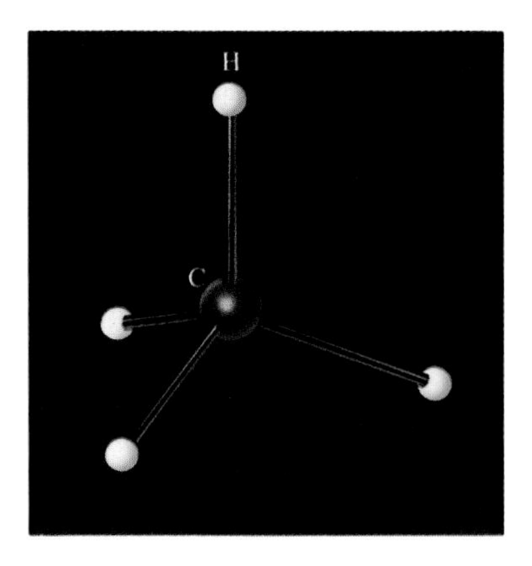

central carbon atom is *not* in the same plane as any set of three hydrogen atoms. In CH_4, all of the bond angles are the same, and the four hydrogen atoms are in equivalent environments. In the picture, the top hydrogen atom in the tetrahedron may look different, but any of the hydrogen atoms in the molecule can be rotated to this position and the picture would appear identical.

The electron-pair arrangement for atoms with five pairs is called trigonal bipyramidal. As shown below for PF_5, this shape can be considered to be the combination of a linear [F(1)—P—F(5)] perpendicular to a trigonal planar [PF(2) F(3) F(4)] arrangement. In contrast to the other electron-pair arrangements, the trigonal bipyramid has two types of bond angles and positions in which to place bonded atoms.

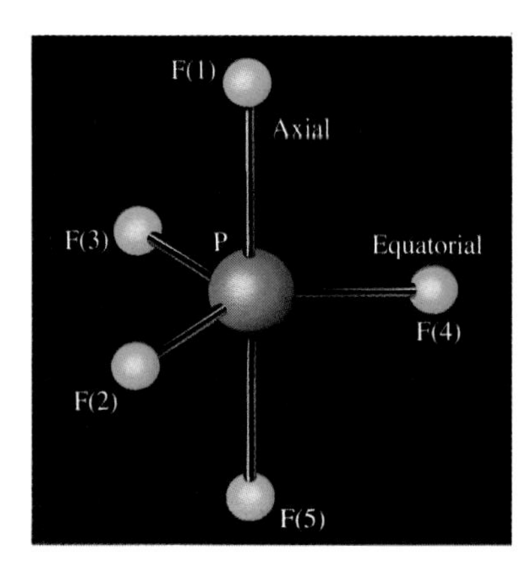

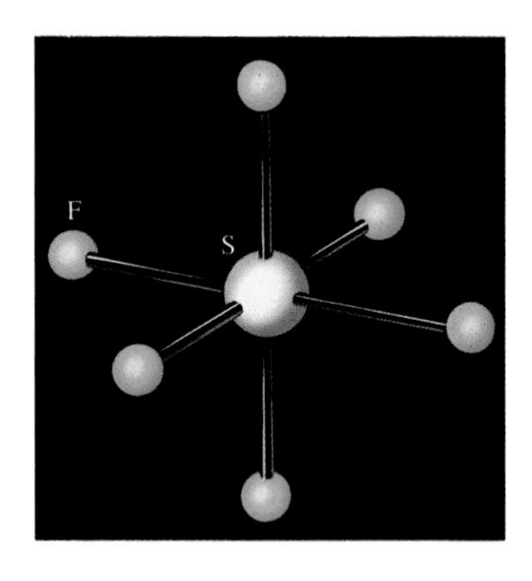

The F(1) and F(5) atoms are in the *axial* positions, and their bonds make a 90° angle with the plane occupied by the phosphorus atom and the F(2), F(3), and F(4) atoms. The F(2), F(3), and F(4) atoms are in the *equatorial* positions, and their bonds make an angle of 120° with each other. The F(1)—P—F(5) angle is, of course, 180°. The axial and equatorial positions are different, and we would expect different properties for the axial fluorine atoms than for the equatorial ones. For example, the P—F bond distances for F(1) and F(5) are equal, but different from the three equivalent distances to F(2), F(3), and F(4).

The final electron-pair arrangement, for an atom with six pairs of electrons, is octahedral as in SF_6. The bond angles in this case are 90° and 180°. In this geometry, all six positions are equivalent.

Example 9.1 **Shapes of Molecules**

What is the electron-pair arrangement for the central carbon atom and what are the bond angles between the atoms in the molecules (a) CCl_4 and (b) CO_2?

Solution

(a) First, write the Lewis structure for CCl_4. The electron-pair arrangement can be predicted from it.

```
        ..
       :Cl:
        |
        C
     .. / | \ ..
   :Cl  |  Cl:
     ''  :Cl: ''
         ..
```

The central carbon atom has four bonded atoms and no lone pairs. The electron-pair arrangement, based on four pairs of electrons, is tetrahedral. The Cl—C—Cl bond angles are all 109.5°.

(b) The Lewis structure of CO_2 is

:Ö=C=Ö:

The central carbon atom is bonded to two oxygen atoms and has no lone pairs. The electron-pair arrangement is linear. Each C=O double bond is counted the same as a single bond. The O—C—O bond angle is 180°.

Understanding What is the electron-pair arrangement for phosphorus, and what are the Cl—P—Cl bond angles in PCl_5?

Answer: The electron-pair arrangement is trigonal bipyramidal. There are three types of bond angles: 90° from the axial to equatorial, 120° among the equatorial, and 180° from the axial to axial chlorine atoms.

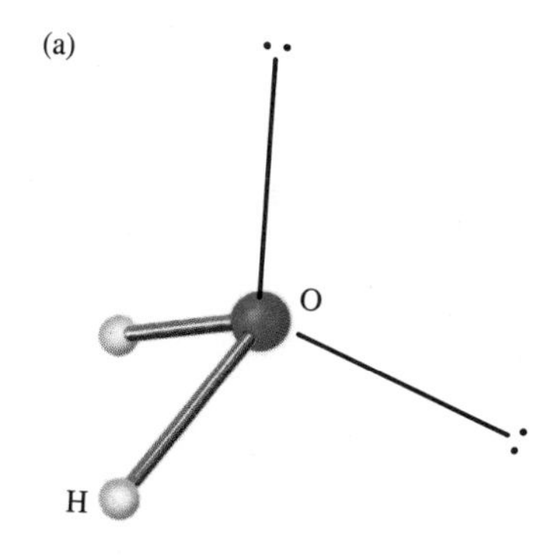

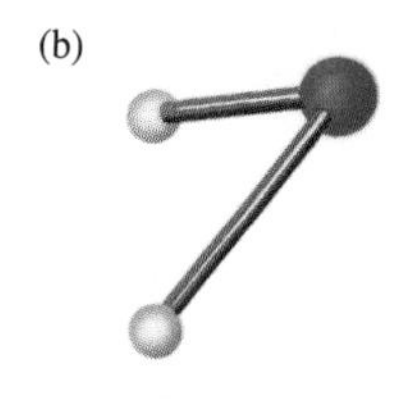

Figure 9.2 The electron-pair arrangement and molecular shape of water (a) The electron-pair arrangement of water is tetrahedral, with two lone pairs and two bond pairs. (b) The molecular geometry of water is bent or V-shaped.

Central Atoms That Have Lone Pairs

In all of the examples shown so far, the central atom has had no unshared pairs of electrons. If there are unshared pairs on the central atom, the electron-pair arrangement is determined by the sum of the numbers of bonded atoms and lone pairs. For example, the Lewis structure of water is

H—Ö—H (with two lone pairs on O)

There is a total of four electron pairs (two lone pairs and two bonded atoms) around the oxygen atom. The *electron-pair arrangement* is tetrahedral and the predicted H—O—H angle is 109.5°. For central atoms with lone pairs, the *molecular shape* will *not* be the same as the electron-pair arrangement. The **molecular shape** describes the arrangement of the atoms, not the lone pairs. The molecular shape of water is bent or V-shaped (Figure 9.2). The bent geometry is predicted by the tetrahedral electron-pair arrangement. The lone pairs are not ignored in assigning molecular shape, because the H—O—H bond angle is deduced from the tetrahedral electron-pair arrangement that counted both the bonded atoms and the lone pairs.

Rule 1 of the VSEPR model predicts a bond angle in water of 109.5°, but the *measured* bond angle is 104.5°. The small deviation is explained by a difference in repulsions between bonding pairs and lone pairs. Each type of electron pair exerts a different repulsion on other electron pairs. Bonding pairs are spread out over the two bonded atoms, whereas lone pairs reside only on the central atom. Thus, lone pairs have greater repulsive forces than do the bonding pairs. This leads to a second rule of VSEPR theory.

VSEPR Rule 2: Lone pair–lone pair repulsion is greater than lone pair–bond pair repulsion, which in turn is greater than bond pair–bond pair repulsion.

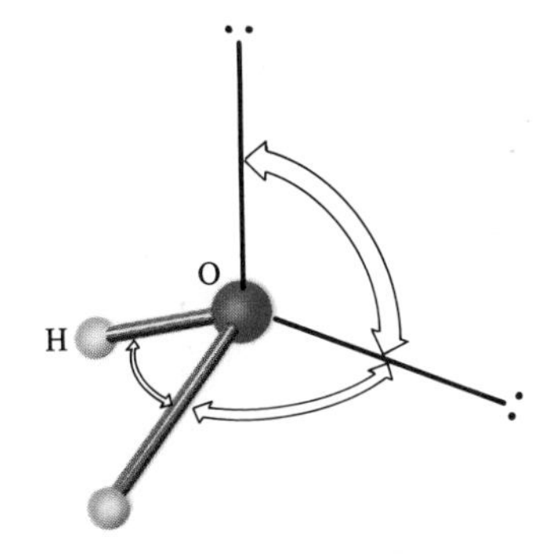

Figure 9.3 Repulsion of electron pairs in water Lone pair–lone pair repulsion is greater than lone pair–bond pair and bond pair–bond pair repulsions, causing the H—O—H angle in H_2O to be 104.5°, slightly smaller than the 109.5° tetrahedral angle.

For water, the larger lone pair–lone pair repulsion will increase the lone pair separation, forcing the bonding pairs to move closer together (Figure 9.3).

An analogous effect is observed with ammonia, NH_3. The Lewis structure shows there are four pairs of clectrons around nitrogen.

The number of the lone pairs and bonded atoms determines the electron-pair arrangement about any central atom.

```
   ..
H—N—H
  |
  H
```

The electron-pair arrangement, as predicted by VSEPR theory, is tetrahedral, and the expected bond angles are again 109.5°. In this case, the larger lone pair–bond pair repulsions cause a small decrease in the H—N—H bond angles (Figure 9.4). The measured angles are 107.3°.

Although the electron-pair arrangement of NH_3 is tetrahedral, the *molecular shape* is trigonal pyramidal (Figure 9.5). Note how this shape is different from that of BF_3. The electron-pair arrangement for BF_3 is trigonal planar (three pairs of electrons), and all four atoms are in the same plane. In NH_3, the nitrogen is not in the plane of the hydrogen atoms. The difference in geometry between BF_3 and NH_3 is caused by the lone pair on nitrogen.

In molecules with lone pairs, the molecular geometry differs from the electron-pair arrangement.

Figure 9.6 shows a variety of cases for central atoms with both bonding and lone pairs of electrons. Each type of arrangement is designated as $\mathbf{AB_nE_m}$, where A is the central atom, B designates the bonded atoms, and E represents the lone pairs. In all cases, the bond angles are determined from the *total* number of bonded atoms plus lone pairs.

Rule 2 can be used to predict the locations of the lone pairs in certain of the cases in Figure 9.6. In the Lewis structure for SF_4, there are four bond pairs and one lone pair (AB_4E).

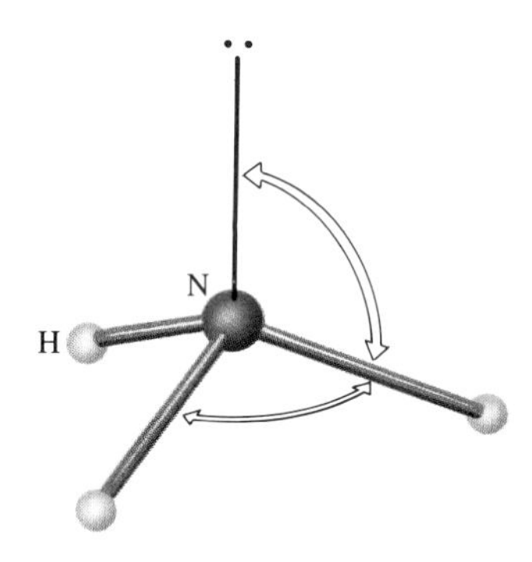

Figure 9.4 Repulsion of electron pairs in ammonia Lone pair–bond pair repulsion is greater than bond pair–bond pair repulsion, causing the H—N—H angles in NH_3 to be 107.3°, slightly smaller than the 109.5° tetrahedral angle.

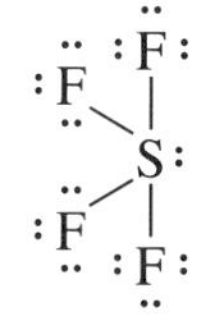

The electron-pair arrangement is trigonal bipyramidal. Four of the positions in the trigonal bipyramid will be occupied by fluorine atoms, and the fifth by the lone pair. By rule 2, the lone pair should be placed to minimize the lone pair–bond pair repulsions. Two molecular shapes are possible, one in which

(a)

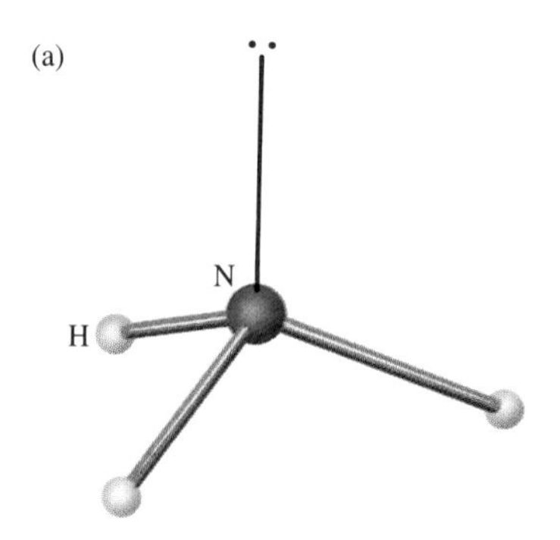

(b)

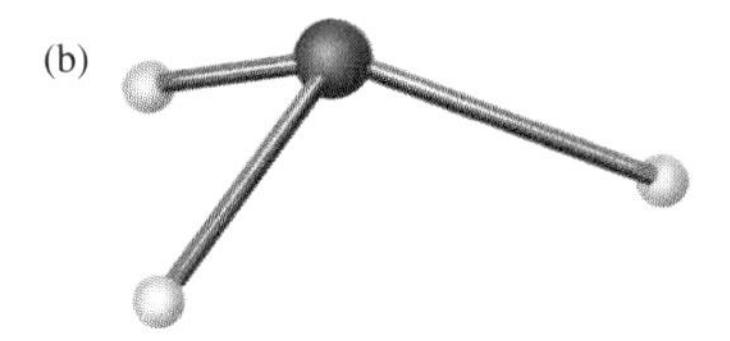

Figure 9.5 The electron-pair arrangement and molecular shape of ammonia (a) The electron-pair arrangement of ammonia is tetrahedral, with one lone pair and three bond pairs. (b) The molecular geometry of ammonia is trigonal pyramidal.

Type of molecule	Number of bonded atoms	Number of lone pairs	Electron pair arrangement	Molecular geometry	Bond angles	Examples
AB_2E	2	1	 Trigonal planar	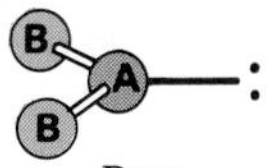 Bent	120°	SO_2, $SnCl_2$
AB_3E	3	1	 Tetrahedral	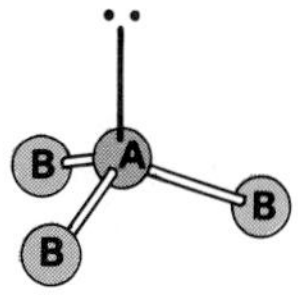 Trigonal pyramidal	109°	NH_3
AB_2E_2	2	2	 Tetrahedral	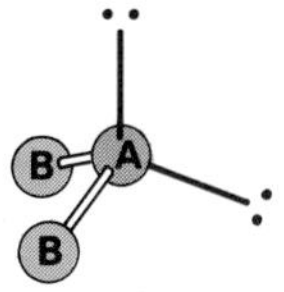 Bent	109°	H_2O
AB_4E	4	1	 Trigonal bipyramidal	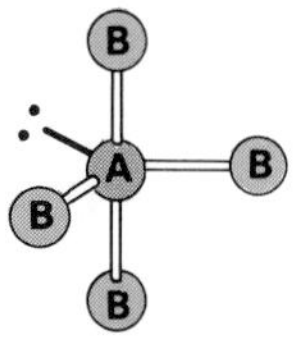 See-saw	90°, 120°, 180°	SF_4, XeO_2F_2
AB_3E_2	3	2	 Trigonal bipyramidal	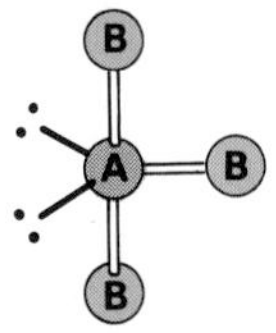 T-shaped	90°, 180°	ClF_3
AB_2E_3	2	3	 Trigonal bipyramidal	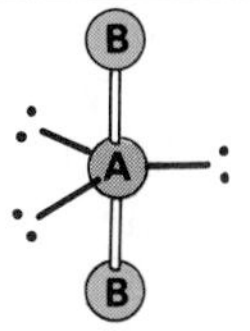 Linear	180°	XeF_2, I_3^-
AB_5E	5	1	 Octahedral	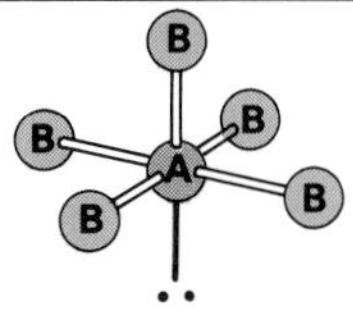 Square pyramidal	90°, 180°	ClF_5, $XeOF_4$
AB_4E_2	4	2	 Octahedral	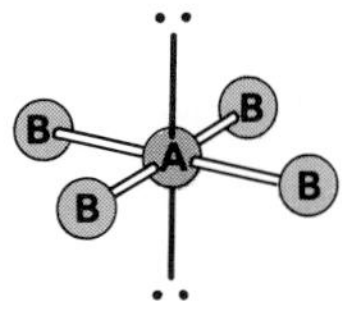 Square planar	90°, 180°	XeF_4, ICl_4^-

Figure 9.6 **Molecular shapes expected for molecules with lone pairs.**

the lone pair is located in an axial position (**A**) and the other in which the lone pair is in an equatorial position (**B**).

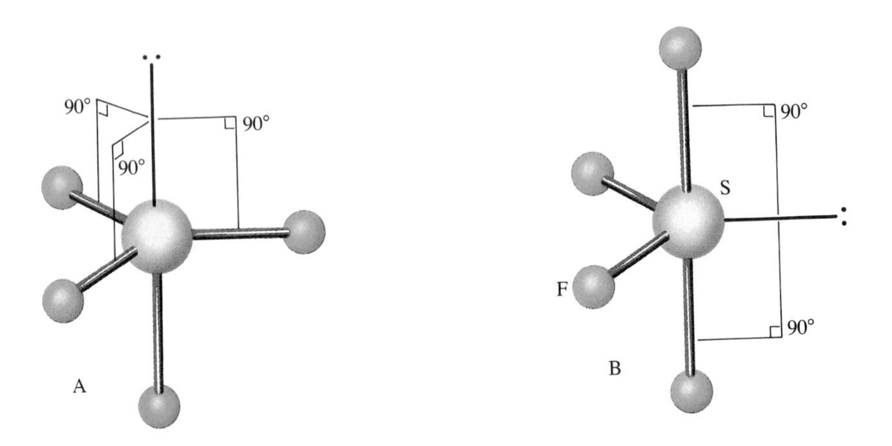

These two *molecular* geometries are different: **A** is a trigonal pyramid and **B** is an irregular shape sometimes referred to as a see-saw geometry (the axial fluorines are the seats and the equatorial fluorines the support legs). Experiment has shown that **B** is the correct geometry. This result is consistent with rule 2 because the repulsion from 90° interactions is greater than that from interactions at larger angles. In structure **A**, the lone pair has three 90° interactions with S—F bond pairs (the angle might increase a little because of the larger size of the lone pair), whereas the lone pair in **B** has only two 90° interactions. In **B** there are also two 120° lone pair–bond pair interactions, but because of the large angle, these interactions are not as important as the 90° interactions. Geometry **B** is favored because it has a smaller number of unfavorable lone pair–bond pair interactions at 90°.

For the same reason, the placement of lone pairs into equatorial sites in the trigonal bipyramidal electron-pair arrangement is always favored over lone pairs in the axial locations. For example, in the structure of ClF_3 (an AB_3E_2 molecule) both lone pairs are located in the equatorial positions, leading to a T-shaped molecular geometry.

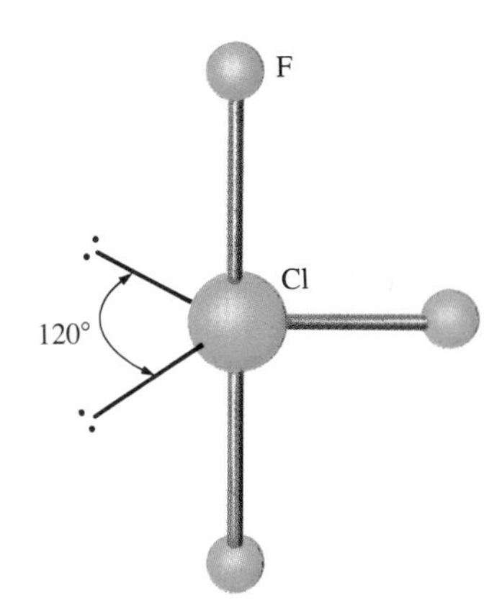

In this geometry the lone pairs are well separated at 120°, and the number of 90° lone pair–bond pair interactions is minimized. The structure that minimizes unfavorable 90° interactions will always be favored.

The structure that minimizes the number of 90° interactions is favored.

We should consider one other entry in Figure 9.6. Xenon tetrafluoride, XeF_4, has four bond pairs and two lone pairs (AB_4E_2). For six pairs of electrons the electron-pair arrangement is octahedral. The first lone pair can go in any of the six equivalent positions in the octahedron, but the second can either be adjacent to the first (**C**) or away from it (**D**).

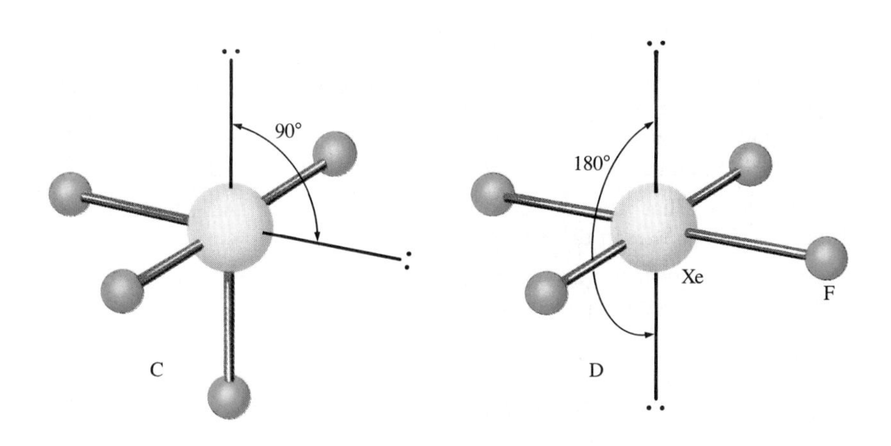

In **C** the lone pairs are at an angle of 90°, whereas in **D** they are at 180°. The more stable structure **D** minimizes the lone pair–lone pair repulsion. This predicted square planar molecular geometry is observed experimentally.

Example 9.2 Shapes of Molecules

What are the electron-pair arrangements, the molecular shapes, and the bond angles for (a) BrF_5, (b) ClNO, and (c) SO_2?

Solution

(a) The Lewis structure is

The central bromine atom has five bonded atoms and one lone pair for a total of six; the electron-pair arrangement is octahedral and the F—Br—F bond angles are 90° and 180°. The molecular shape is square pyramidal.

(b) The Lewis structure is

The central nitrogen atom is bonded to two other atoms and has one lone pair; the electron-pair arrangement is trigonal planar. The Cl—N—O bond angle is about 120° (we expect it to be slightly less than 120° because of the greater lone pair–bond pair repulsions) and the molecule is V-shaped.

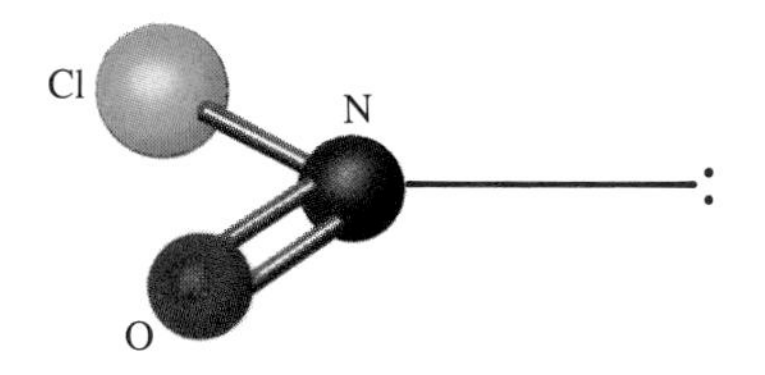

Structure of ClNO.

(c) The Lewis structure of SO_2 is

$$:\ddot{O}=\ddot{S}-\ddot{O}:$$

Two other resonance forms can be written, but the shape can be determined from any resonance structure. The central sulfur atom is bonded to two other atoms and has one lone pair; the electron-pair arrangement is trigonal planar. The O—S—O bond angle is about 120° and the molecule is V-shaped.

Understanding What are the electron-pair arrangement, the shape, and the Cl—I—Cl bond angles in the ICl_4^- ion?

Answer: The electron-pair arrangement is octahedral, the Cl—I—Cl bond angles are 90° and 180°, and the molecular shape is square planar.

Geometry of Molecules with Multiple Central Atoms

Most molecules have more than one atom for which the VSEPR model can be used to assign geometry (in all the preceding cases, only one atom in the molecules was bonded to more than one other atom). In cases with more than one central atom, we simply apply the VSEPR rules to each central atom. For CH_3CN, the Lewis structure is

$$\begin{array}{c} \text{H} \\ | \\ \text{H—C(1)—C(2)}\equiv\text{N:} \\ | \\ \text{H} \end{array}$$

The C(1) atom forms four bonds and has no lone pairs, so its electron-pair arrangement is tetrahedral. The H—C(1)—H and H—C(1)—C(2) bond angles should be about 109°. The C(2) atom makes two bonds (one single bond and one triple bond) and has no lone pairs, so its electron-pair arrangement is linear. The C(1)—C(2)—N bond angle is therefore 180°. Figure 9.7 is a drawing of this structure.

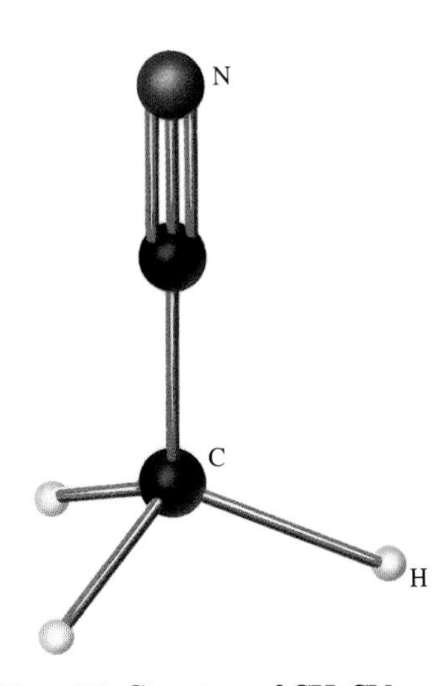

Figure 9.7 Structure of CH_3CN

Example 9.3 Shapes of Molecules

What is the geometry about each carbon atom in ethylene, C_2H_4?

Solution First we need the Lewis structure:

$$\begin{array}{ccc} \text{H} & & \text{H} \\ & \text{C=C} & \\ \text{H} & & \text{H} \end{array}$$

We need to assign a geometry to each carbon atom, but both are clearly the same. Each carbon is bonded to three other atoms and has no lone pairs; the electron-pair arrangement for three pairs is trigonal planar. The H—C—H and H—C=C bond angles are approximately 120°.

Understanding What is the geometry about each carbon atom in acetylene, HC≡CH?

Answer: Linear

Note that the VSEPR model does not predict how the geometry about one central atom will be oriented with respect to others in the molecule. For example, is ethylene completely planar or are the planes of the two CH_2 groups about the carbon atoms perpendicular or at some other angle (Figure 9.8)? The VSEPR model does not answer this question. A bonding model will be developed in Section 9.4 that does explain the overall shape of ethylene.

The geometry about each central atom is determined separately by applying the VSEPR model.

9.2 Polarity of Molecules

Objective

- To predict the polarity of a molecule from bond polarities and molecular shape

The knowledge of the shapes of molecules gives us important information about their physical and chemical properties. One such physical property is the polarity of a molecule. A **polar molecule** contains an unequal distribution of charge and thus has a dipole moment. We can use the differences in electronegativity between elements to predict whether *polar bonds* are present in a molecule. For diatomic molecules, if the bond is polar then the molecule is polar. For all other molecules, we need to know the molecular shape to be able to predict the polarity.

Consider the molecule CO_2. The Lewis structure is

$$:\ddot{O}=C=\ddot{O}:$$

From the VSEPR model we know that this is a linear molecule. Since oxygen

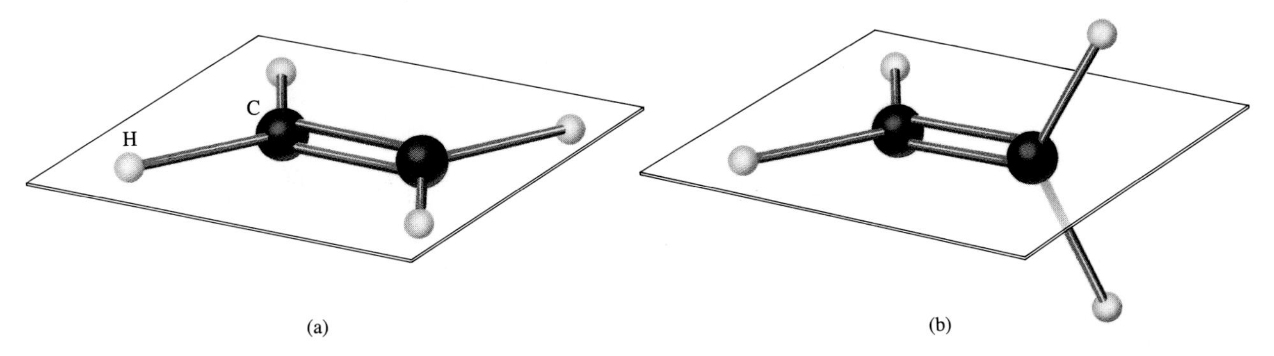

Figure 9.8 Two possible shapes of C_2H_4 Ethylene could exist in (a) planar or (b) nonplanar overall geometry. The VSEPR model does not predict how the bonds about two central atoms will be arranged.

is more electronegative than carbon, the bonds are polarized as follows:

$$\overset{\delta-}{\mathrm{O}}=\overset{\delta+}{\mathrm{C}}=\overset{\delta-}{\mathrm{O}}$$

Both bond polarity and molecular shape must be known to predict the polarity of a molecule.

Even though CO_2 has polar bonds, experiment shows that this molecule is nonpolar (has no dipole moment). To understand how this is possible, consider the arrows we would use to indicate the dipole moment of each bond.

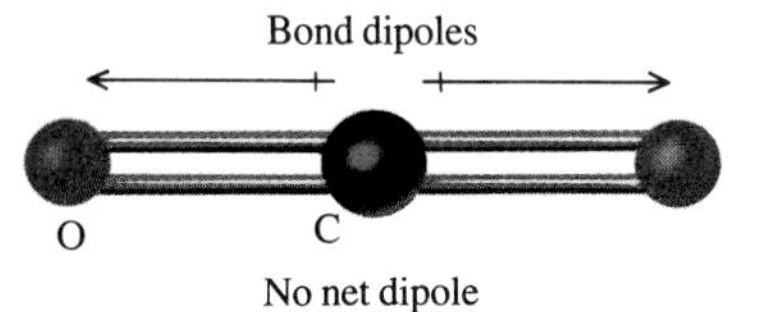

Arrows are used to help emphasize that dipole moments are vector quantities; they have both direction and magnitude. The arrow points toward the negative end of the bond (direction), and the length is proportional to the charge separation (magnitude). The polarity of the molecule is the sum of the dipole moments generated by the bonds. For CO_2, the vectors used to represent the dipole moments are of equal length. Because of the linear geometry, the vectors point in exactly opposite directions and cancel. We predict that CO_2 is a nonpolar molecule that contains polar bonds, a prediction verified by experiment.

It is useful to consider two similar molecules to help clarify these ideas. First, consider SO_2. As shown in Example 9.2(c), the O—S—O bond angle is about 120°.

$$\ddot{\mathrm{O}}=\ddot{\mathrm{S}}-\ddot{\mathrm{O}}$$

The oxygen atoms are more electronegative than sulfur, so the polarity is

$$^{\delta-}\mathrm{O}=\mathrm{S}^{\delta+}=\mathrm{O}^{\delta-}$$

In this case, the dipole moments do *not* cancel, but would sum as

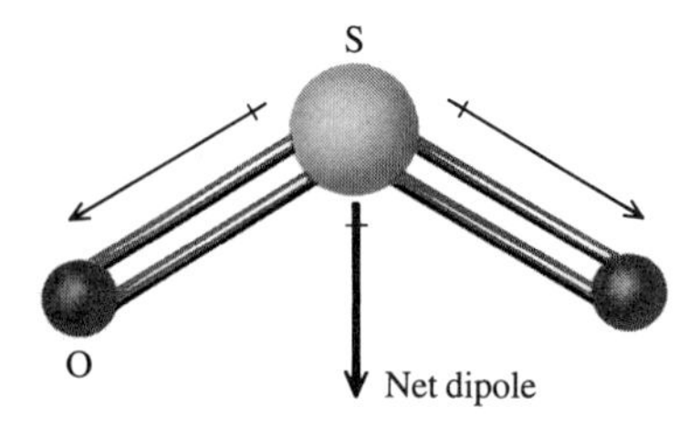

Only in molecules with linear geometry will two equal bond dipole moments cancel. Thus, SO_2 is a polar molecule; it differs from CO_2 because it has a different shape due to the lone pair on sulfur. *Molecules with lone pairs of electrons on a central atom are generally polar.*

Second, let us consider OCS. This molecule has a Lewis structure analogous to that of CO_2, but the two bonds have different polarities. In this case, the electronegativities of sulfur and carbon are about the same, so the only important bond dipole is along the C—O bond.

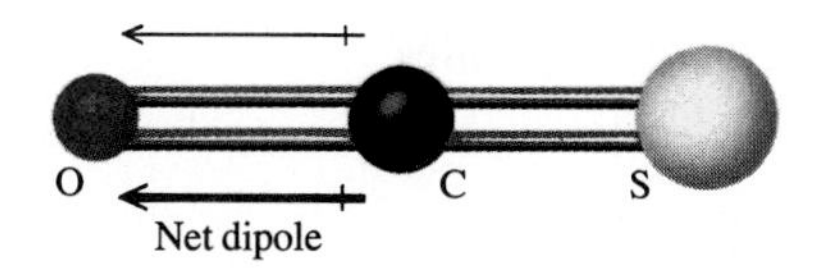

The molecule is linear, the same as CO_2, but OCS is a polar molecule because the bond dipole moments are not equal and therefore do not cancel each other.

Two other examples of nonpolar molecules with polar bonds are BCl_3 and CCl_4. The cancellation of the bond dipoles is not as clear as with CO_2, but in the trigonal planar geometry of BCl_3 and the tetrahedral geometry of CCl_4 the dipole moments of the equally polar bonds also cancel exactly.

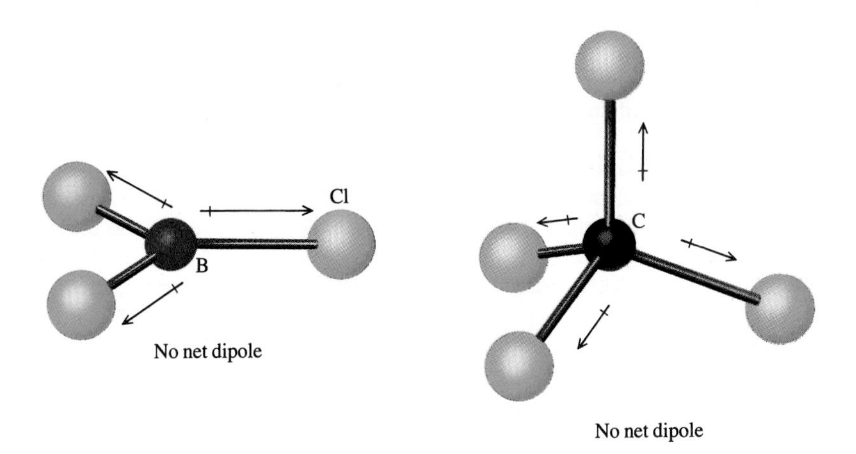

If we replace one of the chlorine atoms with an atom of another element, the new molecule is polar. For example, both BCl_2F and $HCCl_3$ are polar molecules.

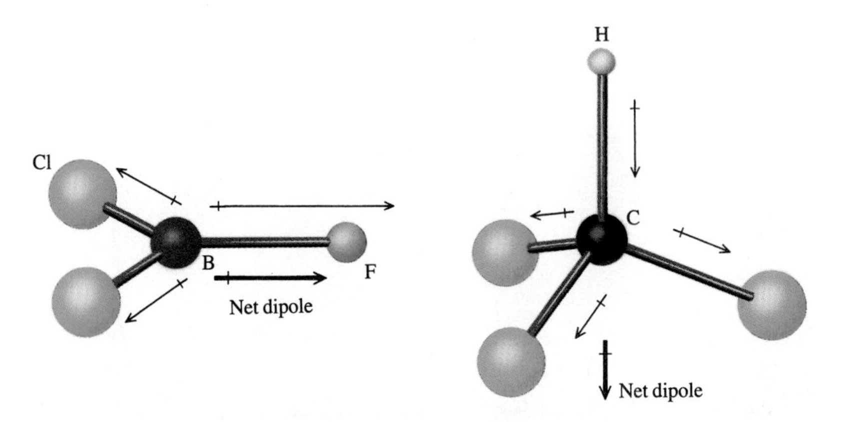

Only identical bond dipoles cancel completely for these geometries. *Nonpolar molecules that contain polar bonds are observed when there are no lone pairs on the central atom, and all of the atoms bonded to the central atom are identical.*

A molecule with polar bonds is nonpolar if the geometry causes the bond polarities to cancel.

Example 9.4 Polarity of Molecules

Classify the following molecules as polar or nonpolar: (a) CH_4, (b) HCN, (c) H_2O, (d) XeF_4.

Solution

(a) As shown in Figure 9.1, CH_4 is a tetrahedral molecule. The four bond dipole moments (which point toward carbon) are equal and cancel because of the regular geometry. The molecule is nonpolar.

(b) The structure of HCN is

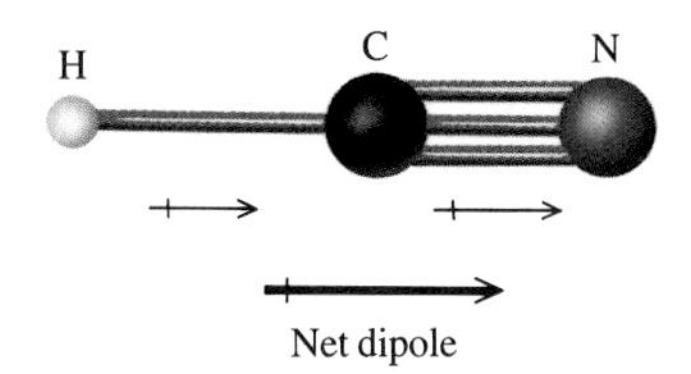

This linear molecule is polar, with a dipole moment that is the sum of the two bond dipoles.

(c) The oxygen atom in water has two lone pairs and is bonded to two hydrogen atoms; the electron-pair arrangement is tetrahedral and the measured H—O—H bond angle is 104.5°. The bond dipoles do not cancel and the molecule is polar (Figure 9.9).

(d) The structure of XeF_4, based on an octahedral electron-pair arrangement, is

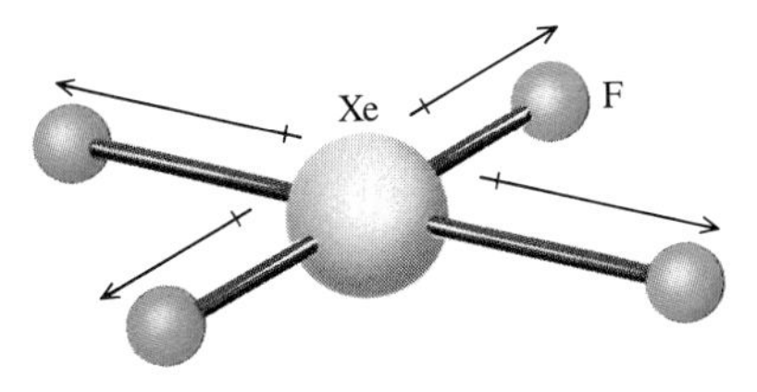

The bond dipoles of the four fluorine atoms cancel because there are two pairs oriented at 180°, and the molecule is nonpolar. XeF_4 is a rare example of a

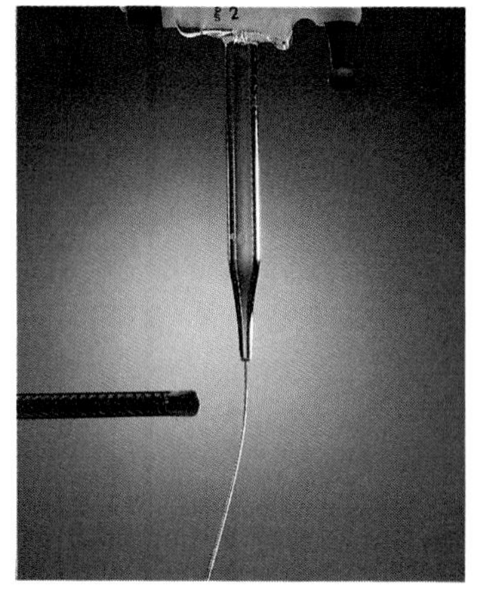

(a)

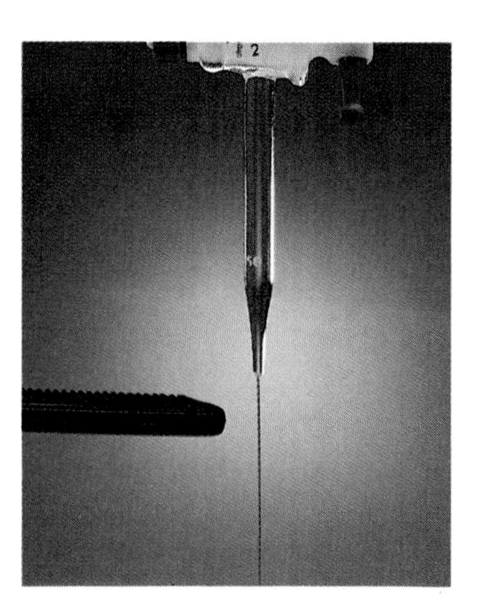

(b)

Figure 9.9 Polarity of molecules (a) A stream of water is deflected by a charged comb because the compound is polar. (b) A stream of CCl_4 is not deflected by the charged comb because the compound is nonpolar.

nonpolar molecule with lone pairs on the central atom, because the lone pairs, as well as the fluorine atoms, are oriented at 180°.

Understanding Is PF_3 polar or nonpolar?

Answer: Polar

Example 9.5 Polarity of Molecules

An experiment shows that the molecule PF_2Cl_3 is nonpolar. What is the molecular shape and what is the arrangement of the atoms in that shape?

Solution This molecule has five bonding pairs and no lone pairs about phosphorus, so the electron-pair arrangement is trigonal bipyramidal. Because the axial and equatorial sites are different, there are three possible arrangements of the fluorine and chlorine atoms:

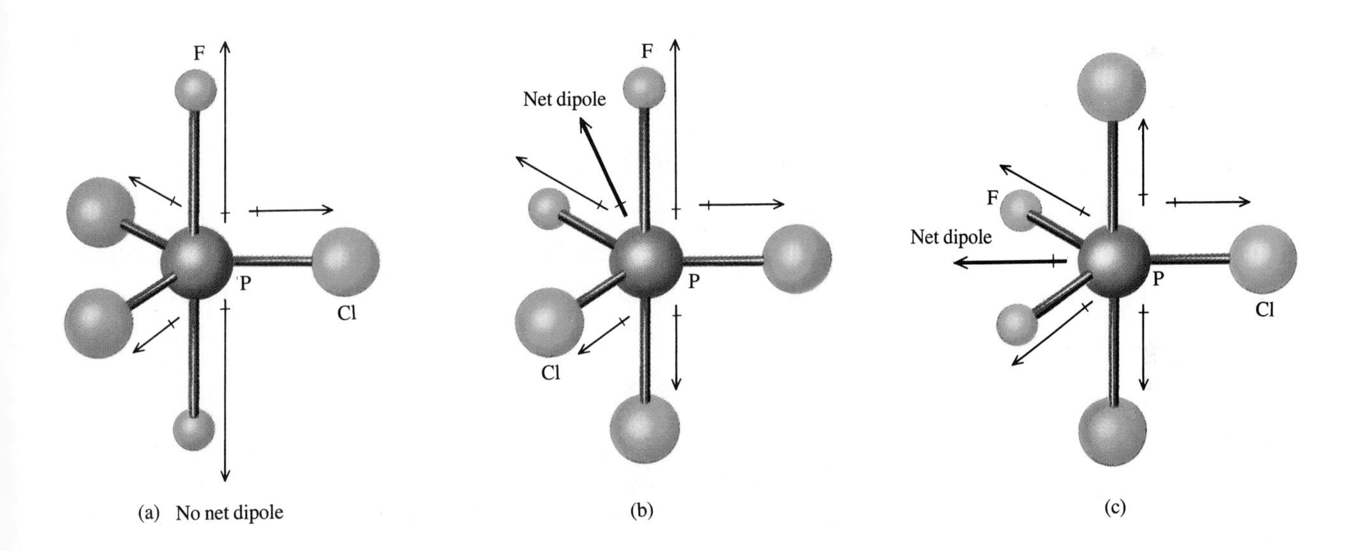

(a) No net dipole (b) (c)

The two fluorine atoms can both be in the axial positions (a) or both in the equatorial positions (c), or there can be one in each kind of position (b). For a, the dipole moments of the axial P—F bonds will cancel (this is the same as linear geometry) and the dipole moments of the equatorial P—Cl bonds will also cancel (this is a trigonal planar geometry). The molecule is nonpolar in this arrangement. The bond dipole moments will not cancel in the arrangements shown in b or c. In b, neither the axial nor equatorial bonds have equal dipole moments that could cancel. In c, the dipole moments of the axial P—Cl bonds will cancel, but the bond dipole moments in the equatorial plane will not cancel. Thus, the arrangement in a is correct if experiment has shown that the molecule is nonpolar. This example is one of the few cases of a nonpolar molecule that does not place the same type of atom in all sites of the polyhedron, but it does place the same atoms in the two sets of *equivalent* sites that occur in a trigonal bipyramid.

Understanding If the hypothetical molecule $SF_2Cl_2Br_2$ is nonpolar, what is its shape and what is the arrangement of the atoms?

Answer: Octahedral with the arrangement

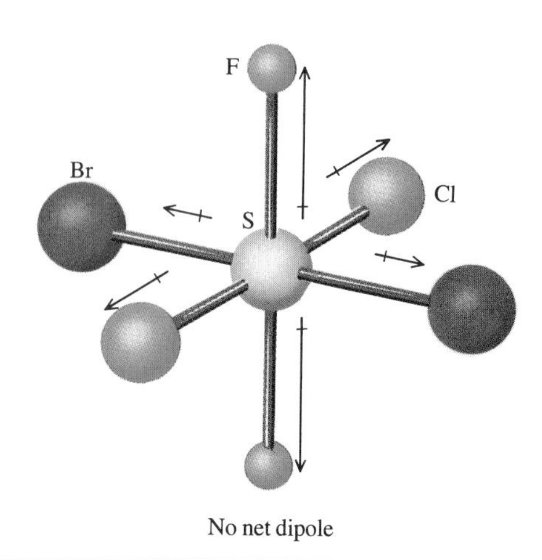

9.3 Valence Bond Theory

Objectives

- To identify the orbitals used to form the bonds in any specific molecule
- To assign the hybrid orbitals used by a central atom to form bonds and hold lone pairs

Lewis structures and VSEPR theory are very useful in describing the shapes of molecules. A more detailed bonding description includes the orbitals being used by the atoms to make the chemical bonds. Covalent bonds are formed from overlap of **valence orbitals,** orbitals in the valence shell, on each atom involved in the bond. **Valence bond theory** describes bonds as being formed by atoms sharing valence electrons in overlapping valence orbitals.

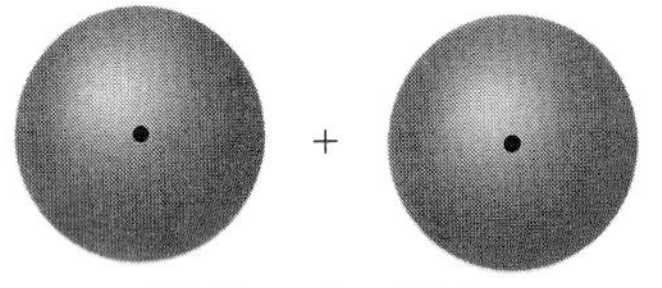

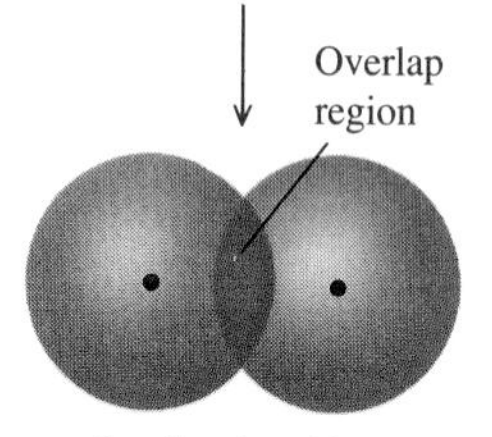

Figure 9.10 Orbitals used for the bond in H_2 The bond in H_2 is formed by overlap of a 1*s* orbital on each hydrogen atom, each containing one electron.

Section 8.3 described the orbitals used in forming the bond in H_2. As shown in Figure 9.10, the 1*s* orbitals of the two hydrogen atoms overlap. The two electrons must pair because they are in overlapping orbitals. The electron density of the pair is concentrated between the two hydrogen nuclei. Each hydrogen atom uses one valence-shell orbital and one valence electron to form the bond.

In valence bond theory the simple overlap of valence orbitals explains the bonding. Each shared pair of electrons must use one valence orbital on each of the bonded atoms. We need to consider the *orbitals available for bonding* and the *number of valence electrons* that each atom provides for bonding. To form the bond in HF, we know that the hydrogen atom will use a 1*s* orbital, but the energy level diagram of fluorine is needed to determine which orbital it uses. Only the energy levels of the valence electrons of fluorine are considered, because the inner 1*s* orbital is very low in energy and close to the nucleus, and thus is unavailable for bond formation.

2p ↑↓ ↑↓ ↑

2s ↑↓

F

Three of the valence orbitals are filled, but one of the $2p$ orbitals contains a single electron. This partially filled orbital overlaps with the hydrogen atom $1s$ orbital to form the covalent bond (Figure 9.11).

In valence bond theory, bonds are formed by atoms sharing two electrons in overlapping atomic orbitals.

Note three features of this bonding description. First, the orientation of the bond is along the axis of the p orbital. This orientation maximizes the overlap of the orbitals while minimizing the repulsive interaction of the nuclei of the bonded atoms. Second, each of the atomic orbitals used to form the bond is occupied by one electron in the separate atoms. Third, the three filled valence orbitals on fluorine are the lone pairs on fluorine in the Lewis structure of HF.

H—F: (with lone pairs)

Thus, this valence bond theory description of the bonding in HF is really just an elaboration of Lewis theory in which we consider the orbitals that are used to form the covalent bond and those that hold the lone pairs.

Example 9.6 Orbital Description of Covalent Bonds

Identify the atomic orbitals that are used in the formation of the covalent bond in Cl_2 and the atomic orbitals that hold the lone pairs shown in the Lewis structure.

Solution The energy-level diagram for the valence electrons of chlorine is

3p ↑↓ ↑↓ ↑

3s ↑↓

Cl

Each chlorine atom has a single $3p$ orbital containing one electron. The covalent bond is formed from the overlap of these half-filled $3p$ orbitals on each of the chlorine atoms. The three lone pairs on each chlorine atom in the Lewis structure are those paired electrons in the $3s$ and the two filled $3p$ valence orbitals in the separate chlorine atoms.

Understanding Identify the atomic orbitals that are used to form the covalent bond in HCl and the atomic orbitals that accommodate the lone pairs.

Answer: A $1s$ orbital on hydrogen overlaps with a $3p$ orbital on chlorine. The three lone pairs on chlorine are in the $3s$ and two of the $3p$ valence orbitals.

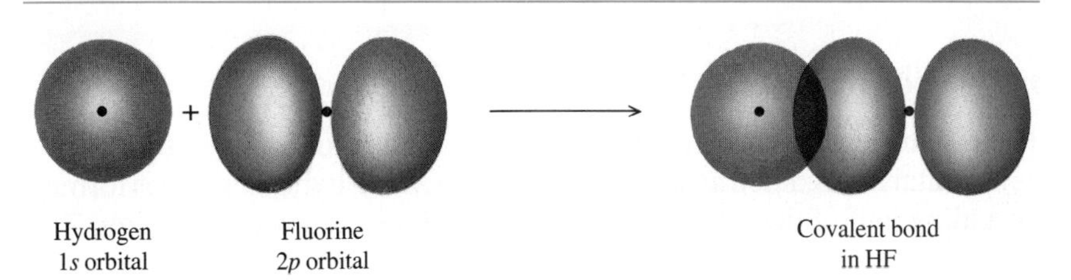

Figure 9.11 Covalent bond in HF The bond in HF is formed from overlap of a $1s$ orbital on the hydrogen atom with a p orbital on fluorine.

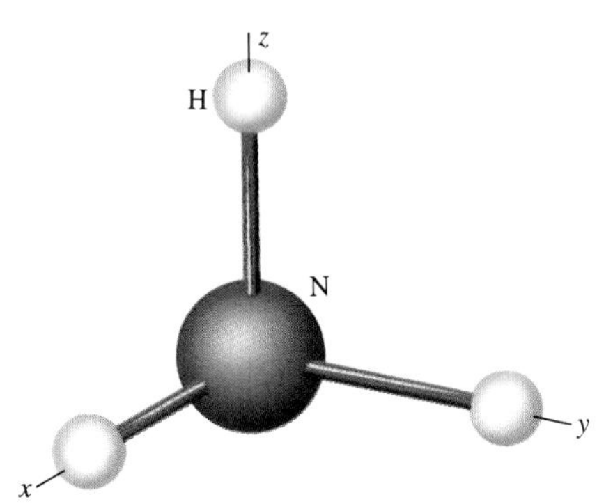

Figure 9.12 Bonding of three hydrogen atoms with three *p* orbitals If three bonds were made by the three *p* orbitals of a central atom, the bond angles would all be 90°.

Hybridization of Atomic Orbitals

Experiment shows that the *s*, *p*, and *d* valence orbitals and the ground state arrangement of electrons in atoms cannot always be used to describe the bonding in molecules. For example, the H—N—H bond angles in ammonia, NH_3, are all 107° and the three bonds are identical in strength. The valence orbitals available to form bonds for nitrogen atom are the 2*s* and three 2*p* orbitals. There is an electron in each of the 2*p* orbitals. These electrons are well-suited to form the three bonds.

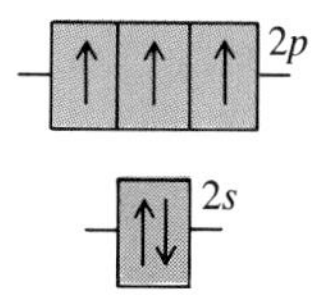

These three *p* orbitals are oriented at 90° angles with respect to each other; if bonds were formed directly by these orbitals, there would be three 90° H—N—H bond angles (Figure 9.12). This predicted result is at odds with the experimental observations, indicating that the orbitals used by atoms in molecules to form bonds are not necessarily the same as the orbitals of the free atom.

These problems can be solved by remembering that each orbital is a mathematical expression that describes the electron as a wave. We can mathematically mix (or average) two or more of these wave functions that describe the electron, and produce an equal number of wave functions that have different shapes and orientations. The details of such mixing are left for more advanced courses, but we can use the idea to understand molecular shape. **Hybrid orbitals** are orbitals obtained by mixing two or more atomic orbitals on the same atom. The new hybrid orbitals will have different shapes and directional properties than the orbitals used in constructing them. However, the total number of orbitals is the same before and after hybridization.

Hybrid orbitals are needed to explain the shapes of many molecules.

The formation of hybrid orbitals is based on the mathematical combining of the orbitals, and it is really not very different from the averaging of numbers. A simple analogy is the mixing of different colors of paint to obtain a new color. For example, we could make a pink-colored paint by mixing a cup of red and a cup of white paint (Figure 9.13). If we place the new mixture back into the cups, we still have *two cups* of paint that contain everything that was in the original two cups, but the color is now an average of the originals. If we wanted a darker color of paint, we could mix two cups of red with one of white, producing *three cups* of a darker-colored paint. The mixing of orbitals is analogous; the energy and directional characteristics of the new hybrid orbitals will be decided by the type and number of atomic orbitals used in the mixing. The number of new hybrid orbitals will be the same as the number of orbitals used to form them.

sp Hybrid Orbitals

The mixing of one *s* orbital with one *p* orbital leads to the formation of two new orbitals designated as *sp* hybrid orbitals. One *sp* hybrid orbital is formed by adding half of the *p* wave function and half of the *s* wave function. The other *sp* hybrid orbital is formed by subtracting half of the *p* wave function from half of the *s* wave function. The results of this mixing can be seen

Figure 9.13 Mixing paint as an analogy to the formation of hybrid orbitals The mathematical mixing of atomic orbitals to form hybrid orbitals is similar to mixing different colors of paint to obtain a new color. After the paint is mixed, there are still the same number of cups of paint, but the color has changed.

pictorially (Figure 9.14). The *sign* of the wave function's amplitude in one lobe of a *p* orbital is opposite to the sign in the other lobe (indicated by shading in the figure). When the *p* orbital is added to the *s* orbital, there is reinforcement of the wave amplitude on one side of the nucleus and cancellation on the other, forming an *sp* hybrid orbital that has one large lobe and

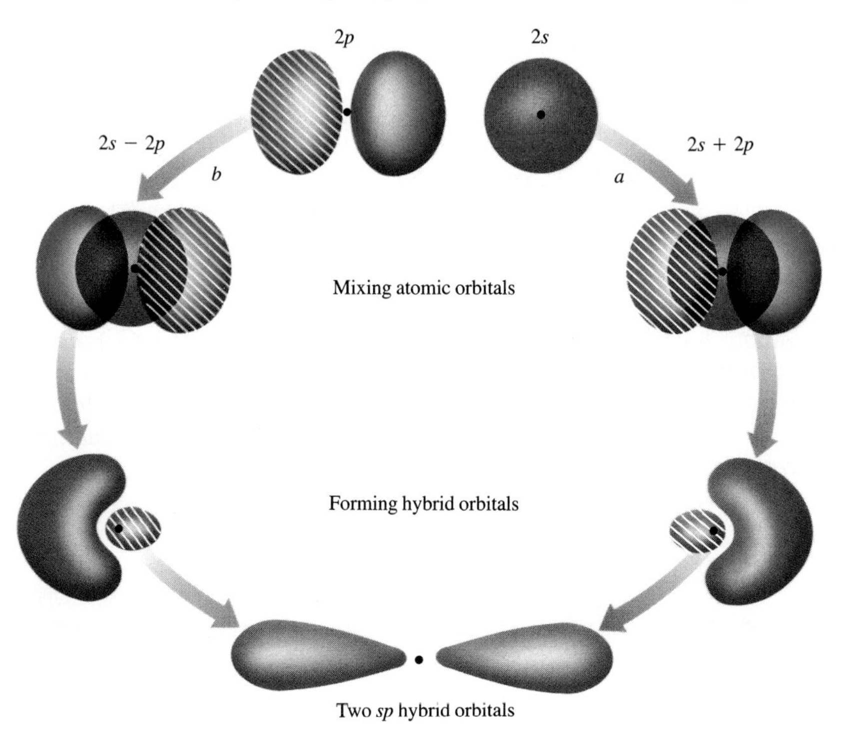

Figure 9.14 *sp* Hybrid orbital The *sp* hybrid orbitals are formed from the addition, *a*, and subtraction, *b*, of one *s* and one *p* atomic orbital. In *a*, the amplitudes of the wave functions reinforce on the right side of the nucleus and in *b* on the left side. The bottom drawing shows the two new hybrid orbitals about the nucleus, with the shape of the orbitals elongated and the smaller lobes omitted for clarity.

one small lobe, as pictured. The other *sp* hybrid orbital is formed by subtraction of the *p* orbital (this reverses the signs of the amplitude in the two lobes) from the *s* orbital, causing reinforcement on the opposite side of the nucleus. Both hybrid orbitals have the same shape, but their large lobes are oriented at 180° with respect to each other.

Beryllium chloride, $BeCl_2$, is an example of a molecule in which the bonding is best described by *sp* hybrid orbitals on the Be atom. The bonding in $BeCl_2$ can be pictured by first considering the energy-level diagram of beryllium.

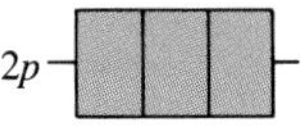

Experiment shows that the molecule is nonpolar, so the bonds in $BeCl_2$ must be equivalent and oriented at 180°. The bonding is explained by the formation of two *sp* hybrid orbitals on the beryllium atom. Note that the energy of the two new hybrid orbitals is the average of the energies of the *s* and *p* orbitals from which they were formed. Following Hund's rule, the two electrons are placed in the *sp* hybrid orbitals with the same spin.

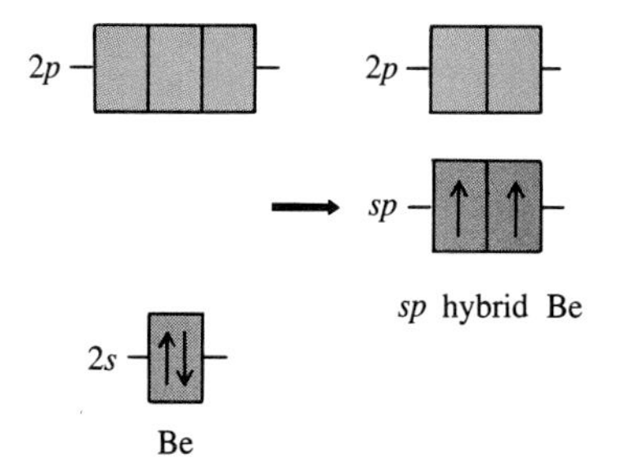

The two bonds in $BeCl_2$ are made by overlap of a $3p$ orbital on each chlorine with an *sp* hybrid orbital on beryllium (Figure 9.15). The two unhybridized *p* orbitals on beryllium remain vacant in the molecule. Note that the electronic state of two electrons in the *sp* hybrid orbitals is higher in energy than the $2s^2$ ground state configuration, but the energy required for this promotion is much less than is released when the two Be—Cl bonds form.

It is important to remember that two equivalent *sp* hybrid orbitals form bonds with a bond angle of 180°. *The sp hybrid orbitals describe the bonding on central atoms that have 180° bond angles.*

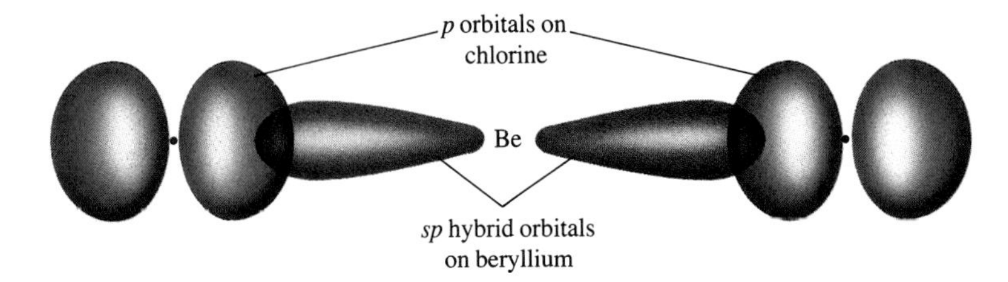

Figure 9.15 Bonding in $BeCl_2$ The bonds in $BeCl_2$ are made from the overlap of two *sp* hybrid orbitals on the beryllium atom with $3p$ orbitals on the two chlorine atoms.

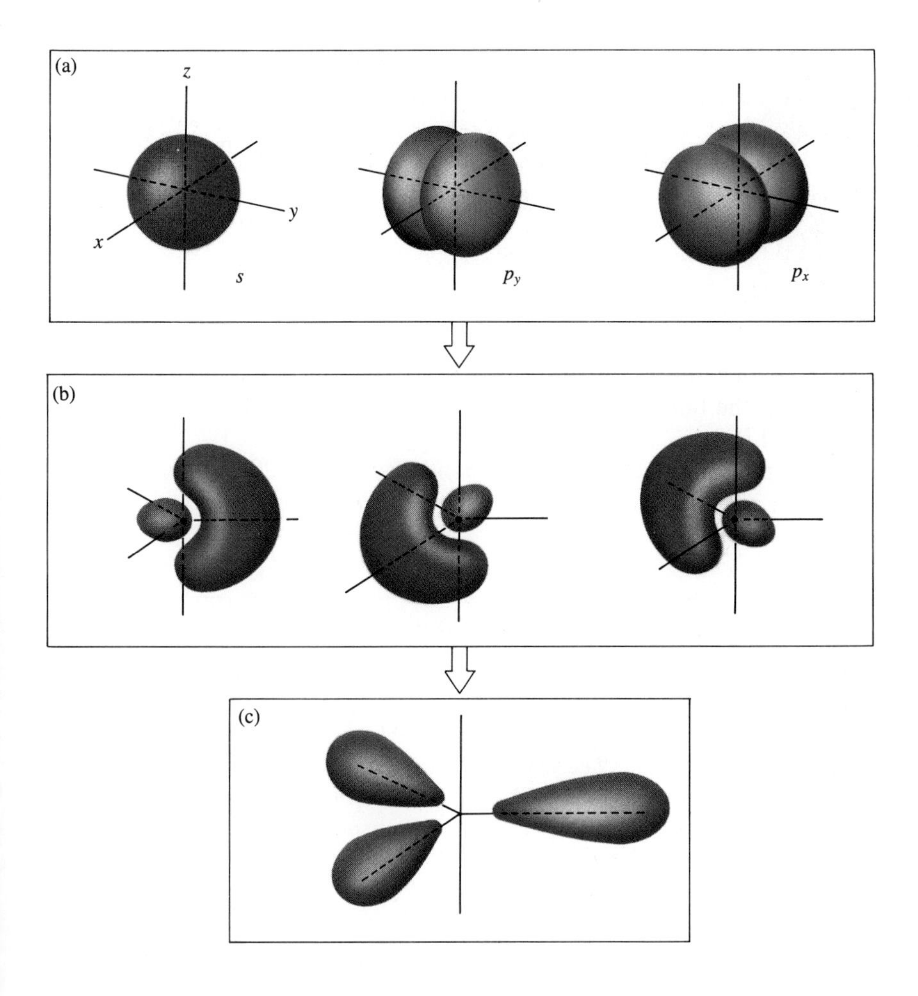

Figure 9.16 ***sp*^{2} Hybrid orbitals** The mixing of an *s* and two *p* orbitals (a) leads to the formation of three sp^2 hybrid orbitals (b). They point toward the corners of an equilateral triangle and are located in the same plane as the two *p* orbitals from which they were formed. In (c) the orbitals are elongated and the small lobes omitted for clarity.

sp^2 Hybrid Orbitals

The combining of one *s* orbital with two *p* orbitals leads to the formation of three sp^2 hybrid orbitals. Each of these orbitals has one large and one small lobe similar to those of *sp* hybrid orbitals. They lie in the same plane (the plane containing the two *p* orbitals from which they were formed) and point toward the corners of an equilateral triangle (Figure 9.16).

The bonding in BF_3 is best described by sp^2 hybridization of the boron atom. As predicted by the VSEPR model, experiment shows that BF_3 is a planar molecule with equal F—B—F bond angles of 120° and equal B—F bond lengths. The energy-level diagram of boron has two electrons in the 2*s* orbital and one in a 2*p* orbital. Hybridization of the *s* and two of the *p* orbitals produces three sp^2 hybrid orbitals, each containing one electron.

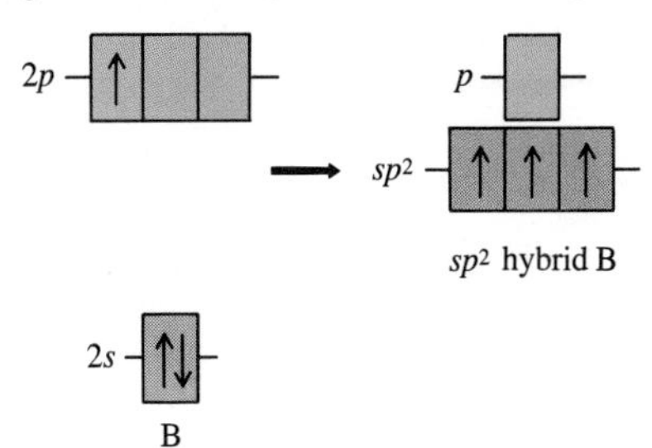

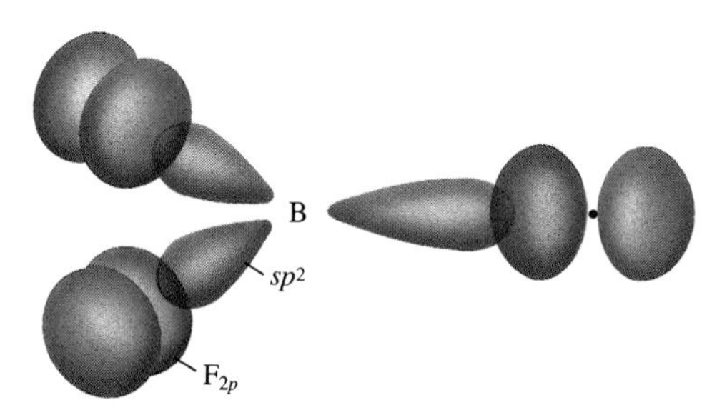

Figure 9.17 Bonding in BF_3 The bonds in BF_3 are made from the overlap of three sp^2 hybrid orbitals on the boron atom with $2p$ orbitals on the three fluorine atoms.

The bonds in BF_3 are formed by overlap of each of the three sp^2 hybrid orbitals on the boron atom with a p orbital on one of the fluorine atoms (Figure 9.17).

The sp^2 hybrid orbitals produce bonds that make 120° angles with each other. *The sp^2 hybrid orbitals describe the bonding on central atoms that have 120° bond angles.*

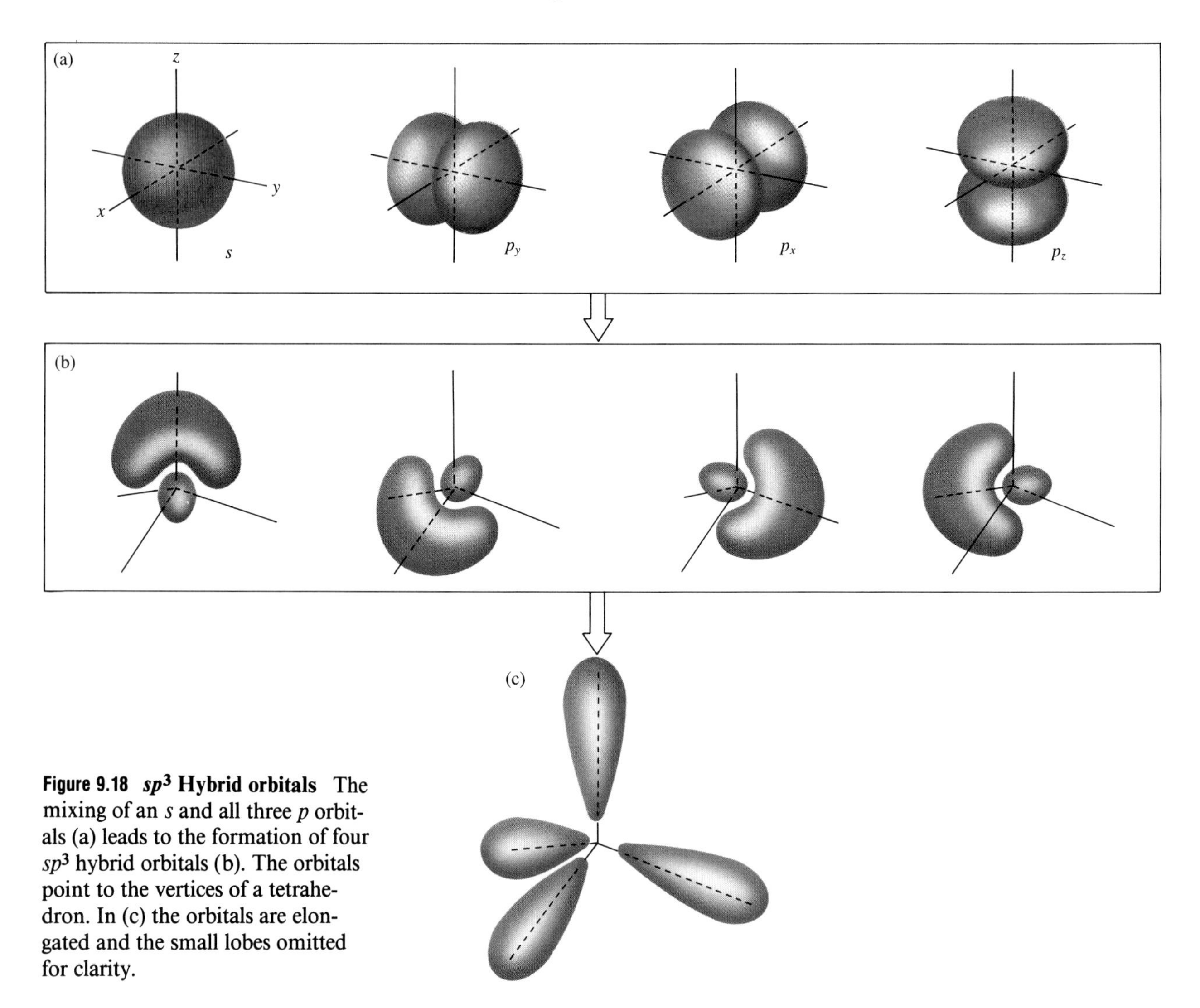

Figure 9.18 sp^3 Hybrid orbitals The mixing of an s and all three p orbitals (a) leads to the formation of four sp^3 hybrid orbitals (b). The orbitals point to the vertices of a tetrahedron. In (c) the orbitals are elongated and the small lobes omitted for clarity.

sp^3 Hybrid Orbitals

The mixing of the *s* and all three of the *p* orbitals yields four sp^3 hybrid orbitals. Again, each of these orbitals has a large and a small lobe. The large lobes point at the corners of a tetrahedron (Figure 9.18).

Methane (CH_4) is a molecule in which the central atom has sp^3 hybrid orbitals. All four valence orbitals are sp^3 hybridized, making four equivalent orbitals directed at angles of 109.5°. Each hybrid orbital contains one of the four valence electrons on the carbon atom.

2p ↑ ↑ → sp^3 ↑ ↑ ↑ ↑

sp^3 hybrid C

2s ↑↓

C

The bonds in methane are formed by overlap of each of the four sp^3 hybrid orbitals on the carbon atom with a 1*s* orbital on one of the hydrogen atoms (Figure 9.19).

The sp^3 hybrid orbitals describe the bonding on central atoms that have bond angles of approximately 109°.

Another example of a molecule with a tetrahedral electron-pair arrangement is water. The bonding in water is different from the bonding in methane because not all of the valence electrons in water are used to make bonds. The bond angles in water are about 109°, so sp^3 hybridization of the orbitals on oxygen is necessary.

2p ↑↓ ↑ ↑ → sp^3 ↑↓ ↑↓ ↑ ↑

sp^3 hybrid O

2s ↑↓

O

Two of the hybrid orbitals on oxygen are filled and two are occupied by one electron. The two half-filled hybrid orbitals are used to form the O—H bonds by overlap with the 1*s* orbitals on the hydrogen atoms (Figure 9.20). The two filled hybrid orbitals contain the lone pairs shown in the Lewis structure of water. One hybrid orbital is needed on the oxygen atom for each bonding pair and each lone pair of electrons.

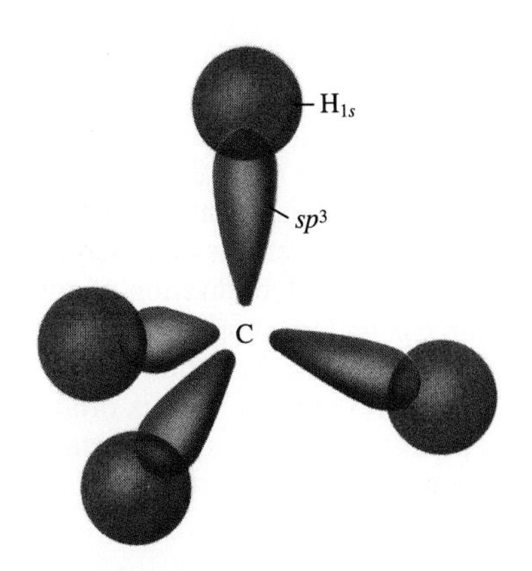

Figure 9.19 Bonding in CH_4 The bonds in CH_4 are made from the overlap of four sp^3 hybrid orbitals on the carbon atom with 1*s* orbitals on the four hydrogen atoms.

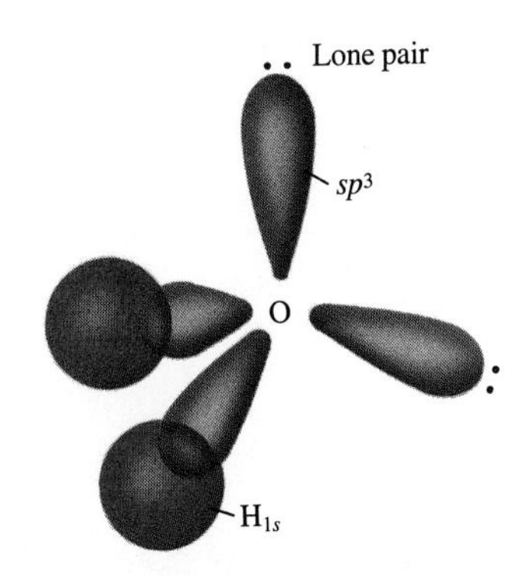

Figure 9.20 Bonding in water The bonds in H_2O are made from the overlap of two sp^3 hybrid orbitals on the oxygen atom with 1*s* orbitals on the hydrogen atoms. The lone pairs on oxygen occupy the remaining two sp^3 orbitals.

Hybridization Involving *d* Orbitals

The number of hybrid orbitals of any given type is equal to the number of atomic orbitals from which they are formed. The hybridization of a central atom must provide a hybrid orbital for each of its lone pairs and each atom that is bonded to it. Note that the number of hybrid orbitals on any central

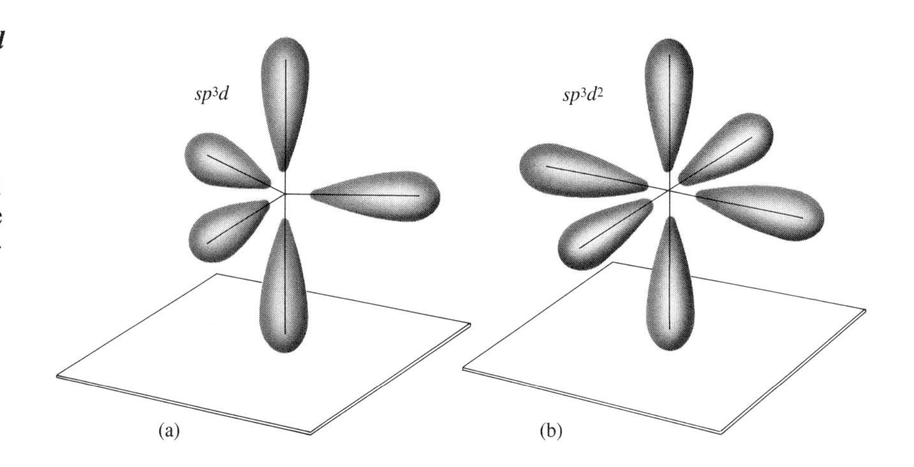

Figure 9.21 Hybridization involving *d* orbitals (a) Hybridization of one *s*, three *p*, and the appropriate *d* orbital leads to the formation of five sp^3d hybrid orbitals. The five orbitals point to the vertices of a trigonal bipyramid. (b) Hybridization of one *s*, three *p*, and two appropriate *d* orbitals leads to the formation of six sp^3d^2 hybrid orbitals. The six orbitals point to the vertices of an octahedron.

atom and the electron-pair arrangement counted for the VSEPR model are exactly the same. For electron rich molecules with five pairs of electrons around a central atom, such as the phosphorus atom in PF_5, five atomic orbitals are needed to make five hybrid orbitals. This bonding is described by sp^3d hybridization, which generates five hybrid orbitals pointing toward the vertices of a trigonal bipyramid (Figure 9.21). For molecules with six electron pairs around a central atom, such as the sulfur atom in SF_6, sp^3d^2 hybridization is used. The six sp^3d^2 hybrid orbitals point toward the vertices of an octahedron (Figure 9.21).

Example 9.7 Hybridization

Describe the hybridization of the central atom and the bonding in (a) BCl_3, (b) NH_3, (c) IF_5.

Solution

(a) The correct set of hybrid orbitals to be used in any bonding situation is determined by the bond angles about the central atom. The Lewis structure of BCl_3 has three bonded atoms and no lone pairs on the central boron atom, so the electron-pair arrangement is trigonal planar with 120° bond angles. In order to make bonds at 120°, the boron atom is sp^2 hybridized with one electron in each hybrid orbital. The three bonds are made from overlap of these sp^2 orbitals on boron with a $3p$ orbital on each of the chlorine atoms (Figure 9.22).

(b) The electron-pair arrangement of ammonia is tetrahedral. The 109° bond angles of the tetrahedron (measured bond angles are 107°) are formed by sp^3 hybrid orbitals on the nitrogen atom. Three of the four hybrid orbitals hold one electron and overlap with the $1s$ orbitals on the hydrogen atoms. The fourth sp^3 hybrid orbital holds the lone pair (Figure 9.22).

(c) The electron-pair arrangement of IF_5 is octahedral. The iodine atom uses sp^3d^2 hybrid orbitals. The F—I—F bond angles are about 90°, and each bond is formed from overlap of an sp^3d^2 hybrid orbital on iodine with a *p* orbital on a fluorine atom. The remaining hybrid orbital holds a lone pair (Figure 9.22).

Understanding Describe the bonding in SF_4.

Answer: The four S—F bonds are made from overlap of an sp^3d orbital on sulfur with a $2p$ orbital on each fluorine atom. The remaining hybrid orbital on sulfur contains a lone pair.

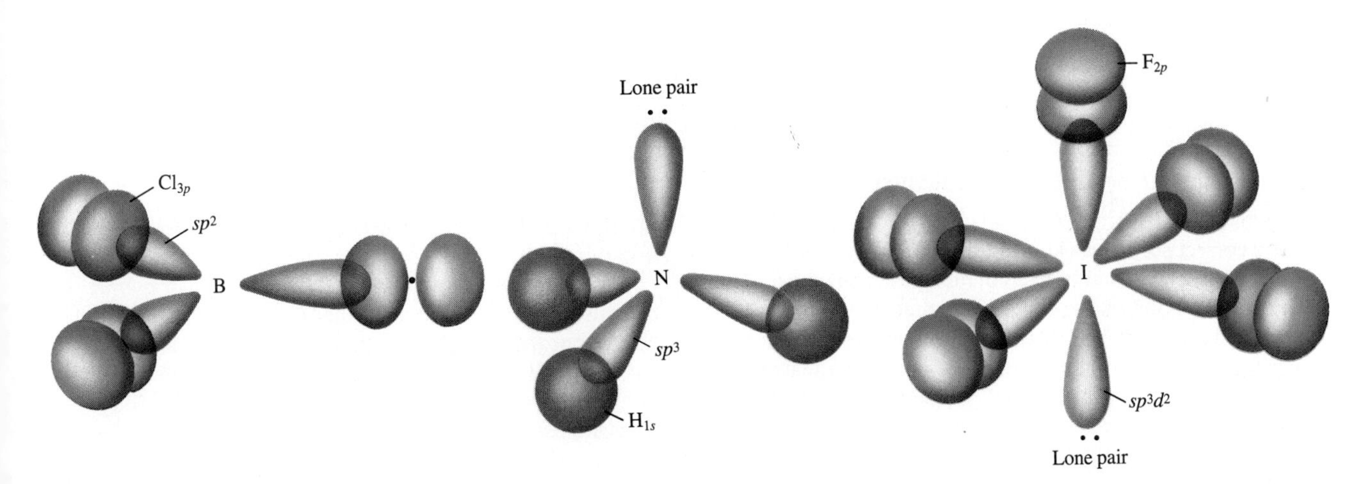

Figure 9.22 Bonding pictures of BCl_3, NH_3, and IF_5

Coordinate Covalent Bonds

Hybrid orbitals can also be used to describe coordinate covalent bonds. The coordinate covalent bond is formed by overlap of a filled orbital on one atom with an empty orbital on another atom. When ammonia reacts with BCl_3, the lone pair on nitrogen provides both electrons to form a coordinate covalent bond with boron. The electrons on nitrogen are in an sp^3 hybrid orbital; the unused orbital on boron in BCl_3 is a *p* orbital (BCl_3 is described by sp^2 hybridization, so there is an empty *p* orbital in the valence shell). As the coordinate covalent bond forms, the electron-pair arrangement of boron changes from trigonal planar to tetrahedral, changing the hybridization from sp^2 to sp^3. In forming the coordinate covalent bond, the hybridization on the donor atom, nitrogen, does not change, but the hybridization on the acceptor atom does (Figure 9.23).

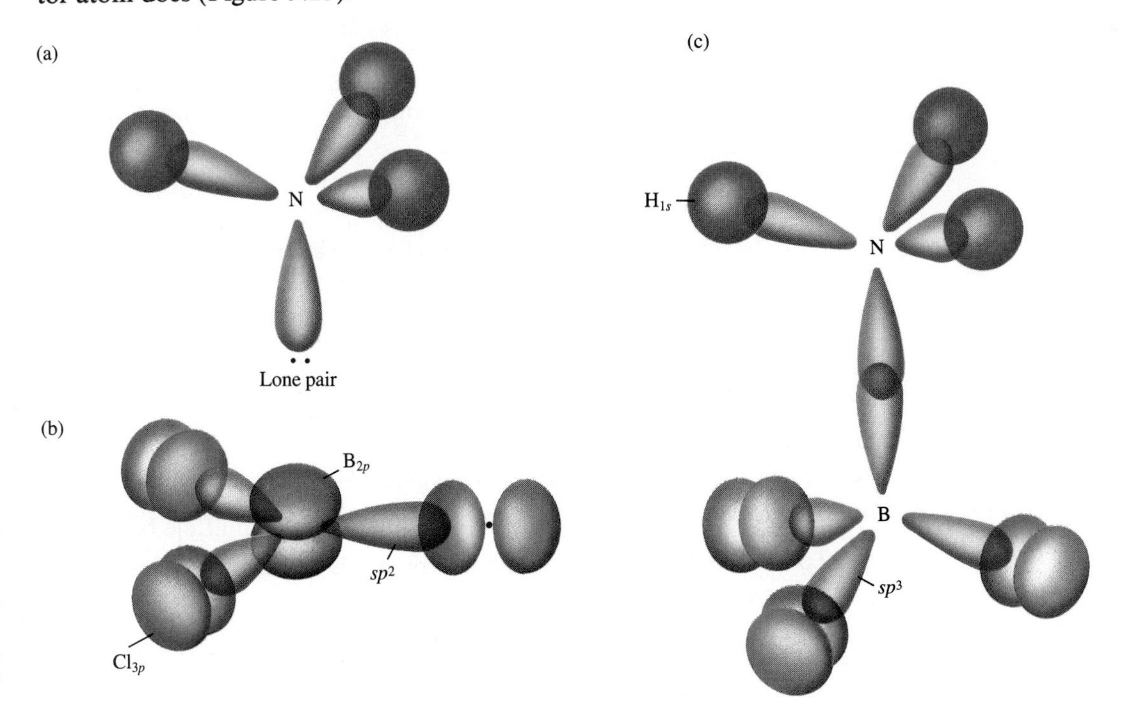

Figure 9.23 Hybrid orbitals forming a coordinate covalent bond The lone pair on the nitrogen atom in NH_3 (a) can donate two electrons to the electron-deficient boron atom in BCl_3 (b) to form a coordinate covalent bond (c).

INSIGHTS into CHEMISTRY

A Closer View

Experimental Bond Angles Explained by Hybridization Theory

The correct set of hybrid orbitals used in any bonding situation is determined by the bond angles about the central atom. The bond angles can be measured experimentally or predicted from VSEPR theory. The higher the percentage of *p* versus *s* orbitals used in the bonding, the closer the bond angle approaches 90°. In the three types of *s-p* hybrid orbitals, the bond angles decrease from 180° to 120° to 109.5° as the fraction of *p* orbitals increases from sp ($\frac{1}{2}$ *p*) to sp^2 ($\frac{2}{3}$ *p*) to sp^3 ($\frac{3}{4}$ *p*).

The hybridization of oxygen in H_2O is described as sp^3 because the electron-pair arrangement is tetrahedral. Experiment shows that the H—O—H bond angle is actually 104.5°, certainly close to 109°, but not exactly equal to it. The bonding description for water can be improved by using hybrid orbitals that have slightly more *p* character to form the O—H bonds, leaving the lone pairs in orbitals with slightly more *s* character. A larger fraction of *p* orbitals in the hybrid orbitals that form the bonds reduces the bond angles. The same type of hybrid orbitals must be used to form the equivalent O—H bonds, but the orbitals used to form the bonds do not have to be the same as the lone pair orbitals. Hybrid orbitals with fractional values of *s* and *p* orbital contributions are used in order to obtain the correct bond angles.

In certain cases, VSEPR theory does not correctly predict bond angles. An example is PH_3, for which the electron-pair arrangement is predicted to be tetrahedral; experiment shows that the H—P—H bond angles are really 93.7°. In order to make bonds at this angle, the phosphorus atom must use nearly pure *p* orbitals. This leaves the lone pair in an orbital that is mostly *s* in character. Hybridization theory cannot predict this bond angle, but can be used to interpret the bonding.

Example 9.8 Hybrid Orbitals and Coordinate Covalent Bonding

Describe the bonding in PF_5 and in PF_6^-, the ion formed in the reaction of PF_5 with F^-.

Solution The electron-pair arrangement of PF_5 is trigonal bipyramidal. The hybridization of phosphorus is sp^3d and there are several empty *d* orbitals available to accept additional electrons. The coordinate covalent bond in PF_6^- is formed by donation of a lone pair from the fluoride anion. The hybridization on phosphorus changes from sp^3d to sp^3d^2.

Understanding Describe the bonding in BF_3 and in BF_4^-, the ion formed in the reaction of BF_3 with F^-.

Answer: The boron atom is sp^2 hybridized in BF_3 and changes to sp^3 hybridization in BF_4^-.

9.4 Multiple Bonds

Objectives

- To determine the orbitals used in the formation of sigma and pi bonds
- To describe the bonding in molecules that contain multiple bonds
- To use the properties of sigma and pi bonds to describe *cis* and *trans* isomers

In all of the examples of bonding that are shown in Section 9.3, the atoms are connected by single bonds. The bonds are formed by overlap of *s* or *p* orbitals or hybrid orbitals oriented along the bond (Figure 9.24). These bonds, formed from the head-to-head overlap of atomic or hybrid orbitals, are called **sigma bonds** (σ), bonds in which the shared pair of electrons is concentrated around the line joining the nuclei of the bonded atoms.

Sigma bonds are formed by orbitals directed toward each other.

Double Bonds

Many molecules have double bonds; an important example is ethylene, C_2H_4.

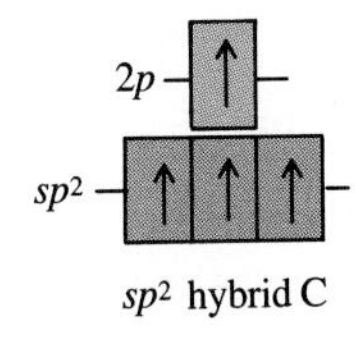

The electron-pair arrangement of each carbon atom is trigonal planar, indicating that the hybridization is sp^2. The energy-level diagram for an sp^2 hybridized carbon atom that will make four bonds is

$2p$ ↑

sp^2 ↑ ↑ ↑

sp^2 hybrid C

There are three sp^2 hybrid orbitals and one p orbital, *each containing one electron,* on each carbon atom. Two of the hybrid orbitals on each carbon atom overlap with the $1s$ orbitals on the hydrogen atoms to make four of the six bonds shown in the Lewis structure. One of the two C—C bonds is formed from overlap of two hybrid orbitals (Figure 9.25). These are all sigma bonds.

After the sp^2 hybrid orbitals are used to form these five sigma bonds, there remains one valence p orbital containing a single electron on each carbon atom. Both of these p orbitals are perpendicular to the plane of the hybrid orbitals. These two p orbitals can form a bond by *sideways* overlap

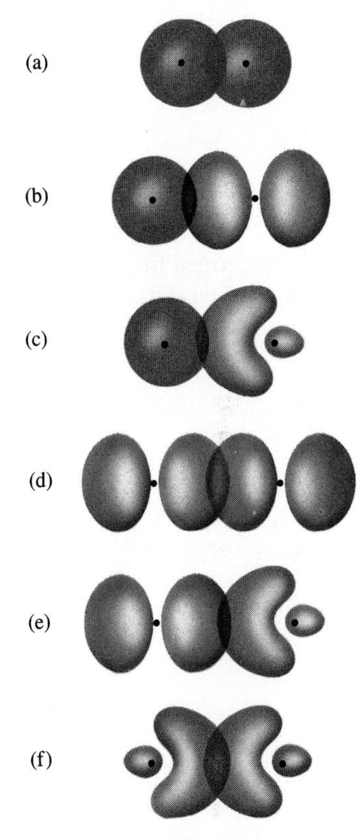

Figure 9.24 Sigma bonds (a) Overlap of s orbitals. (b) Overlap of s and p orbitals. (c) Overlap of s and hybrid orbitals. (d) Head-on overlap of p orbitals. (e) Overlap of p and hybrid orbitals. (f) Overlap of hybrid orbitals.

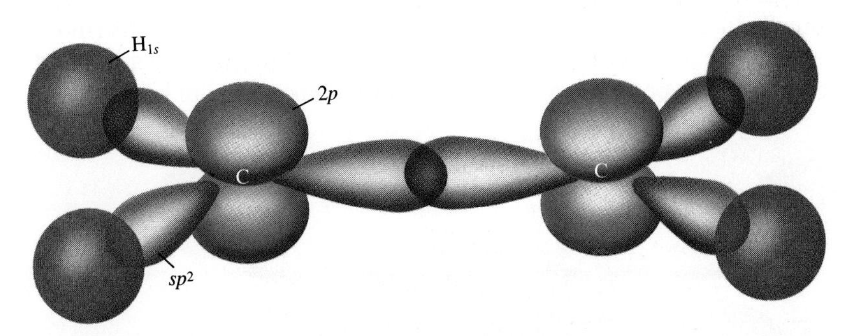

Figure 9.25 Sigma bonds in ethylene The C—H sigma bonds in ethylene are formed from overlap of sp^2 hybrid orbitals on the carbon atoms with $1s$ orbitals on the hydrogen atoms. The C—C sigma bond is formed from overlap of carbon sp^2 hybrid orbitals.

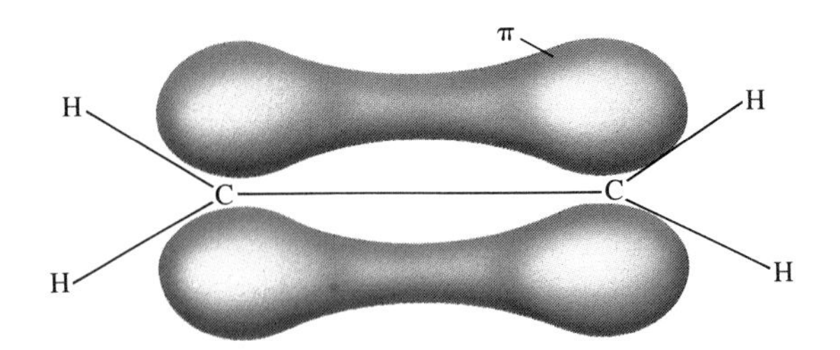

Figure 9.26 Pi bond The sideways overlap of two p orbitals forms a pi bond. The electron density in this *one* pi bond is concentrated above and below a line joining the bonded atoms.

(Figure 9.26). This type of bond is called a **pi bond** (π), a bond that places electron density above and below the line joining the bonded atoms.

Pi bonds are formed from the sideways overlap of p orbitals.

The sideways overlap of two p orbitals, each containing one electron, forms *one* pi bond. Half of the electron density in the bond is above a line joining the bonded atoms, and half is below. Pi bonds are expected to have the electron density in two regions of space because the p orbitals that make up the pi bond have electron density above and below the nuclei. Figure 9.27 shows the complete bonding in ethylene, which includes the five sigma bonds and the pi bond.

Molecular Geometry of Ethylene

Earlier, VSEPR theory was used to predict the geometry about each carbon atom in ethylene. We concluded that the bonding of each carbon atom was trigonal planar, but we could not predict the shape of the whole molecule. An important question is how the planes of the two CH_2 groups are oriented with respect to each other. The complete description of the bonding orbitals as shown in Figure 9.27 predicts that all of the atoms in the molecule will be in the same plane. The molecule is planar because the p orbitals that overlap to form the pi bond are perpendicular to the sigma bonding plane. The maximum overlap of these p orbitals occurs if the molecule is planar. When the two planes are oriented perpendicularly, the overlap of these two orbitals is zero (Figure 9.28).

Note that the 90° rotation of one end of the ethylene molecule in going from the arrangement shown in structure **A** to that shown in **B** completely breaks the pi bond, but does not change the C—C sigma bond at all. This is an important difference between sigma and pi bonds. Atoms or groups of atoms held together by sigma bonds can rotate about the bonds (in compounds with noncyclic structures), but those held together by a pi bond as well as a sigma bond cannot rotate without breaking the pi bond.

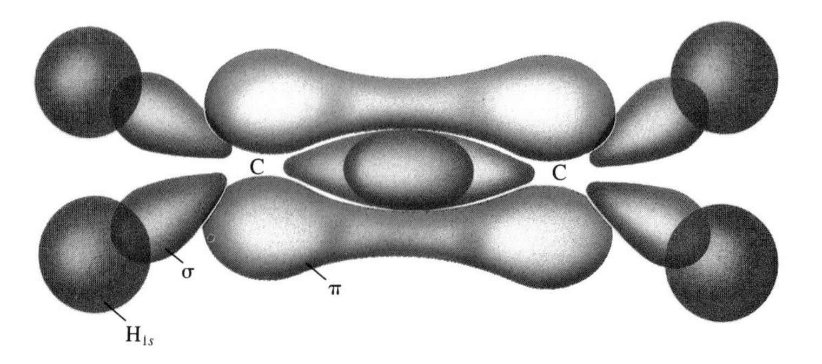

Figure 9.27 Total bonding picture of ethylene Ethylene has five sigma bonds and one pi bond.

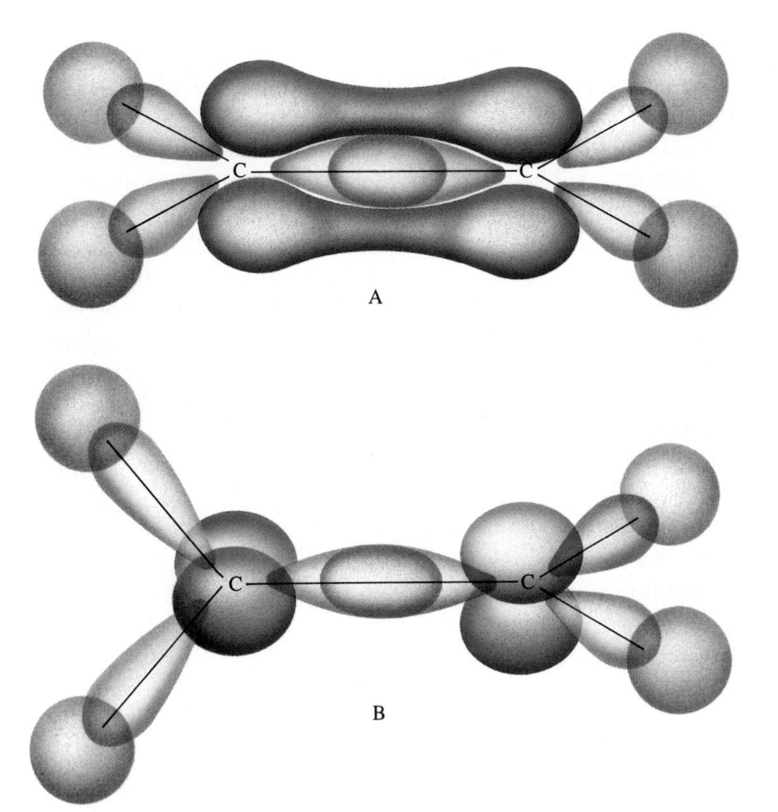

Figure 9.28 Overlap of *p* orbitals Orientation **A** shows ethylene as a planar molecule with overlap of the *p* orbitals to form the pi bond. Rotation of the plane of one CH_2 group by 90° yields **B**, in which there is no overlap of the *p* orbitals and the pi bond is broken. Orientation **A** is the correct shape.

Isomers

There are two different compounds that have the formula HClC=CHCl, each of which has different physical and chemical properties. For example, one form boils at 47.5 °C while the other boils at 60.3 °C. Structurally, the two forms differ in the orientation of the chlorine and hydrogen atoms in these planar molecules (Figure 9.29).

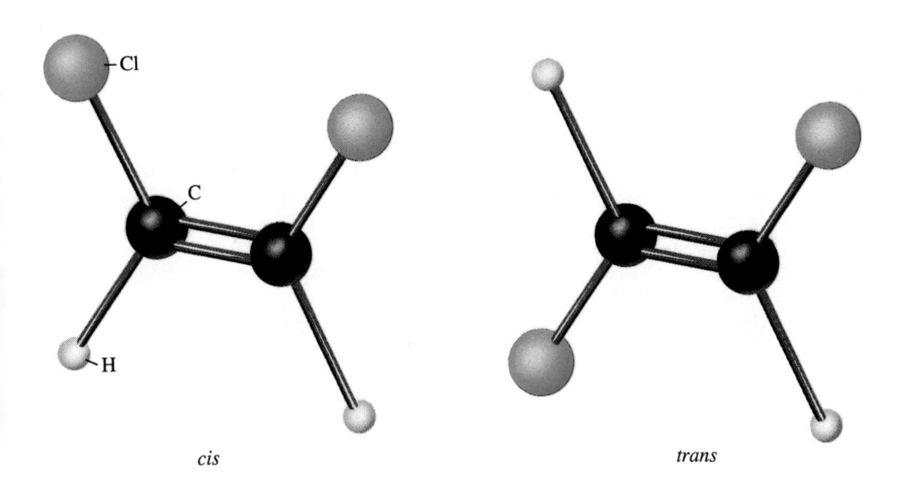

Figure 9.29 *Cis* and *trans* isomers of HClC=CHCl

In the form labeled *cis*, the chlorine atoms are both on the same side of the planar molecule; in the *trans* form they are on opposite sides. **Isomers** are different compounds with the same molecular formula but with different structures. The *cis* and *trans* isomers of the molecules shown in Figure 9.29 do not easily interconvert. The interconversion of the isomers requires the rotation of one H—C—Cl group about the C=C bond. Such a rotation must pass through 90° where the overlap of the two *p* orbitals that form the pi bond is zero; in other words the pi bond must be broken. From the table of bond energies of C=C and C—C bonds in Section 8.8, breaking the pi bond requires 611 kJ/mol − 348 kJ/mol = 263 kJ/mol. This considerable energy is not generally available in molecules at normal temperatures. In contrast to pi bonds, rotation about single bonds (sigma bonds) does not change the overlap of the atomic orbitals, so rotation about single bonds is a relatively low energy process. Thus, there is only one compound with the formula $H_2ClC—CH_2Cl$.

Rotation around pi bonds is not allowed but rotation around sigma bonds is allowed.

A third isomer of those shown in Figure 9.29 is $Cl_2C=CH_2$ (boiling point = 32 °C). To convert either of the isomers in Figure 9.29 into this compound, sigma bonds must be broken and reformed in a different way. The $Cl_2C=CH_2$ isomer does not exist in separate *cis* and *trans* forms. We will consider isomers in greater detail in Chapter 18.

Bonding in Formaldehyde

Formaldehyde, CH_2O, is another compound that contains a double bond.

```
  :O:
   ||
H—C—H
```

The electron-pair arrangement at carbon is trigonal planar; the hybridization is sp^2. As in ethylene, two of the hybrid orbitals on carbon overlap with $1s$ orbitals on the hydrogen atoms to form the C—H sigma bonds. The third sp^2 hybrid orbital points at the oxygen to make a C—O sigma bond. The orbitals available for the oxygen atom to form bonds are

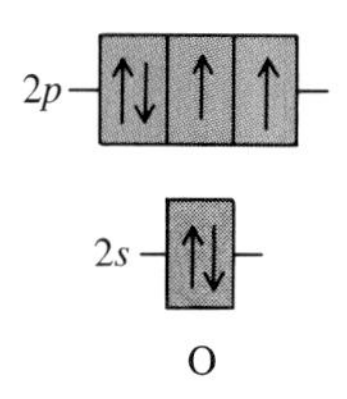

A *p* orbital occupied by one electron and directed at the carbon atom overlaps with the remaining sp^2 orbital on carbon to form the C—O sigma bond. The second C—O bond is a pi bond formed from the overlap of the remaining *p* orbital on carbon with the second *p* orbital on the oxygen atom that is occupied by one electron (Figure 9.30).

Note that we cannot determine the hybridization of the oxygen atom since it is bonded to only one other atom. In the absence of bond angles that indicate the hybridization of orbitals on an atom, we shall use unhybridized atomic orbitals to describe the bonding.

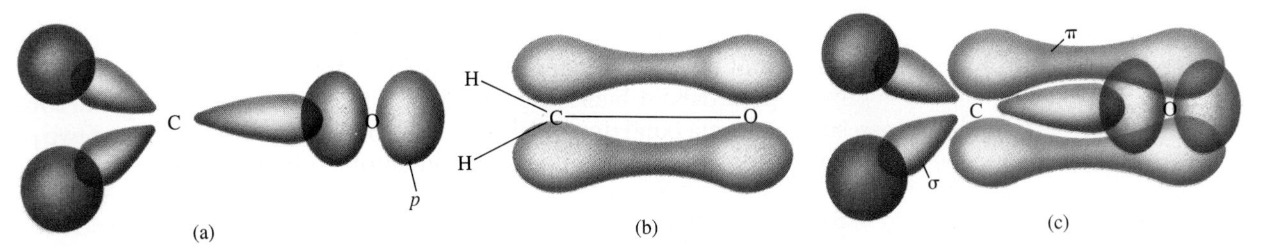

Figure 9.30 Bonding in formaldehyde The sp^2 hybridized carbon atom in formaldehyde forms three sigma bonds (a) and one pi bond (b) to give the overall bonding scheme (c).

Triple Bonds

The Lewis structure of acetylene, C_2H_2, shows that it contains a triple bond.

$$H-C\equiv C-H$$

The linear geometry of each carbon atom indicates *sp* hybridization; the energy level diagram for an *sp* hybridized carbon atom that forms four bonds is

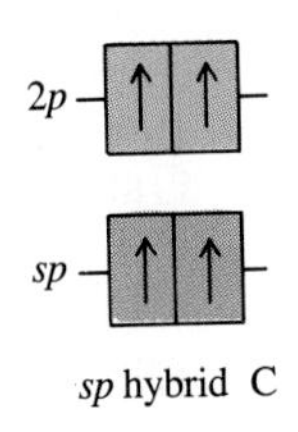

The *sp* hybrid orbitals form two C—H sigma bonds by overlap with the 1*s* orbitals on the two hydrogen atoms. The first of the three C—C bonds, the sigma bond, results from the overlap of two *sp* hybrid orbitals, one on each carbon atom (Figure 9.31a).

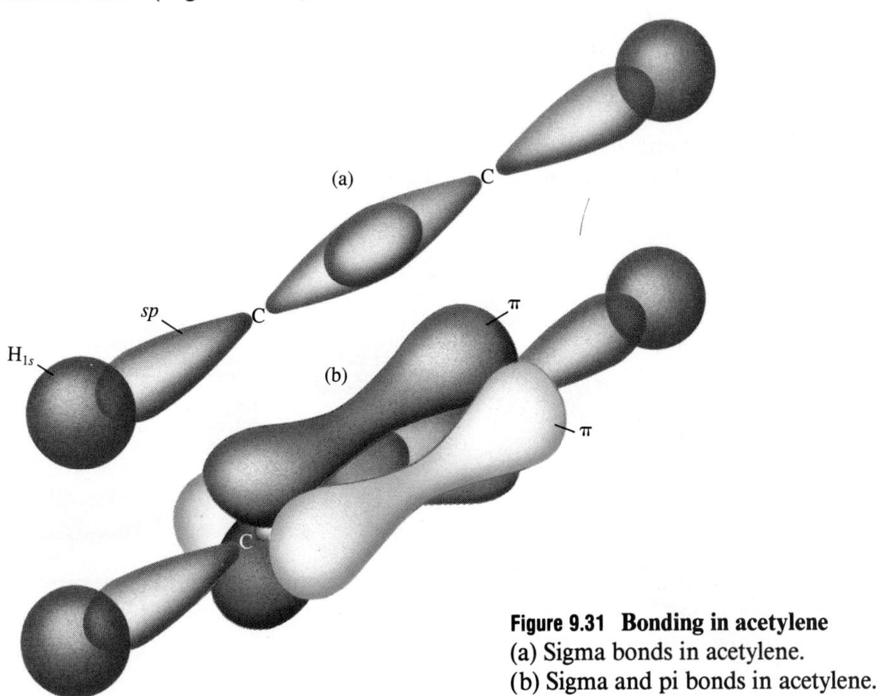

Figure 9.31 Bonding in acetylene
(a) Sigma bonds in acetylene.
(b) Sigma and pi bonds in acetylene.

In addition to the hybrid orbitals, there are two singly occupied orbitals on each of the carbon atoms. Two pi bonds are formed from the sideways overlaps of these orbitals (Figure 9.31b). The triple bond in acetylene consists of a sigma bond (formed by the overlap of *sp* hybrid orbitals) and two pi bonds (formed from the sideways overlap of *p* orbitals).

Summary of Bonding

1. When atoms are connected by a single bond, the bond is a sigma bond. Sigma bonds are formed from overlap of *s* or *p* orbitals or hybrid orbitals oriented along the axis of the bond.
2. When atoms are connected by a multiple bond, one bond is a sigma bond and the second and third bonds are pi bonds, formed by the sideways overlap of *p* orbitals.

Bonding in Benzene

Hybrid orbitals explain the bonding in molecules for which more than one resonance form can be written. An interesting example is benzene, C_6H_6. This molecule is planar, all C—C—C and C—C—H bond angles are 120°, and all C—C bond distances are equal. In Section 8.6, it was pointed out that the two equivalent resonance forms of benzene are frequently written as a single structure by using a dashed circle.

The arrangement about each of the carbon atoms is trigonal planar; the hybridization for each is sp^2. Two of the hybrid orbitals on each carbon atom

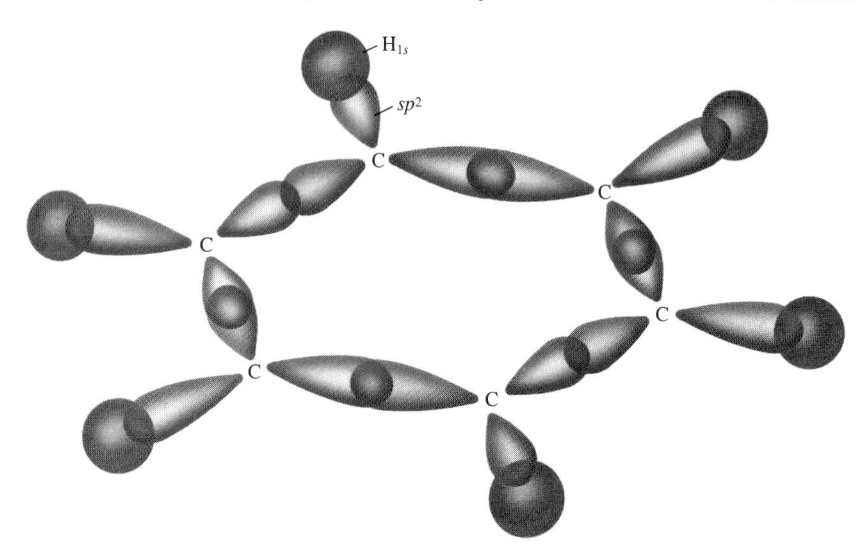

Figure 9.32 Sigma bonds of benzene Each carbon atom in benzene is sp^2 hybridized, using one hybrid orbital to form its C—H bond and the other two to form its two C—C sigma bonds.

are used to make C—C sigma bonds and the third is used to make the C—H sigma bond (Figure 9.32).

Each sp^2 hybridized carbon atom has one electron in a *p* valence orbital that is perpendicular to the plane of the sigma bonds (Figure 9.33a). These *p* orbitals can overlap to form pi bonds in two different arrangements that correspond to the two resonance forms shown above, but the best representation analogous to the "dashed circle" structure is equal overlap of the six *p* orbitals to form a large pi electron cloud that includes all six pi electrons (Figure 9.33b).

The pi bonds in benzene are formed by the sideways overlap of six carbon *p* orbitals.

In summary, the C—C sigma bonds in benzene are formed from the overlap of sp^2 hybrid orbitals on adjacent carbon atoms, and the C—H bonds from overlap of sp^2 hybrid orbitals on carbon with 1*s* orbitals on the hydrogen atoms. The three pi bonds are made from sideways overlap of *p* orbitals. The three pi bonds can be formed by two different combinations of

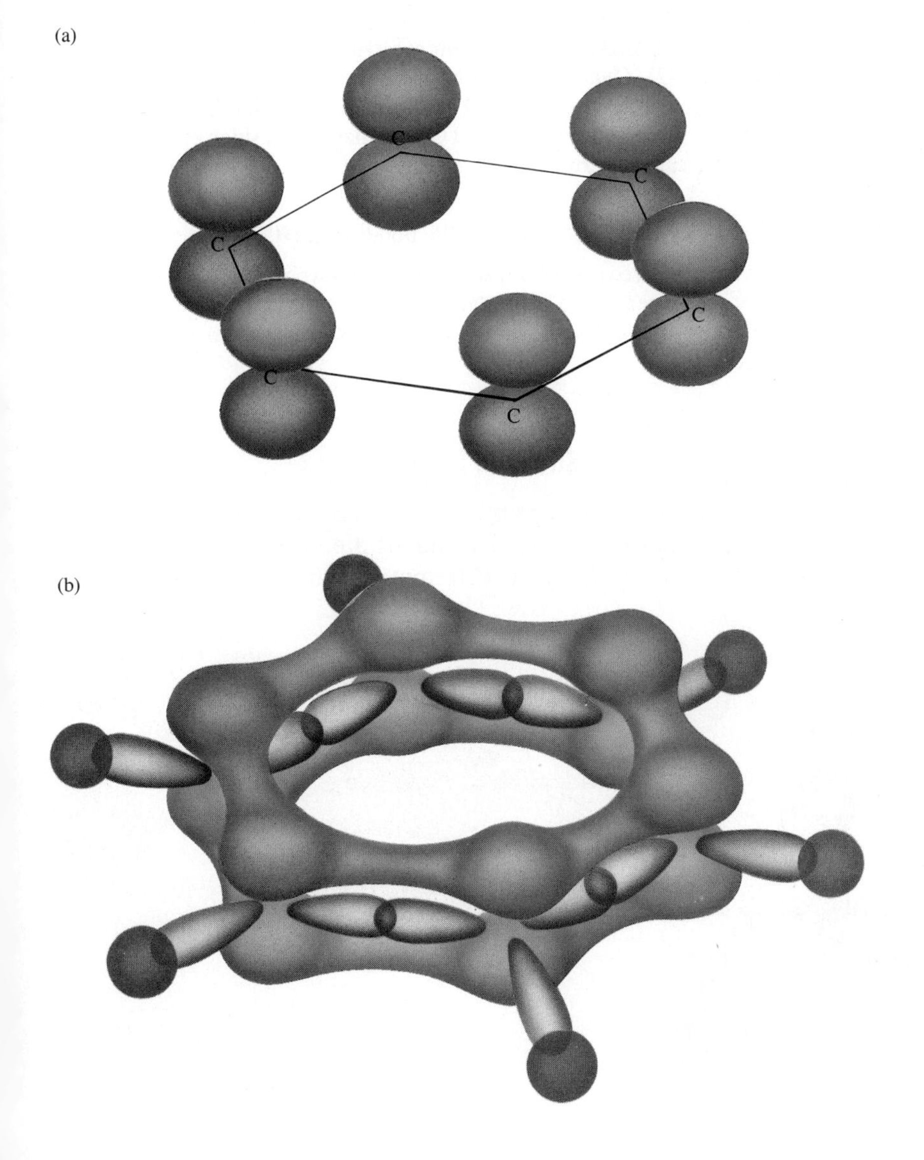

Figure 9.33 Pi bonding in benzene (a) The *p* orbital on each carbon atom contains one electron. (b) These six orbitals overlap in benzene to form a large pi bond.

overlap of the p orbitals, but the best representation is a large pi cloud containing all six electrons.

Example 9.9 Bonding Descriptions

Describe completely the bonds in propylene, $CH_3CH{=}CH_2$. Identify the hybridization of each atom and the nature (σ or π) of each bond.

Solution The Lewis structure is

```
    H   H        H
    |   |       /
H — C — C  ═  C(3)
    |(1)  (2)   \
    H            H
```

The electron-pair arrangement of C(1) is tetrahedral; the bond angles about C(1) are approximately 109° and its hybridization is sp^3. The geometry around both C(2) and C(3) is trigonal planar; the hybridizations are sp^2. The C—H bonds are formed by overlap of these hybrid orbitals with $1s$ orbitals on the hydrogen atoms. The C(1)—C(2) bond is formed by overlap of an sp^3 hybrid orbital on C(1) with an sp^2 hybrid orbital on C(2). One of the two bonds between C(2) and C(3) is formed by overlap of an sp^2 orbital on each. Thus, eight of the nine bonds are sigma bonds. The second bond between C(2) and C(3) is made from sideways overlap of a p orbital on each carbon atom, forming a pi bond. These bonds are shown in Figure 9.34.

Understanding Describe completely all of the bonding in HCN.

Answer: The sp hybridized carbon atom forms two sigma bonds by overlapping with a $1s$ orbital on hydrogen and a p orbital on the nitrogen atom. The carbon and nitrogen atoms form two pi bonds from sideways overlap of the two unhybridized p orbitals on each atom.

Summary of Bonding and Structure Models

Over the past two chapters a number of models and theories have been developed that describe how covalent bonds are formed and how the arrangement of valence electrons can determine the shapes of compounds. To

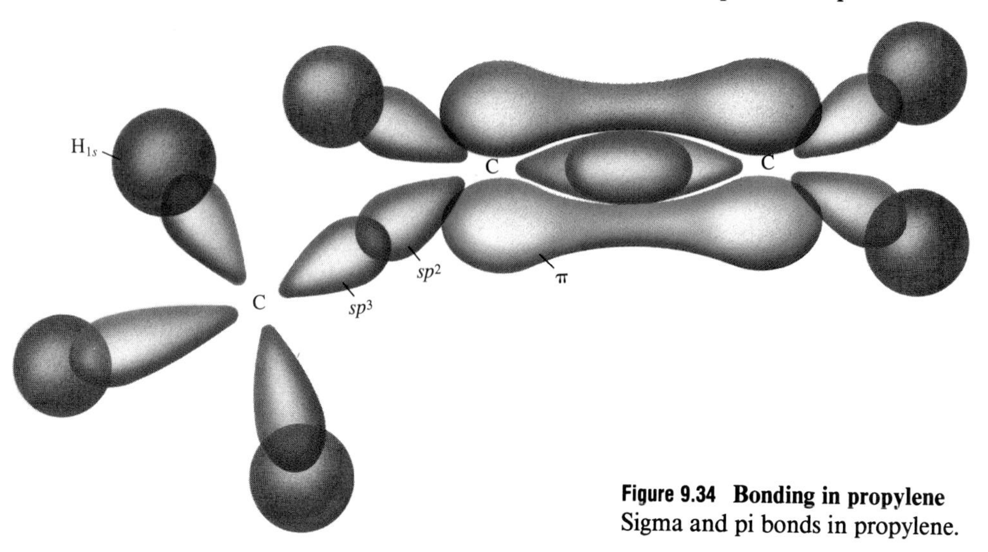

Figure 9.34 Bonding in propylene
Sigma and pi bonds in propylene.

describe the bonding and shape of a compound, the connectivity must be known and the number of valence electrons must be calculated. Generally the Lewis structure describes the bonding quite well and can be used with the VSEPR model to predict the geometry about the central atoms of the molecule. The polarity of the molecule can be determined from the electronegativities of the elements and the shape of the molecule. Valence bond theory allows you to determine which orbitals are used to make each bond and to hold the lone pairs. Frequently, hybrid orbitals are used to form the bonds in order to correctly explain the molecule's shape. Valence bond theory also introduces sigma and pi bonds, which are necessary to account for some properties of molecules that contain multiple bonds.

Example 9.10 Bonding Summary

For (a) CO_2 and (b) CH_2CHOCH_3, write the connectivity, Lewis structure, electron-pair arrangement, hybridization of each central atom, and polarity.

Solution

(a)

Connectivity	Lewis Structure	Electron-Pair Arrangement
O—C—O	$:\ddot{O}=C=\ddot{O}:$	linear

Hybridization of Each Central Atom	Polarity
sp	nonpolar

(b)

Connectivity

```
H     H     H
 \    |     |
  C—C—O—C—H
 /          |
H           H
```

Lewis Structure

```
H     H       H
 \    |  ..   |
  C=C—O—C—H
 /       ..   |
H             H
```

Electron-Pair Arrangement

CH_2, CH = trigonal planar
O, CH_3 = tetrahedral

Hybridization of Each Central Atom

CH_2, CH = sp^2
O, CH_3 = sp^3

Polarity

polar

9.5 Molecular Orbitals: Homonuclear Diatomic Molecules

Objectives

- To write the molecular orbital diagrams for homonuclear diatomic molecules and ions of the first and second period elements
- To determine the bond order and the number of unpaired electrons in diatomic species from the molecular orbital diagrams

Our models of bonding, Lewis structures and valence bond theory, are similar in that bonds are formed when an electron pair is shared between two adjacent atoms. A different approach is to describe the electron arrangement using wave functions (orbitals) that allow the valence shell electrons to be spread over all of the atoms in the molecule rather than shared between only two atoms. The wave functions and energy level diagram that describe the

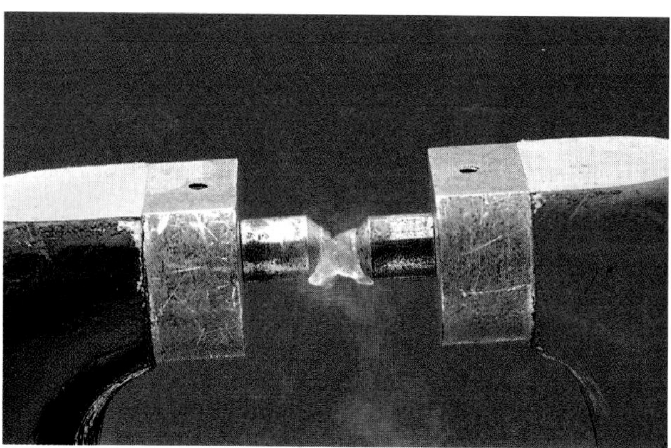

Figure 9.35 The behavior of liquid nitrogen and oxygen in a magnetic field Compounds such as liquid nitrogen, which contain only paired electrons, are not attracted by a magnetic field (left). In contrast, liquid oxygen remains suspended between the poles of a magnet (right), indicating that it is paramagnetic because it contains unpaired electrons.

entire molecule are analogous to those used in describing the electrons in atoms. **Molecular orbital theory** is a model that treats bonding as delocalized over the entire molecule rather than limiting the sharing of electrons to two atoms. Molecular orbital theory is needed because valence bond theory cannot explain certain unusual bonding situations and in some cases gives answers that are inconsistent with experimental results.

A surprisingly simple molecule for which the Lewis structure gives an incorrect bonding description is O_2. The Lewis structure of O_2 is

$$:\ddot{O}=\ddot{O}:$$

The experimentally determined O—O bond distance in this molecule is consistent with a double bond, but measurements show that O_2 is paramagnetic (attracted into a magnetic field) and contains two unpaired electrons (Figure 9.35). The Lewis structure of O_2 does not predict two unpaired electrons; all of the electrons are present as pairs.

Molecular orbital theory describes the bonding in a molecule by using a series of molecular orbitals. A **molecular orbital** is a wave function of an electron in a molecule. It differs from an atomic orbital in that molecular orbitals may extend over the entire molecule rather than being restricted to one individual atom. Molecular orbitals are described by mathematically combining the valence shell atomic orbitals of all the atoms in a molecule. The mathematical operation used to form molecular orbitals is similar to that used to form hybrid orbitals. The difference is that hybrid orbitals are formed by combining valence orbitals on the *same* atom, whereas molecular orbitals are formed from combinations of valence orbitals on *different* atoms. Just as with hybridization, the number of molecular orbitals that are formed must equal the number of atomic orbitals used to make them. As with atomic orbitals, we are interested in both the energy and the shape of the electron cloud for each molecular orbital.

The Hydrogen Molecule

The simplest molecule to consider is H_2, since there are only two 1*s* valence orbitals available to form the molecular orbitals. The molecular orbitals are formed by the addition and subtraction of the two atomic wave functions (Figure 9.36).

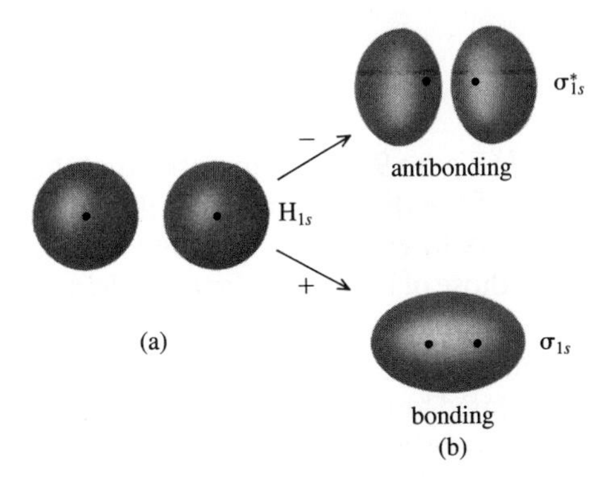

Figure 9.36 Molecular orbitals for H_2 Atomic orbitals are added and subtracted (a) to form bonding and antibonding molecular orbitals (b).

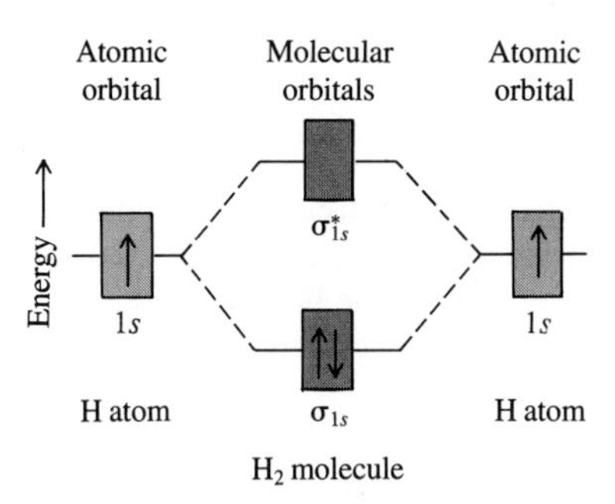

Figure 9.37 Molecular orbital diagram of H_2 The diagram for H_2 shows the starting atomic orbitals on the outside for reference and the molecular orbitals in the middle. The two electrons in the H_2 molecule are placed in the bonding molecular orbital.

Addition of the two atomic wave functions produces one of the two new molecular orbitals. The reinforcement of the two wave functions forms a large amplitude (large electron probability) directly between the two nuclei. A **bonding molecular orbital** is one that concentrates the electron density between the atoms in the molecule. An electron in this orbital is mainly located between the two nuclei and will be attracted by both, so the molecular orbital is lower in energy than the atomic orbitals from which it formed. The bonding molecular orbital in hydrogen is labeled σ_{1s} because it is symmetric about the line joining the two nuclei (the definition of a sigma bond) and is formed from 1*s* atomic orbitals.

Subtraction of the 1*s* atomic wave function on one hydrogen atom from that on the other produces the second molecular wave function. The interference of the wave functions of opposite sign causes the electron probability to be very low between the nuclei. An **antibonding molecular orbital** is one that reduces the electron density in the region between the atoms in the molecule. The lower electron density between the two nuclei than is present in the isolated atoms produces a less stable (higher in energy) wave function than the separate atomic wave function from which it is made. An asterisk is used to designate an antibonding molecular orbital, σ^*_{1s}. The increase in energy of an electron in an antibonding orbital is approximately the same as the decrease in energy of an electron in the corresponding bonding molecular orbital. Although less stable than the atomic orbitals from which it is formed, an antibonding orbital can hold two electrons just like any orbital.

Molecular orbitals are formed from combinations of atomic orbitals on different atoms.

The filling of molecular orbitals follows the same procedure used to fill atomic orbitals. The aufbau principle, the Pauli exclusion principle, and Hund's rule all apply. For H_2, the two electrons are placed in the σ_{1s} orbital with opposing spins (Figure 9.37). The electron configuration for the *molecule* H_2 is represented by $(\sigma_{1s})^2$.

Molecular orbital theory defines *bond order* by the following equation:

$$\text{bond order} = \tfrac{1}{2}[\text{number of electrons in bonding orbitals} - \text{number of electrons in antibonding orbitals}]$$

The bond order for H_2 is one $[\frac{1}{2}(2 - 0) = 1]$, the same as that found using the Lewis structure.

The bond order is one half the number of electrons in bonding orbitals minus one half the number of electrons in antibonding orbitals.

The He_2 Molecule

The importance of the antibonding orbital is evident if the hypothetical molecule He_2 is considered. The molecular energy-level diagram for He_2 is qualitatively the same as that for H_2. A total of four electrons, two from the valence shell of each helium atom, must now occupy the molecular orbitals.

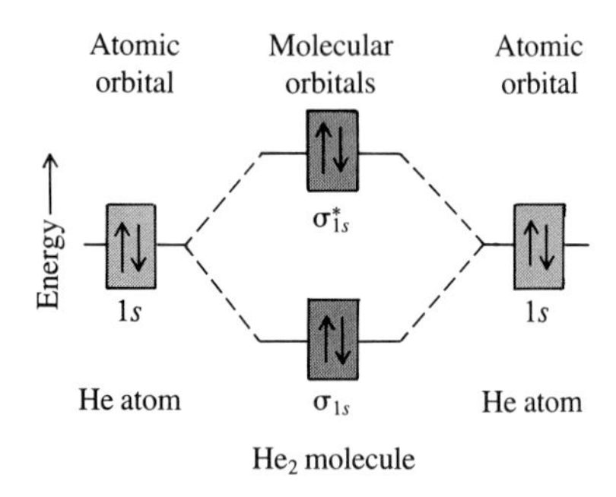

Figure 9.38 Molecular orbital diagram of He_2 The four electrons in He_2 fill both the bonding and antibonding molecular orbitals.

The electron configuration is $(\sigma_{1s})^2(\sigma^*_{1s})^2$ (Figure 9.38). Remember that the 1s atomic orbitals shown on either side of the diagram have combined to form the molecular orbitals and cannot hold electrons. They are on the diagram to show their energies relative to those of the molecular orbitals, and show which atomic orbitals were used to form the molecular orbitals.

The bond order in the He_2 molecule is $\frac{1}{2}[2-2]=0$. All of the energy that was gained (in comparison to the isolated atoms) by filling the bonding orbital is cancelled in filling the antibonding orbital. In fact, the completely filled bonding and antibonding orbitals are slightly less stable than the filled atomic orbitals from which they arise. Molecular orbital theory predicts that the isolated atoms are a little more stable than He_2, so this molecule will not form. We already know that atoms of He are stable and unreactive.

A molecule is less stable than its isolated atoms if the number of electrons in bonding orbitals is equal to the number of electrons in antibonding orbitals.

Example 9.11 Bonding in He_2^+

Write the electron configuration for He_2^+, and calculate the bond order. Predict whether or not this species will be stable.

Solution The molecular orbital diagram will be the same as for H_2 and He_2. The electron configuration is $(\sigma_{1s})^2(\sigma^*_{1s})^1$ and the bond order is $\frac{1}{2}[2-1]=\frac{1}{2}$. The He—He bond in this cation would be weak, but the presence of a bond shows that this species should exist. The He_2^+ ion has been detected experimentally and has the properties predicted by molecular orbital theory.

Understanding Write the electron configuration for H_2^+ and calculate the bond order. Will this species be stable?

Answer: $(\sigma_{1s})^1$, bond order $=\frac{1}{2}$, the species is stable

Second Period Diatomic Molecules

A **homonuclear diatomic molecule** contains two atoms of the same element. Several of the nonmetallic elements in the second period exist as homonuclear diatomic molecules (N_2, O_2, and F_2). Molecular orbital theory is very useful in describing the bonding in these well-known molecules and other

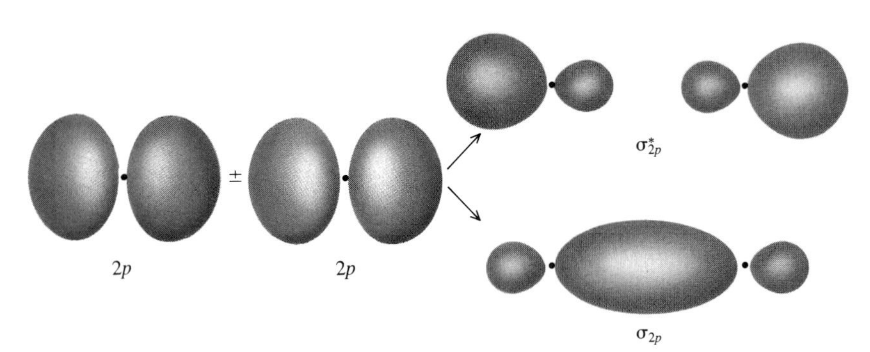

Figure 9.39 Sigma molecular orbitals from *p* atomic orbitals The combination of the *p* orbitals oriented along the line joining the nuclei yields a σ_{2p} bonding molecular orbital when the signs of the amplitude reinforce and a σ^*_{2p} antibonding molecular orbital when the signs of the amplitude interfere. Both orbitals can hold two electrons.

less common molecules such as Li_2, B_2, and C_2. In the second period, the 2*s* and 2*p* atomic orbitals comprise the valence shell; molecular orbitals are derived from these atomic orbitals. Since each atom has four valence shell orbitals (*s*, p_x, p_y, and p_z), there are a total of eight molecular orbitals in each diatomic molecule.

For a second period diatomic molecule, the 2*s* orbitals on the two atoms mix to form σ_{2s} and σ^*_{2s} molecular orbitals analogous to those formed for H_2. The *p* orbitals combine to yield two different types of molecular orbitals. The first set of molecular orbitals arises from the combination of a *p* orbital on one atom with a *p* orbital on the other atom, oriented along the line joining the nuclei (Figure 9.39). The combination yields a σ_{2p} bonding molecular orbital when the sigma of the amplitude reinforce and a σ^*_{2p} antibonding molecular orbital when the signs of the amplitude interfere. As before, the combination of two atomic orbitals forms two molecular orbitals.

The remaining *p* orbitals, two on each atom, are perpendicular to the internuclear line and combine as shown in Figure 9.40. Each pair of *p* orbitals combines to form one bonding and one antibonding molecular orbital. These are pi molecular orbitals because the region of overlap lies both above and below the line joining the nuclei. The two *p* orbitals on each atom combine with those on the other atom to form two bonding π_{2p} molec-

Figure 9.40 Pi molecular orbitals from *p* atomic orbitals Sideways combination of *p* orbitals yields two π_{2p} bonding molecular orbitals and two π^*_{2p} antibonding molecular orbitals.

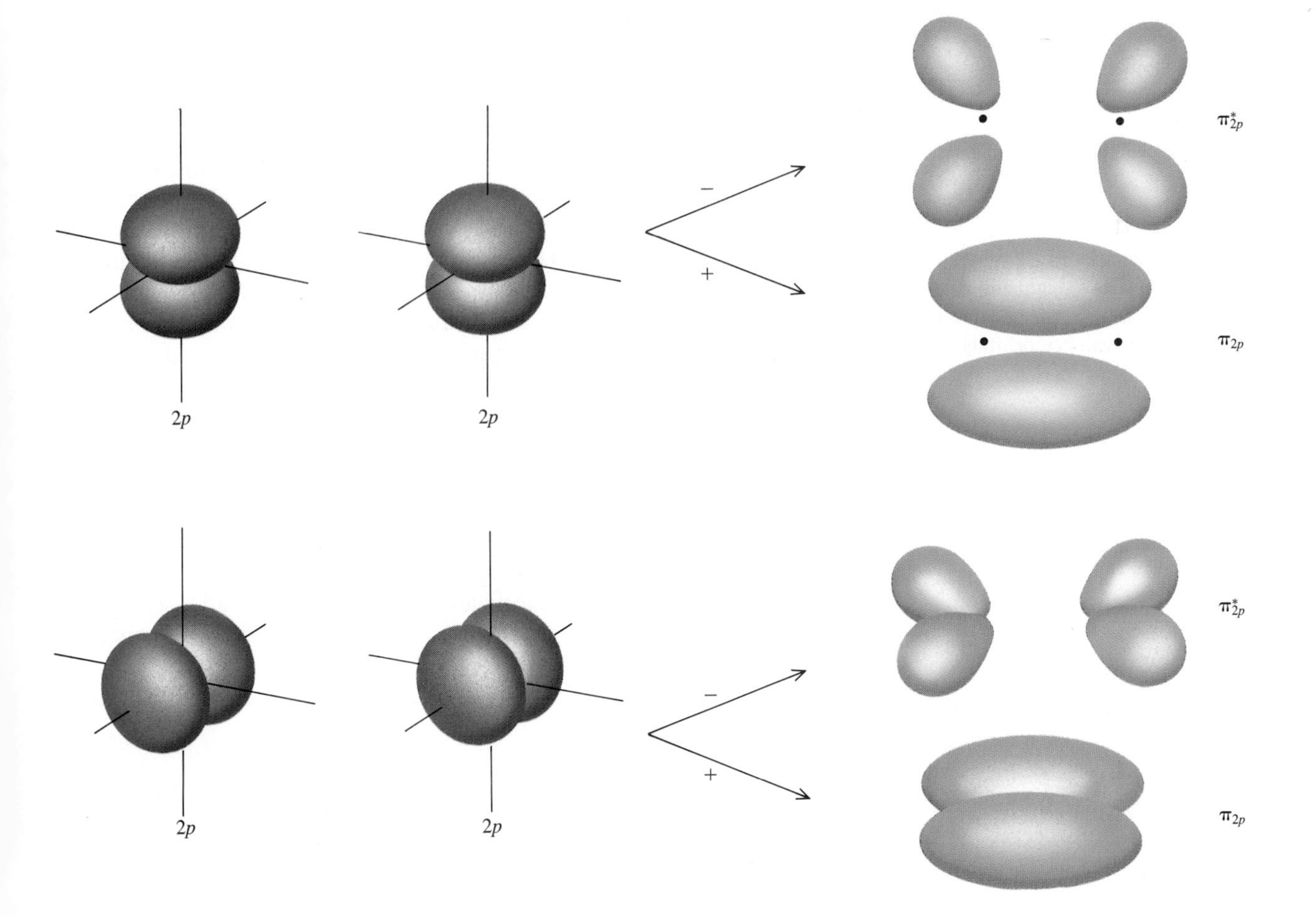

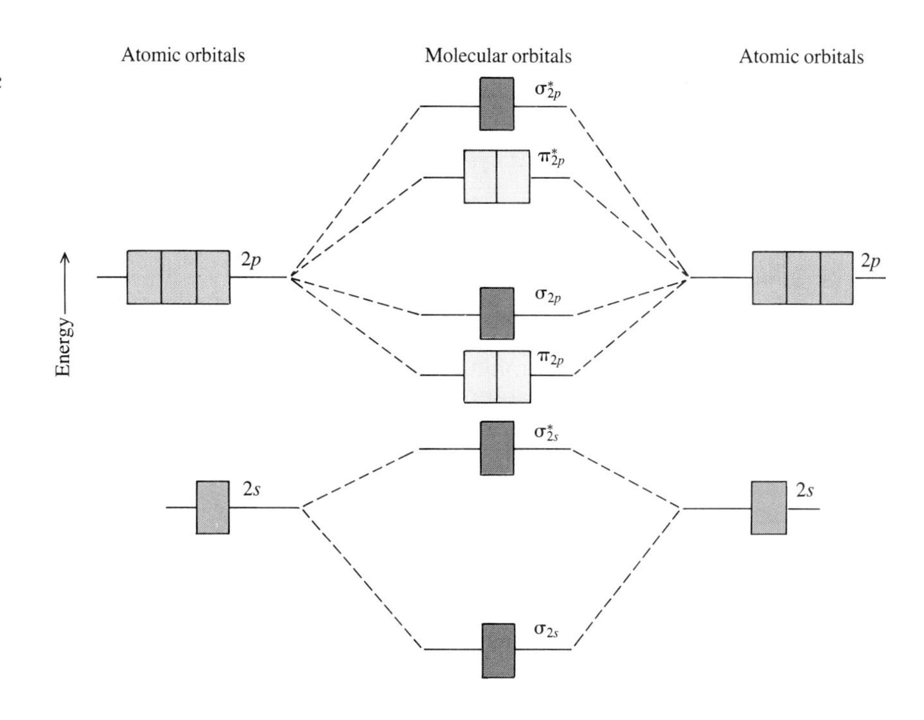

Figure 9.41 Molecular orbital diagram for the second period diatomic molecules This diagram is correct for Li_2 through N_2. For O_2 and F_2 the energy order of the σ_{2p} and π_{2p} are reversed, but since these are filled orbitals for these two molecules, this inversion does not affect the bond order or magnetic properties.

ular orbitals and two π^*_{2p} antibonding molecular orbitals. The two π_{2p} molecular orbitals are perpendicular to each other and have exactly equal energies. The two π^*_{2p} molecular orbitals are also perpendicular to each other and have equal energies, which are higher than the energy of the π_{2p} orbitals. Overall, four *p* type atomic orbitals form two π bonding molecular orbitals at one energy and two π^* antibonding molecular orbitals at a higher energy.

Diatomic molecules of elements in the second row have eight molecular orbitals formed from the combination of the four valence orbitals on each atom.

The complete molecular orbital diagram for a second period diatomic molecule is shown in Figure 9.41. Both the bonding and antibonding molecular orbitals that arise from the combination of the 2*s* atomic orbitals are lower in energy than the molecular orbitals arising from combination of the 2*p* atomic orbitals, because the 2*s* atomic orbitals are lower in energy than the 2*p* atomic orbitals. Also, the π_{2p} orbitals are lower in energy than the σ_{2p} orbitals. This order can be explained by noting that the σ_{2s} and σ_{2p} orbitals are located in the same region of space, principally between the two nuclei. The two orbitals interact, decreasing the energy of the σ_{2s} orbital and increasing the energy of the σ_{2p}. Similar interactions decrease the energy of the σ^*_{2s} and increase the energy of σ^*_{2p}. These interactions decrease across the period, and there is evidence that σ_{2p} is lower in energy than π_{2p} for O_2 and F_2. For simplicity, we will use the diagram in Figure 9.41 for all cases because these energy levels are filled for O_2 and F_2 and do not greatly influence the molecular properties.

Electron Configuration of N_2

The electron configuration of N_2 can be determined by placing the correct number of valence electrons in the molecular orbital diagram (Figure 9.42), using the general principles (aufbau, Hund's rule, Pauli exclusion) that were used to fill atomic orbitals.

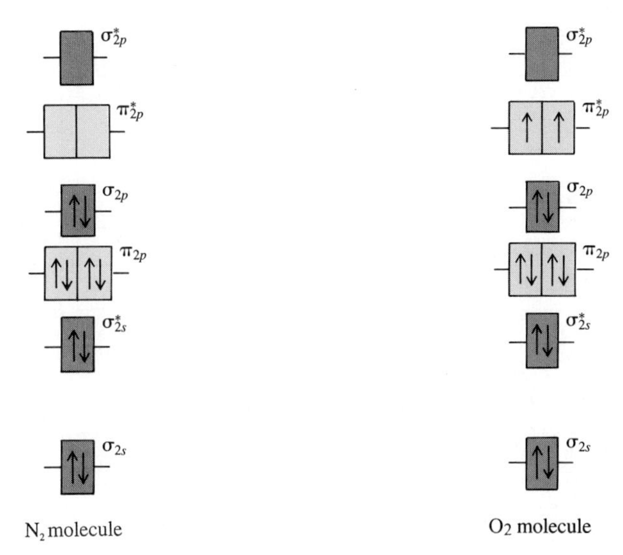

Figure 9.42 Molecular orbital diagram for N_2 N_2 places ten electrons in the molecular orbital diagram for homonuclear diatomic molecules. The diagram predicts a bond order of 3 and no unpaired electrons.

Figure 9.43 Molecular orbital diagram for O_2 O_2 places twelve electrons in the molecular orbital diagram for homonuclear diatomic molecules. The diagram predicts a bond order of 2 and two unpaired electrons.

The electron configuration is $(\sigma_{2s})^2(\sigma^*_{2s})^2(\pi_{2p})^4(\sigma_{2p})^2$. There are eight valence-shell electrons in bonding molecular orbitals and two in antibonding orbitals, so the predicted bond order is

$$\text{bond order} = \tfrac{1}{2}[8 - 2] = 3$$

The four electrons in σ_{2s} and σ^*_{2s} orbitals contribute essentially no net bonding. The *two* filled π_{2p} orbitals and the filled σ_{2p} orbital yield the three bonds. This bond description is consistent with the bond order we get from the Lewis structure, :N≡N:. The lone pairs on the nitrogen atoms correspond to the electron pairs in the filled σ_{2s} and σ^*_{2s} molecular orbitals.

Electron Configuration of O_2

As noted earlier, valence bond theory does not account for the experimental fact that O_2 has two unpaired electrons and a double bond between the oxygen atoms. In contrast, molecular orbital theory correctly predicts this behavior. The molecular orbital diagram for the 12 valence electrons of O_2 is shown in Figure 9.43.

The energy-level diagram now has two electrons occupying the degenerate (of equal energy) π^*_{2p} levels. Following Hund's rule, these electrons are placed in separate orbitals with parallel spins. The predicted bond order is

$$\text{bond order} = \tfrac{1}{2}[8 - 4] = 2$$

Thus molecular orbital theory correctly accounts for the experimental observations of a double bond and the presence of *two unpaired electrons* in O_2.

Molecular orbital theory correctly predicts that O_2 will have two unpaired electrons.

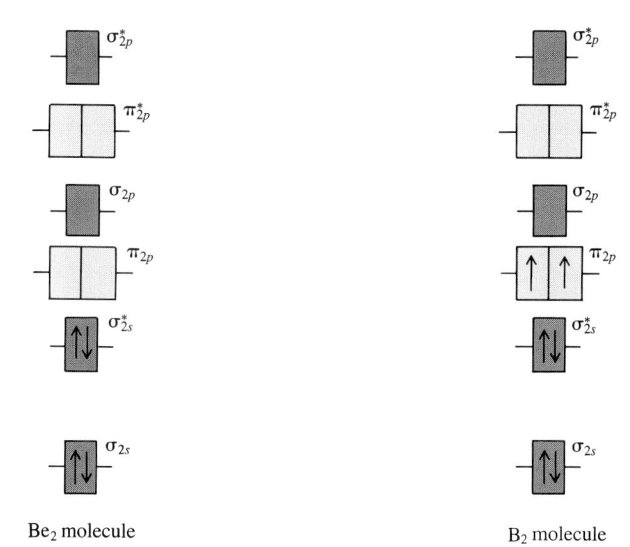

Figure 9.44 Molecular orbital diagram for Be_2 Be_2 places four electrons in the molecular orbital diagram for homonuclear diatomic molecules. The diagram predicts a bond order of zero, so the molecule will not form.

Figure 9.45 Molecular orbital diagram for B_2 B_2 places six electrons in the molecular orbital diagram for homonuclear diatomic molecules. The diagram predicts a bond order of 1 and two unpaired electrons.

Example 9.12 Molecular Orbitals

Write the molecular orbital diagram and the electron configuration for (a) Be_2 and (b) B_2. Predict the bond order and number of unpaired electrons for each molecule.

Solution

(a) Be_2 has four valence electrons. The molecular orbital diagram is shown in Figure 9.44.

The electron configuration is $(\sigma_{2s})^2\ (\sigma^*_{2s})^2$. The bond order is $\frac{1}{2}[2 - 2] = 0$. Because the bond order is 0, this molecule is not expected to exist.

(b) B_2 has six valence electrons. The molecular orbital diagram is shown in Figure 9.45.

The electron configuration is $(\sigma_{2s})^2(\sigma^*_{2s})^2(\pi_{2p})^2$. The bond order is $\frac{1}{2}[4 - 2] = 1$ and the molecule has two unpaired electrons. Experiments have confirmed both of these predictions, although the molecule B_2 exists only at high temperature and low pressure (other elemental forms of boron are more stable). This is another case like O_2 in which the Lewis structure, B≡B, does not agree with the results of experimental measurements. For the B_2 molecule, neither the bond order nor the paramagnetism is correctly predicted by the Lewis structure.

Understanding Use molecular orbital theory to predict the bond order and number of unpaired electrons for C_2.

Answer: Bond order = 2, no unpaired electrons

Summary of Second Row Homonuclear Diatomic Molecules

Table 9.1 summarizes the electron configurations of the second period homonuclear diatomic molecules along with important physical data.

Table 9.1 Molecular Orbital Electron Configurations, Bond Order, and Physical Data for Second Period Homonuclear Diatomic Molecules

Species	Electron Configuration	Predicted Bond Order	Measured Bond Energy (kJ/mol)	Number of Unpaired Electrons
Li_2	$(\sigma_{2s})^2$	1	105	0
Be_2	$(\sigma_{2s})^2(\sigma_{2s}^*)^2$	0	unstable	
B_2	$(\sigma_{2s})^2(\sigma_{2s}^*)^2(\pi_{2p})^2$	1	290	2
C_2	$(\sigma_{2s})^2(\sigma_{2s}^*)^2(\pi_{2p})^4$	2	620	0
N_2	$(\sigma_{2s})^2(\sigma_{2s}^*)^2(\pi_{2p})^4(\sigma_{2p})^2$	3	946	0
O_2	$(\sigma_{2s})^2(\sigma_{2s}^*)^2(\pi_{2p})^4(\sigma_{2p})^2(\pi_{2p}^*)^2$	2	498	2
F_2	$(\sigma_{2s})^2(\sigma_{2s}^*)^2(\pi_{2p})^4(\sigma_{2p})^2(\pi_{2p}^*)^4$	1	159	0

9.6 Heteronuclear Diatomic Molecules and Delocalized Molecular Orbitals

Objectives

- To construct the molecular orbital diagram for heteronuclear diatomic molecules
- To describe the formation of delocalized molecular orbitals

In the previous section, molecular orbital diagrams were constructed for species in which the two atoms that combined were identical. The molecular orbital diagrams for species made up of two different atoms are similar to those for homonuclear molecules, but with important differences. Molecular orbitals can also be constructed for molecules containing more than two atoms. In these molecules, the differences between valence bond theory and molecular orbital theory become more evident. In valence bond theory only two adjacent atoms in the molecule can share a pair of electrons. Such bonds are called localized bonds. This limitation is not present in molecular orbital theory, in which a single orbital may form from atomic orbitals on three or more atoms in the molecule, producing *delocalized bonds.* A **delocalized molecular orbital** is a wave function of an electron in a molecule that is spread over *more* than two atoms.

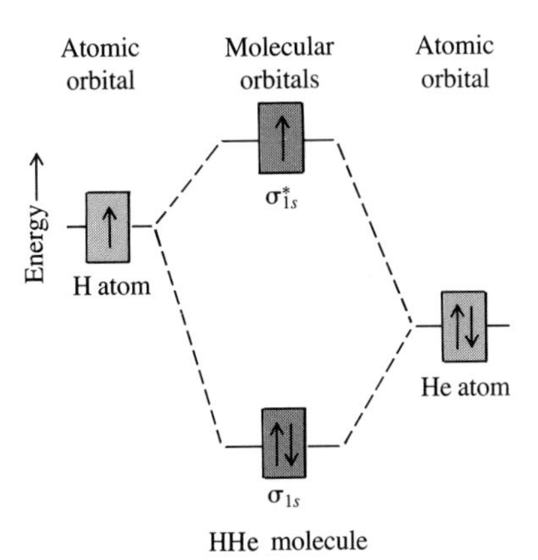

Figure 9.46 Molecular orbital diagram for HHe The atomic orbitals used to form the molecular orbitals for the HHe molecule are different in energy. The molecule has three electrons, a bond order of 0.5, and one unpaired electron.

HHe Molecule

Consider HHe, a molecule that has been observed in the gas phase. The molecular orbitals for HHe are formed from the $1s$ orbitals of the H and He atoms. The energy of the $1s$ orbital on the helium atom is lower than that of the hydrogen atom, as shown in Figure 9.46. This energy difference was first noted in Chapter 7, when trends in ionization energies within a period were presented. Since the Z_{eff} increases from left to right, the atomic orbitals become more stable from left to right in any period. Two molecular orbitals, σ_{1s} and σ_{1s}^*, are formed. The bonding orbital is lower in energy than the helium $1s$ atomic orbital, and the antibonding orbital is higher in energy than the hydrogen $1s$ orbital. Following the aufbau procedure, two of the

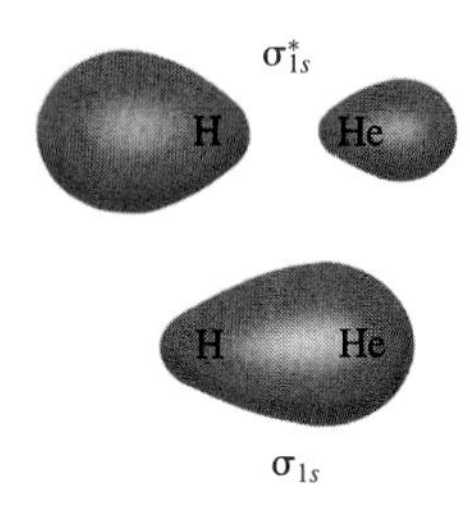

Figure 9.47 Molecular orbitals for HHe The molecular orbitals for HHe are not symmetrical. The bonding molecular orbital has more electron density near the helium atom, and the antibonding orbital has more electron density near the hydrogen atom.

three electrons are placed in the σ_{1s} orbital and one in the σ^*_{1s} orbital. The bond order is 0.5.

In Figure 9.37, the molecular orbital diagram of H_2, each atomic orbital contributes equally (the mathematical mixing uses the wave function of both orbitals equally) to the bonding and antibonding molecular orbitals. In the case of the HHe orbitals, the wave function of the helium atom 1*s* orbital contributes a larger fraction to the bonding molecular orbital than does the hydrogen atom 1*s* orbital. The hydrogen atom contributes more to the σ^*_{1s} orbital. In general, a molecular orbital uses a larger fraction of the atomic orbital that is closer to it in energy. Because of these differing contributions, the molecular orbitals in HHe are not symmetric as was observed with H_2 (Figure 9.36). The amplitude of the σ_{1s} orbital is greater around the helium atom, and the amplitude of the σ^*_{1s} orbital is greater around the hydrogen atom (Figure 9.47).

Second Row Heteronuclear Diatomic Molecules

A **heteronuclear diatomic molecule** contains one atom of each of two different elements. Carbon monoxide, CO, and nitrogen monoxide, NO, are two common compounds of second row elements that exist as heteronuclear diatomic molecules. The molecular orbital energy-level diagrams for these molecules are similar to those for the homonuclear diatomic molecules. In both cases the same valence shell atomic orbitals (2*s* and 2*p*) are available to form molecular orbitals, so we expect the formation of two σ orbitals, two σ^* orbitals, two degenerate π orbitals, and a pair of degenerate π^* orbitals. The diagram differs from the homonuclear case because the energies of the

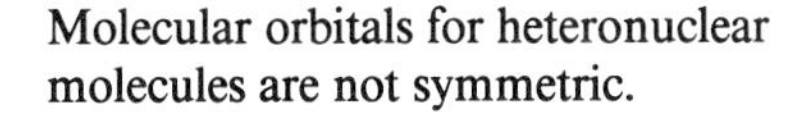

Molecular orbitals for heteronuclear molecules are not symmetric.

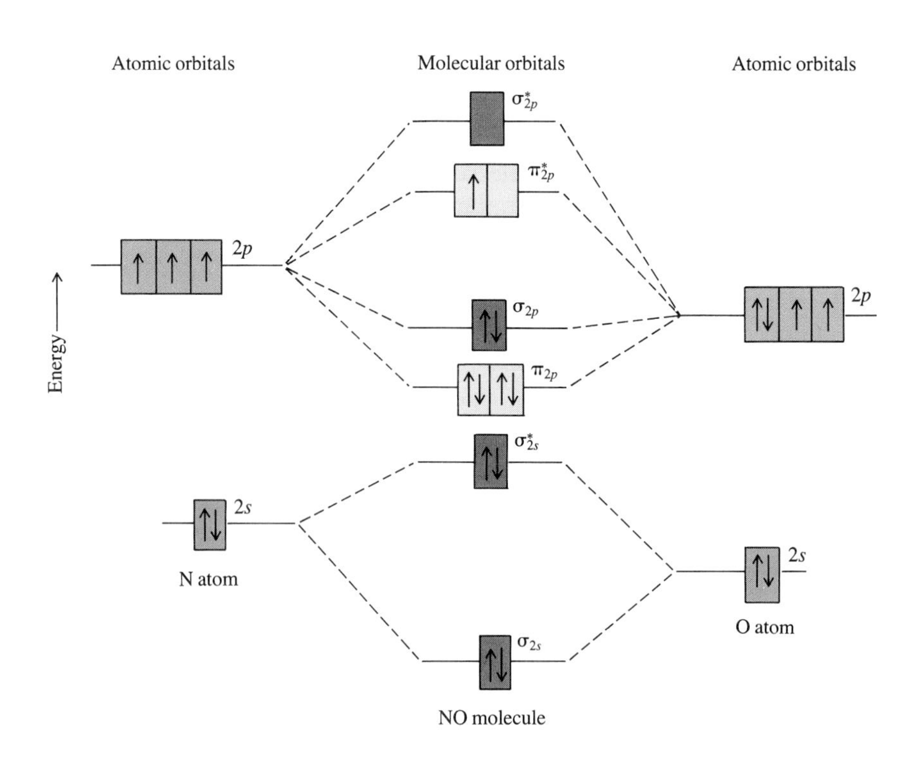

Figure 9.48 Molecular orbital diagram for NO The molecular orbital diagram for NO predicts a bond order of 2.5, and predicts that the molecule is paramagnetic with one unpaired electron.

atomic orbitals on the two atoms are no longer the same. Elements to the right of the period have orbitals of lower energy. A diagram similar to that in Figure 9.41 is used for heteronuclear diatomic molecules such as NO that are formed from nearby elements in the periodic table, but the diagram is modified to indicate that the atomic orbitals used to form the molecular orbitals are of different energies (Figure 9.48).

The electron configuration of NO is $(\sigma_{2s})^2(\sigma_{2s}^*)^2(\pi_{2p})^4(\sigma_{2p})^2(\pi_{2p}^*)^1$. The 11 valence electrons of NO fill the σ_{2p} orbital completely and place one electron in the π_{2p}^* orbitals. The bond order is $\frac{1}{2}[8 - 3] = 2.5$, and one unpaired electron is expected. Experiment shows that both of these predictions are correct.

It is interesting to compare NO with NO^+. The molecular orbital configuration for NO^+ is $(\sigma_{2s})^2(\sigma_{2s}^*)^2(\pi_{2p})^4(\sigma_{2p})^2$. Removal of an electron from NO causes the bond order to increase to 3 because the electron that is removed comes from an antibonding orbital. It has been shown experimentally that the bond length in NO^+ is 9 pm shorter than in NO. Because NO^+ and N_2 are isoelectronic, the molecular orbital electron configuration is the same for both.

Diatomic molecules formed from elements that are close to each other in the periodic table have molecular orbital diagrams that resemble those of homonuclear diatomic molecules.

Example 9.13 Molecular Orbitals

Write the molecular orbital electron configuration and give the bond order of CN^-.

Solution The cyanide anion has 10 valence electrons. The electron configuration is $(\sigma_{2s})^2(\sigma_{2s}^*)^2(\pi_{2p})^4(\sigma_{2p})^2$. The bond order is $\frac{1}{2}[8 - 2] = 3$. This ion is also isoelectronic with N_2 and NO^+.

Understanding Write the molecular orbital electron configuration and give the bond order of NO^-. Is the species diamagnetic or paramagnetic?

Answer: $(\sigma_{2s})^2(\sigma_{2s}^*)^2(\pi_{2p})^4(\sigma_{2p})^2(\pi_{2p}^*)^2$; the bond order is two, and the ion is paramagnetic.

Molecular Orbital Diagram for LiF

The molecular orbital diagram in Figure 9.48 is correct only for molecules formed by second row elements close together in the periodic table, for which the energies of the atomic orbitals of the two elements are fairly similar. For heteronuclear diatomic molecules formed from elements with very different orbital energies, different diagrams are needed. The interaction of overlapping orbitals decreases as the energy difference of the atomic orbitals increases. If there is a large energy difference between the atomic orbitals being used to form the molecular orbitals, the bonding orbital is only slightly more stable than the separate atomic orbitals.

The molecular orbital diagram of LiF is shown in Figure 9.49. The relatively high energy 2*p* orbitals on lithium are not shown. The energy separation of the atomic orbitals in LiF is so large that the lower energy molecular orbitals are almost pure fluorine atomic orbitals. There is just a weak interaction between the lithium 2*s* orbital and the fluorine 2*p* orbital directed toward lithium. The 2*p* orbital on fluorine is used rather than the 2*s* because the 2*p* is closer in energy to the lithium 2*s* orbital. When the eight valence electrons for LiF are placed in this diagram, they are all in molecular

(Text continues on p. 360.)

INSIGHTS into CHEMISTRY

A Closer View

Atomic Orbitals Overlap to Form Delocalized Molecular Orbitals

Some of the advantages of the delocalized approach of molecular orbital theory can be demonstrated by considering BeH_2. The Lewis structure for BeH_2 is

H—Be—H

The lines that connect the beryllium atom to the two hydrogen atoms represent the sharing of pairs of electrons by two adjacent atoms. This valence bond picture of the molecule, in conjunction with the valence-shell electron-pair repulsion model, proved to be useful in predicting that this molecule has a linear geometry. In forming the two equivalent sigma bonds, the $2s$ orbital and one $2p$ orbital of the beryllium atom combine to form two *sp* hybrid orbitals. Each bond is formed by overlap of a hydrogen atom $1s$ orbital and an *sp* hybrid orbital on beryllium, resulting in two bonds.

The molecular orbital description also begins with the linear arrangement of atoms. The molecular orbitals are formed from the valence-shell atomic orbitals of the beryllium atom ($2s$, $2p_x$, $2p_y$, and $2p_z$) and the two $1s$ orbitals on the hydrogen atoms. The $2s$ orbital on the beryllium atom overlaps equally with each of the hydrogen $1s$ orbitals as shown in the following diagram. These orbitals combine to give a sigma bonding orbital and a sigma antibonding orbital. An electron pair in the bonding molecular orbital contributes equally to the bonding of both the hydrogen atoms, and is therefore a delocalized molecular orbital.

One of the $2p$ atomic orbitals of the beryllium atom is directed along the molecular axis, and also overlaps with the orbitals of the two hydrogen atoms. Since the sign of the amplitude of the *p* wave function is opposite in the two lobes, for both the bonding and the antibonding orbital the sign of the molecular wave function on the first hydrogen atom is opposite that in the second hydrogen atom, as shown. Once again a sigma bonding orbital and a sigma antibonding orbital result from the negative combinations.

Since the two remaining *p* orbitals of the beryllium atom are perpendicular to the Be—H bonds, there is no net overlap with the hydrogen $1s$ orbitals, so they do not contribute to the bonding in any way. These orbitals are the same as the starting $2p$ orbitals on the beryllium atom and are called *nonbonding* (nb) orbitals.

The molecular orbital energy-level diagram for BeH_2 is shown on the next page. On the left side are the four valence-shell atomic orbitals of the beryllium atom, and on the right side are the $1s$ orbitals of the two hydrogen atoms. The molecular orbitals are in the center. The four electrons in the BeH_2 molecule fill the two σ bonding molecular orbitals. The total number of bonds is $\frac{1}{2}$[bonding electrons − antibonding electrons] = 2. Both bonding electron pairs in the molecule contribute equally to the bonding of each hydrogen atom to the beryllium. Each of the pairs is delocalized over all three atoms in the molecule.

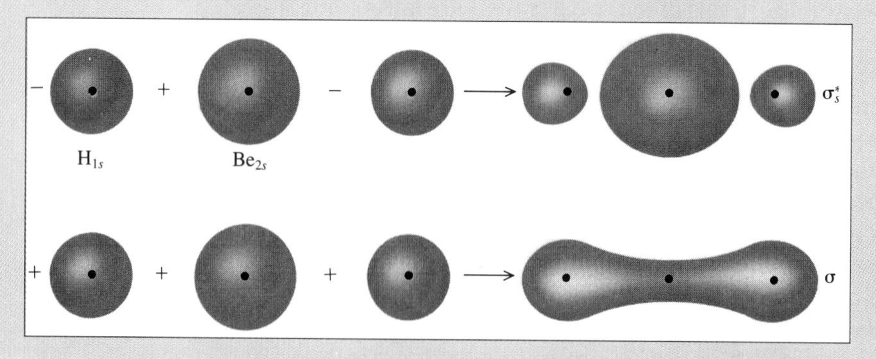

Sigma molecular orbitals from interaction of beryllium 2s and hydrogen 1s atomic orbitals Both bonding and antibonding orbitals are formed from the interaction of the beryllium 2s orbital with the 1s orbitals of the two hydrogen atoms. The new molecular orbitals are delocalized over all three atoms.

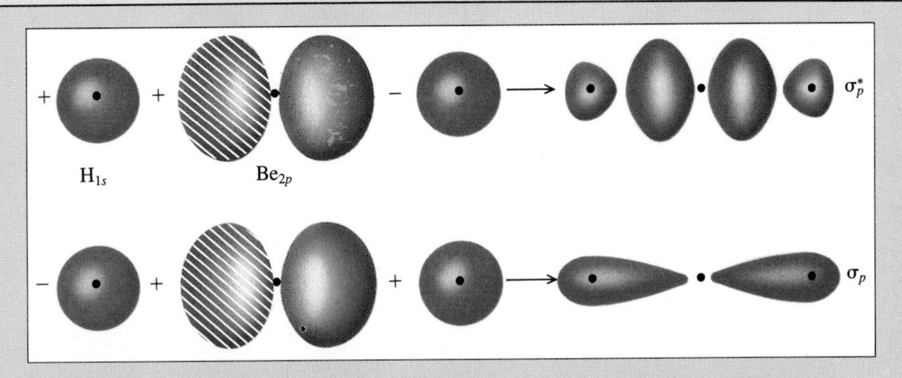

Sigma molecular orbitals from interaction of beryllium 2*p* and hydrogen 1*s* atomic orbitals Both a bonding and antibonding orbital are formed from the interaction of the beryllium 2*p* orbital with the 1*s* orbitals of the two hydrogen atoms. The new molecular orbitals are delocalized over all three atoms. The shading indicates the areas where the *p* orbital wave function's amplitude has a negative sign.

It is important to recognize that both the valence bond theory and the molecular orbital theory produce the same distribution of the electrons in the molecule. In both pictures the four electrons are present as two pairs, with a net of one pair contributing to the bonding of each hydrogen atom. They differ in that both pairs are delocalized and are involved in the bonding of each hydrogen atom in the molecular orbital model, while the valence bond picture has localized bonds.

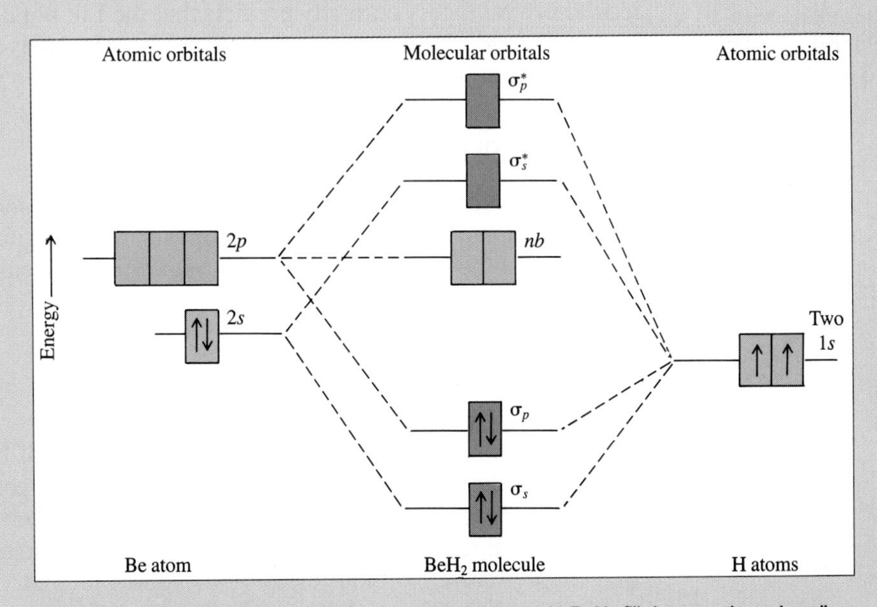

Molecular orbital diagram for BeH_2 The four electrons in BeH_2 fill the two sigma bonding molecular orbitals.

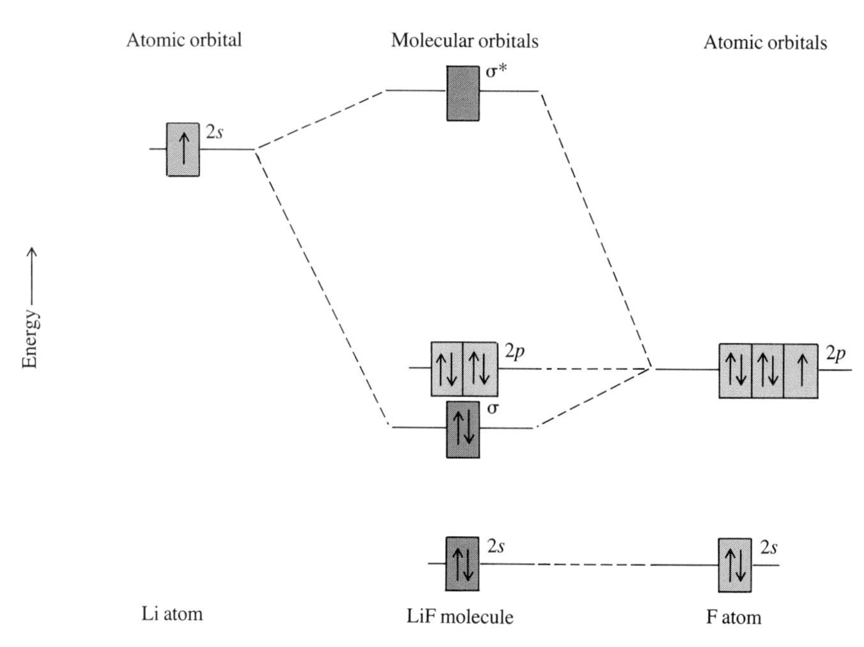

Figure 9.49 The molecular orbital diagram for LiF The energy of the 2*s* orbital on lithium is much higher than those of the 2*s* and 2*p* orbitals on fluorine. The eight valence electrons are in molecular orbitals that are nearly the same as the atomic orbitals on fluorine.

The molecular orbital diagram for LiF places all of the valence electrons around the fluorine, predicting that the compound is ionic.

orbitals that are essentially the same as the atomic orbitals of fluorine. The eight electrons are located mainly around the fluorine. In other words, molecular orbital theory correctly predicts that the LiF bond is ionic since the valence electron in the lithium atom is nearly completely transferred to the fluorine atom.

Delocalized Pi Bonding

In using Lewis diagrams to describe the bonding of some molecules and ions, we drew resonance structures to account for the experimentally observed properties of the species. For example, the ozone molecule, O_3, was represented by two resonance configurations.

A single Lewis structure cannot account for the equal bond lengths and strengths of the two O—O bonds. This problem is a direct result of the fact that only localized bonds are shown in conventional Lewis diagrams. Molecular orbital theory overcomes this problem, since the molecular orbitals may involve atomic orbitals on all of the atoms present.

Valence bond theory describes the bond that changes location in the two resonance forms of ozone as a pi bond. This pi bond is formed from three *p*

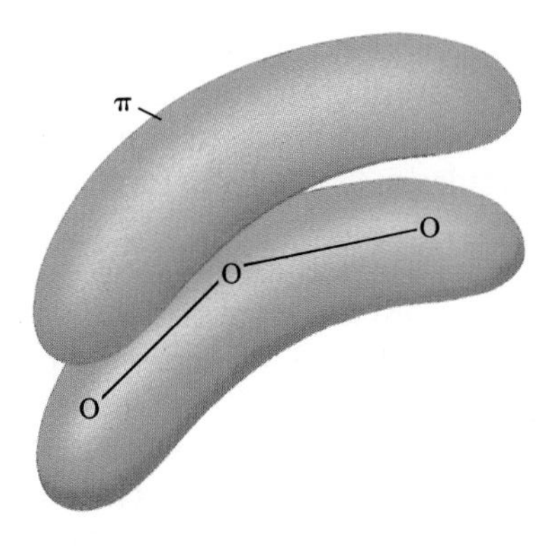

Figure 9.50 Pi bonding molecular orbital for ozone The three p orbitals perpendicular to the O_3 plane interact to form a delocalized pi bonding molecular orbital.

orbitals that are perpendicular to the molecular plane of O_3 (remember that π bonds are made from sideways overlap of p orbitals). The p orbital on the central oxygen atom overlaps with a p orbital on the oxygen atom on its left in the first resonance structure, and with a p orbital on the oxygen atom on its right in the second resonance structure. In molecular orbital theory, all three of these p orbitals can interact to form one large molecular orbital that is spread out over all three nuclei (Figure 9.50).

The complete molecular orbital treatment of O_3 and most other molecules is not appropriate at this point. The energy-level diagrams can become quite complicated, even for small molecules. While molecular orbital theory is being used more and more frequently to describe the bonding and distribution of electrons in molecules, it does not provide a simple means of anticipating molecular geometries. Valence bond theory and Lewis structures are generally more useful in predicting molecular shapes. Each of the models for covalent bonding is valuable in different ways for describing the observed properties of molecules. One very important advantage of molecular orbital theory is that it produces energy-level diagrams that allow the interpretation of the electromagnetic spectra of molecules, just as the energy-level diagrams for atoms can account for atomic spectra.

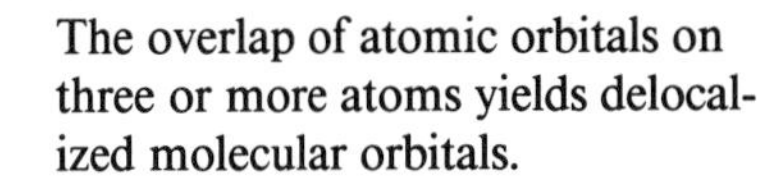

The overlap of atomic orbitals on three or more atoms yields delocalized molecular orbitals.

Summary

The *valence-shell electron-pair repulsion (VSEPR)* model can be used to predict the shapes of many molecules and ions. In this model, we use a Lewis structure to count the number of lone pairs plus the number of bonded atoms for each central atom (one that is bonded to at least two other atoms). This sum is used to determine the *electron pair arrangement*, which is the arrangement adopted by the valence-shell electron pairs to maximize the distance between them. The electron-pair arrangements (and bond angles) for given numbers of electron pairs are: linear for two pairs (180°); trigonal planar for three pairs (120°); tetrahedral for four pairs (109.5°); trigonal bipyramidal for five pairs (both 90° and 120°); and octahedral for six pairs (90°). Because lone pairs are not considered part of the molecular geometry, the electron-pair arrangements for molecules with lone pairs on the central atom are not the same as the molecular geometries, but the bond angles are approximately the same in both. The rule that lone pairs exert greater repulsive forces than bond pairs is useful in explaining details of the shapes of molecules with lone pairs.

The *polarity* of a molecule can be determined from its shape and the polarities of its bonds. A molecule in which the same type of atom occupies each vertex of the electron-pair arrangement is nonpolar, even though it may contain polar bonds, because the individual bond dipoles cancel. Most other molecules are polar, especially those with lone pairs on the central atom.

Valence bond theory is an extension of Lewis theory in which the orbitals that are used to form bonds are specified. *Hybrid orbitals,* orbitals that are mixtures of *s* and *p* and even *d* orbitals, are frequently used in this description of bonding. The type of hybrid orbitals to be used is determined from the bond angles as follows: *sp* hybrid orbitals form bonds at 180°, sp^2 hybrid orbitals form bonds at 120°, and sp^3 hybrid orbitals form bonds at 109.5°. In addition, sp^3d hybrid orbitals are used to explain trigonal bipyramidal geometry and sp^3d^2 hybrid orbitals to explain octahedral geometry.

The bonds shown in Lewis structures can be described as *sigma* or *pi* bonds. Sigma bonds are formed from end-on overlap of *s*, *p*, or hybrid orbitals; pi bonds are formed from sideways overlap of *p* orbitals. The orientation needed to form the pi bond explains the planar geometry of molecules such as ethylene and its derivatives.

Molecular orbital theory describes bonding as being delocalized over the entire molecule. Molecular orbitals are formed by combinations of appropriate atomic orbitals. The molecular orbitals are either *bonding molecular orbitals* or *antibonding molecular orbitals.* The electron configuration of a molecule is determined in the same manner as for atoms, by adding the appropriate number of electrons to the molecular orbital diagram. The molecular orbital description of bonding accurately predicts the bond order and number of unpaired electrons in molecules.

Chapter Terms

Electron-pair arrangement—the positions of the valence-shell electron pairs that minimizes their repulsion from each other. *(9.1)*

Molecular shape—a description of the positions of the atoms, not the lone pairs. *(9.1)*

Polar molecule—a molecule that contains an unequal distribution of charge and thus a dipole moment. *(9.2)*

Valence orbital—an orbital in the valence shell. *(9.3)*

Valence bond theory—a theory that describes bonds as being formed by atoms sharing valence electrons in overlapping valence orbitals. *(9.3)*

Hybrid orbitals—orbitals obtained by mixing two or more atomic orbitals on the same atom. *(9.3)*

Sigma bond (σ)—a bond in which the shared pair of electrons is concentrated around the line joining the nuclei of the bonded atoms. *(9.4)*

Pi bond (π)—a bond that places electron density above and below the line joining the bonded atoms. It is formed from sideways overlap of *p* orbitals. *(9.4)*

Isomers—different compounds with the same molecular formula but with different structures *(9.4)*

Molecular orbital theory—a model that treats bonding as delocalized over the entire molecule. *(9.5)*

Molecular orbital—a wave function of an electron in a molecule. *(9.5)*

Bonding molecular orbital—an orbital that concentrates the electron density between the atoms in the molecule. *(9.5)*

Antibonding molecular orbital—an orbital that reduces the electron density in the region between the atoms in the molecule. *(9.5)*

Homonuclear diatomic molecule—a molecule formed by two atoms of the same element. *(9.5)*

Delocalized molecular orbital—a molecular orbital that involves atomic orbitals on more than two atoms. *(9.6)*

Heteronuclear diatomic molecule—a molecule formed by one atom of each of two different elements. *(9.6)*

Exercises

Exercises designated with color have answers in Appendix J.

VSEPR Model

9.1 What is the basic premise of VSEPR theory that is used to assign the electron-pair arrangement? State carefully how the electron-pair arrangement is determined.

9.2 What electron-pair arrangement is expected for a central atom that has
(a) three bonded atoms and no lone pairs
(b) two bonded atoms and two lone pairs
(c) four bonded atoms and no lone pairs
(d) four bonded atoms and one lone pair?

9.3 What electron-pair arrangement is expected for a central atom that has

(a) two bonded atoms and one lone pair
(b) three bonded atoms and two lone pairs
(c) four bonded atoms and two lone pairs
(d) five bonded atoms and one lone pair?

9.4 Use VSEPR theory to predict the molecular shape of (a) CF_4 (b) CS_2 (c) AsF_5 (d) F_2CO (e) NH_4^+.

9.5 Use VSEPR theory to predict the molecular shape of (a) BeF_2 (b) SF_6 (c) SiH_4 (d) ClCN (e) BeF_3^-.

9.6 Give the electron-pair arrangement and the molecular shape of
(a) SeO_2 (b) N_2O (N is the central atom)
(c) H_3O^+ (d) IF_5 (e) SCl_4.

9.7 Give the electron-pair arrangement and the shape of (a) XeO_2 (b) I_3^- (c) NO_2^- (d) PCl_5 (e) $AlCl_3$.

9.8 In each pair of species, indicate which species has the smaller bond angles. Explain your answer.
(a) BCl_3 or NCl_3 (b) OF_2 or SF_6.

9.9 In each pair of species, indicate which species has the smaller bond angles. Explain your answer.
(a) SO_4^{2-} or $AlBr_3$ (b) CCl_4 or BeI_2.

9.10 Write the Lewis structure for each of the following molecules. Indicate all of the bond angles, as predicted by VSEPR theory. Deduce the connectivity from the way the formula is written.
(a) H_3CCCH (b) Br_2CCH_2 (c) H_3CNH_2.

9.11 Write the Lewis structure for each of the following molecules. Indicate all of the bond angles, as predicted by VSEPR theory. Deduce the connectivity from the way the formula is written.
(a) $ClC(O)NH_2$ (oxygen bonded only to the carbon atom)
(b) $HOCH_2CH_2OH$
(c) NCCN

9.12 Use VSEPR theory to determine the bond angles around each central atom in the following molecules. Note that the drawing may incorrectly picture the angles and is not a Lewis structure.

```
       H  O
       |  |
(a) H—C—C—H
       |
       H

       H           H
       |          /
(b) H—C—O—C—C
       |     |    \
       H     H     H

(c) F—Xe—F
```

9.13 Use VSEPR theory to determine the bond angles around each central atom in the following molecules. Note that the drawing may incorrectly picture the angles and is not a Lewis structure.

```
     H     H  H     H
      \    |  |    /
(a)    C—C—C—C
      /          \
     H            H

                 H
                 |
(b) H—C—C—C—H
                 |
                 H

     Cl
     |
(c)  P—Cl
     |
     Cl
```

Polarity of Molecules

9.14 Give an example of a nonpolar molecule that contains polar bonds. Show the polarity of the bonds with arrows and show how these bond dipoles cancel.

9.15 In Exercise 9.4 the shapes of (a) CF_4, (b) CS_2, (c) AsF_5, and (d) F_2CO were determined. Are each of these molecules polar or nonpolar?

9.16 In Exercise 9.5 the shapes of (a) BeF_2, (b) SF_6, (c) SiH_4, and (d) ClCN were determined. Is each of these molecules polar or nonpolar?

9.17 Of the molecules (a) SeO_2, (b) N_2O (N is the central atom), (c) IF_5, and (d) SCl_4, which are polar? For the polar molecules, draw the molecular structure of each and indicate the bond dipoles and the molecular dipole moment.

9.18 Of the molecules (a) XeO_2, (b) I_2, (c) NO, and (d) PCl_5, which are polar? For the polar molecules, draw the molecular structure of each and indicate the bond dipoles and the molecular dipole moment.

9.19 Of the molecules (a) BCl_3, (b) OF_2, and (c) SF_6, which are polar? For the polar molecules, draw the molecular structure of each and indicate the bond dipoles and the molecular dipole moment.

9.20 Of the molecules (a) NCl_3, (b) CBr_4, and (c) BeI_2, which are polar? For the polar molecules, draw the molecular structure of each and indicate the bond dipoles and the molecular dipole moment.

Hybrid Orbitals

9.21 What atomic orbitals overlap to form the bonds in HI?

9.22 What atomic orbitals overlap to form the bonds in FCl?

9.23 Why are hybrid orbitals needed to explain the bonding in NH_3?

9.24 What hybridization of the central atom would yield bond angles of all adjacent atoms of
(a) 120° (b) 90° (c) 180°?

9.25 What is the hybridization of the central atom that has the electron-pair arrangement of
(a) a tetrahedron
(b) a trigonal bipyramid
(c) an octahedron?

9.26 What hybrid orbitals on the central atom are used to make the bonds in
(a) CF_4 b) $SbCl_6^-$ (c) AsF_5
(d) SiH_4 (e) NH_4^+?

9.27 What hybrid orbitals on the central atom are used to make the bonds in
(a) SeH_2 (b) SCl_2 (c) H_3O^+
(d) IF_5 (e) SCl_4?

9.28 What hybrid orbitals on the central atom are used to make the bonds in
(a) N_2O (b) $SnCl_2$ (c) I_3^- (d) SeO_2?

9.29 What hybrid orbitals are used by the nitrogen atom in
(a) $HNCl_2$ (b) NO_3^- (c) N_2H_2?

9.30 What hybrid orbitals are used by the carbon atom in
(a) CO_3^{2-} (b) CH_2F_2 (c) H_2CO?

9.31 What types of orbitals are used from each atom to form the bonds in
(a) ClF_3 (b) BBr_3 (c) BeF_2?

9.32 What orbitals are used by selenium and fluorine to form the bonds in SeF_4? What orbital holds the lone pair on selenium?

9.33 Nitrous acid has the connectivity HONO. What are the hybrid orbitals used on the nitrogen atom and the central oxygen atom?

9.34 What orbitals are used to form the bonds and hold the lone pairs in
(a) OF_2 (b) NH_3 (c) BCl_3.

9.35 What are the hybridizations of the carbon atoms in
(a) C_2H_6 (b) C_2H_4 (c) CBr_4?

9.36 What is a coordinate covalent bond and how does it differ from other electron-pair bonds?

9.37 Describe the bonding in BCl_3 and BCl_4^-, the ion formed from the reaction of BCl_3 and Cl^-.

9.38 Describe the bonding in $SbCl_5$ and $SbCl_6^-$, the ion formed from the reaction of $SbCl_5$ and Cl^-.

9.39 Describe the changes in the hybridizations of the boron and nitrogen atoms when a coordinate covalent bond forms between NH_3 and BH_3.

9.40 Explain why it is difficult to determine the hybridization of the chlorine atom in HCl.

9.41 Define a sigma bond and a pi bond. Show how *p* orbitals overlap in a sigma bond and in a pi bond.

9.42 If the *z* axis is defined as the bond axis, draw a picture that shows the overlap of the following pairs of orbitals and indicate whether a sigma or pi bond is formed from:
(a) p_z, p_z (b) p_y, p_y (c) *sp* hybrid formed from p_z and *s* orbitals, p_z.

9.43 If the *z* axis is defined as the bond axis, draw a picture that shows the overlap of the following pairs of orbitals and indicate whether a sigma or pi bond is formed from:
(a) p_x, p_x (b) *s*, p_z (c) sp^2 hybrid formed from p_x, p_z, and *s* orbitals, *s*.

9.44 Identify the orbitals on each atom that are used to form all of the bonds in H_3CCN. How many sigma bonds and pi bonds are formed?

9.45 Sketch the bonds in butadiene, which is pictured below. Label the type of orbitals being used to form each bond and indicate whether the bond is a sigma or a pi bond.

```
H       H   H       H
 \      |   |      /
  C  =  C - C  =  C
 /                 \
H                   H
```

9.46 Sketch the bonds in H_2CNH, label the type of orbital from which each bond is formed, and indicate whether the bond is a sigma or pi bond.

9.47 Sketch the bonds in N_2, label the types of orbitals from which each bond is formed, and indicate whether the bond is a sigma or pi bond.

9.48 Give the hybridization of each central atom in
(a) cyclohexene,

```
        H   H
         \ /     H
          C     /
        /   \  /
  H - C       C - H
      ||      |
  H - C       C - H
        \   /  \
          C     H
         / \
        H   H
```

(b) phosgene, Cl_2CO
(c) glycine, H_2NCH_2COOH

```
   H   H  :O:
   |   |   ||   ..
  :N - C - C  - O - H
   |   |        ..
   H   H
```

9.49 Give the hybridization of each central atom in
(a) CO_2
(b) H_3CCCH
(c) $H_3CC(O)H$,

```
H  :O:
|   ||
H—C—C—H
|
H
```

9.50 Two resonance forms can be written for NO_2^-. In each resonance form, indicate the hybridization used by the central atom.

9.51 Three resonance forms can be written for N_3^-. In each resonance form, indicate the hybridization used by the central atom.

9.52 Predict the hybridization at each central atom in

```
     H  H                   H
     |  |                   |
(a) H—C—N:           (b) H—C—C≡C—H
     |  |                   |
     H  H                   H
```

9.53 Predict the hybridization at each central atom in

```
     H :O: H                   H       H
     |  ||  |                  |   ..  |
(a) H—C—C—C—H          (b) H—C—O—C—H
     |     |                   |   ..  |
     H     H                   H       H
```

9.54 The main compound used for the production of Orlon is acrylonitrile, H_2CCHCN. Draw the Lewis structure of acrylonitrile and indicate the hybridization of each central atom.

9.55 Tetrafluoroethylene, C_2F_4, is used to prepare Teflon. Draw the Lewis structure of tetrafluoroethylene and indicate the hybridization of each carbon atom.

Molecular Orbitals

9.56 Draw the complete molecular orbital diagram, including the electrons, and write the electron configuration for H_2^+. State the bond order and the number of unpaired electrons, if any. Is this a stable species?

9.57 Draw the complete molecular orbital diagram, including the electrons, and write the electron configuration for H_2^-. State the bond order and the number of unpaired electrons, if any. Is this a stable species?

9.58 Draw the complete molecular orbital diagram, including the electrons, for Li_2. State the bond order and the number of unpaired electrons, if any. Is this a stable species?

9.59 Draw the complete molecular orbital diagram, including the electrons, for C_2. State the bond order and the number of unpaired electrons, if any. Is this a stable species?

9.60 Write the molecular orbital electron configuration and determine the bond order and number of unpaired electrons in
(a) C_2^+ (b) N_2^- (c) Be_2^-.

9.61 Use the molecular orbital diagram in Figure 9.41 to predict which species in each pair has the stronger bond: (a) B_2 or B_2^- (b) N_2^- or N_2^+ (c) O_2^{2+} or O_2.

9.62 Identify *two* homonuclear diatomic molecules or ions that both have the following molecular orbital electron configurations. Will these species be stable?
(a) $(\sigma_{2s})^2(\sigma_{2s}^*)^2(\pi_{2p})^4(\sigma_{2p})^2(\pi_{2p}^*)^3$
(b) $(\sigma_{2s})^2(\sigma_{2s}^*)^2(\pi_{2p})^4(\sigma_{2p})^2$
(c) $(\sigma_{2s})^2(\sigma_{2s}^*)^2$

9.63 Assuming that the molecular orbital diagram shown in Figure 9.41 is correct for heteronuclear diatomic molecules containing elements that are close in the periodic table, write both a homonuclear and a heteronuclear diatomic molecule (remember that molecules are neutral) in which both molecules have the given electron configuration.
(a) $(\sigma_{2s})^2(\sigma_{2s}^*)^2(\pi_{2p})^4(\sigma_{2p})^2(\pi_{2p}^*)^2$
(b) $(\sigma_{2s})^2(\sigma_{2s}^*)^2(\pi_{2p})^4$
(c) $(\sigma_{2s})^2(\sigma_{2s}^*)^2(\pi_{2p})^4(\sigma_{2p})^2$

9.64 The molecular orbital diagram for NO shown in Figure 9.48 can be used for the following species. Write the electron configuration of each, indicating the bond order and number of unpaired electrons.
(a) CN (b) CO^- (c) BeB^- (d) BC^+.

9.65 The molecular orbital diagram for NO shown in Figure 9.48 can be used for CO. Draw the complete molecular orbital diagram for CO. What is the C—O bond order?

9.66 The molecular orbital diagram for NO shown in Figure 9.48 can be used for OF^-. Draw the complete molecular orbital diagram for OF^-. What is the O—F bond order?

9.67 Write the molecular orbital electron configuration and give the bond order and number of unpaired electrons for (a) $LiBe^+$, (b) CO^+, (c) CN^-, (d) OF.

9.68 The delocalized bonding picture used for O_3 is also correct for NO_2^-. Draw the delocalized π molecular orbital for NO_2^-.

9.69 Draw the delocalized π orbital for benzene. Clearly indicate the atomic orbitals being used to make the molecular orbital.

Additional Problems

9.70 Write one Lewis structure of N_2O_5 (O_2NONO_2 connectivity). What are the bond angles around the central oxygen atom and the two nitrogen atoms? What is the hybridization of each?

9.71 Pictured below are three isomers of difluorobenzene, $C_6H_4F_2$. Are any of them nonpolar?

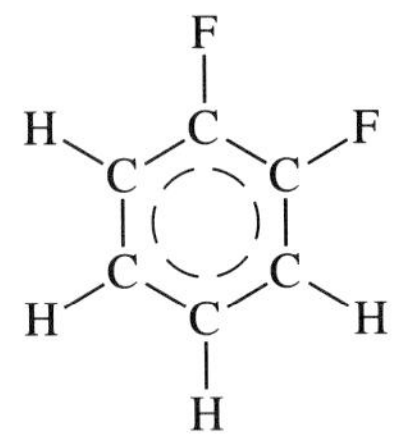

A

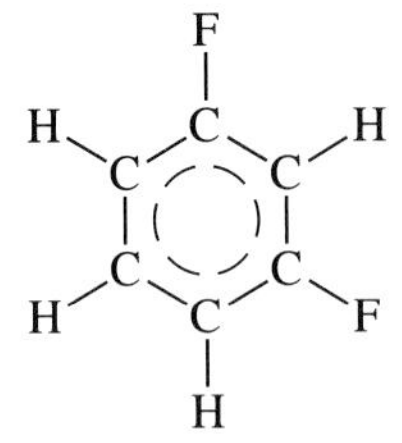

B

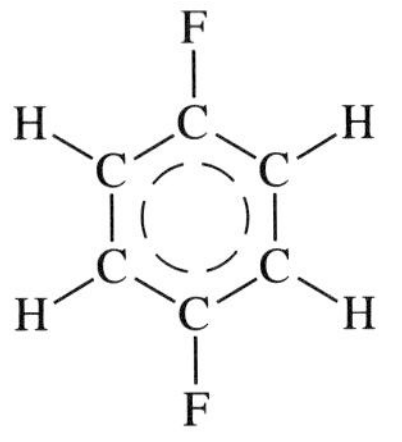

C

9.72 More than 5 billion pounds of ethylene oxide, C_2H_4O, are produced annually. It is used in the production of ethylene glycol, $HOCH_2CH_2OH$, the main component of antifreeze, and acrylonitrile, CH_2CHCN, used in the production of synthetic fibers and other chemicals. It has an interesting cyclic structure.

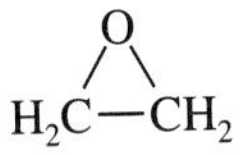

O

$H_2C—CH_2$

Draw the Lewis structures of ethylene oxide, ethylene glycol, and acrylonitrile.

9.73 Aspirin, or acetylsalicylic acid, has the formula $C_9H_8O_4$ and the connectivity

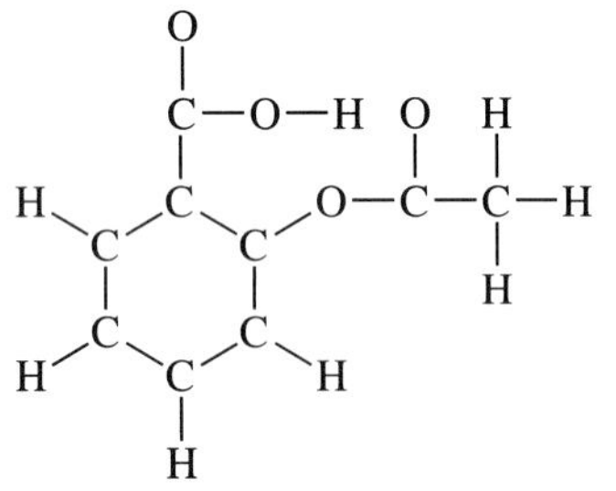

(a) How many sigma bonds and how many pi bonds are in aspirin?

(b) What is the hybridization about the CO_2H carbon atom?

(c) What is the hybridization about the carbon atom in the benzene-like ring that is bonded to an oxygen atom? Also, what is the hybridization of this oxygen atom?

9.74 Recently the structure of an amine compound, NR_3 (R = organic group, with very large R-substituents) has been determined to have C—N—C bond angles of 119.2°. It is believed that the bond angles of about 109° expected from VSEPR theory are not observed because of the large substituents bonded to the nitrogen atom. Given this large bond angle, what type of orbitals are used on the nitrogen atom to make the N—C sigma bonds and in what type of orbital is the lone pair located?

9.75 Aspartame is a compound that is 200 times sweeter than sugar and is used extensively (under the trade name NutraSweet) in diet soft drinks. The connectivity of the atoms in aspartame is

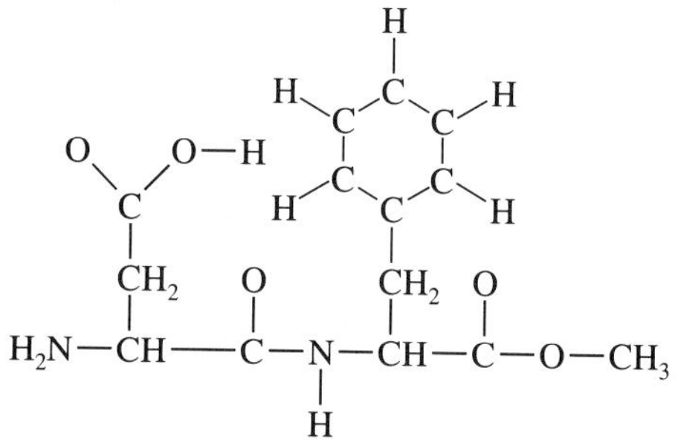

(a) How many sigma and how many pi bonds are in aspartame?

(b) What is the hybridization about each carbon atom that forms a double bond with an oxygen atom?

(c) What is the hybridization about each nitrogen atom?

9.76 Phosgene, $COCl_2$, is a highly toxic gas that was used in combat during World War I. It is an important intermediate in the preparation of a number of organic compounds, but must be handled with extreme care. Given that carbon is the central atom in phosgene, what is the Lewis structure, electron-pair arrangement, hybridization of the carbon atom, and the polarity?

9.77 Calcium cyanamide, CaNCN, is used both to kill weeds and as a fertilizer. For the NCN^{2-} ion, what is the Lewis structure, electron-pair arrangement, and the hybridization of the carbon atom?

9.78 Histidine is an essential amino acid that is used by the body to form proteins. The Lewis structure of histidine is shown below. What are the approximate values for the bond angles 1 through 5 indicated on the structure?

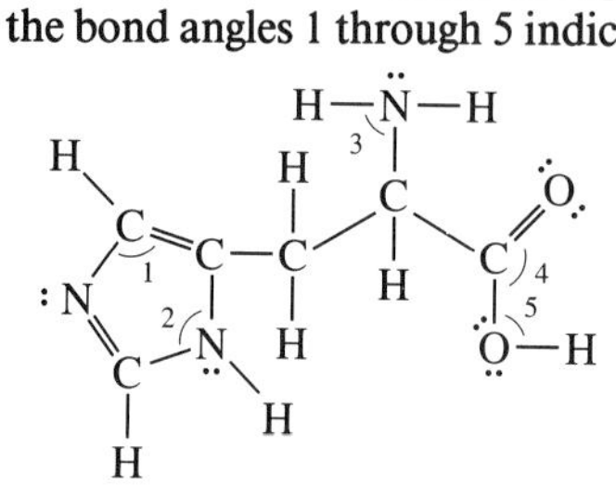

9.79 Formamide, $HCONH_2$, can be prepared at high pressures from carbon monoxide and ammonia, and is used as an industrial solvent. Two resonance forms (one with formal charge) can be written for formamide. Write both resonance structures and predict the bond angles about the carbon and nitrogen atoms *for each resonance form.* Are they the same? Describe how the experimental determination of the H—N—H bond angle could be used to indicate which resonance form is more important.

9.80 Draw the molecular orbital diagrams for NO^- and NO^+. Compare the bond orders in these two ions.

9.81 Ionization energies can be determined for molecules as well as atoms. Draw the molecular orbital diagrams for NO and CO, and predict which will have the lower ionization energy.

Chapter Ten 10

Liquids and Solids

The physical properties of gases were presented in some detail in Chapter 5. Except under unusual conditions, a single relationship—the ideal gas law—describes the properties of all gases quite well. The kinetic molecular theory offers an explanation for this common behavior of gases. The physical properties of the other two states of matter, liquids and solids, are more difficult to describe. The kinetic molecular theory, with some modification, accounts for the characteristics of changes of state, in which a substance in one state is converted into another state. However, there is no single equation of state like the ideal gas law that applies to all substances in the liquid and solid states. As we shall see in this chapter, the greater variations of properties in the liquid and solid states arise because the molecules are close together, and thus the strength of the forces of attraction between the molecules becomes important.

In this chapter we examine the different kinds of attractions that exist between molecules, since they play a key role in determining the conditions under which a substance is a solid, a liquid, or a gas. We shall present the properties of substances that are related to the strengths of these intermolecular attractions. The structures of solids on the atomic and molecular scale are discussed, and the experimental means of determining these structures are also presented. Finally, we consider the characteristics of changes of state and introduce the idea of dynamic equilibrium. Table 10.1 summarizes the characteristic properties of the three states of matter.

The kinetic molecular theory accounts for both the low density and high compressibility of gases by assuming that nearly all of the volume of a sample is empty space. In the *condensed states,* liquids and solids, the matter occu-

◀ Water, like most substances, can exist in the solid, liquid, and gaseous states.

Table 10.1 Characteristic Properties of Gases, Liquids, and Solids

State	Volume and Shape of Sample	Density	Compressibility
gas	assumes the volume and shape of container	low	easily compressed
liquid	has definite volume, assumes shape of container	high	nearly incompressible
solid	has both definite volume and shape	high	nearly incompressible

pies a very large fraction of the sample volume, causing the characteristic high densities and low compressibilities that we observe. A solid differs from a liquid in that it is rigid (a sample maintains a definite shape), while the liquid is fluid and adapts to the shape of the container. These properties of the condensed phases suggest that the molecules attract each other. To begin the consideration of the liquid and solid states, we discuss the nature of the attractive forces that cause molecules to stay close together.

10.1 Intermolecular Attractions

Objectives

- To distinguish between intermolecular and intramolecular forces
- To explain why the strengths of intermolecular forces determine the temperatures over which each of the three states is stable
- To use boiling points as an indication of the strengths of intermolecular attractions
- To identify the kinds of intermolecular attractions that are important for a given substance

Intermolecular forces are the attractive forces that exist between molecules. Intermolecular forces cause molecules to stay close to each other in the liquid and solid states. These forces are the same attractions that were introduced in Chapter 5 to account for deviations of gases from ideal behavior. It is important to avoid confusing *intermolecular forces* with *intramolecular forces,* the forces that hold the atoms together in molecules. When a liquid or a solid vaporizes (becomes a gas) the intermolecular attractions must be overcome, but the forces within the molecules remain intact.

Intramolecular forces hold atoms together in molecules; intermolecular forces are those between molecules.

The existence of the liquid and solid states is ample evidence that atoms and molecules do attract each other. Experimental measurements have shown that molecules must be quite close together before intermolecular attractions become an important factor. In all cases, the origin of intermolecular forces is the attraction between electrical charges of opposite sign.

Kinetic Molecular Theory and Intermolecular Forces

The physical state of any sample of matter depends upon the strengths of the intermolecular attractions and the average kinetic energy of the molecules. The intermolecular attractions are important only at short distances (a few molecular diameters), so they are significant only in the condensed states and in gases at very high pressures. The strengths of the intermolecular attractions do not change much with the temperature, except indirectly, because intermolecular distances change with temperature. On the other hand, the average kinetic energy of molecules is proportional to the absolute temperature (refer to the discussion of the kinetic molecular theory in Chapter 5). Furthermore, according to the kinetic molecular theory, there is a distribution of kinetic energies among the molecules; at any instant some molecules have kinetic energies greater than the average and others have kinetic energies less than the average.

In solids, the energy of the attractions between molecules is quite large compared to the average kinetic energy of molecules at that temperature, and almost none of the molecules have enough energy to overcome the attractions. In this situation the molecules vibrate, but rarely move far from their nearest neighbors. The molecules adopt an orderly arrangement that maximizes the energy of attraction, leading to the rigid characteristics of a solid. When the average kinetic energy is considerably greater than the attractive energy between molecules, there is no tendency for the molecules to stick together and each molecule behaves independently of all the others. The result is the completely random behavior of molecules in the gaseous state. In the liquid state, the forces of attraction are large enough to keep the molecules close together, but small enough that the more energetic molecules are able to move about.

This model predicts that substances that can remain solid at higher temperatures must have stronger intermolecular forces of attraction. Molecules of these substances must have a high average kinetic energy (a high temperature) before they can exist as gases. The melting and boiling points of substances are therefore good indicators of the strength of the intermolecular forces that exist between the molecules.

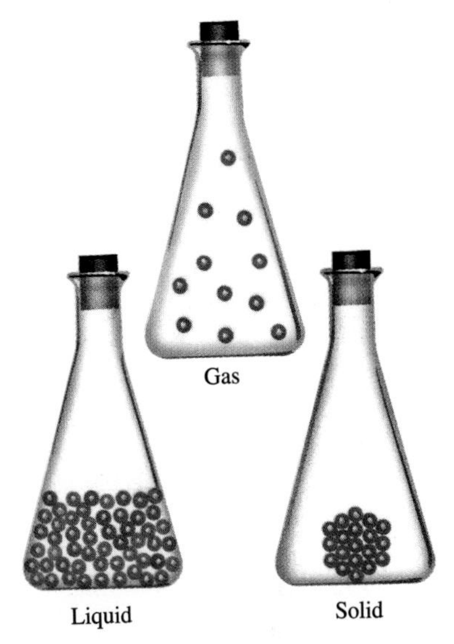

Molecular view of the states of matter In a gas the molecules are so far apart that each one moves independently of the others. In the liquid the molecules are close together but may move about relative to each other. In the solid state the molecules are held together in a regular arrangement.

The stronger the intermolecular attractions in a substance, the higher its boiling point will be.

Example 10.1 Intermolecular Forces and Boiling Points

The boiling points for the hydrogen compounds of some group VIA elements are: H_2O, 100 °C; H_2S, −61.8 °C; H_2Se, −42 °C. Arrange these compounds in order of increasing strength of the intermolecular forces.

Solution The boiling points of substances increase as the intermolecular forces of attraction get stronger, so the correct order is from the lowest to the highest boiling point.

$$H_2S < H_2Se < H_2O$$

Understanding The boiling points for some of the noble gases are: Ar, −185.7 °C; Kr, −152.9 °C; Ne, −245.9 °C. Arrange these elements in order of increasing strength of the intermolecular attractions.

Answer: Ne < Ar < Kr

In addition to the melting and boiling points, there are several other properties that also reflect the relative strengths of intermolecular attractions. These will be discussed after we examine the origin of these attractions.

Dipole-Dipole Attractions

The attractions between electrical charges of opposite sign are very important in nature. In Chapter 8 the stability of solid ionic compounds was shown to arise mainly from the electrostatic attraction between the ions. In covalent compounds, partial positive and negative charges are produced on atoms by the unequal sharing of electron pairs. In many cases the arrangements of these polar bonds (see Chapter 9) cause the molecules to have a dipole moment. Such polar molecules are drawn to one another, since the negative end of one molecule attracts the positive end of another one in accord with Coulomb's law. **Dipole-dipole attractions** are the intermolecular forces that arise from simple electrostatic attractions between the molecular dipoles. For example, the difference in the boiling points of the polar ICl molecule (97.4 °C) and that of Br_2 (58.8 °C) is explained by dipole-dipole attractions. The Br_2 molecule is nonpolar, but the dipoles of the ICl arrange to maximize the attractive interactions, similar to that shown in Figure 10.1. In general, the larger the dipole moment of a molecule, the stronger is the dipole-dipole attraction. Because the charges involved are smaller, the dipole-dipole attractions are much weaker than the attractions between ions or those that result from covalent bond formation between atoms.

Dipole-dipole attractions are present only in substances that contain polar bonds.

Dispersion Forces

Dipole-dipole attractions contribute to the intermolecular attractions in many substances. However, nonpolar molecules also attract each other. Nonpolar molecular substances such as argon (Ar), nitrogen (N_2), and chlorine (Cl_2) all condense to liquids and solids at low temperatures, and therefore attractive forces between the molecules must also exist in nonpolar substances. Table 10.2 shows the boiling points of several series of related nonpolar substances. In each of these series, there is a trend of increasing boiling points, and therefore an increase in the strengths of the intermolecular attractions, as the molar masses of the compounds increase.

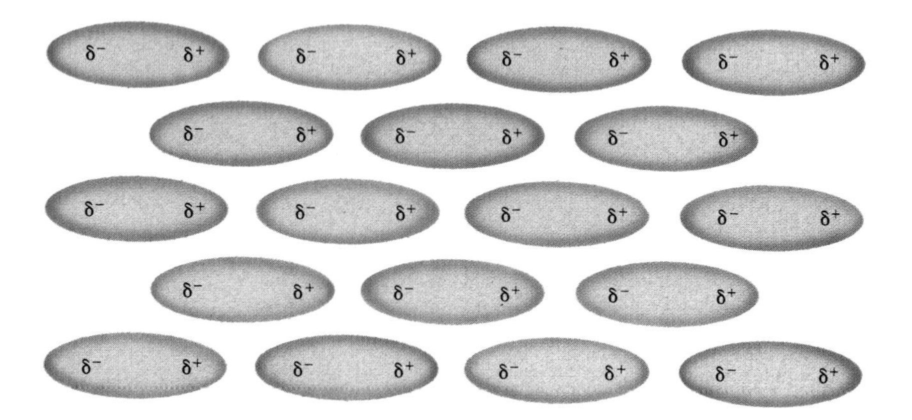

Figure 10.1 Dipole-dipole attractions In a solid, polar molecules pack together so that the positive end of one molecule is close to the negative end of other molecules.

Table 10.2 Boiling Points of Some Nonpolar Substances

Substance	Molar Mass	Boiling Point (°C)	Substance	Molar Mass	Boiling Point (°C)
Halogens			**Group IV hydrides**		
fluorine (F_2)	38	−188	methane (CH_4)	16	−184
chlorine (Cl_2)	71	−34.6	silane (SiH_4)	32	−111.8
bromine (Br_2)	160	58.8	germane (GeH_4)	77	−90.0
iodine (I_2)	254	184.4	stannane (SnH_4)	123	−52
Boron halides			**Noble gases**		
BF_3	68	−101	helium	4	−268.9
BCl_3	117	12.5	neon	20	−245.9
BBr_3	250	90.1	argon	40	−185.7
BI_3	392	210	krypton	84	−152.9
			xenon	131	−107.1

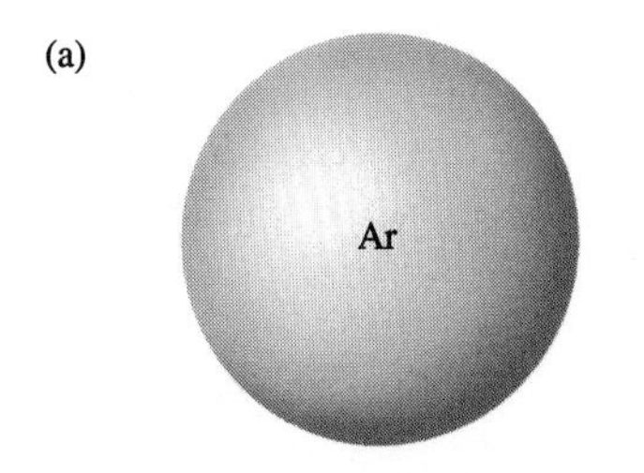

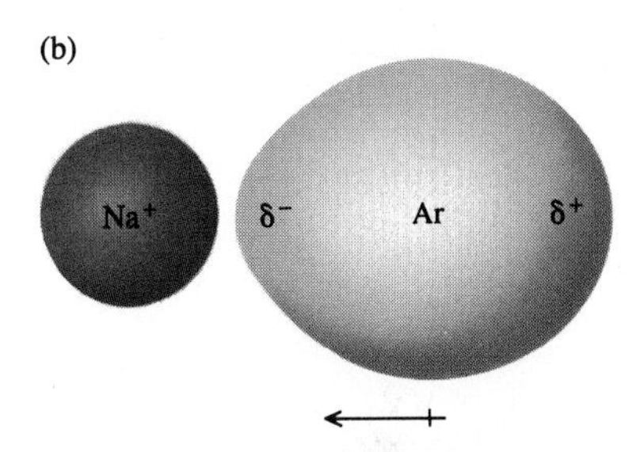

Figure 10.2 Ion–induced dipole attraction (a) The electron cloud of an argon atom is spherical, so the isolated atom has no dipole moment. (b) The electron cloud of an argon atom is distorted by the presence of a nearby ion, producing an induced dipole moment in the nonpolar atom.

To gain some insight into the nature of these attractions, consider what happens when an ion with a 1+ charge comes close to an argon atom. The argon atom consists of a nucleus (charge = +18) surrounded by a cloud of 18 electrons. An argon atom that is very far from other molecules and ions has a spherical distribution of the electrons and no dipole moment. As an ion of 1+ charge comes close to the argon atom, it attracts the outer electrons toward itself, distorting the electron cloud and causing the argon atom to have a dipole moment (see Figure 10.2). Such an **induced dipole moment** is caused by the presence of an electrical charge close to an otherwise nonpolar molecule. In a similar manner a polar molecule, like HCl, can induce a dipole moment in an argon atom by distorting the electron cloud (See Figure 10.3). A small electrostatic attraction exists between the ion and the induced dipole, or between the permanent dipole and the induced dipole.

In each of these cases, there are permanent charges that distort the electron cloud of a nearby nonpolar molecule. However, there are no permanent charges present when all of the molecules are nonpolar. How can such molecules attract each other? *Averaged over time,* there is no net dipole moment in a nonpolar molecule, but the electrons in the nonpolar molecule are constantly in motion. If it were possible to "freeze" their positions at any instant in time, it is almost certain that a nonsymmetric charge distribution would exist within the molecule. An **instantaneous dipole moment** is the result of a very slightly unequal charge distribution within a molecule caused by the motion of the electrons. The rapid motion of the electrons in the molecule means that this instantaneous dipole will be pointed in a different direction or gone a fraction of a microsecond later. At almost any instant, though, the molecule possesses a dipole moment.

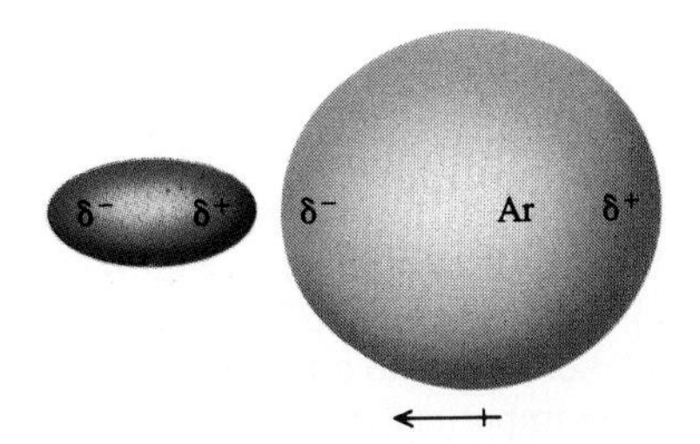

Figure 10.3 Dipole–induced dipole interaction Dipolar molecules can also induce a dipole moment in a nonpolar species like argon. Both the positive and negative ends of the dipole will distort the electron cloud, resulting in a dipole–induced dipole attraction.

The very small charges of an instantaneous dipole moment in one nonpolar molecule can induce a dipole moment in a nearby nonpolar molecule, causing the two molecules to attract each other weakly. These instantaneous dipole–induced dipole attractions are called **London dispersion forces,** named after Fritz London (1900–1954), a German physicist who developed this model to explain the intermolecular attractions that exist between non-

Table 10.3 Boiling Points of Some Hydrogen Halides

Compound	Dipole–Dipole Forces	Boiling Point (°C)	London Forces
hydrogen chloride	|	−85	|
hydrogen bromide	decrease	−67	increase
hydrogen iodide	↓	−36	↓

polar molecules. **Polarizability** refers to the ease with which the electron cloud of a molecule can be distorted by a nearby charge. The greater the polarizability of a molecule, the larger the induced dipole moment and the size of the electrostatic attraction. In general, polarizability increases with the size of the electron cloud. The series of substances shown in Table 10.2 reflect the effect of increasing the size of the electron cloud on the polarizability of molecules. For molecules of similar shape, the greater the number of electrons in a molecule, the more polarizable it becomes. In all four families of compounds shown in Table 10.2, there is a regular increase in the boiling points as the sizes of the electron clouds increase.

Dispersion forces contribute to the attractions between all molecules. Even in molecules with dipole moments, most of the energy of intermolecular attraction arises from the dispersion forces. For Br_2 and ICl (with similar molecular weights and the same number of electrons) considered earlier, the difference in boiling points of 38.6 °C (58.8 °C and 97.4 °C) is attributed to dipole–dipole attraction. For comparison, the boiling points of both substances are more than 300 °C higher than that of H_2 (−252.8 °C), the lightest molecule and therefore a molecule with very weak dispersion forces. The energy of attraction between the molecules of ICl is caused mainly by the dispersion forces, with dipole–dipole attractions making a relatively small contribution.

The boiling points of the heavier hydrogen halides (Table 10.3) increase in the order HCl < HBr < HI. For these hydrogen halides the dipole moments (and therefore the dipole–dipole attractions) decrease with the difference between the electronegativities of the bonded atoms, HCl > HBr > HI. The strengths of the dispersion forces increase as the number of electrons in the molecule increase, HCl < HBr < HI. The observed trend in the boiling points correlates with the increase in the size of the dispersion forces. In most similar situations where the strengths of the London dispersion forces and dipole–dipole attractions predict different trends in boiling points, the dispersion forces dominate. All of the attractive forces discussed (dipole–dipole, dipole–induced dipole, and instantaneous dipole–induced dipole) are collectively called **van der Waals forces.**

London dispersion forces contribute to the intermolecular forces between all molecules and are generally more important than dipole–dipole attractions.

Hydrogen Bonding

The boiling points for the hydrides of the group IVA elements (Table 10.2) increase regularly from $CH_4 < SiH_4 < GeH_4 < SnH_4$ with the increasing strength of the dispersion forces. However, examination of the boiling points of the hydrides of groups VA, VIA, and VIIA reveals that the first compound

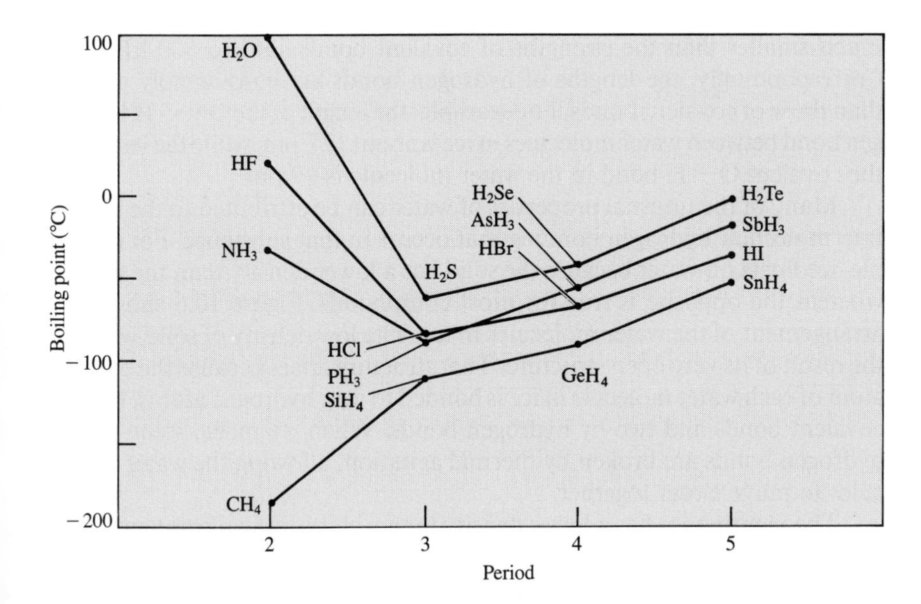

Figure 10.4 The boiling points of the hydrogen compounds of the group IVA, VA, VIA, and VIIA elements In all but the group IVA compounds, the unusually high boiling point of the lightest member of the series is attributed to hydrogen bonding.

in each of these series has an unexpectedly high boiling point, as shown in Figure 10.4. The abnormally strong intermolecular forces of attraction that are reflected in the boiling points of ammonia (NH_3), water (H_2O), and hydrogen fluoride (HF) are attributed to hydrogen bonding. **Hydrogen bonding** is a particularly strong intermolecular attraction between a hydrogen atom, that is bonded to a small, highly electronegative atom, and a lone pair of electrons on fluorine, oxygen, or nitrogen. In Figure 10.5 the hydrogen bonds between molecules are shown as dotted lines.

Hydrogen bonding requires very polar bonds to hydrogen, and atoms that have unshared pairs of electrons, usually N, O, or F.

Hydrogen bonds are very strong (4 to 30 kJ/mol) compared to other dipole–dipole attractions (less than 1 kJ/mol). The energies of hydrogen bonds, while much greater than those of other intermolecular forces, are still

```
-----O—H
     |
     H
     :         H
     :         |
  H—O-----H—O-------
     |         :
     H         :
```

Water

```
   H        H
   |        |
H—N-----H—N-----
   |        |
   H        H
```

Ammonia

```
H—F-----H—F-----
```

Hydrogen fluoride

Figure 10.5 Hydrogen Bonds Hydrogen bonds between molecules of H_2O, NH_3, and HF are shown as dotted lines.

Ice floats in water because the hydrogen bonding in the solid leaves large spaces between the molecules. After melting, the orderly arrangement of the hydrogen bonds is destroyed, allowing the molecules to move closer together.

much smaller than the strengths of covalent bonds (140 to 600 kJ/mol). Correspondingly, the lengths of hydrogen bonds are considerably greater than those of covalent bonds. For example, the length of the O · · · H hydrogen bond between water molecules in ice is about 177 pm, while the length of the covalent O—H bond in the water molecule is 99 pm.

Many of the unusual properties of water can be attributed to the strong intermolecular hydrogen bonding that occurs in that substance. For example, ice floats on water because the solid has a lower density than the liquid, whereas the opposite is true for most compounds. Figure 10.6 shows the arrangement of the water molecules in ice. The low density of solid water is the result of its very open structure. That structure arises because the oxygen atom of each water molecule in ice is bonded to four hydrogen atoms, two by covalent bonds and two by hydrogen bonds. When ice melts, some of the hydrogen bonds are broken by thermal agitation, allowing the water molecules to move closer together.

The fact that ice has a lower density than water has significant environmental consequences. As the temperature of the air decreases and cools lakes, a layer of ice forms on the surface. Because the solid water is less dense than the liquid, it remains on the surface rather than sinking to the bottom of the lake. The surface layer of solid insulates the liquid below it, providing an environment in which aquatic life can survive. If the ice sank to the bottom of the lakes, their temperatures would be much lower, and in colder climates some lakes would freeze solid.

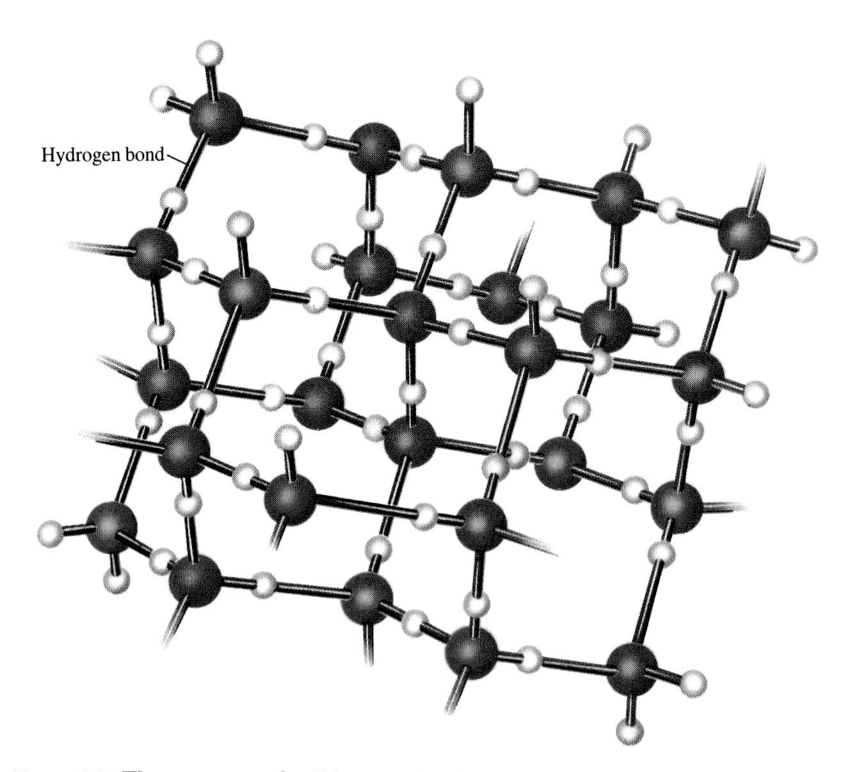

Figure 10.6 The structure of solid water Each oxygen atom is covalently bonded to two hydrogen atoms and hydrogen bonded to two other hydrogen atoms. Because of hydrogen bonding there are large vacant spaces within the solid.

Example 10.2 Nature of Intermolecular Forces and Their Strengths

Identify the intermolecular forces of attraction for the molecules in each part, and determine which substance has the stronger forces of attraction.
(a) CF_4, CCl_4 (b) CH_3OH, CH_3Cl (c) ClF, BrCl

Solution

(a) Both of these molecules have a symmetric tetrahedral structure, so neither of them possesses a dipole moment. The only intermolecular forces possible in both substances are the London dispersion forces that are present in all substances. The larger number of electrons in CCl_4 means that the dispersion forces should be stronger in CCl_4 than in CF_4.
(b) London dispersion forces and dipole–dipole attractions are possible for both of these substances, but only the CH_3OH molecules have strong hydrogen-bonding attractions. Since hydrogen bonding is much stronger than other kinds of intermolecular attraction, CH_3OH is expected to have stronger intermolecular forces than are present in CH_3Cl.
(c) Both molecules possess dipole moments. The ClF molecule has a larger dipole moment than BrCl, based on the electronegativity differences between the bonded atoms. The dispersion forces between BrCl molecules are greater than those between the ClF molecules, because of the larger size of the atoms in BrCl. In cases where the changes in the strengths of dispersion forces and dipole–dipole attractions are opposite, the observed trend of increasing attraction is determined by the dispersion forces. Therefore, the BrCl exhibits stronger intermolecular forces of attraction than does ClF. The boiling points of BrCl (5 °C) and ClF (−100.8 °C) support this conclusion.

Understanding Compare the origins and relative strengths of intermolecular attractions in SF_4 and SeF_4.

Answer: Both London dispersion forces and dipole–dipole attractions increase in strength from SF_4 to SeF_4, so the attractions are stronger in SeF_4.

10.2 Properties of Liquids

Objective

- To understand how the strengths of intermolecular forces influence the surface tension, viscosity, and enthalpy of vaporization of liquids

The liquid state exhibits some similarities to both the solid and gaseous states. Liquids have high densities and incompressibilities like solids. On the molecular scale, we attribute these properties of a liquid to the intermolecular attractions holding the molecules close together so they occupy most of the volume of a sample. Liquids are like gases in that they are fluid and adopt the shape of the container that they occupy. Microscopically, the fluidity of liquids arises because the molecules are able to move about and lack the long-range order that is found in crystalline solids.

Properties of Liquids and Intermolecular Forces

The boiling point of a substance is only one property that can be related to the relative strength of intermolecular forces of attraction; there are several others. Some of the properties of liquids that are related to intermolecular attractions are discussed in the next few paragraphs.

Figure 10.7 Liquid drops The attraction of molecules for each other in a liquid produces surface tension, which causes small drops of a liquid to adopt a spherical shape.

Table 10.4 Boiling Points and Molar Enthalpies of Vaporization of Selected Substances

Substance	Boiling Point (°C)	ΔH_{vap} (kJ/mol)
argon	−185.7	6.1
hydrogen chloride	−83.7	15.1
carbon dioxide	−78.3	16.1
butane (C_4H_{10})	−0.6	22.3
carbon disulfide	46.3	26.9
water	100.0	44.0

Enthalpy of Vaporization

In Chapter 4 we saw the need to include the physical states of all substances in a thermochemical equation, because there is an enthalpy change associated with any change in the physical state of matter. The **molar enthalpy of vaporization, ΔH°_{vap}**, is the enthalpy change that accompanies the conversion of one mole of a substance from the liquid state to the gas state at constant temperature. The enthalpy of vaporization of a liquid is the energy needed to separate the molecules by overcoming the intermolecular attractions. Thus the stronger the intermolecular attractions in a substance, the larger the enthalpy of vaporization. Table 10.4 presents the boiling points and enthalpies of vaporization for several substances. The correlation of these enthalpies with the boiling points and strength of the intermolecular attractions is readily seen.

Surface Tension

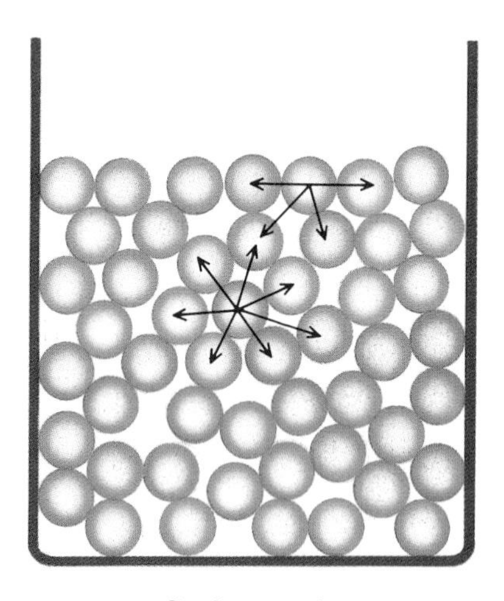

Figure 10.8 Surface tension A molecule in the interior of the liquid is attracted equally in all directions by the surrounding molecules. A surface molecule has unbalanced forces of attraction toward the interior of the liquid, resulting in surface tension.

The intermolecular forces of attraction cause small drops of any liquid to assume a spherical shape (Figure 10.7). The intermolecular attractions for a molecule in the interior of a liquid sample and those for a molecule on the surface of the sample are represented by arrows in Figure 10.8. A molecule in the interior is attracted by its neighbors in all directions. In contrast, a surface molecule has no neighbors above it, so there will be a net force that attracts it toward the interior of the liquid. The unbalanced forces on the surface molecules cause a liquid to adopt a shape that has the smallest surface area possible for a fixed volume, namely a sphere.

Increasing the surface area of a liquid requires an expenditure of energy, because the number of surface molecules increases, and each molecule on the surface has fewer neighboring molecules to attract it. **Surface tension** is the quantity of energy required to increase the surface area of a liquid, and has SI units of joules/m^2. Liquids with strong intermolecular forces of attraction have high surface tensions.

Capillary Action

Capillary action, which causes water to rise in a small diameter glass tube as shown in Figure 10.9a, is another property of liquids that results from intermolecular attractions. There are two kinds of intermolecular attractions that

Table 10.5 Surface Tension and Viscosity of Liquids at 20 °C

Liquid	Surface Tension (J/m^2)	Viscosity ($N\ s/m^2$)
acetone (C_3H_6O)	0.0237	0.327×10^{-3}
chloroform ($CHCl_3$)	0.0271	0.580×10^{-3}
benzene (C_6H_6)	0.0289	0.652×10^{-3}
glycerol ($CH_2OHCHOHCH_2OH$)	0.0634	1.49×10^{-3}
water (H_2O)	0.0730	1.005×10^{-3}
mercury (Hg)	0.487	1.55×10^{-3}

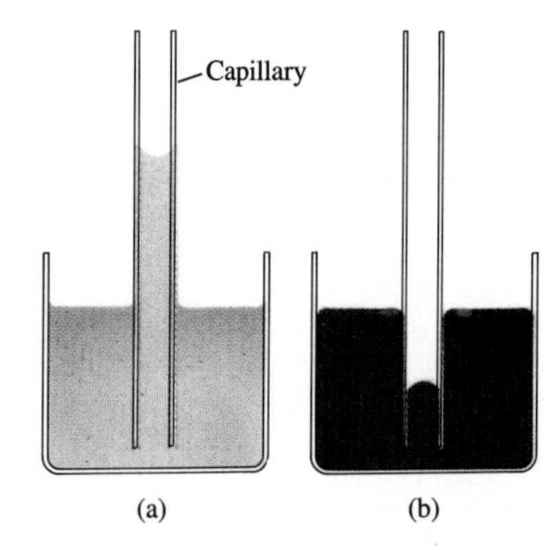

Figure 10.9 Capillary action The meniscus, or curved surface of a liquid, results from a balance between the cohesive and adhesive forces. (a) The curvature of the water surface results from cohesive forces that are greater than adhesive forces. The water rises in the capillary because of the unbalanced forces. (b) Mercury has cohesive forces that are stronger than the adhesive forces to the glass, resulting in a depression of the liquid inside the capillary, and an upward curvature of the liquid surface at the glass-liquid interface.

contribute to this phenomenon. **Cohesive forces** result from the attraction of molecules for other molecules of the same substance. **Adhesive forces** are the attractions that molecules of one substance exert on those of a different substance. The water rises in the capillary tube because the adhesive forces between the water and the glass are quite strong (Figure 10.9a). Since the glass (largely SiO_2) has many polar sites on its surface, the adhesive attractions between these polar sites and the dipoles of the water molecules are sufficiently strong to draw the liquid up the tube against the force of gravity. For liquid mercury (Figure 10.9b) the cohesive forces are greater than the adhesive forces, so the surface of the liquid inside the tube is actually depressed. The upward or downward curvature of a liquid surface is called the *meniscus*. The direction of curvature depends upon the relative strengths of the adhesive and cohesive forces. Capillary action is one of the factors that contribute to the ability of plants to get water out of the ground.

Viscosity

We have all observed differences between syrup and water when they are poured. The syrup flows much more slowly than does water, reflecting the difference between the viscosities of these two liquids. **Viscosity** is the resistance of a fluid to flow, and is another property that is related to the forces of intermolecular attraction. The stronger the intermolecular forces between the molecules in a liquid, the more viscous the liquid becomes. Other factors are also important in determining the viscosity of a liquid, such as the structure, size, and shape of the molecules. Although the surface tension of water is considerably greater than that of glycerol (Table 10.5), glycerol has a higher viscosity. We attribute the higher viscosity of glycerol to the elongated shape of the polar glycerol molecules, which allows them to become entangled and slows their flow.

As intermolecular forces increase in strength, the boiling point, enthalpy of vaporization, surface tension, and viscosity all increase.

10.3 The Solid State

Objectives

- To distinguish between crystalline and amorphous solids
- To describe the characteristics of molecular, covalent network, ionic, and metallic solids

- To determine the number of ions or atoms present in a cubic unit cell of a solid
- To calculate the density of a crystalline solid from the dimensions and the contents of a unit cell
- To know the relation between the cubic closest packing and face-centered cubic crystal structures

An amorphous solid does not have the regular, repeating arrangement of units that is found in crystalline materials.

Rigidity is a characteristic of most solids. Unlike the gaseous and liquid states, a sample of a solid has a definite shape. This rigidity means that the energy of the intermolecular attractions is much larger than the kinetic energy of the individual molecules. We shall place solids into two categories, crystalline solids and amorphous solids. A **crystalline solid** is a solid in which the units that make up the substance are arranged in a very regular repeating pattern. If the relative positions of a few units in a crystalline solid are known, the locations of all the other particles in the sample can be accurately predicted. An **amorphous solid** lacks the long range order of a crystalline solid. Many amorphous solids consist of large molecules that cause the liquid state to become very viscous as the temperature is lowered. As the temperature of the liquid decreases, the large molecules move so slowly that they cannot arrange themselves into the pattern present in the crystalline state. Glasses and most plastics, such as polyethylene, are examples of amorphous solids. In this section we restrict the consideration of the solid state to crystalline solids.

Types of Crystals

We can classify crystalline solids according to the nature of the forces that hold the units together in a regular arrangement. The kinds of forces used are usually referred to as crystal forces.

Molecular Solids

Molecular solids are usually soft and have low melting points.

Molecular solids consist of small molecules held together by van der Waals forces and/or hydrogen bonds. Physical properties such as the melting points and hardnesses of molecular solids such as H_2O, CO_2, and I_2 vary considerably, depending on the strengths of the intermolecular interactions. Because intermolecular forces are weak in comparison to ionic or covalent bonds, molecular solids are usually rather soft substances. Many small molecules exist in the liquid or gaseous states at room temperature, and need to be cooled in order to form the solid state.

Covalent Network Solids

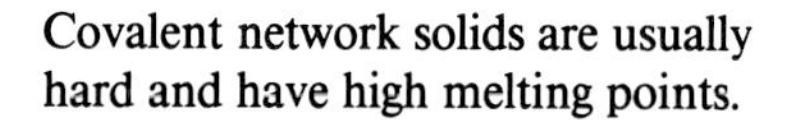
Covalent network solids are usually hard and have high melting points.

In **covalent network solids** the atoms are held together by covalent bonds. Diamond is an example of a covalent network solid. In the diamond structure of elemental carbon, each atom is covalently bonded to four other carbon atoms at the corners of a tetrahedron, as shown in Figure 10.10a. Because the atoms are held together in three dimensions by strong covalent bonds, covalent network solids usually have very high melting points and are hard materials. Diamond is the hardest substance known.

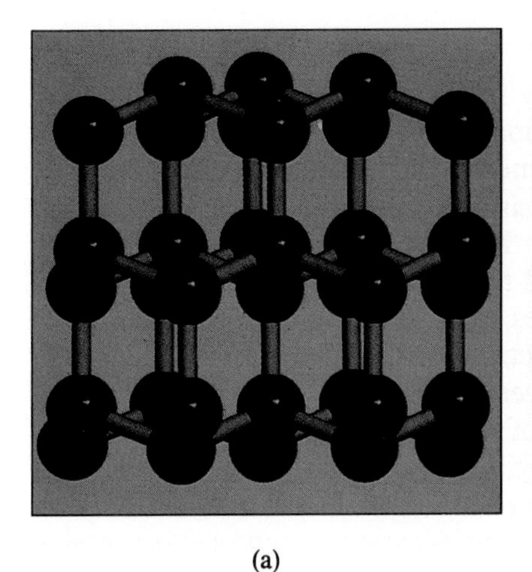

(a)

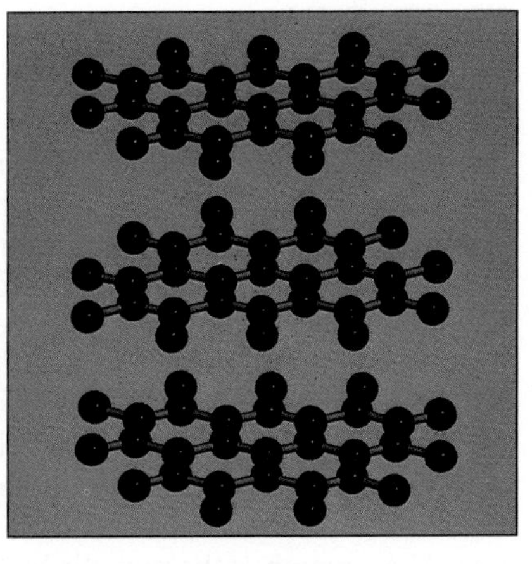

(b)

Figure 10.10 Two allotropes of carbon (a) In the diamond crystal each carbon atom is covalently bonded to four other atoms at the corners of a tetrahedron. Boron nitride and silicon carbide also form covalent network crystals with the diamond structure. (b) Graphite consists of layers of covalently bonded atoms of carbon, with van der Waals forces holding the sheets together.

Graphite is another crystalline form of carbon, but it has both covalent bonding and van der Waals forces as the crystal forces. In graphite (Figure 10.10b), sp^2 hybridized carbon atoms form sheets of atoms held together by covalent bonds. The covalent bonds hold the solid together in only the two dimensions of the sheets, while the weaker van der Waals forces hold one sheet to another. In contrast to diamond, graphite is soft because the weak van der Waals attractions allow the two-dimensional layers to slip past each other. We use the term **allotropes** to refer to two or more molecular or crystalline forms of an element in the same physical state that exhibit different chemical and physical properties. Diamond and graphite are allotropes, and there are many other allotropic forms of carbon that will be discussed in Chapter 19. Sulfur, phosphorus, and tin are other elements that have allotropic forms in the solid state, and O_2 and O_3 are examples of gaseous allotropes.

Other examples of covalent crystals are the many minerals that have extensive networks of covalently bonded silicon and oxygen atoms. The structure of quartz, SiO_2 has each Si atom bonded to four oxygen atoms at the corners of a tetrahedron, and each oxygen atom is bonded to two silicon atoms.

Quartz Quartz, a crystalline form of silicon dioxide, is held together by Si—O—Si covalent bonds. Many gems and minerals contain extensive covalent bonding of silicon and oxygen atoms.

Ionic Solids

An **ionic solid** consists of oppositely charged ions that are held together by electrostatic attractions. The electrostatic attractions that hold the ions in place in the crystal of an ionic solid are very strong, comparable in strength to covalent bonds. Ionic compounds characteristically have high melting points and are relatively hard and brittle. A unique property of ionic compounds is that they do not conduct an electric current in the solid state, because the ions are held firmly in place, but that the liquids are excellent electrical conductors because the ions are then free to move. Many binary compounds formed by the reactions of metallic elements with nonmetals, such as NaCl and MgO, form ionic solids.

Ionic solids are generally brittle and have high melting points.

Metallic Solids

Metallic solids have a luster and large thermal and electrical conductivities; a special kind of bonding, metallic bonding, is needed to account for these properties.

Metallic solids are solids formed by metal atoms. The metallic elements form crystalline solids that exhibit many unique properties, such as high thermal and electrical conductivity, metallic luster, and malleability (that is, they can be reshaped by pressure or hammering). A special kind of bonding, called metallic bonding, accounts for these properties. One relatively simple model for metallic bonding is called the electron sea model. Since metals have relatively low ionization energies, the valence shell electrons are lost by the metal atoms and form a sea of electrons with the metal ions embedded in it. The electrons in the "sea" are very mobile and account for the high thermal and electrical conductivity of solid metals. In Chapter 19 an alternate description of metallic bonding, based on quantum mechanics and called band theory, is outlined and applied to both metals and semiconductors. The strength of the metallic bond varies greatly, as reflected in the wide range of melting and boiling points observed for various metals. For example, mercury (Hg) boils at 356.6 °C and rhenium (Re) has a boiling point of 5650 °C.

Gold leaf Gold leaf is made by hammering out bars of the metal into very thin sheets. The dome of the state capital building in Atlanta, Georgia is covered with five pounds of gold leaf.

Crystal Structure

Much of what we know about the bond angles and bond lengths in molecules and the sizes of atoms and ions have come from study of the crystalline solid state. While a single crystal contains an extremely large number of atomic-sized particles, it is possible to describe their arrangement from the positions of only a few of the particles within the solid. In a crystalline solid, the particles (atoms, ions, or molecules) are arranged in a regular geometric pattern so that the attractive forces are at a maximum. This pattern in three-dimensional space can be described by referring to a set of mathematical points, each of which has the same environment (that is, the same angles and distances to its nearest neighbors). Each point is called a **lattice point,** and the geometric arrangement of the lattice points is called the *crystal lattice.* The crystal lattice, along with the arrangement of physical particles with respect to the lattice points, is called the **crystal structure.**

The **unit cell** is a small regular geometric figure needed to define the repeating pattern of lattice points. A single crystal can be viewed as the packing together of the unit cells like boxes, as shown in Figure 10.11.

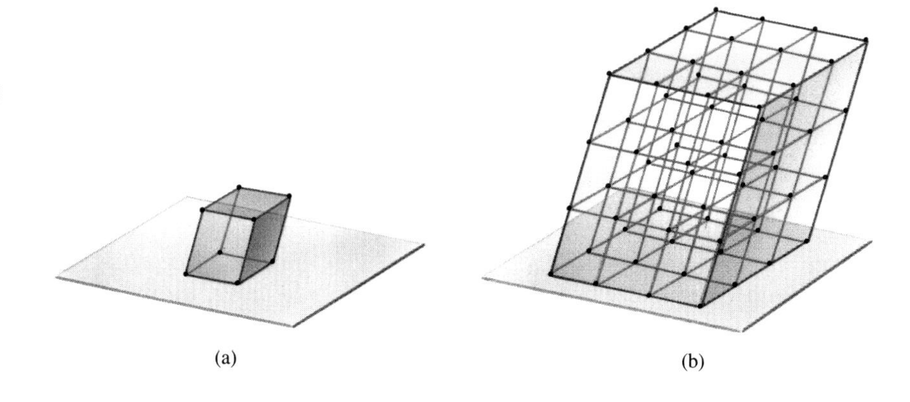

Figure 10.11 The unit cell The black dots represent the lattice points, which are usually the centers of atoms or molecules. (a) The unit cell consists of a small number of lattice points needed to define the arrangement in the crystal. (b) The crystal structure of the solid. In this structure each lattice point is at the corner of eight different unit cells.

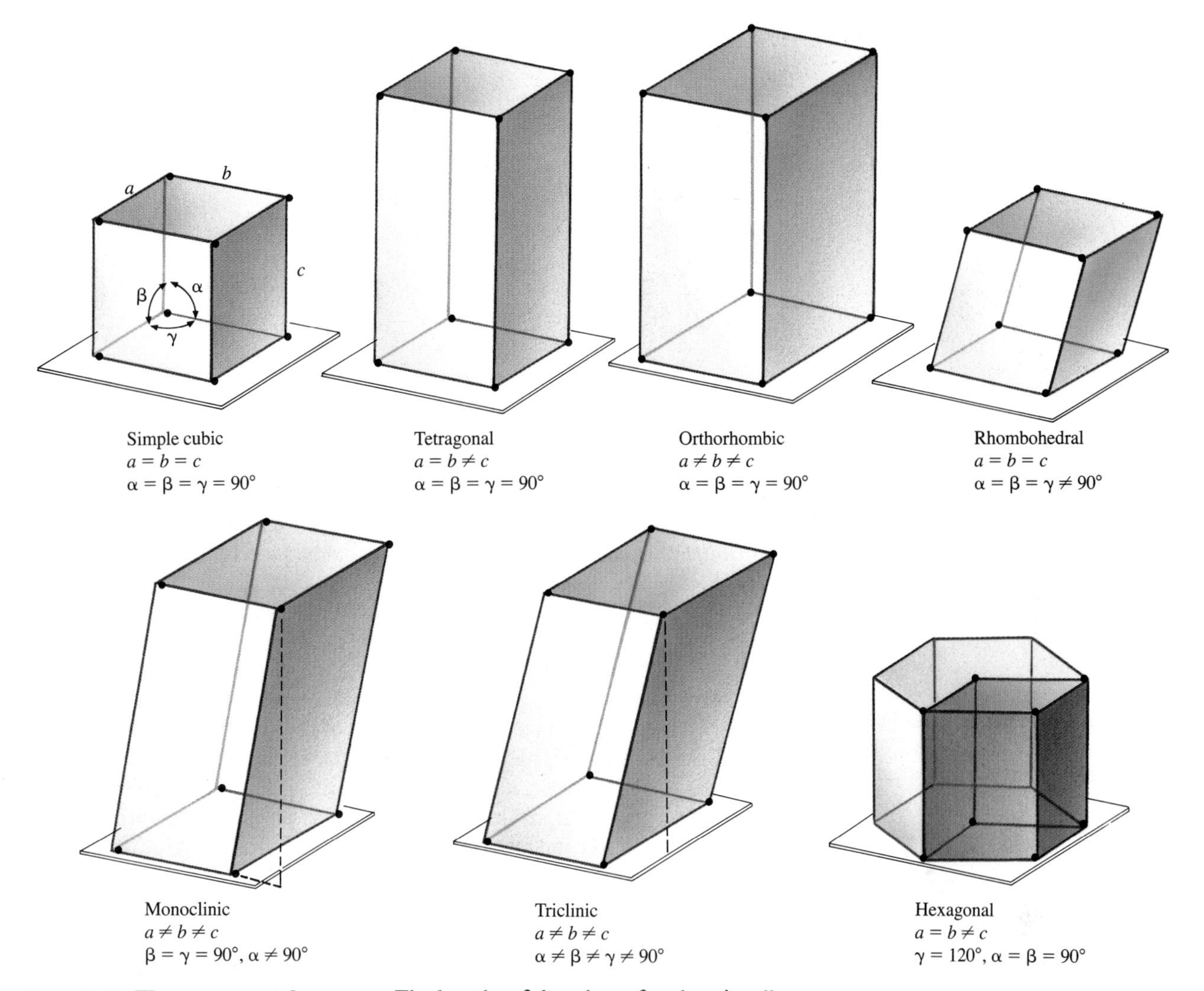

Figure 10.12 The seven crystal systems The lengths of the edges of each unit cell are represented by a, b and c. The angle between edges b and c is α, the angle between edges a and c is β, and that between edges a and b is γ.

Because of the restrictions of geometry, only a limited number of crystal lattices are possible. These can be grouped into seven systems, defined by the relative lengths of the edges and the angles between the edges of the unit cell. Figure 10.12 shows the seven crystal systems. For simplicity, we shall limit our consideration of crystal lattices to the **cubic system** ($a = b = c$ and $\alpha = \beta = \gamma = 90°$).

Crystalline solids have planar faces and distinctive geometric shapes that are determined by the crystal systems to which they belong. For example, sodium chloride (Figure 10.13) forms crystals that are perfect cubes, and the unit cell for that compound belongs to the cubic system. In addition to the cubic shape, octahedral and rectangular shapes are often observed in substances that crystallize with unit cells that belong to the cubic system.

There are three kinds of cubic unit cells, defined by the locations of the lattice points within the unit cell. To keep the discussion simple, we shall assume that each lattice point in the unit cell is occupied by a spherical atom.

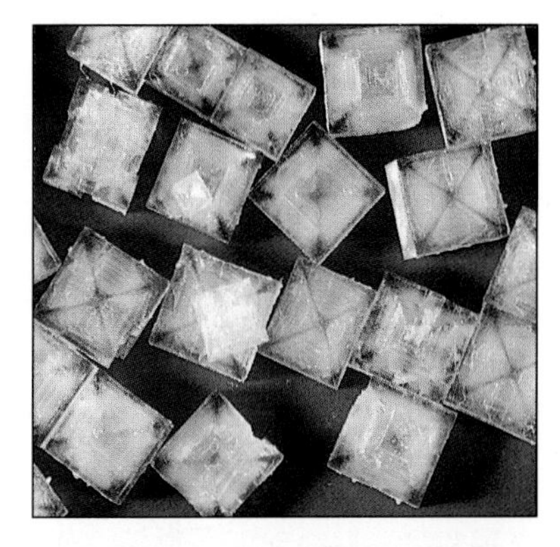

Figure 10.13 Sodium chloride Crystals of sodium chloride have a cubic shape.

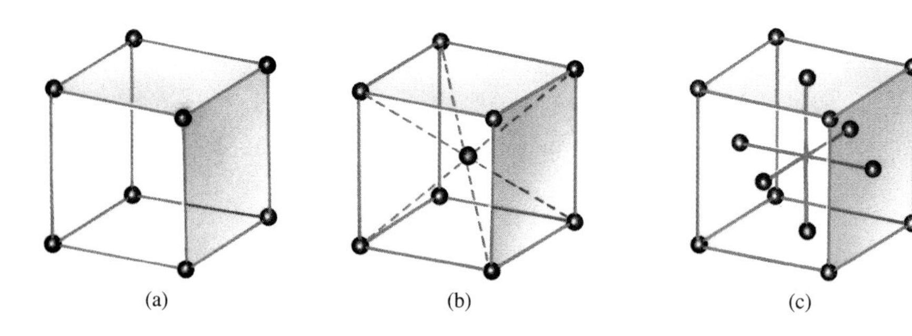

Figure 10.14 The three cubic unit cells The cubic unit cells are (a) simple cubic, (b) body-centered cubic (BCC), and (c) face-centered cubic (FCC).

In each of these cubic structures there is an atom centered at each of the eight corners of the cube. In the **simple** or **primitive cubic** unit cell, only these eight atoms are present. In the **body-centered cubic** (BCC) unit cell, there is an additional atom at the center of the cube. In the **face-centered cubic** (FCC) unit cell, in addition to the eight corner atoms, there is an atom at the center of each of the six faces of the cube. The three cubic unit cells are illustrated in Figure 10.14.

In counting the number of atoms or ions in a unit cell, only a fraction of each corner, edge, and face atom is counted.

Because the entire crystal consists only of repetitions of the unit cell, the intrinsic properties of a unit cell (such as elemental composition and density) must be consistent with the bulk properties of the substance. Many calculations of these properties require that we count the number of atoms or ions contained within the unit cell. Counting the number of atoms in a unit cell is more complicated than it first appears, because we must recognize that the atoms on the corners and faces of the unit cell are shared by the adjacent cells. As shown in Figure 10.15, each corner of a unit cell is actually part of eight cells in the lattice, so only one eighth of each atom at a corner is in each unit cell. The eight corner atoms in one cell therefore contribute a total of one atom ($8 \times \frac{1}{8}$) to that cell. Figure 10.15 also shows that two unit cells share the atoms that are centered on the faces of a cubic unit cell. Therefore, one half of each face atom is contained in a single unit cell, for a total of three face atoms ($6 \times \frac{1}{2}$) in the face-centered cubic unit cell.

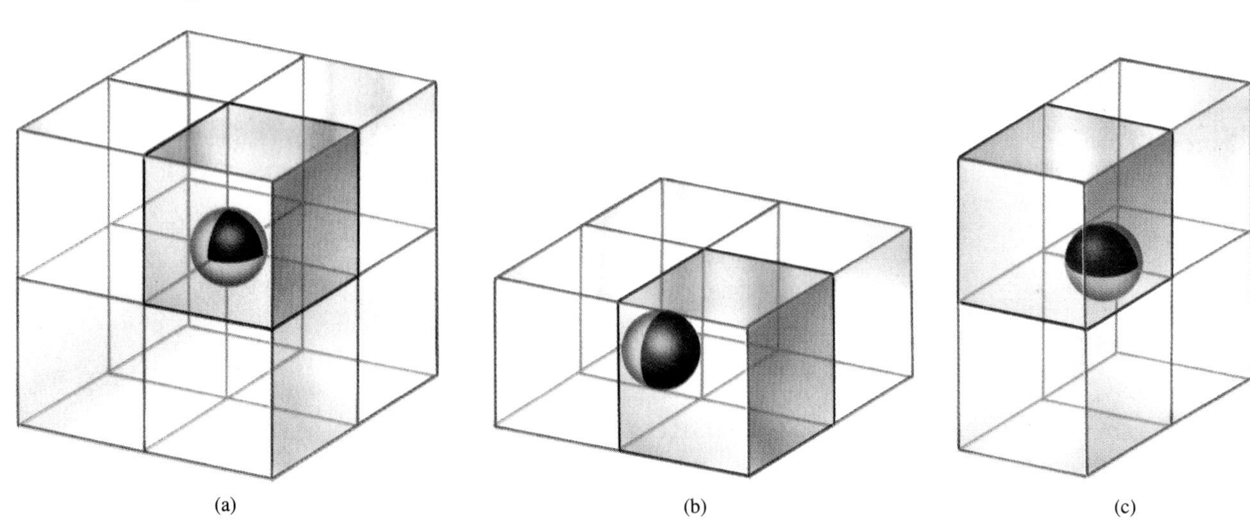

Figure 10.15 Sharing of atoms by unit cells (a) Eight unit cells share each corner atom in any cubic array; (b) two cells share a face atom; (c) four cells share an atom located on the edge of the unit cell.

Bulk properties of crystalline solids, such as compound stoichiometry and the density, can be found from the crystal structure and the dimensions of a unit cell. The experimentally measured densities of most metal samples differ slightly from those calculated from the unit cell because of imperfections and impurities in the crystals. Nevertheless, the experimental and calculated densities are sufficiently close to confirm that the correct unit cell has been identified. The calculation of the density of a metal from the crystal structure and cell dimensions is illustrated in Example 10.3.

Example 10.3 Calculation of Density from Crystal Data

Molybdenum crystallizes in a body-centered cubic array of atoms. The edge of the unit cell is 0.314 nm long. Calculate the density of this metal.

Solution The density of a substance is the mass per unit volume. In the body-centered cubic cell there are atoms at the corners and in the center of the cube, so the number of atoms in the unit cell is

$$8 \text{ corners} \times \left(\frac{1 \text{ atom}}{8 \text{ corner}}\right) + 1 \text{ atom at center of cell} = 2 \text{ atoms}$$

The mass of the two atoms in the unit cell is found from the molar mass of molybdenum and Avogadro's number.

$$\text{mass} = (2 \text{ atoms Mo})\left(\frac{1 \text{ mole}}{6.022 \times 10^{23} \text{ atoms}}\right)\left(\frac{95.94 \text{ g}}{1 \text{ mole}}\right) = 3.186 \times 10^{-22} \text{ g Mo}$$

The volume of the cubic unit cell is the length of the edge raised to the third power. This volume should be expressed in cm^3, to obtain the units usually used for the density of solids.

$$\text{volume} = \left(0.314 \text{ nm} \times \frac{100 \text{ cm}}{1 \times 10^9 \text{ nm}}\right)^3 = 3.10 \times 10^{-23} \text{ cm}^3$$

The calculation of the density is completed by combining the mass and volume.

$$\text{density} = \frac{\text{mass}}{\text{volume}} = \frac{3.19 \times 10^{-22} \text{ g}}{3.10 \times 10^{-23} \text{ cm}^3} = 10.3 \frac{\text{g}}{\text{cm}^3}$$

The measured density of metallic molybdenum is 10.2 g cm^{-3}.

Understanding Copper crystallizes in a face-centered cubic array of atoms in which the unit cell's edge is 0.362 nm long. What is the calculated density of copper?

Answer: 8.90 g cm^{-3}

Some atomic properties are also based on crystal structures. For example, the radii of metal atoms are generally determined from their crystal structures, as illustrated in Example 10.4.

Example 10.4 Calculation of a Metal Atom Radius from Crystal Data

Nickel crystallizes in a face-centered cubic array of atoms, and the length of the unit cell edge is 351 pm. What is the radius of the nickel atom?

Solution On one face of this cell there are five nickel atoms—one at each corner and one in the center of the face. Since we assume that each atom is a sphere, the three atoms along the face diagonal must be in contact with each other. The length of the face diagonal is four times the radius of a nickel atom.

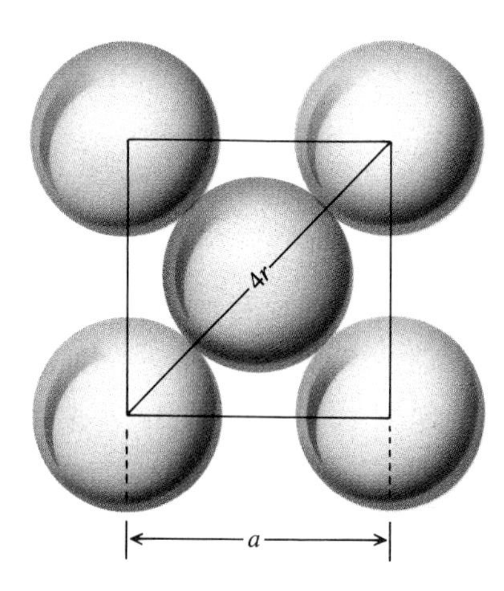

length of diagonal $= 4 \times r$

Since the diagonal of the face of the unit cell is the hypotenuse of a right triangle in which each of the legs is equal to the length of the edge of the unit cell *(a)*, we can use the Pythagorean theorem (the sum of the squares of the legs of a right triangle is equal to the square of the hypotenuse) to find the length of the diagonal in terms of the edge length.

length of diagonal $= \sqrt{a^2 + a^2} = a\sqrt{2}$

Equating the two expressions for the length of the diagonal gives

$4 \times r = a\sqrt{2}$

We substitute the length of the cell edge and solve the equation for the radius.

$$r = \frac{(351 \text{ pm})\sqrt{2}}{4} = 124 \text{ pm}$$

Understanding Use the data in Example 10.3 to calculate the atomic radius (in pm) of the molybdenum atom. In a body-centered cubic structure the atoms along the cube diagonal are in contact, and the cube diagonal $= a\sqrt{3}$.

Answer: $r_{\text{Mo}} = 136$ pm

Closest Packing Structures

In Example 10.4 the atoms were viewed as spheres that were in contact with each other. In many metals and other solid monatomic elements, the atoms are packed together in the most efficient manner, called **closest packing,** in which there is a minimum volume of empty space. A familiar example of closest packing is the stacking of oranges in a grocery store display. The closest packing arrangement of spheres in a single layer is shown in Figure 10.16a. Each sphere is in contact with six other spheres, leaving small trian-

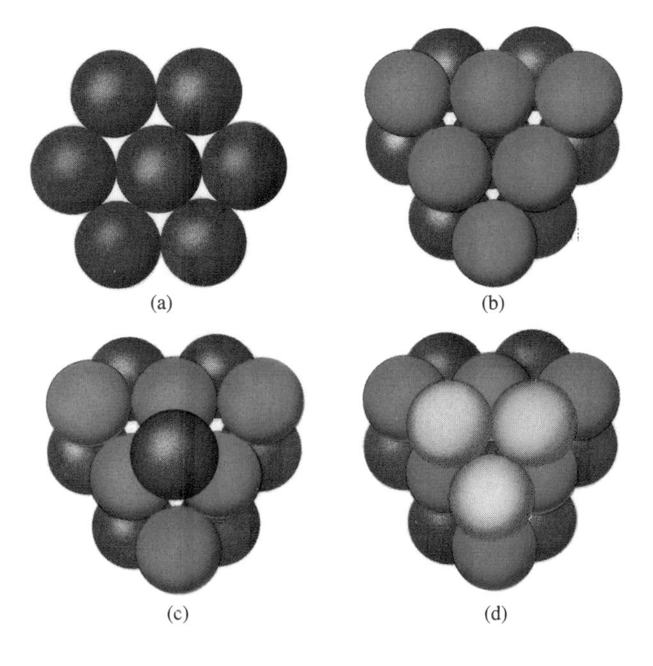

Figure 10.16 Closest packing arrays (a) Within a single layer each sphere is in contact with six others, producing triangular holes between them. (b) Spheres in the second layer cover half of the triangular holes. There are three spheres in the second layer in contact with each sphere in the first layer. (c) In the hexagonal closest packing array, the spheres in the third layer are directly above those in the first layer. (d) Placing the spheres in the third layer over the uncovered triangular holes in the first layer produces the cubic closest packing array.

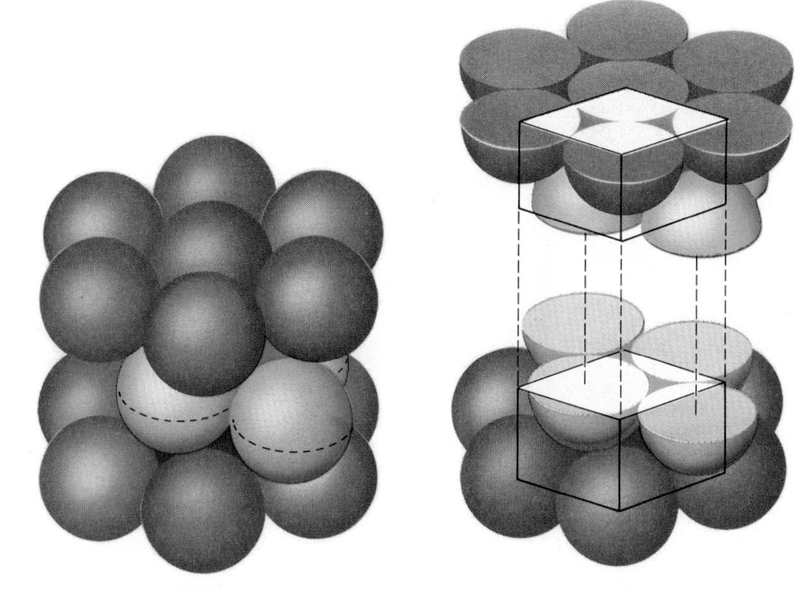

Figure 10.17 Hexagonal closest packing On the left are three closest packing layers. On the right the top and bottom halves are separated to show the parts of the spheres (shown in white) in each layer that lie within the unit cell. The boundaries of the unit cell are outlined with solid black lines.

gular holes between them. If a second layer of spheres is added, they will drop into the indentations over these triangular holes. However, as can be seen in Figure 10.16b, atoms in the second layer cover only half of the triangular holes in the first layer. Each sphere in the first layer is in contact with three of the spheres in the second layer. Spheres in the third layer can have either of two possible arrangements. In Figure 10.16c each sphere in the third layer is directly above a sphere in the first layer. This kind of stacking of the layers is referred to as ABA stacking, and it produces the **hexagonal closest packing** structure.

Placing the third layer of spheres over the triangular holes in the second layer that are not occupied by the spheres in the first layer produces the **cubic closest packing** structure (Figure 10.16d). This arrangement of layers is called ABC stacking. In both of the closest packing arrays, each sphere is in contact with twelve other spheres: six in the same layer and three in each of the layers above and below. The number of nearest neighbors is called the **coordination number.** The coordination number of twelve in the closest packing arrays is the largest possible, and results in the smallest volume of free space between the spheres. The spheres occupy 74% of the total volume of the structure in both of the closest packing arrays.

Figure 10.17 shows the unit cell for the hexagonal closest packing arrangement. The direction of stacking the layers coincides with the *c* axis of the hexagonal unit cell (see Figure 10.12). There are two spheres in each of these unit cells. The eight corners are centered on atoms in the A layers, and contribute one atom to the unit cell. Three atoms in the B layer are each partially in the unit cell and contribute a second atom. Figure 10.18 shows the relation of the cubic closest packing layers to the cubic unit cell. The arrow along the diagonal of the unit cell is perpendicular to the layers shown in Figure 10.16d. The colors are the same as those used in Figure 10.16 to represent the A, B, and C layers. Two of the corners come from the A layers,

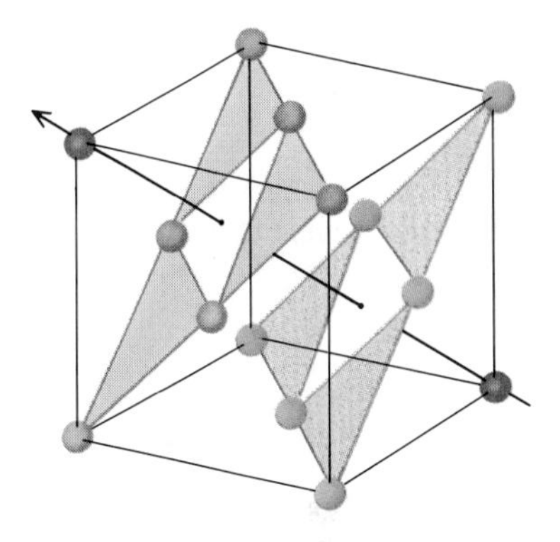

Figure 10.18 Cubic closest packing The drawing shows the relation of the face-centered cubic cell to the cubic closest packing arrangement. The color-coding of the closest packing layers is the same as in Figure 10.16(d). Two diagonally opposite corners of the FCC unit cell come from A layers. The arrow shows the direction of the stacking of the closest packing layers.

The coordination number is the number of neighbors in contact with a single atom.

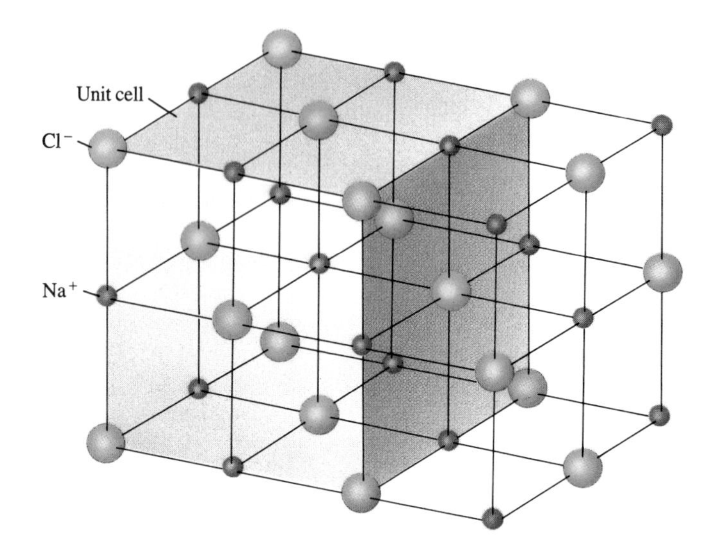

Figure 10.19 Crystal structure of NaCl A portion of the sodium chloride crystal. The face-centered cubic unit cell is shaded and has chloride ions at the corners and in the faces of the unit cell. The sodium ions occupy the center of the cell and each of the 12 edges.

and portions of six spheres in each of the B and C layers are present in the unit cell. From Figure 10.18, we see that the cubic closest packing array is the same as the face-centered cubic cell.

Cubic closest packing results in a face-centered cubic unit cell.

Ionic Crystal Structures

In ionic crystals there are two kinds of particles, cations and anions. The unit cell is usually described in terms of the arrangement of only the cations or only the anions, with the oppositely charged ions occupying specific locations within the cell. The type of crystal lattice that will form is determined by the relative charges and sizes of the anions and cations. In an ionic crystal the ratio of the number of cations to anions in the unit cell must be consistent with the formula of the compound. Figure 10.19 shows a portion of a sodium chloride crystal; it identifies the unit cell as a face-centered cubic array of chloride ions, with a sodium ion at the center of each of the twelve edges of the cube and one in the center of the cell. Figure 10.20 shows a space-filling representation of a unit cell of sodium chloride. In Example 10.5 we verify that the numbers of cations and anions are equal in one unit cell of sodium chloride.

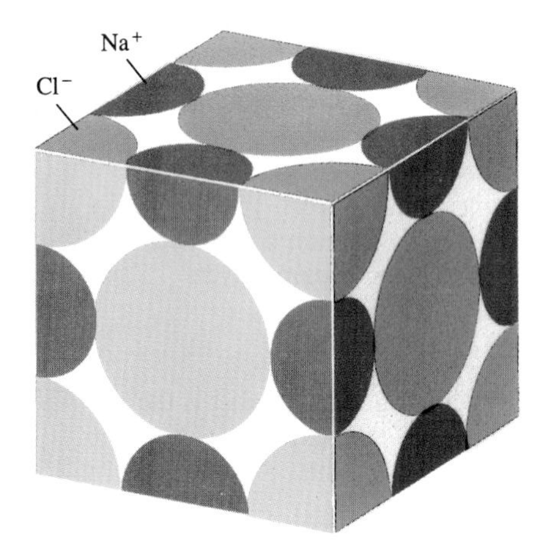

Figure 10.20 A unit cell of NaCl A space-filling model of a single unit cell of sodium chloride. Only the portions of the ions that occupy this unit cell are shown.

Example 10.5 Determining the Number of Ions in a Unit Cell

Determine the number of sodium ions and chloride ions present in the face-centered cubic unit cell of sodium chloride.

Solution In this unit cell (see Figure 10.20) all of the Cl^- ions are shared by other unit cells. Only one eighth of each chloride ion on the corners is in this unit cell, and two different unit cells share each of the six ions in the centers of faces.

$$8\ \text{corners} \times \left(\frac{1}{8}\frac{Cl^-}{\text{corner}}\right) = 1\ Cl^-$$

$$6\ \text{faces} \times \left(\frac{1}{2}\frac{Cl^-}{\text{face}}\right) = 3\ Cl^-$$

There are 1 + 3 = 4 Cl^- ions in each unit cell. Four different unit cells share each of the 12 sodium ions along the edges of the cube, so one quarter of each of them is present in this unit cell. In addition, there is one Na^+ ion at the center of the unit cell that is not shared with any other cell.

$$1\ Na^+ + 12\ \text{edges} \times \left(\frac{1\ Na^+}{4\ \text{edge}}\right) = 4\ Na^+$$

The unit cell of the sodium chloride contains four Na^+ ions and four Cl^- ions, giving the one-to-one ratio of cations to anions that the formula of the compound demands.

Understanding Cesium chloride crystallizes in a simple cubic array of Cs^+ ions, with one Cl^- at the center of the unit cell. How many ions of each type are present in the unit cell?

Answer: 1 Cs^+ and 1 Cl^-.

10.4 X-Ray Diffraction

Objectives

- To know how x-ray diffraction data are used to determine structures
- To use the Bragg equation to relate the angle of diffraction and wavelength of x rays to the distances in a crystal

The experimental determination of crystal structures by x ray diffraction has been one of the most important discoveries of the twentieth century. X rays are short-wavelength electromagnetic radiation. When this radiation is directed at a crystal, the atoms scatter the radiation in all directions, so each atom in the crystal behaves as if it were a small source of x rays. When these x rays come together they interfere with each other. If the waves are in phase (the maximum and minimum amplitudes occur at the same places), *constructive interference* occurs and a more intense wave results. When the separate waves are exactly out of phase, *destructive interference* occurs; that is, the waves cancel each other, and no radiation is observed. Constructive and destructive interferences are illustrated in Figure 10.21.

In 1913 William H. Bragg and his son William L. Bragg interpreted the diffraction of a narrow beam of x rays by crystals. They found that the

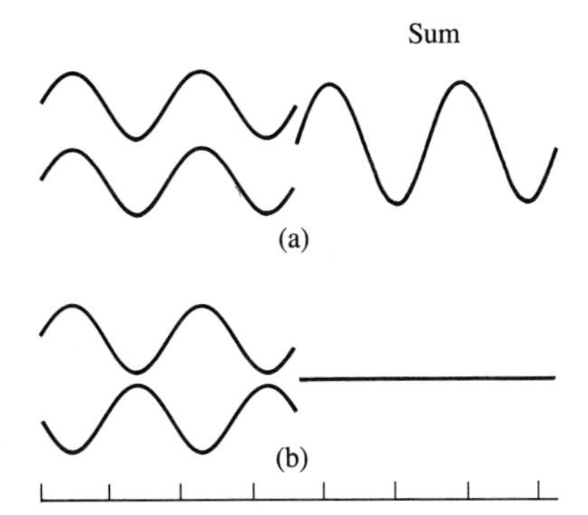

Figure 10.21 Interference of waves (a) Constructive interference occurs when two in-phase waves combine to produce a wave of greater amplitude. (b) Destructive interference results from the combination of two equal waves of electromagnetic radiation that are exactly out of phase and results in no amplitude.

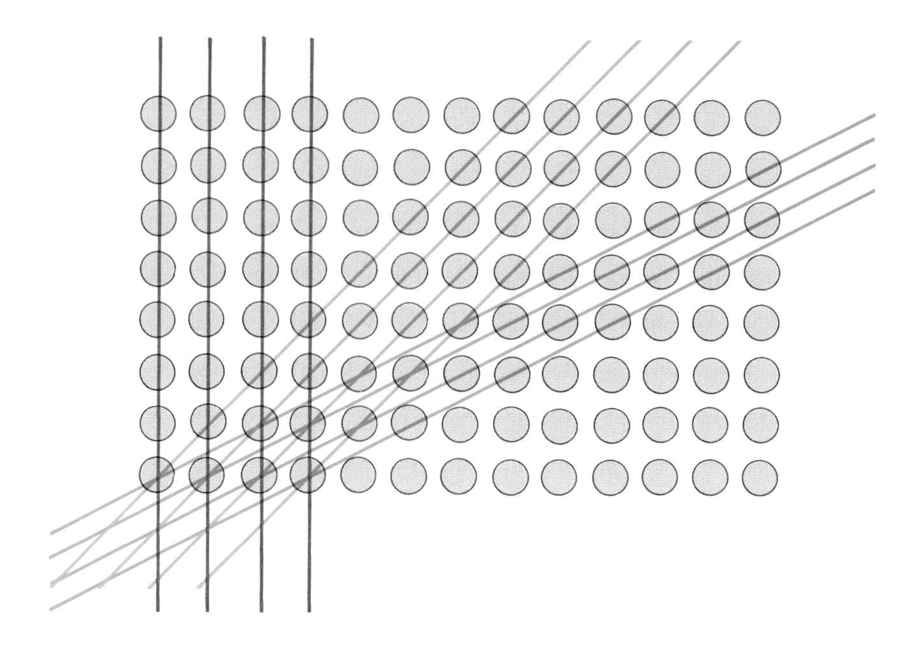

Figure 10.22 Spacing of layers in crystals Within the single layer of atoms shown, there are many different interlayer spacings. Three of these spacings are shown by the red, blue, and green lines. There are characteristic angles of diffraction for each of these distances.

The Bragg equation is used to calculate the distances between identical layers of atoms in crystalline solids.

locations of the dots on the photographic film could be predicted by using a very simple equation that we now call the **Bragg equation:**

$$n\lambda = 2d \sin \theta \qquad [10.1]$$

where λ is the wavelength of the x rays used, d is the distance between layers of atoms in the crystal, θ is the angle between the beam of x rays and the layer of atoms, and n is a whole positive number called the order. There are many different distances between layers of atoms in a crystal (as shown in two dimensions in Figure 10.22), giving rise to many angles of diffraction. The Bragg equation is useful only when the spacing between layers and the wavelength of the radiation are the same order of magnitude.

Example 10.6 Distances in Crystals from the Bragg Equation

A crystal is found to diffract 154-pm x rays at an angle of 19.3°. Assuming that $n = 1$ in the Bragg equation, calculate the distance (in pm) between the layers of atoms that give rise to this diffraction.

Solution We must solve the Bragg equation for the distance between the layers of atoms:

$$d = \frac{n\lambda}{2 \sin \theta}$$

Substitution of the wavelength of the x rays and the angle of diffraction into the equation gives the distance between the layers of atoms.

$$d = \frac{(1)(154 \text{ pm})}{2 \sin 19.3} = \frac{154}{(2)(0.3305)} = 233 \text{ pm}$$

Understanding Using the same x-ray source, a different crystal produces a diffraction at an angle of 12.3°. Using $n = 1$, calculate the distance between the layers of atoms that produce this diffraction.

Answer: 361 pm

Example 10.7 Bragg Equation

The distance determined in Example 10.6 is the length of the unit cell edge, a, in a cubic lattice. A second diffraction arises from the layers of atoms along the face diagonals of the unit cell. From Figure 10.22, we can see that the distance between these layers of atoms is 1/2 the diagonal of the face, $\sqrt{2}/2 \times a$ or $a/\sqrt{2}$. Calculate the angle of diffraction for these layers of atoms.

Solution The distance between these layers of atoms is

$$d = \frac{a}{\sqrt{2}} = \frac{233 \text{ pm}}{1.414} = 165 \text{ pm}$$

Solve the Bragg equation for $\sin\theta$, and substitute this distance and the wavelength of the x rays into it.

$$\sin\theta = \frac{n\lambda}{2d} = \frac{(1)(154 \text{ pm})}{(2)(165 \text{ pm})} = 0.467$$

The angle of diffraction, θ, is found to be 27.8°.

Understanding What is the angle of the first order diffraction ($n = 1$) found from layers of atoms that are 210 pm apart, for x rays with a wavelength of 166 pm?

Answer: 23.2°

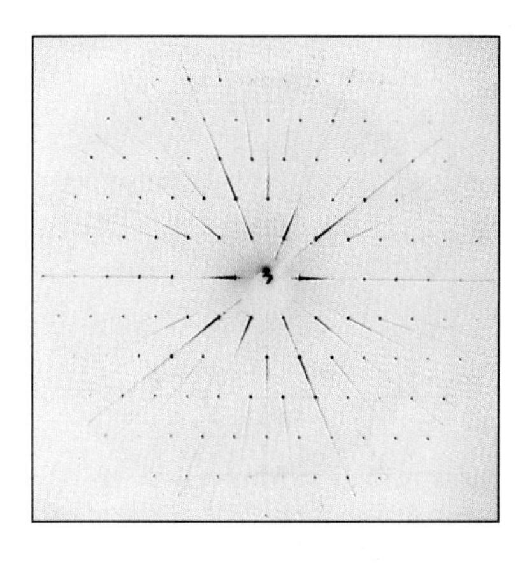

X ray diffraction pattern A photographic film exposed to x rays that have been diffracted by a crystal. Each of the black spots corresponds to a distance between layers of atoms in the crystal.

In a modern x-ray crystallography laboratory the intensities of the x rays are measured at thousands of angles. High-speed computers analyze the data to determine accurately the locations of the atoms in the crystal. Today the technique is applied to determine the structures of molecules that contain hundreds and even thousands of atoms.

10.5 Phase Changes

Objectives

- To describe phase changes as equilibrium processes
- To relate enthalpy of phase changes and the critical temperature to the relative strengths of the intermolecular attractions
- To describe heating curves and their relation to the heat capacities and enthalpies of phase transitions.

We have now examined the nature of each of the physical states in some detail. In this section we turn our attention to another important aspect of the states of matter, the changes that occur when a substance is converted from one physical state into another. From everyday experience we expect a substance to go from the solid to the liquid and then to the gaseous state as the temperature increases. In the earlier sections of this chapter, we noted that this behavior is determined by the strengths of intermolecular attractions and by the average kinetic energy of the molecules. In the course of considering phase changes, we shall introduce the idea of dynamic equilibrium, a concept that is the theme of several of the remaining chapters of this text.

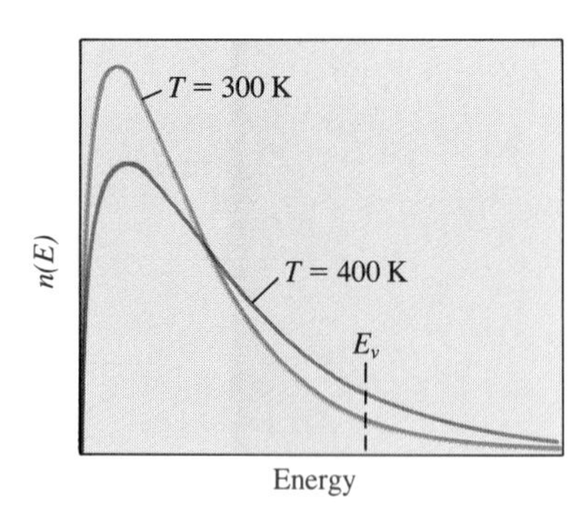

Figure 10.23 The Maxwell-Boltzmann distribution The distribution of kinetic energy for molecules at 300 and 400 K. The fraction of molecules that have an energy greater than E_v is larger at the higher temperature.

Liquid-Vapor Equilibrium

In Chapter 5 we saw that the kinetic molecular theory offers an explanation for the behavior of gases. The average kinetic energy of molecules is proportional to the absolute temperature of the sample, but there is a wide range of energies possessed by the individual molecules. Some molecules have much less energy than the average value, while others have much more. The distribution of the kinetic energy of identical molecules, called the Maxwell-Boltzmann distribution, is shown in Figure 10.23 for two different temperatures.

The Maxwell-Boltzmann energy distribution also applies to molecules in the liquid state and is an important factor in understanding the evaporation process. To escape from a liquid, a molecule must have a kinetic energy that is sufficient to overcome the forces of attraction from the other molecules in the sample, which is shown as E_v in Figure 10.23. For any temperature at which the liquid state is stable, only a small fraction of the molecules possess enough energy to **evaporate** or **vaporize,** that is, to escape from the surface of the liquid.

Vapor Pressure

Suppose a sample of a liquid is put into an evacuated vessel (pressure = 0) held at a fixed temperature. Some of the molecules (those with enough kinetic energy) will evaporate, causing an increase in the pressure within the vessel (see Figure 10.24). If the temperature is held constant, the rate of evaporation is also constant (as long as the surface area of the liquid does not change) because the fraction of the molecules with enough energy to escape from the liquid does not change. However, as the concentration of the molecules in the gas state builds up, some of them collide with the liquid surface and rejoin it. **Condensation** is the conversion of a gas to a liquid. The higher the concentration (pressure) of the molecules in the gas state, the larger is the rate of condensation, simply because there are more gaseous molecules that collide with the surface of the liquid. The rates of evaporation

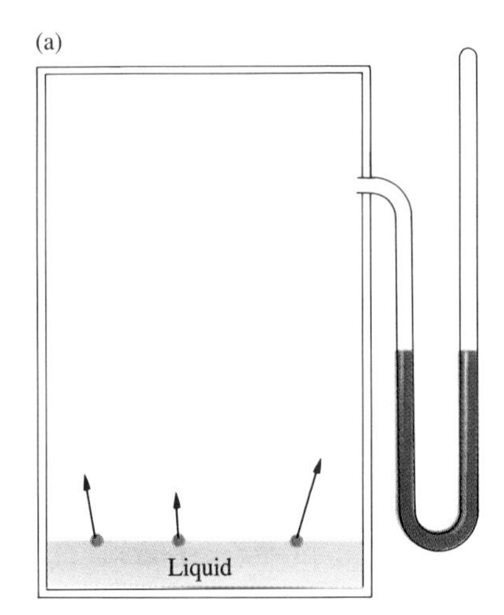

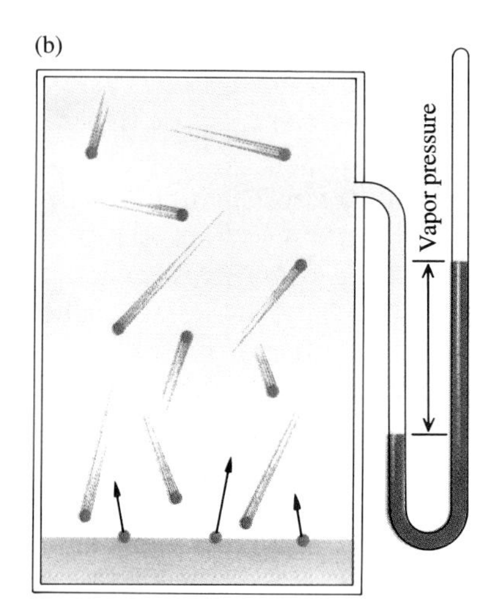

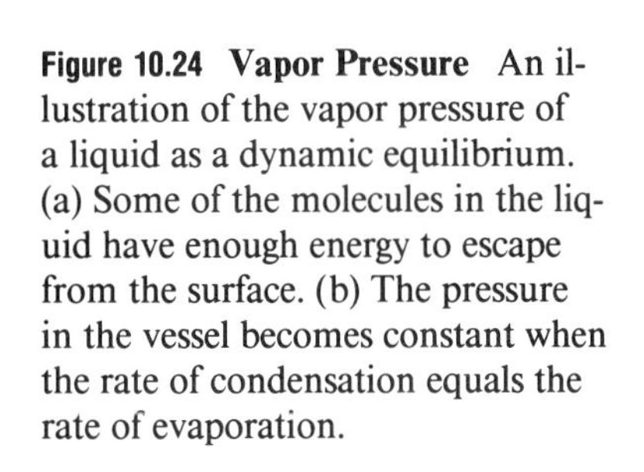

Figure 10.24 Vapor Pressure An illustration of the vapor pressure of a liquid as a dynamic equilibrium. (a) Some of the molecules in the liquid have enough energy to escape from the surface. (b) The pressure in the vessel becomes constant when the rate of condensation equals the rate of evaporation.

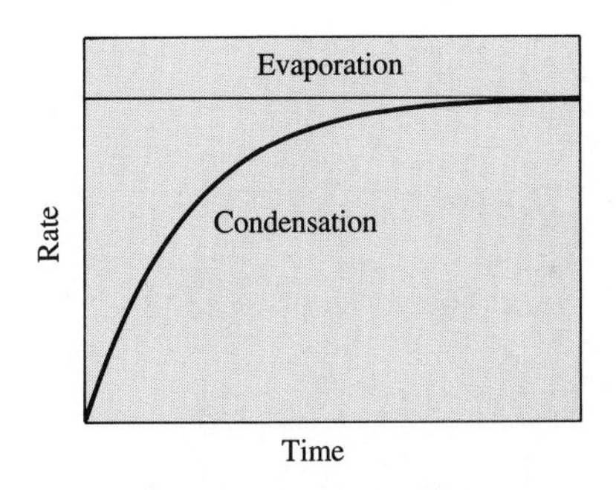

Figure 10.25 Rates of evaporation and condensation At a constant temperature the rate of evaporation is a constant. As the number of molecules in the gas phase increases, the rate of condensation approaches and finally equals that of evaporation.

and condensation (when a liquid is placed into an evacuated vessel) are shown in Figure 10.25. After a relatively short time the number of molecules that condense per second is equal to the number of molecules that evaporate per second. When the rate of condensation becomes equal to the rates of evaporation, the pressure in the vessel no longer changes. The constant pressure achieved is called the **vapor pressure** of the liquid.

A state of **dynamic equilibrium** is one in which two opposing changes occur at equal rates, so no *net* change is apparent. In any system at equilibrium, it is important to recognize that both opposing processes continue to occur, but since the rates are equal, no net change is observed. In a chemical equation, we indicate an equilibrium process by adding a second arrow that points toward the reactants. The equilibrium of liquid and gaseous diethyl ether is represented in the equation

Dynamic equilibrium is established when the rates of opposing processes are equal.

$$(C_2H_5)_2O(\ell) \rightleftharpoons (C_2H_5)_2O(g)$$

In this example the two opposing processes are vaporization and condensation.

The value of E_v in Figure 10.23 is the minimum kinetic energy a molecule must have to escape from the liquid, and it depends on the strength of the intermolecular attractions. As the temperature increases, the energy distribution curve shifts, so the number of molecules with enough energy to evaporate increases. Therefore, the rate of evaporation and the equilibrium vapor pressure of the liquid must increase with increasing temperature. Figure 10.26 shows the vapor pressures of several liquids as functions of temperature. Any point on one of these lines represents a combination of temperature and pressure at which the liquid and the gaseous states of the substance are in equilibrium. Each liquid has a characteristic vapor pressure curve that depends on the strength of its intermolecular attractions. Since the vapor pressure of diethyl ether is higher than that of water at all temperatures, the intermolecular attractions in diethyl ether must be weaker.

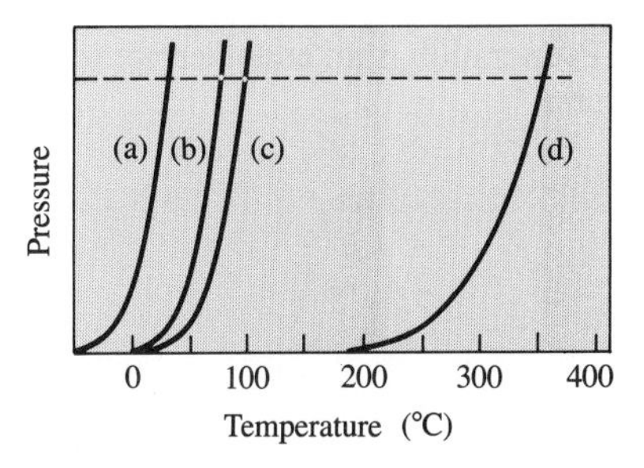

Figure 10.26 Vapor pressure curves The vapor pressure as a function of temperature for (a) diethyl ether, (b) ethanol, (c) water, and (d) mercury. The horizontal dotted line at a pressure of 1 atmosphere intersects each vapor pressure curve at the normal boiling point of the liquid.

The **boiling point** of a liquid is the temperature at which the vapor pressure is equal to the external applied pressure. The boiling point is a function of the external applied pressure corresponding to every point on the liquid-gas equilibrium line. At the boiling point, bubbles of vapor form below the surface of the liquid as heat is added to the sample. The **normal boiling point** of a liquid is the temperature at which its equilibrium vapor pressure is equal to 1 atmosphere. The horizontal dotted line in Figure 10.26 intersects each of the equilibrium vapor pressure curves at the normal boiling point of the corresponding liquid.

The vapor pressure and the rate of evaporation are small and the boiling point is high for a liquid with strong intermolecular forces of attraction.

(a) (b)

Cooling by evaporation (a) The evaporation of perspiration cools people. (b) Dogs and other animals that cannot sweat are cooled by breathing more rapidly. In both cases the enthalpy of vaporization of water is used to remove excess heat.

Evaporation is an endothermic process.

Vaporization is an endothermic process, because the most energetic molecules escape from the liquid, leaving behind molecules with a lower average kinetic energy (a lower temperature). Unless heat is added to the sample, a liquid cools as it evaporates. The cooling effect of vaporization is used in nature to control the body temperature of warm-blooded animals. When excess heat is produced by exercise or by warm surroundings, it is removed from the human body by the evaporation of perspiration. When the humidity is high (a high partial pressure of water vapor in the air), the efficiency of the cooling decreases. Condensation is exothermic, since it is the reverse of the evaporation process. The heat released in the condensation of water from the humid air partially cancels the heat absorbed by evaporation.

Critical Temperature and Pressure

Substances with strong intermolecular forces of attraction have high critical temperatures.

All substances at sufficiently high temperatures can no longer exist in the liquid state. The **critical temperature** is the maximum temperature at which a substance can exist in the liquid state. Above its critical temperature, no matter how high the applied pressure, a substance has only one phase that completely occupies the volume of the vessel. The **critical pressure** is the minimum pressure needed to cause the liquid state to exist up to the critical temperature. The single phase present above the critical temperature is usually referred to as a gas, since it occupies the entire volume of the container, but its density at the critical pressure or higher is often comparable to those found in the condensed states. Like the boiling point, enthalpy of vaporization, vapor pressure, and many other properties, the critical temperature is also directly related to the strength of the intermolecular attractions of a substance. The single phase that exists above the critical temperature and pressure is sometimes called a **supercritical fluid.**

INSIGHTS into CHEMISTRY

Science, Society, and Technology

Beyond the Critical Point: Supercritical Fluids

The change in phase from liquid to vapor or from solid to liquid depends upon the temperature and the pressure applied to a system. For every substance a temperature exists above which the substance cannot be liquefied, no matter how great a pressure is applied; this temperature is called the critical temperature. The pressure required at the critical temperature to liquefy the substance is called the critical pressure. The critical temperature and pressure mark the termination of the liquid-vapor phase equilibrium line at a point called the critical point. The critical temperatures and pressures for some common substances are listed below.

The Critical Temperatures and Pressures of Some Common Substances

Compound	Critical Temperature (°C)	Critical Pressure (atm)
Hydrogen, H_2	−239.9	12.8
Nitrogen, N_2	−147	33.5
Argon, Ar	−122.5	48.0
Oxygen, O_2	−118.4	50.1
Methane, CH_4	−82.1	45.8
Carbon dioxide, CO_2	31	72.8
Ammonia, NH_3	132.2	111.5
Sulfur dioxide, SO_2	157.8	77.7
Water, H_2O	374.1	218.3

When a substance is subjected to temperatures and pressures that are above the critical point, a fluid phase that has properties *intermediate* between those of a liquid or a gas is produced. Like a gas, the fluid has a low viscosity and expands to fill its container. The density, however, approximates that of a liquid. This intermediate phase has come to be called a *supercritical fluid.* The relatively high, liquid-like density gives the fluid good solvent properties, while the low, gas-like viscosity gives the fluid good penetrating ability into the material being dissolved. These solvent properties can be increased or decreased by changing the density of the fluid by altering the temperature or the pressure, thus allowing a single supercritical fluid to dissolve a wide range of components of differing solubilities. This unique combination of properties has given supercritical fluids one of their most important scientific and industrial uses—as an extremely versatile solvent in a process called *supercritical fluid extraction.*

Although the solvent properties of supercritical fluids have been known for over 100 years, their use in laboratory and commercial separation processes has become significant only in recent years. The continually shrinking list of environmentally safe solvents, the rising costs of energy, and the increasing need for greater solvent selectivity have been the major reasons for the increased growth in the use of the process. Carbon dioxide is a particularly attractive solvent for use in the food and flavor industries. Its supercritical properties occur at relatively mild conditions and it is nontoxic, noncarcinogenic, nonflammable, and relatively inexpensive. The extracted material is readily recovered by simply reducing the pressure below the critical pressure, thus allowing the carbon dioxide to escape as a gas to be recycled. Commercial decaffeination of coffee and tea using supercritical carbon dioxide has been in operation since 1977. Other applications include the removal of cooking oils from foods like potato chips and the extraction of flavors and fragrances from citrus oils.

Research into other uses of supercritical fluids indicates many widespread applications. Recent studies seem to suggest that supercritical water may hold promise as a detoxifying agent for both solid and liquid wastes. Supercritical fluid chromatography is another area of potential application, especially in process analytical chemistry.

Liquid-Solid Equilibrium

The changes of a substance from liquid to solid (freezing) and from solid to liquid (melting or fusion) are also opposing changes that lead to a dynamic equilibrium. This equilibrium process for benzene, C_6H_6, is represented by the equation

$$C_6H_6(s) \underset{\text{freezing}}{\overset{\text{melting}}{\rightleftharpoons}} C_6H_6(\ell)$$

The **normal melting point** of a substance is the temperature at which the solid and liquid phases are in equilibrium at one atmosphere pressure. Associated with melting is a **molar enthalpy of fusion,** the enthalpy change that occurs when one mole of solid is converted into liquid at a constant temperature. For any substance the enthalpy of fusion is considerably smaller than the enthalpy of vaporization. In the vaporization process energy must be provided to separate the molecules completely. In forming the liquid from the solid, the molecules still remain quite close together, so only a small fraction of the attractive energy between the molecules must be provided. The energy required for melting is used to overcome a small part of the intermolecular attractions, giving the molecules greater freedom of motion.

Heating and Cooling Curves

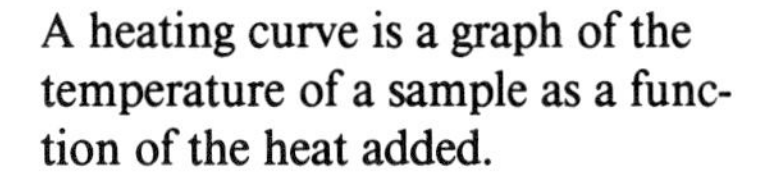

A heating curve is a graph of the temperature of a sample as a function of the heat added.

Adding heat to a solid sample at one atmosphere of pressure usually produces the liquid phase and then the gas phase, in that order. Figure 10.27 shows the heating curve (a plot of temperature as a function of heat added) for water at one atmosphere. At the left side of the graph the solid phase is present at $-40\,^{\circ}C$. As heat is added at a constant rate, the temperature of the solid rises at a constant rate that depends on the heat capacity of the solid. This part of the curve ends abruptly when the melting point is reached. At the melting point the temperature remains constant as long as both the solid and liquid phases are present. All of the heat that is added at the melting point is used to overcome the attractive forces between the molecules in the solid, and release them into the liquid state. For a well stirred mixture of ice and water, the temperature is $0\,^{\circ}C$, no matter how much or how little ice is present.

Once all of the solid has been converted into liquid, the temperature rises again at a constant rate which reflects the heat capacity of the liquid. When the temperature reaches the normal boiling point of the liquid, the temperature again stays constant, because the added heat is used to overcome intermolecular attractions as the molecules move far apart from each other in the vapor phase. As soon as enough heat has been added to vaporize the sample completely, the temperature rises again at a constant rate that depends on the heat capacity of the vapor. The observed lengths of the constant-temperature portions on the heating curve at the melting and boiling points are proportional to the enthalpies of fusion and vaporization, respectively.

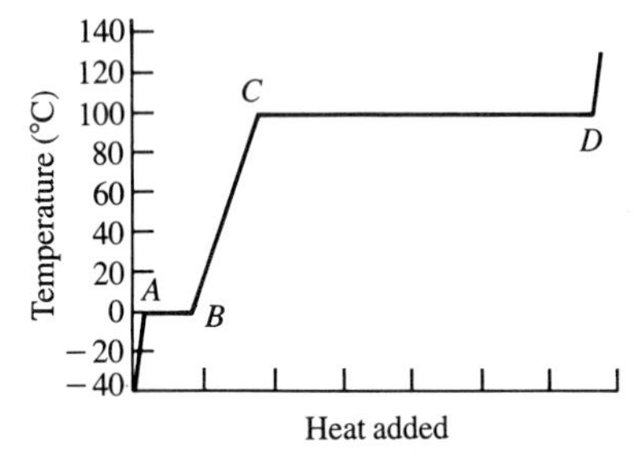

Figure 10.27 A heating curve When heat is added to a typical substance, water, the temperature rises until the melting point of the solid is reached at point A. Between points A and B the temperature remains constant as the solid is converted completely into liquid. Further addition of heat raises the temperature until the boiling point is reached at point C. From point C to D the heat that is added converts the liquid to gas. Further heating raises the temperature of the gaseous sample.

Removing heat from a gaseous sample of this same substance retraces the heating curve. When heat is removed rapidly, we sometimes observe **supercooling,** the cooling of the liquid below its normal freezing point without forming solid. A supercooled liquid is in an unstable state, and stirring or adding a small crystal of the substance causes the rapid formation of the solid, with an abrupt increase of the temperature to the normal freezing point. Supercooling occurs because the freezing process is a cooperative phenomenon—one that requires circumstances involving several independent molecules. The crystallization process requires several low-energy molecules of the substance to come together and establish the arrangement found in the crystal lattice. In a supercooled liquid, a larger number of low-energy molecules are present at the lower temperature, which increases the probability they will come together in the correct arrangement. Once the molecular arrangement of the solid has been established, other low-energy molecules rapidly join the solid phase, releasing heat that rapidly raises the temperature to the freezing point.

Solid-Gas Equilibrium

In the solid state a few surface molecules always have sufficient kinetic energy to overcome the intermolecular attractions and escape into the gas phase. **Sublimation** is the direct conversion of a substance in the solid state into the gaseous state. The reverse of sublimation, called **deposition,** is the conversion of the gas directly into the solid state. The opposing changes of sublimation and deposition lead to a state of dynamic equilibrium. Figure 10.28 shows all three reversible phase transitions on an enthalpy diagram. From the enthalpy diagram and Hess's law, we see that the enthalpy of sublimation is the sum of the enthalpy of fusion and the enthalpy of vaporization when all three enthalpy changes occur at the same temperature. For most solids, such as metals and ionic network solids, at normal temperatures the strengths of the intermolecular attractions are so large that the vapor pressure is too small to measure. Increasing the temperature of these solids first causes them to melt; then the vapor pressure of the liquid increases with temperature until the substance boils. One common solid substance that sublimes but does not melt at a pressure of one atmosphere is carbon dioxide. In calculating the heat absorbed or released when a sample undergoes a temperature change that also includes phase changes, the heat capacities of each phase and the enthalpy changes for the phase transitions must be used. Values of these quantities for water and mercury are given in Table 10.6, and their use in energy calculations is illustrated in Example 10.8.

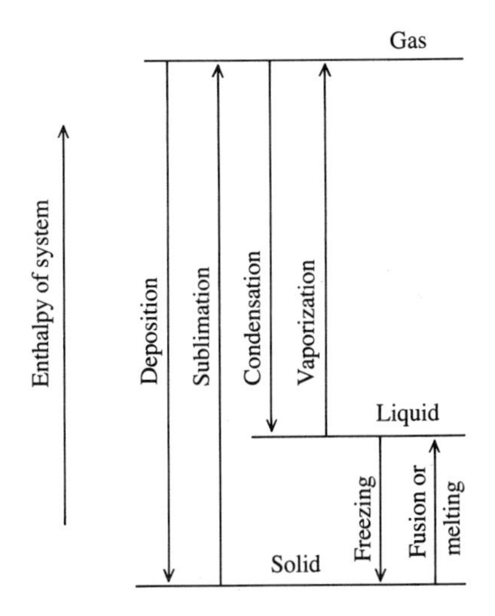

Figure 10.28 Enthalpy diagram for phase changes All of the phase changes shown are reversible at the appropriate combinations of pressure and temperature. For most substances the enthalpy of vaporization is considerably larger than the enthalpy of fusion.

Example 10.8 Enthalpy Calculations

A steam burn is much more severe than one received from hot water. Calculate the heat flow (in kilojoules) when 0.500 mol of liquid water at 100 °C is cooled to the normal body temperature of 37 °C, and compare it to heat flow in by cooling the same amount of steam at 100 °C to 37 °C.

Table 10.6 Some Thermal Properties of Water and Mercury

Property	Water	Mercury
molar heat capacities (J/mol °C)		
solid	37.3	28.2
liquid	75.4	27.7
gas	36.3	20.9
ΔH_{fusion} (kJ/mol)	6.01	2.37
$\Delta H_{\text{vaporization}}$ (kJ/mol)	40.6	59.3
$\Delta H_{\text{sublimation}}$ (kJ/mol)*	50.9	64.3

* The enthalpy of sublimation is different from the sum of the enthalpies of fusion and vaporization, because each is measured at a different temperature. The enthalpies are each measured at the normal temperature of the phase transitions.

Solid iodine has a sufficiently high vapor pressure at 100 °C that the violet color of the vapor can easily be seen. Volatile solids like iodine are often purified by sublimation.

Solution The heat capacity for liquid water given in Table 10.6 is the heat that must be removed from one mole of water to lower its temperature by one degree Celsius. Therefore the heat flow is given by

$$q = C\,\Delta t\,n$$

where q is the heat, C is the molar heat capacity, Δt is the change in temperature, and n is the number of moles of water in the sample (Section 4.5). Substitute the values given in the problem into this equation.

$$q_1 = \frac{75.4\ \text{J}}{\text{mol °C}}(37 - 100)\text{°C}\,(0.500\ \text{mol}) = -2.4 \times 10^3\ \text{J}$$

Converting this into the desired units, 2.4 kJ of heat is released by cooling the liquid.

When the initial sample is steam at 100 °C, the gas first condenses to the liquid, which then cools. First calculate the heat flow during the phase change, using the enthalpy of condensation (the negative of the enthalpy of vaporization):

$$q_2 = 0.500\ \text{mol}\left(\frac{-40.6\ \text{kJ}}{\text{mol}}\right) = -20.3\ \text{kJ}$$

Once the gas has condensed to liquid at 100 °C, it must cool to the final temperature of 37 °C, releasing an additional 2.4 kJ. The total heat flow for the sample of steam is

$$q = q_1 + q_2 = -20.3 - 2.4 = -22.7\ \text{kJ}$$

In cooling from 100 °C to 37 °C the steam releases almost 10 times the energy that liquid water releases when cooled over the same temperature range, so it is not surprising that steam burns are much more severe than those produced by the hot liquid.

Understanding Using the data in Table 10.6, calculate how much heat is needed to convert 20.0 grams of solid mercury at −50 °C to liquid mercury at 20 °C. The normal melting point of mercury is −39 °C.

Answer: 430 joules

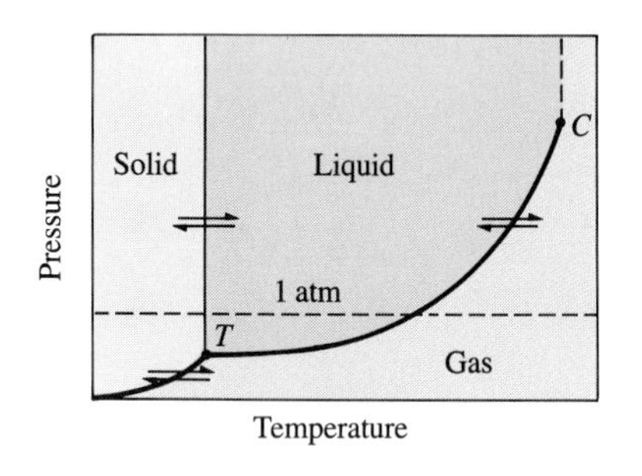

Figure 10.29 A typical phase diagram The lines divide the graph into three regions in which only one phase is present. Two phases may be in equilibrium at any point on the lines. The dashed horizontal line at 1 atm pressure intersects the solid-liquid equilibrium line at the normal freezing (melting) point and the liquid-gas line at the normal boiling point. All three phases are present at the triple point, labeled as T on the graph. The critical point is labeled C.

10.6 Phase Diagrams

Objectives

- To identify the various phases that are stable at any point on a phase diagram
- To use the principle of Le Chatelier to predict equilibrium phase changes
- To relate the sign of the slope for the solid-liquid equilibrium line to the densities of the two states, using the principle of Le Chatelier
- To construct the main features of a heating or cooling curve at constant pressure from the phase diagram

The physical state of a substance reflects the strength of the intermolecular attractions. In Section 10.5, we saw that the stable phase also depends upon the temperature and pressure. A **phase diagram** is a graph of pressure versus temperature that shows the region of stability for each of the physical states. A typical phase diagram is shown in Figure 10.29. The three line segments show the combinations of pressure and temperature at which any two phases exist in equilibrium. The line that separates the liquid from the gas is the vapor pressure curve, examples of which were given for several liquids in Figure 10.26. The line segments in Figure 10.29 divide the diagram into three regions. In each region only one phase is present, but along the lines the two phases are in equilibrium.

The **triple point** is the unique combination of pressure and temperature at which all three phases—the solid, the liquid, and the gas—exist in equilibrium. The triple point occurs where the solid-liquid, solid-gas, and liquid-gas equilibrium lines meet. At the triple point a liquid both boils and freezes at the same temperature and pressure. For water the triple point occurs at 0.0098 °C and a pressure of 4.58 torr. The triple point of carbon dioxide is at −56.4 °C and 5.11 atmospheres. Since the liquid phase of carbon dioxide is stable only at pressures greater than the triple point pressure, this substance does not melt at one atmosphere, but instead sublimes directly to a gas.

The liquid-gas equilibrium line ends at the critical point. Above the critical temperature only one phase exists. The vertical dotted line at the critical temperature does not represent a phase equilibrium. Generally the single fluid phase at pressures greater than the critical pressure is considered to be a liquid below the critical temperature and a gas above the critical temperature.

Example 10.9 Interpreting Phase Diagrams

Using the phase diagram to the right, identify the phase or phases present at each of the lettered points A, C, and E.

Solution First label each of the areas with the phase present in that region of the graph. This is easily done by following any line of constant pressure that is above the triple point, such as the line at $P = 1$ atm. As the temperature increases (left to right), the stable phase changes from the solid to the liquid to the gas in that order. We note that point A is at the intersection of all three phases, the triple point. Therefore at A the solid, liquid, and gas phases are all present. Point C is in the region that contains only solid. Point E is on the line that separates the liquid from the gas, so both of those phases are present in equilibrium.

Understanding What phase or phases are present in equilibrium at points B, D, and F on the phase diagram?

Answer: B—liquid only, D—gas only, F—solid and gas

Phase diagram referenced in Examples 9 and 10

The heating or cooling curve for a substance can be deduced from its phase diagram. A horizontal line at constant pressure intersects the solid-liquid line at the freezing point, and the liquid-vapor line at the boiling point. If the constant pressure is lower than the triple point pressure, the substance sublimes, and the liquid phase is never observed.

Example 10.10 Heating Curves from Phase Diagrams

Using the phase diagram in Example 10.9, sketch the heating curve expected at pressures of (a) 1 atm and (b) 0.20 atm.

Solution

(a) A heating curve is a graph of temperature *vs.* heat added. Starting at a temperature below the melting point of this substance at 1 atm pressure, addition of heat raises the temperature of the sample uniformly, until the melting point is reached. At that temperature, continued addition of heat is used to convert the solid to liquid at a constant temperature. Once the solid is completely melted, the temperature again rises uniformly until the boiling point is reached. The temperature remains constant while the liquid is completely vaporized, at which point the temperature rises again. The resulting heating curve is shown in the graph at the left as line (a).

(b) At a pressure of 0.20 atm the liquid phase is never stable. Thus the temperature of the solid increases with the addition of heat, until the sublimation point is reached (the temperature at which the constant pressure line intersects the solid-gas equilibrium line). The temperature does not change with further addition of heat until the solid has been converted completely to gas. The temperature then rises as the added heat increases the temperature of the gas. The horizontal portion of the heating curve at 0.20 atm is approximately the same as the sum of the two horizontal portions of the heating line at 1 atm. The lengths of the horizontal portions are arbitrary, since the enthalpies of fusion and vaporization of the substance are not given. For most substances the enthalpy of vaporization is considerably larger than the enthalpy of fusion, and this is reflected in the heating curve.

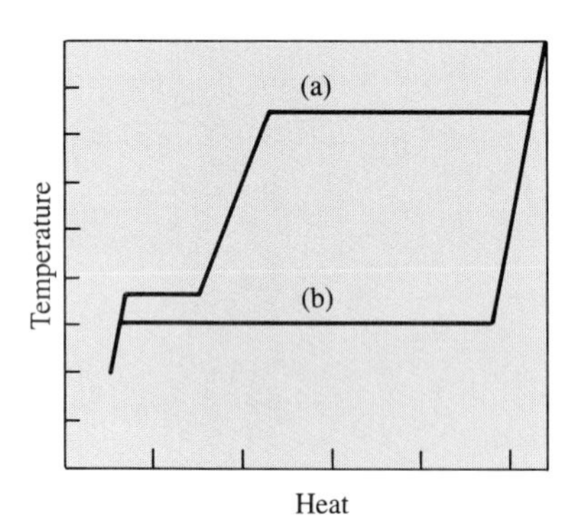

Heating curves from Example 10.10 (a) The heating curve at 1 atm. (b) The heating curve at 0.20 atm. The two lines are the same below the sublimation temperature at 0.20 atm, and above the boiling point at 1 atm.

The Principle of Le Chatelier

The vapor pressure of a liquid changes with the temperature. What can we expect if the volume of an equilibrium mixture of liquid and vapor is changed? Suppose a sample consists of liquid diethyl ether in equilibrium with its vapor at 25 °C, represented by the equation

$$(C_2H_5)_2O(\ell) \rightleftharpoons (C_2H_5)_2O(g)$$

The equilibrium vapor pressure of diethyl ether at 25 °C is 530 torr. Now, we abruptly decrease the volume of the vessel, causing the pressure to increase to 760 torr. Immediately after the volume change the system is no longer at equilibrium, since the higher pressure increases the rate of condensation without changing the rate of evaporation. The resulting net condensation of the vapor reduces the pressure of the diethyl ether in the gas phase. If the temperature is held constant, just enough diethyl ether vapor condenses to restore the original pressure of 530 torr. Figure 10.30 presents the sequence of events.

The condensation of the liquid diethyl ether is an exothermic process. If we do not allow heat to leave the system when the volume is decreased, the energy released by condensation will raise the temperature of the system, and the vapor pressure of the liquid will increase. Thus, when equilibrium is restored both the temperature and the pressure are higher than the initial values of 25 °C and 530 torr. The new equilibrium pressure and temperature depend upon the relative quantities of the two phases initially present.

Whether or not the temperature changes when the volume is decreased, some of the vapor condenses to liquid, which is represented by the chemical equation

$$(C_2H_5)_2O(g) \longrightarrow (C_2H_5)_2O(\ell)$$

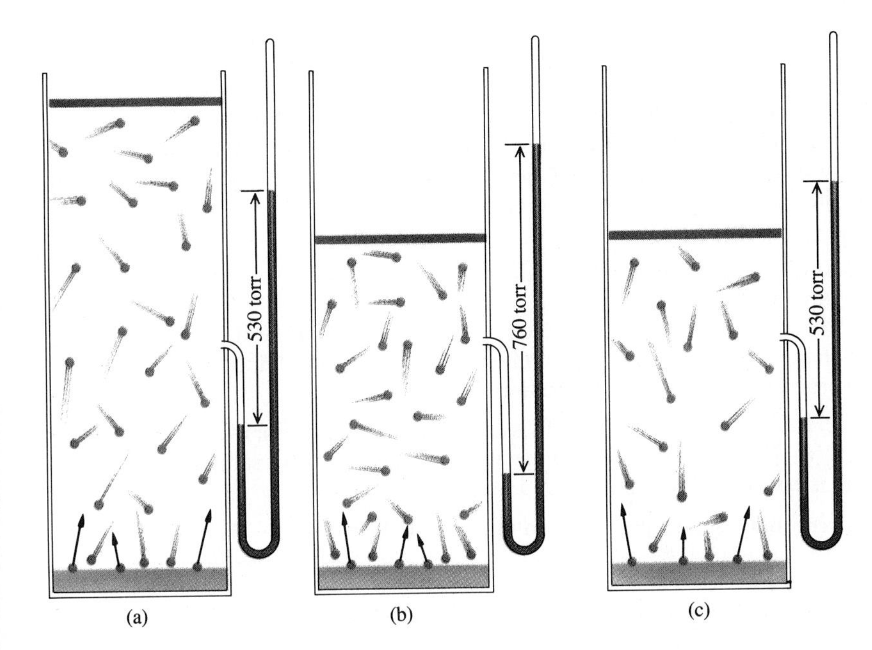

Figure 10.30 Effect of a change in volume on an equilibrium (a) Diethyl ether vapor in equilibrium with liquid at 25 °C. (b) An abrupt decrease in volume temporarily increases the pressure to 760 torr from the equilibrium pressure of 530 torr. (c) Enough of the vapor condenses at constant temperature to restore the original pressure of 530 torr.

These observations illustrate the principle of Le Chatelier. The **principle of Le Chatelier** states that a change to a system at equilibrium causes a shift that reduces the effect of the change.

We can interpret the previous example using the principle of Le Chatelier. The change that was made to the equilibrium system was the decrease in the volume. The system responded by net condensation of gas to form the liquid phase, which occupies a smaller volume. When applying the principle of Le Chatelier to equilibria between phases, you need to consider only three variables that are involved in a shift in the equilibrium: volume, pressure, and temperature. These variables are examined in Example 10.11.

Decreasing the volume of a gas in equilibrium with its liquid causes more gas molecules to join the liquid phase, where they occupy a smaller volume.

Example 10.11 Phase Changes and the Principle of Le Chatelier

Solid iodine at 97.5 °C, where the vapor pressure is 40 torr, is in equilibrium with 1.00 liter of iodine vapor. The volume of the enclosed space is suddenly increased to 2.00 liter.

(a) What is the pressure in the vessel immediately after the volume change?
(b) To restore equilibrium, will deposition or sublimation occur?
(c) If the temperature is held constant, what is the pressure in the vessel when equilibrium is restored?
(d) If no heat is added or removed, in which direction have the temperature and pressure changed when equilibrium is restored?

Solution

(a) From Boyle's law, we know that the product PV is a constant for the vapor, so

$$P_1V_1 = P_2V_2$$

where

$P_1 = 40$ torr $P_2 = ?$

$V_1 = 1.00$ L $V_2 = 2.00$ L

We substitute these values into Boyle's law and solve for P_2.

The pressure exerted on the blades of the skates melts the ice, providing lubrication between the ice and skate blades.

$$P_2 = \frac{P_1V_1}{V_2} = \frac{40 \text{ torr} \times 1.00 \text{ L}}{2.00 \text{ L}} = 20 \text{ torr}$$

(b) The change imposed on the system lowered the pressure of the vapor. The principle of Le Chatelier predicts that the pressure must therefore increase to restore equilibrium, so sublimation will occur. The net change is represented by the equation

$$I_2(s) \longrightarrow I_2(g)$$

(c) At constant temperature enough iodine must sublime to restore the initial pressure of 40 torr.

(d) Since sublimation is an endothermic process, heat must be absorbed as the solid changes to vapor, and the sample cools. At the lower temperature, the vapor pressure of iodine will also be lower.

Understanding A sample of a liquid is in equilibrium with vapor in a sealed container at 25 °C. What happens to the temperature and pressure of the sample when equilibrium is restored after removing 25 J of heat?

Answer: Both the temperature and pressure are lower.

The principle of Le Chatelier can be used to determine the effect of a change in pressure on the temperature at which solid and liquid are in equilibrium. If the pressure is increased on a sample of solid in equilibrium with liquid, a change must occur that reduces the pressure. Therefore, the phase that occupies a smaller volume (the one with the higher density) forms as a result of the pressure increase. For most substances, the density of the liquid phase is less than that of the solid, so the increase in pressure causes some of the liquid to freeze. Freezing is an exothermic process, so the energy released in the freezing of the liquid will raise the temperature. Since neither the solid nor the liquid phase is very compressible, a very small decrease in volume is needed to restore the equilibrium. Very little of the liquid must freeze to produce the necessary volume change, so only a small temperature increase is produced. This means that the graph of pressure versus temperature for the solid-liquid equilibrium line is nearly vertical with a positive slope, as shown in Figure 10.31.

Water is one of the few substances for which the density of the solid is less than the density of the liquid. When the pressure is increased on water in equilibrium with ice, the result is that some of the solid melts, producing a smaller volume of the sample. Since melting is an endothermic process, the temperature of the sample is lower when equilibrium is re-established. The slope of the ice-water equilibrium line is negative. Because ice melts when the pressure is increased, a wire can pass through a block of ice without cutting it in two pieces. Since the temperature of the liquid formed under pressure is less than 0 °C, the water freezes again above the wire.

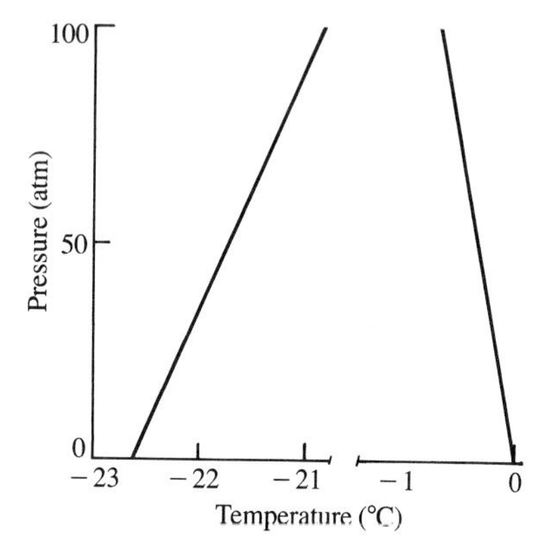

Figure 10.31 Dependence of freezing point on pressure The liquid-solid equilibrium is nearly a vertical line on a graph of pressure versus temperature. The line on the left is typical of the pressure dependence of the melting point of most substances, for which the density of the liquid is less than the density of the solid. For water and a few other substances the melting point is lowered by an increase in pressure, because the density of the liquid is greater than the density of the solid (line on right). The freezing point of water decreases about 0.074 °C for a change in pressure of 10 atmospheres near 0 °C.

INSIGHTS into CHEMISTRY

Science, Society, and Technology

High Heat and Pressure Turn Black Graphite into Sparkling Diamonds

The element carbon exists in several allotropic forms, two of which are graphite and diamond. At room temperature and 1 atm pressure, graphite is the more stable form of the element, so it is not surprising that this form of the element is quite abundant and inexpensive. Diamonds are relatively rare and expensive, so a process that converts graphite into diamonds could be quite profitable. Graphite has a density of 2.26 g/cm^3, and the density of diamonds is 3.51 g/cm^3. The principle of Le Chatelier suggests that the denser form of the element (diamond) should be favored by high pressures. In 1955 General Electric developed a process to make commercial grade diamonds at temperatures of 2000 to 3000 K and pressures near 125,000 atmospheres. Today, more than 40% of the commercial diamonds are synthetic. Gem quality diamonds can also be made by the GE process, but the cost is prohibitively high.

Diamonds and graphite Application of the principle of Le Chatelier contributed to the development of a process to convert graphite into diamonds.

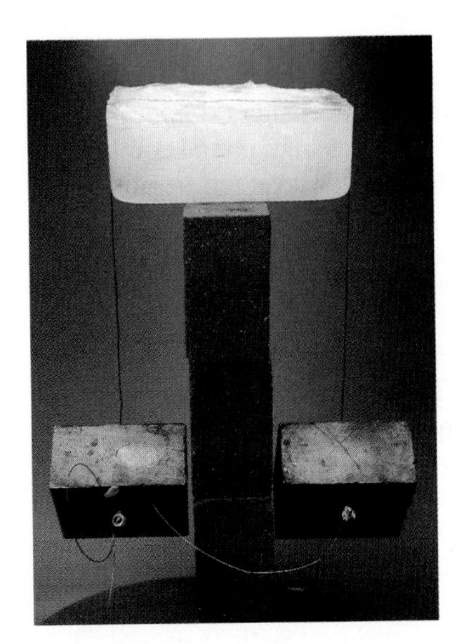

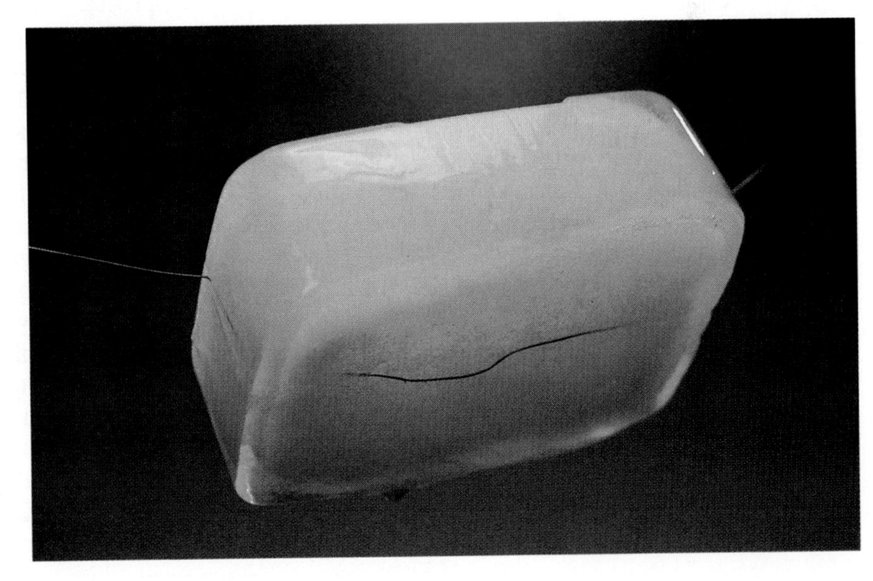

A wire that has weights attached to the ends will slowly pass through a block of ice without cutting it into two pieces. The wire exerts a pressure on the ice immediately under it, causing it to melt. The water flows around the wire where the pressure is lower, and the solid is reformed.

Summary

The solid and liquid phases, or *condensed phases,* do not have a single law to describe them, as do gases. In the solid state the units (atoms, molecules, ions) that make up the substance are held rigidly in place and have long-range order. In the liquid state the units are close together, but have sufficient kinetic energy to move relative to each other. The strengths of *intermolecular attractions* influence and often determine many of the physical properties of substances. Among the properties that may be used to indicate the relative strengths of intermolecular attractions are the boiling points, the melting points, the enthalpies of fusion and vaporization, and the surface tension, viscosity, and vapor pressure of liquids.

In molecular substances there are three main kinds of intermolecular attractions: *London dispersion forces, dipole–dipole attractions,* and *hydrogen bonding.* Dispersion forces and dipole–dipole attractions are collectively called van der Waals forces. London dispersion forces contribute to the attraction of all atoms, molecules, and ions, and depend on the *polarizability* of the species. Hydrogen bonding is considerably stronger than either dispersion forces or dipole–dipole attractions; it occurs only between a molecule that contains hydrogen covalently bonded to a very electronegative element and a molecule with a lone pair of electrons on N, O, or F.

Variations in the strengths of intermolecular forces are related to the properties of molecular substances. Substances with strong intermolecular forces will have higher boiling points, *enthalpies of vaporization, surface tensions,* and *viscosities.* The appearance of the meniscus of a liquid in a glass tube is determined by whether the *cohesive forces* are stronger or weaker than the *adhesive forces.*

The type of attractions between the particles in a crystal is one way to categorize solids. In *molecular solids* the units are molecules held together by van der Waals forces and hydrogen bonds. Molecular solids usually have low melting points. *Covalent network solids* do not contain small discrete molecules, but have a framework of covalent bonds throughout the crystal. These substances are very high melting and are often very hard. *Ionic solids* are held together in the crystal by the strong attractions that exist between the oppositely charged ions. *Metallic solids* can be viewed as metal ions held together by a "sea" of mobile electrons that extends through the entire crystal. These metallic bonds cover a wide range of strengths, as reflected by the melting points of metals that range from −39 °C for mercury to 3410 °C for tungsten.

The structural features of a crystalline solid can be defined by a *unit cell,* the small part of the crystal that can reproduce the three-dimensional arrangement of the particles by simple displacements. There are a total of seven crystal systems. Many solids, particularly metals, are represented by closest packing arrays, in which each particle has a *coordination number* of twelve. The composition and density of a unit cell must be the same as bulk properties of the substance. The crystal structure and unit cell dimensions are used to find the radii of atoms. The *Bragg equation* is used to determine the spacing between layers of atoms from an x ray diffraction pattern. By analyzing the intensities at the angles of diffraction maxima, each atom in a unit cell can be located.

Phase changes are all dynamic equilibrium processes occurring at constant temperature and pressure. *Melting* (solid to liquid), boiling or *vaporization* (liquid to gas), and *sublimation* (solid to gas) are all endothermic processes because energy is used to overcome the intermolecular forces of attraction. From Hess's law, the reverse processes of *freezing, condensation,* and *deposition* are all exothermic changes. *Vapor pressure* is the partial pressure of a gas in dynamic equilibrium with its liquid or solid at a constant temperature. The vapor pressure of any substance increases with increasing temperature.

A *phase diagram* is a graph of pressure versus temperature that shows the combinations of pressure and temperature at which the solid, liquid, and gas phases are in equilibrium. The equilibrium lines separate the regions in which only one of the physical states is stable. At one point, the *triple point,* all three phases are in equilibrium with each other. The *principle of Le Chatelier* makes predictions about the changes that occur in phase equilibria as a result of changes in the pressure, volume, temperature, and heat in a system.

Chapter Terms

Intermolecular forces—the attractive forces that exist between molecules. *(10.1)*

Dipole-dipole attractions—the intermolecular forces that arise from electrostatic attractions between the molecular dipoles. *(10.1)*

Induced dipole moment—a dipole caused by the presence of an electrical charge close to an otherwise nonpolar molecule. *(10.1)*

Instantaneous dipole moment—a dipole moment that results from unequal charge distribution within a molecule caused by the motion of the electrons. *(10.1)*

London dispersion forces—the instantaneous dipole–induced dipole attractions that explain the attractions between nonpolar molecules. *(10.1)*

Polarizability—the ease with which the electron cloud of a molecule can be distorted. *(10.1)*

van der Waals forces—the weak intermolecular forces between small molecules. The term includes both dipole–dipole attractions and London dispersion forces. *(10.1)*

Hydrogen bonding—a particularly strong intermolecular attraction between a hydrogen atom that is bonded to a highly electronegative atom and a lone pair of electrons on N, O, or F. *(10.1)*

Molar enthalpy of vaporization—the enthalpy change that accompanies the conversion of one mole of a substance from the liquid to the gas phase at constant temperature. This process is always endothermic. *(10.2)*

Surface tension—the quantity of energy required to increase the surface area of a liquid; it has SI units of joules/m^2. *(10.2)*

Cohesive forces—the attraction of molecules for other molecules of the same substance. *(10.2)*

Adhesive forces—the attractions that molecules of one substance exert on those of a different substance. *(10.2)*

Viscosity—the resistance of a fluid to flow; it is one of the properties related to the forces of intermolecular attraction. *(10.2)*

Crystalline solid—a solid in which the units that make up the substance are arranged in a very regular repeating pattern. *(10.3)*

Amorphous solid—a solid that lacks the long-range order of a crystalline solid. *(10.3)*

Molecular solid—a solid that consists of small covalently bonded molecules, held together by van der Waals forces or hydrogen bonding. *(10.3)*

Covalent network solid—a solid that consists of atoms held together in the crystal by covalent bonds. *(10.3)*

Allotropes—structurally distinct forms of an element in the same physical state that exhibit different chemical and physical properties. *(10.3)*

Ionic solid—a solid that consists of oppositely charged ions that are held together by electrostatic attractions. *(10.3)*

Metallic solid—a solid formed by metal atoms. In metals the outer electrons occupy highly delocalized orbitals that extend throughout the crystal and account for the electrical and thermal conductivity. *(10.3)*

Lattice point—the point in space that has the same geometric environment as every other lattice point in a crystal lattice. *(10.3)*

Crystal structure—the geometric arrangement of particles (atoms, ions, or molecules) in a crystalline solid. *(10.3)*

Unit cell—a small geometric figure needed to define the pattern of lattice points in the entire crystal. *(10.3)*

Crystal system—one of the seven possible forms of unit cells, defined by the lengths of the edges (a, b, c) and the angles between the edges (α, β, γ) of the unit cell. *(10.3)*

Cubic system—the crystal system in which $a = b = c$, and $\alpha = \beta = \gamma = 90°$. *(10.3)*

Simple cubic—the cubic structure with identical atoms located only at the corners of the unit cell. *(10.3)*

Body-centered cubic (BCC)—the cubic structure with identical atoms at the corners and the center of the unit cell. *(10.3)*

Face-centered cubic (FCC)—the cubic structure with identical atoms at the corners and in the center of each of the six square faces. The FCC and the cubic closest packing arrays are identical. *(10.3)*

Closest packing array—the most efficient packing of spheres, which has the smallest empty space (26%). Each sphere is in contact with 12 other spheres. *(10.3)*

Hexagonal closest packing—the closest packing array with an ABA . . . stacking pattern of the layers. *(10.3)*

Cubic closest packing—the closest packing array with an ABC . . . stacking pattern of the layers. *(10.3)*

Coordination number—the number of nearest neighbors an atom, ion or molecule has in a crystalline solid. The largest possible coordination number of uniform-sized spheres is 12, and it occurs in both of the closest packing arrays. *(10.3)*

Bragg equation—the relation between the angle of diffraction of x rays and the distance between layers of atoms in a crystalline solid:

$$n\lambda = 2d \sin \theta$$

where λ is the wavelength of the x rays used, d is the distance between layers of atoms in the crystal, θ is the angle between the x ray beam and the layers of atoms, and n is a whole positive number called the order. *(10.4)*

Evaporation—the escape of molecules from the liquid state into the gas phase. *(10.5)*

Vaporization—the same as evaporation. *(10.5)*

Condensation—the conversion of a gas to a liquid. *(10.5)*

Vapor pressure—the partial pressure of a substance that is in equilibrium with one of the condensed phases of that substance. *(10.5)*

Dynamic equilibrium—a situation in which two opposing changes occur at equal rates, so no *net* change is apparent. *(10.5)*

Boiling point—the temperature at which the vapor pressure of a liquid is equal to the applied pressure. *(10.5)*

Normal boiling point—the boiling point at a pressure of one atmosphere. *(10.5)*

Critical temperature—the maximum temperature at which a substance can exist in the liquid phase. *(10.5)*

Critical pressure—the minimum pressure needed to cause the liquid state to exist up to the critical temperature. *(10.5)*

Normal melting point—the temperature at which the solid and liquid phases are in equilibrium at one atmosphere pressure. Melting points change very little with pressure. *(10.5)*

Molar enthalpy of fusion—the enthalpy change that accompanies the conversion of one mole of a substance from the solid to the liquid phase at constant temperature. This change is always endothermic. *(10.5)*

Supercooling—lowering the temperature of a liquid below its freezing point without forming a solid. This is an unstable condition. *(10.5)*

Sublimation—the direct conversion of a solid to a gas without first changing to a liquid. *(10.5)*

Deposition—the direct conversion of a gas to a solid. Deposition is the reverse of sublimation. *(10.5)*

Molar enthalpy of sublimation—the enthalpy change that accompanies the sublimation of one mole of a substance. *(10.5)*

Phase diagram—a graph of pressure versus temperature that shows the regions of stability for each phase (solid, liquid, and gas). The lines between the phases represent conditions of temperature and pressure at which two or more phases are in equilibrium. *(10.6)*

Triple point—the unique combination of pressure and temperature at which all three phases (solid, liquid, and gas) exist in equilibrium. *(10.6)*

Supercritical fluid—the single phase that exists above the critical temperature and pressure. *(10.6)*

Principle of Le Chatelier—a change to a system at equilibrium causes a shift in the position of the equilibrium that reduces the effect of the change. *(10.6)*

Exercises

Exercises designated with color have answers in Appendix J.

Intermolecular Forces

10.1 Identify the kinds of intermolecular forces (London dispersion, dipole–dipole, hydrogen bonding) that are the most important in each of the following substances.
(a) methane (CH_4) (b) methanol (CH_3OH) (c) chloroform ($CHCl_3$) (d) benzene (C_6H_6) (e) ammonia (NH_3) (f) sulfur dioxide (SO_2)

10.2 Identify the kinds of intermolecular forces (London dispersion, dipole-dipole, hydrogen bonding) that are the most important in each of the following substances.
(a) propane (C_3H_8)
(b) ethylene glycol ($HO(CH_2)OH$)
(c) cyclohexane (C_6H_{12})
(d) phosphine oxide (PH_3O)
(e) nitrogen monoxide (NO)
(f) hydroxylamine (NH_2OH)

10.3 In each part, select the substance that has the higher boiling point, based upon the relative strengths of the intermolecular attractions.
(a) C_3H_8 and CH_4 (b) I_2 and ICl (c) H_2S and H_2Te (d) H_2Se and H_2O (e) CH_2Cl_2 and CH_3Cl (f) NOF and NOCl

10.4 In each part, select the substance that has the higher boiling point, based upon the relative strengths of the intermolecular attractions.
(a) C_2H_4 and CH_4 (b) Cl_2 and ClF (c) S_2F_2 and S_2Cl_2 (d) NH_3 and PH_3 (e) CH_3I and CHI_3 (f) BBr_2I and $BBrI_2$

10.5 The types of intermolecular forces in the solid, liquid, and gas phases are the same. What determines which of these phases is stable at a given temperature?

10.6 The density of liquid sulfur dioxide is 1.43 g/mL and that of the gas is 0.00293 g/mL at STP. Account for the large difference between these two densities.

10.7 For most substances the density of the liquid state is less than that of the solid state. How does this affect the energy of the intermolecular attractions?

10.8 Identify the types of all of the intermolecular forces that cause each of the following gases to condense to a liquid.
(a) N_2O (b) CH_4 (c) NH_3 (d) SO_2

10.9 Identify the types of all of the intermolecular forces that cause each of the following gases to condense to a liquid.
(a) F_2 (b) BF_3 (c) HF (d) NO_2

10.10 From the graphs in Figure 10.4, is there any evidence that hydrogen bonding occurs for any elements other than nitrogen, oxygen, and fluorine? Explain.

10.11 Explain why the effect of hydrogen bonding on the boiling point is considerably greater for water than for ammonia and hydrogen fluoride. (Hint: The formation of a hydrogen bond requires both a positively charged hydrogen atom, and unshared pairs of electrons on the electronegative atom.)

Properties of Liquids

10.12 Explain why the viscosity of a liquid may not be consistent with the enthalpy of vaporization and the boiling point as a measure of the strength of intermolecular forces.

10.13 Water forms beads on the painted surfaces of new cars. After a car is exposed to the weather for a long period of time, water spreads out into a thin film on the same painted surface. What has happened to the paint that causes this change?

10.14 State how each of the following properties changes with increasing strength of the intermolecular forces.
(a) enthalpy of fusion
(b) melting point
(c) surface tension
(d) viscosity
(e) enthalpy of vaporization
(f) boiling point

10.15 Trouton's rule states that the enthalpy of vaporization for a substance divided by its normal boiling point on the Kelvin scale is approximately 88 joules/kelvin. Is this in agreement with the expected behavior of these properties as intermolecular forces change? Explain.

10.16 Explain why the surface of the HCl(aq) liquid in a buret is curved rather than flat.

10.17 Arrange the liquids ethanol (C_2H_5OH), glycerol ($CH_2(OH)CHOHCH_2(OH)$), and ethylene glycol ($(CH_2OH)_2$) in order of their expected decreasing viscosity.

10.18 Select the liquid in each part that has the larger expected enthalpy of vaporization.
(a) C_3H_8 and CH_4
(b) I_2 and ICl
(c) S_2Cl_2 and S_2F_2
(d) H_2Se and H_2O
(e) CH_2Cl_2 and CH_3Cl
(f) NOF and NOCl

10.19 Select the liquid in each part that has the larger expected enthalpy of vaporization.
(a) C_2H_4 and CH_4
(b) Cl_2 and ClF
(c) H_2S and H_2Te
(d) NH_3 and PH_3
(e) CHI_3 and CH_3I
(f) BBr_2I and $BBrI_2$

10.20 The compounds ethanol (C_2H_5OH) and dimethyl ether ($CH_3{-}O{-}CH_3$) have the same molecular formula. Which of these compounds is expected to have the higher surface tension?

Crystal Forces

10.21 How does an amorphous solid differ from a crystalline solid?

10.22 An amorphous solid can sometimes be converted into a crystalline solid by a process called annealing. Annealing consists of heating the substance to a temperature just below the melting point and then cooling it very slowly. Explain why this process helps produce a crystalline solid.

10.23 Sometimes amorphous solids are referred to as supercooled liquids. In what ways are amorphous solids similar to liquids?

10.24 Identify the kinds of forces that are most important in holding the particles together in crystalline solid samples of the following substances.
(a) H_2O
(b) C_6H_6
(c) $CaCl_2$
(d) SiO_2
(e) Fe
(f) SiC

10.25 Identify the kinds of forces that are most important in holding the particles together in crystalline solid samples of the following substances.
(a) Kr
(b) HF
(c) K_2O
(d) CO_2
(e) Zn
(f) NH_3

10.26 Arrange the following substances in order of increasing boiling points: CO_2, KCl, H_2O, N_2, CaO.

10.27 Arrange the following substances in order of increasing boiling points: He, SiC, NO_2, NaBr, BaO.

The Cubic Unit Cell

10.28 Ammonium chloride consists of a simple cubic array of Cl^- ions, with an NH_4^+ ion at the center of the unit cell. Calculate the number of ammonium ions and chloride ions in each unit cell.

10.29 The coordination number of uniformly sized spheres in a cubic closest packing array (FCC) is 12. What is the coordination number of each atom in
(a) a simple cubic lattice
(b) a body-centered cubic lattice?

10.30 How many atoms of tungsten are present in each unit cell of that metal, if it crystallizes in a body-centered cubic array of atoms?

10.31 Silver crystallizes in a cubic closest packing (FCC) array of atoms. How many silver atoms are present in each unit cell?

10.32 Nickel crystallizes in a face-centered cubic array of atoms in which the length of the unit cell's edge is 351 pm. Calculate the density of this metal.

10.33 Avogadro's number can be calculated quite accurately from the density of a solid and crystallographic data. Europium (molar mass = 151.96 g/mol) has a body-centered cubic array of atoms in which the edge of the unit cell is 458.27 pm. The density of the metal is 5.243 g/cm^3. Use these data to calculate Avogadro's number. (Hint: Calculate the number of atoms/cm^3 from the crystallographic data, and the volume of one mole of the element from the molar mass and density. Combine these values to find the atoms/mol, or Avogadro's number.)

10.34 Argon is a solid below −189 °C, with a face-centered cubic array of atoms and a density of 1.65 g/cm^3.
(a) What is the length of the edge of the unit cell?
(b) Assuming the atoms are spheres in contact along the face diagonal, what is the radius of an argon atom?

10.35 Calcium fluoride (fluorite) has a unit cell with Ca^{2+} ions in a face-centered cubic arrangement. How many fluoride ions are present in each unit cell? (Remember that the ratio of fluoride ions to calcium ions must be the same as in the formula of the compound.)

10.36 In the rutile structure of TiO_2, the unit cell has a titanium ion in the center of the unit cell and a titanium ion at each of the eight corners. How many oxide ions are present in each unit cell? (Remember that the ratio of oxide ions to titanium ions must be the same as in the formula of the compound.)

10.37 Lithium hydride (LiH) has the sodium chloride structure, and the length of the edge of the unit cell is 4.086×10^{-8} cm. Calculate the density of this solid.

10.38 Cesium iodide crystallizes as a simple cubic array of I^- ions with a cesium ion in the center of the cell. The edge of this unit cell is 445 pm. Calculate the density of CsI.

10.39 Chromium has a cubic crystal structure in which the edge of the unit cell is 288 pm. If the density of chromium is 7.20 g/cm^3, how many chromium atoms are in a unit cell? In which of the cubic unit cells does chromium crystallize?

10.40 Palladium has a cubic crystal structure in which the edge of the unit cell is 389 pm. If the density of palladium is 12.02 g/cm^3, how many palladium atoms are in a unit cell? In which of the cubic unit cells does palladium crystallize?

10.41 The edge of the face-centered cubic unit cell of copper is 362 pm. Assume that the copper atoms are spheres that are in contact along the diagonal of the face of the cube. Find the radius of a copper atom.

X-Ray Diffraction

10.42 Which of the following materials should produce well-defined x-ray diffraction patterns?
(a) KBr (b) liquid H_2O (c) glass (d) a sugar crystal

10.43 At what angle does the first-order diffraction occur from layers of atoms that are 293 pm apart, using x rays with a wavelength of 154 pm?

10.44 What is the wavelength of x rays if the first-order diffraction from layers of atoms that are spaced at 232 pm occurs at an angle of 13.4°?

10.45 X rays with a wavelength of 1.790×10^{-10} m are diffracted at an angle of 14.7° by a crystal of KI.
(a) What is the spacing between layers in the crystal that gives rise to this angle of diffraction? (Assume $n = 1$.)
(b) KI has the same crystal structure as NaCl, and the distance calculated in part (a) is one half the length of the cube edge. Calculate the density of KI.

Phase Changes and Vapor Pressure

10.46 How does the vapor pressure of a liquid change for each of the following changes?
(a) The surface area of the liquid is increased from 1 cm^2 to 10 cm^2 by pouring the liquid into a container with a larger diameter.
(b) Enough heat is added to the sample to raise the temperature by 5 °C.
(c) The volume of a closed container of liquid in equilibrium with its gas is decreased at constant temperature.

10.47 "A liquid stops evaporating when the equilibrium vapor pressure is reached." What is wrong with this statement?

10.48 A 1.50-g sample of water is placed in an evacuated 1.00-L container at 30 °C.
(a) Calculate the pressure in the container if all of the water is vaporized. (Assume the ideal gas law, $PV = nRT$.)
(b) What is the vapor pressure of water at 30 °C?
(c) What mass of water actually evaporates?

10.49 A 1.50-g sample of methanol (CH_3OH) is placed in an evacuated 1.00-L container at 30 °C.
(a) Calculate the pressure in the container if all of the methanol is vaporized. (Assume the ideal gas law, $PV = nRT$.)
(b) The vapor pressure of methanol at 30 °C is 158 torr. What mass of methanol actually evaporates?

10.50 Why does water have a lower vapor pressure at 25 °C than dimethyl ether? Which of these two liquids has the larger enthalpy of vaporization?

10.51 Why is greater cooling achieved by perspiring when the wind is blowing than in calm air?

10.52 How does the humidity affect the efficiency of cooling by perspiration? Explain.

10.53 Each of two glasses contains 200 g of water. In one glass 10 g of the water is ice, while in the other glass 90 g of the water is ice. Which sample is colder, assuming both are at equilibrium?

10.54 How much heat is absorbed by a 10.0-g sample of water in going from ice at −10 °C to liquid water at 95 °C? Use the physical properties of water given in Table 10.6.

10.55 How much heat is absorbed by a 15.0-g sample of water in going from liquid at 10 °C to steam at 105 °C at a pressure of 1.00 atm? Use the physical properties of water given in Table 10.6.

10.56 For a typical substance, arrange the enthalpies of sublimation, fusion, and vaporization in order of increasing value.

Phase Diagrams

10.57 The melting point of iodine is 113.5 °C and its normal boiling point is 184.3 °C. The triple point occurs at 92.3 torr and 113.4 °C, and the critical point is 512 °C at 112 atmospheres. Sketch the phase diagram for iodine.

10.58 The heat of sublimation of iodine is 60.2 kJ/mol and its enthalpy of vaporization is 45.5 kJ/mol. What is the enthalpy of fusion of iodine?

10.59 Use the phase diagram shown with this problem to answer each of the following.
(a) Label each region of the diagram with the phase that is present.
(b) Identify the phase or phases present at each of the points A, B, C, D, and E.

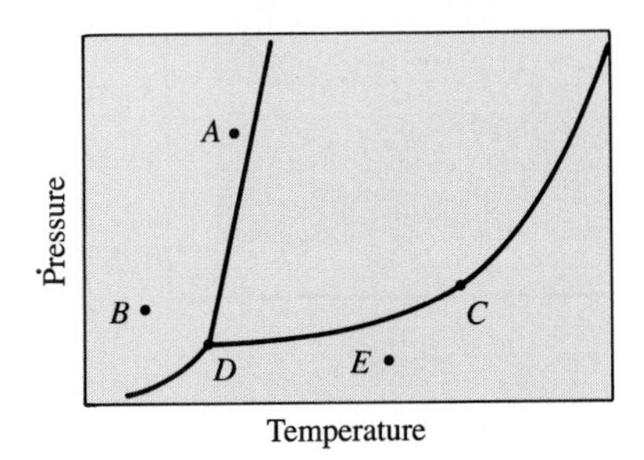

10.60 Answer these questions by using the phase diagram in Exercise 59.
(a) Sketch the heating curve expected when heat is added to the sample at constant pressure, starting at point B.
(b) Starting at point A, describe what happens if the pressure is lowered at constant temperature.
(c) What does the positive slope of the solid-liquid equilibrium line tell you about this substance?

10.61 Use the principle of Le Chatelier to determine the changes that occur when each of the following changes is made to a system at equilibrium.
(a) The pressure is increased abruptly at constant temperature on a solid in equilibrium with its vapor.
(b) The volume of the container is increased at constant temperature on a liquid in equilibrium with its gas.
(c) A solid is in equilibrium with its vapor. How will the temperature of the sample change when the volume of the container is increased if heat is neither added nor removed? When equilibrium is restored, how has the pressure changed from its initial value?

10.62 Use the principle of Le Chatelier to determine the changes that occur when each of the following changes is made to an equilibrium system.
(a) Heat is added to a sample of solid in equilibrium with its gas at constant pressure.
(b) The pressure is increased abruptly on an equilibrium mixture of water and ice.
(c) The pressure is increased on an equilibrium mixture of a liquid and its vapor at constant temperature.

Additional Exercises

10.63 Which of the following pairs are allotropes?
(a) oxygen gas (O_2) and ozone gas (O_3)
(b) liquid oxygen (O_2) and oxygen gas (O_2)
(c) amorphous silicon dioxide and quartz (crystalline silicon dioxide)

10.64 Even when the atmospheric temperature is above normal body temperature, sweating is effective in cooling a person. Explain how this is accomplished, since heat flows from a warmer object to a cooler one.

10.65 There are two allotropic forms of the element tin, called gray tin and white tin. The density of gray tin is 5.75 g/cm^3 and that of white tin is 7.28 g/cm^3. Use the principle of Le Chatelier to determine which allotropic form of tin is favored by high pressure.

10.66 The cooling process in a refrigeration unit involves reducing the pressure above a liquid called the refrigerant. The pressure reduction causes vaporization, which cools the remaining liquid. A common refrigerant is Freon-11 (CCl_3F), which has a boiling point of 23.8 °C, a specific heat for the liquid of 0.870 J g^{-1} °C^{-1}, and an enthalpy of vaporization of 180 J g^{-1}.
(a) How much energy must be removed from 10.0 g of liquid Freon-11 to cool it from its boiling point to 0 °C?
(b) What mass of Freon-11 must vaporize in order to remove the amount of heat calculated in part (a)?

Solutions

Chapter Eleven

11

Solutions play a very important role in chemistry and in the world around us. Most reactions we observe occur in solution. The fuels we use, the air we breathe, and the water we drink are all solutions. The fluids in our bodies are complex solutions that distribute nutrients and oxygen throughout the body. Alloys are generally solid solutions of two or more metals with desirable properties that are not exhibited by the pure metals.

This chapter will focus mainly on liquid solutions because they are so important in experimental chemistry. Particular emphasis is placed on aqueous solutions, because water is the most commonly used solvent and one that is important in biological solutions. Several terms related to solutions were introduced and defined in Chapter 4, and it may be helpful to review these definitions before proceeding with this chapter.

11.1 Solution Concentration

Objectives

- To express the concentration of a solution in several different ways
- To convert between different concentration units

The composition of a solution is expressed as concentration. There are many different units for concentration, but all of them express the composition of the solution as the quantity of the solute that is present in a fixed quantity of the solution or solvent. We shall review concentration units that were used earlier and add some others in this section.

"Salting" the road helps to melt snow and ice.

Concentration Units

Molarity (M)

In Chapter 4 *molarity* was defined as the number of moles of solute in one liter of solution, and it is the preferred concentration unit when performing stoichiometry calculations. This way of expressing concentrations was used extensively in the stoichiometry calculations presented in Chapter 4.

Mole Fraction (χ)

The mole fraction of a substance in a solution is the number of moles of that particular component divided by the total number of moles of all substances present.

$$\chi_A = \frac{\text{moles of A}}{\text{moles of A} + \text{moles of B} + \text{moles of C} + \ldots}$$

Mole fraction was used in Chapter 5 to calculate the partial pressure of one component of gas in a gaseous mixture.

Mass Percent Composition

The concentration of solute may be expressed as a percentage of the mass of the solution, by using the equation

$$\text{mass percent solute} = \frac{\text{grams of solute}}{\text{grams of solution}} \times 100\%$$

The percentage gives the number of grams of solute present in 100 g of solution. The mass percent concentration unit is not widely used by chemists, but is still encountered frequently in pharmacies and appears on the labels of many products that are sold as solutions (Figure 11.1). Mass percent composition is a commonly used concentration unit.

Figure 11.1 Common solutions The concentrations of many solutions sold in grocery stores and pharmacies are expressed as percent by mass.

Example 11.1 Mass Percent Concentration

What is the mass percent concentration of a solution prepared by dissolving 5.00 g of NaCl in 200. g of water?

Solution The mass of the solute and the mass of the solution are needed to find the percent concentration. The mass of solute is given, and the mass of the solution is the sum of the masses of the components, 205 g. The concentration is:

$$\text{mass percent NaCl} = \frac{5.00 \text{ g NaCl}}{205 \text{ g soln}} \times 100\% = 2.44\% \text{ NaCl}$$

Understanding 25.0 g of a solution contains 2.00 g of glucose. Express the concentration in mass percent of glucose.

Answer: 8.00%

Molality (*m*)

The **molality *(m)*** of a solution is defined as

$$\text{molality} = \frac{\text{moles of solute}}{\text{kg of solvent}}$$

This is the only common unit of concentration in which the denominator expresses a quantity of *solvent* rather than *solution.* Even though the names are very similar, do not confuse the definitions of molality and molarity. Molarity is the moles of solute per liter of *solution.*

Example 11.2 Molal Concentration of Solution

What is the molality of the solution in Example 11.1 (5.00 g of NaCl in 200. g of water)?

Solution A flow diagram showing the steps in this calculation is

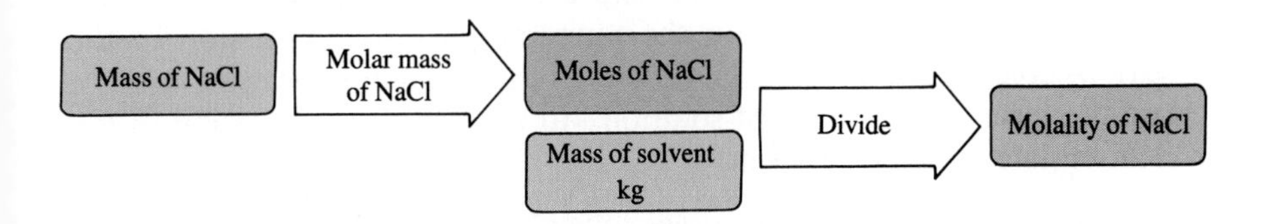

The quantity of the solute, NaCl, must be expressed in moles.

$$\text{moles NaCl} = 5.00 \text{ g NaCl}\left(\frac{1 \text{ mol}}{58.44 \text{ g}}\right) = 8.56 \times 10^{-2} \text{ mol NaCl}$$

The denominator in molality is the mass of the solvent in kilograms. Since this mass is given as 200. g, or 0.200 kg, the molality of the solution is

$$\text{molality NaCl} = \frac{8.56 \times 10^{-2} \text{ mol NaCl}}{0.200 \text{ kg solvent}} = 0.428\ m \text{ NaCl}$$

Understanding What is the molality of a solution that contains 2.00 g of glucose in 25.0 g of solvent? (molar mass of glucose = 180 g/mol)

Answer: 0.444 molal

Example 11.3 Calculations Involving Molality

What mass of acetic acid (CH_3CO_2H, molar mass = 60.05 g/mol) must be dissolved in 250. g of water to produce a 0.150 *m* solution?

Solution This problem is similar to conversions performed in Chapter 4, except the desired molality is used as a conversion factor. The following flow diagram outlines the solution to this problem.

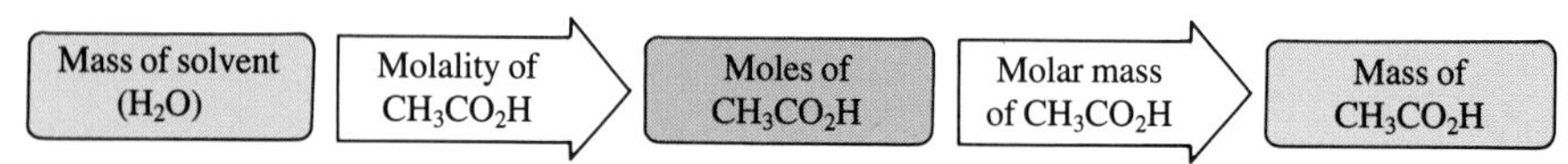

The desired molality is used as a conversion factor, to find the number of moles of solute in 250 g (= 0.250 kg) of the solvent.

$$\text{moles } CH_3CO_2H = 0.250 \text{ kg } H_2O\left(\frac{0.150 \text{ mol } CH_3CO_2H}{1 \text{ kg } H_2O}\right)$$
$$= 0.0375 \text{ mol } CH_3CO_2H$$

The molar mass of the acetic acid is the proper conversion factor for expressing this amount of the compound in grams.

$$\text{mass } CH_3CO_2H = 0.0375 \text{ mol } CH_3CO_2H\left(\frac{60.05 \text{ g } CH_3CO_2H}{1 \text{ mol } CH_3CO_2H}\right)$$
$$= 2.25 \text{ g } CH_3CO_2H$$

Understanding How many grams of water must be added to 4.00 g of urea ($CO(NH_2)_2$, molar mass = 60.06 g/mol) to produce a 0.250 *m* solution of the compound?

Answer: 266 g of water

Conversion Among Concentration Units

Often it is necessary to convert from one concentration unit to another. All of the concentration units are fractions, with the quantity of the solute in the numerator and the quantity of the solution (or solvent) in the denominator. They differ only in the units used to express these two quantities. Table 11.1 summarizes the units used for the numerator and denominator for the various concentration units. In Table 11.1 and subsequent calculations we use the term "moles total of solution" to refer to the summation in the denominator of mole fraction.

$$\text{moles total of solution} = \text{moles solvent} + \text{moles solute}$$

When converting from one concentration unit to another, always write the units of the numerator and denominator.

When performing conversions from one concentration unit to another, always write the units in the form of a fraction. Then the units of the

Table 11.1 Units for Concentration Conversions

Concentration Unit	Numerator Units (solute in each case)	Denominator
mass percent	grams	100 g solution
molarity	moles	1 L solution
molality	moles	1 kg solvent
mole fraction	moles	1 mol total of solution

numerator and denominator can be changed separately. This method of conversion is illustrated by changing a mole fraction of 0.024 NaCl in water into the molal concentration of NaCl. We start by writing the given concentration as a fraction.

$$\text{mole fraction NaCl} = \frac{0.024 \text{ mol NaCl}}{1 \text{ mol total of solution}}$$

To change the concentration to units of molality, no change is needed in the units of the numerator, but we must replace the denominator with kg of solvent. First we find the number of moles of *solvent.*

$$\begin{aligned}\text{mol solvent} &= \text{mol total of solution} - \text{mol solute}\\ &= 1 \text{ mol total of solution} - 0.024 \text{ mol NaCl}\\ &= 0.976 \text{ mol solvent}\end{aligned}$$

This amount of the solvent (water) must be expressed as kilograms, so

$$\begin{aligned}\text{mass water} &= 0.976 \cancel{\text{mol}}\ H_2O \left(\frac{18.02 \cancel{\text{g}}}{1 \cancel{\text{mol}}}\right)\left(\frac{1 \text{ kg}}{1000 \cancel{\text{g}}}\right)\\ &= 1.76 \times 10^{-2} \text{ kg } H_2O\end{aligned}$$

The molal concentration is then found.

$$\text{molality NaCl} = \frac{0.024 \text{ mol NaCl}}{1.76 \times 10^{-2} \text{ kg } H_2O} = 1.4\ m \text{ NaCl}$$

Some additional sample calculations are given in Examples 11.4 and 11.5.

Example 11.4 Conversion of Concentration Units

Hydrochloric acid is sold as a 36% aqueous solution. What is this concentration in (a) molality and (b) mole fraction?

Solution Writing the concentration as a fraction gives

$$\text{mass percent HCl} = \frac{36 \text{ g HCl}}{100 \text{ g solution}}$$

Since the quantity of solvent will be needed, we calculate it by difference.

$$\text{mass } H_2O = 100 \text{ g solution} - 36 \text{ g HCl} = 64 \text{ g } H_2O$$

(a) The steps involved in this conversion of units are shown in the flow diagram

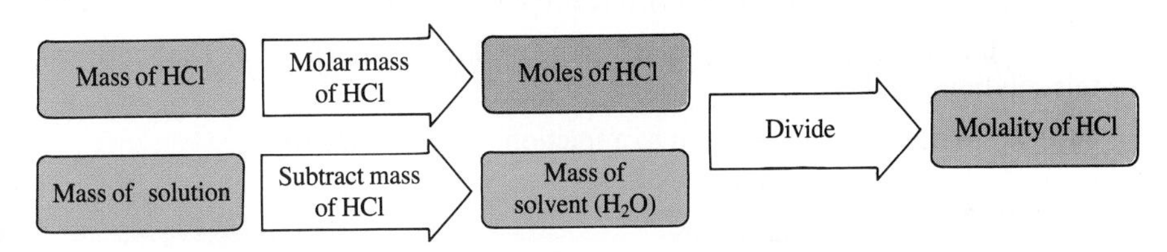

Molality expresses the quantity of solute in moles, so we must express the 36 g of HCl in moles, using its molar mass of 36.5 g/mol:

$$\text{moles HCl} = 36 \cancel{\text{g}} \text{ HCl} \left(\frac{1 \text{ mol}}{36.5 \cancel{\text{g}}}\right) = 0.99 \text{ mol HCl}$$

Since molality is the number of moles of solute per kilogram of solvent, we must express the quantity of water in kg:

$$\text{mass } H_2O = 64 \text{ g } H_2O\left(\frac{1 \text{ kg}}{1000 \text{ g}}\right) = 0.064 \text{ kg } H_2O$$

The conversion is completed by combining these two numbers.

$$\text{molality HCl} = \frac{0.99 \text{ mol HCl}}{0.064 \text{ kg } H_2O} = 15\ m \text{ HCl}$$

(b) When expressing the concentration as the mole fraction of HCl, the numerator is the number of moles of HCl, which has already been calculated in part (a). The denominator is the total number of moles of all substances in the solution. The flow diagram shows the steps in the calculations.

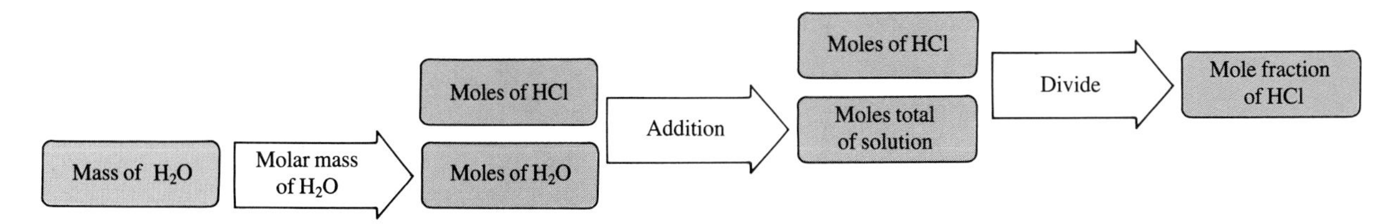

$$\text{mol total of solution} = \text{mol HCl} + \text{mol } H_2O$$

$$\text{mol total of solution} = 0.99 \text{ mol HCl} + 64 \text{ g } H_2O\left(\frac{1 \text{ mol}}{18.0 \text{ g}}\right)$$

$$\text{mol total of solution} = 0.99 + 3.56 = 4.55 \text{ mol solution}$$

The calculation is completed by finding the quotient,

$$\text{mole fraction HCl} = \frac{0.99 \text{ mol HCl}}{4.55 \text{ mol soln}} = 0.22$$

Understanding What are the (a) molality and (b) mole fraction of a 24.5% solution of ammonia (NH_3) in water?

Answer: (a) 19.0 m, (b) 0.256

The density is needed to convert between units of volume and mass.

For some of the conversions it is necessary to know the density of the solution. Density serves as a conversion factor between volume and mass. Any time the quantity of solution must be changed from volume to mass or *vice versa,* the density of the solution must be used in the calculation. This process is illustrated in the next example.

Example 11.5 Conversion of Concentration Units

What is the molarity of an aqueous solution that is 2.00 molal in HCl? The density of this solution is 1.034 g/mL.

Solution Write the concentration as a fraction, including the units.

$$\text{molality HCl} = \frac{2.00 \text{ mol HCl}}{1 \text{ kg solvent}}$$

We can choose a sample that contains exactly 1 kg of the solvent to simplify the calculation. The denominator in the molarity expression is the volume of solution in liters, so we must first find the total mass of the solution in grams, and then convert it into the desired volume in liters. In the chosen sample, the mass of solution is the sum of the masses of solvent and solute. Assuming the mass of the solvent is exactly 1 kg, we calculate

$$\text{total mass} = 1000.0 \text{ g solvent} + 2.00 \cancel{\text{mol}} \text{ HCl} \left(\frac{36.46 \text{ g}}{1 \cancel{\text{mol}}}\right)$$

$$= 1072.9 \text{ g solution}$$

The quantity of solution is expressed as a volume in molarity, so the density of the solution is used as a conversion factor.

$$\text{volume soln} = 1072.9 \cancel{\text{g}} \text{ soln} \left(\frac{1 \cancel{\text{mL}}}{1.034 \cancel{\text{g}}}\right)\left(\frac{1 \text{ L}}{1000 \cancel{\text{mL}}}\right)$$

$$= 1.038 \text{ L soln}$$

Now that the numerator and denominator of the desired concentration unit have been evaluated, all that remains is to find the ratio.

$$\text{molarity HCl} = \frac{2.00 \text{ mol HCl}}{1.038 \text{ L soln}} = 1.93\ M \text{ HCl}$$

Note that the molality and molarity of this solution are numerically quite close to each other. This is usually true for aqueous solutions except for very concentrated ones.

Understanding Find the molarity of a 3.10 molal aqueous ammonia solution that has a density of 0.977 g/mL.

Answer: 2.88 *M*

11.2 Principles of Solubility

Objectives

- To use solute-solvent interactions to predict the relative solubilities of substances in different solvents
- To explain how disorder is a driving force in the mixing of substances

Many chemical reactions do not proceed unless the reactants are dissolved in solution. Indeed, in many reactions the formation of insoluble products serves to cause the reaction to go to completion. Because of the major role that solutions play in chemical reactions, it is helpful to understand the factors that determine whether a given substance will dissolve in a solvent. The processes that occur as a solute dissolves are presented in this section. These processes can be divided into two general components: an enthalpy change due to differences in attractive forces between molecules, and a change in the disorder of the system.

For most solutes there is a limit to the quantity that can dissolve in a fixed volume of any given solvent. When a solid such as $CO(NH_2)_2$ (urea) is stirred with one liter of water, some of it dissolves to form a solution. If enough solute is present, we find that not all of it dissolves, but a maximum, constant concentration of the solution is obtained. Addition of more solute does not change this concentration further; the added solid simply remains undissolved. In this situation a state of dynamic equilibrium has been reached, as represented in Figure 11.2. Molecules or ions continue to enter the liquid phase from the solid, but other molecules or ions of the solute leave the solution at an equal rate, analogous to the solid-liquid phase equilibrium we examined in Chapter 10. The **solubility** is the concentration of solute that

Figure 11.2 Dissolving a solid The solubility of a solute is reached when the rate of dissolution and the rate of crystallization are equal. No net change in the concentration of the solute is observed, no matter how much excess solute is present (or how little, as long as there is some).

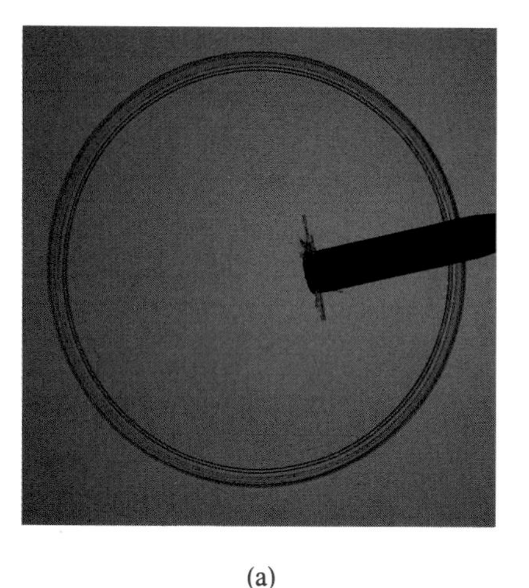

(a)

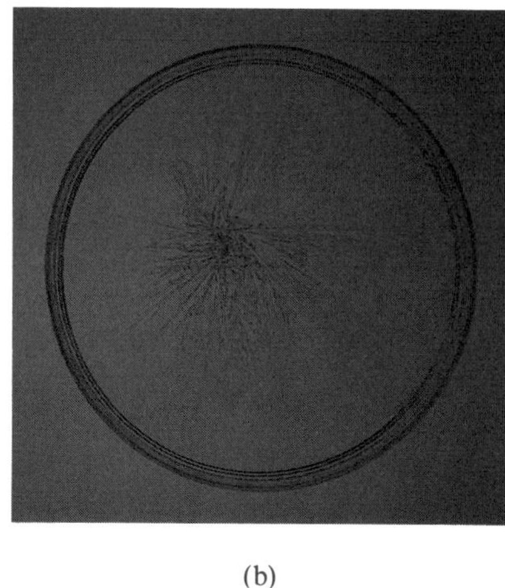

(b)

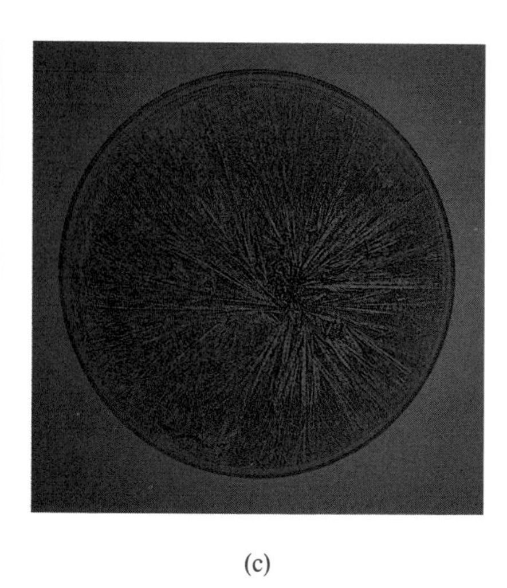

(c)

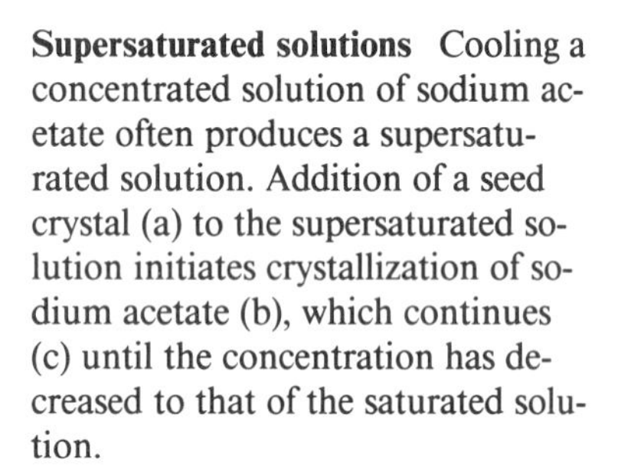

Supersaturated solutions Cooling a concentrated solution of sodium acetate often produces a supersaturated solution. Addition of a seed crystal (a) to the supersaturated solution initiates crystallization of sodium acetate (b), which continues (c) until the concentration has decreased to that of the saturated solution.

exists in equilibrium with an excess of that substance. A **saturated solution** is one that is in equilibrium with an excess of the solute. The concentration of a saturated solution is equal to the solubility. An **unsaturated solution** is one in which the concentration of the solute is less than its solubility.

Under some conditions it is possible to prepare a solution in which the concentration of solute is temporarily greater than its solubility. This is called a **supersaturated solution,** and it is an unstable condition that is analogous to the supercooled liquids discussed in Chapter 10.

The addition of a small quantity of solute to a solution is a simple way to distinguish among unsaturated, saturated, and supersaturated solutions. If the solution is unsaturated, the added solute dissolves, increasing the concentration of the solution. If the solution is saturated, addition of solute produces no change in the concentration of the solution (although the added

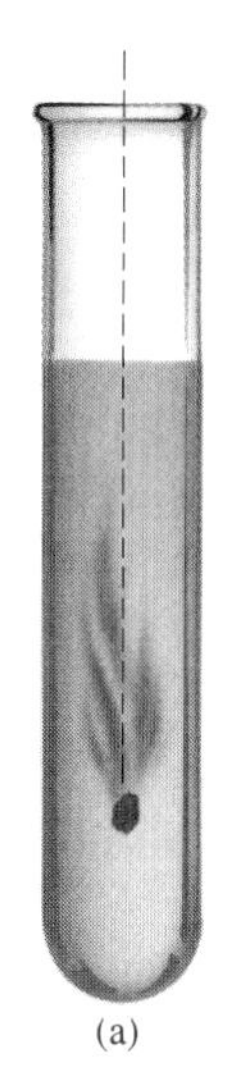

(a)

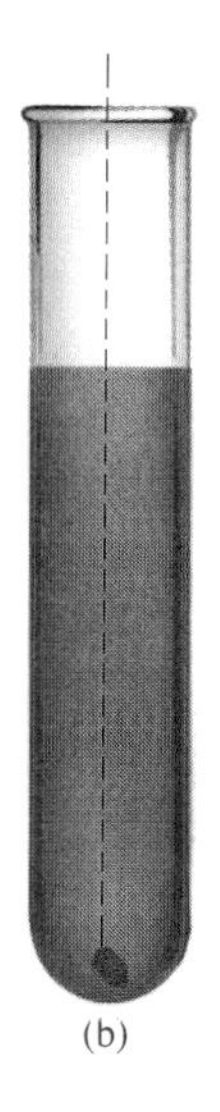

(b)

(c)

Figure 11.3 Solutions (a) When pure solute is added to an unsaturated solution it dissolves. (b) When solute is added to a saturated solution, no more solute dissolves. (c) When solute is added to a supersaturated solution, additional solid forms.

solute also participates in the dynamic equilibrium). When the solution is supersaturated, the addition of pure solute causes the rapid precipitation of additional solute. The precipitation of the solute continues until the solution's concentration has decreased to its equilibrium value—the solubility. These three situations are illustrated in Figure 11.3.

The Solution Process

The solution process is quite complex. Experience shows that some materials are very soluble in water, while others are quite insoluble. Sugar and alcohol readily dissolve in water, while sand and charcoal do not dissolve to any measurable extent. The solubilities of a single substance in different liquids also vary considerably. For example, grease will not dissolve in water, but kerosene can be used to remove grease stains. Figure 11.4 illustrates the difference between the solubilities of alcohol and motor oil in water.

Many factors contribute to solubilities, but two of these, the enthalpy change that accompanies solute–solvent attractions and the change in disorder, are the most important and can provide some insight into the general principles of solubility.

Solute–Solvent Interactions

Most spontaneous processes (those that proceed without outside forces) are accompanied by a decrease in the potential energy of the system. For example, a ball rolls down rather than up a hill. As the ball's height decreases, its potential energy decreases because of the gravitational attraction. In general, we expect that any change accompanied by a large decrease in enthalpy will be spontaneous. In considering whether a liquid and solid will form a solution, we find that the change in enthalpy arises mainly from the changes in the intermolecular attractions.

There are three types of intermolecular forces that are involved in the formation of condensed phase solutions: solute–solute, solvent–solvent, and solute–solvent interactions. In Chapter 10 we described the nature of the intermolecular forces, and emphasized the attractions among molecules of the same substance. The same forces also cause molecules of different

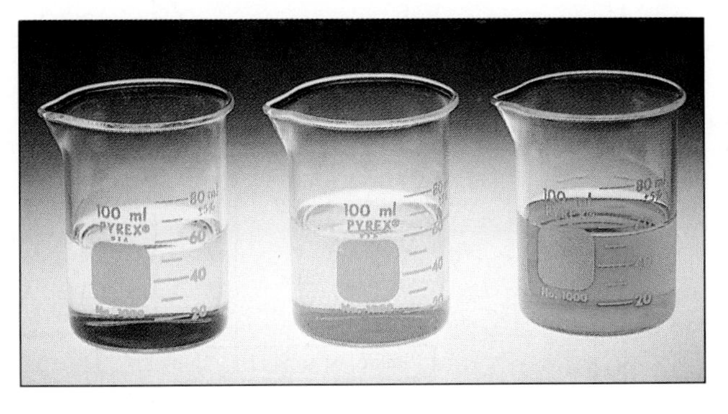

(a)

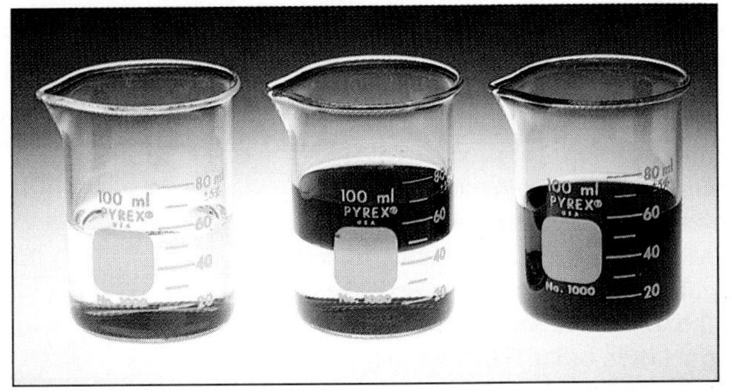

(b)

Figure 11.4 Liquid–liquid solubilities (a) Water and alcohol mix in all proportions. (b) Water and motor oil (dye added for clarity) to any great extent do not dissolve in each other.

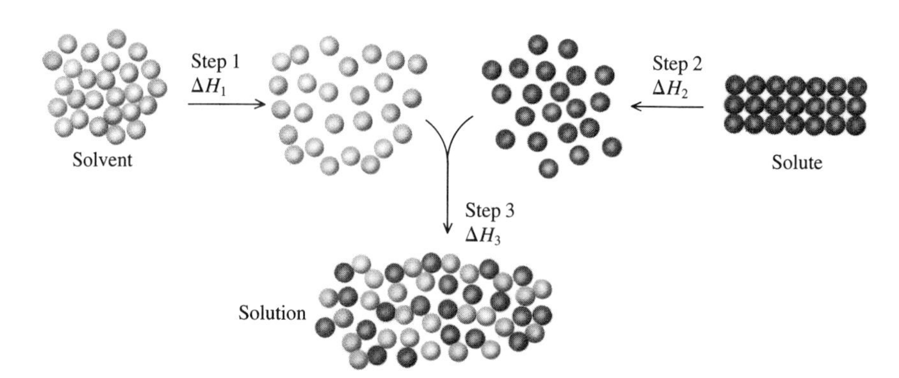

Figure 11.5 Contributions to dissolution A pictorial representation of the factors that contribute to the enthalpy of a solution. Step 1: The solvent molecules move apart. Step 2: The solute molecules are separated. Step 3: The solute and solvent molecules mix. The enthalpy changes in steps 1 and 2 are both endothermic, while the enthalpy change in step 3 is exothermic.

substances to attract each other. Not only do water molecules exert attractive forces on other water molecules, but they also attract molecules of other substances with which they are mixed, such as alcohol. The relative strengths of these different attractive forces play an important role in determining whether or not two substances will form a solution.

The relative strengths of solute-solute, solvent-solvent, and solute-solvent attractions are important in determining if substances will dissolve.

When a solution forms, the molecules of the solvent must move apart to accommodate the solute molecules. Since the solvent molecules attract each other, energy must be expended to separate them. The solute molecules must also be separated from each other to enter the solution, and this process is also endothermic. Because the solvent and solute molecules exert attractive forces on each other, energy will be released when they are brought together. These three steps are shown in Figure 11.5, and the energy changes are graphed on the energy level diagrams of Figure 11.6. The overall enthalpy change that accompanies the formation of the solution, called the *heat of solution,* is simply the sum of the three individual energy changes.

$$\Delta H_{\text{soln}} = \Delta H_1 + \Delta H_2 + \Delta H_3$$

Both endothermic and exothermic heats of solution are observed, as shown in Figure 11.6. When sulfuric acid dissolves in water, the enthalpy

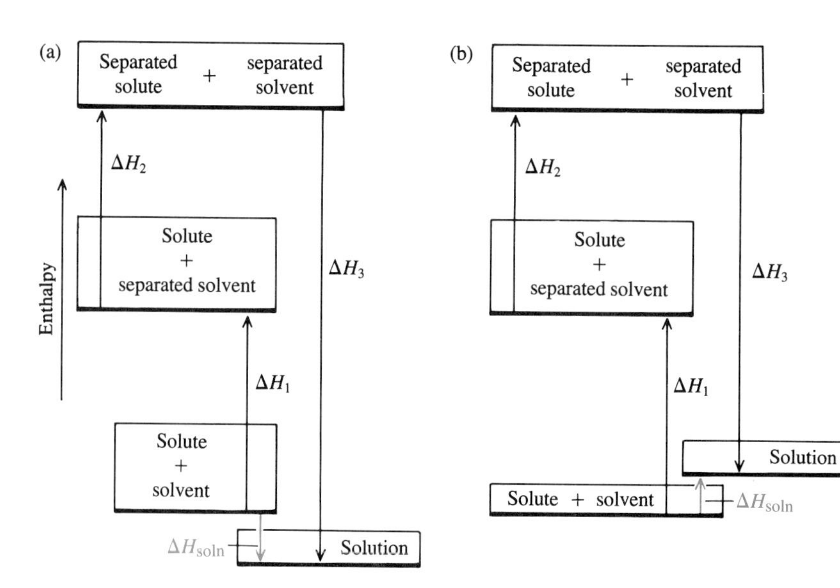

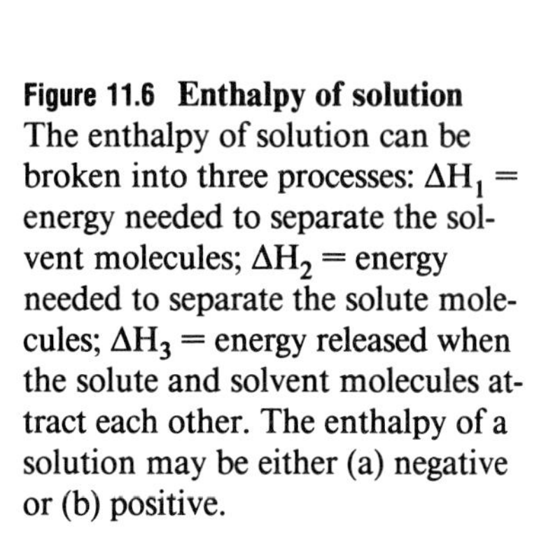

Figure 11.6 Enthalpy of solution The enthalpy of solution can be broken into three processes: ΔH_1 = energy needed to separate the solvent molecules; ΔH_2 = energy needed to separate the solute molecules; ΔH_3 = energy released when the solute and solvent molecules attract each other. The enthalpy of a solution may be either (a) negative or (b) positive.

change is −74.3 kJ mol^{-1}. So much heat is released that spattering can occur when small amounts of water boil. In this case, the energy that is released by the exothermic solute–solvent interactions is greater than that absorbed by the endothermic processes of separating the solvent and solute molecules. In general, substances that have similar properties and thus have similar intermolecular forces will have strong solute–solvent interactions and will tend to form solutions. The statement "like dissolves like" is a simplification that is often used to explain trends in solubility.

Substances in which the intermolecular forces are similar tend to form solutions.

Spontaneity

As noted above, a decrease in enthalpy is an important factor in causing a spontaneous change to occur. Yet many soluble substances dissolve spontaneously, even though the enthalpy of solution is endothermic. There must be another driving force that causes such changes to occur in spite of the unfavorable enthalpy change. To understand how such a process can be spontaneous, let us consider the formation of a solution of two gases from the pure substances.

In Chapter 5 we saw that gases mix spontaneously with each other (diffusion), a process that involves essentially no change in enthalpy. This mixing is illustrated in Figure 11.7. Suppose a container is divided by a removable partition into two compartments of equal volume. On one side of the partition is pure oxygen, and the other compartment contains pure nitrogen, with both gases at the same pressure and temperature. Removing the partition allows the molecules of the gases to mix, and in a relatively short time a uniform solution of the two gases occupies the entire volume. No matter how long we wait, we would not expect the two gases to separate spontaneously into the original arrangement.

Before we removed the barrier, all of the oxygen molecules were confined in the volume to the right of the partition, and the nitrogen molecules were to the left. We could be sure of getting an oxygen molecule if we took any molecule at random from the right compartment. We can no longer be certain of this after the partition has been removed—we might pick either a nitrogen or an oxygen molecule from this half of the container. By removing

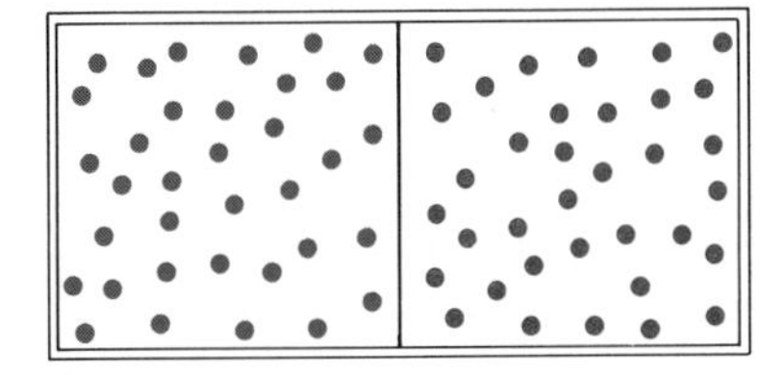

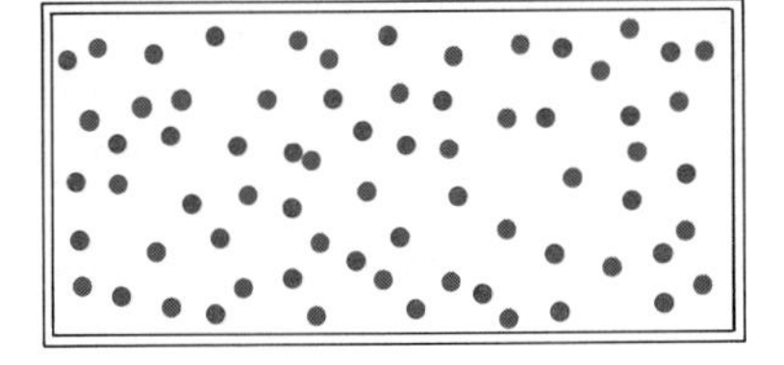

Figure 11.7 Mixing of gases (a) Oxygen and nitrogen gases at the same pressure and temperature are separated by a partition. (b) When the partition is removed there is no change in enthalpy, but the gases mix completely because of the natural tendency toward an increase in disorder.

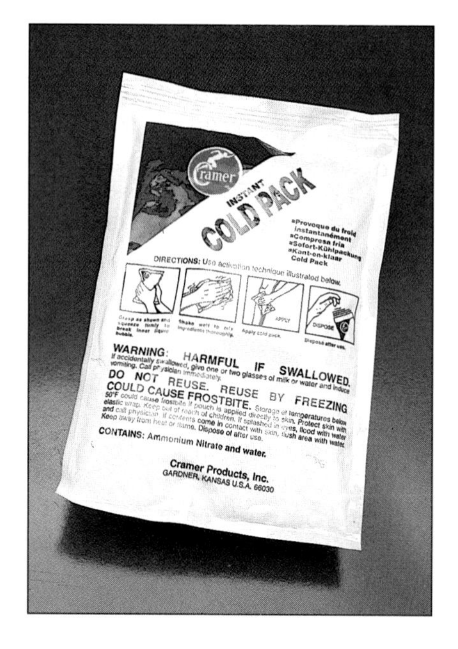

Figure 11.8 Chemical cold packs Cold packs, like the one shown, are used to treat athletic injuries. Squeezing the plastic bag allows ammonium nitrate to mix with water, forming a solution. The endothermic heat of solution produces temperatures below the freezing point of water.

the barrier, we have gone from a more orderly arrangement to a less orderly one. Mixing of gases is one example of a very important principle of nature: *Processes in which disorder increases tend to occur spontaneously.* This natural tendency toward disorder is one of the main driving forces in the formation of solutions. This principle is so important that it is presented in a more formal way in Chapter 15.

An increase in disorder is an important driving force in the formation of solutions.

An increase in disorder can be used to explain the fact that ammonium nitrate will dissolve in water even though the enthalpy of solution is sufficiently endothermic (+26.4 kJ mol^{-1}) that it is used in cold packs (see Figure 11.8). Although the enthalpy change is endothermic in this case, the ammonium nitrate is soluble because there is an increase in disorder when the solution forms. Before dissolving, the ammonium ions and nitrate ions are in the very highly ordered solid state. Once in solution, these ions are free to move independently of one another through the entire volume of the solution. This increase in disorder is more than enough to compensate for the unfavorable change in enthalpy.

Solubility of Molecular Compounds

Hexane (C_6H_{14}) and heptane (C_7H_{16}) are two liquid hydrocarbons. In both of these compounds the intermolecular attractions are the weak London dispersion forces. The same dispersion forces also cause hexane and heptane molecules to attract each other. Since all three of these attractions are very close in energy, it is not surprising that the two liquids mix in any proportion. There is little change in the energy of the attractions, so the increase in disorder upon mixing becomes the controlling factor in the dissolution process.

Now consider the mixing of water with hexane. The intermolecular attractions among the water molecules are dominated by very strong hydrogen bonding interactions. The interactions between water molecules and hexane molecules are the much weaker London dispersion forces. Therefore, the energy needed to break the hydrogen bonding interactions in the dissolution process is much greater than the energy released when the hexane and water molecules attract each other. In this case the increase in the disorder of the mixture is not sufficient to overcome the unfavorable enthalpy change, so very low solubility results. The observed solubility of hexane in water is only 0.14 g per liter.

Consideration of the kinds of intermolecular attractions provides a general guide to relative solubilities, but it does not help us make quantitative predictions with any reliability. Nevertheless, such considerations do allow us to predict the relative solubilities of a substance in two different solvents, as illustrated in Example 11.6.

Example 11.6 Relative Solubilities

Predict the solvent in which the given compound is more soluble:

(a) carbon tetrachloride (CCl_4) in water or hexane (C_6H_{14})
(b) urea ($CO(NH_2)_2$) in water or carbon tetrachloride
(c) iodine (I_2) in benzene (C_6H_6) or water.

Solution

(a) The intermolecular attractions between carbon tetrachloride molecules are dispersion forces, since these molecules are nonpolar. These forces also dominate in hexane, but in water the most important intermolecular forces are the much stronger hydrogen bonding attractions. In both cases the only solute–solvent attractions can be dispersion forces, so the carbon tetrachloride will have a higher solubility in hexane than in water.
(b) In urea the presence of hydrogen atoms that are bonded to the small electronegative nitrogen atoms means that this substance can participate in hydrogen bonding. Therefore, both water and urea can participate in hydrogen bonding, while carbon tetrachloride interacts with other molecules only by the weaker dispersion forces. We expect urea to be more soluble in water than in carbon tetrachloride.
(c) Of the three compounds involved, only water is capable of forming hydrogen bonds. Therefore, iodine should be more soluble in benzene than in water.

Understanding Is methyl alcohol (CH_3OH) more soluble in water or hexane?

Answer: water

Solubility of Ionic Compounds in Water

Water is the most common solvent used to dissolve ionic compounds. The enthalpy changes that occur in the formation of aqueous solutions are an important factor in determining the solubility of ionic substances. If the compound is soluble, the enthalpy of attraction between the ions in the solid must be comparable (within about 50 kJ/mol) to the enthalpy of attractions between the water molecules and the ions in the solution. The forces that hold the ionic solid together arise from the electrostatic attraction between the oppositely charged ions, which are very strong, with energies of 400 kJ/mol or more.

In solution the polar water molecules are attracted by the charges of the ions as shown in Figure 11.9. Several water molecules will be attracted to each ion in solution. The cations attract the negative end of the water dipole, while the positive ends of the water dipoles are drawn to the anions. Experiments indicate that the number of water molecules that surround each ion is between four and ten. The interaction of the ions with the water molecules is called **hydration.** Since many ionic compounds are soluble, we can conclude that enthalpies of hydration must be about the same as the crystal lattice enthalpies in the solid compounds.

When ionic substances dissolve in water, the increase in the disorder of the solute is obvious, since the ions become free to move about. When the water molecules are separated to make room for the ions, there is also an increase in disorder. At the same time, however, hydration of the ions restricts the freedom of some of the solvent molecules, decreasing their disorder. Thus depending on the particular solute and its hydration by the water, the disorder of the solvent might either increase or decrease.

The solubilities of ionic substances are difficult to model; solubilities must be determined by experiment.

As we have seen in this section, there are many factors that enter into the dissolution process: the change in disorder and the strengths of intermolecular attractions in the pure substances, as well as those in the mixture. In view of the complexity of the process, it is not surprising that solubilities are hard to predict, and we must rely on solubility rules such as those given in Chapter 4.

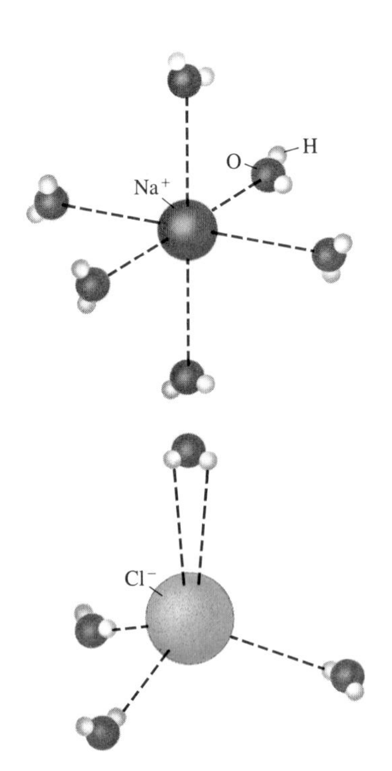

Figure 11.9 Hydration of ions The polar water molecules are attracted to the ions in solution. The positive ends of the water dipole are attracted to the anions, while the cations attract the negative end of the water dipole.

11.3 Effect of Temperature and Pressure on Solubility

Objectives

- To determine the effect of pressure and temperature on solubility
- To relate the sign of the enthalpy of solution to the increase or decrease of solubility with temperature
- To calculate the solubility of gases using Henry's law
- To explain why the solubilities of solids and liquids do not change appreciably with changes in pressure

In the previous section we saw how the characteristics of the solute–solvent interactions influence the solubilities of substances. The solubilities of compounds also depend upon the temperature and pressure. The change of solubility with temperature is used in the purification of many substances, and the effects of temperature and pressure on the solubilities of gases are important to marine organisms that depend on dissolved oxygen. In this section the effect of the temperature and pressure on solubilities is examined.

Effect of Temperature on Solubility

We can prepare a saturated solution by stirring the solvent with an excess of a solid solute. After some time the concentration of the solute no longer increases, and we have a saturated solution. This system is in dynamic equilibrium, with molecules or ions constantly leaving and joining the solid at equal rates. This process can be represented as

$$(\text{solute})_{\text{solid}} \rightleftharpoons (\text{solute})_{\text{solution}}$$

Once equilibrium is established there is no observable change in the concentration of the solution, as long as the temperature and pressure are not changed. The principle of Le Chatelier applies to any system in dynamic equilibrium, so we will use it to establish the influence of temperature on the solubility of a compound.

Heat must be added to a sample of matter to raise its temperature. The addition of heat to a saturated solution in equilibrium with an excess of the solute represents a change to the system. According to the principle of Le Chatelier, a net reaction will occur in the direction that will undo the change. Consider first a substance that has an endothermic enthalpy of solution. Heat is absorbed as the solute dissolves, so it is a "reactant" in the equation.

$$\text{heat} + (\text{solute})_{\text{solid}} \rightleftharpoons (\text{solute})_{\text{solution}}$$

When we add heat to the mixture, the reaction must occur in the forward direction to remove heat, so the quantity of solute in solution will be greater when equilibrium is re-established. In other words, when the enthalpy of solution is positive, the solubility of the solute increases with temperature.

A compound that has a negative enthalpy of solution releases heat in the process, as represented in the equilibrium equation

$$(\text{solute})_{\text{solid}} \rightleftharpoons (\text{solute})_{\text{solution}} + \text{heat}$$

When we add heat to this mixture, the reaction occurs in the reverse direction, reducing the equilibrium concentration of the solute. To summarize, an increase in temperature causes the solubility of a substance to increase if the enthalpy of solution is endothermic and causes the solubility to decrease when the enthalpy of solution is exothermic.

The solubilities of most solids increase as the temperature of the solution increases. The graph in Figure 11.10 shows the solubilities of several ionic compounds as a function of temperature. In general, the more endothermic the enthalpy of solution, the greater is the change in molar solubility with temperature. Note that the solubility of cerium(III) sulfate is seen to *decrease* as the temperature increases, consistent with its exothermic enthalpy of solution.

Unlike solids, the solubilities of nearly all gases in liquids decrease with an increase in temperature, showing that the enthalpy of solution is exothermic. In the gas phase there is almost no energy of attraction between the molecules. There are, however, attractions between the solvent and the solute molecules that result in a negative enthalpy of solution. When tap water is heated, the bubbles that first form are gases from the atmosphere that were dissolved. The bubbles appear because the solubilities of the gases decrease as the temperature is raised. One effect caused by thermal pollution of lakes and streams from factories and power plants is the decrease in the solubilities of gases. The increased temperature of the water can reduce the concentration of dissolved oxygen to a level at which fish cannot survive.

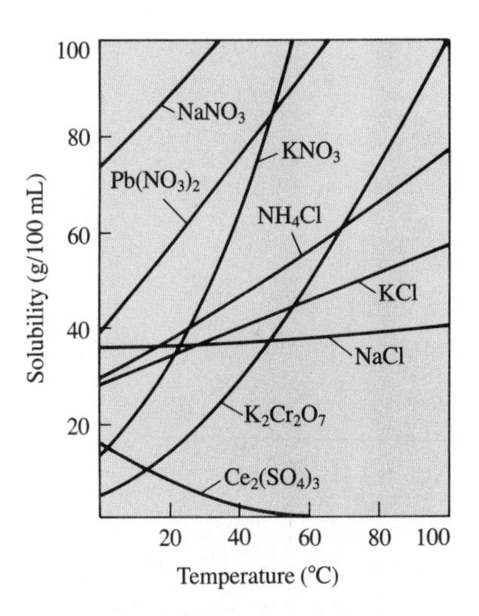

Figure 11.10 Temperature dependence of solubility The solubilities of several ionic compounds in water are shown as a function of temperature. Most ionic compounds have increased solubility at higher temperatures.

As the temperature increases, the solubility increases for any substance with an endothermic enthalpy of solution and decreases for one with an exothermic enthalpy of solution.

The solubilities of gases decrease with increasing temperature.

Effect of Pressure on Solubility

The solubility of a gas in any liquid is very sensitive to the pressure; this is easily understood by applying the principle of Le Chatelier. When we increase the pressure of a gas in contact with a saturated solution, a change

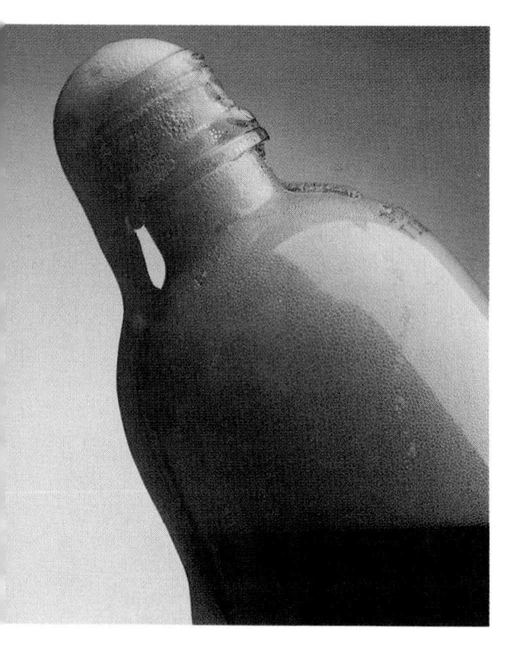

When a warm bottle of carbonated beverage is opened, foaming occurs as the gas escapes from the solution because of the abrupt decrease in CO_2 pressure.

The solubility of a gas in a liquid is directly proportional to its partial pressure.

Table 11.2 Henry's Law Constants for Several Gases in Water

	k (molal/atm)			
Gas	**0 °C**	**20 °C**	**40 °C**	**60 °C**
carbon dioxide	7.60×10^{-2}	3.91×10^{-2}	2.44×10^{-2}	1.63×10^{-2}
ethylene	1.14×10^{-2}	5.60×10^{-3}	3.43×10^{-3}	—
helium	4.22×10^{-4}	3.87×10^{-4}	3.87×10^{-4}	4.10×10^{-4}
nitrogen	1.03×10^{-3}	7.34×10^{-4}	5.55×10^{-4}	4.85×10^{-4}
oxygen	2.21×10^{-3}	1.43×10^{-3}	1.02×10^{-3}	8.71×10^{-4}

must take place that will reduce the pressure of the gas. More gas molecules enter the liquid, thus decreasing the number of moles of gas in the container and increasing the concentration of the gas in solution. By decreasing the amount of gas, the pressure drops—the response predicted by the principle of Le Chatelier. Carbonated beverages are sealed under a high pressure of carbon dioxide. When the container is opened, the sudden drop in pressure causes bubbles of gas to form because of the lower solubility of CO_2 under the new conditions.

At pressures of a few atmospheres or less, it is found that the solubilities of gases obey **Henry's law**—the solubility of a gas is directly proportional to its partial pressure at any given temperature.

$$C = kP \qquad [11.1]$$

Here C is the concentration of the gaseous substance in solution; k is a proportionality constant that is characteristic of the particular solute, solvent, and temperature; and P is the partial pressure of the gaseous solute in contact with the solution. The units of the constant, k, depend upon the units used to express the concentration and the pressure.

Henry's law constants are given for several gases in Table 11.2, with the units of molal/atm. In the table the expected decrease in solubility with increasing temperature is observed (the Henry's law constant for a given system decreases as the temperature increases). These experimentally determined constants are used to calculate the solubilities of gases, as illustrated in Example 11.7.

Example 11.7 Henry's Law Calculation

What is the molal concentration of oxygen in water at 20 °C that has been saturated with air at 1.00 atm? Assume that the mole fraction of oxygen in air is 0.21, and that of nitrogen is 0.79.

Solution Before applying Henry's law we must find the partial pressure of the oxygen in the gas phase, which is simply the mole fraction of oxygen (0.21) times the total pressure (Dalton's law of partial pressure):

$$P_{\text{oxygen}} = 0.21 \times 1.00 \text{ atm} = 0.21 \text{ atm}$$

The Henry's law constant for oxygen at 20 °C (Table 11.2) is 1.43×10^{-3} molal/atm. Using this constant and the partial pressure of oxygen in Equation 11.1, we calculate the concentration in solution.

$$C = kP = \left(1.43 \times 10^{-3} \frac{\text{molal}}{\text{atm}}\right)(0.21 \text{ atm}) = 3.0 \times 10^{-4} \text{ molal}$$

Understanding What is the molal concentration of nitrogen in this same solution?

Answer: 5.8×10^{-4} molal

Unlike gases, the solubilities of liquids and solids in liquids change very little with pressure. For gaseous solutes an increase in pressure is relieved by additional gas dissolving in the liquid, greatly reducing the volume occupied by the system. In terms of Le Chatelier's principle, the equilibrium is shifted toward the smaller volume. When a liquid or solid dissolves in the solvent, there is very little difference between the volume occupied by the solution and the sum of the volumes of the pure substances in the mixture. Thus it requires very large changes in pressure (many atmospheres) to produce even small changes in the solubilities of liquids and solids in liquids.

Pressure has very little effect on the solubilities of solids and liquids.

11.4 Colligative Properties of Solutions

Objectives

- To define and identify the colligative properties of solutions
- To relate the colligative properties to the concentrations of solutions
- To calculate the molar masses of solutes from measurements of colligative properties

Dissolving a solute in a pure liquid changes nearly all of the physical properties of the liquid. **Colligative properties** are those properties of solutions that change in proportion to the concentration of solute particles. They do not depend on the nature of the solute particles, but only on their concentrations. Colligative properties are a means of "counting" the solute particles present in the sample, and give us a way of determining the molar mass of the solute. In this section we examine the origin and uses of several of these colligative properties.

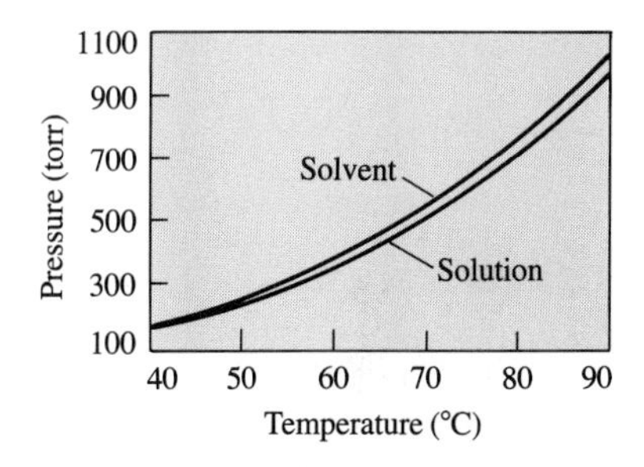

Figure 11.11 Vapor pressure of solutions as a function of temperature The vapor pressure of benzene (upper curve) and that of a benzene solution of a nonvolatile solute (lower curve) are shown. The vapor pressure of the solvent in a solution is always lower than that of the pure solvent at all temperatures.

Vapor Pressure of the Solvent

In Chapter 10 we saw that the equilibrium between a liquid and its vapor gives rise to a characteristic vapor pressure for each substance that depends on the temperature. Experimentally we find that the addition of a nonvolatile solute (one with a very low vapor pressure of its own) to a solvent always reduces the equilibrium vapor pressure below that of the pure solvent. This means that the rate of evaporation of the solvent is reduced and equilibrium is reached with a smaller concentration (partial pressure) of the solvent in the gas phase.

Figure 11.11 shows the temperature dependence of vapor pressure of a pure solvent and that of the solvent when it contains a nonvolatile solute. The vapor pressure of a solution is expressed quantitatively by **Raoult's law:** the vapor pressure (P) of the solvent above a dilute solution is equal to the mole fraction of the solvent (χ) times the vapor pressure of the pure solvent ($P°$):

Raoult's law: The partial pressure of a substance in equilibrium with a solution is equal to its mole fraction in the solution times the vapor pressure of the pure substance.

$$P_{solv} = \chi_{solv} P^{\circ}_{solv} \quad [11.2]$$

By substituting $(1 - \chi_{solute})$ into Equation 11.2 in place of χ_{solv} and rearranging, we see that the *difference* between the vapor pressure of the pure solvent and the vapor pressure of the solution (ΔP) is proportional to the concentration of the solute. In equation form this is

$$\Delta P = P^{\circ}_{solv} - P_{solv} = \chi_{solute} P^{\circ}_{solv} \quad [11.3]$$

For vapor pressure lowering and all other colligative properties, the effect is proportional to the concentration of the solute particles.

Note that Equation 11.3 contains the mole fraction of the *solute,* while in Equation 11.2 the composition of the solution is expressed by the mole fraction of the *solvent.* The lowering of the vapor pressure of the solvent is a colligative property, since it is proportional to the concentration of the solute. Raoult's law applies strictly only to solutions with low concentrations of the solute.

When a nonvolatile solute is added, the concentration of solvent molecules decreases, so the rate of evaporation also decreases (Figure 11.12). In addition, however, there are other thermodynamic considerations beyond the scope of this text that also enter into the lowering of the vapor pressure by solute particles. Raoult's law is used to calculate the vapor pressure above a solution of known concentration, as illustrated in Example 11.8.

Example 11.8 Calculating the Vapor Pressure Above a Solution

Find the partial pressure of water at 25 °C in equilibrium with a 6.0% solution of urea ($CO(NH_2)_2$).

Solution Before we can use Raoult's law to solve this problem, we must first express the concentration of the solute (or solvent) as the mole fraction. A 100.0-g sample of a 6.0% solution of urea would consist of 6.0 g of urea in 94.0 g of water. We first convert these quantities into moles, and then calculate their mole fractions.

$$\text{moles urea} = 6.0\ \text{g urea}\left(\frac{1\ \text{mol}}{60.0\ \text{g}}\right) = 0.10\ \text{mol urea}$$

$$\text{moles water} = 94.0\ \text{g water}\left(\frac{1\ \text{mol}}{18.02\ \text{g}}\right) = 5.22\ \text{mol water}$$

$$\text{moles total of solution} = 0.10 + 5.22 = 5.32\ \text{mol total of solution}$$

$$\chi_{urea} = \frac{0.10\ \text{mol urea}}{5.32\ \text{mol soln}} = 0.019\ \text{urea}$$

$$\chi_{water} = \frac{5.22\ \text{mol water}}{5.32\ \text{mol soln}} = 0.981\ \text{water}$$

The vapor pressure of pure water at 25 °C is 23.77 torr (Appendix E). By substituting the mole fraction of water and its vapor pressure into Equation 11.2, we calculate the vapor pressure above the solution:

$$P_{solv} = \chi_{solv} P^{\circ}_{solv}$$

$$P_{solv} = 0.981 \times 23.77\ \text{torr} = 23.3\ \text{torr}$$

Alternatively, we could have used Equation 11.3 to find the amount by which the vapor pressure is lowered, and subtracted it from the vapor pressure of the pure solvent:

$$\Delta P = \chi_{solute} P^{\circ}_{solv} = 0.019 \times 23.77 \text{ torr} = 0.45 \text{ torr}$$

$$P_{solv} = 23.77 - 0.45 \text{ torr} = 23.32 \text{ torr}$$

As expected, both equations lead to the same result.

Understanding Find the vapor pressure of water above a solution that is 9.0% urea. The temperature is 35 °C, where the vapor pressure of pure water is 42.18 torr.

Answer: 41.0 torr

More commonly, Raoult's law is used to find the mole fraction of solute in a solution of unknown concentration, as shown in Example 11.9.

Example 11.9 Determining the Concentration of Solute by Lowering of Vapor Pressure

At 25 °C the vapor pressure of pure benzene is 93.9 torr. A solution of a nonvolatile solute in benzene has a vapor pressure of 91.5 torr at the same temperature. What is the concentration of the solute, expressed as its mole fraction?

Solution From the data given, the vapor pressure lowering is

$$\Delta P = 93.9 - 91.5 = 2.4 \text{ torr}$$

Substitute this into Raoult's law (Equation 11.3), along with the vapor pressure of the solvent, and solve for the mole fraction of solute.

$$\Delta P = \chi_{solute} P^{\circ}_{solv}$$

$$2.4 \text{ torr} = \chi_{solute}(93.9 \text{ torr})$$

$$\chi_{solute} = \frac{2.4 \text{ torr}}{93.9 \text{ torr}} = 0.026$$

Understanding What is the solute concentration in a benzene solution that has a vapor pressure of 90.6 torr at 25 °C?

Answer: $\chi_{solute} = 0.035$

Very accurate measurements of vapor pressures are required, if accurate concentrations are needed. However, these measurements are difficult to make accurately, so the lowering of vapor pressure is not widely used to determine concentrations. Two other colligative properties—boiling point elevation and freezing point depression—are closely related to the lowering of vapor pressure and are much easier to measure accurately.

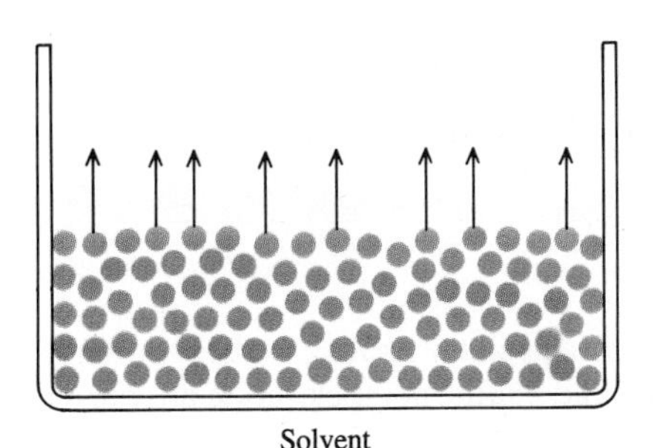

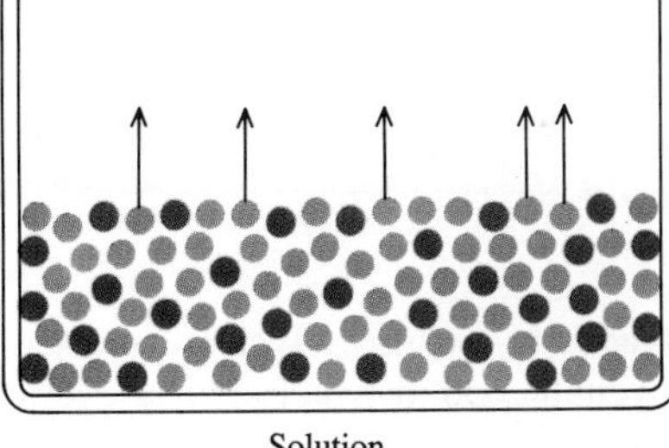

Figure 11.12 Rate of evaporation from solution When a solution is formed, the presence of the solute particles lowers the rate of evaporation of the solvent, thus lowering the equilibrium vapor pressure.

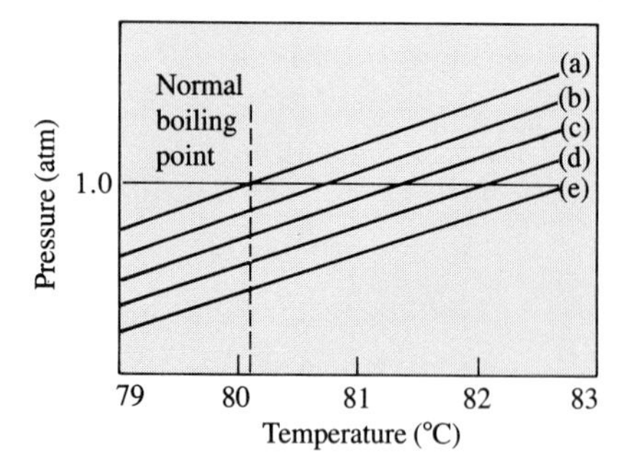

Figure 11.13 The vapor pressure of solutions (a) The vapor pressure of benzene as a function of temperature near its boiling point. The other lines are the vapor pressures of benzene solutions that contain a nonvolatile solute with mole fractions of (b) 0.02, (c) 0.04, (d) 0.06, and (e) 0.08. The intersection of each of these lines with the horizontal line at 1 atmosphere locates the normal boiling points of the solutions.

Boiling Point Elevation

Figure 11.13 shows the vapor pressures of a solvent and several dilute solutions near the normal boiling point of the solvent. At a pressure of one atmosphere (760 torr) the boiling points of the solutions are all higher than that of the pure solvent. Furthermore, the graph in Figure 11.14 shows that the difference between the boiling point of each solution and that of the solvent is proportional to the concentration of the solute particles. When using the **elevation of the boiling point** as a colligative property, the concentration of the solute is usually expressed as molality rather than as mole fraction of the solute. As long as the solution is fairly dilute, molality is proportional to the mole fraction of solute. Expressing the concentration as molality makes it easier to calculate the molar mass of the solute from boiling point elevation data. The effect of the solute concentration on the boiling point can be written in equation form

$$\Delta T_b = (\text{b.p.}_{\text{solution}} - \text{b.p.}_{\text{solvent}}) = mk_b \qquad [11.4]$$

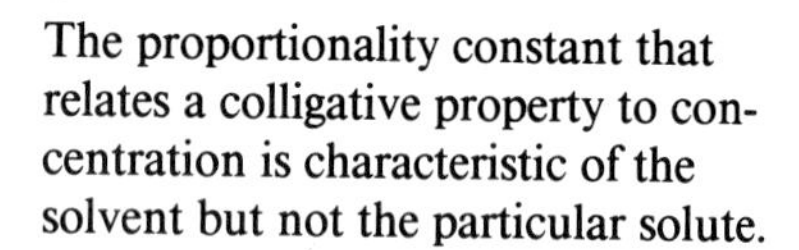

The proportionality constant that relates a colligative property to concentration is characteristic of the solvent but not the particular solute.

where ΔT_b is the elevation of the boiling point, m is the molal concentration of the solute, and k_b is the *boiling point elevation constant,* which depends only on the solvent used. Values of this constant for several solvents are given in Table 11.3.

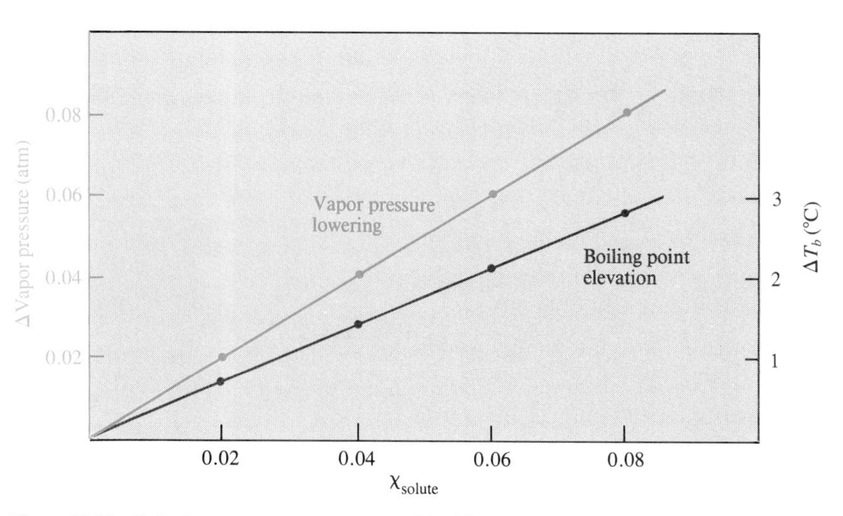

Figure 11.14 Solution vapor pressures and boiling points As long as the solutions are dilute, both the vapor pressure lowering and boiling point elevation are proportional to the concentration.

Table 11.3 Boiling Point Elevation and Freezing Point Depression Constants for Solvents

Solvent	Boiling Point (°C)	k_b (°C/m)	Freezing Point (°C)	k_f (°C/m)
acetic acid	117.90	3.07	16.60	3.90
benzene	80.10	2.53	5.51	4.90
naphthalene	—	—	80.2	6.8
water	100.0	0.512	0.00	1.86

Example 11.10 Calculating the Boiling Point of a Solution

What is the boiling point of a 0.32-molal solution of iodine (I_2) in benzene?

Solution Use the value of k_b given for benzene in Table 11.3 to find the change of the boiling point due to the presence of the solute.

$$\Delta T_b = mk_b$$

$$\Delta T_b = 0.32\ \cancel{m} \times 2.53\ °\text{C}/\cancel{m} = 0.81\ °\text{C}$$

The boiling point of the solution is 0.81 °C higher than that of the pure solvent, so the boiling point of the solution is found by adding ΔT_b to the normal boiling point of the pure solvent.

$$T_b = 80.10 + 0.81 = 80.91\ °\text{C}$$

Understanding What is the boiling point of a 0.60 molal solution of sucrose in water?

Answer: 100.31 °C

Measurements of the boiling point elevation are occasionally used to determine the molar masses of solutes. A solution is prepared that contains accurately known masses of both the solute and solvent. The boiling point elevation is measured experimentally and used to calculate the number of moles of solute in the sample. This number of moles is combined with the sample's mass to find the molar mass of the solute. This calculation is illustrated in Example 11.11.

Example 11.11 Determination of Molar Mass by Boiling Point Elevation of a Solution

A solution is prepared by dissolving 1.00 g of a nonvolatile solute in 15.0 g of acetic acid. The boiling point of this solution is 120.17 °C. Find the molar mass of the solute.

Solution From the boiling point of the solution, and the information given in Table 11.3 for the solvent, acetic acid, the molality can be calculated. The change in the boiling point is

$$\Delta T_b = 120.17 - 117.90 = 2.27\ °\text{C}$$

The molal concentration of the solute is found with Equation 11.4, using the boiling point elevation constant given in Table 11.3.

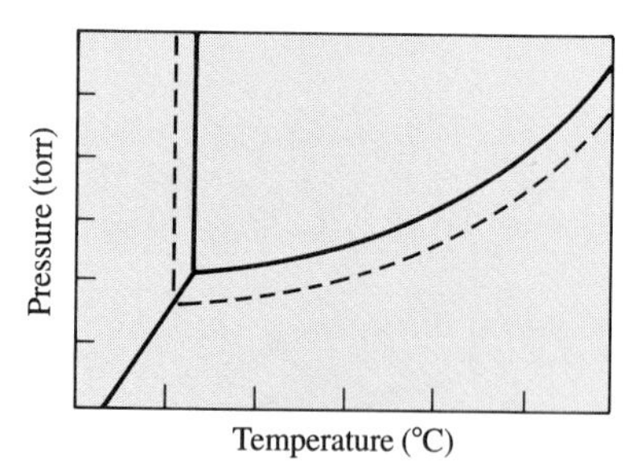

Figure 11.15 Freezing point of a solution The phase diagram shows that the lowering of the vapor pressure of a solution decreases the triple point temperature. As a result, the freezing point of the solution is also lowered. The solid lines show the behavior of the pure solvent, and the dotted lines show the vapor pressure and freezing point of a solution.

$$\Delta T_b = mk_b$$

$$m = \frac{2.27\ °\text{C}}{3.07\ °\text{C/molal}} = 0.739 \text{ molal}$$

The solution contains 0.739 mol of solute for each kilogram of solvent. Since the solution was prepared using 15.0 g of the solvent, we use the molal concentration of the solution as a conversion factor to find the amount of solute present in our sample.

$$\text{moles solute} = 15.0 \text{ g solvent}\left(\frac{1 \text{ kg}}{1000 \text{ g}}\right)\left(\frac{0.739 \text{ mol solute}}{1 \text{ kg solvent}}\right)$$

$$= 1.11 \times 10^{-2} \text{ mol solute}$$

The molar mass is now calculated from the known mass of the sample (1.00 g) and the moles of solute determined in the boiling point elevation experiment.

$$\text{molar mass} = \frac{1.00 \text{ g}}{1.11 \times 10^{-2} \text{ mol}} = 90.1 \text{ g/mol}$$

Understanding Find the molar mass of a nonvolatile solute, if a solution of 1.20 g of the compound dissolved in 20.0 g of benzene has a boiling point of 80.94 °C.

Answer: 1.8×10^2 g/mol

Freezing Point Depression

For boiling point elevation and freezing point depression, the solute concentration is expressed in molality.

The freezing point of a solution is lower than that of the pure solvent. The decrease of the freezing point is related to the lowering of the vapor pressure of the solvent, as shown in Figure 11.15. As the concentration of the solute increases, the triple point temperature decreases, which moves the solid–liquid equilibrium line to lower temperatures. This **freezing point depression** is proportional to the concentration of solute particles, as long as the solution is reasonably dilute. The concentration unit used in freezing point depression experiments is molality, as shown in Equation 11.5.

$$\Delta T_f = mk_f \qquad [11.5]$$

The freezing point depression constant k_f is also a characteristic of the solvent, and independent of the kind of solute particles. The freezing points and the freezing point depression constants for some common solvents are included in Table 11.3. Note the similarity of Equations 11.5 (freezing point depression) and 11.4 (boiling point elevation). The use of freezing point depression data is quite similar to that of boiling point elevation, as illustrated in the following examples.

Example 11.12 Determining the Freezing Point Depression Constant from Experimental Data

Pure ethylene dibromide freezes at 9.97 °C. A solution made by dissolving 0.213 g of ferrocene (molecular formula $Fe(C_5H_5)_2$, molar mass = 186.04 g/mol) in 10.0 g of ethylene dibromide has a freezing point of 8.62 °C. What is the freezing point depression constant for the solvent ethylene dibromide?

Solution From the data given, the freezing point decreased from 9.97 to 8.62 °C, or

$$\Delta T_f = 9.97 - 8.62 = 1.35\ °\text{C}$$

The composition of the solution must be expressed as the molality, which requires converting the quantity of ferrocene to moles, and expressing the quantity of solvent in kg.

$$\text{moles ferrocene} = 0.213 \text{ g ferrocene} \left(\frac{1 \text{ mol}}{186.04 \text{ g}}\right)$$

$$= 1.14 \times 10^{-3} \text{ mol ferrocene}$$

This amount of ferrocene was dissolved in 10.0 g, or 0.0100 kg, of the solvent. The molal concentration of the ferrocene in the solution is

$$\text{molality} = \frac{1.14 \times 10^{-3} \text{ mol ferrocene}}{0.0100 \text{ kg solvent}} = 1.14 \times 10^{-1}\ m$$

The freezing point depression constant is found by rearranging Equation 11.5 and substituting the values we have calculated:

$$k_f = \frac{\Delta T_f}{\text{molal}} = \frac{1.35\ °\text{C}}{1.14 \times 10^{-1} \text{ molal}} = 11.8\ °\text{C/molal}$$

Understanding Benzophenone has a freezing point of 49.00 °C. A 0.450 molal solution of urea in this solvent has a freezing point of 44.59 °C. Find the freezing point depression constant for the solvent benzophenone.

Answer: 9.80 °C/m

Example 11.13 Calculation of Molar Mass from Freezing Point Depression

A solution is prepared by dissolving 0.350 g of an unknown compound in 5.42 g of ethylene dibromide. This solution has a freezing point of 6.51 °C. Using the freezing point depression constant found in Example 11.12, find the molar mass of the solute.

Solution From the freezing point data and the known k_f, the concentration of the solution is found.

$$m = \frac{\Delta T_f}{k_f} = \frac{(9.97 - 6.51)\ °\text{C}}{11.8\ °\text{C/molal}} = \frac{3.46 \text{ molal}}{11.8} = 0.293 \text{ molal}$$

The mass of solvent in the sample is converted to the number of moles of solute by using the measured molality as a conversion factor.

$$\text{mol solute} = 5.42 \text{ g solvent} \left(\frac{1 \text{ kg}}{1000 \text{ g}}\right)\left(\frac{0.293 \text{ mol solute}}{1 \text{ kg solvent}}\right)$$

$$= 1.59 \times 10^{-3} \text{ mol solute}$$

The calculation is completed by combining the amount of solute with its mass, 0.350 g, to find the molar mass of the compound.

$$\text{molar mass} = \frac{0.350 \text{ g in sample}}{1.59 \times 10^{-3} \text{ mol in sample}} = 220.\ \text{g/mol}$$

Understanding A solution of 0.134 g of a compound in 4.76 g of ethylene dibromide has a freezing point of 7.79 °C. Find the molar mass of this solute.

Answer: 152 g/mol

Osmotic Pressure

The final colligative property that we shall discuss is osmotic pressure. Thin layers of certain materials, called **semipermeable membranes,** allow only water and other small molecules to pass through them. **Osmosis** is the

INSIGHTS into CHEMISTRY

A Closer View

Vapor Phase Osmometry Measures Solutes. A Single Drop Yields Useful Information

Another technique used to find the concentration of solute particles is called vapor phase osmometry. The elevation of the boiling point and depression of the freezing point both have the disadvantage that the temperature at which the measurement is made must be near the normal boiling or freezing point of the solvent. Vapor phase osmometry does not have this limitation and may be used at any temperature within the liquid range of the solvent.

The apparatus used in a vapor phase osmometry experiment is shown in the following diagram. It consists of a chamber that is held at a constant temperature by a thermostat. A large reservoir of the pure solvent, at the temperature of the thermostat, establishes a constant partial pressure of the solvent within the chamber. Two small electrical devices, called thermistors, are suspended in the chamber. The electrical resistance of a thermistor changes as the temperature changes, and this apparatus can detect temperature differences as small as 1×10^{-4} °C. A small drop of solvent is suspended on one of the thermistors, and a drop of solution is suspended from the other one. The vapor pressure of the solvent in both drops must be the same when equilibrium is established with the solvent molecules in the vapor phase. This can be true only if the temperature of the drop of solution is higher than that of the drop of the solvent. The origin of this temperature difference can be seen by referring to Figure 11.11. A horizontal line (constant pressure) on the graph of vapor pressure versus temperature intersects the curve for the pure solvent at a lower temperature than it intersects the curve for the solution. The temperature difference between the two thermistors is measured electrically. The instrument is calibrated by using solutions of known concentration.

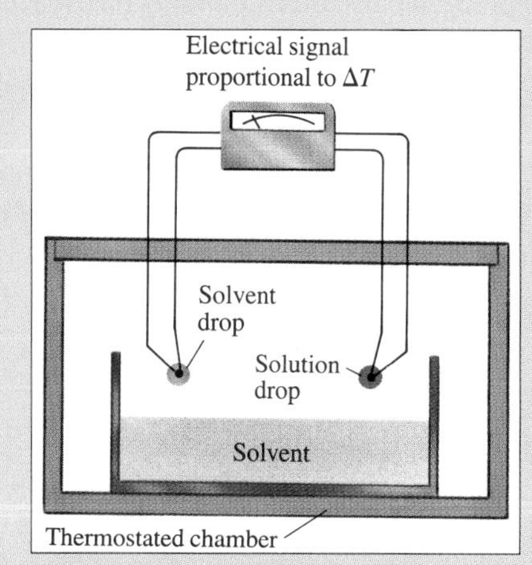

Vapor phase osmometry A schematic diagram of the vapor phase osmometer. When equilibrium with the vapor is reached, the temperature of the drop of solution is higher than that of the solvent. The temperature difference between the two drops is measured electrically, and is proportional to the concentration of solute particles in the solution.

At 25 °C, a 0.05 *m* solution in water has a vapor pressure that is only 0.021 torr lower than that of pure water (23.77 torr). In a vapor phase osmometer, a temperature difference of 1.51×10^{-2} °C is generated. The difference in vapor pressure is almost undetectable, while the temperature difference can be measured to three significant digits.

diffusion of a fluid through a semipermeable membrane. Animal bladders, the skins of fruits and vegetables, and cellophane are examples of semipermeable membranes. Semipermeable membranes are crucial in biological systems, controlling the transport of nutrients and waste products across cell boundaries.

An apparatus can be built, as shown in Figure 11.16, that uses a semipermeable membrane to separate pure water from an aqueous solution. Only the water molecules can pass through the membrane, and they move in both directions. Because the concentration of water is larger in the pure water, more water molecules strike the membrane per second on that side, and more water moves into the solution than leaves it. With time the solution is diluted, and the level of liquid rises in the tube containing the solution, causing a difference in pressure between the two sides of the membrane.

This difference in pressure increases the frequency of water molecules striking the membrane on the solution side, and increases the rate of transfer

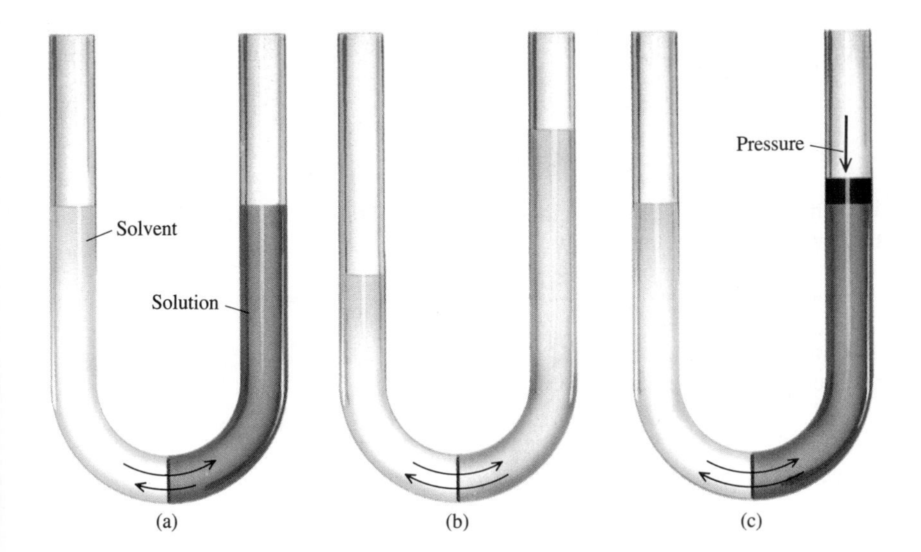

Figure 11.16 Osmotic pressure A semipermeable membrane separates a solution from the pure solvent. (a) Initially there is no pressure difference, so the rate of transport of water is greater into the solution. (b) The level of liquid on the solution side of the membrane has risen to the equilibrium height, and its weight exerts sufficient pressure that the rates of solvent transport are equal in both directions through the membrane. (c) The application of an external pressure to the solution can achieve the same equilibrium situation.

of water back through the membrane to the pure water side. At some point the difference in pressure is sufficient to make the rate of transport in both directions equal, and a state of dynamic equilibrium is achieved. The same state of equilibrium can be achieved by applying a pressure to the solution with a piston, as shown in Figure 11.16c. The **osmotic pressure** of a solution is the pressure difference needed to prevent net transport of solvent across a semipermeable membrane that separates the solution from the pure solvent.

The osmotic pressure of a solution is proportional to the molar concentration of the solute particles in the solution. It has been shown experimentally that the osmotic pressure of a solution can be calculated from an equation similar to the ideal gas law:

$$\Pi = \frac{nRT}{V} = MRT \qquad [11.6]$$

where Π is the osmotic pressure, n is the number of moles of solute present, V is the volume of the solution, R is the ideal gas constant (0.0821 L atm mol^{-1} K^{-1}), and T is the temperature in kelvins. Since n/V is the molar concentration of the solute, M, the right side of the equation is simplified as shown.

Of all the colligative properties, osmotic pressure is the most sensitive. A 0.0100 M solution of sugar in water at 25 °C has an osmotic pressure of 0.245 atm, which corresponds to a column of water 2.53 meters high. The osmotic pressure is so sensitive that it is used to measure the molar masses of very large molecules and substances that are only slightly soluble in water. This procedure is illustrated in Example 11.14.

Osmotic pressure is another colligative property. The concentration of the solute is expressed in units of mol/L (molarity).

Example 11.14 Determining Molar Mass from Osmotic Pressure

Hemoglobin is a large molecule that carries oxygen in human blood. A water solution that contains 0.263 g of hemoglobin (abbreviated here as Hb) in 10.0 mL of solution has an osmotic pressure of 7.51 torr at 25 °C. What is the molar mass of hemoglobin?

Solution We can use the measurement of the osmotic pressure to find the number of moles of solute, which we combine with the mass data given in the problem to find the molar mass. Before using Equation 11.6, we must express the temperature in kelvins and the osmotic pressure in atmospheres.

$$T = 25\ °C + 273 = 298\ K$$

$$\Pi = 7.51\ \text{torr}\left(\frac{1\ \text{atm}}{760\ \text{torr}}\right) = 9.88 \times 10^{-3}\ \text{atm}$$

We solve Equation 11.6 for the molarity of the solution, and substitute the values for the osmotic pressure, temperature, and R.

$$\text{molarity} = \frac{\Pi}{RT} = \frac{9.88 \times 10^{-3}\ \text{atm}}{0.0821\ \text{L atm mol}^{-1}\ \text{K}^{-1} \times 298\ \text{K}} = 4.04 \times 10^{-4}\ \frac{\text{mol Hb}}{\text{L}}$$

The molar concentration is used with the volume of the solution to calculate the number of moles of hemoglobin in the sample.

$$\text{moles Hb} = 10.0\ \text{mL soln}\left(\frac{1\ \text{L}}{1000\ \text{mL}}\right)\left(\frac{4.04 \times 10^{-4}\ \text{mol Hb}}{1\ \text{L soln}}\right)$$

$$= 4.04 \times 10^{-6}\ \text{mol Hb}$$

We combine the mass of the sample with the number of moles of hemoglobin to find the molar mass.

$$\text{molar mass} = \frac{0.263\ \text{g}}{4.04 \times 10^{-6}\ \text{mol}} = 6.51 \times 10^{4}\ \text{g/mol}$$

Understanding A 5.70 mg sample of a protein is dissolved in water to give 1.00 mL of solution. If the osmotic pressure of this solution is 6.52 torr at 20 °C, what is the molar mass of the protein?

Answer: 1.60×10^{4} g/mol

All of the colligative properties fit the relationship

$$\text{property} = \text{solute concentration} \times \text{constant}$$

One colligative property differs from another in the units in which the solute concentration is expressed. In most cases the property is expressed as a difference between its value in the solution and its value in the pure solvent. Table 11.4 matches each of the various colligative properties with the concentration units used and the usual abbreviation for the proportionality constant.

Table 11.4 Concentration Units for Colligative Properties

Property	Symbol	Solute Concentration	Constant
vapor pressure	ΔP	mole fraction	$P°$
boiling point	ΔT_b	molal	k_b
freezing point	ΔT_f	molal	k_f
osmotic pressure	Π	molar	RT

INSIGHTS into CHEMISTRY

Science, Society, and Technology

Colligative Properties Find Application: Reverse Osmosis Desalinizes Sea Water

Water that contains high concentrations of electrolytes is unsuitable for consumption by people. As the population of the world increases, providing fresh water suitable for drinking is increasingly difficult. This situation is particularly true in semi-arid regions, such as the Middle East. Various schemes have been proposed to reduce the concentration of salts in sea water and other natural sources, making it fit for human consumption. One means of purifying water, a process called reverse osmosis, is the subject of continuing research.

Reverse osmosis is the transport of water across a semipermeable membrane, from a solution on one side to pure water on the other side, by applying a pressure greater than the osmotic pressure to the solution. We have seen that no net transport of water across a semipermeable membrane occurs when the osmotic pressure is applied to the solution. If a pressure greater than the osmotic pressure is applied to the solution, then net transport of the water occurs from the solution side to the pure water side. Because the osmotic pressure of natural sea water is fairly high, one of the problems in the design of a reverse osmosis apparatus is to make a membrane that is thin enough to allow rapid transport of the water, yet is strong enough to withstand high pressures without bursting.

Reverse osmosis is used in Saudi Arabia for water purification. In 1991, this water source was threatened when oil from a well in Kuwait was dumped into the Persian Gulf by Iraq. The reverse osmosis plants were shut down because the oil would have ruined the semipermeable membranes in the equipment. The U.S. Navy has developed small portable units for desalinization of sea water for use in life rafts. These manually operated units are capable of producing five liters of drinkable water per hour, which is sufficient to keep several people alive. These units are replacing the bulky containers of fresh water now stored on Navy lifeboats.

11.5 Colligative Properties of Electrolyte Solutions

Objectives

- To predict the ideal van't Hoff factor for solutions of ionic solutes
- To calculate the expected colligative properties for solutions of electrolytes

As data on the colligative properties of solutions were collected, it became apparent that solutions of ionic and molecular solutes behaved differently. This information provided insights into the nature of solutions, particularly the ability of ionic solutes to dissociate into ions. The results and interpretation of such studies are summarized in this section.

Jacobus van't Hoff (1852–1911) and several other scientists noticed that some solutes produce a larger effect on the colligative properties of the solution than was expected. Most of these solutes are **electrolytes,** compounds that separate into ions when they dissolve in water. The **van't Hoff factor,** i, was introduced to account for these deviations. The van't Hoff factor is defined by the equation

$$i = \frac{\text{measured colligative property}}{\text{expected value for a nonelectrolyte}}$$

The equations for the colligative properties, given in Table 11.4, are modified by replacing the concentration of the solute by its product with the van't Hoff factor.

For solutions that involve typical nonelectrolytes, such as urea and sucrose, the van't Hoff factor is 1. For solutions of salts and other electrolytes, i has a value much greater than one. The reason for this becomes

Table 11.5 van't Hoff Factor Values for Electrolytes in Water at 0.05 molal

Compound	ΔT_f(°C)	Measured i	Ideal i
NaCl	0.176	1.9	2.0
HIO_3	0.156	1.7	2.0
$Ca(NO_3)_2$	0.235	2.5	3.0
$MgCl_2$	0.249	2.7	3.0
$AlCl_3$	0.300	3.2	4.0

apparent when we consider what occurs when salts like NaCl and $Ca(NO_3)_2$ dissolve in water; each formula unit dissociates into two or more hydrated ions:

$$NaCl(s) \longrightarrow Na^+(aq) + Cl^-(aq)$$

$$Ca(NO_3)_2(s) \longrightarrow Ca^{2+}(aq) + 2NO_3^-(aq)$$

For each mole of sodium chloride that dissolves, two moles of solute particles (one mole of Na^+ ions and one mole of Cl^- ions) are present in solution. For calcium nitrate, three moles of solute particles (one of Ca^{2+} ions and two of NO_3^- ions) are present in solution per mole of compound that dissolves. Since colligative properties of solutions depend on the concentration of solute particles, we would expect that sodium chloride would have a freezing point depression that is twice that of a nonelectrolyte solution of the same molality. The osmotic pressure of a calcium nitrate solution should be three times that of a nonelectrolyte with the same molarity. On this basis, we would expect the van't Hoff factor to equal the number of ions produced by each formula unit of the compound that dissolves: 2 for NaCl and $MgSO_4$, 3 for $Ca(NO_3)_2$, and so forth.

It is important to remember that colligative properties are proportional to the concentration of solute *particles* in solution.

When electrolyte solutions are very dilute (about 0.01 m or less), we find that the measured colligative properties agree fairly well with those predicted by the ideal value of the van't Hoff factor. At higher concentrations, the observed values for i tend to be smaller than the values expected on the basis of the number of ions produced in solution. The strong electrostatic attractions between oppositely charged ions cause a fraction of them to be held together even in solution and behave as a single particle. This association can partially account for the lower values of i observed for more concentrated solutions. Some typical values of the van't Hoff factor are shown for several electrolytes in Table 11.5.

Example 11.15 Van't Hoff Factor and Colligative Properties

Arrange the following aqueous solutions in order of increasing freezing point, assuming ideal behavior: 0.05 m sucrose, 0.02 m NaCl, 0.01 m $CaCl_2$, 0.03 m HCl.

Solution The freezing point of a solution depends on the molal concentration of solute particles. Since several of these solutes are electrolytes, the concentration of particles is the product of the van't Hoff factor with the molal concentration of the solute. We can summarize this calculation in a table, using an ideal value of the van't Hoff factor.

Compound	Present as	i	m	$i \times m$
sucrose	molecules	1	0.05	0.05
NaCl	$Na^+ + Cl^-$	2	0.02	0.04
$CaCl_2$	$Ca^{2+} + 2Cl^-$	3	0.01	0.03
HCl	$H^+ + Cl^-$	2	0.03	0.06

The lowest freezing point is expected for the solution with the *highest* concentration of solute particles, so the freezing points of these solutions increase in the order

$$0.03\ m\ \text{HCl} < 0.05\ m\ \text{sucrose} < 0.02\ m\ \text{NaCl} < 0.01\ m\ \text{CaCl}_2$$

Understanding Arrange the following solutions in order of increasing osmotic pressure: 0.02 *M* sucrose, 0.02 *M* HNO_3, 0.01 *M* $BaCl_2$.

Answer: 0.02 *M* sucrose < 0.01 *M* $BaCl_2$ < 0.02 *M* HNO_3

11.6 Mixtures of Volatile Substances

Objectives

- To calculate the vapor pressure of each component and the total vapor pressure over an ideal solution
- To interpret positive and negative deviations from Raoult's law in terms of the relative strengths of the intermolecular attractions in the solution

The characteristics of solutions that contain more than one volatile substance are important in the separation and purification of many substances. For example, the separation of crude oil into components such as gasoline, diesel fuel, and asphalt depend upon the different tendencies of compounds to evaporate.

In Section 11.4 we considered the vapor pressure of the solvent over a solution in which the only volatile component was the solvent. There are many solutions in which two or more of the components are volatile—that is, have significant vapor pressures. The vapor phases in equilibrium with these solutions contain both volatile components. In this section we will examine the vapor pressure and composition of the gas in equilibrium with such a mixture.

Consider a solution that contains benzene (C_6H_6) and toluene ($C_6H_5CH_3$). Both of these substances are volatile. At any given temperature, according to Raoult's law, the vapor pressure of the benzene above the solution is given by

$$P_{\text{benzene}} = \chi_{\text{benzene}} P^\circ_{\text{benzene}}$$

Raoult's law can also be used to calculate the partial pressure of toluene above the same solution:

$$P_{\text{toluene}} = \chi_{\text{toluene}} P^\circ_{\text{toluene}}$$

Figure 11.17 shows the vapor pressures of both components as the mole fraction of toluene in the solution varies from 0 to 1. At the left side of the graph the liquid is pure benzene, and on the right side it is pure toluene. The total vapor pressure, also shown in Figure 11.17, is simply the sum of the

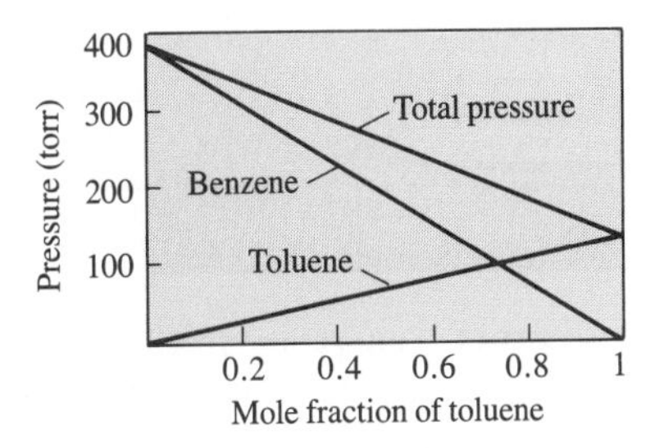

Figure 11.17 Vapor pressure of an ideal solution The graph shows the vapor pressures of mixtures of benzene and toluene as the composition is changed at a constant temperature. Benzene and toluene form an ideal solution, meaning that both liquids obey Raoult's law over the entire range of composition.

vapor pressures of the two compounds. As we expect from the equations, all three of the lines in Figure 11.17 are straight.

We can use Raoult's law to calculate the composition of the vapor above a mixture of two volatile substances. This calculation is illustrated in Example 11.16.

Example 11.16 Composition of Vapor Above a Mixture

At 60 °C the vapor pressure of benzene is 384 torr, and that of toluene is 133 torr. A mixture is made by combining 1.20 mol of toluene with 3.60 mol of benzene. Find:

(a) the mole fraction of toluene in the liquid
(b) the partial pressure of toluene above the liquid
(c) the partial pressure of benzene above the liquid
(d) the total vapor pressure
(e) the mole fraction of toluene in the vapor phase.

Solution

(a) Since the number of moles of each component in the mixture is given in the problem, the mole fraction of toluene is easily found.

$$\chi_{\text{toluene}} = \frac{\text{mol toluene}}{\text{mol toluene} + \text{mol benzene}}$$

$$\chi_{\text{toluene}} = \frac{1.20}{1.20 + 3.60} = 0.250$$

(b) The partial pressure of the toluene is calculated from Raoult's law.

$$P_{\text{toluene}} = \chi_{\text{toluene}}\, P^{\circ}_{\text{toluene}} = 0.250 \times 133 \text{ torr} = 33.2 \text{ torr}$$

(c) Raoult's law is also used to find the vapor pressure of benzene. The vapor pressure of pure benzene is given in the problem, and the mole fraction of the benzene is calculated from that of toluene.

$$\chi_{\text{benzene}} = 1 - \chi_{\text{toluene}} = 1 - 0.250 = 0.750$$

$$P_{\text{benzene}} = 0.750 \times 384 \text{ torr} = 288 \text{ torr}$$

(d) The total vapor pressure above the solution is simply the sum of the vapor pressures of the two components.

$$P_{\text{total}} = 33.2 + 288 = 321 \text{ torr}$$

(e) The composition of the vapor is found from the partial pressures of the components. Recall from Chapter 5 that the partial pressures of gases in a mixture can be used to find the mole fractions of the components. We use the answers to parts (b) and (d) to calculate the mole fraction of toluene in the vapor.

$$\chi_{\text{toluene(g)}} = \frac{33.2 \text{ torr}}{321 \text{ torr}} = 0.103$$

Note that the mole fraction of the less volatile material, in this case toluene, is lower in the vapor phase than in the liquid; thus the mole fraction of the more volatile component, benzene, is greater in the vapor phase than it is in the liquid. In the vapor the mole fraction of benzene is

$$\chi_{\text{benzene}} = 1 - \chi_{\text{toluene}} = 0.897 \text{ benzene}$$

This concentration in the vapor is considerably greater than the 0.750 mole fraction of benzene found in the liquid phase.

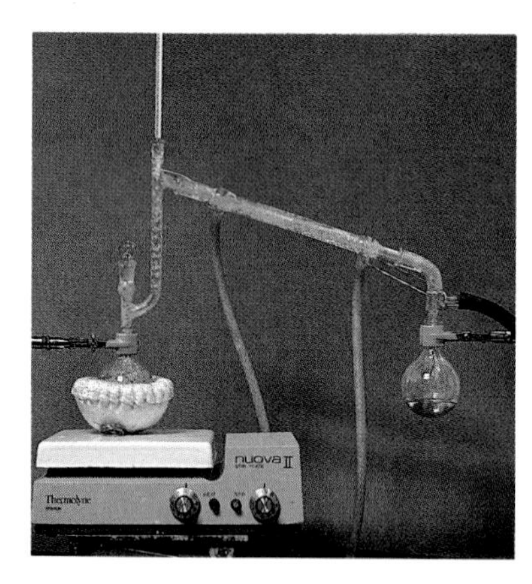

Figure 11.18 Laboratory distillation apparatus A glass column with many indentations (Vigreaux column) is used in the laboratory for fractional distillations. The liquid repeatedly condenses and evaporates as it moves up the column. With each successive evaporation, the vapor becomes richer in the most volatile (lowest boiling) component.

Understanding What is the vapor pressure at 60 °C of a solution with $\chi_{toluene} = 0.10$ and $\chi_{benzene} = 0.90$? Use the data given in this example.

Answer: 359 torr

The vapor over a solution of two volatile components contains a larger fraction of the more volatile substance than does the solution.

The vapor in equilibrium with a mixture of two volatile substances is always richer in the more volatile component. This was pointed out in Example 11.16. While the mole fraction of benzene in the liquid was 0.750, its mole fraction in the vapor was 0.896. As evaporation continues, the composition of the liquid changes, increasing the mole fraction of the less volatile (higher boiling) substance. It is this fact that makes it possible to separate two volatile materials by distillation.

In a fractional distillation the liquid is repeatedly evaporated and condensed as it moves up the distillation column. Figure 11.18 shows a simple fractional distillation apparatus used in the laboratory. Each time the vapor condenses, it produces a mixture that is richer in the more volatile component. As the mole fraction of the more volatile component increases, the boiling point of the mixture decreases, so the temperature of the column decreases from the bottom to the top. By the time the vapor reaches the top of the column, it consists of the more volatile component in a state of purity governed by the number of successive vaporizations and condensations that occurred. Very high purity can be obtained when the column is sufficiently long. A commercial distillation apparatus used in the petroleum industry is shown in Figure 11.19.

In an ideal solution Raoult's law applies to all volatile substances in the solution.

In the data presented in Figure 11.17, we assumed that Raoult's law was obeyed by both of the components over the entire range of composition from pure benzene to pure toluene. An **ideal solution** is one that obeys Raoult's law throughout the entire range of composition. Consider a two-component mixture of compounds A and B. When the strength of the A–B attractions

Figure 11.19 Fractional distillation of crude oil Large-scale distillation equipment is used in the petroleum industry, to separate the hydrocarbons into several fractions with different volatilities.

INSIGHTS into CHEMISTRY

A Closer Look

Chemical Antifreeze: Surviving the Winter

Many species of animals, like turtles and frogs, spiders and beetles, can actually survive the freezing temperatures of winter. But how can this be? As anyone who has refrozen beef knows, freeze-thaw cycles quickly break down the tissue as ice crystals cause cells to burst, explode capillaries, and tear material from bone. Yet some species manage to make it through the winter, in spite of subfreezing temperatures.

The most obvious strategy for creatures who must survive the winter is to avoid freezing temperatures. Many types of turtles and frogs, for example, burrow deep into the floor of ponds and lakes, and are protected from freezing as the ground below the pond remains unfrozen.

Another survival mechanism used by some creatures is familiar to anyone who drives a car: they use an antifreeze compound (like the ethylene or propylene glycol used in automobiles) to avoid freezing. Two basic types of antifreeze are used by animals, proteins and glycols.

The protein antifreeze binds to the "nucleators," the tiny ice crystals that form as the temperature drops, and prevent the growth of large crystals. The formation of large crystals causes the rupture of tissue because the water expands upon freezing. These proteins can prevent the damage from freezing in some insects to temperatures as low as −15 °C.

The animals that rely on glycols are taking direct advantage of the colligative properties that were just studied. By concentrating these low molecular weight compounds in their tissues, the animals lower the freezing point of the fluid. Some of these animals have glycols that total as much as 19% of their body weight, allowing them to withstand temperatures down to −38 °C. Such animals are well-adapted to low temperatures.

A third group of animals survives the cold by living with some freezing. These animals use both antinucleation proteins, as well as a protein that actually encourages ice crystal formation. Thousands of tiny ice crystals develop in the animal, yet they remain extremely small. These creatures also use a variety of compounds—usually sugars—in their blood and extracellular fluids to prevent their cells from collapsing or exploding because of differences in osmotic pressure. Whenever there are large differences in concentrations of solutes on either side of a cell, membrane osmotic pressures can become large enough to burst the membrane. In fact, it is osmotic shock, combined with the action of ice crystals on the muscle fiber, that makes repeated freezing of meat a bad practice.

Animals that can withstand low temperatures have developed highly specialized processes to avoid cell damage and, ultimately, the death of the organism. While at first it might seem that these animals are defeating solution properties through magic, in reality most of these special processes take advantage of the colligative properties of solutions.

are close to the average of the A–A and B–B attractions, the solution behaves ideally. For two very similar liquids such as benzene and toluene, this is very nearly true.

For most mixtures of liquids Raoult's law is strictly obeyed only by very dilute solutions. Usually the intermolecular forces of attraction between two different substances are stronger or weaker than the average of the A–A and B–B attractions. When this is true we find that the straight lines in Figure 11.17 become curved. **Positive deviation** from Raoult's law means that the observed vapor pressure is greater than expected as shown in Figure 11.20, and it occurs when the A–B attractions are weaker than the average of the

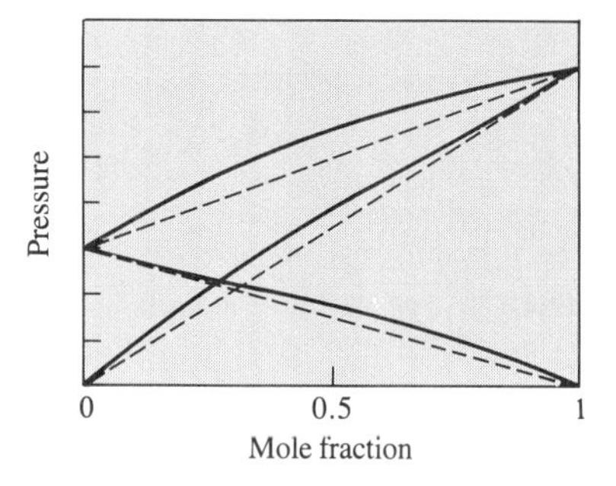

Figure 11.20 Positive deviation from Raoult's law The graph shows the vapor pressure of each component and the total vapor pressure of the mixture as the composition of the liquid changes. The dotted lines represent the behavior of an ideal solution. The solid lines show a positive deviation from Raoult's law.

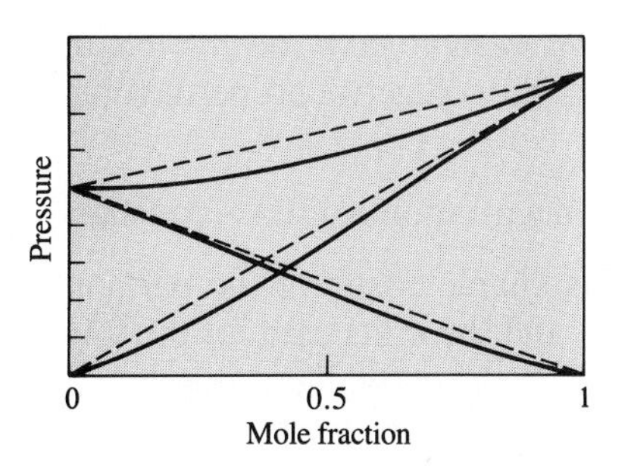

Figure 11.21 Negative deviation from Raoult's law The graph shows the vapor pressure of each component and the total vapor pressure of the mixture as the composition of the liquid changes. The dotted lines represent the behavior of an ideal solution. The solid lines show a negative deviation from Raoult's law.

attractions in the pure components of the mixture. **Negative deviation** from Raoult's law, shown in Figure 11.21, occurs when the intermolecular forces between the dissimilar molecules are stronger than the average of those in the pure substances. Both positive and negative deviations from Raoult's law are observed experimentally for mixtures of different compounds.

Deviations from ideal solution behavior sometimes produce a maximum or minimum in the vapor pressure of the solution, as shown in Figure 11.22. At a maximum or a minimum in the vapor pressure curve for a solution, both the liquid and gas phases have exactly the same composition. When such a mixture boils, both the composition and the boiling point remain constant, so the solution is called a constant boiling mixture. An **azeotrope** (or azeotropic mixture) is a solution that has the composition of the constant boiling mixture. A fractional distillation of substances that form an azeotrope always produces the constant boiling mixture, and no further separation of the components can be achieved at that pressure.

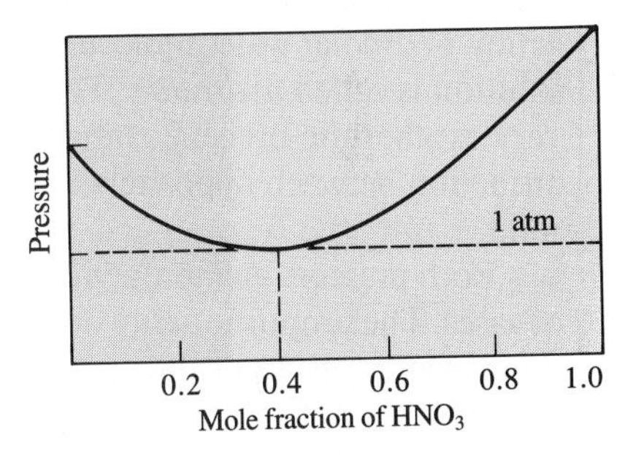

Figure 11.22 Vapor pressure of an azeotrope of nitric acid soution at constant temperature The vapor pressure of water solutions of nitric acid at the boiling point of the azeotropic mixture (120.5° C) is shown as a function of the composition. At the minimum in the vapor pressure curve, the mixture boils at a constant temperature and composition, preventing any further separation of the two substances by distillation.

Ethanol and water form an azeotrope that has a maximum in the vapor pressure curve where χ_{water} is 0.10. The normal boiling point of this azeotrope is 78.17 °C, slightly lower than the 78.4 °C at which pure ethanol boils. The concentrated aqueous nitric acid solution (68 mass percent HNO_3) that is sold for laboratory use is an azeotrope that boils at 120.5 °C, considerably higher than the normal boiling points of either water (100 °C) or nitric acid (86 °C). The nitric acid–water azeotrope results from large negative deviations from Raoult's law. Another azeotrope of some importance is the constant boiling solution of hydrogen chloride in water. At 760 torr, the $HCl-H_2O$ azeotrope boils at 108.6 °C and contains 20.22 mass percent HCl. The concentration of this azeotrope (6.11 *M*) is known with sufficient accuracy that it is used to prepare standard solutions of hydrochloric acid.

Summary

Solutions are homogeneous mixtures of two or more substances, and can occur in the solid, liquid, and gas phases. Solutions are described quantitatively by concentration. Some common units of concentration are mass percent, mole fraction, molarity, and *molality,* defined as the moles of solute per kilogram of solvent. A *saturated solution* is one in which the dissolved and undissolved solute are present in equilibrium. The concentration of a saturated solution is called the *solubility.* Any solution in which the concentration of solute is less than the solubility is *unsaturated.* A *supersaturated solution* is an unstable situation in which the concentration of the solute is greater than the solubility. Unsaturated, saturated, and supersaturated solutions can be distinguished by observing what happens to the solution when a small sample of the pure solute is added to it.

The solution process is complex and depends upon the strengths of solute–solute, solvent–solvent, and solute–solvent attractive forces, as well as the change in disorder that occurs on mixing. While the solution process can be either endothermic or exothermic, dissolution requires that the strengths of the solute–solvent interactions are close to those of the solute–solute and solvent–solvent interactions. While these enthalpy considerations make possible qualitative predictions about the relative solubility of a substance in two different solvents, the dissolution process is too complex to make quantitative predictions. The interaction of dipolar water molecules with ions in aqueous solution is called *hydration.* The hydration of ions is a very exothermic process, comparable to the energy of attraction between oppositely charged ions in an ionic solid.

Both pressure and temperature affect the solubility of gases. The proportionality of the solubility of a gas to the partial pressure of the gas is expressed by *Henry's law.* The solubilities of gases decrease with increasing temperature, but for most solids, both molecular and ionic, solubility gets larger as the temperature of the solution increases. From the principle of Le Chatelier, the sign of the enthalpy of solution determines whether solubility decreases (negative enthalpy change) or increases (positive enthalpy change) with increasing temperature. Pressure has almost no effect on the solubility of liquids and solids.

There are a number of properties of solutions, called *colligative properties,* that depend only on the concentration of solute particles, not on their identity. Four of these were presented: *vapor pressure lowering, boiling point elevation, freezing point depression,* and *osmotic pressure.* All of the colligative properties obey a relationship

$$\Delta(\text{property}) = \text{concentration(solute)} \times \text{constant}$$

where the constant is characteristic of the particular property and solvent, and the units used for the solute concentration depend on the particular property being measured. When the solute is an electrolyte, the *van't Hoff factor, i,* must be used in the equations for the colligative properties. Ideally, the van't Hoff factor is equal to the number of moles of ions produced in solution for each mole of compound dissolved. Electrostatic attractions between the ions cause the experimentally measured values of i to be somewhat lower than the ideal value, for all but very dilute solutions.

When two or more components of a liquid solution are volatile, the vapor pressure of the solution is the sum of the partial pressures of the volatile components. An *ideal solution* is one that obey's Raoult's law over the entire composition range of the mixture. The composition of the vapor in equilibrium with the solution is richest in the most volatile component that is present in the solution. In a fractional distillation, repeated evaporations and condensations are used to separate mixtures of two or more volatile liquids.

Most solutions of volatile substances do not behave ideally, but exhibit either positive or negative deviations from Raoult's law, depending on the strength of the intermolecular attractions between the dissimilar molecules relative to those in the pure components. Many liquids form constant boiling mixtures called *azeotropes,* in which the composition of the vapor and liquid phases are the same. Complete separation of volatile liquids that form an azeotrope cannot be achieved by a fractional distillation.

Chapter Terms

Molality (m)—the concentration of a solute expressed as the number of moles of solute present in one kilogram of solvent. *(11.1)*

Solubility—the concentration of solute that exists in equilibrium with an excess of that substance. *(11.2)*

Saturated solution—a solution that is in equilibrium with an excess of the solute. *(11.2)*

Unsaturated solution—a solution in which the concentration of the solute is less than its solubility. *(11.2)*

Supersaturated solution—a solution in which the concentration of the solute is temporarily greater than its solubility. This is an unstable condition. *(11.2)*

Henry's law—the solubility of a gas is directly proportional to its partial pressure above the solution at any given temperature:

$$C = kP$$

where C is the concentration of the gaseous compound in solution, k is a proportionality constant that is characteristic of the particular solute, solvent, and temperature, and P is the partial pressure of the solute in contact with the solution. *(11.3)*

Colligative properties—those properties of a solution that are proportional to the concentration of solute particles. *(11.4)*

Raoult's law—the vapor pressure (P) of the solvent over a dilute solution is equal to the mole fraction of the solvent (χ) times the vapor pressure of the pure solvent (P°_{solv}):

$$P = \chi_{solvent} P^\circ_{solv}$$

A more useful form of this law is

$$\Delta P = \chi_{solute} P^\circ_{solv}$$

where ΔP is ($P^\circ_{solv} - P_{solution}$). *(11.4)*

Boiling point elevation—a colligative property of solutions of nonvolatile solutes described by the equation

$$\Delta T_b = mk_b$$

where ΔT_b = (boiling point of solution − boiling point of solvent), m is the molal concentration of the solute particles, and k_b is the boiling point elevation constant, which is characteristic of the solvent. *(11.4)*

Freezing point depression—a colligative property of solutions described by the equation

$$\Delta T_f = mk_f$$

where ΔT_f = (freezing point of solvent − freezing point of solution), m is the molal concentration of the solute particles, and k_f is the freezing point depression constant, which is characteristic of the solvent. *(11.4)*

Semipermeable membrane—a thin film of material that allows only water and other small molecules to pass through it. Animal bladders, the skins of fruits and vegetables, and cellophane are examples of semipermeable membranes. *(11.4)*

Osmosis—the diffusion of a fluid through a semipermeable membrane. *(11.4)*

Osmotic pressure—the pressure difference needed to prevent net transport of solvent across a semipermeable membrane that separates a solution from the pure solvent. This is one of the most sensitive colligative properties, and is described by the equation

$$\Pi = MRT$$

where Π is the osmotic pressure, M is the molar concentration of solute particles, R is the ideal gas constant (0.0821 L atm mol^{-1} K^{-1}), and T is the temperature in kelvins. *(11.4)*

Electrolyte—a substance that separates into ions when it dissolves in water. *(11.5)*

van't Hoff factor, *i*—the ratio of the measured colligative property to that expected for a nonelectrolyte.

$$i = \frac{\text{measured colligative property}}{\text{expected value for a nonelectrolyte}}$$

In dilute solutions of electrolytes, the van't Hoff factor approaches the number of ions in solution for each formula unit of the compound that dissolves. *(11.5)*

Reverse osmosis—the transport of solvent across a semipermeable membrane from a solution to pure solvent, caused by applying a pressure greater than the osmotic pressure of the solution. *(11.5)*

Ideal solution—a solution that obeys Raoult's law throughout the entire range of composition. *(11.6)*

Positive deviation (from Raoult's law)—the observed vapor pressure is greater than expected for an ideal solution, which occurs when the attractions between dissimilar molecules are weaker than the average of the attractions in the pure components of the mixture. *(11.6)*

Negative deviation (from Raoult's law)—the observed vapor pressure is less than expected for an ideal solution, which occurs when the attractions between dissimilar molecules are stronger than the average of the attractions in the pure components of the mixture. *(11.6)*

Azeotrope—a solution that has the composition of the constant boiling mixture, for which the compositions of the vapor and liquid phases are identical. *(11.6)*

Exercises

Exercises designated with color have answers in Appendix J.

Solution Concentration

11.1 A solution is prepared by dissolving 25.0 g of $BaCl_2$ in 500. grams of water. Express the concentration of $BaCl_2$ in this solution as
(a) mass percent (b) mole fraction (c) molality.

11.2 A solution contains 1.20 g of benzoic acid ($C_6H_5CO_2H$) in 750.0 g of water. Express the concentration of benzoic acid as
(a) mass percent (b) mole fraction (c) molality.

11.3 How many grams of sodium chloride, NaCl, are present in 350. g of a 3.5% solution?

11.4 How many moles of hydrogen peroxide are present in 250. g of a 3.0% solution?

11.5 Give quantitative directions for preparing 300. g of a 2.50% solution of potassium iodide (KI). What is the molal concentration of this solution?

11.6 Give quantitative directions for preparing 155 g of a

1.00% solution of boric acid (H_3BO_3). What is the molal concentration of this solution?

11.7 What is the mole fraction of bromine (Br_2) in an aqueous 0.10 molal solution?

11.8 What is the molality of silver nitrate in an aqueous 0.10% solution of that compound?

11.9 Rubbing alcohol is a water solution that contains 65% isopropanol (C_3H_7OH). What is the mole fraction of isopropanol in rubbing alcohol?

11.10 A water solution of sodium hypochlorite (NaOCl) is used as a laundry bleach. The concentration of sodium hypochlorite is 0.75 *m*. Express this concentration as a mole fraction.

11.11 Vinegar is a 5.0% solution of acetic acid (CH_3CO_2H) in water. The density of vinegar is 1.0055 g/mL. Express the concentration of acetic acid as
(a) molality (b) molarity (c) mole fraction.

11.12 A 10.0 percent solution of sucrose ($C_{12}H_{22}O_{11}$) in water has a density of 1.038 g/mL. Express the concentration of the sugar as
(a) molality (b) molarity (c) mole fraction.

11.13 A 2.77 *M* NaOH solution in water has a density of 1.109 g/mL. Express the concentration of this solution as
(a) mass percent (b) mole fraction (c) molality.

11.14 A 0.631 *M* H_3PO_4 solution in water has a density of 1.031 g/mL. Express the concentration of this solution as
(a) mass percent (b) mole fraction (c) molality.

11.15 Complete the following table for perchloric acid ($HClO_4$) solutions in water.

Density (g/cm³)	Molality	Molarity	Mass % $HClO_4$	Mole Fraction
(a) 1.060	____	____	10.0	____
(b) 1.011	____	0.2012	____	____
(c) 1.143	2.807	____	____	____
(d) 1.086	____	____	____	0.0284

11.16 Complete the following table for ammonia (NH_3) solutions in water.

Density (g/cm³)	Molality	Molarity	Mass % NH_3	Mole Fraction
(a) 0.973	____	____	6.00	____
(b) 0.936	____	8.80	____	____
(c) 0.950	8.02	____	____	____
(d) 0.969	____	____	____	0.0738

11.17 A solution contains 10.0 g of ethanol (C_2H_5OH), 20.0 g of ethylene glycol ($C_2H_4(OH)_2$), and 90.0 g of water. What is the mole fraction of water in the sample?

11.18 A solution contains 12.0 g of hexane (C_6H_{14}), 20.0 g of octane (C_8H_{18}), and 98.0 g of benzene (C_6H_6). What is the mole fraction of benzene in the solution?

11.19 The density of a 3.75 *M* aqueous sulfuric acid solution in a car battery is 1.225 g/mL. Express the concentration of the solution in molality, mole fraction H_2SO_4, and mass percent of H_2SO_4.

Solubility and Intermolecular Attractions

11.20 Straight chain alcohols ($CH_3(CH_2)_nOH$) that contain more than four carbon atoms have limited solubility in water. Predict how the solubility of these alcohols in water changes as the value of *n* in the formula increases.

11.21 On the basis of the nature of the intermolecular attractions, arrange the following solutes in order of decreasing solubility in carbon tetrachloride (CCl_4): water, ethanol (C_2H_5OH), chloroform ($CHCl_3$).

11.22 Using the intermolecular attractions as a guide, arrange the following solutes in order of increasing solubility in benzene (C_6H_6): hexane (C_6H_{14}), ethanol (C_2H_5OH), water.

11.23 Predict the relative solubility of each compound in the two solvents, on the basis of intermolecular attractions.
(a) Is NaCl more soluble in water or in carbon tetrachloride?
(b) Is I_2 more soluble in water or in benzene (C_6H_6)?
(c) Is ethanol (C_2H_5OH) more soluble in hexane or in water?
(d) Is ethylene glycol [$C_2H_4(OH)_2$] more soluble in ethanol or in benzene (C_6H_6)?

11.24 Predict the relative solubility of each compound in the two solvents, on the basis of intermolecular attractions.
(a) Is Br_2 more soluble in water or in carbon tetrachloride?
(b) Is $CaCl_2$ more soluble in water or in benzene (C_6H_6)?
(c) Is chloroform ($CHCl_3$) more soluble in water or diethyl ether (($C_2H_5)_2O$)?
(d) Is ethylene glycol [$C_2H_4(OH)_2$] more soluble in water or in benzene (C_6H_6)?

11.25 What are the most important types of solute–solvent interactions in each of the following solutions?
(a) CH_3OH in water (c) KBr in water
(b) IBr in CH_3CN (d) argon in water.

11.26 What are the most important types of solute–solvent interactions in each of the following solutions?

(a) $(CH_3)_2CO$ in water
(b) IBr in $CHCl_3$
(c) $CaCl_2$ in water
(d) krypton in CH_3OH.

11.27 Choose the solute in each part that would be more soluble in water. Explain your answer.
(a) Na_2O or CO_2 (b) $TiCl_3$ or $CHCl_3$
(c) C_3H_8 or C_3H_7OH.

11.28 Choose the solute in each part that would be more soluble in hexane (C_6H_{14}). Explain your answer.
(a) $CH_3(CH_2)_{10}OH$ or $CH_3(CH_2)_2OH$
(b) $BaCl_2$ or CCl_4
(c) $Fe(C_5H_5)_2$ (a nonelectrolyte) or $FeCl_2$.

Temperature and Pressure Dependence of Solubility

11.29 The solubilities of most gases decrease as the temperature increases. Is the enthalpy of solution for such gases negative or positive?

11.30 The solubility of potassium chloride in water increases from 34.7 g/100 mL at 20 °C to 56.7 g/100 mL at 100 °C. Is the enthalpy of solution for this compound endothermic or exothermic?

11.31 At 25 °C and 2.0 atm the enthalpy of solution of neon in water is −2.46 kJ/mol and its solubility is 9.07×10^{-4} molal. Would the solubility of neon be greater or less than 9.07×10^{-4} m under each of the following conditions?
(a) 0 °C and 3.0 atm
(b) 25 °C and 1.0 atm
(c) 10 °C and 2.0 atm
(d) 50 °C and 2.0 atm
(e) 50 °C and 1.5 atm

11.32 At 22 °C and 1.0 atm the enthalpy of solution of nitrogen in water is −11.0 kJ/mol and its solubility is 6.68×10^{-4} molal. Would the solubility of nitrogen be greater or less than 6.68×10^{-4} molal under each of the following conditions?
(a) 0 °C and 3.0 atm
(b) 22 °C and 0.75 atm
(c) 10 °C and 1.0 atm
(d) 50 °C and 1.0 atm
(e) 50 °C and 0.5 atm

11.33 The solubility of ethylene (C_2H_4) in water at 20 °C and a pressure of 0.300 atm is 1.27×10^{-4} molal.
(a) Calculate the Henry's law constant for this gas in units of molal/torr.
(b) How many grams of ethylene are dissolved in one kilogram of water at 20 °C if the pressure of the gas is 500 torr?

11.34 The solubility of acetylene (C_2H_2) in water at 20 °C and a pressure of 0.200 atm is 9.38×10^{-3} molal.
(a) Calculate the Henry's law constant for this gas in units of molal/torr.
(b) How many grams of acetylene are dissolved in one kilogram of water at 20 °C if the pressure of the gas is 300 torr?

11.35 Why do carbonated beverages go flat if they are not stored in a tightly sealed container?

11.36 Explain why opening a warm carbonated beverage results in much more frothing than is observed when the container is refrigerated.

11.37 The enthalpy of solution of nitrous oxide (N_2O) in water is −12.0 kJ/mole, and its solubility at 20 °C and 1.00 atm is 0.121 g per 100 g of water.
(a) What is the Henry's law constant (in molal/torr) at 20 °C?
(b) At 10 °C and 1.00 atm, will the Henry's law constant be larger, smaller, or the same as at 20 °C and the same pressure?
(c) Calculate the molal solubility of nitrous oxide in water at 0.500 atm and 20 °C.

11.38 The enthalpy of solution of nitrous oxide (N_2O) in water is −12.0 kJ/mole, and its solubility at 20 °C and 2.00 atm is 0.055 molal. Would you expect the solubility of nitrous oxide to be larger or smaller than 0.055 molal at
(a) 2.00 atm and 0 °C?
(b) 2.00 atm and 40 °C?
(c) 1.00 atm and 20 °C?
(d) 1.00 atm and 50 °C?

11.39 Use the principle of Le Chatelier to explain why the solubilities of solids in a liquid change very little with pressure.

11.40 The solubility of calcium hydroxide in water is 0.165 g per 100 g of water at 20 °C and 0.128 g per 100 g of water at 50 °C. Is the enthalpy of solution for calcium hydroxide endothermic or exothermic?

11.41 From the data presented in Figure 11.10, which has the more positive enthalpy of solution, NaCl or KNO_3? Explain.

11.42 From the data presented in Figure 11.10, which has the more positive enthalpy of solution, NaCl or NH_4Cl? Explain.

Colligative Properties

11.43 Cyclohexane (C_6H_{12}) has a vapor pressure of 99.0 torr at 25 °C. What is the vapor pressure (in torr) of cyclohexane above a solution of 14.0 g of naphthalene ($C_{10}H_8$) in 50 g of cyclohexane at 25 °C?

11.44 The vapor pressure of chloroform ($CHCl_3$) is 360 torr at 40.0 °C. Find the vapor pressure lowering (in torr) produced by dissolving 10.0 g of phenol (C_6H_5OH) in 95.0 g of chloroform. What is the vapor pressure (in torr) of chloroform above the solution?

11.45 A solution contains 1.00 g of sucrose (molar mass = 342.3 g/mol) dissolved in 10.0 g of water. Using the data in Table 11.3, calculate the freezing and boiling points of the solution in degrees Celsius.

11.46 A solution contains 2.00 g of the nonvolatile solute urea (molar mass = 60.06 g/mol) dissolved in 25.0 g of water. Using the data in Table 11.3, calculate the freezing and boiling points of the solution in degrees Celsius.

11.47 The freezing point of cyclohexane is 6.50 °C. A solution that contains 0.500 g of phenol (molar mass = 94.1 g/mol) in 12.0 g of cyclohexane freezes at −2.44 °C. Calculate the freezing point depression constant for cyclohexane.

11.48 Cyclohexane has a normal boiling point of 80.72 °C. The solution in Exercise 47 boils at 81.94 °C. Find the boiling point elevation constant for cyclohexane.

11.49 A 0.500-g sample of nonvolatile, yellow crystalline solid dissolves in 15.0 g of benzene, producing a solution that freezes at 5.03 °C. Use the data in Table 11.3 to find the molar mass of the yellow solid.

11.50 A 0.350-g sample of a nonvolatile compound dissolves in 12.0 g of cyclohexane, producing a solution that freezes at 0.83 °C. Cyclohexane has a freezing point of 6.50 °C and a freezing point depression constant of 20.2 °C/molal. What is the molar mass of the solute?

11.51 Specific samples of aqueous solutions of sucrose and urea both freeze at −0.25 °C. What other properties of these two samples should be the same?

van't Hoff Factor

11.52 List the following aqueous solutions in order of increasing boiling points: 0.02 *m* LiBr, 0.03 *m* sucrose, 0.03 *m* $MgSO_4$, 0.03 *m* $CaCl_2$.

11.53 Arrange the following solutions in order of decreasing osmotic pressure: 0.10 *m* urea, 0.06 *m* NaCl, 0.05 *m* $Ba(NO_3)_2$, 0.06 *m* sucrose.

11.54 A sample of sea water freezes at −2.01 °C. What is the total molality of solute particles? If we assume that all of the solute is NaCl, how many grams of that compound are present in 1 kg of water? (Assume an ideal value for the van't Hoff factor.)

11.55 The saline solution used for intravenous injections contains 8.5 g of NaCl in 1.00 kg of water. Assuming an ideal value for the van't Hoff *i*, calculate the freezing point of this solution.

11.56 A 0.029 *M* solution of K_2SO_4 has an osmotic pressure of 1.79 atm at 25 °C.

(a) Calculate the van't Hoff factor, *i*, for this solution.
(b) Would the van't Hoff factor be larger or smaller for a 0.050 *M* solution of this compound?

Osmotic Pressure

11.57 A solution of 1.00 g of a protein in 20.0 mL of water has an osmotic pressure of 35.2 torr at 298 K. What is the molar mass of the protein?

11.58 A 10.0% solution of an enzyme in water has an osmotic pressure of 13.3 torr at 298 K. What is the molar mass of this enzyme? (Assume that the density of the solution is 1.00 g/mL.)

11.59 An 8.5-g sample of NaCl is dissolved in water to give one liter of solution at 298 K. What is the osmotic pressure of this solution? (Assume an ideal value for the van't Hoff factor.)

11.60 A 0.010 molar solution of sodium chloride is separated from pure water by a semipermeable membrane at 298 K. In which direction will net transport of water occur across the membrane when the applied pressure on the solution is 500 torr?

Solutions of Volatile Compounds

11.61 A mixture contains 15.0 g of hexane (C_6H_{14}) and 20.0 g of heptane (C_7H_{16}). At 40 °C the vapor pressure of hexane is 278 torr, and that of heptane is 92.3 torr. Assume that this is an ideal solution.

(a) What is the mole fraction of each of these substances in the liquid phase?
(b) What are the vapor pressures of hexane and of heptane above the solution?
(c) Find the mole fraction of each substance in the vapor phase.

11.62 A mixture contains 25.0 g of cyclohexane (C_6H_{12}) and 44.0 g of 2-methylpentane (C_6H_{14}). At 35 °C the vapor pressure of cyclohexane is 150 torr, and that of 2-methylpentane is 313 torr. Assume that this is an ideal solution.

(a) What is the mole fraction of each of these substances in the liquid phase?
(b) What are the vapor pressures of cyclohexane and of 2-methylpentane above the solution?
(c) Find the mole fraction of each substance in the vapor phase.

11.63 At a pressure of 760 torr, acetic acid (CH_3COOH, boiling point = 118.1 °C) and 1,1-dibromoethane ($C_2H_4Br_2$, boiling point = 109.5 °C) form an azeotropic mixture boiling at 103.7 °C that is 25% by mass acetic acid. At the boiling point of the azeotrope (103.7 °C) the vapor pressure of pure acetic acid is 471 torr, and that of pure 1,1-dibromoethane is 637 torr.

(a) If the solution had obeyed Raoult's law for both components, calculate the vapor pressure of each component and the total vapor pressure at 103.7 °C.
(b) Compare the answer to part (a) to the actual vapor pressure of the azeotrope. Is the deviation from Raoult's law positive or negative for this azeotropic mixture?
(c) Compare the attractive forces between acetic acid and dibromocthanc to thosc in the two pure substances.

11.64 At a pressure of 760 torr, formic acid (HCO_2H, boiling point = 100.7 °C) and water (H_2O, boiling point = 100.0 °C) form an azeotropic mixture boiling at 107.1 °C that is 77.5% by mass formic acid. At the boiling point of the azeotrope (107.1 °C) the vapor pressure of pure formic acid is 917 torr, and that of pure water is 974 torr.
(a) If the solution had obeyed Raoult's law for both components, calculate the vapor pressure of each component and the total vapor pressure at 107.1 °C.
(b) Compare the answer in part (a) to the actual vapor pressure of the azeotrope. Is the deviation from Raoult's law positive or negative for this azeotropic mixture?
(c) Compare the attractive forces between formic acid and water to those in the two pure substances.

Additional Exercises

11.65 A 10.00-mL sample of a 24.00% solution of ammonium bromide (NH_4Br) required 23.41 mL of 1.200 molar silver nitrate ($AgNO_3$) to react with all of the bromide ion present.
(a) What is the molarity of the ammonium bromide solution?
(b) Use the molarity of the solution to find the mass of ammonium bromide in one liter of this solution.
(c) From the percent concentration and the answer to (b), find the mass of one liter of ammonium bromide solution.
(d) Combine the answer to (c) with the volume of one liter to express the density of the ammonium bromide solution in g/mL.

11.66 At 10 °C the solubility of CO_2 gas in water is 0.240 g per 100 mL of water at a pressure of 1.00 atm. A soft drink is saturated with carbon dioxide at 4.00 atm and sealed.
(a) What mass of carbon dioxide is dissolved in a 12-ounce can of this beverage? (1 ounce = 28.35 mL)
(b) What volume of CO_2(g), measured at STP, is released when the 12-ounce can is left open for several days?

11.67 When 0.030 mol of HCl is dissolved in 100.0 g of benzene, the solution freezes at 4.04 °C. When 0.030 mol of HCl is dissolved in 100.0 g of water, the solution freezes at −1.07 °C. Use the data in Table 11.3.
(a) From the freezing point, calculate the molality of HCl in the benzene solution.
(b) Use the freezing point of the aqueous solution to find the molality of the solute in the water.
(c) Offer an explanation for the different values found in parts (a) and (b).

11.68 A 51.0-mL sample of a gas at 745 torr and 25 °C has a mass of 0.262 g. The entire gas sample dissolves in 12.0 g of water, forming a solution that freezes at −0.61 °C.
(a) Calculate the molar mass of the gas by using the ideal gas law.
(b) Calculate the molar mass from the freezing point of the solution, using the data in Table 11.3.
(c) Offer an explanation for the different values obtained in parts (a) and (b).

11.69 Hemoglobin contains 0.33% Fe by mass. A 0.200-g sample of hemoglobin is dissolved in water to give 10.0 mL of solution, which has an osmotic pressure of 5.5 torr at 25 °C. How many moles of Fe atoms are present in one mole of hemoglobin? (Hint: Calculate the molar mass from the osmotic pressure, and find the mass of iron in one mole of the compound.)

11.70 A benzene solution and a water solution of acetic acid (CH_3COOH, molar mass = 60.05 g/mol) are both 0.50 mass percent acid. The freezing point depression of the benzene solution is 0.205 °C, and that of the aqueous solution is 0.159 °C.
(a) Using the freezing point depression constants in Table 11.3, calculate the molality of the solute in each of the two solutions.
(b) What is the van't Hoff factor for the water solution and the benzene solution from these experiments?
(c) Note the difference and offer an explanation of the experimental results.

PYREX

Chapter Twelve

12

Chemical Equilibrium

Chemists have been studying chemical changes for centuries. When a chemical system is studied, one of the first investigations is to determine whether the reactants will combine with each other to form the desired products. The reactions presented thus far in this book are said to "go to completion," and we assumed that the amount of product formed is determined directly from the amount of the limiting reactant.

Many reactions, however, are not described by "will proceed" or "will not proceed." These other reactions achieve a condition of equilibrium, a state of balance or equality between opposing processes. At equilibrium, the tendency of the reactants to form products is balanced by the tendency of the products to form reactants, so a mixture of reactants and products results. We've already discussed some equilibrium processes in the discussion of phase changes. For example, at 0 °C, the liquid and solid forms of water are in equilibrium.

Knowledge of equilibrium will allow us to determine the extent of reaction, the amounts of products formed, and the amounts of reactants that remain. We can determine the conditions that favor the formation of products, and those that do not. Many important industrial processes are equilibrium reactions, carried out under conditions that provide the largest yield of product at the lowest cost.

This chapter will introduce chemical equilibrium primarily with homogeneous gas-phase reactions. The factors that influence the equilibrium, the response of a system at equilibrium to external changes, and calculations of the amounts of products and reactants are presented. Heterogeneous equilibria are also discussed when we examine the solubility of ionic compounds.

◀ Solid silver chromate forms as two liquids are mixed.

12.1 The Equilibrium Constant

Objectives

- To write the equilibrium constant expression for any chemical reaction
- To calculate the equilibrium constant or reaction quotient from appropriate experimental data
- To convert between equilibrium constants in which the concentrations of gases are expressed in mol/L and those in terms of partial pressures

When we characterize a reaction, one of the most important quantities to measure is the amount of product formed. Relationships between the concentrations of reactants and products have been observed to follow a pattern, which led to the development of a mathematical model that relates equilibrium concentrations of reactants and products. Equilibrium reactions are introduced in this section along with a model that describes the concentration relationships in equilibrium systems. Later sections expand on this material in the further development and uses of these principles.

Not all chemical reactions go to completion. For example, if nitrogen and hydrogen are mixed, some ammonia will form:

$$N_2(g) + 3H_2(g) \longrightarrow 2NH_3(g)$$

The formation of ammonia does not go to completion; the reverse reaction also occurs.

$$2NH_3(g) \longrightarrow 3H_2(g) + N_2(g)$$

The overall reaction reaches a balance when the forward reaction produces as much ammonia (from N_2 and H_2) per second as consumed in the reverse reaction. We say the reaction has achieved **equilibrium** at this point. At equilibrium, the concentrations of nitrogen, hydrogen, and ammonia remain constant because the forward and reverse reactions occur at the same rate. (Rates of phase-change processes were mentioned in Chapter 10, and rates of chemical reactions are discussed in detail in Chapter 16.) Systems at equilibrium are depicted with a double reaction arrow to indicate that both forward and reverse reactions are occurring.

$$N_2(g) + 3H_2(g) \rightleftharpoons 2NH_3(g)$$

The reaction can be tested to see whether it has reached equilibrium by adding some of one of the compounds involved in the reaction. If a little nitrogen is added to the equilibrium system, then additional ammonia will form. If some ammonia is added, then the reaction will proceed in the reverse direction, forming some nitrogen and hydrogen. When the concentrations stop changing, an equilibrium state has again been reached.

Although equilibrium systems are characterized by unchanging concentrations, not all systems with unchanging concentrations are at equilibrium. The changes in some systems occur too slowly to be easily measured, so testing the reaction, by adding a small amount of one of the substances involved, is always a good idea.

Chemists are careful not to try to apply equilibrium methods to reactions that proceed too slowly.

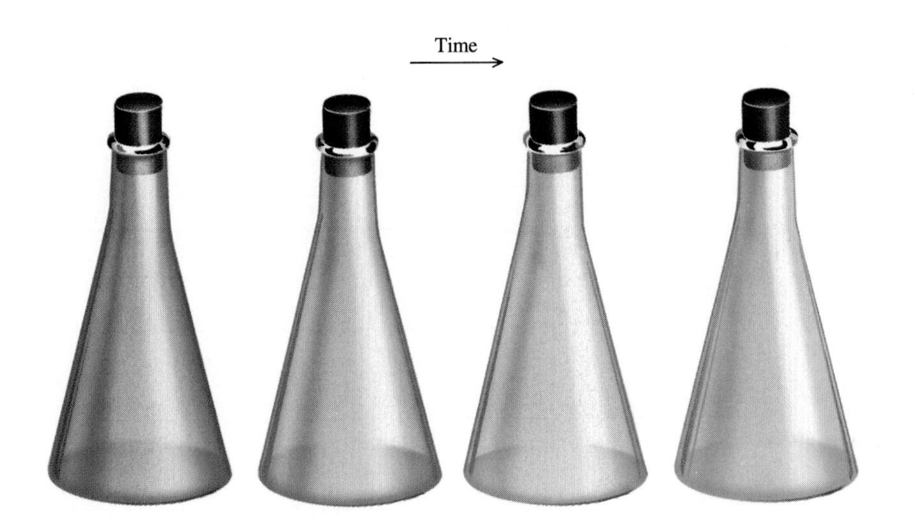

Figure 12.1 Chemical equilibrium A flask filled initially with brown NO_2 becomes lighter by forming the colorless N_2O_4 as the system reaches equilibrium.

Many of the most interesting chemical reactions are equilibrium systems. In this section, we will write the expressions that describe reactions at equilibrium, and study some of the factors that determine the direction a chemical reaction will proceed.

Consider the formation of N_2O_4 from NO_2, a reaction that proceeds at room temperature.

$$2NO_2(g) \rightleftharpoons N_2O_4(g)$$

This reaction is shown in Figure 12.1. If 0.920 g of nitrogen dioxide (equivalent to 0.0200 mol of NO_2) is sealed in a 1.0-L flask, the deep brown color of NO_2 begins to fade quickly as N_2O_4 (a colorless gas) is formed. The concentrations of NO_2 and N_2O_4 change as shown in Figure 12.2.

Initially, the reaction vessel contains only NO_2. During the course of the reaction, NO_2 is consumed and N_2O_4 is formed. Eventually, when the reaction reaches equilibrium, the concentrations of NO_2 and N_2O_4 cease to change. The results of a second experiment, one that starts with 0.920 g of

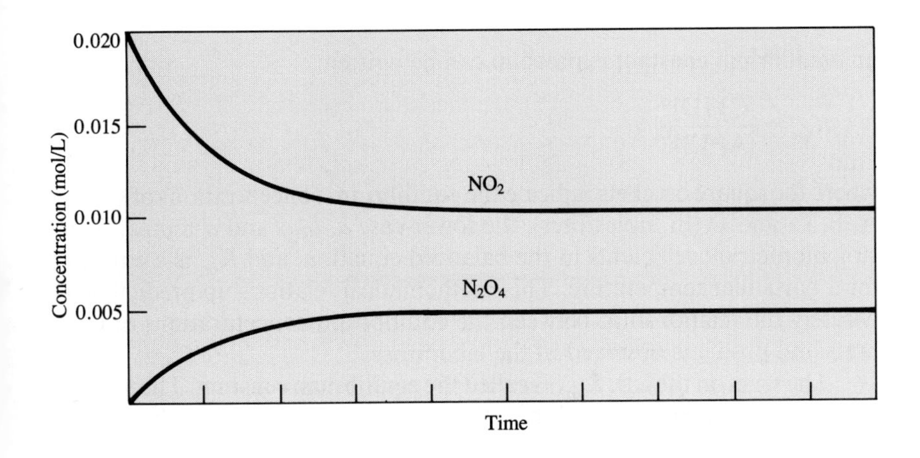

Figure 12.2 The nitrogen dioxide–dinitrogen tetroxide equilibrium Concentrations change as the system reaches equilibrium. Experiment 1: The initial concentrations are 0.0200 M NO_2 and no detectable N_2O_4.

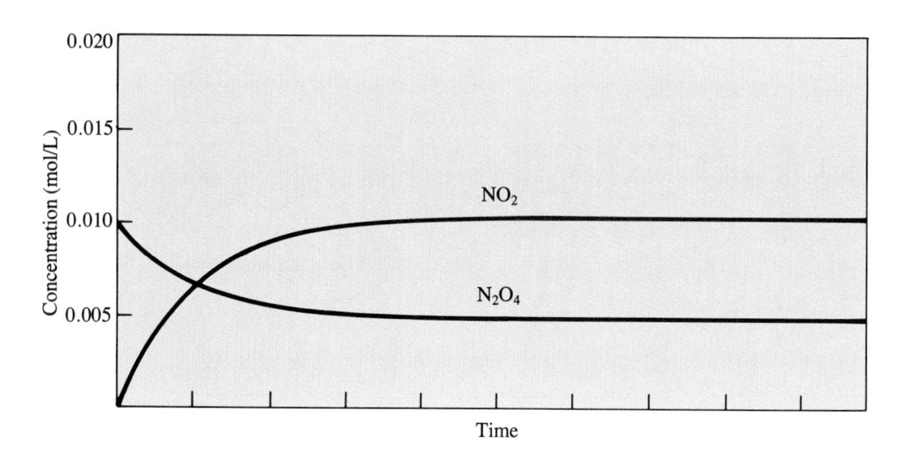

Figure 12.3 The nitrogen dioxide–dinitrogen tetroxide equilibrium Experiment 2: The initial concentrations are 0.0100 M N_2O_4 and no NO_2.

dinitrogen tetroxide (equivalent to 0.0100 mol N_2O_4) sealed in the 1.0-L flask, can also be observed. The results of experiment 2 are shown in Figure 12.3.

The equilibrium concentrations at the end of each experiment are seen to be the same. The reaction will produce the same equilibrium concentrations if initiated with reactants, the equivalent amount of products, or a mixture in which the total mass is the same. Note that the experiments we described are not easily done in the lab since the NO_2 and N_2O_4 are in equilibrium. Experiments that require pure NO_2 or pure N_2O_4 are difficult to perform.

The experiment can be repeated, again, with different starting concentrations. Table 12.1 shows the results of several different experiments.

The relationship between the equilibrium concentrations of NO_2 and N_2O_4 is not obvious. Chemists have studied such equilibrium reactions for many years, and numerous explanations of individual reactions were proposed. However, there was no general treatment until 1864 when two Norwegian scientists, Cato Guldberg (1836–1902) and Peter Waage (1833–1900), proposed the **law of mass action:** For any equilibrium reaction

$$aA + bB \rightleftharpoons cC + dD$$

an **equilibrium constant** expression can be written:

$$K_{eq} = \frac{[C]^c[D]^d}{[A]^a[B]^b} \qquad [12.1]$$

where the square brackets indicate the equilibrium concentrations of species A, B, C, and D (in moles/liter), the lower-case a, b, c, and d represent the stoichiometric coefficients in the balanced equation, and K_{eq} is a constant for a particular temperature. This mathematical relationship predicted accurately the relationships between the equilibrium concentrations of reactants and products observed in the laboratory.

The term on the left, K_{eq}, is called the **equilibrium constant.** The term on the right is generally referred to as the concentration product. The equilibrium constant expression is the product of the concentrations of the prod-

Table 12.1 Initial and Equilibrium Concentrations of Nitrogen Dioxide and Dinitrogen Tetroxide at 317 K

	Initial Concentrations, M		Equilibrium Concentrations, M	
	NO_2	N_2O_4	NO_2	N_2O_4
1	2.00×10^{-2}	0.00	1.03×10^{-2}	4.86×10^{-3}
2	0.00	1.00×10^{-2}	1.03×10^{-2}	4.86×10^{-3}
3	1.00×10^{-2}	1.00×10^{-2}	1.34×10^{-2}	8.29×10^{-3}
4	4.00×10^{-2}	0.00	1.61×10^{-2}	1.19×10^{-2}

ucts, each raised to the power of its stoichiometric coefficient, divided by the product of the concentrations of the reactants, each raised to the power of its stoichiometric coefficient. The value of the equilibrium constant is not affected by the starting concentrations of the species or the order in which the compounds are added, although it is influenced by temperature. We can use the equilibrium constant expression to predict the equilibrium concentrations of species from any starting concentrations.

The equilibrium constant expression for the formation of dinitrogen tetroxide from nitrogen dioxide can be written.

An equilibrium constant expression can be written for any chemical equation.

$$2NO_2(g) \rightleftharpoons N_2O_4(g)$$

$$K_{eq} = \frac{[N_2O_4]}{[NO_2]^2}$$

The equilibrium constant expression can be used together with the experimentally determined equilibrium concentrations to calculate the values of K_{eq} given in the last column of Table 12.2. Within experimental error, the ratio of the equilibrium concentration of N_2O_4 to the square of the equilibrium concentration of NO_2 is nearly constant and independent of the starting concentrations.

Table 12.2 Concentrations of Nitrogen Dioxide and Dinitrogen Tetroxide and the Equilibrium Constant at 317 K

	Initial Concentrations, M		Equilibrium Concentrations, M		
	NO_2	N_2O_4	NO_2	N_2O_4	K_{eq}
1	2.00×10^{-2}	0.00	1.03×10^{-2}	4.86×10^{-3}	45.8
2	0.00	1.00×10^{-2}	1.03×10^{-2}	4.86×10^{-3}	45.8
3	1.00×10^{-2}	1.00×10^{-2}	1.34×10^{-2}	8.29×10^{-3}	46.2
4	4.00×10^{-2}	0.00	1.61×10^{-2}	1.19×10^{-2}	45.9

Example 12.1 Writing the Equilibrium Constant Expression

Write the equilibrium constant expressions for the following reactions:

(a) $N_2(g) + O_2(g) \rightleftharpoons 2NO(g)$
(b) $PCl_5(g) \rightleftharpoons PCl_3(g) + Cl_2(g)$

Solution

(a) According to the law of mass action, the concentration of NO appears in the numerator, raised to the power 2, its coefficient in the chemical equation. The reactants, N_2 and O_2, appear in the denominator, raised to the power 1, their coefficients in the equation.

$$K_{eq} = \frac{[NO]^2}{[N_2][O_2]}$$

(b) The coefficients of all substances are 1, so the law of mass action gives

$$K_{eq} = \frac{[PCl_3][Cl_2]}{[PCl_5]}$$

Understanding Write the equilibrium constant expression for

$$2SO_3(g) \rightleftharpoons 2SO_2(g) + O_2(g)$$

Answer:

$$K_{eq} = \frac{[SO_2]^2[O_2]}{[SO_3]^2}$$

Calculating K_{eq}

The equilibrium constant must be determined by experiment. It is influenced by temperature, so the temperature must be stated.

The equilibrium constant is an experimentally determined quantity that provides information about the concentrations of reactants and products in an equilibrium mixture. It is determined by substituting measured concentrations of all substances present in an equilibrium mixture into the equilibrium constant expression. *The numerical value of K_{eq} depends on temperature, so the reaction temperature must be stated.*

Example 12.2 Calculating K_{eq}

A mixture of nitrogen and hydrogen was introduced into a reaction vessel at 532 °C. After equilibrium was reached, the concentration of nitrogen was 0.079 *M*, that of hydrogen was 0.12 *M*, and that of ammonia was 0.0051 *M*. Calculate the equilibrium constant for

$$N_2(g) + 3H_2(g) \rightleftharpoons 2NH_3(g)$$

Solution Using the chemical equation given, the equilibrium constant expression is

$$K_{eq} = \frac{[NH_3]^2}{[N_2][H_2]^3}$$

Substitute the equilibrium concentrations for all species into the equilibrium constant expression to calculate K_{eq}.

$$K_{eq} = \frac{(0.0051)^2}{(0.079)(0.12)^3}$$

$$K_{eq} = 0.19 \text{ at } 532\ °C$$

Understanding A mixture of phosphorus trichloride and chlorine is placed in a high-pressure reactor at 233 °C, forming phosphorus pentachloride:

$$PCl_3(g) + Cl_2(g) \rightleftharpoons PCl_5(g)$$

At equilibrium the concentrations are: $[PCl_3] = 0.051\ M$; $[Cl_2] = 0.072\ M$; $[PCl_5] = 0.23\ M$. Calculate the equilibrium constant for the reaction.

Answer: $K_{eq} = 63$

Relationships Between Pressure and Concentration

The equilibrium constant would appear to have units, but the common practice is to omit them. The concentrations used for each substance in the equilibrium constant are understood to have units of mol/L. There is no difficulty in accepting these units for solutes in solution, but the concentration of a gas is often expressed in terms of its partial pressure. In fact, the partial pressure of a gas is often more important. Both concentration units (atmospheres or mol/L) can be used for equilibrium calculations.

Consider the dissociation of phosgene, a highly toxic gas:

$$COCl_2(g) \rightleftharpoons CO(g) + Cl_2(g)$$

Two equilibrium constants can be defined for this reaction:

$$K_c = \frac{[CO][Cl_2]}{[COCl_2]}$$

$$K_p = \frac{P_{CO}P_{Cl_2}}{P_{COCl_2}}$$

A subscript describes the equilibrium constant: the c stands for concentration and the p stands for pressure. In equilibrium problems concentrations are expressed in molarity and pressures in atmospheres. Frequently we will use the symbol K_{eq} when we are discussing equilibrium constants in general or when the differences between K_c and K_p aren't important to the discussion.

If concentrations are given, K_c is easier to use. If the problem makes use of partial pressure data, then K_p is generally more convenient. The two equilibrium constants can be interconverted by using an expression that is derived (on the next page) from the ideal gas law:

The ideal gas law forms the basis for relating K_p and K_c.

$$K_p = K_c(RT)^{\Delta n} \qquad [12.2]$$

where Δn is the change in the number of moles of gas. We then calculate Δn from the coefficients in the chemical equation:

$$\Delta n = \text{total number of moles of gas on the product side}$$
$$- \text{ total number of moles of gas on the reactant side}$$

Δn is +1 in the reaction of phosgene dissociating to form carbon monoxide and chlorine.

Conversions between K_p and K_c are illustrated in Example 12.3.

Example 12.3 Convert Between K_p and K_c

Consider the following equilibrium:

$$PCl_3(g) + Cl_2(g) \rightleftharpoons PCl_5(g)$$

If the numerical value of K_p is 1.36 at 499 K, calculate K_c.

Solution K_c is calculated from Equation 12.2.

$$K_p = K_c(RT)^{\Delta n}$$

$$K_c = K_p/(RT)^{\Delta n} \quad \text{or}$$

$$K_c = K_p(RT)^{-\Delta n}$$

INSIGHTS into CHEMISTRY

A Closer View

Conversion Between Pressure and Concentration; Origin Lies in Ideal Gas Law

The ideal gas law describes the relationships between the pressure, temperature, volume, and number of moles of a gas.

$$PV = nRT$$

The concentration of the gas, in mol/L, can be described in terms of its pressure.

$$\frac{n}{V} = \frac{P}{RT}$$

When P is expressed in units of atm, 0.0821 L-atm mol^{-1} K^{-1} is used for R.

We can apply these relationships to the dissociation of phosgene.

$$COCl_2(g) \rightleftharpoons CO(g) + Cl_2(g)$$

The equilibrium concentrations for the reactants and products can be related to the pressures.

$$[COCl_2] = P_{COCl_2}/RT$$

$$[CO] = P_{CO}/RT$$

$$[Cl_2] = P_{Cl_2}/RT$$

These relations can be substituted into the equilibrium constant expression for the phosgene reaction.

$$K_c = \frac{[CO][Cl_2]}{[COCl_2]} = \frac{(P_{CO}/RT)(P_{Cl_2}/RT)}{(P_{COCl_2}/RT)}$$

$$= \frac{P_{CO}P_{Cl_2}}{P_{COCl_2}}\left(\frac{1}{RT}\right)$$

$$K_c = K_p(1/RT)$$

or

$$K_p = K_c(RT)$$

In general

$$K_p = K_c(RT)^{\Delta n}$$

where Δn is the change in the number of moles of gas. We calculate Δn from the balanced equation:

Δn = total number of moles of gas on the product side − total number of moles of gas on the reactant side

Δn is +1 for the decomposition of phosgene into carbon monoxide and chlorine.

First, calculate Δn. There are two moles of gas on the left of the chemical equation, and one mole on the right, so Δn is −1.0 in this case.

Now substitute the numerical values for K_p, R, and T in the equation.

$$K_c = K_p(RT)^{-\Delta n}$$

$$K_c = 1.36 \times (0.0821 \times 499)^{+1}$$

$$K_c = 55.7$$

Understanding If K_c for the reaction of nitrogen and hydrogen to form ammonia,

$$N_2(g) + 3H_2(g) \rightleftharpoons 2NH_3(g)$$

is 3.6×10^8 at 298 K, what is K_p?

Answer: $K_p = 6.0 \times 10^5$

12.2 The Reaction Quotient

Objectives

- To determine the relationship between the expression for the equilibrium constant and the form of the balanced equation

- To define and evaluate Q, the reaction quotient
- To compare Q and K_{eq} to establish in which direction a reaction proceeds when a system is not at equilibrium

A chemical equilibrium can be expressed in different ways. A reaction can be written in the forward direction to emphasize that reactants form products. The reaction might be written in the reverse direction to emphasize that products form reactants under the appropriate conditions. Both reactions express the same general ideas, and the equilibrium constants for both reactions are related.

The equilibrium constant can be used to determine the concentrations of a system at equilibrium, so it follows that we can determine if a given mixture of reactants and products will form more products, more reactants, or is at equilibrium. These topics and their applications to several chemical systems are presented in this section.

Relation of K_{eq} to the Form of the Chemical Equation

The equilibrium constant and its numerical value refer to a particular chemical equation. Since equations can be balanced in a number of ways, we must know the coefficients used in the particular chemical equation. Consider the formation of nitrogen monoxide from nitrogen and oxygen.

$$N_2(g) + O_2(g) \rightleftharpoons 2NO(g)$$

$$K_1 = \frac{[NO]^2}{[N_2][O_2]}$$

Had the equation been written

$$\tfrac{1}{2}N_2(g) + \tfrac{1}{2}O_2(g) \rightleftharpoons NO(g)$$

the equilibrium constant expression would be:

$$K_2 = \frac{[NO]}{[N_2]^{1/2}[O_2]^{1/2}}$$

When the two equilibrium constant expressions are compared, we see that $K_1 = K_2^2$. If each coefficient in the chemical equation is doubled, the equilibrium constant must be squared. If the coefficients are tripled, the equilibrium constant would be cubed. *The value of an equilibrium constant cannot be interpreted unless we know the corresponding chemical equation.*

The equilibrium constant depends on the form of the chemical equation.

If the reaction is written in the reverse direction, then the equilibrium constant is the inverse (negative first power) of the equilibrium constant for the forward reaction.

$$A \rightleftharpoons B \qquad K_{forward} = \frac{[B]}{[A]}$$

$$B \rightleftharpoons A \qquad K_{reverse} = \frac{[A]}{[B]}$$

$$K_{forward} = (K_{reverse})^{-1}$$

$$K_{forward} = 1/K_{reverse}$$

Example 12.4 illustrates how K_{eq} can be determined for different forms of the chemical equation.

Example 12.4 Dependence of K_{eq} on the Form of the Chemical Equation

The reaction of hydrogen and nitrogen to form ammonia can be written in several different ways:

$$(1) \quad N_2(g) + 3H_2(g) \rightleftharpoons 2NH_3(g)$$

$$(2) \quad \tfrac{1}{2}N_2(g) + \tfrac{3}{2}H_2(g) \rightleftharpoons NH_3(g)$$

$$(3) \quad 2NH_3(g) \rightleftharpoons N_2(g) + 3H_2(g)$$

The equilibrium constant, K_c, for Reaction 1 is 0.19 at 532 °C. Write the equilibrium constant expression and calculate K_c for Reaction 2 and Reaction 3.

Solution First write the equilibrium constant expression for each equation.

$$K_1 = \frac{[NH_3]^2}{[N_2][H_2]^3} = 0.19$$

$$K_2 = \frac{[NH_3]}{[N_2]^{1/2}[H_2]^{3/2}}$$

$$K_3 = \frac{[N_2][H_2]^3}{[NH_3]^2}$$

Examination of K_1 and K_2 shows that

$$K_1 = K_2^2$$

This equation can be checked by squaring each term in the equilibrium constant expression for K_2 and verifying that K_2^2 is equal to K_1. Thus,

$$K_2^2 = 0.19$$

Take the square root:

$$K_2 = 0.44$$

Note that when the reaction is halved, the new K is the square root (1/2 power) of the old K.

Similar logic will show that

$$K_3 = 1/K_1$$

$$K_3 = 1/0.19$$

$$K_3 = 5.3$$

All the values of K are valid, and contain equivalent information.

Understanding Use the preceding data to calculate K_c at 532 °C for

$$NH_3(g) \rightleftharpoons \tfrac{1}{2}N_2(g) + \tfrac{3}{2}H_2(g)$$

Answer: $K_c = 2.3$

When chemical equations are added to yield a new equation, K_{eq} for the new reaction is determined by multiplying the equilibrium constants of the component equations. We can calculate the equilibrium constant for the formation of $NO_3(g)$ as follows:

$$(1) \quad 2NO_2(g) \rightleftharpoons N_2O_4(g) \qquad K_1$$

$$(2) \quad N_2O_4(g) + O_2(g) \rightleftharpoons 2NO_3(g) \qquad K_2$$

$$(3) \quad 2NO_2(g) + O_2(g) \rightleftharpoons 2NO_3(g) \qquad K_3$$

INSIGHTS into CHEMISTRY

Science, Technology, and Society

Nitrogen Oxides May Help Form Smog. Solutions to Problems Are Not Clear

Several of the preceding examples have examined the chemistry of the oxides of nitrogen. These compounds are widely studied because they are implicated in the formation of smog.

Scientists are reasonably certain that oxides of nitrogen are important components of smog. These compounds are formed in combustion processes in general, and in internal combustion engines (automobile engines) in particular. Although the combustion of a hydrocarbon might be written as

$$CH_4 + 2O_2 \longrightarrow CO_2 + 2H_2O$$

the actual process is more complicated. Since air, which is about 80% nitrogen, is used for combustion, some of the nitrogen in the air reacts with oxygen in the high-temperature, high-pressure environment of an automobile engine. One result is the formation of various oxides of nitrogen. This family of oxides ranges from N_2O to N_2O_5 and is referred to as NO_x. The chemistry of the oxides of nitrogen is presented in Chapter 19.

Solutions to the smog problem have been proposed. Some people believe that the best solution lies in limiting automobile usage. Others would like to see automobile engines that minimize nitrogen oxide formation by controlling the combustion temperature. Still others would like to see additives developed for gasolines that will limit pollution. There is no single solution that can be labeled "best" at this time; we will probably need several methods. But while the research goes forward, steps that include conservation as well as improved engine design should be followed.

People are often surprised to learn that smog has a past history; many think of smog as a "modern" occurrence. In fact, smog has been present since the erection of the first smokestacks of the Industrial Revolution. Our knowledge of smog, its properties, and its hazards have been increasing, but much of the chemistry of smog is still unknown.

Smog Smog forms from a complex mixture of substances, including combustion products and sunlight.

Automobile exhaust is an important contributor to smog.

Reaction 3 is obtained by adding Reaction 1 to Reaction 2. Likewise, K_3 is the product of K_1 and K_2.

$$K_3 = K_1K_2$$

This relationship can be verified by substituting the concentration expression for each of the three equilibrium constants:

$$K_3 = K_1K_2$$

$$\frac{[NO_3]^2}{[NO_2]^2[O_2]} = \frac{[N_2O_4]}{[NO_2]^2} \times \frac{[NO_3]^2}{[N_2O_4][O_2]} = \frac{[NO_3]^2}{[NO_2]^2[O_2]}$$

Reaction Quotient

The law of mass action not only describes equilibrium systems, but provides important information about systems not yet at equilibrium. The **reaction quotient,** Q, has the same algebraic form as K_{eq}, but the *current concentrations,* rather than the equilibrium concentrations, are used in the calculation. Comparing Q to K_{eq} allows us to predict the direction in which a reaction will proceed to achieve equilibrium.

The reaction quotient, Q, is determined in the same way as K_{eq}, but the concentrations are measured at *any* time. At equilibrium, of course, Q is equal to K_{eq}.

If we examine the general chemical reaction

$$aA + bB \rightleftharpoons cC + dD$$

and use the starting concentrations of a reaction rather than equilibrium concentrations, then the expression is written

$$Q = \frac{[C]^c[D]^d}{[A]^a[B]^b}$$

The numerical value of Q determines the direction in which the net reaction must proceed to reach equilibrium. *The system will always change to make Q closer in value to K_{eq}.*

If $Q < K_{eq}$, the reaction proceeds toward products. If $Q > K_{eq}$, the reaction proceeds toward the reactants.

If Q is less than K_{eq}, the reaction proceeds to the right. The system reacts to increase the concentrations of the products and decrease the concentrations of the reactants, thus increasing Q and bringing it closer in value to that of K_{eq}. Mathematically, the numerator gets larger (since the concentrations of the products increase) and the denominator gets smaller. When Q is greater than K_{eq}, the reaction will proceed toward reactants, to the left.

Suppose a reaction mixture at 532 °C has the following initial concentrations:

$$[NH_3] = 0.10\ M$$
$$[H_2] = 0.20\ M$$
$$[N_2] = 0.30\ M$$

We know (from Example 12.2) that at 532 °C, K_c is 0.19 for

$$N_2(g) + 3H_2(g) \rightleftharpoons 2NH_3(g)$$

We substitute the *starting concentration* of each substance to evaluate the expression for Q.

$$Q = \frac{[NH_3]^2}{[N_2][H_2]^3} = \frac{(0.10)^2}{(0.30)(0.20)^3} = 4.2$$

At equilibrium, at 532 °C, this quotient must be 0.19, so the chemical system will change to decrease this quotient by forming additional N_2 and H_2 from the excess NH_3.

Another way to compare Q to K_c is by a number line, shown in Figure 12.4.

Figure 12.4 Number line representation

Table 12.3 Determination of Direction of Reaction

Experiment	Initial Concentrations, M		Q	Direction of Reaction
	NO_2	N_2O_4		
1	1.00	0.00	0.00	right
2	0.30	0.010	0.11	right
3	0.20	0.018	0.45	equilibrium
4	0.50	0.25	1.0	left
5	0.00	1.0	very large	left

The symbols for Q and K_c are placed on a scale, at positions related to their numerical values. In this example, Q is to the right of K_c, so the system will react to move Q to the left, toward K_c, by forming additional reactants. The direction in which Q moves on the number line is the same direction in which the reaction proceeds.

The same general approach is illustrated in Table 12.3, which presents values of Q for several different mixtures of NO_2 and N_2O_4. The chemical equation is

$$2NO_2(g) \rightleftharpoons N_2O_4(g) \qquad K_c = 0.45 \text{ at } 135\ °C$$

Comparing Q to K_c allows us to predict the direction in which the system responds. Note that any equilibrium mixture must contain some of each species, so if any species is missing, as on lines 1 and 5 of the table, then the reaction will proceed to form the missing substance.

The number line shown in Figure 12.5 is a graphical presentation of the data in Table 12.3. The reaction proceeds to bring Q closer to K_c. The reaction mixtures in experiments 1 and 2 will proceed to the right, to form more N_2O_4. The reaction mixtures in experiments 4 and 5 will proceed to the left, forming more NO_2. The mixture in experiment 3 is at equilibrium.

Figure 12.5 Number line representation of Table 12.3. Q_n refers to Q calculated for Experiment n in Table 12.3.

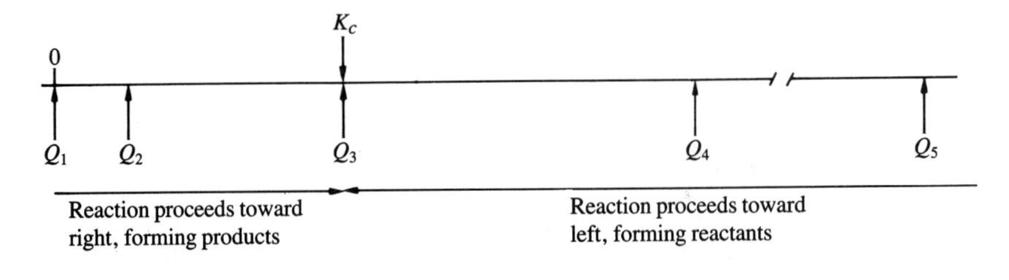

Example 12.5 Determining the Direction of Reaction

If 0.50 mol of NO_2 is mixed with 0.30 mol of N_2O_4 in a 2.0-L flask at 418 K, will the reaction form more NO_2 or form more N_2O_4, or is the system at equilibrium? K_c is 0.32 for the reaction of two moles of NO_2 to form one mole of N_2O_4 at 418 K.

Solution First, write the chemical equation and the equilibrium constant expression:

$$2NO_2(g) \rightleftharpoons N_2O_4(g)$$

$$K_c = \frac{[N_2O_4]}{[NO_2]^2} = 0.32$$

Next, calculate the initial *concentration* of each species:

$$[NO_2] = 0.50 \text{ mol}/2.0 \text{ L} = 0.25\ M$$

$$[N_2O_4] = 0.30 \text{ mol}/2.0 \text{ L} = 0.15\ M$$

Last, calculate Q and compare it to K_{eq}.

$$Q = \frac{0.15}{(0.25)^2} = 2.4$$

Since Q is *greater* than K_{eq}, the reaction will proceed to the left, and more NO_2 will be formed.

Understanding If 0.24 mol of NO_2 is mixed with 0.080 mol of N_2O_4 in a 2.0-L flask at 418 K, will the reaction form more NO_2 or form more N_2O_4, or is the system at equilibrium?

Answer: Q is 2.8, greater than K_{eq}, so the reaction will proceed to the left, forming more NO_2.

12.3 The Principle of Le Chatelier

Objectives

- To determine how the principle of Le Chatelier predicts the response of an equilibrium system when conditions are changed
- To determine how changes in temperature influence the equilibrium system

Henri Louis Le Chatelier

The size of the equilibrium constant helps to determine whether a reaction tends to favor products or reactants. If K_{eq} is large (much greater than 1), the reaction favors the formation of product; a small K_{eq} favors the reactants. Factors that influence the magnitude of K_{eq} will be discussed in this section.

The composition of the system at equilibrium can change if the partial pressures or concentrations of any of the gaseous reactants or products are changed. The direction in which the chemical reaction will proceed is predicted by the principle of Le Chatelier, which we introduced in Chapter 10 in the study of phase equilibria.

Principle of Le Chatelier

The response of a chemical reaction at equilibrium to an external stress was described qualitatively by Henri Louis Le Chatelier (1850–1936). In 1884 he summarized his observations of chemical equilibria:

"Every system in a stable chemical equilibrium submitted to the influence of an exterior force which tends to cause variation in either its temperature or its condensation (pressure, concentration, or number of molecules in the unit of volume) . . . can undergo only those interior

*modifications which, if they occur alone, would produce a change of temperature, or of concentration, of a sign contrary to that resulting from the exterior force."**

Le Chatelier's principle can be restated in modern language: *any change to a chemical reaction at equilibrium causes the reaction to proceed in the direction that will reduce the effect of the change.*

Systems at equilibrium resist change. Shifts in conditions such as concentration, pressure, and temperature cause a reaction to occur in the direction that reduces the impact of the shift. Consider the production of ammonia:

$$N_2(g) + 3H_2(g) \rightleftharpoons 2NH_3(g)$$

If hydrogen were added to an equilibrium mixture of nitrogen, hydrogen, and ammonia, then the system would no longer be at equilibrium and Q would be less than K_{eq}. The equilibrium will be restored as the system reacts to consume some of the added hydrogen. On the other hand, if hydrogen were removed, the system would react to produce more hydrogen (and nitrogen).

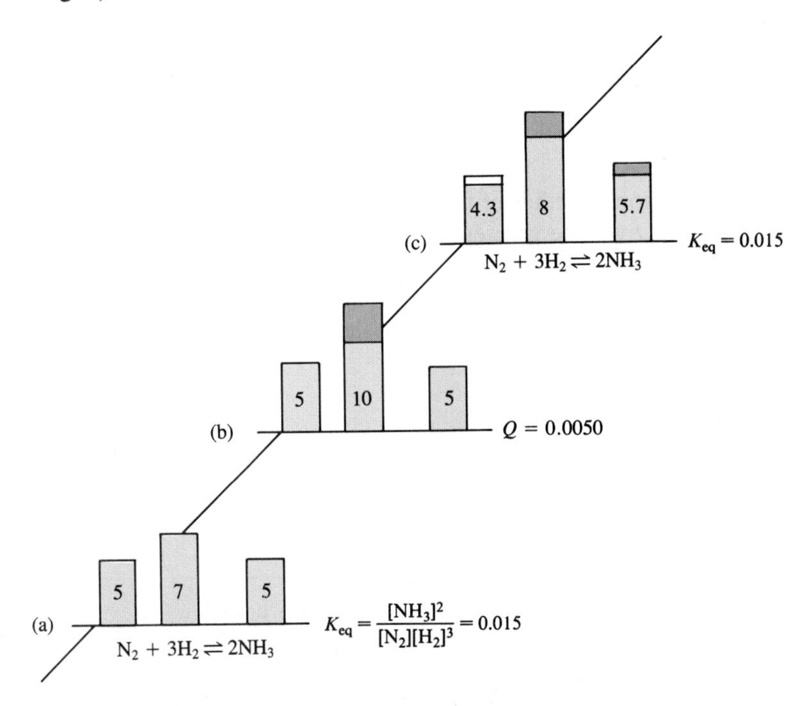

Le Chatelier's principle: The system responds to reduce the change (a) Nitrogen, hydrogen, and ammonia are at equilibrium. (b) The hydrogen concentration is increased. (c) The system reacts to consume some of the added hydrogen. The reaction will decrease the concentration of nitrogen and increase the concentration of ammonia. The systems shown in part (a) and part (c) are at equilibrium. System (b) is not.

The principle of Le Chatelier can be used in a practical way to increase the yield of a reaction. If ammonia were constantly removed from the system, then more ammonia would form, and the reaction would proceed until either the nitrogen or the hydrogen was exhausted. Ammonia is an important industrial chemical, with uses ranging from fertilizer to rocket fuels, so efficient production is an important issue.

* H. M. Leicester and H. S. Klickstein, *A Source Book in Chemistry,* Harvard University Press, Cambridge, Massachusetts, 1963, p. 481.

If the products of a reaction can be removed from the reaction mixture, the system will respond by producing additional products.

Ammonia is easy to liquefy at modest pressures, but nitrogen and hydrogen are not. A reactor can be designed to remove liquid ammonia formed by the reaction of gaseous N_2 and H_2. The system will respond by producing more ammonia, and will continue to produce ammonia until the nitrogen or hydrogen is consumed.

Changes in Concentration or Partial Pressure

When a reaction system is at equilibrium, the concentrations of the species are governed by the law of mass action, and the reaction quotient Q is equal to K_{eq}. If the concentration of any of the species is changed, then Q and K_{eq} are no longer equal, and the reaction will proceed in the direction that consumes the added substance. Scientists often speak colloquially of the "shift in equilibrium" as conditions are changed. The equilibrium constant certainly does not shift—the *composition* changes to produce additional products or reactants so that equilibrium is restored.

Changes in composition can be seen when the reaction of phosgene to form carbon monoxide and chlorine is studied.

$$COCl_2(g) \rightleftharpoons CO(g) + Cl_2(g)$$

We can write the equilibrium constant expression:

$$K_{eq} = \frac{[CO][Cl_2]}{[COCl_2]}$$

If the concentration of either of the products in an equilibrium mixture is increased, then Q will be greater than K_{eq} and the reaction will proceed toward the left, to form more phosgene, and consume some of the added substance.

Example 12.6 demonstrates how a system responds when the equilibrium concentrations are disturbed.

INSIGHTS into CHEMISTRY

Chemical Safety

Poisonous Phosgene Becoming Uncommon As Carbon Tetrachloride Usage Decreases

Phosgene, $COCl_2$, is a deadly gas. It was one of the first chemical warfare agents and was used in poison gas artillery shells in World War I. Phosgene is denser than air, so it would remain in the trenches.

The health hazard is thought to be due to the slow reaction of phosgene with water in the lungs, producing HCl. The lungs begin to fill with water and become inefficient and unable to supply oxygen. Death results within 6 to 24 hours.

Phosgene can be formed by heating some chlorinated compounds, such as carbon tetrachloride (CCl_4), in the presence of oxygen. CCl_4 was once used in fire extinguishers, particularly those used for electrical fires. After chemists determined that phosgene was formed, the use of CCl_4 in fire extinguishers was discontinued.

Carbon tetrachloride was once a popular dry-cleaning fluid and found in several widely used products. It is no longer available to consumers—not only have superior alternatives been found, but CCl_4 is believed to cause cancer. Carbon tetrachloride is still used in many chemistry labs, where access to hoods and waste disposal systems limits the hazards, but even in labs, chemists think of alternatives before using CCl_4.

Example 12.6 Predicting the Direction of Reaction When the System is Disturbed

When oxygen is added to an equilibrium mixture of oxygen, sulfur dioxide, and sulfur trioxide, in which direction will the reaction proceed?

$$SO_2(g) + \tfrac{1}{2}O_2(g) \rightleftharpoons SO_3(g)$$

Solution The equilibrium constant expression is

$$K_{eq} = \frac{[SO_3]}{[SO_2][O_2]^{1/2}}$$

If the concentration of O_2 is increased, then the reaction will proceed in the direction that decreases the concentration of the added substance. Oxygen and SO_2 will be consumed and more SO_3 will be formed.

Another approach to determining the direction of reaction involves comparing Q and K_{eq}. Since the denominator in the mass action expression is increased by the addition of O_2, Q becomes smaller than K_{eq}, and the reaction will proceed to the right.

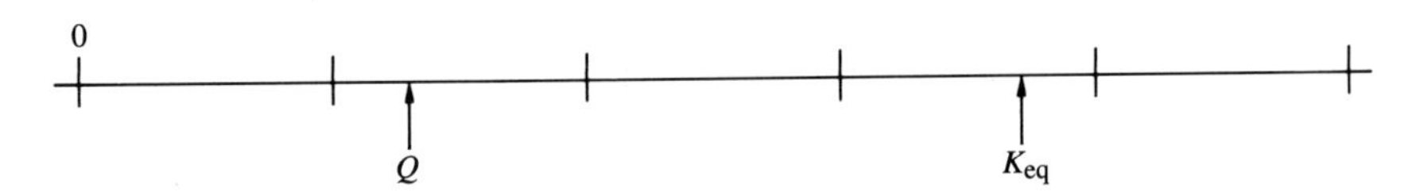

Note that neither K_{eq} nor Q must be evaluated to determine in which direction a reaction proceeds. It will proceed to consume the added substance, or to replace one that has been removed.

Sulfur dioxide is produced when many fossil fuels, particularly coal, are burned—sulfur is a common impurity in these fuels. Coals are classified as "low-sulfur" if the sulfur content is 0.6 to 1%. High-sulfur coals contain up to 4% sulfur. Some of the SO_2 reacts with oxygen to form SO_3, and ultimately H_2SO_4, an important component of acid rain.

Coal Coal is a widely used fuel in the power generation industry. The United States has a 300-year supply of coal.

Understanding In which direction will the preceding reaction proceed if SO_2 is removed from the equilibrium system?

Answer: To the left, to produce more SO_2.

Changes in the pressures of gases cause the same effects as changes in concentration—pressure is just another measure of concentration. A change in the partial pressures of the reacting species can be achieved either by adding (or removing) some of the reactants or products, or by changing the volume of the reaction vessel. The next example demonstrates the effect of changing the volume of the container.

Increasing the partial pressure increases the concentration, and the system responds in the same manner.

Example 12.7 The Response of an Equilibrium System to Changes in Volume

Some PCl_5 is placed in a 10.0-L reaction cylinder. The temperature is raised to 500 K and the following reaction reaches equilibrium.

$$PCl_5(g) \rightleftharpoons PCl_3(g) + Cl_2(g)$$

In which direction will the system react if the volume is decreased to 1.0 L while keeping the temperature constant at 500 K?

Solution To determine the direction in which the reaction will occur, we compare the values of Q and K_{eq}. From the law of mass action, when the system is at equilibrium,

$$K_{eq} = \frac{[PCl_3][Cl_2]}{[PCl_5]}$$

When the volume is reduced from 10.0 L to 1.0 L, the concentration of each species is increased by a factor of 10, since the same number of moles will be present in one tenth the volume. The new concentrations will be:

$$[PCl_5]' = 10\ [PCl_5]$$
$$[PCl_3]' = 10\ [PCl_3]$$
$$[Cl_2]' = 10\ [Cl_2]$$

where the [ ]′ indicates the *new* concentration. The new concentrations can be substituted into the mass action expression:

$$Q = \frac{[PCl_3]'[Cl_2]'}{[PCl_5]'}$$
$$= \frac{10\ [PCl_3] \times 10\ [Cl_2]}{10\ [PCl_5]}$$
$$= 10\ \frac{[PCl_3][Cl_2]}{[PCl_5]}$$
$$Q = 10\ K_{eq}$$

Since Q is greater than K_{eq}, the reaction will proceed to the left, forming more $PCl_5(g)$.

Number line representation

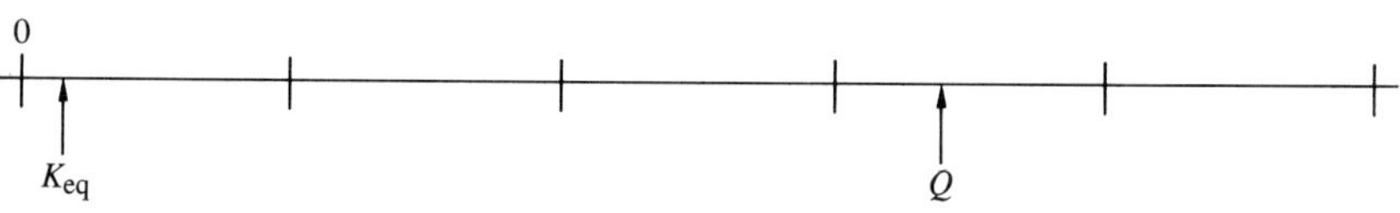

Understanding A gas-phase N_2–H_2–NH_3 system is at equilibrium. In which direction will the reaction ($N_2 + 3H_2 \rightleftharpoons 2NH_3$) proceed if the volume of the container is decreased?

Answer: The system will produce additional NH_3.

To determine the direction of reaction, either use the principle of Le Chatelier to determine the direction in which the system will respond to a change, or calculate Q and compare it to K_{eq}.

Comparing Q and K_{eq} is not the only way to predict how a system will react; the principle of Le Chatelier can be used to determine the direction of reaction on a qualitative basis. In Example 12.7, a chemical system at equilibrium was changed by decreasing the volume. The system will react to counteract the pressure increase by reducing the total number of moles of gas. Since there is one mole of gas on the left side (reactant side) of the chemical equation and two moles of gas on the right, the reaction will proceed to the left, forming additional $PCl_5(g)$ (and reducing the total number of moles of gas in the system). Note that if a reaction has the same number of moles of gas on both sides, then changes in pressure will not cause any net reaction to occur.

Sometimes an inert or nonreactive material is added to a reaction container. As long as these other substances do not affect the partial pressures of the reactants or products, *the pressure of materials other than the reactants or products has no effect on the equilibrium.* The nitrogen dioxide–dinitrogen tetroxide equilibrium can be studied in the presence or absence of other gases and the results will be the same.

INSIGHTS into CHEMISTRY

A Closer View

Decreasing Volume Affects Equilibrium: Colored NO_2 Changes to Colorless N_2O_4

Nitrogen dioxide, NO_2, can be formed by reaction of sulfuric acid with sodium nitrite ($NaNO_2$). The brown NO_2 gas can be withdrawn into a syringe, and then the syringe can be sealed. The gases inside the syringe provide a chemical system that is easy to study. We can use this system to determine the influence of volume on equilibrium.

The equilibrium studied is that between nitrogen dioxide and dinitrogen tetroxide:

$$\underset{\text{brown}}{2NO_2(g)} \rightleftharpoons \underset{\text{colorless}}{N_2O_4(g)}$$

If we decrease the volume by pressing on the syringe plunger, the NO_2 concentration will temporarily be increased. The increased concentration will deepen the color, but the system will quickly react to form additional N_2O_4. In doing so, the system is responding to the change (smaller volume and greater pressure) by producing fewer gas molecules, since one mole of N_2O_4 is formed from two moles of NO_2.

Since N_2O_4 is colorless, the gas in the syringe becomes lighter and closer to the color before the volume was reduced.

If we increase the volume of the syringe by withdrawing the syringe plunger, then the mixture becomes lighter, since the NO_2 concentration decreases. The system will again restore equilibrium by forming more NO_2.

Changing the volume (a) When the volume is decreased, the system acts to form more N_2O_4. (b) When the volume is increased, more NO_2 is formed. (c) The experiment makes a good classroom demonstration.

(a)

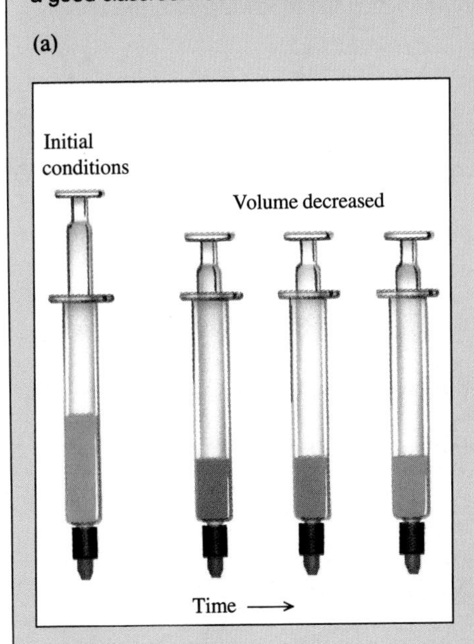

(b)

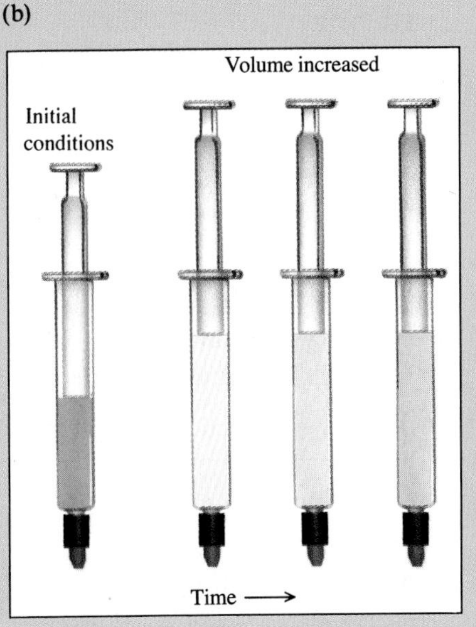

(c)

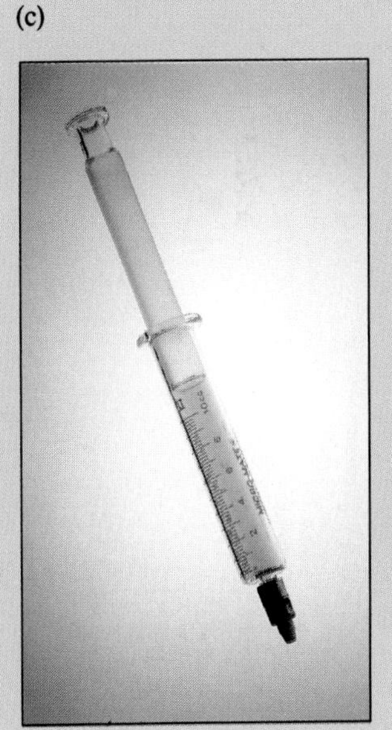

Changes in Temperature

As we saw in Chapter 10, the principle of Le Chatelier predicts the response of an equilibrium system to changes in temperature. If a reaction is exothermic, then adding heat to the system causes the reaction to proceed to the left,

Heating an endothermic reaction increases the concentration of products. Heating an exothermic reaction increases the concentration of reactants.

forming reactants (and consuming the added heat). Heating an endothermic reaction will cause the system to shift to form additional products. *Changing the temperature of a reaction changes the value of K_{eq}, in the direction predicted by the principle of Le Chatelier.*

Sulfur trioxide is produced by the reaction of sulfur dioxide and oxygen.

$$2SO_2(g) + O_2(g) \rightleftharpoons 2SO_3(g)$$

When this reaction is studied at laboratory temperatures, the reaction proceeds toward the formation of SO_3. At high temperatures, such as those found in a furnace, sulfur trioxide decomposes to sulfur dioxide and oxygen.

These results are consistent with the principle of Le Chatelier. The formation of sulfur trioxide is an exothermic reaction; when heated, the reaction will form additional reactants as a reaction occurs to consume the added heat.

The numerical value of K_{eq} changes with temperature, so it is important to specify the temperature when describing a system at equilibrium. The exact dependence of K_{eq} on temperature is presented in Chapter 15; the main point is that K_{eq} changes as predicted by the principle of Le Chatelier.

Example 12.8 The Influence of Temperature Changes on Equilibria

How will an increase in the temperature influence the following equilibria?

(a) $H_2(g) + I_2(g) \rightleftharpoons 2HI(g)$ $\Delta H = +52$ kJ
(b) $N_2(g) + 3H_2(g) \rightleftharpoons 2NH_3(g)$ $\Delta H = -92$ kJ

Solution

(a) The reaction is *endothermic;* the equilibrium constant will increase at higher temperatures.
(b) The reaction is *exothermic;* the equilibrium constant will decrease with increasing temperature.

Understanding ΔH is -112 kJ for

$$CO(g) + Cl_2(g) \rightleftharpoons COCl_2(g)$$

In which direction will the system react if temperature is *decreased?*

Answer: The reaction will shift to the right, forming more phosgene.

12.4 Calculations Using Equilibrium Constants

Objectives

- To employ a systematic approach to chemical equilibria
- To calculate the equilibrium concentrations of species in a chemical reaction

The direction in which a reaction proceeds can be determined by calculating Q from the experimental data, and comparing it to K_{eq}, but the usefulness of the equilibrium constant is not limited to determining the direction of reaction. Once the value of K_{eq} is known, it can be used to find the composition of any mixture of reactants and products at equilibrium. This section will present a systematic approach to help solve equilibrium problems.

Consider the reaction

$$SO_2(g) + NO_2(g) \rightleftharpoons SO_3(g) + NO(g)$$

This reaction was once important in the industrial production of sulfuric acid. The equilibrium constant, K_c, for this reaction is 6.85 at 821 °C. How much SO_3 is formed when 0.50 mol of SO_2 and 0.50 mol of NO_2 are placed in a 1.00-L container at 350 K?

Qualitatively, we can see that as the reaction proceeds, some of the starting materials will be consumed to form products. The amount of products formed and starting materials consumed is unknown; our problem is to calculate it.

Equilibrium problems are best solved by first constructing a table of concentrations. The table has a column for each of the compounds that appear in the reaction. It contains lines for the initial concentration **(i)**, change due to the reaction **(C)**, and the equilibrium concentration **(e)**, and we will call it the iCe table. The upper-case C serves to emphasize that the changes depend on the reaction stoichiometry. We will substitute expressions for the equilibrium concentrations of all species into the equilibrium constant expression, and then solve for the unknown quantity.

Start by writing the blank table:

	SO_2 +	NO_2 ⇌	SO_3 +	NO
initial concentration, M				
Change in concentration, M				
equilibrium concentration, M				

The first step in solving an equilibrium problem is to construct the iCe table.

Next, write the initial concentrations.

	SO_2 +	NO_2 ⇌	SO_3 +	NO
initial concentration, M	0.50	0.50	0.00	0.00

The reaction proceeds to form both SO_3 and NO, but the concentration of each product formed is unknown. We will designate the change in the concentration of SO_2 by $-y$. The − indicates that it is *consumed* rather than formed. The stoichiometric coefficients are used to determine the changes in all the other species in terms of y. The chemical equation tells us that for each mole of SO_2 consumed, we also consume the same amount of NO_2. In addition, we form the same amounts of SO_3 and NO. This information enables us to fill the second row of the iCe table.

	SO_2	+ $NO_2 \rightleftharpoons$	SO_3	+ NO
initial concentration, M	0.50	0.50	0.00	0.00
Change in concentration, M	$-y$	$-y$	$+y$	$+y$
equilibrium concentration, M				

The equilibrium concentration of each compound is the sum of the initial concentration and the change in concentration due to the reaction. We fill the last cell in each column of the table by adding the two cells above it:

	SO_2	+ $NO_2 \rightleftharpoons$	SO_3	+ NO
initial concentration, M	0.50	0.50	0.00	0.00
Change in concentration, M	$-y$	$-y$	$+y$	$+y$
equilibrium concentration, M	$0.50 - y$	$0.50 - y$	y	y

After writing the table, use the information to write an equilibrium constant expression that includes the unknown concentrations.

The final step is to substitute the equilibrium concentrations (the last line in the iCe table) into the equilibrium constant expression. We solve the resulting equation to determine y, the concentration of SO_3.

$$K_c = \frac{[SO_3][NO]}{[SO_2][NO_2]} = 6.85$$

$$\frac{(y)(y)}{(0.50 - y)(0.50 - y)} = 6.85$$

$$\frac{y^2}{(0.50 - y)^2} = 6.85$$

We solve for y by first taking the square root of both sides,

$$\frac{y}{0.50 - y} = 2.617$$

Rearrange,

$$y = 1.3 - 2.617\,y$$

combine terms,

$$3.617\,y = 1.3$$

and solve for y:

$$y = 0.36\ M$$

We can use this number (together with the iCe table) to calculate several quantities:

The concentration of SO_3 formed is y, or 0.36 M.
The concentration of the SO_2 remaining is $0.50 - y$ or 0.14 M.

The equilibrium concentration of NO formed is the same as that of the SO_3, 0.36 *M*.
The equilibrium concentration of NO_2 is 0.14 *M*.

We can and should check our calculations by substituting the equilibrium concentrations into the equilibrium constant expression. If we have done the calculations correctly, we should obtain the value for K_c.

$$K_c = \frac{[SO_3][NO]}{[SO_2][NO_2]}$$

$$= \frac{(0.36)(0.36)}{(0.14)(0.14)}$$

$$= 6.6$$

Equilibrium problems are worked in two steps. First, complete the iCe table; next, solve the equilibrium constant expression. Remember that the upper-case C in iCe reminds us that changes involve reaction stoichiometry —we use the stoichiometric coefficients to determine the changes in the concentrations of all substances. The values in the "C" row must be in the same proportions as the coefficients in the chemical equation.

The next example illustrates a similar problem, but one in which the coefficients in the chemical equation are not all the same.

Example 12.9 Calculating Equilibrium Concentrations

Calculate the equilibrium concentrations of hydrogen and iodine that result when 0.050 mol of HI is sealed in a 2.00-L reaction vessel and heated to 700 °C. The equilibrium constant at 700 °C is 2.2×10^{-2} for

$$2HI(g) \rightleftharpoons H_2(g) + I_2(g)$$

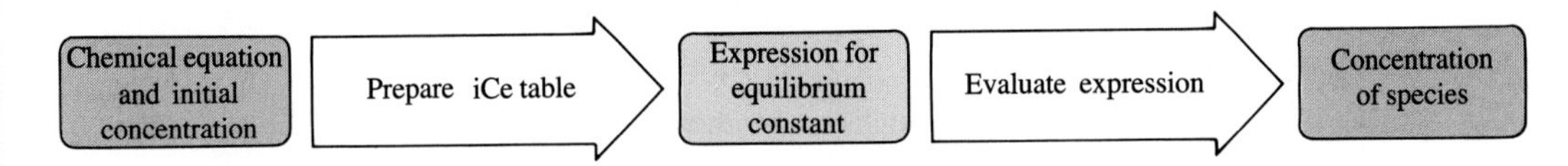

Solution First, write the initial concentrations on line i of the table. The initial concentration of HI is 0.050 mol/2.00 L = 0.025 *M*. Define y as the concentration of H_2 formed. The coefficients of the equation tell us that for each mole of H_2 formed, *one* mole of I_2 will be formed, and *two* moles of HI will be consumed. Write the changes in the concentrations in the C row. A + sign indicates that the species is formed; a − sign indicates that it is consumed. Calculate the equilibrium concentration of each substance from the initial concentration and the change.

	2HI $\rightleftharpoons$	H_2 +	I_2
initial concentration, *M*	0.025	0.00	0.00
Change in concentration, *M*	$-2y$	$+y$	$+y$
equilibrium concentration, *M*	$0.025 - 2y$	y	y

Write the expression for the equilibrium constant, and substitute the concentration expressions from line e.

$$\frac{[H_2][I_2]}{[HI]^2} = K_c$$

$$\frac{(y)(y)}{(0.025 - 2y)^2} = 2.2 \times 10^{-2}$$

Now, solve the equation.

$$\frac{y^2}{(0.025 - 2y)^2} = 2.2 \times 10^{-2}$$

Take the square root of both sides,

$$\frac{y}{0.025 - 2y} = 0.148$$

rearrange,

$$y = 0.148 \times (0.025 - 2y)$$

$$y = 0.0037 - 0.296y$$

combine terms,

$$1.296\ y = 0.0037$$

and solve for y.

$$y = 2.8 \times 10^{-3}$$

The equilibrium concentrations of the species are:

$$[HI] = 0.025 - 2y = 1.9 \times 10^{-2}\ M$$

$$[H_2] = [I_2] = y = 2.8 \times 10^{-3}\ M$$

The answer can be checked by substituting the results of the calculation into the equilibrium constant expression to see whether they reproduce the equilibrium constant.

$$K_c = \frac{[H_2][I_2]}{[HI]^2}$$

$$= (2.8 \times 10^{-3})(2.8 \times 10^{-3})/(1.9 \times 10^{-2})^2$$

$$= 2.2 \times 10^{-2}$$

Understanding At very high temperatures K_c is 0.200 for

$$N_2(g) + O_2(g) \rightleftharpoons 2NO(g)$$

Calculate the equilibrium concentrations of all species if the reaction starts with 5.00×10^{-5} mol of NO in a 10.0-L container.

Answer: $[N_2] = [O_2] = 2.04 \times 10^{-6}\ M$; $[NO] = 0.92 \times 10^{-6}\ M$

Sometimes we must calculate the initial concentrations. Consider the reaction that occurs when 1.00 L of 6.00 M PCl_3 is mixed with 2.00 L of 1.50 M Cl_2 to form PCl_5:

$$PCl_3(g) + Cl_2(g) \rightleftharpoons PCl_5(g)$$

(a)

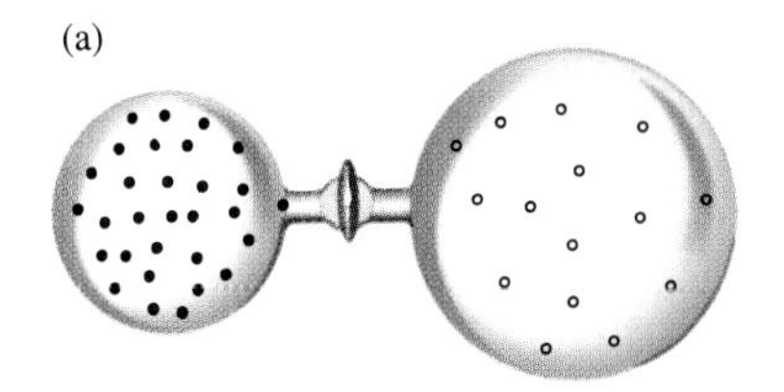

(b)

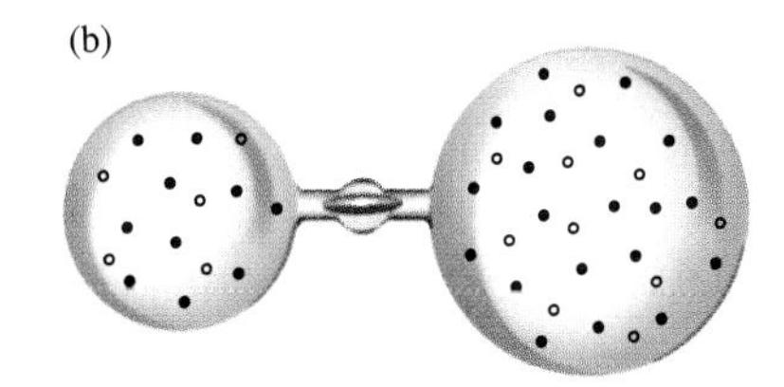

Changes in volume When the valve is opened to mix the reactants, the reaction volume is the sum of the volumes of the original containers. (a) System before mixing: 6.00 M PCl_3 in the left vessel and 1.50 M Cl_2 in the right vessel. (b) After mixing, the concentration of each gas has decreased since the volume has increased.

The reaction vessel is the combination of the two flasks; it has a volume of 3.00 L. The concentration of $PCl_3(g)$ in the reaction vessel is not 6.00 *M*, since the volume changed from 1.00 to 3.00 L. We must calculate the initial concentrations that appear on the first line of the iCe table.

$$[PCl_3] = \frac{1.00\ \cancel{L}\left(\dfrac{6.00\text{ mol } PCl_3}{\cancel{L}}\right)}{3.00\text{ L total volume}} = 2.00\ M$$

$$[Cl_2] = \frac{2.00\ \cancel{L}\left(\dfrac{1.50\text{ mol } Cl_2}{\cancel{L}}\right)}{3.00\text{ L total volume}} = 1.00\ M$$

The next example continues the calculation but requires a different mathematical method to solve the expression for the equilibrium constant.

Example 12.10 Calculating Equilibrium Concentrations

Phosphorus trichloride and chlorine react to form phosphorus pentachloride. At 544 K, K_c is 1.60 for

$$PCl_3(g) + Cl_2(g) \rightleftharpoons PCl_5(g)$$

Calculate the concentration of chlorine when 1.00 L of 6.00 *M* PCl_3 is added to 2.00 L of 1.50 *M* Cl_2 and allowed to reach equilibrium at 544 K.

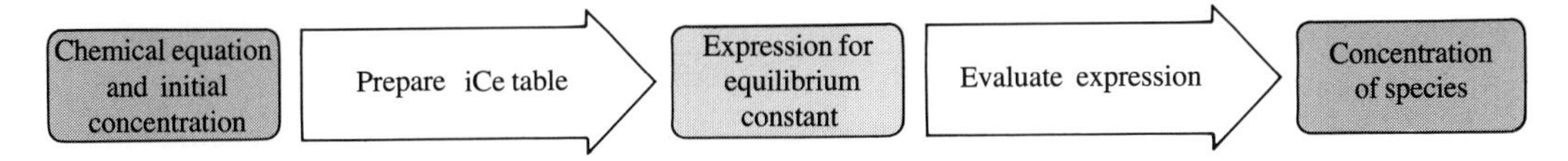

Solution Start with the iCe table. First write the initial concentration, the change, and the equilibrium concentration. The initial concentrations of PCl_3 and Cl_2 were calculated previously.

	PCl_3 +	Cl_2 $\rightleftharpoons$	PCl_5
initial, *M*	2.00	1.00	0.00
Change, *M*	$-y$	$-y$	$+y$
equilibrium, *M*	$2.00 - y$	$1.00 - y$	y

Write the equilibrium constant expression in terms of the equilibrium concentrations from line e of the table.

$$K_c = \frac{[PCl_5]}{[PCl_3][Cl_2]}$$

$$1.60 = \frac{y}{(2.00 - y)(1.00 - y)}$$

Rearrange and combine terms.

$$1.60y^2 - 5.80y + 3.20 = 0$$

This is a *quadratic equation* (one that contains a y^2 term) and can be solved by the *quadratic formula.* Any quadratic equation in the form

$$ay^2 + by + c = 0$$

has the roots

$$y = \frac{-b \pm \sqrt{b^2 - 4ac}}{2a}$$

In this problem

$$y = \frac{+5.80 \pm \sqrt{5.80^2 - 4(1.60)(3.20)}}{2(1.60)}$$

$$y = \frac{+5.80 \pm \sqrt{13.16}}{3.20}$$

$$y = \frac{5.80 \pm 3.63}{3.20}$$

$$y = 0.68 \text{ or } 2.95$$

Mathematically, a quadratic equation has two roots, but a chemical system has only one physically reasonable answer. The last line of the iCe table indicates that $[Cl_2] = 1.00 - y$. In this case, we reject $y = 2.95$, since it predicts a negative concentration for chlorine.

We check the answer by calculating the concentrations of all the species, and substituting these back into the expression for the equilibrium constant to be sure that the results are consistent.

$$[PCl_5] = y = 0.68\ M$$

$$[Cl_2] = 1.00 - y = 0.32\ M$$

$$[PCl_3] = 2.00 - y = 1.32\ M$$

$$K_c = \frac{[PCl_5]}{[PCl_3][Cl_2]} = \frac{0.68}{(1.32)(0.32)} = 1.6$$

This result is in agreement with the original value of 1.60, so we can be confident that the problem has been solved correctly.

Understanding K_c is 0.12 at 1000 K for the dissociation of phosgene:

$$COCl_2(g) \rightleftharpoons CO(g) + Cl_2(g)$$

Calculate the equilibrium concentrations of all species if 2.00 mol of $COCl_2$ is placed in a 5.00-L reactor at 1000 K.

Answer: $[CO] = [Cl_2] = 0.17\ M$; $[COCl_2] = 0.23\ M$

The systematic approach can be used for almost all equilibrium problems.

Equilibrium problems occasionally produce more complicated mathematical equations, but even the most complex problems can be solved. Using the iCe table helps reduce a difficult problem to smaller, more manageable pieces. This process is illustrated in the next example.

Example 12.11 Calculating Equilibrium Concentrations

An industrial process is studied by placing 0.030 mol of sulfuryl chloride, a powerful chemical oxidizer, in a 100-L reactor along with 2.0 mol of SO_2 and 1.0 mol of Cl_2 at 173 °C. At this temperature, K_p is 3.0 for

$$SO_2Cl_2(g) \rightleftharpoons SO_2(g) + Cl_2(g)$$

Calculate the equilibrium concentrations of all species.

Chemical equation and initial concentration → Prepare iCe table → Expression for equilibrium constant → Evaluate expression → Concentration of species

Solution We will fill in the iCe grid to determine the equilibrium concentrations. Write the initial concentrations, and define y as the change in the concentration of SO_2.

The initial concentrations are:

$$[SOCl_2] = 0.030 \text{ mol}/100 \text{ L} = 0.00030\ M$$
$$[SO_2] = 2.0 \text{ mol}/100 \text{ L} = 0.020\ M$$
$$[Cl_2] = 1.0 \text{ mol}/100 \text{ L} = 0.010\ M$$

The stoichiometric coefficients tell us that when y moles of SO_2 are formed, y moles of Cl_2 will also be formed, and y moles of SO_2Cl_2 will be consumed. Write this on the C line of the table. Finally, we calculate the equilibrium concentration by summing the initial concentration and the change in concentration for each column.

	$SO_2Cl_2 \rightleftharpoons$	SO_2 +	Cl_2
initial, M	0.00030	0.020	0.010
Change, M	$-y$	$+y$	$+y$
equilibrium, M	$0.00030 - y$	$0.020 + y$	$0.010 + y$

Write the expression for the equilibrium constant:

$$K_c = \frac{[SO_2][Cl_2]}{[SO_2Cl_2]^2}$$

Since K_p is given, we need to convert K_p to K_c by using Equation 12.2. The change in the number of moles of gases, Δn, is 1 for this reaction. Remember to use kelvins, not °C, for temperature.

$$K_c = K_p(RT)^{-\Delta n} = 3.0 \times (0.082 \times 446)^{-1} = 0.081$$

Substitute the expressions from line e into the equation.

$$K_c = \frac{(0.020 + y)(0.010 + y)}{(0.00030 - y)} = 0.081$$

Solve the equation.

$$K_c = \frac{y^2 + 0.030\ y + 2.0 \times 10^{-4}}{0.00030 - y} = 0.081$$

Gather terms and reduce to a quadratic equation.

$$y^2 + 0.111\ y + 1.757 \times 10^{-4} = 0$$

This is a quadratic equation of the form

$$ay^2 + by + c = 0$$

and has the roots

$$y = -0.11 \quad \text{or} \quad -0.0016$$

Calculate the equilibrium concentrations for each possible root:

	$SO_2Cl_2 \rightleftharpoons$	SO_2 +	Cl_2
equilibrium, M	$0.00030 - y$	$0.020 + y$	$0.010 + y$
$y = -0.11$	0.11	−0.090	−0.10
$y = -0.0016$	0.0019	0.018	0.0084

One of the values of y will produce at least one impossible concentration. In this case, $y = -0.11$ because it would predict negative concentrations for both products.

Check the mathematics by substituting into the equilibrium constant expression.

$$K_c = \frac{[SO_2][Cl_2]}{[SO_2Cl_2]}$$

$$= \frac{(0.018)(0.0084)}{(0.0019)}$$

$$= 0.080$$

The difference between 0.080 (calculated from equilibrium concentrations) and the actual 0.081 is due to roundoff errors in the concentrations.

Understanding K_c for the dissociation of sulfuryl chloride at 173 °C is 0.081. We seal 2.00 mol of SO_2Cl_2 with 1.00 mol each of SO_2 and Cl_2 in a 100-L container. Calculate the concentrations of all species at equilibrium.

Answer: $[SO_2Cl_2] = 6.7 \times 10^{-3}\ M$; $[SO_2] = [Cl_2] = 2.3 \times 10^{-2}\ M$

It is interesting to look at the changes in concentration in Example 12.11 in a different light. The sulfuryl chloride, written as a reactant, increased in concentration and the sulfur dioxide and chloride concentrations decreased. The calculation shows that the reaction will proceed toward the left, but this information was not needed to solve the problem. The iCe table only requires us to define y as a change in concentration. Some calculations will show a positive change, which indicates that the reaction proceeds in the direction it is written; others have a negative change, meaning that the reaction proceeds in the reverse direction.

The iCe table does not require prior knowledge of the direction of the reaction.

12.5 Heterogeneous Equilibria

Objective

- To write expressions for the equilibrium constants that describe heterogeneous equilibria

Heterogeneous equilibria, in which the substances are not all in the same phase, can be treated in the same manner as the homogeneous equilibria that we have already seen. This section will present some of the concepts and calculations that are needed to describe heterogeneous equilibria, many of which have great practical importance.

Concentration of Species

We can use the equilibrium constant and the law of mass action to calculate the equilibrium concentrations of the reactants and products at equilibrium. But the concentrations of some species, such as pure solids and liquids, never change. For example, a liter of sodium chloride (solid) weighs 2.1 kg. It contains 37 moles. The *concentration* of NaCl(s) is 37 mol/L. Clearly, two liters of NaCl(s) contain twice the amount in twice the volume, but at the same concentration. We can exclude solids from our calculations, since their concentrations are constant.

The same argument holds for pure liquids—the concentration of a pure liquid is a constant, related to its density, and does not vary during the course of a chemical reaction. The concentration of a pure solid or liquid is independent of the amount present.

The concentration of a solid or a pure liquid does not change.

A heterogeneous equilibrium system results when calcium carbonate (limestone) is heated in a closed vessel to form calcium oxide (quicklime) and carbon dioxide.

$$CaCO_3(s) \rightleftharpoons CaO(s) + CO_2(g)$$

The equilibrium constant expression is

$$K' = \frac{[CaO][CO_2]}{[CaCO_3]}$$

We are using K' for the equilibrium constant because, as you will see, this equation is just an intermediate step. It can be rewritten as

$$K' \frac{[CaCO_3]}{[CaO]} = [CO_2]$$

Note that the concentrations of $CaCO_3$ and CaO do not change, since they are both solids. Since their concentrations are constant, they are not included in the expression for K_c.

The concentrations of solids and pure liquids are not included in the expression for the equilibrium constant.

$$K_c = [CO_2]$$

If the equilibrium constant is written in terms of pressure, then

$$K_p = P_{CO_2}$$

The equilibrium between a solute in a solution and its solid form is similar to the equilibrium between gases and solids. For example, when silver iodide is added to water, very little actually dissolves. Almost all remains as a solid, in the bottom of the container.

$$AgI(s) \rightleftharpoons Ag^+(aq) + I^-(aq)$$

For this reaction, the equilibrium constant is

$$K' = \frac{[Ag^+(aq)][I^-(aq)]}{[AgI(s)]}$$

$$K = [Ag^+][I^-]$$

Example 12.12 shows the expressions for the equilibrium constants of some heterogeneous systems.

Example 12.12 Equilibrium Constant Expressions for Heterogeneous Equilibria

Write the expressions for both K_p and K_c for the following reactions:

(a) $NaOH(s) + CO_2(g) \rightleftharpoons NaHCO_3(s)$
(b) $CaCl_2(s) + H_2O(g) \rightleftharpoons CaCl_2 \cdot H_2O(s)$
(c) $NH_4Cl(s) \rightleftharpoons HCl(g) + NH_3(g)$

Solution

(a) $K_p = 1/P_{CO_2}$ $\quad K_c = 1/[CO_2]$
(b) $K_p = 1/P_{H_2O}$ $\quad K_c = 1/[H_2O]$
(c) $K_p = P_{HCl}P_{NH_3}$ $\quad K_c = [HCl][NH_3]$

Understanding Write the expressions for K_c and K_p for

$$C(s) + H_2O(g) \rightleftharpoons CO(g) + H_2(g)$$

Answer: $K_p = \dfrac{P_{CO}P_{H_2}}{P_{H_2O}}$, $\quad K_c = \dfrac{[CO][H_2]}{[H_2O]}$

Equilibria of Gases with Solids and Liquids

Phase changes, such as those studied in Chapter 10, can be written as chemical equilibria. We can write a chemical equation for the evaporation of water:

$$H_2O(\ell) \rightleftharpoons H_2O(g)$$

When we write the expression for the equilibrium constant, it is equal to the partial pressure of the water.

$$K_p = P_{H_2O(g)}$$

The liquid water is a pure species, so its concentration is constant and is incorporated into the equilibrium constant.

The pressure of the water vapor reaches the equilibrium vapor pressure. The pressure is relatively small at low temperatures (6.1×10^{-3} atm at 0 °C) but increases to 1.0 atm at 100 °C. The vapor pressure does not depend on the amount of water present, as long as some liquid water is present. We know that the vaporization of water is an endothermic process, so the principle of Le Chatelier tells us that the equilibrium constant and the vapor pressure increase as temperature increases.

Gas-solid equilibria are similar. We can re-examine the equilibrium as limestone (calcium carbonate) is heated to form quicklime (calcium oxide)

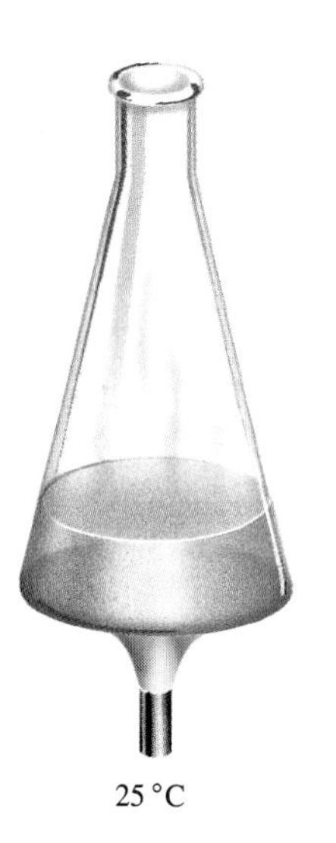

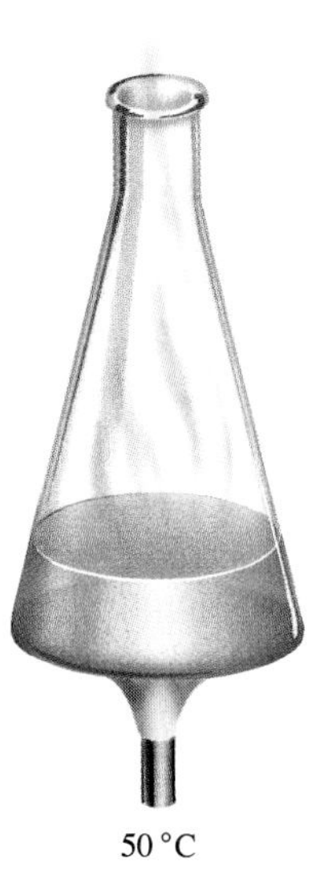

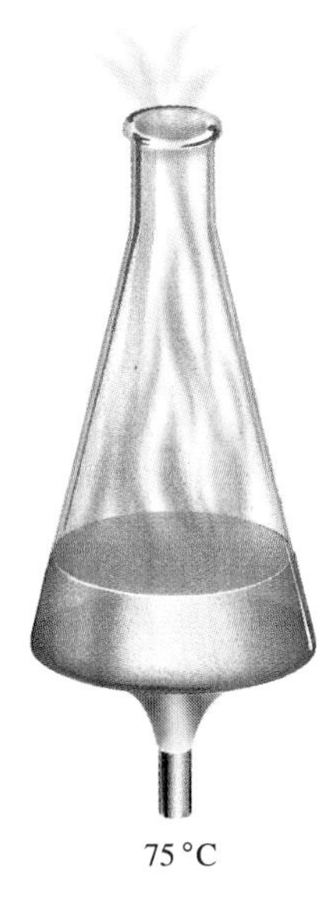

Heating water The concentration of water in the vapor increases as the water is heated.

and carbon dioxide:

$$CaCO_3(s) \rightleftharpoons CaO(s) + CO_2(g)$$

The equilibrium constant expression is

$$K_p = P_{CO_2}$$

Since $CaCO_3$ and CaO are both solids, their concentrations are constant and are not incorporated into the equilibrium constant expression.

Although influenced by temperature, the pressure of CO_2 is independent of the amount of $CaCO_3$ and CaO present, as long as there is some of each solid present. Experimentally, we find the reaction is endothermic, so the principle of Le Chatelier predicts that heating the system will favor the formation of additional CO_2 and CaO.

If any of the substances (CaO, $CaCO_3$, or CO_2) is no longer present, the system is not at equilibrium and it cannot be described by the law of mass action.

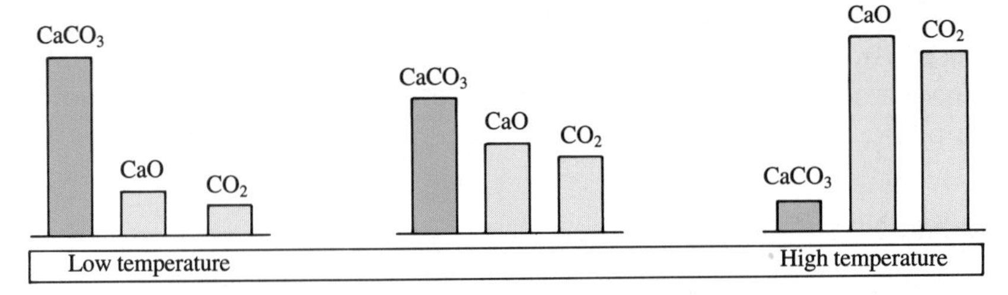

Heating calcium carbonate As calcium carbonate is heated, more calcium oxide and carbon dioxide are produced.

12.6 Solubility Equilibria

Objectives

- To define and write the expression for the solubility product
- To calculate K_{sp} from experimental data
- To calculate the solubilities of slightly soluble salts

The formation of a solid product from reactants in solution is key to many chemical reactions. Most reactions are performed in solution for practical reasons, and if the product is a solid, it is easy to separate from the reactants.

Solubility equilibria, reactions that involve the dissolving and formation of a solid from solution, will be presented in this section. Solubility and precipitation reactions are extensions of the heterogeneous equilibria discussed in the last section.

One important precipitation reaction is the classic test used to determine the presence of silver ions in solution. A few drops of dilute hydrochloric acid solution are added to the test solution; the formation of a white solid (Figure 12.6) indicates the presence of silver.

$$AgNO_3(aq) + HCl(aq) \rightleftharpoons AgCl(s) + HNO_3(aq)$$

The role of the precipitate is often clarified by writing the net ionic equation:

$$Ag^+(aq) + Cl^-(aq) \rightleftharpoons AgCl(s)$$

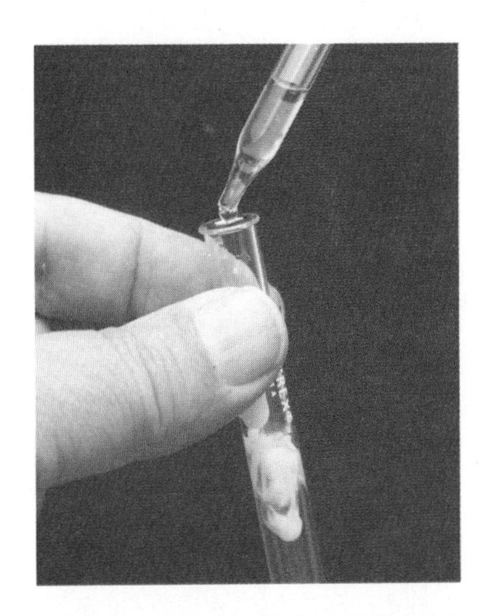

Figure 12.6 The formation of AgCl(s) A characteristic white precipitate forms when a few drops of a hydrochloric acid solution are added to a solution that contains some silver ions. The formation of a white precipitate is not unambiguous evidence of the presence of the silver cation. Mercury and lead cations also form white chloride precipitates, so additional testing may be necessary.

Stalagmites and stalactites These geological features in a cave are formed by naturally occurring precipitation reactions.

The study of solubility equilibria allows us to determine many of the quantitative details of the reaction, including the amount of precipitate formed and the minimum concentration of chloride necessary to form a precipitate.

The Solubility Product

For historical reasons, equilibria that involve precipitation reactions are written as the dissolving of a solid (dissociation into ions), the reverse of the preceding equation:

$$AgCl(s) \rightleftharpoons Ag^+(aq) + Cl^-(aq)$$

We can write the concentration expression for the equilibrium constant:

$$K_{sp} = [Ag^+][Cl^-]$$

The solubility product constant is the equilibrium constant for a solid dissolving and forming ions in solution.

When the chemical reaction is the dissociation of an ionic solid, the equilibrium constant is called the **solubility product constant** and is denoted as K_{sp}. Notice that the solid does not appear in the expression; the concentration of a solid does not change, as discussed in the last section.

Some examples of other solubility equilibria are presented in Example 12.13.

Example 12.13 Expressions for the Solubility Product

Write the chemical equation and expression for the solubility product for each of the following compounds:

(a) $Mg(OH)_2$
(b) LaF_3
(c) $Ca_3(PO_4)_2$

Solution The expression for the solubility product is derived from the law of mass action and the chemical equation that describes dissolving the solid in water.

(a) $Mg(OH)_2(s) \rightleftharpoons Mg^{2+}(aq) + 2OH^-(aq)$

$$K_{sp} = [Mg^{2+}][OH^-]^2$$

(b) $LaF_3(s) \rightleftharpoons La^{3+}(aq) + 3F^-(aq)$

$$K_{sp} = [La^{3+}][F^-]^3$$

(c) $Ca_3(PO_4)_2(s) \rightleftharpoons 3Ca^{2+}(aq) + 2PO_4^{3-}(aq)$

$K_{sp} = [Ca^{2+}]^3[PO_4^{3-}]^2$

Understanding Write the solubility product expression for iron(III) hydroxide.

Answer: $K_{sp} = [Fe^{3+}][OH^-]^3$

The numerical value of the solubility product constant, like all equilibrium constants, is determined experimentally. A tabulation of solubility product constants is given in Appendix F, and some common solubility product constants are repeated in Table 12.4.

The values of some solubility product constants are determined from experiments in which the solubility of a compound is measured. Solubility was defined in Chapter 11. It is the amount of compound that can dissolve in solution, typically measured in mol/L or g/100 mL. Experimental measurements of the concentrations of the dissolved species are sometimes used to determine the solubility product constant, although these calculations are often limited in accuracy. The same electrostatic effects discussed in Chapter 11, when errors in the van't Hoff *i*-parameter were presented, also influence the determination of K_{sp}.

The solubility product constant is generally calculated from experimentally determined solubility data.

Example 12.14 Calculating the Solubility Product Constant

When lead chloride, $PbCl_2$, is added to water, a small amount dissolves. If measurements at 25 °C show that the Pb^{2+} concentration is 1.6×10^{-2} *M*, calculate the value of K_{sp} for $PbCl_2$.

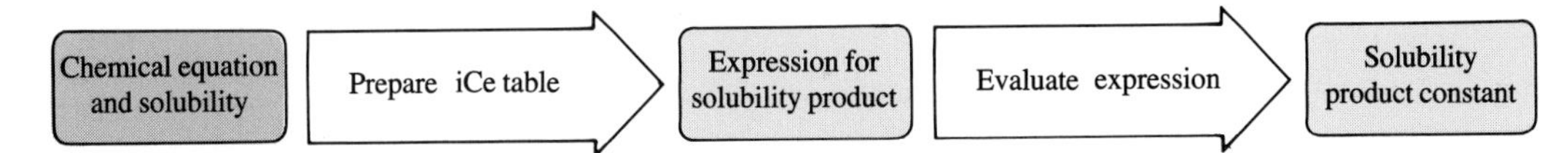

Table 12.4 Solubility Product Constants for Selected Compounds at 25 °C

Compound	Formula	K_{sp}
barium iodate	$Ba(IO_3)_2$	1.5×10^{-9}
barium sulfate	$BaSO_4$	8.7×10^{-11}
calcium fluoride	CaF_2	3.9×10^{-11}
calcium hydroxide	$Ca(OH)_2$	1.3×10^{-6}
calcium phosphate	$Ca_3(PO_4)_2$	1.3×10^{-32}
cerium iodate	$Ce(IO_3)_3$	3.2×10^{-10}
copper(II) hydroxide	$Cu(OH)_2$	1.6×10^{-19}
lanthanum hydroxide	$La(OH)_3$	1.0×10^{-19}
lanthanum iodate	$La(IO_3)_3$	6.2×10^{-12}
lead chloride	$PbCl_2$	1.6×10^{-5}
magnesium hydroxide	$Mg(OH)_2$	8.9×10^{-12}
silver chloride	$AgCl$	1.8×10^{-10}
silver iodate	$AgIO_3$	3.1×10^{-8}
silver iodide	AgI	8.5×10^{-17}
silver sulfate	Ag_2SO_4	1.2×10^{-5}

Solution The problem is made easier if we construct a table. First, write the chemical reaction, the initial concentration, the change in concentration, and the equilibrium concentration of each species.

	$PbCl_2(s)$ $\rightleftharpoons$	$Pb^{2+}(aq)$ +	$2Cl^-(aq)$
initial, *M*	excess	0.0	0.0
Change, *M*	-1.6×10^{-2}	$+1.6 \times 10^{-2}$	$+2(1.6 \times 10^{-2})$
equilibrium, *M*	excess	$+1.6 \times 10^{-2}$	$+3.2 \times 10^{-2}$

We will use the word "excess" when a solid is present, since its concentration is constant as long as the solid is present. Notice that the change in chloride concentration is *twice* the change in lead concentration, because the chemical equation tells us that two chloride ions are formed for each lead ion formed. We next write the expression for the solubility product constant and evaluate it.

$$K_{sp} = [Pb^{2+}][Cl^-]^2$$
$$K_{sp} = (1.6 \times 10^{-2})(3.2 \times 10^{-2})^2$$
$$K_{sp} = 1.6 \times 10^{-5}$$

In the solubility product expression the concentration of chloride ion is doubled *and* it is raised to the second power.

Understanding When silver dichromate dissolves in water, the concentration of silver ion is found to be 8×10^{-3} *M*. Calculate the solubility product for $Ag_2Cr_2O_7$. The chemical equation is

$$Ag_2Cr_2O_7(s) \rightleftharpoons 2Ag^+(aq) + Cr_2O_7^{2-}(aq)$$

Answer: $K_{sp} = 3 \times 10^{-7}$

Calculation of the solubility product constant from the solubility of the substance, rather than concentration of one of the ions, is solved by the same general approach. First, write the chemical equation and the equilibrium constant expression. Then substitute numerical values for the concentration of the anion and cation, as calculated from the solubility. This method is illustrated in Example 12.15.

Example 12.15 Calculating the Solubility Product from Solubility

Calculate the solubility product for silver sulfate, given that the solubility is 0.44 g/100 mL. (Be aware that the many solubility tables, such as those published in the *Handbook of Chemistry and Physics,* express solubility in grams per 100 mL, not moles per liter.)

Solution First, write the chemical equation and the solubility product expression.

$$Ag_2SO_4(s) \rightleftharpoons 2Ag^+(aq) + SO_4^{2-}(aq)$$
$$K_{sp} = [Ag^+]^2[SO_4^{2-}]$$

Next, convert the solubility of silver sulfate from g/100 mL to mol/L. It may be simplest to convert from g/100 mL to g/L to mol/L in two steps. We use the symbol *s* to represent the solubility.

$$s = \frac{0.44 \text{ g } Ag_2SO_4}{100 \text{ mL}}\left(\frac{1000 \text{ mL}}{\text{L}}\right) = \frac{4.4 \text{ g } Ag_2SO_4}{\text{L}}$$

$$s = \frac{4.4 \text{ g } Ag_2SO_4}{\text{L}}\left(\frac{1 \text{ mol } Ag_2SO_4}{311.8 \text{ g } Ag_2SO_4}\right) = 1.4 \times 10^{-2}\ M\ Ag_2SO_4$$

Use the iCe table to calculate the equilibrium concentrations of $Ag^+(aq)$ and $SO_4^{2-}(aq)$.

	$Ag_2SO_4(s) \rightleftharpoons$	$2Ag^+(aq)$ +	$SO_4^{2-}(aq)$
initial, M	excess	0.0	0.0
Change, M	-1.4×10^{-2}	$+2(1.4 \times 10^{-2})$	$+(1.4 \times 10^{-2})$
equilibrium, M	excess	$+2.8 \times 10^{-2}$	$+1.4 \times 10^{-2}$

Last, substitute the concentrations into the expression for the solubility product.

$$K_{sp} = [Ag^+]^2[SO_4^{2-}] = (2.8 \times 10^{-2})^2(1.4 \times 10^{-2})$$

$$K_{sp} = 1.1 \times 10^{-5}$$

Understanding The solubility of $La(IO_3)_3$ is $7 \times 10^{-4}\ M$ at 25 °C. Calculate the solubility product constant.

Answer: $K_{sp} = 6 \times 10^{-12}$

Solubility Calculations

The solubility of a solid can be calculated from the solubility product and the chemical equation for the dissociation of the solid. As in other equilibrium problems, we first set up a table in which we enter the known data. In this particular case, the solubility, which is generally designated by the symbol s, is unknown. The solubility is determined by solving the solubility product constant expression for s. Example 12.16 shows the procedure.

Solubility calculations, although generally simpler than gas-phase equilibrium calculations, can be further simplified by using a systematic approach.

Example 12.16 Calculating the Solubility from the Solubility Product

Given the value of K_{sp} for each compound (see Table 12.4), calculate the solubility of (a) $AgIO_3$ and (b) $Ba(IO_3)_2$.

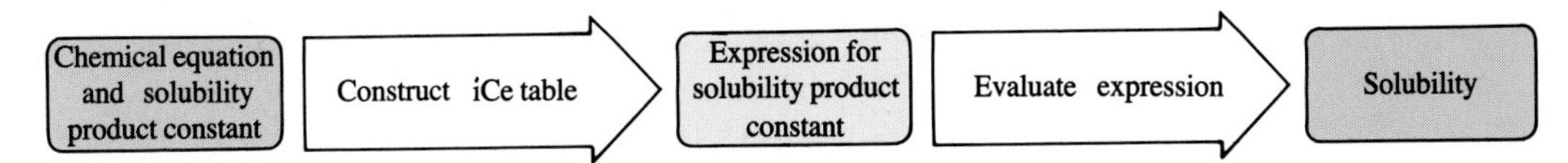

Solution

(a) The equation for dissolution of $AgIO_3$ is

$$AgIO_3(s) \rightleftharpoons Ag^+(aq) + IO_3^-(aq)$$

Let us define s as the solubility, the number of moles of solid that can dissolve in a liter of solution. The solubility is an unknown quantity, and we will fill in a table to derive an equation that can be solved for s. To simplify the problem, we assume we have 1.0 liter of solution.

First, write the initial concentrations. Next, determine the changes. The reaction, of course, is the dissolution of $AgIO_3$; the change in concentration is the unknown, *s*. The equilibrium concentration is determined by summing the initial concentration and the change in concentration.

	$AgIO_3(s) \rightleftharpoons$	$Ag^+(aq)$ +	$IO_3^-(aq)$
initial, *M*	excess	0	0
Change, *M*	$-s$	$+s$	$+s$
equilibrium, *M*	excess	s	s

The last step is to substitute the equilibrium concentrations into the solubility product relationship, and solve for solubility.

$K_{sp} = [Ag^+][IO_3^-]$ (The value of K_{sp} is given in Table 12.4.)

$3.1 \times 10^{-8} = (s)(s) = s^2$

$s = 1.8 \times 10^{-4}\ M$

The solubility of $AgIO_3$ is $1.8 \times 10^{-4}\ M$.

(b) The equation for dissolution of $Ba(IO_3)_2$ is

$Ba(IO_3)_2(s) \rightleftharpoons Ba^{2+}(aq) + 2IO_3^-(aq)$

Again, fill in the grid.

	$Ba(IO_3)_2(s) \rightleftharpoons$	$Ba^{2+}(aq)$ +	$2IO_3^-(aq)$
initial, *M*	excess	0	0
Change, *M*	$-s$	$+s$	$+2s$
equilibrium, *M*	excess	s	$2s$

Note that the stoichiometry tells us that when *s* moles of $Ba(IO_3)_2(s)$ dissolve, *s* moles of $Ba^{2+}(aq)$ and $2s$ moles of $IO_3^-(aq)$ are produced.

After the iCe table is complete, substitute the equilibrium concentrations and the value of the solubility product constant from Table 12.4 into the solubility product expression.

$K_{sp} = [Ba^{2+}][IO_3^-]^2$

$1.5 \times 10^{-9} = (s)(2s)^2 = 4s^3$

$s = 7.2 \times 10^{-4}\ M$

Some calculators do not have a single button that can be used to take a cube root, but almost all have some way to do so. Logarithms (see Appendix A) provide one way to determine a cube root.

Understanding Calculate the solubility of $La(IO_3)_3$.

Answer: $s = 6.9 \times 10^{-4}\ M$

Table 12.5 Solubilities of Selected Iodates

Compound	K_{sp}	Solubility, M
$AgIO_3$	3.1×10^{-8}	1.8×10^{-4}
$Ba(IO_3)_2$	1.5×10^{-9}	7.2×10^{-4}
$La(IO_3)_3$	6.2×10^{-12}	6.9×10^{-4}

Notice that there is no simple proportionality between the value of K_{sp} and the solubility of the compound, as illustrated in Table 12.5. Although the solubility product for $Ba(IO_3)_2$ is smaller than that for $AgIO_3$, the solubility of $Ba(IO_3)_2$ is larger. The solubility depends on both the reaction stoichiometry *and* the solubility product.

The relationship between the expression for the solubility product and the solubility is illustrated in Table 12.6.

You should not try to memorize the table. Instead, note that the coefficients arise from the definition of K_{sp} and from the fact that when a compound dissolves, the concentration of each ion is proportional to the number of ions of that kind in the formula of the compound.

Table 12.6 Relation between Solubility Product, K_{sp}, and Molar Solubility, s

		Solubility of		
Compound	**K_{sp} expression**	**Cation**	**Anion**	**Expression**
AgBr	$K_{sp} = [Ag^+][Br^-]$	s	s	$K_{sp} = s^2$
$CaSO_4$	$K_{sp} = [Ca^{2+}][SO_4^{2-}]$	s	s	$K_{sp} = s^2$
Ag_2SO_4	$K_{sp} = [Ag^+]^2[SO_4^{2-}]$	$2s$	s	$K_{sp} = 4s^3$
PbI_2	$K_{sp} = [Pb^{2+}][I^-]^2$	s	$2s$	$K_{sp} = 4s^3$
LaF_3	$K_{sp} = [La^{3+}][F^-]^3$	s	$3s$	$K_{sp} = 27s^4$
$Ca_3(PO_4)_2$	$K_{sp} = [Ca^{2+}]^3[PO_4^{3-}]^2$	$3s$	$2s$	

12.7 Precipitation and the Common Ion Effect

Objectives

- To predict whether a precipitate will form under specific conditions
- To determine the solubility of a solid in a solution that contains an ion in common with the solid

It is often important to determine whether a precipitate will form when two solutions are mixed. Also important is the dependence of the solubility of a precipitate on the presence of other species in solution. This section will discuss the influence of the concentration of the species in solution, both on the formation and on the solubility of a precipitate.

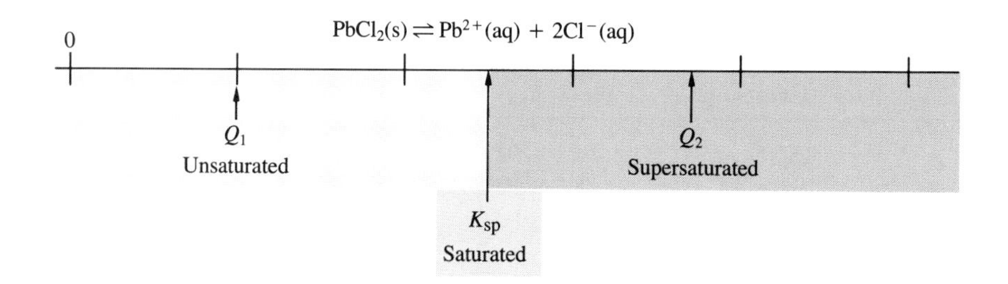

Figure 12.7 Number line representation of solubility

Formation of a Precipitate

We often need to know if a precipitate will form when two solutions are mixed. The mixture, which is not yet at equilibrium, is described by the reaction quotient, Q. If, for example, a solution that contains calcium ions is mixed with a solution that contains phosphate ions, the reaction can be described by a chemical equilibrium:

$$Ca_3(PO_4)_2(s) \rightleftharpoons 3Ca^{2+}(aq) + 2PO_4^{3-}(aq)$$

Solubility equilibria are written for the chemical equation that describes a solid dissolving. If Q is less than K_{sp}, no solid will form.

Even though the reaction under study is the formation of calcium phosphate, it is traditional to write the expression for the dissolution of the solid, the reverse of the reaction that actually occurs, because that is the way that values for the equilibria are tabulated. Our conclusions do not depend on writing the reaction in a particular direction.

The reaction quotient is calculated from the data and the expression

$$Q = [Ca^{2+}]^3[PO_4^{3-}]^2$$

and Q is compared to K_{sp} to determine the direction of reaction. Here Q is called the **ion product,** for obvious reasons. If Q is greater than K_{sp}, the system will react to form solid; if Q is less than K_{sp}, no solid forms, and equilibrium cannot be established.

We can examine the solid-solution equilibrium with the aid of a number line. In Figure 12.7, Q_1 is smaller than K_{sp}; when Q is less than K_{sp} a reaction will occur toward the right and some additional solid will dissolve, forming more ions. Q_2 is greater than K_{sp} and additional precipitation will occur under these conditions—the reaction occurs toward the left. A solution in which Q is less than K_{sp} is unsaturated, a solution in which more material can be dissolved. When Q is equal to K_{sp}, the solution is saturated and no additional solute can be dissolved. Q_2 represents the condition of a supersaturated solution, since Q exceeds K_{sp}. Supersaturated solutions can be prepared, but they are unstable and will eventually form a precipitate.

Supersaturated solutions are unstable; a precipitation reaction will eventually occur.

Example 12.17 illustrates the calculations used to determine if a precipitate will form.

Example 12.17 Formation of a Precipitate

A chemist mixes 100 mL of 0.0050 M NaCl with 200 mL of 0.010 M $Pb(NO_3)_2$. Will lead chloride precipitate?

Solution We need to determine Q and compare it to K_{sp} for

$$PbCl_2(s) \rightleftharpoons Pb^{2+}(aq) + 2Cl^-(aq)$$

To determine Q, the concentrations of Pb^{2+}(aq) and Cl^-(aq) must first be calculated. The concentrations are *not* 0.010 M and 0.0050 M—the substances are diluted by the mixing of the two solutions. The concentrations of Pb^{2+} and Cl^- are found from calculations similar to those illustrated in Chapter 4, where dilution problems were presented. After the two solutions are mixed:

$$[Pb^{2+}] = 6.7 \times 10^{-3}\ M$$

$$[Cl^-] = 1.7 \times 10^{-3}\ M$$

$$Q = [Pb^{2+}][Cl^-]^2$$

$$Q = [6.7 \times 10^{-3}][1.7 \times 10^{-3}]^2$$

$$Q = 1.9 \times 10^{-8}$$

The numerical value for K_{sp} is 1.6×10^{-5}. Since the ion product is less than K_{sp}, no precipitate will form. Only when Q is greater than K_{sp} will precipitation occur.

Understanding If 200 mL of 0.010 M $CaCl_2$(aq) is mixed with 300 mL of 0.150 M NaOH(aq), will $Ca(OH)_2$ precipitate?

Answer: $[Ca^{2+}] = 0.0040\ M$, $[OH^-] = 0.090\ M$, and $Q = 3.2 \times 10^{-5}$, which exceeds $K_{sp}(=1.3 \times 10^{-6})$. A precipitate will form.

The Common Ion Effect

Frequently, we need to calculate the solubility of a precipitate in a solution that already contains one of the same ions that composes the precipitate. These problems illustrate the **common ion effect,** a decrease in solubility of the precipitate in the presence of a solution that contains an ion in common with the precipitate.

The topic can be considered on a qualitative basis by using the principle of Le Chatelier. Let us compare the solubility of silver chloride in water to its solubility in a sodium chloride solution; chloride is the common ion.

$$AgCl(s) \rightleftharpoons Ag^+(aq) + Cl^-(aq)$$

If the solution is initially saturated with silver chloride and the concentration of Cl^-(aq) is increased by adding some sodium chloride, then according to the principle of Le Chatelier, the system will react to form additional AgCl(s), decreasing the solubility of silver chloride.

We can write the reaction quotient as

$$Q = [Ag^+][Cl^-]$$

If the equilibrium is disturbed by increasing the chloride ion concentration, Q becomes greater than K_{sp}, so reaction occurs to consume the added chloride, producing additional AgCl(s).

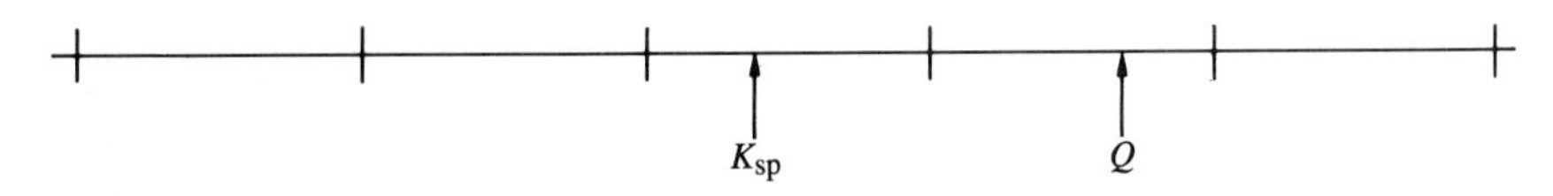

Number line representation of the common ion effect

Comparison of Q and K_{sp} allows us to predict the direction of reaction—we can see that the solubility of a solid is decreased by adding a common ion. Quantitative treatment of the common ion effect is illustrated in Example 12.18.

The solid is *less* soluble in a solution that contains a common ion than it is in water.

INSIGHTS into CHEMISTRY

A Closer View

Selective Precipitation: A Method to Separate, Purify, and Analyze

Selective precipitation, in which one species is precipitated to separate it from another that remains in solution, is often used to isolate and purify substances. It is also quite important in many types of chemical analyses, such as the Mohr titration for determining the concentration of chloride ion in a sample. This titration is performed by adding a silver nitrate solution to the sample to precipitate silver chloride. Potassium chromate is added as an indicator; after nearly all the chloride is precipitated, the next drop of added silver nitrate reacts to form silver chromate, a red solid. We stop the titration at the appearance of the red precipitate.

Ideally, the concentration of chromate should be chosen so that the precipitation of silver chromate starts just after the chloride ion has been precipitated. The proper concentration can be calculated from the following chemical equilibria:

$$AgCl(s) \rightleftharpoons Ag^+(aq) + Cl^-(aq) \qquad K_{sp} = 1.8 \times 10^{-10}$$

$$Ag_2CrO_4(s) \rightleftharpoons 2Ag^+(aq) + CrO_4^{2-}(aq) \qquad K_{sp} = 2.5 \times 10^{-12}$$

After the stoichiometric amount of silver ion has been added and all the chloride has precipitated, the concentration of the silver will be determined by the solubility of the precipitate. (Silver ion is not yet in excess.)

	$AgCl(s) \rightleftharpoons$	$Ag^+(aq)+$	$Cl^-(aq)$
initial, M	excess	0.	0.
Change, M	$-s$	$+s$	$+s$
equilibrium, M	excess	s	s

$$K_{sp} = 1.8 \times 10^{-10}$$

Substitute the expressions for the concentrations of silver and chloride ions into the solubility product expression:

$$K_{sp} = [Ag^+][Cl^-]$$

$$1.8 \times 10^{-10} = s^2$$

$$s = 1.3 \times 10^{-5}\ M$$

Since we know the concentration of the silver ion ($1.3 \times 10^{-5}\ M$) at the equivalence point, we can now calculate the concentration of chromate ion needed to form silver chromate when the silver concentration is $1.3 \times 10^{-5}\ M$.

The solubility product constant for silver chromate is $K_{sp} = 2.5 \times 10^{-12}$, so

$$K_{sp} = [Ag^+]^2[CrO_4^{2-}]$$

Example 12.18 Common Ion Effect

Potassium peroxyborate is a mild household bleach recommended for fabrics that cannot withstand ordinary household bleach (which is sodium hypochlorite, NaOCl). The solubility product for potassium peroxyborate, KBO_3, is 1.3×10^{-2}. Calculate the solubility of KBO_3 (a) in water and (b) in 0.50 M KCl.

Solution

(a) Calculate the solubility of KBO_3 in water.

First, write the chemical equation.

$$KBO_3(s) \rightleftharpoons K^+(aq) + BO_3^-(aq)$$

Next, set up the iCe table.

$$[CrO_4^-] = \frac{K_{sp}}{[Ag^+]^2}$$
$$= \frac{2.5 \times 10^{-12}}{(1.3 \times 10^{-5})^2}$$
$$= 0.015\ M$$

If the concentration of the chromate indicator is 0.015 *M*, the formation of the red precipitate will start as soon as the concentration of silver rises above 1.3×10^{-5} *M*, the point at which all chloride has been consumed.

Other factors actually require that the chromate concentration be somewhat less than 0.015 *M*; a concentration of 0.005 *M* is generally used. The decreased concentration of chromate requires a little excess silver solution (0.03 mL if we are titrating with a 0.1 *M* silver ion solution) beyond the point at which all the chloride is precipitated before a color change occurs.

Notice that even though the solubility product of silver chromate is smaller than K_{sp} for silver chloride, the silver chromate is actually more soluble. Since silver chromate dissociates into three ions and silver chloride into two, we cannot readily compare solubilities from the solubility product alone, as illustrated in Table 12.5.

Mohr titration The solution is initially yellow, the color of the chromate ion, as the white AgCl forms. Near the endpoint, the red silver chromate forms, but it dissipates with swirling. The endpoint is the first permanent red color.

	$KBO_3(s) \rightleftharpoons$	$K^+(aq)$ +	$BO_3^-(aq)$
initial, *M*	excess	0.0	0.0
Change, *M*	$-s$	$+s$	s
equilibrium, *M*	excess	s	s

Substitute into the K_{sp} expression and solve for *s*.

$$K_{sp} = [K^+][BO_3^-]$$
$$1.3 \times 10^{-2} = [s][s]$$
$$s = 1.1 \times 10^{-1}\ M = 0.11\ M$$

(b) Calculate the solubility in 0.50 *M* KCl.

The common ion problem is solved by the same method as part (a), but the starting concentration of potassium ion is 0.50 *M*.

Prepare the iCe table.

	$KBO_3(s)$ ⇌	$K^+(aq)$ +	$BO_3^-(aq)$
initial, *M*	excess	0.50	0.0
Change, *M*	$-s$	$+s$	s
equilibrium, *M*	excess	$0.50 + s$	s

Substitute into the K_{sp} expression:

$$K_{sp} = [K^+][BO_3^-]$$

$$1.3 \times 10^{-2} = [s][0.50 + s]$$

This expression can be expanded to a quadratic equation:

$$s^2 + 0.5s - 1.3 \times 10^{-2} = 0$$

The quadratic formula is used to find the solubility:

$$s = 2.5 \times 10^{-2}\ M = 0.025\ M$$

The solubility of potassium peroxyborate is much lower in the KCl solution than it is in water.

Understanding Calculate the solubility of AgCl in water and in 1.0×10^{-5} *M* NaCl.

Answer: In water, the solubility is 1.3×10^{-5} *M*; in 1.0×10^{-5} *M* KCl the solubility is 9.3×10^{-6} *M*.

Numerical Approximations

The quadratic formula is not always the best method of solving quadratic equations. Consider the problem that arises when we are asked to calculate the solubility of silver iodide ($K_{sp} = 8.5 \times 10^{-17}$) in 0.10 *M* sodium iodide solution. The iCe table is

	$AgI(s)$ ⇌	$Ag^+(aq)$ +	$I^-(aq)$
initial	excess	0	0.10
Change	$-s$	$+s$	$+s$
equilibrium	excess	s	$0.10 + s$

Substitute into the solubility product expression

$$K_{sp} = [Ag^+][I^-]$$

$$8.5 \times 10^{-17} = (s)(0.10 + s)$$

and reduce to a quadratic equation:

$$s^2 + 0.10s - 8.5 \times 10^{-17} = 0$$

If the quadratic formula is used to solve this quadratic equation, a major technical problem becomes apparent. The correct solution requires a calculator or computer that has seventeen digits of accuracy or roundoff error in the calculation becomes significant.

Formulating an approximation is an alternative way to find the roots of a polynomial equation. (Other methods are presented in Appendix A.) Consider the expression for the equilibrium constant that we derived:

$$8.5 \times 10^{-17} = (s)(0.10 + s)$$

Note that s is likely to be much, much less than 0.10; the solubility of silver iodide is 9×10^{-9} *M* in water and will be even smaller in the iodide solution, due to the common ion effect. If s is much less than 0.10, then the sum, $0.10 + s$, is approximately equal to 0.10.

If $s \ll 0.10$ (where $\ll$ means "much less than"), then we can write

When a very small number is added to a larger number, the smaller number can often be ignored.

$$8.5 \times 10^{-17} \approx (s)(0.10)$$

where $\approx$ means "very nearly equal to." This equation is easily solved.

$$s = 8.5 \times 10^{-16}\ M$$

Numerical approximations must be checked.

The result is based on an approximation, so its validity must be checked. Since 8.5×10^{-16} is much less than 0.10, our approximation is valid. Clearly $(0.10 + 8.5 \times 10^{-16})$ doesn't differ significantly from 0.10.

The approximation process, however, needs to be discussed, and we must determine quantitatively what is meant by "much smaller." When solving equilibrium problems, chemists generally neglect the addition or subtraction of one term if the smaller is less than 5% of the larger. The approximation method limits the accuracy of the calculations to about 5%. This figure is a reasonable expectation for the overall accuracy of a solubility product calculation. Other potential sources of error limit the accuracy to about this magnitude.

If the approximation fails, either solve the quadratic equation, or use one of the alternative methods (including a method of *successive* approximations) discussed in Appendix A as well as in the next chapter. The limited solubilities of precipitates makes the simplifying assumption valid for most cases. The next example is typical of most solubility calculations.

Example 12.19 Calculating the Solubility of a Solid in Solution with a Common Ion

What is the solubility of calcium hydroxide in 0.050 *M* sodium hydroxide solution? $K_{sp} = 1.3 \times 10^{-6}$.

Solution

	$Ca(OH)_2(s) \rightleftharpoons$	$Ca^{2+}(aq)$ +	$2OH^-(s)$
initial	excess	0	0.050
Change	$-s$	$+s$	$+2s$
equilibrium	excess	s	$0.050 + 2s$

$$K_{sp} = [Ca^{2+}][OH^-]^2$$
$$1.3 \times 10^{-6} = (s)(0.050 + 2s)^2$$

We make the approximation that if $2s \ll 0.050$, then

$$1.3 \times 10^{-6} \approx (s)(0.050)^2$$
$$s = 5.2 \times 10^{-4}\ M$$

Now check the assumption. Is $2s \ll 0.050$? Since 5% of 0.050 M is $2.5 \times 10^{-3}\ M$, and 10.4×10^{-4} is less than 2.5×10^{-3}, the assumption is valid and we accept the solubility of $5.2 \times 10^{-4}\ M$ as the answer.

Understanding Calculate the solubility of CaF_2 in 0.025 M NaF.

Answer: The solubility is $6.2 \times 10^{-8}\ M$.

Summary

In many reactions the chemical system reaches a state of *equilibrium* in which the reactants and products are present in unchanging concentrations.

The *law of mass action* provides a mathematical basis that is used to evaluate chemical equilibria. An *equilibrium constant expression* can be written, and experimental data are substituted into it and evaluated to determine numerical values for the *equilibrium constant.* The expression for the equilibrium constant can be used to determine the direction of spontaneous reaction. If the *reaction quotient, Q,* is less than K_{eq}, then the reaction will proceed to the right, to form more products.

The *principle of Le Chatelier* predicts how changes in concentration, pressure, volume, and temperature influence a chemical system at equilibrium. If a chemical system at equilibrium is disturbed, it reacts to reduce the effect of the disturbance. If, for example, the concentration of one of the substances is increased, then the reaction will proceed to consume the added substance.

The numerical value of K_{eq} depends on the coefficients in the balanced equation, but if K_{eq} is known for one set of coefficients, K_{eq} can be determined for any other set of coefficients that also balance the equation. Two equilibrium constants, K_c and K_p, are used for systems in which concentrations are given in mol/L and those in which concentrations are expressed as pressures in atmospheres. The relationship between K_p and K_c is derived from the ideal gas law.

The chemical equation, initial composition, and numerical value for K_{eq} can be used to calculate the concentrations of all species in a chemical system at equilibrium. A systematic approach to equilibrium problems (the iCe table) provides the framework for a procedure to solve this type of problem.

In *heterogeneous equilibria,* where the reaction mixture contains more than one phase, the concentrations of pure liquids, solids, and the solvent do not change and do not appear in the expression for the equilibrium constant. Equilibrium expressions contain only the concentrations of the species that change concentration as a result of the chemical reaction.

The solubility product expression is used to describe the process of dissolving a sparingly soluble ionic solid. The numerical value of K_{sp} can be used to determine the solubility of a solid; conversely, the solubility can be used to evaluate K_{sp}.

The expression for the solubility product can also be used to determine whether a precipitate will form. If two solutions are mixed, the concentration of each ion can be calculated and substituted into the solubility product expression. Only if the reaction quotient (ion product) exceeds the solubility product can an equilibrium be established and form a precipitate.

The solubility of a solid is lower in a solution that contains one of the same ions that compose the solid. This effect is called the *common ion effect,* and often reduces the solubility by several orders of magnitude.

Chapter Terms

Equilibrium—a state in which the tendency of reactants to form products is balanced by the tendency of products to form reactants. *(12.1)*

Law of mass action—for any reaction

$$aA + bB \rightleftharpoons cC + dD$$

the equilibrium constant can be calculated from the equilibrium concentrations *(12.1)*:

$$K_{eq} = \frac{[C]^c[D]^d}{[A]^a[B]^b}$$

Equilibrium constant K_{eq}—an experimental measurement of the tendency of a reaction to favor products or reactants. See **law of mass action.** *(12.1)*

Reaction quotient Q—the result of a calculation based on the same expression as the equilibrium constant, except that the concentrations are not restricted to those observed after the system reaches equilibrium. Initial concentrations are often used. *(12.2)*

Principle of Le Chatelier—when a reaction is disturbed from its equilibrium position, it will respond by changing composition to minimize the disturbance. *(12.3)*

Solubility equilibria—those equilibria that describe a solid dissolving in solution, or a solid forming from solution. *(12.6)*

Solubility product constant—the equilibrium constant for the dissolving of a sparingly soluble salt. *(12.6)*

Ion product—the reaction quotient written for dissolution of a slightly soluble solid. *(12.7)*

Common ion effect—the effect on solubility when an ionic compound is dissolved in a solution that contains an ion in common with the solid. *(12.7)*

Exercises

Exercises designated with color have answers in Appendix J.

Gas Phase Equilibrium Expressions

12.1 Describe a physical system that is in equilibrium, and explain how the principles of equilibrium apply to the system.

12.2 Describe a physical system that is *not* in equilibrium, and explain why not.

12.3 Sunlight strikes the upper atmosphere, creating ozone. Is the sunlight-ozone system in equilibrium? Explain your answer.

12.4 Consider a 1.5-V battery. Does the battery represent a system that is at equilibrium? If not, describe the equilibrium status of a battery.

12.5 Write the equilibrium constant expressions (K_p) for the following chemical reactions:

(a) $2H_2O(g) \rightleftharpoons 2H_2(g) + O_2(g)$
(b) $2HCl(g) \rightleftharpoons H_2(g) + Cl_2(g)$
(c) $CO(g) + Cl_2(g) \rightleftharpoons COCl_2(g)$
(d) $2CO(g) + O_2(g) \rightleftharpoons 2CO_2(g)$

12.6 Write the equilibrium constant expressions (K_p) for the following chemical reactions:

(a) $PCl_5(g) \rightleftharpoons PCl_3(g) + Cl_2(g)$
(b) $2NO_2(g) \rightleftharpoons 2NO(g) + O_2(g)$
(c) $2SO_3(g) \rightleftharpoons 2SO_2(g) + O_2(g)$
(d) $H_2(g) + I_2(g) \rightleftharpoons 2HI(g)$

12.7 Write the equilibrium constant expressions (K_c) for the following chemical reactions:

(a) $Cl_2(g) + H_2O(g) \rightleftharpoons 2HCl(g) + \frac{1}{2}O_2(g)$
(b) $2NO_2(g) \rightleftharpoons N_2O_4(g)$
(c) $3O_2(g) \rightleftharpoons 2O_3(g)$
(d) $CO_2(g) \rightleftharpoons CO(g) + \frac{1}{2}O_2(g)$

12.8 Write the equilibrium constant expressions (K_c) for the following chemical reactions:

(a) $HCl(g) + \frac{1}{4}O_2(g) \rightleftharpoons \frac{1}{2}Cl_2(g) + \frac{1}{2}H_2O(g)$
(b) $\frac{1}{2}N_2O_4(g) \rightleftharpoons NO_2(g)$
(c) $N_2O_4(g) \rightleftharpoons N_2(g) + 2O_2(g)$
(d) $\frac{1}{2}O_2(g) + SO_2(g) \rightleftharpoons SO_3(g)$

Calculation of Equilibrium Constant

12.9 Some sulfur dioxide and oxygen are sealed in a reaction vessel to study the formation of sulfur trioxide:

$$2SO_2(g) + O_2(g) \rightleftharpoons 2SO_3(g)$$

(This equilibrium is of interest to scientists who study acid rain because the sulfur trioxide reacts with water to form sulfuric acid.) After equilibrium is achieved, the concentrations are measured: $[SO_2] = 0.015M$; $[O_2] = 0.012\,M$; $[SO_3] = 1.45\,M$. Calculate the equilibrium constant K_c.

12.10 Carbon dioxide is placed in a 5.0-L high-pressure reaction vessel at 1400 °C. After equilibrium is achieved, 4.95 mol of CO_2, 0.050 mol of CO, and 0.025 mol of O_2 are found in the container. Calculate the equilibrium constant K_c for:

$$CO(g) + \tfrac{1}{2}O_2(g) \rightleftharpoons CO_2(g)$$

12.11 To evaluate the equilibrium constant for

$$2NO_2(g) \rightleftharpoons N_2O_4(g)$$

a scientist seals 2.00 mol of nitrogen dioxide in a 2.5-L container. At equilibrium, the reaction vessel is found to contain 1.50 mol of nitrogen dioxide and 0.25 mol of dinitrogen tetroxide. Calculate the equilibrium constant for the reaction at this particular temperature.

12.12 A scientist seals some PCl_5 in a 20.0-L flask. After equilibrium is attained, chemical analysis shows that the flask contains 1.0 mol of $PCl_5(g)$ and 2.0 mol each of $PCl_3(g)$ and $Cl_2(g)$. Calculate the equilibrium constant for

$$PCl_5(g) \rightleftharpoons PCl_3(g) + Cl_2(g)$$

12.13 At 500 K the equilibrium constant is 155 for

$$H_2(g) + I_2(g) \rightleftharpoons 2HI(g)$$

Calculate the equilibrium constant at the same temperature for

$$\tfrac{1}{2}H_2(g) + \tfrac{1}{2}I_2(g) \rightleftharpoons HI(g)$$

12.14 At 3000 K the equilibrium constant 6.1×10^3 for the formation of water:

$$2H_2(g) + O_2(g) \rightleftharpoons 2H_2O(g)$$

Calculate the equilibrium constant at the same temperature for

$$H_2(g) + \tfrac{1}{2}O_2(g) \rightleftharpoons H_2O(g)$$

12.15 Ammonia is placed in a reactor, and the temperature is increased to 745 °C, where some will decompose to nitrogen and hydrogen. The initial concentration of ammonia was 0.0240 *M*. After equilibrium is attained, the concentration of ammonia is 0.0040 *M*. Calculate K_c at 745 °C for

$$2NH_3(g) \rightleftharpoons N_2(g) + 3H_2(g)$$

12.16 The high-temperature reaction of SO_2 with oxygen has been studied. Initially, the reactor contained 0.0076 *M* SO_2, 0.0036 *M* O_2, and no SO_3. After equilibrium was achieved, the SO_2 concentration decreased to 0.0032 *M*. Calculate K_c for

$$2SO_2(g) + O_2(g) \rightleftharpoons 2SO_3(g)$$

12.17 The decomposition of hydrogen iodide has been studied since the beginning of the twentieth century. Some hydrogen iodide is placed in a reactor, which is heated to 322 °C. The reaction was initiated with 1.25 atm of hydrogen iodide; after equilibrium was attained, the total HI pressure had decreased to 1.05 atm. Calculate K_p at this temperature for the reaction

$$2HI(g) \rightleftharpoons H_2(g) + I_2(g)$$

12.18 The formation of phosgene, $COCl_2$, was studied by sealing 0.96 atm of carbon monoxide and 1.02 atm of Cl_2 in a reactor at 682 K. The pressure smoothly dropped from a total pressure of 1.98 atm to 1.22 atm as the system reached equilibrium. Calculate K_p for this reaction.

$$CO(g) + Cl_2(g) \rightleftharpoons COCl_2(g)$$

Determining the Direction of Reaction, Le Chatelier's Principle

12.19 Under what circumstances will changes in the volume of a gaseous system *not* change the equilibrium constant?

12.20 Compare how changes in temperature influence K_{eq} and Q for the endothermic reaction $N_2(g) + O_2(g) \rightleftharpoons NO(g)$.

12.21 The equilibrium constant for the water-gas shift reaction is 5.0 at 469 °C:

$$CO(g) + H_2O(g) \rightleftharpoons CO_2(g) + H_2(g)$$

In which direction will the reaction occur if the following amounts (in moles) of each compound are placed in a 1.0-L container?

	CO	H_2O	CO_2	H_2
(a)	0.50	0.40	0.80	0.90
(b)	0.01	0.02	0.03	0.04
(c)	1.22	1.22	2.78	2.78
(d)	0.61	1.22	1.39	2.39

12.22 The equilibrium constant for the decomposition of hydrogen fluoride is 0.010 at a certain temperature.

$$2HF(g) \rightleftharpoons H_2(g) + F_2(g)$$

Determine the direction of spontaneous reaction if the following amounts in millimoles (1 millimol = 1.0×10^{-3} mol) of each compound are placed in a 1.0-L nickel container. (Glass cannot be used because it would react with HF.)

	HF	H_2	F_2
(a)	0.005	0.020	0.010
(b)	0.020	0.20	0.20
(c)	1.05	0.10	0.090
(d)	2.00	0.050	0.080

12.23 Some ammonia is sealed in a container and allowed to equilibrate at some constant temperature. The reaction is endothermic.

$$2NH_3(g) \rightleftharpoons N_2(g) + 3H_2(g)$$

In which direction will reaction occur
(a) if additional ammonia is added to the system?
(b) if additional hydrogen is added to the system?
(c) if the volume of the container is decreased?
(d) if the temperature is increased?
(e) if argon is added to the system so that the total pressure in the container increases?

12.24 Some sulfur trioxide is sealed in a container and allowed to equilibrate at some constant temperature. The following endothermic reaction occurs:

$$SO_3(g) \rightleftharpoons SO_2(g) + \tfrac{1}{2}O_2(g)$$

In which direction will the reaction proceed
(a) if more sulfur trioxide is added to the system?
(b) if oxygen is removed from the system?
(c) if the volume of the container is increased?
(d) if the temperature is increased?
(e) if argon is added to the container to increase the total pressure?

Conversion between K_p and K_c

12.25 Nitrosyl bromide is formed from nitrogen oxide and bromine:

$$NO(g) + \frac{1}{2}Br_2(g) \rightleftharpoons NOBr(g)$$

K_p for this reaction is 116 at 25 °C. Calculate K_c at this temperature.

12.26 At 3000 K, carbon dioxide dissociates:

$$CO_2(g) \rightleftharpoons CO(g) + \frac{1}{2}O_2(g)$$

If K_p for this reaction is 2.48, calculate K_c.

12.27 Exactly 3.00 mmol of phosphorus pentachloride is sealed in a 5.00-L container at 450 K. Phosphorus trichloride and chlorine are formed:

$$PCl_5(g) \rightleftharpoons PCl_3(g) + Cl_2(g)$$

After equilibrium is established, chemical analysis shows that 1.00 mmol of phosphorus pentachloride is present, the rest having reacted.

(a) Calculate the concentrations of all species.
(b) Calculate K_c.
(c) Calculate K_p.

12.28 If 1.0×10^{-3} mol each of sulfur dioxide and oxygen are sealed in a 1.0-L container at a particular temperature, they react to form sulfur trioxide:

$$2SO_2(g) + O_2(g) \rightleftharpoons 2SO_3(g)$$

At equilibrium at 649 °C, 7.5×10^{-4} mol of oxygen is left.

(a) Calculate the concentrations of all species.
(b) Calculate K_c.
(c) Calculate K_p.

Calculating the Concentration of Species at Equilibrium

12.29 How can you determine whether a system has reached equilibrium?

12.30 A 20.0-L container initially contains 2.00 mol of NO_2. The following reaction occurs:

$$2NO_2(g) \rightleftharpoons N_2O_4(g)$$

At equilibrium, chemical analysis finds that the concentration of $NO_2(g)$ is 0.010 *M*.

(a) Calculate the number of moles of $NO_2(g)$.
(b) Calculate the number of moles of $N_2O_4(g)$ formed, and its concentration.
(c) Calculate K_c.

12.31 If 1.0×10^{-3} mol each of SO_2 and NO_2 are sealed in a 1.0-L flask at 1500 K, they will react to form SO_3 and NO:

$$SO_2(g) + NO_2(g) \rightleftharpoons SO_3(g) + NO(g)$$

The equilibrium constant, K_c, is 1.98 for this reaction. Calculate the concentrations of all species after equilibrium is attained.

12.32 Consider 1.0 mol each of hydrogen and iodine sealed in a 2.0-L flask at 1200 K.

$$H_2(g) + I_2(g) \rightleftharpoons 2HI(g)$$

The equilibrium constant, K_c, is 36 for this reaction. Calculate the concentrations of all species after equilibrium is attained.

12.33 Exactly 2.0 mol each of carbon monoxide and water are sealed in a 4.0-L flask at 1100 K.

$$CO(g) + H_2O(g) \rightleftharpoons CO_2(g) + H_2(g)$$

The equilibrium constant, K_c, is 0.55 for this reaction. Calculate the concentrations of all species after equilibrium is attained.

12.34 Exactly 0.500 mol each of sulfur trioxide and nitrogen monoxide are sealed in a 20.0-L flask.

$$SO_3(g) + NO(g) \rightleftharpoons SO_2(g) + NO_2(g)$$

K_c is 0.50 at 1500 K. Calculate the concentrations of all species after equilibrium is reached.

12.35 Consider 0.200 mol of phosphorus pentachloride sealed in a 2.0-L container at 620 K. The equilibrium constant, K_c, is 0.6 for

$$PCl_5(g) \rightleftharpoons PCl_3(g) + Cl_2(g)$$

Calculate the concentrations of all species after equilibrium has been reached.

12.36 A 2.00-L container at 463 °C contains 0.500 mol of phosgene. The equilibrium constant, K_c, is 4.93×10^{-3} for the following reaction:

$$COCl_2(g) \rightleftharpoons CO(g) + Cl_2(g)$$

Calculate the concentrations of all species after the system reaches equilibrium.

12.37 Consider 0.100 mol each of phosphorus trichloride and chlorine are sealed in a 10.0-L container at 291 °C. The equilibrium constant, K_c, is 8.18 for:

$$PCl_3(g) + Cl_2(g) \rightleftharpoons PCl_5(g)$$

Calculate the concentrations of all species after the system reaches equilibrium.

12.38 Exactly 0.400 mol each of carbon monoxide and chlorine are sealed in a 2.00-L container at 919 K. K_c is 7.52 for:

$$CO(g) + Cl_2(g) \rightleftharpoons COCl_2(g)$$

Calculate the concentrations of all species after the system reaches equilibrium.

Challenging Problems

12.39 Exactly 4 mol of sulfur trioxide is sealed in a 5.0-L container at 1500 K. K_p is 1150 for

$$2SO_3(g) \rightleftharpoons 2SO_2(g) + O_2(g)$$

Calculate the concentrations of all species after the system reaches equilibrium. (*Hint:* This reaction goes essentially to completion.)

12.40 Exactly 0.010 mol of hydrogen iodide is sealed in a 5.00-L container. The temperature is raised to 3000 K. At this temperature, K_p is 0.050 for:

$$2HI(g) \rightleftharpoons H_2(g) + I_2(g)$$

Calculate the equilibrium concentrations of all species.

12.41 Consider 1.00 mol of phosphorus trichloride and 2.00 mol of chlorine sealed in a 3.00-L container at 550 K. The equilibrium constant, K_c, is 8.18 for

$$PCl_3(g) + Cl_2(g) \rightleftharpoons PCl_5(g)$$

Calculate the equilibrium concentrations of all species.

12.42 Calculate the concentrations of all species formed when 1.00 mol of sulfur dioxide and 2.00 mol of chlorine are sealed in a 100-L reactor. The temperature is raised to 400 K where $K_c = 89.3$ for

$$SO_2(g) + Cl_2(g) \rightleftharpoons SO_2Cl_2(g)$$

12.43 Consider a 10.0-L vessel that originally contains 3.00 mol of phosphorus trichloride, 4.00 mol of chlorine, and 5.00 mol of phosphorus pentachloride at 332 °C. K_p is 2.5 for:

$$PCl_3(g) + Cl_2(g) \rightleftharpoons PCl_5(g)$$

Calculate the equilibrium concentrations of all species.

12.44 Exactly 2.00 atm of carbon monoxide and 3.00 atm of hydrogen are placed in a reaction vessel. At a particular temperature K_p is 5.60 for

$$CO(g) + H_2(g) \rightleftharpoons CH_2O(g)$$

Calculate the equilibrium pressures of all species.

12.45 Consider the formation of formaldehyde from carbon monoxide and hydrogen. If a 2.0-L flask of 0.10 M CO is joined to a 5.0-L flask of 0.20 M H_2, and K_c is 0.50 for

$$CO(g) + H_2(g) \rightleftharpoons CH_2O(g)$$

Calculate the equilibrium concentrations of all species.

Heterogeneous Equilibria

12.46 Calculate the partial pressure of $CO_2(g)$, given that K_p is 0.12 (at 1000 K) for:

$$CaCO_3(s) \rightleftharpoons CaO(s) + CO_2(g)$$

12.47 Calculate the partial pressure of $CO_2(g)$, given that K_c is 1.25 (at 1500 K) for:

$$CaCO_3(s) \rightleftharpoons CaO(s) + CO_2(g)$$

Solubility Equilibria

12.48 Solubility is influenced by temperature. Will temperature have the same effect on the solubilities of all solids? Justify your answer.

12.49 Write the solubility product expression for the dissolution of
(a) barium sulfate
(b) silver acetate
(c) copper(I) carbonate
(d) chromium(III) hydroxide

12.50 Write the expression for the solubility product for the dissolution of
(a) magnesium fluoride
(b) calcium phosphate
(c) aluminum carbonate
(d) lanthanum fluoride

12.51 The solubility of silver iodate, $AgIO_3$, is 1.8×10^{-4} M. Calculate the solubility product.

12.52 Silver iodide is sprayed from an airplane by modern "rainmakers" to try to coax rain from promising cloud formations. The AgI crystal provides a "seed" on which the water can begin to condense. Silver iodide satisfies two requirements needed for the formation of water drops. First, the crystals are quite small, and second, the solubility in water is low. If the solubility is 9×10^{-9} M, calculate K_{sp} for AgI.

12.53 Given the following data, calculate the solubility product.
(a) The solubility of barium chromate, $BaCrO_4$ is 1.1×10^{-5} M.
(b) The solubility of cesium permanganate is 0.22 g/100 mL.
(c) The solubility of silver phosphate, Ag_3PO_4, is 4.4×10^{-5} M.
(d) The solubility of silver sulfate, Ag_2SO_4, is 1.4×10^{-2} M.
(e) The solubility of potassium iodate, KIO_3, is 43 g/L.

12.54 Even though barium is toxic, a suspension of barium sulfate is administered to patients who need x rays of the gastrointestinal tract. The barium "milkshake" is safe to drink because the solubility of barium sulfate is so low. Calculate the solubility of barium sulfate in g/L, using the data of Table 12.4.

12.55 The solubility product for copper(II) iodate, $Cu(IO_3)_2$, is 7.4×10^{-8}. Calculate the solubility.

12.56 The solubility product for silver tungstate, Ag_2WO_4, is 5.5×10^{-12}. Calculate the solubility.

12.57 Lead poisoning has been a hazard for centuries. Some scholars believe that the decline of the Roman Empire can be traced, in part, to high levels of lead in water from containers and pipes, and in wine that was stored in lead-glazed containers. If we presume that the typical Roman water supply was saturated with lead carbonate, $PbCO_3$ ($K_{sp} = 1.0 \times 10^{-13}$), how much lead would a Roman ingest in a year if he drank 1 L/day from the container?

Historians feel that lead glazes, principally lead silicate, contributed to the Roman diet. Another potential source was the Roman practice of boiling red wine in lead vessels to get a sweet syrup that was enriched in lead acetate as well as sugar. Problems with lead in water are not restricted to ancient civilizations. Many contemporary municipal water systems include pipes that are soldered with a lead-based solder; people are advised to run the water to flush any lead that dissolved in the water while it was in contact with the solder.

Common Ion Effect

12.58 Calculate the solubility of barium sulfate ($K_{sp} = 8.7 \times 10^{-11}$) in
(a) water
(b) 0.10 M sodium sulfate solution.

12.59 Calculate the solubility of cadmium cyanide, $Cd(CN)_2$ ($K_{sp} = 1.0 \times 10^{-8}$), in
(a) water
(b) 0.050 M sodium cyanide solution.

12.60 Calculate the solubility of copper(II) iodate, $Cu(IO_3)_2$ ($K_{sp} = 7.4 \times 10^{-8}$), in
(a) water
(b) 0.10 M sodium iodate solution.

12.61 Calculate the solubility of copper(I) iodide, CuI ($K_{sp} = 1.1 \times 10^{-12}$), in
(a) water
(b) 0.050 M sodium iodide solution.

Formation of a Precipitate

12.62 Use the solubility product data in Appendix F to determine whether a precipitate will form if the following solutions are mixed:
(a) 10 mL of 0.0010 M $AgNO_3$ and 10 mL of 0.0010 M Na_2SO_4.
(b) 10 mL of 1.0×10^{-6} M iron(II) chloride and 20 mL of 3.0×10^{-4} M barium hydroxide.

12.63 Use the data in Appendix F to determine whether a precipitate forms when the following solutions are mixed:
(a) 5.0 mL of 0.10 M lead nitrate plus 5.0 mL of 0.020 M sodium chloride.
(b) 20.0 mL of 1.0×10^{-6} M magnesium chloride plus 80 mL of 1.0×10^{-6} M potassium fluoride.

12.64 A solution is made by adding 20 mL of 0.010 M potassium bromide to 30 mL of 0.010 M potassium iodide. Concentrated silver nitrate is added to the solution.
(a) Which halide, the bromide or iodide, will precipitate first?
(b) What is the concentration of the *least* soluble halide when the other starts to precipitate?

12.65 A solution that is 0.050 M K_2SO_4 and 0.020 M Na_3PO_4 is mixed with some barium chloride.
(a) Which precipitates first—the barium sulfate or the barium phosphate?
(b) What is the concentration of the anion in the barium compound that precipitated first when the more soluble species begins to precipitate?

12.66 Find the concentration of silver necessary to begin precipitation of AgCl from a solution in which the Cl^- concentration is 7.4×10^{-4} M.

BORAX
100 YEARS
1891
1991
WHITE HOUSE
VINEGAR
PEPSI
ANACIN
ANACIN
FAST PAIN RELIEF

Chapter Thirteen

Solutions of Acids and Bases

Chemists have been classifying materials according to their properties for several hundred years. One classification method is based on the properties of aqueous solutions. Two types of solutions are immediately apparent: acids and bases. Solutions of acids taste sour; lemons taste sour because of the presence of citric acid. (Tasting is neither a safe nor sure way to identify any chemical, but was widely used in the past.) Solutions of acids can neutralize bases and dissolve many metals, such as zinc, cadmium, and iron. Acids cause many organic compounds to change color—the liquid produced when red cabbage is boiled is a good example of a substance that changes color when acid is added (Figure 13.1). Bases, also called alkalis, have different properties. Alkaline solutions taste bitter and feel slippery to the touch. One common base is lye (sodium hydroxide), found in drain cleaners and used in making soap. Bases react with many metal ions to form insoluble precipitates.

Figure 13.1 Cabbage juice The juice produced when red cabbage is boiled changes color when acids or bases are added.

All acids share common properties; so do all bases. Chemists have performed many experiments to determine whether acids have common structural features that are responsible for their behaviors. All of these early experiments (and most of the modern experiments) used water as a solvent. The results of these experiments led the Swedish chemist Svante Arrhenius (1859–1927) to propose a model of an **acid** as a substance that increases the concentration of hydrogen ions when dissolved in water. The Arrhenius model of a **base** was a substance that increases the concentration of hydroxide ions when dissolved in water. These descriptions are the ones we have used since Chapter 3.

An Arrhenius acid increases the hydrogen ion concentration when it is dissolved in water. An Arrhenius base increases the hydroxide ion concentration.

Many commercial products are acidic or basic.

INSIGHTS into CHEMISTRY

Development of Chemistry

Tasting Lab Chemicals Was Once Routine; Now This Dangerous Practice Is Obsolete

The statement about acids tasting sour represents not only a fact about acids, but also a statement about early chemical experiments and chemists. Chemists no longer taste laboratory materials. Not only has the development of modern chemical procedures made such tests obsolete, they are also clearly dangerous. However, chemists once performed these tests routinely. This is a description of one of these tests.

A chemist received a vial of a substance that was connected with a Civil War munitions train. The story of the vial, its contents, and the train appears in Chapter 4. The synthesis of nitroglycerine was a possible use for the vial, and the chemist came across an interesting reference to tasting chemicals while examining documents about the history of nitroglycerine in the United States. An abstract of some experiments by M. Sobrero, originally published in France in *Comptes Rendus,* February 1847, appeared in the *American Journal of Science and Arts,* edited by B. Silliman, B. Silliman, Jr., and James D. Dana, Vol. IV, November 1847.

After mixing the chemicals, "The addition of water precipitates a heavy oily looking liquid, which may be washed in water, dissolved in alcohol and separated by the addition of water. It resembles light yellow olive oil, it is heavier than water, in which it is quite insoluble—it dissolves freely in alcohol and ether, is without smell, and of a sweetish pungent and aromatic flavor. It must be tasted with great caution, as a quantity sufficient to moisten the end of the finger, when applied to the tongue, produces the most unpleasant effects of nausea and headache, which last for hours. No analysis of this compound has been made."

Nitroglycerine is known to most people as a powerful explosive. It is also a very powerful medicine that is prescribed for people with a particular type of heart disease. Nitroglycerine is a *vasodilator,* a compound that causes the blood vessels to relax and expand. Since side effects include headaches and nausea, as noted in the 1847 article, the dosage must be carefully controlled. One method is to dispense the nitroglycerine through the skin by applying a skin patch. Chemists have been able to immobilize nitroglycerine on a fabric patch that is worn like a small bandage. The amount of nitroglycerine delivered to the patient can be controlled and side effects minimized by this convenient delivery system.

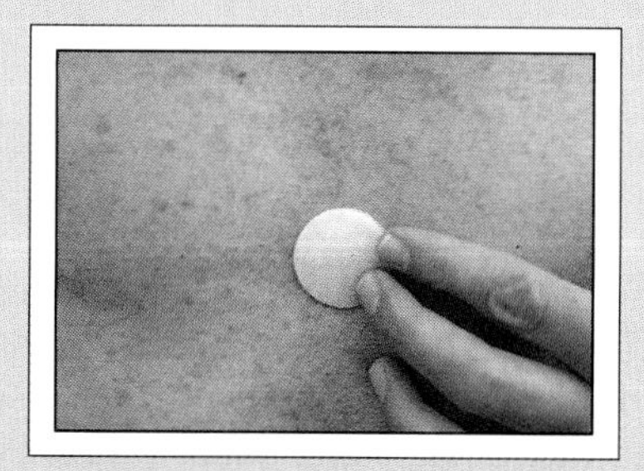

Nitroglycerine skin patch The skin patch is an efficient method of delivering the proper amount of nitroglycerine to the patient.

Other models for describing acids and bases, the Brønsted-Lowry model and the Lewis model, are presented in this chapter. These models extend the definitions of acids and bases. In addition, equilibrium calculations that involve acids and bases will be explained, and some relationships between structures and the acid-base properties of molecules will be examined.

13.1 Brønsted-Lowry Acid-Base Systems

Objectives

- To extend the Arrhenius model to other acid-base systems
- To identify acid-base conjugate pairs

The Arrhenius model of an acid—a substance that increases the hydrogen ion concentration in aqueous solution—while correct, is quite limited. The

Arrhenius definition applies only to aqueous solutions, and does not describe the behavior of substances in other solvents or in gas-phase reactions. The use of a relatively simple concept, classifying substances as acids or bases, was successful in predicting the results of many chemical reactions and the model was extended by Johannes Brønsted (1879–1947) and Thomas Lowry (1874–1936), who independently noted the importance of acid-base behavior in other systems. They recognized that the basis for an acid-base reaction is the transfer of a proton from one species to another, and they proposed a model based on this concept. A **Brønsted-Lowry acid** is defined as a proton donor. A **Brønsted-Lowry base** is defined as a proton acceptor. In this section, the Brønsted-Lowry definitions of acids and bases are discussed, and the application of this model is demonstrated.

A Brønsted-Lowry acid is a proton donor; a Brønsted-Lowry base is a proton acceptor.

It is worthwhile to spend a few moments reviewing some of the terms that are used to describe acids and bases, terms that were first introduced in Chapter 3. A hydrogen ion, H^+, is also referred to as a proton. It is important to remember that the hydrogen ion is always associated with the solvent; "proton" does not refer to the nuclear particle. The designations $H^+(aq)$ and H_3O^+ serve to help recall the role of the solvent. While H^+ is called a hydrogen ion and H_3O^+ is called a hydronium ion, the terms are interchangeable when used to describe solutions in which water is the solvent. This text follows the practice of most chemists and uses the words "hydrogen ion" and the symbol H_3O^+ in describing acid-base equilibria.

$H^+(aq)$, H_3O^+, H^+, hydrogen ion, proton, and hydronium ion are all used interchangeably to describe the same species in water solutions.

The Brønsted-Lowry model is an extension of the Arrhenius model. In water, species that are Brønsted-Lowry acids are also Arrhenius acids, and *vice versa.* But the Arrhenius model does not include gas-phase reactions such as the formation of ammonium chloride:

$$H_3N\!: + H{-}\ddot{\underset{..}{Cl}}\!: \longrightarrow [H_4N]^+ \; [:\ddot{\underset{..}{Cl}}:]^-$$

$$HCl(g) + NH_3(g) \longrightarrow NH_4Cl(s)$$

The hydrogen-chlorine bond in the hydrogen chloride molecule is broken, and the hydrogen ion is transferred to the ammonia molecule; the result is ammonium chloride, an ionic solid. You may have noticed a haze on the glassware (and maybe even the windows) in the laboratory. This film is likely to be ammonium chloride formed by the gas phase reaction of HCl and NH_3, which are often present in the air of chemistry laboratories. The preceding equation shows that ammonia accepts a hydrogen ion, and it is therefore classified as a base by the Brønsted-Lowry model.

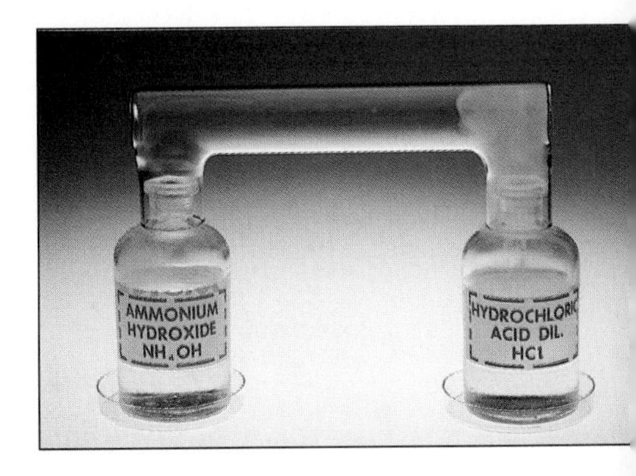

A gas-phase acid-base reaction The white cloud is composed of small $NH_4Cl(s)$ particles formed by the reaction of HCl(g) and $NH_3(g)$. Some containers of aqueous ammonia are still labeled NH_4OH, although $NH_3(aq)$ is preferred.

Acid-Base Conjugate Pairs

When hydrogen fluoride dissolves in water, it transfers a proton, forming hydrogen ion and fluoride ion, F^-, so HF is properly classified as an acid,

$$HF(aq) + H_2O(\ell) \rightarrow H_3O^+(aq) + F^-(aq).$$

Table 13.1 Acid-Base Conjugate Pairs

Acid		Base	
hydrogen chloride	HCl	Cl^-	chloride ion
sulfuric acid	H_2SO_4	HSO_4^-	hydrogen sulfate ion
hydrogen sulfate ion	HSO_4^-	SO_4^{2-}	sulfate ion
acetic acid	CH_3COOH	CH_3COO^-	acetate ion
ammonium ion	NH_4^+	NH_3	ammonia
hydronium ion	H_3O^+	H_2O	water
water	H_2O	OH^-	hydroxide ion

In addition, a proton can bond to F^-, producing HF. The fluoride ion has accepted a proton and behaved as a base.

$$F^-(aq) + H_3O^+(aq) \longrightarrow HF(aq) + H_2O(\ell)$$

When a Brønsted-Lowry acid loses a proton, the species formed is a base because it can accept a proton. The two species are related by the loss and gain of a proton; these species are referred to as an **acid-base conjugate pair**—the acid form of the pair is protonated while the base form has lost the proton. Every Brønsted-Lowry acid has a conjugate base, and every base has a conjugate acid. Several common conjugate pairs are listed in Table 13.1.

Notice that water is listed both in the acid column and in the base column. It can behave either as a proton donor (acid) or as a proton acceptor (base), depending on the species with which it reacts. A substance that can act either as an acid or as a base is called **amphoteric.** (The hydrogen sulfate ion, HSO_4^-, is also amphoteric.) The amphoteric behavior of water is important in understanding the properties of acids and bases in aqueous chemistry and is discussed later.

Water can act as an acid or as a base.

Neutralization Reactions

When an Arrhenius acid reacts with an Arrhenius base, the products are a salt and water.

$$\text{acid} + \text{base} \longrightarrow \text{salt} + \text{water}$$

The Brønsted-Lowry model, as you might guess, is not limited to this type of reaction. *An acid (species 1) will transfer a proton to a base (species 2) to form the conjugate base of species 1 and the conjugate acid of species 2.*

Line 1 in Table 13.2 is the equation for the reaction of hydrofluoric acid with ammonia. The conjugate base of hydrofluoric acid is the fluoride ion, F^-; the conjugate acid of ammonia is the ammonium ion, NH_4^+.

Table 13.2 Representative Acid-Base Neutralization Reactions

	acid$_1$	+	**base$_2$**	⟶	**base$_1$**	+	**acid$_2$**
1	HF	+	NH_3	⟶	F^-	+	NH_4^+
2	HCl	+	H_2O	⟶	Cl^-	+	H_3O^+
3	H_2O	+	NH_2^-	⟶	OH^-	+	NH_3

Lines 2 and 3 illustrate the amphoteric behavior of water. In line 2, water acts as a base; it accepts a hydrogen ion from HCl to form the hydronium ion, H_3O^+ (the conjugate acid of water), and the chloride ion (the conjugate base of HCl). The Brønsted-Lowry definition of a base is a proton acceptor, so water is a base in this example.

$$:\ddot{Cl}-H + :\ddot{O}(H)-H \longrightarrow [:\ddot{Cl}:]^- + [H-\ddot{O}(H)-H]^+$$

$$HCl + H_2O \longrightarrow Cl^- + H_3O^+$$

In line 3, water acts as an acid. It donates a hydrogen ion to the amide ion (NH_2^-), forming ammonia. The hydroxide ion is formed by the transfer of a proton from water to the amide ion. In this example, the amide ion is the base (proton acceptor) and water is the acid (proton donor).

An acid-base neutralization reaction involves a transfer of a proton from the acid to the base.

$$:\ddot{O}(H)-H + :N(H)(H)-H \rightleftharpoons [H-N(H)(H)-H]^+ + [:\ddot{O}-H]^-$$

$$H_2O + NH_3 \rightleftharpoons NH_4^+ + OH^-$$

Example 13.1 Identifying Acid-Base Conjugate Pairs

Identify the conjugate acid-base pairs in the following chemical reactions:

(a) $H_2SO_4(aq) + H_2O(\ell) \longrightarrow HSO_4^-(aq) + H_3O^+(aq)$
(b) $H_2O(\ell) + F^-(aq) \rightleftharpoons OH^-(aq) + HF(aq)$

Solution

(a) The H_2SO_4 loses a proton to form HSO_4^-, so H_2SO_4 is the acid and HSO_4^- is the conjugate base. H_2O accepts a proton, forming H_3O^+.

H_2SO_4/HSO_4^-, H_3O^+/H_2O

(b) Water loses a proton to form OH^-, so H_2O is the acid and OH^- is its conjugate base. The fluoride ion accepts the proton, forming its conjugate acid, HF.

H_2O/OH^-, HF/F^-

Understanding Identify the acid-base conjugate pairs in the following chemical reaction:

$$SO_4^{2-}(aq) + HCl(aq) \longrightarrow HSO_4^-(aq) + Cl^-(aq)$$

Answer: HSO_4^-/SO_4^{2-}, HCl/Cl^-

13.2 Autoionization of Water

Objectives

- To develop the relationship between the hydrogen ion and hydroxide ion concentrations in aqueous solutions
- To define pH and use it to express concentrations

We've seen that water has both acidic and basic properties. In this section, the relationship between the concentrations of the hydrogen ion and the hydroxide ion will be presented.

A water molecule can donate a proton (to form OH^-) or accept a proton (to form H_3O^+), depending on the experimental conditions. Water can, in fact, act both as an acid and a base when it reacts with itself.

$$H_2O(\ell) + H_2O(\ell) \rightleftharpoons H_3O^+(aq) + OH^-(aq)$$

This equation illustrates the **autoionization of water,** which occurs to a small extent. The equilibrium constant for this reaction is

$$K' = \frac{[H_3O^+][OH^-]}{[H_2O]^2}$$

Water is treated like any other pure liquid; its concentration does not change significantly in dilute solutions, so it does not appear in the equilibrium expression.

$$K_w = [H_3O^+][OH^-]$$

The ionization of water is so important to the study of aqueous equilibria that the equilibrium constant is given the special symbol K_w. The expression can be evaluated by measuring the concentrations of H_3O^+ and OH^- in pure water. In fact, we need measure only one, since the stoichiometry of the autoionization reaction requires that $[H_3O^+]$ and $[OH^-]$ must be equal to each other *in pure water.* The results of such experiments, performed at several temperatures, are shown in Table 13.3.

We can see that K_w, like all equilibrium constants, depends on temperature. Since K_w is larger (forward reaction is favored) at higher temperatures, the forward reaction must absorb heat, so the ionization of water must be endothermic. If the temperature is not stated, we will assume a temperature of 25 °C, at which the value for K_w is very close to 1.0×10^{-14}.

The equilibrium constant for the ionization of water is 1.0×10^{-14} at 25 °C.

Table 13.3 Values of the Equilibrium Constant for the Autoionization of Water at Several Temperatures

Temperature (°C)	$K_w = [H_3O^+][OH^-]$
0	0.113×10^{-14}
10	0.292×10^{-14}
15	0.451×10^{-14}
20	0.681×10^{-14}
24	1.000×10^{-14}
25	1.008×10^{-14}
30	1.469×10^{-14}
40	2.919×10^{-14}
50	5.474×10^{-14}
60	9.614×10^{-14}

Reprinted with permission from the *CRC Handbook of Chemistry and Physics,* © 1988.

Calculation of Hydrogen and Hydroxide Ion Concentrations

If we know the concentration of hydrogen ion in a water solution, the expression for K_w can be used to calculate the concentration of hydroxide ion, and *vice versa*. In pure water, we know that $[H_3O^+]$ and $[OH^-]$ are equal to each other. Since

$$K_w = [H_3O^+][OH^-]$$

and $[H_3O^+]$ and $[OH^-]$ are equal, we can write

$$K_w = [H_3O^+]^2$$

At 25 °C, $K_w = 1.0 \times 10^{-14}$, so

$$[H_3O^+]^2 = 1.0 \times 10^{-14}$$

$$[H_3O^+] = 1.0 \times 10^{-7}\ M = [OH^-]$$

The last calculation indicates that the hydrogen ion concentration in pure water is $1.0 \times 10^{-7}\ M$. However, water that is in contact with air dissolves small amounts of carbon dioxide, forming carbonic acid, H_2CO_3. The hydrogen ion concentration of water that is saturated with air is greater than $1.0 \times 10^{-7}\ M$, about $2 \times 10^{-6}\ M$.

If an acid or base is dissolved in the water, the concentrations of hydroxide and hydrogen ions are no longer equal, but the equilibrium constant, K_w, still describes the system. The concentration of one of the ions can be calculated from the equilibrium expression, if the concentration of the other is known. This type of calculation is illustrated in Example 13.2.

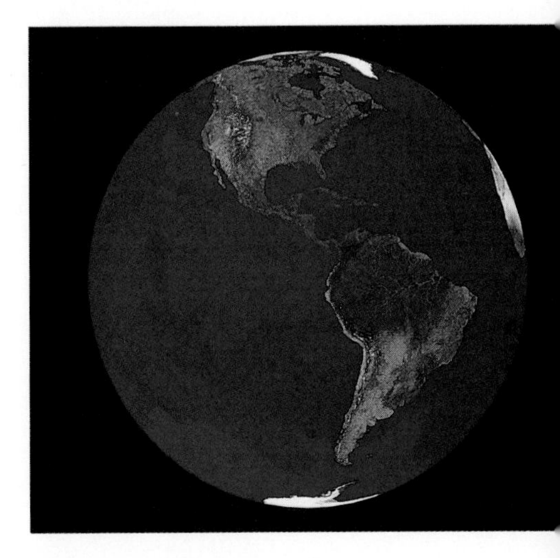

Most of our planet is covered by water. Most of the water, however, is salty.

The product of the hydrogen ion and hydroxide ion concentrations in any aqueous solution is equal to $K_w = 1.0 \times 10^{-14}$ at 25 °C.

Example 13.2 Calculating the Concentration of Hydroxide Ion from the Concentration of Hydrogen Ion

An acid is added to water so that the hydrogen ion concentration is 0.25 *M*. Calculate the hydroxide ion concentration.

Solution We know that

$$K_w = [H_3O^+][OH^-]$$

$$1.0 \times 10^{-14} = (0.25)[OH^-]$$

$$[OH^-] = \frac{1.0 \times 10^{-14}}{0.25}$$

$$[OH^-] = 4.0 \times 10^{-14}\ M$$

Notice that when the hydrogen ion concentration is large, the hydroxide ion concentration is small. The *product* of the two, however, is equal to a constant, K_w, that depends only on temperature.

Understanding Calculate the hydrogen ion concentration in a solution that is $4.0 \times 10^{-5}\ M$ hydroxide ion.

Answer: $[H_3O^+] = 2.5 \times 10^{-10}\ M$

Concentration Scales

In any aqueous solution, the concentration of either the hydrogen ion or the hydroxide ion will be small. After all, the product of the two must be 1.0×10^{-14} (at 25 °C). Powers of ten are awkward to use, so a logarithmic notation, the **pH** scale, has been devised. The notation pH derives from the French *pouvoir hydrogène,* literally translated as hydrogen power. The pH is defined by

$$pH = -\log_{10}[H_3O^+]$$

$$[H_3O^+] = 10^{-pH}$$

We will denote common (base-10) logarithms as "log" instead of $\log_{10}$, simply for convenience. Natural logarithms are denoted "ln" so that they won't be confused with common logarithms. pH measurements are commonly expressed to two decimal places. There are only two significant digits in the H_3O^+ concentration of a solution with pH = 4.77; the 4 serves only to locate the decimal point. Calculations of pH from hydrogen ion concentrations are shown in Example 13.3.

Example 13.3 Calculating pH from Concentration

Calculate the pH of

(a) 0.050 M H_3O^+
(b) 0.12 M OH^-
(c) 2.4 M H_3O^+

Solution

(a) $[H_3O^+] = 0.050\ M$

$$pH = -\log(0.050) = -(-1.30) = 1.30$$

(b) $[OH^-] = 0.12\ M$

$$[H_3O^+] = \frac{K_w}{[OH^-]} = \frac{1.0 \times 10^{-14}}{0.12} = 8.3 \times 10^{-14}\ M$$

$$pH = -\log(8.3 \times 10^{-14}) = 13.08$$

(c) $[H_3O^+] = 2.4\ M$

$$pH = -\log(2.4) = -0.38$$

Understanding Calculate the pH of a solution in which the hydrogen ion concentration is $3.5 \times 10^{-5}\ M$.

Answer: pH = 4.46

Similarly, we can convert from pH to the molarity of H_3O^+ by taking the antilogarithm of the negative of the pH:

$$[H_3O^+] = 10^{-pH}$$

$$[H_3O^+] = \text{invlog}(-pH)$$

where invlog represents taking the antilogarithm, also called the inverse logarithm. These operations are easily performed with a calculator. Appendix A contains instructions on how to perform logarithmic operations on a calculator.

Example 13.4 Calculating the Concentration of Hydrogen Ion from pH

Calculate the hydrogen ion concentration in a solution that has

(a) pH = 3.50
(b) pH = 12.56

Solution

(a) pH = 3.50

$[H_3O^+] = 10^{-3.50}$

$[H_3O^+] = 3.2 \times 10^{-4}\ M$

(b) pH = 12.56

$[H_3O^+] = 10^{-12.56}$

$[H_3O^+] = 2.8 \times 10^{-13}\ M$

Understanding Calculate the hydrogen ion concentration in a solution that has a pH of 4.76.

Answer: $[H_3O^+] = 1.7 \times 10^{-5}$

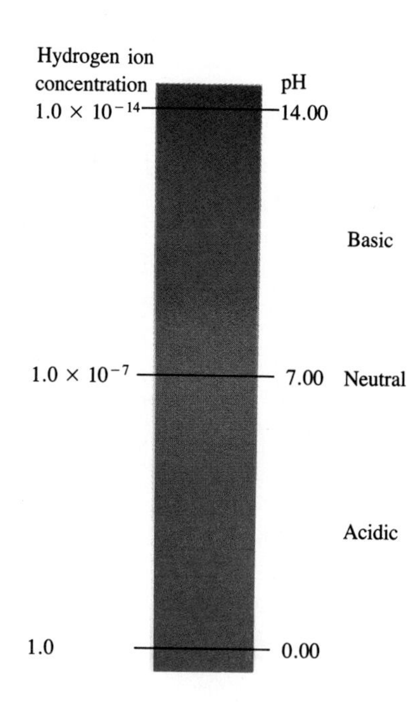

Figure 13.2 The pH scale and acidity

It has become common practice among chemists to use the p-notation for any equilibrium constant or concentration that is very small. Some examples are:

$$pOH = -\log[OH^-]$$

$$pCl = -\log[Cl^-]$$

$$pK_w = -\log(K_w) = 14.00$$

The hydrogen ion concentration in pure water is $1.0 \times 10^{-7}\ M$, so the pH of pure water is 7.00. An acidic solution has a hydrogen ion concentration that is greater than that of pure water, making the pH of an acidic solution less than 7. Alternatively, if some source of hydroxide ion is added to pure water, the solution will be basic and the hydrogen ion concentration will decrease below $1.0 \times 10^{-7}\ M$, so the pH will be greater than 7. These facts are summarized in Figure 13.2.

Most pH measuring devices have calibration markings for the pH range from 0 to 14, as shown in Figure 13.3, but the pH of a solution can be less than 0 or greater than 14. Solutions in which the hydrogen ion concentration exceeds 1 *M* have a negative pH; those solutions in which the hydroxide ion concentration exceeds 1 *M* have a pH concentration greater than 14.

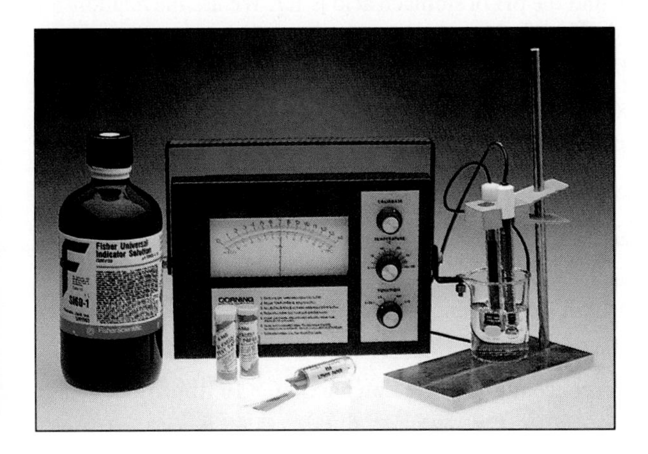

Figure 13.3 Measuring the pH of solutions The pH of a solution can be determined by several different techniques.

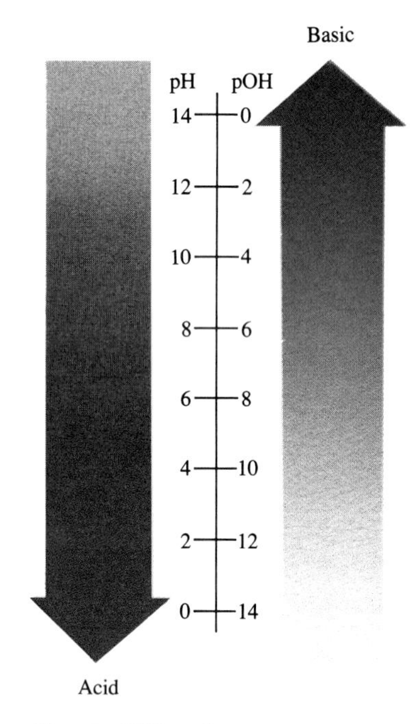

pH and pOH scales

pH + pOH = 14.00
$[H_3O^+][OH^-] = 1.0 \times 10^{-14}$

The Relationship Between pH and pOH

We know that the hydrogen ion concentration can be calculated from the hydroxide ion concentration, and *vice versa.* The key relationship is the equilibrium constant expression that describes the ionization of water.

$$K_w = [H_3O^+][OH^-]$$

A similar expression can be derived for pH and pOH, by taking the negative logarithm of both sides.

$$-\log(K_w) = -\log([H_3O^+][OH^-])$$

Remember that the logarithm of a product can be written as the sum of two logarithms, so

$$-\log(K_w) = -\log[H_3O^+] - \log[OH^-]$$

We have defined the p-function as $-\log_{10}$, or more simply as $-\log$, so we can rewrite

$$pK_w = pH + pOH$$

Since the numerical value for pK_w is 14.00 at 25 °C, the last equation can be rewritten:

$$pH + pOH = 14.00$$

Just as we know that the product of the hydrogen ion and hydroxide ion concentrations is 1.0×10^{-14}, *the sum of the pH and pOH must always be 14.00* (at 25 °C).

Table 13.4 pH Values for Some Common Solutions

Substance	pH
1.00 *M* HCl	0.0
human stomach acid	1.7
lemon juice	2.2
vinegar	2.9
carbonated beverage	3.0
wine	3.5
tomato juice	4.1
black coffee	5.0
urine	6.0
milk	6.9
pure water	7.0
blood	7.4
sea water	8.5
household detergent	9.2
milk of magnesia	10.5
household ammonia	11.9
1.00 *M* NaOH	14.0

Example 13.5 The Hydrogen Ion Concentrations of Common Materials

Refer to Table 13.4 to answer the questions.

(a) What are the hydrogen ion concentration, hydroxide ion concentration, and pOH of stomach acid?
(b) What is the ratio of the hydrogen ion concentration in lemon juice to the hydrogen ion concentration in pure water?

Solution

(a) From Table 13.4, we find the pH of stomach acid is 1.7. We use the relationship between $[H_3O^+]$ and pH to find

$$[H_3O^+] = 10^{-1.7} = 2 \times 10^{-2}\ M$$
$$pOH = 14.00 - pH = 14.00 - 1.7 = 12.3$$
$$[OH^-] = 10^{-12.3} = 5 \times 10^{-13}\ M$$

(b) For lemon juice, the pH is 2.2.

$$[H_3O^+]_{\text{lemon juice}} = 10^{-2.2} = 6 \times 10^{-3}\ M$$
$$[H_3O^+]_{\text{water}} = 1.0 \times 10^{-7}\ M$$

The ratio of the hydrogen ion concentration in lemon juice to that in water is

$$\text{ratio} = 6 \times 10^{-3}\ M / 1 \times 10^{-7}\ M = 6 \times 10^4$$

The hydrogen ion concentration in lemon juice is 6×10^4 times that of pure water.

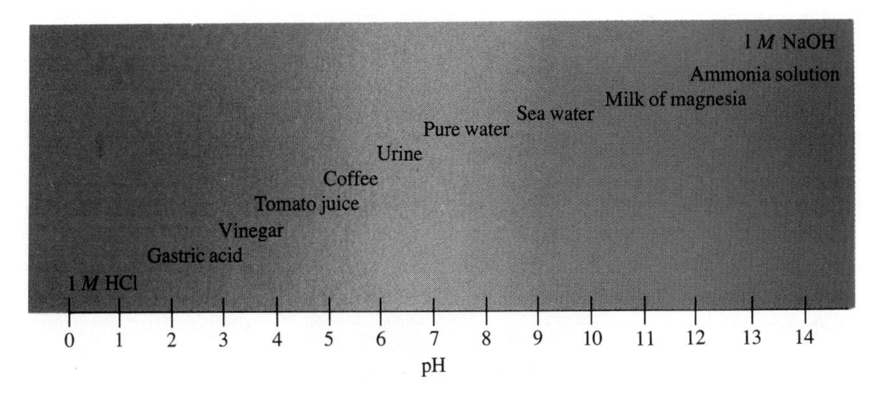

pH values of some common substances.

Understanding What is the ratio of $[OH^-]$ in household ammonia to that in blood?

Answer: Ammonia is 3×10^4 times as basic as blood.

13.3 Strong Acids and Bases

Objectives

- To identify species that are strong acids and bases
- To calculate the concentrations of species, the pH, and the pOH in solutions of strong acids and bases

Compounds that ionize or dissociate completely in water are termed **strong electrolytes;** those that are only partially ionized or dissociated are **weak electrolytes.** Much of the chemistry that occurs in aqueous solution is related to reactions of strong and weak electrolytes. In this section, the acid-base chemistry of these substances is presented, emphasizing the relationships between their concentrations and the pH of their solutions.

When chemists speak of a **strong acid,** they mean one that is completely ionized in solution. An example of a chemical equation that describes a strong acid dissolving in water is

$$HCl(g) + H_2O(\ell) \longrightarrow H_3O^+(aq) + Cl^-(aq)$$

Experiment shows that this reaction goes essentially to completion, and the equilibrium concentration of nonionized HCl is so small that it cannot be detected. We indicate this feature in the chemical equation with a single arrow that points to the right. A weak acid, such as HF, does not ionize completely, so we use a pair of arrows in the chemical equation to indicate equilibrium.

A strong acid ionizes completely in solution; a weak acid does not.

$$HF(g) + H_2O(\ell) \rightleftharpoons H_3O^+(aq) + F^-(aq)$$

It might seem appropriate to use the terms "strong" and "weak" to refer to the chemical reactivity of an acid—that a strong acid would be more reactive than a weak one—but this simply is not true. Hydrofluoric acid, HF, is a weak acid, but it can dissolve glass!

INSIGHTS into CHEMISTRY

Chemical Reactivity

Hydrofluoric Acid
Considered a Weak Acid, Yet Is Highly Reactive—Can Dissolve Glass

Hydrofluoric acid, HF, is a weak acid; it is only partially ionized in solution. The HF molecule is unusually stable in aqueous solution because the hydrogen-fluorine bond is quite strong.

The fact that it is a weak acid does not mean that HF is not highly corrosive and reactive—it is. Chemists who accidentally spill HF on their skin soon discover that HF burns are among the most painful. But perhaps the most startling feature of HF is its reactivity toward glass—HF will actually dissolve glass.

Glass is a very viscous mixture of silicates that are derived from silicon dioxide (SiO_2). When glass is attacked by HF, the following reaction occurs:

$$SiO_2(s) + 4HF(aq) \rightleftharpoons SiF_4(g) + 2H_2O(\ell)$$

This reaction is partially driven by the principle of Le Chatelier. The product SiF_4 is a gas, and it escapes from the system. In general, removing the products from the reaction mixture will drive the reaction toward completion. Hydrofluoric acid and ammonium fluoride are often used to etch glass, to make it opaque or to produce artistic designs.

Glass etching Artists have used HF to dissolve layers of glass, leaving a beautiful object behind.

It is surprising how many fluorides are gases at or near room temperature. The name of the element, fluorine, is derived from *fluere,* Latin for "flow." Common fluoride-containing compounds of industrial importance include SF_6 (a gas at room temperature) and UF_6 (boils at 56 °C).

Table 13.5 Ionization of Strong Acids

hydrochloric	$HCl + H_2O \longrightarrow H_3O^+ + Cl^-$
hydrobromic	$HBr + H_2O \longrightarrow H_3O^+ + Br^-$
hydroiodic	$HI + H_2O \longrightarrow H_3O^+ + I^-$
nitric	$HNO_3 + H_2O \longrightarrow H_3O^+ + NO_3^-$
perchloric	$HClO_4 + H_2O \longrightarrow H_3O^+ + ClO_4^-$
sulfuric	$H_2SO_4 + H_2O \longrightarrow H_3O^+ + HSO_4^-$

There are only six common strong acids; the others are weak.

Only six strong acids are commonly encountered. They are listed in Table 13.5. You should remember these six strong acids. Assume that all the other acids you encounter are weak unless you are told otherwise.

Solutions of Strong Acids

The pH of a solution of a strong acid or base is determined by the concentration of the acid or base.

When a strong acid dissolves in water, the concentrations of the species in the solution can be calculated from the chemical equation and the starting concentrations of the species. Since the acid ionizes completely, these calculations do not require an equilibrium constant.

Example 13.6 Calculating Hydrogen Ion Concentration and pH of Solutions of Strong Acids

Calculate the hydrogen ion concentration and pH of

(a) 0.010 M HNO_3.
(b) a solution prepared by diluting 10.0 mL of 0.50 M $HClO_4$ to 50.0 mL.
(c) a solution prepared by adding 9.67 g of HCl(g) to some water, and then diluting the solution to 500.0 mL.

Solution

(a) Since the ionization of nitric acid is complete

$$HNO_3(aq) + H_2O(\ell) \longrightarrow H_3O^+(aq) + HNO_3^-(aq)$$

we can see that 0.010 M HNO_3 produces 0.010 M hydrogen ion. The pH is 2.00.

(b) A flow diagram can be used to describe the processes.

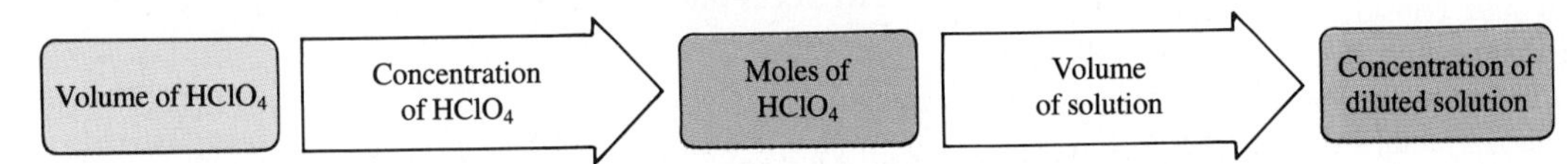

The 0.50 M $HClO_4$ yields 0.50 M H_3O^+ when dissolved in water. This problem is actually one of dilution. First calculate the number of moles of hydrogen ion, and then divide by the total volume of the solution to determine the concentration. The number of moles is calculated from the concentration and volume of perchloric acid added to the solution.

$$\text{number of moles of } H_3O^+ = 0.0100 \text{ L of soln} \frac{0.50 \text{ mol } H_3O^+}{\text{L of soln}}$$

$$= 5.0 \times 10^{-3} \text{ mol of } H_3O^+$$

$$\text{concentration of } H_3O^+ \text{ in final solution} = \frac{5.0 \times 10^{-3} \text{ mol}}{0.0500 \text{ L}} = 0.10\ M$$

$$pH = -\log(0.10) = 1.00$$

(c) When we need to prepare a solution from a gas, one method is to place the container of solvent on a balance, and bubble the gas into it until the mass increases by the appropriate amount. In this problem, we dissolve 9.66 g of HCl in some water, then add additional water until we have 500 mL of solution. First calculate the number of moles of HCl added to solution. Each mole of HCl yields one mole of H_3O^+ since HCl is a strong acid and dissociates completely in water. Then use the amount of H_3O^+ and the final volume to calculate the concentration.

$$\text{moles of HCl} = 9.66 \text{ g HCl} \left(\frac{1 \text{ mol HCl}}{36.46 \text{ g HCl}}\right) = 0.265 \text{ mol HCl}$$

$$[H_3O^+] = \frac{0.265 \text{ mol HCl}}{0.500 \text{ L of solution}} = 0.530\ M$$

$$pH = -\log(0.530) = 0.28$$

Accurate measurements of changes in mass as a gas dissolves in solution are difficult, so a titration is often used to determine the exact concentration.

Understanding Calculate the pH of a solution made by dissolving 1.00 g of HI in enough water to make 250 mL of solution.

Answer: pH = 1.50

Dissolving a gas in solution The mass of the solution increases when a gas dissolves.

INSIGHTS into CHEMISTRY

Chemical Reactivity

Group IA and Soluble Group IIA Hydroxides Are Strong Bases

All the Group IA hydroxides are strong bases. They dissolve in water and dissociate completely. So do the Group IIA hydroxides, but they are not very soluble in water.

The following table lists the solubility of each hydroxide in cold water. Note that some values are missing, and others are not known with good precision because of experimental difficulties.

Most chemists would consider barium hydroxide a strong base, but not magnesium hydroxide. Even though each dissociates fully in water, the use of the term "strong base" is reserved for those metal hydroxides that are also readily soluble in water.

Compound	Solubility, mol/L	Compound	Solubility, mol/L
LiOH	9.2	$Be(OH)_2$	—
NaOH	10.5	$Mg(OH)_2$	0.0002
KOH	19.1	$Ca(OH)_2$	0.02
RbOH	—	$Sr(OH)_2$	0.04
CsOH	16.4	$Ba(OH)_2$	0.2

Data taken from the *Handbook of Chemistry and Physics,* © 1990 by CRC. Used with permission.

Solutions of Strong Bases

The **strong bases** are the soluble compounds that quantitatively produce hydroxide ion when dissolved in water. The most common are metal oxides and hydroxides.

$$NaOH(s) \longrightarrow Na^+(aq) + OH^-(aq)$$

$$Ba(OH)_2(s) \longrightarrow Ba^{2+}(aq) + 2OH^-(aq)$$

$$Li_2O(s) + H_2O(\ell) \longrightarrow 2Li^+(aq) + 2OH^-(aq)$$

The following example illustrates how to calculate the pH of a solution that contains a strong base.

Example 13.7 Calculating Concentrations of Species in Alkaline Solution

Calculate the hydroxide ion concentration, pOH, and pH in a solution made when 1.00 g of barium hydroxide dissolves in enough water to produce 500 mL of solution.

Mass of $Ba(OH)_2$ → [Molar mass of $Ba(OH)_2$] → Moles of $Ba(OH)_2$ → [Coefficients in chemical equation] → Moles of OH^- → [Volume of solution] → Concentration of OH^-, pOH, pH

Solution First, calculate the amount of barium hydroxide from the mass and formula weight.

$$\text{moles of } Ba(OH)_2 = 1.00 \text{ g } Ba(OH)_2 \frac{1 \text{ mol}}{171.3 \text{ g } Ba(OH)_2}$$

$$= 5.84 \times 10^{-3} \text{ mol } Ba(OH)_2$$

Use the chemical equation to determine the number of moles of hydroxide ion produced.

$$Ba(OH)_2 \xrightarrow{H_2O} Ba^{2+} + 2OH^-$$

$$\text{moles of hydroxide} = 5.84 \times 10^{-3}\ \text{mol Ba(OH)}_2 \left(\frac{2\ \text{mol OH}^-}{1\ \text{mol Ba(OH)}_2}\right) = 1.17 \times 10^{-2}\ \text{mol OH}^-$$

The concentration of hydroxide ion is calculated from the amount and volume.

$$[OH^-] = 1.17 \times 10^{-2}\ \text{mol OH}^-/0.500\ \text{L} = 2.34 \times 10^{-2}\ M\ \text{OH}^-$$

$$\text{pOH} = -\log(2.34 \times 10^{-2}) = 1.63$$

$$\text{pH} = 14.00 - \text{pOH} = 14.00 - 1.63 = 12.37$$

Understanding Calculate the pH when 0.010 g of calcium hydroxide is dissolved in enough water to make 100 mL of solution.

Answer: pH = 11.43

Other bases generally are not very soluble or do not dissociate completely. These bases must be described by equilibrium relationships that are discussed in the next section.

13.4 Weak Acids and Bases: Qualitative Aspects

Objectives

- To recognize that the strength of an acid depends on a competition between the solvent and the conjugate base to accept protons
- To illustrate the leveling effect of the solvent

Many important compounds are weak acids or bases. For example, nearly all organic acids and bases are weak, and these compounds are crucial in numerous processes that occur in living systems. One important characteristic of a weak acid is its ability to transfer a proton to a base. In this section, the factors that influence the strengths of acids and bases are discussed and the role of the solvent is examined.

A **weak acid** is one that does not ionize completely when it is dissolved in water. The chemical equation for the ionization of a weak acid is generally written as a Brønsted-Lowry acid-base reaction with water as the base.

$$HF(aq) + H_2O(\ell) \rightleftharpoons H_3O^+(aq) + F^-(aq)$$

The concentration of the solvent does not appear in the equilibrium constant expression, so

$$K_a = \frac{[H_3O^+][F^-]}{[HF]}$$

The subscript a in K_a is a reminder that the constant describes the ionization of an acid to form H_3O^+ and the conjugate base.

We can write a similar equation for the reaction of a weak base, such as ammonia, with water. The solvent functions as a Brønsted-Lowry acid in this reaction.

$$NH_3(aq) + H_2O(\ell) \rightleftharpoons NH_4^+(aq) + OH^-(aq)$$

$$K_b = \frac{[NH_4^+][OH^-]}{[NH_3]}$$

The subscript b indicates the equilibrium constant is for the reaction of a base with water. As usual, the concentration of the solvent is not included in the expression for the equilibrium constant.

The Competition for Protons

Whenever an acid reacts with a base, the products are the conjugate acid of the base and the conjugate base of the acid.

The equilibrium will favor the formation of the weaker acid and the weaker base—the proton will be transferred from the stronger acid to the stronger base, forming the weaker acid and the weaker base.

$$HA(aq) + H_2O(\ell) \rightleftharpoons A^-(aq) + H_3O^+(aq)$$

If K_a is large (much greater than 1), the acid is strong.

If K_a is small, the acid is weak.

When K_a is large (much greater than 1), as in the case of a strong acid, the ionization is favored. The proton is transferred from the stronger acid (HA) to the base (H_2O) to form the weaker acid (H_3O^+). This reaction goes practically to completion when K_a is large. Small values of K_a indicate partial ionization, and thus weak acids.

When we examine the ionization of HCl, we find that the proton may be bonded either to a water molecule or to the chloride ion.

$$HCl(aq) + H_2O(\ell) \longrightarrow H_3O^+(aq) + Cl^-(aq)$$

The strength of an acid is a measure of the competition for protons between water and the conjugate base of the acid.

Since Cl^- is a much weaker base than H_2O, the proton bonds to the water, producing the stoichiometric amount of H_3O^+. The relative strength of an acid results from a competition for protons between the solvent and the conjugate base. Except for a few clearly noted instances, water is the only solvent we will consider in our discussion of acid-base properties.

Weak acids, on the other hand, have conjugate bases that are relatively strong. We can consider the ionization of HF.

$$HF(aq) + H_2O(\ell) \rightleftharpoons H_3O^+(aq) + F^-(aq)$$

Any individual proton will be bonded to a fluoride ion part of the time and to a water molecule the rest of the time. The chemical equation for the ionization of HF describes the competition of two bases, the fluoride ion and water, for the proton. Since the fluoride ion has a stronger attraction for the proton, fewer protons bind to water; therefore, HF(aq) is the predominant species in acid solution at equilibrium.

The Leveling Effect of the Solvent

In an acid-base reaction, a proton is transferred from the stronger acid to a base, forming the weaker acid. One ramification is that *the hydrogen ion is the strongest acid that can exist in water.* Acids that are stronger than H_3O^+ quantitatively transfer their protons to the water to form H_3O^+. We cannot measure any differences in the acidities of the six strong acids in Table 13.5, since they all ionize completely in water. This phenomenon is called the **leveling effect**—the solvent makes the strong acids appear equal, or level, in acidity.

Strong acids appear equal, or level, in strength in aqueous solution, since they ionize completely in water. The conjugate acid of the solvent is the strongest acid that can be present. In water, H_3O^+ is the strongest acid.

Water is a reasonably good base, moderately able to accept protons. We could choose a different solvent that is a poorer base. In this type of solvent system we can differentiate among the six strong acids.

An example of this type of solvent system is pure acetic acid, CH_3COOH. Perchloric acid ionizes completely in this solvent.

$$HClO_4 + CH_3COOH(\ell) \longrightarrow CH_3COOH_2^+ + ClO_4^-$$

However, HCl is a weak acid in this solvent.

$$HCl + CH_3COOH(\ell) \rightleftharpoons CH_3COOH_2^+ + Cl^-$$

Very weak acids such as hydrogen sulfide do not ionize at all in acetic acid solvent.

The strongest base that can exist in solution is the conjugate base of the solvent. If we choose water as a solvent, the conjugate base of water is the hydroxide ion. The oxide ion, O^{2-}, and the amide ion, NH_2^-, are examples of species that are leveled by water. Both react with water to form hydroxide ion in a reaction that goes to completion.

The conjugate base of the solvent is the strongest base that can be present. In water, OH^- is the strongest base.

$$Na_2O(s) + H_2O(\ell) \longrightarrow 2Na^+(aq) + 2OH^-(aq)$$

$$NaNH_2(s) + H_2O(\ell) \longrightarrow Na^+(aq) + OH^-(aq) + NH_3(aq)$$

13.5 Weak Acids and Bases: Quantitative Aspects

Objectives

- To calculate acid ionization constants from experimental data
- To calculate the concentrations of the species present in a solution of a weak acid or base

The concepts and methods that were used in the last chapter to study gaseous equilibria can be applied to solving equilibria among weak acids and bases. The systematic approach simplifies the process.

Expressing the Concentration of an Acid

When a weak acid dissolves in solution, some of the acid molecules transfer a proton to the water. We must be careful when we describe the acid concentration of such a solution. For example, if we prepare a 0.010 *M* solution of acetic acid, the actual concentration of the acetic acid molecules is less than 0.010 *M*, since some lose protons to form hydrogen ions and acetate ions.

$$CH_3COOH(aq) + H_2O(\ell) \rightleftharpoons H_3O^+(aq) + CH_3COO^-(aq)$$

When we speak of the concentration of the acid, we need to specify clearly the species to which we refer. First, we might be speaking of the *nonionized acetic acid* left in the solution. In order to calculate the concentration of the nonionized acid, we must know the starting concentration and

The analytical concentration is the sum of the concentrations of the undissociated acid and all of its conjugate base forms.

the quantity that has ionized. The difference between the two is, of course, the concentration of nonionized acid. The other concentration we might speak of is the *total acetic acid* concentration. The term **analytical concentration** is used to describe the concentration of all the forms of the acid, both the protonated (acetic acid) and the conjugate base, or unprotonated, form (acetate ion). The symbol C_{HA} is used to denote the analytical or total concentration of the weak acid HA. If the analytical concentration of the solution is 0.010 M, the true acetic acid concentration, represented by $[CH_3COOH]$, is somewhat lower.

$$C_{CH_3COOH} = [CH_3COOH] + [CH_3COO^-]$$

When we describe a solution as 0.010 M acetic acid, we are referring to the analytical or total concentration. This solution could be made by dissolving 0.010 mol of acetic acid and diluting to 1.00 L. Although some acetic acid ionizes, the analytical concentration is 0.010 M. The analytical concentration is the most common measure used to describe a solution in which some of the substances are partially or completely ionized.

Determination of K_a for Weak Acids

In addition to the concentration, the ionization constant must also be known to determine the concentrations of the species in a solution of a weak acid. We can write an expression for a general weak acid,

$$HA(aq) + H_2O(\ell) \rightleftharpoons H_3O^+(aq) + A^-(aq)$$

$$K_a = \frac{[H_3O^+][A^-]}{[HA]}$$

and calculate K_a for this equilibrium from experimental measurements of concentration.

One way to determine K_a for the acid is to measure the electrical conductivity of the solution. The conductivity of a solution is proportional to the concentrations of the ions formed in the ionization process. The degree of ionization can be used to determine the ionization constant, as shown in Example 13.8.

Example 13.8 Calculating K_a for a Weak Acid

Picric acid, a weak acid, is dissolved in water to prepare a 0.100 M solution. Conductivity measurements indicate that the picric acid is 83% ionized. Calculate K_a and pK_a.

Solution First, write the chemical equation.

$$HA(aq) + H_2O(\ell) \rightleftharpoons H_3O^+(aq) + A^-(aq)$$

The picric acid is 83% ionized (17% nonionized) so the equilibrium concentrations are

$$[HA] = 17\% \times 0.10\ M = 0.17 \times 0.10 = 0.017\ M$$

$$[A^-] = [H_3O^+] = 83\% \times 0.10\ M = 0.83 \times 0.10 = 0.083\ M$$

INSIGHTS into CHEMISTRY

Chemical Reactivity

Picric Acid Is Unstable; It Can Corrode Storage Bottles and Then Explode

Picric acid, also known as 2,4,6-trinitrophenol, is best known as an explosive rather than as an acid. It is formed when three NO_2 groups (nitro groups) replace three of the hydrogen atoms in phenol, C_6H_5OH. The proton attached to the oxygen atom is the one that ionizes when the compound is dissolved in water.

Picric acid is typical of many explosive compounds. You may recall that oxygen is needed for the combustion of organic compounds such as phenol:

$$C_6H_5OH + 7O_2 \longrightarrow 6CO_2 + 3H_2O$$

Picric acid, however, does not need much external oxygen to burn because oxygen is already available within the compound:

$$(NO_2)_3C_6H_2OH + \tfrac{13}{4}O_2 \longrightarrow 6CO_2 + \tfrac{3}{2}H_2O + \tfrac{3}{2}N_2$$

Notice that 1 mole of picric acid plus 3.25 moles of oxygen produces 9 moles of gases, an enormous expansion of volume. In fact, picric acid can burn, although not completely, without any external oxygen at all, and still produce 8.5 moles of gases.

$$(NO_2)_3C_6H_2OH \longrightarrow \tfrac{11}{2}CO + \tfrac{3}{2}H_2O + \tfrac{3}{2}N_2 + \tfrac{1}{2}C$$

Both reactions are highly exothermic.

Compounds that contain large amounts of oxygen along with carbon and hydrogen are often explosive—they burn quite vigorously, and in a confined space, they explode. Picric acid, however, is even more dangerous than might appear at first glance. It reacts with metals to form metal-picrate salts that are sensitive to shock and can explode when disturbed. Lead picrate, for example, is used to start the rapid reaction of gunpowder in a bullet cartridge.

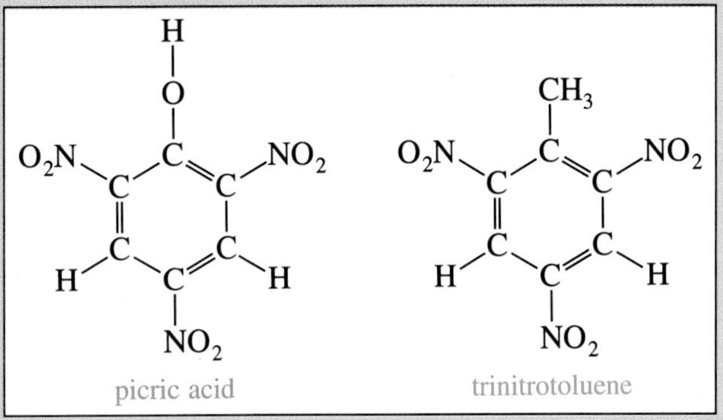

Picric acid has applications in many areas of science and is still found in some stockrooms. Old bottles that have metallic caps can form shock-sensitive picrates and have been known to detonate when moved. If you ever see a corroded picric acid bottle, you would be well advised to call this to the attention of the appropriate authorities. In many areas, bomb-detonation squads have been required to assist in a process that started as a "simple" cleanup of a chemical stockroom.

The structure of picric acid is quite similar to that of another explosive, trinitrotoluene (TNT). TNT is greatly preferred as an explosive, however, since it is more stable and less likely to detonate unexpectedly.

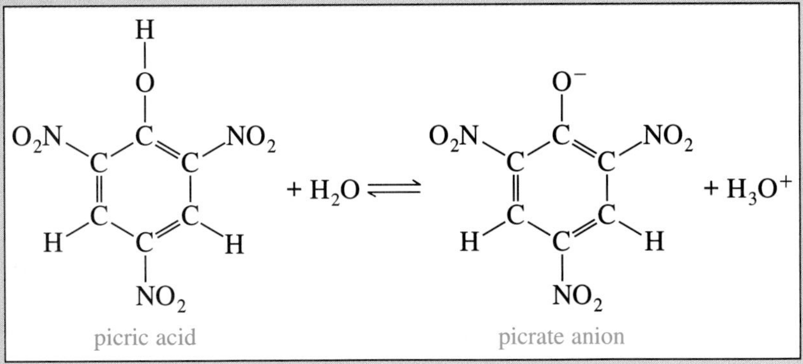

picric acid trinitrotoluene

Picric acid and trinitrotoluene

picric acid $+ H_2O \rightleftharpoons$ picrate anion $+ H_3O^+$

Ionization of picric acid

These values can be substituted into the equilibrium expression

$$K_a = \frac{[A^-][H_3O^+]}{[HA]} = \frac{(0.083)(0.083)}{(0.017)} = 0.40$$

$$pK_a = -\log(0.40) = 0.40$$

Understanding Calculate K_a for hydrazoic acid, HN_3, if a 0.10 M solution is 1.38% ionized.

Answer: $K_a = 1.9 \times 10^{-5}$

The ionization constant for a weak acid can also be determined from experimental measurements of pH. A pH meter like that shown in Figure 13.3 can be used to measure the pH of a solution that contains a known concentration of the weak acid. Example 13.9 illustrates how the ionization constant can be calculated from these measurements.

Example 13.9 Determination of K_a from pH

A 0.100 M solution of chlorobenzoic acid has a pH of 2.50. Calculate K_a.

Solution The iCe table provides a systematic way to approach equilibrium problems. Start by converting the pH to concentration, writing the chemical equation, and filling in the known quantities in the iCe table.

$$[H_3O^+] = 10^{-pH} = 10^{-2.50} = 3.2 \times 10^{-3}\ M$$

	$ClC_6H_4COOH + H_2O \rightleftharpoons$	H_3O^+	$+\ ClC_6H_4COO^-$
initial, M	0.100	0.	0.
Change, M			
equilibrium, M		3.2×10^{-3}	

The stoichiometry of the chemical equation tells us that when $3.2 \times 10^{-3}\ M$ H_3O^+ is formed, the same concentration of chlorobenzoate ion ($ClC_6H_4COO^-$) will be formed, and the concentration of chlorobenzoic acid will decrease by the same amount.

	$ClC_6H_4COOH + H_2O \rightleftharpoons$	H_3O^+	$+\ ClC_6H_4COO^-$
initial, M	0.100	0.	0.
Change, M	-3.2×10^{-3}	$+3.2 \times 10^{-3}$	$+3.2 \times 10^{-3}$
equilibrium, M	9.7×10^{-2}	3.2×10^{-3}	3.2×10^{-3}

The equilibrium values are substituted in the expression for the equilibrium constant.

$$K_a = \frac{[H_3O^+][ClC_6H_4COO^-]}{[ClC_6H_4COOH]} = \frac{(3.2 \times 10^{-3})(3.2 \times 10^{-3})}{(9.7 \times 10^{-2})} = 1.1 \times 10^{-4}$$

Understanding The pH of a 0.100 *M* acetic acid solution is 2.88. Calculate K_a.

Answer: 1.8×10^{-5}

Experimental methods such as determining the fraction ionized from electrical conductivity, or measurement of pH, provide the data needed to tabulate acid ionization constants for many acids. Table 13.6 presents ionization constants for several acids; a more complete listing appears in Appendix F. The table includes the strong acids that are leveled in aqueous solution as well as weak acids that can be differentiated in water.

The strengths of acids and bases are proportional to their ionization constants.

Solutions of Weak Acids

We can use the tabulated value of the acid ionization constant along with the analytical concentration of the acid to calculate the concentrations of the species in a solution of weak acid. These calculations allow us to determine important facts such as the pH of a particular solution.

Determining the Concentrations of Species in a Weak Acid Solution

The tabular approach provides a framework for the systematic solution of weak acid problems. We illustrate this method by calculating the concentrations of the species in a 0.100 *M* acetic acid solution. A flow diagram for a general equilibrium problem follows.

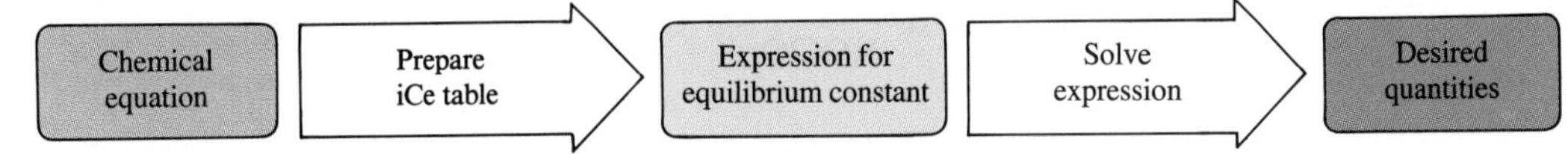

The given concentration, 0.100 *M*, is the analytical concentration of the solution. First, write the chemical equation and iCe table.

	$CH_3COOH + H_2O \rightleftharpoons$	H_3O^+ +	CH_3COO^-
initial, *M*			
Change, *M*			
equilibrium, *M*			

The "initial" line contains the starting concentrations. We know that the initial concentration of acetic acid is 0.100 *M*. The starting concentrations of H_3O^+ and CH_3COO^- are essentially zero, since we begin the problem by assuming acetic acid has not yet ionized.

	$CH_3COOH + H_2O \rightleftharpoons$	H_3O^+ +	CH_3COO^-
initial, *M*	0.100	0.	0.
Change, *M*			
equilibrium, *M*			

Table 13.6 Acid Ionization Constants

Acid	Formula	Conj. Base	K_a at 298 K
perchloric	$HClO_4$	ClO_4^-	100% ionized in water
hydrochloric	HCl	Cl^-	
nitric	HNO_3	NO_3^-	
sulfuric	H_2SO_4	HSO_4^-	
hydrobromic	HBr	Br^-	
hydroiodic	HI	I^-	
hydrogen ion			H_3O^+ is the strongest acid that can exist in water
hydronium ion	H_3O^+	H_2O	
hydrogen sulfate ion	HSO_4^-	SO_4^{2-}	1.2×10^{-2}
chlorous	$HOClO$	ClO_2^-	1.0×10^{-2}
phosphoric	H_3PO_4	$H_2PO_4^-$	7.5×10^{-3}
nitrous	HNO_2	NO_2^-	4.6×10^{-4}
hydrofluoric	HF	F^-	3.5×10^{-4}
formic	$HCOOH$	$HCOO^-$	1.8×10^{-4}
benzoic	C_6H_5COOH	$C_6H_5COO^-$	6.3×10^{-5}
acetic	CH_3COOH	CH_3COO^-	1.8×10^{-5}
hypochlorous	$HOCl$	OCl^-	3.0×10^{-8}
hydrocyanic	HCN	CN^-	7.2×10^{-10}
phenol	C_6H_5OH	$C_6H_5O^-$	1.3×10^{-10}
water	H_2O	OH^-	OH^- is the strongest base that can exist in water

We will define y as the increase in the concentration of the hydrogen ion.

	$CH_3COOH + H_2O \rightleftharpoons$	H_3O^+ +	CH_3COO^-
initial, M	0.100	0.	0.
Change, M	$-y$	$+y$	$+y$
equilibrium, M			

The equilibrium concentration is the sum of the starting concentration and the change in concentration in each column.

	$CH_3COOH + H_2O \rightleftharpoons$	H_3O^+ +	CH_3COO^-
initial, M	0.100	0.	0.
Change, M	$-y$	$+y$	$+y$
equilibrium, M	$0.100 - y$	y	y

The last step is to substitute the equilibrium concentrations into the equilibrium constant expression and solve. The numerical value for K_a is given in Table 13.6.

$$K_a = \frac{[H_3O^+][CH_3COO^-]}{[CH_3COOH]}$$

$$1.8 \times 10^{-5} = \frac{(y)(y)}{0.100 - y}$$

This equation can be solved by approximation. If $y \ll 0.100$ then we can replace $0.100 - y$ with 0.100.

$$1.8 \times 10^{-5} \approx \frac{(y)(y)}{0.100}$$

$$y^2 = 1.8 \times 10^{-6}$$

$$y = 1.3 \times 10^{-3}$$

Check the approximation. Is $1.3 \times 10^{-3} \ll 0.100$? Yes, 1.3×10^{-3} is less than 5% of 0.100, so we can accept the approximation as correct.

Substitute 1.3×10^{-3} for y in the relationships on the last line of the table.

$$[CH_3COOH] = 0.100 - y = 9.9 \times 10^{-2}\ M$$

$$[H_3O^+] = y = 1.3 \times 10^{-3}\ M$$

$$[CH_3COO^-] = y = 1.3 \times 10^{-3}\ M$$

We can see the law of conservation of mass in action. The equilibrium concentration of acetic acid plus the equilibrium concentration of the acetate ion is equal to 0.100 M, the analytical concentration of the acid.

If the numerical approximation fails, there are two approaches that can be used to solve the problem. First, the quadratic equation can be used. Second, the method of successive approximation can be applied; this method is discussed in Appendix A and is utilized later in this text.

Fraction Ionized in Solution

Electrical conductivity measurements can be used to determine the concentration of ions in a solution. The electrical conductivity is directly proportional to the concentrations of the ions, so the concentration-conductivity relationships for strong and weak acids are quite different. Some typical experimental data appear in Figure 13.4.

The conductivity of a strong acid is directly proportional to concentration, since it all dissociates. The conductivity of a weak acid shows a complex dependence on concentration because the fraction ionized depends on concentrations. Weak acids show concentration-conductivity graphs that are curved.

If we know the fraction of a weak acid that has ionized, it is simple to determine the concentrations of the species in solution. For example, we could easily calculate the concentrations of species in a solution of 0.10 *M* weak acid if we are told that it is 25% ionized. The A^- concentration would be 0.025 *M* (25% of the 0.10 *M* analytical concentration of the acid); the concentration of hydrogen ion would also be 0.025 *M*, corresponding to a pH of 1.60; and the concentration of nonionized HA would be 0.075 *M* (75% is not ionized).

For any weak acid we can write the ionization equilibrium, the expression for the analytical concentration, and an expression for α, the **fraction ionized** in solution.

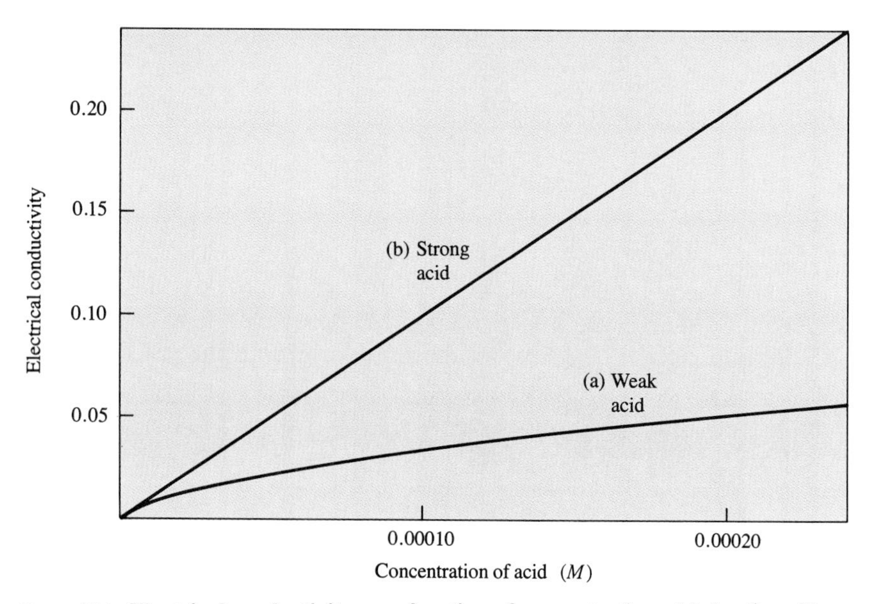

Figure 13.4 Electrical conductivity as a function of concentration (a) Acetic acid, $K_a = 1.8 \times 10^{-5}$. (b) Hydrochloric acid.

$$HA(aq) + H_2O(\ell) \rightleftharpoons H_3O^+(aq) + A^-(aq)$$

$$C_{HA} = [HA] + [A^-]$$

$$\alpha = \frac{[A^-]}{C_{HA}}$$

The calculation of pH is simplified, but unfortunately *the fraction ionized is not a fundamental constant,* but depends on the concentrations of the species in the solution. This dependence is illustrated in the next example, where we see that the fraction ionized increases as the concentration decreases.

Example 13.10 Calculating the Fraction Ionized

Calculate the fraction ionized in 0.10 M and 0.010 M solutions of the weak acid glycine, for which K_a is 1.7×10^{-10}.

Solution

(a) Fraction ionized in 0.10 M glycine:

	$HA(aq) + H_2O(\ell) \rightleftharpoons$	$H_3O^+(aq)$ +	$A^-(aq)$
initial, M	0.10	0.	0.
Change, M	$-y$	$+y$	$+y$
equilibrium, M	$0.10 - y$	y	y

$$K_a = \frac{[H_3O^+][A^-]}{[HA]} = \frac{y^2}{0.10 - y} = 1.7 \times 10^{-10}$$

If $y \ll 0.100$, then

$$\frac{y^2}{0.10} \approx 1.7 \times 10^{-10}$$

$$y = 4.1 \times 10^{-6}$$

Check the assumption: Is $4.1 \times 10^{-6} \ll 0.1$? Yes, so the approximation is valid. The fraction ionized is

$$\alpha = \frac{[A^-]}{C_{HA}} = \frac{4.1 \times 10^{-6}}{0.10} \times 100\% = 4.1 \times 10^{-3}\,\%$$

(b) Fraction ionized in 0.010 M glycine:

	$HA(aq) + H_2O(\ell) \rightleftharpoons$	$H_3O^+(aq) +$	$A^-(aq)$
initial, M	0.010	0.	0.
Change, M	$-y$	$+y$	$+y$
equilibrium, M	$0.010 - y$	y	y

$$K_a = \frac{[H_3O^+][A^-]}{[HA]} = \frac{y^2}{0.010 - y} = 1.7 \times 10^{-10}$$

If $y \ll 0.010$, then

$$\frac{y^2}{0.010} \approx 1.7 \times 10^{-10}$$

$$y = 1.3 \times 10^{-6} \qquad \text{(assumption is valid)}$$

The fraction ionized is

$$\alpha = \frac{[A^-]}{C_{HA}} = \frac{1.3 \times 10^{-6}}{0.010} \times 100\% = 1.3 \times 10^{-2}\ \%$$

Understanding Calculate the fraction ionized in 0.0010 M glycine.

Answer: 4.1×10^{-2} %

The data can be summarized in a table.

Analytical Concentration (M)	$[H_3O^+]$, $[A^-]$ (M)	Fraction Ionized (%)
0.100	4.1×10^{-6}	0.4×10^{-2}
0.0100	1.3×10^{-6}	1.3×10^{-2}
0.00100	4.1×10^{-7}	4.1×10^{-2}

The following graph shows the fraction of glycine ionized as a function of concentration.

Fraction ionized as a function of concentration of glycine

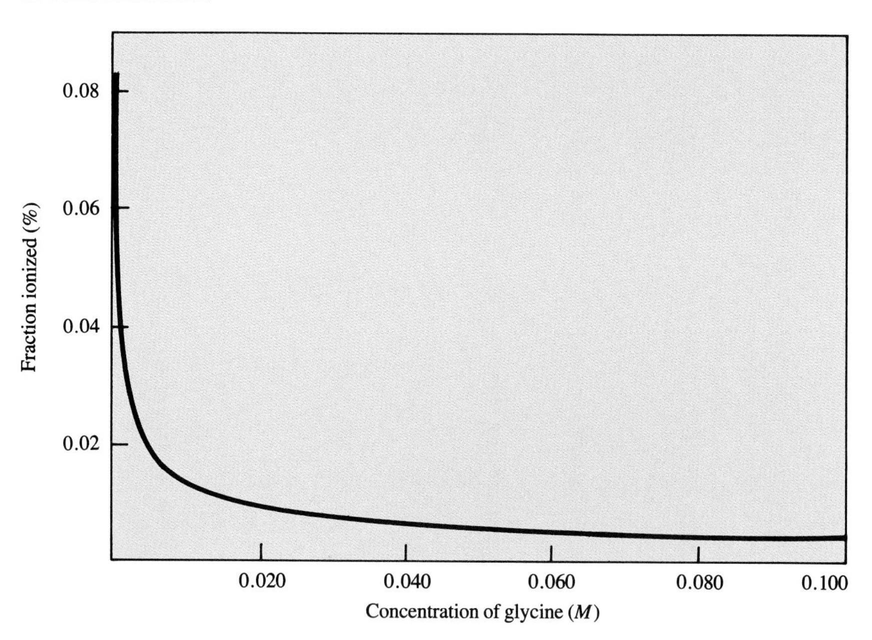

13.6 Solutions of Weak Bases and Salts

Objectives

- To calculate the concentrations of species present in a solution of a weak base

- To determine the relationship between K_b for a base and the K_a of its conjugate acid
- To calculate the pH of the solution obtained when a salt is dissolved in water

Solutions of Weak Bases

The reaction of a weak base with the solvent in aqueous solution can be written as

$$B(aq) + H_2O(\ell) \rightleftharpoons BH^+(aq) + OH^-(aq)$$

The expression for the equilibrium constant is written as usual.

$$K_b = \frac{[BH^+][OH^-]}{[B]}$$

The base, B, is represented as a neutral species in the general equation; it can also have a negative charge, but rarely does a base have a positive charge. Some examples of the ionization of weak bases are:

$$NH_3(aq) + H_2O(\ell) \rightleftharpoons NH_4^+(aq) + OH^-(aq)$$

$$CN^-(aq) + H_2O(\ell) \rightleftharpoons HCN(aq) + OH^-(aq)$$

$$IO_3^-(aq) + H_2O(\ell) \rightleftharpoons HIO_3(aq) + OH^-(aq)$$

Table 13.7 lists values of K_b for several bases.

Calculations for solutions of weak bases are similar to those performed for weak acids. The following example illustrates one such calculation.

The calculation of the concentration of the species formed when a weak base dissolves in water is similar to that done for a weak acid.

Example 13.11 Calculating the pH of a Solution of a Weak Base

Calculate the pH of household ammonia, a 1.44 *M* solution of NH_3. The numerical value of K_b is 1.8×10^{-5} at 25 °C.

Solution Write the chemical equation and the iCe table.

	$NH_3 + H_2O \rightleftharpoons$	NH_4^+ +	OH^-
initial, *M*	1.44	0.	0.
Change, *M*	$-y$	$+y$	$+y$
equilibrium, *M*	$1.44 - y$	y	y

The expressions for the equilibrium concentrations are substituted into the equilibrium expression.

$$K_b = \frac{[NH_4^+][OH^-]}{[NH_3]}$$

$$K_b = \frac{(y)(y)}{1.44 - y} = 1.8 \times 10^{-5}$$

The resulting equation can be solved quickly by approximation. If $y \ll 1.44$ then we can write

$$\frac{y^2}{1.44} \approx 1.8 \times 10^{-5}$$

$$y = 5.1 \times 10^{-3} = [OH^-]$$

Check the approximation. Is $5.1 \times 10^{-3} \ll 1.44$? Yes, we can accept the approximation.

$$[OH^-] = y = 5.1 \times 10^{-3}\ M$$

$$pOH = 2.29$$

$$pH = 14.00 - 2.29 = 11.71$$

Notice how basic the ammonia solution is. Even though ammonia is a relatively weak base, the solution is quite alkaline; its pH is much greater than 7.

Understanding Calculate the pH of 0.20 *M* pyridine. Use Table 13.7 to find K_b.

Answer: The pH is 9.28.

Solutions of Salts

A salt is a compound that contains at least one cation other than H^+ and at least one anion other than OH^-. In principle, any salt can be formed by a neutralization reaction between an acid and a base. When a salt dissolves in water, both an acid and a base are introduced into the solution, so it is not

Table 13.7 Base Ionization Constants

Base		Conjugate Acid	K_b at 298 K
oxide ion	O^{2-}		Reacts quantitatively with water to form hydroxide ion
amide ion	NH_2^-		
hydroxide ion	OH^-	H_2O	Hydroxide is the strongest base that can exist in a water solution
ammonia	NH_3	NH_4^+	1.8×10^{-5}
hydrazine	N_2H_4	$N_2H_5^+$	1.7×10^{-6}
hydroxylamine	NH_2OH	NH_3OH^+	1.1×10^{-8}
pyridine	C_5H_5N	$C_5H_5NH^+$	1.8×10^{-9}

INSIGHTS into CHEMISTRY

Chemical Reactivity

Ammonia Solutions Cut Through Grease and Rinse Clean—Shouldn't be Mixed with Bleach

Aqueous ammonia solutions are common household cleaning materials. An ammonia solution has many properties of alkaline materials, and in particular it reacts with fats and greases to form compounds that can be rinsed away in water. Ammonia is superior to most other alkaline cleaning materials because it is more volatile. If we compare ammonia to sodium hydroxide or trisodium phosphate for washing windows, we see that all are alkaline and will dissolve greases. However, if the rinsing is imperfect and traces of the cleaning solution are left on the window, the water will evaporate and leave the solid sodium salt behind. When ammonia is left on the window, the same general process occurs with one important difference—ammonia is a gas, so after the cleaning and rinsing, all traces of ammonia evaporate.

Household ammonia can react with other commonly used cleaning solutions to produce poisonous gases. Ammonia should *never* be mixed with bleach. In fact, household cleaners contain some of the same compounds that we use in the lab, and should be treated with the same respect. Ammonia (and many other cleaning solutions) should be used only in well-ventilated areas.

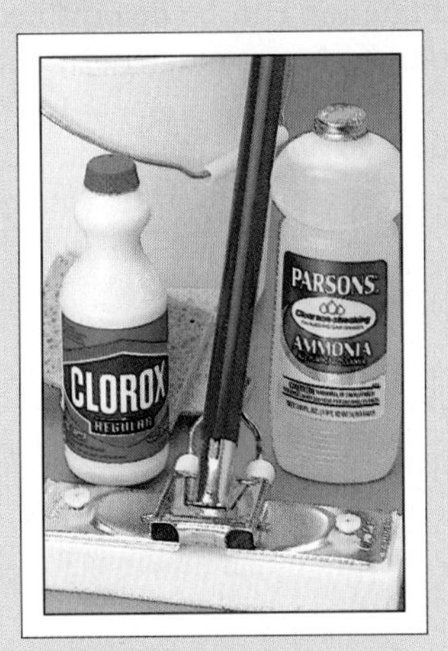

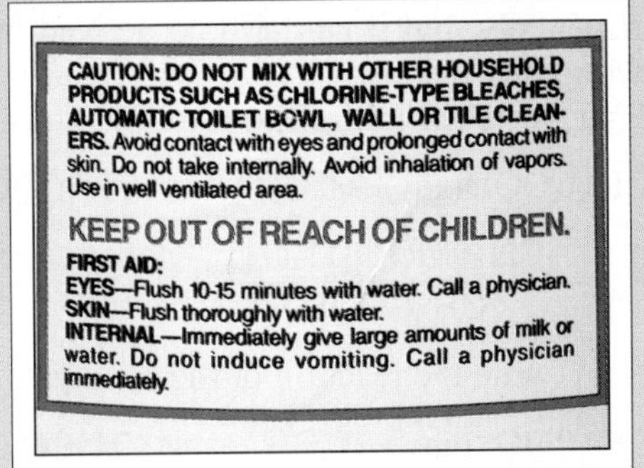

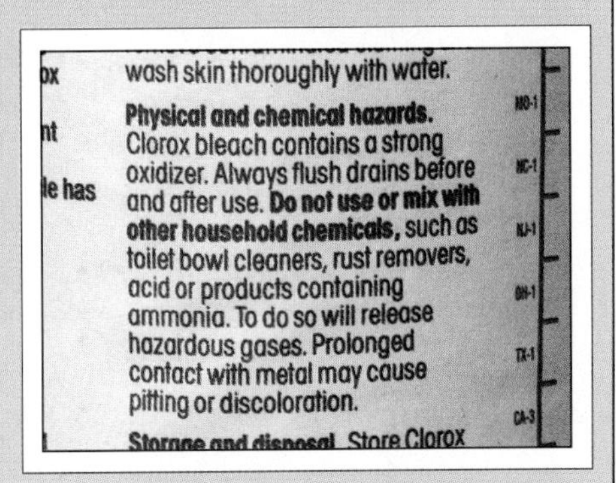

Household cleaning materials Even household chemicals should be treated with respect.

obvious whether the resulting solution will be acidic or basic, or neither. For example, a solution of ammonium chloride is acidic, a sodium chloride solution is neutral, and a sodium fluoride solution is alkaline.

To calculate the pH of a salt solution, we must know the values of K_a and K_b for each substance in solution. Most tables, including those in this text, present the ionization constant for only one form of the acid-base conjugate pairs, usually the neutral form. For example, K_a for HF is tabulated, but K_b for F^- is not. The relationship between K_a and K_b allows us to determine K_b from K_a and *vice versa.* To develop the relationship between K_a and K_b, we will consider acetic acid and acetate ion.

Acetate ion is the conjugate base of acetic acid and can react with water like any other base. Acetate ions can be added to the solution; sodium acetate would be a good source. The net ionic equation for the reaction of acetate ion with water is

$$CH_3COO^-(aq) + H_2O(\ell) \rightleftharpoons CH_3COOH(aq) + OH^-(aq)$$

The acidity of salt solutions The pH of salt solutions varies, depending on the salt.

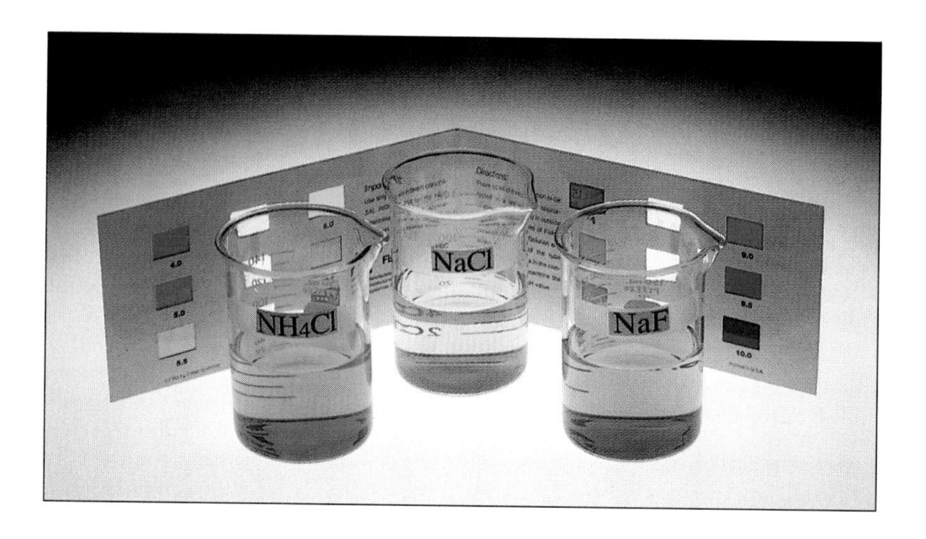

The equilibrium constant, K_b, for the acetate ion can be determined experimentally, but it can also be derived from K_a for acetic acid by the following calculations.

First, write the ionization reaction of acetic acid and K_a.

$$CH_3COOH(aq) + H_2O(\ell) \rightleftharpoons CH_3COO^-(aq) + H_3O^+(aq)$$

$$K_a = \frac{[CH_3COO^-][H_3O^+]}{[CH_3COOH]}$$

Next, write the reaction of the conjugate base with water and K_b.

$$CH_3COO^-(aq) + H_2O(\ell) \rightleftharpoons CH_3COOH(aq) + OH^-(aq)$$

$$K_b = \frac{[CH_3COOH][OH^-]}{[CH_3COO^-]}$$

Last, write the product of K_a and K_b.

$$K_aK_b = \frac{[\cancel{CH_3COO^-}][H_3O^+]}{[\cancel{CH_3COOH}]} \times \frac{[\cancel{CH_3COOH}][OH^-]}{[\cancel{CH_3COO^-}]}$$

After cancelling concentrations we find

$$K_aK_b = [H_3O^+][OH^-]$$

The product of the equilibrium constants for an acid-base conjugate pair is equal to K_w.

or

$$K_aK_b = K_w$$

This equation is valid for any acid-base conjugate pair in water. If we know either K_a or K_b, we can calculate the other. An important consequence of this equation is: *the stronger an acid, the weaker will be its conjugate base.*

Example 13.12 Calculating K_a or K_b for a Conjugate Pair

Use the information in Tables 13.6 and 13.7 to

(a) calculate K_b for acetate ion.
(b) calculate K_a for hydrazinium ion ($N_2H_5^+$).

Solution

(a) Since acetate ion is the conjugate base of acetic acid, use Table 13.6 to find K_a for acetic acid, then calculate K_b for the acetate ion from the relationship between K_a and K_b.

$$K_a K_b = K_w$$

$$K_b = \frac{K_w}{K_a}$$

$$K_b = \frac{1.0 \times 10^{-14}}{1.8 \times 10^{-5}} = 5.6 \times 10^{-10}$$

(b) The equation for the reaction of hydrazinium ion with water is

$$N_2H_5^+(aq) + H_2O(\ell) \rightleftharpoons N_2H_4(aq) + H_3O^+(aq)$$

We cannot find the hydrazinium ion in the list of acids in Table 13.6, so we must use K_b for its conjugate base, hydrazine. The value for K_b for hydrazine is given in Table 13.7 as 1.7×10^{-6}.

$$K_a = \frac{K_w}{K_b}$$

$$K_a = \frac{1.0 \times 10^{-14}}{1.7 \times 10^{-6}} = 5.9 \times 10^{-9}$$

Understanding Calculate K_a for ammonium ion.

Answer: K_a for ammonium $= 5.6 \times 10^{-10}$

We can calculate K_a or K_b, as appropriate, and rank species in order of strength. This is illustrated in Example 13.13.

Example 13.13 Ranking Bases in Order of Strength

Rank the following bases in order of strength: formate ion, cyanide ion, and acetate ion.

Solution We know that $K_b = K_w/K_a$ where K_a is the ionization constant for the conjugate acid.

(a) To calculate K_b for formate ion, first find K_a for formic acid in Table 13.6, and then substitute into the relationship.

$$K_b = \frac{K_w}{K_a} = \frac{1.0 \times 10^{-14}}{1.8 \times 10^{-4}} = 5.6 \times 10^{-11} \text{ (formate)}$$

(b) For CN^-,

$$K_b = \frac{K_w}{K_a} = \frac{1.0 \times 10^{-14}}{7.2 \times 10^{-10}} = 1.4 \times 10^{-5} \text{ (cyanide)}$$

(c) Finally, for the acetate ion

$$K_b = \frac{K_w}{K_a} = \frac{1.0 \times 10^{-14}}{1.8 \times 10^{-5}} = 5.6 \times 10^{-10} \text{ (acetate)}$$

As bases:

cyanide > acetate > formate

Note that we did not really need to calculate the K_b values, since the order of acid strength from the K_a values is

$HCOOH > CH_3COOH > HCN$

We know that the strengths of the conjugate bases must be in reverse order.

Stronger acids have weaker conjugate bases, and *vice versa*.

Understanding Rank the following bases in order of strength: nitrite ion, fluoride ion, benzoate ion.

Answer: benzoate > nitrite > fluoride

The Conjugate Partners of Strong Acids and Bases

The conjugate base of a strong acid is very weak and will have essentially no tendency to remove a proton from water.

$$Cl^-(aq) + H_2O(\ell) \not\longrightarrow HCl(aq) + OH^-(aq)$$

In fact, the reverse of this reaction goes to completion since no HCl(aq) can be detected in the solution. *The conjugate base of a strong acid has no net effect on the pH of a solution.* Examples of these very weak bases are Cl^-, NO_3^-, and ClO_4^-, the anions of strong acids. In any acid-base equilibrium the anion of a strong acid is a spectator ion, one that does not influence the pH of the solution.

The same logic can be used to show that the conjugate acids of strong bases lack acidic behavior in water. The strong bases, ionic compounds such as NaOH and KOH, dissociate completely to form the stoichiometric amount of hydroxide ion. Sodium and potassium ions do not affect the pH of a solution. When considering the acid-base properties of solutions, we treat these ions as spectator ions and do not include them in net ionic equations.

The anions of the strong acids and the cations of the strong bases are spectator ions.

The spectator ions can be remembered; they go hand-in-hand with the strong acids and bases. The anions of the strong acids, Cl^-, Br^-, I^-, NO_3^-, ClO_4^-, and HSO_4^- are spectator ions. The cations associated with the strong bases are also spectator ions. These include Li^+, Na^+, K^+, Rb^+, Cs^+, Ca^{2+}, Sr^{2+}, and Ba^{2+}.

The pH of a Solution of a Salt

We can now consider the pH of a solution made by dissolving a salt in water. The properties of the final solution depend on the relative values of K_a of the cation and K_b of the anion.

If the salt contains just spectator ions, the salt will not affect the pH of the solution. An example is sodium chloride, NaCl. When dissolved in water, neither Na^+ nor Cl^- has acid-base properties, since both HCl and NaOH are strong electrolytes. The pH of the solution is determined by the autoionization of water.

A salt that contains a spectator cation and an anion from a weak acid will produce a basic solution when dissolved. An example is sodium nitrite, $NaNO_2$. When sodium nitrite dissolves in water, it dissociates completely:

$$NaNO_2(aq) \longrightarrow Na^+(aq) + NO_2^-(aq)$$

The sodium cation does not have any acidic properties. The anion is the nitrite ion, which is the conjugate base of nitrous acid. Nitrous acid is a weak acid, so its conjugate base will affect the pH of the solution through the equilibrium

$$NO_2^-(aq) + H_2O(\ell) \rightleftharpoons HNO_2(aq) + OH^-(aq)$$

Table 13.8 Solutions of Salts

Source of		Example		Type of Solution
Cation	Anion			
strong base (KOH)	strong acid (HCl)	K^+	Cl^-	neutral
strong base (NaOH)	weak acid (HF)	Na^+	F^-	basic
weak base (NH_3)	strong acid (HNO_3)	NH_4^+	NO_3^-	acidic
weak base (NH_3)	weak acid (HF)	NH_4^+	F^-	calculations needed

In contrast, NH_4Cl produces NH_4^+ and Cl^- when dissolved in water. The solution will be acidic because the ammonium ion is the conjugate acid of the weak base ammonia, while Cl^- is a spectator ion without acid-base properties. Qualitative descriptions of several salt solutions are presented in Table 13.8.

Chemists frequently refer to the equilibrium constant for the conjugate species as a hydrolysis constant, since it is derived from the reaction of the species with water. ("Hydrolysis" comes from two Greek words meaning "splitting water.") The terminology is more commonly used for the reaction of ions. For example, Fe^{3+}(aq) is acidic, for reasons that we will discuss later, and so K_a for Fe^{3+}(aq) is often termed the hydrolysis constant, because it represents the reaction of Fe^{3+} with water.

The pH of a solution of a salt can be calculated, because we can determine K_a of an acid from K_b of its conjugate base and *vice versa.* Example 13.14 illustrates these calculations.

The pH of a salt solution is determined from the K_b of the anion and the K_a of the cation.

Example 13.14 Calculating the pH of a Salt Solution

Calculate the pH of

(a) a 0.10 *M* sodium nitrate solution.
(b) a 0.050 *M* KCN solution.

Solution

(a) The sodium ion is a weaker acid than water. Nitric acid is strong, so nitrate ion is a weaker base than water. Both the sodium ion and nitrate ion are spectator ions, so the pH is determined by the autoionization of water. The pH is 7.00.
(b) Potassium is a spectator ion, but CN^- is a base; its conjugate acid, HCN, is weak. We must first determine K_b for CN^-. From Table 13.6, we see that K_a for HCN is 7.2×10^{-10}, so

$$K_b \text{ for } CN^- = \frac{K_w}{K_a \text{ for HCN}}$$

$$= \frac{1.0 \times 10^{-14}}{7.2 \times 10^{-10}} = 1.4 \times 10^{-5}$$

We can write the chemical equation and the iCe table.

	$CN^-(aq) + H_2O(\ell) \rightleftharpoons$	$HCN(aq)$ +	$OH^-(aq)$
initial, M	0.050	0.	0.
Change, M	$-y$	$+y$	$+y$
equilibrium, M	$0.050 - y$	y	y

$$K_b = \frac{[HCN][OH^-]}{[CN^-]}$$

$$1.4 \times 10^{-5} = \frac{(y)(y)}{0.050 - y}$$

If $y \ll 0.050$, then

$$1.4 \times 10^{-5} \approx \frac{(y)(y)}{0.050}$$

$$y^2 = 7.0 \times 10^{-7}$$

$$y = 8.4 \times 10^{-4}$$

Check the assumption. Is $8.4 \times 10^{-4} \ll 0.050$? Yes it is, so the assumption is valid.

$$[OH^-] = y = 8.4 \times 10^{-4}$$

$$pOH = -\log[OH^-] = -\log(8.4 \times 10^{-4}) = 3.08$$

$$pH = 14.00 - pOH = 10.92$$

Understanding Calculate the pH of a 0.010 M sodium fluoride solution.

Answer: pH = 7.73

Example 13.15 Calculating the pH of a Solution of a Salt of a Weak Acid or Base

Calculate the pH of a solution that is 0.050 M in ammonium ion.

Solution First, write the chemical equation.

$$NH_4^+(aq) + H_2O(\ell) \rightleftharpoons H_3O^+(aq) + NH_3(aq)$$

Calculate K_a from K_b for ammonia, given in Table 13.7.

$$K_a = \frac{K_w}{K_b} = \frac{1.0 \times 10^{-14}}{1.8 \times 10^{-5}} = 5.6 \times 10^{-10}$$

Now, write the iCe table

	$NH_4^+ + H_2O \rightleftharpoons$	H_3O^+ +	NH_3
initial, M	0.050	0.	0.
Change, M	$-y$	$+y$	$+y$
equilibrium, M	$0.050 - y$	y	y

Substitute into the equilibrium constant expression and solve for y.

$$K_a = \frac{[H_3O^+][NH_3]}{[NH_4^+]}$$

$$5.6 \times 10^{-10} = \frac{(y)(y)}{0.050 - y}$$

If $y \ll 0.050$ then

$$5.6 \times 10^{-10} \approx \frac{y^2}{0.050}$$

$$y^2 = 0.050 \times 5.6 \times 10^{-10} = 2.8 \times 10^{-11}$$

$$y = 5.3 \times 10^{-6}$$

Check the assumption. Is $5.3 \times 10^{-6} \ll 0.050$? Yes, it is. The assumption is good and we can accept the value.

$$[H_3O^+] = 5.3 \times 10^{-6}$$

$$pH = -\log[5.6 \times 10^{-6}] = 5.28$$

Understanding Calculate the pH of 0.020 M pyridinium chloride.

Answer: pH = 3.48

If a salt lacks any spectator ions (both the anion and cation are derived from weak electrolytes), we can tell whether the solution is acidic or basic by comparing K_a to K_b. However, an exact solution requires extensive calculations, and will not be presented here. On a qualitative basis, if $K_a > K_b$, then the solution will be acidic; if $K_b > K_a$, then the solution will be basic.

13.7 Mixtures of Strong and Weak Acids

Objectives

- To calculate the pH of a solution that contains both strong and weak acids or bases
- To determine the pH of a solution that contains a mixture of weak acids or bases with different ionization constants

When there are several different acids in solution, they compete to donate protons to the bases (proton acceptors) present. The solvent always participates in the process. Calculating the concentrations of the various species in solution might seem complicated under these conditions, but in general the opposite is true. As long as the strengths of the acids are quite different (the K_a's differ by at least a factor of 100), we can solve the equilibria sequentially, considering only the strongest acid first. For example, the pH of a solution of HBr (strong) and HF (weak) can be calculated by considering only the HBr. *The contribution by a weak acid toward the pH of a solution is generally negligible in comparison to that of a stronger acid.* Similarly, in solutions of weak acids of different strengths (widely differing K_a values), only the strongest one is important in determining the pH of the solution. The pH of a solution that contains formic acid ($K_a = 1.0 \times 10^{-4}$) and phenol ($K_a = 1.3 \times 10^{-10}$) is determined by calculating the pH of a formic acid solution and disregarding the phenol. If there are neither strong acids nor weak acids present, then the very weakest acid, water, is used to determine the pH of the resulting solution.

A mixture of strong and weak acids can be examined in a qualitative manner by applying the principle of Le Chatelier. The ionization of a weak acid can be written as:

$$HA(aq) + H_2O(\ell) \rightleftharpoons H_3O^+(aq) + A^-(aq)$$

When a strong acid is added, the H_3O^+ concentration increases, and the position of equilibrium will shift to the left, partially consuming the added H_3O^+. We discussed similar systems in connection with the common ion effect, in which the solubility of a precipitate is decreased when one of the ions in the precipitate is added to the solution. The same holds true for acid-base systems. A weak acid produces fewer protons in the presence of a strong acid. Like the solubility calculation, the calculation of pH is often simplified if a strong acid is present.

The influence of a weak acid on the pH of a solution is generally much less than that of a strong acid.

Example 13.16 illustrates the calculation of pH in solutions that contain strong and weak acids and bases as well as a solution that contains two weak acids that differ in acid strengths.

Example 13.16 Calculating the pH of Mixtures of Acids or Bases

Calculate the pH of

(a) a solution that is 0.25 *M* KOH and 1.00 *M* ammonia.
(b) a solution that is 0.40 *M* HCl, 0.20 *M* HBr, 0.10 *M* HCOOH, and 0.20 *M* HF.
(c) a solution that is 0.80 *M* HCOOH and 0.50 *M* HOCl.

Solution

(a) The hydroxide ions contributed by the weak base can be ignored in comparison to those from the strong base.

$$[OH^-] = C_{KOH} = 0.25\ M$$
$$pOH = 0.60$$
$$pH = 13.40$$

(b) The pH will be determined by the strong acids.

$$[H_3O^+] = C_{HCl} + C_{HBr} = 0.40 + 0.20 = 0.60\ M$$
$$pH = 0.22$$

(c) Formic acid, with $K_a = 1.8 \times 10^{-4}$, is a much stronger acid than is hypochlorous acid ($K_a = 3 \times 10^{-8}$). We can consider this to be just a solution of formic acid, and solve for the pH as we do for any other solution of a weak acid.

	$HCOOH + H_2O \rightleftharpoons$	H_3O^+ +	$HCOO^-$
initial, *M*	0.80	0.	0.
Change, *M*	$-y$	$+y$	$+y$
equilibrium, *M*	$0.80 - y$	y	y

$$K_a = \frac{[H_3O^+][HCOO^-]}{[HCOOH]}$$

$$1.8 \times 10^{-4} = \frac{y^2}{0.80 - y}$$

If $0.80 \gg y$ then

$$y^2 \approx 1.8 \times 10^{-4} \times 0.80$$

$$y = 1.2 \times 10^{-2}$$

Check the assumption. Is $1.2 \times 10^{-2} \ll 0.8$? Yes, it is.

$$[H_3O^+] = 1.2 \times 10^{-2}\ M$$

$$pH = 1.92$$

Understanding Calculate the pH of a solution that is 0.050 M HCl and 0.15 M HCOOH.

Answer: pH = 1.30

Our general guidelines, ignoring weaker acids in the presence of stronger ones, applies to mixtures of acids. Titrations, in which acids and bases are mixed, will be treated in detail in the next chapter.

13.8 Influence of Molecular Structure on Acid Strength

Objectives

- To determine how changes in structure influence the acid ionization constants of a series of acids
- To determine the influence of fundamental properties such as size and electronegativity on the strengths of acids

Some acids are strong and others are weak, but we have not yet explored why this is so. In this section we will explore the influence of structure and bonding on the strengths of acids. The ionization of an acid is a complex process influenced by the strength of the bond that holds the proton, the bond polarity, changes in the strengths of other bonds that accompany the loss of the proton, and the solvation of the ions produced in the reaction. We will consider two types of acids, the binary hydrides and the oxyacids, and examine the factors that influence the strengths of these acids.

Binary Hydrides

A compound composed of two elements, one of which is hydrogen, is called a **binary hydride.** Many are acids, and we might expect that the strength of an acid is correlated to its bonding.

When an acid ionizes, the electrons in the H—A bond are retained by the anion, while the hydrogen ion shares electrons from a lone pair on a solvent molecule.

$$\text{H—}\ddot{\text{O}}\text{:(—H)} + \text{H—}\ddot{\underset{\cdot\cdot}{\text{A}}}\text{:} \rightleftharpoons \left[\text{H—}\ddot{\text{O}}\text{(—H)—H}\right]^+ + \left[:\ddot{\underset{\cdot\cdot}{\text{A}}}:\right]^-$$

$$H_2O \quad + \quad HA \quad \rightleftharpoons \quad H_3O^+ \quad + \quad A^-$$

The extent of reaction depends on the relative stabilities of the undissociated acid and the ions in solution. The key factors are the strength of the HA bond and the stability of the A^- ion in solution. A strong H—A bond will be difficult to break, and an unstable A^- anion will be difficult to form.

Bond Strengths

The strengths of the bonds in the hydrogen halides allow us to predict the strengths of the acids:

acid strengths	HF	<	HCl	<	HBr	<	HI
bond energies, kJ/mol	569		431		368		297

The H—X bond strengths predict the trend in acidity down a group of binary hydrides.

The hydrogen-fluorine covalent bond is very strong, the hydrogen-chlorine is weaker, and so forth. This trend in the strength of the bonds is due mainly to increasing size effects. The bond in HF is short and the overlap of orbitals is substantial. In HI, iodine is much larger, so the bond length increases and the overlap decreases as well.

The same arguments hold in other groups, as well. *Within a group in the periodic table, the acidity of the hydrides increases from top to bottom.* The group VIA (16) hydrides follow the same trend:

acidity: $H_2O < H_2S < H_2Se < H_2Te$

Stability of the Anion

A second factor influencing the ionization of HA is the ability of the A atom to accept additional negative charge. *The more electronegative the atom, the more easily it can accommodate additional electron density on the atom.* A more electronegative atom results in a stronger acid.

Next let's examine the changes in acidity of the nonmetal hydrides as we proceed from left to right in the periodic chart, starting with the elements in the second period.

acidity: $CH_4 < NH_3 < H_2O < HF$

Changes in electronegativity difference predict the trend in acidity of binary hydrides going across a period.

In this series of compounds, methane exhibits no acidic properties, ammonia behaves as an acid only in solvents much more basic than water, and hydrofluoric acid is a stronger acid than water. *Within any row of the periodic chart the acidities of the binary hydrides follow the trend expected on the basis of electronegativity, and increase from left to right.*

In many cases, predictions of acid strength based on electronegativity differences are exactly the opposite of predictions based on bond strength. Under these conditions, the experimental evidence indicates that electronegativity trends dominate going across a period, and bond strength is more important down a group.

Oxyacids

Binary hydrides are not the only (or even the most common) compounds that exhibit acidic properties. There are many compounds that contain hydrogen, oxygen, and a third element. These compounds are called **oxyacids,** in which the third element is a nonmetal or a transition metal in a high oxidation state. Oxyacids include the organic acids such as acetic acid (CH_3COOH) as well as carbonic acid (H_2CO_3) and phosphoric acid

(H_3PO_4). These acids have important roles in the processes that occur in living cells. Other examples of oxyacids are HNO_2, HNO_3, H_2CrO_4, H_2SO_4, and $HClO_4$. The hydrogen atoms that ionize in oxyacids are always bonded to an oxygen atom, which in turn is bonded to the third element.

In a series of oxyacids that have the same general structure, *the acidity increases as the electronegativity of the third element increases.* This effect is illustrated by the hypohalous acids, which have the general formula H—O—X, where X = Cl, Br, and I.

	HOCl	HOBr	HOI
electronegativity	3.0	2.8	2.5
K_a	3.0×10^{-8}	2.0×10^{-9}	2.3×10^{-11}

The strengths of the oxyacids, XOH, increase with increasing electronegativity of X.

As the electronegativity of the halogen (X) increases, the halogen atom gains a greater fraction of the shared pair of electrons in the O—X bond. The oxygen atom, in turn, replaces part of this loss by gaining a larger share of the electrons in the O—H bond, so the O—H bond becomes weaker and more polar, increasing the tendency for the H atom to ionize.

The general formula of an oxyacid can be written as $(HO)_mXO_n$. Experiments show that the acid ionization constants depend mainly on the value of n, the number of oxygen atoms that are not bonded to hydrogen atoms. Table 13.9 has several oxyacids grouped in this manner, along with their pK values.

Table 13.9 pK_a Values for Some Oxyacids, $(HO)_mXO_n$

n	Name	Formula	pK_a
0 very weak	hypoiodous acid	(HO)I	10.6
	arsenious acid	$(HO)_3As$	9.2
	hypobromous acid	(HO)Br	8.7
	telluric acid	$(HO)_6Te$	7.7
	hypochlorous acid	(HO)Cl	7.5
1 weak acids	selenious acid	$(HO)_2SeO$	2.5
	arsenic acid	$(HO)_3AsO$	2.3
	phosphoric acid	$(HO)_3PO$	2.1
	chlorous acid	(HO)ClO	2.0
	sulfurous acid	$(HO)_2SO$	1.8
	periodic acid	$(HO)_5IO$	1.6
2 strong acids	iodic acid	$(HO)IO_2$	0.8
	nitric acid	$(HO)NO_2$	(−1)
	selenic acid	$(HO)_2SeO_2$	(−3)
	chloric acid	$(HO)ClO_2$	(−3)
	sulfuric acid	$(HO)_2SO_2$	(−3)
3 very strong	perchloric acid	$(HO)ClO_3$	(−10)

The pK_a values in parentheses are for acids that are completely ionized in water. The K_a values for these acids are estimated from measurements made in a more acidic solvent system.

Table 13.10 Acid Ionization Constants for the Oxyacids of Chlorine

Name	Formula		pK_a
perchloric acid	$HClO_4$	$(HO)ClO_3$	(−10)
chloric acid	$HClO_3$	$(HO)ClO_2$	(−3)
chlorous acid	$HClO_2$	$(HO)ClO$	2.0
hypochlorous acid	$HOCl$	$(HO)Cl$	7.5

There are two factors that explain these observations. First, since oxygen is more electronegative than the central atom, each oxygen atom not bonded to hydrogen will attract electron density from the central atom, which in turn attracts electron density from the O—H bond, making it weaker and easier to ionize in water. This explanation is similar to that given for the relative acid strengths of the hypohalous acids (HOCl, HOBr, and HOI) discussed earlier.

Additional oxygen atoms also help stabilize the anion (conjugate base) formed when the acid ionizes. The charge on the anion will be distributed over the oxygen atoms that are not bonded to a hydrogen atom. The larger the number of such oxygen atoms, the more stable the anion, since the charge is spread out over a larger volume. The oxyacids of chlorine, for example, are HClO, $HClO_2$, $HClO_3$, and $HClO_4$; each forms a conjugate base with a charge of −1. In the hypochlorite ion, ClO^-, this charge resides mainly on the single oxygen atom, while in the perchlorate anion, ClO_4^-, each oxygen atom will have a charge of approximately −1/4. In an ionization reaction, the anion is a product. The more stable the product, the more favored is the reaction. The influence of the number of oxygen atoms on pK_a can be seen in Table 13.10.

ClO^- $$\mathrm{Cl{-}O^{\ominus}}$$

ClO_2^- $$\mathrm{O{=}Cl{-}O^{\ominus} \longleftrightarrow {}^{\ominus}O{-}Cl{=}O}$$

ClO_3^- $$\mathrm{{}^{\ominus}O{-}Cl({=}O){=}O \longleftrightarrow O{=}Cl(-O^{\ominus}){=}O \longleftrightarrow O{=}Cl({=}O){-}O^{\ominus}}$$

ClO_4^- $$\mathrm{O{=}Cl({=}O)_2{-}O^{\ominus} \longleftrightarrow O{=}Cl(-O^{\ominus})({=}O){=}O \longleftrightarrow {}^{\ominus}O{-}Cl({=}O)_2{=}O \longleftrightarrow O{=}Cl({=}O)({-}O^{\ominus}){=}O}$$

In recent years, researchers have been interested in a group of compounds that are related to the oxyacids. As basic knowledge of structural and synthetic chemistry improves, chemists have been able to replace the O—H groups in some acids with fluorine. If one of the O—H groups in sulfuric acid is replaced by fluorine, the compound is called fluorosulfonic acid. The Lewis electron-dot structure of fluorosulfonic acid is very similar to that of perchloric acid.

```
                 :O:
                  ‖
               ..    ..
H2SO4     H—O—S—O—H        sulfuric acid
               ..    ..
                  ‖
                 :O:

              :O:
               ‖
           ..     ..
HFSO4     :F—S—O—H         fluorosulfonic acid
           ..     ..
               ‖
              :O:

              :O:
               ‖
          ..      ..
HClO4     O=Cl—O—H         perchloric acid
          ..      ..
               ‖
              :O:
```

Even though the central atom (sulfur) in fluorosulfonic acid is not as electronegative as the central atom (chlorine) in perchloric acid, fluorosulfonic acid is a stronger acid than perchloric acid, because fluorine is more electronegative than oxygen. Fluorosulfonic acid can replace perchloric acid in many reactions. This change is welcome because there are many organic reactions that require a very strong acid, but perchloric acid cannot be used safely because it reacts explosively with many organic compounds. Fluorosulfonic acid is a useful substitute, since it is both stronger and safer than the perchloric acid that it replaces.

13.9 Lewis Acids and Bases

Objectives

- To define Lewis acids and bases
- To identify Lewis acids, bases, and their salts

Brønsted and Lowry broadened the Arrhenius definition of a base to include species other than OH^-, but they continued to define an acid as a source of protons as did Arrhenius. In 1932, G. N. Lewis expanded the definitions of both acids and bases. Lewis noted that a common feature of all Brønsted bases is the presence of an unshared pair of electrons that is donated to a hydrogen ion. Therefore, he chose to define a **base** as an electron-pair donor. Rather than limiting the definition of an acid to a proton donor, he defined an **acid** as any substance that can accept a pair of electrons, forming a coordinate covalent bond in the process (Table 13.11). His intent was to describe phenomena *other* than proton transfer by the well-understood model used for Brønsted-Lowry acids. Modifying and extending a model to include unstudied species is an important way in which scientific knowledge advances.

Table 13.11 Summary of Acid-Base Definitions

Arrhenius

Acid: increases H_3O^+ concentration when dissolved in water
Base: increases OH^- concentration when dissolved in water
Examples:

$$\underset{\text{acid}}{HF(aq)} + H_2O(\ell) \rightleftharpoons H_3O^+(aq) + F^-(aq)$$

$$\underset{\text{base}}{NH_3(aq)} + H_2O(\ell) \rightleftharpoons NH_4^+(aq) + OH^-(aq)$$

$$\text{neutralization: } H_3O^+(aq) + OH^-(aq) \longrightarrow 2H_2O(\ell)$$

Brønsted-Lowry

Acid: proton donor
Base: proton acceptor
Examples:

$$\underset{\text{acid}}{HCl(g)} + \underset{\text{base}}{NH_3(g)} \rightleftharpoons NH_4Cl(s)$$

$$\underset{\text{acid}}{HClO_4(\ell)} + \underset{\text{base}}{CH_3COOH(\ell)} \rightleftharpoons CH_3COOH_2^+ + ClO_4^-$$

Lewis

Acid: electron-pair acceptor
Base: electron-pair donor
Examples:

$$\underset{\text{acid}}{BF_3(g)} + \underset{\text{base}}{NH_3(g)} \longrightarrow BF_3NH_3(g)$$

$$\underset{\text{acid}}{Cd^{2+}(aq)} + \underset{\text{base}}{4Cl^-(aq)} \rightleftharpoons CdCl_4^{2-}(aq)$$

Lewis Acid-Base Reactions

Crystals of hydrated salts. (Left) $CoCl_2 \cdot 6H_2O$ is a dark red. (Middle) $CuSO_4 \cdot 5H_2O$ is blue. (Right) $NiSO_4 \cdot 6H_2O$ is blue-green.

Arrhenius and Brønsted-Lowry acids and bases are also acids and bases according to the Lewis definitions, but so are many other compounds, and even some atoms and ions. The reaction of HCl with NH_3 in aqueous solution is a typical Brønsted-Lowry reaction.

$$NH_3(aq) + HCl(aq) \longrightarrow NH_4^+(aq) + Cl^-(aq)$$

A proton is transferred from the acid (HCl) to the base (NH_3) in a Brønsted-Lowry acid-base reaction. This reaction is also classified as an acid-base reaction in the Lewis dcfinition, since the ammonia molecule donates its unshared electron pair to the hydrogen ion in forming the ammonium ion.

$$H{-}\overset{H}{\underset{H}{N}}{:} + H{-}\ddot{Cl}{:} \longrightarrow H{-}\overset{H}{\underset{H}{N}}{:} + H^+ + {:}\ddot{Cl}{:}^- \longrightarrow \left[H{-}\overset{H}{\underset{H}{N}}{-}H\right]^- + \left[{:}\ddot{Cl}{:}\right]^+$$

$$NH_3 + HCl \longrightarrow NH_4^+ + Cl^-$$

The reaction of BF_3 with NH_3 is an acid-base reaction according to the model proposed by Lewis, but does not fit the Brønsted-Lowry definition of an acid-base reaction.

Reaction of BF_3 with NH_3

$$F{-}\overset{F}{B}{-}F \;\; {:}\overset{H}{\underset{H}{N}}{-}H \longrightarrow F{-}\overset{F}{\underset{F}{B}} \longleftarrow \overset{H}{\underset{H}{N}}{-}H$$

$$BF_3 + NH_3 \longrightarrow BF_3NH_3$$

The product of a Lewis acid-base reaction is called an **adduct,** derived from the Latin *adductus,* meaning "addition." An adduct is a product that is formed in an addition reaction.

$$\overset{H}{{:}O{-}H} \;\; O{=}C{=}O \longrightarrow \overset{H}{{:}\overset{\oplus}{O}{-}H} \;\; O{=}C{-}O{:}^{\ominus} \longrightarrow \overset{H}{{:}O{:}} \;\; O{=}C{-}O{-}H$$

$$H_2O + CO_2 \longrightarrow H_2CO_3$$

A Lewis acid must have an orbital available to hold the pair of electrons donated by the base.

Lewis recognized that most chemical reactions occur because empty orbitals on one species can be filled by electron pairs from another. Metal ions are Lewis acids, since they have vacant valence shell orbitals and are able to form coordinate covalent bonds with Lewis bases. There is a great deal of experimental evidence to show that metal ions in water solution form such bonds, often with several water molecules. Most metal ions can accept four or six pairs of electrons from Lewis bases. Coordinated solvent molecules are retained in many solid compounds, accounting for the formulas of many hydrated salts. Cobalt(II) chloride hexahydrate ($CoCl_2 \cdot 6H_2O$), copper(II) sulfate pentahydrate ($CuSO_4 \cdot 5H_2O$), and nickel(II) sulfate tetrahydrate ($NiSO_4 \cdot 6H_2O$) are examples of these compounds.

Lewis bases other than water can also form coordinate covalent bonds to metal ions. The very deep blue color that is observed when ammonia is added to a solution of copper(II) ions is a result of the formation of the $Cu(NH_3)_4^{2+}$ ion. The very interesting chemistry of these compounds, called coordination chemistry, is discussed in more detail in Chapter 18.

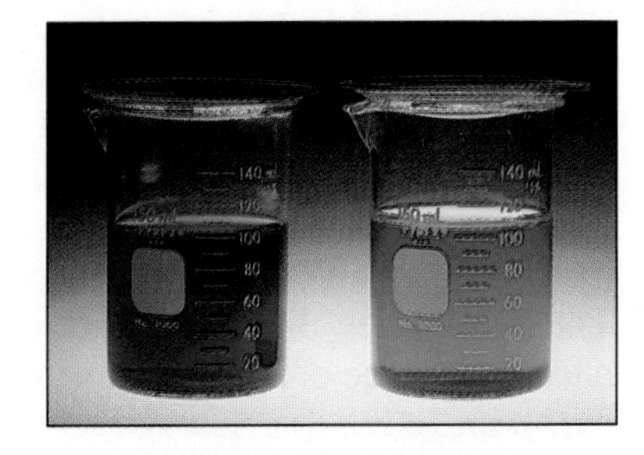

The copper ion exists in water as the light blue $Cu(H_2O)_4^{2+}$ cation. When ammonia is added, the dark blue $Cu(NH_3)_4^{2+}$ cation forms.

The formation of Lewis acid-base adducts can influence the strengths of Brønsted acids. We know that hydrofluoric acid, HF, is a weak Brønsted-Lowry acid in water. The hydrogen fluoride molecule is also a Lewis base, since the fluorine atom has unshared pairs of electrons. When a Lewis acid such as BF_3 is added to a solution of HF, a coordinate covalent bond is formed between the fluorine atom and the boron atom.

$$H{-}\ddot{\underset{..}{F}}{:} + BF_3 \longrightarrow H^+ [F{-}BF_3]^-$$

$$HF + BF_3 \longrightarrow HBF_4$$

The formation of the added bond decreases the strength of the H—F bond, making the resulting compound (tetrafluoroboric acid) a much stronger Brønsted acid than HF. The increase in acidity is large—so large that tetrafluoroboric acid is completely ionized in aqueous solution.

In recent years compounds called super acids have been used in the synthesis of materials that require very acidic conditions. The super acids are extremely strong acids, even stronger than sulfuric and nitric acids. The super acids use the formation of Lewis acid-base adducts to increase the strengths of Brønsted acids. In Section 13.8, fluorosulfonic acid was discussed as a strong acid that ionizes completely in water. It can also be used as a solvent. In pure fluorosulfonic acid, the Brønsted-Lowry acidity is determined by the autoionization equilibrium

$$2HSO_3F \rightleftharpoons H_2SO_3F^+ + SO_3F^-$$

Any compound that will increase the concentration of $H_2SO_3F^+$ will increase the acidity of this solvent. The addition of antimony pentafluoride, SbF_5, to fluorosulfonic acid increases the acidity by a factor of 10,000. The increased acidity is presumably caused by the reaction of the Lewis acid SbF_5 with the fluorosulfonate ion to form the Lewis acid-base adduct $SbF_5(SO_3F)^-$. This reaction decreases the concentration of SO_3F^-, the conjugate base in the autoionization equilibrium

$$SbF_5 + SO_3F^- \rightleftharpoons (SbF_5)(SO_3F)^-$$

The principle of Le Chatelier tells us that a decrease in the concentration of the fluorosulfonate ion will cause an increase in the concentration of $H_2SO_3F^+$. The acid strength of the SbF_5–HSO_3F mixture is so great that it is popularly referred to as "magic acid."

Summary

The Arrhenius definitions of acid and base limited the number of species that could be classified as acids or bases. Brønsted and Lowry extended the definitions as they noted the common feature of proton transfer in acid-base reactions. They defined an acid as a proton donor and a base as a proton acceptor. Further, after a compound has donated a proton, the substance remaining is capable of accepting a proton, so it is a base. The two species, differing in composition by the presence or absence of a proton, form an *acid-base conjugate pair.*

The solvent most commonly used is water, which undergoes an *autoionization* reaction,

$$2H_2O(\ell) \rightleftharpoons H_3O^+(aq) + OH^-(aq)$$

that is described numerically by K_w:

$$K_w = [H_3O^+][OH^-] = 1.0 \times 10^{-14} \text{ at } 25\ °C$$

The product of the concentrations of the hydrogen and hydroxide ions is a constant that depends only on temperature. Concentrations are frequently quite small, and the p-notation is used to describe such solutions.

When a *strong acid or base* is dissolved in water, it ionizes completely, providing the stoichiometric amount of hydrogen ion or hydroxide ion. *Weak acids and bases* do not ionize completely, and the acid or base ionization constant is used to calculate the concentrations of species in solution. The ionization constants are determined experimentally, and the equilibrium constants for an acid-base conjugate pair are related by

$$K_w = K_a K_b$$

When a solution contains a mixture of acids, the concentration of hydrogen ions is determined by the strongest acid as long as the ionization constant is much greater than that of the weaker acids.

The ability of a molecule to donate a proton is influenced by the strength of the bond holding the proton to the acid and the electronegativity of the atoms in the conjugate base. When the acid molecule donates a proton, the resulting conjugate base has a negative charge. The ionization is enhanced if the negatively charged base is more stable. Acids with strongly electronegative atoms in the conjugate bases, or polyatomic species with several oxygen atoms, are stronger than those with less electronegative atoms in the conjugate bases.

Acid-base behavior can be generalized further—G. N. Lewis extended the Brønsted-Lowry model by noting that the most general feature of a base is the presence of an electron pair, and that of an acid is the presence of an empty orbital that can accept the electron pair to form a coordinate covalent bond.

Chapter Terms

Arrhenius acid—a species that increases the hydrogen ion concentration when dissolved in water. *(13.1)*

Arrhenius base—a species that increases the hydroxide ion concentration when dissolved in water. *(13.1)*

Brønsted-Lowry acid—a species that can donate a proton. *(13.1)*

Brønsted-Lowry base—a species that can accept a proton. *(13.1)*

Acid-base conjugate pair—two species that differ by the presence or absence of a proton. The protonated species is the acid; the deprotonated species is the base. *(13.1)*

Amphoteric species—a species that can act either as an acid or as a base. *(13.1)*

Autoionization of water—the equilibrium that describes the ionization of water *(13.2)*:

$$2H_2O(\ell) \rightleftharpoons H_3O^+(aq) + OH^-(aq) \quad K_w = [H_3O^+][OH^-]$$

pH—equal to $-\log[H_3O^+]$. *(13.2)*

Strong electrolyte—a substance that dissociates or ionizes completely in solution. *(13.3)*

Weak electrolyte—a substance that is only partially dissociated or ionized in solution. *(13.3)*

Strong acid—an acid that ionizes completely in solution. *(13.3)*

Strong base—one of the soluble metal oxides or hydroxides that dissociate completely in solution. *(13.3)*

Weak acid—an acid that partially ionizes in solution. *(13.4)*

Weak base—a base that partially ionizes in solution. *(13.4)*

Leveling effect—the effect that makes strong acids (or bases) appear equal or level in strength since they ionize or dissociate completely. *(13.4)*

Analytical concentration—the total concentration of all protonated and unprotonated forms of the same species. *(13.5)*

Fraction ionized—α, concentration of the ionized form divided by the analytical concentration. *(13.5)*

Binary hydride—a compound formed from one element in combination with hydrogen. HF and HCl are binary hydrides. *(13.8)*

Oxyacid—a molecule of the general formula $(OH)_mXO_n$ in which ionization breaks one of the the hydrogen-oxygen bonds. *(13.8)*

Lewis acid—an electron-pair acceptor. *(13.9)*

Lewis base—an electron-pair donor. *(13.9)*

Adduct—the product of a Lewis acid-base addition reaction. *(13.9)*

Exercises

Exercises designated with color have answers in Appendix J.

Acid-Base Conjugate Pairs

13.1 Write the formula and name the conjugate base of (a) nitric acid (b) hydrogen sulfate ion (c) water (d) hydrogen sulfide

13.2 Write the formula and name the conjugate acid of the following species. The information in Tables 13.6 and 13.7 may be helpful.
(a) hydrogen sulfate ion (b) water (c) ammonia (d) pyridine

13.3 Write the formula and name the conjugate acid of the following species. The information in Tables 13.6 and 13.7 may be helpful.
(a) N_2H_4 (b) NO_2^- (c) S^{2-} (d) I^-

13.4 Write the formula and name the conjugate base of (a) HCN (b) HSO_4^- (c) $H_2PO_3^-$ (d) HCO_3^-

13.5 The following reactions occur in aqueous solution. Predict the products and identify the acids and bases (and their conjugate species) in the reaction of
(a) ammonia with hydrochloric acid
(b) hydrogen carbonate ion with nitric acid
(c) formic acid with cyanide ion
(d) acetate ion with water

13.6 The following reactions occur in aqueous solution. Predict the products and identify the acids and bases (and their conjugate species) in the reaction of
(a) NH_3 with CH_3COOH
(b) $N_2H_5^+$ with CO_3^{2-}
(c) H_3O^+ with OH^-
(d) HSO_4^- with $HCOO^-$

Calculation of pH, pOH, $[H^+]$, and $[OH^-]$

13.7 Fill in the following table and indicate whether the solution is acidic, basic, or neutral.

	pH	$[H_3O^+](M)$
(a)	2.00	____
(b)	____	1.04×10^{-3}
(c)	9.84	____
(d)	____	2.00×10^{-1}
(e)	____	9.40×10^{-9}
(f)	11.34	____
(g)	____	4.57×10^{-4}
(h)	4.51	____
(i)	____	6.65×10^{-15}

13.8 Fill in the following table and indicate whether the solution is acidic, basic, or neutral.

	pH	$[H_3O^+](M)$
(a)	2.34	____
(b)	____	1.04×10^{-13}
(c)	−1.09	____
(d)	____	2.12×10^{-11}
(e)	____	7.40×10^{-2}
(f)	13.41	____
(g)	____	7.07×10^{-5}
(h)	9.80	____
(i)	____	0.505

13.9 Fill in the following table and indicate whether the solution is acidic, basic, or neutral.

	pH	$[H_3O^+](M)$	pOH	$[OH^-](M)$
(a)	10.34	____	____	____
(b)	____	____	10.34	____
(c)	____	0.412	____	____
(d)	____	____	____	11.2×10^{-12}

13.10 Fill in the following table and indicate whether the solution is acidic, basic, or neutral.

	pH	$[H_3O^+](M)$	pOH	$[OH^-](M)$
(a)	−1.04	____	____	____
(b)	____	____	0.34	____
(c)	____	1.98×10^{-7}	____	____
(d)	____	____	____	4.42×10^{-2}

13.11 What is the concentration of hydrogen ion and of hydroxide ion in each of the following? (See Table 13.4.)
(a) lemon juice
(b) wine
(c) milk
(d) household ammonia

13.12 What is the concentration of hydrogen ion and of hydroxide ion in each of the following? (See Table 13.4.)
(a) vinegar (b) stomach acid (c) coffee (d) milk

Solutions of Strong Acids and Bases

13.13 Calculate the pH and pOH in solutions of
(a) 0.050 M HCl
(b) 0.024 M KOH
(c) 0.014 M $HClO_4$
(d) 1.05 M NaOH

13.14 Calculate the pH and pOH in solutions of
(a) 0.0045 M $Ba(OH)_2$
(b) 0.080 M HI
(c) 0.030 M $Sr(OH)_2$
(d) 12.3 M HNO_3

13.15 Calculate the pH of a solution obtained by adding exactly 10.0 mL of a 14.8 M KOH solution to 200 mL of water, then adding water until the volume of solution is exactly 250 mL.

13.16 Calculate the pH and pOH in a solution that is made by dissolving 15.0 g of sodium hydroxide in approximately 450 mL of water. The solution initially becomes quite warm, but after it is allowed to return to room temperature, water is added to dilute to precisely 500 mL.

13.17 What are the pH and pOH of a solution prepared by mixing 25 mL of 0.020 M NaOH with 75 mL of 0.010 M HCl?

13.18 What are the pH and pOH of a solution prepared by mixing 35 mL of 0.020 M $Ba(OH)_2$ with 65 mL of 0.010 M HCl?

Challenging Problems: Nonaqueous Solvent Systems

13.19 Pure, liquid ammonia ionizes in a manner similar to that of water.

(a) Write the equilibrium for the autoionization of liquid ammonia.
(b) Identify the conjugate acid and base formed by this reaction.
(c) Is $NaNH_2$ an acid or a base in this solvent?
(d) Is ammonium bromide an acid or a base in this solvent?

13.20 Liquid HF undergoes an autoionization reaction:

$$2HF \rightleftharpoons H_2F^+ + F^-$$

(a) Will KF be an acid or a base in this solvent?
(b) Perchloric acid, $HClO_4$, is a strong acid in liquid HF. Write the chemical equation for the ionization reaction.
(c) Ammonia is a strong base in this solvent. Write the chemical equation for the ionization reaction.
(d) What is the net ionic equation for the neutralization of perchloric acid with ammonia in this solvent?

13.21 Liquid ammonia can be used as a solvent (see Exercise 13.19).
(a) Will ammonium chloride be an acid, a base, or neither in this solvent?
(b) H_2SO_4 is a strong acid in liquid ammonia. Write the chemical equation for the ionization reaction.
(c) Sodium amide, $NaNH_2$, is a base in this solvent. Write the chemical equation for the ionization reaction.
(d) What is the net ionic equation for the neutralization of sulfuric acid with sodium amide in this solvent?

13.22 Pure acetic acid is sometimes used as a solvent for acid-base reactions.
(a) Write the equilibrium for the autoionization of acetic acid.
(b) If a small amount of water were added to the acetic acid solvent, would the water react as an acid or a base? Write the appropriate equation for the reaction.
(c) What is the strongest base that can exist in acetic acid solvent? What compound might you select if you needed to add a strong base to the solution?
(d) What is the strongest acid that can exist in acetic acid solvent? What compound might you select if you needed to add a strong acid to the solution?

13.23 Which of the following solvent systems allows a chemist to differentiate between the acid strengths of HCl and H_2S?
(a) pure acetic acid (b) liquid ammonia (c) water (d) none of the above

13.24 Which of the following solvent systems allows a chemist to differentiate between the base strengths of NH_3 and theobromine, a weak base with $K_b = 1.3 \times 10^{-14}$?
(a) pure acetic acid
(b) liquid ammonia
(c) water
(d) none of the above

Theobromine is found in the cacao bean, and is present in chocolate. Some scientists think it is an appetite suppressant. (Beware: manufactured chocolate also contains sugar and fats, so eating chocolate bars will not result in a weight loss.) Other scientists speculate that it is involved in the "chocolate high" experienced by people who have unusual cravings for chocolate.

Determination of the Dissociation Constant

13.25 When 1.00 g of thiamine hydrochloride (also called vitamin B_1 hydrochloride) is dissolved in water and then diluted to exactly 10.00 mL, the pH of the resulting solution is 4.50. The formula weight of thiamine hydrochloride is 337.28. Calculate K_a for this acid.

13.26 Consider the solution formed when 50.0 mg of butyric acid, C_3H_7COOH, a bad-smelling organic acid (formed when butter turns rancid), is dissolved in water to make 1.00 mL of solution. The pH of the solution is 2.52. Calculate pK_a for butyric acid.

13.27 Lactic acid, $CH_3CH(OH)(COOH)$, is formed in muscles as a by-product of the muscle's contraction. It causes the pain and stiffness common to the weekend athlete. If the pH of a 0.0376 *M* solution is 2.66, what is pK_a for lactic acid?

13.28 If a 0.0100 *M* solution of caproic acid ($C_5H_{11}COOH$), thought to be at least partially responsible for the unique (and generally considered foul) smell of goats, has a pH of 3.43, what is its K_a?

Concentrations of Species in Solutions of Weak Acids

The asterisk indicates a more challenging calculation.

13.29 Use the K_a values in Table 13.6 to calculate the pH of:
(a) 1.25 *M* HOCl
(b) 0.80 *M* HF
(c) 0.14 *M* CH_3COOH
*(d) 0.25 *M* $HClO_2$

13.30 Use the K_a values in Table 13.6 to calculate the pH of:
(a) 0.20 *M* C_6H_5COOH
(b) 1.50 *M* HCOOH
(c) 0.0055 *M* HCN
*(d) 0.075 *M* HNO_2

13.31 Use the K_a values in Table 13.6 to calculate the pH of:
(a) 0.050 *M* HI
(b) 0.20 *M* HF
(c) 0.15 *M* CH_3COOH
*(d) 0.027 *M* C_6H_5COOH

13.32 Use the K_a values in Table 13.6 to calculate the pH of:
(a) 0.33 M HNO_2
(b) 0.016 M C_6H_5OH
(c) 0.15 M HF
*(d) 0.010 M HCOOH

Degree of Ionization

13.33 Phenol (C_6H_5OH, also called carbolic acid) has $pK_a = 9.9$. It is used in preserving body tissues, and it is quite toxic. Calculate the fraction ionized in 0.010 M and 0.0010 M phenol.

13.34 Calculate the fraction of benzoic acid, a useful food preservative, that is ionized in 0.010 M and 0.0010 M solutions.

13.35 A solution is prepared by dissolving 0.121 g of uric acid, $C_5H_3N_4O_3H$ (molar mass = 168 g/mol), and diluting to make exactly 10 mL of solution. Conductivity measurements indicate that the uric acid, known to be monoprotic (that is, only one hydrogen ion dissociates per molecule), is 34% ionized. Calculate K_a for uric acid, which plays an important role in the arthritic disease gout.

13.36 HCN is a deadly gas that has the faint smell of bitter almonds. HCN, formed by the reaction of H_2SO_4 and KCN, is used by some states to execute criminals in the gas chamber. Measurements (carefully) performed on a 0.0050 M solution of HCN indicate that it is 0.038% ionized. Calculate K_a.

13.37 Assuming that the conductivity of an acid solution is proportional to the concentration of H_3O^+, sketch plots of conductivity *vs.* concentration for HCl and HF in the region from 0 to 0.020 M.

13.38 Measurements of conductivity of solutions of two acids, A and B, produced the following data:

	Conductivity of	
Concentration of Acid, g/L	**A**	**B**
0.480	0.0059	0.0067
0.850	0.0078	0.0125
1.220	0.0093	0.0170

Are the acids strong or weak?

Concentrations of Species in Solutions of Weak Bases

13.39 Morphine is a weak base with $K_b = 1.6 \times 10^{-6}$. It is a prescription drug used to deaden pain, and the average dose is 10 mg. Calculate the pH of a 0.0010 M solution.

13.40 Coniine (2-propyl piperidine) is a weak base ($pK_b = 3.1$). It has the formula $C_8H_{17}N$. Calculate the pH of a 0.50 M solution of coniine. Coniine is extracted from the plant *Conium maculatum,* also called hemlock. This harmless-looking relative of the carrot produces a deadly poison, used to execute the Greek philosopher Socrates.

Socrates was a gadfly. He spent a great deal of time demonstrating that many prominent Athenians were more concerned with their own self-interests than that of the society as a whole. He was charged with corrupting the youth, impiety, and disturbing the society. Socrates, who defended himself, was found guilty by the other Athenians. They asked Socrates to recommend his own punishment; he recommended that he be compensated for his work with young people, since he had no other source of income. This suggestion angered his peers, who sentenced him to death.

Socrates was not well-guarded—the Athenians hoped he would escape. But he felt a moral obligation to follow the edict of the state. So he drank the hemlock and died.

Three of Plato's dialogues speak of the events surrounding the death of Socrates. The *Apology* depicts the trial; *Crito* includes Socrates' reasons for choosing death; and *Phaedro* recounts his musings as the coniine took effect.

Calculating K_a from K_b

13.41 Use the data in Table 13.7 to calculate the acid ionization constants for:
(a) ammonium ion
(b) pyridinium ion
(c) hydrazinium ion

13.42 Use the data in Table 13.6 to calculate K_b for:
(a) formate ion
(b) nitrite ion
(c) chlorite ion

Ranking Acids and Bases in Order of Strength

13.43 Rank the species in order of increasing acid strength: HF, HCl, NH_4^+, NH_3

13.44 Rank the species in order of increasing acid strength: NH_4^+, H_2O, HF, HSO_4^-

13.45 Rank the species in order of increasing acid strength: HCl, NH_3, HF, Na^+

13.46 Rank the species in order of increasing acid strength: CH_3COOH, H_2O, HCOOH, F^-

Calculating the pH of a Salt Solution

13.47 Calculate the pH of 1.5 M ammonium chloride.

13.48 Calculate the pH of 0.060 M pyridinium iodide.

13.49 Calculate the pH of 0.25 M potassium nitrite.

13.50 Calculate the pH of 0.010 M sodium acetate.

The pH of a Mixture

13.51 Calculate the pH of a solution that is 0.050 M HCl and 0.15 M HF.

13.52 Calculate the pH of a solution that is 0.20 M CH_3COOH and 0.050 M HI.

13.53 Calculate the pH of a solution that is 0.10 M formic acid and 0.050 M phenol.

13.54 What is the pH of a solution that is 0.10 M acetic acid and 0.20 M HCN?

13.55 What is the pH of a solution that is prepared by mixing 10 mL of 1.0 M NaOH with 100 mL of 0.10 M ammonia?

13.56 What is the pH of a solution prepared when 10.0 g of sodium benzoate is added to 100 mL of 0.10 M KOH?

Evaluating Acids on the Basis of Structure

13.57 Which of the acids is stronger, and why?
(a) H_3AsO_3, H_3AsO_4
(b) PH_3, H_2S
(c) HClO, $HClO_2$

13.58 Which of the acids is stronger, and why?
(a) $HClO_3$, $HClO_4$
(b) HNO_2, HNO_3
(c) GeH_4, AsH_3

Identifying Lewis, Brønsted-Lowry, and Arrhenius Acids

13.59 Consider the following reactions. Determine whether the reaction would be classified as an acid-base reaction in the Arrhenius system, the Brønsted-Lowry system, or the Lewis system.
(a) $HCl(aq) + H_2O(\ell) \longrightarrow H_3O^+(aq) + Cl^-(aq)$
(b) $Zn(OH)_3^-(aq) + OH^-(aq) \rightleftharpoons Zn(OH)_4^{2-}(aq)$
(c) $CO_2(g) + LiOH(s) \longrightarrow LiHCO_3(s)$
(d) $SO_2(g) + H_2O(g) \rightleftharpoons H_2SO_3(g)$

13.60 Consider the following reactions. Determine whether the reaction would be classified as an acid-base reaction in the Arrhenius system, the Brønsted-Lowry system, or the Lewis system.
(a) $HCl(aq) + NH_3(aq) \longrightarrow NH_4Cl(aq)$
(b) $SO_2(g) + NaOH(s) \longrightarrow NaHSO_3(s)$
(c) $LiH(s) + H_2O(\ell) \longrightarrow LiOH(aq) + H_2(g)$
(d) $HSO_4^-(aq) + F^-(aq) \rightleftharpoons HF(aq) + SO_4^{2-}(aq)$

Predicting the Direction of Change

13.61 A solution contains H^+, $HCOO^-$, and HCOOH. In which direction will the pH change if the following substances are added to the solution?
(a) HCl
(b) HCN
(c) CH_3COONa
(d) KBr
(e) H_2O

13.62 Do the following reactions favor the reactants or the products?
(a) $HCl(aq) + NH_3(aq) \rightleftharpoons NH_4Cl(aq)$
(b) $HNO_3(aq) + NaOH(aq) \rightleftharpoons NaNO_3(aq) + H_2O(\ell)$
(c) $2KCl(aq) + Ba(OH)_2(aq) \rightleftharpoons BaCl_2(aq) + 2KOH(aq)$
(d) $HSO_4^-(aq) + NH_3(aq) \rightleftharpoons NH_4^+(aq) + SO_4^{2-}(aq)$

13.63 Do the following reactions favor the reactants or the products?
(a) $NaOH(aq) + KCl(aq) \rightleftharpoons NaCl(aq) + KOH(aq)$
(b) $HF(aq) + NaOH(aq) \rightleftharpoons NaF(aq) + H_2O(\ell)$
(c) $CH_3COONa(aq) + H_2O(\ell) \rightleftharpoons CH_3COOH(aq) + NaOH(aq)$
(d) $NH_4^+(aq) + H_2O(\ell) \rightleftharpoons NH_3(aq) + H_3O^+(aq)$

Additional Problems

13.64 How much 0.083 M HNO_2 must be added to water to make 1.00 L of a solution with a pH of 4.75?

13.65 What volume of 0.10 M acetic acid is needed to prepare 5.0 L of acetic acid solution that has a pH of 4.00?

13.66 What is the mass of benzoic acid that must be dissolved to prepare 1.00 L of a solution that has a pH of 3.50?

13.67 Calculate the volume of 14.3 M HCl that must be used to prepare 100.0 L of a solution that has a pH of 3.50.

13.68 A solution is made by diluting 10.0 mL of concentrated ammonia (28% by weight, density = 0.90 g/mL) to exactly 1 L. Calculate the pH of the solution.

13.69 A solution is made by diluting 25.0 mL of concentrated HCl (37% by weight, density = 1.19 g/mL) to exactly 500 mL. What is the pH of the resulting solution?

13.70 Use the structure and bonding to explain why $HClO_4$ is a stronger acid than H_2SeO_4.

13.71 Element 85, astatine (At), is a radioactive halogen not present in appreciable amounts in nature. The acid HAt can be prepared and compared to the other hydrogen halides.
(a) Explain why you expect HAt to be stronger or weaker than HI.
(b) Propose an experiment that would allow you to compare the acid ionization constants for HAt and HI. The radioactive nature of At can be ignored in answering this question.

Chapter Fourteen

Reactions Between Acids and Bases

The reaction of an acid with a base is one of the most important classes of chemical reactions. Farmers use such reactions to adjust the pH of soil to improve crop yields. Your body uses them to neutralize the chemicals produced each time your heart beats. When an acid and base are mixed, the conjugate partners are formed as the original acid and base are consumed. But we really have not discussed how to determine the pH of the solution in this seemingly complex mixture of acids, bases, and solvent.

The chemistry of acids and bases in solution was discussed in the last chapter. In this chapter, the methods used previously are extended to include solutions that are prepared by mixing solutions of acids with solutions of bases. Also presented is information about indicators, the substances that change color at the end of an acid-base titration. In addition, polyprotic acids, acids that can donate more than one proton to a base, are described. Last, the influence of acids and bases on other equilibria is presented. An important example of this influence is the dependence of the solubility of certain salts on the pH of the solution.

14.1 Titrations of Strong Acids and Bases

Objectives

- To calculate the concentrations of all species present during the titration of a strong acid with a strong base
- To calculate the concentrations of all species present during the titration of a strong base with a strong acid.

Each day, chemists perform thousands of analyses called titrations. A titration is an analysis based on measuring the amount of a substance, called the

◀ The red color formed as methyl red indicator mixes with a solution shows that the solution is acidic.

titrant, that reacts with the **analyte** (the substance whose concentration is being determined). The titrant is generally chosen so it reacts completely with the analyte.

$$\text{analyte} + \text{titrant} \longrightarrow \text{products}$$

Titrations are commonly employed to determine the amount or concentration of a substance as part of a quantitative analysis. Knowing how the pH of the reaction mixture changes as the titrant is added is essential for understanding acid-base titrations and the methods used to detect the completion of the titration reaction. Much of this chapter will be devoted to the construction of acid-base **titration curves,** graphs of pH as a function of the volume of titrant added. The use of titrations and titration calculations were presented in Section 4.3, but are briefly reviewed here.

The Titration of a Strong Acid with a Strong Base

Perhaps the most common titration involves the reaction of an acid with a base. Consider the determination of the concentration of a hydrochloric acid solution, illustrated in Figure 14.1. A known volume of the HCl solution is placed into a flask. Next, an indicator is added that will change color when the hydrochloric acid is consumed. Phenolphthalein is a compound with such properties—it is colorless in acid solution and pink in alkaline solution. A **standard solution** of base (one of accurately known concentration) is added until the color change, or *endpoint,* is reached. The volume of standard base needed to reach the endpoint is recorded, and then the concentration of the original acid is calculated by the procedure presented in Chapter 4 and illustrated in the following example.

Example 14.1 Calculating the Concentration of an Unknown by Titration

In the titration of nitric acid with standard base, the phenolphthalein indicator changes from colorless to pink after 45.12 mL of 0.0980 *M* NaOH has been added to 50.00 mL of the nitric acid. Calculate the molar concentration of the nitric acid.

Solution The following flow diagram summarizes the procedure.

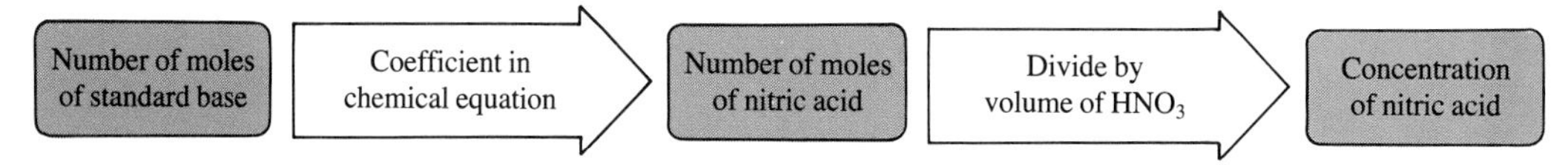

Writing the chemical equation for the reaction is the first step.

$$HNO_3(aq) + NaOH(aq) \longrightarrow NaNO_3(aq) + H_2O(\ell)$$

The number of moles of base is calculated from the concentration and volume added.

$$\text{moles of NaOH} = 0.04512 \text{ L} \left(\frac{0.0980 \text{ mol}}{\text{L of solution}}\right) = 4.42 \times 10^{-3} \text{ mol}$$

$$\text{moles } HNO_3 = 4.42 \times 10^{-3} \text{ mol NaOH} \left(\frac{1 \text{ mol } HNO_3}{1 \text{ mol NaOH}}\right) = 4.42 \times 10^{-3} \text{ mol}$$

$$\text{conc. of } HNO_3 = \frac{4.42 \times 10^{-3} \text{ mol}}{0.05000 \text{ L of } HNO_3} = 0.0884\ M$$

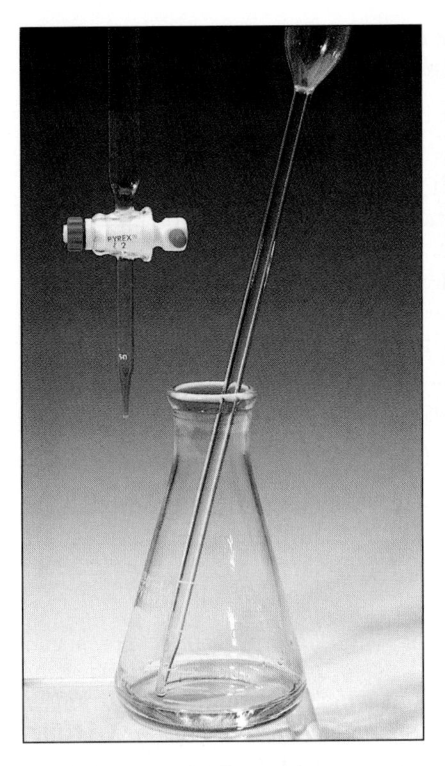

(a)

(b)

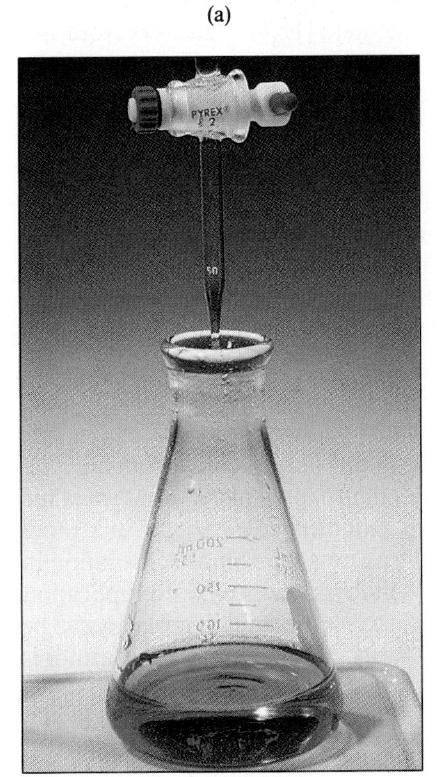

(c)

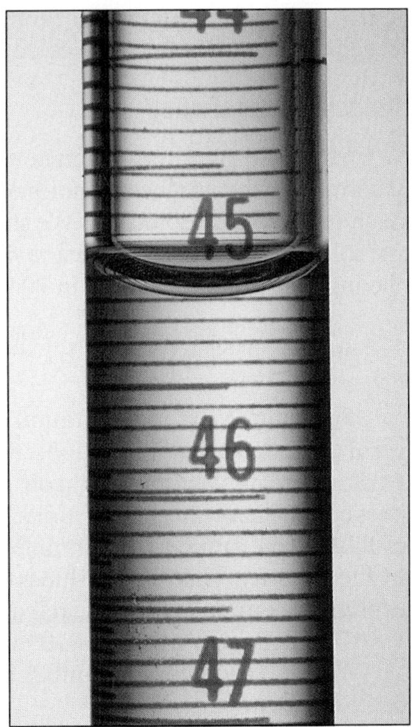

(d)

Figure 14.1 The determination of the concentration of hydrochloric acid by titration (a) A known volume of the acid solution is placed in a flask. (b) Standard base solution is added from a buret. (c) The end of the titration is indicated by a color change. (d) The volume of base solution needed to react with the acid is recorded.

Understanding What is the concentration of KOH if 24.51 mL of 0.110 *M* HCl is needed to titrate 10.00 mL of the KOH solution?

Answer: 0.270 *M*

Once the concentrations of both reactants in the acid-base titration reaction are known, we can calculate how the pH of the solution changed as the titrant was added to the analyte. This procedure is really a limiting reactant stoichiometry calculation, identical to those presented in Chapters 3 and 4.

We can illustrate the procedure with the titration of a strong acid with a standard solution of strong base. Since both the acid and the base are strong electrolytes, the titration reaction is

$$H_3O^+(aq) + OH^-(aq) \longrightarrow 2H_2O(\ell)$$

It is safe to assume that the titration reaction goes to completion if its equilibrium constant is much larger than one. In the case of the reaction between a strong acid and a strong base, K_{eq} is very large.

$$K_{eq} = \frac{1}{[H_3O^+][OH^-]} = \frac{1}{K_w} = 1 \times 10^{14}$$

Before the *equivalence point,* the point at which the amount of titrant is stoichiometrically equal to the amount of analyte, has been reached, H_3O^+ is in excess and the pH is found directly from $[H_3O^+]$. Beyond the equivalence point, OH^- is in excess and we calculate $[H_3O^+]$ and pH from the $[OH^-]$ and the autoionization equilibrium of water. The pH at the equivalence point requires the same calculation as used for the pH of water.

Units of Concentration

A titration is a limiting reactant stoichiometry calculation that involves determining the number of moles of each reactant from the volumes of the analyte and titrant solutions. We showed in Chapter 4 that the molarity of the solution serves as a conversion factor in such calculations. For example, the number of moles of HCl in 20.00 mL of 0.125 *M* HCl solution is

$$\text{moles of HCl} = 0.02000\ \cancel{\text{L}} \left(\frac{0.125\ \text{mol HCl}}{1\ \cancel{\text{L}}}\right) = 2.50 \times 10^{-3}\ \text{mol HCl}$$

The volume and concentration used in this sample conversion are typical of those that occur in titrations. Rather than expressing the amount of the reactants in moles, the **millimole** (1 millimole = 10^{-3} mole) is a more convenient-sized unit for titrations. If we express the volume of solution in milliliters, the number of millimoles of solute is obtained by multiplication by the molarity. This relationship is true because the units of molarity can be expressed as mmoles/mL, as well as moles/L. (Remember that 1 mmol is 1×10^{-3} mol and 1 mL is 1×10^{-3} L.)

mmoles = volume (mL) × concentration (*M*).

We can calculate the number of millimoles of HCl in 20.00 mL of a 0.125 *M* solution:

$$\text{mmoles of HCl} = 20.00\ \cancel{\text{mL}} \left(\frac{0.125\ \text{mmol HCl}}{\cancel{\text{mL}}}\right) = 2.50\ \text{mmol HCl}$$

We can see that the amount of solute is calculated from the volume and molarity. When concentration is expressed in molarity and the volume in liters, we calculate moles of solute. When the volume is in milliliters, we calculate mmoles. Both can be used in stoichiometry problems; sometimes one is more convenient and sometimes the other. For most titrations, including the remaining examples in this chapter, it is more convenient to express amounts in millimoles.

Molarity can be expressed with units of mol/L or mmol/mL; both are equivalent.

Calculation of the Titration Curve

A titration involves adding titrant to the analyte, a process that we call the "titration" reaction. The titration reaction goes to completion so a stoichiometry calculation that starts with the total mmoles of each reactant is used to calculate the amounts of products formed and reactants consumed. To calculate the concentrations of the species in solution, the mmoles of each species must be converted to concentrations, using the total volume of solution.

If we perform a titration in the lab, we might measure the pH after 5 mL of titrant is added. We might make another measurement after an additional 7 mL of titrant is added. We would describe these as the pH after 5 mL and the pH after 12 mL of titrant are added. Volumes are expressed in a cumulative manner.

The titration system differs from the equilibrium system since the addition of titrant causes a stoichiometric reaction to occur. The titration calculation is similar to a limiting-reactant calculation.

A tabular approach, which we call the sRfc table, is helpful in organizing the stoichiometry calculation. The starting number of moles of each reactant is placed on the s-line (s for starting). The changes in the number of moles are on the R-line (R for react), and the final number of moles is shown on the f-line (f for final). The c-line (concentrations) is obtained by dividing the moles of each species on the f-line by the total volume of solution. The process is illustrated in the following calculations.

2.00 mL of 0.250* M *NaOH Added to 20.0 mL of 0.125* M *HCl Since HCl is a strong acid and NaOH is a strong base, the titration reaction is

$$H_3O^+(aq) + OH^-(aq) \longrightarrow 2H_2O(\ell)$$

The first step in calculating a titration curve is a stoichiometry calculation that requires the chemical equation for the titration.

For the s-line, we calculate the number of mmoles of H_3O^+ introduced into the solution from the volume (20.00 mL) and concentration (0.125 M) of the HCl solution. The concentration of hydrogen ions in the HCl solution is

$$\text{conc of } H_3O^+ = \left(\frac{0.125 \text{ mmol HCl}}{\text{mL solution}}\right)\left(\frac{1 \text{ mmol } H_3O^+}{1 \text{ mmol HCl}}\right)$$

$$= 0.125\ M\ H_3O^+$$

The number of mmoles of H_3O^+ is

$$\text{mmol } H_3O^+ = 20.00 \text{ mL}\left(\frac{0.125 \text{ mmol } H_3O^+}{\text{mL}}\right)$$

$$= 2.50 \text{ mmol } H_3O^+$$

The number of mmoles of OH^- is found from the volume (2.00 mL) and concentration (0.250 M) of the NaOH solution.

$$\text{mmol OH}^- = 2.00\ \text{mL}\left(\frac{0.250\ \text{mmol NaOH}}{\text{mL}}\right)\left(\frac{1\ \text{mmol OH}^-}{1\ \text{mmol NaOH}}\right)$$

$$= 0.500\ \text{mmol OH}^-$$

The start line is filled in with this information.

	H_3O^+	+ OH^-	$\longrightarrow 2H_2O$
s, mmol	2.50	0.500	excess
R, mmol	−0.500	−0.500	+1.00

The Reaction line is filled in by recognizing that the OH^- ion is the limiting reactant. The number of moles on the f-line is the sum of the first two lines in the table. Since concentrations are needed to evaluate the pH, the last line in the table, the c-line, is obtained by dividing the number of moles of each species on the f-line by the total volume of the solution, which in this case is

$$V_T = 20.00\ \text{mL} + 2.00\ \text{mL} = 22.00\ \text{mL}$$

The complete table for our stoichiometry calculation is shown below.

	H_3O^+	+ OH^-	$\longrightarrow 2H_2O$
s, mmol	2.50	0.500	excess
R, mmol	−0.500	−0.500	+1.00
f, mmol	2.00	0.	excess
c, M	0.0909	0.	excess

Since the goal is to find the pH of the solution, we need to calculate the pH from the hydrogen ion concentration.

$$\text{pH} = -\log[H_3O^+] = -\log(0.0909) = 1.04$$

If only strong acids or bases remain after the titration step, the pH is determined from their concentrations.

The calculation of the pH of any point before the equivalence point in the titration of a strong acid with a strong base is identical to that shown above.

We next calculate the pH at two addition points on this titration curve: after addition of 10.00 mL of the sodium hydroxide solution and after 20.00 mL of the base has been added.

10.00 mL of 0.250* M *NaOH Added Again, we start with the initial amounts of H_3O^+ and OH^- that have been placed in the titration vessel. These are calculated from the volumes and concentrations of the HCl (20.00 mL of 0.125 M solution) and hydroxide ion (10.00 mL of 0.250 M solution).

$$\text{mmol } H_3O^+ = 20.00 \text{ mL}\left(\frac{0.125 \text{ mmol } H_3O^+}{\text{mL}}\right) = 2.50 \text{ mmol } H_3O^+$$

$$\text{mmol } OH^- = 10.00 \text{ mL}\left(\frac{0.250 \text{ mmol } OH^-}{\text{mL}}\right) = 2.50 \text{ mmol } OH^-$$

These values are placed in the s-line and the sRfc table is completed.

	H_3O^+ +	OH^- ⟶	$2H_2O$
s, mmol	2.50	2.50	excess
R, mmol	−2.50	−2.50	+5.00
f, mmol	0.	0.	excess
c, *M*	0.	0.	excess

Since both the H_3O^+ and OH^- ions have been completely consumed, the titration is at the equivalence point. Since water is the only acid and the only base left in the solution, the hydrogen ion concentration is calculated from the autoionization equilibrium.

$$2H_2O \rightleftharpoons H_3O^+ + OH^- \qquad K_w = 1.0 \times 10^{-14}$$

This equilibrium calculation was presented in Chapter 13, with the result

$$[H_3O^+] = [OH^-] = 1.0 \times 10^{-7}\ M$$

$$pH = -\log[H_3O^+] = 7.00$$

In any strong acid-strong base titration the pH at the equivalence point is 7.00.

The pH at the equivalence point in the reaction of any strong acid with any strong base is 7.00.

20.00 mL of 0.250* M *NaOH Added The starting amount of HCl is 2.50 mmol, as in the last calculation. The amount of hydroxide ion on the s-line of the table is

$$\text{mmol } OH^- = 20.00 \text{ mL}\left(\frac{0.250 \text{ mmol } OH^-}{\text{mL}}\right) = 5.00 \text{ mmol } OH^-$$

The stoichiometry calculation is completed using the sRfc table.

	H_3O^+ +	OH^- ⟶	$2H_2O$
s, mmol	2.50	5.00	excess
R, mmol	−2.50	−2.50	+5.00
f, mmol	0.	2.50	excess
c, *M*	0.	0.0625	excess

The concentration of OH^- on the last line results from dividing the millimoles on the f-line by the total volume of the solution, 40.00 mL. The pH is determined by the excess hydroxide ion that is present.

$$pOH = -\log(0.0625) = 1.20$$

$$pH = 14.00 - pOH = 14.00 - 1.20 = 12.80$$

Hydroxide ion is in excess beyond the equivalence point.

In Example 14.2, additional calculations of points on this titration curve are illustrated.

Example 14.2 Titration of a Strong Acid with a Strong Base

Calculate the pH in the titration of 20.00 mL of 0.125 *M* HCl after addition of (a) 9.60 mL and (b) 10.40 mL of 0.250 *M* NaOH.

Solution In both parts (a) and (b), the source of the H_3O^+ is the HCl solution, and we calculate the amount of H_3O^+ from the volume and concentration of the acid solution.

$$\text{mmol } H_3O^+ = 20.00 \text{ mL}\left(\frac{0.125 \text{ mmol } H_3O^+}{\text{mL}}\right) = 2.50 \text{ mmol } H_3O^+$$

(a) The added OH^- ion is found from the volume and concentration of the standard NaOH solution.

$$\text{mmol } OH^- = 9.60 \text{ mL}\left(\frac{0.250 \text{ mmol } OH^-}{\text{mL}}\right) = 2.40 \text{ mmol } OH^-$$

The starting number of millimoles of each reactant is substituted into the sRfc table, and the limiting reactant stoichiometry calculation completed.

	H_3O^+	+ OH^- ⟶	$2H_2O$
s, mmol	2.50	2.40	excess
R, mmol	−2.40	−2.40	+4.80
f, mmol	0.10	0.	excess
c, *M*	3.4×10^{-3}	0.	excess

The concentration of H_3O^+ is found by dividing 0.10 mmol (the final amount of the f-line) by the total volume of solution, 20.00 + 9.60 mL. Since the $[H_3O^+]$ is known, the pH is found directly.

$$pH = -\log[H_3O^+] = -\log(3.4 \times 10^{-3}) = 2.47$$

(b) The starting amount of hydroxide ion is found from the volume and concentration of the NaOH solution.

$$\text{mmol } OH^- = 10.40 \text{ mL}\left(\frac{0.250 \text{ mmol } OH^-}{\text{mL}}\right) = 2.60 \text{ mmol } OH^-$$

Note that the limiting reactant at this point in the titration is now the H_3O^+. The completed sRfc table shows that the OH^- ions are present in excess in the reaction mixture.

	H_3O^+ +	OH^-	$\longrightarrow$ $2H_2O$
s, mmol	2.50	2.60	excess
R, mmol	−2.50	−2.50	+5.00
f, mmol	0	0.10	excess
c, *M*	0.	3.3×10^{-3}	excess

The pH of a 3.3×10^{-3} *M* solution of hydroxide ion is

$$pOH = 2.48$$

$$pH = 14.00 - pOH = 11.52$$

Understanding Find the pH in this titration, after 12.00 mL of the 0.250 *M* NaOH solution is added.

Answer: pH = 12.19

The Titration Curve

Table 14.1 shows data for the points on the titration curve of 0.125 *M* HCl with 0.250 *M* NaOH. The table includes the results of several calculations in addition to those performed above. The colored shading indicates which species, hydrogen ion or hydroxide ion, was calculated from the titration stoichiometry. The concentration of the other species is calculated from the concentration of the first species and the equilibrium constant for the autoionization of water. Figure 14.2 illustrates how the pH changes during the course of this titration.

pH as a Function of Volume

The **titration curve** (pH as a function of titrant volume) for the data in Table 14.1 is given in Figure 14.3. The use of the titration curve in locating the equivalence point is clear from the graph. The point at which the pH changes most rapidly is the **inflection point** and occurs at the equivalence point.

Figure 14.2 The pH during the course of a titration (a) 0, 5.00, 10.00, and 15.00 mL of 0.10 *M* NaOH added to acid solution that contains Universal Indicator. (b) Color chart for Universal Indicator.

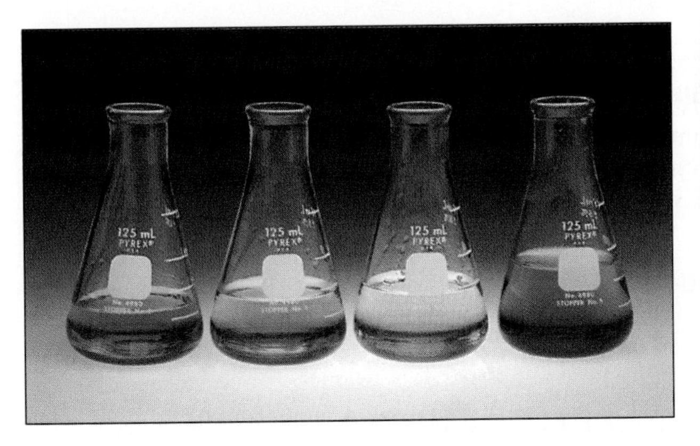

(a)

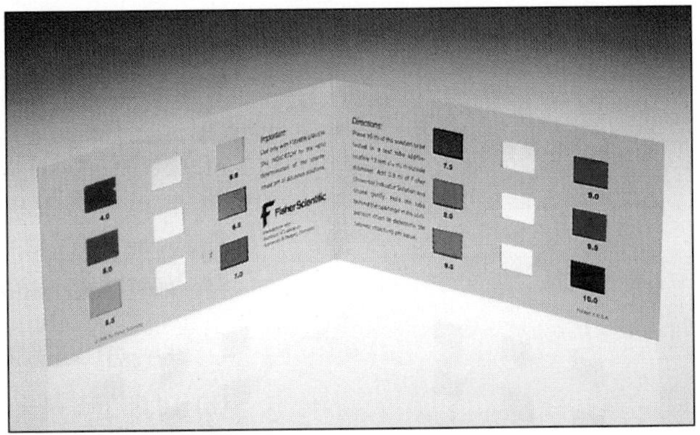

(b)

Table 14.1 Titration of 20.00 mL of 0.125 *M* HCl with 0.250 *M* NaOH

Volume of NaOH Added, mL	$[H_3O^+]$, *M*	$[OH^-]$, *M*	pH
0.00	0.125	8.0×10^{-14}	0.90
2.00	0.0909	1.1×10^{-13}	1.04
5.00	0.0500	2.0×10^{-13}	1.30
9.60	3.4×10^{-3}	3.0×10^{-12}	2.47
10.00	1.0×10^{-7}	1.0×10^{-7}	7.00
10.40	3.0×10^{-12}	3.3×10^{-3}	11.52
11.00	1.2×10^{-12}	8.1×10^{-3}	11.91
12.00	6.4×10^{-13}	0.0156	12.19
20.00	1.6×10^{-13}	0.0625	12.80

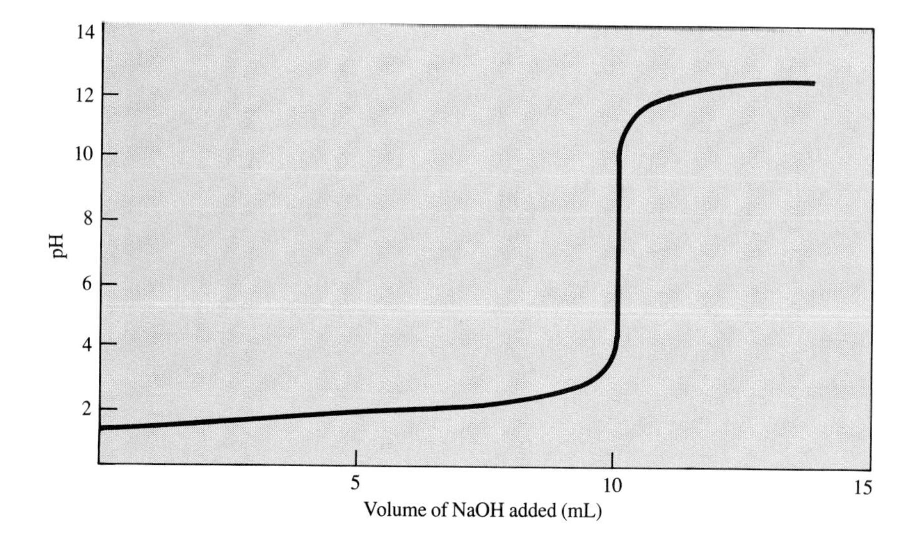

Figure 14.3 pH as a function of the volume of added base

Example 14.3 shows the calculations and titration curve of a strong base with a standard strong acid. The pH is initially high and decreases rapidly at the equivalence point.

Example 14.3 Titration of a Strong Base with a Strong Acid

Calculate the pH values and plot the titration curve for the titration of 50.0 mL of 0.500 *M* KOH with 1.00 *M* HCl. Calculate the pH after 0, 10.00, 24.00, 25.00, 26.00, and 40.00 mL of acid are added.

Solution

(a) 0 mL acid added.

The solution is 0.500 *M* KOH. The OH^- concentration will be 0.500 *M*, so the pOH is 0.30 and the pH is 13.70.

(b) 10.00 mL of acid added.

We started with 50.00 mL of 0.500 *M* KOH. The starting amount of KOH is 50.00 mL × 0.500 mmol/mL = 25.0 mmol. We add 10.00 mL of 1.00 *M* acid to the base. This volume of acid corresponds to 10.0 mmol. The total volume in the titration flask is (50.00 mL base + 10.00 mL acid) = 60.00 mL of solution. The sRfc table can be completed.

	$KOH(aq)$	+ $HCl(aq)$	$\longrightarrow KCl(aq) + H_2O(\ell)$
s, mmol	25.0	10.0	0.
R, mmol	−10.0	−10.0	10.0
f, mmol	15.0	0.	10.0
c, *M*	0.250	0.	0.167

The pH of the solution is determined by the excess of strong base. Since $[OH^-]$ is 0.25 *M*, the pOH is 0.60 and the pH is 13.40.

The remaining calculations are summarized in the following table.

	Amount of Excess					
Added Vol HCl, mL	**HCl, mmol**	**KOH, mmol**	V_T, **mL**	$[H_3O^+]$, *M*	$[OH^-]$, *M*	**pH**
0.	0.	25.0	50.0		0.500	13.70
10.0	0.	15.0	60.0		0.250	13.40
24.0	0.	1.0	74.0		0.014	12.15
25.0	0.	0.	75.0	1.0×10^{-7}	1.0×10^{-7}	7.00
26.0	1.0	0.	76.0	0.013		1.88
40.0	15.0	0.	90.0	0.167		0.78

A plot of the titration curve is shown in Figure 14.4. Notice that the curve drops sharply. Just 1.00 mL before the equivalence point, the solution is still quite basic; 1.00 mL after the equivalence point, the solution is quite acidic. The solution changes from an excess of base to an excess of acid as the titration passes through the equivalence point.

Understanding Calculate the pH after the addition of 0, 5.00, 10.00, and 15.00 mL of 0.200 *M* HCl to 20.00 mL of 0.100 *M* NaOH.

Answer: pH = 13.00, 12.60, 7.00, and 1.54

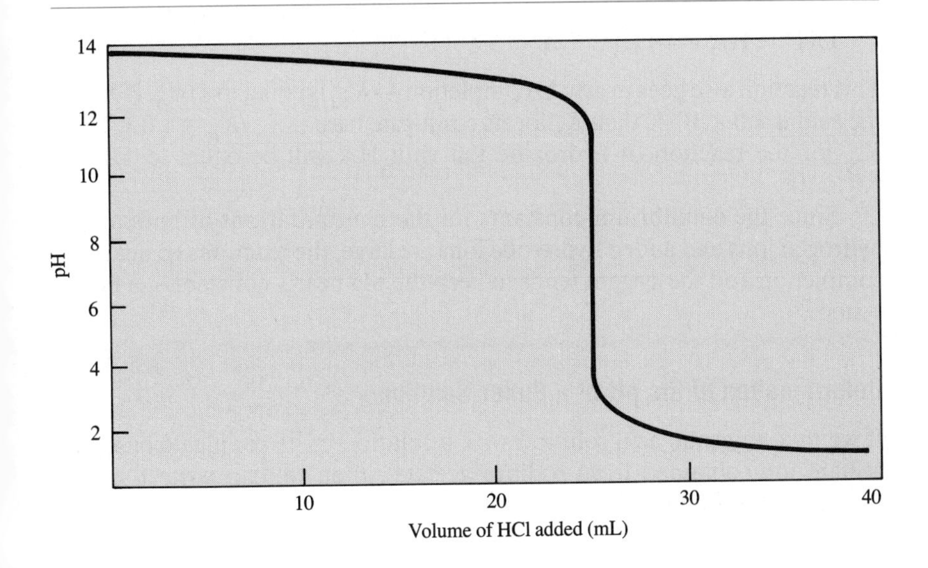

Figure 14.4 Titration curve for the titration of a strong base with a strong acid The graph shows the titration of 50.00 mL of 0.500 *M* KOH with 1.00 *M* HCl.

14.2 Buffers

Objectives

- To determine the pH of a buffer solution containing a weak acid and its conjugate base
- To determine how the addition of small amounts of acids or bases changes the pH of a buffer solution

For the success of many chemical reactions, a constant or nearly constant pH is required. Some reactions must be carried out in a medium that can consume almost completely any small amounts of H_3O^+ or OH^- that are added to the system. We define a **buffer** as a system that resists changes in pH when hydrogen ions or hydroxide ions are added. A buffer system must have a base present to react with any added hydrogen ions, and an acid present to react with any added hydroxide ions.

In this section, the methods used to calculate the pH of a buffer will be presented. These methods will be applied to determine how much the pH of various buffers changes as small amounts of H_3O^+ or OH^- are added to the buffer system.

A buffer solution is made from a weak acid and its conjugate base.

A buffer solution can be prepared from a weak acid and its conjugate base (or a weak base and its conjugate acid). We will consider a buffer made from a generic acid, HA, and its conjugate base, A^-. If hydrogen ions are added to our buffer, they will react with the base:

$$H_3O^+ + A^- \rightleftharpoons H_2O + HA$$

The equilibrium constant for this reaction is quite large. The reaction is the reverse of the ionization of the acid HA, so its equilibrium constant will be the reciprocal of K_a. (You can confirm $K_{eq} = 1/K_a$ by comparing the concentration expressions for both equilibria.) If K_a of the acid is 1×10^{-6}, then the equilibrium constant for the reaction of hydrogen ions with A^-, the conjugate base, is $1 \times 10^{+6}$. A large equilibrium constant means that the reaction will proceed nearly to completion.

Similarly, the weak acid of the buffer will react with any added hydroxide ions:

$$OH^- + HA \rightleftharpoons H_2O + A^-$$

This reaction also goes nearly to completion—K_{eq} is equal to $1/K_b$. If K_a for the acid is 1.0×10^{-6}, then K_b for its conjugate base $= K_w/K_a = 1.0 \times 10^{-8}$. K_{eq} for the reaction of hydroxide ion with HA will be equal to $1/K_b = 1.0 \times 10^8$.

A solution of a weak acid and its conjugate base consumes added H_3O^+ or OH^- to maintain the pH nearly constant.

Since the equilibrium constants for the neutralizations of both added hydrogen ions and added hydroxide ions are large, the reactions go nearly to completion and the system tends to keep the pH nearly constant—a buffer system.

Determination of the pH of a Buffer Solution

If we mix an acetic acid solution with a solution of its conjugate base, the acetate ion (obtained from sodium acetate), then we can write the acid ionization equilibrium:

$$CH_3COOH(aq) + H_2O(\ell) \rightleftharpoons H_3O^+(aq) + CH_3COO^-(aq)$$

Note that $[H_3O^+]$ does not equal $[CH_3COO^-]$ in this equilibrium because both the ionization of acetic acid and the sodium acetate are sources of acetate ion. Regardless, the equilibrium expression is the same:

$$K_a = \frac{[H_3O^+][CH_3COO^-]}{[CH_3COOH]}$$

$$[H_3O^+] = K_a \frac{[CH_3COOH]}{[CH_3COO^-]}$$

By taking the negative logarithm of both sides to convert to pH, we obtain the relationship between the pH of the solution and the concentrations of the acid and its conjugate base:

$$pH = pK_a + \log \frac{[CH_3COO^-]}{[CH_3COOH]}$$

Remember, pK_a is equal to $-\log K_a$.

Buffer solutions are prepared so the concentrations of the acid and its conjugate base are large compared to the concentrations of H_3O^+ and OH^-. Under these circumstances, the degree of ionization is small, and the equilibrium concentrations of the acid and conjugate base (*i.e.*, $[CH_3COOH]$ and $[CH_3COO^-]$) are nearly equal to the starting or analytical concentrations (*i.e.*, C_{CH_3COOH} and $C_{CH_3COO^-}$). Then we can make the approximation

$$pH \approx pK_a + \log \frac{C_b}{C_a} \qquad [14.1]$$

where C_a is the analytical concentration of the acid, and C_b is the analytical concentration of the conjugate base.

The relationship shown in Equation 14.1 was first described by Henderson and almost simultaneously by Hasselbalch, two chemists of the early 20th century. We have honored their research by naming the equation the **Henderson-Hasselbalch equation.**

When the concentrations of the acid and conjugate base are equal, the Henderson-Hasselbalch equation predicts that the pH of the solution will be equal to pK_a. When C_a is equal to C_b, their ratio is 1 and the log of 1 is zero.

$$pH = pK_a + \log(1) = pK_a$$

Since we usually prepare buffer solutions with approximately equal amounts of acid and base (so they can neutralize added hydrogen ions or hydroxide ions with equal effectiveness), the pH of a buffer solution is generally limited to the vicinity of pK_a for the acid.

Most buffers are made from solutions with concentrations in the range from 0.1 to 1.0 *M*, so the Henderson-Hasselbalch equation is a good approximation.

The Henderson-Hasselbalch equation should be used only when the concentrations of both the weak acid and its conjugate base are relatively large. When C_a and C_b are equal, they should be at least 100 times K_a (or K_b for an alkaline buffer) for the approximation to be valid.

The Henderson-Hasselbalch equation and the acid ionization constant expression both contain equivalent information.

$$HA + H_2O \rightleftharpoons H_3O^+ + A^-$$

$$K_a = \frac{[H_3O^+][A^-]}{[HA]}$$

When C_a and C_b are large, then $C_a \approx [HA]$ and $C_b \approx [A^-]$.

$$[H_3O^+] = K\frac{C_a}{C_b}$$

Either method can be used to solve for the pH of a buffer solution.

Example 14.4 Determining the pH of a Buffer Solution

Calculate the pH of a solution that is 0.40 *M* sodium acetate and 0.20 *M* acetic acid. K_a for acetic acid is 1.8×10^{-5}.

Solution One method to solve this problem is to use the Henderson-Hasselbalch equation.

$$pH = pK_a + \log\frac{C_b}{C_a}$$

$$pH = -\log(1.8 \times 10^{-5}) + \log\left(\frac{0.40}{0.20}\right)$$

$$pH = 5.04$$

The expression for the ionization of acetic acid can also be used to calculate the hydrogen ion concentration.

$$CH_3COOH + H_2O \rightleftharpoons H_3O^+ + CH_3COO^-$$

$$K_a = \frac{[H_3O^+][CH_3COO^-]}{[CH_3COOH]}$$

Substitute the numerical value for K_a ($K_a = 1.8 \times 10^{-5}$) and the analytical concentrations for $[CH_3COO^-]$ and $[CH_3COOH]$.

$$1.8 \times 10^{-5} = \frac{[H_3O^+](0.40)}{(0.20)}$$

$$[H_3O^+] = 9.0 \times 10^{-6} \qquad pH = 5.04$$

Understanding Calculate the pH of a solution that is 0.50 *M* hydrofluoric acid and 0.10 *M* sodium fluoride. K_a for HF is 3.5×10^{-4}.

Answer: pH = 2.76

In a buffer solution, the ratio of concentrations (mol/L) is the same as the ratio of moles.

The ratio of *concentrations* of the weak acid and conjugate base is used in the Henderson-Hasselbalch relationship. The concentration ratio is the same as the ratio of the numbers of moles (or mmoles) of acid and base. If we represent the analytical concentrations as

$$C_a = n_a/V$$

$$C_b = n_b/V$$

where n_a is the number of moles of acid and V the volume of solution, then

$$pH = pK_a + \log\frac{n_b/V}{n_a/V} = pK_a + \log\frac{n_b}{n_a}$$

As long as the Henderson-Hasselbalch equation applies, the pH of a buffer made from n_a moles of acid and n_b moles of base does not depend on volume. This concept is illustrated in the next example.

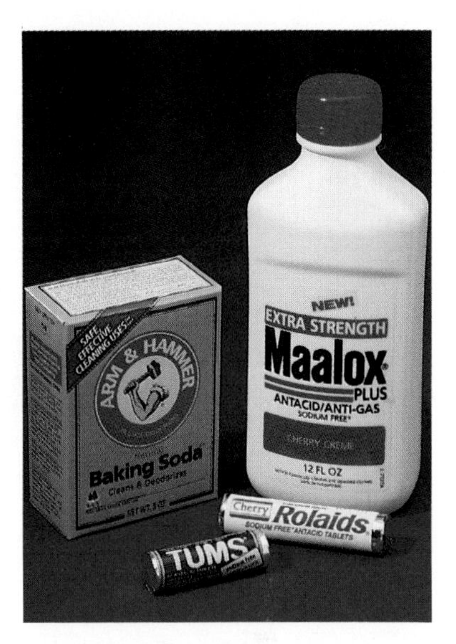

Some common buffers These products are all used to neutralize stomach acid; a buffer is preferred to a base to avoid the bitter taste of the base.

Example 14.5 Determining the Amounts of Acid and Base Needed to Prepare a pH Buffer

How many grams of ammonium chloride must be added to 500.0 mL of 0.32 M NH_3 to prepare a pH 8.50 buffer? K_b for NH_3 is 1.8×10^{-5}.

Solution

We have an acid-base equilibrium, with NH_4^+ as the acid and NH_3 as the base.

$$NH_4^+(aq) + H_2O(\ell) \rightleftharpoons NH_3(aq) + H_3O^+(aq)$$

$$pH = pK_a + \log\frac{C_b}{C_a} = pK_a + \log\frac{n_b}{n_a}$$

For NH_4^+,

$$K_a = \frac{K_w}{K_b} = \frac{1.0 \times 10^{-14}}{1.8 \times 10^{-5}} = 5.6 \times 10^{-10}$$

$$pK_a = 9.25$$

Calculate n_b:

$$n_b = 0.500 \text{ L} \times 0.32\ M = 0.16 \text{ mol}$$

Substitute into the Henderson-Hasselbalch equation:

$$8.50 = 9.25 + \log\frac{0.16}{n_a}$$

$$\log\frac{0.16}{n_a} = 8.50 - 9.25 = -0.75$$

$$\frac{0.16}{n_a} = 10^{-0.75} = 0.18$$

$$n_a = 0.89 \text{ mol}$$

$$\text{mass of } NH_4Cl = 0.89 \text{ mol}\left(\frac{53.49 \text{ g}}{\text{mol } NH_4Cl}\right) = 48 \text{ g } NH_4Cl$$

Understanding How many moles of sodium benzoate must be added to one liter of a 0.022 M solution of benzoic acid ($pK_a = 4.19$) to prepare a liter of pH 4.50 buffer?

Answer: 0.045 mol or 45 mmol

A buffer solution is prepared by mixing a weak acid and its conjugate base in the correct proportions, followed by addition of some concentrated strong acid or base if needed to adjust the pH to the exact value desired.

In general, we prepare buffer solutions that are relatively concentrated. If the buffer is dilute, the Henderson-Hasselbalch equation is no longer accurate and we must solve a quadratic (or even a cubic) equation, as shown in Appendix A.

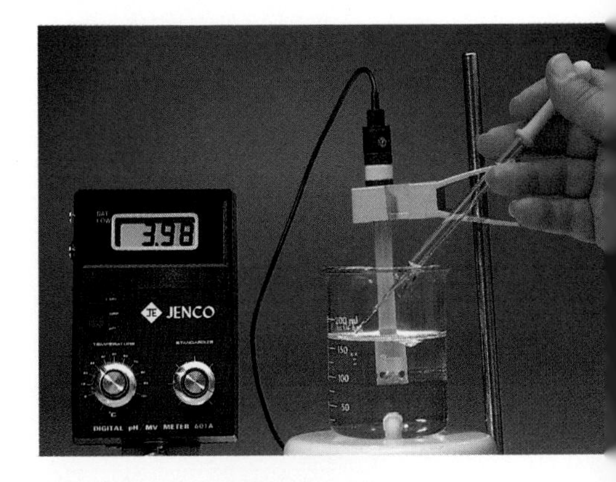

Preparation of a buffer solution The pH of a buffer is adjusted to the exact value by adding small amounts of concentrated solutions of strong acid or base.

The Resistance of a Buffer to Changes in pH

We will compare how the pH changes when we add some acid to two solutions of the same pH—one made from dilute HCl, the other from a mixture of acetic acid and sodium acetate.

Example 14.6 Determining the Change in pH When an Acid Is Added to a Buffer System

Calculate the change in pH observed when 1.00 mL of 0.100 M H_3O^+ solution is added to

(a) 100.0 mL of an HCl solution of pH 4.74.
(b) 100.0 mL of pH 4.74 buffer prepared from 0.100 M acetic acid and 0.100 M sodium acetate. pK_a for acetic acid is 4.74.

Solution (a)

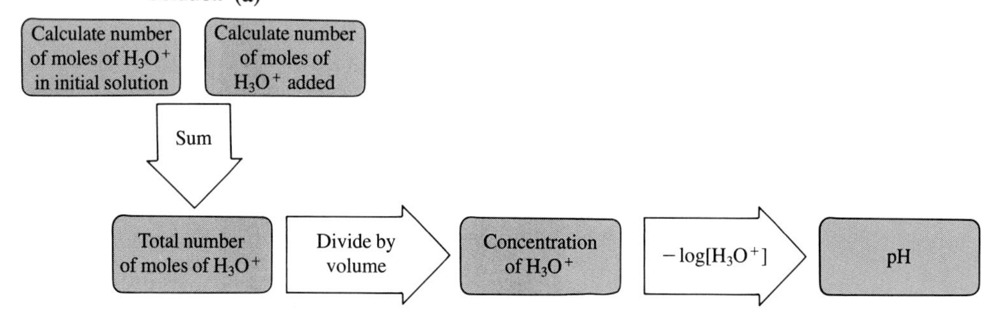

Consider first the 100.0 mL of HCl solution at pH 4.74. The H_3O^+ concentration is calculated from the pH:

$$[H_3O^+] = 10^{-4.74}\ M = 1.8 \times 10^{-5}\ M$$

The number of millimoles of acid in 100 mL of this solution is

$$\text{moles HCl} = \left(\frac{1.8 \times 10^{-5}\ \text{mmol}}{\text{mL of solution}}\right) \times 100.0\ \text{mL} = 1.8 \times 10^{-3}\ \text{mmol}$$

The quantity of additional acid is 1.00 mL of 0.10 M H_3O^+, which contributes

$$\text{added moles } H_3O^+ = \left(\frac{0.100\ \text{mmol}}{\text{mL of solution}}\right) \times 1.00\ \text{mL} = 1.00 \times 10^{-1}\ \text{mmol}$$

The *total* acid in solution is the sum of the two sources:

from the 100 mL of pH 4.74 acid	1.8×10^{-3} mmol
from the 1.00 mL of 0.10 M acid	1.00×10^{-1} mmol
total	1.02×10^{-1} mmol

Now calculate the concentration of the H_3O^+.

$$[H_3O^+] = \frac{1.02 \times 10^{-1}\ \text{mmol}}{101\ \text{mL}} = 1.01 \times 10^{-3}\ M$$

$$\text{pH} = 3.00$$

The pH of the solution decreases by 1.74 pH units when 1.00 mL of 0.100 M acid is added.

(b) When acid (or base) is added to a buffer, the buffer reacts to consume the added material, in accordance with the principle of Le Chatelier. In the acetic acid/acetate ion buffer, the acetate ion will react as a base with the added hydrogen ions.

$$H_3O^+(aq) + CH_3COO^-(aq) \longrightarrow CH_3COOH(aq) + H_2O(\ell)$$

Adding acid to base is, of course, a titration and we can use the sRfc table to determine the concentrations of the products.

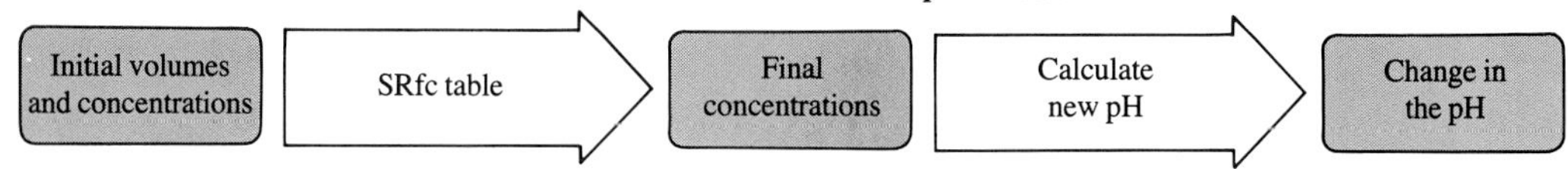

The starting amount of base present in the buffer is calculated from the concentration of acetate ion, 0.100 *M*, and the volume of buffer, 100.0 mL.

$$\text{amount of acetate ion} = 0.100\ M \times 100.0\ \text{mL} = 10.0\ \text{mmol}$$

The starting amount of acetic acid is the same. The amount of hydrogen ion added is 1.00 mL of 0.100 *M* solution, or 0.100 mmol. We set up the table:

	H_3O^+	$+\ CH_3COO^-$	$\longrightarrow CH_3COOH + H_2O$
s, mmol	0.100	10.0	10.0
R, mmol	−0.100	−0.100	+0.10
f, mmol	≈0.	9.9	10.1

We have found the number of millimoles of acetate ion (the base) and acetic acid after the addition of 0.100 mmol of H_3O^+; we next use the Henderson-Hasselbalch equation to calculate the pH.

$$\text{pH} = \text{p}K_a + \log\frac{C_b}{C_a} = 4.74 + \log\frac{9.9}{10.1}$$

$$\text{pH} = 4.73$$

The calculation illustrates how well the buffer resists changes in pH. The pH of the buffer changed by only 0.01 pH unit while the pH of the dilute solution of

Adding acid to buffered and unbuffered solutions (a) When a small amount of acid is added to a solution of HCl whose pH is 4.74 (a), the pH decreases to 3.00 (b). When the same amount of acid is added to the pH 4.74 acetic acid–acetate buffer system (c), the change is much less; the pH decreases by 0.01 to 4.73 (d).

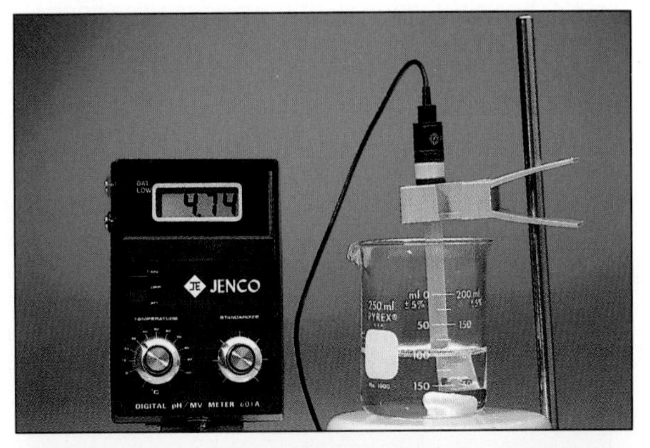

(a)

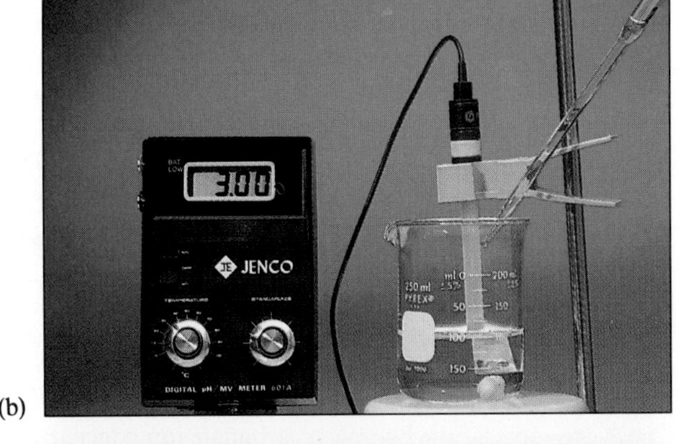

(b)

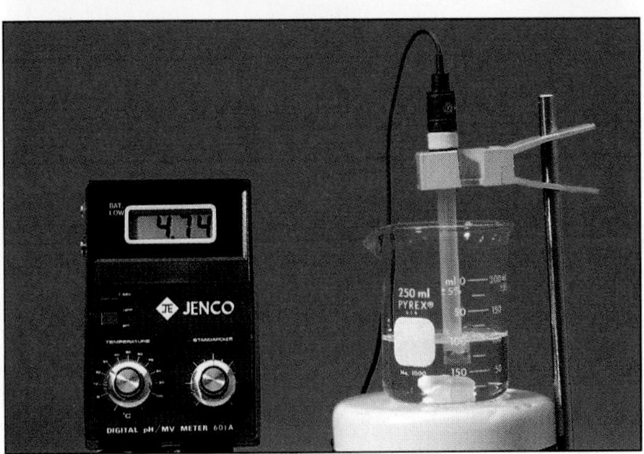

(c)

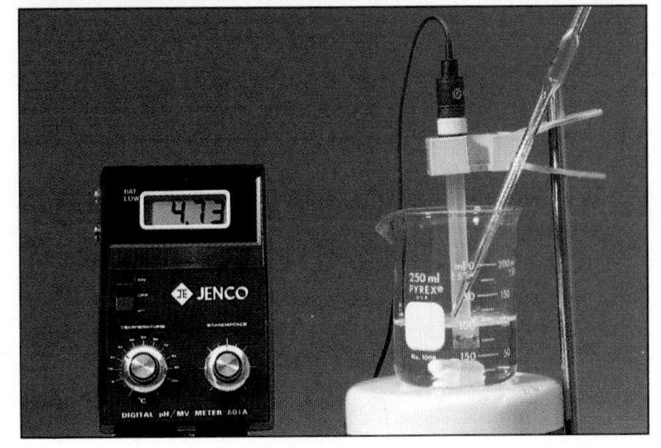

(d)

HCl changed by 1.74 pH units. This calculation helps clarify why chemists prepare buffers from relatively concentrated solutions (0.100 *M*, in this example) of weak acids and their conjugate bases—the concentrated solutions can absorb added hydrogen ions or hydroxide ions without significantly changing pH.

It is helpful to provide a qualitative check when doing these calculations. When acid is added to a solution, the pH must decrease, although the decrease is small in a well-buffered system.

INSIGHTS into CHEMISTRY

Chemical Reactivity

Blood as an Effective Buffer System, Neutralizing Excess Acids or Bases

One of the most important examples of a buffered system is our blood. Not only can it absorb the acids and bases produced in metabolic reactions, it also keeps the pH of blood constant even as we ingest foods with a wide range of acidity. The pH of blood ranges from 7.35 to 7.45 in healthy individuals. If the pH is outside the range of 6.7 to 7.8, the condition is life-threatening and the blood pH must be adjusted immediately.

If the pH of the blood is too low, the condition is called **acidosis;** when the pH is too high, the condition is labeled **alkalosis.**

The principal buffer system in blood is the hydrogen carbonate (bicarbonate) buffer. The hydrogen carbonate ion is amphoteric. It reacts with excess hydrogen ion (either generated or ingested):

$$H_3O^+(aq) + HCO_3^-(aq) \rightleftharpoons H_2CO_3(aq) + H_2O(\ell).$$

It also reacts with excess hydroxide ion:

$$OH^-(aq) + HCO_3^-(aq) \rightleftharpoons CO_3^{2-}(aq) + H_2O(\ell).$$

The buffering capacity depends on the concentration of bicarbonate ion, which is usually about 0.027 *M*. The concentration is kept relatively constant by chemical reactions in the kidneys and lungs. Reactions in the kidneys can increase the hydrogen carbonate ion concentration from dissolved carbonic acid:

$$H_2CO_3(aq) \rightleftharpoons H_3O^+(aq) + HCO_3^-(aq)$$

The excess hydrogen ions are excreted into the urine.

Hydrogen carbonate ion is consumed in the lungs—the blood gives up about 10% of its dissolved carbon dioxide (present as H_2CO_3, HCO_3^-, and CO_3^{2-}) as it passes through the lungs. If the concentration of hydrogen carbonate ion becomes too large, additional $CO_2(g)$ is expelled from the lungs.

The pH of blood is influenced by metabolic processes as well as by respiration. *Metabolic acidosis* is seen in severe diabetics, and temporarily in all individuals after heavy exercise. Exercise generates lactic acid in the muscles. Some of it ionizes ($K_a = 8.4 \times 10^{-4}$), and a large influx of hydrogen ions appears in the bloodstream. *Metabolic alkalosis* is less common and occurs mostly as a side effect of certain drugs that change the concentrations of sodium, potassium, and chloride ions in the blood.

Respiratory acidosis and *alkalosis* result from changes in breathing patterns. Lung diseases and obstructed air passages result in **hypoventilation,** the condition that results when breathing is too shallow and the amount of carbon dioxide removed from the blood by respiration drops. Since the buildup of CO_2 will decrease the pH, hypoventilation results in respiratory acidosis. Anesthesiologists and operating room personnel must be aware of this problem, since most general anesthetics also depress respiration.

Respiratory alkalosis is caused by **hyperventilation,** where the breathing is too deep and frequent. Too much CO_2 is expelled, and the blood becomes alkaline. Patients with this condition are instructed to breathe into a paper bag. The "re-breathing" of exhaled CO_2 will increase its concentration in the blood, making it more acidic. Hyperventilation often results from hysteria or anxiety, possibly induced by studying the material in this chapter.

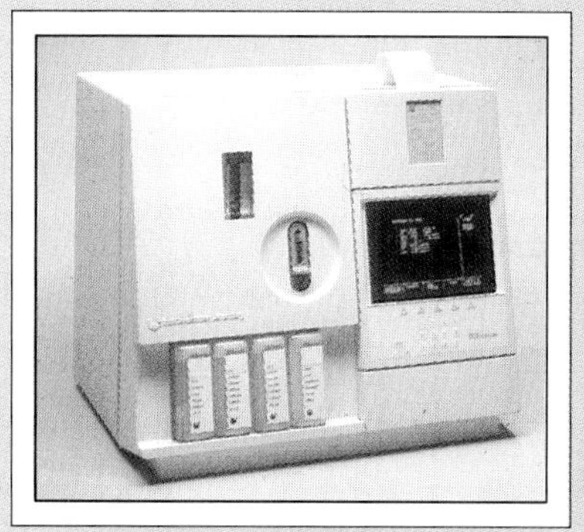

Blood pH and blood gas measurements Hospital labs are well equipped to make measurements of pH and dissolved CO_2 and O_2 in blood to help diagnose diseases.

Understanding Calculate the change in pH observed when 5.00 mL of 0.050 *M* NaOH is added to 100 mL of pH 4.74 acetate buffer (prepared from 0.100 *M* acetic acid and 0.100 *M* sodium acetate).

Answer: The pH increases 0.02 pH units to 4.76.

The **buffer capacity** is defined as the amount of strong acid or base needed to change the pH of one liter of buffer by 1 unit. The buffer capacity is a quantitative measure of the ability of the buffer to resist changes in pH and is directly proportional to the concentrations. A buffer that is 0.30 *M* HF and 0.30 *M* F^- has a capacity that is three times that of a buffer in which the concentrations are 0.10 *M* HF and 0.10 *M* F^- although the pH is the same in both solutions, since the ratio C_{HF}/C_{F^-} is the same.

Buffer solutions have wide applications, but certainly one of the most important is to maintain a constant pH for reactions that involve enzymes. Enzymes are complex proteins that help to metabolize food, store energy, and even synthesize the compounds that are needed to contract muscles, including our heart muscle. If the pH changes very much, most enzymes stop working. Since some enzyme reactions generate hydrogen or hydroxide ions, it is important that these ions be neutralized immediately, or the reaction products will "poison" or retard the reaction.

14.3 The Titration of Weak Acids or Bases

Objectives

- To determine the pH during the course of the titration of a weak acid with strong base
- To determine the pH during the course of the titration of a weak base with strong acid
- To sketch qualitative titration curves for acids of different strengths and concentrations

The titration of a weak acid with strong base is treated in the same way as is a strong acid, using an sRfc table to determine which species are important in the equilibrium calculation. The calculation of pH is a little more complicated, however, since the conjugate acids and bases formed in the titration reaction will influence the final pH.

We will use a two-part process to determine the pH during the course of the titration of a weak acid or base. First, we consider the stoichiometry of the titration reaction, which goes to completion, to find the concentrations of species before considering any equilibria. When we evaluate the results, we will know whether we have a strong or weak acid, strong or weak base, or a buffer solution; we must then choose the method best suited to calculate the pH of the solution. No new equilibrium calculations are involved when we consider titrations of weak acids or bases. The solutions generated in the course of the titration are solutions of strong or weak acids or bases, or buffer solutions; calculations of the pH of all these types of solutions have been illustrated previously.

Most titrations require two calculations. The first is a limiting reactant problem (the titration reaction), and the second involves an equilibrium among species formed by the titration.

We can illustrate the technique by calculating the pH during the titration of 25.00 mL of 0.500 *M* HCOOH (formic acid, $K_a = 1.8 \times 10^{-4}$) with 0.500 *M* NaOH. The pH will be calculated after 0, 12.50, 25.00, and 40.00 mL of base are added.

Before any titrant is added, the system is just a solution of a weak acid.

0 mL of Base Added The solution is 0.500 *M* formic acid—a solution of a weak acid ($K_a = 1.8 \times 10^{-4}$). Since we have not yet added any base, we solve for the pH in the same way as for any weak acid equilibrium system:

	$HCOOH(aq) + H_2O(\ell) \rightleftharpoons$	$H_3O^+(aq)$ +	$HCOO^-(aq)$
i, *M*	0.500	0	0
C, *M*	$-y$	$+y$	$+y$
e, *M*	$0.500 - y$	y	y

The relationships at equilibrium are substituted into the expression for K_a:

$$K_a = \frac{[H_3O^+][HCOO^-]}{[HCOOH]} = 1.8 \times 10^{-4}$$

$$\frac{y^2}{0.500 - y} = 1.8 \times 10^{-4}$$

We solve by approximation. If $y \ll 0.50$ then

$$y^2 \approx 1.8 \times 10^{-4} \times 0.50 = 9.0 \times 10^{-5}$$

$$y = 9.5 \times 10^{-3}$$

After we check that the assumption is valid, and that additional calculations are not needed, we can write

$$[H_3O^+] = 9.5 \times 10^{-3} \quad \text{and} \quad pH = 2.02$$

12.5 mL (Total) of Base Added First, prepare the sRfc table for the titration reaction. The titration is the addition of strong base to a weak acid, a reaction that goes to completion. We start with 25.00 mL of 0.500 *M* formic acid, a total of 12.5 mmol. We have added 12.50 mL of 0.500 *M* base, or 6.25 mmol.

	$HCOOH(aq) +$	$OH^-(aq) \longrightarrow$	$HCOO^-(aq) + H_2O(\ell)$
s, mmol	12.50	6.25	0.
R, mmol	−6.25	−6.25	+6.25
f, mmol	6.25	0.	6.25

The second step is to look at the results of the titration equilibrium and calculate the pH. Is there any strong acid or strong base? No, there is none, because all of the strong base was consumed. Next, determine whether there is any weak acid or base. We have both. There is some formic acid, a weak acid, and some formate ion, its conjugate base. This is a buffer solution. We can calculate the pH from the Henderson-Hasselbalch equation:

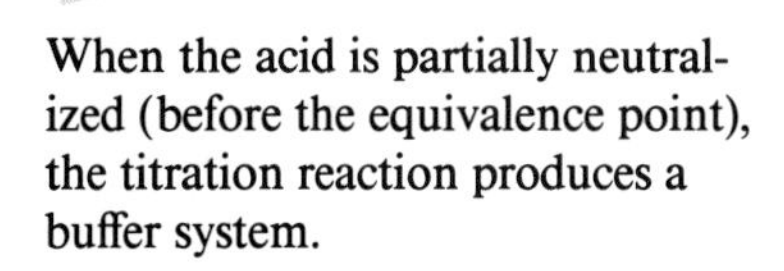

When the acid is partially neutralized (before the equivalence point), the titration reaction produces a buffer system.

$$\mathrm{pH} = \mathrm{p}K_a + \log\frac{n_a}{n_b} = \mathrm{p}K_a + \log\frac{6.25 \text{ mmol}}{6.25 \text{ mmol}}$$

$$= \mathrm{p}K_a + 0 = -\log(1.8 \times 10^{-4}) = 3.74$$

Note that when the concentration of the acid and its conjugate base are equal, pH is equal to $\mathrm{p}K_a$. This situation occurs when exactly half the acid has been neutralized. The pH is equal to the $\mathrm{p}K_a$ halfway to the equivalence point.

In a weak acid titration, the pH is equal to the $\mathrm{p}K_a$ halfway to the equivalence point.

The concentrations of the species are not needed for a buffer calculation, just the number of mmoles.

25.00 mL (Total) of Base Added Start by writing the titration reaction and evaluating the amounts and concentrations of species.

	$HCOOH(aq)$ +	$OH^-(aq) \longrightarrow$	$HCOO^-(aq) + H_2O(\ell)$
s, mmol	12.5	12.5	0.
R, mmol	−12.5	−12.5	+12.5
f, mmol	0.	0.	12.5
c, M	0.	0.	0.250

Enough strong base has been added to exactly neutralize the weak acid, so this condition is the equivalence point. Examine the results of the titration to determine what type of calculation is needed. There is no strong acid or strong base. There is no weak acid (the formic acid has been consumed), *but there is a weak base,* formate ion, produced by the titration. The solution of this kind of equilibrium was shown in Section 13.6, where we used the iCe table to help calculate the pH of a weak base.

A solution of a weak base results at the equivalence point in the titration of a weak acid.

	$HCOO^-(aq) + H_2O(\ell) \rightleftharpoons$	$HCOOH(aq)$ +	$OH^-(aq)$
i, M	0.250	0.	0.
C, M	$-y$	$+y$	$+y$
e, M	$0.250 - y$	y	y

Write the expression for K_b and substitute the relationships from the bottom line of the iCe table into the expression.

$$K_b = \frac{[HCOOH][OH^-]}{[HCOO^-]} = \frac{y^2}{0.250 - y}$$

We calculate K_b from

$$K_b = \frac{K_w}{K_a} = \frac{1.0 \times 10^{-14}}{1.8 \times 10^{-4}} = 5.6 \times 10^{-11}$$

Finally, solve

$$\frac{y^2}{0.250 - y} = 5.6 \times 10^{-11}$$

This is most easily solved by approximation. If $y \ll 0.250$ then

$$y^2 \approx 5.6 \times 10^{-11} \times 0.250$$

$$y = 3.7 \times 10^{-6}$$

After we check the approximation, we can write

$$y = [OH^-] = 3.7 \times 10^{-6}\ M$$

$$pOH = 5.43;\ pH = 8.57$$

The pH at the equivalence point of the titration of a weak acid is greater than 7.

Notice that *in the titration of a weak acid with a strong base, the solution at the equivalence point is alkaline,* not neutral.

40.00 mL (Total) of Base Added Again, start with the titration reaction, and assume it goes to completion. We have added 40.00 mL of 0.500 M NaOH (20.0 mmol of NaOH), so the total volume of the solution is 65.00 mL.

	$HCOOH(aq)$	$+\ OH^-(aq) \longrightarrow$	$HCOO^-(aq) + H_2O(\ell)$
s, mmol	12.5	20.0	0.
R, mmol	−12.5	−12.5	+12.5
f, mmol	0.	7.5	12.5
c, M	0.	0.12	0.192

The pH beyond the equivalence point is determined by the concentration of the excess strong base.

We have an excess of strong base, which determines the pH of the solution.

$$[OH^-] = 0.12\ M$$

$$pOH = 0.92;\ pH = 13.08$$

This calculation is identical to the one performed in the titration of a strong acid with strong base. After we reach the equivalence point, we have excess strong base. It does not matter whether we started with strong acid or weak acid—the acid has been consumed. The weak base can be neglected in comparison to the strong base.

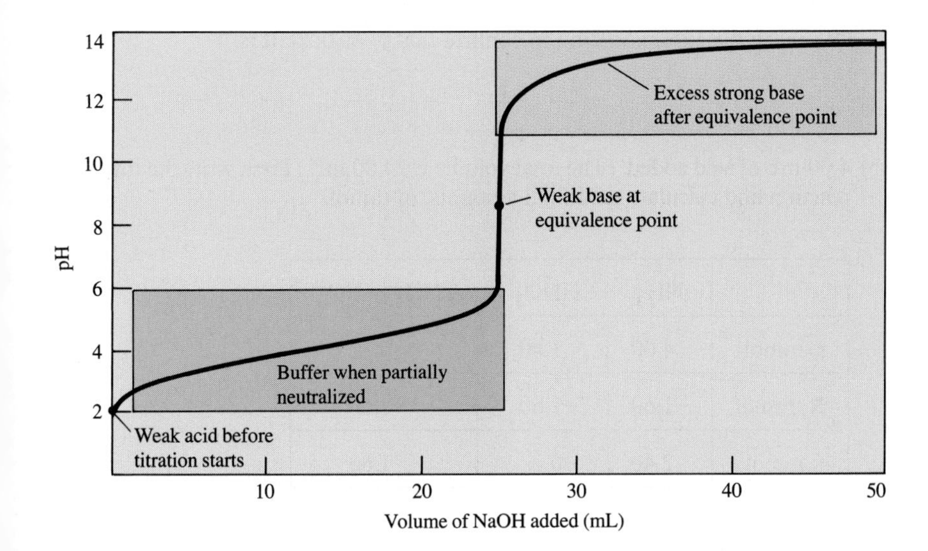

Figure 14.5 Titration curve for the titration of a weak acid with a strong base The graph shows the titration of 25.00 mL of 0.50 M formic acid with 0.500 M sodium hydroxide. The equilibria that determine the pH in the various regions are indicated.

The Titration Curve

We can plot the titration curve for the titration of a weak acid with a strong base. The formic acid titration just examined is shown in Figure 14.5. Notice that we can divide the titration curve into four zones, depending on the type of calculation that we perform.

The titration of a weak base with a strong acid is quite similar to that of a weak acid with a strong base. It is illustrated in Example 14.7.

Example 14.7 Titration of a Weak Base with Strong Acid

Calculate the pH during the titration of 20.00 mL of 0.200 M ammonia ($K_b = 1.8 \times 10^{-5}$) with 0.400 M nitric acid. Calculate the pH after 0, 4.00, 10.00, and 11.00 mL of acid have been added.

Solution

(a) 0 mL of acid added. (The total volume is 20.00 mL.) Before any acid is added, the solution is that of a weak base.

	$NH_3 + H_2O \rightleftharpoons$	NH_4^+ +	OH^-
i, M	0.200	0.	0.
C, M	$-y$	$+y$	$+y$
e, M	$0.200 - y$	y	y

$$K_b = \frac{[NH_4^+][OH^-]}{[NH_3]} = \frac{y^2}{0.20 - y} = 1.8 \times 10^{-5}$$

If $y \ll 0.200$ then

$$y^2 \approx 0.36 \times 10^{-5}$$

$$y = 1.9 \times 10^{-3}\ M$$

We check the approximation to confirm that $y \ll 0.20$. It is.

$$[OH^-] = 1.9 \times 10^{-3}$$

$$pOH = 2.72; \; pH = 11.28$$

(b) 4.00 mL of acid added. (The total volume is 24.00 mL.) First, write the titration reaction and calculate the initial amounts, in mmol.

	NH_3 +	HNO_3 $\longrightarrow$	$NH_4^+ + NO_3^-$
s, mmol	4.00	1.60	0
R, mmol	−1.60	−1.60	+1.60
f, mmol	2.40	0	1.60

After this reaction has gone to completion, the solution contains both the weak base and its conjugate acid, so it is a buffer solution. We calculate pK_a from pK_b and then use the Henderson-Hasselbalch relationship to calculate the pH.

$$pK_a = pK_w - pK_b = 14.00 - 4.74 = 9.26$$

$$pH = pK_a + \log \frac{C_b}{C_a} = 9.26 + \log \frac{2.4}{1.6} = 9.44$$

(c) 10.00 mL of acid added. (The total volume is 30.00 mL.)

	NH_3 +	HNO_3 $\longrightarrow$	$NH_4^+ + NO_3^-$
s, mmol	4.00	4.00	0
R, mmol	−4.00	−4.00	+4.00
f, mmol	0	0	4.00
c, *M*	0	0	0.133

The concentration of the ammonium ion is calculated from the amount formed in the titration (4.00 mmol) and the total volume of solution (30.00 mL). Ammonium ion is a weak acid. K_a for ammonium ion is calculated from K_b for ammonia, and has the value 5.6×10^{-10}.

	$NH_4^+ + H_2O \rightleftharpoons$	H_3O^+ +	NH_3
i, *M*	0.133	0	0
C, *M*	$-y$	$+y$	$+y$
e, *M*	$0.133 - y$	y	y

$$K_a = \frac{[NH_3][H_3O^+]}{[NH_4^+]} = \frac{y^2}{0.133 - y} = 5.6 \times 10^{-10}$$

If $y \ll 0.133$ then

$$y^2 \approx 5.6 \times 10^{-10} \times 0.133 = 7.4 \times 10^{-11}$$

$$y = 8.6 \times 10^{-6}$$

The assumption is checked and verified.

$$[H_3O^+] = 8.6 \times 10^{-6}$$

$$pH = 5.06$$

(d) 11.00 mL of acid added. (The total volume is 31.00 mL.)

	NH_3 +	HNO_3 ⟶	$NH_4^+ + NO_3^-$
s, mmol	4.00	4.40	0
R, mmol	−4.00	−4.00	+4.00
f, mmol	0	0.40	4.00
c, *M*		0.013	0.129

This is a solution in which we have a strong acid (HNO_3) and a weak acid (NH_4^+) in 31.00 mL of solution. We can ignore the weak acid in the presence of strong acid.

$$[H_3O^+] = 0.013\ M$$

$$pH = 1.89$$

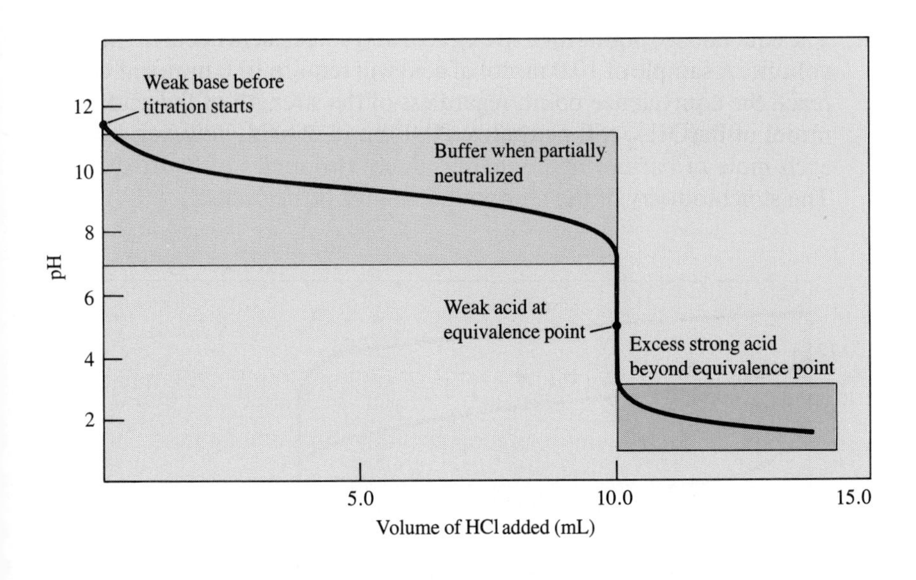

Titration curve Graph of pH as a function of concentration for the titration of 20.00 mL of 0.200 *M* ammonia with 0.400 *M* nitric acid. The equilibria that dominate in the various regions are shown.

Understanding Calculate the pH values after the addition of 0, 10.00, 20.00, and 30.00 mL of 0.500 *M* hydrochloric acid to 20.00 mL of 0.500 *M* methylamine ($K_b = 3.7 \times 10^{-4}$).

Answer: 0 mL, pH = 12.13; 10.00 mL, pH = 10.57; 20.00 mL (equivalence point), pH = 5.58; 30.00 mL, pH = 1.00

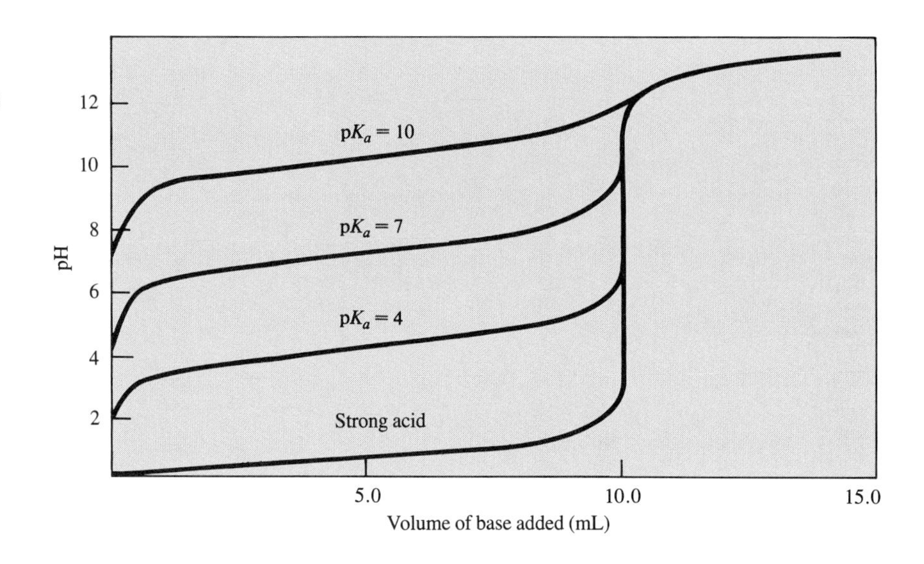

Figure 14.6 Titration curves for acids of different strengths Titration curves for 10.00 mL of 1.00 *M* acids with 1.00 *M* strong base. Note that the change in pH near the equivalence point is greater for the stronger acids.

Shapes of Titration Curves

The shape of an acid-base titration curve is determined by the concentrations and the strengths of the acid and base in the titration. Figure 14.6 shows calculated titration curves for several acids ("strong", $pK_a = 4$, $pK_a = 7$, and $pK_a = 10$), all titrated with strong base.

First, compare the curve of the strong acid with that of the one with $pK_a = 4$. Notice several points:

1. The equivalence points for a strong acid and a weak acid occur at the same volume. A sample of 10.0 mmol of acid will require 10.0 mmol of base to reach the equivalence point, regardless of the strength of the acid. (10.0 mmol of $Ba(OH)_2$ will neutralize 20.0 mmol of acid, however, because each mole of barium hydroxide produces *two* moles of hydroxide ion. The stoichiometry of the compound cannot be neglected.)

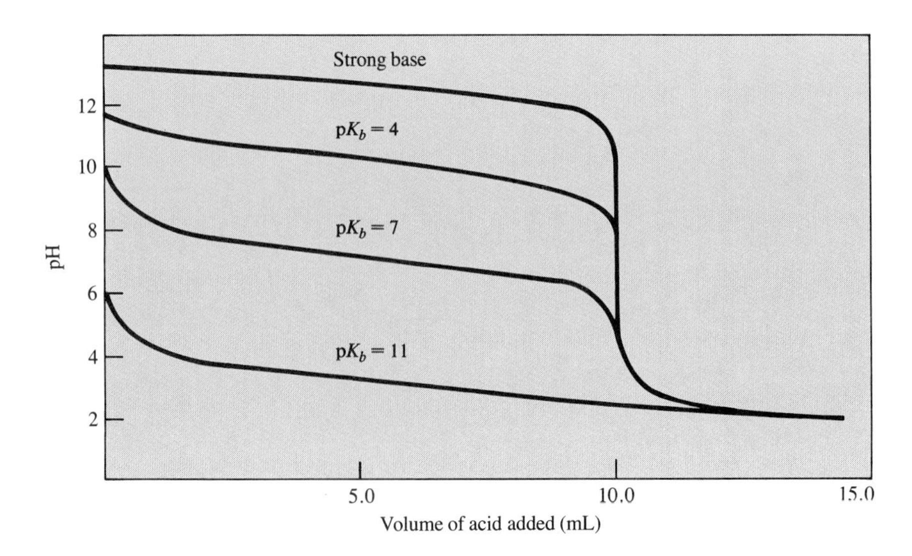

Figure 14.7 Titration curves for bases of different strengths Stronger bases give sharper inflections. A very weak base (such as one with $pK_b = 11$) does not show any inflection.

2. The titration curves for all the acids are indistinguishable beyond the equivalence point. The pH of the solution is determined by the excess hydroxide ion, not the acid.
3. The pH halfway to the equivalence point in the titration of a weak acid is equal to pK_a. The abrupt change in pH near the equivalence point becomes smaller as the acid strength decreases.
4. The pH at the equivalence point is not 7.0, unless both the acid and base are strong. (It is more accurate to say that the pH at the equivalence point is 7.0 only if the acid and base are equal in strength.) As the strength of the acid decreases, the pH at the equivalence point increases.

Figure 14.7 shows curves for the titrations of several bases. The similarity to the curves for acids is clear.

14.4 Indicators

Objectives

- To describe the acid-base chemistry of indicators
- To determine how to choose an indicator and what factors affect the color change of an indicator

We mentioned indicators when we discussed titrations in Section 14.3. The *equivalence point* in a titration occurs when an acid and base are present in stoichiometrically equivalent amounts, while the *endpoint* is the point in the titration at which a color change occurs. The **indicator** is the substance that changes color to signal the end of the titration. If we choose the indicator properly, the endpoint and equivalence point occur simultaneously, within experimental error. In this section we will present some of the properties of indicators and some guidelines for their selection.

We would like our indicator to have several properties:

1. The indicator must change color as a function of pH. The change should be abrupt, rather than drawn out, and the reaction that causes the change in color must occur rapidly. A color change that requires several minutes would not be practical for indicating the equivalence point in a titration.
2. The color change should be easily discerned by eye. A dramatic color change, for example from red to green or from colorless to blue, is easy to detect. A subtle change, say from blue-gray to gray-green, is difficult to detect and to reproduce.
3. The indicator must not change the properties of the solution.

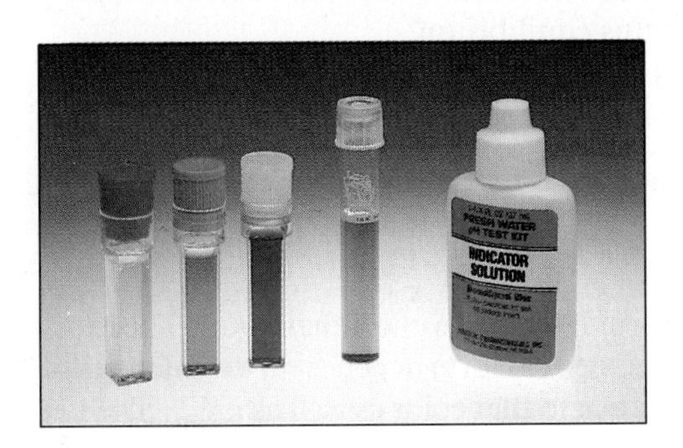

Bromthymol blue at various pH values Bromthymol blue changes from yellow in acid solution to blue in basic solution, with a gray-green color at pH 7. The color of the unknown solution in the tall container matches that of the pH 7 solution. Bromthymol blue is the indicator used in pH testing kits for tropical fish aquariums.

Acid form, colorless

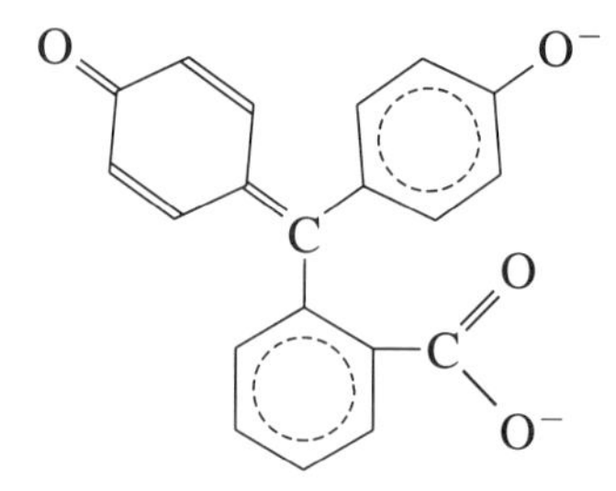

Basic form, pink

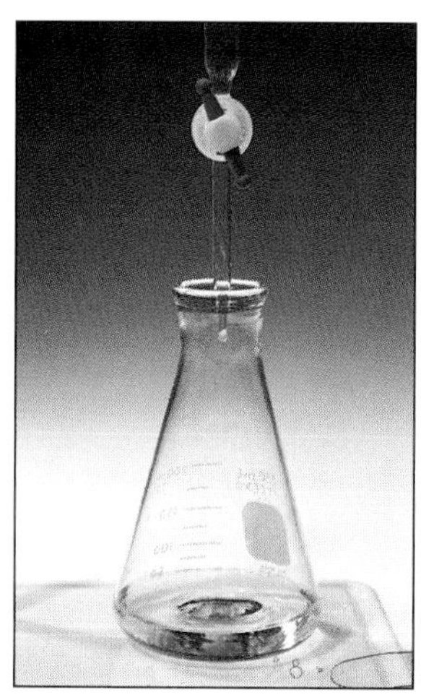

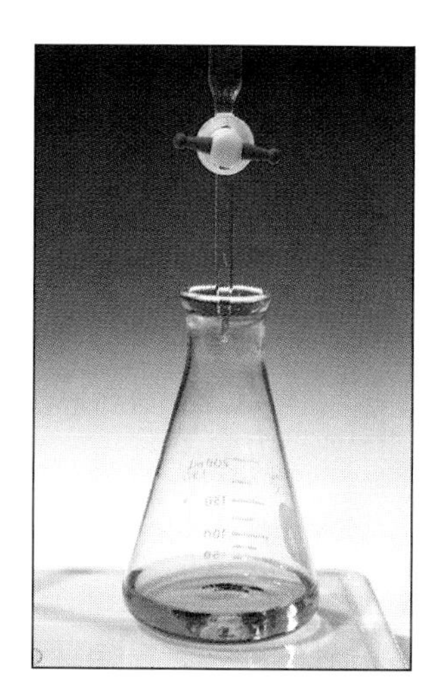

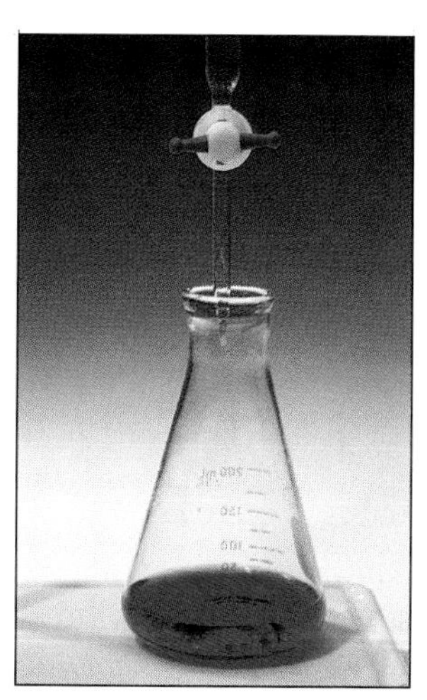

Phenolphthalein The drawings show the structure of the acid and base form of phenolphthalein, an indicator commonly used in the titration of an acid with base. When base is added, the solution turns pink, but the color disappears with swirling. The first permanent pink, shown in the third photo, indicates the endpoint. The fourth photo shows the solution after the base is in excess.

Most common indicators are weak acids or bases. We can represent an indicator molecule in the acid form as HIn, and in the base (deprotonated) form as In^-. The relationships between the two forms of the indicator are the same as for any other weak acid-base conjugate pair, but the indicator is special in that the acid form and base form have different colors, even in very dilute solutions:

$$HIn + H_2O \rightleftharpoons H_3O^+ + In^-$$

$$K_{In} = \frac{[H_3O^+][In^-]}{[HIn]}$$

where K_{In} is the acid ionization constant for the indicator. Most often, pK_{In} $(= -\log K_{In})$ is used to describe this equilibrium.

If the solution is acidic, that is, the pH is much lower than pK_{In}, most of the indicator will be in the acid form. If the pH is much higher than pK_{In}, the base form will predominate.

Choosing the Proper Indicator

An indicator should be chosen to change color at or just beyond the equivalence point.

The indicator will be in the acid form if the pH is lower than pK_{In}. The exact pH at which we observe the color of the acid form depends on the intensity of thc color and the sensitivity of our eye to that color as well as pK_{In}. The pH properties of several indicators are presented in Table 14.2.

Table 14.2 Properties of Several Indicators

Name	Acid Color	Alkaline Color	pH Range	pK_{In}
thymol blue*	red	yellow	1.2–2.8	1.65
methyl orange	red	yellow	3.1–4.4	3.46
methyl red	red	yellow	4.2–6.3	5.00
bromthymol blue	yellow	blue	6.2–7.6	7.30
phenolphthalein	clear	pink	8.3–10.0	8.7
thymol blue*	yellow	blue	8.0–9.6	9.20

* Thymol blue is a diprotic weak acid and has two color changes.

Figure 14.8 shows titration curves for hydrochloric acid and acetic acid with strong base, and the color changes of two indicators. The error is quite large if we use an indicator that changes color *before* the equivalence point, such as methyl red, in the titration of a weak acid. The indicator should be chosen to change color at or slightly after the equivalence point to avoid large errors. When titrating an unknown acid, choose an indicator that changes color in the pH region from 8 to 9; this choice will introduce negligible error for all but the very weakest acids.

The indicator needs to be intensely colored. We do not want to add too much indicator to the solution, since indicators are acids or bases and react with the sample or titrant.

Good technique will often require adding the indicator to water and then adding a drop of titrant to establish the reference color at which the titrations are stopped. Titrations can produce extremely accurate and precise results when performed properly.

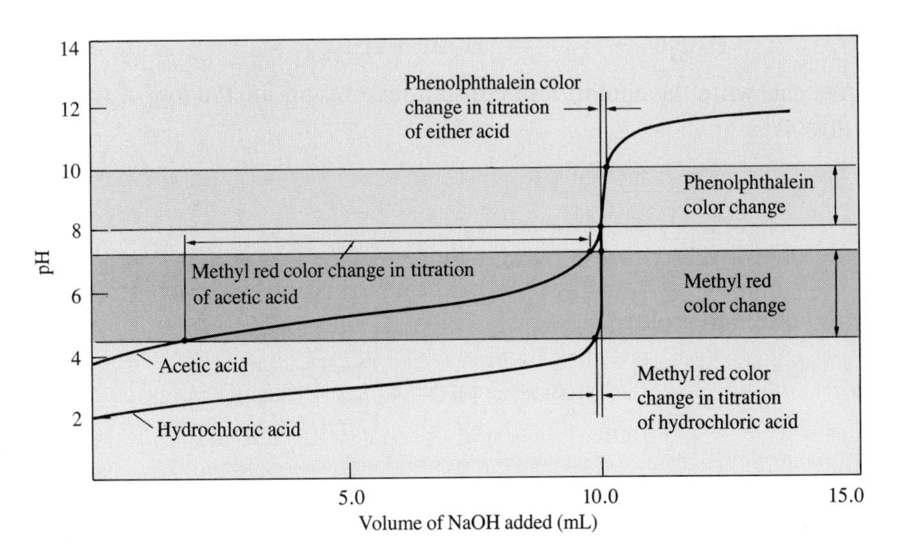

Figure 14.8 Titration curves for strong and weak acids The indicator may not change color exactly at the equivalence point. A substantial error can occur if you choose an indicator that changes color before the equivalence point of a weak acid.

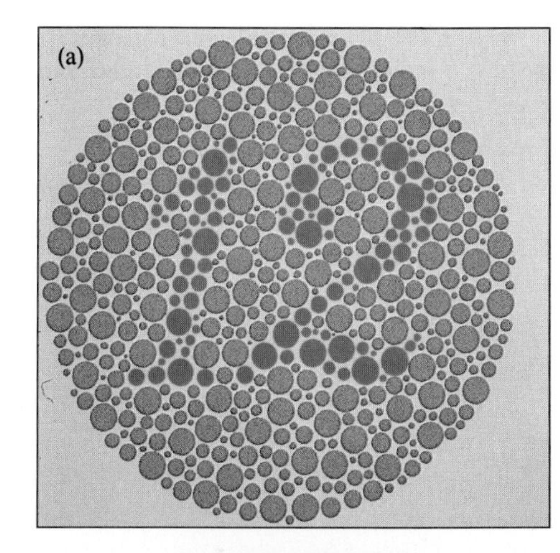

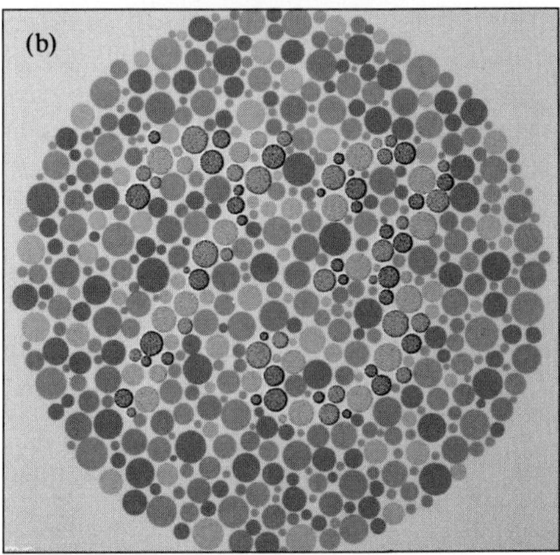

Does everyone see the same?
(a) Both people with normal vision and those with color deficiencies should be able to see the figure 12 in this picture. (b) People with normal vision will see a number here. Individuals with red-green color blindness, the most common deficiency, will see only a pattern of random dots. Color-blind chemists should not despair, because titrations can use an electronic device such as a pH meter to help determine the equivalence point. Several researchers have published plans for a digital pH meter that has an audio output with a computerized voice. These devices can enable vision-impaired students to acquire meaningful laboratory experiences.

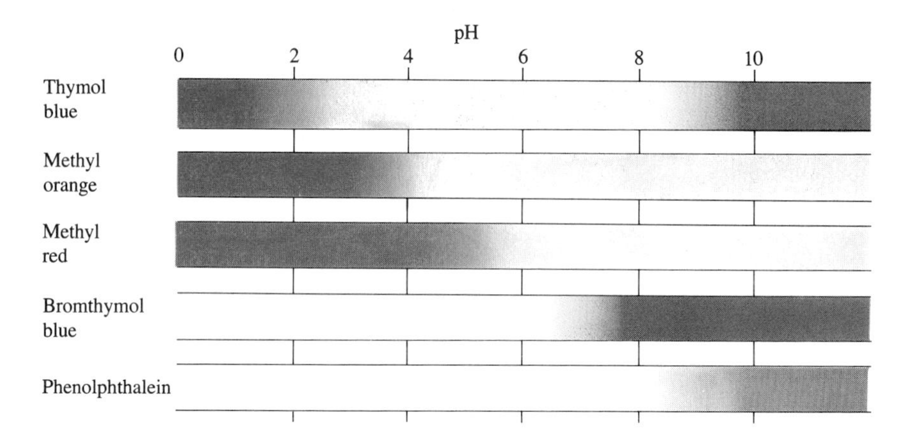

The pH ranges of some common indicators These are the indicators listed in Table 14.2.

14.5 Polyprotic Acids

Objectives

- To calculate the pH of solutions of polyprotic acids
- To calculate K_b from K_a for a polyprotic acid

Some acids, called **polyprotic acids,** can provide more than one proton when they ionize. Two common examples are phosphoric acid, H_3PO_4, and sulfuric acid, H_2SO_4. In this section we will briefly discuss polyprotic acids and methods used to calculate the pH of solutions of polyprotic acids.

Consider the first ionization of the diprotic oxalic acid, $H_2C_2O_4$:

$$\begin{matrix} O{=}C{-}O{-}H \\ | \\ O{=}C{-}O{-}H \end{matrix} + H_2O \rightleftharpoons H_3O^+ + \begin{matrix} O{=}C{-}O{-}H \\ | \\ O{=}C{-}O^- \end{matrix}$$

or

$$H_2C_2O_4 + H_2O \rightleftharpoons H_3O^+ + HC_2O_4^-$$

We can write the equilibrium constant expression for the loss of the first proton.

$$K_{a1} = \frac{[H_3O^+][HC_2O_4^-]}{[H_2C_2O_4]}$$

We use the subscript 1 to indicate the loss of the first proton.

Similarly, we can write the equilibrium constant expression for the loss of the second proton.

$$\begin{matrix} O{=}C{-}O{-}H \\ | \\ O{=}C{-}O^- \end{matrix} + H_2O \rightleftharpoons H_3O^+ + \begin{matrix} O{=}C{-}O^- \\ | \\ O{=}C{-}O^- \end{matrix}$$

or

$$HC_2O_4^- + H_2O \rightleftharpoons H_3O^+ + C_2O_4^{2-}$$

$$K_{a2} = \frac{[H_3O^+][C_2O_4^{2-}]}{[HC_2O_4^-]}$$

The values for K_{a1} and K_{a2} are 5.9×10^{-2} and 6.4×10^{-5}, respectively. We see that K_{a1} is larger than K_{a2}. This situation is normal, because the loss

Table 14.3 Ionization Constants of Polyprotic Acids

Name	Formula	K_{a1}	K_{a2}	K_{a3}
ascorbic	$H_2C_6H_6O_6$	8.0×10^{-5}	1.6×10^{-12}	
carbonic	H_2CO_3	4.3×10^{-7}	5.6×10^{-11}	
citric	$H_3C_6H_5O_7$	7.4×10^{-4}	1.7×10^{-5}	4.0×10^{-7}
malic	$H_2C_4H_4O_5$	4.0×10^{-4}	9.0×10^{-6}	
malonic	$H_2C_3H_2O_4$	1.6×10^{-2}	2.1×10^{-6}	
oxalic	$H_2C_2O_4$	5.9×10^{-2}	6.4×10^{-5}	
sulfurous	H_2SO_3	1.7×10^{-2}	6.4×10^{-8}	
sulfuric	H_2SO_4	strong	1.2×10^{-2}	
phosphoric	H_3PO_4	7.5×10^{-3}	6.2×10^{-8}	2.2×10^{-13}
tartaric	$H_2C_4H_4O_6$	1.0×10^{-3}	4.6×10^{-5}	

Values taken from the CRC *Handbook of Chemistry and Physics,* 59th edition. Copyright CRC, used with permission.

of a proton forms a negatively charged conjugate base, which is also the acid in the second ionization. It is harder to lose another proton, since the ionization process involves separating the positively charged proton from the negative $C_2O_4^{2-}$ species. If the acid were triprotic, it would be harder still to remove the third proton.

Some common polyprotic acids along with their ionization constants are listed in Table 14.3.

When successive K_a values for an acid differ by a factor of 1000 or more, the equilibria of the polyprotic acid can be treated as a series of steps, or successive equilibria. Nearly all of the acid molecules will lose their most easily ionized proton before any of them loses the second one.

We can see that a polyprotic acid system is analogous to that of a mixture of strong and weak acids. We can approach polyprotic acids in the same manner, disregarding the ionization of the weaker species with respect to the stronger, as used for mixtures of strong and weak acids. This approach is illustrated in Example 14.8.

Example 14.8 The pH of a Polyprotic Acid Solution

Calculate the concentrations of all species in 0.500 *M* sulfurous acid.

Solution We obtain the values of K_{a1} and K_{a2} for H_2SO_3 from Table 14.3. Since $K_{a1} \gg K_{a2}$, the dominant equilibrium will be the first ionization:

$$H_2SO_3(aq) + H_2O(\ell) \rightleftharpoons H_3O^+(aq) + HSO_3^-(aq)$$

$$K_{a1} = \frac{[H_3O^+][HSO_3^-]}{[H_2SO_3]} = 1.7 \times 10^{-2}$$

We can set up the iCe table:

	$H_2SO_3 + H_2O \rightleftharpoons$	H_3O^+ +	HSO_3^-
i, *M*	0.500	0.	0.
C, *M*	$-y$	$+y$	$+y$
e, *M*	$0.500 - y$	y	y

Substitute these relationships into the equilibrium constant expression:

$$1.7 \times 10^{-2} = \frac{y^2}{0.500 - y}$$

If $y \ll 0.5$, then

$$1.7 \times 10^{-2} \times 0.500 \approx y^2$$

$$y = 9.2 \times 10^{-2}$$

We check our approximation, and we find that 9.2×10^{-2} is *not* negligible (less than 5%) with respect to 0.500. We have two options: we can solve the quadratic equation or we can make a second approximation. We choose the successive approximation method, explained in more detail in Appendix A.

Rather than assuming that the denominator is 0.500 (neglecting y with respect to 0.500), we make a *second* approximation and assume that y is about equal to 9.2×10^{-2}, the result of our first calculation. We will calculate a new value of y, using $(0.500 - 9.2 \times 10^{-2})$ in the denominator.

$$1.7 \times 10^{-2} = \frac{y^2}{0.500 - 9.2 \times 10^{-2}}$$

$$y = 8.3 \times 10^{-2}$$

To determine whether the second approximation is acceptable, we ask whether it is within 5% of the first one. If the result is between 95% and 105% of our estimate, we consider the approximation acceptable.

Is 8.3×10^{-2} within 5% of 9.2×10^{-2}? No, 95% of 9.2×10^{-2} is 8.74×10^{-2}, so our approximation cannot be accepted. A third iteration is needed, using 8.3×10^{-2} for y.

$$1.7 \times 10^{-2} = \frac{y^2}{0.500 - 8.3 \times 10^{-2}}$$

$$y = 8.4 \times 10^{-2}$$

The third value, 8.4×10^{-2}, is within 5% of 8.3×10^{-2}, so the third iteration can be accepted. Solving the quadratic equation would have yielded the same result.

We finish the iCe table to calculate the concentrations of all the species present:

$$[H_3O^+] = y = 8.4 \times 10^{-2}\ M$$

$$[HSO_3^-] = y = 8.4 \times 10^{-2}\ M$$

$$[H_2SO_3] = 0.500 - y = 0.42\ M$$

We use these concentrations and the equilibrium expression for the ionization of the second proton to solve for the concentration of SO_3^{2-}.

	$HSO_3^- + H_2O \rightleftharpoons$	H_3O^+	$+\ SO_3^{2-}$
i, M	8.4×10^{-2}	8.4×10^{-2}	0.
C, M	$-z$	$+z$	$+z$
e, M	$8.4 \times 10^{-2} - z$	$8.4 \times 10^{-2} + z$	z

$$K_{a2} = \frac{[H_3O^+][SO_3^{2-}]}{[HSO_3^-]} = 6.4 \times 10^{-8}$$

$$6.4 \times 10^{-8} = \frac{(8.4 \times 10^{-2} + z)(z)}{8.4 \times 10^{-2} - z}$$

If $z \ll 8.4 \times 10^{-2}$ then

$$z = [SO_3^{2-}] = 6.4 \times 10^{-8}\ M$$

(The approximation is accepted.)

The second ionization does not change the concentrations of the other species in solution much at all.

Understanding Calculate the pH of 0.033 *M* carbonic acid. (This concentration corresponds to a solution that is saturated with CO_2 at one atmosphere.)

The pH is 3.92.

Determination of K_b from K_a

Polyprotic acids, like other weak acids, have conjugate bases that react with water to form hydroxide ions. Consider carbonic acid, a compound that adds much of the appeal to carbonated beverages. We can write two acid ionization equilibria.

$$H_2CO_3(aq) + H_2O(\ell) \rightleftharpoons H_3O^+(aq) + HCO_3^-(aq) \quad K_{a1} = \frac{[H_3O^+][HCO_3^-]}{[H_2CO_3]}$$

$$HCO_3^-(aq) + H_2O(\ell) \rightleftharpoons H_3O^+(aq) + CO_3^{2-}(aq) \quad K_{a2} = \frac{[H_3O^+][CO_3^{2-}]}{[HCO_3^-]}$$

We can also write two equilibria that describe conjugate base forms of carbonic acid reacting with water.

$$CO_3^{2-}(aq) + H_2O(\ell) \rightleftharpoons HCO_3^-(aq) + OH^-(aq) \quad K_{b1} = \frac{[HCO_3^-][OH^-]}{[CO_3^{2-}]}$$

$$HCO_3^-(aq) + H_2O(\ell) \rightleftharpoons H_2CO_3(aq) + OH^-(aq) \quad K_{b2} = \frac{[H_2CO_3][OH^-]}{[HCO_3^-]}$$

It is important to note that the first base dissociation constant is related to the second acid ionization constant, and so on.

$$K_{a1}K_{b2} = K_w$$

$$K_{a2}K_{b1} = K_w$$

The relationship $K_aK_b = K_w$ holds for acid-base conjugate pairs.

We use the acid ionization constants for carbonic acid from Table 14.3 to determine K_b for carbonate and hydrogen carbonate ion.

$$K_{b1} = \frac{K_w}{K_{a2}} = \frac{1.0 \times 10^{-14}}{5.6 \times 10^{-11}} = 1.8 \times 10^{-4}$$

$$K_{b2} = \frac{K_w}{K_{a1}} = \frac{1.0 \times 10^{-14}}{4.3 \times 10^{-7}} = 2.3 \times 10^{-8}$$

We can see that the hydrogen carbonate ion, HCO_3^-, is an amphoteric species. It can act as an acid:

$$HCO_3^- + H_2O \rightleftharpoons CO_3^{2-} + H_3O^+ \qquad K_{a2} = 5.6 \times 10^{-11}$$

It can also act as a base:

$$HCO_3^- + H_2O \rightleftharpoons H_2CO_3 + OH^- \qquad K_{b2} = 2.3 \times 10^{-8}$$

Determining the pH of solutions of amphoteric species is the topic of the next section.

14.6 Amphoteric Species

Objective

- To estimate the pH in solutions of amphoteric species

Substances that have the properties of both acids and bases are called **amphoteric,** from the Greek *amphoteros,* meaning "both." In this section, amphoteric species and their reactions are described. Methods used to estimate the pH of solutions of amphoteric species are presented.

At the end of the last section, we determined the values of K_a and K_b for hydrogen carbonate ion.

$$K_a = 5.6 \times 10^{-11}$$

$$K_b = 2.3 \times 10^{-8}$$

To estimate the pH of a solution of an amphoteric substance, compare K_a and K_b. If K_a is greater than K_b, the solution will be acidic.

The pH of a solution of hydrogen carbonate ion can be estimated by examining the magnitudes of K_a and K_b. For hydrogen carbonate ion, K_b is larger than K_a. Although it is a better base than an acid, the acidic character of the hydrogen carbonate ion cannot be ignored. A solution of sodium hydrogen carbonate is slightly alkaline; measurements show that the pH of a 0.10 *M* solution is 8.30.

Example 14.9 Amphoteric Acid-Base Systems

Write the equilibria and determine whether the pH of 0.10 *M* sodium hydrogen sulfite is greater than or less than 7.0.

Solution The reaction of HSO_3^- as an acid is:

$$HSO_3^- + H_2O \rightleftharpoons H_3O^+ + SO_3^{2-}$$

$$K_a = K_{a2} = 6.4 \times 10^{-8}$$

As a base:

$$HSO_3^- + H_2O \rightleftharpoons H_2SO_3 + OH^-$$

$$K_b = \frac{K_w}{K_{a1}} = \frac{1.0 \times 10^{-14}}{1.7 \times 10^{-2}} = 5.9 \times 10^{-13}$$

The solution will be acidic because the HSO_3^- ion is a stronger acid than it is a base.

Understanding Write the equilibrium constants for a solution that is 0.10 *M* sodium hydrogen ascorbate, and determine whether the solution is acidic or basic.

Answer: $K_a = 1.6 \times 10^{-12}$; $K_b = 1.25 \times 10^{-10}$. The solution will be slightly basic.

Many amphoteric species are oxides and hydroxides that react with both acids and bases. These materials form complex ions with several (often four) hydroxide ions. (Complex ions are discussed in the next section and in more detail in Chapter 18.) Aluminum hydroxide is a typical amphoteric substance. Solid aluminum hydroxide dissolves in solutions of hydroxide ion, as well as in acids.

$$Al(OH)_3(s) + OH^-(aq) \rightleftharpoons Al(OH)_4^-(aq)$$

$$Al(OH)_3(s) + 3H_3O^+(aq) \rightleftharpoons Al^{3+}(aq) + 6H_2O(\ell)$$

The amphoteric nature of many species provides us with a method to separate substances from mixtures. For example, iron and calcium can be separated from an aluminum ore as part of a purification process. The ore can be added to a strongly alkaline solution. The aluminum will go into solution as the $Al(OH)_4^-$ species, but the iron and calcium form insoluble hydroxides that do not dissolve in alkaline solution.

14.7 Factors That Influence Solubility

Objectives

- To determine how pH influences the solubility of precipitates
- To determine the effect of complex formation on the solubility of precipitates

The solubility of a salt can be influenced by the acid-base properties of the anions and cations from which it is composed. If the anion or cation reacts with hydrogen or hydroxide ion, then the solubility will be influenced by pH. The two important equilibria, the solubility equilibrium and the acid-base reaction, occur at the same time, so we refer to them as simultaneous equilibria.

Salts of Anions of Weak Acids

When a salt is formed from an anion of a weak acid, then the acid-base properties of the anion influence the solubility. We can consider the effect of pH on the solubility of barium fluoride, BaF_2. The solid is a fluoride salt, and fluoride is a base (the conjugate base of HF). We need to consider two steps: (1) the dissolving of the salt and (2) the reaction of the fluoride anion with hydrogen ion:

$$BaF_2(s) \rightleftharpoons Ba^{2+}(aq) + 2F^-(aq)$$

$$F^-(aq) + H_3O^+(aq) \rightleftharpoons HF(aq) + H_2O(aq)$$

The first equation describes the dissociation of a solid. The equilibrium is described mathematically by the solubility product, $K_{sp} = [Ba^{2+}][F^-]^2$. The second equation is the reverse of the dissociation of HF, described quantitatively by $K_{eq} = 1/K_a$.

The two chemical equations can be combined, but the second one must be multiplied by 2 and then added to the first to give a balanced equation for the overall process:

$$BaF_2(s) + 2H_3O^+(aq) \rightleftharpoons Ba^{2+}(aq) + 2HF(aq) + 2H_2O(\ell)$$

We can evaluate the influence of pH qualitatively by applying the principle of Le Chatelier. High concentrations of hydrogen ion (low pH) will force the equilibrium to the right, favoring the dissolution of the solid. This conclusion is general: *the solubility will increase in acidic solutions for any salt whose anion is a weak base.*

When barium fluoride dissolves in an acidic solution, aqueous fluoride ion is formed. But some of the fluoride ions will react with hydrogen ions to form HF(aq). Each time a fluoride ion is converted into a molecule of HF, some additional BaF_2 must dissociate to keep the $[Ba^{2+}][F^-]^2$ ion product equal to K_{sp}. Salts in which one of the ions reacts with hydrogen ion or hydroxide ion, are generally more soluble than predicted by K_{sp} calculations.

The solubility of salts of weak acids increases in acidic solution.

Formation of Complexes

The chemical nature of the cation also influences the solubility of the precipitate. If the cation is a transition metal ion, it can form a compound with a species that can donate electrons. We call the species that donates electrons a **ligand;** the resulting compound is a **complex.** Many common bases, such as ammonia, water, hydroxide ions, and chloride ions, can act as ligands and form complexes; in fact, the formation of a complex is a Lewis acid-base reaction. (Ligands are electron-pair donors so they are Lewis bases.)

Additional discussion of complex formation will be deferred to Chapter 18, but the influence of complex formation on solubility can be described simply by the principle of Le Chatelier. When the cation forms a complex, the concentration of the free (uncomplexed) metal ion in the solution decreases, and additional precipitate will dissolve until the ion product is again equal to K_{sp}. Consider the reaction of silver ions with ammonia to form a complex ion:

The solubility of a precipitate increases in the presence of a complexing agent.

$$Ag^+(aq) + 2NH_3(aq) \rightleftharpoons Ag(NH_3)_2^+(aq)$$

The equilibrium constant for the formation of the complex ion (named the diamminesilver(I) ion) is about 10^{+8}, so the concentration of $Ag^+(aq)$ is much lower in the presence of ammonia.

Complex formation equilibria can be used to our advantage in many chemical analyses and syntheses, for example, determining the presence of silver ions in solution. The first step is to add some chloride to an unknown sample to precipitate the silver:

$$Ag^+(aq) + Cl^-(aq) \rightleftharpoons AgCl(s)$$

If silver is present, we observe the white silver chloride precipitate. Unfortunately, mercury, lead, and thallium also form insoluble white chlorides. To determine whether the precipitate is silver chloride, and not one of the other insoluble chlorides, we add ammonia. If the white precipitate is indeed silver chloride, the diamminesilver complex is formed, and the precipitate dissolves.

$$AgCl(s) + 2NH_3(aq) \rightleftharpoons Ag(NH_3)_2^+(aq) + Cl^-(aq)$$

Other metals, including Cu^{2+}, Zn^{2+}, and Ni^{2+}, also form complexes with ammonia.

Summary

A *titration* is the stepwise addition of a reactant, generally in solution, that reacts with a component of the sample. Among the most common titrations are the neutralization of a strong acid with a strong base and *vice versa.* The pH of the solution can be calculated at any point during the titration by applying basic stoichiometry and equilibrium calculations. Stoichiometry calculations (the sRfc table) are used to determine the concentrations of the species present after the titration reaction goes to completion. In the titration of a strong acid with a strong base, the pH of the solution is determined from the concentration of the excess of strong acid or base. The pH at the *equivalence point* (when the acid and base have been added in stoichiometrically equivalent amounts) is 7.00 for the titration of a strong acid with strong base.

Chemical systems that contain a weak acid and its conjugate base resist changes in pH; such solutions are called *buffers.* The acid reacts to consume any added hydroxide ion; the base reacts with any added hydrogen ion. The pH of a buffer is generally calculated by a relationship first described by Henderson and Hasselbalch:

$$pH = pK_a + \log \frac{C_b}{C_a}$$

where C_b and C_a are the concentrations of the base and acid conjugate pairs.

The pH during the course of the titration of a weak acid with strong base requires two calculations. First, the *analyte* and *titrant* are assumed to react completely. The results of the stoichiometry calculation (the sRfc table) are used to determine the nature of the system (strong or weak acid or base, buffer, or water). Second, the pH of the resulting solution is calculated as for any other solution of weak acids, bases, or buffers. The pH at the equivalence point is greater than 7 for the titration of a weak acid with strong base, and needs to be calculated on an individual basis. The inflection in *titration curves* for systems that include weak acids or weak bases are not as sharp as those for systems that include only strong acids and strong bases.

An *indicator* is often added to the titration flask so that a color change occurs after the analyte has been consumed. The point at which the indicator changes color is called the *endpoint.* If the indicator is chosen to change color at the equivalence point pH, then the endpoint and the equivalence point coincide. If the indicator changes color at a pH that is substantially different from the equivalence point pH, errors may result. Since the titrant is generally a strong acid or strong base, the pH changes quite sharply just beyond the equivalence point, and it is logical to select an indicator that changes color in this pH region (just beyond the equivalence point) in order to minimize errors.

Systems of *polyprotic acids* can sometimes be treated like monoprotic acids. If the first ionization constant is much larger than the second, then the second one can be ignored. If the ionization constants are similar in magnitude, then this approach fails, and simultaneous equilibria must be considered.

Solutions of amphoteric species can be evaluated qualitatively by examining the magnitudes of the acid and base dissociation constants. If the acid ionization constant is larger than K_b for the same ion, then the solution will be acidic. K_b for a polyprotic acid can be calculated from the appropriate acid ionization constant, so the pH of the solution can be estimated.

If the anion of an ionic compound is the conjugate base of a weak acid, then its solubility increases in acidic solutions.

$$MA(s) \rightleftharpoons M^+(aq) + A^-(aq)$$

If the pH is adjusted so that A^- reacts to form HA, the solubility will increase, since more solid dissolves to compensate for the decrease in anion concentration as the weak acid is formed. Similarly, if the cation can form a complex, the solubility of the solid increases in the presence of the complexing agent.

Chapter Terms

Titrant—the substance added to react with the analyte. *(14.1)*

Analyte—the substance whose concentration is being determined. *(14.1)*

Titration curve—a graph of pH as a function of the added volume of titrant. *(14.1)*

Standard solution—one of accurately known concentration. *(14.1)*

Millimole—1×10^{-3} mol. *(14.1)*

Inflection point—the point in a titration curve at which the pH changes most rapidly. *(14.1)*

Buffer—a system that resists changes in pH. Most buffers are solutions that contain a weak acid–weak base conjugate pair. *(14.2)*

Henderson-Hasselbalch equation—generally used to calculate the pH of a buffer solution. *(14.2)*

$$pH = pK_a + \log \frac{C_b}{C_a}$$

Buffer capacity—the amount of strong acid or base needed to change the pH of one liter of a buffer solution by 1 unit. *(14.2)*

Indicator—a substance that changes color to signal the end of a titration. *(14.4)*

Polyprotic acid—an acid that can ionize to produce more than one proton per molecule. *(14.5)*

Amphoteric—a substance that has properties of both acids and bases. *(14.6)*

Ligand—a substance that donates a pair of electrons to form a bond to a transition metal ion. Ligands are Lewis bases. *(14.7)*

Complex—the species formed by the Lewis acid-base reaction of a metal atom or ion with a ligand. *(14.7)*

Exercises

Exercises designated with color have answers in Appendix J.

Titrations

14.1 A titration was used to determine the concentration of a solution of hydrochloric acid. Exactly 20.00 mL of the acid solution was placed in a flask, phenolphthalein was added, and 18.34 mL of 0.0982 *M* NaOH was needed to reach the endpoint. What was the concentration of the hydrochloric acid?

14.2 A pipet was used to deliver 10.00 mL of sulfuric acid to a titration flask. It took 31.77 mL of 0.102 *M* NaOH to neutralize the sulfuric acid completely. Calculate the concentration of the sulfuric acid solution. Assume the reaction is

$$H_2SO_4(aq) + 2NaOH(aq) \rightarrow Na_2SO_4(aq) + 2H_2O(\ell).$$

14.3 Exactly 1.2451 g of a solid white acid was dissolved in water. It was completely neutralized by the addition of 36.69 mL of 0.404 *M* NaOH. Calculate the molar mass of the acid, assuming it to be a monoprotic acid. If additional experiments indicate that the acid is diprotic, what is its molar mass?

14.4 A scientist has synthesized a diprotic organic acid, H_2A, with a molar mass of 124.0 g/mol. The acid must be neutralized (forming the potassium salt) for an important experiment. Calculate the volume of 0.221 *M* KOH that is needed to neutralize 24.93 g of the acid, forming K_2A.

14.5 How much 0.100 *M* NaOH is needed to neutralize completely
(a) 45.00 mL of 0.0500 *M* HCl
(b) 5.00 mL of 0.350 *M* H_2SO_4 (forming Na_2SO_4)
(c) 10.00 mL of 0.100 *M* acetic acid

14.6 How much 0.100 *M* HCl is needed to neutralize completely
(a) 10.0 mL of 0.150 *M* KOH

(b) 250.0 mL of 0.00520 *M* $Ba(OH)_2$
(c) 100.0 mL of 0.100 *M* ammonia

Back Titrations (Challenging Problems)

14.7 A monoprotic organic acid that has a molar mass of 176.1 g/mol is synthesized. Unfortunately, the acid is not completely pure. In addition, it is not very soluble in water. A chemist weighs a 1.8451-g sample of the impure acid and adds it to 100.0 mL of 0.1050 *M* NaOH. The acid is soluble in the NaOH solution and reacts to consume most of the NaOH. The amount of excess NaOH is determined by titration: it takes 3.28 mL of 0.0970 *M* HCl to neutralize the excess NaOH. What is the purity of the original acid?

14.8 A classical test for nitrogen in plant material involves adding some compounds to the plant material to produce ammonia (NH_3, a base) from the nitrogen. The solution is then heated to drive off the ammonia. The ammonia passes through a container of HCl, where it reacts. After all the ammonia has been absorbed, there is still an excess of HCl in the container. The amount remaining after reaction with the ammonia can be determined by titration. The difference between the amount initially added and the amount determined by titration represents the amount that was neutralized by the ammonia.

Exactly 21.34 g of plant material is weighed into the reactor. All the nitrogen is converted to ammonia and collected in 100.0 mL of an HCl solution. If the initial HCl solution concentration is 0.121 *M* and the final solution requires 34.22 mL of 0.118 *M* NaOH to neutralize, calculate the percent nitrogen in the plant material.

Strong Acid-Base Titration Curves

14.9 Calculate the pH in the titration of 100.0 mL of 0.200 *M* HCl with 0.400 *M* NaOH. Calculate the pH after 0, 25.00, 50.00, and 75.00 mL of NaOH have been added. Sketch the titration curve.

14.10 Calculate the pH in the titration of 50.00 mL of 0.250 *M* HNO_3 with 0.500 *M* KOH. Calculate the pH after 0, 12.50, 25.00, and 40.00 mL of KOH have been added. Sketch the titration curve.

14.11 Calculate the pH in the titration of 50.00 mL of 0.100 *M* NaOH with 0.100 *M* HNO_3. Calculate the pH after 0, 25.00, 50.00, and 75.00 mL of nitric acid have been added. Sketch the titration curve.

14.12 Calculate the pH in the titration of 1.00 mL of 0.240 *M* LiOH with 0.200 *M* HNO_3. Calculate the pH after 0, 0.25, 0.50, 1.20, and 1.50 mL of nitric acid have been added. Sketch the titration curve.

The pH of Buffer Solutions

14.13 Calculate the pH of a solution made by
(a) adding 10.0 g of sodium benzoate and 3.00 g of benzoic acid and dissolving in water to make 1.00 L of solution.
(b) adding 25.0 g of sodium acetate to 500.0 mL of 0.30 *M* acetic acid.

14.14 Calculate the pH of a solution made by
(a) adding 15.45 g of potassium fluoride to 100.0 mL of 0.850 *M* HF.
(b) adding 45.00 g of NH_4Cl to 250.0 mL of 0.455 *M* ammonia.

14.15 Calculate the pH that results when the following solutions are mixed:
(a) 10.00 mL of 0.500 *M* sodium acetate and 20.00 mL of 0.350 *M* acetic acid.
(b) 350.0 mL of 0.150 *M* pyridinium chloride and 650.0 mL of 0.450 *M* pyridine.

14.16 Calculate the pH that results when the following solutions are mixed:
(a) 100.0 mL of 0.800 *M* formic acid and 200.0 mL of 0.100 *M* sodium formate;
(b) 300.0 mL of 0.350 *M* ammonia and 200.0 mL of 0.150 *M* ammonium chloride.

14.17 How many grams of sodium acetate must be added to 400.0 mL of 0.500 *M* acetic acid to prepare a pH 4.35 buffer?

14.18 What volume of 0.500 *M* HF must be added to 750 mL of 0.200 *M* sodium fluoride to prepare a buffer of pH 3.95?

Challenging Buffer Problems

14.19 A buffer solution that is 0.100 *M* acetate and 0.200 *M* acetic acid is prepared.
(a) What is the pH of the buffer?
(b) Calculate the initial pH, final pH, and change in pH that result when 1.00 mL of 0.100 *M* HCl is added to 100.0 mL of the buffer.
(c) Calculate the initial pH, final pH, and change in pH that result when 1.00 mL of 0.100 *M* HCl is added to 100.0 mL of water.

14.20 A buffer solution that is 1.00 *M* acetate ion and 0.500 *M* acetic acid is prepared.
(a) Calculate the initial pH, final pH, and change in pH that result when 1.00 mL of 0.100 *M* NaOH is added to 100.0 mL of the buffer.
(b) Calculate the initial pH, final pH, and change in pH that result when 1.00 mL of 0.100 *M* NaOH is added to 100.0 mL of water.

14.21 Calculate the minimum concentrations of acetic acid and sodium acetate that are needed to prepare 100.0 mL of a pH 4.50 buffer whose pH will not change by more than 0.05 units if 1.00 mL of 0.100 *M* strong acid or strong base is added.

14.22 Calculate the minimum concentrations of formic acid and sodium formate that are needed to prepare 500.0 mL of a pH 3.80 buffer whose pH will not change by more than 0.10 units if 1.00 mL of 0.100 *M* strong acid or strong base is added.

14.23 Saccharin is an artificial sweetener that is also a weak acid. It has the formula $C_7H_5NSO_3$ and its pK_a is 11.68. A 12-oz (350-mL) container of diet cola contains 3.0 mg of saccharin and has a pH of 4.50. What are the concentrations of saccharin and the saccharide ion?

14.24 A buffer is made by dissolving 0.0500 mol of acetate and 0.0500 mol of acetic acid in some water. The pH is adjusted to 5.00 by adding small amounts of concentrated acid (HCl) and base (NaOH) as needed, and the solution is then diluted to 1.00 L. What are the concentrations of acetic acid and sodium acetate in the solution?

Titrations of Weak Acids and Bases

14.25 Calculate the pH in the titration of 25.00 mL of 0.200 *M* acetic acid with 0.500 *M* NaOH. Calculate the pH after 0, 5.00, 10.00, and 15.00 mL of base have been added. Sketch the titration curve.

14.26 Calculate the pH in the titration of 20.00 mL of 0.250 *M* formic acid with 0.500 *M* NaOH. Calculate the pH after 0, 5.00, 10.00, and 15.00 mL of base have been added. Sketch the titration curve.

14.27 Calculate the pH during the titration of 10.00 mL of 0.400 *M* hypochlorous acid with 0.400 *M* KOH. Calculate the pH after addition of 0, 50%, 95%, 100%, and 105% of the amount of base needed to reach the equivalence point. Graph the titration curve (pH *vs.* volume of KOH).

14.28 Calculate the pH during the titration of 100.0 mL of 0.230 *M* hydrofluoric acid with 0.500 *M* NaOH. Calculate the pH after addition of 0, 50%, 95%, 100%, and 105% of the amount of base needed to reach the equivalence point. Graph the titration curve (pH *vs.* volume of NaOH).

14.29 Calculate the pH in the titration of 50.00 mL of 0.100 *M* ammonia with 0.100 *M* HCl. Calculate the pH after 0, 25.00, 50.00, and 75.00 mL of acid have been added. Sketch the titration curve.

14.30 Calculate the pH in the titration of 30.00 mL of 0.200 *M* pyridine with 0.200 *M* HCl. Calculate the pH after 0, 15.00, 30.00, and 40.00 mL of acid have been added. Sketch the titration curve.

14.31 Calculate the pH of each of the following solutions:
(a) 10.0 mL of 0.300 *M* HF plus 30.0 mL of 0.100 *M* NaOH
(b) 100.0 mL of 0.250 *M* ammonia plus 50.0 mL of 0.100 *M* HCl
(c) 25.0 mL of 0.200 *M* H_2SO_4 plus 50.0 mL of 0.400 *M* NaOH

14.32 Calculate the pH of each of the following solutions:
(a) 1.00 mL of 0.150 *M* formic acid plus 2.00 mL of 0.100 *M* NaOH
(b) 25.00 mL of 0.250 *M* ammonia plus 5.00 mL of 0.100 *M* HI
(c) 5.00 mL of 0.200 *M* $Ba(OH)_2$ plus 50.00 mL of 0.400 *M* HBr

14.33 What is the concentration of ammonium ion in a pH 9.00 solution that is formed when concentrated NaOH is added to 0.100 *M* NH_4Cl?

Indicators

14.34 A chemist is developing a titration analysis for lactic acid. Lactic acid is a monoprotic acid with $K_a = 8.4 \times 10^{-4}$. Calculate the pH at the equivalence point of a titration of 100 mL of 0.100 *M* lactic acid with 0.500 *M* NaOH. Suggest an indicator from Table 14.2, and explain why you chose it.

14.35 Chloropropionic acid, $ClCH_2CH_2COOH$, is a weak monoprotic acid with $K_a = 7.94 \times 10^{-5}$. Calculate the pH at the equivalence point in a titration of 10.00 mL of 0.100 *M* ascorbic acid with 0.100 *M* KOH. Choose an indicator from Table 14.2 for the titration. Explain your choice.

14.36 A 25.0-mL sample of 1.44 *M* NH_3 is titrated with 1.50 *M* HCl. Calculate the pH at the equivalence point, and choose an indicator from Table 14.2. Justify your choice.

14.37 Exactly 50 mL of a 0.0500 *M* solution of ethylamine, a base with $K_b = 6.5 \times 10^{-4}$, is titrated with 0.100 *M* HNO_3. What is the pH at the equivalence point. Suggest a good indicator from Table 14.2 for this titration, and justify your selection.

Polyprotic Acids

14.38 Write the chemical equilibria and expressions for the equilibrium constants for the ionizations of

(a) oxalic acid
(b) sulfurous acid

14.39 Write the chemical equilibria and expressions for the equilibrium constants for the ionizations of
(a) tartaric acid
(b) malic acid

14.40 Write the equilibrium and calculate K_b for the sulfate ion.

14.41 Write the equilibrium and calculate K_b for the carbonate ion.

14.42 Write the equilibrium and calculate K_b for the citrate ion.

14.43 Write the equilibrium and calculate K_b for the malonate ion.

Identification of Acids from Titration Curves (Challenging Problems)

The titration curves shown in Exercises 14.44 to 14.47 are for the titration of 1.00 mL of 0.100 *M* acid with 0.100 *M* NaOH. The identity of the acid is unknown, but the titration curve often allows the chemist to choose among possible alternatives. Use the data to identify the unknown acids in the titration curves. The acids are chosen from:

Acid	pK_{a1}	pK_{a2}	pK_{a3}
citric	3.13	4.77	6.40
oxalic	1.23	4.19	
malic	3.40	5.04	
phthalic	2.95	5.41	

14.44 Identify the acid.

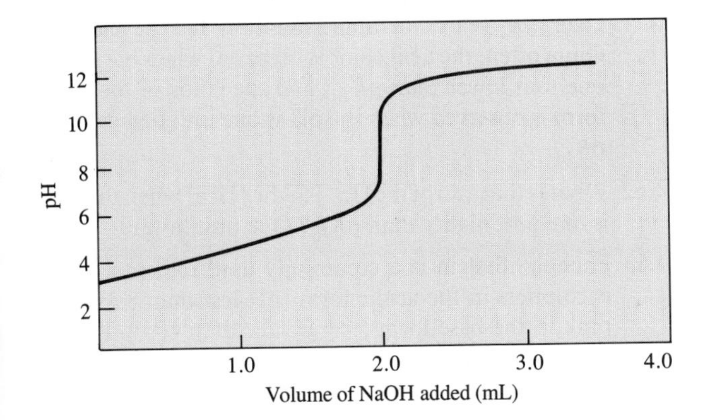

14.45 Identify the acid.

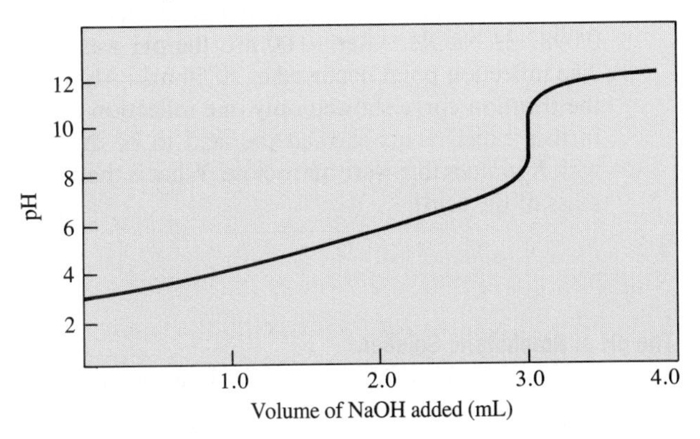

14.46 Identify the acid.

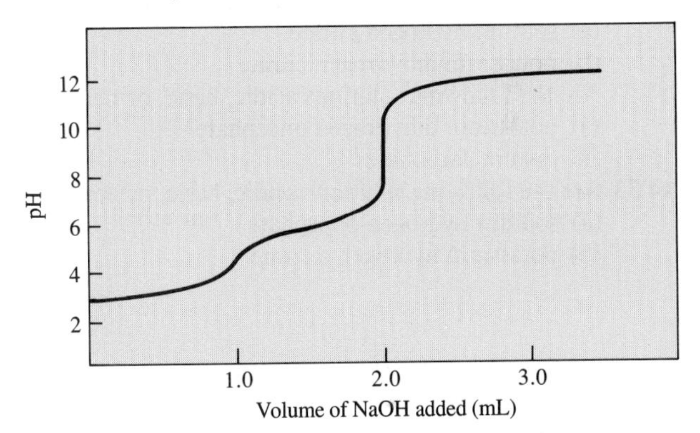

14.47 Identify the acid.

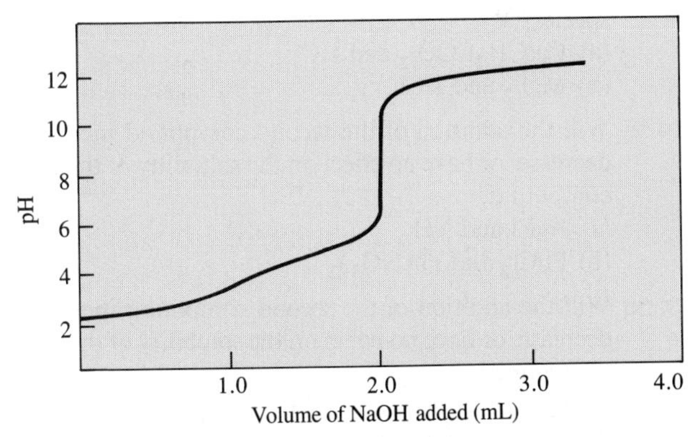

14.48 An unknown acid was titrated and the pH was recorded during the course of the titration. The student weighed 0.3419 g of acid and titrated with 0.0982 *M* NaOH. After 24.35 mL of NaOH was added, the pH was 4.56. At 41.35 mL, the equivalence point was reached. The titration curve showed only one inflection point, so the student assumed that the acid was monoprotic. What were the molar mass and pK_a of the unknown acid?

14.49 An unknown was titrated with standard sodium hydroxide. A sample weighing 0.2469 g was titrated with 0.0982 *M* NaOH. After 10.00 mL the pH was 3.78. The inflection point occurred at 20.54 mL. Although the titration curve showed only one inflection point, further experiments showed the acid to be diprotic with K_a values that were quite close. What is the molar mass of the acid?

The pH of Amphoteric Solutions

14.50 Are the following solutions acidic, basic, or neutral?
(a) sodium hydrogen oxalate
(b) potassium hydrogen malonate

14.51 Are the following solutions acidic, basic, or neutral?
(a) sodium hydrogen citrate
(b) potassium dihydrogen citrate

14.52 Are the following solutions acidic, basic, or neutral?
(a) potassium dihydrogen phosphate
(b) cesium carbonate

14.53 Are the following solutions acidic, basic, or neutral?
(a) sodium hydrogen phosphate
(b) potassium hydrogen tartrate

Solubility as a Function of pH

14.54 Will the addition of the second compound increase, decrease, or have no effect on the solubility of the first species?
(a) $Ca(CH_3COO)_2$ and HCl
(b) MgF_2 and HCl

14.55 Will the addition of the second compound increase, decrease, or have no effect on the solubility of the first compound?
(a) AgCl and NH_3
(b) $PbCl_2$ and $Pb(NO_3)_2$

14.56 Will the addition of the second compound increase, decrease, or have no effect on the solubility of the first compound?
(a) calcium oxalate and HCl
(b) copper sulfate and NH_3

14.57 Will the addition of the second compound increase, decrease, or have no effect on the solubility of the first compound?
(a) zinc hydroxide and NaOH
(b) magnesium phosphate and HNO_3

14.58 What is the pH of a solution that is saturated with magnesium hydroxide? (*Hint:* Find K_{sp} in Appendix F.)

14.59 What is the pH of a solution that is saturated with iron(II) hydroxide? (*Hint:* Find K_{sp} in Appendix F.)

Additional Exercises

14.60 A bottle of concentrated hydroiodic acid is 57% HI by weight and has a density of 1.70 g/mL. A solution of this strong and corrosive acid is made by adding exactly 10.0 mL to some water and diluting to 250.0 mL. If the information on the label is correct, what volume of 0.988 *M* NaOH will be needed to titrate the HI? Suggest an indicator that could be used.

14.61 Concentrated hydrochloric acid is 38% HCl by weight and has a density of 1.19 g/mL. A solution is prepared by measuring 83 mL of the concentrated HCl, adding it to water, and diluting to 1.00 L.
(a) What is the approximate molarity of this solution?
(b) The exact concentration is determined by titration. A 25.00-mL portion of the solution was titrated with 1.04 *M* NaOH. Phenolphthalein changed color after 23.88 mL of the base was added. What is the concentration of the solution? How does the approximate concentration calculated in (a) compare to the exact concentration?

Acid-base indicators have been studied for many years because of their importance in the overall accuracy of an acid-base titration. Acid-base indicators are generally weak acids or bases, and the acid form is a different color than the base form. We can write the chemical equation and equilibrium expression for the ionization of an indicator.

$$\underset{\text{acid form}}{HIn} + H_2O(\ell) \rightleftharpoons H_3O^+ + \underset{\text{base form}}{In^-}$$

$$K_{In} = \frac{[H_3O^+][In^-]}{[HIn]}$$

$$pK_{In} = -\log K_{In}$$

Over the years, an approximation has developed. Quite often, the acid color is observed when the pH is one unit lower than pK_{In}, and the color of the base form is observed when the pH is one unit higher than pK_{In}.

14.62 What is the ratio of the $[In^-]$ to the [HIn] when the pH is one unit higher than pK_{In}? One unit lower?

14.63 Phenolphthalein is a commonly used indicator that is colorless in the acidic form (pH less than 8.3) and pink in the base form (pH greater than 10.0). It is a weak acid with a pK_a of 8.7. What fraction is in the acid form when the acid color is apparent? What fraction is in the base form when the base color is apparent?

14.64 The indicator methyl red is a weak acid with pK_{In} of 5.00. Calculate the pH values at which the indicator will be 1%, 5%, 95%, and 99% in the acid form.

14.65 Use Table 14.2 as a source of data about methyl red. What fraction of the indicator is in the acid form when the acid color is observed? What fraction is in the base form when it is observed?

Chapter Fifteen

15

Chemical Thermodynamics

Our lives are completely dependent on our sources of energy. Our bodies require fuel, so does our society. We need energy from food to breathe, grow, and learn, and our global society requires energy to maintain civilization as we know it. Fuels include milk, bread, and meat as well as wood, coal, and oil; they all provide energy when consumed in chemical reactions. The study of the *energetics* of chemical reactions is called **chemical thermodynamics.**

The word thermodynamics is derived from the Greek words *thermes,* meaning heat, and *dynamikos,* meaning strength. The word dynamic makes us think of motion and energy, so the term thermodynamics is used to describe heat flow. The words apply well to our chemical systems because heat is generally produced or absorbed in the course of a chemical reaction.

One of the major goals of chemists is to develop a model that will help us predict whether or not a reaction will proceed. In chemistry, the word **spontaneous** describes a process that can occur without outside intervention. Our everyday experience tells us that many processes go in a particular direction spontaneously. We know that young organisms grow old; we know that dropped objects fall. If we were shown a movie of broken glass leaping together to form a flask, we would logically assume the film was shown in reverse because glass fragments simply do not spontaneously reform into an unbroken container. If a chemical reaction is spontaneous in one direction, then under the same conditions the reverse reaction is not spontaneous. This statement does not mean that the reverse reaction *cannot* proceed, only that it may require different conditions. Changing temperature, pressure, or con-

The destruction of hundreds of oil wells during the 1991 Iraq/Kuwait conflict not only wasted vast amounts of a nonrenewable energy resource, but produced air pollution with consequences that are not yet known.

Energy from the reaction of hydrogen with oxygen Hydrogen and oxygen can be mixed at room temperature without reaction, but under other conditions the reaction can be energetic and even explosive. The main engines of the Space Shuttle consume liquid hydrogen and oxygen at a rate of 4000 L (1000 gal) per second. The Space Shuttle launch requires about 2×10^{10} kJ of energy (about 1/6 from the hydrogen/oxygen engines and the rest from ammonium perchlorate solid-fuel rockets). If this energy were converted to electricity, it would satisfy the needs of 1 million people for almost a day.

centration can sometimes make the reverse reaction occur. Water, for example, does not spontaneously form ice at room temperature and pressure, but the process is spontaneous at a lower temperature.

Spontaneity does not imply that the process occurs quickly, but rather describes a *capability* to proceed. Oxygen and hydrogen can form water in a spontaneous reaction; indeed, if we supply a spark, the reaction occurs explosively. However, hydrogen and oxygen sealed in a flask will remain uncombined indefinitely. Before we examine details such as the speed of a reaction, we must first determine whether the reaction is *feasible*—that is, whether it will occur at all under the given conditions. We will see that thermodynamics, with its laws and theories, enables us to predict whether a chemical reaction can proceed spontaneously under a certain set of conditions.

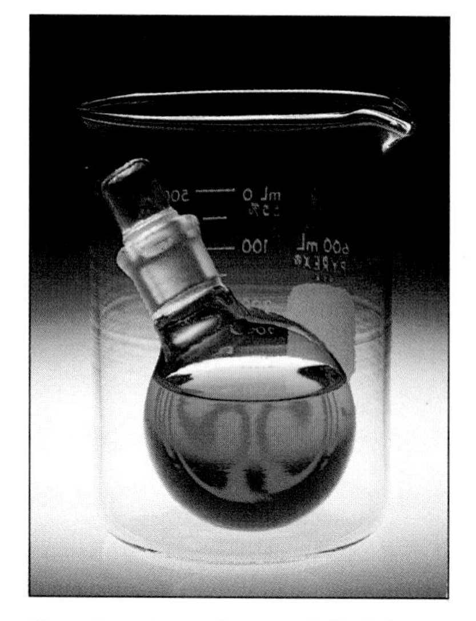

Closed system A capped flask is a closed system.

15.1 Work and Heat

Objectives

- To relate heat, work, and energy
- To calculate work from pressure-volume relationships

Knowledge of the heat absorbed or evolved in a chemical reaction is information of the utmost importance. Chemists use this information not only to adjust conditions to avoid explosions, but also to maximize efficiency. Thermodynamics provides the tools needed to evaluate the heat exchanged by a reaction, to predict the maximum energy that can be produced, and to determine whether or not a proposed reaction is feasible. In this section, we will discuss the heat evolved or absorbed during a reaction and the models used to predict the quantity of that heat.

First, let us repeat some definitions from Chapter 4. The *system* is the matter of interest. It could be a beaker full of solution, the turbines in a generator, or a patient in a hospital.

While several categories of systems have been identified, the systems discussed in this book are generally **closed systems,** those in which the exchange of energy with the surroundings is allowed but not the exchange of matter.

The *surroundings* are all other matter, everything that isn't part of the system. If the system is defined as the particular group of atoms that undergo chemical change in a beaker in the laboratory, then the surroundings include the beaker, the lab bench, the Chemistry Building, and so on. The *universe* is the system *plus* the surroundings.

The experimental conditions needed to characterize the properties of a system, including temperature, pressure, and the amounts and the phases of substances, specify the state of the system. Many thermodynamic properties, called **state functions,** depend only upon the state of the system and not the manner in which the system arrived at the state. Changes in state functions depend only on the initial and final state of the system, not the path by which the change occurs.

Work, Heat, and Energy

Benjamin Thompson (Count Rumford, 1753–1814), a Massachusetts-born scientist, was hired by the King of Bavaria as the chief administrator of the Bavarian army. Among his duties was the supervision of the production of cannons. A cannon was cast as a solid bronze cylinder and then drilled with large, water-cooled drill bits driven by horses. Thompson observed that the water used to cool the drills started to boil about the same time each day. He correctly deduced that the source of the heat was the work done by the horses. A great many years and experiments later, the scientific community accepted the equivalence of heat and work. (The delay in accepting Thompson's work was possibly due to the many enemies he made among scientists of the time.) We now understand that heat and work are two different ways by which energy is transferred.

The power to drill Count Rumford's cannons was supplied by horses. Power houses as pictured here were in common use in the 18th century.

The **energy** of a system is the capacity of the system to perform work. The **internal energy,** E, represents the total energy of the system. It consists of kinetic energy, derived from the motion of the system or its components, and potential energy. Potential energy is derived from the position of the system or the arrangement of particles that form the system. For a chemical system, most of the available potential energy is the energy of the bonds.

Internal energy, E, includes kinetic and potential energy.

Work

Work, the first form of energy transfer that will be discussed, has a specific definition in science. A scientist defines **work** as the application of a force through a distance, measured by the product of forces times distance. Units of force are newtons (kg m s^{-2}) so work has units of newton-meters (N-m) or joules (J); 1 J = 1 N-m. An example of work is lifting a weight against the force of gravity.

When we presented thermochemical equations in Chapter 4, we used the following convention for the algebraic sign of the heat change, ΔH: If the

If the system . . .
performs work on the surroundings, the sign of w is negative.
receives work from the surroundings, the sign of w is positive.
transfers heat to the surroundings, the sign of q is negative.
absorbs heat from the surroundings, the sign of q is positive.

system supplies heat to the surroundings (it is exothermic), then ΔH is negative because the system loses energy. By analogy, when the system performs work on the surroundings, the work w is negative because the internal energy of the system decreases.

PV Work

A system that contains a gas can perform work as the gas expands against an opposing pressure exerted by the surroundings. This type of work is called "*PV* work" because a change in volume (against an external pressure) is the source of the work. An automobile engine, in which the gases formed by the combustion of the fuel push against a piston, is powered by *PV* work. We can show (the derivation is in the following *Insights into Chemistry*) that the work done by the expansion of a gas against a constant pressure is

$$w = -P\,\Delta V \qquad [15.1]$$

where P is the pressure opposing the system.

INSIGHTS into CHEMISTRY

A Closer View

Work Is Performed as a Gas Expands Against External Pressure

A gas that expands against an external pressure performs work on the surroundings. In order to derive the expression for that work, we will consider an ideal gas (the system) confined in a cylinder by a massless, frictionless piston.

The pressure of the gas confined by the piston is P_1. The pressure on the outside is P_{ext}. This external pressure is due to atmospheric pressure.

If the pressures inside and outside the cylinder are not the same, the piston will move, and it will continue to move until the pressures are equal. The final pressure, P_2, is equal to the opposing pressure, P_{ext}.

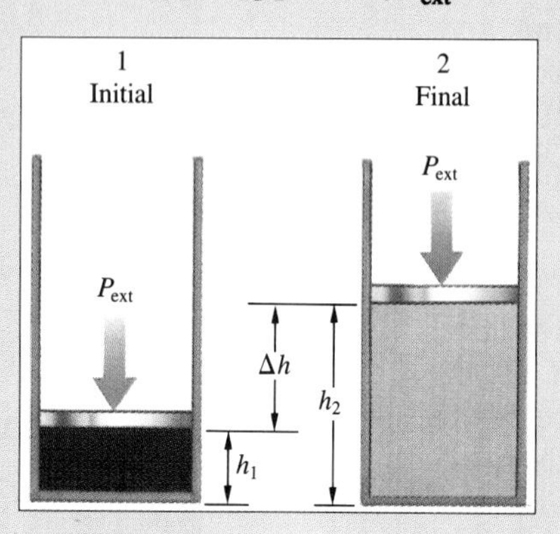

Expansion of a gas.

The work done by the system is defined as

$$\text{work} = \text{force} \times \text{distance}$$

Remember that pressure is defined as force per unit area, so the force that appears in the definition of the work is

$$F = P_{ext} \times A$$

where A is the area of the piston. The distance in the expression for the work is $\Delta h = h_2 - h_1$, the distance that the piston moves. Substitute the force and distance:

$$\text{work} = F \times \Delta h = P_{ext} \times A \times \Delta h$$

To take the next step, we note that the volume of the gas in the cylinder at any time is the area of the piston, A, times the current height of the piston. Therefore,

$$\Delta V = V_2 - V_1 = Ah_2 - Ah_1 = A\,\Delta h$$

Substituting ΔV in place of $A \times \Delta h$ in the work expression,

$$\text{work performed} = P_{ext}\,\Delta V$$

There is one last detail: the algebraic sign of the work. According to our convention, work performed by the system has a negative sign. Since the system performs work when it raises the piston (increases the volume), we must include a negative sign in the final equation so that a negative value for work is associated with a positive ΔV:

$$w = -P_{ext}\,\Delta V$$

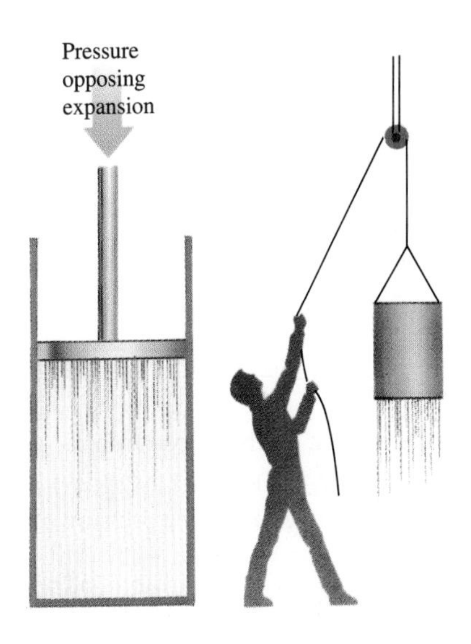

Work A gas (the system) that expands against an opposing pressure is analogous to a person (the system) lifting a weight against gravity. In both cases energy is transferred to the surroundings.

When a gas expands, w *is negative* because the system transfers energy to the surroundings. The expansion of a gas against a pressure is analogous to a person lifting a weight. The sign of the work associated with a contraction is positive, because work is done on the system as the surroundings compress it. Equation 15.1 produces the correct sign for work, since ΔV has a sign. (ΔV is equal to $V_{\text{final}} - V_{\text{initial}}$ and can be positive or negative.)

A gas expanding against an opposing pressure performs work.

When we calculate PV work, we usually obtain units of L-atm. To convert L-atm to joules, the SI unit of work, remember the units of R, the ideal gas constant:

$$R = 0.0821 \text{ L-atm mol}^{-1}\text{ K}^{-1} = 8.314 \text{ J mol}^{-1}\text{ K}^{-1}$$

Cancelling like units yields

$$0.0821 \text{ L-atm} = 8.314 \text{ J}$$

This equivalence can be used to form a conversion factor.

PV work can be expressed in L-atm or in J.

$$\text{conversion factor} = \frac{8.314 \text{ J}}{0.0821 \text{ L-atm}} = \frac{101 \text{ J}}{\text{L-atm}}$$

This conversion factor is used in Example 15.1.

Example 15.1 Pressure-Volume Work

A cylinder contains 45 L of an ideal gas at a pressure of 140 atm. If the gas expands at a constant temperature against an opposing pressure of 1.00 atm, how much work (in joules) is done?

Solution

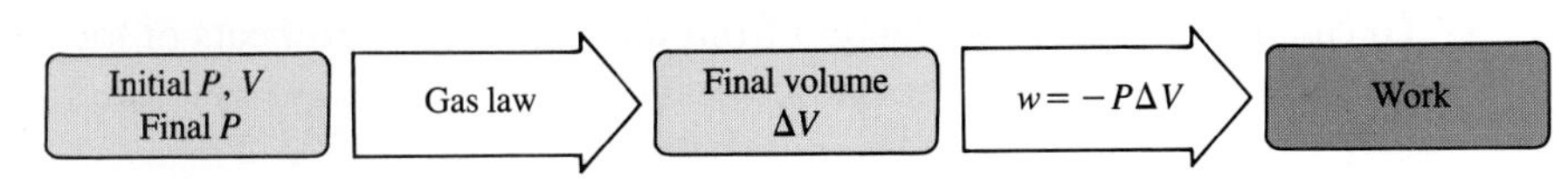

Initial conditions: $V_1 = 45 \text{ L}$, $P_1 = 140 \text{ atm}$

Final conditions: $V_2 = \text{unknown}$, $P_2 = P_{\text{ext}} = 1.00 \text{ atm}$

We use Boyle's law to determine the final volume. Since PV is constant,

$$P_1V_1 = P_2V_2$$

$$V_2 = \frac{P_1V_1}{P_2} = 6300 \text{ L}$$

$$w = -P_{ext}\,\Delta V = -1.00 \text{ atm } (6300 \text{ L} - 45 \text{ L})$$

$$w = -6.26 \times 10^3 \text{ L-atm}$$

Convert the work to joules.

$$w = -6.26 \times 10^3 \cancel{\text{L-atm}} \times 101 \frac{\text{J}}{\cancel{\text{L-atm}}} = -6.32 \times 10^5 \text{ J} = -640 \text{ kJ}$$

The negative sign indicates that the system performs work on the surroundings.

Understanding Calculate the amount of work done when 6300 L of an ideal gas (the system) initially at 1.00 atm is compressed at constant temperature by a pressure of 140 atm to a final volume of 45 L.

Answer: $w = +8.76 \times 10^5$ L-atm $= 8.84 \times 10^4$ kJ. The positive sign indicates that the surroundings do work on the system in order to compress it. Although the changes in the system in Example 15.1 and the "Understanding" are exactly opposite, the work is not.

Heat

A second form of energy transfer is heat. Count Rumford, the cannon-drilling supervisor, recognized that the heat produced by the drill bit was related to the movement of the horses that powered the drill. Heat is a transfer of kinetic energy between the system and the surroundings. Although the system itself is not moving, the atoms and molecules that make up the system are in motion. A transfer of some of this energy to or from the surroundings is the source of the phenomenon that we call heat. Notice that heat implies a *random* motion, while work is a *directed* motion (which includes electrons moving through a wire as well as a gas expanding against a pressure).

Both heat and work are forms of energy.

Heat is generally measured by calorimetry, discussed in Chapter 4. The change in temperature of a substance is related to the heat transferred, so the quantity of heat evolved or absorbed in a process can be determined by experimental measurements of the temperature change.

15.2 The First Law of Thermodynamics

Objectives

- To relate heat, work, energy, and the first law of thermodynamics
- To distinguish between energy and enthalpy
- To calculate the heat of reaction from tabulated standard heats of formation

Centuries of observations led to the **first law of thermodynamics,** also known as the law of conservation of energy: *energy can be neither created nor destroyed.* Scientists cannot explain "why" the first law is observed; we can only marvel that a wondrously complex Nature can also be so simple and direct. In a closed system, matter cannot be exchanged with the surroundings, but heat and work can. Under these conditions the first law can be written in equation form as

$$\Delta E = q + w$$

where ΔE is the change in internal energy of the system, q is the heat flow into the system, and w is the work performed on the system. Heat and work are simply different forms of energy. The three terms in the first law are all expressed in units of joules in the SI system.

Notice two important details. First, thermodynamics is concerned with the *change* in state. Scientists have found no way to evaluate absolute energies; thermodynamicists cannot say "the energy decreases from 2700 J to 2400 J" but can only say "the energy decreases by 300 J in this change." Second, ΔE is a *state function* and independent of the path between the initial and final states. The heat and work, q and w, however, depend on the path, so reactions can be adjusted to give more heat at the expense of work, and *vice versa.*

In Example 15.1 we calculated the work when an ideal gas expanded at a constant temperature against a 1.00 atm pressure.

$$\text{gas sample (45 L, 140 atm, T)} \longrightarrow \text{gas sample (6300 L, 1.00 atm, T)}$$

In the "Understanding" accompanying the example, the gas sample was restored to its original condition by performing work against a pressure of 140 atm.

$$\text{gas sample (6300 L, 1.00 atm)} \longrightarrow \text{gas sample (45 L, 140 atm)}$$

If we consider the expansion and compression together, the initial and final states of the system are the same, and therefore the state function $\Delta E = 0$. The net work in the two changes is

$$w = w_1 + w_2 = -632 \text{ kJ} + 8.84 \times 10^4 \text{ kJ} = 8.78 \times 10^4 \text{ kJ}$$

Even though the internal energy of the system did not change, a large amount of work was done. From the first law (Equation 15.2), the net work done on the system appears as heat transferred to the surroundings.

$$\Delta E = q + w$$

$$q = \Delta E - w = 0 - 8.78 \times 10^4 = -8.78 \times 10^4 \text{ kJ}$$

If a different path had been followed, the work and heat flow would be different, but the change in internal energy, ΔE, would still be zero.

The first law of thermodynamics influences nearly every process we can name. The designers of electrical generating plants know the energy available from a ton of coal (or a ton of uranium or oil). They are responsible for designing the equipment that converts some of the energy of the fuel into useful work, turning the generators. Any energy that is converted to heat is simply lost to the environment. Not only is efficiency a problem, but the heat is a form of pollution. Similar problems are faced by automobile engine designers who strive to maximize efficiency while minimizing thermal (and chemical) pollution.

Combustion of kerosene A large jet can weigh up to 385,000 kg (850,000 lb) and consumes aviation fuel similar to kerosene at the rate of 1200 kg/min (160,000 lb/hr). Jet engines are designed to turn the energy of combustion into work.

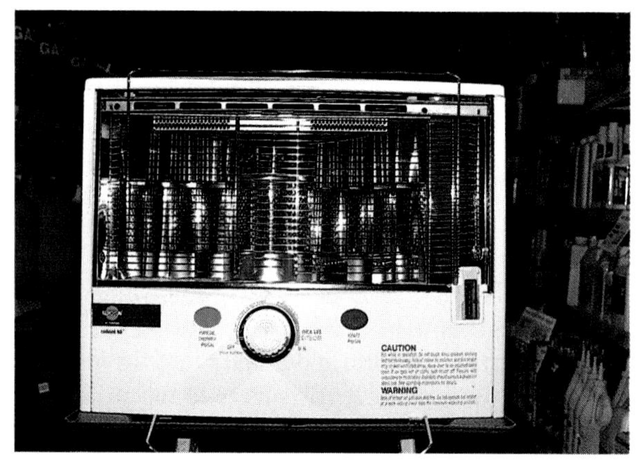

Kerosene stoves are designed to convert the energy of combustion into heat; burn with high thermal efficiency, but a stove that is 99.9% efficient can still produce pollutants in significant quantities. Manufacturers carefully warn customers that carbon monoxide—a colorless, odorless, lethal gas—can be formed in kerosene heaters.

Work, like heat, is *not* a state function. The combustion of a kilogram of kerosene can be used to propel an airplane when used as fuel for jet engines. The kilogram of kerosene can also be burned in a stove, supplying heat but little or no useful work.

Energy and Enthalpy

Most chemical reactions are performed in containers that are open to the atmosphere; the volume of the chemical system is allowed to change, but the pressure remains constant. (Although atmospheric pressure varies with the local weather conditions, the pressure changes very little over the course of a typical experiment, several hours.) If the work is limited to PV work due to a change of volume, and no electrical or other work is performed, a state function called **enthalpy**, H, can be defined.

$$H = E + PV \qquad [15.2]$$

The change in enthalpy for a process that occurs at constant pressure can be written as

$$\Delta H = \Delta E + P\Delta V \qquad [15.3]$$

The heat absorbed at constant pressure, when the only work is PV work, is ΔH.

The change in enthalpy, ΔH, is equal to the heat absorbed by the system under conditions of constant pressure, and under conditions where only PV work is done, the conditions of a typical lab reaction. The change in internal energy, ΔE, is the heat absorbed by the system when the *volume* is constant, as it is inside a heavy-walled, sealed reactor. We generally use the terms "change in enthalpy" and "heat of reaction" interchangeably, since the vast majority of our reactions occur under conditions of constant pressure. A reaction for which ΔH is negative transfers heat to the surroundings and is

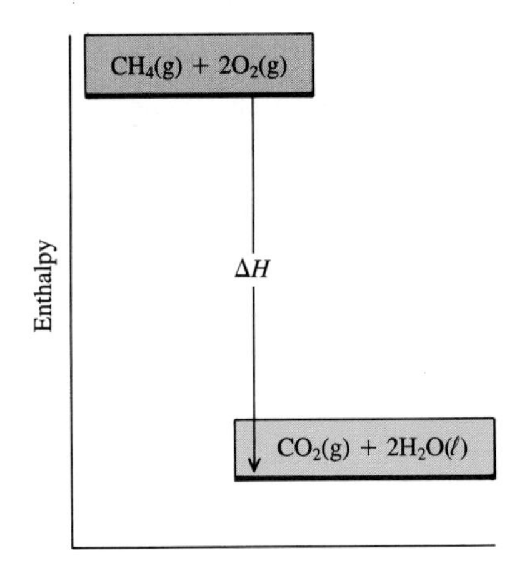

Figure 15.1 Energy level diagram for the combustion of methane
$CH_4(g) + 2O_2(g) \longrightarrow CO_2(g) + 2H_2O(\ell)$

called "exothermic." An endothermic reaction absorbs heat from the surroundings, and ΔH is positive.

The energy level diagram of an exothermic reaction is pictured in Figure 15.1.

Thermochemical Equations

Early scientists hypothesized that ΔH might predict spontaneity; they felt that any reaction in which the enthalpy of the system decreases would be spontaneous. However, later studies disproved this hypothesis by the discovery of some spontaneous reactions in which the enthalpy of the system increases. Still, measuring or calculating enthalpy changes is often crucial because the enthalpy change can change the temperature of the system. Temperature frequently determines whether a reaction will proceed in a controlled manner to form products, whether undesirable side products will be formed, or whether the reaction will proceed out of control. This section presents methods that enable chemists to determine the enthalpy change in a chemical reaction from experimental information about the reactants and products.

Enthalpy changes and thermochemical equations were first discussed in Chapter 4. We saw that ΔH does not depend on the path of the reaction, since E, P, and V are all state functions. We employed this property in Chapter 4, where the additive nature of thermochemical equations was used to determine the enthalpy change for a reaction that cannot be measured directly.

Let us consider a reaction taking place under conditions of constant pressure, and write a thermochemical equation for the reaction.

$$\text{reactants} \longrightarrow \text{products} \qquad \text{heat absorbed} = \Delta H_{\text{reaction}}$$

The overall reaction can also be written as two reactions that proceed through an intermediate state. Although any intermediate state can be chosen, a very convenient one (the reason will become clear later) is the constituent elements in their standard states.

(1)	reactants $\longrightarrow$ elements	heat absorbed $= \Delta H_1$
(2)	elements $\longrightarrow$ products	heat absorbed $= \Delta H_2$
(overall)	reactants $\longrightarrow$ products	heat absorbed $= \Delta H_{reaction}$

Hess's law (Section 4.6) tells us that the total change in enthalpy is the sum of the changes for the two steps:

$$\Delta H_{reaction} = \Delta H_1 + \Delta H_2$$

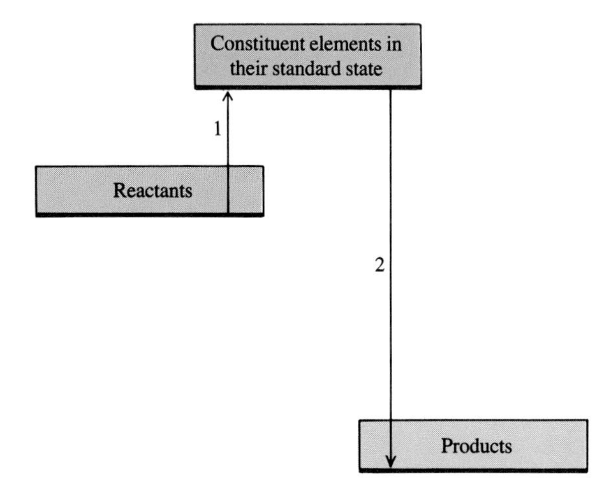

Since enthalpy is a state function, a two-step process provides exactly the same enthalpy change as the direct conversion of reactants into products. The choice of the constituent elements as an intermediate state is arbitrary, but it greatly reduces the number of thermochemical reactions that must be tabulated to provide the information needed for accurate calculations of enthalpy change. These calculations are discussed over the next several pages.

Standard States

The standard state is the pure liquid, pure solid, gas at 1 atm partial pressure, or solute at a concentration of 1 *M*, at a designated temperature and 1 atm pressure.

In any thermochemical equation, the states of all products and reactants must be specified. The reference is the substance at 1 atm pressure and a specified temperature. We call this state the **standard state.** In addition, the standard state of a liquid or solid refers to the pure substance; the standard state of a gas is the gas at a partial pressure of 1 atm, and the standard state of a solute in solution is a concentration of 1 *M*. The reference temperature for nearly all thermodynamic data is 298 K (25 °C), a convention that will be used in this book. Thermodynamic quantities that refer to the standard state are given a superscript °, as in $\Delta H°$. The temperature, if not 298 K, is often indicated as a subscript, *e.g.*, $\Delta H°_{273}$.

Standard Heat of Formation

The reason for choosing the elements as intermediates is to simplify the determination of the enthalpy change in a reaction. The **standard heat of formation** of any substance is the heat absorbed when *one mole* of the substance in its standard state is formed from the most stable form of the elements in their standard states. The equations for some standard enthalpies (heats) of formation are

$$\text{C(graphite)} + \text{O}_2(\text{g}) \longrightarrow \text{CO}_2(\text{g}) \qquad \Delta H_f^\circ[\text{CO}_2(\text{g})]$$

$$\text{C(graphite)} \longrightarrow \text{C(diamond)} \qquad \Delta H_f^\circ[\text{C(diamond)}]$$

$$\text{H}_2(\text{g}) + \tfrac{1}{2}\text{O}_2(\text{g}) \longrightarrow \text{H}_2\text{O}(\ell) \qquad \Delta H_f^\circ[\text{H}_2\text{O}(\ell)]$$

$$\text{Na(s)} + \tfrac{1}{2}\text{N}_2(\text{g}) + \tfrac{3}{2}\text{O}_2(\text{g}) \longrightarrow \text{NaNO}_3(\text{s}) \qquad \Delta H_f^\circ[\text{NaNO}_3(\text{s})]$$

$$\text{O}_2(\text{g}) \longrightarrow \text{O}_2(\text{g}) \qquad \Delta H_f^\circ[\text{O}_2(\text{g})] = 0$$

The enthalpy change for each of these reactions is the standard heat of formation of the single substance that appears on the right side of the arrow. There are several points to note in these equations. (1) There is one mole of a single substance on the product side of each of these reactions. (2) Even though it is impractical to produce sodium nitrate by reaction of the elements at 298 K, it is still possible to evaluate the enthalpy change (the standard heat of formation) for the reaction. (3) Finally, as we see in the last equation, the heat of formation of $O_2(g)$ in its standard state (and indeed all of the elements in their most stable form) is zero, since the equation defining the heat of formation involves no change.

The standard heats of formation for many substances, for example $CO_2(g)$, can be measured directly using a calorimeter, while others, such as that of $NaNO_3(s)$, are calculated using Hess's law with several measured enthalpies of reaction. The standard heats of formation of selected substances appear in Table 15.1, and Appendix G contains a more complete list of ΔH_f° values. Tables of standard enthalpies of formation such as these allow the calculation of enthalpy changes for many chemical reactions.

Table 15.1 Standard Enthalpies of Formation

Compound	Name	ΔH_f°, kJ/mol
$Br_2(\ell)$	bromine	0.
C(diamond)	diamond	1.895
C(graphite)	graphite	0.
$CH_4(g)$	methane	−74.81
$C_2H_6(g)$	ethane	−84.68
$C_3H_8(g)$	propane	−103.85
$C_4H_{10}(g)$	butane	−124.73
$CH_3OH(\ell)$	methyl alcohol	−238.66
$CH_3CH_2OH(\ell)$	ethyl alcohol	−277.69
CO(g)	carbon monoxide	−110.52
$CO_2(g)$	carbon dioxide	−393.51
$H_2(g)$	hydrogen	0.
$H_2O(g)$	water	−241.82
$H_2O(\ell)$	water	−285.83
$N_2(g)$	nitrogen	0.
$NH_3(g)$	ammonia	−46.11
$O_2(g)$	oxygen	0.

The standard heat of reaction can be calculated from standard heats of formation for a typical chemical reaction such as the combustion of methane.

$$CH_4(g) + 2O_2(g) \longrightarrow CO_2(g) + 2H_2O(\ell)$$

The following diagram shows how this reaction can be represented by the two-step process that passes through the intermediate state consisting of the constituent elements in their standard states. Note that the enthalpy change for reaction 1 is the *negative* of the standard heat of formation of the reactants. The enthalpy change for reaction 2 is the heat of formation of the products.

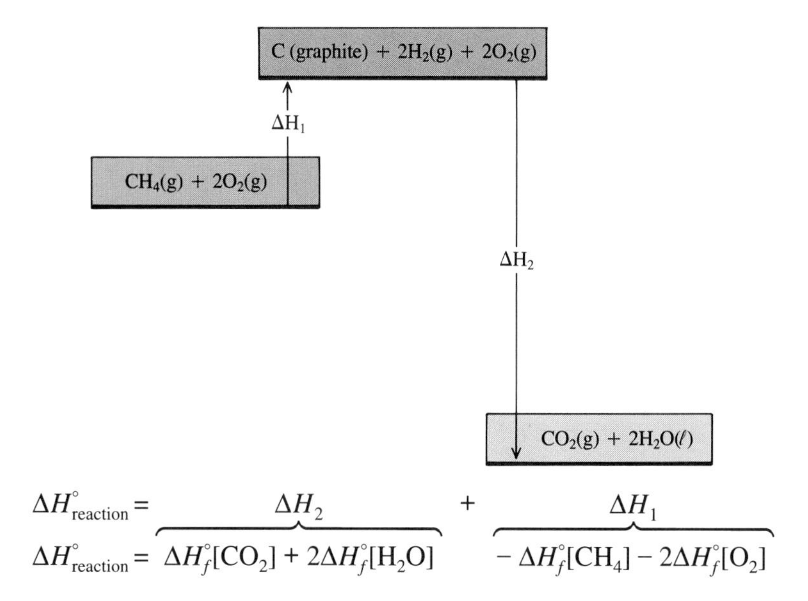

$$\Delta H^\circ_{\text{reaction}} = \overbrace{\Delta H^\circ_f[CO_2] + 2\Delta H^\circ_f[H_2O]}^{\Delta H_2} + \overbrace{- \Delta H^\circ_f[CH_4] - 2\Delta H^\circ_f[O_2]}^{\Delta H_1}$$

For any chemical reaction,

reactants $\longrightarrow$ products

$$\Delta H^\circ_{\text{reaction}} = \Sigma\, n\Delta H^\circ_f[\text{products}] - \Sigma\, n\Delta H^\circ_f[\text{reactants}] \qquad [15.4]$$

where n designates the number of moles of each reactant and product in the chemical equation. For example, two moles of water appear in this chemical equation, so two times the standard heat of formation of water must appear in the thermochemical expression.

Standard heats of formation have units of kJ mol^{-1}, while heats of reaction have units of kJ and depend on the form of the balanced equation, in which the coefficients represent numbers of moles. To calculate the heat generated by combustion of butane, you could use either of the following chemical equations (among others), which describe different amounts of reactants and products.

The coefficients in thermochemical equations refer to moles, so fractional coefficients can be used.

(A) $C_4H_{10}(g) + \frac{13}{2}O_2(g) \longrightarrow 4CO_2(g) + 5H_2O(\ell)$

(B) $2C_4H_{10}(g) + 13O_2(g) \longrightarrow 8CO_2(g) + 10H_2O(\ell)$

The heat of reaction for B will be twice that of A, just as the amounts of the substances are doubled. *The heat of reaction refers to a specific chemical*

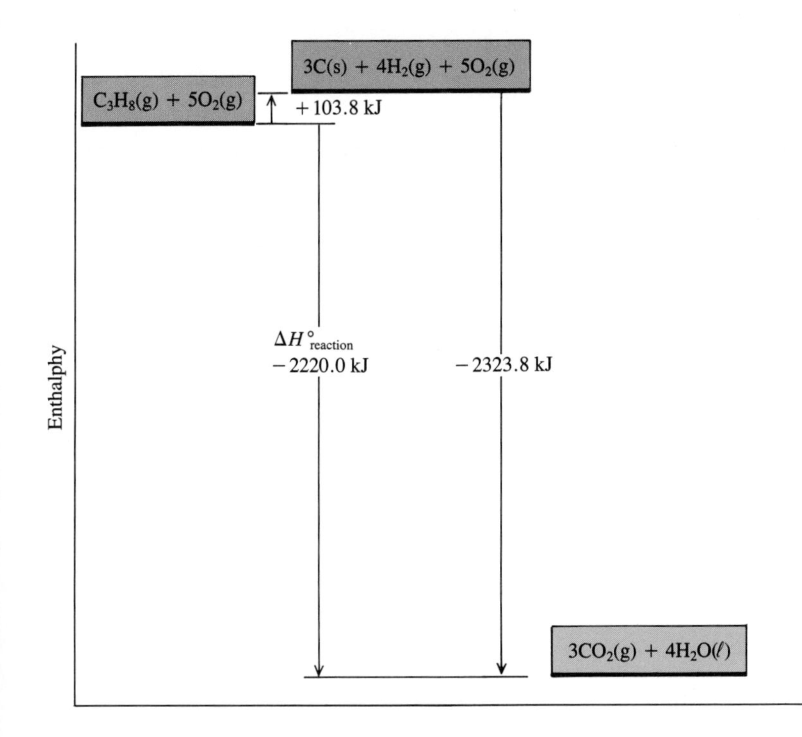

Figure 15.2 Enthalpy diagram for the combustion of propane
$C_3H_8(g) + 5O_2(g) \longrightarrow 3CO_2(g) + 4H_2O(\ell)$

equation. Often, we balance the equation so that the coefficients are based on one mole of the compound of interest.

To calculate a heat of reaction, a chemical equation with a specific set of coefficients must be provided.

The use of standard heats of formation can be illustrated by using the combustion of propane as an example.

$$C_3H_8(g) + 5O_2(g) \longrightarrow 3CO_2(g) + 4H_2O(\ell)$$

The standard enthalpy change, $\Delta H^\circ_{reaction}$, accompanying the complete reaction of a substance with oxygen is often called the **heat of combustion.** The heat of combustion of one mole of propane can be expressed in terms of the standard heats of formation of the reactants and products using Equation 15.4:

$$\Delta H^\circ_{reaction} = \Sigma\, n\Delta H^\circ_f[\text{products}] - \Sigma\, n\Delta H^\circ_f[\text{reactants}]$$

$$\Delta H^\circ_{reaction} = (3\Delta H^\circ_f[CO_2] + 4\Delta H^\circ_f[H_2O]) - (\Delta H^\circ_f[C_3H_8] + 5\Delta H^\circ_f[O_2])$$

Notice that the coefficients of the chemical equation are used. Completing the calculation with values from Table 15.1:

$$\begin{aligned}\Delta H^\circ_{reaction} = {} & [3\ \cancel{\text{mol CO}_2} \times (-393.51\ \text{kJ}/\cancel{\text{mol of CO}_2}) \\ & + 4\ \cancel{\text{mol H}_2\text{O}} \times (-285.83\ \text{kJ}/\cancel{\text{mol of H}_2\text{O}})] \\ & - [1\ \cancel{\text{mol C}_3\text{H}_8} \times (-103.85\ \text{kJ}/\cancel{\text{mol of C}_3\text{H}_8}) \\ & + 5\ \cancel{\text{mol O}_2} \times (0\ \text{kJ}/\cancel{\text{mol of O}_2})]\end{aligned}$$

$$\Delta H^\circ_{reaction} = -2323.85 - (-103.85)\ \text{kJ}$$

$$\Delta H^\circ_{reaction} = -2220.00\ \text{kJ}$$

Note that the units derived for the heat of reaction, kJ, are correct. The burning of propane is illustrated with an enthalpy level diagram in Figure 15.2.

The standard heat of reaction is the sum of the standard heats of formation of the products minus the sum of the standard heats of formation of the reactants.

When propane burns, the propane and oxygen do not first form carbon, hydrogen, and oxygen, which then combine to form water and carbon dioxide. However, since the enthalpy change is a state function and does not depend on the path, the way in which the reaction really occurs has no effect on the way we determine the enthalpy change; we need to know only the initial and final states.

The heat of combustion, a value that is measured in the laboratory, can be used to determine heats of formation. This process is illustrated in Example 15.2.

Example 15.2 Calculating Heat of Formation

Calculate the standard enthalpy of formation for glucose, given the heat of combustion of one mole of glucose to form carbon dioxide and water at 298 K is −2821.8 kJ. Use the data in Table 15.1 for the standard heats of formation of carbon dioxide and water.

Solution

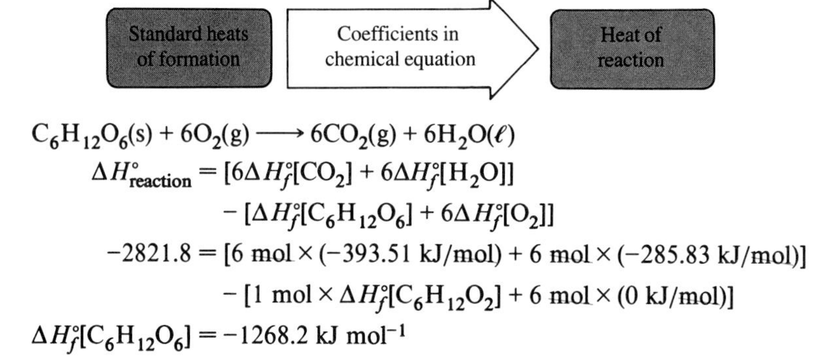

$$C_6H_{12}O_6(s) + 6O_2(g) \longrightarrow 6CO_2(g) + 6H_2O(\ell)$$

$$\Delta H^\circ_{\text{reaction}} = [6\Delta H^\circ_f[CO_2] + 6\Delta H^\circ_f[H_2O]] - [\Delta H^\circ_f[C_6H_{12}O_6] + 6\Delta H^\circ_f[O_2]]$$

$$-2821.8 = [6 \text{ mol} \times (-393.51 \text{ kJ/mol}) + 6 \text{ mol} \times (-285.83 \text{ kJ/mol})] - [1 \text{ mol} \times \Delta H^\circ_f[C_6H_{12}O_2] + 6 \text{ mol} \times (0 \text{ kJ/mol})]$$

$$\Delta H^\circ_f[C_6H_{12}O_6] = -1268.2 \text{ kJ mol}^{-1}$$

Understanding The standard enthalpy change when a mole of liquid ethyl alcohol, CH_3CH_2OH, is burned to form carbon dioxide and liquid water at 298 K is −1366.7 kJ. Calculate the standard heat of formation of ethyl alcohol.

Answer: $-277.8 \text{ kJ mol}^{-1}$

15.3 Entropy

Objectives

- To define entropy and examine its statistical nature
- To apply the second law of thermodynamics to chemical systems
- To compare and contrast entropy increases in the system, surroundings, and universe
- To recognize that absolute entropies can be measured because the third law of thermodynamics defines a zero point

The change in enthalpy, the heat absorbed by the system at constant pressure, often seemed to predict whether or not a reaction could proceed. After all, nearly all spontaneous reactions are accompanied by a release of heat. Unfortunately, enthalpy changes do not always predict spontaneity

INSIGHTS into CHEMISTRY

Science, Society, and Technology

Gasohol: A Blend of Gasoline and Alcohol. Can It Be a Practical Fuel for the Future?

Ethyl alcohol (ethanol) is a popular fuel in some parts of the world, and a common gasoline additive in much of the United States. The heat generated by the combustion of ethyl alcohol is much lower than that from gasoline, based on moles, mass, or even cost (using gasoline and ethanol costs over the last decade). At first glance, it appears illogical to use ethanol as a fuel because it provides less energy for the dollar, but detailed economic analysis provides some interesting information. In countries that have low labor costs, spacious farmland, and a good growing climate, crops can be planted specifically for processing into ethanol for fuel. The costs can then be competitive with those of imported oil. In addition, lighter engines can be used, so fuel mileage might increase when alcohol is used as a fuel.

In the United States, ethanol is often blended with gasoline to make "gasohol." Although the presence of ethanol will decrease the amount of energy available from a gallon of fuel, gasohol proponents note that it increases the burning efficiency, reducing the need for expensive additives. In addition, the ethanol is produced within the same country, and often the same region, so local cash flow (and patriotism, too) are also important.

Some municipalities require the use of gasohol during the winter, when carbon monoxide emissions pose a serious health problem. In summer months, the pollution problem is not as bad, and the gasoline blenders are free to use other additives as dictated by supply and demand.

Gasohol

correctly. Even though most spontaneous reactions show a decrease in enthalpy, there are a few that have a positive ΔH. If barium hydroxide octahydrate is mixed with ammonium chloride, the two white solids form an exceptionally cold, liquid solution.

$$Ba(OH)_2 \cdot 8H_2O(s) + 2NH_4Cl(s) \longrightarrow BaCl_2(aq) + 2NH_3(g) + 10H_2O(\ell)$$

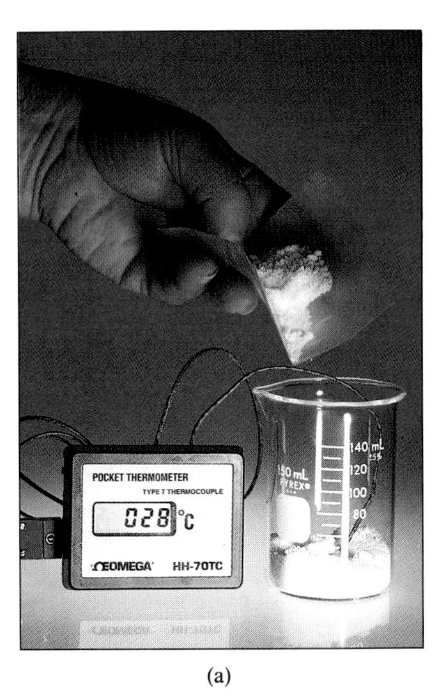

(a)

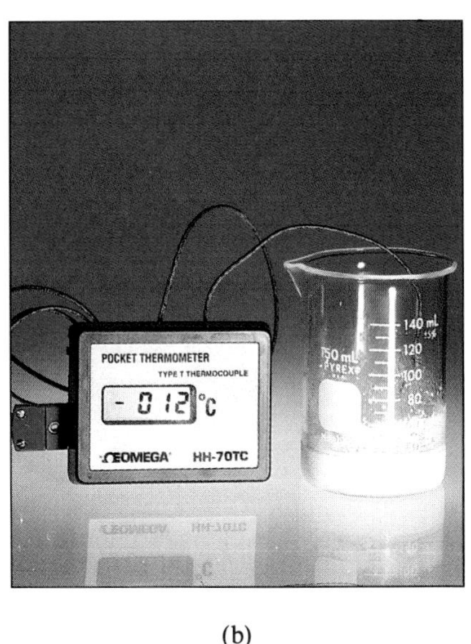

(b)

Figure 15.3 Mixing barium hydroxide and ammonium chloride The system becomes quite cold as the reaction proceeds.

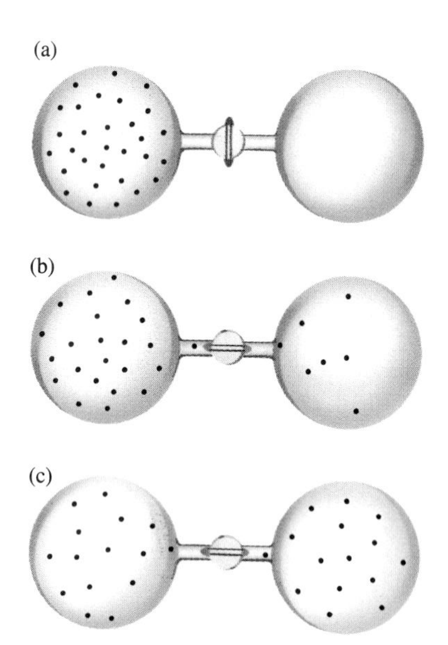

Figure 15.4 Expansion of a gas into a vacuum (a) All the gas is in the container on the left. (b) When the valve is opened, the gas moves into the empty container on the right. (c) At equilibrium, the gas is present at the same concentration in both containers, and disorder has increased.

The products become cold enough to cause frostbite. Figure 15.3 illustrates the temperature changes measured for this spontaneous, endothermic reaction.

Another spontaneous endothermic reaction occurs when ammonium chloride is added to water. As the white, crystalline solid dissolves, the beaker becomes much cooler. Spontaneous endothermic reactions are the basis of "cool-packs" often used by athletes to treat sprains and other injuries, shown in Figure 11.8.

The enthalpy change of a process does not, by itself, predict spontaneity. One other factor must be considered: disorder. Not only do reactions tend to form products with stronger bonds, they also tend to form more disordered products. Disorder plays a measurably important role in predicting whether a reaction will be spontaneous. **Entropy** is the property of a system that describes the amount of disorder. Entropy is a state function. Disorder, chaos, and randomness all describe a state of high entropy. Our observations show that reactions in which the entropy increases are generally favored.

The change in entropy is the sole driving force for some processes. Consider the ideal gas in the apparatus shown in Figure 15.4. When the stopcock is opened, the gas will move spontaneously to occupy the evacuated chamber. There is no change in bonding or intermolecular attractions, so the change in internal energy is zero. There is no work involved (the expansion of a gas against an opposing pressure of 0 atm produces no work), so ΔH is also zero. The process is driven by the change in disorder. In the final state, the system occupies a larger volume, and there are more possible locations for the gas particles, so the entropy has increased.

Practical aspects of entropy Many everyday systems tend toward disorder.

Entropy as Randomness

In the late nineteenth century, Ludwig Boltzmann (1844–1906) showed that the entropy of a system is related to the number of different ways it can be arranged. Boltzmann's work can be illustrated with a deck of playing cards. If a shuffled (randomized) deck has only two cards, there is a 50% chance the deck will be in order, arranged from low to high. If the deck has 13 cards (perhaps all the hearts), the chance of finding the deck in numerical order is small, about one in 6 billion times. If the deck has 52 cards, the chance of it being in order is still smaller.

Chemical systems are described by similar logic. The arrangement of atoms in a solid shows a high degree of order; the entropy is low. On the other hand, there are a large number of ways that atoms can be arranged in a gas, so the entropy of a sample of a gas is much higher than that of the same sample in solid form.

Entropy, given the symbol S, is a state function, so changes in entropy are independent of the path.

$$\Delta S = S_{\text{final}} - S_{\text{initial}}$$

Changes in entropy are related to changes in the randomness of the system. Most changes in entropy can be predicted from a few general concepts.

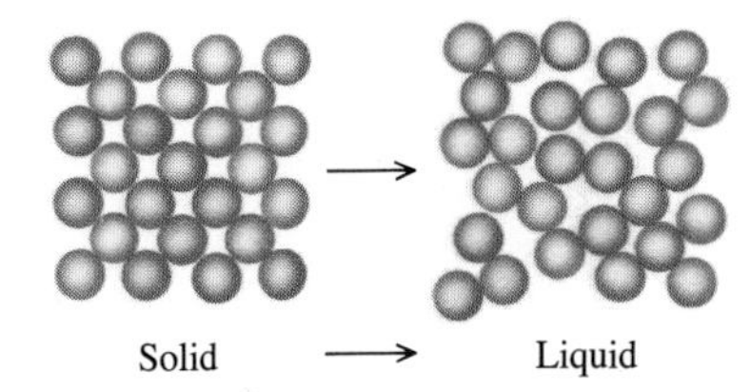

Solids and liquids The liquid form of a substance is much more random than is the solid form.

The Influence of Phase Changes on Entropy

1. *The entropy of a substance increases when the solid forms a liquid and when the liquid forms a gas.* In Chapter 10 we saw that the location of every unit in a crystal can be predicted from the location of the few units that are present in a unit cell. There is a very large degree of order in the solid state. When the solid melts, this order is disrupted and the molecules

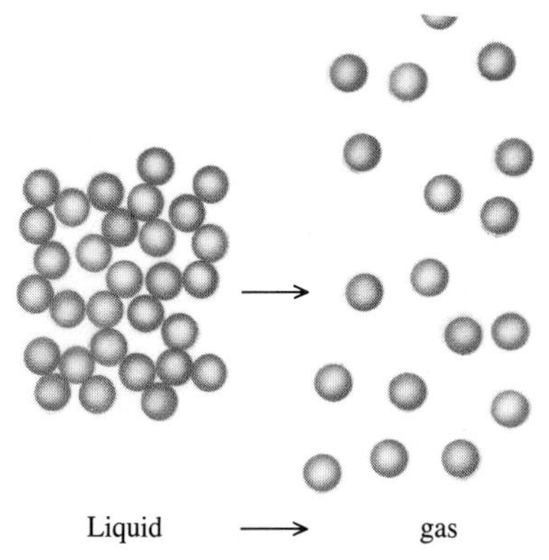

Liquids and gases Liquids are lower in entropy than are gases.

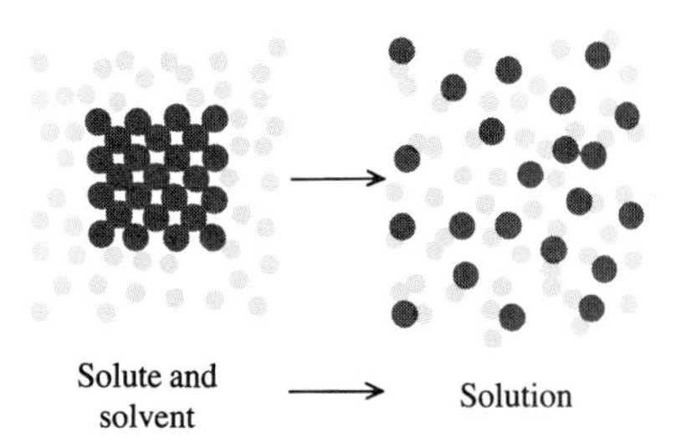

Dissolving a solid in a liquid The entropy of the system increases when the solute dissolves.

are no longer regularly arranged, so the entropy of the liquid is greater. Finally, when the liquid vaporizes, forming a gas, there is a large increase in the volume occupied by the sample. This increase in the volume means that each molecule has many more locations available to it than in the liquid, and so the gas is much more disordered (has higher entropy) than the liquid.

$$S_{\text{solid}} < S_{\text{liquid}} < S_{\text{gas}}$$

In general, the entropy increase on going from the liquid to the gas is considerably greater than the change from the solid to the liquid.

2. *Entropy generally increases when a condensed phase dissolves in a solvent.*

$$C_{12}H_{22}O_{11}(s) \xrightarrow{H_2O} C_{12}H_{22}O_{11}(aq)$$

When sucrose dissolves in water, the sucrose molecules go from a highly ordered arrangement in the solid to a disordered system with much larger volume, the solution. At the same time, the order of the water molecules increases, because of the hydration of the solute molecules. In most cases, the increase in disorder of the solute is greater than the decrease in disorder of the solvent. The solvent does not change phase, starting and remaining a liquid. The entropy change is dominated by the change in the solute from a solid phase to a liquid solution.

3. *Entropy decreases when a gas is dissolved in a solvent.*

$$CO_2(g) \xrightarrow{H_2O} CO_2(aq)$$

When a gas such as carbon dioxide is dissolved in a solvent, the solute goes from the gas phase to the liquid phase, so it becomes substantially less random, mainly because the dissolved molecules are confined to a smaller volume. The change in entropy of the solvent is small, since it is a liquid in the initial and final states. Overall, the entropy decreases.

Quite often the entropy change can be predicted by examining the chemical equation and determining whether the reactant side or the product side contains more moles of gas. Consider the following reaction.

If there are more moles of gases on the product side of the equation than on the reactant side, the entropy generally increases.

$$C_3H_8(g) + 5O_2(g) \longrightarrow 3CO_2(g) + 4H_2O(\ell)$$

The reactant side includes 6 moles of gas; the products have 3 moles of gas. There is a decrease in disorder, so the entropy of the system decreases during this reaction.

Increasing entropy is often associated with increasing volume of the system. This general rule is consistent with the examples we have presented.

Changes in phase are the largest factors influencing changes in entropy.

Second Law of Thermodynamics

One goal throughout this section has been to find a criterion to predict whether a reaction will be spontaneous. The **second law of thermodynamics** provides such a criterion: in any spontaneous process, the entropy of the universe increases. This remarkable statement has been tested against experimental observations and is entirely consistent with laboratory results.

In any spontaneous process, the entropy of the universe increases.

The change in entropy of the universe is the change in entropy of the system plus that of the surroundings.

$$\Delta S_{univ} = \Delta S_{sys} + \Delta S_{surr}$$

The second law of thermodynamics can be interpreted on the basis of probabilities. For any naturally occurring phenomenon, the most probable state is the most random. The second law of thermodynamics states that every spontaneous change is accompanied by an increase in the randomness or disorder of the universe.

If a reaction causes the entropy of the universe to increase, the reaction proceeds spontaneously. If the change in the entropy of the universe is negative, then the *reverse* reaction is spontaneous. If the change in the entropy of the universe is zero, the system is at equilibrium. The equilibrium condition will be discussed at length later.

Spontaneous:	$\Delta S_{univ} > 0$
Equilibrium:	$\Delta S_{univ} = 0$
Nonspontaneous:	$\Delta S_{univ} < 0$ (spontaneous in reverse direction)

If the second law is to be used to determine whether a process is spontaneous, entropy changes in both the system and the surroundings must be measured. It would be much easier if spontaneity could be expressed solely with variables that describe the system, since the surroundings are difficult to characterize.

It is important to note that the second law of thermodynamics refers to the entropy of the *universe,* not that of the system or of the surroundings.

Entropy Changes in the Surroundings

Fortunately, the entropy change in the surroundings *can* be expressed in terms of changes in the state of the system. *The entropy of the surroundings will be affected only by the heat transferred into or out of any closed system.* When heat is added to the surroundings, the kinetic energy increases, the disorder increases, so the entropy increases. At constant pressure (and if the only work is PV work), the heat transferred to the surroundings is equal to $-\Delta H_{sys}$, so ΔS_{surr} will be proportional to $-\Delta H_{sys}$.

The proportionality constant that relates ΔS_{surr} and ΔH_{sys} depends on the temperature of the surroundings. Transferring 10 kJ of heat will have a greater effect on a cold body than on a warm body. For example, change of 100 K from 273 K to 373 K increases the disorder more than does a temperature change from 3000 K to 3100 K. The proportionality constant is $1/T$, where T is the absolute temperature of the surroundings:

$$\Delta S_{surr} = \frac{-\Delta H_{sys}}{T} \qquad [15.5]$$

Equation 15.5 allows us to express the change in the entropy of the surroundings in terms of the change in enthalpy of the system. The units of entropy are energy/temperature, J K^{-1}. On a molar basis, entropy has units of J mol^{-1} K^{-1}.

Notice that the signs follow logical patterns. An exothermic reaction (ΔH negative) transfers heat to the surroundings, and ΔS_{surr} will be positive.

Changes in the Entropy of the System: The Third Law of Thermodynamics

The entropy of the system, unlike energy or enthalpy, is measured on an absolute basis. While we can only measure *changes* in E and H (ΔE or ΔH), we can measure an absolute value of S, the disorder, because there is a perfectly ordered system that can serve as a reference. The **third law of thermodynamics** gives us a reference point: *The entropy of a pure crystalline substance is 0 at a temperature of absolute zero.* There is minimum disorder in the pure crystal at 0 K.

The entropy of a perfect crystalline substance at 0 K is zero.

Appendix G lists entropies for many materials. The table contains the value for each substance in its standard state at 298 K. Standard entropies are designated by the symbol S°_{298}. Notice that pure elements at 298 K *do not* have zero entropy, since they are more disordered than the pure crystalline material would be at absolute zero. The data in the appendix can be used to calculate entropy changes for many reactions, because S is a state function and Hess's law applies:

$$\Delta S^\circ_{\text{reaction}} = \Sigma\, nS^\circ[\text{products}] - \Sigma\, nS^\circ[\text{reactants}]$$

This process is illustrated in Example 15.3.

Example 15.3 Calculating Entropy Changes

Use the standard entropies in Appendix G to calculate the standard entropy change when one mole of propane is burned in oxygen.

$$C_3H_8(g) + 5O_2(g) \longrightarrow 3CO_2(g) + 4H_2O(\ell)$$

Solution

$$\Delta S^\circ = (3S^\circ[CO_2] + 4S^\circ[H_2O(\ell)]) - (S^\circ[C_3H_8] + 5S^\circ[O_2])$$

$$\Delta S^\circ = [3\ \cancel{\text{mol}}\ (213.63\ \text{J}/\cancel{\text{mol}}\text{-K}) + 4\ \cancel{\text{mol}}\ (69.91\ \text{J}/\cancel{\text{mol}}\text{-K})] - [1\ \cancel{\text{mol}}\ (269.91\ \text{J}/\cancel{\text{mol}}\text{-K}) + 5\ \cancel{\text{mol}}\ (205.03\ \text{J}/\cancel{\text{mol}}\text{-K})]$$

$$\Delta S^\circ = -374.53\ \text{J K}^{-1}$$

The primary reason that entropy decreases is that the reactants include 6 moles of gases and the products only 3 moles.

Understanding Calculate the standard entropy change when one mole of water vapor is formed at 25 °C and 1 atm from the liquid.

$$H_2O(\ell) \longrightarrow H_2O(g)$$

Answer: $\Delta S^\circ = 118.81\ \text{J K}^{-1}$

Changes in the Entropy of the Universe

The second law speaks of the *universe,* not the system (or the surroundings). Now that we can calculate entropy changes, let us see what happens to the entropy of the universe in a process. Some of the easiest reactions to analyze are phase changes, like freezing or melting, because no new substances are formed. Consider the melting of one mole of ice.

$$H_2O(s) \longrightarrow H_2O(\ell)$$

The change in enthalpy and entropy of the melting process can be determined from laboratory measurements.

$$\Delta H^\circ_{298} = +6.01 \text{ kJ}$$

$$\Delta S^\circ_{298} = +22.2 \text{ J K}^{-1}$$

The melting of ice is endothermic; when ice melts, the surroundings are cooled. The change in the entropy of the universe can be calculated from

$$\Delta S_{univ} = \Delta S_{surr} + \Delta S_{sys}$$

Since $\Delta S_{surr} = -\Delta H_{sys}/T$, from Equation 15.5, we can write

$$\Delta S_{univ} = \frac{-\Delta H_{sys}}{T} + \Delta S_{sys}$$

At 298 K,

$$\Delta S_{univ} = \frac{-6.01 \times 10^3 \text{ J}}{298 \text{ K}} + 22.2 \text{ J K}^{-1}$$

$$\Delta S_{univ} = -20.2 + 22.2 = +2.0 \text{ J K}^{-1}$$

Since the entropy of the universe increases, the process is spontaneous. Notice also that the reverse process, the freezing of liquid water to form ice, does not occur spontaneously at 298 K.

The melting and boiling processes are examined in more detail in Section 15.4.

15.4 Free Energy

Objectives

- To define free energy and see that the sign of a free energy change corresponds to the direction of spontaneous reaction
- To determine the influence of temperature on free energy

The second law of thermodynamics was modified by J. W. Gibbs (1839–1903), who recognized a way to simplify calculations of spontaneity. Gibbs realized that the second law could be rewritten so that all terms were based on a single state function of the *system,* rather than the entropy change in the universe. He defined a new function, which he called the "free energy." This new function is introduced and its relation to spontaneity is examined in this section.

The universe is defined as the system plus the surroundings, and the entropy change in the universe is defined similarly.

$$\Delta S_{univ} = \Delta S_{surr} + \Delta S_{sys}$$

For a change at constant pressure and temperature, we can substitute $-\Delta H_{sys}/T$ for ΔS_{surr} (Equation 15.5) and drop the subscripts for quantities that refer to the system.

$$\Delta S_{univ} = \frac{-\Delta H}{T} + \Delta S$$

Table 15.2 Free Energy Changes and the Direction of Reaction

$\Delta G < 0$	Forward reaction is spontaneous
$\Delta G = 0$	Reaction at equilibrium
$\Delta G > 0$	Reverse reaction is spontaneous

Multiply by $-T$.

$$-T\,\Delta S_{univ} = \Delta H - T\,\Delta S$$

Gibbs defined a new function, called the **free energy** of the system, and given the symbol, G, as

$$G = H - TS.$$

Then, at constant T,

$$\Delta G = \Delta H - T\,\Delta S \qquad [15.6]$$

Comparing equations, we see that $\Delta G = -T\,\Delta S_{univ}$. Since all the quantities on the right side of Equation 15.6 are state functions, ΔG is also a state function. Like enthalpy and internal energy, the absolute value of G cannot be measured. We can, however, measure changes in G. At constant temperature and pressure, the change in free energy can be calculated from the changes in enthalpy and entropy by using Equation 15.6.

Gibbs realized that ΔG was related to ΔS_{univ} and recognized that *at constant temperature and pressure, any reaction is spontaneous if the free energy decreases.* The free energy is a quantitative indicator of spontaneity that depends only on changes in the system. For any spontaneous reaction, $\Delta G < 0$. Note that when $\Delta G > 0$, the *reverse* reaction is spontaneous. When the system reaches equilibrium, the change in free energy is zero. The direction of spontaneous reaction is summarized in Table 15.2.

A number line may be helpful in visualizing the direction of reaction. A spontaneous reaction will occur to bring ΔG to zero, toward equilibrium.

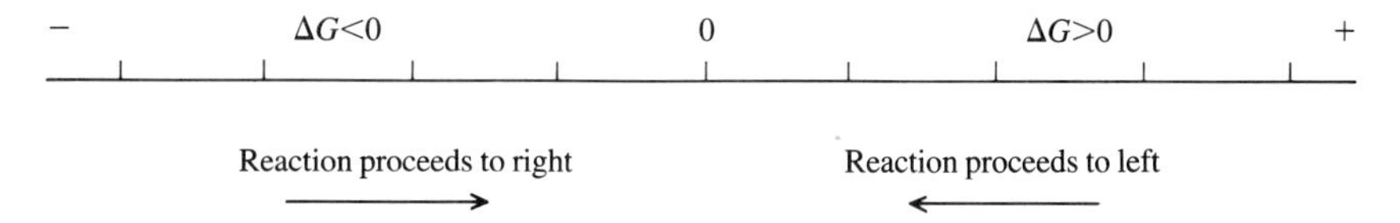

The number line should look somewhat familiar. It is quite similar to the number line used in Chapter 12, where the dependence of the reaction on the equilibrium constant, K_{eq}, and reaction quotient, Q, was presented. The relationships between free energy and equilibrium are explored later in this section.

Just as the change in free energy is related to the enthalpy and entropy, a similar relationship holds for the **standard free energy of formation,** the free energy change during the formation of one mole of a substance in its standard state from the most stable form of the elements in their standard states.

$$\Delta G_f^\circ = \Delta H_f^\circ - T\,\Delta S^\circ \qquad [15.7]$$

Calculations are often simplified when the second law is expressed in terms of ΔG. Changes in standard free energy of a reaction can be calculated from values of $\Delta H°$ and $S°$ by using $\Delta G° = \Delta H° - T\Delta S°$ or by using the standard free energies of formation of the products and reactants. Most tabulations, including Appendix G, provide standard free energies of formation. Like enthalpy, the standard free energy of formation for an element in its standard state is zero. Example 15.4 illustrates the use of tabulated values of standard free energies of formation to determine whether a reaction is spontaneous.

Example 15.4 Determination of $\Delta G°$ for a Reaction

Calculate $\Delta G°$ to determine whether the following reaction will take place spontaneously under standard-state conditions at 298 K:

$$H_2(g) + CO(g) \longrightarrow CH_2O(g)$$

Solution

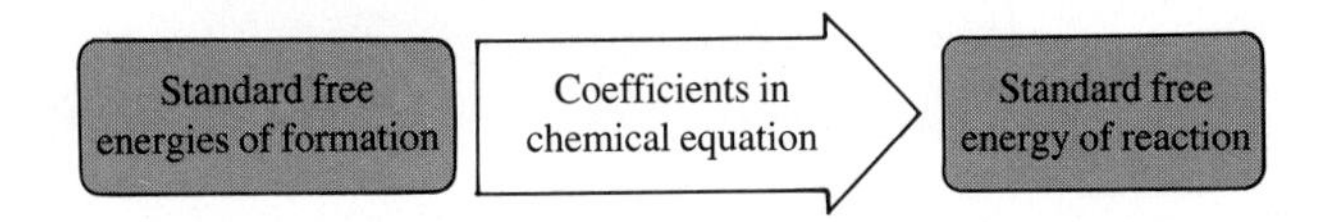

Use the standard free energies of formation from Appendix G.

$$\Delta G°_{reaction} = \Sigma\, n\Delta G°_f[\text{products}] - \Sigma\, n\Delta G°_f[\text{reactants}]$$

$$\Delta G°_{reaction} = (\Delta G°_f[CH_2O]) - (\Delta G°_f[CO] + \Delta G°_f[H_2])$$

$$\Delta G°_{reaction} = -102.55 - (-137.15 + 0)$$

$$\Delta G°_{reaction} = +34.60 \text{ kJ}$$

Since $\Delta G°_{reaction}$ is positive, this reaction is *not* spontaneous at 298 K. Note that a chemist simply cannot look at a chemical equation to determine whether a reaction will be spontaneous. Calculation is required.

Understanding Propene, C_3H_6, is proposed as a starting material in the production of butyraldehyde, C_4H_8O.

$$C_3H_6(g) + CO(g) + H_2(g) \longrightarrow C_4H_8O(\ell)$$

Calculate $\Delta G°_{298}$ to determine whether this reaction is spontaneous under standard-state conditions.

Answer: $\Delta G° = -44.8$ kJ. The reaction is spontaneous. The reaction represents the conversion of propene (often called propylene) into butyraldehyde, also called butanal. Butyraldehyde is an important intermediate in the production of other materials, including polymers and textiles. Propene is derived from natural gas and crude oil and is a *feedstock,* or starting material, in this reaction. The chemical industry produces about 7 billion pounds of butyraldehyde each year.

Even if values for $\Delta G°_f$ are not available, $\Delta G°_f$ can be calculated from $\Delta H°_f$ and $S°$, as illustrated in Example 15.5. This type of calculation provides information that enables us to compare enthalpy and entropy separately to see how each influences the change in free energy.

The free energy change in a reaction can be calculated from tables of $\Delta G°_f$ for the products and reactants, or from tables of $\Delta H°_f$ and $\Delta S°$.

INSIGHTS into CHEMISTRY

Science, Society, and Technology

Nitromethane, A High-Energy Race Car Fuel, Emits Too Many Pollutants for General Use

The following chemical equation describes a possible reaction for the decomposition of nitromethane.

$$4CH_3NO_2(\ell) \longrightarrow 2N_2(g) + 2CO(g) + 2C(s) + 6H_2O(\ell) \qquad \Delta G^\circ = -1639 \text{ kJ}$$

Even though ΔG° is negative, indicating that this reaction is spontaneous, it is a mistake to assume that nitromethane is unstable and prone to explode. The decomposition reaction is extremely slow, and nitromethane is not considered an explosion hazard when handled properly.

Nitromethane will produce even more energy when burned with oxygen. The high energy content of nitromethane makes it a popular fuel for racers who strive to get the highest possible performance from their machines.

When nitromethane is burned with oxygen, the chemical reaction produces nitrogen, carbon dioxide, and water.

$$4CH_3NO_2(\ell) + 3O_2(g) \longrightarrow 2N_2(g) + 4CO_2(g) + 6H_2O(\ell)$$

The free energy change for the oxidation of nitromethane is −2942 kJ, about 80% more than for the decomposition reaction.

Even though nitromethane has favorable burning characteristics, it is not used as a general fuel additive because its combustion can produce some of the nitrogen oxides that have been implicated in the formation of smog.

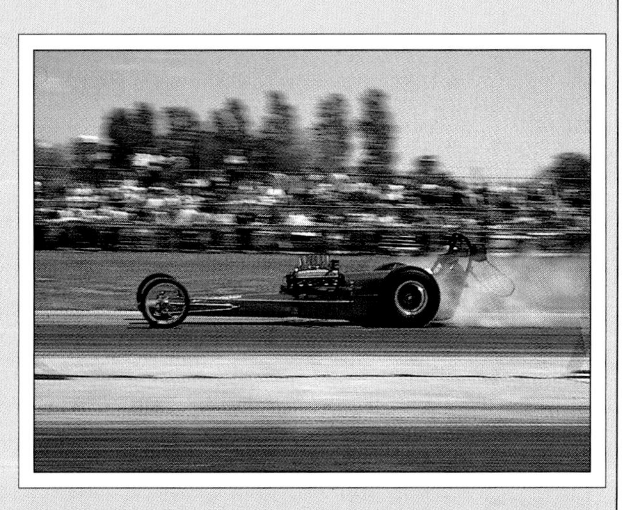

Nitromethane as a fuel A nitromethane-fueled car can accelerate from 0 to 60 mph (100 km/hr) in less than one second. These racing machines have engines that produce more than 2000 horsepower and can complete a quarter-mile (400 m) race in 6.5 seconds.

Example 15.5 Determining ΔG° for a Reaction

Calculate ΔG° for the formation of one mole of HI from H_2 and I_2 in their standard states.

Solution First obtain the thermodynamic data from Appendix G. Since the temperature wasn't specified, we will assume 298 K. We base our calculation on the chemical equation balanced to yield one mole of HI.

	$\frac{1}{2}H_2(g)$ +	$\frac{1}{2}I_2(s)$ →	$HI(g)$
ΔH_f°	0.	0.	26.48 kJ mol^{-1}
S°	130.57	116.14	206.48 J K^{-1} mol^{-1}
ΔG_f°	0.	0.	1.72 kJ mol^{-1}

$$\Delta H^\circ = 1 \text{ mol} (26.48 \text{ kJ mol}^{-1}) = +26.48 \text{ kJ}$$

$$\Delta S^\circ = [1 \text{ mol} (206.48 \text{ J K}^{-1} \text{ mol}^{-1})] - [\tfrac{1}{2} \text{ mol} (130.57 \text{ J K}^{-1} \text{ mol}^{-1}) + \tfrac{1}{2} \text{ mol} (116.14 \text{ J K}^{-1} \text{ mol}^{-1})]$$

$$= +83.13 \text{ J K}^{-1} = 0.08313 \text{ kJ K}^{-1}$$

$$\Delta G^\circ = \Delta H^\circ - T\,\Delta S^\circ$$
$$= +26.48 \text{ kJ} - 298 \text{ K } (0.08313 \text{ kJ K}^{-1})$$
$$= +1.71 \text{ kJ}$$

Alternatively, we could have computed the standard free energy change from ΔG_f° data.

$$\Delta G^\circ = \Sigma\, n\Delta G_f^\circ[\text{products}] - \Sigma\, n\Delta G_f^\circ[\text{reactants}]$$
$$= 1 \text{ mol } (1.72 \text{ kJ mol}^{-1}) - (0 + 0)$$
$$= +1.72 \text{ kJ}$$

The two answers are the same within the uncertainty of the significant figure convention.

Calculating ΔH° and ΔS° separately emphasizes that the formation of hydrogen iodide is a nonspontaneous, endothermic reaction, but that it has a substantial favorable entropy change. This additional information is often useful.

Understanding Compare the free energy of reaction calculated from standard free energies of formation with the free energy calculated from enthalpy and entropy for the decomposition of nitromethane.

$$4CH_3NO_2(\ell) \longrightarrow 2N_2(g) + 2CO(g) + 2C(\text{graphite}) + 6H_2O(\ell)$$

Answer: $\Delta G^\circ = -1639.3$ kJ, $\Delta H^\circ = -1483.6$ kJ, and $\Delta S^\circ = +522.02$ J K^{-1}. Using $\Delta G^\circ = \Delta H^\circ - T\,\Delta S^\circ$, we find $\Delta G^\circ = -1639.2$ kJ.

The Influence of Temperature on Free Energy

The free energy change is strongly influenced by temperature, so the temperature must be specified. We know that a decrease in enthalpy and an increase in entropy both favor spontaneous change. These observations are in agreement with Equation 15.6,

$$\Delta G = \Delta H - T\Delta S$$

because a negative ΔH and a positive ΔS both contribute to a negative ΔG.

Temperature influences the sign of ΔG mainly through the $T\Delta S$ term, although both ΔH° and ΔS° change slightly with changes in the temperature. As long as no phase changes are involved, and the temperature change is not very large (a few hundred kelvin), little error is introduced by assuming the enthalpy and entropy changes are constant. At low temperatures, the reaction is generally dominated by the sign of ΔH; as temperature increases, the $T\Delta S$ contribution becomes increasingly important. Whether an increase in temperature favors spontaneity depends on the sign of ΔS. Since the terms enthalpy and entropy can be positive or negative, any chemical reaction falls into one of four general classes, shown in Table 15.3.

Temperature influences the free energy change primarily through the $T\Delta S$ term.

When ΔH° and ΔS° have the same sign, the enthalpy term determines the sign of ΔG° at low temperature, and at high temperature, the sign of the entropy change determines the sign of ΔG°. When they have opposite signs, increasing temperature does not change the direction of spontaneous reaction. For a large majority of chemical reactions, the signs of ΔH° and ΔS° are the same. A negative ΔH° usually means that there are stronger bonds in the products than in the reactants. This type of reaction usually produces a more

Table 15.3 The Influence of Temperature on the Direction of Spontaneous Reaction

ΔH	ΔS	Temperature	Sign of ΔG	Direction
negative	positive	all	negative	forward
negative	negative	low	negative	forward
negative	negative	high	positive	reverse
positive	positive	low	positive	reverse
positive	positive	high	negative	forward
positive	negative	all	positive	reverse

ordered system, so the entropy of the products is less than that of the reactants.

The next example illustrates a crucial concept, the ability of temperature to influence a reaction. If the thermodynamics of a reaction are understood, it is sometimes possible to change the temperature to achieve spontaneity. The next example also includes the approximation that ΔH_f° and S° are not strongly influenced by temperature. Although the changes are generally small, scientists who need more exact calculations use extended models that predict ΔH_f° and S° at other temperatures, or they use experimental data obtained at these temperatures.

Example 15.6 The Influence of Temperature on a Spontaneous Reaction

Consider the production of methane from carbon monoxide and hydrogen.

$$2CO(g) + 2H_2(g) \longrightarrow CO_2(g) + CH_4(g)$$

Use the following data (from Appendix G) to calculate ΔG°_{298} (ΔG° at 298 K) and ΔG°_{1000}. Assume that ΔH° and ΔS° do not depend on temperature.

	$2CO(g)$	+ $2H_2(g)$ ⟶	$CO_2(g)$	+ $CH_4(g)$	
ΔH_f°	−110.52	0.	−393.51	−74.81	kJ mol^{-1}
S°	197.56	130.57	213.63	186.15	J mol^{-1} K^{-1}
ΔG_f°	−137.15	0.	−394.36	−50.75	kJ mol^{-1}

Solution

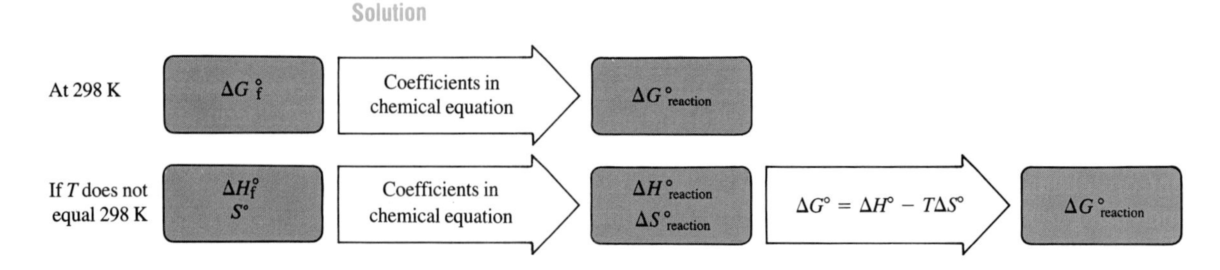

At 298 K,

$$\Delta G^\circ_{298} = (\Delta G_f^\circ[CO_2] + \Delta G_f^\circ[CH_4]) - (2\Delta G_f^\circ[CO] + 2\Delta G_f^\circ[H_2])$$

$$\Delta G^\circ_{298} = [1 \text{ mol} (-394.36 \text{ kJ/mol}) + 1 \text{ mol} (-50.75 \text{ kJ/mol})] - [2 \text{ mol} (-137.15 \text{ kJ/mol}) + 2 \text{ mol} (0 \text{ kJ/mol})]$$

$$\Delta G^\circ_{298} = -170.81 \text{ kJ}$$

so the reaction is spontaneous at 298 K.

At 1000 K, use ΔH° and $T\,\Delta S^\circ$ to calculate ΔG°:

$$\Delta H^\circ = [-393.51 - 74.81] - [2(-110.52)] = -247.28 \text{ kJ}$$

$$\Delta S^\circ = [213.63 + 186.15] - [2(197.56) - 2(130.57)] = -256.48 \text{ J K}^{-1}$$

Convert entropy to kJ K^{-1} and calculate ΔG.

$$\Delta G^\circ_{1000} = \Delta H^\circ - T\,\Delta S^\circ = -247.28 \text{ kJ} - 1000 \text{ K} (-0.25648 \text{ kJ K}^{-1})$$

$$\Delta G^\circ_{1000} = +9.20 \text{ kJ}$$

At 1000 K the reaction is no longer spontaneous.

Understanding Calculate ΔG° at 25 °C and at 300 °C for the synthesis of ammonia under standard conditions to determine whether the reaction is spontaneous at those temperatures.

$$3H_2(g) + N_2(g) \rightleftharpoons 2NH_3(g)$$

Answer: $\Delta G^\circ_{298} = -33.0$ kJ. The reaction is spontaneous. $\Delta G^\circ_{573} = +21.6$ kJ, so the reaction is not spontaneous at the higher temperature.

Phase Transitions

The effect of temperature on phase transitions such as melting and freezing is particularly interesting. Phase transitions also provide a good practical test for the laws of thermodynamics, since the temperature at which a substance melts or boils is relatively easy to measure with high accuracy.

Heat of Fusion

We can use our qualitative knowledge of thermodynamics to predict signs of the entropy and enthalpy changes that occur during a phase transition. The melting (or fusion) of ice is a good transition to study.

$$H_2O(s) \longrightarrow H_2O(\ell)$$

At equilibrium, ΔG° is zero.

$$\Delta G^\circ = 0 = \Delta H^\circ - T\,\Delta S^\circ$$

$$\Delta H^\circ = T\,\Delta S^\circ$$

The sign of ΔH° will be the same as the sign of ΔS°, and since the entropy increases upon going from the ordered solid to the less orderly liquid, ΔS° is positive, so ΔH° is also positive. Melting is always endothermic; ice cools the surroundings as it melts. These conclusions apply to materials other than water: *for any fusion or melting, the standard heat of fusion is positive.*

Heat of Vaporization

Let us compare the enthalpy and entropy of vaporization to the enthalpy and entropy of fusion. The thermodynamic data for the fusion and vaporization processes of several liquids are presented in Table 15.4. The transition from a

Table 15.4 Thermodynamics of Phase Changes

Substance		Molar Mass, g/mol	Fusion[1] (solid → liquid) T_f, K	ΔH°_{fus}, kJ/mol	ΔS°_{fus}, J/mol-K	Vaporization[2] (liquid → gas) T_b, K	ΔH°_{vap}, kJ/mol	ΔS°_{vap}, J/mol-K
H_2O	water	18.0	273	6.01	22.2	373	44.0	118.8
CH_3CH_2OH	ethanol	46.0	156	4.60	29.7	351	37.4	121.3
C_6H_6	benzene	78.1	279	9.83	35.4	353	33.9	96.2

[1] Measured at T_f
[2] Measured at 298 K

ΔS°_{vap} is positive and always larger than ΔS°_{fus} for the same substance.

liquid to a gas creates more disorder and requires more heat than does the melting process.

The standard heats of vaporization shown in Table 15.4 can be calculated from standard heats of formation, as for any other reaction, if the data are available. Appendix G has entries for the standard enthalpy of formation of $H_2O(\ell)$ and $H_2O(g)$, so the heat of the vaporization process can be calculated.

$$H_2O(\ell) \longrightarrow H_2O(g)$$

$$\Delta H^\circ = \Delta H^\circ_f[H_2O(g)] - \Delta H^\circ_f[H_2O(\ell)]$$

$$\Delta H^\circ = 1\ \text{mol}\ (-241.82\ \text{kJ/mol}) - 1\ \text{mol}\ (-285.83\ \text{kJ/mol})$$

$$\Delta H^\circ = +44.01\ \text{kJ to vaporize 1 mol of}\ H_2O$$

The entropy change can be calculated in a similar manner.

The Influence of Temperature on Phase Changes

Experimentally, we know that ice melts at 0 °C. We can calculate the free energy change associated with freezing at −5 °C and +5 °C to determine whether the reaction is spontaneous at these temperatures.

First, write the "chemical" equation.

$$H_2O(s) \longrightarrow H_2O(\ell)$$

Next, calculate ΔG° at −5 °C (268 K).

$$\Delta G^\circ_{268} = \Delta H^\circ - T\Delta S^\circ$$

Remember to divide the entropy by 1000 to convert from J K^{-1} to kJ K^{-1}.

$$\Delta G^\circ_{268} = +6.01\ \text{kJ} - 268\ \text{K} \times 22.0\ \frac{\text{J}}{\text{K}}\left(\frac{1\ \text{kJ}}{1000\ \text{J}}\right)$$

$$\Delta G^\circ_{268} = +6.01\ \text{kJ} - 5.90\ \text{kJ} = +0.11\ \text{kJ}$$

The positive sign for ΔG° indicates that the reaction (the melting of ice) is *not* spontaneous at −5 °C. In fact the reverse reaction, the freezing of water to ice, is spontaneous at −5 °C. Next, calculate ΔG° at +5 °C (278 K).

$$\Delta G^\circ_{278} = \Delta H^\circ - T\,\Delta S^\circ$$

$$= +6.01 \text{ kJ} - 278 \text{ K} \times 22.0 \frac{\text{J}}{\text{K}} \left(\frac{1 \text{ kJ}}{1000 \text{ J}}\right)$$

$$= +6.01 \text{ kJ} - 6.12 \text{ kJ}$$

$$= -0.11 \text{ kJ}$$

The negative sign for the free energy change indicates that the melting of ice *is* spontaneous at 5 °C. Note that these calculations use data for 273 K and apply them at other temperatures, so some (hopefully small) errors are expected.

We can use similar methods to find the temperature at which water will boil. The *normal boiling point* is the temperature at which the vapor pressure of the liquid reaches one atmosphere.

$$H_2O(\ell) \longrightarrow H_2O(g)$$

At the boiling point, the gas (at one atmosphere) is in equilibrium with the liquid, so ΔG is zero.

At the boiling point, the liquid and vapor are in equilibrium, so ΔG is zero.

$$\Delta G^\circ = 0 = \Delta H^\circ - T\,\Delta S^\circ$$

The data for one mole of water, from Table 15.4, can be used to calculate the boiling point.

$$\Delta H^\circ_{\text{vap}} = +44.0 \text{ kJ}$$

$$\Delta S^\circ_{\text{vap}} = 118.8 \text{ J K}^{-1} = 0.1188 \text{ kJ K}^{-1}$$

$$T = \frac{\Delta H^\circ}{\Delta S^\circ} = \frac{44.01 \text{ kJ}}{0.1188 \text{ kJ K}^{-1}}$$

$$T = 370.5 \text{ K} = 97.30 \text{ °C}$$

The true boiling point is 100 °C. The discrepancy is due to the temperature dependence of ΔH° and ΔS°. The values of ΔH° and ΔS° that we used are valid at 298 K and cannot be used at 373 K without sacrificing some accuracy.

15.5 Concentration, Free Energy, and the Equilibrium Constant

Objectives

- To evaluate the effect of concentration on free energy
- To develop the relationship between standard free energy change and the equilibrium constant
- To determine the effect of temperature on the equilibrium constant
- To determine the influence of temperature on vapor pressure

Studies of equilibria led to the observation that some spontaneous reactions can be forced in the reverse direction by adding an overwhelming excess of one or more of the products. This observation was described qualitatively by the principle of Le Chatelier. The purpose of this section is to determine how

INSIGHTS into CHEMISTRY

Science, Society, and Technology

Phase Changes of Glauber's Salt Can Be Used to Heat Your Home

We use energy to power our factories, drive our automobiles, and heat our homes. For the last several centuries, this energy has come from burning fossil fuels, namely, oil, coal, and natural gas. Most homes and workplaces are heated by burning natural gas or oil, or by electrical power that is also generated from fossil fuels. Despite the universal availability of solar power, it is used very little, and only during the day, to provide warmth. However, solar heating systems may be an important energy source in the future.

A desirable energy source for home heating would be completely reusable, cheap, and long-lasting. Ideally, there should be no waste products; even spent systems should be easily and cheaply recycled. Solar heating meets most of these criteria.

Scientists and engineers have been studying solar heating techniques for years. One clever solar heating system was developed by Dr. Maria Telkes, of MIT, in the late 1940s. The home she had built, called the Dover House, took advantage of the large heat of fusion of **Glauber's salt,** sodium sulfate decahydrate or $Na_2SO_4 \cdot 10H_2O$. This substance undergoes a phase transition at a particularly convenient temperature, 32.4 °C. Because of this behavior, Glauber's salt is called a *phase change material.*

$$Na_2SO_4 \cdot 10H_2O(s) \rightleftharpoons Na_2SO_4(aq) + 10H_2O(\ell)$$

Glauber's salt absorbs energy during the day by melting, and it releases that energy during cooler hours as the liquid freezes. The heat of fusion is 80.93 kJ/mol; in comparison, the heat of fusion of water is only 6.01 kJ/mol. Taking into account the difference in density, Glauber's

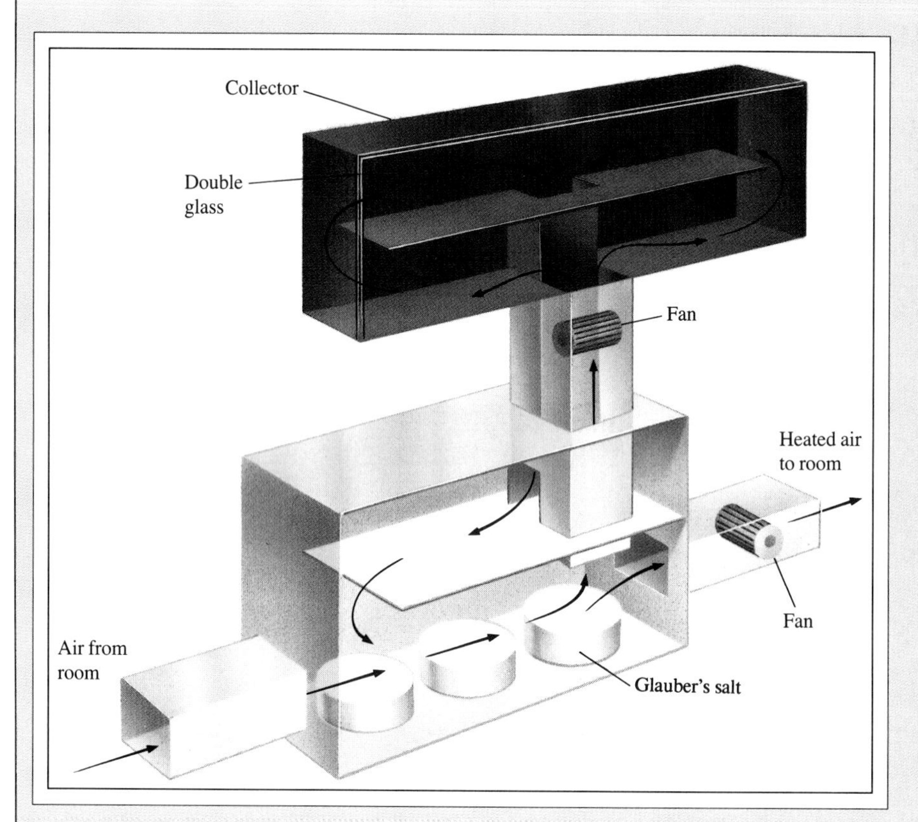

Solar energy collector Solar energy is absorbed during the day by the collector. The hot air melts the Glauber's salts during the day, but the salts freeze at night. The heat evolved as the salts refreeze is used to maintain the house at a comfortable temperature.

salt can store about nine times more heat than the same volume of water. The following table lists a few other phase change materials.

Some Properties of Select Phase Change Materials

Name	Formula	Melting Point, °C	Heat of Fusion	
			kJ/kg	kJ/mol
sodium hydroxide	$NaOH \cdot H_2O$	64	272.1	15.79
sodium acetate	$CH_3COONa \cdot 3H_2O$	58	180.0	24.49
sodium thiosulfate	$Na_2S_2O_3 \cdot 5H_2O$	48	200.5	49.75
disodium hydrogen phosphate	$Na_2HPO_4 \cdot 12H_2O$	36	263.7	94.44
sodium sulfate	$Na_2SO_4 \cdot 10H_2O$	32	251.2	80.93
sodium carbonate	$Na_2CO_3 \cdot 10H_2O$	34	251.2	71.88

The Dover House had iron solar collectors that surrounded cans of Glauber's salt. Air circulated around the collectors and through the Glauber's salt, and was then used to heat the house. Although Dr. Telkes designed the system with enough capacity for seven cold, completely sunless days—which required an enormous volume of the phase change material—it was not long before a period of 11 consecutive cloudy days exhausted the solar heating system. Because of limitations like this one, modern solar houses are always designed with a conventional backup heating system.

Dover House This 50-year-old Massachusetts house was designed to be heated by solar energy.

free energy is influenced by concentration, a detail that has been deferred to this point.

Concentration and Free Energy

The calculations performed so far have assumed that all species are present in their standard states (generally at 298 K). Solids were present as pure materials in the most stable form; liquids as pure liquids; solutions with the solute present at a concentration of 1 *M*; and gases at a partial pressure of 1 atm.

When a substance is *not* in its standard state, its free energy depends on its concentration—higher concentrations will have more positive free energies. Consider a gas that is present at some concentration other than 1.0 atm. Its free energy is given by

$$G = G^\circ + RT \ln P$$

where

G is the free energy of the substance, kJ mol^{-1}
G° is the free energy of the substance in its standard state, kJ mol^{-1}
R is the gas constant, 8.314 J mol^{-1} K^{-1}
T is the temperature, in K
P is the partial pressure of the gas, in atm

The *Insights into Chemistry* on page 629 examines the reaction of nitrogen with hydrogen to form ammonia.

$$N_2(g) + 3H_2(g) \rightleftharpoons 2NH_3(g)$$

The free energy change is given by:

$$\Delta G = \Delta G^\circ + RT \ln \left(\frac{P^2_{NH_3}}{P_{N_2} P^3_{H_2}} \right)$$

Concentration influences the free energy change, since $\Delta G = \Delta G^\circ + RT \ln Q$.

Note that the free energy change in this example includes a term for the standard free energy change and one for the reaction quotient. For any reaction,

$$\Delta G = \Delta G^\circ + RT \ln Q \qquad [15.8]$$

We can now determine ΔG, at any combination of concentrations, from ΔG° and Q. The change in the standard free energy is calculated from tabulated data, such as those in Appendix G; the reaction quotient is determined from the concentrations or partial pressures of the reactants and products present in the experiment. Note that when ΔG° and $RT \ln Q$ are added, they must have the same units. R is generally expressed as 8.314 J mol^{-1} K^{-1}, so $RT \ln Q$ will have units of J mol^{-1}, and ΔG° must be expressed in J, rather than kJ as it appears in tabulations, to be added correctly. A sample calculation that illustrates the influence of concentration on free energy is presented in Example 15.7.

Example 15.7 Determining the Free Energy Change of a Reaction Under Nonstandard Conditions

Calculate the free energy change for the reaction of nitrogen monoxide and bromine to form nitrosyl bromide at 298 K under two sets of conditions.

$$2NO(g) + Br_2(\ell) \longrightarrow 2NOBr(g)$$

(a) The partial pressure of each gas is 1.0 atm.
(b) The partial pressure of NO is 0.10 atm and the partial pressure of NOBr is 2.0 atm.

Solution

(a) Standard-state conditions

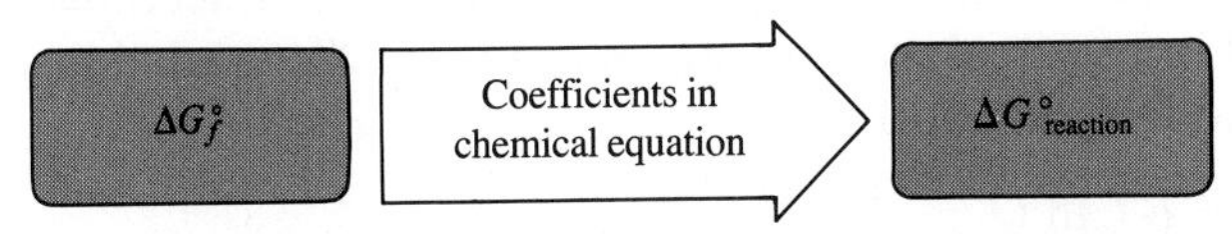

The reactants and products are all in their standard states: gases are present with partial pressures of 1 atm and bromine is present as the liquid. Under these conditions, ΔG will equal ΔG°, which we calculate from standard free energies of formation from Appendix G.

$$\Delta G^\circ = (2\Delta G_f^\circ[NOBr]) - (2\Delta G_f^\circ[NO] + \Delta G_f^\circ[Br_2])$$

$$\Delta G^\circ = [2 \text{ mol } (82.4 \text{ kJ mol}^{-1})] - [2 \text{ mol } (86.55 \text{ kJ mol}^{-1}) + 1 \text{ mol } (0 \text{ kJ mol}^{-1})]$$

$$\Delta G^\circ = -8.3 \text{ kJ}$$

(b) Nonstandard state conditions

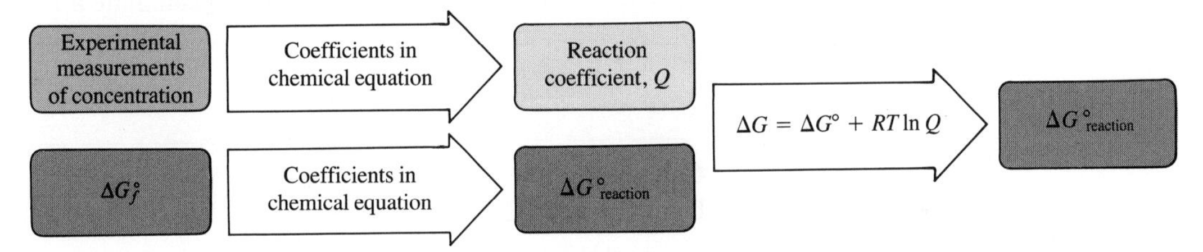

When the substances are not present at standard state conditions, the reaction quotient must be calculated.

$$2NO(g) + Br_2(\ell) \longrightarrow 2NOBr(g)$$

The bromine is present as the liquid, so its concentration is not included in Q.

$$Q = \frac{P^2_{NOBr}}{P^2_{NO}} = \frac{2.0^2}{0.1^2} = 4.0 \times 10^2$$

ΔG is calculated from ΔG° and Q. Convert ΔG° (from part (a)) to J before adding.

$$\Delta G = \Delta G^\circ + RT \ln Q$$

$$\Delta G = -8.3 \text{ kJ} + (8.314 \text{ J mol}^{-1} \text{ K}^{-1})(298 \text{ K}) \ln (4.0 \times 10^2)$$

$$\Delta G = -8.3 \times 10^3 \text{ J} + 1.5 \times 10^4 \text{ J}$$

$$\Delta G = +0.7 \times 10^3 \text{ J} = 0.7 \text{ kJ}$$

Notice how increasing the concentration of the products and decreasing the concentration of a reactant made a reaction that is spontaneous in the forward direction (part (a)) become spontaneous in the reverse direction.

Understanding Calculate the free energy change for the reaction of carbon dioxide and ammonia to form urea at 298 K when all gases are present at 0.10 atm pressures.

$$CO_2(g) + 2NH_3(g) \longrightarrow CO(NH_2)_2(s) + H_2O(\ell)$$

Answer: $\Delta G° = -7.30$ kJ; $Q = 1.0 \times 10^3$; $\Delta G = +9.81$ kJ. This reaction is spontaneous in the standard state, but not when the reactants are present at low pressure.

Equilibrium Constant and Free Energy

A system at equilibrium does not change spontaneously in either direction; at equilibrium ΔG must equal zero and Q is equal to K_{eq}. These relationships can be substituted in Equation 15.8:

$$\Delta G = \Delta G° + RT \ln Q$$

$$0 = \Delta G° + RT \ln K_{eq}$$

$$\Delta G° = -RT \ln K_{eq} \quad [15.9]$$

Equation 15.9 relates the standard free energy change to the temperature and equilibrium constant. If we rearrange this equation, the equilibrium constant can be calculated from the standard free energy change.

$$K_{eq} = e^{-\Delta G°/RT} \quad [15.10]$$

These equations provide the relationships between thermodynamic and equilibrium measurements. Notice particularly that the temperature is an important parameter, one that is discussed in the next section. The following example shows how thermodynamic data can be used to calculate an equilibrium constant.

$\Delta G° = -RT \ln K_{eq}$

$K_{eq} = e^{-\Delta G°/RT}$

Example 15.8 Calculating the Equilibrium Constant from the Free Energy Change

The standard free energy change is −79.9 kJ at 298 K for

$$H_3O^+(aq) + OH^-(aq) \longrightarrow 2H_2O(\ell)$$

Calculate the equilibrium constant for the reaction.

Solution

$$K_{eq} = e^{-\Delta G°/RT}$$

$$\frac{\Delta G°}{RT} = \frac{-79900 \text{ J mol}^{-1}}{(8.314 \text{ J mol}^{-1} \text{ K}^{-1})(298 \text{ K})} = -32.25$$

$$K_{eq} = e^{+32.25}$$

$$K_{eq} = 1.0 \times 10^{14}$$

(The reaction is, of course, the reverse of the autoionization of water.) Notice that the energy was expressed in joules to match the units of R.

Understanding For the formation of formaldehyde,

$$H_2(g) + CO(g) \longrightarrow CH_2O(g)$$

we calculated $\Delta G° = +34.6$ kJ in Example 15.4. Calculate the equilibrium constant at 298 K.

Answer: $K_{eq} = 8.6 \times 10^{-7}$.

INSIGHTS into CHEMISTRY

A Closer Look

Free Energy Changes Are Influenced by Changes in Pressure and Concentration

The first free energy calculations performed have assumed that all species are present in their standard states (generally at 298 K). Solids and liquids were in the pure state; solutions with a solute concentration of 1 *M*; and gases at a partial pressure of 1 atm. Free energy changes, unlike enthalpy changes, are quite sensitive to changes in pressure and concentration. In this *Insights into Chemistry* we investigate the dependence of free energy on these two variables.

To investigate the effect of a change in pressure on the free energy, examine the change in free energy when one mole of an ideal gas undergoes a change in pressure from 1 atm to some other pressure, *P*, at a constant temperature.

$$\text{gas (1 atm)} \longrightarrow \text{gas } (P \text{ atm})$$

$$\Delta G = \Delta H - T\Delta S$$

ΔH for the constant-temperature expansion of an ideal gas is zero, and any change in free energy must arise from the entropy term.

One of the principal factors that contribute to the entropy of a system is the volume it occupies. For example, the positive entropy change observed when most solutes dissolve is due to the increase in the volume going from the pure solid to the solution.

The quantitative relationship between the change in entropy and the change in volume for one mole of an ideal gas is known.

$$\Delta S = R \ln \frac{V_2}{V_1}$$

R is the ideal gas constant (8.314 J mol^{-1} K^{-1}), V_1 is the initial volume of the gas, and V_2 is the final volume. By using Boyle's law (PV = constant), this entropy change can be expressed in terms of the initial (P_1) and final (P_2) pressures of the ideal gas.

$$\Delta S = -R \ln \frac{P_2}{P_1}$$

Therefore, when one mole of an ideal gas changes from its standard state of 1 atm to some other pressure, P, the change in entropy is $-R \ln P$, and the change in free energy is

$$\Delta G = \Delta H - T\Delta S = 0 - T(-R \ln P) = RT \ln P$$

In its initial state, the free energy of the sample is $G°$, since the gas was initially in its standard state. The free energy after the pressure change, G, is therefore

$$G = G° + \Delta G = G° + RT \ln P$$

Note that the pressure *must be expressed in atmospheres* for this equation to be valid.

Next we determine how nonstandard state conditions affect the free energy change for a reaction involving gases. As an example, we take the formation of ammonia from nitrogen and hydrogen, with all species in the gas phase, but not necessarily at partial pressures of 1.0 atm.

$$N_2(g) + 3H_2(g) \longrightarrow 2NH_3(g)$$

The free energy change of the reaction can be calculated from the free energies of the reactants and products.

$$\Delta G = 2G_{NH_3} - G_{N_2} - 3G_{H_2}$$

where the G's with subscripts represent the free energies of one mole of the substance. We use the relationship between free energy and partial pressure to express the free energy of each substance as a function of its standard free energy and pressure and substitute into the previous equation.

$$\Delta G = 2(G°_{NH_3} + RT \ln P_{NH_3}) - (G°_{N_2} + RT \ln P_{N_2}) - 3(G°_{H_2} + RT \ln P_{H_2})$$

$$\Delta G = (2G°_{NH_3} - G°_{N_2} - 3G°_{H_2}) + RT\,(2 \ln P_{NH_3} - \ln P_{N_2} - 3 \ln P_{H_2})$$

$$\Delta G = \Delta G° + RT\,(\ln P^2_{NH_3} - \ln P_{N_2} - \ln P^3_{H_2})$$

$$\Delta G = \Delta G° + RT \ln \frac{P^2_{NH_3}}{P_{N_2} P^3_{H_2}}$$

In the series of equations we rearranged the terms and used the rules for combining logarithms. The combination of the partial pressures is the reaction quotient, Q, for the reaction.

$$\Delta G = \Delta G° + RT \ln Q \qquad [15.8]$$

Table 15.5 The Effect of Changing Concentrations of Species on a Chemical Reaction

Chemical Change	Reaction Quotient	ΔG	Direction Favored
increasing the concentration of the reactants or decreasing the concentration of the products	decrease	decrease	forward
increasing the concentration of the products or decreasing the concentration of the reactants	increase	increase	reverse

If Q is less than K_{eq}, or if ΔG is less than 0, the forward reaction proceeds spontaneously.

Both the free energy change and the equilibrium constant can be used to determine the direction of spontaneous reaction. If the free energy change is negative, the reaction is spontaneous in the forward direction. Likewise, if the reaction quotient Q is less than K_{eq}, the forward reaction occurs spontaneously. To determine whether a reaction is spontaneous, you can use a number line to compare Q with K_{eq}. Alternatively, you can use a number line to display ΔG.

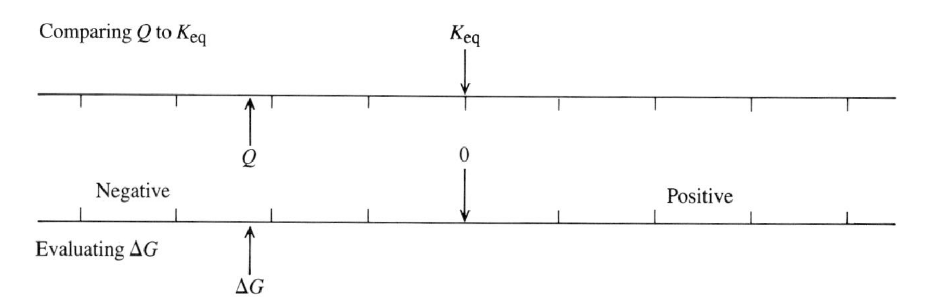

The manner in which the direction of spontaneous reaction changes with changing concentrations is summarized in Table 15.5.

The principle of Le Chatelier predicts the effect of changing concentration on a qualitative basis. Quantitative methods, such as calculating ΔG or comparing Q with K_{eq}, work as well.

Temperature and Equilibrium Constant

Chemical reactions differ widely in their response to changing temperature. For any particular chemical system, the direction of spontaneous reaction is determined by comparing Q to K_{eq}. However, the equilibrium constant itself depends on temperature, as we noted in Equation 15.9:

$$\Delta G^\circ = -RT \ln K_{eq}$$

Since we also have the definition of free energy,

$$\Delta G^\circ = \Delta H^\circ - T\,\Delta S^\circ$$

the effect of increasing temperature on K_{eq} can be determined mathematically. Substituting,

$$\Delta H^\circ - T\,\Delta S^\circ = -RT \ln K_{eq}$$

$$\ln K_{eq} = \frac{\Delta S^\circ}{R} - \frac{\Delta H^\circ}{RT}$$

If we ignore the small temperature dependences of ΔH° and ΔS°, we see that the effect of temperature on K_{eq} depends on the sign of ΔH°. Consider the influence of increasing temperature on the equilibrium constant of an exothermic reaction. Since ΔH° is negative, $\ln K_{eq}$ will decrease at higher temperatures and the equilibrium will favor the reactants. Increasing the temperature of an endothermic reaction (one in which ΔH° is positive) causes the equilibrium constant to increase, so increased temperature favors the products. In summary, the influence of temperature on the equlibrium constant is exactly as predicted by the principle of Le Chatelier: heating an exothermic reaction shifts the equilibrium toward the reactants, in the direction that consumes the added heat. Table 15.6 summarizes the effect of temperature on the equilibrium constant.

The effect of temperature on a reaction can be predicted by the principle of Le Chatelier. Adding heat to an exothermic reaction favors the reactants.

Do not confuse the effect of temperature on free energy with the effect on the equilibrium constant. Temperature influences free energy through the $T\Delta S$ term in $\Delta G = \Delta H - T\,\Delta S$, but it influences the equilibrium constant through the ΔH term.

The Influence of Temperature on Vapor Pressure

When water is heated to 100 °C at a pressure of 1 atm, it boils; at this temperature the vapor pressure of the water is equal to 1.0 atm. At temperatures lower than the boiling point, the vapor pressure is less than 1 atm.

The relationship between temperature and vapor pressure is not difficult to determine.

For the vaporization of water

$$H_2O(\ell) \longrightarrow H_2O(g)$$

the equilibrium constant expression is simply

$$K_{eq} = P_{H_2O}$$

The temperature dependence of the vapor pressure, which equals the equilibrium constant, is given by Equation 15.10.

$$P_{H_2O} = K_{eq} = e^{-\Delta G^\circ/RT}$$

Table 15.6 Influence of Temperature on the Equilibrium Constant

Sign of ΔH	Change in Temperature	Change in K_{eq}
+ (endothermic)	increase	increase
− (exothermic)	increase	decrease

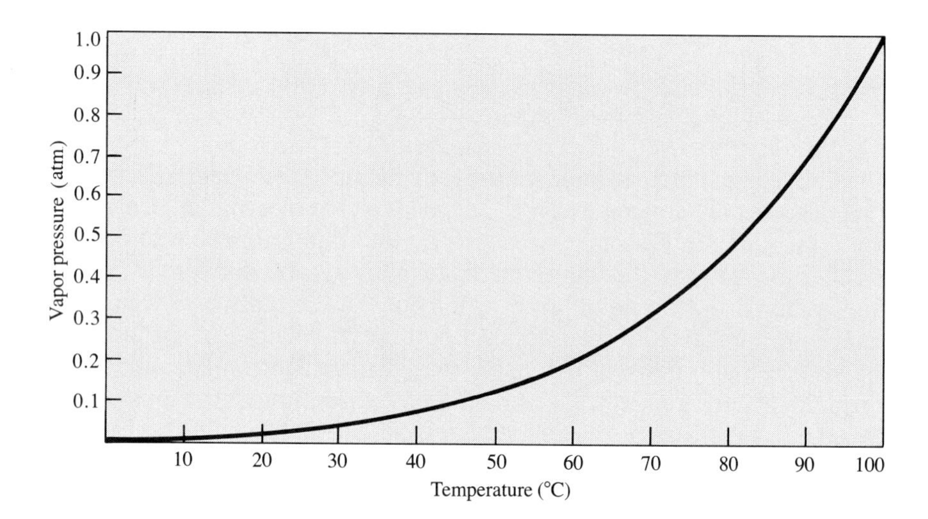

Figure 15.5 Dependence of vapor pressure of water on temperature

The equilibrium constant has an exponential dependence on the reciprocal of temperature. The same relationship is true for the vaporization of any liquid. Figure 15.5 shows the calculated dependence of vapor pressure on temperature, a calculation illustrated in Example 15.9.

Example 15.9 Determining the Vapor Pressure of Water as a Function of Temperature

Calculate the vapor pressure of water at 15 °C.

Solution

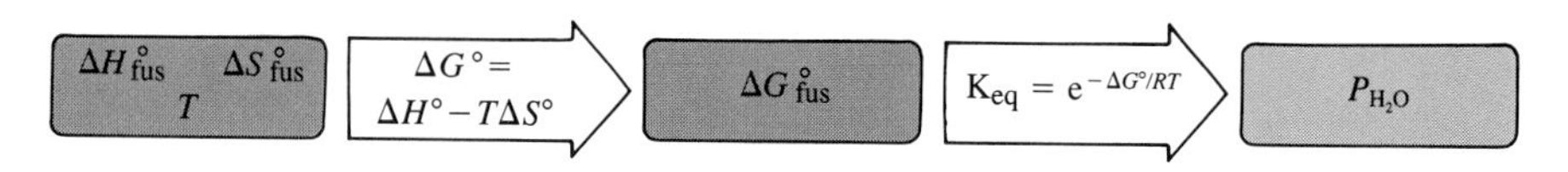

First, write the equation and expression for the equilibrium constant.

$$H_2O(\ell) \longrightarrow H_2O(g) \qquad K_{eq} = P_{H_2O}$$

Next determine $\Delta H°$ and $\Delta S°$ for the process, using the enthalpy and entropy of vaporization. In this case, $\Delta H^\circ_{vap} = +44.01\ \text{kJ mol}^{-1}$ and $\Delta S^\circ_{vap} = +118.8\ \text{J mol}^{-1}\ \text{K}^{-1}$.

Calculate the standard free energy change:

$$\Delta G° = \Delta H° - T\Delta S°$$

$$\Delta G° = 44010\ \text{J} - 288\ \text{K} \times 118.8\ \text{J K}^{-1}$$

$$\Delta G° = 9796\ \text{J}$$

Last, calculate the equilibrium constant at the given temperature.

$$K_{eq} = e^{-\Delta G°/RT}$$

$$\frac{\Delta G°}{RT} = \frac{-9796\ \text{J}}{(8.314\ \text{J mol}^{-1}\ \text{K}^{-1})(298\ \text{K})} = +\,4.09$$

$$K_{eq} = e^{-\Delta G°/RT} = e^{-4.09} = 0.0167$$

$$P_{H_2O} = 0.0167\ \text{atm}$$

Notice that the units of pressure are atmospheres.

Trouton's Law States Constant Relationship Between Heat of Vaporization and Boiling Point

The enthalpy of vaporization is determined by measuring the heat needed to produce a mole of vapor from a mole of liquid. Another way is to measure the vapor pressure at several temperatures. The heat of vaporization depends strongly on the kind of bonding in the liquid; liquids with strong intermolecular forces must absorb more heat when they vaporize than do those with weaker attractions. On the other hand, the entropy of vaporization depends on the change in disorder as a mole of liquid becomes a mole of gas. This increase in disorder is nearly independent of structure and bonding, so *the entropy of vaporization is approximately the same for most substances.*

Measurements by the Irish physicist Frederick Trouton (1863–1922) showed that the entropy of vaporization indeed is nearly the same for most substances. Trouton measured boiling points and heats of vaporization for many compounds, and he noticed that the ratio of the heat of vaporization to the boiling point was almost the same for most of them. A series of similar compounds, such as the organic hydrocarbons shown in the following table, provides consistent data; most other substances have surprisingly similar entropies of vaporization.

Trouton used the data to estimate ΔH°_{vap} for other species by measuring their boiling points and applying the equation

$$\Delta H^\circ_{vap} = T_b \times 88$$

We now understand the basis of this relationship. Write the chemical and thermochemical equations for the vaporization process:

$$\text{liquid} \rightleftharpoons \text{gas}$$

$$\Delta G^\circ_{vap} = \Delta H^\circ_{vap} - T\Delta S^\circ_{vap}$$

At the boiling point, the liquid and gas are in equilibrium, so ΔG° is 0. At equilibrium:

$$\Delta H^\circ_{vap} = T\Delta S^\circ_{vap}$$

$$\Delta H^\circ_{vap}/T = \Delta S^\circ_{vap}$$

Therefore, the ratio of $\Delta H^\circ_{vap}/T_b$ in the table is ΔS°_{vap}. According to Trouton's rule, the "average" value for the entropy of vaporization is 88 J mol^{-1} K^{-1}.

Most chemists would guess that there is a relationship between the boiling temperature and the strength of the forces that hold the liquid together. Trouton's law confirms this hypothesis. The heat of vaporization, which is directly related to the strengths of the intermolecular forces, is directly proportional to the boiling point.

$$\Delta H^\circ_{vap} = \Delta S^\circ_{vap} \times T_b$$

The observation that the entropy of vaporization is about 88 J mol^{-1} K^{-1} for many substances has been studied by chemists in detail. None of the current models of chemical structure and bonding, even the most complex and detailed, can explain this particular value.

Boiling Points and Heats of Vaporization (Determined at T_b) for Several Substances

Compound		ΔH°_{vap} kJ/mol	Boiling Point (T_b)		$\Delta H^\circ_{vap}/T_b$ J mol^{-1} K^{-1}
			°C	K	
methane	CH_4	8.2	−161	112	73
ethane	C_2H_6	13.8	−89	184	75
propane	C_3H_8	19.0	−42	231	82
butane	C_4H_{10}	22.4	0	273	82
hexane	C_6H_{14}	28.8	69	342	84
octane	C_8H_{18}	34.4	126	399	86
decane	$C_{10}H_{22}$	38.8	174	447	87
Average for organic hydrocarbons:					80
bromine	Br_2	32.5	58	331	98
methanol	CH_3OH	35.2	65	338	104
ethanol	C_2H_5OH	38.6	78	351	110
mercury	Hg	59.2	357	630	94
Average used by Trouton:					88

The Great Pyramid at Giza This massive structure was built as a burial vault for the Pharaoh Cheops (Khufu). The construction was done about 4000 years ago, and probably took about 25 years to complete. The energy of 100,000 workers was required to build the pyramids.

Understanding Calculate the vapor pressure of liquid water, in torr, at 0 °C.

Answer: $P = 4.6$ torr. Even at the freezing point, water has a significant vapor pressure. It is possible for ice to evaporate; under conditions of very cold weather (temperatures below 0 °C) and bright sunny skies, ice and snow will evaporate without ever melting.

15.6 Energy Sources and Resources

Objective

- To compare energy production from fossil fuels, nuclear energy, and solar energy

The United States has about 5% of the world's population, but consumes about 30% of the world's energy. In the last 40 years, Americans have consumed more fuels and metal ores than were consumed by all the inhabitants of the earth prior to 1950.

Energy is required to nurture civilization. In years past, energy was measured by the number of people working at a task. The pharaohs of Egypt used the energy of 100,000 laborers to build the pyramids. More than two million workers built the Great Wall of China. The average American consumes about 3×10^8 kJ of energy each year (the equivalent of the work of 200 full-time laborers) to supply food, build housing, fabricate automobiles, and publish chemistry textbooks.

It is common today to speak of the "energy shortage," a concept that is at odds with the first law of thermodynamics: energy cannot be created or destroyed. When we "use" energy, it is transformed into work and heat, which are generally transferred to the surroundings. Rather than an energy shortage, we have a shortage of resources such as oil or gas that are conveniently burned to release the energy.

Fossil Fuels

Fossil fuels are the major source of energy in the world.

The vast majority, about 80%, of the energy that is consumed in the United States is derived from the combustion of coal, natural gas, and petroleum. These materials are called fossil fuels, related to the Latin *fossilis,* meaning "dug from the ground." The fossil fuels originated millions of years ago as living organisms; they have been buried through the ages and have decayed, forming fuels. There are several advantages associated with extracting energy from fossil fuels: they are accessible, portable (particularly good for transportation purposes), and easy to use. Most of the energy is derived from combustion with air, which is widely available and generally free.

Fossil fuels are nonrenewable resources that are being consumed 100,000 times faster than they are formed. The earth's supply of petroleum will be consumed in 170 years at current rates. The major constituents of fossil fuels are hydrocarbons, C_mH_n. Some of the properties of selected fossil fuels are presented in Table 15.7.

The energy provided by fossil fuels is obtained when they are burned. If there is an excess of oxygen, the products are carbon dioxide and water:

$$C_mH_n + (m + n/4)O_2 \longrightarrow mCO_2 + (n/2)H_2O$$

When oxygen is the limiting reactant, some carbon monoxide is produced. If air is used to supply oxygen, some nitrogen compounds are formed as well. Nitrogen can combine with the carbon, hydrogen, and oxygen to form compounds such as $(CN)_2$, called cyanogen, and the nitrogen oxides, NO_x, where x varies from 1/2 in dinitrogen oxide, N_2O, to 2.5 in dinitrogen pentoxide, N_2O_5. If the fossil fuels contain elements other than carbon and hydrogen, then the reaction forms a more complicated mixture of products.

The heat of combustion is about 40 to 50 kJ/g for most fossil fuels. Let us consider the combustion of propane, a component of natural gas:

$$C_3H_8(g) + 5O_2(g) \longrightarrow 3CO_2(g) + 4H_2O(\ell)$$

$$\Delta H^\circ = -2220 \text{ kJ}$$

$$\text{heat evolved} = \frac{2220 \text{ kJ}}{\text{mol } C_3H_8} \times \frac{1 \text{ mol } C_3H_8}{44.10 \text{ g}}$$

$$= 50.3 \text{ kJ/g of propane}$$

The energy released by combustion is used to heat our homes and to move our automobiles. Much more energy, however, is used to power our factories and to manufacture the goods that surround us. The heat of combustion of these fossil fuels is also used to generate alternative forms of energy, mostly electricity.

Three major problems are associated with burning fossil fuels. First, fossil fuels are not renewable; these compounds are available in limited supply. Second, the combustion of fossil fuels produces more than just carbon dioxide and water. There are both impurities in the fuels and complications in the combustion process that lead to air pollution. Third, even if pollution problems were eliminated, the production of large quantities of carbon dioxide is detrimental to the environment. Each of these problems is discussed in the following sections.

Table 15.7 Properties of Some Components of Fossil Fuels

Natural Gas

Composed of hydrocarbons with 1 to 4 carbon atoms. About 70 to 80% of natural gas is CH_4, methane.

Liquids

Petroleum consists of liquid hydrocarbons having from 5 to 12 carbons. Some heavier hydrocarbons (C_{14} to C_{20}) and lighter ones (C_3 and C_4) are dissolved in the liquids.

Solids

Tar and asphalt contain 20 to 40 carbon atoms per hydrocarbon molecule. Coal is even heavier. Coal is a complex and heterogeneous material that has an average molar mass of about 3000 g/mol.

Fossil Fuel Resources

Each year until the mid 1980s, the rate of discovery of fossil fuels exceeded their consumption. That is no longer the case; consumption now exceeds discovery of new deposits. If the current trends continue, the earth's oil supply (about 8×10^{18} kJ, including estimates for undiscovered oil) will be depleted by 2050. Natural gas reserves are about 7×10^{18} kJ and will last 60 years longer than oil.

The most abundant fossil fuel is coal, making up about 90% of the world's fossil fuels. The known supply of coal is about 1.5×10^{20} kJ, sufficient to meet needs for 1500 more years. Many of the processes that utilize petroleum can be modified to burn coal, but certainly not all. Coal differs from petroleum in several important respects. First, coal is a solid and petroleum is a liquid. Coal contains much more complex compounds, many with 100 or more carbon atoms, while petroleum consists of compounds with fewer than 20 carbon atoms. The hydrogen-to-carbon ratio in coal is about 1, so coal would have an empirical formula of CH. Petroleum has more hydrogen, and an approximate empirical formula of CH_2. Fuels with higher H/C ratios burn cleaner, since H_2O has little or no environmental impact.

Fuels with higher H/C ratios generally burn cleaner than fuels with lower H/C ratios.

Many important processes, such as the production of electricity, can be powered by either coal or petroleum. Other processes, especially transportation, involve petroleum liquids almost exclusively. Diesel fuel, aviation fuel, and gasoline are the most common liquid fuels. With an excess of coal and a shortage of petroleum, efforts to change coal into liquefied petroleum-like fuels are an active area of research. In countries that have large coal reserves and lack the cash to import foreign oil, coal liquefaction plants have been built. Coal gasification makes the energy of coal available in a gas form; such plants are now being constructed in several locations.

As the supply of petroleum declines, it will be used more as a source of plastics, fibers, and industrial feedstocks than burned as a source of energy. Mendeleev once noted that burning petroleum made as much sense as burning bank notes, meaning that it had more value than as a fuel.

Air Pollution

Fossil fuels are complex materials and contain elements other than hydrogen and carbon. Some of these elements, particularly nitrogen and sulfur, are implicated in air pollution problems.

When sulfur-containing fossil fuels are burned, some sulfur dioxide is formed. The sulfur dioxide eventually reacts with oxygen to form sulfur trioxide.

$$2SO_2(g) + O_2(g) \longrightarrow 2SO_3(g)$$

Sulfur trioxide is soluble in water and forms sulfuric acid, one of the important components of acid rain, rain with a pH that is often considerably less than 6.

$$SO_3(g) + H_2O(\ell) \longrightarrow H_2SO_4(\ell)$$

The formation of sulfuric acid from sulfur dioxide is actually quite complicated. In fact, this equation represents the sum of a series of individual reactions that can involve light, ozone, and other pollutants such as nitrogen oxide; this complex multistep reaction is quite slow. Most of the sulfur in petroleum is removed in the refining stage, and the burning of refined petroleum products such as gasoline is not the major contributor to acid rain. Crude oil that has a low sulfur content is more expensive than high-sulfur oil, since high-sulfur oils require additional steps to remove the sulfur.

Most of the sulfuric acid in the atmosphere comes from burning coal. Coal is usually not refined, but is used in the same form in which it is mined. Methods for refining coal (including removal of sulfur, called coal desulfurization) are still in the research stages. About 85% of all coal mined is used by utility companies to generate electricity. These power companies prefer to buy low-sulfur coal, but sometimes high-sulfur coal is used, especially if it is less expensive and mined close to the utility. Not only does this practice decrease transportation expenses, it also emphasizes a commitment to buy "local" products. Policies that keep local employment levels high also help please the customers.

The oxides of sulfur and nitrogen have bad odors. To avoid offending their neighbors and to dilute these gases in additional air, most fossil-fuel plants use very tall smokestacks. Recent studies have shown that releasing pollutants high above the ground makes them more likely to react with substances in the upper atmosphere. Sulfur dioxide released well above the ground is more readily oxidized to sulfur trioxide than if it were released at ground level. Many scientists believe that the major cause of acid rain, which can occur hundreds or thousands of miles from the source, is the release of sulfur-containing gases from tall industrial smokestacks. These conclusions are not supported by all investigators, however.

Sulfur compounds are generally removed *after* coal is burned. Chemical reactors called **scrubbers** are used to remove most of the sulfur dioxide from the gases leaving the smokestack of the coal-burning plant. Often the reactor contains a suspension of calcium carbonate (limestone). The exhaust gases are passed through the limestone/water mixture and the SO_2 reacts to form a solid calcium salt. The salt must be stored or buried, but changing the form of a pollutant from a gas to a solid greatly reduces potential hazards.

$$CaCO_3(s) + SO_2(g) \longrightarrow CaSO_3(s) + CO_2(g)$$

Scrubbers are relatively expensive, primarily because they decrease the efficiency of the plant. A well-designed coal plant turns about 37% of the

Power generating plants These smokestacks are several hundred feet tall, so any pollution produced in the process is well on its way into the upper atmosphere.

Greenhouse Tropical plants such as cacti can be grown in a greenhouse even during the winter. In fact, the greenhouse can get too hot, even when the outside temperature is below freezing.

energy from coal into electricity. This value is quite close to the theoretical maximum. Adding scrubbers decreases the efficiency to about 34%.

The Greenhouse Effect

Even if technology were developed to minimize air pollution and if current combustion processes were perfected, the consumption of fossil fuels would still cause a potential environmental problem. The carbon dioxide that is produced from burning acts as an insulator that keeps the earth from radiating heat. This phenomenon is called the **greenhouse effect,** since a blanket of carbon dioxide acts like a greenhouse, letting light energy in during the day but blocking heat losses at night. Computer models indicate that increased levels of atmospheric carbon dioxide will cause a global temperature increase with potentially catastrophic results.

The atmosphere's carbon dioxide concentration has been slowly but surely increasing over the last several centuries. The average temperature is increasing as well, but some increases are expected because the earth has been warming since the last glacial maximum (ice age) about 19,000 years ago. (Geological evidence shows many glacial maxima over the last million years, typically spaced about 40,000 years apart. Most geologists feel that the planet is nearly halfway between ice ages and close in time to its warmest global climate.) Data are needed to determine whether temperatures are higher than expected for this period, but so far there is no clear consensus. Data from Europe indicate a temperature increase, but North American data do not. If there is an increase in temperature beyond values expected for this era, the temperature increase is not worldwide in scope.

Although there has been a significant increase in carbon dioxide concentrations over the last 200 years, there is not enough information to conclude that current carbon dioxide emission is causing the temperature to increase. It should be noted that gases such as chlorofluorocarbons (used in refrigeration systems) and methane are also greenhouse gases. Most of the methane emission to the atmosphere is natural, coming from plants and animals, but a recent study of methane in Eastern Europe showed that leaks from a natural gas distribution system were also significant.

Even if the level of greenhouse gases has not contributed to a temperature increase at this time, it is clear that increasing the carbon dioxide concentration in the atmosphere will ultimately lead to increased temperatures. Many scientists believe that current carbon dioxide production, about 6×10^{12} kg/year, will cause the temperature to increase at the rate of 0.3 °C each decade. These changes are not small; geochemists believe that a change of 2 to 3 °C can trigger a major change in climate.

Much of the carbon dioxide, at least 30%, comes from the burning of timber, principally in South America, as forests are cleared for agricultural purposes.

Nuclear Fuels

Nuclear fission, discussed in Chapter 20, is an important energy source. The energy released during the fission of nuclei such as ^{235}U can ultimately be converted into heat and work. As in all energy generation systems, the use of nuclear power has advantages and disadvantages.

Nuclear reactors are primarily used to generate electricity, although some are used to prepare radioactive materials for use in research and medicine and others are used to fabricate materials for weapons. In 1989, 11% of the world's electrical power (4% of the power in the United States) was generated from 426 nuclear reactors. About 100 additional reactors are under construction, although there have been no new reactors started in the United States since 1978.

The advantages of nuclear energy include a relatively low fuel cost. The fuel cost of nuclear energy includes the cost of mining uranium, purifying it, performing isotopic enrichment, and packaging it in a form suitable for use in a reactor. Other advantages include the generation of power without air pollution or a contribution to the greenhouse effect.

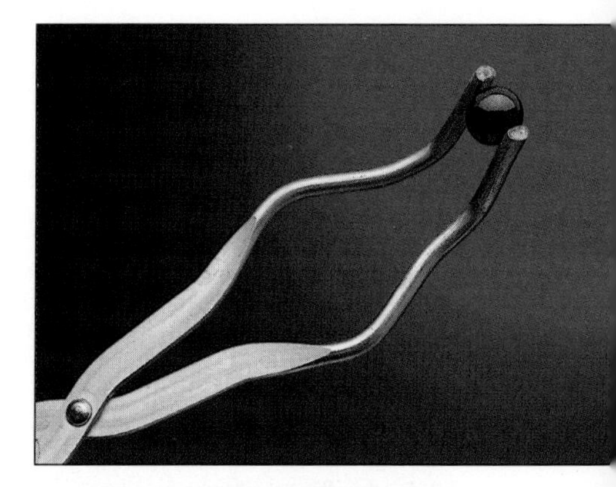

Nuclear waste If 30% of U.S. electricity were produced by nuclear reactors, the waste produced generating the annual electrical energy consumption of one U.S. resident could be stored in a glassy sphere about 1 cm in diameter.

Nuclear power generation is not totally free from pollution, since radioactive wastes are produced. These wastes contain some elements that decay very slowly, so levels of radioactivity remain high for thousands of years. Ideally, radioactive materials should be stored as inert solids. Liquids (or gases) can escape from containers and pollute the environment; solids cannot. Further, the solid must be chemically inert so that air, water, and even strong acids and bases will not attack the radioactive solid. The technology needed to cast radioactive materials into ceramics has been recently developed. The Department of Energy is operating a plant at the Savannah River Site in Aiken, South Carolina, that forms glasses and ceramics from nuclear waste.

In addition to the problems associated with storing radioactive waste, other disadvantages include the costs of dismantling an obsolete nuclear generating plant, and the expenses associated with installing a safety system that is acceptably effective. Although nuclear accidents are rare, the potential damage is enormous. Government safety regulations make the cost of building a nuclear power plant so high that economic choices have steered utilities away from nuclear power.

Many economists believe that when all costs are included, electricity generated from a nuclear reactor is about three times more expensive than that from coal. Regardless, nuclear proponents note that fossil fuel use must be curtailed to avoid the greenhouse effect. If global warming is to be reduced to a tolerable 0.05 °C per decade, it is estimated that an additional 2000 nuclear generating plants must be built in the next 40 years.

Solar Energy

The energy supplied by the sun is a potential replacement for burning fossil fuels. The amount of solar energy that strikes the earth is 1×10^{19} kJ per day. The world's consumption of energy is about 7×10^{14} kJ per day, equivalent to less than 0.01% of the solar energy that strikes the earth. Further, solar energy appears to be free and nonpolluting. Obviously, there must be some problems, or solar-powered facilities would be far more common.

In fact, there are several problems. First, solar energy is not constant. It is available only on clear days, and never at night. Second, solar energy is low in "intensity," and collection devices must be large to collect useful amounts of energy. Still, the energy striking 0.1% of the United States is sufficient to supply the energy needs of the nation, if we could only capture it and use it.

Solar water heater

The most practical solar energy conversion devices are water heaters. Water can be circulated through collectors that are aimed at the sun; the hot

Adobe dwellings These Indian dwellings serve as good examples of passive solar design.

water is stored in an insulated tank. Water heaters require an auxiliary energy source, usually an electrical heating unit, so that hot water is available even when the sun is not shining. Solar heating systems have been used successfully in many countries, but are not widely accepted in the United States. Problems include the need for an extra-strength roof support system, the need for a very large storage tank, and the higher cost of fabricating and installing such a system. Most studies indicate that solar-energy water heaters are less expensive than conventional electric water heaters but more expensive than gas-heated systems, when all costs are considered. These conclusions will change as energy costs shift in the future.

Passive solar energy devices use no machinery or heat-transfer equipment. Some of the earliest solar devices can be seen in the pueblos of some Indians of the southwestern United States, who built dwellings with openings that face south. In the summer, the sun is high in the sky, and the sun's heat is blocked from entering the dwelling. In the winter, the sun is low in the southern sky, and the rooms are flooded with warm sunlight. The same technique can be adapted to modern homes in which the direction and size of windows and the extent of the overhang can be adjusted to minimize summer heating while maximizing solar heating in the winter.

Solar energy cannot yet be used to generate electricity in an economical fashion. Commercial photovoltaic cells, devices that generate electricity directly from light, are about 18% efficient, although 33% efficiency has been achieved in the lab. A solar collector with an area of 40 m^2 on a south-facing roof of the average home in the United States will generate as much electricity as is used by the average family, but the cost would be high. Solar electrical generation is comparable in price to using a small gasoline-powered generator, so solar generators are economically viable in locations that lack power lines.

A promising large-scale use of solar energy would be to force an endothermic reaction to form products. The reaction of methane with water has been considered.

$$CH_4(g) + H_2O(g) \longrightarrow CO(g) + 3H_2(g)$$

This reaction is endothermic and nonspontaneous at room temperature, but proceeds spontaneously at high temperatures. Solar collectors can be used to heat methane and water, forming hydrogen and carbon monoxide. The energy of the sun is then stored as the potential energy of the hydrogen and

Oil well flare The flame is due to the burning of natural gas, a process that wastes energy.

carbon monoxide molecules. The reverse reaction, combustion, proceeds spontaneously and produces heat at times when direct solar energy is unavailable.

Energy Conservation

No summary of energy and its use can be complete without discussing energy conservation measures. Energy has historically been an inexpensive commodity, and people have become accustomed to consuming energy in large quantities. The nation's (and world's) future may depend on a well-designed energy policy that includes fossil fuels, nuclear fuels, and solar energy (particularly passive energy) and that addresses environmental concerns. Such a policy must also include energy conservation.

When energy costs are low, wasting energy is often more economical than building equipment needed to conserve the energy. Thus for years, procedures that wasted energy were common. Oil producers were annoyed when pockets of natural gas were found, since the gas was an explosion hazard, so they simply burned off huge amounts of gas. Buildings and homes were constructed without insulation, since it was cheaper to heat them than to buy effective thermal insulation. Very little thought was given to the fuel efficiency of automobiles; gasoline was a minor expense in the cost of operating a car.

The situation is now changing. Much of the inexpensive domestic petroleum has already been pumped from the ground. As oil wells have gone from 1800 feet deep to 20,000 feet deep, the cost of oil production has skyrocketed.

Many steps have been made toward conserving energy. Automobile efficiencies have almost tripled, most communities require insulation in all new buildings, and industries realize that energy is not the cheap commodity that it once was. There is still room for additional improvement. One well-known scientist recently asked Congress to demand the production of cars that would travel over 30 km per liter of gasoline (80 miles per gallon). In response to the argument that such small, fuel-efficient cars may be unsafe, a prototype car (Volvo LCP-2000) was designed to ensure the safety of all passengers in a 50 km/hr (30 mph) head-on collision. The car was fabricated from modular components for ease in assembly. The prototype test car was rated at 63 miles per gallon in the city and 81 miles per gallon on the highway. The inexpensive construction techniques would offset the expense of the high-technology materials that were used to make the car both strong and light. Such a car would appeal to many consumers, particularly when gasoline prices are skyrocketing.

Energy conservation will have to improve even further. While research in chemistry and technology will continue to address energy issues, the social habits of the population must change. A standard 7000-lumen fluorescent light fixture costs about as much as this textbook and requires 180 W of electrical energy. A high-efficiency fixture (with more reflective paint, better design of the surfaces, but no electrical improvements) costs about twice as much, but requires only 80 W to produce the same 7000 lumens. The more efficient device can pay for itself within one to three years, depending on usage. A 50% reduction in energy consumption is a realistic goal for homes and industries in the next several decades.

A great deal of energy can be saved without high technology devices. For example, the average American car contains 1.7 passengers; increasing this number to 4 passengers would reduce the energy consumption and emissions by about 45%. Wider use of public transportation would result in even greater gains. Changes in habits are important, but an international energy policy that includes economic growth and development, efficiency, and conservation must also be implemented.

Summary

Both heat and work are forms of energy transfer. The *internal energy, E,* of the system includes both kinetic and potential energy. Changes in the internal energy obey the *first law of thermodynamics,* $\Delta E = q + w$, where q is the heat absorbed by the system and w is the work done on the system. If the system performs work on the surroundings, w is negative; if the surroundings do work on the system, then w is positive. Similarly, if the system absorbs heat from the surroundings, q is positive; if the surroundings absorb heat from the system, q is negative.

An example of a system that does work is a gas expanding against a constant opposing pressure. Under these circumstances, $w = -P\,\Delta V$. If a process occurs without a change in the volume of the system, then $P\Delta V$ is zero; the heat absorbed at constant volume is ΔE. If the system is allowed to perform only PV work at constant pressure, then the heat absorbed by the system is ΔH, the change in *enthalpy.* Enthalpy is defined as $H = E + PV$, and it is a *state function.*

The heat absorbed or evolved during a reaction involving substances in their standard states can be calculated: it is the sum of the standard heats of formation of the products minus the sum of the standard heats of formation of the reactants. The *standard heat of formation* of a substance is the heat absorbed when the substance in its standard state is formed from the most stable form of the component elements in their standard states. (The *standard state* is a designated temperature and 1 atm pressure.) A chemical equation must be provided before you can calculate the heat of reaction, since the heat evolved or absorbed depends on the amounts of the reactants present.

Reactions are driven by both changes in *enthalpy* and changes in *entropy,* a measure of disorder. The largest factor that influences entropy changes is a change in phase, because gases are more disordered than liquids, and liquids are more disordered than solids. If there are more moles of gases on the product side of the equation, cntropy increases.

The *second law of thermodynamics* states that, for any spontaneous process, the entropy of the universe increases. It is important to note that the second law of thermodynamics refers to the entropy of the *universe,* not that of the system or the surroundings. Although absolute energy and enthalpy measurements cannot be performed, entropy can be measured since there is a reference point. The entropy of a pure crystalline substance is zero at 0 K, as stated by the *third law of thermodynamics.*

Free energy, G, defined as $G = H - TS$, is a state function. For a change at constant temperature and pressure, $\Delta G = \Delta H - T\Delta S$. ΔG is negative for any spontaneous process and zero when the system is at equilibrium.

Changes in ΔG° can be calculated from tables of ΔG_f° for the products and reactants or from tables of ΔH_f° and S°. The change in free energy is mainly influenced by temperature through the $T\,\Delta S$ term.

Concentration influences the free energy change since

$$\Delta G = \Delta G^\circ + RT \ln Q$$

If Q, the reaction quotient, is less than K_{eq}, then ΔG is less than 0, and the forward reaction proceeds spontaneously. When the system reaches equilibrium, then ΔG is 0 and Q is equal to K_{eq}, so ΔG° can be related to the equilibrium constant:

$$\Delta G^\circ = -RT \ln K_{eq}$$

$$K_{eq} = e^{-\Delta G^\circ/RT}$$

The effect of temperature on the equilibrium constant can be predicted by the principle of Le Chatelier. Heating an exothermic reaction favors the reactants.

When phase changes such as vaporization are studied, the change in entropy can be determined. Since the entropy of a liquid is less than that of a gas, ΔS°_{vap} is positive and always larger than ΔS°_{fus} for the same substance. At the boiling point, the liquid and vapor are in equilibrium, so ΔG is zero. At this temperature,

$$\Delta H^\circ_{vap} = T_b \, \Delta S^\circ_{vap}$$

The standard enthalpy of vaporization is always positive—vaporization is an endothermic reaction. The same is true for melting.

The dominant source of energy in the world is fossil fuels. Although these fuels are convenient, air pollution results when they are burned. Some pollutants, such as the components that lead to acid rain, can be removed. However, combustion of fossil fuels generates carbon dioxide, which may be another form of pollution. Carbon dioxide is one of the gases that can cause the *greenhouse effect.* These gases prevent heat from leaving the earth and may result in global warming. Other technologies, including solar energy, may be more important in the future, but now they are not widely used.

Chapter Terms

Chemical thermodynamics—the study of the energetics of chemical reactions. *(15.1)*

Spontaneous—able to occur without outside intervention. *(15.1)*

Closed system—one that is allowed to exchange energy with surroundings, but does not exchange matter. *(15.1)*

State function—a thermodynamic property that depends only upon the state of the system and not on the manner in which the system arrived at the state. *(15.1)*

Energy—the ability of the system to perform work. *(15.1)*

Internal energy—the total energy of the system. *(15.1)*

Work—the product of force times distance. *(15.1)*

Heat of combustion—the enthalpy change when a compound is burned. *(15.2)*

First law of thermodynamics—energy can be neither created nor destroyed. In equation form: $\Delta E = q + w$. *(15.2)*

Enthalpy—heat content, $H = E + PV$. It is a state function. *(15.2)*

Standard state—the most stable form of a substance at the designated temperature, usually 298 K, and 1 atm pressure. *(15.2)*

Standard heat of formation—the heat absorbed when one mole of a substance in its standard state is formed from the most stable forms of the elements in their standard states. *(15.2)*

Entropy—the property of a system that describes the amount of disorder. It is a state function. *(15.3)*

Second law of thermodynamics—in any spontaneous process, the entropy of the universe will increase. *(15.3)*

Third law of thermodynamics—the entropy of a perfect crystalline substance is 0 at a temperature of absolute zero. *(15.3)*

Free energy—$G = H - TS$. It is a state function. *(15.4)*

Standard free energy of formation—the free energy change during the formation of one mole of a substance in its standard state from the most stable form of the elements in their standard states. *(15.4)*

Scrubbers—chemical reactors used to clean gases leaving the smokestack of an industrial plant. *(15.6)*

Greenhouse effect—the effect of a blanket of gases that insulates the earth to prevent it from radiating heat. *(15.6)*

Passive solar energy device—one that uses no machinery or heat-transfer equipment. *(15.6)*

Exercises

Exercises designated with color have answers in Appendix J.

Evaluation of Work

15.1 How is the sign of w, work, defined?

15.2 How is the sign of q, heat, defined?

15.3 Identify the sign of the work when a fuel-oxygen mixture (the system) is burned, propelling an automobile (part of the surroundings).

15.4 What is the sign of the work when a refrigerator compresses a gas (the system) to a liquid during the refrigeration cycle?

15.5 When a rocket is launched, the burning gases are the source of the motion. If the system is the rocket (including fuel), what is the sign of the work?

15.6 The following processes occur at constant pressure. State the sign of w (consider only PV work from ideal gases) for

(a) $CaCO_3(s) + H_2SO_4(aq) \longrightarrow CaSO_4(aq) + H_2O(\ell) + CO_2(g)$

(b) $HCl(g) + NH_3(g) \longrightarrow NH_4Cl(s)$

15.7 The following processes occur at constant pressure.

State the sign of w (consider only PV work from ideal gases) for
(a) $Fe_2S_3(s) + 6HNO_3(aq) \longrightarrow 2Fe(NO_3)_2(aq) + 3H_2S(g)$
(b) $CH_4(g) + 2O_2(g) \longrightarrow CO_2(g) + 2H_2O(\ell)$

15.8 Calculate w for the following reactions that occur at 298 K and 1 atm pressure. Consider only PV work from the change in volume of gas, and assume that the gases are ideal and the chemical equation represents amounts in moles.
(a) $CO_2(g) + NaOH(s) \longrightarrow NaHCO_3(s)$
(b) $2O_3(g) \longrightarrow 3O_2(g)$

15.9 Calculate w for the following reactions at 298 K and 1 atm pressure. Consider only PV work from the change in volume of gas and assume that the gases are ideal and the chemical equation represents amounts in moles.
(a) $2Na(s) + 2H_2O(\ell) \longrightarrow 2NaOH(s) + H_2(g)$
(b) $2Fe_2O_3(s) + 3C(s) \longrightarrow 4Fe(s) + 3CO_2(g)$

15.10 Calculate w for the following reactions that occur at 298 K and 1 atm pressure. Consider only PV work from the change in volume of gas and assume that the gases are ideal and the chemical equation represents amounts in moles.
(a) $Ni(CO)_4(g) \longrightarrow Ni(s) + 4CO(g)$
(b) $2NO(g) \longrightarrow N_2(g) + O_2(g)$

15.11 A 125-L cylinder contains an ideal gas at a pressure of 100 atm. If the gas is allowed to expand against an opposing pressure of 0.0 atm, how much work (in kJ) will be done?

15.12 A 220-L cylinder contains an ideal gas at a pressure of 150 atm. If the gas is allowed to expand against an opposing pressure of 1.0 atm, how much work (in kJ) will be done?

15.13 A balloon contains 2.0 L of helium at 1.10 atm. Calculate the work done (in J) if the gas expands against atmospheric pressure of 754 torr.

First Law of Thermodynamics

15.14 A reaction produces 4.5 L of a gas at 0.94 atm and consumes 4.35 kJ of heat (endothermic). Calculate ΔE, the change in internal energy of the system.

15.15 A reaction at 1.02 atm consumes 3.5 L of a gas and evolves 2.71 kJ of heat (exothermic). Calculate ΔE, the change in internal energy of the system.

(15.16–15.18) When an ideal gas expands at constant temperature (isothermal conditions), ΔE is zero because the kinetic energy of a gas is related to its temperature (which is unchanged for the isothermal case) and the potential energy is related only to bonding and structure, which is unchanged in an expansion. Consider 1.00 L of a gas initially at 9.00 atm and 15 °C.

15.16 Calculate q and w if the gas expands isothermally against an opposing pressure of 1.00 atm to its final volume.

15.17 Calculate q and w if the gas expands isothermally first against an opposing pressure of 3.00 atm, and then against an opposing pressure of 1.00 atm.

15.18 Consider the results of the single-step expansion and the two-step expansion. How does a three-step isothermal expansion (first 3.00 atm, then 2.00 atm, finally 1.00 atm opposing pressure) compare to the two- and one-step expansions?

15.19 A tank of compressed gas contains about 220 cubic feet (at STP) of gas at an approximate pressure of 2000 psi. The amount of work that can be done when this gas expands is equivalent to the destructive force of about $\frac{1}{4}$ lb of an explosive such as dynamite. Calculate the work (in J) performed when this gas expands against an opposing pressure of 1.0 atm.

A description of an accident was presented in the *National Safety Council Chemical Section Newsletter* in May 1969. It was reprinted in 1976 in the *Journal of Chemical Education* **53**(6), 373 (1976). It is used here with permission of the copyright holder.

"Six 220 cubic foot cylinders, part of a fire-extinguishing system, had been moved away from their wall supports to allow painters to complete the painting of the area. While moving them back into position, it was noted that one cylinder was leaking, having been damaged earlier. The painter had the cylinder leaning against his shoulder, and was attempting to scoot it across the floor.

"At this time, the valve separated from the cylinder, and the man suddenly found himself with a jet-propelled 215 lb piece of steel. He wrestled it to the floor, but was unable to hold it. The cylinder scooted across the floor, hitting another cylinder, knocking it over, and bending its valve. The cylinder then turned 90° to the right and travelled 20 feet where it struck a painter's scaffold, causing a painter to fall 7 feet to the floor and break his leg. After spinning around several times, the cylinder travelled back to its approximate starting point where it struck a wall.

"At this point, the cylinder turned 90° to the left and took off, chasing an electrician in front of it. It crashed into the end wall, 40 feet away, breaking loose four concrete blocks. It turned again 90° to the right, scooted through a door opening, still pursuing the electrician. The electrician ducked into the next door opening, and the cylinder continued its travel in a straight line for another 60 feet, where it fell into a truck well, striking the truck well door. The balance of the cylinder pressure was released as the cylinder spun harmlessly around in the truck well area."

The description of the cylinder incident describes a cylinder that was well less than $\frac{1}{2}$ full. A full cylinder might have done much, much more damage. If you should see unanchored cylinders, please call them to the attention of the proper authorities.

Enthalpy

15.20 Explain the difference between internal energy and enthalpy.

15.21 What is the difference between the heat of reaction and the heat of formation?

15.22 Write the chemical equation for the reaction that defines the standard heat of formation of
(a) $HF(g)$
(b) $H_2SO_4(\ell)$
(c) $Al(OH)_3(s)$
(d) $NaKSO_4(s)$

15.23 Write the chemical equation for the reaction that defines the standard heat of formation of
(a) $CH_3COOH(\ell)$
(b) $H_3PO_4(\ell)$
(c) C(diamond)
(d) $CaCl_2 \cdot 2H_2O(s)$

15.24 Use the standard enthalpies of formation to calculate the enthalpy change for each of the following reactions at 298 K and 1 atm. Label each as endothermic or exothermic.
(a) The fermentation of glucose to ethyl alcohol and carbon dioxide

$$C_6H_{12}O_6(s) \longrightarrow 2CH_3CH_2OH(\ell) + 2CO_2(g)$$

(b) The combustion of normal (straight-chain) butane

$$n\text{-}C_4H_{10}(g) + \tfrac{13}{2}O_2(g) \longrightarrow 4CO_2(g) + 5H_2O(\ell)$$

15.25 Use the standard enthalpies of formation from Appendix G to calculate the enthalpy change for each of the following reactions at 298 K and 1 atm. Label each as endothermic or exothermic.
(a) The photosynthesis of glucose

$$6CO_2(g) + 6H_2O(\ell) \longrightarrow C_6H_{12}O_6(s) + 6O_2(g)$$

(b) The reduction of iron(III) oxide with carbon

$$2Fe_2O_3(s) + 3C(s) \longrightarrow 4Fe(s) + 3CO_2(g)$$

15.26 Use the standard enthalpies of formation from Appendix G to calculate the enthalpy change for each of the following reactions at 298 K and 1 atm. Label each as endothermic or exothermic.
(a) $NaOH(s) + CO_2(g) \longrightarrow NaHCO_3(s)$
(b) $2SO_2(g) + O_2(g) \longrightarrow 2SO_3(g)$

15.27 Use the standard enthalpies of formation from Appendix G to calculate the enthalpy change for each of the following reactions at 298 K and 1 atm. Label each as endothermic or exothermic.
(a) $NH_4NO_3(s) \longrightarrow N_2O(g) + 2H_2O(\ell)$
(b) $CO(g) + H_2O(\ell) \longrightarrow HCOOH(\ell)$

15.28 Use the standard enthalpies of formation from Appendix G to calculate the enthalpy change for each of the following reactions at 298 K and 1 atm. Label each as endothermic or exothermic.
(a) C(graphite) $\longrightarrow$ C(diamond)
(b) $H_2O(\ell) \longrightarrow H_2O(g)$

15.29 Use the standard enthalpies of formation from Appendix G to calculate the enthalpy change for each of the following reactions at 298 K and 1 atm. Label each as endothermic or exothermic.
(a) $H_2O(g) \longrightarrow H_2O(\ell)$
(b) $AlCl_3(s) + 3H_2O(\ell) \longrightarrow Al(OH)_3(s) + 3HCl(g)$

15.30 Calculate $\Delta H°$ when a 28-g sample of glucose, $C_6H_{12}O_6(s)$, is burned to form $CO_2(g)$ and $H_2O(\ell)$ in a reaction at constant pressure and 298 K?

15.31 Calculate the amount of heat evolved or absorbed when a 0.1045-g sample of benzene, $C_6H_6(\ell)$, is burned in excess oxygen to form $CO_2(g)$ and $H_2O(\ell)$ in a reaction at constant pressure and 298 K.

Entropy Changes

15.32 What is the sign of the entropy change for each of the following processes? The system is underlined. State any assumptions that you make.
(a) A test-tube is dropped on the floor and shatters.
(b) A new deck of cards is shuffled.
(c) Steel is manufactured from iron ore and carbon.
(d) A wooden fence rots.

15.33 What is the sign of the entropy change for each of the following processes? The system is underlined. State any assumptions that you make.
(a) Glass is made from sand.
(b) Hydrogen and oxygen combine to form water.
(c) An artisan makes a patterned rug from yarn.
(d) Wood and oxygen are burned to provide heat.

15.34 Predict the sign (the system is underlined) for the entropy change that occurs when
(a) Dry ice ($CO_2(s)$) melts
(b) Water freezes
(c) Gasoline evaporates

15.35 Predict the sign (the system is underlined) for the entropy change that occurs when
(a) Water evaporates
(b) Butter melts
(c) Gold freezes

15.36 Predict the sign and estimate the magnitude of the entropy change, and then use the data in Appendix G to calculate the standard entropy change, for:
(a) $CO(g) + 2H_2(g) \longrightarrow CH_3OH(\ell)$

(b) $3H_2(g) + N_2(g) \longrightarrow 2NH_3(g)$
(c) $CH_4(g) + 2O_2(g) \longrightarrow CO_2(g) + 2H_2O(\ell)$
(d) $CO(g) + H_2O(g) \longrightarrow CO_2(g) + H_2(g)$

15.37 Predict the sign and estimate the magnitude of the entropy change, and then use the data in Appendix G to calculate the standard entropy change, for:
(a) $C(s) + H_2O(g) \longrightarrow CO(g) + H_2(g)$
(b) $2NO(g) + O_2(g) \longrightarrow 2NO_2(g)$
(c) $NaCl(s) \longrightarrow Na^+(aq) + Cl^-(aq)$
(d) $C_5H_{12}(g) + 8O_2(g) \longrightarrow 5CO_2(g) + 6H_2O(\ell)$

Second Law of Thermodynamics

15.38 Calculate $\Delta G°$ to determine whether the reaction is spontaneous with all species present in their standard states at 298 K.
(a) $Fe_2O_3(s) + 2Al(s) \longrightarrow Al_2O_3(s) + 2Fe(s)$
(b) $CO(g) + 2H_2(g) \longrightarrow CH_3OH(\ell)$

15.39 Calculate $\Delta G°$ to determine whether the reaction is spontaneous with all species present in their standard states at 298 K.
(a) $4NH_3(g) + N_2(g) \longrightarrow 3N_2H_4(\ell)$
(b) $2H_2O_2(\ell) \longrightarrow 2H_2O(\ell) + O_2(g)$

15.40 Calculate $\Delta G°$ to determine whether the reaction is spontaneous at 298 K with all species present in their standard states.
(a) $Zn(s) + H_2SO_4(\ell) \longrightarrow ZnSO_4(s) + H_2(g)$
(b) $Cu(s) + H_2SO_4(\ell) \longrightarrow CuSO_4(s) + H_2(g)$

15.41 Calculate $\Delta G°$ to determine whether the reaction is spontaneous at 298 K with all species present in their standard states.
(a) $2NO(g) + O_2(g) \longrightarrow 2NO_2(g)$
(b) $CO(g) + Cl_2(g) \longrightarrow COCl_2(g)$

15.42 Assume all substances are present in their standard states at 298 K and 1 atm pressure, and the only work is *PV* work. Using the data in Appendix G, calculate the change in entropy in the system, surroundings, and universe for

$$H_2(g) + CuO(s) \longrightarrow H_2O(\ell) + Cu(s)$$

15.43 Assume all substances are present in their standard states at 298 K and 1 atm pressure, and the only work is *PV* work. Using the data in Appendix G, calculate the change in entropy in the system, surroundings, and universe at 298 K for

$$2SO_2(g) + O_2(g) \longrightarrow 2SO_3(g)$$

15.44 Explain why absolute enthalpies and energies cannot be measured, and only changes can be determined.

15.45 Explain why absolute entropies *can* be measured.

Free Energy

15.46 Use the data in Appendix G to calculate the change in the standard free energy for each of the following reactions. Note whether the direction of spontaneous reaction is consistent with the sign of the enthalpy change, entropy change, or both.
(a) $H_2(g) + \frac{1}{2}O_2(g) \longrightarrow H_2O(\ell)$ at 25 and 90 °C
(b) $CO(g) + 2H_2(g) \longrightarrow CH_3OH(\ell)$ at −20 and 50 °C

15.47 Use the data in Appendix G to calculate the change in the standard free energy for each of the following reactions. Note whether the direction of spontaneous reaction is consistent with the sign of the enthalpy change, entropy change, or both.
(a) $2H_2O(g) + 2Cl_2(g) \longrightarrow 4HCl(g) + O_2(g)$
(b) $2CO_2(g) + 4H_2O(\ell) \longrightarrow 2CH_3OH(\ell) + 3O_2(g)$

15.48 Use the data in Appendix G to calculate the change in the standard free energy for each of the following reactions. Note whether the direction of spontaneous reaction is consistent with the sign of the enthalpy change, entropy change, or both.
(a) $CH_3COOH(\ell) + NaOH(s) \longrightarrow Na^+(aq) + CH_3COO^-(aq) + H_2O(\ell)$
(b) $AgNO_3(s) + Cl^-(aq) \longrightarrow AgCl(s) + NO_3^-(aq)$

15.49 Use the data in Appendix G to calculate the change in the standard free energy for each of the following reactions. Note whether the direction of spontaneous reaction is consistent with the sign of the enthalpy change, entropy change, or both.
(a) $Fe_2O_3(s) + 3Cu(s) \longrightarrow 3CuO(s) + 2Fe(s)$
(b) $2NH_3(g) \longrightarrow N_2H_4(\ell) + H_2(g)$

Temperature Dependence of Free Energy

(15.50–15.59) Use the data in Appendix G. Ignore any temperature dependences of ΔH and ΔS, and assume the reactants and products are present in their standard states at 298 K unless specified otherwise.

15.50 What is the sign of the standard free energy change at low temperatures and at high temperatures for the combustion of acetaldehyde?

$$CH_3CHO(\ell) + \tfrac{5}{2}O_2(g) \longrightarrow 2CO_2(g) + 2H_2O(\ell)$$

15.51 What is the sign of the standard free energy change at low temperatures and at high temperatures for the explosive decomposition of TNT? Use your knowledge of TNT and the chemical equation, particularly the phases, to answer this question.

$$2C_7H_5N_3O_6(s) \longrightarrow 3N_2(g) + 5H_2O(\ell) + 7C(s) + 7CO(g)$$

15.52 What is the sign of the standard free energy change at low temperatures and at high temperatures for the synthesis of ammonia?

$$3H_2(g) + N_2(g) \longrightarrow 2NH_3(g)$$

15.53 What is the sign of the standard free energy change at low temperatures and at high temperatures for the decomposition of phosgene?

$$COCl_2(g) \longrightarrow CO(g) + Cl_2(g)$$

15.54 Over what range of temperatures is each of the following reactions spontaneous?
(a) $CO(g) + Cl_2(g) \longrightarrow COCl_2(g)$
(b) $NO(g) + \frac{1}{2}O_2(g) \longrightarrow NO_2(g)$

15.55 Over what range of temperatures is each of the following reactions spontaneous?
(a) $2POCl_3(g) \longrightarrow 2PCl_3(g) + O_2(g)$
(b) $PbO(s) + CO_2(g) \longrightarrow PbCO_3(s)$

15.56 Over what range of temperatures is each of the following reactions spontaneous?
(a) A reaction with $\Delta H° = +53.4$ kJ and $\Delta S° = +112.4$ J K^{-1}
(b) A reaction with $\Delta H° = -29.4$ kJ and $\Delta S° = -91.2$ J K^{-1}

15.57 Over what range of temperatures is each of the following reactions spontaneous?
(a) A reaction with $\Delta H° = -53.4$ kJ and $\Delta S° = -112.4$ J K^{-1}
(b) A reaction with $\Delta H° = +29.4$ kJ and $\Delta S° = +91.2$ J K^{-1}

15.58 Calculate $\Delta G°$ at 400 K and 600 K for
(a) $NO_2(g) + N_2O(g) \longrightarrow 3NO(g)$
(b) $2NH_3(g) \longrightarrow N_2H_4(\ell) + H_2(g)$

15.59 Calculate $\Delta G°$ at 300 K and 390 K for
(a) $BaO(s) + CO_2(g) \longrightarrow BaCO_3(s)$
(b) $CH_3COOH(\ell) \longrightarrow CH_4(\ell) + CO_2(g)$

15.60 A chemical reaction that proceeds spontaneously at low temperatures but proceeds in the reverse direction at elevated temperatures is needed. What are the thermodynamics (signs of $\Delta G°$, $\Delta H°$, and $\Delta S°$) of such a reaction?

15.61 A chemical reaction that proceeds spontaneously at high temperatures but proceeds in the reverse direction at low temperatures is needed. What are the thermodynamics (signs of $\Delta G°$, $\Delta H°$, and $\Delta S°$) of such a reaction?

Challenging Free Energy Problems

15.62 Identify the incorrect statements and provide correct answers. The questions refer to the formation of one mole of methanol (CH_3OH) from carbon monoxide and hydrogen (all at 1 atm pressure and in the gas phase).
(a) The spontaneous direction of the reaction depends entirely on $\Delta H°$.
(b) The reaction is spontaneous at 298 K.
(c) The reaction will proceed spontaneously at all temperatures above 298 K.
(d) The reaction will proceed spontaneously at all temperatures below 298 K.
(e) The equilibrium constant will increase with increasing temperature.

15.63 Identify the incorrect statements and provide correct answers. The questions refer to the formation of formic acid (HCOOH) from carbon monoxide and water (all at 1 atm pressure and in the gas phase).
(a) The direction of the spontaneous reaction depends entirely on $\Delta H°$.
(b) The reaction is spontaneous at 298 K.
(c) The reaction will proceed spontaneously at all temperatures above 298 K.
(d) The reaction will proceed spontaneously at all temperatures below 298 K.
(e) The equilibrium constant will increase with increasing temperatures.

15.64 The equilibrium constant for a reaction decreases as temperature increases. Explain how this observation is used to determine the sign of either $\Delta H°$ or $\Delta S°$.

15.65 The free energy for a reaction decreases as temperature increases. Explain how this observation is used to determine the sign of either $\Delta H°$ or $\Delta S°$.

Phase Transitions

15.66 Write chemical equations for (a) fusion, (b) vaporization, and (c) sublimation.

15.67 Write chemical equations for (a) freezing, (b) condensation, and (c) deposition.

15.68 Use the thermodynamic data in Appendix G to determine whether the vaporization of methanol is spontaneous under the stated conditions. State any assumptions made.

$$CH_3OH(\ell) \longrightarrow CH_3OH(g) \quad \text{at 80 °C and 1 atm}$$

15.69 Use the thermodynamic data in Appendix G to determine whether the condensation of nitromethane is spontaneous under the stated conditions. State any assumption you make.

$$CH_3NO_2(g) \longrightarrow CH_3NO_2(\ell) \quad \text{at 40 °C and 1 atm}$$

15.70 Use the thermodynamic data in Appendix G to calculate the normal boiling point of methanol, assuming $\Delta H°$ and $\Delta S°$ do not change with temperature. Compare with the experimentally measured boiling point.

15.71 Use the thermodynamic data in Appendix G to calculate the normal boiling point of nitromethane, assuming $\Delta H°$ and $\Delta S°$ do not change with temperature. Compare with the experimentally measured boiling point.

Challenging Phase Equilibria

(15.72–15.73) Vapor pressure is related to K_{eq}, which is related to $\Delta G°$ and to $\Delta H°$ and $T\Delta S°$. Use these relationships to answer Exercises 15.72 and 15.73.

15.72 The vapor pressure of a substance has been measured as a function of temperature. Calculate the standard enthalpy and entropy of vaporization for this substance.

Temperature, °C	Vapor Pressure, torr
−115.7	1
−93.3	10
−76.2	40
−62.7	100
−37.8	400
−23.7	760

15.73 The vapor pressure of a substance has been measured as a function of temperature. Calculate the standard enthalpy and entropy of vaporization for this substance.

Temperature, °C	Vapor Pressure, torr
−26.1	1
2.4	10
23.8	40
39.5	100
67.8	400
82.5	760

Influence of Concentration on Free Energy

15.74 Calculate the free energy of formation for the following substances under the stated conditions.
(a) $PCl_3(g)$ 0.2 atm, 438 K
(b) $PCl_5(g)$ 0.1 atm, 438 K
(c) $Cl_2(g)$ 2 atm, 438 K

15.75 Calculate the free energy change and determine whether the following reactions are spontaneous as written under the stated conditions.
(a) $PCl_5(0.1\ atm) \longrightarrow PCl_3(0.2\ atm) + Cl_2(2\ atm)$ at 438 K
(b) $H_2(g) + F_2(g) \longrightarrow 2HF(g)$ at 298 K and 1 atm of each substance

15.76 Write the expression and use the standard free energy change to calculate the value of the equilibrium constant for

$$PCl_5(g) \rightleftharpoons PCl_3(g) + Cl_2(g)$$

(a) at 25 °C
(b) at 250 °C

15.77 Write the expression and use the data in Appendix G to calculate the value of the equilibrium constant for

$$2SO_2(g) + O_2(g) \longrightarrow 2SO_3(g)$$

(a) at 25 °C
(b) at 250 °C

(15.78–15.81) We generally assume that $\Delta H°$ and $\Delta S°$ do not change with temperature. This approximation would not be valid over the whole range of temperatures encountered in the next several problems.

15.78 Calculate the changes in free energy and the equilibrium constant with temperature for an endothermic reaction with $\Delta H° = +10$ kJ and an entropy change of $+100\ J\ K^{-1}$. Calculate $\Delta G°$ and K_{eq} at 10, 100, and 1000 K.

15.79 Calculate the changes in the standard free energy and the equilibrium constant with temperature for an endothermic reaction with $\Delta H° = +10$ kJ and an entropy change of $-100\ J\ K^{-1}$. Calculate $\Delta G°$ and K_{eq} at 10, 100, and 1000 K.

15.80 Calculate the changes in the standard free energy and the equilibrium constant with temperature for an exothermic reaction with $\Delta H° = -10$ kJ and an entropy change of $+100\ J\ K^{-1}$. Calculate $\Delta G°$ and K_{eq} at 10, 100, and 1000 K.

15.81 Calculate the changes in the standard free energy and the equilibrium constant with temperature for an exothermic reaction with $\Delta H° = -10$ kJ and an entropy change of $-100\ J\ K^{-1}$. Calculate $\Delta G°$ and K_{eq} at 10, 100, and 1000 K.

15.82 Calculate ΔG at 303 °C.

$$CO(2\ atm) + Cl_2(1\ atm) \rightleftharpoons COCl_2(0.1\ atm)$$

15.83 Calculate ΔG at 37°C.

$$N_2O(1\ atm) + H_2(0.4\ atm) \rightleftharpoons N_2(1\ atm) + H_2O(\ell)$$

15.84 State how temperature will influence the equilibrium constant for:
(a) $N_2O(g) + H_2(g) \rightleftharpoons N_2(g) + H_2O(\ell)$
(b) $CO(g) + Cl_2(g) \rightleftharpoons COCl_2(g)$
(c) $CO(g) + H_2O(g) \rightleftharpoons CO_2(g) + H_2(g)$
(d) $PCl_5(g) \rightleftharpoons PCl_3(g) + Cl_2(g)$
(e) $2SO_2(g) + O_2(g) \rightleftharpoons 2SO_3(g)$

15.85 Calculate the standard free energy change and note whether the reaction is spontaneous under the stated conditions:
(a) $CS_2(\ell) \longrightarrow CS_2(g)$ 15 °C and 1 atm
(b) $CCl_4(g) \longrightarrow CCl_4(\ell)$ 92 °C and 1 atm
(c) $C_6H_6(g) \longrightarrow C_6H_6(\ell)$ 54 °C and 1 atm

Vapor Pressure and Temperature

15.86 Calculate the vapor pressure of the following at the stated conditions:

(a) $CS_2(\ell)$ at 5 °C
(b) $CCl_4(\ell)$ at 29 °C
(c) $C_6H_6(\ell)$ at 45 °C

15.87 Calculate the vapor pressure of the following at the stated conditions:
(a) $CH_3OH(\ell)$ at 58 °C
(b) $CH_3CH_2OH(\ell)$ at 29 °C
(c) $Hg(\ell)$ at 45 °C

Energy Resources

15.88 List the factors that influence the cost of coal.

15.89 List the factors that influence the cost of gasoline.

15.90 What factors influence the cost of heating water with solar energy?

15.91 Why is gasoline generally considered a "cleaner" fuel than coal?

15.92 In one part of the United States, the electricity cost is $\$2.0 \times 10^{-8}$/J. The cost of bottled gas is $\$1.0 \times 10^{-8}$/J. Is it logical to say that an electric water heater will cost twice as much to operate as one that uses bottled gas?

15.93 Calculate the standard free energy change when SO_3 is formed from SO_2 and O_2. Why is sulfur trioxide an important substance to study?

15.94 If a lake has become acidified to the point that plant and animal growth is affected, could limestone ($CaCO_3$) be added to neutralize the acid? Describe the chemical, biological, environmental, and economic ramifications of such a process.

15.95 Name some practical energy-saving measures that you might implement in your home. Why haven't these been implemented?

15.96 List some practical energy-saving measures that could be implemented in the Chemistry Building.

15.97 Do you think that the Chemistry Building should:
(a) recirculate air to conserve energy? (A recirculation system is the norm in domestic heating—house air is taken from air-return ducts, reheated or cooled, and recirculated.)
(b) discard air for health and safety? (The air in most chemistry labs is exhausted by hoods and replaced with fresh air that must be heated or chilled to room temperature. This procedure is wasteful of energy.)

Additional Problems

15.98 A compound is 82.7% carbon and 17.3% hydrogen. It has a heat of vaporization of 38.6 J/g. Determine the molecular formula by applying Trouton's rule. The boiling point is −0.5 °C.

15.99 A volatile organic hydrocarbon is found to contain 85.6% carbon; the remainder is hydrogen. It has a heat of vaporization of 476 J/g and a boiling point of −33 °C. Determine the molecular formula by applying Trouton's rule.

15.100 The equilibrium constant for the formation of phosgene is measured at two different temperatures.

$$CO(g) + Cl_2(g) \longrightarrow COCl_2(g)$$

At 506 °C, $K_{eq} = 1.3$; at 530 °C, $K_{eq} = 0.78$. Calculate $\Delta H°$ and $\Delta S°$ for this reaction. Under standard state conditions, over what temperature range is the reaction spontaneous?

15.101 The equilibrium constant for

$$CO(g) + H_2O(g) \longrightarrow CO_2(g) + H_2(g)$$

is measured at several temperatures.

Temperature, K	K_{eq}
400	1.4×10^3
450	3.6×10^2
489	1.5×10^2
502	1.2×10^2
522	8.0×10^1

Determine $\Delta H°$ and $\Delta S°$ for this reaction. Under standard state conditions, over what temperature range is the reaction spontaneous?

15.102 Engineers who study the design of nuclear power plants have to worry about high-temperature reactions since it is possible for water to decompose. Under what conditions might this reaction occur spontaneously?

$$2H_2O(g) \longrightarrow 2H_2(g) + O_2(g)$$

15.103 If the decomposition of water is spontaneous, would engineers have to worry about an oxygen/hydrogen explosion? Justify your answer.

15.104 The standard free energy change for the "thermite" reaction is −840.1 kJ.

$$2Al(s) + Fe_2O_3(s) \longrightarrow Al_2O_3(s) + 2Fe(s)$$

Calculate K_{eq} for this reaction. Explain why K_{eq} is difficult to determine with your calculator. Look at the phases of the reactants and products, and speculate on what an equilibrium constant tells us about this reaction.

Chemical Kinetics

Chapter Sixteen

16

Two issues are important in the study of chemical reactions. First, we must determine whether a reaction can occur. The calculations of thermodynamics, discussed in Chapter 15, can be used to predict whether or not a reaction is spontaneous. For example, calculations will show that the formation of graphite from diamond is a spontaneous process, since ΔG is less than 0.

$$C(diamond) \longrightarrow C(graphite)$$

Experience tells us that diamonds do not decompose to graphite, even over many years. This observation does not contradict the second law of thermodynamics, but its explanation focuses on another aspect of chemistry, the speed of a reaction. Even if a reaction can proceed spontaneously to form the product, it may be too slow (or too fast) for any practical purpose. Under normal conditions, the conversion of diamonds to graphite is too slow to be measured. The study of the rates of chemical reactions is called **chemical kinetics,** derived from the Greek *kinetikos,* meaning "putting in motion."

The study of kinetics is inherently an experimental science; many factors influence the rate of a reaction. For example, most people realize that iron will ultimately rust in the presence of oxygen, a complex series of chemical reactions that is approximated by:

$$4Fe(s) + 3O_2(g) + xH_2O(\ell) \longrightarrow 2Fe_2O_3 \cdot xH_2O(s)$$

Rusting can occur quickly—perhaps you've observed a garden tool that was left out during a day or two of rain. However, the iron sword of Mohammed misri is still bright and shiny, about 300 years after it was forged, because it was stored under conditions for which rusting is a very slow process. The sword is shown on the next page.

Reactions occur when reactant molecules strike each other and interact to form the molecules of the products. Chemists study various reactions in

◀ Chemical reactions occur over a period of time.

(a)

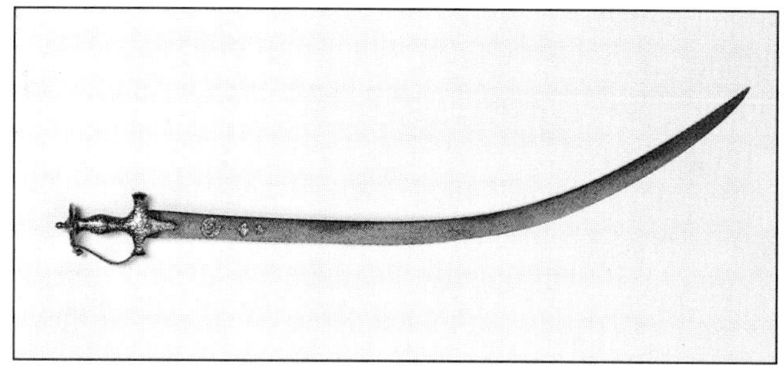

(b)

Rusting and preserving tools. (a) Garden tools can rust in a day or two. The speed depends on the weather and the composition of the steel (iron plus other elements) used to make the tool. Stainless steel (an iron alloy that has about 18% Cr and 8% Ni added) is very slow to rust. (b) This Egyptian sword is about 300 years old and shows no sign of rust; it was stored under conditions for which rusting is a very slow process.

an effort to determine the exact processes (which molecules collide, how many, under what conditions, etc.) that occur. This information is used to improve the production of materials, minimize pollution, and increase the energy efficiency of manufacturing processes.

This chapter first presents the experimental approaches for measuring and classifying reaction rates. The results of the experiments are discussed, and the treatment of experimental data is presented. The currently accepted theories that explain the rates of reaction are next presented, and the role of catalysts in increasing reaction rate is examined. A discussion of reaction mechanisms concludes this chapter.

16.1 The Rate of Reaction and Rate Laws

Objectives

- To define reaction rate
- To calculate the rate of reaction from experimental data
- To express the rate of reaction in terms of changes in the concentrations of reactants and products

The rates of reactions and the factors that influence the rates are crucial components of our knowledge of chemical reactions. The study of reaction rates is important to the synthesis of chemicals and the design of reactors, and provides fundamental insight into chemical reactivity as it applies to biological, geological, and industrial processes.

The Rate of a Reaction

Rate is change per unit time. Although we talk about crime rates and the rate of inflation, the most common use of the word "rate" is to describe a moving object quantitatively. The rate of movement of a car is expressed in miles per hour and represents the change in location over a period of time. For chemical reactions, a rate is measured in terms of a change in concentration per unit time.

$$\text{rate} = \frac{\Delta c}{\Delta t}$$

Figure 16.1 shows a chemical reaction that has an initial product concentration of 0.034 M when the measurement starts (time = 0). The concentration has increased to 0.083 M after 15.1 seconds. The rate at which the concentration increases can be determined from the data.

$$\text{rate of increase} = \frac{0.083\ M - 0.034\ M}{15.1\ \text{s} - 0\ \text{s}} = 3.2 \times 10^{-3}\ M\ \text{s}^{-1}$$
$$= 3.2 \times 10^{-3}\ \text{mol L}^{-1}\ \text{s}^{-1}$$

This simple example illustrates an important point—the units used to express reaction rates are mol L^{-1} s^{-1} (although mol L^{-1} min^{-1} or mol L^{-1} hr^{-1} might be used for slower reactions).

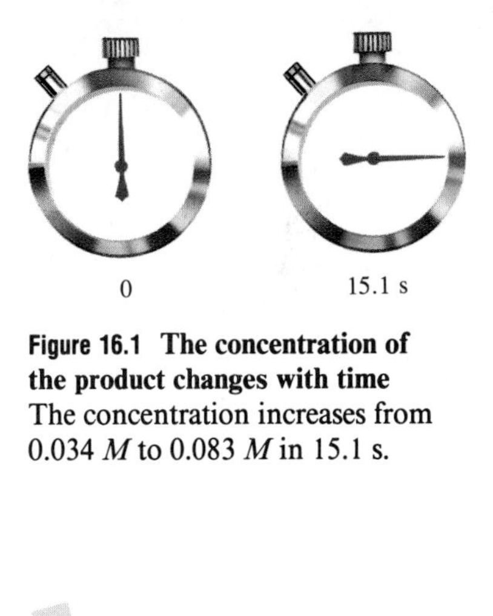

Figure 16.1 The concentration of the product changes with time The concentration increases from 0.034 M to 0.083 M in 15.1 s.

Instantaneous and Average Rates

The measurement of the rate for the reaction illustrated in Figure 16.1 provides an average rate involving the net change in concentration in the first 15.1 s. Often, a more detailed knowledge of the rate is needed, so we make several measurements at different times.

This process can be illustrated with a study of the changes in the concentration during the course of a reaction of the type

$$\text{reactant} \longrightarrow \text{product}$$

In this reaction, the reactant decreases in concentration (Δ[reactant] is negative) and the product increases in concentration (Δ[product] is positive). For this simple reaction, the rate is the rate at which the product appears. The negative of the rate of disappearance of the reactant could be used as well. This definition of reaction rate will be modified for more complex reactions, because reaction stoichiometry will be included.

$$\text{rate} = \frac{\Delta[\text{product}]}{\Delta t} = \frac{-\Delta[\text{reactant}]}{\Delta t}$$

It is important to remember that *the rate of reaction is always expressed as a positive number.*

By convention the rate of reaction is always positive.

It is often convenient to follow the course of a reaction by measuring the concentration of the reactant, rather than the product. Figure 16.2 shows a

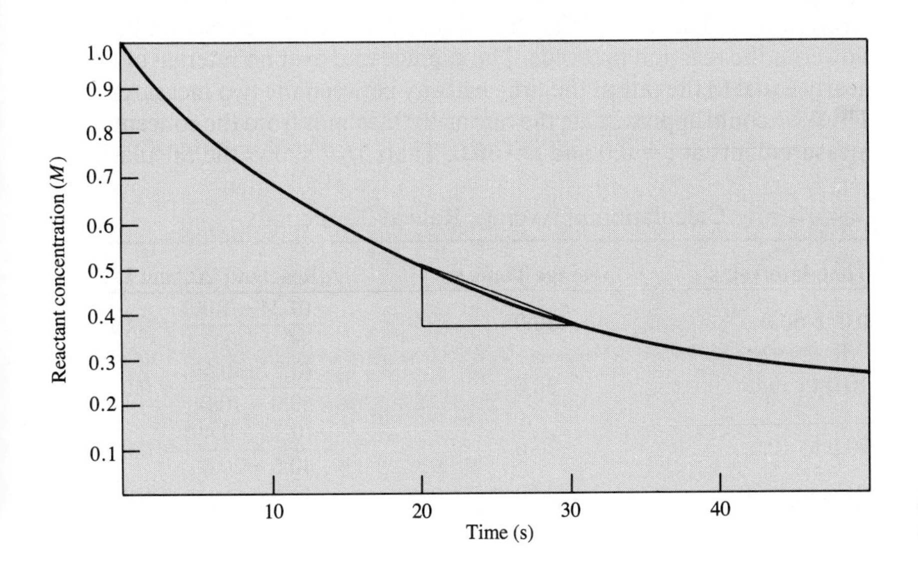

Figure 16.2 The decrease in reactant concentration as a function of time

Table 16.1 The Time Dependence of Concentration

t, s	[Reactant], *M*	Δ*t*, s	Δ[Reactant], *M*	Rate of Decrease of [Reactant] mol L^{-1} s^{-1}
0.0	1.0			
5.0		(10.0 − 0.0) = 10	(0.69 − 1.00) = −0.31	0.031
10.0	0.69			
15.0		(20.0 − 10.0) = 10	(0.49 − 0.69) = −0.20	0.020
20.0	0.49			
25.0		(30.0 − 20.0) = 10	(0.38 − 0.49) = −0.11	0.011
30.0	0.38			
35.0		(40.0 − 30.0) = 10	(0.31 − 0.38) = −0.07	0.007
40.0	0.31			
45.0		(50.0 − 40.0) = 10	(0.27 − 0.31) = −0.04	0.004
50.0	0.27			
55.0		(60.0 − 50.0) = 10	(0.24 − 0.27) = −0.03	0.003
60.0	0.24			

plot of the concentration of the reactant over a period of time. The reactant concentration decreases regularly with time. The average rate of reaction over any time period can be determined from the difference in concentrations divided by the difference between the measurement times. Let us use the data of Figure 16.2 to calculate the average rate of reaction in the interval between 20 and 30 seconds. The reactant concentration decreases from 0.49 *M* (after 20 s) to 0.38 *M* after 30 s. Since the disappearance of the reactant is observed, a negative sign is needed:

$$\text{rate of decrease} = \frac{-(0.38 - 0.49)\ M}{(30 - 20)\ \text{s}} = 0.011\ \text{mol L}^{-1}\ \text{s}^{-1}$$

We should note that this rate is the average rate over the 10-second interval between 20 and 30 s.

Table 16.1 presents the data of Figure 16.2 and the calculated rates of reaction in tabular form. We observe that the rate of the reaction becomes slower as the reaction proceeds. The average rate over an interval of time is nearly equal to the rate at the time halfway between the two measurements. Thus, we could approximate the rate at 30.0 seconds from the concentration measurements at $t = 0.0$ and $t = 60.0$. Table 16.2 shows the calculation of

Table 16.2 Calculation of Average Rate at 30 Seconds

Time Interval, s	Average Time, s	−Δ[Reactant]/Δ*t*, mol L^{-1} s^{-1}
0.0 to 60.0	30.0	$\frac{-(0.24 - 1.00)}{60.0 - 0.0} = 0.013$
10.0 to 50.0	30.0	$\frac{-(0.27 - 0.69)}{50.0 - 10.0} = 0.010$
20.0 to 40.0	30.0	$\frac{-(0.31 - 0.49)}{40.0 - 20.0} = 0.0090$
tangent at 30	30.0	0.0089

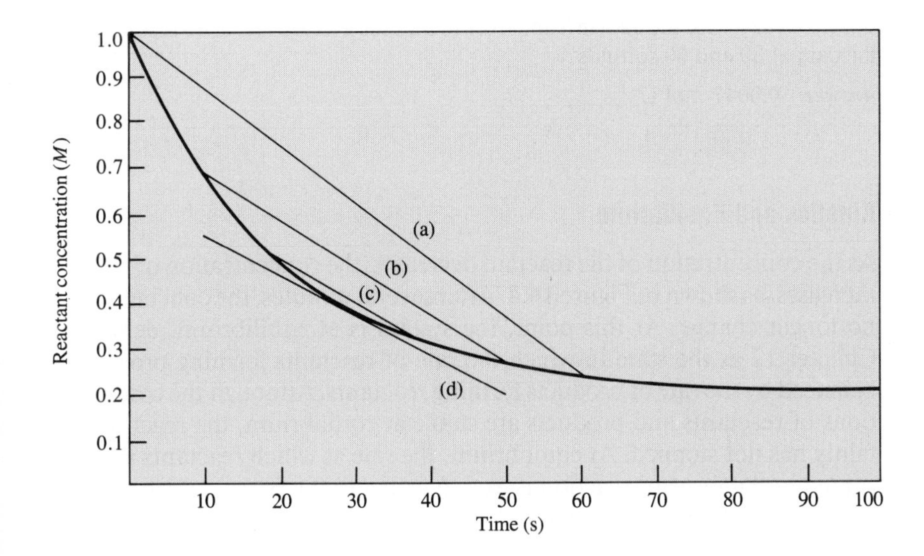

Figure 16.3 Average rates and instantaneous rates The concentration of the reactant decreases as a function of time. The slopes of the straight lines labeled (a) through (c) are the average rates calculated for different intervals: (a) 0 to 60 seconds, (b) 10 to 50 seconds, (c) 20 to 40 seconds. Line (d) is the tangent to the curve at 30 seconds, and its slope is called the instantaneous rate at 30 seconds.

the average rate at 30 seconds for this and several other intervals. The rates calculated in Table 16.2 are shown by the straight lines in Figure 16.3, along with the tangent to the curve at 30 seconds. As the interval decreases, the slopes of the lines approach the **instantaneous rate,** which is the slope of the tangent to the curve.

A simple analogy to instantaneous and average rates is found in driving an automobile. Explaining that it has taken 4 hours for a car to cover 200 miles, an average rate of 50 mph, is not going to help defend an accused speeder whose instantaneous speed was measured at 75 mph.

Example 16.1 Average and Instantaneous Rates

Use the data in Table 16.1 to calculate the average rate of disappearance of the reactant at 35 seconds, using the intervals (a) 10 to 60 seconds and (b) 20 to 50 seconds. (c) Including the rate calculated in Table 16.1, which of the average rates is closest to the instantaneous rate at 35 seconds?

Solution

(a) From Table 16.1 the concentration of reactant at 10 seconds is 0.69 M, and that at 60 seconds is 0.24 M. The negative of the change in concentration divided by the change in time is the average rate we seek.

$$\text{average rate (10 to 60)} = \frac{-(0.24 - 0.69)}{60 - 10} = 0.0090 \text{ mol L}^{-1}\text{ s}^{-1}$$

(b) The same calculation is repeated using the concentrations at 20 seconds (0.49 M) and 50 seconds (0.27 M).

$$\text{average rate (20 to 50)} = \frac{-(0.27 - 0.49)}{50 - 20} = 0.0073 \text{ mol L}^{-1}\text{ s}^{-1}$$

(c) Since the time interval of 10 seconds used to calculate the rate in Table 16.1 was the smallest one, that rate, 0.007 mol L^{-1} s^{-1}, is closest to the instantaneous rate.

Understanding Use the data in Table 16.1 to find the average rate at 45 seconds from the data at 30 and 60 seconds.

Answer: 0.0047 mol L^{-1} s^{-1}

Kinetics and Equilibrium

At equilibrium, the rate at which reactants form products is equal to the rate at which products form reactants.

As the concentration of the reactant decreases, the concentration of product increases, as shown in Figure 16.4. After several minutes, the concentrations no longer change. At this point, the system is at equilibrium, defined in Chapter 12 as the state in which the rate of reactants forming products is balanced by the rate of products forming reactants. Although the concentrations of reactants and products are stable at equilibrium, the reaction certainly has not stopped. At equilibrium, the rate at which reactants are consumed in the forward reaction is exactly *equal* to the rate at which they are formed in the reverse reaction.

$$\text{reactants} \rightleftharpoons \text{products}$$

Rate and Reaction Stoichiometry

The rate of a reaction does not depend on which species is measured.

The rate at which the reactants are consumed and the rate at which the products are formed depend on the stoichiometry of the reaction. Let's consider the dissociation of hydrogen bromide at high temperatures to illustrate the relationship between stoichiometry and rate.

$$2HBr(g) \longrightarrow H_2(g) + Br_2(g)$$

For every two moles of HBr that react, one mole of each product is formed. The same two-to-one stoichiometry must also apply to the rates of change of reactants and products. For example, if the concentration of HBr is decreasing at a rate of 0.50 mol L^{-1} s^{-1} at a particular instant, the rate of change of hydrogen at that time can be calculated from the stoichiometric equivalencies.

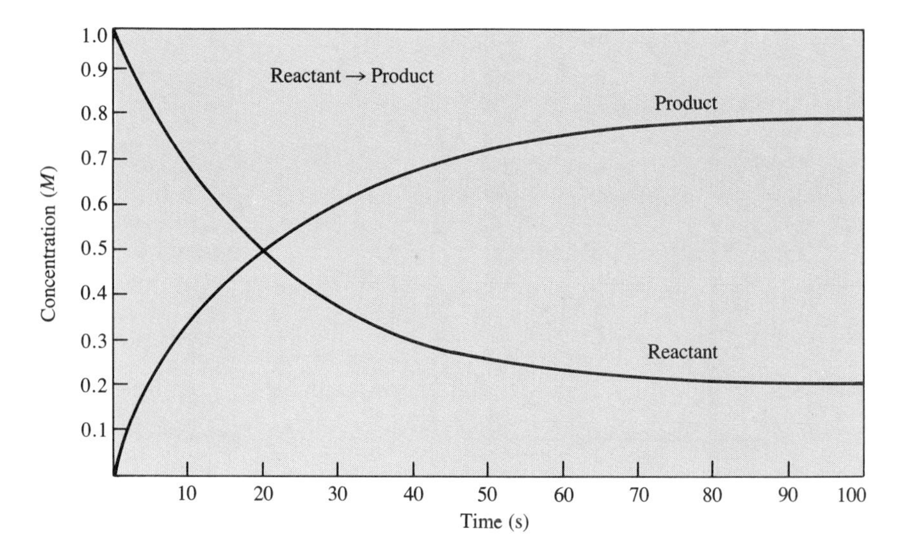

Figure 16.4 Changes in concentration of reactant and product with time After several minutes, the concentrations cease to change.

$$\frac{\Delta[H_2]}{\Delta t} = \frac{-\Delta[HBr]}{\Delta t}\left(\frac{1 \text{ mol } H_2}{2 \text{ mol HBr}}\right)$$

$$= 0.50 \text{ mol HBr L}^{-1}\text{ s}^{-1} \times \frac{1 \text{ mol } H_2}{2 \text{ mol HBr}}$$

$$= 0.25 \text{ mol } H_2 \text{ L}^{-1}\text{ s}^{-1}$$

The rate at which a chemical reaction proceeds should not depend on which species is measured. Therefore, by convention we define the **rate of reaction** as the ratio of the rate of change of any substance to its coefficient in the chemical equation. For the hydrogen bromide reaction,

$$\text{rate of reaction} = \frac{\Delta[Br_2]}{\Delta t} = \frac{\Delta[H_2]}{\Delta t} = \frac{-\Delta[HBr]}{2\Delta t}$$

These concepts are illustrated in Example 16.2.

Example 16.2 Expressing the Rate of Reaction

Consider the formation of ammonia:

$$N_2(g) + 3H_2(g) \longrightarrow 2NH_3(g)$$

(a) If the ammonia concentration is increasing at a rate of 0.024 mol L^{-1} s^{-1}, what is the rate of disappearance of hydrogen?
(b) What is the rate of the reaction?

Solution

(a) rate of disappearance of hydrogen $= \frac{-\Delta[H_2]}{\Delta t}$

$$\frac{-\Delta[H_2]}{\Delta t} = \frac{\Delta[NH_3]}{\Delta t} \times \frac{3 \text{ mol } H_2}{2 \text{ mol } NH_3}$$

$$\frac{\Delta[H_2]}{\Delta t} = 0.024 \text{ mol L}^{-1}\text{ s}^{-1}\left(\frac{3 \text{ mol } H_2}{2 \text{ mol } NH_3}\right) = 0.036 \text{ mol L}^{-1}\text{ s}^{-1}$$

(b) rate of reaction $= \frac{\Delta[NH_3]}{2\,\Delta t} = \frac{0.024 \text{ mol L}^{-1}\text{ s}^{-1}}{2} = 0.012 \text{ mol L}^{-1}\text{ s}^{-1}$

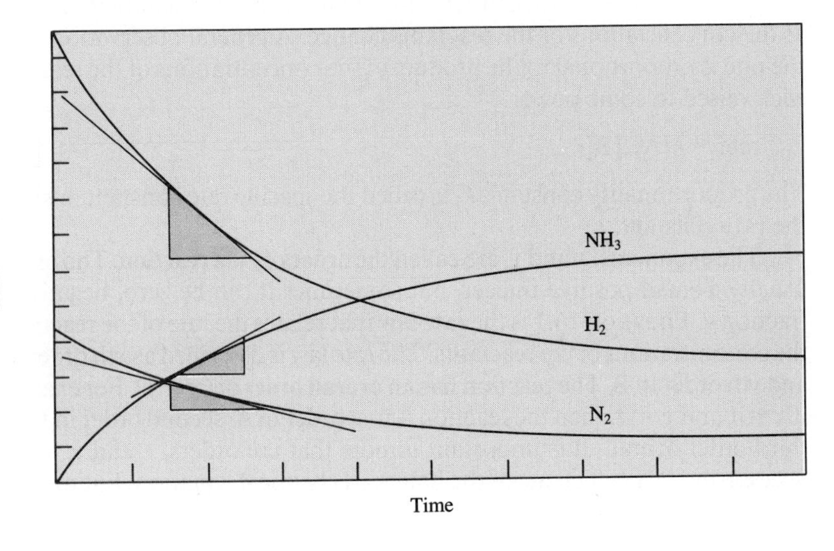

Changes in concentration during the formation of ammonia The rate of reaction is

$$\text{rate} = \frac{-\Delta[N_2]}{\Delta t} = \frac{-\Delta[H_2]}{3\Delta t} = \frac{+\Delta[NH_3]}{2\Delta t}$$

The same answer, of course, is obtained using the rate of change of H_2:

$$\text{rate of reaction} = \frac{-\Delta[H_2]}{3\ \Delta t} = \frac{0.036\ \text{mol L}^{-1}\ \text{s}^{-1}}{3} = 0.012\ \text{mol L}^{-1}\ \text{s}^{-1}$$

Understanding Under different conditions, the rate of disappearance of N_2 was 0.14 mol L^{-1} s^{-1} of N_2. What is the rate of disappearance of hydrogen? What is the rate of reaction?

Answer: The rate of disappearance of hydrogen is 0.42 mol L^{-1} s^{-1}. The rate of the reaction is 0.14 mol L^{-1} s^{-1}.

16.2 Relationships Between Concentration and Rate

Objectives

- To define rate law, rate constant, and reaction order
- To use initial concentrations and initial rates of reactions to determine the rate law and rate constant

The rates of chemical reactions have been studied for quite a long time, and one important observation is that *the rate of reaction is often strongly influenced by the concentrations of the reacting species.* In this section, we present the relationships between rates of reaction and concentrations, as well as some methods used to determine these relationships from experimental data.

Experimental Rate Laws

We will consider a chemical system in which there are two reactants, A and B:

$$aA + bB \rightarrow \text{products}$$

We perform a series of experiments in which the rate of reaction is measured as the concentrations of the reactants change. A general observation is that the rate is proportional to the product of the concentrations of the reactants, each raised to some power.

$$\text{rate} = k[A]^x[B]^y \qquad [16.1]$$

The proportionality constant, k, is called the specific rate constant or simply the **rate constant.**

The exponents, x and y, are called the **orders of the reaction.** The order is usually a small positive integer, but sometimes it can be zero, negative, or fractional. Equation 16.1 is the **rate law** that relates the rate of the reaction to the concentrations of the reactants. The rate law is described as xth order in A and yth order in B. The reaction has an **overall order** of $(x + y)$. For example, if $x = 1$ and $y = 2$, then the reaction is first order in A, second order in B, and third order overall. It is important to note that the orders, x and y, are not necessarily the coefficients of the balanced chemical equation, but numbers that must be determined by experiment.

Measurements of the Initial Rate of Reaction

Before we can write the rate law for any reaction, we must first experimentally measure the reaction order for each substance. One rapid and convenient way of accomplishing this is called the **initial rate method.** We perform several experiments in which the initial concentrations of all substances are accurately known (but different for each experiment). In each experiment the reaction is allowed to proceed for a period short enough that the concentrations of the reactants do not change very much (usually less than 3%) over the time interval. For these small changes, the average rate is very close to the instantaneous rate. The change in concentration of a reactant or product, and the length of the time interval the reaction has proceeded, are used to calculate the rate for each experiment. The initial rate method allows the initial concentration of each substance to be adjusted to a particular value and correlates these concentrations with measurements of rate.

The initial rate method is used to determine the relation between the initial rate of reaction and the initial concentrations of the reactants.

Consider the gas-phase reaction of water with methyl chloride.

$$H_2O(g) + CH_3Cl(g) \longrightarrow CH_3OH(g) + HCl(g)$$

It would be difficult to stop the reaction, collect the HCl, and titrate it every few seconds. Instead, the amount of time needed for the HCl to neutralize a small, but known amount of sodium hydroxide can be used as a measure of the rate at which the HCl is produced. Since the measurement takes place in the initial part of the reaction, the rate corresponds to the initial rate of reaction.

The following data were obtained in the laboratory in an investigation of the gas-phase reaction of water and methyl chloride.

$$H_2O(g) + CH_3Cl(g) \longrightarrow CH_3OH(g) + HCl(g)$$

$$\text{rate} = \frac{\Delta[\text{HCl}]}{\Delta t}$$

	Initial Concentration, *M*		**Initial Rate of Reaction**
Expt.	H_2O	CH_3Cl	**mol L^{-1} s^{-1}**
1	0.010	0.020	3.6×10^{-4}
2	0.020	0.020	14.4×10^{-4}
3	0.030	0.020	32.3×10^{-4}
4	0.010	0.040	7.2×10^{-4}
5	0.010	0.060	10.9×10^{-4}

The data can be used to determine the exponents x and y in the rate law

$$\text{rate} = k[H_2O]^x[CH_3Cl]^y$$

Experiments 1, 2, and 3 all have the same concentration of CH_3Cl and can be used to evaluate the order in H_2O. It is easier to recognize the dependence of rate on concentration if the rate is expressed on a *relative* basis. The relative concentrations of water and the relative rates of reaction in these three experiments are obtained by dividing each concentration by 0.010 (the smallest concentration) and each rate by 3.6×10^{-4} (the slowest rate).

Expt.	Initial Concentration, M $[H_2O]$	Initial Rate of Reaction, mol L^{-1} s^{-1}	Relative Concentration of H_2O	Relative Initial Rate of Reaction
1	0.010	3.6×10^{-4}	1.0	1.00
2	0.020	14.4×10^{-4}	2.0	4.00
3	0.030	32.3×10^{-4}	3.0	8.97

As the relative concentration increases from 1.0 to 2.0 to 3.0, the relative rate of reaction goes from 1.0 to 4.0 to 9.0. This result indicates that the rate of reaction is proportional to the concentration of water raised to the second power. *The reaction is second order in water,* so we can update the rate law to

$$\text{rate} = k[H_2O]^2[CH_3Cl]^y$$

In experiments 1, 4, and 5 only the initial concentration of CH_3Cl changes. The following table includes both the actual and relative concentrations and rates.

Expt.	Initial Concentration, M $[CH_3Cl]$	Initial Rate of Reaction, mol L^{-1} s^{-1}	Relative Concentration of $CHCl_3$	Relative Initial Rate of Reaction
1	0.020	3.6×10^{-4}	1.0	1.00
4	0.040	7.2×10^{-4}	2.0	2.00
5	0.060	10.9×10^{-4}	3.0	3.03

Since the relative rate is the same as the relative concentration in each experiment, the order in methyl chloride is 1, and the rate law is

$$\text{reaction rate} = k[H_2O]^2[CH_3Cl]$$

The final step in the solution of this problem is to evaluate the rate constant, k. We solve the rate law for k:

$$k = \frac{\text{reaction rate}}{[H_2O]^2[CH_3Cl]}$$

We can select any of the experiments; we will choose the data from experiment 3.

$$k = \frac{32.3 \times 10^{-4}\ M\ s^{-1}}{(0.030)^2\ M^2(0.020)\ M} = 1.8 \times 10^2\ L^2\ mol^{-2}\ s^{-1}$$

The complete rate law is

$$\text{reaction rate} = 1.8 \times 10^2\ L^2\ mol^{-2}\ s^{-1}[H_2O]^2[CH_3Cl]$$

Checking the rate constant verifies that the rate law and orders are correct, as well as checking the numerical calculation of the rate constant.

We need to check our work, and calculating a second rate constant from the data of another experiment is a good way. If the rate constant found from both experiments is the same, we are confident that the orders in water and methyl chloride are correct; results that differ by more than experimental uncertainty indicate that an error in determining the order has been made.

Example 16.3 Determining the Rate Law from Initial Rate Data

When methyl bromide reacts with hydroxide ion in solution, methyl alcohol and bromide ion are formed.

$$CH_3Br + OH^- \longrightarrow CH_3OH + Br^-$$

Determine the rate law and evaluate the rate constant from the experimental data.

	Initial Concentration, *M*		
Expt.	**$[CH_3Br]$**	**$[OH^-]$**	**Initial Rate, $mol\ L^{-1}\ s^{-1}$**
1	0.050	0.010	2.4×10^{-3}
2	0.080	0.020	7.7×10^{-3}
3	0.080	0.010	3.8×10^{-3}

Solution

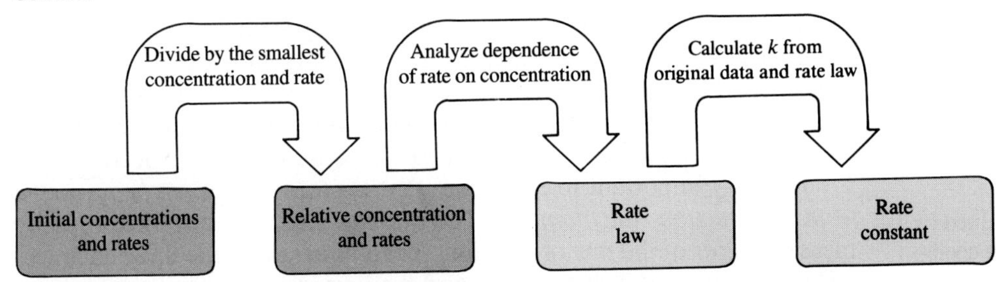

To obtain relative concentrations and rates, divide the CH_3Br concentrations by 0.050; divide the OH^- concentrations by 0.010; and divide the reaction rates by 2.4×10^{-3}.

	Relative Concentration		
Expt.	**$[CH_3Br]$**	**$[OH^-]$**	**Relative Rate**
1	1.0	1.0	1.0
2	1.6	2.0	3.2
3	1.6	1.0	1.6

Experiments 1 and 3 (constant hydroxide ion concentration) show that when the concentration of CH_3Br increases by a factor of 1.6, the rate increases by the same factor. We can conclude that the reaction is first order in methyl bromide. Experiments 2 and 3 (constant concentration of methyl bromide) show that when the concentration of hydroxide doubles, so does the rate. Thus, the reaction is also first order in hydroxide ion. The rate law can now be written.

$$\text{rate} = k[CH_3Br][OH^-]$$

The reaction is second order overall—first order in methyl bromide and first order in hydroxide.

Use the original data for Experiment 1 from the first table (*not* the relative data of the second table) to calculate the rate constant:

$$k = \frac{\text{rate}}{[CH_3Br][OH^-]}$$

$$k = \frac{2.4 \times 10^{-3}\ mol\ L^{-1}\ s^{-1}}{(0.05 \times 0.01)\ mol^2\ L^{-2}}$$

$$k = 4.8\ L\ mol^{-1}\ s^{-1}$$

This value can be confirmed using the data from Experiment 2.

Understanding The results of a similar experiment for the reaction of *t*-butyl bromide, $(CH_3)_3CBr$, with hydroxide ion are given in the following table. Calculate the rate law and the rate constant.

$$(CH_3)_3CBr + OH^- \longrightarrow (CH_3)COH + Br^-$$

Expt.	**Initial Concentration, *M*** $[(CH_3)_3CBr]$	$[OH^-]$	**Initial Rate, mol L^{-1} s^{-1}**
1	0.050	0.010	4.1×10^{-8}
2	0.080	0.020	6.6×10^{-8}
3	0.080	0.010	6.6×10^{-8}

Answer: rate = $k[(CH_3)_3CBr]$; $k = 8.2 \times 10^{-7}\ s^{-1}$. Notice that the rate is zero order in $[OH^-]$. The rate of reaction does not depend on the hydroxide ion concentration.

The reactions shown in Example 16.3 are representative of some interesting organic substitution reactions. Experiments show that the first reaction (methyl bromide with hydroxide) is second order, but that *t*-butyl bromide reacts with hydroxide with first order kinetics. The differences arise because the reactions proceed by different steps, as discussed in Section 16.6.

The rate law must be determined by experiment and not from the coefficients of the chemical equation.

It is very important to emphasize that *the rate law cannot be predicted from the reaction stoichiometry.* The rate law can be determined only by measurements of the rate of reaction. As an illustration, the three chemical systems mentioned in this section and their rate laws are:

$$H_2O + CH_3Cl \longrightarrow CH_3OH + HCl \qquad \text{rate} = k[H_2O]^2[CH_3Cl]$$

$$CH_3Br + OH^- \longrightarrow CH_3OH + Br^- \qquad \text{rate} = k[CH_3Br][OH^-]$$

$$(CH_3)_3CBr + OH^- \longrightarrow (CH_3)COH + Br^- \qquad \text{rate} = k[(CH_3)_3CBr]$$

16.3 Dependence of Concentrations on Time

Objectives

- To compare the concentration-time behavior for reactions with different rate laws
- To use experimental concentration and time data to determine the rate law
- To evaluate the rate constants for zero, first, and second order reactions from laboratory data
- To describe the relationships between reaction rate, rate constant, and half-life
- To calculate the concentration-time behavior from the rate law and rate constant or half-life

The experimental data shown in Figure 16.4 indicate that the rate of reaction changes over the course of the reaction. The concentrations of the reactants decrease, the concentrations of the products increase, and both cease to change when the system reaches equilibrium. We observe that the rate of reaction decreases with time, reaching zero at equilibrium.

In Section 16.2, we used the dependence of the initial rate of reaction on concentration to determine the rate law. Another way to determine the rate law is to examine how the concentration of a reactant changes with time during the course of a *single* experiment. Reactions of different orders behave quite differently. The units of the rate constant also depend on the order of the reaction.

Zero Order Rate Laws

Some reactions show rates that are independent of the concentrations of the reactants, and obey a zero order rate law.

reaction rate $= k$

The units of a zero order rate constant are mol L^{-1} s^{-1}.

The rate constant, k, has the same units as the reaction rate—mol L^{-1} s^{-1}. One example of a zero order reaction is the oxidation of ethyl alcohol in the body. Studies have shown that most people take about five hours to metabolize one ounce of pure alcohol (the amount in two 12-ounce beers, two mixed drinks, or two 5-ounce glasses of wine). After a person consumes several drinks, the alcohol concentration decreases at a constant rate until it is completely consumed; one drink takes 2.5 hours, two drinks take 5 hours, and so forth, for the last of the alcohol to be removed from the body. Figure 16.5 shows graphical representations of a zero order reaction. The graph of concentration *vs.* time (Figure 16.5a) is a straight line that ends when all of the reactant has been consumed. The reaction rate, shown in Figure 16.5b, is a horizontal line that abruptly drops to zero when the reactant has been completely consumed. Zero order rate laws are seldom encountered, except for reactions involving enzymes, an interesting class of reactions that is discussed later in this chapter.

When the graph of concentration vs. time is a straight line, the reaction is zero order.

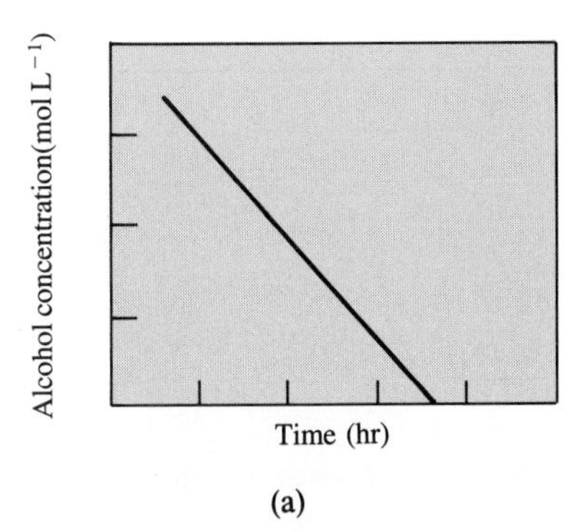

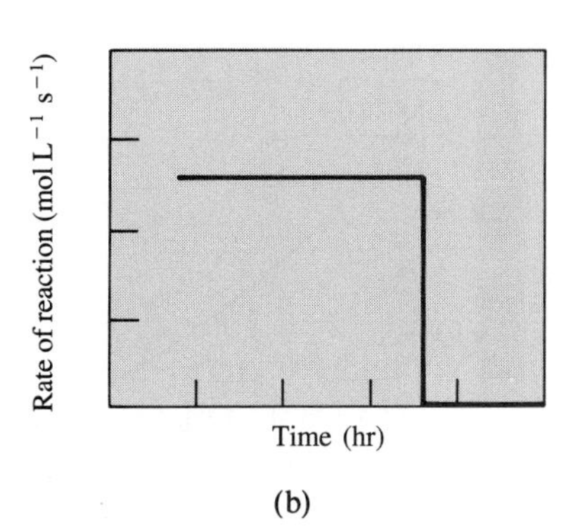

Figure 16.5 The metabolism of alcohol in the body (a) The alcohol concentration in the bloodstream decreases linearly until all the alcohol is metabolized. (b) As long as any alcohol remains present, its rate of disappearance remains constant. One characteristic of zero order reactions is an abrupt drop in the rate when the reactant has been consumed.

First Order Rate Laws

When a reaction is first order in a reactant, A, the rate is proportional to the concentration of the reactant.

$$A \longrightarrow \text{product}$$

$$\text{reaction rate} = \frac{-\Delta[A]}{\Delta t} = k[A] \qquad [16.2]$$

The units of a first order rate constant are generally s^{-1}.

The rate constant in this case will have units of s^{-1}.

Time-Dependent Behavior of Concentration

Equation 16.2 represents a first order reaction. This particular form of the equation is called the **differential form of the rate law** because it relates *differences* in concentration and time ($\Delta[A]/\Delta t$) to the concentrations of the species. Calculus can be used to express the same equation in another way. Equation 16.3 is called the **integral form of the rate law** because concentrations, rather than *changes* in concentration, appear in the equation.

$$[A] = [A]_0 e^{-kt} \qquad [16.3]$$

An exponential decay of reactant is found for first order kinetics.

The concentration of A at any time is designated by [A]; $[A]_0$ is the initial concentration or concentration when $t = 0$; and e is the base of the natural logarithms, approximately 2.718. Equation 16.3 describes an *exponential decay.* The exponential decrease of reactant concentration is illustrated in Figure 16.6.

Another way to express the integral form of the first order rate law results when we take the natural logarithm of both sides of Equation 16.3:

$$\ln[A] = -kt + \ln[A]_0 \qquad [16.4]$$

For a first order reaction, a plot of ln[A] as a function of t is a straight line with a slope of $-k$ and an intercept of $\ln[A]_0$ as shown in Figure 6.7. Please refer to the material in Appendix A if you need to review topics such as slope and intercept.

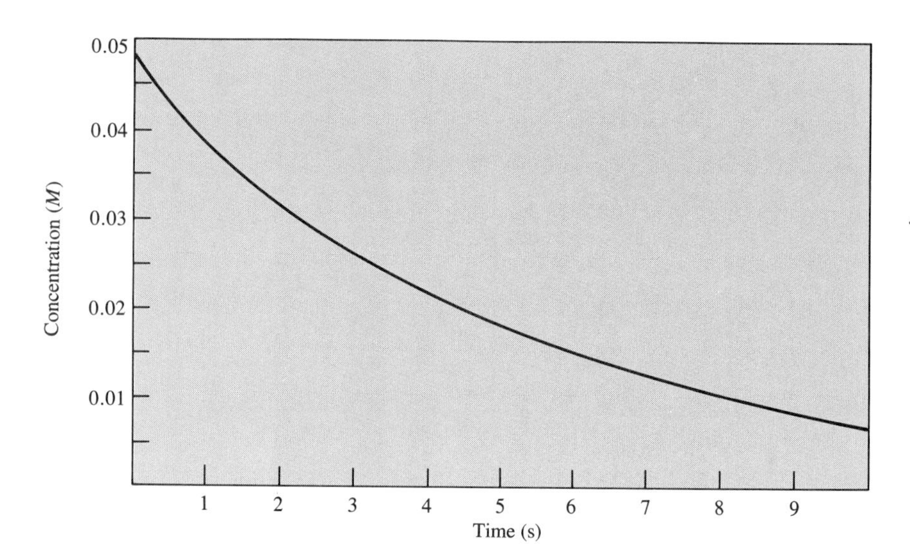

Figure 16.6 Decrease in the concentration of a reactant for a system showing first order kinetics

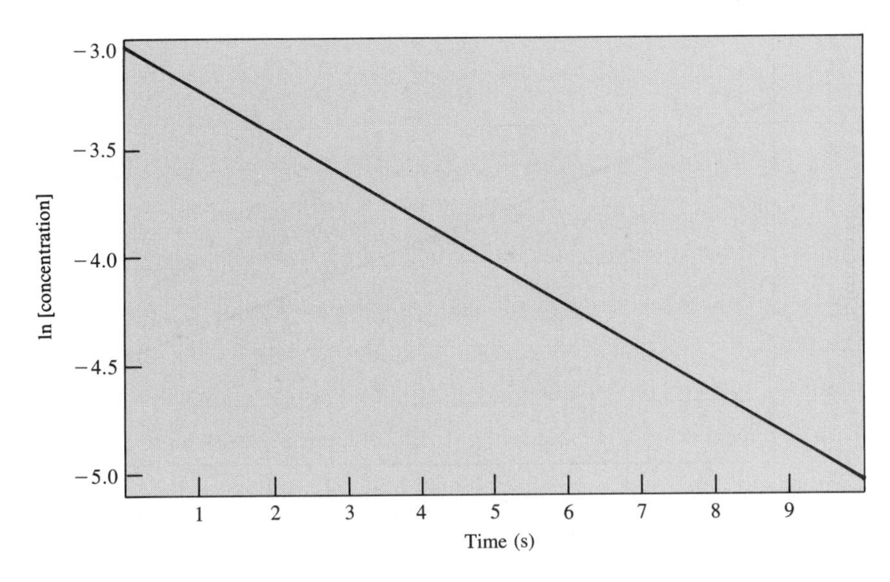

Figure 16.7 The natural logarithm of concentration plotted against time The concentrations are the same as those plotted in Figure 16.6.

When chemists study a new reaction to determine its rate law (reaction order), they generally construct a plot of the natural logarithm of concentration as a function of time. If the graph is a straight line, the reaction is first order. If this plot is not a straight line, then the reaction is not a simple first order reaction and the investigation will continue.

The next example shows how to determine the rate constant from concentration-time data.

If a plot of ln [concentration] *vs.* time is a straight line, then the system is described by first order kinetics.

Example 16.4 Determining a First Order Rate Law and Rate Constant

Determine the rate law (order and rate constant) from the following data, obtained in a research laboratory that studies fast reactions. The reaction can be assumed to be

reactant ⟶ product

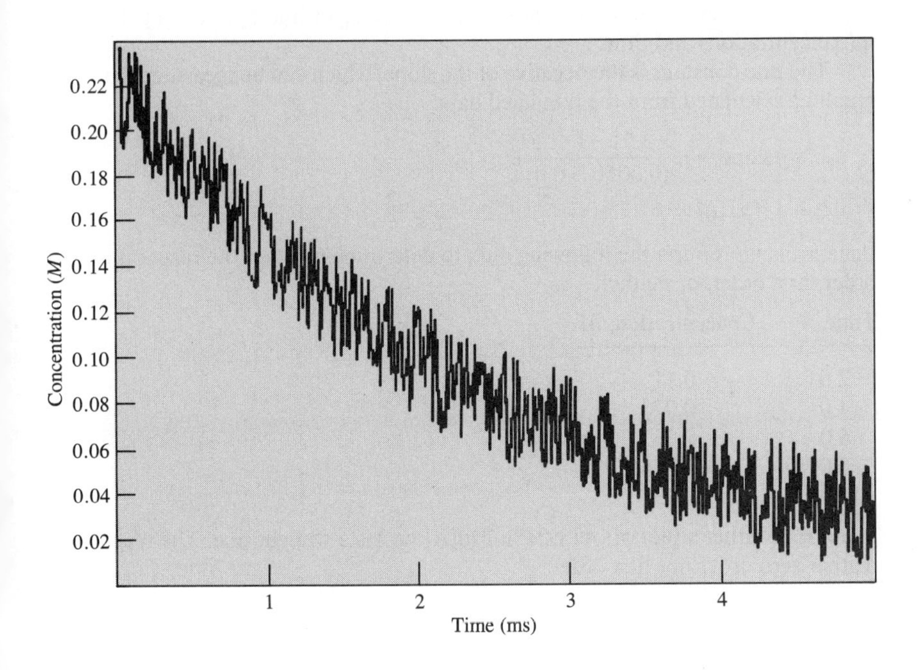

Data from a stopped-flow spectrophotometer This instrument is designed to measure the rates of reactions that take place in milliseconds. The lack of a smooth line is due to noise (a form of experimental uncertainty) that is present along with the signal.

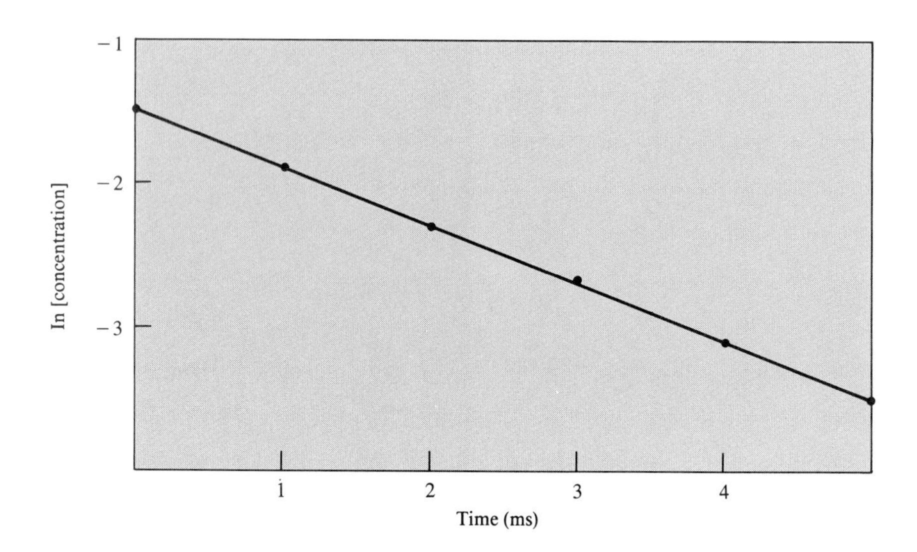

Natural logarithm of concentration as a function of time The data show a straight-line relationship. Although not every point lies on the line, the reason is the lack of precision in the measurement (due to noise) and not curvature in the data.

Solution First, determine the rate law for the reaction. The graph of the data is curved, so the reaction is *not* zero order. Since the data resemble those expected for first order kinetics, start by tabulating the experimental data and the natural logarithm of the concentration.

Time, s	Concentration, *M*	ln(concentration)
0.0	0.22	−1.51
0.001	0.15	−1.90
0.002	0.10	−2.30
0.003	0.070	−2.66
0.004	0.045	−3.10
0.005	0.030	−3.51

Prepare a graph of ln(concentration) against time.

A first order reaction is indicated by the straight-line relationship between ln(concentration) and time.

The rate constant is the negative of the slope, which can be measured from the graph or calculated from the tabulated data:

$$k = -\text{slope} = -\frac{-3.51 - (-1.51)}{(0.0050 - 0.0)\text{ s}}$$

$$k = 4.0 \times 10^2 \text{ s}^{-1}$$

Understanding Graph the following data to determine whether the reaction is zero order, first order, or neither.

Time, s	Concentration, *M*
2.0	0.11
4.0	0.072
6.0	0.055
8.0	0.044
10.0	0.037

Answer: Neither a plot of [A] *vs.* t nor ln[A] *vs.* t is a straight line. The reaction is neither zero order nor first order.

The integrated rate law incorporates a large amount of information. Example 16.5 illustrates how this form of rate law is used to predict the concentration of a reactant at any particular time.

Example 16.5 Concentration-Time Relationships in a First Order Reaction

The degradation of the pesticide fenvalerate in the environment is found to be first order with a rate constant of $3.9 \times 10^{-7}\ s^{-1}$. An accidental discharge of 100 kg of fenvalerate into a holding pond results in a fenvalerate concentration of $1.3 \times 10^{-5}\ M$. Calculate the concentration left after one month (2.6×10^{6} s).

Solution Since the decay is first order

$$\ln[A] = \ln[A]_0 - kt$$
$$= \ln(1.3 \times 10^{-5}) - 3.9 \times 10^{-7} \times 2.6 \times 10^{6}$$
$$= -12.26$$
$$[A] = 4.7 \times 10^{-6}\ M$$

Fenvalerate is a member of a class of compounds called "pyrethrins." These compounds were originally isolated from plants (including chrysanthemums) that exhibited a tendency to repel insects.

Understanding What is the concentration of fenvalerate in the pond one year (3.15×10^{7} s) after the spill?

Answer: $6.0 \times 10^{-11}\ M$

Chrysanthemums These chrysanthemums are the source of a class of natural insecticides called pyrethrins.

Although molarity was used in the last example, the decrease in the mass of fenvalerate could have been used as well. The reason that either unit can be used becomes clear if the first order rate equation is rearranged.

$$\frac{[A]}{[A]_0} = e^{-kt}$$

The units used to express $[A]$ and $[A]_0$ are not important as long as both quantities are expressed with the same unit. Beware—this fortunate circumstance is true only for a first order rate law!

The next example illustrates how the time needed to reach specified concentration can be calculated.

Example 16.6 Time-Concentration Relationships in a First Order Reaction

Dinitrogen pentoxide decomposes to nitrogen dioxide and oxygen. When the reaction takes place in carbon tetrachloride (CCl_4), both nitrogen oxides are soluble but the oxygen escapes as a gas.

$$N_2O_5(soln) \longrightarrow 2NO_2(soln) + \tfrac{1}{2}O_2(g)$$

The rate of reaction ($-\Delta[N_2O_5]/\Delta t$) can be measured by monitoring the volume of gas that is produced by the reaction.

The rate law is found to be first order in N_2O_5 with a rate constant of $8.1 \times 10^{-5}\ s^{-1}$ at 303 K. If the initial concentration of $[N_2O_5]$ is 0.032 M, how long will it take for its concentration to decrease to 0.015 M?

Solution The problem involves concentration-time behavior so the integrated rate equation will be used. It is probably easiest to solve if we use the logarithmic form of the integrated first order rate law.

$$\ln[N_2O_5] = \ln[N_2O_5]_0 - kt$$
$$\ln(0.015) = \ln(0.032) - 8.1 \times 10^{-5}\, t$$
$$-4.20 = -3.44 - 8.1 \times 10^{-5}\, t$$
$$t = 9.4 \times 10^3 \text{ s}$$

It takes about $2\frac{1}{2}$ hours for $[N_2O_5]$ to reach 0.015 *M*.

Understanding In this experiment, how long does it take for the NO_2 concentration to rise to 0.016 *M*?

Answer: 3.6×10^3 s. Remember, the rate of reaction is $\Delta[NO_2]/2\Delta t$.

Half-Life

The rate constant, k, is one way to describe the speed of a reaction. A large value for k implies a fast reaction. Another way to describe the speed of the reaction is by the **half-life,** designated $t_{1/2}$, which is the time needed for the concentration of a reactant to decrease to one-half its original value. A short half-life indicates a rapid reaction.

Let $[A]_0$ be the initial concentration (at time 0). Then $[A]_0/2$ is the concentration when t is equal to $t_{1/2}$. Assume that this reaction is first order in A, so that Equation 16.3 applies:

$$[A] = [A]_0\, e^{-kt}$$
$$\frac{[A]_0}{2} = [A]_0\, e^{-kt_{1/2}}$$

Divide both sides by $[A]_0$:

$$\frac{1}{2} = e^{-kt_{1/2}}$$

and take the natural logarithm of both sides:

$$\ln(1/2) = -kt_{1/2}$$

Evaluate the logarithm and solve for $t_{1/2}$:

$$-0.693 = -kt_{1/2}$$
$$t_{1/2} = \frac{0.693}{k} \qquad [16.5]$$

Note that *the half-life of a first order reaction is independent of the concentration of the reactant.* The time needed for the reactant to decrease to 50% of its initial concentration depends only on the rate constant. (This behavior is not observed for reactions of order other than first.)

Figure 16.8 shows the decrease in the concentration of the reactant in a first order reaction. The time needed to decrease from 1.0 to 0.5 *M* is the same as the time needed to decrease from 0.25 to 0.125 *M*. A constant half-life is a mark of a first order reaction.

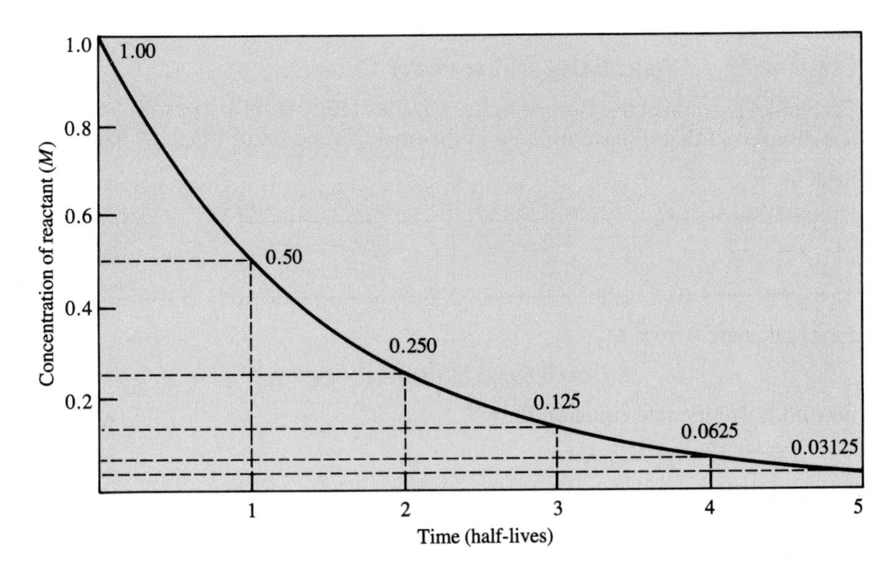

Figure 16.8 Half-lives of a first order reaction The horizontal axis is time, but expressed in half-lives rather than in seconds.

Nuclear decays are typical of first order kinetics. Plutonium-239, an isotope produced by nuclear reactors, is an extreme radioactive hazard with a half-life of 24,000 years. Given that the world's electrical power reactors are producing about 20,000 kg of ^{239}Pu per year, we can calculate how long it would take for one year's total production of ^{239}Pu to decay to 1.0 kg. (One kg is still an extremely large quantity of plutonium, given its extraordinary toxicity. Even one kilogram must be stored with extreme care.)

First, calculate the rate constant from $t_{1/2}$.

$$t_{1/2} = \frac{0.693}{k}$$

$$24{,}000 = \frac{0.693}{k}$$

$$k = 2.9 \times 10^{-5}\ \text{yr}^{-1}$$

Then calculate the time needed for 20,000 kg of plutonium to decay to 1.0 kg:

$$\ln\,[\text{Pu}] = \ln\,[\text{Pu}]_0 - kt$$

$$\ln\,[\text{Pu}] - \ln\,[\text{Pu}]_0 = -kt$$

$$\ln \frac{[\text{Pu}]}{[\text{Pu}]_0} = -2.9 \times 10^{-5}\,t$$

$$\ln \frac{1.0\ \text{kg}}{20{,}000\ \text{kg}} = -2.9 \times 10^{-5}\,t$$

$$\ln\,(5.0 \times 10^{-5}) = -2.9 \times 10^{-5}\,t$$

$$-9.90 = -2.9 \times 10^{-5}\,t$$

$$t = 3.4 \times 10^{5}\ \text{yr}$$

Example 16.7 Calculating a First-Order Decay

Carbon-14 is a radioactive isotope with a half-life of 5730 years. If a particular sample has decayed so that it has only 2.5% of the original amount of ^{14}C, how old is it?

Solution

Half-life → $k = \frac{0.693}{t_{1/2}}$ → Rate constant → Equation for first-order kinetics → Time needed for concentration to decrease

First, calculate k from $t_{1/2}$.

$$k = 0.693/5730 = 1.21 \times 10^{-4}\ \text{yr}^{-1}$$

Second, solve the rate equation for t.

$$\ln [^{14}C] = \ln [^{14}C]_0 - kt$$

$$\ln \frac{[^{14}C]}{[^{14}C]_0} = -kt$$

$$\ln \frac{0.025}{1.000} = -kt$$

$$\ln 0.025 = -1.21 \times 10^{-4}\, t$$

$$-3.69 = -1.21 \times 10^{-4}\, t$$

$$t = 3.0 \times 10^4\ \text{yr}$$

Understanding The carbon in a piece of wood has 40% of the carbon-14 originally present. How old is the wood?

Answer: 7600 yr

The last two examples use nuclear decays to illustrate first order kinetics. While these decays illustrate the concepts well, there are also a great many chemical reactions that show first order kinetics. Nuclear chemistry is a very interesting (and timely) subject that is discussed in Chapter 20.

Second Order Rate Laws

For a reaction that is second order, the second order differential rate equation is

$$\text{rate} = \frac{-\Delta[A]}{\Delta t} = k[A]^2 \qquad [16.6]$$

where the chemical reaction is A → product.

The units for a second order rate constant are L mol^{-1} s^{-1}.

The units of a second order rate constant are typically L mol^{-1} s^{-1}. Again, be aware that zero, first, and second order rate constants all have different units. In fact, if the units are known, then so is the reaction order.

Concentration-Time Dependence

If a graph of 1/[A] *vs.* time is a straight line, then the kinetics are second order.

The rate equation can be solved by calculus to get the integrated second order rate law:

$$\frac{1}{[A]} = \frac{1}{[A]_0} + kt \qquad [16.7]$$

INSIGHTS into CHEMISTRY

Science, Society, and Technology

Scientists Use Carbon Dating Techniques. "Ice Man" Found to be Over 5000 Years Old

Carbon-14, an unstable isotope, is formed at a steady rate by cosmic radiation striking the planet. Atmospheric nitrogen reacts with neutrons produced by the high-energy cosmic rays to form ^{14}C, an isotope that has a half-life of 5730 years. The rate of formation (from cosmic radiation) and the rate of decay (by nuclear processes) are in balance in the atmosphere, so the concentration of ^{14}C is constant. Living organisms use CO_2 in energy transport and storage, so the fraction of the carbon that is ^{14}C in the organism is the same as in the atmosphere. When the organism dies, CO_2 exchange ceases, and the decay of ^{14}C in the organism is no longer offset by intake of atmospheric CO_2.

Carbon-14 dating is widely used by archaeologists to determine the ages of samples of wood, cloth, and other organic materials. The technique is well suited for these samples, since the range over which ^{14}C dating is considered accurate (several half-lives) spans the development of civilization. The decay of other isotopes is used to date samples that are older than about 20,000 years.

The Ice Man In 1990, an Italian glacier melted enough to expose the 5,000-year-old remains of a mountaineer. His possessions, particularly his tools, have been able to shed light on the way people lived in the Bronze Age. Carbon-14 dating was used by archaeologists to determine the time frame of their discoveries.

The Ice Man's body is remarkably well-preserved. Scientists hope to obtain some DNA (a molecule that contains genetic information, discussed in Chapter 21) to determine how much humanity has changed in 5000 years.

We can find the rate constant for a second order reaction from a graph of $1/[A]$ *vs.* t. The graph will be a straight line (demonstrating a second order reaction) whose slope is equal to the rate constant, k.

Notice that a reaction's concentration-time behavior identifies its rate law. The straight-line relationships for zero, first, and second order reactions are summarized in Table 16.3.

Table 16.3 Concentration-Time Relationships

Order	Straight-Line Relationship
0	concentration *vs.* time
1	ln[concentration] *vs.* time
2	1/[concentration] *vs.* time

Half-Life

The half-life of a second order reaction can be determined from Equation 16.7 by substituting the concentration $[A]_0/2$ and the time, $t_{1/2}$:

$$kt_{1/2} = \frac{2}{[A]_0} - \frac{1}{[A]_0} = \frac{1}{[A]_0}$$

$$t_{1/2} = \frac{1}{k[A]_0}$$

The half-life of a second order reaction depends on concentration.

The half-life of a second order reaction depends on the starting concentration, so it is seldom useful to measure it. The half-life of a zero order reaction also depends on concentration.

Calculations of the rate of a second order reaction are shown in Example 16.8.

Example 16.8 The Rate of a Second Order Reaction

Nitrogen dioxide decomposes to nitrogen monoxide and oxygen in a second order reaction.

$$NO_2(g) \longrightarrow NO(g) + \tfrac{1}{2}O_2(g)$$

The decrease in concentration of NO_2 is shown as a function of time in the following figure. Use the data to calculate the rate constant for the reaction.

Solution

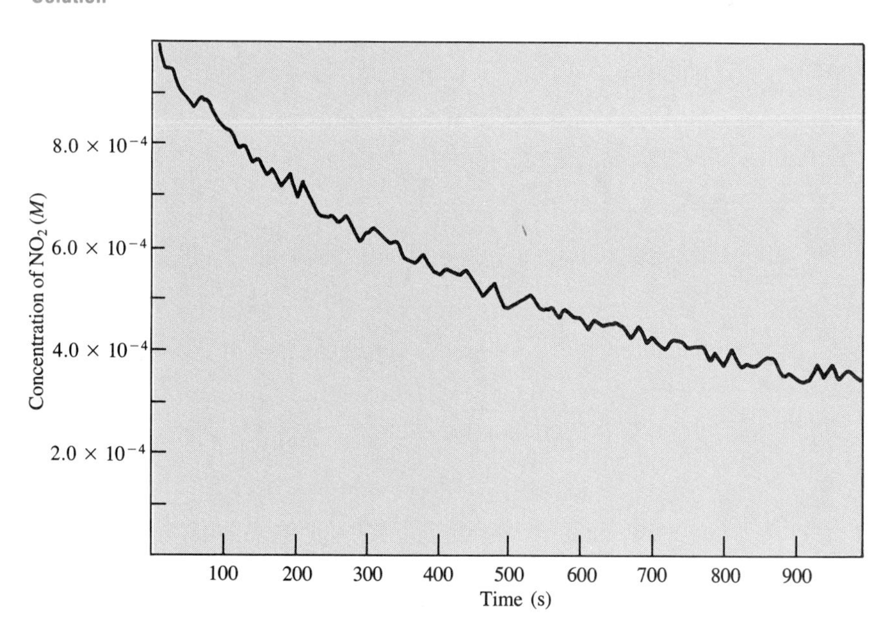

The first two columns in the following table are obtained by reading the data from the graph. The last column in the table is the reciprocal of the concentration.

Time, *s*	Concentration, *M*	1/(concentration), L mol^{-1}
0	0.0010	1000
200	0.00069	1400
400	0.00055	1800
600	0.00046	2200
800	0.00038	2600
1000	0.00033	3000

Plot the reciprocal of concentration against time:

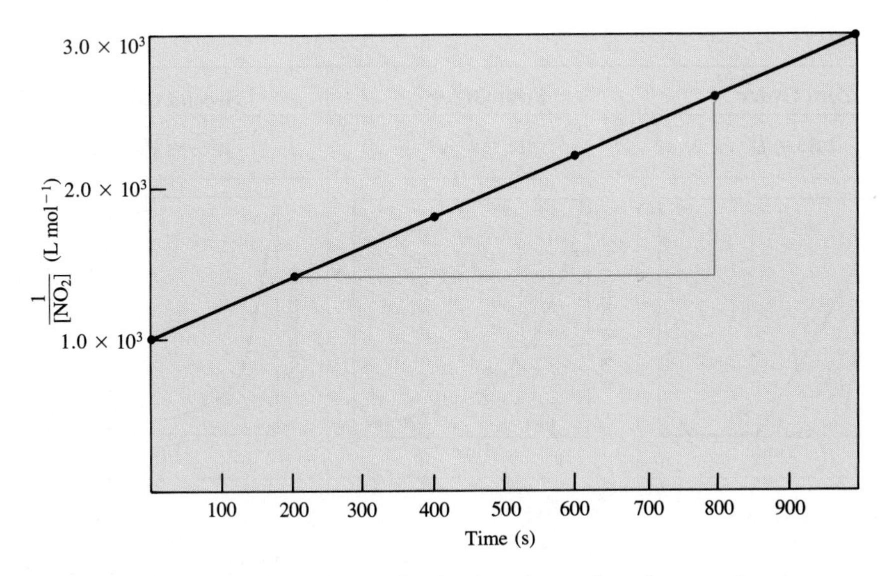

Measure the slope from the graph; the slope is equal to the second order rate constant.

$$k = \text{slope} = \frac{2600 - 1400}{800 - 200}$$

$$k = 2.0 \text{ mol}^{-1} \text{ L s}^{-1}$$

Understanding The decomposition of HI is measured in the laboratory at 117 °C.

$$2HI(g) \longrightarrow H_2(g) + I_2(g)$$

What are the rate law and the rate constant?

Time, s	Concentration of HI, *M*
2000	0.0088
12,000	0.0054
20,000	0.0042
30,000	0.0033

Answer: rate = $k[HI]^2$, $k = 6.8 \times 10^{-3}$ mol^{-1} L s^{-1}

Summary of Rate Laws

It is important to stress that the rate law is determined by experimentation. Many types of experiments can be adapted to provide kinetic information. For the reaction

$$A \longrightarrow \text{products}$$

we can categorize the experimental data in several ways, shown in Table 16.4.

A kinetic study of a reaction is illustrated in Example 16.9.

Example 16.9 Determining the Rate Law and Rate Constant for a Reaction

Determine the rate law and rate constant for the decomposition of the organic compound 1,3-pentadiene, abbreviated 1,3-P.

Table 16.4 Comparison of Rate Laws

	Zero Order	First Order	Second Order
differential rate law	rate = k	rate = k[A]	rate = $k[A]^2$
concentration *vs.* time	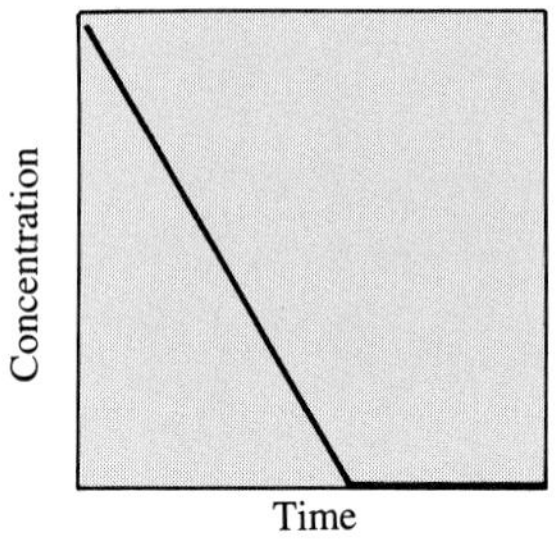 Concentration / Time	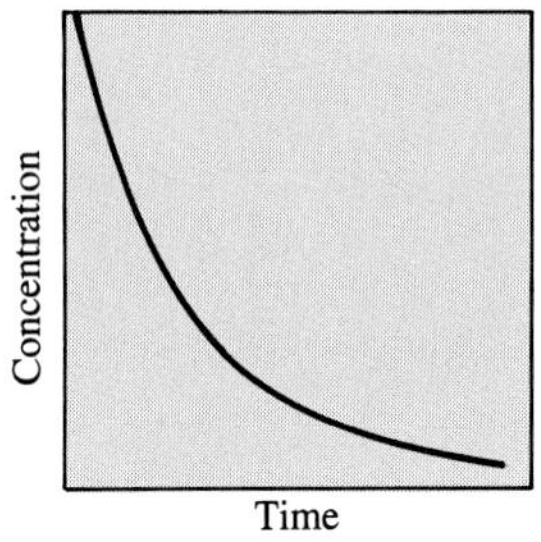 Concentration / Time	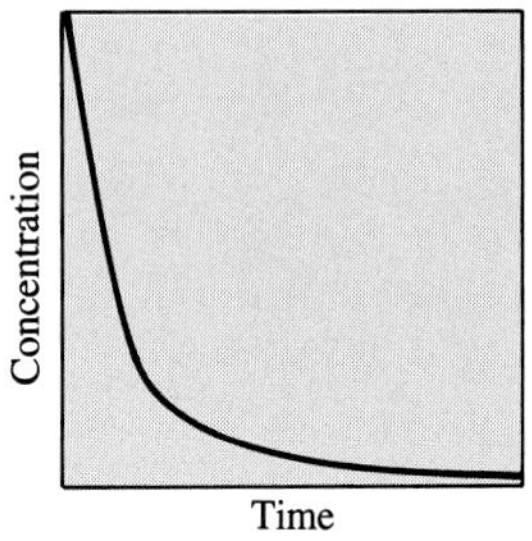 Concentration / Time
integrated rate law	$[A] = [A]_0 - kt$	$[A] = [A]_0e^{-kt}$ or $\ln[A] = \ln[A_0] - kt$	$\frac{1}{[A]} = \frac{1}{[A]_0} + kt$
straight-line plot to determine the rate constant	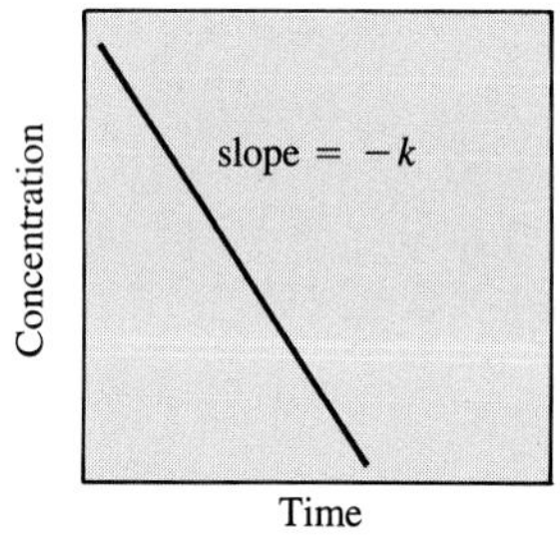 Concentration / Time; slope = $-k$	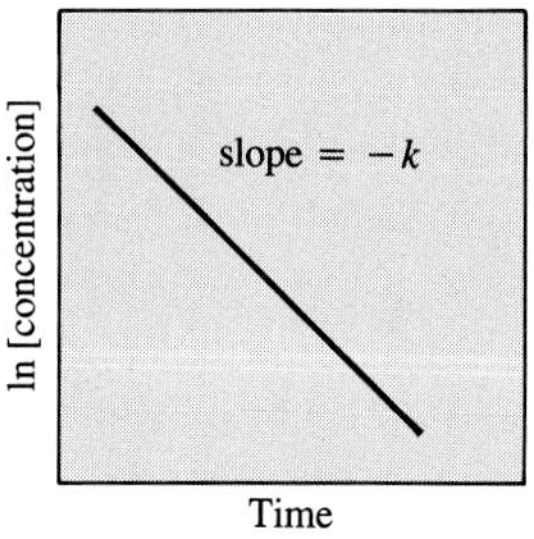 ln [concentration] / Time; slope = $-k$	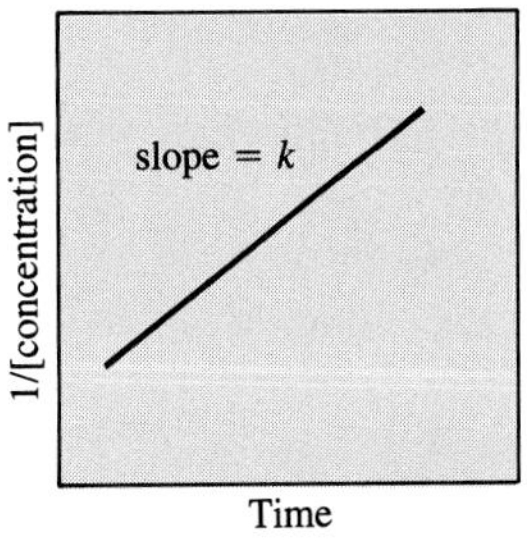 1/[concentration] / Time; slope = k
relative rate *vs.* concentration	[A] M : Rate, mol L^{-1} s^{-1} 1 : 1 2 : 1 3 : 1	[A] M : Rate, mol L^{-1} s^{-1} 1 : 1 2 : 2 3 : 3	[A] M : Rate, mol L^{-1} s^{-1} 1 : 1 2 : 4 3 : 9
log (rate) *vs.* log [conc]	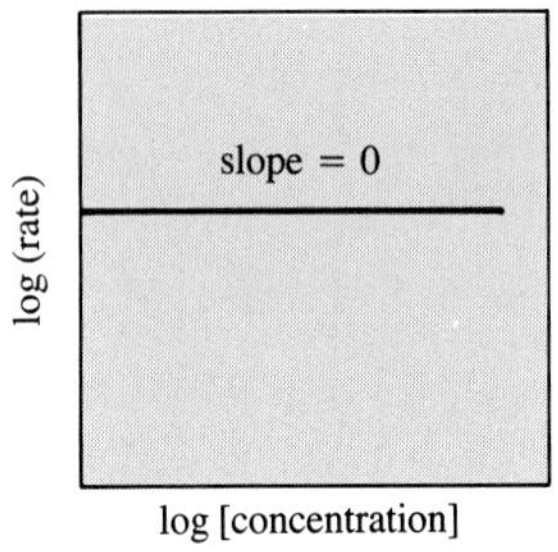 log (rate) / log [concentration]; slope = 0	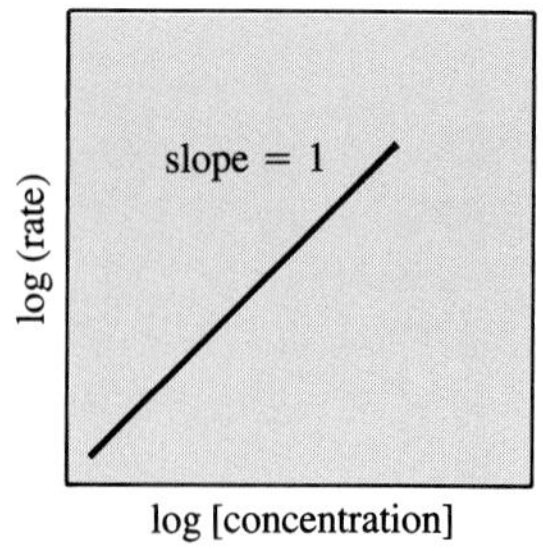 log (rate) / log [concentration]; slope = 1	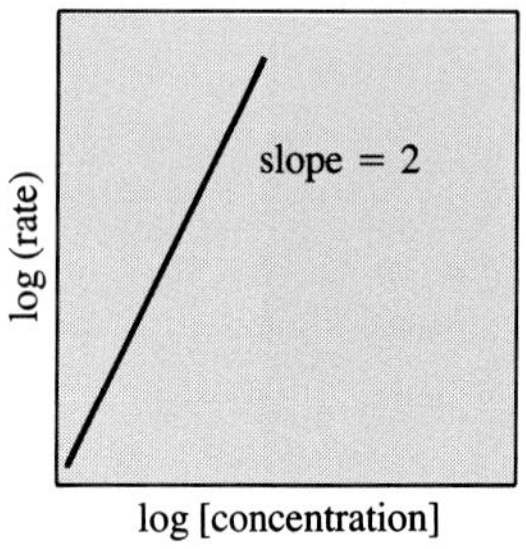 log (rate) / log [concentration]; slope = 2
half-life	$t_{1/2} = \frac{[A]_0}{2k}$	$t_{1/2} = \frac{0.693}{k}$	$t_{1/2} = \frac{1}{k[A_0]}$
units of k, rate constant	mol L^{-1} s^{-1}	s^{-1}	L mol^{-1} s^{-1}

1,3-P ⟶ product

The concentration of 1,3-P was measured as a function of time. The data follow.

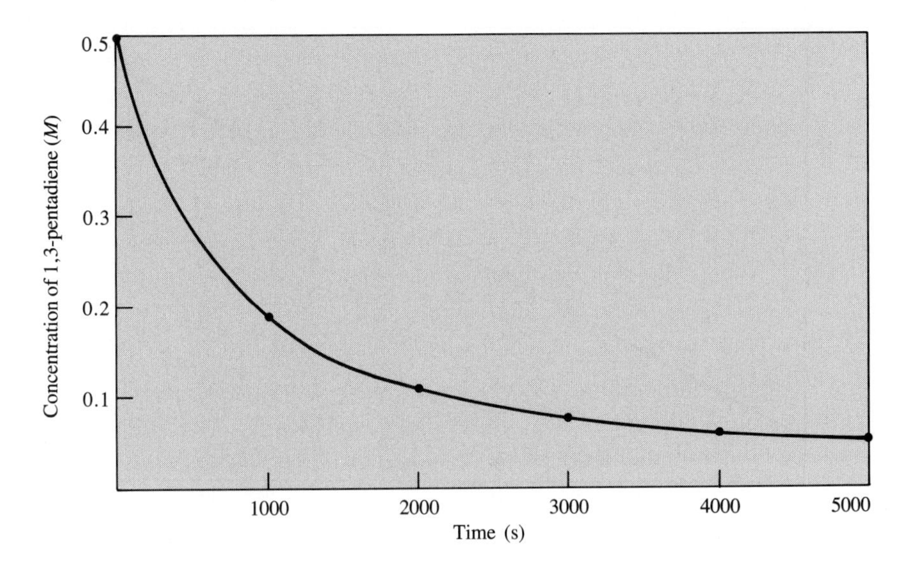

Decrease in the concentration of 1,3-pentadiene with time

Time, s	[1,3-P], M	ln [1,3-P]	1/[1,3-P]
0.	0.480	−0.734	2.08
1000.	0.179	−1.720	5.59
2000.	0.110	−2.207	9.09
3000.	0.0795	−2.532	12.6
4000.	0.0622	−2.777	16.1
5000.	0.0510	−2.976	19.6

Solution First, determine the reaction order. Begin by testing whether a plot of [1,3-P] *vs.* time is a straight line, indicating a zero order reaction. The original data are shown in a plot of [1,3-P] *vs.* time, and it is immediately obvious that it is not a straight line.

Next, to check for first order kinetics, plot ln[1,3-P] *vs.* time.

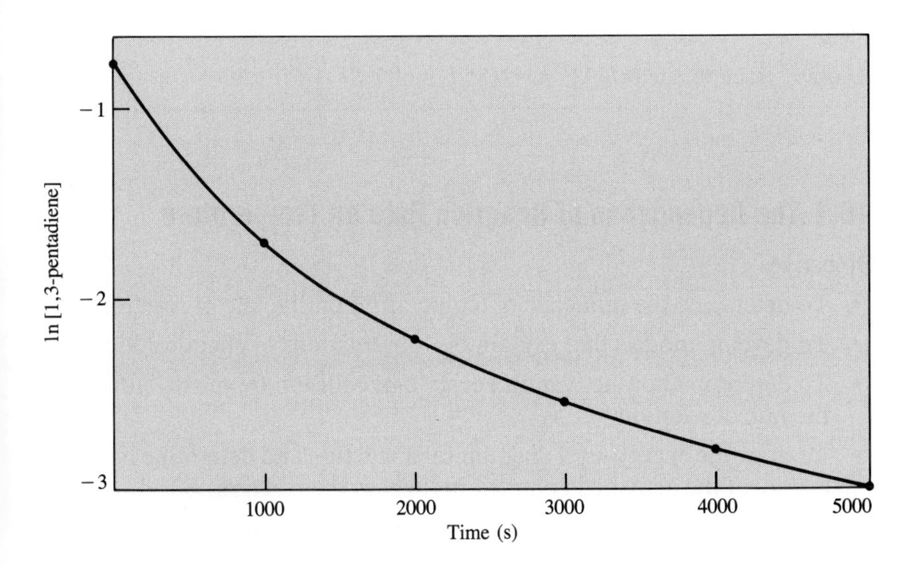

Graph of ln[1,3-P] *vs.* time

The curvature indicates that the reaction is not first order. A plot of 1/[1,3-P] *vs.* time is needed.

Graph of 1/[1,3-P] *vs.* time

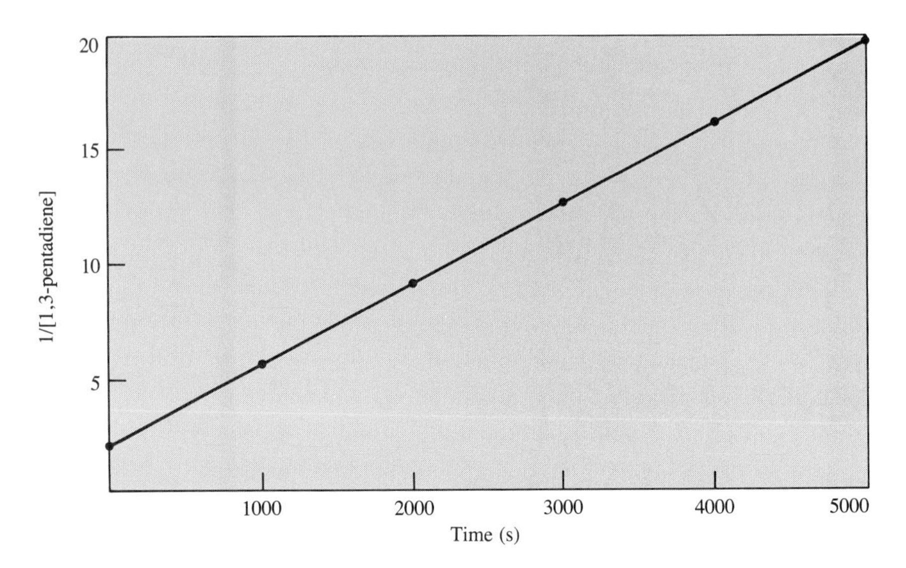

The straight line obtained for the last graph indicates that the reaction is second order.

rate = $k[1,3\text{-P}]^2$

The rate constant is evaluated from the slope of the graph.

$k = 3.5 \times 10^{-3}$ L mol^{-1} s^{-1}

Understanding Determine the rate law and rate constant for the reaction

A ⟶ products

Time, s	[A], *M*
0.0	0.43
2.0	0.25
4.0	0.17
6.0	0.13
8.0	0.11

Answer: reaction rate = $k[A]^2$, $k = 0.85$ L mol^{-1} s^{-1}

16.4 The Dependence of Reaction Rate on Temperature

Objectives

- To determine the influence of temperature on the rate of reaction
- To develop models that explain our observations of chemical kinetics
- To determine how activation energy and collision frequency influence the rate of reaction
- To draw the energy-level diagram for a reaction and determine how the energy of the activated complex influences the reaction
- To evaluate experimental data to measure the activation energy

Nearly all reactions go faster when heated, a fact that influences our everyday life as well as our chemistry. We know that plants grow faster in warm weather than in cold, and we know that higher temperatures cook food more quickly. Note that heating a reaction does not guarantee that *more* products will be formed, only that the reaction will reach equilibrium more quickly.

The temperature dependence of the reaction rate provides important information about the way a chemical reaction proceeds. Careful measurements of the relationship between temperature and reaction rate have provided insight into the inner workings of many chemical reactions. In this section, strong evidence will be presented to show that collisions between molecules (or atoms) are necessary for reactions to occur, and knowledge of the manner in which temperature influences the rate of reaction helps to characterize these collisions.

Influence of Temperature on Rate Constant

Experimental data show that the rates of most reactions increase dramatically with temperature. As an example, the reaction of nitrogen oxide with ozone has been studied and found to be a second order reaction.

$$NO(g) + O_3(g) \longrightarrow NO_2(g) + O_2(g)$$

$$\text{rate} = k[NO][O_3]$$

The temperature dependence of the reaction rate is the result of a change in the value of the rate constant, as shown in Table 16.5. Note that the order of the reaction does not change with temperature.

Table 16.5 Temperature Dependence of the Rate Constant

$NO(g) + O_3(g) \rightarrow NO_2(g) + O_2(g)$

Temperature, K	Rate Constant, $L\ mol^{-1}\ s^{-1}$
200	0.32×10^8
250	1.0×10^8
300	2.2×10^8
350	3.8×10^8

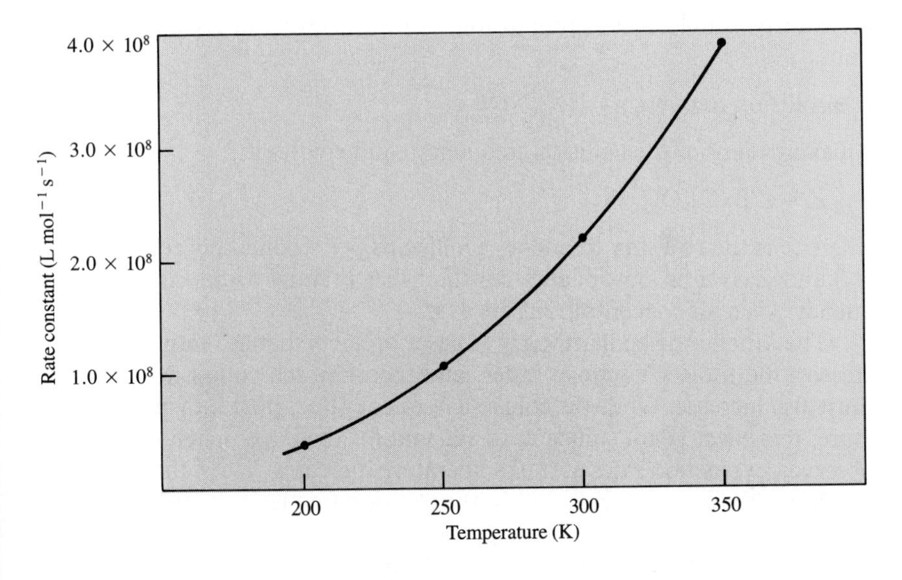

Dependence of rate constant on temperature

The shape of this graph is very similar to that of the vapor pressure *vs.* temperature graphs shown in Figures 10.26 and 15.5. Further, the similarity in shapes arises from similar reasons. The vapor-pressure–temperature relationship was explained by two important principles: the liquid molecules are in constant motion, and there is a force that binds them together in the liquid state. Some of the molecules in a liquid travel on a path toward the surface. If these molecules have enough energy to overcome the intermolecular forces, then they escape to the gas phase. When the liquid is heated, the vapor pressure increases because the molecules move more quickly and a larger fraction are able to escape into the gas phase.

In the following discussion we will apply this type of model, one based on molecular motion, to chemical kinetics. The key parameters will be the kinetic energy of the molecules, and the energy changes that must occur as reactants form products.

Collision Theory

Collision theory is a model that explains the rates of reactions in terms of the collision frequency.

In the collision theory of reaction rates, gas-phase chemical reactions are interpreted on the molecular scale. A basic assumption of collision theory is that molecules must collide in order to react. The kinetic molecular theory of gases is used to interpret these collisions.

We will continue to consider the reaction of nitrogen monoxide and ozone. This reaction has been studied by many researchers because it has an important role in several atmospheric processes such as the production of smog and the formation of an ozone "hole."

$$NO(g) + O_3(g) \longrightarrow NO_2(g) + O_2(g)$$

The **collision frequency,** Z, is the number of collisions per second: Z depends on the concentrations of the gases. For example, if the concentration of O_3 is doubled, the number of collisions between O_3 and NO molecules will double. Doubling the concentration of NO will also double the number of collisions between O_3 and NO molecules. In general, the collision frequency between two molecules is proportional to the product of their concentrations.

$$\text{collision frequency} \propto [O_3][NO]$$

An expression for the collision frequency can be written:

$$Z = Z'[NO][O_3] \qquad [16.8]$$

where Z is the collision frequency (collisions per second between NO and O_3) and Z' is a proportionality constant that is equal to the collision frequency when all concentrations are 1 M.

The kinetic molecular theory of gases predicts that as temperature increases, the molecules move faster, and therefore the collision frequency must also increase. While the collision frequency does increase with temperature, this effect is not sufficient to account for the large increases that are observed in reaction rates as the temperature increases. When the temperature increases from 200 K to 350 K, the collision frequency between the NO and O_3 molecules goes up by about 30%, but the observed reaction rate increases by more than 1000%, as calculated from Table 16.5.

The collision frequencies calculated from the kinetic theory of gases are generally 10^3 to 10^9 times the experimentally observed reaction rates. Clearly, not every collision results in a chemical reaction. Chemical reactions must involve factors in addition to the collision frequency. Further, these factors must be able to account for the large change in rates observed as the temperature changes.

Cold-blooded animals Temperature and metabolism both play important roles for cold-blooded animals. In such animals, metabolic processes speed up when the external temperature increases. They must eat much more during hot weather than during cold, to supply increased amounts of energy. During very cold weather, these animals are often dormant and use very little energy.

Activation Energy

In 1888, Svante Arrhenius (1859–1927) expanded the simple collision model to include the possibility that not all collisions result in the formation of the products. According to this model, if a collision is to be productive, the molecules must collide with enough energy to rearrange the bonds. When the energy of the collision is too small, the molecules simply bounce off each other. The **activation energy,** E_a, is the minimum collision energy required for a reaction to occur. These features of chemical reactions have analogs in the evaporation process in which only molecules with sufficient energy to overcome the intermolecular forces can escape from the liquid phase.

Only collisions with enough energy to rearrange the bonds will result in the formation of products.

The Activated Complex

The energy changes in a reaction are often displayed on an energy-level diagram. The vertical axis is potential energy; the horizontal axis, called the **reaction coordinate,** is a relative scale that begins with the reactants and ends with the products. The energy-level diagram for the reaction of nitrogen dioxide with ozone is shown in Figure 16.9. The overall reaction is exothermic, but the energy-level diagram indicates that the reactants have a "hill" or barrier to climb before products can be formed. The height of the barrier is given by E_a, the activation energy. In the terminology of chemical kinetics, this least-stable arrangement of atoms is called the **activated complex** or transition state. Since the activated complex is the least-stable arrangement, its concentration is extremely small and virtually undetectable.

Another way of looking at the activation energy is as the energy needed to form the activated complex from the reactants. The instability of the activated complex is stressed by the asterisk in $[NO—O_3]^*$. The activated complex can break apart to form the products or to reform reactants.

Reactions with high activation energies are slower than reactions with low activation energies, if all other factors are the same. The activation energy for the gas-phase reactions of NO range from 9.6 kJ/mol (for the reaction with ozone) up to 82 kJ/mol (for the reaction with Cl_2 to form NOCl).

The activation energy is the energy needed to form the activated complex from the reactants.

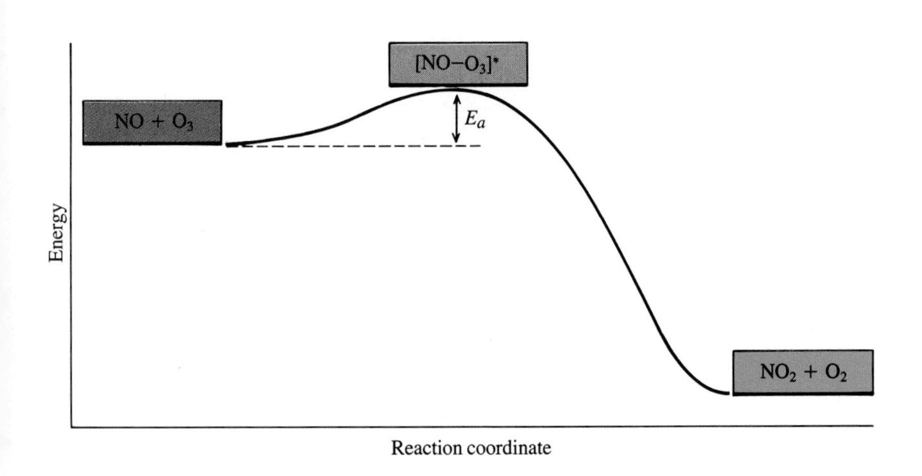

Figure 16.9 Energy-level diagram The activation energy for this reaction is 9.6 kJ/mol.

The Influence of Temperature on Kinetic Energy

The dependence of the rate of a reaction on temperature is strongly influenced by the size of the activation energy. The number of molecules with kinetic energies large enough to initiate a reaction is related to the temperature, as shown in Figure 16.10.

The fraction of collisions with energy in excess of E_a can be derived from the kinetic theory of gases; it is given by

$$f_r = e^{-E_a/RT} \qquad [16.9]$$

The fraction f_r is a number between 0 and 1. For example, if f_r is equal to 0.05, then the energies of 5% of the collisions exceed E_a, the activation energy.

The number of collisions that exceed E_a grows exponentially with increasing temperature.

Equation 16.9 shows one important influence of temperature: *as the temperature increases, the number of collisions with energies that exceed the activation energy grows exponentially.*

We can now predict a rate constant for the second order reaction in which NO and O_3 molecules collide to form NO_2 and O_2. The rate of reaction should be equal to the rate of collision times the fraction of collisions with energies that exceed the activation energy.

$$\text{rate} = \underset{\text{(collision frequency)}}{Z} \times \underset{\text{(fraction exceeding } E_a)}{e^{-E_a/RT}}$$

We know that the collision frequency depends on the concentrations of the colliding species, as shown by Equation 16.8.

$$Z = Z'\,[\text{NO}][\text{O}_3]$$

The rate of reaction is also described by the experimental rate law.

$$\text{rate} = k[\text{NO}][\text{O}_3]$$

Now, the experimental observation can be equated to the rate predicted by collision theory.

$$\text{rate} = k[\text{NO}][\text{O}_3] = Z'[\text{NO}][\text{O}_3] \times e^{-E_a/RT}$$

We solve for k, the rate constant.

$$k = Z'\,e^{-E_a/RT}$$

While this equation produces the correct temperature dependence of rate constants, it still predicts rates much faster than those observed in the laboratory. One additional factor is needed.

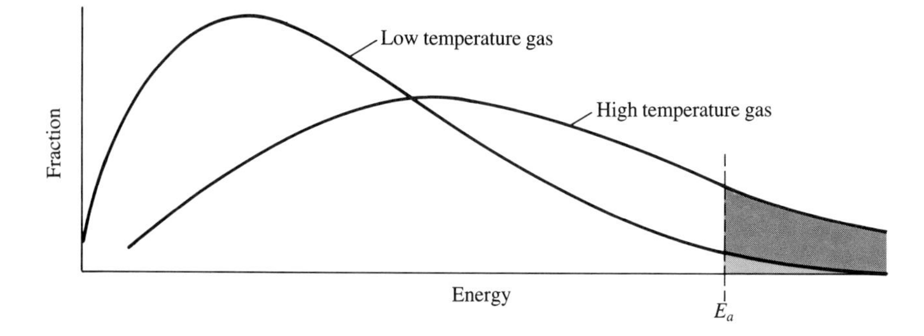

Figure 16.10 Energy distribution in gas molecules A much larger fraction of molecules have energies in excess of E_a at high temperatures than at low temperatures.

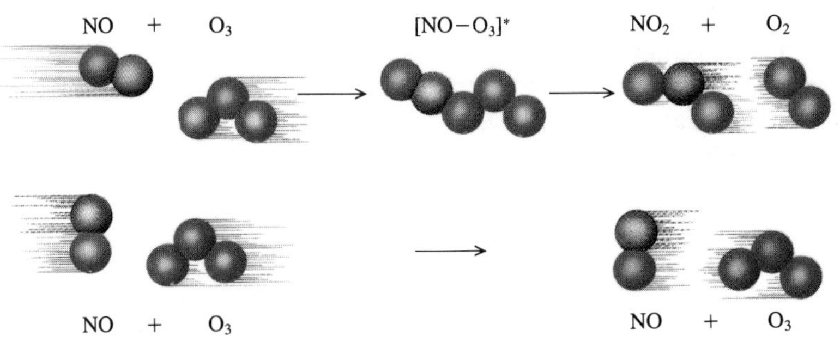

Figure 16.11 Orientation of reactants Not every collision places the reactants in the correct orientation to form products.

The Steric Factor

It seems reasonable that not all collisions with energies greater than E_a will always result in reaction. Some possible collisions between nitrogen oxide and ozone are depicted in Figure 16.11. Since nitrogen dioxide is a bent molecule, the formation of a nitrogen-oxygen bond from the second collision shown is unlikely. The orientation of the atoms in the first collision is favorable for the formation of the O—N—O bent molecule.

Not all collisions with energies that exceed E_a are productive. The geometry of collisions must be considered as well.

Conditions such as these are represented in the predicted rate law by an additional factor, *p*, the **steric factor,** which has a value between zero and one. ("Steric" means "related to the spatial arrangement of atoms.")

$$\text{rate} = \underset{\begin{pmatrix}\text{steric}\\ \text{factor}\end{pmatrix}}{p} \times \underset{\begin{pmatrix}\text{collision}\\ \text{frequency}\end{pmatrix}}{Z'[A][B]} \times \underset{\begin{pmatrix}\text{fraction}\\ \text{exceeding } E_a\end{pmatrix}}{e^{-E_a/RT}}$$

The steric factor and collision frequency terms can be collected in a single factor, called the **pre-exponential term,** which is given the symbol *A*.

$$\text{rate} = Ae^{-E_a/RT}[NO][O_3]$$

Another way to express the same result is to write the expression for the rate constant.

$$k = Ae^{-E_a/RT} \qquad [16.10]$$

Equation 16.10 is called the **Arrhenius equation.**

The Arrhenius equation relates the rate constant, activation energy, and temperature.

Evaluation of Activation Energy and the Pre-exponential Term

The Arrhenius equation fits the observed temperature dependences for a wide range of chemical reactions. The Arrhenius equation can be used in conjunction with experimental measurements of the rate constant at different temperatures to determine the activation energy. If we know E_a for a reaction, we will be able to determine how much faster (or slower) it will proceed as the temperature is changed. To see how to determine E_a from experimental measurments, we start with Equation 16.10.

$$k = Ae^{-E_a/RT}$$

Take the natural logarithm of both sides.

$$\ln k = \ln A - \frac{E_a}{RT}$$

Table 16.6 The Dependence of the Rate Constant on Temperature

	$2NO_2(g) \rightarrow 2NO(g) + O_2(g)$		
T, K	*k*, L mol^{-1} s^{-1}	ln *k*	1/*T*
500	0.003	−5.8	0.00200
550	0.037	−3.30	0.00182
600	0.291	−1.234	0.00167
650	1.66	0.507	0.00154
700	7.39	2.000	0.00143

Rewrite this equation to show the temperature dependence:

$$\ln k = \ln A - \frac{E_a}{R}\left(\frac{1}{T}\right)$$

A plot of ln k versus $1/T$ will have a slope of $-E_a/R$ and an intercept of ln A.

We will illustrate the determination of the activation energy from experimental data for the decomposition of NO_2,

$$2NO_2(g) \longrightarrow 2NO(g) + O_2(g)$$

This reaction is found to obey a second order rate law.

$$\text{rate} = k[NO_2]^2$$

Table 16.6 presents measurements of the rate constant at several temperatures. A graph of ln k *vs.* $1/T$ based on the data from Table 16.6 is shown in Figure 16.12.

Figure 16.12 is an example of an **Arrhenius plot.** The activation energy is determined by measuring the slope of the graph, which is equal to $-E_a/R$.

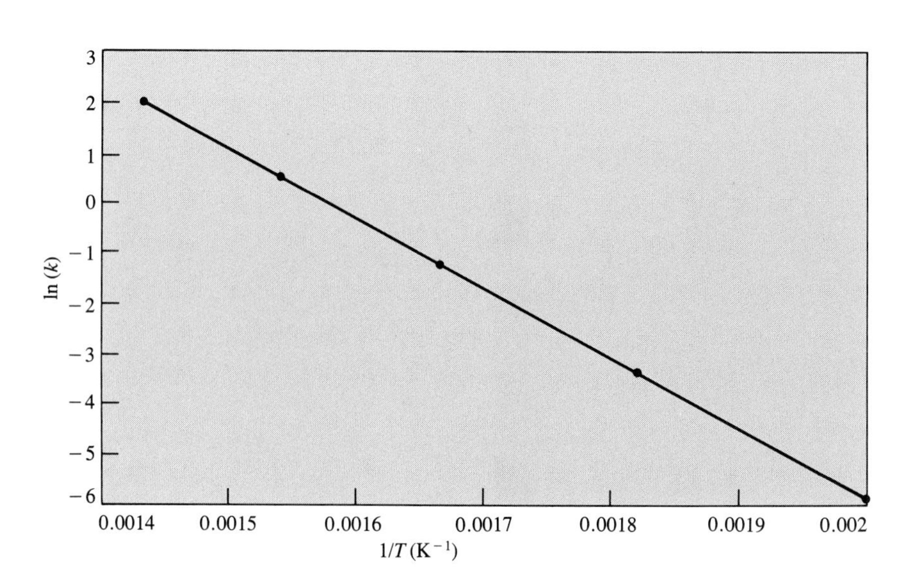

Since only two experiments are needed to calculate the slope, an equation can be derived that relates the activation energy to the rates or rate constants at two temperatures.

$$\ln\left(\frac{k_1}{k_2}\right) = -\frac{E_a}{R}\left(\frac{1}{T_1} - \frac{1}{T_2}\right) \qquad [16.11]$$

Measuring the rate or rate constant at two temperatures provides enough information to evaluate the activation energy.

The next example demonstrates how this equation can be used to determine activation energy from rate measurements at two temperatures.

Example 16.10 Determining the Activation Energy from the Temperature Dependence of Rate Constant

Consider the decomposition of nitrogen dioxide.

$$2NO_2(g) \longrightarrow 2NO(g) + O_2(g)$$

At 650 K, the rate constant is 1.66; at 700 K it is 7.39. Use these rate constants to determine the activation energy.

Solution Substitute the data in Equation 16.11:

$$\ln\left(\frac{k_1}{k_2}\right) = -\frac{E_a}{R}\left(\frac{1}{T_1} - \frac{1}{T_2}\right)$$

$$\ln\left(\frac{1.66}{7.39}\right) = -\frac{E_a}{8.314}\left(\frac{1}{650} - \frac{1}{700}\right)$$

$$E_a = 1.13 \times 10^5 \text{ J mol}^{-1} = 113 \text{ kJ mol}^{-1}$$

The units of E_a must be the same as those of RT (because E_a/RT is dimensionless). When R is 8.314 J mol^{-1} K and T is expressed in K, then E_a has units of J mol^{-1}.

Understanding Calculate the activation energy in kJ/mol for a reaction that has $k = 1.0 \times 10^8$ at 250 K and $k = 3.8 \times 10^8$ at 350 K.

Answer: $E_a = 9.7$ kJ mol^{-1}

The activation energy is determined from the dependence of the natural logarithm of rate constant (or rate) on the reciprocal of temperature. The method used in the last example used two points to determine the activation energy; the precision of the determination is increased if more points are available. When the rate constant is known at more than two temperatures, the activation energy can be determined from the slope of the ln k *vs.* $1/T$ graph, as shown previously. This type of data treatment has the effect of averaging random experimental error, and it also allows the scientist to see whether the result of an individual experiment lies farther from the line than might be expected from random error. When all the data in Table 16.6 are used to prepare a graph of ln k *vs* $1/T$, the slope of the line gives a value of 114 kJ/mol for E_a.

When the rate or rate constant has been measured at several temperatures, the activation energy is calculated from the slope of the graph of ln k *vs.* $1/T$.

Example 16.11 Determining the Activation Energy from the Graph of ln k *vs.* $1/T$

Use the following Arrhenius plot to determine the activation energy and the pre-exponential term.

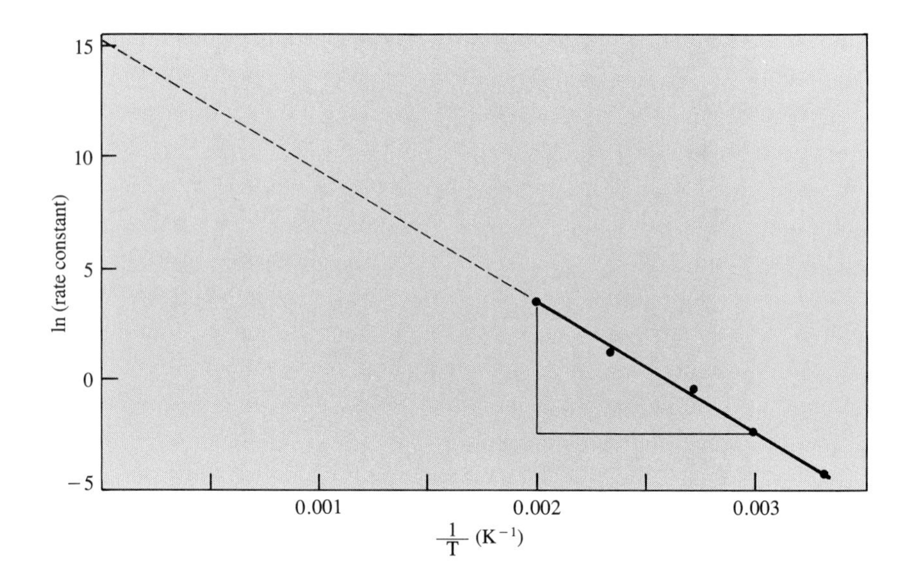

Solution Analysis of the experimental data provides a slope and intercept. The slope is −5600 K and the intercept is 15.7.

Recall that we wrote the Arrhenius equation in the form

$$\ln k = \ln A - \frac{E_a}{R}\left(\frac{1}{T}\right)$$

A plot of ln k *vs.* $1/T$ will have a slope of $-E_a/R$ and an intercept of ln A, so

$$\text{slope} = -5600 \text{ K} = -E_a/R$$

$$E_a = -5600 \text{ K} \, (-8.314 \text{ J mol}^{-1}\text{K}^{-1}) = 4.7 \times 10^4 \text{ J/mol}$$

$$E_a = 47 \text{ kJ/mol}$$

$$\text{intercept} = \ln A = 15.7$$

$$A = e^{15.7} = 6.6 \times 10^6$$

Notice that drawing the best straight line through the collection of points will provide a more accurate answer than using any pair of experimental data points.

A rule of thumb that is often used assumes a "typical" reaction rate doubles with a 10 °C increase from room temperature. Most chemists remember "rule-of-thumb" generalizations like this—they can be handy. For example, food that is stored at 20 °C will spoil about twice as fast as food stored at 10 °C.

Example 16.12 Determining the Activation Energy from the Temperature Dependence of Rate

A reaction rate doubles when the temperature is raised 10 °C from 298 to 308 K. What is the activation energy?

Solution We will use Equation 16.11,

$$\ln\left(\frac{k_1}{k_2}\right) = -\frac{E_a}{R}\left(\frac{1}{T_1} - \frac{1}{T_2}\right)$$

Skilled Cooks Manipulate Time and Temperature; Desired Results Come from Combination of Factors

Temperature plays an important role in cooking food. Cooks once called an oven of 325 °F a "slow" oven, 375 °F a "medium" oven, and 425 °F a "fast" oven, describing well the kinetic effects of temperature. A good cook can adjust temperatures and times to fit the situation.

In fact, there are two possible approaches to cooking. Most people use a kinetic method to cook a beef roast by placing it in a 375 °F oven for several hours, until the internal temperature of the roast reaches 140 °F (for rare beef) to 160 °F (for well-done meat). However, it can also be cooked by allowing the roast to reach an equilibrium with the oven temperature. If rare beef is desired, a roast can be placed in a 140 °F oven for several hours. The temperature will not rise above 140 °F, so the roast will remain rare, regardless of the length of time it spends in the oven. People who use this method must first "brown" the roast by applying high temperature for a short time to obtain the traditional crusty appearance. The initial high temperature also ensures that any surface bacteria will be killed.

Low-temperature cooking is used to make jerky, a dried meat. Jerky was traditionally sun-dried, but now low-temperature ovens are also used.

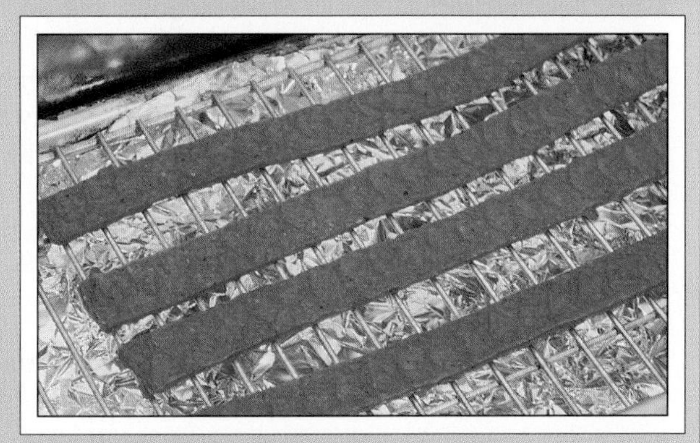

Low temperatures and long cooking times can be substituted for high temperatures and short cooking times. (Recipes adapted from *Stocking Up. How to Preserve the Foods You Grow Naturally.* By the staff of *Organic Gardening and Farming.*® Edited by Carol Hupping Stoner. Emmaus, PA: Rodale Press, © 1977.

Jerky

Jerky is meat that is cut in narrow strips and allowed to dry and become somewhat leathery. Traditionally, it is made from beef, venison, or other game meat. Jerky should be made from lean meat only, as fatty meat will not dry out as thoroughly or keep well. It may be dried over coals, in a smokehouse, or in a warm oven. To prepare lean meat for jerky, cut the meat into strips about ½ inch thick and 2 inches wide and about 4 to 5 inches long.

Brining. Mix together 1½ cups of pickling salt and 1 gallon of clean cold water. Soak the meat strips in this mixture for one to two days. Drain. The meat can then be dried outdoors over a low fire. The coals should be just warm so that they dry, not cook, the meat.

Dry-salting. Weigh the meat, and use 1 teaspoon of salt for each pound of meat. Season with pepper, garlic, and dried herbs mixed in a blender and rubbed into the meat. Arrange meat on racks in a warm 150° to 175° oven. Do not let the temperature rise above 175°—lower is actually better. Turn the meat at least once during the drying process, which should take about four to six hours. The salted strips may also be dried over warm coals. This method should dry the meat in about 24 hours.

The meat is done when it bends before breaking (let cool before testing). Once the meat is cool and dry, wrap it in moisture-proof material, and either freeze or refrigerate, or store in a lidded container in a cool, dry place.

We will let T_1 be 298 and T_2 be 308 K. The ratio of the relative rates of reaction will be the same as the ratio of the rate constants, and the ratio of the rates is 1 to 2. Thus,

$$\ln\left(\frac{1}{2}\right) = -\frac{E_a}{8.314}\left(\frac{1}{298} - \frac{1}{308}\right)$$

$$E_a = 5.3 \times 10^4 \text{ J mol}^{-1} = 53 \text{ kJ mol}^{-1}$$

Understanding The rate of a reaction triples when the temperature changes by 15 °C from 298 to 313 K. Calculate the activation energy for the reaction.

Answer $E_a = 5.7 \times 10^4 \text{ J mol}^{-1} = 57 \text{ kJ mol}^{-1}$

Reactions with large activation energies show a greater dependence on temperature than do reactions with small activation energies.

Reactions with high activation energies are influenced more by temperature changes than are reactions with low activation energies. Consider how changing the temperature from 15 °C to 35 °C will change the rates of two reactions, one with an activation energy of 40 kJ/mol and the other with $E_a = 80$ kJ/mol. For the reaction with $E_a = 40$ kJ/mol, the rate constant at 35 °C will be about three times that at 15 °C. For the reaction with an activation energy of 80 kJ/mol, the rate at 35 °C will be almost nine times faster than the rate at 15 °C.

16.5 Catalysis

Objectives

- To define catalysis and identify heterogeneous, homogeneous, and enzymatic catalysts
- To determine how catalysts influence chemical reactions and to draw energy-level diagrams for catalyzed and uncatalyzed reactions

To increase the rate of a reaction, the frequency of productive collisions must increase. Two different methods can accomplish this goal. The first is to increase the collision frequency, generally by increasing the temperature. Although simple, this approach is not always successful, because unfavorable changes in the value of the equilibrium constant and undesirable side reactions often occur at higher temperatures.

A second approach is to make more of the collisions productive. If the activation energy can be decreased, or if the steric factor can be increased, then the rate of reaction will increase. A **catalyst** is a substance that increases the rate of reaction, but is not consumed in the reaction. A catalyst is inti-

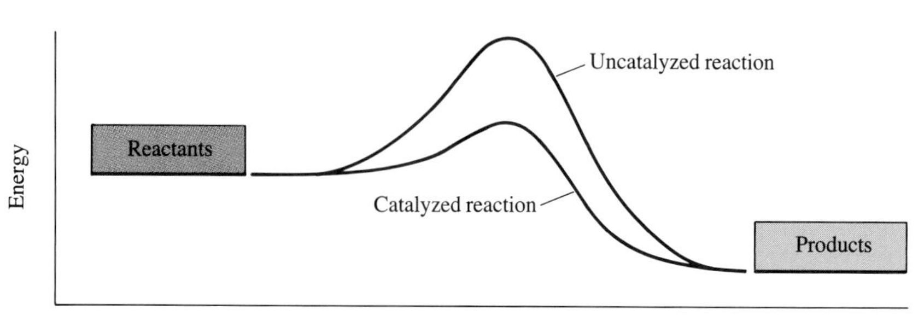

Figure 16.13 Energy-level diagram for a chemical reaction

mately involved in the course of the chemical reaction, making and breaking bonds as the reactants form products, but the catalyst does not undergo a permanent change.

A catalyzed reaction proceeds by a different set of steps than does an uncatalyzed reaction. The catalyzed reaction generally has a lower activation energy, and thus a higher reaction rate at any given temperature. Figure 16.13 shows energy-level diagrams for the same reaction in the presence and absence of a catalyst.

Catalysts increase the rate of reaction by decreasing the activation energy.

Homogeneous Catalysis

A **homogeneous catalyst** is one that is present in the same phase as the reactants. The acid-catalyzed decomposition of formic acid, shown in Figure 16.14, is an example of homogeneous catalysis.

A homogeneous catalyst is in the same phase as the reactants.

In the absence of a catalyst, the reaction involves movement of a hydrogen atom from a carbon atom to an oxygen atom. The energy requirement is high, and the reaction proceeds slowly.

In the acid-catalyzed reaction, a hydrogen ion from the solution bonds to the oxygen atom of formic acid to make $HCOOH_2^+$. The carbon-oxygen

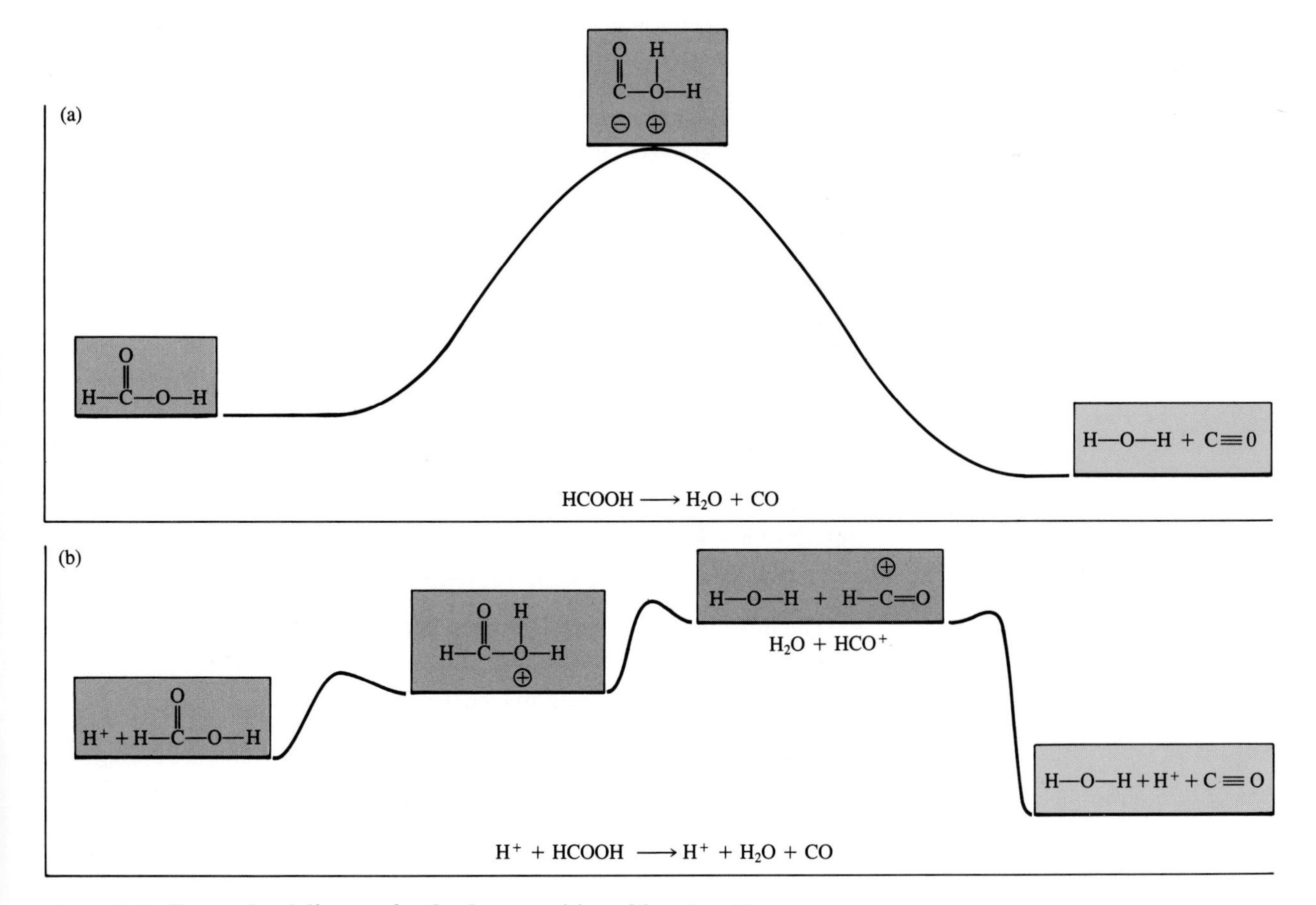

Figure 16.14 Energy-level diagram for the decomposition of formic acid

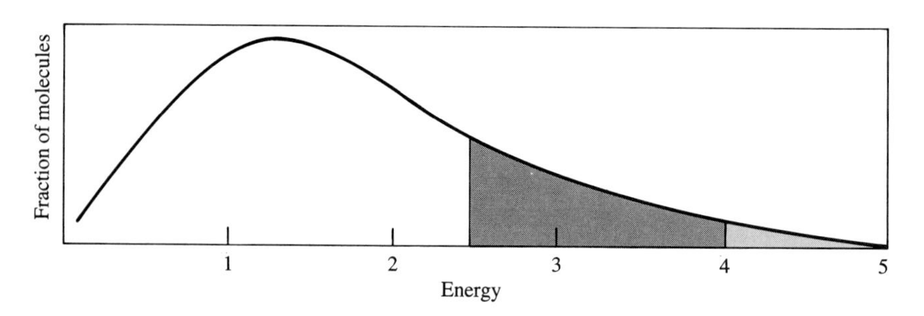

Figure 16.15 Fraction of molecules with enough energy to form the activated complex The activation energy for the catalyzed reaction is 2.5 units. Many more collisions exceed this activation energy than exceed the activation energy for the uncatalyzed reaction, 4.0 units.

bond breaks, forming HCO^+ and H_2O. The H^+ ion is immediately lost from HCO^+ back to the solution.

$$HCOOH + H^+ \longrightarrow [HCOOH_2]^* \longrightarrow H_2O + HCO^+ \longrightarrow H_2O + CO + H^+$$

The hydrogen ion is the catalyst. It is present in the activated complex, but is returned to solution after the activated complex breaks into the products. The catalyst is not consumed in the overall reaction even though it is a part of the activated complex.

The acid-catalyzed, multistep conversion of formic acid to CO and H_2O has a lower activation energy than does the one-step reaction. The addition of a catalyst changes only the activation energy, E_a, and *not* $\Delta E°$, $\Delta H°$, and $\Delta G°$, since these are state functions that depend only on the initial and final states. Figure 16.15 shows that the fraction of molecules that have enough energy to form the activated complex is higher for catalyzed than for uncatalyzed reactions.

More collisions exceed the (lower) activation energy of a catalyzed reaction, so the reaction rate increases.

Another example of homogeneous catalysis is illustrated by the effect of bromide ions on the decomposition of hydrogen peroxide.

$$2H_2O_2(aq) \longrightarrow 2H_2O(\ell) + O_2(g)$$

The reaction is thought to occur in two steps. First, bromide ion reacts with hydrogen peroxide to form bromine.

$$H_2O_2(aq) + 2Br^-(aq) + 2H^+(aq) \longrightarrow Br_2(aq) + 2H_2O(\ell)$$

The bromine reacts with hydrogen peroxide in the second step to form oxygen.

$$H_2O_2(aq) + Br_2(aq) \longrightarrow 2Br^-(aq) + 2H^+(aq) + O_2(g)$$

The concentration of bromine builds up with time, and violent bubbling (the generation of oxygen gas) is observed. When the reaction is complete, all the bromine has been converted back to bromide ion. The sum of the two reactions gives the overall reaction. Bromide ion is a true catalyst—it increases the rate of reaction but is not consumed.

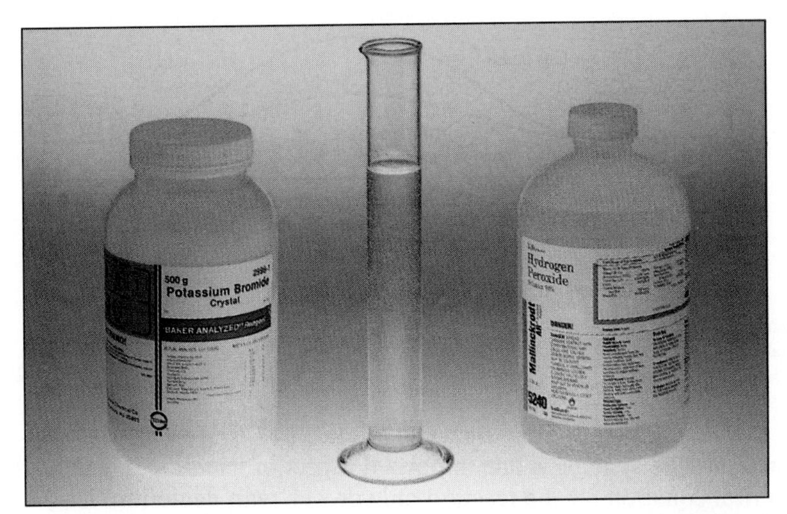

Bromide-catalyzed decomposition of hydrogen peroxide Shortly after a bromide salt such as sodium or potassium bromide is added to a hydrogen peroxide solution, the brown color of bromine is seen, and bubbles of oxygen form. The brown color fades at the conclusion of the reaction.

Heterogeneous Catalysis

Platinum, palladium, nickel, and iron metal surfaces are commonly used to catalyze many reactions, particularly those that involve small gas molecules. A **heterogeneous catalyst** is one that is in a different phase than are the reactants. The formation of methanol, CH_3OH, from hydrogen and carbon monoxide can be catalyzed by metal surfaces.

Metals are frequently used as heterogeneous catalysts, particularly for gas-phase reactions.

$$2H_2(g) + CO(g) \longrightarrow CH_3OH(g)$$

The uncatalyzed formation of methanol has a high activation energy, and high temperatures are needed if the reaction is to proceed at a satisfactory rate. Unfortunately, the reaction is exothermic, so the principle of Le Chatelier indicates that the high temperatures decrease the equilibrium concentration of methanol.

Instead, the formation of methanol from hydrogen and carbon monoxide is catalyzed by a platinum surface. The way in which the catalyst takes part in this reaction is not completely known, but scientists have approached the problem by studying the metal-catalyzed hydrogen-deuterium reaction. (Studying simple systems with the idea of extrapolating the knowledge to

Figure 16.16 Energy-level diagram for the reaction of hydrogen with deuterium

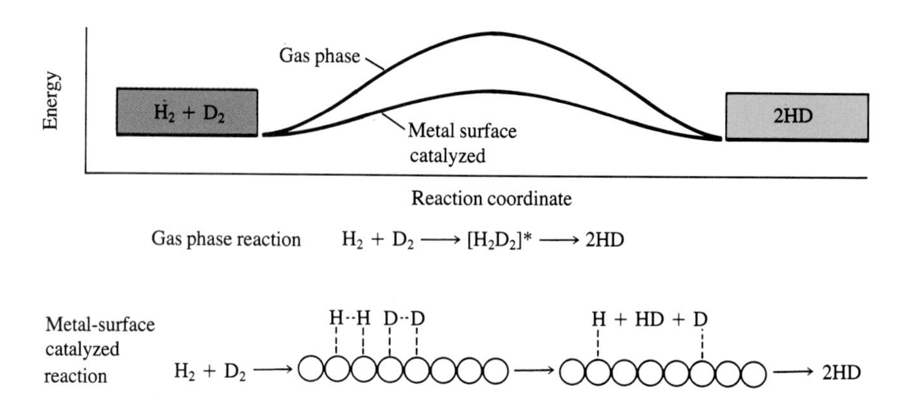

more complex systems is one important aspect of scientific research.) The chemical equation for the reaction of hydrogen with deuterium is

$$H_2(g) + D_2(g) \longrightarrow 2HD(g)$$

In the catalyzed reaction, a hydrogen or deuterium molecule forms a weak bond to the metal surface, and the H—H or D—D bond weakens. If a dissimilar molecule bonds to the surface nearby, the product can form. This process is illustrated schematically in Figure 16.16; many scientists believe that this bond-weakening process explains why so many gas-phase reactions (such as the production of methanol from H_2 and CO) proceed rapidly at certain metal surfaces.

When a substance adheres to the surface of a solid, we call the process **adsorption.** The adsorption of hydrogen and deuterium to the surface of the catalyst is the first step in the reaction. Adsorption should not be confused with *absorption,* in which substances are incorporated into the interior of the material, not just at the surface. One common example is the absorption of liquid by a sponge.

Heterogeneous catalysis MnO_2-catalyzed decomposition of hydrogen peroxide occurs when the solid MNO_2 is added to a solution of H_2O_2.

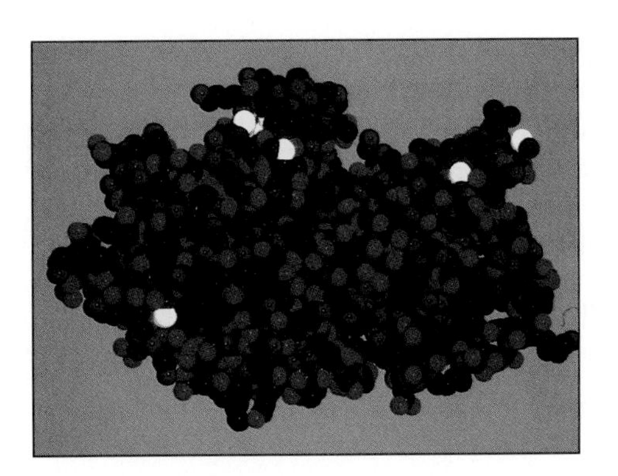

Structure of an enzyme This is a model of the enzyme called alcohol dehydrogenase. Its structure was determined several years ago. Chemists are trying to match the functions and actions of enzymes with their structures. One step in such determinations is to determine the mechanism by which the enzyme catalyzes a particular reaction.

Enzyme Catalysis

Enzymes are large molecules that catalyze specific biochemical reactions. They are often named after their functions—the reactions that they catalyze. Alcohol dehydrogenase, ADH, is an enzyme that catalyzes the removal of two hydrogen atoms from ethyl alcohol, C_2H_5OH, forming acetaldehyde. The way enzymes work is discussed in the next section, using ADH as an example.

It is believed that some enzymes function as catalysts by increasing the value of the steric factor. The enzyme interacts with the reactant molecules in a way that places them in the correct geometry to form the products.

Enzymes have many practical uses. When a textile company recently found that their machinery was becoming clogged with starch that was used in the manufacture of a fabric, they used a cleaning solution that included the enzyme amylase to digest the starch.

On a more familiar level, have you ever wondered how the liquid center is placed inside of candies? The answer lies in an interesting blend of kinetics and equilibrium. Most liquid-filled candies are made by blending sucrose and water, forming a paste, and then coating the paste with chocolate. If these were the only ingredients, the candy would have a "gritty," somewhat crystalline center. The manufacturer, however, blends an enzyme called

Enzymatic decomposition of hydrogen peroxide Fresh liver contains an enzyme that catalyzes the decomposition of hydrogen peroxide. Have you noticed that the photographs show homogeneous, heterogeneous, and enzymatic catalysts for the decomposition of hydrogen peroxide? Chemists often have several methods available for a given task, and they must weigh the advantages and disadvantages of each method.

Liquid-filled candies

"invertase" into the sugar-water paste. This enzyme breaks the 12-carbon sucrose molecule into two 6-carbon sugars. Since these six-carbon sugars are more soluble in water than is the 12-carbon sugar, the paste liquefies. By the time the candies are sold, they are filled with a sweet liquid. Although liquid-filled candies have been available for years, only recently have the details of the process been understood on a scientific basis.

One important aspect of enzyme-catalyzed kinetics is presented in the *Insights into Chemistry* at the end of Section 16.6. This material describes the metabolism of alcohol and the effect of alcohol on the brain.

16.6 Reaction Mechanisms

Objectives

- To divide a chemical reaction into elementary processes
- To define the rate-limiting step of a reaction and determine its molecularity
- To predict the experimental rate law given the chemical equation for the rate-limiting step
- To differentiate among several possible reaction mechanisms by examining experimental rate data

The **mechanism** of a reaction is the sequence of molecular-level steps that lead from reactants to products. In some reactions a single collision, perhaps even with the wall of the container, is all that is needed; in others, several collisions are needed and **intermediates,** compounds that are produced in one step and then consumed in another, are formed. Scientists strive to determine the mechanisms of reactions to learn the order in which bonds are broken, formed, and rearranged during the course of the reaction. Mechanistic studies can lead to improved reactions, better catalysts, and more cost-effective production of useful materials.

Elementary Steps

The stoichiometry of a reaction does not determine its mechanism. The burning of propane, C_3H_8, is a good example to study.

$$C_3H_8(g) + 5O_2(g) \longrightarrow 3CO_2(g) + 4H_2O(g)$$

The reaction, as written, could be explained by a single collision in which one propane and five oxygen molecules strike each other simultaneously. Common sense tells us that it is extremely unlikely that six molecules will collide at the same time, with enough energy and with all reactants in the proper positions to form the products. Instead, the overall reaction is a series of **elementary reactions**—single molecular events that are summed to provide the overall reaction. An **elementary step** is an equation that describes an actual molecular-level event.

Elementary reactions describe the molecular events that add up to the overall reaction.

An overall reaction is described by the chemical equation

$$A + B \longrightarrow C + D$$

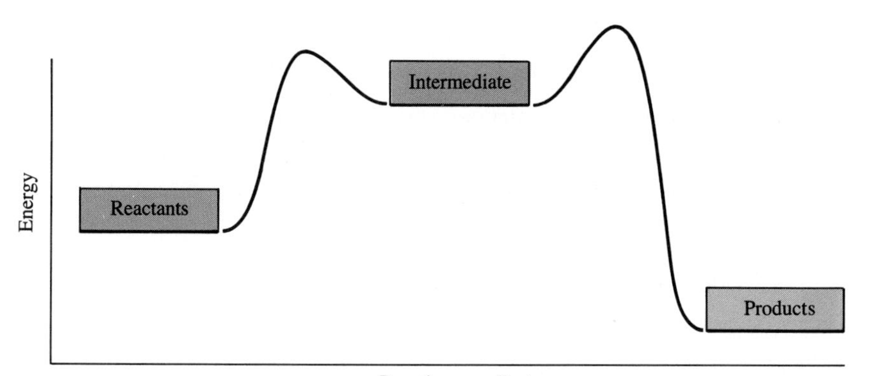

Energy-level diagram for a reaction involving an intermediate The diagram shows two peaks that represent the formation of activated complexes. The valley between them represents the intermediate, which is more stable than the activated complexes but less stable than the reactants or products.

This reaction might occur through a two-step mechanism.

$$\begin{array}{r} A \longrightarrow M + C \\ M + B \longrightarrow D \\ \hline M + A + B \longrightarrow C + D + M \end{array}$$

M appears on both sides of the chemical equations, so it can be removed from the net equation, just as we would remove a spectator ion:

$$A + B \longrightarrow C + D$$

The substance "M" is an intermediate. It is produced in one step and consumed in a later one, so it is not one of the reaction products. The two elementary reactions constitute the mechanism of the overall reaction.

A mechanism is composed of one or more elementary steps.

An intermediate differs from the activated complex. While the activated complex occurs at the maximum in the energy-level diagram, an intermediate is at a shallow minimum.

As an example of a multistep reaction, ozone is known to decompose to oxygen.

$$2O_3(g) \longrightarrow 3O_2(g)$$

One mechanism that has been proposed involves two steps.

$$\begin{array}{r} O_3 \longrightarrow O_2 + O \\ O + O_3 \longrightarrow 2O_2 \\ \hline 2O_3 \longrightarrow 3O_2 \end{array}$$

In this mechanism, atomic oxygen is an intermediate. It is produced in the first step and consumed in the second, so it is never observed among the products of the overall reaction.

Example 16.13

The gas-phase reaction of nitrogen dioxide with fluorine proceeds to form nitrosyl fluoride, NO_2F.

$$2NO_2(g) + F_2(g) \longrightarrow 2NO_2F(g)$$

Determine whether the following mechanism provides the correct overall stoichiometry and identify intermediates, if any.

$$NO_2(g) + F_2(g) \longrightarrow NO_2F(g) + F(g)$$
$$NO_2(g) + F(g) \longrightarrow NO_2F(g)$$

Solution Sum the proposed elementary steps and eliminate atomic fluorine from both sides:

$$\begin{array}{rl} NO_2(g) + F_2(g) & \longrightarrow NO_2F(g) + F(g) \\ NO_2(g) + F(g) & \longrightarrow NO_2F(g) \\ \hline 2NO_2(g) + F_2(g) + \not{F}(g) & \longrightarrow 2NO_2F(g) + \not{F}(g) \\ 2NO_2(g) + F_2(g) & \longrightarrow 2NO_2F(g) \end{array}$$

The two steps can be summed to provide the overall reaction. Atomic fluorine, F, is an intermediate.

Understanding Evaluate the proposed mechanism (all substances are gases) to determine the overall stoichiometry. Identify intermediates.

$$NO_2 + NO_2 \longrightarrow NO_3 + NO$$
$$NO_3 + CO \longrightarrow NO_2 + CO_2$$

Answer: The overall stoichiometry is $NO_2 + CO \longrightarrow NO + CO_2$. Nitrogen trioxide, NO_3, is an intermediate.

In certain cases, the mechanism is quite complicated. The gas phase decomposition of dinitrogen pentoxide is an example of a three-step mechanism.

$$2N_2O_5(g) \longrightarrow 4NO_2(g) + O_2(g)$$

Experimental results indicate that the reaction mechanism may be:

$$N_2O_5 \longrightarrow NO_2 + NO_3$$
$$NO_2 + NO_3 \longrightarrow NO_2 + O_2 + NO$$
$$NO + NO_3 \longrightarrow 2NO_2$$

The three steps can be added, but the first step must be multiplied by 2.

$$\begin{array}{rl} 2N_2O_5 & \longrightarrow 2NO_2 + 2NO_3 \\ NO_2 + NO_3 & \longrightarrow NO_2 + O_2 + NO \\ NO + NO_3 & \longrightarrow 2NO_2 \\ \hline 2N_2O_5 + NO_2 + 2NO_3 + NO & \longrightarrow 5NO_2 + 2NO_3 + O_2 + NO \\ 2N_2O_5 & \longrightarrow 4NO_2 + O_2 \end{array}$$

Rate Laws for Elementary Reactions

Since an elementary step describes a molecular collision, the rate law for an *elementary step* (unlike that of the overall reaction) can be written from the stoichiometry of the step.

$$iA + jB \longrightarrow \text{products}$$
$$\text{rate} = k[A]^i[B]^j$$

A rate law can be written from the stoichiometry of an elementary step, but not from the overall reaction stoichiometry.

For an elementary step, the relationship between rate and concentration can be determined from the number of reactants because the rate will be proportional to the concentrations of the colliding species. The rate for an overall reaction cannot be determined from the overall stoichiometry, since the overall reaction generally consists of several elementary steps.

The number of species that are involved in a single elementary step is called its **molecularity.** When an elementary step involves the spontaneous decomposition of a single molecule, it is called a **unimolecular step.** A *first order rate law describes the kinetics of a unimolecular step.*

The molecularity refers to the number of species that collide in a single elementary step.

If the process involves the collision of two species, the step is **bimolecular** and is described by *second order kinetics.* An elementary step involving the collision of three species is **termolecular,** and the process would show *third order kinetics.* Consideration of simple probability leads to the conclusion that collisions involving four (or more) species are extremely rare—a scientist would never propose such an elementary step. In fact, even termolecular reactions are extremely uncommon.

Example 16.14 illustrates how the rate law of an elementary step can be determined.

Example 16.14

Write the expected rate law and molecularity for the following elementary reactions.

(a) $HCl \longrightarrow H + Cl$
(b) $NO_2 + NO_2 \longrightarrow N_2O_4$
(c) $NO + NO_2 + O_2 \longrightarrow NO_2 + NO_3$

Solution The rate law for an elementary reaction is written directly from its stoichiometry. The molecularity is the same as the overall order of the reaction.

(a) rate $= k[HCl]$ unimolecular
(b) rate $= k[NO_2]^2$ bimolecular
(c) rate $= k[NO][NO_2][O_2]$ termolecular

Rate-limiting processes The number of cars that can travel along the highway is determined by the rate-limiting step; here it is the speed of the drivers paying tolls.

Rate-Limiting Steps

Most reactions involve more than one elementary step. When one step is much slower than any of the others, the overall rate is determined by the slowest step, which is called the **rate-limiting step.** Rate-limiting processes are not unique to chemical kinetics. Perhaps you've driven down a highway that has been narrowed to one lane by construction work. If one car can pass through the construction area every ten seconds, then the overall rate of travel is limited to six cars per minute. It really doesn't matter what the speed limit is, or how many lanes of traffic are available in areas that are not under construction. The overall traffic flow is limited to the flow set by the slowest step.

Reaction products cannot be formed at rates faster than the rate of the slowest elementary step.

Consider the reaction of NO with F_2 to form ONF:

$$2NO(g) + F_2(g) \longrightarrow 2ONF(g)$$

If we assume that the reaction occurs in a single step by collision of two molecules of NO with one of F_2, then the elementary step would be termolecular, and the rate law would be third order.

$$\text{rate} = k[NO]^2[F_2]$$

Experiments show that the rate-limiting step is *not* termolecular, but bimolecular, because the rate law is second order:

$$\text{rate} = k[\text{NO}][\text{F}_2]$$

The termolecular mechanism must be rejected because it does not agree with the experimental data. An alternative, two-step mechanism has been proposed.

$$\text{NO} + \text{F}_2 \longrightarrow \text{ONF} + \text{F} \qquad \text{rate}_1 = k_1[\text{NO}][\text{F}_2] \qquad \text{slow}$$

$$\text{NO} + \text{F} \longrightarrow \text{ONF} \qquad \text{rate}_2 = k_2[\text{NO}][\text{F}] \qquad \text{fast}$$

Note that the sum of these two steps provides the correct overall stoichiometry. As with any multiple-step mechanism, the rate will be limited by the slowest elementary step, which in this case is the first step. The rate-limiting step is bimolecular, involving a collision between NO and F_2, giving the observed second order rate law.

Note that the chemistry of the proposed mechanism is relatively logical. The atomic fluorine (produced in the first step) is known from other experiments to be highly reactive and short-lived. The reactions of atomic fluorine are expected to be fast, so the first step is likely to be slower than the second. The proposed mechanism thus has the first step as the rate-determining step. Last, note that scientists say "the proposed mechanism is *consistent* with the experimental data." There may be several mechanisms that are consistent with the observed data, and selecting among them often requires many additional experiments and a great deal of knowledge about chemical reactions.

The proposed mechanism must be consistent with the observed rate law and overall stoichiometry.

Example 16.15 illustrates another mechanistic analysis.

Example 16.15 Evaluating a Proposed Mechanism

Nitrogen dioxide can react with carbon monoxide to form carbon dioxide and nitrogen monoxide:

$$\text{NO}_2 + \text{CO} \longrightarrow \text{NO} + \text{CO}_2$$

The rate law, found by experiment, is second order.

$$\text{rate} = k[\text{NO}_2]^2$$

Evaluate the following mechanism to determine whether it is consistent with experiment.

$$\text{NO}_2 + \text{NO}_2 \longrightarrow \text{NO}_3 + \text{NO} \qquad \text{slow}$$

$$\text{NO}_3 + \text{CO} \longrightarrow \text{NO}_2 + \text{CO}_2 \qquad \text{fast}$$

Solution From the slow step, we deduce that the rate law will be

$$\text{rate} = k[\text{NO}_2]^2$$

which agrees with the experimental rate law. When we examine the stoichiometry, we see that the two steps can be summed to provide the overall reaction. The two-step mechanism is consistent with both the experimental rate law and the stoichiometry. We cannot be sure, however, that the proposed mechanism is correct.

Understanding For the same reaction

$$\text{NO}_2 + \text{CO} \longrightarrow \text{NO} + \text{CO}_2$$

is the two step mechanism

$$NO_2 + NO_2 \longrightarrow N_2O_4 \qquad \text{slow}$$
$$N_2O_4 + CO \longrightarrow NO + NO_2 + CO_2 \qquad \text{fast}$$

the correct description?

Answer: We cannot say that the mechanism is *correct,* but the mechanism is *consistent* with the experimental data.

The previous examples help clarify reasons to study chemical kinetics. Kinetic studies are necessary to identify the species that collide during the rate-limiting step, which is important information about the mechanism of a reaction. These studies are used as chemists try to find ways to speed up a slow reaction (or to tame a reaction that is dangerously fast) as well as to increase their knowledge of chemical reactions.

Kinetic studies provide fundamental information about the collisions that comprise the mechanism of a reaction.

When a potential mechanism is examined to determine whether it is plausible, two factors are initially considered. First, the rate-limiting step must be consistent with the observed rate law. Second, the sum of all the steps must provide the observed stoichiometry. (Many other factors, including reasonable intermediates and analogy to known reactions, also help to determine which mechanism the chemist proposes.)

Complex Reaction Mechanisms

A mechanism in which the rate-limiting step is *not* the first step adds another detail. We can consider a general chemical reaction.

$$A \longrightarrow B$$

The proposed mechanism has two elementary steps.

$$A \longrightarrow \text{intermediates} \qquad \text{rate}_1 = k_1[A]$$
$$\text{intermediates} \longrightarrow B \qquad \text{rate}_2 = k_2[\text{intermediates}]$$

If the rate is limited by the first step, then the rate law for the overall reaction will be first order in [A]. However, if the second step is the slower, then the reaction rate will depend on the intermediates. Since intermediates are often unstable and their concentrations are difficult to measure, *rate laws are not written in terms of an intermediate.* The solution is to express the intermediate concentration in terms of the stable species, if possible. These methods are illustrated in the next section.

Rate laws do not include the concentrations of unstable intermediates.

Reactions with a Rapid Equilibrium Step

Many multistep reactions contain steps that are rapid and reversible (in other words, equilibrium steps) prior to the rate-limiting step. The reaction of nitrogen monoxide with hydrogen is typical.

$$2NO(g) + 2H_2(g) \longrightarrow N_2(g) + 2H_2O(g)$$

Experiment shows that this reaction is described by a third order rate law.

$$\text{rate} = k[NO]^2[H_2]$$

The following mechanism has been proposed:

$2NO \rightleftharpoons N_2O_2$	fast equilibrium	$K_{eq} = [N_2O_2]/[NO]^2$
$N_2O_2 + H_2 \longrightarrow N_2O + H_2O$	slow	rate $= k_2[N_2O_2][H_2]$
$N_2O + H_2 \longrightarrow N_2 + H_2O$	fast	rate $= k_3[N_2O][H_2]$

The rate of reaction is limited by the rate of the slowest step, step 2.

$$\text{rate} = k_2[N_2O_2][H_2]$$

Although N_2O_2 does not have a measurable concentration in the reaction mixture, its concentration can be calculated from the equilibrium expression.

$$K_{eq} = \frac{[N_2O_2]}{[NO]^2}$$

$$[N_2O_2] = K_{eq}[NO]^2$$

This expression for $[N_2O_2]$ can be substituted into the rate expression.

$$\text{rate} = k_2K_{eq}[NO]^2[H_2] = k[NO]^2[H_2]$$

The proposed mechanism is consistent with the observed rate law and stoichiometry. One reason to choose this mechanism instead of one with a termolecular collision step is that three-body collisions are quite rare. The hypothesis might be tested by seeing whether N_2O_2 is produced during the reaction.

The Hydrogen-Iodine Reaction

Occasionally scientists are able to probe reaction intermediates. The intermediates of the reaction between hydrogen and iodine have been studied. The overall stoichiometry and rate law are:

$$H_2(g) + I_2(g) \longrightarrow 2HI(g)$$

$$\text{rate} = k[H_2][I_2]$$

For years, scientists assumed that the reaction proceeded in one bimolecular step, since a simple two-body collision could explain the kinetics and the stoichiometry. An alternative mechanism could also be proposed.

$I_2 \rightleftharpoons 2I$	fast equilibrium	$K_{eq1} = [I]^2/[I_2]$
$I + H_2 \rightleftharpoons H_2I$	fast equilibrium	$K_{eq2} = [H_2I]/[I][H_2]$
$H_2I + I \longrightarrow 2HI$	slow	rate $= k_3[H_2I][I]$
$H_2 + I_2 \longrightarrow 2HI$		

The rate of the reaction is determined by the rate of the slow step.

$$\text{rate} = k_3[H_2I][I]$$

Unfortunately, both H_2I and I are intermediates. The concentration $[H_2I]$ can be determined from the second equilibrium:

$$[H_2I] = K_{eq2}[I][H_2]$$

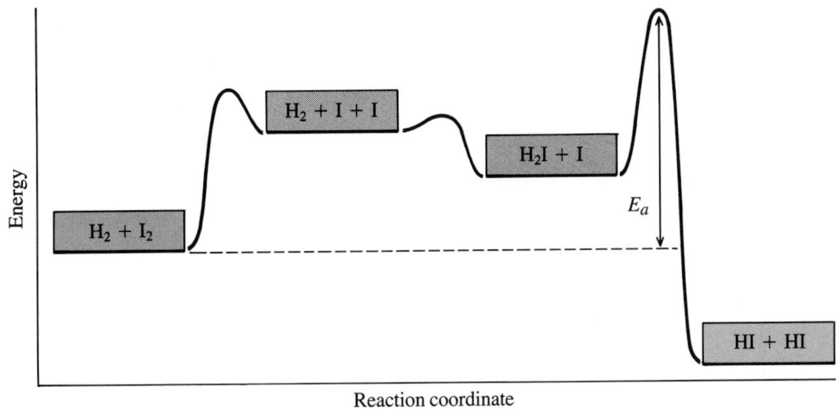

Energy-level diagram for $H_2(g) + I_2(g) \rightarrow 2HI(g)$

This expression for the concentration of H_2I can be substituted into the rate equation derived from the slow step.

$$\text{rate} = k_3K_{eq2}[H_2][I]^2$$

The value of $[I]^2$ can be determined from the first equilibrium:

$$[I]^2 = K_{eq1}[I_2]$$

This relationship can be substituted into the rate equation.

$$\text{rate} = k_3K_{eq1}K_{eq2}[H_2][I_2] = k[H_2][I_2]$$

Both proposed mechanisms have rate laws that are first order in hydrogen and iodine (and thus consistent with the experimental rate law), and both are consistent with the stoichiometry. Further experimentation is needed before either may be ruled out.

Ultraviolet light is known to cause I_2 to dissociate and form 2I. If the hydrogen-iodine reaction proceeds by a single bimolecular collision, the rate will not change appreciably when the reaction mixture is exposed to intense UV light—the I_2 concentration might change from 0.0100 to 0.0099 *M*. On the other hand, the concentration of the intermediate, I, might go from 1.0×10^{-6} *M* to 1.0×10^{-4} *M*, a factor of 100-fold. When the reaction mixture was irradiated, the rate increased dramatically, and in proportion to the quantity of UV radiation used. These results are consistent with a multistep mechanism involving iodine atoms as an intermediate.

Enzyme Catalysis

Alcohol dehydrogenase, ADH, is an enzyme that catalyzes the oxidation of ethyl alcohol to acetaldehyde and other species.

$$CH_3CH_2OH \xrightarrow{ADH} CH_3CHO + 2\text{ H-containing products}$$

The reaction is thought to follow the **Michaelis-Menten mechanism,** illustrated in the following equations. Let E represent the enzyme and S represent the *substrate,* the compound on which the enzyme acts. In this particular example, E is alcohol dehydrogenase and S is ethanol. The first step is a rapid equilibrium in which E and S form a complex.

$$E + S \rightleftharpoons ES \qquad K_{eq} = \frac{[ES]}{[E][S]}$$

The Michaelis-Menten mechanism involves a fast equilibrium to form an enzyme-substrate complex, followed by a slow dissociation of the complex into the products.

The second step is the formation of the product and enzyme from the enzyme-substrate complex.

$$ES \longrightarrow E + P \qquad \text{rate} = k[ES]$$

Note that the enzyme is "recycled"—it is a catalyst that is not consumed in the reaction, so it can act on another substrate molecule.

The last two equations can be combined to write the rate law.

$$\text{rate} = k\,K_{eq}\,[E][S] \qquad [16.12]$$

Equation 16.12 indicates that the rate of reaction can be limited by either the enzyme concentration *or* the substrate concentration, whichever is smaller.

When the enzyme concentration is high, the reaction rate is limited by the substrate concentration, so the rate is directly proportional to the concentration of the substrate. When the substrate concentration is high, the rate is limited by the enzyme concentration.

Kinetics and Equilibrium

When a chemical reaction has reached equilibrium, the reaction on the molecular level does not stop, even though there is no longer any change in the concentrations of the substances in the mixture. Consider the equilibrium established between NO, Cl_2, and NOCl.

$$2NO(g) + Cl_2(g) \rightleftharpoons 2NOCl(g)$$

$$K_{eq} = \frac{[NOCl]^2}{[NO]^2[Cl_2]}$$

The rate law for the reaction of nitrogen oxide with chlorine has been determined.

$$\text{forward rate} = k_f[NO]^2[Cl_2]$$

The subscript on the rate constant, k_f, indicates that it is the rate constant for the *forward* reaction. The rate law for the reverse reaction is also known:

$$\text{reverse rate} = k_r[NOCl]^2$$

A system at equilibrium is one in which the rate of the forward reaction is equal to the rate of the reverse reaction.

The fact that there is no net change in the concentrations of the substances when equilibrium is reached means that *the rate of the forward reaction must be equal to the rate of the reverse reaction.* Therefore, at equilibrium the two rates may be equated, and the equilibrium concentrations are used in the following expression:

$$k_f[NO]^2[Cl_2] = k_r[NOCl]^2$$

This equation can be rearranged so the rate constants are on one side and the concentrations are on thc othcr.

$$\frac{k_f}{k_r} = \frac{[NOCl]^2}{[NO]^2[Cl_2]} = K_{eq}$$

The result of this calculation is generally true for any reversible reaction regardless of the mechanism of the forward or reverse reaction.

$$K_{eq} = \frac{k_f}{k_r} \qquad [16.13]$$

There are several important conclusions that can be reached from Equation 16.13. First, the kinetic treatment of equilibrium, in which equilibrium is defined as the condition in which the forward and reverse reactions rates are equal, gives the same result as the experimentally determined law of mass action.

Another conclusion that follows from Equation 16.13 is that any catalyst that speeds the forward reaction must also be a catalyst for the reverse reaction, because the value of K_{eq} is not changed by a catalyst.

The Influence of Temperature on the Equilibrium Constant

The principle of Le Chatelier tells us the direction in which a reaction occurs in response to a change in temperature, but does not provide an explanation. The explanation can be derived from a kinetic approach to equilibrium. In Section 16.4 we saw that reactions with larger activation energies are affected more by changes in temperature, described quantitatively by Equation 16.11. Examination of Figure 16.17 shows that for an exothermic reaction, the reverse reaction has a higher activation energy than the forward reaction. Thus, the reverse reaction rate will increase more than the forward reaction rate when the temperature increases.

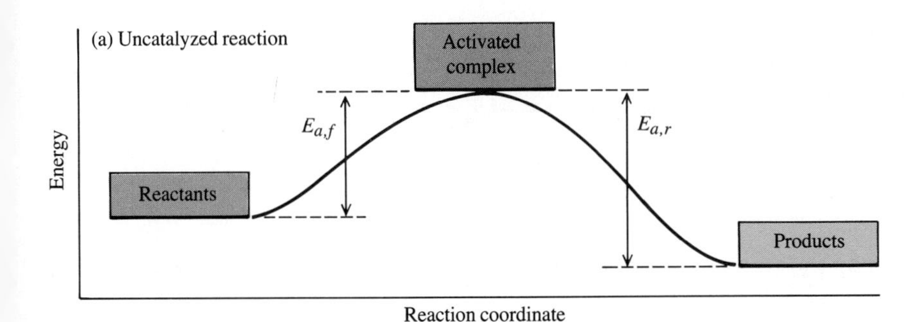

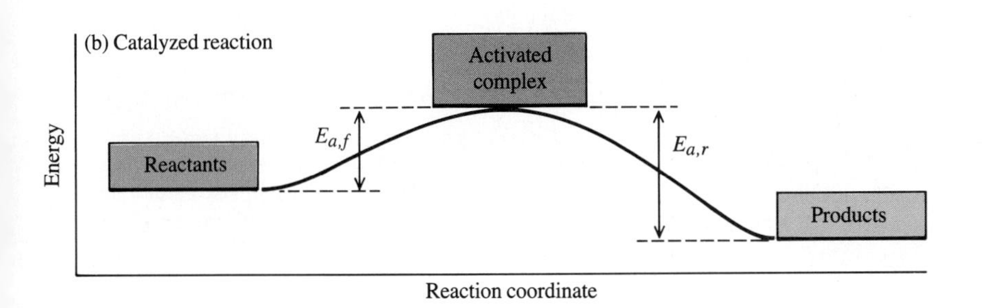

Figure 16.17 Energy-level diagram for catalyzed and uncatalyzed reactions $E_{a,f}$ is the activation energy for the forward reaction, and $E_{a,r}$ is the activation energy for the reverse reaction.

Metabolism of Alcohol Is a Kinetic Process: Time Needed to Restore Reflexes and Judgment

People have been fermenting grain to make ethyl alcohol (or ethanol, or sometimes just "alcohol") for thousands of years. In fact, some anthropologists believe that beer-making was one reason that ancient hunters became farmers! Ethanol is produced by the action of yeast on sugar, a process called fermentation, in which carbon dioxide and alcohol are produced. The sugar comes from barley, rice, or wheat in beer; grape juice in wine; and many varied sources for liquor. All liquor is made by using a distillation process to increase the concentration of alcohol in the product. Twelve ounces of beer, five ounces of wine, and one ounce of distilled liquor all contain about half an ounce of ethyl alcohol.

The metabolism of ethanol is well understood. About 20% of the amount ingested is absorbed into the bloodstream through small veins and arteries in the stomach wall. The rest is absorbed in the small intestine, and ultimately distributed throughout the body by the bloodstream. The rate of alcohol absorption depends on the stomach's contents. If alcohol is consumed as part of a meal, then it will remain in the stomach along with the food for several hours. If alcohol is consumed on an empty stomach, it will enter the bloodstream more quickly.

Some alcohol is excreted in sweat, breath, and urine, but most is eventually converted to carbon dioxide and water in the liver. In the first step of the oxidation of ethanol, the alcohol is changed into a highly toxic substance acetaldehyde, by an enzyme called alcohol dehydrogenase, ADH.

$$CH_3CH_2OH \longrightarrow CH_3CHO + 2H$$

The two hydrogen atoms that are removed from the ethyl alcohol are transferred to a large molecule called NAD. The acetaldehyde reacts further and ultimately produces CO_2 and H_2O. Metabolism and excretion of the alcohol is the *only* way someone can become sober after drinking. Breathing pure oxygen, drinking black coffee, and eating special herbal preparations have no effect.

The reaction rate for ethanol oxidation by alcohol dehydrogenase is proportional to the concentrations of both ADH and ethanol.

$$\text{rate} = k[\text{ADH}][C_2H_5OH]$$

When the concentration of ethanol is much greater than the concentration of the enzyme, the rate of reaction is limited by the concentration of the enzyme. Most enzyme-catalyzed reactions have these rate characteristics because the enzyme becomes saturated with substrate; every available enzyme molecule is bound to a substrate molecule. Increasing the ethanol concentration beyond this concentration has no effect on the overall rate of acetaldehyde production, so the rate law becomes

$$\text{rate} = k'[\text{ADH}]$$

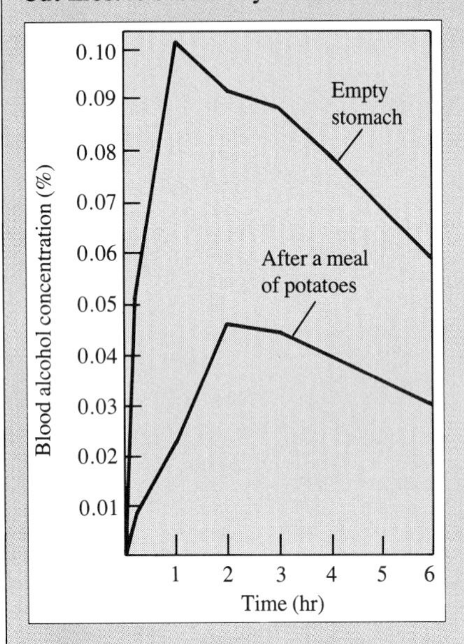

Alcohol absorption The rate of alcohol absorption depends on several factors. The same amount of alcohol consumed on an empty stomach is absorbed more quickly than when consumed along with a meal.

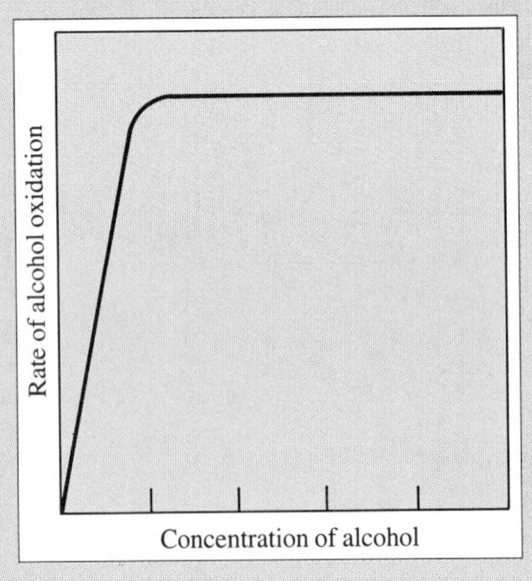

Rate of alcohol oxidation *vs.* alcohol concentration in blood When the concentration of ethanol is much higher than that of the enzyme, as it would be after as little as one drink, the rate of reaction is constant.

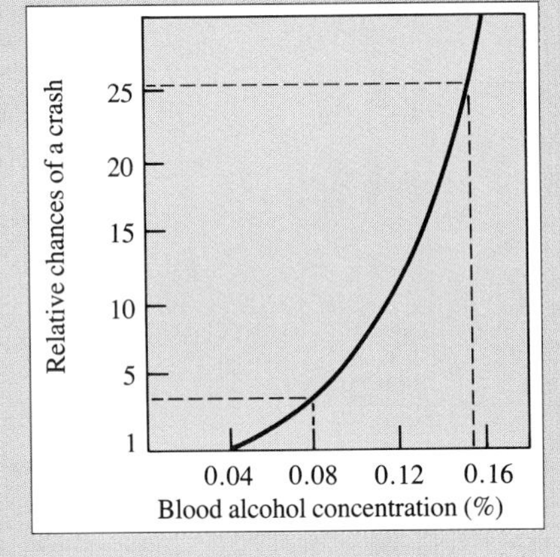

The chances of a motor vehicle accident and blood alcohol levels The chances of an accident are about 4 times higher than normal when the blood alcohol concentration is 0.08%. The risks increase to 25 times normal at a blood alcohol concentration of 0.15%.

The dominant effect of alcohol on the body occurs in the brain—it is a central nervous system depressant. Alcohol diffuses from the outside of the brain toward the center, causing profound changes along the way. As the alcohol concentration in the blood (and brain) increases, judgment is lost first, then the voluntary reflexes, and finally the involuntary reflexes.

People with high alcohol levels are unable to operate automobiles properly. Many people think that loss of reflexes is the main problem, but the loss of judgment is also an important factor. Drinking and driving simply do not mix.

Most states use a blood-alcohol test to determine whether someone operating a motor vehicle is impaired by alcohol. Some laws require the courts to consider an individual impaired when the blood-alcohol concentration is greater than 0.04%; other laws set the threshold concentration at 0.10%.

The following nomograph can be used to determine an approximate blood-alcohol level from the number of drinks consumed by a person and his or her weight. Notice that the predicted concentrations must be corrected for the constant (zero order) rate at which alcohol is metabolized.

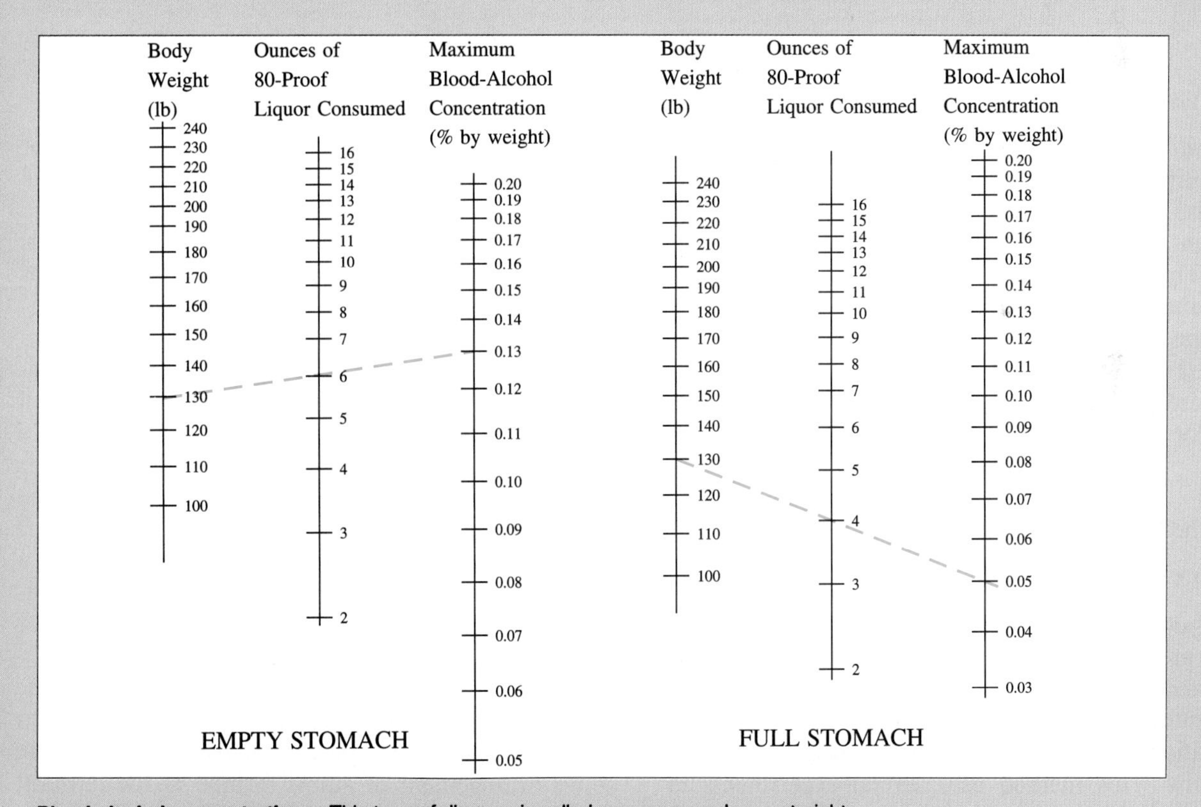

Blood alcohol concentrations This type of diagram is called a nomogram. Lay a straightedge across your weight and the amount of alcohol consumed. (One beer, one glass of wine, and one mixed drink are all equivalent.) The point at which the line crosses the column on the right is the maximum concentration of alcohol in your blood, in units of percent. Subtract 0.015% for each hour that has elapsed since the start of drinking to predict an approximate blood alcohol concentration.

The dotted lines show that a 150-lb person who consumes six drinks on an empty stomach will have a maximum blood alcohol concentration of about 0.13%. A 130-lb person who consumes four drinks along with a meal will have a maximum blood alcohol concentration of about 0.05%. The alcohol will be metabolized at a rate of 0.015% per hour, so it will take a little longer than 3 hours for the second person's alcohol consumption to be metabolized.

For an endothermic reaction, the opposite holds true; increasing the temperature will increase the rate of the forward reaction (which has higher activation energy) more than that of the reverse reaction.

The equilibrium constant is the ratio of the forward rate constant to the reverse rate constant:

$$K_{eq} = \frac{k_f}{k_r}$$

The equilibrium constant for an exothermic reaction decreases with temperature; increasing temperature increases the rate of formation of the reactants more than that of the products. *The equilibrium constant for an endothermic reaction increases with temperature.* The kinetic model provides results that are consistent with the experiment, but the kinetic model offers an improvement over Le Chatelier's principle, becase kinetics both *predicts* the effect of temperature on the equilibrium constant and *explains* these effects.

Summary

The *rate of reaction* is measured by the change in concentration of a product per unit time. The rate, which can also be expressed in terms of the disappearance of a reactant, is determined by evaluation of experimental data. The rate of reaction can be measured over a specified time interval or the *instantaneous rate* can be determined at a particular time. The rate of a reaction is defined so it does not depend on the species that is being measured.

The *rate law* for the reaction

$$A + B \longrightarrow \text{products}$$

is of the form:

$$\text{rate} = k[A]^x[B]^y$$

and is determined from experiment. The exponents x and y are called the *orders* of the reaction with respect to A and B. Since the rate of reaction is influenced by concentration, the use of *initial rates* and initial concentrations often simplifies the determination of the rate law. This method uses several experiments in which the initial concentrations of reactants are varied, the initial rate is measured, and the changes in concentrations are correlated with observed rates.

The rate law can also be determined by measuring the change in the concentration of a reactant as a function of time. Zero order, first order, and second order reactions all show different concentration-time relationships. Plots of concentration, ln[concentration] and 1/[concentration] are straight lines for zero, first, and second order reactions, respectively. The units of the *rate constant* differ for each reaction as well. The *half-life* is another measure of the rate of reaction. For a first order reaction, the half-life is a fundamental quantity that is related to the rate constant. For zero and second order reactions, the half-life depends on concentration.

The rate of reaction is strongly influenced by temperature. The effects of temperature are consistent with a model in which the reactants must collide with a minimum energy called the *activation energy* in order for a reaction to proceed. The activation energy can be determined by measuring the rates of reaction at two or more different temperatures. The rate of reaction is a function of the *collision frequency,* the *activation energy,* and a *steric factor* that includes the geometry of the molecular collisions.

If an energy-level diagram is drawn for a particular reaction, the species with an energy that corresponds to the activation energy is the *activated complex,* the least stable arrangement of atoms formed as the reactant molecules collide to form the products.

Catalysts are substances that alter the path of the reaction, generally by lowering the activation energy. Catalysts are not consumed in the formation of the product. Catalysts may be *homogeneous* (the catalyst is in the same phase as the reactants) or *heterogeneous. Enzymes* are biological molecules that act as catalysts.

The *mechanism* of a reaction can be divided into a series of *elementary steps.* The *molecularity* describes the number of species that collide in any particular elementary step. If one step is much slower than any of the others, the slow step limits the rate of reaction. Reaction mechanisms must be consistent with both the experimentally determined rate law and the observed stoichiometry. Often, more than one mechanism can be proposed for a reaction, each of which is consistent with experimental data.

Chapter Terms

Chemical kinetics—the study of the rates of chemical reactions. *(16.1)*

Rate—change per unit time. *(16.1)*

Instantaneous rate—the slope of the tangent to the concentration-time graph at any given point, rather than an average over a time interval. *(16.1)*

Rate of reaction—ratio of the rate of change of any substance to its coefficient in the chemical equation. *(16.1)*

Rate constant—k, the proportionality constant in the rate law, equal to the rate of reaction when the reactants are present at 1.0 M concentrations. *(16.2)*

Order of the reaction—the power to which the concentration of a species is raised in the rate law. *(16.2)*

Rate law—the relationship between rate and concentrations. It is of the form rate = $k[A]^x[B]^y$. *(16.2)*

Overall order—the sum of the orders for all substances that appear in the rate law. *(16.2)*

Initial rate method—determination of the rate law by measuring the initial rate of reaction in several experiments in which the initial concentrations of the reactants are varied. *(16.2)*

Differential form of the rate law—an expression that relates *differences* in concentration and time ($\Delta[A]/\Delta t$) to the concentrations of the species. *(16.3)*

Integral form of the rate law—an equation that provides concentration-time relationships. *(16.3)*

Half-life—the time needed for the concentration of a reactant to decrease to one-half its original value. *(16.3)*

Collision frequency—the number of collisions per second. *(16.4)*

Activation energy—the minimum collision energy required for a reaction to occur; equal to the difference in energy between the reactants and the activated complex. *(16.4)*

Reaction coordinate—a relative scale on an energy-level diagram that begins with the reactants and progresses to the products. *(16.4)*

Activated complex—an unstable arrangement of atoms at the highest energy point on the reaction coordinate. *(16.4)*

Steric factor—a term in the predicted rate law that expresses the need for the correct orientation of reactants when they collide in order to form the activated complex. *(16.4)*

Pre-exponential term—the product of the steric factor and the collision frequency, collected in a single term. *(16.4)*

Arrhenius equation—the relationship between rate constant and temperature. *(16.4)*

$$k = Ae^{-E_a/RT}$$

Arrhenius plot—a graph of the natural logarithm of rate constant against 1/temperature. *(16.4)*

Catalyst—a substance that increases the rate of reaction, but is not consumed in the reaction. *(16.5)*

Homogeneous catalyst—a catalyst that is in the same phase as the reactants. *(16.5)*

Heterogeneous catalyst—a catalyst that is in a different phase than are the reactants. *(16.5)*

Adsorption—the process by which a substance adheres to the surface of a solid. *(16.5)*

Enzyme—a biological compound that catalyzes a specific biochemical reaction. *(16.5)*

Mechanism—the sequence of molecular-level steps that lead from reactants to products. *(16.6)*

Intermediate—a compound that is produced in one step of the reaction mechanism and then consumed in a subsequent one. Intermediates are not found among the reactants or products. *(16.6)*

Elementary reactions—single molecular events that are summed to provide the overall reaction. *(16.6)*

Elementary step—an equation that describes a molecular-level interaction or collision. *(16.6)*

Molecularity—the number of reactant species that are involved in a single elementary step. *(16.6)*

Unimolecular step—an elementary step that involves the spontaneous decomposition of a single molecule. *(16.6)*

Bimolecular step—an elementary step that involves the collision of two species *(16.6)*

Termolecular step—an elementary step that involves the collision of three species. *(16.6)*

Rate-limiting step—an elementary step that is much slower than any of the others in the mechanism. *(16.6)*

Michaelis-Menten mechanism—an enzyme catalysis mechanism in which an enzyme-substrate complex is formed in a reversible step, followed by a rate-limiting formation of products from the complex. *(16.6)*

Exercises

Exercises designated with numbers in color have answers in Appendix J. Many of the numerical exercises use laboratory data and have some experimental error, so your answer might not agree exactly with the answer in Appendix J.

Rate of Reaction

16.1 List the factors that affect the rate of reaction.

16.2 Oxalic acid can decompose to formic acid and carbon dioxide.

$$HOOC{-}COOH(g) \longrightarrow HCOOH(g) + CO_2(g)$$

The decrease in the concentration of oxalic acid can be measured. The data follow.

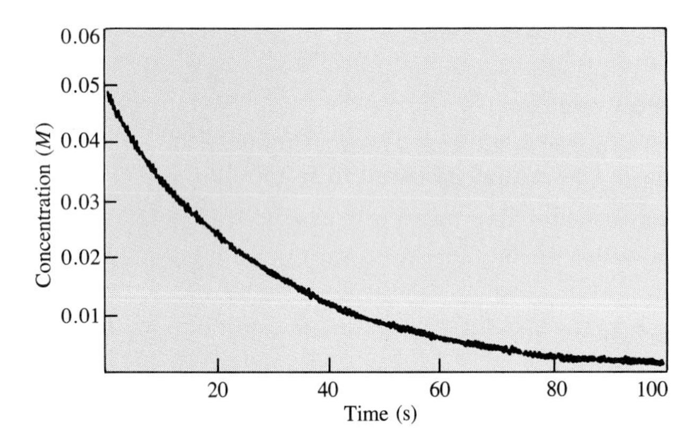

(a) Define the rate of reaction and write an expression for rate in terms of a changing concentration.
(b) Calculate the average rate of reaction between 10 and 30 seconds.
(c) Calculate the instantaneous rate of reaction after 20 seconds.
(d) Calculate the initial rate of reaction.
(e) Calculate the instantaneous rate of formation of CO_2 40 s after the start of the reaction.

16.3 Cyclobutane can decompose to form ethylene:

$$C_4H_8(g) \longrightarrow 2\ C_2H_4(g)$$

The decrease in the cyclobutane concentration can be measured by mass spectrometry (graph follows).
(a) Define the rate of reaction and write an expression for rate in terms of a changing concentration.
(b) Calculate the average rate of reaction between 10 and 30 seconds.
(c) Calculate the instantaneous rate of reaction after 20 seconds.
(d) Calculate the initial rate of reaction.
(e) Calculate the rate of formation of ethylene 40 s after the start of the reaction.

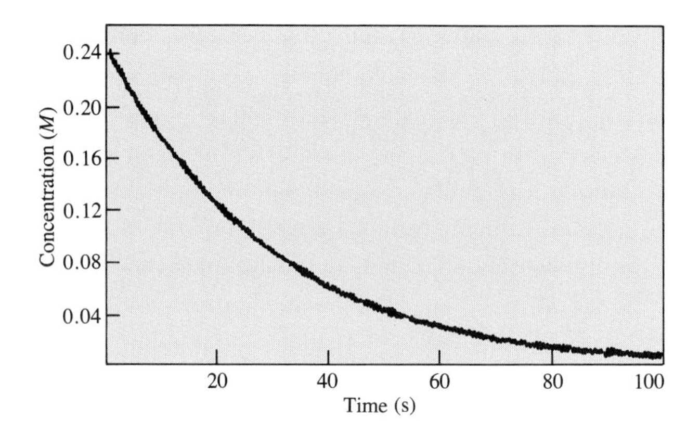

16.4 Nitrogen monoxide reacts with chlorine to form nitrosyl chloride.

$$NO(g) + \tfrac{1}{2}Cl_2(g) \longrightarrow NOCl(g)$$

The figure shows the increase in nitrosyl chloride concentration under appropriate experimental conditions. The concentration of nitrosyl chloride actually starts at 0, although this fact may be difficult to see in the figure.

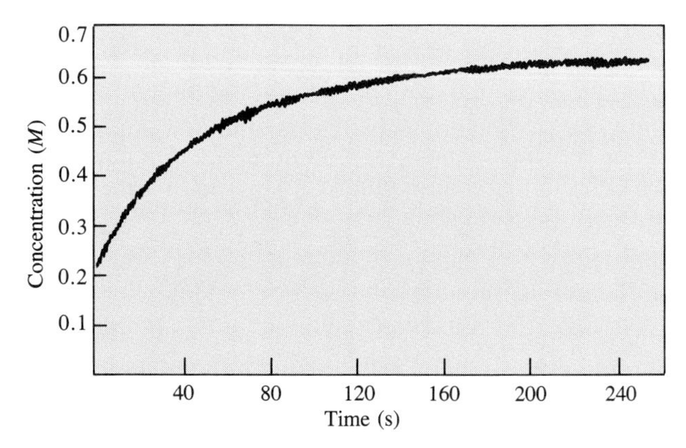

(a) Define the rate of reaction and write an expression for rate in terms of a changing concentration.
(b) Calculate the average rate of reaction between 40 and 120 seconds.
(c) Calculate the instantaneous rate of reaction after 80 seconds.
(d) Calculate the rate of consumption of chlorine 60 s after the start of the reaction.

16.5 Hydrogen iodide can be formed from hydrogen and iodine:

$$H_2(g) + I_2(g) \longrightarrow 2HI(g)$$

The following figure shows the increase in hydrogen iodide concentration under appropriate experimental conditions.

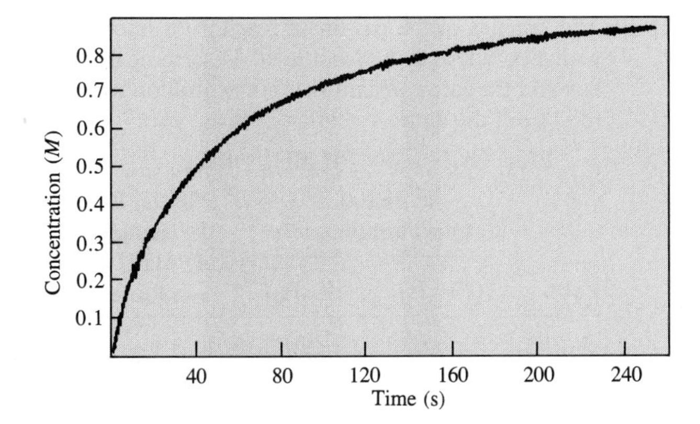

(a) Define the rate of reaction and write an expression for rate in terms of a changing concentration.
(b) Calculate the average rate of reaction between 20 and 60 seconds.
(c) Calculate the instantaneous rate of reaction after 40 seconds.
(d) Calculate the initial rate of reaction.
(e) Calculate the rate of consumption of hydrogen 60 s after the start of the reaction.

Reaction Rate and Stoichiometry

16.6 Define the rate of reaction in terms of changing concentrations for

$$aA + bB \longrightarrow cD + dD$$

16.7 Under certain conditions biphenyl, $C_{12}H_{10}$, can be produced by the reaction of cyclohexane, C_6H_{12}:

$$2C_6H_{12} \longrightarrow C_{12}H_{10} + 7H_2$$

The following table represents part of the data obtained in the kinetics experiment:

Time, s	$[C_6H_{12}]$	$[C_{12}H_{10}]$	$[H_2]$
0.0	0.200	0.000	0.000
1.00	0.159	0.021	—
2.00	0.132	—	—
3.00	—	0.044	—

(a) Fill in the missing concentrations.
(b) Calculate the rate of reaction at 1.5 s.

16.8 Dinitrogen tetroxide decomposes to nitrogen dioxide under laboratory conditions.

$$N_2O_4(g) \longrightarrow 2NO_2(g)$$

The following table represents part of the data obtained in the kinetics experiment:

Time, μs	$[N_2O_4]$	$[NO_2]$
0.000	0.050	0.000
20.00	0.033	—
40.00	—	0.050
60.00	0.020	—

(a) Fill in the missing concentrations.
(b) Calculate the rate of reaction at 30 μs.

16.9 For the reaction

$$NO(g) + \tfrac{1}{2}O_2(g) \longrightarrow NO_2(g)$$

if the nitrogen dioxide concentration increases at a rate of 0.026 mol L^{-1} s^{-1}, what is the rate of disappearance of oxygen? What is the rate of the reaction?

16.10 For the reaction

$$2NO(g) + Cl_2(g) \longrightarrow 2NOCl(g)$$

the NOCl concentration increases at a rate of 0.030 mol L^{-1} s^{-1} at a particular temperature. Calculate the rate of disappearance of chlorine at this time. What is the rate of the reaction?

16.11 For the reaction

$$3N_2O(g) + C_2H_2(g) \longrightarrow 3N_2(g) + 2CO(g) + H_2O(g)$$

if water is produced at the rate of 0.10 mol L^{-1} s^{-1}, what are the rates of production of the other species? What is the rate of the reaction?

16.12 The chromium(III) species reacts with hydrogen peroxide in aqueous solution to form the chromate ion.

$$2CrO_2^- + 3H_2O_2 + 2OH^- \longrightarrow 2CrO_4^{2-} + 4H_2O$$

If the chromate ion is produced at an instantaneous rate of 0.0050 mol L^{-1} s^{-1}, what are the rates at which the other species change concentration? What is the rate of reaction?

16.13 In aqueous solution the permanganate ion reacts with nitrous acid to form Mn^{2+} and nitrate ions.

$$2MnO_4^- + 5HNO_2 + H^+ \longrightarrow 2Mn^{2+} + 5NO_3^- + 3H_2O$$

If the permanganate ion concentration is decreasing at a rate of 0.012 mol L s^{-1}, calculate the rates at which the other concentrations change.

Initial Rate of Reaction

16.14 Use the experimental data to determine the rate law and the rate constant for the gas-phase reaction of nitrogen monoxide with hydrogen.

$$2NO + 2H_2 \longrightarrow N_2 + 2H_2O$$

Expt.	Initial Concentration, M $[NO]$	$[H_2]$	Initial Rate of Reaction $(-\Delta[NO]/2\Delta t)$, mol L^{-1} s^{-1}
1	0.10	0.10	3.8
2	0.10	0.20	7.7
3	0.10	0.30	11.4
4	0.20	0.10	15.3
5	0.30	0.10	34.2

16.15 Use the following experimental data to determine the rate law and rate constant for the gas-phase reaction of nitrogen monoxide and chlorine.

$$2NO(g) + Cl_2(g) \longrightarrow 2NOCl(g)$$

Expt.	Initial Concentration, *M* [NO]	[Cl_2]	Initial Rate of Reaction ($-\Delta[Cl_2]/\Delta t$), mol L^{-1} s^{-1}
1	0.010	0.010	2.5
2	0.010	0.020	5.1
3	0.020	0.010	5.2
4	0.020	0.020	10.3

16.16 Use the following experimental data to determine the rate law and rate constant for the gas-phase reaction of nitrogen oxide and hydrogen.

$$2NO + H_2 \longrightarrow N_2O + H_2O$$

Expt.	Initial Concentration, *M* [NO]	[H_2]	Initial Rate of Reaction ($-\Delta[H_2]/\Delta t$), mol L^{-1} s^{-1}
1	0.12	0.10	0.051
2	0.12	0.20	0.100
3	0.25	0.30	0.313

16.17 Use the following experimental data to determine the rate law and rate constant for the reaction of hydrogen iodide with ethyl iodide.

$$HI + C_2H_5I \longrightarrow C_2H_6 + I_2$$

Expt.	Initial Concentration, *M* [HI]	[C_2H_5I]	Initial Rate of Reaction ($\Delta[I_2]/\Delta t$), mol L^{-1} s^{-1}
1	0.053	0.23	3.7×10^{-5}
2	0.106	0.23	7.4×10^{-5}
3	0.106	0.46	14.8×10^{-5}

16.18 The gas-phase reaction of nitrogen dioxide with ozone was studied. The initial rate method was used to study the reaction at 15 °C. Determine the rate law and the rate constant from the data.

$$NO_2 + O_3 \longrightarrow NO_3 + O_2$$

Expt.	Initial Concentration, *M* [NO_2]	[O_3]	Initial Rate of Reaction ($\Delta[O_2]/\Delta t$), mol L^{-1} s^{-1}
1	2.0×10^{-6}	2.0×10^{-6}	2.1×10^{-7}
2	3.0×10^{-6}	2.0×10^{-6}	3.1×10^{-7}
3	4.0×10^{-6}	3.0×10^{-6}	6.2×10^{-7}
4	4.0×10^{-6}	4.0×10^{-6}	8.3×10^{-7}

16.19 The kinetics of the gas-phase reaction of phosphine with diborane at 0 °C was studied. Determine the rate law and the rate constant (use concentration units of torr) from the data.

$$PH_3 + B_2H_6 \longrightarrow PH_3BH_3 + BH_3$$

Expt.	Initial Concentration, torr [PH_3]	[B_2H_6]	Initial Rate of Reaction ($-\Delta[PH_3]/\Delta t$), torr s^{-1}
1	1.2	1.2	1.9×10^{-3}
2	3.0	1.2	4.7×10^{-3}
3	4.0	1.2	6.4×10^{-3}
4	4.0	3.0	1.6×10^{-2}

Challenging Initial Rate Problems

16.20 The kinetics of the gas-phase formation of nitrosyl bromide was studied. Determine the rate law and the rate constant (use concentration units of atm) from the data.

$$2NO + Br_2 \longrightarrow 2NOBr$$

Expt.	Initial Concentration, atm [NO]	[Br_2]	Initial Rate of Reaction ($-\Delta[NO]/2\Delta t$) atm s^{-1}
1	1.1×10^{-5}	1.2×10^{-5}	0.37
2	1.9×10^{-5}	1.3×10^{-5}	0.69
3	3.5×10^{-5}	1.2×10^{-5}	1.2
4	4.0×10^{-5}	3.0×10^{-5}	3.4

16.21 Use the following experimental data to determine the rate law and rate constant for formation of phosgene.

$$CO + Cl_2 \longrightarrow COCl_2$$

Expt.	Initial Concentration, *M* [CO]	[Cl_2]	Initial Rate of Reaction, mol L^{-1} s^{-1}
1	0.053	0.23	3.7×10^{-5}
2	0.106	0.23	7.4×10^{-5}
3	0.106	0.46	10.4×10^{-5}

16.22 The reactant in a first order reaction decreased in concentration from 0.451 *M* to 0.235 *M* in 131 s. How long will it take to decrease from 0.235 *M* to 0.100 *M*?

16.23 Ethyl chloride decomposes to form ethylene and hydrogen chloride at 437 K.

$$C_2H_5Cl(g) \longrightarrow C_2H_4(g) + HCl(g)$$

The reaction is monitored by measuring the time needed for the hydrogen chloride to react with a known amount of base. Use the ideal gas law to convert pressures to mol/L, given that the volume of the rcaction container is 4.0 L.

Initial Pressure of C_2H_5Cl, torr	Mass of NaOH, g	Time Needed for Reaction, s
131.	0.0321	11.4
160.	0.0399	11.6
172.	0.0336	9.1

(a) What is the rate law?
(b) What is the rate constant?

Determination of Rate Law from Concentration-Time Behavior

(16.24–16.30) Assume the chemical reaction is

$$\text{Reactants} \rightarrow \text{Products.}$$

16.24 Determine the rate constant and order from the concentration-time dependence.

Time, s	[Reactant], *M*
0	0.250
1	0.216
2	0.182
3	0.148
4	0.114
5	0.080

16.25 Determine the rate constant and order from the concentration-time dependence.

Time, s	[Reactant], *M*
0	0.0451
2	0.0421
5	0.0376
9	0.0316
15	0.0226

16.26 Determine the rate constant and order from the concentration-time dependence.

Time, s	[Reactant], *M*
0.001	0.220
0.002	0.140
0.003	0.080
0.004	0.050
0.005	0.030

16.27 Determine the rate constant and order from the concentration-time dependence.

Time, s	[Reactant], *M*
0	0.0350
10	0.0223
20	0.0142
50	0.0037
70	0.0015

16.28 Nitrosyl chloride decomposes to nitrogen monoxide and chlorine at elevated temperatures. Determine the rate constant and order from the concentration-time dependence.

$$NOCl(g) \longrightarrow NO(g) + \tfrac{1}{2}Cl_2(g)$$

Time, s	[NOCl], *M*
0	0.100
30	0.064
60	0.047
100	0.035
200	0.021
300	0.015
400	0.012

16.29 Determine the rate constant and order from the concentration-time dependence.

Time, s	[Reactant], *M*
0.00	0.5073
20.00	0.3247
40.00	0.2506
60.00	0.1951
80.00	0.1629
100.00	0.1405
150.00	0.1029

16.30 Determine the rate of reaction and order from the concentration-time dependence.

Time, s	[Reactant], *M*
0.00	0.7014
10.00	0.4534
20.00	0.3304
40.00	0.2181
60.00	0.1638
80.00	0.1297
100.00	0.1084

Half-Lives of Chemical Reactions

16.31 Explain why half-lives are not normally used to describe reactions other than first order.

16.32 Derive an expression for the half-life of a reaction with a half-order rate law from the integrated rate law

$$[A]^{1/2} = [A]_0^{1/2} - \tfrac{1}{2}kt$$

16.33 Derive an expression for the half-life of a reaction with a third order rate law

$$\frac{1}{[A]^2} - \frac{1}{[A]_0^2} = 2kt$$

16.34 When formic acid is heated, it decomposes to hydrogen and carbon dioxide in a first order decay:

$$HCOOH(g) \longrightarrow CO_2(g) + H_2(g)$$

At 550 °C, the half-life of formic acid is 24.5 min.

(a) What is the rate constant, and what are its units?
(b) How many seconds are needed for formic acid, initially 0.15 *M*, to decrease to 0.015 *M*?

16.35 Cyclobutane decomposes to form ethylene:

$$C_4H_8(g) \longrightarrow 2C_2H_4(g)$$

The rate constant for the first order reaction is 3.05×10^{-2} s^{-1} at 525 °C.
(a) What is the half-life, in seconds, of cyclobutane at 525 °C?
(b) How many seconds are needed for the partial pressure of cyclobutane to decrease to 1 torr from an initial pressure of 760 torr?

16.36 Calculate the half-life of a first order reaction if the concentration of the substance decreases from 0.012 *M* to 0.0082 *M* in 66.2 s.

16.37 Calculate the half-life of a first order reaction if 2.1 s after the reaction starts, the concentration of the reactant is 0.155 *M* and drops to 0.104 *M* 15.0 s after the reaction starts.

Expressing Rates in Units of Pressure

16.38 The decomposition of ozone is a second order reaction with a rate constant of 30.6 atm^{-1} s^{-1} at 95 °C.

$$2O_3(g) \longrightarrow 3O_2(g)$$

If ozone is originally present at a partial pressure of 21 torr, calculate the length of time needed for the ozone pressure to decrease to 1.0 torr.

16.39 Consider the second order decomposition of nitrosyl chloride.

$$2NOCl(g) \longrightarrow 2NO(g) + Cl_2(g)$$

At 450 K the rate constant is 15.4 atm^{-1} s^{-1}.
(a) How much time is needed for NOCl originally at a concentration of 0.0044 *M* to decay to 0.0022 *M*?
(b) How much time is needed for NOCl originally at a partial pressure of 44 torr to decay to 22 torr?

Influence of Temperature on Rate Constant

16.40 Define activation energy and explain how it influences the rate of reaction.

16.41 Explain why a collision must exceed a minimum energy in order for a product to form.

16.42 The number of collisions per second that have energies exceeding the activation energy can be determined. Explain how this determination is or is not useful in predicting the rate of a chemical reaction.

16.43 Explain the effect of doubling the concentration of a reactant on the collision frequency.

16.44 The following data were obtained by studying the change in rate constant as a function of temperature:

Rate Constant, L mol^{-1} s^{-1}	Temperature, K
0.36×10^6	500
3.7×10^6	550
27×10^6	600

Calculate the activation energy and the pre-exponential term.

16.45 The decomposition of formic acid (see Exercise 16.34) was measured at several temperatures. The temperature dependence of the first order rate constant is:

T, K	*k*, s^{-1}
800	0.00027
825	0.00049
850	0.00086
875	0.00143
900	0.00234
925	0.00372

Calculate the activation energy, in kJ, and the pre-exponential term.

16.46 A reaction rate doubles when the temperature is raised 15 °C from 298 to 313 K. What is the activation energy?

16.47 A reaction rate triples when the temperature is raised 20 °C from 15 to 35 °C. What is the activation energy?

16.48 The activation energy for the isomerization of cyclopropane to propene is 274 kJ mol^{-1}. By what factor will the rate of reaction increase as the temperature is increased from 500 to 550 °C?

16.49 The activation energy for the decomposition of cyclobutane (C_4H_8) to ethylene (C_2H_4) is 261 kJ mol^{-1}. If the system is producing ethylene at the rate of 0.043 g s^{-1} at 500 °C, what will be the rate if the temperature is increased to 600 °C?

16.50 Explain the influence of temperature on an uncatalyzed and catalyzed (lower E_a) endothermic reaction.

16.51 Explain how temperature influences an uncatalyzed and catalyzed (lower E_a) exothermic reaction.

16.52 Consider the results of an experiment in which nitrogen dioxide reacts with ozone at two different temperatures, 13 °C and 29 °C.

$$NO_2(g) + O_3(g) \longrightarrow NO_3(g) + O_2(g)$$

If the activation energy is 29 kJ/mol, by what factor will the rate constant increase with this temperature change?

Challenging Problems

16.53 Reaction A has an activation energy of 30 kJ mol^{-1}; reaction B has an activation energy of 40 kJ mol^{-1}. The ratio of their rates is called R:

$$R = \frac{\text{rate of reaction A}}{\text{rate of reaction B}}$$

If the two reactions proceed at the same rate at 25 °C, what is the value of R at 35 °C?

16.54 Two reactions have activation energies of 45 and 40 kJ mol^{-1}, respectively. Which reaction will show the greater increase in rate with an increase in temperature?

16.55 The following data were obtained for the decomposition of nitrogen dioxide.

$$2NO_2(g) \longrightarrow 2NO(g) + O_2(g)$$

	Pressure of NO_2, torr	
Time, s	310 K	315 K
0	24.0	24.0
1	18.1	15.2
2	13.7	9.7
3	10.3	6.1
4	7.8	3.9
5	5.9	2.5
6	4.5	1.6
7	3.4	1.0
8	2.6	0.6
9	1.9	0.4
10	1.5	0.3

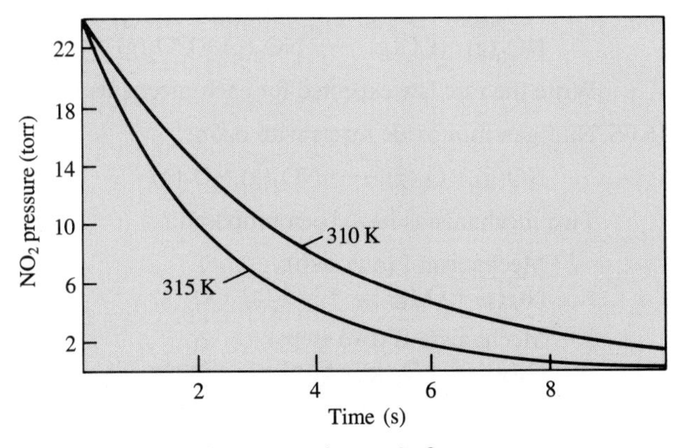

(a) What is the reaction order?
(b) What is the rate constant at each temperature?
(c) What is the activation energy?

Catalysis

16.56 Draw energy-level diagrams for catalyzed and uncatalyzed exothermic reactions. Label the activation energy.

16.57 Draw energy-level diagrams for catalyzed and uncatalyzed endothermic reactions. Label the activation energy.

16.58 The enzyme catalase lowers the activation energy for the decomposition of hydrogen peroxide from 72 to 28 kJ/mol. Calculate the factor by which the rate of reaction will increase at 298 K, assuming that everything else is unchanged. Sketch an approximate energy-level diagram, using the data in Appendix G and assuming that energy is equal to enthalpy.

$$H_2O_2(aq) \longrightarrow H_2O(\ell) + \tfrac{1}{2}O_2(g)$$

16.59 If a catalyst lowers the activation energy of a reaction from 50 kJ/mol to 25 kJ/mol, by what factor will the rate of reaction increase at 15 °C? Assuming that the reaction is exothermic and the products are 40 kJ lower in energy than the reactants, sketch an approximate energy-level diagram.

16.60 Use the following energy-level diagram to explain how an increase in temperature affects the rate of the forward reaction, rate of the reverse reaction, and value of the equilibrium constant for

$$NO + N_2O \longrightarrow N_2 + NO_2$$

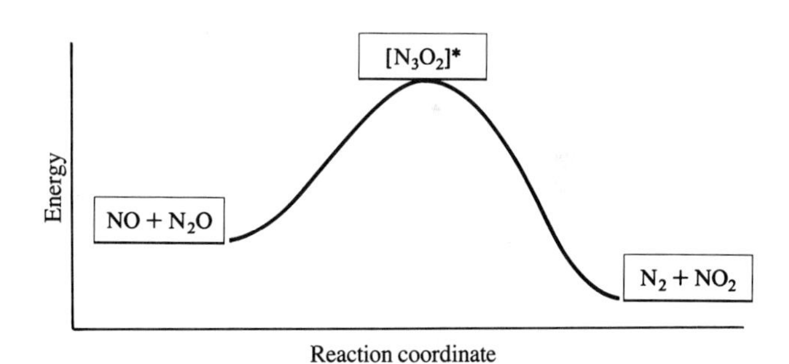

16.61 Use the following energy-level diagram to explain how an increase in temperature affects the rate of the forward reaction, rate of the reverse reaction, and value of the equilibrium constant for

$$2NO_2 \longrightarrow 2NO + O_2$$

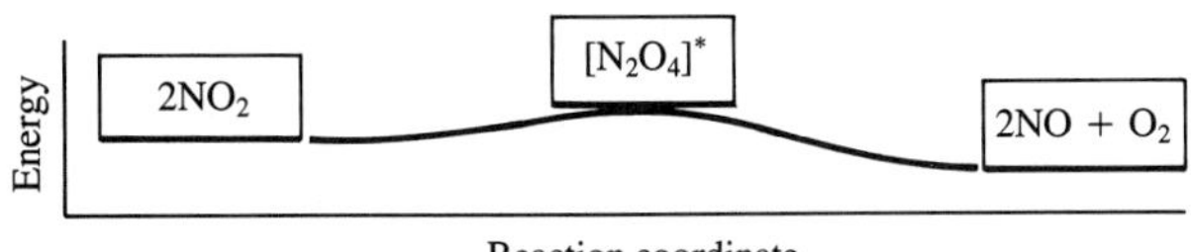

16.62 Use the following energy-level diagram to explain how an increase in temperature affects the rate of the forward reaction, rate of the reverse reaction, and value of the equilibrium constant for

$$NOCl + NO_2 \longrightarrow NO + NO_2Cl$$

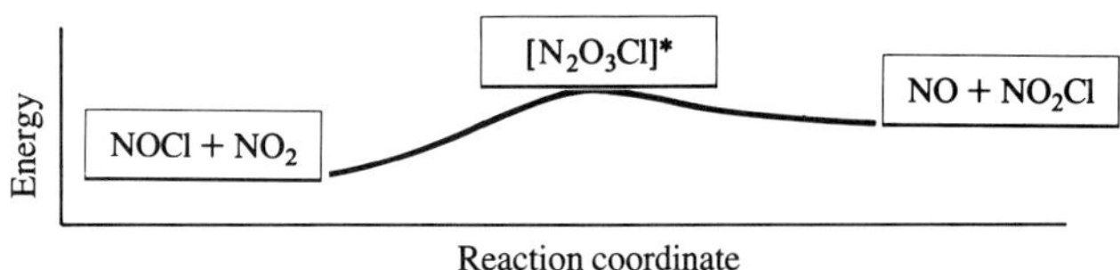

16.63 Use the following energy-level diagram to explain how an increase in temperature affects the rate of the forward reaction, rate of the reverse reaction, and value of the equilibrium constant for

$$NO + CO_2 \longrightarrow NO_2 + CO$$

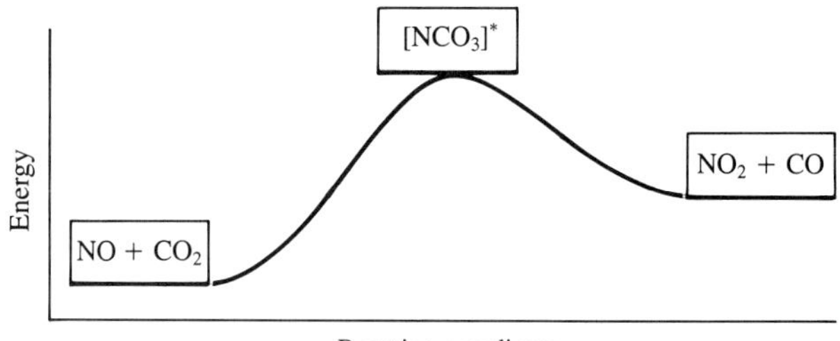

Reaction Mechanisms

16.64 Define an elementary step.

16.65 Explain why elementary reactions can be used to predict the rate law but the overall stoichiometry cannot.

16.66 Sum the following elementary steps to determine the overall stoichiometry of the reaction.

$$NO_2(g) + NO_2(g) \longrightarrow NO_3(g) + NO(g)$$
$$NO_3(g) + CO(g) \longrightarrow NO_2(g) + CO_2(g)$$

16.67 Sum the following elementary steps to determine the overall stoichiometry of the reaction.

$$Cl_2(g) \longrightarrow 2Cl(g)$$
$$Cl(g) + CO(g) \longrightarrow COCl(g)$$
$$COCl(g) + Cl(g) \longrightarrow COCl_2(g)$$

16.68 Write the rate law for each of the following elementary reactions.
(a) $HCl \rightarrow H + Cl$
(b) $H_2 + Cl \rightarrow HCl + H$
(c) $2NO_2 \rightarrow N_2O_4$

16.69 Write the rate law for each of the following elementary reactions.
(a) $H + Br \rightarrow HBr$
(b) $CH_3Br + OH^- \rightarrow CH_3OH + Br^-$
(c) $NO + NO_2 + O_2 \rightarrow NO_2 + NO_3$

16.70 Write the rate law for each of the following elementary reactions.
(a) $C_2H_5Cl \rightarrow C_2H_4 + HCl$
(b) $NO + O_3 \rightarrow NO_2 + O_2$
(c) $HI + C_2H_5I \rightarrow C_2H_6 + I_2$

16.71 Write the rate law for each of the following elementary reactions.
(a) $NO + NO_2Cl \rightarrow NO_2 + NOCl$
(b) $NO_2 + SO_2 \rightarrow NO + SO_3$
(c) $N_2O_4 \rightarrow 2NO_2$

16.72 Nitryl chloride, NO_2Cl, decomposes to NO_2 and Cl_2 with first order kinetics. The following mechanism has been proposed. Identify the rate-limiting step.

$$NO_2Cl \longrightarrow NO_2 + Cl$$
$$NO_2Cl + Cl \longrightarrow NO_2 + Cl_2$$

16.73 Nitrogen dioxide can react with ozone to form dinitrogen pentoxide and oxygen.

$$2NO_2(g) + O_3(g) \longrightarrow N_2O_5(g) + O_2(g)$$
$$\text{rate} = k[NO_2][O_3]$$

A two-step mechanism has been proposed. Which is the rate-limiting step?

$$NO_2 + O_3 \longrightarrow NO_3 + O_2$$
$$NO_3 + NO_2 \longrightarrow N_2O_5$$

16.74 Nitrogen dioxide reacts with carbon monoxide to form carbon dioxide and nitrogen monoxide:

$$NO_2(g) + CO(g) \longrightarrow CO_2(g) + NO(g)$$

Two mechanisms are proposed:

Mechanism I (one step):
$NO_2(g) + CO(g) \longrightarrow CO_2(g) + NO(g)$

Mechanism II (two steps):
$NO_2(g) + NO_2(g) \longrightarrow NO_3(g) + NO(g)$ slow
$NO_3(g) + CO(g) \longrightarrow NO_2(g) + CO_2(g)$ fast

Write the rate law expected for each mechanism.

16.75 Nitrogen monoxide reacts with ozone:

$$NO(g) + O_3(g) \longrightarrow NO_2(g) + O_2(g)$$

Two mechanisms have been proposed:

Mechanism I (one step):
$NO(g) + O_3(g) \longrightarrow NO_2(g) + O_2(g)$

Mechanism II (two steps):
$O_3(g) \longrightarrow O_2(g) + O(g)$ slow
$NO(g) + O(g) \longrightarrow NO_2(g)$ fast

Write the rate law expected for each mechanism.

Reactions with a Fast Equilibrium Step

16.76 The gas-phase reaction of nitrogen monoxide with chlorine proceeds to form nitrosyl chloride.

$$2NO(g) + Cl_2(g) \longrightarrow 2NOCl(g)$$
$$\text{rate} = k[NO]^2[Cl_2]$$

Evaluate the following proposed mechanism to determine whether it is consistent with the experimental results and identify intermediates, if any.

$$2NO(g) \rightleftharpoons N_2O_2(g) \quad \text{fast equilibrium,} \quad K_{eq} = \frac{[N_2O_2]}{[NO]^2}$$

$$N_2O_2(g) + Cl_2(g) \longrightarrow 2NOCl(g) \quad \text{slow (rate-limiting) step}$$

16.77 Is either of the following mechanisms consistent with the given rate law?

$$2NO_2 + O_3 \longrightarrow N_2O_5 + O_2$$

$$\text{rate} = k[NO_2][O_3]$$

(a) $NO_2 + NO_2 \rightleftharpoons N_2O_4$ fast equilibrium
$N_2O_4 + O_3 \longrightarrow N_2O_5 + O_2$ slow

(b) $NO_2 + O_3 \longrightarrow NO_3 + O_2$ slow
$NO_3 + NO_2 \longrightarrow N_2O_5$ fast

16.78 When methyl bromide reacts with hydroxide ion, methyl alcohol and bromide ion are formed:

$$CH_3Br + OH^- \longrightarrow CH_3OH + Br^-$$

Consider the two mechanisms below and write the expected rate law for each mechanism.

(a) Two-step mechanism with rate limited by dissociation of methyl bromide.

$$CH_3Br \xrightarrow{\text{slow}} H_3C^+ + Br^-$$

$$H_3C^+ + OH^- \xrightarrow{\text{fast}} CH_3OH$$

(b) Formation of a transition state followed by fast rearrangement.

$$OH^- + CH_3Br \longrightarrow CH_3OH + Br^-$$

A famous experiment was performed to compare the kinetics of the reaction of *t*-butyl bromide and hydroxide ion to that of methyl bromide and hydroxide ion (Exercise 16.78). The hydrogen atoms of methyl bromide are all replaced with CH_3 groups in *t*-butyl bromide:

(structures: methyl bromide — C bonded to H, Br, H, H; *t*-butyl bromide — C bonded to CH_3, Br, CH_3, CH_3)

methyl bromide *t*-butyl bromide

The experimental results indicate that the reaction of methyl bromide is second order but the reaction of *t*-butyl bromide is first order:

$$CH_3CBr + OH^- \longrightarrow CH_3OH + Br^-$$

$$\text{rate} = k[CH_3Br][OH^-]$$

$$(CH_3)_3CBr + OH^- \longrightarrow (CH_3)_3COH + Br^-$$

$$\text{rate} = k[(CH_3)_3CBr]$$

The reaction of methyl bromide is consistent with mechanism (b) in Exercise 16.78. However, the reaction of *t*-butyl bromide *cannot* occur by mechanism (b) because there simply isn't enough room to place five relatively large groups (the hydroxide ion, the bromide ion, and three bulky methyl groups, rather than hydrogen atoms) around a carbon center.

$$OH^- + (CH_3)_3CBr \not\longrightarrow \left[HO\cdots C(CH_3)_3\cdots Br\right]^- \longrightarrow HOC(CH_3)_3 + Br^-$$

Mechanism (a) is proposed instead:

$$(CH_3)_3CBr \longrightarrow (CH_3)_3C^+ + Br^- \quad \text{slow}$$

$$(CH_3)_3C^+ + OH^- \longrightarrow (CH_3)_3COH \quad \text{fast}$$

Mechanism (a) is termed a *first order nucleophilic substitution,* or S_N1 reaction. ("Nucleophile" is a traditional word used to describe basic species such as hydroxide ion. Such species react with protons, thus they "love nuclei.") The mechanism is consistent with the observed rate law and the stoichiometry.

The presumed mechanism for the reaction of methyl bromide is mechanism (b):

$$OH^- + CH_3Br \xrightarrow{\text{slow}} \left[HO\cdots CH_3\cdots Br\right]^- \xrightarrow{\text{fast}} HOCH_3 + Br^-$$

This mechanism is called a second order nucleophilic substitution, or S_N2 reaction.

Additional Problems

16.79 In water the reaction of ethyl acetate with hydroxide ion forms the acetate ion and ethyl alcohol.

$$CH_3COOC_2H_5 + OH^- \longrightarrow CH_3COO^- + C_2H_5OH$$

$$\text{rate} = k[CH_3COOC_2H_5][OH^-]$$

$$k = 6.5 \text{ L mol}^{-1} \text{ min}^{-1}$$

If the initial concentrations of ethyl acetate and hydroxide ion are both 0.010 *M*, what is the reaction's half-life? How much time is needed for the pH to reach 10.5?

16.80 Hydrogen peroxide decomposition is catalyzed by any of several ions in solution.

$$H_2O_2(aq) \longrightarrow \tfrac{1}{2}O_2(g) + H_2O(\ell)$$

The hydrogen peroxide concentration is measured as a function of time. The following data are obtained.

Time, s	$[H_2O_2]$, *M*
0	0.0334
10	0.0300
20	0.0283
30	0.0249
40	0.0198
50	0.0164
60	0.0130

(a) What is the order?
(b) What is the rate law of the decomposition?

16.81 The decomposition of *p*-toluene sulfinic acid (pTSA) was studied by Kice and Bowers (*J. Am. Chem. Soc,* **84**:605, 1962). The stoichiometry was found to be

$$3\text{pTSA} \longrightarrow A + B + H_2O$$

The pTSA concentration was measured as a function of time:

Time, s	[pTSA], *M*
0	0.100
900	0.0863
1800	0.0752
2700	0.0640
3600	0.0568
7200	0.0387
10800	0.0297
18000	0.0196

Determine the rate law and reaction order.

16.82 The mechanism for the reaction of nitrogen monoxide with oxygen has been studied.

$$2NO(g) + O_2(g) \longrightarrow 2NO_2(g)$$

The mechanism is believed to occur in two steps.

$$2NO \rightleftharpoons N_2O_2 \quad \text{fast equilibrium}$$

$$N_2O_2 + O_2 \longrightarrow 2NO_2 \quad \text{slow}$$

(a) What rate law would be observed if this mechanism is correct?
(b) The reaction is unusual because the rate *decreases* when temperature increases. Given the fact that the reaction is exothermic, propose an explanation for these interesting results. *Hint:* Predict the effect of temperature on each of the two steps.

16.83 In 1926, Hinshelwood and Green studied the reaction of nitrogen monoxide and hydrogen.

$$2NO(g) + 2H_2(g) \longrightarrow N_2(g) + 2H_2O(g)$$

They measured the rate of reaction as a function of the total pressure at 1099 K.

P_{H_2}, torr	P_{NO}, torr	Initial Rate, torr s^{-1}
289	400	0.160
205	400	0.110
147	400	0.079
400	359	0.150
400	300	0.103
400	152	0.025

(a) What is the rate law for the reaction?
(b) Use the data from the first experiment to calculate the rate constant at 1099 K, using concentration units of torr.
(c) The *relative* rate of the reaction changed with temperature, as shown in the following data.

T, K	*k*, relative
956	1.00
984	2.34
1024	5.15
1061	10.9
1099	18.8

Calculate the activation energy, in kJ mol^{-1}, for the reaction.

16.84 When formic acid is heated, it decomposes to hydrogen and carbon dioxide in a first order decay:

$$HCOOH(g) \longrightarrow CO_2(g) + H_2(g)$$

The rate of reaction is monitored by measuring the total pressure in the reaction container.

Time, s	*P*, torr
0	220
50	324
100	379
150	408
200	423
250	431
300	435

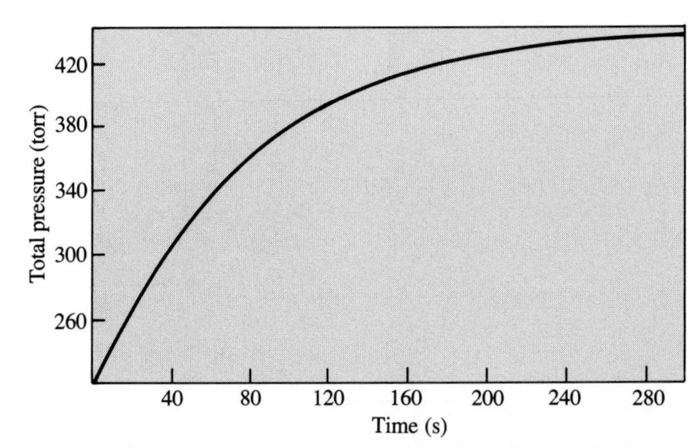

Calculate the rate constant and half-life, in s, for the reaction. At the start of the reaction (time = 0) only formic acid is present. *Hint:* Find the partial pressure of formic acid (use Dalton's law of partial pressure and the reaction stoichiometry to find P_{HCOOH} at each time.

16.85 Acetaldehyde decomposes to methane and CO:

$$CH_3CHO(g) \longrightarrow CH_4(g) + CO(g)$$

Some acetaldehyde was added to a reactor, and the pressure of the reaction mixture was measured during the course of the reaction. The following results were obtained at 791 K;

T, s	P_{total}, torr
0.00	10.0
0.10	11.6
0.20	12.8
0.50	14.9
1.00	16.6
2.00	17.9
5.00	19.0
10.00	19.5

The initial pressure is due to acetaldehyde. See the "hint" in Exercise 16.84.

(a) What is the rate law for the reaction?

(b) Calculate the rate constant at 791 K, using concentration units of torr.

Electrochemistry

Chapter Seventeen

Much of our current understanding of the structure of matter has come from studying chemical reactions and how they are related to electricity. The relationship between chemical reactions and electricity has also been very important in the development of modern technology. The power for flashlights, radios, electric watches, pacemakers, calculators, and starter motors for automobiles comes from chemical reactions. Many of the materials essential to our society, most notably aluminum and chlorine, are produced by electrical methods. **Electrochemistry** is the study of the relation between chemical reactions and electricity, and is the subject of this chapter.

The basis for all electrochemical processes is the transfer of electrons from one substance to another. It is necessary to examine electron transfer, or oxidation-reduction reactions, before beginning our study of electrochemical processes.

17.1 Oxidation States

Objectives

- To define oxidation and reduction
- To assign oxidation states to the elements in a compound or ion, using the Lewis structure of the species
- To apply the rules for assigning oxidation states to atoms in chemical species

If we are to study the relationship between chemical reactions and electricity, we must be able to identify the gain and loss of electrons by chemical species.

Battery-powered cars provide transportation with less pollution than do cars with internal combustion engines.

In this section the chemical changes that result from electron transfer are described.

Basic Definitions

Oxygen reacts directly with nearly all of the elements, and with a wide variety of compounds. We have seen numerous examples of the reactivity of oxygen in earlier chapters. Originally the term "oxidation" referred to any reaction in which oxygen combined with another substance. A reaction that removed oxygen from a compound was called a "reduction." The equation

$$4Fe(s) + 3O_2(g) \longrightarrow 2Fe_2O_3(s)$$

describes the oxidation of iron by oxygen, while the reaction

$$Fe_2O_3(s) + 3C(s) \longrightarrow 2Fe(s) + 3CO(g)$$

describes the reduction of iron oxide (iron ore) by carbon.

In the oxidation reaction, the change of elemental iron to iron oxide involves the transfer of electrons from iron to oxygen. We define **oxidation** as the loss of electrons by an atom or other chemical species. Many reactions that do not involve oxygen as a reactant fit into this classification. For example, all of the following equations represent reactions that involve the oxidation of calcium.

$$2Ca(s) + O_2(g) \longrightarrow 2CaO(s)$$

$$Ca(s) + Cl_2(g) \longrightarrow CaCl_2(s)$$

$$Ca(s) + 2HBr(g) \longrightarrow CaBr_2(s) + H_2(g)$$

Since all three calcium-containing products are ionic compounds, in each of these reactions a calcium atom loses two electrons to produce a Ca^{2+} ion.

The current definition of **reduction** is the *gain* of electrons by any chemical species. In the three preceding equations, the elements oxygen and chlorine, and the hydrogen ion are reduced. Every reaction that involves the oxidation of one substance must involve the reduction of a second substance. An **oxidation-reduction reaction,** or **redox reaction,** is one in which electrons are transferred from one species to another. In a redox reaction at least one species must be oxidized, and at least one species is reduced.

Oxidation involves the loss of electrons, and reduction involves the gain of electrons.

Redox reactions are often written as two half-reactions to emphasize that these are electron transfer reactions. In a **half-reaction** either the oxidation or the reduction portion of a redox reaction is given, showing the electrons explicitly. The oxidation half-reaction has electrons on the product side of the equation; the reduction half-reaction has electrons on the reactant side. In the three preceding equations, elemental calcium is oxidized to Ca^{2+} ions. The oxidation half-reaction in all of these equations is

$$Ca \longrightarrow Ca^{2+} + 2e^-$$

The reduction half-reaction is different in each of these equations. They are

$$O_2 + 4e^- \longrightarrow 2O^{2-}$$

$$Cl_2 + 2e^- \longrightarrow 2Cl^-$$

$$2HBr + 2e^- \longrightarrow 2Br^- + H_2$$

Just as in any chemical equation, the charges and the numbers of atoms of each element must be balanced in all half-reactions.

In any oxidation-reduction reaction the **oxidizing agent** is the substance that causes the oxidation of another species by accepting electrons from it. The oxidizing agent is always reduced. The **reducing agent** is the substance that transfers electrons to a second species, which is reduced. The reducing agent loses electrons and therefore must be oxidized. In the reaction of calcium with chlorine, the Ca is the reducing agent and the Cl_2 is the oxidizing agent.

In everyday life we encounter many oxidation-reduction reactions; in fact, life itself depends on the energy produced by many complex redox reactions. Household bleaches function by oxidizing stains to soluble or colorless products. The chlorine added to swimming pools to keep them clean kills bacteria and algae by oxidizing them. Most of the elements are isolated from their naturally occurring compounds by oxidation-reduction reactions. These represent only a few examples of this important class of chemical reactions.

Oxidation States

In oxidation-reduction reactions in which an uncombined element becomes part of an ionic compound, the loss and gain of electrons is quite obvious. In reactions that involve only covalently bonded substances, the loss and gain of electrons by the elements is more difficult to recognize. For example, in the reaction

$$H_2(g) + F_2(g) \longrightarrow 2HF(g)$$

the product, HF, is partially covalent and partially ionic. In F_2 the two atoms share equally a single pair of electrons. In HF, the more electronegative fluorine atom attracts the shared pair of electrons more strongly than does the hydrogen atom, so each of the fluorine atoms has gained electron density in the course of this reaction. To help resolve cases of partial electron transfer and to keep track of electrons in redox reactions, chemists have devised the bookkeeping concept of oxidation state. The **oxidation state** of an element in a covalently bonded species is defined as the charge it would possess if the

Lewis structure	*Assignment of electrons*	
H : H	2 [H·]	
oxidation state	0	
$:\ddot{F}:\ddot{F}:$	2 [$:\ddot{F}\cdot$]	
oxidation state	0	
$H:\ddot{F}:$	$[H]^+$	$[:\ddot{F}:]^-$
oxidation state	+1	−1

The oxidation state is the charge the atom has when the shared electrons are assigned to the more electronegative atom.

When determining oxidation states of atoms in species with covalent bonds, assign the shared electrons to the more electronegative atom.

shared electrons were completely transferred to the more electronegative atom. Electron pairs shared by atoms of the same element are divided equally. The oxidation state of a monatomic ion is simply the charge on the ion. The oxidation state of hydrogen is 0 in H_2 and +1 in HF, and that of fluorine is 0 in F_2 and −1 in HF. In the reaction of H_2 and F_2, each hydrogen atom loses one electron (is oxidized), while each fluorine atom gains an electron (is reduced).

This method of assigning electrons is somewhat artificial, and is simply another kind of "electron bookkeeping." Do not make the mistake of interpreting oxidation states as representing the actual charges that exist on the atoms in covalent molecules.

An element is oxidized if its oxidation state increases, and it is reduced if its oxidation state decreases in a chemical reaction. An oxidation-reduction reaction can be restated as a reaction in which some of the elements undergo a change in oxidation state. The assignment of the oxidation states of elements in covalent compounds can always be made from the Lewis structures, as illustrated in Example 17.1.

Example 17.1 Oxidation States from Lewis Structures

Draw the Lewis structure and assign the oxidation state of each atom in (a) CO_2, (b) NO_3^-, (c) H_2O_2.

Solution

(a) The Lewis structure of carbon dioxide is

$$\ddot{\underset{\cdot\cdot}{O}}=C=\ddot{\underset{\cdot\cdot}{O}}$$

Since oxygen is more electronegative than carbon (Table 8.8), the shared pairs of electrons are assigned to the oxygen atoms. The Lewis symbols that result from this assignment are

$$C^{4+} \quad \text{and} \quad 2\,[:\ddot{\underset{\cdot\cdot}{O}}:]^{2-}$$

A total of eight valence shell electrons (the two shared pairs and the two unshared pairs) are assigned to each oxygen atom, and none are assigned to the carbon atom. An oxygen atom with eight valence shell electrons has a charge of 2−, since oxygen is a group VIA element; a carbon atom with no valence shell electrons has a charge of 4+. Therefore the oxidation state of the oxygen atoms is −2, and that of the carbon atom is +4.

(b) One of the resonance configurations of the nitrate ion is

$$\left[\begin{array}{c} :\ddot{O}: \\ \quad\diagdown \\ \qquad N=\ddot{O}: \\ \quad\diagup \\ :\ddot{O}: \end{array}\right]^{-} \Longrightarrow [N]^{5+} \text{ and } 3\left[:\ddot{\underset{\cdot\cdot}{O}}:\right]^{2-}$$

The shared pairs of electrons are assigned to the oxygen atoms, because they are more electronegative than the nitrogen. A total of eight valence shell electrons are assigned to each oxygen atom, two more than are present in the neutral atom, giving an oxidation state of −2. A nitrogen atom (group VA) with no valence shell electrons has a 5+ charge, or an oxidation state of +5. Note that it does not matter which of the three resonance structures for the nitrate ion is used for assigning the oxidation states of the atoms.

(c) Hydrogen peroxide has the Lewis structure

$$H-\ddot{\underset{\cdot\cdot}{O}}-\ddot{\underset{\cdot\cdot}{O}}-H$$

The electron pairs shared with the hydrogen atoms are assigned to the more electronegative oxygen atoms. We assign one electron from the O—O bond to each of the oxygen atoms; electrons shared by like atoms are divided equally. Thus the oxidation state of the hydrogen is +1 (no valence shell electrons), and the oxygen has an oxidation state of −1 (seven assigned electrons, one more than needed for a neutral oxygen atom). The Lewis symbols for the individual atoms are

2 [H]$^+$ and 2 [:Ö·]$^-$

Understanding What is the oxidation state of each atom in HCN?

Answer: H is +1, C is +2, and N is −3

Often it is not necessary to construct the Lewis diagram for a molecule or polyatomic ion in order to assign the correct oxidation states of the atoms. Usually a few rules, based on the electronegativities of the elements, can be applied to determine the correct oxidation states of the atoms.

Rules for Assigning Oxidation States

1. The oxidation state is zero for any element in its free state, that is, when it is not combined with a different element.
2. The oxidation state of a monatomic ion is the electrical charge on the ion. All group IA elements form ions with a single positive charge, group IIA elements form 2+ ions, and the halogens form 1− ions; the oxidation states are thus +1, +2, and −1, respectively.
3. Fluorine always has an oxidation state of −1 in its compounds. The other halogens (Group VIIA) have an oxidation state of −1, unless they are combined with a more electronegative halogen or oxygen.
4. Hydrogen has an oxidation state of +1, except when it is combined with a less electronegative element (the metallic elements and boron), in which case its oxidation state is −1.
5. The oxidation state of oxygen is −2, except when it is bonded to fluorine (where it may be +1 or +2), and in substances that contain an O—O bond (peroxides), where it has an oxidation state of −1.
6. The sum of the oxidation states of all the atoms in a molecule or ion is equal to the charge on the molecule or ion.

These rules should always be applied in the order given. Example 17.2 illustrates the application of these rules.

Oxidation states may usually be assigned by applying the rules, and do not require writing a Lewis structure.

Example 17.2 Assigning Oxidation States by Rules

Assign the oxidation states to all elements in (a) K_2S, (b) NH_3, (c) BaO_2, (d) $Cr_2O_7^{2-}$.

Solution

(a) Potassium in group IA has an oxidation state of +1 (rule 2). Since the sum of oxidation states must equal the charge of zero (rule 6), the oxidation state of S must be −2 (which agrees with the expected charge on a monatomic ion of a group VIA element).
(b) Applying rule 4, the oxidation state of H is +1, so that of N is −3 (rule 6).
(c) The oxidation state of the group IIA element Ba is +2 (rule 2), so the oxidation state of the oxygen atoms must be −1 (rule 6). This assignment correctly sug-

gests that this compound is a peroxide that contains an O—O bond in the anion O_2^{2-}.

(d) The oxidation state of oxygen is usually −2 (rule 5); since the dichromate ion has a charge of 2−, the oxidation state of chromium is +6 (rule 6).

Understanding Assign the oxidation states to the elements in ClO_3^-

Answer: Cl is +5, and O is −2

The application of the rules will sometimes result in fractional oxidation states. For example, applying these rules to the azide ion, N_3^-, the oxidation state of nitrogen is −1/3. This is an average oxidation state of the three nitrogen atoms. Using the Lewis structure $[\ddot{N}{=}N{=}\ddot{N}]^-$, we assign an oxidation state of −1 to each of the terminal nitrogen atoms, and +1 is assigned to the center atom. For most applications of oxidation states the average value is acceptable.

17.2 Balancing Oxidation-Reduction Equations

Objective

- To balance oxidation-reduction reactions, using the ion-electron method, for both acidic and basic solutions

Many redox equations can be balanced by inspection.

Any reaction in which the oxidation state of an element changes is called an oxidation-reduction or redox reaction. Many redox reactions are simple enough that the equation can be balanced by inspection. For example,

$$2Na(s) + Cl_2(g) \longrightarrow 2NaCl(s)$$

$$CH_4(g) + 2O_2(g) \longrightarrow CO_2(g) + 2H_2O(\ell)$$

$$Zn(s) + Cu^{2+}(aq) \longrightarrow Zn^{2+}(aq) + Cu(s)$$

The balancing of redox equations can become quite complicated in other cases, and special techniques are used. The **ion-electron method** of balancing redox reactions emphasizes the fact that these are electron transfer reactions. In the balanced equation, the number of electrons lost by one substance must be gained by a second substance.

The ion-electron method for balancing redox equations provides a systematic approach. The steps in this procedure must be followed in the order given.

1. Use the change in oxidation states of the elements to identify the species that are oxidized and reduced. Write two skeleton net ionic half-reactions, one involving the species that is oxidized, and the other for the species that is reduced.
2. Balance each of the half-reactions separately.
   a. Adjust the coefficients of the reactants and products, so that all elements except oxygen and hydrogen are in balance.
   b. Balance the element oxygen by adding water, H_2O, to either the reactant or product side of the equation.

c. Add enough hydrogen ions (H^+) to one side of the equation to balance the hydrogen atoms.*
d. Add the number of electrons needed to make the net charge the same on both sides of the equation. In the oxidation half-reaction the electrons appear on the right side, while in the reduction half-reaction the electrons appear on the left side.

3. If necessary, multiply the two half-reactions by whole numbers to make the number of electrons produced in the oxidation equal to those consumed in the reduction. Add the two half-reactions, eliminating the electrons.

As an example, consider balancing the expression

$$Fe^{2+}(aq) + MnO_4^-(aq) \longrightarrow Fe^{3+}(aq) + Mn^{2+}(aq)$$

This is an oxidation-reduction reaction, since the oxidation state of iron changes from +2 to +3, and the oxidation state of the manganese changes from +7 to +2. We divide the reaction into the two skeleton half-reactions

$$Fe^{2+}(aq) \longrightarrow Fe^{3+}(aq)$$
$$MnO_4^-(aq) \longrightarrow Mn^{2+}(aq)$$

First we consider the steps required to balance the iron half-reaction. Since the iron is already balanced and there are no oxygen or hydrogen atoms to balance, the half-reaction is completed by addition of a single electron to the product side of the equation:

$$Fe^{2+}(aq) \longrightarrow Fe^{3+}(aq) + e^-$$

We balance the manganese half-reaction as follows. First, balance all elements except oxygen and hydrogen. Since Mn is already balanced with one atom of the element on each side of the half-reaction, we add four molecules of water on the right side of the equation to obtain an oxygen balance.

$$MnO_4^-(aq) \longrightarrow Mn^{2+}(aq) + 4H_2O(\ell)$$

We must add $8H^+$ to the reactant side of the manganese half-reaction to balance the hydrogen atoms present on the product side.

$$MnO_4^-(aq) + 8H^+(aq) \longrightarrow Mn^{2+}(aq) + 4H_2O(\ell)$$

Now all of the elements are balanced and five electrons are added to the left side of the half-reaction, giving a net charge of 2+ on each side.

$$MnO_4^-(aq) + 8H^+(aq) + 5e^- \longrightarrow Mn^{2+}(aq) + 4H_2O(\ell)$$

Since only one electron appears in the oxidation half-reaction of the iron(II), and the reduction of the MnO_4^- consumes five electrons, we multiply the iron half-reaction by 5 before adding to obtain the balanced equation.

$$\begin{array}{rl} 5Fe^{2+}(aq) & \longrightarrow 5Fe^{3+}(aq) + \cancel{5e^-} \\ MnO_4^-(aq) + 8H^+(aq) + \cancel{5e^-} & \longrightarrow Mn^{2+}(aq) + 4H_2O(\ell) \\ \hline 5Fe^{2+}(aq) + MnO_4^-(aq) + 8H^+(aq) & \longrightarrow \\ & 5Fe^{3+}(aq) + Mn^{2+}(aq) + 4H_2O(\ell) \end{array}$$

The ion-electron method for balancing redox equations provides a systematic way to obtain the correct equation, even for very complex reactions.

* This step is only correct if the reaction occurs in an acidic solution. We will discuss redox reactions that occur in basic solutions later in this section.

Example 17.3 provides some additional examples of the ion-electron method for balancing redox reactions.

Example 17.3 Ion-Electron Method of Balancing Redox Equations

Complete and balance the following oxidation-reduction reactions.

(a) $Zn(s) + NO_3^-(aq) \longrightarrow Zn^{2+}(aq) + NO(g)$
(b) $H_2O_2(aq) + Cr_2O_7^{2-} \longrightarrow Cr^{3+}(aq) + O_2(g)$

Solution

(a) The two skeleton half-reactions, with the Zn and N balanced, are

$$Zn(s) \longrightarrow Zn^{2+}(aq)$$
$$NO_3^-(aq) \longrightarrow NO(g)$$

In the second half-reaction, water must be added to balance the oxygen.

$$NO_3^-(aq) \longrightarrow NO(g) + 2H_2O(\ell)$$

Now add $4H^+$ to balance the hydrogen atoms on the product side of this half-reaction.

$$NO_3^-(aq) + 4H^+(aq) \longrightarrow NO(g) + 2H_2O(\ell)$$

Now all elements are balanced in both half-reactions, and charge balance is obtained by adding electrons to each of them.

$$Zn(s) \longrightarrow Zn^{2+}(aq) + 2e^-$$
$$NO_3^-(aq) + 4H^+(aq) + 3e^- \longrightarrow NO(g) + 2H_2O(\ell)$$

Multiply the zinc half-reaction by 3, and the nitrogen half-reaction by 2, to obtain six electrons in each half-reaction, and add to get the balanced equation.

$$3Zn(s) \longrightarrow 3Zn^{2+}(aq) + 6e^-$$
$$2NO_3^-(aq) + 8H^+(aq) + 6e^- \longrightarrow 2NO(g) + 4H_2O(\ell)$$

$$3Zn(s) + 2NO_3^-(aq) + 8H^+(aq) \longrightarrow 3Zn^{2+}(aq) + 2NO(g) + 4H_2O(\ell)$$

It is always wise to check the final equation, by confirming that all of the elements and the charge are properly balanced.

	Reactant Side		Product Side
Zn	3	=	3
N	2	=	2
O	2×3	=	$2 \times 1 + 4 \times 1$
H	8	=	4×2
charge	$2 \times (1-) + 8 \times (1+)$	=	$3 \times (2+)$

(b) The elements oxygen and chromium change oxidation states, so we divide the equation into two half-reactions, one containing hydrogen peroxide and oxygen, and the other containing the chromium species.

$$H_2O_2(aq) \longrightarrow O_2(g)$$
$$Cr_2O_7^{2-}(aq) \longrightarrow Cr^{3+}(aq)$$

In the first of these oxygen is already balanced, so hydrogen ions are added to balance that element.

$$H_2O_2(aq) \longrightarrow O_2(g) + 2H^+(aq)$$

Adding two electrons to the product side balances the first half-reaction.

$$H_2O_2(aq) \longrightarrow O_2(g) + 2H^+(aq) + 2e^-$$

For the second half-reaction, a coefficient of two on the product side is needed to balance chromium.

$$Cr_2O_7^{2-}(aq) \longrightarrow 2Cr^{3+}(aq)$$

Adding seven water molecules to the product side balances the oxygen.

$$Cr_2O_7^{2-}(aq) \longrightarrow 2Cr^{3+}(aq) + 7H_2O(\ell)$$

Add fourteen hydrogen ions and then six electrons to the reactant side to complete the balancing of this half-reaction.

$$Cr_2O_7^{2-}(aq) + 14H^+(aq) + 6e^- \longrightarrow 2Cr^{3+}(aq) + 7H_2O(\ell)$$

The oxidation half-reaction of the hydrogen peroxide must occur three times to produce the six electrons consumed in the reduction half-reaction. The final step is the addition of the two half-reactions.

$$3H_2O_2(aq) \longrightarrow 3O_2(g) + 6H^+(aq) + 6e^-$$
$$Cr_2O_7^{2-}(aq) + 14H^+(aq) + 6e^- \longrightarrow 2Cr^{3+}(aq) + 7H_2O(\ell)$$

$$3H_2O_2(aq) + Cr_2O_7^{2-}(aq) + 8H^+ \longrightarrow 2Cr^{3+}(aq) + 3O_2(g) + 7H_2O(\ell)$$

Only the net change in the number of hydrogen ions is shown in the final equation. A check of the elemental and charge balances confirms that the equation is correctly balanced.

	Reactant Side		Product Side
Cr	2	=	2
O	$3 \times 2 + 1 \times 7$	=	$3 \times 2 + 7 \times 1$
H	$3 \times 2 + 8 \times 1$	=	7×2
charge	$1 \times (2-) + 8 \times (1+)$	=	$2 \times (3+)$

Understanding Complete and balance the redox reaction

$$V^{3+}(aq) + Ce^{4+}(aq) \longrightarrow VO_2^+(aq) + Ce^{3+}(aq)$$

Answer: $V^{3+}(aq) + 2Ce^{4+}(aq) + 2H_2O(\ell) \longrightarrow VO_2^+(aq) + 2Ce^{3+}(aq) + 4H^+(aq)$

Balancing Redox Reactions in Basic Solutions

In all the previous examples, the redox reactions were assumed to take place in acidic solution. Very often the products of redox reactions are different in acidic and basic solutions. For example, the reduction of MnO_4^- usually produces Mn^{2+} in acidic solution, while MnO_2 is a common product in basic solution.

When a reaction occurs in a basic solution, hydrogen ions should not be shown in the final equation, since any hydrogen ions react immediately with the excess hydroxide ions present in the base, forming water. The equations for oxidation-reduction reactions that occur in basic solution can be balanced by using the ion-electron method with only slight modifications. Follow the procedure given earlier. If H^+ appears in the final equation, add the equations

H^+ ions cannot appear in the equation for a reaction that occurs in basic solution.

$$H^+ + OH^- \longrightarrow H_2O \qquad \text{or} \qquad H_2O \longrightarrow H^+ + OH^-$$

a sufficient number of times to eliminate the hydrogen ions. For example, hypochlorite ion oxidizes CrO(s) to CrO_4^{2-} in basic solution, forming chloride ions. The two half-reactions obtained from the ion-electron method as outlined earlier are

$$CrO(s) + 3H_2O \longrightarrow CrO_4^{2-} + 6H^+ + 4e^-$$
$$ClO^- + 2H^+ + 2e^- \longrightarrow Cl^- + H_2O$$

Multiply the reduction half-reaction by two and add to get the balanced equation.

$$CrO(s) + 2ClO^- + H_2O \longrightarrow CrO_4^{2-} + 2Cl^- + 2H^+$$

Since the reaction occurs in basic solution, the hydroxide ions present react with the hydrogen ions formed, producing water.

$$2H^+ + 2OH^- \longrightarrow 2H_2O$$

We obtain the final equation by adding the two previous equations.

$$CrO(s) + 2ClO^- + 2OH^- \longrightarrow CrO_4^{2-} + 2Cl^- + H_2O$$

It should be emphasized that this replacement of hydrogen ions with hydroxide ions is a simple device to arrive at the correctly balanced equation but in no way reflects the way in which the reaction actually occurs in solution. The replacement of hydrogen ions by hydroxide ions can also be performed on the half-reactions separately. We illustrate using the half-reactions in this example.

$$\begin{array}{rl} CrO(s) + 3H_2O & \longrightarrow CrO_4^{2-} + 6H^+ + 4e^- \\ 6H^+ + 6OH^- & \longrightarrow 6H_2O \\ \hline CrO(s) + 6OH^- & \longrightarrow CrO_4^{2-} + 3H_2O + 4e^- \end{array}$$

$$\begin{array}{rl} ClO^- + 2H^+ + 2e^- & \longrightarrow Cl^- + H_2O \\ 2H_2O & \longrightarrow 2H^+ + 2OH^- \\ \hline ClO^- + H_2O + 2e^- & \longrightarrow Cl^- + 2OH^- \end{array}$$

When these two half-reactions are combined, exactly the same balanced equation is obtained. When half-reactions are tabulated (see Section 17.4), the presence of OH^- or H^+ in the half-reaction shows whether it is applicable under acidic or basic conditions.

Example 17.4 Balancing Redox Equations in Basic Solution

Use the ion-electron method to balance the following oxidation reduction reaction that occurs in basic solution.

$$MnO_4^- + Br^- \longrightarrow MnO_2(s) + BrO_3^-$$

Solution Divide into the half-reactions and balance all elements except H and O.

$$MnO_4^- \longrightarrow MnO_2(s)$$
$$Br^- \longrightarrow BrO_3^-$$

Add water to balance oxygen, and then H^+ to balance hydrogen.

$$MnO_4^- + 4H^+ \longrightarrow MnO_2(s) + 2H_2O$$
$$Br^- + 3H_2O \longrightarrow BrO_3^- + 6H^+$$

Complete the half-reactions by adding electrons to balance the charge.

$$MnO_4^- + 4H^+ + 3e^- \longrightarrow MnO_2(s) + 2H_2O$$

$$Br^- + 3H_2O \longrightarrow BrO_3^- + 6H^+ + 6e^-$$

Multiply the first half-reaction by 5, and the second by 3, so that the electrons cancel; we then add the half-reactions

$$2MnO_4^- + Br^- + 2H^+ \longrightarrow 2MnO_2(s) + BrO_3^- + H_2O$$

To eliminate the hydrogen ions, we add the equation

$$2H_2O \longrightarrow 2H^+ + 2OH^-$$

The final equation is

$$2MnO_4^- + Br^- + H_2O \longrightarrow 2MnO_2(s) + BrO_3^- + 2OH^-$$

Understanding $CrO_2^- + BrO_4^- \longrightarrow BrO_3^- + CrO_4^{2-}$

What is the balanced equation for this reaction in basic solution?

Answer: $2CrO_2^- + 3BrO_4^- + 2OH^- \longrightarrow 3BrO_3^- + 2CrO_4^{2-} + H_2O$

Oxidation-Reduction Titrations

Many oxidation-reduction reactions are particularly suitable for use in titrations. Titrations require reactions that proceed rapidly to completion. The equilibrium constants for many redox reactions are very large. For example, the reaction between iron(II) and permanganate ion in acid solution

$$5Fe^{2+} + MnO_4^- + 8H^+ \longrightarrow 5Fe^{3+} + Mn^{2+} + 4H_2O$$

has an equilibrium constant of 10^{61}! The permanganate ion has another advantage as a titrant, since it has an intense purple color that is used to detect the endpoint.

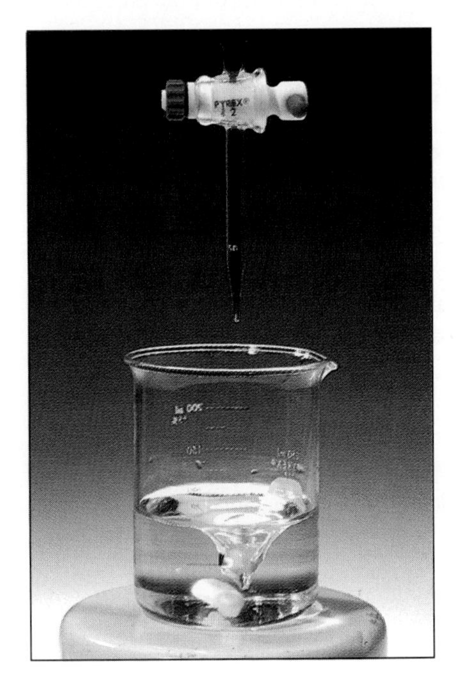

Potassium permanganate titration In an oxidation-reduction titration with potassium permanganate, the reactant also serves as the indicator. The endpoint of the titration is signaled by a faint pink color in the reaction vessel. The color can be detected at extremely low concentrations of the permanganate ion.

Example 17.5 Titration Using $KMnO_4$

A 1.225 g sample of iron ore is dissolved in acid and treated to convert all of the metal to iron(II). The titration of this sample in acid solution with 0.0180 *M* $KMnO_4$ solution requires 45.30 mL to reach the endpoint. What is the mass percentage of iron in the sample of ore?

Solution In order to find the mass percentage of iron in the sample, we must solve a reaction stoichiometry problem to find the mass of iron. The strategy we will employ is shown in the diagram.

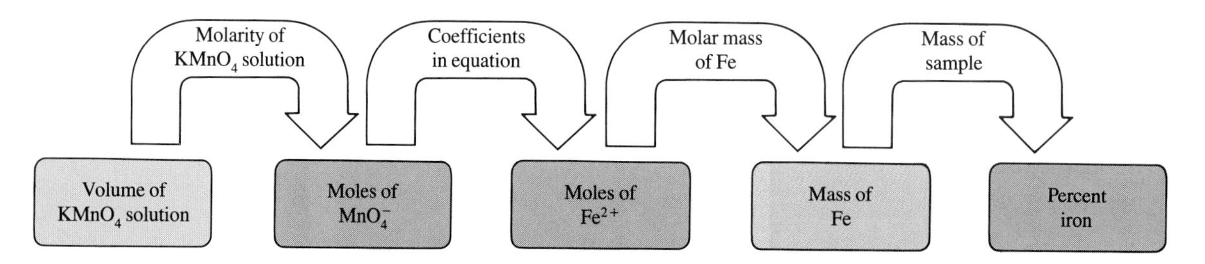

As in all reaction stoichiometry problems, a balanced equation is needed. The reaction occurring in this example is described by the equation

$$5Fe^{2+} + MnO_4^- + 8H^+ \longrightarrow 5Fe^{3+} + Mn^{2+} + 4H_2O$$

The first step is the calculation of the number of moles of the known reactant, in this case the permanganate ion.

$$\text{moles } MnO_4^- = 45.30 \text{ mL } KMnO_4 \left(\frac{1 \text{ L}}{1000 \text{ mL}}\right)\left(\frac{0.0180 \text{ mol}}{1 \text{ L}}\right)$$
$$= 8.15 \times 10^{-4} \text{ mol } MnO_4^-$$

The reaction stoichiometry is used to find the number of moles of iron. Since 5 mol Fe^{2+} reacts with 1 mol $KMnO_4$,

$$\text{moles of Fe} = 8.15 \times 10^{-4} \text{ mol } MnO_4^- \left(\frac{5 \text{ mol Fe}}{1 \text{ mol } MnO_4^-}\right) = 4.08 \times 10^{-3} \text{ mol Fe}$$

The quantity of iron must be expressed in grams for calculating the mass percentage in the sample, so we multiply by the molar mass of iron:

$$\text{mass of Fe} = 4.08 \times 10^{-3} \text{ mol Fe} \left(\frac{55.85 \text{ g}}{1 \text{ mol}}\right) = 0.228 \text{ g Fe}$$

The percentage of iron in the sample is calculated from the mass of iron, as determined in the titration, and the mass of the sample.

$$\text{percentage of Fe} = \frac{0.228 \text{ g Fe}}{1.225 \text{ g sample}} \times 100\% = 18.6\% \text{ Fe}$$

Understanding A 20.00 mL sample of an iron(II) solution requires 24.30 mL of a 0.0192 *M* $KMnO_4$ solution to react completely. What is the molar concentration of iron(II) in the sample solution?

Answer: 0.117 *M* iron(II)

17.3 Voltaic Cells

Objectives

- To sketch a voltaic cell from the balanced redox equation, labelling the direction of electron flow, the positive and negative electrodes, and the direction ions must move through the salt bridge
- To sketch half-cells that involve metal/metal ion, metal ion/metal ion, gas/ion, and metal/metal salt redox couples
- To identify cells that require a salt bridge and those that do not

A **voltaic cell,** also called a **galvanic cell,** is an apparatus that produces electrical energy directly from the chemical energy released in a redox reaction. These cells are used daily in many familiar devices such as batteries, but they have other, less obvious uses as well. They are used in the laboratory as sensors to measure concentrations in solution and are also an important source of thermodynamic data. In this section, the basics of voltaic cells are presented, with emphasis on their relation to redox reactions.

When metallic zinc is placed in a solution of copper(II) sulfate, a spontaneous chemical reaction occurs (Figure 17.1). The zinc reacts and the blue color of copper ions in the aqueous solution gradually fades as finely divided red solid (copper metal) is deposited. The reaction that accounts for these observations is

$$Zn(s) + Cu^{2+}(aq) \longrightarrow Zn^{2+}(aq) + Cu(s)$$

Under the conditions used, this spontaneous chemical change releases energy in the form of heat. By changing the physical arrangement of the reactants it is possible to convert the available energy directly into electrical

Figure 17.1 The Zn/Cu^{2+} reaction Metallic zinc reacts with a solution of blue copper(II) sulfate, forming metallic copper and a colorless zinc sulfate solution.

work. The net chemical change that occurs can be represented by the two half-reactions

$$Zn(s) \longrightarrow Zn^{2+}(aq) + 2e^-$$

$$Cu^{2+}(aq) + 2e^- \longrightarrow Cu(s)$$

As long as the copper ions contact the zinc metal, the electron transfer takes place directly, and energy is released in the form of heat. If the reactants can be separated physically, so that the transferred electrons pass through an electrical conductor, the resulting electrical current can do work. In the zinc-copper voltaic cell, a bar of metallic zinc is immersed in a beaker containing $ZnSO_4$ solution. A separate beaker contains a copper sulfate solution and a bar of copper metal. A wire connects the two pieces of metal in the separate beakers, providing a path for the electrons released by the zinc to reach the copper ions in the other beaker.

There is one thing wrong with this experimental design. As electrons are lost by the zinc, the contents of that beaker would gain a net positive charge, and the contents of the other beaker would acquire a negative charge as the positive copper ions are removed by reduction. For the reaction to occur, electrical neutrality must be maintained. This is accomplished by connecting the two solutions with a **salt bridge,** a tube that contains an electrolyte solution. Anions move through the salt bridge from the copper sulfate solution to the zinc sulfate solution, while cations move through the bridge in the opposite direction, and both solutions remain electrically neutral.

In the voltaic cell shown in Figure 17.2, the pieces of zinc and copper are called **electrodes,** and they provide electrical contacts through which the electrons leave and enter the solutions. The voltaic cell consists of one container in which the oxidation half-reaction occurs, and a second container in which the reduction half-reaction takes place. The complete cell

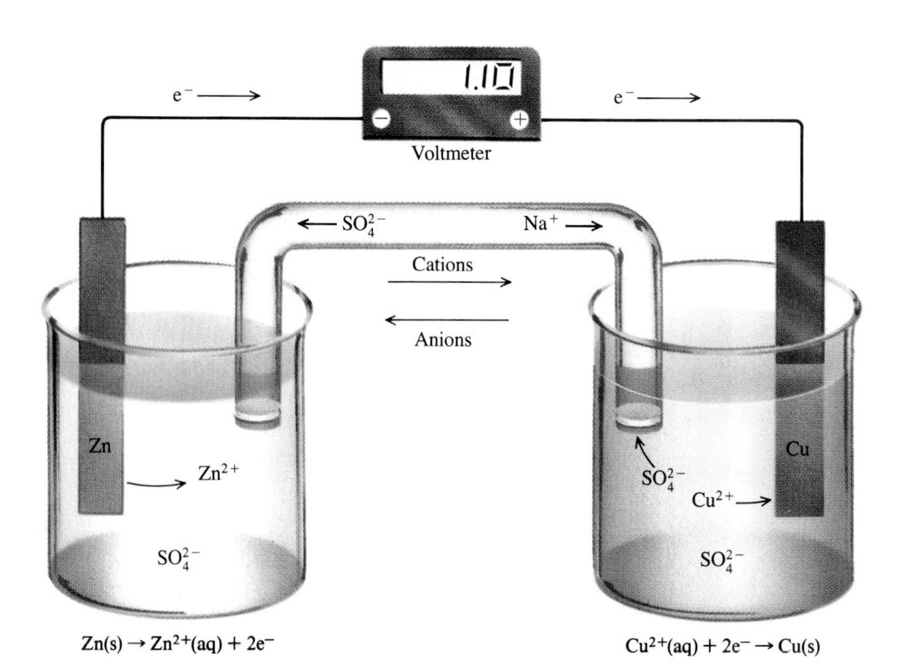

Figure 17.2 The zinc/copper voltaic cell A schematic diagram of the zinc/copper cell described in the text.

consists of these two **half-cells,** plus the external circuit and the salt bridge. The electrons flow through the external circuit from the electrode where oxidation occurs to the one where reduction takes place. The electrode where oxidation takes place is the negative electrode of the voltaic cell, and the other electrode is the positive electrode. The meaning of positive and negative electrodes will be presented in more detail in Section 17.4.

Electrons flow through the external circuit, from the negative electrode, where oxidation occurs, to the positive electrode, where they are used up in a reduction half-reaction.

Example 17.6 Voltaic Cells

The cell shown in Figure 17.3 is made up of two half-cells; one consists of silver metal in a silver nitrate solution and the other is a piece of copper metal immersed in a copper(II) nitrate solution. The half-cells are connected by a salt bridge that contains a sodium nitrate solution. The electrons flow in the direction shown in the diagram.

(a) Write the half-reaction that occurs in each half-cell.
(b) Write the equation for the overall chemical change that takes place.
(c) Indicate the direction in which the nitrate ions flow through the salt bridge.

Solution

(a) Since Figure 17.3 indicates that the electrons flow from the copper electrode to the silver electrode, oxidation of the copper metal occurs, producing electrons. The half-reaction is

$$Cu(s) \longrightarrow Cu^{2+}(aq) + 2e^-$$

A reduction must occur in the other half-cell that contains silver metal and silver ions:

$$Ag^+(aq) + e^- \longrightarrow Ag(s)$$

(b) The overall reaction must be balanced, so multiply the silver half-reaction by 2 before adding it to the copper half-reaction. The net ionic equation for the chemical reaction that takes place in this voltaic cell is

$$Cu(s) + 2Ag^+(aq) \longrightarrow Cu^{2+}(aq) + 2Ag(s)$$

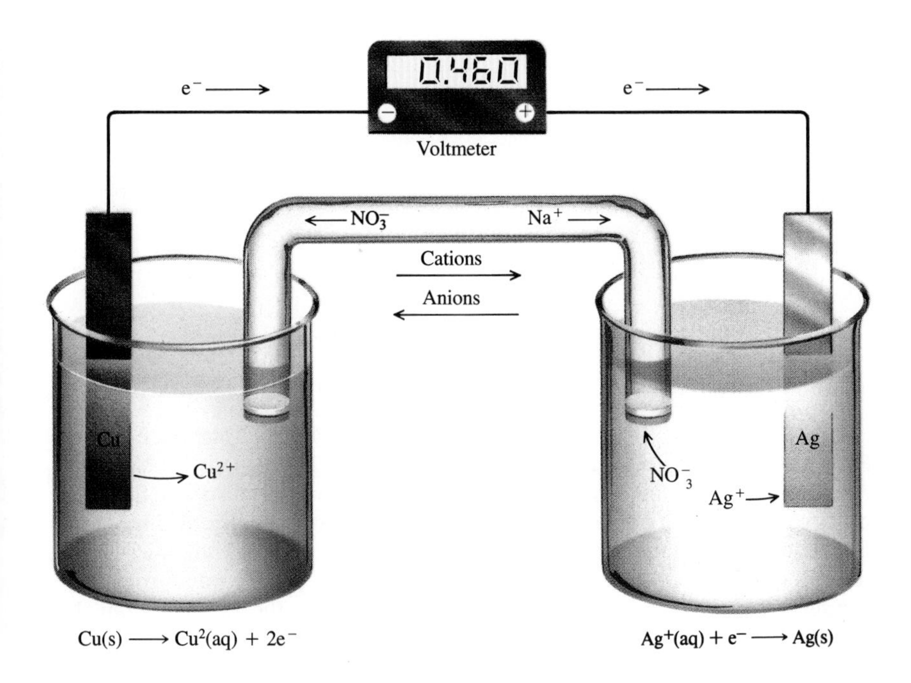

Figure 17.3 The copper/silver cell The schematic diagram of the cell in Example 17.6 shows the positive and negative electrodes and the direction of electron flow.

(c) To keep the contents of the half-cells neutral, nitrate anions must leave the silver half-cell, where cations are removed by reduction, and flow into the copper half-cell, balancing the charge of the Cu^{2+} ions that are produced. There will also be a flow of sodium ions through the salt bridge in the opposite direction.

Other Types of Electrodes and Half-Cells

All of the half-cells described so far consisted of a metal and one of its salts in solution. The metal serves as the electrode in these examples. A number of other redox half-reactions do not involve a metal as the reduced species in the half-reaction. For example, consider the half-reaction

$$Fe^{3+}(aq) + e^- \longrightarrow Fe^{2+}(aq)$$

In a half-cell that involves this half-reaction (or its reverse), a solid conductor of electrons is placed into the solution. The electrical conductor is connected to the electrode in the other half-cell by a wire. The material used to make the electrode in such a half-cell must not be easily oxidized; it is referred to as an **inert electrode.** Ions in the solution can transfer electrons to or from this inert electrical conductor. Inert metals such as platinum or gold, are often chosen as inert electrodes because they are very difficult to oxidize. Graphite is another substance that is used for inert electrodes, because it is a good electrical conductor and does not readily react with either oxidizing or re-

When neither the oxidized nor the reduced form of a couple is a solid conductor, an inert electrode is used to provide an electrical contact to the solution.

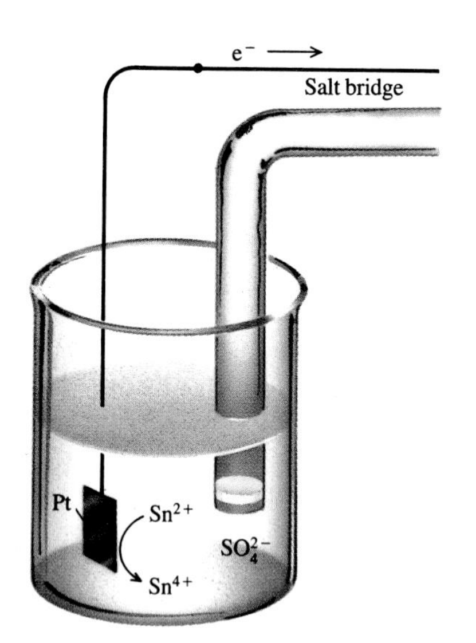

Figure 17.4 The Sn^{4+}/Sn^{2+} half-cell A half-cell in which Sn^{2+} ions are oxidized to Sn^{4+} at a platinum surface. Each Sn^{2+} ion in the solution releases two electrons to the metallic conductor, forming a Sn^{4+} ion in solution.

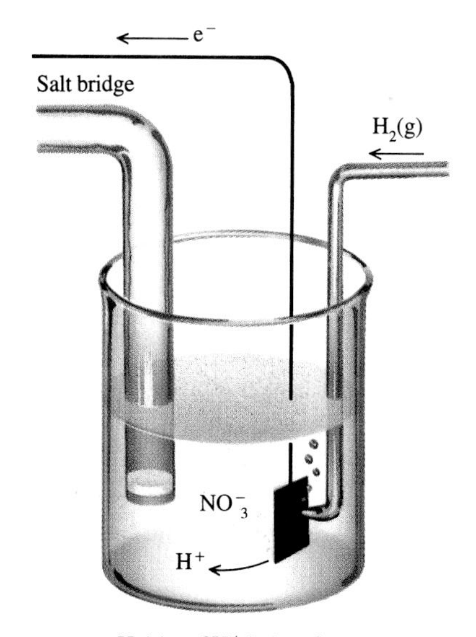

Figure 17.5 The hydrogen electrode Gaseous hydrogen is bubbled over a platinum metal surface that is in contact with a nitric acid solution. A molecule of H_2 is oxidized to two hydrogen ions by transferring its electrons to the inert platinum electrode.

ducing agents in aqueous solutions. A schematic diagram for a typical half-cell involving two ions in aqueous solution is shown in Figure 17.4.

Another type of redox couple, or half-reaction, that uses an inert electrode for electrical contact with the solution involves the oxidation or reduction of a gaseous element to ions in solution. Two such couples are

$$2H^+(aq) + 2e^- \longrightarrow H_2(g)$$

$$Cl_2(g) + 2e^- \longrightarrow 2Cl^-(aq)$$

In these half-cells the gas is bubbled over the surface of the inert electrode, which is in contact with a solution containing the ions of that element. The hydrogen electrode, shown in Figure 17.5, is an example.

Another kind of electrode involves a metal and a slightly soluble salt of that metal. An example of this type of electrode is the calomel electrode. A mixture of mercury(I) chloride (Hg_2Cl_2, called calomel) and liquid mercury is placed in electrical contact with pure mercury. The solution in the half-cell contains a soluble chloride salt such as potassium chloride, and a platinum wire is used as an inert electrical contact to the liquid mercury. (Copper cannot be used for the electrical contact because it dissolves in the mercury.) The reduction half-reaction for this electrode is

$$Hg_2Cl_2(s) + 2e^- \longrightarrow 2Hg(\ell) + 2Cl^-(aq)$$

A schematic diagram of a calomel electrode is shown in Figure 17.6.

In some voltaic cells a salt bridge is not necessary to prevent the direct reaction between the oxidizing and reducing agents. One such example is shown in Figure 17.7, in which the overall reaction is

$$2AgCl(s) + H_2(g) \longrightarrow 2Ag(s) + 2H^+(aq) + 2Cl^-(aq)$$

The spontaneous flow of electrons occurs from the hydrogen electrode to the silver/silver chloride electrode. Since neither the silver chloride nor the gas-

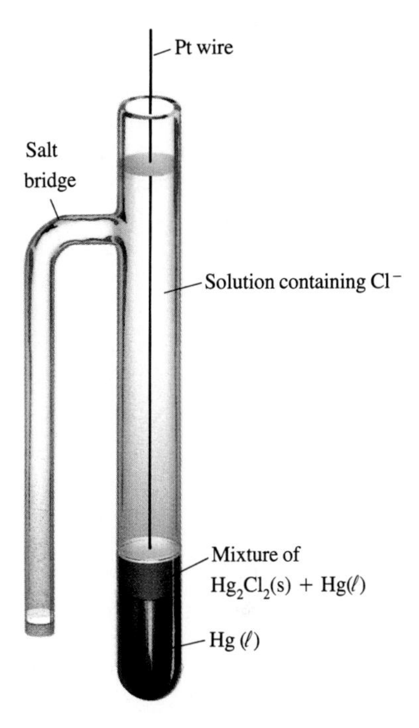

Figure 17.6 The calomel electrode In this electrode, mercury(I) chloride and metallic mercury are the oxidized and reduced forms, respectively.

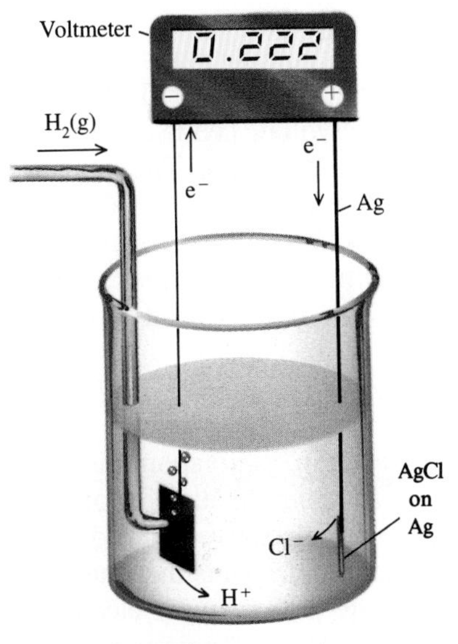

Figure 17.7 A hydrogen–silver/silver chloride cell Schematic diagram of a cell in which AgCl is reduced to silver by hydrogen. A salt bridge is not needed in this cell because the reactants (AgCl(s) and $H_2(g)$) cannot come in direct contact with each other.

When neither of the reactants in the voltaic cell is present in solution, a single electrolyte solution may be used, eliminating the need for a salt bridge.

eous hydrogen is present in the solution phase, they cannot react directly because they are physically separated. Therefore, both electrodes can be immersed in the same electrolyte solution, eliminating the need for a salt bridge.

17.4 Voltages of Voltaic Cells

Objectives

- To calculate the voltage of a standard cell from tabulated standard reduction potentials
- To identify the electrode that has the positive potential
- To use standard reduction potentials to determine the direction of spontaneous reaction under standard conditions

A net force must be applied to a stationary object to produce motion. Water flows through a pipe only if there is a difference in pressure (force per unit area) at the two ends of the pipe. The pressure serves as the driving force to move the water. Similarly, the movement of electrons through a wire requires an unbalanced electrical force. The **electromotive force** or **emf** is the electrical driving force that pushes the electrons generated in the oxidation half-reaction of a voltaic cell toward the electrode where the reduction occurs. In this section, the relation between the chemical reactions and the electrical driving force is presented.

The greater the potential energy difference between the electrons at the two electrodes, the larger is the emf. Since electrons have a negative charge, their potential energy is greater at the negative electrode. The electromotive force of a voltaic cell is an intensive property analogous to pressure. The **volt** (V) is the SI unit for emf. A difference of one volt in emf causes a charge of one coulomb to acquire an energy of one joule.

$$1 \text{ V} = 1 \text{ joule/coulomb} = 1 \text{ J/C}$$

The emf of a voltaic cell is easily measured by using a voltmeter. The potential energy difference between the electrodes of a voltaic cell is commonly referred to as the **potential** of the cell, and is designated as E_{cell}. The **standard potential** of a cell, E°_{cell}, refers to the voltage that is measured when all of the reactants and products in the redox reaction are in their standard states; *i.e.*, solids, liquids, and gases in the pure state at 1 atmosphere of pressure, and solutes present at a concentration of 1 molar. The standard cell shown in Figure 17.2, with the reaction

$$Zn(s) + Cu^{2+}(aq, 1\ M) \longrightarrow Cu(s) + Zn^{2+}(aq, 1\ M) \quad [17.1]$$

has a potential of +1.10 V. Whenever any cell reaction occurs spontaneously as written, the voltage of the cell is positive. The spontaneous flow of electrons always occurs from the negative electrode to the positive electrode. For the zinc-copper cell, the positive sign of the potential means that the electrons flow spontaneously from the zinc electrode to the copper electrode. The copper electrode is at a positive voltage with respect to the zinc electrode.

When a cell voltage is positive, the reaction proceeds, spontaneously as written.

If we replace the zinc half-cell with a silver/silver ion standard half-cell, the reaction that occurs spontaneously is

$$Cu(s) + 2Ag^+(aq,\ 1\ M) \longrightarrow 2Ag(s) + Cu^{2+}(aq,\ 1\ M) \qquad [17.2]$$

and the cell potential is +0.46 V (the silver electrode is positive).

A voltaic cell that combines the standard zinc half-cell with the standard silver half-cell has a potential of 1.56 V with the silver electrode positive, so the spontaneous cell reaction is

$$Zn(s) + 2Ag^+(aq,\ 1\ M) \longrightarrow 2Ag(s) + Zn^{2+}(aq,\ 1\ M) \qquad [17.3]$$

Notice that Equation 17.3 is the sum of Equations 17.1 and 17.2, and its potential is the sum of the potentials of the other two cells. Measurements of the potentials of many cells have confirmed that *cell potentials are additive.*

Cell potentials are additive. In a voltaic cell, reduction always occurs at the positive electrode.

Example 17.7 Adding Cell Potentials

A standard hydrogen half-cell (H_2/H^+) is connected to a standard copper half-cell. The potential of the cell is 0.34 V, with the copper electrode positive.

(a) Write the spontaneous cell reaction.
(b) Determine the potential and write the spontaneous reaction that occurs when the standard hydrogen electrode is used to form a cell with the standard silver half-cell. Indicate which electrode in this cell is positive.

Solution

(a) Electrons flow toward the positive electrode where reduction must occur, so copper ions are reduced to copper. Oxidation must occur in the hydrogen half-cell. We can write the half-reactions occurring in the half-cells as

$$H_2(g,\ 1\ atm) \longrightarrow 2H^+(1\ M) + 2e^-$$
$$Cu^{2+}(1\ M) + 2e^- \longrightarrow Cu(s)$$

The overall spontaneous reaction is

$$Cu^{2+}(1\ M) + H_2(g) \longrightarrow 2H^+(1\ M) + Cu(s) \qquad E° = +0.34\ V$$

(b) The spontaneous reaction that occurs in the copper-silver cell with a potential of +0.46 V is given in Equation 17.2. Addition of this equation to the one found in part (a) gives the equation for the desired cell reaction.

$$Cu(s) + 2Ag^+(1\ M) \longrightarrow Cu^{2+}(1\ M) + 2Ag(s) \qquad E° = +0.46\ V$$
$$Cu^{2+}(1\ M) + H_2(g) \longrightarrow 2H^+(1\ M) + Cu(s) \qquad E° = +0.34\ V$$

$$2Ag^+(1\ M) + H_2(g) \longrightarrow 2Ag(s) + 2H^+(1\ M)$$

We add these potentials to find that of the cell

$$E° = +0.34 + 0.46 = +0.80\ V$$

Since reduction occurs in the silver half-cell, this is the positive electrode in the voltaic cell.

Understanding The equation and potential of the standard nickel-hydrogen cell are

$$Ni(s) + 2H^+(1\ M) \longrightarrow Ni^{2+}(1\ M) + H_2(g) \qquad E° = +0.25\ V$$

Determine the voltage and the spontaneous cell reaction when a nickel half-cell is used with a copper half-cell. (Hint: use the results of part (a) of this example.)

Answer: $Ni(s) + Cu^{2+}(1\ M) \longrightarrow Ni^{2+}(1\ M) + Cu(s) \qquad E° = +0.59\ V$

Standard Potentials for Half-Reactions

Just as in the case of the thermodynamic state functions enthalpy (H) and free energy (G), absolute values of electrode potentials cannot be determined experimentally. Only the *difference in potential* between two half-reactions can be measured experimentally. In Example 17.7 we saw that the potential of a cell can be calculated from the measured potentials of other cells, using the additivity property of cell voltages. By extending this idea one step further, it is possible to assign standard potentials to half-reactions, by arbitrarily defining the standard potential of one particular half-reaction. The half-reaction chosen as a reference point, and assigned a standard voltage of zero, is the reduction of hydrogen ions to hydrogen.

$$2H^{+}(1\ M) + 2e^{-} \longrightarrow H_2(g,\ 1\ atm)$$

The standard potential of this reduction is exactly 0 V, *by definition.*

We can construct a cell that combines the standard hydrogen half-cell with any other standard half-cell and measure its voltage. Since we have chosen the potential of the standard hydrogen half-cell to be zero, the observed voltage of the cell is attributed to the reaction in the second half-cell. In tabulating these **standard reduction potentials** or **electrode potentials,** certain conventions are followed.

1. The half-reactions are always written as reductions; *i.e.,* electrons appear as reactants in the half-reaction.
2. A negative sign is given to the reduction potential for the half-reaction if the hydrogen electrode in the cell is positive.

In a table of standard reduction potentials, the potentials are measured relative to the standard hydrogen electrode.

Table 17.1 is a list of standard reduction potentials at 25 °C for some common half-reactions. In tables of standard potentials it is common to arrange the reduction half-reactions in order of increasing potential. Another list of standard reduction potentials is given in Appendix H in alphabetical order of the element undergoing reduction.

In Table 17.1 the oxidizing agents appear on the left side of the half-reaction and the reducing agents are on the right side. The voltage given in the table is the potential of the given half-reaction coupled in a cell with a standard hydrogen electrode.

Uses of Standard Reduction Potentials

Each of the voltages in Table 17.1 represents the potential of the given half-cell reaction measured against a standard hydrogen half-cell. Because cell voltages are additive, the standard voltage of any cell can be calculated from a table of standard reduction potentials. In the spontaneous cell reaction, the half-reaction with the more negative $E°$ proceeds in the reverse direction, and is the negative electrode of the voltaic cell. For example, the potential of a standard iodine/iodide half-cell is +0.54 V measured against the standard hydrogen couple. In a standard cell involving these two couples, the standard hydrogen half-reaction proceeds spontaneously as an oxidation. The spontaneous oxidation-reduction reaction in the cell is

$$I_2(s) + H_2(g) \longrightarrow 2I^{-}(aq) + 2H^{+}(aq) \qquad E° = +0.54\ V$$

Table 17.1 Standard Reduction Potentials at 25 °C

Reduction Half-Reaction			*E*°, V
$Li^+(aq) + e^-$	$\longrightarrow$	$Li(s)$	−3.045
$Ba^{2+}(aq) + 2e^-$	$\longrightarrow$	$Ba(s)$	−2.90
$Na^+(aq) + e^-$	$\longrightarrow$	$Na(s)$	−2.714
$Mg^{2+}(aq) + 2e^-$	$\longrightarrow$	$Mg(s)$	−2.37
$Al^{3+}(aq) + 3e^-$	$\longrightarrow$	$Al(s)$	−1.66
$2H_2O(\ell) + 2e^-$	$\longrightarrow$	$H_2(g) + 2OH^-(aq)$	−0.83
$Zn^{2+}(aq) + 2e^-$	$\longrightarrow$	$Zn(s)$	−0.76
$Fe^{2+}(aq) + 2e^-$	$\longrightarrow$	$Fe(s)$	−0.44
$Cr^{3+}(aq) + e^-$	$\longrightarrow$	$Cr^{2+}(aq)$	−0.41
$Ni^{2+}(aq) + 2e^-$	$\longrightarrow$	$Ni(s)$	−0.25
$Pb^{2+}(aq) + 2e^-$	$\longrightarrow$	$Pb(s)$	−0.126
$2H^+(aq) + 2e^-$	$\longrightarrow$	$H_2(g)$	0.000...
$Sn^{4+}(aq) + 2e^-$	$\longrightarrow$	$Sn^{2+}(aq)$	0.15
$AgCl(s) + e^-$	$\longrightarrow$	$Ag(s) + Cl^-(aq)$	0.222
$Cu^{2+}(aq) + 2e^-$	$\longrightarrow$	$Cu(s)$	0.34
$O_2(g) + 2H_2O(\ell) + 4e^-$	$\longrightarrow$	$4OH^-(aq)$	0.40
$I_2(s) + 2e^-$	$\longrightarrow$	$2I^-(aq)$	0.54
$Fe^{3+}(aq) + e^-$	$\longrightarrow$	$Fe^{2+}(aq)$	0.77
$Ag^+(aq) + e^-$	$\longrightarrow$	$Ag(s)$	0.80
$NO_3^-(aq) + 4H^+(aq) + 3e^-$	$\longrightarrow$	$NO(g) + 2H_2O(\ell)$	0.96
$Br_2(\ell) + 2e^-$	$\longrightarrow$	$2Br^-(aq)$	1.06
$O_2(g) + 4H^+(aq) + 4e^-$	$\longrightarrow$	$2H_2O(\ell)$	1.23
$Cl_2(g) + 2e^-$	$\longrightarrow$	$2Cl^-(aq)$	1.36
$MnO_4^-(aq) + 8H^+(aq) + 5e^-$	$\longrightarrow$	$Mn^{2+}(aq) + 4H_2O(\ell)$	1.51
$Ce^{4+}(aq) + e^-$	$\longrightarrow$	$Ce^{3+}(aq)$	1.61
$F_2(g) + 2e^-$	$\longrightarrow$	$2F^-(aq)$	2.87

The positive sign of the cell voltage means that the reaction proceeds spontaneously in the direction written.

$Ni^{2+}(aq) + 2e^-$	$\longrightarrow$	$Ni(s)$	−0.25
$2H^+(aq) + 2e^-$	$\longrightarrow$	$H_2(g)$	0.000
$Cu^{2+}(aq) + 2e^-$	$\longrightarrow$	$Cu(s)$	0.34
$I_2(s) + 2e^-$	$\longrightarrow$	$2\,I^-(aq)$	0.54
$Fe^{3+}(aq) + e^-$	$\longrightarrow$	$Fe^{2+}(aq)$	0.77
$Ag^+(aq) + e^-$	$\longrightarrow$	$Ag(s)$	0.80

The colored arrows indicate the direction in which the half-reactions proceed in the spontaneous cell reaction.

The cell that uses the Ni^{2+}/Ni couple and the hydrogen couple has a potential of 0.25 V. Since the standard reduction potential for the nickel couple is more negative than that for the hydrogen couple, the tabulated half-reaction for nickel is reversed, giving the spontaneous cell reaction

$$Ni(s) + 2H^+(aq) \longrightarrow Ni^{2+}(aq) + H_2(g) \qquad E° = 0.25 \text{ V}$$

The hydrogen half-cell is the positive electrode, the electrode at which reduction occurs.

$Ni^{2+}(aq) + 2e^-$	$\longrightarrow$	$Ni(s)$	− 0.25
$2H^+(aq) + 2e^-$	$\longrightarrow$	$H_2(g)$	0.000
$Cu^{2+}(aq) + 2e^-$	$\longrightarrow$	$Cu(s)$	0.34
$I_2(s) + 2e^-$	$\longrightarrow$	$2I^-(aq)$	0.54
$Fe^{3+}(aq) + e^-$	$\longrightarrow$	$Fe^{2+}(aq)$	0.77

The arrow shows that the nickel couple proceeds in the reverse direction, producing a cell reaction that has a positive potential of +0.25 V.

The standard potential of an oxidation half-reaction is the negative of that for the reduction half-reaction.

Since potentials are additive, any two standard half-cell potentials can be combined to calculate the voltage of a standard cell. Of course, in the cell one of the half-reactions must proceed as an oxidation, *in the direction opposite to that shown in the table.* Whenever we reverse half-reaction, we change the sign of its standard potential. Let's calculate the standard voltage of a cell involving the silver and nickel half-reactions. From Table 17.1, the two couples involved are

$$Ni^{2+}(aq) + 2e^- \longrightarrow Ni(s) \qquad E° = -0.25 \text{ V}$$

$$Ag^+(aq) + e^- \longrightarrow Ag(s) \qquad E° = +0.80 \text{ V}$$

One of these half-reactions must occur as an oxidation. Reversing the half-reaction with the more negative reduction potential always gives a positive voltage for the cell. In this example, the nickel couple is reversed, with a change in sign of its standard potential.

$$Ni(s) \longrightarrow Ni^{2+}(aq) + 2e^- \qquad E° = +0.25 \text{ V}$$

$$2Ag^+(aq) + 2e^- \longrightarrow 2Ag(s) \qquad E° = +0.80 \text{ V}$$

$$Ni(s) + 2Ag^+(aq) \longrightarrow Ni^{2+}(aq) + 2Ag(s)$$

The voltage of this spontaneous reaction is

$$E° = +0.25 + 0.80 = +1.05 \text{ V}$$

$Ni^{2+}(aq) + 2e^-$	$\longrightarrow$	$Ni(s)$	− 0.25
$2H^+(aq) + 2e^-$	$\longrightarrow$	$H_2(g)$	0.000
$Cu^{2+}(aq) + 2e^-$	$\longrightarrow$	$Cu(s)$	0.34
$I_2(s) + 2e^-$	$\longrightarrow$	$2I^-(aq)$	0.54
$Fe^{3+}(aq) + e^-$	$\longrightarrow$	$Fe^{2+}(aq)$	0.77
$Ag^+(aq) + e^-$	$\longrightarrow$	$Ag(s)$	0.80

The arrow shows that the nickel couple proceeds in the reverse direction, producing a cell reaction that has a positive potential of + 0.25 + 0.80 v = +1.05 V.

Even though we multiply the coefficients in the silver half-reaction by two in obtaining the balanced redox equation, the potential for that couple is not doubled. The reason for this can be understood by considering the two half-reactions

$$Ag^+ + e^- \longrightarrow Ag(s)$$

$$2Ag^+ + 2e^- \longrightarrow 2Ag(s)$$

The energy change in the second half-reaction is twice that in the first one, but the charge transferred is also doubled, from one mole of electronic charge to two moles. A voltage is the energy released *per coulomb of charge transferred,* so the energy change per charge transferred is identical for both half-reactions. The energy change and the charge transferred are both *extensive properties,* while the cell voltage is an *intensive property.* Whenever two half-reactions are combined to give a balanced redox equation, the standard voltages are added, disregarding any constant multipliers. This additivity of potentials for half-reactions is true only when the resulting equation is a balanced redox reaction. Other situations will not be considered in this text.

If $E°$ is positive the reaction proceeds spontaneously in the forward direction under standard conditions.

The sign of the standard potential for any redox reaction is used to determine the direction in which that reaction proceeds spontaneously under standard conditions. A positive sign for the voltage means the reaction proceeds spontaneously as written. If the sign of the voltage is negative, the reaction proceeds spontaneously in the reverse direction. In a table of standard reduction potentials arranged in order of increasing voltages, the half-reaction that is higher in the table proceeds spontaneously in the reverse direction, that is, as an oxidation. The half-reaction that is closer to the bottom of the table occurs as written, that is, as a reduction.

The substance that is oxidized in a redox equation appears on the right side of the half-reaction in a table of standard reduction potentials.

The table of standard reduction potentials helps us determine whether species can react with each other. Consider the possible reaction of hydrogen ions with two metals, copper and zinc. To determine whether or not a reaction occurs spontaneously under standard conditions, we calculate the standard potential using hydrogen ions and the metal as reactants.

$$\begin{array}{llr} 2H^+(aq) + 2e^- & \longrightarrow H_2(g) & E° = 0.00 \text{ V} \\ Cu(s) & \longrightarrow Cu^{2+}(aq) + 2e^- & E° = -0.34 \text{ V} \\ \hline 2H^+(aq) + Cu(s) & \longrightarrow Cu^{2+}(aq) + H_2(g) & E° = -0.34 \text{ V} \end{array}$$

For zinc

$$\begin{array}{llr} 2H^+(aq) + 2e^- & \longrightarrow H_2(g) & E° = 0.00 \text{ V} \\ Zn(s) & \longrightarrow Zn^{2+}(aq) + 2e^- & E° = 0.76 \text{ V} \\ \hline 2H^+(aq) + Zn(s) & \longrightarrow Zn^{2+}(aq) + H_2(g) & E° = 0.76 \text{ V} \end{array}$$

These simple calculations tell us that zinc can be oxidized by hydrogen ions, producing H_2 gas, since the reaction has a positive potential as written. The negative sign for the voltage of the reaction of copper with hydrogen ions means that acid cannot dissolve the copper by oxidizing it. It is no wonder that water pipes are usually made of copper.

We can predict the direction of spontaneous reaction for any redox process under standard conditions from a table of standard reduction potentials such as that given Table 17.1. This process is illustrated in Example 17.8.

Example 17.8 Using Standard Potentials to Predict Spontaneity

For each of the following reactions, calculate the standard potential, and state the direction in which the reaction proceeds spontaneously.

(a) $2Al(s) + 6H^+(aq) \longrightarrow 2Al^{3+}(aq) + 3H_2(g)$
(b) $2Fe^{2+}(aq) + Cu^{2+}(aq) \longrightarrow 2Fe^{3+}(aq) + Cu(s)$
(c) $2Cr^{2+}(aq) + Br_2(\ell) \longrightarrow 2Cr^{3+}(aq) + 2Br^-(aq)$

Solution

(a) In this reaction the aluminum is oxidized (its oxidation state increases from 0 to +3), so the aluminum half-reaction in the table of *reduction* potentials must be reversed. Therefore, Al(s) will be found on the right side of the half-reaction in the table. The tabulated voltage for

$$Al^{3+}(aq) + 3e^- \longrightarrow Al(s)$$

is −1.66 V, so for the reverse process the potential is +1.66 V. The reduction of hydrogen ions has a potential of 0.00 V, so under standard conditions the voltage of the equation as written is

$$E° = 1.66 + 0.00 = 1.66 \text{ V}$$

Since the calculated voltage is positive, the reaction is spontaneous in the direction shown in the equation.

(b) Since Fe^{2+} is oxidized, we look for this species on the right-hand side of the half-reactions in the table, and find a voltage of +0.77 V for the reduction of Fe^{3+}. The oxidation of Fe^{2+} therefore has a potential of −0.77 V. The copper is reduced (its oxidation state changes from +2 to 0), and the potential of +0.34 V is obtained from the table. Combining the potentials for the two half-reactions gives

$$E° = 0.34 - 0.77 = -0.43 \text{ V}$$

The negative sign of the potential shows that this reaction proceeds spontaneously in the reverse direction.

(c) From Table 7.1, the oxidation of Cr^{2+} has a potential of +0.41 V, and the reduction of Br_2 under standard conditions has a potential of +1.06 V. The potential of this reaction is

$$E° = 0.41 + 1.06 = 1.47 \text{ V}$$

so the reaction proceeds spontaneously as written.

Understanding For the equation

$$3Ag(s) + NO_3^-(aq) + 4H^+(aq) \longrightarrow NO(g) + 2H_2O(\ell) + 3Ag^+(aq)$$

calculate the standard potential, and state whether the reaction is spontaneous as written or spontaneous in the reverse direction.

Answer: +0.16 V; spontaneous as written

17.5 Cell Potentials, ΔG, and K

Objectives

- To relate $E°$, $\Delta G°$, and K_{eq}
- To calculate the equilibrium constant for a reaction from the standard potentials of voltaic cells

The spontaneity of a redox reaction is related to the sign of the voltage of a voltaic cell based on that reaction. If the potential is positive, the reaction proceeds spontaneously as written under standard conditions. In Chapter 15, a negative value of the Gibbs free energy, ΔG, was shown to indicate spontaneity of any chemical reaction. In this section, the relationships among free energy changes, equilibrium constants, and standard cell potentials are developed and used.

Since both the sign of the cell potential and that of the free energy indicate the spontaneity of reactions, we might expect to find a simple relationship between these two quantities. Indeed there is such a relationship; it is

$$\Delta G = -nFE \qquad [17.4]$$

When all substances are in their standard states, both a positive standard potential and a negative standard free energy change indicate that the reaction is spontaneous in the direction written.

where n is the number of moles of electrons transferred in the redox equation, and F is the **faraday,** the negative electrical charge on one mole of electrons. The value of the faraday is

$$1 \text{ F} = 96{,}485 \frac{\text{coulombs}}{\text{mole e}^-}$$

Three significant figures are sufficient for our purposes, so a value of 9.65×10^4 C/mol is used in calculations.

Note that in Equation 17.4 the number of electrons that are transferred must be included. In using Equation 17.4, if E is expressed in volts (J/C), then ΔG will be in joules.

Example 17.9 Standard Free Energy Change from Standard Cell Potential

Use the standard reduction potentials in Table 17.1 to calculate the standard free energy change for the reaction

$$2Fe^{3+}(aq) + 2I^-(aq) \longrightarrow 2Fe^{2+}(aq) + I_2(s)$$

Solution From the table of standard reduction potentials, the two half-reactions involved in this equation are

$$2Fe^{3+}(aq) + 2e^- \longrightarrow 2Fe^{2+}(aq) \qquad E^\circ = +0.77 \text{ V}$$

$$2I^-(aq) \longrightarrow I_2(s) + 2e^- \qquad E^\circ = -0.54 \text{ V}$$

so the voltage for the oxidation of iodide by iron(III) is

$$E^\circ = +0.77 - 0.54 = +0.23 \text{ V}$$

In the chemical equation two electrons are transferred, so we substitute $n = 2$ and the voltage of +0.23 V into Equation 17.4.

$$\Delta G^\circ = -nFE^\circ = -(2 \text{ mol e}^-)\left(9.65 \times 10^4 \frac{\text{C}}{\text{mol e}^-}\right)\left(+0.23 \frac{\text{J}}{\text{C}}\right)$$

$$\Delta G^\circ = -4.4 \times 10^4 \text{ J} = -44 \text{ kJ}$$

In showing the units of the quantities, we used the equivalent unit J/C instead of volts for the cell potential. This redox reaction shows that iron(III) iodide cannot be prepared in aqueous solution, since the spontaneous redox reaction between the ions consumes both Fe^{3+} and I^-.

Understanding From the standard reduction potentials in Table 17.1, find the standard free energy change for the reaction

$$Ni(s) + Cl_2(g) \longrightarrow Ni^{2+}(aq) + 2Cl^-(aq)$$

Answer: $\Delta G^\circ = -311$ kJ

Relation of E° to K_{eq}

As we showed in Section 15.4, the relationship between the equilibrium constant for a chemical reaction and its standard free energy change is

$$\Delta G^\circ = -RT \ln K_{eq} \quad [15.9]$$

Combining this equation with Equation 17.4 provides a direct relationship between the standard cell potential and the equilibrium constant for any redox reaction:

The value of the equilibrium constant for a redox reaction is related to the standard potential of the reaction by the equation

$E^\circ = (2.303RT/nF) \log K_{eq}$.

$$-nFE^\circ = -RT \ln K_{eq}$$

$$E^\circ = \frac{RT}{nF} \ln K_{eq} = \frac{2.303RT}{nF} \log K_{eq} \quad [17.5]$$

As shown in Equation 17.5, we can substitute 2.303 log K_{eq} for ln K_{eq}. At 25 °C (298 K) the value of $2.303RT/F$ is 0.0591 V mol e⁻.

Example 17.10 Calculate K_{eq} from E°

Determine the equilibrium constant for the reaction

$$Sn^{2+}(aq) + Ni(s) \rightleftharpoons Sn(s) + Ni^{2+}(aq) \qquad E^\circ = +0.11 \text{ V}$$

Solution The standard voltage of the reaction is given, and n is 2. Rearrange Equation 17.5 to solve for log K_{eq}, and substitute the values of n, E°, and $2.303RT/F$.

$$\log K_{eq} = \frac{nFE^\circ}{2.303RT} = \frac{2 \times 0.11}{0.0591} = 3.7$$

$$K_{eq} = 10^{3.7} = 5 \times 10^3$$

Notice that a relatively small potential (0.11 V) corresponds to a fairly large equilibrium constant (5000).

Understanding Find the equilibrium constant for the reaction

$$Sn^{4+}(aq) + U^{4+}(aq) + 2H_2O(\ell) \rightleftharpoons UO_2^{2+}(aq) + Sn^{2+}(aq) + 4H^+(aq)$$

E° for the reaction is −0.176 V.

Answer: $K_{eq} = 1.1 \times 10^{-6}$

17.6 Dependence of Voltage on Concentration — The Nernst Equation

Objectives

- To find the voltage of a cell under nonstandard conditions of concentration
- To use the Nernst equation to find the potential of an electrode under nonstandard conditions

Just as the free energy change for a reaction depends on concentration, the voltage must also change when concentrations of reactants and products change. In this section, the dependence of potentials for redox reactions and half-reactions is developed.

In Chapter 15 we saw that

$$\Delta G = \Delta G^\circ + RT \ln Q \quad [15.8]$$

where Q is the reaction quotient. (Remember, the reaction quotient has the same form as the equilibrium constant of the chemical reaction, but contains nonequilibrium values for the concentrations.) We replace the free energy changes in this equation with $-nFE$ and $-nFE°$ respectively.

$$-nFE = -nFE° + RT \ln Q$$

The equation obtained by dividing by $-nF$ is called the **Nernst equation.**

$$E = E° - \frac{RT}{nF} \ln Q = E° - \frac{2.303RT}{nF} \log Q \qquad [17.6]$$

Since the value of $2.303RT/F$ is 0.0591 V mol at 25 °C, the temperature we used for the standard potentials, the Nernst equation is rewritten as

$$E = E° - \frac{0.0591}{n} \log Q$$

This equation shows how the voltage of any cell changes when the concentrations of reactants and products vary. When using Equation 17.6, the value of Q must be calculated by expressing the concentrations of substances in solution as molarity, and the concentrations of gases as their partial pressures in atmospheres. We illustrate the use of the Nernst equation in Example 17.11.

Example 17.11 Cell Voltage and Concentration

A voltaic cell consists of a half-cell of zinc metal in a solution with $[Zn^{2+}] = 2.0\ M$, and a half-cell of copper metal immersed in a solution containing Cu^{2+} at a concentration of $3.6 \times 10^{-4}\ M$. What is the voltage of this cell at 298 K?

Solution Since zinc is above copper in Table 17.1, the cell reaction is the oxidation of zinc metal by copper(II).

$$Zn(s) + Cu^{2+}(aq) \longrightarrow Zn^{2+}(aq) + Cu(s)$$

From the standard reduction potentials (Table 17.1), calculate the standard cell potential.

$$E° = 0.76 + 0.34 = +1.10\ V$$

In the cell reaction, the concentration expression for Q is

$$Q = \frac{[Zn^{2+}]}{[Cu^{2+}]}$$

When this concentration expression is substituted for Q in the Nernst equation, we obtain

$$E = E° - \frac{0.0591}{n} \log \frac{[Zn^{2+}]}{[Cu^{2+}]}$$

We substitute the concentrations given, and 2 for the electrons transferred.

$$E = 1.10 - \frac{0.0591}{2} \log \frac{2.0}{3.6 \times 10^{-4}} = 0.99\ V$$

Note that E is less than $E°$, because the reaction quotient has a value greater than one.

Understanding What is the voltage of the zinc-copper cell when $[Zn^{2+}] = 1.0 \times 10^{-3}\ M$ and $[Cu^{2+}] = 0.500\ M$?

Answer: $E = 1.18$ V

The potential of an electrode under nonstandard conditions can be calculated by applying the Nernst equation to the half-cell reaction. For example, we can calculate the reduction potential of a silver/silver chloride electrode that is immersed in a 1.5×10^{-3} *M* sodium chloride solution. The reduction half-reaction is

$$AgCl(s) + e^- \longrightarrow Ag(s) + Cl^-(aq) \qquad E° = 0.222\ V$$

For the silver/silver chloride half-reaction the only soluble species is the Cl^- ion on the product side, so the Nernst equation at 25 °C is

$$E = E° - \frac{0.0591}{n} \log [Cl^-] = +0.222 - \frac{0.0591}{1} \log [Cl^-]$$

After we substitute 1.5×10^{-3} *M* for the concentration of Cl^-, the electrode potential is found.

$$E = +0.222 - \frac{0.0591}{1} \log (1.5 \times 10^{-3}) = +0.389\ V$$

This is the potential of the Ag/AgCl electrode under the conditions given, when it is combined in a cell with the standard hydrogen half-cell.

Example 17.12 Electrode Potential under Nonstandard Conditions

Find the reduction potential of a hydrogen electrode when the hydrogen pressure is 1.5 atm, and $[H^+] = 4.7 \times 10^{-3}$ *M*.

Solution From the table of standard reduction potentials we get

$$2H^+(aq) + 2e^- \longrightarrow H_2(g) \qquad E° = +0.000\ V$$

In the reaction quotient the concentration of hydrogen is expressed as its partial pressure (in atmospheres), and the concentration of hydrogen ions is squared, because its coefficient in the half-reaction is 2. Thus the Nernst equation is

$$E = +0.000 - \frac{0.0591}{2} \log \frac{P_{H_2}}{[H^+]^2}$$

Substitution of the partial pressure of hydrogen and the concentration of hydrogen ions yields

$$E = +0.000 - \frac{0.0591}{2} \log \frac{1.5}{(4.7 \times 10^{-3})^2} = -0.143\ V$$

Understanding What is the potential of a platinum wire that is immersed in a solution containing $[Sn^{2+}] = 2.5 \times 10^{-2}$ *M* and $[Sn^{4+}] = 7.5 \times 10^{-4}$ *M*?

Answer: $E = +0.10$ V

17.7 Applications of Voltaic Cells in Chemistry

Objectives

- To measure the concentrations of species in solution, using cell potentials
- To determine equilibrium constants for non-redox reactions from the potentials of voltaic cells

There are many applications of voltaic cells in the chemistry laboratory. The ease with which cell voltages can be measured and the electrical signals they produce make them very desirable for automation and direct interfacing with computers. Cells have been designed to measure the concentrations of species in solution. These devices are in widespread use, ranging from pH meters to electrodes that measure the concentration of sugar in the blood of a person with diabetes. In addition, voltaic cells can be designed to determine the equilibrium constants of systems that are not redox reactions.

Measurement of the Concentrations of Ions in Solution

The concentration dependence of cell voltages provides a convenient means of measuring the concentrations of species in solution. An **indicating electrode** is one whose potential depends on the concentration of a particular species in solution. Voltage measurements provide a rapid, easily automated means of determining concentrations. Furthermore, the potential of an electrode can measure much lower concentrations than can be found by techniques such as titration.

We illustrate the analytical use of cell potentials by using the Ag/AgCl half-cell to measure the concentration of chloride ions in a test solution. The concentration dependence of the potential of a silver/silver chloride electrode, calculated using the Nernst equation, is shown in Figure 17.8. The silver/silver chloride electrode is combined with a half-cell called a **reference electrode,** which has an accurately known potential that is independent of the composition of the test solution. The measured voltage of the cell can then be used to calculate the concentration of chloride ions in the test solution.

The potential of an electrode provides a rapid means of measuring the concentration of a species in solution.

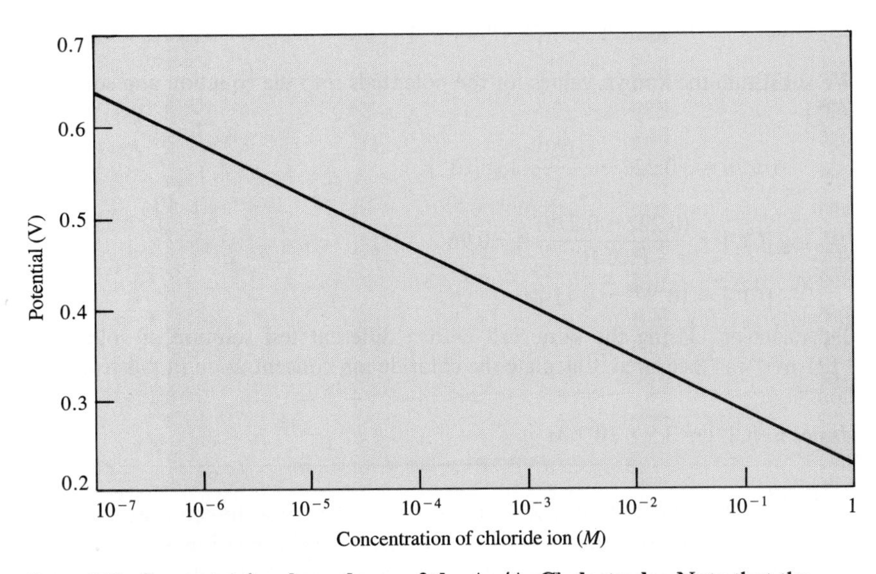

Figure 17.8 Concentration dependence of the Ag/AgCl electrode Note that the horizontal axis in this graph is log [Cl^-]. The potential of a silver/silver chloride electrode changes linearly with the logarithm of the concentration of chloride ion.

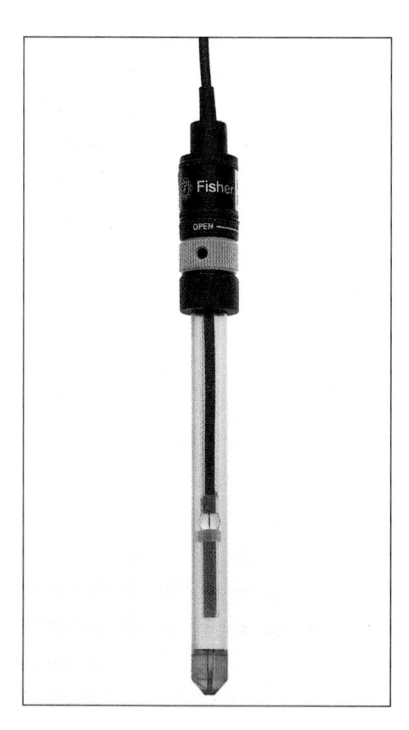

Figure 17.9 A saturated calomel electrode The saturated calomel electrode consists of the mercury/mercury(I) chloride couple in a saturated solution of potassium chloride. It is used as a reference electrode. The tip of the probe has a built-in salt bridge that connects this half-cell to the test solution.

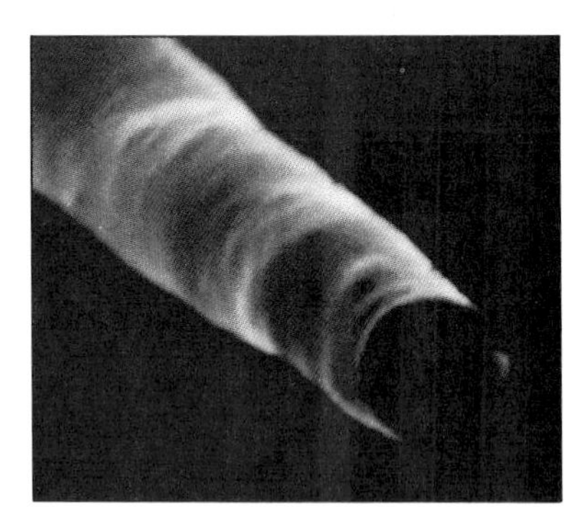

Figure 17.10 Miniature electrode The scanning electron micrograph shows one of the smallest electrodes that has been produced. It is about 400 nm in diameter and is coated with a polymer film. The electrode was designed to monitor chemical changes in and around a single nerve cell.

One of the most common reference electrodes is the saturated calomel electrode, which consists of a mercury/mercury(I) chloride electrode in a saturated solution of potassium chloride.

$$Hg_2Cl_2(s) + 2e^- \longrightarrow 2Hg(\ell) + 2Cl^-(\text{saturated KCl})$$

The saturated calomel electrode has a potential of +0.2458 V at 25 °C. Figure 17.9 shows a commercial saturated calomel electrode.

Probes are available that contain both the reference and test electrodes and are small enough to insert in a blood vessel of a test animal. Miniature voltaic cells employing these electrodes are used to monitor the level of electrolytes in living subjects on a real time basis (Figure 17.10).

Example 17.13 Determining Concentration from Cell Voltage

A probe consisting of a silver/silver chloride electrode and a saturated calomel half-cell is placed in a test solution. The cell voltage is +33 mV (Ag/AgCl electrode positive). What is the concentration of chloride ion in the solution?

Solution Since the silver/silver chloride electrode is positive, the oxidation occurs in the calomel electrode, so the voltage of the cell is the potential of the Ag/AgCl electrode minus the potential of the SCE (saturated calomel electrode).

$$E_{cell} = E_{Ag/AgCl} - E_{SCE} = E_{Ag/AgCl} - 0.2458 \text{ V}$$

Substituting the measured cell voltage of +33 mV, or 0.033 V, and solving for the potential of the Ag/AgCl electrode, we get

$$E_{Ag/AgCl} = 0.033 + 0.2458 = +0.279 \text{ V}$$

The half-reaction for the silver/silver chloride couple is

$$AgCl(s) + e^- \longrightarrow Ag(s) + Cl^-(aq)$$

so the Nernst equation for the Ag/AgCl electrode ($E° = +0.222$ V) is

$$E_{Ag/AgCl} = E°_{Ag/AgCl} - \frac{0.0591}{1} \log [Cl^-]$$

We substitute the known values for the potentials into the equation and solve for $[Cl^-]$.

$$+0.279 = +0.222 - \frac{0.0591}{1} \log [Cl^-]$$

$$\log [Cl^-] = \frac{(0.222 - 0.279)}{0.0591} = -0.96$$

$$[Cl^-] = 10^{-0.96} = 0.11 \ M$$

Understanding Using the same cell with a different test solution, a voltage of +121 mV was measured. Calculate the chloride ion concentration in this test solution.

Answer: $[Cl^-] = 3.5 \times 10^{-3} \ M$

The **pH meter** is the most common example of a device that uses cell voltages to determine concentrations of ions in solution (Figure 17.11). The probe that is connected to the meter consists of a reference electrode and an electrode that obeys the Nernst equation for hydrogen ions. The potential of the cell is related to the hydrogen ion concentration by the equation

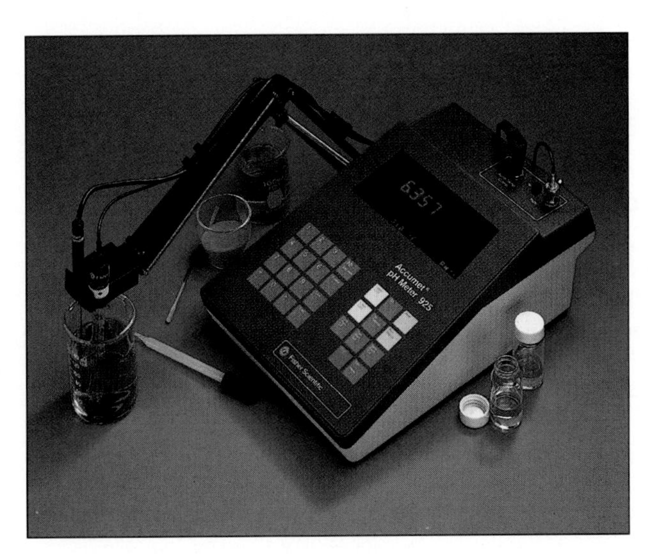

Figure 17.11 A pH meter A pH meter provides a direct reading of the pH of a solution, by measuring the voltage of a cell. The voltage changes by 59.1 mV for a change of one unit of pH.

$$E = k - 0.0591 \log\left(\frac{1}{[H^+]}\right) = k - 0.0591 \text{ pH}$$

where k is a constant that depends upon the particular electrodes used in the measurement. The meter itself is a voltmeter, with a scale marked in pH units rather than volts. A change of one pH unit causes the potential of the electrode to change by 0.0591 V.

For many pH-sensitive electrodes, the value of k changes slightly with time. A control is provided on the pH meter that is adjusted to produce a direct reading of the pH. In an experiment using a pH meter, the electrodes are first placed in a standard buffer solution of known pH, and the meter is adjusted to give the correct reading for the pH. The voltage produced by the cell when the electrodes are placed in a test solution provides a direct reading of the pH.

Determining Equilibrium Constants

In Section 17.5 the standard potential of a voltaic cell was used to determine the equilibrium constant for the redox reaction that occurred in the cell. We can also design cells for determining the values of equilibrium constants for systems that are not redox reactions, such as solubility product constants. Consider the solubility equilibrium of silver bromide.

$$AgBr(s) \rightleftharpoons Ag^+(aq) + Br^-(aq) \qquad K_{sp} = [Ag^+][Br^-]$$

A cell can be constructed that has this reaction as the cell reaction. One half-cell consists of a piece of silver in a solution of $AgNO_3$ of known concentration. The second half-cell is a silver electrode in a solution of NaBr that is saturated with AgBr. The two half-reactions, both written as reductions, are

$$Ag^+(aq) + e^- \longrightarrow Ag(s) \qquad E^\circ = +0.80 \text{ V}$$

$$AgBr(s) + e^- \longrightarrow Ag(s) + Br^-(aq) \qquad E^\circ = +0.071 \text{ V}$$

By reversing the first of these and then adding the two half-reactions, we obtain the equation for dissolving silver bromide.

$$AgBr(s) \rightleftharpoons Ag^+(aq) + Br^-(aq) \qquad E° = -0.73\ V$$

The equilibrium constant for this cell reaction can be found by using the relationship (derived from Equation 17.5)

$$2.303RT \log K = nFE° \qquad [17.7]$$

All of the quantities in Equation 17.7 are known, except for the equilibrium constant. Solve for log K, and then substitute the values that are known.

$$\log K = \frac{nFE°}{2.303RT} = \frac{1(9.65 \times 10^4)(-0.73)}{2.303(8.314)(298)} = -12.3$$

$$K = 10^{-12.3} = 5 \times 10^{-13}$$

The equilibrium constant for the cell reaction is K_{sp} for silver bromide. Thus we find that

$$K_{sp}(AgBr) = 5 \times 10^{-13}$$

Many of the solubility product constants that were used in Section 12.6 have been determined by using the voltages of voltaic cells.

17.8 Commercial Voltaic Cells

Objectives

- To describe the features that make voltaic cells useful as portable energy sources
- To describe the chemical reactions that occur in common commercial cells
- To describe the advantages and disadvantages of fuel cells as an alternative to the combustion of hydrocarbon fuels

Voltaic cells offer a very convenient source of portable energy. The cells described in the previous sections are not very useful as sources of electrical energy, because they are fragile, involve liquids that can spill, and usually have a very high internal resistance. In cells that have high internal resistance much of the chemical energy released is converted into useless heat within the cell rather than into electricity. As a result only small currents can be produced by these cells, and they are suitable mainly for determining thermodynamic quantities such as $\Delta G°$ and equilibrium constants. The design of a voltaic cell that can provide high currents for short times often requires a great deal of ingenuity.

A **battery** consists of two or more voltaic cells, with the negative electrode of one cell connected to the positive electrode of the next cell. The voltage produced by a battery is the sum of the voltages of the individual cells. Although technically incorrect, single voltaic cells are sometimes referred to as batteries.

The Lead Storage Battery

The lead storage battery has been used in automobiles since 1915 to provide the energy to operate the starter motor. The most common batteries produce either 12 or 6 volts, depending on whether six or three individual cells are combined. In each cell the oxidation reaction involves the conversion of lead to lead sulfate, and the reduction reaction produces lead sulfate from lead dioxide. The electrolyte solution in the cell is sulfuric acid. The half-reactions are

$$Pb(s) + HSO_4^- \longrightarrow PbSO_4(s) + H^+ + 2e^- \qquad E° = +0.356 \text{ V}$$

$$PbO_2(s) + HSO_4^- + 3H^+ + 2e^- \longrightarrow PbSO_4(s) + 2H_2O \qquad E° = +1.685 \text{ V}$$

The overall cell reaction is

$$Pb(s) + PbO_2(s) + 2HSO_4^- + 2H^+ \longrightarrow 2PbSO_4(s) + 2H_2O \qquad E° = 2.041 \text{ V}$$

No salt bridge is needed to separate the oxidizing agent and reducing agent, since they are both solids and cannot come in direct contact with each other (Figure 17.12). Without a salt bridge the internal resistance of the cell is very low, allowing the large currents needed for an energy supply. The Nernst equation for the cell is

$$E = E° - \frac{0.0591}{2} \log \frac{1}{[H^+]^2[HSO_4^-]^2}$$

As electrical energy is drawn from the cell, the voltage decreases because the sulfuric acid is consumed as it produces lead sulfate. The electrodes are designed so the lead sulfate that is formed adheres to the surface; this feature allows the battery to be recharged. In an automobile, some of the mechanical energy produced by the engine is converted into an electric current, which reverses the cell reaction by electrolysis (see Section 17.9). The battery can usually be discharged and recharged several thousand times before failure of a cell occurs.

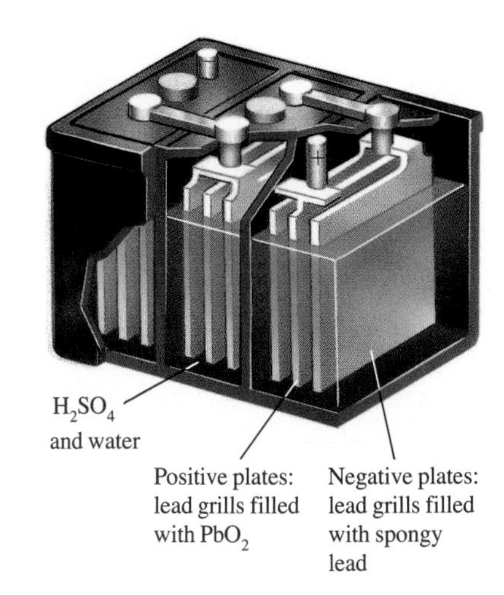

Figure 17.12 The lead storage cell The design of the lead storage cell keeps the solid products of the reaction in contact with the electrodes, one of the conditions necessary in a rechargeable battery. In a 6-V battery, three of the cells are connected together in series.

Dry Cells

The common dry cell is used to power flashlights, watches, calculators, and many other small portable devices. The acid version of this cell was patented by the French chemist George Leclanché in 1866. The cell consists of a zinc case that serves as the negative electrode, and a carbon rod in the center that serves as an inert positive electrode (Figure 17.13). The space between the electrodes contains a moist paste of MnO_2, NH_4Cl, and carbon. The electrode reactions are quite complex, but are generally represented by

$$Zn(s) \longrightarrow Zn^{2+}(aq) + 2e^-$$

$$2MnO_2(s) + 2NH_4^+(aq) + 2e^- \longrightarrow Mn_2O_3(s) + 2NH_3(aq) + H_2O(\ell)$$

The Leclanché cell produces a voltage of about 1.56 V when it is new. As the cell is used, the concentrations of the soluble products (Zn^{2+} and NH_3) increase, causing the voltage produced by the cell to decrease gradually.

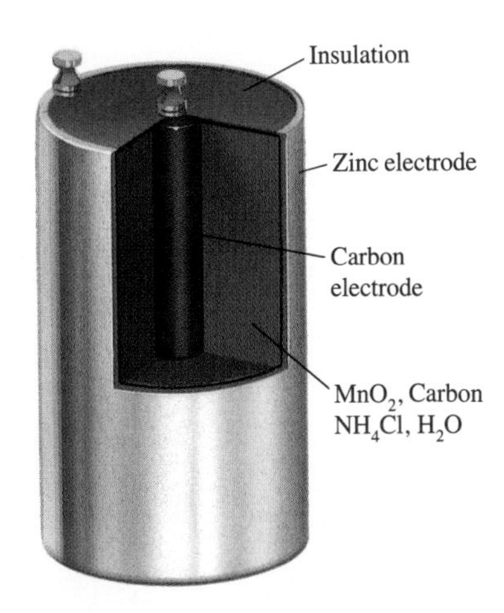

Figure 17.13 The LeClanché dry cell The initial voltage is about 1.56 V, but it decreases as the cell discharges.

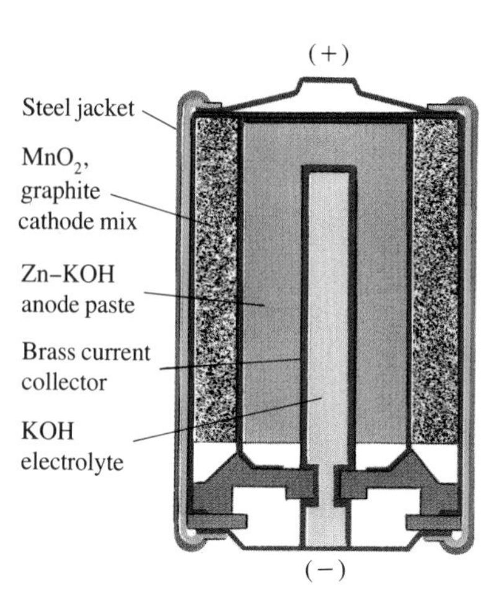

Figure 17.14 The alkaline LeClanché cell The alkaline cell has a voltage of 1.54 V that remains constant throughout the useful life of the cell.

An alkaline version of this same cell replaces NH_4Cl with KOH as the electrolyte (Figure 17.14). Under alkaline conditions the half-reactions can be represented by

$$Zn(s) + 2OH^-(aq) \longrightarrow ZnO(s) + H_2O(\ell) + 2e^-$$

$$2MnO_2(s) + H_2O(\ell) + 2e^- \longrightarrow Mn_2O_3(s) + 2OH^-(aq)$$

The overall cell reaction is

$$Zn(s) + 2MnO_2(s) \longrightarrow ZnO(s) + Mn_2O_3(s)$$

Unlike the acid cell, the alkaline cell keeps a constant voltage as it discharges. Since the reactants and products are all solids, the value of the reaction quotient in the Nernst equation is 1, so the cell voltage does not change. Although they are more expensive to produce, alkaline cells have a longer useful lifetime than do acid cells. In both the acid and alkaline cells, the outer zinc case is consumed. Toward the end of the useful life of a Leclanché cell, holes may develop in the casing, exposing the surroundings to the rather corrosive contents. Modern designs of dry cells no longer use the zinc electrode as the outer casing, so leaking is seldom a problem.

Many other dry cells have been developed in the past few decades. The mercury cell, for example, uses the cell reaction

$$Zn(s) + HgO(s) \longrightarrow ZnO(s) + Hg(\ell) \qquad E° = 1.35 \text{ V}$$

Since all of the reactants and products are in their standard states, the potential of this cell remains constant at 1.35 V, until one of the reactants is completely consumed. This feature makes mercury cells useful in applications in which a constant voltage is essential.

None of the dry cells we have mentioned can be recharged, so they must be replaced frequently. The nickel-cadmium cell is a dry cell that can be recharged, and for that reason has become popular for use in battery-operated tools that require moderately large power. As in the lead storage cell, the products of the cell reaction adhere to the electrodes, and the cell reaction can be reversed by electrolysis. The oxidized nickel species is quite complex and not fully understood, but the electrode reactions are approximated by

$$Cd(s) + 2OH^-(aq) \longrightarrow Cd(OH)_2(s) + 2e^-$$

$$NiO(OH)(s) + H_2O(\ell) + e^- \longrightarrow Ni(OH)_2(s) + OH^-(aq)$$

Fuel Cells

A **fuel cell** is a voltaic cell in which the reactants are continuously supplied, and generally the products of the cell reaction are continuously removed. A great deal of research has been done in recent years to develop practical and economical fuel cells. A fuel cell using the combustion reaction of hydrogen has been developed by the U.S. space program. The electrode reactions are

$$H_2(g) + 2OH^-(aq) \longrightarrow 2H_2O(\ell) + 2e^-$$

$$O_2(g) + 2H_2O(\ell) + 4e^- \longrightarrow 4OH^-(aq)$$

Power plants now burn hydrocarbon fuels to produce steam for turbines that turn electrical generators. Only about 35% of the total energy release is electricity. A fuel cell that consumes methane or other hydrocarbons and oxygen could convert a much larger fraction of the available chemical energy into electricity than is obtained by combustion of conventional fuels like petroleum products and coal. While such cells have been available for a number of years, they are still too expensive to compete economically with the present methods for energy production.

17.9 Electrolysis

Objective

- To predict the reactions that occur in the electrolysis of aqueous solutions, using the standard potentials for half-reactions and considering the effects of overvoltage

Voltaic cells produce electrical energy directly from spontaneous oxidation-reduction reactions. It is also possible to use electrical energy to cause a chemical reaction to occur, by a process called electrolysis. **Electrolysis** is the passage of an electric current through an electrolyte, causing an otherwise nonspontaneous oxidation-reduction reaction to occur. Electrolysis is a very powerful method that is used in the production of very strong oxidizing and reducing agents. For example, the passage of a current through a molten mixture of KHF_2 and anhydrous HF causes the HF to decompose into the elements.

$$2HF(\ell) \longrightarrow H_2(g) + F_2(g)$$

The reaction occurs as written only if energy is supplied; without the consumption of electrical energy, the reverse chemical reaction is spontaneous.

An **electrolytic cell** consists of two electrodes in a molten salt or an electrolyte solution. A battery or other voltage source is attached across the two electrodes, as shown in Figure 17.15. The battery serves as an electron pump, drawing electrons in at the positive electrode and forcing them out at the negative electrode. A reduction half-reaction occurs at the electrode that is attached to the negative terminal of the battery, using the electrons pro-

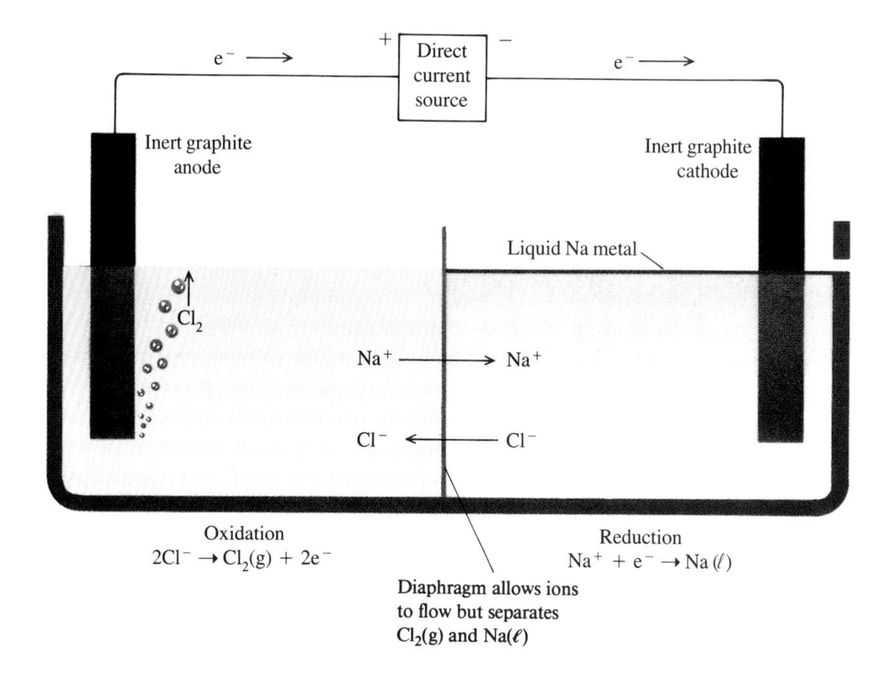

Figure 17.15 An electrolysis cell Schematic diagram of an electrolysis cell in which molten sodium chloride is decomposed into the elements. The sodium metal floats on the molten salt and is removed as a liquid.

vided by the battery. The electrons that enter the battery at its positive terminal must be obtained from an oxidation half-reaction in the electrolysis cell. In any cell, the **anode** is the electrode where oxidation occurs, and the **cathode** is the electrode at which reduction occurs. Thus, in the case of the electrolysis of molten sodium chloride depicted in Figure 17.15, the processes at the two electrodes are

$$2Cl^-(\ell) \longrightarrow Cl_2(g) + 2e^- \qquad \text{(anode)}$$

$$Na^+(\ell) + e^- \longrightarrow Na(\ell) \qquad \text{(cathode)}$$

Electrolysis in Aqueous Solutions

Electrolysis of aqueous electrolyte solutions is complicated by the fact that water can be oxidized and reduced in addition to (or instead of) the ions in solution:

$$2H_2O(\ell) + 2e^- \longrightarrow H_2(g) + 2OH^-(aq) \qquad \text{(reduction)}$$

$$2H_2O(\ell) \longrightarrow 4H^+(aq) + O_2(g) + 4e^- \qquad \text{(oxidation)}$$

Also, many metal electrodes are oxidized to ions in solution.

It is usually necessary to consider more than one oxidation and one reduction in determining the expected products of an electrolysis in aqueous solution. In an electrolysis, the most easily oxidized and most easily reduced species react at the anode and cathode, respectively. The standard potentials for half-reactions are used to determine the reaction that occurs at each electrode during electrolysis. In the electrolysis of aqueous sodium fluoride, the possible oxidations are

$$2F^-(aq) \longrightarrow F_2(g) + 2e^- \qquad E^\circ = -2.87 \text{ V}$$

$$2H_2O(\ell) \longrightarrow 4H^+(aq) + O_2(g) + 4e^- \qquad E^\circ = -1.23 \text{ V}$$

The most easily oxidized species in an electrolysis is the one with the most positive oxidation potential (the negative of the reduction potential). The more positive potential for the oxidation of water (by 1.64 V) means that this is the oxidation half-reaction that should occur in the electrolysis.

Two reductions are also possible in the electrolysis of aqueous sodium fluoride. Both the cation, Na^+, and water might be reduced.

$$Na^+(aq) + e^- \longrightarrow Na(s) \qquad E^\circ = -2.71 \text{ V}$$

$$2H_2O(\ell) + 2e^- \longrightarrow H_2(g) + 2OH^-(aq) \qquad E^\circ = -0.83 \text{ V}$$

The species that react in an electrolysis cell are those that are easiest to oxidize and reduce.

The most easily reduced species in an electrolysis is the one with the highest (most positive) reduction potential. Water is much more easily reduced than sodium ions, since its reduction potential is more positive by 1.88 V.

The net chemical reaction that occurs in the electrolysis of sodium fluoride solution is the decomposition of water into the elements.

$$2H_2O(\ell) \longrightarrow 2H_2(g) + O_2(g)$$

The sodium fluoride undergoes no chemical change—it simply serves as an electrolyte that allows electrical currents to flow through the solution.

Overvoltage

When a sodium chloride solution is subjected to electrolysis, water is reduced to hydrogen at the cathode, just as we predicted for NaF. At the anode, the oxidation of chloride ions must be considered, in addition to the oxidation of water. The half-reactions, along with their potentials, are

$$2Cl^-(aq) \longrightarrow Cl_2(g) + 2e^- \qquad E^\circ = -1.36 \text{ V}$$

$$2H_2O(\ell) \longrightarrow 4H^+(aq) + O_2(g) + 4e^- \qquad E^\circ = -1.23 \text{ V}$$

The more positive potential for the oxidation of water indicates that gaseous oxygen should be the product formed at the anode. However, the voltage required to cause a reaction at an electrode is sometimes considerably greater than the calculated potential. The **overvoltage** is the difference between the calculated potential and the potential needed to cause reaction. The overvoltage is probably caused by slow electron transfer at the electrode surface and is similar to an activation energy. The size of the overvoltage is very dependent on the electrode material and the particular electrode reaction. The effect of overvoltage on the potential of a half-reaction is to make the required applied potential more negative, regardless of whether it is an oxidation or a reduction. Overvoltages must be measured experimentally, and they are often quite high for reactions that produce gases.

The overvoltages for both the oxidation and reduction of water are unusually high, and at most inert electrodes they are about 0.40 V. Because of the overvoltage, the effective potential needed for the oxidation of water to oxygen is approximately −1.6 V, considerably more negative than for the oxidation of chloride ions. When overvoltage effects are considered, the actual electrolysis reaction in a sodium chloride solution is

$$2Cl^-(aq) + 2H_2O(\ell) \longrightarrow H_2(g) + Cl_2(g) + 2OH^-(aq)$$

The electrolysis of brine (a concentrated sodium chloride solution) is used in the commercial manufacture of chlorine, and sodium hydroxide is formed as a by-product of the electrolysis.

Electrolysis Using Active Electrodes

In the previous examples of electrolysis, we assumed that the electrodes were inert and merely served as conductors of electrons needed in the oxidation and reduction of the species in solution. When the metal is sufficiently reactive, it may be oxidized to its ions in solution. For example, metallic copper is more easily oxidized than is water.

In predicting the products of the electrolysis of an aqueous solution, you must consider the decomposition of water, the overvoltage, and the possible oxidation of the metal of the electrode.

$$Cu(s) \longrightarrow Cu^{2+}(aq) + 2e^- \qquad E° = -0.34 \text{ V}$$

If copper were used as the anode in the electrolysis of an aqueous solution, the electrode would dissolve, forming an aqueous solution of Cu^{2+} ions.

Example 17.14 Predicting Electrolysis Products in Water Solution

Using standard potentials, predict the electrolysis reaction that occurs, and the standard potential of the electrolysis reaction, for each of the following electrolyte solutions:

(a) a solution of nickel sulfate using inert electrodes
(b) a solution of magnesium chloride using inert electrodes
(c) a solution of 1 *M* HCl using copper electrodes

Solution

(a) Since neither sulfate nor Ni^{2+} ions can be oxidized easily, the anode reaction must be the oxidation of water.

$$2H_2O(\ell) \longrightarrow 4H^+(aq) + O_2(g) + 4e^- \qquad E° = -1.23 \text{ V}$$

At the cathode either water ($E° = -0.83$ V) or Ni^{2+} ($E° = -0.25$ V) could be reduced. Since the potential for reducing nickel ions is more positive, the cathode reaction is

$$Ni^{2+}(aq) + 2e^- \longrightarrow Ni(s) \qquad E° = -0.25 \text{ V}$$

The cell reaction expected in this electrolysis is

$$2Ni^{2+}(aq) + 2H_2O(\ell) \longrightarrow 2Ni(s) + 4H^+(aq) + O_2(g)$$

The standard potential of the reaction is

$$E° = -1.23 - 0.25 = -1.48 \text{ V}$$

This indicates that at least 1.48 V from an external source is needed to cause the reaction.

(b) The oxidation of water and Cl^- must be considered at the anode. The oxidation potentials are close enough that the overvoltage of water allows chlorine to form.

$$2Cl^-(aq) \longrightarrow Cl_2(g) + 2e^- \qquad E° = -1.36 \text{ V}$$

At the cathode the possible reactions are

$$2H_2O(\ell) + 2e^- \longrightarrow H_2(g) + 2OH^-(aq) \qquad E° = -0.83 \text{ V}$$

$$Mg^{2+}(aq) + 2e^- \longrightarrow Mg(s) \qquad E° = -2.37 \text{ V}$$

The magnesium potential is sufficiently negative that even with a hydrogen overvoltage of 0.4 V, the reduction of water is still favored. The overall reaction expected is

$$2Cl^-(aq) + 2H_2O(\ell) \longrightarrow H_2(g) + Cl_2(g) + 2OH^-(aq)$$

$$E° = -1.36 - 0.83 = -2.19 \text{ V}$$

The reaction does not proceed until the applied voltage exceeds 2.19 V. As the reaction proceeds, the hydroxide ion concentration increases and could become high enough to precipitate the Mg^{2+} ions from solution as $Mg(OH)_2$.

(c) At the anode, in addition to the oxidation of water and chloride ion, the metallic copper in the electrode might be oxidized. The three half-reactions are

$$2Cl^-(aq) \longrightarrow Cl_2(g) + 2e^- \qquad E° = -1.36 \text{ V}$$

$$2H_2O(\ell) \longrightarrow 4H^+(aq) + O_2(g) + 4e^- \qquad E° = -1.23 \text{ V}$$

$$Cu(s) \longrightarrow Cu^{2+}(aq) + 2e^- \qquad E° = -0.34 \text{ V}$$

The most favorable reaction (the one with the least negative potential) is by far the oxidation of the copper in the electrode. At the cathode the reduction is

$$2H^+(aq) + 2e^- \longrightarrow H_2(g) \qquad E° = 0.00 \text{ V}$$

The overall cell reaction is

$$Cu(s) + 2H^+(aq) \longrightarrow Cu^{2+}(aq) + H_2(g)$$

$$E° = -0.34 + 0.00 = -0.34 \text{ V}$$

Once again, a product of the cell reaction, Cu^{2+}, will accumulate in the electrolyte. Since Cu^{2+} is more easily reduced than H^+ ions, after a period of time the cathode reaction becomes

$$Cu^{2+}(aq) + 2e^- \longrightarrow Cu(s)$$

Understanding Predict the oxidation-reduction reaction that occurs in the electrolysis of a solution of $ZnBr_2$.

Answer: $2H_2O(\ell) + 2Br^-(aq) \longrightarrow H_2(g) + Br_2(\ell) + 2OH^-(aq)$

The standard voltage calculated for each of the electrolysis reactions in Example 17.14 is negative. All of these reactions are spontaneous in the reverse direction under standard conditions. Before current will flow in these electrolysis cells, a voltage greater than the $E°$ for the reaction must be applied. In actual practice, a voltage considerably greater than the $E°$ must be applied to overcome the electrical resistance of the cell and any overvoltages that may exist.

17.10 Quantitative Aspects of Electrolysis

Objectives

- To relate the electrical charge to the amount of products formed
- To perform stoichiometry calculations involving electrolysis reactions
- To calculate the energy consumed in electrolysis processes from the applied voltage, current, and time of electrolysis

Much of our current understanding of chemistry is based on the quantitative aspects of electrochemical processes. The lifetime of batteries and the cost of manufacturing chemicals by electrolysis depend upon these quantitative relationships. In this section we examine the quantitative relationships between electrical charge and energy to the electrochemical reactions that occur.

When we perform an electrolysis, the amount of electricity that is used determines the quantity of products formed. Consider the nickel produced by electrolysis of a nickel sulfate solution. The half-reaction occurring at the cathode is

$$Ni^{2+}(aq) + 2e^- \longrightarrow Ni(s)$$

One mole of nickel metal (58.69 g) is produced for each two moles of electrons that pass through the solution. Experimentally, we measure the electric current in amperes, and the length of time the current flows, rather than the number of moles of electrons. The SI unit of charge is the coulomb, which is one ampere-second. Therefore, the total electrical charge that passes in an electrolysis is simply the product of the current and the time.

$$\text{charge (coulombs)} = \text{current (amperes)} \times \text{time (seconds)}$$

The product of current and time is the stoichiometric quantity that determines how much product is formed by an electrolysis reaction.

The total charge in an electrolysis serves as the basis for stoichiometry calculations. In performing these stoichiometry calculations it is convenient to replace the unit of ampere with its equivalent of coulombs/second. For example, if a constant current of 0.200 A passes through a nickel sulfate solution for 30.0 minutes, the total charge is the current times the time, or

$$\text{charge} = 30.0\ \cancel{\text{min}} \left(\frac{60\ \text{s}}{\cancel{\text{min}}}\right)\left(\frac{0.200\ \text{C}}{\cancel{\text{s}}}\right) = 3.60 \times 10^2\ \text{coulombs}$$

This total charge is converted to the number of moles of electrons, using the faraday (9.65×10^4 C/mol) as the conversion factor.

$$\text{moles of } e^- = 3.60 \times 10^2\ \cancel{\text{C}} \left(\frac{1\ \text{mol } e^-}{9.65 \times 10^4\ \cancel{\text{C}}}\right) = 3.73 \times 10^{-3}\ \text{mol } e^-$$

From the coefficients in the half-reaction, one mole of nickel is deposited at the cathode for each two moles of electrons, so the mass of nickel produced is

$$\text{mass Ni} = 3.73 \times 10^{-3}\ \cancel{\text{mol } e^-} \left(\frac{1\ \cancel{\text{mol Ni}}}{2\ \cancel{\text{mol } e^-}}\right)\left(\frac{58.69\ \text{g}}{1\ \cancel{\text{mol Ni}}}\right) = 0.109\ \text{g Ni}$$

Example 17.15 Stoichiometry of Electrolysis

A constant current of 0.500 A passes through a silver nitrate solution for 90.0 minutes. What mass of silver metal is deposited at the cathode?

Solution The flow diagram for solving this problem is similar to those for other stoichiometry problems.

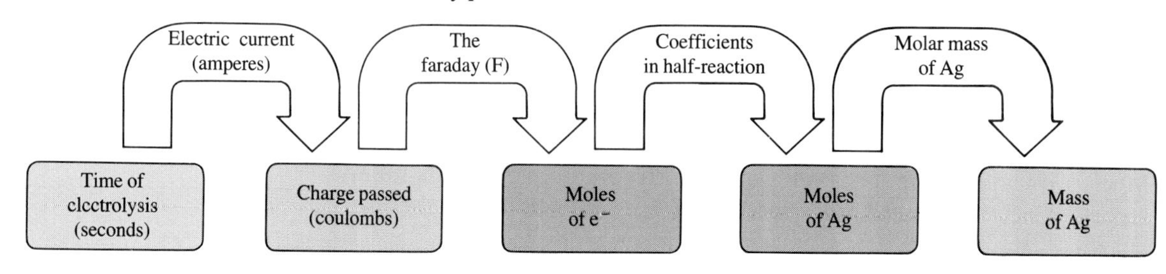

The product of the time and the current is the total charge used, which we want to express as the equivalent number of moles of electrons.

$$\text{mole of e}^- = 90.0\ \cancel{\text{min}}\left(\frac{60\ \cancel{\text{s}}}{\cancel{\text{min}}}\right)\left(\frac{0.500\ \cancel{\text{C}}}{\cancel{\text{s}}}\right)\left(\frac{1\ \text{mol e}^-}{9.65\times10^4\ \cancel{\text{C}}}\right)$$
$$= 2.80\times10^{-2}\ \text{mol e}^-$$

From the half-reaction for the reduction of silver ions

$$Ag^+(aq) + e^- \longrightarrow Ag(s)$$

we see that one mole of Ag is deposited for each mole of electrons, and we complete the calculation using the molar mass of silver.

$$\text{mass Ag} = 2.80\times10^{-2}\ \cancel{\text{mol e}^-}\left(\frac{1\ \cancel{\text{mol Ag}}}{1\ \cancel{\text{mol e}^-}}\right)\left(\frac{107.87\ \text{g}}{\cancel{\text{mol Ag}}}\right) = 3.02\ \text{g Ag}$$

Understanding The anode reaction in this electrolysis cell is the oxidation of water to $O_2(g)$. What volume of $O_2(g)$, measured at STP, was produced?

Answer: 0.157 L of O_2

Energy Considerations

The electrical energy consumed in an electrolysis reaction is the total charge passed multiplied by the applied voltage,

$$E = Q \times V_{\text{applied}} \qquad [17.8]$$

where Q is the total charge passed through the cell in coulombs. The cost of an electrolysis reaction is greatly affected by the applied voltage. If the electrolysis cell has a high electrical resistance, a larger voltage must be applied to obtain a practical current, and the cost of the process increases. The extra energy expended in overcoming the resistance of the cell is largely converted into heat. The design of the electrolysis cell is a very important consideration in the economics of electrolysis processes. Electrical energy is usually expressed in kilowatt-hours. A watt is 1 joule/second, so 1 kw-hr is

$$1\ \text{kw-hr} = \left(1000\ \frac{\text{J}}{\cancel{\text{s}}}\right)(1\ \cancel{\text{hr}})\left(\frac{60\ \cancel{\text{min}}}{\cancel{\text{hr}}}\right)\left(\frac{60\ \cancel{\text{s}}}{\cancel{\text{min}}}\right) = 3.60\times10^6\ \text{joules}$$

Note that the conversion factor between the kilowatt-hour and the joule is an exact number, since all of the factors are defined quantities. Therefore, when this factor is used in calculations, it does not limit the number of significant figures in the result.

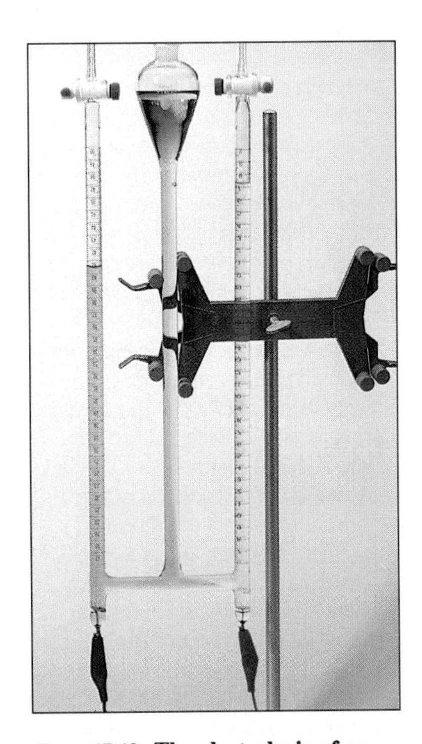

Figure 17.16 The electrolysis of sodium sulfate solution An indicator that changes from blue to yellow in acid was added to the solution, and shows that hydrogen ions are produced in the anode compartment.

$$2H_2O(\ell) \longrightarrow O_2(g) + 4H^+(aq) + 4e^-$$

The electrolysis products, hydrogen and oxygen, are formed in a volume ratio of 2:1, as expected for the electrolysis reaction that occurs.

$$2H_2O(\ell) \longrightarrow 2H_2(g) + O_2(g)$$

Example 17.16 Energy Consumption in Electrolysis

Calculate the energy consumed per mole of hydrogen produced when a sodium sulfate solution is electrolyzed (Figure 17.16), using an applied potential of 4.20 V.

Solution The cathode reaction that produces the hydrogen gas is

$$2H_2O(\ell) + 2e^- \longrightarrow H_2(g) + 2OH^-(aq)$$

The formation of one mole of $H_2(g)$ consumes two moles of electrons, so the total charge that must pass through the electrolysis cell is

$$Q = 1\ \cancel{\text{mol H}_2}\left(\frac{2\ \cancel{\text{mol e}^-}}{1\ \cancel{\text{mol H}_2}}\right)\left(\frac{9.65\times10^4\ \text{C}}{1\ \cancel{\text{mol e}^-}}\right) = 1.93\times10^5\ \text{C}$$

INSIGHTS into CHEMISTRY

The Development of Chemistry

Michael Faraday—The Unassuming Genius

Michael Faraday, an English chemist and physicist, is regarded by many as the greatest experimental scientist of the 1800s. Despite the limitation of having almost no mathematical training, he made an astonishing number of discoveries in both chemistry and physics. He developed methods for liquefying gases such as carbon dioxide and chlorine, using high pressure and low temperature. He discovered benzene and prepared the first compounds of chlorine and carbon. He also determined the empirical formula for naphthalene and prepared various sulfonic acids.

However, it is for his contributions in electricity and magnetism that he is most remembered. He succeeded in making almost all of the basic discoveries on which modern uses of electricity depend. He formulated the quantitative relationships involved during electrolysis that bear his name. He is also credited with inventing the electric motor, electric generator, and electric transformer, as well as discovering electromagnetic induction. Faraday was also the first to observe that the plane of polarized light is rotated in a magnetic field. It was Faraday who gave us the terms anode, cathode, anion, cation, electrode, and electrolyte. He proposed the concept of lines of force to explain the effect of magnets and current-carrying wires, thus foreshadowing field theory. There can be little doubt why Albert Einstein, who kept a picture of Faraday on the wall of his study, considered Faraday as one of the greatest scientists of all time.

The story of Michael Faraday's life reads like a Charles Dickens novel. Faraday was born in Newington, Surrey, England, on September 22, 1791. One of ten children of a poor blacksmith, he received only a minimal education and became an apprentice to a bookbinder in London at the age of 14. This position proved to be a fortunate occurrence, for his employer allowed him to read the books in the shop and even encouraged him to study. Faraday became fascinated by articles on electricity and chemistry and even carried out some scientific experiments with his limited resources. He attended lectures at the City Philosophical Society and there received a basic knowledge of science.

Michael Faraday (1791–1867)

The energy consumed, using a voltage of 4.20 volts, is

$$\text{energy} = 1.93 \times 10^5 \text{ C} \times 4.20 \text{ J/C} = 8.11 \times 10^5 \text{ joules}$$

Expressed in kw-hr this energy is $8.11 \times 10^5/3.60 \times 10^6$, or 0.225 kw-hr.

Understanding How much energy is consumed (in kw-hr) when 0.500 kg of magnesium is formed by electrolysis of molten $MgCl_2$, using an electrolysis voltage of 4.80 V?

Answer: 5.29 kw-hr

When he finished his apprenticeship in 1812, Faraday fully expected to devote himself to the bookbinding trade rather than to science. That same year, he received tickets from a customer to attend a series of lectures at the Royal Institution by Sir Humphrey Davy, then England's leading scientist. Faraday took careful notes at each lecture, later illustrating them with his own diagrams. He ended with 386 pages of notes, which he bound in leather and sent to Davy in the hope of getting a job at the Royal Institution. Davy was enormously impressed by the gesture, but he did not immediately respond with an offer. When Davy later fired his assistant for brawling, he remembered Faraday's gesture and offered him the position of assistant and bottlewasher. Although the pay was less than he was making as a bookbinder, Faraday eagerly accepted the position in 1813. Almost at once, Davy left England for a two-year grand tour of Europe, taking Faraday along as secretary and valet. The trip gave Faraday a chance to broaden his scientific education and the opportunity to meet many of the leading scientists of the day. When he returned to London, Faraday was determined to start his own research at the Royal Institution. Faraday proved to be particularly adept at experimental work, both assisting Davy and conducting his own original research into the fields of electrolysis and the liquefaction of gases. He continued his scientific apprenticeship, and in 1825 he replaced Davy as director of the laboratory. As the signs of Faraday's genius became ever more evident, his reputation began to rival and then to eclipse that of his former employer. Davy grew openly bitter and resentful of Faraday, and cast the only negative vote on Faraday's election to the Royal Society. There is no evidence that Faraday, who had become a deeply religious man, ever exhibited any anger or resentment toward Davy.

In 1826, Faraday—in much the same fashion as Davy—began giving lectures at the Royal Institution. He proved to be a gifted teacher and lecturer; Charles Dickens and Prince Albert, the husband of Queen Victoria, were among Faraday's many admirers. Perhaps because of his own background, Faraday believed in the importance of scientific education for the young, and he initiated the practice of including special Christmas lectures on science for schoolchildren. One of these lectures, "The Chemical History of a Candle," proved immensely popular and is still in print today.

Because of his religious beliefs and his own sense of modesty, Faraday did not believe in any outward displays of vanity. In 1857, when he was offered the presidency of the Royal Society, he declined, and he also declined the offer of a knighthood. He politely accepted the many honors, medals, and degrees that his discoveries brought to him in the name of science. His ideals and beliefs also led him to refuse a governmental request to investigate the possibilities of preparing large quantities of poison gas for use on the battlefield during the Crimean War. His decision helped to postpone the use of this horrible weapon on a large scale for almost 60 years. He also turned down numerous chances to make more money, his real motivation being his love of science and discovery.

In 1867, after a period of declining health, Faraday died at Hampton Court, a residence provided for him by Queen Victoria. Ever modest, he had requested that he be buried under the simplest of gravestones. Of course, the true monuments to Faraday are the many uses of his discoveries in today's world. In science, his many achievements are recognized by the use of his name—the farad is the SI unit of capacitance, while the faraday is the quantity of electricity required to provide one mole of electrons.

17.11 Electrolysis in Industry

Objectives

- To describe important applications of electrolysis, such as chemical synthesis, electroplating, and electrorefining
- To perform quantitative calculations for industrial electrolysis processes

Several industrial processes use electrolysis to isolate elements from their naturally occurring compounds. In fact, very reactive metals and nonmetals must be isolated by electrolysis, because no other methods are available. The recycling of many substances, such as aluminum and copper, depends upon electrolysis for their purification and recovery. In addition, decorative and protective layers of metals such as silver and gold are placed on objects by electrolysis. Some important examples of these uses of electrolysis are described in this section.

Isolation of Elements

Very strong reducing agents like the reactive metals must be produced by electrolysis rather than by use of chemical reducing agents.

Only the very unreactive metals occur naturally in their uncombined form. The rest must be extracted from their ores by reduction. For the very reactive metals electrolysis is often the most economical (and sometimes the only) method of reduction. Similarly, the most reactive nonmetals must be isolated from compounds in nature by an electrolytic oxidation.

Aluminum

Although aluminum is the third most abundant element in the earth's crust, it was not until 1886 that Charles Hall developed a practical means of isolating the metal from its compounds. Before the discovery of the Hall process, aluminum was made by the reduction of aluminum chloride with sodium metal. In the mid-19th century aluminum was considered a precious metal and was used mainly in jewelry. Aluminum metal has a low density and is resistant to corrosion. Aluminum's resistance to corrosion, in spite of its very positive oxidation potential ($E^\circ_{ox} = +1.66$ V), is attributed to the formation of a very thin transparent layer of aluminum oxide on the surface of the metal that protects it from further oxidation by air. The low density of the metal makes it a desirable structural material in airplanes and automobiles where low weight is important.

In the Hall process, aluminum oxide ore is mixed with cryolite, Na_3AlF_6, to produce a mixture that melts at about 980 °C, considerably lower than the melting point of the oxide alone. Carbon electrodes are used

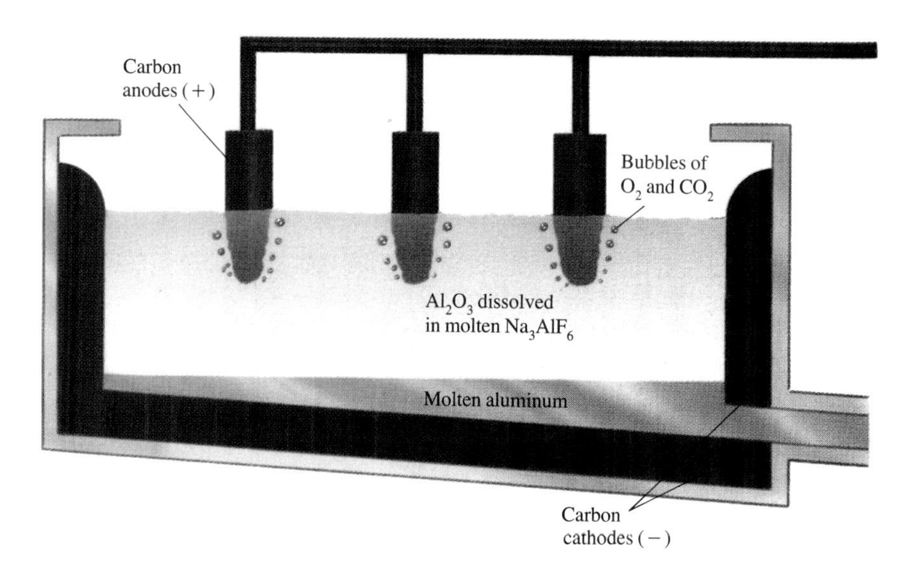

Figure 17.17 Electrolysis to produce aluminum Aluminum is manufactured by the electrolysis of molten salts at carbon electrodes. The anode reaction produces oxygen, part of which reacts with the carbon in the electrode at the cell's high temperature to form CO_2.

in the electrolysis cell, which operates at about 4.2 V. Molten aluminum is formed at the cathode, and oxygen is the principal product formed at the anode, along with some carbon dioxide. Figure 17.17 shows a schematic diagram of the equipment needed for the Hall process.

Example 17.17 Electrical Energy in Electrolysis Processes

Calculate the electrical energy needed to produce 1000 pounds of aluminum by electrolysis of aluminum oxide at a voltage of 4.25 V. Express your answer in kilowatt-hours (1 kw-hr = 3.60×10^6 J).

Solution The energy consumed in an electrolysis is the product of the applied voltage and the total charge that is used. The half-reaction in this example can be represented by

$$Al^{3+} + 3e^- \longrightarrow Al$$

Three moles of electrons, or three faradays of charge, are needed to form one mole of aluminum metal. Thus the total charge needed for 1000 pounds of aluminum is

$$Q = 1000 \text{ lb Al}\left(\frac{454 \text{ g}}{\text{lb}}\right)\left(\frac{1 \text{ mol Al}}{27.0 \text{ g}}\right)\left(\frac{3 \text{ mol e}^-}{\text{mol Al}}\right)\left(\frac{9.65 \times 10^4 \text{ C}}{\text{mol e}^-}\right)$$

$$= 4.87 \times 10^9 \text{ C}$$

In the electrolysis, 4.25 volts was used, so the electrical energy consumed is

$$E = Q \times V = 4.87 \times 10^9 \text{ C}(4.25 \text{ J/C}) = 2.07 \times 10^{10} \text{ J}$$

Using the conversion factor given in the question, we express this energy in kw-hr.

$$E = 2.07 \times 10^{10} \text{ J}\left(\frac{1 \text{ kw-hr}}{3.60 \times 10^6 \text{ J}}\right) = 5.75 \times 10^3 \text{ kw-hr}$$

In addition to the electrical energy used to cause the reaction, energy is required to melt the salts and maintain the electrolysis cell at a temperature above 980 °C. Much of this heat is due to the electrical resistance of the cell.

Sodium

Because sodium is so reactive ($E^\circ_{oxidation} = 2.71$ V) that it reduces water, it must be produced by electrolysis of molten sodium chloride. The boiling point of sodium is lower than the melting point of sodium chloride, so calcium chloride is added to lower the melting point of the electrolyte. Since the density of liquid sodium is less than that of the molten salt, it floats to the surface, where it is removed. The anode reaction produces chlorine, which is also sold as a by-product. This cell was shown in Figure 17.15.

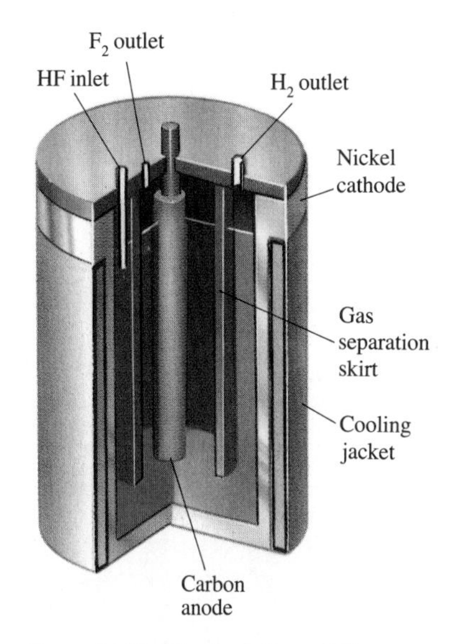

Figure 17.18 Electrolytic cell for producing fluorine This electrolysis cell is designed to isolate the fluorine from the HF/KF electrolyte. A barrier keeps the H_2 separated from the F_2 to prevent the explosive reaction of these gases.

Fluorine

Fluorine is one of the strongest known chemical oxidizing agents, so it must be produced by electrolysis. Since fluorine would rapidly oxidize water, a molten mixture of hydrogen fluoride and potassium fluoride is used as the electrolyte in its manufacture. The electrolysis cell (Figure 17.18) is made from nickel, one of the few metals that is not attacked by hydrogen fluoride. The anode is made from carbon. The hydrogen formed at the cathode must

There are no oxidizing agents strong enough to produce fluorine by a chemical reaction, so electrolysis must be used.

not come in contact with the fluorine produced, since these two elements react explosively with each other. The two half-reactions are

$$\text{Cathode:}\quad 2HF + 2e^- \longrightarrow H_2 + 2F^-$$

$$\text{Anode:}\quad 2HF \longrightarrow F_2 + 2H^+ + 2e^-$$

Chlorine

Because of its reactivity, it is impractical to produce chlorine using chemical oxidizing agents. While some chlorine is manufactured as a by-product in the electrolytic production of sodium and other reactive metals, most of it is prepared by the electrolysis of concentrated sodium chloride solutions (brine). The overall reaction that occurs during the electrolysis is

$$2NaCl(aq) + 2H_2O(\ell) \longrightarrow Cl_2(g) + H_2(g) + 2NaOH(aq)$$

The by-products, hydrogen and sodium hydroxide, are also sold. The cells must be designed to prevent mixing of the chlorine with the other products of the electrolysis. The chlorine reacts with aqueous sodium hydroxide to form sodium hypochlorite (NaOCl) in solution, and also reacts explosively with hydrogen.

Mercury cathodes have been used for years since the hydrogen overvoltage is extremely high, and instead of reducing water and forming H_2, the reduction of sodium ions to sodium amalgam occurs. (Sodium amalgam is a solution of sodium in the liquid mercury.)

$$Na^+(aq) + e^- \longrightarrow Na(Hg)$$

The sodium amalgam is removed from the electrolysis cell and treated with water in a separate container, forming hydrogen and aqueous sodium hydroxide, and the mercury is recycled. The danger of contaminating the environment with mercury outweighs the advantages of the mercury electrodes, so their use in the commercial manufacture of chlorine has been almost completely abandoned.

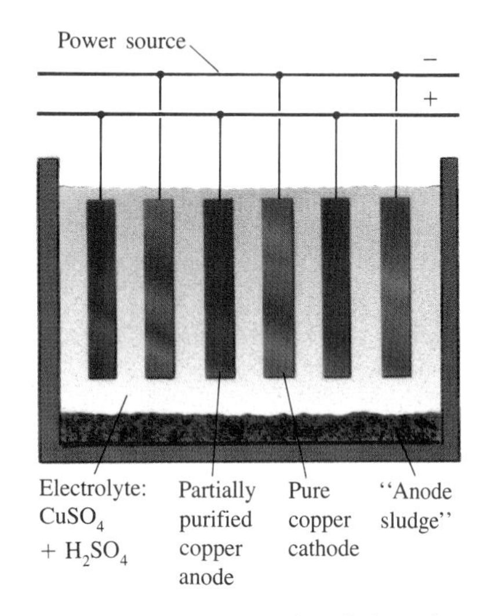

Figure 17.19 Electrolytic refining of copper The impure copper in the anode is dissolved and deposited at the cathode. The sludge is a source of precious metals such as silver, platinum, and gold.

Electrorefining of Copper

Electrorefining is an electrolysis that dissolves an impure metal anode and deposits the pure metal on the cathode. Electrorefining is used to purify and recycle copper. The anode of the electrolysis cell is made of impure copper, and pure copper is used as the cathode (Figure 17.19). Both electrodes are immersed in a solution of copper sulfate. The copper is oxidized to Cu^{2+} at the anode, while Cu^{2+} ions are reduced to the metal and deposited at the cathode. Since the anode and cathode reactions are the reverse of each other, only a very small voltage is needed to cause a current to flow. Metal impurities that are more reactive than copper are oxidized at the anode, but are not reduced at the cathode by the low voltage that is applied across the cell. The impurities in the anode that are less reactive than copper fall from the electrode as the copper dissolves, and form a sludge in the bottom of the electrolysis cell. This sludge, which often contains metals such as silver, gold, and platinum, can be further treated to recover these precious metals as by-products of the copper refining.

INSIGHTS into CHEMISTRY

Science, Society, and Technology

Chemists Balance Time and Energy Input When Using Electrolysis to Make Chemicals

Electrolysis is an important method for the manufacture of many substances. Cost is an important factor in all manufacturing processes. Two of the cost factors in an electrolysis process are closely related: the electrical energy consumed and the time required (time is money). In Example 17.16 we found the amount of electrical energy needed to produce a mole of hydrogen gas when 4.20 V is used in the electrolysis. The quantity of electrical energy consumed can be reduced if the applied voltage in Equation 17.8 is made smaller. The voltage applied to the electrolysis cell is used in three ways.

1. A portion of it is needed to cause the reaction to occur. The net chemical reaction is

$$2H_2O(\ell) \longrightarrow 2H_2(g) + O_2(g) \qquad E° = -1.23 \text{ V}$$

   Thus 1.23 V of the applied potential is needed to cause the reaction to take place.
2. Another part of the voltage is used for the overvoltage at the electrodes. In the decomposition of water, both electrode reactions have a sizeable overvoltage. If we assume the overvoltage is about 0.40 V at each electrode, this accounts for another 0.8 V of the applied voltage.
3. The remainder of the applied voltage is used to overcome the electrical resistance of the electrolysis cell. The voltage across an electrical resistance is given by Ohm's law,

$$E = IR$$

   where I is the electrical current in amperes, and R is the resistance in ohms. The energy produced from the IR product is released as heat.

We can now modify Equation 17.8 by replacing V_{applied} with the sum of the three components.

$$V_{\text{applied}} = -E° + E_{\text{overvoltage}} + IR$$

We can also replace the total charge, Q, in Equation 17.8 with its equivalent, the product of current with time. When these substitutions are made, Equation 17.8 becomes

$$\text{energy} = It(-E° + E_{\text{overvoltage}} + IR)$$

No current flows through the cell until the applied voltage is greater than $(-E° + E_{\text{overvoltage}})$. As the applied voltage increases beyond that value, the current increases, and thus the length of time needed for production of one mole of hydrogen gas decreases. The following graph shows the relationship between the electrical energy consumed and the time required to produce one mole of hydrogen by electrolysis, for three different cell resistances.

As can be seen from the graph, the smaller the electrical resistance of the cell, the lower is the electrical energy consumed. Thus, in designing electrolysis cells, it is very important to reduce the electrical resistance as much as possible. In the electrolysis of solutions, the concentration of electrolyte, the distance between the electrodes, and the area of the electrodes all affect the electrical resistance.

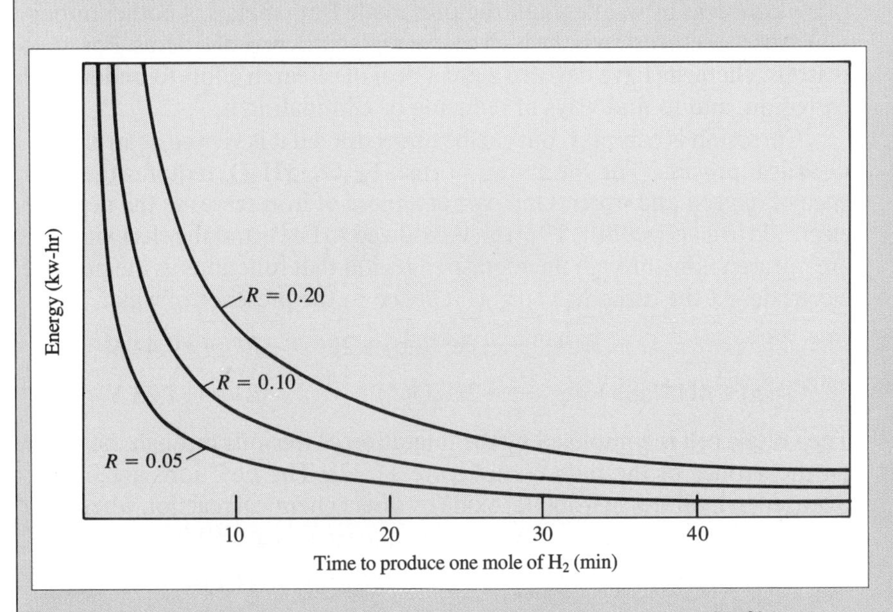

The energy consumed as a function of the time required to produce one mole of hydrogen in the electrolysis of water. For any given time of electrolysis, the lower the resistance of the electrolysis cell, the less electrical energy must be consumed.

Figure 17.20 Electroplating Electrolysis is used to deposit decorative and protective films of metals on the surface of objects.

Electroplating

The electrolytic deposition of a thin metal film on the surface of a metallic object is called **electroplating.** Chromium is often plated onto iron and steel surfaces to improve the appearance and protect the object from corrosion. Objects are often plated with thin layers of silver or gold for appearance. In the electroplating process, the object being plated is used as the cathode, and the source of the metal in the coating is usually a complex metal ion in solution (Figure 17.20).

17.12 Corrosion

Objectives

- To describe corrosion as an electrochemical process
- To explain how acidity and electrolyte concentration contribute to corrosion
- To explain the electrochemistry of anodic and cathodic protection

Corrosion is the oxidation of a metal to produce compounds of the metal through interaction with its environment. The most familiar and most costly example of corrosion is the rusting of iron and its alloys. Rust consists of hydrated forms of iron(III) oxide. The green coating commonly seen on bronze statues (Figure 17.21) is another example of corrosion. The green color is caused by copper(II) compounds formed from the corrosion of the copper in bronze. In addition to being unsightly, corrosion can lead to severe safety concerns by weakening structures made from metals. Another important aspect of corrosion is the high cost of replacing corroded items. For these reasons, chemists have devoted a great deal of research effort to understand corrosion, and to find ways of reducing or eliminating it.

Figure 17.21 The Statue of Liberty The green color of the Statue of Liberty was caused by the oxidation of copper to copper(II) compounds, primarily oxides and carbonates. Before the recent restoration of the statue, corrosion had created holes in the outer surface and greatly weakened the steel frame.

Corrosion is complex, but can be understood if it is viewed as an electrochemical process. The formation of rust, $Fe_2O_3 \cdot xH_2O$, requires the presence of oxygen and water. One part of a piece of iron serves as the negative electrode in a voltaic cell. The iron is oxidized to Fe^{2+}, and the electrons that are released flow through the metal to a region that functions as the positive electrode. At the cathode, $O_2(g)$ is reduced in the presence of water.

$$Fe(s) \longrightarrow Fe^{2+}(aq) + 2e^- \qquad E^\circ = +0.44\ V$$

$$O_2(g) + 4H^+(aq) + 4e^- \longrightarrow 2H_2O(\ell) \qquad E^\circ = +1.23\ V$$

The voltaic cell is completed by the migration of the ions through the water on the surface of the metal (see Figure 17.22). The Fe^{2+} ions are further oxidized to hydrates of iron(III) oxide by direct chemical reaction when they migrate to the surface of the water, where oxygen is available.

$$4Fe^{2+}(aq) + O_2(g) + (4 + 2x)H_2O(\ell) \longrightarrow 2Fe_2O_3 \cdot xH_2O(s) + 8H^+(aq)$$

The actual potential of the reduction half-reaction depends upon the acidity of the water solution, as shown in Figure 17.23. As the concentration of hydrogen ions increases in the water solution, the electrochemical driving

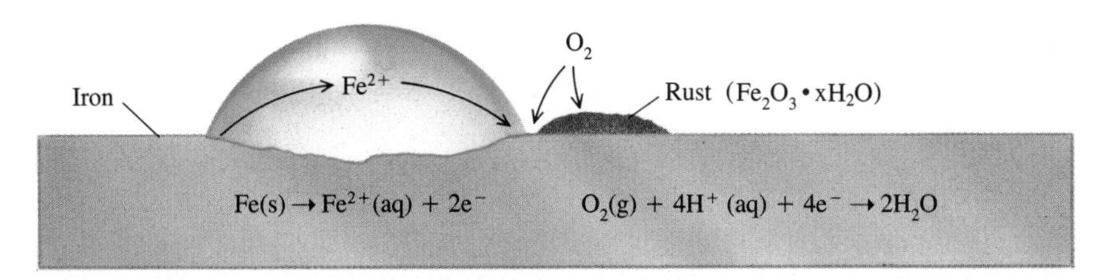

Figure 17.22 The rusting of iron Iron in contact with water forms an anode where the metal is oxidized to $Fe^{2+}(aq)$. The metal in contact with oxygen and water acts as a cathode where oxygen is reduced. The $Fe^{2+}(aq)$ ions are further oxidized to rust ($Fe_2O_3 \cdot xH_2O$) when they migrate to the surface of the water, where there is an abundant supply of oxygen.

force (indicated by the electrode potential) increases. Sulfur oxides in the atmosphere dissolve to produce acidic solutions and tend to enhance the corrosion of metals. Not only is the potential of the electrochemical cell increased, but protective oxide films are more soluble in the acidic water solution. The higher concentration of ions in acid rain also contributes to corrosion, by allowing larger currents to flow through the aqueous solution.

Corrosion is an electrochemical process, with the oxidation and reduction occurring at different sites on the metal surface.

The concentration of electrolytes in the aqueous solution also affects the rate at which metals corrode. The presence of chloride and other ions increases the electrical conductivity of the solution, and the larger currents that result speed the rate of deterioration of the metal. Because sea water has a high salt concentration, rusting and corrosion of metals occurs much more rapidly in waterfront locations. High electrolyte concentrations cause the bodies of automobiles to rust out more quickly in areas where salt is used on roads to melt snow.

The oxidation of the corroded metal usually occurs at a site that is remote from the source of oxygen. When a car is damaged, the metal exposed at the break in the coating of paint often remains bright, since the reduction of O_2 occurs at this point. Several weeks later the paint blisters from the

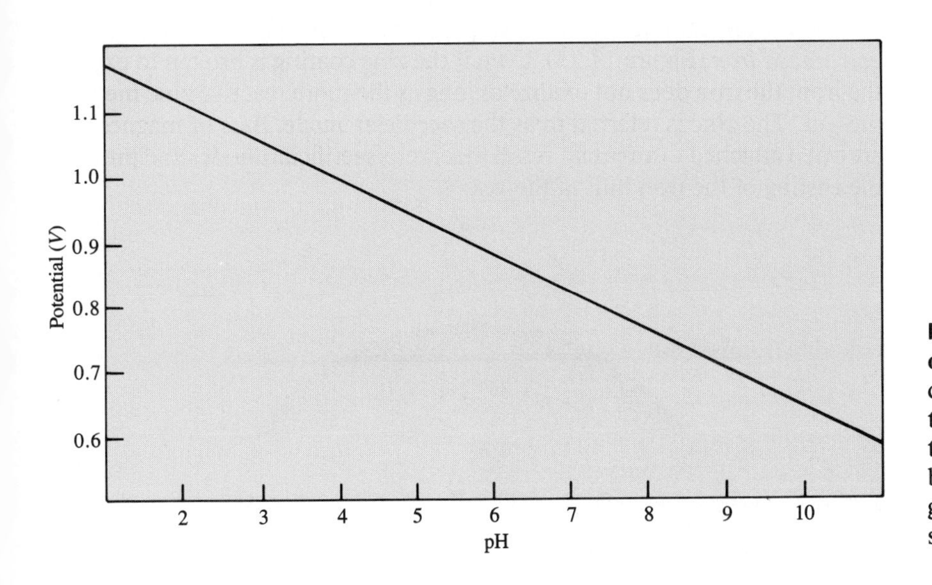

Figure 17.23 The reduction of oxygen The potential for the reduction of oxygen in aqueous solution depends on the pH. The potential increases as the solution becomes more acidic, providing a greater driving force for the corrosion of metals.

Figure 17.24 Corrosion A break in the painted surface of an automobile exposes the steel to air and water. Often the paint blisters as rust forms under it. The oxidation of the metal occurs away from the point where the paint was originally broken.

deterioration of the metal a few centimeters from the exposed surface (Figure 17.24), where an impurity supplies a site for oxidation of the metal.

Protection from Corrosion

Corrosion limits the useful lifetime of many products, so a great deal of effort has been expended to find ways to inhibit or prevent corrosion. Since most corrosion requires the presence of water and oxygen in direct contact with the metal, one obvious way to reduce it is to apply an impervious coating to the metal surface. The painting of automobiles, in addition to improving appearance, reduces rusting. Plating of iron with chromium also protects auto parts from rusting. Even though the oxidation potential of Cr ($E^\circ_{oxidation} = +0.91$ V) indicates that this metal should be quite reactive, it forms a thin surface film of oxide that protects the chromium underneath from further oxidation. Even a very small break in the chromium plating allows the iron alloys underneath to corrode.

Another popular method for inhibiting corrosion is a process called anodic protection. In **anodic protection** the metal is intentionally oxidized under carefully controlled conditions to form a thin, adhering layer of oxide on the surface of the metal. The treatment of iron with aqueous sodium chromate forms a layer of Fe(III) and Cr(III) oxides that protects the iron from contact with oxygen and water:

$$2Fe(s) + 2Na_2CrO_4(aq) + 2H_2O(\ell) \longrightarrow Fe_2O_3(s) + Cr_2O_3(s) + 4NaOH(aq)$$

Anodic protection produces a thin, impervious layer of oxide on the surface of the metal, which stops further oxidation.

The protective oxide layer is sometimes produced by electrolytic oxidation. Aluminum is *anodized* by coating it with an impervious layer of aluminum oxide by electrolysis.

A metal may be protected from corrosion by placing it in contact with a more reactive metal.

Another way to protect metals from corrosion is to force the metal to behave as a cathode in an electrochemical cell. In **cathodic protection** a second, more reactive metal is placed in electrical contact with the metal object being protected from corrosion. The more reactive metal will behave as the anode in the electrochemical cell, thus forcing the other metal to function as the cathode. Iron that has been coated with a layer of zinc is called *galvanized iron* (Figure 17.25). Even if the zinc coating is broken to expose the iron, the iron does not oxidize as long as the more reactive zinc metal is present. The zinc is referred to as the **sacrificial anode.** Bars of magnesium are often attached to an ocean vessel to serve as sacrificial anodes and prevent the rusting of the iron hull of the vessel.

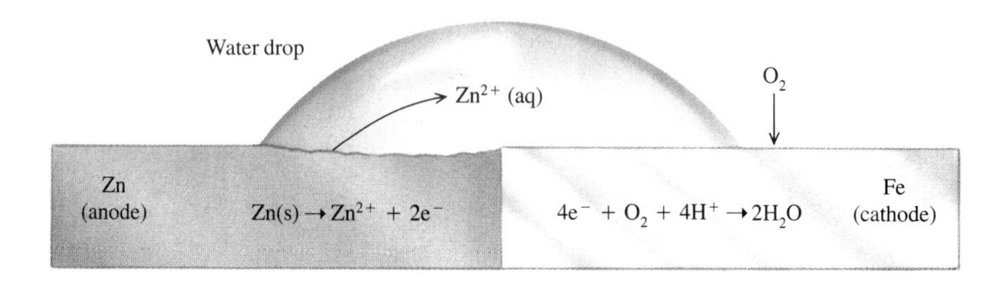

Figure 17.25 Cathodic protection Contact with a more reactive metal protects iron from rusting. Galvanized iron has a coating of zinc. The oxidation of the zinc to Zn^{2+} takes place more readily than the oxidation of Fe to Fe^{2+} or Fe^{3+}, so the iron does not rust.

Summary

Electrochemistry is the study of the relationship between chemical reactions and electricity. All electrochemical processes involve oxidation-reduction or *redox* reactions (electron transfer reactions). In a redox reaction, the substance that loses electrons is oxidized, and is called the *reducing agent.* The *oxidizing agent* is the substance that gains electrons and is reduced. The assignment of *oxidation states* to the elements in substances is useful in recognizing redox processes. The oxidation states of the atoms may be determined from the Lewis structures, by assigning shared electrons to the more electronegative atom participating in a covalent bond. The oxidation state is obtained by subtracting the number of electrons that are assigned to each atom in the structure from the number of valence shell electrons in the neutral atom. The average oxidation state of each element can be determined by using the rules given in Section 17.1.

The *ion-electron method* for balancing oxidation-reduction reactions first divides the chemical expression into two *half-reactions,* each of which is balanced separately. The resulting oxidation and reduction half-reactions are then added, after each is multiplied by a constant so that the number of electrons released in the oxidation is equal to the number consumed by the reduction. The ion-electron method may be used for redox reactions that occur in either acidic or basic solutions.

The chemical energy released in a spontaneous redox reaction can be directly converted into electrical energy by using a *voltaic cell.* A voltaic cell consists of two *half-cells* that are often connected through a *salt bridge.* The salt bridge allows ion transport between the half-cells, but keeps the oxidizing and reducing agents from reacting directly. The word *electrode* is used to refer either to the solid electrical conductor in a half-cell or to the entire half-cell. The half-reactions in a voltaic cell may involve a metal that is oxidized to an ion, two ions of the same metal in different oxidation states, a gas and its ion in solution, or a metal and a slightly soluble salt of the metal ion.

The *electromotive force (emf),* measured in volts, is an intensive property that measures the electrical driving force moving the electrons through an external circuit, from the negative electrode to the positive electrode of a voltaic cell. The voltage of a cell is the energy produced when one coulomb of electronic charge moves from the negative electrode to the positive electrode. The additivity of cell potentials allows us to assign a voltage to any half-reaction, by defining the standard voltage of the hydrogen half-reaction as zero. The standard potentials of reduction half-reactions are tabulated, and may be used to calculate the voltage of any redox reaction under standard conditions. When the voltage of a redox reaction is positive, the reaction proceeds spontaneously as written. A negative voltage shows that the reverse reaction is spontaneous.

The standard free energy change for a redox reaction is related to the standard potential by the equation

$$\Delta G^\circ = -nFE^\circ$$

where n is the number of electrons transferred, and F is the *faraday,* which is the negative charge on one mole of electrons (96,485 coulombs). The standard potential is also related to the equilibrium constant for the cell reaction by the relationship

$$\log K_{eq} = \frac{nFE^\circ}{2.303RT}$$

Thus standard cell potentials provide a means of determining free energy changes and equilibrium constants.

The potentials of redox reactions depend on the concentrations of the reactants and products. The concentration dependence of potentials is given by the *Nernst equation:*

$$E = E^\circ - \frac{2.303RT}{nF} \log Q$$

where Q is the reaction quotient. The factor $2.303RT/F$ has a value of 0.0591 V mol at 25 °C. The Nernst equation may be applied to balanced redox reactions or half-reactions. The Nernst equation is the basis for very easy and rapid methods for measuring concentrations of species in solution. The cell used to measure concentrations consists of a *reference electrode,* which has a constant voltage, and an *indicating electrode* whose voltage changes with the composition of the test solution. Some of the more important uses of voltaic cells in the laboratory are the determination of equilibrium constants and the measurement of the concentrations of species in solution (particularly hydrogen ions).

Voltaic cells and batteries also provide portable energy sources for radios, watches, flashlights, cordless electric tools, calculators, etc. Among the more important of these cells and batteries are the *lead storage cell,* the *LeClanché* dry cell, and a variety of other cells, such as the nickel-cadmium cell. *Fuel cells,* which convert the chemical energy of combustion reactions directly into electrical energy, are also being investigated as a means of conserving our natural resources.

While voltaic cells produce electrical energy from chemical reactions, *electrolysis* consumes electrical energy to cause nonspontaneous chemical reactions to occur. Passing an electrical current through an electrolysis cell causes redox reactions to occur at the electrodes. In the electrolysis of aqueous solutions, the oxidation and reduction of water, solutes present and the electrodes themselves must all be considered in determining the expected products. Because of *overvoltage,* oxidizing agents stronger than oxygen and reducing agents stronger than hydrogen may be generated by the electrolysis of an aqueous solution.

The amount of products formed in an electrolysis is proportional to the total charge (the product of current and time) that is passed through the electrolyte. The negative charge on one mole of electrons is used in stoichiometric calculations involving electrolysis reactions. Under standard conditions, the voltage that must be applied to an electrolysis cell has to be greater than the negative of the standard voltage of the electrolysis reaction. The cell design and the rate of formation of the products are both factors that contribute to the actual voltage that must be applied to the cell. The quantity of electrical energy consumed in an electrolysis is the product of the charge and the applied voltage.

Many strong oxidizing agents (*e.g.,* fluorine and chlorine) and reducing agents (the alkali metals, magnesium, and aluminum, for example) are isolated from their naturally occurring compounds by electrolysis. In addition, the purification and recycling of some metals often involve electrolysis methods. The production of protective and decorative coatings by *electroplating* is another commercial application of electrolysis.

Corrosion of metals is an electrochemical process in which oxygen combines with the metal. Corrosion can be reduced by *anodic protection,* which produces a protective oxide film on the surface of the metal. In *cathodic protection* a more reactive metal is used as a *sacrificial anode* to prevent oxidation of the protected metallic object.

Chapter Terms

Electrochemistry—the study of the relation between chemical reactions and electricity. *(17.1)*

Oxidation—the loss of electrons by an element, a compound, or an ion. *(17.1)*

Reduction—the gain of electrons by an element, a compound, or an ion. *(17.1)*

Oxidation-reduction reaction—a reaction in which electrons are transferred from one species to another. *(17.1)*

Redox reaction—an abbreviation for an oxidation-reduction reaction. *(17.1)*

Half-reaction—an equation describing either the oxidation or the reduction portion of a redox reaction, with the electrons explicitly shown. An oxidation half-reaction has electrons on the product side of the equation. The electrons are on the reactant side of the equation in a reduction half-reaction. *(17.1)*

Oxidizing agent—the reactant that is reduced in a redox reaction. *(17.1)*

Reducing agent—the reactant that is oxidized in a redox reaction. *(17.1)*

Oxidation state—the numerical equivalent of the charge of monatomic ions or the charge an atom would possess if the shared electrons in each covalent bond were assigned to the more electronegative atom. This is also called the oxidation number. *(17.1)*

Ion-electron method—a method of balancing oxidation-reduction reactions by first balancing two half-reactions and then combining them to obtain the chemical equation. *(17.2)*

Voltaic cell—an apparatus that converts the chemical energy produced by a redox reaction directly into electrical energy. This is also called a galvanic cell. *(17.3)*

Half-cell—the compartment in a voltaic cell in which either the reduction or oxidation half-reaction occurs. *(17.3)*

Salt bridge—an electrolyte medium that allows the transport of ions between the two half-cells of a voltaic cell. *(17.3)*

Electrode—(1) a metal or other electrical conductor that connects an electrochemical cell to the external circuit; (2) a half-cell. *(17.3)*

Inert electrode—a solid electrical conductor, usually an unreactive metal or graphite, that is neither oxidized nor reduced in the reaction of a voltaic cell. It serves only as a means of electrical contact to the solution. *(17.3)*

Electromotive force (emf)—the electrical driving force that pushes the electrons released in the oxidation half-cell (the negative electrode) through the external circuit into the reduction half-cell (the positive electrode). The emf is an intensive property of a cell and is measured in volts. *(17.4)*

Volt (V)—the SI unit for emf, which is defined as the potential difference required to increase the energy of a charge of 1 coulomb by 1 joule. *(17.4)*

$$1 \text{ V} = 1 \text{ joule/coulomb} = 1 \text{ J/C}$$

Cell potential—the voltage difference between the two electrodes in a voltaic cell. *(17.4)*

Standard cell potential (E°_{cell})—the voltage of a voltaic cell when all of the reactants and products are in their standard states (pure solids, liquids, or gases at 1 atmosphere pressure; solutes at 1 *M* concentration). *(17.4)*

Standard reduction potential—the standard voltage of a reduction half-reaction, based on a voltage of zero for the standard hydrogen electrode. *(17.4)*

Faraday (F)—the negative electrical charge on one mole of electrons, 96,485 C. *(17.5)*

Nernst equation—the equation that describes the dependence of cell voltage on the concentrations of reactants and products. *(17.6)*

$$E = E^\circ - \frac{RT}{nF} \ln Q = E^\circ - \frac{2.303RT}{nF} \log Q$$

Indicating electrode—an electrode whose potential depends upon the concentration of a particular species in solution. *(17.7)*

Reference electrode—a half-cell with an accurately known potential that is independent of the composition of the test solution. *(17.7)*

pH meter—a voltmeter that is scaled to read the pH of a solution from the measured potential of a voltaic cell. *(17.7)*

Battery—two or more voltaic cells, with the negative electrode of each cell connected to the positive electrode of the next cell. *(17.8)*

Fuel cell—a voltaic cell in which the reactants are continuously supplied, and generally the products of the redox reaction are continuously removed. *(17.8)*

Electrolysis—an otherwise nonspontaneous oxidation-reduction reaction that is caused by the passage of an electric current. *(17.9)*

Electrolytic cell—an apparatus that consists of two electrodes immersed in a molten salt or electrolyte solution, which is used to carry out an electrolysis. *(17.9)*

Anode—the electrode in a cell at which oxidation occurs. *(17.9)*

Cathode—the electrode in a cell at which reduction occurs. *(17.9)*

Overvoltage—the difference between the standard potential and the potential needed to cause an electrolysis reaction. *(17.9)*

Electrorefining—an electrolytic process that converts an impure metal anode into a very pure sample of the metal at the cathode. *(17.11)*

Electroplating—the electrolytic deposition of a thin uniform metal film on the surface of a metallic object. *(17.11)*

Corrosion—the oxidation of a metal to produce compounds of the metal through interaction with its environment. *(17.12)*

Anodic protection—reducing corrosion by the intentional oxidation of a metal under carefully controlled conditions, to form a thin, adhering layer of oxide on the surface of the metal. The protective layer can be generated either chemically or electrochemically. *(17.12)*

Cathodic protection—protection of a metallic object from corrosion by placing it in electrical contact with a second, more reactive metal. *(17.12)*

Sacrificial anode—a piece of active metal placed in electrical contact with a less reactive metal. *(17.12)*

Exercises

Exercises designated with color have answers in Appendix J.

Assigning Oxidation States

17.1 Construct a Lewis structure and assign the oxidation states of all atoms in each molecule or ion.

(a) ClO_3^-
(b) PF_3
(c) CO
(d) N_2
(e) $B(OH)_3$
(f) IF_4^-

17.2 Construct a Lewis structure and assign the oxidation states of all atoms in each molecule or ion.

(a) NO_3^-
(b) NO_2^-
(c) NH_4^+
(d) Br_2
(e) H_2SO_4
(f) CO_2

17.3 In each part assign the oxidation states of all elements, using the rules given in Section 17.1.

(a) ZrO_2
(b) FeO
(c) $Ca(NO_3)_2$
(d) PF_5
(e) Na_2CrO_4
(f) NO_2^-
(g) BaO_2
(h) F_2
(i) Sn^{2+}

17.4 In each part assign the oxidation states of all elements, using the rules given in Section 17.1.

(a) $KMnO_4$
(b) H_2O
(c) Cl_2
(d) NO_2
(e) CrO_2^-
(f) $Co(NO_3)_3$
(g) $CaCO_3$
(h) $HBrO_4$
(i) Fe^{3+}

17.5 In each part assign the oxidation states of all elements, using the rules given in Section 17.1.
(a) KHF_2
(b) H_2Se
(c) NaO_2
(d) NO
(e) BO_2^-
(f) $Cr(NO_3)_3$
(g) CaC_2O_4
(h) $Ba(ClO_4)_2$
(i) Tl^{3+}

Oxidation Reduction Reactions

17.6 Complete and balance each half-reaction in acid solution, and identify it as an oxidation or a reduction.
(a) $Cr^{3+}(aq) \longrightarrow Cr(s)$
(b) $I^-(aq) \longrightarrow I_2(aq)$
(c) $NO_2^-(aq) \longrightarrow NO_3^-(aq)$
(d) $Fe^{2+}(aq) \longrightarrow Fe^{3+}(aq)$
(e) $Cr_2O_7^{2-}(aq) \longrightarrow Cr^{3+}(aq)$
(f) $VO_2^+(aq) \longrightarrow V^{3+}(aq)$

17.7 Complete and balance each half-reaction in acid solution, and identify it as an oxidation or a reduction.
(a) $UO_2^{2+}(aq) \longrightarrow U^{4+}(aq)$
(b) $Zn(s) \longrightarrow Zn^{2+}(aq)$
(c) $IO_3^-(aq) \longrightarrow I^-(aq)$
(d) $N_2O_4(g) \longrightarrow NO_3^-(aq)$
(e) $Mn^{3+}(aq) \longrightarrow MnO_4^-(aq)$
(f) $HOCl(aq) \longrightarrow ClO_3^-(aq)$

17.8 Using the ion-electron method, complete and balance the following redox reactions in acid solution.
(a) $Sn(s) + Fe^{3+}(aq) \longrightarrow Sn^{2+}(aq) + Fe^{2+}(aq)$
(b) $HAsO_3^{2-}(aq) + I_2(aq) \longrightarrow H_2AsO_4^-(aq) + I^-(aq)$
(c) $Cu(s) + Ag^+(aq) \longrightarrow Cu^{2+}(aq) + Ag(s)$
(d) $MnO_4^-(aq) + H_2C_2O_4(aq) \longrightarrow Mn^{2+}(aq) + CO_2(g)$
(e) $Cl_2(g) + Br^-(aq) \longrightarrow Cl^-(aq) + Br_2(\ell)$
(f) $Cu(s) + NO_3^-(aq) \longrightarrow NO(g) + Cu^{2+}(aq)$
(g) $VO^{2+}(aq) + Cr_2O_7^{2-}(aq) \longrightarrow Cr^{3+}(aq) + VO_2^+(aq)$

17.9 Using the ion-electron method, complete and balance the following redox reactions in acid solution.
(a) $Fe(s) + Ag^+(aq) \longrightarrow Ag(s) + Fe^{2+}(aq)$
(b) $I_2(aq) + S_2O_3^{2-}(aq) \longrightarrow I^-(aq) + S_4O_6^{2-}(aq)$
(c) $MnO_4^-(aq) + Fe^{2+}(aq) \longrightarrow Fe^{3+}(aq) + Mn^{2+}(aq)$
(d) $Zn(s) + NO_3^-(aq) \longrightarrow Zn^{2+}(aq) + N_2(g)$
(e) $IO_3^-(aq) + I^-(aq) \longrightarrow I_2(aq)$
(f) $Ce^{4+}(aq) + Cl^-(aq) \longrightarrow Cl_2(aq) + Ce^{3+}(aq)$
(g) $Cr_2O_7^{2-}(aq) + HSO_3^-(aq) \longrightarrow Cr^{3+}(aq) + SO_4^{2-}(aq)$

17.10 Use the ion-electron method to balance each of the redox reactions in basic solution.
(a) $Al(s) + ClO^-(aq) \longrightarrow Al(OH)_4^-(aq) + Cl^-(aq)$
(b) $MnO_4^-(aq) + SO_3^{2-}(aq) \longrightarrow MnO_2(s) + SO_4^{2-}(aq)$
(c) $Zn(s) + NO_3^-(aq) \longrightarrow Zn(OH)_4^{2-}(aq) + NH_3(aq)$
(d) $ClO^-(aq) + CrO_2^-(aq) \longrightarrow Cl^-(aq) + CrO_4^{2-}(aq)$
(e) $Br_2(aq) \longrightarrow Br^-(aq) + BrO_3^-(aq)$
(f) $H_2O_2(aq) + N_2H_4(aq) \longrightarrow N_2(g) + H_2O(\ell)$

17.11 Use the ion-electron method to balance each of the redox reactions in basic solution.
(a) $Cl_2(aq) \longrightarrow Cl^-(aq) + ClO_3^-(aq)$
(b) $MnO_4^-(aq) + I^-(aq) \longrightarrow IO_3^-(aq) + MnO_2(s)$
(c) $ClO_3^-(aq) + CN^-(aq) \longrightarrow Cl^-(aq) + CNO^-(aq)$
(d) $PH_3(g) + CrO_4^{2-}(aq) \longrightarrow CrO_2^-(aq) + P_4(s)$
(e) $F_2(g) + H_2O(\ell) \longrightarrow F^-(aq) + O_2(g)$
(f) $H_2O_2(aq) + Cr(OH)_3(s) \longrightarrow CrO_4^{2-}(aq)$

17.12 The titration of a 0.103 g sample of sodium oxalate ($Na_2C_2O_4$) in acid solution required 24.30 mL of potassium permanganate solution to reach the endpoint. In this reaction the oxalate ion ($C_2O_4^{2-}$) is oxidized to carbon dioxide, and the permanganate ion (MnO_4^-) is reduced to Mn^{2+}. What is the molarity of the $KMnO_4$ solution?

17.13 A sample containing H_3AsO_3 is dissolved in acid solution and titrated with 0.0501 M I_3^- solution, requiring 31.05 mL to reach the endpoint. In this redox reaction the arsenic is converted to H_3AsO_4 and the triiodide ion is reduced to iodide ion. How many grams of H_3AsO_3 are present in the sample?

Voltaic Cells

17.14 A cell is based on the reaction

$$Zn(s) + Ni^{2+}(aq) \longrightarrow Zn^{2+}(aq) + Ni(s)$$

Sketch the cell, and label the positive and negative electrodes, the direction of electron flow in the external circuit, and the direction that cations and anions flow through the salt bridge. Write the half-reaction that occurs at each electrode.

17.15 A cell is based on the reaction

$$Pb(s) + 2Ag^+(aq) \longrightarrow Pb^{2+}(aq) + 2Ag(s)$$

Sketch the cell, and label the positive and negative electrodes, the direction of electron flow in the external circuit, and the direction that cations and anions flow through the salt bridge. Write the half-reaction that occurs at each electrode.

17.16 A platinum wire is in contact with a mixture of mercury and solid mercury (I) chloride (Hg_2Cl_2) in a beaker containing 1 M KCl solution. A salt bridge connects this half-cell to a beaker that contains a copper electrode immersed in 1 M $CuSO_4$ solution. Electrons flow from the platinum electrode to the copper electrode.
(a) Write balanced half-reactions for the two electrodes.
(b) Write the equation for the cell reaction.
(c) Which electrode is positive?
(d) Would direct reaction occur if both electrodes were placed in a container containing an aqueous solution of 1 M $CuSO_4$ and 1 M KCl?

17.17 Two electrodes are immersed in a 1 M HBr solution. One of the electrodes is a silver wire coated with a deposit of AgBr. The second electrode is a platinum wire in electrical contact with a mixture of metallic mercury and Hg_2Br_2. Oxidation occurs at the Ag/AgBr electrode.
(a) Write balanced half-reactions for the two electrodes.
(b) Write the equation for the cell reaction.
(c) In which direction do the electrons flow in the external circuit? Which electrode is positive?
(d) Why is a salt bridge unnecessary in this cell?

Standard Potentials

17.18 Use the data from Table 17.1 to calculate the standard potentials of the cells described in Exercises 17.14 and 17.15.

17.19 Use the data from Table 17.1 to calculate the standard voltage of the cell based on each of the following reactions. In each case state whether the reaction proceeds spontaneously as written or in the reverse direction under standard state conditions.
(a) $Fe^{2+}(aq) + Ag^+(aq) \longrightarrow Fe^{3+}(aq) + Ag(s)$
(b) $Cu(s) + 2AgCl(s) \longrightarrow Cu^{2+}(aq) + 2Ag(s) + 2Cl^-(aq)$
(c) $Ni(s) + Br_2(\ell) \longrightarrow Ni^{2+}(aq) + 2Br^-(aq)$

17.20 Use the data from the table of standard reduction potentials in Appendix H to calculate the standard voltage of the cell based on each of the following reactions. In each case state whether the reaction proceeds spontaneously as written or spontaneously in the reverse direction under standard state conditions.
(a) $H_2(g) + Cl_2(g) \longrightarrow 2H^+(aq) + 2Cl^-(aq)$
(b) $Al^{3+}(aq) + 3Cr^{2+}(aq) \longrightarrow Al(s) + 3Cr^{3+}(aq)$
(c) $Sn^{2+}(aq) + 2Fe^{3+}(aq) \longrightarrow Sn^{4+}(aq) + 2Fe^{2+}(aq)$

17.21 Balance the equations and use the data from Appendix H to determine the standard potential of each of the following reactions. Comment on the direction of spontaneous reaction and the stability of the metal in acid solution.
(a) $H^+(aq) + Cu(s) \longrightarrow Cu^{2+}(aq) + H_2(g)$
(b) $H^+(aq) + Fe(s) \longrightarrow Fe^{2+}(aq) + H_2(g)$
(c) $H^+(aq) + Zn(s) \longrightarrow Zn^{2+}(aq) + H_2(g)$
(d) $H^+(aq) + Ni(s) \longrightarrow Ni^{2+}(aq) + H_2(g)$
(e) $H^+(aq) + Cr(s) \longrightarrow Cr^{2+}(aq) + H_2(g)$

17.22 Balance the equations and use the data from Appendix H to determine the standard potential of each of the following reactions. Comment on the direction of spontaneous reaction and the stability of the metal in water.
(a) $H_2O(\ell) + Cu(s) \longrightarrow Cu^{2+}(aq) + H_2(g) + OH^-(aq)$
(b) $H_2O(\ell) + Na(s) \longrightarrow Na^+(aq) + H_2(g) + OH^-(aq)$
(c) $H_2O(\ell) + Al(s) \longrightarrow Al^{3+}(aq) + H_2(g) + OH^-(aq)$
(d) Is the use of aluminum pans consistent with the answer to (c). If not, explain.

17.23 The standard potential of the reaction

$$Ag^+(aq) + Eu^{2+}(aq) \longrightarrow Ag(s) + Eu^{3+}(aq)$$

is $E° = +1.23$ V. Use the standard potential of the silver couple to find the standard reduction potential for the europium couple. Which electrode is the positive electrode in this cell? Which is the negative electrode?

17.24 The standard potential for the reaction

$$UO_2^{2+}(aq) + Pb(s) + 4H^+(aq) \longrightarrow U^{4+}(aq) + Pb^{2+}(aq) + 2H_2O(\ell)$$

is $E° = +0.460$ V. Use the standard potential of the lead couple to find the standard reduction potential of the uranium couple. Which electrode is the positive electrode in this cell? Which is the negative electrode?

17.25 A half-cell that consists of a silver wire in 1.00 M $AgNO_3$ solution is connected by a salt bridge to a 1.00 M thallium(I) acetate solution that contains a metallic Tl electrode. The voltage of the cell is 1.136 V, with the silver as the positive electrode.
(a) Write the half-reactions and overall equation for the spontaneous chemical reaction.
(b) Use the standard potential of the silver couple, with the voltage of the cell, to calculate the standard reduction potential for the thallium half-reaction.

17.26 A half-cell that consists of a copper wire in 1.00 M $Cu(NO_3)_2$ solution is connected by a salt bridge to a solution that is 1 M in both Pu^{3+} and Pu^{4+} and contains an inert metal electrode. The voltage of the cell is 0.642 V, with the copper as the negative electrode.
(a) Write the half-reactions and overall equation for the spontaneous chemical reaction.
(b) Use the standard potential of the copper couple, with the voltage of the cell, to calculate the standard reduction potential for the plutonium half-reaction.

$\Delta G°$, $E°$, and K

17.27 For each of the given equations, find $E°$ from the standard potentials, and calculate $\Delta G°$ and the equilibrium constant for the reaction. State whether the reaction is spontaneous as written or spontaneous in the reverse direction under standard conditions.
(a) $Zn(s) + Fe^{2+}(aq) \longrightarrow Zn^{2+}(aq) + Fe(s)$
(b) $AgCl(s) + Fe^{2+}(aq) \longrightarrow Ag(s) + Fe^{3+}(aq) + Cl^-(aq)$
(c) $Br_2(\ell) + 2Cl^-(aq) \longrightarrow Cl_2(g) + 2Br^-(aq)$

17.28 For each of the given equations, find $E°$ from the standard potentials, and calculate $\Delta G°$ and the equilibrium constant for the reaction. State whether the

reaction is spontaneous as written or spontaneous in the reverse direction under standard conditions.
(a) $Cu^{2+}(aq) + Ni(s) \longrightarrow Cu(s) + Ni^{2+}(aq)$
(b) $2Ag(s) + Cl_2(g) \longrightarrow 2AgCl(s)$
(c) $Cl_2(g) + 2I^-(aq) \longrightarrow 2Cl^-(aq) + I_2(s)$

17.29 For the equation

$$2Fe(CN)_6^{3-}(aq) + 2I^-(aq) \longrightarrow 2Fe(CN)_6^{4-}(aq) + I_2(s)$$

$\Delta G° = -34.3$ kJ. Find the standard potential for this reaction.

17.30 For the equation

$$2VO^{2+}(aq) + Br_2(\ell) + 2H_2O(\ell) \longrightarrow 2VO_2^+(aq) + 2Br^-(aq) + 4H^+(aq)$$

the equilibrium constant is 1.58×10^2. Calculate $\Delta G°$ and $E°$ for this reaction.

The Nernst Equation

17.31 Find the voltage for each of the cells in Exercise 17.27 when the concentrations of the soluble species are as follows:
(a) $[Fe^{2+}] = 0.05\ M$, $[Zn^{2+}] = 1.0 \times 10^{-3}\ M$
(b) $[Fe^{2+}] = 0.20\ M$, $[Fe^{3+}] = 0.010\ M$, $[Cl^-] = 4.0 \times 10^{-3}\ M$
(c) $[Br^-] = 3.5 \times 10^{-3}\ M$, $[Cl^-] = 0.10\ M$, $P_{Cl_2} =$ 0.50 atm

17.32 Find the voltage for each of the cells in Exercise 17.28 when the concentrations of the soluble species are as follows:
(a) $[Cu^{2+}] = 0.050\ M$, $[Ni^{2+}] = 1.40\ M$
(b) $P_{Cl_2} = 320$ torr
(c) $[I^-] = 0.0010\ M$, $P_{Cl_2} = 0.300$ atm, $[Cl^-] =$ 0.60 M

17.33 What is the voltage of the half-reaction

$$Fe^{3+} + e^- \longrightarrow Fe^{2+}$$

when the concentrations in solution are $[Fe^{3+}] =$ 0.033 M and $[Fe^{2+}] = 0.0025\ M$?

17.34 What is the reduction potential of the hydrogen electrode if the pressure of gaseous hydrogen is 2.5 atm in a solution of pH 6.00?

17.35 A cell consists of a silver/silver chloride electrode and a reference electrode with a constant potential. The voltage of this cell is 345 mV when placed in a 0.200 M NaCl solution (the silver electrode is positive). What voltage is measured when the same cell is placed in a $3.0 \times 10^{-3}\ M$ HCl solution?

17.36 A cell consists of a lead electrode and a reference electrode with a constant potential. The voltage of this cell is 53 mV when placed in a 0.100 M $Pb(NO_3)_2$ solution (the lead electrode is positive). What voltage is measured when the same cell is placed in a saturated lead chloride solution, in which $[Pb^{2+}]$ is $1.6 \times 10^{-2}\ M$?

17.37 Using the cell in Exercise 17.35, a voltage of 0.425 V was measured when the cell was placed in a solution of unknown chloride ion concentration. What is the molar concentration of chloride ions in this test solution?

17.38 The cell in Exercise 17.36 produced a voltage of 0.010 V in an unknown test solution. What is the concentration of lead ions in the unknown solution?

Applications of Cells

17.39 Given the electrode potentials:

$$PbSO_4(s) + 2e^- \longrightarrow Pb(s) + SO_4^{2-}(aq)\ E° = -0.356\ V$$
$$Pb^{2+}(aq) + 2e^- \longrightarrow Pb(s)\ E° = -0.126\ V$$

(a) Use these electrode potentials to calculate the value of the solubility product constant for $PbSO_4$.
(b) Describe a cell in which the cell reaction (or the reverse of the cell reaction) is

$$PbSO_4(s) \longrightarrow Pb^{2+}(aq) + SO_4^{2-}(aq)$$

17.40 The solubility product constant for solid $Ba(IO_3)_2$ is 1.5×10^{-9}. Use the standard reduction potential of $Ba^{2+}(aq)$ to find the standard potential for the half-reaction

$$Ba(IO_3)_2(s) + 2e^- \longrightarrow Ba(s) + 2IO_3^-(aq)$$

(*Hint:* Find $\Delta G°$ for both the solubility equilibrium and the reduction half-reaction for Ba^{2+}, and add the reactions. Use the $\Delta G°$ for the sum reaction to find $E°$.)

17.41 Use the standard reduction potentials in Table 17.1 to find
(a) a reducing agent that will reduce Cu^{2+} but not Pb^{2+}
(b) an oxidizing agent that will react with Cu but not Fe^{2+}
(c) a metal ion that can reduce Fe^{3+} to Fe^{2+}

17.42 Use the standard reduction potentials in Table 17.1 to find
(a) a metal ion that reduces Ni^{2+}
(b) a metal ion that can oxidize Cu
(c) a metal ion that is reduced by Cr^{2+} but not H_2

Practical Cells for Energy

17.43 (a) Use standard reduction potentials to calculate the voltage of a nickel-cadmium cell that uses a basic electrolyte. The nickel is reduced from NiO(OH)(s) to

$Ni(OH)_2(s)$ and cadmium metal is oxidized to $Cd(OH)_2(s)$. (b) Will the voltage of this cell change as the cell is discharged? Explain.

17.44 Use standard reduction potentials to calculate the voltage of a silver-zinc cell that uses a basic electrolyte. The silver is reduced from Ag_2O to Ag, and zinc metal is oxidized to $Zn(OH)_2$. (b) Will the voltage of this cell change as the cell is discharged? Explain.

17.45 A possible reaction for a fuel cell is

$$CH_4(g) + 2O_2(g) \longrightarrow CO_2(g) + H_2O(\ell)$$

(a) Write the oxidation and reduction half-reactions that occur, assuming a basic electrolyte.
(b) Use standard free energies of formation from Appendix G to calculate $\Delta G°$ for a methane/oxygen cell. From the standard free energy change, calculate the voltage this cell could produce under standard conditions.
(c) Calculate the number of kilojoules of electrical energy produced when 1 g of the hydrocarbon is consumed.

17.46 A possible reaction for a fuel cell is

$$C_3H_8(g) + 5O_2(g) \longrightarrow 3CO_2(g) + 4H_2O(\ell)$$

(a) Write the oxidation and reduction half-reactions that occur, assuming a basic electrolyte.
(b) Use standard free energies of formation to calculate $\Delta G°$ for a propane/oxygen cell. From the standard free energy change, calculate the voltage this cell could produce under standard conditions.
(c) Calculate the number of kilojoules of electrical energy produced when 1 g of the hydrocarbon is consumed.

Electrolysis Reactions

17.47 Write the half-reactions and the chemical equations for the reactions that occur in the electrolysis of
(a) molten $CaCl_2$ using inert electrodes
(b) a saturated solution of magnesium sulfate solution using inert electrodes
(c) a nickel(II) bromide solution using nickel electrodes

17.48 Write the half-reactions and the balanced chemical equations for the reactions that occur in the electrolysis of
(a) a zinc chloride solution using zinc electrodes
(b) a sodium iodide solution using inert electrodes
(c) a calcium bromide solution using inert electrodes

17.49 A solution contains the ions H^+, Cu^{2+}, Ca^{2+}, and Ni^{2+}, each at a concentration of 1.0 *M*.
(a) Which of these ions would be reduced first at the cathode during an electrolysis?
(b) After the first ion has been completely removed by electrolysis, which is the second ion that will be reduced?
(c) Which, if any, of these ions cannot be reduced by the electrolysis of the aqueous solution?

17.50 A solution contains the ions H^+, Ag^+, Pb^{2+}, and Ba^{2+}, each at a concentration of 1.0 *M*.
(a) Which of these ions would be reduced first at the cathode during an electrolysis?
(b) After the first ion has been completely removed by electrolysis, which is the second ion that will be reduced?
(c) Which, if any, of these ions cannot be reduced by the electrolysis of the aqueous solution?

17.51 The commercial production of magnesium is accomplished by electrolysis of molten $MgCl_2$.
(a) Why isn't electrolysis of an aqueous solution of $MgCl_2$ used in this process?
(b) What are the anode and cathode half-reactions in the electrolysis of molten $MgCl_2$?

17.52 The commercial production of fluorine uses a mixture of molten potassium fluoride and anhydrous hydrogen fluoride as the electrolyte. The products of the electrolysis are hydrogen and fluorine.
(a) Why is the potassium fluoride necessary, since it is not involved in the redox reaction that occurs?
(b) What products would be formed if the hydrogen fluoride was not present?

Quantitative Aspects of Electrolysis

17.53 How many coulombs of charge are needed to accomplish the following conversions by electrolysis?
(a) Form 0.50 mole of Ca from $CaCl_2$.
(b) Produce 3.0 g Al by electrolysis of Al_2O_3.
(c) Form 0.52 g O_2 by electrolysis of an aqueous Na_2SO_4 solution.
(d) Make 1.0 L of gaseous H_2 at STP by electrolysis of water.

17.54 How many coulombs of charge are needed to accomplish the following conversions by electrolysis?
(a) Produce 0.50 mole of Al by electrolysis of Al_2O_3.
(b) Reduce all of the Cu^{2+} in 100 mL of 0.20 *M* $Cu(NO_3)_2$.
(c) Make 10.0 g of Cl_2 by electrolysis of molten NaCl.
(d) Deposit 0.32 g of silver from an aqueous $AgNO_3$ solution.

17.55 What mass of cadmium is deposited from a $CdCl_2$ solution by passing a current of 1.50 A for 38.0 minutes?

17.56 Find the mass of hydrogen produced by electrolysis of hydrochloric acid for 59.0 minutes using a current of 0.500 A.

17.57 How many minutes are needed to deposit 15.0 g of copper from a Cu^{2+} solution, using a current of 3.00 A?

17.58 One way that has been used to determine the quantity of charge that passes through an electrolysis cell is the silver coulometer. The same current is passed through two electrolysis cells in series, so the total charge through both cells is identical. If 0.158 g of silver is deposited in one cell, how many moles of Sn^{4+} are reduced to Sn^{2+} in the other cell?

17.59 How many seconds are needed to deposit 0.100 g of silver from a solution containing Ag^+, using a current of 250 mA?

17.60 An object with a surface area of 100 cm^2 is gold-plated. The source of the gold is a solution of $Au(CN)_4^-$. How many minutes does it take to cover the object with gold to a thickness of 0.0020 mm by using a current of 0.500 A? The density of gold is 19.3 g cm^{-3}.

Commercial Applications of Electrolysis

17.61 The electrolysis of molten Al_2O_3 at 980 °C is used to produce metallic aluminum. A current of 980 A at 4.85 V is used in the electrolysis cell.
(a) What is the rate of formation of aluminum in kilograms per hour?
(b) How much electrical energy is consumed to produce 1 kg of aluminum?
(c) Why isn't the electrolysis of an aqueous solution of aluminum chloride used to manufacture this metal?

17.62 (a) Find the current that must be used to produce 500 g of copper per hour by the electrorefining process.
(b) If 0.100 V is needed to maintain this current, how many kilowatt-hours of electrical energy are consumed in the refining of the 500 g of copper?

Corrosion

17.63 The electrochemical processes that occur in the corrosion of iron are represented by the half-reactions

$$Fe(s) \longrightarrow Fe^{2+}(aq) + 2e^-$$

$$O_2(g) + 4H^+(aq) + 4e^- \longrightarrow 2H_2O(\ell)$$

(a) Write the overall reaction.
(b) What is the standard potential for the overall chemical reaction?
(c) Natural water has a pH of about 5.9, and air is 21 mole percent oxygen. If the concentration of iron(II) in the water is 5×10^{-5} *M*, what is the potential of the corrosion reaction in the presence of air and natural water at 1 atm pressure and 298 K?

17.64 Would it be possible to use metallic sodium for cathodic protection of the iron hull of an ocean vessel? Explain.

17.65 Magnesium is used in the cathodic protection of metal coffins that are guaranteed to last a century without corroding. If the average current produced by the electrochemical cell is 2.0 mA, how much magnesium (in pounds) is needed to protect the coffin for 100 years?

17.66 An aluminum bulkhead in a swimming pool collapsed. To strengthen the bulkhead, stainless steel (mostly iron) braces were bolted to the aluminum. Within a few months the bulkhead collapsed again, and showed extreme corrosion of the aluminum close to the steel bolts. Explain the electrochemical processes that occurred in the reinforced bulkhead.

Additional Exercises

17.67 Sphalerite is zinc sulfide, a naturally occurring mineral, from which zinc metal is produced. The ore is first roasted in oxygen to form zinc oxide and sulfur dioxide, followed by the reaction of metal oxide with carbon.
(a) Write balanced equations for both of these reactions.
(b) For each reaction, identify the element that is oxidized and the one that is reduced.

17.68 The alkaline zinc-mercury cell was described in Section 17.8, in which the cell reaction is

$$Zn(s) + HgO(s) \longrightarrow Hg(\ell) + ZnO(s) \qquad E^\circ = 1.35 \text{ V}$$

(a) What is the standard free energy change for this reaction?
(b) The standard free energy change in a voltaic cell is the maximum electrical energy that the cell can produce. If a zinc-mercury cell consumes 1.00 g of mercury oxide, what is the maximum electrical energy that could be produced?
(c) For how many hours could a 10 mA-current be produced by a mercury cell in which the limiting reactant is 3.50 g of mercury oxide?

17.69 In the analytical technique called electrogravimetry, electrolysis is used to separate the analyte from a solution, by depositing it on an inert electrode. The electrode is weighed before and after the experiment to find the mass that was deposited. A 0.122 g-sample of a copper-zinc alloy was treated with concentrated sulfuric acid to produce a solution containing copper(II) and zinc(II) sulfates. The platinum cathode used in the electrolysis of this solution increased in mass by 0.073 g after exhaustive electrolysis.
(a) Which metal was deposited on the cathode during

the electrolysis? Write the balanced equation for the electrolysis reaction.

(b) What was the mass percent of zinc in the alloy sample?

(c) After the metal was completely deposited, electrolysis was continued. Write the balanced equation for the reaction that took place. Will this second reaction interfere in the analysis?

17.70 The solubility product constant for PbC_2O_4 is 8.5×10^{-10}, and the standard reduction potential of the Pb^{2+}/Pb couple is −0.126 V.

(a) Find the standard potential of the half-reaction

$$PbC_2O_4(s) + 2e^- \longrightarrow Pb(s) + C_2O_4^{2-}(aq)$$

(*Hint:* the desired half-reaction is the sum of the equations for the solubility product and the reduction of Pb^{2+}. Find $\Delta G°$ for these two reactions and add them to find $\Delta G°$ for their sum. Convert the $\Delta G°$ into the potential of the desired half-reaction.)

(b) Calculate the voltage of the PbC_2O_4/Pb electrode in 0.025 M solution of $Na_2C_2O_4$.

17.71 The solubility product constant for $Pb(IO_3)_2$ is 2.6×10^{-13}, and the standard reduction potential of the Pb^{2+}/Pb couple is −0.126 V.

(a) Find the standard potential of the half-reaction

$$Pb(IO_3)_2(s) + 2e^- \longrightarrow Pb(s) + 2IO_3^-(aq)$$

(*Hint:* the desired half-reaction is the sum of the equations for the solubility product and the reduction of Pb^{2+}. Find $\Delta G°$ for these two reactions and add them to find $\Delta G°$ for their sum. Convert the $\Delta G°$ into the potential of the desired half-reaction.)

(b) Calculate the voltage of the $Pb(IO_3)_2/Pb$ electrode in 3.5×10^{-3} M solution of $NaIO_3$.

17.72 The acid-base titration curves discussed in Chapter 14 can be determined using a pH meter, which measures the voltage of a cell that is comprised of a reference electrode and an indicating electrode that responds to the hydrogen ion concentration in solution. Assume that the cell potential, in volts, follows the equation

$$E_{cell} = k - 0.059 \text{ pH}$$

The potential of the cell is 135 mV at the start of a titration of a 0.032 M solution of a monoprotic strong acid with a strong base. What is the potential of this cell when the equivalence point is reached?

17.73 Use the standard reduction potentials in Appendix H to answer the following questions.

(a) What products, if any, are formed when $KClO_3$ and KCl are mixed in an acid solution?

(b) Which is a stronger oxidizing agent in acid solution, Fe^{3+} or Cr^{3+}?

(c) Which is a stronger reducing agent, $Fe(CN)_6^{4-}$ or Fe^{2+}?

17.74 Write the chemical equation, and calculate the standard potentials for each of the following, using the data in Appendix H.

(a) Fe^{3+} is added to $H_2SO_3(aq)$.

(b) Iron(II) is titrated with $K_2Cr_2O_7$.

(c) Nitrous acid (HNO_2) is added to $Fe^{2+}(aq)$.

Chapter Eighteen 18

Metallurgy, Transition Metals, and Coordination Chemistry

The majority of the elements are metals, and many of them have played key roles in the development of civilization. In our modern society metals are used as electrical conductors, structural materials, and reducing agents. Among the representative elements, sodium and other group IA elements are used as strong reducing agents, aluminum and magnesium are used in lightweight alloys, and lead is used in the manufacture of storage batteries. We have already discussed the chemistry of the group IA and IIA metals in Chapter 7.

With a few exceptions (Au, Ag, Ru, Rh, Pd, Ir, and Pt), all of the metals occur in nature as compounds, so extensive processing is needed to prepare the free elements. In the first section of this chapter, we discuss the isolation of the metals from their naturally occurring ores.

The transition elements and their compounds are important in all aspects of our society. Iron and its alloys are widely used structural materials. Copper is used in electrical wiring, silver compounds are used in photographic processes, titanium is used in light-weight alloys, and its compounds are used as paint pigments, gold has long been a standard of value and beauty, and platinum is an important catalyst. In addition, the transition elements play vital roles in the chemistry of biological systems. Iron, copper, cobalt, zinc, molybdenum, and many other metals are essential for many biological functions.

◀ Metallurgy has been practiced for many centuries.

Following the discussion of the isolation of metals from their ores in Section 18.1, the remainder of this chapter is devoted to the chemistry of the transition elements, with particular attention to the structure, bonding, and characteristic properties of coordination compounds.

18.1 Metallurgy

Objectives

- To state the principal goals of the pretreatment of ores in the metallurgical process
- To recognize and give examples of chemical and physical pretreatment of ores
- To describe and write equations for the chemical changes that occur in the reduction of iron ores
- To provide examples of metals that are isolated by electrolysis, reduction with more active metals, reduction with carbon, and thermal decomposition of compounds
- To list and give examples of the common techniques used to purify metals

With only a few exceptions, metals do not occur in nature as the free elements. The common ores for several elements are listed in Table 18.1. **Metallurgy** is the science of extracting metals from their ores, purifying and preparing them for practical use. Metallurgy is among the oldest chemical processes known, dating from before 1000 BC. Since the metals in ores are in positive oxidation states, the isolation of the elements involves chemical reductions. However, the overall processes involve several steps, including

1. Pretreatment of the ore to concentrate the valuable material and to convert it into compounds suitable for reduction
2. Reduction of the concentrated ore to the element
3. Purification of the metal

Table 18.1 Composition of Some Typical Metallic Ores

Type of Ore	Examples
oxide	Fe_2O_3, Fe_3O_4, Al_2O_3, SnO_2
sulfide	PbS, ZnS, FeS_2, HgS, Cu_2S
chloride	$NaCl$, KCl
carbonate	$FeCO_3$, $CaCO_3$, $MgCO_3$
sulfate	$BaSO_4$, $CaSO_4 \cdot 2H_2O$
silicate	$Be_3Al_2Si_6O_{18}$, $Al_2(Si_2O_8)(OH)_4$
free metal	Au, Ag, Pt, Pd, Rh, Ir, Ru

Pretreatment of Ores

The ores that are found in nature are usually complex mixtures of several minerals. Some minerals contain the desired element but are often mixed with others that interfere with the isolation of the metal. The ore is usually pulverized as the first step. Differences in physical and chemical properties then make it possible to concentrate the minerals containing the metal and separate them from impurities before attempting to extract the metal.

Panning for gold Separation of gold from other materials in the ore takes advantage of the higher density of the metal.

Concentration

Several physical processes are used to concentrate ores. A procedure used in the recovery of gold takes advantage of the high density of the metal. When gold ore is stirred under a flow of water, the less dense materials, often silicates, are suspended and carried away with the water, while the metallic gold settles to the bottom of the container. This method was used by the prospectors in the 1849 gold rush in California when they panned for gold.

Another process, often used to concentrate sulfide ores (PbS and ZnS, for example) is called flotation. The ore is added to a mixture of water and oil with other additives. The resulting mixture is then stirred while a stream of air is bubbled through it. The metal sulfides become coated with the oil and are carried to the surface as a foam, while the unwanted material, called gangue, settles to the bottom of the container and is later discarded.

Flotation is an example of concentrating ores by using differences in physical properties.

Chemical processes for concentrating ores vary greatly, since they depend on the properties of the compounds containing the desired metals. As

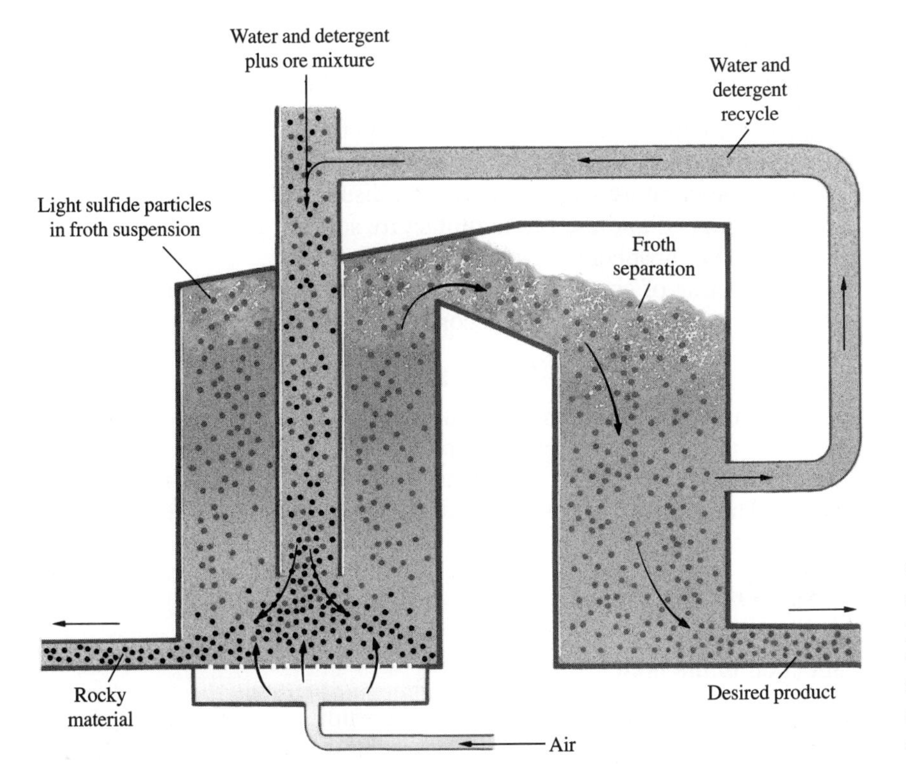

Ore concentration by flotation A schematic diagram of the flotation process used to concentrate sulfide ores. The oil adheres to the metal sulfides and forms a foam with the air bubbles, which carry the desired compounds to the surface.

Aluminum ores are concentrated by using the differences in chemical properties of aluminum oxide and iron(III) oxide.

an example, the principal ore of aluminum is bauxite (hydrated aluminum oxide, $Al_2O_3 \cdot xH_2O$), which usually contains iron oxide as a major impurity. Treatment of the ore with aqueous sodium hydroxide dissolves the amphoteric Al_2O_3, leaving the other metal oxides as solids.

$$Al_2O_3(s) + 2OH^-(aq) + 3H_2O(\ell) \longrightarrow 2Al(OH)_4^-(aq)$$

After the solution is separated from the solid residue, acid or carbon dioxide is added to precipitate $Al(OH)_3$, which is heated to produce the purified oxide.

$$Al(OH)_4^-(aq) + H^+(aq) \longrightarrow Al(OH)_3(s) + H_2O(\ell)$$

$$2Al(OH)_3(s) \xrightarrow{\text{heat}} Al_2O_3(s) + 3H_2O(g)$$

Roasting

Another process in the pretreatment of some ores, called **roasting,** consists of heating the material at a temperature below its melting point, usually in the presence of air, to convert the ore into a chemical form more suitable for the reduction step. The dehydration of aluminum hydroxide shown in the previous paragraph could be classified as roasting. More commonly, the roasting step converts sulfide and carbonate ores into oxides. Galena (PbS), pyrite (FeS_2), and sphalerite (ZnS) are all converted to the metal oxides by roasting the minerals in air.

$$2ZnS(s) + 3O_2(g) \longrightarrow 2ZnO(s) + 2SO_2(g)$$

$$2PbS(s) + 3O_2(g) \longrightarrow 2PbO(s) + 2SO_2(g)$$

$$3FeS_2(s) + 8O_2(g) \longrightarrow Fe_3O_4(s) + 6SO_2(g)$$

Sulfide ores are roasted to form the metal oxides, which can often be reduced by carbon, a relatively inexpensive reducing agent.

The conversion from sulfides to oxides allows the use of carbon in the later reduction to the metal, because the carbon compound that is formed, carbon dioxide, is much more stable than carbon disulfide. In modern operations the SO_2 is trapped and used to manufacture sulfuric acid. Any SO_2 that is vented to the atmosphere is a serious pollutant.

The roasting of cinnabar (HgS) produces the free element, since mercury compounds are very easily decomposed to the metal.

$$HgS(s) + O_2(g) \longrightarrow Hg(\ell) + SO_2(g)$$

Extreme care must be taken in the production and handling of mercury, because of that element's very high toxicity.

Carbonates decompose to the metal oxide and carbon dioxide upon heating. The roasting of limestone ($CaCO_3$), for example, results in the chemical reaction

$$CaCO_3(s) \longrightarrow CaO(s) + CO_2(g)$$

Reduction to the Metal

Different reduction methods are chosen, depending on the reactivity of the metal. The very reactive metals, including most of the alkali and alkaline earth elements, must be reduced by electrolysis of molten salts, because there

Table 18.2 Reduction Methods for Several Metals

Metal	Method of Reduction
Li, Na, Mg, Ca, Al	Electrolytic reduction of molten salts
Cr, Mn, Ti, Ta	Reduction of oxides by more active metals
Fe, Zn, Pb	Reduction of oxides by carbon and CO
Hg, Au, Ag, Cu	The uncombined metal occurs in nature or is produced by roasting the sulfides

are no inexpensive chemical reducing agents that are strong enough to accomplish the task. The displacement of metals from the oxides by reaction with a more reactive metal is used when carbon or carbon monoxide are not strong enough reducing agents. For example, the ore pyrolusite, MnO_2, is reduced using aluminum at high temperature.

$$3MnO_2(s) + 4Al(\ell) \longrightarrow 3Mn(\ell) + 2Al_2O_3(s)$$

Whenever possible, carbon is used as the reducing agent, because it is fairly abundant and inexpensive. The least reactive metals occur in nature as the uncombined elements or are released during the roasting of the ore. Table 18.2 shows the reduction methods used for isolating several of the common metals.

Iron is by far the most widely produced metal, and it is typically isolated from oxide ores by using carbon as the reducing agent. The reduction is carried out in a blast furnace, shown in Figure 18.1. The iron ore, mixed with coke (mainly carbon) and limestone ($CaCO_3$), is added continuously to the

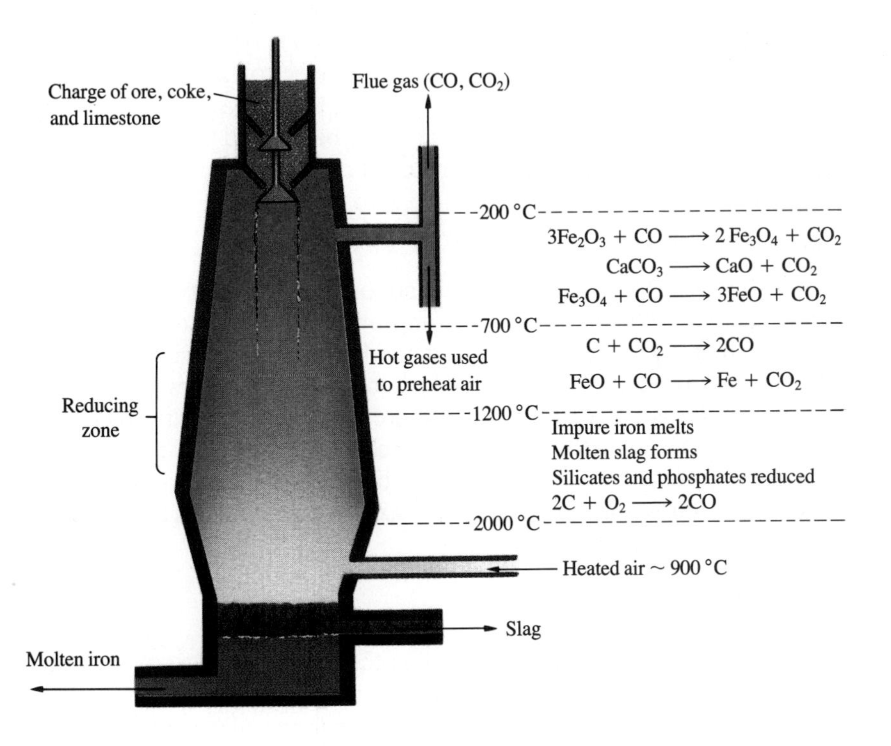

Figure 18.1 The blast furnace A schematic diagram of the blast furnace. The iron ore, coke, and calcium carbonate are added at the top of the furnace, and molten iron and slag are removed at the bottom.

top of the furnace. Hot air is fed into the furnace at the bottom. Several chemical reactions are involved in the overall process that occurs in the blast furnace. The oxygen in the air combines with the carbon, forming carbon monoxide. This exothermic reaction maintains the high temperatures needed in the furnace.

$$2C(s) + O_2(g) \longrightarrow 2CO(g) \qquad \Delta H° = -221.0 \text{ kJ}$$

Carbon monoxide is formed rather than carbon dioxide because oxygen is the limiting reactant. The hot gases rise through the furnace, mixing with the ore. Carbon monoxide is the principal reducing agent, leading to the production of molten iron.

$$Fe_2O_3(s) + 3CO(g) \longrightarrow 2Fe(\ell) + 3CO_2(g)$$

The limestone decomposes at the high temperatures in the furnace to form carbon dioxide and calcium oxide. The calcium oxide combines with impurities in the ore, mostly aluminum oxide and silicon dioxide, to form molten silicates and aluminosilicates.

$$CaCO_3(s) \longrightarrow CaO(s) + CO_2(g)$$

$$CaO(s) + SiO_2(s) \longrightarrow CaSiO_3(\ell)$$

$$CaO(s) + Al_2O_3(s) \longrightarrow Ca(AlO_2)_2(\ell)$$

Purification of metals is achieved by separations based on either chemical or physical properties.

The mixture of calcium silicate and calcium aluminate is called slag, and it is a liquid at the temperatures inside the blast furnace. The liquid iron and slag fall to the bottom of the furnace, forming two separate layers as shown in Figure 18.1. These are drawn off separately, and the iron is cast into bars, called *pig iron.* The pig iron is fairly impure, containing up to 5% carbon and silicon, manganese, and several other impurities, so this crude material must be further purified before it can be used in modern applications.

The Hall process for production of aluminum is typical of commercial processes that use electrolytic reduction. The electrolysis process was described in Section 17.11. Because of its cost, electrolysis is used only to produce the more reactive metals that cannot be isolated from their ores by using carbon or other inexpensive chemical reducing agents.

Purification of Metals

The method used to purify a metal depends on the chemical properties of the particular metal, the nature of the common impurities, and the degree of purity needed. For many metals, particularly iron, more desirable physical and chemical properties are obtained when they are mixed with other elements. An *alloy* is a mixture of two or more elements that has the properties of a metal. Thus in the production of steel, an alloy of iron, the complete removal of chromium and carbon from the crude pig iron is not required. Some of the more common techniques used to purify metals are described briefly in the following paragraphs.

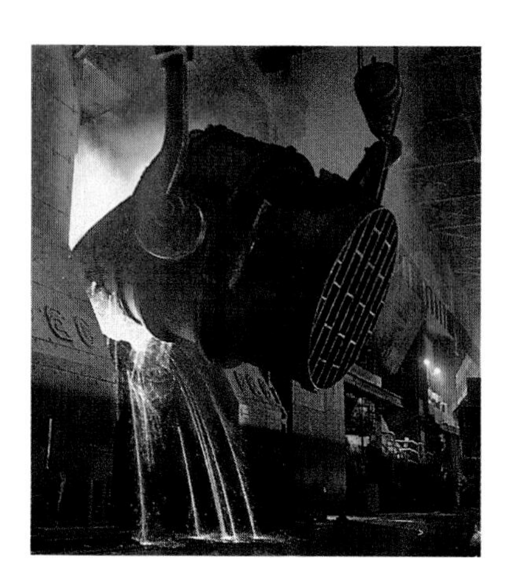

Purified molten iron is poured into molds.

Distillation

Several metals are sufficiently volatile that they can be purified by fractional distillation. Mercury, zinc, and magnesium are all refined (purified) by distillation. For the more reactive metals, the absence of oxygen is essential during the high temperature distillation, so the separation is carried out under an inert atmosphere or in a vacuum. Vacuum distillation offers a second advantage, since it lowers the temperature needed for the distillation.

The Mond process for purifying nickel is an interesting example that uses volatility as a method of separation. The impure nickel is treated with carbon monoxide at about 70 °C to form the compound $Ni(CO)_4$, tetracarbonylnickel(0), which has a boiling point of 43 °C. After the gas is separated from the solid impurities in the nickel, the mixture of CO and $Ni(CO)_4$ is heated to a higher temperature at which the nickel compound decomposes, yielding the nonvolatile metal.

$$Ni(CO)_4(g) \xrightarrow{200\ °C} Ni(s) + 4CO(g)$$

The carbon monoxide released in this decomposition is recycled through the reactor.

Electrolysis for purification Plates of copper produced in the electrorefining of the metal. Impure copper is dissolved at the anode and deposited in much higher purity at the cathode.

Electrolysis

The electrorefining of copper was described in Section 17.11. Other metals that are purified by electrorefining are cobalt, lead, and plutonium.

Zone Refining

When extremely high purity is needed, the process of zone refining can be used. This technique is described in Section 19.4 for the purification of the nonmetal silicon. Because zone refining is a relatively expensive process, and is effective only in removing small amounts of impurities, it is used when a very high purity (99.99+%) of the element is essential.

Refining of Iron and Manufacturing of Steel

Before iron is used, the pig iron produced in the blast furnace must be further refined to remove most of the nonmetal impurities, which include silicon, sulfur, phosphorus, and relatively large amounts of carbon. The *basic oxygen process* is the most widely used process to manufacture steel from the impure pig iron. In a typical operation the vessel is charged with 75 tons of molten iron and 10 tons of limestone. Oxygen gas, sometimes diluted with argon, is forced into the bottom of the converter. The nonmetal impurities are rapidly converted into oxides in exothermic reactions. The heat released by these oxidations maintains the temperature of the mixture well above the melting point of the iron. The oxides of silicon and phosphorus combine with calcium oxide to form a slag. The entire process takes about 20 minutes; the converter is then tilted to pour the molten metal into molds. Some of the important chemical reactions involved are

$$C + O_2 \longrightarrow CO_2$$

$$S + O_2 \longrightarrow SO_2$$

$$Si + O_2 \longrightarrow SiO_2$$

$$CaO + SiO_2 \longrightarrow CaSiO_3$$

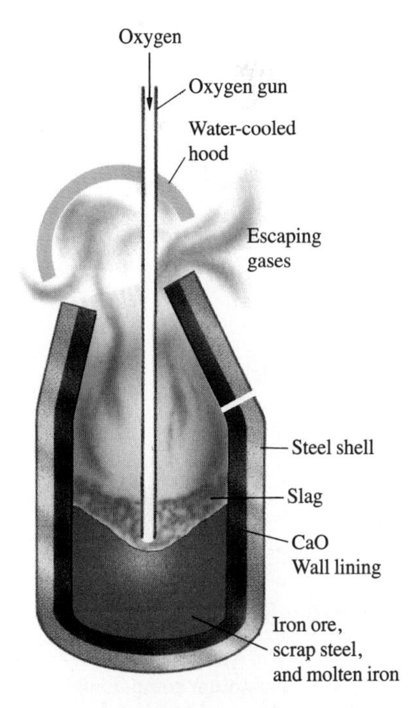

The oxygen furnance Pig iron is further purified in an oxygen furnace. At high temperatures, oxygen removes most of the nonmetal impurities.

18.2 Properties of the Transition Elements

Objectives

- To relate the atomic properties of the transition elements, such as atomic radius and ionization energy, to their relative positions in the periodic table
- To describe the effect of the lanthanide contraction on the sizes and ionization energies of the transition elements
- To use the model of metallic bonding in explaining the trends in physical properties (melting points, boiling points, hardness) observed across a transition series

Transition elements have partially filled *d* orbitals in the metal or one of its oxidation states, and include all of the elements in groups IIIB through IB.

The **transition elements** are characterized by having a partially filled *d* subshell in the metal atom or one of its oxidation states. These elements are found in the center of the periodic table, in the nine columns from group IIIB through group IB (3 to 11). The elements in Group IIB (12), the zinc family (Zn, Cd, and Hg), are not transition metals as we have defined them. Even though they are found in the "*d*-block" of the periodic table, their *d* orbitals are always completely filled. In contrast, the elements La (atomic number 57) and Lu (atomic number 71) do fit the definition of transition elements, since they both contain a single *d* electron in addition to the filled 6*s* subshell. Sometimes the lanthanides and actinides, the two rows of 14 elements across the bottom of the periodic table, are included in the definition of transition elements. We have chosen to refer to these elements that have partially filled *f* orbitals as the *inner transition elements*.

Figure 18.2 Vanadium compounds All of the oxidation states of vanadium are colored. From left to right are V^{2+} (violet), V^{3+} (green), VO^{2+} (blue), and VO_2^+ (yellow).

The transition elements exhibit several properties that distinguish them from the representative elements. Unlike the representative elements, all of the transition elements have high melting points. With the exception of the group IIIB elements, each of the transition elements can have more than one stable oxidation state in its compounds. Transition metal compounds are generally colored, and the color depends on the oxidation state of the metal, as well as on the other elements in the compound. For example, Figure 18.2 shows solutions that contain compounds of vanadium in four different oxidation states, each of which has a different color.

In this section we shall examine several of the properties of the transition metals, and relate them to the atomic properties.

Melting Points and Boiling Points

Table 18.3 lists the melting points and boiling points for the transition elements, and the melting points for the fourth period transition elements are shown graphically in Figure 18.3. These properties reach a maximum in group VB, VIB, or VIIB, suggesting that the metallic bonds are strongest in those groups. If we assume that only the $(n-1)d$ and ns orbitals are used by the transition metals in the formation of the metallic bonds, then a metal atom can form a maximum of one bond for each singly occupied *s* and *d* orbital. The number of electrons available on each atom to form bonds increases from three for the group IIIB (3) elements, up to six in group VIB (6). In group VIIB, one of the six available orbitals must be occupied by two electrons and is no longer available to form a metal-metal bond, which

Table 18.3 The Melting and Boiling Points of the Transition Elements, in °C

		Fourth Period			Fifth Period			Sixth Period		
			m.p.	b.p.		m.p.	b.p.		m.p.	b.p.
IIIB	(3)	Sc	1541	2831	Y	1522	3338	La	921	3457
IVB	(4)	Ti	1660	3287	Zr	1852	4377	Hf	2227	4602
VB	(5)	V	1890	3380	Nb	2468	4742	Ta	2996	5425
VIB	(6)	Cr	1857	2672	Mo	2617	4612	W	3410	5660
VIIB	(7)	Mn	1244	1962	Tc	2172	4877	Re	3180	5627
VIIIB	(8)	Fe	1535	2750	Ru	2310	3900	Os	3045	5027
VIIIB	(9)	Co	1495	2870	Rh	1966	3727	Ir	2410	4130
VIIIB	(10)	Ni	1453	2732	Pd	1552	2970	Pt	1772	3827
IB	(11)	Cu	1083	2567	Ag	962	2212	Au	1064	2807

results in fewer bonds per metal atom. With each successive group after that, the number of doubly occupied orbitals increases by one, causing a further reduction in the number of bonds each metal atom may form.

It should be noted that all of the metals crystallize in body-centered cubic, cubic closest-packing, or hexagonal closest-packing arrays, and thus each atom has either eight or twelve nearest neighbors. Since the transition metals can form a maximum of six bonds, the electrons must occupy highly delocalized orbitals, causing the high electrical conductivity observed for metals.

Several other properties, such as heats of fusion and hardness, reflect the strength of the bonding in the transition elements. It is the strong metallic

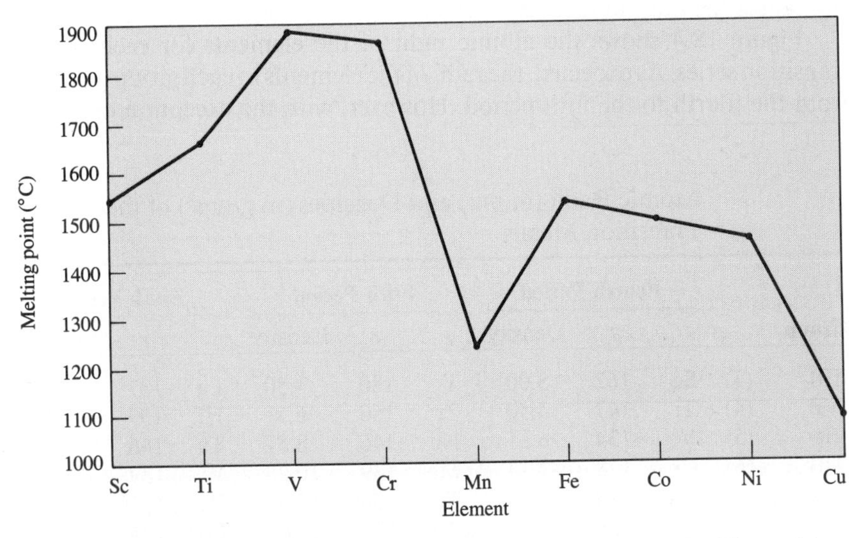

Figure 18.3 The melting points of the fourth period transition metals The melting points of the transition metals are highest in groups VB and VIB, consistent with the model of metallic bonding presented in the text. From the graph it can be seen that other factors also contribute to the variation of the melting points.

bonding that gives the transition elements the properties that make them particularly desirable as structural materials.

Atomic and Ionic Radii

The trends in the chemical properties of the transition metals in each group parallel the changes in atomic and ionic radii. The atomic radius of each transition element is given in Table 18.4. In the fourth period the atomic radius decreases fairly rapidly through the element chromium, in group VIB. After chromium, the radius decreases much more slowly, and actually increases from nickel to copper. This pattern is repeated for the elements in the fifth and sixth periods. The general trend of decreasing radius from left to right within a period is the same as that observed with the representative elements, but the changes in size are not as great for the transition elements.

It is the outermost electrons (those with the highest principal quantum number) and the effective nuclear charge to which they are subject that determine the sizes of the atoms (see Section 7.2). Unlike the representative elements, the electrons added to the transition elements do not enter the highest occupied principal shell, but the d subshell with a smaller value of n. Since the added d electrons in the transition elements are closer to the nucleus, they are very effective in shielding the outer s electrons (the ones that determine the size of the metal atom) from the nuclear charge. Therefore, the effective nuclear charge experienced by the outer s electrons increases quite slowly across the transition elements, resulting in a relatively small decrease in the radius of the atoms. The strength of the metallic bonding also contributes to the atomic size. Up through group VIB the metallic bonds increase in strength, reducing the distance between nearest neighbors in the solid, and thus contribute to the more rapid decrease in the radius. Toward the end of the "d block," the decrease in metal-metal bonding causes the radii to increase slightly.

The radii of transition metals decrease slowly from left to right through group VIIIB and then increase slightly.

Figure 18.4 shows the atomic radii of the elements for each of the transition series. As expected, the radii of the elements in each group increase from the fourth to the fifth period. However, with the exception of group

Table 18.4 Atomic Radii (in pm) and Densities (in g/cm³) of the Transition Metals

Group		Fourth Period			Fifth Period			Sixth Period		
			r	Density		r	Density		r	Density
IIIB	(3)	Sc	162	3.00	Y	180	4.50	La	187	6.17
IVB	(4)	Ti	147	4.50	Zr	160	6.51	Hf	159	13.28
VB	(5)	V	134	6.11	Nb	146	8.57	Ta	146	16.65
VIB	(6)	Cr	128	7.14	Mo	139	10.28	W	139	19.30
VIIB	(7)	Mn	127	7.43	Tc	136	11.5	Re	137	21.00
VIIIB	(8)	Fe	126	7.87	Ru	134	12.41	Os	135	22.57
VIIIB	(9)	Co	125	8.90	Rh	134	12.39	Ir	136	22.61
VIIIB	(10)	Ni	124	8.91	Pd	137	12.0	Pt	139	21.41
IB	(11)	Cu	128	8.95	Ag	144	10.49	Au	144	19.32

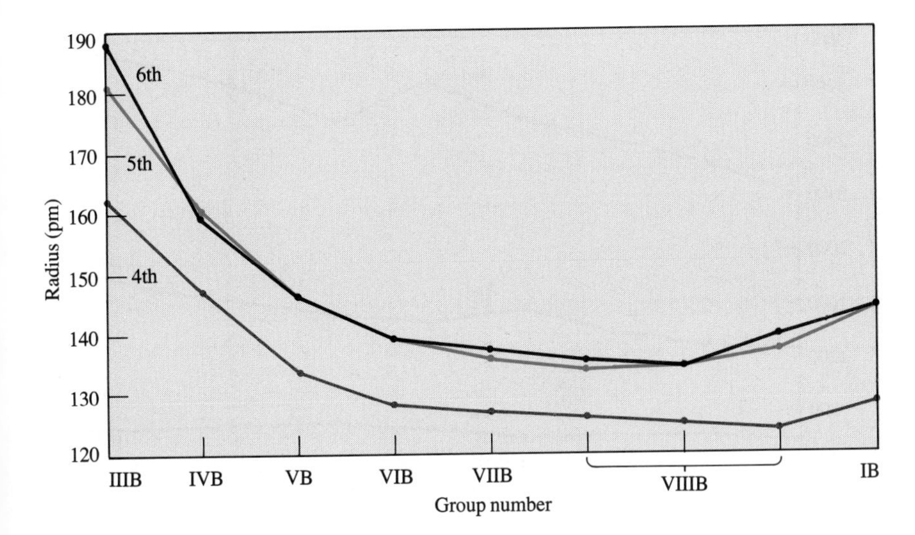

Figure 18.4 Atomic radii of the transition elements In the three transition series, the atomic radii decrease from left to right, with a small increase near the end of the series. With the exception of group IIIB, the atoms in each group of the 4*d* and 5*d* transition series have nearly identical radii because of the lanthanide contraction.

The lanthanide contraction causes the sizes and chemical properties of the fifth and sixth period transition elements to be very similar.

IIIB, the radii of the fifth and sixth period elements in each group are nearly identical. This can be explained by the **lanthanide contraction.** Between the elements lanthanum and hafnium are the 14 elements (the lanthanides) in which the 4*f* subshell is filled. There is a very small decrease in the radius of each of the 14 successive elements in the inner transition series, caused by a slight increase in the effective nuclear charge for each successive element. The contraction in size over the entire 14 elements is sufficient to reduce the sizes of the sixth period transition metal atoms to nearly the same radii as those in the fifth period.

The radii of the positively charged transition metal ions, as expected, are smaller than the radii of the parent metal atoms, and for a given charge they decrease slightly from left to right within any period. The ionic radii of the fifth and sixth row transition elements, just like the atomic radii, are almost the same within any group because of the lanthanide contraction.

The trends in the chemical properties of the transition metals in each group parallel the changes in the radii within each group. There is usually a much larger difference between the chemical behavior of the first and second elements in a transition metal group than is observed between the second and third elements. Because of the lanthanide contraction, zirconium and hafnium in group IVB have nearly the same atomic radii, and their chemistries are nearly identical, but quite different from that of titanium. The chemical behavior of zirconium and hafnium is so similar that the two elements are very difficult to separate.

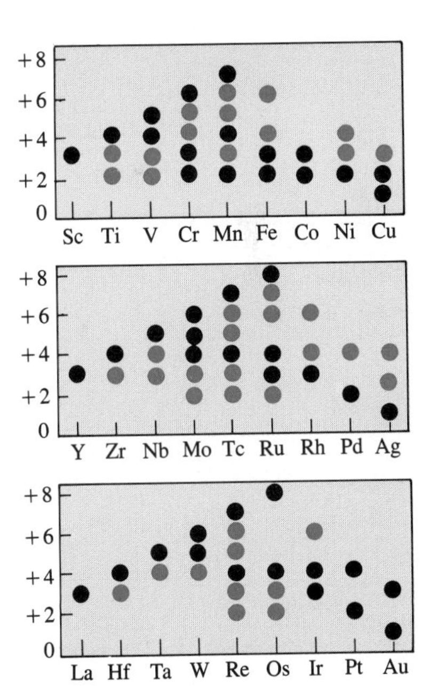

Figure 18.5 Stable oxidation states of the transition metals The stable nonzero oxidation states of the transition metals are shown. The more common oxidation states are indicated by black circles. With very few exceptions, negative oxidation states are not observed for these metallic elements.

Oxidation States and Ionization Energies

With the exception of the group IIIB elements, each of the transition metals has at least two stable oxidation states. The full range of oxidation states for the transition elements is shown in Figure 18.5. The maximum positive oxidation state is equal to the group number for groups IIIB through VIIB (Mn) and the first group VIIIB elements in the fifth and sixth periods (Ru and Os). Note that the group number for these metals equals the total

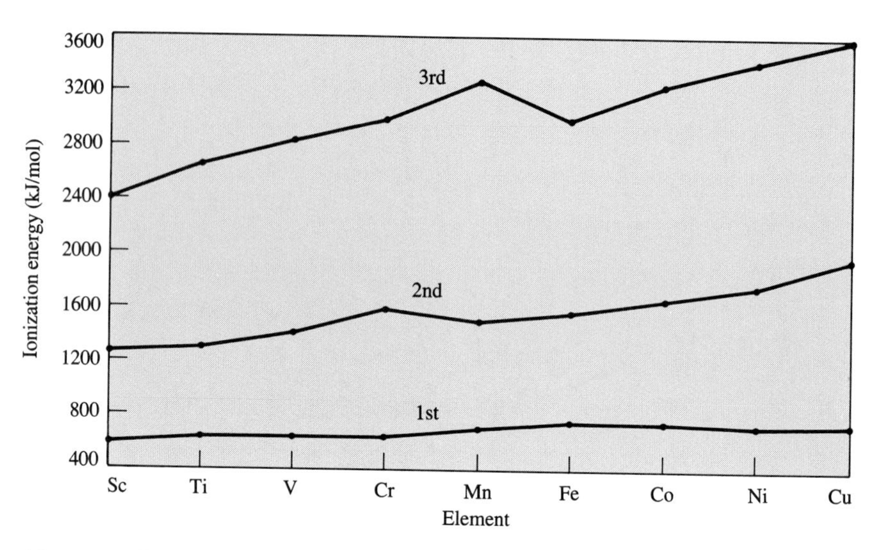

Figure 18.6 Ionization energies of transition elements The first, second, and third ionization energies of the fourth period transition metals.

number of valence-shell electrons (the *ns* and $(n-1)d$ electrons) in the element. To the right of the elements Mn, Ru, and Os in each period, the maximum positive oxidation state observed for each successive element decreases. Generally the highest oxidation state of each of the transition elements is observed only when the metals are combined with the very electronegative elements oxygen and fluorine.

All but the highest oxidation state of the transition metals are characterized by partially filled *d* orbitals. The availability of *d* orbitals in these compounds causes them to exhibit a wide range of colors, and leads to interesting magnetic properties. Both of these topics will be discussed in detail later in this chapter. The large number of oxidation states exhibited by the transition elements leads to a very extensive oxidation-reduction chemistry.

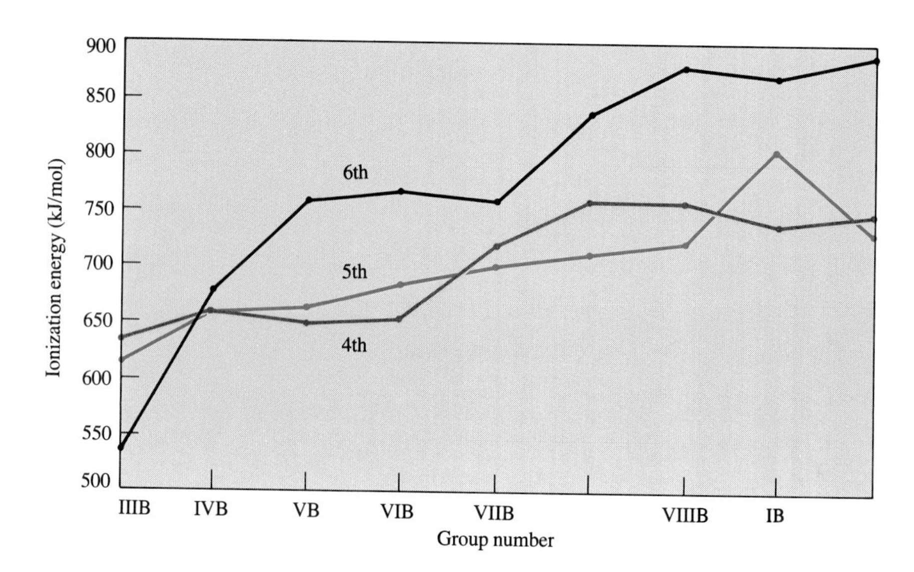

Figure 18.7 First ionization energies For the transition elements there is an increase in the ionization energies within each group. The large increase in ionization energy between the fifth and sixth period elements is partially a result of the lanthanide contraction.

The first, second, and third ionization energies for the fourth period transition metals are shown graphically in Figure 18.6. Each of these lines shows a general increasing trend from left to right, with the difference between the successive ionization energies getting larger for each successive element. In contrast to the behavior of the representative elements, the first ionization energy increases from the top to the bottom within each periodic group. For the transition elements, as can be seen in Figure 18.7, the change in ionization energy from the fourth to the fifth period is very small, and may increase or decrease. The larger increase in the ionization energy that is observed from the fifth to the sixth period is at least partially explained by the lanthanide contraction.

18.3 Chemistry of Selected Transition Elements

Objectives

- To write equations for some important reactions that involve the elements chromium, iron, and copper
- To write equations for the reactions involved in the isolation of chromium from ores

The chemistry of the transition metals is quite varied, producing many colored compounds that range in chemical activity from strong oxidizing agents to strong reducing agents. Many of the compounds of transition metals in the lower oxidation states (up to +3) are ionic, although others—such as anhydrous $FeCl_3$—are covalent. In high oxidation states the transition metals typically form oxy-anions, which usually have the same formulas as those of the representative elements that share the same group number. For example, manganese (group VIIB) and chlorine (group VIIA) form the MnO_4^- and ClO_4^- anions where they are in the +7 oxidation state. In this section the chemistries of the elements chromium, iron, and copper are described briefly, as examples of typical transition metals.

Chromium

Although chromium is a relatively rare element, it is very important to us because of its desirable properties. The metal is very bright and corrosion-resistant. A thin impervious layer of oxide forms on the surface of the metal, protecting it from corrosion. Automobile bumpers that are made of steel are usually coated with a layer of chromium, for appearance as well as protection of the iron from corrosion. Chromium is also used in iron alloys (for example, in stainless steels) to improve the physical properties of the metal, such as hardness and resistance to corrosion.

Chromite, $FeCr_2O_4$, is the principal chromium ore. Reduction of this mineral by carbon produces an alloy of iron called ferrochrome. The chromium in iron alloys is added as ferrochrome, without first separating the two metals. To produce pure chromium, the chromite ore is subjected to extensive chemical pretreatment. The chromite is first heated in the presence of limestone ($CaCO_3$) and air. The oxygen in the air reacts with the chromium to form calcium chromate.

Figure 18.8 Ammonium dichromate decomposition The exothermic decomposition of ammonium dichromate to chromium(III) oxide is one step in the pretreatment of chromite ore before the isolation of chromium. The decomposition was once used in demonstrations of a "chemical volcano" but it has been abandoned for safety reasons.

$$4FeCr_2O_4 + 7O_2 + 8CaCO_3 \longrightarrow 8CaCrO_4 + 8CO_2 + 2Fe_2O_3$$

The mixture is dissolved in water to separate the chromate from the insoluble iron(III) oxide. Ammonium dichromate, $(NH_4)_2Cr_2O_7$, precipitates from the solution after acidification and addition of ammonium chloride.

$$2CrO_4^{2-}(aq) + 2H^+(aq) \longrightarrow Cr_2O_7^{2-}(aq) + H_2O(\ell)$$

$$2NH_4^+(aq) + Cr_2O_7^{2-}(aq) \longrightarrow (NH_4)_2Cr_2O_7(s)$$

The solid ammonium dichromate is heated to cause it to decompose to chromium(III) oxide and gaseous products (Figure 18.8).

$$(NH_4)_2Cr_2O_7(s) \longrightarrow Cr_2O_3(s) + N_2(g) + 4H_2O(g)$$

The final step is the thermal reduction of the chromium(III) oxide with aluminum.

$$Cr_2O_3(\ell) + 2Al(s) \longrightarrow Al_2O_3(s) + 2Cr(\ell)$$

The last reaction is so exothermic that molten chromium is produced.

Most of the compounds of chromium contain the element in the +2, +3, or +6 oxidation state. The most stable compounds of chromium contain Cr(III). The metal dissolves slowly in hydrochloric or sulfuric acid, liberating hydrogen and forming bright blue solutions that contain Cr^{2+} ions.

$$Cr(s) + 2H^+(aq) \longrightarrow Cr^{2+}(aq) + H_2(g) \qquad E° = +0.91 \text{ V}$$

In spite of the favorable potential for its oxidation by hydrogen ions, chromium does not dissolve in nitric acid, because a film of oxide is formed by reaction with the nitrate ions, protecting the metal from attack by the hydrogen ions.

Solutions of chromium(II) are oxidized to chromium(III) unless the solution is protected from air.

$$4Cr^{2+}(aq) + O_2(g) + 4H^+(aq) \longrightarrow 4Cr^{3+}(aq) + 2H_2O(\ell)$$

Solutions of chromium(III) nitrate are purple. The colors of chromium(III) compounds, as with most transition compounds, often depend upon other species present (Figure 18.9). For example, solutions of chro-

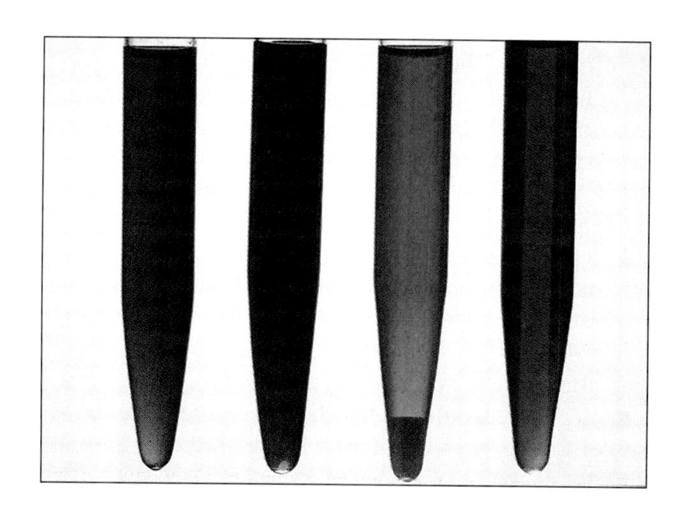

Figure 18.9 Aqueous solutions of Cr(II) and Cr(III) On the left is chromium (II) in sulfuric acid, followed by chromium (III) nitrate (purple), chromium (III) chloride (green), and chromium (III) in an ammonia solution.

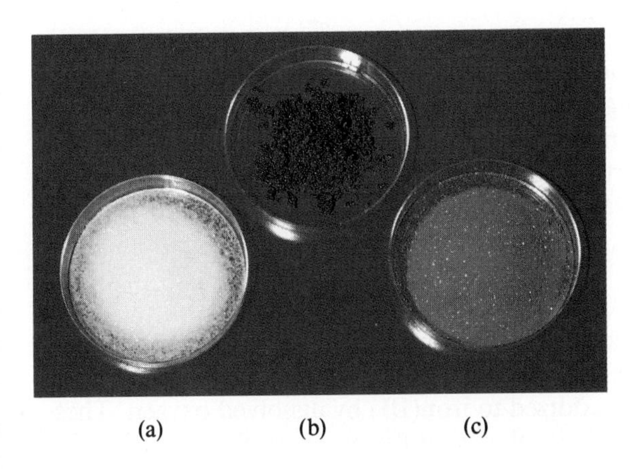

(a) (b) (c)

Figure 18.10 Chromium(VI) compounds From left to right are samples of potassium chromate, chromium trioxide, and potassium dichromate. Each is a different color, but all contain Cr(VI). Chromium(VI) compounds have been shown to be carcinogenic, so proper safety precautions should be taken.

mium(III) chloride are green, and chromium(III) in ammonia solution is red. Addition of base to Cr^{3+} forms a dark green precipitate of $Cr(OH)_3$, which is easily dehydrated to green Cr_2O_3. Chromium(III) hydroxide is amphoteric, and will dissolve in an excess of strong base, forming a green solution containing anionic species. The origin of these different colors is explained in Section 18.6.

In basic solution, chromium(III) is easily oxidized by the hydrogen peroxide ion and other mild oxidizing agents to the yellow chromate ion, CrO_4^{2-}, which contains the metal in the +6 oxidation state.

$$2Cr(OH)_4^-(aq) + 3HO_2^-(aq) \longrightarrow \underset{\text{chromate}}{2CrO_4^{2-}(aq)} + OH^-(aq) + 5H_2O(\ell)$$

Addition of acid to a chromate solution produces the orange-red dichromate ion, which also contains Cr(VI).

$$2CrO_4^{2-}(aq) + 2H^+(aq) \longrightarrow \underset{\text{dichromate}}{Cr_2O_7^{2-}(aq)} + H_2O(\ell)$$

In acid solution, chromium(VI) is a strong oxidizing agent, as indicated by its very positive standard reduction potential.

$$Cr_2O_7^{2-} + 14H^+(aq) + 6e^- \longrightarrow 2Cr^{3+}(aq) + 7H_2O(\ell) \quad E° = +1.33 \text{ V}$$

Many oxidation-reduction titrations use potassium dichromate solution as the titrant.

When a solution of sodium or potassium dichromate is treated with concentrated sulfuric acid, a bright red precipitate of CrO_3 is formed (Figure 18.10). This oxide melts at a relatively low temperature (197 °C), consistent with the covalent (rather than ionic) bonding expected for metallic elements in high oxidation states. Above its melting point CrO_3 loses oxygen readily, forming Cr_2O_3.

The chromium is in the +3 or +6 oxidation states in most of the common compounds of that element.

Iron

Metallic iron has been known since antiquity, and it has played a major role in the development of modern technology. The desirable properties of iron and its alloys are essential in manufacturing most of the machinery we use.

Iron is the fourth most abundant element in the earth's crust (after oxygen, silicon, and aluminum) and occurs mostly in the form of oxides. It is isolated from its ores by reducing the oxides with carbon (see Section 18.1). Its compounds are found in many biological systems, most importantly in the hemoglobin that transports oxygen in the blood.

Metallic iron dissolves in hydrochloric or sulfuric acid to produce iron(II) ions.

$$Fe(s) + 2H^+(aq) \longrightarrow Fe^{2+}(aq) + H_2(g) \qquad E° = +0.44 \text{ V}$$

Aqueous solutions of iron(II) compounds must be protected from the air, or oxygen oxidizes the metal ion to iron(III).

The $Fe(H_2O)_6^{2+}$ ion in solution has a very pale green color. It is a moderately good reducing agent, as it is oxidized to iron(III). Solutions of iron(II), when exposed to air, are slowly oxidized to iron(III) by dissolved oxygen. These solutions usually have a yellow color from the presence of hydroxide complexes, which form when one or more of the water molecules attached to the Fe^{3+} ion loses a proton:

$$Fe(H_2O)_6^{3+}(aq) + H_2O(\ell) \rightleftharpoons Fe(H_2O)_5OH^{2+}(aq) + H_3O^+(aq)$$

In strongly acidic solutions, the position of the equilibrium is shifted toward the reactant side, converting the yellow hydroxide complex to the $Fe(H_2O)_6^{3+}$ ion, which is nearly colorless. As the pH of an iron(III) solution is raised by the addition of base, the solution darkens as polymeric ions are formed. At a pH below 7, a finely divided, gelatinous, dark red-brown precipitate of hydrated iron(III) hydroxide forms.

The color of iron(III) in water solution depends upon the anions that are present, many of which form complexes with the metal ion (Figure 18.11). Thiocyanate ions produce a red color, chloride ions cause a yellow color, and the bromide complexes are brown. The red color of iron(III) in the presence of thiocyanate ions is used in both the qualitative and quantitative analysis for iron.

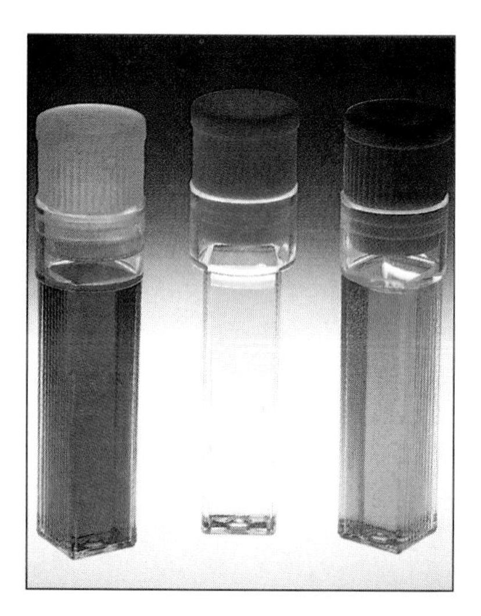

Figure 18.11 Solutions containing iron(III) The color of iron(III) in solution depends upon the anions present. From left to right, the photograph shows iron (III) in solutions containing thiocyanate, chloride, and bromide ions.

Copper

Unlike chromium and iron, metallic copper will not displace hydrogen ions from acids, as indicated by the standard potential of the reaction

$$Cu(s) + 2H^+(aq) \longrightarrow Cu^{2+}(aq) + H_2(g) \qquad E° = -0.34 \text{ V}$$

Copper is not oxidized by hydrogen ions.

Hot, concentrated sulfuric acid and nitric acid both dissolve the metal, producing solutions that contain the Cu^{2+} ion. In both cases, the oxidizing agent is the anion. Sulfuric acid is reduced to sulfurous acid in the process

$$Cu(s) + 2H_2SO_4(aq) \longrightarrow CuSO_4(aq) + H_2SO_3(aq) + H_2O(\ell)$$

The reduction product formed by nitric acid as it oxidizes copper depends on the concentration of the acid, the temperature, and other experimental conditions. At concentrations of about 1 molar nitric acid, the principal nitrogen products are NO and NO_2.

$$3Cu(s) + 8HNO_3(aq) \longrightarrow 3Cu(NO_3)_2(aq) + 2NO(g) + 4H_2O(\ell)$$

$$Cu(s) + 4HNO_3(aq) \longrightarrow Cu(NO_3)_2(aq) + 2NO_2(g) + 2H_2O(\ell)$$

Several copper(I) compounds are also known. However, in aqueous solution the Cu^+ ion is not stable, and decomposes to the metal and copper(II).

INSIGHTS into CHEMISTRY

Chemical Reactivity

Copper Dissolves in Acid; Experiments are Essential

The reaction between copper and nitric acid has been used as a demonstration of several different aspects of chemistry. The reaction is pictured in the photograph; the addition of concentrated nitric acid causes the copper to dissolve, forming a green-colored solution that contains copper(II) nitrate and produces clouds of brown-colored noxious NO_2.

The table of standard reduction potentials can be used to determine that hydrogen ions cannot dissolve copper.

$$Cu(s) + 2H^+(aq) \longrightarrow Cu^{2+}(aq) + H_2(g) \qquad E° = -0.34 \text{ V}$$

When nitric acid dissolves copper, the oxidizing agent is the nitrate ion. The standard potential is +0.47 V for

$$Cu(s) + 2NO_3^-(aq) + 4H^+(aq) \longrightarrow Cu^{2+}(aq) + 2NO_2(g) + 2H_2O(\ell)$$

The green color of the solution is caused by a combination of the blue $Cu^{2+}(aq)$ with the yellow color of the dissolved NO_2.

This reaction has been observed by many people, but one of the most entertaining descriptions was provided by the American chemist and educator Ira Remson (1846–1927), and quoted by F. H. Getman in the *Journal of Chemical Education,* 1940, pp. 9–10. His description serves also to remind us that chemistry is an experimental science, and the theories and hypotheses are based on observations of laboratory results.

"While reading a textbook of chemistry I came upon the statement, 'nitric acid acts upon copper.' I was getting tired of reading such absurd stuff and I was determined to see what this meant. Copper was more or less familiar to me, for copper cents were then in use. I had seen a bottle marked nitric acid on a table in the doctor's office where I was then 'doing time.' I did not know its peculiarities, but the spirit of adventure was upon me. Having nitric acid and copper, I had only to learn what the words 'act upon' meant. The statement 'nitric acid acts upon copper' would be something more than mere words. All was still. In the interest of knowledge I was even willing to sacrifice one of the few copper cents then in my possession. I put one of them on the table, opened the bottle marked nitric acid, poured some of the liquid on the copper and prepared to make an observation. But what was this wonderful thing which I beheld? The cent was already changed and it was no small change either. A green-blue liquid foamed and fumed over the cent and over the table. The air in the neighborhood of the performance became colored dark red. A great colored cloud arose. This was disagreeable and suffocating. How should I stop this? I tried to get rid of the objectionable mess by picking it up and throwing it out of the window. I learned another fact. Nitric acid not only acts upon copper, but it acts upon fingers. The pain led to another unpremeditated experiment. I drew my fingers across my trousers and another fact was discovered. Nitric acid acts upon trousers. Taking everything into consideration, that was the most impressive experiment and relatively probably the most costly experiment I have ever performed. . . . It was a revelation to me. It resulted in a desire on my part to learn more about that remarkable kind of action. Plainly, the only way to learn about it was to see its results, to experiment, to work in a laboratory."

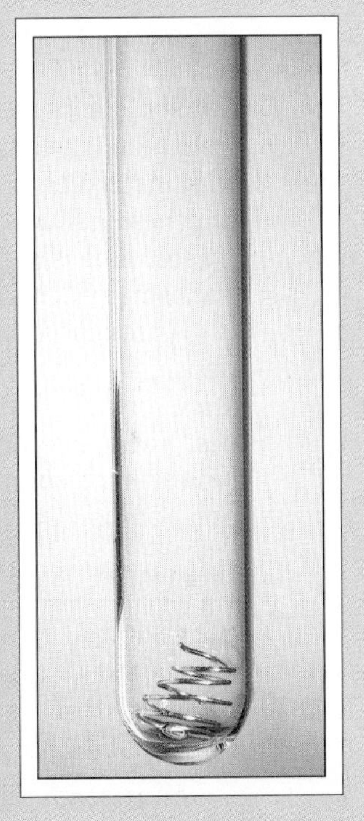

(a)

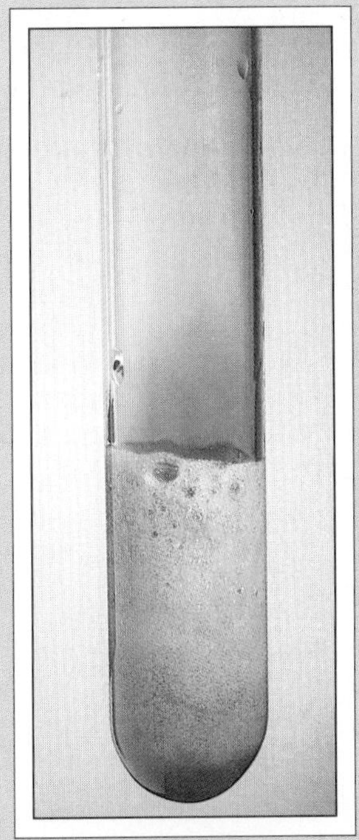

(b)

(a) Nitric acid acts upon copper (b) to form brown NO_2 gas and a solution of copper (II) ions.

$$2Cu^{+}(aq) \longrightarrow Cu(s) + Cu^{2+}(aq) \qquad E^\circ = +0.37\ V$$

The copper(I) halides are fairly insoluble compounds, formed by reducing copper(II) in solution with the appropriate halide ion present. Iodide ion serves as both the reducing agent and the anion in the solid copper(I) salt when it is added to a solution of Cu^{2+}.

$$2Cu^{2+}(aq) + 4I^{-}(aq) \longrightarrow 2CuI(s) + I_2(aq)$$

The reaction of copper(II) with iodide ion can be used as a qualitative test for the presence of copper in solution.

18.4 Coordination Compounds — Structure and Nomenclature

Objectives

- To describe the structures of coordination complexes
- To identify the geometric arrangements for complexes containing 2, 4, and 6 ligands
- To use the standard conventions for writing the formulas of coordination compounds
- To assign the systematic name for a coordination compound, given the formula
- To give the formula of a coordination compound, given its systematic name

Much of the chemistry of the transition metals is that of compounds formed from metal cations acting as Lewis acids, forming coordinate covalent bonds with molecules or ions that donate unshared pairs of electrons. These Lewis acid-base adducts play an important role in the development of photographic film, electroplating, the action of metals in enzymes, and the oxygen transport systems in animals.

Throughout the nineteenth century, many highly colored compounds of transition metals were made. Often, several of these compounds had the same formula, but exhibited quite different physical and chemical properties. Various compounds containing cobalt(III) chloride and ammonia are typical; such complex compounds are now more properly called coordination compounds. The following list contains some examples.

Formula	Color	Early name
$CoCl_3 \cdot 6NH_3$	yellow	Luteo complex
$CoCl_3 \cdot 5NH_3$	purple	Purpureo complex
$CoCl_3 \cdot 4NH_3$	green	Praeseo complex
$CoCl_3 \cdot 4NH_3$	violet	Violeo complex

Nineteenth century chemists devoted much effort attempting to explain the large number of isomers that are found in compounds of transition elements. In 1890, the Swiss chemist Alfred Werner proposed an explanation of how the stoichiometric quantities of small neutral molecules such as ammonia are incorporated into metal compounds. Werner devoted his professional life to developing and proving his theory of coordination com-

Table 18.5 Shapes of Transition Metal Complexes

Coordination Number	Geometric Arrangement of Ligands	Examples
2	linear	Cu(I), Ag(I), Au(I)
4	tetrahedral	Co(II), Ni(II), Mn(II)
4	square planar	Cu(II), Ni(II), Pt(II), Au(III)
6	octahedral	Fe(II), Fe(III), Cr(III), Co(II), Co(III), Ni(II), Mn(II), Mn(III), Ti(III), Pt(IV)

pounds, and he is recognized today as the father of modern coordination chemistry. This section presents the terminology and essential features of Werner's proposals about the structure of transition metal compounds.

Each transition metal ion in a compound is a Lewis acid capable of forming bonds with a number of Lewis bases. In the terminology of coordination chemistry, the Lewis bases bonded to the metal ion are called **ligands,** and they may be either neutral molecules or anions. For a species to behave as a ligand, it must have an unshared pair of electrons that can be donated to an empty orbital of the metal ion, forming a coordinate covalent bond. The **coordination number** is the number of donor atoms that are bonded to the metal ion. A **coordination compound** or **complex** is one that contains a metal ion bound to ligands by Lewis acid-base interactions.

Bonds in coordination complexes are formed using electron pairs donated to the central metal ion. The Lewis base donating the electron pair is called a ligand.

Coordination Number

The number of bonds between the ligands and a metal ion in a coordination complex ranges from two to nine, depending on the nature of the particular metal ion and the ligands that are present. The most common coordination numbers are four and six. Table 18.5 shows some typical coordination numbers for several transition metal ions, and the geometric arrangement of the ligands for each coordination number. Both tetrahedral and square planar shapes are found for a coordination number of four. Several of the transition metal ions exhibit more than one coordination number or arrangement in different compounds. Structures for examples of each of the geometric arrangements are shown in Figure 18.12.

Ligands

Neutral molecules and anions that have unshared pairs of electrons function as ligands. The metal ion-ligand bond is best described as a coordinate-covalent bond, with the ligand providing the shared pair of electrons. Most ligands donate only one pair of electrons to the metal atom and are called **monodentate** ligands. The use of "dentate" in describing ligands is derived from the Latin word for teeth. Examples of monodentate ligands are given in Table 18.6.

Some molecules and ions have more than one atom with unshared pairs of electrons, which can form bonds to the same metal atom. These ligands

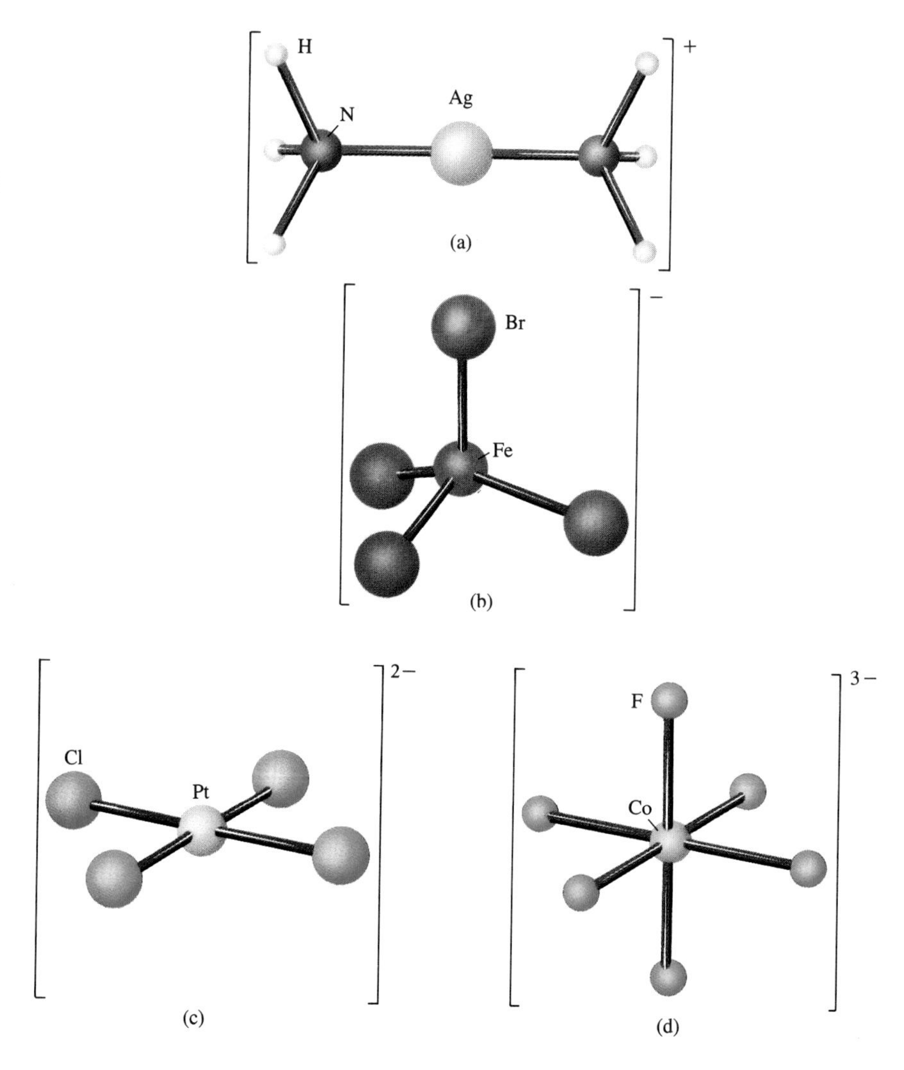

Figure 18.12 Structural formulas of complexes The structural formulas for common geometric arrangements found in complexes: (a) linear complex, (b) tetrahedral complex, (c) square planar complex, and (d) octahedral complex.

are called **chelating ligands,** and the complexes formed are called **chelates.** The word "chelate" comes from the Greek word for claw; to the originator of the term, a chelating ligand looked like a crab using both of its claws to attach itself to the metal ion. Ligands that donate two pairs of electrons to the same metal ion are called bidentate, those that donate three pairs are tridentate, and so forth. As a group, chelating ligands are referred to as **polydentate ligands.** Examples of several polydentate ligands are also shown in Table 18.6. In the hexadentate ligand, ethylenediaminetetraacetic acid (EDTA), two of the bonds to the metal use electron pairs from the nitrogen atoms, and the other four shared pairs are provided by oxygen atoms. EDTA forms especially stable complexes with most metal ions. The addition of EDTA to many food products increases shelf life, because it combines with transition metal ions that catalyze the decomposition of food. It is also used as a water softener, mostly to remove calcium and magnesium ions from hard water.

A chelate is a coordination compound in which two atoms in a single ligand each donate a pair of electrons to the same metal ion.

Table 18.6 Some Common Ligands

Type	Examples
monodentate	H_2O, NH_3, CN^-, NO_2^-, OH^-, SCN^-, X^- (halides), CO, pyridine
bidentate	oxalate, ethylenediamine oxalate ethylenediamine
tridentate	diethylenetriamine, 1,3,5-triaminocyclohexane diethylenetriamine 1,3,5 triaminocyclohexane
polydentate	ethylenediaminetetraacetic acid EDTA

Formulas of Coordination Compounds

Prior to the development of the modern view of coordination compounds, the formulas of coordination compounds were written in forms such as $CoCl_3 \cdot 6NH_3$, showing that the compound contained six molecules of ammonia for each metal atom or ion present. Today we enclose the metal ion and its coordinated ligands inside square brackets, with other ions of the compound outside the brackets. Thus, the formula of the cobalt(III) chloride compound containing six ammonia ligands is now written as $[Co(NH_3)_6]Cl_3$. In this compound the six ammonia molecules are covalently bonded to the cobalt ion to form a complex cation, which combines through ionic interactions with the three chloride ions. Uncoordinated ions are often referred to as counterions. For a complex that is an anion, its

Atoms joined by a Lewis acid-base interaction are said to be "coordinated."

In the formula of a coordination compound, square brackets enclose the formula of the coordination complex.

formula is also enclosed in brackets preceded by the cations that are present, for example, $K_3[Fe(CN)_6]$. Here the six cyanide ions are covalently bonded to the Fe^{3+} ion, forming the complex anion $[Fe(CN)_6]^{3-}$, which is ionically bonded to the three potassium ions.

Example 18.1 Writing the Formulas of Coordination Compounds

Write the formula for each of the following coordination compounds.

(a) $CoCl_3 \cdot 5NH_3$, in which only one of the chloride ions is coordinated to the metal.
(b) $CoCl_3 \cdot 4NH_3$, in which two chloride ions are coordinated to the metal.
(c) The potassium salt of a complex containing chromium(III) coordinated to five CN^- ions and one CO molecule. Note that the CO ligand is a neutral molecule, carbon monoxide.

Solution

(a) The complex, consisting of the Co^{3+} ion coordinated to the five ammonia molecules and one Cl^-, is enclosed by square brackets, and precedes the two uncoordinated chloride anions.

$$[Co(NH_3)_5Cl]Cl_2$$

(b) In this compound the metal, four ammonia molecules, and two of the Cl^- ions constitute the complex ion, which is enclosed in square brackets. The formula is $[Co(NH_3)_4Cl_2]Cl$.
(c) The complex ion, enclosed in square brackets, has a charge of 2−, which is the sum of the charges on the chromium ion and five CN^- ions. There must be two K^+ ions in the compound, which precede the formula of the complex anion.

$$K_2[Cr(CN)_5CO]$$

Understanding Write the formula of the coordination compound that contains platinum(II) coordinated to four ammonia molecules, and platinum(IV) coordinated by six chloride ions.

Answer: $[Pt(NH_3)_4][PtCl_6]$

In structural formulas, bidentate ligands, such as ethylenediamine, are usually represented as a curved line connecting the two donor atoms as shown in Figure 18.13b. This convention greatly simplifies the representation of these structures. Curved lines connecting the donor atoms are also used to represent any polydentate ligand.

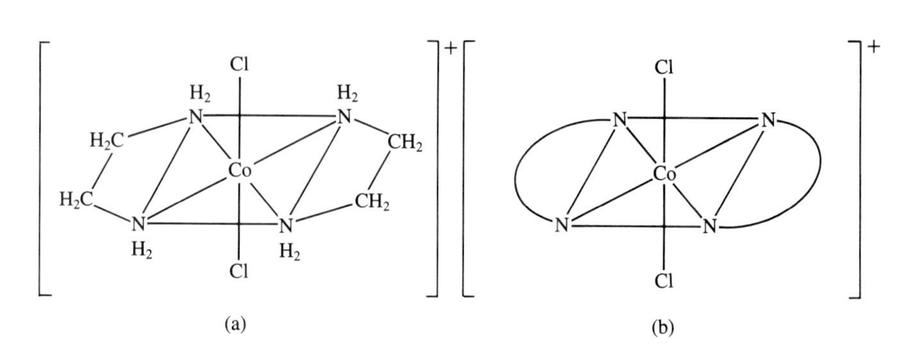

Figure 18.13 Representation of chelating ligands (a) The structural formula of $[Co(en)_2NH_3Cl]^{2+}$, showing all of the atoms in the ethylenediamine (en) ligand. (b) The same structure, representing the en ligand with a curved line connecting the donor atoms. The lines connecting the nitrogens are not bonds, but are placed there to help emphasize that the geometry is octahedral.

Naming of Coordination Compounds

The naming of compounds that contain metal complexes is a part of the International Union of Pure and Applied Chemistry (IUPAC) system of nomenclature. The following rules are an expansion of the simple nomenclature given in Section 2.4.

1. In naming a salt, the cation name is given first followed by the name of the anion as a separate word. The same rule is used for naming any ionic compound, whether or not it contains a complex ion.
2. In naming a coordination complex, the ligands are named first in alphabetical order, and the metal in the complex is named last, followed by the oxidation state of the metal as a Roman numeral in parentheses. The name of the complex is written as one word.
3. The names of the anionic ligands are changed to end in the letter *o*. In naming the neutral ligands, the name of the molecule is usually unchanged. Four common neutral ligands are exceptions to this rule: *aqua* is used for water, *ammine* is used for ammonia, *nitrosyl* is used for NO, and *carbonyl* is used for carbon monoxide. The names used for some of the commonly encountered ligands are given in Table 18.7.
4. The number of times each ligand occurs in a complex is indicated by a prefix: (1) mono (usually omitted), (2) di, (3) tri, (4) tetra, (5) penta, (6) hexa. When there are numeric prefixes in the name of a ligand, its name is usually enclosed in parentheses, and the numerical prefixes are changed to *bis, tris, tetrakis,* and *pentakis* for 2, 3, 4, and 5 ligands, respectively.
5. For cationic and neutral complexes the name of the metal is used. When

Table 18.7 Names of Some Common Ligands

Ligand	Name Used in Coordination Complex
Anions	
bromide, Br^-	bromo
carbonate, CO_3^{2-}	carbonato
chloride, Cl^-	chloro
cyanide, CN^-	cyano
hydroxide, OH^-	hydroxo
nitrite, NO_2^-	nitrito, nitro
oxalate, $C_2O_4^{2-}$	oxalato
oxide, O^{2-}	oxo
sulfate, SO_4^{2-}	sulfato
thiocyanate, SCN^-	thiocyanato
Neutral molecules	
ammonia, NH_3	ammine
carbon monoxide, CO	carbonyl
ethylenediamine, $NH_2CH_2CH_2NH_2$	ethylenediamine, (en)
nitrogen oxide, NO	nitrosyl
pyridine, C_5H_5N	pyridine
water, H_2O	aqua

Table 18.8 Names Used for Metals in Anionic Complexes

Metal	Name Used in Coordination Complex
chromium, Cr	chromate
cobalt, Co	cobaltate
copper, Cu	cuprate
gold, Au	aurate
iron, Fe	ferrate
manganese, Mn	manganate
mercury, Hg	mercurate

the complex is an anion, the ending *-ate* is attached to the name of the metal. For a metal whose symbol is based on a Latin name for the element, the Latin stem is used when naming anionic complexes. (Mercury is one exception to this rule—mercurate is used rather than hydrargentate). See Table 18.8.

The application of these rules is illustrated in Examples 18.2 and 18.3.

Example 18.2 Naming of Coordination Compounds

Give the IUPAC name for each of the following compounds: (a) $K_3[Fe(CN)_6]$, (b) $[Co(NH_3)_5Br]Br_2$, (c) $[Cr(NH_3)_2(en)Cl_2]Cl$, (en-ethylenediamine) (d) $[Ni(CO)_4]$.

Solution

(a) The complex is an anion so it is named second, and it will end in *ate*. The charge of each cyanide ion is 1−, and that of the complex ion is 3− (to balance the charge of 3+ on the three potassium ions). The oxidation state of the iron is therefore +3. The name of the compound is potassium hexacyanoferrate(III).

(b) The cobalt is in the +3 oxidation state, since the charge of the complex is 2+, the charge of the bromide ion ligand is 1−, and ammonia is a neutral molecule. The name of the complex is given first because it is the cation in the compound. Using *ammine* for the ammonia ligands, the name of the compound is then pentaamminebromocobalt(III) bromide.

(c) Since ammonia and ethylenediamine are both neutral ligands and the charge of a chloride ion is 1−, charge balance is used to determine that the chromium is in the +3 oxidation state. Naming the ligands in alphabetical order, and using the numerical prefixes, gives the name diamminedichloro(ethylenediamine)chromium(III) chloride.

(d) Carbon monoxide is a neutral ligand, so the nickel is in the 0 oxidation state. The IUPAC name is tetracarbonylnickel(0).

Understanding What is the name of $K_2[RuCl_5(H_2O)]$?

Answer: potassium aquapentachlororuthenate(III)

Example 18.3 Formulas from the Names of Coordination Compounds.

Write the formula of each of the following compounds:

(a) sodium aquapentacyanocobaltate(III)
(b) dichlorobis(ethylenediamine)chromium(III) nitrate

(c) triamminetrichlorocobalt(III)
(d) ammonium diaquadioxalatoferrate(II)

Solution

(a) The complex anion has a net charge of $5(-1) + (+3) = 2-$, from the sum of the oxidation state of the cobalt and the charges on the cyanide ions. There must be two sodium ions to produce a neutral compound. The formula is therefore $Na_2[Co(CN)_5(H_2O)]$.
(b) From the charge of the chloride ion (1−) and the oxidation state of the chromium (+3), the complex cation has a charge of 1+, which must be balanced by a single NO_3^- ion in the compound. The formula of the complex is enclosed in square brackets to give the formula $[CrCl_2(en)_2]NO_3$. The common abbreviation "en" for the ethylenediamine ligand has been used in this formula.
(c) The complex is neutral, containing three chloride ions, three molecules of ammonia, and a Co^{3+} ion. The formula is written as $[Co(NH_3)_3Cl_3]$.
(d) The anion consists of two oxalate ions ($C_2O_4^{2-}$), two water molecules, and an Fe^{2+} ion. The net charge of the anion is thus 2−, which combines with two NH_4^+ ions to give a neutral compound. The formula is $(NH_4)_2[Fe(C_2O_4)_2(H_2O)_2]$.

Understanding What is the formula of hexaamminecobalt(III) hexacyanoferrate(III)?

Answer: $[Co(NH_3)_6][Fe(CN)_6]$

18.5 Isomers

Objectives

- To identify and distinguish structural isomers and stereoisomers
- To recognize and classify structural isomers as coordination isomers or linkage isomers
- To identify examples of geometric isomers and optical isomers
- To use the terminology introduced for describing the different isomers (*cis-* and *trans-*, *mer-* and *fac-*, chiral, enantiomers, racemic mixture, optical isomers)

In Section 9.4, *isomers* were defined as different compounds with the same chemical formula. Experimentally, isomers are recognized as different compounds because they differ in one or more of their physical and chemical properties. Coordination compounds exhibit a wide variety of isomers, as a result of the fairly large number of atoms present in the species. Figure 18.14 outlines the categories of isomerism that are used to classify coordination compounds.

Structural Isomers

Structural isomers contain the same numbers and kinds of atoms, but differ in the bonds that are present. In coordination compounds there are two kinds of structural isomerism: coordination isomerism and linkage isomerism.

Coordination isomers contain the same numbers and kinds of atoms but have different Lewis bases directly bonded to the metal ion. An anion can function either as a counterion or as one of the ligands. The compounds

Figure 18.14 Categories of isomerism This block diagram shows how the various kinds of isomerism exhibited by coordination compounds are related.

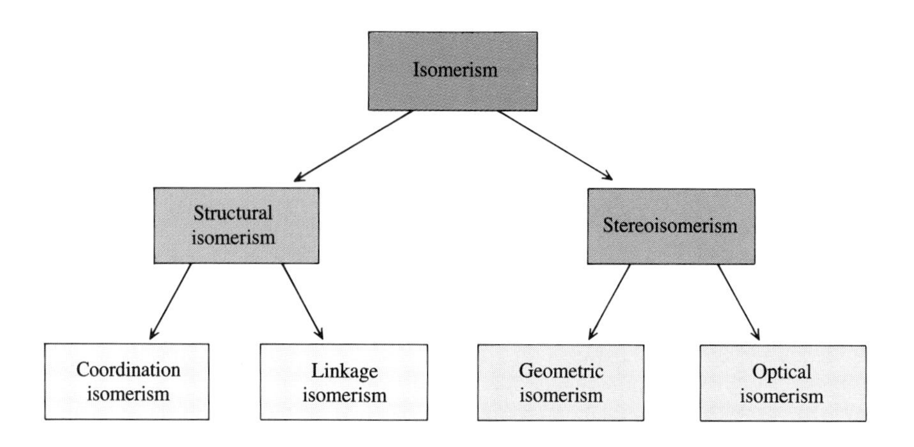

$[Co(NH_3)_4Cl_2]Br$ and $[Co(NH_3)_4ClBr]Cl$ are *structural* isomers because the same atoms are present. They are *coordination* isomers because the bromide ion in the second compound is coordinated to the metal atom, replacing one of the coordinated chloride ions in the first compound. Coordination isomerism can also occur in compounds that contain coordination complexes as both the cation and anion in a salt, as in the compounds $[Cr(NH_3)_6][Co(C_2O_4)_3]$ and $[Co(NH_3)_6][Cr(C_2O_4)_3]$. These two compounds have exactly the same atoms present, but the oxalate ions and ammonia molecules are coordinated to different metal ions in the two compounds.

A second type of structural isomerism is **linkage isomerism,** in which the same ligand is coordinated to the metal through a different donor atom. The most frequently encountered example of this occurs with the ligand NO_2^-, the nitrite ion. All three of the atoms in this anion have unshared pairs of electrons that can form bonds to a metal ion center. The structures of two linkage isomers with the nitrite ion as a ligand are shown in Figure 18.15. In one isomer the metal is bonded to the nitrogen atom, and in the other isomer the metal is bonded to an oxygen atom. In the case of NO_2^-, the ligand is given a different name, depending on the atom that forms the bond to the metal ion. The N-bonded ligand is called nitro and the O-bonded ligand is called nitrito. The names of the complexes shown in Figure 18.15 are (a) pentamminenitrocobalt(III) and (b) pentamminenitritocobalt(III). These names are not part of the IUPAC nomenclature system, but are still widely used by coordination chemists.

Structural isomers, both coordination and linkage, have different connectivity.

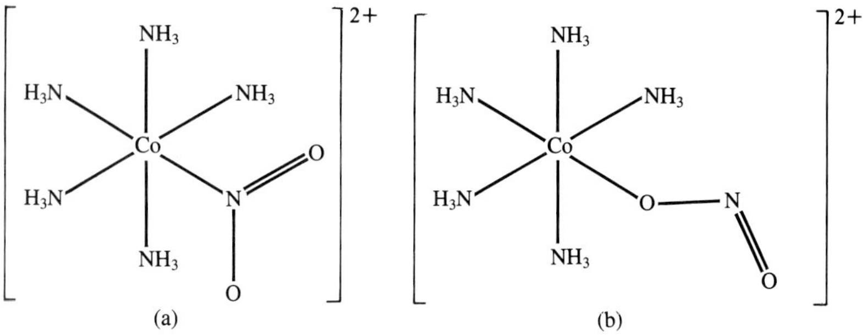

Figure 18.15 The linkage isomers of $(Co(NH_3)_5NO_2]^{2+}$ Two linkage isomers of the NO_2^- ligand are possible. (a) The attachment to the metal ion is through the nitrogen atom, and this is called the *nitro* complex. (b) The bond is formed by donation of the electron pair of one of the oxygen atoms of the ligand, and this is called the *nitrito* complex.

(a) (b)

Another ligand that shows linkage isomerism is the thiocyanate ion, NCS^-. The ion can coordinate to a metal through either the nitrogen or the sulfur atom. In the IUPAC nomenclature system, linkage isomers are distinguished by preceding the name of the ligand by the symbol of the coordinated atom. The name of the complex $[Pt(NH_3)_3SCN]^+$ in which the sulfur atom is bonded to the platinum is called triammine-S-thiocyanatoplatinum(II). If the thiocyanate ion were bonded through the nitrogen atom, this complex would be called triammine-N-thiocyanatoplatinum(II). The IUPAC names for the two complexes shown in Figure 18.15 are (a) pentammine-N-nitritocobalt(III) and (b) pentammine-O-nitritocobalt(III).

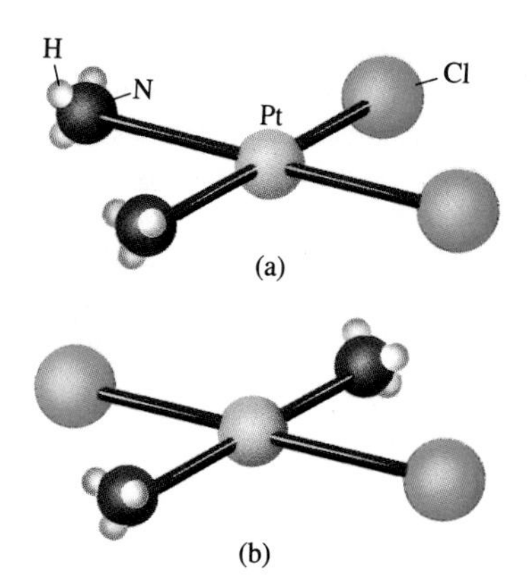

Figure 18.16 Geometric isomers The structures of (a) the *cis*-isomer and (b) the *trans*-isomer of diammine-dichloroplatinum(II). The two isomers differ in the Cl—Pt—Cl bond angles.

Stereoisomers

Stereoisomers have the same bonds present in both molecules, but differ in the arrangement of the atoms in space. There are two categories of stereoisomers: geometric isomers and optical isomers.

Geometric Isomers

Geometric isomers have the same numbers and kinds of bonds, but differ in the relative positions of the ligands. The compound diamminedichloroplatinum(II), $[Pt(NH_3)_2Cl_2]$, is a square planar complex. The two geometric isomers shown in Figure 18.16 differ in the Cl—Pt—Cl bond angle. In the *cis*-isomer this angle is 90°, and in the *trans*-isomer the angle is 180°. *Cis*-refers to the isomer in which the two like ligands are on the same side of the metal atom, and *trans*- describes the complex in which the like ligands are on opposite sides of the metal. This change in the arrangement of the ligands about the platinum atom changes the properties of the compounds markedly. Some of the properties of the two geometric isomers are given in Table 18.9. While *cis*-diamminedichloroplatinum(II) is a highly effective antitumor agent, the *trans*-isomer shows virtually no therapeutic activity. Both isomers are quite toxic, so careful control of the dosage in cancer chemotherapy is needed.

Geometric isomerism is also observed in octahedral complexes that have the general formulas MA_4B_2 and MA_3B_3, where A and B represent different ligands. Examples of these kinds of geometric isomers are illustrated in Figure 18.17. In the MA_4B_2 formula the *cis*- and *trans*- designations are used. While *cis*- and *trans*- are sometimes used to name the two isomers of the type MA_3B_3, they are usually identified as *fac*- (facial) and *mer*- (meridional) isomers, respectively. In the *fac*-isomer the three like

Table 18.9 Some Properties of Geometric Isomers of Diamminedichloroplatinum(II)

Property	*cis*-Isomer	*trans*-Isomer
density	3.738 g/cm^3	3.746 g/cm^3
color	bright yellow	pale yellow
solubility	0.2523 g/100 g H_2O	0.0366 g/100 g H_2O
synthesis	$[PtCl_4]^{2-} + 2NH_3 \longrightarrow [Pt(NH_3)_2Cl_2] + 2Cl^-$	$[Pt(NH_3)_4]^{2+} + 2Cl^- \longrightarrow [Pt(NH_3)_2Cl_2] + 2NH_3$

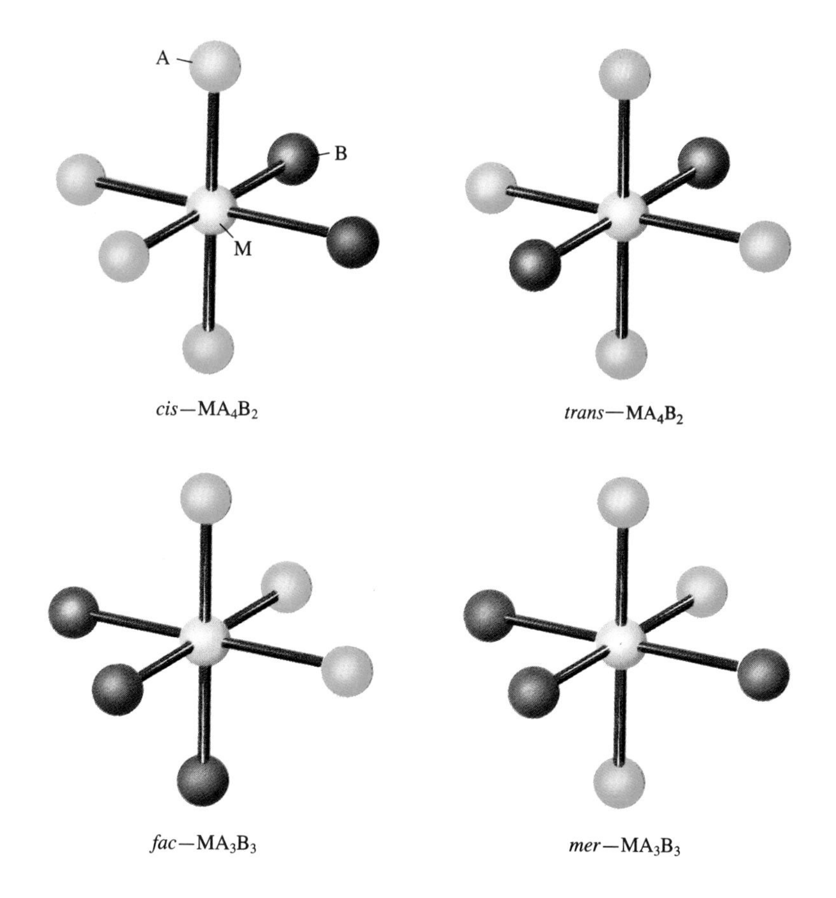

Figure 18.17 Geometric isomers in octahedral complexes At the top are the *cis*- and *trans*-isomers of an octahedral complex having the general formula MA_4B_2. The *fac*- and *mer*-isomers of an octahedral complex having the general formula MA_3B_3 are shown at the bottom of the figure.

Geometric isomers contain the same bonded atoms, but differ in the arrangement of the ligands around the metal ion.

ligands are at the corners of one of the triangular faces of the octahedron, while in the *mer*-isomer two of the like ligands are at opposite vertices of the octahedron.

Example 18.4 Geometric Isomers of Coordination Complexes

How many geometric isomers exist for the octahedral complex $[Ru(H_2O)_2Cl_4]^-$? Name each isomer.

Solution There are two water molecules and four chloride ligands in the complex. Therefore, it should have two geometric isomers. In one the O—Ru—O angle is 90° and in the other this angle is 180°. The isomer with both water molecules on the same side of the metal (90° angle) is named *cis*-diaquatetrachlororuthenate(III), and the other isomer is *trans*-diaquatetrachlororuthenate(III).

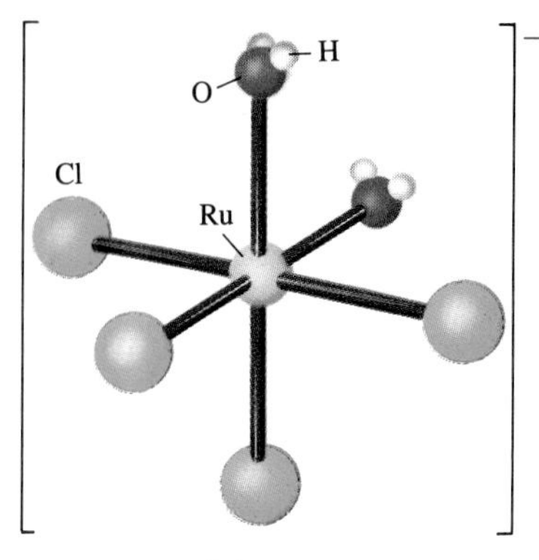

cis-isomer

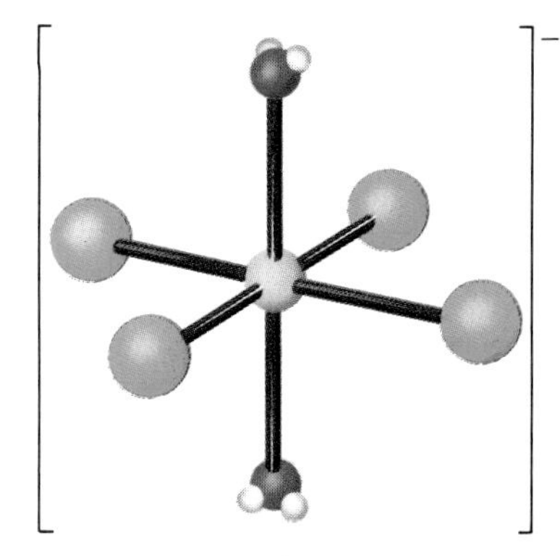

trans-isomer

Understanding Show the structure and name all geometric isomers of the square complex $[Pt(NH_3)_2BrCl]$.

Answer: Both isomers are named diamminebromochloroplatinum(II) but differ in the prefix. The prefix used for each isomer is shown below its structure.

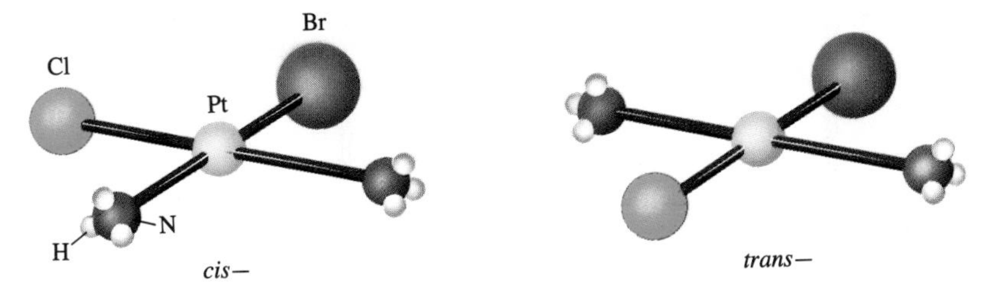

Optical Isomers

A second class of stereoisomerism is called optical isomerism, because of the effect the isomers have on polarized light. **Optical isomers** are molecules or ions that differ in the way they rotate plane polarized light. Normal light consists of electromagnetic radiation in which the electric field oscillates randomly in all directions perpendicular to the direction of the beam. In **polarized light** the electric field oscillates in a single plane, as shown in Figure 18.18. When polarized light passes through a sample of an optical isomer, the plane of polarization of the light is rotated, as shown in Figure 18.19.

Chiral molecules or ions are those with mirror image structures that cannot be superimposed. Only chiral molecules are optically active, that is, can rotate plane polarized light. A familiar example of objects that have nonsuperimposable mirror images are a person's right and left hands. The reflection of a left hand in a mirror is identical to the right hand (see Figure 18.20). No matter how a right hand is twisted and turned, it can never be superimposed on the left hand.

Enantiomers are the chiral molecules that are nonsuperimposable mirror images of each other. For example, octahedral complexes that contain three bidentate ligands exist as enantiomers. The mirror image relationship for the complex ion tris(ethylenediamine)cobalt(III) is shown in Figure 18.21. No matter how the mirror images are turned, they are not superimposable. One enantiomer rotates plane polarized light in a clockwise direction, and its mirror image rotates the light by the same amount in a counterclockwise direction.

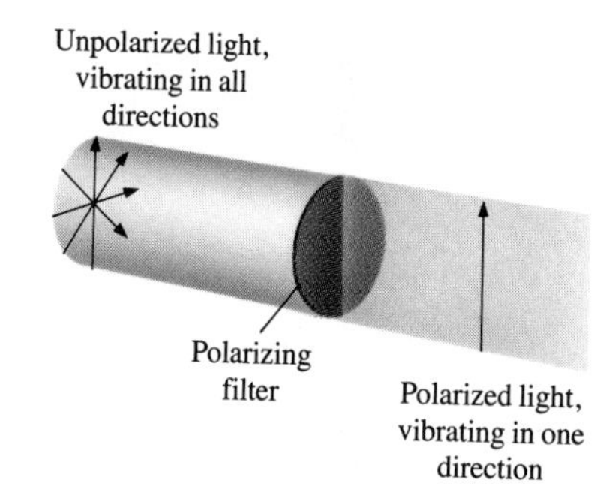

Figure 18.18 Polarizing light The electric field in unpolarized light oscillates in all directions perpendicular to the direction in which the light travels. Only light that is oscillating in a single plane is transmitted by a polarizing filter.

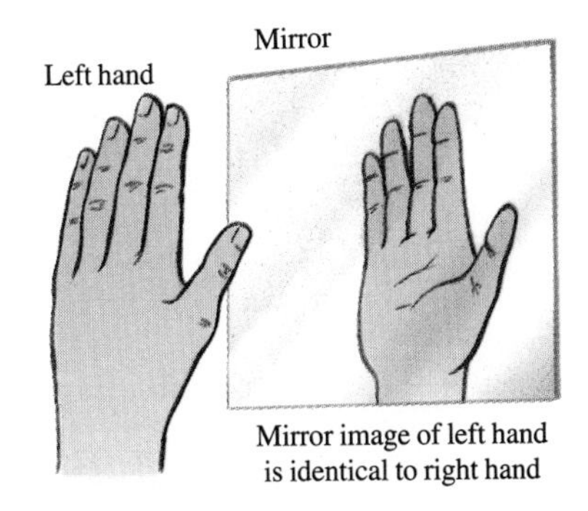

Figure 18.20 Nonsuperimposable mirror images The mirror image of a left hand is the same as a right hand; a right hand and a left hand cannot be superimposed.

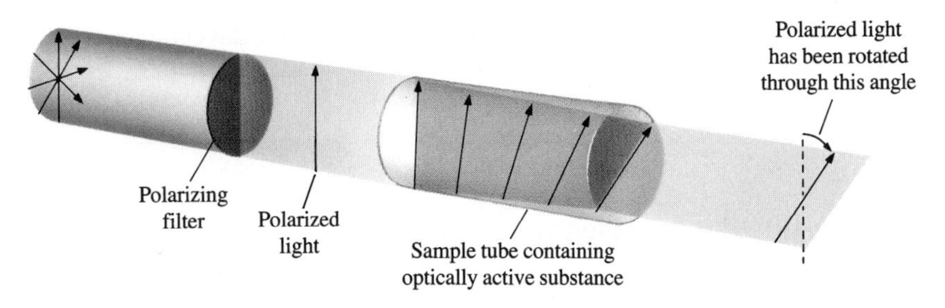

Figure 18.19 Rotation of polarized light When plane polarized light passes through a sample containing an optical isomer, the plane of polarization is rotated.

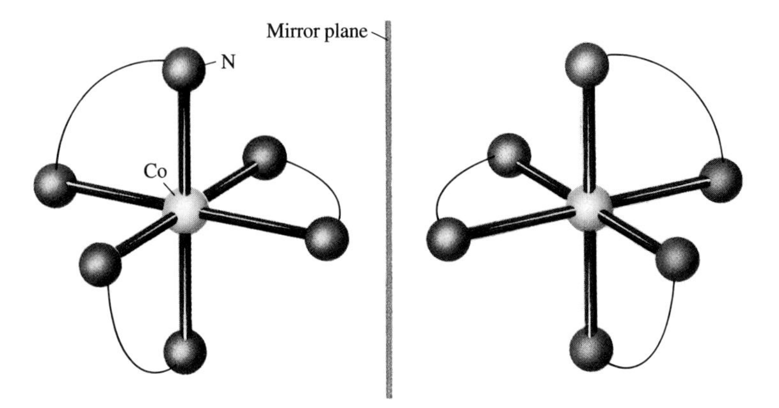

Figure 18.21 The enantiomers of $[Co(en)_3]^{3+}$ The mirror images of the tris(ethylenediamine)cobalt(III) ion cannot be superimposed, and therefore they constitute a pair of enantiomers.

Enantiomeric isomers are mirror image structures that cannot be superimposed on each other; each rotates plane polarized light in the opposite direction.

The synthesis of a chiral compound usually results in a mixture of equal quantities of the two enantiomers, called a **racemic mixture.** A racemic mixture produces no net rotation of plane polarized light, because the presence of equal numbers of molecules of the two enantiomers results in cancellation of the rotations.

Example 18.5 Identifying Isomers of Octahedral Complexes

Identify all the isomers of dichlorobis(ethylenediamine)chromium(III).

Solution Two geometric isomers for this complex are possible, in which the two chloride ligands are in either the *cis-* or the *trans-* positions. Comparison of the mirror images of these geometric isomers shows that the *cis-* complex exists as two enantiomers. The structures of the three possible isomers are shown below.

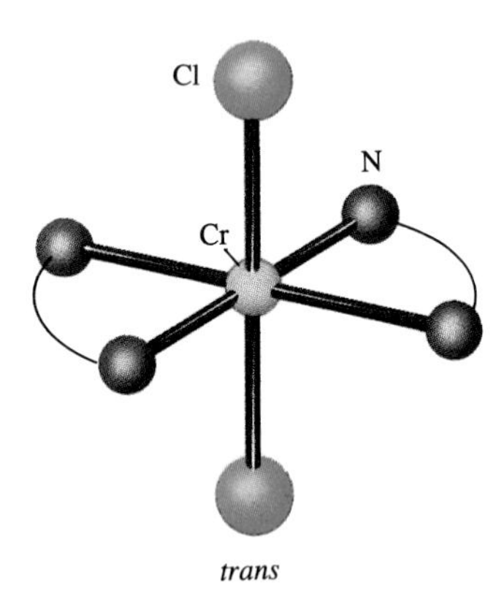

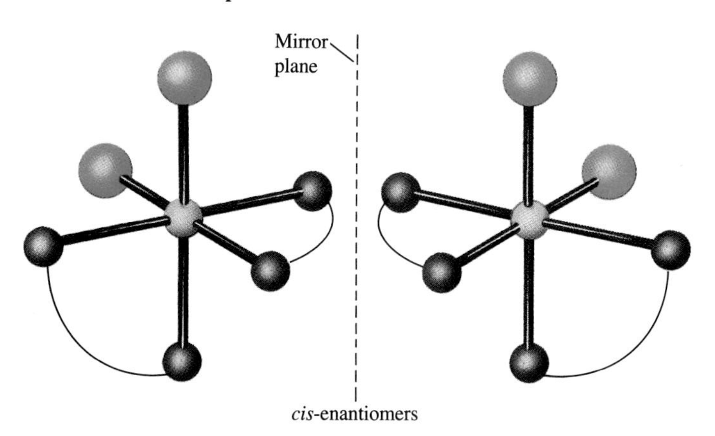

Understanding Will optical isomers be possible for a square planar complex of platinum(II), in which four different ligands are coordinated to the metal?

Answer: No

18.6 Bonding in Coordination Complexes

Objectives

- To explain the color and magnetic properties of transition metal complexes using crystal field theory

- To predict the magnetic properties of transition metal complexes from the position of the ligands in the spectrochemical series

Now that we have some knowledge of the structure of coordination complexes, it is time to turn our attention to the forces that hold the ligands and the metal ion together. At the same time, we must address some of the features of transition metal complexes that are unique to this class of compounds. It has already been noted that the compounds of transition elements are often highly colored. Not only do the colors of transition metal compounds change with the oxidation state of the metal, but the observed color is also affected by the particular ligands that are bound to the metal ion (see Figure 18.22). The magnetic properties of transition metal compounds are also somewhat unusual, when compared to the magnetic properties of compounds of the representative elements.

A successful bonding model for transition metal complexes must account for their colors and magnetic properties. There are currently several such models, but none of these is entirely satisfactory. In this section we present the crystal field model because of its simplicity and its ability to account easily for the colors and magnetic properties of transition metal complexes. Before we begin our discussion of crystal field theory, a brief review of color and magnetism is in order.

Figure 18.22 Nickel(II) compounds The colors of transition metal complexes change in the presence of different ligands. From left to right, nickel(II) is coordinate to water, dimethylglyoxime, and ammonia.

Color and Magnetism

The interaction of electromagnetic radiation (light) with matter was discussed in earlier chapters. What is perceived as color by our eyes is the result of electromagnetic radiation over the small range of wavelengths from about 400 to 700 nm. The colors of visible light are shown in the spectrum of Figure 18.23. The energy of the photons at any given wavelength is given by Planck's relationship

$$E = h\nu = hc/\lambda$$

The energy of a photon of red light ($\lambda = 700$ nm) is lower than the energy of a photon of violet light ($\lambda = 400$ nm). The observed colors of coordination

Figure 18.23 The visible spectrum of light Electromagnetic radiation with wavelengths from 400 nm to 700 nm is called visible light. This is the only light detected by the human eye.

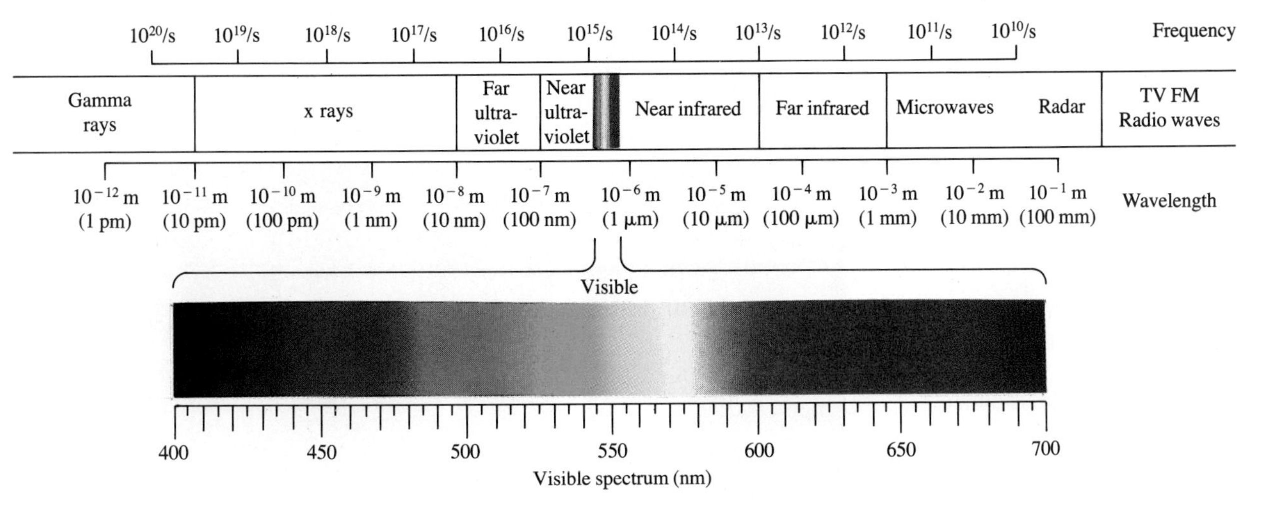

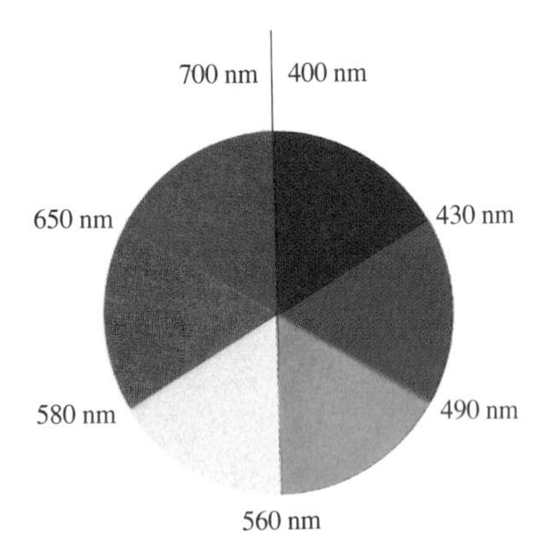

Figure 18.24 The artist's color wheel The artist's color wheel arranges the six colors in order of wavelength around a circle. Complementary colors are located opposite each other. The presence of yellow light and the absence of violet light (the complementary color of yellow) both appear yellow to the human eye.

compounds are related to the allowed energy states of electrons in the coordination complexes.

When all wavelengths of visible light strike our eyes, white light is observed. A sample that equally absorbs all wavelengths of visible light appears black. If a sample of matter absorbs part of the visible light, only some of the colors strike the eye and the object appears colored. When only yellow light strikes the eye, we see a yellow color. A yellow color is also observed if all wavelengths are present except violet light. The colors violet and yellow are referred to as *complementary.* Complementary colors can be identified on the artist's color wheel shown in Figure 18.24. Complementary colors are located opposite each other on the color wheel. The observed colors of most transition metal complexes arise because the sample absorbs the light with the complementary color. Thus the blue color of many copper(II) compounds is caused by the absorption of orange light with wavelengths near 600 nm. The **absorption spectrum** of a substance is a graph of the quantity of light absorbed by the sample as a function of wavelength.

Paramagnetism is associated with unpaired electrons (see Chapter 6), and the measured magnetic properties of a sample are used to determine the number of unpaired electrons in compounds. While paramagnetism is relatively rare in compounds of representative elements, it is quite commonly observed in transition metal compounds. Also, different complexes of the same transition metal ion can have different numbers of unpaired electrons. For example, the $[Co(NH_3)_6]^{3+}$ ion contains no unpaired electrons, while there are four unpaired electrons in the $[CoF_6]^{3-}$ complex. Any successful model for transition metal complexes must adequately account for this kind of magnetic behavior.

Crystal Field Theory

In the early 1930s the colors and the magnetic properties of crystalline solids that contain transition metal ions were explained by physicists, using a purely ionic model called crystal field theory. It was more than 20 years later that chemists recognized its usefulness in explaining the properties of transition metal complexes in solution.

Crystal field theory assumes that the interaction between the ligands and the metal ion in a complex is electrostatic. When the ligands are anions, such as chloride or cyanide, a strong attractive force is exerted on them by the positive charge of the metal ion. Neutral ligands such as water and ammonia are polar molecules, and the negative end of the dipole is strongly attracted by the positive charge of the metal ion. In this case the attractive forces are of

Colors of solid transition metal compounds Crystal field theory accounts for the colors of gemstones. The red color of rubies is due to chromium(III), and sapphires are blue because of iron(II), iron(III), and titanium(IV).

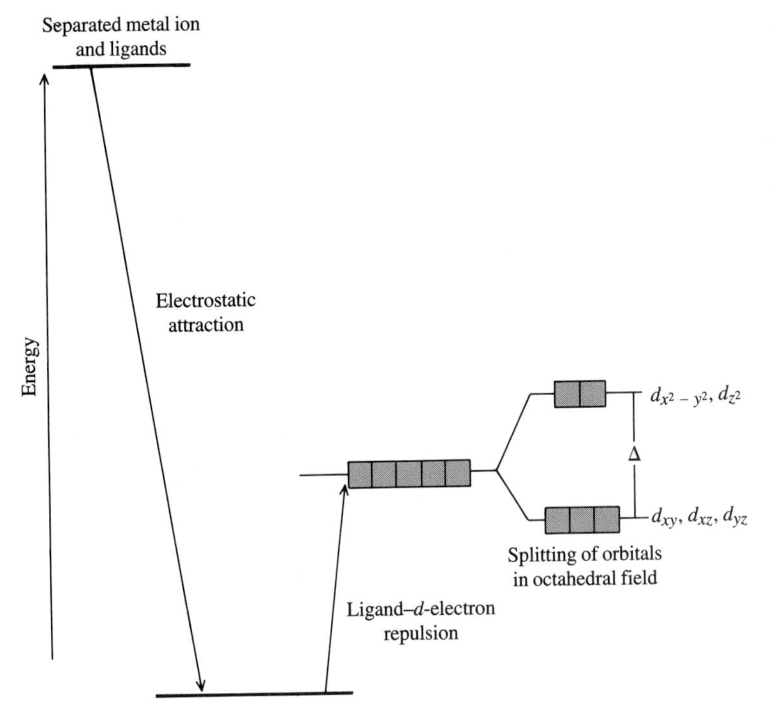

Figure 18.25 Energy changes in complex ion formation On the left, the large decrease in energy results from the attraction of the negative charge of the ligands by the positive transition metal ion. In the center, the repulsion of the *d* electrons of the metals by the negative charges of the ligands causes these orbitals to increase in energy. On the right, the separation of the *d* orbitals into two-fold and threefold degenerate sets by the octahedral arrangement of the ligands is shown.

the ion-dipole type. In either case the electrostatic forces cause a decrease in the energy of the system as the ligands are drawn close to the metal ion. At the same time the negative charges of the ligands repel each other, so they will stay as far apart from each other as possible. We can describe an octahedral arrangement by having the six ligands approach the metal ion along the *x*, *y*, and *z* axes in a Cartesian coordinate system. The left side of Figure 18.25 shows the decrease in the energy resulting from the attraction of the negative ligands by the positive metal ion.

The crystal field theory can be applied to any coordination number and geometry.

The valence shell electrons in transition metal ions occupy the *d* subshell (the *s* electrons have ionized). While there is a net positive charge on the metal ion, the negative charges of the ligands repel the outer *d* electrons, causing them to be less stable (of higher energy) in the complex than they are in the absence of the ligands. The destabilizing of the *d* electrons by the charges of the ligands is shown as an increase in energy in the center of Figure 18.25. However, the electrons in the different *d* orbitals do not experience the same repulsions by the ligands. Figure 18.26 shows the contours for the five *d* orbitals. Three of these orbitals have the lobes of high electron density directed diagonally between the *x*, *y* and *z* axes (d_{xy}, d_{yz}, and d_{xz}), and away

Figure 18.26 The degeneracy of the *d* orbitals Electrons in three of the *d* orbitals (d_{xy}, d_{yz}, and d_{xz}) are further from the locations of the ligands than are those in the other two orbitals ($d_{x^2-y^2}$ and d_{z^2}). The locations of the negatively charged ligands are shown as red dots on the *x*, *y*, and *z* axes.

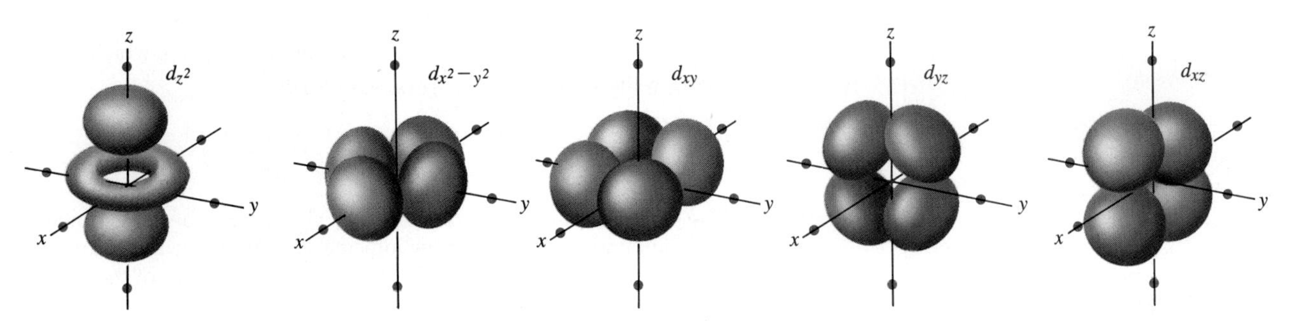

When a transition metal ion is surrounded by ligands, some d orbitals point at the ligands, and others point between them, so the d orbitals are no longer equal in energy.

from the locations of the negative charges of the ligands. The other two d orbitals ($d_{x^2-y^2}$ and d_{z^2}) are directed along the Cartesian axes, and consequently point directly at the negative charges of the ligands. As a result, the five d orbitals are no longer degenerate (equal in energy) in the presence of the six ligands—three of them are lower in energy than are the other two. The separation of the d orbitals into a doubly degenerate and a triply degenerate set is shown on the right side of Figure 18.25.

The **crystal field splitting** is the separation in energy between the two sets of d orbitals, caused by the unequal repulsion of the d electrons of the metal by the negative charges of the ligands that are arranged octahedrally about the central metal ion. The crystal field splitting is represented by Δ, the Greek letter "delta." Figure 18.27 shows the way this energy separation is usually represented in orbital energy level diagrams for a metal ion in an octahedral crystal field. As we shall see, the energy separation of the d orbitals by the crystal field offers a simple explanation of the colors and the magnetic properties of transition metal complexes.

Visible Spectra of Complex Ions

The color of a transition metal ion is caused by the absorption of visible light when an electron is moved from one of the lower-energy d orbitals to one of the higher-energy d orbitals.

The size of Δ, the energy difference of the d orbitals, for most transition metal complexes is within the energy range of visible light (2.8×10^{-19} J to 5.0×10^{-19} J per photon, or 170 kJ/mol to 300 kJ/mol). The absorption of light that moves an electron from one of the lower energy d orbitals to one of the higher energy d orbitals occurs in the visible region of the electromagnetic spectrum and produces the colors observed for transition metal complexes. Figure 18.27 represents the electron transition for the single d electron in the red $Ti(H_2O)_6^{3+}$ ion, along with the observed spectrum of the complex.

Example 18.6 Calculating Δ for the $Ti(H_2O)_6^{3+}$ Ion

The maximum absorption of visible light by the $Ti(H_2O)_6^{3+}$ complex occurs at 510 nm. Express Δ in kJ/mol.

Solution From the energy-level diagram in Figure 18.27, Δ, the crystal field splitting, is the same as the energy of a photon of light with a wavelength of 510 nm. Using Planck's relationship between the energy and wavelength of a photon, we obtain

$$\Delta = h\nu = \frac{hc}{\lambda} = \frac{6.63 \times 10^{-34}\ \text{J s} \times 3.00 \times 10^{8}\ \text{m s}^{-1}}{510 \times 10^{-9}\ \text{m}} = 3.90 \times 10^{-19}\ \text{J}$$

This is the energy of a single photon, which we must multiply by Avogadro's number to obtain the answer on a molar scale.

$$\Delta = 3.90 \times 10^{-19}\ \text{J} \left(\frac{6.02 \times 10^{23}}{\text{mol}}\right)\left(\frac{1\ \text{kJ}}{1000\ \text{J}}\right) = 235\ \text{kJ/mol}$$

Understanding The TiF_6^{3-} ion has an absorption maximum at 590 nm, which corresponds to Δ. Express the crystal field splitting, Δ, in kJ/mol.

Answer: 203 kJ/mol

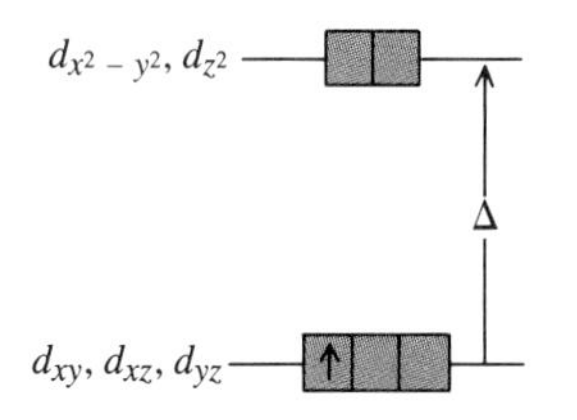

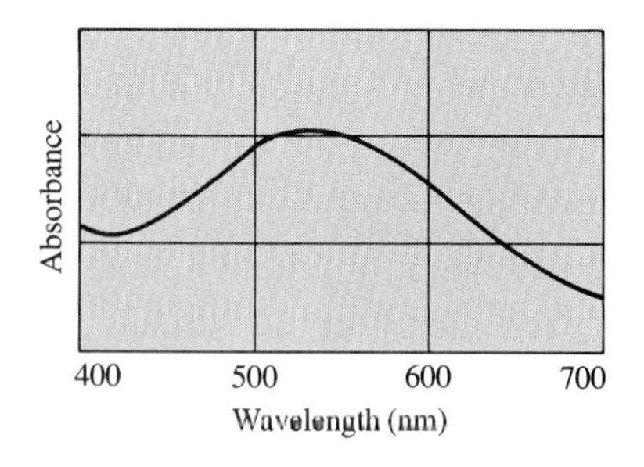

Figure 18.27 The crystal field energy level diagram The energies of the d orbitals in the $Ti(H_2O)_6^{3+}$ ion are shown. The arrow indicates the electronic transition that gives rise to the absorption of visible light by that ion. The spectrum of the ion is shown below the energy diagram.

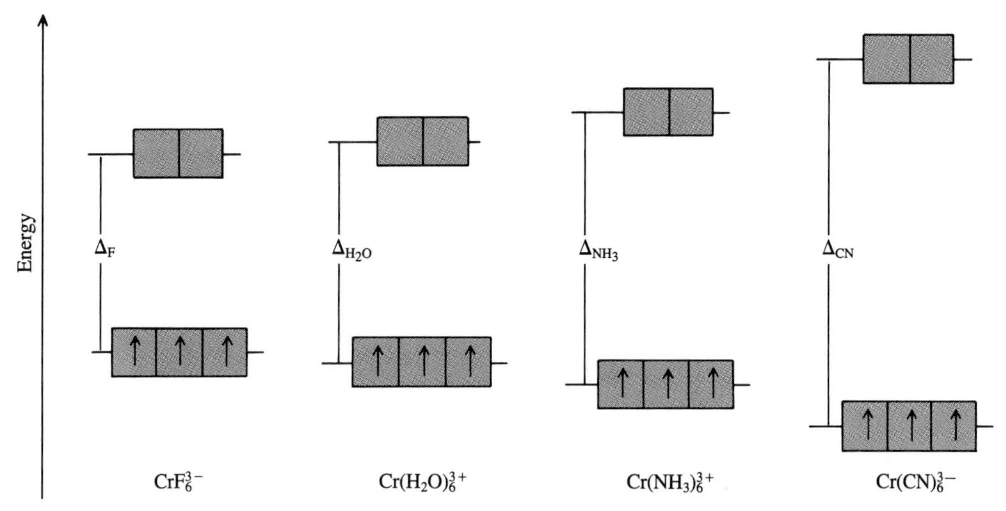

Figure 18.28 The spectrochemical series The effect of changing ligands on Δ for a series of Cr(III) complexes. In the energy level diagrams, Δ increases from left to right as the ligands are changed from fluoride to water to ammonia to cyanide.

When more than one electron occupies the *d* orbitals of the metal, the relation between the value of Δ and the energy of the absorption bands in the visible spectrum is more complicated. For many of these situations more than one absorption band is observed.

The value of the crystal field splitting, Δ, is calculated from the observed spectrum of a transition metal complex. The spectra of many transition metal complexes have been observed, and several generalizations about the resulting values of Δ have been noted. Regardless of the metal ion used, it is found that as the ligands are changed, the size of Δ increases in the order

$$I^- < Cl^- < OH^- < F^- < H_2O < NH_3 < en < NO_2^- < CN^- < CO$$

This arrangement of the ligands in order of increasing Δ that they cause is called the **spectrochemical series.** More complete spectrochemical series, which contain many additional ions and molecules, have been compiled. In Figure 18.28 we show the change of Δ with different ligands for a series of chromium(III) complexes.

The charge on the metal ion also influences the size of Δ. In general Δ is greater for a metal ion with a higher positive charge. Thus in the water complexes of Fe^{2+} and Fe^{3+} the value of Δ increases 40% from 120 kJ/mol to 168 kJ/mol. Furthermore, within any periodic group Δ increases from top to bottom. Thus, the size of Δ increases from the 3*d* to the 4*d* to the 5*d* transition series. Table 18.10 presents a comparison of the values of Δ for the group VIIIB(9) complexes of Co^{3+}, Rh^{3+}, and Ir^{3+} with several different ligands.

Table 18.10 Δ in kJ/mol for Selected Complexes

	Metal Ion		
Ligand	Co^{3+}	Rh^{3+}	Ir^{3+}
Cl^-	—	243	299
H_2O	218	323	—
NH_3	274	408	490
en	278	414	495
CN^-	401	544	—

The size of Δ depends on the particular ligand, the charge of the metal ion, and the period in which the metal is found.

The Electron Configurations of Complexes

The valence-shell electron configuration of a transition metal atom in its ground state is usually $ns^2(n-1)d^m$, where *n* is the value of the principal quantum number, and *m* is the number of electrons in the occupied *d* subshell. When a transition element ionizes, *the ns electrons are lost before*

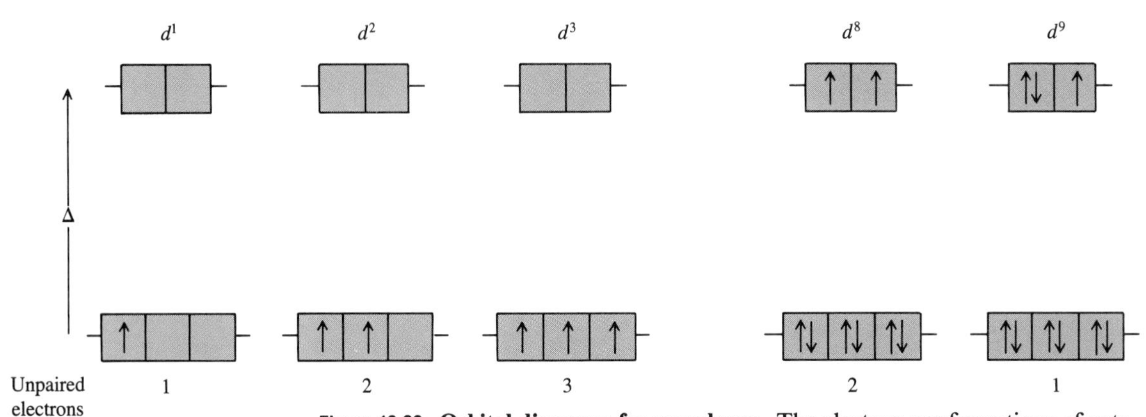

Figure 18.29 Orbital diagrams for complexes The electron configurations of octahedral complexes containing one, two, three, eight, and nine d electrons are the same, regardless of the size of Δ. The number of unpaired electrons is given at the bottom of the figure.

All of the valence-shell electrons in a transition metal ion occupy the d orbitals.

any electrons are removed from the (n − 1)d subshell (see Section 7.3). Thus, in a transition metal ion *all* of the valence-shell electrons are in the $(n-1)d$ subshell. The electrons in the d orbitals of a complex ion obey Hund's rule of maximum multiplicity; they remain unpaired with one electron in each orbital as long as possible. The d^1, d^2, and d^3 configurations have one, two, and three unpaired electrons, respectively, as shown in Figure 18.29. For the d^3 electron configuration in a complex ion, each d orbital in the lower-energy degenerate set contains a single unpaired electron.

The addition of the fourth electron into the lower energy orbitals produces a pair of electrons. There is an energy price that is paid to do this, since the electrons that share the same orbital are close together and repel each other more strongly than do the electrons occupying different orbitals. The **electron pairing energy,** P, is the additional energy required for two electrons to occupy the same orbital, compared to the two electrons singly occupying separate degenerate orbitals. It is this pairing energy that causes the electrons in a degenerate level to occupy separate orbitals as far as is possible (Hund's rule).

When the crystal field splitting is large, the electrons fill the lower energy d orbitals, forming a low spin complex.

Instead of forming a pair of electrons, the fourth electron could be placed in one of the two d orbitals at an energy Δ above the others. The two possible arrangements of the four electrons in the d orbitals are shown on the left side of Figure 18.30. The relative sizes of Δ and P determine which of these configurations is preferred in a given complex. If $P > \Delta$, the fourth electron enters the higher energy d orbital, and the complex contains four unpaired electrons. When $P < \Delta$, a pair of electrons is formed in the lower energy d orbitals, and the complex has only two unpaired electrons. Depending on the size of Δ (which depends on the nature of the ligands), both of these situations are observed for d^4 metal complexes; the MnF_6^{3-} complex contains four unpaired electrons, but the $Mn(CN)_6^{3-}$ complex has only two unpaired electrons. A **high spin** or **weak field** complex occurs when $P > \Delta$. A **low spin** or **strong field** complex occurs when $P < \Delta$. Note that a small value of Δ means that the ligand field is *weak* and results in a *high* spin state. A large Δ produces a *strong* ligand field and the complex is in a *low* spin state.

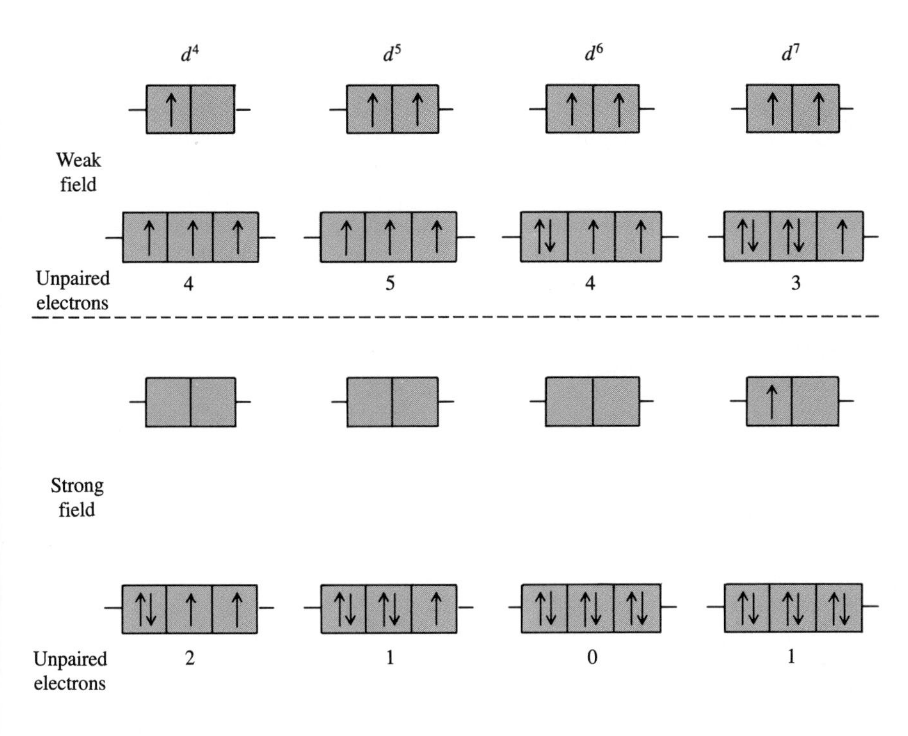

Figure 18.30 The *d* electron configurations in octahedral complexes Two electron configurations are possible for octahedral complexes that contain four to seven *d* electrons. The top row shows the weak field (high spin) complexes, and the bottom row shows the strong field (low spin) complexes.

There are two possible spin states for octahedral complexes that contain four, five, six, or seven *d* electrons.

Two spin states are possible for the d^4 to d^7 electron configurations, depending on the strength of the crystal field. The different spin states and occupied orbitals for all of these are shown in Figure 18.30. Once eight or more electrons are present in the *d* subshell (see Figure 18.29), the number of unpaired electrons is the same regardless of the strength of the crystal field.

The factors that influence the size of Δ determine the number of unpaired electrons in complexes that have four through seven *d* electrons. For example, the CoF_6^{3-} ion contains four unpaired electrons, while $Co(NH_3)_6^{3+}$ contains no unpaired electrons, because ammonia is higher in the spectrochemical series, and its Δ is larger than the pairing energy. Since Δ increases with the charge on the metal ion, low spin complexes are more common for complexes of 3+ metal ions than for 2+ ions. Because Δ is larger for the transition elements in the 4*d* and 5*d* series than for the 3*d* series, nearly all complexes of the heavier transition metals are low spin.

Example 18.7 Magnetic Properties of Transition Metal Complexes

Classify each of the following octahedral complexes as high spin or low spin, and predict the number of unpaired electrons in each. (a) FeF_6^{3-}, (b) $Cr(CN)_6^{4-}$, (c) $Mn(H_2O)_6^{2+}$, (d) $RhCl_6^{3-}$.

Solution

(a) The fluoride ion produces a weak crystal field (see the spectrochemical series); consequently we expect a weak field complex for a 3*d* transition metal ion. The Fe^{3+} ion has a d^5 electron configuration, so the high spin complex should contain five unpaired electrons (Figure 18.30).

(b) This is a complex of Cr^{2+}, which has the d^4 configuration. From its position in the spectrochemical series, the CN^- ion produces one of the strongest crystal field splittings, so a low spin complex is expected. All four of the electrons are in the lower-energy *d* orbitals, so there are two unpaired electrons.

(c) Water is a moderately weak field ligand. With the low ionic charge of the Mn^{2+} ion, a high spin complex is anticipated. There are five *d* electrons in the complex, and each singly occupies a *d* orbital, so all five electrons are unpaired.

(d) The chloride ligand occurs very early in the spectrochemical series. However, Rh is a metal in the 5*d* transition series, so all of its complexes are expected to be low spin. The d^6 configuration of Rh^{3+} forms a complex in which all of the electrons are paired in the lower-energy orbitals.

Understanding Two complexes of iron(II), $Fe(H_2O)_6^{2+}$ and $Fe(CN)_6^{4-}$, have different numbers of unpaired electrons. How many unpaired electrons are present in each of the complexes?

Answer: $Fe(H_2O)_6^{2+}$ has 4 unpaired electrons. $Fe(CN)_6^{4-}$ has no unpaired electrons.

Complexes of Other Shapes

The crystal field theory also applies to tetrahedral and square planar complexes, the other two arrangements most commonly observed in transition metal complexes.

Let us first examine a tetrahedral complex of a transition metal ion. One way of visualizing this arrangement of ligands about the metal is to place the metal at the center of a cube, with the Cartesian axes passing through the centers of the cube's faces. The four ligands then are located at diagonally opposite corners of the cube, as shown in Figure 18.31. With this orientation, three of the *d* orbitals (d_{xy}, d_{xz}, and d_{yz}) have the lobes of electron density directed at the centers of the twelve edges of the cube (four lobes for each of the *d* orbitals). The other two *d* orbitals ($d_{x^2-y^2}$ and d_{z^2}) are directed at the centers of the faces of the cube. The centers of the edges of the cube are closer to the ligands ($a/2$) than are the centers of the faces ($a/\sqrt{2}$). As a result, electrons in three of the *d* orbitals experience a stronger repulsion than do those in the other two. Just as in the case of an octahedral complex, the difference in the repulsion energy causes the *d* orbitals to split into a threefold degenerate set and a twofold degenerate set. In the case of the tetrahedral complex, however, the stronger repulsions occur with the electrons in the set of three orbitals, so these are of higher energy. The result of this analysis is

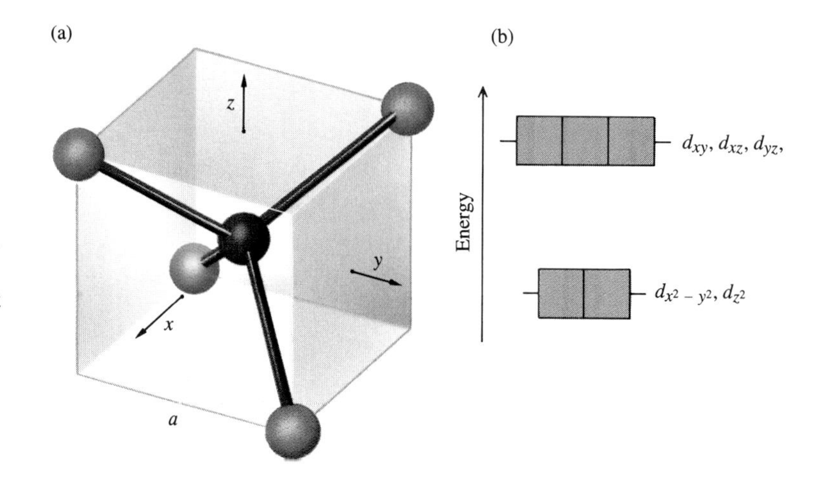

Figure 18.31 Tetrahedral complexes (a) A tetrahedral complex can be represented by placing the ligands at alternating corners of a cube, with the metal ion at the center of the cube. The length of the edge of the cube is *a*. The *x*, *y*, and *z* axes pass through the centers of the cube's faces. The electrons in the three *d* orbitals (d_{xy}, d_{yz}, and d_{xz}) with lobes directed at the centers of the cube's edges are closer to the ligands than are those in the other two orbitals ($d_{x^2-y^2}$ and d_{z^2}), which point at the centers of the cube's faces. (b) The unequal replusions cause splitting of the orbitals into a twofold degenerate set and a higher energy threefold degenerate set.

that the crystal field energy-level diagram for a tetrahedral complex is inverted from that for an octahedral complex, with two orbitals at lower energy and three orbitals at higher energy, as shown in Figure 18.31b.

A quantitative treatment of the ligand field produced by a tetrahedral arrangement of four ligands about a metal ion shows that the crystal field splitting, Δ, is 4/9 of the Δ for an octahedral complex using the same ligands and metal ion. Experimental measurements confirm that Δ for tetrahedral complexes is about half that observed in similar octahedral complexes. As a consequence of the much smaller Δ in the tetrahedral arrangement, nearly all tetrahedral complexes are high spin. Generally, it is safe to assume that a tetrahedral arrangement means that a weak field complex is formed.

The crystal field energy-level diagram for a metal surrounded by ligands in a tetrahedral arrangement is inverted from that produced by an octahedral arrangement.

Square planar geometry is observed almost exclusively in complexes where the metal ion has a d^8 electron configuration. In these d^8 complexes the electrons are invariably paired, suggesting a strong field configuration. The square planar geometry is usually visualized as a distortion of an octahedral complex, produced by removing the two ligands along the z axis. As the negative ligands are removed from the z axis, the electrons in the d_{z^2} orbital become more stable than those in the $d_{x^2-y^2}$ because of the reduced electrostatic repulsion. The electrons in the d_{yz} and d_{xz} also experience a greater reduction in the repulsions (as the ligands on the z axis are removed) than do those in the d_{xy} orbital. The resulting energy-level diagram for the d electrons in a square planar complex is shown in Figure 18.32. In this

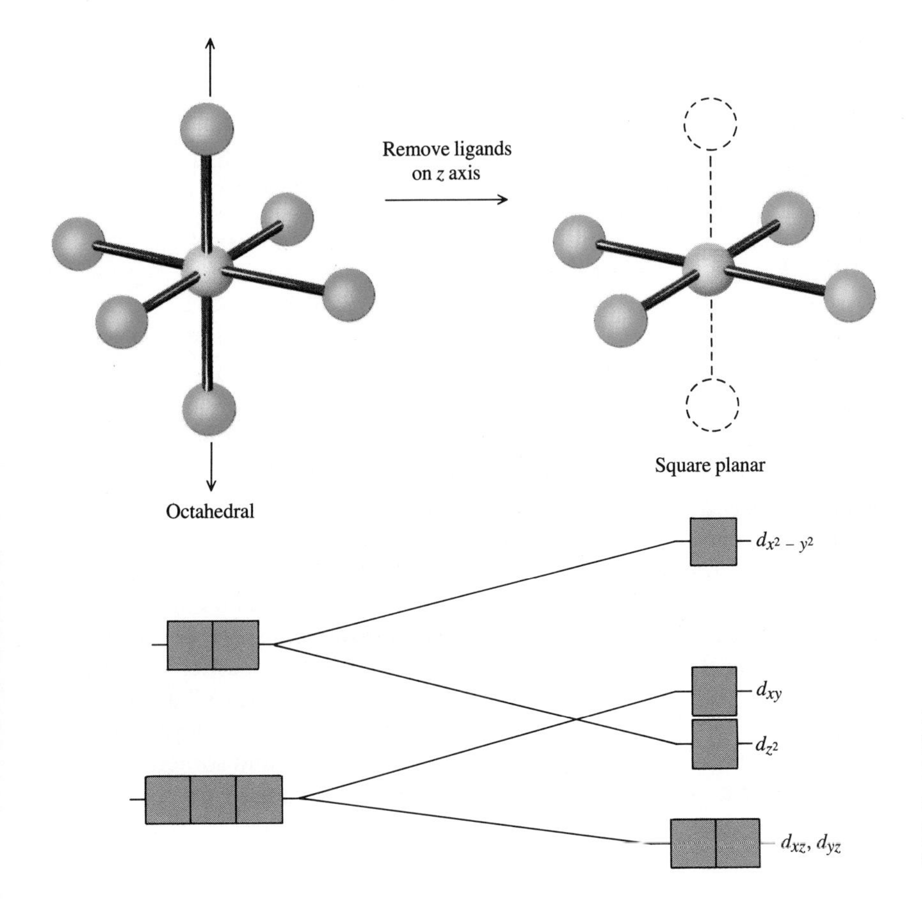

Figure 18.32 Crystal field diagram for a square complex The change in the relative energies of the d orbitals is shown as an octahedral complex is converted into a square planar complex by removing the two ligands along the z axis.

arrangement, four of the *d* orbitals are much lower in energy than the fifth one. Metal ions that contain eight *d* electrons favor a square planar arrangement, because there are just enough electrons to fill the four lowest energy orbitals.

Example 18.8 Electron Configurations of Four-Coordinate Complexes

Predict the geometry, *d* orbital energy-level diagram, and number of unpaired electrons in (a) $Ni(CN)_4^{2-}$ and (b) $FeBr_4^-$.

Solution

(a) Nickel is in the +2 oxidation state and therefore has a d^8 electron configuration. Furthermore, the cyanide ion is a strong field ligand. Under these conditions, a square planar complex is expected. The energy-level diagram for the *d* orbitals is that shown at the right in Figure 18.32, with a pair of electrons in each of the orbitals except the $d_{x^2-y^2}$. There are no unpaired electrons in the complex, so it is diamagnetic.

(b) This complex is formed from $Fe^{3+}(d^5)$ and Br^- ions. The small size of the 3+ metal ion and the relatively large bromide ligands favor the formation of a tetrahedral complex. The *d* orbital energy level diagram expected for a tetrahedral complex is

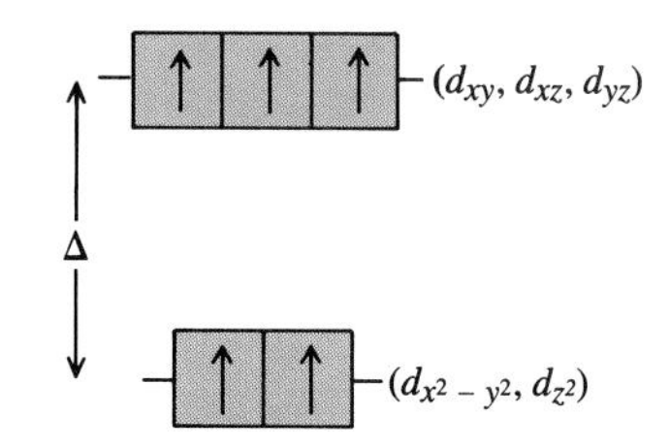

The crystal field splitting for tetrahedral complexes is small, so the high spin complex is expected. Therefore, one electron is present in each of the *d* orbitals for a total of five unpaired electrons.

Summary

The majority of the elements are metals, and the ability to isolate these elements from naturally occurring compounds has been an important factor in the development of our modern technological society. *Metallurgy,* the science of extracting metals from their ores, may involve chemical reactions in all the stages of pretreatment, concentration, reduction, and purification. Concentration of ores is accomplished by both physical and chemical means. The *roasting* of ores converts naturally occurring compounds into chemical forms more suitable for reduction.

The isolation of metals from their compounds involves a chemical reduction. The agent used for reduction depends on the reactivity of the metal being isolated. Very reactive metals, such as sodium and aluminum, are prepared by electrolysis. With other metals, including chromium and titanium, displacement by more reactive metals is chosen for isolating the metals from ores. For economic reasons carbon and carbon monoxide are the reducing agents of choice in the isolation of metals such as iron and zinc.

The final purification of metals is highly dependent on the chemical and physical properties of both the metal and the impurities present, as well as the purity needed in the final application. Often alloys, which are mixtures of two or more metals (and sometimes small quantities of other elements, such as carbon, silicon, and phosphorus), have more desirable properties than pure metals. Distillation, electrolysis, and zone refining are three of the processes used for purifying metals. In the oxygen furnace, used to purify pig iron, several chemical reactions occur.

The *transition elements* are those metals that have partially filled *d* orbitals in at least one of their oxidation states, and generally exhibit more than one stable oxidation state. Consequently, much of the chemistry of these elements involves oxidation-reduction reactions. High melting points and boiling points are characteristic of the transition metals and indicate the presence of very strong metallic bonds.

The assumption that the *ns* and $(n - 1)d$ electrons and orbitals are used to form highly delocalized bonds in these metals is consistent with the increase in melting points within each period up to group VIB or VIIB, followed by decreasing melting points through the remainder of the transition series. There is a small increase in the effective nuclear charge within a period as the atomic number increases and electrons are added to the *d* subshell, causing a small decrease in atomic radius from left to right across the transition elements. The decrease in radius is smaller for the transition metals than for the representative elements. The *lanthanide contraction* causes the sizes of the transition metal atoms in the sixth period to be nearly the same as those of the elements directly above them in the fifth period. Consequently there is usually a much larger difference between the chemical behavior of the first and second elements in a transition metal group than is observed between the second and third elements.

All of the transition elements exhibit stable oxidation states of +2, +3, or both, as well as higher and sometimes lower states. The maximum oxidation state achieved by the transition elements is equal to the number of *s* and *d* electrons present in the valence shell through manganese (+7) in the first transition series, and ruthenium and osmium (+8) in the second and third rows. Following the maximum in positive oxidation state in group VIIB or VIIIB, the highest observed oxidation state decreases through the remaining groups. The lanthanide contraction is also evident in the ionization energies of the elements in each group of the transition elements. The first and second elements in each group have very similar ionization energies, while the ionization energy of the third element in the group is markedly higher.

A large part of the chemistry of the transition elements involves *coordination complexes.* In many compounds, transition metal ions behave as Lewis acids, forming coordinate covalent bonds with several *ligands,* molecules or ions that donate unshared electron pairs to the metal ion. The most commonly observed *coordination numbers* in transition metal complexes and their geometric shapes are two (linear), four (square planar or tetrahedral), and six (octahedral). Ligands are classified as *monodentate, bidentate, tridentate,* and so forth, by the number of atoms (one, two, three, . . .) in the ligand that bond to a metal ion. *Chelating ligands* have two or more donor atoms that coordinate to the same metal atom. In writing the formula of a coordination compound, the metal ion and its coordinated ligands are enclosed in square brackets.

Several kinds of isomers are possible in coordination compounds. These fall into two main categories, *structural isomers* and *stereoisomers.* Structural isomers are further subdivided into *coordination isomers,* which differ in the ligands coordinated to the metal, and *linkage isomers,* in which the same ligand is bonded to the metal ion through different donor atoms. *Geometric isomers,* a kind of stereoisomer, contain the same ligands with different geometric arrangements about the metal ion. Stereoisomers also include *optical isomers,* which rotate the plane of polarized light. This behavior is characteristic of *chiral molecules,* which have nonsuperimposable mirror image structures. The nonsuperimposable mirror images are called *enantiomers,* and they rotate the plane of polarized light in opposite directions. A *racemic mixture* is an equimolar mixture of enantiomers, which produces no net rotation of polarized light.

Crystal field theory is a model for the bonding in coordination complexes. It explains the magnetic properties and visible spectra that are so characteristic of these species. The arrangement of ligands surrounding the transition metal ion in a regular geometric pattern removes the degeneracy of the partially occupied *d* orbitals. In an octahedral complex, three of the *d* orbitals have a lower energy than the other two *d* orbitals. The energy difference between the two sets of *d* orbitals is called the *crystal field splitting,* Δ, and it corresponds to the energy of light in the visible region of the electromagnetic spectrum. The absorption of light in the visible spectrum moves an electron from a lower energy *d* orbital to a higher one, producing the observed color of coordination compounds. The *spectrochemical series* arranges ligands in order of the values of Δ they produce. When the ligands produce a small splitting of the *d* orbitals, the complexes are called *weak field. Strong field complexes* have a large energy separation of the *d* orbitals. When a strong field is produced by the ligands, the electrons in the complex form pairs in the lower energy *d* orbitals before any enter the higher energy *d* orbitals. The electrons in a weak field complex singly occupy all five *d* orbitals, before pairing occurs in the lower energy *d* orbitals. Weak field complexes are called *high spin* and strong field complexes are called *low spin.* Crystal field theory is also successful in accounting for the spectral and magnetic properties of other arrangements of ligands, such as tetrahedral and square planar complexes.

Chapter Terms

Metallurgy—the science of extracting metals from their ores, purifying and preparing them for practical use. *(18.1)*

Roasting—pretreatment of ores by heating the material below its melting point, usually in the presence of air, to convert the ore into a chemical form more suitable for the reduction step. *(18.1)*

Transition elements—those elements characterized by having partially filled *d* orbitals in the metal or at least one of its oxidation states. These elements are found in the center of the periodic chart, in the nine columns from group IIIB through group IB (3 to 11). *(18.2)*

Lanthanide contraction—the small decrease in the radii of the lanthanides as the 4*f* subshell is filled, causing the transition elements of the fifth and sixth periods to have nearly identical radii within each group in group IVB and beyond. *(18.2)*

Ligand—an anion or molecule that functions as a Lewis base in forming one or more coordinate covalent bonds to metal ions. *(18.4)*

Coordination number—the number of donor atoms bonded to a single metal ion in a coordination complex. *(18.4)*

Coordination compound or **complex**—a species that contains a metal ion bound to ligands by Lewis acid-base interactions. *(18.4)*

Monodentate ligand—a ligand that donates one pair of electrons to the metal in a coordination complex. *(18.4)*

Chelating ligand—a ligand that simultaneously bonds to the same metal ion through two or more different atoms. *(18.4)*

Chelate—a coordination complex that contains one or more chelating ligands. *(18.4)*

Polydentate ligand—a ligand having more than one atom that forms a coordinate covalent bond with a metal ion. Numerical prefixes (bi-, tri-, tetra-, etc.) are combined with "dentate" to designate the number of donor atoms in the ligand. *(18.4)*

Structural isomers—compounds that contain the same numbers and kinds of atoms, but differ in the bonds that are present. *(18.5)*

Coordination isomers—compounds that contain the same numbers and kinds of atoms but have different Lewis bases directly bonded to the metal ion; *e.g.*, $[Co(NH_3)_4Cl_2]Br$ and $[Co(NH_3)_4ClBr]Cl$ are coordination isomers. *(18.5)*

Linkage isomers—compounds in which the same ligand is coordinated to the metal ion through one of two possible donor atoms; *e.g.*, NO_2^- may form a bond to the metal ion through one of the oxygen atoms (nitrito) or the nitrogen atom (nitro). *(18.5)*

Stereoisomers—molecules or ions that have the same bonds but are different in the arrangement of the atoms in space. *(18.5)*

Geometric isomers—stereoisomers that have the same number and kind of bonds, but differ in the relative positions of the atoms. *(18.5)*

Optical isomers—stereoisomers that rotate the plane of polarized light. *(18.5)*

Polarized light—electromagnetic radiation that has the electric field oscillating in a single plane. *(18.5)*

Chiral molecule or ion—a molecule or ion that has a mirror image structure that cannot be superimposed on the original. *(18.5)*

Enantiomers—molecules or ions that are nonsuperimposable mirror images of each other. *(18.5)*

Racemic mixture—an equimolar mixture of enantiomers that produces no net rotation of the plane of polarized light. *(18.5)*

Absorption spectrum—a graph of the quantity of light a sample absorbs as a function of wavelength. *(18.6)*

Crystal field splitting (Δ)—the separation in energy between the *d* orbitals caused by the electric field produced by the arrangement of the ligands about the central metal ion. *(18.6)*

Spectrochemical series—the arrangement of ligands in order of increasing size of the crystal field splitting that they cause. *(18.6)*

Electron pairing energy (*P*)—the additional energy required for two electrons to occupy the same orbital, compared to the two electrons singly occupying separate degenerate orbitals. *(18.6)*

High spin complex—a coordination complex in which each of the five *d* orbitals is occupied by a single electron, before two electrons occupy one of the low energy orbitals. High spin complexes are observed when $P > \Delta$. *(18.6)*

Low spin complex—a coordination complex in which the lower energy *d* orbitals are completely filled with two electrons each, before electrons enter the higher energy *d* orbitals. Low spin complexes are observed when $P < \Delta$. *(18.6)*

Strong field complex—a complex in which the crystal field splitting is greater than the pairing energy; a low spin complex. *(18.6)*

Weak field complex—a complex in which the crystal field splitting is less than the pairing energy; a high spin complex. *(18.6)*

Exercises

Exercises designated with color have answers in Appendix J.

Metallurgy

18.1 List three goals in the pretreatment of ores. Give an example of each of these.

18.2 For each of the following compounds, give the preferred method of reduction used to isolate the metal.
(a) LiCl
(b) Fe_2O_3
(c) Al_2O_3
(d) HgS

18.3 List and describe three methods used to purify metals once they have been reduced.

18.4 Write equations for the principal reactions involved in the reduction of Fe_2O_3 in a blast furnace.

Transition elements

18.5 Define the transition elements.

18.6 (a) Which element in the fourth period has one or more 3*d* electrons in the free element, but none in any of its common oxidation states?
(b) Which elements appear in the "*d* block" of the periodic table, but do not meet the definition of a transition element?
(c) Is actinium ($Z = 89$) a transition element? Explain.

18.7 Which of the following elements are transition metals?
(a) Fe
(b) Ba
(c) Hg
(d) Mo
(e) La
(f) Pd

18.8 What distinguishes a transition element from a representative element?

Properties of Transition Elements

18.9 In each part select the transition element that has the higher melting point.
(a) Cr, Co
(b) Ti, Hf
(c) Nb, V
(d) Y, W

18.10 In each part select the transition element that has the higher melting point.
(a) Cr, Cu
(b) Fe, Os
(c) Cr, V
(d) La, W

18.11 Based on the general trends in metallic bonding, which transition metal should have the highest heat of fusion?

18.12 Why do the atomic radii of the transition elements within a period decrease more rapidly from group IIIB through group VIB than through the rest of the transition elements in that period?

18.13 The ratio of the density of tantalum to that of niobium is 1.94, nearly identical to the 1.95 ratio of their atomic weights. Explain how this is a result of the lanthanide contraction.

18.14 Arrange the following transition metal atoms in order of decreasing atomic radius. V, Co, Nb, W

18.15 Arrange the following transition metal atoms in order of decreasing atomic radius. Fe, Mo, Hf, Ta

18.16 What is the maximum positive oxidation state expected for each of the following?
(a) Ti
(b) W
(c) Ta
(d) Re

18.17 What is the maximum positive oxidation state expected for each of the following?
(a) Cr
(b) Zr
(c) Y
(d) Tc

18.18 Only the group IB transition elements form simple compounds in which the oxidation state of the metal is +1. For all of the other transition elements the lowest positive oxidation state is +2. What common feature in the electron configuration of the transition elements contributes to this fact?

18.19 Use the information in Figure 18.6 to explain why the +3 oxidation state becomes less common for the elements near the end of the transition series.

18.20 In each part select the element that has the higher first ionization energy.
(a) Ti or Mn
(b) V or Ta
(c) Ru or Rh
(d) Mo or Os

18.21 In each part select the element that has the higher first ionization energy.
(a) Zr or Tc
(b) Mo or W
(c) Fe or Pt
(d) Mn or Co

The Chemistry of Typical Transition Elements

18.22 Write chemical equations for the following processes.
(a) The reduction of the mineral chromite ($FeCr_2O_4$) with carbon.
(b) The reduction of chromium(III) oxide with aluminum.
(c) The reaction of Cr^{2+}(aq) with dissolved oxygen from the air.
(d) The oxidation of the ion $Cr(OH)_4^-$ by hypochlorite ion in basic solution.
(e) The titration of iron(II) with $Cr_2O_7^{2-}$ in acid solution.

18.23 Write chemical equations for the following processes.
(a) The preparation of CrO_3(s) from $K_2Cr_2O_7$(s).
(b) Equations that illustrate the amphoteric nature of Cr_2O_3.
(c) The oxidation of Cr^{2+} by $Cr_2O_7^{2-}$ in acid solution.
(d) Forming a solution of CrO_4^{2-} from Cr^{3+} in aqueous solution.

18.24 Give the electron configurations of Fe, Fe^{2+}, and Fe^{3+}.

18.25 The hydrated iron(III) ion $[Fe(H_2O)_6]^{3+}$ is a Brønsted acid. Write an equation that shows this ion reacting as an acid.

18.26 What is the oxidizing agent that converts Cu(s) into $CuSO_4$(aq) when the metal dissolves in hot sulfuric acid? Write a chemical equation for this reaction.

Formulas and Names of Coordination Compounds

18.27 Write the formula for each of the coordination compounds described.
(a) Chromium(III) chloride, in which one Cl^- and five water molecules are coordinated to the metal.
(b) $CrCl_3 \cdot 4NH_3$, which contains two coordinated chloride ions.
(c) The potassium salt of the coordination complex containing six CN^- ions and Fe(III).

18.28 Write the formula for each of the coordination compounds described.
(a) A coordination compound containing two complex ions of Co(III), one that contains six CN^- ions, and the other containing three ethylenediamine molecules. (You may use the abbreviation en for the ethylenediamine molecule.)
(b) A coordination compound derived from platinum(II) nitrate, in which four ammonia molecules are coordinated to the transition metal ion.
(c) The sodium salt of the complex formed from Rh(III), five Cl^- ions, and one water molecule.

18.29 Name each of the following compounds.
(a) $[Pt(NH_3)_2Cl_2]$
(b) $[Co(en)_2(NO_2)_2]NO_3$ (en = ethylenediamine)
(c) $K_3[RhCl_6]$
(d) $[Pt(NH_3)_4][PtCl_4]$
(e) $[Cr(CO)_6]$

18.30 Name each of the following compounds.
(a) $[Fe(CO)_5]$
(b) $K_2[Cr(CN)_5NO]$
(c) $[Ru(NH_3)_5Cl]Cl_2$
(d) $[Co(dien)Br_3]$ (dien = diethylenetriamine)
(e) $[Cr(NH_3)_6][Cr(CN)_6]$

18.31 Write the formula of each of the following ions or compounds.
(a) pentaaquachlorochromium(III) chloride
(b) tetraamminedinitrorhodium(III) bromide
(c) dichlorobis(ethylenediamine)ruthenium(III)
(d) diaquatetrachlororhodate(III)
(e) triamminetribromoplatinum(IV)

18.32 Write the formula of each of the following ions or compounds.
(a) hexaaquachromium(III) hexacyanoferrate(III)
(b) bromochlorobis(ethylenediamine)cobalt(III)
(c) carbonylpentacyanocobaltate(III)
(d) (diethylenetriamine)trinitro chromium(III) (abbreviate the neutral ligand with "dien")
(e) pentaaquathiocyanatoiron(III)

Isomers in Coordination Compounds

18.33 Draw the structures and name the geometric isomers of tetraaquadibromochromium(III).

18.34 Draw the structures and name the geometric isomers of triamminetrichlorocobalt(III).

18.35 Use the following list of complexes in answering each part. There may be more than one correct choice for each part.
(1) $[Co(NH_3)_3Cl_3]$
(2) $[Co(en)_2Br_2]^+$
(3) $\{Cr(H_2O)_2Cl_2Br_2]^-$
(4) $[Pt(NH_3)_3SCN]^+$
(5) $[Cr(C_2O_4)_3]^{3-}$
(a) Which complex has three isomers, two of which are enantiomers?
(b) Which complex might have linkage isomers?
(c) Which complexes cannot form optically active isomers?
(d) Identify the complex that has the largest number of possible isomers. Show the structures for all of the isomers.

18.36 For each complex give the structures of all isomers and identify the kind(s) of isomerism that is illustrated.
(a) $[Co(NH_3)_5SCN]^{2+}$
(b) $[Pd(NH_3)_2Br_2]$
(c) $[Cr(C_2O_4)_3]^{3-}$
(d) $[Rh(en)_2(H_2O)_2]^{3+}$
(e) $[Mn(CN)_4(OH)(CO)]^{3-}$

18.37 Draw the structures of
(a) *mer*-triamminetribromorhodium(III)
(b) *trans*-dinitrobromochloroplatinum(II)

18.38 Which of the following ligands could display linkage isomerism? N_3^-, SCN^-, NO_2^-, NCO^-, ethylenediamine.

18.39 What structural feature is used to determine whether a compound can exist as optical isomers?

18.40 What is a racemic mixture?

18.41 What physical property is different for two enantiomers?

Crystal Field Theory

18.42 Find Δ in kJ/mol for an octahedral Ti^{3+} complex that has an absorption maximum in the visible spectrum at 450 nm. (Ti^{3+} has a single *d* electron, so Δ = energy of the absorption maximum.)

18.43 In each pair of complexes, select the one that has the larger value of the crystal field splitting.
(a) $[Co(NH_3)_6]^{3+}$ and $[Co(CN)_6]^{3-}$
(b) $[Cr(H_2O)_6]^{2+}$ and $[Cr(H_2O)_6]^{3+}$
(c) $[Fe(H_2O)_6]^{2+}$ and $[Ru(H_2O)_6]^{2+}$
(d) $[CrF_6]^{3-}$ and $[Cr(H_2O)_6]^{3+}$

18.44 In each pair of complexes, select the one that absorbs light at a shorter wavelength.
(a) $[Rh(NH_3)_6]^{3+}$ and $[Rh(CN)_6]^{3-}$
(b) $[Fe(H_2O)_6]^{2+}$ and $[Fe(H_2O)_6]^{3+}$
(c) $[Co(H_2O)_6]^{3+}$ and $[Rh(H_2O)_6]^{3+}$
(d) $[TiF_6]^{3-}$ and $[Ti(H_2O)_6]^{3+}$

18.45 For each *d* electron configuration, state the number of unpaired electrons expected in octahedral complexes. Give an example complex for each case. (Two answers are possible for some of these cases.)
(a) d^2 (c) d^6
(b) d^4 (d) d^8

18.46 For each *d* electron configuration, state the number of unpaired electrons expected in octahedral complexes. Give an example complex for each case. (Two answers are possible for some of these cases.)
(a) d^3 (c) d^7
(b) d^5 (d) d^9

18.47 For each of the following octahedral complexes, give the number of unpaired electrons expected.
(a) $[CrCl_6]^{3-}$ (d) $[Rh(H_2O)_6]^{3+}$
(b) $[Co(CN)_5(H_2O)]^{2-}$ (e) $[V(H_2O)_6]^{3+}$
(c) $[Mn(H_2O)_6]^{2+}$

18.48 For each of the following octahedral complexes, give the number of unpaired electrons expected.
(a) $[MnF_6]^{3-}$ (d) $[Fe(H_2O)_6]^{2+}$
(b) $[Fe(CN)_6]^{3-}$ (e) $[Ni(H_2O)_6]^{2+}$
(c) $[Re(H_2O)_6]^{2+}$

18.49 Ni(II) forms the complex $NiBr_4^{2-}$, which contains two unpaired electrons. Which of the four coordinate geometries discussed in this chapter is most likely in this complex?

18.50 In each part give the number of unpaired electrons present in the two complexes. In each part both complexes contain the same number of *d* electrons, but one is high spin and the other is low spin.
(a) $[Cr(H_2O)_6]^{2+}$ and $[Mn(CN)_6]^{3-}$
(b) $[Fe(H_2O)_6]^{2+}$ and $[Ru(H_2O)_6]^{2+}$
(c) $[Co(H_2O)_6]^{2+}$ and $[Co(CN)_5(H_2O)]^{3-}$

18.51 Show that the high spin complex for the d^5 electron configuration is favored when $P > \Delta$.

18.52 Recently a low spin tetrahedral complex of cobalt(III) has been found in which the ligands are large hydrocarbon groups bonded to the metal through carbon atoms. Construct a crystal field energy level diagram containing the metal *d* electrons of this complex. How many unpaired electrons are expected in this complex?

18.53 Give the number of unpaired electrons and the geometry expected for each of the following four-coordinate complexes.
(a) $[Au(CN)_4]^-$ (c) $[Pd(NH_3)_4]^{2+}$
(b) $[CoCl_4]^-$

18.54 Give the number of unpaired electrons and the geometry expected for each of the following four-coordinate complexes.
(a) $[Ni(CN)_4]^{2-}$ (c) $[Pt(NH_3)_4]^{2+}$
(b) $[FeCl_4]^{2-}$

30ml
KIMAX
USA
NO. 14000

Chapter Nineteen 19

The Chemistry of Hydrogen, Elements in Groups IIIA Through VIA, and the Noble Gases

The atomic properties that influence the chemistry of the representative elements, such as electronegativity, atomic and ionic radii, and ionization energy, have been discussed in detail in earlier chapters. In this chapter, the properties and chemical reactivity of hydrogen, the elements in groups IIIA through VIA, and the noble gases will be surveyed. Chapter 7 presented a similar survey of groups IA, IIA, and VIIA.

19.1 General Trends

Objective

- To discuss how and why the properties of the second period elements are different from those of other elements in their groups

The nonmetallic elements, with the exception of hydrogen, are located in the upper right portion of the periodic table. The chemistry of these elements is controlled by relatively high electronegativities and ionization energies. The nonmetals form ionic compounds with metals, and they form covalent compounds with each other. Within any group, there is usually a large

◀ Gallium reacts vigorously with bromine.

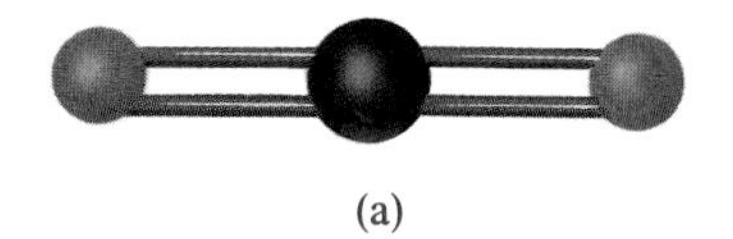

(a)

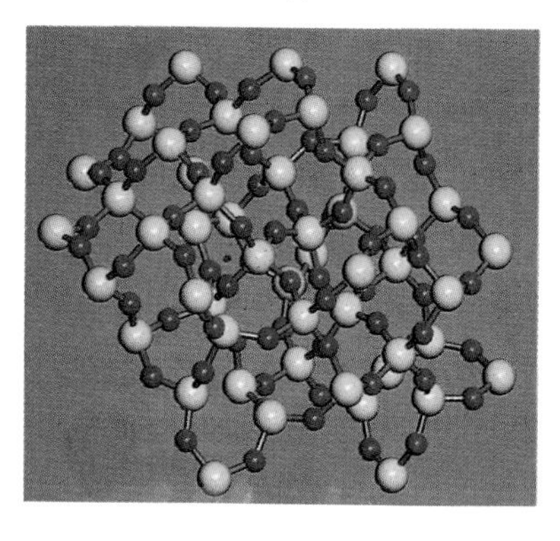

(b)

Figure 19.1 Structures of CO_2 and SiO_2 (a) Carbon dioxide is a triatomic molecule containing two sigma and two pi bonds. (b) Silicon dioxide exists as a covalent solid with each silicon (green) forming sigma bonds to four oxygen atoms.

difference between the chemistry of the second period element and that of the remaining members of the group. High electronegativity is particularly important for the nonmetallic elements of the second period. Fluorine, of course, is the most electronegative element, and oxygen is second. The nonmetallic elements of the second period have the smallest radius in each of their respective groups (helium is the smallest noble gas).

An important consequence of the small sizes of the elements in the second period is the tendency to form strong π bonds from the sideways overlap of p orbitals. Elements of the third and higher periods are too large to form strong π bonds from extensive p orbital overlap. These heavier elements tend to form two σ bonds rather than one σ and one π bond because of the weakness of the π bond. This point is demonstrated by comparison of the structures of CO_2 and SiO_2. Carbon dioxide is a molecular compound with strong π bonding, whereas SiO_2 is a covalent solid in which each silicon atom forms four sigma bonds to four bridging oxygen atoms in an extended array (Figure 19.1). It is a more stable bonding situation for silicon to form four σ bonds rather than the two σ and two π bonds formed by carbon in CO_2. The lower entropy of the polymeric structure for SiO_2 is more than compensated by the greater bond energy.

The difference in the importance of π bonding is also noticeable in the elemental forms of nitrogen and phosphorus. The stable form of nitrogen is N_2, a molecule containing a triple bond (one σ and two π bonds). The simplest form of phosphorus is P_4, in which the atoms are arranged at the vertices of a tetrahedron. In this arrangement, each phosphorus atom forms three sigma bonds. The other stable forms of elemental phosphorus also contain only σ bonds.

Another major difference of the second period elements is that they do not form compounds in which the Lewis structures would place more than eight electrons around a central atom. For example, in group VIA only one type of compound is formed from the combination of one oxygen atom with fluorine atoms, OF_2, but three compounds (two that are electron-rich) can form from the combination of one sulfur atom with fluorine atoms, SF_2, SF_4, and SF_6 (Figure 19.2). As a group VIA element from the second period, oxygen has six valence electrons and four valence orbitals and thus can share only two additional electrons with fluorine atoms. The oxygen atom in OF_2 is sp^3 hybridized. The valence shell of sulfur is the $n = 3$ level, and the

Figure 19.2 Structures of group VIA fluorides Oxygen forms only one fluoride, (a) OF_2, whereas sulfur forms three fluorides, (b) SF_2, (c) SF_4, and (d) SF_6.

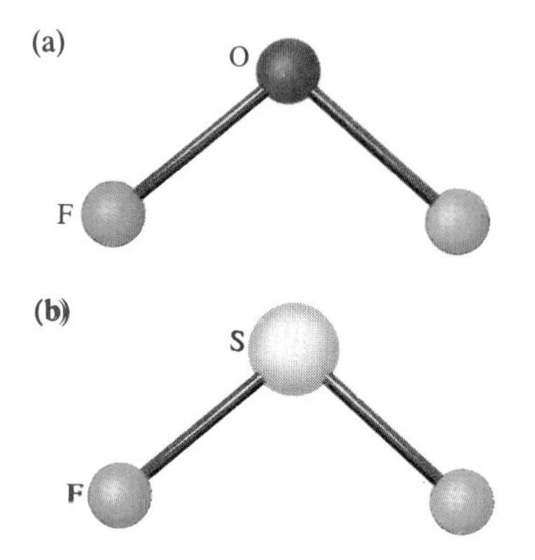

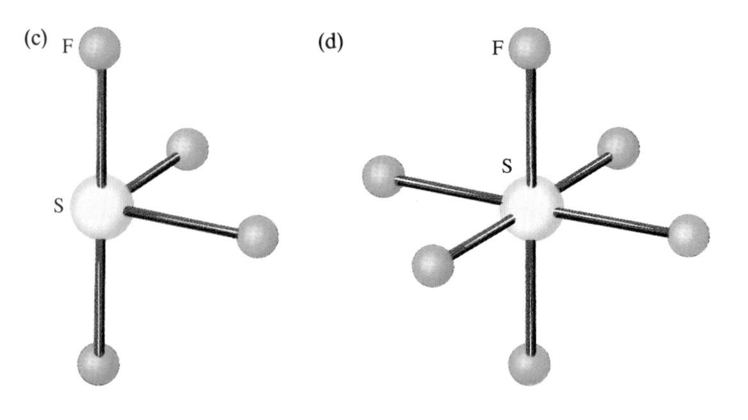

valence orbitals can include the $3d$ as well as $3s$ and $3p$ orbitals. The six valence electrons can be used to form up to six electron pair bonds. In SF_2 the sulfur uses sp^3 hybrid orbitals, in SF_4 sp^3d hybrid orbitals, and in SF_6, sp^3d^2 hybrid orbitals.

Another important general trend for the elements in groups IIIA through VIA is that the metallic character of the elements increases going down the group. This trend is expected on the basis of the decreases in ionization energies and electronegativities observed going down a group. The metals, of course, tend to form cations and react in a way that is very different from the nonmetals. The elements between the metals and nonmetals, the metalloids, have intermediate properties.

The chemistry of the second period elements is very different from that of other elements in their groups because of their relatively small size and high electronegativity, and the availability of only four valence orbitals.

19.2 Hydrogen

Objective

- To describe the properties, sources, and important uses of hydrogen

Although it is generally listed in group IA (1) on a periodic table, hydrogen should probably be considered in a group by itself. It is not surprising that the lightest element has unique properties. Hydrogen can lose an electron to form a proton ($H^+(aq)$ in water); in combination with electropositive metals, it can gain an electron to form the hydride ion, H^-, which has the electron configuration of the noble gas helium. With an electronegativity near the middle of the scale (2.1), hydrogen can also form strong covalent bonds with the other nonmetallic elements.

Hydrogen is the most abundant element in the universe. It is the nuclear fuel consumed by the sun in producing energy. In contrast, hydrogen makes up only 0.87% of the mass of the earth's crust. There are three isotopes of hydrogen: ^1H (99.985% abundant) and ^2H (0.015% abundant, frequently called deuterium) are stable, and ^3H (frequently called tritium) is radioactive and very rare. The molecular form of the element is H_2, a tasteless and odorless gas. Because it is nonpolar and has a very low molecular weight, H_2 has very weak intermolecular forces and boils at −253 °C. The gas has a very low density compared to air, and it was used to lift the dirigible *Graf Zeppelin,* a transatlantic passenger liner, in service during the early 1930s. Unfortunately, it was also used in the *Hindenburg,* which caught fire and exploded upon landing at Lakehurst, New Jersey, in 1937 (Figure 19.3). The Hindenburg exploded because hydrogen reacts explosively when mixed with oxygen in the presence of a source of ignition such as a spark. Thus, hydrogen gas and reactions that produce it must be handled carefully.

Figure 19.3 ***Hindenburg* explosion** The hydrogen gas in the airship *Hindenburg* caught fire and exploded while docking at Lakehurst, New Jersey, in 1937.

Although this famous explosion would indicate that hydrogen is very reactive, reactions of H_2 with most substances at room temperature are relatively difficult to initiate, mainly because of the strong H—H bond (bond energy of 436 kJ/mol). It does react with the highly electropositive group IA metals and calcium, strontium, and barium to form ionic hydrides.

$$2Na(s) + H_2(g) \longrightarrow 2NaH(s)$$

$$Ba(s) + H_2(g) \longrightarrow BaH_2(s)$$

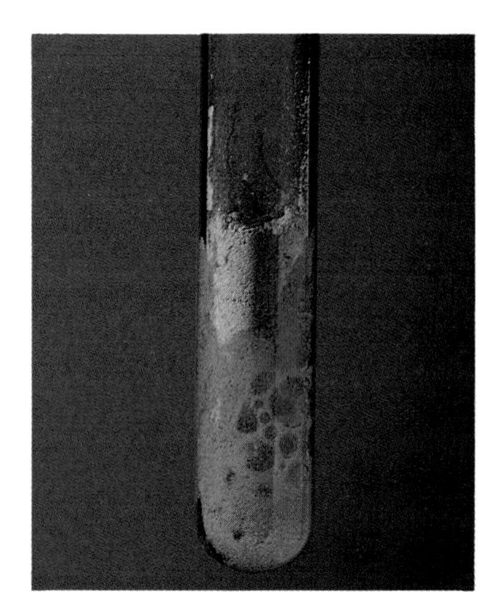

Figure 19.4 Reaction of calcium hydride and water Calcium hydride reacts vigorously with water, producing hydrogen gas.

The hydride ion, H^-, in these salts is a very strong Lewis base and reacts vigorously with water (Figure 19.4).

$$CaH_2(s) + 2H_2O(\ell) \longrightarrow Ca(OH)_2(aq) + 2H_2(g)$$

This reaction of CaH_2 can be viewed as an oxidation-reduction reaction in which the hydride is the reducing agent. Metal hydrides are often used when a very strong reducing agent is needed ($H_2 + 2e^- \rightarrow 2H^-$ $E° = -2.23$ V)

Hydrogen forms covalent hydrides with nonmetals, but reacts rapidly only with oxygen, fluorine, and chlorine. Reactions with the other halogens and nitrogen are slow.

Sources of Hydrogen

Hydrogen can be produced by reactions of either hydrocarbons (mainly CH_4) or carbon with water.

The preparation of small amounts of hydrogen is generally accomplished by the reaction of hydrochloric or sulfuric acid and zinc (Figure 19.5). However, these reactants are too expensive for industrial use.

$$Zn(s) + 2HCl(aq) \longrightarrow H_2(g) + ZnCl_2(aq)$$

Hydrogen is a very important industrial chemical. At the present time, the major commercial source of hydrogen is the high-temperature reaction of methane and steam.

$$CH_4(g) + H_2O(g) \longrightarrow 3H_2(g) + CO(g)$$

Hydrogen is also formed in the reaction of red-hot carbon (from coal) with steam:

$$C(s) + H_2O(g) \longrightarrow H_2(g) + CO(g)$$

The equimolar mixture of hydrogen and carbon monoxide formed in this reaction is known as *water gas,* and it was used extensively in the late 19th and early 20th centuries as a fuel in much the same way that natural gas is used today. Natural gas is safer because the carbon monoxide in water gas is very toxic.

The carbon monoxide produced in either of these reactions can react further with water to produce additional hydrogen.

$$CO(g) + H_2O(g) \longrightarrow H_2(g) + CO_2(g)$$

This reaction is known as the **water gas shift reaction.** Note that in the synthesis of hydrogen from either methane or carbon followed by the water gas shift reaction, large amounts of carbon dioxide are produced. Carbon dioxide, although not poisonous, is a "greenhouse" gas that may lead to global warming (Section 15.6).

Hydrogen is also formed by the electrolysis of water, the other product being oxygen. Unfortunately, this clean method of preparing hydrogen is expensive because of the high cost of electricity. If a method can be developed to produce hydrogen inexpensively from water (sunlight or wind would be good sources to supply the energy), it would be an extremely clean energy source because the only product of its combustion is water. A large-scale economic system based on hydrogen as an energy source is still a dream for the future.

Figure 19.5 Reaction of zinc and HCl Zinc metal reacts with HCl(aq) to produce hydrogen gas.

Uses of Hydrogen

The largest single commercial use of hydrogen is in the synthesis of ammonia by the Haber process.

$$N_2(g) + 3H_2(g) \longrightarrow 2NH_3(g)$$

More than 15 million tons of ammonia are prepared in the United States annually. Most of the ammonia is used as fertilizer, either directly or after conversion into other compounds.

Hydrogen is used in the syntheses of ammonia and methanol.

Hydrogen is also employed for the synthesis of methanol.

$$2H_2(g) + CO(g) \longrightarrow CH_3OH(\ell)$$

Methanol is used as a solvent and an additive in gasoline. Recently, a process was developed to convert methanol into gasoline. Thus coal, a source of carbon, can be converted into water gas and then into methanol, and the methanol can then be converted into gasoline. The synthesis of methanol requires a 2:1 ratio of H_2 to CO, and the additional H_2 needed can be obtained from the water gas shift reaction. Since coal is abundant in the United States, this process could be used to replace oil as an energy source. At present, the cost of the gasoline made from coal is not competitive with gasoline refined from crude oil, and there are also a number of environmental problems (buildup of CO_2 gas in the atmosphere and problems with coal mining). Nevertheless, the conversion of coal to a liquid hydrocarbon fuel may be important in the future.

Hydrogen is also used to hydrogenate some of the double bonds in vegetable oils.

$$\sim\overset{\overset{H}{|}}{C}=\overset{\overset{H}{|}}{C}\sim + H_2 \longrightarrow \sim\underset{\underset{H}{|}}{\overset{\overset{H}{|}}{C}}-\underset{\underset{H}{|}}{\overset{\overset{H}{|}}{C}}\sim$$

The reaction converts liquid vegetable oils into margarine or solid cooking fats. Vegetable oils are known as *unsaturated* fats, while the solid oils are partially *saturated.* Although the solid oils have characteristics that make them desirable for use in the preparation of foods, it has been shown that eating large amounts of saturated fats can cause health problems.

19.3 The Chemistry of Group IIIA (13) Elements

Objectives

- To discuss the inert pair effect
- To describe the isolation, purification, and fundamental chemistry of boron and aluminum

IIB	IIIA	IVA
	5 B Boron	6 C Carbon
	13 Al Aluminum	14 Si Silicon
30 Zn Zinc	31 Ga Gallium	32 Ge Germanium
48 Cd Cadmium	49 In Indium	50 Sn Tin
80 Hg Mercury	81 Tl Thallium	82 Pb Lead

Group IIIA (13) elements have the valence electron configuration ns^2np^1. They generally form compounds in which the element has an oxidation number of +3, but heavier members of the family, especially thallium, also form the +1 oxidation state in many compounds. The trend for the heavier

The heavier members of groups IIIA to VA form some compounds in which the pair of valence *s* electrons is not used for bonding.

members of the group to use only the *p* valence electron(s) in forming compounds is general and is observed in groups IVA and VA also. This trend is called the **inert pair effect,** the tendency for the heavier members of groups IIIA to VA not to use the pair of valence *s* electrons for bonding. The origins of the inert pair effect are complicated, but it does not arise from simple differences in ionization energies. The main reason for the effect appears to be that these large elements form weaker bonds, thus reducing the tendency to use all valence electrons in bonding.

All of the elements of the group are metals with the exception of boron, which is a metalloid. They occur in nature generally as oxides. In contrast to Groups IA and IIA, these elements become less reactive toward the bottom of the group.

(a)

(b)

Figure 19.6 Borax (a) View of a borax mine in the Mojave desert. (b) Borax has a variety of uses in the home.

Boron

Boron is extremely rare in the earth's crust. It occurs in nature in combination with oxygen. Although the element is rare, the boron-containing mineral borax is found in high concentrations in the Mojave desert in California (Figure 19.6). The formula of borax was generally written as $Na_2B_4O_7 \cdot 10H_2O$, until its structure was found to be $Na_2B_4O_5(OH)_4 \cdot 8H_2O$. It has been mined for years for use as a water softener (it precipitates Ca^{2+} and Mg^{2+}) and in cleaning products (it forms weakly basic solutions of pH $\approx$ 9).

The pure element is difficult to prepare. A form of low-purity boron can be prepared by the reduction of B_2O_3 with magnesium.

$$B_2O_3(s) + 3Mg(s) \longrightarrow 2B(s) + 3MgO(s)$$

High-purity boron can be prepared by the high-temperature reduction of BBr_3 by hydrogen in the presence of a solid catalyst.

$$2BBr_3(g) + 3H_2(g) \longrightarrow 2B(s) + 6HBr(g)$$

Boron exists in a number of allotropic forms that all contain an unusual arrangement of the boron atoms, an **icosahedron,** which is a regular polyhedron with 20 faces and 12 vertices (Figure 19.7). In the various allotropic forms, the icosahedra are connected differently, but all have extended bonding arrangements between the icosahedra. As a result of this very stable arrangement, several of the elemental forms of boron produce very hard crystals.

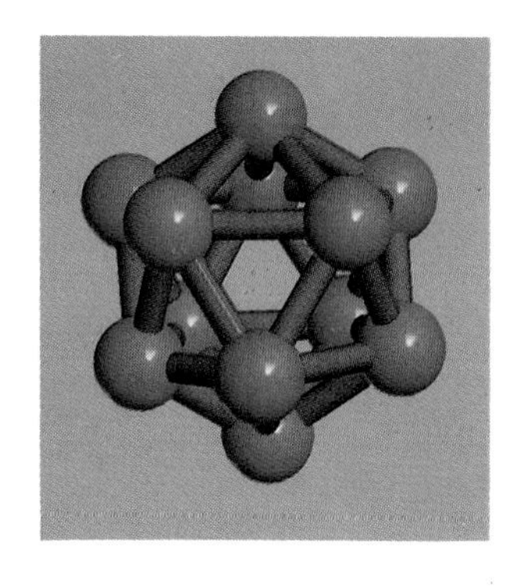

Figure 19.7 Icosahedron of boron atoms The elemental forms of boron contain icosahedral arrays of atoms.

Treatment of purified borax with sulfuric acid produces boric acid, H_3BO_3. Its formula is frequently written as $B(OH)_3$ to reflect its molecular structure. In the solid it exists as sheets containing trigonal planar $B(OH)_3$ groups held together by hydrogen bonds to oxygen atoms in other boric acid molecules. The electron-deficient, sp^2 hybridized boron atom in boric acid acts as a Lewis acid that reacts with water to form weakly acidic solutions containing the $B(OH)_4^-$ ion. These solutions have antiseptic qualities and are used as an eyewash.

$$B(OH)_3 + :OH_2 \rightleftharpoons B(OH)_4^- + H^+(aq)$$

$$K_a = 7.3 \times 10^{-10}$$

Heating boric acid causes its dehydration to B_2O_3. This oxide is used extensively in the manufacture of borosilicate glass (Section 19.4) and in the preparation of elemental boron.

The boron trihalides are interesting nonpolar molecules in which the boron atom is also sp^2 hybridized. Although there are only six electrons about the boron atom in these BX_3 derivatives (and boric acid), it is believed that the lone pairs on the halogen or oxygen atoms interact with the empty p orbital on boron to help stabilize the compounds. The boron trihalides also act as Lewis acids to form adducts with neutral donor molecules such as NH_3.

$$BCl_3 + :NH_3 \longrightarrow Cl_3B \leftarrow NH_3$$

They can also react with anionic donors to form ions such as BF_4^-. In these adducts, the boron atom is tetrahedral and uses sp^3 hybrid orbitals.

$$BF_3(aq) + HF(aq) \longrightarrow H^+(aq) + BF_4^-(aq)$$

Another important example of a tetrahedral anion of boron is BH_4^-. Sodium borohydride, $NaBH_4$, is a reducing agent used in industry and in research laboratories.

Boron Hydrides

Boron forms an extremely interesting series of binary compounds with hydrogen. The simplest member of the series might be expected to be BH_3, in which boron uses each of its three valence electrons to form a sigma bond with each of the three hydrogen atoms. In fact, BH_3 does not exist at room temperature, but dimerizes to form diborane, B_2H_6. At high temperatures, BH_3 can be observed in the gaseous state by mass spectrometry. Diborane is prepared by the reaction of $NaBH_4$ with I_2 and has the interesting structure shown in Figure 19.8.

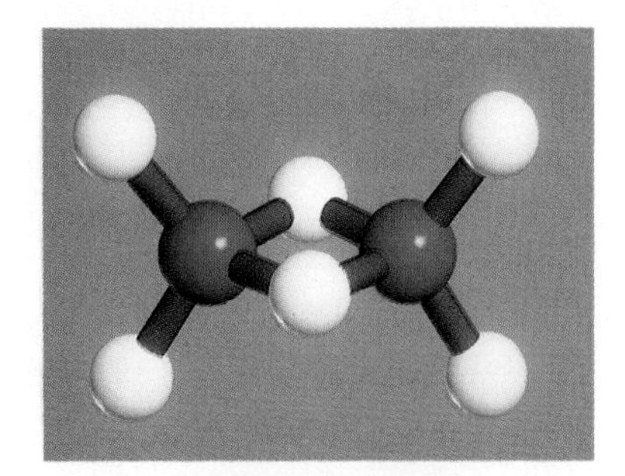

Figure 19.8 Structure of B_2H_6 The structure of B_2H_6 has two three-center B—H—B bonds and four normal two-center B—H bonds.

$$2NaBH_4 + I_2 \longrightarrow B_2H_6 + 2NaI + H_2$$

If BH_3 were a monomer, it would have an empty valence orbital and only six valence electrons about boron. In order to use all four valence

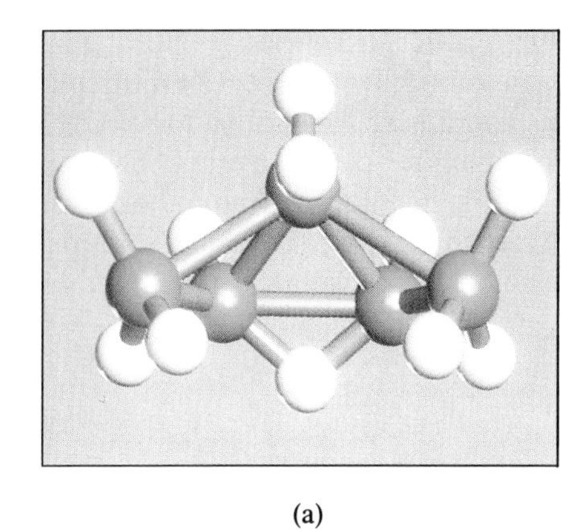

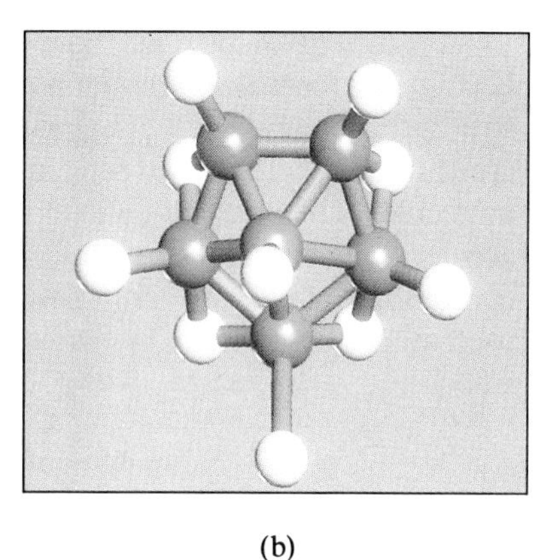

(a) (b)

Figure 19.9 Structure of B_5H_{11} and B_6H_{10} Both B_5H_{11} (a) and B_6H_{10} (b) contain three-center bonds, and the basic structures can be viewed as part of the icosahedron pictured in Figure 19.7.

orbitals and attain an octet of electrons, it forms the dimer. The bonding in this dimer can be explained as two sp^3 hybridized boron atoms, each of which forms two normal two-center, two-electron bonds with two of the hydrogen atoms and two *three-center, two-electron bonds* with the other boron atom and the bridging hydrogen atoms. In a three-center bond, three orbitals (one from each atom) overlap to form a bond. Just as with normal two-center bonds, this three-center bond contains two electrons. Thus, each boron atom uses all four valence orbitals and attains an octet of electrons.

Delocalized three-center bonds are used to describe the bonding in the boron hydrides.

Because three-center bonding allows boron to attain an octet of electrons, it often occurs in boron compounds. Figure 19.9 shows the structures of two other boron hydrides, B_5H_{11} and B_6H_{10}. In these compounds, three-center bonds can also be formed by three adjacent boron atoms. It is interesting to note that the basic arrangement of the boron atoms in these two compounds, as with most of the boron hydrides, consists of pieces of the icosahedron pictured in Figure 19.7. For example, the arrangement of the boron atoms in B_6H_{10} is the same as that of the top six atoms in the icosahedron in Figure 19.7.

These boron hydrides ignite spontaneously in air, giving a green flame. Because of the high enthalpy of combustion of the boron hydrides (shown in the following equation), they were considered as possible rocket fuels, but they are expensive to produce, the reactions are hard to control, and the resulting B_2O_3 would cause damage to the engines.

$$B_2H_6(g) + 3O_2(g) \longrightarrow B_2O_3(s) + 3H_2O(g) \qquad \Delta H = -2034 \text{ kJ}$$

Figure 19.10 Flame test for boron compounds Compounds containing boron burn with a green flame.

Aluminum

Aluminum is the most abundant metal and the third most abundant element in the earth's crust. Metallic aluminum, generally as an alloy with silicon and copper or magnesium, is an important structural material in aircraft because of its low density. Aluminum protects itself from corrosion by forming a thin, strongly adhering, protective coating of the inert oxide Al_2O_3.

$$4Al(s) + 3O_2(g) \longrightarrow 2Al_2O_3(s)$$

The formation of Al_2O_3 from the elements is so exothermic that powdered aluminum is used as a solid rocket fuel. Aluminum is also a good conductor

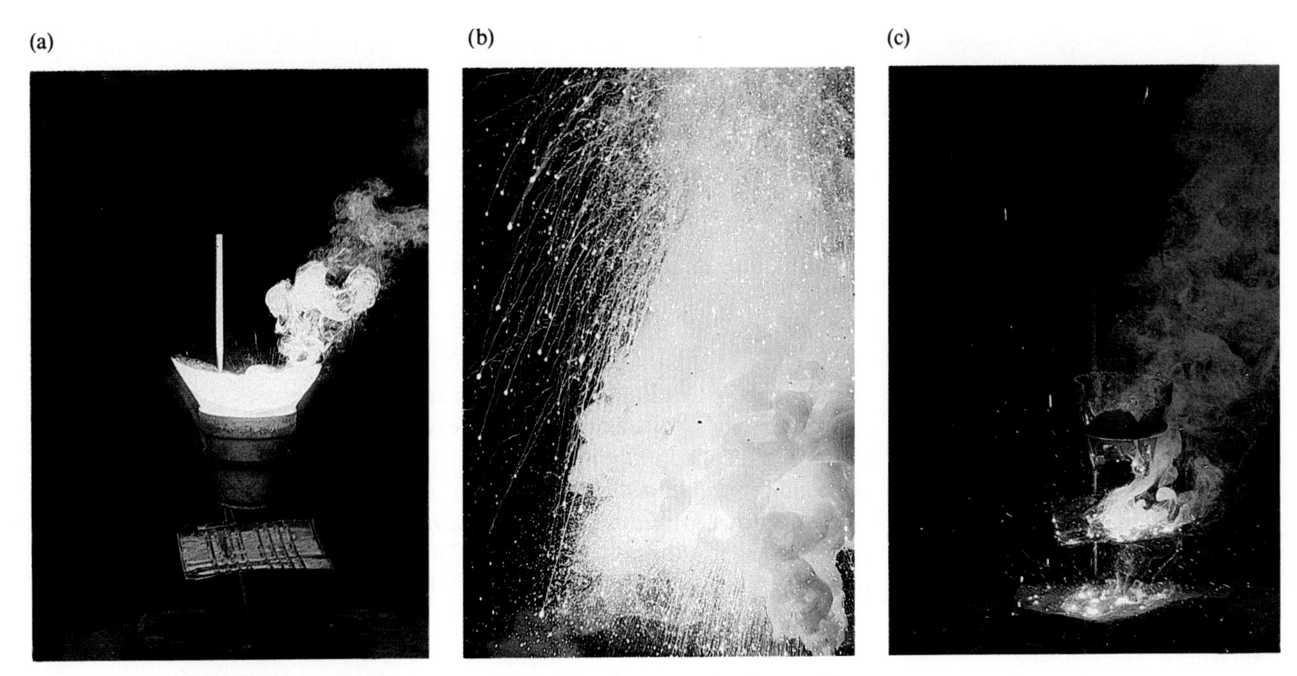

Figure 19.11 Thermite reaction The reaction of aluminum metal with iron oxide is extremely exothermic, producing molten iron.

of electricity and is used in overhead power lines, since it is less dense than the more highly conducting copper.

Aluminum is isolated from the ore bauxite, a hydrated oxide ($Al_2O_3 \cdot xH_2O$), by electrolysis, in a process that was developed by Charles Hall just after his graduation from Oberlin College (Section 17.11). Bauxite ores contain small amounts of Fe_2O_3 and SiO_2 that must be separated prior to electrolysis. The basis for this separation, known as the Bayer process, takes advantage of the amphoteric properties of Al_2O_3. The Al_2O_3 and SiO_2 are dissolved in strong base, and the solid Fe_2O_3 is removed by filtration. The solution is then acidified and the aluminum oxide precipitates, leaving the silicon in solution as silicates (silicates are discussed in Section 19.4).

Aluminum reacts with transition metal oxides in extremely exothermic reactions. The reaction with iron oxide is known as the *thermite reaction* (Figure 19.11).

$$2Al(s) + Fe_2O_3(s) \longrightarrow Al_2O_3(s) + 2Fe(\ell) \qquad \Delta H = -852\ \text{kJ}$$

The thermite reaction is so exothermic that it produces molten iron that can be used to weld iron and steel.

Aluminum oxide has many important industrial uses. It exists in a number of different forms. The α-alumina form is a very hard substance known as corundum, which is used as an abrasive. Several gemstones are clear crystals of α-alumina that contain metal ion impurities, as shown for ruby in Figure 19.12.

A less dense and more reactive form of Al_2O_3 is γ-alumina, which is used as a support in chromatographic separations, and as a heterogeneous catalyst

Figure 19.12 Aluminum oxides Bauxite is a white hydrated ore of aluminum oxide from which aluminum metal is isolated. A ruby is α-alumina colored red by Cr^{3+} impurities.

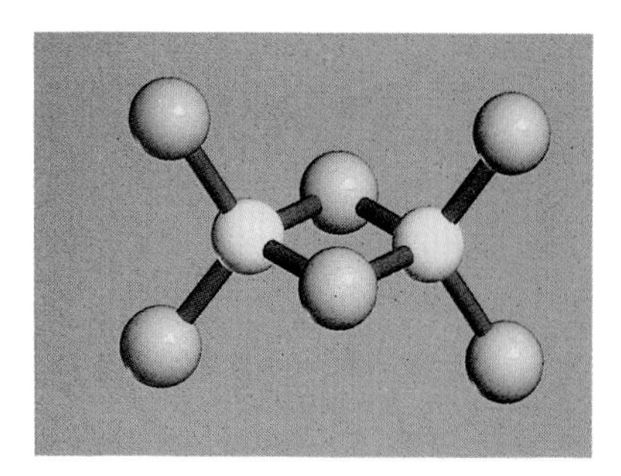

Figure 19.13 Structure of Al_2Cl_6 Al_2Cl_6 is a dimer in the gas phase.

or catalyst support for many chemical reactions. Since heterogeneous catalysis occurs on the surface of a solid, a high surface-to-volume ratio makes γ-alumina desirable in these applications.

Mixing Al_2O_3 with sulfuric acid produces aluminum sulfate, which is used to strengthen paper.

$$Al_2O_3(s) + 3H_2SO_4(aq) \longrightarrow Al_2(SO_4)_3(aq) + 3H_2O(\ell)$$

Alumina also dissolves in base to form the aluminate anion.

$$Al_2O_3(s) + 2OH^-(aq) + 3H_2O(\ell) \longrightarrow 2[Al(OH)_4]^-(aq)$$

A mixture of $Al_2(SO_4)_3$ and $Na[Al(OH)_4]$ produces insoluble $Al(OH)_3$, which upon precipitation is used to remove impurities from water. The gelatinous $Al(OH)_3$ adsorbs (the process by which a substance adheres to the surface of a solid, Section 16.5) dissolved impurities, and also carries small suspended solid particles along with it as it precipitates.

Another aluminum compound familiar to many people is the antiperspirant "aluminum chlorhydrate." This compound is actually aluminum hydroxychloride, $Al_2(OH)_5Cl \cdot 2H_2O$. It is an astringent; that is, it contracts the pores of the skin.

Aluminum chloride, $AlCl_3$, has a solid-phase structure in which aluminum is six coordinate. It sublimes at elevated temperatures to form molecular Al_2Cl_6. The bromide Al_2Br_6 and iodide Al_2I_6 are dimers in the solid as well as in the gas phase. The dimers have structures similar to that of B_2H_6 (Figure 19.13).

Although the structures of Al_2Cl_6 and B_2H_6 are similar, the three-center bonding description used for B_2H_6 is not needed for Al_2Cl_6. The bonding in Al_2Cl_6 can be viewed as simply a Lewis acid–Lewis base interaction of a lone pair on the bridging chlorine atoms of each $AlCl_3$ unit with the empty orbital on each aluminum atom. This is a very common type of bonding interaction in metal halides.

Gallium, Indium, and Thallium

Gallium and indium are not very abundant but are becoming increasingly important because gallium arsenide (GaAs) and indium phosphide (InP) are very useful semiconductor materials. Metallic gallium is unique in having a very large liquid range, from 30 to 2403 °C, and it will melt in your hand (Figure 19.14).

Figure 19.14 Gallium Gallium will melt at body temperature.

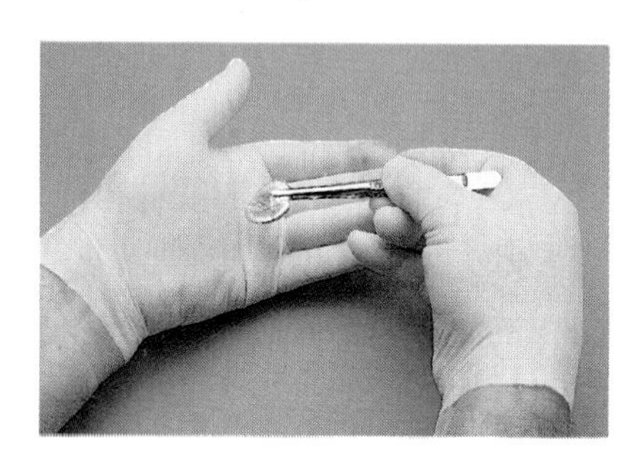

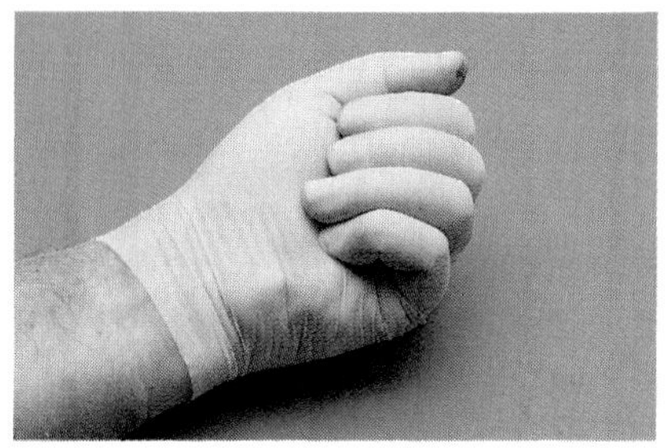

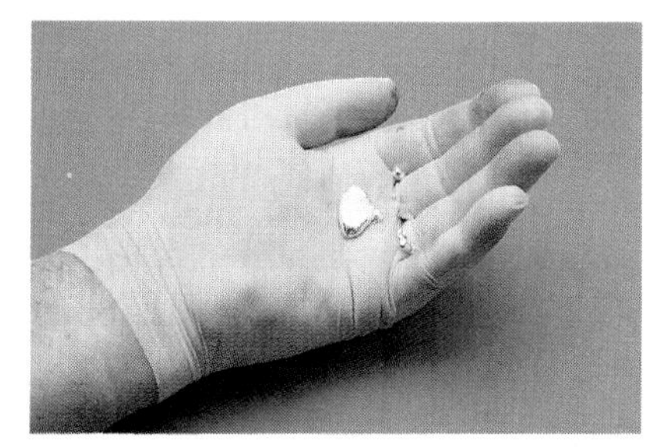

All three elements are isolated as by-products of the refining of other metals. Gallium is recovered from the refining of aluminum, indium from the purification of zinc, and thallium from the smelting of lead.

IIIA	IVA	VA
5 B Boron	6 C Carbon	7 N Nitrogen
13 Al Aluminum	14 Si Silicon	15 P Phosphorus
31 Ga Gallium	32 Ge Germanium	33 As Arsenic
49 In Indium	50 Sn Tin	51 Sb Antimony
81 Tl Thallium	82 Pb Lead	83 Bi Bismuth

19.4 The Chemistry of Group IVA (14) Elements

Objectives

- To describe the bonding and properties of three allotropic forms of carbon
- To discuss the occurrence and chemistry of silicon

Group IVA (14) elements have the electron configuration ns^2np^2. With four valence electrons and four valence orbitals, these elements generally attain an octet by forming four covalent bonds. As with the group IIIA elements, the heavier members of the group—tin and especially lead—exhibit the *inert pair effect* and form compounds in the +2 oxidation state as well as in the +4 oxidation state. The elements span the entire range of properties from carbon, a typical nonmetal, to metallic lead. Silicon and germanium are metalloids. Tin exists in two allotropic forms—white and gray tin. White tin is a metal and is the form used in plating "tin cans," but gray tin is a nonmetallic form that is quite brittle and is not an electrical conductor.

Carbon

Carbon is distributed widely in the earth's crust, mostly as the calcium and magnesium salts of the carbonate ion, CO_3^{2-}. The matter of living organisms contains a high percentage of carbon, as do the fossil fuels: oil, coal, and natural gas. Millions of compounds of carbon have been isolated from plants and animals or synthesized in laboratories. An introduction to the chemistry of carbon compounds is presented in Chapter 21.

Carbon is also found as the free element. Graphite, composed of layers of sheets of sp^2 hybridized carbon atoms (see Figure 10.10b), is the stable form of the element at room temperature and pressure. Graphite has a high melting point and is used to make molds for casting metals. It is a reasonable conductor of electricity and is used as an electrode material in many industrial electrolytic processes, such as the production of aluminum. It is also the "lead" in lead pencils (no elemental lead is present). Recently, new high-strength and lightweight materials have been prepared from graphite fibers mixed with plastics. These "composite" materials have found uses ranging from the shell of the Stealth fighter plane to high-quality sports equipment (Figure 19.15).

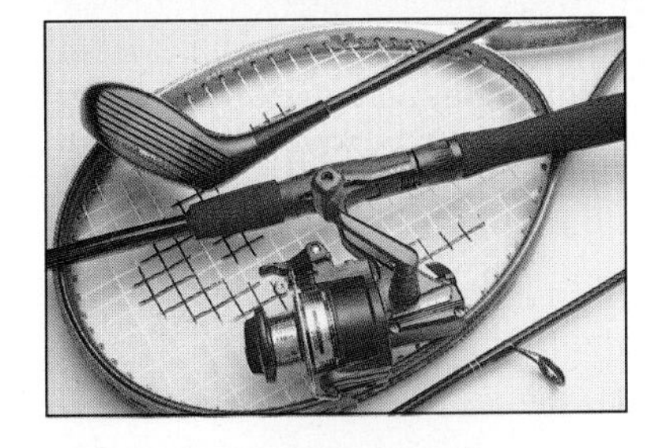

Figure 19.15 Uses of graphite Graphite composites are used to make stealth fighter planes and sports equipment because they are strong and very lightweight.

Charcoal and carbon black are finely divided forms of graphite. Charcoal has a large surface area per unit volume (particularly "activated charcoal," formed by heating charcoal with steam or CO_2). It is used in the purification of water and other liquids and gases because it efficiently ad-

INSIGHTS into CHEMISTRY

A Closer View

Buckminsterfullerene is a Geodesic Molecule That's Tough, Pliable, and Full of Potential

Buckminsterfullerene is a spherical cluster of 60 carbon atoms arranged in a series of 5- and 6-membered rings to form a soccer ball shape (see Figure 19.17). This compound's unusual name comes from the American architect Buckminster Fuller, who designed geodesic dome structures with a similar shape.

Buckminsterfullerene, frequently called buckyball, is an allotrope of carbon. Until its discovery at Rice University in 1985 by H. W. Kroto (visiting from Sussex University), R. Smalley, and their colleagues, there were only two common allotropes of carbon: graphite and diamond. Buckminsterfullerene is a black powdery material that can be dissolved in solvents such as benzene, forming deep magenta solutions.

Buckminsterfullerene is a surprisingly tough and resilient molecule. It can be accelerated to 15,000 miles per hour and slammed against steel surfaces without damage, a property unknown for other molecular particles. Buckyballs can be compressed to less than 70% of their initial volume without destroying the carbon cage. Buckminsterfullerene is also stable toward heating; one method of purification is to sublime the crude material at 600 °C under vacuum. Unlike diamond and graphite, which exist as crystalline network solids, buckminsterfullerene is a discrete molecule.

Like many discoveries in science, the isolation of buckminsterfullerene was serendipitous. The scientists who first recognized its existence were looking for something different. These researchers were examining carbon clusters and chains in an effort to explain why certain wavelengths of light from stars are absorbed as the light passes through interstellar dust clouds. They vaporized graphite from disks with a laser in an attempt to simulate conditions that might be found in these dust clouds. The questions they were trying to answer were in the area of

Geodesic dome Buckminster Fuller proposed the geodesic dome as an excellent design for sturdy buildings. Shown here is the Spaceship Earth in Future World at the Epcot Center.

Graphite and diamond are two allotropes of carbon in which the carbon atoms are sp^2 and sp^3 hybridized, respectively.

sorbs impurities (Figure 19.16). Carbon black and other amorphous forms of carbon are used to reinforce rubber; this is the reason that tires are black.

Diamond is an extremely hard allotrope of carbon formed at high pressures and temperatures, in which the carbon atoms are sp^3 hybridized (see Figure 10.10a). Diamond is the more stable form thermodynamically at high pressures because diamond is more dense than graphite. In fact, graphite can be converted into diamond by applying high pressures and temperatures.

astrophysics, but what they found was a new type of molecule and an avalanche of new questions, some of which are just now being answered.

Today, fairly large quantities of buckyballs can be made by vaporizing graphite rods in a helium atmosphere, using a high-current electric arc. Helium is used because it is a noble gas and thus does not react with the graphite; if air were present, the hot carbon gas would react with nitrogen and oxygen rather than condense to form buckyballs. As the graphite rods are consumed, soot is formed and up to 20% of this soot is buckminsterfullerene. In addition to C_{60}, a whole series of additional allotropes of carbon called "fullerenes," such as C_{70}, also form in the soot. The buckyballs and the other carbon clusters are separated from the soot by solvent extraction, and the pure C_{60} is separated by column chromatography. Even though modest amounts of this material can be made in laboratories, it is still fairly difficult to obtain. New methods of preparation are being developed that will increase the availability of buckyballs.

Scientists are actively studying buckminsterfullerene to learn about its properties and potential uses. Buckyballs can be mixed with alkali metals such as potassium to make superconductors. Because the molecule is spherical, it has been suggested that C_{60} might be used as a lubricant in which the buckyballs would act as molecular ball bearings. Fibers made up of other fullerenes, many of which have the shapes of tubes or rods, could make lightweight composite materials of great strength. Such materials would have uses in the construction of lightweight cars and aircraft. Whatever is done with buckminsterfullerene eventually, it will be keeping chemists and engineers busy and excited for many years to come.

Buckminsterfullerene Buckminsterfullerene dissolves in benzene, forming a deep magenta solution.

Reactor for preparing fullerenes The fullerenes are prepared by vaporizing graphite rods using a high-current electric arc.

Because diamonds are extremely hard, they are used in cutting and drilling tools. Larger, more perfect crystals of diamonds are, of course, valued for their beauty.

An interesting new allotrope of carbon has recently been isolated from the carbon dust formed by heating graphite at high temperatures. This allotrope, with the formula C_{60}, is named buckminsterfullerene. Its shape is similar to the surface of a soccer ball (its nickname is "buckyball"), as shown

"Buckyball," C_{60}, is an allotrope of carbon whose shape is similar to the surface of a soccer ball. It is the parent compound of a whole series of fullerenes.

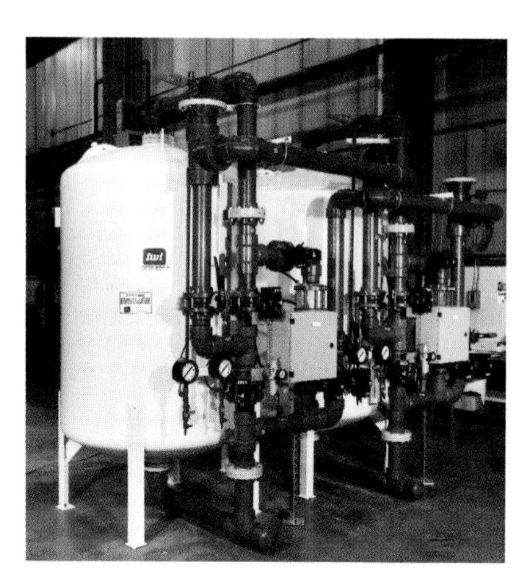

Figure 19.16 Activated charcoal Activated charcoal is used as part of this automatic water purification system because it adsorbs impurities.

in Figure 19.17. Buckyball is just the first of a whole series of new allotropes of carbon, called fullerenes, that can be formed by heating graphite. Scientists are studying the properties and potential uses of these unusual species.

Silicon

Silicon is the second most abundant element on earth. It is not found as the free element in nature, but in minerals called **silicates,** compounds containing silicon combined with oxygen and various metals. There are many different types of silicates with different ratios of silicon to oxygen, but in all forms the silicon atoms are in the +4 oxidation state and are tetrahedrally bonded to oxygen atoms (Figure 19.18).

Silicon is found in nature in combination with oxygen.

Silicon is also found as silica, SiO_2, which is the main component of common beach sand. We have already seen (Figure 19.1) that silica is a network solid with silicon in the center of a tetrahedral arrangement of oxygen atoms. It does not form simple molecules analogous to CO_2 because third period elements do not form strong π bonds. Silica is used to prepare glass, an amorphous solid (Section 10.3). To make glass, the silica is melted and the liquid melt is cooled rapidly. The rapid cooling prevents the forma-

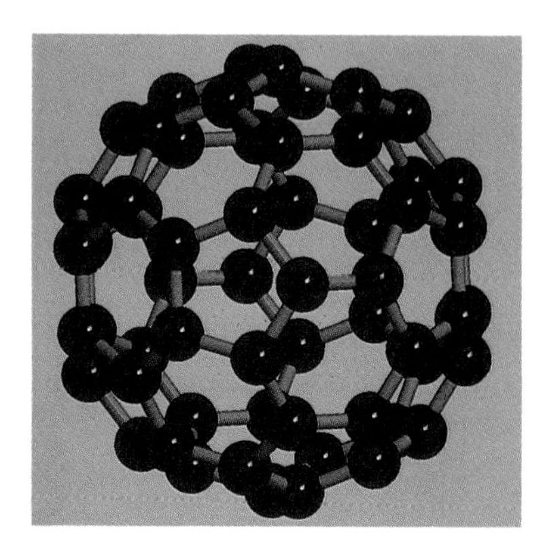

Figure 19.17 Buckminsterfullerene Buckminsterfullerene is an allotrope of carbon that was just recently discovered.

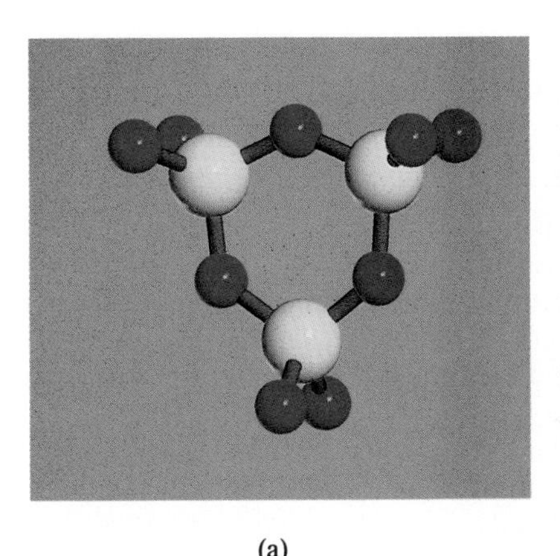

(a)

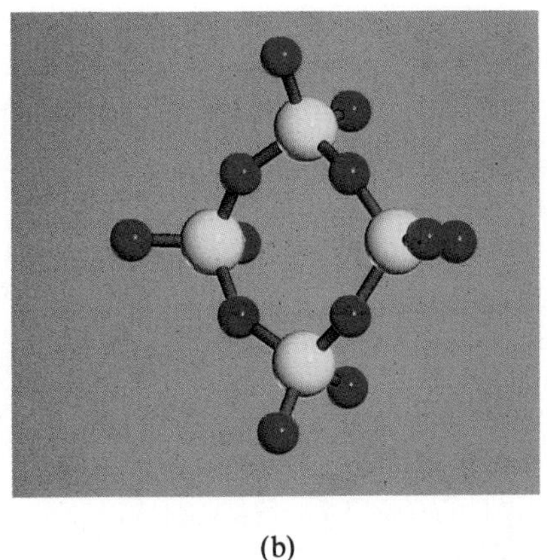

(b)

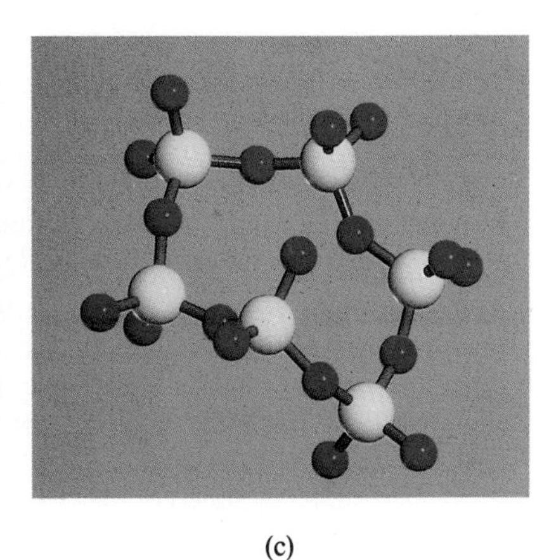

(c)

Figure 19.18 Silicates The structures of three silicate anions, (a) $Si_3O_9^{6-}$, (b) $Si_4O_{12}^{8-}$, (c) $Si_6O_{18}^{12-}$.

tion of crystalline silica. Additives such as Na_2CO_3, B_2O_3 (to form Pyrex glass), and K_2O (to make an especially hard glass) are added to the melt to change the appearance and properties of the glass (Figure 19.19).

Glass is just one example of a group of materials known as **ceramics,** nonmetallic, solid materials that are hard, resistant to heat, and chemically inert. Most ceramics are formed from silicates. Clays, which are mainly aluminosilicates (kaolinite, $Al_2Si_2O_5(OH)_4$, for example), have been used for more than 5000 years to prepare pottery and dishes.

Glass, clays, and cement are important materials based on silicon compounds.

Cement is also mainly an aluminosilicate material. It is formed by heating a mixture of clay and limestone ($CaCO_3$) in a kiln at 1400 to 1600 °C to produce small lumps called "clinkers," which are ground with some gypsum ($CaSO_4 \cdot 2H_2O$) into a powder. Concrete is formed by adding water to a mixture of cement and sand or gravel. This mixture slowly hardens (sets) through a complicated series of reactions that are not fully understood. As with glass, the properties of cement can be varied by using different additives and heating procedures.

Figure 19.19 Glass Normal plate glass is formed by addition of Na_2O and CaO to the silica melt. The beaker is Pyrex glass, formed by addition of B_2O_3 and has a low thermal expansion. Blue-colored glass is formed by addition of cobalt (II) compounds.

INSIGHTS into CHEMISTRY

Science, Society, and Technology

Properties of Glass Changed by Additives. Hubble Space Telescope Requires Unstable Mirrors

The volume of a solid changes very little when it is heated, but these small changes can be significant. Consider what happens when a glass container is heated and rapidly cooled. Since glass is a poor conductor of heat, temperature differences cause different parts of the object to change volume at different rates. This leads to stress, which often results in the glass breaking into pieces.

Some glasses are specially formulated to change volume only slightly with temperature. The borosilicate glass used to make the beakers and flasks used in the chemistry laboratory are examples of such materials. These containers can be heated with a Bunsen burner and then placed on the cold laboratory bench without shattering. One common brand of borosilicate glass is Pyrex, a trademark of the Corning Glass Company.

Pure silica glass (silica, SiO_2) has even better thermal properties than borosilicate. It can be heated to over 1000 K and then placed in liquid nitrogen at 77 K, without shattering. Although silica glass has good thermal properties, compounds such as sodium oxide (soda ash, Na_2O) and calcium oxide (lime, CaO) are added to silica to lower its melting point. The lower melting point decreases the cost of producing glass objects, because the ovens used to melt the glass last longer at the lower temperatures and the energy costs are also lower. Although costs are lower, these additives degrade the tolerance of the glass to rapid changes in temperature.

Material scientists characterize glasses by their coefficients of linear expansion; the most common units are ppm/K, the relative change in length (in parts-per-million, or microinches of change per inch of material) per degree change in temperature. Some common glasses and their coefficients of thermal expansion are shown in the following table.

Glass	Compounds Added to Silica	Coefficient of Linear Expansion, ppm/K
soda lime	Na_2O, CaO	5
borosilicate	Na_2O, B_2O_3	3.25
silica glass	—	0.8

One group of scientists who are concerned with the thermal properties of glasses are the astronomers. Telescope lenses and mirrors are fabricated from dimensionally stable glasses that do not change size or shape much as the temperature changes. A mirror or lens that changes shape will render the telescope useless.

The Hubble Space Telescope is an orbiting observatory that was launched in 1989. For a telescope in space, the effects of extreme temperature changes are very important. The primary mirror is 94 inches (2.4 m) in diameter and was ground to a particular shape. The mirror was designed to be within 10 nm (4×10^{-7} inches) of the specific shape *while in the microgravity environment of space.* (Correcting for forces due to the earth's gravity during the fabrication process was a difficult and time-consuming procedure.) Even small fluctuations in shape cannot be tolerated in the glass of the mirrors, so they were not only constructed from Ultra Low Expansion glass, but they are thermostated in the telescope.

Unfortunately, the primary mirror is flawed; the mirror is shaped very slightly too flat. Five pairs of error-correcting mirrors, each about the size of a postage stamp, are being fabricated and are to be fitted to the telescope in 1993. Scientists feel confident that the telescope will operate as planned after the correction.

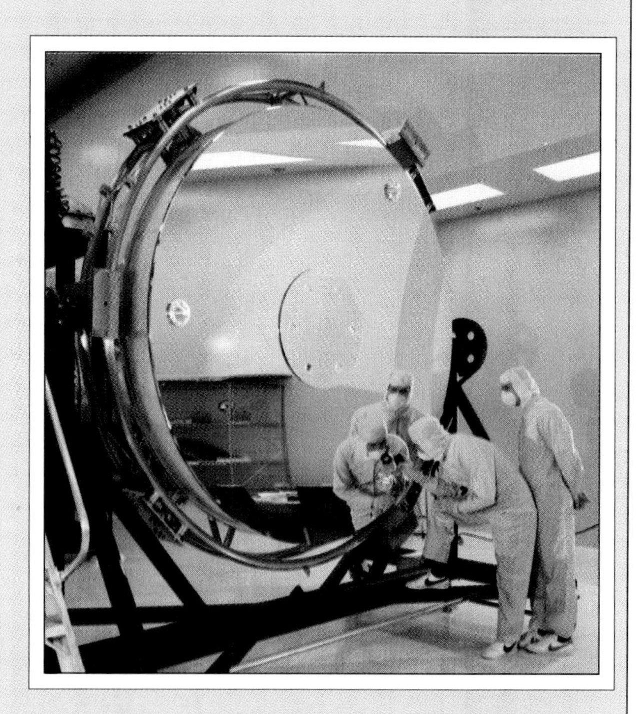

Mirror for Hubble Space Telescope

Figure 19.20 Chrysotile asbestos

Another important silicate is *asbestos* (Figure 19.20). Asbestos is a term used for a family of silicates with fibrous properties. Asbestos is a good heat insulator and does not burn, so it has been used extensively to insulate buildings and ships. Unfortunately, it has been shown that a rarely used form of asbestos, amphibole crocidolite, can be very dangerous to lung tissue, and use of asbestos as an insulating material has been banned. The asbestos in most buildings in the United States is *not* amphibole crocidolite, but chrysotile. Most experts do not think chrysotile presents a serious health hazard.

Preparations of Silicon

Silicon is used by the thousands of tons in alloys of iron and aluminum, in silicone polymers, and (in highly purified form) in solid-state electronic components. In the preparation of the element, SiO_2 is reduced by carbon in an electric arc furnace. In this procedure, the SiO_2 must be kept in excess to prevent formation of SiC. Highly purified silicon is prepared by treating this silicon with chlorine gas to form the volatile tetrachloride. Silicon tetrachloride can be purified by repeated distillation, followed by reduction with magnesium or hydrogen to recover the element.

$$SiO_2(\ell) + 2C(s) \xrightarrow{3000\ ^\circ C} Si(\ell) + 2CO(g)$$

$$Si(s) + 2Cl_2(g) \longrightarrow SiCl_4(\ell)$$

$$SiCl_4(g) + 2Mg(s) \longrightarrow Si(s) + 2MgCl_2(s)$$

Rods of silicon are then further purified by zone refining. In this process, a thin band at one end of a silicon rod is melted, and the heat source is slowly moved toward the other end. As the heat source is moved, the impurities stay in the molten silicon zone, leaving behind high-purity silicon (Figure 19.21).

(a)

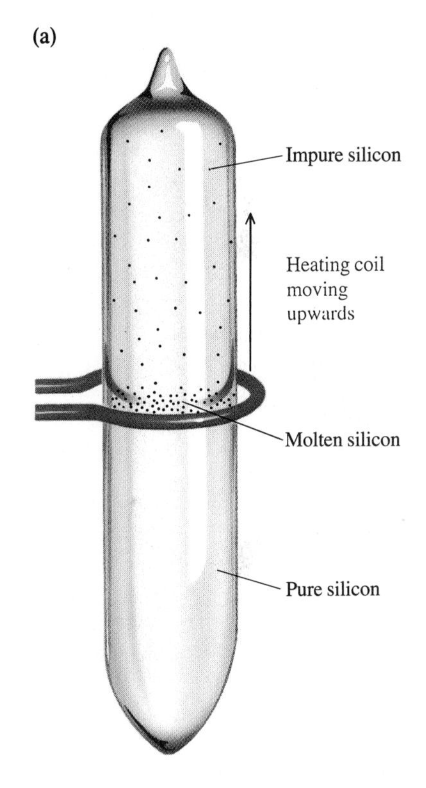

(b)

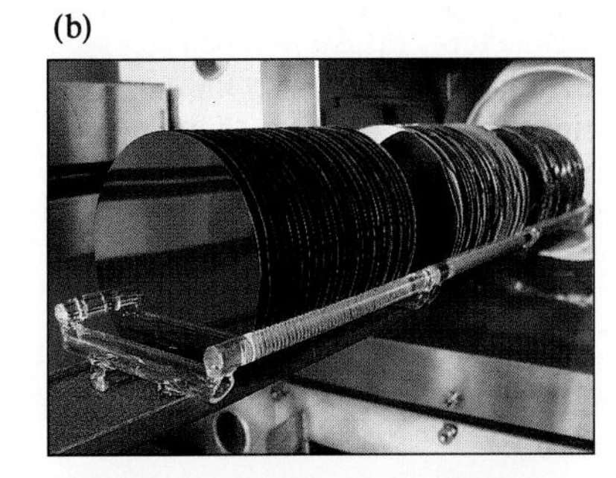

Figure 19.21 Zone refining of silicon (a) Silicon is purified by zone refining. (b) The very pure rods are cut into wafers for use in the production of computer chips.

Semiconductors

The bonding in solid metals or metalloids can be described as arising from the overlap of many orbitals to form energy bands that extend throughout the solid. For example, the 3*s* orbitals in a piece of sodium that contains one

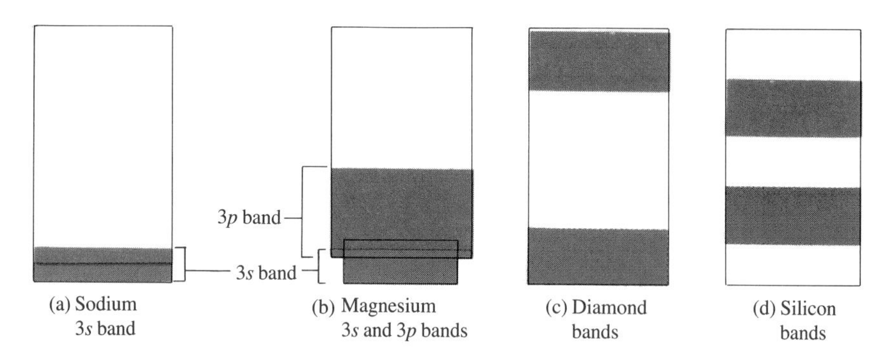

Figure 19.22 Bands of orbitals (a) The band of metal orbitals (red and blue) in sodium formed from overlap of 3*s* atomic orbitals is half-filled (orbitals filled with electrons are indicated by blue). (b) The 3*s* band in magnesium overlaps the 3*p* band. (c) In diamond, the filled band is well-separated in energy from the conduction band. (d) In silicon, the filled band is close in energy to the conduction band.

mole of atoms overlap to form a band of one mole of orbitals (in the terms of molecular orbital theory, one-half mole of bonding and one-half mole of antibonding orbitals) closely spaced in energy. Since for sodium, one mole of electrons is available to fill this band, the orbitals (called a 3*s* band because it arises from the overlap of 3*s* atomic orbitals) are half filled (Figure 19.22a). Sodium can conduct an electrical current because electrons can easily move through the empty, but low-energy, orbitals of the 3*s* band. A **conduction band** is a partially or completely empty band of orbitals that can conduct an electrical current. Magnesium is also a conductor even though this 3*s* band by itself would be filled (one mole of magnesium atoms has two moles of electrons). In magnesium, the band that arises from the 3*p* orbitals overlaps the 3*s* band (Figure 19.22b), and electrons can move through this 3*p* band, which is now the conduction band.

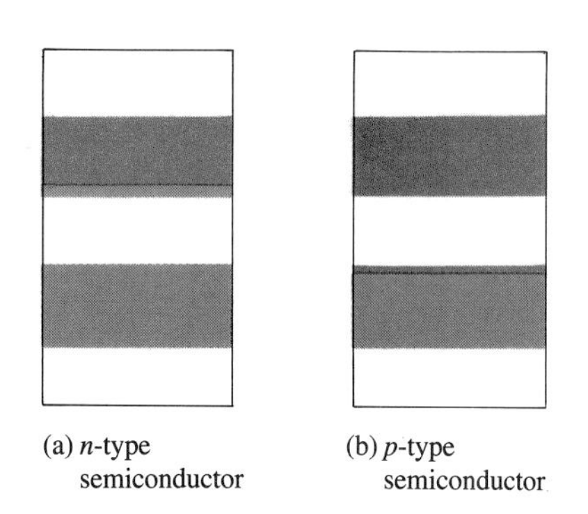

Figure 19.23 Doping of semiconductors (a) An *n*-doped semiconductor is formed by the substitution of impurity atoms that contain more valence electrons. (b) A *p*-doped semiconductor is formed by the substitution of impurity atoms that contain fewer valence electrons than present in the pure substance.

The properties of a nonmetallic solid such as diamond are very different because there is a large energy gap **(band gap)** between the filled band and the conduction band. It is an insulator; the energy gap is too large for the electrons to move into the conduction band (Figure 19.22c) so no empty, low-energy orbitals are available to conduct an electric current.

Pure silicon has the same structure as diamond, but the solid is not nearly as hard as diamond because the Si—Si bonds are much weaker. Because of the weaker bonds, the energy gap between the filled band and the empty band is much smaller in silicon than in diamond (Figure 19.22d). In this situation, a few electrons can cross the band gap because of thermal energy, making silicon a weak conductor of electricity.

The conductivity of silicon can be increased by adding trace quantities of a selected impurity, a process called *doping.* We can dope very pure silicon with a small quantity of an element that has five valence electrons (one more than silicon), such as phosphorus. The extra electrons enter the conduction band, increasing the conductivity of the solid (Figure 19.23a). The doping atoms add negative charge carriers (electrons) to the silicon, so the result is called an *n*-type semiconductor.

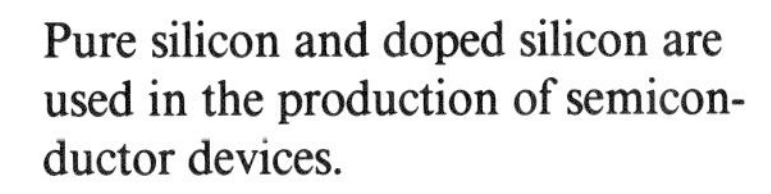

Pure silicon and doped silicon are used in the production of semiconductor devices.

In contrast, we can dope the silicon with an element that has three valence electrons (one less than silicon), such as boron. Each of these atoms leaves a "hole" in the lower-energy band, again increasing the conductivity (Figure 19.23b) by allowing an electron to move into the "hole" and leaving a "hole" at its previous location. Because the entity that appears to move is a positive "hole" (the absence of an electron), this kind of doped silicon is called a *p*-type semiconductor. Note that the extreme sensitivity of the electrical conductivity of silicon and other semiconductors to even trace amounts of impurities makes it very important that these materials are carefully purified.

Germanium, Tin, and Lead

Germanium was once important in the semiconductor industry, but it has given way to silicon, which retains its desirable properties better at higher temperatures. It is recovered from the flue dust produced in the processing of zinc ore, and it may be purified by zone refining. Germanium forms gray-white crystals that have the same structure as silicon, and it has similar properties.

Both tin and lead were among the first metals isolated by humans. Tin is found as the oxide in the mineral cassiterite, SnO_2 (Figure 19.24). The metal can easily be isolated by reduction with charcoal.

$$SnO_2(s) + C(s) \longrightarrow Sn(s) + CO_2(g)$$

The metallic form of tin is soft, and is used in low-melting alloys such as solder and pewter. Tin became very important in the development of early civilizations when it was alloyed with copper to form bronze. Bronze is much harder than either copper or tin. Tin is also used as a coating for other metals such as iron, since it does not react with air and water. An interesting use of tin is in the production of "plate" glass. The molten glass is poured onto the surface of molten tin; upon cooling, the glass surface is so smooth that it does not need to be polished.

Lead was used extensively by the Romans for eating utensils and for plumbing. It occurs naturally as lead(II) sulfide, PbS, in the mineral galena (Figure 19.24). Practically all common lead compounds contain the element in the +2 oxidation state. The pure metal is obtained by first converting the sulfide to the oxide by reaction with oxygen (roasting), followed by reduction of the oxide with carbon and carbon monoxide.

$$2PbS(s) + 3O_2(g) \longrightarrow 2PbO(s) + 2SO_2(g)$$

$$PbO(s) + C(s) \longrightarrow Pb(\ell) + CO(g)$$

$$PbO(s) + CO(g) \longrightarrow Pb(\ell) + CO_2(g)$$

The main use of lead today is in the electrodes in lead storage batteries. In addition, lead shields are used to absorb high-energy radiation such as x rays. Lead was used extensively as an additive to gasoline in the form of tetraethyllead, $(C_2H_5)_4Pb$, but this use has been curtailed because of harmful environmental and physiological effects of the metal. The use of lead compounds in paints has also been significantly reduced for the same reasons. These reductions in the use of lead compounds have resulted in an 85%

(a)

(b)

Figure 19.24 Lead and tin minerals Tin is found in the mineral cassiterite (a) and lead in galena (b).

IVA	VA	VIA
6 C Carbon	7 N Nitrogen	8 O Oxygen
14 Si Silicon	15 P Phosphorus	16 S Sulfur
32 Ge Germanium	33 As Arsenic	34 Se Selenium
50 Sn Tin	51 Sb Antimony	52 Te Tellurium
82 Pb Lead	83 Bi Bismuth	84 Po Polonium

decrease in lead levels in the air of urban areas during the period from 1981 to 1990.

19.5 The Chemistry of Group VA (15) Elements

Objectives

- To discuss the properties of the compounds of nitrogen and phosphorus with hydrogen and oxygen
- To compare the elemental forms of nitrogen and phosphorus

The elements in group VA (15) have five valence electrons with the electron configuration ns^2np^3. These elements generally form compounds with three covalent bonds, leaving one lone pair on the central atom.

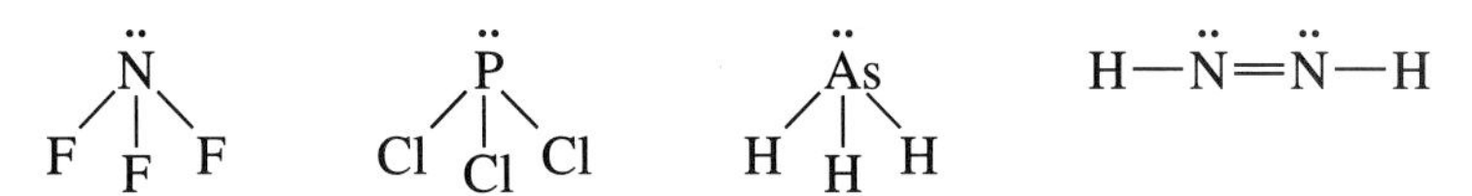

The first three elements in the group can also gain three electrons to form 3− anions in compounds with the highly electropositive group IA and IIA metals. The properties of the elements in group VA range from the nonmetals nitrogen and phosphorus to bismuth, an element that is a metal. As with other groups, the chemistry of the lightest member, nitrogen, is unique because it can form compounds with strong multiple bonds. In contrast to nitrogen, the heavier members of the group form numerous electron-rich species such as PF_5 and AsF_6^-.

Nitrogen

Nitrogen comprises 78% (by volume) of the atmosphere, and it is easily isolated by fractional distillation of liquid air (boiling point = −196 °C). The liquid is used extensively as a low-temperature coolant, and the gas is used to protect foods and reactive chemicals from oxidation by oxygen in the air. The quantity of nitrogen produced by the chemical industry is second only to sulfuric acid.

Nitrogen is relatively nonreactive because it exists as nonpolar diatomic molecules containing a strong triple bond (945 kJ/mol). However, it will react with the active metals lithium and magnesium to form ionic nitrides.

$$6Li(s) + N_2(g) \longrightarrow 2Li_3N(s)$$

$$3Mg(s) + N_2(g) \longrightarrow Mg_3N_2(s)$$

Ammonia

Nitrogen also reacts with hydrogen gas to form ammonia, but only under special conditions. This reaction is known as nitrogen *fixation,* the combination of the element with another element. Nitrogen is fixed by bacteria that live in the roots of leguminous plants such as soybeans and alfalfa. Nitrogen compounds are needed for plant life (as well as all other living organisms). Frequently farmers "rotate" alfalfa and soybeans with crops such as corn and wheat that do not fix nitrogen. Other natural sources of fixed nitrogen are guano (the excrement of bats found concentrated in caves, and of sea-birds found on isolated islands) and the minerals saltpeter, KNO_3, and Chile saltpeter, $NaNO_3$. In modern farming, the nitrogen needed for the growth of crops such as corn and wheat is frequently supplied by addition of commercial fertilizers.

At the turn of this century there was tremendous demand for fertilizer as well as for nitrates used to prepare explosives such as TNT. In attempting to find a way to meet this need, the German chemist Fritz Haber carried out considerable research and concluded that the direct synthesis of ammonia from its elements was practical.

$$N_2(g) + 3H_2(g) \rightleftharpoons 2NH_3(g) \qquad \Delta H^\circ = -92 \text{ kJ}$$

Ammonia, NH_3, is prepared from its elements at high temperatures and pressures.

The reaction requires high pressures (200 atm or more) and high temperatures for the efficient production of ammonia from N_2 and H_2. The principle of Le Chatelier predicts that high pressures will favor formation of product because four volumes of reactant gas are converted into two volumes of product. This reaction has a very high activation energy because of the strong bond in N_2, and the **Haber process** is carried out at elevated temperatures (380 to 450 °C) in the presence of an iron catalyst. The high temperatures are unfavorable for the position of the equilibrium, but are necessary for the reaction to proceed at a reasonable rate. The temperature selected for the reaction is a compromise between the rate and the position of the equilibrium.

Ammonia production The Haber process is used to produce ammonia from elemental nitrogen and hydrogen.

Ammonia is a colorless gas that condenses at −33 °C. The gas has a pungent odor, the smell in "smelling salts." It dissolves in water, forming a solution that is often called "ammonium hydroxide," although NH_4OH has never been isolated. Like water, liquid ammonia undergoes autoionization, but the equilibrium constant at −50 °C is only about 10^{-33}.

$$2NH_3 \rightleftharpoons NH_4^+ + NH_2^- \qquad K \approx 10^{-33}$$

The other important hydride of nitrogen is hydrazine, N_2H_4. Hydrazine is a liquid with a variety of industrial uses. One use of hydrazine is to remove oxygen from water that is used in boilers of electrical generating plants.

$$N_2H_4(aq) + O_2(aq) \longrightarrow N_2(g) + 2H_2O(\ell)$$

Dissolved oxygen causes rapid deterioration of metal pipes at the high temperatures and pressures used in boilers. The reaction of derivatives of hydrazine with N_2O_4 is quite violent, and this reaction was used to propel the Apollo lunar lander rockets.

Nitrogen Oxides

Nitrogen forms numerous oxides, six of which are shown in Table 19.1.

Dinitrogen oxide (N_2O, sometimes called nitrous oxide), is a nontoxic gas that is used as an anesthetic (laughing gas). An interesting commercial use of dinitrogen oxide is as the propellent in cans of whipped cream. It is prepared by the thermal decomposition of ammonium nitrate.

$$NH_4NO_3(s) \xrightarrow{200\ ^\circ C} N_2O(g) + 2H_2O(g)$$

Table 19.1 Nitrogen Oxides

Name	Formula
dinitrogen oxide	N_2O
nitrogen monoxide	NO
dinitrogen trioxide	N_2O_3
nitrogen dioxide	NO_2
dinitrogen tetroxide	N_2O_4
dinitrogen pentoxide	N_2O_5

Nitrogen monoxide (NO, sometimes called nitric oxide) is a gas at room temperature. It has an odd number of electrons and was used in Chapter 8 as an example of a molecule that does not satisfy the octet rule. In the solid state, NO dimerizes to form N_2O_2, thus pairing all the electrons.

Nitrogen monoxide reacts readily with oxygen to form nitrogen dioxide.

$$2NO(g) + O_2(g) \longrightarrow 2NO_2(g)$$

Nitrogen monoxide is a major contributor to air pollution in the lower atmosphere. It forms in the engines of automobiles and jet-propelled planes by the direct combination of $O_2(g)$ and $N_2(g)$. When formed by jets in the upper atmosphere, it can lead to the destruction of ozone, O_3.

$$NO(g) + O_3(g) \longrightarrow NO_2(g) + O_2(g)$$

$$NO_2(g) + O_3(g) \longrightarrow NO(g) + 2O_2(g)$$

The NO is regenerated in these reactions (it is a catalyst), so each NO molecule can destroy a large amount of ozone.

Nitrogen dioxide, NO_2, is also an odd-electron species; the odd electron is located mainly on the nitrogen atom. In the gas phase, NO_2 is in equilibrium with dinitrogen tetroxide, N_2O_4, and exists as N_2O_4 in the solid state. Nitrogen dioxide is formed in the reaction of many metals, such as copper, with concentrated nitric acid (Figure 19.25).

$$Cu(s) + 4HNO_3(aq) \longrightarrow Cu(NO_3)_2(aq) + 2H_2O(\ell) + 2NO_2(g)$$

Nitrogen dioxide is a red-brown gas that is very toxic. It reacts with water to form nitric acid, HNO_3, and nitrous acid, HNO_2.

$$2NO_2(g) + H_2O(\ell) \longrightarrow HNO_2(aq) + HNO_3(aq)$$

Dinitrogen oxide is a nonreactive gas, whereas both nitrogen monoxide and nitrogen dioxide are reactive, toxic gases that contribute to air pollution and acid rain.

The nitrous acid is then easily oxidized to nitric acid by oxygen. The reaction of oxygen with NO to form NO_2 and the reaction of NO_2 with water make both of these odd-electron molecules major contributors to the formation of acid rain.

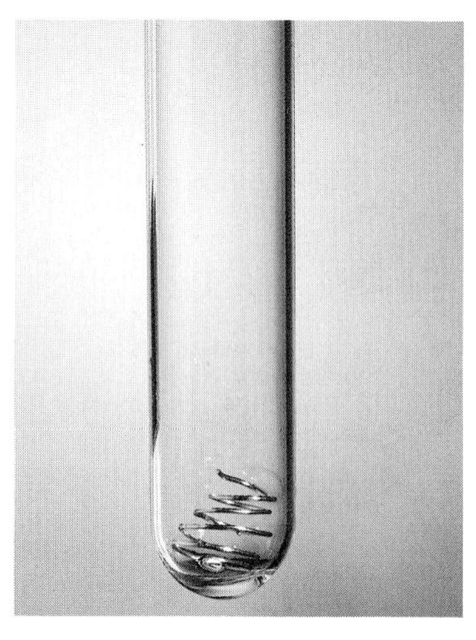

(a)

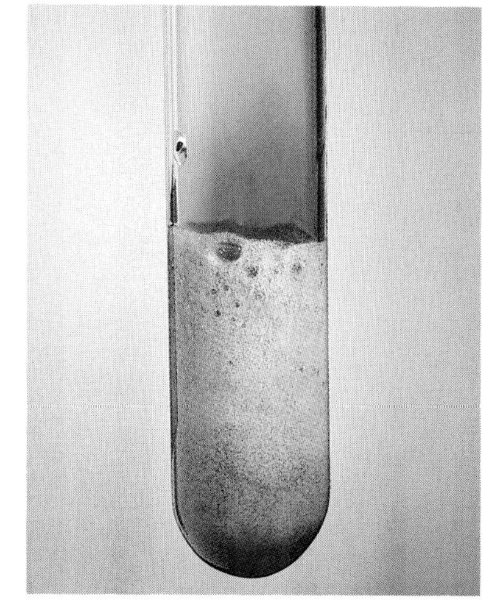

(b)

Figure 19.25 Nitrogen dioxide Copper metal (a) reacts with nitric acid, producing off brown NO_2 gas (b).

Nitric Acid

Nitric acid is an important industrial product. It can be formed in the laboratory by the reaction of $NaNO_3$ and sulfuric acid.

$$2NaNO_3(s) + H_2SO_4(aq) \longrightarrow 2HNO_3(aq) + Na_2SO_4(aq)$$

Commercially, most of this acid is produced by the **Ostwald process.** In the first step of this process, ammonia (formed by the Haber process) is oxidized by oxygen over a platinum catalyst.

$$4NH_3(g) + 5O_2(g) \xrightarrow{Pt} 4NO(g) + 6H_2O(g)$$

In the second step the NO is oxidized with $O_2(g)$ to give $NO_2(g)$ as previously described, and in the final step the $NO_2(g)$ is mixed with water to form nitric acid. The NO that is also formed in this reaction is recycled through the process.

$$2NO(g) + O_2(g) \longrightarrow 2NO_2(g)$$

$$3NO_2(g) + H_2O(\ell) \longrightarrow 2HNO_3(aq) + NO(g)$$

Nitric acid is an unstable, colorless liquid usually used as a 70% solution in water. It frequently has a yellow color because of the presence of NO_2 formed upon exposure to light. It is a strong oxidizing agent as well as a strong acid. It oxidizes all metals, with the exception of the noble metals gold, iridium, platinum, and rhodium. The largest use of HNO_3 is in the production of ammonium nitrate, NH_4NO_3, by reaction with ammonia.

$$NH_3(g) + HNO_3(aq) \longrightarrow NH_4NO_3(aq)$$

The ammonium nitrate is used widely as a fertilizer.

Phosphorus

Phosphorus is found in many minerals in the form of the tetrahedral PO_4^{3-} ion or related species. The element exists in a number of allotropic forms, none of which are analogous to N_2 because of the tendency for elements of the third period to form sigma rather than pi bonds. The most common form is white phosphorus, P_4, which is prepared by the reaction of calcium phosphate with coke and sand at high temperatures.

$$2Ca_3(PO_4)_2(s) + 6SiO_2(s) + 10C(s) \longrightarrow P_4(g) + 6CaSiO_3(\ell) + 10CO(g)$$

The gaseous P_4 condenses from this reaction as white phosphorus, which has a tetrahedral arrangement of phosphorus atoms (Figure 19.26). White phos-

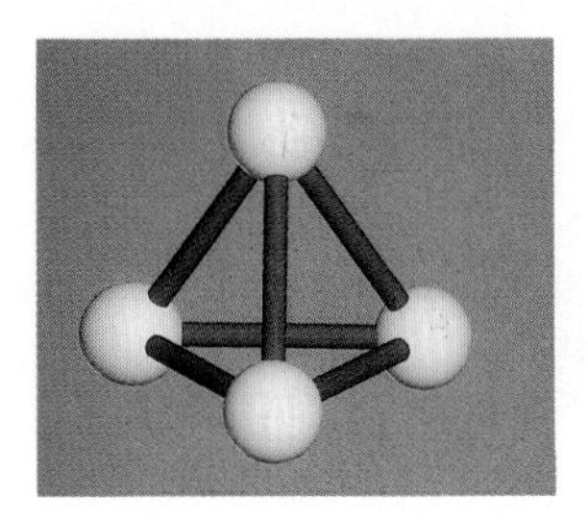

Figure 19.26 Structure of white phosphorus White phosphorus contains tetrahedral P_4 molecules in which each phosphorus atom makes three sigma bonds.

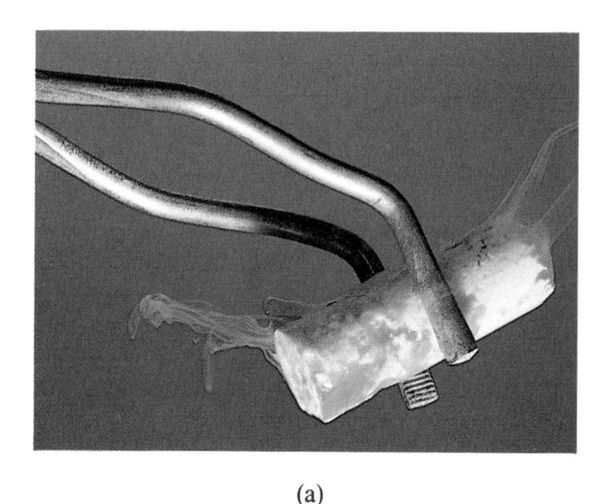

(a)

(b)

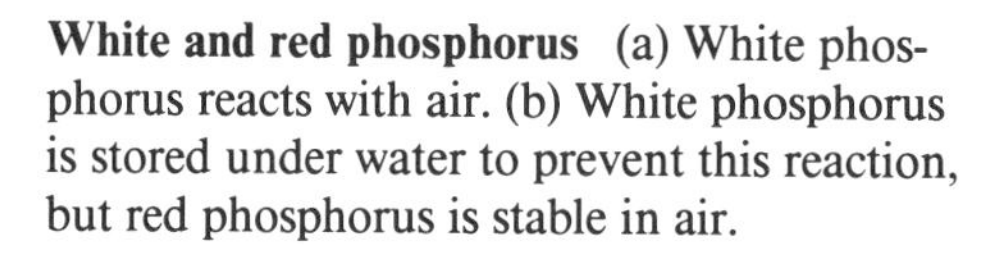

White and red phosphorus (a) White phosphorus reacts with air. (b) White phosphorus is stored under water to prevent this reaction, but red phosphorus is stable in air.

White phosphorus consists of reactive P_4 molecules, whereas red phosphorus is a less reactive polymer of P_4 units.

phorus is a very toxic material that burns when exposed to the air. This form is usually stored under water to protect it from air. Heating white phosphorus in the absence of air converts it to a second allotropic form called red phosphorus, which is stable in air. This form is believed to have a polymeric structure of linked P_4 units.

Industrially, most of the white phosphorus is oxidized to P_4O_{10}, which is mixed with water to give phosphoric acid.

$$P_4(s) + 5O_2(g) \longrightarrow P_4O_{10}(s)$$

$$P_4O_{10}(s) + 6H_2O(\ell) \longrightarrow 4H_3PO_4(aq)$$

The second reaction is quite rapid and complete, making $P_4O_{10}(s)$ a useful drying agent for gases and liquids. Pure H_3PO_4 is a solid that melts at 42 °C. It is very **hygroscopic**—that is, it absorbs water vapor from the air—and is generally sold as a water solution. It is used in very dilute solution (0.01 to 0.05%) to give a tart taste to carbonated drinks. Phosphoric acid can also be prepared by the reaction of $Ca_3(PO_4)_2$ with sulfuric acid.

$$Ca_3(PO_4)_2(s) + 3H_2SO_4(aq) \longrightarrow 2H_3PO_4(aq) + 3CaSO_4(s)$$

Another oxide of phosphorus, P_4O_6, can be prepared by the controlled oxidation of white phosphorus. Hydrolysis of P_4O_6 produces phosphorous acid, H_3PO_3:

$$P_4(s) + 3O_2(g) \longrightarrow P_4O_6(s)$$

$$P_4O_6(s) + 6H_2O(\ell) \longrightarrow 4H_3PO_3(aq)$$

Phosphorus is essential to life. In biological systems it is generally found in the form of phosphates, compounds that contain the PO_4^{3-} ion. Calcium phosphates are major constituents of human bones and teeth. Phosphorus is important for the growth of plants and, along with nitrogen, is a component of fertilizers. The minerals found in nature, such as $Ca_5(PO_4)_3F$ (fluorapatite), are insoluble in water and are converted into more soluble compounds for use as fertilizers by treatment with sulfuric acid or phosphoric acid.

Nitrogen and phosphorus compounds are produced by industry for use as fertilizers.

Phosphine

The stable hydride of phosphorus, PH_3, is called phosphine. This highly poisonous gas boils at −88 °C. Its boiling point is 55 °C lower than that of ammonia because there are no hydrogen bonding forces with PH_3.

The structure of PH_3 differs from that of ammonia in that the H—P—H bond angles are only 93.7°, compared to the 107.3° angles in NH_3 (Figure 19.27). Clearly, the bond angles in PH_3 are not accurately predicted by the VSEPR model. According to valence bond theory, the nearly 90° H—P—H bond angles suggest that the bonds are formed from almost pure 3*p* orbitals from phosphorus, leaving the 3*s* orbital to accommodate the lone pair of electrons.

Phosphine is prepared from the reaction between white phosphorus and aqueous NaOH.

$$P_4(s) + 3NaOH(aq) + 3H_2O(\ell) \longrightarrow 3NaH_2PO_2(aq) + PH_3(g)$$

In spite of its toxicity, phosphine is an important starting material in manufacturing a major flame-proofing material used on cotton cloth.

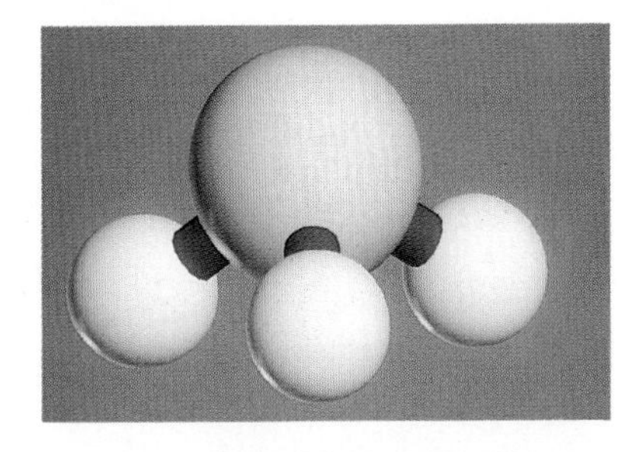

(a)

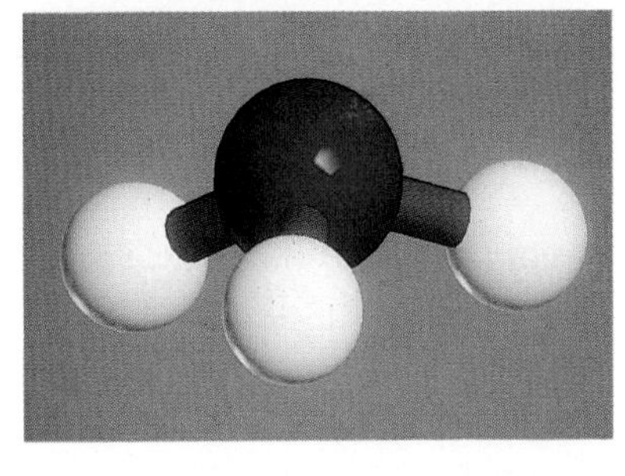

(b)

Figure 19.27 Structures of NH_3 and PH_3 The bond angles in phosphine (a) are much smaller than those in ammonia (b).

Arsenic, Antimony, and Bismuth

The heavier group VA elements occur in nature as sulfides—As_2S_3, Sb_2S_3, and Bi_2S_3. Arsenic and antimony are metalloids, and their chemistry resembles that of phosphorus. The +3 oxidation state becomes increasingly more stable than +5 for the heavier elements in the group. For example, burning the elements in excess air leads to the formation of As_4O_6 and Sb_4O_6, rather than the P_4O_{10} formed by phosphorus. As_4O_{10} can be prepared, but only when strong oxidizing agents are used. The oxide of bismuth, Bi_2O_3, is basic and will dissolve in acid solution, as is expected for the oxide of a metal.

$$Bi_2O_3(s) + 2H^+(aq) \longrightarrow 2BiO^+(aq) + H_2O$$

19.6 The Chemistry of Group VIA (16) Elements

Objectives

- To describe the allotropic forms of oxygen and sulfur and know their properties
- To know the chemistry of compounds containing both oxygen and sulfur.

VA	VIA	VIIA
7 N Nitrogen	8 O Oxygen	9 F Fluorine
15 P Phosphorus	16 S Sulfur	17 Cl Chlorine
33 As Arsenic	34 Se Selenium	35 Br Bromine
51 Sb Antimony	52 Te Tellurium	53 I Iodine
83 Bi Bismuth	84 Po Polonium	85 At Astatine

The elements in group VIA (16) have six valence electrons with the electron configuration ns^2np^4. These elements generally form two covalent bonds and have two lone pairs on the central atom. As in the other groups, the heavier elements also form electron-rich compounds, such as SF_4 and SF_6.

Oxygen

Oxygen is the most abundant element on our planet; it makes up about half of the earth's crust and is present in both air (21% O_2 by volume) and water. On the earth's surface, oxygen is mostly found combined with a variety of other elements. It occurs mainly as water in the oceans and lakes, and combined with silicon in silica and the silicates.

The stable allotrope of oxygen is $O_2(g)$, but a second allotrope, $O_3(g)$, is very important in the upper atmosphere.

The most common allotrope of oxygen is O_2, a pale blue gas at room temperature. The liquid has a light blue color, and boils at −183 °C. The other important allotrope is ozone, O_3. Ozone is a very reactive gas that causes the pungent odor noticed during electrical storms. The solid and liquid phases (boiling point = −112 °C) of ozone are very unstable and decompose explosively. Although ozone is a pollutant in the lower atmosphere, its presence in the upper atmosphere is extremely important to us because it absorbs much of the ultraviolet light coming from the sun. There is concern at present that a number of synthetic compounds, mainly the chlorofluorocarbons (Freons or CFCs), are destroying the ozone layer. CFCs consume ozone in a complicated series of reactions, such as these for trichlorofluoromethane, $CFCl_3$:

Reaction of metals with oxygen At high temperatures, iron reacts with the oxygen present in air.

$$CFCl_3(g) \xrightarrow[\text{light}]{\text{ultraviolet}} CFCl_2(g) + Cl(g)$$

$$Cl(g) + O_3(g) \longrightarrow ClO(g) + O_2(g)$$

$$O_3(g) \longrightarrow O(g) + O_2(g)$$

$$ClO(g) + O(g) \longrightarrow Cl(g) + O_2(g)$$

Note that the last three reactions constitute a cycle showing that Cl atoms are a catalyst, so each molecule of $CFCl_3$ can destroy a large number of ozone molecules. Without ozone as a natural filter for ultraviolet light, the incidence of skin cancer might increase.

Molecular oxygen can be prepared on a small scale by heating potassium chlorate, using MnO_2 as a catalyst.

$$2KClO_3(s) \xrightarrow[150\ °C]{MnO_2} 2KCl(s) + 3O_2(g)$$

On an industrial scale, oxygen, like molecular nitrogen, is recovered from the fractional distillation of liquid air. The largest industrial consumption of O_2 is in the production of steel (see Section 18.1). It is also used in the oxidation of hydrocarbons, in the treatment of wastewater, in medicine, and in rocket engines.

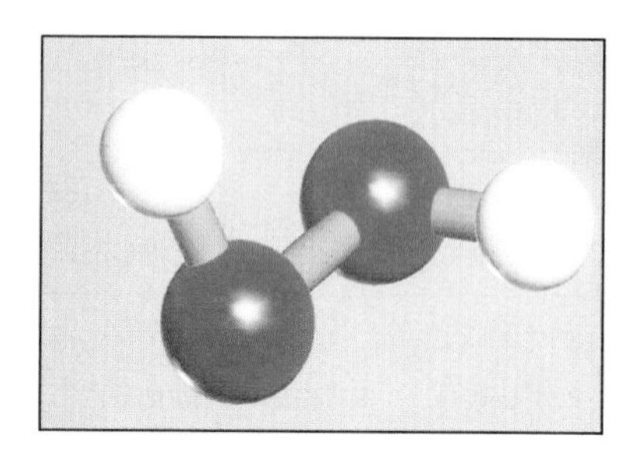

Figure 19.28 Structure of hydrogen peroxide Hydrogen peroxide has a folded structure with much smaller H—O—O angles than predicted by VSEPR theory.

Most elements will react directly with molecular oxygen. All elements except helium, neon, argon, and possibly krypton form binary compounds with oxygen. Oxygen forms strong covalent bonds to most of the nonmetals and many of the transition metals. In its ionic compounds, the small size and 2− charge of the oxide ion produce very stable ionic structures.

Changes in the electronegativity of the elements across the periodic table lead to dramatic changes in the properties of the binary compounds of oxygen. Oxides of the metals on the left side of the table are generally high melting ionic solids, and react to form basic solutions in water. Oxides of the other metals and metalloids are also generally solids, but they are less ionic and may be amphoteric in water. Oxides of the nonmetals are generally covalent compounds that form acidic solutions by reaction with water.

In addition to water, hydrogen forms a second compound with oxygen, hydrogen peroxide, H_2O_2. Hydrogen peroxide has a freezing point very close to that of water, −0.41 °C, but boils at 150 °C and has a much higher density, 1.4 g/mL. It has the structure shown in Figure 19.28.

Hydrogen peroxide is unstable, particularly when pure, and its decomposition is catalyzed by many metals and even by glass.

$$2H_2O_2(\ell) \longrightarrow 2H_2O(\ell) + O_2(g)$$

The 2% to 3% solutions of hydrogen peroxide that are sold as a germicide are safe to handle, but solutions with concentrations above 20% can be very dangerous. Note from the decomposition reaction that when hydrogen peroxide is used as a germicide it gives off $O_2(g)$. This "foaming" helps clean the wound. Hydrogen peroxide is also a bleach that is used extensively in industry and for hair coloring. An important reason for its widespread use is that the products of the reaction, $O_2(g)$ and $H_2O(\ell)$, are nonpollutants.

Sulfur

Sulfur is abundant in nature. It occurs as the free element, as well as in sulfides and sulfates. Sulfur deposits are generally found underground and are brought to the surface by the **Frasch process.** In this process, the sulfur is melted with superheated steam and pumped to the surface by air pressure (Figures 19.29 and 19.30).

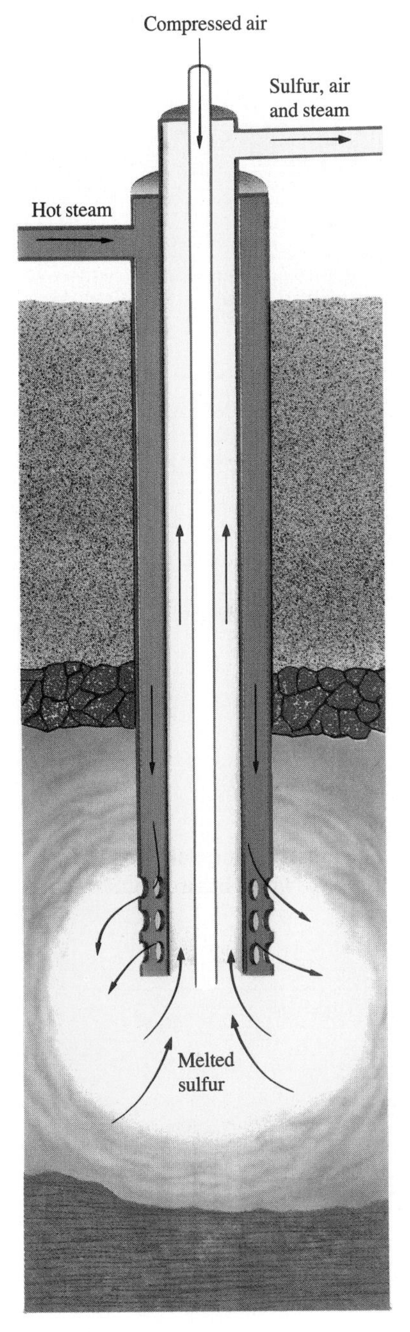

Figure 19.29 Frasch process Underground deposits of sulfur are melted with hot steam and forced to the surface with compressed air.

Figure 19.30 Sulfur A hot mixture of sulfur and water is pushed to the surface by the Frasch process.

Figure 19.31 Coal-fueled power plant The sulfur emitted from the burning of coal (mainly as SO_2) is removed from the flue gas just before the gas goes to the stacks. Notice how clean the stack gases appear in this photograph compared to the photograph on page 637.

Another major source of sulfur (and sulfur compounds) is as a by-product in the purification of fossil fuels. Both crude oil and natural gas are contaminated by H_2S that must be removed before the fuels are burned, to prevent air pollution. As a pollution control measure, sulfur dioxide is removed from the flue gas produced when coal is burned (Figure 19.31). Sulfur dioxide is also recovered from the roasting of metal sulfide ores in the refining of many metals.

$$2PbS(s) + 3O_2(g) \longrightarrow 2PbO(s) + 2SO_2(g)$$

The most common allotrope of sulfur is S_8, which has a cyclic structure.

Sulfur exists in many allotropic forms. The yellow solid that forms around some volcanic steam vents in the earth consists of orthorhombic sulfur, a form that contains molecules of S_8 rings (Figure 19.32).

When orthorhombic sulfur is heated above its melting point of 113 °C, the rings break and the short chains link together to form longer chains. Depending on the temperature and the rate of heating or cooling, many different forms of sulfur can occur. If molten sulfur is cooled rapidly, for instance by pouring it into water (Figure 19.33), the result is a rubbery form called *plastic sulfur,* which contains long chains of sulfur atoms. At room temperature, plastic sulfur reverts to orthorhombic sulfur.

The most important hydride of sulfur is hydrogen sulfide, H_2S. Analogous to the comparison of ammonia and phosphine, H_2S has a lower boiling point (−60 °C) than water because of the lack of strong hydrogen bonding forces. It is extremely toxic and has the unpleasant odor of rotten eggs. Although easily detected by its characteristic smell at low concentrations, it can actually deaden the olfactory nerve at higher concentrations. Large

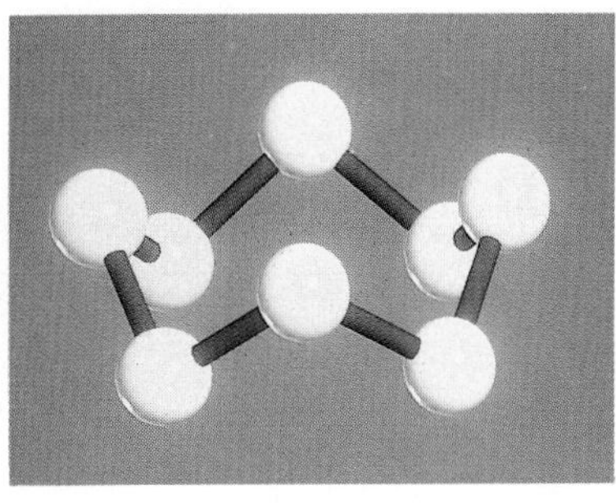

(a) (b)

Figure 19.32 Orthorhombic sulfur (a) Sulfur deposits can be found around natural steam vents in the earth. (b) This allotropic form consists of eight-membered rings of sulfur atoms.

volumes of H_2S are emitted by volcanic activity, and at these elevated temperatures it reacts with oxygen to produce SO_2.

$$2H_2S(g) + 3O_2(g) \longrightarrow 2SO_2(g) + 2H_2O(g)$$

Compounds of Oxygen and Sulfur

The most common oxide of sulfur is sulfur dioxide, SO_2. Sulfur dioxide is a toxic gas (boiling point = −10 °C) that has a choking odor. It is produced industrially on a large scale by the burning of sulfur and as a by-product in the roasting of sulfide ores. Unfortunately, much of the SO_2 that is produced, mainly from the burning of coal, is not trapped and is emitted into the air. The SO_2 from this source, along with SO_2 produced naturally, reacts with water to contribute to acid rain.

$$SO_2(g) + H_2O(\ell) \longrightarrow H_2SO_3(aq)$$

Much of the SO_2 that is produced (both industrially and in the atmosphere) is converted to sulfur trioxide, SO_3, by reaction with oxygen.

$$2SO_2(g) + O_2(g) \xrightarrow{\text{catalyst}} 2SO_3(g)$$

This reaction is very slow but is catalyzed by a variety of solids, such as dust particles in the presence of sunlight, and by metal ions, such as Fe^{3+}, dissolved in water. The SO_3 thus produced reacts with water to form sulfuric acid, the major contributor to acid rain.

Figure 19.33 Plastic sulfur A rubbery form of sulfur, called *plastic sulfur,* which contains long chains of sulfur atoms, is formed upon rapid cooling of hot sulfur.

$$SO_3(g) + H_2O(\ell) \longrightarrow H_2SO_4(aq)$$

The SO_3 is a volatile liquid (boiling point = 45 °C) that is extremely reactive. It has a triangular structure in the gas phase, but in the solid or liquid

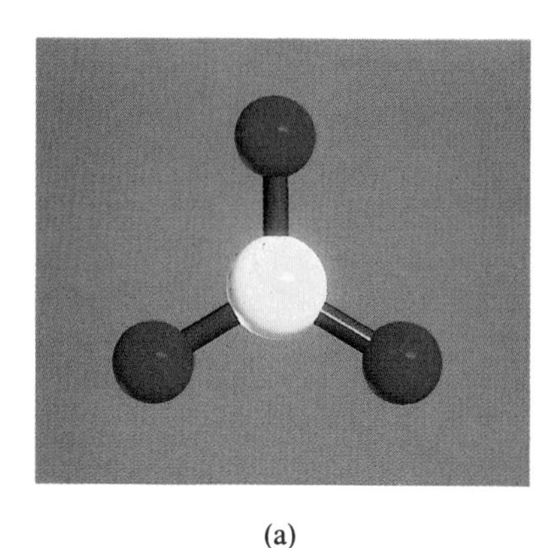

(a)

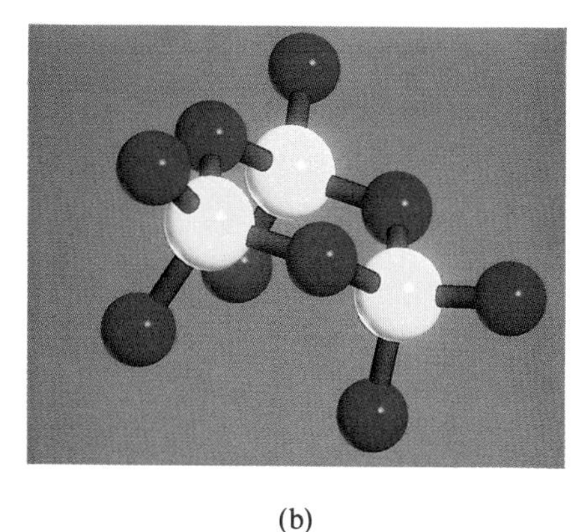

(b)

Figure 19.34 SO_3 and S_3O_9 (a) SO_3 has a trigonal planar structure. (b) S_3O_9 is a cyclic trimer.

phase it exists in various polymeric forms, such as the cyclic trimer S_3O_9 (Figure 19.34).

Sulfuric acid, H_2SO_4, is commercially produced in larger quantity than any other compound. The main industrial method of preparing sulfuric acid is the **contact process.** In this process, the reaction of SO_2 with O_2 is catalyzed by V_2O_5; the SO_3 product is trapped by dissolving it in concentrated H_2SO_4, which is then diluted by mixing with additional water. The resulting acid is generally sold as a solution that is 96% to 98% H_2SO_4 by weight.

Sulfuric acid is a strong, corrosive acid that must be handled carefully. Dilution of the concentrated acid with water produces a large amount of heat. When diluting sulfuric acid, remember the rule to *add acid to water,* and do it slowly with stirring. If you make the mistake of adding water to sulfuric acid, the heat that is released can cause droplets of the water to boil and spatter the concentrated acid out of its container. Even when the dilution is done properly, you should wear protective clothing and a full face

Figure 19.35 Reaction of H_2SO_4 and sugar Sulfuric acid reacts with sugar, removing the water and leaving only a column of carbon.

shield. The affinity of concentrated sulfuric acid for water is so high that it is frequently used to remove water from gases.

A dramatic demonstration of the ability of H_2SO_4 to act as a dehydrating agent is its reaction with sugar to form carbon and hydrated sulfuric acid (Figure 19.35).

$$C_{12}H_{22}O_{11}(s) + 11H_2SO_4(\ell) \longrightarrow 12C(s) + 11H_2SO_4 \cdot H_2O(\ell)$$

In addition to being a strong acid and dehydrating agent, H_2SO_4 is also an oxidizing agent. While it has only mild oxidizing properties at room temperature, it is much more reactive at higher temperatures.

More sulfuric acid is commerically produced than any other compound. It is used as an acid, an oxidizing agent, and a dehydrating agent.

The main use of sulfuric acid is in the conversion of insoluble phosphorus minerals into soluble phosphate fertilizers, as outlined in the previous section. It is also used in the refining of petroleum and in the synthesis of many chemicals.

Selenium and Tellurium

Selenium and tellurium are rare elements whose chemistry resembles that of sulfur. Selenium exists in a number of allotropic forms, including an Se_8 ring form. Another form of selenium and the only crystalline form of tellurium are composed of spiral chains of atoms. The last element of the family, polonium, is radioactive and exists in only trace amounts. Selenium is used in the manufacture of red glass. Selenium is recovered as a by-product in the roasting of certain ores. Although a toxic element, selenium has been found recently as a trace element in the body. This is an interesting case in which trace amounts of an element are beneficial whereas larger amounts are toxic.

19.7 The Noble Gases

Objective

- To describe the properties and chemical behavior of the noble gas elements

The elements in group VIIIA (18), of course, have a noble gas electron configuration, ns^2np^6 ($1s^2$ for helium). As expected, they are very unreactive and exist as monatomic gases. Because of their lack of reactivity, these gases were once called the "inert gases," but over the last 30 years compounds of krypton and especially xenon have been prepared. Although there are no known compounds of helium, neon, or argon, these gases have a number of important uses.

VIIA	VIIIA
	2 He Helium
9 F Fluorine	10 Ne Neon
17 Cl Chlorine	18 Ar Argon
35 Br Bromine	36 Kr Krypton
53 I Iodine	54 Xe Xenon
85 At Astatine	86 Rn Radon

Helium, the second most abundant element in the universe, is found in very low concentration in our atmosphere because the gravitational pull of the earth is too weak to prevent it from escaping to outer space. It is found in considerable concentrations in the United States in certain natural gas deposits, where it forms from the α-particle decay (He^{2+}) of radioactive elements. Helium has a density much lower than that of air, and it is used to float balloons and lighter-than-air ships.

Liquid helium, with a boiling point of 4 K, is used as a coolant for low-temperature experiments and for devices such as the superconducting

The Goodyear blimp Helium is used as the gas inside a blimp to make it lighter than air.

magnets used in magnetic resonance imaging, an important new diagnostic technique in the health field (Figure 19.36).

A helium-oxygen mixture is often used instead of a nitrogen-oxygen mixture for deep-sea diving and in spacecraft. At the high pressures encountered during a deep dive, the concentration of dissolved nitrogen in the blood increases and can cause narcosis. This problem is avoided by using helium rather than nitrogen as the diluting gas for oxygen.

Neon is used as the gas in "neon" lights. The red color of the light can be changed by mixing in some argon or mercury vapors. Argon is the third most abundant gas in dry air (0.93%); along with neon, krypton, and xenon, it is produced by the distillation of liquid air. Argon has a higher density than air and is used as the inert atmosphere in electric light bulbs and in welding to protect metals from oxygen.

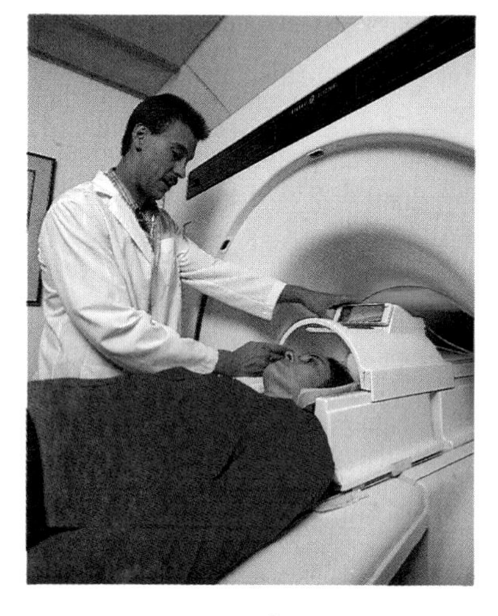

Figure 19.36 Magnetic resonance imaging Magnetic resonance imaging (MRI) is an important technique for the detection of medical problems. The superconducting magnet used in this equipment is cooled by liquid helium.

Krypton and Xenon

In 1962, Neil Bartlett prepared the compound $[O_2^+][PtF_6^-]$ from the reaction of $O_2(g)$ with PtF_6. Realizing that the ionization energy of xenon is similar to that of O_2, he carried out a similar reaction with xenon and isolated $[Xe^+][PtF_6^-]$, the first compound of an "inert gas." While in retrospect his decision to carry out this reaction may seem obvious, at the time it was extraordinary for a scientist to even try a reaction with an element thought to be completely nonreactive. Bartlett's success demonstrates that it is important to question and test accepted scientific theories. Soon after this discovery, workers at the Argonne National Laboratory demonstrated that fluorine will also oxidize xenon at elevated temperatures, leading to the covalent compound XeF_4.

$$Xe(g) + 2F_2(g) \xrightarrow{400\ °C} XeF_4(s)$$

By carefully controlling the conditions, XeF_2 and XeF_6 can also be prepared. A number of oxides (XeO_3, an explosive compound, and XeO_4) and mixed

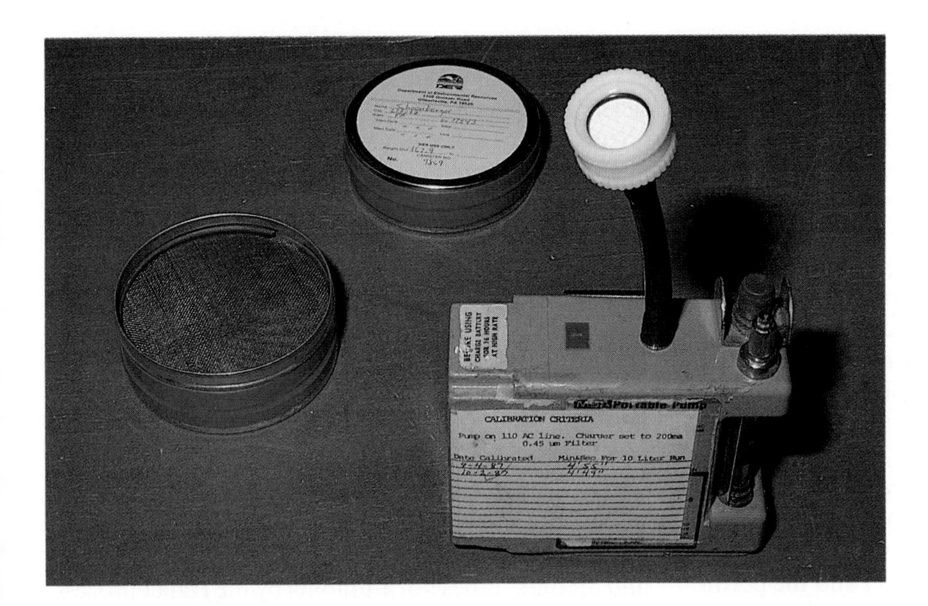

Figure 19.37 Radon testing A number of testing devices are available to determine whether radon is building up in the basement of a home.

fluoro-oxides (XeF_2O_3 and XeF_4O_2) have also been prepared. Sodium perxenate, Na_4XeO_6, forms in the reaction of XeO_3 with O_3 in aqueous NaOH solution. Both XeO_3 and Na_4XeO_6 are very strong oxidizing agents. Although krypton is not as reactive, KrF_2 and KrF_4 have been prepared.

Although a few compounds of krypton and xenon are known, the group VIIIA elements are very unreactive.

Radon

All isotopes of radon are radioactive and are difficult to work with in the laboratory. Radon has received considerable publicity in recent years because it forms in the radioactive decay of an isotope of uranium, natural deposits of which are found in many parts of North America. Because radon is a gaseous element, it can escape from the soil into the air. In well-insulated houses with little ventilation, the gas can build up to levels that can cause a considerable health hazard. Since it is impossible to remove the source of the radon, the problem is best solved by ventilation. In most cases, a small exhaust fan in the basement is sufficient to prevent buildup of the gas. Houses in areas that are known to have this problem should be tested for radon (Figure 19.37).

Summary

The chemistry of the nonmetallic elements of the second period is dominated by their small atomic size and relatively high electronegativity. These second period elements readily form both sigma and pi bonds, whereas elements in the later periods tend to form only sigma bonds because pi overlap of *p* orbitals in these larger elements produces relatively weak bonds. For example, the elemental form of nitrogen is N_2, a compound with one sigma and two pi bonds, whereas the simplest form of phosphorus is P_4 in which each phosphorus atom makes three sigma bonds. The metallic character of the elements increases down each group, as does the stability of the lower oxidation states.

Hydrogen is generally listed in group IA on a periodic table, but is really in a group by itself. Hydrogen can lose an electron to form a proton ($H^+(aq)$ in water) or gain an electron to form the hydride ion, H^-. It also forms strong covalent bonds with other nonmetals.

Hydrogen is produced on a large scale by the high-temperature reaction of methane and water, and also by the reaction of carbon and water. It combines with nitrogen to form ammonia by the *Haber process.* It also can be used to prepare methanol by reaction with carbon monoxide.

Group IIIA elements generally form compounds in which the element has an oxidation number of +3. Heavier members of the family, especially thallium, also form the +1 oxidation state in some compounds. The trend for the heavier members of a group not to use the pair of valence *s* electrons is also observed in groups IVA and VA and is called the *inert pair effect.* The elemental forms of boron contain an unusual arrangement of the boron atoms, an *icosahedron.* The types of compounds formed by boron are dominated by the fact that boron has only three valence electrons. Electron-deficient compounds such as boric acid and the boron halides react with neutral and anionic Lewis bases to form compounds with an octet of electrons at boron. In the boron hydrides, the boron atoms make four bonds by forming *three-center, two-electron bonds.*

Aluminum is the most abundant metal in the earth's crust. Because of its low density, its alloys with silicon, copper, and magnesium are used extensively as structural materials. Aluminum metal is protected from corrosion by the formation of a thin, strongly adhering protective coating of inert aluminum oxide, Al_2O_3. Aluminum oxide exists in a number of forms and has many important industrial uses. Corundum, or α-alumina, is a very hard substance that is used as an abrasive. Crystals of α-alumina that contain metal ion impurities are well-known gems, such as rubies and sapphires. Another form, γ-alumina, is used as a support in chromatographic separations and as a catalyst or catalyst support for many chemical reactions. Gallium and indium are becoming increasingly important because gallium arsenide (GaAs) and indium phosphide (InP) are useful semiconductor materials.

Elements in group IVA have four valence electrons, and generally attain an octet by forming four covalent bonds. The heavier members of the group, tin and especially lead, exhibit the *inert pair effect* and form the +2 oxidation state in some compounds. Millions of compounds of carbon have been isolated from living organisms or synthesized in laboratories. Graphite and diamond are both elemental forms of carbon. The former is composed of layers consisting of sp^2 hybridized carbon atoms, and the latter is an extremely hard allotrope in which the carbon atoms are sp^3 hybridized. Silicon is the second most abundant element in the earth's crust and is found in nature in minerals called *silicates* and silica, SiO_2. In these minerals, the silicon atom is surrounded by a tetrahedral arrangement of the oxygen atoms. Glass is formed by melting and rapidly cooling a mixture of silica and various additives that modify the properties of the glass. Silicon is used in common alloys and to make silicone polymers. Pure silicon is used in the fabrication of solid-state electronic components. Both tin and lead were among the first metals isolated by humans.

Elements in group VA have five valence electrons and generally form three covalent bonds. Elemental nitrogen is easily isolated by fractional distillation of liquid air, and is used extensively by the chemical industry as a low-temperature coolant, as an inert gas to protect materials from reactions with oxygen in the air, and for the synthesis of ammonia by the Haber process. Hydrazine (N_2H_4) is another compound of nitrogen and hydrogen. Nitrogen forms a series of oxides that have interesting properties and structures. Two of these oxides, NO and NO_2, are major sources of air pollution and acid rain. Nitric acid, HNO_3, is made from ammonia by reaction with oxygen and water. The major compounds prepared from phosphorus are phosphoric acid, H_3PO_4, used in carbonated drinks, and various phosphates (compounds of PO_4^{3-}), used as fertilizers.

Group VIA elements, especially oxygen, generally form two covalent bonds and have two lone pairs on the central atom. Oxygen is the most abundant element on our planet, and is found in the atmosphere, in water in the oceans, and in the earth's crust as oxygen compounds of silicon such as silica and the silicates. Oxygen combines with nearly every other element on the periodic table. It forms ionic oxides with metals that form basic solutions in water, and covalent oxides with nonmetals that form acidic solutions in water. Sulfur is found as the element in nature and is recovered by the Frasch process. It is also isolated as H_2S or SO_2 from contaminants in crude oil and natural gas, or from the combustion of coal. Most of the sulfur is converted to sulfuric acid, H_2SO_4, the world's largest-volume industrial compound.

The elements in group VIIIA are called the noble gases because they are not very reactive and exist as monatomic gases. While these elements are nonreactive, compounds of both krypton and xenon are known. The compounds of xenon include three fluorides, XeF_2, XeF_4, and XeF_6, and a number of oxides and mixed fluoro-oxides.

Chapter Terms

Water gas shift reaction—the reaction of carbon monoxide and steam to produce hydrogen and carbon dioxide. *(19.2)*

Inert pair effect—the tendency for the heavier members of groups IIIA through VA not to use the valence *s* electrons for bonding. *(19.3)*

Icosahedron—a regular polyhedron with 20 faces and 12 vertices. *(19.3)*

Silicates—compounds containing silicon combined with oxygen and various metals. *(19.4)*

Ceramics—nonmetallic, solid materials that are hard, resistant to heat, and chemically inert. *(19.4)*

Conduction band—a partially or completely empty band of orbitals in solids that results in high electrical conductivity. *(19.4)*

Band gap—the difference in energy between the highest filled band of orbitals and the conduction band in solids. *(19.4)*

Haber process—the direct synthesis of ammonia from nitrogen and hydrogen. *(19.5)*

Ostwald process—a multi-step process for the production of nitric acid from ammonia, oxygen, and water. *(19.5)*

Hygroscopic—having the ability to absorb water vapor from the air. *(19.5)*

Frasch process—a method for melting underground deposits of sulfur and bringing it to the surface. *(19.6)*

Contact process—a process for the production of sulfuric acid from sulfur dioxide, oxygen, and water. *(19.6)*

Exercises

Exercises designated with color have answers in Appendix J.

General Trends

19.1 Discuss the factors that cause the chemistry of the elements in the second period to be very different from that of the elements in the same group in later periods.

19.2 Why do elements of the second period form stronger π bonds than elements of the third period? Give a specific example of structural differences between the elemental forms of elements from the same group that can be explained by this difference.

19.3 Compare the electronegativities and ionization energies of metals and nonmetals.

19.4 Are the following elements metals, metalloids, or nonmetals?
(a) carbon (c) chlorine
(b) tin (d) silicon

19.5 Are the following elements metals, metalloids, or nonmetals?
(a) gallium (c) arsenic
(b) nitrogen (d) indium

19.6 From the elements nitrogen, silicon, and gallium, pick the ones with the most and least metallic properties. Explain your choices.

19.7 From the elements silicon, germanium, and tin, pick the ones with the most and least metallic properties. Explain your choices.

19.8 Why does sulfur form electron-rich compounds such as SF_6 whereas oxygen does not?

Hydrogen

19.9 Three different bonding modes are found for hydrogen. Describe and give a specific example of each.

19.10 Describe the bonding of the hydrogen in (a) KH, (b) HCl, (c) H_2.

19.11 Why does H_2 have such a low boiling point (−253 °C)?

19.12 List the symbols of the three isotopes of hydrogen and the approximate abundance of each.

19.13 Why is helium rather than hydrogen used today as the gas in blimps?

19.14 Write the equation for the reaction of NaH and water. What mass of NaH is needed to prepare a liter of hydrogen gas at 25 °C and one atmosphere of pressure?

19.15 Write the equation for the reaction of zinc metal with hydrochloric acid. What mass of zinc metal is needed to prepare a liter of hydrogen gas at 25 °C and one atmosphere of pressure, assuming excess HCl?

19.16 Give two important industrial preparations for H_2.

19.17 Write the equation for the water gas shift reaction.

19.18 What is the most important industrial use of H_2? Write the equation for this use.

19.19 Write a series of equations that show how coal can be converted into methanol. Be careful to prepare the correct amount of H_2 needed to form the methanol.

19.20 What is the main difference between saturated and unsaturated vegetable oils?

Group IIIA

19.21 What is the inert pair effect? How does it affect the chemistry of group IIIA?

19.22 Explain why the +1 oxidation state is more stable for thallium than for aluminum.

19.23 Classify the group IIIA elements as nonmetals, metals, or metalloids.

19.24 What is the unusual structural feature found in the elemental forms of boron?

19.25 Draw the structure of B_2H_6 and describe the bonding in this molecule. What is the hybridization at the boron atoms?

19.26 What is a three-center, two-electron bond?

19.27 Because of the high reactivity of the boron hydrides with oxygen, they were considered as possible solid rocket fuels. Write the equation for the reaction of oxygen with $B_{10}H_{14}$ (a solid at room temperature).

19.28 Write the equation and describe the changes in hybridization of the boron atom in the reaction between BCl_3 and NH_3.

19.29 Describe the bonding in BCl_3. What is the hybridization at the boron atom? Is the boron really electron deficient?

19.30 Explain why aluminum, a reactive metal, can be used in airplanes and on the exterior of houses, where it is exposed to the oxygen in the air.

19.31 Describe the Hall process for the production of aluminum from the mineral bauxite. How is energy saved by recycling aluminum rather than preparing it by the Hall process?

19.32 What is the thermite reaction and why can it be used to weld steel?

19.33 Describe the composition of a ruby.

19.34 How is γ-alumina used to purify water?

19.35 Draw the structure of Al_2Cl_6. Compare the bonding in Al_2Cl_6 and B_2H_6.

19.36 The oxide Ga_2O_3 is amphoteric. Write an equation for its reactions, if any, with HCl and NaOH.

Group IVA

19.37 Explain why group IVA elements, especially carbon and silicon, are ideally suited to make four electron-pair bonds.

19.38 State three commercial uses for graphite.

19.39 What is meant by the term adsorption, and why does activated charcoal have high adsorption qualities?

19.40 What are the hybridizations of silicon and carbon in SiO_2 and CO_2?

19.41 Draw the structure of silica, SiO_2, and compare it to the structure of CO_2. Why are the structures so different?

19.42 How is silicon purified for use in the electronics industry?

19.43 How does doping silicon with phosphorus change its conducting properties?

19.44 Describe a *p*-type semiconductor based on silicon.

19.45 Describe an *n*-type semiconductor based on silicon.

19.46 What is the hybridization of silicon in $SiCl_4$? Is this compound polar or nonpolar?

19.47 What is the hybridization of silicon in silicates?

19.48 What mineral is mined for the production of lead? Describe the process for obtaining the metal from this mineral.

Group VA

19.49 What are the structures of the most common allotropic forms of nitrogen and phosphorus? Explain why they are so different.

19.50 Write the Lewis structure of P_4.

19.51 Nitrogen is the compound isolated in second-largest quantity in the chemical industry. How is it isolated?

19.52 What is meant by the *fixation* of nitrogen with hydrogen?

19.53 Describe the Haber synthesis of ammonia. Be sure to comment on the positive and negative effects that the principle of Le Chatelier has on the production of ammonia by this process.

19.54 Write the Lewis structures of NO_2 and N_2O_3. What is the hybridization of the nitrogen atoms in each compound?

19.55 Write the Lewis structures of N_2O and N_2O_4. What is the hybridization of the nitrogen atoms in each?

19.56 Write equations for the following reactions:
(a) The reaction between magnesium and nitrogen.
(b) The preparation of P_4O_{10}.
(c) The reaction of nitrogen dioxide with water.

19.57 Write equations for the following reactions:
(a) The preparation of dinitrogen oxide.
(b) The reaction of hydrazine with oxygen.
(c) The reaction between P_4O_{10} and water.

19.58 Discuss the structure of $(NO_2)_x$ in both the gas and solid phases.

19.59 How is nitric acid prepared by the Ostwald process?

19.60 Write the equation for the oxidation of ammonia to nitrogen monoxide (the first step in the Ostwald process). What mass of ammonia is needed to produce 25 kg of nitrogen monoxide?

19.61 Write the equation for the preparation of ammonium nitrate from NH_3. What mass of ammonia is needed to produce 5.22 kg of ammonium nitrate?

19.62 Even though phosphates are found widely in most soils, fertilizers are used to supply additional phosphates. What is the problem with the "natural phosphates" and how is it overcome in commercial fertilizers?

19.63 Compare the boiling points of NH_3 and PH_3 and give an explanation for the difference.

Group VIA

19.64 Identify the two most abundant elements on our planet, and give their principal occurrences.

19.65 Briefly outline the important physical and chemical properties of the two main allotropes of oxygen.

19.66 Indicate which elements form binary compounds with oxygen.

19.67 Classify as acidic, basic, or amphoteric the oxides of the metals, the nonmetals, and the metalloids.

19.68 Oxygen can be prepared in the laboratory by heating $KClO_3$ in the presence of MnO_2. Write the equation and determine the mass of $KClO_3$ needed to produce 0.50 L of O_2 gas at 27 °C and 755 torr pressure.

19.69 Write the equation for the roasting of lead sulfide. What volume of SO_2 gas measured at STP is produced from the roasting of 1.0×10^2 g of lead sulfide?

19.70 Where is sulfur found in nature, and how is it recovered from each of these sources?

19.71 Describe two allotropic forms of sulfur.

19.72 How is sulfuric acid prepared industrially?

19.73 Describe the three main types of reactions that sulfuric acid will undergo.

Group VIII

19.74 Why are elements in group VIIIA expected to be monomeric and relatively nonreactive?

19.75 What realization led Bartlett to prepare the first compound of a noble gas?

19.76 Draw the Lewis structures and assign the shapes of XeF_2 and XeO_3. Is either of these compounds polar?

19.77 Draw the Lewis structures and assign the shapes of XeF_4 and XeO_4. Is either of these compounds polar?

19.78 What is the main source of radon in homes?

Nuclear Chemistry

Chapter Twenty 20

Throughout the previous chapters our focus has been on chemical reactions. In chemical reactions, changes in the arrangement of the electrons in ions, atoms, and molecules explain the observed changes. The nuclei of the atoms do not change in chemical reactions, but simply provide the mass of the atom and positive charges that determine the energies of the electrons in atoms and molecules.

Nuclei do change, though, and the changes in the nuclei of atoms have become increasingly important in the 20th century, so no survey of chemistry is complete without considering nuclear chemistry. In this chapter we discuss the processes, both natural and induced, that involve changes in the nuclei of atoms. We present important uses of nuclear processes and the new problems created for society by the nuclear age. We shall emphasize the role of chemistry in the uses, discoveries, and possible solutions to problems arising from nuclear processes.

A brief review of the nucleus of the atom is necessary before we begin the discussion of nuclear changes. The nucleus of an atom consists of protons and neutrons, held together in a very small fraction of the volume occupied by the atom. The radius of an atom is about 10^{-8} cm, while that of a nucleus is between 10^{-13} and 10^{-12} cm. Since nearly all the mass of the atom is that of the protons and neutrons, the density of nuclear matter is very great, about 10^{13} to 10^{14} g/cm^3. An object the size of a pea that was made of nuclear matter would have a mass of more than one million *tons!*

Most of the naturally occurring elements are mixtures of several isotopes, atoms with the same number of protons but different numbers of neutrons. The term **nuclide** refers to the nucleus of a particular isotope. **Nucleon** is used to refer to a proton or a neutron. The atomic number, Z, is the number of protons in a nucleus. The mass number, A, is the number of

◀ The eerie blue glow emitted by the reactor is due to radiation from high-energy beta particles.

nucleons, or the sum of the number of protons and neutrons in a particular nuclide. A particular nuclide of an atom is represented by

$$^{A}_{Z}X$$

where X is the chemical symbol for the element (H, He, Fe), the pre-subscript, Z, is the atomic number, and the pre-superscript, A, is the mass number. Using this notation, the three naturally occurring isotopes of oxygen (oxygen-16, oxygen-17, and oxygen-18) have the complete symbols

$$^{16}_{8}O \qquad ^{17}_{8}O \qquad ^{18}_{8}O$$

20.1 Nuclear Stability and Radioactivity

Objectives

- To relate the number of protons and neutrons to the stability of a nuclide
- To classify radioactive decays
- To write balanced nuclear equations for radioactive decays
- To understand the uses of isotopes in the study of chemical reactions
- To use radioactive decay data to determine the ages of samples of matter

In Chapter 2 we mentioned that some nuclei undergo spontaneous radioactive decay, but the majority of the naturally occurring elements are mixtures of stable isotopes of the elements. A **stable isotope** is one that does not spontaneously decompose into a different nuclide. Figure 20.1 shows a graph of the number of neutrons (A − Z) versus the number of protons (Z) for all the 279 known stable nuclides. A number of features of this band of nuclear stability are worthy of note.

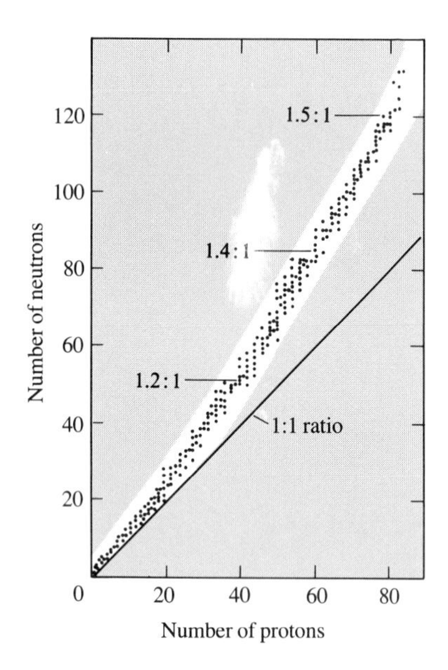

Figure 20.1 The zone of stability of nuclides The dots represent the combinations of neutrons and protons in the known stable isotopes. As the number of protons increases, the ratio of neutrons to protons needed to produce a stable nuclide also increases.

1. With two exceptions, hydrogen-1 and helium-3, the number of neutrons is equal to or greater than the number of protons in the stable nuclides.
2. In the elements with low atomic numbers, the numbers of neutrons and protons in stable nuclei are very nearly equal. Above an atomic number of about twenty, the ratio of neutrons to protons gradually increases to a maximum of about 1.5 : 1.
3. Nuclear stability is greater for nuclides that contain even numbers of protons, neutrons, or both (see Table 20.1).
4. There are certain numbers of protons and neutrons that confer unusual stability to the nuclides. These numbers, called **magic numbers,** are 2, 8, 20, 26, 28, 50, 82, and 126. There is an analogy between the magic numbers for the nucleus and the unusual electronic stability found for the noble gases (2, 10, 18, 36, 54, and 86 electrons).
5. The zone of stability marked in Figure 20.1 contains all of the stable nuclides. However, not all of the nuclides within this band are stable. For example, argon ($Z = 18$) has stable isotopes with mass numbers of 36, 38, and 40, while those with mass numbers of 37 and 39 are unstable.
6. None of the elements beyond bismuth ($Z = 83$) have any stable isotopes. Two other elements, technetium, Tc ($Z = 43$), and promethium, Pm ($Z = 61$), also have no stable isotopes.

Table 20.1 Numbers of Protons and Neutrons in the Stable Nuclides

Number of Protons	Number of Neutrons	Number of Stable Nuclides	Examples
even	even	168	$^{12}_{6}C$, $^{72}_{32}Ge$
even	odd	57	$^{47}_{22}Ti$, $^{67}_{30}Zn$
odd	even	50	$^{35}_{17}Cl$, $^{63}_{29}Cu$
odd	odd	4	$^{2}_{1}H$, $^{14}_{7}N$

These and other observations concerning the properties of atomic nuclei have helped nuclear physicists develop a quantum theory for nuclear particles. There are specific allowed energy states that the nucleons may have, which account for the magic numbers and the emissions that are produced by unstable nuclides.

Radioactivity

Radioactivity is the emission of small particles and electromagnetic radiation by the nucleus.

Some nuclei are *radioactive;* that is, they spontaneously emit small particles and high-energy electromagnetic radiation, forming more stable nuclei. By studying the effect of an electric field on the rays emitted by radioactive material (see Figure 20.2), Ernest Rutherford and Paul Villar identified three kinds of emissions from unstable nuclei. The first three letters of the Greek alphabet are used to refer to the different kinds of radiation (alpha, α; beta, β; and gamma, γ rays). The quantitative studies of nuclear radiation showed that the **alpha particles** are high-energy helium-4 nuclei, **beta particles** consist of high-energy electrons that originate from the nucleus, and **gamma rays** are very short wavelength (high energy) electromagnetic radiation. The origin of the gamma rays is similar to that of the light observed in an atomic emission spectrum. An atom in an excited electronic state can return to the ground state by emission of a photon of light (Section 6.2). Similarly, when a

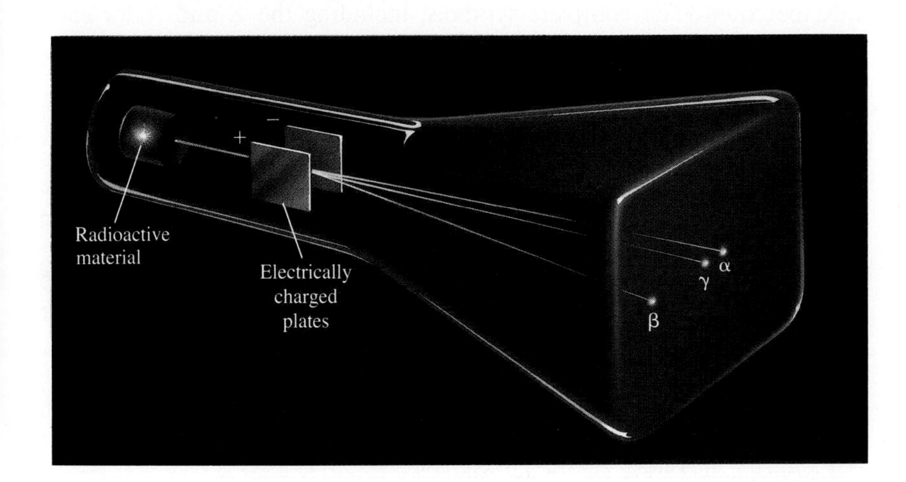

Figure 20.2 Electrical deflection of nuclear radiation The emissions from a radioactive material are classified as α, β, and γ rays depending on their mass and behavior in an electric field. Alpha rays are deflected toward the negative plate; beta rays are attracted toward the positive plate; and gamma rays are not deflected. More quantitative study shows that the alpha rays are helium-4 nuclei, beta rays are electrons, and gamma rays are high energy electromagnetic radiation.

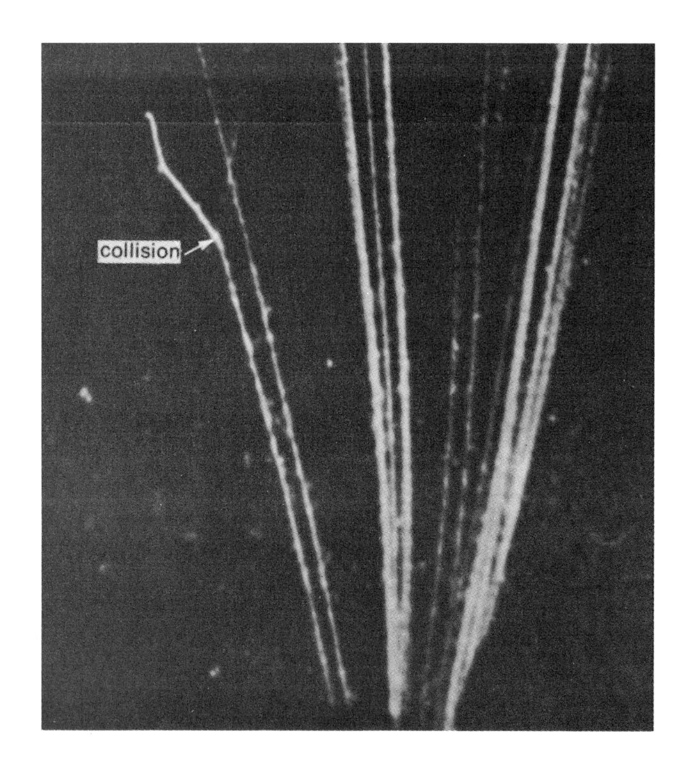

Cloud chamber When radiation passes through a supersaturated gas, the ionization produced causes small droplets of liquid to form, leaving a visible trail. The white lines are the trails left by alpha particles. On the left side, the trail divides as the result of a nuclear reaction that occurred between an alpha particle and a nitrogen atom. The cloud chamber has been used to study radiation since the original one was built by C. T. R. Wilson in 1911.

radioactive nuclide decomposes, it will leave some of the nucleons in excited nuclear states. Gamma rays carry away the energy released as the nucleons return to the ground-state nuclear configuration.

Table 20.2 identifies the particles that make up the nucleus and the kinds of rays that are emitted when unstable nuclides decompose. The mass numbers and atomic numbers used for these particles in writing nuclear equations are included in the table.

A **nuclear equation** is used to describe the radioactive decay of unstable nuclides. A nuclear equation, just as a chemical equation, must be balanced. In a nuclear equation the sum of the mass numbers (mass balance) and the sum of the atomic numbers (charge balance) on the two sides of the equation must be equal. The complete symbols, including the Z and A for each

Table 20.2 Nuclear Particles and Those Involved in Radioactive Decays

Particle	Charge	Mass	Symbol(s)
proton	1	1	${}^{1}_{1}H$, ${}^{1}_{1}p$
neutron	0	1	${}^{1}_{0}n$
alpha particle	2	4	${}^{4}_{2}He^{2+}$, ${}^{4}_{2}\alpha$
beta particle	−1	0	${}^{0}_{-1}e$, ${}^{0}_{-1}\beta$
gamma ray	0	0	${}^{0}_{0}\gamma$
positron (β^{+})	1	0	${}^{0}_{1}e$, ${}^{0}_{1}\beta$

particle, are used. For example, uranium-238 emits an alpha particle to form thorium-234. The nuclear equation is

$$^{238}_{92}U \longrightarrow ^{234}_{90}Th + ^{4}_{2}He$$

In this equation the 2+ charge of the alpha particle was not included, because in nuclear equations there is no need to keep track of the charges on the particles. The equation shows that the neutral uranium atom contains 92 electrons, and the neutral thorium atom contains 90 electrons, so the atom loses two electrons when the nuclear decay occurs. The two electrons freed by a thorium atom are eventually captured by an alpha particle when it slows down, forming a neutral helium-4 atom. A charge balance is always met in any nuclear reaction, so it is not necessary to complicate the nuclear equations with ionic charges.

Any nuclear decay can be represented by a nuclear equation in which mass numbers and atomic numbers of the particles are balanced.

Carbon-14 is a radioactive isotope formed in the outer fringes of the atmosphere by a nuclear reaction (see Section 20.2) of nitrogen-14 with neutrons from outer space. Carbon-14 emits a beta particle when it decomposes.

$$^{14}_{6}C \longrightarrow ^{14}_{7}N + ^{0}_{-1}\beta$$

Note that we use 0 and −1, respectively, for the mass and charge of the beta particle, representing those quantities relative to the values of 1 that are possessed by the proton. The net effect of beta emission is to convert a neutron in the original nuclide into a proton in the product nucleus.

$$^{1}_{0}n \longrightarrow ^{1}_{1}p + ^{0}_{-1}\beta$$

It is important to recognize that the nucleus does not contain electrons, even though some unstable nuclides emit them (as beta particles). The beta particle is created at the instant of its emission from the nucleus. Beta emission lowers the neutron-to-proton ratio in the nucleus, so this type of decay occurs for isotopes that are above or near the top side of the zone of stability shown in Figure 20.1.

Beta decay occurs when the neutron-to-proton ratio in a nucleus is too large for stability, and increases the atomic number of the atom by 1.

Two types of radioactive decays related to beta emission are observed—positron emission (or β^+ emission) and electron capture. Both of these processes have the net effect of converting a proton in the original nuclide into a neutron in the product nuclide.

$$^{1}_{1}p \longrightarrow ^{1}_{0}n + ^{0}_{1}\beta$$

$$^{1}_{1}p + ^{0}_{-1}e \longrightarrow ^{1}_{0}n$$

A **positron** is a particle that has properties identical to those of the electron, except that its charge is positive. Potassium-40 is a radioactive isotope that decomposes by positron emission. The nuclear equation for this decay is

$$^{40}_{19}K \longrightarrow ^{40}_{18}Ar + ^{0}_{1}\beta$$

The positron is an example of antimatter. When a positron collides with its antiparticle, the electron, the entire mass of both particles is converted into two photons of electromagnetic radiation (gamma rays), in a process called **annihilation.**

$$^{0}_{1}\beta + ^{0}_{-1}\beta \longrightarrow 2\,^{0}_{0}\gamma$$

Some unstable nuclides can capture one of the electrons in the atom, usually from the 1*s* subshell, converting one of the protons in the nucleus into a neutron. For example,

$$^{44}_{22}\text{Ti} + ^{0}_{-1}\text{e} \longrightarrow ^{44}_{21}\text{Sc}$$

When **electron capture** occurs, the product atom is in an excited *electronic* state, since the capture of an electron leaves a vacancy in the 1*s* subshell. The atom returns to its ground electronic state by the emission of x rays. The electromagnetic radiation observed from electron capture is called x rays, rather than gamma rays, because it arises from the electronic transitions outside the nucleus. Generally gamma radiation has higher energy than x rays, but there is some overlap of the energy ranges of these two sources of electromagnetic radiation.

Both positron emission and electron capture increase the neutron-to-proton ratio in the nucleus.

Both positron emission and electron capture occur with nuclides in which the neutron-to-proton ratio is below or near the bottom edge of the zone of stability shown in Figure 20.1. All nuclides that decay by positron emission can also decay by electron capture, but the reverse is not true.

Example 20.1 Balancing Nuclear Equations for Radioactive Decays

Write nuclear equations for the following radioactive decays: (a) beta decay of radium-228, (b) alpha decay of polonium-210, (c) electron capture by lanthanum-137.

Solution

(a) From the periodic table we find that the atomic number of Ra is 88. The beta particle has A = 0 and $Z = -1$. We can write part of the nuclear equation as

$$^{228}_{88}\text{Ra} \longrightarrow ? + ^{0}_{-1}\beta$$

The missing nuclide must have $Z = 89$ and $A = 228$, to balance the atomic numbers and mass numbers, respectively. The element with $Z = 89$ is actinium, so the complete equation is

$$^{228}_{88}\text{Ra} \longrightarrow ^{228}_{89}\text{Ac} + ^{0}_{-1}\beta$$

(b) Polonium has an atomic number of 84, and Z for the alpha particle is 2, so the product nucleus has $Z = 82$, which is the element lead, Pb. The product nucleus must have a mass number that is four less than the unstable polonium nuclide, so $A = 206$. The nuclear equation is

$$^{210}_{84}\text{Po} \longrightarrow ^{206}_{82}\text{Pb} + ^{4}_{2}\text{He}$$

(c) Lanthanum has $Z = 57$, so we can write the partial equation as

$$^{137}_{57}\text{La} + ^{0}_{-1}\text{e} \longrightarrow ?$$

The product nucleus has $A = 137 + 0 = 137$ and $Z = 57 - 1 = 56$. The element with an atomic number of 56 is Ba, so the complete equation is

$$^{137}_{57}\text{La} + ^{0}_{-1}\text{e} \longrightarrow ^{137}_{56}\text{Ba}$$

Understanding Write the nuclear equation that forms tin-117 by a positron emission.

Answer: $^{117}_{51}\text{Sb} \longrightarrow ^{117}_{50}\text{Sn} + ^{0}_{1}\beta$

Use of Isotopes as Tracers

The chemical behavior of different isotopes of the same element is nearly identical. When we write a chemical equation, the mass numbers of the elements are not included, since the same reaction occurs for all of the different isotopes. Chemists use this fact to follow the fate of particular atoms in chemical reactions. Consider the oxidation of hydrogen peroxide by permanganate ion in an acidic water solution. The chemical equation for this decomposition is

$$5H_2O_2(aq) + 2MnO_4^-(aq) + 6H^+(aq) \longrightarrow 2Mn^{2+}(aq) + 5O_2(g) + 8H_2O(\ell)$$

There are three possible sources of the elemental oxygen produced in the reaction mixture: the water that is the solvent, permanganate ion, and hydrogen peroxide. From the balanced equation it is not possible to know the source of the atoms in the oxygen gas. We can find out where the oxygen atoms came from, by making the hydrogen peroxide from ^{18}O. When the specially prepared hydrogen peroxide is used, we find that all of the gaseous oxygen contains only oxygen-18. From this experimental observation we can be sure that all of the atoms in the gaseous oxygen originated from the hydrogen peroxide.

In the experiment just described, the ^{18}O was used as a tracer, and provides information about the mechanism of the chemical reaction. We conclude that the O—O bond in the hydrogen peroxide was not broken when the chemical change occurred. The use of radioactive isotopes as tracers is an extremely important tool for following particular atoms in a chemical reaction. Since radioactivity is easily detected, the use of unstable isotopes as tracers is very convenient.

An example of the use of a radioactive tracer is the study of the reduction of $[Co(NH_3)_5Cl]^{2+}$ complex by Cr(II) in water solution. The overall reaction is

$$Co(NH_3)_5Cl^{2+}(aq) + Cr(H_2O)_6^{2+}(aq) + 5H_3O^+(aq) \longrightarrow Co(H_2O)_6^{2+}(aq) + Cr(H_2O)_5Cl^{2+}(aq) + 5NH_4^+(aq)$$

The reaction is first order in both the Cr(II) and the cobalt complex. There are several mechanisms that would be consistent with the observed rate law. A sample of the cobalt compound containing radioactive chlorine (^{36}Cl) as a tracer is prepared. The cobalt complex reacts with Cr(II) in a solution that contains free nonradioactive chloride ions. After separation of the reaction products, all of the chloride ions bound to the chromium are found to be radioactive and the free chloride ions are not. Therefore, any satisfactory mechanism must form a bond between the chromium and the chlorine, before the cobalt-chlorine bond breaks. If this were not so, the chloride ions released by the cobalt would mix with those in the solution, so some of the radioactivity would be found in the free chloride ions.

Radioactive tracers can provide valuable information about reaction mechanisms.

Rates of Radioactive Decays

The decay of an unstable nucleus obeys a first order rate law.

$$\text{rate} = \frac{-\Delta N}{\Delta t} = kN \qquad [20.1]$$

The rate is the number of disintegrations per unit time, k is the decay constant, and N is the number of radioactive nuclei present in the sample.

This is identical to the rate law for a first order chemical process, but there is one important difference. The constant k in a nuclear decay does not change with the temperature, whereas the specific rate constant for a chemical reaction does change. Therefore, it is not necessary to control the temperature when measuring the rate of a radioactive decay.

With radioactive isotopes, it is common to express the rate of decay by the half-life (the length of time it takes for one half of the atoms in the sample to undergo nuclear disintegration) instead of the specific rate constant k. As we saw in Chapter 16, the half-life of a first order process is related to the rate constant by the equation

$$t_{1/2} = \frac{0.693}{k} \qquad [20.2]$$

The half-lives for known radioactive nuclides cover a very wide range, from fractions of seconds up to more than 10^{15} years. For reasonably short-lived isotopes, the following integrated rate law can be used to determine the half-life for the radioactive decay.

The half-life of a short-lived radioactive isotope can be determined by measuring the decay rate at two different times.

$$\ln\left(\frac{N}{N_0}\right) = -kt = -\frac{0.693}{t_{1/2}}\,t \qquad [20.3]$$

In Equation 20.3, N is the number of radioactive atoms present at time t, when N_0 atoms were present at $t = 0$.

Suppose a sample containing iodine-131 has a decay rate of 3153 disintegrations per minute at $t = 0$. The decay rate of this sample 52.5 hours later is 2613 disintegrations per minute. What is the half-life of ^{131}I? The solution to this problem involves both Equations 20.1 and 20.3. The disintegration rate of the sample is proportional to the number of radioactive atoms present at the time of measurement (Equation 20.1), so the ratio N/N_0 in Equation 20.3 may be replaced by R/R_0, (rate/rate$_0$), which is 2613/3153. Solving for the half-life, we find

$$\ln\left(\frac{2613}{3153}\right) = -\frac{0.693}{t_{1/2}}\,52.5\text{ hr}$$

$$t_{1/2} = -\frac{0.693 \times 52.5\text{ hr}}{\ln(0.8287)} = 194\text{ hr}$$

Example 20.2 Finding Half-Life from Decay Data

The nuclide ^{31}Si is a radioactive isotope that disintegrates by beta decay.

$$^{31}_{14}Si \longrightarrow {}^{31}_{15}P + {}^{0}_{-1}\beta$$

A sample of silicon-31 is found to produce 251 disintegrations per second. Exactly 3.00 hours later the same sample produces 113 disintegrations per second. What is the half-life of ^{31}Si?

Solution Since the disintegration rate at any time is proportional to the number of radioactive atoms,

$$\frac{N}{N_0} = \frac{R}{R_0} = \frac{113}{251} = 0.450$$

Substitute this value and the time, 3.00 hours, into Equation 20.3, and solve for $t_{1/2}$.

$$\ln(0.450) = -\frac{0.693}{t_{1/2}} 3.00 \text{ hr}$$

$$t_{1/2} = -\frac{0.693 \times 3.00 \text{ hr}}{-0.798} = 2.60 \text{ hr}$$

This value for the half-life is reasonable, since slightly more than one half of the atoms decayed in 3.00 hours.

Understanding The radioactive decay of a sample containing ^{41}Ar produces 555 disintegrations per minute. The rate decreases to 314 disintegrations per minute exactly 90.0 minutes later. Calculate the half-life of ^{41}Ar.

Answer: 110 min

The integrated rate law cannot be used to determine the half-life of a very long-lived isotope such as ^{36}Cl, which has a half-life of 3.1×10^5 years. The change in the decay rate of such an isotope is too small to measure accurately in any reasonable period of time. We can use Equation 20.1 to measure the half-life of an isotope that decays very slowly, as illustrated in Example 20.3.

The half-life of a long-lived isotope is determined from the disintegration rate of a sample containing a known number of radioactive atoms.

Example 20.3 Determining Half-Life from Decay Rate

A sample of CO_2 was prepared from carbon that had been enriched in ^{14}C. The mass spectrum of the CO_2 shows that a 0.100-gram sample contains 6.23×10^{-12} mol of ^{14}C. The absolute disintegration rate of the 0.100-g sample is 863 disintegrations per minute. Express the half-life of ^{14}C in years.

Solution From the disintegration rate of the sample, we know how many atoms undergo decay each minute. Equation 20.1 may be used to find the decay constant k, after finding the number of ^{14}C atoms present in the sample. The number of moles of ^{14}C in this sample is used to find the number of atoms.

$$6.23 \times 10^{-12} \text{ mol} \left(\frac{6.02 \times 10^{23} \text{ atoms } ^{14}\text{C}}{1 \text{ mol}}\right) = 3.75 \times 10^{12} \text{ atoms } ^{14}\text{C}$$

Equation 20.1 is used to find k.

$$\text{rate} = \frac{863 \text{ atoms}}{1 \text{ min}} = k \times 3.75 \times 10^{12} \text{ atoms}$$

$$k = \frac{863 \text{ atoms min}^{-1}}{3.75 \times 10^{12} \text{ atoms}} = 2.30 \times 10^{-10} \text{ min}^{-1}$$

The half-life is obtained from the rate constant by using Equation 20.2.

$$t_{1/2} = \frac{0.693}{2.30 \times 10^{-10} \text{ min}^{-1}} = 3.01 \times 10^9 \text{ min}$$

This value is easily converted into years.

$$t_{1/2} = 3.01 \times 10^9 \text{ min} \left(\frac{1 \text{ hr}}{60 \text{ min}}\right)\left(\frac{1 \text{ day}}{24 \text{ hr}}\right)\left(\frac{1 \text{ year}}{365 \text{ day}}\right) = 5.73 \times 10^3 \text{ years}$$

Understanding A sample that contains 4.50×10^{14} atoms of a radioactive isotope decays at a rate of 503 disintegrations per minute. What is the half-life (in years) of this isotope?

Answer: 1.18×10^6 years

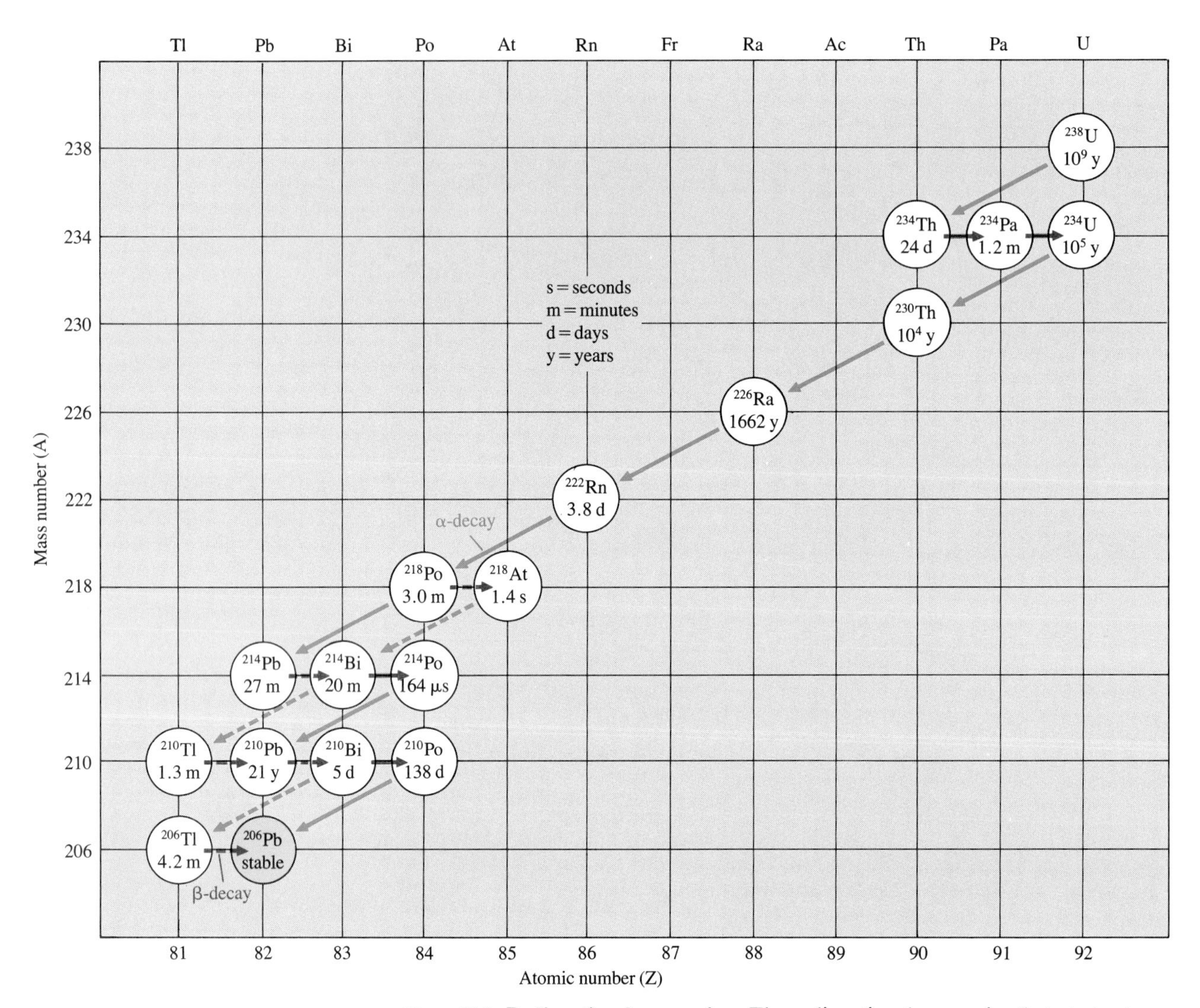

Figure 20.3 Radioactive decay series The radioactive decay series that starts at uranium-238 and concludes with lead-206 is shown. The half-life of each atom is given in the circles. The blue arrows are alpha decays, and red arrows represent beta decays. Note that the slowest decay rate is that of ^{238}U, so nearly all of the uranium that has decayed will be present in the sample as ^{206}Pb.

Radioactive Series

In a number of cases the radioactive decay of a nuclide produces another unstable nuclide that also undergoes decay. With the heavy elements, several decays occur in sequence until a stable nuclide is produced. One such series of decays begins with ^{238}U, finally producing the stable isotope ^{206}Pb after several α and β decays have occurred. The ^{238}U series is shown in Figure 20.3. We can determine the number of α and β decays that occur in such a series if the parent nuclide and the final stable isotope are known, as shown in Example 20.4.

From the change in mass number and atomic number in a decay series, we can calculate the number of α and β decays that occur.

Example 20.4 Analysis of a Decay Series

Determine the number of alpha and beta decays needed to change ^{238}U into ^{206}Pb.

Solution For each alpha particle emitted, the mass number decreases by 4, while a beta decay does not change the mass number. In the present series the change in mass number is

$$\Delta A = 206 - 238 = -32$$

To balance the mass numbers in the decay series, a total of eight (32/4) alpha particles must be emitted.

Once the number of alpha particles is known, we turn our attention to the change in Z. In the present series

$$\Delta Z = 82 - 92 = -10$$

Each alpha particle emitted must reduce the atomic number by two, while a beta emission increases the atomic number by one. We can express this in an equation

$$\Delta Z = \text{number of beta} - 2 \times \text{number of alpha}$$

$$-10 = \text{number of beta} - 2 \times 8$$

$$\text{number of beta} = 16 - 10 = 6$$

The nuclear equation for the overall change is

$$^{238}_{92}U \longrightarrow {}^{206}_{82}Pb + 8\,^{4}_{2}He + 6\,^{0}_{-1}\beta$$

Of course, it is impossible to determine from the given data the order in which the eight alpha and six beta decays occur. Counting the number of alpha and beta decays in Figure 20.3 confirms that our solution is correct.

Understanding Another natural decay series begins with ^{235}U, finally producing the stable isotope ^{207}Pb. Find the number of alpha and beta decays that occurred in this transformation.

Answer: $7\ \alpha$ and $4\ \beta$

Dating Artifacts by Radioactivity

The constant half-lives of radioactive isotopes have been widely used by archaeologists and geologists to determine the ages of objects. One of these techniques, radiocarbon dating, is useful in determining the age of organic matter. Atmospheric carbon dioxide contains a very small amount of ^{14}C, a

The Shroud of Turin This piece of linen was purported to be the burial shroud of Jesus Christ, with an image of his crucified body. On the left is part of the shroud as it appears to the eye; the negative on the right has been enhanced to bring out the image more clearly. In recent years several scientific studies of the shroud have been conducted, among which was ^{14}C dating. The dating experiments showed that the flax from which the linen was made had been grown between 1260 and 1390 AD. On the basis of these and other results, there is no possibility that the shroud existed at the time of Christ.

beta emitter with a half-life of 5730 years (see Example 20.3). A sample of atmospheric carbon dioxide produces about 15.3 disintegrations per minute per gram of carbon. Plants and animals incorporate this radioactive isotope within their cells during their lifetimes. When an organism dies, it no longer exchanges carbon with the atmosphere, and the ratio of $^{14}C/^{12}C$ decreases because of the radioactive decay. If we assume that the ^{14}C content of the atmosphere has remained constant for the past several thousand years, then the age of a sample can be determined, as illustrated in Example 20.5.

Example 20.5 Radiocarbon Dating

An artifact found in an ancient Egyptian tomb produced 11.8 disintegrations of ^{14}C per minute per gram of carbon in the sample. Estimate the age of this sample, assuming its original radioactivity was 15.3 disintegrations per minute per gram of carbon.

Solution Since the rate of disintegration of carbon-14 is known at two different times, Equation 20.3 may be used with the known half-life of the unstable isotope.

$$\ln\left(\frac{11.8}{15.3}\right) = -\frac{0.693}{5730 \text{ years}} t$$

$$t = -\frac{5730 \text{ years} \times \ln(0.771)}{0.693} = 2.15 \times 10^3 \text{ years}$$

Understanding What is the age of a bone found by an archaeologist, if its carbon-14 activity was 2.9 disintegrations per minute per gram of carbon?

Answer: 1.4×10^4 years

The ages of some uranium ore deposits have been estimated by another dating technique. We have seen (Example 20.4) that ^{238}U decays, finally producing ^{206}Pb, a stable isotope. The slowest decay in this series is that of ^{238}U with a half-life of 4.51×10^9 years (see Figure 20.3), so nearly all of the uranium that has decayed is present in the sample as ^{206}Pb. Assuming no lead was present in the sample when the deposit formed, the ratio of $^{206}Pb/^{238}U$ can be used to determine the age of the mineral deposit. In using this technique, a number of corrections must be made, and tests run to verify the assumptions. The use of this method is illustrated in Example 20.6.

The half-life of radioactive nuclides provide a means of determining the age of a sample.

Example 20.6 Using Pb/U Ratio for Dating

The mass spectrum of a rock sample shows that it contains 4.40 grams of ^{238}U and 1.21 grams of ^{206}Pb. Assuming that all of the lead was produced from radioactive decay of uranium-238, how old is the rock? (The half-life of ^{238}U is 4.5×10^9 years.)

Solution We must first calculate the original mass of the ^{238}U in the sample by assuming that each 206 g of lead now present in the sample required the decay of 238 g of uranium. The original mass of uranium-238 was

$$\text{original mass} = 4.40 \text{ g U} + 1.21 \text{ g Pb}\left(\frac{238 \text{ g U}}{206 \text{ g Pb}}\right) = 5.80 \text{ g U}$$

Thus during the lifetime of the rock, the original 5.80 g of the uranium decreased to 4.40 grams today. Using these quantities in the integrated rate law (Equation 20.3), along with the half-life of ^{238}U, we can solve for the age of the rock.

$$\ln\left(\frac{4.40}{5.80}\right) = -\frac{0.693}{4.5 \times 10^9 \text{ yr}} t$$

$$t = -\frac{4.5 \times 10^9 \text{ yr}}{0.693}(-0.276) = 1.8 \times 10^9 \text{ years}$$

Understanding The ratio of $^{40}Ar/^{40}K$ can also be used for dating of minerals, since ^{40}Ar is a stable nuclide.

$$^{40}_{19}K \longrightarrow {}^{40}_{18}Ar + {}^{0}_{1}\beta \qquad t_{1/2} = 1.28 \times 10^9 \text{ years}$$

Analysis of a rock showed that the atom ratio of $^{40}Ar/^{40}K$ was 0.44. What is the age of the sample?

Answer: 6.7×10^8 years

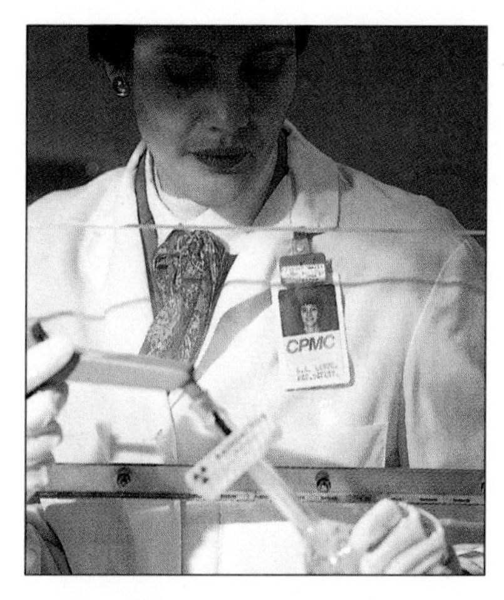

A film badge Film badges like the one shown are worn by workers exposed to radiation. The badges are periodically developed to determine the amount of radiation exposure of the worker.

Detection of Radioactivity

Nuclear radiation can be detected in a number of ways. All of these methods depend on the ability of high-energy particles to cause ionization of atoms and molecules. Historically, the exposure of photographic film by radioactive materials was the first means used to detect radioactivity, and it is still used by people in the nuclear industry. The workers wear film badges that are periodically developed to measure their exposure to radiation.

The Geiger counter, shown schematically in Figure 20.4, is an electrical device that is used to measure radiation. Most of the portable devices used by radiation safety officers work on the same principle. A Geiger counter consists of a metal tube with an insulated wire in the center. The tube is filled with a gas at low pressure, and a high voltage is applied between the center wire and the outer tube. Alpha and beta particles enter the tube through a thin mica window and ionize the gas inside the tube. The ions and the electrons cause the gas to conduct electricity for a brief instant, and a pulse of electric current is amplified and detected. One electrical pulse is produced for each alpha or beta particle that enters the tube.

The detection and measurement of radioactivity depend on the radiation ionizing atoms and molecules.

Some substances emit visible light as a result of the ionization caused by radioactivity. In a scintillation counter, shown in Figure 20.5, each particle of ionizing radiation produces a small burst of light, which is converted to an electrical signal by the photoelectric effect (see Section 6.1). In many scintillation counters each radioactive particle produces an electrical signal that is proportional to the energy of the particle that caused it.

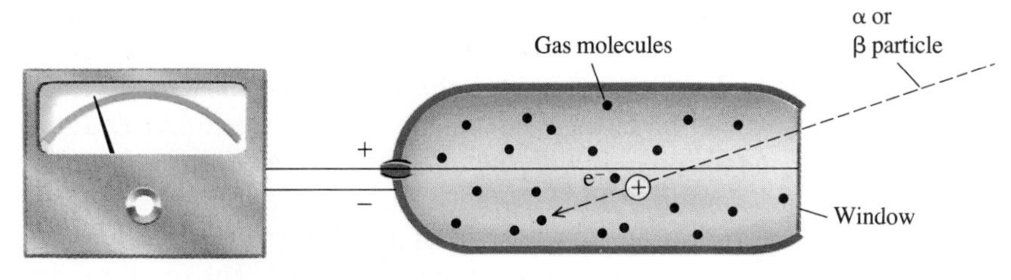

Figure 20.4 Schematic representation of a Geiger counter Radiation enters the Geiger tube and produces electrons and ions in the gas. A pulse of current is produced for each alpha or beta particle that enters the tube.

Figure 20.5 A scintillation counter Each radioactive particle causes a burst of light, which is converted to an electrical signal that is amplified. The electrical signal is often proportional to the energy of the radiation.

20.2 Nuclear Reactions

Objective

- To describe nuclear transmutation by writing a nuclear equation

Natural spontaneous radioactive decay is only one kind of nuclear reaction that can occur. Most of the radioactive isotopes of the elements do not occur naturally on earth. In 1919 Ernest Rutherford observed that the bombardment of nitrogen with alpha particles produced protons. The nuclear reaction that occurred was

$$^{14}_{7}N + ^{4}_{2}He \longrightarrow ^{17}_{8}O + ^{1}_{1}p$$

Another nuclear reaction resulted in the discovery of the neutron in 1932 by James Chadwick, when he bombarded beryllium with alpha particles.

$$^{9}_{4}Be + ^{4}_{2}He \longrightarrow ^{12}_{6}C + ^{1}_{0}n$$

These two reactions are examples of **nuclear transmutations,** reactions in which two particles or nuclei produce elements or isotopes that are different from the reactant species.

Although the first transmutation of elements involved the reactions of alpha particles with other nuclei, this is usually a very difficult change to achieve. Little attraction exists among nuclear particles when the nucleons are much more than 10^{-12} cm apart. To reach this small distance, the alpha particle must have a very high energy to overcome the electrostatic repulsion by the nucleus, since both the alpha particle and the nucleus have positive charges. The energy required to get an alpha particle close enough to the beryllium nucleus to react is about 10^{-12} J, which corresponds to about 6×10^{8} kJ/mol, an enormous energy.

For the elements with high atomic numbers, much higher energies are needed to induce nuclear reactions by bombardment with positively charged particles. Several instruments have been made that accelerate protons, alpha particles, and other small nuclei to these high energies. In one of these, the cyclotron, an alternating voltage is applied between two hollow D-shaped electrodes. The particles are accelerated by the electric field when they are in the gap between the two electrodes. A magnetic field perpendicular to the electrodes causes the ions to follow a spiral of increasing radius as their velocities increase. A schematic diagram for a cyclotron is shown in Figure 20.6.

Another device used to accelerate positive ions, the linear accelerator, is shown schematically in Figure 20.7. Positive ions are introduced into the first tube. The voltage accelerates the ions between the first and second tubes. By the time these ions leave the second tube, the voltage has been reversed and they are accelerated further between the second and third tubes. Each successive tube is longer than the previous one, so the particle arrives at the next gap just in time to coincide with the maximum accelerating voltage. In 1958 the linear accelerator at the Lawrence Radiation Laboratory in California was used to produce the first sample of nobelium ($Z = 102$) by bombarding a sample of curium with high-energy ^{12}C nuclei. The reaction that occurred was

Nuclear reactions may be induced by high-energy charged particles produced by an accelerator.

$$^{246}_{96}Cm + ^{12}_{6}C \longrightarrow ^{254}_{102}No + 4^{1}_{0}n$$

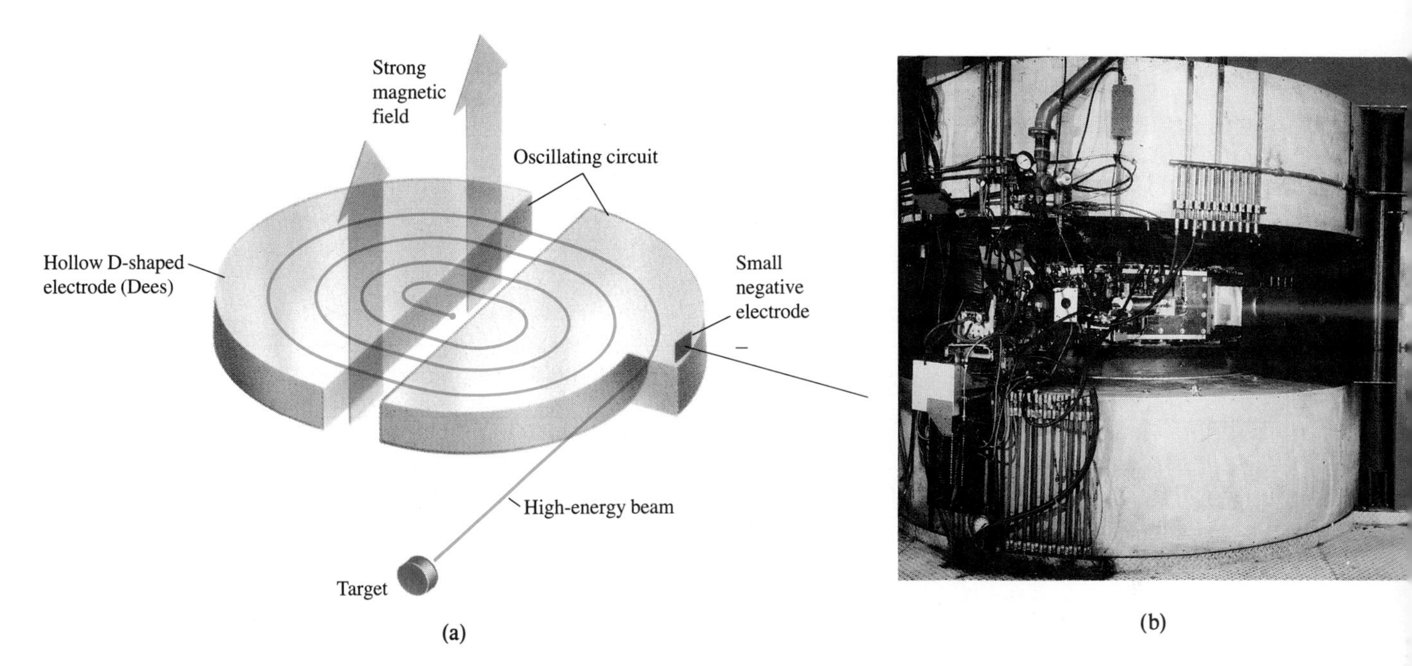

Figure 20.6 The cyclotron The cyclotron is used to accelerate protons and other small, positively charged particles to very high energies. An alternating voltage is applied to the two electrodes, while a magnetic field is used to produce the spiral path of the accelerated particles. (a) A schematic diagram of the cyclotron. (b) A modern cyclotron. The beam of high energy protons produced is blue.

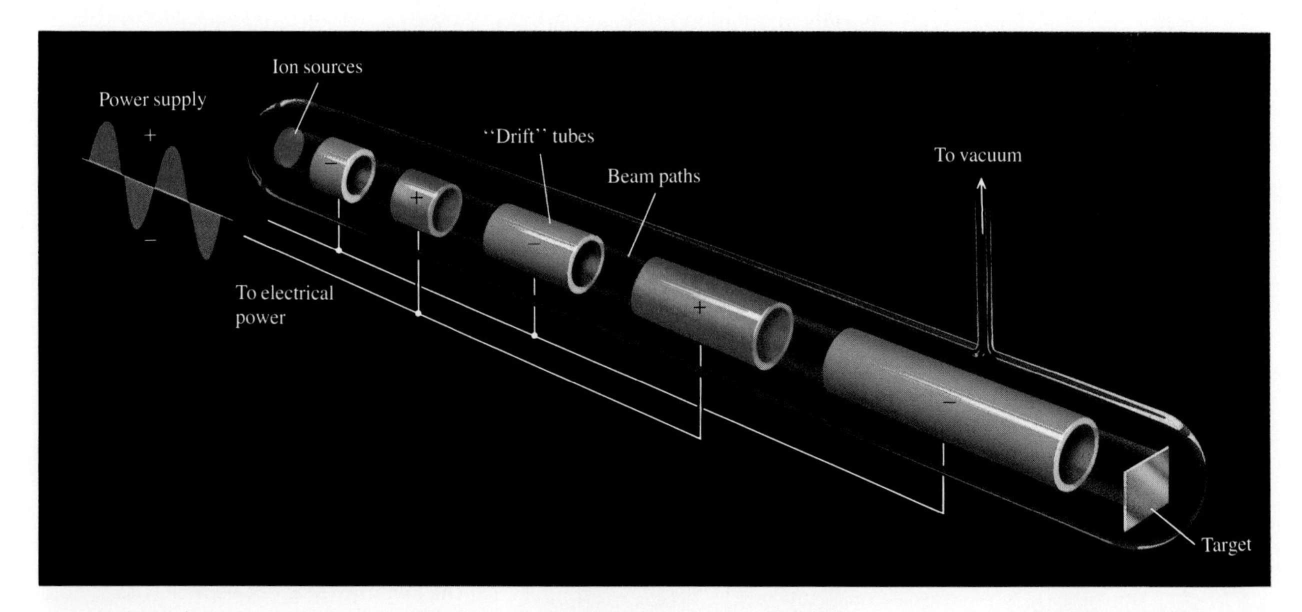

Figure 20.7 The linear accelerator Positive ions are introduced into the first tube. By the time these ions leave the first tube, the voltage has been reversed and they are accelerated further between the first and second tubes. Very high ion energies are achieved by repeating this process several times. The electrical power consumption by a linear accelerator is so great that its use is often restricted to avoid peak hours.

When neutrons are used as the bombarding particles, there is no electrostatic repulsion barrier, so nuclear transmutations occur at much lower energies. Essentially all nuclides react with neutrons, so most radioactive elements are produced by the reaction of neutrons with nuclei. Some examples of neutron-induced nuclear reactions are

$$^{35}_{17}\text{Cl} + ^{1}_{0}\text{n} \longrightarrow ^{36}_{17}\text{Cl} + ^{0}_{0}\gamma$$

INSIGHTS into CHEMISTRY

A Closer View

Quarks: These Subatomic Particles Hold the Key to Understanding Nuclear Forces

For most chemists it is sufficient to consider the proton, neutron, and electron as the basic components of atoms. Theoretical and nuclear physicists who are interested in explanations of the origin of nuclear forces have found evidence for a vast number of other subatomic particles. Many of these particles were initially proposed by theoreticians, as they devised models for the structure of matter. The theoretical models provided the impetus for the design of some new experiments, in which very high energy particles collide. Special accelerators are needed to produce particles with energies much greater than are obtained with a conventional cyclotron or linear accelerator. The aerial photograph shows the largest accelerator operating in the United States.

Today scientists believe that nuclear particles are composed of *quarks.* Six types of quarks have been identified: *up* and *down; charmed* and *strange; top* and *bottom.* While investigations continue, most scientists believe there are no undiscovered types of quarks.

The heavier nuclear particles, protons and neutrons, are made from *up* and *down* quarks. *Up* quarks have a charge equal to +2/3; *down* quarks have a charge equal to −1/3. In this model a proton is composed of two *up* quarks and one *down* quark, so the net charge on a proton is +1. A neutron contains two *down* quarks and one *up* quark, giving a net charge of zero. Other particles such as neutrinos, mesons, and muons can also be described as combinations of quarks.

Quarks and the Composition of Protons and Neutrons

Proton		Neutron	
Type of Quark	Charge	Type of Quark	Charge
up	+2/3	*down*	−1/3
up	+2/3	*down*	−1/3
down	−1/3	*up*	+2/3
Net charge	+1		0

The Fermi National Accelerator This particle accelerator is a distant relative of the cyclotron and was dedicated in 1978, in Batavia, Illinois. The circumference of this proton accelerator is about 4 miles, and it can produce particle energies of 10^9 eV (10^{11} kJ mol).

$$^{10}_{5}B + ^{1}_{0}n \longrightarrow ^{7}_{3}Li + ^{4}_{2}He$$

$$^{14}_{7}N + ^{1}_{0}n \longrightarrow ^{14}_{6}C + ^{1}_{1}H$$

The last reaction is quite important, since it is the natural source of the ^{14}C that is used in radiocarbon dating (Section 20.1). Example 20.7 illustrates the use of nuclear equations to describe the reactions of small particles with other nuclei.

Current knowledge of quarks and other subatomic particles reveals an impressive symmetry. The effects of five of the six quarks have been seen experimentally. (The particle itself is not observed, but the result of the interaction of the particle with other matter or with energy can be detected.) The effect of the *top* quark has not yet been seen. Even the largest particle accelerator in the world, the CERN collider in Switzerland, does not produce particles with enough energy to provide direct evidence for its existence. However, calculations provide a good estimate of the energy that is needed to find the *top* quark, and such an accelerator is under construction. The superconducting supercollider (SSC) will be located near Dallas, Texas. The total cost will be about 8 billion U.S. dollars; much of it is to be contributed by countries other than the United States. The detection of the *top* quark is one of the most important objectives of the SSC's research. More accurate measurement of the properties of known subatomic particles is another major goal of the facility.

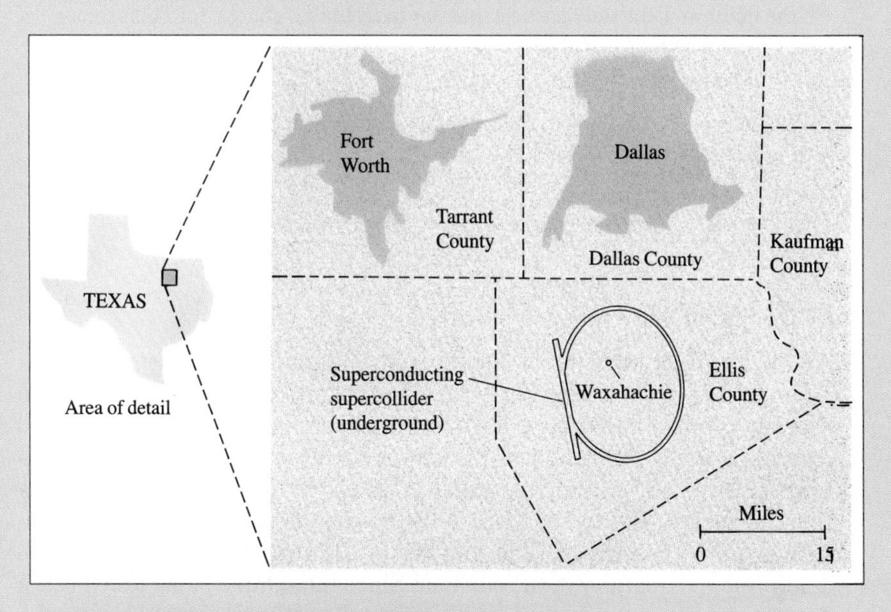

The superconducting supercollider This accelerator is under construction near Waxahachie, Texas. Its circumference is 53 miles, and it will produce collisions with energies up to 4×10^{12} eV.

Example 20.7

In each part, write a complete balanced nuclear equation.

(a) A neutron transforms ^{7}Be into ^{7}Li.
(b) A neutron reacts with ^{17}O to form ^{14}C.
(c) ^{69}Ga reacts with a neutron and emits a γ ray.
(d) ^{238}U is bombarded with ^{16}O ions, forming ^{250}Fm.

Solution

(a) On the reactant side of the equation, the sum of the mass numbers is $7 + 1 = 8$, and that of the atomic numbers is $4 + 0 = 4$. The missing product must have a mass number of 1 and an atomic number of 1, to have a balanced nuclear equation. The missing product is a proton (hydrogen ion).

$$^{7}_{4}\text{Be} + ^{1}_{0}\text{n} \longrightarrow ^{7}_{3}\text{Li} + ^{1}_{1}\text{H}$$

(b) To balance the mass numbers in the equation, the other product must have a mass number of $17 + 1 - 14 = 4$. The atomic number is $8 + 0 - 6 = 2$. The particle with A = 4 and $Z = 2$ is an alpha particle, so the complete equation is

$$^{17}_{8}\text{O} + ^{1}_{0}\text{n} \longrightarrow ^{14}_{6}\text{C} + ^{4}_{2}\text{He}$$

(c) For the gamma ray, both the mass number and the atomic number are equal to 0. Thus, the product isotope has the same atomic number as the reacting nucleus, and is a gallium atom ($Z = 31$); the mass number increases by one ($Z = 70$). The balanced nuclear equation is

$$^{69}_{31}\text{Ga} + ^{1}_{0}\text{n} \longrightarrow ^{70}_{31}\text{Ga} + ^{0}_{0}\gamma$$

(d) Balancing the mass numbers in the partial equation shows that 4 mass units are missing from the product side, while the atomic numbers are in balance. Since the neutron is the only particle that has mass but no charge, four neutrons must be produced in this reaction.

$$^{238}_{92}\text{U} + ^{16}_{8}\text{O} \longrightarrow ^{250}_{100}\text{Fm} + 4^{1}_{0}\text{n}$$

Understanding Write the balanced nuclear reaction for the bombardment of ^{56}Fe with protons to form ^{56}Co.

Answer: $^{56}_{26}\text{Fe} + ^{1}_{1}\text{p} \longrightarrow ^{56}_{27}\text{Co} + ^{1}_{0}\text{n}$

Activation Analysis

An interesting application of nuclear transmutations in chemistry is a technique known as **activation analysis:** the exposure of a sample to a neutron beam to produce radioactive isotopes, which are then used to identify and determine the composition of the sample. After exposure to the neutron beam, the radiation from the sample is measured. From the kind of radiation, its energy, and the half-life of the decay, the presence of specific elements can be confirmed. The number of disintegrations is then used to determine the quantity of the element. Since the sample being tested is not destroyed (only a very few atoms are activated), activation analysis is particularly desirable for analyzing valuable objects such as paintings. In addition, the method is very sensitive, so it is a powerful technique for detecting trace impurities in a sample. As little as 10^{-15} g of some elements can be detected with neutron activation analysis.

Neutron activation analysis is a very sensitive method that uses nuclear transmutation to identify the elements present in a sample.

One recent application of neutron activation analysis occurred in 1991. The body of Zachary Taylor was exhumed and subjected to analysis for arsenic, because some historians thought his death may have been caused by arsenic poisoning. Small samples of his hair and fingernails were irradiated with neutrons at the Oak Ridge National Laboratory. The nuclear reaction that is induced by neutrons is

$$^{75}_{33}\text{As} + ^{1}_{0}\text{n} \longrightarrow ^{76}_{33}\text{As} + ^{0}_{0}\gamma$$

The unstable ^{76}As emits a beta particle with a half-life of 26.3 hours. The conjecture that Zachary Taylor had died of arsenic poisoning was disproved by the absence of detectable quantities of arsenic in the samples studied by neutron activation analysis.

Less commonly, irradiation with other particles, such as deuterons and protons, has been used for activation analysis. One procedure for analysis of carbon in steel uses the reaction

$$^{12}_{6}\text{C} + ^{2}_{1}\text{H} \longrightarrow ^{13}_{7}\text{N} + ^{1}_{0}\text{n}$$

The ^{13}N decays by positron emission with a half-life of 10 minutes, and is easily measured.

Zachary Taylor In 1991 the body of Zachary Taylor (1784–1850), the twelfth President of the United States, was exhumed. A 2.7-mg sample of his hair and a 12.3-mg sample of fingernails were subjected to neutron activation analysis to determine the presence of arsenic. On the basis of the results of this analysis, it was concluded that he had not died from arsenic poisoning as some historians suspected.

20.3 Nuclear Binding Energy

Objectives

- To calculate the mass defect of a nuclide
- To relate nuclear binding energy to the mass defect

The energy released by nuclear weapons and in nuclear power plants is testimony to the tremendous energies that are involved in the formation and disintegration of the nuclei of atoms. Einstein's theory of relativity shows that mass is equivalent to energy through the equation

$$E = mc^2 \qquad [20.4]$$

where c is the speed of light (2.9979×10^8 m/s), m is the mass in kilograms, and the energy E is in joules. According to Einstein's relationship, for any exothermic process the products have a smaller mass than the reactants. For any chemical change, the difference in mass that is predicted is much smaller than the uncertainty in the measurement of the mass. For example, consider the combustion of methane

$$CH_4(g) + 2O_2(g) \longrightarrow CO_2(g) + 2H_2O(\ell) \qquad \Delta H^\circ = -890 \text{ kJ}$$

The mass change we calculate is less than 1×10^{-8} g, when 80 g of reactants are converted into the products. Thus, the law of conservation of mass for any ordinary chemical change is valid within experimental accuracy.

When nuclear changes take place, the energy changes are much larger and measurable differences in mass take place.

The amount of energy required to keep the protons and neutrons bound together in any nuclide, overcoming the coulombic repulsions among the positive charges, is called the **nuclear binding energy.** The source of this

Table 20.3 Masses of the Hydrogen-1 Atom and the Neutron

Particle	Symbol	Mass	
		u	grams
hydrogen-1 atom*	$^{1}_{1}H$	1.007825	1.67353×10^{-24}
neutron	$^{1}_{0}n$	1.008665	1.67496×10^{-24}

* Since the measured mass of an atom includes Z electrons, we must use the mass of the hydrogen-1 atom (one proton and one electron) to calculate the mass defect.

Nuclear binding energy is large enough to be detected as a difference in mass between the nucleons and the nucleus.

energy is the conversion of some of the mass of the nucleons (Equation 20.4). Suppose we measure the mass of an atom, and then compare it to the masses of an equivalent number of hydrogen-1 atoms and neutrons (Table 20.3). In every case (except the hydrogen-1 atom itself), the mass of the atom is less than the mass of the individual particles in it. The difference is known as the **mass defect.** The larger the mass defect, the greater is the nuclear binding energy.

Using the information in Table 20.3, the mass defect of ^{4}He is calculated from its atomic mass of 4.002602 u as follows:

$$\begin{aligned}
&\text{mass of } 2\ {}^{1}\text{H} = 2 \times 1.007825 &&= 2.015650 \text{ u} \\
&\text{mass of } 2\ {}^{1}\text{n} = 2 \times 1.008665 &&= \underline{2.017330 \text{ u}} \\
&\text{sum of nucleons} &&= 4.032980 \text{ u} \\
&\text{mass of } {}^{4}\text{He} &&= \underline{4.002602 \text{ u}} \\
&\text{difference (mass defect)} &&= 0.030378 \text{ u}
\end{aligned}$$

The mass defect is used to calculate the nuclear binding energy.

The difference in mass is used to find the nuclear binding energy:

$$E = \Delta mc^2 \qquad [20.5]$$

Example 20.8 Calculating Nuclear Binding Energy from Mass Defect

What is the energy change when one mole of ^{4}He is made from fundamental particles?

Solution The mass defect of one mole of ^{4}He is 0.030378 g, or 3.0378×10^{-5} kg. Using this molar mass defect in Equation 20.5, we get

$$E = (3.0378 \times 10^{-5}\ \text{kg})(2.9979 \times 10^{8}\ \text{m/s})^2 = 2.730 \times 10^{12}\ \text{J}$$

This energy is roughly ten million times the energy of a typical covalent bond. A mass of 54 *tons* of natural gas would have to be burned to produce the energy that is released when one mole (4 g) of helium is formed from the nucleons.

Understanding The mass defect for ^{16}O is 0.1369 u. Express the nuclear binding energy for this nucleus in kJ/mol.

Answer: 1.230×10^{10} kJ/mol

Nuclear binding energies are extremely large, so a different energy unit, the megaelectron volt (MeV), has traditionally been used to express these energies:

$$1\ \text{MeV} = 1.602 \times 10^{-13}\ \text{J}$$

A convenient equivalency for the direct conversion of atomic mass units into MeV is

$$1 \text{ u} = 931.5 \text{ MeV}$$

The nuclear binding energy for the ^4He nucleus, expressed in MeV, is

$$\Delta E = 0.030378 \text{ u} \left(\frac{931.5 \text{ MeV}}{1 \text{ u}}\right) = 28.30 \text{ MeV}$$

In any nuclear change, the number of nucleons is conserved, so the binding energies of nuclei are usually compared by calculating the binding energy per nucleon.

In any nuclear transformation, the number of nucleons (protons plus neutrons) does not change.

Example 20.9 Calculating Binding Energy per Nucleon

The atomic mass of ^{40}Ca is 39.96259 u. What is the binding energy per nucleon (in MeV) in this atom?

Solution From the atomic and mass numbers of this nuclide, we know that the nucleus contains 20 neutrons and 20 protons. The mass defect is calculated as

$$\Delta m = (20 \times 1.007825) + (20 \times 1.008665) - 39.96259 = 0.36721 \text{ u}$$

The binding energy per nucleon is then found by converting this mass defect into MeV, and dividing by the 40 nucleons that compose the nucleus.

$$\text{binding energy} = \left(\frac{0.36721 \text{ u}}{40 \text{ nucleons}}\right)\left(\frac{931.5 \text{ MeV}}{1 \text{ u}}\right) = 8.551 \frac{\text{MeV}}{\text{nucleon}}$$

Understanding What is the binding energy per nucleon (in MeV) for ^{16}O, which has a mass defect of 0.1369 u?

Answer: 7.970 MeV/nucleon

Nuclear explosion The explosion of a thermonuclear bomb is dramatic testimony to the vast energy stored in the atomic nucleus.

Since the number of nucleons does not change in nuclear reactions, the binding energy per nucleon is a measure of nuclear stability. Figure 20.8 shows a graph of the binding energy per nucleon versus mass number. This plot shows that the most stable nuclei are in the region of ^{56}Fe. The nuclides with mass numbers greater than 56 and those less than 56 have smaller binding energies per nucleon. This curve suggests that the nucleus is a possible source of great energy, if nuclei of intermediate mass are formed from

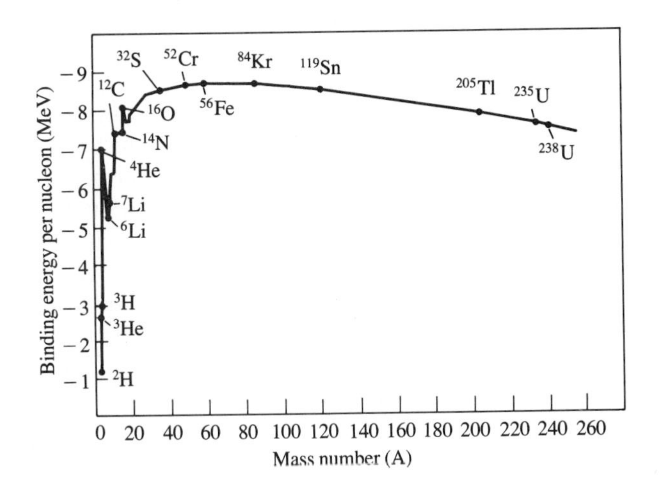

Figure 20.8 Binding energy per nucleon The binding energy per nucleon is shown as a function of the mass number of the nuclides. The nuclei at the top of the curve are the most stable.

either lighter or heavier nuclides. The following section explains the nuclear processes that take advantage of these changes in binding energy per nucleon to release very large quantities of energy.

20.4 Fission and Fusion

Objectives

- To interpret the energy released in nuclear fission and fusion reactions from the nuclear binding energy curve in Figure 20.8
- To distinguish among critical, subcritical, and supercritical conditions for fission
- To describe a typical fusion reaction with a nuclear equation

Fission Reactions

In 1938 Otto Hahn and Fritz Strassman detected the presence of radioactive barium among the products formed when uranium was bombarded with neutrons. Other studies led to the conclusion that ^{235}U had captured a neutron to form ^{236}U, which split into two smaller nuclei, according to the nuclear equation

$$^{236}_{92}U \longrightarrow ^{141}_{56}Ba + ^{92}_{36}Kr + 3^{1}_{0}n \quad [20.6]$$

This was the first example of **nuclear fission,** the splitting of a heavy nucleus into two nuclei of comparable size. The energy released in this fission is about 6.8×10^7 kJ/gram of uranium.

On August 2, 1939, Albert Einstein, at the request of several other scientists, sent a letter to President Franklin Roosevelt. In that letter he urged that the United States begin research to develop a bomb based upon the energy released from nuclear fission, so that Germany would not develop one first. As a result of Einstein's letter, the Manhattan Project was begun for the purpose of developing the atomic bomb. The success of this top-secret research effort was demonstrated when the first fission bomb was tested on July 16, 1945, in New Mexico. On August 6 and 9 of that year similar bombs were used by the United States against Japan. These are the only two thermonuclear devices that have ever been used in war.

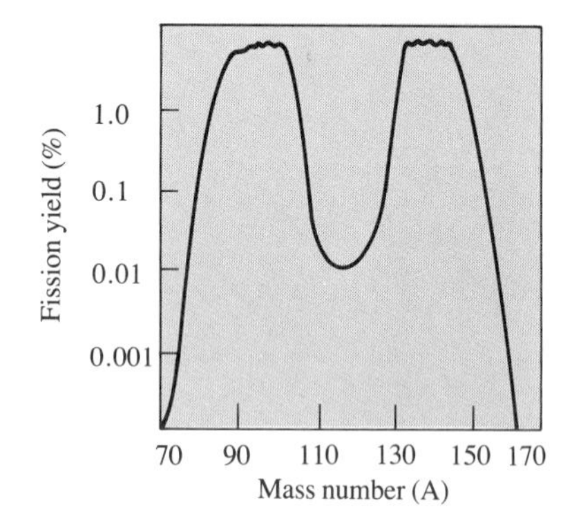

Figure 20.9 Fission yield curve The distribution of fission products from the neutron-induced fission of ^{235}U. If two fragments of equal size were formed, they should have mass numbers of about 116. Most of the fission products fall in the two mass ranges of 84 to 105 and 130 to 150.

When fission of a heavy element occurs, many products are formed. More than 370 product nuclides, with mass numbers ranging from 72 to 161, have been observed from the neutron-induced fission of ^{235}U. Figure 20.9 shows a fission yield curve for this reaction. For reasons not fully understood, fission is unsymmetrical, with the mass number of one product nuclide about 1.4 times the mass number of the other nuclide. The determination of fission yield curves requires the chemical separation of the products, by procedures not greatly different from those used in the qualitative analysis exercises in most general chemistry laboratories.

Each fission event releases about 200 MeV, equivalent to 80 million kilojoules per gram of ^{235}U consumed. By way of contrast, the energy released by the detonation of TNT, a very powerful chemical explosive, is only 16 kilojoules per gram, so a nuclear fission produces about 5 million times as

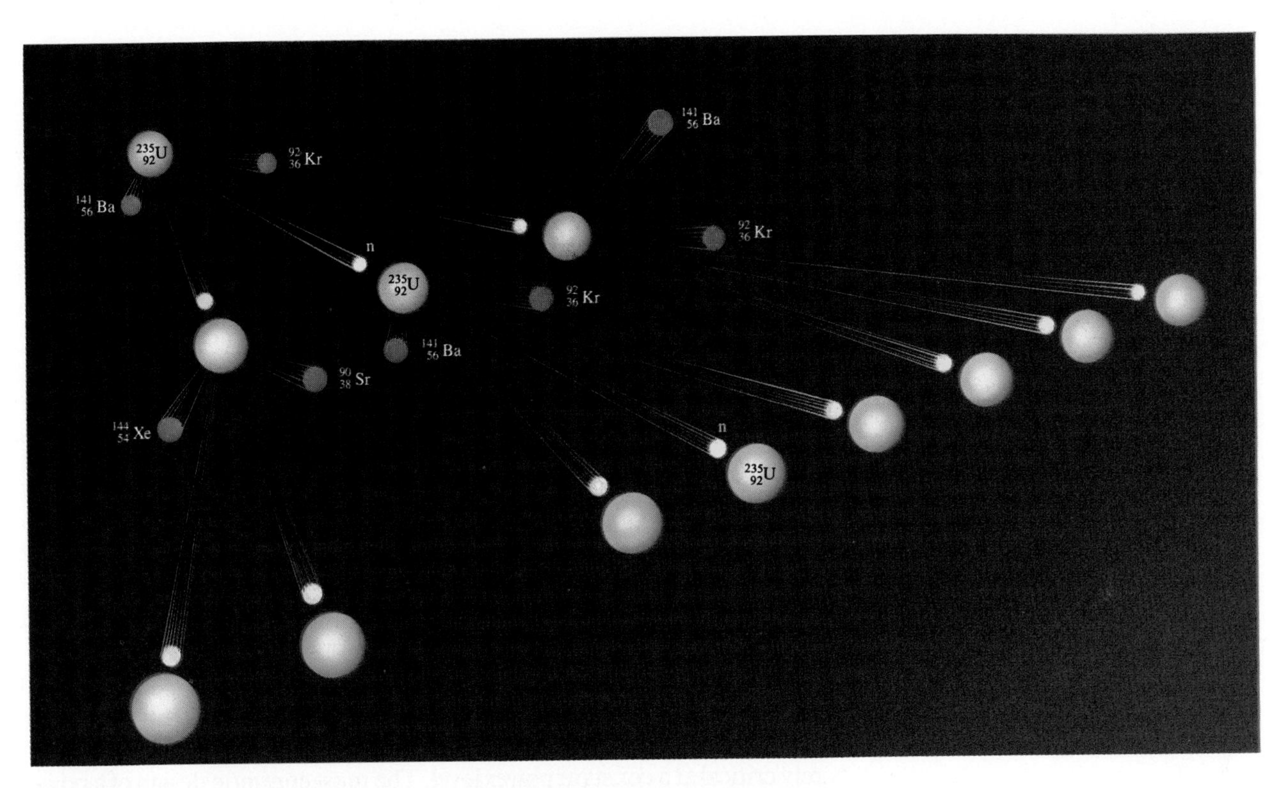

Figure 20.10 The chain reaction of ^{235}U A single neutron induces a fission that produces three neutrons that induce fissions to produce eight more neutrons, and so forth. In the 20th generation of this chain, 3.5 billion fissions occur. This rapidly expanding chain of events led to the awesome explosion that occurred over Japan on August 6, 1945.

much energy per gram of fuel consumed as does a chemical reaction. Often the energies of nuclear explosives are expressed in megatons (millions of tons) of chemical explosives.

Nuclear fission releases about 5 million times the energy of a chemical reaction.

On an average, each nuclear fission of ^{235}U produces 2.5 neutrons, each of which may induce another fission if it is absorbed by another ^{235}U atom. Thus a single fission event may start a **chain reaction,** leading to many more nuclear fissions. If each of the neutrons formed initiates another fission, then a rapidly expanding chain of events occurs, as illustrated in Figure 20.10.

The diagram in Figure 20.10 assumes that every neutron produced by a fission induces another fission. If the size of the sample is small, many of the neutrons will escape from the sample before reacting with another nucleus to initiate another fission. The chain reaction is said to be *critical* when, on an average, exactly one of the neutrons produced by one fission is absorbed to induce a second fission, thus causing a self-sustaining reaction. The reaction is *supercritical* if each fission event induces more than one fission, and is *subcritical* when less than one fission is caused by each fission that occurs. The **critical mass** is the size of the sample of fissionable matter needed for the

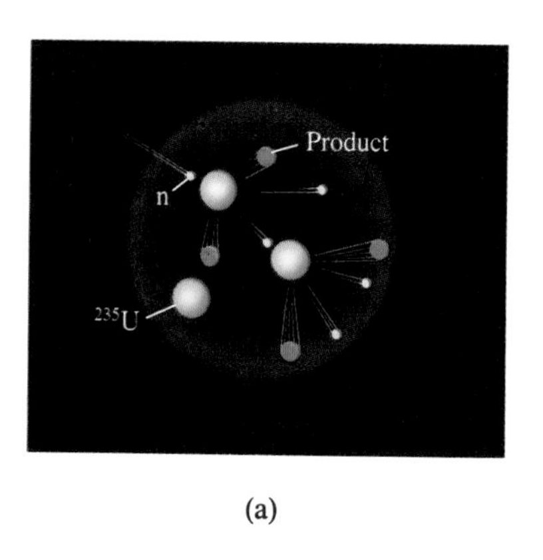

(a)

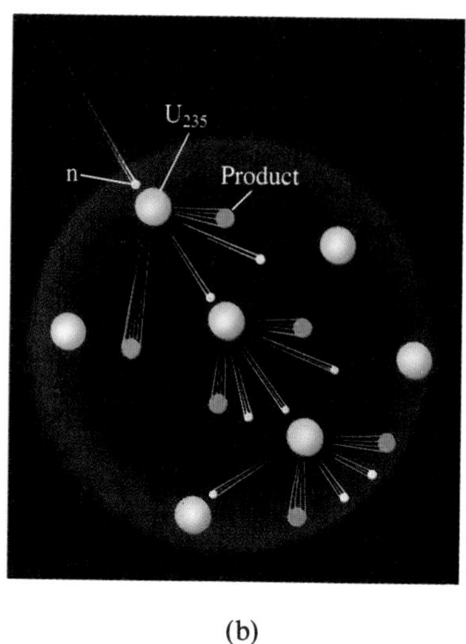

(b)

Figure 20.11 Subcritical and critical nuclear fission (a) When most of the neutrons produced by the fission events escape from the sample, the chain reaction is subcritical and is not self-sustaining. (b) The chain reaction is critical when each fission induces exactly one additional fission.

chain reaction to be critical. Figure 20.11 schematically illustrates the conditions of subcritical and critical nuclear fission.

In the first nuclear bombs, two subcritical masses of fissionable material were forced together to form a supercritical mass (see Figure 20.12). In a nuclear power reactor, the chain reaction is carefully controlled to remain barely critical at a constant power level. The most common design of a power reactor is shown in Figure 20.13. The fuel rods are made from uranium oxide that has been enriched in ^{235}U. These fuel rods are shaped so that nearly all of the neutrons produced by fission in one fuel rod escape into the surrounding water. The heavy water, D_2O, that fills the space between the fuel rods is the *moderator,* a material that slows down the high-energy neutrons. Slow neutrons are more efficiently captured by fissionable nuclei than are those with high energy. Some of these neutrons reach other fuel rods in the reactor core, where they produce additional fissions. The control rods consist of a material, such as cadmium or boron, that absorbs neutrons very efficiently. When the control rods are inserted all the way into the core, the reactor is subcritical, because almost all of the neutrons are absorbed by the control rods and are not available to initiate another fission. During operation of the reactor, the control rods are used to ensure that the reactor does not become supercritical. In a power reactor, the energy released by the fission events produces steam that is used to operate turbines that produce electricity.

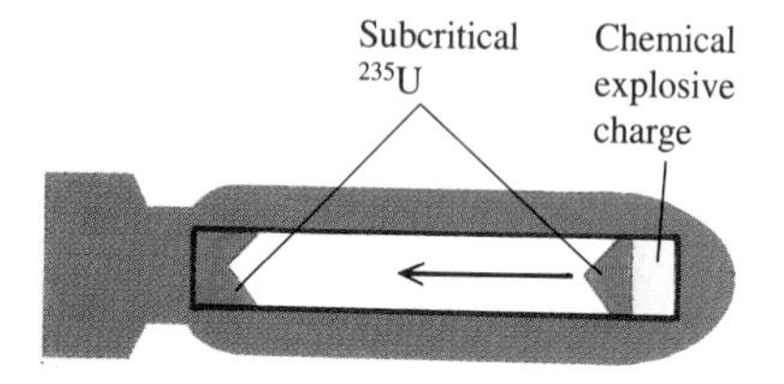

Figure 20.12 Schematic of a fission bomb Two subcritical masses of fissionable material are forced together by chemical explosives to initiate the nuclear explosion.

Natural uranium is mostly ^{238}U (99.3%), and the remaining 0.7% of the atoms are ^{235}U. Since only the ^{235}U undergoes neutron-induced fission, reactor fuels are enriched to about 3.5% in the lighter isotope by gaseous diffusion of UF_6 (see Section 5.8). A typical nuclear reactor that supplies 1000 mW of electricity consumes more than a ton of ^{235}U in one year of operation. There are more than 340 such nuclear reactors in operation today.

At one time the small abundance of ^{235}U led to concerns that fissionable material might be depleted in a relatively short time. The **breeder reactor,** which generates more fissionable atoms than it consumes, provides a means

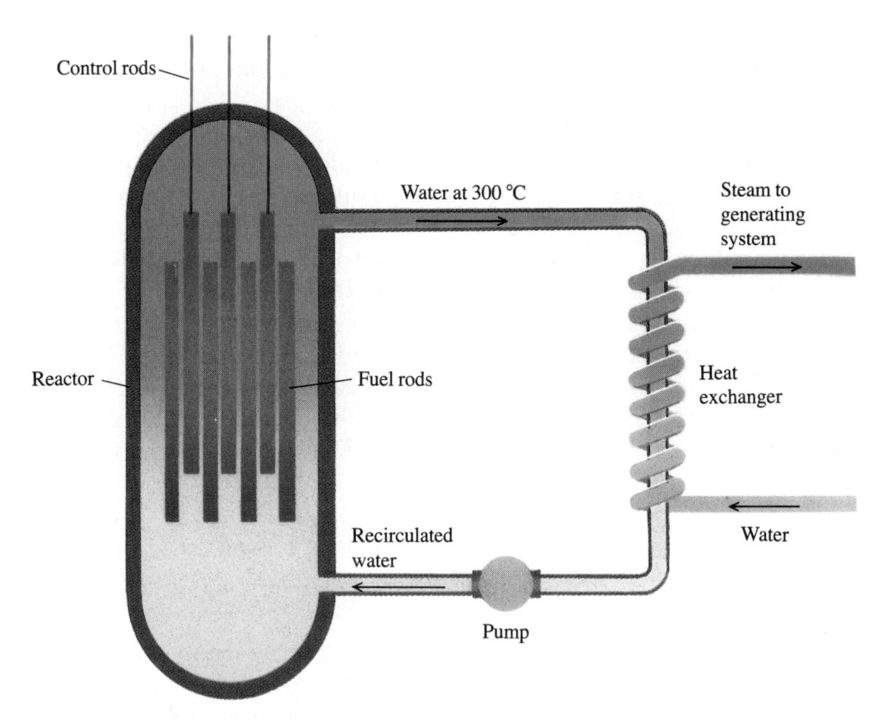

Figure 20.13 Schematic diagram of a nuclear reactor During operation the control rods absorb sufficient numbers of neutrons to maintain the chain reaction at a critical level. The output of the reactor must be constantly monitored, to assure that the reactor does not become supercritical.

of increasing the supply of nuclear fuel by a factor of 100 or more. While ^{235}U is the only naturally occurring nucleus that undergoes fission, many of the man-made elements also undergo fission when they absorb neutrons. One of these, ^{239}Pu, is generated from ^{238}U in the following nuclear changes:

$$^{1}_{0}n + ^{238}_{92}U \longrightarrow ^{239}_{92}U \xrightarrow[\beta]{24\text{ min}} ^{239}_{93}Np \xrightarrow[\beta]{2.3\text{ d}} ^{239}_{94}Pu$$

A breeder reactor is designed so that a fraction of the neutrons produced by fission is captured by ^{238}U to generate ^{239}Pu, which is isolated and used to fuel other nuclear reactors. Thus, the nonfissionable ^{238}U nuclei are converted into the fissionable ^{239}Pu.

Nuclear Power and Safety

It may appear that fission reactors offer an abundant supply of the energy needed by modern society, but there are serious concerns related to nuclear power generation. Compared to the generation of electricity by burning fossil fuels, nuclear power plants are quite "clean." Nuclear power plants do not introduce ash, smoke, sulfur oxides, nitrogen oxides, and carbon dioxide into the atmosphere, as do conventional power plants. As we have seen in previous chapters, these waste products from conventional fuels cause pollution of the environment and make substantial contributions to such problems as acid rain and the greenhouse effect.

Nuclear power, however, is not without undesirable side products. The nuclides produced from fission reactions all have neutron-to-proton ratios that are quite high, and lie above the band of nuclear stability (Figure 20.1). Consider the products of the fission reaction given in Equation 20.6, ^{141}Ba and ^{92}Kr, which have n/p ratios of 1.52 and 1.56, respectively. The only stable nuclide with a mass number of 141 (^{141}Pr) has a neutron-to-proton ratio of 1.39, and the two stable nuclides with mass number 92, ^{92}Mo and ^{92}Zr, have neutron-to-proton ratios of 1.19 and 1.30 respectively. The high neutron-to-proton ratios of fission products mean that nearly all of them are radioactive, decaying by a series of beta emissions toward the zone of stability in Figure 20.1. The beta decay chains from ^{92}Kr and ^{141}Ba, along with the half-lives of the nuclides, are

$$^{92}_{36}Kr \xrightarrow{3.0\ s} {}^{92}_{37}Rb \xrightarrow{5.3\ s} {}^{92}_{38}Sr \xrightarrow{2.7\ h} {}^{92}_{39}Y \xrightarrow{3.5\ h} {}^{92}_{40}Zr$$

$$^{141}_{56}Ba \xrightarrow{18\ m} {}^{141}_{57}La \xrightarrow{3.9\ h} {}^{141}_{58}Ce \xrightarrow{33\ d} {}^{141}_{59}Pr$$

Thus the waste products from nuclear reactors are highly radioactive. Because the half-lives of many other fission products are fairly long (for example, ^{90}Sr has $t_{1/2} = 28.1$ years), the fission product waste from nuclear reactors remains dangerously radioactive for years and even centuries. A solution to the long-term problem of safe storage of the radioactive products from reactors must be found. At present, the most promising solution to nuclear waste disposal is the incorporation of the radioactive products into glasses or ceramics to immobilize them, followed by burial deep under ground.

Furthermore, many people fear disastrous accidents in the operation of nuclear reactors, which could release radioactive nuclides into the surroundings. Two incidents of reactor accidents have received wide publicity. In 1979 a cooling system failure led to the release of radioactive gases into the atmosphere at the Three Mile Island reactor in Pennsylvania. A second and more serious accident occurred in a nuclear reactor at Chernobyl in the Soviet Union in 1986. Subsequent investigation of both of these incidents showed that they resulted from avoidable operator errors. Concerns about the safety of nuclear reactors and the disposal of the radioactive waste they generate has deterred the construction of additional nuclear power reactors in the United States for the past several years.

All means of meeting the demand for power by our modern technological society involve a certain amount of risk, in the form of environmental contamination and potential loss of life. The advantages and disadvantages of nuclear power, fossil fuels, solar power, and other energy sources must be considered carefully and rationally in planning for our future energy needs.

Nuclear Fusion

The binding energy curve in Figure 20.8 suggests that **nuclear fusion,** the combination of two light nuclei to form a larger one, is also a source of energy. Fusion is a much cleaner nuclear process than fission, since it does not directly form radioactive products. The neutron-to-proton ratio of the stable elements is essentially constant at 1, up to Ca. Fusion reactions are the source of the energy emitted by the sun and other stars. It is estimated that the sun is about 73% hydrogen and 26% helium. Some of the fusion reactions believed to occur constantly in the sun are:

$$^{1}_{1}\text{H} + ^{1}_{1}\text{H} \longrightarrow ^{2}_{1}\text{H} + ^{0}_{1}\beta \qquad \Delta E = -9.9 \times 10^{7}\ \text{kJ/mol}$$

$$^{1}_{1}\text{H} + ^{2}_{1}\text{H} \longrightarrow ^{3}_{2}\text{He} \qquad \Delta E = -5.2 \times 10^{8}\ \text{kJ/mol}$$

$$^{3}_{2}\text{He} + ^{1}_{1}\text{H} \longrightarrow ^{4}_{2}\text{He} + ^{0}_{1}\beta \qquad \Delta E = -1.9 \times 10^{9}\ \text{kJ/mol}$$

Before a fusion reaction can occur, however, the energy barrier created by electrostatic repulsion must be overcome. It is estimated that temperatures in the range of 10^6 to 10^7 K are needed before fusion reactions can occur. At these high temperatures, all matter exists as a **plasma,** a mixture of free electrons and completely ionized atoms. In the fusion or hydrogen bomb, a fission explosion is used to raise the temperature of light nuclei and to initiate the fusion reaction.

This explosive release of nuclear energy is of no use as an energy source for generation of electrical power. If fusion is to become a useful source of energy, some means of controlled release of the nuclear energy must be developed. Investigations related to the development of a controlled fusion reactor have been in progress in the United States for more than 40 years. The confinement of matter at high pressures and temperatures in excess of 10^7 K is achieved by strong magnetic fields. While fusion has been achieved, the power consumed to produce the conditions needed is more than that yielded by the fusion.

The United States, Russia, Japan, and the European Community are jointly participating in the design of an experimental international thermonuclear reactor, capable of producing 1000 megawatts of thermal power. The magnitude of this project, begun in 1988, is such that operation of the experimental reactor will not begin before the year 2002. If the experimental reactor is successful, it will be many years later before fusion reactors will be in use to generate power.

20.5 Biological Effects of Radiation

Objective

- To describe the factors that influence the effect of radiation on the human body

There is great concern among the general population about the negative effects of radiation on the functions of biological systems, especially the human body. The consideration of radiation effects is complicated by several factors. The different kinds of radiation—alpha particles, beta particles, gamma rays and x rays, and neutrons—have different characteristics in their interaction with matter. Furthermore, the radiation effects depend on whether the source of the radiation is inside or outside the body, and whether the entire body or only part of it is exposed to the radiation. For these reasons the established radiation safety standards involve many qualitative judgments and assumptions.

The SI unit for radioactivity is the *bequerel* (*Bq*), defined as an activity of one disintegration per second. However, the biological damage produced by radiation originating outside the body depends on the depth of penetration of the particles and their energies, as well as the number of particles. Alpha particles are generally stopped by the skin and can produce very little internal damage. Beta radiation can penetrate only about 1 cm below the skin. Gamma rays and x rays are the most penetrating forms of radiation, with a range of 30 cm or more, and can therefore produce damage throughout the interior of the body.

The biological effects of radiation depend on the quantity and type of radiation.

The biological damage caused by radiation is the result of ionization of molecules as the energy is transferred to matter. The physical effects of radiation on matter depend on the amount of energy transferred to the molecules, which is commonly referred to as the radiation dose. The *rad* is defined as the quantity of radiation that transfers 1×10^{-2} joules of energy per kilogram of matter. It has been found that the damage to biological tissues per rad differs, depending on the kind of radiation. For example, one rad of alpha radiation is more damaging than one rad of beta particles. In measuring the biological effects of radiation, it is necessary to multiply the radiation dose by a quality factor, Q, which depends on the type of radiation, as well as other factors. The value of Q is approximately 1 for beta and gamma radiation and 10 for alpha particles. The *effective dose* is the product of the quality factor times the number of rads of radiation.

$$\text{effective dose (in rems)} = Q \times \text{dose (in rads)}$$

The unit for effective dose is called the *rem,* an abbreviation for "roentgen equivalent man." It should be noted that Q is quite variable and depends on

Table 20.4 Estimated Annual Effective Dose in U.S. Population

Source	Dose (millirem)
Natural sources	
cosmic rays	28
terrestrial	28
in the body	39
inhaled radon	~ 200
Human activity	
occupational	0.9
nuclear fuel	0.05
consumer products	5–13
environmental sources	0.06
medical	
diagnostic x rays	39
nuclear medicine	14
total	360

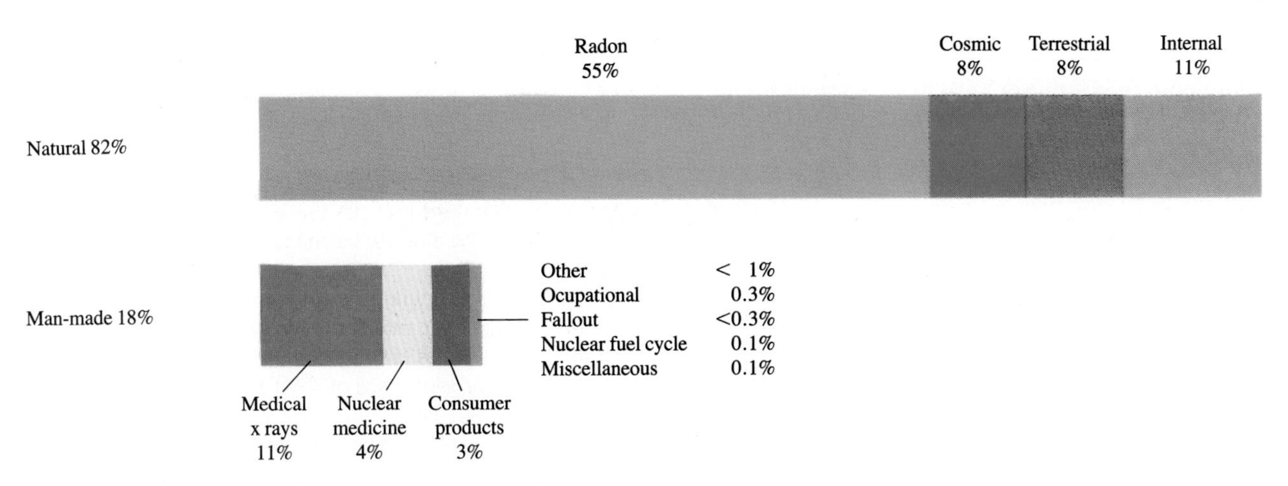

Figure 20.14 **Average radiation exposure** The sources of average radiation exposure of the U.S. population, expressed in percentages. Note that the exposure from natural sources far exceeds that from man-made origins.

the rate of the dosage, the kind of tissue that is exposed, and the total dose received. Table 20.4 presents the sources of radiation and their contributions to the average effective dose of the U.S. population. These data are shown graphically in Figure 20.14, as percentages of the average total exposure.

It should be emphasized that these figures are very approximate, since they depend a great deal upon geographical location, altitude, and occupation. For example, the cosmic ray exposure of a person on a commercial jet liner is well above the average dose from that source at sea level. The exposure to radon depends on the concentration of uranium in the ground, which varies considerably from one location to another.

Radon

As Figure 20.14 shows, a majority of the average radiation exposure in the United States comes from radon, the heaviest of the noble gases. The source of ^{222}Rn in the environment is the decay chain of ^{238}U. Since radon is a noble gas, it escapes from the ground easily and enters the atmosphere. It undergoes alpha decay with a 3.82-day half-life, producing ^{218}Po. If this decay occurs when the radon is in a person's lungs, the polonium (group VIA) remains in the tissue. Within a few days the ^{218}Po that is trapped within the body emits two more alpha particles and two beta particles, causing further radiation damage. Home radon test kits (see Figure 19.37) are sold in some parts of the country, because of the great public awareness and the potential dangers of that gas when it accumulates in homes. The radon generally enters homes through the basement, so one of the most effective means of eliminating the radiation danger is to add ventilation fans in the basement.

Most of the radiation to which the population is exposed comes from natural sources.

(Text continues on p. 892.)

Radiopharmaceuticals Zero in on Target Organs and Enable Images that Improve Health Care Delivery

Dr. Edward A. Deutsch
Mallinckrodt Medical, Inc.

"Nuclear medicine" is listed in Table 20.4 as one of the sources of radiation exposure generated by human activity. In fact, nuclear medicine is a widely practiced discipline that can provide doctors with crucial diagnostic information that is often not attainable by any other means. More than one out of every three patients who stays overnight in a hospital is studied by a diagnostic nuclear medicine procedure. In addition, many nuclear medicine departments also treat patients with radioisotopes which emit beta particles. As a notable example, former President Bush had his overactive thyroid glands treated with radioactive ^{131}I-iodide to reduce the level of hormone production.

In the practice of diagnostic nuclear medicine, a specific chemical form of a gamma-ray-emitting isotope is injected into a patient with the goal of having the isotope localize in a target organ. The chemical species into which the radioisotope is incorporated is known as a radiopharmaceutical. The distribution of the radiopharmaceutical in the patient is then monitored with a gamma-ray camera, which can provide either two-dimensional (planar) or three-dimensional (single-photon emission computed tomography, or SPECT) images. Visualization of the target organ provides information about both its structure and function. While other imaging techniques such as x-ray CT scanning or magnetic resonance imaging (MRI) can provide excellent information about organ *structure,* they are not well suited to assessing organ *function.* Since nuclear medicine imaging tracks how an organ processes a given chemical species (i.e., the radiopharmaceutical), it can provide unique information about the physiological and biochemical state of that organ.

As a simple example, let's consider ^{123}I-iodide. Iodine-123 is a gamma-ray-emitting isotope suitable for use in nuclear medicine, while the chemical species iodide is known to be taken up by the normal thyroid gland. X-ray CT and MRI can tell us the shape of the thyroid gland and might be able to identify the presence of a nodule on the gland but would not be able to tell us if the nodule performs the biological function of taking up iodide. A nuclear medicine scan with ^{123}I-iodide allows this diagnosis, and even more subtle *biochemical* diagnoses to be made quite easily.

In the case of iodide, we are fortunate that the thyroid gland naturally accumulates this chemical species (i.e., iodide exhibits organ specificity for the thyroid). In order to image other organs, we must create radiopharmaceuticals that will exhibit specificity for the target organs. This is essentially a chemical challenge—how can one manipulate the chemistry of an element in order to incorporate it into species that exhibit organ specificity?

Most modern radiopharmaceuticals are based on the isotope technetium-99m (the "m" stands for *metastable,* designating a particular nuclear energy state). Fully 85% of all the diagnostic nuclear medicine studies conducted in the United States employ some chemical form of this isotope. The preeminence of ^{99m}Tc in nuclear medicine stems from three facts.

First, the nuclear properties of ^{99m}Tc are ideally suited to the requirements of diagnostic nuclear medicine. The gamma-ray emission of ^{99m}Tc has an energy of 141 KeV, which is energetic enough to escape from the body easily, but yet is low enough to be captured efficiently by standard scintillation counters and easily stopped by lead shielding. The decay of ^{99m}Tc releases no beta or alpha particles that would generate a high radiation dose to the patient. The physical half-life of ^{99m}Tc is six hours, which is long enough to provide adequate time for the preparation of the radiopharmaceutical and subsequent imaging of the patient but is short enough to minimize the radiation dose to the patient.

Second, ^{99m}Tc is readily available to all clinics and hospitals *via* the commercial ^{99}Mo/^{99m}Tc generator. This generator obviates the need for an on-site nuclear reactor or particle accelerator and thus makes ^{99m}Tc one of the cheapest medical isotopes available. In this generator, ^{99}Mo as the dinegative molybdate anion $[MoO_4]^{-2}$ is sorbed onto an alumina column. The "parent" ^{99}Mo undergoes beta decay with a half-life of 66 hours, which is sufficiently long to allow the generators to be shipped around the world. The "daughter" produced by this beta decay is the mononegative pertechnetate anion $[TcO_4]^-$, which can be simply eluted from the alumina column by washing with sterile 0.15 M NaCl solution. Note that the beta particle emitted from ^{99}Mo carries off the negative charge which is lost when dinegative $[^{99}MoO_4]^{2-}$ is converted into mononegative $[^{99m}TcO_4]^-$.

Third, and probably most important, the element technetium exhibits a wide-ranging and intricate chemistry. Since technetium is a transition metal which is located in the center of the periodic table, it can exist in several oxidation states (from -1 to $+7$); many of these oxidation states exhibit diverse chemistries that allow ^{99m}Tc to be incorporated into a variety of chemical forms. Thus, many classes of chemical species can be constructed about a ^{99m}Tc core, and these chemical species can be engineered to exhibit reasonable organ specificity. One chemical form of ^{99m}Tc can be used to image the bones, another to image the liver, another to image the brain, and so on. It has been the development of the inorganic chemistry of technetium coordination complexes that has paved the way for the generation of a host of new, organ-specific ^{99m}Tc radiopharmaceuticals.

Technetium chemistry is a relatively new discipline. All isotopes of this element (Z = 43) are unstable on a geological time scale, and thus, throughout the early history of chemistry, no technetium was available for study. But with the advent of nuclear power, the relatively stable isotope ^{99g}Tc (where "g" designates *ground state*), a fission product, became available in kilogram quantities. Since about 1979, inorganic chemists have utilized ^{99g}Tc for studies of fundamental aspects of technetium chemistry in the anticipation that the knowledge gained by these basic studies might have direct applications in nuclear medicine. The resulting interactions between inorganic chemistry and nuclear medicine have led to the development of new classes of clinically useful ^{99m}Tc radiopharmaceuticals. Of these, two of the most significant are ^{99m}Tc agents for heart and for brain imaging.

Heart-imaging agents Heart disease is the leading cause of death in the United States. Nearly 1 million Americans die of it each year—38% of all deaths. The number of deaths could be dramatically reduced by screening large populations for evidence of heart disease. In effect, the use of a ^{99m}Tc heart-imaging agent allows a physician to determine which regions of the heart receive sufficient, insufficient, or no blood flow. This information is used to categorize the patient's disease state and prescribe appropriate therapy.

On the basis of the known heart uptake of both inorganic and organic cations, it was proposed in 1980 that cationic complexes of ^{99m}Tc might accumulate in the heart to an extent sufficient to provide gamma-ray images. In 1981 it was first demonstrated that cationic technetium(III) complexes provide acceptable heart images in mongrel dogs. These early studies set the tone and general procedure for developing new technetium complexes as potential heart-imaging agents. First, the complex is synthesized and characterized by the use of macroscopic amounts of ^{99g}Tc; then the preparative chemistry is translated to the very dilute solutions of technetium provided by the ^{99}Mo/^{99m}Tc generator.

Subsequent studies established that the early technetium(III) cations failed as heart-imaging agents, but by 1986 over 14 different cationic ^{99m}Tc complexes had been evaluated in humans, and one of these was sufficiently effective to warrant commercial production. This particular agent is an octahedrally coordinated technetium(I) complex in which the six ligands are carbon-bonded isonitriles: $:C{\equiv}N{-}R$ (R = functional group). Since the ligands are neutral, the technetium(I) complex has a unipositive charge: $[^{99m}Tc(C{\equiv}N{-}R)_6]^+$.

Figure 1 shows a planar, whole-body human image obtained after a normal volunteer has been injected with a typical cationic ^{99m}Tc agent. The heart is visualized in the upper middle of the chest, while the initial activity

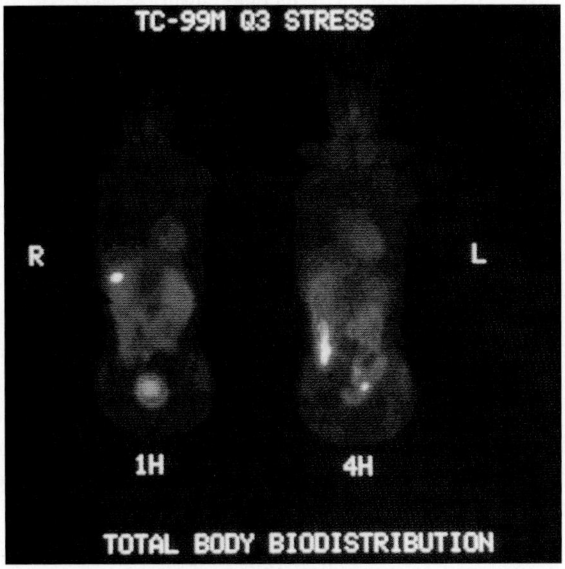

Figure 1 Images obtained with a typical ^{99m}Tc imaging agent.

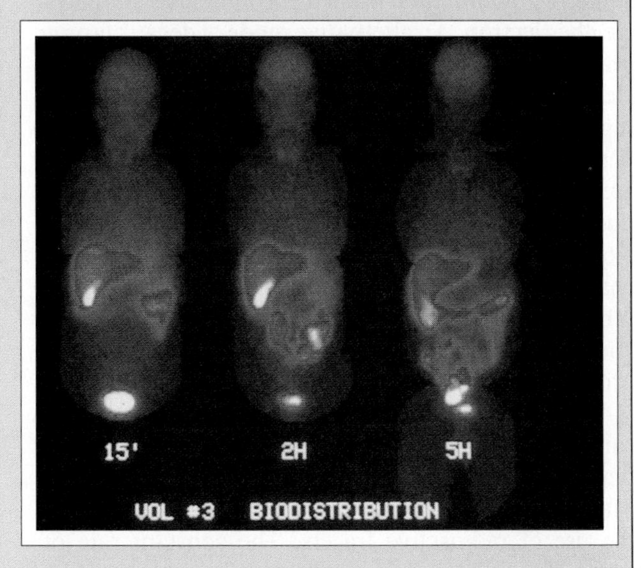

Figure 2 Images obtained with a typical brain-imaging agent.

INSIGHTS into CHEMISTRY

Science, Society, and Technology (Continued)

present in the liver is seen to have cleared through the gallbladder and into the intestines. (Large amounts of radioactivity in the liver would interfere with heart imaging.) The heart activity does not clear with time, so that images can be obtained several hours after injection of the agent into a patient. Thus, a suspected heart attack victim can be injected with the agent upon arrival in the emergency room, and the actual imaging can be delayed until the patient has been treated and stabilized.

Brain-imaging agents Cerebrovascular disease is a major cause of death and disability in the United States. About 500,000 new stroke cases and more than 200,000 deaths attributable to stroke are reported annually. New medical therapies are available for the treatment of stroke victims, but for these therapies to be effective, the presence and extent of a circulation defect must be detected as early as possible. To accomplish this feat, a radiopharmaceutical must traverse the blood-brain barrier (BBB), a tightly connected lining of cells which excludes molecules that are large, charged, or hydrophilic. Moreover, for the radiopharmaceutical to provide the most useful type of diagnostic information, it must remain immobile in the brain long enough to allow three-dimensional SPECT imaging.

In the last few years several ^{99m}Tc brain-imaging radiopharmaceuticals have been developed, and one of them, ^{99m}Tc HM-PAO, is now commercially available. This agent is a small, neutral and lipophilic molecule that can cross through the BBB. Once in the brain, it undergoes rapid decomposition into an unidentified but relatively hydrophilic product that cannot cross the BBB back into the bloodstream. Figure 2 shows a planar, whole-body human image obtained after a normal volunteer has been injected with a typical ^{99m}Tc brain-imaging agent. The brain is readily visualized by the 4 to 6% of the injected activity that is trapped within it, while the bulk of the radioactivity is seen to have been excreted through the liver and ultimately the intestine. Compare this image with that in Figure 1 where no radioactivity is taken up by the brain because the injected radiopharmaceutical is cationic and thus cannot cross the BBB.

Summary

While many isotopes of the elements are stable, there are many more that are radioactive. With few exceptions the unstable nuclides are transformed into more stable ones by emitting an alpha particle, beta particle, gamma ray, or positron (also called a β^+ particle). The positron is an example of *antimatter,* which is consumed by *annihilation* when it encounters its antiparticle, the electron. An alternative mode of decay of some unstable nuclides is *electron capture,* in which the nucleus absorbs one of the electrons of the atom, converting a proton into a neutron inside the nucleus. A number of factors that contribute to the stability of nuclei are the neutron-to-proton ratio, *magic numbers,* even numbers of protons, and even numbers of neutrons. Any radioactive decay, as well as other nuclear changes, can be described by a *nuclear equation,* which is balanced when the sums of both the atomic numbers and mass numbers are the same on the reactant and product sides.

Because of the ease with which radioactive isotopes can be detected, they allow us to study the fate of particular atoms in chemical and biological systems. Many advances in chemistry have been made through the use of radioactive tracers. The constant rate of decay of radioisotopes is usually expressed as a half-life, and has been used to determine the age of objects. Among the most common means of detecting and measuring radioactivity are photographic film, Geiger counters, and scintillation counters.

Most radioactive isotopes do not occur in nature, but some have been produced by bombarding nuclei with high-energy ions to induce nuclear transformations. This technique was used to produce the first samples of many of the elements beyond uranium in the periodic table. Many other radioisotopes have been made by using neutrons as the bombarding particles. *Neutron activation analysis* is a powerful technique to perform a chemical analysis without destroying the sample.

The energy changes that accompany many nuclear transformations are large enough to produce measurable changes in mass in accordance with Einstein's relationship, $\Delta E = \Delta mc^2$. The *mass defect,* which results from the *nuclear binding energy,* is the difference between the mass of the constituent nucleons and that of the nuclide. The possibility of nuclear fission and nuclear fusion was indicated by comparison of the nuclear binding energies per nucleon that are observed for stable nuclides.

Many heavy nuclides, ^{235}U and ^{239}Pu in particular, undergo *fission* when the nucleus absorbs a neutron. Fission reactions are usually chain reactions, which lead to devastating explosions when they proceed under *supercritical* conditions. When the fission process is controlled at barely *critical* conditions in a nuclear reactor, it provides a useful source of energy. Nuclear reactors are now used to generate a significant fraction of electrical energy. *Breeder reactors* are also used to generate electricity, and are designed to produce more fissionable matter than is consumed. Concerns about the safety of nuclear reactors and the disposal of the radioactive waste they generate has deterred the construction of additional nuclear power reactors in the United States.

While nuclear *fusion* has been achieved as an uncontrolled explosion, after more than 40 years of research the development of a controlled fusion reactor that generates useful amounts of energy is still several decades away from being achieved. The containment of the reacting nuclides at the pressures and temperatures needed (several million kelvins) for fusion to occur has proved to be a formidable task.

Radiation damage to biological tissue is caused mainly by the ionization of molecules as the radioactive particles give up their energy. The SI unit for radioactivity is the *bequerel,* and the energy transferred to matter, called radiation dose, is measured in *rads.* An estimate of the radiation damage to biological tissue must also include a quality factor, Q, for different kinds of radiation, the kind of tissue, and several other factors. The effective radiation dose (measured in *rems*) is the product of the quality factor and the radiation dose (in rads). Current estimates indicate that the majority of radiation exposure of the U.S. population comes from natural sources, particularly from radon.

Chapter Terms

Nuclide—the nucleus of a particular isotope. *(20.1)*

Nucleon—a particle that is present in the nucleus of an atom; both protons and neutrons are called nucleons. *(20.1)*

Stable isotope—an atom with a nuclide that does not spontaneously decompose into one or more different nuclides. *(20.1)*

Magic numbers—certain numbers of protons and neutrons that confer unusual stability to the nuclide. These numbers are 2, 8, 20, 26, 28, 50, 82, and 126. *(20.1)*

Radioactive atom—a nuclide that spontaneously emits small particles or high-energy electromagnetic radiation or both, forming a more stable nuclide. *(20.1)*

Alpha particle—a high-energy He^{2+} ion produced by the radioactive decay of some atoms. *(20.1)*

Beta particle—a high-energy electron emitted by the nucleus of some radioactive atoms. *(20.1)*

Gamma ray—high-energy electromagnetic radiation originating from the radioactive decay of unstable nuclides. *(20.1)*

Nuclear equation—describes the changes in a nuclear reaction in shorthand notation. Nuclear equations are balanced by conservation of atomic number and mass number of the particles on the reactant and product sides of the equation. *(20.1)*

Positron—a positively charged electron that is produced in the radioactive decay of some unstable isotopes; the antiparticle of the electron. *(20.1)*

Annihilation—the process in which two antiparticles collide and their mass is converted into electromagnetic energy. *(20.1)*

Electron capture—a mode of decay for some unstable nuclides, in which the nucleus captures an electron, converting a proton in the nucleus into a neutron. *(20.1)*

Nuclear transmutation—a nuclear reaction that involves two or more particles or nuclei as reactants and produces elements or isotopes that are different from the reactants. *(20.2)*

Activation analysis—the exposure of a sample to a neutron beam, producing radioactive isotopes which are used to identify the elements present and determine the composition of the sample. *(20.2)*

Nuclear binding energy—the energy of attraction among the protons and neutrons in a nuclide. This energy is calculated from the mass defect by using $E = \Delta mc^2$ where Δm is the mass defect of the nucleus in kg/mol, c is the speed of light, 3.00×10^8 m/s, and E is in J/mol. *(20.3)*

Mass defect—the difference between the mass of the atom and the mass of the equivalent number of ^1H atoms and neutrons; a measure of the energy holding the nucleons together in the nucleus. *(20.3)*

Nuclear fission—the splitting of a heavy nucleus into two nuclei of comparable size. *(20.4)*

Nuclear chain reaction—a fission event, initiated by the absorption of a neutron, that produces more neutrons than are consumed, allowing a self-sustained nuclear reaction. *(20.4)*

Critical mass—the minimum size of the sample of fissionable matter needed for the chain reaction to be self-sustained; *i.e.,* on an average at least one neutron formed by a fission is used to induce a second fission. *(20.4)*

Breeder reactor—a fission reactor that generates more fissionable atoms than are consumed. *(20.4)*

Nuclear fusion—the combination of two light nuclides to form a heavier one. *(20.4)*

Plasma—an electrically neutral gas at very high temperature, that consists of highly charged ions and electrons. *(20.4)*

Exercises

Exercises designated with color have answers in Appendix J.

Isotopes and Notations

20.1 Fill in the missing entries in the table.

Symbol	Z	A	# protons	# neutrons
$^{40}_{20}Ca$				
	15	31		
			50	68
		239	93	

20.2 Fill in the missing entries in the table.

Symbol	Z	A	# protons	# neutrons
$^{23}_{11}Na$				
	45	103		
			32	38
		234	90	

Nuclear Stability

20.3 Calculate the neutron-to-proton ratio (to 2 decimal places) for each of the following stable isotopes.
(a) $^{17}_{8}O$
(b) $^{39}_{19}K$
(c) $^{75}_{33}As$
(d) $^{118}_{50}Sn$
(e) $^{196}_{78}Pt$

20.4 Calculate the neutron-to-proton ratio (to 2 decimal places) for each of the following stable isotopes.
(a) $^{12}_{6}C$
(b) $^{40}_{20}Ca$
(c) $^{90}_{40}Zr$
(d) $^{138}_{56}Ba$
(e) $^{208}_{82}Pb$

20.5 Which of the following isotopes lie within the band of nuclear stability shown in Figure 20.1?
(a) $^{16}_{8}O$
(b) $^{37}_{17}Cl$
(c) $^{68}_{29}Cu$
(d) $^{239}_{94}Pu$
(e) $^{88}_{41}Nb$

20.6 Which of the following isotopes lie within the band of nuclear stability shown in Figure 20.1?
(a) $^{17}_{8}O$
(b) $^{93}_{38}Sr$
(c) $^{67}_{30}Zn$
(d) $^{233}_{92}U$
(e) $^{28}_{12}Mg$

Radioactive Decay

20.7 Does the neutron-to-proton ratio increase, decrease, or remain unchanged when each of the following radioactive decays occurs? (a) beta decay, (b) positron decay, (c) gamma decay, (d) alpha decay.

20.8 A nuclide that lies above the band of stability shown in Figure 20.1 has too large a neutron-to-proton ratio to be stable. Which of the natural decay processes for such a nuclide will move it toward the band of stability?

20.9 A nuclide that lies below the band of stability shown in Figure 20.1 has too small a neutron-to-proton ratio to be stable. Which of the natural decay processes for such a nuclide will move it toward the band of stability?

20.10 Radioactive decay always moves the nuclide toward the band of stability. From the mode of decay for each of the following radioactive nuclides, state whether it lies above or below the band of stability. (a) ^{32}Si, beta decay; (b) ^{44}Ti, electron capture; (c) ^{52}Mn, positron decay.

20.11 Why are the high-energy photons that accompany an electron capture decay called x rays rather than gamma rays?

20.12 Most alpha decays and beta decays are accompanied by the emission of gamma rays as well. What is the reason for this?

20.13 Write the balanced nuclear equation for each of the following nuclear decays. Show the mass and atomic numbers for each of the particles in the equation.
(a) beta decay of carbon-14
(b) alpha decay of thorium-234
(c) positron decay of potassium-40
(d) electron capture decay of lead-198

20.14 Write the balanced nuclear equation for each of the following nuclear decays. Show the mass and atomic numbers for each of the particles in the equation.
(a) alpha decay of bismuth-201
(b) positron decay of iridium-184
(c) electron capture of lanthanum-135
(d) beta decay of bromine-80

20.15 Identify the missing particles by balancing the mass and atomic numbers in each of the nuclear decay equations.
(a) $^{67}_{31}Ga + ^{0}_{-1}e \longrightarrow$ ______
(b) ______ $\longrightarrow ^{211}_{85}At + ^{4}_{2}He$
(c) $^{67}_{29}Cu \longrightarrow ^{67}_{30}Zn +$ ______
(d) ______ $\longrightarrow ^{124}_{53}I + ^{0}_{1}\beta$
(e) $^{227}_{90}Th \longrightarrow$ ______ $+ ^{4}_{2}He$

20.16 Identify the missing particles by balancing the mass

and atomic numbers in each of the nuclear decay equations.

(a) ____ $\longrightarrow {}^{223}_{88}Ra + {}^{4}_{2}He$

(b) ${}^{22}_{11}Na + {}^{0}_{-1}e \longrightarrow$ ____

(c) ${}^{232}_{90}Th \longrightarrow$ ____ $+ {}^{4}_{2}He$

(d) ${}^{72}_{31}Ga \longrightarrow {}^{72}_{32}Ge +$ ____

(e) ${}^{60}_{29}Cu \longrightarrow {}^{60}_{28}Ni +$ ____

20.17 When acetic acid (A) reacts with methyl alcohol (B) in acidic solution, methyl acetate (C) is formed, as shown in the equation

$$\underset{A}{CH_3{-}\overset{\overset{\large O}{\|}}{C}{-}O{-}H} + \underset{B}{CH_3{-}O{-}H} \longrightarrow \underset{C}{CH_3{-}\overset{\overset{\large O}{\|}}{C}{-}O{-}CH_3} + H_2O$$

Describe a tracer experiment using ^{18}O that would determine whether the oxygen atom bonded to the two carbon atoms in the ester originated in the alcohol or in the acid. (Experimentally it has been shown that the oxygen is originally in the methyl alcohol.)

Half-Life and Decay Rates

20.18 ^{130}I decays by emission of beta particles to form stable ^{130}Xe. A sample of iodine-130 was recorded as having 1245 disintegrations per second at 11:00 A.M. on June 23. At 9:32 A.M. on June 24 the same sample had a disintegration rate of 350 disintegrations per second. What is the half-life in hours for the beta decay of ^{130}I?

20.19 The radioactive decay of a sample containing ^{121}Te produced 865 disintegrations per minute. Exactly 7.00 days later the rate of decay was found to be 650 disintegrations per minute. Calculate the half-life in days for the decay of ^{121}Te.

20.20 A sample that contains 3.75×10^{13} atoms of a beta emitter has a disintegration rate of 382 disintegrations per minute. What is the half-life (in years) for this isotope?

20.21 By mass spectral analysis, a sample of strontium is known to contain 2.64×10^{10} atoms of ^{90}Sr as the only radioactive element. The absolute disintegration rate of this sample is measured as 1238 disintegrations per minute.

(a) Calculate the half-life (in years) of ^{90}Sr.

(b) How long will it take for the disintegration rate of this sample to drop to 1000 disintegrations per minute?

Radioactive Decay Series

20.22 Determine the number of α particles and β particles that are emitted in the decay series that starts with ^{239}Pu and finally produces ^{207}Pb.

20.23 How many α and β particles are emitted in the decay series that begins with ^{237}Np and finally produces the stable isotope ^{209}Bi?

Radioactive Dating

20.24 A sample of uranium ore contains 6.73 mg of ^{238}U and 3.22 mg of ^{206}Pb. Assuming all of the ^{206}Pb arose from decay of the ^{238}U, and the half-life of ^{238}U is 4.51×10^9 years, determine the age of the ore.

20.25 A tree was cut down and used to make a statue 2300 years ago. What fraction of the carbon-14 present originally remains today? ($t_{1/2}$ of ^{14}C is 5730 years.)

20.26 The ratio of ^{87}Rb to ^{87}Sr can be used for dating of geological samples, assuming all of the strontium came from the beta decay of rubidium ($t_{1/2} = 5.0 \times 10^{10}$ years). What is the age of a rock sample if $^{87}Sr/^{87}Rb = 0.051$?

20.27 Would carbon-14 dating be useful in determining whether a sample of paper was 20 or 50 years old?

20.28 In a living organism the decay of ^{14}C produces 15.3 disintegrations per minute per gram of carbon. What percentage of the atoms of carbon in the biosphere are ^{14}C?

20.29 What property of the products of radioactive decay is used to detect and measure the presence of the radiation?

Nuclear Reactions

20.30 Complete and balance the following nuclear equations.

(a) ${}^{54}_{26}Fe + {}^{4}_{2}He \longrightarrow 2{}^{1}_{1}H +$ ____

(b) ____ $+ {}^{1}_{0}n \longrightarrow {}^{24}_{11}Na + {}^{4}_{2}He$

(c) ${}^{238}_{92}U + {}^{16}_{8}O \longrightarrow$ ____ $+ 5{}^{1}_{0}n$

(d) ${}^{96}_{42}Mo +$ ____ $\longrightarrow {}^{97}_{43}Tc + {}^{1}_{0}n$

(e) ${}^{250}_{98}Cf + {}^{11}_{5}B \longrightarrow$ ____ $+ 5{}^{1}_{0}n$

20.31 Complete and balance the following nuclear equations.

(a) ${}^{249}_{98}Cf + {}^{10}_{5}B \longrightarrow {}^{257}_{103}Lr +$ ____

(b) ${}^{14}_{7}N + {}^{1}_{1}H \longrightarrow {}^{4}_{2}He +$ ____

(c) ${}^{238}_{92}U + {}^{1}_{0}n \longrightarrow$ ____ $+ {}^{0}_{-1}\beta$

(d) ${}^{6}_{3}Li + {}^{1}_{0}n \longrightarrow {}^{4}_{2}He +$ ____

(e) ____ $+ {}^{4}_{2}He \longrightarrow {}^{12}_{6}C + {}^{1}_{0}n$

20.32 Why is it necessary to accelerate charged particles to very high energies to cause nuclear reactions with the heavier elements?

20.33 Why are low-energy neutrons able to react with nearly all nuclides?

Nuclear Binding Energy

20.34 The molar mass of ^{19}F is 18.9984 g/mol.
(a) Use the masses of hydrogen-1 and the neutron given in Table 20.3 to calculate the mass defect of this nuclide in grams per mole.
(b) Express the total nuclear binding energy in MeV for this nuclide.
(c) Calculate the binding energy per nucleon for ^{19}F.

20.35 The molar mass of ^{31}P is 30.9738 g/mol.
(a) Use the masses of hydrogen-1 and the neutron given in Table 20.3 to calculate the mass defect of this nuclide in grams per mole.
(b) Express the total nuclear binding energy in MeV for this nuclide.
(c) Calculate the binding energy per nucleon for ^{31}P.

20.36 The atomic masses of three isotopes of aluminum are:

^{26}Al	25.9869
^{27}Al	26.9815
^{28}Al	27.9819

(a) Calculate the binding energy per nucleon in MeV for each of these isotopes of aluminum.
(b) Two of these nuclides are radioactive and the third one is stable. Based on your answer to part (a), which isotopes are unstable?

20.37 The atomic masses of three isotopes of phosphorus are:

^{30}P	29.9783
^{31}P	30.9738
^{32}P	31.9739

(a) Calculate the binding energy per nucleon in MeV for each of these isotopes of phosphorus.
(b) Two of these nuclides are radioactive and the third one is stable. Based on your answer to part (a), which isotopes are unstable?

Nuclear Fission and Fusion

20.38 Explain why most fission products formed in a nuclear reactor decay by beta emission rather than undergoing another kind of nuclear decay.

20.39 When ^{239}Pu is used in a nuclear reactor, one of the fission events that occurs is

$$\underset{}{{}^1_0n} + \underset{239.052\ u}{{}^{239}_{94}Pu} \longrightarrow \underset{95.916\ u}{{}^{96}_{39}Y} + \underset{139.917\ u}{{}^{140}_{55}Cu} + 4\,{}^1_0n$$

The atomic mass of each atom is given below its symbol in the equation.
(a) Calculate the change in mass, expressed in atomic mass units, that accompanies this fission event.
(b) Convert the mass loss into joules per fission.
(c) Find the energy released when 1.00 g of plutonium undergoes this particular fission.

20.40 When ^{239}Pu is used in a nuclear reactor, one of the fission events that occurs is

$${}^1_0n + \underset{239.052\ u}{{}^{239}_{94}Pu} \longrightarrow \underset{97.913\ u}{{}^{98}_{40}Zr} + \underset{138.919\ u}{{}^{139}_{54}Xe} + 3\,{}^1_0n$$

The atomic mass of each atom is given below its symbol in the equation.
(a) Calculate the change in mass, expressed in atomic mass units, that accompanies this fission event.
(b) Convert the mass loss into joules per fission.
(c) Find the energy released when 1.00 g of plutonium undergoes this particular fission.

20.41 What is the role of the moderator in a nuclear reactor? the control rods?

20.42 One kilogram of high-grade coal produces about 2.8×10^4 kJ of energy when it is burned. Fission of one mole of ^{235}U releases 1.9×10^{10} kJ.
(a) Calculate the number of metric tons (1 metric ton = 1 megagram) of coal needed to produce the same amount of energy as the fission of 1 kg of uranium.
(b) How many metric tons of sulfur dioxide (a major source of acid rain) are produced from the burning of the coal in part (a), if the coal is 0.90 % by mass sulfur?

20.43 One of the fission products that causes major concern is ^{90}Sr, since it is incorporated into milk and other high-calcium foods. ^{90}Sr undergoes beta decay with a half-life of 28.1 years. What fraction of the ^{90}Sr that was formed in the detonation of the first fission bombs in August 1945 is still present in the environment in 1993?

20.44 One of the promising reactions for a controlled fusion reactor consumes deuterium (^2H) and tritium (^3H) in the reaction

$$\underset{2.0140}{{}^2_1H} + \underset{3.0161}{{}^3_1H} \longrightarrow \underset{4.0026}{{}^4_2He} + \underset{1.008665}{{}^1_0n}$$

The atomic mass of each particle is given below the equation. Calculate the energy released by this fusion reaction per gram of helium formed, and compare it to the energy generated by the fission of one gram of ^{235}U (8×10^7 kJ/g).

Biological Effects of Radiation

20.45 Compare the general penetrating abilities of α, β, and γ radiation. Why are alpha particles produced inside the body particularly dangerous?

20.46 A person's exposure to radiation can depend very

much on occupation. List several occupations that might result in greater exposure to radiation than the average exposure of the U.S. population. Explain your choices.

Additional Exercises

20.47 Gaseous diffusion is used to enrich natural uranium in ^{235}U. Use Graham's law (Section 5.8) to calculate the ratio of enrichment of ^{235}U by a single diffusion of UF_6. The atomic masses of the two principal isotopes of uranium are 235.0493 and 238.0508, for ^{235}U and ^{238}U, respectively, and that of fluorine is 18.9984.

20.48 Cadmium is used in the control rods of fission reactors because it absorbs neutrons very efficiently. Most of the neutrons are absorbed by ^{113}Cd to produce ^{114}Cd and a gamma ray. The atomic masses of these two isotopes of cadmium are $^{113}Cd = 112.9044$ u and $^{114}Cd = 113.9034$ u. Calculate the energy of the gamma ray emitted in this nuclear transformation, expressing your answer in MeV. What is the wavelength of the gamma ray?

20.49 ^{128}I decays by both β^- and β^+ decay. What are the product nuclides from each of these decays?

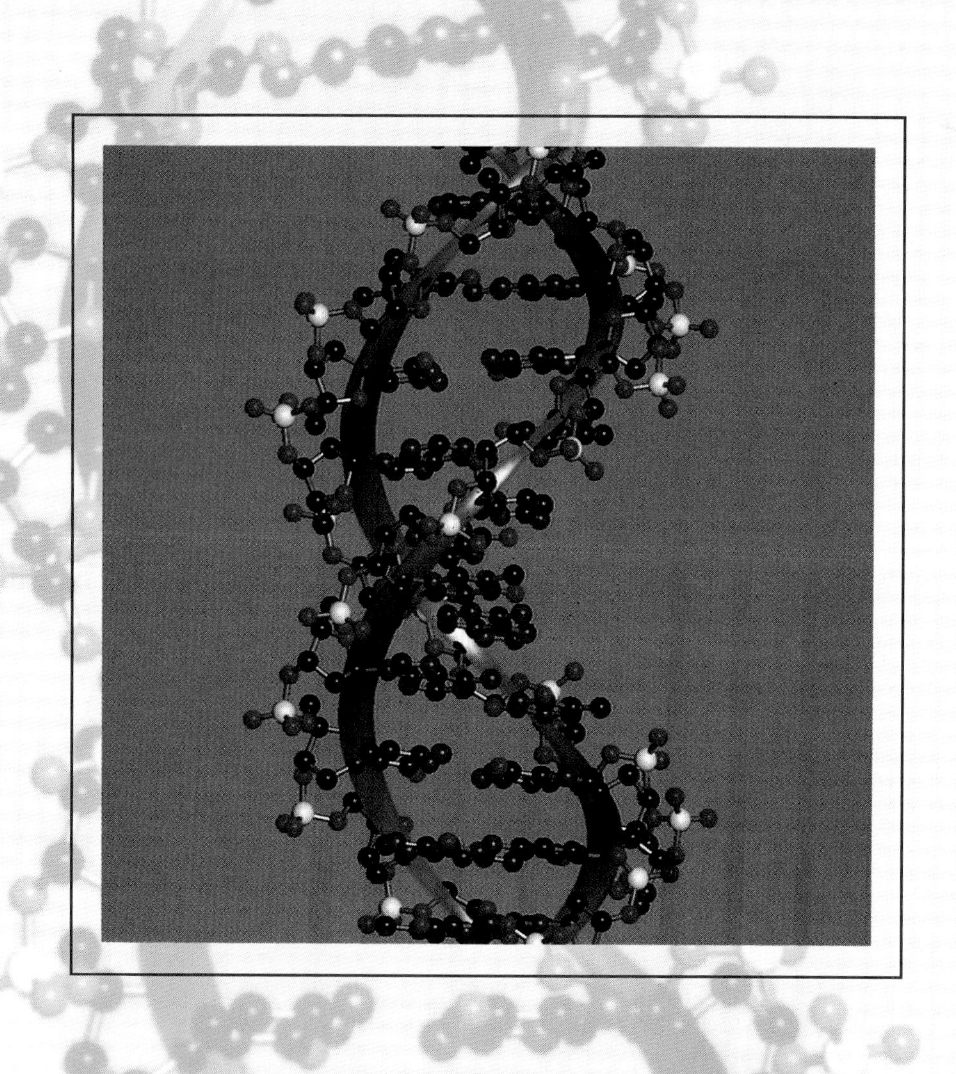

Chapter Twenty-one

21

Organic Chemistry and Biochemistry

Organic chemistry is the study of carbon-containing compounds. Compounds that contain carbon atoms have a rich and diverse chemistry, and well over six million different organic compounds have been characterized. Originally, substances produced by living systems (organic) were differentiated from all others (inorganic). It was believed that living systems had a "vital force" that produced special compounds. However, in 1828 the German chemist Friedrich Wöhler demonstrated that urea, an organic compound first isolated from urine, could be prepared by the reaction of three inorganic compounds, lead(II) cyanate, ammonia, and water.

$$Pb(OCN)_2 + 2NH_3 + 2H_2O \longrightarrow \underset{\text{urea}}{2(NH_2)_2CO} + Pb(OH)_2$$

Although organic compounds can be produced from nonliving systems, the name is used for carbon-containing compounds in which carbon forms bonds to itself, to hydrogen, and to other nonmetals such as nitrogen, oxygen, and sulfur. The many carbon-containing minerals, such as carbonate salts, are not considered organic compounds. **Biochemistry** is the study of the chemistry of systems in living organisms.

Why is the chemistry of carbon so extensive? The primary reason is that carbon atoms make strong bonds to each other, forming chains or rings, a process known as **catenation.** Carbon atoms are ideally suited to form four strong covalent bonds because they have four valence orbitals and four valence electrons. In addition to forming strong carbon-carbon bonds, car-

◀ DNA has a double helix structure and provides the basic genetic information that determines who we are.

bon atoms make strong bonds to hydrogen, nitrogen, oxygen and sulfur. Also, as a second period element, carbon readily forms multiple bonds with itself and with nitrogen and oxygen.

21.1 Alkanes

Objectives

- To identify and name linear and cyclic alkanes
- To write the structural isomers of alkanes
- To name the important substituent groups
- To describe the shapes of the two important conformers of cyclohexane
- To write some reactions of alkanes

Hydrocarbons are compounds that contain only the elements hydrogen and carbon. An **alkane** is a hydrocarbon that contains no multiple bonds or rings. Alkanes are also known as **saturated hydrocarbons** because each carbon atom makes bonds to four other atoms, the maximum number possible. The simplest hydrocarbon is methane, CH_4. Using valence bond theory to describe the bonding, each C—H bond in this tetrahedral molecule is described as formed from the overlap of an sp^3 hybridized orbital on the carbon atom with the $1s$ orbital on a hydrogen atom (Figure 21.1).

The alkane with two carbon atoms is ethane, C_2H_6. As with all alkanes, the carbon atoms in ethane are sp^3 hybridized. Three of these hybrid orbitals on each carbon atom overlap with $1s$ orbitals on the hydrogen atoms, and the fourth is used to form the C—C bond. The third member of the family is propane, C_3H_8.

There are a number of ways to draw the structures of these molecules (Figure 21.2). While convenient to write, the commonly used structural formulas do not properly show the shapes of these molecules. When you look at the molecular formula or a structural formula of an alkane, remember that the geometry around each carbon atom is tetrahedral, so the carbon chain in propane forms a bent or "V" shape.

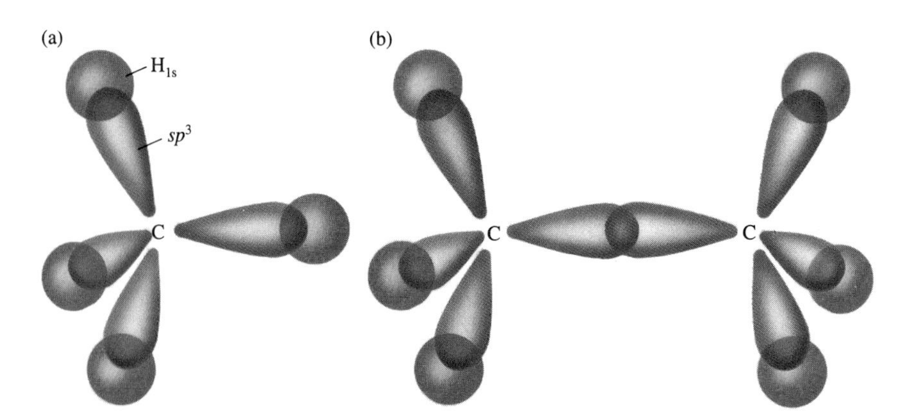

Figure 21.1 The bonding in ethane and methane Carbon atoms form covalent bonds from sp^3 hybridized orbitals in (a) methane and (b) ethane.

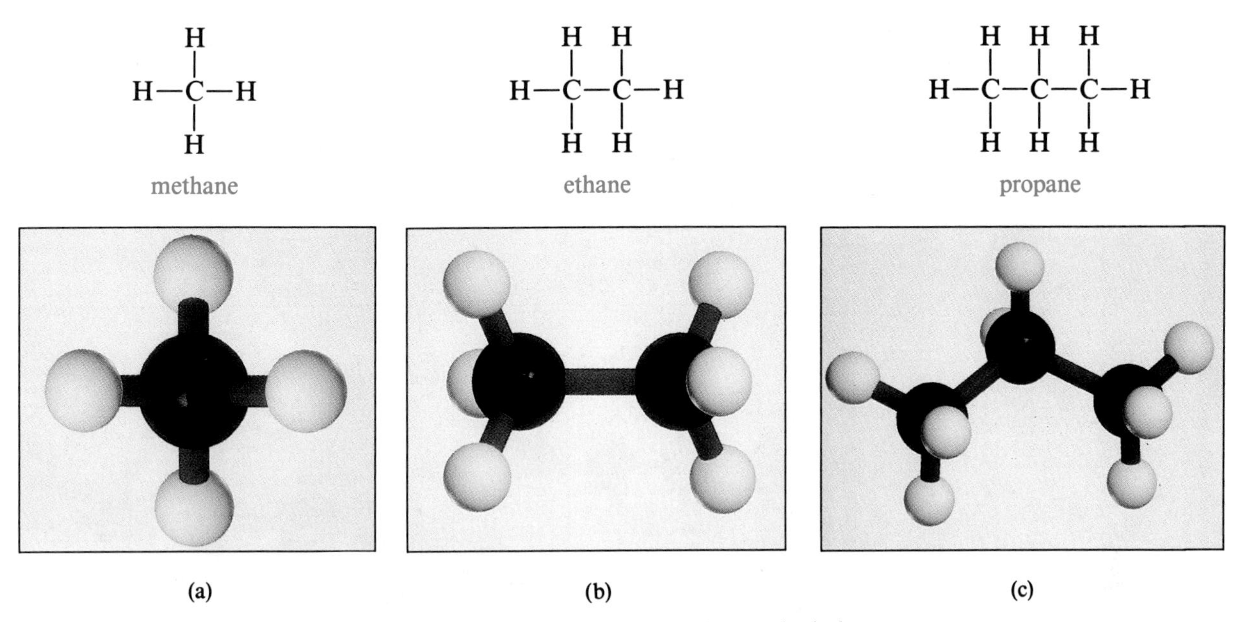

(a) (b) (c)

Figure 21.2 Structures of (a) methane, (b) ethane, and (c) propane The tetrahedral shape of the carbon atoms in alkanes makes the actual arrangement of the atoms in space (computer-generated figures at the bottom) very different from those shown by structural formulas (top).

Note that from methane to ethane to propane, the chemical formula is increased by one CH_2 group. In general, the alkanes have the formula C_nH_{2n+2}, where $n = 1, 2, \ldots$. Table 21.1 lists the names of the first ten straight-chain members (isomers with all of the carbon atoms in a continuous chain) of the alkane family along with their boiling points. As expected, the boiling points increase with increasing molecular weight because increasing the chain length increases the polarizability of the alkane (see Section 10.1), leading to stronger intermolecular forces of the London dispersion type.

Table 21.1 The First Ten Straight-Chain Alkanes

Name	Formula	Normal Boiling Point (°C)	Alkyl Group	
methane	CH_4	−162	methyl	CH_3—
ethane	C_2H_6	−89	ethyl	C_2H_5—
propane	C_3H_8	−42	propyl	C_3H_7—
n-butane	C_4H_{10}	0	butyl	C_4H_9—
n-pentane	C_5H_{12}	36	pentyl	C_5H_{11}—
n-hexane	C_6H_{14}	69	hexyl	C_6H_{13}—
n-heptane	C_7H_{16}	98	heptyl	C_7H_{15}—
n-octane	C_8H_{18}	126	octyl	C_8H_{17}—
n-nonane	C_9H_{20}	151	nonyl	C_9H_{19}—
n-decane	$C_{10}H_{22}$	174	decyl	$C_{10}H_{21}$—

Figure 21.3 Structures of *n*-butane and isobutane There are two structural isomers of butane that differ in the connectivity of the carbon chain.

```
    H  H  H  H
    |  |  |  |
 H—C—C—C—C—H
    |  |  |  |
    H  H  H  H
```

n-butane

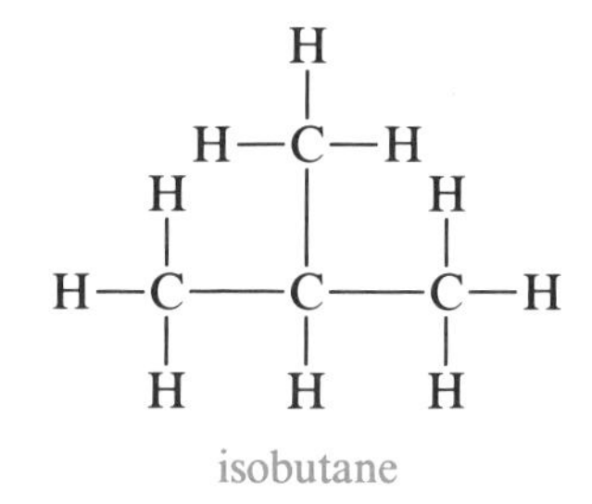

isobutane

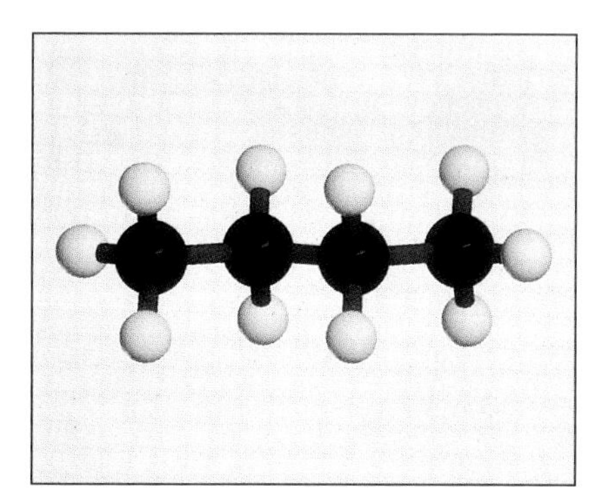

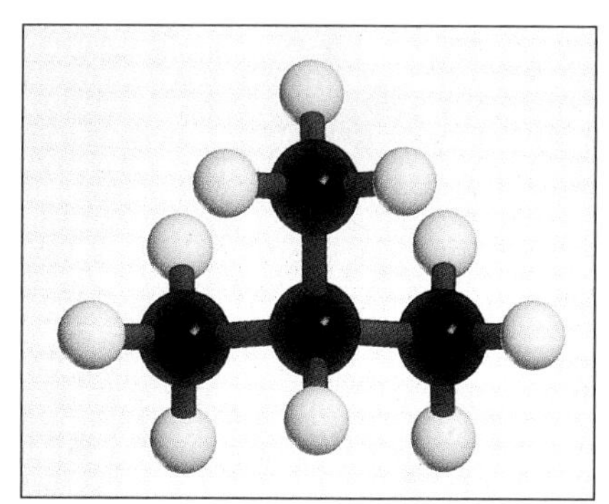

Structural Isomers

There are two hydrocarbons with the formula C_4H_{10}. The carbon atoms can be arranged in an unbranched chain (called straight-chain even though the chain is puckered) or arranged in a branched chain (Figure 21.3). The straight-chain isomer is called normal butane (*n*-butane) and the branched isomer is called isobutane. These two alkanes are *structural isomers,* molecules that have the same molecular formula but differ in the connectivity or arrangement of the bonds. They have different physical and chemical properties. For example, the boiling point of isobutane (−12 °C) is lower than that of *n*-butane (0 °C). The boiling points are different because the more linear shape of *n*-butane allows the chains to line up next to each other and exert slightly stronger intermolecular forces. In comparison, the more spherical shape of isobutane limits the contact between molecules, so its intermolecular forces are slightly weaker.

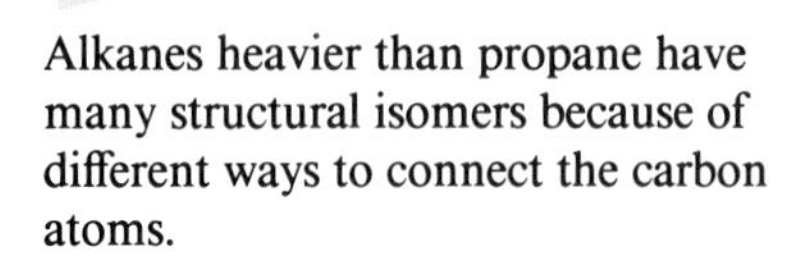

Alkanes heavier than propane have many structural isomers because of different ways to connect the carbon atoms.

The number of structural isomers that are possible for an alkane increases rapidly as the number of carbon atoms increases. Example 21.1 shows the isomers of C_5H_{12}, pentane.

Example 21.1 Isomers of Pentane

Draw structural formulas of all of the possible structural isomers of the pentanes, C_5H_{12}.

Solution One isomer, *n*-pentane, has all of the carbon atoms in a linear chain. A second isomer, isopentane, is formed by branching the chain once at the second carbon atom in the chain. A third isomer, neopentane, is formed by branching the chain twice at the second carbon atom. No other connectivity patterns can be drawn in which each carbon atom has four bonds and each hydrogen atom has one bond; therefore, only these three isomers are possible.

```
     H   H   H   H   H
     |   |   |   |   |
 H — C — C — C — C — C — H
     |   |   |   |   |
     H   H   H   H   H
```

n-pentane

```
          H
          |
      H — C — H
  H       |       H   H
  |       |       |   |
H—C ——— C ——— C — C — H
  |       |       |   |
  H       H       H   H
```

isopentane

```
          H
          |
      H — C — H
  H       |       H
  |       |       |
H—C ——— C ——— C — H
  |       |       |
  H       |       H
      H — C — H
          |
          H
```

neopentane

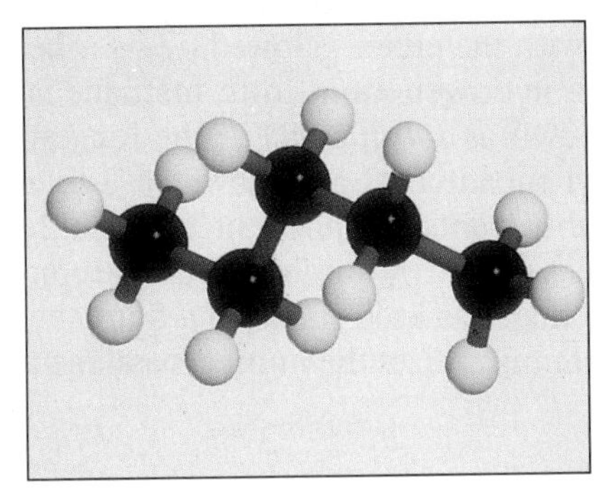

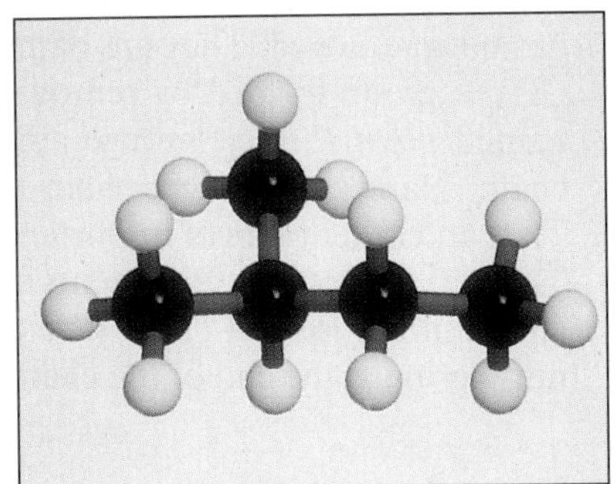

Note that each isomer has a different pattern of connections among the carbon atoms. Simply changing the position of a group by rotation of the drawing or rotation about a single bond does *not* lead to additional isomers. All four of the following drawings are different representations of isopentane. They all have the same connectivity and thus are *not* isomers.

```
          H
          |
      H — C — H
  H       |       H   H
  |       |       |   |
H—C ——— C ——— C — C — H
  |       |       |   |
  H       H       H   H
```

```
  H   H       H       H
  |   |       |       |
H—C — C ——— C ——— C — H
  |   |       |       |
  H   H       |       H
          H — C — H
              |
              H
```

```
  H       H       H   H
  |       |       |   |
H—C ——— C ——— C — C — H
  |       |       |   |
  H       |       H   H
      H — C — H
          |
          H
```

```
              H
              |
          H — C — H
  H   H       |       H
  |   |       |       |
H—C — C ——— C ——— C — H
  |   |       |       |
  H   H       H       H
```

Alkane Nomenclature

Given the large number of organic compounds, the International Union of Pure and Applied Chemistry (IUPAC) has established an extensive system for naming compounds. A few rules will allow the naming of alkanes.

1. Alkanes are named by using the suffix *-ane* with the appropriate prefix (*pro-* for three, *but-* for four; after that, numerical prefixes are used—*pent-* for five, *hex-* for six, and so forth, as shown in Table 21.1) to name the *longest* continuous chain of carbon atoms. For example, the alkane

$$\begin{array}{l} CH_3CH_2CH_2\underset{\displaystyle CH_3}{\underset{|}{C}}HCH_2CH_2CH_2CH_3 \end{array}$$

   is named as a substituted octane.
2. Carbon chains that branch from the longest continuous chain are named as alkyl groups. **Alkyl groups** are alkanes from which one hydrogen atom has been removed. They are named with the prefix followed by *-yl*. The CH_3— group formed by removing a hydrogen atom from methane is named *methyl*. Alkyl groups are known as *substituents* of the longest chain. Names of straight-chain alkyl substituents are shown in Table 21.1, and other types of important substituents are shown in Table 21.2.
3. The positions of the substituents are designated by numbering the carbon atoms in the longest chain (the one that was used in step 1). Start the numbering at the end of the chain that minimizes the numbers assigned

Table 21.2 Important Substituents

Name	Formula	Name	Formula
fluoro	—F	phenyl	$-C_6H_5$ (benzene ring)
chloro	—Cl	vinyl	$-CH{=}CH_2$
bromo	—Br	isopropyl	$-\overset{\displaystyle CH_3}{\underset{\displaystyle CH_3}{C}}-H$
iodo	—I	isobutyl	$-CH_2-\overset{\displaystyle CH_3}{\underset{\displaystyle CH_3}{C}}-H$
nitro	$-NO_2$	*sec*-butyl	$-\overset{\displaystyle CH_3}{\underset{\displaystyle CH_2-CH_3}{C}}-H$
amino	$-NH_2$	*tert*-butyl	$-\overset{\displaystyle CH_3}{\underset{\displaystyle CH_3}{C}}-CH_3$

to the carbon atoms to which the substituents are bonded. For example, if we number the octane shown in step 1 from left to right, the methyl group is attached at position 4 (**A**); if we numbered it from right to left, the methyl group would be attached at position 5 (**B**). The numbering scheme in **A** is correct because it minimizes the position number of the substituent.

$$\begin{array}{l} \scriptstyle 1\;\;2\;\;3\;\;4\;5\;\;6\;\;7\;\;8 \\ CH_3CH_2CH_2\underset{\displaystyle |\atop \displaystyle CH_3}{C}HCH_2CH_2CH_2CH_3 \end{array} \qquad \begin{array}{l} \scriptstyle 8\;\;7\;\;6\;\;5\;4\;\;3\;\;2\;\;1 \\ CH_3CH_2CH_2\underset{\displaystyle |\atop \displaystyle CH_3}{C}HCH_2CH_2CH_2CH_3 \end{array}$$

A B

The position of the substituent is indicated by a number before its name (the number and substituent name are separated by a dash). This number and substituent name are placed before the name of the longest chain. For cases with more than one substituent, the various substituents are listed in alphabetical order. If there is more than one substituent of the same type, the prefixes *di* (2), *tri* (3), and *tetra* (4) are used and the position numbers are separated by commas. The preceding compound is 4-methyloctane.

An alkane is named with the prefix that designates the number of carbon atoms in the longest chain, followed by the suffix *-ane*.

Example 21.2 Naming Substituted Alkanes

Name the following compounds.

(a) $CH_3CH_2CH_2\underset{\displaystyle |\atop \displaystyle CH_3}{C}HCH_2CH_3$

(b) $CH_3\underset{\displaystyle |\atop \displaystyle Br}{C}HCH_2\underset{\displaystyle |\atop \displaystyle CH_2CH_3}{C}HCH_2CH_2CH_3$

(c) $CH_3\overset{\displaystyle NO_2\atop \displaystyle |}{C}HCH_2\overset{\displaystyle CH_3\atop \displaystyle |}{\underset{\displaystyle |\atop \displaystyle CH_3}{C}}CH_2CH_2CH_2CH_3$

Solution

(a) The longest chain contains six carbon atoms, so it is a hexane.

$$\begin{array}{l} \scriptstyle 6\;\;5\;\;4\;\;3\;2\;\;1 \\ CH_3CH_2CH_2\underset{\displaystyle |\atop \displaystyle CH_3}{C}HCH_2CH_3 \end{array}$$

In this case, numbering from right to left yields a lower number for the methyl substituent, the 3-position. The name is 3-methylhexane.

(b) The longest chain is a heptane.

$$\begin{array}{l} \scriptstyle 1\;\;2\;3\;\;\;4\;5\;\;6\;\;7 \\ CH_3\underset{\displaystyle |\atop \displaystyle Br}{C}HCH_2\underset{\displaystyle |\atop \displaystyle CH_2CH_3}{C}HCH_2CH_2CH_3 \end{array}$$

The substituents are at positions 2 and 4. The name is 2-bromo-4-ethylheptane.

(c) The longest chain is an octane. The nitro substituent is at the 2-position and the two methyl substituents are both at the 4-position. Numbering the chain in the other direction would give higher position numbers, 5 and 7. The name is 4,4-dimethyl-2-nitrooctane.

Understanding Name the following substituted alkane.

$$CH_3CH_2CCl_2CH_2CH_2CH_2CH_3$$

Answer: 3,3-dichloroheptane

Cycloalkanes

A **cycloalkane** is a saturated hydrocarbon that contains a ring of carbon atoms. The general formula of cycloalkanes that contain one ring and no substituents is C_nH_{2n} where $n = 3, 4, \ldots$. The first four simple cycloalkanes are pictured in Figure 21.4. As shown, these compounds are frequently drawn as polygons, with each corner representing a carbon atom with the appropriate number of hydrogen atoms attached for each carbon atom to make four bonds.

We know that a carbon atom making bonds to four other atoms generally has a tetrahedral shape with approximately 109° bond angles. In cyclopropane and cyclobutane, the small rings force C—C—C bond angles that are much smaller than this value. The resulting *ring strain* causes these two molecules to be less stable and consequently more reactive than other hydrocarbons. For larger rings, the carbon-carbon bond angles can be approximately 109° only if the rings are not planar. For cyclohexane, the carbon atoms can achieve approximately 109° bond angles by two arrangements, a "chair" form and a "boat" form, as shown in Figure 21.5. These two arrangements are called **conformers,** different arrangements of atoms caused

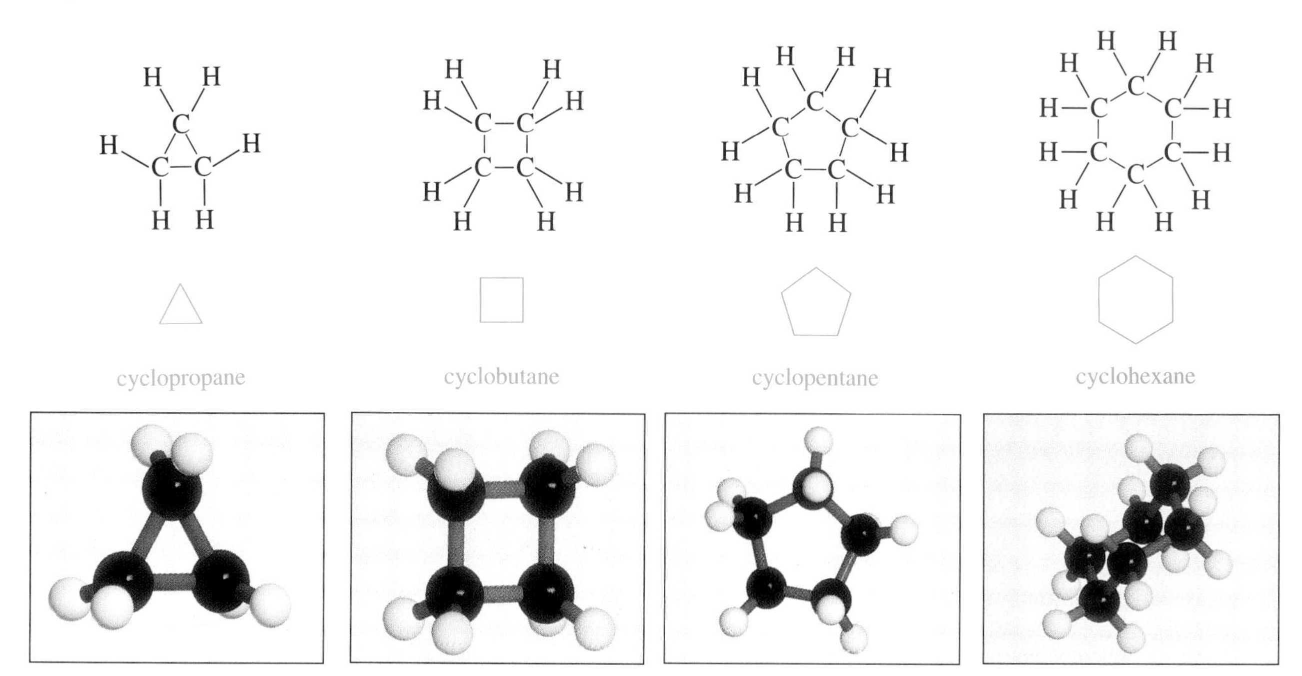

Figure 21.4 Cycloalkanes The structural formula of a cycloalkane (top) is frequently written as the polygon of the carbon atom framework (middle). The computer generated figures (bottom) show the spatial arrangements of the atoms.

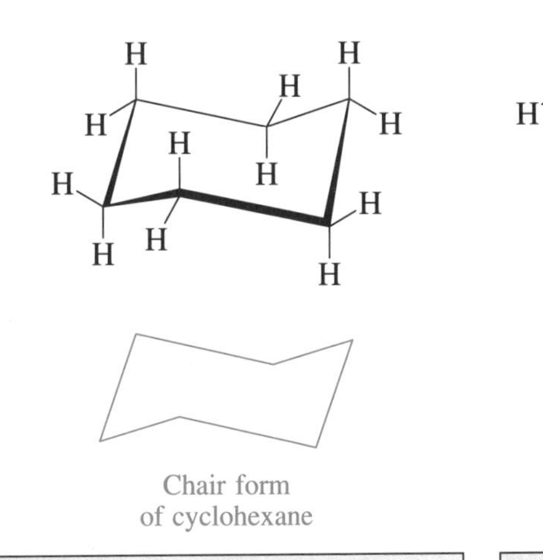

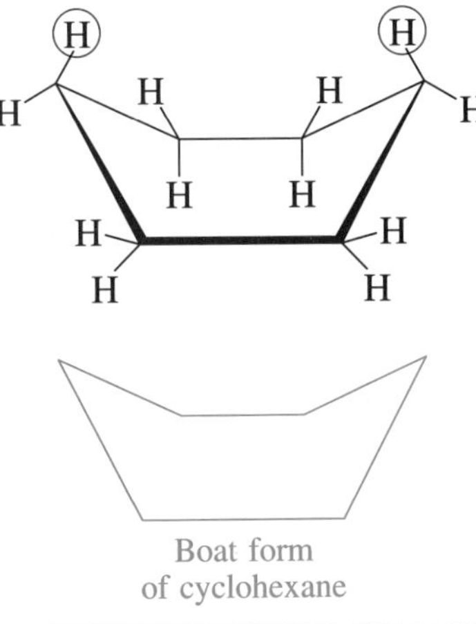

Figure 21.5 Boat and chair forms of cyclohexane The chair form of cyclohexane is favored over the boat form because repulsive interactions between the hydrogen atoms are reduced in this arrangement. The line drawings at the top are frequently written as just the bent polygons of the carbon framework, as indicated in the middle of the figure. The computer-generated figures at the bottom represent the spatial arrangements of the atoms.

by rotations about single bonds. The two conformers of cyclohexane can interconvert by rotations about the C—C bonds without breaking them, and they are in rapid equilibrium with each other at room temperature. The "chair" arrangement is favored because in the "boat" form several pairs of the hydrogen atoms are fairly close to each other, especially the pair at the inside top of the structure (colored red in the top drawing of Figure 21.5). The closeness of these hydrogen atoms causes repulsions that make the boat form higher in energy than the chair form. The properties of many biological molecules are determined by the orientations of the six-membered rings.

There are two stable conformers of cyclohexane, with the chair form being more stable than the boat form.

Cycloalkanes are named like alkanes, adding the prefix *cyclo* to the Greek prefix that designates the size of the ring.

Example 21.3 Names of Cycloalkanes

Name the following two compounds.

(a) Br, CH_3 (b) NH_2, CH_2CH_3

Solution

(a) The ring is numbered starting at the bromine substituent because that substituent will be named first in alphabetical order. In order to minimize the number locating the other substituent, the numbering sequence is chosen so that the methyl group is at position 3.

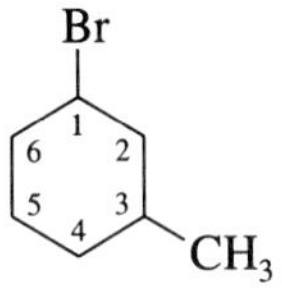

Br

1 2 3 4 5 6

CH_3

The name is 1-bromo-3-methylcyclohexane.

(b) The substituents are at the 1 and 2 positions. The name is 1-amino-2-ethylcyclobutane.

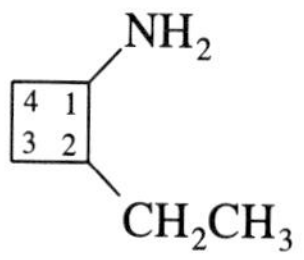

NH_2

4 1 3 2

CH_2CH_3

Understanding Name the following compound.

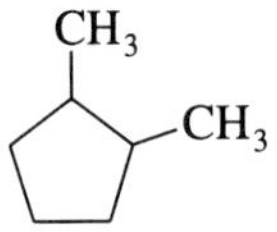

CH_3

CH_3

Answer: 1,2-dimethylcyclopentane.

Reactions of Alkanes

In general, alkanes are not very reactive because only relatively strong C—C (348 kJ/mol) and C—H (414 kJ/mol) bonds are present. Despite this stability, alkanes are important fuels because they react at high temperatures with oxygen in a combustion reaction. We know that the reaction of a hydrocarbon with oxygen is highly exothermic and yields carbon dioxide and water. For example, methane is the main component of the natural gas that is used to heat many homes.

$$CH_4(g) + 2O_2(g) \longrightarrow CO_2(g) + 2H_2O(\ell) \qquad \Delta H = -890 \text{ kJ/mol}$$

Another important fuel, gasoline, is a complex mixture of compounds, the main components of which are alkanes that contain between 5 and 11 carbon atoms. The combustion of gasoline in an automobile engine is carried out with a limited amount of oxygen, so CO as well as CO_2 is a product.

Alkanes will also react with halogens in the presence of light or an appropriate catalyst, leading to replacement of one or more of the hydrogen atoms, a **substitution reaction.**

$$CH_3CH_3 + Cl_2 \xrightarrow{h\nu} CH_3CH_2Cl + HCl$$

$$CH_3CH_2Cl + Cl_2 \xrightarrow{h\nu} CH_3CHCl_2 \text{ or } CH_2ClCH_2Cl + HCl$$

In the second reaction, either of two isomers can form, 1,1-dichloroethane or 1,2-dichloroethane. Chlorinated hydrocarbons such as these are important as cleaners (dry-cleaning solvents), as insecticides, and as intermediates in the production of a number of chemical products.

The mechanism of these substitution reactions is complex. The light is needed to provide the energy to break the bond in Cl_2.

$$Cl_2 \xrightarrow{h\nu} 2Cl$$

The chlorine atoms have an unpaired electron and are reactive enough to break C—H bonds. Two products are formed in the second reaction because these highly reactive chlorine atoms react with any of the C—H bonds.

21.2 Alkenes and Alkynes

Objectives

- To identify and name unsaturated compounds
- To list the possible isomers of alkenes, alkynes, and aromatic compounds
- To write some reactions of alkenes, alkynes, and aromatic compounds

Alkenes

An **unsaturated hydrocarbon** contains one or more multiple carbon-carbon bonds. **Alkenes** are unsaturated hydrocarbons that contain carbon-carbon double bonds. The general formula of an alkene containing one double bond is C_nH_{2n} where $n = 2, 3 \ldots$. The two simplest alkenes are ethene, C_2H_4, and propene, C_3H_6 (Figure 21.6). The older "common" names for these compounds are ethylene and propylene. Each carbon atom involved in

ethene (ethylene)

propene (propylene)

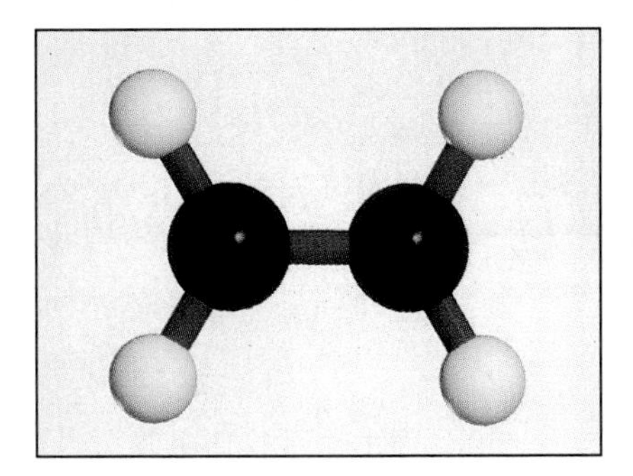

(a)

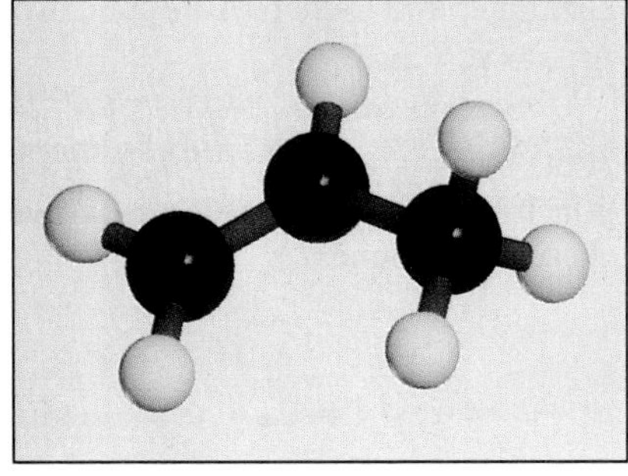

(b)

Figure 21.6 Structures of (a) ethene and (b) propene In the computer-generated figures, the C—C double bonds are shorter than the C—C single bond. Typical C—C bond distances are 154 pm for a single bond and 133 pm for a double bond.

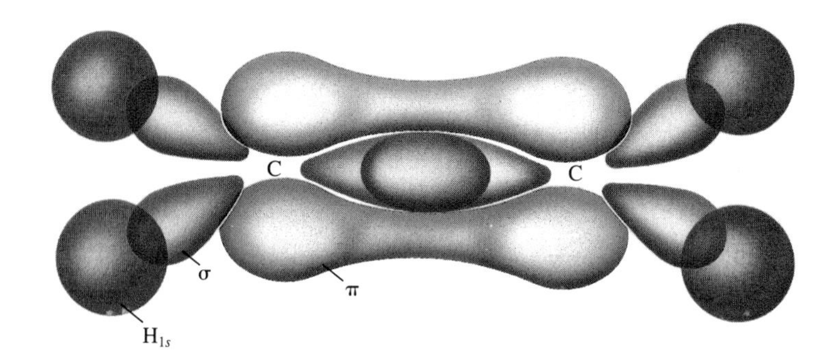

Figure 21.7 Bonding in ethene The sigma bonds in ethene are formed from sp^2 hybrid orbitals on the carbon atoms, and the pi bond is formed from sideways overlap of a *p* orbital on each.

forming the double bond uses sp^2 hybridized orbitals to form three sigma bonds, and a *p* orbital to form a pi bond (Figure 21.7).

An alkene contains at least one carbon-carbon double bond. It is named in the same way as an alkane by replacing the *-ane* suffix with *-ene.*

Alkenes are named by the same system as alkanes, but replacing the *-ane* suffix with *-ene.* The position of the double bond is indicated by the number in the chain of the first carbon atom involved in the multiple bonding. The chain is numbered so as to minimize the number given to the double bond, and this numbering takes precedence over the number of any substituent. In an older system of nomenclature, ethene was known as ethylene, and this name is still common. The second member of the alkene family, C_3H_6, is called propene and not "1-propene" because only one location of the double bond is possible in this molecule.

The third member of the alkene family, C_4H_8, has four possible isomers (Figure 21.8). Two isomers differ in the position of the double bond. The double bond can be between the first two carbon atoms in the chain, $CH_2{=}CHCH_2CH_3$ (1-butene), or between the second and third carbon atoms, $CH_3CH{=}CHCH_3$ (2-butene). Because rotation about a double bond would break the pi bond, there are two isomers of 2-butene: the *cis* isomer has both methyl groups on the same side of the double bond, and the *trans* isomer has them on opposite sides. *Cis* and *trans* isomers are *geometric isomers,* isomers with the same connectivity and kind of bonds between atoms, but differing in the arrangement of their atoms in space (Section 18.5).

Geometric *cis* and *trans* isomers exist for alkenes in which both alkene carbon atoms have two different substituents.

The final alkene isomer with the formula C_4H_8 is the branched isomer 2-methylpropene, $CH_2{=}C(CH_3)_2$. *Cis* and *trans* isomers of 2-methylpropene are not possible. *Cis* and *trans* isomers of any alkene are not possible if one of the carbon atoms involved in the double bond has two substituents that are identical. The 1-butene, the two geometric isomers of 2-butene, and 2-methylpropene are structural isomers.

Example 21.4 *Cis-Trans* Isomers

Which of the following substituted alkenes have *cis* and *trans* isomers? Give the name of each compound.

(a) $(Cl)(Cl)C{=}C(H)(CH_3)$ — Cl (top left), Cl (bottom left), H (top right), CH_3 (bottom right)

(b) $(F)(Cl)C{=}C(F)(Cl)$ — F (top left), Cl (bottom left), F (top right), Cl (bottom right)

(c) $(CH_3)(H)C{=}C(Br)(H)$ — CH_3 (top left), H (bottom left), Br (top right), H (bottom right)

1-butene

cis-2-butene

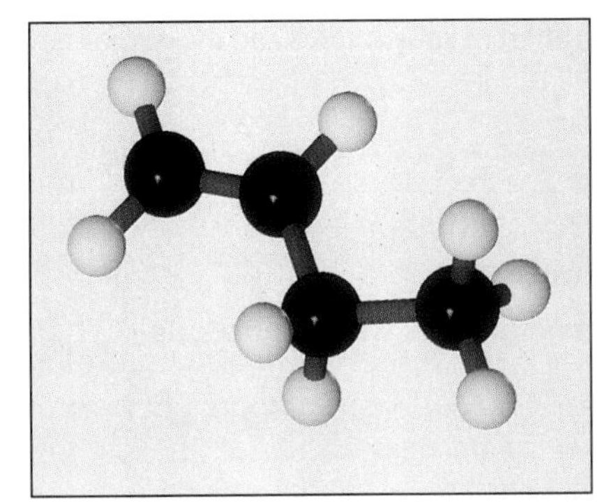

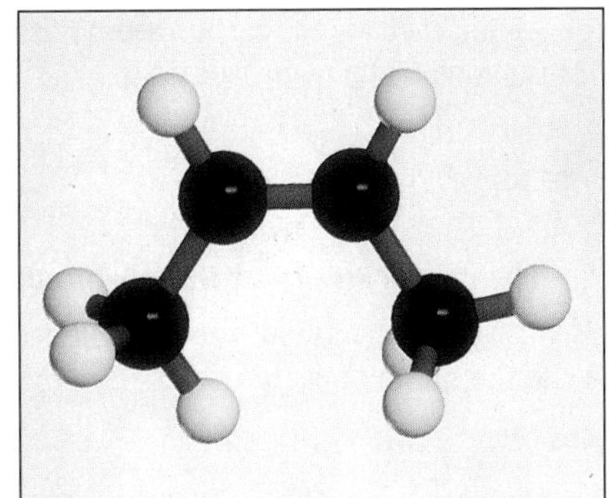

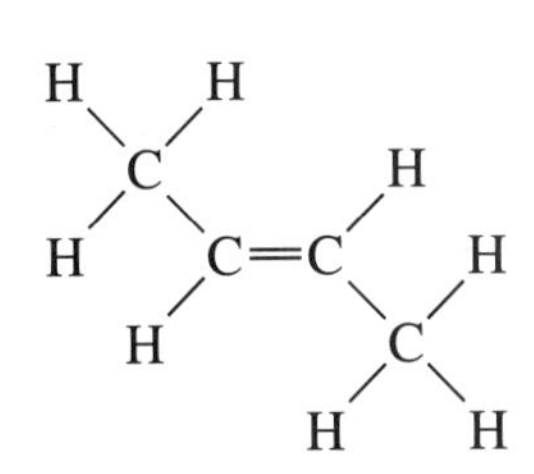

trans-2-butene

2-methylpropene

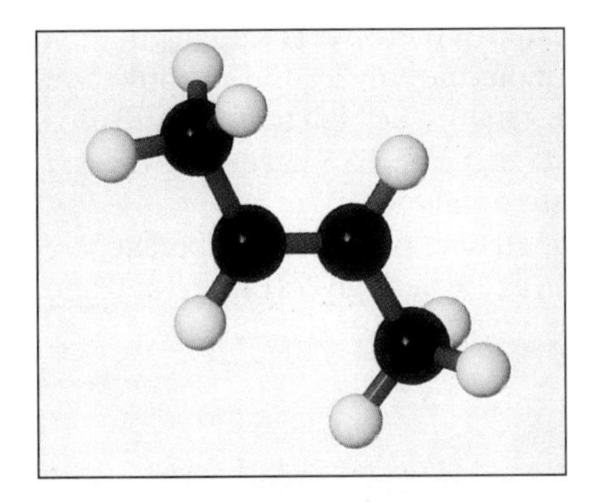

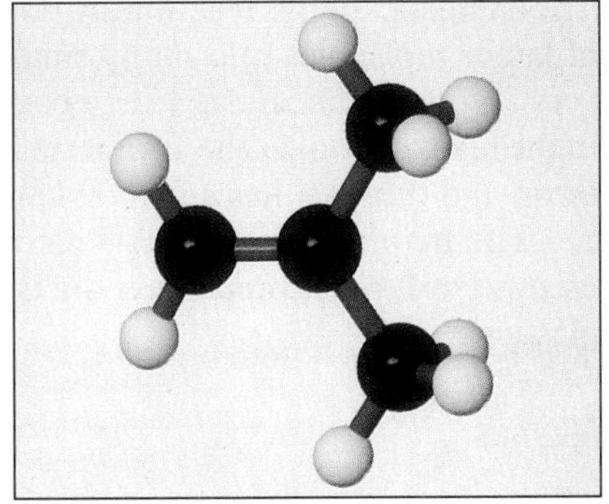

Figure 21.8 Structures of the four alkene isomers of C_4H_8 In the computer-generated figures, the double bonds are the shorter C—C distances.

Solution

(a) Two of the substituents on one of the alkene carbon atoms are the same—there are no *cis-trans* isomers. This compound is 1,1-dichloropropene.

(b) Each alkene carbon atom has two different substituents. The isomer shown is a *cis* isomer. The *trans* isomer is

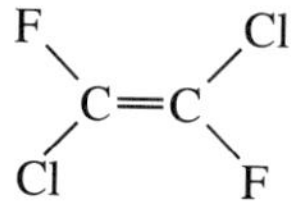

```
F       Cl
 \     /
  C=C
 /     \
Cl      F
```

The names are *cis-* or *trans*-1,2-dichloro-1,2-difluoroethene.

(c) Again, each alkene carbon atom has two different substituents and the *cis* isomer is shown. The *trans* isomer is

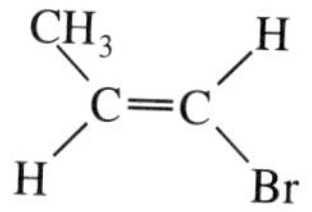

```
CH3      H
  \     /
   C=C
  /     \
 H       Br
```

The names are *cis-* or *trans*-1-bromopropene.

Understanding Draw all the possible isomers of alkenes with the formula C_5H_{10}. Write the name of each.

Answer:

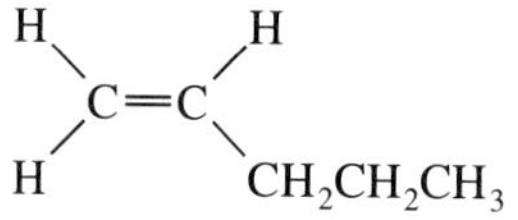

```
H       H                CH3       CH2CH3          CH3       H
 \     /                   \      /                  \      /
  C=C                       C=C                       C=C
 /     \                   /     \                   /     \
H       CH2CH2CH3         H       H                 H       CH2CH3
   1-pentene                cis-2-pentene              trans-2-pentene

H       CH3              CH3       H               H        H
 \     /                   \      /                 \      /
  C=C                       C=C                      C=C
 /     \                   /     \                  /     \
H       CH2CH3           CH3      CH3              H       CH(CH3)2
 2-methyl-1-butene        2-methyl-2-butene         3-methyl-1-butene
```

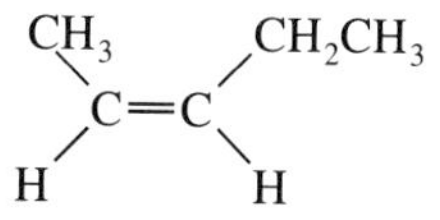

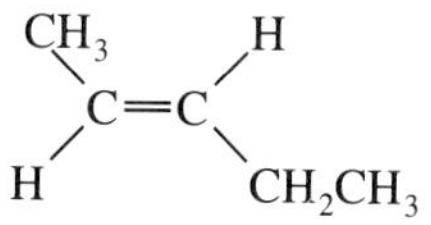

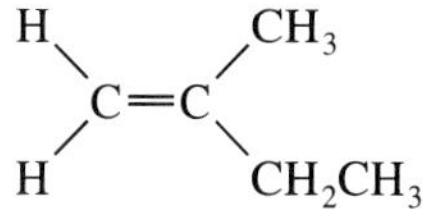

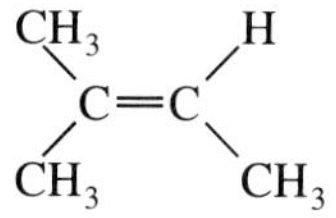

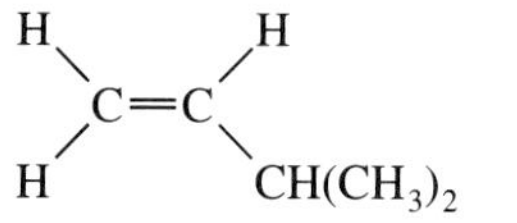

Alkenes can be prepared from alkanes by the high-temperature reaction known as thermal cracking, an important industrial process.

$$CH_3(CH_2)_nCH_3 \xrightarrow[\text{steam}]{900\ ^\circ C} H_2 + CH_4 + CH_2{=}CH_2 + CH_2{=}CHCH_3 + CH_2{=}CHCH_2CH_3, \text{ etc.}$$

$n = 0–6$

The energetics of this reaction are dominated by entropy. The fragmentation of larger molecules into many smaller molecules makes the entropy term very favorable (*i.e.*, it has a large positive value). The high temperature used in the reaction makes the $T\Delta S$ term in $\Delta G = \Delta H - T\Delta S$ larger than the ΔH term, and thus ΔG has a favorable negative value.

The main products of this reaction, ethene, propene, and butene, are used extensively to prepare products such as ethylene glycol (antifreeze) and plastics.

Alkynes

Alkynes are unsaturated hydrocarbons that contain carbon-carbon triple bonds. The general formula of an alkyne with one triple bond is C_nH_{2n-2}. Alkynes are named like alkenes, but a *-yne* suffix is used instead of *-ene*. The alkynes for $n = 2$ to 4 are shown in Figure 21.9.

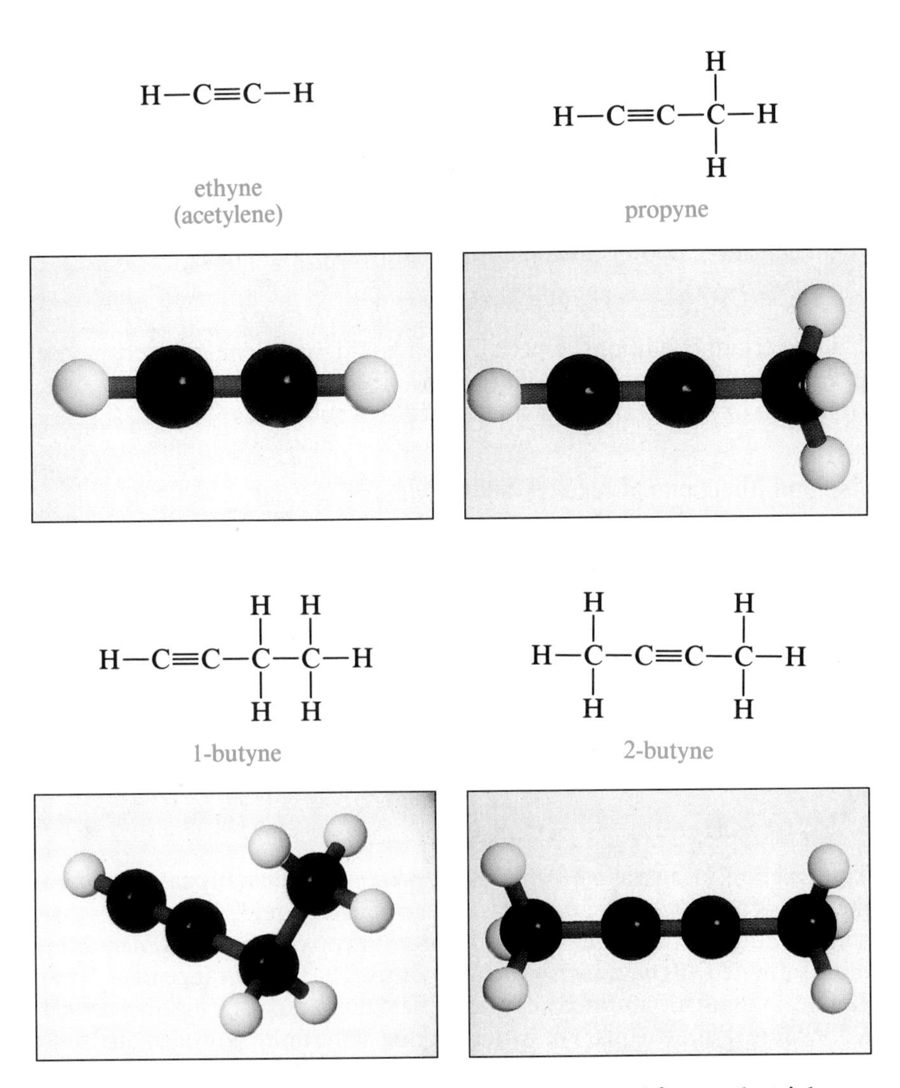

Figure 21.9 Structures of alkynes In the computer-generated figures, the triple bonds are the shorter C—C distances. A typical C—C triple bond length is 120 pm.

Each alkyne carbon atom uses *sp* hybridized orbitals to form two sigma bonds and two *p* orbitals to form two pi bonds. Because of the linear arrangement of the —C≡C— group, there are no *cis-trans* isomers for alkynes.

The simplest alkyne is ethyne, which is more commonly called by its older name, acetylene. Acetylene was once used as a starting material for synthesizing a variety of compounds such as acetic acid, but better routes to these compounds have now been developed that use ethene as the starting material. The industrial preparation of acetylene is based on an extremely high-temperature reaction of methane, again a reaction that is helped by a large and favorable $T\Delta S$ term.

$$2CH_4 \xrightarrow[1200\ °C]{steam} HC{\equiv}CH + 3H_2$$

Alkynes are named in the same way as alkenes, except that an *-yne* suffix is used instead of *-ene*. Like alkenes, alkynes have structural isomers; however, the linear —C≡C— bonding does not allow *cis-trans* isomers.

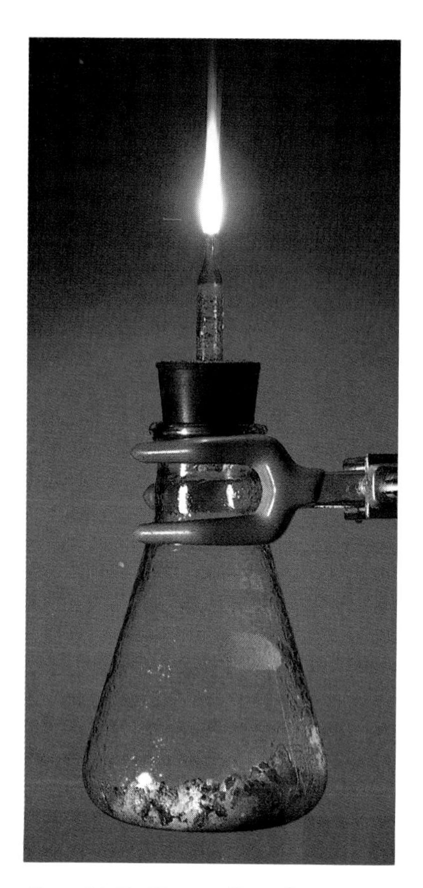

Figure 21.10 Preparation of acetylene Acetylene gas is prepared from the reaction of calcium carbide and water. The gas will burn in air if ignited.

In the laboratory, acetylene is prepared from the reaction of calcium carbide, CaC_2, and water (Figure 21.10).

$$CaC_2(s) + 2H_2O(\ell) \longrightarrow C_2H_2(g) + Ca(OH)_2(aq)$$

Acetylene is one of the few hydrocarbons that has a positive standard free energy of formation (209 kJ/mol). The decomposition reaction into hydrogen and carbon is thus thermodynamically favorable.

$$HC{\equiv}CH(g) \longrightarrow H_2(g) + 2C(s)$$

Under certain conditions, especially as a liquid, acetylene can decompose explosively. The combustion of acetylene produces very high temperatures, so that an oxygen/acetylene flame is used to weld metals (Figure 21.11).

Addition Reactions of Alkenes and Alkynes

Unsaturated hydrocarbon compounds are more reactive than saturated hydrocarbons because of the double or triple bonds. One common type of reaction is an **addition reaction,** the combination of two or more substances to form one new substance. For unsaturated hydrocarbons, small molecules such as hydrogen and HCl can add across the multiple bond. For example, hydrogen can add across the double bond of an alkene to form an alkane. In this reaction, each carbon atom changes its hybridization from sp^2 to sp^3, and one hydrogen atom attaches to each carbon atom.

$$CH_2{=}CH_2 + H_2 \xrightarrow{\text{catalyst}} CH_3CH_3$$

This reaction, known as a *hydrogenation reaction,* is generally catalyzed by a metal such as nickel. The reaction is used to convert liquid vegetable oils that contain double bonds into margarine or solid cooking fats. In general, the more saturated oils have higher melting points. The fixed arrangement of the double bonds in unsaturated oils makes them unable to pack together as well as the more saturated oils. The better packing in the more saturated oils leads

Figure 21.11 Oxyacetylene torch The reaction of acetylene and oxygen produces very high temperatures and is used to weld metals.

to stronger intermolecular forces, and thus to higher melting points. Studies have shown that unsaturated oils are better to consume for health reasons.

Halogens will also add across double and triple bonds. Either one or two moles of halogen can be added per mole of alkyne, depending on the conditions of the reaction.

The pi bonds make alkenes and alkynes more reactive than alkanes.

$$HC{\equiv}CH \xrightarrow{Br_2} CHBr{=}CHBr \xrightarrow{Br_2} CHBr_2CHBr_2$$

Aromatic Hydrocarbons

Aromatic hydrocarbons contain one or more benzene rings. Remember from Section 9.4 that benzene, C_6H_6, has a planar cyclic structure in which all of the carbon atoms are sp^2 hybridized (Figure 21.12). The bonding can be represented by two resonance forms, which are frequently indicated by using a dashed line or circle to indicate the delocalized pi bond formed by the unhybridized *p* orbitals. The delocalized structure is often written as a simple hexagon with the dashed circle in the middle.

A large number of compounds can be prepared by replacing one or more of the hydrogen atoms with some other group or groups. These compounds are named as substituted benzenes, although many aromatic compounds also have historical, nonsystematic names. Toluene, a common solvent, has

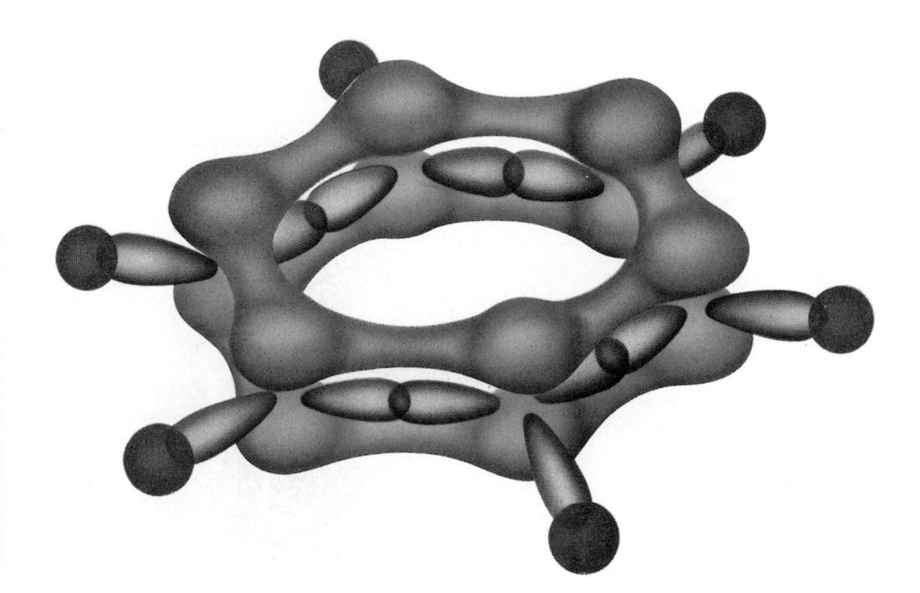

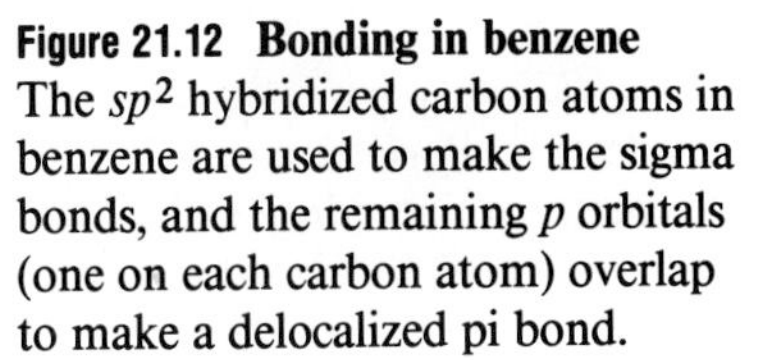

Figure 21.12 Bonding in benzene The sp^2 hybridized carbon atoms in benzene are used to make the sigma bonds, and the remaining *p* orbitals (one on each carbon atom) overlap to make a delocalized pi bond.

one methyl substituent on the ring. The systematic name, methylbenzene, is rarely used.

toluene

The second substituent on the benzene ring can take any of three different positions. Counting the carbon atom in the ring with the first substituent as position 1, a second substituent at position 2 is named *ortho* (*o*), one at position 3 is named *meta* (*m*), and one at position 4 is named *para* (*p*). (There are two ortho and meta positions, but, as always in naming compounds, the numbers are kept as low as possible.) The three isomers of dimethyl-substituted benzenes are known as xylenes, and they can also be named as substituted benzenes (Figure 21.13). Both toluene and the xylenes

Figure 21.13 Structures of the xylenes

1,2-dimethylbenzene
(*ortho*-xylene)

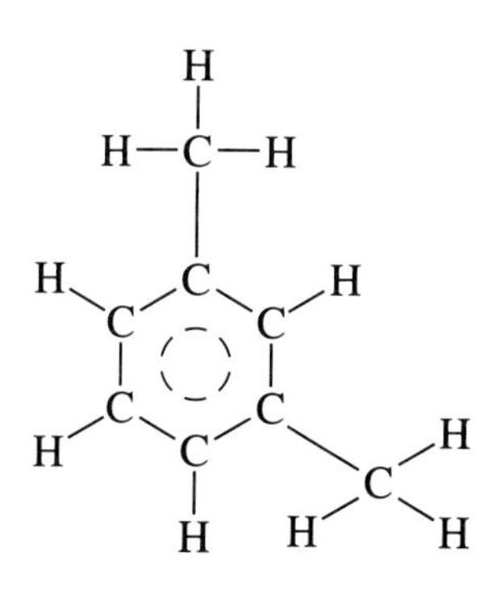

1,3-dimethylbenzene
(*meta*-xylene)

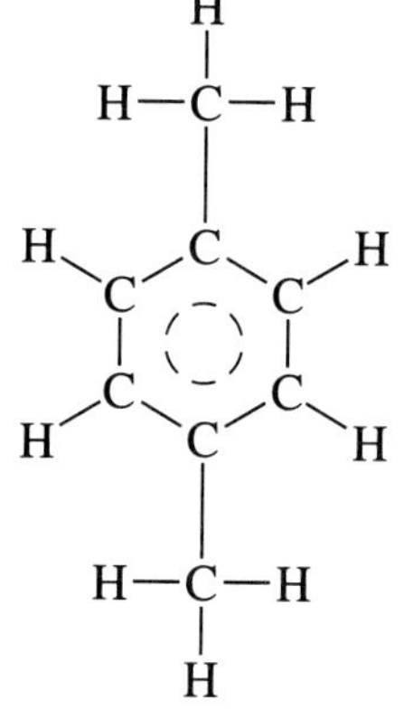

1,4-dimethylbenzene
(*para*-xylene)

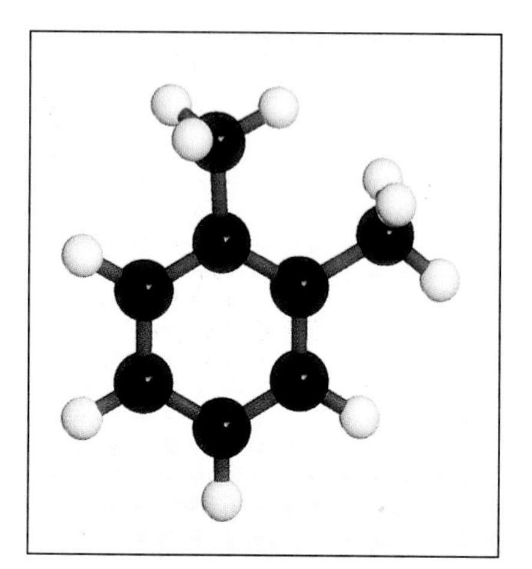

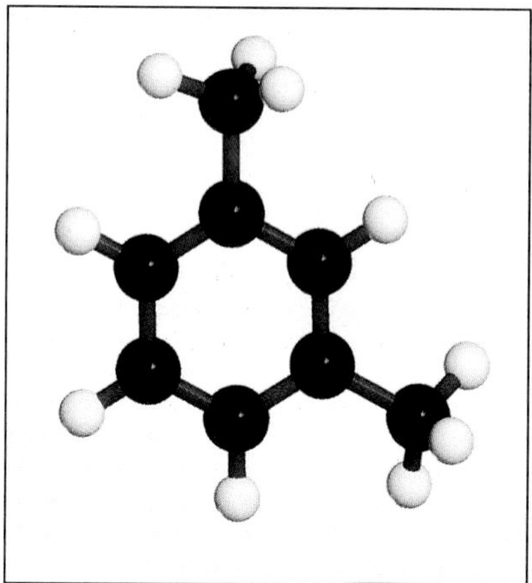

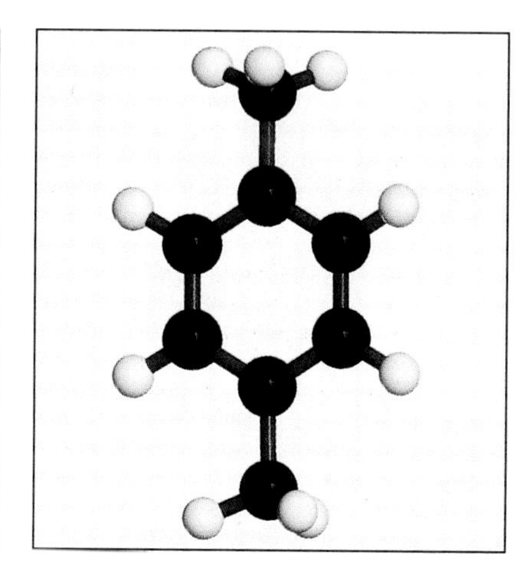

are used as industrial solvents and in paints. They are much less toxic than is benzene itself.

Many aromatic compounds with three or more substituents are also known. These are named as substituted benzenes, using numbers to indicate the position of the substituents; they may sometimes be named from the nonsystematic base name. The ring is numbered so as to minimize the numbers of the alphabetically listed substituents. When an aromatic ring is considered as a substituent, for instance on an alkyl chain, it is known as an **aryl group.**

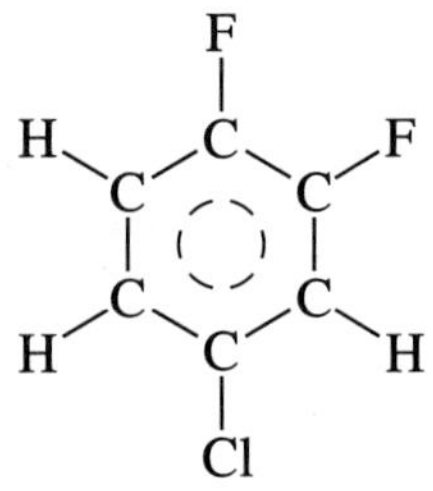

4-chloro-1,2-difluorobenzene

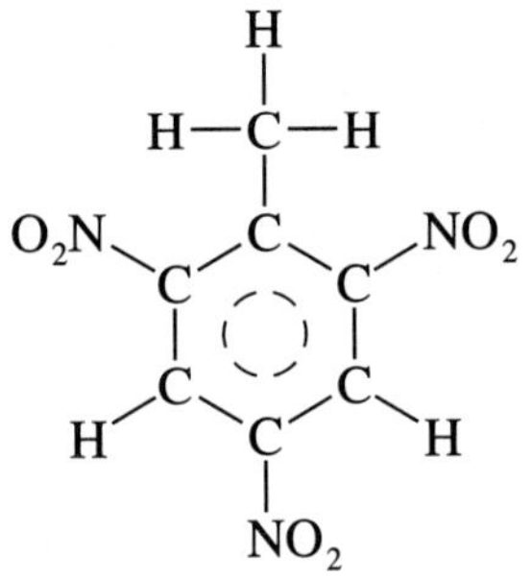

2,4,6-trinitrotoluene (TNT)

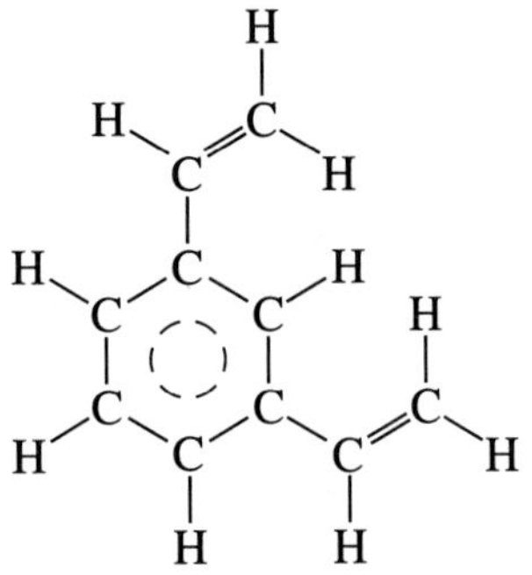

1,3-divinylbenzene

Example 21.5 Naming Substituted Benzenes

Name the following compounds.

(a)

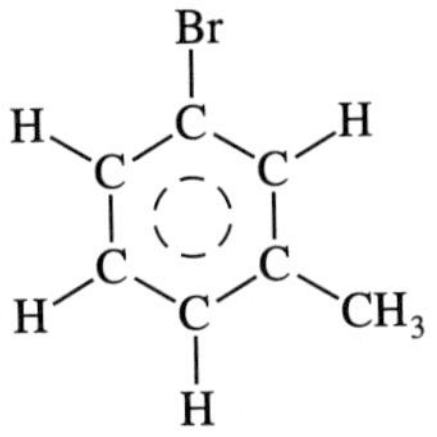

(b)

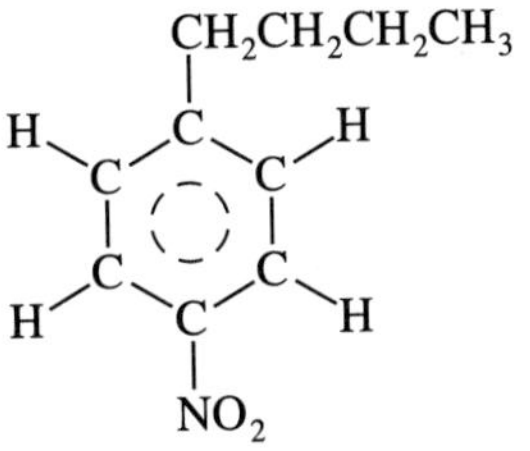

Solution

(a) The ring is numbered starting with the first group to be listed alphabetically, the bromine-substituted carbon. Thus, its systematic name is 1-bromo-3-methylbenzene. The compound can also be named as a substituted toluene: *m*-bromotoluene or 3-bromotoluene.

(b) This compound is 1-butyl-4-nitrobenzene.

Understanding Draw the structure of 1,2-dibromo-4-methylbenzene.

Answer:

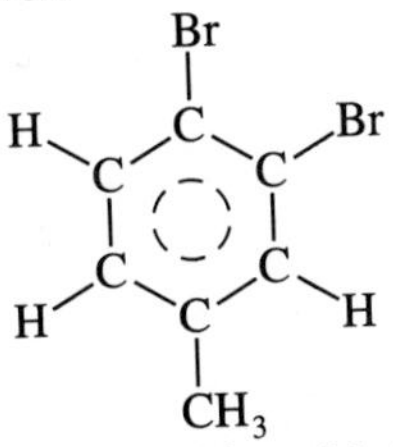

This compound could also be named 3,4-dibromotoluene.

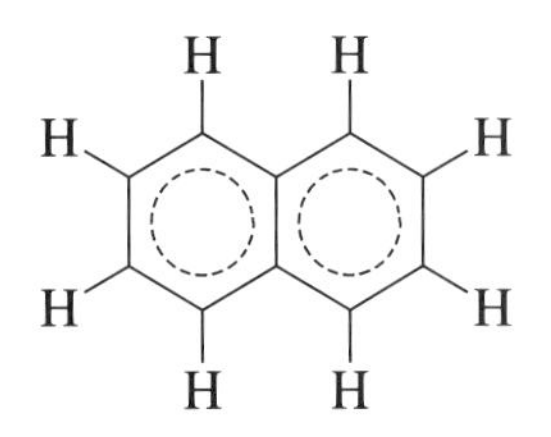

naphthalene

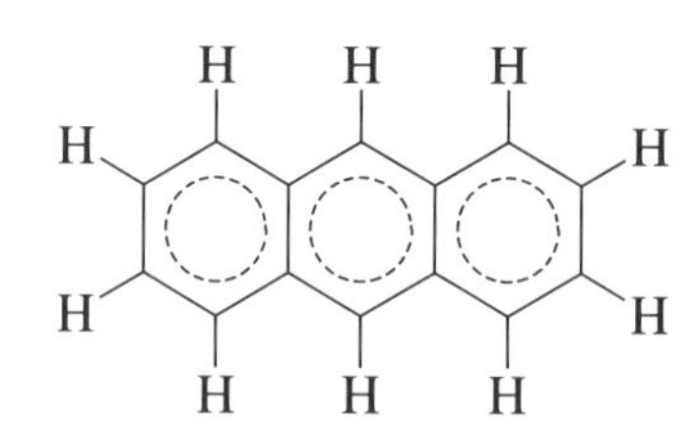

anthracene

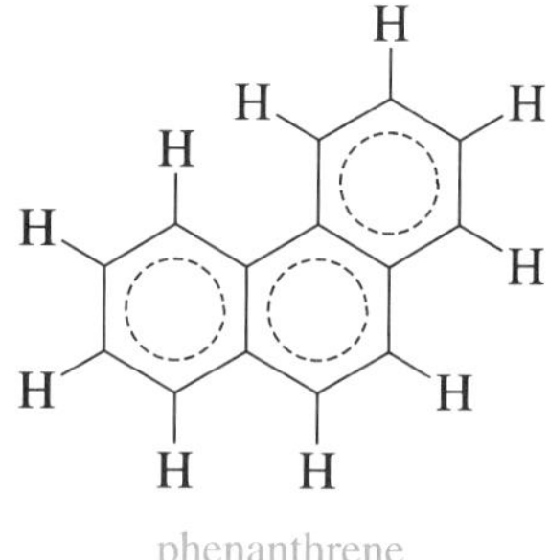

phenanthrene

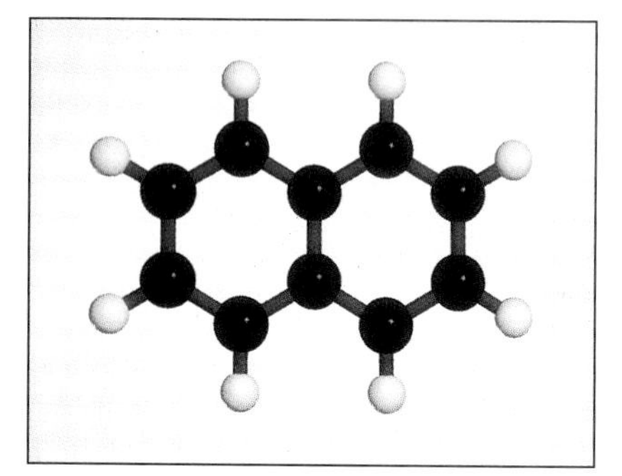

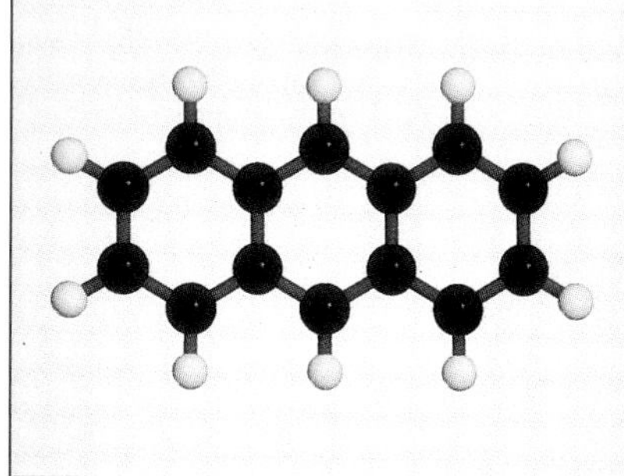

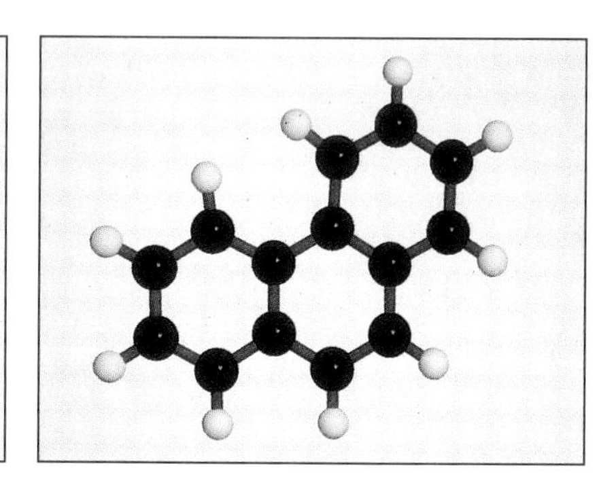

Figure 21.14 Fused ring aromatic hydrocarbons

Benzene rings can be fused together at two adjacent positions to form more complex aromatic compounds known as polycyclic aromatic hydrocarbons (Figure 21.14). The polycyclic aromatic hydrocarbons are planar because the aromatic rings are fused together with sp^2 hybridized carbon atoms. One such compound, naphthalene, is the white compound that gives the characteristic smell to mothballs. Many of these polycyclic hydrocarbons and their derivatives are formed by heating soft coal in the absence of oxygen. Coke (mainly carbon), used as the reducing agent in the steel industry, is also formed in this process.

Aromatic compounds generally undergo substitution reactions rather than addition reactions.

The reactions of aromatic hydrocarbons are different from those of other unsaturated hydrocarbons, in that addition reactions are not favored. An addition reaction would destroy the delocalized pi bonding of the aromatic ring. Instead, aromatic compounds undergo substitution reactions, frequently in the presence of a catalyst.

$$C_6H_6 + Cl_2 \xrightarrow{FeCl_3} C_6H_5Cl + HCl$$

21.3 Functional Groups

Objectives

- To identify the important functional groups
- To name organic compounds containing functional groups
- To identify chiral carbon atoms

$$\begin{array}{ccc} & \text{H} & \\ & | & \\ \text{H}- & \text{C} & -\text{OH} \\ & | & \\ & \text{H} & \end{array}$$

methanol

$$\begin{array}{ccccc} & \text{H} & & \text{H} & \\ & | & & | & \\ \text{H}- & \text{C} & - & \text{C} & -\text{OH} \\ & | & & | & \\ & \text{H} & & \text{H} & \end{array}$$

ethanol

$$\begin{array}{ccccccc} & \text{H} & & \text{OH} & & \text{H} & \\ & | & & | & & | & \\ \text{H}- & \text{C} & - & \text{C} & - & \text{C} & -\text{H} \\ & | & & | & & | & \\ & \text{H} & & \text{H} & & \text{H} & \end{array}$$

2-propanol (isopropanol)

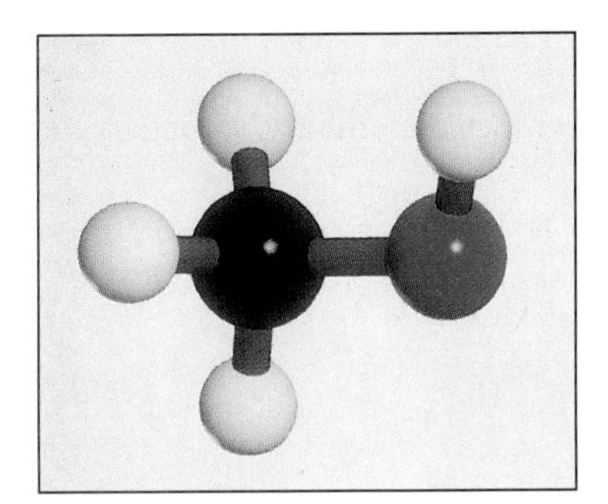

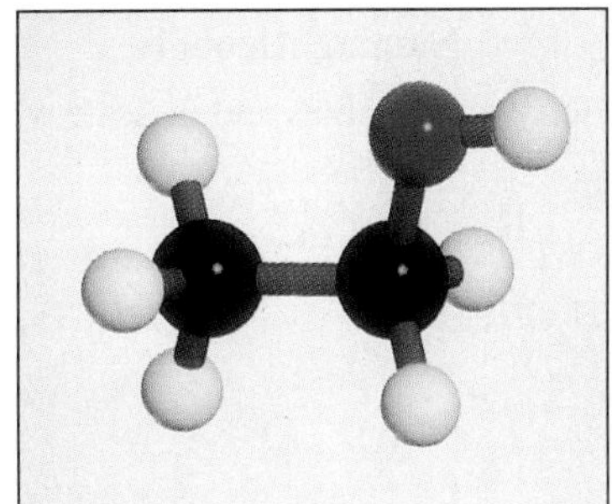

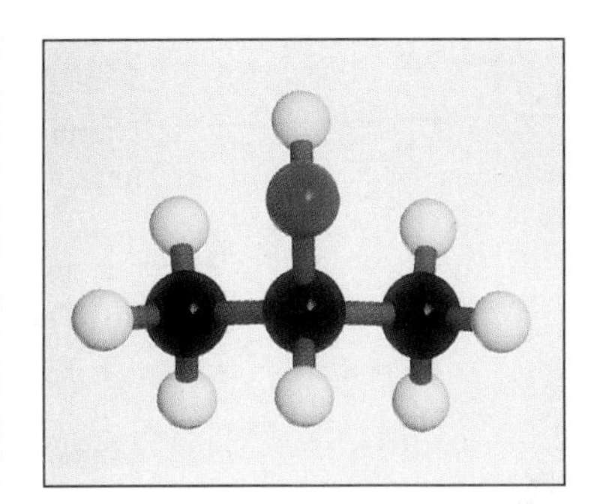

Figure 21.15 Structures of methanol, ethanol, and isopropanol

A **functional group** is an atom or small group of atoms in a molecule that undergoes characteristic reactions. Carbon-carbon double and triple bonds are functional groups. For example, it is expected that propene and 1-butene will react similarly because both contain the $CH_2{=}CH{-}$ (vinyl) functional group. In this section we will focus on functional groups containing oxygen or nitrogen atoms.

Alcohols

An **alcohol** contains the —OH *(hydroxyl)* functional group. Alcohols can be thought of as being derived from water by replacing one of the hydrogen atoms with an alkyl or substituted alkyl group. Like water, alcohols have strong intermolecular hydrogen bonding forces. As a result, alcohols tend to have higher boiling points than do hydrocarbons of approximately equal molecular weight. For example, the boiling point of butane is 0 °C, whereas the boiling point for an alcohol of similar molecular weight, $CH_3CH_2CH_2OH$, is 97 °C.

Alcohols containing fairly short alkyl substituents mix freely with water, but as the hydrocarbon portion increases in size the water solubility drops significantly. The three most common alcohols are methanol, ethanol, and isopropanol (Figure 21.15).

The name of an alcohol is based on the longest carbon chain to which the —OH group is attached, with the suffix *-ol* added. The chain is numbered beginning at the end nearer the hydroxyl group. The systematic name for isopropanol is 2-propanol. Other examples are

Alcohols contain the —OH functional group and are named with the suffix *-ol*.

$$CH_3CH_2CH_2OH \qquad \begin{array}{c} \qquad\qquad\quad CH_3 \\ \qquad\qquad\quad | \\ CH_3CH_2CH_2COH \\ \qquad\qquad\quad | \\ \qquad\qquad\quad H \end{array} \qquad \begin{array}{c} \quad\; CH_3 \\ \quad\; | \\ CH_3COH \\ \quad\; | \\ \quad\; CH_3 \end{array}$$

1-propanol (primary)   2-pentanol (secondary)   2-methyl-2-propanol (tertiary)

Alcohols are classified as *primary* when the —OH group is bonded to a carbon atom that is bonded to no more than one other carbon atom, *secondary* when the —OH group is bonded to a carbon atom that is bonded to two other carbon atoms, and *tertiary* when the —OH group is bonded to a carbon atom that is bonded to three other carbon atoms. It is important to distinguish these three types of alcohols because they react differently.

Example 21.6 Naming Alcohols

Name the following alcohols and indicate whether they are primary, secondary, or tertiary.

(a) 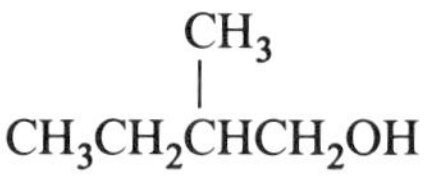 $$\begin{array}{l} \qquad\qquad CH_3 \\ \qquad\qquad\ \ | \\ CH_3CH_2CHCH_2OH \end{array}$$

(b) 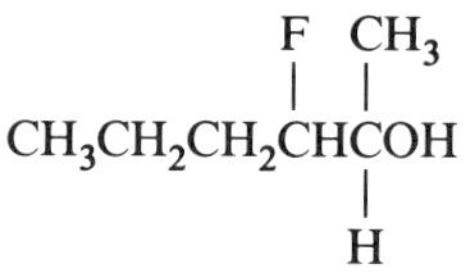 $$\begin{array}{l} \qquad\qquad\qquad\quad F\ \ CH_3 \\ \qquad\qquad\qquad\quad |\ \ \ \ | \\ CH_3CH_2CH_2CHCOH \\ \qquad\qquad\qquad\qquad\ \ | \\ \qquad\qquad\qquad\qquad\ H \end{array}$$

Solution

(a) The longest chain containing the hydroxyl group contains four carbon atoms. The hydroxyl group is at the 1-position, and the additional methyl group that is not in this chain is at the 2-position. The name is 2-methyl-1-butanol. This is a primary alcohol.

(b) All six of the carbon atoms are in the chain. The hydroxyl group is at the 2-position and the fluorine atom is at the 3-position. The name is 3-fluoro-2-hexanol. This is a secondary alcohol.

Name the following alcohol and indicate whether it is primary, secondary, or tertiary.

$$\begin{array}{l} \qquad\qquad CH_3 \\ \qquad\qquad\ \ | \\ CH_3CH_2COH \\ \qquad\qquad\ \ | \\ \qquad\qquad CH_2CH_3 \end{array}$$

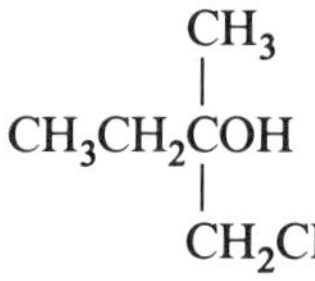

Answer: 3-methyl-3-pentanol, tertiary

Ethanol, often referred to as grain alcohol, is the alcohol in beer, wine, and other alcoholic beverages. It is formed by the fermentation of sugars and starches, which is catalyzed by yeasts.

$$\underset{\text{sugar}}{C_6H_{12}O_6} \xrightarrow{\text{yeast}} \underset{\text{ethanol}}{2CH_3CH_2OH} + 2CO_2$$

Oxygen must be excluded in this reaction or acetic acid, CH_3CO_2H, will also form. The carbon dioxide gas produced in the reaction helps to protect the reaction from the oxygen in air in the early stages of fermentation of grapes to form wine (Figure 21.16).

Only a small fraction of the ethanol used industrially is produced by fermentation. One exception is in Brazil, which has a major national effort to prepare ethanol from sugarcane to supply much of its energy needs. Most ethanol is prepared commercially by the acid-catalyzed addition reaction of ethene and water.

$$CH_2{=}CH_2 + H_2O \xrightarrow{H_2SO_4} CH_3CH_2OH$$

Ethanol is used as an intermediate for the production of other compounds and as a solvent. It is being used increasingly as an additive (5 to 10%) in gasoline (gasohol).

Methanol is another important alcohol. It was initially prepared by heating wood in the absence of air (to prevent combustion) and became known as wood alcohol. Methanol is very toxic to humans, causing blindness and death when ingested. In order to prevent the consumption of ethanol intended for commercial use, a small amount of methanol or other alcohol is sometimes added. This toxic mixture is known as *denatured alcohol,* and it should not be consumed.

Methanol is prepared industrially by the reaction of carbon monoxide and hydrogen, a reaction that is catalyzed by zinc oxide.

$$CO + 2H_2 \xrightarrow{400\ ^\circ C,\ ZnO} CH_3OH$$

While this reaction seems very simple, chemists continue to devote a considerable amount of research—some of which has been very successful—to lowering the temperature needed. Methanol is widely used as an industrial solvent. Recently, there has been interest in using methanol as a starting material for a variety of compounds because carbon monoxide and hydrogen can be prepared from coal and water. The increased use of methanol as a starting material would reduce the demand for crude oil, the present source of many products in the chemical industry.

The compound ethylene glycol, CH_2OHCH_2OH, is sold as antifreeze. This alcohol is added to automobile radiators to prevent the water from freezing at low temperatures and bursting the engine block. It is a *diol,* a compound containing two hydroxyl groups.

Figure 21.16 Fermentation of grapes The fermentation of grapes is so vigorous in its early stages that large amounts of CO_2 gas are produced.

Phenols

A **phenol** is a compound in which a hydroxyl group is substituted for a hydrogen atom on an aromatic ring. The name "*phenol*" also is used for the parent compound of this type, C_6H_5OH. Phenols are considered a separate class of compounds from alcohols because phenols are much more acidic. For example, the reaction of phenol and sodium hydroxide will go to completion, whereas the same reaction with cyclohexanol does not occur to any appreciable extent.

$$C_6H_5OH + NaOH \longrightarrow C_6H_5O^-Na^+ + H_2O$$

$$C_6H_{11}OH + NaOH \longrightarrow \text{no reaction}$$

INSIGHTS into CHEMISTRY

Science, Society, and Technology

Reaction of Dichromate Ion with Alcohol Is Basis for Breath Alcohol Test

Every court in this country considers driving a privilege, not a right. When a person drives a car (or other motorized vehicle), he or she pledges not to drive under the influence of alcohol. The local police authorities have the right (maybe even the obligation) to check whether a person who is driving erratically is under the influence of alcohol.

Most municipalities use a two-part test to determine if a driver is under the influence of alcohol. First, the officer will conduct a "field sobriety test," such as asking the suspect to walk a straight line or recite the alphabet. Second, a chemical analysis is employed. One common system uses the reaction of the dichromate ion, $Cr_2O_7^{2-}$, with ethyl alcohol to determine the concentration of alcohol in a suspect's breath.

The suspect breathes into an instrument that monitors pressure (to make sure that the suspect is cooperating) and routes the last portion of breath through an ampule that contains potassium dichromate in sulfuric acid. If the sample contains ethyl alcohol, the following reaction occurs:

$$CH_3CH_2OH + 2Cr_2O_7^{2-} + 16H^+ \longrightarrow 2CO_2 + 4Cr^{3+} + 11H_2O$$

The reaction takes about a minute to go to completion. The earlier breath analyzers (1950s-era devices) used an oven to heat the reaction, but modern instruments add a trace of a silver catalyst to ensure that the reaction proceeds within the time allotted to the analysis.

The dichromate solution is yellow before the analysis begins, but as $Cr_2O_7^{2-}$ is consumed, the yellow color fades. The alcohol concentration is measured by comparing the color of the ampule that contains breath to one that does not, the "blank." The loss of color is proportional to the amount of dichromate that has been used, a property that is relatively easy to measure.

Many phenols have industrial applications; phenol itself is used in the manufacture of resins and dyes, and 4-methylphenol (*p*-cresol) has been used as a disinfectant.

```
          OH                       OH
          |                        |
   H      C      H          H      C      H
    \   /   \   /            \   /   \   /
      C       C                C       C
      |  ( )  |                |  ( )  |
      C       C                C       C
    /   \   /   \            /   \   /   \
   H      C      H          H      C      H
          |                        |
          H                        CH3
       phenol                   p-cresol
```

The analysis of breath is not without controversy. Questions have been raised, and the techniques have been modified in response. Here are some of the most common questions voiced about breath analysis.

1. How does the instrument respond to someone who has just used an alcohol-containing mouthwash?

Operating procedures require at least a 20-minute observation time under police supervision. This period is sufficient to avoid effects of "mouth" alcohol. Similarly, police will ask a suspect to remove dental appliances or anything that could be a source of "mouth" alcohol.

2. What about other organic compounds in the breath? Won't the instrument respond to them?

Very few organic compounds can be expected in breath. The only substance that is occasionally found is acetone, which occurs in the breath of diabetics. The oxidation of acetone (or other similar organic compounds) will consume dichromate, causing an error. The conditions of the test, however, are adjusted so that the only common compound that reacts *during the time of the analysis* is a primary alcohol. The silver catalyst and the timed reaction discriminate against acetone and other organic substances.

3. Can the instrument give a wrong answer?

Yes. There can be operator errors or instrument malfunctions. Many of these problems would be seen in the calibration step when a synthetic breath sample of known alcohol concentration was introdued into the instrument. Any reading other than the expected one indicates a problem. The procedure is not perfect. For example, errors can cancel each other during the calibration step but not during the analysis step. In general, the procedure is accurate, but it relies heavily on a well-trained operator.

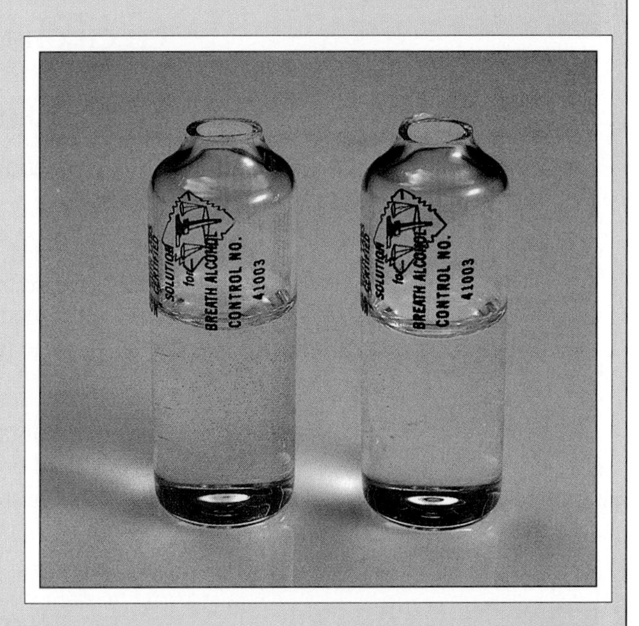

Ampules from a breathalyzer unit The ampule is yellow from $Cr_2O_7^{2-}$ at the start of the analysis and becomes clear as the alcohol reacts after a positive test.

Ethers

An **ether** contains a C—O—C functional group. An ether can be viewed as being formed from water by the replacement of both hydrogen atoms with carbon-based substituents. Unlike water and alcohols, the ether functional group cannot form hydrogen bonds by itself (there are lone pairs on oxygen but no hydrogen atoms with highly polarized bonds). Ethers have lower boiling points than do alcohols of similar molecular weight because of the lack of hydrogen bonding.

Ethers contain the C—O—C functional group and are named by the substituents, followed by the word *ether*.

Most simple ethers are named by giving the two substituents on the oxygen atom followed by the word *ether.* Diethyl ether, $CH_3CH_2—O—CH_2CH_3$, is the "ether" that is a well-known anesthetic. More recently, divinyl ether ($CH_2=CH—O—CH=CH_2$) and methyl propyl ether ($CH_3—O—CH_2CH_2CH_3$) have been used because they have fewer side effects. The boiling point of diethyl ether is quite low, 35 °C, only a few degrees above room temperature. This low boiling point means that a room in which diethyl ether is being used has a substantial buildup of the highly flammable vapors, and care must be taken to avoid any source of a spark or flame. A potential danger in storing ethers is that they react slowly with oxygen to form peroxides (compounds with a —OO— functional group). Peroxides have a tendency to explode violently.

A **condensation reaction** is a reaction that joins two molecules while producing a small molecule such as water. An ether can be prepared by an acid-catalyzed condensation reaction of two alcohol molecules.

$$CH_3CH_2OH + HOCH_2CH_3 \xrightarrow{H_2SO_4} CH_3CH_2OCH_2CH_3 + H_2O$$

Aldehydes and Ketones

The C═O functional group is called the **carbonyl group.** An **aldehyde** contains the $—\overset{\overset{\displaystyle O}{\|}}{C}—H$ functional group (Figure 21.17). The systematic names for aldehydes are based on naming the longest carbon chain containing the C═O group and adding the suffix *-al.* The carbonyl carbon atom is counted in the base name for the chain.

Aldehydes contain the $—\overset{\overset{\displaystyle O}{\|}}{C}—H$ functional group and are named with the suffix *-al.*

A **ketone** contains the $R—\overset{\overset{\displaystyle O}{\|}}{C}—R'$ functional group where the R and R′ groups are alkyl or aryl substituents (Figure 21.18). Ketones are frequently named by the method used for ethers; that is, by giving the names of the two

$H—\overset{\overset{\displaystyle O}{\|}}{C}—H$

methanal
(formaldehyde)

$H—\underset{\underset{\displaystyle H}{|}}{\overset{\overset{\displaystyle H}{|}}{C}}—\overset{\overset{\displaystyle O}{\|}}{C}—H$

ethanal
(acetaldehyde)

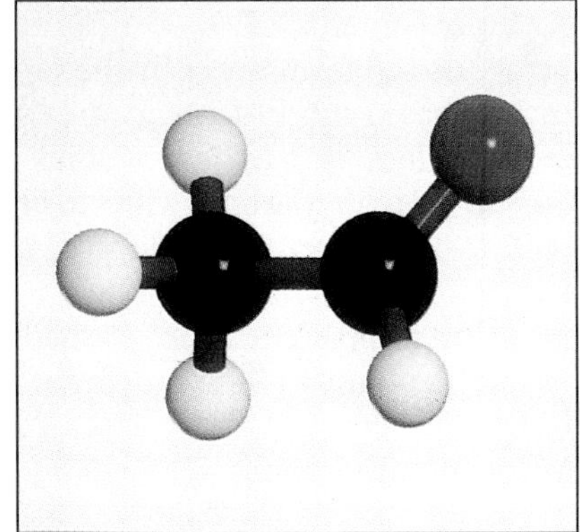

Figure 21.17 Aldehydes

$$\begin{array}{c} \text{H} \quad \text{O} \quad \text{H} \\ | \quad \| \quad | \\ \text{H}-\text{C}-\text{C}-\text{C}-\text{H} \\ | \qquad\quad | \\ \text{H} \qquad\quad \text{H} \end{array}$$

propanone
(acetone)

$$\begin{array}{c} \text{H} \quad \text{O} \quad \text{H} \quad \text{H} \\ | \quad \| \quad | \quad | \\ \text{H}-\text{C}-\text{C}-\text{C}-\text{C}-\text{H} \\ | \qquad\quad | \quad | \\ \text{H} \qquad\quad \text{H} \quad \text{H} \end{array}$$

2-butanone
(methyl ethyl ketone)

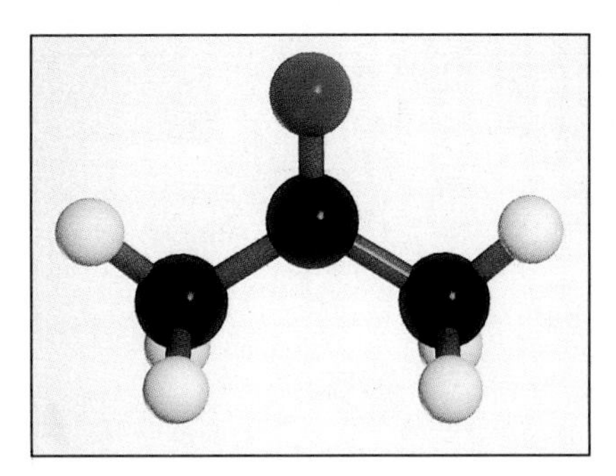

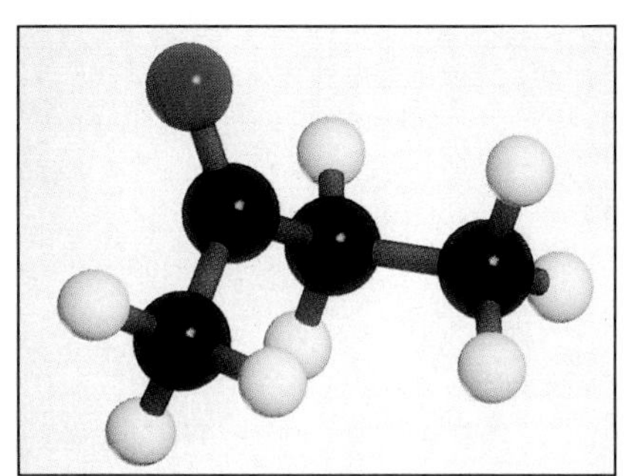

Figure 21.18 Ketones In the computer-generated figures, the double bonds are represented by the short C—O distances.

substituents followed by the word *ketone.* However, the systematic names for ketones are based on naming the longest carbon chain containing the C=O group and adding the suffix *-one.* A number is used to indicate the position of the carbonyl group. Propanone, the simplest ketone, has the carbonyl group at the 2-position and is better known by its common name, acetone. Acetone is widely used as a solvent. The carbonyl group makes acetone polar so that it is soluble in water as well as being a good solvent for many organic substances. Methyl ethyl ketone is also used extensively as a solvent. Both ketones have the important property of being relatively nontoxic compounds.

Ketones contain the $-\overset{\overset{\displaystyle O}{\|}}{C}-$ functional group and are named with the suffix *-one.*

Aldehydes and ketones may be prepared by the mild oxidation of alcohols. The oxidation of a primary alcohol yields an aldehyde, and the oxidation of a secondary alcohol yields a ketone.

$$CH_3CH_2OH + \tfrac{1}{2}O_2 \longrightarrow CH_3-\overset{\overset{\displaystyle O}{\|}}{C}-H + H_2O$$

$$CH_3-\underset{\underset{\displaystyle H}{|}}{\overset{\overset{\displaystyle OH}{|}}{C}}-CH_3 + \tfrac{1}{2}O_2 \longrightarrow CH_3-\overset{\overset{\displaystyle O}{\|}}{C}-CH_3 + H_2O$$

Carboxylic Acids and Esters

A **carboxylic acid** contains the $-\overset{\overset{\displaystyle O}{\|}}{C}OH$ (or CO_2H) functional group, the carboxyl group. The carboxyl group is the combination of a carbonyl and a hydroxyl group. Compounds that contain the carboxyl group are acids because the electron-withdrawing carbonyl group makes the O—H bond more

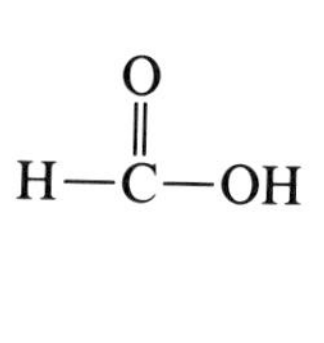

methanoic acid
(formic acid)

ethanoic acid
(acetic acid)

butanoic acid
(butyric acid)

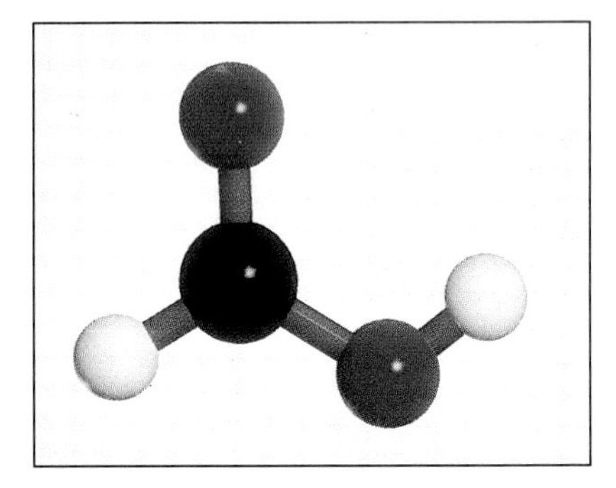

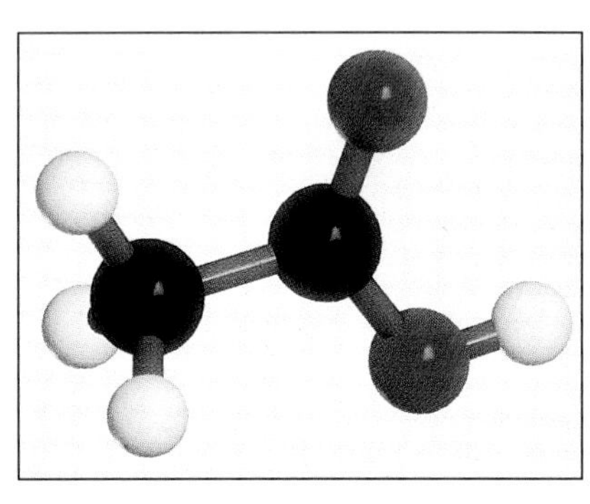

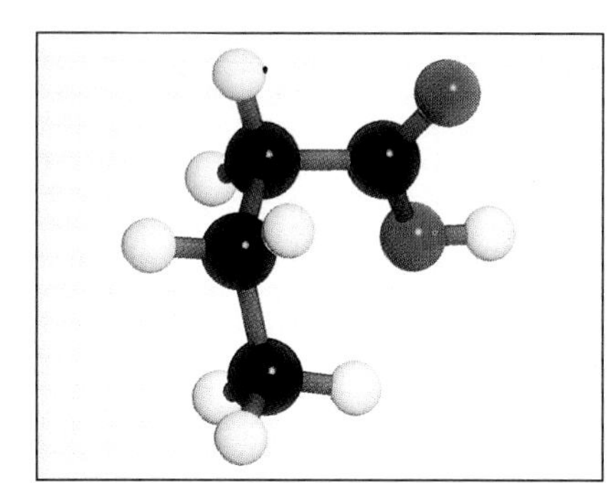

Figure 21.19 Carboxylic acids In the computer-generated figures, the double bonds are the short C—O distances.

polar than in alcohols, so the ionization of a proton is more likely. Ionization of the proton leaves the **carboxylate group,** $—CO_2^-$, an anion that is stabilized by two resonance structures.

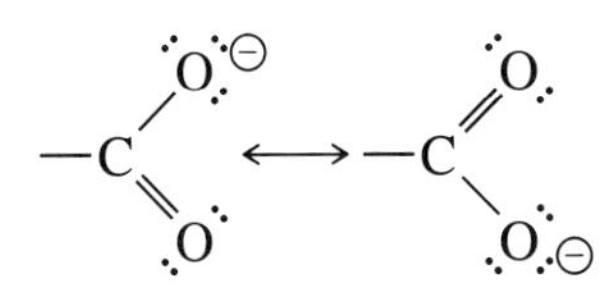

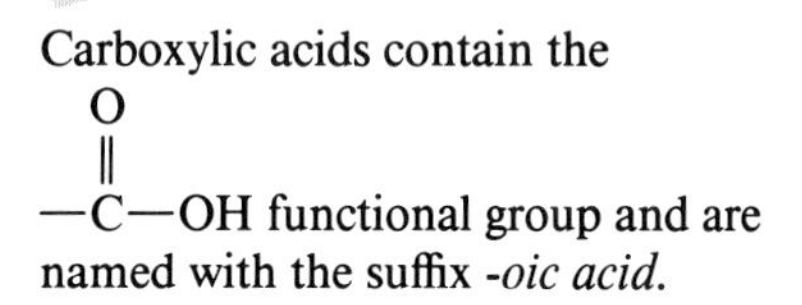

Carboxylic acids contain the —C(=O)—OH functional group and are named with the suffix *-oic acid.*

Still, carboxylic acids are fairly weak acids compared to H_2SO_4 or HCl.

The systematic method for naming carboxylic acids is based on the longest chain attached to the carboxylic acid group, with the suffix *-oic acid.* The carboxyl carbon atom is counted as part of the base name for the chain.

Carboxylic acids have been known for centuries. The more common ones have nonsystematic names based on their natural sources (Figure 21.19). The simplest acid, methanoic acid, was first isolated from ants, so its common name is formic acid (from the Latin *formica,* meaning "ant"). Ethanoic acid, known as acetic acid, is responsible for the sour taste in vinegar. Butanoic acid, also called butyric acid, is the compound that gives rancid butter its unpleasant odor.

Carboxylic acids are prepared from the oxidation of primary alcohols or aldehydes. A more powerful oxidizing agent is used in these reactions than is used in the oxidation of alcohols to form aldehydes and ketones. Milder oxidizing agents in the presence of a catalyst can also be used.

$$CH_3CH_2OH \xrightarrow{KMnO_4} CH_3—\overset{\displaystyle O}{\overset{\|}{C}}—OH$$

$$CH_3—\overset{\displaystyle O}{\overset{\|}{C}}—H + \tfrac{1}{2}O_2 \xrightarrow{Mn^{2+}} CH_3—\overset{\displaystyle O}{\overset{\|}{C}}—OH$$

INSIGHTS into CHEMISTRY

A Closer View

Soap's Unique Chemical Structure Enables it to Dissolve Oil into Water

It is difficult to wash an oil spot out of clothing with plain water because oil is a hydrocarbon that does not dissolve in water. Oil and water actually repel one another, so that oil, in the presence of water, will adhere even more strongly to clothing. The addition of soap or detergent to water changes the situation; soapy water can dissolve oil from clothing and rinse it away. What is special about the structure of soaps that makes them effective cleaning agents for oils and greases?

Most soaps are soluble sodium or potassium salts of carboxylic acids. The most common commercial soap is sodium stearate, $NaC_{18}H_{35}O_2$. It dissolves in water, forming the sodium and stearate ions. Even though most of the stearate ion is a hydrocarbon chain, it dissolves in water because of the carboxylate group. The carboxylate end is called *hydrophilic* (water-loving) and the hydrocarbon tail is called *hydrophobic* (water-fearing).

It is the long hydrocarbon chains of the stearate anions that dissolve the oils and greases. If water containing dissolved soap is mixed with oil, the hydrocarbon chains strongly attract the oil, while the ionic ends keep the soap dissolved into water. The oil spot is broken up into small droplets and dispersed into the water. The "tails" of many soap anions are needed to remove each oil droplet.

While the sodium salt of stearate ions and the anions of other soaps are soluble in water, the calcium and magnesium salts are not. Hard water contains these metal cations, so the anions precipitate, reducing the efficiency of the soap for dissolving oils. "Bathtub ring" is an example of the precipitation of soap by hard water. Thus, soaps will not clean well in hard water until most of the metal cations have been precipitated by reacting with the soap. In recent years, this problem has been solved by replacing soaps with detergents, generally compounds with long hydrophobic tails and the charged sulfate group such as sodium dodecyl sulfate, $NaCH_3(CH_2)_{11}OSO_3$. The calcium and magnesium salts of detergents generally remain soluble in water.

```
H     H   H   H   H   H   H   H   H   H   H   H   H   H   H   H   H
 \    |   |   |   |   |   |   |   |   |   |   |   |   |   |   |   |    O
H—C — C — C — C — C — C — C — C — C — C — C — C — C — C — C — C — C — C//
 /    |   |   |   |   |   |   |   |   |   |   |   |   |   |   |   |     \
H     H   H   H   H   H   H   H   H   H   H   H   H   H   H   H   H      O⁻
```

hydrophobic — hydrophilic

stearate anion

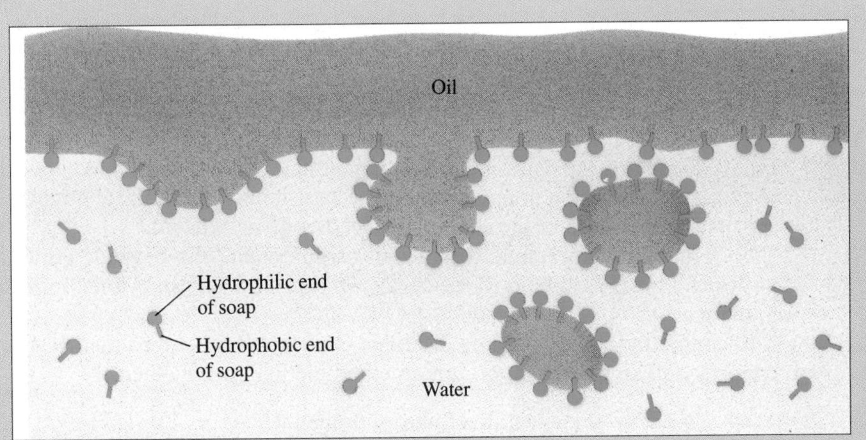

The hydrocarbon chain is hydrophobic. It penetrates the oil deposit and pulls small droplets into the water. The hydrophilic end of the soap keeps the droplets suspended in the water.

Esters contain the $-\overset{O}{\overset{\|}{C}}-OR$ functional group.

An **ester** contains the $-\overset{O}{\overset{\|}{C}}OR$ functional group, where R is an organic substituent such as methyl or phenyl. Esters are prepared by the acid-catalyzed condensation of a carboxylic acid and an alcohol. The OR portion of the ester originates with the alcohol.

$$CH_3\overset{O}{\overset{\|}{C}}-OH + HOCH_3 \longrightarrow CH_3\overset{O}{\overset{\|}{C}}-OCH_3 + H_2O$$

An ester is named by combining the name of the parent alcohol with the name of the parent acid, modified by replacing the *-ic* with the suffix *-ate*. The systematic name for the ester shown in the preceding equation is methyl ethanoate. It is more commonly called methyl acetate. In general, esters formed from the $CH_3C(O)O-$ group are called acetates. An important ester for industrial applications is vinyl acetate.

$$CH_3\overset{O}{\overset{\|}{C}}-OCH{=}CH_2$$

vinyl acetate

In contrast to many carboxylic acids, esters generally have pleasant odors and also contribute to the flavors of fruits.

Example 21.7 Names of Organic Compounds with Functional Groups

Name the following compounds.

(a) $CH_3CH_2\underset{CH_2CH_3}{\underset{|}{C}}H\overset{O}{\overset{\|}{C}}-OH$ (b) $CH_3CH_2\overset{O}{\overset{\|}{C}}CH_2Br$ (c) $C_6H_5-O-CH_3$

Solution

(a) The carboxyl functional group is part of a four-carbon chain (regardless of which CH_3CH_2- group is counted, both chains are the same length). There is also a five-carbon atom chain, but it does *not* contain the carbon atom of the carboxyl group. Therefore, the name is 2-ethylbutanoic acid. Note that it is not necessary to indicate the position of the acid group because it must always be at position 1, the first carbon atom of the chain on which the name is based.

(b) This compound is a ketone with a four-carbon chain, the bromide group at position 1, and the carbonyl group at position 2. The name is 1-bromo-2-butanone.

(c) This compound has a benzene substituent (a phenyl group) and a methyl group connected to an oxygen atom. It is an ether: methyl phenyl ether.

Understanding Draw the structure of phenyl propanoate.

Answer:

$$CH_3CH_2\overset{O}{\overset{\|}{C}}-OC_6H_5$$

Amines and Amides

An **amine** is a derivative of ammonia in which one or more hydrogen atoms are replaced with an organic substituent (R). Amines are related to ammonia in the same way that alcohols and ethers are related to water. Amines are named by alphabetically listing the names of the substituents attached to the nitrogen atom, followed by the suffix *-amine.* Amines with the formula RNH_2 are *primary amines* (methylamine and aniline), R_2NH are *secondary amines* (diethylamine), and R_3N are *tertiary amines* (triethylamine). Primary and secondary amines, like ammonia and alcohols, can form intermolecular hydrogen bonds and thus generally have fairly high boiling points.

Amines are derivatives of ammonia.

CH_3NH_2

methylamine

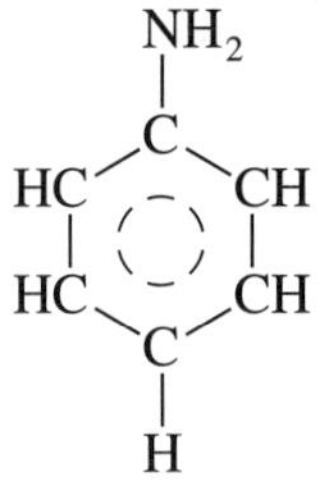

$C_6H_5NH_2$

aniline

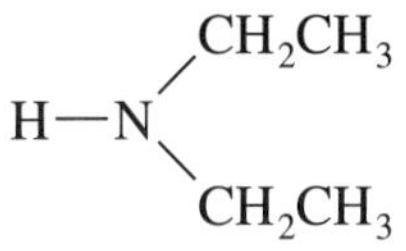

$H-N(CH_2CH_3)_2$

diethylamine

$CH_3CH_2N(CH_2CH_3)_2$

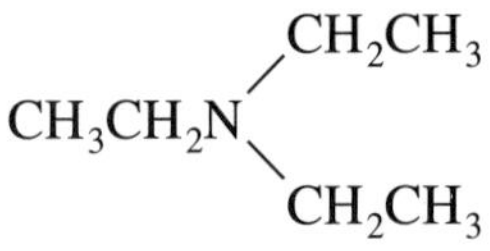

triethylamine

An **amide** contains the $-\overset{O}{\overset{\|}{C}}-NR_2$ functional group. An amide is formed in a condensation reaction of a primary or secondary amine or ammonia with a carboxylic acid.

$$CH_3\overset{O}{\overset{\|}{C}}-OH + HNR_2 \longrightarrow CH_3\overset{O}{\overset{\|}{C}}-NR_2 + H_2O$$

The amide functional group is an important link in the backbones of protein molecules (Section 21.5).

Amides contain the $-\overset{O}{\overset{\|}{C}}-NR_2$ functional group.

Review of Functional Groups

Table 21.3 on the next page lists the important functional groups that have been covered in the previous two sections.

Amino Acids and Chirality at Carbon

An **amino acid** contains a carboxyl group ($-CO_2H$) and an amine group ($-NH_2$). Glycine is the simplest amino acid.

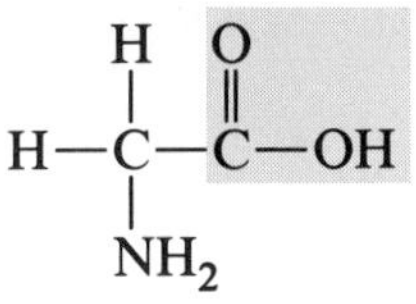

$H-CH(NH_2)-\overset{O}{\overset{\|}{C}}-OH$

glycine

$CH_3-CH(NH_2)-\overset{O}{\overset{\|}{C}}-OH$

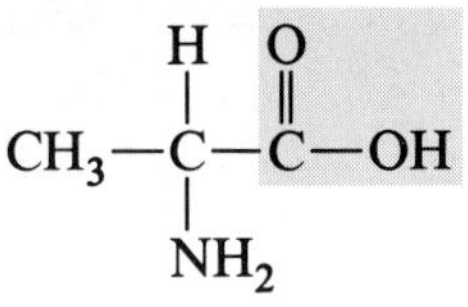

alanine

Table 21.3 Important Functional Groups

Functional Group	Name	Example
C=C	alkene	$H_2C{=}CH_2$ (ethylene)
C≡C	alkyne	$HC{\equiv}CH$ (acetylene)
—OH	alcohol (attached to alkyl group)	CH_3CH_2OH (ethanol)
—OH	phenol (attached to aryl group)	C_6H_5OH (phenol)
C—O—C	ether	$CH_3CH_2OCH_2CH_3$ (diethyl ether)
$-\overset{O}{\overset{\Vert}{C}}-H$	aldehyde	$CH_3\overset{O}{\overset{\Vert}{C}}-H$ (acetaldehyde or ethanal)
$-\overset{O}{\overset{\Vert}{C}}-$	ketone	$CH_3\overset{O}{\overset{\Vert}{C}}CH_3$ (acetone or propanone)
$-\overset{O}{\overset{\Vert}{C}}-OH$	carboxylic acid	$CH_3\overset{O}{\overset{\Vert}{C}}-OH$ (acetic acid or ethanoic acid)
$-\overset{O}{\overset{\Vert}{C}}-O-R$ (R = alkyl, aryl)	ester	$CH_3\overset{O}{\overset{\Vert}{C}}-O-CH_2CH_3$ (ethyl acetate or ethyl ethanoate)
$-NR_2$ (R = H, alkyl, aryl)	amine	$(CH_3CH_2)_2NH$ (diethylamine)
$-\overset{O}{\overset{\Vert}{C}}-NR_2$ (R = H, alkyl, aryl)	amide	$CH_3\overset{O}{\overset{\Vert}{C}}-NH_2$ (acetamide or ethanamide)

Glycine and alanine are called alpha amino acids, indicating that the acid and amine functional groups are attached to the same carbon atom. α-Amino acids are the building blocks of proteins.

A carbon atom that is bonded to four different substituents is a chiral center.

As outlined in Section 18.4, *chiral* molecules or ions are those whose mirror image structures cannot be superimposed on the original structure. A carbon atom that is bonded to four different substituents is chiral. As with most α-amino acids, alanine is chiral because the central carbon atom is bonded to four different groups, in this case —H, $-CH_3$, $-NH_2$, and —COOH. Chiral carbon atoms are called *asymmetric* centers. As shown in

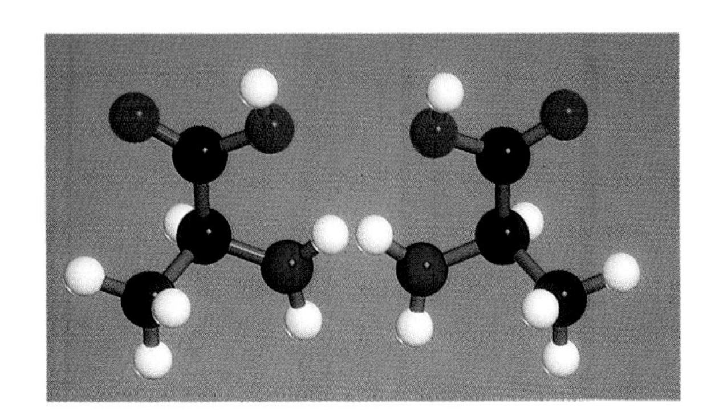

Alanine mirror images Computer drawings of L-alanine (left) and D-alanine (right).

Figure 21.20, the mirror image of alanine is not superimposable on the original. The form on the left is designated as L-alanine and that on the right as D-alanine to distinguish the arrangements of the groups.

While the difference between the two forms of alanine and other molecules containing a chiral carbon atom may seem small, it is crucial in biological reactions. Nearly all the amino acids that occur in living organisms are L-amino acids. The D-amino acids are not biologically equivalent.

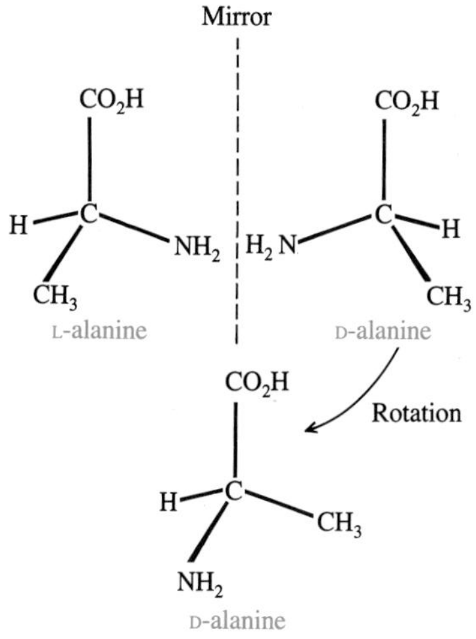

Mirror

CO_2H CO_2H

H—C—NH_2 H_2N—C—H

CH_3 CH_3

L-alanine D-alanine

Rotation

CO_2H

H—C—CH_3

NH_2

D-alanine

Figure 21.20 D and L isomers of alanine L-Alanine and its mirror image, D-alanine, are not superimposable. Rotation of the structure on the right, so the orientation of the C—H bond matches that of the structure on the left, places the CH_3 and NH_2 substituents in reversed positions.

21.4 Synthetic Organic Polymers

Objective

- To describe two common methods for the formation of polymers

A **polymer** is a very large molecule formed by the repeated bonding together of many smaller units *(monomers)*. Organic polymers are formed by joining together many small organic molecules. They may be synthetic polymers prepared by the chemical industry, or natural polymers made by living systems (most of these will be covered in the next three sections). Many synthetic polymers with different chemical and physical properties can be prepared.

Chain-Growth Polymers

A **chain-growth polymer** (also called an **addition polymer)** is a polymer chain formed from monomeric units with no loss of atoms. The simplest are **homopolymers,** polymers formed from the combination of many units of a single monomer compound. An example of a homopolymer is polyethylene, formed from a large number of ethylene molecules in the presence of a catalyst.

Chain-growth or addition polymers form from monomeric units with no loss of atoms.

$$n\mathrm{CH_2{=}CH_2} \xrightarrow{\text{catalyst}} \begin{array}{c} \mathrm{H\ \ H\ \ H\ \ H\ \ H\ \ H\ \ H\ \ H\ \ H} \\ -\mathrm{C{-}C{-}C{-}C{-}C{-}C{-}C{-}C{-}C}- \\ \mathrm{H\ \ H\ \ H\ \ H\ \ H\ \ H\ \ H\ \ H\ \ H} \end{array} \text{ or } \left(\begin{array}{c} \mathrm{H\ \ H} \\ -\mathrm{C{-}C}- \\ \mathrm{H\ \ H} \end{array} \right)_n$$

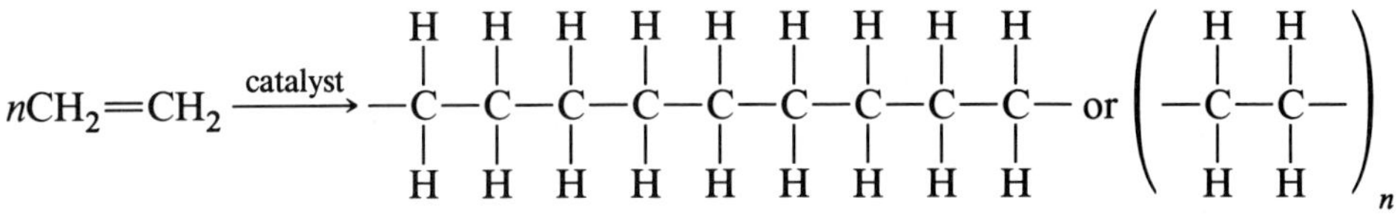

polyethylene

In this case the repeating unit of the polymer is simply a CH_2CH_2 group. In the preparation of polyethylene, not all of the chains are completely linear as shown in the preceding equation. Depending on the conditions of the reaction, some branching can occur.

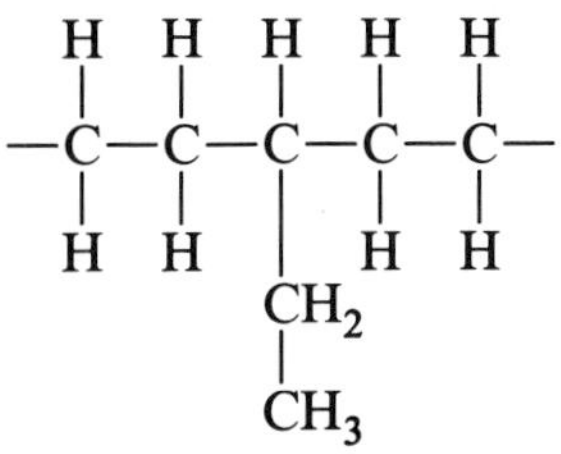

$$\begin{array}{c} \mathrm{H\ \ H\ \ H\ \ H\ \ H} \\ -\mathrm{C{-}C{-}C{-}C{-}C}- \\ \mathrm{H\ \ H\ \ \ |\ \ \ H\ \ H} \\ \mathrm{CH_2} \\ | \\ \mathrm{CH_3} \end{array}$$

Two forms of polyethylene are commercially available. Low density polyethylene has some branching in the chains, whereas in high density polyeth-

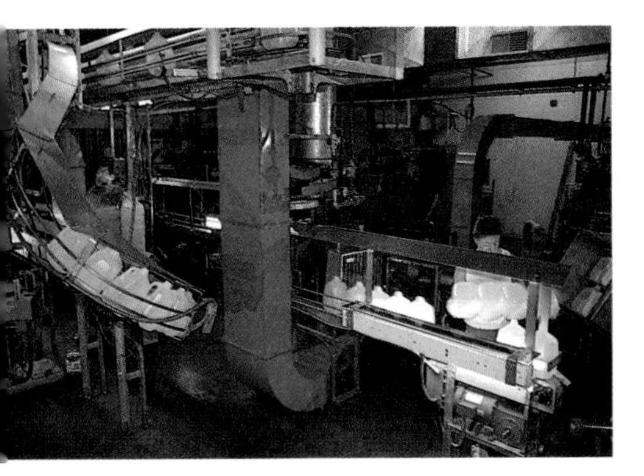

Figure 21.21 Polyethylene bottles High-density polyethylene is used to make containers for milk.

ylene the chains are nearly all linear (the more uniform linear chains can pack better, producing a denser solid). The high density polymer is harder and has greater strength, but the low density polymer is more transparent. Low density polyethylene is used as a film to wrap food and other products, and the high density polymer is used in piping, toys, and bottles (Figure 21.21).

The properties of the polymer can be further altered by starting with substituted alkenes. The chain-growth polymerization of propene (common name propylene) yields polypropylene, with the repeating unit of a CH_2CHCH_3 group.

$$n\mathrm{CH_2{=}CHCH_3} \xrightarrow{\text{catalyst}} \left(\begin{array}{cc} \mathrm{H} & \mathrm{H} \\ | & | \\ -\mathrm{C}- & \mathrm{C}- \\ | & | \\ \mathrm{H} & \mathrm{CH_3} \end{array} \right)_n$$

polypropylene

The formation of a polymer from a substituted ethylene is complicated by the fact that different arrangements of the substituents along the chain are possible, in addition to differences in the branching of the chains. Depending on the conditions of the polypropylene polymerization reaction, products ranging from a soft rubbery material to a hard solid can be formed. Much research has been carried out in this area in order to produce materials having a variety of desirable physical properties.

The chain-growth polymerization of tetrafluoroethylene produces the polymer known as Teflon.

$$n\mathrm{CF_2{=}CF_2} \xrightarrow{\text{catalyst}} \left(\begin{array}{cc} \mathrm{F} & \mathrm{F} \\ | & | \\ -\mathrm{C}- & \mathrm{C}- \\ | & | \\ \mathrm{F} & \mathrm{F} \end{array} \right)_n$$

Teflon

Teflon is an extremely unreactive material that is used for nonsticking cooking utensils, for coatings on valves, and as insulation around wires (Figure 21.22).

Figure 21.22 Teflon-insulated wire Teflon is extruded over a copper wire. Teflon is a good electrical and thermal insulator, so a much thinner layer of insulation can be used. The thinner wires are frequently used when older buildings are rewired, perhaps for computer networks, because a conduit can hold more Teflon-insulated wires than wires with conventional insulation.

Natural and Synthetic Rubbers

Natural rubber is a polymer secreted as a liquid by rubber trees. It is a polymer of isoprene, $CH_2{=}C(CH_3){-}CH{=}CH_2$. In the polymerization of a monomer containing two double bonds (a *diene*), the backbone of the polymer can have either a *cis* or a *trans* arrangement about the remaining double bond.

$$\left(\begin{array}{cccccc} \mathrm{CH_3} & & \mathrm{H} & \mathrm{CH_3} & & \mathrm{H} \\ & \mathrm{C{=}C} & & & \mathrm{C{=}C} & \\ -\mathrm{CH_2} & & \mathrm{CH_2{-}CH_2} & & & \mathrm{CH_2}- \end{array} \right)_n$$

natural rubber (all *cis*)

$$\left(\begin{array}{ccccccc} -CH_2 & & H & CH_3 & & CH_2- \\ & C{=}C & & & C{=}C & \\ CH_3 & & CH_2-CH_2 & & & H \end{array}\right)_n$$

gutta-percha (all *trans*)

The isomer generally found in nature has a *cis* configuration in the backbone of the chain, which leads to a material that is elastic. The *trans* isomer is hard and nonelastic, but it found a few specialty uses, such as in hockey pucks and the covers of golf balls before being replaced by modern polymers. Both isoprene isomers can also be produced synthetically.

Natural rubber is soft and can be pulled apart easily. Charles Goodyear developed a process known as *vulcanization,* which improves the properties of rubber. In vulcanization, the polymer is heated with a small amount of sulfur, and the polymer chains become interconnected with C—S—S—C bridges (Figure 21.23). These bridges prevent the individual chains from sliding over each other, leading to a harder and more elastic polymer. This bridging of the individual chains in the polymer is known as *cross-linking.*

Soft polymers can be converted to more useful materials by cross-linking the polymer chains.

The development of synthetic polymers was spurred when the disruptions caused by World War II made it difficult to import natural rubber. An example of a process developed at that time is the polymerization of $CH_2{=}CCl{-}CH{=}CH_2$ to the polymer with *trans* double bonds to produce neoprene, a compound with characteristics somewhat similar to those of natural rubbers.

$$2n\,CH_2{=}\overset{\overset{\displaystyle Cl}{|}}{C}{-}CH{=}CH_2 \longrightarrow$$

$$\left(\begin{array}{ccccccc} -CH_2 & & H & Cl & & CH_2- \\ & C{=}C & & & C{=}C & \\ Cl & & CH_2-CH_2 & & & H \end{array}\right)_n$$

neoprene

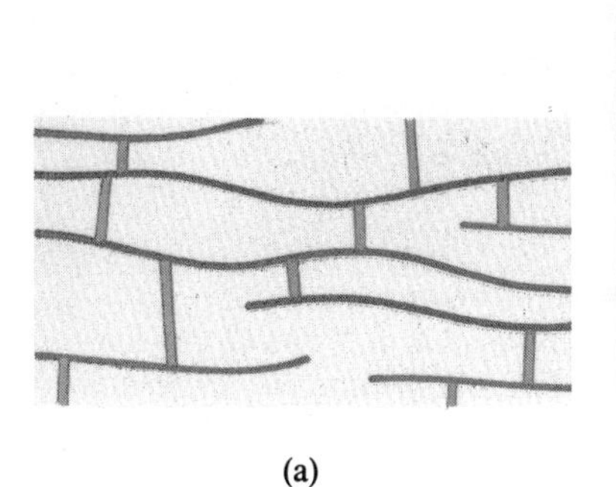

(a)

(b)

(c)

Figure 21.23 Vulcanized rubber Cross-linking the polymer chains (a) prevents them from coming apart. This process converts soft natural rubber (b) into a harder material (c).

Copolymers

The chain-growth polymers discussed so far are *homopolymers,* ones formed from one type of monomer. **Copolymers** are polymers formed from the combination of any units of more than one type of monomer. The plastic film Saran, used to wrap food and packages, is made from chloroethene (vinyl chloride) and 1,1-dichloroethene.

$$2n\mathrm{CH_2{=}CH(Cl)} + 2n\mathrm{CH_2{=}CCl_2} \longrightarrow \left(-\mathrm{CH_2{-}CH(Cl){-}CH_2{-}CCl_2{-}CH_2{-}CH(Cl){-}CH_2{-}CCl_2}-\right)_n$$

Saran

In the formation of Saran shown here, the two groups that form the polymer alternate in the backbone of the chain. In practice, some imperfections in the alternation occur.

Step-Growth or Condensation Polymers

Step-growth or condensation polymers form from monomeric units with the elimination of small molecules.

A **step-growth** or **condensation polymer** is a polymer formed by a reaction that eliminates a small molecule each time a monomer is linked to the polymer chain. Nylon 66 is an example; it results from the condensation of

Saran Wrap Saran is a copolymer made from chloroethene and 1,1-dichloroethene. Because the film is clear and sticks to itself and to smooth surfaces, it is useful as a food covering.

Formation of Nylon 66 Nylon forms at the interface of hexamethylenediamine (dissolved in water) and a derivative of adipic acid (adipyl chloride, dissolved in hexane).

hexamethylenediamine and adipic acid. In this polymerization reaction, an *amide* linkage forms with the elimination of water.

$$n\,\mathrm{HO{-}\overset{O}{\overset{\|}{C}}(CH_2)_4\overset{O}{\overset{\|}{C}}{-}OH} + n\,\mathrm{H{-}N(H)(CH_2)_6N(H){-}H} \longrightarrow$$

adipic acid hexamethylenediamine

$$\left(\mathrm{-\overset{O}{\overset{\|}{C}}(CH_2)_4\overset{O}{\overset{\|}{C}}-\overset{H}{\overset{|}{N}}(CH_2)_6\overset{H}{\overset{|}{N}}-} \right)_n + 2n\mathrm{H_2O}$$

nylon 66

Dacron Dacron was used to cover the wings and body of the Gossamer Albatross, the first human-powered plane to cross the English Channel. Dacron was chosen because it is lightweight and durable.

Polyesters are formed from the condensation of diacids and dialcohols. An example is Dacron, an extremely strong but lightweight material that is used in clothing and as a tire cord. It also has been used in surgery to help support weak or damaged arteries. It is formed from terephthalic acid and ethylene glycol.

$$n\,\mathrm{HO{-}\overset{O}{\overset{\|}{C}}{-}C_6H_4{-}\overset{O}{\overset{\|}{C}}{-}OH} + n\,\mathrm{HO{-}CH_2CH_2{-}OH} \longrightarrow$$

terephthalic acid ethylene glycol

$$\left(\mathrm{-\overset{O}{\overset{\|}{C}}-C_6H_4-\overset{O}{\overset{\|}{C}}-O-CH_2CH_2O-} \right)_n + 2n\mathrm{H_2O}$$

Dacron

21.5 Proteins

Objective

- To describe the structure of proteins

A variety of polymeric compounds are synthesized by living systems. **Proteins,** polymers of amino acids, are one of three important classes of polymers found in living systems. Proteins account for about 15% of the mass of the human body and perform many functions. Some proteins are used for structural purposes, such as in the skin, muscles, and cartilage. Others can act as catalysts for important biological reactions, as hormones to regulate reactions, and as antibodies to help fight disease.

Proteins form as a result of the step-growth polymerization of α-amino acids. Because each amino acid contains both an amine and an acid group, they can polymerize by forming amide linkages. This method of polymerization is similar to the formation of nylon, except that nature may use any of 20 different amino acids, each containing a different R group (see Table 21.4), to form the chains in proteins.

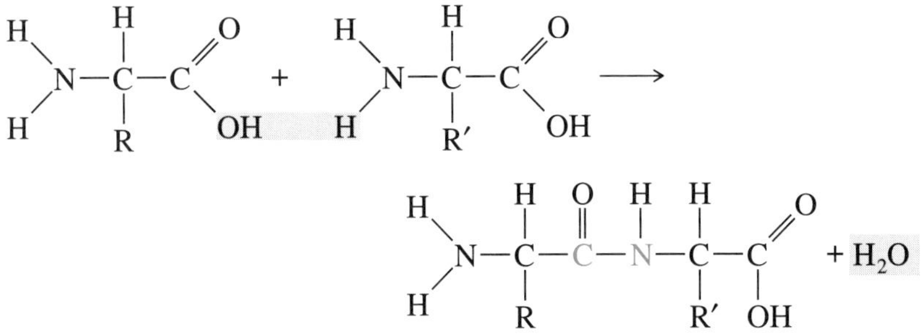

$$H_2N-\underset{\substack{|\\R}}{\overset{\substack{H\\|}}{C}}-\overset{\substack{O\\\|}}{C}-OH + H_2N-\underset{\substack{|\\R'}}{\overset{\substack{H\\|}}{C}}-\overset{\substack{O\\\|}}{C}-OH \longrightarrow H_2N-\underset{\substack{|\\R}}{\overset{\substack{H\\|}}{C}}-\overset{\substack{O\\\|}}{C}-\overset{\substack{H\\|}}{N}-\underset{\substack{|\\R'}}{\overset{\substack{H\\|}}{C}}-\overset{\substack{O\\\|}}{C}-OH + H_2O$$

The new molecule formed from two or more amino acids is called a **peptide,** and the new amide linkage is called a *peptide bond. Polypeptides* contain many amino acids, and extremely long ones are termed *proteins.*

With 20 different amino acids available to form a polymer, nature has the potential to synthesize a wide variety of proteins. One way to classify the various amino acids is to note that some have R groups that are polar, such as the alcohol group in serine, whereas others are nonpolar, such as the methyl group in alanine. Another useful classification involves whether the R group is acidic, basic, or neutral. The types and order in which R groups occur in proteins greatly influence their properties.

Protein Structure

The primary structure of a protein is the sequence of amino acids in the polypeptide chain.

The structures of proteins are complex and can be described at four different levels. The **primary structure** is the sequence of amino acids in the polypeptide chain. The primary structure can be represented as a picture of all the atoms or by using the abbreviations given in Table 21.4 (see Figure 21.24). When we write the primary structure using the three-letter abbreviations, we assume that the amino terminal end of each amino acid, and thus of the whole peptide, is on the left and the carboxylic acid end is on the right. This convention is important because the dipeptide ser-val is a different compound from val-ser.

The protein chains can assume different shapes. The **secondary structure** describes the shape of the polypeptide chain, which is determined by the types of hydrogen bonds made by the amide portion of the chains. The

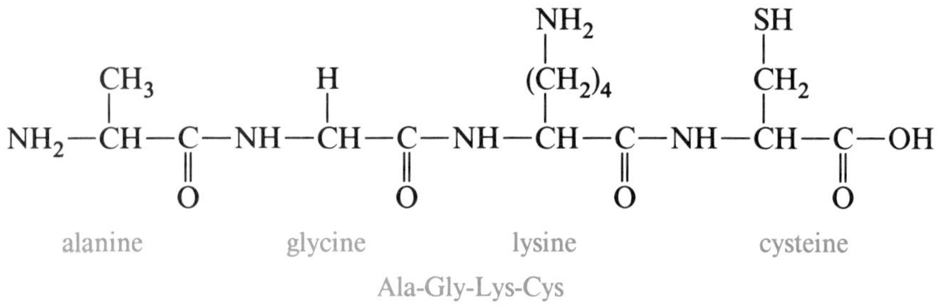

$$NH_2-\underset{\text{alanine}}{\overset{\substack{CH_3\\|}}{CH}}-\overset{}{\underset{\substack{\|\\O}}{C}}-NH-\underset{\text{glycine}}{\overset{\substack{H\\|}}{CH}}-\underset{\substack{\|\\O}}{C}-NH-\underset{\text{lysine}}{\overset{\substack{NH_2\\|\\(CH_2)_4\\|}}{CH}}-\underset{\substack{\|\\O}}{C}-NH-\underset{\text{cysteine}}{\overset{\substack{SH\\|\\CH_2\\|}}{CH}}-\underset{\substack{\|\\O}}{C}-OH$$

Ala-Gly-Lys-Cys

Figure 21.24 The primary chain of a polypeptide The primary sequence of a peptide is the order of amino acids in the polypeptide chain. This structure can be written using the structural formula or the shorthand notation for each amino acid.

Table 21.4 The 20 α-Amino Acids in Proteins

Nonpolar R Groups

glycine (gly)

valine (val)

alanine (ala)

isoleucine (ile)

proline (pro)

tryptophan (trp)

phenylalanine (phe)

methionine (met)

leucine (leu)

Polar R Groups

serine (ser)

cysteine (cys)

asparagine (asn)

tyrosine (tyr)

threonine (thr)

histidine (his)

glutamine (gln)

glutamic acid (glu)

aspartic acid (asp)

lysine (lys)

arginine (arg)

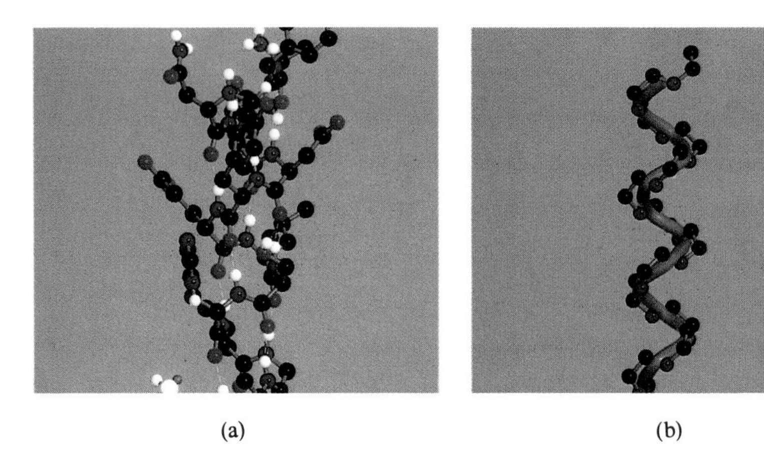

Figure 21.25 **α-Helix** (a) Hydrogen bonding between amine groups in the same region of a polypeptide chain leads to the formation of the α-helix structure. (b) A projection containing only the atoms in the polypeptide chain shows the spiral clearly.

hydrogen atoms of the polar N—H bonds can form hydrogen bonds to the amide oxygen atoms that are in the same chain or to those in other chains. An **α-helix** is the spiral structure adopted by a peptide chain that is held together by hydrogen bonding between amide groups in the same region of the chain (Figure 21.25). The coiled structure of an α-helix is the main arrangement in structural proteins that need to be able to stretch, such as those in hair and skin.

The secondary structure determines the shape of the polypeptide chain.

Hydrogen bonding between amide groups in different polypeptide chains or in different sections of the same chain leads to the formation of a **β-pleated sheet** structure (Figure 21.26). In a β-pleated sheet structure, the peptide chain is nearly fully extended, in contrast to the coiled structure of the α-helix. This structure is found in fibers that are strong but do not readily stretch, such as silk.

The types of R groups in the amino acids within the chain control whether a segment of the protein forms an α-helix or a β-pleated sheet. Bulky R groups favor the spiraling chain of the helix, because the groups are well separated in this structure. In contrast, bulky R groups will disrupt the structure of a β-pleated sheet. In general, any section of a polypeptide chain has only one of these two types of structures. In addition, a typical polypeptide chain has random coil sections that have no organized structure.

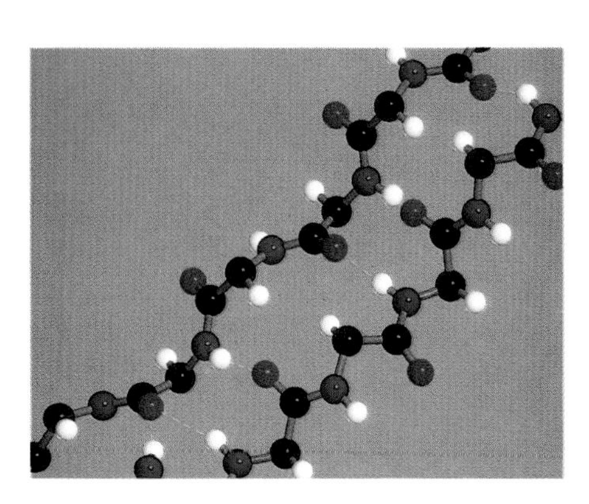

Figure 21.26 **β-pleated sheet** Hydrogen bonding between different sections of a polypeptide chain or different chains leads to β-pleated sheets. Only a part of the extended sheet is shown.

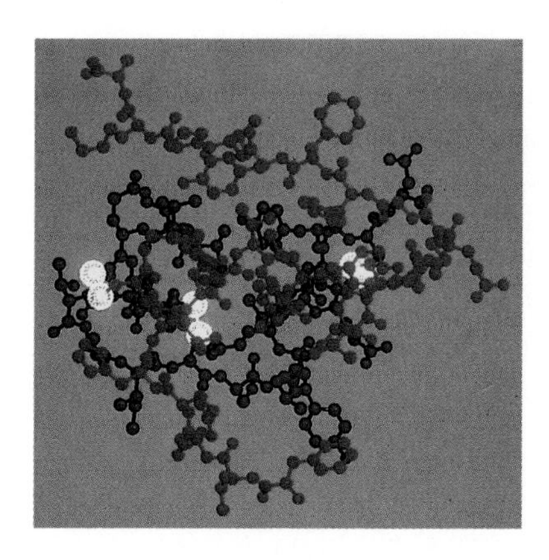

Figure 21.27 Tertiary structure of insulin The three disulfide linkages (yellow) are a controlling factor in the arrangement of the tertiary structure of insulin. Insulin is formed from two peptide chains held together by two S—S bonds. A third S—S bond is one chain.

The **tertiary structure** is the overall three-dimensional arrangement of the protein. Tertiary structure is determined by a variety of factors such as dipole-dipole, dispersion, and hydrogen bonding forces within the protein. Also, covalent bonds between different chains or with another part of the same chain contribute to the tertiary structure. Disulfide bridges, formed by oxidation of —SH groups in cysteine, are particularly important.

The tertiary structure is the overall three-dimensional arrangement of the protein.

$$\text{—SH} + \text{HS—} \xrightarrow{[O]} \text{—S—S—} + H_2O$$

Insulin, a hormone that regulates glucose utilization, contains three disulfide linkages. Insulin is formed from two peptide chains held together by two S—S bonds, while a third S—S bond exists within the shorter chain (Figure 21.27).

Finally, some proteins have **quaternary structure,** which describes the orientation of different polypeptide chains with respect to each other. The classic example is hemoglobin, which consists of four polypeptide chains (Figure 21.28).

The quaternary structure describes the orientation of different polypeptide chains within the protein.

Figure 21.28 Hemoglobin The structure of hemoglobin consists of four interacting polypeptide chains, shown here with two colored green and two colored blue. Each polypeptide chain coordinates an iron ion (red disk) that is used for the transport of oxygen.

Denaturation is the loss of the structural organization of a protein. Mild heating or the presence of metal ions such as mercury or lead will cause many proteins to lose part or all of their structure. One example is the changes that take place when an egg is boiled. On heating, the protein albumin is converted from a soluble protein (clear egg white) to an insoluble denatured protein (a white opaque gel). Heavy metals disrupt protein structure by breaking the disulfide bridges and bonding to the sulfide functional groups.

21.6 Carbohydrates

Objective

- To describe the structures of carbohydrates

Carbohydrates, polyhydroxyl aldehydes or ketones or substances that react with water to yield such compounds, are the second important class of polymers in living systems. The name arises from the fact that most carbohydrates have the empirical formula $C_n(H_2O)_m$. There are a wide variety of carbohydrates, including sugars, starches, and cellulose. Sugar and starch are common food sources, and cellulose is the main structural material of plants. Carbohydrates form in plants from water and CO_2 in a process catalyzed by chlorophyll and activated by sunlight.

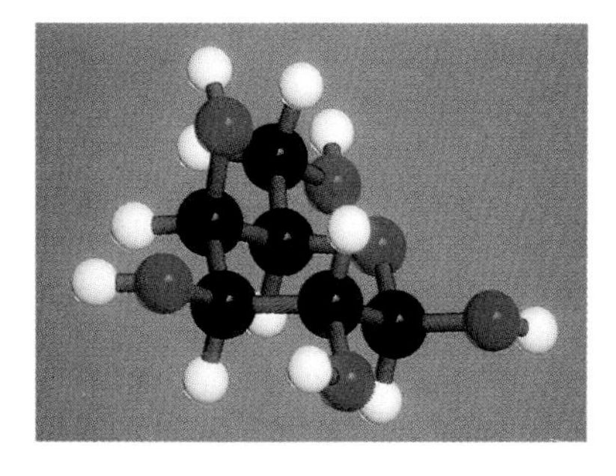

β-galactose

$$mH_2O + nCO_2 \xrightarrow[\text{sunlight}]{\text{chlorophyll}} C_n(H_2O)_m + nO_2$$

This overall reaction, known as photosynthesis, is endothermic; the energy that is consumed is supplied by the sunlight.

Monosaccharides

Monosaccharides are the basic units of carbohydrates, having the general formula $(CH_2O)_x$ where $x = 3$ to 8. Monosaccharides are simple sugars such as glucose, $C_6H_{12}O_6$ (Figure 21.29).

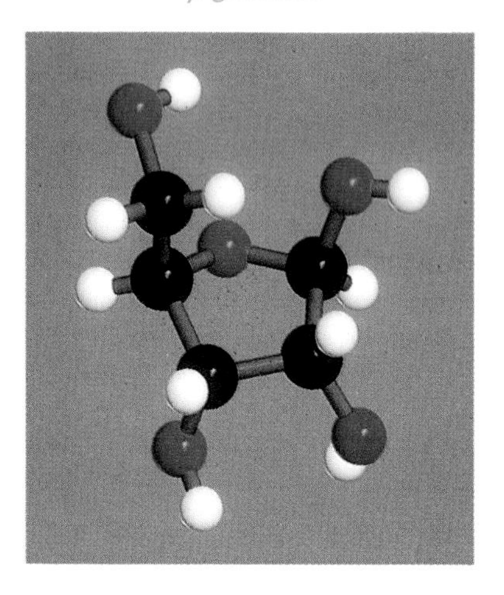

Ribose

As pictured, glucose exists in water in two different cyclic forms. Both are six-membered rings that adopt a "chair" conformation. In this conformation, there are two locations for the hydrogen atoms and hydroxyl groups on the ring carbon atoms. The bonds to hydrogen atoms can be parallel to the rough plane formed by the carbon atoms in the ring (equatorial) or perpendicular to this plain (axial). Bulky functional groups prefer the equatorial sites because axial groups can be located too close to other atoms in the molecule. The difference between the two forms of glucose is that in the β-form all of the hydroxyl groups are in equatorial positions, whereas in the α-form the hydroxyl group at C1 is in an axial position. The two forms interconvert in solution, but the β-form is favored because the larger hydroxyl group is placed in the equatorial position. There are other isomers of glucose, such as β-galactose, in which the hydroxyl group at C4 is in an axial position. There are also other monosaccharides, such as ribose ($C_5H_{10}O_5$), that have rings of different sizes.

Axial H
4 CH₂OH
C 5 O
Equatorial HO C H
HO 3 H 2 C 1
C C—H
H OH OH

α-glucose

H
4 CH₂OH
C 5 O
HO C H
HO 3 H 2 C 1
C C—OH
H OH H

β-glucose

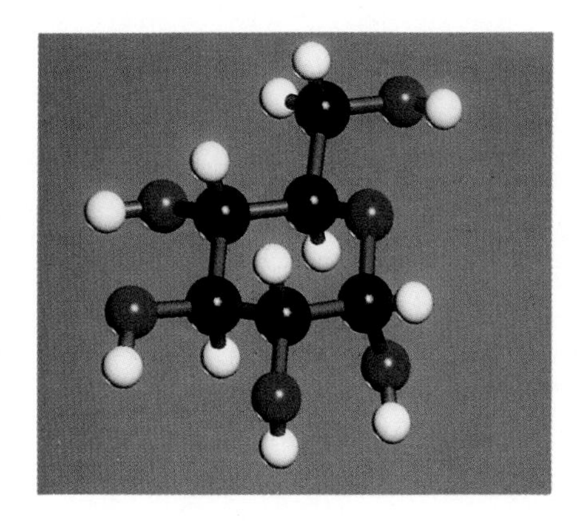

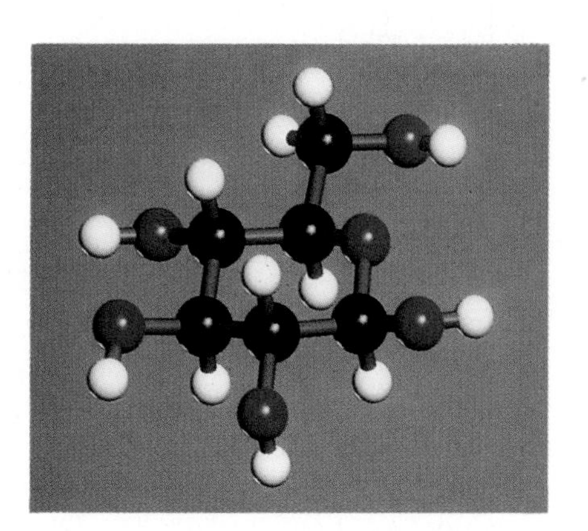

Figure 21.29 α- and β-Glucose The two isomers of glucose differ by the arrangement of the —OH and —H substituents at the number 1 carbon atoms.

Disaccharides

Two monosaccharides can condense with the elimination of water to yield a **disaccharide.** For example, the combination of α-glucose with fructose (a component of honey) yields sucrose, the main constituent of cane sugar. This condensation reaction is reversed in digestion, and the monosaccharides formed can then be used as an energy source.

H
HO— CH_2OH O
H H
H —H
HO— OH OH
H

α-glucose

\+

CH_2OH
O H
HO H HO CH_2OH
OH H

fructose

⟶

H
HO— CH_2OH O
H H
H H
HO— OH CH_2OH
H O
O— H
H HO CH_2OH
OH H

sucrose

\+ H_2O

Polysaccharides

A **polysaccharide** is a step-growth polymer of monosaccharides. Well-known polysaccharides are cellulose and starch, both polymers of glucose. Cellulose contains long chains of β-glucose units joined as shown in Figure 21.30.

Polysaccharides form by the step-growth polymerization of monosaccharides.

Figure 21.30 Polysaccharides
(a) Cellulose is a polymer of glucose. The glucose units are joined through oxygen bridges at the β (equatorial) positions. (b) Starch is also a polymer of glucose, but the glucose units are joined through oxygen bridges at the α (axial) positions of carbon atom 1.

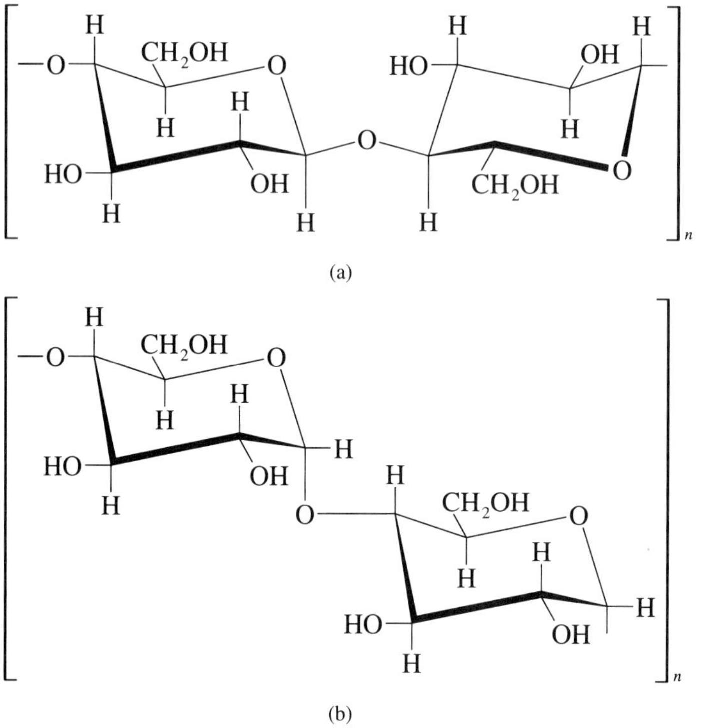

The strands of cellulose can form strong hydrogen bonds, leading to strong fibers such as found in cotton. Cellulose is the main structural component of plants. Humans do not have the ability to digest cellulose, although cows and some other species do. Thus cellulose can be used as a source of glucose for those animals, but it passes through the digestive systems of humans essentially unchanged (cellulose is popularly known as "fiber" in food).

On the other hand, starch (also a polymer of glucose) is digested by the human system, and is a main source of energy. In starch, the polymer linkages are through the α-positions (there is more than one type of starch). The structure of starch is not as suitable for hydrogen bonding interactions and thus it does not form strong fibers like cellulose.

21.7 Nucleic Acids

Objective

- To describe the structures of nucleic acids

Nucleic acids are the third important type of polymer found in living systems. Nucleic acids are responsible for directing the syntheses of the various proteins needed for the existence of each species. The building blocks of these polymers are nucleotides. A **nucleotide** is composed of three units, a five-carbon sugar, a base (a cyclic amine), and a phosphate group.

1. *Sugar* The sugars ribose and deoxyribose are used to make nucleotides.

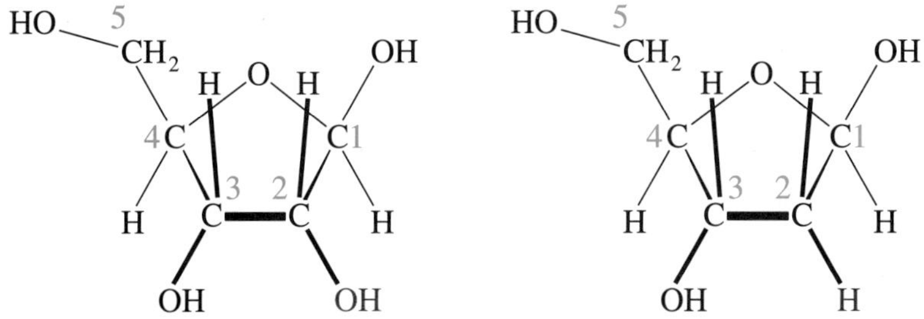

ribose

deoxyribose

2. *Base* Five cyclic amine bases are used to form nucleotides as shown in Figure 21.31. Each of these bases has been given a single-letter abbreviation (the first letter of its name), much like the three-letter abbreviation given to the amino acids used to form proteins.
3. *Phosphate* The third component is derived from phosphoric acid, H_3PO_4.

The nucleotide is formed by condensation reactions of the sugar, phosphoric acid, and the base.

A nucleotide is composed of a five-carbon sugar, a base, and a phosphate group.

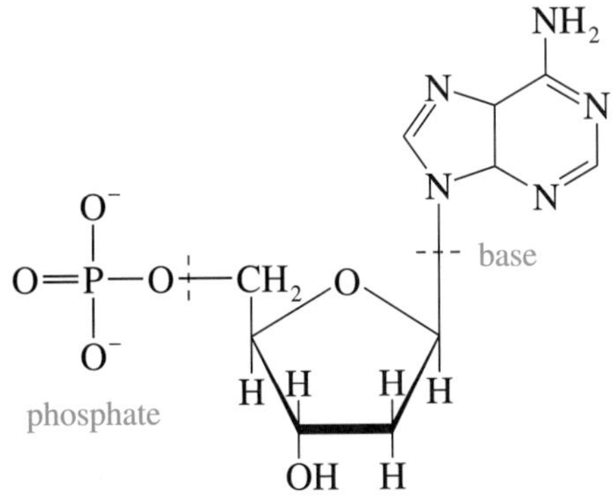

phosphate

base

sugar

A **nucleic acid** consists of a chain of nucleotides. The phosphates bond to carbon 3 of one sugar and to carbon 5 of the next, forming a sugar-phosphate backbone (Figure 21.32). The base is attached at carbon 1 of the sugar.

There are two types of nucleic acids. **Deoxyribonucleic acids (DNA)** are polymers of nucleotide units located inside the chromosomes. DNA is the molecule that stores genetic information. **Ribonucleic acids (RNA)** are polymers of nucleotide units located outside the chromosomes. They transfer genetic information from DNA and direct the syntheses of proteins. The sugar in RNA is ribose, and that in DNA is deoxyribose. The bases adenine, guanine, and cytosine are found in both DNA and RNA, but uracil is found only in RNA and thymine only in DNA.

The primary structure of a nucleic acid is determined by the sequence of its bases. It is the ordering of the bases that stores information in DNA and RNA. Each three-base sequence signals the incorporation of a particular amino acid into a protein and is called a *codon*. For example, the sequence UCU calls for the incorporation of serine.

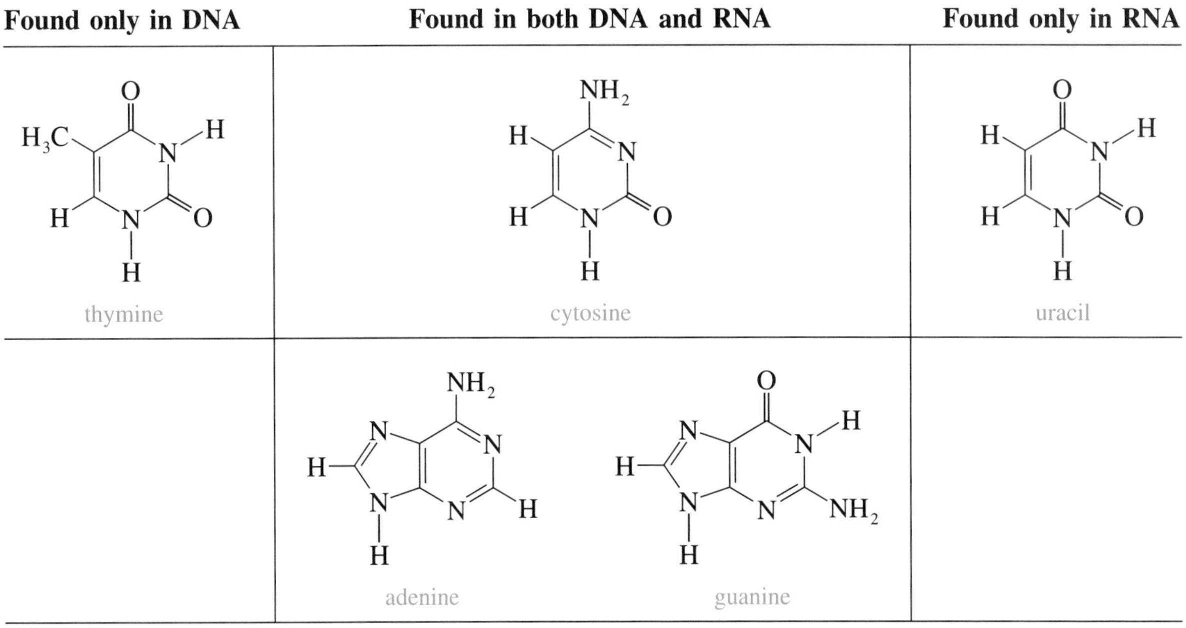

Found only in DNA	Found in both DNA and RNA		Found only in RNA
thymine	cytosine		uracil
	adenine	guanine	

Figure 21.31 The five bases in DNA and RNA

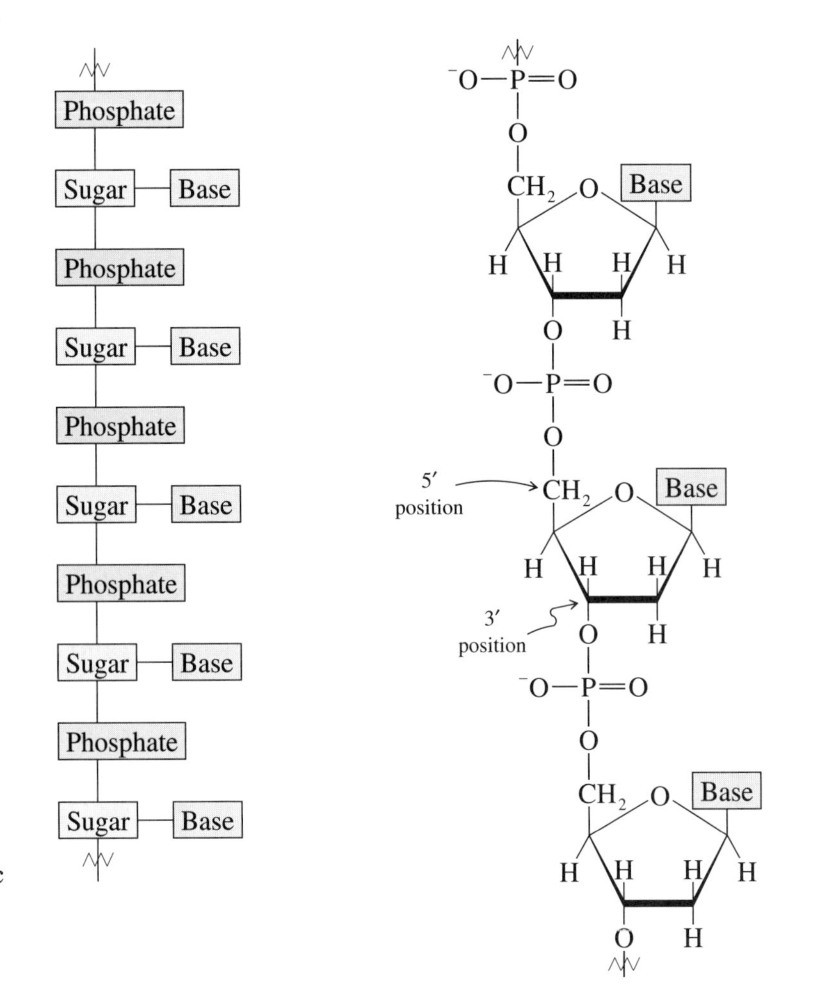

Figure 21.32 Nucleic acids Nucleic acids are polymers of nucleotides. The picture on the right shows a fragment of DNA.

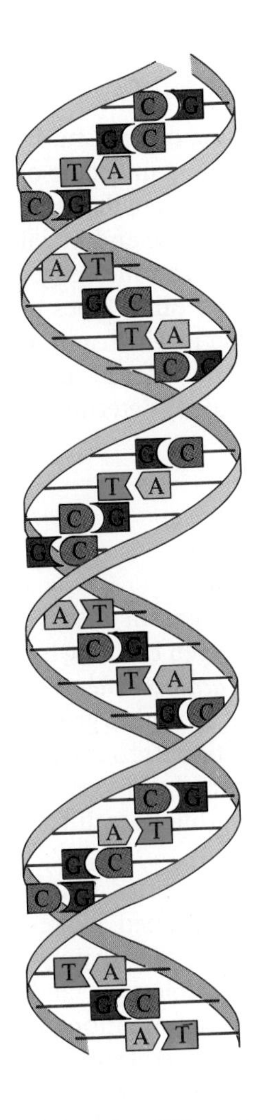

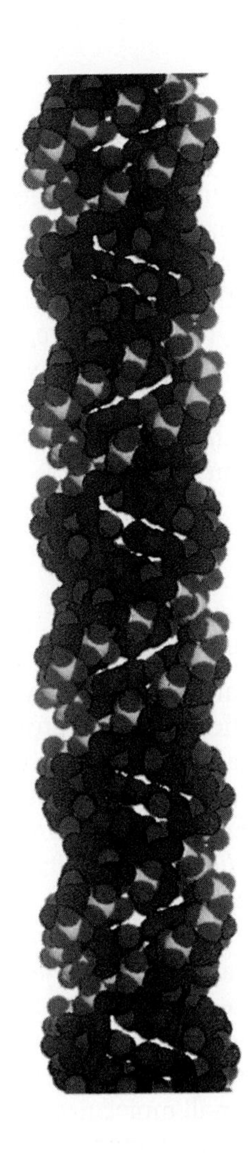

Figure 21.33 The structure of DNA The two strands of DNA are held together by specific adenine-thymine (A—T) and guanine-cytosine (G—C) hydrogen bonding interactions.

Secondary Structure of DNA

Careful analysis of the makeup of DNA reveals that in any sample of DNA there is the same molar amount of adenine base as there is thymine base, and the same amount of guanine base as there is cytosine base, but the ratio of the two pairs of bases varies considerably from one type of DNA to another. Using this information and available information from x-ray crystallographic studies, Watson and Crick proposed in 1953 that the secondary structure of DNA is a double helix of two entwined nucleic acid strands (Figure 21.33). The double helix is held together by very specific hydrogen bonding interactions in which the adenine bases on one strand interact *only* with thymine bases on the second strand, and the guanine bases interact *only* with cytosine bases (Figure 21.33). This base pairing explains why the A/T and G/C ratios are 1. It also explains how cell division produces two identical cells. The double helix unwinds and each strand creates a new matching strand to form two new *identical* double helices.

DNA is composed of two nucleic acid chains held in a double helix by specific hydrogen bonding interactions.

Protein Synthesis

Nucleic acids control the assembly of proteins. The factors that distinguish species and differentiate individuals of each species are determined by the types of proteins that are synthesized. Protein synthesis does not take place in the nucleus of the cell where the DNA is located, but in the cytoplasm. In order to transfer the genetic information stored in the DNA, a molecule of RNA (called messenger RNA) is synthesized by the DNA, much in the same way as DNA reproduces, except that in RNA the sugar is ribose and uracil replaces thymine as the base that interacts with adenine. The messenger RNA travels to the cytoplasm where, with the help of two other types of RNA, *transfer* RNA and *ribosomal* RNA, it directs the order in which amino acids are incorporated into newly synthesized proteins. The order is determined by the order of the bases in the RNA, which was originally determined by the order of bases in DNA.

Summary

A study of carbon-containing compounds is called *organic chemistry.* The simplest organic compounds are *hydrocarbons,* compounds containing only the elements hydrogen and carbon. If they have no multiple bonds, they are *saturated* and are known as *alkanes.* Saturated hydrocarbons that contain rings are known as *cycloalkanes.* Hydrocarbons containing double bonds are *alkenes,* and those containing triple bonds are known as *alkynes.* A series of rules have been established by the IUPAC for naming these compounds and their substituted derivatives, although many nonsystematic names are still in use. A hydrocarbon frequently exists as a number of *isomers.*

Alkanes are not very reactive, but the pi bonds in alkenes and alkynes are good locations for *addition reactions. Aromatic hydrocarbons* are also unsaturated but rarely undergo addition reactions because of the stability of benzene (aromatic) rings. Aromatic hydrocarbons undergo substitution reactions.

There are a number of *functional groups* that occur in many organic compounds. Important groups containing elements other than carbon and hydrogen are found in *alcohols* (—OH), *phenols* (aromatic alcohols), *ethers* (C—O—C), *aldehydes* (—C(=O)H), *ketones* (—C(=O)R), *carboxylic acids* (—C(=O)OH), *esters* (—C(=O)OR), *amines,* ($—NR_2$), and *amides* ($—C(=O)NR_2$). The properties of organic molecules containing these functional groups are mainly determined by the degree of unsaturation, the polarity, and whether hydrogen bonding by the functional group is possible. An *amino acid* contains a carboxyl group ($—CO_2H$) and an amino group ($—NH_2$). In α-amino acids these functional groups are bonded to the same carbon atom. This carbon atom is a *chiral* center if the other two bonded groups are different.

Small organic molecules can be converted into a variety of *polymers,* large molecules formed by repeated bonding of *monomer* units. *Chain-growth* polymers form with the loss of no other molecules, and *step-growth* polymers form with the elimination of small molecules. Some polymers are formed from one type of monomer (homopolymers), and some from two or more types of monomers (copolymers). *Cross-linking* can lead to the formation of harder and stronger polymers.

Organic molecules can be formed in living systems or synthesized. The study of systems in living organisms is *biochemistry. Proteins* are polymers of α-amino acids, and perform many functions in biological systems. They are synthesized in a controlled manner by the nucleic acids DNA and RNA. *Nucleic acids* are polymers of *nucleotide* units, which are composed of a sugar, a cyclic amine base, and a phosphate group. The ordering of the amino acids in proteins is controlled by the ordering of the nucleotides in the nucleic acids. The arrangement in space of proteins helps to determine their properties. *Polysaccharides* are polymers of *monosaccharides* (simple sugars) that are used by living organisms for structural support and as energy sources.

Chapter Terms

Organic chemistry—the study of carbon-containing compounds. *(21.1)*

Biochemistry—the study of the chemistry of systems in living organisms. *(21.1)*

Catenation—the formation of chains or rings of like atoms. *(21.1)*

Hydrocarbon—a compound that contains only the elements hydrogen and carbon. *(21.1)*

Alkane—a hydrocarbon that contains no multiple bonds or rings. *(21.1)*

Saturated hydrocarbon—a hydrocarbon with no multiple bonds. *(21.1)*

Alkyl group—an alkane from which one hydrogen atom has been removed. *(21.1)*

Cycloalkane—a saturated hydrocarbon that contains a ring of carbon atoms. *(21.1)*

Conformers—different arrangements of atoms in a molecule that are caused by rotations about single bonds. *(21.1)*

Substitution reaction—a reaction in which one atom or group of atoms in a molecule is replaced by a different atom or group of atoms *(21.1)*

Unsaturated hydrocarbon—a hydrocarbon that contains one or more multiple carbon-carbon bonds. *(21.2)*

Alkene—an unsaturated hydrocarbon that contains one or more carbon-carbon double bonds. *(21.2)*

Alkyne—an unsaturated hydrocarbon that contains one or more carbon-carbon triple bonds. *(21.2)*

Addition reaction—the combination of two or more substances to form one new substance. *(21.2)*

Aromatic hydrocarbon—a compound that contains one or more benzene rings. *(21.2)*

Aryl group—an aromatic ring that is being considered as a substituent. *(21.2)*

Functional group—an atom or small group of atoms in a molecule that undergoes characteristic reactions. *(21.3)*

Alcohol—an organic compound that contains the —OH *(hydroxyl)* functional group. *(21.3)*

Phenol—an organic compound in which a hydroxyl group is substituted for a hydrogen atom on an aromatic ring. The parent compound of this type, C_6H_5OH, is called phenol. *(21.3)*

Ether—an organic compound that contains a C—O—C functional group. *(21.3)*

Condensation reaction—a reaction that joins two molecules with the elimination of a small molecule such as water. *(21.3)*

Carbonyl group—the C=O functional group. *(21.3)*

Aldehyde—an organic compound that contains the $-\overset{\overset{\displaystyle O}{\|}}{C}-H$ functional group. *(21.3)*

Ketone—an organic compound that contains the $R-\overset{\overset{\displaystyle O}{\|}}{C}-R'$ functional group where the R and R′ groups are alkyl or aryl substituents. *(21.3)*

Carboxylic acid—an organic compound that contains the $-CO_2H$ functional group, a carboxyl group. *(21.3)*

Carboxylate group—the $-CO_2^-$ functional group. *(21.3)*

Ester—an organic compound that contains the $-CO_2R$ functional group where R is an organic substituent. *(21.3)*

Amine—a derivative of ammonia in which one or more of the hydrogen atoms is replaced with an organic substituent (R). Amines with the formula RNH_2 are *primary amines,* R_2NH are *secondary amines,* and R_3N are *tertiary amines.* *(21.3)*

Amide—an organic compound that contains the $-C(O)NR_2$ functional group where R is an organic substituent or hydrogen. *(21.3)*

Amino acid—an organic compound that contains both a carboxyl group ($-CO_2H$) and an amino group ($-NH_2$). *(21.3)*

Polymer—a large molecule formed by the repeated bonding together of many smaller units *(monomers). (21.4)*

Chain-growth polymer (also called **addition polymer**)—a polymer formed from monomeric units with no loss of atoms. *(21.4)*

Homopolymer—a polymer formed from the combination of many units of a single monomer compound. *(21.4)*

Copolymer—a polymer formed from the combination of many units of more than one type of monomer. *(21.4)*

Step-growth polymer (also called **condensation polymer**)—a polymer formed by a reaction that eliminates a small molecule each time a monomer is linked to the polymer chain. *(21.4)*

Polyester—a polymer formed from the condensation of a dicarboxylic acid and a dialcohol. *(21.4)*

Protein—a polymer of amino acids. *(21.5)*

Peptide—a small polymer of amino acids. *(21.5)*

Primary structure—the *sequence* of amino acids in the polypeptide chain. *(21.5)*

Secondary structure—the shape of the polypeptide chain. *(21.5)*

α-Helix—the spiral structure adopted by a peptide chain that is held together by hydrogen bonding between amide groups in the same region of the chain. *(21.5)*

β-Pleated sheet—the flattened structure adopted by a protein that is held together by hydrogen bonding between amide groups in different chains or in different sections of the same chain. *(21.5)*

Tertiary structure—the overall three-dimensional arrangement of the protein. *(21.5)*

Quaternary structure—the orientation of different polypeptide chains with respect to each other. *(21.5)*

Denaturation—the lose of structural organization of a protein. *(21.5)*

Carbohydrate—a polyhydroxyl aldehyde or ketone, or substances that react with water to yield such a compound. *(21.6)*

Monosaccharide—the basic unit of carbohydrates, having the general formula $(CH_2O)_x$ where $x = 3$ to 8. *(21.6)*

Disaccharide—the product from the condensation of two monosaccharides with the elimination of water. *(21.6)*

Polysaccharide—a step-growth polymer of monosaccharides. *(21.6)*

Nucleotide—the building block of DNA and RNA; each nucleotide is composed of three units, a five-carbon sugar, a base (a cyclic amine), and a phosphate group. *(21.7)*

Nucleic acid—a chain of nucleotides. *(21.7)*

Deoxyribonucleic acid (DNA)—a polymer of nucleotide units located inside the chromosomes. DNA is the molecule that stores genetic information. *(21.7)*

Ribonucleic acid (RNA)—a polymer of nucleotide units located outside the chromosomes. RNA transmits genetic information and directs the assembly of proteins. *(21.7)*

Exercises

Exercises designated with color have answers in Appendix J.

Alkanes

21.1 Explain the differences between organic chemistry and biochemistry.

21.2 Explain what is special about the element carbon that causes it to have such an extensive chemistry.

21.3 Why are alkanes also called saturated hydrocarbons? What is the hybridization of the carbon atoms in alkanes?

21.4 Which of the following molecules are alkanes?
(a) C_8H_{16}
(b) C_6H_{14}
(c) C_4H_9F
(d) C_9H_{20}

21.5 Write the molecular formula of an alkane containing 12 carbon atoms.

21.6 Show a structural formula for the straight-chain isomer of C_5H_{12}. What is the name of the alkyl group formed by removing a hydrogen atom from one of the terminal carbon atoms?

21.7 Show the structural formula for C_3H_8. What is the name of the alkyl group formed by removing a hydrogen atom from the center carbon atom?

21.8 Draw the structural formulas for the isomers of hexane and name each isomer.

21.9 Draw the structural isomers of dimethylcyclohexane. Remember that one of the substituents will always be at position 1.

21.10 Write the structural formula for each of the following alkanes:
(a) 2-methylhexane
(b) 3,3-dichloroheptane
(c) 2-methyl-3-phenyloctane
(d) 1,1-diethylcyclohexane

21.11 Write the structural formula for each of the following alkanes:
(a) 1-bromobutane
(b) 2-methyl-3-nitropentane
(c) 2,2-dimethylhexane
(d) 1-chloro-1-methylcyclopentane

21.12 Name the following compounds:

(a) $CH_2(F)—CH(CH_3)—CH_2—CH_2—CH_3$

(b) $CH_3—CH(CH_2CH_3)—CH_2—CH_2—CH_3$

(c) $CH_3—C(CH_3)_2—CH_2—CH_3$

(d) cyclobutane ring bearing Cl and $CH_2—CH_3$ on adjacent carbons

21.13 Name the following compounds:

(a) $CH_3—CH(CH_3)—CH_2—CH(Br)—CH_2—CH_3$

(b) cyclohexane ring bearing CH_3

(c) 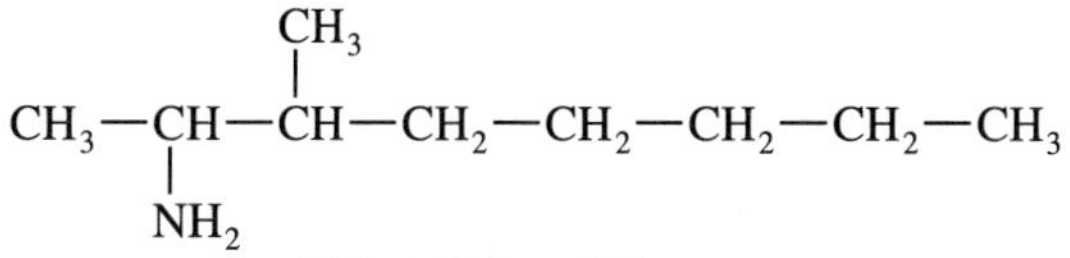 $CH_3-CH(NH_2)-CH(CH_3)-CH_2-CH_2-CH_2-CH_2-CH_3$

(d) 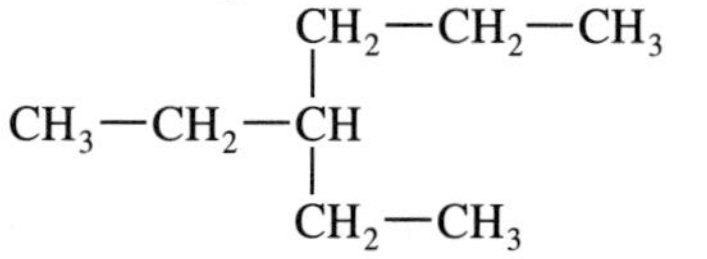 $CH_3-CH_2-CH(CH_2-CH_2-CH_3)(CH_2-CH_3)$

21.14 Draw 1,1,3,3-tetramethylcyclohexane in the boat and chair configurations. Which is more stable and why?

Alkenes and Alkynes

21.15 Show an example of an alkene that will exist as both *cis* and *trans* isomers and another that cannot.

21.16 Write the general formulas for a noncyclic alkene and alkyne.

21.17 Name the following unsaturated compounds:

(a) 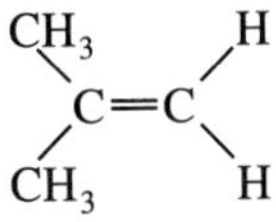 $(CH_3)_2C=CH_2$ (two CH_3 on one carbon, two H on the other)

(b) 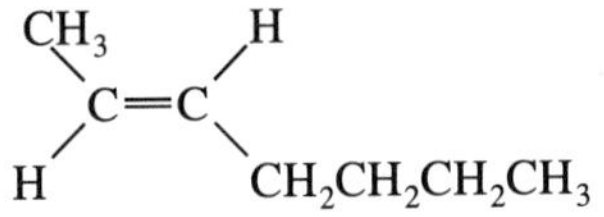 $CH_3(H)C=C(H)CH_2CH_2CH_2CH_3$ (CH_3 and H on one carbon; H and $CH_2CH_2CH_2CH_3$ on the other)

(c) $CH_3CH_2CH_2(H)C=CF_2$ ($CH_3CH_2CH_2$ and H on one carbon; F and F on the other)

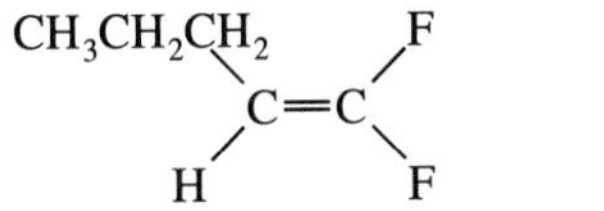

(d) $CH_3C{\equiv}CCH_2Br$

21.18 Name the following unsaturated compounds:

(a) 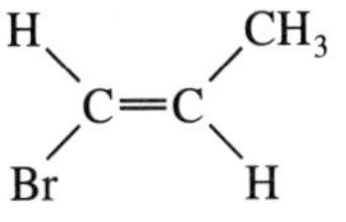 H, Br on one carbon; CH_3, H on the other: $H(Br)C=C(CH_3)H$

(b) 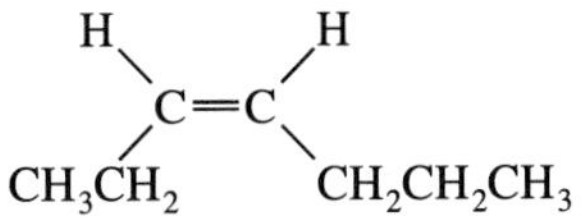 H, CH_3CH_2 on one carbon; H, $CH_2CH_2CH_3$ on the other: $H(CH_3CH_2)C=C(H)CH_2CH_2CH_3$

(c) $CH_3C{\equiv}CCH_2CH_2F$

(d) 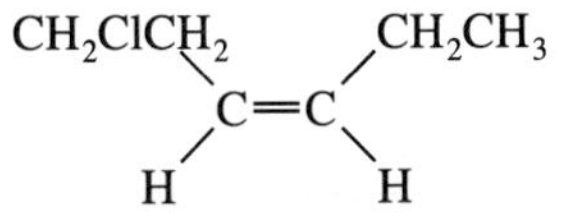 CH_2ClCH_2, H on one carbon; CH_2CH_3, H on the other: $CH_2ClCH_2(H)C=C(H)CH_2CH_3$

21.19 Which of the two compounds pictured below exists as *cis* and *trans* isomers?

$(H)(H)C=C(CH_3)(CH_3)$ $\qquad$ $(H)(CH_3)C=C(CH_3)(H)$

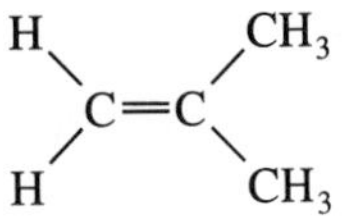

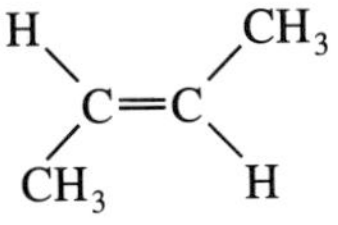

21.20 Which of the two compounds pictured below exists as *cis* and *trans* isomers?

$(Br)(Br)C=C(H)(Cl)$ $\qquad$ $(Cl)(Br)C=C(Br)(H)$

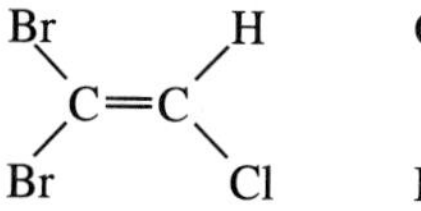

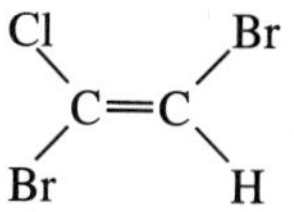

21.21 Draw all possible isomers for the substituted alkene C_3H_5F.

21.22 Draw all possible isomers for the substituted alkene C_4H_7Cl that have the chlorine atom attached to a double-bonded carbon atom.

21.23 Write the structural formula for each of the following unsaturated compounds:
(a) 2-bromo-1-hexene
(b) *cis*-4-nitro-2-pentene
(c) 1,2-dichloro-3-hexyne
(d) 1,1-difluoro-3-chloro-1-heptene

21.24 Write the structural formula for each of the following unsaturated compounds:
(a) 2,3-dimethyl-1-pentene
(b) 1-methylcyclohexene
(c) *cis*-1-chloro-2-butene
(d) 4-methyl-1-hexene

21.25 Write the structural formula and name the product of the reaction of 2-hexene with Br_2.

21.26 Complete the following equations:

(a) $CH_2{=}CH(CH_3) + H_2 \longrightarrow$

(b) $CH{\equiv}CCH_2CH_3 + 2Cl_2 \longrightarrow$

(c) 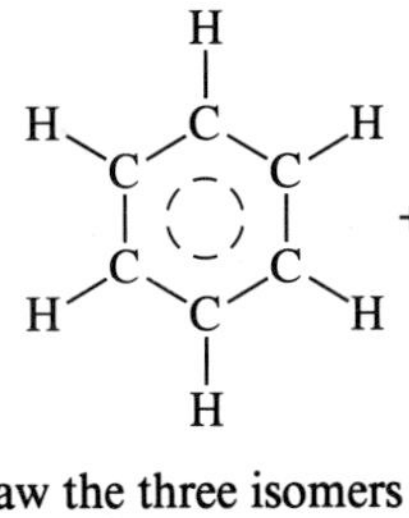 C_6H_6 (benzene ring, each C bonded to H) $+ Cl_2 \xrightarrow{FeCl_3}$

21.27 Draw the three isomers of dimethylbenzene.

Functional Groups

21.28 Name the following functional groups:

(a) 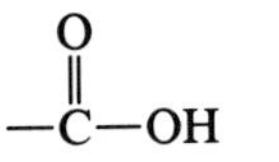 $-C(=O)-OH$

(b) 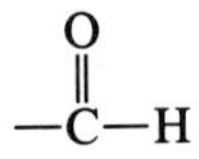 $-C(=O)-H$

(c) $C-O-C$

21.29 Name the following functional groups:

(a) 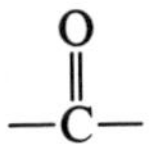 $-C(=O)-$

(b) $-\overset{\overset{\displaystyle O}{\|}}{C}-O-R$

(c) $-\overset{\overset{\displaystyle O}{\|}}{C}-\overset{\overset{\displaystyle H}{|}}{N}-$

21.30 State which compound in each of the following pairs is expected to have the higher boiling point. Explain your answer.

(a) $CH_3CH_2OCH_2CH_3$ or $CH_3CH_2\overset{\overset{\displaystyle OH}{|}}{C}=O$

(b) $CH_3CH_2CH_2CH_2NH_2$ or $CH_3CH_2OCH_3$

(c) $CH_3CH_2C{\equiv}CF$ or $CH_2{=}CH_2$

21.31 State which compound from each of the following pairs is expected to have the higher boiling point. Explain your answer.

(a) CH_3OCH_3 or $CH_3CH_2CH_2CH{=}CH_2$

(b) $CH_3\overset{\overset{\displaystyle O}{\|}}{C}NH_2$ or $CH_3C{\equiv}CCH_3$

(c) $CH_3CH_2CH_2CH_2OH$ or $CH_3\overset{\overset{\displaystyle O}{\|}}{C}OCH_3$

21.32 Write the structural formula and name the organic product expected from the acid-catalyzed condensation reaction of CH_3OH.

21.33 Write the structural formula and name the organic product expected from the mild air oxidation of CH_3OH.

21.34 Write the structural formula and name the organic product expected from the acid-catalyzed condensation reaction of CH_3CO_2H and CH_3CH_2OH.

21.35 Write the structural formula and name the organic product expected from the condensation reaction of CH_3CO_2H and CH_3NH_2.

21.36 Draw the structure of the following molecules:
(a) 1-butanol
(b) 3-methyl-2-pentanone
(c) methyl acetate
(d) ethyl phenyl amine

21.37 Draw the structure of each of the following molecules:
(a) ethyl vinyl ether
(b) 2-bromopropanal
(c) pentanoic acid
(d) 3-fluorophenol

21.38 Name the following compounds:

(a) $FCH_2CH_2CH_2OH$

(b) $CH_3CH_2\overset{\overset{\displaystyle O}{\|}}{C}OH$

(c) $CH_3\overset{\overset{\displaystyle O}{\|}}{C}O\overset{\overset{\displaystyle CH_3}{|}}{C}HCH_3$

(d) $CH_3CH_2CH_2NH_2$

21.39 Name the following compounds:

(a) $CH_3CH_2CH{=}CH_2$

(b) $CH_3CH_2CH_2\overset{\overset{\displaystyle O}{\|}}{C}OCH_2CH_3$

(c) $CH_3CH_2O\overset{\overset{\displaystyle CH_3}{|}}{C}HCH_3$

(d) $CH_3CH_2CH_2CH_2\overset{\overset{\displaystyle O}{\|}}{C}NH_2$

21.40 Indicate which of the following compounds have a chiral or asymmetric center. Mark the chiral center in those molecules.

(a) $CH_3CH_2CH_2CH_2CH_3$

(b) $H-\underset{\underset{\displaystyle NH_2}{|}}{\overset{\overset{\displaystyle H}{|}}{C}}-\overset{\overset{\displaystyle O}{\|}}{C}-OH$

(c) $CH_3\underset{\underset{\displaystyle NH_2}{|}}{\overset{\overset{\displaystyle H}{|}}{C}}-\overset{\overset{\displaystyle O}{\|}}{C}-OH$

21.41 Indicate which of the following compounds have a chiral or asymmetric center. Mark the center in the chiral molecules.

(a) $CH_3\underset{\underset{\displaystyle H}{|}}{\overset{\overset{\displaystyle F}{|}}{C}}CH_2CH_3$

(b) $CH_3\underset{\underset{\displaystyle NH_2}{|}}{\overset{\overset{\displaystyle CH_3}{|}}{C}}-\overset{\overset{\displaystyle O}{\|}}{C}-OH$

(c) $CH_3\underset{\underset{\displaystyle F}{|}}{\overset{\overset{\displaystyle CH{=}CH_2}{|}}{C}}-NH_2$

Polymers

21.42 Describe and give a specific example of the formation of the following:
(a) A chain-growth polymer
(b) A homopolymer
(c) A step-growth polymer

21.43 Polyvinyl chloride, a chemically resistant polymer used in house siding and floor tiles, is a chain-growth polymer of chloroethene (vinyl chloride). Draw the repeating unit of polyvinyl chloride.

21.44 Polystyrene, a polymer used for thermal insulation (coolers) and toys, is a chain-growth polymer of phenylethylene (styrene). Draw the repeating unit of polystyrene.

21.45 What is the monomer that is used to form polyacrylonitrile, a chain-growth polymer used in carpets, as formulated below?

$$\left(-\underset{\text{H}}{\overset{\text{H}}{\text{C}}}-\underset{\text{CN}}{\overset{\text{H}}{\text{C}}}- \right)_n$$

21.46 Viton is a strong, flexible chain-growth copolymer used in gaskets. It is formed from 1,1-difluoroethylene and hexafluoropropene. Draw the repeating unit.

21.47 Tires are frequently made from SBR, the trade name of the chain-growth copolymer made from styrene ($CH_2{=}CHC_6H_5$) and butadiene ($CH_2{=}CHCH{=}CH_2$). Draw the repeating unit.

21.48 Explain what is meant by cross-linking a polymer. How does it affect the properties of the polymer?

21.49 Kevlar, a step-growth polymer used in bulletproof vests, is formulated below. What is the monomeric unit used to form Kevlar?

$$\left(-\overset{\text{H}}{\text{N}}(C_6H_4)\overset{\text{O}}{\overset{\|}{\text{C}}}- \right)_n$$

21.50 Lexan is a step-growth polymer with high-impact strength formed from the following monomers with the elimination of phenol. Draw the copolymer.

$$HO-C_6H_4-C(CH_3)_2-C_6H_4-OH \qquad (C_6H_5O)_2CO$$

Proteins

21.51 Write the general reaction for the combination of two amino acids to form a peptide.

21.52 Describe the four levels of structure for a protein.

21.53 Describe the bonding in an α-helix and a β-pleated sheet. What type of structure within the protein do these arrangements refer to?

21.54 Draw the structures of the two dipeptides formed from serine and valine.

21.55 Draw the structures of two of the tripeptides that can be formed from alanine, glycine, and cysteine.

Carbohydrates

21.56 Draw the two forms of glucose, clearly indicating how they are different.

21.57 Draw a disaccharide formed from β-galactose and ribose.

21.58 Lactose (milk sugar) is the disaccharide formed from β-galactose and β-glucose. Draw the structure.

21.59 Describe the difference between cellulose and starch. How are they treated differently by the human digestive system?

Nucleic Acids

21.60 Discuss the different functions of DNA and RNA.

21.61 Describe the three components of a nucleotide. Be specific about the components used in DNA and those in RNA.

21.62 Draw and name the five principal bases used in DNA and RNA. Be specific about the bases used in DNA and those in RNA.

21.63 Describe the Watson and Crick model of DNA. How does this model account for the experimental fact that the A/T and G/C ratios are 1?

21.64 Describe how both DNA and RNA are used in the synthesis of a specific protein.

21.65 Draw the nucleotides containing the following bases and sugars: (a) adenine and ribose, (b) cytosine and deoxyribose, (c) uracil and ribose.

Additional Exercises

21.66 Write the structural formula for each of the following molecules:
(a) *cis*-2-bromo-3-hexene
(b) 2-nitrophenol

21.67 Indicate the functional groups in each of the following compounds.
(a) aspirin

$$C_6H_4(\overset{\text{O}}{\overset{\|}{\text{C}}}-OH)(O-\underset{\text{O}}{\underset{\|}{\text{C}}}-CH_3)$$

(b) estrone

(H_3C, O, H, H, H, HO — steroid ring structure)

(c) butacetin, an analgesic (pain-killing) agent

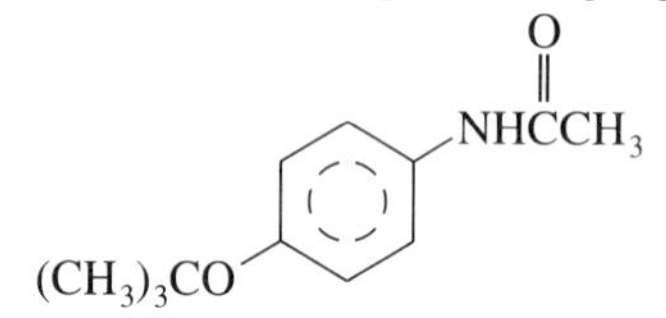

21.68 Explain why ethanol dissolves freely in water whereas its isomer, dimethyl ether, is only slightly soluble in water.

21.69 Draw the structure of an α-amino acid with a chiral center.

21.70 Write the structures of two monomers that you think might be used to make an interesting copolymer. Draw the repeating unit of the copolymer.

21.71 Explain the correlation between the three-letter base pairs in a codon and the order of α-amino acids in proteins.

Appendix A

Math Procedures

The study of chemistry requires certain mathematical skills. This appendix briefly reviews some of the mathematical operations that are used in the course, although it is not intended to replace a mathematics textbook.

A.1 Electronic Calculators

A modern scientific calculator is an important tool that will help you in your study of chemistry. Your calculator should be able to express numbers in exponential notation (explained in the next section), and perform operations such as natural and base-10 logarithms, antilogarithms (raising *e* or 10 to a power), roots, and raising a number to any power.

In order to use a calculator efficiently, you should learn to *chain* calculations together. When evaluating an expression such as

$$\frac{(1.202 \times 0.850) - 0.0307}{0.576}$$

it is easiest to multiply 1.202 by 0.850 and leave the result (1.0217) on the calculator display, subtract 0.0307 (to get 0.991), and divide by 0.576 to get 1.7204861, the final result. On most algebraic notation calculators, it is not necessary to press the [=] key after the first multiplication to obtain 1.0217 on the display. The preceding calculation can be performed by these keystrokes:

1.202 [×] .85 [−] .0307 [=] [÷] .576 [=]

display reads [1.0217] [0.991] [1.7204861]

On most algebraic calculators, if the [=] is not pressed before the [÷], only 0.0307 will be divided by 0.576, and the final result will be incorrect (0.96840 . . .).

Note: the calculator does not determine the correct number of significant figures. You must truncate the answer as necessary, according to the rules described in Chapter 1.

A.2 Exponential Notation

The numbers of science range from very large to very small. We frequently use *exponential notation* (or scientific notation) to express a number as a product of a *digit term* and an *exponential term.* The digit term is a number between 1 and 10. The exponential term represents 10 raised to a whole number power. For example, the number 2468 is expressed as 2.468×1000 or 2.468×10^3 in exponential notation.

$10000 = 1 \times 10^4$	$13579 = 1.3579 \times 10^4$
$1000 = 1 \times 10^3$	$1357 = 1.357 \times 10^3$
$100 = 1 \times 10^2$	$135 = 1.35 \times 10^2$
$10 = 1 \times 10^1$	$13 = 1.3 \times 10^1$
$1 = 1 \times 10^0$	
$0.1 = 1 \times 10^{-1}$	$0.13579 = 1.3579 \times 10^{-1}$
$0.01 = 1 \times 10^{-2}$	$0.01357 = 1.357 \times 10^{-2}$
$0.001 = 1 \times 10^{-3}$	$0.00135 = 1.35 \times 10^{-3}$
$0.0001 = 1 \times 10^{-4}$	$0.00013 = 1.3 \times 10^{-4}$

The exponential term locates the decimal point. The exponent is the number of places that the decimal point is shifted while going from the original number to the digit term. A positive exponent indicates that the decimal point is shifted to the left, and a negative exponent indicates that the decimal point is shifted to the right. Most calculators have a mode in which any number entered or calculated is displayed in exponential notation. In this mode the display shows the digit part of the number on the left, followed by a gap and the power of ten at the right side of the display.

Addition and Subtraction

To add or subract two numbers, convert the numbers to the same power of ten and add or subtract the digit terms.

$$1.23 \times 10^{-2} + 4.5 \times 10^{-3} = 1.23 \times 10^{-2} + 0.45 \times 10^{-2} = 1.68 \times 10^{-2}$$

Multiplication

Multiply the digit terms and add the exponents. Shift the decimal place, if necessary, so the result is written as a digit term between 1 and 10 times the exponential term. The exponents are simply added in this calculation.

$$1.23 \times 10^{-2} \times 4.5 \times 10^{-3} = (1.23)(4.5) \times 10^{-5} = 5.535 \times 10^{-5} = 5.5 \times 10^{-5}$$

Modern calculators perform calculations with high accuracy and can display numbers with more digits than can be justified by the data. The concept of significant digits is discussed in Chapter 1.

Division

Divide the digit terms and subtract the exponents.

$$\frac{4.03 \times 10^2}{1.24 \times 10^{-3}} = \frac{4.03}{1.24} \times 10^{2-(-3)} = \frac{4.03}{1.24} \times 10^5 = 3.25 \times 10^5$$

Powers and Roots

To raise a number to a power, raise the digit term to the power and multiply the exponent by the power.

$$(4.0 \times 10^{-2})^3 = (4.0)^3 \times 10^{(-2 \times 3)} = 64 \times 10^{-6} = 6.4 \times 10^{-5}$$

Many calculators have an $\boxed{x^y}$ (or $\boxed{y^x}$) operation. If you enter 4.0×10^{-2}, press $\boxed{x^y}$, and then enter 3 and press $\boxed{=}$, you should see 6.4×10^{-5}.

A root is a fractional power—the square root is equivalent to raising a number to the 1/2 power, a cube root is the 1/3 power, etc. Many calculators have a separate button for the square root operation but not for other roots. Treating a root as a fractional power will work, as will using logarithms, discussed in the next section.

A.3 Logarithms

A logarithm is the power to which you must raise a base number to obtain the desired number. In this text we use two types of logarithms, common logarithms (log), for which the base is 10, and natural logarithms (ln), for which the base is e (2.71828 . . .). Important logarithmic relationships are:

$$\log x = y \text{ where } x = 10^y$$

$$\ln x = z \text{ where } x = e^z$$

The common logarithms and natural logarithms are related by the equation

$$\ln x = (\ln 10)(\log x) \approx 2.303 \log x$$

The common logarithm of a number that is a power of 10 is always a whole number.

Number	Exponential Notation	Common Logarithm
100	10^2	2
10	10^1	1
1	10^0	0
0.1	10^{-1}	−1
0.01	10^{-2}	−2

The logarithm of a number that is not an integral power of ten is found by using a calculator or table of logarithms. The common logarithm of 45 is 1.65. We can use an estimate to check this value—since 45 is part way between 10 and 100, we expect the logarithm to be part way between 1 and 2. An alternative way of expressing the relationship is

$$\log 45 = 1.65 \qquad \text{or} \qquad 45 = 10^{1.65}$$

When you take the logarithm of a number, the number of *decimal places* in the result is equal to the number of significant *digits* in the original quantity.

To find the log on the calculator, enter 45, then depress the $\boxed{\text{log}}$ key to obtain 1.65.

The second kind of logarithm is the natural logarithm. The base of natural logarithms is $e \approx 2.7183$. Natural logarithms make some operations more convenient. To find the natural logarithm of 45, enter 45, then depress the [ln] key to obtain 3.81.

$$\ln 45 = 3.81 \qquad \text{or} \qquad 45 = e^{3.81}$$

Antilogarithms

Frequently, the logarithm is given, and the equivalent number must be found. For example, we might have to determine the number whose common logarithm is −7.82. To calculate the antilogarithm, we must raise 10 (the base) to the −7.82 power. On most calculators, enter −7.82 and then depress the [10^x] button (often labeled inv log) to obtain 1.5136×10^{-8}. Since there are two decimal places in the logarithm, there are two significant figures in the number.

$$10^{-7.82} = 1.5 \times 10^{-8}$$

The process of finding the antilogarithm of a natural logarithm is called exponentiation, since it involves the exponential function. To find the natural antilogarithm of −7.82, first enter −7.82 and then depress [e^x] (often called exp or inv ln) to obtain 4.0162×10^{-4}.

$$e^{-7.82} = 4.0 \times 10^{-4}$$

Different calculators have different labels on their keys, so refer to your user's manual for the proper key to press for each function.

Operations Using Logarithms

Since logarithms are exponents, adding logarithms is the same operation as multiplying the numbers that they represent. Subtracting corresponds to division. The type of the logarithm, natural or common, doesn't matter in these operations.

$$\log (xy) = \log (x) + \log(y)$$

$$\log (x/y) = \log(x) - \log(y)$$

$$\log (x^y) = y \log (x)$$

$$\log \sqrt[y]{x} = \log (x^{1/y}) = (1/y) \log (x)$$

Consider raising 4.300 to the fifth power. We will find the logarithm, multiply by the power, and then take the antilogarithm.

$$\log 4.300 = 0.63347$$

$$\log 4.300^5 = 5 \times \log 4.300 = 5 \times 0.63347 = 3.16735$$

$$10^{3.16735} = 1470 = 4.300^5$$

Similarly, we can find the cube root of 4.300 by taking the logarithm, dividing by 3, and then taking the antilogarithm. Either common or natural logarithms can be used—natural logarithms will be used just to offer a comparison with the last calculation.

$$\ln 4.300 = 1.45866$$

$$\ln 4.300^{1/3} = \tfrac{1}{3} \times \ln 4.300 = \tfrac{1}{3} \times 1.45862 = 0.4862$$

$$e^{0.4862} = 1.626 = 4.300^{1/3}.$$

A.4 Graphs

In many situations, the value of one quantity depends on the value of another. It is often easier to understand the relationship between two quantities by looking at a graph that displays the value of one quantity against the value of the other. Graphs also help us to predict trends and to answer "what-if" problems.

Let us consider the relationship between the amount of some product formed in a reaction and the temperature of the reaction. The experiment is designed so that we change the temperature and then measure the amount of product at that temperature. The horizontal axis (x-axis) is used for the quantity that can be controlled or adjusted; in this example, temperature is placed on the horizontal axis. The vertical axis (y-axis) is used for the quantity that responds to changes in the quantity on the x-axis, in this case the amount of product formed. We can say that amount of product *depends on* temperature; often the quantity on the y-axis is called the *dependent variable* and the quantity on the x-axis is the *independent variable.* The experimental data are placed on the graph and connected with a smooth line.

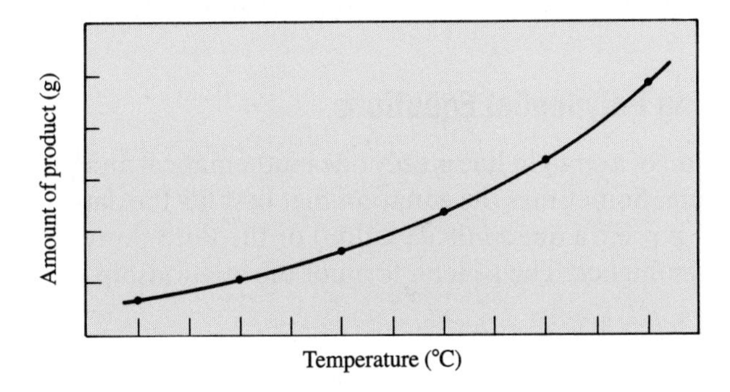

Figure A.1

The line connecting the points on a graph can assume many different shapes, but one of the most important is the straight line. A straight-line or linear relationship can be described by an equation such as

$$y = 3x + 2$$

This relationship predicts the value of y for a given value of x. For example, when x is adjusted to have a value of 4, then y will be 14. The general equation of a straight line is

$$y = mx + b$$

The quantity b is called the y-intercept, and it provides the value of y when x is equal to 0. The quantity m is called the *slope* of the line, and it provides information about the degree of slant. Lines with positive values of

m slant up toward the right, a line with a value of zero for the slope is level, and lines with negative values of the slope slant down toward the right. The slope is evaluated by measuring the change in the dependent variable (Δy) for a given change in the independent variable (Δx). We represent the *change* in any variable by using a Greek delta, Δ, which should be understood to mean (final value − initial value).

$$m = \frac{\Delta y}{\Delta x}$$

The slope of a line representing a measured quantity must have units. If, for example, the x-axis is time in seconds and the y-axis is temperature in °C, then the slope will have units of °C/s.

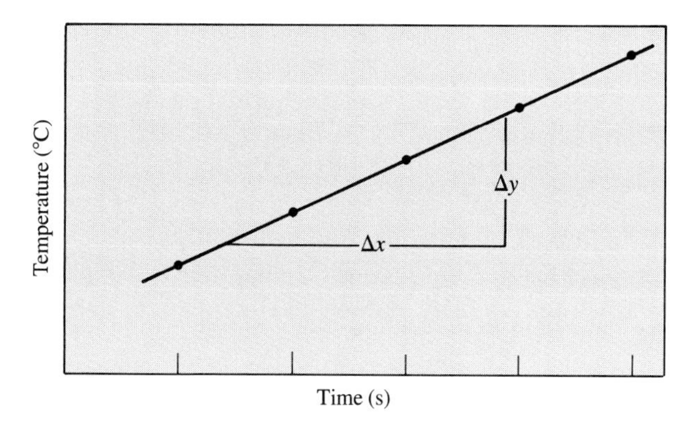

Figure A.2

A.5 Solving Polynomial Equations

The equation of a straight line is only one mathematical model that is used to describe data. Sometimes the equation that best fits the data has x raised to the second power (a quadratic equation) or the third power (a cubic equation) or even higher. The general form of the relationship is

$$y = \beta_0 + \beta_1 x + \beta_2 x^2 + \beta_3 x^3 + \ . \ . \ .$$

This relationship is called a *polynomial equation.* In this section, some of the methods used to solve these equations will be presented. When we speak of "solving" an equation, we generally mean finding a *root,* which is a value of x for which y is equal to zero. The number of roots of a polynomial equation cannot be larger than the largest power of x in the equation, but may be smaller.

Quadratic Equations

A quadratic equation contains x raised to the second power. It has the general form

$$y = ax^2 + bx + c$$

where a, b, and c are constants. To find the roots of the equation, we need to find the values of x for which y is zero.

$$ax^2 + bx + c = 0$$

A formula, called the quadratic formula, enables us to determine the roots of the quadratic equation:

$$x = \frac{-b \pm \sqrt{b^2 - 4ac}}{2a}$$

A quadratic equation has *two* roots, one calculated from the addition and the other from the subtraction process. Both represent values of x for which y is equal to zero; the text (Chapters 12 through 14) includes ways to determine which of the two values should be used in a particular situation.

There is an alternate solution to the quadratic formula that has some advantages. For example, if the quadratic formula and a calculator are used to solve the equation

$$x^2 + 0.100x - 8.0 \times 10^{-17} = 0$$

the roots found are 0 and −0.100. The smaller root is not exactly equal to 0, but its very small value has been lost in the rounding errors of the calculator. An alternate way to find the roots of a quadratic equation avoids this rounding error.

We define

$$Q = \frac{b + (\text{signb})\sqrt{b^2 - 4ac}}{2}$$

where a, b, and c have the same meaning as in the quadratic formula, and (signb) is the sign of b: +1 if b is positive, and −1 if b is negative. The two roots of the quadratic equation are then

$$x = \frac{-c}{Q} \text{ and } x = \frac{-Q}{a}$$

We can apply this to the preceding quadratic equation, where $a = 1$, $b = 0.100$, and $c = -8.0 \times 10^{-17}$:

$$Q = \frac{0.100 + (+1)\sqrt{0.100^2 - 4(1)(-8.0 \times 10^{-17})}}{2} = 0.100$$

The two roots are

$$x = \frac{-c}{Q} = 8.0 \times 10^{-16} \text{ and } x = \frac{-Q}{a} = -0.100$$

There are chemical situations in which the small number is the physically meaningful solution. By using this alternate solution to the quadratic equation, the desired solution is not lost in rounding errors.

Other ways can be used to solve for the root of a quadratic equation. Since these other methods are also applicable to any polynomial equations, they are discussed in the next section.

Higher-Order Polynomial Equations

There are several ways to determine the roots of a higher-order polynomial equation. We use a quadratic and a cubic (third-order) equation as examples, but the methods discussed can be applied to any polynomial.

One way to estimate the root is by graphing the function and analyzing the graph to find the root. Remember that a root is the value of x for which $y = 0$.

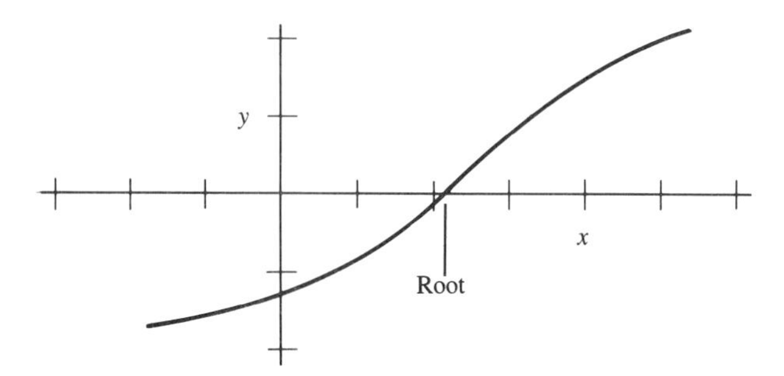

Figure A.3

Exact Solutions

When the equation is simple, we can find an exact, or analytical, solution. We have presented the formula for the roots of a quadratic equation; the formula to find the roots of higher-order equations is much more complex. Further, mathematicians can prove that for equations of order five and higher, there is no formula that can be used. In general, the roots of equations of order greater than two are found by numerical methods, as described next.

Numerical Solutions

The root of an equation is a number. In the study of chemistry, the method used to find the root of an equation is not important; what is important is the number itself and its interpretation. All numerical methods are similar in that they require an initial estimate of the root, and they require a test for the desired accuracy of the approximation to the root.

Successive Approximation

Often quadratic equations can be solved without using the quadratic formula, by a method called *successive approximation.* We can illustrate the method by using an equation from the solubility equilibrium considered in Chapter 12. We will solve a cubic equation (a quadratic would be solved in similar fashion) by successive approximation.

$$1.0 \times 10^{-8} = s(5.0 \times 10^{-3} + 2s)^2$$

From the origin of this equation there is reason to believe that $2s$ is quite a bit smaller than 5.0×10^{-3}, so we make the approximation that s in the sum is 0. In using successive approximations, *the approximate value is substituted only in sums and differences.* With this approximation, the equation becomes

$$1.0 \times 10^{-8} = s(5.0 \times 10^{-3})^2$$

which is easily solved for s.

$$s = \frac{1.0 \times 10^{-8}}{(5.0 \times 10^{-3})^2} = 4.0 \times 10^{-4}$$

This is a better approximation for s than is 0, so we substitute it into the sum in the original equation.

$$1.0 \times 10^{-8} = s(5.0 \times 10^{-3} + 2s)^2$$

$$1.0 \times 10^{-8} = s(5.0 \times 10^{-3} + 2(4.0 \times 10^{-4}))^2 = s(5.8 \times 10^{-3})^2$$

We now solve this equation for the third approximation for s:

$$s = \frac{1.0 \times 10^{-8}}{(5.8 \times 10^{-3})^2} = 3.0 \times 10^{-4}$$

3.0×10^{-4} is a better approximation for s, and we repeat the substitution into the sum in the equation one more time.

$$5.0 \times 10^{-3} + 2s = 5.0 \times 10^{-3} + 2(3.0 \times 10^{-4}) = 5.6 \times 10^{-3}$$

$$s = \frac{1.0 \times 10^{-8}}{(5.6 \times 10^{-3})^2} = 3.2 \times 10^{-4}$$

We accept this value of $s = 3.2 \times 10^{-4}$ as the answer, since the sum we calculate to the correct number of significant figures $(5.0 \times 10^{-3} + 2s)$ would not change from the previous iteration. In the chemical calculations where successive approximations are used, if successive approximations are within 5% of each other, we consider the answer satisfactory.

Interval Halving

When we know that the desired root of a polynomial lies between two values, we can approach it by a method of halving. In an acid ionization equilibrium problem we obtain the equation

$$1.3 \times 10^{-3} = \frac{x^2}{0.0500 - x}$$

From the chemistry involved, we know that the desired value of x must lie between 0 and 0.0500. First expand the equation into the standard form of a quadratic equation.

$$f(x) = x^2 + 1.3 \times 10^{-3}\,x - 6.5 \times 10^{-5} = 0$$

We evaluate $f(x)$ for the two limits (0 and 0.0500) and for the average of these two values.

$$f(0) = -6.5 \times 10^{-5}$$

$$f(0.0500) = 2.5 \times 10^{-3}$$

$$f(0.0250) = 5.9 \times 10^{-4}$$

Since the sign of the function changes between $x = 0$ and $x = 0.0250$, the desired root must lie in that interval (remember that the function is equal to 0 when x is equal to the root), so we use the average of the two numbers, 0.0125, for the next evaluation of the function.

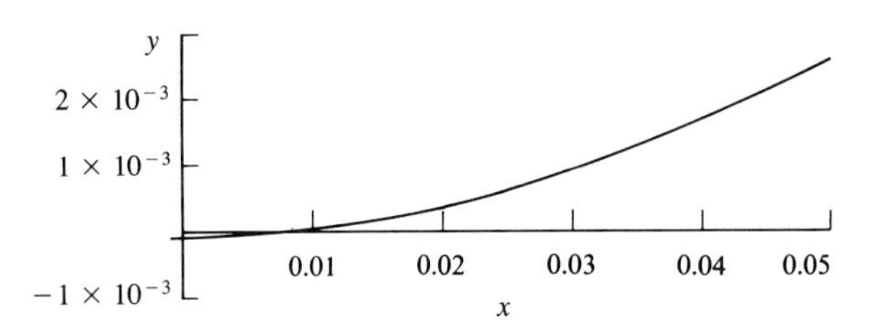

Figure A.4

$$f(0.0125) = 2.6 \times 10^{-4}$$

This value of x is also larger than the root since the function has a positive value. The following table summarizes the rest of the procedure.

lower x	upper x	middle x	$f(x)$(at middle x)	root between
0.00	5.00×10^{-2}	2.50×10^{-2}	5.9×10^{-4}	lower and middle
0.00	2.50×10^{-2}	1.25×10^{-2}	1.1×10^{-4}	lower and middle
0.00	1.25×10^{-2}	6.25×10^{-3}	-1.8×10^{-5}	middle and upper
6.25×10^{-3}	1.25×10^{-2}	9.38×10^{-3}	3.5×10^{-5}	lower and middle
6.25×10^{-3}	9.38×10^{-3}	7.82×10^{-3}	6.3×10^{-6}	lower and middle
6.25×10^{-3}	7.82×10^{-3}	7.04×10^{-3}	-6.3×10^{-6}	middle and upper
7.04×10^{-3}	7.82×10^{-3}	7.43×10^{-3}	-1.4×10^{-7}	middle and upper

We can accept 7.43×10^{-3} as the root. Note how close the function is to zero ($f(x) = -1.4 \times 10^{-7}$) when x is equal to 7.43×10^{-3}.

The Newton-Raphson Method

Probably the fastest method for finding the roots of a polynomial equation is the Newton-Raphson method. We will illustrate this method by finding a root (a value of x for which $y = 0$) of the equation

$$y = 4x^3 - 3x^2 + 2x - 1$$

The graph of this equation shows that there is a root at about 0.6, so we'll use that value as our first estimate for x. The value of y at $x = 0.6$ is

$$y = 4(0.6)^3 - 3(0.6)^2 + 2(0.6) - 1 = -0.016$$

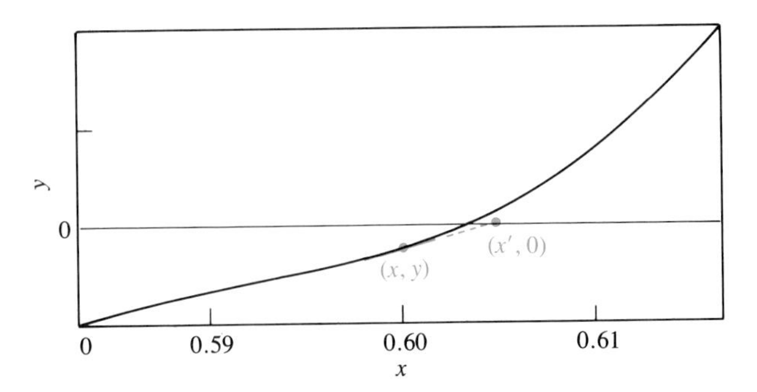

Figure A.5

To find the second estimate for x, we determine the slope of the function at the first estimate and draw a straight line with that slope from the first estimate to the x-axis. The point at which the extrapolated line crosses the x-axis is x', the second estimate for the root.

Notice that the line of slope m passes through points with coordinates (x, y) and $(x', 0)$.

$$\text{slope} = m = \frac{\Delta y}{\Delta x} = \frac{0 - y}{x' - x}$$

This relationship can be rearranged to solve for x':

$$x' = x - \frac{y}{m}$$

These relationships can be applied to the particular example. In this particular example, $x = 0.6$, and $y = -0.016$. We will find the slope by evaluating the derivative of the function. (If you do not know how to obtain the derivative, the slope is found by graphing the function and drawing the tangent.)

$$\text{slope} = m = 12x^2 - 6x + 2$$
$$= 12(0.6)^2 - 6(0.6) + 2 = 2.72$$

Now, solve for x'

$$x' = x - \frac{y}{m} = 0.6 - \frac{-0.016}{2.72} = 0.606$$

This is our second estimate of the root of the function. The value of the function when $x = 0.606$ is $y = 4.72 \times 10^{-4}$, which is much closer to zero than the previous estimate.

We can repeat this procedure until the estimates converge to the desired accuracy. (In fact, in this example the third estimate is the same as the second one to three significant figures.) The Newton-Raphson method is easily automated (programmed into a computer or calculator), and it is one of the best overall methods of solving for roots of polynomial equations.

Appendix B

Selected Physical Constants*

Quantity	Symbol	Value	Units
atmosphere (standard)	atm	$1.013\,25 \times 10^{5}$	Pa
atomic mass unit	u	$1.660\,540 \times 10^{-27}$	kg
Avogadro's number	N_A	$6.022\,14 \times 10^{23}$	particles mol^{-1}
Bohr radius	a_0	$5.291\,8 \times 10^{-11}$	m
Boltzmann constant	k	$1.380\,66 \times 10^{-23}$	$J\ K^{-1}$
charge-to-mass ratio of electron	e/m_e	$-1.758\,820 \times 10^{11}$	$C\ kg^{-1}$
elementary charge	e	$1.602\,177 \times 10^{-19}$	C
electron mass	m_e	$9.109\,39 \times 10^{-31}$	kg
faraday	F	$9.648\,53 \times 10^{4}$	C/mol e^-
gas constant	R	0.082 058	$L\ atm\ mol^{-1}\ K^{-1}$
		8.314 5	$J\ mol^{-1}\ K^{-1}$
neutron mass	m_n	$1.674\,929 \times 10^{-27}$	kg
Planck's constant	h	$6.626\,08 \times 10^{-34}$	J s
proton mass	m_p	$1.672\,623 \times 10^{-27}$	kg
Rydberg constant	R_h	$1.096\,78 \times 10^{7}$	m^{-1}
		$1.096\,78 \times 10^{5}$	cm^{-1}
speed of light	c	$2.997\,924\,58 \times 10^{8}$	$m\ s^{-1}$
(in vacuum)		$1.862\,81 \times 10^{5}$	$mi\ s^{-1}$

The uncertainty is less than $\pm$ 1 in the last place shown.

* E. R. Cohen and B. N. Taylor, *J. Phys. Chem. Ref. Data,* **17**, 1797 (1988).

Appendix C

Unit Conversion Factors

The metric system was implemented by the French National Assembly in 1790 and has been modified many times. The International System of Units, or *le Système International* (SI), represents an extension of the metric system and is used widely by scientists. It was adopted by the 11th General Conference of Weights and Measures in 1960. It has been modified since that time.

The SI defines seven base units, and defines other units based on these seven. The seven base units are shown in Table C.1.

Table C.1 SI Base Units

Physical Quantity	Name of Unit	Symbol
length	meter	m
mass	kilogram	kg
time	second	s
temperature	kelvin	K
amount of substance	mole	mol
electric current	ampere	A
luminous intensity	candela	cd

Decimal fractions and multiples of metric and SI units are designated by the prefixes listed in Table C.2. Those most commonly used in chemistry are underlined.

Table C.2 Metric and SI Prefixes

Factor	Prefix	Symbol	Factor	Prefix	Symbol
10^{12}	tera	T	10^{-1}	deci	d
10^{9}	giga	G	10^{-2}	<u>centi</u>	c
10^{6}	mega	M	10^{-3}	<u>milli</u>	m
10^{3}	<u>kilo</u>	k	10^{-6}	<u>micro</u>	μ
10^{2}	hecto	h	10^{-9}	<u>nano</u>	n
10^{1}	deka	da	10^{-12}	<u>pico</u>	p
			10^{-15}	femto	f
			10^{-18}	atto	a

All other units are derived from the seven base units. Table C.3 shows important derived units.

Table C.3 Derived SI Units

Physical Quantity	Name of Unit	Symbol	Definition
area	square meter	m^2	
volume	cubic meter	m^3	
density	kilogram per cubic meter	kg/m^3	
force	newton	N	$kg\ m/s^2$
pressure	pascal	Pa	N/m^2
energy	joule	J	$kg\ m^2/s^2$
electric charge	coulomb	C	A s
electric potential difference	volt	V	J/(A s)

Important Equalities

Many different units of measure are used. The following table gives important equalities among these units.

Common Units of Mass and Weight

1 pound = 453.59 grams = 0.45359 kilogram
1 kilogram = 1000 grams = 2.205 pounds
1 gram = 1000 milligrams = 6.022×10^{23} atomic mass units
1 atomic mass unit = 1.660540×10^{-27} kilograms
1 short ton = 2000 pounds = 907.2 kilograms
1 metric ton = 1000 kilograms = 2205 pounds

Common Units of Length

1 inch = 2.54 centimeters (exactly) = 0.0254 meter
1 foot = 12 inches = 30.48 centimeters
1 mile = 5280 feet = 1.609 kilometers
1 yard = 36 inches = 0.9144 meter
1 meter = 100 centimeters = 39.37 inches = 3.281 feet = 1.094 yards
1 kilometer = 1000 meters = 1094 yards = 0.6214 mile
1 picometer = 100 Ångstrom = 1.0×10^{-10} centimeter
= 1.0×10^{-12} meter = 3.937×10^{-11} inch

Common Units of Volume

1 quart = 0.94635 liter
1 liter = 1 cubic decimeter = 1000 cubic centimeters = 0.001 cubic meter
1 milliliter = 1 cubic centimeter = 0.001 liter = 1.0567×10^{-3} quart
1 cubic foot = 28.317 liters = 29.922 quarts = 7.480 gallons

Common Units of Pressure

1 atmosphere = 760 torr = 1.01325×10^5 pascals = 14.70 pounds per square inch
1 bar = 10^5 pascals
1 torr = 1 millimeter of mercury
1 pascal = 1 kg/m s^2 = 1 N/m^2

Common Units of Energy

1 thermochemical calorie* = 4.184 joules
= 4.129×10^{-2} liter-atmospheres
= 2.611×10^{19} electron volts
1 electron volt = 1.6022×10^{-19} joule = 96.486 kJ/mol
1 liter-atmosphere = 24.217 calories = 101.321 joules
1 British thermal unit = 1055.06 joules = 252.2 calories

* The amount of heat required to raise the temperature of one gram of water from 14.5 °C to 15.5 °C.

Appendix D

Names of Ions

Common Ions

Cations	Anions
1+	**1−**
ammonium (NH_4^+)	acetate (CH_3COO^-)
cesium (Cs^+)	bromide (Br^-)
copper(I) (Cu^+)	chlorate (ClO_3^-)
hydrogen (H^+, H_3O^+)	chloride (Cl^-)
lithium (Li^+)	cyanide (CN^-)
potassium (K^+)	dihydrogen phosphate ($H_2PO_4^-$)
silver (Ag^+)	fluoride (F^-)
sodium (Na^+)	hydride (H^-)
thallium(I) (Tl^+)	hydrogen carbonate (HCO_3^-)
	hydrogen sulfate (HSO_4^-)
2+	hydrogen sulfite (HSO_3^-)
barium (Ba^{2+})	hydroxide (OH^-)
cadmium (Cd^{2+})	hypochlorite (ClO^-)
calcium (Ca^{2+})	iodide (I^-)
chromium(II) (Cr^{2+})	nitrate (NO_3^-)
cobalt(II) (Co^{2+})	nitrite (NO_2^-)
copper(II) (Cu^{2+})	perchlorate (ClO_4^-)
iron(II) (Fe^{2+})	permanganate (MnO_4^-)
lead(II) (Pb^{2+})	thiocyanate (SCN^-)
magnesium (Mg^{2+})	
manganese(II) (Mn^{2+})	**2−**
mercury(I) (Hg_2^{2+})	carbonate (CO_3^{2-})
mercury(II) (Hg^{2+})	chromate (CrO_4^{2-})
strontium (Sr^{2+})	dichromate ($Cr_2O_7^{2-}$)
nickel (Ni^{2+})	hydrogen phosphate (HPO_4^{2-})
tin(II) (Sn^{2+})	oxalate ($C_2O_4^{2-}$)
zinc (Zn^{2+})	oxide (O^{2-})
	peroxide (O_2^{2-})
3+	sulfide (S^{2-})
aluminum (Al^{3+})	sulfate (SO_4^{2-})
chromium(III) (Cr^{3+})	sulfite (SO_3^{2-})
cobalt (III) (Co^{3+})	
iron(III) (Fe^{3+})	**3−**
lanthanum(La^{3+})	arsenate (AsO_4^{3-})
thallium(III) (Tl^{3+})	phosphate (PO_4^{3-})
titanium(III) (Ti^{3+})	
vanadium(III) (V^{3+})	

Oxyanions and Their Corresponding Oxyacids

Oxyanions		**Oxyacids**	
CO_3^{2-}	carbonate ion	H_2CO_3	carbonic acid
NO_2^-	nitrite ion	HNO_2	nitrous acid
NO_3^-	nitrate ion	HNO_3	nitric acid
PO_4^{3-}	phosphate ion	H_3PO_4	phosphoric acid
SO_3^{2-}	sulfite ion	H_2SO_3	sulfurous acid
SO_4^{2-}	sulfate ion	H_2SO_4	sulfuric acid
ClO^-	hypochlorite ion	$HClO$	hypochlorous acid
ClO_2^-	chlorite ion	$HClO_2$	chlorous acid
ClO_3^-	chlorate ion	$HClO_3$	chloric acid
ClO_4^-	perchlorate ion	$HClO_4$	perchloric acid

Appendix E

Properties of Water*

Physical Properties of Water

molar mass	18.01528	g mol^{-1}
melting point (1 atm)	0.00	°C
boiling point (1 atm)	100.00	°C
triple point temperature	0.0098	°C
triple point pressure	4.58	torr
triple point density (ℓ)	0.99978	g cm^{-3}
triple point density (g)	4.885	g L^{-1}
critical temperature	374.1	°C
critical pressure	218.3	atm
critical density	0.322	g cm^{-3}
maximum density	0.99995	g cm^{-3}
temperature of maximum density	4.0	°C
specific heat (at 15 °C)	4.184	J g^{-1} K^{-1}
heat of fusion (at 0 °C)	6.01	kJ mol^{-1}
entropy of fusion (at 0 °C)	22.2	J mol^{-1} K^{-1}
heat of vaporization (at 25 °C)	44.0	kJ mol^{-1}
entropy of vaporization (at 25 °C)	118.8	J mol^{-1} K^{-1}
freezing point depression constant	1.86	° molal^{-1}
boiling point elevation constant	0.512	° molal^{-1}

Dependence of Water Density, Vapor Pressure, and pK_w (−log of ionization constant) on Temperature

T °C	Density g cm^{-3}	Vapor Pressure torr	pK_w
0	0.99984	4.585	14.9435
10	0.99970	9.212	14.5346
20	0.99821	17.542	14.1669
25	0.99707	23.769	13.9965
30	0.99565	31.844	13.8330
40	0.99222	55.365	13.5348
50	0.98803	92.588	13.2617
60	0.98320	149.50	13.0171
70	0.97778	233.84	
80	0.97182	355.33	
90	0.96535	525.92	
100	0.95840	760.00	

* Reprinted with permission from the *CRC Handbook of Chemistry and Physics,* 1991. Copyright CRC Press, Boca Raton, FL.

Appendix F

Solubility Products, Acids, and Bases*

Solubility Product Constants at 25 °C

Substance	Formula	Solubility Product
aluminum phosphate	$AlPO_4$	9.8×10^{-21}
barium carbonate	$BaCO_3$	2.6×10^{-9}
barium chromate	$BaCrO_4$	1.2×10^{-10}
barium fluoride	BaF_2	1.8×10^{-7}
barium iodate	$Ba(IO_3)_2$	1.5×10^{-9}
barium phosphate	$Ba_3(PO_4)_2$	6.0×10^{-39}
barium sulfate	$BaSO_4$	8.7×10^{-11}
bismuth arsenate	$BiAsO_4$	4.4×10^{-10}
cadmium arsenate	$Cd_3(AsO_4)_2$	2.2×10^{-33}
cadmium carbonate	$CdCO_3$	6.2×10^{-12}
cadmium cyanide	$Cd(CN)_2$	1.0×10^{-8}
cadmium fluoride	CdF_2	6.4×10^{-3}
cadmium hydroxide	$Cd(OH)_2$	5.3×10^{-15}
cadmium iodate	$Cd(IO_3)_2$	2.5×10^{-8}
cadmium phosphate	$Cd_3(PO_4)_2$	2.5×10^{-33}
calcium carbonate	$CaCO_3$	5.0×10^{-9}
calcium fluoride	CaF_2	3.9×10^{-11}
calcium hydroxide	$Ca(OH)_2$	1.3×10^{-6}
calcium iodate	$Ca(IO_3)_2$	6.5×10^{-6}
calcium phosphate	$Ca_3(PO_4)_2$	1.3×10^{-32}
calcium sulfate	$CaSO_4$	7.1×10^{-5}
cerium(III) iodate	$Ce(IO_3)_3$	3.2×10^{-10}
cobalt(II) arsenate	$Co_3(AsO_4)_2$	6.8×10^{-29}
cobalt(II) phosphate	$Co_3(PO_4)_2$	2.1×10^{-33}
copper(I) bromide	$CuBr$	6.3×10^{-9}
copper(I) chloride	$CuCl$	1.7×10^{-7}
copper(I) iodide	CuI	1.1×10^{-12}
copper(I) thiocyanate	$CuSCN$	1.8×10^{-13}
copper(II) arsenate	$Cu_3(AsO_4)_2$	7.9×10^{-36}
copper(II) hydroxide	$Cu(OH)_2$	1.6×10^{-19}
copper(II) iodate	$Cu(IO_3)_2$	7.4×10^{-8}
copper(II) oxalate	CuC_2O_4	4.4×10^{-10}
copper(II) phosphate	$Cu_3(PO_4)_2$	1.4×10^{-37}

Solubility Product Constants at 25 °C *Continued*

Substance	Formula	Solubility Product
iron(II) carbonate	$FeCO_3$	3.1×10^{-11}
iron(II) fluoride	FeF_2	2.4×10^{-6}
iron(II) hydroxide	$Fe(OH)_2$	1.6×10^{-14}
iron(III) hydroxide	$Fe(OH)_3$	2.6×10^{-39}
iron(III) phosphate	$FePO_4$	9.9×10^{-29}
lanthanum hydroxide	$La(OH)_3$	1.0×10^{-19}
lanthanum iodate	$La(IO_3)_3$	6.2×10^{-12}
lead bromide	$PbBr_2$	6.6×10^{-6}
lead carbonate	$PbCO_3$	1.0×10^{-13}
lead chloride	$PbCl_2$	1.6×10^{-5}
lead fluoride	PbF_2	7.1×10^{-7}
lead hydroxide	$Pb(OH)_2$	1.4×10^{-20}
lead iodate	$Pb(IO_3)_2$	3.7×10^{-13}
lead iodide	PbI_2	8.5×10^{-9}
lead oxalate	PbC_2O_4	8.5×10^{-10}
lead sulfate	$PbSO_4$	1.8×10^{-8}
lead thiocyanate	$Pb(SCN)_2$	2.1×10^{-5}
lithium carbonate	Li_2CO_3	8.2×10^{-4}
magnesium carbonate	$MgCO_3$	6.8×10^{-6}
magnesium fluoride	MgF_2	7.4×10^{-11}
magnesium hydroxide	$Mg(OH)_2$	8.9×10^{-12}
magnesium phosphate	$Mg_3(PO_4)_2$	9.9×10^{-25}
manganese(II) carbonate	$MnCO_3$	2.2×10^{-11}
manganese(II) hydroxide	$Mn(OH)_2$	2.1×10^{-13}
manganese(II) iodate	$Mn(IO_3)_2$	4.4×10^{-7}
mercury(I) bromide	Hg_2Br_2	6.4×10^{-23}
mercury(I) carbonate	Hg_2CO_3	3.7×10^{-17}
mercury(I) chloride	Hg_2Cl_2	1.5×10^{-18}
mercury(I) fluoride	Hg_2F_2	3.1×10^{-6}
mercury(I) iodide	Hg_2I_2	5.3×10^{-29}
mercury(I) oxalate	$Hg_2C_2O_4$	1.8×10^{-13}
mercury(I) sulfate	Hg_2SO_4	8.0×10^{-7}
mercury(I) thiocyanate	$Hg_2(SCN)_2$	3.1×10^{-20}
mercury(II) hydroxide	$Hg(OH)_2$	3.1×10^{-26}
mercury(II) iodide	HgI_2	2.8×10^{-29}
nickel(II) carbonate	$NiCO_3$	1.4×10^{-7}
nickel(II) hydroxide	$Ni(OH)_2$	5.5×10^{-16}
nickel(II) iodate	$Ni(IO_3)_2$	4.7×10^{-5}
nickel(II) phosphate	$Ni_3(PO_4)_2$	4.7×10^{-32}
palladium(II) thiocyanate	$Pd(SCN)_2$	4.4×10^{-23}
potassium hexachloroplatinate(IV)	$K_2[PtCl_6]$	7.5×10^{-6}
potassium perchlorate	$KClO_4$	1.1×10^{-2}
silver acetate	$AgC_2H_3O_2$	1.9×10^{-3}
silver arsenate	Ag_3AsO_4	1.0×10^{-22}
silver bromate	$AgBrO_3$	5.3×10^{-5}
silver bromide	$AgBr$	7.7×10^{-13}
silver carbonate	Ag_2CO_3	8.5×10^{-12}
silver chloride	$AgCl$	1.8×10^{-10}
silver chromate	Ag_2CrO_4	2.5×10^{-12}
silver cyanide	$AgCN$	6.0×10^{-17}
silver iodate	$AgIO_3$	3.1×10^{-8}

Solubility Product Constants at 25 °C *Continued*

Substance	Formula	Solubility Product
silver iodide	AgI	8.5×10^{-17}
silver oxalate	$Ag_2C_2O_4$	5.4×10^{-12}
silver phosphate	Ag_3PO_4	1.0×10^{-16}
silver sulfate	Ag_2SO_4	1.2×10^{-5}
silver sulfite	Ag_2SO_3	1.5×10^{-14}
silver thiocyanate	$AgSCN$	1.0×10^{-12}
strontium arsenate	$Sr_3(AsO_4)_2$	4.3×10^{-19}
strontium carbonate	$SrCO_3$	5.6×10^{-10}
strontium fluoride	SrF_2	4.3×10^{-9}
strontium iodate	$Sr(IO_3)_2$	1.1×10^{-7}
strontium sulfate	$SrSO_4$	3.4×10^{-7}
tin(II) hydroxide	$Sn(OH)_2$	5.5×10^{-27}
zinc arsenate	$Zn_3(AsO_4)_2$	3.1×10^{-28}
zinc carbonate	$ZnCO_3$	1.2×10^{-10}
zinc fluoride	ZnF_2	3.0×10^{-2}
zinc hydroxide	$Zn(OH)_2$	6.9×10^{-17}
zinc iodate	$Zn(IO_3)_2$	4.3×10^{-6}

Selected Acids and Their Properties

Properties of Some Common, Commercially Available Acids

Name	Formula	Approximate Weight Percent	Approximate Molarity	Volume Needed to Prepare 1.0 L of a 1 *M* Solution, mL
acetic	CH_3COOH	99.8	17.4	57.55
hydrochloric	HCl	37.2	12.1	82.6
hydrofluoric	HF	49.0	28.9	34.6
nitric	HNO_3	70.4	15.9	62.9
perchloric	$HClO_4$	70.5	11.7	85.5
phosphoric	H_3PO_4	85.5	14.8	67.7
sulfuric	H_2SO_4	96.0	18.0	55.6

Safety Concerns

All acids, particularly when present in concentrated form, are potentially dangerous. They can all cause burns to skin. Always wear goggles, gloves, face shield, and an acid-resistant apron when handling concentrated solutions; even dilute acid solutions need to be treated with respect. (Goggles, gloves, and an apron are *always* appropriate when handling chemicals.) When you prepare an acid solution, remember to *add acid to water.* Adding water to acid can cause localized heating that spatters the acid out of its container. You should also stir while you are adding the acid.

Particular safety problems are associated with the behavior of some of the acids:

HCl	Strong fumes, handle in hood only.
HF	Dissolves glass. Use polymer beakers and bottles. Avoid skin contact.
HNO_3	Strong fumes, handle in hood only. Powerful oxidant. Avoid contact with organic materials.
$HClO_4$	Powerful oxidant. Can form explosive compounds with organic materials.
H_2SO_4	Reacts violently with water.

Selected Bases and Their Properties

Properties of Some Common, Commercially Available Bases

Name	Formula	Approximate Weight Percent	Approximate Molarity	Volume Needed to Prepare 1.0 L of a 1 *M* Solution, mL
ammonia	NH_3	28.0	14.5	69.0
sodium hydroxide	NaOH	50.5	19.4	51.5
potassium hydroxide	KOH	45.0	11.7	85.5

Safety Concerns

All bases, particularly when present in concentrated form, are potentially dangerous. They are especially hazardous if spilled into eyes. Bases require extensive washing; at least 15 minutes is needed, even for dilute solutions of bases. Always wear goggles, gloves, and an apron when handling concentrated solutions. Even if a dilute solution of a base contacts the eye, 15 minutes of washing is needed.

Ionization Constants of Selected Weak Acids at 25 °C

Acid	Formula	K_{a1}	K_{a2}	K_{a3}
acetic	CH_3COOH	1.8×10^{-5}		
arsenic	H_3AsO_4	5.6×10^{-3}	1.7×10^{-7}	4.0×10^{-12}
arsenious	H_3AsO_3	6.3×10^{-10}		
ascorbic	$H_2C_6H_6O_6$	8.0×10^{-5}	1.6×10^{-12}	
benzoic	C_6H_5COOH	6.3×10^{-5}		
carbonic	H_2CO_3	4.3×10^{-7}	5.6×10^{-11}	
chlorous	$HClO_2$	1.0×10^{-2}		
citric	$H_3C_6H_5O_7$	7.4×10^{-4}	1.7×10^{-5}	4.0×10^{-7}
formic	HCOOH	1.8×10^{-4}		
hydrocyanic	HCN	7.2×10^{-10}		
hydrofluoric	HF	3.5×10^{-4}		
hypobromous	HOBr	2.0×10^{-9}		
hypochlorous	HOCl	3.0×10^{-8}		

Acid	Formula	K_{a1}	K_{a2}	K_{a3}
hypoiodous	HOI	2.3×10^{-11}		
iodic	HIO_3	1.7×10^{-1}		
malic	$H_2C_4H_4O_5$	4.0×10^{-4}	9.0×10^{-6}	
malonic	$H_2C_3H_2O_4$	1.6×10^{-2}	2.1×10^{-6}	
nitrous	HNO_2	4.6×10^{-4}		
oxalic	$H_2C_2O_4$	5.9×10^{-2}	6.4×10^{-5}	
periodic	H_5IO_6	2.3×10^{-2}		
phenol	C_6H_5OH	1.3×10^{-10}		
phosphoric	H_3PO_4	7.5×10^{-3}	6.2×10^{-8}	2.2×10^{-13}
selenious	H_2SeO_3	3.5×10^{-3}	5.0×10^{-8}	
sulfuric	H_2SO_4	strong	1.2×10^{-2}	
sulfurous	H_2SO_3	1.5×10^{-2}	1.0×10^{-7}	

Dissociation Constants of Selected Weak Bases at 25 °C

Base	Formula	K_b
ammonia	NH_3	1.8×10^{-5}
ethylamine	$C_2H_5NH_2$	6.5×10^{-4}
hydrazine	N_2H_4	1.7×10^{-6}
hydroxylamine	NH_2OH	1.1×10^{-8}
morphine	$C_{17}H_{19}NO_3$	1.6×10^{-6}
pyridine	C_5H_5N	1.8×10^{-9}
strychnine	$C_{21}H_{22}N_2O_2$	1.8×10^{-6}
urea	$(NH_2)_2CO$	1.2×10^{-14}

Appendix G

Thermodynamic Constants for Selected Compounds

This table lists standard enthalpies of formation ΔH_f°, standard entropies S°, and standard free energies of formation ΔG_f° for a variety of substances all at 25 °C (298.15 K) and 1 atm. The entries in the table are ordered by group number in the periodic table.

Note that the *solution-phase* entropies are not absolute entropies but are measured relative to the arbitrary standard that S° for $H_3O^+(aq)$ is equal to 0. It is for this reason that some of them are negative.

Most of the thermodynamic data in these tables were taken from the *NBS Tables of Chemical Thermodynamic Properties* (1982). The data for organic compounds $C_nH_m (n > 2)$ were taken from the *Handbook of Chemistry and Physics* (1981).

	Substance	ΔH_f°(25 °C) kJ mol^{-1}	S° (25 °C) J mol^{-1} K^{-1}	ΔG_f°(25 °C) kJ mol^{-1}
	$H(g)$	217.96	114.60	203.26
	$H_2(g)$	0	130.57	0
	$H^+(aq)$	0	0	0
	$H_3O^+(aq)$	−285.83	69.91	−237.18
IA	$Li(s)$	0	29.12	0
	$Li(g)$	159.37	138.66	126.69
	$Li^+(aq)$	−278.49	13.4	−293.31
	$LiH(s)$	−90.54	20.01	−68.37
	$Li_2O(s)$	−597.94	37.57	−561.20
	$LiF(s)$	−615.97	35.65	−587.73
	$LiCl(s)$	−408.61	59.33	−384.39
	$LiBr(s)$	−351.21	74.27	−342.00
	$LiI(s)$	−270.41	86.78	−270.29
	$Na(s)$	0	51.21	0
	$Na(g)$	107.32	153.60	76.79
	$Na^+(aq)$	−240.12	59.0	−261.90
	$Na_2O(s)$	−414.22	75.06	−375.48
	$NaOH(s)$	−425.61	64.46	−379.53
	$NaF(s)$	−573.65	51.46	−543.51

	Substance	ΔH_f°(25 °C) kJ mol^{-1}	S° (25 °C) J mol^{-1} K^{-1}	ΔG_f°(25 °C) kJ mol^{-1}
	NaCl(s)	−411.15	72.13	−384.15
	NaBr(s)	−361.06	86.82	−348.98
	NaI(s)	−287.78	98.53	−286.06
	$NaNO_3$(s)	−467.85	116.52	−367.07
	Na_2S(s)	−364.8	83.7	−349.8
	Na_2SO_4(s)	−1387.08	149.58	−1270.23
	$NaHSO_4$(s)	−1125.5	113.0	−992.9
	Na_2CO_3(s)	−1130.68	134.98	−1044.49
	$NaHCO_3$(s)	−950.81	101.7	−851.1
	K(s)	0	64.18	0
	K(g)	89.24	160.23	60.62
	K^+(aq)	−252.38	102.5	−283.27
	KO_2(s)	−284.93	116.7	−239.4
	K_2O_2(s)	−494.1	102.1	−425.1
	KOH(s)	−424.76	78.9	−379.11
	KF(s)	−567.27	66.57	−537.77
	KCl(s)	−436.75	82.59	−409.16
	$KClO_3$(s)	−397.73	143.1	−296.25
	KBr(s)	−393.80	95.90	−380.66
	KI(s)	−327.90	106.32	−324.89
	$KMnO_4$(s)	−837.2	171.71	−737.7
	K_2CrO_4(s)	−1403.7	200.12	−1295.8
	$K_2Cr_2O_7$(s)	−2061.5	291.2	−1881.9
	Rb(s)	0	76.78	0
	Rb(g)	80.88	169.98	53.09
	Rb^+(aq)	−251.17	121.50	−283.98
	RbCl(s)	−435.35	95.90	−407.82
	RbBr(s)	−394.59	109.96	−381.79
	RbI(s)	−333.80	118.41	−328.86
	Cs(s)	0	85.23	0
	Cs(g)	76.06	175.49	49.15
	Cs^+(aq)	−258.28	133.05	−292.02
	CsF(s)	−553.5	92.80	−525.5
	CsCl(s)	−443.04	101.17	−414.55
	CsBr(s)	−405.81	113.05	−391.41
	CsI(s)	−346.60	123.05	−340.58
IIA	Be(s)	0	9.50	0
	Be(g)	324.3	136.16	286.6
	BeO(s)	−609.6	14.14	−580.3
	Mg(s)	0	32.68	0
	Mg(g)	147.70	148.54	113.13
	Mg^{2+}(aq)	−466.85	−138.1	−454.8
	MgO(s)	−601.70	26.94	−569.45
	$MgCl_2$(s)	−641.32	89.62	−591.82
	$MgSO_4$(s)	−1284.9	91.6	−1170.7
	Ca(s)	0	41.42	0
	Ca(g)	178.2	154.77	144.33
	Ca^{2+}(aq)	−542.83	−53.1	−553.58
	CaH_2(s)	−186.2	42	−147.2
	CaO(s)	−635.09	39.75	−604.05

	Substance	ΔH_f°(25 °C) kJ mol^{-1}	S° (25 °C) J mol^{-1} K^{-1}	ΔG_f°(25 °C) kJ mol^{-1}
	$CaS(s)$	−482.4	56.5	−477.4
	$Ca(OH)_2(s)$	−986.09	83.39	−898.56
	$CaF_2(s)$	−1219.6	68.87	−1167.3
	$CaCl_2(s)$	−795.8	104.6	−748.1
	$CaBr_2(s)$	−682.8	130	−663.6
	$CaI_2(s)$	−533.5	142	−528.9
	$Ca(NO_3)_2(s)$	−938.39	193.3	−743.20
	$CaC_2(s)$	−59.8	69.96	−64.9
	$CaCO_3$(s, calcite)	−1206.92	92.9	−1128.84
	$CaCO_3$(s, aragonite)	−1207.13	88.7	−1127.80
	$CaSO_4(s)$	−1434.11	106.9	−1321.86
	$CaSiO_3(s)$	−1634.94	81.92	−1549.66
	$CaMg(CO_3)_2(s)$	−2326.3	155.18	−2163.4
	$Sr(s)$	0	52.3	0
	$Sr(g)$	164.4	164.51	130.9
	$Sr^{2+}(aq)$	−545.80	−32.6	−559.48
	$SrCl_2(s)$	−828.9	114.85	−781.1
	$SrCO_3(s)$	−1220.0	97.1	−1140.1
	$Ba(s)$	0	62.8	0
	$Ba(g)$	180	170.24	146
	$Ba^{2+}(aq)$	−537.64	9.6	−560.77
	$BaO(s)$	−582.0	70.3	−552.3
	$BaCl_2(s)$	−858.6	123.68	−810.4
	$BaCO_3(s)$	−1216.3	112.1	−1137.6
	$BaSO_4(s)$	−1473.2	132.2	−1362.3
IIIB	$Sc(s)$	0	34.64	0
	$Sc(g)$	377.8	174.68	336.06
	$Sc^{3+}(aq)$	−614.2	−255	−586.6
IVB	$Ti(s)$	0	30.63	0
	$Ti(g)$	469.9	180.19	425.1
	$TiO_2(s)$	−944.7	50.33	−889.5
	$TiCl_4(\ell)$	−804.2	252.3	−737.2
	$TiCl_4(g)$	−763.2	354.8	−726.8
VIB	$Cr(s)$	0	23.77	0
	$Cr(g)$	396.6	174.39	351.8
	$Cr_2O_3(s)$	−1139.7	81.2	−1058.1
	$CrO_4^{2-}(aq)$	−881.15	50.21	−727.75
	$Cr_2O_7^{2-}(aq)$	−1490.3	261.9	−1301.1
	$W(s)$	0	32.64	0
	$W(g)$	849.4	173.84	807.1
	$WO_2(s)$	−589.69	50.54	−533.92
	$WO_3(s)$	−842.87	75.90	−764.08
VIIB	$Mn(s)$	0	32.01	0
	$Mn(g)$	280.7	238.5	173.59
	$Mn^{2+}(aq)$	−220.75	−73.6	−228.1
	$MnO(s)$	−385.22	59.71	−362.92

	Substance	ΔH_f°(25 °C) kJ mol^{-1}	S° (25 °C) J mol^{-1} K^{-1}	ΔG_f°(25 °C) kJ mol^{-1}
	MnO_2(s)	−520.03	53.05	−465.17
	MnO_4^-(aq)	−541.4	191.2	−447.2
VIIIB	Fe(s)	0	27.28	0
	Fe(g)	416.3	180.38	370.7
	Fe^{2+}(aq)	−89.1	−137.7	−78.9
	Fe^{3+}(aq)	−48.5	−315.9	−4.7
	FeO(s)	−272	—	—
	Fe_2O_3(s, hematite)	−824.2	87.40	−742.2
	Fe_3O_4(s, magnetite)	−1118.4	146.4	−1015.5
	$Fe(OH)_3$(s)	−569.0	88	−486.6
	FeS(s)	−100.0	60.29	−100.4
	$FeCO_3$(s)	−740.57	93.1	−666.72
	$Fe(CN)_6^{3-}$(aq)	561.9	270.3	729.4
	$Fe(CN)_6^{4-}$(aq)	455.6	95.0	695.1
	Co(s)	0	30.04	0
	Co(g)	424.7	179.41	380.3
	Co^{2+}(aq)	−58.2	−113	−54.4
	Co^{3+}(aq)	92	−305	134
	CoO(s)	−237.94	52.97	−214.22
	$CoCl_2$(s)	−312.5	109.16	−269.8
	Ni(s)	0	29.87	0
	Ni(g)	429.7	182.08	384.5
	Ni^{2+}(aq)	−54.0	−128.9	−45.6
	NiO(s)	−239.7	37.99	−211.7
	Pt(s)	0	41.63	0
	Pt(g)	565.3	192.30	520.5
	$PtCl_6^{2-}$(aq)	−668.2	219.7	−482.7
IB	Cu(s)	0	33.15	0
	Cu(g)	338.32	166.27	298.61
	Cu^+(aq)	71.67	40.6	49.98
	Cu^{2+}(aq)	64.77	−99.6	65.49
	CuO(s)	−157.3	42.63	−129.7
	Cu_2O(s)	−168.6	93.14	−146.0
	CuCl(s)	−137.2	86.2	−119.88
	$CuCl_2$(s)	−220.1	108.07	−175.7
	$CuSO_4$(s)	−771.36	109	−661.9
	$Cu(NH_3)_4^{2+}$(aq)	−348.5	273.6	−111.07
	Ag(s)	0	42.55	0
	Ag(g)	284.55	172.89	245.68
	Ag^+(aq)	105.58	72.68	77.11
	AgCl(s)	−127.07	96.2	−109.81
	$AgNO_3$(s)	−124.39	140.92	−33.48
	$Ag(NH_3)_2^+$(aq)	−111.29	245.2	−17.12
	Au(s)	0	47.40	0
	Au(g)	366.1	180.39	326.3
IIB	Zn(s)	0	41.63	0
	Zn(g)	130.73	160.87	95.18
	Zn^{2+}(aq)	−153.89	−112.1	−147.06

	Substance	ΔH_f°(25 °C) kJ mol^{-1}	S° (25 °C) J mol^{-1} K^{-1}	ΔG_f°(25 °C) kJ mol^{-1}
	ZnO(s)	−348.28	43.64	−318.32
	ZnS(s, sphalerite)	−205.98	57.7	−201.29
	$ZnCl_2$(s)	−415.05	111.46	−369.43
	$ZnSO_4$(s)	−982.8	110.5	−871.5
	$Zn(NH_3)_4^{2+}$(aq)	−533.5	301	−301.9
	Hg(ℓ)	0	76.02	0
	Hg(g)	61.32	174.85	31.85
	HgO(s)	−90.83	70.29	−58.56
	$HgCl_2$(s)	−224.3	146.0	−178.6
	Hg_2Cl_2(s)	−265.22	192.5	−210.78
IIIA	B(s)	0	5.86	0
	B(g)	562.7	153.34	518.8
	B_2H_6(g)	35.6	232.00	86.6
	B_5H_9(g)	73.2	275.81	174.9
	B_2O_3(s)	−1272.77	53.97	−1193.70
	H_3BO_3(s)	−1094.33	88.83	−969.02
	BF_3(g)	−1137.00	254.01	−1120.35
	BF_4^-(aq)	−1574.9	180	−1486.9
	BCl_3(g)	−403.76	289.99	−388.74
	BBr_3(g)	−205.64	324.13	−232.47
	Al(s)	0	28.33	0
	Al(g)	326.4	164.43	285.7
	Al^{3+}(aq)	−531	−321.7	−485
	$Al(OH)_3$(s)	−1287	85.4	−1150.
	Al_2O_3(s)	−1675.7	50.92	−1582.3
	$AlCl_3$(s)	−704.2	110.67	−628.8
	Ga(s)	0	40.88	0
	Ga(g)	277.0	168.95	238.9
	Tl(s)	0	64.18	0
	Tl(g)	182.21	180.85	147.44
IVA	C(s, graphite)	0	5.74	0
	C(s, diamond)	1.895	2.377	2.900
	C(g)	716.682	157.99	671.29
	CH_4(g)	−74.81	186.15	−50.75
	C_2H_2(g)	226.73	200.83	209.20
	C_2H_4(g)	52.26	219.45	68.12
	C_2H_6(g)	−84.68	229.49	−32.89
	C_3H_6(g, propene)	20.41	267	62.74
	C_3H_6(g, cyclopropane)	53.3	237.4	104.39
	C_3H_8(g)	−103.85	269.91	−23.49
	n-C_4H_{10}(g)	−124.73	310.03	−15.71
	i-C_4H_{10}(g)	−131.60	294.64	−17.97
	C_4H_8O(ℓ, butyraldehyde)	−238.7	246.8	−119.2
	n-C_5H_{12}(g)	−146.44	348.40	−8.20
	C_6H_6(g)	82.93	269.2	129.66
	C_6H_6(ℓ)	49.03	172.8	124.50
	CO(g)	−110.52	197.56	−137.15
	CO_2(g)	−393.51	213.63	−394.36
	CO_2(aq)	−413.80	117.6	−385.98

Substance	ΔH_f°(25 °C) kJ mol^{-1}	S° (25 °C) J mol^{-1} K^{-1}	ΔG_f°(25 °C) kJ mol^{-1}
$CS_2(\ell)$	89.70	151.34	65.27
$CS_2(g)$	117.36	237.73	67.15
$H_2CO_3(aq)$	−699.65	187.4	−623.08
$HCO_3^-(aq)$	−691.99	91.2	−586.77
$CO_3^{2-}(aq)$	−677.14	−56.9	−527.81
$HCOOH(\ell)$	−424.72	128.95	−361.42
$HCOOH(g)$	−378.61	248.74	351.00
$HCOOH(aq)$	−425.43	163	−372.3
$HCOO^-(aq)$	−425.55	92	−351.0
$CH_2O(g)$	−108.57	218.66	−102.55
$CH_3OH(\ell)$	−238.66	126.8	−166.35
$CH_3OH(g)$	−200.66	239.70	−162.01
$CH_3OH(aq)$	−245.93	133.1	−175.31
$H_2C_2O_4(s)$	−827.2	—	—
$HC_2O_4^-(aq)$	−818.4	149.4	−698.34
$C_2O_4^{2-}(aq)$	−825.1	45.6	−673.9
$CH_3COOH(\ell)$	−484.5	159.8	−390.0
$CH_3COOH(g)$	−432.25	282.4	−374.1
$CH_3COOH(aq)$	−485.76	178.7	−396.46
$CH_3COO^-(aq)$	−486.01	86.6	−369.31
$CH_3CHO(\ell)$	−192.30	160.2	−128.12
$C_2H_5OH(\ell)$	−277.69	160.7	−174.89
$C_2H_5OH(g)$	−235.10	282.59	−168.57
$C_2H_5OH(aq)$	−288.3	148.5	−181.64
$(CH_3)_2CO(\ell)$	−248.1	200.4	155.4
$CH_3OCH_3(g)$	−184.05	266.27	−112.67
$C_6H_{12}O_6(s)$	−1268	212	−910
$CF_4(g)$	−925	261.50	−879
$CCl_4(\ell)$	−135.44	216.40	−65.28
$CCl_4(g)$	−102.9	309.74	−60.62
$CHCl_3(g)$	−103.14	295.60	−70.37
$COCl_2(g)$	−218.8	283.53	−204.6
$CH_2Cl_2(g)$	−92.47	270.12	−65.90
$CH_3Cl(g)$	−80.83	234.47	−57.40
$CBr_4(s)$	79	357.94	67
$CH_3I(\ell)$	−15.5	163.2	13.4
$HCN(g)$	135.1	201.67	124.7
$HCN(aq)$	107.1	124.7	119.7
$CN^-(aq)$	150.6	94.1	172.4
$CH_3NH_2(g)$	−22.97	243.30	32.09
$CH_3NO_2(\ell)$	−113.1	171.76	−14.52
$CH_3NO_2(g)$	−74.73	274.42	−6.91
$CO(NH_2)_2(s)$	−333.51	104.49	−197.44
Si(s)	0	18.83	0
Si(g)	455.6	167.86	411.3
SiC(s)	−65.3	16.61	−62.8
SiO_2(s, quartz)	−910.94	41.84	−856.67
SiO_2(s, cristobalite)	−909.48	42.68	−855.43
Ge(s)	0	31.09	0
Ge(g)	376.6	335.9	167.79
Sn(s, white)	0	51.55	0
Sn(s, gray)	−2.09	44.14	0.13

	Substance	ΔH_f°(25 °C) kJ mol^{-1}	S° (25 °C) J mol^{-1} K^{-1}	ΔG_f°(25 °C) kJ mol^{-1}
	Sn(g)	302.1	168.38	267.3
	SnO(s)	−285.8	56.5	−256.9
	SnO_2(s)	−580.7	52.3	−519.6
	$Sn(OH)_2$(s)	−561.1	155	−491.7
	Pb(s)	0	64.81	0
	Pb(g)	195.0	161.9	175.26
	Pb^{2+}(aq)	−1.7	10.5	−24.43
	$PbCO_3$(s)	−699.1	130.96	−625.5
	PbO(s)	−218.99	66.5	−188.95
	PbO_2(s)	−277.4	68.6	−217.36
	PbS(s)	−100.4	91.2	−98.7
	PbI_2(s)	−175.48	174.85	−173.64
	$PbSO_4$(s)	−919.94	148.57	−813.21
VA	N_2(g)	0	191.50	0
	N(g)	472.70	153.19	455.58
	NH_3(g)	−46.11	192.34	−16.48
	NH_3(aq)	−80.29	111.3	−26.50
	NH_4^+(aq)	−132.51	113.4	−79.31
	$N_2H_4(\ell)$	50.63	121.21	149.24
	N_2H_4(aq)	34.31	138	128.1
	NO(g)	90.25	210.65	86.55
	NO_2(g)	33.18	239.95	51.29
	NO_2^-(aq)	−104.6	123.0	−32.2
	NO_3^-(aq)	−205.0	146.4	−108.74
	N_2O(g)	82.05	219.74	104.18
	N_2O_4(g)	9.16	304.18	97.82
	N_2O_5(s)	−43.1	178.2	113.8
	NOBr(g)	82.1	273.5	82.4
	HNO_2(g)	−79.5	254.0	−46.0
	$HNO_3(\ell)$	−174.10	155.49	−80.76
	NH_4NO_3(s)	−365.56	151.08	−184.02
	NH_4Cl(s)	−314.43	94.6	−202.97
	$(NH_4)_2SO_4$(s)	−1180.85	220.1	−901.90
	P(s, white, $\frac{1}{4}$ P_4)	0	41.09	0
	P(g)	314.64	163.08	278.28
	P_2(g)	144.3	218.02	103.7
	P_4(g)	58.91	279.87	24.47
	PH_3(g)	5.4	210.12	13.4
	H_3PO_4(s)	−1279.0	110.50	−1119.2
	H_3PO_4(aq)	−1288.34	158.2	−1142.54
	$H_2PO_4^-$(aq)	−1296.29	90.4	−1130.28
	HPO_4^{2-}(aq)	−1292.14	−33.5	−1089.15
	PO_4^{3-}(aq)	−1277.4	−222	−1018.7
	$POCl_3$(g)	−592.7	324.6	−545.2
	PCl_3(g)	−287.0	311.67	−267.8
	PCl_5(g)	−374.9	364.47	−305.0
	As(s, gray)	0	35.1	0
	As(g)	302.5	174.10	261.0
	As_2(g)	222.2	239.3	171.9
	As_4(g)	143.9	314	92.4

	Substance	ΔH_f°(25 °C) kJ mol^{-1}	S° (25 °C) J mol^{-1} K^{-1}	ΔG_f°(25 °C) kJ mol^{-1}
	AsH_3(g)	66.44	222.67	68.91
	As_4O_6(s)	−1313.94	214.2	−1152.53
	Sb(s)	0	45.69	0
	Sb(g)	262.3	180.16	222.1
	Bi(s)	0	56.74	0
	Bi(g)	207.1	186.90	168.2
VIA	O_2(g)	0	205.03	0
	O(g)	249.17	160.95	231.76
	O_3(g)	142.7	238.82	163.2
	OH^-(aq)	−229.99	−10.75	−157.24
	H_2O(ℓ)	−285.83	69.91	−237.18
	H_2O(g)	−241.82	188.72	−228.59
	H_2O_2(ℓ)	−187.78	109.6	−120.42
	H_2O_2(aq)	−191.17	143.9	−134.03
	S(s, rhombic, $\frac{1}{8}S_8$)	0	31.80	0
	S(s, monoclinic)	0.30	32.6	0.096
	S(g)	278.80	167.71	238.28
	S_8(g)	102.30	430.87	49.66
	S^{2-}(aq)	33.1	−14.6	85.8
	H_2S(g)	−20.63	205.68	−33.56
	H_2S(aq)	−39.7	121	−27.83
	HS^-(aq)	−17.6	62.8	12.08
	SO(g)	6.26	221.84	−19.87
	SO_2(g)	−296.83	248.11	−300.19
	SO_3(g)	−395.72	256.65	−371.08
	H_2SO_3(aq)	−608.81	232.2	−537.81
	HSO_3^-(aq)	−626.22	139.7	−527.73
	SO_3^{2-}(aq)	−635.5	−29	−486.5
	H_2SO_4(ℓ)	−813.99	156.90	−690.10
	HSO_4^-(aq)	−887.34	131.8	−755.91
	SO_4^{2-}(aq)	−909.27	20.1	−744.53
	SF_6(g)	−1209	291.71	−1105.4
	Se(s, black)	0	42.44	0
	Se(g)	227.07	176.61	187.06
VIIA	F_2(g)	0	202.67	0
	F(g)	78.99	158.64	61.94
	F^-(aq)	−332.63	−13.8	−278.79
	HF(g)	−271.1	173.67	−273.2
	HF(aq)	−320.08	88.7	−296.82
	Cl_2(g)	0	222.96	0
	Cl(g)	121.68	165.09	105.71
	Cl^-(aq)	−167.16	56.5	−131.23
	HCl(g)	−92.31	186.80	−95.30
	ClO^-(aq)	−107.1	42	−36.8
	ClO_2(g)	102.5	256.73	120.5
	ClO_2^-(aq)	−66.5	101.3	17.2
	ClO_3^-(aq)	−103.97	162.3	−7.95
	ClO_4^-(aq)	−129.33	182.0	−8.52
	Cl_2O(g)	80.3	266.10	97.9

	Substance	ΔH_f°(25 °C) kJ mol^{-1}	S° (25 °C) J mol^{-1} K^{-1}	ΔG_f°(25 °C) kJ mol^{-1}
	HClO(aq)	−120.9	142	−79.9
	ClF_3(g)	−163.2	281.50	−123.0
	Br_2(ℓ)	0	152.23	0
	Br_2(g)	30.91	245.35	3.14
	Br_2(aq)	−2.59	130.5	3.93
	Br(g)	111.88	174.91	82.41
	Br^-(aq)	−121.55	82.4	−103.96
	HBr(g)	−36.40	198.59	−53.43
	BrO_3^-(aq)	−67.07	161.71	18.60
	I_2(s)	0	116.14	0
	I_2(g)	62.44	260.58	19.36
	I_2(aq)	22.6	137.2	16.40
	I(g)	106.84	180.68	70.28
	I^-(aq)	−55.19	111.3	−51.57
	I_3^-(aq)	−51.5	239.3	−51.4
	HI(g)	26.48	206.48	1.72
	ICl(g)	17.78	247.44	−5.44
	IBr(g)	40.84	258.66	3.71
VIIIA	He(g)	0	126.04	0
	Ne(g)	0	146.22	0
	Ar(g)	0	154.73	0
	Kr(g)	0	163.97	0
	Xe(g)	0	169.57	0
	XeF_4(s)	−261.5	—	—

Appendix H

Standard Reduction Potentials at 25 °C

Half-Reaction	$E°$, V
$Ag^+(aq) + e^- \rightleftharpoons Ag(s)$	+0.799
$AgBr(s) + e^- \rightleftharpoons Ag(s) + Br^-(aq)$	+0.095
$AgCl(s) + e^- \rightleftharpoons Ag(s) + Cl^-(aq)$	+0.222
$Ag_2CrO_4(s) + 2e^- \rightleftharpoons 2Ag(s) + CrO_4^{2-}(aq)$	+0.446
$AgI(s) + e^- \rightleftharpoons Ag(s) + I^-(aq)$	−0.151
$Ag_2O(s) + H_2O(\ell) + 2e^- \rightleftharpoons 2Ag(s) + 2OH^-(aq)$	0.342
$Al^{3+}(aq) + 3e^- \rightleftharpoons Al(s)$	−1.66
$H_3AsO_4(aq) + 2H^+(aq) + 2e^- \rightleftharpoons H_3AsO_3(aq) + H_2O(\ell)$	+0.559
$Ba^{2+}(aq) + 2e^- \rightleftharpoons Ba(s)$	−2.90
$Br_2(\ell) + 2e^- \rightleftharpoons 2Br^-(aq)$	+1.06
$BrO_3^-(aq) + 6H^+(aq) + 5e^- \rightleftharpoons \frac{1}{2} Br_2(\ell) + 3H_2O(\ell)$	+1.52
$Ca^{2+}(aq) + 2e^- \rightleftharpoons Ca(s)$	−2.87
$2CO_2(g) + 2H^+(aq) + 2e^- \rightleftharpoons H_2C_2O_4(aq)$	−0.49
$Cd^{2+}(aq) + 2e^- \rightleftharpoons Cd(s)$	−0.403
$Cd(OH)_2(s) + 2e^- \rightleftharpoons Cd(s) + 2OH^-(aq)$	−0.83
$Ce^{4+}(aq) + e^- \rightleftharpoons Ce^{3+}(aq)$	+1.61
$Cl_2(g) + 2e^- \rightleftharpoons 2Cl^-(aq)$	+1.36
$HClO(aq) + H^+(aq) + 2e^- \rightleftharpoons Cl^-(aq) + H_2O(\ell)$	+1.63
$ClO^-(aq) + H_2O(\ell) + 2e^- \rightleftharpoons Cl^-(aq) + 2OH^-(aq)$	+0.89
$ClO_3^-(aq) + 6H^+(aq) + 5e^- \rightleftharpoons \frac{1}{2} Cl_2(g) + 3H_2O(\ell)$	+1.47
$Co^{2+}(aq) + 2e^- \rightleftharpoons Co(s)$	−0.28
$Co^{3+}(aq) + e^- \rightleftharpoons Co^{2+}(aq)$	+1.83
$Cr^{3+}(aq) + e^- \rightleftharpoons Cr^{2+}(aq)$	−0.41
$Cr^{2+}(aq) + 2e^- \rightleftharpoons Cr(s)$	−0.91
$Cr_2O_7^{2-}(aq) + 14H^+(aq) + 6e^- \rightleftharpoons 2Cr^{3+}(aq) + 7H_2O(\ell)$	+1.33
$CrO_4^{2-}(aq) + 4H_2O(\ell) + 3e^- \rightleftharpoons Cr(OH)_3(s) + 5OH^-(aq)$	−0.13
$Cu^{2+}(aq) + 2e^- \rightleftharpoons Cu(s)$	+0.34
$Cu^{2+}(aq) + e^- \rightleftharpoons Cu^+(aq)$	+0.153
$Cu^+(aq) + e^- \rightleftharpoons Cu(s)$	+0.521
$F_2(g) + 2e^- \rightleftharpoons 2F^-(aq)$	+2.87
$Fe^{2+}(aq) + 2e^- \rightleftharpoons Fe(s)$	−0.44
$Fe^{3+}(aq) + e^- \rightleftharpoons Fe^{2+}(aq)$	+0.771
$Fe(CN)_6^{3-}(aq) + e^- \rightleftharpoons Fe(CN)_6^{4-}(aq)$	+0.358
$2H^+(aq) + 2e^- \rightleftharpoons H_2(g)$	0.000
$H_2(g) + 2e^- \rightleftharpoons 2H^-(aq)$	−2.23
$2H_2O(\ell) + 2e^- \rightleftharpoons H_2(g) + 2OH^-(aq)$	−0.83

Half-Reaction	$E°$, V
$HO_2^-(aq) + H_2O(\ell) + 2e^- \rightleftharpoons 3OH^-(aq)$	+0.88
$H_2O_2(aq) + 2H^+(aq) + 2e^- \rightleftharpoons 2H_2O(\ell)$	+1.776
$Hg^{2+}(aq) + 2e^- \rightleftharpoons Hg(\ell)$	+0.85
$2Hg^{2+}(aq) + 2e^- \rightleftharpoons Hg_2^{2+}(aq)$	+0.92
$I_2(s) + 2e^- \rightleftharpoons 2I^-(aq)$	+0.54
$IO_3^-(aq) + 6H^+(aq) + 5e^- \rightleftharpoons \frac{1}{2} I_2(aq) + 3H_2O(\ell)$	+1.195
$K^+(aq) + e^- \rightleftharpoons K(s)$	−2.93
$Li^+(aq) + e^- \rightleftharpoons Li(s)$	−3.04
$Mg^{2+}(aq) + 2e^- \rightleftharpoons Mg(s)$	−2.37
$Mn^{2+}(aq) + 2e^- \rightleftharpoons Mn(s)$	−1.18
$MnO_4^-(aq) + 8H^+(aq) + 5e^- \rightleftharpoons Mn^{2+}(aq) + 4H_2O(\ell)$	+1.51
$MnO_4^-(aq) + 2H_2O(\ell) + 3e^- \rightleftharpoons MnO_2(s) + 4OH^-(aq)$	+0.59
$HNO_2(aq) + H^+(aq) + e^- \rightleftharpoons NO(g) + H_2O(\ell)$	+0.983
$N_2(g) + 2H_2O(\ell) + 4H^+(aq) + 2e^- \rightleftharpoons 2NH_3OH^+(aq)$	−1.87
$N_2(g) + 4H_2O(\ell) + 2e^- \rightleftharpoons 2NH_2OH(aq) + 2OH^-(aq)$	−3.17
$NO_3^-(aq) + 4H^+(aq) + 3e^- \rightleftharpoons NO(g) + 2H_2O(\ell)$	+0.96
$Na^+(aq) + e^- \rightleftharpoons Na(s)$	−2.71
$Ni^{2+}(aq) + 2e^- \rightleftharpoons Ni(s)$	−0.25
$NiO(OH)(s) + H_2O(\ell) + e^- \rightleftharpoons Ni(OH)_2(s) + OH^-(aq)$	+0.52
$O_2(g) + 2H_2O(\ell) + 4e^- \rightleftharpoons 4OH^-(aq)$	+0.40
$O_2(g) + 4H^+(aq) + 4e^- \rightleftharpoons 2H_2O(\ell)$	+1.23
$O_2(g) + 2H^+(aq) + 2e^- \rightleftharpoons H_2O_2(aq)$	+0.68
$Pb^{2+}(aq) + 2e^- \rightleftharpoons Pb(s)$	−0.126
$PbSO_4(s) + H^+(aq) + 2e^- \rightleftharpoons Pb(s) + HSO_4^-(aq)$	−0.356
$PbO_2(s) + HSO_4^-(aq) + 3H^+(aq) + 2e^- \rightleftharpoons PbSO_4(s) + 2H_2O(\ell)$	+1.685
$S(s) + 2H^+(aq) + 2e^- \rightleftharpoons H_2S(g)$	+0.14
$H_2SO_3(aq) + 4H^+(aq) + 4e^- \rightleftharpoons S(s) + 3H_2O(\ell)$	+0.45
$HSO_4^-(aq) + 3H^+(aq) + 2e^- \rightleftharpoons H_2SO_3(s) + H_2O(\ell)$	+0.17
$Sn^{2+}(aq) + 2e^- \rightleftharpoons Sn(s)$	−0.14
$Sn^{4+}(aq) + 2e^- \rightleftharpoons Sn^{2+}(aq)$	+0.15
$VO_2^+(aq) + 2H^+(aq) + e^- \rightleftharpoons VO^{2+}(aq) + H_2O(\ell)$	+1.00
$Zn^{2+}(aq) + 2e^- \rightleftharpoons Zn(s)$	−0.76
$Zn(OH)_2(s) + 2e^- \rightleftharpoons Zn(s) + 2OH^-$	−1.25

Appendix I

Glossary*

α-Helix—the spiral structure adopted by a peptide chain that is held together by hydrogen bonding between amide groups in the same region of the chain. *(21.5)*

Absolute uncertainty—a measure of the expected range in the measurement. *(1.6)*

Absorption spectrum—a graph of the quantity of light a sample absorbs as a function of wavelength. *(18.6)*

Accuracy—the agreement of the measured value to a true or accurately known value of the same quantity. *(1.6)*

Acid—a substance that provides the hydrogen cation in water solution (see also Lewis and Brønsted-Lowry acids). *(3.1)*

Acid-base conjugate pair—two species that differ by the presence or absence of a proton. The protonated species is the acid; the deprotonated species is the base. *(13.1)*

Actinides—the elements in period 7 following actinium (Ac): thorium (Th) to lawrencium (Lr). *(1.3)*

Activated complex—an unstable arrangement of atoms at the highest energy point on the reaction coordinate. *(16.4)*

Activation analysis—the exposure of a sample to a neutron beam producing radioactive isotopes which are used to identify the elements present and determine the composition of the sample. *(20.2)*

Activation energy—the minimum collision energy required for a reaction to occur; equal to the difference in energy between the reactants and the activated complex. *(16.4)*

Actual yield—the amount of product isolated when a chemical reaction is carried out. *(3.4)*

Addition reaction—the combination of two or more substances to form one new substance. *(21.2)*

Adduct—the product of a Lewis acid-base addition reaction. *(13.9)*

Adhesive forces—the attractions that molecules of one substance exert on those of a different substance. *(10.2)*

* The number in parentheses is the number of the section in which the term is defined.

Adsorption—the process by which a substance adheres to the surface of a solid. *(16.5)*

Alcohol—an organic compound that contains the —OH (*hydroxyl*) functional group. *(21.3)*

Aldehyde—an organic compound that contains the $\overset{\text{O}}{\overset{\|}{-\text{C}}}-\text{H}$ functional group. *(21.3)*

Alkali metals—the elements in group IA (1): Li, Na, K, Rb, Cs, Fr. *(1.3)*

Alkaline earth metals—the elements in group IIA (2): Be, Mg, Ca, Sr, Ba, Ra. *(1.3)*

Alkane—a hydrocarbon that contains no multiple bonds or rings. *(21.1)*

Alkene—an unsaturated hydrocarbon that contains one or more carbon-carbon double bonds. *(21.2)*

Alkyl group—an alkane from which one hydrogen atom has been removed. *(21.1)*

Alkyne—an unsaturated hydrocarbon that contains one or more carbon-carbon triple bonds. *(21.2)*

Allotropes—structurally distinct forms of an element in the same physical state that exhibit different chemical and physical properties. *(10.3)*

Alloy—a mixture of a metal and one or more additional elements, often a second metal. *(1.2,7.5)*

Alpha particle—a high-energy helium nucleus produced by the radioactive decay of some atoms. *(20.1)*

Amide—an organic compound that contains the —C(O)NR_2 functional group where R is an organic substituent or hydrogen. *(21.3)*

Amine—a derivative of ammonia in which one or more of the hydrogen atoms is replaced with an organic substituent (R). Amines with the formula RNH_2 are *primary amines,* R_2NH are *secondary amines,* and R_3N are *tertiary amines.* *(21.3)*

Amino acid—an organic compound that contains both a carboxyl group (—CO_2H) and an amino group (—NH_2). *(21.3)*

Amorphous solid—a solid that lacks the long-range order of a crystalline solid. *(10.3)*

Amphoteric species—species that can act as either an acid or as a base. *(13.1,14.6)*

Amplitude—the height of a wave. *(6.1)*

Analyte—a substance whose concentration is being determined. *(14.1)*

Analytical concentration—the total concentration of all protonated and deprotonated forms of the same species. *(13.5)*

Angular momentum quantum number, ℓ—the quantum number that describes the shape of the electron probability wave in an atom. The allowed values of this quantum number are 0 and all positive whole numbers up to $(n - 1)$. *(6.4)*

Anion—a negatively charged ion. *(2.2)*

Annihilation—the process in which two anti-particles collide and their mass is converted into electromagnetic energy. *(20.1)*

Anode—the electrode in a cell at which oxidation occurs. *(17.9)*

Anodic protection—reducing corrosion by the intentional oxidation of a metal under carefully controlled conditions, to form a thin, adhering layer of oxide on the surface of the metal. The protective layer can be generated either chemically or electrochemically. *(17.12)*

Antibonding molecular orbital—an orbital that reduces the electron density in the region between the atoms in the molecule. *(9.5)*

Aromatic hydrocarbon—a compound that contains one or more benzene rings. *(21.2)*

Arrhenius acid—a species that increases the hydrogen ion concentration when dissolved in water. *(13.1)*

Arrhenius base—a species that increases the hydroxide ion concentration when dissolved in water. *(13.1)*

Arrhenius equation—the relationship between rate constant and temperature. *(16.4)*

$$k = Ae^{-E_a/RT}$$

Arrhenius plot—a graph of the natural logarithm of rate constant against 1/temperature. *(16.4)*

Aryl group—an aromatic ring that is being considered as a substituent. *(21.2)*

Atom—the smallest unit of an element that enters into a chemical combination. *(2.1)*

Atomic mass unit—the base unit of a mass scale that defines one unit (u) as one-twelfth the mass of a single atom of ^{12}C. *(3.2)*

Atomic number—the number of protons in the nucleus of an atom. It has the symbol Z. *(2.2)*

Atomic orbital—a wave function of an electron in an atom that has assigned values for all three of the quantum numbers n, ℓ, and m_ℓ. *(6.4)*

Atomic radius—one-half the distance between adjacent atoms of the same element in a molecule. *(7.2)*

Atomic weight or **atomic mass**—the mass in atomic mass units of one atom of an element. It is an average mass that reflects the natural isotopic distribution of the element. *(3.2)*

Autoionization of water—the equilibrium that describes the ionization of water. *(13.2)*

$$2H_2O(\ell) \rightleftharpoons H_3O^+(aq) + OH^-(aq)$$

$$K_W = [H_3O^+][OH^-]$$

Avogadro's law—at constant pressure and temperature the volume of a sample of gas is proportional to the number of moles of gas present ($V = k_3 \times n$). *(5.2)*

Avogadro's number—the number of units in one mole, 6.022×10^{23} units/mol. *(3.2)*

Azeotrope—a solution that has the composition of the constant boiling mixture, in which the compositions of the vapor and liquid phases are identical. *(11.6)*

β-Pleated sheet—the flattened structure adopted by a protein that is held together by hydrogen bonding between amide groups in different chains or in different sections of the same chain. *(21.5)*

Band gap—the difference in energy between the highest filled band of orbitals and the conduction band in solids. *(19.4)*

Base—a substance that provides hydroxide anion in water (see also Lewis and Brønsted-Lowry bases). *(3.1)*

Base units—any of the defined quantities in the SI. *(1.5)*

Battery—two or more voltaic cells with the negative electrode of each cell connected to the positive electrode of the next cell. *(17.8)*

Beta particle—a high-energy electron emitted by the nucleus of some radioactive atoms. *(20.1)*

Bimolecular step—an elementary step that involves the collision of two species. *(16.6)*

Binary compound—a compound composed of only two elements. *(2.4)*

Binary hydride—a compound formed from one element in combination with hydrogen. HF and CH_4 are binary hydrides. *(13.8)*

Biochemistry—the study of the chemistry of systems in living organisms. *(21.1)*

Body centered cubic (BCC)—the cubic structure with identical atoms at the corners and the center of the unit cell. *(10.3)*

Boiling point—the temperature at which the vapor pressure of a liquid is equal to the applied pressure. *(10.5)*

Boiling point elevation—a colligative property of solutions described by the equation

$$\Delta T_b = mk_b$$

where ΔT_b = (b.p. of solution − b.p. of solvent), m is the molal concentration of the solute particles, and k_b is the boiling point elevation constant characteristic of the solvent. *(11.4)*

Bond dissociation energy or **bond energy**—the energy required to break one mole of bonds in a gaseous species. *(8.8)*

Bond length—the distance between the nuclei of two bonded atoms in a molecule. *(8.3)*

Bond order—the number of electron pairs that are shared between two atoms. *(8.3)*

Bonding molecular orbital—an orbital that concentrates the electron density between the atoms in the molecule. *(9.5)*

Bonding pairs—pairs of electrons shared between two atoms. *(8.3)*

Boyle's law—at constant temperature the volume of a gas sample is inversely proportional to the pressure ($P \times V = k_1$). *(5.2)*

Bragg equation—the relation between the angle of diffraction of x rays and the distance between layers of atoms in a crystalline solid.

$$n\lambda = 2d \sin \theta$$

where λ is the wavelength of the x rays used, d is the distance between layers of atoms in the crystal, θ is the angle between the x-ray beam and the layers of atoms, and n is a whole positive number called the order. *(10.4)*

Breeder reactor—a fission reactor that generates more fissionable atoms than are consumed. *(20.4)*

Brønsted-Lowry acid—a species that can donate a proton. *(13.1)*

Brønsted-Lowry base—a species that can accept a proton. *(13.1)*

Buffer—a system that resists changes in pH. Most buffers are solutions that contain a weak acid-base conjugate pair. *(14.2)*

Buffer capacity—the amount of strong acid or base needed to change the pH of one liter of buffer solution by 1 unit. *(14.2)*

Buret—a device calibrated to measure the volume of added liquid in a titration. *(4.3)*

Calorimeter—the device used to measure heat flow. *(4.5)*

Calorimetry—the measurement of heat flow. *(4.5)*

Carbohydrate—a polyhydroxyl aldehyde or ketone, or substances that react with water to yield such a compound. *(21.6)*

Carbonyl group—the C=O functional group. *(21.3)*

Carboxylate group—the $-CO_2^-$ functional group. *(21.3)*

Carboxylic acid—an organic compound that contains the $-CO_2H$ functional group, a carboxyl group. *(21.3)*

Catalyst—a substance that increases the rate of reaction, but is not consumed in the reaction. *(16.5)*

Catenation—the formation of chains or rings of like atoms. *(21.1)*

Cathode—the electrode in a cell at which reduction occurs. *(17.9)*

Cathodic protection—protection of a metallic object from corrosion by placing it in electrical contact with a second, more reactive metal. *(17.12)*

Cation—a positively charged ion. *(2.2)*

Cell potential—the voltage difference between the two electrodes in a voltaic cell. *(17.4)*

Central atom—an atom bonded to two or more other atoms. *(8.3)*

Ceramics—nonmetallic, solid materials that are hard, resistant to heat, and chemically inert. *(19.4)*

Chain-growth polymer or **addition polymer**—a polymer formed from monomeric units with no loss of atoms. *(21.4)*

Change in enthalpy, ΔH—the heat absorbed by the system under constant pressure. *(4.4)*

Charles's law—at constant pressure the volume of a sample of gas is proportional to the absolute temperature ($V = k_2 \times T$). *(5.2)*

Chelate—a coordination complex that contains one or more chelating ligands. *(18.4)*

Chelating ligand—a ligand that simultaneously bonds to the same metal ion through two or more different atoms. *(18.4)*

Chemical bonds—the forces that hold the atoms together in substances. *(8.1)*

Chemical change—a process in which one or more new substances are produced. *(1.2)*

Chemical energy—a form of potential energy derived from the forces that hold the atoms together. *(4.6)*

Chemical equation—an equation that describes the identities and relative amounts of reactants and products in a chemical reaction. *(3.1)*

Chemical expression—an expression that describes the identities of reactants and products in a chemical reaction, but is not necessarily balanced. *(3.1)*

Chemical kinetics—the study of the rates of chemical reactions. *(16.1)*

Chemical nomenclature—the organized system for the naming of compounds. *(2.4)*

Chemical property—the tendency to react and form new substances. *(1.2)*

Chemical thermodynamics—the study of the energetics of chemical reactions. *(15.1)*

Chemistry—the study of matter and its interactions with other matter and with energy. *(1.1)*

Chiral molecule or ion—a molecule or ion that has a mirror image structure that cannot be superimposed on the original. *(18.5)*

Chromatography—a technique that employs two materials, one moving and one stationary, to separate a mixture into its components. *(1.4)*

Closed system—one that is allowed to exchange energy with the surroundings, but does not exchange matter. *(15.1)*

Closest-packing array—the most efficient packing of spheres, which has the smallest empty space (26%). Each sphere is in contact with 12 other spheres. *(10.3)*

Coefficient—the number of units of each substance in the chemical reaction. *(3.1)*

Cohesive forces—the attraction of molecules for other molecules of the same substance. *(10.2)*

Colligative properties—those properties of a solution that are proportional to the concentration of solute particles. *(11.4)*

Collision frequency—the number of collisions per second. *(16.4)*

Combustion reaction—the process of burning. Examples are the reactions of organic compounds with excess oxygen to yield carbon dioxide and water. *(3.1)*

Common ion effect—the effect on solubility when an ionic compound is dissolved in a solution that contains an ion in common with the solid. *(12.7)*

Complete ionic equation—the equation that shows separately all species, both ions and molecules, as they are present in solution. *(4.2)*

Complex—the species formed by the Lewis acid-base reaction of a metal atom or ion with ligands. *(14.7)*

Compound—a substance that can be decomposed into its elements by chemical processes. *(1.2)*

Condensation—the conversion of a gas to a liquid. *(10.5)*

Condensation reaction—a reaction that joins two molecules with the elimination of a small molecule such as water. *(21.3)*

Condensed phase—the solid and liquid states of matter. Any phase that is resistant to volume changes. *(5.1)*

Conduction band—a partially or completely empty band of orbitals in solids that results in high electrical conductivity. *(19.4)*

Conformers—different arrangements of atoms in a molecule that are caused by rotations about single bonds. *(21.1)*

Contact process—a process for the production of sulfuric acid from sulfur dioxide, oxygen, and water. *(19.6)*

Continuum spectrum—a spectrum in which all wavelengths of light are present. This spectrum is characteristic of the light emitted by a heated solid. *(6.2)*

Coordinate covalent bond—a covalent bond in which both electrons of the bonding pair have come from one atom. A coordinate covalent bond is also frequently referred to as a *dative bond. (8.7)*

Coordination compound or complex—a species that contains a metal ion bound to ligands by Lewis acid-base interactions. *(18.4)*

Coordination isomers—compounds that contain the same numbers and kinds of atoms but have different Lewis bases directly bonded to the metal ion; *e.g.*, $[Co(NH_3)_4Cl_2]Br$ and $[Co(NH_3)_4ClBr]Cl$ are coordination isomers. *(18.5)*

Coordination number—the number of nearest neighbors an atom, ion, or molecule has in a crystalline solid. The largest possible coordination number of uniform-sized spheres is 12, and it occurs in both of the closest packing arrays; also the number of donor atoms bonded to a single metal ion in a coordination complex. *(10.3,18.4)*

Copolymer—a polymer formed from the combination of many units of more than one type of monomer. *(21.4)*

Core electrons—the inner shell electrons that are not in the valence shell. *(7.1)*

Corrosion—the oxidation of a metal to produce compounds of the metal through interaction with its environment. *(17.12)*

Coulomb's law—the force between two charged objects is directly proportional to the product of the charges and inversely proportional to the square of the distance between them. *(2.2)*

$$\text{force} = \frac{kQ_1Q_2}{r^2}$$

Covalent bond—the bond that arises from atoms sharing electron pairs. *(8.3)*

Covalent network solid—a solid that consists of atoms held together in the crystal by covalent bonds. *(10.3)*

Critical mass—the minimum size of the sample of fissionable matter needed for the chain reaction to be self-sustained; *i.e.*, on an average at least one neutron formed by a fission is used to induce a second fission. *(20.4)*

Critical pressure—the minimum pressure needed to cause the liquid state to exist up to the critical temperature. *(10.5)*

Critical temperature—the maximum temperature at which a substance can exist in the liquid phase. *(10.5)*

Crystal field splitting, Δ—the separation in energy between the *d* orbitals caused by the electric field produced by the arrangement of ligands about the central metal ion. *(18.6)*

Crystal structure—the geometric arrangement of particles (atoms, ions, or molecules) in a crystalline solid. *(10.3)*

Crystal system—one of the seven possible forms of unit cells, defined by the lengths of the edges (a, b, c) and the angles between the edges (α, β, γ) of the unit cell. *(10.3)*

Crystalline solid—a solid in which the units that make up the substance are arranged in a very regular repeating pattern. *(10.3)*

Cubic closest packing—the closest-packing array with an *ABCABC* . . . stacking pattern of the layers. *(10.3)*

Cubic system—the crystal system in which $a = b = c$, and $\alpha = \beta = \gamma = 90°$. *(10.3)*

Cycloalkane—a saturated hydrocarbon that contains a ring of carbon atoms. *(21.1)*

Dalton's law of partial pressure—the total pressure of a mixture of gases is the sum of the partial pressures of the component gases. *(5.6)*

Degenerate orbitals—all orbitals in an atom that have identically the same energy. In the hydrogen atom and one-electron ions, all wave functions that belong to the same principal shell form a degenerate set. In many-electron atoms, all orbitals in the same subshell are degenerate. *(6.5)*

Delocalized molecular orbital—a molecular orbital that involves atomic orbitals on more than two atoms. *(9.6)*

Denaturation—the loss of structural organization of a protein. *(21.5)*

Density—the ratio of mass to volume. *(1.2)*

Deoxyribonucleic acid (DNA)—a polymer of nucleotide units located inside the chromosomes. DNA is the molecule that stores genetic information. *(21.7)*

Deposition—the direct conversion of a gas to a solid. Deposition is the reverse of sublimation. *(10.5)*

Derived unit—a unit that is composed of combinations of base units. *(1.5)*

Diamagnetic—matter that is repelled by a magnetic field. All electrons are paired in diamagnetic materials. *(6.6)*

Diatomic molecule—a molecule that is composed of two atoms. *(2.1)*

Differential form of the rate law—an expression that relates *differences* in concentration and time ($\Delta[A]/\Delta t$) to the concentrations of the species. *(16.3)*

Diffusion—the mixing of particles due to motion. *(5.8)*

Dipole moment—the magnitudes of separated charges times the distance between the charges. *(8.4)*

Dipole-dipole attractions—the intramolecular forces that arise from electrostatic attractions between the molecular dipoles. *(10.1)*

Disaccharide—the product from the condensation of two monosaccharides with the elimination of water. *(21.6)*

Dissociation—the separation of an ionic solid into individual cations and anions when dissolved in solution (generally water). *(2.3)*

Distillation—the separation of components based on differences in volatility. It consists of two steps: (1) the mixture is heated to convert the liquid into a vapor, its gaseous form, and (2) the vapor is condensed to a liquid in a different container. *(1.4)*

Double bond—the bond formed by sharing two pairs of electrons between two atoms. *(8.3)*

Dual nature of light—electromagnetic radiation (light) is described both as waves and as particles. *(6.1)*

Dynamic equilibrium—a situation in which two opposing changes occur at equal rates, so no *net* change is apparent. *(10.5)*

Effective nuclear charge—the weighted average of the nuclear charge that influences any particular electron in an atom, after correcting for the effect of interelectronic repulsions. *(6.5)*

Effusion—the passage of a gas through a small hole into an evacuated space. *(5.8)*

Electrochemistry—the study of the relation between chemical reactions and electricity. *(17.0)*

Electrode—(1) a metal or other electrical conductor that connects an electrochemical cell to the external circuit; (2) a half-cell. *(17.3)*

Electrolysis—an otherwise nonspontaneous oxidation-reduction reaction that is caused by the passage of an electric current. *(17.9)*

Electrolyte—a substance that separates into ions when it dissolves in water. *(11.5)*

Electrolytic cell—an apparatus that consists of two electrodes immersed in a molten salt or electrolyte solution, which is used to carry out an electrolysis. *(17.9)*

Electromagnetic radiation—oscillating electric and magnetic fields that are both perpendicular to each other and to the direction of motion. *(6.1)*

Electromotive force (emf)—the electrical driving force that pushes the electrons released in the oxidation half-cell (the negative electrode) through the external circuit into the reduction half-cell (the positive electrode). The emf is an intensive property of a cell and is measured in volts. *(17.4)*

Electron—a very small particle that has a mass of 9.11×10^{-31} kg and a charge of -1.602×10^{-19} coulombs. The relative charge is 1− and the relative mass is approximately 0. *(2.2)*

Electron affinity—the energy change that accompanies the addition of an electron to a gaseous atom or ion. *(7.4)*

Electron capture—a mode of decay for some unstable nuclides, in which the nucleus captures an electron, converting a proton in the nucleus into a neutron. *(20.1)*

Electron configuration—a notation to describe the number of electrons in each subshell of an atom or ion; *e.g.*, $1s^2 2s^2 2p^2$ is the electron configuration of the carbon atom. *(6.6)*

Electron-deficient molecule—a molecule for which the Lewis structure has fewer than eight electrons around any atom. *(8.7)*

Electron-pair arrangement—the positions of the valence-shell electron pairs that minimize their repulsion from each other. *(9.1)*

Electron pairing energy, *P*—the additional energy required for two electrons to occupy the same orbital, compared to the two electrons singly occupying separate degenerate orbitals. *(18.6)*

Electron-rich molecule—a molecule that has more than eight electrons about an atom in its Lewis structure. *(8.7)*

Electron spin quantum number, m_s—the quantum number that represents one of the two allowed magnetic states of an electron. The allowed values of m_s are $+\frac{1}{2}$ and $-\frac{1}{2}$ only. *(6.4)*

Electronegativity—a measure of the ability of an atom to attract the shared electrons in a chemical bond. *(8.4)*

Electroplating—the electrolytic deposition of a thin uniform metal film on the surface of a metallic object. *(17.11)*

Electrorefining—an electrolytic process that converts an impure metal anode into a very pure sample of the metal at the cathode. *(17.11)*

Element—a substance that cannot be decomposed into a simpler substance by normal chemical means. *(1.2)*

Elementary reactions—single molecular events that are summed to provide the overall reaction. *(16.6)*

Elementary step—an equation that describes a molecular-level interaction or collision. *(16.6)*

Empirical formula—gives the relative numbers of different elements in a substance, using the smallest whole numbers for subscripts. *(2.3)*

Enantiomers—molecules or ions that are nonsuperimposable mirror images of each other. *(18.5)*

Endothermic reaction—a reaction that absorbs heat from the surroundings. *(4.4)*

Enthalpy, *H*—heat content, $H = E + PV$. *(15.2)*

Entropy, *S*—the property of a system that describes the amount of disorder. It is a state function. *(15.3)*

Enzyme—a biological compound that catalyzes a specific biochemical reaction. *(16.5)*

Equilibrium—a state in which the tendency of reactants to form products is balanced by the tendency of products to form reactants. *(12.1)*

Equilibrium constant—an experimental measurement of the tendency of a reaction to favor products or reactants. See **law of mass action.** *(12.2)*

Equivalence point—the point in a titration at which the reactants have been added in stoichiometrically equivalent amounts. *(4.3)*

Equivalency—a relationship that expresses the same quantity in two different types of units. Equivalent quantities are represented by the symbol $\triangleq$. *(1.5)*

Error—the difference between the measured result and the true value. *(1.6)*

Ester—an organic compound that contains the $—CO_2R$ functional group where R is an organic substituent. *(21.3)*

Ether—an organic compound that contains a C—O—C functional group. *(21.3)*

Evaporation—the escape of molecules from the liquid phase into the gas phase. *(10.5)*

Excited state—an atom in which one or more electrons occupy orbitals that leave lower-energy orbitals partially or completely vacant. *(6.6)*

Exothermic reaction—a reaction that releases heat to the surroundings. *(4.4)*

Exponential notation—a quantity expressed as the product of a number between 1 and 10 multiplied by a power of 10. Also called *scientific notation.* *(1.6)*

Extensive properties—those that depend upon the specific sample that is under observation. Examples are mass and volume. *(1.2)*

Face-centered cubic (FCC)—the cubic unit cell with identical atoms at the corners and in the center of each of the six square faces. The FCC and the cubic closest-packing arrays are identical. *(10.3)*

Faraday, *F*—the negative electrical charge on one mole of electrons, 96,485 C. *(17.5)*

Ferromagnetic—matter in which the magnetic moments of the atoms line up to produce a permanent magnet. *(6.6)*

Filtration—the process of separating a mixture of a solid and liquid by passing it through a porous barrier. *(1.4)*

First law of thermodynamics—energy can be neither created nor destroyed. In equation form: $\Delta E = q + w$. *(15.2)*

Formal charges—charges assigned to each atom in a Lewis structure, obtained by assuming the shared electrons are divided equally between the bonded atoms. *(8.5)*

Formula weight—the sum of the atomic weights of atoms in a formula. *(3.2)*

Fraction ionized, α—the concentration of the ionized form divided by the analytical concentration. *(13.5)*

Frasch process—a method for melting underground deposits of sulfur and bringing them to the surface. *(19.6)*

Free energy, G—$G = H - TS$. It is a state function. *(15.4)*

Freezing point depression—a colligative property of solutions described by the equation

$$\Delta T_f = mk_f$$

where ΔT_f = (f.p. of solvent − f.p. of solution), m is the molal concentration of the solute particles, and k_f is the freezing point depression constant characteristic of the solvent. *(11.4)*

Frequency, ν—the number of waves that pass a fixed point in one second. The SI unit for frequency is s^{-1}, and has been given the name **hertz.** *(6.1)*

Fuel cell—a voltaic cell in which the reactants are continuously supplied, and generally the products of the redox reaction are continuously removed. *(17.8)*

Functional group—an atom or small group of atoms in a molecule that undergoes characteristic reactions. *(21.3)*

Gamma ray—high-energy electromagnetic radiation originating from the radioactive decay of unstable nuclides. *(20.1)*

Gas—a fluid that has no definite shape or volume. *(5.1)*

Geometric isomers—stereoisomers that have the same number and kind of bonds, but differ in the relative positions of the atoms. *(18.5)*

Graham's law—the rate of effusion of gases is inversely proportional to the square root of the molar mass. *(5.8)*

Gravimetric analysis—the selective precipitation of one component of a solution, followed by isolation and weighing of the precipitate, in order to determine the amount of that component. *(4.3)*

Greenhouse effect—the effect of a blanket of gases that insulate the earth and prevent it from radiating heat. *(15.6)*

Ground state—the state of the atom in which the electron configuration has the lowest possible energy. *(6.6)*

Group—a column of the periodic table. *(1.3)*

Haber process—the direct synthesis of ammonia from nitrogen and hydrogen. *(19.5)*

Half-cell—the compartment in a voltaic cell in which either the reduction or oxidation half-reaction occurs. *(17.3)*

Half-life—the time needed for the concentration of a reactant to decrease to one-half its original value. *(16.3)*

Half-reaction—an equation describing either the oxidation or the reduction portion of a redox reaction, with the electrons explicitly shown. An oxidation half-reaction has electrons on the product side of the equation. The electrons are on the reactant side of the equation in a reduction half-reaction. *(17.1)*

Halogens—the elements in group VIIA (17): F, Cl, Br, I, At. *(1.3)*

Heat capacity—the quantity of heat required to raise the temperature of a sample by 1 kelvin (or 1 °C). *(4.5)*

Heat of combustion—the enthalpy change when a compound is burned. *(15.2)*

Heisenberg uncertainty principle—it is not possible to know simultaneously and precisely both the position and momentum of a particle. *(6.2)*

Henderson-Hasselbalch relationship—generally used to calculate the pH of a buffer solution. *(14.2)*

$$\text{pH} = \text{p}K_a + \log \frac{C_b}{C_a}$$

Henry's law—the solubility of a gas is directly proportional to its pressure above the solution at any given temperature:

$$C = k \times P$$

where C is the concentration of the gaseous compound in solution, k is a proportionality constant that is characteristic of the particular solute and temperature, and P is the partial pressure of the solute in contact with the solution. *(11.3)*

Hertz, Hz—the SI unit of frequency, 1 Hz = 1 s^{-1}. *(6.1)*

Hess's law—the change in enthalpy for an equation obtained by adding two or more thermochemical equations is the sum of the enthalpy changes of the equations that have been added. *(4.6)*

Heterogeneous catalyst—a catalyst that is in a different phase from that containing the reactants. *(16.5)*

Heterogeneous mixture—a mixture in which different parts have different properties and composition. *(1.2)*

Heteronuclear diatomic molecule—a molecule formed by one atom of each of two different elements. *(9.6)*

Hexagonal closest packing—the closest-packing array with an *ABABA* . . . stacking pattern of the layers. *(10.3)*

High-spin complex—a coordination complex in which each of the five *d* orbitals is occupied by a single electron, before two electrons occupy one of the low-energy orbitals. High-spin complexes are observed when $P > \Delta$. *(18.6)*

Homogeneous catalyst—a catalyst that is in the same phase as the reactants. *(16.5)*

Homogeneous mixture—a mixture in which all parts of the sample exhibit identical properties. *(1.2)*

Homonuclear diatomic molecule—a molecule formed by two atoms of the same element. *(9.5)*

Homopolymer—a polymer formed from the combination of many units of a single monomer compound. *(21.4)*

Hund's rule—in filling a set of degenerate orbitals, each orbital is occupied by one electron, all with identical spins, before two electrons are placed in the same orbital. *(6.6)*

Hybrid orbitals—orbitals obtained by mixing two or more atomic orbitals on the same atom. *(9.3)*

Hydrocarbon—a compound that contains only the elements hydrogen and carbon. *(21.1)*

Hydrogen bonding—a particularly strong intermolecular attraction between a hydrogen atom that is bonded to a highly electronegative atom and a lone pair of electrons on N, O, or F. *(10.1)*

Hygroscopic—having the ability to absorb water vapor from the air. *(19.5)*

Hypothesis—a possible explanation for observed results. *(1.1)*

Icosahedron—a regular polyhedron with 20 faces and 12 vertices. *(19.3)*

Ideal gas law—the equation that describes the state of a gas, $PV = nRT$, where P = pressure, V = volume, n = number of moles, R = ideal gas constant (0.0821 L atm/mol K), and T = temperature. *(5.2)*

Ideal solution—a solution that obeys Raoult's law throughout the entire range of composition. *(11.6)*

Indicating electrode—an electrode whose potential depends upon the concentration of a particular species in solution. *(17.7)*

Indicator—a compound that changes color at the *end point* of a titration. The indicator should be chosen so that the end point coincides as nearly as possible with the equivalence point. *(4.3, 14.4)*

Induced dipole moment—a dipole caused by the presence of an electrical charge close to an otherwise nonpolar molecule. *(10.1)*

Inert electrode—a solid electrical conductor, usually an unreactive metal or graphite, that is neither oxidized nor reduced in the reaction of a voltaic cell. It serves only as a means of electrical contact to the solution. *(17.3)*

Inert pair effect—the tendency for the heavier members of groups IIIA–VA not to use the valence *s* electrons for bonding. *(19.3)*

Inflection point—the point in a titration curve at which the pH changes most rapidly. *(14.1)*

Initial rate method—determination of the rate law by measuring the initial rate of reaction in several experiments in which the initial concentrations of the reactants are varied. *(16.2)*

Inner transition elements—the lanthanides and the actinides. The inner transition elements are placed at the bottom of the periodic table. *(1.3)*

Instantaneous dipole moment—a dipole moment that results from unequal charge distribution within a molecule caused by the motion of the electrons. *(10.1)*

Instantaneous rate—the slope of the tangent to the concentration-time graph at any given point, rather than an average over a time interval. *(16.1)*

Integral form of the rate law—equation that provides concentration–time relationships. *(16.3)*

Intensive properties—those that are identical in any sample of the substance. Examples are color, density, and melting points. *(1.2)*

Interhalogens—compounds formed from two different halogens. *(7.5)*

Intermediate—a compound that is produced in one step of the reaction mechanism and then consumed in a subsequent one. Intermediates are not found among the reactants or products. *(16.6)*

Intermolecular forces—the attractive forces that exist between molecules. *(10.1)*

Internal energy—the total energy of the system. *(15.1)*

Ion—a charged particle formed by the addition or removal of electrons from an atom or group of atoms. *(2.2)*

Ion product—the reaction quotient written for dissolution of a slightly soluble solid. *(12.7)*

Ion-electron method—a method of balancing oxidation-reduction reactions by first balancing two half-reactions and then combining them to obtain the chemical equation. *(17.2)*

Ionic bonding—the bonding that results from the electrostatic attraction between positively charged cations and negatively charged anions. *(8.2)*

Ionic compound—a compound composed of cations and anions joined to form a neutral species. *(2.3)*

Ionic radius—the measure of the size of an ion in an ionic solid. *(7.2)*

Ionic solid—a solid that consists of oppositely charged ions that are held together by electrostatic attractions. *(10.3)*

Ionization—the separation of a molecular compound into individual cations and anions when dissolved (generally in water). *(3.1)*

Ionization energy—the energy required to remove the highest-energy electron from a gaseous atom or ion in its electronic ground state. *(7.3)*

Isoelectronic series—a group of atoms and ions that have the same number of electrons. *(7.1)*

Isomers—different compounds with the same molecular formula but with different structural formulas. *(9.4)*

Isotopes—atoms of the same element that have different numbers of neutrons. *(2.2)*

Joule, J—the SI unit of heat defined as

$$1\ J = 1\ kg\ m^2/s^2$$

In comparison to the calorie, 1 cal = 4.184 joule. *(4.4)*

Ketone—an organic compound that contains the $R-\overset{\overset{O}{\|}}{C}-R'$ functional group, where the R and R′ groups are alkyl or aryl substituents. *(21.3)*

Kinetic energy—the energy that matter possesses because of its motion. *(4.6)*

Kinetic molecular theory—a model that describes the behavior of gas particles at the atomic or molecular level. *(5.7)*

Lanthanide contraction—the small decrease in the radii of the lanthanides as the 4*f* subshell is filled, causing the transition elements of the fifth and sixth periods to have nearly identical radii within each group in groups IVB and beyond. *(7.2, 18.2)*

Lanthanides—the elements in period 6 between lanthanum (La) and hafnium (Hf): cerium (Ce) through lutetium (Lu). *(1.3)*

Lattice energy—the energy required to separate one mole of an ionic solid into isolated gaseous ions. *(8.2)*

Lattice point—the point in space that has the same geometric environment as every other lattice point in a crystal lattice. *(10.3)*

Law—a statement or equation that summarizes a large number of observations. *(1.1)*

Law of conservation of energy—the total energy of the universe (the system plus the surroundings) is constant during a chemical change. *(4.4)*

Law of conservation of mass—there is no detectable loss or gain in mass when a chemical reaction occurs. *(2.1)*

Law of constant composition—all samples of a pure substance contain the same elements in the same proportions by mass. *(2.1)*

Law of mass action—for any reaction

$$aA + bB \rightleftharpoons cC + dD$$

the equilibrium constant can be calculated from the equilibrium concentrations. *(12.1)*

$$K_{eq} = \frac{[C]^c[D]^d}{[A]^a[B]^b}$$

Law of multiple proportions—if two elements unite to form more than one compound, the masses of one element (in the compound) that combine with a fixed mass of the second element are in a ratio of small whole numbers. *(2.1)*

Leveling effect—the effect that makes strong acids (or bases) appear equal in strength since they ionize or dissociate completely. *(13.4)*

Lewis acid—an electron-pair acceptor. *(13.9)*

Lewis base—an electron-pair donor. *(13.9)*

Lewis electron-dot symbol—the symbol of the element with one dot to represent each valence electron. *(8.1)*

Lewis structure—a representation of covalent bonding, using Lewis symbols, that shows shared electrons as dots or lines between atoms and unshared electrons as dots. *(8.3)*

Ligand—an anion or molecule that functions as a Lewis base in forming one or more coordinate covalent bonds to metal ions. *(14.7, 18.4)*

Limiting reactant—the reactant that is completely consumed in a chemical reaction. The amounts of products that can form and the amounts of the other reactants that can be consumed are determined by the amount of the limiting reactant present. *(3.5)*

Line spectrum—a spectrum that contains light at discrete wavelengths separated by regions where no light is emitted. Line spectra are produced by gaseous atoms of the elements. *(6.2)*

Linkage isomers—compounds in which the same ligand is coordinated to the metal ion through one of two possible donor atoms; *e.g.*, NO_2^- may form a bond to the metal ion through one of the oxygen atoms (nitrito) or the nitrogen atom (nitro). *(18.5)*

Liquid—a fluid that has a fixed volume but no definite shape. *(5.1)*

London dispersion forces—the instantaneous dipole-induced dipole attractions that explain the attractions between nonpolar molecules. *(10.1)*

Lone or nonbonding pairs—pairs of electrons that are not shared. *(8.3)*

Low-spin complex—a coordination complex in which the lower-energy *d* orbitals are completely filled with two electrons each, before electrons enter the higher-energy *d* orbitals. Low-spin complexes are observed when $P < \Delta$. *(18.6)*

Magic numbers—certain numbers of protons and neutrons that confer unusual stability to the nuclide. These numbers are 2, 8, 20, 26, 28, 50, 82, and 126. *(20.1)*

Magnetic quantum number, m_ℓ—the quantum number that describes the orientation of an electron wave function in an atom. The allowed values of this quantum number depend upon the value of the ℓ quantum number, and may have all integer values from $-\ell$ to $+\ell$. *(6.4)*

Mass—a measure of the quantity of matter in a sample or object. *(1.2)*

Mass defect—the difference between the mass of the atom and the mass of the equivalent number of ^{1}H atoms and neutrons. *(20.3)*

Mass number—the total number of protons and neutrons in an atom. It has the symbol A. *(2.2)*

Matter—anything that has mass and occupies space. *(1.2)*

Mechanism—the sequence of molecular-level steps that lead from reactants to products. *(16.6)*

Metal—a material that has luster and is a good conductor of electricity. Metallic elements are located in the center and left-hand side of the periodic table. *(1.3)*

Metallic solid—a solid formed by metal atoms. In metals the outer electrons occupy highly delocalized orbitals that extend throughout the crystal and account for the electrical and thermal conductivity. *(10.3)*

Metalloid—an element with properties between those of a metal and a nonmetal. The elements along the dividing line between metals and nonmetals in the periodic table. *(1.3)*

Metallurgy—the science of extracting metals from their ores, purifying, and preparing them for practical use. *(18.1)*

Michaelis-Menten mechanism—an enzyme catalysis mechanism in which an enzyme-substrate complex is formed in a reversible step, followed by a rate-limiting formation of products from the complex. *(16.6)*

Millimole—1×10^{-3} mole. *(14.1)*

Mixture—two or more substances that can be separated by taking advantage of different physical properties of the substances. *(1.2)*

Molality, *m*—the unit of concentration of a solute expressed as the moles of solute present in one kilogram of solvent. *(11.1)*

Molar enthalpy of fusion—the enthalpy change that accompanies the conversion of one mole of a substance from the solid to the liquid phase at constant temperature. This change is always endothermic. *(10.5)*

Molar enthalpy of sublimation—the enthalpy change that accompanies the sublimation of one mole of a substance. *(10.5)*

Molar enthalpy of vaporization—the enthalpy change that accompanies the conversion of one mole of a substance from the liquid to the gas phase at constant temperature. This process is always endothermic. *(10.2)*

Molar heat capacity—the heat needed to raise the temperature of one mole of a substance by 1 kelvin. *(4.5)*

Molar mass—the mass in grams of one mole of any substance; numerically the same as the molecular or formula weight. *(3.2)*

Molarity—the concentration unit defined as the number of moles of solute in one liter of solution. *(4.1)*

Mole—the amount of substance that contains as many entities as there are atoms in exactly 12 grams of ^{12}C (the isotope of carbon that has six neutrons). *(3.2)*

Mole fraction—the number of moles of one component of a homogeneous mixture divided by the total number of moles of all substances present in the mixture. *(5.6)*

Molecular formula—gives the number of every type of atom in a molecule. *(2.1)*

Molecular orbital—a wave function of an electron in a molecule. *(9.5)*

Molecular orbital theory—a model that treats bonding as delocalized over the entire molecule. *(9.5)*

Molecular shape—a description of the positions of the atoms, not the lone pairs. *(9.1)*

Molecular solid—a solid that consists of small covalently bonded molecules, held together by van der Waals forces or hydrogen bonding. *(10.3)*

Molecular weight—the sum of the atomic weights of all atoms present in the molecular formula of a molecule. *(3.2)*

Molecularity—the number of reactant species that are involved in a single elementary step. *(16.6)*

Molecule—a combination of atoms joined tightly together so that they behave as a single particle. *(2.1)*

Monatomic ion—an ion formed by the loss or gain of electrons by a single atom. *(2.3)*

Monodentate ligand—a ligand that donates one pair of electrons to the metal in a coordination complex. *(18.4)*

Monosaccharide—the basic unit of carbohydrates, having the general formula $(CH_2O)_x$ where $x = 3$ to 8. *(21.6)*

Negative deviations from Raoult's law—the observed vapor pressure is less than expected for an ideal solution, and occurs when the attractions between dissimilar molecules are stronger than the average of the attractions in the pure components of the mixture. *(11.6)*

Nernst equation—the equation that describes the dependence of cell voltage on the concentrations of reactants and products. *(17.6)*

$$E = E^\circ - \frac{RT}{nF}\ln Q = E^\circ - \frac{2.303RT}{nF}\log Q$$

Net ionic equation—the equation that shows only those species in the reaction that undergo change. *(4.2)*

Neutralization reaction—the reaction of an acid and a base to yield water and a salt. *(3.1)*

Neutron—a small particle that has a mass of 1.675×10^{-27} kg and no charge. The relative mass is 1. *(2.2)*

Noble gases—the elements of group VIIIA. They were formerly known as the "inert gases": He, Ne, Ar, Kr, Xe, Rn. *(1.3)*

Nonmetal—a material that lacks the characteristics of a metal. Nonmetallic elements are in the top right part of the periodic table. *(1.3)*

Normal boiling point—the boiling point at a pressure of one atmosphere. *(10.5)*

Normal melting point—the temperature at which the solid and liquid phases are in equilibrium at one atmosphere pressure. Melting points change very little with pressure. *(10.5)*

Nuclear binding energy—the energy of attraction among the protons and neutrons in a nuclide. This energy is calculated from the mass defect by using

$$E = \Delta mc^2$$

where Δm is the mass defect of the nucleus in kg/mol, c is the speed of light, 3.00×10^8 m/s, and ΔE is in joules/mol. *(20.3)*

Nuclear chain reaction—a fission event, initiated by the absorption of a neutron, that produces more neutrons than are consumed, allowing a self-sustained nuclear reaction. *(20.4)*

Nuclear equation—describes the changes in a nuclear reaction in shorthand notation. Nuclear equations are balanced by conservation of atomic number and mass number of the particles on the reactant and product sides of the equation. *(20.1)*

Nuclear fission—the splitting of a heavy nucleus into two nuclei of comparable size. *(20.4)*

Nuclear fusion—the combination of two light nuclides to form a heavier one. *(20.4)*

Nuclear transmutation—a nuclear reaction that involves two or more particles or nuclei as reactants and produces elements or isotopes that are different from the reactants. *(20.2)*

Nucleic acid—a chain of nucleotides. *(21.7)*

Nucleon—a particle that is present in the nucleus of an atom; both protons and neutrons are called nucleons. *(20.1)*

Nucleotide—the building block of DNA and RNA; each nucleotide is composed of three units: a five-carbon sugar, a base (a cyclic amine), and a phosphate group. *(21.7)*

Nucleus—the small, heavy, positively charged core of an atom. *(2.2)*

Nuclide—the nucleus of a particular isotope. *(20.1)*

Number of significant digits—the number of digits from the first nonzero digit to the last significant digit in the quantity. *(1.6)*

Octet rule—each atom in a molecule shares electrons until it is surrounded by eight valence electrons. The octet rule is most important for elements in the second period. *(8.3)*

Optical isomers—stereoisomers that rotate the plane of polarized light. *(18.5)*

Orbital diagram—a drawing that represents the spins of electrons as "up" and "down" arrows placed in boxes that represent the orbitals. *(6.6)*

Order of the reaction—the power to which the concentration is raised in the rate law. *(16.2)*

Organic chemistry—the study of carbon-containing compounds. *(21.1)*

Organic compound—a compound made up of carbon atoms in combination with other elements such as hydrogen, oxygen, and nitrogen. *(3.1)*

Osmosis—the diffusion of a fluid through a semipermeable membrane. *(11.4)*

Osmotic pressure—the pressure difference needed to prevent net transport of solvent across a semipermeable membrane that separates a solution from the pure solvent. This is one of the most sensitive colligative properties, and is described by the equation

$$\Pi = MRT$$

where Π is the osmotic pressure, M is the molar concentration of solute particles, R is the ideal gas constant (0.0821 L atm mol^{-1} K^{-1}), and T is the temperature in kelvins. *(11.4)*

Ostwald process—a multistep process for the production of nitric acid from ammonia, oxygen, and water. *(19.5)*

Overall equation—the equation that shows all of the reactants and products in undissociated form. *(4.2)*

Overall order—the sum of the orders for all substances that appear in the rate law. *(16.2)*

Overvoltage—the difference between the standard potential and the potential needed to cause an electrolysis reaction. *(17.9)*

Oxidation—the loss of electrons by an element, compound, or ion. *(17.1)*

Oxidation state—the numerical equivalent of the charge of monatomic ions, or the charge an atom would possess if the shared electrons in each covalent bond were assigned to the more electronegative atom. This is also called the *oxidation number*. *(17.1)*

Oxidation-reduction reaction—a reaction in which electrons are transferred from one species to another. *(17.1)*

Oxidizing agent—the reactant that is reduced in a redox reaction. *(17.1)*

Oxyacid—a molecule of the general formula OH_mXO_n in which ionization breaks one of the hydrogen-oxygen bonds. *(13.8)*

Paramagnetic—matter that is attracted by a magnetic field. Only substances that contain unpaired electron spins are paramagnetic. *(6.6)*

Passive solar energy device—one that uses no machinery or heat-transfer equipment. *(15.6)*

Pauli exclusion principle—no two electrons in the same atom can have the same set of all four quantum numbers. *(6.6)*

Peptide—a small polymer of amino acids. *(21.5)*

Percent yield—the actual yield divided by the theoretical yield and multiplied by 100%. *(3.4)*

Period—a row of the periodic table. *(1.3)*

Periodic table—an arrangement of the elements (shown by their symbols) into rows and columns so that elements with similar chemical properties are placed in vertical columns. *(1.3)*

pH—equal to $-\log[H_3O^+]$. *(13.2)*

pH meter—a voltmeter that is scaled to read the pH of a solution from the measured potential of a voltaic cell. *(17.7)*

Phase diagram—a graph of pressure versus temperature that shows the regions of stability for each phase (solid, liquid, and gas). The lines between the phases represent conditions of temperature and pressure at which two or more phases are in equilibrium. *(10.6)*

Phenol—an organic compound in which a hydroxyl group is substituted for a hydrogen atom on an aromatic ring. The parent compound of this type, C_6H_5OH, is called phenol. *(21.3)*

Photoelectric effect—the ejection of electrons from a solid by the absorption of a photon of light. *(6.1)*

Photon—a particle of light that possesses an energy of $h\nu$. *(6.1)*

Physical property—one that can be observed without changing the substances present in the sample. *(1.2)*

Pi (π) bond—a bond that places electron density above and below the line joining the bonded atoms. It is formed from sideways overlap of *p* orbitals. *(9.4)*

Pipet—a device calibrated either to deliver or to contain a specific volume of liquid; it is used to measure accurately a fixed volume of solution. *(4.1)*

Planck's constant, *h*—the proportionality constant that relates a quantum of energy to the frequency of the radiation absorbed or emitted. $h = 6.63 \times 10^{-34}$ J s. *(6.1)*

Planck's equation—$\Delta E = h\nu$ *(6.1)*

Plasma—an electrically neutral gas at very high temperature that consists of highly charged ions and electrons. *(5.1, 20.4)*

Polar bond—a covalent bond in which the bonding electrons are not equally shared by the two atoms. *(8.4)*

Polar molecule—a molecule that contains an unequal distribution of charge and thus a dipole moment. *(9.2)*

Polarizability—the ease with which the electron cloud of a molecule can be distorted. *(10.1)*

Polarized light—electromagnetic radiation that has the electric field oscillating in a single plane. *(18.5)*

Polyatomic ion—charged species made up of more than one atom. *(2.3)*

Polydentate ligand—a ligand having more than one atom that forms a coordinate covalent bond with a metal ion. Numerical prefixes (bi, tri, tetra, etc.) are combined with "dentate" to designate the number of donor atoms in the ligand. *(18.4)*

Polyester—a polymer formed from the condensation of a dicarboxylic acid and a dialcohol. *(21.4)*

Polymer—a large molecule formed by the repeated bonding together of many smaller units (*monomers*). *(21.4)*

Polyprotic acid—an acid that can ionize to produce more than one proton per molecule. *(14.5)*

Polysaccharide—a step-growth polymer of monosaccharides. *(21.6)*

Positive deviations from Raoult's law—the observed vapor pressure is greater than expected for an ideal solution, and occurs when the attractions between dissimilar molecules are weaker than the average of the attractions in the pure components of the mixture. *(11.6)*

Positron—a positively charged electron that is produced in the radioactive decay of some unstable isotopes; the antiparticle of the electron. *(20.1)*

Potential energy—the energy derived from the position or condition of matter. *(4.6)*

Pre-exponential term—the product of the steric factor and collision frequency, collected in a single term. *(16.4)*

Precipitation reaction—the formation of an insoluble product or products from the reaction of soluble reactants. *(4.2)*

Precision—agreement among repeated measurements. *(1.6)*

Pressure—the force per unit area exerted on a surface. *(5.1)*

Primary structure—the sequence of amino acids in the polypeptide chain. *(21.5)*

Principal quantum number, *n*—the quantum number that contains information about the distance of an electron from the nucleus. It may have any positive integer values, and it affects the energy of the electron. The value determines the energy of an electron in the hydrogen atom. *(6.4)*

Principal shell—the set of all wave functions in an atom that have the same value of *n*. *(6.4)*

Principle of Le Chatelier—a change to a system at equilibrium causes a shift in the position of the equilibrium that reduces the effect of the change. *(10.6)*

Product—a substance that is formed in a chemical reaction. *(3.1)*

Property—anything that can be observed or measured. *(1.2)*

Protein—a polymer of amino acids. *(21.5)*

Proton—a small particle that has a mass of 1.673×10^{-27} kg and a charge of $+1.602 \times 10^{-19}$ coulombs. The absolute charge is the same as that of the electron but opposite in sign. The relative charge is 1+, and the relative mass is 1. *(2.2)*

Pseudo-noble gas electron configurations—electron configurations of the type [noble gas] $(n-1)d^{10}$. *(7.3)*

Quantum—the smallest quantity of energy that is absorbed or emitted by matter as electromagnetic radiation. *(6.1)*

Quantum numbers—numbers that describe the characteristics of wave functions, and are analogous to coordinates that describe the location of a particle. In the hydrogen atom, four quantum numbers are needed to describe the wave function of the electron. *(6.4)*

Quaternary structure—the orientation of different polypeptide chains with respect to each other. *(21.5)*

Racemic mixture—an equimolar mixture of enantiomers that produces no net rotation of the plane of polarized light. *(18.5)*

Radioactive atom—a nuclide that spontaneously emits small particles or high-energy electromagnetic radiation or both, forming a more stable nuclide. *(20.1)*

Radioactivity—the spontaneous nuclear reaction that transforms a relatively unstable nucleus into a more stable nucleus and a small particle and energy. *(2.2)*

Raoult's law—the vapor pressure (P) of the solvent over a dilute solution is equal to the mole fraction of the solvent (χ) times the vapor pressure of the pure solvent ($P°$).

$$P = \chi_{\text{solvent}} P°$$

A more useful form of this law is

$$\Delta P = \chi_{\text{solute}} P°$$

where ΔP is $(P° - P_{\text{solution}})$. *(11.4)*

Rate—change per unit time. *(16.1)*

Rate constant, *k*—the proportionality constant in the rate law, equal to the rate of reaction when the reactants are present at 1.0 M concentrations. *(16.2)*

Rate law—the relationship between rate and concentrations. It is of the form: rate = $k[A]^x[B]^y$. *(16.2)*

Rate of reaction—the ratio of the rate of change of any substance to its coefficient in the chemical equation. *(16.1)*

Rate-limiting step—a step that is much slower than any of the others in the mechanism. *(16.6)*

Reactant—a substance that is consumed in a chemical reaction. *(3.1)*

Reaction coordinate—a relative scale on an energy-level diagram that begins with the reactants and progresses to the products. *(16.4)*

Reaction quotient—the result of a calculation based on the same expression as the equilibrium constant, except that the concentrations are not restricted to those observed after the system reaches equilibrium. Initial concentrations are often used. *(12.2)*

Redox reaction—an abbreviation for an oxidation-reduction reaction. *(17.1)*

Reducing agent—the reactant that is oxidized in a redox reaction. *(17.1)*

Reduction—the gain of electrons by an element, compound, or ion. *(17.1)*

Reference electrode—a half-cell with an accurately known potential that is independent of the composition of the test solution. *(17.7)*

Relative uncertainty—the ratio of the absolute uncertainty of a number to the number itself, often expressed as a percentage. *(1.6)*

Representative elements—the elements in groups labeled A (IA–VIIIA) or groups 1–2 and 13–18. *(1.3)*

Resonance—the use of two or more Lewis structures that differ only in the distribution of the valence electrons in representing the bonding in a species. *(8.6)*

Reverse osmosis—the transport of water across a semipermeable membrane from a solution to pure water on the other side of the membrane, by applying a pressure greater than the osmotic pressure to the solution. *(11.5)*

Ribonucleic acid (RNA)—a polymer of nucleotide units located outside the chromosomes. RNA transmits genetic information and directs the assembly of proteins. *(21.7)*

Roasting—pretreatment of ore by heating the material below its melting point, usually in the presence of air, to convert the ore into a chemical form more suitable for the reduction step. *(18.1)*

Root mean square (rms) speed u_{rms}—the square root of the average squared speeds of a collection of particles. *(5.7)*

Rydberg equation—the equation that predicts the wavelengths of the lines in the hydrogen atom spectrum.

$$\frac{1}{\lambda} = R_h\left(\frac{1}{n_1^2} - \frac{1}{n_2^2}\right)$$

where n_1 and n_2 are whole positive numbers, with $n_1 < n_2$, and R_h is a constant, called the Rydberg constant, which has a value of 1.097×10^7 m^{-1}. *(6.2)*

Sacrificial anode—a piece of active metal placed in electrical contact with a less reactive metal. *(17.12)*

Salt—a compound made up of the cation from a base and the anion from an acid. *(3.1)*

Salt bridge—an electrolyte medium that allows the transport of ions between the two half-cells of a voltaic cell. *(17.3)*

Saturated hydrocarbon—a hydrocarbon with no multiple bonds. *(21.1)*

Saturated solution—a solution that is in equilibrium with an excess of the solute. *(11.2)*

Scientific method—investigations that are guided by theory. *(1.1)*

Scrubbers—chemical reactors used to clean gases leaving the smokestack of an industrial plant. *(15.6)*

Second law of thermodynamics—in any spontaneous process, the entropy of the universe will increase. *(15.3)*

Secondary structure—the shape of the polypeptide chain. *(21.5)*

Semiconductor—a weak conductor of electricity. *(1.3)*

Semipermeable membranes—thin films of materials which allow only water and other small molecules to pass through them. Animal bladders, the skins of fruits and vegetables, and cellophane are examples of semipermeable membranes. *(11.4)*

Sigma (σ) bond—a bond in which the shared pair of electrons is concentrated around the line joining the nuclei of the bonded atoms. *(9.4)*

Significant digit—all digits in a measurement from the first nonzero digit through the first digit that is uncertain. Also called "significant figure." *(1.6)*

Silicates—compounds containing silicon combined with oxygen and various metals. *(19.4)*

Simple cubic—the cubic structure with identical atoms located at the corners of the unit cell. *(10.3)*

Single bond—the bond formed by sharing one pair of electrons between two atoms. *(8.3)*

Skeleton structure—the drawing that shows which atoms are bonded to each other in a molecule. *(8.3)*

Solid—state of matter with a fixed shape and volume. *(5.1)*

Solubility—the concentration of solute that exists in equilibrium with an excess of that substance. *(11.2)*

Solubility equilibrium—an equilibrium that describes a solid dissolving in solution, or a solid forming from solution. *(12.6)*

Solubility-product constant—the equilibrium constant for the dissolving of a sparingly soluble salt. *(12.6)*

Solute—any substance being dissolved to form a solution. Solutes are usually present in lesser quantity than the solvent. *(4.1)*

Solution—another name for a homogeneous mixture. *(1.2)*

Solvent—the substance that has the same physical state as the solution. It is generally the component present in largest quantity; in aqueous solutions the solvent is water. *(4.1)*

Specific heat—the heat needed to raise the temperature of one gram of a substance by 1 kelvin; it has the units of J/g-K (or J/g-°C). *(4.5)*

Spectator ions—ions present in solution that do not undergo change. *(4.2)*

Spectrochemical series—the arrangement of ligands in order of increasing size of the crystal field splitting that they cause. *(18.6)*

Spectrum—a graph of the intensity of light as a function of the wavelength or frequency. *(6.2)*

Spontaneous—able to occur without outside intervention. *(15.1)*

Stable isotope—an atom with a nuclide that does not spontaneously decompose into one or more different nuclides. *(20.1)*

Standard cell potential, E°_{cell}—the voltage of a voltaic cell when all of the reactants and products are in their standard states (pure solids, liquids, or gases at 1 atmosphere pressure; solutes at 1 M concentration). *(17.4)*

Standard free energy of formation—the free energy change during the formation of one mole of a substance in its standard state from the most stable form of the elements in their standard states. *(15.4)*

Standard heat of formation—the heat absorbed when one mole of a substance in its standard state is formed from the most stable forms of the elements in their standard states. *(15.2)*

Standard reduction potential—the standard voltage of a reduction half-reaction based on a voltage of zero for the standard hydrogen electrode. *(17.4)*

Standard solution—a solution of accurately known concentration. *(4.3, 14.1)*

Standard state—the most stable form of a substance at the designated temperature, usually 298 K, and 1 atm pressure. *(15.2)*

Standard temperature and pressure (STP)—the conditions of 273 K (or 0°C) and 1.00 atm. *(5.2)*

State function—a thermodynamic property that depends only upon the state of the system and not the manner in which the system arrived at the state. *(4.6, 15.1)*

Step-growth polymer or **condensation polymer**—a polymer formed by a reaction that eliminates a small molecule each time a monomer is linked to the polymer chain. *(21.4)*

Stereoisomers—molecules or ions that have the same bonds but are different in the arrangement of the atoms in space. *(18.5)*

Steric factor—a term in the predicted rate law that expresses the need for the correct orientation of reactants when they collide in order to form the activated complex. *(16.4)*

Stoichiometry—the study of quantitative relationships involving substances and their reactions. *(3.1)*

Strong acid—an acid that ionizes completely in solution. *(13.3)*

Strong base—one of the soluble metal oxides or hydroxides that dissociates completely in solution. *(13.3)*

Strong electrolyte—a substance that dissociates or ionizes completely in solution. *(13.3)*

Strong field complex—a complex in which the crystal field splitting is greater than the pairing energy; a low-spin complex. *(18.6)*

Structural formula—indicates how the atoms are connected in the molecule. *(2.1)*

Structural isomers—compounds that contain the same numbers and kinds of atoms, but differ in the bonds that are present. *(18.5)*

Sublimation—the direct conversion of a solid to a gas without first changing to a liquid. *(10.5)*

Subshell—all the possible wave functions of electrons in an atom that have the same values for both the n and ℓ quantum numbers. Each principal shell consists of n subshells. *(6.4)*

Substance—matter that cannot be separated into component parts by a physical process. *(1.2)*

Substitution reaction—a reaction in which one atom or group of atoms in a molecule is replaced by a different atom or group of atoms. *(21.1)*

Supercooling—lowering the temperature of a liquid below its freezing point without forming a solid. This is an unstable condition. *(10.5)*

Supercritical fluid—the single phase that exists above the critical temperature and pressure. *(10.6)*

Supersaturated solution—a solution in which the concentration of the solute is temporarily greater than its solubility. This is an unstable condition. *(11.2)*

Surface tension—the quantity of energy required to increase the surface area of a liquid; it has SI units of joules/m^2. *(10.2)*

Surroundings—all matter other than the system in a chemical reaction. *(4.4)*

Symbol—an abbreviation for an element that consists of one or two letters usually related to the name of the element. *(1.2)*

System—the matter of interest in a chemical reaction. *(4.4)*

Termolecular step—an elementary step that involves the collision of three species. *(16.6)*

Tertiary structure—the overall three-dimensional arrangement of the protein. *(21.5)*

Theoretical yield—the maximum quantity of product that can be obtained in a chemical reaction, based on the amounts of starting materials. *(3.4)*

Theory—an explanation of the laws of nature. *(1.1)*

Thermochemical equation—a chemical equation that includes heat flow from the system to the surroundings. *(4.4)*

Thermochemistry—the study of the heat flow that accompanies chemical reactions. *(4.4)*

Third law of thermodynamics—the entropy of a perfect crystalline substance is 0 at a temperature of absolute zero. *(15.3)*

Threshold frequency, ν_0—the lowest frequency of light that can eject an electron from a solid. *(6.1)*

Titrant—the substance added to react with the analyte. *(14.1)*

Titration—a procedure for the determination of the quantity of one substance by the addition of a measured amount of a second substance. *(4.3)*

Titration curve—a graph of pH as a function of the added volume of titrant. *(14.1)*

Transition elements—those elements characterized by having partially filled d orbitals in the metal or at least one of its oxidation states. These elements are found in the center of the periodic chart, in the nine columns from group IIIB through group IB (3 to 11). *(1.3, 18.2)*

Triple bond—the bond formed by sharing three pairs of electrons between two atoms. *(8.3)*

Triple point—the unique combination of pressure and temperature at which all three phases (solid, liquid, and gas) exist in equilibrium. *(10.6)*

Uncertainty—related to precision. Measurements of high precision have small uncertainty. *(1.6)*

Unimolecular step—an elementary step that involves the spontaneous decomposition of a single molecule. *(16.6)*

Unit cell—a small geometric figure needed to define the pattern of lattice points in the entire crystal. *(10.3)*

Unit conversion factor—a fraction in which the numerator and the denominator express the same quantity in different units. *(1.5)*

Units—standards used for quantitative comparison between measurements of the same type of quantity. *(1.5)*

Unsaturated hydrocarbon—a hydrocarbon that contains one or more multiple carbon-carbon bonds. *(21.2)*

Unsaturated solution—a solution in which the concentration of the solute is less than its solubility. *(11.2)*

Valence bond theory—a theory that describes bonds as being formed by atoms sharing valence electrons in overlapping valence orbitals. *(9.3)*

Valence electrons—electrons that occupy the valence shell orbitals. *(7.1)*

Valence orbital—an orbital in the valence shell. *(9.3)*

Valence shell—those orbitals in the atom of the highest occupied principal level, and the orbitals with *partially filled sublevels* of lower principal quantum number. *(7.1)*

van der Waals equation—a description of the behavior of real gases, based on correcting the ideal gas law for particle size and attractive forces:

$$\left(P + \frac{an^2}{V^2}\right)(V - nb) = nRT$$

where a and b are constants experimentally determined for each gas. *(5.9)*

van der Waals forces—the weak intermolecular forces between small molecules. The term includes both dipole-dipole attractions and London dispersion forces. *(10.1)*

van't Hoff factor, *i*—the ratio of the measured colligative property to that expected for a nonelectrolyte.

$$i = \frac{\text{measured colligative property}}{\text{expected value for a nonelectrolyte}}$$

In dilute solutions of strong electrolytes the van't Hoff factor approaches the number of ions in solution for each formula unit of the compound that dissolves. *(11.5)*

Vapor pressure—the partial pressure of a substance that is in equilibrium with one of the condensed phases of that substance. *(10.5)*

Vaporization—the same as evaporation. *(10.5)*

Viscosity—the resistance of a fluid to flow; it is one of the properties that is related to the forces of intermolecular attraction. *(10.2)*

Visible light—electromagnetic radiation in the wavelength range from 400 to 700 nm. Light in this wavelength range is seen by the human eye. *(6.1)*

Volt, V—the SI unit for emf, which is defined as the potential difference required to increase the energy of a charge of 1 coulomb by 1 joule. *(17.4)*

$$1 \text{ V} = 1 \text{ joule/coulomb} = 1 \text{ J/C}$$

Voltaic cell—an apparatus that converts the chemical energy produced by a redox reaction directly into electrical energy. This is also called a galvanic cell. *(17.3)*

Volumetric analysis—a quantitative determination in which the volume of a solution or substance is measured. *(4.3)*

Volumetric flask—a container calibrated to hold an accurately known volume of liquid. *(4.1)*

Water gas shift reaction—the reaction of carbon monoxide and steam to produce hydrogen and carbon dioxide. *(19.2)*

Wave—a periodic disturbance in a medium or in space which is described by specifying its amplitude, speed, wavelength, and frequency. *(6.1)*

Wave function, ψ—the equation of the wave that describes the location and energy of a particle in the wave model of matter. *(6.3)*

Wavelength, λ—the distance from one peak of a wave to the next. In the SI system wavelength is measured in meters. *(6.1)*

Weak acid—an acid that partially ionizes in solution. *(13.4)*

Weak base—a base that partially ionizes in solution. *(13.4)*

Weak electrolyte—a substance that is only partially dissociated or ionized in solution. *(13.3)*

Weak field complex—a complex in which the crystal field splitting is less than the pairing energy; a high-spin complex. *(18.6)*

Weight—the force of attraction between two objects. The weight of an object will change from one location to another but the mass is always the same. *(1.2)*

Work—the product of force times distance. *(15.1)*

Work function, $h\nu_0$—the minimum energy of light needed to eject an electron from a solid. *(6.1)*

Appendix J

Answers to Selected Exercises

Note: Numerical answers can differ by one or two in the last digit because of rounding differences in intermediate steps.

Chapter 1

1.8 (a) intensive, physical (b) intensive, physical (c) intensive, chemical (d) extensive, physical (e) intensive, physical

1.10 (a) chemical (b) physical (c) physical (d) chemical (e) physical (f) physical (g) chemical (h) physical

1.12 pale, yellow (physical) corrosive (chemical) gas (physical) that reacts (chemical) . . . burn in fluorine (chemical) with a bright flame (chemical). Small amounts prevent cavities (chemical). The melting point (physical) . . . and boiling point (physical) . . . forms compounds (chemical).

1.16 (a) homogeneous mixture (b) heterogeneous mixture (c) element (d) heterogeneous mixture (e) homogeneous mixture

1.18 (a) Ag (b) C (c) S (d) Br (e) He

1.20 group = column and row = period on periodic table
roman = arabic when arabic is 7 or less
roman = VIII when arabic is 8, 9, or 10
roman = (arabic − 10) when arabic is 11 or greater

1.22 (a) Zr, Hf (b) S, Se, Te, or Po (c) Cl, Br, I, or At (d) Be, Mg, Ca, or Sr (e) He, Ne, Kr, Xe, or Rn (f) P, As, Sb, or Bi

1.24 (a) representative (b) transition (c) representative (d) inner transition (e) representative

1.26 (a) Na, sodium (b) Cl, chlorine (c) Ra, radium (d) Ne, neon

1.28 Sr and Ra should have similar properties

1.32 (a) kg (b) m (c) K (d) s

1.34 0.9321 mi

1.36 9.51 qt

1.38 (a) $T_C = (T_F - 32)\dfrac{1\ ^\circ\text{C}}{1.8\ ^\circ\text{F}}$

(b) $T_F = T_K\dfrac{1.8\ ^\circ\text{F}}{1\ \text{K}} - 459.67\ ^\circ\text{F}$

(c) $T_K = (T_F - 32)\dfrac{1\ ^\circ\text{C}}{1.8\ ^\circ\text{F}} + 273.15\ \text{K}$

1.40 −268.94 °C, −452.09 °F

1.42 2.58 g/cm^3

1.44 2.84 L

1.46 2.59×10^4 g

1.50 (a) 5 (b) 2 (c) 2 (d) 3 (e) 6 (f) 5

1.52 (a) 2.16×10^2
(b) 3.7×10^{-2}
(c) 0.224
(d) 0.4
(e) 21.9
(f) 1.98×10^3

1.54 (a) 0.377
(b) 7.5
(c) 36.42

1.58 761 mi/hr

1.60 5.88×10^{12} mi

1.62 47.3 m^2

1.64 1.24×10^2 ft^2

1.66 19.1 g/cm^3

Chapter 2

2.2 The ratio of sulfur that combines with 1 g of oxygen is $\frac{3}{2}$.

2.12 (a) $^{15}_{7}N$ (b) $^{70}_{31}Ga$ (c) $^{40}_{18}Ar$

2.14 (a) 33 protons, 46 neutrons, 33 electrons
(b) 23 protons, 28 neutrons, 23 electrons
(c) 52 protons, 76 neutrons, 52 electrons

2.16 (a) $^{16}_{8}O^{2-}$ (b) $^{79}_{34}Se^{2-}$ (c) $^{59}_{28}Ni^{2+}$

2.18

symbol	$^{23}Na^+$	$^{40}Ca^{2+}$	$^{81}Br^{1-}$	$^{128}Te^{2-}$
atomic number	11	20	35	52
mass number	23	40	81	128
charge	1+	2+	1−	2−
number protons	11	20	35	52
number electrons	10	18	36	54
number neutrons	12	20	46	76

2.24 (a) I^- (b) Mg^{2+} (c) O^{2-} (d) Na^+
2.26 (a) CaS (b) Mg_3N_2 (c) FeF_2
2.28 (a) $CaCl_2$ (b) Rb_2S (c) Li_3N (d) Y_2Se_3
2.30 (a) OH^- (b) ClO_3^- (c) MnO_4^-
2.32 (a) $Mg(NO_2)_2$ (b) Li_3PO_4 (c) $Ba(CN)_2$
2.34 (a) lithium iodide (b) magnesium nitride (c) sodium phosphate (d) barium perchlorate
2.36 (a) cobalt(III) chloride (b) iron(II) sulfate (c) copper(II) oxide
2.38 (a) Mn_2S_3 (b) $Fe(CN)_2$ (c) K_2S (d) $HgCl_2$
2.40 (a) SF_4 (b) NCl_3 (c) N_2O_5 (d) ClF_3
2.42 (a) phosphorus pentabromide (b) selenium dioxide (c) diboron tetrachloride (d) disulfur dichloride
2.44 KCl
2.46 magnesium hydroxide
2.48 (a) $^{35}_{17}Cl^-$ (b) $^{39}_{19}K^+$
2.50 (a) nitrogen monoxide, molecular (b) yttrium sulfate, ionic (c) sodium oxide, ionic (d) nitrogen tribromide, molecular

2.52

symbol	$^{70}Ga^{3+}$	$^{103}Rh^{3+}$	$^{114}In^{1+}$	$^{28}Si^{2-}$
atomic number	31	45	49	14
mass number	70	103	114	28
charge	3+	3+	1+	2−
number protons	31	45	49	14
number electrons	28	42	48	16
number neutrons	39	58	65	14

2.54 (a) $CaCl_2$ (b) CO_2 (c) Fe_2O_3
2.56 (a) $^{23}Na^+$ (b) $^{121}Sb^{3+}$ (c) ^{84}Kr

Chapter 3

3.2 (a) $C_5H_{12} + 8O_2 \longrightarrow 5CO_2 + 6H_2O$
(b) $4NH_3 + 3O_2 \longrightarrow 2N_2 + 6H_2O$
(c) $2KOH + H_2SO_4 \longrightarrow K_2SO_4 + 2H_2O$
3.4 (a) $2N_2H_4 + N_2O_4 \longrightarrow 3N_2 + 4H_2O$
(b) $2F_2 + 2H_2O \longrightarrow 4HF + O_2$
(c) $Na_2O + H_2O \longrightarrow 2NaOH$
3.6 (a) $C_6H_{12} + 9O_2 \longrightarrow 6CO_2 + 6H_2O$
(b) $C_4H_8 + 6O_2 \longrightarrow 4CO_2 + 4H_2O$
(c) $C_3H_6O + 4O_2 \longrightarrow 3CO_2 + 3H_2O$
(d) $2C_4H_6O_2 + 9O_2 \longrightarrow 8CO_2 + 6H_2O$
3.8 (a) $2KOH + H_2SO_4 \longrightarrow 2H_2O + K_2SO_4$
(b) $Ca(OH)_2 + 2HCl \longrightarrow 2H_2O + CaCl_2$
(c) $HNO_3 + LiOH \longrightarrow H_2O + LiNO_3$
3.10 $N_2H_4 + 3O_2 \longrightarrow 2NO_2 + 2H_2O$
3.12 $S_8 + 4Cl_2 \longrightarrow 4S_2Cl_2$
3.14 69.7 u, gallium
3.16 (a) 0.293 mol S (b) 89 g Al (c) 1.8×10^3 g Cl
3.18 (a) 40.00 g/mol (b) 28.05 g/mol (c) 58.32 g/mol
3.20 (a) 0.183 mol C_6H_6 (b) 1.72 g SiH_4
3.22 (a) 399.0 g/mol (b) 21 g (c) 0.817 mol (d) 1.1×10^3 atoms C
3.24 (a) 0.013 mol K_2SO_4 (b) 0.039 mol $C_8H_{12}N_4$ (c) 0.0383 mol $Fe(C_5H_5)_2$
3.26 (a) 695 g N_2O_4 (b) 1.0×10^3 g $CaCl_2$ (c) 12 g CO
3.28 (a) 1.61×10^2 g NO_2 (b) 2.1×10^{24} molecules NO_2 (c) 2.1×10^{24} N atoms, 4.2×10^{24} O atoms
3.30 (a) 85.6% C, 14.4% H (b) 52.9% C, 5.9% H, 41.2% N (c) 69.94% Fe, 30.06% O
3.32 (a) 42.9% C, 2.1 g C
3.34 (a) 1.18 g C (b) 1.89 g C (c) 6.99×10^{-3} g C
3.36 CH_2
3.38 C_2H_3N
3.40 C_4H_8O
3.42 $FeCl_2$
3.44 $TiC_{10}H_{10}$
3.46 $C_3H_4O_2$
3.48 $C_4H_6NO_2$
3.50 (a) $C_6H_{12}O_3$ (b) $C_{12}H_{16}N_4O_{12}$
3.52 $C_9H_{18}O_3$
3.54 $C_2H_4O_2$
3.56 (a) $2C_3H_6 + 9O_2 \longrightarrow 6CO_2 + 6H_2O$ (b) 7.69 g CO_2
3.58 3.24 g Cl_2
3.60 118 g Al
3.62 21%
3.64 0.97 g Li_2O
3.66 5.6 g NH_3
3.68 2.76 g Ag
3.70 37.4%
3.72 (a) 45.3% (b) 9.7 g NaOH
3.74 572 tons of 10% ore
3.76 $x = 5$
3.78 $C_{17}H_{19}NO_3$
3.80 56.9%
3.82 0.022 g H_2

Chapter 4

4.2 0.587 *M*
4.4 51.0 g $AgNO_3$

4.6 0.49 L
4.8 0.025 *M*
4.10 (a) 4.4×10^{-2} *M* (b) 1.8×10^{-2} *M*
4.12 (a) 0.10 mol HNO_3 (b) 3.2 mol HNO_3
4.14 (a) 252 g HCl (b) 1.3×10^2 g KCl
4.16 19 g KSCN
4.18 3.00×10^{-3} *M*
4.20 5.2 g NaOH
4.22 0.90 L
4.24 0.42 *M*
4.26 0.93 g $BaSO_4$
4.28 3.2×10^{-2} *M* $Ba(OH)_2$
4.30 a, c, d dissolve
4.32 (a) $Mg^{2+}(aq) + 2OH^-(aq) \longrightarrow Mg(OH)_2(s)$
(b) all are soluble
(c) $Ba^{2+}(aq) + CO_3^{2-}(aq) \longrightarrow BaCO_3(s)$
4.34 Pb^{2+}
4.36 Ca^{2+}
4.38 Add Na_2SO_4 to test for Hg^{2+}. If precipitate forms, Hg^{2+} is present. If no precipitate forms, Hg^{2+} is absent, and Na_2CO_3 can be added to test for Ca^{2+}.
4.40 $AgNO_3(aq) + NaBr(aq) \longrightarrow AgBr(s) + NaNO_3(aq)$
$Ag^+(aq) + NO_3^-(aq) + Na^+(aq) + Br^-(aq) \longrightarrow AgBr(s) + Na^+(aq) + NO_3^-(aq)$
$Ag^+(aq) + Br^-(aq) \longrightarrow AgBr(s)$
21 g AgBr
4.42 Solid is $CaCO_3$; 0.15 *M* $CaCl_2$
4.44 15.5 g $PbSO_4$
4.46 $(NH_4)_2SO_4(aq) + BaCl_2(aq) \longrightarrow BaSO_4(s) + 2NH_4Cl(aq)$ 17 g $BaSO_4$
4.48 0.73 *M*
4.50 23.2% calcium
4.52 (a) released from the system (b) −98.3 kJ
4.54 14.1 kJ
4.56 −193 kJ
4.58 −478 kJ
4.60 20.8 kJ
4.62 0.211 kJ
4.64 −2.2 kJ; endothermic
4.66 2.4 kJ; exothermic
4.68 1.4×10^2 kJ
4.70 -1.09×10^4 kJ
4.72 −113 kJ
4.74 0.0440 g CH_4
4.76 −312 kJ
4.78 −348.0 kJ
4.80 55 mL
4.82 12.9 g total
4.84 2.91 *M* in Cl^-
4.86 0.299%
4.88 0.100 L
4.90 −97 kJ

Chapter 5

5.6 (a) 0.439 atm
(b) 3.893×10^{-2} atm
(c) 1.8×10^3 torr
5.8 (a) 318 K (b) 245 K (c) 503 K
5.10 2.52 atm
5.12 52.8 mL
5.14 87.2 mL
5.16 4.38×10^4 ft^3
5.18 474 mL
5.20 3.88×10^{-2} mol
5.22 12.2 mol
5.24 2.02×10^3 L
5.26 52.7 g/mol
5.28 54.0 g/mol
5.30 30.3 mL
5.32 1.31 L O_2
5.34 0.444 L H_2
5.36 1.06 L N_2
5.38 2.44 L SO_2, 3.66 L O_2
5.40 5.55 atm
5.42 6.06 atm
5.44 17.6 atm
5.46 $P_{H_2} = 5.60$ atm, $P_T = 7.13$ atm
5.48 730 torr O_2
5.50 1.76 atm H_2
5.54 $O_2 < N_2 < Ne$
5.56 Ar < Ne (25°) < Ne (100°)
5.58 679 m/s
5.60 28.0 g/mol
5.64 rate He/rate Ne = 2.245
5.66 16 mL
5.68 116 g/mol
5.72 (a) O_2 (100 °C)
(b) N_2 (−100 °C)
(c) Ar (1 atm)
5.74 ideal = 208 atm, van der Waals = 218 atm
5.76 27 psi
5.78 64 atm
5.80 5.0×10^3 L
5.82 6.04 g
5.84 786 g

Chapter 6

6.2 $c_{560} = c_{720} = 3.00 \times 10^8$ m/s
$\nu_{560} = 5.36 \times 10^{14}\ s^{-1}$, $\nu_{720} = 4.17 \times 10^{14}\ s^{-1}$

6.4 $\lambda = 5.00 \times 10^{-6}$ m, infrared

6.6 $\lambda = 1.21$ pm

6.8 (a) 4.08×10^{-19} J
(b) 2.45×10^{18} photons/s

6.10 $\Delta E = 3.37 \times 10^{-19}$ J

6.12 (a) $\lambda = 259$ nm, ultraviolet (b) no

6.14 (a) 6.2×10^{12} *e*/s (b) 6.2×10^{12} photons/s

6.16 292 nm, UV

6.18 Only lines that originate from $n = 1$ are seen in the absorption spectrum.

6.20 $\lambda = 3.32 \times 10^{-10}$ m; the same as $2\pi r$

6.22 (a) 1.66×10^{-34} m (b) 3.58×10^{-38} m
(c) 1.47×10^{-10} m

6.24 2.65×10^3 m/s

6.26 (a) 6*p* (b) 3*s* (c) 5*d* (d) 4*s* (e) not allowed, $\ell > (n - 1)$

6.28

	(a)	(b)	(c)	(d)	(e)
n	3	5	7	4	2
ℓ	1	2	0	3	0

6.30 $2s < 3s = 3p_x = 3p_y < 4s = 4p_z = 4d_{xy}$

6.32 (a) 2*p* (b) not allowed, $\ell > (n - 1)$
(c) 3*s* (d) not allowed, $m_\ell > \ell$ (e) 3*d*
(f) 5*s*

6.44 (a) $2s < 2p < 3p < 3d < 5p$
(b) $1s < 2s < 2p < 3s < 3d < 4d$
(c) $1s < 2s < 2p < 3s < 3p < 3d < 4p$

6.46 (a) Be: [⇅] 1*s* [⇅] 2*s*
Ne: [⇅] 1*s* [⇅] 2*s* [⇅|⇅|⇅] 2*p*
(b) N: [⇅] 1*s* [⇅] 2*s* [↑|↑|↑] 2*p*
(c) H: [↑] 1*s*
Li: [⇅] 1*s* [↑] 2*s*
B: [⇅] 1*s* [⇅] 2*s* [↑] 2*p*
F: [⇅] 1*s* [⇅] 2*s* [⇅|⇅|↑] 2*p*

6.48 Ne, Be

6.50 (a) 1*s* (b) 2*s* (c) 2*p* (d) 2*p*

6.52 [↑|↑|] 2*p*
[⇅] 2*s*
[⇅] 1*s*

Quantum numbers for electrons in C:

electron	n	ℓ	m_ℓ	m_s
1	1	0	0	+1/2
2	1	0	0	−1/2
3	2	0	0	+1/2
4	2	0	0	−1/2
5	2	1	−1	+1/2
6	2	1	0	+1/2

6.54 $n = 2, \ell = 0, m_\ell = 0, m_s = 1/2$

6.56 1.875×10^3 nm, 1.282×10^3 nm

6.58 $\Delta x = 1 \times 10^{-37}$ m; it is not significant

6.60 Li, $1s^22s^1$, [⇅] 1*s* [↑] 2*s*
F, $1s^22s^22p^5$, [⇅] 1*s* [⇅] 2*s* [⇅|⇅|↑] 2*p*
O, $1s^22s^22p^4$, [⇅] 1*s* [⇅] 2*s* [⇅|↑|↑] 2*p*

6.62 (a) $n = 3$ to $n = 2$, $\Delta E = 3.02 \times 10^{-19}$ J
$n = 33$ to $n = 32$, $\Delta E = 1.27 \times 10^{-22}$ J
(b) 2.18×10^{-18} J
(c) There is a limit to the energy difference, since the $n = \infty$ level corresponds to the electron being infinitely far from the nucleus.
(d) The energy associated with removing an electron from an atom in the gas phase is known as the ionization energy of the atom; see Section 7.3. The energy that was calculated in part (c) corresponds to removal of the electron from one atom of H. The value that is commonly tabulated for the ionization energy of H corresponds to the energy that is needed to remove the electron from one mole of H atoms, namely

$$\left(2.18 \times 10^{-18} \frac{\text{J}}{\text{atom}}\right)\left(6.022 \times 10^{23} \frac{\text{atom}}{\text{mol}}\right) = 1.314 \times 10^6 \frac{\text{J}}{\text{mol}} \text{ or } 1314 \frac{\text{kJ}}{\text{mol}}$$

6.64 (a) 7.28×10^6 m/s (b) 3.97×10^3 m/s, $u_{\text{rms}} = 2.73 \times 10^3$ m/s

Chapter 7

7.4 (a) $1s^22s^22p^63s^23p^64s^23d^{10}4p^2 = [\text{Ar}]4s^23d^{10}4p^2$
[Ar] [⇅] 4*s* [⇅|⇅|⇅|⇅|⇅] 3*d* [↑|↑|] 4*p*
(b) $1s^22s^22p^63s^23p^4 = [\text{Ne}]3s^23p^4$
[Ne] [⇅] 3*s* [⇅|↑|↑] 3*p*

(c) $1s^22s^22p^63s^23p^64s^23d^{10}4p^65s^1 = [Kr]5s^1$

[Kr] [↑] 5s

7.6 (a) Al (b) Na (c) Mn

7.8 Br

7.10 (a) $1s^22s^22p^63s^23p^64s^2 = [Ar]4s^2$

[Ar] [⇅] 4s

(b) $1s^22s^22p^63s^23p^64s^23d^{10}4p^65s^24d^5 = [Kr]5s^24d^5$

[Kr] [⇅] 5s [↑|↑|↑|↑|↑] 4d

(c) $[Kr]5s^24d^{10}5p^1$

[Kr] [⇅] 5s [⇅|⇅|⇅|⇅|⇅] 4d [↑| |] 5p

7.12 (a) 1 (b) 2 (c) 0

7.14 (a) $3s^23p^1$ (b) $6s^1$ (c) $4s^24p^3$

7.16 IIA

7.18 (a) ns^1 (b) ns^2np^2 (c) ns^2np^5

7.22 (a) [Ar] (b) $[Ar]3d^5$ (c) $[Ar]4s^23d^{10}$

7.24 (a) Si^+ (b) Mg^+ (c) Fe^+

7.26 (a) 0 (b) 2 (c) 0

7.28 $Fe^{3+} = [Ar]3d^5$ $Cr^{3+} = [Ar]3d^3$

7.30 As^-, Br^+

7.32 Li^{3+}, B^{3+}, N^{3+}

7.34 manganese

7.36 (a) Na (b) O^{2-} (c) Ni^{2+}

7.38 (a) $O < B < Li$ (b) $N < C < Si$
(c) $S < As < Sn$

7.40 (a) $Be^{2+} < Be < Li$ (b) $Cl < S < S^{2-}$
(c) $N < C < Si$

7.46 (a) $Cl > Si$ (b) $Na > Rb$ (c) $F^- > O^{2-}$

7.48 (a) $Cl > Ge$ (b) $F > B$ (c) $Al^{3+} > Na^+$

7.50 (a) $O^{2-} < O < F$ (b) $Si < C < N$
(c) $Sr < Ru < Te$

7.56 Boron. In both cases a valence 2*s* electron is being removed, but B^+ has one more proton in its nucleus and a smaller size.

7.72
3*p* ⇅ ⇅ ⇅
3*s* ⇅
2*p* ⇅ ⇅ ⇅
2*s* ⇅
1*s* ⇅

7.74 [Kr] $4d^8$. The exception with palladium atom does not influence the electron configuration of the 2+ ion.

7.76 CrO_2, 2 unpaired electrons

7.78 0.468 L

7.80 $Fe^{2+} = [Ar]\ 3d^6$ $Fe^{3+} = [Ar]\ 3d^5$

7.82 size $Ca^{2+} < K^+ < S^{2-}$
ionization energy $S^{2-} < K^+ < Ca^{2+}$

Chapter 8

8.2 (a) Na· (b) :F̤̈· (c) :Ö:$^{2-}$ (d) Mg^{2+}

8.4 (a) beryllium
(b) carbon
(c) boron

8.6 (a) ·Ba· + 2:B̈r: ⟶ Ba^{2+} + 2:B̈r:$^-$
(b) 2K· + ·S̈· ⟶ $2K^+$ + :S̈:$^{2-}$

8.10 (a) Cl^- is smaller than Br^-, thus the ion separation in LiCl is smaller.
(b) O^{2-} has a greater charge than F^-.

8.12 (a) Li_2O, Na_2S; Li_2O has the greater lattice energy because of the smaller sizes of the ions.
(b) KCl, MgF_2; MgF_2 has the greater lattice energy because of the higher charge, smaller size of Mg^{2+}, and the smaller size of F^-.

8.14 (a) $Ca^{2+} < K^+ < S^{2-}$
(b) $Na^+ < F^- < O^{2-}$

8.18 (a) CF_4 (b) NI_3 (c) Cl_2O

8.20 (a) H—S̈—H (two lone pairs on S)

dots are lone pairs
lines are bond pairs

(b) 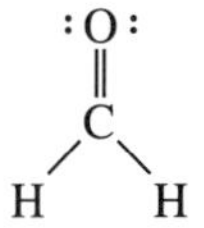 :Ö: double-bonded (‖) to C, with C bonded to H and H

dots are lone pairs
lines are bond pairs

(c) 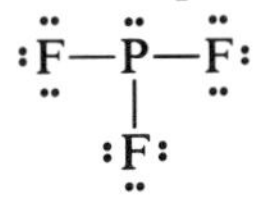 :F̈—P̈—F̈: with :F̈: bonded below P

dots are lone pairs
lines are bond pairs

8.22 (a) H—S bond order = 1
(b) C=O bond order = 2
C—H bond order = 1
(c) P—F bond order = 1

8.24 (a) 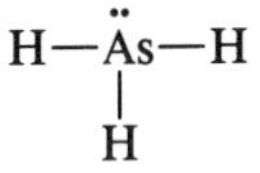 H—Äs—H with H bonded below As (b) [:N≡O:]$^+$

(c) 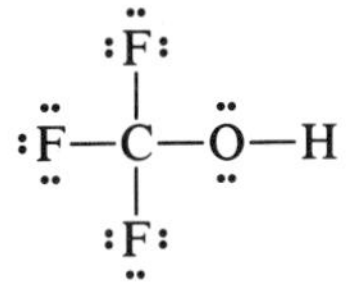 :F̈—C—Ö—H, with :F̈: bonded above C and :F̈: bonded below C

8.26

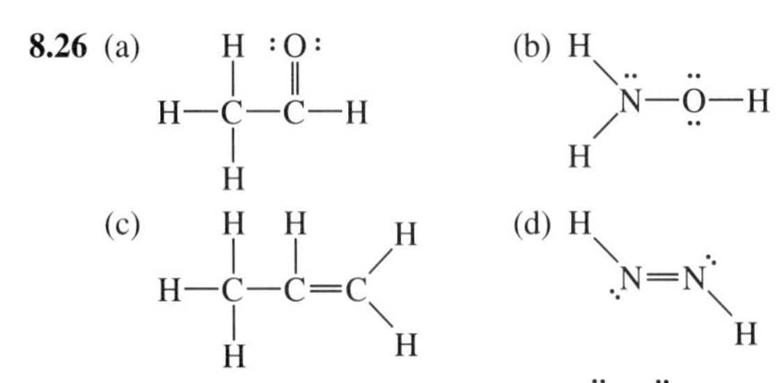

8.28 (a) H—C≡C—H (b) H—O—Cl: (c) (d)

8.32 (a) Br (b) Cl (c) N (d) O (e) Ge

8.34 (a) C—O (b) Ge—C (c) O—H (d) B—C

8.36 BrF

8.40 The first Lewis structure is better because it does not require formal charges.

8.42 (a) Both contribute equally.

(b) :Cl—C≡N: ⟷ Cl=C=N: ⟷ :Cl≡C—N:

The first Lewis structure is much better.

8.44 (a) The first resonance structure is favored because the −1 formal charge is on the more electronegative element.

(b) O=C=N: ⟷ :O—C≡N: ⟷ :O≡C—N:

The second structure is favored because the −1 formal charge is on the more electronegative element.

8.46 (a) :N=N=N: ⟷ :N≡N—N: ⟷ :N—N≡N:

The first structure is best because it avoids a −2 formal charge on one atom.

(b) All resonance forms are of equal energy.

8.48

8.50 (a) electron-rich

(b) As written, this is electron-deficient.

(c) There are additional resonance forms for this odd-electron molecule that place the odd electron on oxygen, but the two shown are favored because the +1 formal charge is on the less electronegative element.

8.54 (a) (b)

8.56 (a) :Ö—Cl—Ö: ⟷ :Ö—Cl=Ö ⟷ Ö=Cl—Ö: ⟷ Ö=Cl=Ö (formal charges: ⊖ ⊕ ⊖; ⊖; ⊖; ⊖)

(b) $^{\ominus}$:O—S$^{\oplus}$(—O$^{\ominus}$)—Ö—H ⟷ O=S(—O$^{\ominus}$)—Ö—H ⟷ $^{\ominus}$O—S(=O)—Ö—H ⟷ $^{\ominus}$O—S(—O$^{\ominus}$)=O$^{\oplus}$—H

8.58 (a) 1167 kJ (b) 2056 kJ

8.60 (a) −482 kJ (b) −554 kJ

8.62 (a) −1287 kJ (b) 7 kJ

8.66 :S$^{\oplus}$—N$^{\ominus}$, N—S (ring) ⟷ :S=N:, :N—S: (ring)

8.68 $F_4S=CH_2$ (:F: groups on S, S=C, C bonded to two H)

The Lewis structure places a double bond between S and C.

8.70 :N=N=Ö ⟷ :N≡N—Ö: (formal charges ⊖ ⊕; ⊕ ⊖)

:N=O=N: ⟷ :N≡O—N: ⟷ :N—O≡N: (formal charges ⊖ +2 ⊖; +2 −2; −2 +2)

The N—O—N arrangement has a higher formal charge, making the N—N—O arrangement more favorable.

8.72 :Br—N=Ö N—O bond is more polar.

8.74 −136 kJ

Chapter 9

9.2 (a) trigonal planar (b) tetrahedral (c) tetrahedral (d) trigonal bipyramidal

9.4 (a) tetrahedral (b) linear (c) trigonal bipyramidal (d) trigonal planar (e) tetrahedral

9.6 (a) trigonal planar, bent (b) linear, linear (c) tetrahedral, trigonal pyramidal (d) octahedral, square pyramidal (e) trigonal bipyramidal, see-saw

9.8 (a) NCl_3, angle based on tetrahedron (109°) rather than trigonal planar (120°) for BCl_3.
(b) SF_6, 90° angles based on octahedron compared to 109° for OF_2 based on tetrahedron.

9.10 (a) $H_3C—C≡C—H$ 109°, 180° 180° (b) Br(Br)C=C(H)H all angles 120°

(c) $H_3C—NH_2$ (H—C—N:) all angles 109°

9.12 (a) H—C(H)(H)—C(=O:)—H 109° 120°

(b)

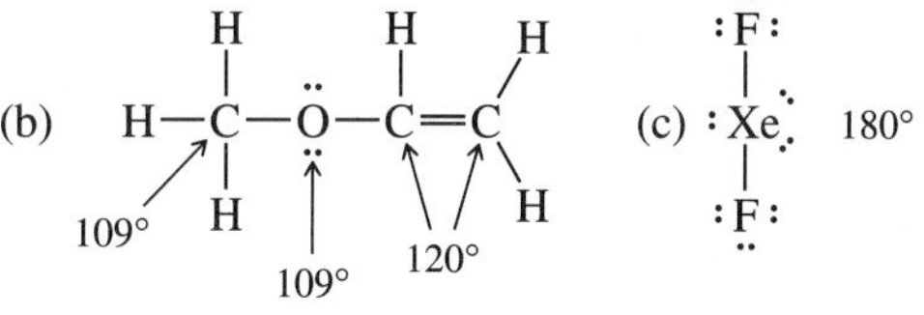

(c) :F:—Xe—:F: 180°

9.16 (a) nonpolar (b) nonpolar (c) nonpolar (d) polar

9.18 (a) polar (Xe, O, O) molecular (b) nonpolar (c) N=O both bond and molecular dipole (d) nonpolar

9.20 (a) polar (N, Cl, Cl, Cl) molecular dipole goes from N down between Cl atoms (b) nonpolar (c) nonpolar

9.22 $3p$ on Cl with $2p$ on F

9.24 (a) sp^2 (b) sp^3d^2 (c) sp

9.26 (a) sp^3 (b) sp^3d^2 (c) sp^3d (d) sp^3 (e) sp^3

9.28 (a) sp (b) sp^2 (c) sp^3d (d) sp^2

9.30 (a) sp^2 (b) sp^3 (c) sp^2

9.32 sp^3d for both bonds and lone pair on Se, $2p$ orbital on F

9.34 (a) sp^3, O; $2p$, F (b) sp^3, N; $1s$, H (c) sp^2, B; $3p$, Cl

9.38 $SbCl_5$—sp^3d hybrid orbitals on Sb overlapping $3p$ on Cl
$[SbCl_6]^-$—sp^3d^2 hybrid orbitals on Sb overlapping $3p$ on Cl

9.42 (a) 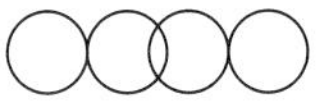 (b) 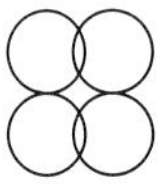 (c) 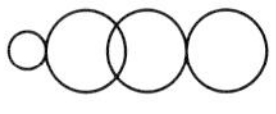

sigma bond pi bond sigma bond

9.44 5 sigma and 2 pi bonds
9.48 (a) two carbon atoms involved in the double bond are sp^2, all others are sp^3 (b) sp^2 (c) N sp^3; CH_2 sp^3; C of CO_2H sp^2; O of COH sp^3
9.50 sp^2 in both
9.52 (a) C sp^3, N sp^3 (b) CH_3 sp^3, other two C sp

9.54 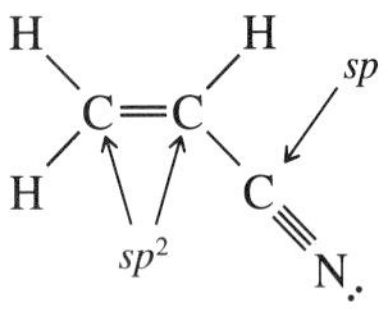

9.56 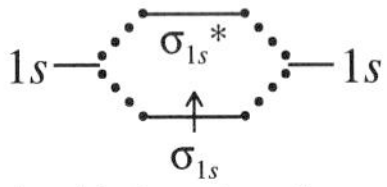

$(\sigma_{1s})^1$, bond order = $\frac{1}{2}$, one unpaired electron, stable

9.58 bond order = 1, no unpaired electrons, stable
9.60 (a) $(\sigma_{2s})^2(\sigma^*_{2s})^2(\pi_{2p})^3$, bond order = $1\frac{1}{2}$, one unpaired electron
(b) $(\sigma_{2s})^2(\sigma^*_{2s})^2(\pi_{2p})^4(\sigma_{2p})^2(\pi^*_{2p})^1$, bond order = $2\frac{1}{2}$, one unpaired electron
(c) $(\sigma_{2s})^2(\sigma^*_{2s})^2(\pi_{2p})^1$, bond order = $\frac{1}{2}$, one unpaired electron
9.62 (a) F_2^+, O_2^-, stable (b) N_2, O_2^{2+}, C_2^{2-}, stable
(c) Be_2, Li_2^{2-}, B_2^{2+} unstable
9.64 (a) $(\sigma_{2s})^2(\sigma^*_{2s})^2(\pi_{2p})^4(\sigma_{2p})^1$, bond order = $2\frac{1}{2}$, one unpaired electron
(b) $(\sigma_{2s})^2(\sigma^*_{2s})^2(\pi_{2p})^4(\sigma_{2p})^2$, $(\pi^*_{2p})^1$, bond order = $2\frac{1}{2}$, one unpaired electron
(c) $(\sigma_{2s})^2(\sigma^*_{2s})^2(\pi_{2p})^2$, bond order = 1, two unpaired electrons
(d) $(\sigma_{2s})^2(\sigma^*_{2s})^2(\pi_{2p})^2$, bond order = 1, two unpaired electrons

9.66 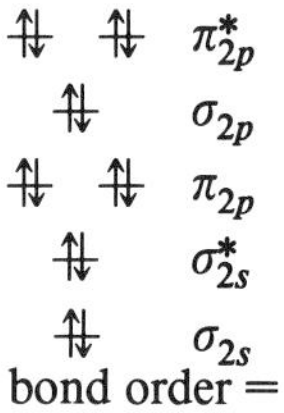

bond order = 1

9.70 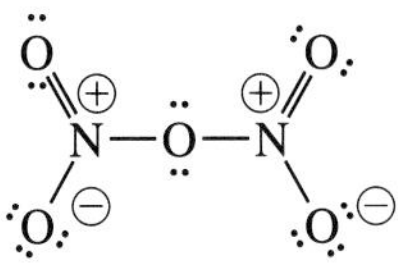

Central O has 109° bond angles, sp^3.
N atoms have 120° bond angles, sp^2.

9.72 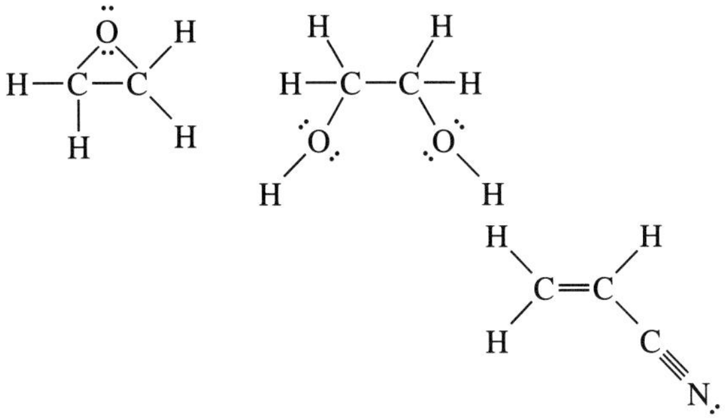

9.74 N uses sp^2 orbitals to make the bonds, and the lone pair is in a p orbital.

9.76

Lewis Structure	Electron Pair Arrangement
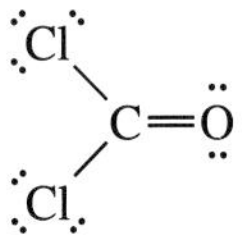	trigonal planar
Hybridization	**Polarity**
sp^2	polar

9.78 1 120°
2 109°
3 109°
4 120°
5 109°
9.80 Bond order for NO^- = 2, for NO^+ = 3.

Chapter 10

10.2 (a) London dispersion (b) hydrogen bonding (c) London dispersion (d) dipole-dipole, London dispersion (e) London dispersion (f) hydrogen bonding
10.4 (a) C_2H_4 (b) Cl_2 (c) S_2Cl_2 (d) NH_3 (e) CHI_3 (f) $BBrI_2$
10.8 (a) London dispersion, dipole-dipole (b) London dispersion (c) London dispersion, hydrogen bonding (d) London dispersion, dipole-dipole
10.14 (a)–(f) all increase

10.16 Adhesive forces to glass are greater than cohesive forces in the liquid.
10.18 (a) C_3H_8 (b) I_2 (c) S_2Cl_2 (d) H_2O (e) CH_2Cl_2 (f) NOCl
10.20 Because of OH group, C_2H_5OH can hydrogen bond, so it will have a higher surface tension.
10.24 (a) hydrogen bonding (b) London dispersion (c) ionic attraction (d) covalent bonds (e) metallic bonds (f) covalent bonds
10.26 $N_2 < CO_2 < H_2O < KCl < CaO$
10.28 one NH_4^+ and one Cl^- per unit cell
10.30 2 W atoms/cell
10.32 9.02 g/cm^3
10.34 (a) 544 pm (b) 192 pm
10.36 4 O^{2-} ions/cell
10.38 4.90 g/cm^3
10.40 4 Pd atoms/cell; FCC
10.42 (a) and (d)
10.44 108 pm
10.46 (a) no change (b) increases (c) no change
10.48 (a) 2.07 atm (b) 4.19×10^{-2} atm (c) 3.04×10^{-2} g
10.54 7.53 kJ
10.56 $\Delta H_f < \Delta H_v < \Delta H_s$
10.58 $\Delta H_f = 14.7$ kJ/mol
10.60 (a)

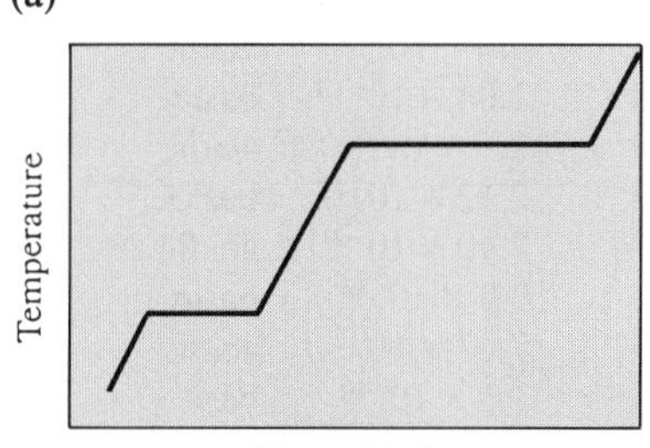

(b) First the solid melts to liquid; after liquid forms, at a lower pressure, it vaporizes.
(c) The solid is denser than liquid.
10.62 (a) Solid sublimes.
(b) Some ice melts and the temperature decreases.
(c) Vapor condenses to restore original pressure or until all of the vapor has condensed.
10.66 (a) 207 J (b) 1.15 g

Chapter 11

11.2 (a) 0.160% (b) 2.36×10^{-4} (c) 0.0131 m
11.4 0.22 mol H_2O_2
11.6 Add water to 1.55 g of H_3BO_3, to give a total mass of 155 g of sol'n; molality = 0.163 m
11.8 5.9×10^{-3} m $AgNO_3$
11.10 1.3×10^{-2}
11.12 (a) 0.324 m (b) 0.303 M (c) 5.82×10^{-3}
11.14 (a) 6.00% H_3PO_4 (b) 1.16×10^{-2} (c) 0.651 m
11.16

	Molality	Molarity	Mass % NH_3	Mole Fraction
(a)	3.75	3.43	6.00	6.33×10^{-2}
(b)	11.2	8.80	16.0	0.168
(c)	8.02	6.70	12.0	0.126
(d)	4.42	3.98	7.00	0.0738

11.18 0.801
11.20 As n increases, solubility decreases.
11.22 $H_2O < C_2H_5OH < C_6H_{14}$
11.24 (a) CCl_4 (b) water (c) ether (d) water
11.26 (a) hydrogen bonding (b) dipole-dipole and dispersion (c) ion hydration (d) dispersion
11.28 (a) $CH_3(CH_2)_{10}OH$ (b) CCl_4 (c) $Fe(C_5H_5)_2$
11.30 endothermic
11.32 (a) higher (b) lower (c) higher (d) lower (e) lower
11.34 (a) 6.17×10^{-5} m/torr (b) 0.482 g/kg H_2O
11.38 (a) higher (b) lower (c) lower (d) lower
11.40 $\Delta H_{sol'n}$ is exothermic.
11.42 NH_4Cl, the molar solubility increases more rapidly with temperature.
11.44 $\Delta P = 42.5$ torr, $P = 318$ torr
11.46 b.p. = 100.68 °C, f.p. = −2.47 °C
11.48 2.76 °C/m
11.50 $M = 104$ g/mol
11.52 sucrose < LiBr < $MgSO_4$ < $CaCl_2$
11.54 total $m = 1.08$ m; 31.6 g/kg of water
11.56 (a) $i = 2.5$ (b) at high concentration, i should be smaller
11.58 1.40×10^5 g/mol
11.60 The net transport occurs into the pure water ($\pi = 372$ torr and applied pressure = 500 torr).
11.62 (a) $\chi_{cyclohex} = 0.368$; $\chi_{Me} = 0.632$
(b) $P_{cyclohex} = 55.2$ torr; $P_{Me} = 198$ torr
(c) $\chi_{cyclohex} = 0.218$; $\chi_{Me} = 0.782$
11.64 (a) $P_{HCO_2H} = 525$ torr; $P_{H_2O} = 416$ torr; $P_T = 941$ torr
(b) actual $P = 760$ torr; negative deviation from Raoult's law
(c) The attractions between formic acid and water are stronger than the average.
11.66 (a) 3.27 g CO_2 (b) 1.66 L
11.68 (a) 128 g/mol in gas
(b) 66 g/mol in solution
(c) the gas dissociates into two ions when it dissolves

11.70 (a) $m(\text{benzene}) = 4.18 \times 10^{-2}\ m$; $m(H_2O) = 8.55 \times 10^{-2}\ m$
(b) if HOAc is a nonelectrolyte, $m = 8.4 \times 10^{-2}\ m$

$$i(\text{benzene}) = \frac{4.18 \times 10^{-2}}{8.4 \times 10^{-2}} = 0.50; \quad i(H_2O) = \frac{8.55 \times 10^{-2}}{8.4 \times 10^{-2}} = 1.0$$

(c) In benzene the molecule dimerizes as $(CH_3CO_2H)_2$; it remains monomeric in water.

Chapter 12

12.6 (a) $K_p = \frac{P_{Cl_2}P_{PCl_3}}{P_{PCl_5}}$ (b) $K_p = \frac{(P_{NO})^2 P_{O_2}}{(P_{NO_2})^2}$
(c) $K_p = \frac{(P_{SO_2})^2 P_{O_2}}{(P_{SO_3})^2}$ (d) $K_p = \frac{(P_{HI})^2}{P_{H_2}P_{I_2}}$

12.8 (a) $K_c = \frac{[Cl_2]^{1/2}[H_2O]^{1/2}}{[HCl][O_2]^{1/4}}$ (b) $K_c = \frac{[NO_2]}{[N_2O_4]^{1/2}}$
(c) $K_c = \frac{[N_2][O_2]^2}{[N_2O_4]}$ (d) $K_c = \frac{[SO_3]}{[SO_2][O_2]^{1/2}}$

12.10 $K_c = 1.4 \times 10^3$
12.12 $K_c = 0.20$
12.14 $K_c = 78$
12.16 $K_c = 1.4 \times 10^3$
12.18 $K_p = 15$
12.22 (a) left (b) left (c) right (d) right
12.24 (a) right (b) right (c) right (d) right (e) no change
12.26 $K_c = 0.158$
12.28 (a) $[SO_2] = 5.0 \times 10^{-4}$, $[O_2] = 7.5 \times 10^{-4}$, $[SO_3] = 5.0 \times 10^{-4}$ (b) $K_c = 1.3 \times 10^3$ (c) $K_p = 17$
12.30 (a) 0.20 mol NO_2 (b) 0.90 mol N_2O_4 $[N_2O_4] = 4.5 \times 10^{-2}\ M$ (c) $K_c = 4.5 \times 10^2$
12.32 $[H_2] = [I_2] = 0.12\ M$; $[HI] = 0.76\ M$
12.34 $[SO_3] = [NO] = 0.015\ M$; $[SO_2] = [NO_2] = 0.010\ M$
12.36 $[COCl_2] = 0.217\ M$; $[CO] = [Cl_2] = 0.0327\ M$
12.38 $[CO] = [Cl_2] = 0.110\ M$; $[COCl_2] = 0.0904\ M$
12.40 $[H_2] = [I_2] = 3.1 \times 10^{-4}\ M$; $[HI] = 1.4 \times 10^{-3}\ M$
12.42 $[SO_2] = 0.00438\ M$; $[Cl_2] = 0.0144\ M$; $[SO_2Cl_2] = 0.00562\ M$
12.44 $P_{CO} = 0.25$ atm; $P_{H_2} = 1.25$ atm; $P_{CH_2O} = 1.75$ atm
12.46 0.12 atm
12.50 $K_{sp} =$ (a) $[Mg^{2+}][F^-]^2$ (b) $[Ca^{2+}]^3[PO_4^{3-}]^2$ (c) $[Al^{3+}]^2[CO_3^{2-}]^3$ (d) $[La^{3+}][F^-]^3$
12.52 $K_{sp} = 8 \times 10^{-17}$
12.54 2.2×10^{-3} g/L
12.56 $1.1 \times 10^{-4}\ M$
12.58 (a) $9.3 \times 10^{-6}\ M$ (b) $8.7 \times 10^{-10}\ M$
12.60 (a) $2.6 \times 10^{-3}\ M$ (b) $7.4 \times 10^{-6}\ M$
12.62 (a) no precipitate (b) precipitate forms
12.64 (a) AgI (b) $[I^-] = 4.5 \times 10^{-7}\ M$
12.66 $[Ag^+] = 2.4 \times 10^{-7}\ M$

Chapter 13

13.2 (a) H_2SO_4, sulfuric acid (b) H_3O^+ or H^+, hydronium ion or hydrogen ion (c) NH_4^+, ammonium ion (d) $C_5H_5NH^+$, pyridinium ion
13.4 (a) CN^-, cyanide ion (b) SO_4^{2-}, sulfate ion (c) HPO_3^{2-}, hydrogen phosphite ion (d) CO_3^{2-}, carbonate ion
13.6 (a) $NH_3 + CH_3COOH \longrightarrow NH_4^+ + CH_3COO^-$ (base, acid, acid, base)
(b) $N_2H_5^+ + CO_3^{2-} \longrightarrow N_2H_4 + HCO_3^-$ (acid, base, base, acid)
(c) $H_3O^+ + OH^- \longrightarrow 2H_2O$ (acid, base, acid and base)
(d) $HSO_4^- + HCOO^- \longrightarrow SO_4^{2-} + HCOOH$ (acid, base, base, acid)

13.8

	pH	$[H_3O^+]$	
(a)	2.34	4.6×10^{-3}	acidic
(b)	12.98	1.04×10^{-13}	basic
(c)	−1.09	12.	acidic
(d)	10.67	2.12×10^{-11}	basic
(e)	1.13	7.40×10^{-2}	acidic
(f)	13.41	3.9×10^{-14}	basic
(g)	4.15	7.07×10^{-5}	acidic
(h)	9.80	1.6×10^{-10}	basic
(i)	0.30	0.50	acidic

13.10

	pH	$[H_3O^+]$, M	pOH	$[OH^-]$, M	
(a)	−1.04	11.	15.04	9.1×10^{-16}	acidic
(b)	13.66	2.2×10^{-14}	0.34	0.46	basic
(c)	6.70	1.98×10^{-7}	7.30	5.0×10^{-8}	acidic
(d)	12.65	2.3×10^{-13}	1.35	4.4×10^{-2}	basic

13.12

	pH	$[H_3O^+]$, M	$[OH^-]$, M
(a)	2.9	1×10^{-3}	1×10^{-11}
(b)	1.7	2×10^{-2}	5×10^{-13}
(c)	5.0	1×10^{-5}	1×10^{-9}
(d)	6.9	1×10^{-7}	1×10^{-7}

13.14

	pH	pOH
(a)	11.95	2.05
(b)	1.10	12.90
(c)	12.78	1.22
(d)	−1.09	15.09

13.16 pH = 13.88; pOH = 0.12
13.18 pH = 11.88; pOH = 2.12
13.20 (a) KF is a base
(b) $HClO_4 + HF \longrightarrow H_2F^+ + ClO_4^-$
(c) $NH_3 + HF \longrightarrow NH_4^+ + F^-$
(d) $H^+ + NH_3 \longrightarrow NH_4^+$
13.22 (a) $2CH_3COOH \rightleftharpoons CH_3COOH_2^+ + CH_3COO^-$
(b) base: $H_2O + CH_3COOH \longrightarrow H_3O^+ + CH_3COO^-$
(c) CH_3COO^-; CH_3COONa is a good source for acetate ion
(d) $CH_3COOH_2^+$; $HClO_4$
13.24 (a) and (c)
13.26 4.80
13.28 1.4×10^5
13.30 (a) 2.46 (b) 1.80 (c) 5.70 (d) 2.23
13.32 (a) 1.92 (b) 5.85 (c) 2.14 (d) 2.89
13.34 0.010 *M*—7.7%; 0.0010 *M*—22% ionized
13.36 7.2×10^{-10}
13.38 A is weak; B is strong.
13.40 12.30
13.42 (a) 5.6×10^{-11} (b) 2.2×10^{-11}
(c) 1.0×10^{-12}
13.44 $H_2O < NH_4^+ < HF < HSO_4^-$
13.46 F^- cannot act as a Brønsted acid;
$H_2O < CH_3COOH < HCOOH$.
13.48 3.24
13.50 8.37
13.52 1.30
13.54 2.89
13.56 13.00
13.58 (a) $HClO_4$ (b) HNO_3 (c) AsH_3
13.60 (a) Arrhenius, Brønsted, and Lewis
(b) Lewis (c) Arrhenius, Brønsted, and Lewis (d) Arrhenius, Brønsted, and Lewis
13.62 (a) products (b) products (c) no reaction (d) products
13.64 0.23 mL
13.66 0.23 g
13.68 11.20

Chapter 14

14.2 0.162 *M* H_2SO_4
14.4 1.82 L
14.6 (a) 15.0 mL (b) 26.0 mL (c) 100 mL
14.8 0.53% N
14.10 0 mL, pH = 0.60; 12.50 mL, pH = 1.00; 25.00 mL, pH = 7.00; 40.00 mL, pH = 12.93
14.12 0 mL, pH = 13.38; 0.25 mL, pH = 13.18; 0.50 mL, pH = 12.97; 1.20 mL, pH = 7.00; 1.50 mL, pH = 1.62
14.14 (a) 3.95 (b) 8.39
14.16 (a) 3.14 (b) 9.80
14.18 97 mL
14.20 (a) initial pH = 5.04, final pH = 5.04, $\Delta pH < 0.01$ (b) initial pH = 7.00, final pH = 11.00, $\Delta pH = +4.00$
14.22 $[HCOO^-] = 1.89 \times 10^{-3}$, $[HCOOH] = 1.65 \times 10^{-3}$
14.24 $[CH_3COOH] = 0.036\ M$; $[CH_3COO^-] = 0.064\ M$
14.26 0 mL, pH = 2.17; 5.0 mL, pH = 3.74; 10.0 mL, pH = 8.48; 15.0 mL, pH = 12.85
14.28 0%, pH = 2.05; 50%, pH = 3.46; 95%, pH = 4.74; 100%, pH = 8.33; 105%, pH = 11.89
14.30 0 mL, pH = 9.28; 15.0 mL, pH = 5.25; 30.0 mL, pH = 3.12; 40.0 mL, pH = 1.54. Pyridine is too weak a base to show a clear inflection point.
14.32 (a) 12.22 (b) 10.32 (c) 0.48
14.34 pH = 8.00—use phenolphthalein
14.36 pH = 4.69—use methyl red
14.38 (a) $H_2C_2O_4 + H_2O \rightleftharpoons H_3O^+ + HC_2O_4^-$

$$K_{a1} = \frac{[H_3O^+][HC_2O_4^-]}{[H_2C_2O_4]}$$

$HC_2O_4^- + H_2O \rightleftharpoons H_3O^+ + C_2O_4^{2-}$

$$K_{a2} = \frac{[H_3O^+][C_2O_4^{2-}]}{[HC_2O_4^-]}$$

(b) $H_2SO_3 + H_2O \rightleftharpoons H_3O^+ = HSO_3^-$

$$K_{a1} = \frac{[H_3O^+][HSO_3^-]}{[H_2SO_3]}$$

$HSO_3^- + H_2O \rightleftharpoons H_3O^+ + SO_3^{2-}$

$$K_{a2} = \frac{[H_3O^+][SO_3^{2-}]}{[HSO_3^-]}$$

14.40 $SO_4^{2-} + H_2O \rightleftharpoons HSO_4^- + OH^-$

$$K_b = \frac{[HSO_4^-][OH^-]}{[SO_4^{2-}]} = 8.3 \times 10^{-13}$$

14.42 $C_6H_5O_7^{3-} + H_2O \rightleftharpoons HC_6H_5O_7^{2-} + OH^-$

$$K_b = \frac{[HC_6H_5O_7^{2-}][OH^-]}{[C_6H_5O_7^{3-}]} = 2.5 \times 10^{-8}$$

14.44 citric acid
14.46 oxalic acid
14.48 $M = 84.2$ g/mol; $pK_a = 4.40$
14.50 (a) acid (b) acid
14.52 (a) acid (b) base
14.54 (a) *s* increases (b) *s* increases
14.56 (a) *s* increases (b) *s* increases
14.58 pH = 10.42
14.60 77 mL NaOH. Phenolphthalein is a good choice of indicator.

14.62 $\Delta pH = 1$; $[In]/[HIn] = 10$; $\Delta pH = -1$; $[In]/[HIn] = 0.10$

14.64 1% acid, pH = 7.00; 5% acid, pH = 6.28; 95% acid, pH = 3.72; 99% acid, pH = 3.00

Chapter 15

15.6 (a) w = negative (b) w = positive

15.8 (a) $w = +24.5$ L-atm (b) $w = -24.5$ L-atm

15.10 (a) $w = -73.4$ L-atm (b) $w = 0$

15.12 $w = -3.3 \times 10^3$ kJ

15.14 $\Delta E = +3.92$ kJ

15.16 $\Delta E = 0$; $w = -8.00$ L-atm $= -0.810 \times 10^2$ kJ
$q = +8.00$ L-atm $= +0.810 \times 10^2$ kJ

15.18 $w_1 = -6.00$ L-atm; $w_2 = -3.00$ L-atm; $w_3 = -4.50$ L-atm; $w_{total} = -13.50$ L-atm. $q_{total} = +13.50$ L-atm. The 3-step expansion generates more work, and consumes more heat, than either the 1- or 2-step expansion.

15.22 (a) $\frac{1}{2}H_2(g) + \frac{1}{2}F_2(g) \longrightarrow HF(g)$
(b) $H_2(g) + S(s, \text{rhombic}) + 2O_2(g) \longrightarrow H_2SO_4(\ell)$ ($\frac{1}{8}S_8$ (s, rhombic) also correct)
(c) $Al(s) + \frac{3}{2}O_2(g) + \frac{3}{2}H_2(g) \longrightarrow Al(OH)_3(s)$
(d) $Na(s) + K(s) + S(s, \text{rhombic}) + 2O_2(g) \longrightarrow NaKSO_4(s)$ ($\frac{1}{8}S_8$ (s, rhombic) also correct)

15.24 (a) $\Delta H = -74$ kJ exothermic
(b) $\Delta H = -2878.46$ kJ exothermic

15.26 (a) $\Delta H = -131.69$ kJ exothermic
(b) $\Delta H = -197.78$ kJ exothermic

15.28 (a) $\Delta H = +1.895$ kJ endothermic
(b) $\Delta H = +44.01$ kJ endothermic

15.30 $\Delta H = -4.4 \times 10^2$ kJ

15.32 (a) + (b) + (c) − (d) +

15.34 (a) + (b) − (c) +

15.36 (a) $\Delta S = -331.9$ J/K (b) −198.53 J/K
(c) −242.76 J/K (d) −42.08 J/K

15.38 (a) $\Delta G = -840.1$ kJ spontaneous
(b) $\Delta G = -29.2$ kJ spontaneous

15.40 (a) $\Delta G = -181.4$ kJ spontaneous
(b) $\Delta G = +28.2$ kJ nonspontaneous

15.42 $\Delta S_{sys} = -70$ J/K; $\Delta S_{surr} = 431$ J/K; $\Delta S_{univ} = +361$ J/K

15.46 (a) at 25 °C, $\Delta G° = -237.18$ kJ spontaneous
at 90 °C, $\Delta G° = -226.60$ kJ spontaneous
(b) at −20 °C, $\Delta G° = -44.2$ kJ spontaneous
at 50 °C, $\Delta G° = -20.9$ kJ spontaneous

15.48 (a) $\Delta G° = -99.2$ kJ, consistent with the sign of $\Delta H°$ but not $\Delta S°$ (b) $\Delta G° = -54.0$ kJ, consistent with the sign of both $\Delta H°$ and $\Delta S°$

15.50 $\Delta H°$ is negative and $\Delta S°$ is negative. At low temperatures, $\Delta G°$ is negative; at high temperatures, $\Delta G°$ is positive.

15.52 $\Delta H°$ is negative and $\Delta S°$ is negative. At low temperatures, $\Delta G°$ is negative; at high temperatures, $\Delta G°$ is positive.

15.54 (a) spontaneous when $T < 791$ K (b) spontaneous when $T < 779$ K

15.56 (a) spontaneous when $T > 475$ K (b) spontaneous when $T < 322$ K

15.58 (a) at 400 K, $\Delta G° = +86.6$ kJ; at 600 K, $\Delta G° = +52.2$ kJ (b) at 400 K, $\Delta G° = +196$ kJ; at 600 K, $\Delta G° = +223$ kJ

15.60 $\Delta H°$ is negative and $\Delta S°$ is negative, so $\Delta G°$ is negative at low temperatures and positive at high temperatures.

15.62 (a) incorrect—the direction depends on the sign of $\Delta G°$ (b) incorrect—reaction is not spontaneous (c) and (d) incorrect—the reaction is nonspontaneous above 245 K (e) correct

15.64 The dependence of K_{eq} on T provides information about the sign of $\Delta H°$; K_{eq} will become smaller when an exothermic reaction is heated.

15.66 (a) fusion: solid $\longrightarrow$ liquid
(b) vaporization: liquid $\longrightarrow$ gas
(c) sublimation: solid $\longrightarrow$ gas

15.68 At 80 °C, $\Delta G° = -1.85$ kJ, so the vaporization is spontaneous assuming that tabulated values for $\Delta H_f°$ and $S°$ at 25 °C are correct at 80 °C.

15.70 Calculated boiling point from $\Delta H°$ and $\Delta S°$ is 63.4 °C, compared to the experimental value of 65.0 °C.

15.72 $\Delta S°_{vap} = +150$ J mol^{-1} K^{-1}, $\Delta H°_{vap} = 23.5$ kJ mol^{-1}

15.74 (a) $\Delta G = -34.7$ kJ (b) $\Delta G = +72.1$ kJ

15.76 (a) $K_{eq} = P_{PCl_3}P_{Cl_2}/P_{PCl_5} = 3.0 \times 10^{-7}$ at 25 °C
(b) $K_{eq} = 1.3$ at 250 °C

15.78 (a) at 10 K, $\Delta G° = +9.0$ kJ and $K_{eq} = 2 \times 10^{-48}$
(b) at 100 K, $\Delta G° = 0$ kJ and $K_{eq} = 1.0$
(c) at 1000 K, $\Delta G° = -90.0$ kJ and $K_{eq} = 5.0 \times 10^4$

15.80 at 10 K, $\Delta G = -11.0$ kJ and $K_{eq} = 2.9 \times 10^{57}$
at 100 K, $\Delta G = -20.0$ kJ and $K_{eq} = 2.8 \times 10^{10}$
at 1000 K, $\Delta G = -110$ kJ and $K_{eq} = 5.6 \times 10^5$

15.82 $\Delta G = -81.8$ kJ

15.84 Increasing the temperature will decrease K_{eq} for (a), (b), (c), and (e) since $\Delta H°$ is negative; increasing the temperature will increase K_{eq} for (d) since $\Delta H°$ is positive.

15.86 (a) 0.21 atm (b) 0.18 atm (c) 0.29 atm

15.98 C_4H_{10}

15.100 reaction spontaneous for $T < 790$ K

Chapter 16

16.2 (a) rate = $-\Delta$[HOOC—COOH]/Δt = Δ[HCOOH]/Δt = $\Delta[CO_2]/\Delta t$
(b) rate = 9.0×10^{-4} mol L^{-1} s^{-1}
(c) rate = 8.0×10^{-4} mol L^{-1} s^{-1}
(d) rate = 2.0×10^{-3} mol L^{-1} s^{-1}
(e) rate = 4.1×10^{-4} mol L^{-1} s^{-1}

16.4 (a) rate = Δ[NOCl]/Δt
(b) rate = 1.9×10^{-3} mol L^{-1} s^{-1}
(c) rate = 1.6×10^{-3} mol L^{-1} s^{-1}
(d) rate = 9×10^{-4} mol L^{-1} s^{-1}; 4.5×10^{4} mol L^{-1} s^{-1}

16.8 (a)

Time, μs	$[N_2O_4]$	$[NO_2]$
0.0		
20.00		0.034
40.00	0.025	
60.00		0.060

(b) rate = 3.6×10^{-4} mol L^{-1} s^{-1}

16.10 $\Delta[Cl_2]/\Delta t = 0.015$ mol L^{-1} s^{-1} = rate of reaction

16.12 $\Delta[CrO_2^-]/\Delta t = -0.0050$ mol L^{-1} s^{-1}
$\Delta[H_2O_2]/\Delta t = -0.0075$ mol L^{-1} s^{-1}
$\Delta[OH^-]/\Delta t = -0.0050$ mol L^{-1} s^{-1}
$\Delta[H_2O]/\Delta t = +0.0100$ mol L^{-1} s^{-1}
rate = 0.0025 mol L^{-1} s^{-1}

16.14 rate = $3.8 \times 10^3[NO]^2[H_2]$
16.16 rate = $4.2[NO][H_2]$
16.18 rate = $5.2 \times 10^4[NO_2][O_3]$
16.20 rate = $2.8 \times 10^9[NO][Br_2]$
16.22 172 s
16.24 zero order, $k = 3.4 \times 10^{-2}$ mol L^{-1} s^{-1}
16.26 first order, $k = 501$ s^{-1}
16.28 second order, $k = 0.186$ L mol^{-1} s^{-1}
16.30 second order, $k = 0.080$ L mol^{-1} s^{-1}
16.34 (a) 0.023 min^{-1} (b) 4.88×10^3 s
16.36 121 s
16.38 24 s
16.44 $E_a = 109$ kJ mol^{-1}; $A = 7.83 \times 10^{16}$
16.46 35.8 kJ mol^{-1}
16.48 13 times faster
16.52 The rate constant will increase by 91%.
16.54 The reaction with 45 kJ activation energy will show the greatest change with a change in temperature.
16.58 5.2×10^7
16.60 The reaction is exothermic and E_a is quite large, so increasing the temperature will strongly favor the reactants.
16.62 The reaction is slightly endothermic and E_a is small, so increasing the temperature will slightly favor the products.
16.66 $NO_2 + CO \longrightarrow CO_2 + NO$
16.68 (a) k[HCl] (b) $k[H_2][Cl]$ (c) $k[NO_2]^2$
16.70 (a) $k[C_2H_5Cl]$ (b) $k[NO][O_3]$
(c) $k[HI][C_2H_5I]$
16.72 The first step is the rate-limiting step.
16.74 I rate = $k[NO_2][CO]$ II rate = $k[NO_2]^2$
16.76 rate = $K[N_2O_2][Cl_2] = kK_{eq}[NO]^2[Cl_2]$; N_2O_2 is an intermediate; mechanism is consistent with the rate law
16.78 (a) rate = $k[CH_3Br]$ (b) rate = $k[CH_3Br][OH^-]$
16.80 zero order, rate = 3.4×10^{-4} mol L^{-1} s^{-1}
16.82 (a) rate = $k[NO]^2[O_2]$ (b) The first reaction is exothermic, so increasing the temperature will force the reaction back toward the reactants, so the rate at which the products are formed will decrease.
16.84 $k = 0.0124$ s^{-1}, $t_{1/2} = 55.9$ s

Chapter 17

17.2 (a) oxidation state of N = +5, O = −2
(b) oxidation state of N = +3, O = −2
(c) oxidation state of N = −3, H = +1
(d) oxidation state of Br = 0
(e) oxidation state of S = +6, O = −2, H = +1
(f) oxidation state of C = +4, O = −2

17.4 (a) oxidation state of K = +1, Mn = +7, O = −2
(b) oxidation state of H = +1, O = −2
(c) oxidation state of Cl = 0
(d) oxidation state of N = +4, O = −2
(e) oxidation state of Cr = +3, O = −2
(f) oxidation state of Co = +3, N = +5, O = −2
(g) oxidation state of Ca = +2, C = +4, O = −2
(h) oxidation state of H = +1, Br = +7, O = −2
(i) oxidation state of Fe = +3

17.6 (a) $Cr^{3+} + 3e^- \longrightarrow Cr$ reduction
(b) $2I^- \longrightarrow I_2 + 2e^-$ oxidation
(c) $NO_2^- + H_2O \longrightarrow NO_3^- + 2H^+ + 2e^-$ oxidation
(d) $Fe^{2+} \longrightarrow Fe^{3+} + e^-$ oxidation
(e) $Cr_2O_7^{2-} + 14H^+ + 6e^- \longrightarrow 2Cr^{3+} + 7H_2O$ reduction
(f) $VO_2^+ + 4H^+ + 2e^- \longrightarrow V^{3+} + 2H_2O$ reduction

17.8 (a) $Sn + 2Fe^{3+} \longrightarrow Sn^{2+} + 2Fe^{2+}$
(b) $HAsO_3^{2-} + I_2 + H_2O \longrightarrow H_2AsO_4^- + 2I^- + H^+$
(c) $Cu + 2Ag^+ \longrightarrow Cu^{2+} + 2Ag$
(d) $2MnO_4^- + 5H_2C_2O_4 + 6H^+ \longrightarrow 2Mn^{2+} + 10CO_2 + 8H_2O$
(e) $Cl_2 + 2Br^- \longrightarrow 2Cl^- + Br_2$
(f) $3Cu + 2NO_3^- + 8H^+ \longrightarrow 3Cu^{2+} + 2NO + 4H_2O$
(g) $6VO^{2+} + Cr_2O_7^{2-} + 2H^+ \longrightarrow 6VO_2^+ + 2Cr^{3+} + H_2O$

17.10 (a) $2Al + 3ClO^- + 2OH^- + 3H_2O \longrightarrow 2Al(OH)_4^- + 3Cl^-$
(b) $2MnO_4^- + 3SO_3^{2-} + H_2O \longrightarrow 2MnO_2 + 3SO_4^{2-} + 2OH^-$

(c) $4Zn + NO_3^- + 7OH^- + 6H_2O \longrightarrow 4Zn(OH)_4^{2-} + NH_3$
(d) $3ClO^- + 2CrO_2^- + 2OH^- \longrightarrow 3Cl^- + 2CrO_4^{2-} + H_2O$
(e) $3Br_2 + 6OH^- \longrightarrow 5Br^- + BrO_3^- + 3H_2O$
(f) $2H_2O_2 + N_2H_4 \longrightarrow N_2 + 4H_2O$

17.12 1.26×10^{-2} M $KMnO_4$

17.16 (a) $2Hg + 2Cl^- \longrightarrow Hg_2Cl_2 + 2e^-$
$Cu^{2+} + 2e^- \longrightarrow Cu$
(b) $2Hg + Cu^{2+} + 2Cl^- \longrightarrow Hg_2Cl_2 + Cu$
(c) Cu
(d) yes

17.18 (a) $E° = +0.51$ V;
$Zn(s) + Ni^{2+}(aq) \longrightarrow Zn^{2+}(aq) + Ni(s)$
(b) $E° = +0.93$ V;
$Pb(s) + 2Ag^+(aq) \longrightarrow Pb^{2+}(aq) + 2Ag(s)$

17.20 (a) +1.36 V spontaneous forward
(b) −1.25 V spontaneous reverse
(c) +0.62 V spontaneous forward

17.22 (a) $Cu + 2H_2O \longrightarrow Cu^{2+} + H_2 + 2OH^-$
$E° = -1.17$ V
spontaneous reverse, does not react with water
(b) $2Na + 2H_2O \longrightarrow 2Na^+ + H_2 + 2OH^-$
$E° = +1.88$ V
spontaneous forward, reacts with water
(c) $2Al + 6H_2O \longrightarrow 2Al^{3+} + 3H_2 + 6OH^-$
$E° = +0.83$ V
spontaneous forward, reacts with water
(d) surface oxide film protects the metal from contacting water

17.24 $E° = +0.334$ V Pt positive electrode, Pb negative electrode

17.26 (a) $Cu \longrightarrow Cu^{2+} + 2e^-$ $Pu^{4+} + e^- \longrightarrow Pu^{3+}$
$Cu + 2Pu^{4+} \longrightarrow Cu^{2+} + 2Pu^{3+}$
(b) $E°$ for $Pu^{4+} + e^- \longrightarrow Pu^{3+}$ is +0.98 V

17.28 (a) $E° = +0.59$ V spontaneous forward
$\Delta G° = -1.1 \times 10^2$ kJ
$K_{eq} = 1.0 \times 10^{20}$
(b) $E° = +1.14$ V spontaneous forward
$\Delta G° = -2.20 \times 10^2$ kJ
$K_{eq} = 4 \times 10^{38}$
(c) $E° = +0.82$ V spontaneous forward
$\Delta G° = -1.6 \times 10^2$ kJ
$K_{eq} = 1 \times 10^{28}$

17.30 $E° = +0.0650$ V $\Delta G° = -12.5$ kJ

17.32 (a) +0.55 V (b) +1.13 V (c) +0.64 V

17.34 −0.37 V

17.36 +0.030 V

17.38 3.4×10^{-3} M

17.40 −3.16 V

17.44 (a) +1.59 V (b) No, all reactants and products are solids so $Q = 1$.

17.46 (a) $5O_2 + 10H_2O + 20e^- \longrightarrow 20OH^-$
$C_3H_8 + 20OH^- \longrightarrow 3CO_2 + 14H_2O + 20e^-$
(b) $\Delta G° = -2108$ kJ/mol $E° = +1.09$ V
(c) 47.8 kJ

17.48 (a) $Pb^{2+}(aq) + 2e^- \longrightarrow Pb(s)$
$Pb(s) \longrightarrow Pb^{2+}(aq) + 2e^-$
No net reaction
(b) $2H^+(aq) + 2e^- \longrightarrow H_2(g)$
$2I^-(aq) \longrightarrow I_2(s) + 2e^-$
$2H^+(aq) + 2I^-(aq) \longrightarrow H_2(g) + I_2(s)$
(c) $Cu^{2+}(aq) + 2e^- \longrightarrow Cu(s)$
$2Br^-(aq) \longrightarrow Br_2(\ell) + 2e^-$
$Cu^{2+}(aq) + 2Br^-(aq) \longrightarrow Cu(s) + Br_2(\ell)$

17.50 (a) Ag^+ (b) Pb^{2+} (because of hydrogen overvoltage) (c) Ba^{2+}

17.52 (a) It provides charge carriers in the electrolyte.
(b) K and F_2

17.54 (a) 1.4×10^5 C (b) 3.9×10^3 C (c) 2.72×10^4 C (d) 2.9×10^2 C

17.56 1.85×10^{-2} g

17.58 7.32×10^{-4} mol Sn^{4+}

17.60 19 min

17.62 (a) 422 A (b) 4.22×10^{-2} kw-hr

17.68 (a) −260 kJ (b) 1.20 kJ (c) 86.6 h

17.70 (a) −0.394 V (b) −0.347 V

17.72 $k = 0.223$ V $E_{cell} = -0.191$ V

17.74 (a) $2Fe^{3+} + H_2SO_3 + H_2O \longrightarrow 2Fe^{2+} + HSO_4^- + 3H^+$
$E° = +0.60$ V
(b) $6Fe^{2+} + Cr_2O_7^{2-} + 14H^+ \longrightarrow 6Fe^{3+} + 2Cr^{3+} + 7H_2O$
$E° = +0.56$ V
(c) $Fe^{2+} + HNO_2 + H^+ \longrightarrow Fe^{3+} + NO + H_2O$
$E° = +0.212$ V

Chapter 18

18.6 (a) Sc (b) group IIB (c) Yes, it has a partially filled d electron level in the atom

18.10 (a) Cr (b) Os (c) Cr (d) W

18.12 Metallic bond strength, as well as Z_{eff}, is increasing for these elements.

18.14 Nb > W > V > Co

18.16 (a) +4 (b) +6 (c) +5 (d) +7

18.18 Ionization of the two ns electrons

18.20 (a) Mn (b) Ta (c) Rh (d) Os

18.22 (a) $FeCr_2O_4(s) + 4C(s) \longrightarrow Fe(\ell) + 2Cr(\ell) + 4CO(g)$

(b) $Cr_2O_3(s) + 2Al(\ell) \longrightarrow 2Cr(\ell) + Al_2O_3(s)$

(c) $4Cr^{2+}(aq) + O_2(g) + 4H^+(aq) \longrightarrow 4Cr^{3+}(aq) + 2H_2O(\ell)$

(d) $2Cr(OH)_4^-(aq) + 3ClO^-(aq) + 2OH^-(aq) \longrightarrow 2CrO_4{}^{2-}(aq) + 3Cl^-(aq) + 5H_2O(\ell)$

(e) $6Fe^{2+}(aq) + Cr_2O_7^{2-}(aq) + 14H^+(aq) \longrightarrow 6Fe^{3+}(aq) + 2Cr^{3+}(aq) + 7H_2O(\ell)$

18.24 Fe $1s^22s^22p^63s^23p^64s^24d^6 = [Ar]4s^23d^6$

Fe^{2+} $[Ar]3d^6$

Fe^{3+} $[Ar]3d^5$

18.26 H_2SO_4 is the oxidizing agent;

$Cu(s) + 2H_2SO_4(aq) \longrightarrow CuSO_4(aq) + SO_2(g) + 2H_2O(\ell)$

18.28 (a) $[Co(en)_3][Co(CN)_6]$ (b)$[Pt(NH_3)_4](NO_3)_2$

(c) $Na_2[RhCl_5(H_2O)]$

18.30 (a) pentacarbonyliron(0)

(b) potassium pentacyanonitrosylchromate(III)

(c) pentaamminechlororuthenium(III) chloride

(d) tribromo(diethylenetriamine)cobalt(III)

(e) hexaamminechromium(III) hexacyanochromate(III)

18.32 (a) $[Cr(H_2O)_6][Fe(CN)_6]$

(b) $[Co(en)_2ClBr]^+$

(c) $[CoCO(CN)_5]^{2-}$

(d) $[Cr(NO_2)_3(dien)]$

(e) $[Fe(H_2O)_5(SCN)]^{2+}$

18.34

fac-triamminetrichlorocobalt (III) *mer*-triamminetrichlorocobalt (III)

18.36 (a)

linkage isomers

(b)

trans- *cis-*

geometric isomers

(c)

optical isomers

(d)

cis- (two enantiomers) *trans-*

optical isomers

geometric isomers

(e)

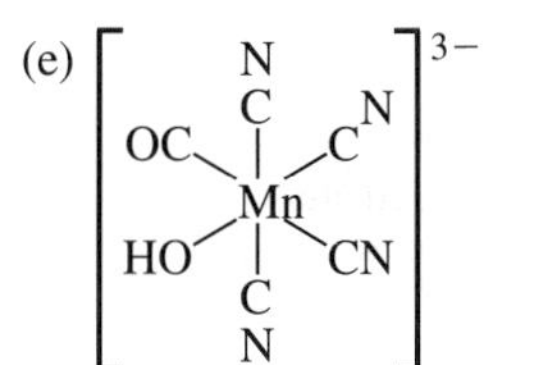

geometric isomers

18.38 SCN^-, NO_2^-, NCO^- can have linkage isomers
18.42 $\Delta = 266$ kJ/mol
18.44 (a) $[Rh(CN)_6]^{3-}$
(b) $[Fe(H_2O)_6]^{3+}$
(c) $[Rh(H_2O)_6]^{3+}$
(d) $[Ti(H_2O)_6]^{3+}$
18.46 (a) 3,$[Cr(H_2O)_6]^{3+}$
(b) 1,$[Mn(CN)_6]^{4-}$ 5,$[Mn(H_2O)_6]^{2+}$
(c) 1,$[Rh(H_2O)_6]^{2+}$ 3,$[Co(H_2O)_6]^{2+}$
(d) 1,$[Cu(H_2O)_6]^{2+}$
18.48 (a) 4 (b) 1 (c) 1 (d) 4 (e) 2
18.50 (a) $[Cr(H_2O)_6]^{2+}$, 4 $[Mn(CN)_6]^{3-}$, 2
(b) $[Fe(H_2O)_6]^{2+}$, 4 $[Ru(H_2O)_6]^{2+}$, 0
(c) $[Co(H_2O)_6]^{2+}$, 3 $[Co(CN)_5H_2O]^{3-}$, 1
18.52

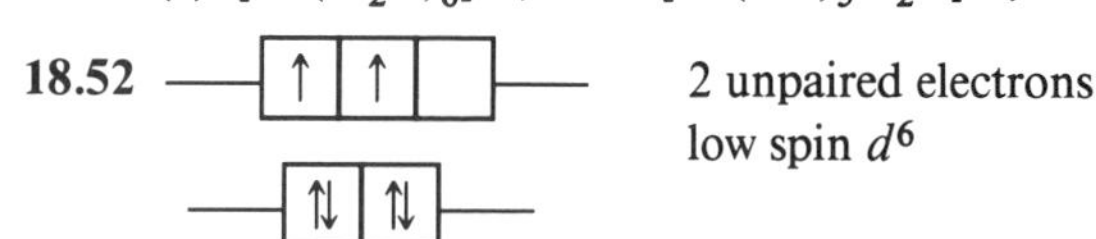

2 unpaired electrons
low spin d^6

18.54 (a) 0, square planar
(b) 4, tetrahedral
(c) 0, square planar

Chapter 19

19.4 (a) nonmetal (b) metal (c) nonmetal
(d) metalloid
19.14 $NaH(s) + H_2O(\ell) \longrightarrow NaOH(aq) + H_2(g)$
0.981 g NaH
19.28 $BCl_3 + NH_3 \longrightarrow Cl_3B \longleftarrow NH_3$
sp^2 hybrid boron in BCl_3, sp^3 hybrid boron in product
19.40 Si, sp^3; C, sp
19.46 sp^3, nonpolar
19.50
19.54

sp^2 N in all cases

19.56 (a) $3Mg(s) + N_2(g) \longrightarrow Mg_3N_2(s)$
(b) $P_4(s) + 5O_2(g) \longrightarrow P_4O_{10}(s)$
(c) $2NO_2(g) + H_2O(\ell) \longrightarrow HNO_2(aq) + HNO_3(aq)$
19.60 $4NH_3(g) + 5O_2(g) \longrightarrow 4NO(g) + 6H_2O(g)$
1.4×10^4 g NH_3
19.68 $2KClO_3(s) \longrightarrow 2KCl(s) + 3O_2(g)$
1.6 g $KClO_3$
19.76 linear, nonpolar

trigonal pyramidal, polar

Chapter 20

20.2

Symbol	Z	A	#p	#n
$^{23}_{11}Na$	11	23	11	12
$^{103}_{45}Rh$	45	103	45	58
$^{70}_{32}Ge$	32	70	32	38
$^{234}_{90}Th$	90	234	90	144

20.4 (a) 1.00 (b) 1.00 (c) 1.25 (d) 1.46
(e) 1.54
20.6 (a) and (c) are stable
20.8 β decay
20.10 (a) above (b) below (c) below
20.12 The product nuclei are produced in excited states.
20.14 (a) $^{201}_{83}Bi \longrightarrow {}^{197}_{81}Tl + {}^{4}_{2}He$
(b) $^{184}_{77}Ir \longrightarrow {}^{184}_{76}Os + {}^{0}_{1}\beta$
(c) $^{135}_{57}La + {}^{0}_{-1}e \longrightarrow {}^{135}_{56}Ba$
(d) $^{80}_{35}Br \longrightarrow {}^{80}_{36}Kr + {}^{0}_{-1}\beta$
20.16 (a) $^{227}_{90}Th \longrightarrow {}^{223}_{88}Ra + {}^{4}_{2}He$
(b) $^{22}_{11}Na + {}^{0}_{-1}e \longrightarrow {}^{22}_{10}Ne$
(c) $^{232}_{90}Th \longrightarrow {}^{228}_{88}Ra + {}^{4}_{2}He$
(d) $^{72}_{31}Ga \longrightarrow {}^{72}_{32}Ge + {}^{0}_{-1}\beta$
(e) $^{60}_{29}Cu \longrightarrow {}^{60}_{28}Ni + {}^{0}_{1}\beta$
20.18 12.3 hr
20.20 1.29×10^5 years
20.22 8α, 4β
20.24 2.86×10^9 years
20.26 3.6×10^9 years
20.28 1.3×10^{-10}%
20.30 (a) $^{54}_{26}Fe + {}^{4}_{2}He \longrightarrow 2\,{}^{1}_{1}H + {}^{56}_{26}Fe$
(b) $^{27}_{13}Al + {}^{1}_{0}n \longrightarrow {}^{24}_{11}Na + {}^{4}_{2}He$
(c) $^{238}_{92}U + {}^{16}_{8}O \longrightarrow {}^{249}_{100}Fm + 5\,{}^{1}_{0}n$
(d) $^{96}_{42}Mo + {}^{2}_{1}H \longrightarrow {}^{97}_{43}Te + {}^{1}_{0}n$
(e) $^{250}_{98}Cf + {}^{11}_{5}B \longrightarrow 5\,{}^{1}_{0}n + {}^{256}_{103}Lr$
20.32 High energy is needed to overcome electrostatic repulsion.
20.34 (a) $\Delta m = 0.1587$ g/mol (b) 147.8 Mev
(c) 7.779 Mev/nucleon

20.36 (a) $^{26}Al = 8.150$ MeV/nuc., $^{27}Al = 8.333$ MeV/nuc., $^{28}Al = 8.310$ MeV/nuc.
(b) ^{26}Al and ^{28}Al are unstable

20.38 n/p ratio is too large, so β decay moves the product nucleus closer to the zone of stability

20.40 (a) $\Delta m = 0.203$ u (b) 3.03×10^{-11} J
(c) 7.63×10^{7} kJ/g

20.42 (a) 2.9×10^{3} metric tons of coal
(b) 52 metric tons of SO_2 produced

20.44 4.2×10^{8} kJ/g-He; 5 times the energy of fission of 1.0 g of ^{235}U

20.48 9.0 Mev, 0.14 pm

Chapter 21

21.4 b and d

21.6
```
    H   H   H   H   H
    |   |   |   |   |
H — C — C — C — C — C — H
    |   |   |   |   |
    H   H   H   H   H
```
alkyl group = *n*-pentyl

21.8
```
    H   H   H   H   H   H
    |   |   |   |   |   |
H — C — C — C — C — C — C — H
    |   |   |   |   |   |
    H   H   H   H   H   H
```
n-hexane

```
                H
                |
            H — C — H
    H   H   H   |   H
    |   |   |   |   |
H — C — C — C — C — C — H
    |   |   |   |   |
    H   H   H   H   H
```
2-methylpentane

```
            H
            |
        H — C — H
    H   H   |   H   H
    |   |   |   |   |
H — C — C — C — C — C — H
    |   |   |   |   |
    H   H   H   H   H
```
3-methylpentane

```
            H
            |
        H — C — H
    H   H   |   H
    |   |   |   |
H — C — C — C — C — H
    |   |   |   |
    H   H   |   H
        H — C — H
            |
            H
```
2,2-dimcthylbutane

```
            H
            |
        H — C — H
    H   H   |   H
    |   |   |   |
H — C — C — C — C — H
    |   |   |   |
    H   |   H   H
    H — C — H
        |
        H
```
2,3-dimethylbutane

21.10 (a)
```
        H
        |
    H — C — H
    H   |   H   H   H   H
    |   |   |   |   |   |
H — C — C — C — C — C — C — H
    |   |   |   |   |   |
    H   H   H   H   H   H
```

(b)
```
    H   H   Cl  H   H   H   H
    |   |   |   |   |   |   |
H — C — C — C — C — C — C — C — H
    |   |   |   |   |   |   |
    H   H   Cl  H   H   H   H
```

(c)
```
        H
        |
    H — C — H
    H   |   H   H   H   H   H   H
    |   |   |   |   |   |   |   |
H — C — C — C — C — C — C — C — C — H
    |   |   |   |   |   |   |   |
    H   H   |   H   H   H   H   H
            C6H5 (phenyl ring bearing 5 H)
```

(d)
```
H   H       H   H
 \ /         \ /
  C    H H    C
 / \  /   \  / \
H   C       C   H
     \     /
  H   \   /   H
H  C ——C—— C  H
H  |        |  H
H  C        C  H
H    \    /    H
       C
      / \
     H   H
```

21.12 (a) 1-fluoro-2-methylpentane
(b) 3-methylhexane
(c) 2,2-dimethylbutane
(d) 1-chloro-2-ethylcyclobutane

21.16 alkene, C_nH_{2n}; alkyne, C_nH_{2n-2}

21.18 (a) *trans*-1-bromopropene (b) *cis*-3-heptene
(c) 5-fluoro-2-pentyne (d) *cis*-1-chloro-3-hexene

21.20 (Cl)(Br)C=C(Br)(H) — Cl and Br on the left carbon; Br (upper) and H (lower) on the right carbon

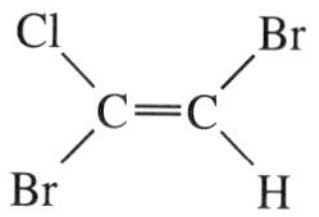

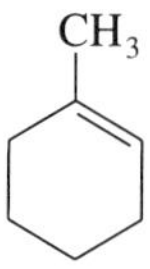

21.22 CH_3(H)C=C(Cl)(CH_3)   H(CH_3)C=C(Cl)(CH_3)

H(Cl)C=C(CH_2CH_3)(H)   H(Cl)C=C(H)(CH_2CH_3)

H(H)C=C(CH_2CH_3)(Cl)   Cl(H)C=C(CH_3)(CH_3)

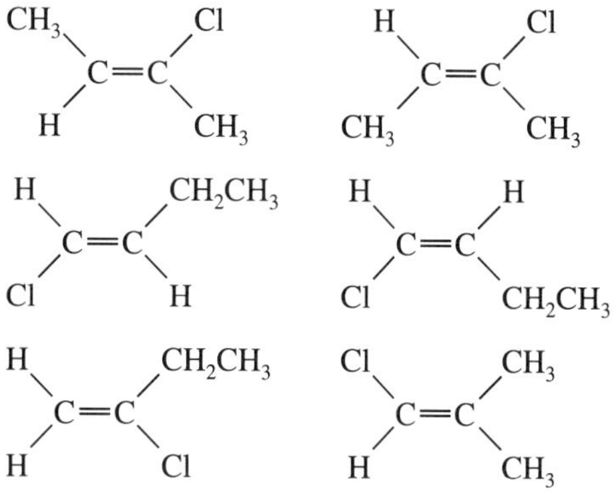

21.24 (a) 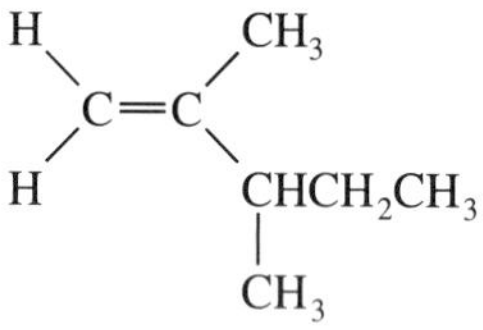 H(H)C=C(CH_3)($\underset{\displaystyle CH_3}{CH}CH_2CH_3$)

(b) 1-methylcyclohexene ring with CH_3 on the double-bonded carbon

(c) H(CH_3)C=C(H)(CH_2Cl)

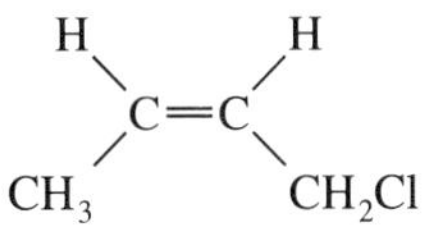

(d) 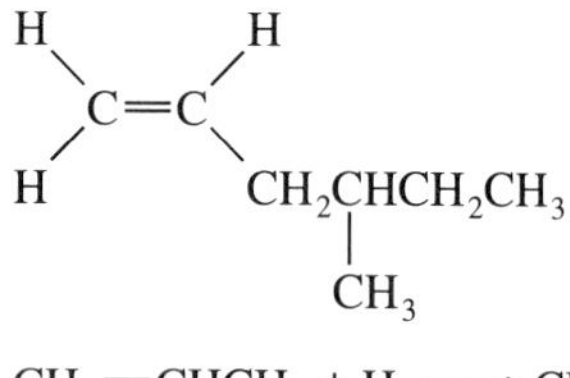 H(H)C=C(H)($CH_2\underset{\displaystyle CH_3}{CH}CH_2CH_3$)

21.26 (a) $CH_2{=}CHCH_3 + H_2 \longrightarrow CH_3CH_2CH_3$
(b) $CH{\equiv}CCH_2CH_3 + 2Cl_2 \longrightarrow CHCl_2CCl_2CH_2CH_3$
(c) 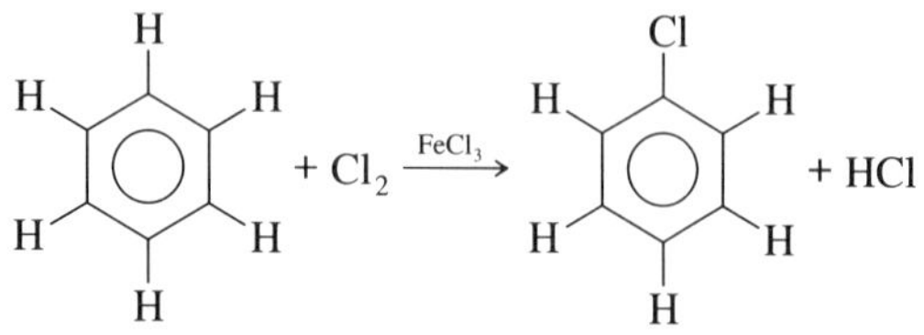 $C_6H_6 + Cl_2 \xrightarrow{FeCl_3} C_6H_5Cl + HCl$ (benzene ring with six H; product ring with Cl replacing one H)

21.28 (a) carboxylic acid (b) aldehyde (c) ether

21.30 (a) $CH_3CH_2\overset{\displaystyle OH}{C}{=}O$, it has hydrogen bonding forces
(b) $CH_3CH_2CH_2CH_2NH_2$, it has hydrogen bonding forces
(c) $CH_3CH_2C{\equiv}CF$, it is larger and more polarizable and thus has higher London dispersion forces

21.32 CH_3OCH_3, dimethylether

21.34 $CH_3\overset{\displaystyle O}{\overset{\|}{C}}{-}OCH_2CH_3$
ethyl acetate (ethyl ethanoate)

21.36 (a) $CH_3CH_2CH_2CH_2OH$
(b) $CH_3\overset{\displaystyle O}{\overset{\|}{C}}{-}\overset{\displaystyle CH_3}{CH}CH_2CH_3$
(c) $CH_3\overset{\displaystyle O}{\overset{\|}{C}}{-}OCH_3$
(d) 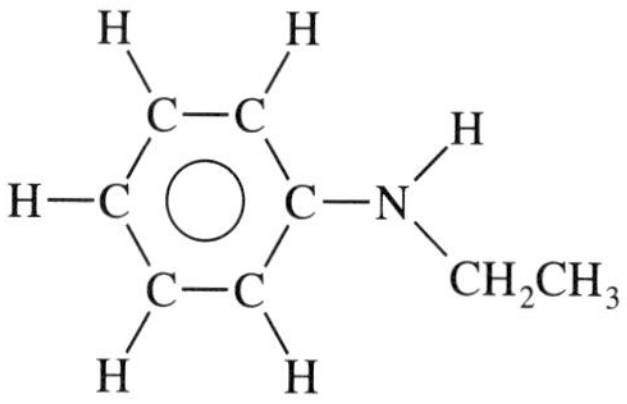 benzene ring (H on five carbons) bonded to $-N(H)(CH_2CH_3)$

21.38 (a) 3-fluoro-1-propanol (b) propanoic acid
(c) isopropyl acetate (d) propyl amine

21.40 (c) only, the chiral center is the middle carbon atom in the three-carbon atom chain.

21.44 $\left(\begin{array}{cc} C_6H_5 & H \\ | & | \\ C{-} & {-}C \\ | & | \\ H & H \end{array} \right)_n$

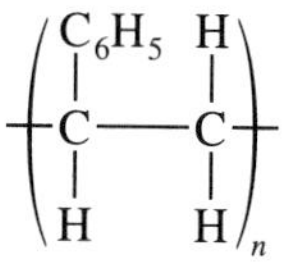

21.46 $\left(\begin{array}{cccc} F & H & F & F \\ | & | & | & | \\ C{-} & C{-} & C{-} & C \\ | & | & | & | \\ F & H & F & CF_3 \end{array} \right)_n$

21.50 $\left(O{-}C_6H_4{-}\underset{\displaystyle CH_3}{\overset{\displaystyle CH_3}{C}}{-}C_6H_4{-}O{-}\overset{\displaystyle O}{\overset{\|}{C}} \right)_n$

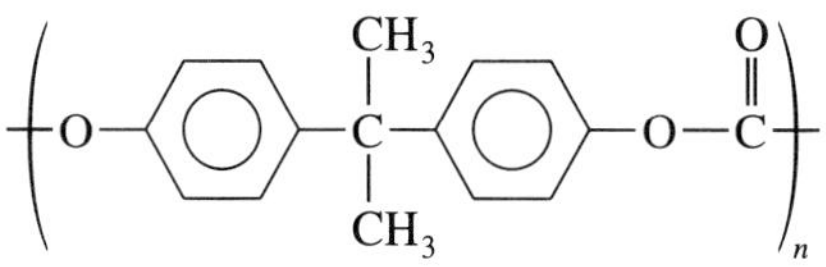

21.66 (a) 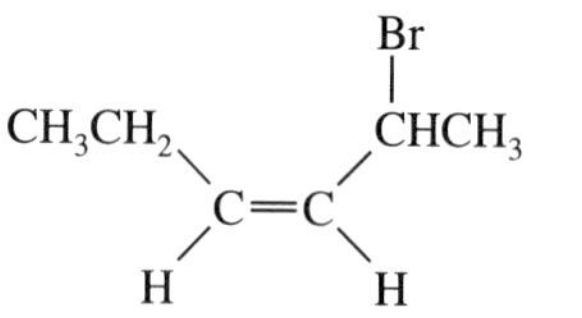 CH_3CH_2(H)C=C(H)($\overset{\displaystyle Br}{CH}CH_3$)
(b) 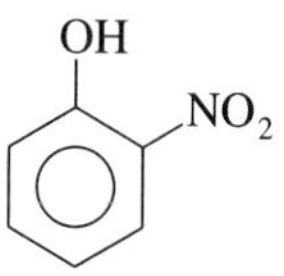 phenol ring with OH and adjacent NO_2

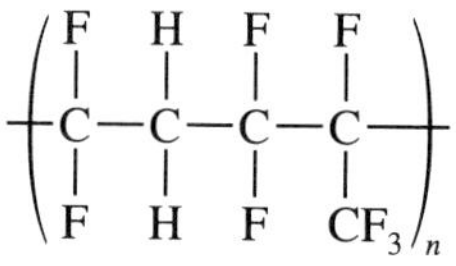

Illustration Credits

Photographs not credited below were taken by Larry Cameron.

Chapter 1

Unnum. figure p. 3: Scala/Art Resource, NY; 1.3 (Moon, Earth, Jupiter): NASA; unnum. figure p. 7 (left): Yoav Levy/ Phototake, NYC; unnum. figure p. 7 (center): Michael S. Yamashita/The Stock Shop; unnum. figure p. 11 (top): The Granger Collection, NY; unnum. figures p. 13 (top and bottom): Quesada/Burke Studios: unnum. figure p. 15 (left); Novosti/Science Photo Library/Photo Researchers, Inc.; unnum. figure p. 15 (right): AIP Niels Bohr Library; unnum. figure p. 15 (center): from *Annalen der Chemie und Pharmacie,* VII, Supplementary volume from 1872; 1.12: Courtesy of Ashland Oil, Inc.; unnum. figure p. 35: Courtesy of L.S. Starrett Company.

Chapter 2

Chapter Opener: Bruce C. Schardt, from cover of *Science,* Vol. 243, 24 February © 1989 by the AASS; unnum. figure p. 50: The Bettmann Archive; unnum. figure p. 51: Philippe Plailly, Science Source/Photo Researchers, Inc.; unnum. figure p. 52: The British Library; unnum. figure p. 53: Science Museum Library; unnum. figure p. 71: Dr. Jeremy Burgess/ Science Photo Library/Photo Researchers, Inc.

Chapter 3

Unnum. figure p. 86: P. Turner/The Image Bank; unnum. figure p. 87 (right): Whitney Lane/The Image Bank; unnum. figure p. 88: David Weintraub/Science Source/Photo Researchers, Inc.; unnum. figures p. 89: the Dean of Lincoln; unnum. figure p. 91: Leon Lewandowski; unnum. figure p. 92: The Granger Collection, NY; unnum. figure p. 103 (bottom): Courtesy of Fisons Instruments; unnum. figure p. 106: The Granger Collection, NY; unnum. figures p. 111: Courtesy of The Doe Run Company, Herculaneum, MO.

Chapter 4

4.7: Courtesy of CPAC, Inc., Leicester, NY; unnum. figure p. 138: Ted Spiegal/Black Star; unnum. figure p. 152 (left): Whitney Lane/The Image Bank; unnum. figure p. 152 (right): David R. Frazier Photolibrary.

Chapter 5

Chapter Opener: Bernard Assett/Photo Researchers, Inc.; 5.1: Courtesy of Bethlehem Steel; unnum. figure p. 168 (left): H. Wendler/The Image Bank; unnum. figure p. 168 (right): Scott R. Goode; unnum. figure p. 175: James A. Marshall, University of South Carolina; unnum. figures pp. 176 and 182: National Center for Atmospheric Research/University Corp. for Atmospheric Research/National Science Foundation; 5.21: Oak Ridge National Laboratories.

Chapter 6

Chapter Opener: W. Rivelli/The Image Bank; unnum. figure p. 202 (top): Uniphoto, Inc.; unnum. figure p. 202 (bottom): Don King/The Image Bank; unnum. figure p. 213:

Chapter 7

Unnum. figure p. 263: Courtesy of Instrumentation Laboratory/Lexington, MA.

Chapter 8

Unnum. figure p. 276: Paul Silverman/Fundamental Photographs, NY; unnum. figure p. 290: UPI/The Bettmann Archive; unnum. figure p. 293: Keith Kent/Science Photo Library/Photo Researchers, Inc.; unnum. figure p. 296: IBM

Research/Peter Arnold, Inc.; unnum. figure p. 301: Dr. E. R. Degginger; unnum. figure p. 305: Will McIntyre/Photo Researchers, Inc.

Chapter 10

Chapter Opener: Farrell Grehan/Photo Researchers, Inc.; unnum. figure p. 375: Emil Muench/Photo Researchers, Inc.; 10.7: Hans Pfetschinger/Peter Arnold, Inc.; unnum. figure p. 381: Charles D. Winters; unnum. figure p. 382: Ron Sherman; 10.13: Dr. E. R. Degginger; unnum. figure p. 391: Dr. M. B. Hursthouse/Science Photo Library/Photo Researchers, Inc.; unnum. figure p. 394 (left): Michael P. Gedomski/Photo Researchers, Inc.; unnum. figure p. 394 (right): David R. Frazier Photolibrary; unnum. figure p. 402: Craig Blankenhorn/Black Star; unnum. figure p. 403 (top right): Kristen Brochmann, Fundamental Photographs, NY; unnum. figures p. 403 (bottom left, bottom right): Charles D. Winters.

Chapter 11

Chapter Opener: Thomas Hovland/Grant Heilman, Inc.; unnum. figure p. 426: Kristen Brochmann/Fundamental Photographs, NY; 11.19: Courtesy of Ashland Oil, Inc.; 11.18: J. W. Morgenthaler.

Chapter 12

Unnum. figure p. 461 (left): Tom McHugh/Photo Researchers, Inc.; unnum. figure p. 461 (right): Comstock, Inc.; unnum. figure p. 464: Oesper Collection in the History of Chemistry, University of Cincinnati; unnum figure p. 467: Phil Degginger; unnum. figure p. 482: R. Rowan/Photo Researchers, Inc.

Chapter 13

Unnum. figure p. 502: Van Bucher/Photo Researchers, Inc.; unnum. figure p. 507: Tom Van Sant/Geosphere Project, Santa Monica/Science Photo Library/Photo Researchers, Inc.; unnum. figure p. 512: The Corning Museum of Glass.

Chapter 14

Unnum. figure p. 568: Courtesy of Instrumentation Laboratory, Lexington, MA; unnum. figures p. 579 (top and bottom): Adam Hart-Davis/Science Photo Library/Photo Researchers, Inc.

Chapter 15

Chapter Opener: Peter Menzel; unnum. figure p. 596 (upper left): NASA; unnum. figure p. 597: The Bettmann Archive; unnum. figure p. 602 (left): David O. Hill/Photo Researchers, Inc.; unnum. figure p. 602 (right): David R. Frazier Photolibrary; unnum. figure p. 609 (right): David R. Frazier Photolibrary; unnum. figure p. 609 (left): Gary Milburn/Tom Stack & Associates; unnum. figure p. 611 (left): Schuster/Superstock; unnum. figure p. 618: J. Barry O'Rourke/The Stock Market; unnum. figure p. 624: Redrawn from Bruce Anderson, *The Solar Home Book,* p. 24 R.A.K. Publishing, Co., 1976; unnum. figure p. 625: Eleanor Raymond, Architect, supplied by Doris Cole; unnum. figure p. 634: Roland Birke/Peter Arnold, Inc.; unnum. figure p. 637: Lionel Delevingue/Phototake, NYC; unnum. figures p. 638 (top and bottom): Michael S. Thompson/Comstock, Inc.; unnum. figure p. 639 (bottom): Larry Mulvehill/Photo Researchers, Inc.; unnum. figure p. 640 (top): Jeff Hunter/The Image Bank; unnum. figure p. 640 (bottom): Co Rentmeester/The Image Bank.

Chapter 16

Unnum. figure p. 652 (left): Tony Freeman/PhotoEdit; unnum. figure p. 652 (right): The Metropolitan Museum of Art, Bequest of George C. Stone, 1935; unnum. figure p. 667: Alan Pitcairn/Grant Heilman, Inc.; unnum. figure p. 671: Paul Hanny/Gamma Liaison; unnum. figures p. 678: Jeff Lepore/Photo Researchers, Inc.; unnum. figure p. 695: David Maenza/The Image Bank; unnum. figures pp. 702 and 703: redrawn from Saferstein, R., *Criminalistics: An Introduction to Forensic Science,* 3rd ed., Prentice-Hall, Inc., 1987.

Chapter 17

Chapter Opener: Courtesy of General Motors; 17.9: Courtesy of Fisher Scientific; 17.10: courtesy of A. G. Ewing, The Pennsylvania State University, *Analytical Chemistry,* Vol. 64, 1 July, 1992, p. 1373; 17.11: Courtesy of Fisher Scientific; unnum. figure p. 758: Oesper Collection in the History of Chemistry, University of Cincinnati; 17.20: Courtesy of Reed and Barton Silversmiths; 17.21: Andy Levin/Photo Researchers, Inc.; 17.24: Runk/Schoenberger/Grant Heilman, Inc.

Chapter 18

Chapter Opener: David Hiser/The Image Bank; unnum. figure p. 779: Robert Lebeck/The Image Bank; unnum. figure p. 782: Gerhard Gscheidle/The Image Bank: unnum. figure p. 783: Courtesy of Bethlehem Steel; 18.2: Richard Megna/

Fundamental Photographs, NY; unnum. figures p. 808: Runk/Schoenberger/Grant Heilman, Inc.

Chapter 19

19.3: UPI/Bettmann Archive; 19.6(a): Rich McIntyre/Tom Stack & Associates; 19.10: Tom Pantages; 19.11 Charles D. Winters; 19.12 (top): Paul Silverman/Fundamental Photographs, NY; 19.15: Randy Jolly/Comstock, Inc.; unnum. figure p. 834: F. Stuart Westmorland/Tom Stack & Associates; unnum. figure p. 835 (right): Courtesy of Dr. R. E. Smalley/Rice University; 19.16: Courtesy of U.S. Filter/IWT; unnum. figure p. 838: Courtesy of Hughes-Danbury Optical Systems; 19.20: Runk/Schoenberger/Grant Heilman, Inc.; 19.21: Dick Luria/Science Source/Photo Researchers, Inc.; 19.24(a): Renee Purse/Photo Researchers, Inc.; 19.24(b): Dr. E. R. Degginger; unnum. figure p. 843: Larry Lee/Westlight; 19.30: Nathan Benn/Woodfin Camp & Associates, Inc.; 19.31: Barrie Rokeach/The Image Bank; 19.32(a): Comstock, Inc.; unnum. figure p. 854 (top): Courtesy of the Goodyear Tire and Rubber Company; 19.36: Alvis Upitis/The Image Bank.

Chapter 20

Chapter Opener: Bruce Roberts/Photo Researchers, Inc.; unnum. figure p. 864: from Brescia, Mehlman, Pellegrini and Stambler, *Chemistry: A Modern Introduction,* 2nd ed., Saunders College Publishing, 1978; unnum. figure p. 871: Santi Visali/The Image Bank; unnum. figure p. 873: Yoav Levy/Phototake, NYC; 20.5: Courtesy of Beckman Instruments, Inc.; 20.6(b): Courtesy of Argonne National Laboratory; unnum. figure p. 876: Courtesy of Fermilab; unnum. figure p. 879: The Bettmann Archive; unnum. figure p. 881: Courtesy of the Atomic Museum; unnum. figures p. 891: Instituto H San Raffaele, Milano.

Chapter 21

21.11: Phil Degginger; 21.16: Rod Westwood/The Image Bank; unnum. figure p. 922: Blair Seitz/Photo Researchers, Inc.; unnum. figure p. 923: Dr. E. R. Degginger; 21.21: David R. Frazier Photolibrary; 21.22: Courtesy of DuPont; 21.23(b): Peter Frey/The Image Bank; 21.23(c): Gregory Heisler/The Image Bank; unnum. figure p. 935: UPI/Bettmann Newsphotos; 21.28: from M. F. Perutz, *The Hemoglobin Molecule,* Scientific American, © 1964, All rights reserved.

Index

Selected Physical Constants*

Quantity	Symbol	Value	Units
Atmosphere (standard)	atm	101 325.	Pa
Atomic mass unit	u	$1.660\,540 \times 10^{-27}$	kg
Avogadro's number	N_A	$6.022\,14 \times 10^{23}$	particles mol^{-1}
Bohr radius	a_o	$5.291\,8 \times 10^{-11}$	m
Boltzmann constant	k	$1.380\,66 \times 10^{-23}$	J K^{-1}
Charge-to-mass ratio of electron	e/m_e	$-1.758\,820 \times 10^{11}$	C kg^{-1}
Elementary charge	e	$1.602\,177 \times 10^{-19}$	C
Electron mass	m_e	$9.109\,39 \times 10^{-31}$	kg
Faraday constant	F	96 485.3	C/mol e^-
Gas constant	R	0.082 058 8.314 5	L atm mol^{-1} K^{-1} J mol^{-1} K^{-1}
Neutron mass	m_n	$1.674\,929 \times 10^{-27}$	kg
Planck's constant	h	$6.626\,08 \times 10^{-34}$	J s
Proton mass	m_p	$1.672\,623 \times 10^{-27}$	kg
Rydberg constant	R_h	$1.096\,78 \times 10^{7}$ $1.096\,78 \times 10^{5}$	m^{-1} cm^{-1}
Speed of light (in vacuum)	c	299 792 458. 186 281	m s^{-1} mi s^{-1}

The uncertainty is less than 1 in the last place shown.

* E. R. Cohen and B. N. Taylor, *J. Phys. Chem. Ref. Data,* **17**, 1797 (1988).

Relations and Conversions in the SI and English Systems

SI Units	English Units	SI–English
length		
1 km = 10^3 m	1 ft = 12 in	in = 2.54 cm
1 cm = 10^{-2} m	1 yd = 3 ft	1 m = 39.37 in
1 mm = 10^{-3} m	1 mile = 5280 ft	
1 nm = 10^{-9} m		
volume		
1 m^3 = 10^6 cm^3	1 gal = 4 qt	1 L = 1.057 qt
1 cm^3 = 1 mL	1 qt = 57.75 in^3	1 qt = 0.946 L
mass		
1 kg = 10^3 g	1 lb = 16 oz	1 lb = 453.6 g
1 mg = 10^{-3} g	1 ton = 2000 lb	